P9-DOF-086

THE CONDENSED CHEMICAL DICTIONARY

Sixth Edition

Completely revised and enlarged by

ARTHUR and ELIZABETH ROSE

State College, Pa.

Formerly Directed by

FRANCIS M. TURNER

1889-1952

REINHOLD PUBLISHING CORPORATION

New York

CHAPMAN & HALL, LTD., LONDON

Second Printing 1962

Library of Congress Catalog Card Number: 61-14790

Photocopy set by
APPLIED SCIENCE LABORATORIES
State College, Pa.

Printed in the United States of America
by The Guinn Company Inc.

Table of Contents

Publisher's Preface

With the appearance of the Sixth Edition, "The Condensed Chemical Dictionary" rounds out 42 years of continuous service to the chemical and process industries and to the many thousands of people whose work and interests bring them into contact with the chemicals of commerce. The First Edition appeared in 1919. The need for such a reference source became apparent to the late Francis M. Turner, who was at that time Vice President of the Chemical Catalog Company, as a result of the great influx of questions about chemicals and materials from users of the "Chemical Engineering Catalog," which had been started several years previously.

Many of the features of the Dictionary were new and original contributions to the reference literature; most of these have been retained throughout succeeding editions, including the present one. The material presented is specifically "tailored" to the needs of the busy chemist, while at the same time it provides ready answers to the more commonplace questions of the non-scientist.

Until the time of his death, Francis Turner was the editorial director of the Dictionary. The Second and Third Editions were prepared with the capable assistance of Thomas C. Gregory. The Fourth Edition underwent considerable expansion, particularly in regard to inclusion of a large number of definitions dealing with chemical terminology, as distinct from materials and substances. At this time Arthur and Elizabeth Rose of State College, Pennsylvania, assumed the burden of editorship. They have done a most competent and effective job in expanding the scope of the book and improving its organization and accuracy. Updating, and addition of new material has been in progress almost continuously since 1951, resulting in appearance of the Fifth Edition in 1956 and this present Sixth Edition in 1961.

It is no exaggeration to state that "The Condensed Chemical Dictionary" has earned for itself a recognized position in the literature of chemistry—a position which the publishers hope will not only be maintained but improved throughout the years.

August, 1961 — Reinhold Publishing Corporation

Numerical List of Manufacturers

1. Stauffer Chemical Company, 380 Madison Ave., New York 17, N.Y.
3. Abbott Laboratories, North Chicago, Ill.
4. Rhode Island Laboratories, Inc., 100 Pulaski St., West Warwick, R.I.
5. The Firestone Tire and Rubber Co., Akron 17, Ohio
7. Cluett, Peabody & Co., Inc., 530 Fifth Ave., New York 36, N.Y.
8. Scientific Oil Compounding Co., Inc., Cunilate Div., 1637-55 S. Kilbourn Ave., Chicago 23, Ill.
9. Knoll Pharmaceutical Co., Orange, N.J.
10. Van Dyk & Company, Inc., Main and William Sts., Belleville 9, N.J.
11. Koppers Company, Inc., Koppers Bldg., Pittsburgh 19, Pa.
12. Sindar Corporation, 321 W. 44 St., New York 36, N.Y.
13. The Formica Company, 4614 Spring Grove Ave., Cincinnati 32, Ohio
14. Resistoflex Corporation, Roseland, N.J.
15. Armour Industrial Chemical Co., Div. of Armour and Co., 110 N. Wacker Dr., Chicago 6, Ill.
16. Arthur H. Thomas Company, P.O. Box 779, Philadelphia 5, Pa.
17. Heresite & Chemical Co., Manitowoc, Wis.
18. Central Scientific Co., Subsidiary of Cenco Instruments Corp., 1700 Irving Park Rd., Chicago 13, Ill.
19. Fries Bros. Inc., P.O. Box 8, Carlstadt, N.J.
20. Corning Glass Works, Corning, N.Y.
21. Neville Chemical Company, Neville Island, Pittsburgh 25, Pa.
22. UBS Chemical Company, Div. of A. E. Staley Mfg. Co., 491 Main St., Cambridge 42, Mass.
23. Rohm & Haas Company, Washington Square, Philadelphia 5, Pa.
24. Wyeth Laboratories, Inc., P.O. Box 8299, Philadelphia 1, Pa.
25. Pennsylvania Refining Company, Butler, Pa.
27. Thiokol Chemical Corp., Chemical Div., 780 N. Clinton Ave., Trenton 7, N.J.
28. E. I. du Pont de Nemours & Co., Inc., Wilmington 98, Del.
29. Enjay Chemical Co., Div. of Humble Oil & Refining Co., 15 W. 51 St., New York 19, N.Y.
30. Corn Products Sales Co., Corn Products Div., 717 Fifth Ave., New York 22, N.Y.
31. Handy & Harman, 82 Fulton St., New York 38, N.Y.
33. The Penetone Company, Tenafly, N.J.
35. Firestone Plastics Co., P.O. Box 690, Pottstown, Pa.
36. Reichhold Chemicals, Inc., 525 N. Broadway, White Plains, N.Y.
37. Rayonier, Inc., 161 E. 42 St., New York 17, N.Y.
38. Sierra Talc Company, Box 390, South Pasadena, Calif.
40. Virginia-Carolina Chemical Corporation, Richmond, Va.
41. Atlas Mineral Products Co., Mertztown, Pa.
42. Warwick Chemical Co., Div. of Sun Chemical Corp., Wood River Junction, R.I.
45. Sonneborn Chemical and Refining Corp., 300 Park Avenue South, New York 10, N.Y.
46. Acheson Colloids Co., Port Huron, Mich.
47. The Duriron Co., Inc., Dayton, Ohio
48. Crown Zellerbach Corp., Chemical Products Div., Camas, Wash.
49. W. A. Cleary Corp., New Brunswick, N.J.
50. Allied Chemical Corp., General Chemical Div., 40 Rector St., New York 6, N.Y.
51. Esso Standard Div. of Humble Oil & Refining Co., 15 W. 51 St., New York 19, N.Y.
52. Rubber Corporation of America, New South Road, Hicksville, N.Y.
53. National Starch and Chemical Corp., Box 50, Grand Central Post Ofc., New York 17, N.Y.
54. Carrier Corp., Carrier Pkwy., Syracuse 1, N.Y.
55. Food Machinery & Chemical Corp., 161 E. 42 St., New York 17, N.Y.
56. Canadian Industries Ltd., Box 10, Montreal, Quebec, Canada
57. American Cyanamid Co., 30 Rockefeller Plaza, New York 20, N.Y.
58. Monsanto Chemical Co., 800 N. Lindbergh Blvd., St. Louis 66, Mo.
60. Cerro Sales Corp., Subsidiary of Cerro Corp., 300 Park Ave., New York 22, N.Y.
61. Shawinigan Resins Corporation, Springfield 1, Mass.
62. Hooker Chemical Corporation, Durez Plastics Div., 8 Walck Rd., North Tonawanda, N.Y.
63. The Richardson Company, Melrose Park, Ill.

64. Spencer Kellogg & Sons, Inc., Buffalo 5, N.Y.
65. The Borden Chemical Co., Div. of The Borden Co., 350 Madison Ave., New York 17, N.Y.
67. Climax Molybdenum Co., 1270 Avenue of the Americas, New York 20, N.Y.
69. R. T. Vanderbilt Co., Inc., 230 Park Ave., New York 17, N.Y.
70. G. D. Searle & Co., P.O. Box 5110, Chicago 80, Ill.
71. Smith, Kline & French Laboratories, 1530 Spring Garden St., Philadelphia 1, Pa.
72. Hanson-Van Winkle-Munning Co., Matawan, N.J.
73. Glyco Chemicals, Div. of Chas. L. Huisking & Co., Inc., 417 Fifth Ave., New York 16, N.Y.
74. Nuodex Products Co., Div. of Heyden Newport Chemical Corp., Elizabeth, N.J.
75. American LaFrance Corporation, Elmira, N.Y.
76. Ceresit Corp., 3227 S. Shields Ave., Chicago 16, Ill.
77. American Celcure Wood Preserving Corp., 1074 E. Eighth St., Jacksonville 6, Fla.
78. Jacques Wolf & Co., Subsidiary of Nopco Chemical Co., 60 Park Pl., Newark 1, N.J.
79. Newport Industries Company, Div. of Heyden Newport Chemical Corp., 342 Madison Ave., New York 17, N.Y.
80. Visking Company, Div. of Union Carbide Corp., 6733 W. 65 St., Chicago 38, Ill.
81. Wood Ridge Chemical Corp., Park Place East, Wood-Ridge, N.J.
82. Graphite Metallizing Corp., 1050 Nepperhan Ave., Yonkers, N.Y.
83. Synthetic Chemicals, Inc., 335 McLean Blvd., Paterson 4, N.J.
84. Olin Mathieson Chemical Corp., Chemicals Div., 10 Light St., Baltimore 3, Md.
85. Shulton, Inc., 697 Route 46, Clifton, N.J.
86. Colgate-Palmolive Co., 105 Hudson St., Jersey City 2, N.J.
88. American Potash & Chemical Corp., 300 W. Sixth St., Los Angeles 54, Calif.
89. Atlas Powder Co., Wilmington 99, Del.
90. Brazil Oiticica, Inc., 80 Broad St., New York 4, N.Y.
91. Schwarz BioResearch, Inc., 230 Washington St., Mount Vernon, N.Y.
92. Masonite Corporation, 111 W. Washington St., Chicago 2, Ill.
93. Tennessee Corporation, 617-29 Grant Bldg., Atlanta 1, Ga.
94. The C. P. Hall Co., 414-418 S. Broadway, Akron 8, Ohio
96. H. N. Hartwell & Son, Inc., 947 Park Square Bldg., Boston 16, Mass.
97. Chesebrough-Pond's Inc., 485 Lexington Ave., New York 17, N.Y.
98. Floridin Company, P.O. Box 989, Tallahassee, Fla.
99. Minerals & Chemicals Philipp Corporation, 20 Essex Turnpike, Menlo Park, N.J.
100. Eli Lilly and Company, Indianapolis 6, Ind.
101. Commerce Chemical Corp., Div. of Commerce Oil Corp., 1271 Ave. of the Americas, New York 20, N.Y.
103. Arthur S. Hoyt Co., P.O. Box 24, Hicksville, N.Y.
104. Witco Chemical Co., Inc., 122 E. 42 St., New York 17, N.Y.
105. Polak's Frutal Works, Inc., Middletown, N.Y.
107. American Norit Co., Inc., 6301 Glidden Way, Jacksonville 8, Fla.
108. Hagan Chemicals & Controls, Inc., P.O. Box 1346, Pittsburgh 30, Pa.
109. Arapahoe Chemicals, Inc., 2855 Walnut St., Boulder, Colo.
110. United Carbon Co., Inc., 410 Park Ave., New York 22, N.Y.
111. American Manganese Bronze Co., Holmesburg, Philadelphia 36, Pa.
112. The Chemstrand Corporation, 350 Fifth Ave., New York 1, N.Y.
114. Pabst Brewing Company, Merchandise Mart, Chicago 54, Ill.
115. Eastman Kodak Co., 343 State St., Rochester 4, N.Y.
116. Allis-Chalmers Mfg. Co., Box 512, Milwaukee 1, Wis.
117. Alox Corp., P.O. Box 556, Niagara Falls, N.Y.
118. California Industrial Minerals Co., P.O. Box 1666, Fresno 17, Calif.
119. B. F. Goodrich Chemical Co., Div. of The B. F. Goodrich Co., 3135 Euclid Ave., Cleveland 15, Ohio
121. Marathon, Div. of American Can Co., Menasha, Wis.
122. G & A Laboratories, Inc., Box 1217, Savannah, Ga.
123. Merck & C., Inc., Rahway, N.J.
124. Marine Colloids, Inc., 24 State St., New York 4, N.Y.
125. Shell Chemical Co., Div. of Shell Oil Co., 50 W. 50 St., New York 20, N.Y.
126. The Pacific Lumber Co., 100 Bush Ave., San Francisco 4, Calif.
128. Bareco Wax Co., Div. of Petrolite Corp., 917 Enterprise Bldg., Tulsa 3, Okla.
129. Weyerhaeuser Timber Co., P.O. Box 1645, Tacoma 1, Wash.
133. Columbian Carbon Co., 380 Madison Ave., New York 17, N.Y.
134. The Harshaw Chemical Co., 1945 E. 97 St., Cleveland 6, Ohio
135. Illinois Zinc Co., Div. of Hydrometals, Inc., 230 N. Michigan Ave., Chicago 1, Ill.
136. The Atlantic Refining Co., Chemicals Div., 2600 S. Broad St., Philadelphia 1, Pa.
137. Jefferson Chemical Co., Inc., Box 4128, N. Austin Station, Austin, Texas
138. Heyden Newport Chemical Corp., Heyden Chemical Div., 342 Madison Ave., New York 17, N.Y.
139. The Moores Lime Company, Springfield, Ohio
140. Pennsylvania Industrial Chemical Corp., 120 State St., Clairton, Pa.
141. The Sherwin-Williams Co., Pigment, Color & Chemical Div., 260 Madison Ave., New York 16, N.Y.
142. Enthone, Inc., Subsidiary of American Smelting & Refining Co., Box 1900, New Haven 8, Conn.
144. Air Reduction Chemical & Carbide Co., 150 E. 42 St., New York 17, N.Y.
145. Lithcote Corporation, 5000 W. Lake St., Melrose Park, Ill.

147. Chipman Chemical Co., Inc., Bound Brook, N.J.
148. National Lead Co., DeLore Div., Mississippi River and River Des Peres, St. Louis 11, Mo.
149. Dow Corning Corp., Midland Mich.
150. Plant Protection, Ltd., Chipman Chemical Co., Inc., Bound Brook, N.J.
151. California Chemical Co., Oronite Div., 200 Bush St., San Francisco 20, Calif.
152. Swift & Company, Union Stock Yards, Chicago 9, Ill.
154. Wallace & Tiernan Incorporated, Lucidol Div., 1740 Military Rd., Buffalo 5, N.Y.
155. Wilbur B. Driver Company, Newark, N.J.
157. American Molasses Co., The Nulomoline Div., 120 Wall St., New York 5, N.Y.
158. Minnesota Mining and Mfg. Co., Chemical Div., 900 Bush Ave., St. Paul 6, Minn.
159. Royce Chemical Co., Carlton Hill, N.J.
160. Simplex Wire & Cable Co., Cambridge 39, Mass.
161. H. I. Thompson Fiber Glass Co., 1733 Cordova St., Los Angeles 7, Calif.
162. Winthrop Laboratories, 1450 Broadway, New York 18, N.Y.
164. Rumford Chemical Works, Rumford 16, R.I.
165. Apex Chemical Co., Inc., 200 S. First St., Elizabethport 1, N.J.
166. Hoskins Manufacturing Co., 4445 Lawton Ave., Detroit 8, Mich.
167. Kendall Refining Company, Bradford, Pa.
168. Lake States Yeast & Chemical Div. of St. Regis Paper Co., Rhinelander, Wis.
169. The LaMotte Chemical Products Co., Chestertown, Md.
170. Morningstar-Paisley, Inc., 630 W. 51 St., New York 19, N.Y.
171. Protexol Corporation, Kenilworth, N.J.
172. Victor Chemical Works, Div. of Stauffer Chemical Co., 155 N. Wacker Dr., Chicago 6, Ill.
173. Wallerstein Company, Div. of Baxter Laboratories, Inc., Wallerstein Sq., Mariners Harbor, Staten Island 3, N.Y.
175. Allied Chemical Corp., Plastics Div., 40 Rector St., New York 6, N.Y.
176. American Products Mfg. Co., 8127-33 Oleander St., New Orleans 18, La.
177. Columbia-Southern Chemical Corp., One Gateway Center, Pittsburgh 22, Pa.
179. The General Tire & Rubber Co., Chemical Div., Akron 9, Ohio
181. Chemagro Corp., 220 E. 42 St., New York 17, N.Y.
182. Nalco Chemical Company, 6216 W. 66 Pl., Chicago 38, Ill.
183. Natural Gas Odorizing Co., Inc., Box 1645, Houston 1, Tex.
184. The Permutit Company, Div. of Pfaudler Permutit Inc., 50 W. 44 St., New York 36, N.Y.
186. Gunk Laboratories, Inc., 5829 W. 66 St., Chicago 38, Ill.
188. Fritzsche Brothers, Inc., Port Authority Bldg., 76 Ninth Ave., New York 11, N.Y.
189. Wallace & Tiernan Incorporated, Harchem Div., Box 178, Newark 1, N.J.
190. Hoffmann-La Roche Inc., Nutley 10, N.J.
191. Owens-Corning Fiberglas Corporation, Toledo 1, Ohio
192. U.S. Industrial Chemicals Co., Div. of National Distillers and Chemical Corp., 99 Park Ave., New York 16, N.Y.
194. Stoner-Mudge Co., Div. of American-Marietta Co., 2000 Westhall St., Pittsburgh 33, Pa.
195. Standard Dry Wall Products, Inc., New Eagle, Pa.
196. The American Agricultural Chemical Co., Chemical Sales Div., 100 Church St., New York 7, N.Y.
197. Allied Chemical Corp., Nitrogen Div., 40 Rector St., New York 6, N.Y.
199. The Alpha-Molykote Corp., 65 Harvard Ave., Stamford, Conn.
200. APCO Oil Corp., Liberty Bank Bldg., Oklahoma City 2, Okla.
201. Philadelphia Quartz Co., Public Ledger Bldg., Independence Sq., Philadelphia 6, Pa.
202. The Baker Castor Oil Co., Subsidiary of National Lead Co., Bayonne, N.J.
203. Wyandotte Chemicals Corporation, Wyandotte, Mich.
204. Pennsalt Chemicals Corp., 3 Penn Center, Philadelphia 2, Pa.
205. A.C. Horn Co., Bldg. Matls. Div. East, 750 Third Ave., New York 17, N.Y.
206. Imperial Chemical Industries Ltd., Imperial Chemical House, Millbank, London, S.W. 1, England
207. Imperial Chemical (Pharmaceuticals) Ltd., Imperial Chemical House, Millbank, London, S.W. 1, England
209. Michigan-Standard Alloy Casting Div., Consolidated Foundries & Mfg. Corp., 1999 Guoin St., Detroit 7, Mich.
210. Ionac Chemical Co., Div. of Pfaudler Permutit Inc., Birmingham, N.J.
212. Miles Chemical Co., Div. of Miles Laboratories, Inc., Elkhart, Ind.
214. Union Carbide Corporation, 270 Park Ave., New York 17, N.Y.
216. Amoco Chemicals Corp., 130 E. Randolph Dr., Chicago 1, Ill.
217. Filtrol Corp., 3250 E. Washington Blvd., Los Angeles 23, Calif.
218. Great Lakes Carbon Corp., 333 N. Michigan Ave., Chicago 1, Ill.
219. Geigy Industrial Chemicals, Div. of Geigy Chemical Corp., P.O. Box 430, Yonkers, N.Y.
220. Thermal American Fused Quartz Co., 1820 Salem St., Dover, N.J.
221. Archer-Daniels-Midland Co., 700 Investors Bldg., Minneapolis 2, Minn.
223. C.J. Osborn Company, 1301 W. Blancke St., Linden, N.J.
224. The Quaker Oats Co., Merchandise Mart Plaza, Chicago 54, Ill.
225. Bromine Producers Co., Adrian, Mich.
226. Aluminum Co. of America, 1501 Alcoa Bldg., Pittsburgh 19, Pa.
227. Givaudan-Delawanna, Inc., 321 W. 44 St., New York 36, N.Y.
228. West Virginia Pulp and Paper, Industrial Chemical Sales Div., 230 Park Ave., New York 17, N.Y.
229. West Virginia Pulp and Paper, Polychemicals Div., Charleston, S.C.

230. Advance Solvents & Chemical Div. of Carlisle Chemical Works, Inc., New Brunswick, N.J.
231. Brown Company, 650 Main St., Berlin, N.H.
232. L. B. Holliday & Co., Ltd., Huddersfield, England
233. The Dow Chemical Company, Midland, Mich.
235. National Rosin Oil Products, Inc., 1270 Avenue of the Americas, New York 20, N.Y.
236. Baroid Div., National Lead Co., P.O. Box 1675, Houston 1, Tex.
238. Chemical Development Corp., Danvers, Mass.
239. Clay-Adams, Inc., 141 E. 25 St., New York 10, N.Y.
241. Davison Chemical Co., Div. of W. R. Grace & Co., Baltimore 3, Md.
242. Emery Industries, Inc. Carew Tower, Cincinnati 2, Ohio
243. Allied Chemical Corp., National Aniline Div., 40 Rector St., New York 6, N.Y.
244. Diamond Alkali Co., Union Commerce Bldg., Cleveland 14, Ohio
245. General Electric Co., 1 River Rd., Schenectady 5, N.Y.
247. Johns-Manville, 22 E. 40 St., New York 16, N.Y.
248. U.S. Rubber Corp., Naugatuck Chemical Div., Naugatuck, Conn.
249. Norton Company, 551 New Bond St., Worcester 6, Mass.
250. Foote Mineral Co., 18 W. Chelten Ave., Philadelphia 44, Pa.
251. Republic Steel Corporation, Republic Bldg., Cleveland 1, Ohio
252. Arizona Chemical Co., 30 Rockefeller Plaza, New York 20, N.Y.
253. California Chemical Co., Ortho Div., Lucas St. & Ortho Way, Richmond, Calif.
255. Lukens Steel Company, Coatesville, Pa.
256. Eastman Chemical Products, Inc., Subsidiary of Eastman Kodak Co., 260 Madison Ave., New York 16, N.Y.
258. Morton-Withers Chemical Co., Div. Chas. Pfizer & Co., Inc., 2110 High Point Rd., Greensboro, N.C.
259. General Mills, Inc., Chemical Div., S. Kensington Rd., Kankakee, Ill.
260. Kessler Chemical Co., Inc., State Road & Cottman Ave., Philadelphia 35, Pa.
261. American Viscose Corp., P.O. Box 455, Marcus Hook, Pa.
263. Scott Bader & Co. Ltd., Wollaston, Wellingborough, Northamptonshire, England
265. The Goodyear Tire & Rubber Co., Akron 16, Ohio
266. Hercules Powder Company, Inc., 910 Market St., Wilmington 99, Del.
267. Kenrich Petrochemicals, Inc., 57-02 48 St., Maspeth 78, N.Y.
268. The New Jersey Zinc Co., 160 Front St., New York 38, N.Y.
269. Worthington Corporation, Harrison, N.J.
270. The Mearl Corporation, 41 E. 42 St., New York 17, N.Y.
271. American Lignite Products Co., Ione, Calif.
272. Ames Co., Inc., Elkhart, Ind.
274. Beckman Instruments, Inc., 2500 Fullerton Rd., Fullerton, Calif.
275. Cabot Corp., 125 High Street., Boston 10, Mass.
276. Shawinigan Chemicals Limited, P.O. Box 6072, Montreal, Quebec, Canada
277. Xylos Rubber Co., Div. of The Firestone Tire & Rubber Co., Akron 1, Ohio
278. Firestone Synthetic Rubber & Latex Co., 381 W. Wilbeth Rd., Akron 1, Ohio
280. The Carborundum Co., P.O. Box 337, Niagara Falls, N.Y.
281. Continental-Diamond Fibre Corp., Subsidiary of The Budd Co., Newark, Del.
282. Wallace & Tiernan Incorporated, Box 178, Newark 1, N.J.
283. The International Nickel Co., Inc., Huntington Alloy Products Div., Huntington 17, W.Va.
284. West Chemical Products, Inc., West Disinfecting Div., 42-16 West St., Long Island City 1, N.Y.
285. J. M. Huber Corporation, 630 Third Ave., New York 17, N.Y.
287. Beaunit Mills, Inc., Elizabethton, Tenn.
288. Metal & Thermit Corporation, Rahway, N.J.
289. The Lockrey Co., Lubricants Div., Southampton, N.Y.
292. Allied Chemical Corp., Solvay Process Div., P.O. Box 271, Syracuse 1, N.Y.
293. Interchemical Corporation, 67 W. 44 St., New York 36, N.Y.
294. The Glidden Co., Chemicals Div., Hammond, Ind.
296. The Glidden Co., Chemicals Div., 3901 Hawkins Point Rd., Baltimore 26, Md.
299. Chas. Pfizer & Co., 800 Second Ave., New York 17, N.Y.
300. Arkansas Co., Inc., P.O. Box 210, Newark 1, N.J.
301. Burroughs Wellcome & Co. (U.S.A.) Inc., One Scarsdale Rd., Tuckahoe, N.Y.
302. Bee Chemical Company, Logo Div., 2700 E. 170 St., Lansing, Ill.
303. Phillips Petroleum Co., Bartlesville, Okla.
304. National Lead Co., New Products Development, 105 York St., Brooklyn 1, N.Y.
305. Ciba Pharmaceutical Products Inc., Summit, N.J.
306. Hooker Chemical Corporation, Niagara Falls, N.Y.
307. General Aniline & Film Corp., Dyestuff and Chemical Div., 435 Hudson St., New York 14, N.Y.
308. Westinghouse Electric Corporation, East Pittsburgh, Pa.
309. Nopco Chemical Company, 60 Park Pl., Newark 1, N.J.
311. W. R. Grace & Co., Grace Chemical Div., Box 4915, Memphis 7, Tenn.
312. Evans Chemetics, Inc., 250 E. 43 St., New York 17, N.Y.
313. Ethyl Corp., 100 Park Ave. Bldg., New York 17, N.Y.
314. Parr Instrument Co., Inc. 211 53 St., Moline, Ill.

315. Lederle Laboratories, Div. of American Cyanamid Co., Pearl River, N.Y.
317. Bolidens Gruvaktiebolag, Sturegatan 22, Stockholm, Sweden
318. The Aktivin Div., Heyden Newport Chemical Corp., 342 Madison Ave., New York 17, N.Y.
319. Commercial Solvents Corp., 260 Madison Ave., New York 16, N.Y.
321. Schering Corporation, 60 Orange St., Bloomfield, N.J.
322. Kelco Company, 75 Terminal Ave., Clark, N.J.
323. Pittsburgh Chemical Co., A Subsidiary of Pittsburgh Coke & Chemical Co., Grant Bldg., Pittsburgh 19, Pa.
324. Anaconda American Brass Company, Waterbury 20, Conn.
325. Arnold, Hoffman & Co., Inc., 55 Canal St., Providence 1, R.I.
326. U.S. Stoneware Co., Tallmadge Rd., Tallmadge, Ohio
327. The Upjohn Company, 7171 Portage Rd., Kalamazoo, Mich.
328. Onyx Chemical Corporation, 190 Warren St., Jersey City 2, N.J.
329. Mallinckrodt Chemical Works, Second and Mallinckrodt St., St. Louis 7, Mo.
330. Parke, Davis & Company, Joseph Campau Ave. at the River, Detroit 32, Mich.
331. Socony Mobil Oil Co., Inc., 150 E. 42 St., New York 17, N.Y.
332. The National Filter Media Corp., 1717 Dixwell Ave., New Haven 14, Conn.
333. Maas & Waldstein Co., 2121 McCarter Highway, Newark 4, N.J.
334. A. R. Maas Chemical Co., Div. of Stauffer Chemical Co., 4570 Ardine St., South Gate, Calif.
336. Titanium Pigment Corp., Subsidiary of National Lead Co., 111 Broadway, New York 6, New York.
337. National Lead Co., Titanium Alloy Mfg. Div., 111 Broadway, New York 6, N.Y.
341. Pyrene Manufacturing Co., 560 Belmont Ave., Newark 8, N.J.
342. S. B. Penick & Company, 100 Church St., New York 8, N.Y.
343. Parker Rust Proof Company, 2177 E. Milwaukee Ave., Detroit 11, Mich.
344. Mineral Pigments Corp., Muirkirk, Md.
345. W. A. Hammond Drierite Co., Xenia, Ohio
346. Lehn & Fink Products Corp., 192 Bloomfield Ave., Bloomfield, N.J.
347. Kennametal Inc., Latrobe, Pa.
348. Hynson, Westcott & Dunning, Inc., Baltimore 1, Md.
349. Haveg Industries, Inc., 900 Greenbank Rd., Wilmington 8, Del.
350. Driver-Harris Co., Harrison, N.J.
351. The Celotex Corporation, 120 S. LaSalle St., Chicago 3, Ill.
352. Celanese Chemical Co., A Div. of Celanese Corp. of America, 180 Madison Ave., New York 16, N.Y.
353. Catalin Corp. of America, One Park Ave., New York 16, N.Y.
354. Beacon Manufacturing Co., 180 Madison Ave., New York 16, N.Y.
355. Wisconsin Alumni Research Foundation, P.O. Box 2217, Madison 5, Wis.
400. Aceto Chemical Co., Inc., 40-40 Lawrence St., Flushing 54, N.Y.
401. Morton Chemical Company, 110 N. Wacker Dr., Chicago 6, Ill.
402. Calcium Carbonate Co., 520 S. Fourth St., Quincy, Ill.
403. Werner G. Smith, Inc., Chemical Div., 1730 Train Ave., Cleveland 13, Ohio
404. Austenal Co., Div. of Howe Sound Co., 224 E. 39 St., New York 16, N.Y.
405. MONA Industries, Inc., Chemical Div., Paterson 17, N.J.
406. Meer Corporation, 318 W. 46 St., New York 36, N.Y.
407. Ampco Metal, Inc., P.O. Box 2004, Milwaukee 1, Wis.
408. The Chas. Taylor Sons Co., Subsidiary of National Lead Co., P.O. Box 58, Annex Station, Cincinnati 14, Ohio
409. Skelly Oil Co., 605 W. 47 St., Kansas City 41, Mo.
410. A Gross and Co., 295 Madison Ave., New York 17, N.Y.
411. Southern Electrical Co., Metals Div., P.O. Box 989, Chattanooga, Tenn.
412. E. R. Squibb & Sons, 745 Fifth Ave., New York 22, N.Y.
413. Olin Mathieson Chemical Corp., Organics Div., East Alton, Ill.
414. Flamort Chemical Co., 746 Natoma St., San Francisco 3, Calif.
415. Robeco Chemicals, Inc., 25 E. 26 St., New York 10, N.Y.
416. Universal Oil Products Co., 30 Algonquin Rd., Des Plaines, Ill.
417. The Inerto Co., 1489 Folsom St., San Francisco 3, Calif.
418. Roussel Corporation, 155 E. 44 St., New York 17, N.Y.
419. Cadet Chemical Corp., Lockport-Olcott Rd., Burt, N.Y.
420. Union Bag-Camp Paper Corp., Chemical Products Div., 233 Broadway, New York 7, N.Y.
421. Celanese Polymer Co., Div. of Celanese Corp. of America, 180 Madison Ave., New York 16, N.Y.
422. Verona Pharma Chemical Corp., P.O. Box 385, Union, N.J.
423. Insul-Mastic Ink, A Subsidiary of Pittsburgh Chemical Co., Grant Bldg., Pittsburgh 19, Pa.
424. S. C. Johnson & Son, Inc., Racine, Wis.
425. Applied Science Laboratories, Inc., 140 N. Barnard St., State College, Pa.
426. Michigan Chemical Corporation, Saint Louis, Mich.
427. Alberene Stone Corp., 386 Park Ave., South, New York 16, N.Y.
428. Cowles Chemical Co., 7016 Euclid Ave., Cleveland 3, Ohio
429. AviSun Corp., 1345 Chestnut St., Philadelphia 7, Pa.
430. Fine Organics, Inc., 205 Main St., Lodi, N.J.
431. James G. Biddle Co., 1316 Arch St., Philadelphia 7, Pa.

432. Southern Clays, Inc., 33 Rector St., New York 6, N.Y.
433. American Metal Climax, Inc., 61 Broadway, New York 6, N.Y.
434. National Lead Co., Evans Lead Div., Box 1467, Charleston 25, W.Va.
435. New Wrinkle, Inc., 1771 Springfield St., Dayton 3, Ohio
436. Pennsylvania Glass Sand Corp., 375 Park Ave., New York 22, N.Y.
437. Pioneers, Inc., 2411 Grove St., Oakland 12, Calif.
438. Allied Chemical Corp., Harmon Colors, 40 Rector St., New York 6, N.Y.
439. National Lead Co., Nuclear Metals Div., 111 Broadway, New York 6, N.Y.
440. BASF Incorporated, 375 Park Ave., New York 22, N.Y.
441. Pacific Coast Borax Co., Div. of Borax Consolidated, Ltd., 630 Shatto Pl., Los Angeles 5, Calif.
442. Harbison-Carborundum Corp., P.O. Box 337, Niagara Falls, N.Y.
443. Ciba Company, Inc., Pigments Div., Fair Lawn, N.J.
444. Coors Porcelain Co., 600 Ninth St., Golden, Colo.
445. Devcon Corp., Danvers, Mass.
446. Harbison-Walker Refractories Co., Garber Research Center, P.O. Box 98037, Pittsburgh 27, Pa.
447. Calcium Aluminate Cement Corp., 104 E. 40 St., New York 16, N.Y.
448. The Glidden Co., Paint Div., 900 Union Commerce Bldg., Cleveland 14, Ohio
449. Ultra Chemical Works, Inc., Div. of Witco Chemical Co., Inc., 2 Wood St., Paterson, N.J.
450. Hostachem Corporation, 270 Sheffield St., Mountainside, N.J.
451. The Stepan Chemical Company, 427 W. Randolph St., Chicago 6, Ill.
452. Metalsalts Corporation, 200 Wagaraw Rd., Hawthorne, N.J.

Alphabetical List of Manufacturers

	Mfr. No.
Abbott Laboratories	3
Aceto Chemical Co., Inc.	400
Acheson Colloids Co.	46
Advance Solvents & Chemical	230
Air Reduction Chemical & Carbide Co.	144
Aktivin Div., Heyden Newport	318
Alberene Stone Corp.	427
Allied Chem., General Chemical Div.	50
Allied Chem., Harmon Colors	438
Allied Chem., National Aniline Div.	243
Allied Chem., Nitrogen Div.	197
Allied Chem., Plastics Div.	175
Allied Chem., Solvay Process Div.	292
Allis-Chalmers Mfg. Co.	116
Alox Corp.	117
Alpha-Molykote Corp.	199
Aluminum Co. of America	226
American Agricultural Chemical Co.	196
American Celcure Wood Preserving Corp.	77
American Cyanamid Co.	57
American LaFrance Corporation	75
American Lignite Products Co.	271
American Manganese Bronze Co.	111
American Metal Climax, Inc.	433
American Molasses Company	157
American Norit Co., Inc.	107
American Potash & Chemical Corp.	88
American Products Mfg. Co.	176
American Viscose Corp.	261
Ames Co., Inc.	272
Amoco Chemicals Corp.	216
Ampco Metal, Inc.	407
Anaconda American Brass Company	324
APCO Oil Corp.	200
Apex Chemical Co., Inc.	165
Applied Science Laboratories, Inc.	425

	Mfr. No.
Arapahoe Chemicals, Inc.	109
Archer-Daniels-Midland Co.	221
Arizona Chemical Co.	252
Arkansas Co., Inc.	300
Armour Industrial Chemical Co.	15
Arnold, Hoffman & Co., Inc.	325
Atlantic Refining Co.	136
Atlas Mineral Products Co.	41
Atlas Powder Co.	89
Austenal Co.	404
AviSun Corp.	429
Baker Castor Oil Co.	202
Bareco Wax Co.	128
Baroid Div., National Lead Co.	236
BASF Incorporated	440
Beacon Manufacturing Co.	354
Beaunit Mills, Inc.	287
Beckman Instruments, Inc.	274
Bee Chemical Company	302
James G. Biddle Co.	431
Bolidens Gruvaktiebolag	317
Borden Chemical Co.	65
Brazil Oiticica, Inc.	90
Bromine Producers Co.	225
Brown Company	231
Burroughs Wellcome & Co. (U.S.A.) Inc.	301
Cabot Corp.	275
Cadet Chemical Corp.	419
Calcium Aluminate Cement Corp.	447
Calcium Carbonate Co.	402
California Chemical Co., Oronite Div.	151
California Chemical Co., Ortho Div.	253
California Industrial Minerals Co.	118
Canadian Industries, Ltd.	56
Carborundum Co.	280
Carrier Corp.	54

Interstate Commerce Commission Shipping Regulations

In the Condensed Chemical Dictionary many entries include a separate line or section entitled "Shipping Regulations." This states briefly the category of hazard involved for definitely dangerous materials, and the corresponding shipping label. Where no information on shipping regulations is given, or where the statement "Shipping regulations: None" appears, and the shipper has any reason to believe that hazards may exist, we would advise getting in touch with the Bureau of Explosives, New York, or other authorities indicated in the following paragraphs, for information concerning regulations that may possibly have been put in effect since the compilation of this edition of the Dictionary.

The categories of dangerous materials, and their labels, as set up by the Interstate Commerce Commission and the Bureau of Explosives, are indicated in abbreviated terms in the Condensed Chemical Dictionary, as follows:

1. Flammable gas or flammable compressed gas.
   Red gas label.
2. Flammable liquid. Red label.
3. Flammable solid. Yellow label.
4. Oxidizing material. Yellow label.
5. Corrosive liquid. White label.
6. Radioactive material. Red or blue label.
7. Nonflammable compressed gas. Green label.
8. Poison. Poison label.
   Class A. Poison gas or liquid.
   Class B. Poisonous liquid or solid.
9. Explosives.

The Interstate Commerce Commission regulations also include detailed requirements on packaging and containers, on maximum quantities per container and per shipment, and on classes of materials as distinct from individual materials. The Commission's regulations are published under the title: "Regulations for transportation of explosives and other dangerous articles by land and water, in rail freight service, and by motor vehicle (highway), and water, including specifications for shipping containers." Copies of these regulations may be obtained under various plans on application to The Bureau of

Explosives, 63 Vesey Street, New York 7, N.Y., or from the Superintendent of Documents, or other tariff agents. These regulations are amended periodically as new data are obtained and such amendments are described in supplements issued from time to time by the Bureau of Explosives.

The Interstate Commerce Commission regulations are binding upon all common carriers by land or water, including transport by motor vehicles using the public highways.

Regulations as to transportation of dangerous cargo by water are prescribed by the U.S. Coast Guard as per Title 46, Code of Federal Regulations Parts 146 through 149, published by the U.S. Government Printing Office, and obtainable from the Superintendent of Documents, Washington D.C. The National Cargo Bureau Inc., 99 John St., New York 38, N.Y., provides information and recommendations regarding cargo stowage.

Rulings as to what products, and in what quantities, may be transported by air are made by the U.S. Civil Aeronautics Board but the packaging and labeling of such air shipments is subject to the regulations of the Interstate Commerce Commission as interpreted by the Bureau of Explosives.

The Board of Transport Commissioners for Canada has adopted regulations identical with those of the U.S. Interstate Commerce Commission as interpreted by the Bureau of Explosives, so consequently the information on this subject in this book applies generally to Canada as well as the United States.

Various states, municipalities, and other public agencies such as those responsible for tunnels, etc., have special regulations and laws of various types.

Interstate Commerce Commission. Regulations for the transportation of explosives and other dangerous articles are prescribed by the Interstate Commerce Commission under the authority of the federal law, Act of Congress, approved June 25, 1948.

Section 835 of this law states, "The Interstate Commerce Commission shall formulate regulations for the safe transportation within the limits of the jurisdiction of the United States of explosives and other dangerous articles ... which shall be binding upon all common carriers engaged in interstate or foreign commerce, and upon all shippers.

"The Commission ... may make changes ... made desirable by new information or altered conditions.

"Such regulations shall be in accord with the best known practicable means for securing safety in transit, covering the packing, marking, loading, handling while in transit, and the precautions necessary to determine whether the material when offered is in proper condition to transport....

"The Interstate Commerce Commission may utilize the services of the Bureau for the Safe Transportation of Explosives and Other Dangerous Articles, and may avail itself of the advice and assistance of any department, commission, or board of the Government."

Bureau of Explosives. The Bureau of Explosives was organized by the American Railway Association in 1906 to secure safety in the transportation of explosives. Its membership consists of practically all of the railroads and express companies in the United States and Canada, and a large number of steamship companies, as well as manufacturers of explosives and other dangerous articles, manufacturers of shipping containers for such articles, and trade associations comprised of such manufacturers. The Bureau of Explosives endeavors, through its corps of inspectors, its chemical laboratory and its headquarters to educate shippers' and carriers' employees in the requirements of the regulations, and to secure their enforcement.

Method of Procedure. Regulations are ordinarily prepared by the Bureau of Explosives after consultation with shippers and others interested and are then submitted to the Interstate Commerce Commission by the Bureau of Explosives for approval and publication. These regulations and the Act of September 1, 1948, place upon the shipper of a dangerous article the duty of properly preparing a shipment for transportation, knowing its characteristics, and instructing his employees in the requirements of the regulations and securing compliance therewith. If a shipper has no knowledge of the dangerous characteristics of the material which he desires to ship, he should confer with the Bureau of Explosives and, if necessary, submit a sample for examination and report.

Interpretation of the Regulations and Their Use in the Condensed Chemical Dictionary. Although these regulations have been drawn with great care, nevertheless, in their interpretation, differences of opinion may arise even among experts. Many factors must be considered in determining whether a substance is hazardous. A small amount of an oil in a bottle in a laboratory may not present any fire hazard, whereas a 50,000-gallon tank of the same oil may.

The editors and publishers of this book believe that the interpretation given is the most comprehensive and accurate attempt made to guide shippers and other handlers of materials in bulk to ship their product in accordance with the law.

Nevertheless, these interpretations are not offered, nor must they be construed, as official publications either of the Interstate Commerce Commission or of the Bureau of Explosives. Accordingly, the obligation to definitely determine the hazardous character of a commodity rests with the shipper.

Editors' Statement

The editors take pleasure in acknowledging the assistance of the many persons who helped make the task of this latest revision possible. Appreciation is expressed to the managements of the several hundred companies whose products are listed and to their chemists, engineers, and the sales, library, patent and legal representatives who provided commercial information not otherwise available. Especially helpful scientific assistance was given by Warren W. Miller, W. S. Hodgkiss, C. H. Jeglum, H. B. Comstock, H. A. R. Zehrlaut, H. J. Poel, D. F. King, B. A. Cohrssen, R. E. Maizell, D. M. C. Reilly, E. A. Carpovich and T. C. George.

The highly essential and complex tasks of screening, assembling, confirming, editing, cross checking and proofreading were done with the help of Jo N. Hays, Virginia B. Hetrick, Jeanne B. O'Brian, and Ursula Miller. The majority of the typing for photoreproduction was done by Beverly Lightner, Shirley Kempl and Marilyn Lent.

Suggestions for further improvement will be welcomed.

Elizabeth Rose

Arthur Rose

Safety Information ∿ Warnings and Labels

M.C.A. Warning Labels. To achieve uniform and more adequate labeling of hazardous chemical products the Labels and Precautionary Information (LAPI) Committee of the Manufacturing Chemists' Association, Inc., Washington, D.C., has prepared a manual entitled "Warning Labels" for the benefit and guidance of its members. In this dictionary we have quoted the permissive Danger, Warning, and Caution labels advised (1956 edition) and information as to their use.

Individual statutes, regulations or ordinances may require that particular information be included in a label or that a specific label be affixed to a container. In each case, the requirements of these laws should be studied. The warning labels suggested in this manual should be used in addition to, or in combination with, any label required by law.

Federal statutes and regulations affecting the labeling of chemical materials include Federal Caustic Poison Act and regulations; Federal Insecticide, Fungicide, and Rodenticide Act and regulations; Federal Food, Drug and Cosmetic Act and regulations; as well as the Interstate Commerce Commission Regulations for Transportation of Explosives and Other Dangerous Articles and others. Copies are obtainable at nominal cost from the Superintendent of Documents, Washington, D.C., or directly from the agencies that administer the laws.

State and local governments frequently regulate chemical labeling through statutes, ordinances, and regulations affecting poisons; insecticides, fungicides, rodenticides and herbicides; foods, drugs, and cosmetics; agricultural and horticultural materials; the practice of pharmacy; and other subjects. Copies are usually obtainable from state or local Departments of Health, Agriculture, Pharmacy, or other regulatory agencies.

The safety information given does not purport to indicate relative degrees of hazard of a product or the manner in which it may prove to be hazardous. The omission of such warning does not mean that the substances are harmless, especially if they are improperly handled. This book is not intended to furnish complete toxicity data on chemical substances. Such information should be obtained from the supplier or manufacturer of each product.

Trademarks

The constantly increasing development and use of trademarks, trade names and brand names to designate chemical products has played an important part in chemistry and its related fields.

For this reason, the publisher has incorporated in this book the descriptions of many chemical specialties and products which have come to be identified either officially or unofficially by trademarks, trade names or brand names claimed to connote origin of product, as distinguished from chemical composition.

In such cases the trademark, trade name or brand name has been enclosed in quotation marks, followed by a description of the product and a reference by number to the alleged or reputed manufacturer of the product so described. While every reasonable effort for accuracy has been made in such cases, the absence of trademark, trade name or brand name designation does not indicate that proprietary rights may not exist in the word involved. Neither the editors nor the publisher assume any responsibility for the accuracy of any such description or for the validity or ownership of any such trademark, trade name or brand name.

Reinhold Publishing Corporation

Explanation of Arrangement

Articles are in a straight alphabetical arrangement. This is strictly according to the complete title of the article, written as if it were one word, with exceptions noted below. For example, the article on **waxes** would follow **wax distillate; woodbine** precedes **wood copper,** etc.

Exceptions to the alphabetical arrangement are the usual organic prefixes: ortho-, meta-, para-, alpha-, beta-, etc., sec-, tert-, sym-, as-, uns-, cis-, trans-, d-(dextro-), l-(levo-), n-(normal), N-(attachment to the nitrogen atom), C-(attachment to the carbon atom), O-(attachment to the oxygen atom), and all numbers denoting structure.

Thus alpha is disregarded inside as well as outside the word, so **dimethyl-**alpha**-naphthylamine** is filed as **dimethylnaphthylamine** and follows **dimethylmethane.** Of course, where names are identical except for prefixes, they are filed accordingly, so ortho-, meta-, and para-**dichlorobenzenes** are filed in the order meta-, ortho-, and para-.

The prefixes iso-, di-, tri-, tetra-, cyclo-, and bis- are considered integral parts of the names, so **dimethylamine** is under D, and **isobutane** under I. Similarly, the prefixes for inorganic salts are considered part of the name, as **sodium orthophosphate,** filed as "s-o-d-i-u-m-o-r."

There is no grouping of articles under the larger classes such as the alcohols or oils. However, minor class terms are often inverted, as **starch, chlorinated; starch, permanent.** Where an article is headed by a two-name title, as **starch, permanent,** no attempt has been made to cross reference it under the second name, **permanent starch.**

Entries containing numbers only are listed after the Z entries. Otherwise numbers are ignored in making the alphabetical arrangement. Thus "**3M Brand**" is alphabetized as "m-b-r-a-n-d," "**PE-16-Clear**" is alphabetized as "p-e-c-l-e," "**#20 Hot Galvanizing Flux**" is alphabetized as "h-o-t-g-a," and "**O & W Compound**" is alphabetized as "o-w-c-o-m."

Thus, alpha is disregarded [illegible] as well as outside the [illegible] dimethylamine is filed as [illegible] methylmethanine [illegible] In cases where [illegible] are filed in the order [illegible] and para-.

A. Abbreviation for angstrom (q. v.).

A. Formerly symbol for argon, now Ar.

"A-1." [58] Trade name for thiocarbanilide, (q. v.).

"A-32." [58] Trade name for a rubber accelerator, a reaction product of butyraldehyde and butylidene aniline, of complex chemical structure.
Properties: Red-yellow to orange-brown oily liquid, slightly turbid when cold; sp. gr. 0.98; flash point approx. 175°F; moisture, no separation on standing; soluble in benzene, chloroform, acetone, and solvent naphtha.
Containers: 150-, 400-lb steel drums.
Uses: Rubber accelerator; often useful as an activator of thiazole-type accelerators. Excellent for hard rubber.

"A-100." [58] Trade name for a reaction product of butyraldehyde, acetaldehyde, and aniline.
Properties: Dark reddish-brown oily liquid; sp. gr. 1.04; flash point approx. 185°F; moisture, no separation on standing; soluble in benzene, chloroform, acetone, and solvent naphtha.
Containers: 150-lb steel drums.
Uses: Accelerator for hard rubber vulcanizates, either natural or synthetic.

AA. Abbreviation for allyl alcohol; also for adenylic acid.

"AA" Oil. [202] Trademark for a cold pressed #1 castor oil having light color, high purity and low acidity, meeting U.S.P. and Federal Specification JJJ-C-86 (Grade 1) requirements.

"AA Quality." [196] Trademark for a line of fertilizers of various compositions, types and grades; also used for ground phosphate rock.

abaca (Manila hemp). The strongest of all vegetable fibers, obtained from the leaves of Musa textilis, a tree of the banana family. The fibers are 4-8 ft long, light in weight, soft, lustrous, nearly white in color, and do not swell or stiffen when wet.
Sources: Philippines, Central America, Sumatra.
Grades: Sold in 18 grades based on color and length.
Uses: Heavy cordage and twine, especially for marine use; manila paper; fine tissue paper.
Shipping regulations: None.*

"Abalyn." [266] Trademark for a pale liquid resin, a methyl ester of rosin used as a solvent, penetrant, and plasticizer.

"Abasin." [162] Trademark for acetylcarbromal (q. v.).

abelmoschus. See ambrette seed.

Abel's reagent. An etching reagent used in the microanalysis of carbon steels. It consists of a 10% solution of chromic acid (CrO_3).

abies bark. The bark of firs and spruces, used in tanning.

abietates. Salts of abietic acid. Abietic acid is the most abundant acid in the mixture obtained by treating rosin with acetic acid. If the crude acid is employed the metal abietates are identical with resinates, although theoretically an abietate is a definite compound derived from abietic acid, $C_{19}H_{29}COOH$.

abietic acid. (abietinic acid; sylvic acid) $C_{19}H_{29}COOH$ (having a phenanthrene ring system). A major active ingredient of rosin, where it occurs with other acids of closely related structure and properties, i.e., the resin acids. The term abietic acid is often applied to these mixtures, separation of which is difficult and not achieved in technical grade material.
Properties: Yellowish resinous powder; m.p. 172-175°C; optical rotation −106°; soluble in alcohol, ether, chloroform, and benzene; insoluble in water.
Derivation: Rosin, colophony, pine resin; tall oil.
Method of purification: Crystallization.
Grades: Technical.
Containers: Wooden kegs; drums; multiwall paper sacks.
Uses: Resinates of heavy metals as varnish driers; esters in lacquers and varnishes; fermentation (lactic and butyric acid ferments' growth promoter, preventive of raw material infection and decomposition); also in manufacture of soaps.
Shipping regulations: None.*

abietic acid, ethyl ester. See ethyl abietate.

abietinic acid. See abietic acid.

"Abitol." [266] Trademark for a colorless, tacky, very viscous liquid; mixture of tetra-, di-, and dehydroabietyl alcohols made from rosin.
Properties: Viscous, pale, sticky liquid; hydroxyl value, approximately 5%; acid

*See "I.C.C. Shipping Regulations," page xiii.
Reference numbers refer to name of manufacturer. See "List of Manufacturers," page v.

number, about 0.3; color (Lovibond, 50 mm tube) 0.5 amber; sp. gr. (20/20°C) 1.007; refractive index 1.528 (20°C); viscosity (200°C) 30 poises, (270°C) 1.85 poises, (340°C) 0.15 poises. Soluble in and miscible with a wide variety of organic materials.

Uses: Suggested uses are plasticizers, tackifiers, resins, and adhesive modifiers.

ablation compounds (as used in space technology). Materials used as coatings on rockets and missiles for the purpose of cooling their outer surfaces. For instance, a base of magnesium oxide can be protected by a coating of iron, nickel and chromium oxides. These evaporate preferentially up to 4500°F, leaving a thin layer of molten magnesium oxide over the sintered base.

A-bomb. See atomic bomb.

"Abopon." [73] Brand name for proprietary product. Sodium boro-phosphate complex. Water-white viscous liquid.

Properties: Sp. gr. (25°C) 1.68; pH (10% solution) 7.7. Soluble in water, diethylene glycol, glycerin; insoluble in ethyl alcohol, methyl alcohol, toluol, mineral spirits, mineral oil, vegetable oil.

Containers: 1-gal can (14 lbs); 5-gal can (70 lbs); 55-gal drum (800 lbs).

Uses: Paper (flameproofing agent, adhesive, sizing, stiffening, glazing, weighting); textiles (scouring); cosmetics (replaces gums in finger-waving solutions); polishes (abrasive and pigment suspensions); paints, lacquers, etc. (sealer for porous surfaces prior to painting); insulation (flameproofing agent for cotton coatings for insulating wire. Poor conductor of heat and electricity and can be used as a binder for other insulated materials). Suggested for water inks, metal cleaning, suspending and binding agent for coloring materials for ceramics and glass enamels.

abradants. See abrasives.

abrasives (abradants). Substances used to wear off or grind objects in order to give them the desired size, shape, or finish. The smoothness of the finish produced by an abrasive depends upon the size or coarseness of the grains. The principal factors in abrasive power, in order of importance, are hardness, brittleness, and refractoriness. The hardness of the abrasive determines what materials can be ground with it since the grains will not cut or scratch anything harder than themselves (see Mohs' scale). Brittleness, with the resulting fracture of the abrasive grains under stress, presents fresh, sharp surfaces, increasing the efficiency of grinding. Refractoriness of the abrasive grain is its resistance to deterioration under the high, local temperatures produced during grinding; it is often less important than the strength and temperature resistance of the bonding agent used to hold the abrasive grains together. The principal types of bonds are: silicate (waterglass); vitreous or ceramic; shellac; rubber; resinoid (especially alkyd resin); and metallic (made by powder metallurgy). As examples of abrasives, see also boron carbide, corundum, diamond, and silicon carbide.

absinthe. Green liqueur containing oils of wormwood, angelica, anise and marjoram. Toxic. Its manufacture is prohibited by law in the United States.

absinthe oil. See wormwood oil.

absinthin $C_{30}H_{40}O_8$. A glycoside from absinthium.

Properties: Lustrous needles; very bitter taste. M.p. 68°C; slightly soluble in water and petroleum ether; soluble in alcohol, ether, benzene, and chloroform.

Derivation: Extraction from absinthium.

Containers: Glass bottles.

Uses: Medicine; flavoring.

absinthium (wormwood). Leaves and tops of wormwood plant, Artemisia absinthium. Also used for the plant itself and for the oil distilled from the leaves and tops.

Occurrence: Europe, northern and western Asia, and Africa; cultivated in United States.

Grades: Technical.

Containers: Bags; boxes.

Uses: Manufacture of absinthin and absinthe; essential oils (raw material); flavor in beverages and condiments; perfumery (aromatic waters and lotions); medicine.

Shipping regulations: None.*

absolute (as in perfumery). See concrete (2).

absolute alcohol. Expression for dehydrated ethyl alcohol, at least 99% pure.

absolute temperature. The fundamental temperature scale used in theoretical physics and chemistry, and in certain engineering calculations such as the change in volume of a gas with temperature. Absolute temperatures are expressed either in degrees Kelvin or in degrees Rankine, corresponding respectively to the centigrade and Fahrenheit scales. Degrees Kelvin are obtained by adding 273 to the centigrade temperature, while degrees Rankine are obtained by adding 460 to the Fahrenheit temperature. The nearest practical approach to the absolute zero is the melting point of helium which is below −272°C.

"Abson." [119] Trade name for a line of acrylonitrile-butadiene-styrene (ABS) resins and compounds. Tough, dimensionally stable over a wide temperature range, chemical resistant and light weight. Used in automotive, appliance and business machine industries.

absorption. Most commonly means the taking up of a gas or vapor by a liquid (physical or physiochemical absorption), or the taking up of energy (heat, light, x-rays) by any material (spectral absorption).

Ammonia is separated from coal gas by passage through water or sulfuric acid

*See "I.C.C. Shipping Regulations," page xiii.

Reference numbers refer to name of manufacturer. See "List of Manufacturers," page v.

solutions which take up (absorb, dissolve) the ammonia. As in this example, chemical combination often accompanies absorption, but there are very many instances where the process is entirely physical. The occlusion of hydrogen by certain metals (palladium) is usually termed absorption.

Absorption should be distinguished from adsorption, in that the latter is a surface phenomenon, i.e., the material taken up is distributed over the surface of the adsorbing material. In absorption the material taken up is distributed throughout the body of the absorbent.

Spectral absorption is illustrated when certain wave lengths of sunlight are absorbed by ordinary glass, thereby warming it slightly (the light energy is changed to heat) and depriving the transmitted light of certain wavelengths, particularly those of the ultraviolet spectrum. Transparent liquids are often identified and analyzed by passing a beam of light through them and noting the extent of absorption of original light.

absorption oils (scrubbing oil; wash oil). Generally refer to a moderately high boiling oil distilled from petroleum (i.e., a gas oil) or coal tar, and used for separating desired gases or vapors by dissolving them from some mixture. Thus the vapors of natural gasoline are separated from certain natural gases by passage up a tower through which a stream of an absorption oil is passed; and benzene, toluene, and xylene are recovered from coal gas by a similar procedure.

ABS resins. Acrylonitrile-butadiene-styrene resins (q.v.).

Abyssinian gold
1. (Talmi gold). Brass having a thin facing of gold applied by rolling; used for costume jewelry.
2. A yellow or gold-colored aluminum bronze containing 5-10% aluminum, the remainder being copper.

Ac. Symbol for actinium.

AC. Abbreviation for allyl chloride.

acacia. See arabic, gum.

acacia bark. Bark of acacia tree, used as an astringent because of its tannin content.

acacia cavenia. See cassie oil.

acacia farnesiana. See cassie oil.

acajou nut. See semecarpus nut.

acaricide. A substance, such as tetraethylpyrophosphate, having the power to kill acarids, i.e., mites and ticks.

acaroid resin. See accroides gum.

"Accel." [123] Trademark for a lactic acid starter culture for use in food processing.

accelerator
1. A substance which accelerates the vulcanization of rubber or permits vulcanization at lower temperature, thus reducing time and cost of manufacture and improving the finished products. Accelerators comprise various organic compounds of nitrogen and sulfur, among the most important of which are diphenylguanidine, hexamethylenetetramine, mercaptobenzothiazole, tetramethyl- and tetraethylthiuram disulfides, thiocarbanilide, and zinc dimethyldithiocarbamate. A few inorganic types are still used to a minor extent, e.g., antimony pentasulfide, calcium oxide, magnesium oxide, and zinc oxide.
2. See particle accelerator.

Accelerator "8." [28] Trademark for formaldehyde-para-toluidine.
Properties: White powder.
Containers: Drums (100 lbs, net).
Use: To accelerate and improve the vulcanization of natural rubber.

"Accelerator 49." [57] A proprietary product. Di-substituted guanidine half way between diphenylguanidine and di-ortho-tolylguanidine in accelerating strength. Used as a primary accelerator and as an activator for other primary accelerators.

Accelerator "552." [28] Trademark for piperidinium pentamethylene dithiocarbamate $\underbrace{CH_2(CH_2)_4}NC(S)S\underbrace{NH_2(CH_2)_4C}H_2$. A cream-colored crystalline powder; sp. gr. 1.20; melts with decomposition at not lower than 167°C.
Use: To accelerate vulcanization of natural and synthetic rubber and latex compounds; as peptizer or plasticizer for neoprene.
Containers: 125-lb drums.

Accelerator "808." [28] Trademark for butyraldehyde-aniline condensation product.
Properties: Amber liquid.
Containers: Drums (250 lbs, net).
Use: To accelerate and improve the vulcanization of natural and synthetic rubber and latex compounds.

Accelerator "833." [28] Trademark for butyraldehyde-monobutylamine condensation product. Reddish-amber liquid; sp. gr. 0.86.
Containers: 225-lb drums.
Use: To accelerate and improve the vulcanization of natural and synthetic rubber, especially for neoprene type cements.

"Accelerator B." [58] Brand name of a rubber accelerator consisting of 2 parts "Thiurad" and 1 part "Thiotax," the ratio commonly used for butyl rubber stocks.
Containers: 150-lb fiber drums.

"Accobond" 3900 Cellulosic Film Resin. [57] Trademark. Aqueous cationic melamine-formaldehyde resin syrup soluble in all proportions in water.
Use: As an agent for bonding coatings and printings to cellulosic films.

"Accobrite" Rosin Sizes. [57] Trademark for a series of pale or light colored rosin sizes for the paper industry.

"Accocel" 741 Dispersant. [57] Trademark for a complex sulfonic acid condensate used in controlling pitch troubles in paper making, and dispersion of inks, pigments, and asphalt.

"Accomite." [57] Trademark for a blasting agent.

"Acco" Rosin Sizes. [57] Trademark for a series of dry and liquid rosin sizes for the paper industry.

"Accosperse." [57] Trademark for a series of aqueous dispersions of chemically manufactured pigments. These dispersions, shipped in polyethylene-lined containers of 30-gal capacity, find use in latex paints and other aqueous systems.

"Accostrength." [57] Trademark for a synthetic water-soluble polymer used to improve the dry strength of paper.

"Acco" Streptomycin D. [57] A proprietary product, contains not less than 45% active streptomycin as the sulfate salt.
Properties: A light tan powder; soluble in water.
Containers: 12.5-lb can.
Uses: For agricultural usage in the control of plant diseases.

accroides gum (black-boy gum; xanthorrhea resin; acaroid resin; Botany Bay gum).
Properties: Red or yellow gum; soluble in alcohol.
Derivation: A resin obtained from several species of the xanthorrhea tree (Australia grass tree).
Occurrence: Australia.
Grades: Technical.
Containers: Bags.
Uses: Varnishes; lacquers; rosin substitute; leather (finishing agent); sealing wax compositions; paper (finishing agent); toilet soaps; medicine.
Shipping regulations: None. *

accuracy. The extent to which a measured or enumerated value differs from the true value, the true value being assumed or accepted on the basis of independent evidence. See also precision.

"Ace Alkali." [292] Trademark for a solid alkaline composition containing predominantly caustic alkali. Used for machine washing of bottles and other heavy duty cleansing.

acecoline. See acetylcholine chloride.

"Acele." [28] Trademark for a cellulose acetate fiber available in natural and color-sealed normal tenacity continuous filament yarns in various deniers and lusters.
Properties: Sp. gr. 1.32; tensile strength (psi), 18,000-24,000; break elongation 28%; moisture regain 6%; soluble in glacial acetic acid, acetone, acetonitrile, butyrolactone, dimethyl formamide; dioxane-1,4.
Containers: Tubes and cones in cases; beams.
Uses: Textiles.

acenaphthene (naphthyleneethylene; ethylenenaphthalene) $C_{10}H_6(CH_2)_2$.
Properties: White needles; sp. gr. 1.024 (99/4°C); freezing point 93.6°C; b.p. 277.5°C; refractive index (100°C) 1.6048. Soluble in hot alcohol; insoluble in water.
Derivation: From coal-tar.
Grades: Technical; 98%.
Containers: Wooden barrels or fiber drums.
Uses: Dye intermediates; chemicals; pharmaceuticals; insecticide; fungicide; plastics; horticulture.
Shipping regulations: None. *

1,2-acenaphthenedione. See acenaphthenequinone.

acenaphthenequinone (1,2-acenaphthenedione) $C_{10}H_6(CO)_2$.
Properties: Yellow needles; m.p. 261-263°C; insoluble in water; soluble in alcohol.
Derivation: By oxidizing acenaphthene, using glacial acetic acid and sodium or potassium dichromate.
Grades: Technical.
Use: Dye synthesis.

acetal (diethylacetal; 1,1-diethoxyethane; ethylidenediethyl ether) $CH_3CH(OC_2H_5)_2$.
Properties: Colorless, volatile liquid; agreeable odor; nutty after-taste. Stable to alkalies but readily decomposed by dilute acids. Forms a constant boiling mixture with ethyl alcohol. Soluble in alcohol, and ether; sparingly soluble in water.
Constants: Sp. gr. 0.831; b.p. 103-104°C; vapor pressure 20.0 mm (20°C); flash point (closed cup) 37°F; specific heat 0.520; refractive index 1.38193 (20°C); wt (lbs/gal) 6.89. Typical specifications: Acetal 97% min.; boiling range 97-112°C; color water-white; metals none; chlorides, sulfates none; water none; sp. gr. 0.826-0.830 at 20°C; wt/gal 6.89 lbs at 20°C.
Derivation: By the partial oxidation of ethyl alcohol, the acetaldehyde first forming condensing with the alcohol.
Grades: Technical.
Containers: Nonreturnable: 1-gal cans, net wt: 6 lbs; 5-gal cans, net wt: 30 lbs; 55-gal drums, net wt: 365 lbs. Returnable: 5-gal carboys, net wt: 30 lbs; 12-gal carboys, net wt: 75 lbs.
Danger: Extremely flammable.
Uses: Medicine; solvent; cosmetics; organic synthesis; perfumes.
Shipping regulations: Flammable liquid. Red label. *

acetaldehyde (acetic aldehyde; aldehyde; ethanal; ethyl aldehyde) CH_3CHO.
Properties: Colorless, flammable liquid; pungent, fruity odor. Sp. gr. 0.783 (18/4°C); b.p. 20.2°C; m.p. −123.5°C; vapor pressure 740.0 mm (20°C); flash point −40°F (open cup); specific heat 0.650; refractive index 1.3316 (20°C); wt. 6.50 lbs/gal (20°C); miscible with water, alcohol, ether, benzene, gasoline, solvent naphtha, toluene, xylene, turpentine and acetone.

*See "I.C.C. Shipping Regulations," page xiii.
Reference numbers refer to name of manufacturer. See "List of Manufacturers," page v.

Derivation: (a) Oxidation of ethyl alcohol vapor over platinum black or other catalyst; (b) direct oxidation of propane and butane; (c) hydration of acetylene by means of mercuric sulfate or ferric sulfate catalysts, or by high-pressure reaction with an alcohol; (d) by-product in the fermentation production of ethyl alcohol; oxidation of ethylene over a platinum catalyst.
Grades: Technical; 99%.
Containers: 5-, 10-, 55-, and 110-gal steel drums; 10,000-gal tank cars.
Uses: (in approximate order of volume): Acetic acid; n-butyl alcohol; acetic anhydride (other than that obtained from acetic acid); 2-ethylhexanol; pentaerythritol; 2-methyl-5-ethylpyridine; chloral; phenol and urea condensation products; intermediate for drugs, perfumes, photographic agents.
Danger! Extremely flammable. May form explosive peroxides under air pressure. MCA warning label.
Shipping regulations: Flammable liquid. Red label. *

acetaldehyde ammonia. See aldehyde ammonia.

acetaldehyde cyanohydrin. See lactonitrile.

acetaldol. See aldol.

acetal resins. A type of polymers or synthetic resins obtained by controlled polymerization of formaldehyde (CH_2O) to obtain a linear molecule of the type $-O-CH_2-O-CH_2-O-CH_2-$. Single molecules may have over 1000 $-CH_2O-$ units. The molecule has no side chains so that dense crystals are formed causing acetal resins to be much more like metals than any other resin. These polymers are comparatively hard, rigid, strong, tough and resilient (not brittle), dimensionally stable under exposure to moisture and heat, resistant to chemicals and solvents, resistant to flexing and creep, and have a high gloss and slippery low friction surface.
Typical Properties: Sp. gr. 1.425; thermal conductivity 0.13 Btu/hr/sq ft/°F/ft; coefficient of thermal expansion 4.5×10^{-5} per °F; specific heat 0.35 Btu/lb/°F; water absorption 0.41%/24 hrs; flammability 1.1 in/min; tensile strength 10,000 psi; elongation 15%; hardness (Rockwell) R120; impact strength (notched) 1.4 ft lb/in; flexural strength 14,100 psi; shear strength 9,500 psi.
Uses: In appliance parts; gears; bushings; bearings; movie projector and typewriter parts. See also "Delrin."

acetamide (acetic acid amine, ethanamide) CH_3CONH_2.
Properties: Colorless deliquescent crystals. Mousy odor. Soluble in water and alcohol; slightly soluble in ether.
Constants: Sp. gr. 1.159; m.p. 82°C; b.p. 223°C; refractive index 1.4274 (78.3°C).
Typical specifications (technical grade): Acetamide 99% min; free acid, acetic 0.3% max; chlorides none; sulfates none; color grayish; odor slight, mousy; m.p. 77-79°C.
Typical Specifications (C.P. odorless grade): Acetamide 99.5-99.9%; free acid, acetic trace; chlorides none; sulfates none; color white; odor none; m.p. 79-81°C; nonvolatile 0.04% max.
Derivation: By the interaction of ethyl acetate and ammonium hydroxide.
Method of purification: Crystallization.
Grades: Technical; C.P. odorless.
Containers: Nonreturnable: 5-, 25-, 50-, 100-lb fiber cartons; 200-lb fiber containers. All net weight.
Uses: Organic synthesis (reactant, solvent, peroxide stabilizer); general solvent; lacquers; explosives; soldering flux; hygroscopic agent; wetting agent; penetrating agent.
Shipping regulations: None. *

acetamido-. Prefix for CH_3CONH-. Also called acetamino- or acetylamino-.

5-acetamido-8-amino-2-naphthalenesulfonic acid. $C_{10}H_5(SO_3H)(NH_2)(CH_3CONH)$.
Properties: Reddish-brown paste containing approximately 40% solids.
Grade: Technical.
Use: Intermediate.

para-**acetamidobenzenesulfonyl chloride.** See para-acetylaminobenzenesulfonyl chloride.

para-**acetamidobenzoic acid.** See para-acetylaminobenzoic acid.

acetamidocyanoacetic ester. See ethyl acetamidocyanoacetate.

3-acetamido-4-hydroxybenzenearsonic acid. See acetylaminohydroxyphenylarsonic acid.

8-acetamido-2-naphthalenesulfonic acid magnesium salt $[C_{10}H_6(CH_3CONH)(SO_3)]_2Mg$.
Properties: Brownish-gray paste containing approximately 80% solids.
Grade: Technical.
Use: Intermediate.

para-**acetamidophenol.** See para-acetylaminophenol.

5-acetamido-1,3,4-thiadiazole-2-sulfonamide. See acetazolamide.

2-acetamidothiophene $C_4H_3S(CH_3CONH)$.
Properties: Fine tan crystals; practically odorless; stable; m.p. 158-160°C.
Use: Organic synthesis.

3-acetamido-2,4,6-triiodobenzoic acid. See acetrizoic acid.

"Acetamine." [28] Trademark for a group of azo dyes and developers made for application to acetate yarn, and especially suited to the coloration of nylon.

acetamino-. See acetamido-.

acetaminophen. See para-acetylaminophenol.

para-**acetaminophenetol.** See acetophenetidin.

para-**acetaminophenyl allyl ether** (allyl para-acetylaminophenolate; allyl para-acetaminophenolate) $C_6H_4(NHCOCH_3)(C_3H_5O)$.

*See "I.C.C. Shipping Regulations," page xiii.
Reference numbers refer to name of manufacturer. See "List of Manufacturers," page v.

Properties: Scales or plates; m. p. 94°C; soluble in alcohol; less so in water.

acetanilide (N-phenylacetamide; antifebrin) $C_6H_5NH(COCH_3)$.

Properties: White, shining crystalline scales or white, crystalline powder; odorless; stable in air; slightly burning taste; sp. gr. 1.2105; m.p. 114-116°C; b.p. 305°C; soluble in hot water, alcohol, ether, chloroform, acetone, glycerol and benzene.

Typical specifications: Melting range 112-114°C, ash 0.05% max; completely soluble in 95% ethyl alcohol.

Derivation: By the acetylation of aniline with glacial acetic acid.

Method of purification: Crystallization.

Grades: Technical; C.P.

Containers: 1-lb cartons; 150-, 200-lb barrels; bottles; fiber drums; multiwall paper sacks.

Use: Medicine; rubber accelerator; preservative for hydrogen peroxide; stabilizer for cellulose ester "dopes" and lacquers; manufacture of intermediates (para-nitroaniline, para-nitroacetanilide, para-phenylenediamine); synthetic camphor; pharmaceutical chemicals; dyestuffs; precursor in penicillin manufacture.

Shipping regulations: None. *

acetanisidine. See methacetin.

acetanisole. See para-methoxyacetophenone.

acetarsone. See acetylaminohydroxyphenylarsonic acid.

acetate

1. A compound derived from acetic acid, CH_3COOH, by replacing the acid hydrogen by a metal or a radical, so that the resulting compound contains the acetate radical or group (CH_3COO-). See ethyl acetate, copper acetate, etc.

2. Generic name for a manufactured fiber in which the fiber-forming substance is cellulose acetate. Where not less than 92% of the hydroxyl groups are acetylated, the term triacetate may be used as a generic description of the fiber (Federal Trade Commission). This fiber was formerly called "acetate rayon" or "acetate silk." The term "rayon" may not now be used for this type fiber.

acetate C-8. See n-octyl acetate.

acetate C-9. See nonyl acetate.

acetate C-10. See decyl acetate.

acetate C-11. See undecylenyl acetate.

acetate C-12. See dodecyl acetate.

acetate dyes. These are for the most part insoluble azo or anthraquinone dyes that have been highly dispersed to make them capable of penetrating and dyeing acetate fibers. A second class is insoluble amino azo dyes that are made water soluble by treatment with formaldehyde and bisulfite. After absorption by the fiber the resulting sulfonic acids hydrolyze and regenerate the insoluble dyes.

acetate fiber. See acetate (2).

acetate film. A durable, highly transparent film with nondeforming characteristics, produced from cellulose acetate resin. It is grease-, oil-, dust-, and air-proof and hygienic.

Available forms: Rolls and cut-to-size sheets.

Uses: Principally for laminations; foliation; document preservation; pressure sensitive tape; magnetic sound recording tape; window cartons and envelopes; packaging of items, such as textiles, paper specialties, tomatoes, avocados, mushrooms and for bottle overwraps.

acetate green. A chrome-green pigment with yellowish blue tone in which the yellow is made from lead acetate.

acetate of lime. Commercial term for calcium acetate and sometimes applied particularly to the calcium acetate made from pyroligneous acid and milk of lime. There are a brown and gray acetate of lime. For further data see calcium acetate.

acetate process. A process for making cellulose acetate resin or fiber by treating cellulose (wood pulp or cotton linters) with acetic acid, acetic anhydride, and sulfuric acid as catalyst. The cellulose is fully acetylated (three acetate groups per glucose unit) and at the same time the sulfuric acid causes appreciable degradation of the cellulose polymer so that the product contains only 200-300 glucose units per polymer chain. At this point in the process the cellulose acetate ordinarily is partially hydrolyzed by the addition of water until an average of 2-2.5 acetate groups per glucose unit remain. This product is thermoplastic and soluble in acetone. Fibers are produced by forcing an acetone solution through orifices of the spinneret into a stream of warm air, which evaporates the solvent. Fibers are also produced in a similar manner from cellulose triacetate, which is insoluble in acetone but soluble in methylene chloride. See "Arnel."

acetazolamide (5-acetamido-1,3,4-thiadiazole-2-sulfonamide) $(CH_3CONH)C_2N_2S(SO_2NH_2)$.

Properties: White to faintly yellowish white crystalline powder; odorless; m.p. 258°C; slightly soluble in water; very slightly soluble in alcohol.

Grades: U.S.P. XVI.

Use: Medicine.

acetazolamide sodium. $C_4H_5N_4NaO_3S_2$.

Properties: Soluble in water; solution (1 in 10) has a pH range 9.0-10.0.

Grades: U.S.P. XVI.

Use: Medicine.

acethydrazidepyridinium chloride. See Girard's "P" reagent.

acetic acid (ethanoic acid, vinegar acid, methanecarboxylic acid) CH_3COOH. Glacial acetic acid is the term for the pure compound, in distinction to the frequently encountered water solutions known as acetic acid. Vinegar is a dilute acetic acid.

Properties: Clear, colorless, acid liquid; very pungent odor. M.p. 16.63°C; b.p. 118°C (765 mm), 80°C (202 mm); sp. gr. 1.0492 (20/4°C); wt/gal (20°C) 8.64 lbs; viscosity (20°C) 1.22 cps; flash point (open cup) 110°F; refractive index 1.3715 (20°C). Miscible with water, alcohol, glycerin, and ether; insoluble in carbon disulfide.

Derivation: (a) By oxidation of acetaldehyde with air at 60-80°C in the presence of manganous acetate or cobalt acetate; (b) by bacterial oxidation of dilute ethyl alcohol; (c) recovery from pyroligneous acid by solvent extraction with ether or ethyl acetate (Coahran process), by absorption (Suida process), or by azeotropic distillation (Othmer process); (d) by catalytic combination of methanol and carbon monoxide at 700 atmospheres and 350°C; (e) by liquid-phase oxidation of butane, with a cobalt or manganese catalyst.

Grades: U.S.P. XVI (glacial and dilute); C.P.; technical (80; 99.5%); commercial (6, 28, 30, 36, 56, 60, 70, 80, and 99.5%); N.F. XI (diluted).

Containers: 5-lb bottles; 6-, 13-gal carboys; 8.5-, 13-, 55-, 100-gal barrels and drums; tankcars.

Uses: Manufacture of acetic anhydride and acetate esters, especially vinyl acetate; very widely used as an acid, solvent, and reagent in the production of rubber, plastics, acetate fibers, pharmaceuticals, dyes, insecticides, photographic chemicals, etc; textile printing.

Caution! Causes severe burns. MCA warning label.

acetic acid amine. See acetamide.

acetic acid, glacial. See acetic acid.

acetic aldehyde. See acetaldehyde.

acetic anhydride (acetyl oxide; acetic oxide) $(CH_3CO)_2O$.

Properties: Colorless, very mobile, strongly refractive liquid; very strong acetic odor; sp. gr. 1.0830 (20/20°C); b.p. 139.9°C; f.p. −73.1°C; flash point 150°F; wt/gal (20°C) 9.01 lbs. Miscible with alcohol, ether, and acetic acid; soluble in cold water; decomposes in hot water to form acetic acid.

Typical specifications: Sp. gr. 1.080-1.085 (20/20°C); color not darker than a .00001 N iodine solution; purity not less than 96.0%; chlorides none; sulfates none; phosphates none; nitrates none; heavy metals none; paraffin none; $KMnO_4$ test 2 cc. shall not reduce more than 0.1 cc. of 1 N $KMnO_4$; sulfur compounds not more than .009%; average weight 9.01 lbs/gal (20°C).

Derivation: (a) By oxidation of acetaldehyde with air in the presence of manganous acetate or cobalt acetate; (b) by reaction of acetylene and glacial acetic acid in the presence of mercuric sulfate to form ethylidene diacetate, which is subsequently decomposed by distillation over sodium pyrophosphate or zinc chloride, forming acetic anhydride and acetaldehyde; (c) from ketene and glacial acetic acid.

Grades: C.P., technical (75, 85, 90-95%).

Containers: Various bottles; 5-gal carboys; 55-gal aluminum drums; tankcars.

Uses: Mainly for cellulose acetate fibers and plastics; dehydrating and acetylating agent in production of pharmaceuticals, dyes, perfumes, explosives, etc.

Caution! Causes severe burns. Vapor harmful. MCA warning label.

acetic ester. See ethyl acetate.

acetic ether. See ethyl acetate.

acetic oxide. See acetic anhydride.

acetin (monoacetin; glyceryl monoacetate) $C_3H_5(OH)_2OOCCH_3$. Acetin may also refer to glyceryl di- or triacetate, also known as diacetin and triactin (q.v.).

Properties: Colorless, thick liquid; hygroscopic; sp. gr. 1.206 (20/4°C); b.p. 158°C (165 mm); 130°C (3 mm); soluble in water, alcohol; slightly soluble in ether; insoluble in benzene.

Derivation: By heating glycerol and strong acetic acid, distilling off the weak acetic acid formed and again heating with strong acetic acid and distilling.

Method of purification: Rectification.

Impurities: Uncombined acetic acid.

Grades: Technical.

Containers: 500-gm bottles; 1-, 10-, 50-lb tins; 500-lb drums.

Uses: Gelatinizing smokeless powders; preparing noncongealing dynamites; production of dinitroacetyl glycerin; tanning; solvent for basic dyes, indulin dyes, Perkin's violet dye.

Shipping regulations: None.*

acetoacetanilide (acetylacetanilide) $CH_3COCH_2CONHC_6H_5$.

Properties: White, crystalline solid; m.p. 85°C. Resembles ethyl acetoacetate in chemical reactivity. Slightly soluble in water, soluble in dilute sodium hydroxide, alcohol, ether, acids, chloroform, and hot benzene.

Typical specifications: M.p. 83.0-85.0°C; iron none; aniline none; solubility, 2 g soluble in 100 cc. of 0.5% NaOH; color above solution not darker than 0.00003 N iodine; appearance white, flaky solid; purity not less than 97.0% by the CO_2 evolution method; density 25 lbs/cu ft.

Derivation: By reacting ethyl acetoacetate with aniline, eliminating ethyl alcohol. Acetoacetanilide may also be prepared from aniline and diketene.

Grades: Technical.

Containers: Fiber drums. Net content: 3, 15, 125 lbs. Paper-lined wooden barrels, net weight 175 lbs.

Uses: Organic synthesis; dyestuffs (intermediate in the manufacture of the dry

*See "I.C.C. Shipping Regulations," page xiii.

Reference numbers refer to name of manufacturer. See "List of Manufacturers," page v.

colors generally referred to as Hansa and benzidine yellows).
Fire hazard: None.
Shipping regulations: None.*

acetoacet-ortho-anisidide $CH_3COCH_2CONHC_6H_4OCH_3$.
Properties: White crystalline powder; m.p. 86.6°C; sp. gr. 1.1320 (86.6/20°C); flash point (open cup) 325°F.
Containers: 1-gal cans; 5-, 55-gal drums.
Uses: Intermediate for azo pigments.

acetoacet-ortho-chloranilide $CH_3COCH_2CONHC_6H_4Cl$.
Properties: White crystalline powder; m.p. 107°C; b.p. decomposes; sp. gr. 1.1920 (107/20°C); flash point (open cup) 350°F. Almost insoluble in water.
Containers: 1-gal cans; 5-, 55-gal drums.
Use: Intermediate for azo pigments.

acetoacet-para-chloranilide $CH_3COCH_2CONHC_6H_4Cl$.
Properties: White crystalline powder; m.p. 133°C; b.p. decomposes; flash point (open cup) 320°F. Very slightly soluble in water.
Containers: 200-lb drums.
Use: Intermediate for azo pigments.

acetoacetic acid (acetylacetic acid; diacetic acid; acetone carboxylic acid) CH_3COCH_2COOH.
Properties: Colorless oily liquid; soluble in water, alcohol, and ether; decomposes below 100°C into acetone and carbon dioxide.
Uses: Organic syntheses.

acetoacetic ester. See ethyl acetoacetate.

acetoacet-para-phenetidide $CH_3COCH_2CONHC_6H_4OCH_2CH_3$.
Properties: Crystalline powder; m.p. 108.5°C; b.p. decomposes; sp. gr. 1.0378 (108.5/20°C); flash point (open cup) 325°F.
Containers: 1-gal cans; 5-, 55-gal drums.
Use: Intermediate for azo pigments.

acetoacet-ortho-toluidide $CH_3COCH_2CONHC_6H_4CH_3$.
Properties: Fine, white granular powder; m.p. 106°C; slightly soluble in water.
Grades: Technical.
Containers: Paper-lined wooden barrels. Net weight 250 lbs.
Uses: Intermediate in the manufacture of Hansa and benzidine yellows.

acetoacet-meta-xylidide $(CH_3)_2C_6H_3NHCOCH_2COCH_3$.
Properties: White to light yellow crystalline solid; m.p. 89-90°C; sp. gr. (20°C) 1.238; solubility in water (25°C) 0.5%.
Use: Organic synthesis.

acetoaminosalol. See para-acetylaminophenyl salicylate.

para-**acetoanisol.** See para-methoxyacetophenone.

acetobromal. See diethylbromoacetamide.

aceto-caustin. See trichloroacetic acid.

acetocinnamone. See benzylidene acetone.

acetoglycerides. Term commonly used to refer to acetylated monoglycerides although commercial acetoglycerides will contain di- and tri-glycerides. See acetostearin.

acetoin. See acetylmethylcarbinol.

acetol (acetonyl alcohol; hydroxyacetone; acetyl carbinol; pyruvic alcohol). CH_3COCH_2OH.
Properties: Colorless liquid; sp. gr. 1.0824 at 20/20°C; b.p. 146°C; m.p. −17°C. Soluble in water, alcohol, and ether.
Derivation: (a) By action of potassium acetate or potassium formate on a solution of bromo- or chloroacetone in dry methanol; (b) by bacterial fermentation of propylene glycol.
Grades: Technical.
Use: Solvent for nitrocellulose.

acetoluidide. See acetyl ortho-, or para-toluidine.

acetomeroctol $CH_3COOHgC_6H_3(OH)C(CH_3)_2CH_2C(CH_3)_3$ (2-Acetoxymercuri-4-(1,1,3,3-tetramethylbutyl) phenol.
Properties: White solid; m.p. 155-157°C; freely soluble in alcohol; soluble in ether or chloroform; sparingly soluble in benzene; practically insoluble in water.
Grade: N.N.D.
Use: Medicine.

acetone (dimethylketone; ketopropane; pyroacetic ether; 2-propanone) CH_3COCH_3.
Properties: Colorless liquid; characteristic odor; flammable. M.p. −94.3°C; b.p. 56.1°C; refractive index (20°C) 1.3591; sp. gr. (15°C) 0.7972; wt/gal (15°C) 6.64 lbs; flash point (open cup) 15°F. Miscible with water, alcohol, ether, chloroform and most oils.
Derivation: (a) Oxidation of isopropyl alcohol with catalyst such as brass or copper (in one variation, hydrogen peroxide is produced commercially); (b) fermentation of carbohydrates by bacterial organism; (c) oxidation of cumene; (d) vapor-phase oxidation of butane.
Grades: Technical; C.P.; N.F. XI; electronic.
Containers: 1-gal cans; 1-, 5-, 30-, 54-gal drums; 350-lb barrels; 8000-gal tank cars; tank trucks.
Uses (in approximate order of volume): Synthesis of acetic anhydride; derivatives such as diacetone alcohol; mesityl oxide, etc.; solvent for cellulose acetate; solvent in paints, lacquers, and adhesives; absorbent for acetylene; general solvent uses; epoxy resins; fibers; pharmaceuticals; rubber antioxidants; very pure grade in electronics industry to dry and cleanse parts.
Danger! Extremely flammable. MCA warning label.
Shipping regulations: Flammable liquid. Red label.*

*See "I.C.C. Shipping Regulations," page xiii.
Reference numbers refer to name of manufacturer. See "List of Manufacturers," page v.

acetone-bromoform. See tribromo-tert-butyl alcohol.

acetone carboxylic acid. See acetoacetic acid.

acetone chloroform. See chlorobutanol.

acetone cyanohydrin (alpha-hydroxyisobutyronitrile; 2-methyl-lactonitrile) $(CH_3)_2COHCN$.

Properties: Colorless liquid; b.p. 82°C (23 mm); m.p. −20°C; density 0.932 (19°C); refractive index n 20/D 1.3996; flash point 165°F; soluble in water, alcohol, and ether. Distillation not recommended because of decomposition to hydrocyanic acid and acetone. Insoluble in petroleum ether.

Derivation: By condensing acetone with hydrocyanic acid.

Grades: Technical (97-98% pure).

Containers: 6-gal carboys; 380-lb drums.

Uses: Insecticides; intermediate for organic synthesis, especially methyl methacrylate.

Shipping regulations: Poison, class B. Poison label.*

acetonedicarboxylic acid. See beta-ketoglutaric acid.

acetone oxime. See acetoxime.

acetone sodium bisulfite. See sodium acetone bisulfite.

acetonitrile (methyl cyanide) CH_3CN.

Properties: Colorless, limpid liquid; aromatic odor; poisonous; sp. gr. 0.783; m.p. −41°C; b.p. 82°C. Soluble in water and alcohol.

Typical specifications: Sp. gr. 0.782-0.785 (20°C); boiling range 80-82°C; purity 99% (min).

Derivation: By heating acetamide with glacial acetic acid; from dimethyl sulfate and sodium cyanide.

Grades: Technical.

Containers: Drums; tanks.

Uses: Organic synthesis of vitamin B pharmaceuticals and others; perfumes; extracts; denaturant; purification of a variety of chemicals; specialized solvent, especially for extractive distillation; crystallization medium; fiber synthesis.

Shipping regulations: Poison, Class B. Poison label.*

acetonyl acetone (1,2-diacetylethane; hexanedione-2,5; 2,5-diketohexane) $CH_3COCH_2CH_2COCH_3$.

Properties: Colorless liquid. Soluble in water; sp. gr. 0.9734 (20/20°C); b.p. (760 mm) 192.2°C; vapor pressure 0.43 mm at 20°C; freezing point −5.4°C; flash point 185°F; wt 8.1 lbs/gal (20°C); xylene nitrocellulose dilution ratio 1.8.

Typical specifications: Sp. gr. 0.971-0.976 at 20/20°C; boiling range 185-195°C (760 mm); acidity not more than 0.02% (as acetic).

Derivation: By-product in the production of acetaldehyde from acetylene.

Grades: Technical.

Containers: 1-gal cans; 5-, 55-gal drums. Net content: 8, 40, 430 lbs.

Caution! Volatile solvent; avoid prolonged breathing of vapor; use with adequate ventilation.

Uses: Solvent for cellulose acetate; roll-coating inks; tanning agent; lacquers; stains.

acetonyl alcohol. See acetol.

3-(alpha-acetonylbenzyl)-4-hydroxycoumarin. See warfarin.

acetophenetidin (para-acetylphenetidin; para-acetaminophenetol; phenacetin; para-ethoxyacetanilide) $CH_3CONHC_6H_4OC_2H_5$.

Properties: White crystals or powder; odorless and stable in air. Soluble in alcohol, chloroform and ether; slightly soluble in water; has slightly bitter taste.

Constants: M.p. 135°C.

Derivation: By the interaction of para-phenetidin and glacial acetic acid, or of ethyl bromide and para-acetaminophenol.

Method of purification: Crystallization.

Grades: Technical; U.S.P. XVI.

Containers: 100-, 200-, 1000-lb drums.

Uses: Medicine.

Shipping regulations: None.*

acetophenone (phenyl methyl ketone; hypnone; acetylbenzene) $C_6H_5COCH_3$.

Properties: Colorless liquid with sweet, pungent odor and taste. B.p. 201.7°C; f.p. 19.7°C; sp. gr. (20/20°C) 1.030; wt/gal (20°C) 8.75 lbs; refractive index (20°C) 1.5363; flash point (Cleveland open cup) 180°F. Slightly soluble in water; soluble in organic solvents.

Derivation: (a) Friedel-Crafts process with benzene and acetic anhydride or acetyl chloride; (b) by-product from the oxidation of cumene to produce acetone and phenol; (c) oxidation of ethyl benzene.

Method of purification: Distillation and crystallization.

Grades: Technical.

Containers: Glass bottles; 1-gal cans; 5-, 55-gal drums; tank cars; tank trucks.

Uses: Perfumery; tear gas, by chlorination; solvent; intermediate for pharmaceuticals, resins, etc.

Shipping regulations: None.*

acetopropionic acid. See levulinic acid.

"Acetoquat." [400] Trade name for a series of quaternary ammonium salts.

"Acetoquat BZA." Benzalkonium chloride.

"Acetoquat CDAC." Cetyldimethylbenzylammonium chloride. See cetalkonium chloride.

"Acetoquat CPB." Cetylpyridinium bromide.

"Acetoquat CPCX." Cetylpyridinium chloride.

"Acetoquat CTAB." Cetyltrimethylammonium bromide.

"Aceto-Slip." [400] Trade name for a refined oleamide (q.v.).

acetostearin
$CH_3(CH_2)_{16}COOCH_2CHOHCH_2OOCCH_3$.

*See "I.C.C. Shipping Regulations," page xiii.
Reference numbers refer to name of manufacturer. See "List of Manufacturers," page v.

Acetylated glyceryl monostearate. It is a solid with the peculiar combination of flexibility and nongreasiness. Derived from glyceryl monostearate or mixed glycerides by acetylation with acetic anhydride. Suggested uses are as a superior protective coating for food and as a plasticizer.

acetotoluidide. See acetyl-ortho- or -para-toluidine.

acetoxime (acetone oxime; 2-propanone oxime) $(CH_3)_2CNOH$.
Properties: Colorless crystals; both basic and acidic in properties. Chloral-like odor. Fairly readily hydrolyzed by dilute acids. Soluble in alcohols, ethers, and water.
Constants: Sp. gr. 0.97 (20/20°C); b.p. 136.3°C; m.p. 61°C.
Grades: Technical.
Uses: Organic synthesis (intermediate); solvent for cellulose ethers; primer for Diesel fuels.

ortho-**acetoxybenzoic acid.** See acetylsalicylic acid.

10-acetoxy-1-hydroxy-cis**-7-hexadecene.** See gyplure.

12-acetoxy-1-hydroxy-cis**-9-octadecene.** See gyplure.

4(para**-acetoxyphenyl)-2-butanone.** See Q-lure.

acetozone. See acetylbenzoyl peroxide.

acetrizoic acid (3-acetamido-2,4,6-triodobenzoic acid) $CH_3CONHC_6HI_3COOH$.
Properties: White, odorless powder. Soluble in alcohol and in solutions of alkali hydroxides; slightly soluble in ether; very slightly soluble in chloroform; practically insoluble in benzene. M.p. (dec) 278-283°C.
Use: Medicine.

aceturic acid (N-acetylglycine; acetylaminoacetic acid; acetylglycocoll) $CH_3CONHCH_2COOH$.
Properties: Long needles; m.p. 206-208°C; soluble in water and alcohol; slightly soluble in acetone, chloroform, glacial acetic acid; practically insoluble in ether, benzene. Forms stable salts with organic bases.
Use: Medicine.

acetylacetanilide. See acetoacetanilide.

acetylacetic acid. See acetoacetic acid.

acetylacetone $CH_3COCH_2OCCH_3$ (diacetylmethane; pentanedione-2,4).
Properties: Very mobile, colorless liquid. Unpleasant odor. When cooled, solidifies to lustrous, pearly spangles. The liquid is affected by light. It turns a brownish color and there is formation of resinous products. B.p. 140.5°C (760 mm); sp. gr. (20/20°C) 0.9753; wt 8.1 lbs/gal; coefficient of expansion 0.00105 (20°C); vapor pressure 7.0 mm (20°C); freezing point −23.5°C; viscosity 0.0058 poise (20°C). Soluble in water (acidified by hydrochloric acid); fairly soluble in plain water; soluble in alcohol, chloroform, ether, benzene, acetone, and glacial acetic.
Derivation: By condensing ethyl acetate with acetone.
Containers: 5-gal cans; 55-gal drums.
Uses: Solvent for cellulose acetate; intermediate; metal chelates.

acetylamino-. See acetamido-.

acetylaminoacetic acid. See aceturic acid.

para-**acetylaminobenzaldehyde thiosemicarbazone.** See thiosemicarbazone.

para-**acetylaminobenzenesulfonyl chloride** (para-acetamidobenezenesulfonyl chloride; N-acetylsulfanilyl chloride) $(CH_3CONH)C_6H_4(SO_2Cl)$.
Properties: Light tan to brownish powder or fine crystals. M.p. 149°C; soluble in benzene, chloroform, and ether.
Containers: 150-lb steel drums.
Use: As an intermediate in the manufacture of sulfa drugs.

ortho-**acetylaminobenzoic acid.** See acetylanthranilic acid.

para-**acetylaminobenzoic acid** (para-acetamidobenzoic acid) $CH_3CONHC_6H_4COOH$.
Properties: Needle-like crystals. Soluble in alcohol; slightly soluble in water.
Constants: M.p. 256.5°C (dec).
Derivation: Oxidation of acetyl-para-toluidine by potassium permanganate.
Grades: Technical.
Use: Chemical (intermediate).

acetylaminohydroxyphenylarsonic acid (acetarsone; N-acetyl-4-hydroxy-meta-arsanilic acid; 3-acetamido-4-hydroxybenzenearsonic acid) $(HO)(CH_3CONH)C_6H_3AsO(OH)_2$.
Properties: White to slightly yellow powder containing 27% arsenic; odorless; slight acid taste; soluble in alkali and alkali carbonate solutions; slightly soluble in water; insoluble in alcohol.
Grades: N.F. XI.
Use: Medicine.
Shipping regulations: None.*

1-acetylamino-8-naphthol-3,6-disulfonic acid (acetyl H-acid) $CH_3CONHC_{10}H_4OH(SO_3H)_2$.
Properties: Slightly soluble in water.
Derivation: Acetylation of H acid.
Grades: Technical.
Uses: Dyestuffs (reds).

para-**acetylaminophenol** (APAP; N-acetyl-para-aminophenol; acetaminophen; para-acetamidophenol; para-hydroxyacetanilide) $CH_3CONHC_6H_4OH$.
Properties: Crystals; odorless; slightly bitter taste; sp. gr. 1.293 (21/4°C); m.p. 168°C; slightly soluble in water and ether; soluble in alcohol; pH saturated aqueous solution 5.5-6.5.
Derivation: Interaction of para-aminophenol and an aqueous solution of acetic anhydride.
Grade: N.F. XI.

Uses: Chemical intermediate in making pharmaceuticals; stabilizer for hydrogen peroxide; medicine.

para-**acetylaminophenyl salicylate** (acetoaminosalol; para-acetylaminosalol; phenetsal) $C_6H_4(NHCOCH_3)OOCC_6H_4OH$.
Properties: Fine, white, crystalline scales; odorless, tasteless. Soluble in alcohol, ether, benzene, dilute solutions of the alkalies and hot water; insoluble in light hydrocarbon solvents; decomposed by strong alkalies. M.p. 187-188°C.
Derivation: By reducing para-nitrophenol salicylate to para-aminophenol salicylate and acetylating the latter.
Method of purification: Recrystallization.
Grades: Technical; pure.
Containers: Tins; glass bottles.
Use: Medicine.
Shipping regulations: None.*

para-**acetylaminosalol.** See para-acetylaminophenyl salicylate.

para-**acetylanisole.** See para-methoxyacetophenone.

acetylanthranilic acid (ortho-acetylaminobenzoic acid) $CH_3CONHC_6H_4COOH$.
Properties: Needles, plates, rhombic crystals (crystallized in glacial acetic acid); m.p. 185°C; slightly soluble in water; soluble in hot alcohol, ether and benzene.
Derivation: By oxidation of ortho-acetyltoluidine with potassium permanganate in the presence of magnesium sulfate or potassium chloride.
Grades: Technical.
Uses: Chemical (organic synthesis, anthranilic acid).

acetylation. Introduction of an acetyl radical (CH_3CO-) into the molecule of an organic compound having OH or NH_2 groups. The usual reagents for this purpose are acetic anhydride or acetyl chloride. Thus ordinary ethyl alcohol C_2H_5OH may be converted to $C_2H_5OCOCH_3$ (ethyl acetate). Cellulose is similarly converted to cellulose acetate by treatment with a mixture containing acetic anhydride. Acetylation is commonly used to determine the number of hydroxyl groups in fats and oils (see acetyl value).

acetylbenzene. See acetophenone.

acetylbenzoyl peroxide (acetozone; benzozone) $C_6H_5CO \cdot O_2 \cdot OCCH_3$.
Properties: White crystals; decomposed by water, alkaloids, organic matter and some organic solvents; decomposes slowly and evaporates when gently heated, and instantaneously (possibly explosively) if quickly heated, ground or compressed. The commercial product is mixed with a neutral drying powder and contains 50% acetylbenzoyl peroxide; m.p. 36.6°C; b.p. 130°C (19 mm); moderately soluble in ether, chloroform, carbon tetrachloride and water; slightly soluble in mineral oils and alcohol.
Uses: Medicine (active germicide); disinfectant.
Shipping regulations: In solution, oxidizing material. Yellow label. Solid, not accepted.*

acetyl bromide CH_3COBr.
Properties: Colorless, fuming liquid; turns yellow in air; reacts violently with water or alcohol; fumes irritate the eyes; soluble in ether, chloroform and benzene.
Constants: B.p. 81°C; m.p. −96°C; sp. gr. 1.663 (16/4°C).
Derivation: By the interaction of acetic acid and phosphorus pentabromide.
Grades: Technical.
Containers: Metal bottles; iron drums.
Uses: Organic synthesis; manufacture of dyes.
Shipping regulations: Corrosive liquid. White label.*

N-acetyl-N-bromodiethylacetyl urea. See acetylcarbromal.

alpha-**acetylbutyrolactone** (alpha-acetyl-gamma-hydroxybutyric acid, gamma-lactone) $C_6H_8O_3$.
Properties: Liquid with ester-like odor. Sp. gr. 1.18-1.19 (20°C); b.p. (30 mm) 142-144°C; soluble in water.
Derivation: Prepared from sodium acetoacetate and ethylene oxide in absolute alcohol.
Toxicity: Avoid swallowing and skin contact.
Use: Intermediate.

acetyl carbinol. See acetol.

acetylcarbromal (N-acetyl-N-bromodiethylacetylurea) $(C_2H_5)_2CBrCONHCONHCOCH_3$.
Properties: Crystals, slightly bitter taste; m.p. 109°C. Slightly soluble in water; freely soluble in alcohol, and ethyl acetate.
Use: Medicine.

acetyl chloride CH_3COCl.
Properties: Colorless, flammable, highly refractive, fuming liquid; strong odor; irritating to eyes; sp. gr. 1.1051; m.p. −112°C; b.p. 51-52°C; soluble in ether, acetone, acetic acid; violent reaction with water and alcohol.
Derivation: By mixing glacial acetic acid and phosphorus trichloride in the cold and heating a short time to drive off hydrochloric acid. The acetylchloride is then distilled.
Containers: Iron drums; 110-lb carboys. Protect from moisture.
Uses: Organic preparations (acetylating agent); dyestuffs; pharmaceuticals.
Danger! Flammable, causes severe burns. MCA warning label.
Shipping regulations: Corrosive liquid. White label.*

acetylcholine (acetylethanoltrimethylammonium hydroxide) $CH_3COOCH_2CH_2N(CH_3)_3OH$. A derivative of choline important in the mechanism of nerve action. The enzyme cholinesterase hydrolyzes acetylcholine into comparatively inactive choline and acetic acid and is

*See "I.C.C. Shipping Regulations," page xiii.
Reference numbers refer to name of manufacturer. See "List of Manufacturers," page v.

necessary in the body to prevent acetylcholine poisoning. See nerve gases.
Properties: White, crystalline powder; hygroscopic; soluble in water.
Use: Medicine.

acetylcholine bromide $(CH_3)_3NBr(CH_2)_2OCOCH_3$.
Properties: Colorless, hygroscopic crystals; m.p. 143°C; very soluble in water; soluble in alcohol; decomposes in hot water; insoluble in ether.
Derivation: Reaction of choline bromide and acetic anhydride.
Use: Medicine.

acetylcholine chloride (acecoline) $(CH_3)_3NCl(CH_2)_2O\cdot COCH_3$.
Properties: Colorless hygroscopic crystals; m.p. 149-152°C; odorless; very soluble in water and alcohol; decomposed by hot water or alkalies; insoluble in ether.
Derivation: From reaction of choline chloride and acetic anhydride.
Use: Medicine.

acetylcholinesterase. See cholinesterase.

alpha-**acetyldigitoxin (anhydrous)** $C_{43}H_{66}O_{14}$.
Properties: Crystals; sparingly soluble in water, ether, petroleum ether; soluble in most organic solvents.
Derivation: Obtained by enzymatic hydrolysis of digilanide A.
Grade: N.N.D.
Use: Medicine.

acetylene (ethine; ethyne) HC:CH.
Properties: Colorless gas; ethereal odor; highly flammable; explosive when compressed or mixed with air in certain proportions; toxic when inhaled; forms explosive compounds with copper and silver; sp. gr. 0.91 (air = 1); m.p. −81.8°C (890 mm); b.p. −84°C (760 mm); soluble in alcohol, acetone, and water.
Derivation: (a) By the action of water on calcium carbide; (b) cracking of petroleum hydrocarbons by the BASF process, or the Wulff process (q.v.); (c) by the Montecatini process, the partial oxidation of natural gas.
Grades: Technical, containing 98% acetylene and not more than 0.05% by volume of phosphine or hydrogen sulfide; 99.5%.
Containers: Steel cylinders. Much acetylene is also delivered by pipe line.
Uses (in approximate order of volume): Welding and cutting of metals; vinyl and vinylidene chlorides; acrylonitrile; general illuminating purposes; per- and trichloroethylene; chlorinated rubber industry; acetic acid, acetaldehyde and acetate monomers; vinylacetylene for neoprene; miscellaneous chemicals, including those of the Reppe process (q.v.).
Danger! Flammable gas.
Shipping regulations: Flammable compressed gas. Red gas label.*

acetylene black. The carbon black resulting from incomplete combustion or thermal decomposition of acetylene gas.
Properties: High liquid adsorption, retention of high bulk volume, purity and electrical conductivity. Has property of instilling this conductivity into plastics, rubbers, and other materials.
Containers: Bags; cases.
Uses: Manufacture of dry cell batteries and conductive rubber and plastics; component of explosives; as a filler in natural rubber and in thermal and sound insulation, as a gloss suppressor in the paint industry, as a carburizing agent in hardening of steel and as a pigment in special printing inks.

acetylene dichloride. See sym-dichloroethylene.

acetylene hydrocarbons. Unsaturated hydrocarbons of the homologous series having the empirical formula C_nH_{2n-2} and a structural formula containing a triple bond.

acetylene tetrabromide (Muthmann's liquid; sym-tetrabromoethane) $CHBr_2CHBr_2$.
Properties: Yellowish liquid. Soluble in alcohol and ether; insoluble in water.
Constants: Sp. gr. 2.98 to 3.00; b.p. 239-242°C (760 mm) with decomposition; 151°C (54 mm); m.p. 0.1°C; refractive index 1.638.
Derivation: By the interaction of acetylene and bromine, and subsequent distillation.
Method of purification: Rectification.
Grades: Technical.
Containers: 1- and 5-lb bottles; 10-lb metal cans; 55-gal steel drums.
Uses: Separating minerals by specific gravity; solvent for fats, oils, and waxes; fluid in liquid gauges; solvent in microscopy.
Shipping regulations: None.*

acetylene tetrachloride. See sym-tetrachloroethane.

acetylenogen. See calcium carbide.

N-**acetyl ethanolamine** (hydroxyethyl acetamide) $CH_3CONHC_2H_4OH$.
Properties: Brown viscous liquid, soluble in alcohol, ether, and water; sp. gr. 1.122 (20/20°C); boiling range 150-152°C (5 mm), decomposes (10 mm); lbs/gal 9.34 (20°C); refractive index n 20/D 1.4730; flash point (open cup) 350°F; f.p. 15.8°C.
Grades: Technical (75% solution in water).
Uses: Plasticizer for polyvinyl alcohol and for cellulosic and proteinoid substances; humectant for paper products, glues, cork and inks; high-boiling solvent used in compounding fountain-pen inks; textile conditioner.

acetylethanoltrimethylammonium hydroxide. See acetyl choline.

acetyl eugenol. See eugenol acetate.

acetylformic acid. See pyruvic acid.

N-**acetylglycine.** See aceturic acid.

acetylglycocoll. See aceturic acid.

acetyl H-acid. See 1-acetylamino-8-naphthol-3,6-disulfonic acid.

N-**acetyl-4-hydroxy**-meta-**arsanilic acid.** See

*See "I.C.C. Shipping Regulations," page xiii.
Reference numbers refer to name of manufacturer. See "List of Manufacturers," page v.

acetylaminohydroxyphenylarsonic acid.

alpha-**acetyl**-gamma-**hydroxybutyric acid,** gamma-**lactone.** See alpha-acetyl butyrolactone.

acetyl iodide CH_3COI.
Properties: Colorless, transparent, fuming liquid, turning brown on exposure to air or moisture; soluble in ether and benzene; decomposed by water and alcohol.
Constants: Sp. gr. 1.98; b.p. 105-108°C.
Derivation: By the interaction of acetic acid, iodine, and phosphorus.
Method of purification: Distillation.
Grades: Technical.
Containers: Glass bottles.
Uses: Organic synthesis.
Shipping regulations: Corrosive liquid. White label.*

acetylisoeugenol (isoeugenol acetate) $C_6H_3(CHCHCH_3)(OCH_3)(OCOCH_3)$.
Properties: White crystals; spicy clove-like odor; congealing point 77.0°; soluble 1 part in 27 parts of 95% alcohol.
Method of purification: Crystallization.
Grades: Technical.
Containers: Tin cans.
Uses: Perfumery, particularly for carnation-type odors.
Shipping regulations: None.*

acetyl ketene. See diketene.

acetylmethylcarbinol (acetoin; 3-hydroxy-2-butanone; dimethylketol) $CH_3COCHOHCH_3$.
Properties: Slightly yellow liquid or crystalline solid (dimer); oxidizes gradually to diacetyl on exposure to air; sp. gr. 1.016; b.p. 140-148°C; m.p. 15°C; soluble in alcohol; miscible with water in all proportions; slightly soluble in ether.
Derivation: Reduction of diacetyl.
Grades: Technical.
Containers: 10-, 100-gm, 1-, 5-lb glass bottles; 2-lb polyethylene-lined cartons.
Uses: Aroma carrier; preparation of flavors and essences.
Shipping regulations: None.*

acetyl-beta-**methylcholine chloride.** See methacholine chloride.

acetyl-1,4-naphthylenediamine-6- and 7-sulfonic acids $C_{10}H_5NH_2NHCH_3COSO_3H$.
Properties: Needle-like crystals; sparingly soluble in water.
Derivation: Acetylation of 1-naphthylamine-6-and 7-sulfonic acids followed by (a) nitration in the presence of concentrated sulfuric acid, and (b) reduction.
Grades: Technical.
Uses: Organic synthesis; dyestuffs.

acetyl oxide. See acetic anhydride.

acetyl peroxide (diacetyl peroxide) $(CH_3CO)_2O_2$.
Properties: Colorless crystals; m.p. 30°C; explodes on heating; slightly soluble in cold water; soluble in alcohol and ether. Marketed as a 25% solution in dimethyl phthalate; flash point (open cup) 113°F; f.p. (approx.) −8°C; sp. gr. (20°C) 1.18.
Containers: 65-lb carboys.
Uses: Initiator and catalyst for resins.
Shipping regulations: Solution: Oxidizing material. Yellow label. Solid: Not accepted.*

acetylphenetidin. See acetophenetidin.

acetyl phenol. See phenyl acetate.

N-**acetyl**-para-**phenylenediamine.** See para-aminoacetanilide.

acetylphenylhydrazine (hydracetin) $C_6H_5(NH)_2C_2H_3O$.
Properties: Colorless, inodorous crystals; m.p. 128°C; soluble in alcohol; moderately soluble in water and ether.
Derivation: Replacement of hydrogen in hydrazine by phenyl and acetyl radicals.
Containers: Fiber drums; steel drums; multiwall paper sacks.
Uses: In medicine; organic synthesis; stabilizer (transformer oils, vegetable oils, coal-carbonization spirits, lubricating oils, petroleum oils, shale oils).
Shipping regulations: None.*

N^1-acetyl-N^4-phthalylsulfanilamide. See phthalylsulfacetamide.

acetyl propionyl (2,3-pentanedione; methyl ethyl glyoxal; methyl ethyl diketone) $CH_3COCOCH_2CH_3$.
Properties: Yellow liquid; m.p. −52°C; b.p. 106-110°C; sp. gr. (15/4°C) 0.955-0.959; partly soluble in water.
Use: Flavors of butterscotch and chocolate type.

4-acetyl resorcinol (2,4-dihydroxyacetophenone) $C_6H_3(OH)_2COCH_3$.
Properties: Light tan crystals; m.p. 146-148°C. High absorptivity in ultra violet light region. Slightly soluble in water; soluble in most organic solvents except benzene and chloroform.
Uses: Light stabilizer for plastics; dye intermediate; fungicide; plant growth promoter.

acetylsalicylic acid (aspirin; ortho-acetoxybenzoic acid) $C_2H_3O_2C_6H_4CO_2H$.
Properties: White crystals or white, crystalline powder. Odorless or has faint odor. Stable in dry air; slowly hydrolyzes in moist air to salicylic and acetic acids. Slightly soluble in water; soluble in alcohol, chloroform, and ether; less soluble in absolute ether. Dissolves with decomposition in solutions of alkali hydroxides and carbonates. M.p. 132-136°C; b.p. 140°C (dec.).
Derivation: Action of acetic anhydride on salicylic acid.
Method of purification: Crystallization.
Impurities: Salicylic acid.
Grades: Technical; U.S.P. XVI.
Containers: 25-lb boxes; 25-, 100-, 250-lb drums.
Use: Medicine.
Shipping regulations: None.*

acetylsalol. See phenylacetylsalicylate.

acetyl sulfamethoxypyridazine (N-acetyl-N-(6-methoxy-3-pyridazinyl)sulfanilamide;

3-(N-acetyl-sulfanilamido)-6-methoxy-pyridazine) $C_{13}H_{14}N_4O_4S$.
Properties: Crystals; tasteless. Decomposes 186-187°C.
Grade: N. N. D.
Use: Medicine.

N-acetylsulfanilamide. See sulfacetamide.

N-acetylsulfanilamide sodium. See sulfacetamide sodium.

N-acetylsulfanilyl chloride. See para-acetylaminobenzenesulfonyl chloride.

acetylsulfisoxazole
$NH_2C_6H_4SO_2N(COCH_3)C_3NO(CH_3)_2$.
Properties: White to slightly off-white, crystalline solid with slight characteristic odor; m.p. 192-195°C. Practically insoluble in water; slightly soluble in alcohol, chloroform, and ether.
Grade: U.S.P. XVI.
Use: Medicine.

acetyltannic acid (tannyl acetate; diacetyltannin) $C_{14}H_8O_9(COCH_3)_2$.
Properties: Yellowish-white or grayish-white powder, darkening on exposure to light; soluble in ethyl acetate, aqueous solution of sodium borate or sodium phosphate; slightly soluble in water and alcohol.
Derivation: Heating tannin with acetic anhydride in presence of glacial acetic acid.
Use: Medicine.
Shipping regulations: None.*

acetyltannin. See acetyltannic acid.

2-acetylthiophene (methyl 2-thienyl ketone) $CH_3COC_4H_3S$.
Properties: Colored oil; m.p. 10-11°C; b.p. 213.5°C; very slightly soluble in ether.
Use: Organic intermediate.

acetyl-ortho-toluidine (ortho-acetotoluidide) $CH_3CONHC_6H_4CH_3$.
Properties: Colorless crystals; m.p. 110°C; b.p. 296°C; sp. gr. 1.168 (15°C); soluble in alcohol, ether, benzene, chloroform, glacial acetic acid; slightly soluble in cold water; insoluble in hot water.
Derivation: By boiling glacial acetic acid with ortho-toluidine and distilling the product.
Grades: Technical.
Use: Organic synthesis.

acetyl-para-toluidine (para-acetotoluidide) $CH_3CONHC_6H_4CH_3$.
Properties: Colorless needles; m.p. 153°C; b.p. 307°C; density 1.212 (15/4°C); slightly soluble in water; soluble in alcohol, ether, ethyl acetate, glacial acetic acid.
Grades: Technical.
Containers: Wooden barrels or fiber drums.
Uses: Dyes.
Shipping regulations: None.*

acetyl triallyl citrate
$CH_3COOC_3H_4(COOCH_2CH:CH_2)_3$.
Properties: Liquid; boiling range 142-143°C (0.2 mm); sp. gr. 1.140 (20°C); refractive index n 25/D 1.4665; flash point 171-174°C.
Use: Cross linker for polyesters; monomer for polymerization.

acetyl tributyl citrate
$CH_3COOC_3H_4(COOC_4H_9)_3$.
Properties: Colorless, odorless liquid. Distillation range (1 mm) 172-174°C; pour point −75°F; sp. gr. (25°C) 1.046; wt/gal (25°C) 8.74 lbs; refractive index (25°C) 1.4408; viscosity (25°C) 42.7 cps. Insoluble in water.
Derivation: Esterification and acetylation of citric acid.
Grades: Technical.
Containers: Metal drums; tanks.
Uses: Plasticizer for vinyl resins, especially for food packaging.

acetyl triethyl citrate
$CH_3COOC_3H_4(COOC_2H_5)_3$.
Properties: Colorless, odorless liquid. Distillation range (1 mm) 131-132°C; pour point −45°F; sp. gr. (25°C) 1.135; wt/gal (25°C) 9.47 lbs; refractive index (25°C) 1.4386; viscosity (25°C) 53.7 cps. Slightly soluble in water.
Derivation: Esterification and acetylation of citric acid.
Grades: Technical.
Containers: Metal drums; cans.
Uses: Plasticizer for cellulosics, particularly ethyl cellulose.
Shipping regulations: None.*

acetyl tri-2-ethylhexyl citrate $C_{32}H_{58}O_8$.
Properties: Liquid; b.p. 225°C (1 mm); insoluble in water.
Grades: Technical.
Use: Low-volatility plasticizer for the vinyls.

acetyl-dℓ-tryptophan. See tryptophan.

acetyl value. The number of milligrams of KOH required for neutralization of acetic acid obtained by the saponification of one gram of acetylated fat or oil sample. Acetylation is carried out by boiling the sample with an equal amount of acetic anhydride, washing and drying. Saponification values on the acetylated and on untreated fat are determined. From the results the acetyl value is calculated. It is a measure of the number of free hydroxyl groups in the fat or oil.

achillea (milfoil; yarrow). The flowering tops of the perennial herb, Achillea millefolium.
Occurrence: Europe, America.
Chief constituents: Volatile oil; tannin; achillein.
Use: Medicine.

"**Achrocidin.**" [57] Trademark for a tetracycline-antihistamine-analgesic compound.

"**Achromycin.**" [57, 315] Trademark for tetracycline, $C_{22}H_{24}N_2O_8$ (q.v.).

"**Aciban.**" [57] Trademark for an antacid.

acicular. Needle-shaped; used in describing crystals or the particles in powders.

acid. In common everyday language an acid is

*See "I.C.C. Shipping Regulations," page xiii.
Reference numbers refer to name of manufacturer. See "List of Manufacturers," page v.

any one of a large class of chemical substances which have one or more of the following properties: sour taste, ability to make litmus dye turn red, and to cause other indicator dyes to change to characteristic colors, ability to react with and dissolve certain metals to form salts, ability to react with bases or alkalies to form salts. Chemists have long defined an acid as a compound that contains the element hydrogen in its formula, this hydrogen being replaceable by metals to form salts. A later definition of acid, entirely consistent with the preceding one, is as a substance whose molecules yield hydrogen ions (H^+) in water or under other suitable conditions. These definitions are adequate to explain the chemistry of most well known and common acids. Extensive chemical research has shown that for a complete understanding it is desirable to define an acid as a substance that produces the positive ion of solvent in which the acid is dissolved, or as any substance (molecule or ion) which will release a proton or hydrogen ion. Thus an acid is a proton donor. These more general definitions include the simpler concepts mentioned at first, and also explain many reactions not otherwise easily understood. The names of Brönsted and Lewis are associated with the latter definitions.

It should also be mentioned that some acid oxides (oxides of nonmetals) are sometimes improperly referred to as acids, because these acid oxides react in water to form acids. Thus phosphoric oxide, P_2O_5, is sometimes called phosphoric acid, and As_2O_3 is commonly called arsenic acid. See also base; alkali; neutralization; pH.

1, 2, 4 acid. See 1-amino-2-naphthol-4-sulfonic acid.

acid amide. See amide.

acid ammonium tartrate. See ammonium bitartrate.

acid ammonium valerate. See ammonium valerate.

acid anhydride. An oxide of a nonmetallic element or an organic radical which is capable of forming an acid when united with water, or which can be formed by the abstraction of water from the acid molecule, or which can unite with basic oxides to form salts.

acid calcium phosphate. See calcium phosphate, monobasic.

acid dyes. These are usually azo, triarylmethane or anthraquinone dyes with acid substituents such as nitro-, carboxy-, or sulfonic acid. They are most frequently applied in acid solution to wool and silk, and no doubt combine with the basic groups of the proteins of those animal fibers. Orange II (C.I. 151), black 10 B (C.I. 246), and acid alizarine blue B (C.I. 2054) are examples.

acid ethylsulfate. See ethylsulfuric acid.

acid glaucine blue. See peacock blue.

acidic oxides. The oxides of nonmetals, e. g. SO_2, CO_2, P_2O_5, SO_3, which form acids when combined with water. See also acid anhydrides.

acidimetry. The determination of the concentration of acid solutions or of the quantity of acid in a sample or mixture. This is usually done by titration with a solution of base of known strength (standard solution) and an indicator is used to establish the end point. See also pH.

acid lining. Silica brick lining used in Bessemer or open hearth furnaces.

acid liquor, sulfite. See sulfite acid liquor.

acid magnesium citrate. See magnesium citrate, dibasic.

acid magnesium phosphate. See magnesium phosphate, monobasic.

acid methyl sulfate. See methylsulfuric acid.

acid of sugar. Oxalic acid (q. v.).

acidogen nitrate. See urea nitrate.

"Acidolene." [244] Trademark for a series of sulfonated oils made from neatsfoot, sperm, cod, fish and coconut oil.

Containers: Non-returnable steel drums averaging 400-425 lbs net.

Uses: Primarily used by the leather industry and referred to as fatliquors. Also used wherever an oil emulsifiable in water is needed for plasticizing or softening.

acid open hearth slag. See slag.

acid phosphatase. An enzyme found in blood serum which catalyzes the liberation of inorganic phosphate from phosphate esters. Optimum pH 5; is less active than alkaline phosphatase.

Uses: Biochemical research.

acid phosphate. An acid salt of phosphoric acid such as NaH_2PO_4, $CaHPO_4$, etc. Also used to refer specifically to calcium phosphate monobasic, $Ca(H_2PO_4)_2$, or superphosphate of lime.

acid potassium oxalate. See potassium binoxalate.

acid potassium sulfate. See potassium bisulfate.

acids. See under specific title, e. g., sulfuric acid, not acid, sulfuric.

"Acidulin." [100] Trademark for glutamic acid hydrochloride (q. v.).

acid value. The number of milligrams of potassium hydroxide neutralized by the free acids present in one gram of oil. The determination is done by titrating the sample in hot 95% ethyl alcohol and using phenolphthalein as an indicator.

"Acintene." [252] Trademark for a group of pinene, terpene, and turpentine products.

"Acintene" A: alpha-Pinene. Close boiling solvent.
"Acintene" B: beta-Pinene. Used for terpene resins, synthetic pine oil.
"Acintene" DP: Dipentene. A solvent used in reclaimed rubber manufacture and rubber compounding.
"Acintene" N: Liquid terpene polymer. Used as a vehicle constituent for printing inks; as a plasticizer for resins and rubber; caulking compounds; mastic tile adhesives.
"Acintene" L: Turpentine still residue. For use in printing inks and resins; as a rubber softener; as a replacement for pine tar.
"Acintene" P: Solvent grade refined sulfate wood turpentine. Meets ASTM and Federal Specifications. Used in the manufacture of shoe polishes, wax preparations, cleaning compounds, and as oil additives.

"Acintol." [252] Trademark for a series of tall oils and tall oil derivatives.
"Acintol" C: Crude tall oil, a mixture of rosin and fatty acids containing a small amount of unsaponifiable matter.
"Acintol" FA-1 and FA-2 Fractionated Tall Oils: Tall oil fatty acids. Acintol FA-1 has a Gardner color of 8, 4% rosin acids and 4% unsaponifiables. Acintol FA-1 Special has a Gardner color of 4+, 3.8% rosin acids and 1.3% unsaponifiables. Acintol FA-2 has a Gardner color of 3+, 1.0% rosin acids and 1.0% unsaponifiables. Acintol FA-3 has a Gardner color of 2+, 0.6% rosin acids and 0.7% unsaponifiables.
"Acintol" D Distilled Tall Oil: Fractionated tall oil characterized by light color, uniform composition, high acid and saponification numbers, and high linoleic acid content.
"Acintol" H Tall Oil Heads: A by-product of tall oil fractionation. High saturated fatty acid content and a low rosin content.
"Acintol" R Tall Oil Rosin: A rosin suitable for many applications where wood and gum rosins have been used.
"Acintol" P Tall Oil Pitch: Light-colored semisolid product of uniform composition.

"ACL." [58] Trademark for bleaching compounds used in bleaches, cleansers, sanitizers, etc.
"ACL" 85 is trichloroisocyanuric acid (90% available chlorine).
"ACL" 70 is dichloroisocyanuric acid (70% available chlorine).
"ACL" 60 is sodium dichloroisocyanurate (61% available chlorine).
"ACL" 59 is potassium dichloroisocyanurate (60% available chlorine).
Containers: Fiber drums with polyethylene liners.

"Aclar." [50] Trademark for a series of fluorohalocarbon films. Retain useful properties from −320°F to +390°F. Used in packaging applications, where a transparent, vapor and/or gas barrier are required. Used in electronic and electrical applications because of insulating and heat resistant properties. Extreme chemical resistance and ability to seal make it useful as a tank lining, etc.

"Acofor." [79] Trade name for distilled tall oil fatty acids.
Properties: Sp. gr. 0.905 (25°C); refractive index (20°C) 1.471; flash point (open cup) 367°F; acid number 192; saponification number 194; unsaponifiable matter 2.5%.
Containers: 55-gal drums; tank cars.
Uses: Paint and varnish; inks; soaps; disinfectants; textile oils; core oils, etc. See also "Aconew Extra," "Aconew 500."

acoin. See di-para-anisyl-para-phenetyl-guanidine hydrochloride.

"Acolin." [79] Trade name for distilled tall oil, specially processed to improve color and odor.
Properties: Sp. gr. 0.947 (25°C); refractive index 1.488 (20°C); flash point 370°F (open cup); acid number 178; saponification number 182; iodine value 143.
Containers: 55-gal drums; tank cars.
Uses: Paint and varnish; inks; soaps; disinfectants; textile oils; core oils.

"Aconew 500." [79] Trade name for distilled tall oil fatty acids.
Properties: Sp. gr. 0.898 (25°C); refractive index 1.468; acid number 195; saponification number 197; iodine value 128; unsaponifiable matter 1.5%.
Containers: 55-gal drums; tank cars.
Uses: Paint and varnish; inks; soaps; disinfectants; textile oils; core oils, etc. See also "Aconew Extra" and "Acofor."

"Aconew Extra." [79] Trademark for distilled tall oil fatty acids.
Properties: Sp. gr. (25/25°C) 0.898; refractive index 1.468 (20°C); flash point (open cup) 370°F; acid number 195; saponification number 197; unsaponifiable matter 1.5%.
Containers: 55-gal drums; tank cars.
Uses: Paint and varnish; inks; soaps; disinfectants; textile oils; core oils. See also "Aconew 500," "Acosix" and "Acofor."

aconite (monkshood; wolfsbane; friar's cowl).
Derivation: Dried tuberous root or leaves of the perennial herbaceous plant Aconitum napellus. Poisonous!
Occurrence: Mountainous regions of Europe, Asia and North America.
Grades: Technical.
Containers: Burlap bags and bales.
Uses: Medicine.
Shipping regulations: None.*

aconitic acid (propene-1,2,3 tricarboxylic acid) $H(COOH)C:C(COOH)CH_2(COOH)$.
Properties: White to yellowish crystalline solid; m.p. (about) 195°C with decomposition; soluble in water and alcohol.
Derivation: (a) By dehydration of citric acid with sulfuric acid; (b) extraction from sugar cane bagasse, Aconitum napellus and other natural sources.
Uses: Preparation of plasticizers and wetting agents; antioxidant; organic syntheses.

*See "I.C.C. Shipping Regulations," page xiii.
Reference numbers refer to name of manufacturer. See "List of Manufacturers," page v.

aconitine $C_{34}H_{49}NO_{11}$.
Properties: White crystalline alkaloid; feeble bitter taste; intensely poisonous! Soluble in alcohol, ether, benzene, and chloroform; very slightly soluble in water.
Constants: M. p. 204°C; specific rotation, in chloroform, +17.3°.
Derivation: By extraction and crystallization from the root of Aconitum napellus.
Containers: 1/8-, 1/4-, 1-oz vials, 5-, 10- and 15-grain vials.
Uses: Medicine, in form of the base or as the hydrobromide, hydrochloride, nitrate.
Shipping regulations: None. *
Salts obtained by interaction of the acid and alkaloid:
Hydrobromide $C_{34}H_{49}NO_{11} \cdot HBr \cdot 2\frac{1}{2} H_2O$. M. p. (dec.) 180°C; specific rotation −30.5°.
Hydrochloride $C_{34}H_{49}NO_{11} \cdot HCl \cdot 3\frac{1}{2} H_2O$. M. p. (dec.) 171°C; specific rotation −31.3°.
Nitrate $C_{34}H_{49}NO_{11} \cdot HNO_3$. M. p. (dec.) about 200°C; specific rotation (2% aqueous solution) −35°.

aconitine hydrobromide. See aconitine.

aconitine hydrochloride. See aconitine.

aconitine nitrate. See aconitine.

acorn sugar. See quercitol.

"Acosix." [79] Trademark for distilled tall oil.
Properties: Sp. gr. 0.936 (25°C); refractive index 1.484 (20°C); flash point (open cup) 367°F; acid number 190; saponification number 192; iodine number 150; unsaponifiable matter 2.0%.
Containers: 55-gal drums; tank cars.
Uses: Paint and varnish; inks; soaps; disinfectants; textile oils; core oils.
See also "Aconew Extra", "Acosix 700" and "Acolin."

"Acosix 700." [79] Trade name for distilled tall oil.
Properties: Sp. gr. 0.936 (25°C); refractive index 1.484 (20°C); flash point (open cup) 367°C; acid number 190; saponification number 192; iodine number 150; unsaponifiable matter 2%.
Containers: 55-gal drums; tank cars.
Uses: Paint and varnish; inks; soaps; disinfectants; textile oils; core oils, etc.
See also "Acosix" and "Acolin."

"A-C" Polyethylenes. [175] Trademark for a line of low molecular weight polyethylenes.
Properties: Translucent white; tasteless; excellent electrical properties; abrasion resistant; resistant to water and most chemicals; sp. gr. 0.92; slightly soluble in turpentine, petroleum naphtha, xylene, and toluene at room temperature; soluble in xylene, toluene, trichloroethylene, turpentine, and mineral oils at 180°F; practically insoluble in water; slightly soluble in methyl acetate, acetone, and ethanol up to the boiling point of these solvents.
Grades: Nos. 6 and 6A: Average molecular weight, 2000; m. p. 219-26°F; hardness 3-5; viscosity 180 cps (140°C).
No. 7: Average molecular weight 2000; m. p. 223-30°F; hardness 2-3; viscosity 220 cps (140°C).
No. 615: Average molecular weight 5000; m. p. 224-32°F; hardness 3-4; viscosity 4000 cps (140°C).
Nos. 617 and 617A: Average molecular weight 1500; m. p. 210-17°F; hardness 6-9; viscosity 100 cps (140°C).
No. G-201: Average molecular weight 2000; m. p. 201-8°F; hardness 6-9; viscosity 230 cps (140°C).
No. 629: M. p. 213-21°F; acid number 14-17; hardness 3-6; viscosity 160 cps (140°C).
Containers: 50-lb multiwall paper bags.
Uses: In coating containers and paper; in printing inks, paints and paste polishes; as rubber lubricants and mold release agents; emulsifiable grade (No. 629) in various liquid polishes, textile finishes, and paper sizes.

acraldehyde. See acrolein.

"Acrawax." [73] Brand name for proprietary product. Modified fatty acid ester.
Properties: Hard, light-brown, synthetic wax. Good luster. Soluble (hot) in ethyl alcohol, toluol, butyl acetate and turpentine. Partly soluble in mineral oil and mineral spirits. Insoluble in water. Blends with carnauba wax, rosin, shellac, stearic acid, estergum, etc.
Constants: Sp. gr. (24°C) 1.04; m. p. 95-97°C; flash point 230°C (open cup).
Containers: 1-gal can (8 lbs); 5-gal can (40 lbs); 55-gal drum (400 lbs).
Uses: For the manufacture of polishes, dental waxes, record waxes, wax coatings, etc.

"Acrawax" B. [73] Brand name for a proprietary product. Claimed to be a modified fatty acid ester.
Properties: A hard, brown wax of good luster. Soluble (hot) in mineral spirits, alcohol, toluol, butyl acetate, mineral oil. Insoluble in water. This product differs from "Acrawax" in that it is compatible with paraffin wax and forms gels with mineral spirits and kerosene.
Constants: Sp. gr. 0.955 (25°C); flash point 235°C (open cup); m. p. 81° to 84°C.
Containers: 1-gal, 5-gal cans; 55-gal drums. Net weight: 8, 50, 400 lbs.
Uses: Flatting agent for paints, enamels and varnishes.

"Acrawax" C. [73] Brand name for proprietary product.
Properties: Hard, brown synthetic wax, having a good luster. Soluble (hot) in toluol, mineral spirits, mineral oil, vegetable oil and turpentine. Insoluble in water and isopropanol. Blends with paraffin wax, carnauba wax, candelilla, rosin. This product is of interest where a high-melting wax which is not brittle is desired.
Constants: M. p. 140-142°C; flash point 283°C (open cup); sp. gr. 0.975 (25°C).
Containers: 1-gal can (8 lbs); 5-gal can

*See "I.C.C. Shipping Regulations," page xiii.
Reference numbers refer to name of manufacturer. See "List of Manufacturers," page v.

(40 lbs); 55-gal drum (400 lbs).
Uses: In the manufacture of polishes, electrical insulation, waterproofing, record waxes, dental waxes and special wax combinations. Because of its high flash point, "Acrawax" C can be used for many purposes where ordinary waxes are unsuitable due to fire hazards.

acridine $C_{13}H_9N$ (tricyclic).
Properties: Small colorless needles. When the dust or vapor is inhaled it causes violent sneezing; solutions of acridine and salts irritate the skin. Soluble in alcohol, ether or carbon disulfide; sparingly soluble in hot water.
Constants: Sublimes at 100°C; m.p. 111°C; b.p. above 360°C.
Derivation: (a) By extraction with dilute sulfuric acid from the anthracene fraction from coal tar and adding potassium dichromate. The acridine chromate precipitated is recrystallized, treated with ammonia and recrystallized. (b) It has been obtained synthetically by a number of processes.
Uses: Manufacture of dyes; derivatives, especially acriflavine, proflavine; analytical reagent.
Shipping regulations: None.*

acriflavine (neutral acriflavine; euflavine; trypaflavine neutral) $C_{14}H_{14}N_3Cl$. A mixture of 3,6-diamino-10-methylacridinium chloride and 3,6-diaminoacridine.
Properties: Brownish or orange, odorless, granular powder. Soluble in 3 parts of water; incompletely soluble in alcohol; nearly insoluble in ether and chloroform; the aqueous solutions fluoresce green on dilution.
Derivation: Synthetic.
Method of purification: Crystallization.
Containers: Amber glass bottles.
Uses: Antiseptic and bacteriostat.
Shipping regulations: None.*

acriflavine hydrochloride (trypaflavine; flavine) $C_{14}H_{15}N_3Cl_2$. A mixture, as above.
Properties: Brownish-red, odorless, crystalline powder. Soluble in 3 parts of water; somewhat soluble in alcohol; nearly insoluble in ether and chloroform. Aqueous solutions are dark-red in color and fluoresce green on dilution. An aqueous solution (1:250) is distinctly acid.
Derivation: Synthetic.
Method of purification: Crystallization.
Containers: Amber glass bottles.
Uses: Antiseptic and bacteriostat.
Shipping regulations: None.*

acriflavine, neutral. See acriflavine.

"Acrilan." [112] Trademark for a synthetic fiber made from acrylonitrile.
Properties: Tensile strength (psi) 37,000-40,000; elongation 36%; sp. gr. 1.17; moisture regain 1.2% (70°F, 65% relative humidity); shrinks 5% at 485°F. Resistant to mineral acids; fair to good resistance to weak alkalies. Insoluble in alcohol, acetone, benzene, carbon tetrachloride, and petroleum ether; soluble in dimethyl sulfoxide, maleic anhydride, ethylene carbonate, nitriles, and nitrophenols.
Derivation: A solution of polymerized acrylonitrile is forced through minute holes of a spinneret, the solvent is removed, and the resulting fiber is stretched.
Uses: Woven and knitted clothing fabrics; carpets; drapes; upholstery; electrical insulation; laminates.

"Acriviolet." [243] Trademark for dye mixture used as oral antiseptic.

acroleic acid. See acrylic acid.

acrolein (propenal; acrylic aldehyde; allyl aldehyde; acraldehyde) CH_2CHCHO.
Properties: Colorless or yellowish liquid; flammable; disagreeable choking odor; violent action on the eyes; poisonous! Soluble in water, alcohol and ether. Polymerizes readily unless inhibitor (hydroquinone) is added. Very reactive.
Constants: B.p. 52.7°C; m.p. —87.0°C; sp. gr. (20/20°C) 0.8427; wt/gal (20°C) 7.03 lbs; flash point (Cleveland open cup) below 0°F.
Derivation: (a) By the oxidation of allyl alcohol or propylene; (b) by heating glycerol with magnesium sulfate; (c) from propylene by use of special bismuth-phosphorus-molybdenum catalyst.
Method of purification: Rectification.
Grades: Technical.
Containers: 1-gal cans; 5-, 55-gal drums; tanks.
Uses: War gas (lachrymator); resins; intermediate for polyurethane and polyester resins, methionine, pharmaceuticals.
Fire hazard: Dangerous.
Shipping regulations: Flammable liquid. Red label.*

acrolein dimer (2-formyl-3,4-dihydro-2H-pyran) $\underline{OCH{:}CHCH_2CH_2}CHCHO$.
Properties: Liquid; sp. gr. 1.0775 (20°C); b.p. 151.3°C; freezing point —100°C; flash point (open cup) 118°F; soluble in water.
Containers: 55-gal drum.
Uses: Intermediate for resins, pharmaceuticals, dyestuffs.

"Acronal." [440] Trademark for an aqueous dispersion of polyacrylates or copolymers of acrylates with other vinyl compounds. Solids content between 40 and 50%. At room temperature dry to films which vary from soft and very tacky to relatively hard and block resistant.
Uses: Coating, impregnating and laminating paper, fabrics, non-wovens, fibers, etc.; adhesives raw materials and binders for leatherboard.

"Acronize." [57] Trademark for chlortetracycline.

"Acronol." [206] Brand name for a line of basic dyes for cotton, rayon, silk, and paper.

acrylamide $CH_2CHCONH_2$.
Properties: Colorless, odorless crystals; m.p. 84.5°C; b.p. (25 mm) 125°C; sp. gr.

1.122 (30°C); soluble in water, alcohol, acetone; insoluble in benzene, heptane. The solid is stable at room temperature but may polymerize violently on melting. Monomer is toxic.
Derivation: Hydration of acrylonitrile with sulfuric acid (84.5%) and neutralization.
Grade: Technical (approximately 97% pure).
Containers: Fiber drums.
Uses: Synthesis of dyes, etc.; polymers or copolymers as plastics, adhesives, paper and textile sizes, soil conditioning agents.
Shipping regulations: None.*

acrylate resins (acrylic resins). Thermoplastic polymers or copolymers of acrylic acid, methacrylic acid, esters of these acids, or acrylonitrile. The latter, and the methyl and ethyl esters are the most frequently used starting materials. These colorless monomer liquid esters polymerize readily in the presence of light, heat, or catalysts such as benzoyl peroxide, and must always be stored or shipped with inhibitor present to avoid spontaneous and explosive polymerization. The resins range from soft, sticky semifluid materials to hard solids. Molding powders are granular solids, often cube cut or minute spherical granules. The polymethacrylates are harder than the corresponding polyacrylates and the methyl esters form harder resins than the ethyl or butyl esters. Polymethylmethacrylate, well known examples of which are "Lucite" and "Plexiglas," is outstanding for its clarity and transparency, which lead to important uses. An almost unique property is that of carrying light around corners and reflecting it out the edges of a piece of cast polymer. Additional characteristics of the acrylate resins are low water absorption, low specific gravity, good shock resistance and high dielectric strength. These resins are stable to outdoor weathering, and also chemically stable at moderate temperatures, and resistant to action of water, aqueous salt solutions, and to moderate concentrations of acid and base. The acrylates are soluble in aromatic hydrocarbons, chlorinated hydrocarbons, esters, and ketones. They are compatible with many plasticizers. Insoluble, thermosetting resins can be obtained as cross-linked polymers by addition of acrylic anhydride, glycol esters of acrylic or methacrylic acid, or acrylamide.
Uses: Aircraft canopies and windows; automotive instrument dials, horn buttons, tail light lenses; comb and brush backs, lighting fixtures, pen and pencil barrels; refrigerator parts, costume jewelry and decorative ware and trim; binder in rocket propellants; contact lenses and optical parts; dentures; surgical instruments; safety glass; protective coatings including lacquers, paints, and finishes; adhesives; plasticizers and modifiers for vinyls and other resins; lubricating oil additives; textile and leather finishes and coatings; display fixtures and signs; aerosol snow. The synthetic fibers "Orlon," "Acrilan," "Dynel," and "Vinyon N" are also acrylate resins or acrylics since they are copolymers of acrylonitrile. Acrylic rubbers including acrylonitrile rubber are also of this general class.
See also polyacrylic acid.

acrylic acid (acroleic acid; ethylenecarboxylic acid; vinylformic acid; propenoic acid) $H_2C{:}CHCOOH$.
Properties: Colorless, corrosive liquid; acrid odor. Polymerizes readily. Miscible with water, alcohol and ether.
Constants: B.p. 140.9°C; m.p. 12.1°C; sp. gr. (20/20°C) 1.052; vapor pressure (20°C) 3.1 mm; wt/gal (20°C) 8.6 lbs; refractive index (20°C) 1.4224; flash point (open cup) 130°F.
Derivation: (a) Oxidation of acrolein; (b) hydrolysis of acrylonitrile; (c) molecular rearrangement of beta-propiolactone; (d) by Reppe process from acetylene, carbon monoxide and water.
Grades: Technical (esterification and polymerization grades); glacial (97%).
Containers: Bottles; drums; tank cars.
Uses: Intermediate; as monomer for acrylate resins.

acrylic acid resins. See acrylate resins.

acrylic aldehyde. See acrolein.

acrylic esters. See acrylate resins.

acrylic fiber. Generic name for a manufactured fiber in which the fiber-forming substance is any long chain synthetic polymer composed of at least 85% by weight of acrylonitrile units $-CH_2CH(CN)-$ (Federal Trade Commission). See "Acrilan," "Orlon," "Creslan," "Zefran."

acrylic resin molding powders. See acrylate resins.

acrylic resins. Same as acrylate resins.

acrylic rubber. A synthetic rubber made at least in part from acrylonitrile, the outstanding example being acrylonitrile rubber.

"Acryloid" Coating Resins. [23] Trademark for acrylic ester polymers in organic solvent solutions or 100% solids form, water-white and transparent. Films range from very hard to very soft.
Use: Exceptionally resistant surface coatings, such as heat-resistant and fumeproof enamels, vinyl and plastic printing, fluorescent coatings, clear and pigmented coatings on metals.

"Acryloid" Modifiers. [23] Trademark for thermoplastic acrylic polymers in powder form. Various grades facilitate processing or improve physical properties of rigid or semi-rigid poly(vinyl chloride) formulations.

"Acryloid" Oil Additives. [23] Trademark for acrylic polymers supplied in special oil solution or in diester lubricant.
Use: Viscosity-index improvement, pour-

*See "I.C.C. Shipping Regulations," page xiii.
Reference numbers refer to name of manufacturer. See "List of Manufacturers," page v.

point depression of lubricating oils and hydraulic fluids, sludge dispersancy in lubricating and fuel oils.

acrylonitrile (propenenitrile; vinyl cyanide) $H_2C{:}CHCN$.

Properties: Colorless, mobile liquid; mild odor; fumes are toxic; freezing range −83° to −84°C; b.p. 77.3-77.4°C (760 mm); sp. gr. 0.8004 (25°C); flash point (Tag open cup) 32°F. Soluble in all common organic solvents; partially miscible with water.

Derivation: (a) Addition of hydrogen cyanide to acetylene in the presence of cuprous chloride catalyst; (b) catalytic dehydration of ethylene cyanohydrin; (c) from propylene, air and ammonia.

Grade: Over 99% pure.

Containers: 5-gal cans; 55-gal drums; tank cars; tank trucks.

Uses: Synthetic rubber; plastics; synthetic fibers ("Acrilan," "Dynel," and "Orlon"); organic synthesis; grain fumigant; paper industry.

Danger! Vapor hazardous. Flammable.

Shipping regulations: Flammable liquid. Red label.*

acrylonitrile-butadiene rubber. See acrylonitrile rubber.

acrylonitrile-butadiene-styrene resins (ABS resins). Thermoplastic resins with uniform molecular structure, high impact strength, high heat distortion strength, good electrical and low-temperature properties; resistant to action of most solvents, oils, and chemicals.

"Acrylon" Rubber. [65] Trademark for a group of synthetic acrylic rubbers outstanding in resistance to oil, grease, ozone, and oxidation.

Uses: Gaskets and rubber parts for contact with oils and diester lubricants.

"Acrysol." [23]

"Acrysol" Thickeners. Trademark for aqueous solutions of sodium polyacrylate or other polymeric acrylic salts. Viscous solutions, powerful thickening action on rubber latices and polymer emulsions. ASE-type is alkali-soluble acrylic emulsion, forms clear, viscous solution of thickener upon addition of alkali.

Use: Thickeners in paints, fabric coatings and backings, adhesives.

"Acrysol" Warp Sizes. Trademark for polyacrylic acid and copolymer products in aqueous solutions or dispersions. Some grades are solutions of sodium polyacrylate.

Use: Warp size for synthetic fibers, cotton and rayon; modifier of starch sizes.

ACS. Abbreviation for American Chemical Society.

"Actamer." [58] Trademark for bithionol (q.v.).

"Actane." [142] Trademark for fluoride-containing additive supplied in powder form for acid pickling solutions to dissolve siliceous films on metals as well as to assist in etching aluminum and other metals.

ACTH (adrenocorticotropin; corticotropin). A protein consisting essentially of a single chain of amino acids; 19- and 23-unit amino acid chains have been synthesized. The 23-unit substance is 100% biologically active. ACTH is one of the hormones secreted by the anterior lobe of the pituitary gland. It stimulates an increase in the secretion of the adrenal cortical steroid hormones.

Properties: White powder; freely soluble in water; soluble in 60-70% alcohol or acetone. Solutions are stable to heat. Molecular weight is approximately 3500.

Source: Extracted from whole pituitary glands of swine, sheep and ox. Normally isolated from swine.

Units: Pure ACTH has 150-200 potency units per mg.

Uses: Medicine; biochemical research; nutrition; veterinary medicine.

"Actifed." [301] Trademark for a combination of triprolidine hydrochloride and pseudoephedrine hydrochloride (q.v.).

"Actidil." [301] Trademark for triprolidine hydrochloride (q.v.).

"Acti-Dione." [327] Trademark for the crystalline antibiotic cycloheximide ($C_{15}H_{23}NO_4$) isolated from cultures of Streptomyces griseus.

Properties: Very soluble in chloroform, methanol and acetone; moderately soluble in 99% isopropanol, n-butanol, and amyl acetate; very slightly soluble in carbon tetrachloride and the saturated hydrocarbons; solubility in water, 2%; the solutions are stable at pH 3 to 7 for several weeks; rapidly destroyed at pH above 8.5; m.p. 114°-118°C.

Uses: May be added as the crystalline material to bacteriological media to facilitate the isolation or counting of bacteria in the presence of yeasts or molds; as "Actispray" for cherry leaf spot, cedar-apple rust, and cedar-hawthorn rust; as "Acti-Dione Ferrated"; "Acti-Dione"; and "Acti-Dione-Thiram" for turf diseases; "Acti-Dione" for powdery mildew on roses and other ornamentals; and as "Acti-Dione" (concentrate and ready-to-use) for white pine blister rust.

Hazards: Should be handled with caution as it is a potent skin irritant in high concentration or long exposure.

"Actidip." [273] Trademark for a mildly alkaline powder having the property of refining crystal size in phosphate baths when used just prior to the phosphate. Commercially available in 100-lb fiber drums.

actinides. See actinide elements.

actinide elements (actinide series). The term applied to the group of chemical elements of increasing atomic number, starting with actinium and extending through the recently discovered element number 103. The names, chemical symbols, and atomic

*See "I.C.C. Shipping Regulations," page xiii.

Reference numbers refer to name of manufacturer. See "List of Manufacturers," page v.

numbers of the members of the series are: actinium, Ac, 89; thorium, Th, 90; protoactinium, Pa, 91; uranium, U, 92; neptunium, Np, 93; plutonium, Pu, 94; americium, Am, 95; curium, Cu, 96; berkelium, Bk, 97; californium, Cf, 98; einsteinium, Es, 99; fermium, Fm, 100; mendelevium, Md, 101; nobelium, No, 102. Element 103, discovered in 1961, tentatively named lawrencium, is expected to be the last member of the series. These names and symbols are those approved in the 1957 Report of the IUPAC. The symbols Ei and Mv formerly in use have been superseded.

The name actinide is given to the series by analogy with the lanthanide series of elements, the rare earths, since the successive differences between members of the series are due to a similar change in their atomic structure. Chemically, the early members of the actinide series show a greater variability in their reactions and kinds of compounds formed than do the lanthanides, but the later members have the same close similarity to each other.

All of these elements are unstable and undergo radioactive decay. Only two of the series, uranium and thorium, occur naturally in any significant quantity. The longest lived isotopes of protoactinium and actinium are members of the decay chain that starts with uranium-235. Numerous other isotopes of these elements have been prepared by synthesis through nuclear reactions carried out in nuclear reactors or accelerators.

The name transuranic has frequently been given to the elements with atomic number greater than 92. They were first made by the methods of heavy element build-up starting with the then heaviest known element uranium. While many isotopes of the elements from 93 to 101 are known, none have half lives long enough to allow the accumulation of any significant quantity, with the notable exception of plutonium which is made in quantity for nuclear weapons. Many show the property of fission by means of thermal neutrons, and some of the heavier ones undergo fission spontaneously. The general instability increases as the elements have higher atomic numbers, and all of the known isotopes of the heaviest have very short half-lives, making the problem of the synthesis of the as yet unknown members more and more difficult. There is much interest in making number 104, which should not be an actinide, but be similar to hafnium in its chemistry, and the confirmation of this point would check the theory.

actinide series. See actinide elements.

actinium Ac. A radioactive element found in nature as a constituent of all uranium ores, one ton of pure pitchblende containing 0.15 mg of actinium. Actinium has an atomic number of 89 and is classified as the first member of the actinide series of elements. The longest lived isotope, Ac-227, decays with emission of alpha and beta particles and has a half-life of 21.7 years. The most important source of actinium is pile neutron bombardment of radium. Actinium is similar chemically to the more basic of the rare earth elements and may be precipitated with them as the oxalate. Except for the sulfides, the compounds of actinium are colorless. Compounds of the types $Ac(OH)_3$, Ac_2O_3, $AcCl_3$, etc., have been prepared.

actinium decay series (actinium series). A series of little known radioactive elements, of which natural actinium is the best known and most stable member. These elements are produced in the radioactive decay of uranium 235 into actinium 227 and of the latter into lead of atomic weight 207. The lead is the stable end product of the series.

actinium series. See actinium decay series.

actinolite $Ca_2(Mg, Fe)_5Si_8O_{22}(OH)_2$. A natural hydroxy-calcium-magnesium-iron silicate.
Properties: Color green; luster vitreous to silky; fibrous to granular; fibers brittle.
Constants: Sp. gr. 2.3-3.2; hardness 5-6.
Occurrence: Canada, United States, Europe.
Uses: A minor asbestos mineral; building material.

actinomycin. A family of antibiotics produced by Streptomyces; reported to be active against E. coli, other bacteria, fungi and to have cytostatic and radiomimetic activity. There are many forms of actinomycin and two of commercial importance include A and C. Actinomycin has been suggested for use in cancer.

"Actispray." [327] See "Acti-Dione."

activated alumina. A highly porous and granular form of aluminum oxide having preferential adsorptive capacity for moisture from gases, vapors, and some liquids. When saturated, it can be revived or reactivated by the application of heat within the temperature range of 350-600°F to drive off the moisture. The cycle of adsorption and reactivation can be repeated many times. Available in granules ranging in size from a powder to pieces approximating 1 ½ inches. The 8 to 14 and ¼" to 8 mesh are the more commonly used sizes. The average weight approximates 50 lbs/cu ft.

Activated alumina is also effective for the removal of oil vapor from commercial gases, such as oxygen, hydrogen, carbon dioxide, natural gas, etc., introduced as lubrication during compression.

It is also used as a catalyst or catalyst carrier; to remove fluorides from drinking water; and in chromatography.

activated carbon. See active carbon.

activated charcoal. See active carbon.

activated clays. Clays whose adsorbent character or bleaching action has been enhanced by treatment with acid. Bentonite clay is most frequently treated in this fashion.

activated sludge. See sewage sludge.

activation.
1. Process of heating an adsorbent such as charcoal (see adsorption). This is usually done in the presence of steam or other inert gas in order to condition the adsorbent so that it will be able to exert its powers of adsorption to a maximum degree. The same procedure is used to regenerate or reactivate an adsorbent after it has been in use for a period of time. See, for example, active carbon.
2. Physico-chemical concept for a process which increases the internal energy of molecules and so allows them to undergo chemical reaction when otherwise they would not do so.

activation analysis. An extremely sensitive and specific technique for identifying and measuring very small amounts of various elements. A sample is exposed to neutrons in a nuclear reactor and from the characteristics of the induced radiation, the trace elements present are identified. The technique is particularly useful when concentrations of the elements are too small to be measured by ordinary means. Trace elements have thus been determined in drugs, fertilizers, foods, fuels, glass, minerals, dusts, water, toxicants, etc.

activator.
1. A substance which increases the effectiveness of a rubber vulcanization accelerator. Zinc oxide is the most important example but zinc laurate, zinc stearate, and litharge are also used.
2. A substance which is required in trace quantities to impart luminescence to certain crystals. Silver and copper are activators for zinc sulfide and cadmium sulfide pigments.

active amyl alcohol. See 2-methyl-1-butanol.

active carbon (activated carbon; activated charcoal). Any form of carbon characterized by high adsorptive capacity for gases, vapors and colloidal solids. The carbon, char or charcoal is produced by destructive distillation of wood, peat, lignite, nut shells, bones, vegetable or other carbonaceous matter, but must usually be "activated" to develop adsorptive power. Activation is usually achieved by heating to high temperatures (800-900°C) with steam or carbon dioxide, which brings about a porous particle structure. In some cases hygroscopic substances, such as zinc chloride and/or phosphoric acid or sodium sulfate, are added prior to the destructive distillation or activation, to increase adsorptive capacity. The carbon content of active carbons ranges from about 10% for bone charcoal to 98% for some wood chars. The internal surface area of active carbon has been estimated to be about 3600 square feet per gram. The density ranges from 0.08 to nearly 0.5. The chief uses are: bone charcoal, coal and chars from light woods and lignites are suitable for adsorbing color and other impurities from liquids. Chars from medium dense woods and shells are suitable for general vapor and solvent recovery. Only chars of considerable density and structural hardness, especially prepared for such use, are satisfactory for gas-mask and toxic-vapor adsorption.
Grades: Various; N. F. XI.
Shipping regulations: (As activated charcoal) flammable solid. Yellow label required for express only.*

activity. The number of atoms, or a quantity proportional to the number of atoms, of a radioactive substance decaying per unit time. The unit of activity is the curie. The specific activity is the activity per unit weight of the radioactive material. Frequently the term activity will be used synonomously with radioactivity.

activity series. See electromotive series.

"Acto." [51] Trademark for refined petroleum sodium sulfonate. Used as an oil-soluble emulsifier and surface-active agent.

"Actol." [51] Trademark for a naphthenic base crankcase oil for use where low cost is prime factor.

ADA. Abbreviation for acetonedicarboxylic acid. See beta-ketoglutaric acid.

"Adakane." [221] Trademark for a line of saturated long chain hydrocarbons. Available in various grades for specific applications in lubricants, solvents, inert carriers for organic compounds, liquid radiation shield, denaturant, substitute for purified kerosene.

"Adalin." [162] Trademark for carbromal or bromodiethylacetylurea (q. v.).

adamantine spar. See corundum.

adamsite. See phenarsazine chloride.

"Addition Agent 774." [28]
Properties: Pale amber, mildly alkaline water solution of organic wetting agents; faint amine odor.
Containers: 9-lb (1-gal) bottles; 125-lb fiber drums.
Uses: As antipitting agent in "Coppralyte" copper plating; as antifume agent in "Durobrite" zinc plating.

addition polymers. Polymers formed by the direct addition or combination of the monomer molecules with one another, without the formation of low molecular weight byproducts such as water.

adducts. See inclusion complexes.

"Adecene." [221] Trademark for a line of fatty olefins. Available in various grades for applications in viscosity index improvers, pour-point depressants, detergents, lead scavengers, leather treating, textile and paper chemicals, adhesives, plastics, polymers and protective coatings.

"Ademol." [412] Trademark for flumethiazide (q. v.).

*See "I.C.C. Shipping Regulations," page xiii.
Reference numbers refer to name of manufacturer. See "List of Manufacturers," page v.

Aden gum. See arabic gum.

adeniin. $C_{19}H_{28}O_8$.
Properties: Light yellow, amorphous powder m. p. 84-85°C; dextrorotatory; soluble in alcohol; insoluble in water.
Derivation: From the juice of Adenium honghel.
Use: Medicine.

adenine (6-aminopurine) $C_5H_5N_5$. A purine found in ribonucleic acids and deoxyribonucleic acids, nucleosides, nucleotides, and many important coenzymes.
Properties: White, odorless microcrystalline powder with sharp salty taste. M. p. 360-5°C (dec.). Very slightly soluble in cold water; soluble in boiling water, acids and alkalies; slightly soluble in alcohol; insoluble in ether and chloroform. Aqueous solutions are neutral.
Derivation: By extraction from tea; by synthesis from uric acid; prepared from yeast ribonucleic acid.
Use: Medicine and biochemical research.

adenine hydrochloride $C_5H_5N_5 \cdot HCl \cdot \frac{1}{2} H_2O$.
Properties and uses similar to adenine.

adenine riboside. See adenosine.

adenine hemisulfate $(C_5H_5N_5)_2 \cdot H_2SO_4 \cdot 2H_2O$.
Properties and uses similar to adenine.

adenohypophyseal luteotropin. See luteotropin.

adenosine (adenine riboside; 9-beta-D-ribofuranosyladenine) $C_{10}H_{13}N_5O_4$. The nucleoside composed of adenine and ribose.
Properties: White, crystalline, odorless powder with mild, saline or bitter taste. M. p. 229°C. Quite soluble in hot water; practically insoluble in alcohol.
Derivation: Isolation following hydrolysis of yeast nucleic acid.
Use: Biochemical research.

adenosine diphosphate (5'-adenylphosphoric acid; ADP; adenosine diphosphoric acid) $C_{10}H_{15}N_5O_{10}P_2$. A nucleotide of great importance in the maintenance of life. It is found in all living cells and is important in the storage of energy for chemical reactions.
Derivation: (a) From adenosine triphosphate by hydrolysis with the enzyme adenosinetriphosphatase from lobster or rabbit muscle; (b) By yeast phosphorylation of adenosine.
Use: Biochemical research.
Commercially available as the sodium or barium salt.

adenosinediphosphoric acid. See adenosine diphosphate.

adenosine monophosphate. See adenylic acid.

adenosine-3'-phosphoric acid. See adenylic acid.

adenosine-5'-phosphoric acid. See adenylic acid.

adenosine triphosphate (5'-adenyldiphosphoric acid; ATP) $C_{10}H_{16}N_5O_{13}P_3$. A nucleotide of great importance to the body. It serves as a source of energy for many chemical reactions in the body, especially those reactions associated with muscular activity.
Properties: White amorphous powder; odorless; very faint sour taste. Very soluble in water; insoluble in alcohol, ether, and organic solvents; stable in acidic solutions; decomposes in alkaline solution.
Derivation: Isolation from muscle tissue; yeast phosphorylation of adenosine.
Use: Experimental work in physiology and biochemistry.
Commercially available as the disodium, dipotassium, and dibarium salts.

5'-adenyldiphosphoric acid. See adenosine triphosphate.

adenylic acid (adenosine monophosphate; AA; adenosine phosphoric acid; AMP). The monophosphoric ester of adenosine; i. e., the nucleotide containing adenine, D-ribose and phosphoric acid. The phosphate may be esterified to either the 2, 3, or 5 carbon of ribose yielding adenosine-2'-phosphate, adenosine-3'-phosphate, and adenosine-5'-phosphate, respectively. Yeast adenylic acid, previously believed to be adenosine-3'-phosphate, is now thought to be a mixture of the C2' and C3' compounds. Muscle adenylic acid (A5MP) has been shown to be adenosine-5'-phosphate. Adenylic acid is a constituent of many important coenzymes.
Properties (muscle adenylic acid): Crystalline solid; m. p. 196-200°C. Readily soluble in boiling water. Gives only traces of furfural when boiled with 20% hydrochloric acid.
(yeast adenylic acid monohydrate): Long crystalline rods. Decomposes 195°C. Anhydrous form decomposes at 208°C. Almost insoluble in cold water; slightly soluble in boiling water. Gives quantitative yield of furfural when distilled with 20% HCl.
Derivation: Yeast adenylic acid by precipitation from yeast nucleic acid. Muscle adenylic acid by precipitation from tissues; by hydrolysis of ATP with barium hydroxide; by enzymatic phosphorylation of adenosine.
Uses: Medicine and biochemical research.

5'-adenylphosphoric acid. See adenosine diphosphate.

adeps. See lard.

"Ad-Flex." 265 Trademark for vulcanizable synthetic rubber produced by the polymerization of diolefin or diolefin/mono-olefin mixture further modified by addition of low molecular weight mercaptans.

adhesion
1. The phenomenon of holding surfaces together with an adhesive (q. v.).
2. The phenomenon of the sticking of two surfaces together due to molecular attraction for each other.

adhesive. A substance capable of holding materials together by surface attachment.

Note: Adhesive is the general term and includes among others cement (pyroxylin and rubber), glue, mucilage and paste. All of these terms are loosely used interchangeably. Various descriptive adjectives are applied to the term adhesive to indicate certain characteristics as follows:
Physical form, that is, liquid adhesive, tape adhesive.
Chemical type, that is, silicate adhesive, resin adhesive.
Materials bonded, that is, paper adhesive, metal-plastic adhesive, can label adhesive.
Conditions of use, that is, hot-setting adhesive. (ASTM definition: ASTM D 907-52).
See also dextrin; furan resin; sodium silicate.

"Adhevia." [206] Brand name for proprietary adhesion agents for tar and bitumen.

adiabatic. A process, condition, or operation during which there is no gain or loss of heat from the surroundings.

adiphenine hydrochloride (diethylaminoethyl diphenylacetate hydrochloride) $(C_6H_5)_2CHCOOC_2H_4N(C_2H_5)_2 \cdot HCl$.
Properties: Crystals; m.p. 113°C. Soluble in water; almost insoluble in alcohol and ether.
Use: Medicine.

adipic acid (hexanedioic acid; 1,4-butanedicarboxylic acid) $COOH(CH_2)_4COOH$.
Properties: White, crystalline solid. M.p. 152°C; b.p. (100 mm) 265°C; sp. gr. (20/4°C) 1.360; flash point (closed cup) 385°F. Slightly soluble in water; soluble in alcohol and acetone. Relatively stable.
Derivation: Oxidation of cyclohexane, cyclohexanol, or cyclohexanone with air or nitric acid.
Grades: Technical.
Containers: Glass bottles; tins; 50-lb multiwall paper bags; drums.
Uses: Manufacture of nylon and polyurethane foams; preparation of esters for use as plasticizers and lubricants; ingredient of baking powders; insecticides; adhesives.
Shipping regulations: None.*

adipocerite. Synonym for hatchettite (q.v.).

"Adipol" 10A Plasticizer. [55] Trade name for diisooctyl adipate (q.v.).

"Adipol" BCA Plasticizer. [55] Trademark for dibutoxyethyl adipate (q.v.).

"Adipol" DIBA Plasticizer. [55] Trade name for diisobutyl adipate (q.v.).

"Adipol" 2EH Plasticizer. [55] Trademark for di-2-ethylhexyl adipate (q.v.).

"Adipol" ODY Plasticizer. [55] Trademark for n-octyl n-decyl adipate (q.v.).

"Adipol" XX Plasticizer. [55] Trademark for di-isodecyl adipate (q.v.).

adiponitrile $NC(CH_2)_4CN$. An intermediate in the manufacture of nylon 66. M.p. 1-3°C. It can be hydrolyzed to adipic acid or hydrogenated to form hexamethylenediamine. Industrially, there are two important synthetic methods for adiponitrile: (1) from furfural, through the intermediate furan, tetrahydrofuran, and 1,4-dichlorobutane; and (2) from butadiene via 1,4-dicyanobutene-2.

"Adiprene" C. [28] Trademark for a urethane rubber. Reaction product of a diisocyanate and a polyalkylene ether glycol.
Properties: Amber colored solid in stick form; sp. gr. 1.07.
Containers: 50-lb drums.
Use: A synthetic rubber for products that must have exceptional abrasion resistance, high tensile strength and low temperature brittle point.

"Adiprene" L-100. [28] Trademark for a liquid urethane rubber. Reaction product of a diisocyanate and a polyalkylene ether glycol.
Properties: Honey-colored liquid; viscosity 14,000-19,000 cps at 30°C; sp. gr. 1.06.
Containers: 450-lb drums.
Uses: A fluid elastomer which can be converted into a tough elastic solid rubber having excellent resistance to abrasion, oxidation and ozone.

"Adiprene" L-167. [28] Trademark for a liquid urethane rubber. Reaction product of a diisocyanate and a polyalkylene ether glycol.
Properties: Honey-colored liquid; viscosity 6000-8000 cps at 30°C; sp. gr. 1.06.
Containers: 450-lb drums.
Use: A type of "Adiprene" L-100 which will give harder vulcanizates.

"Admex." [221] Trademark for a line of plasticizers consisting variously of epoxidized soybean oil, monomeric esters and polyesters. Typical member:
"Admex" 710, epoxidized soybean oil.
Properties (approximate): A light colored liquid; sp. gr. 0.994 (25/25°C); 8.25 lbs/gal (25°C); insoluble in water; viscosity 400 cps (25°C).
Uses: Plasticizers for polyvinyl chloride and copolymers, cellulosics and other natural and synthetic resins.

admiralty metal. A non-ferrous alloy containing 70-73% copper, 0.75-1.20% tin, and the remainder zinc. It offers good resistance to dilute acids and alkalies, sea water, and moist sulfurous atmospheres. Sp. gr. (20°C) 8.53; liquidus temperature 935°C; solidus temperature 900°C.
Uses: Condenser, evaporator, and heat exchanger tubes, plates, and ferrules.

adobe clay. See brick clay.

"Adogen." [221] Trademark for a line of fatty nitrogen chemicals including amines, amides, nitriles and quaternary ammonium compounds. Available in various grades for specific applications in fabric softeners, ore separation, detergents, petroleum additives, corrosion inhibitors, bactericides, printing inks, anti-block and slip agents, water proofing formulations and chemical intermediates.

"Adol." [221] Trademark for a line of industrial fatty alcohols, saturated and mono-unsaturated. Available in a variety of grades for specific applications in detergents, vinyl plastics, shampoos, corrosion inhibitors, agricultural chemicals and synthetic waxes.

ADP. See adenosine diphosphate.

adraganthin. See bassorin.

adrenal cortical hormones. See corticoid hormones.

"Adrenalin(e)." [330] Trademark for epinephrine (q. v.).

adrenocorticotropic hormone. See ACTH.

"Adroyd." [330] Trade name for oxymetholone (q. v.).

adsorbent. A substance which has the ability of condensing or holding other substances on its surface. Active carbon, activated alumina, and silica gel are examples.

"Adsorbol." [38] Trademark for a montmorillonite clay produced in both activated and natural form in both granular and in 200-mesh powder. Used to remove odors and colors in oils, fats, fatty acids and similar substances.

adsorption. The taking up of a gas, vapor or dissolved material on the surface of a solid. All solids have on their surface some adsorbed material. Certain finely divided materials, particularly active carbon (q. v.) and silica gel, can adsorb relatively large quantities of other materials. The process of adsorption must be distinguished from absorption.

adsorption indicator. A substance used in analytical chemistry to detect the presence of a slight excess of another substance or ion in solution as the result of a color produced by adsorption of the indicator on a precipitate present in the solution. Thus a precipitate of silver chloride will turn red in a solution containing even a minute excess of silver ion (silver nitrate solution), if fluorescein is present. In this example, fluorescein is the adsorption indicator.

"Advacar." [230] Trademark for a series of water-dispersible paint driers.

"Advacide." [230] Trademark for a series of fungicides and mildewcides for paints, etc., wood and fabric preservatives.

"Advasol." [230] Trademark for a series of liquid soluble paint driers of a linoresinate base. Supplied in high metal concentrations: lead 24%, calcium 4%, cobalt 6%, manganese 6%, iron 6%, zinc 5%.

"Advawax." [230] Trademark for a line of wax modifiers.

"Advawax P." A master batch of high molecular weight polybutene and paraffin wax designed to facilitate incorporation of the rubbery polymer into additional wax.

"Advawax M." A master batch of high molecular weight polybutene and amber colored microcrystalline wax; designed to facilitate incorporation of the rubbery polymer into additional wax.

"Advawax 280." A hard, high melting point synthetic wax for use in lacquers, varnishes, as a plastic lubricant and anti-blocking agent, as an extender or substitute for carnauba wax.

"Advawet." [230] Trademark for a series of wetting agents for emulsion paints, latex paints and vinyl plastisols.

"AD-X2." [437] Trademark for a treated mixture of salts including sodium sulfate and magnesium sulfate. Used as additive to lead storage batteries to reduce effects of sulfation.

aerate. To impregnate or saturate a material (usually a liquid) with air, or some similar gas. This is usually achieved by bubbling the air through the liquid, or by spraying the liquid into air.

"Aero" Brand Cyanide. [57] Trademark for calcium cyanide sold in the form of black flakes and used primarily for leaching of gold and silver ores.

"Aerobrite." [57] Trademark for a metal processing aid.

"Aerocarb." [57] Brand name for mixtures containing salts of sodium, potassium, and barium used as a molten bath for case-hardening and heat treatment of steel.

Containers: Steel drums.

Shipping regulations: Poison Class B. Poison label.*

"Aerocase." [57] Brand name for mixtures of calcium chloride and sodium chloride, used as a molten bath activated with calcium cyanide, to case-harden and heat-treat steel.

Containers: Steel drums.

Shipping regulations: Calcium cyanide component, Poison, Class B. Poison label.*

"Aero" Catalysts. [57] Trademark for a line of various special catalysts used mostly in the manufacture of ammonia and sulfuric acid.

"Aerocat" Cracking Catalysts. [57] Trademark for a line of synthetic silica-alumina catalysts used for refining of petroleum in fluid catalytic cracking units.

"Aero" Cyanamid. [57] Brand name.

Properties: Gray to black dry material of various finenesses.

Containers: 50- and 100-lb bags.

Uses: For direct application to soil; formulating complete fertilizers and special uses such as leaf removal, vine killing, wood control and disease control.

"Aero" Cyanate. [57] Brand name for product containing 92% potassium cyanate.

Properties: White flakes.

Containers: 5-lb can; 40-lb pail; 100-lb drum.

Uses: Weed control in onions. Weed control

in other crops being investigated.
Caution! Avoid breathing dust or spray mist. Avoid prolonged or repeated contact with skin. MCA warning label.

"Aerofat." [57] Trademark for a flotation agent.

"Aerofloat." [57] Trademark for aryl dithiophosphoric acids (liquids) and alkyl dithiophosphoric acid salts (solids) used as mining flotation reagents to promote frothing properties.

"Aerofloc" Promoters. [57] Trademark for a group of synthetic water soluble polymers used as flocculating agents to improve solid-liquid separations by thickening and filtration.

"Aeroform" Catalysts. [57] Trademark for a line of alumina-based catalysts used for the upgrading of gasoline in either fixed bed or fluid reforming units.

"Aerofroth" Frothers. [57] Trademark for a group of surface-active agents used primarily as foaming agents or frothers in flotation processing of ores and minerals.

"Aeroheat" Neutral Salts. [57] Trademark for various mixtures containing alkaline chlorides, carbonates, nitrates and nitrites. Used in the form of molten baths in the hardening, quenching, annealing and tempering of ferrous and non-ferrous metals.

"Aero" Hydrocyanic Acid Discoids. [57] Trademark for proprietary product containing hydrocyanic acid absorbed in cellulose discs.
Properties: Off-white cellulose discs which evolve hydrocyanic acid gas.
Containers: 1-lb can; 2 ½-lb can.
Use: For killing household and shipboard pests.
Shipping regulations: Class B poison. Poison label.*

"Aerolin." [57] Trademark for bating for use in the treatment of hides in the manufacture of leather.

"Aero" Liquid Hydrocyanic Acid. [57] Brand name for a proprietary product; hydrocyanic acid in liquid form of 96-98% commercial purity.
Containers: Steel cylinders. Drums for citrus fumigation in California only.
Uses: For fumigation of mills, warehouses and food-processing plants; fumigation of food and tobacco in chambers; also citrus fumigation.

"Aerolube." [57] Trademark for a line of additives for motor and industrial oils to improve oxidation stability, wear, and other properties.

"Aeromet." [57] Trademark for metallurgical additive in steel making where nitrogen content is desired; also applied to iron and steel for desulfurization.
Containers: Bulk freight cars and in moisture-resistant paper bags in fiber drums.

"Aeromine" Promoters. [57] Trademark for a group of cationic flotation reagents used in floth flotation processing of ores and minerals, primarily silica and silicates.

"Aero" Nitriding Compound. [57] Trademark for cyanide-carbonate mixture used for cyanide nitriding of steels.
Containers: Steel drums.

"Aeronox." [57] Trademark for a line of antioxidants for use in industrial oils and lubricants.

"Aero-Phos." [57] Proprietary product; finely ground Florida natural phosphate, 31%, 33%, 35% P_2O_5.
Properties: Light tan material generally 85% through a 200-mesh screen.
Containers: Bulk freight car and 100-lb bags.
Use: For direct application to the soil as a fertilizer.

"Aeroprills." [57] Proprietary product; ammonium nitrate fertilizer containing 33.5% nitrogen (16.75% nitrate nitrogen; 16.75% ammonia nitrogen).
Properties: Light cream colored pellets of small uniform size.
Containers: 80 or 100 lb. bags.
Uses: For direct application and in fertilizer mixtures.

"Aerosize" Sizing Compounds. [57] Trade name for a series of aqueous high free rosin emulsions used for treatment of paper and paper board to impart water, ink, and lactic acid resistance. May be used to give a surface sizing effect as on floor coverings.

aerosol.
1. Colloidal suspension of liquid or solid particles in a gas. For example, the insecticide DDT may be dissolved in "Freon" and kept under pressure. When released the solvent vaporizes rapidly leaving an aerosol of DDT. The "smog" causing so much trouble in Los Angeles and other industrial regions is an aerosol. In chemical warfare aerosols play an important part as screening smokes, etc.
2. Term also used for aerosol bomb, a container filled with liquefied gas and dissolved or suspended ingredients which can be dispersed as a spray or aerosol.

aerosol propellants. A compressed gas or vapor whose expansion carries another substance or mixture from a container when a valve is opened and pressure is thus released. Most non-food aerosol propellants (for insecticides, paints, perfumes), are fluorocarbons, as, for example, dichlorodifluoromethane. Compressed nitrogen and propane are used for aerosol shaving creams and toothpastes, and compressed nitrous oxide, carbon dioxide, and nitrogen are used for foods, as in aerosol whipped cream.

"Aerosol" Wetting Agents. [57] Brand name for neutral wetting agents obtainable in various types. They do not decompose on heating and are soluble in practically all nonaqueous media as well as in water.

*See "I.C.C. Shipping Regulations," page xiii.
Reference numbers refer to name of manufacturer. See "List of Manufacturers," page v.

"Aerosol" OT 100%: A nearly pure (99.5-100%) nontoxic, nonvolatile, practically odorless ester of a sulfonated dicarboxylic acid which resembles paraffin in appearance, is somewhat soft and plastic, practically nonhygroscopic and is supplied in pellet form. Solubility in water: 1.5 g/100 cc (25°C); 5.5 g/100 cc (70°C).

"Aerosol" OT Aqueous: A somewhat opaque, slightly gelatinous 10% (or 25%) dispersion of "Aerosol" OT 100% in water.

"Aerosol" OT Clear: A water-white, free-pouring liquid consisting of "Aerosol" OT 100% in 70%, 25% and 10% concentrations in water and a mutual solvent.

"Aerosol" OT-B: Free flowing white powder containing 85% "Aerosol" OT and 15% sodium benzoate.

"Aerosol" OS: Light tan colored powder containing approximately 80% active ingredient, the remainder being sodium sulfate.

"Aerosol" AY: A nonvolatile, practically odorless substance similar in appearance to "Aerosol" OT but markedly less plastic. Solubility in water: 39.2 g/100 cc (25°C), 50.2 g/100 cc (70°C).

"Aerosol" MA: A product similar to "Aerosol" AY. Solubility in water: 34.3 g/100 cc (25°C), 44.7 g/100 cc (70°C).

"Aerosol" 18: Light cream colored paste containing 35% active ingredient, 65% water. Readily dispensable in water, yielding a pearly, opalescent dispersion that becomes clear when heated. Of value as a detergent for fabrics in hard water.

"Aerosol" C-61: A cationic surface active agent.

Uses: Wetting agent; antiseptics; cosmetics; detergents; inks; cleaning; degreasing metals; case-hardening processes for metals; paints; varnishes, and lacquers; paper; printing inks; shampoos; toilet preparations; acid treatment of oil wells; embalming fluids; manufacture of crystalline materials; settling operations; breaking mineral oil emulsions; wall-paper removers; shoe cleaners and polishes.

"Aerosporin." [301] Trademark for polymyxin B, an antibiotic used in medicine.

"Aerospray." [57] Trademark for a resin emulsion to prevent dust loss, rain erosion or coal fines.

"Aerotex" Accelerator 187. [57] A buffered inorganic salt catalyst. Accelerator for melamine and urea formaldehyde resins, particularly where a rapid rate of cure is required.

"Aerotex" Accelerator AS. [57] An acid salt catalyst used primarily to cure "Permel" Resin B.

"Aerotex" Accelerator MX. [57] A buffered inorganic salt catalyst for melamine formaldehyde resins. Increases the efficiency of melamine formaldehyde resins.

"Aerotex" Accelerator S. [57] A catalyst for use with silicone water repellent emulsions only.

"Aerotex" Accelerator UTX. [57] A non-ammoniacal catalyst for urea and melamine formaldehyde resins. Resins catalyzed with this product polymerize at a moderate uniform rate which minimizes overcuring.

"Aerotex" Buffer DCY. [57] An organic, nitrogen containing compound which is added to thermosetting resin baths to prevent amine type or formaldehyde odors in resin-treated fabrics.

"Aerotex" Cream 450. [57] Trademark. A urea formaldehyde resin used as an anticreasing agent, a durable resin finish, a finishing agent, an antislip finishing agent and for shrinkage control.

"Aerotex" CSN. [57] Trademark for a 25% solids aqueous solution of a cationic organic compound. Pale yellow liquid; pH 6-7; imparts antistatic properties to acrylic, acetate, polyester, nylon, wool, and resin-treated cellulose fibers and fabrics.

"Aerotex" Fire Retardant NDC. [57] A water-soluble phosphate mixture. This product is a non-durable fire retardant for cellulosic textiles. It gives a non-hygroscopic finish resistant to tensile strength loss when exposed to high temperatures.

"Aerotex" Fire Retardant NDS. [57] A water-soluble phosphate mixture. This product is a non-durable fire retardant finish for textiles which produces a somewhat softer hand than "Aerotex" Fire Retardant NDC. It may be applied by padding or spraying.

"Aerotex" Resins. [57] Trademark for a group of synthetic resins used in finishing textile fabrics.

"Aerotex" Resin 110: A resin emulsion used for finishing cotton and as a sizing for hat felts.

"Aerotex" Resin 120: An oil-in-water type of emulsion of a thermoplastic resin. Useful as a durable finish for cellulosic textiles to increase the abrasion resistance, improve hand, and produce a durable finish.

"Aerotex" Resin 133: A thermoplastic copolymer used in creaseproofing formulations with melamine formaldehyde resins to modify the hand and increase the wrinkle recovery properties and the tensile strength of textiles.

"Aerotex" Resin 159: A dispersion of a thermoplastic resin used primarily for stiffening textile fabrics.

"Aerotex" Resin 180: A urea formaldehyde resin used as a bodying agent and also to insolubilize starch finishes.

"Aerotex" Resin 801: A modified urea formaldehyde resin used for creaseproofing, shrinkage control and increasing the body of rayon and cotton fabrics. This product has excellent stability.

"Aerotex" Resin 802: A modified urea formaldehyde resin used for producing durable body, wrinkle recovery and shrinkage control on rayon fabrics. It is a durable finishing agent which produces an anti-slip finish on all fibers.

"Aerotex" Resin 803: A modified urea formaldehyde resin which gives a very stiff finish on all fibers. It produces a stiff, wrinkle resistant finish on rayon.

"Aerotex" Resin 7513: A resin emulsion used as a finish for cotton and synthetics; also useful for building body of finish, along with "Aerotex" Resin M3.

"Aerotex" Resin M3: A melamine formaldehyde resin used for textile treatment to achieve crease resistance, shrinkage control, durable glazed finishes. Also used as a gasfading inhibitor for acetate dyes.

"Aerotex" Resin P-114: A resin dissolved in mineral spirits used in resin bonded pigment applications as a pigment binder.

"Aerotex" Resin P-116: A solution of a resin in mineral spirits. This product is used to prepare a printing emulsion for resin bonded pigments.

"Aerotex" Resin P-117: A solution of a resin in xylol. This product is used to prepare a printing emulsion for resin bonded pigments.

"Aerotex" Resin P-200: A solvent soluble modified melamine formaldehyde resin used in resin bonded pigment applications to improve scrub resistance.

"Aerotex" Resin UM: A melamine formaldehyde resin used for producing crease resistance, shrinkage control and durable mechanical finishes with a soft hand to textile fabrics.

"Aerotex" Softener H. [57] Trademark. A durable synthetic type textile softener of mixed cationic and anionic agents. Especially suited for use in conjunction with most of the synthetic resin processes because of its compatibility, being equally efficient on wool, cotton, or synthetic fibers.

"Aerotex" Softener W. [57] An anionic emulsion which acts as a softening agent and improves the luster and sewability of fabrics. Used as a plasticizing agent for "Aerotex" Resin 159.

"Aerotex" Syrup 55. [57] A modified urea-formaldehyde resin for use with other thermosetting resins or alone to produce durable stiffness on textile fabrics.

"Aerotex" Syrup 250 (Conc.). [57] Trademark. A urea formaldehyde resin used as an antislip finishing agent and for durable finishes on textiles.

"Aerotex" Thickener 37. [57] A water-soluble polymer which is used as a thickener and as an additive to warp sizes to improve the abrasion resistance, flexibility and size adhesion.

"Aerotex" Water Repellent S. [57] This product is a silicone emulsion for producing water repellency on textile fabrics.

aesculin. See esculin.

AET. See aminoethylisothiuronium bromide hydrobromide.

"Afaxin." [162] Trademark for synthetic oleovitamin A.

affinity. Chemical affinity is loosely used to express the tendency for a mixture of substances to react chemically. The free energy decrease is a quantitative measure of chemical affinity.

African pepper. See capsicum.

"Afrox." [89] Trademark for a foaming agent used to remove water and cuttings from drill holes in drilling oil wells by air or gas methods.

after-chromed dyes. Those which are improved in color or fastness by treatment with sodium dichromate, copper sulfate or similar materials, after the fabrics are dyed.

Ag. Symbol for silver.

agalite. A variety of talc (q. v.) from New Jersey. Used in paper coatings.

agalmatolite. A variety of pyrophyllite (q. v.).

agar. See agar-agar.

agar-agar. (Japanese, Bengal, Ceylon, or Chinese isinglass or gelatin; macassargum; Layor caranga; sometimes erroneously called vegetable glue or gelatin).

Properties: Consists in part, at least, of a mixed sulfate of calcium and a polysaccharide containing galactose. Dried mucilaginous substance extracted from various species of Gelidium and Algae. Unground, in thin, translucent membranous pieces; ground, pale, buff powder. Soluble in boiling water; insoluble in cold water; swells slowly in water. Insoluble in organic solvents.

Occurrence: Pacific and Indian Oceans; Japan Sea; also off southeastern U. S. coast.

Grades: Technical; U. S. P. XVI.

Containers: Multiwall paper sacks; fiber drums; bales.

Uses: Culture medium in bacteriology; sizing for silk; adhesives; substitute for gelatin; pharmaceutical preparations; photography (ingredient of sensitized emulsions); ingredient of vegetarian foods; sausage casing manufacture; food stuffs (thickening agent in milk, cream, ice cream, etc., substitute for white of egg); medicine; as coating or paste base in many specialties.

Shipping regulations: None.*

agaric (touchwood; spunk; tinder; larch agaric; German fungus). The dried fruit body of a fungus, Polyporus officinalis.

Properties: Light colored, spongy irregular-shaped pieces; faint odor; disagreeable taste.

Chief constituent: Agaric acid.

Occurrence: Upon European larch and various coniferous trees in western United States and Canada; commercially, principally from larch forests near Archangel.

Use: Medicine.

agaric acid (agaricinic acid; agaricic acid) $C_{19}H_{36}OH(COOH)_3 \cdot 1\frac{1}{2} H_2O$.

Properties: White, microcrystalline powder;

*See "I.C.C. Shipping Regulations," page xiii.
Reference numbers refer to name of manufacturer. See "List of Manufacturers," page v.

odorless; tasteless; m.p. 141.5°C with decomposition. Soluble in hot water, alcohol, and alkali solutions; barely soluble in cold water, ether, and chloroform.
Derivation: From agaric (q.v.).
Uses: Medicine; synthesis (esters and salts).

agaricic acid. See agaric acid.

agaricin. A preparation containing active agaric acid, obtained from agaric by alcoholic extraction.
Use: In medicine (similar to agaric).

agaricinic acid. See agaric acid.

agate. A form of native silica or quartz (q.v.) essentially a variegated chalcedony. The colors are either banded, irregularly clouded, or, due to visible impurities as in moss agate which has brown moss-like or dendritic forms, as of manganese oxide, distributed through the mass. The bands differ in porosity and may be colored artificially.
Uses: Gems; pestles and mortars; burnishers or polishers for gold workers and bookbinders; textile rollers; pivot supports for balances and magnetic needles; balls for ball mills, spatulas for mixing corrosive substances.

agave fibers. Hard, strong fibers obtained from various species of Agave, especially sisal (q.v.). They are used in rope, cordage, and sacking but are not as strong or water-resistant as abaca. See also cantala; henequen; istle.

age. See Fermi age.

age-resisters. See antioxidants.

"Agerite Alba." [69] Trade name for a proprietary product, para-benzyloxyphenol.
Properties: Light powder; sp. gr. 1.26; faint pleasant odor; none in rubber stocks; m.p. 115-120°C; nontoxic. Discoloration, essentially none in indirect sunlight; minimum in direct sunlight. Very slightly soluble in water; practically insoluble in petroleum hydrocarbons; soluble in benzene and alkalies.
Uses: Antioxidant in white and light-colored rubber products.

"Agerite DPPD." [69] Trademark for a proprietary product, N,N'-diphenyl-para-phenylenediamine.
Properties: Dark blue to black powder; sp. gr. 1.28 ± .03; melting range 145-152°C; moderately soluble in acetone, benzene, carbon disulfide, chloroform; slightly soluble in alcohol, petroleum hydrocarbons; insoluble in water.
Uses: Flex-crack resisting rubber antioxidant.

"Agerite Gel." [69] Trademark for a proprietary product, ditolylamines with a selected petroleum wax.
Properties: Grayish-white, soft, waxy solid; sp. gr. 1.01; odor faint; m.p. about 65°C; nontoxic. Insoluble in water; soluble in gasoline, chloroform, benzene, and carbon disulfide.
Use: Antioxidant in practically any rubber stock, especially sponge rubber.

"Agerite Hipar." [69] Trademark for a proprietary product, phenyl-beta-naphthylamine, plus isopropoxydiphenylamine, plus diphenyl-para-phenylenediamine.
Properties: Gray-brown soft powder; sp. gr. 1.19; odor, slight, none in rubber stocks; m.p. 75-90°C; nontoxic. Insoluble in water; slightly soluble in petroleum hydrocarbons; soluble in benzene.
Uses: Antioxidant in tire treads, rubber soling, and other goods where flex-cracking resistance is important.

"Agerite HP." [69] Trademark for a proprietary blend of phenyl-beta-naphthylamine and N,N'-diphenyl-para-phenylenediamine.
Properties: Gray to brown powder; sp. gr. 1.21 ± .03; melting range 89-96°C; soluble in acetone, benzene, chloroform, carbon disulfide; moderately soluble in alcohol, gasoline; insoluble in water.
Uses: Flex-crack resisting antioxidant.

"Agerite Iso." [69] Trademark for a proprietary product, para-isopropoxydiphenylamine.
Properties: Tan to gray flakes; sp. gr. 1.15 ± .03; melting range 80-86°C; very soluble in acetone, benzene, chloroform, carbon disulfide; soluble in alcohol; moderately soluble in gasoline, petroleum ether, insoluble in water.
Uses: Antioxidant for tires.

"Agerite Powder." [69] Trademark for phenyl-beta-naphthylamine.
$C_{10}H_7NHC_6H_5$.
Properties: Gray; odor very faint; m.p. 106-107°C; sp. gr. 1.19; nontoxic in ordinary handling. Insoluble in water; moderately soluble in gasoline, alcohol; soluble in chloroform, benzol, acetone, and carbon disulfide.
Use: Antioxidant in tire treads, inner tubes, wire insulation, mechanicals, and footwear.

"Agerite Resin." [69] Trademark for aldol-α-naphthylamine. $C_{10}H_7NCHCH_2CHOHCH_3$.
Properties: Cherry red resin; odor characteristic; m.p. 80-100°C; sp. gr. 1.16. Insoluble in water, gasoline; sparingly soluble in alcohol; soluble in chloroform, benzene, and carbon disulfide.
Use: Antioxidant in tire carcasses, tubes, uncured tape, black soles.

"Agerite Resin D." [69] Trademark for a polymerized trimethyldihydroquinoline.
Properties: Reddish-brown, brittle resin in small pellets. Odor faint, not objectionable; sp. gr. 1.08; m.p. 60-90°C; nontoxic. Insoluble in water and petroleum hydrocarbons. Easily soluble in benzene.
Use: Antioxidant in rubber products such as tire carcasses, inner tubes, wire insulation, belting, brake lining, air-bags, and steam hose.

"Agerite Spar." [69] Trademark of a proprietary product, mixed mono-, di-, and tristyrena-

ted phenols.
Properties: Light straw colored liquid; sp. gr. 1.08 ± .02; very soluble in petroleum hydrocarbons, alcohol, esters; insoluble in water.
Uses: General purpose, reasonably nonstaining and nondiscoloring rubber antioxidant.

"Agerite Stalite." [69] Trademark of a proprietary product, mixture of mono-, and dioctylated diphenylamines.
Properties: Reddish brown, viscous liquid; sp. gr. 0.99 ± .02; very soluble in alcohol, benzene, carbon disulfide, gasoline; insoluble in water.
Uses: General purpose antioxidant for all elastomers.

"Agerite Stalite S." [69] Trademark of a proprietary product, a purified mixture of mono- and dioctylated diphenylamines.
Properties: Light tan, friable powder; sp. gr. 0.97 ± .03; melting range 89-103°C; very soluble in alcohol, benzene, carbon disulfide, gasoline.
Uses: General purpose antioxidant in all rubbers.

"Agerite Superflex." [69] Trademark for a proprietary diphenylamine-acetone reaction product.
Properties: Dark brown liquid; sp. gr. 1.10 ± .02; very soluble in acetone, benzene, chloroform, carbon disulfide; slightly soluble in gasoline; insoluble in water.
Uses: Flex-crack resisting antioxidant for dark rubber, such as tires and heavy duty mechanicals.

"Agerite Superlite." [69] Trademark for a proprietary product, a mixture of polybutylated bisphenol A.
Properties: Amber liquid; sp. gr. 0.945-0.965; very soluble in benzene, chloroform, gasoline; insoluble in water.
Uses: Essentially nondiscoloring, nonstaining, rubber antioxidant.

"Agerite White." [69] Trademark for sym-di-beta-naphthyl-para-phenylenediamine. $C_{10}H_7NHC_6H_4NHC_{10}H_7$.
Properties: Grayish-white; odor very faint; m.p. 230-235°C; sp. gr. 1.20; nontoxic in ordinary handling. Insoluble in water; moderately soluble in chloroform, acetone, carbon disulfide, benzene, and gasoline.
Uses: Preventive of deterioration caused by copper and as an all-purpose antioxidant in pure gum stocks, thread, bands, bathing caps, proofing, stocks containing factice; acid-cured stocks and latex compounds.

agglutination. Aggregation of suspended bacteria into lumps that usually settle out. Sometimes applied to other similar suspended particles but usually applied to the clumping together of bacteria under the action of antitoxins.

aglucone. The nonsugar-like portion of a glucoside molecule. See glycosides.

aglycones. The nonsugar hydrolytic products of glycosides. See glycosides.

"Agrico." [196] Trademark for a line of fertilizers of various compositions, types and grades.

agricultural chemicals. Chemicals used to increase the productivity and quality of farm crops. They include fertilizers, soil conditioners, fungicides, insecticides, and weed-killing agents.

"Agrilon." [65] Trademark for synthetic polyelectrolytes.

"Agrimul." [309] Trademark for an insecticide emulsifier.

"Agri-Mycin." [299] Trademark for an agricultural preparation containing streptomycin.

"Agrinite." [196] Trademark for an organic fertilizer.

"Agri-Phos." [196] Trademark for a superphosphate fertilizer.

"Agri-Strep." [123] Trademark for agricultural antibiotics.

"Agritracin." [342] Trademark for antibiotic sprays and dusts to combat plant diseases.

"Agrox." [150] Trademark for a mercurial seed disinfectant, containing 6.70% phenyl mercury urea.
Containers: 14-oz and 3-lb tins; 50- and 100-lb drums.
Uses: On cereal crops, cotton, rice, and flax for protection against bunt, smut, seed rot, and seed blight.

"Agrozyme." [123] Trademark for an enzyme preparation for use in animal and poultry feed.

ague tree. See sassafras bark.

"Ahco" 1250. [325] Trademark for a polyvinyl acetate emulsion. Produces durable, stiff finish on textiles. Used for backfilling, bodying, binding.

"Ahco" Assistant 100. [325] Trademark for a water soluble polymer. Stripping assistant for vat dyed fibers.

"Ahco" Base Oil 90. [325] Trademark for an oleic acid-ethylene oxide condensate. Emulsifier for mineral oil, kerosene, fatty acid esters.

"Ahcobond S." [325] Borated resin used as a water soluble resin binder and plasticizer for textile sizing.

"Ahco" DD-50. [325] Trademark for a blend of mineral oil with sulfonated oils and esters. Used as textile softener and shrink-proofing oil.

"Ahcofix" C. [325] Resinous condensation product; a water-colored syrupy liquid used as a fixation product for direct dyes.

"Ahcols." [325] A series of sulfated oils for dyeing and finishing assistants.

"Ahcoquinone." [325] Brand name for a series of acid and mordant dyestuffs.

"Ahcotex" W-100. [325] Trademark for emulsifiable mineral oil. Lubricant for wool rawstock and cotton yarns.

"Ahcovat." [325] Brand name for series of anthraquinonoid vat dyestuffs.

"Ahcovat" Solubles. [325] Brand name for a series of leuco vat esters.

"Ahcovels" A, G. [325] Fatty carbamides. Cationic substantive softening agents.

"Ahcovels" E, F, R, RM. [325] Fatty carbamide salts. Anionic substantive softening agents.

"Ahcowet." [325] Trade name for a series of synthetic surface-active agents.

"Ahcowet" 129. Sulfated fatty acid ester. Anionic wetting and non-foaming agent for continuous high-speed vat dyebaths.

"Ahcowet" ANS. Alkylaryl sulfonate. Anionic detergent and penetrant in 50% aqueous solution used as dye bath assistant for sulfur and direct colors, "boil off" assistant, carbonizing agent.

"Ahcowet" N. Ethylene oxide condensate. Non-ionic general purpose detergent and wetting agent in 25% aqueous solution.

"Ahcowet" RS. Sulfated ester. Anionic wetting and rewetting agent for textile processing. 65% aqueous solution.

"Ahcowet" SDS. Sulfonated alkyl ester. Anionic general-purpose surface active agent in 65% aqueous solution. For general wetting and for emulsification problems.

"Ahcowet" VL. Ethylene oxide condensate. Non-ionic dispersant and retardant for vat dyes used as a dispersant for acetate dyes. 20% aqueous solution.

A. I. Ch. E. Abbreviation for American Institute of Chemical Engineers.

air. A mixture of gases, the composition of which varies with altitude and other conditions at the collection point. A rule of thumb composition for dry clean air is 79.1% nitrogen and 20.9% oxygen by volume, or 76.8% nitrogen, 23.2% oxygen by weight. An average composition at surface altitudes is

	By volume %	By weight %
Nitrogen.......	78.09	75.54
Oxygen	20.93	23.14
Argon	0.93	1.27

Pure air also contains very small percentages of the rare gases neon, krypton, xenon and helium. Air as ordinarily encountered also always contains appreciable quantities of water vapor and carbon dioxide. See also liquid air.

Shipping regulations: Compressed air: Nonflammable gas. Green label. *

air classification. The separation of solid particles according to weight and/or size, by suspension in and settling from an air stream of appropriate velocity.

"Aircoflex." [144] Trademark for plasticizers tailored to modify homopolymer and copolymer emulsion systems.

airfloat clays. Clays of fine state of subdivision as the result of separation by an air process, after grinding.

"Air-Flo Green." [147] Brand name for a mosquito larvicide containing copper meta-arsenite.

Containers: 100-lb drums; granular 35-lb bags.

air gas. Infrequently used term for a gas obtained by blowing air over a bed of incandescent solid fuel. An air gas is also obtained by blowing air through layers of the very volatile petroleum distillates; i. e., those having gravities from 80-90° Bé.

air hardening steel. See steel, self hardening.

ajava oil. See ajowan oil.

"Ajax." [172] Trade name for a hydrated monocalcium phosphate.

Properties: Brilliant white, free-flowing granular material, especially sieved to eliminate both coarse and extremely fine particles. Purity complies with all Food and Drug laws.

Containers: 100-lb paper bags.

Uses: Manufacture of household baking powder, phosphated flours, and prepared mixes; frits for vitreous enamels.

ajowan oil (ajava oil; ptychotis oil).

Properties: Colorless or brownish-colored essential oil; strong thymol odor; sharp burning taste. Soluble in alcohol and ether.

Chief known constituents: Thymol, thymene, pinene, para-cymene, dipentene and terpinene.

Constants: Sp. gr. 0.910-0.930; optical rotation, 0° to +5°; refractive index n 20/D 1.485-1.510.

Derivation: By distillation of the fruits of Carum copticum. Native to India, Egypt, Persia, and Afghanistan.

Uses: Medicine, perfumery, and formerly for preparation of thymol.

Shipping regulations: None. *

"Akroflex" C. [28] Trademark for a rubber antioxidant containing 35% diphenyl-para-phenylenediamine $C_6H_4(NHC_6H_5)_2$, and 65% phenyl-alpha-naphthylamine ("Neozone"A).

Properties: Dark gray waxy pellets; sp. gr. 1.23; f.p. not lower than 75°C.

Use: To improve the aging and service life of natural and synthetic rubbers; anti-crosslinking agent for SBR (styrene butadiene rubber).

"Akroflex" CD contains 35% diphenyl-para-phenylenediamine and 65% phenyl-beta-naphthylamine ("Neozone" D). Dark gray pellets; sp. gr. 1.22.

"Aktivin." [318] Proprietary product. Technical grade of para-toluene-sulfo-sodiumchloramide with 20% available chlorine: $CH_3C_6H_4SO_2NClNa \cdot 3H_2O$.

Properties: White powder soluble in water with practically neutral reaction. Not

*See "I.C.C. Shipping Regulations," page xiii.
Reference numbers refer to name of manufacturer. See "List of Manufacturers," page v.

soluble in organic solvents. Stable; a 10% solution in water loses 10% of its available chlorine after 2 hours boiling. Oxidizing and disinfecting agent. Not toxic in ordinary sense. Compatible and stable with alkaline compounds such as caustic soda, soda ash, soda phosphates, silicates, etc.; not compatible with ammonia and ammonium compounds and acids.
Types: "Aktivin," "Aktivin" S, "Aktivin" S Special.
Containers: 25-, 50-, 100-lb drums; 250-lb barrels.
Uses: Disinfecting agent for technical purposes, i.e., in breweries, etc. Oxidizing and bleaching agent in textile manufacturing, starch solubilizing in preparation of starch size and finish for textiles.

"Aktone." [285] Trademark for a modified urea activator for thiazole and thiuram accelerators for GR-S, cold rubber, and natural rubber. White flakes; m.p. 100°C; sp. gr. 1.32; stable to heat and storage.
Containers: 50-lb multiwall bags.

"Akweons." [152] Trade name for a series of wetting agent-emulsifier type acid additives to control acid fumes, corrosion and flash rusting in acid pickling operations.
Containers: 5 and 55 gallons.

"Akwilizer." [152] Trade name for brominated cottonseed oil used primarily to balance the specific gravity of the low specific gravity flavoring oils used in the citrus drink industry.
Properties: Sp. gr. 1.33 ± .005; color is light amber; practically odorless and tasteless.
Containers: 1-gal jugs; 55-gal drums.

"AK-33X." [313] See "Ethyl," Antiknock Compound-TEL-Motor 33 Mix.

Al. Symbol for aluminum.

alabamine. Obsolete name for astatine.

alabaster. See gypsum.

"alacreatine." See N-amidinoalanine.

"Alamac." [259] Trademark for the acetic acid salts of the "Alamines."
Derivation: From primary aliphatic amines of coconut, tallow, and other fatty acids.
Grades: Technical and distilled.
Containers: 1-, 5-gal pails; 55-gal steel drums.
Uses: Mineral flotation; biostats; emulsifiers; corrosion inhibitors; pigment treating; surface active agents.

"Alamide." [259] Trade name for a series of high molecular weight, aliphatic amides, such as palmitamide and stearamide, produced by reacting ammonia with fatty acids.
Uses: Intermediates for durable water repellents and finishes for textiles and paper; mold release and antiblocking agents.

"Alamine." [259] Trademark for a series of high molecular weight, primary, secondary and tertiary aliphatic amines. A wide range of chain lengths and degrees of unsaturation are available.
Derivation: From fatty acids.
Grades: Technical and distilled.
Containers: 1-, 5-gal pails; 55-gal steel drums; tank cars; tank trucks.
Uses: Corrosion inhibitors; flotation agents; petroleum product additives; surface active agents; emulsifiers; biostats; hard rubber additives; plastics additives; pigment treating; chemical intermediates.

"Alanap." [248] Trademark for N-1-naphthylphthalamic acid.
Properties: Crystalline solid; m.p. 185°C; almost insoluble in water; slightly soluble in acetone, benzene, and ethanol; not stable in solutions above pH 9.5 nor at temperatures in excess of 180°C; non-corrosive.
Uses: Selective, pre-emergence herbicide for weed control on cucurbits, asparagus, peanuts, nursery stock and turf (crabgrass control). A solution of the sodium salt, "Alanap-3," is used on cotton and soybeans. Also available in a granular form, "Alanap 10G."
Hazards: Do not store near seeds or fertilizers; relatively nontoxic to warmblooded animals but normal precautions in handling should be taken.

alanine (alpha-alanine; alpha-aminopropionic acid; 2-aminopropanoic acid) $CH_3CH(NH_2)COOH$. A naturally occurring nonessential amino acid.
Properties: Colorless crystals; soluble in water; slightly soluble in alcohol; insoluble in ether; optically active.
DL-alanine, m.p. 295°C with decomposition; sublimes at 200°C.
L(+)-alanine, m.p. 297°C with decomposition.
D(-)-alanine, m.p. 295°C with decomposition.
L(+)-alanine hydrochloride: prisms; m.p. 204°C with decomposition.
L(+)-alanine, N-acetyl; crystals; m.p. 116°C.
L(+)-alanine, N-benzoyl; crystals; m.p. 152-154°C.
Derivation: Hydrolysis of protein (silk, gelatin, zein); organic synthesis.
Grades: Reagent; technical.
Containers: Drums (DL-form).
Uses: Microbiological investigations; studies of amino acid metabolism; biochemical research.

beta-**alanine** (3-aminopropanoic acid; beta-aminopropionic acid). $NH_2CH_2CH_2COOH$. The only known naturally occurring amino acid with the amino group in the beta position (see amino acid).
Properties: White prisms; m.p. 198°C with decomposition. Soluble in water; pH (50% solution) 6.0-7.3); slightly soluble in alcohol; insoluble in ether. Hydrochloride: plates and leaflets; m.p. 122.5°C; platinichloride: yellow leaflets; m.p. 210°C (dec).
Derivation: Hydrolysis of protein; by the addition of ammonia to beta-propiolactone, and by other processes based on the reaction of ammonia with acrylonitrile, etc.

Containers: Fiber drums.
Uses: Biochemical studies; organic synthesis; in the production of calcium pantothenate (q. v.); buffer in electroplating.

alant starch. See inulin.

beta-**alanylhistidine.** See carnosine.

alapurin. See lanolin, anhydrous.

"Alathon." [28] Trademark for polyethylene resin (q. v.), including linear varieties.

"Albacer." [73] Brand name for proprietary product; a fatty acid ester.
Properties: White, hard synthetic wax of high luster. Soluble (hot) in naphtha, turpentine, toluol, mineral oil, vegetable oil, carbon tetrachloride, butyl acetate. Insoluble in water, ethyl alcohol, methyl alcohol. This product, when melted, changes to a heavy-bodied, non-gelatinous liquid which sets very slowly.
Constants: M.p. 95-96°C; sp. gr. (25°C) 0.968; saponification value 180-185.
Containers: 1-gal can (8 lbs); 5-gal can (50 lbs); 55-gal drum (430 lbs).
Uses: Compatible with synthetic resins, mineral oils, vegetable oils and other waxes.

albahaca oil. See tolu oil.

"Albaoil." [244] Trademark for series of sulfonated castor oils varying in percentage of oil.
Containers: Non-returnable steel drums averaging 400-425 lbs net.
Uses: By leather manufacturers as fatliquor and as a plasticizer in finishes; also in the textile and other industries.

albargin. A silver drug, containing 14.8-15% silver.
Properties: Yellow, shining powder, sometimes coarse. Acts slowly on albumin, incompatible with tannin and chlorides. Slowly reduced by light. Soluble in water; odorless.
Containers: Amber-colored bottles.
Use: Medicine.
Shipping regulations: None.*

"Albasol." [309] Trademark for a series of chemicals suitable for rendering mineral oils emulsifiable.

"Alberene Stone." [427] Trademark for a natural talc type rock widely used for chemical laboratory benches and similar uses because it is highly homogeneous, very dense and tough, and highly resistant to acid and alkali. Some grades are serpentine rather than talc. Density 185 lbs/cu ft; fusion point 2200°F; water absorption 0.07-0.10% by weight in 48 hours; dielectric strength 37-70 kilovolts; transverse physical strength 3000 (talc) to 5000 (serpentine) psi.

albertite. A natural, jet black, brittle hydrocarbon, an asphaltic pyrobitumen, distinguished from other varieties of bitumen by being almost insoluble in carbon disulfide, and infusible. Hardness 2; sp. gr. about 1.1; fracture conchoidal.

"Albigen." [440] Trademark for a water-soluble polymer used in the textile industry for stripping vat and other dyestuffs. Has no affinity for the fiber; promotes the stripping effect of alkaline hydrosulfite solutions.
Uses: Prevents staining of whites in the wet treatment of fabrics dyed or printed with direct dyestuffs.

"Albone." [28] Trademark for technical grade hydrogen peroxide.
"Albone" 35: H_2O_2 content 35% by weight; active oxygen 16.5% by weight; wt/gal 9.44 lb at 20°C; density 1.133 g/cc at 20°C; pH (Beckman meter) 3.4-3.8.
"Albone" 50: H_2O_2 content 50% by weight; active oxygen 23.5% by weight; wt/gal 9.98 lb at 20°C; density 1.196 g/cc at 20°C; pH (Beckman meter) 2.4-2.7.
Containers: 27.5-gal drums; tank trucks; 4000-, 6000- and 8000-gal tank cars.

albumen. Egg white. See also albumin.

albumen glue. See glue.

albumin. A simple naturally occurring protein soluble in water, coagulated by heat, may be thrown out of solution by saturation with ammonium sulfate, hydrolyzes only to alpha-amino acids or their derivatives. Found in egg white (ovalbumin), in blood (serum albumin), in milk (lactalbumin).

albumin, blood. (albumin, serum).
Properties: Brown amorphous lumps soluble in water and alcohol.
Derivation: Ox-blood is allowed to coagulate and the serum separated by centrifuging. The decanted liquor is filtered, decolorized and subsequently evaporated.
Grades: Technical, light and dark.
Containers: 100-, 332-lb drums; 300-lb barrels.
Uses: Photographic papers; textile printing; dye preparations; foodstuffs; sweetmeats; baked products; medicine; albuminate preparations; clarifying agent; leather adhesive; glue; in pesticides.
Shipping regulations: None.*

albumin colors. Colors used in textile printing and so called because they are mechanically held to the fiber through the agency of coagulated albumin.

albumin, egg (ovalbumin). Principal protein found in egg white. Is sold either as liquid or solid.
Preparation: Fresh white separated from the yolk, diluted with water, beaten to froth and subsequently filtered and evaporated.
Grades: Technical; edible.
Containers: 150-, 200-, 225-lb cases.
Uses: Leather industry; foodstuffs; clarifying agent; photography; adhesive for cork liners of bottle caps; manufacture of varnishes; ivory substitutes; fixing certain dyes; sugar refining.
Shipping regulations: None.*

albumin, milk (whey albumin; lactalbumin). Casein is coagulated from milk by rennet

*See "I.C.C. Shipping Regulations," page xiii.
Reference numbers refer to name of manufacturer. See "List of Manufacturers," page v.

or by dilute acids, filtered and dried.
Grades: Technical.
Containers: Wooden kegs.
Uses: Adhesives; varnishes; ivory substitutes.
Shipping regulations: None. *

albuminoids. See scleroproteins.

albumin, serum (technical). See albumin, blood.

albumin, serum (normal human).
Properties: Light yellow to cream colored lumps; practically odorless. About 96% of the total protein is albumin.
Derivation: Fractionation of human blood and careful drying.
Grades: U.S.P. XVI. May be dispensed in solution.
Use: Medicine.

albumin tannate (tannin albuminate).
Properties: Yellowish-white powder; odorless. Contains about 50% tannin. Decomposed by aqueous solutions of alkali hydroxides and carbonates. Slowly soluble in artificial gastric juice; almost insoluble in water, alcohol, chloroform, and ether.
Derivation: Interaction of dilute aqueous solutions of egg albumin and tannic acid.
Use: Medicine.
Shipping regulations: None. *

albumin, whey. See albumin, milk.

"Albusol." [329] Trademark for a stabilized, egg-albumin solution used in photo-engraving and photo-lithographic processes to replace dry albumin or eggs.

ALCA. Abbreviation for American Leather Chemists' Association.

alcanin. See alkanna.

"Alcian." [206] Brand name for a proprietary line of dyestuffs.

"Alcoa." [226] Trademark for a line of aluminum products (ingot, sheet, wire, tubing, castings, powder, foil, extrusions, collapsible tubes, electrical conductors, closures and screw machine products), aluminum compounds and fluorine compounds.

"Alcoblaks." [133] A series of carbon black dispersions in various alcohols. Used for pigmenting specialty coatings.

alcohol. The term as used in common parlance applies to ethyl alcohol. Chemically, alcohol is a generic term applied to a series of hydroxyl (OH) compounds, the simplest of which has the general formula $C_nH_{2n+1}OH$. There may also be alcohols of the less saturated hydrocarbons. Alcohols may be mono-, di- or tri-hydric, etc., according to the number of hydroxyl radicals (OH) they contain, and primary, secondary or tertiary, according to the position of the hydroxyl radical in the formula. Alcohols having many hydroxyl radicals are called polyhydric alcohols, polyalcohols, or polyols. The phenol series of compounds are sometimes known as aromatic alcohols, but are not comparable in properties. Among the various alcohols covered in this book are: ethyl, allyl, amyl, benzyl, butyl, cetyl, isobutyl, isopropyl, phenylethyl, propyl. Alternative names for these compounds use the ending -ol, thus, methanol, octanol, glycerol. In general the ending -ol in the name of an organic compound signifies the presence of an OH radical, and alcoholic properties are to be expected, although they may be markedly modified by other elements present in the molecule.
Uses: Many alcohols are especially useful as solvents, and as fairly simple hydrocarbon derivatives, are much used in synthesizing other derivatives.

alcohol, absolute. See ethyl alcohol.

alcohol, C-7. See 1-heptanol.

alcohol, C-8. See 1-octanol.

alcohol, C-9. See nonyl alcohol.

alcohol, C-10. See 1-decanol.

alcohol, C-11. See 1-undecanol and undecylenic alcohol.

alcohol, C-12. See lauryl alcohol.

alcohol, C-16. See cetyl alcohol.

alcohol, caustic. See sodium ethylate.

alcohol, dehydrated. See ethyl alcohol.

alcohol dehydrogenase. An enzyme found in animal and plant tissue which acts upon ethyl alcohol and other alcohols producing acetaldehyde and other aldehydes.
Use: Biochemical research.

alcohol, denaturants. See denaturants for alcohol.

alcohol, denatured. See denatured alcohol.

alcohol, grain. Synonym for ethyl alcohol, and applying to that made from grain.

alcohol, industrial. Ninety-five per cent alcohol which has been made unfit for beverage use with a difficultly separable material. See denatured alcohol.

"Alcond DX-100." [72] Trade name for a dry acid containing a blend of inorganic salts for desmutting or deoxidizing solution for aluminum and etched aluminum. Used prior to anodizing or application of conversion coatings; also to prepare aluminum surfaces for spotwelding where extremely low surface resistance is required.

"Aldactone." [70] Trademark for a brand of spironolactone, 3-(3-oxo-7 alpha-acetylthio-17 beta-hydroxy-4-androsten-17 alpha-yl) propionic acid-gamma-lactone, $C_{24}H_{32}O_4S$, m.p. 135°C.
Use: Medicine.

"Aldaromes." [188] Trademark for compositions of aromatic chemicals and essential oils used in embalming fluids and sprays to mask or cover the unpleasant odor of formalde-

*See "I.C.C. Shipping Regulations," page xiii.
Reference numbers refer to name of manufacturer. See "List of Manufacturers," page v.

hyde. Will not decompose when in contact with formaldehyde; give clear solutions in 40% formaldehyde. Not water-soluble.

aldehyde. A generic term applied to a class of organic compounds containing the group R-CHO and holding an intermediate position between the alcohols and the acids derived from the alcohols by oxidation:

$CH_3 \cdot CH_2 \cdot OH$ $CH_3 \cdot CHO$ $CH_3 \cdot CO \cdot OH$
ethyl alcohol acetaldehyde acetic acid

Acetaldehyde is commonly called simply aldehyde. Among the various aldehydes covered in the book are: formaldehyde and anisic, cinnamic, isobutyl, decyl, propyl, pyruvic, etc., aldehydes.

aldehyde ammonia (acetaldehyde ammonia; 1-aminoethanol) $CH_3CHOHNH_2$.
Properties: White crystalline solid; stable in closed containers, resinifies on long exposure to air. Very soluble in water and alcohol.
Constants: M.p. 97°C; b.p. 110°C (partly decomposed).
Derivation: Action of acetaldehyde on ammonia.
Methods of purification: Crystallization.
Grades: Technical.
Containers: 5-, 10-, 25-, 100-lb drums.
Uses: Accelerator for vulcanization of thread rubber; organic synthesis (source of acetaldehyde and ammonia).
Shipping regulations: None.*

aldehyde, C-7. See heptanal.

aldehyde, C-8. See octanal.

aldehyde, C-9. See nonanal.

aldehyde, C-10. See decanal.

aldehyde, C-11. See undecylenic aldehyde.

aldehyde, C-12 lauric. See lauryl aldehyde.

aldehyde, C-12 MNA. See methyl nonyl acetaldehyde.

aldehyde, C-14 (so-called). See gamma-undecalactone.

aldehyde, C-16 (so-called). See ethyl methyl phenyl glycidate.

aldehyde, C-18 (so-called). See gamma-nonyl lactone.

aldehyde collidine. See methylethylpyridine.

aldehydine. See 2-methyl-5-ethylpyridine.

"Aldo 25." [73] Trademark (propylene glycol stearate, self-emulsifying).
Properties: White, wax-like solid; sp. gr. 0.94; m.p. 43-46°C; pH (3% dispersion) 9.2-10.0. Dispersible in hot water; soluble (hot) in alcohols and hydrocarbons.
Containers: 8-, 50-lb containers; 450-lb drums.
Uses: Edible emulsifying agent in baking and nongreasy cosmetic creams and ointments; emulsifying agent for oils, solvents and waxes where slightly alkaline cream-like products are required.

"Aldo 28." [73] Trademark (glyceryl monostearate, self-emulsifying).
Properties: Yellow, bead form; faint odor; m.p. 57-61°C; pH (3% aqueous dispersion at 25°C) 9.0-10.0; sp. gr. (25/20°C) 0.97. Soluble in alcohol, hydrocarbons, mineral oils, vegetable oils; dispersible in hot water.
Containers: 8-, 50- and 400-lb containers.
Uses: Suggested as emulsifying agent for edible and pharmaceutical emulsions, greaseless creams. Protective coatings for edible hygroscopic powders, etc. Addition to shortenings, cooking oils, fats, etc., to improve whiteness and smoothness of the finished material.

"Aldo 33." [73] Trademark (glyceryl monostearate, neutral grade).
Properties: White, bead form; faint odor; m.p. 57-61°C; sp. gr. (25/25°C) 0.97. Insoluble in water; completely soluble in hot alcohol, hydrocarbons, mineral oil, vegetable oils.
Containers: 8-, 50-lb containers; 400-lb drums.
Uses: Suggested as addition in baked goods, bread, cake to improve shelf life, texture and assured volume. In creamed icings, candy emulsifier, water-in-oil emulsifying agent, ice cream emulsifier, plasticizer for the polyvinyl acetals, shortenings, cooking oils, fats, etc., to improve smoothness and color. As a synthetic wax for waterproofing, insulation, polishes, dental waxes, etc.

"Aldo 40." [73] Trademark (glyceryl oleo stearate).
Properties: White, plastic, soft-solid. Soluble (hot) in alcohols and hydrocarbons; dispersible (in high concentrations) in hot water.
Containers: 425-lb drums; 8-, 40-lb containers.
Uses: Edible emulsifier for food products; antistaling and softening agent for bread.

"Aldocet." [73] Trademark for acetostearin.
Properties: Solid, white wax; tasteless, bland odor; sp. gr. 0.97 (25°C); m.p. 36-46°C; free fatty acid 5% max; saponification value 345-362; iodine value 5 max; soluble in certain proportions in methanol, ethanol, toluol, and naphtha; when heated it is soluble in vegetable or mineral oils, acetone, and ethyl acetate; non-hygroscopic.
Containers: 425-lb (net) drums.
Uses: Coatings for cheese, dried fruits, vegetables, meat and fish; plasticizer for chewing gum and brittle waxes; special industrial and textile fiber coatings; replacement in certain applications for cocoa butter, beeswax, partially hydrogenated fats, etc.

aldol (acetaldol; oxybutyric aldehyde; beta-hydroxybutyraldehyde) $CH_3CHOHCH_2CHO$.
Properties: Clear water-white to pale yellow syrupy liquid. Decomposes into crotonaldehyde and water on distillation under atmospheric pressure. Miscible with

*See "I.C.C. Shipping Regulations," page xiii.
Reference numbers refer to name of manufacturer. See "List of Manufacturers," page v.

water, alcohol, ether, organic solvents.
Constants: Sp. gr. 1.1098 (15.6/4°C); b.p. 83°C (20 mm); vapor pressure <0.1 mm (20°C); specific heat 0.737; weight 9.17 lbs/gal (20°C). Flash point 181.4°F; freezing point below 0°C.
Derivation: By condensation of acetaldehyde.
Grades: Technical (98%).
Containers: 1-, 5-gal cans; 5-, 10-, 55-, 100-gal drums; tank cars.
Uses: Synthesis of rubber accelerators and age resisters; perfumery; engraving; ore flotation; solvent for pyroxylin and many organic substances; solvent mixtures for cellulose acetate; fungicides; organic synthesis; printer's rollers; cadmium plating; dyes; drugs; dyeing assistant; synthetic resins.
Shipping regulations: None.*

aldolase (zymohexase). An enzyme present in muscle involved in glycogenolysis and anaerobic glycolysis. It catalyzes production of dihydroxyacetone, phosphate and phosphoglyceric aldehyde from fructose 1,6-diphosphate.
Use: Biochemical research.

aldosterone (electrocortin) $C_{21}H_{28}O_5$. An adrenal cortical steroid hormone which is the most powerful mineralocorticoid. It is probably the chief regulator of sodium, potassium, and chloride metabolism, approximately 30 times as active as deoxycorticosterone.
Properties: Crystals; m.p. 108-112°C.
Derivation: Isolated from adrenals; has been synthesized.
Use: Medicine.

"Aldox." [204] Trademark for an acidic powdered compound used to deoxidize aluminum prior to spot welding or to desmut aluminum subsequent to etching. Commercially available in 400-lb fiber drums.

aldrin (HHDN) $C_{12}H_8Cl_6$. The assigned common name for an insecticidal product containing from 22 to 95% of HHDN or 1,2,3,4,10,10-hexachloro-1,4,4a,5,8,8a-hexahydro-1,4,5,8-endo-exodimethanonaphthalene. Accepted as generic name by Ent. Soc. See also dieldrin and endrin.
Properties: Brown to white crystalline solid; m.p. 104-105.5°C; insoluble in water; soluble in most organic solvents. Not affected by alkalies or dilute acids; compatible with most fertilizers, herbicides, fungicides, and insecticides. May be formulated as emulsifiable concentrate, wettable powder or dust.
Grade: Technical.
Containers: Fiber drums.
Uses: Insecticide.
MCA warning labels (technical grade and formulations, 10% and over):
Warning! Poisonous by skin contact or inhalation; rapidly absorbed through the skin; spray mist extremely hazardous. (formulations less than 10%): Caution! May be fatal by skin contact, inhalation or swallowing. (Similar labels required by U.S.D.A.).
Shipping regulations: Liquid formulations containing more than 60% aldrin and dry formulations containing more than 65% aldrin: Class B Poison. Poison label.*

"Alert." [51] Trademark for a white, high melting, adhesive lubricant for textile twister rings; also suitable for food handling machinery, etc.

"Alert 80, 101." [151] Trademark for mercaptan type gas odorants, recommended for odorization of natural gas.

"Alevaire." [162] Trademark for superinone.

"Alfane." [41] Trade name for an acid-, solvent-resistant, synthetic-resin cement of the epoxy type used as a mortar cement up to temperatures of 200°F.

Alfin catalysts. Catalysts obtained from alkali alcoholates derived from a secondary alcohol which is a methyl carbinol and olefins possessing the grouping $-CH=CH-CH_2-$, which may even be part of a ring as in toluene.

The interaction of the alkali alcoholate, sodium isopropoxide, with the olefin halide, allyl chloride, gives a slurry of sodium chloride on which sodium isopropoxide and allyl sodium are adsorbed. This slurry is a typical Alfin catalyst used to convert olefins such as ethylene, propylene and butenes into polyolefin polymers.

"Alflorone." [123] Trade name for fludrocortisone (9-alpha-fluorohydrocortisone).

"Alfrax." [280] Trademark for refractory products composed principally of electrically fused aluminum oxide grain. Available as bonded refractories and refractory cements.
Properties: High refractoriness; strength at elevated temperatures; chemical stability and low thermal expansion.
Bonded Refractories.
"Alfrax" K. Approximately 90% crystalline aluminum oxide. The thermal conductivity curve is a straight line between 20 Btu/sq ft/in. thickness/°F/hr at 1000°F and 25 Btu at 2600°F. Porosity averages 20%, permeability is very low.
Uses: Brick and special shapes for use in furnace and kiln constructions; tile for furnace or kiln hearths; muffle plates for continuous tunnel kilns.
"Alfrax" B. Approximately 85% aluminum oxide. The thermal conductivity is 10 Btu/sq ft/in thickness/°F/hr. Porosity averages 30%, permeability higher than "Alfrax" K refractories.
Uses: Bricks and shapes for oil and gas fired combustion chamber linings.
"Alfrax" BI. Light weight insulating refractories having a large percentage of sealed pores. The thermal conductivity is 7 Btu/sq ft/in. thickness/°F/hr. Weight of standard 9 inch straight brick, 4.5 lbs.
Uses: Linings of furnaces and kilns both electric and fuel fired. Will withstand

temperatures of 3300°F and can be used as primary linings except where slagging conditions prevail.

Refractory Cements.

No. 20 is a coarse plastic material, No. 22 is somewhat finer, and No. 25 is considerably finer. Used in laying "Alfrax" brick and in ramming monolithic linings for nonferrous melting furnaces.

"Algaecide." [108] A pelletized, chlorine releasing biocide.

Containers: 14-oz bottles; 20-lb cans.

Uses: To kill algae or slimy bacteria growths in cooling tower systems and evaporative condensers.

algaroth powder. See antimony oxychloride.

algarroba bean. See carob seed.

"Algeeclear." [282] Trademark for copper sulfate compounded with an organic dispersing agent; contains 67% active ingredient and 33% copper complexing agent.

Properties: Finely ground powder; average bulk density 67 lb/cu ft; sp. gr. 1.1.

Containers: 25-lb multiwall bags; 5-lb packages in 30-lb cartons; 250-lb drums.

Uses: Algicide for swimming pools, ponds, lakes, reservoirs, cooling towers.

Caution! Keep away from nose and mouth. Do not breathe the dust.

"Algepon." [300] Trademark for series of dyeing, stripping and discharge printing assistants for various applications in textile processing. Several types are each formulated for a specific function:

"Algepon" AK: A quaternary compound used for the removal of pigment colors from textiles.

"Algepon" LN: A nonionic compound used as a leveling and retarding agent for dyeing blends of natural and synthetic fibers.

"Algepon" PD: A complex quaternary compound for use in discharge printing of difficultly dischargeable pigment colors such as the phthalocyanines; incorporated into the conventional printing pastes containing gums, hydrosulfite, etc.

"Algepon" VA: A cationic compound used as a leveling and retarding agent in the dyeing of vat colors. Also effective for stripping vat and naphthol dyes.

"Algepon" XR: A highly efficient stripping assistant for vat dyes.

algicide. Chemical agent added to water to destroy algae. Copper sulfate is commonly employed as an algicide for large water systems.

"Algimaster." [426] Trademark for a composition containing the active ingredients alkyl quaternary ammonium bromides, organic, polyamine, amine hydrobromides.

Containers: 1-qt polyethylene bottles.

Uses: Control of algae in swimming pool water.

Shipping regulations: None.*

algin. Term referring to alginic acid and its derivatives (i.e., sodium, ammonium, potassium, propylene glycol alginate, etc.). Alginic acid is $(C_6H_8O_6)_n$, a polyuronic acid composed of beta-D-mannuronic acid residues linked so that the carboxyl group of each unit is free, while the aldehyde group is shielded by a glycosidic linkage.

Properties: White to dark brown powder, possessing marked hydrophilic colloidal properties. Soluble in water; insoluble in organic solvents. Properties of algin products vary with their source and method of manufacture, usually lying within the following approximate limits:

pH of water solutions	3.5-10
viscosity of 1% by weight water solution	1.0-100 poises
moisture content	5-20%
particle size	10-200 mesh
color	white to dark brown

Derivation: Extracted from brown seaweed or kelps, chiefly from the giant kelp, Macrocystis pyrifera, on the Pacific coast, and from the horsetail kelp, Laminaria digitata, on the Atlantic coast. In refined grades of algin the natural cellulose present is removed by filtration, and the product is bleached and purified. Technical grades may contain some cellulose and be unbleached.

Grades: Refined (food); technical (commercial); N.F. XI (sodium alginate).

Containers: Drums; fiber containers; 1- and 5-lb bottles; multiwall paper sacks; 50-, 100-, 200-, 700-lb drums.

Uses: In general, wherever a hydrophilic colloid possessing marked suspending, thickening, emulsifying, stabilizing, and waterholding properties is required. Specific uses are in the food industry; tooth paste; cosmetics; pharmaceuticals; textile sizing; coatings; waterproofing agent for concrete; boiler water treatment; oil-well drilling muds; adhesives; fibers; storage of gasoline as a solid.

Shipping regulations: None.*

algin fibers. Filaments or threads composed of metal alginates.

Derivation: By projecting a fine stream of an aqueous solution of alkali alginate into a bath containing a salt of a metal which forms an insoluble alginate.

Properties: Sodium alginate fibers are water-soluble, but calcium alginate fibers become swollen and are dissolved in alkaline solution. Beryllium fibers (colorless) and the chromium fibers (grayish to green to purple) are alkali resistant and possess an affinity for mordant dyes.

Uses: Sodium or calcium alginate threads are used as a support in weaving fine woollen threads and can be completely dissolved in a suitable alkaline solution leaving the pure wool fabric. Beryllium alginate fibers may be used in fireproof fabric for public buildings. Calcium alginate fibers may be spun into silk substitutes.

alginic acid. See algin.

alginoid iron. See ferric alginate.

algiron. See ferric alginate.

"Algol." [307] Brand name of proprietary line of vat dyestuffs.

"Algosol." [307] Brand name of proprietary line of water-soluble leuco derivatives of vat dyestuffs.

"Algrain." [319] Trademark for a highly refined grain alcohol.
Use: Pharmaceuticals, perfumes, toilet waters, flavoring extracts.

"Algran." [147] Brand name for free-flowing granular insecticide containing 20 or 25% aldrin.
Containers: 50-lb bags.
Uses: For control of soil and turf-inhabiting insects.

"Alidase." [70] Trademark for hyaluronidase.

"Alipal." [307] Trademark for a series of anionic surfactants.
"Alipal" CO-433. Sodium salt of sulfate ester of an alkyl phenoxypolyoxyethylene ethanol; 28% active.
Properties: Amber liquid; soluble in water, ethylene glycol; insoluble in carbon tetrachloride, corn oil, kerosene, mineral oil, xylene; stable in hard water, alkaline solutions and in mild acid.
Uses: Surfactant with foaming, wetting, and dispersing properties; detergent base for high foaming formulations, and shampoos, scrub soaps and hard surface detergents; lime soap dispersant.
"Alipal" CO-436. Ammonium salt of sulfate ester of an alkyl phenoxypolyoxyethylene ethanol; 58% active.
Properties: Amber liquid; soluble in water, ethylene glycol; stable in hard water and in mild acid solutions.
Uses: Detergent with good wetting, foaming, and dispersing properties; detergent base in high foaming liquid formulations; surfactant in cosmetic preparations; emulsifier for copolymerization of vinyls; anti-static agent for natural and synthetic fibers.

aliphatic. Organic compounds whose molecules do not have their carbon atoms arranged in a ring structure. This category therefore includes all the paraffin hydrocarbons and their saturated and unsaturated derivatives of all types.

"Aliphatic Ester Sulfate." [328] Brand name for sodium fatty acid ester sulfate product that serves as a dyeing assistant and a softener in textile processing.

"Aliquat." [259] Trademark for a series of aliphatic quaternary salts.
Derivation: Primary, secondary, and tertiary aliphatic amines and diamines. Aliphatic chain lengths from C_8 to C_{18} available.
Containers: 1-, 5-gal pails; 55-gal steel drums; tank trucks; tank cars.
Uses: Textile softeners and conditioners; foam rubber stabilizers; biostats and emulsifiers; clay and talc treating; antistatic agents; chemical intermediates; corrosion inhibitors.

aliquot. A part which is a definite fraction of a whole; as, aliquot samples for testing or analysis.

"Alite." [326] Trademark for a series of sintered metallic oxides.

"Alitrile." [259] Trade name for a series of high molecular weight, aliphatic nitriles produced from fatty acids from tallow, hydrogenated tallow, coconut and soybean oils, and palmitic and stearic acids.
Uses: Lubricating oil additives; plasticizers.

alizarin (1, 2-dihydroxyanthraquinone; anthraquinonic acid) $C_6H_4(CO)_2C_6H_2(OH)_2$. Parent form of many dyes and pigments.
Properties: Orange-red crystals; brownish-yellow powder. Soluble in alcohol and ether; sparingly soluble in water. C.I. No. 1027.
Constants: M.p. 289°C; b.p. 430°C (sublimable).
Derivation: Anthracene is oxidized to anthraquinone, the sulfonic acid of which is then fused with caustic soda and potassium chlorate, the melt is run into hot water and the alizarin precipitated with hydrochloric acid. Occurs naturally in madder root.
Grades: Technical.
Containers: Wooden barrels; kegs; fiber containers.
Uses: Manufacture of dyes; production of lakes; both dye and intermediate.
Shipping regulations: None.*

alizarin assistant. The same as Turkey red oil (q. v.).

alizarin dyes. A name sometimes used synonymously with mordant dyes, since many of the latter are related chemically to alizarin.

alizarin oil. The same as Turkey red oil (q. v.).

alizarin red. See madder lake.

alizarin yellow R (para-nitrobenzene azosalicylate sodium salt) $O_2NC_6H_4NNC_6H_3OHCOONa$.
Properties: Yellow brown powder; soluble in water.
Use: Acid base indicator in pH range 10.1 (yellow) to 12.0 (violet) (see indicators); also a biological stain.

"Alizarol." [243] Trademark of acid mordant dyes.

"Alizurol." [206] Brand name for the proprietary dyestuff "Alizurol Green 2YS," which is an alizarin type dye used in textile printing.

"Alkalate." [244] Trademark for a compound consisting of sodium sesquisilicate base intimately combined with sodium phosphate.
Properties: White, crystalline granular product; soluble in water; total Na_2O content 38.2%; per cent of total Na_2O in active form 32.4%; per cent of total Na_2O in inactive form 5.8%; compounded to balance the effect of bicarbonates and soil conditions.
Containers: 125-lb plywood drums; 350-lb

*See "I.C.C. Shipping Regulations," page xiii.
Reference numbers refer to name of manufacturer. See "List of Manufacturers," page v.

fiber drums; 350-lb wooden barrels.
Uses: Laundries, dairies and creameries. For general free rinse cleaning in high bicarbonate areas.

alkali. A term applied to the hydroxides and carbonates of the alkali metals and the radical ammonium.

In commerce the term is used to designate the main products of the alkali industry and embraces the hydroxide and carbonates of sodium (or combinations thereof) and the hydroxide and carbonate of potassium; such products being known as caustic soda, soda ash, bicarbonate ("bicarb") of soda, sal soda (washing soda), caustic potash, modified sodas (neutral sodas), causticized ash, etc. For additional data on these see under each item.

The term is also applied more generally to any strong base in aqueous solution, i. e., a substance which gives a high concentration of hydroxyl ion when dissolved in water. Calcium oxide and barium hydroxide are included in this sense.

See also base.

alkali blue. A class name for a series of pigment dry powders prepared by the phenylation of para-rosaniline or fuchsine, followed by drowning in hydrochloric acid, washing and sulfonating. Alkali blue, on a weight basis, has the highest tinting strength of all blue pigments, approximately 5 to 6 times greater than iron blue and 50% stronger than phthalocyanine blue.

This pigment class also has excellent resistance to bleed in water, acid, oils, fats and waxes but bleeds profusely in alcohol and lacquer thinner. Has poor stability in alkali and soap and fair permanency in both fulltone and tint.
Containers: 250-lb barrels.
Uses: Printing inks; some inside paints.

alkali blue, flushed. The chemical type described under alkali blue, but marketed as an ink. Alkali blue presscake is "flushed" with vehicle to replace the water in the pulp and provide for a commercially acceptable ink.
Containers: 125-, 250-lb barrels.
Uses: Printing inks, toning of black inks.

alkali cellulose. The product formed by steeping wood pulp with sodium hydroxide, this being the first step in the manufacture of viscose rayon, and other cellulose derivatives.

alkali lignin. Material recovered by allowing alkaline solutions to act upon wood while under pressure, with subsequent treatment with acid.

alkali metals. Metals comprising Group 1A of the periodic system; potassium, sodium, lithium, rubidium, cesium, and francium. Except for francium (q. v.), the alkali metals are all soft, silvery metals, which may be readily fused and volatilized, the melting and boiling points becoming lower with increasing atomic weight. The specific gravity increases with, but less rapidly than, the atomic weight, the atomic volume therefore becoming greater as the series is ascended. They are the most strongly electro-positive of the metals. They react vigorously, at times violently with water; within the group itself the basicity increases with atomic weight, that of cesium being the greatest. All burn readily in air and are very reactive.

alkalimetry. The measurement of the concentration of bases or of the amount of free base present in a solution by titration or some other means of analysis.

alkaline earth metals. Divalent metals of Group II of periodic table; beryllium, magnesium, calcium, strontium, barium, and radium.

alkaline earths. The oxides of the alkaline earth metals.

alkaloids. A group of basic nitrogenous organic compounds of vegetable origin which exhibit a powerful toxic action on the human or animal system. They are usually derivatives of the nitrogen ring compounds: pyridine, quinoline, isoquinoline, pyrrole, and are designated by the ending -ine. Though some are liquids, they are usually colorless, crystalline solids, having a bitter taste, which combine with acids without elimination of water. They are soluble in alcohol; insoluble, or sparingly soluble in water. Examples are atropine, morphine, nicotine, quinine, cocaine and strychnine.

"Alkamortar H-W." [446] Trade name for a ceramic air-setting mortar, resistant to caustic solutions of moderate concentrations at atmospheric temperatures.
Uses: Bonding acid resistant brick and in limited installations when alkaline conditions prevail.

alkane.
1. General term for a saturated aliphatic hydrocarbon; a paraffin hydrocarbon. Its formula is C_nH_{2n+2}. Examples range from methane, CH_4, through eicosane, $C_{20}H_{42}$, and higher homologs.
2. A trade designation for dodecylbenzene (q. v.).

"Alkane." [151] Trademark for a colorless liquid alkyl aromatic synthesized from petroleum hydrocarbons; used as raw material for both household and industrial detergents.

alkanesulfonic acid, mixed RSO_3H (R is methyl, ethyl, propyl, mixed). Trade designation for a mixture of methane-, ethane-, and propane-sulfonic acids. A strong nonoxidizing, nonsulfonating liquid acid which is thermally stable at moderately elevated temperatures.
Properties: Light amber liquid with sour odor; very corrosive; miscible with water, and saturated fatty acids.
Typical specifications: M.p. below —40°C; b. p. 120-140° (1 mm); sp. gr. 1. 38 (20°C);

pH (1% solution) 1.15.
Containers: 5-, 25-, 70-lb carboys.
Uses: Catalyst; intermediate; reaction medium.

alkanet. See alkanna.

alkanna. (alkanet; orcanette; anchusa; dyer's alkanet). Root of an herbaceous perennial plant Alkanna (Anchusa) tinctoria.
Occurrence: Asia Minor, southeastern Europe.
Containers: Bags.
Use: For its coloring principle, alkannin (alcannin; anchusin) which is used for coloring pharmaceuticals, oils, leathers, and in an indicator paper (red for acids, green-blue for alkalies).

alkannin. See alkanna.

"Alkanol." [28] Trademark for a group of surface active agents which, by reducing surface tension, promote wetting, rewetting, penetration, dispersion, emulsification, or detergency.

alkanolamines. Compounds such as ethanolamine, $HOCH_2CH_2NH_2$, or triethanolamine, $(HOCH_2CH_2)_3N$, in which nitrogen is attached directly to the carbon of an alkyl alcohol.

"Alkarb." [88] Trademark for an alkali metal carbonate mixture.
Properties: Melts over a range, becoming completely liquid at 850°C; hygroscopic.
Derivation: From lepidolite lithium ore.
Typical analysis: Potassium carbonate 70.4%; rubidium carbonate 23.1%; cesium carbonate 2.2%; sodium carbonate 2.9%; lithium carbonate 1.1%.
Containers: 100-lb multiwall paper bags; bulk hopper cars.
Uses: Glass and enamel raw material; carbon dioxide absorbent; detergents and heat treating salts.
Hazard: May cause skin irritation.

"Alkar" Process. [416] Patented process for catalytic alkylation of a variety of aromatic hydrocarbons with numerous olefins to yield the corresponding alkylaromatic in either monoalkylated or polyalkylated form. The olefin reactant may be present in low concentration in feed stream. Alkylaromatic products yields are essentially quantitative and characterized by high purity.
Uses: Production of high purity ethylbenzene from benzene and ethylene, but not limited to this application.

"Alkaterge." [319] Trade name for oil-soluble surface active agents.
Grades and Properties:
"Alkaterge-A." Color, Gardner (1933) 7; m.p. −41°C; sp. gr. (25/25°C) 0.883; refractive index (25°C) 1.4631; flash point (COC) 325°F; solubility in water (ml/100 ml) 0.005.
"Alkaterge-C." Color, Gardner (1933) 15; m.p. −40°C; sp. gr. (25/25°C) 0.925; flash point, (COC) 400°F; solubility in water (ml/100 ml) 0.005.
"Alkaterge-E." Color, Gardner (1933) 7; m.p. −50°C; sp. gr. (25/25°C) 0.924; refractive index (25°C) 1.4738; flash point (COC) 395°F; solubility in water (ml/100 ml) 0.4.
"Alkaterge-T." Color, Gardner (1933) 7; m.p. 59°C; flash point (COC) none; solubility in water (ml/100 ml) 0.01.
Uses: Auxiliary emulsification agents; antifoam agents; dispersing agents; spreading agents; pigment-grinding assistants; acid acceptors and in numerous other applications.

alkavervir. Mixture of alkaloids obtained by selective extraction of Veratum viride, N.F., with various organic solvents and selective precipitation from acidic and basic solutions.
Properties: Light yellow powder which provokes sneezing. Freely soluble in alcohol and acetone; practically insoluble in water.
Grade: N.N.D.
Use: Medicine.

"Alkazene." [233] Trademark applied to a series of alkyl and halogenated alkylbenzenes, useful as solvents, gauge fluids, etc.

alkermes. See kermes.

"Alkolene." [244] Trademark for series of slightly alkaline oils emulsified with soap.
Containers: Non-returnable steel drums averaging 400-425 lbs net.
Uses: Used in the leather industry.

"Alkophos." [58] Trade name for aluminum acid phosphates in liquid (Grades C and E) and solid (Grades B and D) form.
Containers: B, 200-lb barrels; C, 150-lb carboys, 600-lb lacquer-lined drums; D, 400-lb barrels; E, 650-lb lacquer-lined drums.
Uses: Bonding agents for refractories, for high temperature cements and for refractory paints.

"Alkor." [41] Trade name for a synthetic, furan-type resin cement which is acid- and alkali-proof and used as a mortar cement where temperatures do not exceed 380°F.

beta-**alkoxy proprionitrile.** See beta-isopropoxy-proprionitrile.

"Alkron." [88] Trademark for anti-dusting, wettable powders and emulsifiable solutions containing parathion in various concentrations.

"Alk-Tri." [233] Trademark for trichloroethylene (q.v.).

alkyd resins. Alkyd resins in general are made by the union of dibasic acids or anhydrides—usually phthalic anhydride, with a polybasic alcohol such as glycerine. Modification is accomplished by use of other anhydrides (maleic), dibasic acids, glycols, polyols and also other substances. The most common of the latter are various natural oils or the acids derived from them. Use of linseed oil or linoleic acid or similar drying oil materials produces an oxidizing alkyd, while use of essentially saturated oils and their derivatives produces non-oxidizing types. Hard resin types are produced by using rosin or similar resins as modifying agents. Their retention of initial appearance after long exposure to severe weather, heat resistance, color retention, toughness, adhesion, flexibility, and ease of application, explain the extensive uses. They can be made to close specification of viscosity, acid number and color.

These resins have value in protective and decorative coatings for metals, wood, paper, textiles, in adhesives, printing inks rubber compounding and floor coverings. They are also used as vehicles in architectural, automotive and industrial finishes, water-thinned paints, nitrocellulose lacquers, urea- and melamine-formaldehyde enamels, aluminum vehicles, bulletin colors, grinding vehicles and marine paints.

alkyl. A paraffin hydrocarbon radical which may be represented as derived from an alkane by dropping one hydrogen from the formula. Examples are ethyl, C_2H_5-; propyl, $CH_3CH_2CH_2$-; isopropyl, $(CH_3)_2CH$-; etc. Corresponding aromatic radicals are known as aryl.

alkyl aryl polyethyleneglycol ethers. See isooctylphenoxypolyoxyethylene ethanol for a typical example of this class of compound. They are used as surface-active agents, as in detergents.

alkylaryl sulfonates. Class name for a widely used group of anionic synthetic detergents. The aryl portion is usually benzene, and the alkyl portion often dodecyl or decyl. Typical are sodium tridecylbenzene sulfonate, and sodium dodecylbenzene sulfonate. See the latter for further details.

alkylate. Generic term, particularly in the oil industry, applied to the product of an alkylation process. (See "Alkylation Process, HF;" alkylation process, sulfuric acid.) Alkylate (consisting of branched paraffin hydrocarbons with high octane number) generally is blended in varying proportions with other hydrocarbon mixtures also boiling in the gasoline boiling range to produce military and civilian aviation gasolines and motor fuels of commerce. (See also detergent alkylate.)

alkylation. A process used in petroleum refining which consists in causing the chemical combination of isoparaffin hydrocarbons with olefins. For example, isobutane combines with ethylene to give neohexane or, 2,2-dimethylbutane. A pressure of 3000 psi and a temperature of 900-1000°F are required. Sulfuric acid and hydrofluoric acid are often used as catalysts. The products of alkylation (alkylates) are usually branched paraffin hydrocarbons with high octane number, and therefore of value in preparing automobile and airplane fuels.

"Alkylation Process, HF." [416] Patented process employing virtually anhydrous liquid hydrogen fluoride as catalyst in combining an isoparaffin with an olefin to produce an "alkylate" product composed of the corresponding branched-chain paraffins. Used extensively by oil industry to produce very high octane gasoline blending component, usually from isobutane and butylenes (sometimes the olefin is a C_3-C_4 mixture, less frequently propylene only or amylenes only), the alkylate being a mixture of branched-chain paraffins, generally octanes. Apparatus for continuous catalyst regeneration is part of process equipment.

alkylation process, sulfuric acid. Sulfuric acid also is used as an alkylation catalyst. It requires a considerably different and more complex arrangement of apparatus than hydrogen fluoride alkylation, since refrigeration is required. "Spent" acid (88-92% H_2SO_4) must be continuously withdrawn for reconstitution in a separate acid plant and make-up acid (98% H_2SO_4) added.

alkyldimethylbenzylammonium chloride. See, for example, benzalkonium chloride.

alkyl fluophosphates. See, for example, diisopropyl fluophosphate.

allanite (orthite)
$(Ca, Ce, La)_2(Al, Fe)_3(OH)(SiO_4)_3$. A natural complex silicate of calcium, aluminum, iron, the cerium metals (see rare earth metals), and yttrium metals in smaller quantities. Occurs in igneous rocks.
Properties: Color black to brown or grayish; pitchy, submetallic luster; streak greenish gray to brown.
Constants: Sp. gr. 3.5-4.2, hardness 5.5-6.
Occurrence: Greenland, Sweden, U.S.S.R., Madagascar, New York, Connecticut, New Jersey, North Carolina, Texas, Virginia.
Use: Source of cerium.

allantoin (glyoxyldiureid; 5-ureidohydantoin)
$C_4H_6N_4O_3$. The end product of purine metabolism in mammals other than man and other primates; it results from the oxidation of uric acid.
Properties: White to colorless, odorless, tasteless powder or crystals; m.p. 230-236°C (dec.). One gram is soluble in 190 cc water or 500 cc alcohol; readily soluble in alkalies.
Preparation: Produced by oxidation of uric acid. Also present in tobacco seeds, sugar beets, wheat sprouts, and in mammalian excretions.
Grades: C.P.
Containers: Glass bottles.
Use: Medicine.
Shipping regulations: None.*

allene (propadiene; dimethylenemethane)
$H_2C{:}C{:}CH_2$.
Properties: Gas; unstable; m.p. −146°C; b.p. −32°C.
Containers: Cylinders.
Use: Intermediate.

*See "I.C.C. Shipping Regulations," page xiii.
Reference numbers refer to name of manufacturer. See "List of Manufacturers," page v.

"Allercur." [299] Trademark for clemizole hydrochloride.

allethrin $C_{19}H_{26}O_3$ (2-allyl-4-hydroxy-3-methyl-2-cyclopenten-1-one ester of chrysanthemummonocarboxylic acid). A synthetic insecticide structurally similar to pyrethrin and used in the same manner. Accepted by Ent. Soc. as generic name. For other synthetic analogs see barthrin, cyclethrin, ethythrin, furethrin. Pyrethrin I differs in having a 2,4-pentadienyl group in place of the allyl of allethrin.

Properties: Clear, amber colored, viscous liquid. Sp. gr. (20/20°C) 1.005-1.015; refractive index (20°C) 1.5040. Insoluble in water; incompatible with alkalies; soluble in alcohol, carbon tetrachloride, kerosine, and nitromethane.

Derivation: Synthetically (glycerine, acetylene, and ethyl acetoacetate are the major raw materials).

Grades: 90% technical (about 90% pure, with 10% of isomers or related compounds); 20% technical; 2.5% technical.

Containers: Drums.

Use: Insecticide.

Shipping regulations: None. *

"Allexcel." [342] Trademark for allethrin-containing insecticidal concentrates.

allicin ($C_6H_{10}OS_2$). An antibacterial substance extracted from garlic (allium).

Properties: Colorless oily liquid; irritating to the skin; sharp garlic odor; unstable, decomposing rapidly when heated; slightly soluble in water; very soluble in alcohol, benzene and ether.

Use: Medicine.

allium (garlic). The fresh bulb of the perennial plant Allium sativum.

Uses: As condiment, in medicine, and as source of garlic oil.

allo-. A prefix designating the more stable of two geometrical isomers (Chemical Abstracts).

allomaleic acid. See fumaric acid.

allo-ocimene $(CH_3)_2C(CH)_3CCH_3CHCH_3$.

Properties: Clear, almost colorless liquid. Boiling range (5-95%) 89-91°C (20 mm); sp. gr. (15/15°C) 0.824; refractive index (20°C) 1.5278. Polymerizes and oxidizes readily.

Uses: Component of varnishes and a variety of polymers.

allophanamide. See biuret.

allo-threonine. See threonine.

allotrope (allotropic form). One of several possible forms of an element, e.g., carbon may occur as diamond, carbon black and graphite.

alloxan (mesoxalylurea) $C_4H_2O_4N_2 \cdot H_2O$ and $4H_2O$.

Properties: White crystals, become pink on exposure to air; colorless, aqueous solution imparts pink color to skin; sometimes explodes if bottled; m.p. 170°C (dec.); soluble in water and alcohol.

Derivation: Oxidation of uric acid in acid solution.

Uses: Organic synthesis; experimental work in nutrition, physiology, and biochemistry; cosmetics.

alloxazine $C_{10}H_6O_2N_4$.

Properties: Greyish-green powder; decomposes above 300°C; insoluble in water, alcohol, and ether; soluble in aqueous alkalies.

Derivation: Occurs naturally in some plant pigments.

Uses: Intermediate in preparation of pharmaceuticals and dyes.

"750 Alloy." [166] An iron-chromium-aluminum resistance material. For electric heating elements up to 2050°F.

"815 Alloy." [166] An iron-chromium-aluminum resistance material. For electric heating elements up to 2150°F.

"875 Alloy." [166] An iron-chromium-aluminum resistance material for electric heating elements up to 2350°F.

"Alloy 815-R." [166] An iron-chromium-aluminum material. This alloy is resistant to corrosion, has high strength and ductility and low temperature coefficient; used for a wide range of precision resistor applications.

alloys. A solid or liquid mixture of two or more metals; or of one or more metals with certain nonmetallic elements by fusing the components. The properties of alloys are often greatly different from those of the component metals, making them more satisfactory for many uses than any pure metals. The composition and uses of various alloys are listed under their specific names.

alloy steels. Steel possessing distinctive properties depending on the presence of some element or elements other than carbon or on the presence of these elements and carbon. Low alloy steels are those having less than 5% of such added constituents.

allspice. See pimenta.

allspice oil. See pimenta oil.

allyl para-**acetaminophenolate.** See para-acetaminophenyl allyl ether.

allyl acetone $CH_2CHCH_2CH_2COCH_3$.

Properties: Colorless liquid; sp. gr. (20/4°C) 0.846; wt/gal (20°C) 6.99 lbs.

Containers: 5-gal steel drums; 55-gal resin steel drums.

Uses: Intermediate in pharmaceutical synthesis; perfume materials; fungicides; insecticides; fine chemicals.

allyl para-**acetylaminophenolate.** See para-acetaminophenyl allyl ether.

allyl alcohol (AA; 2-propen-1-ol; propenyl alcohol) CH_2CHCH_2OH.

Properties: Colorless liquid with pungent mustard-like odor; irritating to the eyes; poisonous! B.p. 96.9°C; m.p. −129°C;

*See "I.C.C. Shipping Regulations," page xiii.
Reference numbers refer to name of manufacturer. See "List of Manufacturers," page v.

sp. gr. (20/4°C) 0.8520; wt/gal (20°C) 7.11 lbs; refractive index n 20/D 1.4131; flash point (Tag open cup) 90°F. Miscible with water, alcohol, chloroform, ether, petroleum ether.
Derivation: (a) Hydrolysis of allyl chloride (from propylene) with dilute caustic; (b) isomerization of propylene oxide over lithium phosphate catalyst at 230-270°C; (c) dehydration of propylene glycol.
Method of purification: Rectification.
Grades: Technical.
Containers: 1-, 5-gal pails; 55-gal drums; tank cars.
Uses: Preparation of esters for use in resins and plasticizers; intermediate for pharmaceuticals and other organic chemicals; manufacture of glycerol and acrolein; military poison gas; herbicide.
Danger! Extremely hazardous liquid and vapor. Rapidly absorbed through skin. Flammable. MCA warning label.
Shipping regulations: Poison, class B. Poison label. *

allyl aldehyde. See acrolein.

allyl amine (2-propenyl amine) $C_3H_5NH_2$.
Properties: Colorless to light yellow liquid; strong ammoniacal odor; attacks rubber and cork; causes sneezing and lachrymation; b.p. 55-58°C; sp. gr. 0.759-0.761 (20/20°C); refractive index n 22/D 1.4194. Soluble in water, alcohol, ether, and chloroform.
Grades: C.P.; technical.
Containers: 1-, 5-, 10-lb bottles; 5-, 30-, 55-gal drums; tank cars.
Uses: Pharmaceutical intermediate (e.g., mercury diuretics); possible use as resin intermediate.

allylbarbituric acid. N.F.X name for 5-allyl-5-isobutylbarbituric acid (q.v.). 5-Allylbarbituric acid is $C_7H_8N_2O_3$, solid, m.p. 164-166°C. Insoluble in water; soluble in alcohols. Used as an intermediate in pharmaceuticals, tobacco flavors, resins, etc.

allyl bromide (3-bromopropene; bromoallylene) C_3H_5Br.
Properties: Colorless to light yellow liquid; irritating, unpleasant odor. Caution! Lachrymator; burns skin. Flammable. Sp. gr. 1.398 (20/4°C); m.p. −119°C; b.p. 71.3°C; refractive index 1.4654; soluble in alcohol, ether, chloroform, carbon tetrachloride, carbon disulfide; very slightly soluble in water.
Grades: Technically pure (95% minimum purity by bromine titration).
Containers: Returnable 5-gal carboys, 55-lbs net, 92 lbs gross.
Uses: Organic synthesis; preparation of resins and perfume intermediates.
Shipping regulations: Flammable liquid. Red label. *

5-allyl-5-sec-butylbarbituric acid. See talbutal.

allyl caproate (allyl hexanoate; 2-propenyl hexanoate) $CH_3(CH_2)_4COOCH_2CHCH_2$.
Properties: Colorless to pale yellow liquid; pineapple odor. Insoluble in water; soluble in 1 volume of 80% alcohol.
Constants: Sp. gr. 0.885-0.888; refractive index 1.424-1.426; b.p. 186-188°C.
Grades: Technical.
Containers: Glass bottles; demijohns; aluminum containers.
Uses: Perfumery; flavors.
Shipping regulations: None. *

N-allyl ortho-(carboxymethoxy)benzamide $HOOCCH_2OC_6H_4CONHCH_2CH{:}CH_2$.
Properties: Buff-colored crystals; relatively insoluble in water; m.p. 115-123.5°C.

allyl chloride (3-chloropropene; alpha-chloropropylene; AC; chloroallylene) CH_2CHCH_2Cl.
Properties: Colorless liquid with unpleasant pungent odor; b.p. 45.0°C; f.p. −134.5°C; sp. gr. (20/4°C) 0.9382; wt/gal (20°C) 7.83 lbs; refractive index n 20/D 1.416; flash point (Tag open cup) −20°F. Insoluble in water; miscible with alcohol, chloroform, ether, and petroleum ether.
Derivation: By gas-phase direct chlorination of propylene at 15 psi and 400-500°C.
Method of purification: Distillation.
Grades: Technical.
Containers: 1-, 5-gal pails; 55-gal drums; tank cars.
Uses: Preparation of allyl alcohol and other allyl derivatives; thermosetting resins for varnishes, plastics, adhesives; synthesis of pharmaceuticals and insecticides.
Danger! Vapor harmful. Flammable. MCA warning label.
Shipping regulations: Flammable liquid. Red label. *

allyl chlorocarbonate (allyl chloroformate) C_3H_5OOCCl.
Shipping regulations: Corrosive liquid. White label. *

allyl chloroformate. See allyl chlorocarbonate.

allyl cyanide (3-butenenitrile; vinylacetonitrile) $CH_2{:}CHCH_2CN$.
Properties: Liquid; agreeable onion-like odor; sp. gr. 0.8341; m.p. −87°C; b.p. 119°C. Slightly soluble in water.
Derivation: Prepared by treating dry cuprous cyanide with allyl bromide.
Source: Found in some mustard oils.
Use: Cross-linking agent in polymerization.

allyl-9,10-epoxystearate $CH_2{:}CHCH_2OCO(CH_2)_7CHOHC(CH_2)_7CH_3$.
Properties: Sp. gr. 0.9217; b.p. decomposes; freezing point 4°C. Insoluble in water.

allyl hexanoate. See allyl caproate.

2-allyl-4-hydroxy-3-methyl-2-cyclopenten-1-one ester of chrysanthemummonocarboxylic acid. See allethrin.

allylidene diacetate $CH_2{:}CHCH(OOCCH_3)_2$.
Properties: Colorless liquid; sp. gr. (20/20°C) 1.0749; lbs/gal (20°C) 8.945;

b. p. (50 mm) 107°C; vapor pressure (20°C) 0.25; freezing point −37.6°C; viscosity 2.8 cps (20°C); flash point (open cup) 180°F; soluble in water (1.8% by weight).

Uses: Chemical intermediate of primary interest as a source of acrolein, and as a monomer for copolymerization to form resinous materials.

5-allyl-5-isobutylbarbituric acid $C_{11}H_{16}N_2O_3$. Allylbarbituric acid in N. F. X.

Properties: White crystalline powder; odorless; slightly bitter taste; soluble in alcohol, ether and chloroform; almost insoluble in water; m. p. 138-139°C.

Use: Medicine.

5-allyl-5-isopropyl-barbituric acid (allylisopropylmalonylurea; aprobarbital) $C_{10}H_{14}O_3N_2$.

Properties: White crystalline powder; odorless; slightly bitter taste; m. p. 140-141.5°C; insoluble in water and paraffin hydrocarbons; soluble in alcohol, chloroform, and ether.

Grades: N. F. XI.

Use: Medicine.

5-allyl-5-isopropyl barbituric sodium (sodium allyl isopropyl barbiturate; aprobarbital sodium) $C_{10}H_{13}O_3N_2Na$.

Properties: White crystalline powder; odorless; slightly bitter taste; hygroscopic; soluble in water; insoluble in alcohol and ether.

Use: Medicine.

allylisopropylmalonylurea. See 5-allyl-5-isopropylbarbituric acid.

allyl isosulfocyanate. See allyl isothiocyanate.

allyl isothiocyanate (allyl isosulfocyanate; mustard oil, artificial; 2-propenyl isothiocyanate) C_3H_5NCS.

Properties: Colorless to light yellow, oily liquid; pungent and irritating mustard odor; will cause blistering of skin; harmful to lungs; sp. gr. 1.013-1.016 (25/25°C); b. p. 152°C; refractive index n 25/D 1.5271; optically inactive; neutral to moistened litmus in freshly distilled alcoholic solution. Soluble in alcohol (suffers deterioration by interaction during storage), ether, carbon disulfide; slightly soluble in water.

Derivation: Obtained by distilling sodium thiocyanate and allyl chloride, or by distillation of dried ripe seeds of Brassica nigra L. (black mustard seed).

Containers: 5-lb bottles (cases containing 10 x 5 lb bottles), 1-, 2-lb bottles.

Uses: Military poison gas; flavoring ingredient in food; counter-irritant in ointments and mustard plasters.

allyl mercaptan (allyl sulfhydrate; 2-propene-1-thiol) CH_2CHCH_2SH.

Properties: Water-white liquid (darkens on standing); strong garlic odor; sp. gr. 0.925 (23/4°C); b. p. 67-68° (90°C); insoluble in water; soluble in ether and alcohol.

Grades: C.P.

Containers: Bottles.

Uses: Flavoring agent of garlic type; pharmaceutical intermediate; rubber vulcanization accelerator intermediate.

Shipping regulations: None. *

4-allyl-1,2-methylenedioxybenzene. See safrole.

N-allylnormorphine hydrochloride. See nalorphine hydrochloride.

allyl resins. A special class of vinyl resins derived from esters of allyl alcohol and dibasic acids. Common monomers are allyl diglycol carbonate, also known as diethylene glycol bis(allyl carbonate), diallyl phthalate, diallyl isophthalate, and diallyl maleate. Polymerization occurs through the unsaturated allyl double bond to form thermosetting resins which are highly resistant to chemicals, moisture, abrasion, and heat. The allyl resins are used as laminating adhesives since they cure without the application of heat or pressure. They are also used in varnishes and molding compositions.

N-allylsalicylamide $HOC_6H_4CONHCH_2CH:CH_2$.

Properties: Normally a sticky solid, light yellow to light brown; m. p. 48-52°C.

allyl sulfhydrate. See allyl mercaptan.

allyl sulfide (diallyl sulfide; thioallyl ether) $(CH_2CHCH_2)_2S$.

Properties: Colorless liquid with garlic odor. B. p. 139°C; sp. gr. (27/4°C) 0.888; refractive index n 27/D 1.4877. Insoluble in water; miscible with alcohol, ether, chloroform, and carbon tetrachloride.

Grades: Technical.

Containers: Bottles.

Uses: Component of artificial oil of garlic.

Shipping regulations: None. *

allylsulfocarbamide. See allylthiourea.

allylsulfourea. See allylthiourea.

allylthiourea (allylsulfocarbamide; rhodalline; thiosinamine; allylsulfourea). $C_3H_5NHCSNH_2$.

Properties: White crystalline solid; slight garlic odor; bitter taste; sp. gr. 1.22; m. p. 78°C. Toxic! Soluble in water, ether, and solutions of borax, benzoates, urethane; insoluble in benzene; slightly soluble in 70% alcohol.

Derivation: Made by warming a mixture of equal parts of allyl isothiocyanate and absolute alcohol with an equal amount of 30% ammonia.

Grades: Technical (95% min purity); commercial (90% purity, approx).

Containers: Bottles; fiber drums, 25 lbs net, 28 lbs gross; steel drums; bottles (technically pure grade).

Uses: Medicine; corrosion inhibitor; organic synthesis.

Shipping regulations: None. *

allyltrichlorosilane $CH_2CHCH_2SiCl_3$.

Properties: Colorless liquid; b. p. 117.5°C; sp. gr. 1.217 (27°C); refractive index

n 20/D 1.487; flash point (Cleveland open cup) 95°F. Readily hydrolyzed by moisture, with the liberation of hydrochloric acid; polymerizes easily.
Derivation: Reaction of allyl chloride with silicon (copper catalyst).
Grades: Technical.
Use: Intermediate for silicones; glass fiber finishes.
Shipping regulations: Corrosive liquid. White label. *

almond, bitter (amygdala amara; bitter almond). Ripe seed of Prunus amygdalus var. amara; chief constituents 35-50% fixed oil, about 3% amygdalin.
Occurrence: Italy, Spain and southern France.
Grades: Technical.
Containers: Barrels; burlap bags.
Uses: Preparation of amygdalin; recovery of the essential oil; flavoring compounds.
Shipping regulations: None. *

almond meal. Residue obtained after expressing oil from almonds.
Grades: Technical.
Containers: 200-lb barrels.
Uses: Cosmetics; manufacturing bitter almond water; perfume base; cooking; confectionery.
Shipping regulations: None. *

almond oil, bitter (amygdala amara oil). An essential oil derived by steam distillation from the meal or seeds of Prunus amygdalus, Stokes (Amygdalus communis, L.) the almond tree, or Prunus armeniaca, L., the apricot tree. Only a small amount of the commercial oil is derived from almonds. Bitter almond oil may be deprived of its hydrocyanic acid. It changes to a crystalline mass on exposure to the air, hence should be kept in tightly closed completely filled containers.
Grades: (a) Containing hydrocyanic acid; (b) hydrocyanic acid free.
Properties: (a) Colorless oil subsequently becoming yellow; bitter almond (or benzaldehyde) odor. Toxic; smell cautiously! Hydrocyanic acid content may run up to 11%. (b) Colorless oil subsequently becoming yellow. Oxidizes more rapidly than the oil containing hydrocyanic acid. Miscible with alcohol, ether and oils. Slightly soluble in water.
Chief known constituents: (a) Benzaldehyde (may be 95%); hydrocyanic acid, benzaldehyde cyanhydrin.
Constants: (a) Sp. gr. 1.038-1.060 (25/25°C); optical rotation about +0.167°; refractive index n 20/D 1.5442; (b) Sp. gr. 1.050-1.055 (15°C); b.p. 179°C; optically inactive; refractive index 1.542-1.546.
Adulteration: Nitrobenzene, synthetic benzaldehyde.
Containers: 1-, 5-, 10-lb bottles; 25-lb tins.
Uses: Flavor for cosmetics and medicines; perfumes. May be used as flavor in foods only if free from hydrocyanic acid.

almond oil, expressed. See almond oil, sweet.

almond oil, sweet (almond oil, expressed). A fixed nondrying oil. Note that sweet almond oil is the expressed oil; bitter almond oil is the steam-distilled oil.
Properties: Yellowish oily liquid; almost odorless; bland taste; sp. gr. 0.910-0.915 (25/25°C); clear at −10°C, congeals near −20°C; saponification number 190-200; iodine number 95-105. Soluble in ether, chloroform, benzene; slightly soluble in alcohol; insoluble in water.
Derivation: Expressed from the seed of Prunus amygdalus.
Chief constituents: Oleic, linoleic, myristic, palmitic acids.
Grades: Technical; N.F. XI.
Containers: Iron drums; tins; glass bottles.
Uses: Lubricant for delicate mechanisms; medicine (emollient); cosmetic creams.
Shipping regulations: None. *

almond, sweet (amygdala dulcis; Jordan almond). Ripe seed of Prunus amygdalus var. dulcis.
Constituents: About 50% fixed oil, proteids, emulsin, sugar.
Occurrence: Europe and California.
Uses: Food, medicine (nutrient, demulcent); source of sweet almond oil.

aloe
Properties: Orange-brown to blackish-brown opaque, resin-like masses; saffron-like odor; strongly bitter taste. Also sold as dark yellow powder.
Derivation: Dried juice of leaves of Aloe perryi and other species of Aloe.
Principal varieties: Socotrine (east coast of Africa, and Arabia); Barbados or Curacao (West Indies); Cape (South Africa).
Grades: Technical; U.S.P. XVI.
Containers: Cases; kegs; bags.
Use: Medicine.
Shipping regulations: None. *

aloe-emodin (rhabarberone) $C_{15}H_{10}O_5$. 1,8-dihydroxy-3-hydroxymethylanthraquinone. An isomer of emodin.
Properties: Orange needles. M.p. 223-224°C; sublimes readily. Soluble in hot alcohol, benzene, ether, ammonia water, and sulfuric acid.
Occurrence: Free and as glycoside in aloe, senna, and rhubarb.
Use: Medicine.

aloe oil. An oil obtained from Socotrine aloe.

aloin (barbaloin). A mixture of active principles obtained from aloe. Varies in properties according to variety of aloe used.
Properties: Yellow crystals with bitter taste; darkens on exposure to air; odorless or slight odor of aloe; soluble in acetone, alkalies, water, and alcohol; slightly soluble in ether.
Grades: Technical; N.F. XI.
Containers: Barrels; drums; kegs.
Uses: Medicine; proprietary laxatives; electroplating baths; stimulates fermentation.
Shipping regulations: None*

alon. A cellulose acetate fiber originated in Japan which is spun as viscose rayon from an aqueous solution and then acetylated in the fiber form.

"Alox" Compounds. [117] Trademark for mixtures of organic acids, oxyacids, and esters derived from oxidation of petroleum hydrocarbons. Derivatives such as methyl esters and metallic soaps are available.
Uses: Corrosion inhibitors; temporary protective coatings; lubricity additives; upper cylinder lubricant additives; blown rapeseed oil replacement in marine engine lubricants.
Shipping regulations: According to flash points. *

"Aloxite." [280] A trademark for aluminum oxide made by fusing materials high in alumina, such as bauxite, and for articles made therefrom.
Properties: Abrasive aluminum oxide made by electric furnace fusion of refined alumina is a white crystalline material; hardness, 9.03 in Mohs' scale; sp. gr. 3.92-3.96. Abrasive aluminum oxide made by the electric furnace fusion and treatment of bauxite varies in color from light brown to deep brown; hardness 9.07 in Mohs' scale; sp. gr. 3.94-4.00. Claimed to be a very good electrical insulator, fairly good heat conductor; highly refractory.
Containers: Multiwall paper sacks.
Uses: Abrasive grains and powders; grinding wheels; stones; razor hones; refractory cements; filter plates and tubes; diffuser plates and tubes; porous undergrain plates and coated abrasive products.

"Alpco." [271] Trademark for a series of high melting point mineral waxes and resins, used for carbon paper inks, polishes, paper, plasticizers, surfactants, dispersants, casting waxes and surface coatings.

"Alperox." C. [154] Trademark for technical lauroyl peroxide (q. v.).

alpha. A prefix denoting the position of a substituting group of atoms (radical) in the main group of a compound. The Greek letters alpha, beta, gamma, etc., are usually not identical with the International Union of Chemistry numbering system, 1, 2, 3, etc., since they do not start from the same carbon atom. Thus acids start lettering from the carboxylic (COOH) group: CH_3CH_2COOH propionic acid, $CH_3CHClCOOH$ alpha-chloropropionic acid, CH_2ClCH_2COOH beta-chloropropionic acid. (The International systematic names are propanoic acid, 2-chloropropanoic acid and 3-chloropropanoic acid). Alpha, beta, etc. are also used to designate attachment to the side chain of a ring compound. Thus alpha-chlorotoluene, (benzyl chloride) $C_6H_5CH_2Cl$, may be distinguished from ortho-chlorotoluene, $CH_3C_6H_4Cl$. Specific alpha compounds will be found under the name of the compound, viz: beta-alanine. See alanine.

"Alphalin." [100] Trademark for vitamin A, U.S.P. (q. v.).

"Alphanol." [307] Trademark for line of acid dyestuffs.

alpha particle. The nucleus of the helium atom, having mass about four times that of the hydrogen atom, and a doubly positive charge compared with the unit positive atomic charge on the proton. Produced in the normal disintegration of radium, and in many other nuclear changes. Used as a bombarding particle in cyclotron and similar nuclear experimentations.

alphaprodine hydrochloride (prisilidene hydrochloride) $C_{16}H_{23}NO_2 \cdot HCl$. 1,3-dimethyl-4-phenyl-4-piperidyl propionate hydrochloride.
Properties: White, crystalline bitter powder; aminelike odor; m.p. 218-220°C; stable to air, light and heat. Freely soluble in alcohol, chloroform, and water; very slightly soluble in ether; pH (1% solution) 4.5-5.2.
Grade: N.N.D.
Use: Medicine.

"Alphazurine." [243] Trademark of triphenylmethane acid blues.

"Alrowet" D-65. [219] Trademark for sodium dioctyl sulfosuccinate $C_8H_{17}OOCCH_2$-$CH(SO_3Na)COOC_8H_{17}$. Miscible with water and soluble in most organic solvents.
Uses: Dispersant for paints, inks, polishes; drycleaning aid; polymerization emulsifier; textile penetrant and dyeing assistant.

alseroxylon. Fat-soluble alkaloidal fraction of the purified extract of Rauwolfia serpentina.
Grade: N.N.D.
Use: Medicine.

"Altax." [69] Trademark for a proprietary benzothiazyl disulfide.
Properties: Cream to light yellow powder; sp. gr. 1.51 ± .03; moderately soluble in benzene, carbon disulfide, chloroform; insoluble in water, dilute caustic, gasoline.
Uses: Primary accelerator in natural and nitrile rubber and SBR; plasticizer and vulcanization retarder in neoprene Type G; cure modifier in neoprene Type W; oxidation cure activator in butyl. For extruded and molded goods, tire and tubes, wire and cable, sponge.

"Altex." [206] Brand name of proprietary line of union dyestuffs.

althea (marshmallow). Dried root of Althea officinalis, deprived of brown corky layer.
Occurrence: Europe; United States.
Grades: Technical; N.F. XI.
Containers: Bags.
Use: Medicine.
Shipping regulations: None. *

althein. See asparagine.

alum. Refers to hydrated double sulfates of aluminum and univalent metals such as potassium, sodium or ammonium. These usually crystallize easily from solutions

*See "I.C.C. Shipping Regulations," page xiii.
Reference numbers refer to name of manufacturer. See "List of Manufacturers," page v.

containing proper proportions of the dissolved sulfates, the compositions being $K_2SO_4 \cdot Al_2(SO_4)_3 \cdot 24H_2O$, $(NH_4)_2SO_4 \cdot Al_2(SO_4)_3 \cdot 24H_2O$, etc. These are distinguished as potash alum, ammonium alum. The formulas are sometimes written as $KAl(SO_4)_2 \cdot 12H_2O$. The term alum is also applied to analogous compounds of other trivalent metals such as iron and chromium. Thus $(NH_2)_2SO_4 \cdot Fe_2(SO_4)_3 \cdot 24H_2O$ is ammonium ferric alum. Aluminum sulfate itself, or its hydrates are sometimes incorrectly referred to as alums, especially in the paper industry.

Alum, N.F. XI, refers either to ammonium alum or potassium alum.

alum, ammonia. See aluminum-ammonium sulfate.

alum, burnt (alum, exsiccated) $AlNH_4(SO_4)_2$ or $AlK(SO_4)_2$. Aluminum-ammonium sulfate or aluminum-potassium sulfate heated just sufficiently to drive off the water of crystallization.

Properties: White, odorless powder; sweetish, astringent taste. Soluble in hot and slowly soluble in cold water; insoluble in alcohol.

Grades: N.F. XI.

Use: Medicine.

alum, chrome. See chromium-potassium sulfate.

alum, chrome ammonium. See chromium-ammonium sulfate.

"Alumel." [166] Trademark product. An alloy consisting of about 94% nickel, with small, carefully controlled amounts of silicon, aluminum and manganese. It is chiefly used as the negative element in "Chromel-Alumel" thermocouples and lead wire.

"Alumex." [285] Trademark for a china clay; particle size 55-60% minus 2 microns, 20-25% plus 5 microns; oil absorption 30 cc/100 grams clay; pH 4.5-5.5.

alum, exsiccated. See alum, burnt.

alumina. See aluminum oxide.

alumina, activated. See activated alumina.

"Alumina Ceramics, High Strength." [444] Available as dense (AD) or porous (AP) varieties.

"AD"-85: 85% Al_2O_3. Tensile strength 17,000 psi (76°F); 8,000-9,000 psi (2000°F); sp. gr. 3.40; color, white; sp. heat 0.18 Btu/lb; thermal conductivity 92 Btu/hr/sq ft/°F/in. at 72°F, 50 at 600°F; thermal coefficient of expansion 5.68×10^{-6}/°C (25-200°C).

"AD"-94: 94% Al_2O_3. Tensile strength 25,000-27,000 psi (76°F); 9,000-10,000 psi (2000°F); sp. gr. 3.61; color, white; sp. heat 0.188 Btu/lb; thermal conductivity 140 Btu/hr/sq ft/°F/in. at 72°F, 74 at 600°F; thermal coefficient of expansion 6.67×10^{-6}/°C (25-200°C).

"AD"-99: 99% Al_2O_3. Tensile strength 34,000-35,000 psi (76°F); 21,000-22,000 psi (2000°F); sp. gr. 3.90; color, white; sp. heat 0.2 Btu/lb; thermal conductivity 202 Btu/hr/sq ft/°F/in. at 72°F, 108 at 600°F; thermal coefficient of expansion 6.7×10^{-6}/°C (25-200°C).

"AP-100": 100% Al_2O_3. Sp. gr. 3.1; color, white; thermal conductivity 105 Btu/hr/sq ft/°F/in. at 72°F, 57 at 600°F

alumina, fused. See "Alundum."

alumina gel. See aluminum hydroxide, gelatinous.

alumina trihydrate (aluminum hydroxide; aluminum hydrate; hydrated alumina; hydrated aluminum oxide) $Al_2O_3 \cdot 3H_2O$ or $Al(OH)_3$.

Properties: White crystalline powder, balls or granules; sp. gr. 2.42; insoluble in water; soluble in mineral acids and caustic soda.

Derivation: From bauxite; the ore is dissolved in strong caustic and aluminum hydroxide precipitated from the sodium aluminate solution by neutralization (as with carbon dioxide) or by autoprecipitation (Bayer process).

Grades: Technical; C.P.

Containers: Bags; drums; tonnage lots.

Uses: Glass, ceramics, iron-free aluminum and aluminum salts; manufacture of activated alumina; base for organic lakes. Finely divided form (0.1-0.6 microns) used for rubber reinforcing pigment, paper coating, filler, cosmetics.

Shipping regulations: None.*

aluminite (websterite) $Al_2O_3 \cdot SO_3 \cdot 9H_2O$. A natural hydrous aluminum sulfate.

Properties: Color white; streak white; luster dull or earthy; sp. gr. 1.66; hardness 1-2. Usually occurs in white chalky masses in clay beds.

Occurrence: Germany; England.

Uses: Tanning; water proofing; dyeing; paper making; water purification.

aluminium. See aluminum.

aluminosilicates. Compounds of aluminum silicate with metal oxides or other radicals. Used as catalysts in the refining of petroleum; to soften water. See also zeolites.

aluminum (aluminium) Al. Element of atomic number 13, of group III of the periodic system. A silvery, ductile metal. Most abundant metal in earth's crust.

Properties: Sp. gr. 2.708; m.p. 660°C; b.p. 1800°C. Soluble in strong acids and alkalies; insoluble in water. Pure (99+%) aluminum is resistant to ordinary corrosion and most acids; is attacked by caustic alkalies, the halogens and their acids.

Derivation: Purified aluminum oxide (q.v.) is made from bauxite. Aluminum metal is then made by the electrolysis of the oxide in a bath of molten cryolite. The original process was called the Hall process (q.v.) and various modifications of this are now in use. A recent process

*See "I.C.C. Shipping Regulations," page xiii.

Reference numbers refer to name of manufacturer. See "List of Manufacturers," page v.

involves an extra step in which aluminum chloride vapors are passed over the partially reduced aluminum oxide in a reversible process which continually generates the free metal.
Forms available: Ingots; pigs; innumerable structural shapes; foil; flakes; powder (technical and U.S.P. XVI grade). A superpure grade (99.99%) is made by further electrolytic reduction of the commercial grade. Foamed aluminum is now made by mixing zirconium hydride with molten aluminum.
Uses (in approximate order of volume): Military (aircraft, ships, etc.); building materials; consumer durable goods; transportation; machinery and equipment; construction (electric power construction); destructive uses; electrical and communications equipment; containers and packaging; exports; chemicals; photography; miscellaneous.
Powder: Usual metal powder uses; protective ointment; solid fuel propellant.
Superpure grade: Electronics; catalyst in petroleum refining; jewelry; roofing; car trim; cladding for aluminum alloys for aircraft.
Flakes: As insulation for liquid fuels.
Caution: Aluminum powder forms flammable and explosive mixtures with air.

aluminum acetate (waterproofing salts). (1) Normal $Al(C_2H_3O_2)_3$; (2) basic $Al(C_2H_3O_2)_2OH$.
Properties: (a) Known only in solution; (b) amorphous, white powder; insoluble in water.
Derivation: By the interaction of aluminum hydroxide and acetic acid. The product is recovered by crystallization.
Grades: Technical; C.P.; also sold in solution; U.S.P. XVI (solution).
Containers: Wooden barrels; boxes; drums; carboys (solutions).
Uses: Waterproofing cloth; fireproofing fabrics; mordant in textile dyeing; preparation of lakes; embalming fluids; medicine; calico printing.
Shipping regulations: None.*
See also mordant rouge.

aluminum acetotartrate. Consists of approximately 70% basic aluminum acetate and 30% tartaric acid.
Properties: Colorless crystals or white powder; slight acetic odor; astringent, acidulous taste; slowly soluble in cold water; insoluble in alcohol.
Use: Medicine.

aluminum acetylacetonate $Al(C_5H_7O_2)_3$.
Properties: Solid; m.p. 189°C; b.p. 315°C. Soluble in benzene and alcohol.
Uses: Deposition of aluminum; catalyst.

aluminum acetylsalicylate (aluminum aspirin) $[C_6H_4(OCOCH_3)(COO)]_2AlOH$.
Properties: White to offwhite granules or powder; odorless or slight odor; m.p., decomposes. Insoluble in water and organic solvents; soluble with decomposition in alkali, hydroxides and carbonates.
Derivation: Reaction of aluminum hydroxide with acetylsalicylic acid.
Grade: N.F. XI.
Uses: Medicine.

aluminum alkyls (aluminum trialkyls). Catalysts used in the Ziegler process.
See triethylaluminum and triisobutylaluminum.

aluminum alloys. Alloys based on aluminum containing variable amounts of manganese, silicon, copper, magnesium, lead, bismuth, nickel, chromium, zinc, and titanium. A wide range of uses and properties is possible. Alloys may be obtained for casting or working, heat-treatable or non-heat-treatable, with a wide range of strength and corrosion resistance, machinability and weldability.

aluminum aminoacetate, basic. See dihydroxyaluminum aminoacetate.

aluminum-ammonium chloride (ammonium-aluminum chloride) $AlCl_3 \cdot NH_4Cl$.
Properties: White crystals; soluble in water. M.p. 304°C.
Uses: Used in treatment of furs.

aluminum ammonium sulfate (alum, ammonia; ammonium alum; alum N.F.) $Al_2(SO_4)_3(NH_4)_2SO_4 \cdot 24H_2O$ or $AlNH_4(SO_4)_2 \cdot 12H_2O$.
Properties: Colorless crystals; odorless; strong astringent taste. Soluble in water, glycerine; insoluble in alcohol.
Constants: Sp. gr. 1.645; m.p. 94.5°C; b.p., loses $20H_2O$ at 120°C.
Derivation: By crystallization from a mixture of ammonium and aluminum sulfates.
Method of purification: Recrystallization.
Grades: Technical, lump, ground, powdered; C.P.; N.F. XI.
Containers: 100-lb bags; 100-, 250-, 350-, 360-lb drums.
Uses: Medicine; mordant in dyeing; water and sewage purification; sizing paper; tanning; clarifying agent; ingredient in baking powder.
Shipping regulations: None.*

aluminum anodized. The resistance of aluminum to abrasion may be increased by anodic treatment in certain electrolytes, usually sulfuric acid, chromic acid, or oxalic acid. The coating appears to be amorphous aluminum oxide. Anodic coatings may be colored by impregnation with organic dyes or mineral pigments.

aluminum aspirin. See aluminum acetylsalicylate.

aluminum borate $2Al_2O_3 \cdot B_2O_3 \cdot 3H_2O$ (approx.).
Properties: White, granular powder; decomposed by water.
Derivation: By the interaction of aluminum hydroxide and boric acid.
Method of purification: Crystallization.
Grades: Technical; C.P.
Containers: 1-lb bottles; wooden barrels.
Use: Glass and ceramic industries.
Shipping regulations: None.*

aluminum boride AlB_{12}.
Properties: Powder; apparent bulk density (fully settled): light, 0.6-0.8 g/cc; dense, 1.2-1.4 g/cc.
Uses: Neutron absorption applications.

aluminum borohydride $Al(BH_4)_3$.
Properties: Volatile liquid; b.p. 44.5°C; m.p. —64.5°C. Reacts vigorously with water to liberate hydrogen; ignites spontaneously in air.
Derivation: (1) By reaction of sodium borohydride and aluminum chloride in the presence of a small amount of tributyl phosphate; (2) by reaction of aluminum trimethyl and diborane.
Uses: Intermediate in organic synthesis; jet fuel additive.

aluminum brass. An alloy containing 76% copper, 21.5 to 22.25% zinc, 1.75 to 2.50% aluminum. Sp. gr. 8.33; mean coefficient of thermal expansion (32-212°F) 0.97 to 1.02×10^{-5}; thermal conductivity (at 20°C) 58 $Btu/ft^2/ft/°F/hr$; melting range 1710 to 1780°F; specific heat at 20°C 0.09 cal/g/°C. Available cold rolled, drawn, tubes. Methods of fabrication include deep drawing, flanging, riveting, and brazing. Typical physical properties of an 1.0 in. diameter annealed tube with 0.065 in. wall: tensile strength 60,000 psi; yield point 27,000 psi; per cent elongation in 2 in. 55%; machining quality 30 based on free cutting brass as 100.
Uses: Condenser, evaporator, and heat exchanger tubes and ferrules.

aluminum bromide (a) $AlBr_3$; (b) $AlBr_3 \cdot 6H_2O$.
Properties: White to yellowish, deliquescent crystals; exists as double molecules Al_2Br_6 in the vapor; soluble in water, alcohol, carbon disulfide or ether. The anhydrous form reacts violently with water, liberating hydrogen bromide.
Constants: (a) Sp. gr. 3.01; m.p. 97.5°C; b.p. 265°C; (b) sp. gr. 2.54; m.p. 93°C (decomposes).
Derivation: (a) By passing bromine over heated aluminum; (b) the reaction of hydrobromic acid with aluminum hydroxide.
Method of purification: Crystallization.
Grades: Technical; C.P.
Containers: 1-, 2-, 4-, 8-lb glass jars; air-tight drums.
Uses: Bromination, alkylation, and isomerization catalyst in organic synthesis.

aluminum bronze. A nonferrous alloy containing 88-96.1% copper, 2.3-10.5% aluminum, and small amounts of iron and tin. Sometimes small amounts of other additives are present such as nickel, manganese or zinc. These alloys are characterized by high strength, ductility, hardness and resistance to shock and fatigue. Sp. gr. 7.50-8.19. Chemically there is good resistance to dilute hot or cold solutions of sulfuric acid, acetic acid in the absence of air, hot or cold sodium hydroxide up to 50%, and sea water.

aluminum bronze powder (gold bronze powder).
Composition: Alloy of 90% copper and 10% aluminum reduced from leaf form to powder and polished mechanically and coated with stearic acid.
Grades: Litho, moulding, printing-ink, and radiator.
Uses: As a pigment in paints and inks.

aluminum n-butoxide $Al(OC_4H_9)_3$.
Properties: Yellow to white crystalline solid; m.p. 101.5°C (pure) and 88-96°C (commercial); density 1.0251 (20°C); b.p. 290-310°C (30 mm). Soluble in aromatic, aliphatic and chlorinated hydrocarbons.
Containers: 6-gal open head container.
Uses: Ester exchange catalyst; defoamer ingredient; hydrophobic agent intermediate.

aluminum carbide Al_4C_3.
Properties: Yellow crystals or powder; decomposes in water with liberation of methane.
Constants: Sp. gr. 2.36. Stable to 1400°C.
Derivation: By heating aluminum oxide and coke in an electric furnace.
Grades: Technical.
Containers: Iron drums.
Uses: Generating methane; catalyst; metallurgy; drying agent.
Fire hazard: Dangerous! Keep dry!

aluminum carbonate. A basic carbonate of variable composition; formula sometimes given as $Al_2O_3 \cdot CO_2$. White lumps or powder, insoluble in water, dissolves in hot hydrochloric acid or sulfuric acid. Formerly used as mild astringent, styptic.
Normal aluminum carbonate $Al_2(CO_3)_3$ is not known as an individual compound that can be isolated.

Aluminum Chelates. [134] Available as:
Aluminum Chelate BEA-1: Chemically modified aluminum secondary butoxide.
Properties: Pale yellow liquid; sp. gr. 1.030 (21°C); aluminum content 8.9-9.1%.
Aluminum Chelate PEA-1: Chemically modified aluminum isopropylate.
Properties: Pale yellow liquid; sp. gr. 1.035 (25°C); aluminum content 9.3-10.0%.
Aluminum Chelate PEA-2: Chemically modified aluminum isopropoxide.
Properties: Pale yellow; soluble in aromatic, aliphatic and chlorinated hydrocarbons; aluminum content 7.8-7.9%.
Uses: Curing of epoxy, phenolic, castor oil alkyds and high molecular weight polymers which are hydroxyl or carboxyl bearing; textile hydrophobing, in solvent based systems; adhesion promotion.

aluminum chloride (a) $AlCl_3$ (anhydrous); (b) $AlCl_3 \cdot 6H_2O$.
Properties: Yellowish-white, granular crystals or powder; (a) sp. gr. 2.44; m.p. 190°C (2.5 atm.); sublimes readily; the vapor consists of double molecules, Al_2Cl_6; (b) sp.gr. 2.40; m.p. decomposes. Soluble in water, ether and alcohol. The anhydrous form combines violently with

water.
Derivation: (a) (1) By reaction of chlorine with molten aluminum; (2) by reaction of alumina or bauxite with coke and chlorine at about 1500°F. (b) By crystallizing the anhydrous form from hydrochloric acid solution.
Impurities: Ferric chloride; titanium chloride; silicon chloride; basic aluminum chloride.
Grades: Technical anhydrous; technical crystals; 50% solution; C. P. crystals; U.S.P. XVI (hydrate).
Containers: Drums; tank cars.
Uses: Petroleum refining and catalysis (Friedels-Craft catalyst); polymerization catalyst for synthetic rubber, plastics, and lubricants; metallurgy; dyes; aluminum chemicals; external medicine.
Warning! Causes burns. Contact with water or moist air liberates hydrochloric acid gas. MCA warning label (anhydrous aluminum chloride).

aluminum chloride solution 32° Bé. Special grade of a solution low in iron content, having an acid reaction but containing no free acid. Typical analysis: $Al_2Cl_6 \cdot 12H_2O$, 50.3%; Al_2O_3, 10.6%; Fe, 0.005%.
Containers: Carboys; tank cars; tank trucks.
Uses: Carbonizing fine wool such as piece goods, raw stock or shoddy, resulting in an exceptionally white, lofty wool; carbonizing colored goods where it has very little stripping action; soap manufacture for treatment of glycerin lyes where chloride of alumina is more desirable than sulfate.

aluminum citrate $AlC_6H_5O_7$; usually contains about 10% water.
Properties: Odorless, fine white powder. Very soluble in water; slowly in cold water but rapidly in warm.
Derivation: Hydrated aluminum oxide dissolved in citric acid.
Containers: Glass bottles.
Use: Medicine.
Shipping regulations: None. *

aluminum diformate (aluminum formate, basic) $Al(OH)(CHO_2)_2 \cdot H_2O$.
Properties: White or gray powder. Soluble in water.
Derivation: Aluminum hydroxide is dissolved in formic acid and spray-dried. Solutions are also prepared by treating aluminum sulfate with formic acid, followed by lime.
Grades: Technical solutions (12-20°Bé).
Containers: Carboys.
Uses: Waterproofing; mordanting; antiperspirant compounds.

aluminum dihydroxy glycinate. See dihydroxy aluminum aminoacetate.

aluminum distearate $Al(OH)(C_{18}H_{35}O_2)_2$.
Properties: White powder; m.p. 145°C; sp. gr. 1.009. Insoluble in water, alcohol, ether. Forms gel with aliphatic and aromatic hydrocarbons.
Uses: The most commonly used aluminum stearate. Thickener in paints, inks, and greases; water repellent; lubricant in plastics and ropes; in cement production.

aluminum ethylhexoate (aluminum octoate). A metallic salt of 2-ethylhexoic acid, used as a paint additive.

aluminum fluoride anhydrous AlF_3.
Properties: White crystals. Sublimes about 1260°C without melting; sp. gr. 2.882. Slightly soluble in water; insoluble in most organic solvents.
Derivation: Solution of alumina trihydrate in aqueous hydrofluoric acid followed by crystallization and calcination to remove water.
Grades: Technical (82-93% AlF_3).
Containers: Multiwall paper sacks; barrels; fiber drums.
Uses: In the production of aluminum to lower the melting point and increase the conductivity of the electrolyte; as a flux in ceramic glazes and enamels.

aluminum fluoride hydrate $AlF_3 \cdot 3\frac{1}{2} H_2O$.
Properties: White crystalline powder. Insoluble in water.
Derivation: Action of hydrofluoric acid on alumina hydrate and subsequent recovery by crystallization.
Method of purification: Crystallization.
Grades: Technical; C. P.
Containers: 1-lb bottles; barrels.
Uses: Ceramics (production of white enamel); repressant of alcoholic side fermentations.

aluminum fluosilicate (aluminum silicofluoride) $Al_2(SiF_6)_3$.
Properties: White powder. Slowly soluble in cold water; readily soluble in hot water.
Grades: Technical.
Uses: Artificial gems, enamels, glass.

aluminum formate. See aluminum triformate, and aluminum diformate.

aluminum formate, basic. See aluminum diformate.

aluminum formate, normal. See aluminum triformate.

aluminum formoacetate
$Al(OH)(OOCH)(OOCCH_3)$. White powder; soluble in water and alcohol; used in textile water repellents.

aluminum glycinate, basic. See dihydroxy aluminum aminoacetate.

aluminum hydrate. See alumina trihydrate.

aluminum hydroxide. See alumina trihydrate.

aluminum hydroxide, anhydrous. See alumina trihydrate.

aluminum hydroxide, gelatinous (hydrous aluminum oxide; alumina gel)
$Al_2O_3 \cdot xH_2O$.
Properties: White, gelatinous precipitate. Constants variable with the composition; sp. gr. about 2.4. Insoluble in water and alcohol; soluble in acid and alkali.
Derivation: By treating a solution of aluminum sulfate or chloride with caustic

soda, sodium carbonate or ammonia; by precipitation from sodium aluminate solution by seeding or acidifying (carbon dioxide is commonly used).
Grades: Technical; C. P.; U. S. P. XVI.
Containers: Fiber drums.
Uses: Dyeing mordant; water purification; waterproofing fabrics; manufacture of lake; filtering medium; chemicals (manufacture of aluminum salts); lubricating compositions; manufacture of glassware; sizing paper; ceramic glaze; medicine.

aluminum hydroxystearate $Al(OH)[OOC(CH_2)_{10}CHOH(CH_2)_5CH_3]_2$.
Properties: White powder; m. p. 155°C; sp. gr. 1.045. Less soluble in nonpolar compounds than other aluminum stearates, and more soluble in polar compounds.
Uses: Waterproofing of leather and cements; lubricant for plastics and ropes; ingredient for paint and inks.

aluminum iodide AlI_3 (anhydrous).
Properties: Brown-black crystalline pieces (white when pure); m. p. 191°C; b. p. 385°C; sp. gr. 3.9825. Soluble in water with violent reaction; soluble in alcohol, ether, carbon disulfide.
Derivation: By heating aluminum and iodine in a sealed tube.
Method of purification: Crystallization.
Grades: Technical.
Containers: Bottles.
Uses: Organic synthesis.

aluminum isopropoxide. See aluminum isopropylate.

aluminum isopropylate (aluminum isopropoxide) $Al(OC_3H_7)_3$.
Properties: White solid; sp. gr. 1.035(20°C); b. p. 135-145°C (10-20 mm); 125-130°C (4-6 mm); m. p. 118-118.5°C. Anhydrous. Soluble in alcohol, benzene; decomposes in water.
Grades: Distilled (purity approximately 100%).
Containers: 1-, 2-, 7 ½-, 40-lb cans.
Uses: Very efficient dehydrating agent; reducing agent; cross-linking agent; pharmaceutical intermediate; antiperspirant; soaps.
Shipping regulations: None.*

aluminum metaphosphate $Al(PO_3)_3$.
Properties: White powder; insoluble in water; m. p. approx. 1537°C.
Containers: 100-lb drums; 275-lb barrels.
Uses: As a constituent of glazes, enamels and glasses, and as a high temperature insulating cement.

aluminum monobasic stearate. See aluminum monostearate.

aluminum monostearate (aluminum monobasic stearate) $Al(OH)_2[OOC(CH_2)_{16}CH_3]$.
Properties: Fine, white to yellowish white powder; faint characteristic odor; m. p. 155°C; sp. gr. 1.020. Insoluble in water, alcohol and ether. Forms a gel with aliphatic and aromatic hydrocarbons.
Derivation: Mixing solutions of a soluble aluminum salt and sodium stearate.
Grades: U. S. P. XVI.
Uses: Medicine; manufacture of paints, inks, greases, waxes; drier in protective coatings; thickening for lubricating oils; waterproofing; gloss producer; stabilizer for plastics.

aluminum naphthenate.
Properties: Yellow substance of rubbery consistency with high thickening power.
Derivation: Made by addition of solution of aluminum salt to aqueous solution of alkali naphthenate.
Uses: Paint and varnish drier and bodying agent; detergent in lube oils; the solution in organic solvents has been proposed for insecticides and siccatives.

aluminum beta-naphtholdisulfonate $Al_2[C_{10}H_5(OH)(SO_3)_2]_3$.
Properties: White powder, darkens slightly on exposure to air. Its solutions are incompatible with albumin, gelatin solutions, silver salts, soluble nitrates, caustic alkalies, alkaline carbonates; soluble in cold water and glycerol; slightly soluble in alcohol. Coagulates albumin, but soluble in excess.
Use: Medicine.
Shipping regulations: None.*

aluminum nitrate $Al(NO_3)_3 \cdot 9H_2O$.
Properties: White crystals. Soluble in cold water; decomposes in hot water. Soluble in alcohol and acetone.
Constants: M. p. 73°C; decomposes at 134°C.
Derivation: Formed by the action of nitric acid on aluminum and crystallization.
Method of purification: Recrystallization.
Grades: Technical; C. P.; 99.75%.
Containers: 1-, 5-lb bottles; multiwall paper sacks; drums.
Uses: Textiles (mordant in printing with alizarin red); leather tanning and finishing; manufacture of electrical incandescent filaments; catalysts in petroleum refining; nucleonics.
Shipping regulations: Oxidizing material. Yellow label.*

aluminum octoate. See aluminum ethylhexoate.

aluminum oleate $Al(C_{18}H_{33}O_2)_3$.
Properties: Yellowish-white viscous mass. Insoluble in water; soluble in alcohol, benzene, ether, oil and turpentine.
Derivation: By heating aluminum hydroxide, water and oleic acid. The resultant mixture is filtered and dried.
Grades: Technical.
Containers: Barrels.
Uses: Waterproofing; drier for paints, etc.; thickener for lubricating oils; medicine; as lacquer for metals; as lubricant for plastics.
Shipping regulations: None.*

aluminum oxide (alumina) Al_2O_3. The mineral corundum is natural aluminum oxide, and emery, ruby, and sapphire are impure crystalline varieties. The mixed mineral bauxite is a hydrated aluminum oxide with

a somewhat variable proportion of water and the oxides of iron, silicon and titanium as the principal impurities. This is the source of commercial aluminum oxide used in the production of metallic aluminum.
Properties: These vary according to the method of preparation. White powder, balls or lumps of various mesh. Sp. gr. 3.4-4.0; m.p. 2030°C; insoluble in water; difficultly soluble in mineral acids and strong alkali. The special properties and uses of corundum and hydrated forms of aluminum oxide are described under their individual headings.
Derivation: (a) Leaching of bauxite with caustic soda to separate the alumina from iron oxide, silica and other insoluble matter, followed by precipitation of a hydrated aluminum oxide by hydrolysis and seeding of the solution. The alumina hydrate is then washed, filtered and calcined to remove water and obtain the anhydrous oxide. See Derivation under alumina trihydrate; also Bayer process. (b) Coal mine waste waters are used to obtain aluminum sulfate, which is then reduced to alumina.
Grades: Technical; C.P.; fibers; high purity; fused.
Containers: Multiwall paper sacks; 50-lb drums; 100-lb barrels.
Uses: The major use is for production of aluminum, but other important uses are for manufacture of abrasives, refractories, ceramics, electrical insulators, catalyst and catalyst supports; crucibles and laboratory wares; for absorbing gases and water vapors; chromatographic analysis; fluxes.

aluminum oxide, hydrous. See aluminum hydroxide, gelatinous.

aluminum palmitate $Al(OH)_2(C_{16}H_{31}O_2)$.
Properties: White powder; m.p. 200°C; sp. gr. 1.072. Insoluble in alcohol and acetone; forms gel with hydrocarbons.
Derivation: By heating aluminum hydroxide and palmitic acid and water. The resultant mixture is filtered and dried.
Grades: Technical.
Containers: 50-lb paper bags; cartons.
Uses: Waterproofing leather, paper, textiles; thickening for lubricating oils; thickening or suspending agent in paints and inks; production of high gloss on leather and paper; ingredient of varnishes; lubricant for plastics.

aluminum penicillin G. See penicillin.

aluminum phenolsulfonate (aluminum sulfocarbolate) $Al_2(C_6H_4OHSO_3)_6$.
Properties: Reddish-white powder with slight phenol odor. Strongly astringent taste. Soluble in water and alcohol, glycerin.
Use: Medicine.
Shipping regulations: None.*

aluminum phosphate $AlPO_4$.
Properties: White crystals. Insoluble in water and alcohol, soluble in acids and alkalies.
Constants: Sp. gr. 2.566; m.p. 1500°C.
Derivation: By the interaction of solutions of aluminum sulfate and sodium phosphate.
Grades: Technical; C.P.
Containers: 100-lb bags, drums; 100-, 200-, 700-lb barrels.
Uses: Ceramics, dental cements; cosmetics; paints and varnishes; pharmaceuticals; pulp and paper industry.
Shipping regulations: None.*

aluminum potassium sulfate (potash alum; alum, N.F.; potassium alum) $Al_2(SO_4)_3 \cdot K_2SO_4 \cdot 24H_2O$, sometimes written $AlK(SO_4)_2 \cdot 12H_2O$.
Properties: White odorless crystals having an astringent taste; sp. gr. 1.75; m.p. 92°C; b.p., loses $18H_2O$ at 64.5°C; sublimes 400°C; soluble in water; insoluble in alcohol. Solutions in water are acid.
Derivation: (a) Alunite is roasted in reverberatory furnaces. The melt is leached and the salt recovered by crystallization. (b) Also derived by crystallization from a solution made by dissolving aluminum sulfate and potassium sulfate.
Grades: Technical; lump; ground; powdered; N.F. XI.
Containers: Bags and drums.
Uses: Medicine; dyeing (mordant); baking powder; textiles; paper; matches; leather; paints; catalyst in ammonia synthesis; tanning agent; waterproofing agent; purification of water; aluminum salts.
Shipping regulations: None.*

aluminum resinate.
Properties: Brown mass. Insoluble in water; soluble in oils.
Derivation: By heating soluble aluminum salts and rosin.
Grades: Technical (fused, precipitated).
Containers: Fused: 300-lb barrels. Precipitated: drums.
Uses: Drier for varnishes.
Caution! Combustible.

aluminum-rubidium sulfate (rubidium alum) $AlRb(SO_4)_2 \cdot 12H_2O$.
Properties: Colorless crystals. Soluble in water (hot); insoluble in alcohol.
Constants: Sp. gr. 1.867; m.p. 99°C.

aluminum salicylate $Al(C_6H_4OHCOO)_3$.
Properties: Reddish-white powder. Odorless. Soluble in dilute alkalies; insoluble in water, alcohol. Decomposed by acid.
Use: Medicine.
Shipping regulations: None.*

aluminum silicate pigments (ASP). The aluminum silicates form pigments which are both organophilic (compatible with organic solvents and substances) and hydrophobic.

aluminum silicates. Varying proportions of Al_2O_3 and SiO_2. Occur naturally in clays (q.v.). The only stable compound is synthetic (see mullite). See also aluminosilicates.
Uses: Glass; ceramics; pigment and filler for paints, printing inks, rubber and

plastics; where formed by precipitation with aluminum sulfate from soluble silicate solutions, are used for the beater sizing of paper, coating of pigments, decolorizing mineral oils.

aluminum silicofluoride. See aluminum fluosilicate.

aluminum soaps. See such materials as aluminum oleate, aluminum palmitate, aluminum resinate, etc.

aluminum-sodium chloride (sodium-aluminum chloride) $AlCl_3 \cdot NaCl$.
Properties: White or yellowish crystalline powder; hygroscopic. Soluble in water. M. p. 185°C.
Grades: Technical.
Uses: Leather industry; aluminum manufacture.

aluminum-sodium sulfate (SAS; sodium-aluminum sulfate; soda alum; alum, porous) $Al_2(SO_4)_3 \cdot Na_2SO_4 \cdot 24H_2O$.
Properties: Colorless crystals; saline, astringent taste; effloresces in air. Soluble in water; insoluble in alcohol.
Constants: Sp. gr. 1.675; m. p. 61°C.
Derivation: By heating a solution of aluminum sulfate and adding sodium chloride. The solution is allowed to cool, with constant stirring. The alum meal deposited is washed with water and centrifuged.
Method of purification: Recrystallization.
Grades: Pure crystals; technical; C. P.
Containers: 1-, 5-lb bottles; bulk in cars; 200-lb bags; 350-, 400-lb barrels.
Uses: Textiles (mordant, waterproofing); dry colors; ceramics; tanning; paper size precipitant; matches; inks; engraving; sugar refining; water purification; medicine; confectionery; baking powders.
Shipping regulations: None. *

aluminum stearate (aluminum tristearate) $Al(C_{18}H_{35}O_2)_3$ (approx.).
Properties: White powder; sp. gr. 1.070; m. p. 115°C. Insoluble in water, alcohol, ether; soluble in alkali, petroleum, turpentine oil. Forms gel with aliphatic and aromatic hydrocarbons.
Derivation: Reaction of aluminum salts with stearic acid.
Grade: Technical.
Containers: 25-, 50-lb bags; cartons; drums.
Uses: Paint and varnish drier; greases; waterproofing agent; cement additive; lubricants; cutting compounds; flatting agent; cosmetics and pharmaceuticals; additive for chewing gums.
Shipping regulations: None. *

aluminum subgallate $Al_4(C_7H_2O_5)_3 \cdot 4H_2O$.
Properties: Brown powder; soluble in dilute acids; insoluble in water.
Derivation: Interaction of solutions of alum and sodium gallate.

aluminum sulfate (known in the trade as alum; pearl alum; pickle alum; cake alum; filter alum; paper makers' alum; patent alum). (a) $Al_2(SO_4)_3$; (b) $Al_2(SO_4)_3 \cdot 18H_2O$.
Properties: White crystals; soluble in water (has a sweet taste); insoluble in alcohol. Stable in air.
Constants: Sp. gr. (a) 2.71, (b) 1.62; m. p. (a) decomposes at 770°C, (b) decomposes at 86.5°C.
Derivation: (a) By treating pure kaolin or aluminum hydroxide or bauxite with sulfuric acid. The insoluble silicic acid is removed by filtration and the sulfate is obtained by crystallization; (b) similarly from waste coal-mining shale and sulfuric acid.
Grades: Iron-free; technical; C. P.; U. S. P. XVI.
Containers: Bags; fiber drums; multiwall paper sacks; bulk in carloads; solution in rubberlined tank cars or trucks.
Uses: Sizing paper; lakes; alums; mordant for dyeing; water purification; fireproofing cloth; tannage of white leather; catalyst in manufacture of ethane; waterproofing agent for concrete; clarifying agent for fats and oils; ingredient of lubricating compositions; manufacture of satin white; deodorizer and decolorizer in petroleum refinery processes; precipitating agent in sewage-treating plants.
Shipping regulations: None. *

aluminum sulfide Al_2S_3.
Properties: Yellowish-gray lumps. Odor of hydrogen sulfide gas. In moist air, decomposes and forms a gray powder; decomposed by water.
Constants: Sp. gr. 2.02; m. p. 1100°C.
Grades: Technical.
Use: Preparation of hydrogen sulfide gas.

aluminum sulfocarbolate. See aluminum phenolsulfonate.

aluminum sulfocyanate. See aluminum thiocyanate.

aluminum thiocyanate (aluminum sulfocyanate) $Al(SCN)_3$.
Properties: Yellowish powder. Soluble in water; insoluble in alcohol and ether.
Grades: Technical.
Containers: Iron drums.
Uses: Textile industry; manufacturing pottery.
Shipping regulations: None. *

aluminum trialkyls. See aluminum alkyls.

aluminum triformate (aluminum formate, normal) $Al(HCOO)_3 \cdot 3H_2O$.
Properties: White, finely divided crystalline powder which has thixotropic properties when damp. At normal temperatures and under average humidity conditions it is quite stable. Solutions for technical purposes are stabilized by the addition of a small amount of aluminum hydroxide or calcium carbonate. Mildly acid in reaction. Contains 12.5% aluminum. Soluble in hot water; slightly soluble in cold water (in making solutions for technical purposes, a water temperature of from 70-100°C is said to be best; agitation should be continuously maintained during the dissolving process).
Grades: Powder; 15°, 20° Be. solutions.

*See "I.C.C. Shipping Regulations," page xiii.
Reference numbers refer to name of manufacturer. See "List of Manufacturers," page v.

Containers: Barrels; kegs; 13-gal carboys.
Uses: Textile (delustering rayon, mordanting, waterproofing, after-treatment of dyeings); paper (sizing); fur dyeing (mordant); and medicine.

aluminum triricinoleate $Al(C_{17}H_{32}OHCOO)_3$.
Properties: Yellowish to brown plastic mass. Limited solubility in most organic solvents. M. p. 95°C.
Derivation: Castor oil.
Uses: Gelling agent; waterproofing; solvent-resistant lubricants.

aluminum tristearate. See aluminum stearate.

alum meal. Crystals of aluminum potassium sulfate containing small amounts of iron.

alum, N. F. May be either aluminum ammonium sulfate or aluminum potassium sulfate.

"Alumon." [169] Trademark for aurintricarboxylic acid (ammonium salt) used in the colorimetric determination of aluminum.

alum, papermakers'. See aluminum sulfate.

alum, pearl. Specially prepared aluminum sulfate for the paper making industry.

alum, pickle. Aluminum sulfate prepared to meet specifications of packers and preservers.

alum, porous. See aluminum sodium sulfate.

alum, potash. See aluminum-potassium sulfate.

alum root. See geranium.

alum, rubidium. See aluminum-rubidium sulfate.

alum schist. See alum shale.

alum shale (alum schist; alum slate). A clay containing iron pyrites and aluminum silicate; a source of alum.

alum slate. See alum shale.

alum, soda. See aluminum sodium sulfate.

alum stone. See alunite.

"Alundum." [249] Trademark for a line of fused-alumina refractory and abrasive products.
Properties: Sp. gr. 3.93-4.01; fusion point 2000-2050°C depending on purity; coefficient of thermal expansion 0.0000072; mean specific heat (25-100°C) 0.1827-0.200 g cal/g/°C.
Grades and Uses: Available as grains, cements, and refractory shapes.
Grains: Available in all standard mesh sizes for refractory purposes and for increasing alumina content in ceramic bodies. High purity 38500 and 38900 "Alundum" grain is available for electrical insulation in radio and television tube industry. Two degrees of purity are: regular "Alundum" 95% Al_2O_3 and small amounts titania, iron, and silica; No. 38 "Alundum," white, containing 99% Al_2O_3 and traces of soda, iron, titania, and silica.
Cements: Clay-bonded mixtures with maturing temperatures from 600-1000°C for imbedding electrical resistors, metal melting applications, refractory brick setting, etc.
Refractory shapes: Bonded fused-alumina shapes of great variety for very high temperature work both commercial and experimental; high heat conductivity; chemically stable; non-conductive electrically. Can be used at temperatures up to 1800°C depending upon service conditions. Used for all types of furnaces, bricks, plates, muffles, tunnel kilns, enameling furnaces, tubes, laboratory ware.

alunite (alum stone) $KAl_3(OH)_6(SO_4)_3$. A naturally occurring basic potassium aluminum sulfate, usually found with volcanic and other igneous rocks.
Properties: Color white, gray or reddish; streak white; luster vitreous to pearly. Often impure from admixture of silica or clayey material. Sodium may replace most of the potassium to form natroalunite. Sp. gr. 2.6-2.8; hardness 4.
Occurrence: Utah, Arizona, California, Colorado, Nevada, Washington; Italy; Australia.
Use: Production of alum potassium compounds, millstones, metallic aluminum; decolorizing and deodorizing agent; fertilizer.

alunogen $Al_2(SO_4)_3 \cdot 18H_2O$. A natural aluminum sulfate.
Properties: Color white, yellowish or pinkish; streak white; luster vitreous or silky; sp. gr. 1.6-1.8; hardness 1.5-2. Usually occurs as a fibrous crust on aluminum-bearing shales and other aluminous rocks.
Occurrence: New Mexico, Colorado; Europe.

"Alurate." [190] Trademark for 5-allyl-5-isopropylbarbituric acid (q. v.).

"Alusite." [446] Trade name for a 70% alumina brick with relatively low porosity and a PCE value of 38. Good resistance to mechanical abrasion and penetration by molten slags. Used in rotary lime, lime sludge and dolomite kilns, soaking pit curb walls, gas regenerators, non-ferrous metallurgical refining furnaces, in reverberatory and brass melting furnaces, and various lead furnaces.

"Alvar." [276] Trademark for polyvinyl acetal resin by the reaction of acetaldehyde with polyvinyl alcohol. Hard, tough, heat stable resin; good adhesive strength.
Grades: As slightly yellowish free-flowing granules, 7-70, 15-70, 15-80, 5-80, varying in molecular weight and degree of acetyl hydrolysis.
Uses: To impart gloss, toughness and adhesion to cellulose nitrate films for wash primers; solution adhesives; preservation of archeological artifacts; as molding resin.

"Alwax" Sizes. [57] Trademark for series of aqueous emulsions of paraffin and microcrystalline waxes.
Uses: In paper industry to impart resistance

to water, lactic acid, blood serum, and organic liquids. Improves paper pliability, gives smoother surface, added scuff resistance to container board and improved printing qualities. Coatings are plasticized, more water repellent, and smoother surfaced. These products may be acid- or alkaline-stable types.

Am. Symbol for americium.

amalgam. Any alloy of mercury with one or more other metals. An amalgam may be either solid or liquid. Also a native alloy of silver and mercury. Native gold amalgams are also known.

"Amanol" Nitrogen Solutions. [57] Trademark for nitrogen fertilizers containing 41% and 49% nitrogen in the form of aqua ammonia and ammonium nitrate.
Containers: Bulk tank cars or trucks.
Uses: For direct applications and in fertilizer mixtures.

amaranth $C_{20}H_{11}N_2O_{10}Na_3S_3$. A coal tar azo dye, C.I. No. 184.
Properties: Dark red brown powder; soluble in water; very slightly soluble in alcohol.
Derivation: From naphthionic and R acids.
Grade: U.S.P. XVI.
Containers: Glass bottles.
Use: As a red dye.

amaroid. A class of bitter principles found in plants, as ceroxylin; pyrethrosin.

amatol. An explosive mixture of ammonium nitrate and T.N.T. The 50-50 mixture can be melted and poured for filling small shells; the 80% ammonium nitrate mixture is granular and has been used for filling large shells or bombs.
Shipping regulations: Explosive, class A. High Explosive label.*

"Amax." [69] Trademark for a proprietary product, N-oxydiethylene benzothiazyl 2-sulfenamide.
Properties: Tan flakes; sp. gr. 1.37 ± .03; melting range 70-90°C; very soluble in benzene, chloroform, methanol.
Uses: Primary accelerator for natural rubber and SBR; in belting, camelback and tires.

"Amax No. 1." [69] Trademark of a proprietary mixture of N-oxydiethylene benzothiazyl-2-sulfenamide and benzothiazyl disulfide.
Properties: Light tan flakes; sp. gr. 1.40 ± .03; melting range 70-90°C; very soluble in benzene and chloroform; moderately soluble in acetone and methanol.
Uses: Primary accelerator for natural rubber and SBR in belting, camelback and tires.

amazonite (amazon stone). A green gem stone which is a variety of feldspar. The most valuable kinds are apple green in color. The lighter green kinds and those with streaks and specks of yellowish-white or red are of no commercial importance.
Occurence: United States (North Carolina, Virginia, Colorado); Russia.

amazon stone. See amazonite.

amber (fossil resin). A fossil resin derived from an extinct variety of pine which flourished in the Tertiary Period. Strand amber comes from stratum partly below sea level. It is obtained either by dredging or is picked up off the beach after it has been uncovered and washed out by the action of the waves. Baltic amber is mined.
Properties: It is pale yellow, grading to brown or reddish-brown and it varies from transparent to opaque. The transparent variety contains up to 4% and the opaque variety up to 8% succinic acid, hence the name succinite. Sp. gr. 1.07-1.09; hardness 2-2.5.
Occurrence: Strand: Sicily; Holland; England and other European shores. Baltic: Mined in East and West Prussia.
Uses: Beads, ornaments, mouthpieces for pipes, cigarette holders; varnish; dye and drug manufacture.

ambergris.
Properties: Irregular, gray, grayish-brown or black, streaked or mottled, waxy, opaque masses; peculiar odor. Insoluble in water; soluble in alcohol, chloroform, ether, fats, and oils. M.p. about 60°C; sp. gr. 0.80-0.92.
Chief constituents: Cholesterol, ambrein, benzoic acid.
Derivation: Morbid concretion from the intestinal tract of the sperm whale. Its discovery is accidental.
Grades: Technical.
Containers: Wooden boxes; bottles.
Uses: Drugs; perfume (fixative). (Obsolescent, being replaced by synthetic fixatives.)
Shipping regulations: None.*

"Amberlac." [23]
"Amberlac" Coating Resins. Trademark for modified alkyd-type resins for quick-drying lacquers. Some grades are resin-modified; some oxidizing and baking grades.
Use: Metal primers; bottle cap coatings; food can coatings; appliance coatings.
"Amberlac" 165. Trademark for synthetic water-soluble polymer. Colorless additive for film formers; improves scuff resistance; prevents sticking of thermoplastic coatings.
Use: High gloss paper coatings; coating of book covers, decorative papers, paperboard.

"Amberlite." [23]
"Amberlite" Ion Exchange Resins. Trademark for insoluble crosslinked polymers of various types in minute bead form. Strong acid, weak acid, strong base, and weak base forms, each having various grades differing in exchange capacity and porosity, for removing simple and complex cations and anions from aqueous and non-aqueous solutions. Reversible in action, can be regenerated and used repeatedly.
Use: Water conditioning (softening and complete deionization); recovery and concentration of metals, antibiotics, vitamins,

*See "I.C.C. Shipping Regulations," page xiii.
Reference numbers refer to name of manufacturer. See "List of Manufacturers," page v.

organic bases; catalysis; decolorization of sugar; manufacture of chemicals.

"Amberlite" Laboratory Grade Exchangers. Trademark for ion exchange resins made to meet special laboratory needs. High-chemical-purity grades for chromatographic and other analytical work, manufactured to stringent specifications with respect to certain metallic impurities, moisture content and particle size. Less expensive grades, closely controlled with respect to porosity and particle size, for laboratory work not requiring high chemical purity.

"Amberlite" Liquid Exchangers. LA series. Trademark for high boiling amines, oil soluble with ion exchange properties.
Uses: Extraction of metals and separation of electrolytes.

"Amberlite" Mixed-Bed Exchangers. See "Monobed."

"Amberlite" Nuclear-Grade Exchangers. Trademark for strong acid and base ion exchange resins manufactured to meet the exacting specifications required for treating water in the primary loop of water-cooled nuclear reactors.

"Amberlite" Pharmaceutical Grade Exchangers. Trademark for ion exchange resins used for prolonged release of drugs, tablet disintegrants, taste masking, diagnostic tests, and therapeutic agents.

amber mica. See phlogopite.

amber oil.
Properties: Colorless or yellow-brown, thin, liquid, volatile oil; darkens with age; empyreumatic, balsamic odor. Soluble in alcohol, ether, chloroform, carbon disulfide, and fixed oils. It is dextrorotatory.
Chief constituents: Phenols; terpenes.
Constants: Sp. gr. 0.85-0.97; depending on purity.
Derivation: From amber, by destructive distillation and redistillation.
Method of purification: Rectification.
Grades: Crude; rectified.
Containers: Tins; glass bottles.
Use: Medicine.
Shipping regulations: None. *

"Amberol." [23] Trademark for maleic-resin and resin-modified and unmodified phenol-formaldehyde-type and polymers in solid form. Various grades differ in solubility, viscosity in solution, color, hardness, and reactivity. React with various oils to produce fast drying, high gloss protective coatings and vehicles for printing inks.
Uses: Varnishes; enamels; can liners; nitrocellulose sanding sealers; printing inks; tackifying and vulcanization of butyl rubber.

amber seed. See ambrette seed.

"Ambiflo." [233] Trademark for polyglycol lubricants.

"Ambitrol." [233] Trademark for a freeze depressant concentrate based on ethylene glycol (q. v.) and containing inhibitors.

amblygonite $Li(AlF)PO_4$ or $AlPO_4 \cdot LiF$. A natural fluophosphate of aluminum and lithium.
Properties: White to pale greenish-, bluish-, yellowish-, grayish- or brownish-white; white streak; vitreous to pearly luster. Contains 10.1% lithia, sometimes with partial replacement by sodium.
Constants: Sp. gr. 3.01-3.09; hardness 6.
Occurrence: California, Maine, Connecticut, South Dakota; Germany; Norway; France.
Use: An important source of lithium; used in glazes and coatings.

"Ambodryl" Hydrochloride. [330] Trademark for bromodiphenhydramine hydrochloride [beta-(para-bromobenzhydryloxy) ethyl dimethylamine hydrochloride] $C_{17}H_{21}BrNO \cdot HCl$.
Properties: White to off-white odorless amorphous or crystalline powder possessing a bitter taste; melting range 148-150°C. Soluble in water and alcohol; insoluble in ether.
Use: Medicine.

"Ambraloy-927." [324] Trade name for an aluminum brass alloy, manufactured under U.S. Patent No. 2,003,685, containing arsenic as an inhibitor to increase its resistance to dezincification. Its nominal composition is copper 77%, zinc 20.96%, aluminum 2%, arsenic 0.04%. Used principally as a condenser tube alloy in power plants, marine service and oil refineries where cooling water is salt or brackish, and where it is resistant to the impinging action of turbulently flowing sea water containing air bubbles.

ambrette seed (amber seed; muskmallow; abelmoschus; musk seed). Seeds of Hibiscus abelmoschus, an evergreen shrub.
Occurrence: Egypt; India; tropical America.
Uses: Manufacture of ambrette seed oil.

ambrette-seed oil, liquid.
Properties: Yellow oil of musk-like odor, freed of the odorless fatty acid. Soluble in 2.5-8 vols. and more of 80% alcohol.
Constants: Sp. gr. 0.905-0.917 at 15°C; optical rotation slightly dextrogyrate, up to +1°20' or rarely levogyrate up to -2°24'; index of refraction 1.474-1.480; acid value 0.8-2.5; ester value 137-190.
Chief known constituents: Ambrettolide, farnesol.
Derivation: By distilling ambrette seed.
Use: Perfumery.
Shipping regulations: None. *
See also ambrette-seed oil, ordinary.

ambrette-seed oil, ordinary.
Properties: Concrete mass resembling orris oil. Sp. gr. 0.891 (40°C), 0.883 (50°C); acid value 75-140; ester value 70-130.
Derivation: By distilling ambrette seed.
See also ambrette-seed oil, liquid.
Shipping regulations: None. *

ambrettolide (hexadecen-6-olide) $C_{16}H_{28}O_2$.

Properties: Colorless liquid, having very powerful musk-like odor.
Constants: Sp. gr. 0.955-0.957; refractive index 1.480-1.481.
Occurrence: Found in ambrette-seed oil.
Use: In perfumery, as a fixative.
Shipping regulations: None.*

ambutonium bromide $C_{20}H_{27}BrN_2O$ (3-carbamoyl-3,3-diphenylpropyl)ethyldimethylammonium bromide.
Properties: Crystals; m.p. 228-229°C (decomposes).
Use: Medicine.

"Amcrom." [67] Trademark for copper-chromium alloys in cast or wrought form.

American ashes. See potassium carbonate.

"American" Dynamite. [57] Trademark. Based on U.S. Bureau of Mines specification, it is a line of high-density permissible explosives, approved for mining of coal where gassy and dusty conditions are encountered.

American hellebore. See veratrum viride.

American Indian hemp. See apocynum.

American process for zinc oxide pigment. See Wetherill process.

American saffron. See carthamus.

American spikenard. See aralia.

American valerian. See cypripedium.

American vermilion. See chrome red.

americium Am. A synthetic radioactive element of atomic number 95, first prepared as the 241 isotope by neutron bombardment of plutonium in a nuclear reactor. See actinide elements.

Metallic americium is obtained by first preparing americium trifluoride which is reduced with barium vapor at about 1100°C. The metal is silvery white and tarnishes slowly in air; m.p. above 850°C; sp. gr. 11.7. Americium forms compounds of the types AmO_2, Am_2O_3, $AmCl_3$, etc.

amethopterin. See methotrexate.

amethyst. A purple or bluish-colored form of native silica or quartz (q.v.).
Use: Gem stone.
For oriental amethyst see corundum.

amide (acid amide). An organic compound containing the structural group $-CONH_2$, and closely related to the organic acids with the grouping COOH. Common examples are acetamide CH_3CONH_2, and urea $CO(NH_2)_2$.

Amido is the adjective form of amide, but is also used loosely as a synonym for amino. See amine.

N-amidinoalanine (dl-alpha-guanidinopropionic acid; "alacreatine")
$NH_2C(:NH)NHCH(CH_3)COOH$.
Derivation: Prepared by reacting thiourea with ethyl bromide yielding ethyl isothiourea hydrobromide which is added with alkali to alanine; the desired product splits out with ethyl mercaptan.
Use: Biochemical research.

amidophenols. See meta-, ortho-, and para-aminophenol.

amidopyrine. See aminopyrine.

amidothiolactic acid. See cysteine.

amidourea hydrochloride. See semicarbazide hydrochloride.

amination. The process of making an amine (RNH_2). The methods commonly used are (a) reduction of a nitro compound and (b) action of ammonia on a chloro-, hydroxy-, or sulfonic acid compound.

amine. A class of organic compounds of nitrogen that may be considered as derived from ammonia (NH_3) by replacing one or more of the hydrogen atoms by organic radicals, such as CH_3 or C_6H_5, as in methyl amine and aniline. The former is a gas at ordinary temperature and pressure, but other amines are liquids or solids. All amines are basic in nature, and will usually combine readily with hydrochloric or other strong acids to form salts.

amine 220. [214] $C_{17}H_{33}\overline{CNC_2H_4N}C_2H_4OH$.
Properties: Sp. gr. 0.9330 (20/20°C); lbs/gal 7.77 (20°C); b.p. 235°C (1 mm); flash point 465°F.
Containers: 1-gal can; 5- and 55-gal drums (7.5, 35, 420 lbs).
Uses: Demulsifier used particularly in the recovery of tar from water-gas process emulsions. A powerful cationic wetting agent. Useful in flotation processes involving siliceous minerals and the formation of emulsions and dispersions under acidic conditions.

amine absorption process. See Girbotol absorption process.

"Amine C, O and S." [219]
$CH_3(CH_2)_nC_3H_4N_2C_2H_4OH$. Trade names for high molecular weight imidazolines soluble in organic solvents but sparingly soluble in water at alkaline pHs. React with acids such as hydrochloric to form water-soluble positively charged colloids with unusual tolerance for electrolyte.
Amine C: undecyl; m.p. 32°C.
Amine O: heptadecenyl; m.p. 5°C.
Amine S: heptadecyl; m.p. 42°C.
Uses: Fungicides; antistatic treatment for upholstery; water repellent treatment for plaster; detergents; rust preventive oils.

"Amine D." [266] Brand name for a technical grade of dehydroabietylamine.
Properties: Pale yellow viscous liquid; oil soluble.
Uses: Production of bactericides; fungicides; corrosion inhibitors; asphalt additives; flotation agents.

"Amine D Acetate." [266] Brand name for acetic acid salt of "Amine D."
Properties: Amber, water-soluble, surface active product.

*See "I.C.C. Shipping Regulations," page xiii.
Reference numbers refer to name of manufacturer. See "List of Manufacturers," page v.

Uses: Bactericide; fungicide; preservative for alkaline dispersions of proteins.

"Amine D Ethylene Oxide Adduct." [266] See "Polyrad."

para-**aminoacetanilide** (N-acetyl-para-phenylenediamine) $NH_2C_6H_4NHOCCH_3$.
Properties: Colorless crystals, sometimes reddish; soluble in alcohol and ether, slightly soluble in water. M.p. 162°C; b.p. 267°C.
Derivation: Acetylization of para-phenylenediamine.
Method of purification: Crystallization.
Grades: Technical.
Containers: Drums.
Uses: Intermediates; azo dyes.
Shipping regulations: None.*

aminoacetic acid. See glycine.

aminoacetyl-para-phenetidine acetate. See phenocoll acetate.

aminoacetyl-para-phenetidine hydrochloride. See phenocoll hydrochloride.

D-amino acid oxidase. See amino oxidase.

L-amino acid oxidase. See amino oxidase.

amino acids. Organic acids containing an amino group (e.g., $CH_3CH_2NH_2COOH$, alanine). All naturally occurring amino acids are alpha amino acids (i.e., with the $-NH_2$ group attached to the carbon atom next to the -COOH group) except beta-alanine. Amino acids can be obtained by the hydrolysis of a protein or by organic synthesis. An essential amino acid is one which is required for growth and which cannot be synthesized by the body. Essential amino acids in the diet of man are

isoleucine	phenylalanine
leucine	threonine
lysine	tryptophan
methionine	valine

In addition, arginine and histidine are essential in the diet of the rat. Historically "essential amino acids" referred to those amino acids essential for the rat. Amino acids may be classified as neutral, acidic or basic by the ratio of acidic (carboxyl) groups to basic (amino) groups. See individual amino acids.

5-aminoacridine. See 9-aminoacridine.

9-aminoacridine (5-aminoacridine) $C_{13}H_{10}N_2$.
Properties: Sulfur-yellow crystals; m.p. 241°C; moderately strong base. Freely soluble in alcohol; slightly soluble in chloroform, toluene, pyridine; soluble in acetone.
Use: Medicine.

9-aminoacridine hydrochloride $C_{13}H_{10}N_2 \cdot HCl$.
Properties: Pale-yellow crystalline powder; odorless; bitter taste; stable to light, does not stain tissues; soluble in water (1:300).
Use: Medicine (topical antiseptic).

aminoamylene glycol. See 2-amino-2-ethyl-1,3-propanediol.

ortho-**aminoanisole.** See ortho-anisidine.

para-**aminoanisole.** See para-anisidine.

aminoanthraquinone $C_6H_4(CO)_2C_6H_3NH_2$ (tricyclic) (a) 1-amino; (b) 2-amino.
Properties: (a) Red, iridescent needles. (b) Red or orange-brown needles. Soluble in alcohol, chloroform, benzene, and acetone; insoluble in water.
Constants: M.p. (a) 252°C; (b) 302°C; b.p. sublime (both a and b).
Derivation: By reduction of nitroanthraquinones, or by the substitution of the amino radical direct for the sulfonic acids.
Method of purification: Crystallization.
Grades: Technical.
Containers: Wooden barrels.
Use: Dye intermediates.
Shipping regulations: None.*

4-aminoantipyrine (4-amino-1,5-dimethyl-2-phenyl-3-pyrazolone; 1,5-dimethyl-2-phenyl-4-aminopyrazolone) $C_{11}H_{13}N_3O$.
Properties: M.p. 107-109°C.
Uses: Medicine; analytical reagent.

2-amino-4-arsenosophenol hydrochloride (oxophenarsine hydrochloride) $C_6H_3(AsO)(OH)NH_2 \cdot HCl$.
Properties: A white or nearly white hygroscopic powder; soluble in water, solution of alkali hydroxides and carbonates and in dilute mineral acids. Solutions darken on exposure to air.
Derivation: Reduction of the arsonic acid derivative. It is usually marketed with buffering agents and suitable substances to make its solution physiologically compatible with human blood.
Grade: N.F. XI.
Use: Medicine.

para-**aminoazobenzene** (aniline yellow; phenylazoaniline) $C_6H_5NNC_6H_4NH_2$.
Properties: Yellow to tan crystals. Soluble in alcohol and ether; slightly soluble in water.
Constants: M.p. 126-128°C; b.p. above 360°C.
Derivation: (a) Diazoaminobenzene when heated with aniline hydrochloride yields aminoazobenzene. The aniline hydrochloride acts as a catalyzer and does not enter into the reaction. (b) Diazotization of a solution of aniline and aniline hydrochloride with hydrochloric acid and insufficient sodium nitrite to make the diazo compound.
Method of purification: Crystallization.
Grades: Technical.
Containers: Barrels; kegs; fiber containers.
Uses: Dyes (chrysoidine dyes, induline dyes, solid yellow and acid yellow dyes); coloring spirit varnishes (hydrochloride).
Shipping regulations: None.*

aminoazobenzenedisulfonic acid $C_6H_4(SO_3H)NNC_6H_3NH_2(SO_3H)$.
Properties: Bright, violet needles. Soluble in hot water, and alcohol; insoluble in ether.
Derivation: By heating either aminoazobenzene hydrochloride or aminobenzene-mono-

sulfonic acid with fuming sulfuric acid.
Grades: Technical.
Uses: Synthesis of dyes; wool dyeing.

para-**aminoazobenzene hydrochloride** (aminoazobenzene salt) $C_6H_5NNC_6H_4NH_2 \cdot HCl$.
Properties: Steel-blue crystals. Soluble in alcohol; slightly soluble in water.
Derivation: By passing dry hydrogen chloride gas into a solution of aminoazobenzene.
Method of purification: Crystallization.
Grades: Technical.
Containers: Barrels; kegs.
Uses: Dyes; coloring lacquers; intermediate.
Shipping regulations: None.*

aminoazobenzenemonosulfonic acid $NH_2C_6H_4NNC_6H_4SO_3H$.
Properties: Yellowish-white, microscopic needles. Barely soluble in water; almost insoluble in alcohol, ether and chloroform.
Derivation: By sulfonating aminoazobenzene.
Grades: Technical.
Use: Dyestuff manufacture.

aminoazobenzene-beta-**naphthol** (Sudan III; benzene-azo-para-benzene-azo-beta-naphthol) $C_6H_5NNC_6H_4NNC_{10}H_6OH$.
Properties: Brown powder; m.p. 195°C; insoluble in water; soluble in alcohol and oils.
Derivation: By heating aminoazobenzene and beta-naphthol.
Method of purification: Crystallization from alcohol.
Grades: Technical.
Containers: Tins; glass bottles.
Uses: Coloring oils red; biological stain.

aminoazobenzene salt. See para-aminoazobenzene hydrochloride.

ortho-**aminoazotoluene** (2-amino-5-azotoluene; toluazotoluidine) $CH_3C_6H_4N_2C_6H_3NH_2CH_3$.
Properties: Reddish-brown to yellow crystals; soluble in alcohol, ether, oils, and fats; slightly soluble in water.
Constants: M.p. given variously from 99-117°C.
Derivation: From ortho-toluidine by treatment with nitrite and hydrochloric acid.
Method of purification: Crystallization.
Grades: Technical.
Containers: Kegs and barrels.
Uses: Dyes; medicine.
Shipping regulations: None.*

aminoazotoluene hydrochloride $CH_3C_6H_4N_2C_6H_3CH_3NH_2 \cdot HCl$. There are four isomers.
Properties: White crystals; soluble in water, alcohol, and ether.
Derivation: By the interaction of aminoazotoluene and dry hydrogen chloride gas.
Method of purification: Crystallization.
Grades: Technical.
Containers: Barrels; kegs.
Uses: Organic synthesis.
Shipping regulations: None.*

6-para-(para-**aminobenzamido)benzamido-1-naphthol-3-sulfonic acid** $H_2NC_6H_4CONHC_6H_4CONHC_{10}H_5(OH)(SO_3H)$.
Properties: Gray paste containing approx. 35% solids.
Grade: Technical.
Use: Intermediate.

aminobenzene. See aniline.

para-**aminobenzenearsonic acid.** See arsanilic acid.

2-amino-para-**benzenedisulfonic acid** (aniline-2,5-disulfonic acid) $C_6H_3NH_2(SO_3H)_2 \cdot 4H_2O$.
Properties: Crystals; very soluble in water and alcohol.
Derivation: Boiling sodium salt of 4-chloro-3-nitrobenzene sulfonate with sodium sulfite, resulting in formation of sodium 2-nitrobenzene disulfonate, which is reduced with iron and acetic acid to aniline-2,5-disulfonic acid.
Use: Intermediate.

4-amino-meta-**benzenedisulfonic acid** (aniline-2,4-disulfonic acid) $C_6H_3NH_2(SO_3H)_2 \cdot 2H_2O$.
Properties: Needles decomposing when heated over 120°C; very soluble in water and alcohol.
Derivation: By heating sulfanilic acid with fuming sulfuric acid at 170-180°C.
Use: Dye intermediate.

para-**aminobenzenesulfonamide.** See sulfanilamide.

meta-**aminobenzenesulfonic acid.** See metanilic acid.

para-**aminobenzenesulfonic acid.** See sulfanilic acid.

meta-**aminobenzoic acid** $C_6H_4NH_2CO_2H$.
Properties: Yellowish or reddish crystals; sublimes easily; sweet taste. Slightly soluble in water, alcohol, and ether.
Constants: Sp. gr. 1.51 (4°C); m.p. 173°C to 174°C.
Derivation: By the reduction of meta-nitrobenzoic acid.
Method of purification: Crystallization.
Grades: Technical.
Containers: Wooden kegs; fiber containers; multiwall paper sacks.
Use: Dye intermediate.
Shipping regulations: None.*

ortho-**aminobenzoic acid.** See anthranilic acid.

para-**aminobenzoic acid** (PABA) $NH_2C_6H_4CO_2H$. Required by many organisms as a vitamin for growth; is active in neutralizing the antibacteriostatic effect of some sulfamide drugs. Deficiency symptoms are achromotrichia in rats and a failure of growth in chicks.
Properties: Light buff odorless crystals; white when pure; discolors on exposure to light and air; m.p. 186-187°C. Sparingly soluble in cold water and dilute hydrochloric acid; slightly soluble in ether, benzene, chloroform; soluble in hot water, ethyl acetate, glacial acetic acid, warm glycerol; freely soluble in solutions of alkali hydroxides or carbonates; insoluble in petroleum ether. Unstable to ferric salts and oxidizing agents.
Derivation: Synthetic product from reduction

of para-nitrobenzoic acid. Commercially available as the calcium, potassium, and sodium salts.
Food source: Widely distributed in nature. Yeast is an especially good source.
Grades: Technical; N. F. XI.
Containers: Bottles; 25-, 100-lb drums.
Uses: Dyes and pharmaceuticals; nutrition.
Shipping regulations: None. *

meta-**aminobenzotrifluoride** $CF_3C_6H_4NH_2$.
Properties: Colorless to oily yellow liquid.
Grade: Technical (88% min); purified (98% min).
Use: Pharmaceutical intermediate.

para-**aminobenzoyldiethylaminoethanol base.** See procaine base.

para-**aminobenzoyldiethylaminoethanol hydrochloride.** See procaine hydrochloride.

para-**aminobenzoyldiethylaminoethanol nitrate.** See procaine nitrate.

ortho-**aminobenzoylformic acid.** See isatin.

N-(para-**aminobenzoyl)glycine.** See para-aminohippuric acid.

ortho-**aminobiphenyl** (ortho-phenylaniline; ortho-biphenylamine) $C_6H_5C_6H_4NH_2$.
Properties: Colorless or purplish crystals; m. p. 49.3°C; b. p. 299°C; slightly soluble in water.
Derivation: Reduction of ortho-nitrobiphenyl.
Grades: Technical.
Containers: 475-lb drums; tank cars.
Uses: Intermediate for organic synthesis (carbazole); resins; synthetic rubbers; solvent.
Caution: Toxic. Avoid breathing vapors and skin contact.

2-amino-1-butanol $CH_3CH_2CHNH_2CH_2OH$.
Properties: Colorless liquid; sp. gr. 0.944 at 20/20°C; m. p. −2°C; b. p. (760 mm) 178°C, (10 mm) 79-80°C; vapor pressure 0.5 mm at 20°C (est'd); flash point 164°F; wt 7.85 lbs/gal (20°C); pH (0.1M aqueous solution) 11.11; refractive index 1.453 at 20°C. Completely miscible in water at 20°C; soluble in alcohols; corrosive to copper, brass, aluminum.
Containers: 1-gal cans; 5- and 55-gal drums.
Uses: Emulsifying agent (in soap form) for oils; fats, and waxes; absorbent for acidic gases; chemical synthesis.
Caution! Slightly toxic; mildly alkaline; avoid repeated skin contact.

aminobutylene glycol. See 2-amino-2-methyl-1,3-propanediol.

alpha-**aminocaproic acid.** See norleucine.

aminocaproic lactam. See caprolactam.

meta-**aminochlorobenzene.** See meta-chloroaniline.

ortho-**aminochlorobenzene.** See ortho-chloroaniline.

para-**aminochlorobenzene.** See para-chloroaniline.

2-amino-4-chlorophenol (para-chloro-ortho-aminophenol) $C_6H_3OHNH_2Cl$.
Properties: Light brown crystals; m. p. 138°C (decomposes). Soluble in dilute mineral acid or dilute alkaline solution.
Derivation: Reduction of para-chloro-ortho-nitrophenol.
Method of purification: Recrystallization.
Containers: Fiber drums.
Grade: 96% pure (min.).
Use: Intermediate.
Shipping regulations: None. *

2-amino-4-chlorotoluene [5-chloro-2-methylaniline (NH_2=1); 4-chloro-ortho-toluidine (CH_3=1)] $ClNH_2C_6H_3CH_3$.
Properties: Off-white solid or light brown oil which tends to darken on storage; m. p. 20-22°C.
Grade: Fused.
Containers: Barrels; 40-gal steel drums.
Use: Intermediate.

2-amino-6-chlorotoluene [6-chloro-ortho-toluidine (CH_3=1); 3-chloro-2-methylaniline (NH_2=1)] $ClNH_2C_6H_3CH_3$.
Properties: Liquid; m. p. 0-2°C.
Grade: Technical.
Container: Drums.
Use: Intermediate.

4-amino-2-chlorotoluene [2-chloro-para-toluidine (CH_3=1); 3-chloro-4-methylaniline (NH_2=1)] $ClNH_2C_6H_3CH_3$.
Properties: Liquid; m. p. 21-24°C.
Grade: Technical.
Container: Drums.
Use: Intermediate.

meta-**amino**-para-**cresol methyl ether.** See cresidine.

aminocyclohexane. See cyclohexylamine.

L-amino dehydrogenase. See amino oxidase.

para-**aminodiethylaniline** (N, N-diethyl-para-phenylene diamine; diethylaminoaniline) $(C_2H_5)_2NC_6H_4NH_2$.
Properties: Liquid; b. p. 260-2°C; insoluble in water; soluble in alcohol and ether.
Derivation: Treatment of diethylaniline with nitrous acid and subsequent reduction.
Use: Dye intermediate.

para-**aminodiethylaniline hydrochloride** $C_{10}H_{16}N_2 \cdot HCl$.
Properties: Colorless needles; soluble in water, alcohol; insoluble in ether.
Containers: 1-, 5-lb bottles; 25-, 50-, 100-lb drums.
Use: Color photography.

2-amino-4,6-dihydroxypteridine. See xanthopterin.

3-amino-7-dimethylamino-2-methylphenazothionium chloride. See tolonium chloride.

para-**aminodimethylaniline** (dimethylaminoaniline; dimethyl-para-phenylenediamine) $(CH_3)_2NC_6H_4NH_2$.
Properties: Colorless, asbestos-like, long needles; stable in air when pure. If impure, the crystals liquefy. Soluble in water, alcohol and benzene.
Constants: M. p. 41°C; b. p. 257°C.

*See "I.C.C. Shipping Regulations," page xiii.
Reference numbers refer to name of manufacturer. See "List of Manufacturers," page v.

Derivation: By reduction of para-nitrosodimethylaniline with zinc dust and hydrochloric acid. The aminodimethylaniline is not isolated, but the solution is worked up.
Method of purification: Recrystallization from mixture of benzene and ligroin.
Grades: Technical.
Containers: Kegs; 1- and 5-lb glass bottles; special metal cans or drums.
Uses: Base for production of methylene blue; photo-developer; reagent for detection of hydrogen sulfide; reagent for cellulose; organic synthesis; reagent for certain bacteria.
Shipping regulations: None. *

aminodimethylanilinethiosulfonic acid $(CH_3)_2NC_6H_3NH_2SSO_3H$.
Properties: Prismatic crystals. Soluble in hot water; less so in cold water.
Constants: M. p. 193-204° (decomposes).
Derivation: From para-nitrosodimethylaniline by reacting with sodium thiosulfate and weak acids.
Grades: Technical.
Use: Synthesis of dyestuffs.

aminodimethylbenzene. See xylidine.

1-amino-2,3-dimethylbenzene. See 2,3-xylidine.

1-amino-2,4-dimethylbenzene. See 2,4-xylidine.

1-amino-2,5-dimethylbenzene. See 2,5-xylidine.

1-amino-2,6-dimethylbenzene. See 2,6-xylidine.

4-amino-1,5-dimethyl-2-phenyl-3-pyrazolone. See 4-aminoantipyrine.

2-amino-4,6-dimethylpyridine $(CH_3)_2C_5H_2NNH_2$.
Properties: B. p. 235.3°C; freezing point 65.2-68.5°C. Soluble in water.
Derivation: Prepared from 2-aminopyridine.
Grade: 95% (minimum).
Use: Organic intermediate.

2-amino-4,6-dinitrophenol. See picramic acid.

para-**aminodiphenylamine** (N-phenyl-para-phenylene diamine) $NH_2C_6H_4NHC_6H_5$.
Properties: Colorless needles; m. p. 75°C; slightly soluble in water; soluble in absolute alcohol and ether.
Derivation: Reduction of the coupling product of diazotized sulfanilic acid and diphenylamine.
Grade: Technical.
Containers: Drums.
Use: Dye intermediate.

para-**aminodiphenylamine hydrochloride** $C_6H_5NHC_6H_4NH_2 \cdot HCl$.
Properties: Green powder; soluble in hot water, hot alcohol; insoluble in ether.

1,4,2-aminodiphenylaminesulfonic acid (2-aniline-5-aminobenzenesulfonic acid) $C_6H_5NHC_6H_3NH_2SO_3H$.
Properties: Needle-like crystals. Barely soluble in water.
Derivation: From para-nitrodiphenylamine-ortho-sulfonic acid by reduction with iron and hydrochloric acid.
Grades: Technical.
Use: Synthesis of dyestuffs.

amino dithioformic acid. See dithiocarbamic acid.

aminoethane. See ethylamine.

2-aminoethanesulfonic acid. See taurine.

2-aminoethanethiol. See cysteamine.

1-aminoethanol. See aldehyde ammonia.

2-aminoethanol. See ethanolamine.

alpha-**(1-aminoethyl) benzyl alcohol hydrochloride.** See phenylpropanolamine hydrochloride.

aminoethylethanolamine. See hydroxyethylethylene diamine.

4-aminoethylglyoxaline. See histamine.

beta-**aminoethylisothiuronium bromide hydrobromide** (AET) $C_3H_9N_3S \cdot 2HBr$.
Properties: Crystals; hygroscopic; m. p. 194-195°C.
Derivation: Thiourea is refluxed with 2-bromoethylamine hydrobromide in isopropanol.
Use: Suggested as a prophylactic in radiation sickness; enzyme activator; free radical detoxifier.

para-beta-**aminoethylphenol.** See tyramine.

N-aminoethylpiperazine $H_2NC_2H_4\underline{NCH_2CH_2NHCH_2CH_2}$. High boiling triamine combining a primary, secondary and tertiary amine in one molecule.
Properties: Liquid; sp. gr. 0.9837; b. p. 222.0°C; flash point 200.0°F; freezing point 17.6°C. Soluble in water.
Containers: 55-gal drums; tank cars.
Uses: Epoxy curing agent; intermediate for pharmaceuticals, anthelmintics, surface active agents, synthetic fibers.

2-amino-2-ethyl-1,3-propanediol (amino-amylene glycol) $CH_2OHC(C_2H_5)NH_2CH_2OH$.
Properties: Colorless liquid, white crystalline solid. Completely miscible in water at 20°C. Corrosive to copper, brass, aluminum.
Constants: Sp. gr. 1.099 at 20°C/20°C; b. p. (10 mm) 152-153°C; m. p. 37.5-38.5°C; wt. 9.15 lbs/gal (20°C); refractive index 1.490 at 20°C; flash point 166°F; pH (0.1M aqueous solution) 10.82.
Containers: 5- and 55-gal drums.
Uses: Emulsifying agent (in soap form) for oils, fats, and waxes; absorbent for acidic gases; chemical synthesis.
Caution! Slightly toxic; mildly alkaline; avoid repeated skin contact.

2-aminoethylsulfuric acid $NH_2CH_2CH_2OSO_3H$.
Properties: White noncorrosive crystalline powder; m. p. 274-280°C; sinters at 274°C and darkens without complete melting at 280°C; sp. gr. 1.782; bulk density 1.007 g/cc. Soluble in water; insoluble in most organic solvents; pH of 1% aqueous solution 4.0 (20°C), 5% aqueous solution 3.3 (39°C).

Use: Organic synthesis of ethyleneimine and various other compounds.

4-aminofolic acid. See aminopterin.

aminoform. See hexamethylenetetramine.

amino-G acid. See 2-naphthylamine-6,8-disulfonic acid.

alpha-**aminoglutaric acid.** See glutamic acid.

alpha-**aminoglutaric acid hydrochloride.** See glutamic acid hydrochloride.

aminoglutethimide [alpha-(para-aminophenyl)-alpha-ethylglutarimide]. Amino analog of glutethimide.
Use: Medicine.

amino-4-guanidovaleric acid. See arginine.

2-aminoheptane (1-methylhexylamine; tuaminoheptane) $CH_3(CH_2)_4CH(NH_2)CH_3$.
Properties: Volatile, colorless to pale yellow liquid; amine odor. Boiling range 138.5-142.5°C; refractive index 1.4150-1.4200 (n 25/D); sp. gr. 0.7600-0.7625 (25/25°C). Freely soluble in alcohol, benzene, chloroform, and ether; slightly soluble in water; pH (1% solution) 11.45.
Use: Medicine.

para-**aminohippuric acid** [N-(para-aminobenzoyl) glycine; PAHA] $NH_2C_6H_4CONHCH_2COOH$.
Properties: White, crystalline powder. Discolors on exposure to light. Slightly soluble in water, alcohol, and most organic solvents. Very soluble in dilute hydrochloric acid and alkalies. Forms a water soluble sodium salt. M.p. 197-199°C.
Grade: N.F. XI.
Uses: Medical diagnostic reagent; intermediate.

aminohydroxybenzoic acids. See aminosalicylic acids.

alpha-**amino**-beta-**hydroxybutyric acid.** See threonine.

alpha-**amino**-beta-para-**hydroxyphenylpropionic acid.** See tyrosine.

alpha-**amino**-beta-**hydroxypropionic acid.** See serine.

alpha-**amino**-beta-**imidazolepropionic acid.** See histidine.

dℓ-alpha-**aminoisobutyric acid** (2-amino-2-methylpropanoic acid) $(CH_3)_2C(NH_2)COOH$.
Properties: White crystals; sweetish taste. M.p. 335°C; sublimes; freely soluble in water; insoluble in alcohol and ether.
Derivation: (a) Treatment of acetone with ammonium cyanide; (b) by heating dimethylhydantoin with hydrochloric acid.

alpha-**aminoisocaproic acid.** See leucine.

alpha-**aminoisovaleric acid.** See valine.

D-4-amino-3-isoxazolidone (cycloserine; oxamycin) $C_3H_6N_2O_2$ or $\underline{NHCOCH(NH_2)CH_2O}$.

A soluble, crystalline, broad spectrum antibiotic; highly stable in alkaline solutions; somewhat unstable in neutral or acid solutions.
Derivation: (a) Submerged aerobic fermentation; (b) synthesis from serine.
Grade: N.N.D.
Use: Medicine.

amino-J acid. See 2-naphthylamine-5,7-disulfonic acid.

aminomercuric chloride. See mercury, ammoniated.

para-**(4-amino-3-methoxyphenylazo) benzenesulfonic acid** $H_2NC_6H_3(OCH_3)NNC_6H_4SO_3H$.
Properties: Maroon paste containing approximately 38% solids.
Grade: Technical.
Use: Intermediate.

para-**aminomethylbenzenesulfonamide hydrochloride.** See maphenide hydrochloride.

ℓ-alpha-**(aminomethyl)-3,4-dihydroxybenzyl alcohol.** See levarterenol.

4-amino-4'-methyldiphenylamine-2-sulfonic acid (aminotoluidinobenzene sulfonic acid) $CH_3C_6H_4NHC_6H_3NH_2SO_3H$.
Properties: Light to dark gray paste with characteristic odor.
Use: Intermediate.

6-amino-3-methyl-1-(2-methylallyl)uracil $C_9H_{13}N_3O_2$.
Properties: White crystalline powder; odorless with mild, slightly bitter taste; m.p. 170-176°C. Slightly soluble in water, with increasing solubility as water is heated; freely soluble in alcohol; insoluble in ether.
Grade: N.F. XI.
Use: Medicine.

2-amino-3-methylpentanoic acid. See isoleucine.

2-amino-2-methyl-1,3-propanediol (AMPD; aminobutylene glycol; butanediolamine) $CH_2OHC(CH_3)NH_2CH_2OH$. Corrosive to copper, brass, aluminum.
Properties: White crystalline solid. Solubility in water: 250 g/100 cc at 20°C, also soluble in alcohol.
Constants: M.p. 109-111°C; b.p. (10 mm) 151-152°C; pH (0.1M aqueous soln) 10.78.
Containers: 1-, 5-, 51-gal fiberpaks.
Uses: Emulsifying agent (in soap form) for oils, fats, and waxes; absorbent for acidic gases; chemical synthesis; cosmetics.
Caution: Slightly toxic; mildly alkaline; avoid repeated skin contact.

2-amino-2-methylpropanoic acid. See dℓ-alpha-aminoisobutyric acid.

2-amino-2-methyl-1-propanol (isobutanolamine; AMP) $CH_3C(CH_3)NH_2CH_2OH$.
Properties: Colorless liquid, or white crystalline solid. Completely miscible in water at 20°C.
Constants: Sp. gr. 0.934 at 20/20°C; b.p. (760 mm) 165°C, (10 mm) 67.4°C; m.p. 30-31°C; vapor pressure 1 mm at 20°C (est'd); flash point 153°F; wt. 7.76 lbs per gal (20°C); refractive index 1.449 at 20°C; pH (0.1M aqueous solution) 11.27.
Containers: 5- and 55-gal drums. May

solidify in drum in cool weather.
Uses: Emulsifying agent (in soap form) for oils, fats, and waxes; absorbent for acidic gases; chemical synthesis.
Caution: Slightly toxic; mildly alkaline; avoid repeated skin contact.

2-amino-3-methylpyridine
N:CNH$_2$CCH$_3$:CHCH:CH
Properties: B. p. 221. 1°C; freezing point 29. 5-33. 3°C; soluble in water.
Derivation: From 2-aminopyridine.
Grade: 98% (minimum).
Use: Intermediate.

2-amino-4-methylpyridine (2-amino-4-picoline) N:CNH$_2$CH:CCH$_3$CH:CH
Properties: Crystals; b.p. 230.9°C (760 mm), 115-117°C (11 mm); freezing point 96-99.0°C. Sublimes on slow heating; soluble in water, lower alcohols, dimethylformamide, coal tar bases; slightly soluble in petroleum ether, aliphatic hydrocarbons.
Derivation: Prepared from 2-aminopyridine.
Grade: 95% (minimum).
Uses: Intermediate; medicine.

2-amino-5-methylpyridine
N:CNH$_2$CH:CHCCH$_3$:CH.
Properties: B. p. 227. 1°C; freezing point 76. 6°C; soluble in water.
Derivation: Prepared from 2-aminopyridine.
Grade: 95% (minimum).
Use: Intermediate.

2-amino-6-methylpyridine
N:CNH$_2$CH:CHCH:CCH$_3$.
Properties: B. p. 214. 4°C; freezing point 43. 7°C; soluble in water.
Derivation: Prepared from 2-aminopyridine.
Use: Intermediate.

alpha-**amino**-beta-**methylvaleric acid.** See isoleucine.

alpha-**amino**-gamma-**methylvaleric acid.** See leucine.

3-amino-1,5-naphthalenedisulfonic acid. Preferred name for 2-naphthylamine-4, 8-disulfonic acid.

6-amino-1,3-naphthalenedisulfonic acid. Preferred name for 2-naphthylamine-5, 7-disulfonic acid.

7-amino-1,3-naphthalenedisulfonic acid. Preferred name for 2-naphthylamine-6, 8-disulfonic acid.

2-amino-1-naphthalenesulfonic acid. Preferred name for 2-naphthylamine-1-sulfonic acid.

4-amino-1-naphthalenesulfonic acid. See naphthionic acid.

5-amino-1-naphthalenesulfonic acid. Preferred name for 1-naphthylamine-5-sulfonic acid.

5-amino-2-naphthalenesulfonic acid. Preferred name for 1-naphthylamine-6-sulfonic acid.

6-amino-2-naphthalenesulfonic acid. Preferred name for 2-naphthylamine-6-sulfonic acid.

8-amino-1-naphthalenesulfonic acid. Preferred name for 1-naphthylamine-8-sulfonic acid.

8-amino-2-naphthalenesulfonic acid. Preferred name for 1-naphthylamine-7-sulfonic acid.

8-amino-1,3,6-naphthalenetrisulfonic acid. Preferred name for 1-naphthylamine-3, 6, 8-trisulfonic acid.

1-amino-2-naphthol-3,6-disulfonic acid.
$C_{10}H_4NH_2OH(SO_3H)_2$.
Properties: Needles. Soluble in water.
Derivation: By the action of hydrochloric acid and stannous chloride upon 1-benzene-azo-2-naphthol-3, 6-disulfonic acid.
Grades: Technical.
Containers: Wooden barrels; fiber containers.
Use: Photography (in the form of the sodium salt as a developer).

1-amino-8-naphthol-2,4-disulfonic acid. See 8-amino-1-naphthol-5, 7-disulfonic acid.

1-amino-8-naphthol-3,5-disulfonic acid (B acid; 8-amino-1-naphthol-4, 6-disulfonic acid) $C_{10}H_4NH_2OH(SO_3H)_2$.
Derivation: By sulfonating 1-amino-8-naphthol-3-sulfonic acid.
Grades: Technical.
Containers: Wooden barrels.
Use: Dye intermediate.

1-amino-8-naphthol-3,6-disulfonic acid (H acid; 8-amino-1-naphthol-3, 6-disulfonic acid) $C_{10}H_4NH_2OH(SO_3H)_2$.
Properties: Gray powder. Soluble in water, alcohol and ether.
Derivation: From alpha-naphthylamine trisulfonic acid by soda fusion.
Method of purification: Crystallization.
Grades: Technical, 80%, 85%.
Containers: 300-, 385-lb wooden barrels.
Use: Dye intermediate.
Shipping regulations: None. *

1-amino-8-naphthol-4,6-disulfonic acid (K acid; 8-amino-1-naphthol-3, 5-disulfonic acid) $C_{10}H_4NH_2OH(SO_3H)_2$.
Derivation: By soda fusion of a naphthylamine trisulfonic acid.
Grades: Technical.
Containers: Wooden barrels.
Use: Dye intermediate.

2-amino-8-naphthol-3,6-disulfonic acid (2R acid; RR acid; 7-amino-1-naphthol-3, 6-disulfonic acid) $C_{10}H_4NH_2OH(SO_3H)_2$.
Properties: Soluble in water.
Derivation: By soda fusion of a naphthylamine trisulfonic acid.
Grades: Technical.
Containers: Wooden barrels.
Use: Dye intermediate.

7-amino-1-naphthol-3,6-disulfonic acid. Preferred name for 2-amino-8-naphthol-3, 6-disulfonic acid.

8-amino-1-naphthol-3,5-disulfonic acid. Preferred name for 1-amino-8-naphthol-4, 6-disulfonic acid.

8-amino-1-naphthol-3,6-disulfonic acid. Preferred name for 1-amino-8-naphthol-3, 6-disulfonic acid.

*See "I. C. C. Shipping Regulations," page xiii.
Reference numbers refer to name of manufacturer. See "List of Manufacturers," page v.

8-amino-1-naphthol-4,6-disulfonic acid. Preferred name for 1-amino-8-naphthol-3,5-disulfonic acid.

8-amino-1-naphthol-5,7-disulfonic acid (Chicago acid; SS acid; 1-amino-8-naphthol-2,4-disulfonic acid) $C_{10}H_4NH_2OH(SO_3H)_2$.
Properties: Gray paste; white when pure. Soluble in water and sodium hydroxide solution..
Derivation: Fusion of 1,8-naphthosultam-2,4-disulfonic acid with caustic potash.
Method of purification: Recrystallization from water.
Grade: Technical.
Containers: Wooden barrels.
Use: Dye intermediate.
Shipping regulations: None. *

1-amino-2-naphthol-4-sulfonic acid (1,2,4 acid) $C_{10}H_5NH_2OHSO_3H$.
Properties: Pinkish-white to gray needles. Soluble in hot water; almost insoluble in cold water.
Derivation: Beta-naphthol is changed to the 1-nitroso-beta-naphthol which is treated with sodium bisulfite. Upon acidification the free sulfurous acid effects simultaneous reduction and sulfonation.
Method of purification: By washing the crude acid paste with water.
Grades: Technical.
Containers: Glass bottles; 500- to 800-lb wooden barrels.
Uses: Aniline dyes; analysis (determination of phosphates and calcium).
Shipping regulations: None. *

1-amino-2-naphthol-6-sulfonic acid $C_{10}H_5NH_2OHSO_3H$.
Properties: Needles or prisms. Slightly soluble in hot water, alcohol; insoluble in ether.
Derivation: By reducing 1,2,6-nitrosonaphtholsulfonic acid.
Grades: Technical.
Uses: Photography (making sodium salt used as a developer); dye intermediate.

1-amino-2-naphthol-6-sulfonic acid, sodium salt. See eikonogen.

1-amino-5-naphthol-7-sulfonic acid (M acid; 5-amino-1-naphthol-3-sulfonic acid) $C_{10}H_5NH_2OHSO_3H$.
Properties: Gray needles. Slightly soluble in cold water; soluble in hot water and alcohol.
Grades: Technical.
Containers: Wooden barrels.
Use: Dye intermediate.

1-amino-8-naphthol-4-sulfonic acid (S acid; 8-amino-1-naphthol-5-sulfonic acid) $C_{10}H_5NH_2OHSO_3H$.
Properties: Gray; white when pure. Slightly soluble in water; insoluble in alcohol and ether.
Derivation: Fusion of 1-naphthylamine-4,8-disulfonic acid with caustic soda.
Method of purification: Precipitation from a dilute solution of its sodium salt.
Grades: Technical.
Containers: Wooden barrels.
Use: Dye intermediate.
Shipping regulations: None. *

1-amino-8-naphthol-5-sulfonic acid $C_{10}H_5NH_2OHSO_3H$.
Properties: Small needles; slightly soluble in water.
Derivation: Fusion of 1-naphthylamine-5,8-disulfonic acid with caustic soda.
Grades: Technical.
Containers: Wooden barrels.
Use: Dye intermediate.

2-amino-1-naphthol-4-sulfonic acid $C_{10}H_5NH_2OHSO_3H$.
Properties: Needles; white; insoluble in water, alcohol, ether, benzene.
Derivation: By heating 2-nitroso-1-naphthol with sodium sulfite.
Grades: Technical.
Containers: Wooden barrels.
Use: Dye intermediate.

2-amino-3-naphthol-6-sulfonic acid (R acid) $C_{10}H_5NH_2OHSO_3H$.
Properties: Needles; slightly soluble in water.
Derivation: Fusion of 2-naphthylamine-3,6-disulfonic acid with caustic soda.
Grades: Technical.
Containers: Wooden barrels.
Use: Dyes.

2-amino-5-naphthol-7-sulfonic acid (J acid; 6-amino-1-naphthol-3-sulfonic acid) $C_{10}H_5NH_2OHSO_3H$.
Properties: Gray; white when pure. Soluble in hot water; sparingly soluble in cold water.
Derivation: Fusion of beta-naphthylamine-5,7-disulfonic acid with caustic.
Method of purification: Recrystallization from hot water.
Grades: Technical.
Containers: Wooden barrels.
Use: Dyes.

2-amino-5-naphthol-7-sulfonic acid, sodium salt $C_{10}H_5NH_2OHSO_3Na$.
Properties: Gray to brown paste with characteristic odor.
Grade: 50% (minimum).
Use: Dye intermediate.

2-amino-8-naphthol-6-sulfonic acid. See 7-amino-1-naphthol-3-sulfonic acid.

5-amino-1-naphthol-3-sulfonic acid. Preferred name for 1-amino-5-naphthol-7-sulfonic acid.

6-amino-1-naphthol-3-sulfonic acid. Preferred name for 2-amino-5-naphthol-7-sulfonic acid.

7-amino-1-naphthol-3-sulfonic acid (gamma acid; 2-amino-8-naphthol-6-sulfonic acid) $C_{10}H_5NH_2OHSO_3H$.
Properties: White crystals. Soluble in alcohol and ether; very slightly soluble in water.
Derivation: By heating caustic soda and

*See "I.C.C. Shipping Regulations," page xiii.
Reference numbers refer to name of manufacturer. See "List of Manufacturers," page v.

2-naphthylamine-6,8-disulfonic acid in an autoclave.
Method of purification: Crystallization.
Grades: Technical.
Containers: 50-, 100-lb kegs; 200-, 300-lb barrels.
Use: Azo dye intermediate.
Shipping regulations: None.*

8-amino-1-naphthol-5-sulfonic acid. Preferred name for 1-amino-8-naphthol-4-sulfonic acid.

2-amino-5-nitrothiazole $C_3H_3N_3O_2S$.
Properties: Greenish yellow to orange-yellow fluffy powder; slightly bitter taste. Decomposes 202°C; very sparingly soluble in water; soluble in dilute acids.
Use: Veterinary medicine.

amino oxidase (L-amino acid oxidase; D-amino acid oxidase; L-amino dehydrogenase). An enzyme which catalyzes the deamination of amino acids by dehydrogenation to keto acids and ammonia. Two types are recognized, acting on the D- and L-amino acids. Recent emphasis has been on characterization of the D-amino oxidase, which is known to contain a flavin as coenzyme. Both types are found in animal tissue, especially in liver and kidney, as well as in snake venom and certain bacteria.

2-amino-6-oxypurine. See guanine.

aminopentamide sulfate (4-dimethylamino-2, 2-diphenylvaleramide sulfate) $H_2NOCC(C_6H_5)_2CH_2CH(CH_3)N(CH_3)_2 \cdot \frac{1}{2} H_2SO_4$.
Properties: White, odorless, crystalline powder. Freely soluble in alcohol and water; pH range of 2.5% solution is 1.3-2.2.
Grade: N.N.D.
Use: Medicine.

1-aminopentane. See n-amylamine.

aminophenacetin acetate. See phenocoll acetate.

aminophenacetin hydrochloride. See phenocoll hydrochloride.

2-aminophenetole. See ortho-phenetidine.

4-aminophenetole. See para-phenetidine.

meta-**aminophenol** (meta-hydroxyaniline) $C_6H_4NH_2OH$.
Properties: White crystals; m.p. 122°C; soluble in water, alcohol, and ether.
Derivation: Fusion of meta-sulfanilic acid with caustic soda and subsequent extraction of the melt with ether.
Method of purification: Recrystallization.
Containers: Drums.
Uses: Dye intermediate: intermediate for para-amino-salicylic acid.

ortho-**aminophenol** (ortho-hydroxyaniline) $C_6H_4NH_2OH$.
Properties: White crystals; turn brown with age; m.p. 170°C; sublimes on further heating. Slightly soluble in water, alcohol, benzene; freely soluble in ether.
Derivation: By reduction of ortho-nitrophenol mixed with aqueous ammonia by means of a stream of hydrogen sulfide.
Method of purification: Recrystallization.
Grades: Technical.
Containers: 1- and 5-lb glass bottles; fiber cans and drums.
Uses: Dyeing furs and hair; dye intermediate for azo and sulfur dyes.

para-**aminophenol** (para-hydroxyaniline) $C_6H_4NH_2OH$.
Properties: White or reddish yellow crystals; turn violet on exposure to light; m.p. 184°C with decomposition; soluble in water and alcohol.
Derivation: (a) By reduction of para-nitrophenol with iron filings and hydrochloric acid. (b) By electrolytic reduction of nitrobenzene in concentrated sulfuric acid and treatment with an alkali to free the base.
Method of purification: Recrystallization.
Grades: Technical; photographic.
Containers: Glass bottles; 100-lb wooden barrels; 100-lb net fiber drums.
Uses: Dyeing textiles, hair, furs, feathers; photographic developer; pharmaceuticals; antioxidants; oil additives.

4-amino-1-phenol-2,6-disulfonic acid $C_6H_2OHNH_2(SO_3H)_2$.
Properties: Fine needles. Soluble in water; slightly soluble in alcohol; insoluble in ether.
Derivation: By the action of sulfur dioxide on para-nitrophenol.
Grades: Technical.
Containers: Wooden barrels.
Uses: Dyes.

ortho-**aminophenol hydrochloride** $C_6H_7NO \cdot HCl$.
Properties: White crystals changing to gray on exposure to light; m.p. 208-210°C; soluble in alcohol and water.
Uses: Dyeing fur and hair; dye intermediate.

para-**aminophenol hydrochloride** $C_6H_7NO \cdot HCl$.
Properties: Light-brown to white needles; m.p. 306°C with decomposition. Soluble in water and alcohol.
Derivation: By neutralizing para-aminophenol with hydrochloric acid.
Method of purification: Recrystallization.
Grades: Technical; photographic.
Containers: Glass bottles; 100-lbs net fiber drums.
Uses: Dyes; photographic chemicals.
Shipping regulations: None.*

aminophenolsulfonic acid III. See 3-amino-1-phenol-4-sulfonic acid.

aminophenolsulfonic acid IV. See 3-amino-1-phenol-6-sulfonic acid.

2-amino-1-phenol-4-sulfonic acid (ortho-aminophenol-para-sulfonic acid) $C_6H_3OHNH_2SO_3H$.
Properties: Brown crystals. Fairly soluble in hot water; very soluble in alkaline solution. No melting point. Decomposes on heating.
Derivation: (a) Sulfonation and nitration of chlorobenzene followed by hydrolysis to phenol with caustic soda with subsequent

*See "I.C.C. Shipping Regulations," page xiii.
Reference numbers refer to name of manufacturer. See "List of Manufacturers," page v.

reduction by sodium sulfide. (b) Sulfonation of ortho-aminophenol. (c) Sulfonation of phenol followed by nitration and reduction.
Method of purification: Recrystallization.
Grades: Technical, 97% crystals; 10% solution in sodium sulfide.
Containers: 100-lb kegs; 250-lb wooden barrels.
Use: Intermediate for Schultz dyes Nos. 154, 155, 156, and 157.
Shipping regulations: None. *

3-amino-1-phenol-4-sulfonic acid (aminophenolsulfonic acid III) $C_6H_3OHNH_2SO_3H$.
Properties: Needles or plates. Slightly soluble in cold water (sodium salt more so).
Derivation: By heating together (in a water bath) concentrated sulfuric acid and meta-aminophenol-disulfonic acid.
Grades: Technical.
Containers: Wooden barrels.
Use: Dyes.

3-amino-1-phenol-6-sulfonic acid (aminophenolsulfonic acid IV) $C_6H_3OHNH_2SO_3H$.
Properties: Needles or plates. Soluble in water.
Derivation: (a) By soda fusion of aniline disulfonic acid; (b) by sulfonating metanilic acid.
Grades: Technical.
Containers: Wooden barrels.
Use: Dyes.

ortho-**aminophenol**-para-**sulfonic acid.** See 2-amino-1-phenol-4-sulfonic acid.

para-**aminophenylarsonic acid.** See arsanilic acid.

1-amino-2-phenylethane. See beta-phenylethylamine.

alpha-(para-**aminophenyl**)-alpha-**ethylglutarimide.** See aminoglutethimide.

ortho-**aminophenylglyoxalic lactime.** See isatin.

para-**aminophenylmercaptoacetic acid** $NH_2C_6H_4SCH_2CO_2H$.
Properties: M.p. 186-187°C; insoluble in water, alcohol, benzene, chloroform; soluble in aqueous acid or alkali solutions.
Uses: Synthetic intermediate for dyes and pharmaceuticals.

2-(para-**aminophenyl)-6-methylbenzothiazole.** See dehydrothio-para-toluidine.

meta-**aminophenyl methyl carbinol** $NH_2C_6H_4CH(OH)CH_3$.
Properties: Solid; sp. gr. 1.12 (density, g/ml); b.p. 217.3°C (100 mm); freezing point 66.4°C; soluble in water.
Uses: Carrier for dyeing synthetic fibers; intermediate for perfume chemicals and pharmaceuticals.

1-(meta-**aminophenyl)-3-methyl-5-pyrazolone** $NH_2C_6H_4\underline{NNC(CH_3)CH_2C}O$.
Properties: Light tan paste containing approximately 45% solids.
Grade: Technical.
Use: Intermediate.

alpha-**amino**-beta-**phenylpropionic acid.** See phenylalanine.

aminophylline (theophylline ethylenediamine) $C_{16}H_{24}N_{10}O_4 \cdot xH_2O$. Contains 78 to 83.5% anhydrous theophylline; 12.8 to 14.1% ethylenediamine.
Properties: White or slightly yellowish granules or powder with slight ammoniacal odor and a bitter taste. Soluble in water; insoluble in alcohol and ether. Solutions are alkaline to litmus.
Grade: U.S.P. XVI.
Containers: Bottles; 25-, 100-lb drums.
Use: Medicine.

aminopicolines. See aminomethylpyridines.

amino plastics. See amino resins.

2-aminopropane. See isopropylamine.

2-aminopropanoic acid. See alanine.

3-aminopropanoic acid. See beta-alanine.

2-amino-1-propanol (2-aminopropyl alcohol; beta-propanolamine) $CH_3CH(NH_2)CH_2OH$.
Properties: Colorless to pale yellow liquid. Both *l* and d*l* forms are available. d*l*-form: fishy odor; boils at 173-176°C; freely soluble in water, alcohol, ether. *l*-form: refractive index 1.4480-1.4495 at 25°C; distillation range about 114°C at 100 mm.

3-aminopropanol (propanolamine) $H_2NCH_2CH_2CH_2OH$.
Properties: Colorless liquid; m.p. 12.4°C; b.p. 184-186°C (760 mm), 168°C (500 mm); flash point (Tag open cup) 175°F; sp. gr. 0.9786 (30°C). Miscible with alcohol, water, acetone and chloroform.
Grade: 99% pure.
Uses: Organic intermediate.

alpha-**aminopropionic acid.** See alanine.

beta-**aminopropionic acid.** See beta-alanine.

2-aminopropyl alcohol. See 2-amino-1-propanol.

N-aminopropyl morpholine $\underline{CH_2CH_2OCH_2CH_2N}C_3H_6NH_2$.
Properties: Sp. gr. 0.9872 (20/20°C); b.p. 224.5°C; f.p. 220°F; freezing point −15°C; soluble in water.
Use: Fiber synthesis.

para-**(2-aminopropyl) phenol.** See hydroxyamphetamine.

para-**(2-aminopropyl) phenol hydrobromide.** See hydroxyamphetamine hydrobromide.

aminopropyltriethoxysilane.
Properties: Liquid; b.p. 217°C; sp. gr. 0.94 (25°C).
Uses: Sizing of glass fibers for making laminates.

aminopterin (4-aminofolic acid; aminopteroylglutamic acid) $C_{19}H_{20}N_8O_5 \cdot 2H_2O$. Differs slightly in structure from folic acid (q.v.) and antagonizes the utilization of folic acid by the body.

*See "I.C.C. Shipping Regulations," page xiii.
Reference numbers refer to name of manufacturer. See "List of Manufacturers," page v.

Properties: Occurs as clusters of yellow needles which are soluble in aqueous sodium hydroxide solutions.
Grades: The sodium salt (aminopterin sodium) is listed in N. N. D.
Uses: Medicine; rodenticide.

6-aminopurine. See adenine.

2-aminopyridine (alpha-pyridylamine) $C_5H_4NNH_2$.
Properties: White leaflets or large, colorless crystals; freezing point 58.1°C; b.p. 210.6°C; soluble in water, alcohol, benzene, ether, hot petroleum ether.
Use: Intermediate for antihistamines, other drugs and organics.

3-aminopyridine (beta-pyridylamine) $C_5H_4NNH_2$.
Properties: White crystals; m.p. 64°C; b.p. 250-252°C; soluble in water, alcohol, benzene, ether.
Use: Intermediate in preparation of drugs and dyestuffs.

4-aminopyridine $C_5H_4NNH_2$.
Properties: Crystals; m.p. 158.9°C; b.p. 273.5°C; soluble in water.
Derivation: From 2-aminopyridine.
Grade: 95% (minimum).
Use: Intermediate.

aminopyrine (dimethylaminoantipyrine; amidopyrine) $C_{13}H_{17}N_3O$. 4-Dimethylamino-2,3-dimethyl-1-phenyl-3-pyrazolin-5-one.
Properties: Colorless crystalline powder, almost tasteless, odorless; m.p. 107-109°C. Moderately soluble in water; soluble in alcohol, chloroform, ether.
Derivation: By reducing isonitrosoantipyrine with zinc and methylating with methyl iodide.
Containers: 25-, 100-lb drums.
Use: Medicine.
Shipping regulations: None.*

amino resins (amino plastics). A large class of thermosetting resins made by the reaction of an amine with an aldehyde. The only aldehyde in commercial use is formaldehyde, and the most important amines are urea and melamine. These resins are used in molding, adhesives, laminating, textile finishes, protective coatings, and paper manufacture. Less important applications include leather treatment, binders for fabrics, and foundry sands, graphite resistors, plaster of paris fortification, foam structures and ion exchange resins. In general the resins are hard and brittle and are used with fillers.

In commercial practice the reaction between the amine and formaldehyde is carried only part way to completion by the resin manufacturer, and then heat, with or without a catalyst, is used in the final applications to cause thermosetting, i.e., the formation of the hard infusible resin. Many variations of these resins arise by changing the proportions of amine to formaldehyde and by having solvents or chemically reactive materials present during resin formation.

Aniline and thiourea resins are no longer of importance. Ethylene urea resins are of increasing importance for textile finishing.

4-aminosalicylic acid (PASA; PAS; para-aminosalicylic acid; 4-amino-2-hydroxybenzoic acid) $NH_2C_6H_3(OH)COOH$.
Properties: White or nearly white bulky powder. Odorless or has slight acetous odor. M.p. 150-151°C (dec). Affected by light and air. Soluble in sodium bicarbonate and phosphoric acid; somewhat soluble in alcohol; slightly soluble in ether or acetone; practically insoluble in benzene and water; pH (saturated aqueous solution) 3.0-3.7.
Derivation: From meta-aminophenol and potassium bicarbonate solution under pressure.
Grade: U.S.P. XVI.
Containers: Drums.
Use: Medicine.

5-aminosalicylic acid (meta-aminosalicylic acid; 5-amino-2-hydroxybenzoic acid) $NH_2C_6H_3(OH)COOH$.
Properties: White crystals; sometimes pinkish; m.p. 260-280°C with decomposition; soluble in hot water or alcohol.
Derivation: (a) From the corresponding nitrosalicylic acid by reduction. (b) By reducing the azo dye, benzeneazosalicylic acid.
Method of purification: Recrystallization from water.
Grades: Technical.
Containers: Drums.
Uses: Dyes; intermediates; manufacture of transfer paper.
Shipping regulations: None.*

5-aminosalicylic acid hydrochloride $C_6H_3COOH(OH)NH_2 \cdot HCl$.
Properties: Grayish-white crystals. Soluble in water, alcohol and ether.
Derivation: By the reduction of nitrosalicylic acid with zinc and hydrochloric acid.
Method of purification: Crystallization.
Grades: Technical.
Containers: Glass bottles; tins.
Uses: Medicine; manufacture of transfer paper.
Shipping regulations: None.*

meta-**aminosalicylic acid.** See 5-aminosalicylic acid.

para-**aminosalicylic acid.** See 4-aminosalicylic acid.

alpha-**aminosuccinamic acid.** See asparagine.

aminosuccinic acid. See aspartic acid.

2-aminothiazole (2-thiazylamine) $C_3H_4N_2S$.
Properties: Light yellow crystals; m.p. 90°C; distills at 3 mm without decomposition. Slightly soluble in cold water, alcohol and ether; soluble in hot water and dilute mineral acids.

*See "I.C.C. Shipping Regulations," page xiii.
Reference numbers refer to name of manufacturer. See "List of Manufacturers," page v.

Derivation: Chlorination of vinyl acetate and condensation with thiourea.
Use: Intermediate in synthesis of sulfathiazole.

alpha-**amino-beta-thiolpropionic acid.** See cysteine.

aminothiourea. See thiosemicarbazide.

aminotoluene. See meta-, ortho-, or para-toluidine. See also benzylamine.

4-amino-meta-toluenesulfonic acid (SO_3H=1). See ortho-toluidine-metasulfonic acid (CH_3=1).

5-amino-ortho-toluenesulfonic acid (SO_3H=1). See para-toluidine-ortho-sulfonic acid (CH_3=1).

aminotoluidinobenzene sulfonic acid. See 4-amino-4'-methyldiphenylamine-2-sulfonic acid.

aminotrate phosphate. See trolnitrate phosphate.

"Aminotriazole." [57] A proprietary grade of 3-amino-1,2,4-triazole sold as a weed killer. Particularly effective for control of noxious perennial weeds, defoliation, and plant growth inhibition.
Properties: White crystalline solid; m.p. 150-153°C; soluble in water; slightly soluble in ethanol; insoluble in ether or acetone.
Uses: As a weed killer, defoliant, and plant growth inhibitor. Particularly effective for control of perennial weeds such as poison ivy, poison oak, Canada thistle, quack grass, and aquatic weeds. Also effective as a defoliant for cotton and other plants with additional advantage of inhibiting regrowth of new foliage.

3-(3-amino-2,4,6-triiodophenyl)-2-ethyl propanoic acid. See iopanoic acid.

aminourea hydrochloride. See semicarbazide hydrochloride.

"Aminox." [248] Trademark for a low-temperature reaction product of diphenylamine and acetone.
Properties: A light-tan powder; sp. gr. 1.13; m.p. 85-95°C; soluble in acetone, benzol, and ethylene dichloride; insoluble in water and gasoline.
Use: Antioxidant for light-colored rubber stocks; tire carcass, inner tubes, footwear, heels, soles, proofing sundries, and wire insulation.

aminoxylene. See xylidine.

aminoxylol. See xylidine.

"Amite." [67] Trademark for copper-titanium alloys in cast or wrought forms.

"Amizyme." [114] Trade name for a series of enzyme preparations, high in dextrinizing or starch-liquefying properties. Available in tablet, powdered or liquid form.
Derivation: Produced by growing pure microbial cultures on select media.
Uses: Conversion of starch; used in paper coatings, adhesives, and textile sizes.
Shipping regulations: None. *

"Ammate." [28] Trademark for non-selective herbicide (weed and brush killer) based on ammonium sulfamate. Supplied in 95% dry crystalline formulations and also in aqueous solution.
Containers: Tank car lots; 40-lb drums and 60-lb bags.
Use: For control of poison ivy and woody plants.

ammelide. A compound related to melamine; $H_2N\overline{CNC(OH)NC(OH)}N$.

ammeline. A compound related to melamine; $H_2N\overline{CNC(NH_2)NC(OH)}N$.

ammeline resins. Stated to be ceramic-like materials subliming at about 1700°F. Made by action of phosphorus pentoxide on ammeline or melamine.

ammines. Coordination compounds formed by the union of ammonia with a metallic substance, in such a way that the nitrogen atoms are linked directly to the metal. See cobaltammines and coordination compound. Note the distinction from amines, in which the nitrogen is attached directly to the carbon atom.

ammonia NH_3. This term is properly applied only to the pure gas or compressed or cooled liquid of the indicated composition, but the term is also regularly used for solutions of this material in water, i.e., aqua ammonia. The following discussion relates to the pure substance NH_3.
Properties: Colorless gas with characteristic pungent odor; lighter than air; b.p. (colorless liquid) −33.5°C; freezing point −77.7°C; easily liquefied by pressure alone. Vapor pressure of liquid is 8.5 atm at 20°C; sp. volume (70°F) 22.7 cu ft/lb; sp. gr. of liquid 0.77 (0°C), 0.6819 (at b.p.); very soluble in water, alcohol, and ether.
Derivation: Formerly almost entirely as a byproduct of the destructive distillation of coal, but most ammonia is now produced by direct combination of nitrogen and hydrogen gases. (See ammonia synthesis.)
Grades: Commercial 99.5%; refrigerant 99.97%.
Containers: 50-, 100-, 150-lb steel cylinders; 50,000-lb tank cars; tank trucks.
Uses: Refrigeration; fertilizer, as ingredient of mixtures or applied directly in liquid form; chemical manufacture, e.g., nitric acid, or as neutralizing agent, etc.; in rubber vulcanization; water treatment; nitriding of steel; oil refining; extracting certain metals from ores; solvent and reaction medium in organic synthesis; yeast nutrient; sulfite paper pulp process; explosives; pesticide; rocket fuel. Cracked into its constituent gases, it is a protective atmosphere for bright annealing, and a source of hydrogen; also used in powder metallurgy and brazing.

Warning! Hazardous liquid and gas under pressure. Liquid causes burns. Gas extremely irritating. MCA warning label.
Shipping regulations: Nonflammable gas. Green gas label.*

ammonia alum. See aluminum-ammonium sulfate.

ammonia, anhydrous. Ammonia that is free of water, i.e., the pure dry gas NH_3, but much more frequently the pure dry liquid ammonia of commerce. See ammonia.

ammonia, aqua. See ammonium hydroxide.

ammoniac (ammoniac gum).
Properties: Irregular, rounded tears; peculiar odor; sweetish-bitter, somewhat acrid taste; sp. gr. 1.207; partly soluble in water (forming emulsions) and alcohol.
Constituents: Resin, volatile oil, salicylic acid, gum.
Derivation: Gum resins from stems of perennial herb, Dorema ammoniacum.
Occurrence: Persia, northern India and southern Siberia.
Grades: Technical.
Containers: Bags.
Uses: Medicine; as ingredient of porcelain cements; perfumes.
Shipping regulations: None.*

ammoniacal liquor. See ammonia liquor.

ammoniac gum. See ammoniac.

ammonia, household. Name applied in trade to dilute ammonium hydroxide frequently containing small amounts of various detergent additives.

ammonia, liquid. Ammonia (NH_3) is regularly handled in commerce and industry in the form of a liquid, kept under relatively low pressure in steel cylinders or tanks. See ammonia.

ammonia liquor (gas liquor; ammoniacal liquor).
Derivation: A condensed watery solution obtained in the destructive distillation of a bituminous coal in gas or coke manufacture, composed of ammonia and ammonium compounds, and containing hydrogen sulfide and cyanogen.
Uses: Production of anhydrous ammonia, aqua ammonia (ammonium hydroxide), ammonium sulfate and other ammonium salts; as a source of ammonia in the Solvay process for producing soda sah.
Grades: Technical.
Containers: Tank cars.
Shipping regulations: None.*

ammonia-soda process. See Solvay process.

ammonia solution, U.S.P. XVI. Specifications covering two concentrations of aqueous ammonium hydroxide solutions. Diluted ammonia solution contains not less than 9 g or more than 10 g of ammonia (NH_3) per 100 ml. Strong ammonia solution contains not less than 27% or more than 30% by weight of ammonia (NH_3). See ammonium hydroxide.

ammonia synthesis. The direct combination of nitrogen and hydrogen gases at high temperature and pressure in the presence of a catalyst to produce ammonia. Pressures range up to 1000 atmospheres and temperatures up to 700°C, so that special chromium alloy steels and reaction vessel designs are required. The hydrogen is often obtained by reaction of natural gas with steam at high temperature. Hydrogen is also obtained by the water gas process, or by oxidation of natural gas to carbon monoxide and hydrogen, or, recently, as a by-product of certain petroleum refinery processes. Nitrogen is usually supplied by mixing enough air with the hydrogen stream so that after combustion all oxygen is removed and hydrogen and nitrogen are left in proper proportions.

The carbon monoxide usually present is converted to dioxide by reaction with steam over an iron oxide catalyst. The carbon dioxide is then absorbed in water under pressure or in a 20% solution of ethanolamine, which can be continuously regenerated. Various special means are used for removing traces of carbon monoxide since it is a catalyst poison. The original operating process was the Haber process, but a variety of methods have since been developed for obtaining nitrogen and hydrogen, purifying them, and operating the catalytic conversion unit. These are now mostly of historic interest only, but include Haber-Bosch, Claude, Casale, Fauser and Mont Cenis (q. v.).

ammoniated mercury chloride. See mercury, ammoniated.

ammoniated mercury nitrate. See mercurous nitrate, ammoniated.

ammoniated ruthenium oxychloride. See ruthenium red.

ammoniated superphosphate. Fertilizer produced by mixing ammonia with superphosphate in the ratio of 5 parts to 100.

ammonia water. A solution of the gas ammonia (NH_3) in water, also known as ammonium hydroxide (q. v.).

ammonio-cupric sulfate. See copper sulfate ammoniate.

ammonio-ferric oxalate. See ferric-ammonium oxalate.

ammonio-ferric sulfate. See ferric-ammonium sulfate.

ammonio-formaldehyde. See hexamethylenetetramine.

ammonium acetate $NH_4(C_2H_3O_2)$.
Properties: White, deliquescent, crystalline mass. Soluble in water, alcohol.
Constants: M.p. 114°C; sp. gr. 1.073.
Derivation: By the interaction of glacial acetic acid and ammonia gas.
Method of purification: Crystallization.
Grades: Technical; C.P.
Containers: Drums.

*See "I.C.C. Shipping Regulations," page xiii.
Reference numbers refer to name of manufacturer. See "List of Manufacturers," page v.

Uses: Reagent in analytical chemistry; drugs; textile dyeing; preserving meats; foam rubbers; vinyl plastics; explosives.
Shipping regulations: None. *

ammonium acid camphorate. See ammonium camphorate.

ammonium acid carbonate. See ammonium bicarbonate.

ammonium acid fluoride. See ammonium bifluoride.

ammonium acid phosphate. See ammonium phosphate, monobasic.

ammonium alum. See aluminum ammonium sulfate.

ammonium-aluminum chloride. See aluminum-ammonium chloride.

ammonium arsenate $(NH_4)_2HAsO_4$.
Properties: White crystals or powder efflorescing in air with loss of ammonia; sp. gr. 1.99; poisonous! Soluble in water; decomposes in hot water.
Derivation: Interaction of arsenic acid with ammonia.
Grades: C. P.
Containers: 1-lb bottles.
Use: Medicine.
Shipping regulations: Class B poison. Poison label.*

ammonium atreolate.
Properties: Black, syrupy liquid. Bitter taste. Soluble in water and alcohol; miscible with lard, glycerin, petrolatum and wool fat; insoluble in chloroform.
Derivation: Aqueous solution which contains the ammonia salts of a mixture of organic acids. These organic acids are obtained by the action of sulfuric acid on certain petroleum distillates.
Use: Medicine.

ammonium benzenesulfonate (ammonium sulfonate) $C_6H_5SO_3NH_4$. An ashless sulfonate, marketed as a 35% solution in kerosene.

ammonium benzoate $C_6H_5COONH_4$.
Properties: White crystals or powder. Soluble in water, alcohol and glycerol.
Constants: Decomposes at 198°C; sp. gr. 1.260; sublimes at 160°C.
Derivation: By the action of ammonium hydroxide on benzoic acid with subsequent crystallization.
Method of purification: Recrystallization.
Grades: Technical.
Containers: Bottles; fiber drums.
Uses: Medicine; food-preservative; latex and adhesive.
Shipping regulations: None. *

ammonium biborate. See ammonium borate.

ammonium bicamphorate. See ammonium camphorate.

ammonium bicarbonate (ammonium acid carbonate; ammonium hydrogen carbonate) NH_4HCO_3.
Properties: White crystals. Soluble in water; insoluble in alcohol.
Constants: Sp. gr. 1.586; m.p. decomposes at 36° to 60°C.
Derivation: By heating ammonium hydroxide with an excess of carbon dioxide, and evaporation.
Method of purification: Recrystallization.
Impurities: Ammonium carbonate.
Grades: Technical; C. P.; food grade.
Containers: 100-lb drums; carloads.
Uses: Production of ammonium salts; dyes; substitute for yeast in baking; ingredient of fire extinguishing compounds; pharmaceuticals; degreasing textiles; inflater for rubber.
Shipping regulations: None. *

ammonium bichromate. See ammonium dichromate.

ammonium bifluoride (ammonium acid fluoride) $(NH_4)FHF$.
Properties: White crystals, deliquescent; sp. gr. 1.211; decomposed by heat; poisonous; soluble in cold water and alcohol; decomposes in hot water.
Derivation: Action of ammonium hydroxide on hydrofluoric acid with subsequent crystallization.
Method of purification: Recrystallization.
Grades: Technical.
Containers: Drums.
Uses: Ceramics; chemical reagent; etching glass (white acid); sterilizer for brewery, dairy and other equipment; electroplating; processing beryllium.
Shipping regulations: None. *

ammonium binoxalate $(NH_4)HC_2O_4 \cdot H_2O$.
Properties: Colorless crystals. Soluble in water.
Constants: Sp. gr. 1.556; decomposes on heating.
Derivation: Action of ammonium hydroxide on oxalic acid with subsequent crystallization.
Grades: Technical; pure.
Containers: Glass bottles; 25-, 100-lb drums.
Uses: Analytical chemistry; removing ink stains from fabrics.
Shipping regulations: None. *

ammonium biphosphate. See ammonium phosphate, monobasic.

ammonium bisulfate NH_4HSO_4.
Properties: White crystals, deliquescent; sp. gr. 1.78; decomposes at 120°C. Soluble in water; forms strongly acid solution.
Grades: C. P.
Containers: 1-, 5-lb bottles.
Use: Medicine.
Shipping regulations: None. *

ammonium bisulfide. See ammonium sulfide.

ammonium bisulfite NH_4HSO_3.
Properties: White crystals easily decomposed by heat. Soluble in water and alcohol.
Grades: C. P.; 45-47% solution.
Containers: 1-, 5-lb bottles; tanks.
Use: Medicine; preservative.

ammonium bitartrate (acid ammonium tartrate) $(NH_4)HC_4H_4O_6$.
Properties: White crystals. Soluble in water, acids and alkalies; insoluble in alcohol.
Constants: Sp. gr. 1.636.
Derivation: By the action of ammonium hydroxide on tartaric acid.
Method of purification: Crystallization.
Grades: Technical; C.P.
Containers: 1-lb bottles; barrels; multiwall paper sacks.
Use: Baking powder.
Shipping regulations: None.*

ammonium borate (ammonium biborate) $NH_4HB_4O_7 \cdot 3H_2O$.
Properties: Colorless crystals; efflorescent with loss of ammonia. Soluble in water.
Constants: Sp. gr. 2.38-2.95.
Derivation: By the action of ammonium hydroxide on boric acid with subsequent crystallization.
Method of purification: Recrystallization.
Grades: Technical; C.P.
Containers: Glass bottles; 100-lb drums.
Uses: Medicine; fireproofing compounds; electrical condensers.
Shipping regulations: None.*

ammonium bromide NH_4Br.
Properties: Colorless, odorless crystals or a yellowish white crystalline powder; soluble in water and alcohol; somewhat hygroscopic.
Constants: Sp. gr. 2.43; m.p., sublimes.
Derivation: Action of hydrobromic acid on ammonium hydroxide with subsequent crystallization.
Method of purification: Recrystallization.
Grades: Technical; pure; C.P.; N.F. XI.
Containers: 25-, 50-, 150-lb drums; 400-lb barrels.
Uses: Precipitating silver salts for photographic plates; medicine (for its bromide ion); analytical chemistry; process engraving; lithographic work; textile finishing; and fire retardant.
Shipping regulations: None.*

ammonium-cadmium bromide. See cadmium-ammonium bromide.

ammonium camphorate (ammonium bicamphorate; ammonium acid camphorate) $NH_4HC_{10}H_{14}O \cdot 3H_2O$.
Properties: White crystalline powder. Soluble in water.
Use: Medicine.
Shipping regulations: None.*

ammonium carbamate $NH_4CO_2NH_2$.
Properties: White, crystalline rhombic powder; exceedingly volatile; the "anhydride" of ammonium carbonate; forms urea on heating. Soluble in water and alcohol.
Derivation: By the interaction of dry ammonia gas and carbon dioxide. Is recovered from gas liquor with ammonia and ammonium carbonate.
Grades: Technical.
Containers: Iron drums.
Use: Fertilizer.
Shipping regulations: None.*

ammonium carbazotate. See ammonium picrate.

ammonium carbonate (crystal ammonia; ammonium sesquicarbonate; hartshorn). A mixture of ammonium acid carbonate and ammonium carbamate, $(NH_4)HCO_3 \cdot (NH_4)(NH_2)CO_2$.
Properties: Colorless crystal plates or white powder; unstable in air, being converted into the bicarbonate. Strong odor of ammonia, sharp ammoniacal taste. Soluble in water; decomposes in hot water, yielding ammonia and carbon dioxide.
Derivation: Ammonium salts are heated with calcium carbonate.
Method of purification: Sublimation.
Grades: Technical; lumps; cubes; powder; C.P.; U.S.P. XVI.
Containers: Bottles; 5-, 10-, 25-, 100-lb kegs; 250-, 375-, 500-lb barrels; drums.
Uses: Ammonium salts; medicine; baking powders; smelling salts; rubber manufacture; manufacture of casein colors, casein glues and other adhesives; tanning; cleansing powders; fire extinguishing compounds; pharmaceuticals; textiles (mordant, washing fabrics); fermentation accelerator in wine manufacture; also, organic chemicals, ceramics, in washing wool.
Shipping regulations: None.*

ammonium-cesium bromide. See cesium ammonium bromide.

ammonium-cesium-rubidium bromide. See cesium-rubidium-ammonium bromide.

ammonium chloride (sal ammoniac; ammonium muriate) NH_4Cl.
Properties: White crystals; cool, saline taste; somewhat hygroscopic. Soluble in water and glycerol; slightly soluble in alcohol.
Constants: Sublimes 350°C; sp. gr. 1.54.
Derivation: (a) As a by-product of the ammonia-soda process (see soda ash and sodium bicarbonate); (b) reaction of ammonium sulfate and sodium chloride solutions.
Method of purification: Crystallization.
Grades: Technical (lumps or granulated); C.P.; U.S.P. XVI.
Containers: Barrels; multiwall paper sacks.
Uses: Dry batteries; mordant (dyeing and printing); soldering flux; manufacture of various ammonia compounds, etc.; dyes; fertilizer; tanning; manufacture of rust cement for pipe joints; pickling agent in zinc coating and tinning; electroplating; candle manufacture; washing powders; snow treatment; resins and adhesives of urea-formaldehyde; medicine.
Shipping regulations: None.*

ammonium chloroplatinate. See platinic-ammonium chloride.

ammonium chloroplatinite. See platinous-ammonium chloride.

ammonium chromate $(NH_4)_2CrO_4$.
Properties: Yellow crystals; soluble in cold water; insoluble in alcohol.
Constants: Sp. gr. 1.866; m.p., decomposes.
Derivation: By the addition of ammonium hydroxide to a solution of ammonium bichromate; recovery by crystallization.
Method of purification: Recrystallization.
Impurities: Dichromates.
Grades: Technical; C.P.
Containers: 1-, 5-lb bottles; barrels; boxes.
Uses: Mordant in dyeing; photography (sensitizer for gelatin coatings); reagent in analytical chemistry; catalyst; corrosion inhibitor.
Shipping regulations: None.*

ammonium chrome alum. See chromium ammonium sulfate.

ammonium-chromium sulfate. See chromium-ammonium sulfate.

ammonium citrate, dibasic $(NH_4)_2HC_6H_5O_7$.
Properties: White granules; soluble in water; very slightly soluble in alcohol.
Preparation: From citric acid by partial neutralization.
Grades: Pure; reagent.
Containers: Bottles; fiber drums; kegs; barrels.
Uses: Pharmaceuticals; rustproofing; cotton printing; plasticizer; analytically in determination of phosphate in fertilizer.
Shipping regulations: None.*

ammonium-cobalt sulfate. See cobaltous ammonium sulfate.

ammonium-copper chloride. See copper-ammonium chloride.

ammonium-cupric chloride. See copper-ammonium chloride.

ammonium decaborate. See ammonium pentaborate.

ammonium dichromate (ammonium bichromate) $(NH_4)_2Cr_2O_7$.
Properties: Orange needles; soluble in water; insoluble in alcohol. May react explosively with certain organic compounds.
Constants: Sp. gr. 2.152 (25°C); m.p., decomposes with slight heating.
Derivation: Action of chromic acid on ammonium hydroxide with subsequent crystallization.
Method of purification: Recrystallization.
Grades: Technical; C.P.
Containers: Various bottles and cartons; drums; multiwall paper sacks.
Uses: Mordant for dyeing; manufacturing of alizarin; chrome alum; smokeless sporting powders; oil purification; pickling; candle wicks; leather tanning; synthetic perfumes; photography; process engraving; lithographic work; chromic oxide; pyrotechnics.
Warning! Flammable; harmful dust may cause rash or external ulcers. MCA warning label.
Shipping regulations: Flammable solid. Yellow label.*

ammonium dimethyldithiocarbamate $(CH_3)_2NCS_2NH_4$.
Grades: 42% solution in water.
Containers: 5-gal cans, 55-gal drums.
Use: Fungicide.

ammonium ethyl phosphate solution.
Properties: Water-white to yellow-tinged liquid of ammoniacal odor; sp. gr. 1.23 at 25°C; wt/gal 10.37 lbs; pH (75% soln) 7.0-7.2; viscosity (25°C), 60 cps.
Chemical constitution: 67% monoethyl ammonium phosphate; 33% diethyl ammonium phosphate. In thin film is liquid at relative humidities above 40%; not stable to heat; when subjected to elevated temperatures for any length of time as a solution or while on any article, will liberate ammonia with a reduction in pH; at temperatures below 15°C crystals will form in the solution and will dissolve with a rise in temperature or on dilution.
Standard Grade: 75% solution in water.
Chief constituents: Mixture of mono- and diethyl ammonium phosphates.
Containers: 500-lb drums.
Use: Flameproofing agent for paper, textiles, wood and other combustible materials.

ammonium fluoride NH_4F.
Properties: White crystals; sp. gr. 1.31; decomposed by heat; soluble in cold water. Poisonous!
Derivation: Interaction of ammonium hydroxide and hydrofluoric acid with subsequent crystallization.
Method of purification: Recrystallization.
Grades: Technical; C.P.
Containers: 1-lb waxed bottles; 250-lb barrels.
Uses: Fluorides; analytical chemistry; in agriculture for improving sandy soils; antiseptic in brewing; etching glass; textile mordant; wood preservation.
Shipping regulations: None.*

ammonium fluosilicate (ammonium silicofluoride; cryptohalite) $(NH_4)_2SiF_6$.
Properties: White, crystalline powder; sp. gr. 2.01; soluble in alcohol and water.
Containers: Drums.
Uses: Laundry sours; moth proofing; disinfectant in brewing industry; glass etching; light metal casting; electroplating.

ammonium gluconate $NH_4C_6H_{11}O_7$.
Properties: White powder; soluble in water; insoluble in alcohol.
Constants: Optical rotation +11.6° (in water).
Preparation: From gluconic acid by neutralization with ammonia.
Grades: Pure; technical.
Containers: Bottles; fiber drums; kegs; barrels.
Use: In foods as an emulsifying agent for cheese and salad dressings.

ammonium hexanitratocerate. See ceric ammonium nitrate.

ammonium hydrate. See ammonium hydroxide.

*See "I.C.C. Shipping Regulations," page xiii.
Reference numbers refer to name of manufacturer. See "List of Manufacturers," page v.

ammonium hydrogen carbonate. See ammonium bicarbonate.

ammonium hydrosulfide. See ammonium sulfide.

ammonium hydroxide (ammonia solution; aqua ammonia; ammonium hydrate) NH_4OH.
Properties: Colorless liquid; strong characteristic odor. Concentrations of solutions range up to about 30% ammonia.
Derivation: Ammonia gas is dissolved in water.
Grades: Technical; C. P.; 16°; 20°; 26°; U. S. P. XVI (strong and diluted).
Containers: 1-, 4-lb bottles; 5-, 10-gal carboys; 55-, 110-gal drums; 8,000-gal tank cars.
Uses: Textiles; manufacture of rayon, rubber, fertilizers; refrigeration; plastics (condensation agent for resins, accelerator in phenol condensation process); photography (development of latent images); pharmaceutical (smelling salts, aromatic spirits of ammonia, various preparations); ammonia soaps; lubricants; fireproofing wood; ink manufacture; explosives; ceramics; ammonium compounds; saponifying fats and oils; organic synthesis; detergent.
Warning! Liquid causes burns; vapor ex tremely irritating. MCA warning label.
Shipping regulations: None. *

ammonium hypophosphite $NH_4H_2PO_2$.
Properties: Colorless crystals or white powder; decomposes when heated, with evolution of phosphine which ignites spontaneously. Soluble in water and alcohol.
Derivation: Neutralizing hypophosphorous acid with ammonium hydroxide.
Grade: 98% pure.
Use: Medicine.
Shipping regulations: None. *

ammonium hyposulfite. See ammonium thiosulfate.

ammonium ichthosulfonate. See ichthammol.

ammonium iodide NH_4I.
Properties: White crystals or white granular powder; soluble in water or alcohol.
Constants: Sp. gr. 2.56; m. p., sublimes with decomposition.
Derivation: Action of ammonium hydroxide on hydriodic acid with subsequent crystallization.
Method of purification: Recrystallization.
Grades: Technical; C. P.
Containers: 1-oz, 1/4-, 1-, 5-, 7-lb bottles; 25-lb jars.
Uses: Iodides; medicine; photography.
Caution: Store away from light.
Shipping regulations: None. *

ammonium-iron tartrate. See ferric-ammonium tartrate.

ammonium lactate $NH_4C_3H_5O_3$.
Properties: Colorless to yellowish syrupy liquid. Slight odor of ammonia. Decomposes when hot. Sp. gr. 1.19-1.21 (15°C). Soluble in water and alcohol.
Grades: Technical.
Containers: 1-, 5-lb bottles.
Uses: Electroplating, and in leather tanning and finishing.

ammonium laurate, anhydrous $C_{11}H_{23}COONH_4$.
Properties: Tan colored, wax-like material, free from ammonia odor. Soluble in ethyl alcohol, methyl alcohol, cottonseed oil, and mineral oil. Soluble (hot) in naphtha, toluene, and vegetable oil.
Constants: Sp. gr. (25°C) 0.88; pH (5% dispersion) 7.6-7.8; m. p. 42-56°C; neut. value 120-125.
Grades: Technical.
Containers: 1-gal cans (7lbs); 5-gal cans (45 lbs); 55-gal drums (360 lbs).
Use: Emulsifying agent for the production of oil-in-water emulsions with a high oil content; cosmetics.

ammonium lignin sulfonate. See lignin sulfonates.

ammonium linoleate ($C_{17}H_{31}COONH_4$).
Properties: Yellow-colored paste with an ammoniacal odor. Soluble in water, ethyl alcohol, methyl alcohol. Emulsifies in naphtha, toluene, mineral oil, and vegetable oil.
Constants: Sp. gr. 1.1; pH (5% dispersion) 9.5-9.8; total solids 82%.
Grades: Technical, 80%.
Containers: 1-, 5-, 55-gal drums.
Uses: Emulsifying agent for oils, waxes, and hydrocarbon solvents for industrial purposes; surface tension reducer; detergent; water-repellent finishes.

ammonium-magnesium sulfate. See magnesium-ammonium sulfate.

ammonium metavanadate (ammonium vanadate) NH_4VO_3.
Properties: White crystals; insoluble in saturated ammonium chloride solution; slightly soluble in cold water.
Constants: Sp. gr. 2.326; m. p., breaks up at 210°C.
Derivation: Alkali solutions of V_2O_5 and precipitation with ammonium chloride.
Method of purification: Recrystallization.
Grades: Technical; C. P.
Containers: Technical: compressed paper drums; C. P.: glass bottles.
Uses: Raw material for catalyst as V_2O_5; dyes; varnishes; drier for paints and inks; photography.

ammonium molybdate (molybdic acid 85%) $(NH_4)_6Mo_7O_{24} \cdot 4H_2O$. The reagent grade contains 85% MoO_3.
Properties: White crystalline powder; soluble in water; insoluble in alcohol.
Constants: Sp. gr. 2.38-2.95; m. p. decomposes.
Derivation: Dissolving molybdenum trioxide in aqueous ammonia.
Method of purification: Recrystallization.
Grades: Technical; C. P.; reagent (molybdic acid 85%, A. C. S.).
Containers: Glass bottles; boxes; drums; barrels.
Uses: Analytical reagent; pigments; catalyst

for dehydrogenation and desulfurization in petroleum and coal technology; production of molybdenum metal; sources of molybdate ions.
Shipping regulations: None.*

ammonium muriate. See ammonium chloride.

ammonium-nickel chloride. See nickel-ammonium chloride.

ammonium-nickel sulfate. See nickel-ammonium sulfate.

ammonium nitrate (Norway saltpeter) NH_4NO_3.
Properties: Colorless crystals; explosive! but not readily detonated. Soluble in water, alcohol, and alkalies.
Constants: Sp. gr. 1.725; m.p. 169.6°C; b.p., decomposes at 210°C.
Derivation: By the action of ammonia vapor on nitric acid.
Method of purification: Crystallization.
Grades: Usually expressed in percent of nitrogen, as 20.5% N; 33.5% N. FGAN is a fertilizer grade, prilled and usually coated with kieselguhr.
Containers: Bags; carloads.
Uses: Fertilizer; explosives; pyrotechnics; weedicides and insecticides; manufacture of nitrous oxide; absorbent for nitrogen oxides; ingredient of freezing mixtures; oxidizer in solid rocket propellants; nutrient for antibiotics and yeast; catalyst.
Fire hazard: Dangerous. Has been the cause of large serious explosions.
Shipping regulations: Oxidizing material. Yellow label. Very special packaging regulations.*

ammonium nitroso-beta**-phenyl hydroxylamine.** See cupferron.

ammonium oleate (approximately) $C_{17}H_{33}COONH_4$. An ammonium soap.
Properties: Yellow to brownish, ointment-like mass; ammonia odor; decomposes on heating. Soluble in water and alcohol.
Containers: Barrels.
Uses: Emulsifying agent; cosmetics.

ammonium oxalate $(NH_4)_2C_2O_4 \cdot H_2O$.
Properties: Colorless crystals; poisonous! soluble in water. Sp. gr. 1.502; decomposed by heat.
Derivation: Interaction of ammonium hydroxide and oxalic acid and subsequent crystallization.
Method of purification: Recrystallization.
Grades: Technical; pure; C.P.
Containers: 200-, 250-lb drums.
Uses: Analytical chemistry; safety explosives; manufacture of oxalates; rust and scale removal from metals.
Shipping regulations: None.*

ammonium paratungstate. See ammonium tungstate.

ammonium pentaborate (ammonium decaborate) $(NH_4)_2B_{10}O_{16} \cdot 8H_2O$.
Properties: Crystals; soluble in water.
Containers: 100-lb bags.
Use: Intermediate for boron chemicals; as a "control" in atomic submarines.

ammonium perchlorate (AP; APC) NH_4ClO_4.
Properties: White crystals; explosive! Soluble in water. Sp. gr. 1.95; m.p., decomposes on heating.
Derivation: By the interaction of ammonium hydroxide and perchloric acid or from sodium chlorate. Recovery by crystallization.
Method of purification: Recrystallization.
Grades: Technical; C.P.
Containers: Glass bottles; drums.
Uses: Explosives; pyrotechnics; analytical chemistry; etching and engraving agent; as oxidizer in solid propellants.
Fire hazard! Dangerous; oxidizing material; may explode in a fire.
Shipping regulations: Oxidizing material. Yellow label.*

ammonium permanganate.
Shipping regulations: Oxidizing material. Yellow label.*

ammonium persulfate $(NH_4)_2S_2O_8$.
Properties: White crystals; strong oxidizing agent; soluble in water. Sp. gr. 1.98; m.p., decomposes.
Derivation: Electrolysis of a concentrated solution of ammonium sulfate. Recovered by crystallization.
Method of purification: Recrystallization.
Grades: Technical; pure; C.P.
Containers: Bottles; drums.
Uses: Oxidizing agent; bleaching agent; photography; oxidizing copper; electroplating; manufacture of other persulfates; deodorizing and bleaching oils; aniline dyes; preserving food; depolarizer in batteries; washing infected yeast.
Caution! Strong oxidizing agent.

ammonium phosphate. See ammonium phosphate, dibasic; ammonium phosphate, hemibasic; ammonium phosphate, monobasic.

ammonium phosphate, dibasic (ammonium phosphate, secondary; diammonium hydrogen phosphate; diammonium phosphate) $(NH_4)_2HPO_4$.
Properties: White crystals or powder; sp. gr. 1.619; mildly alkaline in reaction; soluble in water; insoluble in alcohol.
Derivation: Interaction of ammonium hydroxide and phosphoric acid in proper proportions.
Method of purification: Recrystallization.
Grades: Technical; C.P.; dentifrice grade; highly purified, for phosphors.
Containers: 1-, 5-lb bottles; 200-lb bags; 350-lb barrels; drums.
Uses: Medicine (ingredient in compound syrups); in the impregnation of wood, paper, and textiles to render them nonflammable; to prevent afterglow in matches, and smoking of candlewicks; fertilizer; in plant nutrient solutions; manufacture of yeast, vinegar, yeast foods, and bread improvers; flux for soldering tin, copper, brass, zinc; purifying sugar; in ammoniated dentifrices; to make halophosphate phosphors.

*See "I.C.C. Shipping Regulations," page xiii.
Reference numbers refer to name of manufacturer. See "List of Manufacturers," page v.

ammonium phosphate, hemibasic $NH_4H_2PO_4 \cdot H_3PO_4$.
Properties: White crystalline material; somewhat hygroscopic. Strongly acid in reaction. Soluble in water.
Containers: Barrels and kegs with moisture-proof liners; 100-lb bags.
Uses: Nutrient for truck gardens; yeast food; buffer for adjustment of pH values; metal cleaning.

ammonium phosphate, monobasic (ammonium acid phosphate; ammonium biphosphate; ammonium phosphate, primary) $NH_4H_2PO_4$.
Properties: Brilliant, white crystals or powder. Mildly acid in reaction. Moderately soluble in water; sp. gr. 1.803.
Derivation: Interaction of phosphoric acid and ammonia in proper proportions.
Grades: Technical; C. P.
Containers: Barrels; kegs; multiwall paper sacks.
Uses: Fertilizer; in the impregnation of wood, paper, and textiles to render them nonflammable; to prevent afterglow in matches, and smoking of candlewicks; in plant nutrient solutions; manufacture of yeast, vinegar, yeast foods, and bread improvers; medicine.

ammonium phosphate, primary. See ammonium phosphate, monobasic.

ammonium phosphate, secondary. See ammonium phosphate, dibasic.

ammonium phosphite (neutral ammonium phosphite) $(NH_4)_2HPO_3 \cdot H_2O$.
Properties: Colorless, crystalline mass. Hygroscopic; keep tightly closed. Soluble in water.
Grades: Technical.
Use: Chemical (reducing agent).

ammonium phosphite, neutral. See ammonium phosphite.

ammonium phosphomolybdate $(NH_4)_3PO_4 \cdot 12MoO_3 \cdot 3H_2O$.
Properties: Yellow crystalline powder; soluble in alkali; insoluble in alcohol and acids; very slightly soluble in water.
Derivation: By the interaction of ammonium molybdate and phosphoric and nitric acids.
Method of purification: Crystallization.
Grades: Technical; C. P.
Containers: Tins; glass bottles.
Use: Reagent for alkaloids.
Shipping regulations: None. *

ammonium phosphotungstate (ammonium phosphowolframate) $2(NH_4)_3PO_4 \cdot 24WO_3 \cdot xH_2O$.
Properties: White powder. Soluble in alkali; insoluble in acid; slightly soluble in water.
Derivation: By the interaction of ammonium tungstate, ammonium phosphate and nitric acid.
Method of purification: Crystallization.
Grades: Technical.
Containers: Glass bottles; tins.
Use: Chemical reagent.
Shipping regulations: None. *

ammonium phosphowolframate. See ammonium phosphotungstate.

ammonium picrate (ammonium carbazotate; ammonium picronitrate) $C_6H_2(NO_2)_3ONH_4$.
Properties: Yellow crystals; highly explosive! Sp. gr. 1.72; m. p., decomposes; soluble in water and alcohol.
Derivation: By the action of ammonium hydroxide on picric acid with subsequent crystallization.
Method of purification: Recrystallization.
Grades: Technical.
Containers: Wooden kegs or boxes free from metal nails or screws.
Use: Explosives; medicine.
Fire hazard: Dangerous; easily exploded by heat or shock.
Shipping regulations: Explosive, class A. High explosive label. *

ammonium picronitrate. See ammonium picrate.

ammonium polysulfide $(NH_4)_2S_x$.
Properties: Known only in solution; yellow, unstable; having hydrogen sulfide odor; decomposed by acids with deposition of sulfur; poisonous.
Derivation: Passing hydrogen sulfide into 28% ammonium hydroxide and dissolving an excess of sulfur in the resulting solution.
Uses: Analytical reagent; insecticide spray.

ammonium reineckate. See reinecke salt.

ammonium rhodanide. See ammonium thiocyanate.

ammonium ricinoleate ($C_{17}H_{32}OHCOONH_4$).
Properties: White paste.
Grades: Technical.
Uses: Detergent; emulsifying agent.

ammonium salicylate $C_6H_4OHCOONH_4$.
Properties: Colorless crystals and white powder with pink tinge; odorless; stable in dry air, but affected by light. Soluble in water and alcohol.
Derivation: By the action of ammonium hydroxide on salicylic acid.
Method of purification: Crystallization.
Grades: Technical.
Containers: 5-, 25-lb cartons; 25-, 50-, 100-lb drums; 200-lb barrels.
Use: Medicine.
Shipping regulations: None. *

ammonium salts. The salts formed on direct union of ammonia or neutralization of ammonium hydroxide with acids. They are in general white salts and soluble in water. Usually decomposed by heat into ammonia and the corresponding acid, which may also be decomposed. All ammonium salts liberate ammonia (NH_3) when heated with a strong base, e. g., sodium hydroxide or calcium hydroxide.

ammonium selenate $(NH_4)_2SeO_4$.
Properties: Colorless crystals; sp. gr. 2.194; soluble in water; insoluble in alcohol.
Use: Mothproofing agent.

ammonium selenite $(NH_4)_2SeO_3 \cdot H_2O$.
Properties: Colorless or slightly reddish crystals. Caution! Keep away from dust or light! Soluble in water.
Grades: Technical.
Uses: Analysis (test for alkaloids); glass industry (red glass).

ammonium sesquicarbonate. See ammonium carbonate.

ammonium silicofluoride. See ammonium fluosilicate.

ammonium soaps. Product resulting from the reaction of an appropriate fatty acid with ammonium hydroxide. The resulting soaps have an appreciable vapor pressure of ammonia and decompose on continual exposure leaving the fatty acid residue. Usually not sold as detergents but find application in the manufacture of toilet preparations and emulsions.

ammonium-sodium sulfate. See sodium-ammonium sulfate.

ammonium stearate $C_{17}H_{35}COONH_4$.
Properties: Tan colored, wax-like solid, free from ammonia odor. Dispensible in hot water. Soluble (hot) in toluene; partially soluble (hot) in butyl acetate and ethyl alcohol.
Constants: Sp. gr. (22°C) 0.89; pH (3% dispersion) 7.6; m.p. 73-75°C; neutralization value 70-80.
Grades: Available as anhydrous solid or as paste.
Containers: 1-gal can (7 ½ lbs); 5-gal can (45 lbs); 55-gal drum (360 lbs).
Uses: In the manufacture of vanishing creams, brushless shaving creams, and other cosmetic products. Integral waterproofing of cements, concrete, stucco, etc.

ammonium sulfamate $NH_4OSO_2NH_2$.
Properties: White crystalline solid; m.p. 130°C, decomposes at 160°C; acidified solutions, when enclosed and heated, may explode. Soluble in water.
Derivation: Hydrolysis of the reaction product obtained when urea is treated with fuming sulfuric acid.
Grades: Technical; C.P.
Containers: 50- and 100-lb fiber drums; 350-lb wooden barrels.
Uses: Flameproofing agent for textiles and certain grades of paper; weed killer; electroplating; generation of nitrous oxide gas.

ammonium sulfate $(NH_4)_2SO_4$.
Properties: Brownish-gray to white crystals according to degree of purity. Soluble in water; insoluble in alcohol and acetone. Sp. gr. 1.77; m.p. 513°C with decomposition.
Derivation: (a) The ammoniacal vapors from the destructive distillation of coal are led into sulfuric acid, followed by crystallization and drying. This forms the crude ammonium sulfate of commerce. (b) Ammonia derived from atmospheric nitrogen by fixation is neutralized with sulfuric acid. (c) Ammonia and carbon dioxide are passed into a slurry of gypsum, converting the latter to insoluble calcium carbonate and a solution of ammonium sulfate, which is filtered and crystallized. The ammonia is sometimes obtained from refinery wastes or natural gas. (d) European process uses sulfurous gases (from smelters and roasters) absorbed in organic base like xylidine. This is oxidized by air and decomposed by ammonia.
Method of purifying: Recrystallization or sublimation.
Grades: Commercial; technical; C.P.; enzyme grade (no heavy metals).
Containers: 1-, 5-lb bottles; 25-lb boxes; 100-lb kegs; 200-lb bags; 300-, 400-lb barrels; carload; multiwall paper sacks.
Uses: Fertilizer; water treatment; fermentation; charging electric batteries; soldering liquids; galvanizing iron; candle manufacture; fireproofing compositions; viscose rayon; tanning.
Shipping regulations: None. *

ammonium sulfide $(NH_4)_2S$. The true sulfide is stable only in the absence of moisture and below 0°C. The ammonium sulfide of commerce is largely ammonium bisulfide or hydrosulfide, NH_4HS.
Properties: Yellow crystals. Soluble in water, alcohol, and alkalies. M.p., decomposes.
Derivation: By the interaction of ammonium hydroxide and hydrogen sulfide.
Method of purification: Crystallization.
Grades: Technical; C.P.; liquid, 40-44%.
Containers: Iron drums; tins; glass bottles; tanks.
Uses: Textile industry; photography (developers); coloring brasses, bronzes; iron control in soda ash production.
Shipping regulations: None. *

ammonium sulfite $(NH_4)_2SO_3 \cdot H_2O$.
Properties: Colorless crystals; acrid, sulfurous taste. Hygroscopic; sublimes at 150°C with decomposition; soluble in water. Caution! Keep well stoppered! Sp. gr. 1.41.
Containers: Multiwall paper sacks; 25-, 100-, 175-lb drums.
Grades: Technical; C.P.
Uses: Chemical (intermediates, reducing agent); medicine; permanent wave solutions; photography.

ammonium sulfocyanate. See ammonium thiocyanate.

ammonium sulfocyanide. See ammonium thiocyanate.

ammonium sulfoichthyolate. See ichthammol.

ammonium sulfonate. See ammonium benzenesulfonate.

ammonium sulforicinoleate.
Properties: Yellow liquid; soluble in alcohol.
Grades: Technical.
Containers: 400-lb barrels.
Use: Medicine; furniture polish.

*See "I.C.C. Shipping Regulations," page xiii.
Reference numbers refer to name of manufacturer. See "List of Manufacturers," page v.

ammonium tartrate $(NH_4)_2C_4H_4O_6$.
Properties: White crystals; soluble in water and alcohol. Sp. gr. 1.601; decomposes on heating.
Derivation: By the action of tartaric acid on ammonium hydroxide with subsequent crystallization.
Method of purification: Recrystallization.
Grades: Technical; C. P.
Containers: 1-lb bottles; barrels; boxes.
Uses: Textile industry; medicine.
Shipping regulations: None. *

ammonium tetrathiocyanodiammono-chromate. See Reinecke salt.

ammonium thiocyanate (ammonium sulfocyanide; ammonium rhodanide; ammonium sulfocyanate) NH_4SCN.
Properties: Colorless, deliquescent crystals; soluble in water, alcohol, acetone, and ammonia.
Constants: Sp. gr. 1.3057; m.p. 149.6°C, decomposes at 170°C.
Derivation: By boiling an aqueous solution of ammonium cyanide with sulfur or polysulfides, or by the reaction of ammonia and carbon disulfide.
Method of purification: Crystallization.
Grades: Technical; C. P.; 50-60% solution.
Containers: 50-, 100-lb bags; plastic-lined kegs; drums; solution in tank cars.
Uses: Analytical chemistry; chemicals (thiourea, guanidine sulfocyanate); fertilizers; photography; ingredient of freezing solutions, especially liquid rocket propellants; coating zinc grayish-black; textiles; weed killer and defoliant; adhesives; curing resins; pickling iron and steel; electroplating; temporary soil sterilizer; polymerization catalyst; separator of zirconium and hafnium, of gold and iron.
Shipping regulations: None. *

ammonium thioglycolate $HSCH_2COONH_4$.
Containers: 55-gal drums, commercial grade.
Use: Hair-waving formulations.

ammonium thiosulfate (ammonium hyposulfite) $(NH_4)_2S_2O_3$.
Properties: White crystals, decomposed by heat; very soluble in water; pH of 60% solution 6.5-7.0.
Impurities: Ammonium sulfite, 1.0% max; heavy metals, trace.
Grades: Pure crystals (97%); 60% photographic solution.
Containers: 1-, 5-lb bottles; drums.
Uses: Photographic fixing agent; analytical reagent; fungicide; reducing agent; brightener in silver plating baths; cleaning compounds for zinc-base die-cast metals; hair waving formulas; fog screens.
Shipping regulations: None. *

ammonium tungstate (ammonium wolframate; ammonium paratungstate) $(NH_4)_6W_7O_{24}\cdot 6H_2O$.
Properties: White crystals; soluble in water; insoluble in alcohol.
Derivation: Interaction of ammonium hydroxide and tungstic acid with subsequent crystallization.
Method of purification: Recrystallization.
Grades: Technical; C. P.
Containers: Wooden kegs.
Uses: Preparation of ammonium phosphotungstate and other tungsten compounds.
Shipping regulations: None. *

ammonium-uranium carbonate. See uranyl ammonium carbonate.

ammonium-uranium fluoride. See uranium-ammonium fluoride.

ammonium valerate (ammonium valerianate; acid ammonium valerate) $NH_4C_5H_9O_2\cdot 2HC_5H_9O_2$.
Properties: Colorless, very deliquescent crystals; pungent unpleasant odor (as of valeric acid); sharp, sweet taste; acid reaction. Keep well stoppered! Soluble in water, alcohol, and ether. When saturated aqueous solution is diluted with 5 vols. of water, free valeric acid separates out in an oily liquid.
Derivation: Interaction of ammonium hydroxide and valeric acid with subsequent crystallization.
Method of purification: Recrystallization.
Grades: Technical.
Containers: 1-, 5-lb glass bottles.
Uses: Medicine; imitation butter flavors.
Shipping regulations: None. *

ammonium valerianate. See ammonium valerate.

ammonium vanadate. See ammonium metavanadate.

ammonium wolframate. See ammonium tungstate.

ammonium-zinc sulfate. See zinc-ammonium sulfate.

ammonium zirconifluoride. See zirconium-ammonium fluoride.

ammonium zirconyl carbonate solution $(NH_4)_3ZrOH(CO_3)_3\cdot 2H_2O$, in water.
Properties: A clear solution; sp. gr. 1.238 (24°C). Stable up to about 60°C; decomposes in dilute acids, alkalies.
Containers: 500-lb drums.
Uses: Ingredient in water repellents for textiles and leather; catalyst; pharmaceutical preparations; ingredient in latex emulsion paints; preparation of zirconium metal and alloys; dyestuff industry.

"Ammonyx 27." [328] Trade name for tallow trimethyl ammonium chloride.
Uses: Germicide; de-emulsifier for foam rubber production.

"Ammonyx" 2194. [328] Trade name for ditallow dimethyl ammonium methyl sulfate, 75% active paste. Used as a softener.

amobarbital (5-ethyl-5-isoamylbarbituric acid) $C_{11}H_{18}N_2O_3$.
Properties: White, crystalline powder; odorless with bitter taste. M.p. 156-158°C; solutions are acid to litmus. Very slightly soluble in water; soluble in alcohol,

*See "I.C.C. Shipping Regulations," page xiii.
Reference numbers refer to name of manufacturer. See "List of Manufacturers," page v.

ether, chloroform, and in solutions of fixed alkali hydroxides and carbonates.
Grade: U. S. P. XVI.
Use: Medicine.

amobarbital sodium $C_{11}H_{17}N_2NaO_3$. The sodium salt of amobarbital.
Properties: White, friable, granular powder; odorless with bitter taste; hygroscopic. Solutions are alkaline to litmus and phenolphthalein. Very soluble in water; soluble in alcohol; practically insoluble in ether and chloroform.
Grade: U. S. P. XVI.
Use: Medicine.

amodiaquine hydrochloride $ClC_9H_5NNHC_6H_3(OH)CH_2N(C_2H_5)_2 \cdot 2HCl \cdot 2H_2O$. A synthetic antimalarial agent; 7-chloro-4-(3'-diethylaminomethyl-4'-hydroxyanilino) quinoline dihydrochloride dihydrate.
Properties: Yellow, odorless, bitter, crystalline solid. M. p. 150-160°C (dec). Soluble in water; sparingly soluble in alcohol; very slightly soluble in benzene, chloroform, and ether; pH (1% solution) 4.0-4.8.
Grade: U. S. P. XVI.
Use: Medicine.

amolanone hydrochloride $C_{20}H_{23}NO_2 \cdot HCl$. [3-(beta-Diethylaminoethyl)-3-phenyl-2-benzofuranone hydrochloride]
Properties: Crystals; m. p. 152-153°C; soluble in water.
Grade: N. N. D.
Use: Medicine.

amorphous. Noncrystalline, having no determinable form or crystalline structure, e. g., glass.

amosite. A variety of asbestos, and of the mineral anthophyllite, containing a minimum of magnesium. Resistant to heat and used in insulation and filters.

AMP.
1. Abbreviation for 2-amino-2-methyl-1-propanol.
2. Abbreviation for adenosine monophosphate. See adenylic acid.

A5MP. Abbreviation for adenosine-5-monophosphoric acid. See adenylic acid (muscle adenylic acid).

"Ampco." [407] Trademark for a series of aluminum-iron-copper alloys containing 6-15% aluminum, 1.5-5.25% iron, balance copper. Available as sand, shell and centrifugal castings, extrusions, forgings, rolled sheet and plate, welding electrodes and filler rod. Resistant to fatigue, corrosion, erosion, wear and cavitation-pitting. Used for brushings, bearings, gears, slides, etc.

"Ampco-Brāz." [407] Trademark for a series of copper-zinc alloy filler rod for brazing and braze-welding with the oxyacetylene process.

"Ampcoflex." [41] Trade name for synthetic-resin, normal-impact rigid sheet and also pipe and fittings of Type I unplasticized polyvinyl chloride used to fabricate structures where optimum corrosion resistance is desired.

"Ampcolite." [41] Trade name for synthetic-resin, high-impact pipe and fittings of the styrene copolymer type used to handle normal corrosive fluids.

"Ampcoloy." [407] Trademark for a series of industrial copper alloys including low iron-aluminum bronzes, nickel-aluminum bronzes, tin bronzes, manganese bronzes, leaded bronzes, beryllium coppers and high conductivity alloys.

"Ampco-Trode." [407] Trademark for a series of aluminum-bronze, arc-welding electrodes and filler rod, containing 9.0-15.0% aluminum, 1.0-5.0% iron, balance copper, for joining like or dissimilar metals and overlaying surfaces resistant to wear, corrosion, erosion and cavitation-pitting.

AMPD. Abbreviation for 2-amino-2-methyl-1,3-propanediol.

ampere-second. See coulomb.

"Amphedroxyn Hydrochloride." [100] Trademark for methamphetamine hydrochloride (q. v.).

amphetamine (1-phenyl-2-aminopropane; methylphenethylamine) $C_6H_5CH_2CH(NH_2)CH_3$.
Properties: Colorless, volatile liquid; characteristic strong odor and slightly burning taste; b. p. 200-203°C (dec); soluble in alcohol and ether; slightly soluble in water.
Derivation: By synthesis.
Grades: Dextro-, dextrolevo-.
Containers: Glass bottles.
Use: Medicine.
Shipping regulations: None. *

amphetamine phosphate, dibasic (1-phenyl-2-aminopropane phosphate) $(C_9H_{13}N)_2 \cdot H_3PO_4$. Both the racemic and dextro- forms are used.
Properties: White, odorless crystalline powder, with a slightly bitter taste; soluble in water; slightly soluble in alcohol; insoluble in ether; pH (solution 1 in 20) 7-8.5; specific rotation of dextro-salt (25°C) +20 to +23°.
Grades: N. F. XI (racemic or $d\ell$-; dextro-).
Containers: Glass bottles (dextro-); drums (racemic).
Use: Medicine.

amphetamine phosphate, monobasic (1-phenyl-2-aminopropane phosphate) $C_9H_{13}N \cdot H_3PO_4$. Both the racemic and dextro- forms are used.
Properties: White, odorless crystalline powder with bitter taste; sinters at 150°; becomes amorphous with further heating; decomposes at 300°C. Soluble in water (dextro- is less soluble than the racemic); slightly soluble in alcohol; practically insoluble in benzene, chloroform, and ether; pH (10% solution) about 4.6. Specific rotation of dextro-salt (25°C) +15 to +19°.
Grade: N. F. XI (racemic or $d\ell$-; dextro-).

*See "I. C. C. Shipping Regulations," page xiii.
Reference numbers refer to name of manufacturer. See "List of Manufacturers," page v.

Containers: Glass bottles (dextro-); drums (racemic).
Use: Medicine.

amphetamine sulfate (1-phenyl-2-aminopropane sulfate)
$(C_9H_{13}N)_2 \cdot H_2SO_4$. Both the racemic and dextro- forms are used.
Properties: White, odorless powder; bitter taste; soluble in water; slightly soluble in alcohol; insoluble in ether. Specific rotation of dextro-salt (25°C) + 20 to + 23.5°; pH of solutions 5 to 6.
Derivation: By synthesis.
Grades: U.S.P. XVI (racemic or $d\ell$ -; dextro-).
Containers: Glass bottles (dextro-); drums (racemic).
Use: Medicine.
Shipping regulations: None.*

amphibole asbestos. See asbestos.

amphiboles. A group of silicate minerals with similar physical properties, chemical composition and atomic structure. The group is characterized by the presence of hydroxyl, by a silica:oxygen ratio of 4:11, and by a fibrous or prismatic cleavage. Amphiboles are common in igneous and metamorphic rocks. The commercially useful members of the group are:

anthophyllite	$(Mg, Fe)_7(Si_4O_{11})_2(OH)_2$
tremolite	$Ca_2Mg_5(Si_4O_{11})_2(OH)_2$
actinolite	$Ca_2(Mg, Fe)_5(Si_4O_{11})_2(OH)_2$
crocidolite	$Na_2Fe_5(Si_4O_{11})_2(OH)_2$.

Uses: As asbestos minerals.

amphoteric. An amphoteric compound has the capacity of behaving either as an acid or base; thus aluminum hydroxide neutralizes acids with the formation of aluminum salts: $Al(OH)_3 + 3HCl = AlCl_3 + 3H_2O$, and also dissolves in strongly basic solutions to form aluminates: $Al(OH)_3 + 3NaOH = Na_3AlO_3 + 3H_2O$.

amphotericin B. A polyene antifungal antibiotic.
Properties: Pale yellow semicrystalline powder; m.p., gradual decomposition above 170°C. Insoluble in water; slightly soluble in methanol, propylene glycol, dimethylformamide and dimethylacetamide; somewhat more soluble in dimethylsulfoxide.
Derivation: Produced by fermentation with Streptomyces nodosus. Commercially available as a desoxycholate complex.
Grade: N.N.D.
Use: Medicine.

"Amplus Improved." [299] Trademark for a preparation containing d-amphetamine, hydroxyzine hydrochloride, vitamins and minerals designed to assist in the reduction of weight.

"Amprol." [123] Trademark for amprolium for use as a coccidiostat.

amprotropine phosphate (phosphate of the d,l-tropic acid ester of 3-diethylamino-2,2-dimethyl-1-propanol) $C_{18}H_{29}NO_3 \cdot H_3PO_4$.
Properties: Bitter crystals. M.p. 142-144°C. Soluble in water, slightly soluble in alcohol.
Use: Antispasmodic.

"Amprozyme." [78] Trademark for a series of enzyme products of bacterial origin used for desizing textiles and removal of blood and albuminous stains prior to dry cleaning, for removal of gelatin from used x-ray and photographic film, and in the manufacture of glue and various adhesives.

"Ampvar." [41] Trade name for synthetic-resin metal conditioner of the vinyl-phosphoric acid-zinc chromate type used to prepare metal surfaces for the application of corrosion-proof coatings.

"Amsulf." [67] Trademark for copper-sulfur alloy castings in the form of wirebars, cakes, billets and the like.

"Amūno." [123] Trademark for a mothproofing compound.

amydricaine hydrochloride $C_{16}H_{26}N_2O_2 \cdot HCl$.
2-Benzoxy-2-dimethylaminomethyl-1-dimethylaminobutane hydrochloride.
Properties: Colorless, hygroscopic crystals; bitter taste. Decomposes about 170°C; soluble in water, alcohol and chloroform; insoluble in ether.
Use: Medicine.

amygdala amara. See almond, bitter.

amygdala amara oil. See almond oil, bitter.

amygdala dulcis. See almond, sweet.

amygdalase. See emulsin.

amygdalic acid. See mandelic acid.

amygdalin (mandelonitrile beta-gentiobioside; amygdaloside) $C_6H_5CHCNOC_{12}H_{21}O_{10}$. A glycoside found in bitter almonds.
Properties: White crystals; bitter taste. Anhydrous form m.p. 214-216°; soluble in water and alcohol; insoluble in ether.
Use: Medicine.

amygdaloside. See amygdalin.

amygdophenine (mandelyl-para-phenetidine; phenetidine amygdalate)
$C_6H_4OC_2H_5NHCOCHOHC_6H_5$.
Properties: Light, crystalline, grayish-white powder. Difficultly soluble in water.
Constants: M.p. 140.5°C.
Derivation: By the action of mandelic acid upon para-phenetidine in the presence of dehydrating agents.
Use: Medicine.
Shipping regulations: None.*

amyl. The five-carbon aliphatic radical C_5H_{11}-, also known as pentyl. Eight isomeric arrangements (exclusive of optical isomers) are possible. In addition to this theoretical source of confusion, the amyl compounds occur (as in fusel oil), or are formed (as from the petroleum pentanes) as mixtures of several isomers, and since their boiling points are close and their other properties similar, it is neither easy nor usually necessary to purify them. As used in this

*See "I.C.C. Shipping Regulations," page xiii.
Reference numbers refer to name of manufacturer. See "List of Manufacturers," page v.

dictionary, amyl means a mixture of isomers, unless a specific isomer is designated. Several entries are under isoamyl. See also amyl alcohols.

amyl acetate (amylacetic ester; banana oil; pear oil) $CH_3COOC_5H_{11}$. Commercial amyl acetate is a mixture of isomers, the composition and properties depending upon the grade and derivation. The principal isomers are isoamyl, normal-, and secondary-amyl acetates (q. v.).

Derivation: Esterification of amyl alcohol (often fusel oil) with acetic acid and a small amount of sulfuric acid as catalyst.

Method of purification: Rectification.

Grades: Commercial (85-88%), flash point 63-70°F; high test (85-88%), flash point 84°F; technical (90-95%), flash point 79°F; pure (95-99%), flash point 77°F; special antibiotic grade. Amyl acetate is also sold by original source, as from fusel oil, pentane, or Oxo process.

Containers: 1-gal bottles; 1-, 5-, 30-, 54-gal drums; tank cars.

Uses: Solvent for lacquers and paints; extraction of penicillin; manufacture of photographic film, waterproofing compounds, artificial leather, artificial pearls; flavoring agent; printing and finishing textile fabrics.

Fire hazard: Keep lights and fire away. MCA warning label.

Shipping regulations: Flammable liquid. Red label (not required for high test grade).*

n-amyl acetate $CH_3COOCH_2CH_2CH_2CH_2CH_3$.

Properties: Colorless liquid; b. p. 148.4°C; m. p. −70.8°C; sp. gr. (20/20°C) 0.879; wt/gal (20°C) 7.22 lbs; flash point (closed cup) 77°F. Very slightly soluble in water; miscible with alcohol and ether.

Derivation: Esterification of n-amyl alcohol with acetic acid.

Containers and Uses: See amyl acetate.

Fire hazard: Keep lights and fire away. MCA warning label.

Shipping regulations: Flammable liquid. Red label.*

sec-amyl acetate $CH_3CO_2C_5H_{11}$.

Properties: Colorless liquid.

Typical specifications: Distillation range 123-145°C; odor mild, nonresidual; purity, ester content as amyl acetate, 85-88%; sp. gr. 0.862-0.866 (20/20°C); flash point 89°F (approximate); wt/gal (20°C) 7.19 lbs (approximate).

Derivation: Esterification of sec-amyl alcohol and acetic acid.

Grades: Technical.

Containers: 1-gal cans; 5-, 55-gal steel drums; tank cars.

Uses: Solvent for nitrocellulose and ethyl cellulose; airplane dopes; artificial leather; celluloid products; cements; coated paper; lacquers; lacquer thinners; leather finishes; linoleum; nail enamels; patent leather; plastic wood; textile sizing and printing compounds; washable wallpaper.

Fire hazard: Combustible but not flammable—flash point over 80°F; use with adequate ventilation.

Shipping regulations: Requires no red caution label.

amylacetic ester. See amyl acetate.

amyl alcohol (amyl hydrate). Eight isomers of amyl alcohol, $C_5H_{11}OH$, are possible (exclusive of several optical isomers) and six are offered commercially. In addition, definite mixtures of the isomers are sold under a variety of names (unfortunately some of them identical with the names of the pure isomers) as well as fusel oil (q. v.), a natural fermentation product. For descriptive data on the pure isomers, see

(1) n-amyl alcohol, primary
$CH_3CH_2CH_2CH_2CH_2OH$

(2) 2-methyl-1-butanol (active amyl alcohol, from fusel oil)
$CH_3CH_2CH(CH_3)CH_2OH$

(3) isoamyl alcohol, primary
$(CH_3)_2CHCH_2CH_2OH$

(4) 2-pentanol
$CH_3CH_2CH_2CHOHCH_3$

(5) 3-pentanol
$CH_3CH_2CHOHCH_2CH_3$

(6) tert-amyl alcohol
$(CH_3)_2COHCH_2CH_3$.

The other two isomers, not described in detail, are

(7) sec-isoamyl alcohol
$(CH_3)_2CHCHOHCH_3$

(8) 2,2-dimethyl-1-propanol
$(CH_3)_3CCH_2OH$.

(1), (2), (3) and (8) are primary alcohols, (4), (5) and (7) are secondary alcohols, and (6) is a tertiary alcohol. (1), (4) and (5) are normal, and (2), (3), (6), (7) and (8) are branched chain compounds. (2), (4) and (7) are asymmetric, and have optically active forms.

amyl alcohol, fermentation. See fusel oil.

amyl alcohol, primary. A mixture of primary amyl alcohols made from normal butenes by the Oxo process is sold under this name. It consists of 60% 1-pentanol, 35% 2-methyl-1-butanol, and 5% 3-methyl-1-butanol. Used as a solvent.

n-amyl alcohol, primary (1-pentanol; n-butyl carbinol) $CH_3(CH_2)_4OH$.

Properties: Colorless liquid; mild odor; b. p. 137.8°C; freezing point −78.9°C; sp. gr. (20/20°C) 0.8240; wt/gal (20°C) 6.9 lbs; flash point (open cup) 135°F. Slightly soluble in water; miscible with alcohol and ether.

Derivation: Fractional distillation of the mixed alcohols resulting from the chlorination and alkaline hydrolysis of pentane.

Grades: Technical; C. P.

Containers: 55-gal drums; tank cars.

Uses: Raw material for certain pharmaceutical preparations; organic synthesis.

Shipping regulations: None.*

amyl alcohol, primary, active. See 2-methyl-1-butanol. Ordinary active amyl alcohol.

n-sec-amyl alcohol. See 2-, and 3-pentanol.

sec-amyl alcohol, active. See 2-pentanol.

tert-amyl alcohol (dimethyl ethyl carbinol; 2-methyl-2-butanol; amylene hydrate; tert-pentanol) $(CH_3)_2COHCH_2CH_3$.
Properties: Colorless liquid having a camphoraceous odor and burning taste; sp. gr. 0.81 (20/20°C); freezing point −11.9°C; b.p. 101.8°C; refractive index 1.4052 (20°C); wt/gal 6.76 lbs; flash point (open cup) 70°F. Slightly soluble in water; miscible with alcohol and ether; solutions neutral to litmus.
Derivation: Fractional distillation of the mixed alcohols resulting from the chlorination and alkaline hydrolysis of pentanes.
Grades: Technical; C.P.; U.S.P. XVI.
Containers: 1-gal can (approximate net contents 6.5 lbs); 5-gal can (approximate net contents 34 lbs); 55-gal drum (approximate net contents 350 lbs).
Uses: Solvent; flotation agent; organic synthesis; medicine.
Shipping regulations: Flammable liquid. Red label.*

amyl aldehyde. See n-valeraldehyde.

n-amylamine (pentylamine; 1-aminopentane) $C_5H_{11}NH_2$.
Properties: Liquid; sp. gr. 0.75 (20/20°C); m.p. −55.0°C; b.p. 104.4°C; flash point 40°F. Soluble in water, alcohol and ether.
Derivation: From the reaction of ammonia and amyl chloride, which gives a mixture of mono-, di-, and triamyl amines.
Grade: Technical.
Containers: 5-gal cans; 55-gal drums; tank cars.
Uses: Chemical intermediate; dyestuffs; rubber chemicals; insecticides; synthetic detergents; flotation agents; corrosion inhibitors; solvent; gasoline additive; pharmaceuticals.
Shipping regulations: Flammable liquid. Red label.*

amylase. A class of enzymes which convert starch into sugars. Fungal and bacterial amylases, from specific fungi and bacteria, have been suggested for commercial fermentation processes. See also amylopsin, diastase, and ptyalin.
Uses: Textile desizing; conversion of starch to glucose sugar in syrups (especially corn syrups); in baking to improve crumb softness and add shelf life (other baking advantages); in dry cleaning to attack food spots and similar stains.

amyl benzoate. See isoamyl benzoate.

amyl butyrate. See isoamyl butyrate.

tert-amyl carbamate (amylene carbamate) $C_2H_5C(CH_3)_2CO_2NH_2$.
Properties: White crystals with camphor odor. M.p. 83-86°C; soluble in alcohol and ether; slightly soluble in water.
Use: Medicine.

amyl carbinol. See 1-hexanol.

n-amyl chloride (1-chloropentane) $CH_3(CH_2)_3CH_2Cl$.
Properties: Colorless liquid. B.p. 107.8°C; freezing point −99°C; sp. gr. 0.883 (20/4°C); refractive index n 20/D 1.4128; flash point 54°F. Miscible with alcohol and ether; insoluble in water.
Derivation: (a) Distillation of amyl alcohol with salt and sulfuric acid; (b) addition of hydrogen chloride to alpha-amylene.
Grades: Technical.
Use: Chemical intermediate.
Shipping regulations: Flammable liquid. Red label.*

amyl chlorides, mixed. $C_5H_{11}Cl$.
Properties: Colorless liquid; sp. gr. (20°C) 0.88; 95% distills between 85° and 109°C; wt/gal 7.33 lbs; refractive index (20°C) 1.406; water solubility negligible; water azeotrope at 77-82°C, 90% $C_5H_{11}Cl$ (approximate); miscible with alcohol and ether.
Constituents: 1-chloropentane, b.p. 107.8°C; 2-chloropentane, b.p. 96.7°C; 3-chloropentane, b.p. 97.3°C; 1-chloro-2-methylbutane, b.p. 99.9°C; 1-chloro-3-methylbutane, b.p. 98.8°C; 3-chloro-2-methylbutane, b.p. 93.0°C; and 2-chloro-2-methylbutane, b.p. 86.0°C.
Derivation: Vapor phase chlorination of mixed normal pentane and isopentane.
Containers: Drums; tank cars.
Use: Synthesis of other amyl compounds; solvent; rotogravure ink vehicles; rubber cements; soil fumigation.
Fire hazard: Flash point (open cup) 34°F.
Shipping regulations: Flammable liquid. Red label.*

alpha-amyl cinnamic aldehyde (jasmine aldehyde; "Buxine") $C_6H_5CH{:}C(CHO)C_5H_{11}$.
Properties: Clear, yellow, oily liquid. Jasmine-like odor. Aldehyde content 98-100%. Soluble in 4 volumes of 80% alcohol.
Constants: Sp. gr. 0.962 to 0.966; refractive index 1.551 to 1.557.
Method of purification: Distillation.
Grades: Technical.
Containers: Glass bottles; demijohns; tin cans; aluminum containers; tin-lined or stainless steel drums.
Use: Perfumery, particularly for jasmine notes.
Shipping regulations: None.*

alpha-n-amylene (1-pentene; propylethylene) $CH_3CH_2CH_2CH{:}CH_2$.
Properties: Colorless liquid. Soluble in alcohol; insoluble in water.
Constants: B.p. 30°C; m.p. −165°C; refractive index (20°C) 1.3714; sp. gr. 0.6410 (20°C).
Derivation: Natural gasoline.
Grades: Technical.
Uses: Organic synthesis; blending agent for high octane motor fuel.

beta-n-amylene (sym-methylethylethylene; 2-pentene) $CH_3CH_2CH{:}CHCH_3$. Properties below are for the trans-isomer.
Properties: Colorless liquid. Flammable, Soluble in alcohol; insoluble in water.
Constants: B.p. 36.4°C; m.p. −139°C; sp. gr. 0.6482 (20°C); refractive index

1.3793 (20°C).
Derivation: Natural gasoline.
Grades: Technical.
Use: Organic synthesis.

amylene carbamate. See tert-amyl carbamate.

amylene dichlorides. See dichloropentane.

amylene hydrate. See tert-amyl alcohol.

amylenes, mixed C_5H_{10}. A mixture of several different amylenes. The chief constituents are sym-methylethylethylene (beta-n-amylene) and trimethylethylene (3-methyl-2-butene). Small proportions of uns-methylethylethylene and propylethylene (alpha-n-amylene) are also present. Pentane may be present in the limits set by the specifications.
Typical specifications: Color, water-white; sp. gr. 0.66 (20°C); pentane content less than 5.0%; sulfur content none; initial b.p. not below 32°C; not less than 90% boils below 45°C; final b.p. not above 60°C; wt/gal 5.50 lbs.
Containers: 1-gal can (approximate net contents 5 lbs); 5-gal can (approximate net contents 27 lbs); 55-gal drum (approximate net contents 282 lbs); tank car capacity (approximate net contents 8,000 gals).
Use: Organic synthesis.
Fire hazard: Flash point below 80°F.
Shipping regulations: Flammable liquid. Red label.*

amyl ether (amyl oxide; diamyl ether) $(C_5H_{11})_2O$. A mixture principally of normal- and iso-amyl ethers.
Properties: Yellowish liquid; unpleasant odor. Soluble in alcohol and ether; insoluble in water.
Typical specifications: Sp. gr. 0.78-0.81; flash point (open cup) 135°F; solidification point below −75°C; refractive index (20°C) 1.42; water azeotrope (96-98°C) 41% amyl ether (approx.); initial b.p. not below 165°C; not less than 95% boils below 200°C; final b.p. not above 210°C; wt/gal 6.61 lbs.
Derivation: By distilling a mixture of amyl alcohol and amyl chloride.
Grades: Technical.
Containers: 1-, 5-gal cans; 55-gal drums.
Use: Solvent.

amyl formate $HCOOC_5H_{11}$.
Properties: Anhydrous, colorless liquid composed of a mixture of isomeric amyl formates with the isoamyl formate in predominance. Plum-like odor. Less odoriferous and more energetic solvent than amyl acetate. It also has both a lower boiling point and a greater speed of evaporation. Miscible with oils, hydrocarbons, alcohols, ketones.
Constants: Sp. gr. 0.880 to 0.885; b.p. 123.5°C; flash point 80°F.
Grades: Technical.
Uses: Solvent for cellulose esters, resins; solvent mixtures; films and coatings; celluloid substitutes; perfume for leather.
Shipping regulations: Flammable liquid. Red label.*

n-**amyl furoate** (amyl pyromucate) $C_4H_3OCO_2C_5H_{11}$.
Properties: Colorless oil, decomposes on standing.
Constants: Sp. gr. 1.0335 (20/4°C); b.p. 233°C. Insoluble in water; soluble in alcohol.
Derivation: By the usual esterification methods from furoic acid.
Uses: Perfumes; lacquers.

amyl hydrate. See amyl alcohol.

amyl hydride. See pentane.

amyl hydrosulfide. See amyl mercaptan.

"Amyliq." [173] Trademark for a bacterial amylase preparation which permits conversion of starch to form superior sizings, coatings and adhesives from starch for use in the paper trade.

amyl laurate $C_{11}H_{23}COOC_5H_{11}$.
Properties: Sp. gr. at 20°C, 0.860; boiling range 290-330°C; color, very light straw; odor, faintly alcoholic. Flash point 300°F.
Uses: Plasticizer and solvent.

amyl mercaptan (amyl hydrosulfide; amyl sulfhydrate; pentanethiol) $C_5H_{11}SH$. A mixture of isomers having an extremely noticeable odor.
Typical specifications: Water-white to light yellow liquid; sp. gr. (20°C) 0.83-0.84; mercaptan content at least 90.0%; initial b.p. not below 104.0°C; not less than 95% boils below 125°C; final b.p. not above 130°C; wt/gal 6.99 lbs; refractive index (20°C) 1.4406. Flash point 65°F (open cup).
Derivation: Mixing amyl bromide and potassium hydrosulfide in alcohol.
Containers: 1-, 5-gal cans; 55-gal drums.
Uses: Synthesis of organic sulfur compounds; chief constituent of odorant used in gas lines to locate leaks.
Shipping regulations: Flammable liquid. Red label.*

tert-**amyl mercaptan** $(CH_3)_2CSH(C_2H_5)$. (2-Methyl-2-butanethiol).
Properties: Boiling range 95-119°C; sp. gr. 0.828 (60/60°F); refractive index 1.438 (20/D); flash point −1°C.
Grades: 95%.
Containers: Bottles and drums.
Use: Odorant.
Shipping regulations: Flammable liquid. Red label.*

amyl naphthalene $C_{10}H_7C_5H_{11}$.
Properties: Liquid; sp. gr. (20°C) 0.965; refractive index (20°C) 1.573; vapor pressure (20°C) < 0.01 mm; b.p. 279-330°C; m.p. −60°C; flash point, 124°C; insoluble in water.
Use: Plasticizer.

amyl nitrate (mixed isomers) $C_5H_{11}NO_3$.
Properties: Sp. gr. at 20°C, 0.990; boiling range 145-156°C; color, water-white; odor, ethereal. Flash point 118°F.

amyl nitrates, primary. See "Ethyl," Diesel Ignition Improver.

amyl nitrite (isoamyl nitrite) $(CH_3)_2CHCH_2CH_2NO_2$.
Properties: A clear, yellowish liquid, of peculiar ethereal, fruity odor and pungent, aromatic taste. It is volatile even at low temperatures and is flammable. Soluble in alcohol; almost insoluble in water.
Constants: Sp. gr. 0.865-0.875 at 25°C; b.p. 96-99°C.
Derivation: By the interaction of amyl alcohol and nitrous acid.
Grades: U.S.P. XVI (95% min); technical.
Containers: Dark amber glass bottles.
Uses: Medicine; perfumes; diazonium compounds.
Fire hazard: Dangerous. Mixture with air explodes if ignited.
Shipping regulations: Flammable liquid. Red label. *

amylodextrin. See soluble starch.

amyloid. Used to refer to various gelatinous and filmlike or parchment-like materials resulting from the action of sulfuric acid and water on cellulose or starch.

amyl oleate $C_{17}H_{33}COOC_5H_{11}$.
Properties: Sp. gr. at 20°C, 0.862; boiling range 200 to 240°C at 20 mm; light straw color; faint alcoholic odor. Flash point 366°F.
Uses: Solvent and plasticizer.
Shipping regulations: None. *

"Amylon." [53] Trade name for a high-amylose starch derived from high-amylose corn.

amylopectin. The outer, almost insoluble portion of starch granules. It is a hexosan, a polymer of glucose, and is a branched molecule of many glucose units. It stains violet with iodine and forms a paste with water.

amylopsin (animal diastase). The starch-digesting enzyme of pancreatic juice, the most powerful enzyme of the digestive tract. It is an amylase which converts starches through the soluble-starch stage to various dextrins and maltose. It acts in neutral, slightly acid and slightly alkaline environments with an optimum pH of 6.3-7.2. It requires the presence of certain negative ions for activation.
Use: Biochemical research.

amylose. The inner, relatively soluble portion of starch granules. Amylose is a hexosan, a polymer of glucose, and consists of long straight chains of glucose units joined by a 1,4-glycosidic linkage. It stains blue with iodine.

amyl oxide. See amyl ether.

ortho-sec-**amyl phenol** $C_5H_{11}C_6H_4OH$.
Typical specification: Clear, straw-colored solid; sp. gr. (30/30°C) 0.955-0.971; initial b.p. not below 235.0°C; final b.p. not above 250.0°C; wt/gal 8.0 lbs; very slightly soluble in water; soluble in oil and organic solvents. Flash point (open cup) 200°F.
Containers: 1-gal cans; 5-, 55-gal drums; tank cars.
Uses: Dispersing and mixing agent for paint pastes; antiskinning agent for paint, varnish and oleoresinous enamels; organic synthesis.

ortho-tert-**amyl phenol** $(CH_3)_2C_2H_5CC_6H_4OH$.
Typical specifications: Pale yellow liquid, sp. gr. (30°C) 0.96-0.97; initial b.p. not below 233°C; final b.p. not above 245°C; wt/gal 8.12 lbs; slightly soluble in water; soluble in oil and organic solvents. Flash point (open cup) 219°F.
Containers: 1-gal cans; 5-, 55-gal drums.
Uses: Dispersing and mixing agent for paint pastes, antiskinning agent for paint, varnish and oleoresinous enamels; organic synthesis.

para-tert-**amyl phenol** $(CH_3)_2C_2H_5CC_6H_4OH$.
Properties: White crystals; irritating to skin, handle with care; m.p. 93°C; b.p. 265-267°C (138°C at 15 mm); slightly soluble in water; soluble in alcohol and ether.
Containers: Drums.
Uses: Manufacture of oil-soluble resins; plasticizer; germicide; fumigant; organic synthesis.

para-tert-**amylphenyl acetate** $C_5H_{11}C_6H_4OOCCH_3$.
Properties: Sp. gr. (20°C) 0.996; boiling range 253-272°C; color, water-white; odor, fruity. Flash point 240°F.
Uses: Perfumes; flavorings.
Shipping regulations: None. *

amyl propionate $CH_3CH_2COOC_5H_{11}$. Probably the isoamyl isomer.
Properties: Colorless, high-boiling, liquid; apple-like odor; sp. gr. (20/20°C) 0.869-0.873; wt/gal (20°C) 7.25 lbs (approximate); distillation range 135-175°C; miscible with most organic solvents.
Derivation: By reacting fusel oil (amyl alcohol) with propionic acid in the presence of sulfuric acid as a catalyst, followed by neutralization, drying and distillation.
Grades: Technical.
Containers: 1-gal cans; 5-, 55-gal steel drums.
Uses: Perfumes; lacquers; flavors.
Shipping regulations: None. *

amyl pyromucate. See n-amyl furoate.

amyl salicylate. See isoamyl salicylate.

amyl stearate $CH_3(CH_2)_{16}COOC_5H_{11}$.
Properties: Liquid; sp. gr. (20°C) 0.860; boiling range 230-270°C at 30 mm; color, light straw; odor, faintly alcoholic; flash point 368°F.
Uses: Solvent and plasticizer.
Shipping regulations: None. *

amyl sulfhydrate. See amyl mercaptan.

amyl sulfide. See diamyl sulfide.

amyltrichlorosilane $C_5H_{11}SiCl_3$. A mixture of isomers.
Properties: Colorless to yellow liquid. B.p.

168°C; sp. gr. 1.137 (25/25°C); refractive index n 20/D 1.4152; flash point (Cleveland open cup) 145°F. Readily hydrolyzed by moisture with the liberation of hydrochloric acid.
Derivation: By Grignard reaction of silicon tetrachloride and amylmagnesium chloride.
Grades: Technical.
Containers: 1-, 4-, 9-lb bottles; drums.
Use: Intermediate for silicones.
Shipping regulations: Corrosive liquid. White label. *

amylum. See starch.

amyl valerate. See isoamyl valerate.

amyl valerianate. See isoamyl valerate.

amyris oil. See sandalwood oil, West Indies.

"Amytal." [100] Trademark for amobarbital, U.S.P.

"Amytal Sodium." [100] Trademark for amobarbital sodium, U.S.P.

"Amzirc." [67] Trademark for zirconium-containing copper-base alloys.

anabasine (neonicotine; 2-(3-pyridyl)piperidine) $C_{10}H_{14}N_2$. A naturally occurring alkaloid.
Properties: Colorless liquid; darkens on exposure to air. B.p. 105°C; freezing point 9°C; sp. gr. 1.046 (20/20°C); refractive index n 20/D 1.5430. Miscible with water; soluble in alcohol and ether.
Derivation: (a) Extraction from Anabasis aphylla and Nicotiana glauca; (b) synthetic.
Use: Insecticide.

anacardium gum. See cashew gum.

"Anaesthesin." [162] Trademark for ethyl para-aminobenzoate.

analcime. See analcite.

analcite (analcime) $Na_2O \cdot Al_2O_3 \cdot 4SiO_2 \cdot 2H_2O$. A mineral; one of the zeolites (q.v.).
Properties: Colorless, white; sometimes greenish-grayish, yellowish or reddish white; hardness 5-5.5; sp. gr. 2.22-2.29.
Occurrence: Europe; United States: Nova Scotia.

analogs, structural. See antimetabolites.

anatase (octahedrite). A natural crystallized form of titanium dioxide (q.v.).

"Anattene." [342] Trademark for annatto derivatives prepared for use in coloring foodstuffs, i.e., cheeses, oranges.

"Anavenol." [207] Trademark of a solution containing 20% w/v beta-naphthoxyethanol and 5% w/v thialbarbitone used as veterinary anesthetic. See "Kemithal."

anchusa. See alkanna.

anchusin. See alkanna.

andalusite Al_2OSiO_4. A natural silicate of aluminum.
Properties: Gray, greenish, reddish, or bluish in color. Streak white or uncolored. Luster vitreous.
Constants: Sp. gr. 3.1-3.2; hardness, 7-7.5.
Occurrence: Massachusetts, Connecticut, California, Nevada; U.S.S.R.; Switzerland; Spain; South Africa; Australia.
Uses: Gem stones; a constituent of sillimanite refractories; spark plug insulators; laboratory ware; superrefractories.

"Andok." [51] Trademark for a high quality grease for lubricating antifriction bearings. Correct grades, easily applied by hand or gun, are available for various speed and temperature conditions.

androgen. A general term for male sex hormones. These hormones cause the development of the secondary male sex characteristics such as the deepening of the voice and the growth of facial hair. The androgenic hormones are steroids and are synthesized in the body by the testis, the cortex of the adrenal gland, and, to slight extent, by the ovary. The international unit (I.U.) is the androgenic activity of 0.1 mg androsterone (q.v.).
Use: Medicine.

androsterone. $C_{19}H_{30}O_2$. An androgenic steroid; metabolic product of testosterone (q.v.). The international unit (I.U.) of androgenic activity is defined as 0.1 mg androsterone.
Properties: Crystalline solid; m.p. 185-185.5°C; sublimes in high vacuum; dextrorotatory in solution; not precipitated by digitonin; practically insoluble in water; soluble in most organic solvents.
Derivation: Isolation from male urine; synthesis from cholesterol.
Use: Medicine; biochemical research.

"Anectine." [301] Trademark for succinylcholine chloride (diacetylcholine chloride), used in anesthesia and electroshock therapy.

anemonin (pulsatilla camphor) $C_{10}H_8O_4$.
Properties: Yellowish-white crystals; m.p. 157-158°C; insoluble in water; soluble in hot alcohol.
Derivation: Separation from the volatile oil of pulsatilla.
Use: Medicine.
Shipping regulations: None. *

anesthesia ether. See ether.

"Anesthesin." [3] Trademark for ethyl-para-aminobenzoate (q.v.).

anethole (anise camphor; para-methoxy-propenylbenzene; para-propenylanisole) $CH_3CH{:}CHC_6H_4OCH_3$.
Properties: White crystals with sweet taste; tending to melt to liquid at warm room temperature; odor characteristic of oil of anise, suggestive of licorice. Affected by light. Soluble in 8 volumes of 80% alcohol, 1 volume of 90% alcohol. Almost immiscible with water.
Constants: Sp. gr. 0.983-0.987; refractive index 1.557-1.561; optical rotation, 0.08°; m.p. 22-23°C; distillation range 234-237°C.
Derivation: By crystallization from anise or fennel oils; synthetically from para-cresol.
Method of purification: Rectification.

Grades: U.S.P. XVI; technical.
Containers: Glass bottles; demijohns; tin cans; aluminum containers; stainless steel drums.
Uses: Perfumes, particularly for dentifrices; flavors; synthesis of anisic aldehyde; for licorice candies; color photography (sensitizer in color-bleaching process); in microscopy.
Shipping regulations: None.*

anethum (dill; dill seed).
Derivation: Fruit of garden dill, Anethum graveolens.
Occurrence: Asia Minor and Europe; cultivated in United States.
Grades: Technical.
Containers: Bags.
Uses: Medicine (aromatic); condiment.
Shipping regulations: None.*

anethum oil. See dill oil.

angelica (garden angelica). Root and seed of the biennial herb, genus Angelica.
Occurrence: Europe; Asia; eastern United States.
Grades: Technical.
Containers: Bags; bales.
Uses: Medicine (aromatic); candy; food ingredient; source of angelica root oil; rectification of alcohol and distilled liquors.
Shipping regulations: None.*

angelica root oil, European.
Properties: Essential oil. A limpid liquid, colorless when freshly distilled, becoming yellowish to brownish on exposure to light and air. Strong aromatic odor; spicy taste. Soluble in 0.5-6 vols. and more of 90% alcohol (sometimes with slight turbidity).
Chief known constituents: Phellandrene; valeric acid.
Constants: Sp. gr. 0.853-0.918; optical rotation +16° to +41°; refractive index 1.477-1.488; acid value to 3.8; ester value 12-37, after distillation 51-75.
Derivation: Distilled from the roots of Angelica archangelica.
Uses: Medicine; preparation of liqueurs; perfumery.
Shipping regulations: None.*

angelica root oil Japanese.
Properties: Essential oil. Odor suggestive of musk. Odor is stronger and more persistent than that of European oil.
Constants: Sp. gr. 0.905-0.908 (15°C); optical rotation −1°40'; refractive index 1.49110; acid value 10.6; ester value 40.
Derivation: Distilled from the root of either Angelica refracta, Fr. Schmidt or Angelica anomala, Lall.
Use: Medicine.
Shipping regulations: None.*

angelica seed oil.
Properties: Pale yellow essential oil; darkens with age; resembles in odor the oil from the root, but is much finer.
Chief known constituents: Phellandrene; valeric acid.
Constants: Sp. gr. 0.851-0.890; optical rotation +11° to +13°30'; refractive index 1.486-1.489; acid value up to 2.9; ester value 13-30. Soluble in 5-9 vols. of 90% alcohol (occasionally with opalescence and turbidity).
Derivation: Distilled from the seeds of Angelica archangelica.
Uses: Medicine; preparation of liqueurs; perfumery.
Shipping regulations: None.*

angelic acid (2-methyl-2-butenoic acid; alpha-methyl-crotonic acid) $CH_3CH{:}C(CH_3)COOH$. The cis isomer of tiglic acid.
Properties: Colorless needles or prismatic crystals; spicy odor. Soluble in alcohol, ether, and hot water.
Constants: Sp. gr. 0.9539 (76/4°C); m.p. 45°C; b.p. 185°C; refractive index n 47/D 1.4434.
Derivation: From the root of Angelica archangelica or from the oil of Anthemis nobilis by distillation.
Method of purification: Crystallization.
Grades: Technical.
Containers: Tins.
Uses: Medicine; flavoring extracts.
Shipping regulations: None.*

angel red. See iron oxide reds.

angiotensin (angiotonin; hypertensin). A peptide found in the blood, important in its effect on blood pressure. Both a decapeptide and an octapeptide are known. Their amino acid sequences, and hence the complete structures, have been decided.

angiotonin. See angiotensin.

anglesite $PbSO_4$. A natural lead sulfate.
Properties: White, gray, yellow, blue or green in color; luster, adamantine to vitreous; slowly soluble in nitric acid.
Derivation: Contains 73.6% PbO. Formed by the oxidation of galena (q.v.) and found wherever exposed deposits of galena occur.
Constants: Sp. gr. 6.12-6.39; hardness 3.
Occurrence: United States; Canada; Mexico; Chile; Europe; Australia; Siberia.
Use: An ore of lead.
See also lead sulfate.

angostura (carony bark; cusparia bark). Bark of tree, Galipea officinalis or G. cusparia.
Occurrence: Northern South America and West Indies.
Containers: Bags.
Use: Medicine (aromatic bitter).
Shipping regulations: None.*

angostura bark oil.
Properties: Light yellow-colored essential oil becoming darker on exposure to the air; aromatic odor and taste. Soluble in 9 vols. of 90% alcohol (with turbidity).
Chief known constituents: Galipol; cadinene; galipine; pinene.
Constants: Sp. gr. 0.928-0.96 (15°C); optical rotation −7°30' to −50°; refractive index 1.50744; acid value 1.8; ester value 5.5, after acetylation 35.7.

Derivation: Distilled from the bark of Galipea cusparia, St. Hil.
Uses: Medicine; preparation of liqueurs and bitters.
Shipping regulations: None. *

angstrom. A unit of length almost one one-hundred-millionth (10^{-8}) centimeter. The angstrom (abbreviated A.) is now defined in terms of the wave length of the red line of cadmium (6438.4696 A.).

anhaline. See hordenine.

anhalonidine $C_{12}H_{17}NO_3$ (7,8-dimethoxy-8-hydroxy-1-methyl-1-,2,3,4-tetrahydroisoquinoline). An alkaloid.
Properties: White crystals; poisonous! M.p. 160°C; soluble in water, alcohol, and chloroform; slightly soluble in ether.
Derivation: By extraction from mescal buttons.
Method of purification: Crystallization.
Grades: Technical.
Containers: Tins; glass bottles.
Use: Medicine.
Shipping regulations: None. *

anhalonine $C_{12}H_{15}NO_3$. An alkaloid.
Properties: White crystals; poisonous! Soluble in alcohol, ether, water and chloroform.
Constants: M.p. 85°C; b.p. 140°C (0.02 mm); specific rotation (methanol solution) −63.8° (25°C).
Derivation: By extraction and subsequent crystallization from seed of mescal buttons.
Method of purification: Recrystallization.
Grades: Technical.
Containers: Tins; glass bottles.
Use: Medicine.
Shipping regulations: None. *

"Anhydrex." [160] Trademark for a proprietary product. Wires or cables with moisture-resistant insulation; suitable for submarine installation; excellent aging properties; low mechanical water absorption; high electrical stability on soaking in water.

anhydrite $CaSO_4$. A natural calcium sulfate usually occurring as compact granular masses of white, gray-bluish or brick-red color and resembling marble in appearance; luster, pearly to vitreous. Found mostly as layers in gypsum and halite deposits. Differs from gypsum in hardness and lack of hydration.
Properties of powdered anhydrite: Hygroscopic, tasteless, odorless free-flowing powder; density 65.0 lbs/cu ft.
Occurrence: United States: Nova Scotia; Europe.
Containers: 5-ply asphalt laminated bags, 100-lbs net; carloads.
Uses: Substitute for gypsum in the manufacture of cement; sometimes employed in agriculture as land plaster; ornamental stone; drying agent; insecticide fillers; manufacture of refrigerant gases.

anhydroecgonine (ecgonidine)$C_9H_{13}NO_2$. An alkaloid.
Properties: White crystals; poisonous! Soluble in water and alcohol.
Constants: M.p. (*dl*-isomer) 226-230°C, with decomposition; (*l*-isomer) 235°C, with decomposition.
Derivation: Obtained from ecgonine.
Method of purification: Crystallization.
Grades: Technical.
Containers: Tins; glass bottles.
Use: Medicine.
Shipping regulations: None. *

anhydroecgonine hydrochloride $C_9H_{13}NO_2 \cdot HCl$.
Properties: White crystals; poisonous! Soluble in water and alcohol.
Constants: M.p. 240-241°C.
Derivation: By the action of hydrochloric acid on anhydroecgonine.
Method of purification: Crystallization.
Grades: Technical.
Containers: Glass bottles.
Use: Medicine.
Shipping regulations: None. *

anhydro-formaldehyde-para-toluidine. See formaldehyde-para-toluidine.

anhydroglucose. Term applied to the unit of molecular structure of cellulose
-OCHCH(CH_2OH)OCHCHOHCHOH
and differing from the composition of glucose by abstraction of the elements of water. The term is also applied to material of essentially the same composition, resulting from oxidation of cellulose by agents such as nitrogen dioxide.

anhydroglucuronic acid. Oxidation product of cellulose, when NO_2 is used as the oxidizing agent. See cellulose, oxidized.

anhydrohydroxyprogesterone. See ethisterone.

"Anhydrol." [214] Trademark for a proprietary gasoline-free solvent composed of 100 gal S.D. 1 ethanol denatured with 10 gal isopropanol (90%) and 1 gal methyl isobutyl ketone.
Properties of anhydrous grade: B.p. (760 mm) 75.5-80.5°C; sp. gr. 0.7895-0.7935 (20/20°C); lb/gal 6.6 (20°C); flash point 54°F.
Grades: Anhydrous and 190 proof.
Containers: 1-gal can; 5-and 55-gal drums; tank cars up to 10,000 gals.
Uses: Solvent in manufacture of printing inks, textile dyestuff solutions, wick deodorants, window cleaners, synthetic detergents, aircraft de-icing fluids, and as a solvent in the photographic industry.
Shipping regulations: Flammable liquid. Red label. *

"Anhydroprene." [160] Trademark for a proprietary product. A wire or cable insulated with "Anhydrex" moisture-resistant insulation and protected with a thin neoprene jacket having high tear and abrasion resistance. Suitable for use underground in ducts, in conduit or racked on walls.

anhydrous aluminum chloride. See aluminum chloride.

*See "I.C.C. Shipping Regulations," page xiii.
Reference numbers refer to name of manufacturer. See "List of Manufacturers," page v.

anhydrous borax. Borax glass. See sodium borate.

anhydrous salt. A dry salt; one which does not contain water either adsorbed on its surface or combined as water of crystallization.

"Anhydrox." [236] Brand name for a compound to prevent or overcome anhydrite or gypsum contamination in drilling mud, by pretreatment of the mud to remove calcium and sulfate ions.

"A" Nickel. [283] Trademark for wrought commercially pure nickel. An "electronic grade" is used in vacuum tubes and similar electronic applications. Low Carbon "A" nickel has a maximum carbon content of 0.02%. Used for applications where the temperature exceeds 600°F and for deep drawing.

anileridine hydrochloride $C_{22}H_{28}N_2O_2 \cdot 2HCl$
Ethyl 1-(4-aminophenethyl)-4-phenylisonipecotate dihydrochloride.
Properties: Crystals; decomposition 280-287°C; freely soluble in water; soluble in alcohol, methanol.
Grade: N. N. D.
Use: Medicine.

anileridine phosphate $C_{22}H_{28}N_2O_2 \cdot H_3PO_4$
Ethyl 1-(4-aminophenethyl)-4-phenylisonipecotate phosphate.
Grade: N. N. D.
Use: Medicine.

aniline (aniline oil; phenylamine; aminobenzene) $C_6H_5NH_2$. One of the most important of the organic bases; the parent substance for many dyes and drugs.
Properties: Colorless oily liquid; characteristic odor and taste; rapidly becomes brown on exposure to air and light; poisonous! Soluble in alcohol, ether, and benzene; slightly soluble in water.
Constants: Sp. gr. 1.0235; m.p. −6.2°C; b.p. 184.4°C; wt/gal (20°C) 8.52 lbs; refractive index n 20/D 1.5863; flash point (open cup) 195°F.
Derivation: (a) Reduction of nitrobenzene with iron filings or borings and 30% hydrochloric acid as catalyst; aniline is recovered by distillation. (b) Reaction of chlorobenzene and aqueous ammonia in the presence of cuprous oxide catalyst at 200°C and 800 psi. (c) By catalytic vapor-phase reduction of nitrobenzene with hydrogen.
Grades: Commercial; C. P.
Containers: 1-lb bottles; 40-, 85-, 500-, 900-lb drums; 60,000-lb tank cars.
Uses (in approximate order of volume): Rubber accelerators and anti-oxidants; dyes and intermediates; veterinary pharmaceuticals; drugs; photographic chemicals (hydroquinone); explosives; rocket fuel; petroleum refining.
Caution: Rapidly absorbed through skin; liquid and vapor hazardous. MCA warning label.
Shipping regulations: Class B poison. Poison label.*

aniline acetate (phenylamine acetate) $C_6H_5NH_2 \cdot CH_3COOH$.
Properties: Colorless liquid; becomes dark with age; on standing or heating is converted gradually to acetanilide; sp. gr. 1.070-1.072; miscible with water and alcohol.
Derivation: Combination of acetic acid and aniline.
Grades: Technical.
Use: Organic synthesis.

2-aniline-5-aminobenzenesulfonic acid. See 1,4,2-aminodiphenylamine-sulfonic acid.

aniline black. A black color developed on cotton and other textiles from a bath containing aniline hydrochloride, an oxidizing agent (usually chromic acid) and a catalyzer (usually a vanadium or copper salt).

aniline chloride. See aniline hydrochloride.

aniline, N, N-dimethyl. See N, N-dimethylaniline.

aniline-2,4-disulfonic acid. See 4-amino-metabenzenedisulfonic acid.

aniline-2,5-disulfonic acid. See 2-amino-parabenzenedisulfonic acid.

aniline dyes. A large class of synthetic dyes made from intermediates based upon, or made from, aniline.

aniline-formaldehyde resins. See amino resins.

aniline hydrochloride (aniline salt; aniline chloride) $C_6H_5NH_2 \cdot HCl$.
Properties: White crystals; commercial article frequently greenish in appearance, darkens in light and air. Soluble in water, alcohol, and ether.
Constants: Sp. gr. 1.2215; m.p. 198°C; b.p. 245°C.
Derivation: (a) By passing a current of dry hydrochloric acid gas into an ethereal solution of aniline; (b) neutralizing aniline at 100°C with concentrated hydrochloric acid and subsequent crystallization.
Method of purification: Recrystallization.
Grades: Technical; C. P.
Containers: 1-, 5-lb bottles; 180-, 225-lb barrels; drums.
Uses: Dyes; intermediates; dyeing and printing aniline black.
Shipping regulations: None.*

aniline inks. Fast-drying printing inks used on kraft paper, cotton fabric, cellophane, polyethylene, etc. The name arises from the fact that original inks for this purpose were solutions of coal tar dyes in organic solvents. Modern inks usually employ pigments rather than dyes and are of two types: spirit inks, containing organic solvent as the vehicle, and emulsion inks, in which water is the main vehicle.

1-aniline-2-methylanthraquinone
$C_6H_5NHC_{14}H_6O_2CH_3$.
Properties: Chocolate-brown to deep-red crystals. Soluble in sulfuric acid and organic solvents such as alcohol, nitroben-

*See "I.C.C. Shipping Regulations," page xiii.
Reference numbers refer to name of manufacturer. See "List of Manufacturers," page v.

zene, and mono- and dichlorobenzenes.
Constants: Melts above 200°C.
Derivation: From 1-chloro-2-methylanthraquinone and aniline in the presence of catalysts of copper salts and acid-binding agents like sodium carbonate, etc.
Method of purification: Crystallization from high boiling organic solvents such as nitrobenzene or from the halogenated benzenes.
Grades: Technical.
Containers: Wooden barrels.
Use: Dyes.
Shipping regulations: None. *

aniline oil. See aniline.

aniline resins. A type of amino resins (q. v.).

aniline salt. See aniline hydrochloride.

para-**anilinesulfonic acid.** See sulfanilic acid.

aniline yellow. See para-aminoazobenzene.

6-anilino-1-naphthol-3-sulfonic acid. Preferred name for phenyl-2-amino-5-naphthol-7-sulfonic acid.

7-anilino-1-naphthol-3-sulfonic acid. Preferred name for phenyl-2-amino-8-naphthol-6-sulfonic acid.

anilino-phenol. See para-hydroxydiphenylamine.

anilipyrine. See anilpyrine.

anilpyrine (anilipyrine).
Properties: White, crystalline powder; m.p. 75°C; soluble in water.
Derivation: By fusing antipyrine with acetanilid.
Use: Medicine.

animal black. Forms of more or less pure and finely divided carbon made by calcination of animal bones or ivory. Used as pigments, decolorizing, purifying and refining agents. Bone black, drop black and ivory black are varieties of animal black.

animal cellulose. See tunicine.

animal char. See animal charcoal.

animal charcoal (animal char). Same as animal black.

animal diastase. See amylopsin.

animal oil. See bone oil.

animal starch. See glycogen.

animal tankage. See tankage.

animé (animi). Any of several resins but especially soft copal. Sometimes applied specifically to Zanzibar gum (q. v.), a variety of copal.

animi. A variation of animé.

anion. An ion having a negative charge; anions in a liquid subjected to electric potential collect at the positive pole or anode. Examples are hydroxide, OH^-; carbonate, $CO_3^=$; phosphate, $PO_4^{\equiv}$.

anion exchange. See ion exchange.

anionic detergents. See detergents, synthetic.

ortho-**anisaldehyde** (ortho-methoxybenzaldehyde; ortho-anisic aldehyde) $C_6H_4(OCH_3)CHO$.
Properties: White to light tan solid; burned, slightly phenolic odor; b.p. 238°C; m.p. (2 crystalline forms) 38-39°C and 3°C; sp. gr. (liquid) 1.1274 (25/25°C); (solid) 1.258 (25/25°C); refractive index (n 20/D) 1.5608; flash point 244°F; slightly soluble in water.
Grade: 95% (min.).
Containers: 1-lb bottles; 1- and 5-gal cans; 55-gal drums.
Use: Intermediate.

para-**anisaldehyde** (aubepine; para-anisic aldehyde; para-methoxybenzaldehyde) $C_6H_4(OCH_3)CHO$.
Properties: Colorless to pale yellow liquid, having odor of hawthorn. Soluble in 5 volumes of 50% alcohol.
Constants: Sp. gr. 1.119-1.122; refractive index 1.570-1.572; m.p. 0°C; b.p. 248°C; refractive index (n 13/D) 1.5764.
Derivation: Obtained from anethole or anisole by oxidation.
Method of purification: Distillation.
Grades: Liquid and crystals, latter being the disulfite compound.
Containers: Glass bottles; demijohns; stainless steel or tin-lined drums.
Uses: Perfumery; intermediate for antihistamines.
Shipping regulations: None. *

anise (anise seed). Fruit of Pimpinella anisum.
Occurrence: Western Asia, Egypt; cultivated in southern Europe, India and North America.
Grades: Spanish; Mexican; Syrian.
Containers: Bags.
Uses: Manufacture of anise oil; condiment; flavor; medicine.
Shipping regulations: None. *

anise alcohol. See anisic alcohol.

anise camphor. See anethole.

aniseed oil. See anise oil.

anise oil (anise seed oil; aniseed oil).
Properties: A colorless, thick liquid having the identifying characteristic of solidifying to a crystalline mass at about 15°C. Characteristic odor; very sweet taste. Soluble in 1.5 to 3 vols. of 90% alcohol.
Chief constituents: Anethole (90%); methylchavicol; anise ketone; acetaldehyde.
Constants: Sp. gr. 0.978-0.988; optical rotation −2° to +1°; n(20°C) 1.5530-1.5600.
Derivation: By distillation of the seeds of Pimpinella anisum or Illicium verum.
Grades: U.S.P. XVI; Chinese; Russian.
Uses: As a source of anethole; medicine; when terpene-free, in perfumery; and to flavor liqueurs.
Shipping regulations: None. *

anise seed. See anise.

anise-seed oil. See anise oil.

anisic acid (para-methoxybenzoic acid) $CH_3OC_6H_4COOH$.
Properties: White crystals or powder; sp. gr. 1.385 (4°C); m.p. 184°C; b.p. 275-280°C; soluble in alcohol and ether; almost insoluble in water.
Derivation: Oxidation of anethole.
Containers: 100-lb drums.
Use: Medicine; repellent and ovicide.

anisic alcohol (anisyl alcohol; anise alcohol; para-methoxybenzyl alcohol) $CH_3OC_6H_4CH_2OH$.
Properties: Colorless liquid, having a floral odor suggesting hawthorn. Soluble in 1 volume of 50% alcohol.
Constants: Sp. gr. 1.111-1.114; congealing point 24°; refractive index 1.541-1.545; boiling range 255-265°.
Derivation: Obtained from anisic aldehyde, by reduction.
Method of purification: Distillation.
Containers: Glass bottles; demijohns; tin cans; aluminum containers; tin-lined or stainless steel drums.
Use: In perfumery, for light floral odors.
Shipping regulations: None.*

anisic aldehyde. See anisaldehyde.

ortho-**anisidine** (ortho-methoxyaniline; ortho-aminoanisole) $CH_3OC_6H_4NH_2$.
Properties: Reddish or yellowish colored oil, becomes brownish on exposure to air; volatile with steam; sp. gr. 1.097 (20°C); b.p. 225°C; m.p. 5°C; soluble in dilute mineral acid, alcohol, and ether; insoluble in water.
Derivation: (a) Reduction of ortho-nitroanisole with tin (or iron) and hydrochloric acid; (b) heating ortho-aminophenol with potassium methyl sulfate.
Method of purification: Steam distillation.
Grades: 99% (1% maximum moisture).
Containers: 55-, 110-gal drums; tank cars.
Use: Intermediate for azo dyes and for guaiacol.
Warning! Hazardous liquid and vapor; absorbed through skin. MCA warning label!
Shipping regulations: None.*

para-**anisidine** (para-methoxyaniline; para-aminoanisole) $CH_3OC_6H_4NH_2$.
Properties: Fused, crystalline mass; crystallizing point 57.2°C min; sp. gr. 1.089 (55/55°C); b.p. 242°C; soluble in hot water, alcohol, and ether.
Derivation: (a) Reduction of para-nitroanisole with iron filings and hydrochloric acid; (b) methylation of para-aminophenol.
Grades: Technical.
Containers: 500-, 800-lb drums.
Uses: As a component of various azo dyestuffs and as an intermediate in chemical synthesis.
Warning! Hazardous solid. Absorbed through skin. MCA warning label.
Shipping regulations: None.*

anisindione (2-para-anisyl-1,3-indandione) $C_{16}H_{12}O_3$.
Properties: Pale yellow crystals; m.p. 156-157°C.
Use: Medicine.

anisoin $CH_3OC_6H_4COCH(OH)C_6H_4OCH_3$.
Properties: White to yellow powder; sweet, cinnamon-like odor; m.p. 111-113°C.

anisole (methylphenyl ether; methoxybenzene) $C_6H_5OCH_3$.
Properties: Colorless liquid; agreeable, aromatic odor; soluble in alcohol and ether; insoluble in water.
Constants: Sp. gr. 0.999 (15/15°C); m.p. —37.8°C; b.p. 155°C; refractive index (n 20/D) 1.5150-1.5170.
Derivation: From sodium phenate and methyl chloride; heating phenol with methyl alcohol.
Containers: 5-lb tins; 60-lb cases; drums.
Uses: Solvent; perfumery; vermicide; intermediate.

anisoyl chloride $CH_3OC_6H_4COCl$.
Properties: Clear crystals or amber colored liquid. M.p. 22°; b.p. 262-263°. Soluble in acetone and benzene; decomposed by water or alcohol.
Containers: 500-lb drums.
Uses: Intermediate for dyes and medicines.
Shipping regulations: Corrosive liquid. White label.*

anisyl acetate (para-methoxybenzyl acetate) $CH_3OC_6H_4CH_2OCOCH_3$.
Properties: Colorless liquid, having a lilac-type odor. Soluble in 4 vols. of 60% alcohol.
Constants: Sp. gr. 1.104-1.107; refractive index 1.514-1.516.
Derivation: Reaction of anisic alcohol with acetic anhydride, using sulphocamphoric acid as catalyst.
Method of purification: Distillation.
Grades: Technical.
Containers: Glass bottles; demijohns; tin cans; aluminum containers; stainless steel or tin-lined drums.
Use: Perfumery.
Shipping regulations: None.*

anisyl alcohol. See anisic alcohol.

anisyl formate (para-methoxybenzyl formate) $CH_3OC_6H_4CH_2OCOH$.
Properties: Colorless liquid, with floral-lilac odor. Soluble in 5.5 vols. of 70% alcohol.
Constants: Sp. gr. 1.139-1.141; refractive index 1.522-1.524.
Method of purification: Distillation.
Grades: Technical.
Containers: Glass bottles; demijohns; tin cans; aluminum containers; stainless steel or tin-lined drums.
Use: Perfumery.
Shipping regulations: None.*

annatto (annotta; arnotta; bixin; butter color). Vegetable dyestuff containing coloring principle called bixin.
Properties: Soluble in alcohol, ether, and

oils.

Derivation: From the seeds of Bixa orellana.

Occurrence: South America; West Indies; India.

Grades: Spanish; Brazilian; French; technical.

Containers: 100-lb boxes.

Uses: Coloring foodstuffs (dairy products); dyeing orange yellow on cotton and silk (not fast); coloring wood stains and varnishes.

annealing. The process of maintaining a material such as glass or metal at a specified range of temperature for a specified period of time, also the process of gradually cooling such a material at a predetermined rate. The objective is to remove strains resulting from previous operations, and thereby to eliminate brittleness and to give a tougher, stronger and more enduring material.

annotta. See annatto.

"Ano." [307] Brand name of a line of dyestuffs used for the coloring of anodized aluminum.

anode. The positive terminal of an electrical source to which electrons and negatively charged ions travel. See also cathode.

anode mud. Residue obtained from the bottom of a copper or other plating bath. In the electrolytic refining of copper the anode mud contains the relatively inert metals platinum, silver and gold and is usually collected and treated for the recovery of these metals and other rare elements.

anodizing. The production of a protective oxide film on aluminum or other light metals by passing a high voltage electric current through a bath in which the metal is suspended. The metal serves as the anode. The bath usually contains sulfuric, chromic, or oxalic acid.

anona oil. See ylang-ylang oil.

anorthosite. A rock consisting principally of soda-lime feldspar, with minor quantities of iron-magnesium silicates and other minerals.

Occurrence: Wyoming, Colorado, New York; Labrador.

Use: A possible low grade ore of aluminum, successfully used in pilot plant recovery operations.

"Anozinc." [288] Trademark for chemical compounds and compositions for use in anodizing metals, both in dry form and in the form of aqueous solutions.

ANPO. Abbreviation for alpha-naphthylphenyloxazole (q.v.).

"Ansol" M. [192] Proprietary product. Said to be substantially anhydrous denatured alcohol to which has been added small amounts of ester and hydrocarbons.

Properties: Water-white; flammable. Keep lights and fire away.

Constants: Acidity, free acid as acetic, not more than 0.02%; distillation range, below 70°C, none; below 80°C, not less than 90%; above 90°C, none. Mild, non-residual odor; sp. gr. 0.796-0.800 at 20/20°C; flash point 52°F (approx.); water solubility (25°C) 100 cc solvent dissolves 44 cc water.

Grades: Technical.

Containers: 1-gal cans; 36,000-lb (gross) drum cars; 6,000- and 8,000-gal tank cars.

Uses: Solvent for many resins not soluble in regular alcohol; solvent for alcohol-soluble nitrocellulose; antiblushing agent; substitute for alcohol and (less) high-boiling solvents; airplane dopes; artificial leather dopes; Bakelite cements; celluloid products; celluloid softeners; cleaning compounds; coated paper; lacquers; lacquer thinners; leather dopes; linoleum; nitrocellulose cements; resin solutions; spirit varnishes; textile finishes.

Fire hazard: Flammable - flash point under 80°F.

Shipping regulations: Flammable liquid. Red label.*

"Ansol" PR. [192] Proprietary product. Said to be compounded from anhydrous denatured alcohol, esters, and hydrocarbons. It contains a considerably larger percentage of esters than "Ansol" M.

Constants: Flash point 54°F (approx.). Dilution ratio (nitrocellulose solution method): with toluol 3.7, with petroleum naphtha 1.1. Water solubility (25°C): 100 cc solvent dissolves 61.5 cc water. Mild, nonresidual odor; sp. gr. 0.841-0.846 (20/20°C).

Grades: Technical.

Containers: 1-gal cans; 5-gal drums to 36,000-lb (gross) drum cars; 6,000- and 8,000-gal tank cars.

Uses: Solvent for nitrocellulose and resins; airplane dopes; artificial leather; celluloid; celluloid softener; cleaning fluids; coated transparent paper; coated wall paper; decalcomanias; lacquers; leather dopes; linoleum; nitrocellulose cements; nitrocellulose solution; oilcloth; paint removers; plastic wood; ribbons; spirit varnishes.

Fire hazard: Flammable - flash point under 80°F.

Shipping regulations: Flammable liquid. Red label.*

"Ansolysen" Tartrate. [24] Trademark for pentolinium tartrate [pentamethylene-1,5-bis(1'-methylpyrrollidinium bitartrate)].

Use: Medicine.

"Anstac-2M." [238] Trade name for an antistatic and cleaning agent for plastics, such as methyl methacrylates, vinyls, and polystyrenes.

Containers: 1-qt bottles; 1-, 5-gal drums.

Uses: In aircraft, sign, novelty, electrical, photographic, optical, and other industries.

"Antaron FC-34." [307] Trademark for a high foaming, water soluble, amphoteric surfactant with soap-like qualities; a complex fatty amido compound; 40% active.

Properties: Amber, viscous liquid; soluble in water; aqueous solutions are stable to

strong alkalies and acids, hard water and high concentrations of electrolytes.
Uses: Fulling agent and detergent for woolen and worsted fabrics, effective under neutral, acid and alkaline conditions; recommended for use in bubble baths, detergents; in soaps for dedusting purposes.

"Antarox G-100." [307] Trademark for a nonionic surfactant which has cationic properties in acid media and is nonionic in alkaline systems; an alkyl polyoxyethylene glycol amide; 100% active.
Properties: Dark brown, viscous liquid; soluble in water, ethanol, ethylene glycol; insoluble in mineral oil; stable in 10% sulfuric acid; stable to alkali.
Uses: Viscose spin bath additive, preventing accumulation of sludge in pipe lines, reels, etc.; prevents clogging of the spinnerets; in manufacture of cellophane, prevents deposits forming on extrusion slits and rollers which cause surface scratches; in the steel industry, used to obtain cleaner sheets in the final wash of the reduced sheet during the cold reduction process.

antazoline hydrochloride $C_{17}H_{19}N_3 \cdot HCl$. 2-(N-benzylanilinomethyl)-2-imidazoline hydrochloride, or $C_6H_5CH_2N(C_6H_5)CH_2C_3H_5N_2 \cdot HCl$.
Properties: White, odorless crystalline powder with bitter taste. M.p. 237-241° (dec). Sparingly soluble in alcohol and water; practically insoluble in benzene and ether.
Use: Medicine.

antazoline phosphate $C_{17}H_{19}N_3 \cdot H_3PO_4$, or $C_6H_5CH_2N(C_6H_5)CH_2C_3H_5N_2 \cdot H_3PO_4$.
Properties: White, odorless crystalline powder with bitter taste. M.p. 194-198° (dec). Soluble in water; sparingly soluble in methanol; practically insoluble in benzene and ether; pH (2% solution) about 4.5.
Grade: N.F. XI.
Use: Medicine.

"Ant-B-Gon." [253] Brand name for an ant bait containing sodium arsenite.

"Antepar." [301] Trademark for piperazine citrate, an anthelmintic.

anthemidis oil. See chamomile oil, Roman.

anthion. See potassium persulfate.

"Anthomine." [300] Trademark for a dyeing assistant primarily for use in wool dyeing to eliminate tippiness, achieve uniformity, and impart softness and antistatic properties.

anthophyllite $(Mg,Fe)_7Si_8O_{22}(OH)_2$. A natural magnesium-iron silicate, usually occurring in metamorphic rocks; a member of the amphibole group (q.v.).
Properties: Gray to various shades of green or brown. Streak uncolored or gray. Luster vitreous to pearly. Sp. gr. 2.85-3.2; hardness 5.5-6.
Occurrence: North Carolina, New York, Pennsylvania, Massachusetts; Greenland; Norway.
Uses: Asbestos filters and paint filler.

anthracene (anthracin; green oil) $C_6H_4(CH_2)C_6H_4$. A tricyclic hydrocarbon.
Properties: Colorless crystals with blue fluorescence. Soluble in alcohol and ether; insoluble in water.
Constants: Sp. gr. 1.25 (27/4°C); m.p. 217°C; b.p. 340°C.
Derivation: (a) By salting out from crude anthracene oil, and draining. The crude salts are purified by pressing and finally, by the use of various solvents, phenanthrene and carbazole are removed; (b) by distilling crude anthracene oil with alkali carbonate in iron retorts, the distillate containing only anthracene and phenanthrene. The latter is removed by carbon disulfide.
Method of purification: By sublimation with superheated steam, or by crystallization from benzene followed by sublimation; for very pure crystals, by zone melting of solid anthracene.
Impurities: Phenanthrene, carbazole and chrysene.
Grades: Commercial (90 to 95%); pure crystals.
Containers: 300-lb bags; 100-, 600-lb drums; casks.
Uses: Dyes; alizarin; phenanthrene; carbazole; anthraquinone; calico printing; also as component of smoke screens; as scintillation counter crystals.

anthracene oil. A coal-tar fraction boiling in the range 270-360°C, used as a source of anthracene and similar aromatics. It is also used as a wood preservative.

anthracin. See anthracene.

anthracite (hard coal). A variety of coal containing 86-98% fixed carbon, and usually with a brilliant luster and conchoidal fracture. It burns with a short blue flame and gives off little smoke or odor.
Occurrence: Pennsylvania, Virginia; U.S.S.R.; Europe; Korea.
Uses: Household fuel; industrial fuel; metallurgy; manufacture of producer gas and water gas.

anthragallic acid. See anthragallol.

anthragallol (1,2,3-trihydroxyanthraquinone; anthragallic acid) $C_{14}H_5(OH)_3O_2$.
Properties: Brown powder. Soluble in alcohol, ether, glacial acetic acid; slightly soluble in water and chloroform.
Constants: Sublimes at 290°C; m.p. 312-313°C.
Derivation: Obtained as a product of the reaction of benzoic, gallic, and sulfuric acids.
Method of purification: Crystallization.
Grades: Technical.
Containers: Wooden kegs.
Use: Dyeing.
Shipping regulations: None.*

"Anthragen." [307] Trademark for a line of lake colors. Used for printing inks, wallpaper, coated paper, paint, rubber, and organic plastics.

*See "I.C.C. Shipping Regulations," page xiii.
Reference numbers refer to name of manufacturer. See "List of Manufacturers," page v.

"Anthralan." [307] Trademark of proprietary line of acid dyestuffs. Used on wool.

anthralin (1,8,9-anthratriol; 1,8-dihydroxyanthranol) $C_{14}H_{10}O_3$.
Properties: Odorless, tasteless, crystalline, yellowish brown powder. M.p. 176-181°C. Filtrate from water suspension is neutral to litmus. Soluble in chloroform, acetone, benzene, and in solutions of alkali hydroxide; slightly soluble in alcohol, ether, and glacial acetic acid; insoluble in water.
Derivation: By catalytic reduction of 1,8-dihydroxyanthraquinone with hydrogen at high pressure.
Grade: U.S.P. XVI (95%).
Use: Medicine.

anthranilamide $C_6H_4CO(NH_2)_2$.
Properties: Tan crystalline powder; m.p. 108°C.
Grade: 98% (min).

anthranilic acid (ortho-aminobenzoic acid) $C_6H_4(NH_2)(CO_2H)$.
Properties: Yellowish crystals; sweetish taste; soluble in hot water, alcohol, and ether.
Constants: M.p. 144-146°C. Sublimes.
Derivation: By the treatment of phthalimide with an alkaline hypobromite solution.
Method of purification: Crystallization.
Grades: Technical (95-98%); 99% or better.
Containers: 10-, 25-lb tins (in cases); 100-, 175-, 225-lb barrels; 50-, 100-, 150-lb drums.
Uses: Manufacture of dyes, drugs, perfumes and pharmaceuticals.
Shipping regulations: None.*

anthranol (9-hydroxyanthracene) $C_{14}H_9OH$.
Properties: Crystals, m.p. 120°C; soluble in organic solvents with a blue fluorescence. Changes in solution to anthrone.
Use: Dyes.

anthranone. See anthrone.

"Anthrapole." [300] Trademark for a group of dye carriers or assistants, for use in dyeing polyester fibers and blends. Active ingredients are aromatic esters, chlorinated hydrocarbons and phenol derivatives. Emulsifying agents are incorporated.

anthrapurpurin (1,2,7-trihydroxyanthraquinone; isopurpurin; purpurin red) $C_{14}H_5O_2(OH)_3$.
Properties: Orange-yellow, crystalline needles. Soluble in alcohol and alkalies; slightly soluble in ether and hot water; very slightly soluble in chloroform and benzene.
Constants: M.p. 369°C; b.p. 462°C.
Derivation: By fusion of anthraquinonedisulfonic acid with caustic soda and potassium chlorate; the melt is run into hot water and the anthrapurpurin precipitated by hydrochloric acid.
Grades: Technical; pure.
Containers: Kegs; wooden barrels.
Uses: Dyeing; organic synthesis.
Shipping regulations: None.*

anthraquinone $C_6H_4(CO)_2C_6H_4$.
Properties: Yellow needles. Soluble in alcohol, ether, and acetone; insoluble in water.
Constants: Sp. gr. 1.419-1.438; m.p. 286°C; b.p. 379-381°C; flash point (closed cup) 365°F.
Derivation: (a) By oxidizing anthracene with alkali bichromate, or electrolytically; (b) heating phthalic anhydride and benzene in the presence of aluminum chloride and dehydrating the product; (c) by direct oxidation of napthalene in a fixed bed catalytic converter.
Method of purification: Sublimation.
Grades: Sublimed; 30% paste (sold on 100% basis); electrical; 99.5%.
Containers: Bags; drums.
Uses: Intermediate for dyes and organics; organic inhibitor; bird repellent for seeds.
Shipping regulations: None.*

anthraquinone-1,5-and-1,8-disulfonic acids (rho acid, chi acid respectively) $C_{14}H_8O_8S_2$.
Properties: In their pure state, trace yellow to white. The technical variety is grayish-white. Soluble in water and strong sulfuric acid. The 1,8-isomer is much more soluble than the 1,5-isomer.
Constants: The 1,5-disulfonic acid melts with decomposition at 310-311°C. The 1,8-isomeric form melts with decomposition at 293-294°C.
Derivation: Anthraquinone is sulfonated with strong oleum in the presence of mercury or mercuric oxide to a mixture of the 1,5- and 1,8-disulfonic acids which are separated by fractional crystallization.
Method of purification: Fractional crystallization from strong sulfuric acid or in form of their alkali salts from either acid or alkaline solutions.
Grades: Technical.
Containers: Wooden barrels.
Use: Dyes.
Shipping regulations: None.*

anthraquinone dyes. Dyes whose molecular structure is based on anthraquinone ($C_6H_4(CO)_2C_6H_4$). The chromophore groups are >C=O and >C=C<; the benzene ring structure is important in the development of color. Color index numbers range from 1027 to 1175. These dyes are acid or mordant dyes respectively when OH or HSO_3 groups are present. Those anthraquinone dyes that can be reduced to an alkaline soluble leuco (vat) derivative that has affinity for fibers, and which can be reoxidized to the dye, are known as anthraquinone vat dyes. They are largely used on cotton, rayon, and silk, and have excellent properties of color and fastness.

anthraquinone-2-sodium sulfonate (silver salt) $C_{14}H_7O_2SO_3Na \cdot H_2O$.
Properties: Silvery leaflets. Soluble in water; insoluble in alcohol and ether.
Derivation: From anthraquinone by sulfonating with an equal weight of 45-50% oleum and heating up to 160°C, diluting, neutralizing with caustic soda and evapora-

ting to crystallization.
Method of purification: Crystallization from water.
Grades: Technical.
Containers: Wooden barrels.
Use: Dyes.
Shipping regulations: None.*

anthraquinonic acid. See alizarin.

anthrarobin (leuco-alizarin, 3,4-dihydroxy-anthranol; deoxyalizarin) $C_{14}H_{10}O_3$.
Properties: Yellowish-brown to dark brown crystals; m.p. 208°; soluble in alcohol, chloroform and ether; slightly soluble in water.
Derivation: Reduction of alizarin with ammonia and zinc dust.
Uses: Medicine; substitute for chysarobin.

anthrarufin. See 1,5-dihydroxyanthraquinone.

anthrasol (colorless coal-tar). Thin, mobile, light yellow oil obtained from pitch by distillation and from coal tar by freeing it from bases by acid treatment.

1,8,9-anthratriol. See anthralin.

anthraxylon. A constituent of coal, which originated from lignin and woody tissue. It is associated with the vitrain structure.

anthrone (anthranone; 9,10-dihydro-9-oxo-anthracene) $C_{14}H_{10}O$. The keto, more stable form of anthranol.
Properties: Colorless needles; m.p. 156°C; insoluble in water; soluble in alcohol, benzene, and hot sodium hydroxide.
Derivation: Reduction of anthraquinone with tin and hydrochloric acid.
Use: Rapid determination of sugar in body fluids, and of animal starch in liver tissue; general reagent for carbohydrates; organic synthesis.

antibiotic. A chemical substance produced by microorganisms that has the capacity, in dilute solutions, to inhibit the growth of other microorganisms or destroy them. Vuillemin, in 1889, first defined the words, antibiosis and antibiotic, as pertaining to the injurious effects of one organism upon another. These effects were later found to be due largely to the production of specific chemical substances. In 1942, Waksman designated these substances as antibiotics.

Antibiotics are produced by aerobic spore-forming (tyrothricin, bacitracin, polymyxin) and non-spore-forming bacteria (pyocyanase), by filamentous fungi (penicillin) and higher or mushroom fungi (polyporin, clitocybin) and by actinomycetes (actinomycin, streptomycin, chloramphenicol, tetracyclines). Antibiotic-like substances are produced by higher plants (quinine, emetine, tomatin) and animals (lysozyme).

More than 500 antibiotics are now known. Only about 20 of these have found extensive application in the treatment of infectious diseases in man and animals. Since the discovery of penicillin, the most important new antibiotics were found to be produced by actinomycetes. The manufacture of antibiotics has now grown to a great pharmaceutical industry, sales amounting to more than a billion dollars annually in the United States alone.

The antibiotics differ greatly in their physical properties, chemical composition, antimicrobial activities (antibiotic spectrum), toxicity and usefulness as therapeutic agents. Some are active only upon bacteria, others upon fungi, still others upon bacteria and fungi; some are active on viruses, some on protozoa, and some are also active on neoplasms.

Antibiotics can now alleviate most bacterial infections. They are finding extensive applications in veterinary medicine and in the treatment of certain bacterial and fungus diseases of plants (fire blights of fruit trees and blights of beans, tomatoes, peppers, and tobacco). Antibiotics also possess certain growth-promoting properties and are being used extensively in the feeding of nonherbivorous animals. They also find application in the preservation of biological materials (bull semen, virus preparations).

antichlor. A term used in the bleaching, dyeing and wood pulp industries to designate any product which serves to neutralize and remove hypochlorite or free chlorine after the bleaching operations. For many years the trade considered the term as synonymous with sodium thiosulfate, but it may equally be applied to sodium disulfite or any other product used for the purpose.

anticoagulin. An inhibitor of the power of blood to coagulate.

anti-diuretic hormone. See vasopressin.

antienzyme. A substance present in the substrate which restricts or negates the catalytic activity of the enzyme on that substrate.

antifebrin. See acetanilide.

"Antifoam A." [149] Trademark for a silicone defoamer used to prevent or suppress foams in a wide variety of aqueous and nonaqueous systems. Generally effective at concentrations in the range of 1 to 200 ppm. Physiologically harmless, it is permissible in food processing up to 10 parts per million. Also available as a water-dilutable emulsion.

anti-foam agents. See defoaming agents.

"Antifoams 60, 66 and SF-96." [245] Trade names for silicone emulsions and fluids designed for the prevention or suppression of foam in aqueous and nonaqueous systems.
Uses: "Antifoam 60" is recommended for use in aqueous systems and makes possible the increases of capacity in kettles by its antifoaming action. Used in adhesive manufacture, pulp and paper manufacture and coating, textile finishing, fermentation processes and metal reclaiming. "Antifoam 66" is used in non-aqueous systems to eliminate foaming in cooking of phenolic

*See "I.C.C. Shipping Regulations," page xiii.
Reference numbers refer to name of manufacturer. See "List of Manufacturers," page v.

resins, antibiotic fermentation, resin polymerization, esterification of vegetable oils, and the manufacture of paints and printing inks. "Antifoam SF-96" is used for defoaming of high-detergency motor oils, petroleum crudes, and other applications where the presence of silica is objectionable.

"Antifoams" C-1 and HP. [108] C-1 is a light-colored, water soluble liquid; HP a dry, powdered, water dispersable blend containing organic dispersing agents and sodium sulfite.

Containers: C-1: 5-gal cans and 55-gal drums. HP: 60 and 180-lb net wt fiber drums.

Uses: Foaming and carryover control in steam boilers. Improves steam purity; disperses sludge; minimizes blowdown.

"Anti-Fume S Solution." [28] A durable protective agent for inhibiting the fading of dyed acetate by atmospheric gases.

antiglobulin. An agent used to coagulate globulin.

antigorite. A variety of serpentine asbestos characterized by thin lamellar or plate-like structure.

antihistamines. Synthetic substances whose presence in minute amounts prevents or counteracts the action of excess histamine formed in body tissues as the result of allergic reactions, or of other circumstances. These compounds are usually complex amines of various types, and also have other physiological effects and medical uses. For examples, see chlorpheniramine maleate, dimenhydrinate, diphenhydramine hydrochloride, pheniramine maleate, pyrilamine maleate, thonzylamine hydrochloride, tripelennamine hydrochloride.

antiknock compounds. Substances added to gasoline to prevent explosive combustion in the engine (knocking). The most familiar and widely used of these is tetraethyl lead, $Pb(C_2H_5)_4$. The term antiknock compound is sometimes incorrectly used as though restricted to this one compound or to closely similar compounds.

antiknock gasoline. Gasoline having a high knock rating (octane number, q. v.) due to presence of tetraethyl lead, benzene, or branched chain hydrocarbons.

"Antilac." [165] Trademark for liquid antimony lactate containing 15% available antimony oxide. Completely soluble in cold water. Recommended as a replacement for technical tartar emetic. See also "Mordantine."

antimatter. See antiparticle.

antimetabolites. Substances structurally analogous to essential metabolites (any such as nucleic acid, proteins, enzymes, etc., essential to the synthesis of new cell substance in a plant or animal organism) which interfere with or prevent growth of an organism, or simply cause it to starve. The sulfa drugs and antibiotics probably work in this way.

antimonial glass. See antimony glass.

antimonial lead alloys (hard lead). Lead containing from about 6 to 28% antimony. Common grades are as follows: (a) 15% antimony; resistant to sulfuric acid; used in type metal; (b) national stock pile specification; 10.7-11.3% antimony; (c) battery grids; 5-11% antimony; (d) battery terminals; 4% antimony; (e) cable sheaths; 1% antimony.

antimonic. The variation of the name antimony used for compounds in which the antimony has a valence of five, as antimony pentachloride, pentasulfide, etc.

antimonic acid. See antimony pentoxide.

antimonic anhydride. See antimony pentoxide.

antimonine. See antimony lactate.

antimonite. See stibnite.

antimonious. See antimonous.

antimonous (antimonious). The variation of the name antimony used for compounds in which the antimony has a valence of three, as in antimony tribromide, antimony trichloride, antimony trioxide, antimony trisulfide.

antimony Sb. Element of atomic number 51 of group V of the periodic system. The symbol Sb comes from the Latin name stibium. Under the name regulus of antimony or antimony regulus, it was one of the earliest known elements. Two forms are known, the ordinary stable metallic form (beta), and an unstable yellow variety (alpha), which can be obtained during the electrolysis of antimony trichloride. As the alpha antimony is deposited on the electrode, it forms a solid solution in the antimony chloride. When this solution is scratched or heated, metallic antimony and clouds of antimony chloride form instantaneously, giving rise to the designation "explosive antimony."

Properties (beta or stable metallic form): Silver white, lustrous, hard, brittle metal. Sp. gr. 6.68; m.p. 630°C; b.p. 1380°C, hardness 3-3.5. Soluble in hot concentrated sulfuric acid. Insoluble in dilute acids. Not acted upon by air at room temperature.

Ores: Stibnite, antimony ocher, valentinite, livingstonite, and jamesonite are most common. Mexico, Bolivia, Yugoslavia, China, Algeria, and South Africa are major producers.

Derivation: Roasting stibnite in air to remove sulfur and obtain the oxide. This is then mixed with carbon and heated. U.S. production includes much antimony recovered as a by-product from antimony-bearing lead and silver ores.

Grades: Technical; lump; ground; powdered. Purity usually 99.0-99.9%; high purity (impurities less than 10 ppm).

Containers: 224-lb cases; 55-lb bars or pigs.
Uses: For hardening lead, particularly in lead storage batteries and cable sheaths; bearing metal, type metal, pewter, Britannia metal, and in making antimony compounds.
See also antimony black.

antimony 124. Radioactive antimony of mass number 124.
Properties: Half-life, 60 days; radiation, beta and gamma.
Uses: As a tracer (q. v.) especially in solid state studies, and marker of interfaces between products in pipe lines; the gamma ray has the proper energy to eject neutrons from beryllium. Convenient portable neutron sources, which may be reactivated in a nuclear reactor, are made by such an irradiation of an antimony pellet encased in a beryllium shell.
Shipping regulations: Class D poison, radioactive material. Red label. *

antimony arsenate.
Properties: Heavy, white powder. Poisonous! Insoluble in alcohol, water.
Derivation: By precipitating a solution of tarter emetic with arsenic. Consists approximately of 40% Sb_2O_3, 20% As_2O_5 and H_2O.
Grades: Technical.
Use: Medicine.
Shipping regulations: Poison, Class B. Poison label. *

antimony arsenite.
Properties: Fine, white powder. Insoluble in alcohol, water or dilute acids. Soluble in solutions of alkali hydroxides. Poisonous!
Derivation: Mixture of equal parts of arsenic trioxide and antimony trioxide.
Grades: Technical.
Use: Medicine.
Shipping regulations: Poison, Class B. Poison label. *

antimony black. Metallic antimony in the form of a fine powder produced by electrolysis or chemical action on an antimony salt solution. Used as a bronzing pigment for metals and plaster casts. The term antimony black is also used to refer to antimony sulfide.

antimony bloom. See antimony trioxide.

antimony blue. See antimony yellow.

antimony bromide. See antimony tribromide.

antimony, butter. See antimony trichloride.

antimony, caustic. See antimony trichloride.

antimony chloride. See antimony trichloride.

antimony chloride, basic. See antimony oxychloride.

antimony fluoride. See antimony trifluoride.

antimony glance. See stibnite.

antimony glass (vitreous antimony; antimonial glass). A vitrified product of variable composition obtained by partial roasting and subsequent fusion of antimony trisulfide. Transparent dark ruby-red mass.
Use: For tinting glass and porcelain yellow. Now seldom used.

antimony gray. See stibnite.

antimony iodide. See antimony triiodide.

antimonyl. The radical or group SbO, which occurs commonly in formulas of antimony compounds. Thus, SbOCl is often named antimonyl chloride, and numerous other antimony compounds are sometimes named in a similar manner.

antimony lactate (antimonine) $Sb(C_3H_5O_3)_3$.
Properties: Tan-colored mass. Soluble in water.
Derivation: By the interaction of antimony hydroxide and lactic acid.
Grades: Technical.
Containers: 500-lb barrels.
Uses: Mordant; textile industry.
Shipping regulations: None. *

antimonyl chloride. See antimony oxychloride.

antimony needles. See antimony trisulfide.

antimony orange. See antimony trisulfide.

antimony oxide. See antimony trioxide.

antimony oxychloride (algaroth powder; antimony chloride, basic; antimonyl chloride) SbOCl.
Properties: White crystalline powder; m. p. 170°C (decomposes); soluble in hydrochloric acid and alkali tartrate solutions; insoluble in alcohol, ether, and water.
Derivation: By the interaction of water and antimony chloride.
Method of purification: Crystallization.
Grades: Technical; C. P.
Containers: Wooden kegs.
Uses: Antimony salts; smoke-producing substance; medicine; flame proofing textiles.
Shipping regulations: None. *

antimony pentachloride (antimony perchloride) $SbCl_5$.
Properties: Reddish-yellow, oily liquid. Offensive odor. Hygroscopic. Caustic! Fumes in moist air. Solidifies by absorption of moisture. Decomposed by excess water into hydrochloric acid and antimony pentoxide. Soluble in an aqueous solution of tartaric acid, in hydrochloric acid, and chloroform.
Constants: M. p. 2.8°C; sp. gr. 2.34; b. p. 92°C (30 mm).
Derivation: Action of chlorine on antimony powder.
Containers: 15-gal drums.
Uses: Analysis (testing for alkaloids and cesium); dyeing; intermediates; as chlorine carrier in organic chlorinations.
Shipping regulations: Corrosive liquid. White label. *

antimony pentafluoride SbF_5.
Properties: Liquid; sp. gr. 2.99 (23°C);

*See "I.C.C. Shipping Regulations," page xiii.
Reference numbers refer to name of manufacturer. See "List of Manufacturers," page v.

m.p. 7°C; b.p. 149.5°C; hydrolyzed by water; soluble in potassium fluoride.
Derivation: Antimony pentachloride and anhydrous hydrogen fluoride.
Use: Catalyst and/or source of fluorine in fluorination reactions.
Shipping regulations: Corrosive liquid. White label. *

antimony pentasulfide (antimony red; antimony persulfide; antimony sulfide golden) Sb_2S_5.
Properties: Orange yellow powder; odorless; insoluble in water; soluble in concentrated HCl with evolution of hydrogen sulfide; soluble in alkali.
Uses: Red pigment for oil or water color; vulcanizing and coloring rubber.

antimony pentoxide (antimonic anhydride; antimonic acid; stibic anhydride) Sb_2O_5.
Properties: White or yellowish powder; sp. gr. 5.6; m.p. 450°C; loses oxygen above 300°C; insoluble in water; soluble in strong bases forming antimonates; insoluble in acids except concentrated hydrochloric.
Derivation: Action of concentrated nitric acid on the metal or the trioxide.
Use: Preparation of antimonates and other antimony compounds.

antimony perchloride. See antimony pentachloride.

antimony persulfide. See antimony pentasulfide.

antimony pigment (white). See antimony white.

antimony potassium tartrate (tartar emetic; potassium antimonyl tartrate; tartrated antimony) $K(SbO)C_4H_4O_6 \cdot \frac{1}{2} H_2O$.
Properties: Transparent, odorless crystals, efflorescing on exposure to air, or white powder; sweetish, metallic taste; poisonous! Sp. gr. 2.6; at 100°C loses all its water. Soluble in water, glycerol; insoluble in alcohol. Aqueous solution is slightly acid.
Derivation: By heating antimony trioxide with a solution of potassium bitartrate and subsequent crystallization.
Grades: Technical; crystals; powdered. C.P.; U.S.P. XVI.
Containers: 25-, 50-, 100-, 250-lb drums; 425-, 625-lb barrels.
Warning! May be fatal if swallowed. MCA warning label.
Uses: Textile and leather mordant; medicine (emetic); perfumery; insecticide.
Shipping regulations: None. *

antimony red. See antimony pentasulfide.

antimony regulus. See antimony.

antimony salt (deHaens salt). Mixture of antimony trifluoride and either sodium fluoride or ammonium sulfate.
Properties: White crystals; soluble in water; poisonous!
Grades: Technical; crystalline.
Use: Dyeing and printing textiles.

antimony sodiate.
See sodium antimonate.

antimony sodium tartrate (sodium antimonyl tartrate $Na(SbO)C_4H_4O_6$.
Properties: Hygroscopic white crystals or powder; sweet taste. Soluble in water; insoluble in alcohol.
Derivation: By interaction of antimony trioxide and sodium acid tartrate.
Use: Medicine (emetic).

antimony sodium thioglycollate
$C_4H_4O_4NaS_2Sb$.
Properties: White or pink powder; odorless or with a faint mercaptan odor; freely soluble in water; insoluble in alcohol.
Grades: Technical.
Use: Medicine.

antimony sulfate (antimony trisulfate) $Sb_2(SO_4)_3$.
Properties: White powder or lumps. Deliquescent. Poisonous! Decomposes in water. Sp. gr. 3.62 (4°C).
Derivation: By the action of sulfuric acid on antimony trioxide and subsequent crystallization.
Grades: C.P.; technical.
Use: Explosives.

antimony sulfide. See antimony trisulfide and antimony pentasulfide.

antimony sulfide golden. See antimony pentasulfide.

antimony sulfuret. See antimony trisulfide.

antimony tribromide (antimony bromide) $SbBr_3$.
Properties: Yellow, deliquescent, crystalline mass. Poisonous! Soluble in carbon disulfide, hydrobromic acid, hydrochloric acid, ammonia. Decomposed by water.
Constants: Sp. gr. 4.148; m.p. 96.6°C; b.p. 280°C.
Grades: Technical.
Containers: Glass bottles.
Uses: Analytical chemistry; mordant; staining iron and copper articles; manufacturing antimony salts.

antimony trichloride (antimonous chloride; antimony chloride; butter of antimony; caustic antimony; mineral butter) $SbCl_3$.
Properties: Colorless, transparent, crystalline mass. Very hygroscopic. Fumes slightly in air. Corrosive! Soluble in alcohol, benzene, carbon disulfide, chloroform, ether, acetone, acids; with water forms antimony oxychloride. Butter of antimony is a clear strongly caustic liquid with an acid reaction. Poisonous!
Constants: Sp. gr. 3.14; b.p. 223.5°C; m.p. 73.2°C.
Derivation: By the interaction of chlorine and antimony or by dissolving antimony sulfide in hydrochloric acid.
Grades: Technical; C.P.
Containers: Crystals: bottles; pails. Liquid: bottles; jugs; demijohns; carboys.
Uses: Antimony salts; bronzing iron; mordant; manufacturing lakes; coloring zinc black; catalyst in organic synthesis; pharmaceuticals (drug, manufacture of

tartar emetic); fireproofing textiles.
Danger! Causes severe burns; vapor hazardous. MCA warning label.

antimony trifluoride (antimony fluoride) SbF_3.
Properties: White to gray crystals, hygroscopic. M.p. 292°C; sp. gr. 4.58; soluble in water. Poisonous!
Uses: Porcelain; pottery; dyeing.

antimony triiodide (antimony iodide) SbI_3.
Properties: Red crystals. Volatile at high temperatures. Decomposed by water. Poisonous! Soluble in carbon disulfide, hydrochloric acid, and solution of potassium iodide; insoluble in alcohol and chloroform, decomposes in water with precipitation of oxyiodide.
Constants: Sp. gr. 4.768; m.p. 167°C; b.p. 401°C.
Derivation: Action of iodine on antimony.
Use: Medicine.

antimony trioxide (antimony white; flowers of antimony; antimony bloom) Sb_2O_3.
Properties: White, odorless, crystalline powder. Sp. gr. 5.67; m.p. 655°C. Insoluble in water; soluble in concentrated hydrochloric and sulfuric acids, strong alkalies; amphoteric.
Derivation: Burning antimony in air; adding ammonium hydroxide to antimony chloride; acidifying a solution of an antimonite.
Grades: Technical; C.P.
Containers: 50-, 100-lb bags; 500-lb barrels.
Uses: Opacifying white enamels; flameproofing and flame-retardant for textiles, tentage, paper, plastics and paints; paint pigments; glass manufacture; infrared transparent glass; mordant; medicine.

antimony trisulfate. See antimony sulfate.

antimony trisulfide (sulfuret of antimony; antimony orange; black antimony; antimony needles; antimony sulfide) Sb_2S_3.
Properties: (a) Black crystals; (b) orange-red crystals. Insoluble in water; soluble in concentrated hydrochloric acid, and sulfide solutions.
Constants: Sp. gr. 4.562; m.p. 546°C.
Derivation: (a) Occurs in nature as black crystalline stibnite (q.v.). (b) As precipitated from solutions of salts of antimony, the trisulfide is an orange-red precipitate, which is filtered, dried and ground. See also antimony vermilion.
Grades: Technical.
Containers: 250-, 350-, 500-lb barrels; bags.
Uses: Pigment; antimony salts; pyrotechnics; matches; percussion pellets for cartridges; ruby glass; refining gold from silver and copper; metallic antimony; veterinary surgery; rubber pigment; fireproofing fabrics and paper.
Shipping regulations: None. *

antimony vermillion. A red trisulfide of antimony formed by the action of hydrogen sulfide on an antimony salt solution. Used as a pigment.

antimony violet. See antimony yellow.

antimony white (antimony pigment, white). A durable paint pigment especially valuable as a flame retardant. Formed in flues and dust chambers of antimony roasting furnaces. It is antimony trioxide.

antimony yellow.
1. A pigment produced by slow oxidation of antimony sulfide. Various shades, (as antimony blue, antimony violet) are obtained by admixture of metal oxides or other mineral compounds.
2. Synonym for lead antimonate.

antimycin A ($C_{28}H_{40}O_9N_2$). An antibiotic substance said to have strong fungicidal properties.
Properties: Crystals. M.p. 139-140°C; soluble in alcohol, ether, acetone, and chloroform; slightly soluble in benzene, carbon tetrachloride, and petroleum ether; insoluble in water.
Derivation: From Streptomyces.
Use: Active against a large group of fungi, but in general not against bacteria; suggested as insecticide and miticide.

antioxidant. Any of a class of compounds added to vulcanized rubber, gasoline, natural fats and oils, soaps, and other substances to retard oxidation, deterioration, and rancidity. Rubber antioxidants are commonly of an aromatic amine type, such as di-beta-naphthyl-para-phenylenediamine and phenyl-beta-naphthylamine, and are added in quantities approximately 1%. Many of the antioxidants used in food products, petroleum oils, as well as rubber, are substituted phenolic compounds Some examples are: butylated hydroxyanisole, di-tert-butyl-para-cresol, and propyl gallate. Food antioxidants are effective in very low concentrations (not more than 0.01% in animal fats) and not only retard rancidity but protect the nutritional value by minimizing the breakdown of vitamins and essential fatty acids. Sequestering agents, such as citric and phosphoric acids, are frequently employed in antioxidant mixtures to nullify the harmful effect of traces of metallic impurities.

"Antioxidant No. 29." [28] Di-tert-butyl-para-cresol. A colorless to pale yellow crystalline solid or finely divided solid. Bulk density, untapped, 0.61 g/cc; sp. gr. 1.04.
Containers: 100-lb fiber drums.
Uses: To retard gum formation and TEL precipitation in motor and aviation gasolines and to inhibit the oxidation of turbine and electrical oils. Soluble in most gasolines at concentrations up to 40%. Concentration required 5 to 40 lbs/1000 bbls.

"Antioxidant 425." [57] A proprietary name for 2,2'-methylene-bis(4-ethyl-6-tertiary-butylphenol). A powerful antioxidant for use in white and light colored rubber products where no discoloration or staining can be tolerated.

"Antioxidant 2246." [57] Brand name for 2,2'-methylenebis (4-methyl-6-tert-butyl-

*See "I.C.C. Shipping Regulations," page xiii.
Reference numbers refer to name of manufacturer. See "List of Manufacturers," page v.

phenol).
Properties: Colorless, crystalline solid; m.p. 131.2°C; sp. gr. 1.074 (30°C); soluble in acetone, benzene, chloroform, dioxane, ethanol, ethyl acetate. Insoluble in water.
Uses: Oxidation inhibitor for use in rubber, polyethylene, waxes, etc. Polymerization inhibitor in chemical processes.

"Antioxidant B." [243] Trade name for a proprietary antioxidant and antiskinning agent.
Properties: Water-white to pale straw liquid, containing 100% active ingredient; sp. gr. 0.921; flash point 69°C, Cleveland open cup; slight odor.
Containers: 15- and 55-gal specially lined drums.
Uses: Antiskinning agent for paint, enamel, varnish, pigment dispersions, ink, baking finishes, and hot dip and hot spray compositions.

"Antioxidant D." [243] Trade name for a proprietary antioxidant and antiskinning agent.
Properties: Off-white powder containing 4.7% inert material to impart free flowing properties; melting point (dried) 86°C; flash point 234°F; slight odor; soluble to various degrees in V.M.&P. naphtha, turpentine, toluene, mineral spirits, tung oil, and linseed oil.
Containers: 10-, 100- and 250-lb drums.
Uses: Antiskinning agent for paint, enamel, varnish, pigment dispersions, ink, baking finishes, and hot dip and hot spray compositions.

antiozidants. See antiozonants.

antiozonants (antiozidants). Substances used to reverse or prevent the severe oxidizing action of ozone. Their most common use is to prevent the harmful action of the ozone in the air on rubbers, both natural and synthetic. Among antiozonant materials used are petroleum waxes, both amorphous and microcrystalline, secondary aromatic amines such as N,N-diphenyl-para-phenylenediamine, quinoline, and furane derivatives.

antiparticle (antimatter). Modern theories of physics, particularly the symmetry laws, assume that for each kind of particle of which matter is made an antiparticle exists. This idea is of most importance in relation to fundamental particles (q.v.). The most important property of the antiparticle is that it will, on collision with its symmetric particle, be annihilated along with its partner and give rise to radiant energy. If a given particle has electromagnetic properties such as charge and magnetic moment, the antiparticle has the same magnitudes of these properties, but reversed in sign. The antiparticle always has the same mass as its symmetry partner, but the two will differ in the kinds and probabilities of the reactions they can undergo.

antipyonin. Neutral sodium tetraborate, used medicinally.

antipyrine (phenazone; phenyldimethylisopyrazolone). $C_{11}H_{12}N_2O$.
Properties: Colorless crystals; fine white crystalline powder; odorless; slightly bitter taste; sp. gr. 1.19; m.p. 110-113°C; b.p. 319°C; soluble in water, alcohol, and chloroform; slightly soluble in ether.
Derivation: By the condensation of methylphenylhydrazine and ethyl acetoacetate.
Method of purification: Crystallization.
Grades: Technical; N.F. XI.
Containers: 25-, 100-, 200-, 250-lb drums; barrels.
Uses: Medicine; an analytical reagent for nitrous acid, nitric acid, and iodine number.
Shipping regulations: None.*

antipyrine acetylsalicylate $C_{11}H_{12}N_2OC_6H_4(COOH)OCOCH_3$. Not to be confused with antipyrine salicylacetate.
Properties: White crystalline powder; acetous odor. Soluble in alcohol and warm water; sparingly soluble in cold water.
Constants: M.p. 63-65°C.
Derivation: Combination of antipyrine and acetylsalicylic acid.
Use: Medicine.
Shipping regulations: None.*

antipyrine amygdalate. See antipyrine mandelate.

antipyrine chloral hydrate. See chloral hydrate antipyrine.

antipyrine mandelate (antipyrine amygdalate) $C_{11}H_{12}N_2OC_6H_5CHOHCOOH$.
Properties: White powder. Soluble in water, alcohol, and ether.
Constants: Fusing point 52-53°C.
Use: Medicine.

antipyrine salicylacetate $C_{11}H_{12}N_2O\cdot C_6H_4(COOH)OCH_2COOH$. Antipyrine ortho-(carboxymethoxy) benzoate. Not to be confused with antipyrine acetylsalicylate.
Properties: Crystals with bitter, acid taste. M.p. 149-150°C; soluble in alcohol; slightly soluble in water.
Derivation: An equimolar mixture of antipyrine and ortho-(carboxymethoxy)benzoic acid.
Use: Medicine.

antipyrine salicylate $C_{11}H_{12}N_2O\cdot C_6H_4COOH(OH)$.
Properties: White, coarse, crystalline powder, odorless, tasteless, sometimes causes vomiting. Soluble in alcohol and benzene.
Constants: M.p. 91.5°C.
Derivation: By the action of salicylic acid upon antipyrine, either at 100°C or in solution.
Containers: Drums; barrels.
Use: Medicine.
Shipping regulations: None.*

antiseptics. Substances applied to humans or animals which inhibit or stop the growth of microorganisms without necessarily destroying them. Such substances may also often be used as disinfectants, which are applied to inanimate objects, but concentrations and other conditions are different. Some substances commonly used as antiseptics are: alcohol; boric acid and borates; certain dyes, as acriflavine; certain essential oil derivatives, as menthol; hydrogen peroxide; hypochlorites; iodine; mercuric chloride; and phenol. Many of these are corrosive and poisonous and should be used with great caution. Among the newer antiseptics are hexachlorophene and some quaternary ammonium compounds.

See also disinfectants; sanitizers.

antistatic agents. Materials which reduce static electrical charges on textiles, wax polishes, resins, and paper products. Such charges are often built up by friction and cause difficulty in handling as well as creating a fire hazard. The antistat allows the charge to leak off, usually by retaining enough moisture to provide good electrical conduction by way of molecularly held water. Most textile antistats are nondurable (will wash out), but durable agents are available. The field is largely given over to trademarked items. High molecular weight fatty alcohols and medium-sized polymers have been mentioned.

"Antistine." [305] Trademark for antazoline, an antihistaminic drug. Available as the hydrochloride or phosphate.

"Antivert." [299] Trademark for a combination drug containing meclizine hydrochloride and nicotinic acid; used in medicine.

antlerite $Cu_3(OH)_4SO_4$. A natural basic sulfate of copper. Found in the oxidized portions of copper deposits.

Properties: Color green; hardness 3.5-4; sp. gr. 3.9; luster vitreous.

Occurrence: Chile.

Use: An ore of copper.

ant oil, artificial. See furfural.

"Antox." [28] Trademark for rubber antioxidant made from condensation product of butyraldehyde-aniline.

Properties: Amber liquid.

Containers: Drums (275 lbs net).

Use: To improve the aging and service life of rubber and synthetic rubber.

"Antrenyl." [305] Trademark for oxyphenonium bromide (q. v.).

"Antron." [28] Trademark for a trilobal multifilament nylon for textile fibers in the form of continuous filament yarns.

"Antrycide." [207] Trademark for quinapyramine sulfate [4-amino-6-(2'-amino-6'-methyl-4'-pyrimidylamino)-2-methylquinoline 1, 1' dimetho(methylsulfate)]. White crystalline powder for veterinary medicine.

"Antrypol." [207] Trademark for suramin (the symmetrical urea of the sodium salt of meta-benzoyl-meta-amino-para-methylbenzoyl-1-aminonaphthalene-4, 6, 8-trisulfonic acid.

Use: Medicine.

ants, artificial oil of. See furfural.

ANTU. Abbreviation for alpha-naphthylthiourea.

Antwerp blue. Applied loosely to any of a number of varieties of iron blue pigments, usually containing considerable extender such as alumina.

AP. Abbreviation for ammonium perchlorate.

"AP 30." [233] Trademark for an anionic flocculant.

APAP. Abbreviation for acetyl-para-aminophenol. See para-acetylaminophenol.

"Apasol W-1345." [78] Trademark for a completely sulfonated ester of a high molecular weight alcohol.

Properties: Clear oil, soluble in water, having penetrating, suspending, leveling, and softening properties.

Uses: Textile operations, such as desizing, kier boiling, pasting dyestuffs, dyeing, soaping, dispersion of acetate printing colors, etc.

apatite $Ca_5(F, Cl, OH)(PO_4)_3$. A natural calcium phosphate usually containing fluorine; sometimes with chlorine, hydroxyl, or carbonate substituting for part or all of the fluorine. The corresponding minerals are known as fluorapatite, chlorapatite, hydroxyapatite, carbapatite. May contain magnesium, manganese, or iron. Collophanite is a finely divided crystalline variety which makes up the bulk of phosphate rock.

Properties: Color variable; sp. gr. 3.1-3.2; hardness 5; luster vitreous to greasy; frequently in hexagonal crystals.

Occurrence: Maine, New Hampshire, Virginia, Massachusetts, New York, Connecticut, Pennsylvania, New Jersey, North Carolina, California; U.S.S.R; Canada; Europe.

Uses: Source of phosphorus and phosphoric acid; manufacture of fertilizers; gems.

APC. Abbreviation for ammonium perchlorate.

"APCO." [200] Trademark for a series of petroleum solvents including "Petrolene," "Troluoil," mineral spirits, etc., as well as the following:

"Deodorized APCO 125":

Properties: Water-white; initial boiling point 326°F, 95% distils at 395°F; sp. gr. 0.774 (60°F); flash point (TCC) 118°F; neutral odor, nonresidual.

Uses: In odorless brushing enamels; fly spray and polish manufacture; some dry cleaning units; and in degreasing, extraction and cleaning operations.

"APCO 140":

Properties: Water-white; initial boiling point 363°F, 95% distils at 403°F; sp. gr. 0.795 (60°F); flash point (TCC) 140°F; mild, nonresidual odor.

Use: In dry cleaning industry.

"Deodorized APCO 140":
Properties: Water-white; neutral odor; boiling range 366-405°F; sp. gr. 0.781 (60°F); wt/gal 6.50 lbs (60°F); flash point 140°F.
Containers: Drums; tank cars; tank wagons.
Uses: Dry cleaning solvent; thinner for low odor brushing enamels.
"APCO 360":
Properties: Water-white; boiling range 310-353°F; sp. gr. 0.779 (60°F); wt/gal 6.49 lbs (60°F); flash point 102°F.
Containers: Drums; tank cars; tank wagons.
Uses: Paint, varnish and enamel thinner; metal cleaning and degreasing.
"APCO 467":
Properties: Water-white; initial boiling point 410°F; 95% distils at 465-470°F; sp. gr. 0.809 (60°F); flash point (TCC) 175°F; mild nonresidual odor.
Uses: In flat wall paint; polish; insecticides; degreasing operations; and as a coolant in milling magnesium and other metals.
"Deodorized APCO 467":
Properties: Approximately the same as "APCO 467" except for the neutral odor.
Uses: In paint; varnish; insecticides; polishes; cleaners; and inks.
"Apco Wood Treating and Weed Treating Solvent": Trade name for a petroleum distillate.
Properties: Straw color; boiling range 425-635°F; wt/gal 7.39 lbs (60°F); flash point, open cup, 245°F.
Containers: Tank cars.

"Apco Inkol No. 0." [200] Trademark for a petroleum solvent.
Properties: Light color; boiling range 470-505°F; sp. gr. 0.824 (60°F); flash point 240°F (open cup).
Use: Solvent for heat set and flash dry printing inks.

"Apcolene." [200] Trademark for a petroleum solvent.
Properties: Water-white; initial boiling point 195-204°F; 95% distils at 236-248°F; sp. gr. 0.739 (60°F); flash point (TCC) 25°F; mild, nonresidual odor.
Use: Preservative wash for lithographic blankets, rubber rolls, and rubber printing plates.
Shipping regulations: Flammable liquid. Red label. *

"Apcoseal." [200] Trademark for a chemical dispersion of a tough and resilient asphalt and mineral filler.
Properties: Flash point (P.M. closed cup) 119°F; thermal conductivity or K factor 1.26; density of material before application 1.05; density of cured coating 0.96; remains flexible at low temperatures and does not flow at high temperatures.
Uses: Weather proofing or corrosion resistant coating; heat or sound insulating coating.

"Apcothinner." [200] Trademark for a petroleum solvent.
Properties: Water-white; initial boiling point 240-248°F; 95% distils at 278-288°F; sp. gr. 0.763 (60°F); flash point (TCC) 52°F; mild, nonresidual odor.
Use: In paints; varnishes; enamels; roto ink; and for degreasing hides.
Shipping regulations: Flammable liquid. Red label. *

aphrodine. See yohimbine.

"Aphrosol." [206] Brand name for a proprietary foaming agent used for fire fighting and for the production of foamed cements and aerated concrete.

API. Abbreviation for American Petroleum Institute.

"Apiezon." [431] Trademark for a line of hydrocarbon oils, greases and waxes that are produced by molecular distillation and characterized by very low vapor pressures and good thermal stability. Vapor pressures of various grades range from 10^{-5} to 10^{-11}mm Hg at room temperature. Some of these materials are liquids at room temperature, others have melting points up to 125°C. Used as lubricants and seals in high vacuum equipment and operations.

apiol (parsley camphor; 1-allyl-2,5-dimethoxy-3,4-methylenedioxybenzene) $C_{12}H_{14}O_4$. Dimethoxymethylene ether of allyltetrahydroxybenzene.
Properties: White, crystalline solid with faint parsley odor; m.p. 29.5°C; b.p. 294°C; refractive index n 20/D 1.536-1.538; soluble in alcohol, ether, and fixed oils; insoluble in water.
Derivation: Separation from the volatile oil of parsley.
Use: Medicine.
Shipping regulations: None. *

aplite. A fine-grained variety of granite consisting mostly of quartz and feldspar.
Occurrence: Virginia.
Use: Source of alumina for glass, pottery, porcelain and enamel-ware.

apo-. A prefix denoting formation from, or relationship to, another compound.

APO. See tris(1-aziridinyl)phosphine oxide.

apoatropine $C_{17}H_{21}NO_2$.
Properties: White crystalline, poisonous alkaloid! Soluble in alcohol, ether, benzene, chloroform, and dilute acids; slightly soluble in water.
Constants: M.p. 60-62°C.
Derivation: Obtained from atropine by dehydration.
Method of purification: Crystallization.
Grades: Technical.
Containers: Cans; glass bottles.
Use: Medicine.
Shipping regulations: None. *

apoatropine hydrochloride $C_{17}H_{21}NO_2 \cdot HCl$.
Properties: Colorless crystals; poisonous! Soluble in water and alcohol; slightly soluble in ether.
Constants: M.p. 239°C.
Derivation: By the action of hydrochloric acid

*See "I.C.C. Shipping Regulations," page xiii.
Reference numbers refer to name of manufacturer. See "List of Manufacturers," page v.

on apoatropine.
Method of purification: Crystallization.
Grades: Technical.
Containers: Glass bottles.
Use: Medicine.
Shipping regulations: None.*

apocodeine $C_{18}H_{19}NO_2$.
Properties: White crystalline, poisonous alkaloid! M.p. 124°C with decomposition. Soluble in alcohol and ether; very slightly soluble in water.
Derivation: Obtained from codeine by fusion with oxalic or metaphosphoric acid.
Method of purification: Crystallization.
Grades: Technical.
Containers: Tins; glass bottles.
Use: Medicine.
Shipping regulations: None.*

apocynum (Canadian hemp; American Indian hemp; Indian physic; black Indian hemp; dogbane).
Derivation: Dried rhizome and roots of Apocynum cannabinum.
Occurrence: United States.
Grades: Technical.
Containers: Sacks.
Use: Medicine.
Shipping regulations: None.*

apomorphine $C_{17}H_{17}NO_2$.
Properties: White crystalline alkaloid; poisonous! Oxidizes rapidly in air and becomes green. Decomposes 195°C. Soluble in alcohol, acetone, chloroform; slightly soluble in water, benzene, and ether.
Derivation: From morphine by extraction of one molecule of water.
Method of purification: Crystallization.
Containers: 15-grain vials; ⅛-, 1-oz bottles; 5-oz tins.
Use: Medicine.
Shipping regulations: None.*

apomorphine hydrochloride $C_{17}H_{17}NO_2 \cdot HCl \cdot \frac{1}{2}H_2O$.
Properties: White crystalline alkaloid; poisonous! Bitter taste, turns green on exposure to light and air; odorless. Soluble in water and alcohol; slightly soluble in chloroform and ether.
Constants: M.p. 200-210°C.
Derivation: Obtained by the action of hydrochloric acid on apomorphine.
Method of purification: Crystallization.
Grades: U.S.P. XVI.
Containers: 15-grain vials; ⅛-, 1-oz bottles; 5-oz tins.
Use: Medicine.
Shipping regulations: None.*

apparent density. See density.

apple acid. See malic acid.

apple essence. See isoamyl valerate.

apple oil. See isoamyl valerate.

apple, Peru. See stramonium.

"App-L-Set." [233] Brand name for a proprietary product. Plant growth control agent to prevent preharvest drop of certain fruits.

"Appramine." [42] Proprietary product. Cationic fatty amides.
Properties: Light yellow paste; disperses readily in water at temperatures above 60°C.
Containers: 44-gal fiber container.
Use: Durable softening agent for all types of textile fibers.

"Appretole." [42] Proprietary product. Anionic fatty amide dispersion.
Properties: Cream colored paste; disperses readily in water above 60°C.
Containers: 55-gal steel drums.
Use: Softener for cotton and rayon textile fabrics. Effects are durable to washing and dry cleaning; does not affect light fastness of dyestuffs.

approx. Abbreviation for approximate or approximately.

"Apresoline." [305] Trademark for hydralazine hydrochloride; used in medicine.

apricot kernel oil. See persic oil.

aprobarbital. See 5-allyl-5-isopropylbarbituric acid.

aprobarbital sodium. See 5-allyl-5-isopropylbarbituric sodium.

"APW." [244] Proprietary compound consisting of a balanced blend of buffered alkalies and a surface active agent.
Properties: Soluble in water; total Na_2O content 39.9%; per cent of total Na_2O in active form 24.0%; per cent of total Na_2O in inactive form 15.9%; white, granular dedusted mechanical mix.
Containers: 5-lb cans (9/case); 10-lb cans (4/case); 125-lb plywood drum; 325-lb wooden barrel.
Shipping regulations: None.*

aqua. Water.

aqua ammonia. See ammonium hydroxide.

"Aquablak." [133] Trademark for carbon black and bone black aqueous dispersions used in latex paints, latex compounding, paper coatings, leather finishing, etc.
Containers: 50-, 300- and 500-lb drums; available in many types, including:
"Aquablak" B: All-purpose utility black; anionic type.
"Aquablak" K: All-purpose utility black; nonanionic type.
"Aquablak" M: Blue-tone gray black for tinting.
"Aquablak" 15: Special black for electrical conductivity.
"Aquablak" 41: Jet-black.

"Aquadag." [46] Trademark for a concentrated colloidal dispersion of pure electric-furnace graphite in water.
Properties: Paste consistency; solids content 22%; average particle size 0.5 micron; max. particle size 4 microns; sp. gr. 1.121; b.p. 100°C; completely miscible with water.
Uses: General industrial applications

including metalworking operations at elevated temperatures, conductive films in electrical and electronic applications, dry-film lubrication, impregnation, and graphic-arts opaquing; meets U. S. Army Specification 2-130.

"Aquadow." [233] Trademark for an aqueous ammonia solution used for fertilizing purposes.

aqua fortis. See nitric acid.

"Aquagel." [236] Brand name for a proprietary product. A gel-forming colloidal bentonite clay used in drilling muds.

"Aqualin." [125] Trademark for an herbicide which is a formulation containing 85% acrolein as the principal constituent for the control of submersed and floating weeds and algae in irrigation canals, ditches, drains, ponds and other bodies of water.

Properties: Colorless, lachrymatory liquid that is highly volatile!; b. p. 52°C; miscible in lower alcohols, ether, hydrocarbons, acetone, and benzene; moderately soluble in water at 68°F; may polymerize with violence under some conditions!

Containers: 16-gauge unlined steel drum containing 370 lbs.

Danger! Extremely flammable and irritating vapor and liquid. Poisonous by inhalation, skin contact or swallowing.

Shipping regulations: Flammable liquid. Red label. *

aquamarine. See beryl.

"Aquaness." [89] Trademark for chemicals used in oil production and refining.

"Aquapel." [266] Trade name for alkylketene dimers used as paper size and for treatment of textiles.

"Aqua Phos Kil 6." [55] Trademark for insecticide used on fruits, vegetables, cotton. Active ingredient, 61% parathion (q. v.).

"Aquaprint." [293] Trademark for a resin-bonded pigment color for printing on textiles. The vehicle, an oil-in-water emulsion, contains a water-insoluble binder which adheres to the fibers and anchors the color permanently to the cloth.

aqua regia (nitrohydrochloric acid; nitromuriatic acid; chloronitrous acid; chloroazotic acid; chlorazotic acid).

Derivation: A mixture of nitric and hydrochloric acids, usually one part of nitric acid and three or four parts of hydrochloric acid.

Properties: Fuming yellow, corrosive, suffocating, volatile liquid.

Grades: Technical.

Containers: Glass bottles.

Uses: Metallurgy; testing metals; dissolving metals (platinum, gold, etc.).

Fire hazard: Dangerous. *

Shipping regulations: Corrosive liquid. White label. Legal label name: nitrohydrochloric acid. *

"Aquaresin." [73] Brand name for proprietary product; glycol bori-borate.

Properties: Water-white viscous liquid which is nondrying. Odorless. Soluble in water, methyl alcohol, glycerine, and diethylene glycol; insoluble in ethyl alcohol and toluene. Sp. gr. (25°C) 1.375; pH (5% dispersion) 8.0.

Containers: 1-gal cans (11 lbs); 5-gal cans (55 lbs); 55-gal drums (625 lbs).

Uses: Textile lubricant and softener; softener or plasticizer for glues, gelatine, gums; adhesive for cellophane, glassine; sealing joints in systems carrying oils, hydrocarbons; fire retardant for treating paper, leather, textiles; prevention of caking in pigments in water suspensions.

"Aquaresin" G. B. [73] Brand name for proprietary product, modified glyceryl borate.

Properties: Water-white viscous liquid. Will not crystallize at any dilution. Odorless, noncorrosive, nontoxic. Soluble in water, methyl alcohol, glycerine, and diethylene glycol. Insoluble in ethyl alcohol, toluene, mineral spirits, mineral oil, vegetable oil.

Containers: 1-gal cans (11 lbs); 5-gal cans (55 lbs); 55-gal drums (675 lbs).

Uses: Textile lubricant and softener; flexibilizer, softener, or plasticizer for glues, gelatine, gums. Cosmetic preparations, astringent lotions; adhesive for cellophane, glassine; sealing joints in systems carrying oils, hydrocarbons; fire retardant for treating paper, leather, textiles.

"Aquarex." [28] Trademark for a line of rubber latex stabilizers and mold lubricants. Available as:

"Aquarex" D. Sodium salts of sulfate mono-esters of a mixture of higher fatty alcohols consisting chiefly of the lauryl and myristyl derivatives of the type RSO_4Na. White powder; sp. gr. 1.33.

Uses: To improve the stability of latex mixtures and to prevent coagulation due to mechanical friction; in water solution as a mold lubricant.

Containers: 100-lb drums.

"Aquarex" G Surface Active Agent. Water solution of sodium alkyl sulfonate of the type RSO_3Na. Clear amber liquid; sp. gr. 1.09.

Use: In emulsion polymerization of synthetic elastomers and resins.

Containers: 425-lb drums.

"Aquarex" L Mold Release Agent. Ammonium salts of alkyl acid phosphates of the type $RHPO_3ONH_4$. Amber viscous paste; sp. gr. 1.01.

Use: As a corrosion inhibiting mold lubricant.

Containers: 400-lb drums.

"Aquarex" MDL Surface Active Agent or Mold Release Agent. Water paste of sodium salts of sulfate mono-esters of a mixture of higher fatty alcohols consisting chiefly of the lauryl and myristyl derivatives. Amber paste; sp. gr. 1.08.

Uses: A stabilizing and wetting agent for latex mixtures; in water solution as a mold lubricant.

Containers: 375-lb drums.
"Aquarex" ME Surface Active or Mold Release Agent. Sodium salts of sulfate mono-esters of a mixture of higher fatty alcohols consisting chiefly of the lauryl and myristyl derivatives. Creamy white powder; sp. gr. 1.06.
Use: A stabilizing and wetting agent for latex mixtures and in water solution as a mold lubricant.
Containers: 150-lb drums.
"Aquarex" NS Surface Active Agent. Water solution of C-cetyl-betaine $(CH_3)_3N^+CH(C_{16}H_{33})COO^-$. Brown slightly cloudy liquid; sp. gr. 1.05.
Use: As a stabilizer for latex mixtures. Due to its amphoteric properties, it functions as a stabilizer in both acid and alkaline dispersions.
Containers: 450-lb drums.
"Aquarex" SMO Surface Active Agent. 33% water solution of mono-sodium salt of sulfated methyl oleate $CH_3(CH_2)_8CH(SO_4Na)(CH_2)_7COOCH_3$. Reddish amber liquid; sp. gr. 1.08.
Use: A surface conditioning agent for rubber latex and neoprene latex mixtures.
Containers: 450-lb drums.
"Aquarex" WAQ Surface Active or Mold Release Agent. A water solution of the sodium salts of the sulfate mono-esters of a mixture of higher fatty alcohols consisting chiefly of the lauryl and myristyl derivatives (see "Aquarex" D). Amber viscous liquid; sp. gr. 1.04.
Use: A stabilizing and wetting agent for latex mixtures; diluted, as a mold lubricant.
Containers: 400-lb drums.

"Aquarol." [300] Trademark for water repellents of the wax-multivalent metal salt type for textiles. Non-durable or renewable.

"Aquasol." [57] Trademark for a highly sulfonated castor oil.

"Aquatar." [323] Trademark for a coal tar emulsion maintenance coating.

"Aquazinc." [354] Trade name for an aqueous dispersion of zinc stearate containing a wetting agent and designed to replace powdered zinc stearate for many applications in order to eliminate the dust, fire hazard, and other difficulties encountered with the dry material. It will volatilize at or below 100°C.
Uses: In manufacture of butyl rubber, neoprene adhesives, various types of rubber latex and molded goods.

Ar. Symbol for argon; official since 1957.

"A.R." [329] Trademark for chemical products for laboratory and industrial use which are specially produced and controlled to meet critical purity and uniformity requirements.

arabic gum. A commercial term for acacia gum, the dried gummy exudate from the stems of Acacia Senegal or related African species of Acacia.
Properties: Thin flakes, powder, granules or angular fragments; color white to yellowish white, almost odorless, and have a mucilaginous taste. Completely soluble in hot and cold water, yielding a viscous solution of mucilage; insoluble in alcohol. The aqueous solution is acid to litmus.
Grades: U.S.P. XVI; a great many varieties named according to color and source, as Senegal, Kordofan, Morocco, Cape, Aden, suakin, white Senaar. See also wattle gum, an Australian variety.
Containers: Bags; multiwall paper sacks; barrels.
Uses: Pharmaceuticals; adhesives; inks; textile printing; cosmetics; food preparation; in general, as a thickening agent and colloidal stabilizer.
Shipping regulations: None.*

arabinose (pectinose; pectin sugar; gum sugar) $C_5H_{10}O_5$.
Properties: White crystals. Three varieties (optical isomers) are known but differ very slightly in most properties. Soluble in water and glycerine; insoluble in alcohol and ether. M.p. 158.5°C; sp. gr. 1.585 (20/4°C).
Derivation: (a) From calcium dextrogluconate and hydrogen peroxide; (b) by boiling vegetable gum with dilute sulfuric acid.
Method of purification: Crystallization.
Grades: Technical.
Containers: Glass bottles.
Uses: Medicine; as a culture medium.
Commercially available as D- and L-arabinose.

"Aracar." [51] Trademark for railroad car journal oils for both plain and roller bearings. Having good viscosity index, they permit operation over wide temperature ranges without waste grab or hot boxes.

arachic acid. See arachidic acid.

arachic alcohol. See arachidyl alcohol.

arachidic acid (arachic acid; eicosanoic acid) $CH_3(CH_2)_{18}COOH$. A widely distributed but minor component of the fats of peanut oils and related plant species.
Properties: Shining, white, crystalline leaflets. Soluble in ether; slightly soluble in alcohol; insoluble in water. M.p. 75.4°C; sp. gr. 0.8240 (100/4°C); b.p. 205°C (1 mm), 328°C (760 mm) (decomposes); refractive index 1.4250 (100°C).
Derivation: From peanut oil.
Grades: Technical; 99%.
Containers: Glass bottles; 50-lb bags.
Uses: Organic synthesis; lubricating greases, waxes, and plastics.
Shipping regulations: None.*

arachidonic acid
$CH_3(CH_2)_4(CH{:}CHCH_2)_4(CH_2)_2COOH$. A C_{20} unsaturated fatty acid. It is considered essential for good health.
Properties: Liquid; m.p. −49.5°C; iodine value 333.50.
Source: Liver; brain; lecithin.
Use: Medicine; biochemical research.

*See "I.C.C. Shipping Regulations," page xiii.
Reference numbers refer to name of manufacturer. See "List of Manufacturers," page v.

arachidyl alcohol (1-eicosanol; arachic alcohol) $CH_3(CH_2)_{18}CH_2OH$. A long-chain saturated fatty alcohol, much like stearyl alcohol.
Properties: M.p. 71°C; b.p. 220°C (3 mm).
Derivation: Reduction of arachidic acid.
Uses: Lubricants, rubber, plastics, textiles.

arachidyl behenyl amine RNH_2, in which R equals principally the saturated straight chain radicals $C_{20}H_{41}$- and $C_{22}H_{43}$-.
Properties: White solid; soluble in most organic solvents.
Suggested uses: Surface active agents; ore flotation; corrosion inhibitor; intermediate for bactericides, detergents, lube oil additives, etc.

arachin (arachine). A protein from peanut, a globulin containing the following amino acids in representable amounts: arginine, histidine, lysine, cystine. A yellow green syrup. Soluble in water and alcohol; insoluble in ether.

arachine. See arachin.

arachis oil. See peanut.

aragonite $CaCO_3$. A natural carbonate of calcium with the same formula as calcite (q. v.) but differing in stability, hardness, specific gravity, and other properties.
Properties: Colorless, white, yellow, various other pale colors; hardness, 3.5-4; sp. gr. 2.9-3. May show luminescence. Occurs in crystals, pisolites, the pearly layer of shells, and in flos ferri (a colloidal variety with siderite).
Occurrence: United States and Europe.

"Aralen" Phosphate. [162] Trademark for chloroquin phosphate.

aralia (American spikenard; spikenard; spignet; pettymorrel; spice berry). Dried rhizome and roots of perennial herb, Aralia racemosa.
Occurrence: Northeastern United States.
Grades: N.F. XI.
Use: Medicine.

"Aramite." [248] Trademark for 2-(para-tert-butylphenoxy) isopropyl 2-chloroethyl sulfite $(CH_3)_3CC_6H_4OCH_2CH(CH_3)OSOOC_2H_4Cl$.
Properties: Clear light colored oil; sp. gr. 1.148-1.152 (20°C); b.p. 175°C (0.1 mm); very soluble in common organic solvents; insoluble in water; non-corrosive.
Grades: Technical (90% min.); wettable powder; emulsifiable concentrates; dusts.
Uses: As an acaricide on a wide variety of non-edible crops and ornamentals.
Hazards: Harmful if absorbed through the skin or ingested. Do not get in eyes; avoid contact with skin or clothing; store in cool, dry place.

"Aranox." [248] Trademark for para-(para-tolylsulfonylamido)-diphenylamine, $CH_3C_6H_4SO_2NHC_6H_4NHC_6H_5$.
Properties: Grey powder; sp. gr. 1.32; m.p. 135°C min; soluble in acetone, benzol and ethylene dichloride; insoluble in gasoline and cold water; slightly soluble in hot water or hot alkaline solutions which extract it from thin rubber sheets after considerable exposure.
Uses: Antioxidant for balloon fabrics, proofing, clothing and light-colored sundries.

"Aranthol." [9] Trademark for methamoctol (2-methyl-amino-6-hydroxy-6-methylheptane) employed in medicine as the mucate and hydrochloride salts.

"Arapen." [51] Trademark for high quality lime-base greases of greenish black color developed for all types of pressure gun fittings. Made with a high viscosity base oil, they adhere sufficiently to reciprocating parts so they will not be thrown off.

araroba (goa powder). A brownish-yellow odorless powder obtained from cavities in the trunk of the Brazilian tree, Andira araroba; it is the source of chrysarobin.

"Arasan." [28] Trademark for seed disinfectants based on thiram (tetramethyl thiuramdisulfide). Formulated for dry applications and with wetting and sticking agents for slurry. Also formulated with thiram plus methoxychlor.
Use: For the treatment of corn, sorghums, peanuts, soybeans, and vegetable seeds.

"Arazate." [248] Trademark for zinc dibenzyl dithiocarbamate, $Zn[SCSN(C_7H_7)_2]_2$.
Properties: White powder; sp. gr. 1.41; melting range 165-175°C; moderately soluble in benzol and ethylene dichloride; insoluble in acetone, gasoline, and water.
Uses: Accelerator for latex, dispersions, and cements.

arbor vitae. See thuja.

arbor vitae oil. See thuja oil.

arbutin (ursin, hydroquinone glucose; arbutoside) $C_{12}H_{16}O_7 \cdot H_2O$.
Properties: Long colorless silky needles with bitter taste; m.p. 195-200°(anhydrous); loses water about 70°C; soluble in water and alcohol; insoluble in ether, chloroform, and carbon disulfide.
Derivation: A glucoside found in the leaves of the cranberry and blueberry and in the roots, trunks and leaves of most pear species. Pure arbutin has also been prepared synthetically from acetobromoglucose and hydroquinone in presence of alkali.
Use: Medicine.

arbutoside. See arbutin.

archil. See orchil.

"Arctic Crystal" Soap. [86] Trademark for neutral sodium soap, 42°C titer. Available in flakes or granules.
Uses: Detergent; fulling; kier-boiling; scouring; soaping.

"Arctic Syntex." [86] Trademark for a series of detergents. Available as:
"Arctic Syntex" 036; polyoxyethylated nonylphenol, 100% active.
"Arctic Syntex" HD; alkylarylsulfonate built

*See "I.C.C. Shipping Regulations," page xiii.
Reference numbers refer to name of manufacturer. See "List of Manufacturers," page v.

with polyphosphates.

"Arctic Syntex" M Beads; alkylarylsulfonate plus polyphosphates.

"Arctic Syntex" M Liquid; modified alkylarylsulfonate (22% active ingredient).

Uses: Emulsifying agents; kier-boiling; wetting; dispensant; scouring; acid fulling; dyeing and carbonizing assistant.

"Arcturus Red." [141] Trade name for azo red pigments derived from beta-naphthol.

Properties: Yellow shade of red; very transparent, fair resistance to light, good resistance to heat; non-bleeding in water and organic solvents.

Uses: Printing inks, especially flexographic inks for foil printing, rubber, plastics.

"Ardil." [206] Trademark for a brand of regenerated protein fiber derived from groundnuts (peanuts).

arecaidine (arecaine; 1,2,5,6-tetrahydro-1-methylnicotinic acid) $C_7H_{11}O_2N \cdot H_2O$. An alkaloid derived from betel nuts (Areca catechu).

Properties: White crystals, optically inactive; decomposes 232° when anhydrous; freely soluble in water; slightly soluble in organic solvents.

arecaidine methyl ester. See arecoline.

arecaine. See arecaidine.

areca nut (betel nut, Pinang). Fruit of Areca catechu. The seeds are hard and heavy, round-conical, depressed at the base, externally brown, mottled with fawn spots. The fresh seeds have a faint cheese-like odor; the powder has an astringent, slightly bitter taste.

Occurrence: East Indies.

Grades: Technical; N.F. XI (dispensed as powder).

Containers: Bags; barrels.

Uses: Medicine; chewed by the natives (mixed with lime) for its stimulating effect.

Shipping regulations: None.*

arecolidine $C_8H_{13}O_2N$. An alkaloid derived from betel nuts (Areca catechu). Isomeric with arecoline.

Properties: White crystals; hygroscopic; m.p. 105°C; sublimes 110°C; soluble in water, alcohol, acetone, ether.

arecoline (arecaidine methyl ester; methyl 1,2,5,6-tetrahydro-1-methylnicotinate; methyl arecaidinate) $C_8H_{13}O_2N$. Alkaloid obtained from Areca catechu or betel nut.

Properties: Colorless, odorless, oily liquid; strongly alkaline; optically inactive; b.p. 209°C; sp. gr. 2.02; soluble in water, alcohol, chloroform, ether; volatile with steam.

Use: Medicine.

arecoline hydrobromide $C_8H_{13}NO_2 \cdot HBr$.

Properties: White crystals or white crystalline powder; odorless; bitter taste; optically inactive; affected by light. Soluble in water and alcohol. M.p. 170-175°C.

Derivation: By the action of hydrobromic acid on arecoline.

Method of purification: Crystallization.

Containers: Vials; bottles; tins.

Grades: Technical; N.F. XI.

Use: Medicine.

Shipping regulations: None.*

"Areskap." [58] Trademark for sodium butyl-ortho-phenylphenolsulfonate in liquid (250-lb barrels) or dry powder form (115-lb drums).

Uses: Wetting, penetrating and spreading agent for insecticides, embalming fluids.

"Aresket." [58] Trademark for sodium butylbiphenylsulfonate in dry powder form.

Containers: 115-lb drums.

Uses: Wetting, penetrating, and spreading agent.

"Aresklene." [58] Trademark for disodium dibutyl-ortho-phenylphenoldisulfonate.

"Arfonad" Camphorsulfonate. [190] Trademark for a brand of trimethaphan camphorsulfonate (q.v.).

argentic. Adjective meaning divalent silver, as in argentic fluoride, AgF_2. The monovalent term is argentous, as AgF.

argentite (silver glance) Ag_2S. Lead-gray to black or blackish-gray mineral, streak same color, metallic luster. A natural silver sulfide. Found both primary and as a product of secondary enrichment. Contains 87.1% silver. Differs from other soft black minerals in cutting like wax. Soluble in nitric acid.

Constants: Sp. gr. 7.2-7.36; hardness 2-2.5.

Occurrence: Nevada, Colorado, Montana; Mexico; Chile; Germany; Canada.

Use: An important ore of silver.

argentous. See argentic.

argentum. The Latin name for silver, hence the symbol Ag in chemical nomenclature.

argilla. See kaolin.

argillaceous hematite (ironstone clay). A variety of natural ferric oxide containing an appreciable portion of clay or sand as impurity. A hard brown to deep red mineral, with submetallic to nonmetallic luster and a red streak. See hematite, red.

argillaceous limestone. A limestone with appreciable clay as impurity. Certain varieties are useful as raw material for cement manufacture and are called cement rock (q.v.).

arginase. An enzyme producing ornithine and urea by splitting arginine. It is found in liver.

Use: Biochemical research.

arginine (guanidine aminovaleric acid; amino-4-guanidovaleric acid) $NHC(NH_2)NH(CH_2)_3CH(NH_2)COOH$. An essential amino acid for rats, occurring naturally in the L(+) form.

Properties: Prisms from water containing 2 molecules of H_2O, anhydrous plates from alcohol solution; dehydrates at 105°C;

*See "I.C.C. Shipping Regulations," page xiii.
Reference numbers refer to name of manufacturer. See "List of Manufacturers," page v.

decomposes at 244°C; sparingly soluble in alcohol; insoluble in ether.
Derivation: Widely found in animal and plant proteins. It is precipitated as the flavianate from gelatin hydrolysate in industry.
Containers: Drums.
Uses: Biochemical research; medicine; pharmaceuticals.

arginine hydrochloride
$NHC(NH_2)NH(CH_2)_3CH(NH_2)COOH \cdot HCl$.
Properties: Platelets and prisms; sinters at 218°C, solidifies again at 225°C, decomposes at 235°C. Soluble in water; slightly soluble in hot alcohol.
Grade: N.N.D.
Use: Medicine.

"Argo." [30] Trademark for a brand of chemical products made from corn.

argols. A by-product of the wine industry containing crude potassium acid tartrate (cream of tartar). It consists of a crystalline crust found on the sides of wine vats in which the grape juice has fermented. White argol is the deposit of the white grape and red argol is the deposit of the red grape. Argols are similar to wine lees (q.v.) in their chemical nature, but are purer since they contain less sediment. They contain from about 50 to 85% of potassium acid tartrate and 6-12% calcium tartrate.
Containers: 1-, 5-lb cans; bags.
Uses: Manufacture of tartaric acid and tartrates; fertilizer ingredient; manufacture of malt vinegar; mordant in dyeing.
Shipping regulations: None. *

argon Ar. Element of atomic number 18, zero group of the periodic system. Occurs to the extent of 0.94% by volume in the atmosphere.
Properties: Colorless, inert, monatomic gas, does not combine chemically with any element; m.p. −189.3°C; b.p. −185.8°C; specific volume (70°F) 9.7 cu ft/lb. Slightly soluble in water.
Derivation: (a) By fractional distillation of liquid air. (b) By the treatment of atmospheric nitrogen with certain metals such as magnesium and calcium to form nitrides. (c) Recovery from natural gas oxidation bottoms stream in ammonia plant.
Method of purification: (a) Highly purified argon is obtained by passing the gas through a bed of titanium at 850°C. (b) Synthetic zeolite molecular sieves separate oxygen from argon to give high-purity gas.
Grades: Technical; highest purity (99.995%).
Containers: Steel cylinders (technical); hermetically sealed glass flasks (highest purity).
Uses (in approximate order of volume): Inert gas shield in arc welding; aircraft; electric lamps (also used with neon in "neon" lights); electronics; titanium and zirconium refining; flushing molten metals (particularly aluminum) to eliminate porosity in castings. Its use as an inert gas or atmosphere in miscellaneous applications is growing as fast as new production can take care of the demand.
Shipping regulations: Nonflammable compressed gas. Green gas label. *

"Aridex." [28] Trademark for a line of non-durable water repellents for textiles, including both aqueous and solvent types.

"Aridye." [293] Trademark for a patented product and process for printing colors on textiles using permanent and insoluble pigments suspended in an organic vehicle, into which water is emulsified to give printing consistency. The vehicle contains a water-insoluble binder which adheres to the cloth and anchors the color permanently to the fibers.

"Ariperm." [300] Trademark for a gas-fading inhibitor for application to dyed acetate fabrics. Protects the dyes from fading due to acid gases in the atmosphere.

"Aristocort." [315] Trademark for triamcinolone (q.v.).

aristoquin (diquinine carbonate; quinine carbonate) $CO(OC_{20}H_{23}ON_2)_2$.
Properties: White or pinkish, amorphous powder containing 96% quinine; tasteless; m.p. 189°C; decomposed by acids and alkalies with liberation of quinine. Soluble in alcohol, chloroform, glycerol and dilute acids; insoluble in water, ether.
Use: Medicine.
Shipping regulations: None. *

"Arizole." [252] Trademark for a group of terpene products, including anethole U.S.P., anethole Technical, and sulfate pine oil. Used in perfumes, soaps, etc.

"Arko." [300] Trademark for a cationic finishing agent used to soften fabrics made from wool, acetate, rayon, and synthetic fibers.

"Arkolene." [300] Trademark for wetting agents of the alkylarylsulfonate type, sodium or ammonium salts. Used in acid carbonizing of wool and other wet processing of textiles where quick wetting or penetration is required.

"Arkolube." [300] Trademark for textile lubricating and softening agents, based on silicones, polyethylenes, and waxes. Used with textile resins to minimize loss of tear strength and abrasion resistance, and to improve sewability.

"Arko Stat-Ex." [300] Trademark for antistatic agent for use on synthetic fibers.

"Arkotan." [300] Trademark for a synthetic tanning agent for leather. Ammonium salt of a naphthalenesulfonic acid complex.

"Arlacel." [89] Trademark for each of a series of non-ionic emulsifiers for use in cosmetics and pharmaceuticals. They are fatty acid partial esters of polyols or polyol anhydrides.

"Arlex." [89] Trademark for non-crystallizing industrial humectant solution, containing

83% solids consisting of sorbitol and related polyhydric materials.
Use: For flexibilizing and moisture-conditioning in industrial applications, including tobacco, glue compositions, cellulose products, etc.

"Armacs." [15] Trademark for a group of amine acetate salts derived from primary amines, secondary amines and diamines. The alkyl groups range from C_8 to C_{18} in chain length.
Containers: 55-gal non-returnable drums.
Uses: Anti-static agents on resin and plastic surfaces; bactericides; algicides; corrosion inhibitors in oil pipelines; mineral flotation; pigment dispersants; emulsifiers.

"Armalon." [28] Trademark for TFE-fluorocarbon fiber felt and also for TFE-fluorocarbon resin coated glass fabrics, tapes, and laminates.

The felt is made in various thicknesses either as porous material or impregnated with TFE resin. The latter is used for gaskets, the plain felt for filter purposes.

The TFE coated fabrics, etc., are used for electrical insulation, for antistick and antifriction conveyor belts and covers in the food and packaging industries, for diaphragms and gaskets requiring unusual resistance to chemicals and to high and low temperatures.

"Armeens." [15] Trademark for a series of high molecular weight aliphatic amines, primary, secondary and tertiary.
Derivation: From natural fats and oils. Available in chain lengths ranging from C_8 to C_{18} of 90% purity of a single homolog, and certain mixtures of these obtained from overall distilled coco, soya, tallow, etc., sources.
Grades: Technical and distilled.
Containers: 55-gal drums; 320-lb fiber drums for certain items.
Uses: As chemical intermediates for textile agents, flotation agents, asphalt additives, emulsifiers, wetting agents, rubber chemicals, metal working lubricants, plastics, fuel oil additives, motor oil additives, catalysts for polyurethane foams.

Armenian bole. See iron oxide reds.

"Armids." [15] Trademark for a class of high-melting, wax-like amides derived from fatty acids. They vary in form from material that can be flaked (like stearic acid) to a soft paste, and are soluble in ketones, esters, alcohol, turpentine, fatty acids, fats and mineral spirits, but insoluble in water.
Containers: Flaked and powdered products packed in 50-lb multiwall paper bags; lump products packed in 100-lb cloth bags.
Use: As antiblocking agents in plastic formulations; as emulsion additives; solvents; printing ink; antiscratch agents; dye solubilizers and in organic synthesis.
See also "Armowax."

"Armofos." [334] Trademark for sodium tripolyphosphate, anhydrous (sodium triphosphate; pentasodium triphosphate) $Na_5P_3O_{10}$.
Properties: White granules or powder; soluble in water; non-deliquescent; pH (1% solution) 9.7; improves the sudsing and detergent properties of soaps and synthetic detergents; prevents rust stains.
Grades: Technical; food processing.
Containers: 100-lb multiwall paper bags; bulk hopper cars.
Uses: Sequestering agent for iron, calcium and magnesium ions; soap builder; detergent mixtures; deflocculator in drilling muds; paper, ceramics and textiles.

armoracia radix (horse-radish root).
Occurrence: Cultivated everywhere.
Derivation: The fresh root of Cochlearia armoracia, containing volatile oil similar to oil of mustard.

armored concrete. See concrete.

"Armowax." [15] Trademark for a synthetic amide wax derived from fatty acids.
Properties: M.p. 130°C; amide, 90%; color, FAC 35.
Uses: In wax mixtures, powder metal lubricant, and steel wire drawing.
See also "Armids."

Armstrong's acid. See naphthalene-1,5-disulfonic acid.

arnatto. See annatto.

Arnaudon's green. See chromic phosphate.

"Arneels." [15] Trademark for a series of high molecular weight aliphatic nitriles. Available in chain lengths ranging from C_8 to C_{18} and certain naturally occurring mixtures of these.
Properties: Consistency ranges from liquids freezing at −36°C to solids melting at 40°C. Good thermal and chemical stability.
Containers: 55-gal non-returnable drums.
Uses: Excellent softening agents and plasticizers for nitrile type rubbers, vinyl plastics; dye solubilizers for inks; metal wetting agents.

"Arnel." [352] Trademark for an acetate fiber made from cellulose triacetate. It has a higher melting point and is less soluble than cellulose acetate.
Properties: Staple or continuous filament. Tensile strength (psi) 20,000-26,000; elongation 22-28%; sp. gr. 1.3; moisture regain 3.2%; m.p. approximately 300°C; soluble in methylene chloride, glacial acetic acid, chloroform; swollen by acetone and trichloroethylene.
Uses: Knitted and woven fabrics for wearing apparel, alone and blended with other fibers; laundry pads; electrical insulation; laminated papers.

arnica flowers (mountain tobacco).
Derivation: Dried flowers and heads of Arnica montana.
Occurrence: Northern Europe; Asia; North America.

Grades: Technical; N.F. XI.
Containers: Sacks.
Use: Medicine.
Shipping regulations: None.*

arnica flowers oil.
Properties: Reddish-yellow to brown-colored essential oil; strong aromatic odor and taste. The consistency of the oil varies; sometimes it is a butter-like mass which melts between 20° and 30°C to a brownish liquid. Soluble in alcohol with great difficulty. Solutions in absolute alcohol are clear at first, later become turbid. Soluble in ether, benzene, chloroform.
Constants: Sp. gr. 0.8905-0.9029; acid value 62.6-127.3; ester value 22.7-32.2.
Derivation: By distillation of the flowers of Arnica montana, L.
Use: Medicine.
Shipping regulations: None.*

arnica root oil.
Properties: Light yellow-colored essential oil, darkens with age; pungent aromatic taste; radish odor.
Constants: Sp. gr. 0.982-1.00 at 15°C; optical rotation +0°25' to −2°; refractive index 1.507-1.508; acid value 4-10; ester value 60-100. Soluble in 7-12 volumes of 80% alcohol, occasionally with turbidity.
Derivation: From the dried root of Arnica montana, L.
Use: Medicine (liniments, etc.).
Shipping regulations: None.*

"Aroclor." [58] Trademark for a series of polychlorinated polyphenyls available as liquids, resins, or solids.
Properties: Water-white mobile liquids and pale yellow viscous oils to light amber resins and opaque crystalline solids. Insoluble in water; thermoplastic; nondrying; stable on long heating at 150°C; electrically nonconducting; not affected by boiling with sodium hydroxide solution.
The light oils can be distilled at atmospheric pressure without appreciable decomposition. The viscous oils and resins distil readily under vacuum, are odorless and tasteless at ordinary temperatures, have adhesive properties; do not support combustion when heated alone above 360°C; are easily soluble in most common organic solvents and drying oils. Hard crystalline materials are, in general, less soluble.
Uses: In protective coatings; as plasticizers and extenders; as sealers in water-proofing compounds and putty; in asphaltic materials, printing inks, waxes, synthetic adhesives. Liquid "Aroclors" are used as dielectrics; hydraulic fluids; in thermostats; in cutting oils; as extreme pressure lubricants; as high temperature lubricants; as grinding fluids; and as a heat transfer medium. Solid "Aroclors" are used to impregnate carbon resistors, as sealers, or impregnating agents for electrical apparatus.

"Aroflo." [285] Proprietary brand name for a carbon black having long flow properties.
Properties: Sp. gr. 1.72; bulk density 12 lbs cu ft; fluffy powder; color (Nigrometer) 84; particle diameter 27 millimicrons; oil absorption 85 lbs/100 lbs carbon black.
Containers: 25-lb bags and 50-lb cartons.
Uses: A new, long-flow carbon black for printing inks and paints, with low oil absorption, fast drying, a dense color and bluish undertone.

"Arogen" (GPF). [285] Proprietary brand name for a general purpose furnace carbon black.
Properties: Sp. gr. 1.77; free-flowing pellets; bulk density 26 lbs/cu ft; particle diameter 60 millimicrons; pH (approx.) 9.3; ash 0.50% max; 99.9% through 325 mesh screen; color (Nigrometer) 96.
Containers: 50-lb paper bags.
Uses: As a reinforcing ingredient for compounding in natural and most synthetic rubbers, contributing to abrasion resistance, good tensile and tear strength. As a black coloring agent in rubber, paper, plastics, paint and ink.

aromatic nucleus. The six-carbon ring characteristic of the molecules of all organic compounds of the benzene and related series, or the condensed six-carbon rings of naphthalene, anthracene, phenanthrene, etc.

aromatization. See hydroforming process.

"Aromex." [285] Proprietary name for a series of furnace carbon blacks; "Aromex HAF," "Aromex CF" and "Aromex ISAF."
Properties: Sp. gr. 1.77; free-flowing pellets; bulk density 26 lbs/cu ft; particle diameter 30 (HAF), 27 (CF) and 25 (ISAF) millimicrons; pH (approx.) 9.3 (HAF), 9.3 (CF) and 8.0 (ISAF); ash 0.75% max.; 99.9% through 325 mesh screen.
Containers: 50-lb paper bags or bulk.
Uses: As reinforcing ingredients for compounding in natural and most synthetic rubbers, contributing to abrasion resistance, good tensile and tear strength. As black coloring agents in rubber, paper, plastics, paint and ink. "Aromex CF" contributes electrical conducting properties to most rubber compounds. "Aromex ISAF" contributes abrasion resistance.

Arosorb process. Separation of benzene, toluene and xylenes from mixtures containing saturated naphthenic and aliphatic hydrocarbons, by selective adsorption on silica gel.

"Arothane." [221] Trademark for a line of urethane resins.

"Arovel" (FEF). [285] Proprietary brand name for fast extrusion furnace carbon black.
Properties: Sp. gr. 1.77; free-flowing pellets; bulk density 26 lbs/cu ft; particle diameter 55 millimicrons; pH (approx.) 8.4; ash 0.50% max; 99.9% through 325 mesh screen; color (Nigrometer) 92-93.
Containers: 50-lb paper bags.
Uses: As a reinforcing ingredient in natural

and most synthetic rubbers, contributing to processing of unvulcanized compounds, and to abrasion resistance, tensile and tear strength; as a black coloring agent.

"Arox." [51] Trademark for specially compounded fluid lubricants for all air tools except those requiring hand-packing with grease. Used for both wet and dry operations and under all temperature conditions.

"Arquads." [15] Trademark for a series of quaternary ammonium salts containing one or more alkyl groups ranging in chain lengths from C_8 to C_{18}. These are water-soluble cationic surface chemicals. Certain oil-soluble and water-dispersible "Arquads" are also available.
Containers: Openhead lined drums.
Uses: Sanitizing and cleansing agents; dispersants for fillers in rubber compounding; emulsifying agents; germicides; textile softeners.

Arrhenius equation. An empirical expression relating to the increase of the rate of chemical reaction with rise in temperature. Mathematically:

$$\frac{d \ln k}{dT} = \frac{A}{RT^2}$$

in which k is the specific reaction velocity, T is the absolute temperature, A is a constant usually referred to as the energy of activation of the reaction and R is the gas law constant.

arrowroot. A starch obtained from the roots of several varieties of plants belonging to the genus Maranta.
Grades: Bermuda; St. Vincent; domestic.
Containers: Bags; barrels.
Uses: Food; sizing; laundry; adhesives; face powder; starches; baking and confectionery industries.
Shipping regulations: None.*

"Arrow TX" (EPC). [285] Proprietary brand name for a medium processing channel carbon black.
Properties: Sp. gr. 1.77; free-flowing pellets; bulk density 25 lbs/cu ft; particle diameter 27 millimicrons; pH 4.1-4.5; ash 0.05% max.; 99.9% through 325 mesh screen; color (Nigrometer) 84-85.
Containers: 50-lb paper bags or bulk.
Uses: As a reinforcing ingredient for compounding in natural and most synthetic rubbers, contributing to abrasion resistance, tensile and tear strength; as a black coloring agent in rubber, paper, plastics, paint and ink.

arrow wood. See eunonymus.

arsacetin (sodium acetylarsanilate; sodium para-acetylaminophenylarsonate) $CH_3CONHC_6H_4AsO(OH)ONa$.
Properties: White, crystalline powder; odorless, tasteless, free of arsenous or arsenic acid; solutions will admit of thorough sterilization. Poisonous! May produce optic atrophy! Soluble in cold, but more so in warm water.
Use: Medicine.

arsanilic acid (atoxylic acid; para-aminobenzenearsonic acid; para-aminophenylarsonic acid; arsenic acid anilide) $C_6H_4NH_2 \cdot AsO(OH)_2$.
Properties: White, crystalline powder; practically odorless; poisonous! Soluble in hot water, amyl alcohol and alkaline carbonate solutions; slightly soluble in cold water, alcohol, and acetic acid; insoluble in acetone, benzene, chloroform, and ether.
Constants: M.p. 232°C.
Derivation: By condensing aniline with arsenic acid, removing the excess of aniline by steam distillation in alkaline solution and setting the acid free by hydrochloric acid.
Method of purification: Conversion into the sodium salt, boiling with animal charcoal, crystallizing the sodium salt and setting the acid free by dilute hydrochloric acid.
Grades: Technical; pure.
Containers: Tins; glass bottles.
Uses: Arsanilates; starting-point for the manufacture of arsenical medicinal compounds, such as arsphenamine, etc.

arsenic As. Element of atomic number 33 of group V of the periodic system. (The word "arsenic" is also frequently used to refer to arsenic trioxide.)
Arsenic is a silver gray or tin-white brittle, crystalline metal that turns black in air. The black form is sometimes encountered as a powder (see arsenic, black). Arsenic also exists in an allotropic form that is a yellow powder (very poisonous!), or a brown or gray powder. Only the metallic form is of commercial importance.
Occurrence: The main commercial source is the flue dust of copper and lead smelters. Arsenic occurs as enargite, realgar, loellingite, arsenopyrite, and danaite (a cobalt-bearing iron arsenic sulfide). The native element is sometimes found. Other minerals are orpiment, mimetite, niccolite, scorodite, smaltite, and sperrylite.
Properties (of metallic form): Sp. gr. 5.72, but commercial material ranges from 5.6-5.9; m.p. 814° (at 36 atm); sublimes at 615°C; appreciably volatile at 100°C; hardness 3.5 Mohs; insoluble in water; soluble in nitric acid.
Derivation: Arsenic trioxide is heated with charcoal.
Grades: Technical; refined 99%; crude 95-98%; high or electronic purity (impurities less than 10 ppm).
Containers: Bulk; 500-lb barrels; 100-lb casks.
Uses: Hardening lead shot; certain copper alloys (bronzes, speculum metal); lead base alloys for battery grids, bearings, cable sheaths, anodes; certain brasses for high temperature use; copper alloy boiler tubes; in medicine in the form of an amalgam; certain solders; diodes.

End uses of arsenic and its compounds, in approximate order of volume: insecticides; weed killers; glass industry; crabgrass controls; cotton desiccants; feed additives; miscellaneous.
Shipping regulations: Poison, class B. Poison label. *

beta-**arsenic**. See arsenic, black.

arsenic acid $H_3AsO_4 \cdot \frac{1}{2}H_2O$.
Properties: White, translucent crystals; poisonous! Soluble in water, alcohol, alkali, glycerin.
Constants: Sp. gr. 2-2.5; m.p. 35.5°C; b.p. loses water at 160°C.
Derivation: By digestion of arsenic with nitric acid.
Grades: Pure; technical; C. P.
Containers: Glass bottles; barrels.
Uses: Manufacture of arsenates, arsenical insecticides; rarely medicinally; glass making; wood treating process.
Shipping regulations: Poison, class B. Poison label. *
See also arsenic pentoxide.

"Arsenic Acid 75." [147] Brand name for arsenic acid for cotton desiccation.
Containers: 5- and 30-gal drums.

arsenic acid anilide. See arsanilic acid.

arsenical Babbitt. Bearing metal with up to 3% arsenic, but usually in the range 0.1 to 1.4%. The arsenic minimizes softening at higher temperatures, promotes fatigue strength. Such bearings are extensively used in automobile and diesel engines.

arsenical nickel. See niccolite.

arsenical pyrites. See arsenopyrite.

arsenicals, liquid. General term referring to solutions or mixtures containing arsenic compounds and intended for use as weed killers.
Warning! Poisonous. May be fatal if swallowed. MCA warning label.
Shipping regulations: Poison, class B. Poison label. *

arsenic anhydride. See arsenic pentoxide.

arsenic, black (beta-arsenic). A form of arsenic obtained by the sublimation of arsenic in a tube of hydrogen; sp. gr. 4.2 (20°C); insoluble in water; soluble in nitric acid, hot alcohol.

arsenic bromide. See arsenic tribromide.

arsenic, butter of. See arsenic trichloride.

arsenic chloride. See arsenic trichloride.

arsenic disulfide (red orpiment; arsenic monosulfide; ruby arsenic; realgar; red arsenic glass; red arsenic sulfide; red arsenic) As_2S_2 or AsS.
Properties: Orange-red powder; poisonous! Soluble in acids and alkalies; insoluble in water.
Constants: Sp. gr. 3.4-3.6; m.p. 307°C.
Derivation: By roasting arsenopyrite and iron pyrites and sublimation.
Grades: Technical.
Containers: Steel drums.
Uses: Leather industry, depilatory agent; paint pigment; shot manufacture; pyrotechnics; calico dyeing and printing; rodenticide.
Shipping regulations: Poison, class B. Poison label. Legal label name: arsenic sulfide. *

arsenic glass, red. See arsenic disulfide.

arsenic hydride. See arsine.

arsenic iodide (arsenous iodide; arsenious iodide; arsenic triiodide) AsI_3.
Properties: Orange-red shining crystalline scales or powder; poisonous!; unstable in sunlight or moisture; sp. gr. 4.70 (25°C); m.p. 146°C; sublimes when heated slowly. Soluble in alcohol, ether, carbon disulfide, chloroform, and benzene; soluble in water with hydrolysis.
Derivation: By the direct union of arsenic and iodine.
Method of purification: Recrystallization.
Grades: Technical.
Containers: Glass bottles.
Uses: Analytical chemistry; medicine.
Shipping regulations: Poison, class B. Poison label. *

arsenic monosulfide. See arsenic disulfide.

arsenic oxide. See arsenic trioxide; arsenic pentoxide.

arsenic pentasulfide As_2S_5.
Properties: Yellow or orange powder. Poisonous! Soluble in nitric acid and alkalies; insoluble in water; decomposes to sulfur and the trisulfide when heated.
Derivation: By the decomposition of sulfoarsenates or by precipitating arsenic acid in a hydrochloric acid solution with hydrogen sulfide. It is filtered, then dried.
Grades: Technical.
Containers: Barrels; boxes.
Uses: Paint pigments; blue fire; Bengal lights.
Shipping regulations: Poison, class B. Poison label. *

arsenic pentoxide (arsenic oxide; arsenic anhydride; arsenic acid) As_2O_5.
Properties: White, amorphous solid; deliquescent. Poisonous! Forms arsenic acid in water. Soluble in water, alcohol.
Constants: Sp. gr. 4.086; m.p. 315°C.
Derivation: By action of oxidizing agent such as nitric acid on arsenious oxide.
Grades: Various as to purity.
Containers: 400-lb drums; cartons; boxes; tins.
Uses: Arsenates; insecticides; dyeing and printing.
Shipping regulations: Poison, class B. Poison label. *

arsenic red. See arsenic disulfide.

arsenic, ruby. See arsenic disulfide.

arsenic sulfide, red. See arsenic disulfide.

arsenic sulfide, yellow. See arsenic trisulfide; orpiment.

arsenic tersulfide. See arsenic trisulfide.

arsenic tribromide (arsenic bromide; arsenious bromide; arsenous bromide) $AsBr_3$.
Properties: Yellowish-white hygroscopic crystals; poisonous! Sp. gr. 3.54 (25°C); m.p. 33°C; b.p. 221°C. Decomposed by water.
Derivation: By the direct union of arsenic and bromine.
Uses: Analytical chemistry; medicine.
Shipping regulations: Poison, class B. Poison label. Legal label name: arsenic bromide. *

arsenic trichloride (arsenic chloride; arsenious chloride; arsenous chloride; butter of arsenic; caustic arsenic chloride; fuming liquid arsenic) $AsCl_3$.
Properties: Colorless or pale yellow, oily liquid. Soluble in concentrated hydrochloric acid and most organic liquids; decomposed by water. Fumes in moist air. Poisonous!
Constants: B.p. 130.5°C; m.p. −18°C; sp. gr. 2.163 (14/4°C); flash point, none.
Derivation: (a) By action of chlorine on arsenic; (b) by distillation of arsenic trioxide with concentrated hydrochloric acid.
Grades: Technical.
Containers: Bottles; 20-, 55-gal drums.
Uses: Intermediate for organic arsenicals (pharmaceuticals, insecticides); poison gases; ceramics.
Shipping regulations: Poison, class B. Poison label. *

arsenic triiodide. See arsenic iodide.

arsenic trioxide (arsenious acid; white arsenic; arsenious oxide; "arsenic," arsenous anhydride) As_2O_3.
Properties: White amorphous, odorless, tasteless powder; poisonous! Slightly soluble in water; soluble in acids and alkalies; freely soluble in glycerin.
Constants: Sp. gr. 3.865; sublimes 193°C.
Derivation: By roasting arsenopyrite and recovery of the arsenic trioxide by sublimation.
Method of purification: Sublimation.
Grades: Technical; 99%; offgrade (96-99%); C.P.; N.F. XI; electronically pure.
Containers: 5-, 50-gal drums; barrels; carloads.
Uses: Manufacture of pigments, glass, shot and bullets; insecticides; rat poison; cattle dip; weed killer; hide preservative; medicine; manufacture of other arsenic compounds; ceramic enamels; aniline colors; mixed with soda ash for boiler compound; textile mordant; sterilizing agent in water purification; wood preservative.
Danger! May be fatal if swallowed. MCA warning label.
Shipping regulations: Poison, class B. Poison label. *

arsenic trisulfide (arsenious sulfide; arsenous sulfide; arsenic tersulfide) As_2S_3.
Properties: Yellow crystals or powder, changes to a red form at 170°C; sp. gr. 3.43; m.p. 300°C; insoluble in water and hydrochloric acid; dissolves in alkaline sulfide solutions and in nitric acid; poisonous!
Derivation: Occurs in nature as the mineral orpiment (q.v.). May be precipitated from arsenious acid solution by the action of hydrogen sulfide.
Grades: Technical.
Uses: Pigment; reducing agent; medicine; fibers to be used in fiber optics.
Shipping regulations: Poison, class B. Poison label. Legal label name: arsenic sulfide. *

arsenic white. See arsenic trioxide.

arsenic yellow. See arsenic trisulfide (pigment); orpiment.

arsenious acid. See arsenic trioxide.

arsenious anhydride. See arsenic trioxide.

arsenious bromide. See arsenic tribromide.

arsenious chloride. See arsenic trichloride.

arsenious iodide. See arsenic iodide.

arsenious oxide. See arsenic trioxide.

arsenious sulfide. See arsenic trisulfide.

arseniuretted hydrogen. See arsine.

arsenobenzene. See arsphenamine.

arsenopyrite (mispickel; arsenical pyrites) $FeS_2 \cdot FeAs_2$.
Properties: Silver-white to gray mineral with grayish-black streak and metallic luster. Found in mineral veins associated with ores of silver, lead, cobalt, tin, zinc. It sometimes is highly auriferous. Soluble in nitric acid. Sp. gr. 5.89 to 6.2; hardness 5.5 to 6.0.
Occurrence: United States (California, Alaska); Canada; Brazil; Australia; Bolivia; England; Germany.
Uses: Ore of arsenic and gold.

arsenous anhydride. See arsenic trioxide.

arsenous bromide. See arsenic tribromide.

arsenous chloride. See arsenic trichloride.

arsenous iodide. See arsenic iodide.

arsenous oxide. See arsenic trioxide.

arsenous sulfide. See arsenic trisulfide.

arsenous sulfide, yellow. See arsenic trisulfide.

arsine (arsenic hydride; arseniuretted hydrogen) AsH_3.
Properties: Colorless gas; extremely poisonous! M.p. −113.5; b.p. −55°C; decomposes 230°C; soluble in water; slightly soluble in alcohol, alkalies.
Derivation: By the action of sulfuric acid on metallic zinc mixed with arsenic compounds.
Grades: Technical.
Containers: Steel cylinders.

Use: Organic synthesis; military poison gas.
Shipping regulations: Poison, class A. Poison gas label. *

arsphenamine (3,3'-diamino-4,4'-dihydroxyarsenobenzene dihydrochloride; arsenobenzene; 606; Ehrlich 606). Originally patented as "Salvarsan." $NH_2(OH)C_6H_3AsAsC_6H_3(OH)NH_2 \cdot 2HCl \cdot 2H_2O$.
Properties: Light yellow, hygroscopic, poisonous powder; odorless or slight odor. Contains not less than 30% arsenic. Oxidized by exposure to air, it becomes more toxic and darker. Soluble in water, alcohol, glycerol and sodium hydroxide solution; slightly soluble in chloroform and ether.
Derivation: By reducing 3-nitro-4-hydroxyphenylarsonic acid with sodium hydrosulfite.
Containers: Ampules; sealed glass tubes.
Use: Medicine.
Shipping regulations: None. *

arsthinol $C_{11}H_{14}NO_3S_2As$ Cyclic 3-hydroxypropylene ester of 3-acetamido-4-hydroxydithiobenzenearsonous acid; 2-(3'-acetamido-4'-hydroxyphenyl)-1,3-dithia-2-arsacyclopentone-4-methanol.
Properties: White, odorless, microcrystalline powder; sparingly soluble in alcohol; very slightly soluble in water.
Grade: N.N.D.
Use: Medicine.

"Artane" Hydrochloride. [57] Trademark for trihexphenidyl hydrochloride [3-(1-piperidyl)-1-phenyl-1-cyclohexyl-1-propanol hydrochloride] $C_{20}H_{31}NO \cdot HCl$.
Properties: White crystalline powder; m.p. 248-252°C; solubility less than 1 g/100 cc water at 25°C; slightly soluble in alcohol; pH (saturated aqueous solution) 6.3; decomposes 258.8°C.
Use: Medicine.

artemisia absinthium oil. See wormwood oil.

artemisia maritima. See worm-seed oil, Levant.

ℓ-arterenol. See levarterenol.

"Artic." [28] Trademark for refrigeration grade 99.5+% pure methyl chloride.
Properties: Boiling range (95%) —24.6° to —23.6°C; wt/gal 7.68 lb at 20°C.
Containers: 100-lb cylinders; multiunit tank cars of 15 (1,300 lb each) tanks.

artificial cinnabar. See mercuric sulfide, red.

artificial gum. See dextrin.

artificial ivory. See ivory, artificial.

artificial malachite. See copper carbonate.

artificial oil of neroli. See methyl anthranilate.

aryl compounds Those whose molecules have the ring structure characteristic of benzene, naphthalene, phenanthrene, anthracene, etc., i.e., either the six-carbon ring of benzene or the condensed six-carbon rings of the other aromatic derivatives. For example, an aryl radical might be phenyl, C_6H_5-; benzyl, $C_6H_5CH_2-$; naphthylene, $C_{10}H_6=$; etc.

As. Symbol for arsenic.

as-. Abbreviation for asymmetrical; same as uns- (q.v.).

"AS-15." [299] Trademark for an agricultural preparation containing streptomycin.

"ASA." [100] Trademark for acetylsalicylic acid, U.S.P.

"ASA." [243] Trade name for a proprietary antiskinning agent.
Properties: Water-white to pale straw liquid, containing 100% active ingredient; sp. gr. (80°F) 0.908; flash point 66°C (Cleveland open cup).
Containers: 5-, 15-, and 55-gal specially lined drums.
Uses: Antiskinning agent for paint, enamel, varnish and ink.

asafetida (devil's dung; food of the gods). A soft brown gum resin obtained by incising the living roots of Ferula fetida and other species of Ferula.
Properties: Very obnoxious odor; bitter, acrid taste.
Occurrence: Native in Tibet; Persia; Turkistan.
Grades: Technical; lump; powdered.
Containers: Cans; barrels.
Uses: Medicine; proprietary remedies; condiment.
Shipping regulations: None. *

asafetida oil.
Properties: Colorless to yellow or brown essential oil. Disagreeable odor suggestive of onions or garlic. Soluble in alcohol, ether, chloroform, benzene.
Constants: Sp. gr. 0.915-0.993 at 15.5°C; optical rotation +10°58' to —17°3'; refractive index 1.4942-1.5259; sulfur content 8.9-31.4%.
Derivation: By distillation of gum asafetida.

asar. See asarum.

asarabacca oil. See asarum europeum oil.

asarum (Canada snake root; wild ginger; asar). The dried rhizomes and roots of Asarum canadense or Asarum europeum, a low perennial herb.
Occurrence: Canada to North Carolina and Kansas.
Chief constituents: An acrid bitter resin and aromatic volatile oil. (See asarum oil).
Uses: Medicine; source of oil.

asarum canadense oil (Canada snake root oil; wild ginger oil; Canadian asarabacca oil).
Properties: Yellowish-brown volatile oil; agreeable, strong aromatic odor and taste; soluble in 70-80% alcohol.
Chief known constituents: Linalool; pinene; borneol; terpineol; geraniol; methyl eugenol.
Derivation: Distilled from the rhizome and roots of Asarum canadense.

Containers: Tins; glass bottles.
Use: Perfumery.
Shipping regulations: None. *

asarum europeum oil (asarabacca oil).
Properties: Thick, brownish liquid; sweetish aromatic odor; pepper-like burning taste. Soluble in alcohol, ether, chloroform and benzene.
Chief known constituents: Asarone, methyl eugenol.
Constants: Sp. gr. 1.015-1.068; optical rotation undetermined.
Derivation: Distilled from rhizome and roots of Asarum europeum.
Method of purification: Rectification.
Grades: Technical.
Containers: Tins; glass bottles.
Use: Perfumery.
Shipping regulations: None. *

asbestos (amphibole; chrysotile; stone flax; earth flax; mountain cork; serpentine). A group of impure magnesium silicate minerals which occur in fibrous form.
Color: White, gray, green, brown.
Two types are used:
1. Serpentine asbestos is the mineral chrysotile, a magnesium silicate. The fibers are strong and flexible so that spinning is possible with the longer fibers. Used for fireproofing and insulating purposes, in the form of textiles; brake lining, clutch facings; gaskets; building materials and electrical and heat insulation.
2. Amphibole asbestos includes the minerals tremolite, actinolite, amosite, crocidolite, and anthophyllite, which are various silicates of magnesium, iron, calcium, and sodium. The fibers are generally brittle and cannot be spun but are more resistant to chemicals and to heat than serpentine asbestos. Used for chemical filters, electrical insulation, and paint filler.

Occurrence: Vermont, Arizona, Georgia, North Carolina; Canada; U.S.S.R.; Africa; China; Italy.
Containers: 100-lb bags; car loads.

"Asbestos TA." [349] An acid-digested asbestos of medium length fiber. The solubility in boiling 15% hydrochloric acid for three hours is not greater than 1.3%. It has a low pH with high purity and uniformity. It is used as a filler and reinforcement in the plastics and missiles industry where it excels in corrosion or heat resistant applications.

"Asbestos TN." [349] The same as TA except instead of medium fibers, the short fibers are used.

ascaridole $C_{10}H_{16}O_2$ 1,4-Peroxido-para-menthene-2.
Properties: A liquid, naturally occurring peroxide; explodes violently on heating to about 130°C or when treated with organic acids. B.p. 84°C (5 mm); sp. gr. 1.011 (13/15°C); refractive index n 20/D 1.4743.
Derivation: By vacuum distillation of chenopodium oil.
Grades: Technical.
Uses: Initiator in polymerizations; medicine.

"Ascarite." [16] A proprietary sodium hydrate asbestos absorbent.
Grades: Mesh 8-20, 20-30.
Containers: 1-lb bottles.
Uses: For the absorption of carbon dioxide in the determination of carbon in iron and steel and in carbon-hydrogen determinations, particularly in quantitative organic microanalysis, and in analysis of respiratory gases; 60 grams will absorb 10-15 g of carbon dioxide.
Shipping regulations: Postal regulations restrict mailing.

asclepias (pleurisy root; butterfly weed). Dried root of perennial herb, Asclepias tuberosa.
Occurrence: Ontario to Minnesota.
Use: Medicine.

ascorbic acid (L-ascorbic acid; vitamin C) $\overline{OCOCOH{:}COHCH}CHOHCH_2OH$. A dietary factor which must be present in the diet of man to prevent scurvy. It cures scurvy and increases resistance to infection. Ascorbic acid is involved in the following metabolic systems:
1. Oxidation of phenylalanine and tyrosine via para-hydroxyphenyl-pyruvate;
2. hydroxylation of aromatic compounds;
3. conversion of folacin to citrovorum factor.

Properties: White crystals (usually plates, sometimes needles); m.p. 192°C; soluble in water; slightly soluble in alcohol; insoluble in ether, chloroform, benzene, petroleum ether, oils, and fats; stable to air when dry.
Sources: Food source: citrus fruits, tomatoes, potatoes, green leafy vegetables, other fruits and vegetables.
Commercial source: synthetic product made by a series of reactions from D-glucose.
Units: One international unit is equivalent to 0.05 mg of L-ascorbic acid.
Grades: U.S.P. XVI.
Containers: Glass bottles; fiber cans; multiwall paper sacks; drums.
Uses: Medicine; nutrition; antioxidant and preservative in foodstuffs; reducing agent in analytical chemistry. The ferric and calcium salts are available for biochemical research.

ascorbic acid oxidase. An enzyme found in plant tissue which acts upon ascorbic acid in the presence of oxygen to produce dehydroascorbic acid.
Use: Biochemical research.

ascorbyl palmitate $C_{22}H_{38}O_7$.
Properties: M.p. 116-117°C; soluble in ether, chloroform, alcohol, animal and vegetable oils; slightly soluble in water.
Uses: Antioxidant for fats and oils; source of vitamin C; stabilizer; emulsifier.
Derivation: Palmitic and L-ascorbic acids.

"Aseptoform." [19] Trademark for esters of para-hydroxybenzoic acid, such as

"Aseptoform" methyl, "Aseptoform" propyl, "Aseptoform" butyl.

"A.S.F." [299] Trademark for a preparation containing water-soluble vitamins; used in medicine.

ash. Mineral content of a product that remains after the product has been burned.

ash, causticized. See causticized ash.

askarel. A generic descriptive name for synthetic nonflammable electrical insulating (dielectric) material which when decomposed by the electric arc evolves only non-explosive gases or gaseous mixtures. Examples are chlorinated aromatic derivatives, particularly pentachlorodiphenyl and trichlorobenzene, but also including pentachlorodiphenyl oxide, pentachlorophenylbenzoate, hexachlorodiphenylmethane, pentachlorodiphenyl ketone, and pentachloroethylbenzene.
Use: Nonflammable liquids for transformer insulation.

A.S.M. Abbreviation for American Society for Metals.

"ASP." [99] Trademark for aluminum silicate pigments; selected particle size, white crystals that have been process engineered to remove moisture, sand, mica, and water-soluble salts. These nonhygroscopic products are essentially inert and insoluble under normal use conditions. Certain grades are available with surface treatment or modifications for special properties in aqueous or organic systems.
Typical analysis (volatile free basis):
ASP (unmodified): SiO_2, 45.4%; Al_2O_3, 38.8%; Fe_2O_3, 0.3%; TiO_2, 1.5%; CaO, 0.1%; Na_2O, 0.1%; K_2O, trace; LOI, 13.8%.
ASP 102, 106, 600 and 602 are used for rubber applications. They have very high whiteness and excellent extrusion properties. ASP 400 and 900, because of their "anti-penetration" characteristics, make effective blocking agents in adhesives for corrugated board, laminated fiber board, paper bag seams, paper tubes, cores, and proprietary adhesives.

asparagic acid. See aspartic acid.

asparagine (alpha-aminosuccinamic acid; beta-asparagine; althein; aspartamic acid; aspartamide)
$NH_2COCH_2CH(NH_2)COOH$. The beta amide of aspartic acid, a non-essential amino acid, existing in the D(+)- and L(−)-isomeric forms as well as the DL-racemic mixture. L(−)-asparagine is the most common form.
Properties L(−)-asparagine monohydrate: White crystals; m.p. 234-235°C; acid to litmus; nearly insoluble in ethanol, methanol, ether, and benzene; soluble in acids and alkalies.
Derivation: Occurs in most young plants, especially Leguminosae.
Uses: Biochemical research; preparation of culture media; medicine.

beta-asparagine. See asparagine.

asparaginic acid. See aspartic acid.

aspartamic acid. See asparagine.

aspartamide. See asparagine.

aspartic acid (asparaginic acid; asparagic acid; aminosuccinic acid)
$COOHCH_2CH(NH_2)COOH$. A naturally occurring nonessential amino acid. The common form is L(+)-aspartic acid.
Properties: Colorless crystals; soluble in water; insoluble in alcohol and ether; optically active:
DL-aspartic acid: M.p. 278-280°C with decomposition; sp. gr. 1.663 (12/12°C)
L(+)-aspartic acid: M.p. 251°C
D(−)-aspartic acid: M.p. 269-271°C with decomposition; sp. gr. 1.6613.
Source: Young sugar cane; sugar beet molasses.
Derivation: Hydrolysis of protein; reaction of ammonia with diethyl fumarate.
Uses: Biological and clinical studies; preparation of culture media; organic intermediate; dietary supplement.
Available commercially as D(−)- and L(+)-aspartic acid.

aspergillic acid $C_{12}H_{20}N_2O_2$. Dibutylhydroxypiperazone. An antibiotic substance from strains of Aspergillus flavus.
Properties: M.p. 97°C; insoluble in cold water; soluble in common organic solvents and dilute acids. Hydrochloride melts at 178°C and is soluble in water.

asphalt (asphaltum; earth pitch; Trinidad pitch; Judean pitch; Jew's pitch; mineral pitch; petroleum asphalt). Black to dark-brown solid or semi-solid material which gradually liquefies when heated; its predominating constituents are bitumens, which occur in the solid or semi-solid form in nature, or are obtained by refining petroleum, or are combinations of the bitumens mentioned with each other or with petroleum or derivatives thereof.
Properties: Black solid, with a dull luster; streak black to brown; solubility in carbon disulfide 69% (native), 99% (petroleum). Flash point 350°F (native), 450°F (petroleum).
Derivation: Found native, or obtained as a residue from the distillation of crude petroleum.
Occurrence: United States; Trinidad; Venezuela; Cuba; Canada; Europe.
Containers: 1-gal friction top pails; 25- and 50-lb cans; 400-lb seamless drums; tank trucks; tank cars.
Uses: Paving, roofing, waterproofing material; paints; oil-base drilling muds; dielectrics; rubber blends; fungicides; fuel, including solid propellants for rockets.
Shipping regulations: None.*
See also asphaltite; bitumen; asphaltic pyrobitumen; rock asphalt; gilsonite.

asphalt, blown. See blown asphalt.

*See "I.C.C. Shipping Regulations," page xiii.
Reference numbers refer to name of manufacturer. See "List of Manufacturers," page v.

asphalt, condensed. See blown asphalt.

asphalt, cut-back. A liquid petroleum product, produced by fluxing an asphaltic base with suitable distillates. (A. S. T. M. definition).
Properties: Flash point (open cup) 100°F. Solubility of residue from distillation in carbon tetrachloride 99.5%.
Use: Road surfaces.

asphalt, emulsified. A suspension or emulsion of ordinary asphalt in water. Such emulsions have attained wide use because, unlike straight asphalt, they do not have to be heated to be applied. The suspended asphalt is spread on in the usual way; and after the water has evaporated, the asphalt hardens into a continuous mass.
Containers: Steel drums and tank cars or trucks.
Uses: Highways; cement waterproofing; roofing compounds; and the like.

asphaltic pyrobitumen. A natural bitumen distinguished by infusibility and low solubility in carbon disulfide. Species include elaterite, wurzilite, albertite, impsonite, and pyrobituminous shales.

asphaltite. A native bituminous mixture similar to asphalt, but more difficultly fusible, and less soluble in carbon disulfide. Materials coming under this classification are gilsonite, glance pitch, grahamite, Egyptian asphalt, and Syrian asphalt.

asphalt, liquid. See residual oils.

asphalt, oxidized. See blown asphalt.

asphalt paint. Asphaltic base in a volatile solvent with or without drying oils, resins, fillers, and pigments.

asphalt, sludge. See sludge asphalt.

asphalt, sulfurized. See sulfurized asphalts.

asphaltum. See asphalt.

asphaltum oil. See residual oils.

aspic oil. See lavender-spike oil.

aspidium (male fern; shield fern).
Derivation: Dried rhizome of Dryopteris flixmas.
Occurrence: North America; northern Asia; Europe; northern Africa.
Grades: Technical; U. S. P. XVI.
Containers: Bags.
Use: Medicine.
Shipping regulations: None. *

aspidosperma. See quebracho.

aspidospermine $C_{22}H_{30}O_2N_2$. An alkaloid.
Properties: White to brownish-yellow crystalline alkaloid.
Constants: M.p. 208°C; b.p. 220°C (1 to 2 mm); sublimes at 180°C under reduced pressure. Soluble in fats and fixed oils; sparingly soluble in absolute alcohol and ether. Its sulfate and hydrochloride are soluble in water.
Use: Medicine.

aspirin. See acetylsalicylic acid.

aspirin, soluble. The calcium, lithium, or sodium salts of aspirin.

assay. A method of analysis, or the results of an analysis, or the process of performing the analysis. Formerly applied to ores and metals, but now more generally used.

association. A chemical reaction, generally caused by a weaker class of chemical bonding forces, but stronger than the forces that hold liquids and solids together, more specific and usually reversible. It may be between molecules of the same kind, for example the dimerization of acetic acid in nonaqueous solvents, or between molecules, or ions and molecules, of unlike kind, as is represented by the formation of ammonia complexes of many metal ions. The simple and the larger molecules are usually in equilibrium with one another with appreciable concentrations of each present. Polymerization is almost synonymous but usually refers to processes in which the larger molecules are not easily reconverted to the simpler form.

A-stage resin (resol). The first stage of condensation in the formation of phenol-formaldehyde and similar resins. At this stage the resin is fusible, i. e., it becomes fluid on heating, and is completely soluble in alcohol. Its molecules have not become cross-linked. This form of resin is used for laminating and impregnating paper and fabrics. On heating further, the resin is converted to the infusible, insoluble, thermoset cross-linked form. See B-stage resin (intermediate) and C-stage resin.

astatine At. Element of atomic number 85, a member of the halogen family of elements (group VII). It is reported to occur naturally in very small quantities from the branched beta decay in the three naturally occurring radioactive decay series. Astatine is also formed by the alpha decay of francium and has been prepared synthetically by bombarding bismuth with alpha particles. Astatine 210 is the longest-lived isotope, having a half-line of 8.3 hours. Thus its properties are difficult to evaluate. Astatine, although more metallic than the other halogens, has halogen-like characteristics, being easily volatilized and easily reduced to the free element in solution. It is similar to iodine in that it is concentrated in the thyroid gland.

"Asterol" Dihydrochloride. [190] Trademark for a brand of diamthazole dihydrochloride (q. v.), an antifungal agent.

asthma weed. See lobelia.

ASTM. Abbreviation for American Society for Testing Materials, a national technical society organized for "the promotion of knowledge of the materials of engineering, and the standardization of specifications and the methods of testing." The society

*See "I. C. C. Shipping Regulations," page xiii.
Reference numbers refer to name of manufacturer. See "List of Manufacturers," page v.

publishes standards for all types of engineering materials in its Book of ASTM Standards, which is frequently revised to keep it up to date.

"Astol" A. [206] Brand name for a proprietary scouring and desizing agent consisting of a stable emulsion of trichloroethylene suitable for use on delicate materials.

"Aston 123." [328] Trade name for a thermosetting polyamine, 20% active.
Use: Durable anti-static agent.

"Astracel." [307] Fast dyes for printing acetate rayon.

"Astrol." [307] A fast alizarine direct blue.

asymmetry. A molecular arrangement in which a particular carbon atom is joined to four different groups. See optical isomerism.

At. Symbol for astatine.

"Atabrine" Hydrochloride. [162] Trademark for the drug, quinacrine hydrochloride.

atacamite $Cu_2Cl(OH)_3$. Naturally occurring basic copper chloride.
Properties: Color, various shades of green; luster, adamantine to vitreous; hardness 3-3.5; sp. gr. 3.75-3.94. Found in the oxidized portions of copper deposits.
Occurrence: Chile; Bolivia; Mexico; Australia; Arizona.
Use: A minor ore of copper.

atactic. A type of polymer molecule in which substituent groups or atoms are arranged randomly above and below the backbone chain of atoms, when the latter are arranged so as all to be in the same plane. See polymer, stereospecific.

"Atarax." [299] Trademark for hydroxyzine hydrochloride.

"Ataraxoid." [299] Trademark for a combination drug containing prednisolone and hydroxyzine hydrochloride.

ATE. Abbreviation for aluminum triethyl. See triethyl aluminum.

"Atlac." [89] Trademark for a series of polyester resins for use in reinforced plastics.

"Atlacide." [147] Brand name for a non-selective chlorate weed and grass killer.
Containers: 100-lb drums.
Use: As water spray or dry powder to destroy roots of undesirable vegetation.

Atlas "A." [147] Trademark for a sodium arsenite solution.
Containers: 5-, 30- and 55-gal drums.
Properties: Contains 4, 6, 8 or 9 ½ lbs arsenic trioxide per gal; mixes readily in cold water; nonflammable and nonexplosive; also available as a soluble powder (Atlas "A" SP); 5 lbs is equivalent to 1 gal Atlas "A" solution.
Uses: For selective weed control in turf; tree or stump killer; potato vine killer prior to harvesting; controls certain submerged vegetation in ponds and lakes; effective for control of subterranean termites.

Atlas "D" Debarking Compound. [147] Brand name for a sodium arsenite solution.
Containers: 2-gal cans; 5-, 30-, 55-gal drums.
Use: Used during sap-flow season to obtain easy removal of bark from standing timber intended for pulpwood, fence posts and poles. Contains animal repellent.

"Atlastavon." [41] Trade name for synthetic-resin sheet lining of the plasticized vinyl type used to protect steel tanks at temperatures up to 160°F without brick sheathing. Outstanding resistance to oxidizing acids.

"Atlastic 31." [41] Trade name for an asphaltic lining material used as a resilient membrane between concrete and acid brick sheathing.

"Atlastic 77." [41] Trade name for a cold-trowel, bituminous-based compound for joining sewer pipe.

"Atlas W.P. & C." [147] Trademark for a non-corrosive, non-poisonous, non-vaporous, and odorless liquid for cleaning and preserving ships' decks.
Containers: 5-gal drums.

"Atlosol." [89] Trademark for an emulsifier used with oil and water to make drilling fluids for oil production.

"Atlox." [89] Trademark for series of emulsifiers developed for use with agricultural pesticides.

ATM. Abbreviation for aluminum trimethyl. See trimethyl aluminum.

"Atmos." [89] Trademark for a series of mono- and diglyceride emulsifiers used in ice cream and frozen desserts.

atmosphere.
1. Synonym for air (q.v.).
2. A unit of pressure equivalent to 760 mm mercury or 14.696 pounds per square inch absolute (psia).

"Atmul." [89] Trademark for a series of mono- and diglyceride emulsifiers used in baked goods and other food products.

atom. The smallest particle of an element which retains the characteristic properties and behavior of the element, or the smallest particle of an element that can exist by itself or in combination with similar particles of the same or other elements.

atomic bomb (A-bomb). An explosive device containing two or more masses of fissionable substances such as uranium-235 or plutonium, together with a mechanical means of uniting these masses suddenly. The size of the individual masses must be subcritical, i.e., less than the critical mass, but the combined mass is greater (super-critical). Immediately upon uniting the masses an extremely violent explosion occurs due to rapid uncontrolled release of nuclear energy by fission of the U-235 or plutonium. This nuclear reaction is propagated into a chain reaction when a neutron from fission of one atom enters another atom and causes its fission, so that it releases energy and more neutrons, which continue the process. In a large mass few neutrons escape and the chain reaction becomes explosively rapid. It has been proposed that such explosive devices of great power may be used for peaceful purposes such as the clearing of underwater obstacles, or the release of petroleum from oil shales.

atomic energy. Misnomer for nuclear energy (q. v.).

atomic hydrogen welding. A term used to describe a method of welding in which hydrogen gas is passed through an arc between two tungsten electrodes. The arc breaks down the molecules to form atomic hydrogen. The recombination of the atoms to form molecules and the combustion of the molecular hydrogen in atmospheric oxygen produce a flame temperature of 4000-5000°C.

atomic number. The number of unit positive charges carried by the nucleus of an atom, or the number of external electrons outside the nucleus of an atom. With a few exceptions, it is the number of the element in the sequence obtained by arranging the elements in the order of increasing atomic weight.

atomic volume. The atomic weight of an element divided by its density.

atomic weight. The weight of an atom of an element compared with the weight of an atom of the lighter isotope of oxygen taken as standard at 16. This is the definition of the so-called physical atomic weights, which differ very slightly from chemical atomic weights based on the normally occurring mixture of isotopes of oxygen.

atoxylic acid. See arsanilic acid.

ATP. Abbreviation for adenosine triphosphate.

"Atpet." [89] Trademark for a series of emulsifiers used in conjunction with oil as corrosion inhibitors, solubilizers for production of soluble cutting oils, and waterblock removal agents in oil well drilling.

atropamine. See apoatropine.

atropine (daturine) $C_{17}H_{23}NO_3$. An alkaloid obtained from species of Atropa, Datura or Hyoscyamus.
Properties: White crystals or white crystalline powder; extremely poisonous! optically inactive (but usually contains levorotatory hyoscyamine). Soluble in alcohol, ether, chloroform, and glycerol. Slightly soluble in water. M.p. 114-116°C.
Derivation: By extraction from Datura stramonium, or synthetically.
Grades: Technical; N. F. XI.
Containers: Glass bottles; tins.
Use: Medicine; antidote for nerve gases.
Shipping regulations: None.*

atropine methylbromide $C_{17}H_{23}NO_3 \cdot CH_3Br$.
Properties: White crystals; m. p. 222-223°C; very poisonous! Soluble in water and alcohol.
Use: Medicine.

atropine methylnitrate $C_{17}H_{23}NO_3 \cdot CH_3NO_3$.
Properties: White crystalline powder; odorless; m.p. 163°C; very poisonous! soluble in water and alcohol.
Use: Medicine.

atropine oleate. Soluble in benzene, chloroform, ether, and oils; insoluble in water.
Derivation: A 2% solution of atropine in equal parts of olive oil and oleic acid.
Use: Medicine.

atropine sulfate $(C_{17}H_{23}O_3N)_2H_2SO_4 \cdot H_2O$.
Intense poison!
Properties: White, odorless, crystalline powder, efflorescent in dry air; very poisonous! Soluble in water, alcohol, and glycerol; less soluble in chloroform and ether. M.p. given variously at 183-194°C. U.S.P. requires not less than 188°C.
Derivation: By interaction of a solution of atropine in ether and of sulfuric acid in strong alcohol.
Grades: U.S.P. XVI.
Containers: 1-oz bottles; 5- and 15-grain vials.
Uses: Similar to those of atropine (q. v.).

"Attaclay." [99] Trademark for a naturally absorbent material produced from the mineral attapulgite, a clay mined only in southwestern Georgia and northwestern Florida. Activated by specific thermal treatment and ground to fine particle size for use as versatile carriers, diluents, and conditioners for the formulation of insecticides, herbicides, fungicides and other agricultural dusts and wettable powders.
Typical analysis (volatile free basis): SiO_2, 67.0%; Al_2O_3, 12.5%; MgO, 11.0%; Fe_2O_3, 4.0%; CaO, 2.5%; other, 3.0%. Tamped bulk density 25-27 lbs/cu ft; sp. gr. 2.47.

"Attacote." [99] Trademark for a fine particle size, sorptive attapulgus clay product recommended as an anti-caking, anti-agglomerating agent for conditioning ammonium nitrate and sulfate crystals, urea, granular fertilizers, etc.
Properties: Tamped bulk density 18-22 lbs/cu ft; sp. gr. 2.47; average particle size 5.3 microns.
Attacote "C": A modified grade of Attacote that will meet a low specification for ammonia release.

"Attagel." [99] Trademark for colloidal grades of attapulgus clay (hydrous magnesium aluminum silicate).
Properties: Tamped bulk density 19-21 lbs/cu ft (Attagel 20); 30-35 lbs/cu ft (Attagel 30). Sp. gr. 2.36; average particle size

*See "I.C.C. Shipping Regulations," page xiii.
Reference numbers refer to name of manufacturer. See "List of Manufacturers," page v.

0.12-0.14 microns.
Attagel 30: A pulverized colloidal form of attapulgus clay which disperses by the application of shear to produce thickening of aqueous systems.
Attagel 20: Similar except that it has been processed to obtain the best particle size distribution.
Uses: In thickening or gelling of organic or aqueous systems, suspending pigments or other solids in aqueous or organic systems, stabilizing emulsions, or for introducing thixotropic properties into organic or aqueous systems.

attapulgite $(Mg, Al)_5Si_8O_{22}(OH)_4 \cdot 4H_2O$. A clay mineral, the active ingredient in most fuller's earth (q.v.).

"Attapulgus" 150. [99] Trademark for attapulgus drilling clay processed from attapulgite. Used in fresh, salt, gyp, or high temperature systems. For typical analysis, see "Attaclay."

attar of roses. See rose oil.

"Attasorb" LVM. [99] Trademark for a complex hydrated magnesium aluminum silicate; an effective sorptive cleaner and polishing agent, emulsion stabilizer, and film strengthener. For typical analysis, see "Attaclay." Tamped bulk density 15-18 lbs/cu ft; pH 7.5-9.0.

Au. Symbol for gold.

aubepine. See para-anisaldehyde.

Auer metal. See pyrophoric alloy.

Auger effect. The phenomenon of the emission of electrons from atoms during de-excitation from energy levels from which x-rays are expected. When an electron is removed from an inner shell of an atom, the x-ray energy of the transition of outer electrons into the inner shells may be internally converted to the ejection of several of the more loosely bound electrons. These emitted electrons are called Auger electrons.

"Aura." [108] Trademark for a powdered, chlorinated, alkaline polyphosphate detergent for mechanical washing of all tableware, including plastic and chinaware.
Containers: 125-, 350-lb drums.

"Auragreen." [329] Trademark for a special dye mixture used for control of brownpatch disease of turf grass. It has advantage as a grass dye.

auramine $(CH_3)_2NC_6H_4(C:NH)C_6H_4N(CH_3)_2 \cdot HCl$.
Properties: Yellow flakes or powder; soluble in water, alcohol, and ether.
Uses: Yellow dye for paper, textiles, leather; also an antiseptic; fungicide.

aurantiin. See naringin.

"Aurantiol." [227] $C_{18}H_{27}O_3N$. Trademark for hydroxycitronellal-methyl anthranilate Schiff base (methyl N-3,7-dimethyl-7-hydroxyoctylidene-anthranilate).
Properties: Linden-orange flower odor; yellow honey-like viscous liquid; stable; refractive index (n 20/D) 1.5350-1.5460; flash point TCC 206°F. Clearly soluble in 2 parts of 70% alcohol, 1 part of 80% alcohol.
Occurrence: Not found in nature.
Uses: Floral scents.

"Aurasperse." [134] Trade name for water dispersions of pigments for use in latex based paints.

aureolin. See Indian yellow.

"Aureomycin." [315] Trademark for chlortetracycline (q.v.).

"Auric." [28] Trademark for a hydrated ferric oxide brown pigment.

auric bromide. See gold tribromide.

auric chloride. See gold chloride.

auric hydroxide. See gold hydroxide.

auric oxide. See gold oxide.

auric trioxide. See gold oxide.

aurin $(C_6H_4OH)_2CC_6H_4O$.
Properties: Reddish-brown pieces with greenish metallic luster; easily powdered; insoluble in water, benzene, and ether; soluble in alcohol.
Uses: Indicator; dye intermediate.
Note: This material is often confused with para-rosolic acid, from which it differs only slightly in composition.

auripigment. See orpiment.

"Aurofac." [57] Trademark for a feed supplement based on chlortetracycline.

alpha-**auromercaptoacetanilid.** See aurothioglycanide.

aurothioglucose (gold thioglucose) $C_6H_{11}O_5SAu$.
Properties: Yellow to yellow-green powder; almost odorless and tasteless. Soluble in water (decomposes); insoluble in acetone, alcohol, chloroform, and ether; pH of 1% aqueous solution 6.3.
Derivation: Reaction of gold bromide with an aqueous solution of thioglucose and sulfur dioxide.
Grade: N.F. XI.
Use: Medicine.

aurothioglycanide (alpha-auromercaptoacetanilid) $C_6H_5NHCOCH_2SAu$.
Properties: Grayish-yellow powder. Insoluble in acids, bases, benzene, ether, chloroform, and water.
Grade: N.N.D.
Use: Medicine.

aurous bromide. See gold monobromide.

aurous iodide. See gold iodide.

aurous sodium thiosulfate. See gold-sodium thiosulfate.

austenite. A nonmagnetic solid solution of carbon or ferric carbide in gamma iron. Very unstable below its critical temperature, but may be obtained in high carbon steels by rapid quenching from high temperatures. Addition of manganese and nickel lowers critical transition temperature and austenite may be obtained stable at room temperature in proper alloys. Characterized by a face-centered cubic lattice.

austenitic alloys (austenitic steels). Alloys of iron, chromium, nickel noted for their resistance to corrosion.

Australian bark. See wattle bark.

Australian black wattle gum. See wattle gum.

Australian fever bark. See alstonia.

Australian fever tree. See eucalyptus.

Australian native quinine. See alstonia.

Austrian cinnabar. See chrome red.

"Autoset." [65] Trademark for urea-formaldehyde resin for use as a binder in the manufacture of particle board.

autumn crocus. See colchicum.

autunite (calcium uranite; lime uranite) $Ca(UO_2)_2(PO_4)_2 \cdot 10H_2O$. A natural hydrated phosphate of calcium and uranium, found in the oxidized zones of uranium deposits.

Properties: Color, lemon yellow to sulfur yellow, sometimes pale greenish yellow; streak yellowish; one good cleavage; hardness 2.0-2.5; sp. gr. 3.1-3.2; fluorescent in ultraviolet light; radioactive.

Occurrence: New York, North Carolina, New Mexico, South Dakota; Europe.

Use: Minor ore of uranium.

auxenolic acid. See auxin.

auxentriolic acid. See auxin.

auxin.

1. General term for any of a group of natural and synthetic substances referred to as plant hormones which stimulate plant growth. The presence of auxins in plant cells contributes to the ability to bend toward the light. The natural materials are formed in the tips of growing plants, and in root tips, and on the shaded side of growing shoots. The materials occur in small amounts in numerous varieties of plant and animal matter. These are mostly ether-soluble carboxylic acids of the formula $R(CH_2)_xCOOH$, where R is either aromatic or an unsaturated naphthenic group. Phenylpropionic acid, and 3-indoleacetic acid are well-known examples.

2. Specifically auxin A and auxin B are particular naturally occurring members of the above group with formulas $C_{18}H_{32}O_5$ and $C_{18}H_{30}O_4$. The former is also known as auxentriolic acid, and the latter as auxenolic acid. Heteroauxin (3-indoleacetic acid) is a related material.

auxochrome. A radical or group of atoms whose presence is essential in enabling a colored organic substance to be retained on fibers. The best examples are the groups -COOH, $-SO_3H$, -OH and $-NH_2$.

ava-ava. See kava.

aventurine. A form of native quartz (q. v.) spangled with scales of mica, hematite or other mineral; also a kind of glass containing gold-colored spangles of brass or similar material.

Use: In ornaments.

"Avertin." [162] Trademark for tribromoethanol solution and amylene hydrate.

"Av-Formula." [123] Trademark for an animal feed supplement containing antibiotics and vitamins.

avgas. Abbreviation for aviation gasoline.

aviation gasoline. A fuel especially suited for aircraft use, in that it has high volatility, high octane rating, and good stability.

"Avicel." [261] Trademark for crystalline cellulose, a highly purified particulate form of cellulose; molecular weight 30,000-50,000; particle size 10-50 microns; density 1.55 (bulk density 0.3-0.8). Dispersible in water; swells in dilute alkali; insoluble in dilute acid, organic solvents, oils. No caloric value in foods. Forms stable gels with water, for calorie control in foods, and for pharmaceutical creams and lotions. Absorbs oil based and syrupy foods to form free flowing forms for dry mixes. Used in column chromatography, and as raw material for cellulose reactions.

"Avisun." [429] Trade name for polypropylene resin; molecular weight 100,000-500,000; sp. gr. 0.90-0.92; m. p. 347°F. Available in several molding and film grades.

"Avitex." [28] Trademark for a group of textile softeners, lubricants, and antistatic agents. Included are both anionic and cationic types.

"Avitone." [28] Trademark for a line of chemical compounds that are used principally as softening, lubricating, and finishing agents for textiles, leather, and paper.

"Avloclor." [207] Trademark for chloroquine phosphate [7-chloro-4-(4-diethylamino-1-methylbutylamino) quinoline diphosphate]. A synthetic drug.

"Avloprocil." [207] Trademark for procaine penicillin G.

"Avlosulfon." [207] Trademark for dapsone (diaminodiphenylsulfone); used in medicine.

Avogadro, principle of (Avogadro's hypothesis). Equal volumes of different gases under the same conditions of temperature and pressure contain the same number of molecules. An alternative and sometimes more useful statement is that equal molecular quantities of different gases occupy the same volume at the same conditions of temperature and pressure.

"Avolin." [188] $C_6H_4(COOCH_3)_2$. Trademark for a special perfume grade of dimethyl phthalate. Free from odor; used as a diluent for perfume materials.

axerophthol. Synonym for vitamin A_1. See vitamin A.

"Ayr-Trap." [205] An air-entraining admixture for concrete to minimize capillary absorption of moisture and the disintegration caused by the freezing and thawing of such moisture.

azacyclonol hydrochloride (alpha, alpha-diphenyl-4-piperidinemethanol hydrochloride) $(C_6H_5)_2COHC_5H_{10}N \cdot HCl$.
Properties: Small white crystals or crystalline powder. Odorless, stable in air. M.p. 270-281°C (dec.). Slightly soluble in water, alcohol. Insoluble in chloroform, ether, acetone, hexane; pH of 1 in 200 solution 5-7.
Grade: N.F. XI.
Use: Medicine.

azapetine phosphate $C_{17}H_{17}N \cdot H_3PO_4$. 6-Allyl-6,7-dihydro-5H-dibenz[c,e]-azepine phosphate.
Properties: M.p. 210-213°C. Soluble in water.
Grade: N.N.D.
Use: Medicine.

azaserine $C_5H_7N_3O_4$. O-Diazoacetyl-L-serine. $N_2CHCOOCH_2CH(NH_2)COOH$. An antibiotic of importance in cancer research. It has had some success in inhibiting certain types of cancer in mice and rats. It was originally isolated from a Streptomyces, but is also prepared synthetically.

6-azauracil riboside. See 6-azauridine.

6-azauridine (6-azauracil riboside; as-triazine-3,5(2H,4H)dione riboside).
Derivation: Microbiological fermentation.
Use: Suggested as treatment for cancer.

azelaic acid (nonanedioic acid; 1,7-heptanedicarboxylic acid) $CO_2H(CH_2)_7CO_2H$.
Properties: Yellowish to white crystalline powder; m.p. 106°C; b.p. 286.5°C at 100 mm; soluble in hot water, alcohol, and organic solvents.
Derivation: Oxidation of oleic acid by ozone.
Grades: Technical
Containers: Bags; wooden barrels; kegs.
Uses: Organic synthesis; lacquers; production of hydrotropic salts; alkyd resins; polyamides; plasticizers; jet engine lubricants.

azelaoyl chloride $ClOC(CH_2)_7COCl$.
Properties: B.p. 125-130°C (3 mm). Slowly decomposes in cold water; soluble in hydrocarbons and ethers.
Use: Organic synthesis.

azeotrope. See azeotropic mixture.

azeotropic distillation. A type of distillation in which a substance is added to the mixture to be separated in order to form an azeotropic mixture with one or more of the constituents of the original mixture. The azeotrope or azeotropes thus formed will have boiling points different from the boiling points of the original mixture and will permit greater ease of separation.

azeotropic mixture (azeotrope). A liquid mixture of two or more substances which behaves like a single substance in that the vapor produced by partial evaporation of liquid has the same composition as the liquid. The constant boiling mixture exhibits either a maximum or minimum boiling point as compared with that of other mixtures of the same substances.

aziminobenzene. See 1,2,3-benzotriazole.

azine dyes. Dyes which are derived from phenazine $(C_6H_4)N_2(C_6H_4)$ (tricyclic). The chromophore group may be >C=N-; but the cause of color is more probably due to the characteristic unsaturation of the benzene rings. The members of the group are quite varied in application. The "Nigrosines" (Color Index 864 and 865) and "Safranine" (Color Index 841) are examples of this group.

aziridine. See ethyleneimine.

"Azite" 900 Liquefier. [57] Trademark for a nonresinous nitrogenous chemical used in the paper industry primarily for the liquefaction of various coating adhesives.

azlon. Generic name for a manufactured fiber in which the fiber-forming substance is composed of any regenerated naturally occurring protein (Federal Trade Commission). Proteins from corn, peanuts and milk have been used. Azlon fiber has a soft "feel", blends well with other fibers, and is used, in general, like wool.

azobenzene (diphenyldiimide; benzeneazobenzene) $C_6H_5N_2C_6H_5$.
Properties: Yellow or orange crystals; m.p. 68°C; b.p. 297°C; sp.gr. 1.203 (20/4°C); soluble in alcohol and ether; insoluble in water.
Derivation: By reducing nitrobenzene with sodium stannite.
Method of purification: Crystallization.
Grades: Technical
Containers: Wooden kegs or fiber containers.
Use: Manufacture of benzidine and induline dyes, rubber accelerators; fumigant.
Shipping regulations: None.*

azobenzene-para-sulfonic acid $C_{12}H_{10}O_3N_2S$.
Properties: Orange crystals; m.p. 129°C.
Uses: Intermediate and reagent chemicals.

"Azocel." [83] Trade name for a nonionic, stable solubilizer for diazotized, developed, naphthol, acid, and acetate dyes.

azodine (benzeneazonaphthylethylenediamine) $C_{18}H_{18}N_4$.
Properties: Red crystals; m.p. 107-108°C.
Use: Reagent for rapid determination of penicillin in blood, urine and other media.

azo dyes. Those that have the -N=N- group as a chromophore group in their molecular structure and are produced from amino compounds by the processes of diazotization and coupling. Over half of the commercial dyestuffs are in this general category. By varying the chemical composition it is possible to produce acid, basic, direct, or mordant dyes. This general group is subdivided as monoazo, disazo, trisazo, and tetrazo according to the number of -N=N- groups in the molecule. Examples of these dyes are Chrysoidine Y, Bismarck Brown 2R and Direct Green B.

"Azo Fuchsine." [307] Brand name of proprietary line of acid dyestuffs, used on wool. Can also be used on certain kinds of paper and leather.

"Azoguard." [206] Brand name for a proprietary product. Increases the stability of diazo compounds and is a useful addition to printing pastes and padding liquors containing these products.

azoic dyes (azoics). Insoluble azo dyes made on or in the fiber, except direct dyes that are further developed on the fiber. Formerly called developed, ice, or ingrain colors.

"Azomel" A. [206] Brand name for a proprietary product used for pasting and dissolving the "Brenthols." Owing to its stability to hard water and metallic salts, it aids in the production of good rubbing fastness and is particularly recommended when machine dyeing with "Brenthols."

azophenylene. See phenazine.

"Azo Phloxine." [307] Brand name for a proprietary line of acid dyestuffs, used on wool. Can also be used on certain kinds of paper and leather.

"Azopol A." [206] Brand name of a proprietary product. An aqueous solution of an ethylene oxide condensate.
Uses: In the developing bath for the production of azoic dyeing on cotton; it is employed as a stabilizer and as a dispersing agent, to give dyeings of improved rubbing fastness.

"Azosol." [307] Brand name for a proprietary line of dyestuffs; soluble in organic solvents.
Uses: For the coloring of spirit lacquers and spirit inks.

azosulfamide (disodium 2-(4'-sulfamylphenylazo)-7-acetamido-1-hydroxynaphthalene-3,6-disulfonate) $C_{18}H_{14}N_4Na_2O_{10}S_3$.
Properties: Dark red, odorless, tasteless powder. Soluble in water with an intense red color; practically insoluble in organic solvents.
Use: Medicine.

azotic acid. See nitric acid.

azoxytoluidine. See diaminoazoxytoluene.

azulene $C_{16}H_{26}O$. The blue coloring matter of chamomile, wormwood and other essential oils. A terpene.
Properties: Blue, oily liquid; b.p. 170°C; sp. gr. 0.987; insoluble in water.

azure A carbacrylic resin. See azuresin.

azure blue. See cobalt blue.

azuresin (azure A carbacrylic resin).
Properties: Moist, irregular, dark blue or purple granules. Slightly pungent odor.
Derivation: Carbacrylic cation-exchange resin, in reversible combination with 3-amino-7-dimethylaminophenazathionium chloride (azure A dye).
Grade: U.S.P. XVI.
Use: Medicine (diagnostic test).

azurite (copper carbonate, blue; hydrated basic copper carbonate; chessy copper; chessylite) $2CuCO_3 \cdot Cu(OH)_2$. Copper mineral, various shades of azure-blue in color, vitreous almost adamantine luster, light blue streak. Found as an alteration product of chalcopyrite and other sulfide ores of copper in the upper oxidized zones of mineral veins. Contains 69.2% CuO, 25.6% CO_2, balance water. Sp. gr. 3.77-3.83; hardness 3.5-4.
Occurrence: United States; Australia; Europe; U.S.S.R.; Africa.
Uses: Ore of copper; jewelry.

azurmalachite. A blue and green mixture of the minerals azurite and malachite. Sometimes used as a gem.

B

B. Symbol for boron.

Ba. Symbol for barium.

babassu oil. An oil expressed or extracted from a species of Brazilian palm nut (Orbignya speciosa).

Properties: Pale yellow when refined; sp. gr. 0.9240 (15°C); saponification value 246-251; iodine value 14.0-16.3; refractive index n 40/D 1.4500.

Use: It is similar in its properties and uses to coconut oil, being edible. The press cake and meal are used as stock feed, while the oil itself is chiefly used in food, soap, and cosmetic industries

Babbitt, bearing metal. See Babbitt metal.

Babbitt metal (Babbitt, bearing metal). A group of soft alloys used widely for bearings. They have good bonding characteristics with the supporting metal, maintain oil films on their surface, and possess nonseizing, antifriction and bearing properties. Used as cast, machined or preformed bimetallic bearings in the form of a thin coating on a steel base. The main types are lead base Babbitt, lead silver base Babbitt, tin base Babbitt, cadmium base Babbitt, arsenical Babbitt, white metal bearing alloys. Named for Isaac Babbitt, the inventor.

Babcock test. A rapid test for butterfat in milk introduced by Stephen M. Babcock in 1890 and now in world-wide use in the dairy industry.

B acid. See 1-amino-8-naphthol-3,5-disulfonic acid.

bacitracin. An antibiotic, of polypeptide structure, produced by the metabolic processes of Bacillus subtilis. It is effective against many gram-positive bacteria, but is ineffective against most aerobic gram-negative bacteria.

Properties: White to pale buff, hygroscopic powder; odorless or with slight odor; bitter taste. Powder is stable to heat but solutions deteriorate at room temperature. Freely soluble in water; soluble in alcohol, methanol, and glacial acetic acid; insoluble in acetone, chloroform, and ether. Solutions are neutral or slightly acid to litmus.

Units: Defined by comparison with a U.S.P. reference standard.

Grade: U.S.P. XVI, having a potency of about 40 to 50 units per milligram.

Use: Medicine; feed supplement.

background counts. See counts.

bactericide. Any product that will kill bacteria; used especially in killing disease bacteria. Bactericides vary in their potency; thus many bactericides will kill ordinary germs, but may not be effective against anthrax or tetanus or other sporeforming bacteria which are particularly resistant. Germicide is a synonymous term.

bacteriophage. An organism introduced into the body for the purpose of killing bacteria. Synonymous with phagocyte.

bacteriostat. A substance which prevents or slows down the growth of bacteria. Examples are quaternary ammonium salts and hexachlorophene.

baddeleyite (brazilite) ZrO_2. A natural zirconium oxide.

Properties: Color, black, brown, yellow, to colorless; streak white; luster submetallic to vitreous to greasy. M.p. 2500-2950°C. Highly resistant to chemicals. Sp. gr. 5.5-6.0.

Grades: Crude (53%, 73-75%); purified (98%).

Occurrence: Brazil; Ceylon.

Uses: In corrosion- and heat-resistant applications, such as furnace linings and muffles.

"Badger Liquid Soap." [86] Trademark for liquid potash soap (15% soap solution).

Uses: Detergent; degumming; soaping; wool and yarn scouring; fulling.

Badische acid. See 2-naphthylamine-8-sulfonic acid.

bagasse (megass). The crushed cane pulp left after the juice has been expressed from sugar canes.

Uses: The tough fiber is used in insulating buildings; fuel; fertilizer; building board; paper pulp.

Bahama white wood. See canella.

"Bake Aid." [172] Brand name for hydrated monocalcium phosphate.

Properties: White, free-flowing, crystalline powder. Purity complies with all Food and Drug laws.

Containers: 100-lb paper bags.

Use: "Rope" preventive in bread baking.

"Bakelite." [214] Trademark for a series of resins; see "Vinylite."

baker's sugar. See dextrose.

baking finish. A paint or varnish that requires baking at temperatures above 150°F for the

*See "I.C.C. Shipping Regulations," page xiii.
Reference numbers refer to name of manufacturer. See "List of Manufacturers," page v.

development of desired properties. (ASTM definition, ASTM D16-52).

baking powder, calcium acid sulfate. A baking powder containing sodium bicarbonate and mono-calcium phosphate.

baking powder, cream of tartar. A baking powder using as its ingredients sodium bicarbonate, tartaric acid and cream of tartar (potassium bitartrate).

baking powders. Powdered chemicals used as a substitute for yeast in baking for generating carbon dioxide gas in the dough. The components include sodium bicarbonate, one or more substances that become slightly acid when the powder is moistened, and starch to adjust the overall concentration of the active materials.

baking soda. See sodium bicarbonate.

"Bakthane." [23] Trademark for an agricultural insecticide based on the microorganism Bacillus thuringiensis Berliner and supplied as a dust containing viable spores.
Use: Controls imported cabbage worms, cabbage loopers, artichoke plume moths, and tobacco hornworm on certain fruits, vegetables, and field crops.

BAL. Abbreviation for British Anti-Lewisite. See 2, 3-dimercaptopropanol.

balas ruby. A variety of spinel used as a gem stone.

balata.
Properties: A rubber-like hydrocarbon polymer gum obtained from the latex of Mimusops balata, which grows wild in Central and South America. Balata is tough and water resistant but softens in hot water and cannot be vulcanized.
Grades: Ciudad Bolivar block, Manaos block; Surinam sheets; amber sheets.
Uses: Machinery belting; golf ball covers; packing; gaskets.

balata, synthetic. A stereospecific rubber, consisting of the trans isomer of polyisoprene.

"Balco." [155] Trademark for an alloy of 70% nickel and 30% iron.
Properties: Resistivity, 120 ohms per circular mil ft at 20°C; temperature coefficient of resistance, 0.0045 between 0-100°C; strongly magnetic; heat resistant below 1100°F.
Forms: Wire; insulated wire; ribbon.
Uses: Thermometer bulbs; applications which require high temperature coefficients of resistance.

ball clay. General term for those clays that possess good plasticity, strong bonding power, high refractoriness and which burn to a white or cream colored product. These clays are fine grained relatively pure hydrated aluminum silicate and usually are gray, tan or blue but may be nearly white or even black. Used as bonding and plasticizing agents, or chief ingredients of whiteware, porcelains, stoneware, terra cotta, glass refractories, graphite crucibles and porcelain enamels.

"Ball Powder." [413] Trademark for a spherical granular smokeless powder.
Use: Propelling charge for small arms ammunition in military and sporting categories.
Containers: Metal cans, 12-16 oz. net weight; fiber drums 30 and 100 lbs. net weight.
Fire hazard: Dangerous.
Shipping regulations: Explosives. Red label. *

balm of Gilead (Mecca balsam). Resinous juice from twigs of the evergreen tree, Commiphora opobalsamum.
Properties: Color ranges from yellow to green, brownish-red; fragrant odor; insoluble in water; soluble in benzene, acetone, chloroform, carbon disulfide, ether, glacial acetic acid.
Occurrence: Shores of the Red Sea.
Uses: Perfumery; medicine.

balsa. A strong light-weight wood principally from the tree Ochroma logopus SW which grows in Central America. It is available in three weights: aero, max weight 9 lb/cu ft; flotation, max weight 18 lb/cu ft; "heavy," max weight 25 lb/cu ft.
Uses: In insulation, life preservers, protective packaging, airplane construction.

balsam. An aromatic, liquid, resinous substance exuding from a tree or shrub and consisting of a resin mixed with a volatile oil, usually with benzoic or cinnamic acids and their esters. For individual balsams see under their respective names; e.g., balsam copaiba. See copaiba resin.

balsam of fir. See Canada balsam.

"Bamadex." [57] Trademark for dextro-amphetamine and meprobamate.

banana liquid. See banana oil.

banana oil.
1. (banana liquid). A solution of nitrocellulose in amyl acetate or similar solvent, so termed because of the odor.
2. Synonym for amyl acetate.

B and R. Abbreviation for ball and ring, a method of determining the softening point of tars, pitches, and various thermoplastic materials, both natural and synthetic.

banewort. See belladonna.

"Banox." [108] Trademark for a series of dry, powdered, phosphate-type corrosion inhibitors. No. 1 is artificially colored. No. 1-P and WT are colorless.
Containers: 135-lb fiber drums.
Uses: Refrigerator cars; refrigeration brine; cooling towers and small water systems.

"Banthīne." [70] Trademark for methantheline bromide (beta-diethylaminoethylxanthene-

*See "I.C.C. Shipping Regulations," page xiii.
Reference numbers refer to name of manufacturer. See "List of Manufacturers," page v.

9-carboxylate methobromide).
Use: Medicine.

BAP. Abbreviation for benzyl-para-aminophenol.

"Barabond." [236] Brand name for southern type bentonite for use in foundry sands. Useful where higher green strength and lower dry strength are desired than are obtained with Wyoming type bentonites.
Containers: Multiwall paper bags containing 100 lbs.

"Barafos." [236] Brand name for a proprietary product, a polyphosphate compound for the treatment of drilling mud to reduce viscosity and gel strength.
Containers: Asphalt laminated bag containing 100 lbs.

"Baragel." [236] Trademark for a compound of purified bentonite and an organic base; useful as an inorganic gelling agent for lubricating oils to prepare nonmelting greases of general usefulness.
Containers: 50-lb multiwall paper bags.

"Baragel" 24. [236] Trademark for compound similar to "Baragel." Particularly adapted for gelling oils of higher aromatic content.

"Barak." [28] Trademark for dibutyl-ammonium oleate.
Properties: Translucent, light brown liquid.
Containers: Drums (175 lbs net).
Use: To activate accelerators and improve processing of rubber and synthetic rubber.

barbaloin. See aloin.

barbasco root. See cube root.

Barberio's solution. An aqueous solution of sodium nitrite (1:2000).
Use: Analysis (testing for indican in urine).

barberite. A nonferrous alloy containing 88.5% copper, 5% nickel, 5% tin, 1.5% silicon. Sp. gr. 8.80; m.p. 1070°C. It offers good resistance to sulfuric acid in all dilutions up to 60%, sea water, moist sulfurous atmospheres, and mine waters.

barberry (jaundice berry; woodsour; sowberry; pepperidge bush; sour-spine). The bark, root bark and root are commercial products obtained from shrubs of Berberis vulgaris. The root contains berberine, oxycanthine and berbamine.
Containers: Bags.
Use: Source of its alkaloids.

barbital (diethylmalonylurea; diethylbarbituric acid) $C_8H_{12}N_2O_3$.
Properties: White crystals or powder; bitter taste; odorless; stable in air; m.p. 187-192°C; soluble in hot water, alcohol, ether, acetone, and ethyl acetate.
Derivation: By the interaction of diethyl ester of diethylmalonic acid and urea.
Grades: Technical; N.F. XI.
Containers: 1-, 5-lb tins; 25-, 100-lb drums.
Use: Medicine; stabilizer for hydrogen peroxide.

barbital sodium (barbital, soluble) $C_8H_{11}N_2NaO_3$.
Properties: White powder, stable, odorless; bitter taste. Soluble in water (solution alkaline to litmus); slightly soluble in alcohol; insoluble in ether.
Grade: N.F. XI.
Containers: 100-lb drums.
Use: Medicine.

barbital, soluble. See barbital sodium.

barbituric acid (malonylurea, pyrimidinetrione) $\overline{CO(NHCO)_2}CH_2 \cdot 2H_2O$.
Properties: White crystals, efflorescent; odorless; m.p. 245°C with some decomposition; slightly soluble in water and alcohol; soluble in ether. Forms salts with metals.
Derivation: By condensing malonic acid ester with urea.
Grades: Technical.
Containers: 1-, 5-lb glass bottles; tins.
Uses: Preparation of pharmaceutical chemicals; plastics (condensation agent with furfural).
Shipping regulations: None.*

"Barden" Clay. [285] Proprietary brand name for a group of hydrous aluminum silicates (sedimentary kaolins) from South Carolina.
Properties: Sp. gr. 2.60; bulk density, aerated, 18-20 lbs/cu ft, packed, 35-40 lbs/cu ft; creamy white; pH 4.5-5; air-floated; particle size 90% minus 2 microns.
Containers: 50-lb multiwall bags or bulk.
Uses: In pesticides as a carrier for dust bases and wettable powders, and as a secondary diluent in field strength dusts; in boxboard adhesives to impart high viscosity, provide solid glue lines, speed up setting, retard silica migration for smoother board and non-reactive printing surfaces; in flooring and tile adhesives as a filler and viscosity control agent; in fertilizers and prilled urea as an anti-caking agent; in roofing granules; putties; caulking compounds; linoleum.

"Bardol" Rubber Compounding Oil. [175] Trademark for a coal-tar oil with high aromatic hydrocarbon content.
Properties: Dark colored liquid; sp. gr. 1.07-1.12 (25/25°C); distillation at 300°C, 60% max; low viscosity at 40°F.
Containers: 55-gal steel drums; tank trucks; tank cars.
Uses: In rubber compounding as a swelling agent for natural and synthetic elastomers; as a dispersing agent for blacks and mineral fillers; to improve tack in butyl rubber; and to impart low set, low heat build-up and high resilience.
Caution: Avoid contact with skin and inhalation of vapors.

"Bardol" B Rubber Compounding Oil. [175] Trademark for a refined coal-tar distillate.
Properties: Pale yellow to straw-colored liquid; naphthalene-like odor; sp. gr. 1.020-1.045 (15.5/15.5°C); distillation range, 230-300°C; non-toxic in normal usage.
Containers: 55-gal steel drums.

*See "I.C.C. Shipping Regulations," page xiii.
Reference numbers refer to name of manufacturer. See "List of Manufacturers," page v.

Uses: Primary uses as a plasticizer, softener, and reclaiming oil; secondary uses as a reinforcer and tackifier. Pale color permits use in light colored stocks. This solvent-swelling type of softener markedly reduces viscosity of all elastomers and is especially suited for low hardness stocks made with "Thiokol" and neoprene-type elastomers.

barite (barytes; heavy spar; tiff) $BaSO_4$. Natural barium sulfate, the chief source of barium chemicals. Found as masses in sedimentary rocks, in metallic veins, and in residual clays.
Properties: Colorless, white, and various light shades; luster vitreous; one good cleavage.
Constants: Sp. gr. 4.5; hardness 3-3.5.
Occurrence: Arkansas, Missouri, Georgia, Tennessee, Nevada, California, Arizona; Germany; Canada; Brazil; England; U.S.S.R.
Grades and Uses:
(Crude) Unground, massive, product used for production of lithopone and barium chemicals.
(Ground barite) Sold in numerous sizes and degrees of purity. Used in oil well drilling mud; glass making; filler for paper, rubber, oilcloth, linoleum, etc; paint pigment and extender; x-ray apparatus; storage batteries; flux for smelting brass.
Containers: 50-, 100-lb bags; box cars.
Shipping regulations: None.*

barite, artificial. See barium sulfate.

barite, synthetic. See barium sulfate.

barium Ba. Element of atomic number 56, of group II of the periodic system; one of the alkaline earth elements.
Properties: Silver-white, slightly lustrous, somewhat malleable metal. All barium salts except the sulfate are poisonous! Soluble in acids; decomposes water.
Constants: Sp. gr. 3.78; m.p. 850°C.
Occurrence: In combination in nature in the form of barite (q.v.) and witherite (q.v.).
Derivation: Reduction of barium oxide by ferrosilicon in a vacuum at a high temperature.
Grades: Technical (not an article of commerce as yet).
Uses: Alloys; pyrotechnics.
Shipping regulations: None.*

barium acetate $Ba(C_2H_3O_2)_2 \cdot H_2O$.
Properties: White crystals; poisonous! Soluble in water; insoluble in alcohol.
Constants: Sp. gr. 2.02; m.p. decomposes.
Derivation: Acetic acid is added to a solution of barium sulfide. The product is recovered by evaporation and subsequent crystallization.
Method of purification: Recrystallization.
Grades: Technical; C.P.
Containers: 1-, 5-lb bottles; 525-lb casks.
Uses: Chemical reagent; acetates; verdigris; textile mordant.
Shipping regulations: None.*

barium aluminate $3BaO \cdot Al_2O_3$.
Properties: Gray pulverized mass, soluble in water, acids.
Uses: Ceramics; water treatment.

barium bichromate. See barium dichromate.

barium binoxide. See barium peroxide.

barium borotungstate (barium borowolframate) $Ba_9[B(W_2O_7)_6]_2$.
Properties: Large, white crystals. Caution! Effloresces in air. Keep well stoppered! Soluble in water.
Use: Making borotungstates.

barium borowolframate. See barium borotungstate.

barium bromate $Ba(BrO_3)_2 \cdot H_2O$.
Properties: White crystals or crystalline powder. Slightly soluble in water; insoluble in alcohol.
Constants: Sp. gr. 3.820; decomposes at 260°C.
Derivation: By passing bromine into a solution of barium hydroxide, barium bromide and barium bromate being formed which are separated by crystallization.
Method of purification: Recrystallization.
Grades: Pure; reagent.
Containers: Glass bottles; 25-lb tin boxes.
Use: Analytical reagent.
Fire hazard: Dangerous.
Shipping regulations: Oxidizing material. Yellow label.*

barium bromide $BaBr_2 \cdot 2H_2O$.
Properties: Colorless crystals; poisonous! Soluble in water and in alcohol.
Constants: Sp. gr. 3.852; m.p. anhydrous $BaBr_2$ 847°C.
Derivation: By the interaction of barium sulfide and hydrobromic acid, with subsequent crystallization.
Method of purification: Recrystallization.
Grades: Technical; C.P.
Containers: 50-lb cases; glass bottles.
Use: Manufacturing bromides.
Shipping regulations: None.*

barium carbonate $BaCO_3$.
Properties: White powder; found in nature as the mineral witherite; poisonous! Insoluble in water; soluble in acids (not in sulfuric acid).
Constants: Sp. gr. 4.275; m.p. 174°C at 90 atm.
Derivation: (a) Interaction of barium sulfide and soda ash solutions. (b) By passing a current of carbon dioxide gas through the solution of barium sulfide.
Grades: Technical; C.P.
Containers: Bags; drums; barrels.
Uses: Barium salts; rat poison; optical glass; flat wall paint; foundry core compounds; water purification; oil well drilling; ceramics (tile, terra cotta, porcelain); marble substitutes; dyes; enamels for ironware; steel carburizing; rubber; beet sugar; chemical reagent.
Shipping regulations: None.*

barium chlorate $Ba(ClO_3)_2 \cdot H_2O$.
Properties: Colorless prisms or white

powder; poisonous! Soluble in water.
Constants: Sp. gr. 3.179; m.p. 414°C.
Derivation: Electrolysis of barium chloride.
Method of purification: Crystallization.
Grades: Technical; C.P.
Containers: 1-lb bottles; 1-lb cans; drums.
Uses: Pyrotechnics; explosives; textile mordant; manufacture of other chlorates.
Caution: Fire hazard; dangerous; oxidizing material; explosive when in contact with combustible material.
Shipping regulations: Oxidizing material. Yellow label.*

barium chloride $BaCl_2 \cdot 2H_2O$.
Properties: Colorless flat crystals; poisonous! Soluble in water; insoluble in alcohol.
Constants: Sp. gr. 3.097; m.p. 960°C (anhydrous).
Derivation: (a) By the action of hydrochloric acid on barium sulfide with subsequent crystallization. (b) By treating witherite (natural barium carbonate) with hydrochloric acid. (c) By roasting crude barite with coal and calcium chloride. (d) By treating a solution of barium sulfide with calcium chloride.
Method of purification: Recrystallization.
Grades: Technical (crystals or powdered) 99%; crystalline; powdered; C.P.
Containers: 1-, 5-lb bottles; 200-lb bags; 400-lb barrels; 800-, 1000-lb casks; multiwall paper sacks; drums.
Uses: Chemicals (artificial barium sulfate, other barium salts, photographic chemicals); in ceramics to precipitate soluble sulfates as insoluble sulfate of barium; color lakes; leather tanning and finishing; heat-treating ferrous metals; rat and vermin poisons; pigments; drug; purification of sugar juice; textiles (mordant, weighting, calico-printing); water softener; boiler compounds; chemical reagent; lube oil additives.
Warning! May be fatal if swallowed. MCA warning label.
Shipping regulations: None.*

barium chromate (lemon chrome; ultramarine yellow; baryta yellow; Steinbuhl yellow) $BaCrO_4$.
Properties: Heavy, yellow, crystalline powder; poisonous! Soluble in acids; insoluble in water.
Constants: Sp. gr. 4.498.
Derivation: By the interaction of barium chloride and sodium chromate. The precipitate is washed, filtered and dried.
Grades: Technical; C.P.
Containers: 1-lb bottles; wooden barrels; kegs; multiwall paper sacks.
Uses: Safety matches; pigment in paints, ceramics; see also chrome yellow.
Shipping regulations: None.*

barium citrate $Ba_3(C_6H_5O_7)_2 \cdot 2H_2O$.
Properties: Grayish white crystalline powder. Highly insoluble in water. Poisonous!
Grades: Technical.
Containers: 55-gal fiber drums (250 lbs net).
Uses: Manufacture of barium compounds; stabilizer for latex paints.

barium cyanide $Ba(CN)_2$.
Properties: White, crystalline powder; poisonous! Soluble in water and alcohol.
Derivation: By the action of hydrocyanic acid on barium hydroxide with subsequent crystallization.
Method of purification: Recrystallization.
Grades: Technical.
Containers: Steel barrels.
Use: Metallurgy; electroplating.
Shipping regulations: Poison, class B. Poison label.*

barium cyanoplatinite. See platinum(ous) barium cyanide.

barium dichromate (barium bichromate) $BaCr_2O_7 \cdot 2H_2O$.
Properties: Poisonous! Brownish-red needles or crystalline masses. Soluble in acids, decomposed by water.

barium dioxide. See barium peroxide.

barium diphenylamine sulfonate $(C_6H_5NHC_6H_4SO_3)_2Ba$.
Properties: White crystals, soluble in water. Poisonous!
Use: Indicator in oxidation-reduction titrations.

barium dithionate (barium hyposulfate) $BaS_2O_6 \cdot 2H_2O$.
Properties: Colorless crystals. Soluble in water; slightly soluble in alcohol. Poisonous!
Constants: Sp. gr. 4.536.
Derivation: By the action of manganese dithionate on barium hydroxide.

barium diuranate. See uranium-barium oxide.

barium ethylsulfate $Ba(C_2H_5SO_4)_2 \cdot 2H_2O$.
Properties: Colorless crystals; poisonous! Soluble in water and alcohol.
Derivation: By the interaction of barium hydroxide and ethylsulfuric acid.
Method of purification: Crystallization.
Grades: Technical.
Containers: Tins.
Use: Organic preparations.
Shipping regulations: None.*

barium fluoride BaF_2.
Properties: White powder; poisonous! Sparingly soluble in water.
Constants: Sp. gr. 4.828; m.p. 1280°C.
Derivation: By the interaction of barium sulfide and hydrofluoric acid followed by crystallization.
Grades: Technical; C.P.; single pure crystals.
Containers: 1-, 5-lb bottles; wooden barrels.
Uses: Enamels; embalming fluids; crystals for spectroscopy.
Shipping regulations: None.*

barium fluosilicate (barium silicofluoride) $BaSiF_6$.
Properties: White, crystalline powder. Insoluble in water.
Grades: Technical.
Containers: 350-lb barrels.
Uses: Ceramics; insecticide; insecticidal compositions.
Caution! May be fatal if swallowed. MCA label.

barium fructose diphosphate. See fructose diphosphates, calcium and barium salts.

barium glass. A glass in which barium oxide (BaO) replaces part of the calcium oxide of ordinary lime soda glass.

barium hydrate. See barium hydroxide.

barium hydrosulfide $Ba(SH)_2$.
Properties: Yellow crystals. Caution! Keep well stoppered! Soluble in water.

barium hydroxide, octahydrate (barium hydrate; barium octahydrate; caustic baryta) $Ba(OH)_2 \cdot 8H_2O$.
Properties: White powder or crystals; poisonous; absorbs carbon dioxide from air. Keep well stoppered! Soluble in water, alcohol and ether.
Constants: Sp. gr. 1.656; m.p. 78°C, losing its water of crystallization (m.p. anhydrous $Ba(OH)_2$ 408°C).
Derivation: (a) By dissolving barium oxide in water with subsequent crystallization. (b) By precipitation from an aqueous solution of the sulfide by caustic soda. (c) Prepared on a large scale by heating barium sulfide in earthenware retorts into which a current of moist carbonic acid is passed after which superheated steam is passed over the resulting heated carbonate.
Impurities: Iron and calcium in some commercial grades.
Method of purification: Recrystallization.
Grades: Technical (crystals or anhydrous powder); C.P.; ACS reagent.
Containers: 1-, 5-lb bottles; 25-lb boxes; 100-lb kegs; 500-, 700-lb barrels; multiwall paper sacks.
Uses: Organic preparations; barium salts; beet sugar industry (now largely replaced by strontium or lime); refining animal and vegetable oils; analytical chemistry.
Shipping regulations: None.*

barium hydroxide, monohydrate (barium monohydrate) $Ba(OH)_2 \cdot H_2O$.
Properties: White powder. Poisonous!
Containers: 50-lb bags.
Uses: Manufacture of oil and grease additives; barium soaps and chemicals. Used in refining of beet sugar; as alkalizing agent in water softening; sulfate removal agent in treatment of water and brine; in boiler scale removal; dehairing agent; catalyst in manufacture of phenol-formaldehyde resins; insecticide and fungicide; sulfate controlling agent in ceramics; purifying agent for caustic soda and as steel carbonizing agent.

barium hydroxide pentahydrate (barium pentahydrate) $Ba(OH)_2 \cdot 5H_2O$.
Properties: Translucent free-flowing white flakes; density 65 lbs/cu ft (approx.).
Containers: 100-lb paper bags; 400-lb fiber drums.
Uses: Same as the octahydrate, above.

barium hypophosphite $BaH_4(PO_2)_2$.
Properties: White, crystalline powder; odorless; soluble in water; insoluble in alcohol. Poisonous! Used as a medicine.

barium hyposulfate. See barium dithionate.

barium hyposulfite. See barium thiosulfate.

barium iodate $Ba(IO_3)_2$.
Properties: White, crystalline powder. Caution! Poisonous! Slightly soluble in water; insoluble in alcohol.
Constants: Sp. gr. 5.23; m.p., decomposes at 476°C.
Use: Medicine.

barium iodide $BaI_2 \cdot 2H_2O$.
Properties: Colorless crystals; decomposes and reddens on exposure to air; poisonous! Soluble in water; slightly soluble in alcohol.
Constants: Sp. gr. 5.150; m.p., loses $2H_2O$ and melts at 740°C.
Derivation: By the action of hydriodic acid on barium hydroxide or of barium carbonate on ferrous iodide solution.
Method of purification: Crystallization.
Grades: Technical; C.P.
Containers: 1-oz vials; ¼-, 1-, 5-lb bottles.
Use: Preparation of other iodides.
Shipping regulations: None.*

barium manganate (manganese green; Cassel green) $BaMnO_4$.
Properties: Emerald-green powder; poisonous! Insoluble in water; decomposed by acids.
Constants: Sp. gr. 4.85.
Use: Paint pigment.

barium-mercury bromide. See mercuric-barium bromide.

barium-mercury iodide. See mercuric-barium iodide.

barium metaphosphate $Ba(PO_3)_2$.
Properties: White powder; slowly soluble in acids; insoluble in water.
Use: As a constituent of glasses, porcelains and enamels.

barium molybdate $BaMoO_4$.
Properties: White powder. Slightly soluble in acids, water.
Grades: Technical.

barium monohydrate. See barium hydroxide, monohydrate.

barium monosulfide. See barium sulfide.

barium monoxide. See barium oxide.

barium nitrate $Ba(NO_3)_2$.
Properties: Lustrous, white crystals; poisonous! Soluble in water; insoluble in alcohol.
Constants: Sp. gr. 3.244; m.p. 575°C.
Derivation: (a) By the action of nitric acid on barium carbonate, oxide or hydroxide and subsequent crystallization. (b) By adding sodium nitrate (Chile saltpeter) either to a solution of barium chloride or barium sulfide.
Method of purification: Recrystallization.
Grades: Technical; crystals; fused mass or powder; C.P.
Containers: 100-lb kegs; 500-lb barrels; 800- to 900-lb casks; multiwall paper sacks; 550-lb drums.
Uses: Pyrotechnics (gives green light);

*See "I.C.C. Shipping Regulations," page xiii.
Reference numbers refer to name of manufacturer. See "List of Manufacturers," page v.

explosives; drug; chemicals (barium peroxide); ceramic glazes; rodenticide; vacuum tubes.
Warning: Contact with other material may cause fire; may be fatal if swallowed. MCA warning label.
Shipping regulations: Oxidizing material. Yellow label.*

barium nitrite $Ba(NO_2)_2 \cdot H_2O$.
Properties: White to yellowish, crystalline powder. Soluble in alcohol, water.
Constants: Sp. gr. 3.173; decomposed by heat.
Grades: Technical; C.P.
Uses: Diazotization.

barium octahydrate. See barium hydroxide, octahydrate.

barium oxalate $BaC_2O_4 \cdot H_2O$.
Properties: White crystalline powder. Sp. gr. 2.66. Insoluble in water; soluble in dilute nitric or hydrochloric acids.
Grades: Technical; reagent.
Use: Analytical reagent.

barium oxide (barium monoxide; barium protoxide; calcined baryta) BaO.
Properties: White to yellowish-white powder; absorbs carbon dioxide readily from air; keep well stoppered; poisonous! Soluble in acids; reacts violently with water to form the hydroxide.
Constants: Sp. gr. 4.73-5.46; m.p. 1923°C.
Derivation: (a) Fusion of barium sulfate mixed with carbon in an electric furnace. (b) Fluid bed reduction. (c) For special porous grade, low temperature reduction of pure barium carbonate.
Grades: Technical regular grind (208 lbs/cu. ft.); technical fine grind (175 lbs/cu. ft.); porous, carbide free; 97%.
Containers: Pails; drums; bulk unit loads.
Uses: Dehydrating agent; glass industry; other barium salts; in Europe for refining beet sugar.
Caution: May cause skin irritation.
Shipping regulations: None.*

barium pentahydrate. See barium hydroxide pentahydrate.

barium perchlorate $Ba(ClO_4)_2 \cdot 3H_2O$.
Properties: Colorless crystals; explosive if in contact with combustible materials. Soluble in alcohol and water.
Constants: Sp. gr. 2.74; m.p. 505°C.
Grades: Technical.
Use: Drying agent for gases.
Shipping regulations: Oxidizing material. Yellow label.*

barium periodate $Ba_5(IO_6)_2$. White, insoluble powder which is decomposed by dilute sulfuric acid to form periodic acid and barium sulfate. It is prepared by heating barium iodate to redness.

barium permanganate $Ba(MnO_4)_2$.
Properties: Brownish-violet crystals. Soluble in water.
Grades: Technical.
Uses: Strong disinfectant; manufacture of permanganates; strong oxidizing agent.
Shipping regulations: Oxidizing material. Yellow label.*

barium peroxide (barium binoxide; barium dioxide; barium superoxide) BaO_2 or $BaO_2 \cdot 8H_2O$.
Properties: Grayish-white powder; poisonous! Decomposes in water.
Constants: Sp. gr. 4.96; m.p. 450°C; decomposes 800°C.
Derivation: By heating the monoxide in a stream of oxygen or air.
Grades: Technical; reagent.
Containers: 700-lb ret. steel drums; 75-lb lug-covered pails.
Uses: Manufacture of oxygen and hydrogen peroxide; bleaching (textiles and straw hat industry); tracer bullets; primer in combination with aluminum powder in aluminic thermic welding; oxygenated water.
Fire hazard: Dangerous; avoid contact with skin.
Shipping regulations: Oxidizing material. Yellow label.*

barium phosphate, secondary $BaHPO_4$.
Properties: White powder. Soluble in nitric acid (dilute), hydrochloric acid (dilute); slightly soluble in water. Toxic.
Grades: Technical.

barium phosphite $BaHPO_3$.
Properties: White powder. Slightly soluble in water.
Grades: Technical.

barium potassium chromate pigment (Pigment E) $BaK_2(CrO_4)_2$.
Properties: A pale yellow pigment. As compared with other chromate pigments, it has a particularly low chloride and sulfate content and forms stronger, more elastic paint films. Sp. gr. 3.65.
Derivation: By a kiln reaction at 500°C between potassium dichromate and barium carbonate.
Uses: As a component of anticorrosive paints for use on iron, steel, and light metal alloys.

barium protoxide. See barium oxide.

barium pyrophosphate $Ba_2P_2O_7$.
Properties: White powder; soluble in acids and ammonium salts; very slightly soluble in water.

barium rhodanide. See barium thiocyanate.

barium silicide $BaSi_2$ (variable).
Properties: Metallic, gray lumps. Decomposed by moisture.
Uses: Deòxidizing and desulfurizing steel.

barium silicofluoride. See barium fluosilicate.

barium stannate $BaSnO_3 \cdot 3H_2O$.
Properties: A white crystalline powder, sparingly soluble in water, readily soluble in hydrochloric acid.
Use: In the production of special ceramic insulations requiring high dielectric properties.

*See "I.C.C. Shipping Regulations," page xiii.
Reference numbers refer to name of manufacturer. See "List of Manufacturers," page v.

barium stearate $Ba(C_{18}H_{35}O_2)_2$.
Properties: White crystalline solid; insoluble in water or alcohol; m.p. 160°C; sp. gr. 1.145.
Uses: Waterproofing agent; lubricant; packing for bearings; wax compounding; preparation of greases.
Containers: Cartons.

barium sulfate (barium sulfate precipitated; blanc fixe; barite, synthetic or artificial; heavy spar artificial; permanent white; terra ponderosa) $BaSO_4$.
Properties: A fine, white, odorless, tasteless, bulky powder. Practically insoluble in water, organic solvents, and solutions of acids and alkalies; soluble in concentrated sulfuric acid; sp. gr. 4.25-4.5.
Derivation: (a) By treating a solution of a barium salt with sodium sulfate (salt cake). (b) By-product in manufacture of hydrogen peroxide. Barium peroxide is added to a 4% solution of sulfuric acid forming weak hydrogen peroxide and a barium sulfate precipitate. This is washed, dried, and marketed as blanc fixe. (c) Occurs in nature as the mineral barite.
Grades: Technical, dry, pulp, bleached, ground, floated, natural; C.P.; U.S.P. XVI; x-ray.
Containers: 1-, 5-lb bottles; 100-, 250-lb drums; wooden barrels; multiwall paper sacks.
Uses: Pigment for paints; filler and delustrant for textiles, rubber, linoleum, oilcloth, plastics, and lithograph inks; base for lake colors; indicator in x-ray photography; medicine.
Shipping regulations: None.*
See also barite, the natural product.

barium sulfate, precipitated. See barium sulfate.

barium sulfide (barium monosulfide; black ash) BaS.
Properties: Yellowish-green or gray powder or lumps; poisonous! Soluble in water, decomposes to the hydrosulfide.
Constants: Sp. gr. 4.25.
Derivation: Barium sulfate (crude barite) and coal are roasted in a furnace. The melt is lixiviated with hot water, filtered and evaporated.
Impurities: Iron, arsenic.
Grades: Technical.
Containers: 500-lb casks; multiwall paper sacks.
Uses: Depilatory; barium salts including lithopone; vulcanizing; weighting gutta-percha; generating perfectly pure hydrogen sulfide for analytical purposes.
Shipping regulations: None.*

barium sulfite $BaSO_3$.
Properties: White powder, decomposed by heat. Soluble in hydrochloric acid (dilute); insoluble in water.
Grades: Technical; C.P.
Uses: Analysis; paper.

barium sulfocyanide. See barium thiocyanate.

barium superoxide. See barium peroxide.

barium thiocyanate (barium sulfocyanide; barium rhodanide) $Ba(SCN)_2 \cdot 2H_2O$.
Properties: White crystals; poisonous! Soluble in water and in alcohol. Deliquescent.
Derivation: By heating barium hydroxide with ammonium thiocyanate and subsequent crystallization.
Method of purification: Recrystallization.
Grades: Technical.
Containers: Iron drums.
Uses: For making aluminum or potassium thiocyanates; dyeing; photography.
Shipping regulations: None.*

barium thiosulfate (barium hyposulfite) $BaS_2O_3 \cdot H_2O$.
Properties: White, crystalline powder; slightly soluble in water. Insoluble in alcohol. Poisonous!
Constants: Sp. gr. 3.5; decomposed by heat.
Grades: Technical.
Uses: Explosives; luminous paints; matches; varnishes.

barium titanate $BaTiO_3$. Widely used ferroelectric ceramic. Single crystals, either pure or doped with iron, are used in storage devices, dielectric amplifiers, and digital calculators.

barium tungstate (barium wolframate; barium white; tungstate white; wolfram white) $BaWO_4$.
Properties: White powder. Insoluble in water.
Constants: Sp. gr. 5.04.
Uses: Pigment; in x-ray photography for manufacture of intensifying and phosphorescent screens.

barium-uranium oxide. See uranium-barium oxide.

barium white. See barium tungstate.

barium wolframate. See barium tungstate.

barium zirconium silicate $BaZrSiO_5$, a complex of BaO, ZrO_2, SiO_2.
Properties: White compound. Density 118 lbs/cu ft; m.p. 2800°F; insoluble in water, alkalies. Slightly soluble in acids; soluble in hydrofluoric acid.
Containers: 80-lb paper bags; 500-lb drums.
Uses: Production of electrical resistor ceramics; glaze opacifiers; stabilizer for colored ground coat enamels.

barm. See yeast.

barn. A unit of measurement, equal to 10^{-24} square centimeters, for the cross-section target area of the nucleus of an atom. The name arose from colloquial reference in the early stages of nuclear technology to a nucleus as being "as big as a barn."

"Barnesite." [88] Trademark for a special rare earth for instrument lens polishing.

"Baroco." [236] Brand name for a high yield clay, compounded for the preparation of drilling

*See "I.C.C. Shipping Regulations," page xiii.
Reference numbers refer to name of manufacturer. See "List of Manufacturers," page v.

muds for use in formations containing moderate quantities of salt or other oil field electrolytes that may flocculate ordinary drilling muds. Used in preparation of starting muds or workover muds where weights of 9.0 lbs/gal (67 lbs/cu ft) are sufficient.

"Baroid." [236] Brand name for a weighting material manufactured from selected barytes (barium sulfate ore). "Baroid" is added to drilling muds to increase the unit weight of the mud, thus increasing the hydrostatic head on the formations being drilled in deep wells, to prevent the walls of the hole from caving.

barometric pressure. The pressure of the air at a particular point on or above the surface of the earth, due to the weight of the air above. At sea level this pressure is sufficient to support a column of mercury about 29.9 in. in height (760 mm). This is equivalent to a pressure of 14.7 psi or 1 atmosphere. These are average values that have been chosen as standards; thus 29.9 in. or 760 mm of mercury is referred to as standard or normal barometric pressure. The actual barometric pressure varies continually with changes in the weather, and the average value decreases about one inch for each 1000 feet of altitude, up to several thousand feet.

"Baron." [233] Weed-killing composition containing 2-(2,4,5-trichlorophenoxy) ethyl-2,2-dichloropropionate.

"Bar-O-Sil." [304] Trademark for a complex barium silicate vinyl stabilizer.

Properties: Fine white powder; sp. gr. 2.7; refractive index 1.5.

Containers: Fiberboard drums containing 15 and 75 lbs.

Uses: A supplementary vinyl stabilizer used to control plating, hazing, crocking and dry hand in vinyl products.

barrel syrup. See molasses.

barthrin. Synthetic analog of allethrin described as the 6-chloropiperonyl ester of chrysanthemummonocarboxylic acid. Used as insecticide with applications similar to allethrin and other analogs as furethrin, ethythrin and cyclethrin. Accepted as generic name by Ent. Soc.

"Bartyls." [12] Trademark for a series of compounded anti-skinning agents. A blend of chemical ingredients designed to retard film formations in storage and during the printing operation.

baryon. The name of a group of fundamental particles (q.v.). A baryon is a fundamental particle with a mass equal to or greater than the mass of a nucleon, and includes the nucleon and hyperons.

baryta, calcined. See barium oxide.

baryta, caustic. See barium hydroxide.

baryta water. A solution of barium hydroxide.

baryta yellow. See barium chromate.

barytes. See barite.

"Basacryl." [440] Trademark for a series of cationic dyestuffs for the dyeing and printing of polyacrylonitrile fiber. Readily soluble in water.

basalt. A dense to glassy, dark-colored, basic volcanic rock composed essentially of soda-lime feldspar and pyroxene, with or without olivine and with accessory magnetite or ilmenite. An average of 198 analyses of typical basalts gave 49.06% SiO_2, 15.70% Al_2O_3, 5.38% Fe_2O_3, 6.37% FeO, 8.95% CaO, 6.17% MgO, 3.11% Na_2O, 1.52% K_2O, 1.36% TiO_2, 0.31% MnO, 0.45% P_2O_5, 1.62% H_2O. Interpretations of the term are rather indefinite and quarrymen sometimes include any dark-colored fine-grained rock under the term basalt or trap rock.

Uses: Road building; paving blocks; building stone.

basanite. See Lydian stone.

base. In common everyday language, a base is any one of a large class of compounds having one or more of the following properties: bitter taste, slippery feeling in solution, ability to make litmus dye turn blue and to cause other indicator dyes to take on characteristic colors, ability to react with (neutralize) acids to form salts. Included are both hydroxides and oxides of metals. More specific definitions state that a base is a compound whose molecules yield hydroxyl ions (OH^-) in water, or yield the negative ion of a solvent, or finally, that a base is any molecule or ion that can combine with protons or hydrogen ions, i.e., a proton acceptor.

See also acid; alkali; neutralization.

"Basekote." [51] Trademark for light brown, firm-consistency lubricant used as a load-carrying medium on ship-launching ways. It is melted and applied hot to the groundways. On cooling, it forms a hard, tough, extremely adhesive coating.

base metals. Commonly used name for lead, zinc, etc., in contrast to gold and silver.

base oils. See blown oils.

BASF process (formerly called Sachsse process). A process for producing acetylene by burning a mixture of low molecular weight hydrocarbons (as, natural gas) with oxygen to produce a 2700°F temperature. The combustion gases are quickly chilled by scrubbing with water, and the acetylene is separated by distillation and solvent extraction from ethylene, carbon monoxide, hydrogen and other reaction products.

basic. Term used to describe certain compounds to distinguish them because of their more alkaline nature as compared with other compounds of nearly the same name. In this book the term basic is usually ignored in making the alphabetical arrangement. Thus basic beryllium carbonate will be found indexed as beryllium carbonate.

basic cinder. See basic slag.

basic dichromate. See bismuth chromate.

basic lining. A furnace lining containing basic oxides or compounds that decompose under furnace conditions to give basic oxides. The usual basic linings contain calcium and magnesium oxides or carbonates.

"Basicol." [188] Trademark for a series of essential oils intended as replacements for oils of lavender, geranium, lemon, pine, ylang ylang, neroli and orris root.

basic oxides. An oxide which is a base or which forms a hydroxide when combined with water, and/or which will neutralize acidic substances. Basic oxides are all metallic oxides, but there is a great variation in the degree of basiscity. Some basic oxides such as those of sodium, calcium and magnesium combine with water with vigor or with relative ease, and also neutralize all acidic substances rapidly and completely. The oxides of the heavy metals are only weakly basic, do not dissolve or react with water to any extent, and neutralize only the more strongly acidic substances. There is a gradual transition from basic to acidic oxides and certain oxides, as aluminum oxide, show both acidic and basic properties.

basic phosphate. See basic slag.

basic refractories. See basic lining.

basic salts. Compounds that belong both to the category of salts and that of bases because they have in their composition the radicals OH (hydroxide) or O (oxide) as well as the usual positive and negative radicals of normal salts. Among the best examples are bismuth subnitrate, often written $BiONO_3$, and basic copper carbonate $Cu_2(OH)_2CO_3$. Most basic salts are insoluble in water and many are of variable composition.

basic slag (basic cinder; basic phosphate; Thomas metal; Thomas slag; Thomas phosphate; Belgian slag). A slag produced in the conversion of pig iron of high phosphorus content into steel. It contains a variable amount of tricalcium phosphate, calcium silicate, lime and oxides of iron, magnesium and manganese. Used as a fertilizer, being particularly valuable for grazing lands.
See also slag.

basil oils (sweet basil oil).
Properties: Essential oils of which there are two commercial varieties, viz., (a) ordinary sweet basil oil and (b) Réunion oil. Yellow color; odor is aromatic, penetrating, estragon-like (ordinary), camphor-like (Réunion). (a) Soluble in 1 to 2 vols. and more of 80% alcohol (sometimes opalescent or with separation of paraffin). (b) Soluble in 2 to 3 vols. of 80% alcohol; (mostly) in 3 to 7 vols. of 80% alcohol (occasionally with separation of paraffin). Insoluble in water.
Chief known constituents: (a) Methyl chavicol; linalool; cineole. (b) Pinene, cineole, camphor and methyl chavicol.
Constants: (a) Sp. gr. 0.904-0.930 (15°C); optical rotation −6° to −22°; refractive index 1.481-1.495; acid value up to 3.5; ester value 1-12. (b) Sp. gr. 0.945-0.987 (15°C); optical rotation +0°22' to +12°; refractive index 1.51505-1.51753; acid value up to 3; ester value 9-22.
Derivation: From the leaves of several varieties of the sweet basil, Ocimum basilicum, L.
Uses: Flavoring; medicine; perfumery.
Shipping regulations: None.*

bassora gum. A term applied collectively to a group of highly colored gums which are somewhat similar to tragacanth gum. A bassora gum of commercial importance is Indian gum (q.v.).

bassorin (tragacanthin; adraganthin). A slimy nonadhesive mucilaginous residue obtained when tragacanth gum is treated with water several times and filtered. In the presence of alkalies the whole of the gum dissolves.

bastard saffron. See carthamus.

bastnaesite $(Ce, La)(CO_3)F$. A natural fluocarbonate of the cerium group of rare earth metals.
Properties: Color wax-yellow to reddish brown; luster vitreous to greasy; hardness 4.0-4.5; sp. gr. 4.9-5.2.
Occurrence: California, Colorado, New Mexico; Sweden.
Use: Ore of the rare earths.

bastose. A combination of cellulose and lignin occurring in jute fiber.

batch distillation. Distillation in which the entire sample of the material to be distilled, the charge, is placed in the still before the process is begun and product is withdrawn only from the condenser of the apparatus.

batteries. See dry cell; storage battery.

battery acid (electrolyte acid). Sulfuric acid of strengths suitable for use in storage batteries. The product is water-white, odorless and practically free from iron.
Derivation: By diluting high-grade commercial sulfuric acid with distilled water to standard strengths.
Approximate freezing points of electrolyte:

Sp. gr.	Charge of Battery	F.p., °F
1.100	Discharged	+18
1.150	Discharged	+ 5
1.165	Discharged	− 0
1.180	Discharged	− 6
1.200	Half charged	−23
1.225	Half charged	−30
1.250	Half charged	−61
1.280	Full charged	−96

Containers: 126- to 200-lb glass carboys;

*See "I.C.C. Shipping Regulations," page xiii.
Reference numbers refer to name of manufacturer. See "List of Manufacturers," page v.

tank cars.
Use: Storage batteries.
Shipping regulations: Corrosive liquid. White label. *

Baumé (abbreviated as Bé). A term used to designate readings on an arbitrary scale of specific gravities devised by the French chemist Antoine Baumé and used by him in the graduation of hydrometers. The relations to specific gravity (at 60/60°F) are as follows:
°Bé = 145 — 145/sp. gr. for materials heavier than water.
°Bé = 140/sp. gr. — 130 for materials lighter than water.

bauxite. A natural aggregate of aluminum-bearing minerals, more or less impure, in which the aluminum occurs largely as hydrated oxides. It is usually formed by prolonged weathering of aluminous rocks. Contains 30-75% Al_2O_3, 9-31% H_2O, 3-25% Fe_2O_3, 2-9% SiO_2, and 1-3% TiO_2.
Properties: Color white, cream, yellow, brown, gray or red; streak variable; luster dull or earthy; sp. gr. 2-2.55; hardness 1-3; often in pisolitic concretions. Insoluble in water; decomposed by hydrochloric acid.
Occurrence: Arkansas, Alabama, Georgia, Virginia; British Guiana; Surinam; Haiti; France; U.S.S.R.; Hungary.
Uses: Most important ore of aluminum; aluminum chemicals; abrasives; cement; refractories; decolorizing and deodorizing agent; catalysts; filler in rubber, plastics, paints, cosmetics.

bay.
bayberry. There is understandable confusion in popular usage among these terms and their derivatives, but official and traditional usage as to the botanical drugs is clear. Bay is the European laurel, Laurus nobilis (see laurus) and is used for laurel oil (sweet bay oil). Bayberry is the wax myrtle, Myrica cerifera (see myrica), and is used for bayberry wax. However, bayberry oil and bay oil (see myrcia oil) are made from the berries and leaves, respectively, of Pimenta racemosa. Note that myrcia is not a mistake for myrica.

bayberry oil.
Properties: Yellowish-brown essential oil; aromatic odor distinct from that of bay oil; phenol content 73%; other constituents are pinene, chavicol, citrol, eugenol; sp. gr. 0.955-0.990 (15°C); optical rotation —7° 3'. Soluble in 1.5 vol. of 70% alcohol; in 0.5 vol. and more of 80% alcohol.
Derivation: By distillation of the berries of Pimenta racemosa (Pimenta acris). Note that this is not the same plant used for bayberry wax, but is identical with that used for bay oil.
Containers: Tins; glass bottles.
Shipping regulations: None. *

bayberry wax (myrtle wax).
Properties: Green wax; bitter taste; feeble odor; chief constituents: palmitin, palmitic acid, myristin and lauric acid; sp. gr. 0.933 (15°C); saponification value 198-199; iodine value 68-80; solidifying point 25°C; partly soluble in alcohol.
Derivation: By boiling the berries of myrica. The wax on the berries melts and floats.
Grades: Technical.
Containers: Boxes; bags.
Uses: Candles (for pleasant odor when burned); soaps; leather polishing; medicine.
Shipping regulations: None. *

Bayer process. Process for making pure alumina from the crude ore bauxite. The main use of alumina is in the production of metallic aluminum.
Bauxite is mixed with hot concentrated sodium hydroxide, which dissolves the alumina and silica. The silica is precipitated and the dissolved alumina is separated from the solids, diluted, cooled and then is crystallized as aluminum hydroxide. The aluminum hydroxide is calcined to give alumina.

"Bayer 21/199." See O,O-diethyl O-3-chloro-4-methyl-2-oxo-2H-1-benzopyran-7-yl-phosphorothioate.

"Bayer 22555." See para-dimethylaminobenzenediazo sodium sulfonate.

"Bayer 29493." See O,O-dimethyl O-(4-(methylthio)-meta-tolyl)phosphorothioate.

Bayer's acid. See 2-naphthylamine-7-sulfonic acid; also crocein acid.

bayleaf oil. See myrcia oil.

bay oil. See myrcia oil.

"Bayol." [51] Trademark for technical, colorless, "white," mineral oils of low viscosity. They are used primarily for textile fiber lubrication, in the preparation of cloth sizes, etc. "Bayol" 50 is best suited as a lubricant on synthetic fibers during twisting and weaving.

bay rum. Properly, the distillate from a mixture of 200 lbs of dried bay leaves, 65 gals rum, 100 gals of water, a little salt. Now commonly substituted is a mixture of bay oil, orange-peel oil and oil of pimenta.
Uses: Aromatic shaving lotions; alcoholic rub.

"Baytex." [181] Trademark for O,O-dimethyl O-(4-(methylthio)-meta-tolyl)phosphorothioate (q.v.).

"Baytown." [110] Brand name for styrene-butadiene rubber masterbatch which contains carbon black and sometimes processing and extending oils.

BBO. See 2,5-dibiphenylyloxazole.

BBP. Abbreviation for butyl benzyl phthalate.

BCWL. Abbreviation for basic carbonate white lead. See lead carbonate, basic.

Be. Symbol for beryllium.

Bé. Abbreviation for Baumé (q.v.).

"Beacon." [51] Trademark for low temperature,

*See "I.C.C. Shipping Regulations," page xiii.
Reference numbers refer to name of manufacturer. See "List of Manufacturers," page v.

low-torque, bearing greases for aircraft control bearings and instrument bearings subjected to wide temperature variation.

"Beaconol." [354] A line of sulfonated fatty alcohols used as detergents, wetting agents, and dispersing agents in the paper, textile, paint, and rubber industries.

bean oil. See soybean oil.

bean oil, Chinese. See soybean oil.

bearberry bark. See cascara sagrada bark.

bearwood. See cascara sagrada bark.

bebeerine (d-bebeerine; ℓ-bebeerine or curine) $C_{36}H_{38}N_2O_6$. An alkaloid. Two stereoisomers known; the levo form (curine) is related to curare.

Properties: White to yellowish-brown, amorphous powder; levo or alpha: m.p. 214°C; insoluble in water; soluble in acetone, chloroform, dilute acid; slightly soluble in alcohol; dextro or beta: m.p. 142-150°C; soluble in chloroform, benzene, acids; slightly soluble in alcohol.

Derivation: By extraction of the bark of Nectandra rodiae (d-bebeerine) or Pareira brava (curine) and subsequent crystallization.

Method of purification: Recrystallization.

Grades: Technical.

Containers: Tins; glass bottles.

Use: Medicine, in form of sulfate and other salts. The commercial sulfate is a mixture of alkaloids allied to bebeerine.

bebeerine hydrochloride.

Properties: Reddish-brown scales. Soluble in water and alcohol; m.p. 259°C.

Derivation: By the action of hydrochloric acid on bebeerine.

Method of purification: Crystallization.

Grades: Technical.

Containers: Glass bottles.

Use: Medicine.

Shipping regulations: None.*

"Becco." [55] Trademark for a line of chemicals, especially active oxygen chemicals, such as hydrogen peroxide, peracetic acid, calcium peroxide, urea peroxide, etc.

"Beckacite." [36] Modified phenolic and nonphenolic or maleic ester synthetic resins.

Physical properties: Color, different grades range from G to WG (U.S. Department of Agriculture rosin standards); acid number 15-20 in most instances, running as high as 130 in specialty types; melting range 176-310°F, capillary tube method.

Chemical properties: All grades soluble cold in acetates and coal-tar solvents, some requiring heat in the presence of turpentine and drying oils. Imparted resistance to abrasion, water, weather and other reagents, as well as drying speed vary according to type and grade.

Uses: Clear varnishes; oleoresinous enamels; clear and pigmented nitrocellulose lacquers; printing ink vehicles; etc.

"Beckamine." [36] Urea-formaldehyde and melamine resin solutions with water or various blends of toluene, n-butyl alcohol, xylene and ethyl alcohol as the volatile component.

Physical properties: Solids, 39% to 61%; color none; acid number −1 to +7; viscosity A to W (Gardner-Holdt).

Chemical properties: Most important properties imparted to surface coatings are greater hardness, better color and color retention and faster baking.

Uses: Clear and pigmented industrial finishes of the baking type; plywood bonding; agent for improving the wet strength of papers.

"Beckolin." [36] Synthetic drying oils. Two grades available: Light and dark.

Physical properties: Nonvolatile, 100%; color 4-7 (Hellige-Klett); acid number 6-16; viscosity X to Y (Gardner-Holdt).

Chemical properties: Imparts faster drying, longer luster life and increased resistance to weathering.

Uses: Paint, varnish, and enamel products.

"Beckopol." [36] A phenolated copal gum or reacted combination of a phenolic synthetic resin and Congo copal gum.

Physical properties: Color M to K (U.S. Department of Agriculture rosin standards); acid number 40-45; melting range 268-302°F.

Chemical properties: Soluble cold in acetates, coal-tar solvents and turpentine; soluble hot in drying oils. Imparts excellent resistance to weather, water and abrasion, high viscosity; unusual hardness, remarkable toughness and outstanding printproof properties.

Use: Exceptionally hard varnishes, principally of the rubbing and polishing types.

"Beckosol." [36] Alkyd resins in solid and solution form. Over 50 different items representing 16 separate and distinct kinds of base resins of the following types: (1) Drying pure alkyds, (2) nondrying pure alkyds, (3) phenol-modified alkyds, (4) oil- and phenol-modified alkyds, and (5) nonphenolic alkyds.

Physical properties: Solids from 45-100% by weight; solvents mineral spirits, high solvent naphtha, V.M. & P. naphtha, xylene and toluene; oil length, short to extra-long; color 1L to 7 (Hellige-Klett standards); acid number 3-23; viscosity A to Z_2.

Chemical properties: Imparted properties (speed of drying, hardness, color and color retention, durability, etc.) vary with the different types, kinds and grades.

Uses: Paints; lacquers; enamels; industrial and architectural surface coatings of all kinds.

"Bedacryl." [206] Trademark for methacrylic ester polymers used in air drying or stoving lacquers showing no discoloration at high temperatures or yellowing in ultraviolet light.

"Bedafil." [206] Brand name of a proprietary leather filling agent.

"Bedafin" 685. [206] Brand name for a resin used as a plasticizer for "Bedafin" 2001 to produce more flexible finishes. Soluble in aqueous ammonia.

"Bedafin" 2001. [206] Brand name for an alternative to albumen for fixing pigments on fabrics. Soluble in ammonia and other alkalies, e.g., triethanolamine.

"Bedafin" E and F. [206] Brand name for transparent, colorless, thermoplastic resins, soluble in trichlorethylene and mixtures of toluene and methylated spirit. Applied from organic solvents they give stiff finishes with a firm handle. They can be applied with plasticizers to give flexible transparent effects.

"Bedafin" 285X. [206] Brand name for a solution of "Bedafin" 685 in xylene.

"Bedesol." [206] Brand name of a line of synthetic resins used in the paint, varnish and printing-ink industries.

beechwood creosote. See creosote, wood-tar.

beerstone. A deposit occurring during brewing operations on containers and consisting of calcium oxalate and organic material.

"Bee-Sive." [302] Trademark for adhesive compositions containing synthetic polymers; used for bonding materials.

beeswax (cera flava; yellow wax). Wax from the honeycomb of the bee. Beeswax consists largely of myricyl palmitate, cerotic acid and some high-carbon paraffins.

Properties: Yellow to grayish brown solid with honey-like odor and faint characteristic taste. Sp. gr. 0.95; melting range 62-65°C; insoluble in water; slightly soluble in alcohol; soluble in chloroform, ether, and oils.

Derivation: From melting the honeycomb in hot water, straining, and cooling the mixture in molds.

Grades: Technical; crude; refined; N.F. XI. For the white grade, see beeswax, bleached.

Containers: Bags; 100-lb cartons.

Uses: Furniture and floor waxes; adhesives; leather dressings; anatomical specimens; artificial fruit and flowers; transparent and wax paper; process engraving; lithographing; textile sizes and finishes; also uses listed under beeswax, bleached.

beeswax, bleached (white wax, cera alba). Beeswax bleached by sunlight or oxidizing agents such as chromic acid, hydrogen peroxide, or ozone.

Properties: Yellowish-white solid, translucent in thin layers, nearly tasteless. Other properties are those of beeswax.

Grades: U.S.P. XVI.

Containers: Bricks and slabs in 100-lb cartons.

Uses: Ointments; cosmetics; candles.

"Beetle" Resins. [57] Trademark for a series of alkylated urea-formaldehyde condensation products supplied in solution in organic solvents. These resins are thermosetting types generally used with alkyd type resins in surface coatings baked in the range of 250-300°F for 15-60 minute time periods. Available in the following grades:

212-9. Solids 60 wt %; 8.4 lb/gal; special solvent; color (Gardner) 1 max; viscosity (Gardner-Holdt, 25°C) Y-Z_2; hydrocarbon solvent tolerance 1500; acid number (solid resin) 1-4.

216-8. Solids 60 wt %; 8.5 lb/gal; butanol-xylol solvent; color (Gardner) 1 max.; viscosity (Gardner-Holdt, 25°C) S-V; hydrocarbon solvent tolerance 350; acid number (solid resin) 0.5-2.0.

220-8. Solids 50 wt %; 8.3 lb/gal; butanol-xylol solvent; color (Gardner) 1 max; viscosity (Gardner-Holdt, 25°C) X-Z_1; hydrocarbon solvent tolerance 200; acid number (solid resin) 1-4.

227-8. Solids 50 wt %; 8.3 lb/gal; butanol-xylol solvent; color (Gardner) 1 max; viscosity (Gardner-Holdt, 25°C) W-Z; hydrocarbon solvent tolerance 100; acid number (solid resin) 1-4.

230-8. Solids 50 wt %; 8.3 lb/gal; butanol-xylol solvent; color (Gardner) 1 max; viscosity (Gardner-Holdt, 25°C) Q-T; hydrocarbon solvent tolerance 150; acid number (solid resin) 0.5-2.0.

beet molasses. See molasses.

beet sugar. Sugar (sucrose) from special types of beets by extraction with water, refining and evaporation. Chemically identical with cane sugar ($C_{12}H_{22}O_{11}$).

behenic acid (docosanoic acid) $CH_3(CH_2)_{20}COOH$. A saturated fatty acid found as a minor component of the oils of the type of peanut and rapeseed.

Properties: M.p. 80.0°C; b.p. 306°C (60 mm), 265°C (15 mm); sp. gr. 0.8221 (100/4°C); refractive index 1.4270 (100°C).

Derivation: Peanut oil; occurs in ben oil, hydrogenated mustard oil and rapeseed oil.

Containers: 50-lb paper bags.

Grades: Technical; 99%.

Uses: Cosmetics; waxes; plasticizers; chemicals; stabilizers.

behenyl alcohol (1-docosanol) $CH_3(CH_2)_{20}CH_2OH$. A long chain, saturated fatty alcohol, much like stearyl alcohol.

Typical specifications: Titer 62.6°C; boiling range 342-377°C.

Impurities: 15% or more stearyl and arachidyl alcohols.

Derivation: Reduction of behenic acid.

Uses: Synthetic fibers; lubricants.

Belgian slag. See basic slag.

belladonna (deadly nightshade; death's herb; banewort; divale; poison black cherry). An herbaceous perennial bush (Atropa belladonna) of which the leaves and roots are used for their content of hyoscyamine and atropine.

Occurrence: Southern and central Europe; Asia Minor; Algeria; cultivated in North America; England; France.

Grades: Belladonna leaf, U.S.P. XVI; belladonna root, N.F. XI.
Containers: Boxes; bales.
Shipping regulations: None.*

bell metal. An alloy used for casting bells. It is generally composed of about 80% copper and 20% tin. It often contains small amounts of lead and zinc. Sp. gr. 8.7; m.p. about 890°C.

"Belro." [266] Trademark for a dark, acidic, thermoplastic resin.
Uses: Adhesives; battery waxes, coatings; cleaning compounds.

"Bemberg." [287] Trademark for cuprammonium yarn, a continuous spun regenerated cellulose product. Spun in ranges from 40 to 1800 denier in straight yarns and 70 to 5200 denier in novelty yarns. Sp. gr. 1.5; 11% moisture regain; tenacity range 1.8 to 2.4 grams per denier.
Uses: Triple sheers; linings; dress goods; sportswear; underwear; draperies; upholstery.

"BEM Brand." [241] Brand name for mixed fertilizers.

bemegride (3,3-methylethylglutarimide; 3-ethyl-3-methylglutarimide; methetharimide) $C_8H_{13}NO_2$.
Properties: Platelets; m.p. 127°C. Sublimes at 100°C and 2 mm pressure. Soluble in water and acetone.
Grade: N.N.D.
Use: Medicine.

"Bemul." [354] Trademark for a nontoxic, practically odorless emulsifying agent.
Properties: A pure white, edible glycerol monostearate in bead form; m.p. 58-59°C; completely dispersible in hot water; completely soluble in alcohols and hot hydrocarbons.
Uses: In the manufacture of pharmaceuticals, cosmetics, and foodstuffs; as a protective coating for edible hygroscopic powders, tablets, and crystals; as a pour-point depressant for lubricating oils; and in textile sizes; etc.

benactyzine hydrochloride (2-diethylaminoethyl benzilate hydrochloride) $(C_6H_5)_2COHCOO(CH_2)_2N(C_2H_5)_2 \cdot HCl$.
Grade: N.N.D.
Use: Medicine.

"Benadryl" Hydrochloride. [330] Trademark for diphenhydramine hydrochloride.

"Ben-A-Gel." [236, 304] Trademark for a highly beneficiated magnesium montmorillonite, used as a thickening agent.
Properties: Soft, milky-white powder; sp. gr. 2.4.
Containers: 10-, 25-, and 100-lb fiberboard drums.
Uses: Viscosity control and pigment suspension in aqueous systems in water-based paints and cosmetics. "Ben-A-Gel" requires high shear and is used primarily in industrial paints, cosmetics and pharmaceuticals. "Ben-A-Gel EW" is designed for use with low shear mixing equipment.

bench gas. See coal gas.

bendroflumethiazide 3-benzyl-3,4-dihydro-6-(trifluoromethyl)-2H-1,2,4-benzothiadiazine-7-sulfonamide, 1,1-dioxide.
Properties: White crystalline solid; faint rose-like odor; m.p. 216-218°C. Insoluble in water and acid; soluble in dilute alkali.
Grade: Pharmaceutical.
Use: Medicine.

Benedict solution. A water solution of sodium carbonate, copper sulfate and sodium citrate. The blue color changes to a red, orange, or yellow precipitate or suspension in the presence of a reducing sugar such as glucose, and is therefore used in testing for such materials, especially for urinalysis in the treatment of diabetes. See Fehling's solution.

"Benemid." [123] Trademark for probenecid; used in medicine.

Bengal gelatin. See agar-agar.

Bengal isinglass. See agar-agar.

Bengal lights. A mixture of realgar or arsenic disulfide, synthetic, potassium nitrate and sulfur, used in pyrotechnics to make blue light.

beni oil. See sesame oil.

benjamin gum. See benzoin gum, Siam, and benzoin gum, Sumatra.

benne oil. See sesame oil.

"Benodaine." [123] Trademark for piperoxan hydrochloride.

ben oil. A non-drying oil obtained from the seeds of Moringa aptera.
Uses: Food; lubricant for delicate machinery; perfumery; pharmaceuticals.

benoxinate hydrochloride $CH_3(CH_2)_3OC_6H_3(NH_2)CO_2(CH_2)_2N(C_2H_5)_2 \cdot HCl$. beta-Diethylaminoethyl-4-amino-3-n-butoxybenzoate hydrochloride.
Properties: White, odorless, crystalline powder; salty taste. M.p. 157-160°C; stable to air, heat and light. Freely soluble in alcohol, chloroform, and water; insoluble in ether; pH (aqueous solution) 4.5-5.2.
Grade: U.S.P. XVI.
Use: Medicine.

"Benthal" Alkyd Resin Intermediate. [58] Trademark for a material consisting of phthalic anhydride 4.5-5.5%, benzoic acid 94.5-95.5%; light yellow flakes; m.p. 118°C min; characteristic odor.
Containers: 200-lb net wt fiber drums.
Uses: In the resin industry, particularly in alkyds where it reduces the acid value, reduces viscosity, retards yellowing on baking at elevated temperatures and after bodying, improves stability and flow.

"Bentone." [236, 304] Trademark for a group of thickening agents:
"Bentone" 18-C: Alkyl ammonium montmorillonite. Finely divided light-cream powder;

*See "I.C.C. Shipping Regulations," page xiii.
Reference numbers refer to name of manufacturer. See "List of Manufacturers," page v.

sp. gr. 1.85.

Uses: Thickening and suspension agent for polar liquids and liquid mixtures; allows pigment suspension without strong thickening action for aerosol paints.

"Bentone" 27: Organic derivative of a special montmorillonite; finely divided creamy-white powder; sp. gr. 1.7.

Uses: Thickening agent with efficient gelling action over a wide range of organic solvents and finishes; prevents hard settling of suspended pigments or heavy reactive particles, fillers or abrasives; prevents excessive penetration into porous surfaces; provides strong reinforcing action in non-volatile organic compositions.

"Bentone" 34: Dimethyldioctadecyl ammonium bentonite. Finely divided light cream powder; sp. gr. 1.80.

Uses: Gelling agent for a wide variety of organic liquids; permits production of lubricating greases from lubricating oils; used in paints to prevent pigment settling and improve film properties.

"Bentone" 38: Organic derivative of a special montmorillonite. Finely divided creamy-white powder; sp. gr. 1.8.

Uses: Gellant for organic liquids of low and intermediate polarity; promotes particle suspension and penetration control for paints, varnishes, enamels, vinyl dispersions and epoxy and polyester resins.

bentonite (wilkenite; colloidal clay). A clay containing appreciable amounts of the clay mineral montmorillonite, and usually with the ability to swell greatly by absorption of water. Composed principally of aluminum and silicates, usually with some magnesium and iron.

Properties: Color, light yellow or green, cream, pink, gray to black; plastic. Earthy taste; insoluble in water, but swells when added to water. Insoluble in common organic solvents; pH of water suspension between 9 and 10. Some varieties are activated by acid treatment before use.

Occurrence: Wyoming, South Dakota, California, Mississippi, Texas, Arizona; Europe; U.S.S.R.; Canada.

Containers: Paper sacks; mesh bags; railroad cars.

Grades: Various, including U.S.P. XVI.

Uses: Metal casting, oil well drilling mud; decolorizing agent for oils, water, wine, and other products; filler and plasticizer for ceramics, refractories, rubber, soap, paper; for closing leaks in walls and dams; polishes, adhesives; to emulsify asphalts for roads.

"Bentox." [253] Brand name for a type of sulfur product used as insecticide dust.

"Benzahex." [147] Trademark for dusts and sprays containing benzene hexachloride sometimes formulated with DDT and/or sulfur.

Containers: Dusts, 50-lb bags; sprays, 5-, 30-, 55-gal drums; spray powders, 4-, 50-lb bags.

Uses: For control of insects on cotton, corn, certain other crops and livestock. Should not be used on potatoes, peas, beans (after pods are set) and other crops on which it may cause off flavor and/or odor.

benzalacetone. See benzylidene acetone.

benzal chloride. See benzyl dichloride.

benzaldehyde (benzoic aldehyde; synthetic oil of bitter almonds; benzoyl hydride; benzene carbonal) C_6H_5CHO.

Properties: Colorless or yellowish, fragrant, strongly refractive, volatile oil with odor resembling oil of bitter almonds, and burning aromatic taste; miscible with alcohol, ether, fixed and volatile oils; soluble in water.

Constants: Sp. gr. 1.041-1.046; refractive index (20°C) 1.5440-1.5464; m.p. −26°C; solidifies −56°C; b.p. 179.9°C.

Derivation: (a) From benzene, carbon monoxide and hydrogen chloride in the presence of cuprous chloride or aluminum chloride; (b) air oxidation of toluene with uranium or molybdenum oxides as catalysts; (c) chlorination of toluene with subsequent hydrolysis by acid or alkali.

Impurities: Usually chlorine derivatives.

Method of purification: Rectification.

Grades: Technical; F.F.C. (meaning free from chlorine); N.F. XI. Note: The specifications, especially regarding impurities, vary considerably for the grades used for dye manufacture from those used in perfumery.

Containers: 25-, 50-lb tins; 100-lb carboys; 425-, 450-lb drums.

Uses: Organic synthesis, especially of dyes and dye intermediates; solvent for oils, resins, some cellulose ethers, cellulose acetate and nitrate; flavoring compounds; production of synthetic perfumes; manufacture of cinnamic acid, benzoic acid; toilet preparations and soaps; photographic chemicals; baking chemicals; and in medicine.

Shipping regulations: None.*

benzaldehyde green. See malachite green.

benzalkonium chloride. A mixture of alkyl dimethylbenzylammonium chlorides of general formula $C_6H_5CH_2N(CH_3)_2RCl$ in which R is a mixture of the alkyls from C_8H_{17} to $C_{18}H_{37}$. It is a typical quaternary ammonium salt.

Properties: Occurs as a white or yellowish-white, amorphous powder or in gelatinous pieces. It has an aromatic odor and very bitter taste; very soluble in water, alcohol or acetone; almost insoluble in ether; slightly soluble in benzene. Water solutions foam strongly when shaken and are alkaline to litmus.

Grade: U.S.P. XVI.

Uses: Cationic detergent; surface antiseptic; fungicide.

benzamide (benzoylamide) $C_6H_5CONH_2$.

Properties: Colorless crystals; m.p. 130°C; b.p. 288°C; sp. gr. 1.341. Soluble in hot water, hot benzene, alcohol, and ether.

*See "I.C.C. Shipping Regulations," page xiii.

Reference numbers refer to name of manufacturer. See "List of Manufacturers," page v.

Derivation: From benzoyl chloride and ammonia or ammonium carbonate.
Grades: Technical.
Uses: Organic synthesis.

benzamine hydrochloride. See eucaine hydrochloride.

benzamine lactate. See eucaine lactate.

benzaminoacetic acid. See hippuric acid.

benzanilide (benzoylanilide; phenylbenzamide) $C_6H_5NH(COC_6H_5)$.
Properties: White to reddish crystals and powder, closely related to acetanilide, containing benzoyl in place of acetyl radical. Sp. gr. 1.306; m.p. 160-162°C. Soluble in alcohol; insoluble in water; slightly soluble in ether.
Derivation: From benzoic anhydride and aniline with caustic soda.
Method of purification: Crystallization.
Grades: Technical.
Containers: Kegs.
Uses: Medicine; intermediate in the synthesis of dyes, drugs and perfumes.
Shipping regulations: None. *

benzanthrone $C_{17}H_{10}O$, a four ring system.
Properties: Pale yellow needles; soluble in alcohol and other organic solvents. M.p. 170°C.
Derivation: (a) From anthranol and glycerol by condensation by means of sulfuric acid (anthranol is made from anthraquinone); (b) from anthracene in sulfuric acid solution by addition of glycerol and heating to 100-110°C until the anthracene disappears. The reaction mass is then diluted with water, salted out and purified.
Method of purification: Crystallization from toluene.
Grades: Technical.
Containers: Wooden barrels.
Use: Dyes.
Shipping regulations: None. *

benzathine penicillin G. See penicillin.

benzazimide. See 4-ketobenzotriazine.

benzazoline hydrochloride. See tolazoline hydrochloride.

"Benzedrine." [71] Trademark for racemic amphetamine sulfate. See amphetamine sulfate.

benzene (benzol) C_6H_6.
Properties: Clear, colorless, flammable liquid of highly refractive nature; characteristic odor; vapors burn with a very smoky flame; narcotic and toxic; b.p. 80.1°C; m.p. 5.5°C; sp. gr. 0.8790 (20/4°C); wt/gal 7.32 lb; refractive index (n 20/D) 1.50110; flash point (closed cup) 12°F. Miscible with alcohol, ether, acetone, carbon tetrachloride, carbon disulfide, acetic acid; slightly soluble in water.
Derivation: (a) Illuminating gas and coke-oven gas are "scrubbed," by passing through oil which thus becomes saturated with benzene and toluene. The resulting oil is distilled; benzene and toluene are recovered, and then separated by fractional distillation. (b) Coal-tar, after dehydration, is fractionally distilled yielding "light oil." On distilling this, the first runnings contain the crude benzene. This is successively washed with caustic soda, sulfuric acid and water, and again distilled. (c) Extraction from catalytic reforming streams in refining of petroleum. The last two methods are the more important.
Impurities: Toluene, xylene, tarry substances.
Grades: Crude; straw color; motor; industrial pure (2°C); nitration (1°C); thiophene-free.
Containers: 1-, 5-lb bottles; various sizes of tin cans; 55-, 110-gal drums; 8,000-10,000-gal tank cars.
Uses (in approximate order of volume): Styrene; phenol; synthetic detergents; nylon intermediates; aniline; DDT; maleic anhydride; dichlorobenzene; benzene hexachloride; nitrobenzene; diphenyl; insecticides; fumigants; solvent and miscellaneous synthetic uses. Large amounts are produced from petroleum and used in motor fuels without separation from the hydrocarbon mixture.
Danger: Extremely flammable, vapor harmful, poison. MCA warning label.
Shipping regulations: Flammable liquid. Red label. *

benzene azimide. See 1,2,3-benzotriazole.

benzeneazoanilide. See diazoaminobenzene.

benzeneazobenzene. See azobenzene.

benzene-azo-para**-benzene-azo-**beta**-naphthol.** See aminoazobenzene-beta-naphthol.

benzeneazonaphthylamine. See yellow AB.

benzeneazonaphthylethylenediamine. See azodine.

benzene carbonal. See benzaldehyde.

benzenecarbothioic acid. See thiobenzoic acid.

benzene carboxylic acid. See benzoic acid.

benzene dibromide. See dibromobenzene.

benzene-ortho**-dicarboxylic acid.** See phthalic acid.

benzene-para**-dicarboxylic acid.** See terephthalic acid.

benzene hexachloride. Common and accepted designation for a commercial mixture of isomers of 1,2,3,4,5,6-hexachlorocyclohexane (q.v.). Used as an insecticide. See also purified gamma-isomer under lindane.

gamma**-benzene hexachloride.** See lindane.

benzenemonosulfonic acid. See benzenesulfonic acid.

benzenephosphinic acid (phenylphosphinic acid) $C_6H_5H_2PO_2$.
Properties: Colorless crystals; m.p. 82-84°C; sp. gr. 1.376 (29°C). Decomposes at 200°C. Stable in air.

*See "I.C.C. Shipping Regulations," page xiii.
Reference numbers refer to name of manufacturer. See "List of Manufacturers," page v.

Soluble in water, alcohol, acetone. Slightly soluble in ether; insoluble in benzene, hexane, carbon tetrachloride.
Containers: 100-lb fiber drums.
Uses: Antioxidant; intermediate for metallic salt formation; accelerator for organic peroxide catalysts.

benzenephosphonic acid (phenylphosphonic acid) $C_6H_5H_2PO_3$.
Properties: Colorless crystals. M.p. 158°C; sp. gr. 1.475 (4°C); decomposes at 275°C; soluble in water, alcohol, ether, acetone; insoluble in benzenes, hexane, carbon tetrachloride.
Containers: 100-lb fiber drums.
Uses: Intermediate in anti-fouling paint agents; catalyst in organic reactions.

benzene phosphorus dichloride $C_6H_5PCl_2$.
Properties: Highly reactive colorless liquid. M.p. −51°C; b.p. 224.6°C (760 mm); sp. gr. 1.315 (25°C); refractive index 1.5958 (n 25/D). Soluble in common inert organic solvents; fumes in air; hydrolyzes in water. Handle with caution.
Containers: 55-gal stainless steel drums.
Uses: Organic synthesis, for derivation of plasticizers, polymers, antioxidants; oil additives.

benzene phosphorus oxydichloride $C_6H_5POCl_2$.
Properties: Reactive colorless liquid. M.p 3.0°C; b.p. 258°C (760 mm); sp. gr. 1.197 (25°C); refractive index 1.5585 (n 25/D). Soluble in common inert organic solvents; hydrolyzes in water.
Containers: 55-gal nickel drums.
Uses: Organic synthesis, for derivation of plasticizers, polymers, antioxidants, oil additives.

benzene ring. The six-carbon ring present in the molecular structure of benzene and in all the organic compounds derived from benzene by replacing the hydrogen atoms by other atoms or radicals.

benzene series. A series of compounds of hydrogen and carbon all having the general formula C_nH_{2n-6} with n never less than six, e.g., benzene C_6H_6, toluene C_7H_8, xylene C_8H_{10}.

benzenesulfonic acid (phenylsulfonic acid) $C_6H_5SO_3H$.
Properties: Fine, deliquescent needles or large plates; m.p. 65-66°C when anhydrous; with 1.5 molecules water, m.p. is 43-44°C; soluble in water and alcohol; slightly soluble in benzene; insoluble in ether and carbon disulfide.
Derivation: By sulfonating benzene with fuming sulfuric acid.
Method of purification: Crystallization.
Grades: Technical.
Containers: Iron drums.
Uses: Making phenol; resorcinol; for other organic syntheses and as a catalyst.
Shipping regulations: None.*

benzenylaminothiophenol $C_6H_5CNC_6H_4S$.
Properties: Yellow needles; pleasant odor of tea roses and geranium. Soluble in alcohol, ether, carbon disulfide, and dilute hydrochloric acid; insoluble in water. M.p. 115°C; b.p. 360°C.
Method of purification: Crystallization.
Grades: Technical.
Containers: Tins.
Use: Perfumery.
Shipping regulations: None.*

benzenyl trichloride. See benzotrichloride.

benzestrol
$HOC_6H_4CH(C_2H_5)CH(C_2H_5)CH(CH_3)C_6H_4OH$. 4,4'-(1,2-Diethyl-3-methyltrimethylene) diphenol.
Properties: Odorless, white, crystalline powder. M.p. 161-163°C; readily soluble in acetone, alcohol, ether, methanol, and sodium hydroxide solution; soluble in vegetable oils; moderately soluble in glacial acetic acid; slightly soluble in dilute alcohol, benzene, chloroform, and petroleum ether; practically insoluble in water and dilute mineral acids.
Grade: N.F. XI.
Use: Medicine.

benzethonium chloride. A synthetic quaternary ammonium compound, $C_{27}H_{42}ClNO_2 \cdot H_2O$. (Benzyldimethyl{2-[2-(para-1,1,3,3-tetramethylbutylphenoxy) ethoxy] ethyl}ammonium chloride or $(CH_3)_3CCH_2C(CH_3)_2C_6H_4OC_2H_4OC_2H_4N(CH_2C_6H_5)(CH_3)_2Cl \cdot H_2O$.
Properties: Colorless, odorless crystals. Very bitter taste; m.p. 164-166°C. Soluble in water, alcohol, and acetone. Aqueous solution yields flocculent white precipitate with soap solutions.
Grade: U.S.P. XVI.
Use: Antiseptic; cationic detergent.

benzhydrol (benzohydrol; diphenylmethanol; diphenylcarbinol) $(C_6H_5)_2CHOH$.
Properties: Needlelike colorless crystals; m.p. 69°C; b.p. 298°C (748 mm), 176°C (13 mm). Slightly soluble in water, easily soluble in alcohol, ether, chloroform and carbon disulfide; insoluble in ligroin.
Derivation: Reduction of benzophenone with magnesium or zinc dust.
Use: Preparation of other organic compounds, including certain antihistamines.

benzhydryl chloride $(C_6H_5)_2CHCl$.
Properties: A water-white to light straw colored liquid; refractive index 1.596; b.p. 140 °C (3 mm); used for synthesis.

2-(benzhydryloxy)-N,N-dimethylethylamine hydrochloride. See diphenhydramine hydrochloride.

benzidine (benzidine base; para-diaminodiphenyl) $NH_2(C_6H_4)_2NH_2$.
Properties: Grayish-yellow, white or reddish gray crystalline powder; m.p. 127°C, but varies with rate of heating, and is lowered by presence of moisture in sample; b.p. 400°C; soluble in hot water, alcohol, and ether; slightly soluble in water.
Derivation: (a) By reducing nitrobenzene with zinc dust in alkaline solution followed by distillation; (b) by electrolysis of

*See "I.C.C. Shipping Regulations," page xiii.
Reference numbers refer to name of manufacturer. See "List of Manufacturers," page v.

nitrobenzene, followed by distillation; (c) nitration of diphenyl followed by reduction of the product with zinc dust in alkaline solution, with subsequent distillation.
Method of purification: Crystallization.
Grades: Technical (paste, powder 80-85%).
Containers: Paste, 500-lb barrels; powder, 250-lb barrels.
Uses: Organic synthesis; manufacture of dyes, especially of Congo red; detection of blood stains; stain for microscopic work; reagent to determine lignification of wood; as a stiffening agent in rubber compounding.
Warning: Hazardous solid and vapor. Rapidly absorbed through skin. Repeated absorption may result in bladder tumors. MCA warning label.
Shipping regulations: None.*

benzidine base. See benzidine.

benzidinedicarboxylic acid. See diaminodiphenic acid.

benzidine hydrochloride $C_{12}H_{12}N_2 \cdot 2HCl$.
Properties: Crystals; soluble in water and alcohol.
Grades: Reagent; technical.
Containers: Barrels.
Use: Quantitative determination of sulfates; reagent for metals; detection of blood.

benzidine orange. A family of organic azo pigments prepared by coupling the tetrazonium salt of 3,3'-dichlorobenzidine with substituted pyrazolones. These pigments have properties and uses similar to those described under benzidine yellow.

benzidine sulfate (para-diaminodiphenyl sulfate) $C_{12}H_{12}N_2 \cdot H_2SO_4$.
Properties: White crystalline powder. Soluble in ether; sparingly soluble in water, alcohol, dilute acids.
Derivation: Action of sulfuric acid or sodium sulfate on benzidine with subsequent recovery by precipitation.
Method of purification: Crystallization.
Grades: Technical.
Containers: Wooden barrels; kegs.
Use: Organic synthesis.
Shipping regulations: None.*

benzidine yellow. A family of organic azo pigments prepared by coupling the tetrazonium salt of 3,3'-dichlorobenzidine with acetoacetarylides having good brightness, light-fastness and alkali resistance. They are approximately twice as strong as hansa yellow but somewhat poorer in light permanency.
Containers: Barrels.
Uses: Printing ink; linoleum and floor tile; plastics and rubber.

benzil (dibenzoyl) $C_6H_5CO \cdot COC_6H_5$.
Properties: Yellow needles. Soluble in alcohol and ether; insoluble in water. M.p. 95°C; b.p. 346-348°C; sp. gr. 1.521.
Derivation: From benzoin by oxidation with nitric acid.
Method of purification: Crystallization from alcohol.
Grades: Technical.
Containers: Tins.
Use: Organic synthesis.
Shipping regulations: None.*

benzine. See ligroin. Note: Term is misleading and outmoded. Do not confuse with benzene, also known commercially as benzol.

benzoate of soda. See sodium benzoate.

benzocaine. See ethyl-para-aminobenzoate.

benzodihydropyrone (dihydrocoumarin).
Properties: White to light yellow oily liquid with a sweet odor; congeals at 23°C. Insoluble in water; soluble in alcohol, chloroform and ether.
Containers: Glass bottles; cans; drums.
Uses: In perfumery and flavors.
Shipping regulations: None.*

benzodioxine. See piperoxan.

"Benzoform." [307] Brand name of proprietary line of direct dyestuffs for aftertreatment with formaldehyde. Used for the dyeing of cotton and rayon. Characterized by fair fastness to washing, water, and perspiration and good dischargeability.

benzofuran. See coumarone.

benzofuran resin. See coumarone-indene resins.

benzoglycolic acid. See mandelic acid.

benzoguanamine (2,4-diamino-6-phenyl-s-triazine) $C_6H_5C_3N_3(NH_2)_2$.
Properties: Crystals; sp. gr. 1.40 (d 25/4); m.p. 227-228°C. Soluble in alcohol, ether and dilute hydrochloric acid; partially soluble in dimethylformamide; practically insoluble in acetone, chloroform, ethyl acetate; insoluble in water.
Derivation: Prepared from benzonitrile and dicyandiamide in the presence of sodium and liquid ammonia.
Containers: Drums; tank cars.
Uses: Manufacture of thermosetting resins, resin modifiers, chemical intermediate for pesticides, pharmaceuticals and dyestuffs.

benzohydrol. See benzhydrol.

benzoic acid (carboxybenzene; benzene carboxylic acid; phenylformic acid) C_6H_5COOH. It occurs naturally in benzoin gum and some berries.
Properties: White scales or needle crystals with odor of benzoin or benzaldehyde; sp. gr. 1.2659; m.p. 121.25°C; b.p. 249.2°C; sublimes at 100°C; flash point (closed cup) 250°F; soluble in alcohol, ether, chloroform, benzene, carbon disulfide, carbon tetrachloride, and turpentine; slightly soluble in water.
Derivation: (a) Decarboxylation of phthalic acid by steam, in the presence of catalysts; (b) chlorination of toluene to yield benzotrichloride, which is hydrolyzed to benzoic acid; (c) by the direct oxidation of toluene.
Method of purification: Sublimation.
Grades: Technical; C.P.; U.S.P. XVI.

*See "I.C.C. Shipping Regulations," page xiii.
Reference numbers refer to name of manufacturer. See "List of Manufacturers," page v.

Containers: 200-lb barrels; multiwall paper sacks; 25-, 50-, 100-lb drums; tanks.
Uses: Chemicals (benzoates), especially sodium and benzyl benzoates; mordant in calico printing; seasoning tobacco and improving the aroma; flavors; perfumes; dentifrices; medicine (germicide); plasticizer and resin intermediate; food packaging; textiles; dyes.
Shipping regulations: None.*

benzoic aldehyde. See benzaldehyde.

benzoic ether. See ethyl benzoate.

benzoic trichloride. See benzotrichloride.

benzoin (bitter almond oil camphor; phenylbenzoyl carbinol) $C_6H_5CHOHCOC_6H_5$.
Properties: White or yellowish crystals; m.p. 137°C; b.p. 344°C; slightly soluble in water, alcohol, and ether.
Derivation: Condensation of benzaldehyde in potassium cyanide solution.
Use: Antiseptic.
Do not confuse with benzoin gum, of which it is not a constituent.

benzoin gum, Siam (benzoin Siam; Benjamin gum; benzoin resin).
Properties: Almond-shaped, pale, reddish-brown tears; balsamic, vanilla-like odor; aromatic; slightly acrid taste; hard and brittle at ordinary temperature, but softened by heat; soluble in warm alcohol and carbon disulfide; insoluble in water. Not less than 90% is soluble in alcohol (U.S.P.).
Chief constituents: An ethereal oil, benzoic acid, cinnamic acid, vanillin, resins.
Derivation: Balsamic resin from Styrax benzoin and other species.
Occurrence: Siam, Cambodia, Cochin-China.
Grades: Technical; U.S.P. XVI.
Containers: Tins.
Uses: Medicine; perfumery; cosmetics.
Shipping regulations: None.*

benzoin gum, Sumatra (benzoin Sumatra; Benjamin gum; benzoin resin).
Properties: Differs from benzoin gum, Siam, in many respects. The odor is not so strong and it does not melt so easily. Generally contains 12-15% of woody matter. Soluble in warm alcohol and carbon disulfide; insoluble in water. Not less than 75% is soluble in alcohol (U.S.P.).
Chief constituents: Cinnamic acid, benzoic acid, vanillin, resins.
Derivation: Balsamic resin from Styrax benzoin or Sumatra benzoin.
Occurrence: Sumatra, Java and Sunda Islands.
Grades: Technical; U.S.P. XVI.
Containers: Bags; cases.
Uses: Cinnamic acid; varnishes; medicine; cosmetics.
Shipping regulations: None.*

benzoin resin. See benzoin gum, Siam; benzoin gum, Sumatra.

benzoin Siam. See benzoin gum, Siam.

benzoin Sumatra. See benzoin gum, Sumatra.

benzol. See benzene. The term benzol is still used commercially, but is not in favor in modern nomenclature.

benzol 160°. See naphtha, solvent.

benzol black. A carbon black made by incomplete combustion of benzene or benzene-containing liquids.

benzonaphthol (beta-naphthol benzoate; naphthyl benzoate; benzoyl naphthol) $C_6H_5COOC_{10}H_7$.
Properties: White crystalline powder, odorless, tasteless; darkens with age. Soluble in chloroform and in alcohol (more so in hot); almost insoluble in water. M.p. 11°C.
Derivation: Interaction of benzoyl chloride and beta-naphthol.
Use: Medicine.
Shipping regulations: None.*

benzonatate (omega-methoxypoly(ethyleneoxy) ethyl-para-butylaminobenzoate; nonaethyleneglycol monomethyl ether para-n-butylaminobenzoate) $C_4H_9NHC_6H_4COO(CH_2)_2-(OC_2H_4)_nOCH_3$. n=8, average.
Properties: Colorless to faintly yellow oil. Soluble in most organic solvents except aliphatic hydrocarbons.
Grade: N.N.D.
Use: Medicine.

benzonitrile (phenyl cyanide) C_6H_5CN.
Properties: Transparent, colorless oil; odor of essential oil of almonds; viscosity (100°F) 1.054 centistokes; refractive index (n 20/D) 1.5289. Very toxic! Soluble in hot water, alcohol, and ether; insoluble in cold water. Sp. gr. 1.0051; b.p. 190.7°C; m.p. −13.1°C.
Derivation: From benzoic acid by heating with lead thiocyanate.
Method of purification: Distillation.
Grades: Technical.
Containers: Glass bottles; iron drums.
Uses: Organic synthesis; possible uses include production of pharmaceuticals; dyestuffs; intermediate for rubber chemicals; solvent for vinyl resins.

benzophenol. Phenol itself, C_6H_5OH, as distinguished from higher phenols such as the cresols.

benzophenone (diphenylketone) $(C_6H_5)_2CO$.
Properties: White prisms, with sweet, rose-like odor. Soluble in alcohol and ether; insoluble in water. Congealing point −47.5°C; chlorine-free.
Purification: Crystallization from alcohol.
Grades: Free from chlorine.
Containers: Tin cans; fiberboard containers; drums.
Uses: Organic synthesis; perfumery, for floral odors and as fixative; derivatives are used as ultraviolet absorbers.
Shipping regulations: None.*

benzophenone oxide. See xanthone.

benzopurpurin (Eclipse; Eclipse Red). A red substantive dye formed by combining naphthionic acid with diazo compound of ortho-toluidine. (Brownish-red powder.)

Used in dyeing wool and silk; various shades known by combinations of letters and numbers, such as 4B, 10B, etc., placed after the name.

benzopyrene (benzpyrene) $C_{20}H_{12}$. A polynuclear (five-ring) aromatic hydrocarbon, held to be a cause of cancer. Found in coal tar, cigarette smoke, and in the atmosphere as a product of incomplete combustion. Occurs as benzo[a]pyrene (1,2-benzopyrene); also called 3,4-benzpyrene) and benzo[e]pyrene (4,5-benzopyrene).

Properties (benzo[a]pyrene): Yellowish crystals; m.p. 179°C; b.p. 310-312°C (10 mm). Insoluble in water; slightly soluble in alcohol; soluble in benzene, toluene, xylene.

benzopyrone. See coumarin.

benzoquinone. See quinone.

benzosulfimide. See saccharin.

1,2,3-benzotriazole (aziminobenzene; benzene azimide) $C_6H_4NHN_2$.

Properties: White to light tan; odorless, crystalline compound; boiling range 201-204°C (15 mm); very stable toward acids and alkalies, and toward oxidation and reduction. Its basic characteristics are very weak but it forms stable metallic salts. Can exist in 2 tautomeric forms. Soluble in alcohol and benzene; slightly soluble in water.

Containers: 1-lb bottles; 5-, 50-, 100-lb fiber drums.

Uses: Photographic restrainer; as a chemical intermediate; derivatives used as ultraviolet absorbers.

Shipping regulations: None.*

benzotrichloride (toluene trichloride; benzenyl trichloride; benzoic trichloride; phenylchloroform) $C_6H_5CCl_3$.

Properties: Colorless to yellowish liquid; characteristic, penetrating odor. Soluble in alcohol and ether; insoluble in water. Sp. gr. 1.38; b.p. 213-214°C; m.p. −5°C; refractive index (n 19/D) 1.5584.

Derivation: By the chlorination of boiling toluene.

Method of purification: Rectification.

Grades: Technical.

Containers: Iron drums; carboys.

Use: Synthetic dye industry.

Shipping regulations: None.*

benzotrifluoride (toluene trifluoride; trifluoromethyl benzene) $C_6H_5CF_3$.

Properties: Water-white liquid with aromatic odor. B.p. 102.1°C; f.p. −29.1°C; sp. gr. 1.1812 (25/4°C); refractive index (n 20/D) 1.4146. Flash point (closed cup) 54°F. Miscible with alcohol, acetone, benzene, carbon tetrachloride, ether, and n-heptane; decomposes in water.

Containers: 55-gal drums.

Uses: Intermediate for dyes, and pharmaceuticals; as a solvent and dielectric fluid; vulcanizing agent; insecticides.

Shipping regulations: Flammable liquid. Red label.*

trans-beta-**benzoylacrylic acid** $C_6H_5COCH{:}CHCOOH$.

Properties: Straw yellow needles or plates; m.p. 99°C; soluble in most solvents but only slightly soluble in cold water and ligroin.

Containers: Polyethylene-lined fiber drums.

Uses: Ovicide for eggs of the body louse; reagent for characterizing phenols; intermediate in the manufacture of bactericides, insecticides, surface active agents and the upgrading of drying oils.

benzoylamide. See benzamide.

benzoylaminoacetic acid. See hippuric acid.

benzoylanilide. See benzanilide.

benzoyl chloride C_6H_5COCl.

Properties: Transparent, colorless pungent liquid; vapor causes tears. Sp. gr. 1.2188; m.p. −0.5°C; b.p. 197.2°C; refractive index (n 20/D) 1.5536; flash point 72°C. Soluble in ether and carbon disulfide; decomposes in water.

Derivation: (a) Interaction of benzoic acid and sulfuryl chloride; (b) benzotrichloride and water in the presence of zinc chloride; (c) phosphorus tri- or pentachloride and benzoic acid.

Containers: 1-lb bottles; returnable carboys of 50- and 100-lbs net; tank trucks.

Grades: Technical; C.P.

Uses: Medicine; intermediate for introduction of benzoyl groups; intermediate for other organics.

Warning! Causes burns; vapor irritating. MCA warning label.

Shipping regulations: Corrosive liquid. White label.*

benzoylglycin. See hippuric acid.

benzoylglycocoll. See hippuric acid.

benzoyl hydride. See benzaldehyde.

benzoylnaphthol. See benzonaphthol.

benzoyl peroxide $(C_6H_5CO)_2O_2$.

Properties: White, granular, crystalline solid; tasteless; odorless; dangerous, has been known to explode spontaneously. Active oxygen, about 6.5%. Soluble in nearly all organic solvents; slightly soluble in alcohols and vegetable oils; insoluble in water. M.p. 103-105°C (dec).

Grades: Technical, wet or dry.

Containers: 1-lb net fiber containers or polyethylene-lined bags; standard shipping cases contain 5, 25, and 50 containers.

Uses: Bleaching agent for flour, fats, oils, and waxes; polymerization catalyst; drying agent for unsaturated oils; pharmaceutical and cosmetic purposes; rubber compounding; burn out agent for acetate yarns.

Fire hazard: Dangerous.

Shipping regulations: Oxidizing material. Yellow label.*

2-benzoylpyridine $C_6H_5COC_5H_4N$.

Properties: Freezing point 42.7°C; insoluble

*See "I.C.C. Shipping Regulations," page xiii.

Reference numbers refer to name of manufacturer. See "List of Manufacturers," page v.

in water.
Grade: 98% (minimum).
Use: Organic synthesis.

4-benzoylpyridine $C_6H_5COC_5H_4N$.
Properties: Freezing point 71.4°C; insoluble in water.
Grade: 98% (minimum).
Use: Organic synthesis.

benzoylsalicylic acid, methyl ester. See methyl benzoylsalicylate.

benzoylsulfonic imide. See saccharin.

benzozone. See acetylbenzoyl peroxide.

1, 2-benzphenanthrene. See chrysene.

benzpyrene. See benzopyrene.

benzpyrinium bromide
$(CH_3)_2NCOOC_5H_4NCH_2C_6H_5Br$. 1-Benzyl-3-(dimethylcarbamyloxy)-pyridinium bromide.
Properties: White to slightly yellow crystalline powder; almost odorless. M.p. 114-120°C; very soluble in alcohol and water; practically insoluble in ether; pH (1% solution) 4.5-5.5.
Use: Medicine.

benztropine methanesulfonate (3-diphenylmethoxytropane methanesulfonate) $C_{21}H_{25}NO \cdot CH_4O_3S$.
Properties: White, colorless, odorless, slightly hygroscopic, crystalline powder. Very soluble in water, freely soluble in alcohol, and very slightly soluble in ether. Melting range 142-144°C.
Grade: N.F. XI.
Use: Medicine.

benzyl abietate $C_{19}H_{29}COOC_6H_5CH_2$.
Properties: Stable, nonvolatile, viscous liquid which resembles Canada balsam. Soluble in most anhydrous solvents.

benzyl acetate (phenylmethyl acetate) $C_6H_5CH_2OOCCH_3$.
Properties: A water-white liquid with a pleasant flowery odor which should not be stale, flat or sharp in a well-manufactured product. The odor should not change on keeping or on exposure to air or on evaporation. Should give a negative test for chlorine by all qualitative methods. Soluble in alcohol and ether; almost insoluble in water and glycerol. Flash point 102°C. Sp. gr. 1.059-1.062 (15°C); b.p. 212°C; refractive index 1.5015-1.5035.
Derivation: (a) By treating benzyl chloride with acetate of soda in various solvents; (b) by esterification of benzyl alcohol with acetic anhydride or acetic acid.
Method of purification: Distillation.
Grades: Free-from-chlorine grade which should have an ester content of 97% but for which lower grade material is sometimes substituted; technical grade which is not free from chlorine and for which ester content varies considerably.
Containers: 55-gal, 100-lb tinned drums; carboys; cans; bottles.
Uses: Essential ingredient of artificial jasmin and many other flowery perfumes; soap perfume; in some flavors; solvent and high boiler for cellulose acetate and nitrate, natural and synthetic resins; oils; lacquers; dopes; polishes; printing inks; varnish removers.
Shipping regulations: None.*

benzyl alcohol (alpha-hydroxytoluene; phenylcarbinol) $C_6H_5CH_2OH$.
Properties: A water-white liquid with a faint aromatic odor which in the course of time, especially on exposure to air, smells slightly of benzaldehyde. Has a sharp, burning taste. B.p. 206°C; flash point 96°C; sp. gr. 1.040-1.050; refractive index (20°C) 1.5385-1.5405. Somewhat soluble in water; miscible with alcohol, ether and chloroform.
Derivation: (a) By hydrolysis of benzyl chloride; (b) by hydrolysis of other benzyl esters, such as benzyl acetate; (c) by Cannizzaro method from benzaldehyde.
Method of purification: Distillation and chemical treatment.
Grades: Free from chlorine (F.F.C.); technical; N.F. XI.
Containers: 1-, 5-gal cans; 100-gal carboys; 450-lb drums.
Uses: Solvent in perfumery and flavoring materials; intermediate in preparing other benzyl esters and ethers; high-boiling solvent in cellulose derivative products; medicine (local anesthetic). Solvent for: cellulose esters and ethers, benzyl abietate, resins, sulfur; lacquers; films; paint and varnish removers.
Shipping regulations: None.*

benzylamine (aminotoluene) $C_6H_5CH_2NH_2$.
Properties: Colorless liquid; strongly alkaline reaction. Soluble in alcohol, ether, and water. Sp. gr. 0.9813; b.p. 184.5°C; refractive index (n 20/D) 1.540.
Derivation: From benzyl chloride and ammonia.
Method of purification: Distillation.

benzyl-para-aminophenol (BAP) $C_6H_5CH_2NHC_6H_4OH$.
Properties: Light brown, finely ground powder, melts between 84-90°C; 96-99% pure; solubility is 50% in anhydrous methanol, 50% in 95% ethyl alcohol, 0.06% in water; 0.1-0.5% in gasoline, varying with chemical nature of gasoline.
Containers: Iron drums (200 lbs net).
Use: In cracked gasoline, in concentration of 0.001-0.004% by weight, to prevent gum formation.

2-benzylamino-1-propanol $C_6H_5CH_2NHCH(CH_2OH)CH_3$.
Properties: White to yellow solid. Both ℓ and dℓ forms are available. M.p. (dℓ-form) 70-73°C. Specific rotation (ℓ-form) +38° to +44° (1.0% solution in alcohol) at 25°C.

benzylaniline $C_6H_5NHCH_2C_6H_5$.
Properties: Colorless prisms. Soluble in alcohol and ether; insoluble in water. M.p. 33°C; b.p. 310°C.

Derivation: By heating aniline with benzyl chloride.
Method of purification: Crystallization.
Grades: Technical.
Containers: Wooden barrels.
Use: Organic synthesis.
Shipping regulations: None. *

benzylbenzene. See diphenylmethane.

benzyl benzoate $C_6H_5CH_2OOCC_6H_5$.
Properties: A water-white liquid which readily freezes to a solid. Has a sharp, burning taste and a faint aromatic, pleasant odor which should not be harsh or have any trace of benzaldehyde or toluene odor even after long standing. Supercools easily. Insoluble in water and glycerin; soluble in alcohol, chloroform and ether. Sp. gr. 1.119; b.p. 325°C; m.p. 18.8°C; should be free from even traces of chlorine and have less than 0.03% free acid; refractive index 1.568-1.569 (20°C).
Derivation: (a) By a Cannizzaro reaction from benzaldehyde; (b) by esterifying benzyl alcohol with benzoic acid; (c) by treating benzoate of soda with benzyl chloride.
Method of purification: By distillation and crystallization.
Grades: N.F. XI; technical.
Containers: 55-gal drums (tinned); 100-lb drums; cans; bottles; aluminum containers.
Uses: Fixative and solvent for musk in perfumes and flavors; medicine; plasticizer; miticide.
Shipping regulations: None. *

benzyl bichloride. See benzyl dichloride.

benzyl bromide (alpha-bromotoluene) $C_6H_5CH_2Br$.
Properties: Clear, refractive liquid. Pleasant odor. Not easily hydrolyzed. Soluble in alcohol, benzene, ether; insoluble in water. A lachrymator. Sp. gr. 1.438 at 16°C; b.p. 198-199°C; m.p. −3.9°C; vapor density 5.8.
Derivation: (a) Bromination of toluene; (b) interaction of benzyl alcohol and hydrobromic acid.
Grades: Technical.
Uses: Making foaming and frothing agents; organic synthesis.
Shipping regulations: Corrosive liquid. White label. *

benzyl butyrate $C_3H_7COOCH_2C_6H_5$.
Properties: Liquid; heavy fruity odor; b.p. 240°C; density 1.016 (17.5°C); soluble in alcohol.
Grades: Technical.
Uses: Plasticizer; for blending in odorants and flavoring.

benzyl carbinol. See phenethyl alcohol.

benzyl chloride (alpha-chlorotoluene) $C_6H_5CH_2Cl$.
Properties: Colorless liquid; pungent odor. Sp. gr. 1.1027; m.p. −43°C; b.p. 179°C; n 25/D 1.5365; flash point 165°F. Soluble in alcohol and ether; insoluble in water.
Derivation: By passing chlorine over boiling toluene until it has increased 38% in weight. The product is washed with water and separated by fractional distillation.
Method of purification: Redistillation.
Grades: Technical; C.P.; 95%; redistilled.
Forms: Anhydrous; stabilized (with aqueous soda ash solution).
Containers: Anhydrous: 475-lb nickel drums; 100-lb carboys. Stabilized: 475-lb steel drums; tank trucks.
Uses: Dyes; intermediates; benzyl compounds; synthetic tannins; perfumery; pharmaceuticals; manufacture of photographic developer; gasoline gum inhibitors; penicillin precursors; quaternary ammonium compounds.
Warning! Causes burns. MCA warning label.
Shipping regulations: Corrosive liquid. White label. *

benzyl chlorocarbonate (carbobenzoxy chloride; benzyl chloroformate) $C_6H_5CH_2OCOCl$.
Properties: Oily liquid, with lachrymatory properties; acrid odor. Decomposes over 100°C.
Use: Peptide synthesis.
Shipping regulations: Corrosive liquid. White label. *

benzyl chloroformate. See benzyl chlorocarbonate.

ortho-**benzyl**-para-**chlorophenol** $C_6H_5CH_2C_6H_3OHCl$.
Properties: White to light tan or pink flakes; crystallizing point 45°C min; sp. gr. 1.202-1.206 (55/55°C); odor slight phenolic max. Insoluble in water; highly soluble in alcohol and other organic solvents; dispersible in aqueous media with the aid of soaps or synthetic dispersing agents; noncorrosive to most metals or other engineering materials.
Containers: 22- and 55-gal lacquer-lined drums.
Uses: As the active principle, or as an enhancing agent for disinfectants.

benzyl cinnamate (cinnamein) $C_9H_7O_2 \cdot C_7H_7$.
Properties: White crystals; aromatic odor; m.p. 39°C; congealing point min 34°C; b.p. 244°C (25 mm); insoluble in water; soluble in alcohol.
Containers: Cans.
Uses: Perfumery and flavors.

benzyl cyanide (phenylacetic acid nitrile) $C_6H_5CH_2CN$.
Properties: Colorless oily liquid; aromatic odor; soluble in alcohol and ether; insoluble in water. Sp. gr. 1.0157; m.p. −24°C; b.p. 233.5°C; refractive index (n 25/D) 1.5211.
Derivation: By the interaction of benzyl chloride and potassium cyanide.
Method of purification: Distillation.
Grades: Technical.
Containers: Iron drums.
Use: Organic synthesis.

benzyl dichloride (benzylidene chloride; benzyl bichloride; benzal chloride; chlorobenzal) $C_6H_5CHCl_2$.
Properties: Colorless oily liquid; faint aromatic odor; sp. gr. 1.295 (16°C); m.p.

−16.1°C; b.p. 207°C; refractive index 1.5502 (20°C); soluble in alcohol, ether, and in dilute alkali; insoluble in water.
Derivation: By the chlorination of toluene, until two formula weights of chlorine are absorbed, in absence of catalysts but presence of light.
Method of purification: Distillation.
Grades: Technical.
Containers: Carboys.
Uses: Dyes.
Shipping regulations: None.*

N-benzyldiethanolamine $C_6H_5CH_2N(C_2H_4OH)_2$.
Properties: Colorless to light yellow liquid. Sp. gr. 1.073; refractive index 1.5345-1.5375; distilling range 155-165°C (2 mm).
Containers: 200-, 400-lb steel drums.
Uses: Corrosion inhibitor; intermediate.

N-benzyldimethylamine $C_6H_5CH_2N(CH_3)_2$.
Properties: Colorless to light yellow liquid. Sp. gr. 0.894 (27°C); refractive index 1.4985-1.5005 (25°C); b.p. 180-182°C; distilling range 65-68°C (18 mm).
Containers: 200-, 400-lb steel drums.
Uses: Intermediate, especially for quaternary ammonium compounds; a dehydrohalogenating catalyst; corrosion inhibitor; acid neutralizer.

N-benzylethanolamine $C_6H_5CH_2NH(C_2H_4OH)$.
Properties: Colorless to light yellow liquid. Sp. gr. 1.044 (27°C); refractive index 1.5400-1.5430; distillation range 240-255°C (760 mm).
Containers: 200-, 400-lb drums.
Uses: Corrosion inhibitor; intermediate.

benzyl ethyl ether $C_6H_5CH_2OC_2H_5$.
Properties: Colorless, oily liquid; aromatic odor; volatile in steam; insoluble in water; miscible with alcohol and ether. B.p. 185°C; sp. gr. 0.949; refractive index 1.4955 at 20°C.
Derivation: By boiling benzyl chloride with either sodium or potassium ethylate.
Grades: Technical.
Use: Organic synthesis.

benzyl fluoride $C_6H_5CH_2F$.
Properties: Colorless liquid. Forms acicular crystals on prolonged cooling. Sp. gr. 1.022 at 25°C; b.p. 139.8°C (753 mm); m.p. −35°C.
Derivation: By decomposing benzyl trimethylammonium fluoride.
Grades: Technical.
Use: Organic synthesis.

benzyl formate $C_6H_5CH_2OOCH$.
Properties: Colorless liquid; fruity-spicy odor. Resembles benzyl acetate in many respects but differs in its greater volatility. Sp. gr. 1.083-1.087; refractive index, 1.511-1.513; b.p. 203°C; miscible with alcohols, ketones, oils, aromatic, aliphatic and halogenated hydrocarbons; insoluble in water.
Grades: Technical.
Containers: Cans.
Uses: Solvent for nitrocellulose, acetylcellulose, and some cellulose ethers, benzyl abietate, ester gum, copal ester; perfumery; flavors.

benzylidene acetone (benzalacetone; acetocinnamone; methylcinnamyl ketone; methylstyryl ketone) $C_6H_5CHCHCOCH_3$.
Properties: Colorless crystals; odor of coumarin. Soluble in alcohol, ether, benzene, and chloroform; insoluble in water. M.p. 42°C; congealing point 39°C (min); b.p. 260-262°C; refractive index 1.5836 at 46°C; density (15/15°C) 1.0377.
Derivation: By the condensation of benzaldehyde and acetone.
Method of purification: Crystallization.
Grades: Technical.
Containers: Bottles; tin cans; aluminum containers.
Use: Organic synthesis; perfumery (fixative, balsamic odors, artificial essence of sweet pea).
Shipping regulations: None.*

benzylidene chloride. See benzyl dichloride.

2-benzyl-2-imidazoline hydrochloride. See tolazoline hydrochloride.

benzyl iodide $C_6H_5CH_2I$.
Properties: Colorless crystals or liquid; vapors cause tears. Soluble in alcohol, carbon disulfide and ether; insoluble in water. Sp. gr. 1.7335; m.p. 34.1°C; b.p., decomposes.
Derivation: By the interaction of benzyl chloride and hydriodic acid.
Method of purification: Crystallization.

benzyl isoamyl ether. See isoamyl benzyl ether.

benzyl isoeugenol
$CH_3CHCHC_6H_3(OCH_3)OCH_2C_6H_5$.
Properties: White crystalline material, having a light floral odor of the carnation type. Soluble in alcohol and ether. Congealing point, 57°C min.
Purification: Crystallization.
Containers: Cans.
Use: Perfumery, in carnation types and in other florals as a fixative.

N-benzylisopropylamine
$C_6H_5CH_2NH(CH_3CHCH_3)$.
Properties: Colorless to yellow liquid; sp. gr. 0.895 (25°C); refractive index 1.4995-1.5015 (25°C).
Containers: 200-, 400-lb steel drums.
Uses: Rust inhibitor; intermediate.

benzylmethylamine $C_6H_5CH_2NHCH_3$.
Properties: Colorless to light yellow liquid. Sp. gr. 0.936 (25°C); refractive index 1.5185-15220 (25°C); distillation range 183-188°C (760 mm).
Containers: 200-, 400-lb steel drums.
Uses: Organic synthesis.

N-benzyl-N-methylethanolamine
$C_6H_5CH_2NCH_3(C_2H_4OH)$.
Properties: Colorless to light yellow liquid; sp. gr. 1.006 (27°C); refractive index 1.5250-1.5270 (25°C); distillation range 95-105°C (2 mm).
Containers: 200- and 400-lb steel drums.
Uses: Corrosion inhibitor; intermediate.

3-benzyl-4-methyl umbelliferone $C_6H_5CH_2CH_3C_9H_4O_3$.
Properties: Fine tan crystalline powder; m.p. 255°C min; slightly soluble in ethyl alcohol; insoluble in water.
Use: Optical whitening agent; intermediate.

benzyl penicillin potassium. See potassium penicillin G, under penicillin.

benzyl penicillin sodium. See sodium penicillin G, under penicillin.

ortho-**benzylphenol** (2-hydroxydiphenylmethane). $C_6H_5CH_2C_6H_4OH$.
Properties: White crystals; m.p. 52°C; soluble in organic solvents; insoluble in water.
Use: Disinfectant.
Note: Ortho-benzylphenol also exists in an unstable form; m.p. 21°C; b.p. 312°C.

para-**benzylphenol** (4-hydroxydiphenylmethane) $C_6H_5CH_2C_6H_4OH$.
Properties: White crystals from ethyl alcohol; m.p. 84°C; b.p. 320-322°C. Soluble in ethyl alcohol, ether, chloroform, benzene, acetic acid, caustic alkalies; moderately soluble in hot water.
Containers: 100-lb drums.
Uses: Antiseptic and germicide; also for organic synthesis.

benzyl phenylacetate $C_6H_5CH_2COOCH_2C_6H_5$.
Properties: Colorless liquid, with sweet honeylike odor. Soluble in alcohol; sp.gr. 1.097-1.099; refractive index 1.554-1.556.
Uses: Perfumery and flavors.

benzyl propionate $C_2H_5COOCH_2C_6H_5$. A compound similar to benzyl acetate but has a sweeter odor and is more expensive.
Properties: B.p. 220°C; sp.gr. 1.036 (17.5°C); insoluble in water.

2-benzylpyridine $C_6H_5CH_2C_5H_4N$.
Properties: Boiling point (760 mm) 276.8°C; freezing point 13.6°C; density (20°C) 1.061; refractive index (n 20/D) 1.5797. Insoluble in water.

4-benzylpyridine $C_6H_5CH_2C_5H_4N$.
Properties: Boiling point (760 mm) 291.1°C; freezing point 11.6°C; density (20°C) 1.067; refractive index (n 20/D) 1.5825. Insoluble in water.

benzyl rhodanide. See benzyl thiocyanate.

benzyl salicylate $C_6H_4(OH)COOCH_2C_6H_5$.
Properties: Colorless liquid, except at rather cold room temperatures; very faint sweet odor. Soluble in 9 vols. of 90% alcohol.
Constants: Sp.gr. 1.176-1.179; refractive index 1.580-1.581; m.p. min 24°C; b.p. (26 mm) 208°C.
Uses: Perfumery, particularly as a solvent for nitro-musks; and in floral odors as a fixative.

benzyl succinate (dibenzyl succinate) $C_6H_5CH_2OOCCH_2CH_2COOCH_2C_6H_5$.
Properties: White crystalline powder, almost tasteless. Soluble in alcohol, ether, chloroform, also in fixed and volatile oils; insoluble in water. M.p. 45°C.

benzyl sulfide $(CH_2C_6H_5)_2S$.
Properties: Colorless plates. Soluble in alcohol and ether; insoluble in water. Sp. gr. 1.0712; m.p. 49°C.
Derivation: By the action of potassium sulfide on benzyl chloride and subsequent distillation.
Method of purification: Recrystallization.
Grades: Technical.
Containers: Wooden kegs.
Use: Organic synthesis.
Shipping regulations: None.*

benzyl thiocyanate (benzyl rhodanide) $C_6H_5CH_2CNS$.
Properties: Colorless crystals; m.p. 41°C; b.p. 230°C; insoluble in water; soluble in alcohol and ether.
Containers: 525-lb drums.
Use: Insecticide.

benzyltrimethylammonium chloride $C_6H_5CH_2N(CH_3)_3 \cdot Cl$. A quaternary ammonium salt.
Properties: Colorless crystals; stable up to 135°C, above which benzyl chloride and trimethylamine are formed. Readily soluble in water, ethyl alcohol, and butanol; slightly soluble in dibutyl phthalate and tributyl phosphate. Properties of 60% solution: Sp. gr. (20/20°C) 1.07; wt/gal 8.90 lbs; f.p. less than −50°C.
Grades: 60-62% aqueous solution.
Containers: 1-gal bottles; 5-gal carboys; 50-gal wax-lined barrels.
Uses: Solvent for cellulose; catalyst in production of phenolic resins; reclaiming scrap rubber.

4-benzyltrimethyl ammonium methoxide $C_6H_5CH_2(CH_3)_3NOCH_3$. A quaternary ammonium salt.
Properties: Yellow liquid; decomposes on distillation.
Containers: Steel drums.
Uses: Catalyst; organic-soluble strong base.

benzyne C_6H_4. A ring-structured intermediate, in which two adjoining carbons of the ring are attached by triple bonds. It is known to exist in solution in some organic reactions and has been important in explaining known reactions and predicting new ones.

"Beraloy." [155] Trademark for beryllium-copper alloys supplied in two grades: "Beraloy" A (1.80-2.05% beryllium) and "Beraloy" D (1.60-1.80% beryllium) "Beraloy" A meets A.S.T.M. Specifications B-194-51T and B-197-51T.
Properties: High electrical conductivity; high resistance to fatigue; very low hysteresis or drift; easily formed when annealed; high strength and rigidity when heat treated; corrosion resistant.
Forms: Strip, round wire, flat wire.
Uses: Diaphragms; springs; fabrication of lightweight, intricate parts.

berbamine. See berberamine.

berberamine (berbamine) $C_{19}H_{19}NO_3 \cdot 2H_2O$. An alkaloid.

*See "I.C.C. Shipping Regulations," page xiii.
Reference numbers refer to name of manufacturer. See "List of Manufacturers," page v.

Properties: White lumps; m.p. 156°C (anhydrous); soluble in alcohol, ether, dilute acids; slightly soluble in water. Extract of root of Berberis shrub.
Use: Medicine.

berberine $C_{20}H_{17}O_4N \cdot 6H_2O$ or $C_{20}H_{19}O_5N$.
Properties: White to yellow crystals; poisonous alkaloid; m.p. 145°C (anhydrous); soluble in water and alcohol; very slightly soluble in ether. Some important salts of berberine are: berberine disulfate $C_{20}H_{17}O_4N \cdot H_2SO_4$; berberine sulfate $(C_{20}H_{17}O_4N)_2H_2SO_4$; berberine hydrochloride $C_{20}H_{17}O_4N \cdot HCl$. All three are yellow crystalline salts, slightly soluble in water.
Derivation: From the root of Berberis vulgaris or Hydrastis canadensis. The salts of this alkaloid are obtained by the action of the respective acid on the alkaloid.
Method of purification: Crystallization.
Grades: Technical.
Containers: 1-oz vials; 1-lb bottles; cans.
Use: Medicine, in form of alkaloid or its sulfate or hydrochloride.
Shipping regulations: None.*

bergamot oil.
Properties: Brownish-yellow or honey-colored liquid, often colored green by copper content; agreeable odor; bitter taste. Linalyl acetate content 34-45% (usually 34-40%). Soluble in 2 vols. of 90% alcohol; soluble in glacial acetic acid; affected by light.
Chief known constituents: Limonene; linalyl acetate; linalool; terpineol.
Constants: Sp. gr. (25/25°C) 0.875-0.880; optical rotation +8° to +22° (occasionally +5°24' and +24°); refractive index 1.464-1.468; acid value 1-3.5; evaporation residue 4.5-6% (occasionally 4%); acid value of evaporation residue 19-30.
Derivation: By expression from the fruits of Citrus bergamia Risso et Poiteau.
Adulteration: Turpentine, lemon, orange oils; distilled bergamot oil; fatty oils; cedarwood oil; gurjun-balsam oil.
Grade: N.F. XI.
Use: Perfumery.
Shipping regulations: None.*

berkelium Bk. A synthetic radioactive element with atomic number 97 first produced as the 243 isotope by bombarding americium with helium ions in a cyclotron. See actinide elements. The chemical properties of berkelium have been studied by tracer techniques and are similar to those of the other transuranium elements.

Berlin blue. A name applied loosely to any of a number of the varieties of iron blue pigments. See iron blues.

Berlin red. A red pigment consisting, essentially, of red iron oxide. See iron oxide reds, and hematite, red.

bertram. See pyrethrum root.

beryl $Be_3Al_2(SiO_3)_6$, sometimes with replacement of Be by Na, Li, Cs. A natural silicate of beryllium and aluminum. Found in pegmatites.
Properties: Various shades of green, blue, yellow, red, white, or colorless; white streak; vitreous luster. Inert to any reagent at low temperature except hydrofluoric acid. Sp. gr. 2.63-2.8; hardness 7.5-8.
Varieties: Emerald: bright emerald green color due to some chromium; aquamarine: sky blue to greenish blue color; morganite: pink color; heliodor: golden color.
Occurrence: South Dakota, New Hampshire, Colorado; Brazil; Argentina; India; Australia.
Uses: Source of beryllium salts; gem stone; dielectric for spark plugs.

beryllium Be. Element of atomic number 4, group II of the periodic system.
Properties: A hard, gray-white metal. Sp. gr. 1.85; m.p. 1280°C. Active metal at high temperatures. Resistant to oxidation at ordinary temperatures. Similar to aluminum chemically. Once called glucinum. The principal source is beryl, containing 5.3% Be. Soluble in acids and alkali hydroxides. Caution! Beryllium dust is toxic when inhaled or in prolonged external contact with the body.
Derivation: By a large variety of processes, as electrolysis of the double fluoride of beryllium and potassium; reduction of the oxide with carbon in the presence of copper to give a beryllium copper alloy; reduction of beryllium fluoride with magnesium metal; also by a sulfate leaching process. Beryllium as the fabricated metal is obtained by powder metallurgy techniques.
Grades: Technical; pure.
Containers: Drums.
Uses: In copper, nickel and aluminum alloys; production of neutrons; special windows for x-ray tubes; neutron moderator in nuclear energy devices; in certain aircraft components; fuel-element cladding in nuclear reactors; suggested for high energy fuels and missile parts.
Shipping regulations: (Metal powder) Poison, class B. Poison label.*

beryllium acetylacetonate $Be(C_5H_7O_2)_2$. Crystalline powder; slightly soluble in water; resistant to hydrolysis. A chelating nonionizing compound.

beryllium carbonate (basic beryllium carbonate) $(BeO)_5 \cdot CO_2 \cdot 5H_2O$.
Properties: White powder. Variable composition. Soluble in acids; insoluble in water.

beryllium carbonate, basic. See beryllium carbonate.

beryllium chloride $BeCl_2$.
Properties: White or slightly yellow, deliquescent crystals. M.p. 440°C; b.p. 520°C; sp. gr. 1.90. Very soluble in water; soluble in alcohol, benzene, ether, carbon disulfide. Readily hydrolyzed.
Derivation: By passing chlorine over a mixture of beryllium oxide and carbon.

beryllium copper. Alloys, often also containing nickel or cobalt, and having relatively high electrical conductivity, high strength, and high hardness.
Properties: Specific gravity 8.22. Tensile strength of heat treated sheet 175,000 psi, elongation 5% in 2 inches; Brinell hardness 350. Typical analysis: Copper 97.4; beryllium 2.25; nickel 0.35.
Uses: In electrical switch parts; watch springs; optical alloys; valves and parts; springs and diaphragms; shims; cams; and bushings.

beryllium hydrate. See beryllium hydroxide.

beryllium hydroxide (beryllium hydrate) $Be(OH)_2$.
Properties: Amorphous, white powder; decomposed to the oxide by heat; soluble in acids, alkalies; insoluble in water.
Derivation: By precipitation with alkali from pure beryllium acetate.
Grades: Technical.

beryllium metaphosphate $Be(PO_3)_2$.
Properties: White porous powder or granular material; has a high melting point; insoluble in water.
Uses: Raw material for special ceramic compositions; as a catalyst carrier.

beryllium nitrate $Be(NO_3)_2 \cdot 3H_2O$.
Properties: White to faintly yellowish, deliquescent mass; m.p. 60°C; soluble in water, alcohol.
Derivation: By the action of nitric acid on beryllium oxide, with subsequent evaporation and crystallization.
Method of purification: Recrystallization.
Grades: Technical; C.P.
Containers: Glass bottles.
Uses: Chemical reagent; gas mantle hardener.
Fire hazard: Dangerous.
Shipping regulations: Oxidizing material. Yellow label. *

beryllium oxide BeO.
Properties: White amorphous powder. Sp. gr. 3.016; m.p. 2570°C. Hardness 9 (Mohs). The dust is quite poisonous! Soluble in acids and alkalies; insoluble in water.
Derivation: By heating beryllium nitrate or hydroxide.
Grades: Technical; C.P.; pure.
Containers: Bottles.
Uses: Preparation of beryllium compounds; ceramics and refractories; phosphor; is under consideration for use in missiles, nucleonics and electronics because of its light weight, heat resistance, electrical insulating properties, transparency to microwave radiation, near-immunity to nuclear radiation, hardness and thermal conductivity.

beryllium-potassium (potassium-beryllium fluoride) $BeF_2 \cdot 2KF$.
Properties: White, crystalline masses. Soluble in water; insoluble in alcohol. Poisonous!
Grades: Technical.

beryllium-sodium fluoride (sodium-beryllium fluoride) $BeF_2 \cdot 2NaF$. Poisonous!
Properties: White, crystalline mass. Soluble in water. M.p. about 350°C.
Grades: Technical.
Use: Making pure beryllium metal.

beryllium sulfate $BeSO_4 \cdot 4H_2O$.
Properties: Colorless crystals. Soluble in water; insoluble in alcohol. Poisonous! Sp. gr. 1.713; decomposes at 540°C.
Grades: Technical.

beryllonite $NaBePO_4$. A natural sodium beryllium phosphate, sometimes used as a gem stone. Colorless, white or light yellow; sp. gr. 2.84; hardness 5.5-6.

"Be Square." [128] Brand name for a grade of petroleum microcrystalline wax.
Properties: Colors, black, amber or white; m.p. ranges, 190-195°F, 180-185°F, or 170-175°F.
Containers: 10-lb slabs, 8/carton or 168/pallet; 350-lb drums.
Uses: Polishes; crayons; candles; bakery board coating; electrical insulation, as in potting compounds.

Bessemer process. A method for the production of steel which utilizes an air blast through the molten pig iron to remove silicon, carbon, manganese, phosphorus and sulfur as their respective oxides. The gaseous oxides escape, while the solid oxides combine with one another or with an oxide from the converter lining to form an easily fusible slag. This floats on the surface of the molten iron and can easily be removed from it. The method has been used chiefly on irons with a low phosphorus content. Invented in England by Sir Henry Bessemer in 1857 and independently in America by William Kelley the same year, it was the first process to make steel available for construction on a large scale.

beta-. Prefix applied to chemical names to denote the position of a substituent or radical. In this book, beta is not ordinarily used in alphabetizing. See also alpha.

betacaine hydrochloride. See eucaine hydrochloride.

"Betachlor." [203] Trademark for a chlorinated solvent of the dichloroethyl ether type.
Properties: A clear liquid; sp. gr. 1.20-1.23; flash point (COC) 180°F; total chlorine 47.8%.
Containers: 55-gal steel drums.
Uses: Detergent additive; selective extractant; for degreasing; for fumigant formulations; in paint and plastics industry as solvent.
Caution: Avoid prolonged or repeated breathing of vapor and contact with skin. Use with adequate ventilation. Do not take internally.

betahypophamine. See vasopressin.

betaine (lycine; oxyneurine; trimethylglycine) $(CH_3)_3\underline{NCH_2COO}$. An alkaloid.

*See "I.C.C. Shipping Regulations," page xiii.
Reference numbers refer to name of manufacturer. See "List of Manufacturers," page v.

Properties: Colorless, sweet crystals; deliquescent; lose water at 100°C; anhydrous material melts 293°C; soluble in water and alcohol; slightly soluble in ether.
Derivation: Occurs widely in plants, recovered from sugar beets; also made synthetically from chloroacetic acid and trimethylamine.
Use: Medicine, as the hydrochloride.

betaine hydrochloride (lycine hydrochloride) $C_5H_{11}O_2N \cdot HCl$.
Properties: Colorless crystals; m.p. 227-228°C (dec.); soluble in water and alcohol; insoluble in chloroform and ether. Aqueous solutions are strongly acid. Liberates hydrogen chloride at the melting point.
Grades: Technical.
Containers: Drums.
Uses: Source of hydrogen chloride in solders and fluxes; organic synthesis; medicine.

betaine phosphate $C_5H_{11}O_2N \cdot H_3PO_4$.
Properties: White, odorless granules; acid taste; m.p. 198-200°C; very soluble in water.
Grades: Technical.
Uses: Crystalline source of phosphoric acid.

"Betalin" 12. [100] Trademark for cyanocobalamin. See vitamin B_{12}.

"Betalin" S. [100] Trademark for thiamine hydrochloride U.S.P. (q.v.).

"Betanol." [354] Trademark for a group of dispersing agents consisting of high molecular weight esters. They form dispersions which are stable over a wide pH range in the presence of acids and mineral salts. They can be used to prepare both water-in-oil and oil-in-water emulsions.
Uses: In the preparation of cosmetics, pharmaceuticals, textiles, paints, wherever a dispersing agent or wetting agent is needed in the presence of acids or bases.

"Betanox" Special. [248] Trademark for low-temperature reaction product of phenol-beta-naphthylamine and acetone.
Properties: Tan-colored powder; sp. gr. 1.16; m.p. above 120°C; soluble in acetone, benzol, and ethylene dichloride; insoluble in water and gasoline.
Uses: Antioxidant for use in wire insulation, tire treads, carcass, inner tubes, dark-colored footwear, proofing, and mechanicals.

beta particles. Electrons ejected from the nucleus of an atom as a result of certain kinds of radioactive decomposition or nuclear change. These electrons are emitted at very high speeds, sometimes approaching the speed of light. The energies of the emitted electrons are characteristic of the parent element. Beta decay is the kind of radioactive change in which beta particles are emitted. A stream of beta particles is referred to as a beta ray. The high speed electrons produced by man-made accelerator devices such as the betatron should not be referred to as beta particles.

beta ray. The name given to one of the first three recognized kinds of radiations emitted by the nuclei of atoms undergoing spontaneous radioactive transformation. See beta particles.

"Betasol." [57] Trade name for a line of organic wetting agents.

betatron. An electron accelerator which operates on the same principle as an ordinary electric generator. The electrons are accelerated by means of magnetic induction. The betatron consists of a ring-shaped evacuated glass tube called the "doughnut" which is placed between the poles of an electromagnet. The magnet produces in the "hole" of the doughnut a strong magnetic field which in turn induces a voltage within the doughnut itself. Electrons are produced from a heated filament and after preliminary acceleration are injected into the doughnut where the induced voltage increases their energy. They move in a circular path and gain energy in each turn. The electrons may be used for the production of x-rays or may be directed out of the apparatus for other purposes.

"Betaxin." [162] Trademark for thiamine hydrochloride.

betazole hydrochloride $C_5H_9N_3 \cdot 2HCl$.
Properties: White, crystalline, nearly odorless powder; pH of 5% solution 1.5; m.p. not higher than 240°C. Soluble in water; practically insoluble in chloroform.
Grades: U.S.P. XVI.
Use: Medicine.

betel.
1. Dried leaves of Piper betel, a shrub of India, Malay, etc., also cultivated in Madagascar and the West Indies. The material is chewed by natives to blacken teeth and possibly as a narcotic and stimulant.
2. Material from nuts of the betel palm (Areca catechu), the source of betel oil and alkaloids. The latter are used in veterinary medicine.

betel nut. See areca nut.

bethanechol chloride (carbamylmethylcholine chloride) $C_7H_{17}ClN_2O_2$, or $H_2NCOOCH(CH_3)CH_2N(CH_3)_3Cl$.
Properties: Colorless hygroscopic crystals with amine-like odor. M.p. 217-221°C (dec). Very soluble in water; freely soluble in alcohol; practically insoluble in chloroform, benzene, and ether. Stable in air; pH (1% solution) 5.5-6.5.
Derivation: Propylene chlorohydrin is treated with phosgene and then with ammonia in ether. The product is heated with trimethylamine.
Grade: U.S.P. XVI.
Use: Medicine.

Bettendorf's reagent. A reagent used for the detection of arsenic in presence of bismuth and antimony compounds. It consists of a concentrated solution of stannous chloride in fuming hydrochloric acid.

"Better Blend Soda." [203] Trademark for bicarbonate of soda treated with approximately 0.7% tricalcium phosphate. Available in food and rubber grades, both of which have a density of 64.5 lbs/cu ft.

Betterton-Kroll process. A process for obtaining bismuth and purifying desilverized lead that contains bismuth. Metallic calcium or magnesium is added to the molten lead to cause formation of high melting intermetallic compounds with bismuth. These separate as a surface scum and are skimmed off. The excess calcium and magnesium are removed from the lead by use of chlorine gas as mixed molten chlorides of lead or zinc.

Betts process. An electrolytic process in which pure lead is deposited on a thin cathode of pure lead, from an anode containing as much as 10% of silver, gold, bismuth, copper, antimony, arsenic, selenium, and other impurities. The electrolyte is lead fluosilicate and fluosilicic acid. The scrap anodes and the residues of impurities associated with them are either recast into anodes or treated to recover antimonial lead, silver, gold, bismuth, etc.

betula oil. See methyl salicylate.

"Beutene." [248] Trademark for a butyraldehyde-aniline reaction product.
Properties: A reddish-brown, free-flowing liquid; sp. gr. 0.95; soluble in acetone, benzol, and ethylene dichloride; slightly soluble in gasoline; insoluble in water.
Uses: Accelerator for tire treads, carcass, belt friction, inner tubes, mechanicals, hard rubber, molded heels and soles and air-cured footwear.

bev. Abbreviation for billion electron volts. See electron volt.

bevatron. See cyclotron.

BFE. Abbreviation for bromotrifluoroethylene.

BF3-MEA. See boron fluoride monoethylamine.

BFPO. Abbreviation for bis(dimethylamino)-fluorophosphine oxide. See dimefox.

BHA. Abbreviation for butylated hydroxy-anisole.

BHC. Abbreviation for benzene hexachloride. Common and accepted designation for commercial mixtures of isomers of 1,2,3,4,5,6-hexachlorocyclohexane (q.v.) used as insecticide.

bhilawan nut. See semecarpus nut.

BHT. Abbreviation for butylated hydroxy-toluene. See di-tert-butyl-para-cresol.

Bi. Symbol for bismuth.

bi-. Prefix meaning two or twice. A compound not found under bi- should be looked for under bis- or di-, since bi-, bis-, and di- are equivalent prefixes, assigned with slight differences in meaning for particular compounds, or according to customary usage.

biacetyl. See diacetyl.

biacetylenes. See diacetylenes.

Biazzi continuous nitration process. The material to be nitrated and the nitrating acid mixture enter the top of a cylindrical vessel containing an impeller type agitator and cooling coils. The mixture is drawn downwards through the central portion of the cylinder and then forced upwards through the spaces between the cooling coils. Part of the material overflows near the top and passes to an acid separator, which in this process is a flat tank of stainless steel. Separation depends upon centrifugal action. A similar device is used for separation of water, after the water-washing steps.

bibenzyl. See diphenylethane, symmetrical.

bicalcic phosphate. See calcium phosphate, dibasic.

bicarbonate of soda. See sodium bicarbonate.

bicarburetted hydrogen. See ethylene.

Bicheroux process. A process for the production of high-quality plate glass. Molten glass is poured between rollers and fed onto a moving table which delivers the strip to a lehr, where the glass cools slowly while passing between a series of asbestos-covered rollers. After being cut into suitable lengths the glass is ground and polished as individual plates.

bichloroacetic acid. See dichloroacetic acid.

bichromate of soda. See sodium dichromate.

"Bicillin." [24] Trademark for benzathine penicillin G (dibenzylethylenediamine dipenicillin G), a penicillin salt, composed of two molecules of crystalline penicillin G and one of base (dibenzylethylenediamine), which has a penicillin potency of 1200 units per milligram and is stable in aqueous suspension. Used in medicine.

bicyclohexyl (dicyclohexyl) $C_{12}H_{22}$.
Properties: Colorless, mobile liquid with pleasant odor. B.p. 238.5°C; f.p. 1 to 3°C; sp. gr. 0.883 (25/16°C); wt/gal 7.37 lbs; refractive index (n 20/D) 1.480; flash point (open cup) 215°F.
Derivation: Hydrogenation of biphenyl.
Grades: Technical.
Uses: High-boiling solvent and penetrant.

Biebrich red. See scarlet red.

biformin. An antiobiotic produced by the fungus Polyporus biformis, reported to be active against various bacteria and fungi.

bihexyl. See dodecane.

"B-I-K." [248] Trade name for a surface coated urea.
Properties: A fine white powder; sp. gr. 1.32;

melting range 129-134°C; soluble in water. Surface coating not soluble in water, but is soluble in rubber. Slightly soluble in acetone; insoluble in benzene, gasoline and ethylene dichloride.
Uses: Promoter for "Celogen AZ," a nitrogen blowing agent; activator for thiazole accelerators; odor reducer when used with nitrosoamine type blowing agents.

"Bikalith." [88] Trademark for a series of lithium ores including lepidolite, petalite, spodumene and amblygonite. Origin: South Africa. Used in glass-making, ceramics; coatings.

bile acids. Acids which are found in the bile (the secretion of the liver). They are steroids, having an alcoholic group, and a five-carbon-atom side chain terminating in a carboxyl group. Cholic acid is by far the most abundant bile acid in human bile. Others are deoxycholic and lithocholic acids. The bile acids do not occur free in bile but are linked to the amino acids, glycine and taurine. These conjugated acids are water-soluble; their salts are powerful detergents and as such, aid in the absorption of fats from the intestine.

bilifulvin. See bilirubin.

bilirubin (bilifulvin) $C_{32}H_{36}O_6N_4$. Red coloring matter of bile. Chemical structure related to hemoglobin.
Properties: Orange-red powder; m.p. 192°C; soluble in acids, alkalies, chloroform and benzene; insoluble in water; very slightly soluble in alcohol and ether.
Derivation: From bile pigment.
Method of purification: Crystallization.
Grades: Technical.
Containers: Glass bottles; tins.
Use: Analytical chemistry.
Shipping regulations: None.*

bimethyl. See ethane.

binder (paint). See description of liquid "vehicle," under paint.

binitrobenzene. See dinitrobenzene.

binitronaphthalene. See dinitronaphthalene.

binitrotoluene. See dinitrotoluene.

"Biobate." [173] Trademark for an enzymatic preparation for use in bating in the leather and tanning industry.

biochemical oxygen demand (B.O.D.). A standardized means for estimating the degree of contamination of water supplies, especially those which receive contamination from sewage and industrial wastes. It is measured as the quantity of dissolved oxygen (in mg/liter) required during stabilization of the decomposable organic matter by aerobic biochemical action in sewage effluents, polluted water or industrial wastes. Determination of this quantity is accomplished by diluting suitable portions of the sample with water saturated with oxygen and measuring the dissolved oxygen in the mixture both immediately and after a period of incubation, usually five days. See also dissolved oxygen (D.O.) and oxygen consumed (C.O.D.) as related terms.

"Biocides" B-2 and B-7. [108] Biocides (effective against most microorganisms) and exceptionally toxic to sulfate-reducing bacteria even in concentrated brines.
Containers: B-2: 40-lb lined cans; 435-lb drums. B7: 55-gal drums.
Uses: In oilfield water injection systems, producing wells and water disposal systems. Protects against corrosion, cleans up systems by removing scale.

"Biocide" RP. [108] Potent liquid biocide that very quickly kills most types of green algae and other microorganisms. Film forming, gives longer protection.
Containers: 8-oz bottles; 5-gal cans.
Uses: Used to destroy most types of slime bacteria; protects wood from fungi attack. Not for use in swimming pools.

biocytin $C_{16}H_{28}N_4O_4S$ (epsilon-N-biotinyl-L-lysine). A naturally occurring complex of biotin isolated from yeast. Water-soluble crystals; m.p. 228.5°C. It is believed to be an intermediate in the utilization of biotin by animal organisms.

bioflavonoids (vitamin P complex; citrus flavonoid compounds). A group of naturally occurring substances concerned with the maintenance of normal conditions in the walls of the small blood vessels. The bioflavonoids are widely distributed among plants, especially citrus fruits, black currants and rose hips. In commercial methods the rinds of citrus fruits are extracted with aqueous alkalies, hot water and water-miscible organic solvents. The more important bioflavonoids are hesperidin, hesperidin methyl chalcone, naringin, and rutin (q.v.). The name vitamin P has been discarded.

bioluminescence. The emission of energy, as light, from living organisms or products derived from them. Light production in the plant kingdom (bacteria and fungi) is continuous and independent of stimulation, varying only with environmental changes. Among animals (fish, fireflies, protozoa) no light appears until the luminous region is excited in some way - by nerves or directly by mechanical, electrical and chemical stimulation. See luminescence.

"Bionol" A-50. [307] Trademark for a cationic bactericidal agent composed of 50% alkyl dimethyl benzyl ammonium chloride.
Properties: Pale yellow, slightly viscous liquid; sp. gr. 0.98; soluble in water; stable to dilute acids and alkalies.
Uses: Disinfectant, germicide, deodorant; can be combined with nonionic detergents to produce detergent-sanitizers.

"Biopal CVL-10." [307] Brand name of an iodophor, consisting of a solution of 10% available iodine in alkyl phenoxy polyoxyethylene ethanol.

Properties: Brownish-black liquid; readily soluble in water.
Uses: Bactericide, sporocide, fungicide, virucide and protozoacide in hard or soft water and at high or low temperatures. Has both detergent and sanitizing properties.
Stability: "Biopal CVL-10," as well as its aqueous solutions, slowly decreases in activity in the presence of light and should be packed in dark amber bottles or in suitably lined metal containers.
Handling: Must be handled with due caution.

biotin (vitamin H) $C_{10}H_{16}N_2O_3S$. 2'-Keto-3, 4-imidazolido-2-tetrahydrothiophene-n-valeric acid. Biotin, frequently referred to as a member of the vitamin B_2 complex, is necessary for the maintenance of health in animals and for growth of many micro-organisms. The exact metabolic role of biotin is not clear. However, it does influence fat metabolism, decarboxylation and carbon dioxide fixation and deamination of some amino acids. It is closely related metabolically to pantothenic acid and folic acid. A biotin deficiency may be induced by the ingestion of avidin, a raw-egg protein, because of the formation of a nonabsorbable biotin-avidin complex. Biotin is synthesized in the intestinal tract of humans; therefore, normally not required.
Properties: White crystals; m. p. 230-232°C; soluble in water and alcohol; insoluble in petroleum ether, chloroform; stable to heat; stable in neutral or acid solutions; destroyed by strong alkali or oxidizing agents.
Sources: Food sources: egg yolk, kidney, liver, yeast, milk, molasses. Commercial sources: synthetic preparations of biotin or its methyl ester.
Units: Amounts are expressed in milligrams or micrograms of biotin.
Containers: Bottles.
Uses: Medicine; nutrition.

biphenyl. See diphenyl.

ortho-**biphenylamine.** See ortho-aminobiphenyl.

ortho-**biphenyl biguanide** $NH_2(CNHNH)_2C_6H_4C_6H_5 \cdot H_2O$.
Properties: White to faintly pink powder; m. p. above 150°C on dried material; ash not over 0.5%. Soluble in alcohol, "Carbitol" and "Cellosolve;" very slightly soluble in water.
Containers: 50-lb paper bags; 150-lb fiber drums.
Use: Soap antioxidant.

2, 2'-biphenyldicarboxylic acid. See diphenic acid.

bipropargyl. See dipropargyl.

bipropenyl. See 2, 4-hexadiene.

birch oil (birch-tar oil).
Properties: Yellowish-brown liquid; characteristic odor; poisonous! Soluble in alcohol, ether, chloroform, glacial acetic acid, amyl alcohol, benzene, carbon disulfide, oil of turpentine.
Chief known constituents: Phenols, guaiacol, cresol, creosol, and xylenol. Sp. gr. 0.956.
Derivation: Distilled from birch-tar, obtained from the dry distillation of the wood of Betula alba.
Method of purification: Rectification.
Grades: Crude; rectified.
Containers: Iron drums; cans.
Uses: Leather dressing; disinfectant; medicine (external use); in small quantities in perfumery.
Shipping regulations: None.*

birch oil, sweet. See methyl salicylate.

birch-tar oil. See birch oil.

bird pepper. See capsicum.

bis-. Prefix meaning two or twice. A compound not found under bis- should be looked for under bi- or di-, since bi-, bis-, and di- are equivalent prefixes, assigned with slight differences in meaning for particular compounds, or according to customary usage.

1, 3-bis(2-benzothiazolyl-mercaptomethyl)-urea.
Properties: Buff to light tan powder; m. p. 220°C; sp. gr. 1.29.
Use: Rubber accelerator.

para-**bis[2-(5-**para-**biphenylyloxazolyl)]-benzene** (BOPOB) $C_{36}H_{25}O_2N_2$.
Properties: Shiny yellow flakes; m. p. 327-328°C; fluorescence peak 4400A; sparingly soluble in toluene.
Grade: Purified.
Use: Scintillation counter; wave length shifter in liquid scintillators.

bis(2-chloroethoxy)methane. See dichloroethyl formal.

4-{para-**[bis(2-chloroethyl)amino]phenyl}butyric acid.** See chlorambucil.

3, 3-bis(chloromethyl)oxetane. See "Penton."

2, 2-bis(para-**chlorophenyl)-1, 1-dichlorethane.** See TDE.

1, 1-bis(para-**chlorophenyl)ethanol.** See di(para-chlorophenyl)ethanol.

1, 1-bis(para-**chlorophenyl)-2, 2, 2-trichloroethanol** $CCl_3C(C_6H_4Cl)_2OH$. An alcohol analog of DDT. See "Kelthane."

bis[S-(diethoxyphosphinothioyl)mercapto]-methane. See ethion.

bis(diethylthiocarbamyl)disulfide. See tetraethylthiuram disulfide.

bis(diethylthiocarbamyl)sulfide. See tetraethylthiuram sulfide.

bis(dimethylamino)fluorophosphate. A systemic insecticide. See dimefox.

bis(1, 3-dimethyl butyl)amine $[(CH_3)_2CH_2CH_2CH(CH_3)]_2NH$.
Properties: Sp. gr. 0.772-0.778 (20/20°C);

*See "I.C.C. Shipping Regulations," page xiii.
Reference numbers refer to name of manufacturer. See "List of Manufacturers," page v.

distillation range 179.0-205.0°C (760 mm); 6.5 lbs/gal; flash point 160°F.

bis(dimethylthiocarbamyl)disulfide. See tetramethylthiuram disulfide.

bis(dimethylthiocarbamyl)sulfide. See tetramethylthiuram sulfide.

1,3-bisethylaminobutane $C_2H_5NHCH_2CH_2CHNH(C_2H_5)CH_3$. A water white amine, boiling range 179-185°C.

bis(2-ethylhexyl) 2-ethylhexyl phosphonate. See di(2-ethylhexyl) 2-ethylhexyl phosphonate.

bis(2-ethylhexyl) hydrogen phosphate. See di(2-ethylhexyl) hydrogen phosphate.

bis(2-ethylhexyl) phosphite. See di(2-ethylhexyl) phosphite.

2,2-bis(para-ethylphenyl)-1,1-dichloroethane. See 1,1-dichloro-2,2-bis(para-ethylphenyl) ethane.

bisethylxanthogen $(C_2H_5OCSS)_2$.
Properties: Yellow needles; onion-like odor; m.p. 28-32°C. Insoluble in water; freely soluble in benzene, ether, petroleum ethers, oils.
Grades: 58% solution in oil.
Containers: 5-gal cans; 30-gal drums.
Uses: Weed control; rubber vulcanizer; fungicide.

bishydroxycoumarin (dicoumarol) $C_{19}H_{12}O_6$. 3,3'-Methylenebis(4-hydroxycoumarin).
Properties: White or creamy-white, crystalline powder; faint pleasant odor and slightly bitter taste; m.p. 287-293°C. Readily soluble in solutions of fixed alkali hydroxides; slightly soluble in chloroform; practically insoluble in water, alcohol and ether.
Derivation: (a) Originally, extracted from spoiled sweet clover; (b) synthetically from methyl acetylsalicylate, sodium, and formaldehyde.
Grade: U.S.P. XVI.
Use: Medicine.

bis(2-hydroxy-3,5-dichlorophenyl) sulfide. See bithionol.

beta-**bishydroxyethyl sulfide.** See thiodiglycol.

1,8-bis(para-hydroxyphenyl)methane $C_{22}H_{28}O_2$.
Properties: Pale amber crystalline material; m.p. 102°C min; slightly soluble in benzene, methanol, ethanol, and isopropanol; soluble in aqueous alkalies.

4,4-bis(4-hydroxyphenyl)pentanoic acid. See "Diphenolic Acid."

bishydroxyphenyl sulfone. See dihydroxydiphenyl sulfone.

bismanol. An alloy of bismuth and manganese which possesses unusually high magnetic force. It is supposed to be manganese bismuthide. It is prepared by powder metallurgy techniques and is separated magnetically from its constituents. Its coercive force is said to be 3000 oersteds.

Bismarck Brown G $C_{18}H_{20}N_8Cl_2$. Benzene-meta-diazo-bis-meta-phenylenediamine hydrochloride.
Properties: Dark blackish-brown powder; soluble in water and alcohol.
Derivation: Action of nitrous acid on meta-phenylenediamine hydrochloride in aqueous solution.
Use: Dye for wool, silk, leather.

Bismarck Brown R $C_{21}H_{26}N_8Cl_2$. Toluene-2,4-diazo-bis-meta-toluylenediamine hydrochloride.
Properties: Dark brown powder; soluble in water and alcohol.
Derivation: Action of nitrous acid on toluylene diamine.
Use: Dye for wool and leather.

"Bismate." [69] Trademark for proprietary product, bismuth dimethyldithiocarbamate $[(CH_3)_2NC(S)S]_3Bi$.
Properties: Lemon yellow powder; sp. gr 2.04±.03; melts above 230°C. with decomposition; soluble in chloroform; slightly soluble in benzene, carbon disulfide; insoluble in water.
Uses: Primary accelerator for natural rubber and SBR; secondary accelerator with thiazole accelerators in natural rubber and SBR; used in wire, cable, extruded and molded goods.
Also supplied in natural yellow "rodform" or in green colored "rodform."

bismon (bismuth oxide, colloidal).
Properties: Translucent, yellowish or light brown amorphous mass. Contains 22.3% bismuth oxide. Cannot be used hypodermically. Soluble in water to form an opalescent suspension.
Derivation: Interaction of bismuth salts and alkaline solutions of sodium lysalbinate or protalbinate.
Use: Medicine.

bismuth Bi. Element of atomic number 83, of group V of the periodic system.
Properties: Grayish-white, hard, brittle metal, with a reddish tinge; sometimes found native. Soluble in hydrochloric acid (in presence of oxygen), hot concentrated sulfuric acid and nitric acid; insoluble in water.
Constants: Sp. gr. 9.78 at 20°C; m.p. 271°C; b.p. 1460-1480°C; hardness 2 to 2.5. The thermal conductivity, 0.018 cal/sec/cc at 100°C, is less than that of any other metal except mercury.
Source: (1) Metallurgical byproducts (often lead bullion) obtained chiefly from ores of lead, silver, copper and gold, and (2) ores used chiefly for their bismuth and one or two other metals, as tin and tungsten. See also bismuthinite, bismutite, cosalite, and tetradymite.
Derivation: Debismuthizing of lead bullion by (a) fractional crystallization, (b) electrolytic (Betts) refining, or (c) addition of calcium or magnesium (Betterton-Kroll process) which removes bismuth.
Purification: By addition of molten caustic,

zinc, and finally chlorine (to make removable chlorides of the impurities).
Impurities: Lead, iron, copper, arsenic, antimony, selenium.
Forms available: Bars, pieces, lump, powder; 99.5+% pure; high purity (impurities less than 10 ppm).
Uses: Low-melting alloys (see fusible alloys); metallic coating of other metals; heat transfer medium in nuclear power production (suggested); ingredient, with manganese, of alloy forming unusually permanent magnets (see bismanol); in molten-metal continuous dyeing process; bismuth salts.
Shipping regulations: None.*

bismuth alloys. See Tables under fusible alloys.

bismuth ammonium citrate.
Properties: Pearly, shining, transparent scales or white powder; slightly acid, metallic taste; composition varies. Soluble in water; slightly soluble in alcohol.
Derivation: By the interaction of bismuth subnitrate, citric acid and ammonium hydroxide.
Grades: Technical.
Containers: Tins; amber glass bottles.
Use: Medicine.
Shipping regulations: None.*

bismuth bromide (bismuth tribromide) $BiBr_3$.
Properties: Yellow, crystalline powder. Hygroscopic. Decomposed by water with formation of bismuth oxybromide. Soluble in hydrochloric acid (dilute), solutions of potassium iodide, potassium bromide and potassium chloride; insoluble in alcohol. Sp. gr. 5.7; b.p. 453°C; m.p. 218°C.

bismuth bromide, basic. See bismuth oxybromide.

bismuth carbolate. See bismuth phenate.

bismuth carbonate. See bismuth subcarbonate.

bismuth carbonate, basic. See bismuth subcarbonate.

bismuth cerium oxalate.
Properties: White powder. Soluble in hydrochloric acid; insoluble in alcohol and water.
Derivation: Mixture of cerium oxalate and bismuth oxalate.

bismuth cerium salicylate.
Properties: White powder; insoluble in alcohol and water.
Use: Medicine.

bismuth cerium valerate.
Properties: White powder; soluble in mineral acids; insoluble in water.
Use: Medicine.

bismuth chloride (bismuth trichloride) $BiCl_3$.
Properties: White, very deliquescent crystals; volatilized by heat. Soluble in acids; insoluble in alcohol; decomposes in water to the oxychloride. Sp. gr. 4.56; m.p. 227°C; b.p. decomposes at 300°C.
Derivation: By the action of hydrochloric acid on bismuth.
Grades: Technical; C.P.
Containers: 1-, 5-lb bottles; jars.
Use: Bismuth salts.
Shipping regulations: None.*

bismuth chloride, basic. See bismuth oxychloride.

bismuth chromate (basic dichromate) $Bi_2O_3 \cdot 2CrO_3$.
Properties: Orange-red amorphous powder; soluble in alkalies and acids; insoluble in water.
Derivation: By the interaction of bismuth nitrate and potassium chromate.
Grades: Technical.
Containers: Tins; kegs.
Use: Pigment.
Shipping regulations: None.*

bismuth citrate $BiC_6H_5O_7$.
Properties: White powder; soluble in ammonia or alkali citrates; insoluble in water.
Derivation: Boiling bismuth subnitrate with citric acid.
Use: Medicine.
Shipping regulations: None.*

bismuth, cosmetic. See bismuth oxychloride.

bismuth ditannate. See bismuth tannate.

bismuth dithiosalicylate $SC_6H_3OHCOOBiO_2BiO_3 \cdot 2H_2O$.
Properties: Yellowish-gray powder; odorless; tasteless. Contains 72% bismuth oxide. Insoluble in water, alcohol, and ether.

bismuth ethyl camphorate $C_{36}H_{57}BiO_{12}$.
Properties: Solid; faint aromatic odor; m.p. 61-67°C. Insoluble in water; soluble in ether, chloroform and oils.
Derivation: Reaction of sodium ethyl camphorate and bismuth nitrate in glycerin solution.
Use: Medicine.

bismuth gallate, basic. See bismuth subgallate.

bismuth glance. See bismuthinite.

bismuth para-glycolylaminophenylarsonate. See glycobiarsol.

bismuth glycolylarsanilate. See glycobiarsol.

bismuth hydrate. See bismuth hydroxide.

bismuth hydroxide (bismuth hydrate; bismuth oxyhydrate; bismuth trihydroxide; bismuth trihydrate; hydrated bismuth oxide) $Bi(OH)_3$.
Properties: White, amorphous powder. Soluble in acids; insoluble in water. Sp. gr. 4.36.
Derivation: By the action of sodium hydroxide on a solution of a bismuth salt.
Grades: Technical; C.P.
Containers: Glass bottles; tins; drums.
Use: Bismuth salts.
Shipping regulations: None.*

bismuthinite (bismuth glance) Bi_2S_3, may contain Cu or Fe.
Properties: Lead-gray mineral, often with yellow tarnish, metallic luster. Contains 81.2% bismuth, 18.8% sulfur. Soluble in nitric acid. Sp. gr. 6.4-6.5; hardness 2.

Occurrence: Utah; Bolivia; Mexico.
Use: Ore of bismuth.

bismuth iodate $Bi(IO_3)_3$.
Properties: White powder. Slightly soluble in nitric acid; insoluble in acetic acid and water.

bismuth iodide (bismuth triiodide) BiI_3.
Properties: Grayish-black, metallic, glistening crystals. Soluble in alcohol, hydriodic acid and potassium iodide; insoluble in water; decomposes in hot water; sp. gr. 5.65; m.p. 408°C.
Derivation: By the interaction of bismuth and iodine.
Method of purification: Crystallization.
Grades: Technical.
Containers: Glass bottles.
Uses: Analytical chemistry; bismuth oxyiodide.
Shipping regulations: None.*

bismuth iodide, basic. See bismuth oxyiodide.

bismuth lactate $BiH(C_3H_4O_3)_2$.
Properties: White powder; slightly soluble in water and decomposed by it.

bismuth magistery. See bismuth subnitrate.

bismuth nitrate (bismuth ternitrate; bismuth trinitrate) $Bi(NO_3)_3 \cdot 5H_2O$.
Properties: Lustrous, clear, colorless, hygroscopic crystals; acid taste. Soluble in dilute nitric acid, alcohol and acetone; slowly decomposed by water to the subnitrate.
Constants: Sp. gr. 2.78; m.p. 74°C; b.p., decomposes 75-80°C.
Derivation: By the action of nitric acid on bismuth with subsequent recovery by evaporation and crystallization.
Method of purification: Recrystallization.
Grades: Technical; C.P.
Containers: 1-, 5-lb bottles; tins; 250-lb drums; multiwall paper sacks.
Uses: Preparation of other bismuth salts; bismuth luster on tin; luminous paints and enamels; medicine.
Shipping regulations: Oxidizing material. Yellow label.*

bismuth nitrate, basic. See bismuth subnitrate.

bismuth oleate.
Properties: Yellowish-brown, soft, granular mass. Soluble in ether; insoluble in water.
Derivation: A combination of bismuth trioxide and oleic acid.
Grades: Technical.
Containers: Tins.
Use: Medicine.
Shipping regulations: None.*

bismuth oxide. See bismuth trioxide; bismuth tetroxide.

bismuth oxide, colloidal. See bismon.

bismuth oxide, hydrated. See bismuth hydroxide.

bismuth oxybromide (basic bismuth bromide; bismuthyl bromide) BiOBr.
Properties: White powder. Soluble in hydrochloric acid (dilute), nitric acid (dilute); insoluble in alcohol and water; sp. gr. 8.08.
Use: Medicine.

bismuth oxycarbonate. See bismuth subcarbonate.

bismuth oxychloride (bismuth chloride, basic; bismuth subchloride; bismuthyl chloride; cosmetic bismuth; pearl white; flake white) BiOCl.
Properties: White, lustrous crystalline powder. Sp. gr. 7.717. Soluble in acid; insoluble in water.
Derivation: By action of water on bismuth chloride; interaction of dilute nitric acid solution of bismuth nitrate with sodium chloride.
Grades: Technical; C.P.
Containers: 25-lb drums.
Uses: Medicine; face powder; pigment; artificial pearls.
Shipping regulations: None.*

bismuth oxyhydrate. See bismuth hydroxide.

bismuth oxyiodide (basic bismuth iodide; bismuth subiodide) BiOI.
Properties: Red powder. Decomposed by acids and alkalies. Caution! Keep away from light! Sp. gr. 7.82. Insoluble in alcohol, chloroform, water.
Containers: Fiber drums.
Use: Medicine.

bismuth oxynitrate. See bismuth subnitrate.

bismuth pentafluoride BiF_5. Used as a fluorinating agent.

bismuth permanganate $Bi(MnO_4)_3$.
Properties: Black powder. Soluble in acids (dilute).
Use: Medicine.

bismuth phenate (bismuth carbolate; bismuth phenolate; bismuth phenylate; phenolbismuth) $C_6H_5O \cdot Bi(OH)_2$.
Properties: Grayish-white powder; odorless and tasteless. Insoluble in water, alcohol and ether.
Derivation: By the interaction of bismuth nitrate and sodium phenolate.
Grades: Technical, 80% Bi_2O_3.
Containers: Tins.
Use: Medicine.
Shipping regulations: None.*

bismuth phenolate. See bismuth phenate.

bismuth phenolsulfonate (bismuth sulfocarbolate; bismuth sulfophenate; bismuth sulfophenylate).
Properties: Pale, reddish powder. Slightly soluble in water.
Derivation: By the interaction of bismuth hydroxide and phenolsulfonic acid.
Grades: Technical.
Containers: Tins.
Uses: Medicine; antiseptic.
Shipping regulations: None.*

bismuth phenylate. See bismuth phenate.

bismuth phosphate $BiPO_4$.
Properties: White powder. Sp. gr. 6.323.

Soluble in hydrochloric acid, nitric acid; insoluble in alcohol, water.
Use: Medicine.

bismuth potassium iodide $Bi_3 \cdot 4KI$.
Properties: Red crystals. Decomposed by water. Soluble in potassium iodide solution.
Use: Medicine.

bismuth-potassium tartrate. See potassium-bismuth tartrate.

bismuth pyrogallate (helcosol; basic bismuth pyrogallate) $C_6H_3(OH)_2OBiOH$.
Properties: Yellowish-green amorphous powder; odorless; tasteless. Soluble in dilute acid and alkaline solutions; insoluble in water and alcohol.
Derivation: By the action of pyrogallic acid on bismuth carbonate.
Grades: Technical, 60% Bi_2O_3.
Containers: Tins.
Use: Medicine.
Shipping regulations: None.*

bismuth pyrogallate, basic. See bismuth pyrogallate.

bismuth salicylate, basic. See bismuth subsalicylate.

bismuth sodium iodide $BiI_3 \cdot 4NaI$.
Properties: Red crystals. Decomposed by water. Soluble in solution of sodium iodide.
Use: Medicine.

bismuth sodium tartrate.
Properties: White powder. Odorless; tasteless; contains about 73% bismuth. Soluble in water; insoluble in organic solvents.
Use: Medicine.

bismuth sodium triglycollamate
$C_6H_7NNaBiO_7 \cdot 3C_6H_7NNa_2O_6$. A sodium bismuth complex of nitrilotriacetic acid. Contains approximately 18.3% bismuth.
Properties: White, odorless, crystalline powder. Somewhat salty taste. Stable to air or light. Very soluble in water; insoluble in organic solvents; pH (2% solution) 7.0-8.0.
Grade: U.S.P. XVI.
Use: Medicine.

bismuth solders. See data under fusible alloys.

bismuth stannate $Bi_2(SnO_3)_3 \cdot 5H_2O$.
Properties: Light colored crystalline powder. Insoluble in water. Approximate temperature of dehydration is 140°C.
Uses: Component of ceramic capacitators, especially useful with barium titanate.

bismuth subcarbonate (bismuth oxycarbonate; bismuth "carbonate"; bismuth carbonate, basic) $(BiO)_2CO_3$ or $Bi_2O_3 \cdot CO_2$, with one-half H_2O.
Properties: White, odorless powder; tasteless; insoluble in water and alcohol; soluble in nitric or hydrochloric acid with effervescence.
Constants: Sp. gr. 6.86.
Derivation: By adding ammonium carbonate to a solution of a bismuth salt.
Grades: Technical; C.P.; U.S.P. XVI (90% Bi_2O_3 min).
Containers: Up to 250-lb drums.
Uses: Bismuth compounds; face powder; medicine; x-ray work.
Shipping regulations: None.*

bismuth subchloride. See bismuth oxychloride.

bismuth subgallate (basic bismuth gallate) $C_6H_2(OH)_3COOBi(OH)_2$.
Properties: Saffron-yellow powder; odorless and tasteless. Soluble in dilute alkalies; insoluble in water, alcohol and ether. Stable in air, but affected by light.
Derivation: Interaction of bismuth nitrate, glacial acetic acid, and gallic acid in aqueous solution.
Grades: Technical; N.F. XI.
Containers: 1-lb cans; 5- to 200-lb drums.
Use: Medicine.
Shipping regulations: None.*

bismuth subiodide. See bismuth oxyiodide.

bismuth subnitrate (magistery of bismuth; basic bismuth nitrate; Spanish white; flake white; pearl white; bismuth oxynitrate) $4BiNO_3(OH)_2 \cdot BiO(OH)$.
Properties: White, heavy, slightly hygroscopic powder, which shows acid to moistened litmus paper. Soluble in acids; insoluble in water and alcohol.
Constants: Sp. gr. 4.928; m.p. decomposes at 260°C.
Derivation: By adding bismuth nitrate to water, filtering and drying.
Impurities: Arsenic, lead, silver carbonates. First two especially should be watched for, if article is to be used in pharmacy or medicine.
Grades: Technical; C P.; N.F. XI.
Containers: Multiwall paper sacks; 5- to 250-lb drums.
Uses: Bismuth salts; perfumery and cosmetics; ceramic enamels; burning gold on ceramic ware; bismuth luster on metals; pharmaceuticals (preparations, antiseptic and deodorizing compositions, drug; analytical reagent).
Shipping regulations: None.*

bismuth subsalicylate (basic bismuth salicylate) $Bi(C_7H_5O_3)_3Bi_2O_3$.
Properties: White, bulky crystalline powder; tasteless; odorless; soluble in acids and alkalies; insoluble in water, alcohol and ether. Stable in air but affected by light.
Derivation: By treating freshly prepared bismuth hydroxide with salicylic acid.
Grades: Technical; U.S P. XVI.
Containers: 5-lb cans; 25-, 100-lb drums.
Use: Medicine.
Shipping regulations: None.*

bismuth sulfate $Bi_2(SO_4)_3$.
Properties: White needles or powder Contains 68.5% (approx.) bismuth. Sp. gr. 5.08. Soluble in hydrochloric acid (dilute), nitric acid (dilute); insoluble in alcohol, water.
Use: Medicine.

bismuth sulfide Bi_2S_3.
Properties: Blackish-brown powder. Soluble in nitric acid; insoluble in water.

*See "I.C.C. Shipping Regulations," page xiii.
Reference numbers refer to name of manufacturer. See "List of Manufacturers," page v.

Constants: Sp. gr. 7.00-7.81; m.p., decomposes.
Derivation: (a) By melting bismuth and sulfur together. (b) By passing hydrogen sulfide into a solution of a bismuth salt. (c) Occurs as the mineral bismuthinite.
Grades: Technical; C.P.
Containers: Tins.
Use: Bismuth compounds.
Shipping regulations: None.*

bismuth sulfocarbolate. See bismuth phenolsulfonate.

bismuth sulfophenate. See bismuth phenolsulfonate.

bismuth sulfophenylate. See bismuth phenolsulfonate.

bismuth tannate (bismuth ditannate).
Properties: Light brownish-yellow powder containing about 36% bismuth. Insoluble in water and alcohol; soluble in mineral acids.
Derivation: From freshly prepared bismuth hydroxide and tannin.
Use: Medicine.

bismuth telluride (bismuth tritelluride) Bi_2Te_3.
Properties: Gray hexagonal platelets. M.p. 585°C; sp. gr. 7.642.
Derivation: Stoichiometric combination of the elements.
Grades: Ingots, single crystals.
Use: Semiconductors, for thermoelectric cooling and power generation applications.
See also tetradymite.

bismuth ternitrate. See bismuth nitrate.

bismuth tetroxide Bi_2O_4.
Properties: Heavy, yellowish-brown powder. Soluble in acids; insoluble in water.
Constants: Sp. gr. 5.6; m.p. 305°C.
Derivation: By further oxidation of bismuth trioxide.
Grades: Technical; C.P.
Containers: 1-lb bottles; tins.
Use: Bismuth salts.

bismuth tribromide. See bismuth bromide.

bismuth tribromophenate (tribromophenol-bismuth) $Bi_2O_3(C_6H_2Br_3OH)$. (?)
Properties: Yellow, odorless powder. Insoluble in water and alcohol; soluble in dilute hydrochloric acid.
Derivation: By the interaction of bismuth chloride and sodium tribromophenolate.
Grades: Technical.
Containers: Glass bottles; boxes.
Use: Medicine.
Shipping regulations: None.*

bismuth trichloride. See bismuth chloride.

bismuth trihydrate. See bismuth hydroxide.

bismuth trihydroxide. See bismuth hydroxide.

bismuth triiodide. See bismuth iodide.

bismuth trinitrate. See bismuth nitrate.

bismuth trioxide (bismuth oxide; bismuth yellow) Bi_2O_3.
Properties: Heavy, yellow powder. Soluble in acids; insoluble in water.
Constants: Sp. gr. 8.8; m.p. 820°C.
Derivation: Ignition of bismuth nitrate.
Grades: Technical; C.P.
Containers: 1-lb bottles, tins; 100-lb drums.
Uses: Medicine; bismuth salts; ceramic colors.
Shipping regulations: None.*

bismuth tritelluride. See bismuth telluride and tetradymite.

bismuth valerate $Bi(C_5H_9O_2)_3 \cdot 2Bi(OH)_3$.
Properties: White powder. Insoluble in water or alcohol; soluble in dilute acid.
Derivation: Interaction of solutions of bismuth trinitrate and sodium valerate.
Use: Medicine.

bismuth yellow. See bismuth trioxide.

bismuthyl bromide. See bismuth oxybromide.

bismuthyl chloride. See bismuth oxychloride.

bismutite $(BiO)_2CO_3$. A natural carbonate of bismuth, resulting from alteration of bismuth minerals.
Properties: Color yellow, green, gray, black; streak gray; luster vitreous to dull; hardness variable; sp. gr. 6.1-7.7. Soluble in strong hydrochloric acid.
Occurrence: South Carolina, Arizona, California; Europe.
Use: Minor ore of bismuth.

para-bis[2-(5-alpha-napthyloxazolyl)]benzene (NOPON) $C_{32}H_{20}O_2N_2$.
Properties: Crystals; m.p. 215-217°C.
Grade: Purified.
Use: Scintillation counter.

bis(3-nitrophenyl) disulfide. See nitrophenide.

bisphenol A (para,para'-isopropylidenediphenol) $(CH_3)_2C(C_6H_4OH)_2$.
Properties: White flakes with a mild phenolic odor. B.p. 220°C (4 mm); freezing point 153°C; sp. gr. 1.195 (25/25°C); insoluble in water; soluble in alcohol and dilute alkalies; slightly soluble in carbon tetrachloride.
Containers: 50-lb bags; 400-lb metal drums.
Uses: In the manufacture of phenolic, epoxy and polycarbonate resins.

1,4-bis-2-(5-phenyloxazolyl)-benzene (POPOP). $C_{24}H_{17}O_2N_2$.
Properties: Light yellow, cottony needles; m.p. 245-246°C; fluorescence max 4200 A; solubilities g/100 g at 25°C: water 0.00; 95% ethanol 0.00; toluene 0.12; hexane 0.02.
Grade: Purified.
Containers: Glass bottles.
Use: Efficient band-shifter in scintillation counting.

bis(tetrachloroethyl)disulfide $C_4H_2Cl_8S_2$.
Properties: Sp. gr. (23.3°C) 1.785; b.p. 185°C (3 mm). Soluble in benzene, hexane, ethanol.
Uses: Agricultural chemicals; additives.

bis(tri-n-butyltin) oxide $(C_4H_9)_3SnOSn(C_4H_9)_3$.
Properties: Slightly yellow liquid; b.p. 180°C

*See "I.C.C. Shipping Regulations," page xiii.
Reference numbers refer to name of manufacturer. See "List of Manufacturers," page v.

(2 mm); f.p. less than −45°C; sp.gr. 1.17 (25°C); flash point, above 212°F (TCC); viscosity 4.8 centistokes at 25°C; practically insoluble in water; miscible with organic solvents.
Derivation: Hydrolysis of tributyl tin chloride.
Uses: Fungicide and bactericide; forms compound with cellulosic and lignin-containing materials not easily decomposed or dissolved in water.

bis-trichlorosilylbenzene $Cl_3SiC_6H_4SiCl_3$. Colorless liquid; b.p. (30 mm) 168°C.

bis-trichlorosilyl ethane $Cl_3SiCH_2CH_2SiCl_3$. 1,1,1,4,4,4-Hexachloro-1,4-disilabutane.
Properties: Colorless liquid. B.p. 202.9°C; sp.gr. 1.475 (29/29°C); flash point (Cleveland open cup) 190°F. Readily hydrolyzed by moisture, with the liberation of hydrochloric acid.
Derivation: By the reaction of acetylene and trichlorosilane in the presence of a peroxide catalyst.
Grades: Technical.
Use: Intermediate for silicones.
Shipping regulations: Corrosive liquid. White label. *

1,3-bis(trimethylamino)-2-propanol diiodide. See propiodal.

bithionol (bis(2-hydroxy-3,5-dichlorophenyl)-sulfide) $HOCl_2C_6H_2SC_6H_2Cl_2OH$. 2,2'-Thiobis(4,6-dichlorophenol).
Properties: White or grayish-white crystalline powder; m.p. 187°C. Odorless or with slight aromatic or phenolic odor. Insoluble in water; freely soluble in acetone, alcohol and in ether; soluble in chloroform and dilute solutions of fixed alkali hydroxides.
Grade: U.S.P. XVI.
Containers: 50-, 200-lb drums.
Uses: Medicine; deodorant; germicide; fungistat; cosmetics; pharmaceuticals.

"Bitrex." Trademark for benzyldiethyl(2,6-xylylcarbamoylmethyl)ammonium benzoate, produced in Europe, and approved by the Alcohol and Tobacco Tax Division for Specially Denatured Alcohol No. 40. Is said to be more bitter than brucine or quassia, also permitted for SDA No. 40.

bitter almond. See almond, bitter.

bitter almond oil. See almond oil, bitter.

bitter almond oil, synthetic. See benzaldehyde.

bitter almond oil camphor. See benzoin.

bitter apple. See colocynth.

bitter ash. See euonymus or quassia.

bitter cucumber. See colocynth.

bitter gourd. See colocynth.

bittern. The solution of bromides, magnesium, and calcium salts that remains after sodium chloride has been crystallized by concentration of sea water or salt brines.

bitter orange-flower oil. See orange-flower oil.

bitter root. See gentian.

bittersweet. See dulcamara.

bitterwood. See quassia.

"Bitumastic." [11] Trademark for a protective coating made from refined coal tar pitch and fillers. It is used primarily for waterproofing concrete or masonry surfaces.

bitumen. A general term for native asphalt-like hydrocarbons. Bitumens are solid or semisolid, insoluble in water, but largely soluble in carbon disulfide and other organic solvents. They include asphalt, asphaltites, asphaltic pyrobitumens and mineral waxes (q.v.). Bitumen is also used to refer to the components of coal that are soluble in organic solvents.

bituminous coal (soft coal). Designation for a broad class of coals having 46-86% fixed carbon and approximately 20-40% volatile matter. They run about 11,000 or more Btu/lb. Bituminous coals are further classified, in order of increasing volatile matter or decreasing fixed carbon, as: low volatile (semi-bituminous), medium volatile, high volatile. They are also classified as coking and non-coking coals.
Occurrence: Pennsylvania, West Virginia, Illinois, Indiana, Wyoming, Utah.
Uses: Fuel; coke production; manufacture of producer gas, illuminating gas, fuel gas and briquets.

bituminous varnish. See varnish.

biuret (allophanamide; carbamylurea) $NH_2CONHCONH_2 \cdot H_2O$.
Properties: White needles; odorless; m.p. 190°C with decomposition; soluble in water and alcohol; very slightly soluble in ether. Loses water of crystallization at about 110°C.
Derivation: From urea by heat.
Methods of purification: Crystallization.
Grades: Technical.
Containers: Glass bottles.
Use: Analytical chemistry.
Shipping regulations: None. *

bivinyl. See butadiene.

bixin. See annatto.

Bk. Symbol for berkelium.

"B-K B-K." [204] Trademark for concentrated uniform flake alkali for machine bottle washing and heavy duty cleaning. Packed in 340-lb and 130-lb steel drums.

"B-K Bottle Kompound." [204] Trademark. A machine bottle washing compound.

"B-K Chlorine-Bearing Powder." [204] Trademark. Calcium hypochlorite powder containing 50% available chlorine.
Use: As dairy bactericide in plants and on farms, restaurants. A disinfectant and deodorant.

"B-Kleer." [204] Trademark for a heavy duty caustic alkali; contains polyphosphate and wetting agent in fused flake form.

*See "I.C.C. Shipping Regulations," page xiii.
Reference numbers refer to name of manufacturer. See "List of Manufacturers," page v.

Properties: Dustless, uniform, highly soluble and effective in hard water.
Containers: Packed in 350-lb open-head and 100-lb steel drums.
Uses: In machine bottle washing; H. T.S. T. Pasteurizer; vacuum pan cleaning.

"B-K Liquid." [204] Trademark. A bactericide.

"BL-60." [172] Trade name for a modified form of sodium aluminum phosphate specifically designed for use in cakes made with lactylated type shortening.
Containers: Drums.
Uses: Leavening agent in prepared cake mixes.

black. See specific types of blacks such as acetylene black, animal black, benzol black, bone black, carbon black, channel black, drop black, furnace black, gas black, ivory black, lamp black, mineral black, thermal black, vegetable black, vine black. All are more or less impure forms of carbon, characterized by fine state of subdivision and/or relatively large surface area, and are made by thermal decomposition or incomplete combustion of carbon compounds. The term black is also applied to other black powdery materials such as minerals.

black, aniline. See aniline black.

black antimony. See antimony trisulfide.

black ash. 1. The product obtained by heating black liquor in furnaces. The organic material is reduced to carbon. The alkaline components are leached out and used again in papermaking. The carbon may be treated to obtain activated carbon.
2. See barium sulfide.

black balsam. See Peru balsam.

blackband. An earthy carbonate of iron containing considerable carbonaceous matter. An iron ore.

blackberry bark. See rubus.

black boy gum. See accroides gum.

"Black Bull." [423] Trademark for a series of asphalt-gilsonite roof coatings.

black cohosh. See cimicifuga.

black cyanide. Term for mixture containing 45% calcium cyanide, made from calcium cyanamid by heating it with salt and carbon. Formerly the main source of cyanides, but now superseded by the methane-ammonia process for making hydrocyanic acid.

black glass. Manganese or ferric oxides are added to ordinary glass.

black haw. See viburnum prunifolium.

black henbane. See hyoscyamus.

black Indian hemp. See apocynum.

black jack. See sphalerite.

black lead. See graphite.

black liquor. 1. The liquor resulting from cooking pulpwood in an alkaline solution in the soda or sulfate papermaking process. It is a source of lignin and tall oil.
2. Iron acetate liquor.

black mordant. See iron acetate liquor.

black mustard oil. See mustard oil, volatile.

black oil. See residual oils.

"Black-Out." [69] Trademark for a series of proprietary products, modified elastomers in solvent.
"Clear" total solids 11-12%; sp. gr. 0.88 ±.03; flash point 41°F.
"Black" total solids 11-12%; sp. gr. 0.88 ±.03; flash point 41°F
"White" total solids 15-16%; sp. gr. 0.91 ±.03; flash point 41°F.
"Metallic" total solids 18.5-20.5%; sp. gr. 0.89-1.0; flash point 43°F.
Uses: Rubber finish.
Shipping regulations: Red label.*

black oxide of manganese. See pyrolusite.

"Black Pearls." [275] Trade name for a series of pelleted channel carbon blacks for paints, inks, and plastics. Available as "Black Pearls"

A	Longest flow channel carbon black.
S	Long flow channel carbon black.
O	Medium color all-purpose channel black.
2	Standard high color channel black.
46	High color channel black.
70	Medium color channel black.
71	Medium color channel black.
74	Medium color channel black.
80	Medium color channel black.
81	Regular color all-purpose channel black.

black plate. Thin sheet steel obtained by rolling and usually used for containers. It is not coated with any metal but a special lacquer or baked enamel finish is usually applied by the can manufacturer.

black, platinum. See platinum black.

black powder. See gunpowder.

black precipitate. See mercurous nitrate, ammoniated.

black root. See leptandra.

black rouge. See iron oxide, black.

black sampson. See echinacea.

black sand. A deposit of dark minerals with a high specific gravity found in stream beds and on beaches. Magnetite and ilmenite are usually present, and gold, monazite and other minerals are sometimes present.

black smalt. A darker form of smalt. See smalt.

black snake-root. See cimicifuga.

blackstrap. See molasses.

blackstrap molasses. See molasses.

blanc fixe. See barium sulfate.

blanching liquor. A solution of calcium hypochlorite.

"Blancol." [232] Brand name for a series of optical bleaching agents for wool, cotton and synthetic fabrics and paper.

"Blancol." [307] Trademark for an anionic dispersing agent, composed of the sodium salt of a sulfonated naphthalene condensate; 90% active.
Properties: Coarse granules ranging in color from tan to brown; density 1.00-1.05; soluble in water, glycerin, ethylene glycol. Also available as a free-flowing granular powder ("Blancol" N), and as a liquid ("Blancol" W Conc).
Uses: Dispersing agent for pigments, earths, and solids in water; peptizing agent in insecticide formulations; in the paper trade for slime control, preventing coagulation of pitch, reducing two-sidedness, to improve sizing, etc.; in the leather trade as a bleaching, dispersing, leveling, and neutralizing agent.

"Blancophor." [307] Trademark for optical whitening agents.
"Blancophor" FFG. A coumarin derivative.
Properties: Tan powder; density 1.05; soluble in water; compatible with urea or melamine formaldehyde resin finishes; produces blue-violet brilliancy on fabrics.
Uses: Whitening agent for wool, nylon, acetate rayon and mixed fibers.
"Blancophor" HS Brands. Stilbene derivatives. Fine yellowish tan, water-soluble powders used on cellulosic fibers, cotton and rayon fabrics, paper, and in household and industrial detergents.

"Blandol." [45] Trademark for white mineral oil, N.F.
Properties: Sp. gr. 0.850-0.860 (60°F); Saybolt viscosity 80-90 (100°F); odorless and tasteless.
Uses: Pharmaceutical and cosmetic formulations; plasticizers; paper penetrants; foam depressants.

"Blankit I." [307] Trademark for a bleaching and reducing agent composed of a stabilized, buffered hydrosulfite product; 67% active.
Properties: White, fine, slightly granular powder; density 1.23-1.30; soluble in water.
Containers: Must be stored in air-tight, dry containers protected from heat and moisture.
Uses: Bleaching agent for wool, jute, linen and hemp; textile stripping agent.

blast-furnace dust. Dust deposited by or recovered from blast-furnace gases. It contains a variable amount of potash which renders it valuable. The black dust is comparatively poor in potash, the reddish dust deposited in the stoves and boilers is richer and the light-colored dust found at the base of the stacks is the richest.

blast-furnace gas. A by-product gas from the smelting of iron ore. It is obtained by the passage of hot air over the coke in the blast furnaces. A typical gas will analyze 12.9% carbon dioxide, 26.3% carbon monoxide, 3.7% hydrogen, 57.1% nitrogen.
Uses: For heating blast-furnace stoves, and as boiler- or gas-engine fuel.

blast-furnace slag. See slag.

blasting agent. Shall mean any material or mixture, consisting of a fuel and oxidizer, intended for blasting, not otherwise classified as an explosive and in which none of the ingredients are classified as an explosive, provided that the finished product, as mixed and packaged for use or shipment, cannot be detonated by means of a No. 8 test blasting cap when unconfined. (National Fire Protection Association, 1959).
An example is a mixture of 94% fertilizer grade ammonium nitrate (prills) and 6% oil.

blasting gelatin. A type of dynamite containing some nitrocellulose, usually in addition to nitroglycerin.

blasting powders. Black, granular, slow-acting explosives which burn with explosive rapidity when ignited under confinement, but do not detonate. They are known, therefore, as deflagrating powders in distinction from high explosives or detonating powders. All blasting powders are Class A explosives under I.C.C. Regulations.
Uses: Mining; road building; quarries.
Shipping regulations: Explosives. Class A, by freight; not accepted by express.*

"B-L-E." [248] Trademark for high-temperature reaction product of diphenylamine and acetone.
Properties: Dark-brown viscous liquid; sp. gr. 1.087; soluble in acetone, benzol, and ethylene dichloride; insoluble in gasoline and water.
Uses: Antioxidant for tire treads and carcass, inner tubes, and as a general purpose antioxidant.

bleaching agents. As used for paper or textiles, these include hydrogen peroxide (the most common), sodium hypochlorite, sodium peroxide, sodium chlorite, and calcium hypochlorite, among inorganic agents, and many organic chlorine derivatives. See, for example, dichloroisocyanuric acid.

bleaching assistants. Any materials added to bleaching baths to secure more rapid and complete penetration of the bleach or improved regulation of the bleaching action. Typical bleaching assistants include compounds of sulfonated oils and solvents, soluble pine oils, fatty alcohol salts, sodium silicate, sodium phosphate, magnesium sulfate, and borax.

bleaching clays. Clays that possess superior decolorizing characteristics for use in refining of mineral, petroleum, vegetable, and animal oils.

bleaching powder. Any powder used, in solution,

*See "I.C.C. Shipping Regulations," page xiii.
Reference numbers refer to name of manufacturer. See "List of Manufacturers," page v.

for textile bleaching. Specifically, the term refers to chlorinated lime, which is an important industrial agent. Among other powders developed for household use are sodium perborate, and dichlorodimethylhydantoin. See also calcium hypochlorite.

bleach liquor. A solution of calcium hypochlorite (q. v.).

blende. See sphalerite.

"Blendene." [73] Brand name for proprietary product; a terpene-soap composition.
Properties: Oily liquid having a pine odor; dispersible in water, in toluene, mineral spirits, and mineral oil. Soluble (hot) in methyl alcohol, ethyl alcohol, and naphtha; partly soluble in cottonseed oil.
Constants: Sp. gr. (25°C) 0.947; titer below 6°C; pH (10% dispersion) 9.2.
Containers: 1-gal cans (8 lbs); 5-gal cans (40 lbs); 55-gal drums (430 lbs).
Uses: For the manufacture of fluid emulsions, oils, solvents, etc., for water paint base, emulsion sprays, furniture and automobile polishes.

blind coal. A term sometimes used to designate anthracite.

blister copper. Copper (96-99% purity) produced by the reduction and smelting of copper ores. It has a blistered appearance probably caused by gas pockets. It is usually further refined electrolytically.

blistering beetles. See cantharides.

blistering fly. See cantharides.

block polymer. A polymer whose molecule is made up of comparatively long sections that are of one chemical composition, these sections being separated from one another by segments of a different chemical character, as for example, blocks of polyvinyl chloride interspersed with blocks of polyvinyl acetate.

block tin lining. Block tin is a common designation for pure tin. Copper vessels are lined or coated with tin by the application of molten tin upon clean copper with the aid of fluxing. Such coatings are sometimes called hot dippings. Tin is sometimes used for coating lead sheet or lining lead pipe, and owing to the method of fabrication, these articles may be called two-ply metal. Frequently tin is the metal chosen for making, holding, and conveying distilled water and it is used in contact with some chemicals.

blood charcoal (blood char). Made by adding caustic potash to blood, evaporating and calcining the residue.
Grades: Fresh; spent.
Uses: (Fresh) decolorizing; deodorizing and filtering.

blood, dried. A packing house by-product obtained by coagulating animal blood, followed by drying and grinding. It is a brown powder with an odor similar to that of glue and contains about 11.8% nitrogen and 1.2% phosphorus.
Grades: According to nitrogen content expressed as ammonia, as, 16-16.5% ammonia; 14% ammonia.
Containers: Bags.
Uses: Fertilizer, also as an ingredient of proprietary fertilizer compounds; clarifying agent for wines, syrups, etc.; ingredient of patent medicines used for blood diseases; in work on immunity by serum therapy; manufacture of adhesives; feed for hogs and chickens.
Shipping regulations: None. *

blood geranium. See sanguinaria.

blood red. A red pigment consisting essentially of red iron oxide. See iron oxide reds.

blood-root. See sanguinaria.

bloodstone.
1. A variety of chalcedony (q. v.) or quartz (q. v.). Color is dark-green with bright red spots. It is engraved for signet ring stones.
2. This term has also been used for hematite (q. v.).

blowing agents. Substances used to produce foam or sponge rubber through the fact that they undergo decomposition when heated. Thus their incorporation into a soft rubber compound results in its conversion to a foam or sponge as soon as heat is applied during vulcanization. Blowing agents are similarly used in producing plastic foams. See, for example, "Celogen," "Nitrosen."

blown asphalt. Hard, friable solid obtained by blowing air at high temperature through mineral residual oils. They are also known as oxidized asphalts, condensed asphalts, and mineral rubber. See also pitches, artificial.

blown oils (oxidized oils; base oils; thickened oils; polymerized oils).
1. Vegetable and animal oils which have been heated and agitated by a current of air or oxygen. They are partially oxidized, deodorized and polymerized by the treatment, and are increased in density, viscosity and drying power. Important blown oils are castor, linseed, rape, whale and fish oils.
Uses: Paints, varnishes, lubricants, and plasticizers.
2. Mineral oils. See blown asphalt.

blubber oil. See whale oil.

blue asbestos. See crocidolite.

blue copperas. See copper sulfate.

blue gas. See water gas.

blue glass. Cobalt oxide is added to a soda lime glass. Cupric oxide gives a green blue.

blue gum tree. See eucalyptus.

blue, iron. See iron blues.

*See "I.C.C. Shipping Regulations," page xiii.
Reference numbers refer to name of manufacturer. See "List of Manufacturers," page v.

blue, laundry. See laundry blue.

blue lead. See lead sulfate, blue basic.

blue oil. Name given to the heavy oil from crude Scottish shale oil after paraffin has been removed.

blue ointment. One of the ointments containing mercury, white wax, and petrolatum. It contains not less than 29% and not more than 31% of mercury; about 20% white wax, and 40% petrolatum.

blue powder. See zinc.

blueprinting. A process for the reproduction of drawings or printed material in which sensitized paper containing ferric ammonium citrate and potassium ferricyanide is placed under tracing (i.e., transparent) paper imprinted with the lines or data to be reproduced and then exposed to bright light. The ferric ions are reduced to ferrous ions by the effect of the light, and the ferrous ions react with the ferricyanide ions forming Turnbull's blue ($Fe_3[Fe(CN)_6]_2$). Under the black lines on the tracing paper no ferrous ions are reduced and no blue color is produced. Thus the black lines of the tracing paper appear on the sensitized paper as white lines against a blue background.

blueprint paper. Paper dipped in solutions of ammonium ferric citrate and potassium ferricyanide and dried in the dark. Exposure to ultraviolet or sunlight reduces the ferric compound to the ferrous state, forming blue ferrous ferricyanide which is fixed on the paper. Unexposed portions are washed out.

blue, Prussian. See iron blues.

bluestone. A variety of thin bedded or easily cleavable sandstone, usually dark in color, used for flagstone. Quarried in New Jersey, Pennsylvania, and New York. The name has also been applied to other types of blue rocks. Note the distinction from blue stone (q.v.).

blue stone. See copper sulfate.

blue verdigris. See copper acetate, basic.

blue verditer. See Bremen blue.

blue vitriol. See copper sulfate.

blue vitriol, natural. See chalcanthite.

blue, washing. See laundry blue.

blue water gas. See water gas.

"3B" Mercaptan. [204] Trademark for tert-dodecylmercaptan.

Properties: Color, water-white to very pale yellow; sp. gr. (20°C), 0.85-0.86; mercaptan content, 93% min; distillation, 95% between 200-235°C; flash point, 205°F; soluble in most organic solvents; insoluble in water.

Uses: Organic synthesis; modifier in polymerization of diene polymers such as nitrile and styrene rubbers.

BMU. Abbreviation for beta-methyl umbelliferone.

B.O.D. Abbreviation for biochemical oxygen demand; see also dissolved oxygen (D.O.) and oxygen consumed (C.O.D.).

body oil. See whale oil.

"Boerite." [309] Trademark for a series of sulfonated fatty acids having a general use in the industrial arts.

boghead coal. A variety of bituminous or subbituminous coal resembling cannel coal in appearance and behavior during combustion. It is characterized by a high percentage of algal remains and volatile matter. Upon distillation it gives exceptionally high yields of tar and oil. (ASTM definition, ASTM D493-39.)

bog iron ore. See limonite.

bog manganese. See wad.

bog moss. See sphagnum.

"Bogol." [48] Trademark for a crude tall oil.

boiler scale. A rocklike deposit occurring on boiler walls and tubes in which hard water has been heated or evaporated. Consists largely of calcium carbonate or calcium sulfate or similar materials, depending on the mineral content of the water. Boiler scale decreases the rate of heat transfer through the boiler and tube walls resulting in increased heating costs and shortening of boiler life. Most boiler feed water is softened (treated to remove calcium and magnesium ions) before being led into the boiler in order to prevent the formation of boiler scale.

boiling point. The temperature at which the vapor pressure of a liquid is just slightly greater than the total pressures of the surroundings. This temperature may be approximately determined in many cases by noting the temperature at which ebullition first occurs, that is, when bubbles of vapor are formed within the body of the liquid as its temperature is gradually raised. Precise determination of the boiling point is more complicated, and requires special methods because of superheating of the liquid, formation of bubbles of air or other dissolved gases, and for other reasons.

bois de rose oil. See oil bois de rose Brazilian.

bole. A red variety of hematite (naturally occurring ferric oxide) used as a pigment. See iron oxide reds.

boletic acid. See fumaric acid.

"Bolidensalt" BIS. [317] Trade name for zinc chrome arsenate.

Approximate composition: H_3AsO_4, 20%; Na_2HAsO_4, 19%; $Na_2Cr_2O_7 \cdot 2H_2O$, 16%; $ZnSO_4 \cdot 7H_2O$, 43%. The zinc sulfate is to be added at the treating plant.

Containers: Steel drums with wooden crates (220-lbs net).

Uses: Wood preservative used in water

solution in plants working under vacuum and pressure; zinc and chromium arsenates are precipitated in the wood fibers making the treated wood resistant to leaching.

"Bolidensalt" BIS Copperized. [317] Trade name for zinc copper chrome arsenate.
Approximate composition: H_3AsO_4, 20%; Na_2HAsO_4, 19%; $Na_2Cr_2O_7 \cdot 2H_2O$, 16%; $ZnSO_4 \cdot 7H_2O$, 22%; $CuSO_4 \cdot 5H_2O$, 22%. Zinc and copper sulfate to be added at treating plant.
Containers: Steel drums with wooden crates (220-lbs net).
Uses: As listed for "Bolidensalt" BIS.

"Bolidensalt" K33. [317] Trade name for non-ionic copper chrome arsenate.
Approximate composition: As_2O_5, 34%; CrO_3, 26.6%; CuO, 14.8%; H_2O, 24.6%.
Containers: Steel drums with wooden crates (220-lbs net).
Uses: Suitable for cooling towers; sleepers and high-transmission line poles as electrical resistance of wood is increased by salt; effective against all wood destroyers including soft-rot organisms. All ingredients in salt are practically completely fixed in the wood and no soluble by-products are formed.

"Boltaron." [96] Trademark for a line of polyvinyl chloride products used for corrosion resistant piping, ducts, hoods, tanks, machined parts.

bolus alba. See kaolin.

bombardment, nuclear or atomic. The directing of gamma rays or a stream of high energy particles such as electrons, neutrons, protons, etc., against a target element. The projectiles may come from cosmic rays, particle accelerators, nuclear reactors or other sources. When a high energy particle hits the target nucleus, (1) it may bounce off and leave the struck nucleus unchanged; (2) it may be absorbed, sometimes producing a new element of higher atomic number; (3) it may bring about the formation of entirely new particles; or (4) it may cause the nucleus to disintegrate Several new elements and many of the previously unknown subatomic particles (mesons, etc.) first appeared in these reactions.

"Bomber." [244] Trademark for a compound consisting of a balanced blend of buffered alkalies and a surface active agent.
Properties: White, granular, free-flowing mechanical mix; soluble in water; total Na_2O content 43.5%.
Containers: 125-lb plywood drums; 325-lb wooden barrels.
Uses: A general dairy and creamery detergent for heavy duty cleaning in low water-hardness areas having a predominance of magnesium.

BON. Abbreviation for beta-oxynaphthoic acid. See 3-hydroxy-2-naphthoic acid.

"Bonadettes." [299] Trademark for meclizine hydrochloride.

"Bonadoxin." [299] Trademark for a combination drug containing meclizine hydrochloride and pyridoxine hydrochloride (vitamin B_6). Used in medicine.

"Bonadur." [57] A trade designation for a series of acid azo red pigments which are made by diazotizing and coupling a substituted toluidine sulfonic acid with beta-hydroxynaphthoic acid and forming a metallic salt with an appropriate metal compound.
Uses: Inks for cartons, food wrappers and waxed papers, plastics, rubber, floor covering, and metal-decorating lacquers and enamels.

"Bonamine." [299] Trade name for meclizine hydrochloride.

"Bonaril." [233] Trademark for a hydrolyzed polyacrylamide for use in foundry sands.

bond clay. Clays which, in addition to a high refractoriness, have a strong bonding power, thus making them of value for crucibles to melt steel and brass and as a bond for abrasive wheels. They are all fire clays and some are also ball clays.

"Bonderizing." [343] Trademark for a process for furnishing a corrosion-resisting base for paint finishes on steel, aluminum, zinc, and their alloys and diecastings. Also used as an aid in deep drawing on steel and aluminum.

"Bondogen." [69] Trademark for proprietary product. Oil soluble sulfonic acid of high molecular weight with a high boiling hydrophilic alcohol and a paraffin oil.
Properties: Dark mahogany liquid; sp. gr. 0.90-0.92; acid number 40-42.
Uses: Peptizing agent and plasticizer for all elastomers.

bone ash (bone earth). An ash containing from 67 to 85% basic calcium phosphate, 2 to 3% magnesium phosphate, 3 to 10% calcium carbonate, some caustic lime and calcium fluoride.
Derivation: By calcining bones.
Uses: Fertilizer; in the preparation of superphosphates; cleaning and polishing compounds; ceramic products.

bone black (bone char; bone charcoal). Black pigment made by carbonizing bones. Carbon content is usually about 10% unless calcium phosphate and other salts are extracted with acid. It is an inferior black pigment but a superior adsorbent for purification of sugar solutions.
Containers: Multiwall paper sacks; fiber drums.
Uses: Manufacturing blackings and polishes; decolorizing agent and filtering medium; cementation reagent; absorptive medium in gas masks; paint and varnish pigment; clarifying shellac; decolorizing paraffin and sugar; filtering, decolorizing and deodorizing water.

bone brown. Partially charred ivory dust or bones.
Use: As a pigment.

bone char. See bone black.

bone charcoal. See bone black.

bone, dissolved. A ground bone or bone meal which has been treated with sulfuric acid.
Use: As fertilizer material.

bone earth. See bone ash.

bone fat (bone tallow). Fat obtained from animal bones by any of the following methods: (a) by boiling fresh bones in water; (b) by treating with steam under pressure; (c) by extraction with an organic solvent.
Use: Manufacture of candles and cheap soaps.

bone fat pitch. See stearin and fatty acid pitches.

bone, fine. The trade terms "bone meal," "bone dust," and "fine bone" are used to indicate mechanical condition, or fineness of division, and do not refer especially to composition.

bone glue. See glue.

bone meal, raw. A meal produced by drying and grinding animal bones not previously steamed under pressure.
Properties: Contains 4-5% ammonia, 20-25% phosphoric acid, 43-55% bone phosphate.
Use: As fertilizer.

bone meal, steamed. A meal produced by grinding animal bones which have been previously steamed under pressure.
Properties: Contains 2-3% ammonia, 50-55% bone phosphate.
Use: As fertilizer.

bone oil (animal oil; Dippel's oil; hartshorn oil; Jeppel's oil).
Properties: Dark brown, fixed oil; repulsive odor. Soluble in water.
Chief constituents: Hydrocarbons, pyridine bases, and amines.
Constants: Sp. gr. 0.900-0.980.
Derivation: By the destructive distillation of bones or other animal substance. After extraction with benzene or carbon disulfide, they are distilled in iron or clay retorts, the volatile products, consisting of gaseous ammonium salts and bone oil, are condensed and the gases containing the ammonium compounds collected in sulfuric acid. The bone oil and aqueous liquor collected are separated by gravity. The crude bone oil is subjected to fractional distillation. The constituents are numerous, the most important being pyridine.
Method of purification: Rectification.
Grades: Technical.
Containers: Iron drums; tank cars.
Uses: Organic preparations; source of pyrrole; denaturant for alcohol.
Shipping regulations: None. *

bone phosphate (BPL; bone phosphate of lime). Phosphoric acid is found in bones in the form of tribasic calcium phosphate (q. v.).

bone tallow. See bone fat.

bone turquoise. See turquoise.

"Bonine." [299] Trademark for meclizine hydrochloride.

BON red. A class name for a group of organic azo pigments made by coupling beta-hydroxynaphthoic acid to various amines and forming the barium, calcium, strontium or manganese salts. They have bright shades ranging from yellow red to deep maroon. They are characterized by good light resistance, good heat resistance, nonbleeding in vehicles and solvents and good opacity. They are widely used in printing inks, paints, enamels, lacquers, rubber plastics, wall paper, textiles, floor coverings, and crayons.

boort. See diamond.

"Boot." [173] Trademark for a pre-spotting material for use in treating stains prior to dry cleaning so as to facilitate the dry cleaning process. It is a concentrated soapless pre-spotter used on the wet side to soften spots and stains on garments prior to dry cleaning.

BOPOB. See para-bis[2-(5-para-biphenylyloxazolyl)]-benzene.

boracic acid. See boric acid.

boracite $Mg_3B_7O_{13}Cl$. A natural chloride and borate of magnesium, occurring in salt beds.
Properties: Colorless, white to yellowish or greenish; pyroelectric; occurs as crystals or as soft masses. Sp. gr. 2.9-3; hardness 7.
Occurrence: Germany; France; Louisiana.

boral. A "sandwich" consisting of boron carbide crystals in aluminum, with a cladding of commercially pure aluminum. Concentrations of up to 50% boron carbide can be obtained.
Uses: As a shielding material against the passage of thermal neutrons, as in reactor shields; neutron curtains; shutters for thermal curtains; safety rods; containers for fissionable material.

"Boran." [169] Trademark for diaminochrysazin used in the colorimetric determination of boron.

borax (tincal; borax decahydrate) $Na_2B_4O_7 \cdot 10H_2O$. A natural hydrated sodium borate, found in salt lakes and alkali soils. Borax is also used as the commercial name for sodium borate (q. v.).
Properties: Color white to grayish or greenish; luster vitreous to dull; taste sweetish alkaline; sp. gr. 1.7; hardness 2-2.5; readily soluble in water.
Containers: Paper bags; bulk.
Occurrence: Tibet; California, Nevada.
Uses: As raw material for the manufacture of commercial sodium borate. Now largely displaced by kernite.

*See "I.C.C. Shipping Regulations," page xiii.
Reference numbers refer to name of manufacturer. See "List of Manufacturers," page v.

borax, anhydrous. See borax, dehydrated.

borax decahydrate. See borax.

borax, dehydrated (borax, anhydrous) $Na_2B_4O_7$.
Properties: White, free-flowing crystals; hygroscopic; forms partial hydrate in damp air; m.p. 741°C; sp. gr. 2.367; rate of solution, slow.
Grades: Technical (99% $Na_2B_4O_7$); standard; fine granular form.
Containers: 100-lb paper bags; boxcars.
Uses: Manufacture of glass, enamels, and other ceramic products.

borax glass. See sodium borate.

borax pentahydrate $Na_2B_4O_7 \cdot 5H_2O$.
Properties: Begins to lose water of hydration at 122°C; sp. gr. 1.815; stable; free-flowing.
Grades: Crude; technical (99.5% $Na_2B_4O_7 \cdot 5H_2O$).
Containers: 100-lb bags; bulk.
Uses: Weed killer and soil sterilant; same uses as sodium borate.

borazon. A boron nitride formed at very high pressures and temperatures from mixtures of boron and nitrogen, or from ordinary hexagonal boron nitride in the presence of catalysts such as lithium, calcium, magnesium, or their nitrides. Comparable to diamond in hardness, and has cubic arrangement of atoms in its crystal lattice. Density 3.48 g/cc. Reacts extremely slowly with water; dissolves in fused sodium carbonate.

Borcher's metal. A group of alloys of chromium with nickel and cobalt, or of chromium and iron with a small proportion of molybdenum and/or silver or gold. Heat and corrosion resistant.
Uses: For chemical apparatus; crucibles; pyrometer tubes; heat treating or annealing pots.

Bordeaux mixture. A liquid fungicide and insecticide mixture made by adding slaked lime to a copper sulfate solution. It is either made by the user or bought as a powder ready for dissolving. Stabilizing agents are sometimes added to delay settling.

Borden's "38." [65] Trademark for ureaform; plant nutrient containing 38% nitrogen.

boric acid (boracic acid; orthoboric acid) H_3BO_3.
Properties: Colorless, odorless scales or white powder. Stable in air; sp. gr. 1.4347; m.p. 184°C; soluble in water, alcohol, and glycerin.
Derivation: (a) By adding hydrochloric or sulfuric acid to a solution of borax and crystallizing; (b) decomposing boracite with hydrochloric acid; (c) found native in Tuscany. See also sassolite.
Method of purification: Recrystallization.
Grades: U.S.P. XVI; C.P.; technical.
Containers: Bags; drums.
Uses: Chemicals (borates, water-glass manufacture, nitric acid from saltpeter with simultaneous production of borax); ceramic glazes; high grade cements capable of taking high polishes; impregnating wicks of stearin candles; glass pastes, and special glasses; intermediates; laundry starch glazes; leather preparation of hides prior to tanning (leather-dressing compounds); artificial precious stones; metallurgy (welding flux, brazing copper, enamel coatings on iron); fireproofing compositions and linings for safes; pigments (Guignet's green, borated ultramarine); enamel paints; manufacture of imitation hard wood from soft wood; paper glazes; medicine; cosmetics; soaps, textiles (fireproofing, mordant, solvent bleach); fiberglass insulation.
Shipping regulations: None. *

boric acid esters. Trimethyl, tri-n-butyl, tricyclohexyl, tridodecyl, tri-p-cresyl, etc. borates. Compounds which are readily hydrolyzed to boric acid and the respective alcohols.
Properties: Colorless to yellow liquids; b.p. 230-350°C.
Suggested uses: Dehydrating agents; catalysts; sources of boric oxide; special solvents; fire retardants in plastics and paints; plasticizers or adhesion additives to latex paints; ingredients of soldering and brazing fluxes.

boric acid, ortho-. See boric acid.

boric anhydride. See boric oxide.

boric oxide B_2O_3 (boric anhydride, boron oxide).
Properties: Colorless transparent glass or powder; hard and brittle; slightly bitter to taste; sp. gr. 1.83-1.88; b.p. above 1500°C. The material has no melting point since it is a congealed liquid which has never crystallized but is nevertheless harder than many crystalline solids. Soluble in alcohol and acids; slightly soluble in cold water with decomposition; soluble in hot water.
Uses: Production of boron; chemical analysis of silicates; in heat resistant glassware; fire-resistant additive for paints.

"Boriresin." [354] Trademark.
Properties: An amber-colored, water-soluble, viscous synthetic resin, miscible with polyhydric alcohols, and on drying forms a hard, transparent film.
Use: For fireproofing fabrics.

borneol (bornyl alcohol; Malayan camphor) $C_{10}H_{17}OH$. Various isomers exist, but commercial product is described here.
Properties: White, translucent lumps; peculiar camphor-like odor; burning taste; soluble in alcohol and ether; very slightly soluble in water.
Constants: Sp. gr. 1.011; m.p. 203°C minimum; b.p. 212°C.
Derivation: (a) By reduction of ordinary camphor with nascent hydrogen. (b) From the wood of Dryobalanops camphora, a tree

*See "I.C.C. Shipping Regulations," page xiii.
Reference numbers refer to name of manufacturer. See "List of Manufacturers," page v.

found in Sumatra and Borneo—deposited in the trunks in solid crystalline mass. (c) A constituent of certain volatile oils including citronella, thyme, African ginger, Canada snake root, nutmeg.
Grades: Technical.
Containers: 1-, 5-lb tins; 5-, 15-lb boxes; cans.
Uses: Medicine; manufacture of synthetic camphor; perfumery; chemical esters.
Caution: Combustible.
Shipping regulations: None.*

bornite (purple copper ore) Cu_5FeS_4. A natural sulfide of copper and iron, usually found with other copper minerals in igneous rocks and vein deposits.
Properties: Color brownish bronze, tarnishing to variegated purple and blue to black; luster metallic; streak grayish black.
Constants: Sp. gr. 5.07; hardness 3.
Occurrence: Arizona; Montana, Virginia, Utah; Canada; Chile; Peru; Bolivia.
Use: An ore of copper.

bornyl acetate $C_{10}H_{17}OOCCH_3$.
Properties: Colorless liquid at warm room temperature; solidifies to colorless crystals at cooler temperature; piney-camphoraceous odor. Soluble in 3 volumes of 70% alcohol; miscible with 95% alcohol and with ether.
Constants: Sp. gr. 0.980-0.984; refractive index 1.463-1.465; m.p. 29°C.
Derivation: (a) Interaction of borneol and acetic anhydride in the presence of formic acid. (b) Intermediate in making synthetic camphor.
Grades: Technical.
Uses: Perfumery; nitrocellulose solvent and plasticizer.

bornyl alcohol. See borneol.

bornyl chloride (pinene hydrochloride; turpentine camphor; terpene hydrochloride) $C_{10}H_{16} \cdot HCl$.
Properties: White, crystalline mass, resembling camphor; turpentine and camphor odor; m.p. 131-132°C; b.p. 208°C; soluble in alcohol; insoluble in water.
Derivation: By treating pinene with dry hydrochloric acid in the cold.
Grades: Technical.
Containers: Boxes.
Use: Medicine.
Shipping regulations: None.*

bornyl formate $C_{10}H_{17}OOCH$.
Properties: Colorless liquid, having a piney odor. Sp. gr. 1.007-1.009.
Grades: Technical.
Uses: In the perfuming of soaps, disinfectants, and sanitary products.

bornyl isovalerate $C_{10}H_{17}OOC_5H_9$. A constituent of valerian oil.
Properties: Limpid fluid, aromatic, valerian-like odor. Soluble in alcohol and ether; insoluble in water.
Constants: Sp. gr. 0.951 at 20°C; b.p. 255-260°C.
Use: Medicine; essential oil intermediate.

bornyl salicylate $C_{10}H_{17}OCOC_6H_4OH$.
Properties: Brown, oily liquid. Incompatible with alkalies and alkaline salts. Insoluble in water; soluble in alcohol, ether, chloroform, and oils.
Use: Medicine.
Shipping regulations: None.*

boroethane. See diborane.

boron B. Element of atomic number 5, of group III in the periodic system. It is an essential plant nutrient.
Properties: Very soft, brown, amorphous powder; or as crystals; ignites in air. Soluble in concentrated nitric acid and sulfuric acid; insoluble in water, alcohol and ether.
Constants: Sp. gr. 2.45; m.p. 2300°C; hardness of crystals 9.3 (Mohs).
Sources: Borax; kernite; colemanite; ulexite; priceite; boracite; sassolite.
Derivation: (a) Electrolysis of a fused bath of potassium chloride or potassium fluoride, potassium fluoborate, and boric oxide; (b) by heating boric oxide with powdered magnesium; (c) by reduction of boron halides with a gaseous dispersion of molten alkali metal.
Grades: Technical; 99% pure; high purity (electronic).
Uses: Catalytic agent; ceramics and heat resistant glassware (a glass in which boric oxide (B_2O_3) replaces the calcium oxide in ordinary lime soda glass); metallurgy (alloy steels, cementation of iron); thermometers and thermoregulators; controlling agent in uranium graphite piles (in the form of boron steel or boron carbide); scavenger to remove gaseous impurities from molten copper; component of delayed action fuses; semiconductors; abrasives (crystals).
Shipping regulations: None.*

boron alloys. These are usually alloys with iron or manganese, but may be with aluminum, titanium, vanadium, zirconium, manganese, silicon, calcium, or carbon or with two or more of these. Ferroboron usually contains 15-25% boron; manganese boron usually 60-65% manganese. These alloys are used as degasifying and deoxidizing agents, or as a means of increasing the hardenability of steel. Only a few thousandths of a per cent are needed for the latter purpose. The other elements are used for their own beneficial properties, and also to prevent oxidation of boron in the melt.

boronatrocalcite. See ulexite.

boron bromide. See boron tribromide.

boron carbide B_4C. Probably not a true compound, but instead a solution of varying amounts of carbon in a slightly distorted boron lattice.
Properties: Black hard crystals ranking next to diamond in hardness (9.3, Mohs); sp. gr. 2.6; m.p. 2350°C; b.p. 3500°C. Soluble in fused alkali; insoluble in water

and acids. Has high capture cross-section for thermal neutrons.
Derivation: By heating boron oxide with carbon in an electric furnace.
Uses: In powder form as an abrasive and in molded form as an abrasion resister; also as control rods for nuclear reactors.

boron chloride. See boron trichloride.

boron fluoride monoethylamine (BF3-MEA) $BF_3-C_2H_5NH_2$.
Properties: White to pale tan flakes. Sp. gr. 1.38; m.p. 88-90°C. Soluble in furfuryl alcohol, polyglycol, acetone. Releases boron trifluoride above 110°C.
Use: Elevated temperature cure of epoxy resins.

boron fuels. Fuels for air-breathing engines and rockets made up principally of boron hydrides or their compounds with carbon. See also rocket propellants.

boron hydride. A compound of boron and hydrogen. See diborane; pentaborane; decaborane.

boronia oil.
Properties: Dark green oil, having a violet-type odor.
Derivation: Derived by volatile solvent extraction of the flowers of the brown boronia of Western Australia (Boronia megastigma Nees).
Constituents: Beta-ionone, ethyl alcohol, ethyl formate.
Use: Perfumery, mainly in Australia.

boron nitride BN.
Properties: Crystals or powder; m.p. about 3000°C; an electrical nonconductor; resists corrosion at temperatures above 3000°F; somewhat soluble in water. Boron nitride is anisotropic and some properties vary according to method of preparation and crystal form. See borazon. The powder has a hardness of 2 on the Mohs' scale.
Derivation: By heating a mixture of boric acid and tricalcium phosphate or similar materials in an ammonia atmosphere in an electric furnace.
Uses: Refractory; furnace insulation; high temperature lubricant, as in glass molds; component of rectifying tubes; dielectric; chemical equipment parts; molten metal pump parts.

boron phosphide BP. A refractory-like material harder than silicon carbide and inert to corrosion. Oxidizes in air at about 800°C but is more stable in reducing atmospheres.
Derivation: Direct union of boron and phosphorus at about 1000°C in a reducing atmosphere.
Use: Suggested for a transistor operable at high temperatures.

boronotungstic acid. See borotungstic acid.

boron oxide. See boric oxide.

boron phosphate (sometimes called borophosphoric acid) BPO_4.
Properties: White, non-hygroscopic crystals; sp. gr. 1.873. Soluble in water; pH (1% solution) 2.0.
Containers: 400-lb barrels.
Use: Special glasses; ceramics; acid cleaner; especially promising as ceramic when fused with silica.

boron steel. See ferroboron.

boron tribromide (boron bromide) BBr_3.
Properties: Colorless, fuming liquid. Decomposed by alcohol and by water.
Constants: Sp. gr. 2.69 at 15°C; b.p. 90°C; m.p. −46°C.
Grades: Technical.
Use: Catalyst in organic syntheses.

boron trichloride (boron chloride) BCl_3.
Properties: Colorless, fuming liquid. Decomposed by alcohol and by water.
Constants: Sp. gr. 1.35 at 15°C; b.p. 18°C (approximate); m.p. −107°C.
Grades: Technical (99%); C.P. (99.5%).
Containers: Up to 50-lb pressure cylinders; tank cars.
Uses: Catalyst in organic syntheses; source of many boron compounds.
Shipping regulations: Corrosive liquid. White label.*

boron trifluoride BF_3.
Properties: Colorless gas; 2.3 times as dense as air; m.p. −126.8°C; b.p. −101°C; soluble in cold water; hydrolyzes in hot water; decomposes in alcohol. Easily forms double compounds such as that with ether, known as boron trifluoride etherate, or BF-3 ether complex.
Derivation: From boron trichloride and anhydrous hydrogen fluoride or by combination of elements.
Grades: Pure (99% min).
Containers: Steel cylinders.
Uses: Catalyst in organic synthesis; instruments for measuring neutron intensity.
Shipping regulations: Nonflammable gas. Green label.*

borophosphoric acid. See boron phosphate.

borosilicate glass. A silicate glass containing at least 5% boric oxide.

"Boro-Spray." [88] Trademark for a crystalline product consisting chiefly of sodium pentaborate. Used for spray applications to tree fruit and truck crops where boron deficiency is indicated.

borotungstic acid (borowolframic acid; boronotungstic acid) $B_2O_3(WO_3)_9 \cdot 24H_2O$.
Properties: Yellowish liquid. Soluble in water, alcohol. Sp. gr. 3.00.
Derivation: By heating ammonium borotungstate with aqua regia.
Grades: Technical.
Containers: Glass bottles.
Use: Mineralogic assays.
Shipping regulations: None.*

borowolframic acid. See borotungstic acid.

bort. See diamond.

Bosch-Meiser process. A process for producing urea from ammonia and carbon dioxide in the presence of some moisture. Ammonium carbamate is formed first, and then converted to urea.

boson. Also called Bose particle, or Bose-Einstein particle. Consideration of symmetry properties of the wave-mechanical description of systems of particles allow the classification of the particles into two kinds. Bosons are particles with integral or zero spin, and their most important property is that the Pauli exclusion principle does not hold for bosons. The other class of particles are fermions (q. v.).

Botany Bay gum. See accroides gum.

bourbonal. See ethyl vanillin.

bournonite (cogwheel ore) $PbCuSbS_3$. A natural sulfantimonide of lead and copper. Color steel gray to black; metallic luster. Often in wheel-shaped crystals. Sp. gr. 5.8; hardness 2.5-3. Found in metallic veins.

Occurrence: Europe; Mexico, Chile, Bolivia.

Use: Minor ore of copper, lead, and antimony.

boxberry. See gaultheria.

Boyle point. The temperature at which a gas obeys the perfect gas laws of Boyle and Avogadro.

Boyle's Law. The volume of a sample of gas varies inversely with the pressure, if the temperature remains constant. The relation is strictly true only for an imaginary perfect or ideal gas, but the law is satisfactory for practical calculations except when pressures are high, or temperatures are approaching the liquefaction point. Van der Waal's equation is a refinement to take care of the inherent inaccuracy of Boyle's Law.

b. p. Abbreviation for boiling point.

BPL. 1. Abbreviation for bone phosphate of lime. See bone phosphate.
2. Abbreviation for beta-propiolactone.

"BP Pyro No. 5." [172] Trade name for a grade of sodium acid pyrophosphate possessing a controlled slow rate of reaction.

Derivation: Dehydration of NaH_2PO_4.

Containers: Bags.

Uses: Leavening agent for use in the manufacture of canned biscuit doughs; for compounding bakers' baking powder.

"BPR." [55] Trademark for insecticidal material containing varying proportions of pyrethrin, piperonyl butoxide and rotenone; in liquid or dust base.

Br. Symbol for bromine.

BRA. Abbreviation for beta-resorcylic acid.

"Bradosol." [305] Trademark for domiphen bromide, dodecyldimethyl (2-phenoxyethyl) ammonium bromide. Used in medicine.

bran. The coarse, husky outer-coat of wheat, rye, and other cereals, which, after grinding, is separated from the flour by sifting or bolting.

brandy mint. See peppermint.

bran oil. A name sometimes used for furfural.

brasilin (brazilin; pernambuco extract; hypernic extract; Brazilwood extract) $C_{16}H_{14}O_5$. The crystalline colorizing principle of pernambuco and sappan wood.

Properties: White or pale yellow rhombic needles from alcohol; turns orange in air or light. Soluble in water, alcohol, ether and in alkali hydroxide solution with a carmine-red color. Decomposes above 130°C.

Uses: Dyeing red and purple shades of wood, ink, textiles, etc. Recommended also as acid-base indicator, turning yellow in acid and carmine-red in alkali.

brass and bronze. Copper base alloys. Brass is mainly a copper-zinc alloy and bronze mainly copper-tin. However, brass may contain some tin, and bronze, some zinc. Other metals, commonly lead, are added to some varieties to give desired properties. A large variety of compositions is produced for various uses in which such characteristics as corrosion resistance, hardness, tensile strength, color, and machinability are of different importance. In commercial practice the terms brass and bronze may be used without much regard for their original meanings. As a class these alloys are inferior to iron base alloys (steels) in hardness and strength, but superior in workability and resistance to corrosion. Some examples of brasses and bronzes with approximate compositions are:
coinage bronze, 95% Cu, 4% Sn, 1% Zn;
bronze, gun metal, 90% Cu, 10% Sn;
phosphor bronze, 79.7% Cu, 10% Sn, 9.5% Sb, 0.8% P;
commercial bronze, 90% Cu, 10% Zn;
cartridge brass, 67-70% Cu, 30-33% Zn, Pb, Fe;
high brass, 65% Cu, 35% Zn;
muntz metal, 60% Cu, 40% Zn;
red brass, 85% Cu, 15% Zn.

Available forms: Sheet, rod, wire, tubing, etc.

See also: admiralty metal; aluminum bronze; phosphor bronze 30; phosphor bronze 47; phosphor bronze 209; red brass.

brassidic acid $C_{22}H_{42}O_2$ or $CH_3(CH_2)_7CHCH(CH_2)_{11}COOH$. The trans isomer of erucic acid.

Properties: White crystals. M.p. 61-62°C; b.p. 282°C (30 mm); sp. gr. 0.859; refractive index (n 57/D) 1.448. Insoluble in water; slightly soluble in alcohol; soluble in ether.

Derivation: By treating erucic acid with nitrous acid (catalyst).

"Brasslyfe." [302] Trademark for corrosion-preventive compositions used to form a metal protective coating.

"Braze Bonding Agent." [69] Trademark for proprietary product, halogenated rubber derivatives and selected modifiers in solvent.

Properties: Deep red liquid; sp. gr. 1.01 ± .02; total solids 20-22%; flash point 93°F.

Uses: Cement to bond natural rubber, SBR or neoprene to steel.

"Braze Cover Cements." [69] Trademark for a series of proprietary products, specially compounded elastomers in appropriate solvents.

"For Natural Rubber and SBR", black liquid; solvent aliphatic; sp. gr. 0.835 ± .02; flash point 39°F.

"For Neoprene", black liquid; solvent system aromatic-aliphatic; sp. gr. 0.86 ± .02; solids 16.5-18%; flash point 39°F.

"For Butyl", black liquid; solvent aliphatic; sp. gr. 0.835 ± .02; solids 15-17%; flash point 46°F.

Uses: Applied over "Braze Bonding Agent" to improve bond strength and uniformity of adhesion after vulcanization.

Shipping regulations: Red label.*

brazilin. See brasilin.

brazilite. See baddeleyite.

Brazil-nut oil. See castanha oil.

Brazil wax. See carnauba wax.

Brazilwood extract. See brasilin.

"BRC" Hydrocarbons. [175] Trademark for solid coal-tar hydrocarbons. Available in the following grades:

No. 20 Softening point 175-185°F; sp. gr. 1.25-1.32 (25/25°C); carbon disulfide insolubles 15-25%.

No. 21 Softening point 205-220°F; sp. gr. 1.26-1.35 (25/25°C); carbon disulfide insolubles 15-25%.

No. 22 Softening point 205-220°F; sp. gr. 1.26-1.35 (25/25°C); carbon disulfide insolubles 25-35%.

No. 30 Softening point 280-305°F; sp. gr. 1.28-1.36 (25/25°C); carbon disulfide insolubles 25-40%.

Uses: Processing aid for all elastomers and rubber reclaiming; used in mechanicals where hardness and stiffness are requisite.

breeder. Name generally applied to any nuclear chain reactor which is capable of producing more fissionable material than it consumes in its operation. By the use of the breeding process, the stockpile of fissionable materials can be steadily increased since a complete conversion of all fertile material such as ordinary uranium 238 or thorium 232 into fissionable material can be accomplished. The first experimental breeder reactor was put into operation in 1951. If breeding succeeds, twenty pounds of natural uranium will yield 51,800,000 kilowatt hours of electricity—enough to light 25,000 average American homes for a year. Natural uranium contains only about 0.7% of directly fissionable uranium 233 and 235 so that without breeding the twenty pounds of natural uranium can provide a correspondingly smaller fraction of the above power.

"Brellin." [342] Trademark for gibberellic acids as plant growth-stimulating compositions.

Bremen blue (blue verditer; copper blue). Greenish-blue pigment consisting chiefly of copper hydroxide together with some carbonate and oxychloride. These blues are opaque in water, become slightly transparent in oil and lose body. They are soluble in acids and ammonia and are darkened by hydrogen sulfide or sulfur fumes.

bremsstrahlung. The continuous x-ray spectrum. When a charged particle is accelerated it must radiate energy as electromagnetic radiation. Therefore if beta rays are stopped in matter, or when the electron beam of an x-ray tube impinges on the target, part of the energy appears as a continuous spectrum of electromagnetic radiation, the bremsstrahlung, superimposed on the x-ray line spectrum.

"Brentamine." [206] Brand name of proprietary line of amino bases and stabilized, diazotized amino bases used in the production of azoic colors on the fiber by dyeing and printing methods.

"Brenthol." [206] Brand name of proprietary line of phenolic and naphtholic bodies, for use in the production of azoic colors on the fiber by dyeing and printing methods.

"Brentogen." [206] Brand name for a proprietary line of dyestuff powders based on mixtures of stabilized diazo compounds and naphthol arylamides; used in textile printing mainly for direct and certain resist styles giving fast bright shades of high tinctorial value. Assistants used in development of these powders are also given this brand name.

"Brentosyn." [206] Brand name of a proprietary line of azoic coupling components for "Terylene" and other polyester fibers.

"Bretol." [430] Trade name for cetyl dimethyl ethyl ammonium bromide, a quaternary ammonium compound used in dental preparations and soldering fluxes.

breunnerite. See magnesite.

"Brevital." [100] Trademark for methohexital (q. v.).

Brewster process. A method for the extraction of acetic acid from the acid distillate of the destructive distillation of wood. Isopropyl ether is used as the solvent for the acetic acid.

"BRH" 2 Rubber Softener. [175] Trademark for a proprietary asphaltic product.

Properties: Viscous fluid; sp. gr. 1.0 (25/25°C); flash point, 400°F; max. evaporation loss of 1% in 5 hours on heating 50 g at 163°C.

Uses: In compounding friction, adhesive, and electrical tapes because of excellent aging and tack characteristics; as a reclaiming

*See "I.C.C. Shipping Regulations," page xiii.

Reference numbers refer to name of manufacturer. See "List of Manufacturers," page v.

oil for natural rubber scrap, especially in pan process reclaiming.

brick, alumina (high alumina brick). Refractory bricks of higher alumina content than ordinary fire-clay brick. They are made from several alumina materials, such as diaspore, bauxite, kaolin, etc.

See also refractories.

Typical properties of bauxite brick: Refractoriness, cone 36-39; ret. U-load, 50 lbs/sq.in., cone 14; true sp.gr., 3.1-3.4; apparent sp.gr., 1.8-2.1; porosity 30-40%; coefficient thermal expansion (0°-1000°C), 65×10^{-7}.

Analysis of one bauxite brick: silica, 8.82%; iron oxides (Fe_2O_3), 6.30%; alumina, 78.01%; lime, 0.98%; magnesia, 4.41%; titania, 1.16%.

Uses: The use of high alumina brick has become increasingly general in the last few years under certain types of operating conditions where the service is severe. A large use of brick of this type is in the hot zone portion of rotary lime, cement or dolomite kilns as well as in the firing zone of shaft lime kilns. High alumina brick are also used in certain portions of large boiler settings and in ceramic kilns of both the continuous and the periodic types; in brief, they find application under certain types of conditions where the service is very severe.

brick, bauxite. See brick, alumina.

brick, chemical. See chemical stoneware.

brick, chrome. Refractory bricks made from chrome ores, largely the spinel chromite. They contain about 35-44% Cr_2O_3. See also refractories.

Typical properties: Refractoriness, cone 40-42; ret. U-load, 50 lbs/sq. in., cone 15; true sp.gr., 3.1-3.6; apparent sp.gr., 2.3-2.6; porosity 20-30%; coefficient thermal expansion (0°-1000°C) 90×10^{-7}.

Uses: Chrome brick are used to some extent in place of magnesia brick because they are somewhat less expensive. Chrome brick resist the action of both acidic and basic oxides. They are also used as a dividing material to separate refractories of opposite chemical character; for example, in the open-hearth furnace, frequently a course or two of chrome brick is used to separate the silica roof from the magnesia sidewalls, thereby preventing chemical interaction.

brick clay. Usually a relatively impure clay that contains considerable fluxing impurities and burns to a red brick. These clays must mold readily, burn hard at as low a temperature as possible and give minimum loss from warping or cracking during burning. Better grade clays (lower impurities, particularly lower amounts of soluble salts, more uniform color, greater hardness, less porosity, and uniform color on burning) are used for pressed brick or face brick. Two common varieties of brick clay found west of the Mississippi River are adobe which is a calcareous silty clay used for sun dried bricks in the Southwest, and loess, a similar material found throughout the Mississippi Valley.

brick, fire-. The term "fire-brick" is sometimes used to refer to any refractory brick. Its use should more properly be limited to those brick manufactured essentially from refractory fire-clays, and containing up to about 48-50% alumina.

Analysis of one low-grade fire-clay brick: Silica 61-72%, iron oxide (Fe_2O_3) 6.43%, alumina 28.7%, lime 0.46%, magnesia 1.04%, soda 0.05%, potash 0.05%, titania 1.60%.

Analysis of one high-grade fire-clay brick: Silica 53.52%, iron oxide (Fe_2O_3) 2.00%, alumina 41.00%, lime 0.30%, magnesia 0.30%, soda 0.90%, potash 0.20%, titania 1.60%.

Typical properties: Refractoriness, cone 30-35; ret. U-load 50 lbs/sq in, cone 14-20; true sp.gr. 2.5-2.6; apparent sp.gr. 1.9-2.1; porosity (per cent) 20-30; coefficient thermal expansion (0°-1000°C) 40×10^{-7}.

See also refractories.

Uses: Restricting the term to fire-clay brick: Entire linings of iron blast furnaces; linings and checkers of blast-furnace stoves; checker work in other operations as in open-hearth furnaces, glass tanks, etc. Large quantities are used in connection with power plants and industrial boiler stations, cupolas, malleable furnaces, heating furnaces of various kinds, gas producers, water gas sets. A large portion of the lining of rotary cement and lime kilns as well as cement coolers is also built of fire-brick.

brick, high alumina. See brick, alumina.

brick, magnesia (magnesite brick). Refractory bricks consisting essentially of MgO (periclase) with about 15% of other oxides. See also refractories.

Typical properties: Refractoriness, cone 30-40; ret. U-load 50 lbs/sq in, cone 16; true sp.gr. 3.1-3.5; apparent sp.gr. 2.4-2.7; porosity (per cent) 20-30; coefficient thermal expansion (0°-1000°C) 100×10^{-7}.

Uses: Magnesia brick are used wherever the corrosion of basic slags is severe, as for almost the whole of open-hearth steel furnaces, of copper reverberatory furnaces, in soaking pits and in the basic electric steel furnace. Certain factors tend to reduce their application, including higher cost, comparatively lower mechanical strength at elevated temperatures and greater thermal expansion.

brick, magnesite. See brick, magnesia.

"Brickmaster" Periclase. [426] Trademark for magnesium oxide, periclase.

Derivation: Synthetic.
Grades: Kiln run.
Containers: Bags and bulk.
Uses: In the manufacture of refractories.

brick, silica. Refractory bricks consisting essentially of lime-bonded quartzite (q.v.).

The best American brands contain approximately 94-96% SiO_2, with about 2% CaO. See also refractories.

Typical properties: Refractoriness, cone 32-35; ret. U-load 50 lbs/sq. in., cone 18-30; true sp. gr., 2.3-2.5; apparent sp. gr., 1.7-1.9; porosity, 20-30%; coefficient thermal expansion:

Practical users of silica brick give their expansion as being 3/16 to 1/4 in. (1.6-2.1%) per foot. Most of this occurs below 600°C. A sudden acceleration in expansion takes place at about 575°C.

Analysis of one brick: Silica 96.42%; iron oxides (Fe_2O_3) 0.50%; alumina, 0.75%; lime, 2.01%; magnesia, 0.08%; titania, 0.06%.

Uses: Because of their excellent mechanical strength at high temperature, their positive expansion at high heat and the fact that they do not begin to deform except at temperatures within a very few degrees of their actual fusion point, silica brick are extensively used for arch work whenever operation is steady. Thus silica brick are used in the open-hearth roof, in the caps of glass tanks and in the crowns of copper reverberatory furnaces. Silica brick are also used in the by-product coke oven because of their higher themal conductivity as compared with fire-clay brick and their resistance to abrasion; the same factors make them suited for coal-gas retorts. The roofs of electric steel furnaces are usually of silica material. Silica brick, however, are attacked by basic slags and dusts.

brick, zirconia. Refractory bricks containing a high percentage of zirconium oxide.

Typical properties: Refractoriness, cone 39-42; ret. U-load 50 lbs/sq. in., cone 19; true sp. gr., 4.8-5.9; apparent sp. gr., 4.0-4.6; porosity, 15-30%; coefficient thermal expansion (0°-1000°C), 8.4 x 10^{-7}.

Use: In metallurgical furnaces to resist basic slag.

bright stock. Lubricating oil of high viscosity, obtained from residues of petroleum distillation by dewaxing and treatment with fuller's earth or similar material. Sometimes also applied to viscous petroleum distillates.

Use: For blending with neutral oils in preparing automotive engine lubricating oils.

"Brij." [89] Trademark for each of a series of emulsifiers and wetting agents developed for use in emulsions of high alkalinity or acidity. They are polyoxyethylene ethers of higher aliphatic alcohols. Soluble in water and lower alcohols. Insoluble in coal tar hydrocarbons.

Brilliant Croceine
$C_6H_5N_2C_6H_4N_2C_{10}H_4OH(SO_3Na)_2$.

Properties: Light brown powder; cherry red solution in water.

Use: To dye wool and silk red from acid solution, and cotton and paper with aid of a mordant; also used for red lakes.

"Brilliant Toning Red." [141] Trade name for Permanent Red 2B azo pigments derived from beta-hydroxynaphthoic acid.

Properties: Good light resistance; good heat resistance; non-bleeding in water and organic vehicles and solvents.

Grades: Light yellow shade red and medium shade red. Resinated and nonresinated.

Uses: Printing inks; paints; enamels; lacquers; rubber; plastics; wallpaper; textiles; floor coverings; crayons; paper coatings.

brimstone. See sulfur.

brimstone acid. A surfuric acid (q. v.) prepared from sulfur.

briquettes. Bricks and blocks of finely divided material molded into shape by pressure, frequently with a specially chosen binder; e.g., coal dust is mixed with pitch for fuel, ore dust with lime for smelting.

brisance index. A measure of the shattering powder of an explosive. The brisance index is the ratio of the weight of graded sand which is shattered when a charge of the explosive is packed in the sand in a bomb and detonated in a standard manner to the weight of sand shattered by TNT, when detonated in the same manner.

britannia metal. See pewter and white metal.

"Britecast O." [99] Trademark for an attapulgite-based binder and thickener for oil-bound foundry sands. Good permeability and excellent green strength.

Properties: Tamped bulk density, 45.4 lbs/cu. ft.; sp. gr. 1.62 g./cc.; free moisture, 14%.

Containers: 50-lb multiwall bags.

British Anti-Lewisite. See 2,3-dimercaptopropanol.

British gum. See dextrin.

British thermal unit. See Btu.

"Britone Reds." [141] Trade name for resinated type lithol reds. See "Graphic Red."

brittle silver ore. See stephanite.

Brix degree. A measure of the density or concentration of sugar solutions. There are two different Brix scales: (1) Degrees Brix equals percent by weight of sucrose in the solution and is related empirically to the specific gravity. (2) Degrees Brix = 400 – [400/(sp. gr. at 15.6°C)].

brochantite $Cu_4SO_4(OH)_6$. Native basic copper sulfate. Found in the oxidized zones of copper deposits.

Properties: Color green; luster vitreous; sp. gr. 3.8; hardness 3.5-4.

Occurrence: Rumania; U.S.S.R.; Bolivia; Mexico; Chile; Arizona.

Use: Minor ore of copper.

Broenner's acid. See 2-naphthylamine-6-sulfonic acid.

bröggerite. A thorium-bearing variety of uraninite (q. v.).

"Brom 55." [225] Trademark for dibromo-dimethylhydantoin (q. v.).

bromacetone. See bromoacetone.

bromal. See tribromoacetaldehyde.

bromargyrite. See bromyrite.

"Bromat." [430] Trade name for cetyl trimethyl ammonium bromide, a quaternary ammonium compound with high germicidal activity.

bromated camphor. See camphor bromate.

bromauric acid. See bromoauric acid.

bromcamphor. See camphor bromate.

bromcresol green. Tetrabromophenol-meta-cresolsulfonphthalein, an acid base indicator, showing color change from yellow to blue over the range pH 3.8-5.4. Yellow crystals; m.p. 218°C; slightly soluble in water, soluble in alcohol.

bromcresol purple. Dibromo-ortho-cresol-sulfonphthalein, an acid base indicator, changes from yellow to purple between pH 5.2-6.8. Yellow crystals; m.p. 241°C; insoluble in water, soluble in alcohol.

bromelain. See bromelin.

bromelia. See beta-naphthyl ethyl ether.

bromelin (bromelain). A milk-clotting, protein-digesting enzyme. It is precipitated from pineapple juice with alcohol or ammonium sulfate.
Use: Biochemical research.

bromeosin. See eosin.

bromic acid $HBrO_3$.
Properties: Colorless or slightly yellow liquid; turns yellow on exposure; unstable except in very dilute solution; soluble in water.
Constants: Sp. gr. 3.1883; b.p. decomposes at 100°C.
Derivation: Sulfuric acid is added to a solution of barium bromate and the product recovered by subsequent distillation and absorption in water.
Uses: Dyes; intermediates; pharmaceuticals.
Shipping regulations: Corrosive liquid. White label.*

brominated camphor. See camphor bromate.

brominated lime. A bleaching agent prepared by treating lime with bromine.

brominated poppyseed oil. Contains 33% bromine. Used in medicine.

bromine Br. Element of atomic number 35, of group VII of the periodic system; one of the halogens.
Properties: Very dark, reddish-brown liquid; irritating fumes; burns skin. Soluble in alcohol, ether, chloroform, carbon disulfide, potassium bromide; slightly soluble in water.
Constants: B.p. 58.8°C; m.p. −7.3°C; sp. gr. (20/4°C) 3.1193; vapor density (air=1) 3.5; wt/gal (20°C) 25.7 lbs.
Derivation: From sea water and natural brines by oxidation of bromide salts with chlorine.
Method of purification: Distillation.
Grades: Technical; C.P.; 99.8%.
Containers: 1-, 6.5-lb bottles; 10-gal drums; tank cars; tank trucks.
Uses: Principally, in manufacture of ethylene dibromide (component of antiknock mixtures); also, water treatment; intermediate for fumigants (methyl bromide), fire extinguisher fluid, bromide salts and organics (used in pharmaceuticals, dyes, photography, catalysis, extraction of gold); military poison gas; shrinkproofing wool.
Fire hazard: May produce fire on contact with organic matter such as sawdust, excelsior, etc. These fumes or vapors are poisonous.
Danger: Causes severe burns; vapor hazardous. MCA warning label.
Shipping regulations: Corrosive liquid. White label.*

bromine-chloride (chlorine-bromide) BrCl.
Properties: Reddish-yellow, mobile liquid. Volatile! Vapors are irritant to the eyes. Decomposes with evolution of chlorine at 10°C. Caution! Keep well stoppered! Soluble in carbon disulfide, ether, water.
Grades: Technical.
Use: Medicine.

bromine cyanide. See cyanogen bromide.

bromine iodide. See iodine monobromide.

bromine pentafluoride BrF_5.
Properties: A colorless liquid; sp. gr. 2.466 (25°C); m.p. 8.8°C; b.p. 40.5°C; vapor pressure 70°F 7 psia; decomposes in water.
Grade: 98% min.
Containers: Cylinders.
Use: Synthesis; oxidizer in liquid rocket propellants.
Shipping regulations: Corrosive liquid. White label.*

bromine trifluoride BrF_3.
Properties: Colorless to yellow liquid; m.p. 9°C; b.p. 135°C; vapor pressure 70°F 0.15 psia; decomposed violently by water.
Grade: 98% min.
Containers: Cylinders.
Use: Fluorinating agent.
Shipping regulations: Corrosive liquid. White label.*

bromine water. A solution of bromine in water.

"Brominol." [3] Trademark for brominated olive oil.
Properties: Clear, reddish-brown, oily liquid with no taste or odor. Sp. gr. between 1.235 and 1.245 at 25°C. Insoluble in water; soluble in organic solvents.
Uses: Medicine; weighting agent for citrus oils in the production of citrus emulsions for use in soft drinks.

bromisovalum. See bromoisovaleryl urea.

N-bromoacetamide (NBA) $CH_3CONHBr$.

*See "I.C.C. Shipping Regulations," page xiii.
Reference numbers refer to name of manufacturer. See "List of Manufacturers," page v.

Properties: White powder with bromine odor. M. p. 105-108°C. Contains about 57% active bromine; decomposes appreciably at temperatures above 80°F.
Grades: Technical.
Containers: Glass bottles; polyethylene-lined fiber drums.
Warning! Poison.
Uses: Brominating and oxidizing agent in organic synthesis.
Shipping regulations: Poison, Class B. Poison label. *

bromoacetic acid $CH_2BrCOOH$.
Properties: Colorless, deliquescent crystals. Keep from air and moisture. Soluble in water, alcohol and ether.
Constants: M. p. 51°C; b. p. 208°C; sp. gr. 1.93.
Derivation: By heating acetic acid and bromine.
Method of purification: Crystallization from alcohol.
Grades: Technical.
Containers: Barrels; tins.
Use: Organic synthesis.
Shipping regulations: None. *

bromoacetone (bromacetone) $CH_2BrCOCH_3$.
Properties: Colorless liquid when pure; rapidly becomes violet even in absence of air. Powerful irritant and lachrymator. Soluble in acetone, alcohol, benzene, and ether; slightly soluble in water.
Constants: Sp. gr. 1.631 at 0°C; b.p. 136°C (partial decomposition); m. p. —54°C; vapor density 4.75; vapor tension 9 mm (20°C).
Derivation: By treating aqueous acetone with bromine and sodium chlorate at 30-40°C.
Grades: Technical.
Containers: Lead-lined containers.
Uses: Organic synthesis; tear gas.
Shipping regulations: Poison, class A. Poison gas label. * Not accepted by express.

bromoacetone cyanohydrin.
Properties: Colorless liquid. Soluble in alcohol, ether, and water.
Constants: Sp. gr. 1.584 at 13°C; b. p. 94.5°C (5 mm Hg).
Derivation: Interaction of bromoacetone and hydrocyanic acid at (approximate) 0°C.
Grades: Technical.
Use: Organic synthesis.

5-(2-bromoallyl)-5-sec-butylbarbituric acid. See butallylonal.

bromoallylene. See allyl bromide.

bromoauric acid (bromauric acid; gold tribromide, acid) $HAuBr_4 \cdot 5H_2O$.
Properties: Dark, red-brown needle crystals or granular masses; odorless; metallic and acid taste. Stable in air if pure but deliquescent if chloride is present. Soluble in water and alcohol.
Constants: M. p. 27°C.
Derivation: By dissolving auric bromide in hydrobromic acid, concentrating and crystallizing.

bromobenzene (phenyl bromide) C_6H_5Br.
Properties: Heavy, mobile, colorless liquid. Pleasant, characteristic odor.
Constants: Sp. gr. 1.499 (15°C); wt/gal 12.51 lbs; b. p. 156.6°C, 312.9°F; freezing point —30.5°C, —22.9°F; flash point 65°C; refractive index 1.5625. Miscible with most organic solvents; insoluble in water.
Derivation: By bromination of benzene in presence of iron.
Method of purification: Washing with caustic soda, followed by steam distillation.
Grades: Technical; pure.
Containers: 100-, 600-, 1200-lb drums.
Uses: Solvent; motor fuels; top-cylinder compounds; crystallizing solvent; organic synthesis.
Shipping regulations: None. *

para-**bromobenzoic acid** $C_6H_4BrCOOH$.
Properties: Colorless or reddish crystals. Soluble in alcohol and ether; very slightly soluble in water.
Constants: M. p. 254°C.
Derivation: From para-bromotoluene by oxidation.
Method of purification: Crystallization.
Grades: Technical.
Containers: Wooden kegs.
Use: Organic synthesis.
Shipping regulations: None. *

ortho-**bromobenzyl cyanide** (ortho-bromophenyl-acetonitrile; 2-bromo-7-cyanotoluene) $BrC_6H_4CH_2CN$.
Properties: Colorless liquid; sp. gr. 1.519; m. p. 25.5°C; b. p. 242°C (decomposes). Lachrymator! Poisonous!
Shipping regulations: Poison, class C. Tear gas label. *

1-bromobutane. See n-butyl bromide.

2-bromobutane. See sec-butyl bromide.

alpha-**bromobutyric acid** $CH_3CH_2CHBrCOOH$.
Properties: Clear, colorless, oily liquid. Soluble in alcohol and ether; sparingly soluble in water.
Constants: Sp. gr. 1.54; b. p. 181° at 250 mm, 214-217°C with decomposition; m. p. —4°C.
Derivation: By heating bromine and butyric acid.
Method of purification: Distillation.
Grades: Technical.
Containers: Iron drums.
Use: Organic synthesis.
Shipping regulations: None. *

bromocamphor. See camphor bromate.

bromocarnallite. An artificial carnallite (q. v.) in which chlorine is replaced by bromine.

bromochlorodimethylhydantoin.
Properties: Free-flowing white powder; faint halogen odor; m.p. 163-164°C; soluble in benzene, methylene dichloride, chloroform. Active bromine, 33% min.; active chlorine, 14% min.

Grade: Commercial.
Containers: 200-lb fiber drums.
Uses: Germicide and fungicide in treatment of water; disinfectant; halogenating agent; catalyst of ionic type; selective oxidant.

sym-bromochloroethane (ethylene chlorobromide) CH_2BrCH_2Cl.
Properties: Colorless, volatile liquid. Chloroform-like odor; soluble in alcohol and ether; insoluble in water. Sp. gr. 1.70; b.p. 107-108°C; wt/gal 14.9 lbs (0°C); f.p. −16.6°C.
Derivation: By action of bromine and chlorine on ethylene gas.
Method of purification: Distillation.
Grades: Technical.
Containers: 50-gal iron drums, 750-lb each.
Use: Solvent for general purposes and for cellulose esters and ethers; organic synthesis.

bromochloromethane (methylene chlorobromide) $BrCH_2Cl$.
Properties: Clear, colorless, volatile liquid with chloroform-like odor; sp. gr. 1.93 (25°C); b.p. 67°C; m.p. −86.5°C; refractive index (n 25/D) 1.48; nonflammable. Soluble in organic solvents; insoluble in water.
Containers: Drums; tankcars.
Uses: In fire extinguishers; organic synthesis.
Caution! Vapors and decomposition products are hazardous, particularly in unventilated or confined spaces.
Shipping regulations: None.*

bromochlorophosgene COBrCl.
Properties: Liquid; sp. gr. 1.82 (15°C); b.p. 25°C.
Derivation: Interaction of phosgene with either aluminum or boron bromides.

1-bromo-3-chloropropane (trimethylene chlorobromide) $BrCH_2CH_2CH_2Cl$.
Properties: Colorless liquid. Freezing point below −50°C; b.p. 143-145°C (760 mm); sp. gr. 1.594 (25/25°C); lbs/gal 13.27 (25°C); refractive index 1.484 (n 25/D); flash point none. Insoluble in water. Soluble in methanol and ether.
Uses: Organic synthesis; pharmaceuticals.

2-bromo-2-chloro-1,1,1-trifluoroethane. See halothane.

2-bromo-7-cyano-toluene. See bromobenzyl cyanide.

bromocyclopentane. See cyclopentyl bromide.

bromodiethylacetylurea (carbromal) $C(C_2H_5)_2BrCONHCONH_2$.
Properties: White, crystalline powder; odorless; tasteless. Soluble in chloroform, ether, alcohol, concentrated mineral acids and alkali hydroxide solutions; almost insoluble in cold water; slightly soluble in hot water. M.p. 116-117°C.
Derivation: Interaction between bromodiethylacetyl bromide and urea.
Grade: N.F. XI (as carbromal).
Containers: Drums.
Use: Medicine.
Shipping regulations: None.*

3-bromo-5,5-dimethylhydantoin $HNCONBrCOC(CH_3)_2$.
Properties: Solid
Containers: Glass bottles, fiber drums.

1-bromododecane. See lauryl bromide.

bromoethane. See ethyl bromide.

2-bromoethylamine hydrobromide $BrCH_2CH_2NH_2 \cdot HBr$.
Uses: Intermediate; suggested as a soldering flux.

bromoethyl chlorosulfonate $BrCH_2CH_2OSO_2Cl$.
Properties: Caution! Very irritant! B.p. 100-105°C (18 mm).
Derivation: Interaction of sulfuryl chloride and glycol bromohydrin.
Grades: Technical.

para-bromofluorobenzene C_6H_4BrF.
Properties: Colorless liquid; b.p. 151-152°C; freezing point −17.4°C; sp. gr. (15°C) 1.593; refractive index (n 25/D) 1.5245; insoluble in water.
Uses: Intermediate; production of para-fluorophenol.

bromoform (tribromomethane; methenyl tribromide) $CHBr_3$.
Properties: Colorless, heavy liquid; odor and taste similar to those of chloroform. Soluble in alcohol, ether, chloroform, benzene, solvent naphtha, fixed and volatile oils; slightly soluble in water.
Constants: Sp. gr. 2.8887; m.p. 9°C; b.p. 151.2°C; wt/gal 24 lbs; boiling range 150.3-151.2°C; freezing point 9°C; surface tension 41.53 dynes/cm (20°C); dielectric constant 4.5 (20°C); refractive index 1.6005.
Derivation: By heating acetone or ethyl alcohol with bromine and alkali hydroxide, and recovery by distillation. (Similar to acetone process for chloroform.)
Method of purification: Redistillation.
Grades: Technical; pure.
Containers: 1-, 5-lb bottles; bottles in 50-lb cases; 5-gal carboys.
Uses: Intermediate in organic synthesis; geological assaying.
Shipping regulations: None.*

"Bromofume." [88] Trademark for a soil fumigant composition, consisting of ethylene dibromide in volatile solvent.
Use: In the control of wireworms and root-knot nematodes.

1-bromohexane. See n-hexyl bromide.

2-bromoisovaleryl urea (bromisovalum) $(CH_3)_2CHCHBrCONHCONH_2$.
Properties: White needles; slightly bitter in taste. Soluble in hot water, alcohol, and ether. M.p. 147-149°C.
Grades: N.F. XI (as bromoisovalum).
Use: Medicine.
Shipping regulations: None.*

bromol (2,4,6-tribromophenol) $C_6H_2Br_3OH$.
Properties: Soft, white needles; sweet taste; disagreeable penetrating bromine odor. Soluble in alcohol, chloroform, ether, and caustic alkaline solutions; almost insoluble in water.
Constants: M.p. 96°C, sublimes; sp. gr. 2.55 (20/20°C); b.p. 244°C.
Derivation: Action of bromine on phenol.

bromomethane. See methyl bromide.

bromomethyl ethyl ketone $BrCH_2COC_2H_5$.
Properties: Colorless to pale-yellowish liquid. Subject to the action of light. Caution! Very irritant to the eyes. Soluble in alcohol, benzene, ether; insoluble in water.
Constants: Sp. gr. 1.43; b.p. 145-146°C (decomposes).
Derivation: Interaction of sodium bromide and methyl ethyl ketone in the presence of sodium chlorate.
Grades: Technical.
Uses: Organic synthesis.

alpha-**bromonaphthalene** $C_{10}H_7Br$.
Properties: Clear, white liquid; sp. gr. 1.4870; m.p. 6.2°C; b.p. 279°C; refractive index 1.6601; soluble in alcohol, ether, and benzene; slightly soluble in water.
Derivation: By bromination of naphthalene.
Method of purification: Rectification.
Uses: Organic synthesis; microscopy; refractometry.

2-bromopentane $CH_3CH_2CH_2CHBrCH_3$.
Properties: A clear, colorless to yellow-colored liquid. It has a strong odor. Sp. gr. 1.1850 (25/25°C).

meta-**bromophenol** $HO(C_6H_4)Br$.
Properties: Crystals; m.p. 33°C; b.p. 235-236°C; insoluble in water; soluble in alcohol, ether, and alkalies.

ortho-**bromophenol** $HO(C_6H_4)Br$.
Properties: A yellow to red oily liquid; unpleasant odor; sp. gr. 1.5; b.p. 194°C; m.p. 6°C; insoluble in water; soluble in alcohol, ether, and chloroform.

para-**bromophenol** $HO(C_6H_4)Br$.
Properties: Crystals; sp. gr. 1.840 (15°C), 1.5875 (80°C); m.p. 64°C; b.p. 238°C, slightly soluble in water; soluble in alcohol, chloroform, ether, and glacial acetic acid. Used as a disinfectant.

bromophenol blue. Tetrabromophenolsulfonphthalein, an acid-base indicator, showing color change from yellow to purple over the range pH 3.0 to 4.6.

ortho-**bromophenylacetonitrile.** See bromobenzyl cyanide.

2-bromo-4-phenylphenol $C_6H_5C_6H_3BrOH$.
Properties: Light-colored solid with faint characteristic odor; sp. gr. (25/4°C) 1.536; m.p. 93.6-95.6°C; flash point 207°C; b.p. with decomposition (18 mm) 195-200°C. Soluble in alkalies, most organic solvents; insoluble in water.
Use: Germicide.

bromophosgene (carbonyl bromide; carbon oxybromide) $COBr_2$.
Properties: Heavy, colorless liquid. Strong odor. Hydrolyzed by water and is decomposed by light and heat. Caution! Very toxic.
Constants: Sp. gr. 2.5 (approx.) (15°C); b.p. 64-65°C.
Derivation: By the action of sulfuric acid upon carbon tetrabromide.
Grades: Technical.
Uses: Military poison gas (toxic suffocant); making crystal violet-type coloring agents.
Shipping regulations: Poison, class A. Poison gas label.*

bromopicrin (nitrobromoform; tribromonitromethane) CBr_3NO_2.
Properties: Prismatic crystals; decomposes with explosive violence if heated rapidly. Caution! Very irritant! Soluble in alcohol, benzene, and ether; slightly soluble in water. Sp.gr. 2.79 at 18°C; b.p. 127°C (118 mm Hg); m.p. 103°C.
Derivation: Action of picric acid on an aqueous solution of bromine and calcium oxide, followed by distillation under reduced pressure.
Grades: Technical.
Uses: Organic synthesis; military poison gas.

3-bromopropene. See allyl bromide.

alpha-**bromopropionic acid** (2-bromopropionic acid) $CH_3CHBrCOOH$.
Properties: Colorless liquid. Sp. gr. 1.69; m.p. 24.5°C; b.p. 203°C, with decomposition; soluble in water, alcohol, and ether.
Derivation: By heating propionic acid with bromine.
Method of purification: Distillation.

3-bromo-1-propyne. See propargyl bromide.

2-bromopyridine C_5H_4NBr.
Properties: Boiling point 194.8°C (at 760 mm); sp. gr. 1.627 (20°C); refractive index 1.5714 (n 20/D); solubility in 100 g. water 2.08 (20°C).
Use: Synthesis of pyridine compounds.

3-bromopyridine C_5H_4NBr.
Properties: B.p. 174.4°C (760 mm); sp. gr. 1.628 (20/20°C); refractive index 1.5710 (n 20/D). Difficultly soluble in water, readily soluble in common organic solvents.

bromostyrol $C_6H_5CHCHBr$.
Properties: Yellowish liquid, with a strong floral odor. Soluble in 4 volumes of 90% alcohol. Sp.gr. 1.395-1.424; refractive index 1.602-1.608; m.p., min -2°C.
Grades: Technical.
Use: Perfumery.
Shipping regulations: None.*

bromosuccinic acid $HOOCCH_2CHBrCOOH$.
Properties: Colorless crystals; sp. gr. 2.073; m.p. 159-161°C; soluble in water and alcohol; insoluble in ether.
Derivation: By heating bromine and succinic acid.

Method of purification: Recrystallization.
Grades: Technical.
Containers: Wooden casks.
Use: Organic synthesis.
Shipping regulations: None.*
Note: The above are the properties of the *dl* form. Optically active forms are also known.

N-bromosuccinimide (NBS) $(CH_2CO)_2NBr$.
Properties: Fine crystal, white to cream in color; melting range 172-178°C (decomposes). Soluble in carbon tetrachloride.
Use: For controlled low-energy bromination.
Containers: Bottles; fiber drums.
Shipping regulations: None.*
Caution: Respirators are recommended for personnel handling the dry material.

bromothymol blue. Dibromothymolsulfonphthalein, an acid-base indicator, showing color change from yellow to blue over the range pH 6.0-7.6.

alpha-**bromotoluene.** See benzyl bromide.

bromotrichloromethane (trichlorobromomethane) CCl_3Br.
Properties: A clear, colorless heavy liquid with chloroform-like odor. Miscible with many organic liquids; sp. gr. 2.0; b.p. 104°C.
Use: Organic synthesis.

bromotrifluoroethylene (BFE) $BrFC{:}CF_2$.
The name is used both for the monomer and polymers made from it. The polymers are usually clear oils at room temperature and noncracking solids at −65°F. Viscosities and densities can be varied widely. They are typical fluorocarbons, chemically inert, thermally stable, and nonflammable.
Uses (BFE polymers): Flotation fluids for gyros or accelerometers used in inertial guidance systems. BFE polymers can also be used like CFE polymers, but are more expensive.

bromotrifluoromethane $CBrF_3$.
Properties: Colorless gas; f.p. −175°C; b.p. −58°C; nonflammable.
Use: Fire extinguisher.

brompheniramine maleate (2-[para-bromo-alpha-(2-dimethylaminoethyl)benzyl] pyridine maleate) $C_{16}H_{19}BrN_2 \cdot C_4H_4O_4$.
Properties: Crystals. Rather soluble in water; less soluble in alcohol.
Grade: N.N.D.
Use: Medicine.

"Bromsulphalein." [348] Trademark for sulfobromophthalein sodium (q.v.).

"Bromural." [9] Trade name for bromoisovalerylurea (bromisovalum).

"Bromvegol." [342] Trademark for brominated vegetable oils used for weighting soft-drink emulsions.

bromyrite (bromargyrite) AgBr. Natural silver bromide, similar to cerargyrite (q.v.).

bronze. See brass and bronze.

bronze blues. A name applied loosely to any of a number of the varieties of iron-blue pigments.
See iron blues.

bronze orange. See red lake C pigments.

bronzing liquid.
1. A solution of pyroxylin in amyl acetate together with a bronze powder, usually aluminum bronze.
2. Gloss oils and aluminum bronze.
3. Spirit varnishes and aluminum bronze.
In general, any liquid used for bronzing.

brookite TiO_2. A natural crystallized titanium oxide.
Properties: Color ranges from brown to yellow, red, and iron black; rhombic crystals; Mohs' hardness 5.5-6; sp. gr. 3.8-4.1.
Occurrence: Europe; Massachusetts, New York, Arkansas.

broom. See scoparius.

brosylate ester. An ester of para-bromobenzenesulfonic acid.

brown acetate. See calcium acetate.

brown hematite. See limonite.

Brownian movement. An incessant motion of colloidal particles caused by the impact of the molecules of the liquid phase, first noted by Robert Brown, a British botanist. See also colloid.

brown iron ore. See limonite.

brown ironstone clay. See limonite.

brown rock. A type of phosphate rock (q.v.) resulting from weathering of phosphatic limestones. Found in Tennessee and used as raw material for fertilizer.

brown sienna. See sienna.

brown umber. See umber.

"Brozone." [233] A liquid formulation of methyl bromide in solvent and "Trizone"; used as soil fumigant to control weeds, nematodes and fungi.

"BRS" 700 Rubber Softener. [175] Trademark for a refined coal-tar product.
Properties: Dark-colored, viscous liquid; sp. gr. 1.17-1.22 (25/25°C); float test, 40-100 (32°C); carbon disulfide insolubles, 4-10 wt %.
Containers: 55-gal steel drums; tank trucks; tank cars.
Uses: As a softener and extender for natural rubber and synthetic elastomers conferring high tensile strength, good resilience, low heat build-up, and low set to vulcanizates; in manufacture of tires, mechanical goods, automotive items, etc.

"BRT" 3 Coal Tar Saturant. [175] Trademark for a refined coal tar.
Properties: Sp. gr. 1.15-1.20 (25/25°C); specific Engler viscosity, 13-18 (40°C); carbon disulfide insolubles, 4-10 wt %.
Containers: 55-gal drums; tank trucks; tank

cars.
Use: As a saturant for brake linings.

"BRT" 4 Rubber Reclaiming Tar. [175] Trademark for a refined coal-tar product.
Properties: Dark-colored liquid; sp. gr. 1.15-1.20 (25/25°C); specific viscosity, Engler, 15-30 (50°C); carbon disulfide insolubles, 4-10 wt %.
Containers: 55-gal steel drums; tank trucks; tank cars.
Uses: Reclaiming natural and synthetic rubbers.

"BRT" 7 Rubber Softener. [175] Trademark for a rubber-compounding material refined from crude coal-tar.
Properties: Heavy fluid; sp. gr. 1.20-1.25 (25/25°C); specific viscosity, Engler, 6-9 (100°C); carbon disulfide insolubles, 15-20 wt %.
Containers: 55-gal drums.
Use: Rubber softener for easy calendering and tubing of rubber compounds.

brucine (dimethoxystrychnine) $C_{23}H_{26}O_4N_2 \cdot H_2O$ or $2H_2O$.
Properties: White crystalline alkaloid; poisonous! Very bitter taste; loses water at 100°C; m.p. 178°C. Soluble in alcohol, chloroform, and benzene; slightly soluble in water, ether, glycerin, and ethyl acetate. Forms brucine sulfate, hydrochloride, and nitrate (m.p. 230°C).
Derivation: By extraction and subsequent crystallization from nux vomica or ignatia seeds.
Method of purification: Crystallization.
Grades: Technical.
Containers: 1-, 5-, 10-oz vials; cans.
Use: Medicine; denaturing alcohol; lubricant additive; separation of racemic mixtures.
Shipping regulations: Poison, class B. Poison label.*

brucine sulfate $(C_{23}H_{26}O_4N_2)_2H_2SO_4 \cdot 7H_2O$. Bitter white crystalline solid; used for denaturing alcohol, oils, and in medicine.
Containers: Cans.

brucite $Mg(OH)_2$. Natural magnesium hydroxide, usually found as a decomposition product of magnesium silicates, or as an alteration of dolomite. Frequently found with magnesite.
Properties: Color white, gray, greenish; luster pearly or waxy; good micaceous cleavage. Sp. gr. 2.39; hardness 2.5.
Occurrence: Nevada, Washington; Canada.
Uses: Refractories.

bruisewort. See saponaria.

brun rouge. A red pigment obtained by calcining yellow ocher. See ocher.

Brunswick blue. Applied loosely to any of a number of varieties of iron blue pigments, usually containing considerable extender such as barytes.

"BRV" Rubber Softener. [175] Trademark for a heavy high-boiling coal-tar distillate.
Properties: Dark, coal-tar oil; sp. gr. 1.14-1.18 (25/25°C); Engler specific viscosity, 5-10 (50°C); distillation, 26% max at 355°C.
Containers: 55-gal steel drums; tank trucks; tank cars.
Uses: Primary uses as plasticizer, softener, and reclaiming oil; secondary use as dispersing agent. Added directly to rubber; acts as solvent type softener; is highly effective in dispersing blacks; gives good aging properties; used on natural, natural reclaim, GR-S, nitrile, and neoprene rubbers; is especially suited for heat resistant stocks, hard rubber, and as a reclaiming oil for highly loaded natural rubber scrap.

"BryKo." [204] Trademark for a multi-purpose liquid cleaner, non-ionic, not affected by any kind of water. Easy on hands, speedy, powerful, safe on metal, wood, painted surfaces, plastic, glass, and rubber. Packaged in quarts, gallons, and thirty-gallon drums.

"Brymul." [51] Trademark for an emulsifiable grade of cleaner for general use on metals, etc. Contains Stoddard-type solvent.

bryonin. A mixture of a glucoside and an alkaloid obtained from bryonia.
Properties: Yellow, amorphous, bitter powder. Poisonous! Slightly soluble in water; soluble in alcohol; insoluble in ether.

B-stage resin (resitol). The second stage of condensation of phenol-formaldehyde resins (q.v.). Thermoplastic; swells in some liquids but does not dissolve readily. Used in molding powders. See A-stage resin.

"BTC-1100." [328] Trade name for alkyl naphthal ammonium chloride, 100% active powder.
Uses: Disinfectant; deodorant; germicide; fungicide.

Btu (British thermal unit). The quantity of heat required to raise the temperature of one pound of water one degree Fahrenheit (usually from 39 to 40°F). This is the accepted standard for the comparison of heating values of fuels. For example, fuel gases range from 100 (low producer gas) to 3200 (pure butane) Btu per cu ft. The usual standard for a city gas is about 500 Btu.

BTX. Abbreviation for benzene, toluene, xylene.

bubulum oil. See neats-foot oil.

bucco. See buchu.

buchu (bucco; bucku; buku; diosma).
Derivation: Dried leaves of Barosma betulina or other species of Barosma.
Habitat: Southern Africa.
Containers: Wooden boxes.
Grades: Technical.
Use: Medicine.
Shipping regulations: None.*

buchu-leaf oil.
Properties: Dark-colored essential oil of strong, sweetish, mint-like odor; bitter, cooling taste. Solubility in alcohol: One

*See "I.C.C. Shipping Regulations," page xiii.
Reference numbers refer to name of manufacturer. See "List of Manufacturers," page v.

volume dissolves in 3 to 5 volumes of 70% alcohol.
Chief known constituents: Diosphenol, menthone, limonene, dipentene.
Constants: Betulina oil: Sp.gr. 0.937-0.97; optical rotation −14° to −48°; refractive index 1.474-1.478.
Derivation: Distilled from the leaves of Barosma betulina. There are also oils distilled from the leaves of B. crenulata and B. serratifolia.
Shipping regulations: None.*

bucku. See buchu.

buclizine hydrochloride. 1-para-chlorobenzhydryl-4-(para-(tert)-butylbenzyl) piperazine dihydrochloride. Used in medicine.

"Budene." [265] Trademark for a cis-1,4-polybutadiene synthetic rubber.

buffer solution. A solution to which moderate amounts of either a strong acid or base may be added without causing any large change in the pH value (q.v.) of the solution. Such solutions usually contain (a) a weak acid and a salt of the weak acid, (b) a mixture of an acid salt with the normal salt, or (c) a mixture of two acid salts, for example NaH_2PO_4 and Na_2HPO_4.

bufotenin $HOC_8H_5NCH_2CH_2N(CH_3)_2$. 3-(beta-Dimethylaminoethyl)-5-hydroxyindole. A poisonous alkaloid obtained from the skin glands of toads. M.p. 138-140°C. It is used in medical research on mental disorders.

bugbane. See cimicifuga.

"Bug-Geta." [253] Brand name for insecticide bait containing metaldehyde and calcium arsenate.

bugroot. See cimicifuga.

bugwort. See cimicifuga.

buhrstone (burrstone; millstone).
Properties: A hard tough stone used for grinding cereals, cement rock, paint, etc. It is white, gray or creamy in color and is a chalcedonic silica or form of quartz of cellular texture. See quartz and chalcedony.
Occurrence: United States (Alabama, Georgia, Mississippi, New York, North Carolina, Ohio, Pennsylvania, Virginia); Canada; France.

buku. See buchu.

bulbocapnine $C_{19}H_{19}NO_4$.
Properties: White crystalline powder. Soluble in alcohol and ether; insoluble in water. M.p. 199°C.
Derivation: By extraction and subsequent crystallization from the tubers of Corydalis cava.
Grades: Technical.
Containers: Tins; glass bottles.
Use: Medicine.

bunamiodyl $C_3H_7CONHC_6HI_3CH{:}C(C_2H_5)COONa$ 3(3-Butyrylamino-2,4,6-tri-iodophenyl)-2 ethyl sodium acrylate. Used in medicine (radiopaque contrast media, diagnostic aid).

buna-S. Name of German origin used rather generally for synthetic rubbers resulting from emulsion polymerization of butadiene and styrene. See styrene-butadiene rubber.

bunker "C" fuel oil. A heavy residual oil used as fuel by ships, industry, and for large-scale heating installations.

bunker fuel. A general term for residual oils used as fuel on steamships and in industry.

"Bunnatol-G." [354] Trademark for a plasticizer for synthetic and reclaimed rubber. It is insoluble in mineral and vegetable oils, and imparts to rubber a high resistance to greases and oils.

Burgundy mixture. Same as soda Bordeaux (q.v.).

Burgundy pitch.
Properties: An opaque yellowish or reddish-brown hard brittle resin obtained either from the Norway spruce or the European silver fir. It has a sweet and aromatic taste. Soluble in glacial acetic acid, acetone, and alcohol. The resin of Pinus sylvestris is also offered as Burgundy pitch and it can be recognized by the fact that it is soluble in ether, chloroform, and solutions of salts of ammonium or sodium carbonate and borax in which true Burgundy pitch is only partially soluble. Common pitch, rosin and turpentine are agitated with water and also offered as Burgundy pitch.
Containers: 500-lb drums.
Use: In medicine (plasters).
Shipping regulations: None.*

burnable poison. See poison.

burning bush. See euonymus.

burnt lime. See calcium oxide.

burnt sugar. See caramel.

burnt umber. See umber.

burn-up. A term in nuclear technology referring to the utilization, or extent of utilization of nuclear fuel. It may be expressed either as a fraction of the fissionable material that has been used or "burned," or as the amount of energy that has been extracted from the fuel. In the latter case it is frequently quoted as megawatt days per ton.

"Buromin." [108] Trademark for sodium hexametaphosphate in glass plate form for boiler water conditioning.
Containers: 25-lb cloth bags; 100-lb paper bags; 100-lb drums.

"Burosil." [108] Trademark for a granular, alkaline, phosphate-silicate compound used in boiler water conditioning to precipitate calcium and magnesium as a loose sludge.
Containers: 125-lb drums.

burrstone. See buhrstone.

busulfan (1,4-dimethanesulfonoxybutane; tetramethylene bismethanesulfonate) $CH_3SO_2OCH_2(CH_2)_2CH_2OSO_2CH_3$.
Properties: White, crystalline powder; almost odorless; m.p. 115-118°C. Very slightly soluble in water and slightly soluble in alcohol and acetone. Very poisonous!
Grade: U.S.P. XVI.
Use: Medicine.

butabarbital sodium $C_{10}H_{15}N_2O_3Na$. Sodium 5-sec-butyl-5-ethylbarbiturate; sodium 5-ethyl-5-(1-methylpropyl) barbiturate.
Properties: White, bitter powder. Very soluble in water; soluble in alcohol; practically insoluble in benzene and dry ether; pH (1% solution) 9.0-10.2.
Grade: N.F. XI.
Use: Medicine.

butacaine sulfate (3-di-n-butylaminopropyl-para-aminobenzoate sulfate) $(C_{18}H_{30}N_2O_2)_2H_2SO_4$.
Properties: White, odorless, crystalline powder; m.p. 100-103°C; affected by light; rapidly produces numbness when placed upon the tongue; soluble in water; very soluble in warm alcohol and in acetone; slightly soluble in chloroform; insoluble in ether.
Use: Medicine.

"Butacite." [28] Trademark for polyvinyl butyral resin (q.v.) available as soft pliable sheeting. See also polyvinyl acetal resins.

butadiene (butadiene-1,3; vinylethylene; erythene; bivinyl; divinyl B) $H_2C{:}CHHC{:}CH_2$.
Properties: Colorless gas with mild aromatic odor; easily liquefied; flammable; b.p. −4.41°C; sp. gr. 0.6211 (liquid at 20°C); f.p. −108.9°C; flash point −76°C; specific volume 6.9 cu ft/lb (70°F); vapor pressure 17.65 psia (0°C). Soluble in alcohol and ether; insoluble in water. The material polymerizes readily, particularly if oxygen is present, and the commercial material contains an inhibitor to prevent spontaneous polymerization during shipment or storage.
Derivation: (a) Catalytic dehydrogenation of field or refinery butanes and butenes; (b) by-product of ethylene production.
Methods of purification: Extractive distillation in the presence of furfural and absorption in aqueous cuprous ammonium acetate.
Grades: Technical, 98.0%; C.P. 99.0%; instrument 99.4%.
Containers: Cylinders; pressure tank trucks and tank cars.
Uses: Principally in styrene-butadiene and nitrile-butadiene rubbers (SBR, NBR); starting material for adiponitrile (nylon 66); styrene-butadiene latex in paints and as binder in nonwoven fabrics; component in rocket fuels (butadiene-acrylonitrile polymer); as cis-polybutadiene, an extender or substitute for natural rubber, and as trans-polybutadiene, a unique new type of rubber.
Danger! Extremely flammable, may form explosive peroxides on exposure to air. MCA warning label.
Shipping regulations (butadiene, inhibited): Flammable gas. Red gas label.*

butadiene-1,3. See butadiene.

butadiyne. See diacetylenes.

butaldehyde. See butyraldehyde.

butallylonal (5-(2-bromoallyl)-5-sec-butylbarbituric acid) $C_{11}H_{15}BrN_2O_3$.
Properties: Fine, white crystals or crystalline powder; slightly bitter taste; m.p. 130-134°C; soluble in alcohol or ether; slightly soluble in cold water; insoluble in paraffin hydrocarbons.
Use: Medicine.

"Butamer" Process. [416] Patented process for the isomerization of normal butane to isobutane in the presence of hydrogen and a solid, noble metal catalyst of undisclosed composition. By recycling unconverted normal butane, ultimate yield of isobutane on a volumetric basis exceeds 100%.

butamin. See tutocaine.

butanal. See butyraldehyde.

butane (n-butane; butyl hydride) C_4H_{10}.
Properties: Colorless gas; characteristic natural-gas odor; extremely stable; has no corrosive action on metals; does not react with moisture; b.p. −0.5°C; f.p. −138.33°C; condensing pressure (approx) 30 lbs gauge at 90°F; sp. gr. of liquid at 0°C, 0.599; sp. gr. of vapor at 0°C (760 mm) (air=1) 2.07; critical temperature 153.2°C; critical pressure (absolute) 525 psi; explosive limits in air, % by volume, lower 1.9; upper 8.5; heating value (77°F) 3266 Btu/cu ft; specific volume (70°F) 6.4 cu ft/lb.
Grades: Research, 99.99 mole %; pure, 99 mole %; also available in various mixtures with isobutane, propane, pentanes, etc.
Containers: 16-gal returnable cylinder (approximate net content 71 lbs); 28-gal returnable cylinder (approx. net content 122 lbs); tank car (approx. net content 10,000 gal).
Uses: Organic synthesis; raw material for synthetic rubber and high-octane liquid fuels; fuel for household and for many industrial purposes, either alone or in admixture with propane or air; extractant; solvent; refrigerant; standby and enricher gas; propellant in aerosols.
Caution! Flammable gas under pressure.
Shipping regulations: Flammable gas. Red label.*

n-butane. See butane.

butanedial. See succinaldehyde.

1,4-butanedicarboxylic acid. See adipic acid.

butanedioic anhydride. See succinic anhydride.

1,3-butanediol. See 1,3-butylene glycol.

1,4-butanediol. See 1,4-butylene glycol.

2,3-butanediol. See 2,3-butylene glycol.

butanediolamine. See 2-amino-2-methyl-1,3-propanediol.

butanedione. See diacetyl.

2,3-butanedione oxime thiosemicarbazone $CH_3C(NOH)C(CH_3)N_2HCSNH_2$. A test reagent for manganese in very dilute solutions made from dimethylglyoxime and thiosemicarbazide.

butane dioxime. See dimethylglyoxime.

butanenitrile. See butyronitrile.

1-butanethiol. See butyl mercaptan.

2-butanethiol. See sec-butyl mercaptan.

1,2,4-butanetriol $HOCH_2CHOHCH_2CH_2OH$.
Properties: Almost colorless, odorless liquid; low toxicity, completely miscible in water and ethyl alcohol; hygroscopic. B.p. 312°C (extrap); sp. gr. 1.184; refractive index 1.473; flash point 332°F.
Derivation: Reaction of 2-butyne-1,4-diol with water, followed by reduction.
Grades: Technical; nitration.
Containers: Bung-type, lined steel drums.
Uses: Intermediate for alkyd resins and explosives; cellulose plasticizer; emulsifier for cosmetics, inks, finishes, paper, cork, textiles.
Shipping regulations: None.*

butanoic acid. See butyric acid.

1-butanol. See n-butyl alcohol.

2-butanol. See sec-butyl alcohol.

butanoyl chloride. See butyroyl chloride.

"Butaprene." [5] Trademark for synthetic rubbers, latices, and resins comprising copolymers of butadiene with other monomers except those copolymers of butadiene with styrene which are classified as general purpose synthetic rubbers. See "FR-S."

"Butaprene N." [277] Trademark for a series of oil resistant synthetic rubbers composed of copolymers of butadiene and acrylonitrile.
Grades: "Butaprene NH," "Butaprene NXM," "Butaprene NAA," "Butaprene NL," "Butaprene NF" in the order of decreasing acrylonitrile content, decreasing oil resistance, and increasing flexibility at low temperatures.
Containers: 75-lb boxes (milled sheets); 50-lb bales in paper bags.
Uses: Oil and fuel resistant tank linings; hose; mechanical rubber parts; modification of vinyl, styrene and phenolic resins.
Shipping regulations: None.*

"Butaprene N Latex." [278] Trademark for a series of oil resistant rubbers in latex form, consisting of copolymers of butadiene and acrylonitrile. Acrylonitrile content varies from 30 to 40%.
Grades: N-300, N-400.
Containers: 5-gal drums to tank cars.
Uses: Adhesives, paper impregnation, textile and paper coatings, leather finishes, oil resistant coatings, modification of phenolic resin emulsions, etc.
Hazards: Keep from freezing.
Shipping regulations: None.*

"Butaprene PL." [35] Trademark for a complete series of latexes comprising copolymers of butadiene with styrene, acrylonitrile, acrylate esters, etc.
Properties: Air dries; forms continuous films capable of carrying high pigment and/or filler loadings.
Containers: 55-gal drums; tank trucks; tank cars.
Uses: Interior and exterior water-based paints; metal primers; adhesives; paper coating; grease resistant coatings and saturants.
Hazards: Keep from freezing.

"Butaprene SL." [277] Trademark for a series of high styrene-butadiene copolymer resins used as reinforcing agents for rubber.
Grades: "Butaprene SL," "Butaprene SL/AB," "Butaprene SL-1" covering a range of hardness and dispersibility. All grades are supplied as white, friable, resinous crumbs.
Containers: 50-lb paper bags or suitably sized fiber drums.
Use: For imparting stiffness, strength, and abrasion resistance to rubber and synthetic rubber compounds, especially shoe soles, floor tile and similar compounds.
Shipping regulations: None.*

"Butasan." [58] Trademark for zinc dibutyldithiocarbamate (q.v.).
Uses: Accelerator for latex; activator for thiazole type accelerators.

"Butazate." [248] Trademark for zinc dibutyldithiocarbamate (q.v.).
Uses: Accelerator for latex, dispersions, cements and proofing.

butazolidine. See phenylbutazone.

2-butenal. See crotonaldehyde.

butene-1 (ethylethylene; alpha-butylene) $CH_2:CHCH_2CH_3$.
Properties: Colorless gaseous hydrocarbon; b.p. −6.3°C; sp. gr. 0.5951 (20/4°C); f.p. about −185°C. Specific volume 6.7 cu ft/lb (70°F); flash point −80°C; soluble in most organic solvents.
Derivation: Gases containing appreciable content of butene-1, along with other butene and butane hydrocarbons, are obtained by fractional distillation of refinery gas.
Grades: Technical, 95%; C.P. 99.0%.
Containers: Cylinders, tanks.
Uses: Polymer and alkylate gasoline; polybutenes; butadiene; intermediate for C_4 and C_5 oxides, aldehydes and alcohols.
Shipping regulations: Flammable gas. Red gas label.*

cis-**butene-2** (dimethylethylene; beta-butylene; also called the "high-boiling" butene-2). $CH_3CH:CHCH_3$. Cis- and trans-butene-2

*See "I.C.C. Shipping Regulations," page xiii.
Reference numbers refer to name of manufacturer. See "List of Manufacturers," page v.

are geometric structural isomers.
Properties: Colorless, gaseous hydrocarbon; b.p. 3.7°C; sp. gr. 0.6213 (20/4°C) freezing point −139°C; specific volume 6.7 cu ft/lb (70°F); flash point −73°C; soluble in most organic solvents.
Derivation: Gases containing appreciable content of cis-butene-2, along with other butene and butane hydrocarbons, are obtained by fractional distillation of refinery gas.
Grades: Technical 95%; C.P. 99%.
Containers: Cylinders; tanks.
Uses: Solvent; cross-linking agent; polymer gasoline; butadiene synthesis; synthesis of C_4 and C_5 derivatives.
Shipping regulations: Flammable gas. Red gas label.*

trans-butene-2 (dimethylethylene; beta-butylene; also called the "low-boiling" butene-2) $CH_3CH:CHCH_3$. Cis- and trans-butene-2 are geometrical structural isomers.
Physical properties: Colorless, gaseous hydrocarbon; b.p. 0.88°C; freezing point −105.8°C; sp. gr. 0.6042 (20/4°C); specific volume 6.7 cu ft/lb (70°F); flash point −73°C; soluble in organic solvents.
Derivation: Gases containing appreciable content of trans-butene-2, along with other butene and butane hydrocarbons, are obtained by fractional distillation of refinery gas.
Grades: Technical, 95%; C.P., 99.0%.
Containers: Cylinders; tanks.
Uses: Same as for cis-butene-2.
Shipping regulations: Flammable gas. Red gas label.*

2-butene-1,4-diol $HOCH_2CH:CHCH_2OH$.
Properties: Almost colorless, odorless liquid; very soluble in water, ethyl alcohol and acetone, sparingly soluble in benzene. Technical butenediol is predominantly the cis isomer. F.p. range 4.0-7.0°C; b.p. range 232-235°C; sp. gr. 1.067-1.074; refractive index (n 25/D) 1.476-1.478; flash point 263°F.
Derivation: By reduction of 2-butyne-1,4-diol; by high pressure synthesis from acetylene and formaldehyde.
Containers: Lined steel drums.
Uses: Intermediate for alkyd resins, plasticizers, nylon, pharmaceuticals; cross linking agent for synthetic resins; fungicides.
Caution: Primary skin irritant but not sensitizer.
Shipping regulations: None.*

3-butenenitrile. See allyl cyanide.

butenoic acid. See crotonic acid.

"Butesin." [3] Trademark for n-butyl para-aminobenzoate.

"Butesin" picrate. [3] Trademark for di-(butyl-para-aminobenzoate) trinitrophenol. Butamben picrate.
Properties: Yellow powder having a bitter taste; m.p. 110°C; soluble in organic solvents; slightly soluble in water.
Use: Medicine (external application).

butethal (5-butyl-5-ethylbarbituric acid) $C_{10}H_{16}N_2O_3$.
Properties: White crystals or powder; odorless; butter taste; m.p. 124-127°C; fairly soluble in alcohol or ether; practically insoluble in water.
Use: Medicine.

butethamine formate $C_{13}H_{20}N_2O_2 \cdot HCOOH$. 2-(Isobutylamino)ethyl-para-aminobenzoate formate.
Properties: Odorless, white crystals. M.p. 136-139°C; freely soluble in alcohol and water; very slightly soluble in benzene; slightly soluble in chloroform and ether. pH (1% solution) about 6.1.
Grade: N.N.D.
Use: Medicine.

butethamine hydrochloride [2-(Isobutylamino)-ethyl-para-aminobenzoate hydrochloride] $NH_2C_6H_4COOCH_2CH_2NHCH_2CH(CH_3)_2 \cdot HCl$.
Properties: White, odorless, crystals or crystalline powder with bitter taste and local anesthetizing effects on tongue. M.p. 192-196°; sparingly soluble in water; slightly soluble in alcohol and chloroform; very slightly soluble in benzene; practically insoluble in ether. pH (1% solution) about 4.7; stable in air.
Grade: N.F. XI.
Use: Medicine.

"Butoben." [123] Trademark for a chemical preparation for prevention of molding and putrefaction of pastes and certain animal and vegetable substances.

butonate $(CH_3O)_2P(O)CH(CCl_3)OOCC_3H_7$. O,O-Dimethyl 2,2,2-trichloro-1-n-butyryloxyethyl phosphonate.
Properties: Colorless, somewhat oily liquid with slight ester odor; miscible with most organic solvents; stable in neutral or acid aqueous solutions; unstable in aqueous alkali. Sp. gr. 1.3742; refractive index 1.4707; wt/gal 11.5 lbs.
Use: Insecticide.

"Buton" Resins. [29] Trademark for butadiene-styrene copolymers for surface coatings and thermosetting plastics. Characteristics of Buton coatings include chemical resistance, hardness, flexibility, adhesion, abrasion resistance, and gloss range. Cure temperatures range from 70° to 1100°F.
Grades and Uses:
Buton 100 - Basic all-hydrocarbon resin for can coatings, metal sheet primers, and chemical intermediates.
Buton 200 - Polar modification for baked metal primers, tank, and drum linings.
Buton 300 - Higher polarity gives greater compatability and reactivity for resistant lacquers and low-bake coatings.
Buton A-500 - High molecular weight resin. Excellent electricals for laminating, potting and encapsulation.

butopyronoxyl (butyl mesityl oxide) $C_{12}H_{18}O_4$. n-Butyl 3,4-dihydro-2,2-dimethyl-4-oxo-1,2H-pyran-6-carboxylate.
Properties: Yellow to pale reddish-brown liquid with aromatic odor. Reasonably

stable in air; slowly affected by light. Insoluble in water; miscible with alcohol, chloroform, ether, glacial acetic acid. Sp. gr. 1.052-1.060 (25/25°C); refractive index (n 25/D) 1.4745-1.4755. Distilling range 256-270°C.
Derivation: Condensation of mesityl oxide and dibutyl oxalate in the presence of sodium ethoxide.
Grade: Technical.
Use: Insect repellent.

"Butoxone." [147] Brand name for a selective hormone type weedkiller based on 2,4-dichlorophenoxybutyric acid (2,4-DB). Contains 2 lbs. of acid equivalent as the dimethylamine salt.
Containers: 1-gal cans; 5-, 30-, and 55-gal drums.

2-butoxyethanol. See ethylene glycol monobutyl ether.

butoxyethyl laurate.
Properties: Oily liquid, mild odor; sp. gr. (25°C) 0.884; color 100 APHA max; boiling range (4 mm) 160-220°C; flash point (open cup) 320°C; viscosity (25°C) 7 cps; f.p. −7°C.
Use: Plasticizer and solvent.

butoxyethyl oleate.
Properties: Oily liquid, mild odor; sp. gr. (25°C) 0.886; Color 250 APHA max; boiling range (4 mm) 199-233°C; viscosity (25°C) 10 cps; f.p. −37°C.
Uses: Plasticizer and solvent.

1-butoxyethoxy-2-propanol $CH_3CHOHCH_2OC_2H_4OC_4H_9$.
Properties: Sp. gr. 0.9310 (20/20°C); b.p. 230.3°C; freezing point −90°C. Solubility in water is infinite; wt/gal 7.8 lbs; flash point 250°F.
Uses: Solvent; hydraulic fluid components; plasticizer intermediate.

butoxyethyl stearate $C_{17}H_{35}COOC_2H_4OC_4H_9$.
Properties: Sp. gr. (20°C) 0.882; refractive index (25°C) 1.446; vapor pressure (20°C) <0.01 mm; b.p. (4 mm) 210-233°C; m.p. 16.5°C; insoluble in water.
Containers: 1-, 5-, 55-gal drums.
Use: Plasticizer.

para-**butoxyphenol** $HOC_6H_4OC_4H_9$.
Properties: White to faint yellow crystalline powder; m.p. 61-65°C. Soluble in alcohol, acetone, ether, benzene, aqueous alkali; insoluble in water.
Grade: 93% pure.
Containers: Fiber drums.
Use: Synthesis.

4′-butoxy-3-piperidinopropiophenone hydrochloride. See dyclonine.

butter. The fat of milk, obtained from cream of milk, churned into a smooth deep yellow semisolid substance with a sp.gr. of 0.926-0.940. It contains 82.5% fat (see butter fat) as required by the USDA with the remainder largely water containing small amounts of sugar, proteins, vitamins, mineral salts, and coloring matter.

butter color. See annatto.

buttercup yellow. See zinc yellow.

butter fat. The oily portions of the milk of mammals. Composition is largely glycerides of oleic, stearic, and palmitic acids with smaller amounts of the glycerides of butyric, caproic, caprylic and capric acids. Sp. gr. range 0.910-0.914.

butterfly weed. See asclepias.

butter of antimony. See antimony trichloride.

butter of arsenic. See arsenic trichloride.

butter of tin. See stannic chloride.

butter of zinc. See zinc chloride.

butterweed oil. See erigeron oil.

butter yellow. See dimethylaminoazobenzene.

button lac. See shellac.

"Butvar." [61] Trademark for polyvinyl butyral resins. Various types available (B-72A, B-73, B-76, B-90 and B-98) as free flowing powders covering a molecular weight range of 32,000 to 225,000 (weight average), hydroxyl content of about 10 to 20 (as % polyvinyl alcohol), butyral content of 80 to 88 (as % polyvinyl butyral) and viscosity of about 75 to 1570 cps. (10% in 95% ethanol solution at 25°C. by Ostwald Viscometer). Also available as small particle, stable, aqueous dispersions with various amounts and types of plasticizers and dispersants. B R type is an anionic dispersion of 50% solids with 40% plasticizer by weight of resin.
Uses: Coatings (metal, textile, wood, etc.) film; adhesives; sealers; molded materials; strip coatings; insulation and safety glass.

butyl. 1. The radical C_4H_9-. 2. Short or slang for butyl rubber.

n-**butyl acetate** $CH_3COOCH_2CH_2CH_2CH_3$.
Properties: Limpid colorless liquid; fruity odor. Soluble in alcohol, ether, and hydrocarbons; slightly soluble in water.
Constants: Sp. gr. 0.8826 (20/20°C); b.p. 126.3°C; vapor pressure 8.7 mm Hg (20°C); freezing point −75°C; refractive index 1.3951 (20°C); wt/gal 7.29 lbs (20°C); flash point (Tag open cup) 100°F.
Derivation: Esterification and then distillation, after contact of butyl alcohol with acetic acid in the presence of a catalyst such as sulfuric acid.
Method of purification: Distillation.
Grades: Technical; pure.
Containers: 1-gal cans; 5-, 55-gal drums; tank cars.
Uses: Solvent in production of lacquers, lacquer enamels, pyroxylin solutions, leather dope, airplane dope, perfumes, flavoring extracts; solvent for natural gums and synthetic resins.
Caution! Keep away from heat and open flame. Avoid prolonged breathing of vapor. Use with adequate ventilation. Avoid prolonged or repeated contact with skin. MCA warning label.
Shipping regulations: None.*

*See "I.C.C. Shipping Regulations," page xiii.
Reference numbers refer to name of manufacturer. See "List of Manufacturers," page v.

sec-butyl acetate $CH_3COOCH(CH_3)(C_2H_5)$. (2-butanol acetate.)
Properties: Colorless liquid; b.p. 112.2°C; sp.gr. 0.8905 at 0/4°C, 0.870 at 20/4°C; refractive index 1.389 (20°C); wt/gal 7.21 lb; flash point (closed cup) 66°F. Miscible with alcohol and ether; insoluble in water.
Derivation: Esterification of sec-butyl alcohol.
Grades: Technical; pure.
Containers: 1-gal cans; 55-gal drums; tank cars.
Uses: Solvent for nitrocellulose; lacquers; thinners; nail enamels; celluloid products; artificial leather; leather finishes; plastic wood; washable wallpaper.
Shipping regulations: Flammable liquid. Red label.*

tert-butyl acetate $CH_3COOC(CH_3)_3$.
Properties: Colorless liquid, b.p. 96°C; sp.gr. 0.896 (20°C). Insoluble in water; soluble in alcohol and ether.
Use: Suggested as an antiknock agent in gasoline.

butyl acetoacetate
$CH_3COCH_2COOCH_2CH_2CH_2CH_3$.
Properties: Colorless liquid; insoluble in water; soluble in alcohol and ether. Sp.gr. 0.9694 (20/20°C); b.p. (760 mm) 213.9°C; vapor pressure 0.19 mm (20°C); flash point 185°F; wt/gal 8.1 lbs (20°C).
Grades: Technical.
Use: Intermediate in synthesis of metal derivatives, dyestuffs, pharmaceuticals.

butyl acetoxystearate
$CH_3(CH_2)_5CH(CH_3COO)(CH_2)_{10}COOC_4H_9$.
Like butyl acetyl ricinoleate, but the double bond is saturated.
Derivation: From castor oil, butyl alcohol, and acetic anhydride, with hydrogenation.
Uses: Plasticizer; textile oils; adhesives.

butyl acetylene. See 1-hexyne.

butyl acetyl ricinoleate $CH_3(CH_2)_5CH(CH_3CO_2)CH_2(CH)_2(CH_2)_7CO_2C_4H_9$.
Properties: Yellow, oily liquid; mild odor; miscible with most organic solvents. Sp. gr. 0.940 (20/20°C); saponification number 235; f.p., indefinite, becomes cloudy at −32°C, solidifies at −65°C; flash point 230°F; refractive index 1.4614 (20°C); Saybolt viscosity 123 secs at 100°F; wt/gal 7.8 lbs (68°F); practically insoluble in water.
Derivation: From castor oil, butyl alcohol and acetic anhydride.
Grades: Technical.
Containers: 1-gal cans; 5-, 55-gal steel drums; tank cars.
Uses: Plasticizer; emulsifier; lubricant; detergent; protective coatings; special cleansing compounds; quick-breaking emulsions.

N-tert-**butylacrylamide** $H_2C{:}CHCONHC(CH_3)_3$.
Properties: White crystalline solid; m.p. 128-130°C; sp.gr. 1.015 (30°C). Soluble in methanol, ethyl alcohol, chloroform, and acetone.
Uses: Monomer; organic intermediate.

n-**butyl acrylate** $CH_2{:}CHCOOC_4H_9$.
Properties: Colorless liquid; m.p. −64°C; polymerizes readily on heating; vapor pressure (20°C) 3.2 mm; sp.gr. 0.9015 (20/20°C); wt/gal 7.5 lbs (20°C); flash point 120°F; nearly insoluble in water.
Derivation: Reaction of acrylic acid or methyl acrylate with butyl alcohol.
Grades: Technical (inhibited).
Containers: 1-gal cans; 5-, 55-gal drums; tank cars.
Uses: Intermediate in organic synthesis; polymers and copolymers for solvent coatings, adhesives, paints, binders; emulsifier.
See also acrylate resins.
Shipping regulations: None.*

n-**butyl alcohol** (1-butanol; butyric alcohol) $CH_3(CH_2)_2CH_2OH$.
Properties: Colorless liquid; vinous odor. B.p. 117.7°C; f.p. −89.0°C; sp.gr. (20/20°C) 0.8109; wt/gal (20°C) 6.76 lb; refractive index (n 20/D) 1.3993; flash point (Tag open cup) 115°F. Solubility in water (20°C) 7.7 wt %; solubility of water in n-butyl alcohol 20.1%. Miscible with alcohol and ether.
Derivation: (a) Condensation of acetaldehyde to form crotonaldyhyde, which is hydrogenated at 30 psi and 180°C; (b) bacterial fermentation of grain or molasses; (c) by-product in the high-pressure oxidation of butane and propane; (d) by the Fischer-Tropsch process.
Grades: Technical.
Containers: 1-, 5-, 55-gal drums; tank cars.
Uses: Preparation of esters, especially butyl acetate; solvent for resins and coatings; plasticizers; dyeing assistant; hydraulic fluids; detergent formulations; dehydrating agent (by azeotropic distillation); intermediate; "butylated" urea and melamine resins.
Caution! Keep away from heat and open flame. Avoid prolonged breathing of vapor. Use with adequate ventilation. Avoid prolonged or repeated contact with skin. MCA warning label.
Shipping regulations: None.*

sec-**butyl alcohol** (SBA; 2-butanol; methylethylcarbinol) $CH_3CH_2CHOHCH_3$.
Properties: Colorless liquid; strong, pleasant odor. B.p. 99.5°C; m.p. −114.7°C; sp.gr. (20/4°C) 0.808; wt/gal (20°C) 6.74 lbs; refractive index (n 25/D) 1.3949; flash point (closed cup) 75°F. Moderately soluble in water; miscible with alcohol and ether.
Derivation: Absorption of butene, from cracking petroleum or natural gas, in sulfuric acid with subsequent hydrolysis by steam.
Grades: Technical.
Containers: 1-, 5-, 55-gal drums; tankcars.
Uses: Preparation of methyl ethyl ketone; solvent in varnishes, lacquers, and paint removers; organic synthesis.
Caution! Keep away from heat and flame. Avoid prolonged breathing of vapor. Use

*See "I.C.C. Shipping Regulations," page xiii.
Reference numbers refer to name of manufacturer. See "List of Manufacturers," page v.

with adequate ventilation. Avoid prolonged or repeated contact with skin. MCA warning label.
Shipping regulations: Flammable liquid. Red label. *

tert-butyl alcohol (2-methyl-2-propanol; trimethyl carbinol). $(CH_3)_2COHCH_3$.
Properties: Low-melting colorless crystals; camphor odor. M.p. 25.5°C; b.p. 82.9°C; sp.gr. (liquid, 26°C) 0.779; refractive index (n 20/D) 1.3878; flash point (closed cup) 52°F. Miscible with water, alcohol, and ether.
Derivation: Absorption of isobutene, from cracking petroleum or natural gas, in sulfuric acid with subsequent hydrolysis by steam.
Grades: Technical.
Containers: Bottles; barrels; 55-gal drums; tanks.
Uses: Solvent; alcohol denaturant; organic synthesis.
Warning: Flammable; use with adequate ventilation. MCA warning label.
Shipping regulations: Flammable liquid. Red label. *

butyl alcohol, iso-. See isobutyl alcohol.

n-butyl aldehyde. See butyraldehyde.

butyl aldehyde, iso-. See isobutyraldehyde.

n-butylamine $C_4H_9NH_2$.
Properties: Colorless, volatile liquid with amine odor. B.p. 77.1°C; f.p. −49.1°C; sp. gr. (20/20°C) 0.7385; wt/gal (20°C) 6.2 lbs; refractive index (n 20/D) 1.401; flash point (open cup) 10°F. Miscible with water, alcohol, and ether.
Derivation: By reaction of butanol or butyl chloride with ammonia.
Grades: Technical.
Containers: 1-gal cans; 5- and 55-gal drums; tank cars.
Uses: Intermediate for emulsifying agents, pharmaceuticals, insecticides, rubber chemicals, dyes, tanning agents.
Warning: Flammable; may cause skin irritation. MCA warning label.
Shipping regulations: Flammable liquid. Red label. *

sec-butylamine $CH_3CHNH_2C_2H_5$.
Properties: Colorless liquid; sp. gr. (20°C) 0.725; boiling range 63-68°C; refractive index (20°C) 1.395; solidification point −104.5°C; odor, amine; wt/gal (20°C) 6.0 lbs.
Fire hazard: Flash point less than 20°F.
Shipping regulations: Flammable liquid. Red label. *

tert-butylamine $(CH_3)_3CNH_2$.
Properties: Colorless liquid; b.p. 44-46°C; sp. gr. 0.700 (15°C); refractive index (n 18/D) 1.3794; flash point below room temperature. Miscible with water; soluble in common organic solvents.
Grades: Technical.
Containers: Drums; tank cars.
Uses: Intermediate for rubber accelerators, insecticides, fungicides, dyestuffs, pharmaceuticals.
Caution! Avoid prolonged contact with skin; prolonged inhalation.
Shipping regulations: Flammable liquid. Red label. *

n-butyl-para-aminobenzoate $H_2NC_6H_4COOC_4H_9$.
Properties: White, crystalline powder, odorless, tasteless; m.p. 57-59°; b.p. 174°C (8 mm). Soluble in dilute acids, alcohol, chloroform, ether, and fatty oils. Almost insoluble in water.
Grade: N. F. XI.
Use: Medicine (local anesthetic).

butylaminoethanol $C_4H_9NHC_2H_4OH$.
Properties: Liquid; sp. gr. 0.88-0.99 (20/20°C); distillation range 192-210°C (760 mm); wt/gal 7.4 lbs; flash point 170°F.

N-n-butyl aniline $C_6H_5NHC_4H_9$.
Properties: Sp. gr. (20°C) 0.932; boiling range 236-242°C; refractive index 1.534 (20°C); color amber; odor aniline; very soluble in alcohol and ether; insoluble in water.
Fire hazard: Flash point 225°F.
Shipping regulations: None. *

2-tert-butylanthraquinone $C_{18}H_{16}O_2$.
Properties: Yellow powder. M.p. 102-104°C; soluble in alcohol and acetone.
Grades: Technical (98%).
Use: Organic synthesis; manufacture of hydrogen peroxide.

butylated hydroxyanisole (BHA) $(CH_3)_3CC_6H_3OH(OCH_3)$.
Grade: Food inhibitor grade.
Containers: 1-pt and 1-gal glass bottles; net weights 1 and 8.5 lbs; drums.
Uses: Antioxidant for fats and oils; food packaging.

butylated hydroxytoluene. See di-tert-butyl-para-cresol.

n-butylbenzene $C_6H_5C_4H_9$.
Properties: Colorless liquid; b.p. 183.2°C; f.p. −87.9°C; sp. gr. 0.860 (20°C); refractive index (n 20/D) 1.489; flash point 71°C.
Grades: Technical; pure; research.
Uses: Organic synthesis.

sec-butylbenzene (2-phenylbutane) $C_6H_5C(CH_3)(C_2H_5)$.
Properties: Colorless liquid; b.p. 170.65°C; vapor pressure 15 mm Hg (60°C); f.p. −75.68°C; sp. gr. 0.8618 (20/4°C); wt/gal 7.2 lbs (20°C); refractive index (n 20/D) 1.4901; flash point (open cup) 120°F.
Typical specification: Boiling range 160-185°C; sp. gr. 0.865-0.870 (60°F); no copper corrosion.
Containers: 1- and 5-gal cans; 55-gal black iron drums; tank trucks and tank cars.
Use: As a medium-high boiling solvent for coating compositions and organic synthesis.

tert-butylbenzene (2-methyl-2-phenylpropane) $C_6H_5C(CH_3)_3$.
Properties: Colorless liquid; insoluble in water; soluble in alcohol; b.p. 169.1°C; f.p. −57.8°C; sp. gr. 0.866 (20°C); re-

*See "I.C.C. Shipping Regulations," page xiii.
Reference numbers refer to name of manufacturer. See "List of Manufacturers," page v.

fractive index (n 20/D) 1.492; flash point 60°C.
Grades: Technical; pure; research.
Use: Organic synthesis.

butyl benzenesulfonamide (N-n-butyl benzenesulfonamide) $C_6H_5SO_2NHC_4H_9$.
Properties: Liquid; pleasant odor; amber to straw color; sp. gr. 1.148 (25/25°C); refractive index 1.5235 (25°C); b.p. 189-190°C (4.5 mm).
Uses: Synthesis of dyes; pharmaceuticals, and other organic chemicals; in resin manufacturing; and as plasticizer for some synthetic resins.

butyl benzoate (n-butyl benzoate) $C_6H_5COOC_4H_9$.
Properties: Colorless oily liquid; insoluble in water; miscible with alcohol or ether. Sp. gr. 1.00 (20°C); b.p. 247.3°C; m.p. −22°C.
Grades: Technical.
Uses: Solvent for cellulose ether; plasticizer; perfume ingredient.

butyl benzyl phthalate (BBP) $C_6H_4(COO)_2C_4H_9C_7H_7$.
Properties: Clear, oily liquid. Sp. gr. 1.113-1.121 (25/25°C).
Grades: Technical.
Containers: 5- and 55-gal drums; tankcars.
Uses: Plasticizer for polyvinyl and cellulosic resins.

butyl benzyl sebacate $C_4H_9OOC(CH_2)_8COOC_7H_7$. Ester used as plasticizer.
Typical specifications: Light straw color; b.p. (10 mm) 245-285°C; sp. gr. 1.023 (25°C); wt/gal 8.6 lbs. As a plasticizer it combines the desirable properties of dibenzyl sebacate and dibutyl sebacate (q.v.).

butyl borate. See tributyl borate.

n-butyl bromide (1-bromobutane) C_4H_9Br.
Properties: Colorless liquid; sp. gr. (20/20°C) 1.279; b.p. 101.6°C; m.p. −112.4°C. Insoluble in water; soluble in alcohol and ether.
Grade: 99.7%.
Use: Alkylating agent.

sec-butyl bromide (2-bromobutane) $CH_3CHBrCH_2CH_3$.
Properties: Clear, colorless liquid with pleasant odor; boiling point 91.2°C; sp. gr. 1.2425 (25/25°C); refractive index at 25°C 1.4320-1.4344; soluble in alcohol and ether.
Containers: Steel drums.
Use: Synthesis.

butyl butanoate. See n-butyl butyrate.

n-butyl butyrate (butyl butanoate) $CH_3(CH_2)_2COOC_4H_9$.
Properties: Colorless liquid; sp. gr. 0.8721 (20/20°C); refractive index (n 20/D) 1.4059; m.p. −91.5°C; b.p. 165.7°C (736 mm); slightly soluble in water; soluble in alcohol and ether.

butylcarbamoylsulfanilamide. See 1-butyl-3-sulfanilylurea.

n-butyl carbinol. See n-amyl alcohol, primary.

sec-butyl carbinol. See 2-methyl-1-butanol.

butyl "Carbitol." [214] $C_4H_9OCH_2CH_2OCH_2CH_2OH$. Trademark for diethylene glycol monobutyl ether (q.v.).

butyl "Carbitol" acetate. [214] Trademark for diethylene glycol monobutyl ether acetate, $CH_3CO(OC_2H_4)_2OC_4H_9$ (q.v.).

para-tert-butylcatechol (4-tert-butyl-1,2-dihydroxybenzene) $(CH_3)_3CC_6H_3(OH)_2$.
Properties: Colorless crystals; m.p. 56-57°C; sp. gr. 1.049 (60/25°C); b.p. 285°C; soluble in ether, alcohol, and acetone; slightly soluble in water at 80°C.
Containers: Glass bottles; lined drums.
Use: Polymerization inhibitor.
Warning: Causes skin irritation. Causes burns when moist. MCA warning label.
Shipping regulations: None.*

butyl "Cellosolve." [214] Trademark for ethylene glycol monobutyl ether, $CH_2OHCH_2OC_4H_9$ (q.v.).

butyl "Cellosolve" acetate. [214] Trademark for ethylene glycol monobutyl ether acetate (q.v.).

butyl chloral (2,2,3-trichlorobutanal; trichlorobutyraldehyde; crotonchloral) $CH_3CHClCCl_2CHO$.
Properties: Colorless, oily liquid with pungent odor. B.p. 164.5-165.5°C; sp. gr. 1.3956 (20/4°C); refractive index (n 20/D) 1.47554. Forms a crystalline hydrate. Soluble in water; miscible with alcohol and ether. Polymerizes.
Derivation: By the action of chlorine on acetaldehyde or paraldehyde; or from crotonaldehyde, hydrogen chloride, and chlorine.
Use: Medicine.

butyl chloral hydrate (trichlorobutyraldehyde hydrate) $CH_3CHClCCl_2CH(OH)_2$.
Properties: Colorless leaflets; sp. gr. 1.693 (20/4°C); m.p. 78°C. Slightly soluble in water; soluble in alcohol and ether.
Derivation: Action of chlorine on paraldehyde.
Use: Medicine.

n-butyl chloride (1-chlorobutane) $CH_3CH_2CH_2CH_2Cl$ or C_4H_9Cl.
Properties: Colorless liquid. Insoluble in water. Miscible with alcohol and ether. Water is insoluble in butyl chloride.
Constants: Sp. gr. 0.8875 (20/20°C); b.p. 78.6°C; wt/gal 7.35 lbs (20°C); refractive index 2.4015 (20°C); vapor pressure 80.1 mm (20°C); f.p. −122.8°C; viscosity 0.0045 poise (20°C); flash point 15°F (open cup). Extremely flammable!
Typical specifications: (a) Distillation range 71-86°C with 95% distilling between 76 and 79.5°C; purity 99.11%; (b) color water-white; sp. gr. (20°C) 0.88-0.89; water content none; acidity as hydrochloric acid not over 0.01%; not more than 25% boils below 77.5°C; wt/gal 7.35 lbs.
Grades: N.F. XI; technical.

*See "I.C.C. Shipping Regulations," page xiii.
Reference numbers refer to name of manufacturer. See "List of Manufacturers," page v.

Containers: 1- and 5-gal cans; 55-gal drums.
Uses: Organic synthesis (alkylating agent, butyl cellulose); solvent; anthelmintic.
Danger! Extremely flammable; avoid prolonged breathing of vapor. MCA warning label.
Shipping regulations: Flammable liquid. Red label.*

butyl citrate. See tributyl citrate.

tert-butyl-meta-cresol (MBMC) $CH_3C_6H_4OC(CH_3)_3$.
Properties: Clear liquid that solidifies slightly below room temperature; f.p. 23.1°C; b.p. 244°C; sp. gr. 0.922 (80°C). Soluble in organic solvents and aqueous potassium hydroxide.
Containers: 5- and 55-gal drums; tank cars.
Uses: Germicide; disinfectant; synthesis of antioxidants and rubber-processing chemicals; additives to lubricating oils; in synthetic resins; in perfumes as a fixative.

butyl crotonate $CH_3CH{:}CHCOOC_4H_9$.
Properties: Water-white liquid; pleasant, persistent odor.
Constants: Sp. gr. 0.9037 (20/20°C); b.p. 180.5°C; wt/gal 7.52 lbs (20°C); soluble in alcohol and ether; insoluble in water.

butyl cyclohexyl phthalate.
Properties: Clear liquid; very mild, characteristic odor; color (Hazen) 80 max; sp. gr. 1.078; saponification number 369; acidity (as phthalic acid), 0.01% max.; miscible with most organic solvents.
Containers: Drums; tank trucks; tankcars.
Use: Plasticizer for polymers and elastomers.

n-butyldiamylamine $C_4H_9N(C_5H_{11})_2$.
Properties: Sp. gr. (20°C) 0.788; boiling range 229-241°C; color light straw; odor amine; flash point 200°F.
Shipping regulations: None.*

n-butyldichloroarsine $C_4H_9AsCl_2$.
Properties: Oily liquid. Somewhat agreeable odor. Decomposed by water. Caution! Very irritant! B.p. 192-194°C.
Derivation: Interaction of hydrochloric acid and n-butylarsenic acid in the presence of sulfur dioxide.

1-n-butyl-3-(3,4-dichlorophenyl)-1-methylurea (neburon) $Cl_2C_6H_3NHCONCH_3(C_4H_9)$.
Properties: White crystalline solid; m.p. 102°C; very low solubility in water and hydrocarbon solvents. Stable towards oxidation and moisture.
Use: Weed killer.

n-butyl diethanolamine $C_4H_9N(CH_2CH_2OH)_2$.
Properties: Liquid; sp. gr. (20°C) 0.97; b.p. 272°C; color very light straw; odor faint amine; wt/gal 8.08 lbs (20°C); flash point 245°F.
Shipping regulations: None.*

butyl diglycol carbonate (diethylene glycol bis(n-butylcarbonate) $C_{14}H_{26}O_7$.
Properties: Colorless liquid of low volatility; insoluble in water (very stable to hydrolysis by water); widely soluble in organic solvents; compatible with many resins and plastics.
Typical specifications: Sp. gr. 1.07 (20/4°C); boiling range 164-166°C (2 mm); flash point 475°C; Saybolt viscosity 21 cps (20°C); refractive index 1.435 (20°C); evaporation rate 0.59 mg/sq cm/hr (100°C).
Uses: Plasticizer; high-boiling-point solvent and softening agent; manufacture of pharmaceuticals and lubricant compositions.

4-tert-butyl-1,2-dihydroxybenzene. See para-tert-butylcatechol.

5-tert-butyl-4,6-dinitrohemimellitene. See "Musk Tibetene."

2-sec-butyl-4,6-dinitrophenol. See dinitro-ortho-sec-butylphenol.

4-butyl-1,2-diphenyl-3,5-pyrazolidinedione. See phenylbutazone.

"Butyl Eight." [69] Trade name for an ultra-accelerator of the dithiocarbamate type.
Properties: Dark red liquid; odor distinct; sp. gr. 1.01; non-toxic. Partly soluble in water; soluble in acetone, benzene, carbon disulfide, chloroform, and gasoline.
Use: Ultra-accelerator in self-curing naphtha cements for proofing and for self-curing calendered and tubed goods.

butylene. See butene-1; cis-butene-2; trans-butene-2; isobutene.

1,3-butylene glycol (1,3-butanediol) $HOCH_2CH_2CH(OH)CH_3$.
Properties: Practically colorless, odorless liquid; hygroscopic; sp. gr. 1.0059 (20/20°C); 8.4 lbs/gal (20°C); b.p. 207.5°C (760 mm); vapor pressure 0.06 mm (20°C); refractive index 1.4401 (20°C); flash point (Cleveland open cup) 250°F; completely soluble in water and alcohol; slightly soluble in ether.
Derivation: Reduction of aldol.
Containers: Drums; tankcars.
Uses: Polyesters; polyurethanes; surface active agents; plasticizers; humectant; coupling agent.

1,4-butylene glycol (1,4-butanediol; tetramethylene glycol) $HOCH_2CH_2CH_2CH_2OH$.
Properties: Colorless, oily liquid. B.p. 230°C; m.p. 16°C; sp. gr. 1.020 (20/4°C). Miscible with water; soluble in alcohol; slightly soluble in ether.
Derivation: From acetylene and formaldehyde by the Reppe process (high pressure synthesis).
Grades: Technical.
Containers: Drums; tankcars.
Uses: Solvents; humectant; intermediate for plasticizers, pharmaceuticals, polyester and polyurethane resins.
Shipping regulations: None.*

2,3-butylene glycol (2,3-dihydroxybutane; 2,3-butanediol; pseudobutylene glycol; sym-dimethylethylene glycol) $CH_3CHOHCHOHCH_3$.
Properties: Nearly colorless crystalline

solid; hygroscopic; sp. gr. 1.045 (20/20°C); m.p. 23-27°C; b.p. 179-182°C; refractive index 1.438 (20°C); soluble in alcohol and ether; miscible with water in all proportions.
Derivation: From corn sugar by acid hydrolysis; also from fermentation of sugar beet molasses.
Grades: 99%.
Uses: Resins; solvent for dyes; intermediate; blending agent.
Shipping regulations: None.*

butylene glycol, pseudo-. See 2,3-butylene glycol.

butylene oxides C_4H_8O. Mixtures of 1,2- and 2,3-butylene oxides.
Typical specification: Nearly colorless liquid; boiling range (5-95%) 60-67°C; sp. gr. 0.826 (25/25°C); refractive index 1.381; flash point 5°F.
Grades: Technical.
Uses: Chemical intermediate, especially for various polymers.
Shipping regulations: Flammable liquid. Red label.*

1,2-butylene oxide $H_2\underline{COC}HCH_2CH_3$.
Properties: Colorless liquid; sp. gr. 0.8312 (20/20°C); b.p. 63°C; sets to a glass below −150°C; flash point (closed cup) −15°F. Soluble in water and miscible with most organic solvents.
Grades: About 97.5% purity.
Uses: Intermediate, especially for various polymers.
Shipping regulations: Flammable liquid. Red label.*

butyl epoxy stearate.
Properties: Clear, colorless liquid with mild, slightly fatty, slightly fruity odor; sp. gr. (20°C) 0.910; wt/gal 7.59 lbs.
Containers: 1-, 5-gal cans; 55-gal drums; tankcars.
Use: Plasticizer for low temperature flexibility improvement of vinyls.

n-butylethanolamine $C_4H_9NHCH_2CH_2OH$.
Properties: Sp. gr. (20°C) 0.892; boiling range 194-204°C; color water-white; odor, very faint amine type; flash point 170°F.
Shipping regulations: None.*

butyl ether (n-dibutyl ether) $C_4H_9OC_4H_9$.
Properties: Colorless liquid; stable; mild, ethereal odor; sp. gr. 0.7694 (20/20°C); b.p. 142.2°C (760 mm); vapor pressure 4.8 mm (20°C); flash point 100°F; f.p. −95.2°C; latent heat of vaporization 67.8 cal/g at 140.9°C; refractive index 1.3992 (20°C); wt/gal 6.4 lbs (20°C); viscosity 0.0069 poise (20°C).
Typical specification: Sp. gr. 0.7680-0.7710 (20/20°C); boiling range 137-143°C (760 mm); acidity not more than 0.02% (as butyric); miscible with most common organic solvents; immiscible in water.
Grades: Technical.
Containers: 1-gal cans; 5- and 55-gal drums; tankcars.
Uses: Solvent for hydrocarbons, fatty materials; extracting agent; solvent purification; organic synthesis (reaction medium).
Caution: Tends to form explosive peroxides, especially when anhydrous. Avoid prolonged breathing of vapor or contact with skin. MCA warning label.

butylethyl acetaldehyde. See 2-ethylhexaldehyde.

5-butyl-5-ethylbarbituric acid. See butethal.

n-butyl ethyl ether. See ethyl n-butyl ether.

2-butyl-2-ethylpropanediol-1,3. See 2-ethyl-2-butylpropanediol-1,3.

butyl formate $HCOOC_4H_9$.
Properties: Colorless liquid. Sp. gr. 0.885-0.9108; b.p. 107°C; m.p. −90°C; miscible with alcohols, ethers, oils, hydrocarbons, and water.
Grades: Technical.
Uses: Solvent for nitrocellulose, some types of cellulose acetate, many cellulose ethers, many natural and synthetic resins; lacquers; perfumes; organic synthesis (intermediate).

n-butyl furfuryl ether $C_4H_9OCH_2C_4H_3O$.
Properties: Colorless liquid, turning dark on exposure to air; extremely hygroscopic, unstable in presence of moisture. Sp. gr. 0.955 (20/0°C); b.p. 189-190°C (765 mm); refractive index (n 20/D) 1.4522.
Derivation: Action of alpha-furfuryl bromide on butyl alcohol.
Shipping regulations: None.*

n-butyl furoate $C_4H_3OCO_2C_4H_9$.
Properties: Colorless oil; decomposes on standing. Sp. gr. 1.055 (20/4°C); b.p. 83-84°C (1 mm), 118-120°C at 25 mm. Insoluble in water; soluble in alcohol and ether.

butyl glycol phthalate.
Properties: Liquid; b.p. 370°C; soluble in most organic solvents.

butyl hydride. See butane.

tert-butyl hydroperoxide $(CH_3)_3COOH$. A highly reactive peroxy compound.
Properties: Liquid; m.p. −8°C; decomposes at 75°C; sp. gr. 0.896 (20/4°C). Soluble in organic solvents.
Use: Polymerization catalyst.

"tert-Butyl Hydroperoxide-70." [154] $(CH_3)_3COOH$.
Properties (typical): Sp. gr. 0.875 (min.) at 25°C; refractive index 1.394 (min.) at 25°C; f.p. −35°C. Slightly soluble in water; very soluble in esters, alcohols, ketones, aliphatic, aromatic and chlorinated hydrocarbons.
Containers: 1-, 7-, 35-lb polyethylene bottles.
Uses: Polymerization catalyst for vinyl type monomers.
Shipping regulations: Red label.*

tert-butylhydroquinone $C_6H_3(OH)_2C(CH_3)_3$. Chemical intermediate; m.p. 125°C; insoluble in water; soluble in alcohol, acetone and ethyl acetate.

*See "I.C.C. Shipping Regulations," page xiii.
Reference numbers refer to name of manufacturer. See "List of Manufacturers," page v.

butyl isodecyl phthalate.
Properties: Clear liquid; mild, characteristic odor; color (Hazen) 50 max; sp. gr. 0.997 (20/20°C); saponification number 310; acidity (as phthalic acid) 0.01% max.
Containers: Drums; tankcars; tank trucks.
Use: Plasticizer.

tert-butylisopropyl benzene hydroperoxide.
Shipping regulations: Oxidizing material. Yellow label.*

butyl lactate $CH_3CHOHCOOC_4H_9$.
Properties: Water-white, stable, non-toxic liquid. Mild odor. Miscible with many lacquer solvents, diluents, and oils; slightly soluble in water; hydrolyzed in presence of acids and alkalies.
Constants: Sp. gr. 0.974-0.984 (20/20°C); flash point 168°F (Tag open cup); m.p. —43°C; wt per U.S. gal 8.15 lbs (68°F); b.p. 188°C; refractive index 1.4216 (20°C); vapor pressure 0.4 mm Hg (20°C); latent heat of vaporization 77.4 cal/g (20°C).
Typical specifications: Purity not less than 95% ester, by wt; acidity not more than 0.15% calculated as lactic acid; water no turbidity when one volume is mixed with 19 volumes of 60° Bé. gasoline at 20°C; nonvolatile matter not more than 0.01 g when 100 cc are evaporated and heated to constant weight at 120°C; distillation range below 140°C none, between 155 and 195°C not less than 90%, between 187 and 189°C not less than 60%, above 200°C none.
Grades: Technical.
Containers: 1-gal cans; 5- and 55-gal steel drums; tank trucks; tankcars.
Uses: Solvent for nitrocellulose, ethyl cellulose, oils, dyes, natural gums and many synthetic resins; lacquers; varnishes; inks; stencil pastes; antiskinning agent; chemical (intermediate); perfumes; dry-cleaning fluids; adhesives.
Shipping regulations: None.*

N-n-butyl lauramide $C_{11}H_{23}CONHC_4H_9$.
Properties: Boiling range 200-225°C at 2 mm; white solid; odor lauric acid; flash point 375°F.
Shipping regulations: None.*

butyl laurate $C_{11}H_{23}COOC_4H_9$.
Properties: Liquid with sp. gr. (25°C) 0.855; b.p. (5 mm) 130-180°C; m.p. < -10°C; insoluble in water.
Derivation: Alcoholysis and fractionation of coconut oil.
Containers: Drums.
Use: Plasticizer.

n-butyllithium C_4H_9Li.
Properties: Liquid at room temperature; sp. gr. 0.68-0.70. Also described as a powder. B.p. 80-90°C (10^{-4} mm); soluble in most organic solvents. Usually sold as a solution in the C_5 to C_7 hydrocarbons, in which it is quite stable.
Derivation: Reaction of finely dispersed lithium metal with butyl chloride.
Grades: Sold according to percentage butyllithium in the solution.
Containers: Bottles; cylinders (1-100 lbs active ingredient); tankcars; tank wagons; multi-unit ton containers.
Uses: As a catalyst in the stereospecific polymerization of isoprene and butadiene; metalating agent for sterically hindered positions; alkylating agent for other metal organics and pharmaceuticals.

n-butylmagnesium chloride C_4H_9MgCl.
Properties: Flammable liquid; sp. gr. 0.88.
Derivation: From magnesium and butyl chloride.
Grade: Available in solution in ethyl ether.
Containers: Glass bottles; 5-gal drums.
Warning! Flammable.
Use: Grignard reagent, as an alkylating agent.
Shipping regulations: Flammable liquid. Red label.*

n-butyl mercaptan (1-butanethiol) C_4H_9SH.
Properties: A colorless liquid; sp. gr. 0.8412 (20/4°C); refractive index 1.4427 (20/D); flash point 0°C; b.p. 92°C; occurs in the odorous secretion of the skunk. Slightly soluble in water; very soluble in alcohol and ether.
Grades: 95%.
Uses: Intermediate; solvent.
Shipping regulations: Flammable liquid. Red label.*

sec-butyl mercaptan (2-butanethiol) $C_2H_5CH(SH)CH_3$.
Properties: Boiling range 73-89°C; sp. gr. (20/4°C) 0.8288; refractive index (20/D) 1.4363; flash point —15°C.
Grades: 95%.
Containers: Cars and drums.
Shipping regulations: Flammable liquid. Red label.*

tert-butyl mercaptan $(CH_3)_3CSH$.
Properties: Colorless liquid with strong skunk odor; sp. gr. 0.79-0.82 (60/60°F); distillation range 62-67°C; refractive index (20/D) 1.422; flash point —15°F; wt/gal 6.71 lbs.
Containers: 1-qt cans; 1-, 5-, 54-gal drums; tank cars.
Use: Intermediate; gas odorant for detecting leaks.
Shipping regulations: Flammable liquid. Red label.*

butyl mesityl oxide. See butopyronoxyl.

n-butyl methacrylate $H_2C:CCH_3COOC_4H_9$.
Properties: Colorless liquid; b.p. 163-164°C; f.p. below —75°C; sp. gr. 0.895 (25/25°C); flash point (open cup) 130°F; readily polymerized; insoluble in water.
Derivation: Reaction of methacrylic acid or methyl methacrylate with butyl alcohol.
Grades: Technical (inhibited).
Containers: Drums.
Uses: Potting compound; cement for optical glass; polymerizable monomer for resins, solvent coatings, adhesives, oil additives; emulsions for textile, leather and paper finishing. See also acrylate resins.

para-tert-butyl-alpha-methylhydrocinnamaldehyde. See "Lilial."

*See "I.C.C. Shipping Regulations," page xiii.
Reference numbers refer to name of manufacturer. See "List of Manufacturers," page v.

n-butyl myristate $CH_3(CH_2)_{12}COOC_4H_9$. The butyl ester of myristic acid. An oily liquid at room temperature.
Typical specification: Color, water-white; saponification number 193-203; f.p. 1-7°C; boiling range 167-197°C at 5 mm; sp. gr. (25°C) 0.850-0.858; insoluble in water; soluble in acetone, castor oil, chloroform, methanol, mineral oil, and toluene.
Derivation: Alcoholysis of coconut oil with butyl alcohol followed by fractional distillation.
Containers: 1-, 7-, and 35-lb tins; 400-lb drums.
Uses: Plasticizer; lubricant for textiles, paper stencils; cosmetic preparations.

"Butyl Namate." [69] Trademark for an aqueous solution of sodium dibutyldithiocarbamate; 47% minimum assay.
Properties: Pale amber liquid; sp. gr. 1.09 ± .02 at 25°C.
Uses: Ultra accelerator for natural and synthetic latices.

n-butyl nitrate $C_4H_9NO_3$.
Properties: Sp. gr. (20°C) 1.03; b.p. 123°C; color water-white; odor ethereal. Insoluble in water; soluble in alcohol and ether; flash point 97°F.

butyl octadecanoate. See butyl stearate.

butyl octyl phthalate.
Properties: Clear water-white liquid; mild characteristic odor; sp. gr. 0.991-0.997 (20/20°C); saponification number 298-308. Miscible with most organic solvents.
Containers: Drums; tank trucks; tankcars.
Uses: Plasticizer for vinyl resins.

butyl oleate $CH_3(CH_2)_7CH{:}CH(CH_2)_7COOC_4H_9$.
Properties: Light-colored, oleaginous liquid; mild odor; insoluble in water; miscible with alcohol, ether, vegetable and mineral oils.
Constants: Sp. gr. 0.873 (20/20°C); flash point 356°F; iodine value 76.8; f.p. opaque at 12°C, solid at −26.4°C; wt/gal 7.26 lbs (20°C); boiling range 173-227°C (2 mm).
Derivation: Alcoholysis of olein or esterification of oleic acid with butyl alcohol.
Containers: 7-, 35-lb tins; 380-lb drums.
Uses: Plasticizer; solvent; lubricant; water-resisting agent; coating compositions; polishes; water-proofing compounds.
Shipping regulations: None. *

tert-butyl perbenzoate $C_6H_5CO \cdot O_2 \cdot C(CH_3)_3$.
Properties: Liquid; sp. gr. 1.04 (25°C); f.p. 8°C; vapor pressure 0.33 mm (50°C); very soluble in alcohols, esters, ethers, ketones.
Containers: 1-, 8-, 40-lb polyethylene bottles.
Grades: 95% min.
Uses: High temperature catalyst for polymerization of acrylates, styrene, and curing of polyesters; compounding of silicones and polyethylene.
Shipping regulations: Oxidizing material. Yellow label. *

tert-butyl permaleic acid
$(CH_3)_3CCO_2COCH{:}CHCOOH$.
Properties: White crystalline solid; m.p. 114-116°C (dec); slightly soluble in water, cool 5% alkaline solutions and alcohols; moderately soluble in oxygenated organic solvents, polyester monomers; slightly soluble in petroleum ether, carbon tetrachloride, and chloroform; insoluble in benzene.
Grade: 95% pure.
Containers: 1- and 5-lb glass bottles.
Uses: Polymerization catalyst; in bleaching and in pharmaceuticals.

tert-butyl perphthalic acid
$(CH_3)_3CO_2COC_6H_4COOH$.
Properties: White crystalline solid; m.p. 96-99°C; insoluble in water; soluble in cool 5% alkaline solutions and in alcohols; moderately soluble in oxygenated organic solvents, chlorinated hydrocarbons, polyester monomers; slightly soluble in petroleum hydrocarbons.
Grades: 95% pure.
Containers: 1- and 5-lb glass bottles.
Uses: Polymerization catalyst and oxidizing agent.

ortho-tert-butylphenol $(CH_3)_3CC_6H_4OH$.
Properties: Light yellow liquid; freezing point −65°C; density 0.982 (20°C); b.p. 224°C; flash point (open cup) 110°C. Soluble in isopentane, toluene, and ethyl alcohol; insoluble in water.
Uses: Chemical intermediate for synthetic resins, plasticizers, surface-active agents, perfumes, and other products.

para-tert-butyl phenol $(CH_3)_3CC_6H_4OH$.
Properties: White crystalline solid, with a distinctive odor; sp. gr. (crystals) 1.03; sp. gr. (molten) 0.908 (114/4°C); b.p. 239°C; m.p. 98°C.
Grades and Containers: Flake in 50-lb bags; molten in insulated tank cars and tank trucks.
Uses: Plasticizer for cellulose acetate; intermediate for antioxidants, special starches; pour-point depressors and emulsion breakers for petroleum oils and some plastics; synthetic lubricants; insecticides; industrial odorants.

2-(para-tert-butylphenoxy)isopropyl 2-chloroethyl sulfite. See aramite.

n-butyl phenyl ether $C_4H_9OC_6H_5$.
Properties: Sp. gr. (20°C) 0.929; boiling range 202-212°C; color water-white; odor aromatic; flash point 180°F.
Shipping regulations: None. *

4-tert-butylphenyl salicylate
$(CH_3)_3CC_6H_4OOCC_6H_4OH$.
Properties: Off-white, odorless crystals; m.p. 62-64°C; soluble in alcohol, ethyl acetate, toluene; insoluble in water.
Use: Light absorber, best at 290-330 μ.

n-butylphosphoric acid $C_4H_9H_2PO_4$.
Properties: Mobile reddish amber liquid; sp. gr. 1.25 (25°C); insoluble in water; can be neutralized with alkalies or amines to give water-soluble salts.
Purity: 97%, with remainder being orthophosphoric acid and n-butyl alcohol.

*See "I.C.C. Shipping Regulations," page xiii.
Reference numbers refer to name of manufacturer. See "List of Manufacturers," page v.

Uses: Textile and paper processing compounds; catalysts in urea-resin formation; polymerizing agents for resins and oils.

n-butyl propionate $C_2H_5CO_2C_4H_9$.
Properties: Water-white liquid. Apple-like odor. Soluble in alcohol and ether; miscible with all coal-tar and petroleum distillates; very slightly soluble in water.
Constants: Sp. gr. 0.875 (20°C), 0.874 (15.5°C); wt/gal 7.3 lbs; b.p. 146°C (commercial grades boil over a range of 130-150°C due to presence of butyl alcohol and esters); m.p. −89°C; flash point 63°F (approx.); dilution ratio (nitrocellulose solution method) with toluol 2.1, with petroleum naphtha 1.2.
Derivation: Esterification of propionic acid with butyl alcohol, with sulfuric acid as a catalyst.
Method of purification: Fractional distillation.
Grades: Technical (85-90% to 95% ester content).
Containers: 1-gal cans; 5- and 55-gal steel drums; tank cars.
Uses: Solvent for nitrocellulose; retarder in lacquer thinner; lacquers; ingredient of perfumes, flavors.
Fire hazard: Flammable; flash point under 80°F. Keep lights and fire away.
Shipping regulations: Flammable liquid. Red label.*

butyl ricinoleate $C_{17}H_{32}(OH)COOC_4H_9$.
Properties: Yellow to colorless oleaginous liquid; soluble in alcohol and ether; insoluble in water.
Constants: Sp. gr. 0.916 (20/20°C); b.p. approximately 275°C (13 mm); flash point 220°C (428°F); Saybolt viscosity 112 (100°F); freezing point indefinite, slightly opaque at −30°C, and very viscous at −50°C; wt/gal 7.62 lbs (20°C).
Derivation: Castor oil and butyl alcohol.
Containers: 5-gal cans; 55-gal drums.
Uses: Plasticizer; lubricant.
Shipping regulations: None.*

butyl rubber. Synthetic rubber produced by copolymerization of isobutene (approx. 98%) with a small proportion (2%) of isoprene or butadiene. Polymerization is carried out at −50 to −100°C in a liquid hydrocarbon, with aluminum chloride as a catalyst. Its outstanding property compared with other rubbers is impermeability to gases; it is practically the only rubber used in inner tubes and similar gas-retaining applications.

The uncured rubber is tacky, but may be compounded like natural rubber and vulcanized. Butyl rubber has good resistance to chemical attack and aging, even at high temperatures. It has superior vibration insulation characteristics and abrasion resistance, but relatively low tensile strength and poor flame resistance. Tires made entirely of butyl show good friction on wet surfaces and do not squeal. Important uses other than inner tubes are as vibration insulators and electrical insulation for wires. Available as latex (paper coating, textile treating, leather finishing, adhesive formulation), liquid (roof coatings), and in halogenated forms (high temperature hose). Also used as binder fuel in solid rocket propellants.

butyl stearamide. $C_{17}H_{35}CONHC_4H_9$.
Properties: Light straw color; sp. gr. (20/20°C) 0.869; boiling range 195-200°C (2 mm); flash point 430°F; amide odor.
Use: Suggested as a plasticizer and as an intermediate for the synthesis of insecticides, surface-active agents, pharmaceuticals, and textile assistants.

butyl stearate (butyl octodecanoate) $C_{17}H_{35}COOC_4H_9$.
Properties: Colorless, stable, oleaginous material solidifying at about 19°C. Practically odorless, sometimes with faint fatty odor.
Constants: Sp. gr. 0.855-0.860 (25/20°C); m.p. 19.5-20°C; flash point 370°F; wt/gal 7.14 lbs (68°F); refractive index 1.4430 (20°C); b.p. 220-225°C (25 mm). Miscible with mineral and vegetable oils; soluble in alcohol and ether; insoluble in water.
Derivation: Alcoholysis of stearin or esterification of stearic acid with butyl alcohol.
Grades: Technical.
Containers: 7-, 35-lb tins; 213-, 380-lb drums.
Uses: Constituent in polishes, special lubricants and coatings; lubricants for metals, and in textile and molding industries; in wax polishes as the dye solvent; plasticizer for laminated fiber products, rubber hydrochloride, chlorinated rubber, and cable lacquers; in carbon paper; as an emollient in cosmetic and pharmaceutical products; damp-proofer for concrete.
Shipping regulations: None.*

1-butyl-3-sulfanilylurea (N-(butylcarbamoyl) sulfanilamide; N-sulfanilyl-N-butylurea; carbutamide) $H_2NC_6H_4SO_2NHCONH(CH_2)_3CH_3$.
Properties: Crystals; m.p. 144-145°C. Soluble in water (pH 5-8).
Derivation: Prepared from butylurea and sulfanilamide.
Use: Medicine.

butyl titanate. See tetrabutyl titanate.

para-tert-**butyltoluene.** See 1-methyl-4-tert-butylbenzene.

1-butyl-3-para-tolylsulfonylurea. See tolbutamide.

n-**butyltrichlorosilane** $C_4H_9SiCl_3$.
Properties: Colorless liquid; b.p. 148.9°C; sp. gr. 1.1608 (25/25°C); refractive index (n 25/D) 1.4363; flash point (Cleveland open cup) 126°F. Readily hydrolyzed by moisture, with the liberation of hydrochloric acid.
Derivation: By Grignard reaction of silicon tetrachloride and n-butylmagnesium chloride.
Grades: Technical.
Use: Intermediate for silicones.

Shipping regulations: Corrosive liquid. White label.*

tert-butyltrimethylmethane. See hexamethylethane.

N-n-butylurea $C_4H_9HNCONH_2$.
Properties: White solid; decomposes on heating; odor none; m.p. 96°C; soluble in water, alcohol, and ether.
Shipping regulations: None.*

n-butyl vinyl ether. See vinyl n-butyl ether.

butyl xanthate. See xanthic acids.

"Butyl Zimate." [69] Trademark for zinc dibutyldithiocarbamate, $[(C_4H_9)_2NC(S)S]_2Zn$.
Properties: White to cream colored powder; sp. gr. 1.21 ± .03; melting range 104-108°C; Zn content 13-15%; soluble in benzene, carbon disulfide, chloroform, gasoline; insoluble in water, dilute caustic.
Uses: Accelerator in some latex formulations.

butynediol $HOCH_2C{:}CCH_2OH$.
Properties: White orthorhombic crystals; m.p. 58°C; b.p. 238°C; flash point 152°C; refractive index 1.450 (n 25/D); soluble in water, aqueous acids, alcohol and acetone. Insoluble in ether and benzene. Very reactive! Toxic!
Derivation: By high pressure synthesis from acetylene and formaldehyde. A bismuth copper acetylide catalyst has been used.
Grades: Crystalline solid, 97%; aqueous solution, 35%.
Containers: Drums; tank cars; tank trucks.
Uses: Intermediate; corrosion inhibitor; electroplating brightener; defoliant for agricultural crops; polymerization accelerator; cross-linker.

3-butyn-1-ol (beta-ethynyl ethanol) $HC{:}CCH_2CH_2OH$.
Properties: Water-white liquid with characteristic odor; sp. gr. 0.9257 (20/4°C); refractive index 1.4409 (20°C); b.p. (760 mm) 128.9°C; f.p. −63.6°C.
Uses: Preparation of perfume bases, acetylenic esters, plastics, plasticizers, pharmaceuticals, wetting agents, medicinals, and organic synthesis.

"Butyn Sulfate." [3] Trademark for butacaine sulfate (q.v.).

butyraldehyde (butaldehyde; n-butanal; n-butyl aldehyde; butyric aldehyde) $CH_3(CH_2)_2CHO$.
Properties: Water-white liquid; characteristic, pungent, aldehyde odor. Flammable. Sp. gr. 0.8048 (20/20°C); b.p. 75.7°C (760 mm); vapor pressure 91.5 mm (20°C); flash point 20°F; wt/gal 6.7 lbs (20°C); coefficient of expansion 0.00114 (20°C); freezing point −99°C; viscosity 0.0043 poise (20°C). Slightly soluble in water; soluble in alcohol and ether.
Derivation: (a) By the Oxo process: the reaction of propylene with carbon monoxide and hydrogen in the presence of a cobalt catalyst; (b) by dehydrogenating butanol vapors over a catalyst, the butyraldehyde being separated by distillation; (c) by partial reduction of crotonaldehyde.
Grades: Technical (93% min.).
Containers: Drums; tank cars.
Uses: Intermediate.
Warning: Flammable. Avoid prolonged breathing of vapor. Avoid prolonged or repeated contact with skin. MCA warning label.
Shipping regulations: Flammable liquid. Red label.*

butyric acid (n-butyric acid; butanoic acid; ethylacetic acid; propylformic acid). $CH_3CH_2CH_2COOH$.
Properties: Colorless limpid liquid; rancid odor; refractive index 1.3981 (20°C); sp. gr. 0.9583 (20/4°C); m.p. −5.0 to −8°C; b.p. 163.5°C (757 mm), 75°C (25 mm); vapor pressure 0.84 mm (20°C); flash point 170°F; viscosity 1.61 cps (20°C); slightly soluble in water; miscible with alcohol and ether.
Typical specifications: Purity not less than 99%; sp. gr. 0.957-0.961 (20/20°C); boiling range (760 mm) below 158°C none, above 165°C none; esters not more than 0.5%; chlorides none; average wt 7.99 lbs/gal (20°C).
Derivation: Occurs as glyceride in animal milk fats. Produced as a byproduct in hydrocarbon synthesis, by oxidation of n-butyraldehyde, and by butyric fermentation of molasses or starch.
Grades: 90%; 95%; 99%; edible; synthetic; reagent; technical.
Containers: 1-gal cans; 5-, 55-, 110-gal drums; tank cars.
Uses: For synthesis of butyrate ester perfume and flavor ingredients; tanning and deliming; butter making; pharmaceuticals; water purification; varnish; disinfectants; emulsifying agents; for sweetening gasolines.

butyric alcohol. See n-butyl alcohol.

butyric aldehyde. See butyraldehyde.

butyric anhydride $(CH_3CH_2CH_2CO)_2O$.
Properties: Water-white liquid. Hydrolyzes to butyric acid in the presence of water. Sp. gr. 0.9681 (20/20°C); m.p. −75°C; b.p. (760 mm) 199.5°C; vapor pressure 0.3 mm (20°C); flash point 190°F; wt/gal 8.1 lbs (20°C); coefficient of expansion 0.00100 (20°C); freezing point −65.1°C; viscosity 0.0159 poise (20°C); decomposes in water.
Typical specifications: Purity not less than 85%; sp. gr. 0.965-0.970 (20/20°C); color water-white; boiling range (760 mm) below 190°C none, above 200°C none, below 195°C not more than 10%; average wt/gal 8.05 lbs (20°C).
Grades: Technical.
Containers: 1-gal glass jugs; 5- and 12-gal (returnable) glass carboys; 55-gal (returnable) aluminum drums.
Use: Manufacture of various butyrates, drugs, and tanning agents.

*See "I.C.C. Shipping Regulations," page xiii.
Reference numbers refer to name of manufacturer. See "List of Manufacturers," page v.

butyrolactone $\underline{OCH_2CH_2CH_2CO}$.

Properties: Colorless liquid with pleasant odor. B. p. 240°C; m. p. −44°C; sp. gr. 1.144; flash point 209°F. Miscible with water, alcohol and ether.
Derivation: By high pressure synthesis from acetylene and formaldehyde.
Grades: Technical.
Containers: Drums; tank cars.
Uses: Intermediate for butyric and succinic acids; solvent for resins; in paint removers; petroleum processing.

butyrone. See dipropyl ketone.

butyronitrile (propyl cyanide; butanenitrile) $CH_3(CH_2)_2CN$.

Properties: Colorless liquid; sp. gr. (15°C) 0.796; m. p. −112.6°C; b. p. 116-117.7°C. Slightly soluble in water; soluble in alcohol and ether.
Containers: Drums and tank cars.
Uses: Basic material in industrial, chemical and pharmaceutical intermediates and products; poultry medicines.

butyroyl chloride (butyryl chloride; butanoyl chloride) C_3H_7COCl.

Properties: Clear colorless liquid with characteristic pungent acid chloride odor. Reacts with alcohol and water; infinitely miscible with ether. Freezing point −89°C; distillation range 100-110°C; sp. gr. 1.028 (15°C); refractive index 1.4121 (n 20/D).
Containers: 5-, 13-gal glass carboys.
Use: Organic synthesis.

butyryl chloride. See butyroyl chloride.

"Buxine." [227] Trademark for amylcinnamic aldehyde (alpha-n-amyl-beta-phenylacrolein), minimum 97% pure.

Properties: Clear yellow liquid; color No. 15 maximum; strong floral odor, suggesting jasmin and lily. Sp. gr. 0.963-0.968 (25/25°C); refractive index 1.5540-1.5590 (20°C); flash point (Tag closed cup) 221°F. Clearly soluble in 6 parts of 80% alcohol.
Occurrence: Not found in nature.
Uses: In jasmin perfumes; in soap perfumes.

BVE. Abbreviation for butyl vinyl ether. See vinyl n-butyl ether.

"BWH-1." [248] Trademark for a specially selected mixture of plasticizing oils.

Properties: Free-flowing, dark brown liquid; sp. gr. 1.01; soluble in acetone, benzol and ethylene dichloride; insoluble in water and gasoline.
Uses: Rubber plasticizer and reclaiming oil.

"B-X-A." [248] Trade name for a diarylamine-ketone-aldehyde reaction product.

Properties: Brown powder; sp. gr. 1.10; melting range, 85-95°C; store in a cool place. Soluble in acetone, benzol and ethylene dichloride; insoluble in water and gasoline. Used as rubber antioxidant.

"BxDC." [177] Trademark for butoxyethyl diglycol carbonate.

"B. Y." [319] Trademark for dried fermentation solubles used as animal feed supplements.

C

C. Symbol for carbon.

C. Abbreviation for centigrade.

"C_2." [84] Brand name of a proprietary product, a sodium chlorite product that bleaches paper pulp (wood, rag, flax) to high brightness without degradation of the fibers. Also used as a slimicide and biocide in paper machines.

"C-46." [306] Trademark for hexachlorobutadiene.

"C-56." [306] Trademark for hexachlorocyclopentadiene (q. v.).

"C-64." [233] Trademark for a proprietary synthetic aromatic chemical. Colorless, crystalline material; soluble even in diluted alcohol; powerful spicy odor with a slight camphoraceous touch, somewhat reminiscent of patchouli oil.

"C-66 (Dow)." [233] Trademark for a proprietary synthetic aromatic chemical. Colorless liquid; soluble in alcohol and essential oils; fresh neutral floral note.

CA. Abbreviation for cellulose acetate and cortisone acetate.

Ca. Symbol for calcium.

ca. Abbreviation for circa, meaning about or approximately.

C_3A. Abbreviation for tricalcium aluminate, as used in cement. See cement, Portland.

CAB. Abbreviation for cellulose acetate-butyrate.

"Cab-O-Lite" [275] (wollastonite, $CaSiO_3$). Trade name for brilliant white nonmetallic mineral extender pigment for paints; filler for ceramics; mineral filler for polyester resins.

"Cab-O-Sil." [275] Trade name for anhydrous and particulate colloidal silica; reinforcing, thixotropic, thickening and gelling, suspending, flatting, anticaking, antislip agent.
Uses: In plastics, silicones, rubber, paints, varnishes and lacquers, printing inks, adhesives, pharmaceuticals, cosmetics, floor waxes, lubricating oils and greases, sealants, reproduction paper, dusting powders, sulfur and low temperature thermal insulation.

cacao (cocoa). Powder prepared from roasted cured kernels of ripe seed of Theobroma cacao.
Properties: Weak reddish to brown powder, with chocolate-like taste and odor.
Grade: U.S.P. XVI.
Use: Flavoring agent for pharmaceuticals.

cacao beans. Seed of the cacao tree, Theobroma cacao, native to Mexico, West Indies, and South America and cultivated in all tropical countries. The seeds contain from 50 to 57% cacao butter. Used in the manufacture of cocoa, chocolate, cacao butter, and theobromine.

cacao butter (theobroma oil; cocoa butter; cacao oil). The fat obtained from the roasted seed of Theobroma cacao.
Properties: Yellowish-white, brittle solid with chocolate-like taste and odor. Sp. gr. 0.858-0.864 (100/25°C); m.p. 30-35°C; refractive index (n 40/D) 1.4537-1.4585; saponification number 188-195; iodine number 35-43; solidification range 45-50°C. Insoluble in water; slightly soluble in alcohol; soluble in boiling dehydrated alcohol; freely soluble in ether and chloroform.
Derivation: From the cacao bean, by expression, decoction, or extraction by solvent.
Chief constituents: Glycerides of stearic, palmitic, and lauric acids.
Grades: Crude; refined; U.S.P. XVI (as theobroma oil).
Containers: Tins; bags; barrels.
Uses: Confectionery and pharmaceuticals; soaps.
Shipping regulations: None.*

cacao oil. See cacao butter.

C acid. See 2-naphthylamine-4,8-disulfonic acid.

cacodylic acid (dimethylarsinic acid) $(CH_3)_2AsOOH$.
Properties: Colorless, odorless, deliquescent crystals; poisonous! M.p. 200°C; soluble in water, alcohol, and acetic acid; insoluble in ether.
Derivation: (a) By distilling a mixture of arsenic trioxide and potassium acetate and oxidizing the resulting product with mercuric oxide; (b) hydrolysis of dimethyltrihaloarsine.
Grades: C.P.
Containers: 1- and 5-lb glass bottles.
Use: Synthesis of dyes, drugs and perfumes; preparation of cacodylates; making dimethylarsine and derivatives; herbicide.
Shipping regulations: Poison, class B. Poison label.*

"Cadalume L." [72]
Trade name for materials for bright cadmium plating; includes cadmium oxide, sodium cyanide and proprietary brighteners.

*See "I.C.C. Shipping Regulations," page xiii.
Reference numbers refer to name of manufacturer. See "List of Manufacturers," page v.

"Cadalyte." [28] Trademark for a series of cadmium electroplating compounds.

"Cadalyte" Bright Dip. Clear, colorless liquid; corrosive; miscible with water.

"Cadalyte" Brightener. Light brown, free-flowing, fine powder.

"Cadalyte" Maintenance Compound. Buff-colored, free-flowing, fine powder; water-soluble. Contains "Cadalyte" Brightener and sodium cyanide in proportions consumed in cadmium plating.

Containers: 100-lb drums.

Uses: Replenishing "Cadalyte" cadmium plating baths.

"Cadalyte" Plating Salts. Reddish-brown powder consisting of a mixture of sodium cyanide, cadmium oxide, and brighteners; contains approximately 19.2% cadmium. Free-flowing, water-soluble, stable.

Containers: 50-lb drums.

Uses: To prepare cadmium plating baths which deposit cadmium in a bright, ductile, adherent form on iron and steel to prevent rusting and corrosion.

cadaverine (1,5-diaminopentane; pentamethylenediamine) $NH_2(CH_2)_5NH_2$. A ptomaine formed in the decay of animal bodies; relatively nonpoisonous.

Properties: Syrupy colorless fuming liquid; m.p. +9°C; b.p. 178-179°C; soluble in water and alcohol; slightly soluble in ether.

Uses: Preparation of high polymers; intermediate; biological research.

"Caddy." [49] Trade name for a liquid cadmium fungicide used on turf grass.

cade oil (juniper tar; juniper tar oil).

Properties: Thick, clear, dark brown liquid; tarry odor; burning, bitter taste. Chief known constituent: cadinene $C_{15}H_{24}$. Sp.gr. 0.950-1.055; soluble in ether, glacial acetic acid, chloroform; partially soluble in alcohol; very slightly soluble in water.

Derivation: By the dry distillation of the wood of Juniperus oxycedrum.

Method of purification: Rectification.

Grades: Technical; U.S.P. XVI.

Containers: 1-, 5-, 10-lb bottles; 25-lb tins.

Uses: Soap; pharmaceuticals; perfumery (Russian leather type odors).

Shipping regulations: None.*

"Cadminate." [329] Trademark for a cadmium-containing turf fungicide. It is used for the prevention and cure of dollar spot, copper spot and red thread (pink patch). It can be mixed with water and sprayed.

Containers: 1/2-oz paper packets (32/canister); bulk.

cadmium Cd. Element of atomic number 48, of group II of the periodic system.

Properties: Soft, blue-white, ductile, malleable metal, or grayish-white powder. Tarnishes in moist air; metal becomes brittle at 80°C and burns when heated. Toxic. Sp.gr. 8.642 (17°C); m.p. 320.9°C; b.p. 767°C; soluble in acids and in ammonium nitrate solutions; insoluble in water.

Derivation: Occurs chiefly as greenockite (CdS) associated with zinc blende; cadmium content of the blende is usually less than 1%. Cadmium is therefore ordinarily obtained as a by-product in zinc (and lead) production. It is more volatile than zinc and distils over in the first stages of zinc distillation; the resulting high-cadmium spelter is redistilled (often refluxed) and refined by distillation at 800°C in the presence of coal. Cadmium is also recovered electrolytically from zinc-dust residues of electrolytic zinc plants and from lead blast-furnace dust.

Grades: Technical; powder; pure sticks; ingots; slabs; anodes; etc.; high purity (less than 10 ppm impurities).

Uses: Cadmium plating; manufacture of cadmium salts; cadmium-vapor lamps; smoke bombs; small-arms ammunition; white pigment; bearing and low-melting alloys; electric instruments; incandescent light filaments; aluminum solder; substitute for tin in solders; dental amalgams; ceramic coatings; deoxidizer in metallurgy; coloring glass; nickel plating; process engraving; lithography; Weston standard cell; nickel-cadmium storage battery. See also cadmium plating.

cadmium acetate $Cd(CH_3COO)_2 \cdot 3H_2O$.

Properties: White monoclinic crystals; slight odor of acetic acid; sp.gr. 2.01; m.p., loses water at 130°C; soluble in water, alcohol, and ether.

Derivation: Interaction of acetic acid and cadmium oxide.

Grades: Technical.

Uses: Ceramics (iridescent glazes); manufacture of acetates; assistant in dyeing and printing textiles; dentistry; laboratory reagent.

cadmium ammonium bromide (ammonium-cadmium bromide) $CdBr_2 \cdot 4NH_4Br$.

Properties: Colorless crystals. Soluble in alcohol and water.

cadmium-base Babbitt. A bearing metal composed of 95% cadmium and 5% silver. Used for relatively high temperatures but is susceptible to corrosion. See Babbitt metal.

cadmium borotungstate $2CdO \cdot B_2O_3 \cdot 9WO_3 \cdot 18H_2O$.

Properties: Yellow, heavy crystals. Soluble in water. Solution yellow or light brown.

Grades: Technical.

Use: Separating minerals.

cadmium bromate $Cd(BrO_3)_2 \cdot H_2O$.

Properties: White crystals or crystalline powder; sp.gr. 3.758; m.p., decomposes. Soluble in water; insoluble in alcohol.

Derivation: By adding cadmium sulfate to a solution of barium bromate.

Method of purification: Recrystallization.

Grades: Pure; reagent.

Containers: Glass bottles; 25-lb tin boxes.

Uses: Analytical reagent.

Fire hazard: Dangerous, oxidizing agent.

cadmium bromide $CdBr_2$ or $CdBr_2 \cdot 4H_2O$.

Properties: White to yellowish, efflorescent crystalline powder; sp.gr. 5.192; m.p. 568°C; b.p. 963°C; soluble in water, acetone, and alcohol; slightly soluble in ether.

*See "I.C.C. Shipping Regulations," page xiii.

Reference numbers refer to name of manufacturer. See "List of Manufacturers," page v.

Derivation: By heating cadmium in bromine vapor.
Grades: Technical; reagent.
Containers: Bottles; cans; cases.
Uses: Photography; process engraving; lithography.
Shipping regulations: None. *

cadmium carbonate $CdCO_3$.
Properties: White, amorphous powder; sp. gr. 4.258; decomposes below 500°C. Soluble in acids (dilute) and in concentrated solutions of ammonium salts; insoluble in water.
Grades: Reagent.

cadmium chlorate $Cd(ClO_3)_2 \cdot 2H_2O$.
Properties: Colorless, prismatic crystals. Hygroscopic. Caution! Keep well stoppered! Sp. gr. 2.28 (18°C); m.p. 80°C. Soluble in alcohol and water.
Grades: Technical.
Shipping regulations: Oxidizing material. Yellow label. *

cadmium chloride
(a) $CdCl_2$; (b) $CdCl_2 \cdot 2\ 1/2\ H_2O$.
Properties: Small white crystals, odorless; sp. gr. (a) 4.05, (b) 3.327; m.p. (a) 568°C; b.p. (a) 861-954°C. Soluble in water and alcohol.
Derivation: By the action of hydrochloric acid on cadmium with subsequent crystallization.
Method of purification: Recrystallization.
Grades: Technical; reagent.
Containers: 1- and 5-lb bottles; tins; 400-lb drums.
Uses: Preparation of cadmium sulfide; analytical chemistry; photography; dyeing and calico printing; ingredient of electroplating baths; addition to tinning solutions; manufacture of special mirrors; vacuum tube industry.
Shipping regulations: None. *

cadmium-copper alloy. An alloy of copper with 1% cadmium. Sp. gr. 8.94. Has been used for trolley wires.

cadmium ethylhexoate. See soaps, metallic.

cadmium hydrate. See cadmium hydroxide.

cadmium hydroxide (cadmium hydrate) $Cd(OH)_2$.
Properties: White, amorphous powder; sp. gr. 4.79; m.p., loses H_2O (300°C); soluble in ammonium hydroxide and in dilute acids; insoluble in water and alkalies; absorbs carbon dioxide from air.
Derivation: By the action of sodium hydroxide on a cadmium salt solution.
Grades: Technical; C.P.
Containers: Glass bottles; boxes.
Use: Cadmium salts; cadmium plating
Shipping regulations: None. *

cadmium iodate $Cd(IO_3)_2$.
Properties: Fine, white powder; sp. gr. 6.48; m.p., decomposes; slightly soluble in water; soluble in nitric acid or ammonium hydroxide.
Grades: Technical.
Use: Oxidizing agent.

cadmium iodide CdI_2.
Properties: White, flaky, crystals; odorless; becomes yellow on exposure to air and light. Occurs in two allotropic forms. Sp. gr. 5.67 (alpha), and 5.30 (beta); m.p. (alpha) 388°C, m.p. (beta) 404°C; b.p. (alpha) 712°C; soluble in water, alcohol, ether, acetone, and ammonia.
Derivation: By the action of hydriodic acid on cadmium oxide.
Method of purification: Recrystallization.
Grades: Technical; reagent.
Containers: 1-, 5-, 10-lb bottles; 25-lb fiber drums.
Uses: Photography; medicine; process engraving and lithography; analytical chemistry.
Shipping regulations: None. *

cadmium lithopone. See lithopone, cadmium.

cadmium naphthenate. See soaps, metallic.

cadmium nitrate $Cd(NO_3)_2 \cdot 4H_2O$.
Properties: White, amorphous pieces or hygroscopic needles. Keep well stoppered. Soluble in water and alcohol; sp. gr. 2.455; m.p. 59.5°C; b.p. 132°C.
Derivation: By the action of nitric acid on cadmium or cadmium oxide and crystallization.
Grades: Technical; reagent.
Containers: Tins; glass bottles; 400-lb drums.
Uses: Ceramic industry for coloring glass and porcelain; laboratory reagent; manufacture of other cadmium salts.
Fire hazard: Dangerous.
Shipping regulations: Oxidizing material. Yellow label. *

cadmium octoate. See soaps, metallic.

cadmium orange. An impure form of cadmium selenide, used as a pigment.

cadmium oxalate $Cd(COO)_2 \cdot 3H_2O$.
Properties: White, amorphous powder; soluble in dilute acids, ammonium hydroxide; insoluble in alcohol and water; sp. gr. 3.32 (dehydrated).
Grades: Technical.

cadmium oxide (anhydrous cadmium oxide) CdO.
Properties: Yellowish-red or brownish-red to brownish-black powder; sp. gr. 6.95-8.11; soluble in dilute acids and ammonium hydroxide; insoluble in water.
Grades: Technical; C.P.
Containers: 25- and 50-lb packages.
Uses: Addition agent for cadmium-plating baths; pigment in ceramics; chemical catalyst; making cadmium salts.

cadmium oxide, anhydrous. See cadmium oxide.

cadmium phosphate $Cd_3(PO_4)_2 \cdot 2CdHPO_4 \cdot 4H_2O$.
Properties: White powder; sp. gr. 4.06; soluble in dilute acids; insoluble in water.
Grades: Technical.

cadmium plating. The electrodeposition of cadmium on iron wire, steel articles, etc., to make them relatively rust-proof.

cadmium potassium iodide $CdI_2 \cdot 2KI \cdot 2H_2O$.
Properties: White powder, becomes yellowish

*See "I.C.C. Shipping Regulations," page xiii.
Reference numbers refer to name of manufacturer. See "List of Manufacturers," page v.

with age; deliquescent; soluble in water, alcohol, ether, and acid; sp. gr. 3.359; m. p. 76°C, with decomposition.
Derivation: By combining cadmium iodide and potassium iodide in solution, in proportion of their combining weights and subsequent crystallization.
Method of purification: Recrystallization.
Grades: Technical; C. P.
Containers: Tins; glass bottles.
Uses: Analytical chemistry; medicine.
Shipping regulations: None.*

cadmium propionate $Cd(OOCC_2H_5)_2$. A solid; used in scintillation counters.

cadmium reds (cadmium selenide lithopone). Pigments manufactured in a series of shades: light, medium light, medium, deep, and maroon. These colors are prepared by precipitating a highly purified solution of cadmium sulfate with barium sulfide in the presence of selenium, forming co-precipitated pigments, containing cadmium sulfide, cadmium selenide, and barium sulfate, which are furnace treated after precipitation. They are fast to light, insoluble in all vehicles (non-bleeding), have high heat and alkali resistance, have acid fastness, are extremely soft and easily ground, and have low oil absorption and good gloss in enamels. Have a tendency to chalk similar to cadmium yellows, but are sometimes used in automotive finishes. Used in manufacture of lacquers, oil paints, and enamels. The approximate specifications are as follows: Sp. gr. 4.30; wt/solid gal 39.70 lbs; 1 lb bulks 0.0250 gal; fineness (residue on 325-mesh screen) 0.06%; oil absorption 17.50.

cadmium resinate. See soaps, metallic.

cadmium ricinoleate
$Cd[CH_3(CH_2)_5CHOHCH_2CH{:}CH(CH_2)_7CO_2]_2$. A nearly odorless, fine, white powder derived from castor oil.
Properties: M. p. 104°C; sp. gr. 1.11.
Uses: Solutions used to stabilize polyvinyl chloride and copolymers against light and heat.

cadmium salicylate $Cd(C_7H_5O_3)_2 \cdot H_2O$.
Properties: Colorless crystals or white powder; soluble in water, alcohol, ether, and glycerol.
Derivation: Action of salicylic acid on cadmium oxide.
Use: Medicine.
Shipping regulations: None.*

cadmium selenide CdSe. Usually a red powder, but may also occur gray to brown. Sp. gr. 5.81 (15/4°C); m.p. above 1350°C; Insoluble in water; stable at high temperatures. The red powder form is used as a paint pigment that withstands light, acid, alkali, and high temperatures. It also serves to increase abrasion resistance of rubber compounds. See also cadmium reds.

cadmium selenide lithopone. See cadmium reds.

cadmium stearate. See soaps, metallic.

cadmium sulfate (a) $CdSO_4$; (b) $3CdSO_4 \cdot 8H_2O$; (c) $CdSO_4 \cdot 4H_2O$.
Properties: Colorless, odorless crystals. Sp. gr. (a) 4.69; (b) 3.09; (c) 3.05; m. p. (a) 1000°C. Soluble in water; insoluble in alcohol.
Derivation: By the action of dilute sulfuric acid on cadmium or cadmium oxide.
Method of purification: Crystallization.
Grades: Technical; reagent.
Containers: 1-lb bottles; tins; kegs.
Use: Manufacture of normal cadmium electric cells; analytical reagent; medicine; vacuum tube manufacture.
Shipping regulations: None.*

cadmium sulfide (orange cadmium; orient yellow; aurora yellow; jaune brilliant; greenockite; see also cadmium reds and cadmium yellows) CdS.
Properties: Light yellow or orange powder. Sp. gr. 3.9-4.8; sublimes at 980°C; soluble in acids and ammonia; insoluble in water.
Derivation: (a) By passing hydrogen sulfide gas into a solution of a cadmium salt acidified with hydrochloric acid. The precipitate is filtered and dried. (b) Occurs naturally as greenockite (q. v.).
Grades: Technical; N. N. D.; high purity (single crystals).
Containers: 100-lb kegs; 200-lb cases.
Uses: Pigments and inks; ceramic glazes; soap color; pyrotechnics; topical medicine; phosphors; fluorescent screens; scintillation counters; transistor material.
Shipping Regulations: None.*

cadmium tallate. See soaps, metallic.

cadmium telluride CdTe.
Properties: Brownish-black, cubic crystals. Oxidizes on prolonged exposure to moist air. Insoluble in water and mineral acids, except nitric, in which it is soluble with decomposition. M. p. 1041°C; sp. gr. 6.2 (15/4°C).
Derivation: Fusion of the elements; reaction of H_2Te and $CdCl_2$.
Use: Semi-conductors.

cadmium tungstate (cadmium wolframate) $CdWoO_4$.
Properties: White or yellow crystals or powder. Soluble in ammonium hydroxide; alkali cyanides; insoluble in water or dilute acids.
Derivation: By the interaction of cadmium nitrate and ammonium tungstate.
Forms: Single crystal rods; broken crystals (crackle).
Grades: Technical.
Containers: Barrels.
Uses: Fluorescent paint; x-ray screens; scintillation counters; catalyst.
Shipping regulations: None.*

cadmium wolframate. See cadmium tungstate.

cadmium yellows. Cadmium sulfide pigments. The types in most general use are co-precipitated pigments containing barium sulfate. Cadmium yellows are available in a range of shades from a greenish to a reddish

*See "I.C.C. Shipping Regulations," page xiii.
Reference numbers refer to name of manufacturer. See "List of Manufacturers," page v.

yellow or golden shade, and they have good brilliancy, alkali-fastness and do not darken on exposure to light or in the presence of sulfide fumes. In paints and enamels there is some tendency, however, to become chalky or to develop a whitish film or scum on the surface as the paint film weathers. In tints they do not last well on weathering as the yellow seems to go "out"; this is particularly evident in green tints made from combinations of cadmium yellow and some suitable blue or bluish-green. The cadmium yellows have excellent heat resistance which is to be expected since they are furnace-treated pigments. They are used in enamels and also in casein paints.

"Cadmolith." [296] Brand name for a line of yellow and red cadmium-lithopone pigments manufactured in a series of shades—primrose, lemon, golden, and orange—by the coprecipitation of a highly purified solution of cadmium sulfate with barium sulfide forming cadmium sulfide and barium sulfate. The red shades also contain selenium. These pigments are fast to light; insoluble in all vehicles (non-bleeding); heat and alkali resistant; acid-fast. Soft and easily ground.

Typical specifications: Sp. gr. 4.22; wt/solid gal 35.17 lbs; 1 lb bulks 0.02825 gal; fineness (residue on 325-mesh screen) 0.04%; oil absorption 17.69. Used as a pigment in automotive finishes, textile coatings, printing inks, lacquer and rubber.

"Cadox." [419] Trademark for a series of organic peroxide catalysts.

"Cadox" B160. Paste containing 55% benzoyl peroxide with butyl benzyl phthalate.

"Cadox" BC. Purified benzoyl peroxide, 50%, with camphor.

"Cadox" BCP. Finely divided, white, free flowing powder, consisting of about 35% benzoyl peroxide. Safe to handle without any special precautions. The filler is calcium phosphate.

"Cadox" BDP. Paste containing 50% benzoyl peroxide with dibutyl phthalate.

"Cadox" BSA. Powdered benzoyl peroxide, 95%, with stearic acid.

"Cadox" BSD. Thick white paste containing 50% finely milled benzoyl peroxide combined with Dow Corning silicone oil.

"Cadox" BSG. Same with General Electric silicone oil.

"Cadox" BTP. Paste containing 50% benzoyl peroxide with tricresyl phosphate.

"Cadox" MDP. 60% methyl ethyl ketone peroxide in dimethyl phthalate. Assay: 60.0% min. Active oxygen 11.0 ± 0.5%. Packing: 1- and 8-lb. glass bottles; 40-lb non-returnable "JALINER" drums.

"Cadox" TBH. Water-white liquid consisting essentially of 70% tert-butyl hydroperoxide (q.v.) with di-tert-butyl peroxide.

"Cadox" TDP. Thick paste containing 50% 2,4-dichlorobenzoyl peroxide with dibutyl phthalate as a stabilizer. Assay: 50.0 ± 1.0%. Active oxygen 2.10 ± 0.04%. Packing: 1 and 10-lb glass jars; 50-lb drums.

"Cadox" TS-40. Thick paste containing 40% 2,4-dichlorobenzoyl peroxide with silicone fluid.

"Cadox" TS-50. Same with 50% 2,4-dichlorobenzoyl peroxide.

caesium compounds. See corresponding cesium compound.

C_4AF. Abbreviation for tetracalcium aluminoferrate, as used in cement. See cement, Portland.

caffea. See coffee.

caffearine. See trigonelline.

caffeine (theine; methyltheobromine; trimethylxanthine $C_8H_{10}N_4O_2 \cdot H_2O$.

Properties: White, fleecy masses or long flexible, silky, crystals; an alkaloid; loses H_2O at 80°C. Efflorescent in air. M.p. 236.8°C; soluble in chloroform, slightly soluble in water and alcohol, very slightly soluble in ether. Odorless; bitter taste; solutions neutral to litmus.

Derivation: By extraction of coffee, tea, guarana, paraguay tea, or kola nuts; also produced synthetically. Much of the caffeine of commerce is a by-product of decaffeinized coffee manufacture.

Method of purification: Recrystallization.

Grades: Technical; U.S.P. XVI.

Containers: 1-lb bottles; 5-, 10-, 25-lb cans; 100-, 150-lb drums.

Uses: Beverages; in medicine in form of basic alkaloid or as the arsenate, arsenite, benzoate, borocitrate, citrate, diiodide, hydriodate, hydrobromide, hydrochloride, phosphate, phthalate, salicylate, sulfate, sodium benzoate, sodium salicylate, valeriate, etc., because of their solubility in alcohol, water, and ether.

Shipping regulations: None.*

caffeine bromide. See caffeine hydrobromide.

caffeine, citrated. White odorless powder; slightly bitter acid taste; corresponds closely to the true salt with the formula $C_8H_{10}O_2N_4 \cdot C_6H_8O_7$; assay 48.0-52.0% anhydrous caffeine and 48-52% anhydrous citric acid after drying; solubility in water (1:4), complete; soluble in alcohol; 15.0% max. retained on 100 mesh.

Grade: N.F. XI.

Use: Medicine.

caffeine hydrobromide (caffein bromide) $C_8H_{10}O_2N_4 \cdot HBr \cdot 2H_2O$.

Properties: Colorless, efflorescent crystals; become brownish on exposure to air; decompose at 80-100°C. Soluble in water or alcohol (with decomposition).

Use: Medicine.

caffeine sodium benzoate. A mixture of caffeine and sodium benzoate containing 47-50% anhydrous caffeine and 50-53% sodium benzoate.

Properties: White, odorless powder with slightly bitter taste. Slightly soluble in

*See "I.C.C. Shipping Regulations," page xiii.

Reference numbers refer to name of manufacturer. See "List of Manufacturers," page v.

chloroform; soluble in alcohol and water.
Grades: U.S.P. XVI.
Uses: Similar to caffeine, but it is more soluble in water.

cajeputene. See dipentene.

cajeput oil (cajuput oil).
Properties: Essential oil. Crude: Bluish-green to green. Rectified: Colorless or yellowish; pleasant camphor-like odor; aromatic taste which at first imparts a burning sensation and later changes to a cooling one; sp. gr. 0.915-0.932 (15°C); optical rotation 0 to -4°; refractive index (n 20/D) 1.4660-1.4720; soluble in alcohol, ether, benzyl benzoate, fixed oils, diethyl phthalate; slightly soluble in mineral oil; insoluble in glycerin.
Chief known constituents: Eucalyptol, alpha-terpineol, pinene.
Derivation: Distilled from the fresh leaves and twigs of several species of Melaleuca but mostly from those of Melaleuca leucadendron, L.
Containers: Cans.
Uses: Medicine and perfumery.
Shipping regulations: None.*

cajeputol. See eucalyptol.

cajuput oil. See cajeput oil.

cake alum. See aluminum sulfate.

calabar bean. See physostigma.

calabarine. See physostigmine.

"Calade." [108] Trademark for a powdered, alkaline sodium hexametaphosphate compound containing wetting agents and an aluminum corrosion inhibitor.
Containers: 3-lb 9-oz packages (12/case); 100- and 300-lb drums.
Uses: Detergent for dishwashing and general cleaning.

calamine (mineral) (hemimorphite) $Zn_4Si_2O_7(OH)_2 \cdot H_2O$.
Properties: White with delicate blue or green shades, also yellow or brownish-white mineral. Luster, vitreous to pearly. Similar to smithsonite (q.v.) but not identical. Contains 67.5% zinc oxide. Loses water only at a red heat, remains unchanged at 340°C, sp. gr. 3.4-3.5; hardness 4.5-5.0. Pyroelectric.
Occurrence: United States; Europe.
Use: An ore of zinc.

calamine (pharmaceutical) (calamine, prepared). Zinc oxide with a small amount of ferric oxide and containing not less than 98% zinc oxide upon ignition.
Properties: Pink, odorless, almost tasteless powder. Insoluble in water, almost completely soluble in mineral acids.
Grades: U.S.P. XVI.
Containers: Drums.
Use: Medicinal and pharmaceutical products.

calamine, prepared. See calamine (pharmaceutical).

calamintha oil. See marjoram oil.

calamus (sweet flag; calmus; sweet cane; sweet grass).
Derivation: Unpeeled, dried rhizome of Acorus calamus.
Habitat: Europe; North America; western Asia; cultivated in Burma and Ceylon.
Containers: Bales; bags.
Use: Medicine, source of calamus oil.
Shipping regulations: None.*

calamus oil.
Properties: Yellow to brownish-yellow oil.
Chief known constituents: Asarone and eugenol.
Constants: Sp. gr. 0.959-0.970 (15°C); refractive index 1.503-1.510; saponification value 6-20. Slightly soluble in water.
Derivation: By steam distillation of calamus.
Containers: Bottles.
Uses: Preparation of liquers, medicine, perfumery.
Shipping regulations: None.*

"Calaroc." [206] Brand name of a range of resinous syrups, used for textile and paper finishing. They are fixed to the fibers by a short heating treatment, and give a resinous finish of the thermo-setting type.

"Calasec." [206] Brand name of aqueous syrups of ammonium polymethacrylate. These syrups are used as thickening agents for resinous dispersions, for textile-finishing purposes.

"Calatac." [206] Brand name of a range of resins, supplied in the form of aqueous dispersions of a polymer, and used for textile finishing. The resins are of the thermoplastic type.

calaverite $AuTe_2$. One of the gold telluride group of minerals. Corresponds to the same general formula as sylvanite and krennerite (q.v.). Pale bronze-yellow color or tin-white, tarnishing to bronze yellow on exposure. Metallic luster. Contains 40-43% gold, 1-3% silver.
Constants: Sp. gr. 9.0; hardness 2.5.
Occurrence: United States (California, Colorado); Australia; Canada.
Use: Important source of gold.

"Calcene." [177] Trademark for a specially prepared precipitated calcium carbonate for use in rubber compounding.
Properties: Fine particle size (0.1 micron average diameter); low alkalinity; only slight activating effect on cure.
Grades: "Calcene" TM, coated with a rubber-soluble agent; "Calcene" NC, non-coated.
Containers: 50-lb paper bags.
Uses: In manufacture of drug sundries, heels, wire insulation, and miscellaneous molded and mechanical rubber goods.

calcic liver of sulfur. See lime, sulfurated.

calciferol. See vitamin D.

calcimine (kalsomine). Essentially chalk and glue in powdered form ready to mix with water. Used as temporary decoration for interior plastic walls. Will not withstand washing.

calcination. The process of heating a material to a high temperature, but below its fusing point, to cause it to lose moisture or other volatile material or be oxidized or reduced; roasting; e.g., calcining limestone to make lime.

calcined baryta. See barium oxide.

calcined clay. Ball clay or china clay that has been heated until combined water is removed and plastic character of the clay is destroyed.

calcite (calcspar) $CaCO_3$ The most common form of natural calcium carbonate. Dogtooth spar, Iceland spar, nailhead spar, and satin spar are varieties of calcite crystal. It is the essential ingredient of limestone, marble, and chalk (q.v.).
Properties: Colorless, white, and various colors; vitreous to earthy luster; good cleavage in 3 directions. May contain small amounts of magnesium, iron, manganese and zinc. Effervesces in acid.
Constants: Sp. gr. 2.72; hardness 2.
Uses: As a phosphor; Iceland spar is used in optical instruments; limestone, marble, and chalk have a variety of uses.

calcium Ca. Element of atomic number 20, group IIa of the periodic system; one of the alkaline earth elements.
Properties: Moderately soft, white metal; brilliant crystalline surface when freshly cut. Keep dry, in well-stoppered bottles. Soluble in acid; decomposes water, liberating hydrogen gas.
Constants: Sp. gr. 1.578 (15°C); m.p. 810°C, sublimes below its m.p. in vacuum; b.p. 1170°C.
Derivation: By electrolyzing molten calcium chloride.
Forms available: Metallic; crystalline.
Containers: Air-tight tins; glass bottles.
Uses: Manufacture of intermediates; alloys; dehydrating oils; metallurgical purposes; reducing agent in smelting; decarburization and desulfurization of iron and iron alloys; in place of sodium in chemical reactions; getter in vacuum tubes.
Shipping regulations: Flammable solid. Yellow label.*

calcium 45. Radioactive calcium of mass number 45.
Properties: Half-life 152 days; radiation, beta; radiotoxicity, very hazardous.
Derivation: By pile irradiation of calcium carbonate, by neutron bombardment of scandium, or as a by-product of the irradiation of calcium nitrate for the preparation of carbon 14.
Forms available: Calcium chloride in hydrochloric acid solution and solid calcium carbonate.
Uses: As a research aid for studying water purification, calcium exchange in clays, detergency, ion exchange, surface wetting and other surface phenomena, calcium uptake and deposition in bone, soil characteristics as related to soil utilization of fertilizer and crop yield, diffusion of calcium in glass, etc.
Shipping regulations: Poison, class D, radioactive material. Blue label.*

calcium abietate $(C_{20}H_{29}O_2)_2Ca$. Product of the action of lime on rosin or resin acids. See abietates; see also calcium resinate.

calcium acetate (vinegar salts; gray acetate; lime acetate; brown acetate) $Ca(C_2H_3O_2)_2 \cdot H_2O$.
Properties: Brown, gray or white (when pure) powder; amorphous or crystalline; decomposes on heating. Soluble in water; slightly soluble in alcohol.
Derivation: By the action of pyroligneous acid on calcium hydroxide, the solution being filtered and evaporated to dryness, yielding gray acetate of lime.
Grades: Technical (80% basis); reagent; C.P.; pure; brown; gray.
Uses: Medicine; manufacture of acetone, acetic acid, acetates; mordant in dyeing and printing of textiles; stabilizer in resins; additive to calcium soap lubricants.

calcium acetylsalicylate (aspirin, soluble) $Ca(CH_3COOC_6H_4COO)_2 \cdot 2H_2O$.
Properties: White powder. Aqueous solutions are unstable. Soluble in water.
Derivation: (a) Action of acetylsalicylic acid upon calcium carbonate in the presence of a small amount of water. (b) By passing carbon dioxide into an aqueous solution of calcium carbonate and acetylsalicylic acid.
Use: Medicine.

calcium acrylate $(H_2C{:}CHCOO)_2Ca$.
Properties: Free-flowing white powder; soluble in water. Forms a dihydrate which is also a free-flowing powder. Solutions polymerize readily to form a resin, which equilibrates with water.
Uses: Soil stabilization; sealing oil wells; ion exchange; binder for clay products and foundry molds.

calcium alginate ("Calginate").
Properties: A cream-colored, refined powder, having a moisture content of about 10%; insoluble in hot or cold water with slight swelling; pH about neutral; soluble in alkaline solutions; insoluble in organic solvents.
Containers: 10-, 100- and 300-lb drums.
Uses: For industrial and pharmaceutical applications.
Shipping regulations: None.*

calcium aluminate (tricalcium aluminate) $3CaO \cdot Al_2O_3$. A refractory, and an important ingredient of cements, especially of aluminous cement. See cement, Portland and other cement articles.

calcium-para-aminosalicylate $[C_6H_3(NH_2)(OH)COO]_2Ca \cdot 3H_2O$.
Properties: White to cream-colored crystals or powder. Odorless and has alkaline, slightly bitter-sweet taste. Somewhat hygroscopic. Solutions slowly decompose and darken in color. Soluble in water, methanol and acetone; slightly soluble in alcohol.

Containers: Fiber drums.
Grade: N. F. XI.
Use: Medicine.

calcium ammonium nitrate. A homogeneous mixture of about 60% ammonium nitrate and 40% limestone and/or dolomite. It is a fertilizer material containing about 20% nitrogen.
Shipping regulations: Oxidizing material. Yellow label.*

calcium arsenate (tricalcium ortho-arsenate) $Ca_3(AsO_4)_2$.
Properties: White powder. Poisonous! Very slightly soluble in water, soluble in dilute acids.
Derivation: By the interaction of calcium chloride and sodium arsenate.
Grades: Technical; C. P.
Containers: Bottles; bags; 100-lb barrels; multiwall paper sacks.
Uses: Insecticide; germicide.
Warning: Poisonous if swallowed. M. C. A. warning label.
Shipping regulations: Poison, class B. Poison label.*

calcium arsenite $CaAsO_3H$.
Properties: White, granular powder. Poisonous! Insoluble in water; soluble in acids.
Grades: Technical.
Containers: Wooden barrels.
Uses: Germicides; insecticides.
Shipping regulations: Poison, class B. Poison label.*

calcium benzoate $Ca(C_7H_5O_2)_2 \cdot 3H_2O$.
Properties: White powder or crystals, odorless. Keep well stoppered. Soluble in water. Sp. gr. 1.44.
Derivation: Oxidation of toluene followed by precipitation by milk of lime.
Grades: Technical.
Containers: Glass bottles.
Use: Medicine.
Shipping regulations: None.*

calcium bichromate. See calcium dichromate.

calcium biphosphate. See calcium phosphate, monobasic.

calcium bisulfide. See calcium hydrosulfide.

calcium bisulfite $Ca(HSO_3)_2$. Exists only in solution and is really a solution of calcium sulfite in an aqueous sulfur dioxide solution.
Properties: Yellowish liquid with strong sulfur-dioxide odor; sp. gr. 1.06; corrosive to metals. Soluble in water and acids.
Derivation: By the action of sulfur dioxide on calcium hydroxide solution.
Grades: Technical (8° Bé.).
Containers: Iron drums.
Uses: Antichlor in bleaching textiles; paper pulp (dissolving lignin); preservative; bleaching sponges; chromium bisulfite; hydroxylamine salts; germicide; disinfectant.
Shipping regulations: None.*

calcium borocitrate.
Properties: Fine, white powder. Soluble in dilute acids; slightly soluble in water.
Use: Medicine.

calcium bromide $CaBr_2 \cdot 2H_2O$.
Properties: White, granular salt; odorless; sharp saline taste; very deliquescent; m. p. 38°C; b. p. 149-150° (dec). Soluble in water and alcohol; insoluble in chloroform and ether.
Derivation: By the action of hydrobromic acid on calcium oxide, carbonate, or hydroxide and subsequent crystallization.
Method of purification: Recrystallization.
Grades: Technical; C. P.; N. F. XI (84-94% $CaBr_2$).
Containers: 25-, 100-lb jars; 25-, 100-lb drums.
Uses: Manufacturing mineral waters; photography; medicine; dehydrating agent; food preservative; road treatment; freezing mixtures; sizing compounds; wood preservative.
Shipping regulations: None.*

calcium carbide (carbide; acetylenogen) CaC_2.
Properties: Grayish-black, irregular lumps; must be kept dry. Sp. gr. 2.22; m. p. about 2300°C. Decomposes in water, with formation of acetylene gas (flammable) and calcium hydroxide (skin irritant upon prolonged contact).
Derivation: By the interaction of finely pulverized limestone or quicklime with crushed coke or anthracite coal in an electric furnace.
Grades: Technical.
Containers: 2 lbs to 5 tons; metal packages, water and airtight.
Uses: Generation of acetylene gas, for welding, illumination, industrial and synthetic chemical purposes; signal fires; reduction of copper sulfide and metallic oxides; production of calcium cyanamide; manufacture of graphite and hydrogen; electric (dehydrating agent in electrostatic work); desiccated foods (dehydrating agent); synthesis of acetaldehyde and acetic acid; steel hardening.
Fire hazard: Must be kept away from water.
Shipping regulations: None.*

calcium carbonate $CaCO_3$.
Properties of pure calcium carbonate: White powder or colorless crystals, odorless, tasteless; sp. gr. 2.7-2.95; decomposes at 825°C; insoluble in water and alcohol; soluble in acids with evolution of carbon dioxide.
Calcium carbonate is one of the most stable, common, and widely dispersed of materials. It occurs in nature as aragonite, calcite, chalk, limestone, lithographic stone, marble, marl, and travertine.
Powdered calcite or limestone and marble chips are accurate source names, but there is confusion and overlapping among the other names for the powdered varieties. These are the various whites, whitings,

and chalks. Precipitated chalk is necessarily a synthetic substance and if of U. S. P. grade must meet the strictest purity requirements of all forms.

Prepared chalk is made from chalk and may be identical with whiting, but if of N. F. grade it also must meet strict purity requirements. Whiting may be powdered chalk or limestone, although chalk was the traditional source.

Containers: Wooden barrels; multiwall paper sacks; lined box cars.

Uses: See under various forms listed.

calcium carbonate, precipitated. See chalk, precipitated.

calcium carbonate, prepared. See chalk, prepared.

calcium caseinate.

Properties: White or slightly yellow, nearly odorless powder. Insoluble in cold water; forms a milky solution when suspended in water, stirred and heated.

Containers: 100-lb bags.

Uses: Medicine; special foods.

calcium chlorate $Ca(ClO_3)_2 \cdot 2H_2O$.

Properties: White to yellowish crystals. Keep well stoppered. Melts when rapidly heated at 100°C. Soluble in water and alcohol. Hygroscopic. Sp. gr. 2.711.

Derivation: By the action of chlorine on hot calcium hydroxide slurry.

Method of purification: Crystallization.

Grades: Technical; reagent.

Containers: Iron canisters; glass bottles.

Uses: Photography, pyrotechnics; dusting powder to kill poison ivy.

Fire hazard: Dangerous.

Shipping regulations: Oxidizing material. Yellow label.*

calcium chloride (a) $CaCl_2$; (b) $CaCl_2 \cdot H_2O$; (c) $CaCl_2 \cdot 6H_2O$.

Properties: White, deliquescent crystals, granules, lumps or flakes. Keep well closed. Sp. gr. (a) 2.151 (25°C); (c) 1.68 (17°C); m.p. (a) 772°C, (b) 260°C, (c) 30°C; b.p. (a) greater than 1600°C, (c) loses all its water at 200°C. Soluble in water and alcohol.

Derivation: (a) By the action of hydrochloric acid on calcium carbonate and subsequent crystallization. (b) Commercially obtained as a by-product in the Solvay soda and other processes.

Method of purification: Recrystallization.

Grades: C. P.; U. S. P. XVI; technical; solid (73-75%); powder (77-80%); flake (77-80%); liquid (40-45%).

Containers: C. P.: 1-, 5-lb bottles; powder: 400-lb drums; solid: 650-, 655-lb drums; flake: 100-lb bags, 375-, 400-lb drums; liquid: tank cars.

Uses: (a) Road treatment (dust-proofing, thawing snow and ice, binding unpaved surfaces); drilling muds; weighting water for drilling; dustproofing, freezeproofing, and thawing coal, coke, stone, sand, ore; concrete conditioning; paper and pulp industry; fungicides; ballast for weighting implement tires; (a), (b) and (c) are used in general as a cheap drying agent and chemical reagent.

Shipping regulations: None. *

calcium chlorite.

Shipping regulations: Oxidizing material. Yellow label. *

calcium chromate (gelbin; Steinbuhl yellow; yellow ultramarine) $CaCrO_4 \cdot 2H_2O$. M. p., loses water at 200°C. Soluble in dilute acids and alcohol; slightly soluble in water.

Containers: Bags.

Grades: Technical.

Use: Pigment; corrosion inhibitor.

calcium citrate (lime citrate) $Ca_3(C_6H_5O_7)_2 \cdot 4H_2O$. A by-product in the manufacture of citric acid.

Properties: A white odorless powder; loses most of its water at 100°C and all of it at 120°C. Almost insoluble in water; insoluble in alcohol.

Grades: Reagent; technical.

Containers: Glass bottles; 50-lb boxes; 100-lb kegs; barrels.

Use: Medicine; production of citric acid; dietary supplement.

Shipping regulations: None. *

calcium cyanamide (lime nitrogen; cyanamide) $CaCN_2$.

Properties: Gray-black lumps or powder; sp. gr. 1.083. Decomposes in water, liberating ammonia.

Derivation: Calcium carbide is finely powdered and heated in an electric oven, into which nitrogen is passed. The charge remains in the furnace 24 to 26 hrs. It is then removed and any uncombined calcium carbide is leached out.

Grades: Fertilizer, 21% N; industrial.

Containers: 200-lb bags; drums; bulk in cars.

Uses: Fertilizer; weed killer; nitrogen products; hardening iron and steel.

Shipping regulations: None. *

calcium cyanide $Ca(CN)_2$.

Properties: Colorless crystals or white powder; gray-black (technical); decomposes in moist air liberating hydrogen cyanide. Dissolves in water and very weak acid, with liberation of hydrogen cyanide gas.

Grades: Made in different granulations for different uses.

Containers: 4-oz to 100-lb metal containers.

Uses: Killing ants, rats, mice, moles, and similar burrowing insects and rodents; also for fumigating greenhouses, mushroom houses, flour mills, grain, and seed; for fumigating citrus trees under tents for control of scale insects; leaching of gold and silver ores; other cyanides.

Caution: Liberates poisonous gas; avoid contact with skin and breathing gas or dust. MCA warning label.

Shipping regulations: Poison, class B. Poison label.*

calcium cyclamate (calcium cyclohexylsulfamate) $(C_6H_{11}NHSO_3)_2Ca \cdot 2H_2O$.

Properties: White, crystalline, practically odorless powder with very sweet taste. Freely soluble in water (solutions are neutral to litmus); practically insoluble in alcohol, benzene, chloroform and ether; pH (10% solution) 5.5-7.5. Sweetening power approximately 30 times that of sucrose.
Grade: U.S.P. XVI.
Containers: 100-lb drums.
Uses: Sweetening agent in certain soft drinks, and in low-calorie and diabetic diets.

calcium cyclohexenylethylbarbiturate. See cyclobarbital calcium.

calcium cyclohexylsulfamate. See calcium cyclamate.

calcium dehydroacetate $(C_8H_7O_4)_2Ca$. See also dehydroacetic acid.
Properties: White to cream powder. Almost insoluble in water and organic solvents.
Grades: 96% minimum.
Use: Fungicide.

calcium dibromobehenate $(C_{22}H_{41}O_2Br_2)_2Ca$.
Properties: White to yellow powder; odorless; tasteless. Protect from light. Soluble in ether, chloroform, acetone, carbon tetrachloride, and benzene; insoluble in water or alcohol.
Use: Medicine.
Shipping regulations: None.*

calcium dichromate (calcium bichromate) $CaCr_2O_7 \cdot 3H_2O$, or 4.5 H_2O.
Properties: Brownish-red crystals; deliquescent; sp. gr. (4.5 H_2O variety) 2.136. Soluble in water.
Grades: Technical; C.P.
Use: Corrosion inhibitor.

calcium dioxide. See calcium peroxide.

calcium disodium EDTA (calcium disodium ethylenedinitrilotetraacetate; ethylenediaminetetraacetic acid calcium disodium chelate; edathamil calcium disodium) $C_{10}H_{12}N_2O_8CaNa_2 \cdot 2H_2O$.
Properties: White, odorless powder or flakes; soluble in water; insoluble in organic solvents. It acts as a chelating agent for heavy metals.
Grade: N.N.D. (edathamil calcium disodium)
Uses: Medicine; in foods to "complex" trace heavy metals, and as a preservative; antigushing agent in fermented malt beverages.

calcium disodium ethylenedinitrilotetraacetate. See calcium disodium EDTA.

calcium 2,2'-dithiobisdibenzoate $Ca(OOCC_6H_4S)_2$. Off-white to grayish powder.

calcium ethylhexoate. See soaps, metallic.

calcium ferrocyanide $Ca_2Fe(CN)_6 \cdot 12H_2O$.
Properties: Yellow crystals; decompose on heating. Soluble in water; insoluble in alcohol; sp. gr. 1.68.
Derivation: By decomposing ferriferrocyanide with quicklime in a closed vessel with steam. The solution of calcium ferrocyanide is evaporated and recovered by crystallization.
Grades: Technical.
Containers: Wooden barrels.
Use: For the removal of metallic impurities in the manufacture of citric, tartaric, and other acids.
Shipping regulations: None.*

calcium fluoride CaF_2.
Properties: White powder, occurring in nature as fluorite (pure form) or fluorspar (mineral). Reacts with hot concentrated sulfuric acid to liberate hydrofluoric acid. Insoluble in water; sp. gr. 3.18.
Derivation: (a) By powdering pure fluorite or fluorspar; (b) by the interaction of a soluble calcium salt and sodium fluoride.
Grades and Uses: See fluorspar. Single pure crystals of calcium fluoride are also produced, for use in spectroscopy.

calcium fluosilicate (calcium silicofluoride) (a) $CaSiF_6$; (b) $CaSiF_6 \cdot 2H_2O$.
Properties: White, crystalline powder. Sp. gr. (a) 2.662 (17.5°C), (b) 2.254. Very slightly soluble in water.
Derivation: By the action of fluosilicic acid on calcium carbonate and subsequent crystallization.
Method of purification: Recrystallization.
Grades: Technical.
Containers: Wooden barrels; multiwall paper sacks.
Use: Ceramics.
Shipping regulations: None.*

calcium folinate (calcium leucovorin) $C_{20}H_{21}CaN_7O_7 \cdot 5H_2O$. The calcium salt of folinic acid, formerly called the citrovorum factor.
Properties: Yellowish-white or yellow, odorless microcrystalline powder. Very soluble in water. Practically insoluble in alcohol.
Grade: U.S.P. XVI.
Use: Medicine.

calcium fructose diphosphate. See fructose diphosphates, calcium and barium salts.

calcium gluconate $Ca(C_6H_{11}O_7)_2 \cdot H_2O$.
Properties: White, odorless, practically tasteless, fluffy powder or granules. Soluble in hot water; less soluble in cold water; insoluble in alcohol, acetic acid, and other organic solvents; specific rotation (20/D) about +6°. Solutions neutral to litmus.
Derivation: Neutralization of gluconic acid with lime or calcium carbonate.
Method of purification: Crystallization.
Grades: Technical; U.S.P. XVI; special for ampules.
Containers: Cans; fiber drums; barrels.
Uses: Medicine and veterinary medicine.
Shipping regulations: None.*

calcium glutamate. Similar to sodium glutamate (q.v.).

calcium glycerinophosphate. See calcium glycerophosphate.

calcium glycerophosphate (calcium glycerinophosphate) $CaC_3H_7O_2PO_4$.
Properties: White, crystalline powder; odorless; almost tasteless; slightly hygroscopic; decomposes above 170°C; slightly soluble in water; insoluble in alcohol.
Derivation: By esterification of phosphoric acid with glycerol and conversion of glycerophosphoric acid to the calcium salt.
Grades: Technical; pure.
Containers: 1-, 5-lb bottles; 5-, 10-, 25-lb tins; 200-lb barrels.
Uses: Medicine and veterinary medicine; stabilizer for plastics; dietary supplement.
Shipping regulations: None. *

calcium glycolate $(CH_2OHCOO)_2Ca$.
Properties: White solid.
Grades: Technical.
Use: As convenient source of glycolic acid and of the glycolic acid radical in chemical synthesis.

calcium grease. See lubricating greases.

calcium hippurate $Ca(C_9H_8NO_3)_2 \cdot 3H_2O$.
Properties: White crystals or powder; soluble in water; sp. gr. 1.32.
Derivation: Interaction of calcium carbonate with hippuric acid.
Use: Medicine.
Shipping regulations: None. *

calcium hydrate. See calcium hydroxide.

calcium hydride CaH_2.
Properties: Grayish-white lumps or crystals. Acted upon by moist air with formation of calcium hydroxide and evolution of hydrogen. Sp. gr. 1.7; decomposes at 675°C. Decomposed by water, acids, and lower alcohols.
Grades: Technical; C. P.
Containers: Drums.
Uses: Reducing agent; drying agent; analytical reagent in organic chemistry; easily portable source of hydrogen; cleaner for blocked-up oil wells.
Caution: Avoid humidity. Flammable gas forms.

calcium hydrosulfide (calcium bisulfide; calcium sulfhydrate) $Ca(HS)_2 \cdot 6H_2O$.
Properties: Colorless, transparent crystals. Soluble in alcohol and water.
Caution: Keep well stoppered! Decomposes in air (15-18°C).
Use: Leather industry to produce a smoother leather.

calcium hydroxide (calcium hydrate; lime hydrate; caustic lime; slaked lime) $Ca(OH)_2$.
Properties: Soft, white crystalline powder with alkaline, slightly bitter taste. Sp. gr. 2.34; m. p., loses water at 580°C. Very slightly soluble in water; soluble in glycerin, syrup, and acids; insoluble in alcohol. Absorbs carbon dioxide from air.
Derivation: By the action of water on calcium oxide.
Impurities: Calcium carbonate, magnesium salts, iron.
Grades: Technical; chemical lime (insoluble matter under 2%, Mg under 3%); building lime; U. S. P. XVI; C. P.
Containers: Wooden barrels; multiwall paper sacks; bulk.
Uses: Mortar; plasters; cements; calcium salts; causticizing soda; hydrogen; depilatory; lime paints; medicine; agriculture (to "sweeten" acid soil); ammonia recovery in gas manufacture; candle manufacture; disinfectant; water softening; purification of sugar juices; hard-rubber products; water paints; soil stabilizers, petrochemicals.
Shipping regulations: None. *

calcium hypochlorite (calcium oxychloride) $Ca(OCl)_2$.
Properties: White crystalline solid; soluble in water; not hygroscopic; practically clear in water solution. Stable chlorine carrier.
Derivation: The chlorination of a slurry of lime and caustic soda with the subsequent precipitation of calcium hypochlorite dihydrate which is dried under vacuum.
Grades: Commercial (70%); high purity (99.2% available chlorine as calcium hypochlorite).
Containers: Steel drums and cans; plastic and glass bottles.
Uses: Algecide; bactericide; deodorant; disinfectant; fungicide; and bleaching agent.
Shipping regulations: Oxidizing material. Yellow label. *
See also chlorinated lime.

calcium hypophosphite (lime hypophosphite) $Ca(H_2PO_2)_2$.
Properties: Colorless, transparent crystals or white to grayish white crystalline powder; odorless; nauseous bitter taste; soluble in water; insoluble in alcohol. Explosive when triturated or heated with nitrates, chlorates or other oxidants.
Derivation: By boiling lime, water and phosphorus together, with subsequent crystallization from the solution.
Method of purification: Recrystallization.
Grades: Technical.
Containers: 1-lb bottles; cartons; 5-, 10-, 25-, 50-lb tins; drums.
Use: Medicine.
Caution: Evolves spontaneously flammable phosphine when heated above 300°C.
Shipping regulations: None. *

calcium hyposulfite. See calcium thiosulfate.

calcium iodate $Ca(IO_3)_2 \cdot 6H_2O$.
Properties: White crystals or powder; odorless; m. p. 35°C, decomposes. Soluble in water and nitric acid; insoluble in alcohol.
Grades: Technical; C. P.
Uses: Deodorant; medicine; mouth washes; feed additive.

calcium iodide $CaI_2 \cdot 6H_2O$.
Properties: Yellowish-white crystals; deliquescent; decomposes in air by absorption of carbon dioxide. Soluble in water, ethyl alcohol, and amyl alcohol; sp. gr. 2.55;

loses 6 H_2O at 42°C; m.p. 783°C.
Derivation: By the action of hydriodic acid on calcium carbonate.
Method of purification: Crystallization.
Grades: Technical; C.P.
Containers: 1-, 5-, 7-lb bottles; 25-lb jars.
Use: Photography; medicine.
Shipping regulations: None.*

calcium iodobehenate $(C_{22}H_{42}O_2I)_2Ca$.
Properties: White or yellowish powder containing approximately 24% iodine; unctuous to the touch; odorless, or slight fatty odor. Soluble in warm chloroform; only slightly soluble in alcohol and ether; insoluble in water.
Derivation: Combination of erucic and hydriodic acids with a soluble calcium salt.
Use: Medicine; feed additive.
Shipping regulations: None.*

calcium lactate $Ca(C_3H_5O_3)_2 \cdot 5H_2O$.
Properties: White, almost tasteless powder; almost odorless. Soluble in water, practically insoluble in alcohol. Loses H_2O at 120°C.
Derivation: By neutralizing dilute lactic acid with calcium carbonate and evaporating the solution.
Grades: N.F. XI; edible.
Containers: 25-, 100-, 200-lb drums.
Uses: Medicine, veterinary medicine; in the manufacture of foods, beverages.
Shipping regulations: None.*

calcium leucovorin. See calcium folinate.

calcium levulinate
$Ca[CH_3CO(CH_2)_2COO]_2 \cdot 2H_2O$.
Properties: White crystalline or amorphous powder, having a faint odor like burnt sugar; bitter salty taste. Freely soluble in water; slightly soluble in alcohol; insoluble in ether and chloroform; m.p. 119-125°C.
Grades: C.P.; N.N.D.
Containers: Drums.
Use: Medicine; food additive.

calcium lignosulfonate. Used as a binder for non-magnetic ores. See also lignin sulfonates.

calcium linoleate $Ca(C_{18}H_{31}O_2)_2$.
Properties: White, amorphous powder. Soluble in alcohol and ether; insoluble in water.
Derivation: By the interaction of solutions of calcium chloride and sodium linoleate.
Grades: Technical.
Containers: Wooden kegs.
Uses: Waterproofing compounds; emulsifying agent and stabilizer for flat paints, fillers, and enamels.
Shipping regulations: None.*

calcium magnesium aconitate. See dicalcium magnesium aconitate.

calcium magnesium carbonate. See dolomite.

calcium magnesium chloride (magnesium calcium chloride) $CaCl_2 \cdot MgCl_2$.
Properties: White, deliquescent crystals; soluble in water and acids; insoluble in alcohol and ether.
Derivation: (a) A by-product in the salt industry; (b) by the action of hydrochloric acid on dolomite.
Method of purification: Crystallization.
Grades: Technical.
Use: Manufacture of intermediates, dyes, fireproof paints, paper and textile sizing; as preservatives; laboratory reagent; dehydrating starch.
Shipping regulations: None.*

calcium mandelate $Ca(C_8H_7O_3)_2$.
Properties: White, odorless powder. Insoluble in alcohol; slightly soluble in cold water.
Containers: 150-lb drums.
Use: Medicine and pharmaceuticals.

calcium metasilicate. See calcium silicates.

calcium methylate $Ca(OCH_3)_2$. A white powder; used as a catalyst intermediate.

calcium molybdate $CaMoO_4$.
Properties: White, crystalline powder; sp. gr. 4.35; soluble in mineral acids; insoluble in alcohol, ether, or water.
Derivation: By the fusion of calcium oxide and a molybdenum ore.
Grades: Technical.
Containers: Wooden barrels.
Use: Molybdic acid; as alloying agent in the production of iron and steel.
Shipping regulations: None.*

calcium naphthenate.
Properties: Light sticky tenacious mass. Insoluble in water; soluble in ethyl acetate, carbon tetrachloride, gasoline, benzene and ether.
Derivation: Precipitation from aqueous solution of calcium salts and sodium naphthenate.
Containers: Drums.
Uses: Waterproofing compositions; adhesives; wood fillers; grafting waxes; cements; varnishes; color lakes.

calcium beta-naphtholsulfonate
$Ca(C_{10}H_6OHSO_3)_2 \cdot 3H_2O$.
Properties: Pale red powder; odorless. Decomposes at about 50°C; soluble in water and alcohol.
Use: Medicine; brewing.
Shipping regulations: None.*

calcium nitrate (lime nitrate; nitrocalcite; Norge niter; lime saltpeter; Norwegian saltpeter) (a) $Ca(NO_3)_2 \cdot 4H_2O$; (b) $Ca(NO_3)_2$.
Properties: White, deliquescent mass. Soluble in water, alcohol and acetone. Sp. gr. (a) 1.82, (b) 2.36; m.p. (a) 42°C, (b) 561°C.
Derivation: By oxidation of ammonia to nitric acid, followed by neutralization with lime.
Method of purification: Crystallization.
Grades: Technical; pure; C.P.; reagent.
Containers: Wooden kegs; glass bottles.
Uses: Pyrotechnics; explosives; matches; fertilizers; other nitrates; incandescent gas mantles; radio tube manufacture; coagulant for rubber latex.
Shipping regulations: Oxidizing material. Yellow label.*

calcium nitride Ca_3N_2.
Properties: Brown crystals; sp. gr. 2.63 (17°C); m.p. 900°C. Soluble in water with decomposition; soluble in dilute acids; insoluble in absolute alcohol.

calcium nitrite $Ca(NO_2)_2 \cdot H_2O$.
Properties: Colorless or yellowish, brittle lumps. Hygroscopic. Caution! Keep well stoppered! Soluble in water; slightly soluble in alcohol. Sp. gr. 2.23 (34°C, anhydrous); m.p., loses water at 100°C.
Grades: Technical.

calcium octoate. See soaps, metallic.

calcium orthophosphate. See calcium phosphate, tribasic.

calcium orthotungstate. See calcium tungstate.

calcium oxalate CaC_2O_4.
Properties: White, crystalline powder; soluble in dilute hydrochloric acid, dilute nitric acid; insoluble in acetic acid and water. Sp. gr. 2.2.
Grades: Technical; C.P.
Containers: Barrels; bags; kegs.
Use: Making oxalic acid and organic oxalates.

calcium oxide (lime; quicklime; burnt lime; calx; fluxing lime) CaO.
Properties: White, or grayish-white hard lumps, sometimes with a yellowish or brownish tint, due to iron; odorless. Crumbles on exposure to moist air; sp. gr. 3.40; m.p. 2570°C; b.p. 2850°C. Soluble in acid; very slightly soluble in water, uniting to form calcium hydroxide.
Derivation: Calcium carbonate (usually limestone; sometimes oyster shells) is roasted in kilns until all of the carbon dioxide is driven off.
Impurities: Calcium carbonate; magnesium, iron, and aluminum oxides.
Grades: N.F. XI; technical; chemical lime; agricultural lime; building lime.
Containers: 1-, 5-lb cans; 25-, 50-lb kegs; wooden barrels; bags; freight cars; multiwall paper sacks.
Uses (in approximate order of volume): refractory; flux in steel manufacture; construction; pulp and paper; calcium carbide; water treatment; nonferrous metallurgy; glass making; other chemicals; waste treatment; insecticides and fungicides; leather tanning; petroleum refining; food processing, sugar refining; in general, as a cheap industrial alkali.
Caution: Heats upon contact with water or moisture and may cause ignition of organic material.
Shipping regulations: None.*

calcium oxychloride. See calcium hypochlorite.

calcium palmitate $Ca(C_{15}H_{31}CO_2)_2$. White or pale-yellow powder produced by reacting sodium palmitate with a soluble calcium salt. Insoluble in water; soluble in alcohol or ether. Used as waterproofing agent, thickener for lubricating oils, and in the manufacture of solidified oils. Available only as technical grade.

This material is also an important component of the curdy precipitate formed when hard waters act on soap. It is often referred to as a calcium soap.

calcium pantothenate $(C_9H_{16}NO_5)_2Ca$, or $[HOCH_2C(CH_3)_2CH(OH)CONH(CH_2)_2COO]_2Ca$. The calcium salt of pantothenic acid; possessing vitamin activity. It is available in either the dextro- or racemic forms.
Properties: (both forms identical) White, slightly hygroscopic, odorless powder; sweetish taste; stable in air; solutions have a pH of 7-9; soluble in water and glycerol; insoluble in alcohol, chloroform, and ether. M.p. 170-172°C; dec. 195-196°C; specific rotation (5% aqueous solution) + 28.2°(25°C).
Source: Same as pantothenic acid.
Grades: U.S.P. XVI (both forms).
Containers: 1-kilo jars.
Uses: Medicine; nutrition; animal feeds; same biological use as pantothenic acid.

calcium perborate $Ca(BO_3)_2 \cdot 7H_2O$.
Properties: Gray-white lumps or powder. Soluble in acids; also in water with partial decomposition.
Uses: Medicine; as a bleach; in tooth powders.

calcium permanganate $Ca(MnO_4)_2 \cdot 4H_2O$.
Properties: Violet crystals, deliquescent. Keep well stoppered. Soluble in water and ammonia; decomposed by alcohol.
Grades: Technical; pure.
Containers: Glass bottles; tins; wooden barrels.
Uses: Textile industry; sterilizing water; dentistry; disinfectant; deodorizer; an additive (with hydrogen peroxide) in liquid rocket propellants.
Shipping regulations: Oxidizing material Yellow label.*

calcium peroxide (calcium superoxide; calcium dioxide) CaO_2.
Properties: White or yellowish, odorless, almost tasteless powder. Decomposes about 200°C. Practically insoluble in water; soluble in acids with formation of hydrogen peroxide. Available oxygen 22.2% (min. 13.3% in technical grade).
Derivation: By the interaction of solutions of a calcium salt and sodium peroxide, with subsequent crystallization.
Grades: 60-75%.
Containers: 25-, 100-, 200-lb drums.
Uses: Seed disinfectant; dentifrices; dough conditioners; medicine; bleaching of oils; modification of starches; high temperature oxidations.
Fire hazard: Dangerous.
Shipping regulations: Oxidizing material. Yellow label.*

calcium phenolsulfonate (calcium sulfocarbolate; calcium sulfophenate; calcium sulfophenylate) $Ca(C_6H_4OHSO_3)_2 \cdot H_2O$.
Properties: White, crystalline powder; odorless. Soluble in water and alcohol.
Derivation: By the action of phenolsulfonic acid on calcium hydroxide.
Grades: Technical.

Containers: 1-lb bottles; 5-, 25-lb cans; drums.
Uses: Disinfectant; medicine.
Shipping regulations: None.*

calcium phosphate. See calcium phosphate, dibasic; calcium phosphate, monobasic; or calcium phosphate, tribasic.

calcium phosphate, acid. See calcium phosphate, monobasic.

calcium phosphate, antimoniated (James' powder). Mixture of precipitated calcium phosphate and antimony trioxide in the proportion of 2:1.
Properties: Greenish-gray, gritty powder; odorless; tasteless. Insoluble in water.

calcium phosphate, dibasic (dicalcium orthophosphate; bicalcic phosphate; secondary calcium phosphate) $CaHPO_4 \cdot 2H_2O$ and $CaHPO_4$.
Properties: White, tasteless, crystalline powder; odorless; soluble in dilute hydrochloric, nitric, and acetic acids; insoluble in alcohol; slightly soluble in water. (Hydrate) sp. gr. 2.306.
Derivation: Interaction of fluorine-free phosphoric acid with milk of lime.
Grades: Food grade; dentifrice grade; N.F. XI; feed grade, 18 ½ or 21% P.
Containers: Anhydrous: 50-lb bags; 100-, 350-lb drums; hydrate: 60-, 100-lb bags; 100-, 225-lb fiber drums; feed grade: 100-lb bags; bulk.
Uses: Animal feed supplement; food supplement; dentifrice; medicine; glass; fertilizer; stabilizer for plastics.
Shipping regulations: None.*

calcium phosphate, monobasic (calcium biphosphate; acid calcium phosphate; calcium phosphate, primary; monocalcium phosphate) $CaH_4(PO_4)_2 \cdot H_2O$.
Properties: Colorless, pearly scales or powder, deliquescent in air. Soluble in water and acids. Aqueous solutions are acid.
Constants: M.p., loses H_2O at 100°C, decomposes at 200°C; sp. gr. 2.20.
Derivation: By dissolving either dicalcium or tricalcium phosphates in phosphoric acid and allowing the solution to evaporate spontaneously.
Grades: Technical; C.P.
Containers: 100-lb bags; 350-lb drums.
Uses: Leavening agent (baking powders); as a plant food; mineral supplement; stabilizer for plastics; to control pH in malt; glass manufacture.
Shipping regulations: None.*

calcium phosphate, precipitated. See calcium phosphate, tribasic.

calcium phosphate, primary. See calcium phosphate, monobasic.

calcium phosphate, secondary. See calcium phosphate, dibasic.

"Calcium Phosphate SF-52." [172] Trade name for a proprietary product, essentially an anhydrous dicalcium phosphate (see calcium phosphate, dibasic).
Properties: Greyish-white, granular, free-flowing material, designed especially for stock food enrichment. Particularly suitable in animal salt blocks.
Containers: 100-lb paper bags.
Use: Mineral supplement for animal and poultry feeds.

calcium phosphate, tertiary. See calcium phosphate, tribasic.

calcium phosphate, tribasic (calcium orthophosphate; tricalcium phosphate; precipitated calcium phosphate; tricalcium orthophosphate; tricalcic phosphate; tertiary calcium phosphate) $Ca_3(PO_4)_2$. (Sometimes contains some $Ca(OH)_2$). See also bone ash.
Properties: The precipitated product is a white, odorless, tasteless, amorphous powder. Sp. gr. 3.18; m.p. 1670°C. Soluble in acids; insoluble in water, alcohol, and acetic acid. Permanent in air.
Derivation: Found abundantly in nature as phosphate rock, apatite, and phosphorite (q.v.). By the interaction of solutions of calcium chloride and sodium triphosphate with excess of ammonia. By interaction of hydrated lime and phosphoric acid.
Grades: Granular; technical; C.P.; N.F. XI; pure precipitated.
Containers: 60-, 100-lb multiwall paper bags; 100-lb drums.
Uses: Ceramics (porcelains, potteries, enamels, milk glass); calcium acid phosphate; phosphorus and phosphoric acid; polishing powder; cattle foods; clarifying sugar syrups; medicine (food); rubber; mordant (dyeing textiles with Turkey red); fertilizers; dentifrices; stabilizer for plastics.
Shipping regulations: None.*

calcium phosphide (photophor) Ca_3P_2.
Properties: Red-brown crystals or gray granular masses; sp. gr. 2.51 (15°C); m.p. about 1600°C; decomposed by water to form phosphine, which is spontaneously flammable; insoluble in alcohol and ether.
Note: Authorities differ both as to formula and decomposition product. Some claim the formula to be Ca_2P_2 and state that the gas liberated by contact with water is flammable because impurities are present. Apparently calcium phosphide varies in composition according to its method of preparation.
Derivation: By heating calcium phosphate with aluminum or carbon; by passing phosphorus vapors over metallic calcium.
Grades: Technical.
Containers: Iron canisters.
Uses: Signal fires; torpedoes.
Fire hazard: Dangerous.
Shipping regulations: Flammable solid. Yellow label.*

calcium phosphite $CaHPO_3 \cdot 2H_2O$.
Properties: White powder; loses its water at 200-300° (with decomposition). Slightly soluble in water; insoluble in alcohol.

calcium phytate (hexacalcium phytate) $C_6H_6(CaPO_4)_6$.
Derivation: Corn steep liquor.
Containers: 50-lb bags.
Uses: To remove excess metals from wine and vinegar; as a source of calcium in pharmaceuticals; source of phytic acid and its salts.

calcium plumbate Ca_2PbO_4.
Properties: Orange to brown crystalline powder; decomposed by hot water or carbon dioxide; sp. gr. 5.71. Soluble in acids (with decomposition); insoluble in cold water.
Uses: Oxidizing agent; pyrotechnics and safety matches; glass; storage batteries.

calcium propionate $Ca(OOCCH_2CH_3)_2$. (Occurs also with one H_2O.) A white powder soluble in water; slightly soluble in alcohol.
Uses: As an antifungal agent in bread.

calcium propyl arsonate $C_3H_7AsO_3Ca$. Crystals, soluble in water. Used as pre-emergence control of crab grass.

calcium pyrophosphate $Ca_2P_2O_7$.
Properties: White powder. Soluble in dilute hydrochloric and in nitric acids; insoluble in water. Sp. gr. 3.09; m.p. 1230°C.
Containers: 100-lb bags; 400-lb barrels; 100-lb drums.
Uses: Polishing agent in dentifrices; mild abrasive for metal polishing; food supplement.

calcium resinate.
Properties: Yellowish-white, amorphous powder or lumps. Soluble in acid; insoluble in water; soluble in amyl acetate, butyl acetate, ether, amyl alcohol.
Derivation: By boiling calcium hydroxide with rosin and filtering; fusion of hydrated lime and melted rosin.
Grades: Technical.
Containers: Wooden barrels; fiber drums.
Uses: Waterproofing; manufacturing paint driers, porcelains, perfumes, cosmetics, enamels; for fabrics, wood, paper; as amber substitute; tanning leather.
Shipping regulations: Flammable solid. Yellow label. *

calcium rhodanate. See calcium thiocyanate.

calcium ricinoleate $Ca[CH_3(CH_2)_5CHOHCH_2CHCH(CH_2)_7CO_2]_2$. A fine white powder with a slight odor of fatty acids. Derived from castor oil.
Properties: M.p. 98°C; sp. gr. 1.04.
Containers: 50-lb bags.
Uses: Greases and lubricants; nontoxic stabilizer for polyvinyl chloride.

calcium D-saccharate $CaC_6H_8O_8 \cdot 4H_2O$.
Properties: White, crystalline powder; odorless; tasteless; insoluble in water and alcohol; soluble in calcium gluconate solutions.
Derivation: Oxidation of D-gluconic acid and neutralization with lime.
Containers: Fiber drums; cans.
Uses: Medicine.
Shipping regulations: None. *

calcium saccharin $(C_6H_4COSO_2N)_2Ca$. Fine white powder; used as a sugar substitute.

calcium salicylate $Ca(C_7H_5O_3)_2 \cdot 2H_2O$.
Properties: White powder; odorless; tasteless. Loses all its water at 120°C. Soluble in water; insoluble in alcohol.
Grades: Purity 99+%.
Use: Medicine.
Shipping regulations: None. *

calcium sequestration. The use of ordinary sodium hexametaphosphate and similar compounds that function as water softeners by forming complex ions with calcium and magnesium ions, thus preventing the formation of insoluble calcium and magnesium curds or precipitates when mixed with soap.
See also sequestration.

calcium silicates. Some occur naturally (see wollastonite); some are found in mixtures, especially in Portland cement, which has been extensively studied. (See cement, Portland.) Calcium silicate hydrates are used as coatings for clays, as absorbents and decolorizing agents, and as rubber fillers. The following describes calcium metasilicate, $CaSiO_3$.
Properties: White, amorphous powder; sp. gr. 2.9. Insoluble in water.
Containers: Bags.
Uses: Absorbent; antacid.

calcium silicofluoride. See calcium fluosilicate.

calcium stannate $CaSnO_3 \cdot 3H_2O$. White crystalline powder; insoluble in water; approximate temperature of dehydration 350°C.
Uses: Additive in ceramic capacitors; production of ceramic colors.

calcium stearate $Ca(C_{18}H_{35}O_2)_2$.
Properties: White powder; m.p. 150°C. Insoluble in any common solvent.
Derivation: By the interaction of sodium stearate and calcium chloride; then filtration.
Grades: Technical.
Containers: Wooden barrels; fiber drums; cartons; multiwall paper sacks.
Uses: Waterproofing; flatting agent in lacquers; also used in varnish, paints, enamels, plastics; lubricant; emulsions; cements; wax crayons; stabilizer for vinyl resins.
Shipping regulations: None. *

Calcium "Sucaryl." [3] Trademark for calcium cyclamate (q.v.).

calcium succinate $CaC_4H_4O_4 \cdot 3H_2O$.
Properties: Colorless crystals; indices of refraction 1.460 (alpha), 1.540 (beta), 1.610 (gamma). Slightly soluble in water; soluble in dilute acids.
Use: Medicine.

calcium sulfamate $Ca(SO_3NH_2)_2 \cdot 4H_2O$.
Properties: White, crystalline solid. Soluble in water. Aqueous solution is stable on boiling.
Grades: Technical.

Containers: 150-lb fiber drums; 350-lb wooden barrels.
Use: Flameproofing agent for textiles and certain grades of paper.

calcium sulfate (anhydrite) $CaSO_4$. See also gypsum, and terra alba, which are hydrated forms. See also gypsum cements,
Properties: White powder; white crystals with gray, blue, or reddish tinge, or brick-red crystals. Insoluble in water.
Constants: Sp. gr. 2.964; m.p. 1450°C.
Derivation: Found in large quantities in nature; by-product in many industrial operations.
Grades: Technical; C.P.
Containers: Bags; barrels; freight cars; multiwall paper sacks.
Uses: Interior plasters; polishing powder; cements (especially "Keene's cement," also as a retarder); paints (white pigment, filler, drier); paper (size, filler, surface-coating); dyeing and calico printing; metallurgy (reduction of zinc minerals); agriculture (as such and in compounds used to correct soils poor in calcium); drying industrial gases, solids, and many organic liquids.
Shipping regulations: None.*

calcium sulfhydrate. See calcium hydrosulfide.

calcium sulfide (hepar calcis) CaS. See also lime, sulfurated.
Properties: Yellow to light-gray powder with odor of hydrogen sulfide in moist air; unpleasant alkaline taste. Gradually decomposes in moist air or in weak acids. Soluble in acids; slightly soluble in water with partial decomposition; insoluble in alcohol. Sp. gr. 2.8.
Derivation: By strongly heating pulverized calcium sulfate and charcoal.
Grades: Technical.
Containers: 1-lb bottles; steel drums; multiwall paper sacks.
Uses: Luminous paint; medicine; depilatory; preparation of arsenic-free hydrogen sulfide; veterinary medicine.
Shipping regulations: None.*

calcium sulfide, crude. See lime, sulfurated.

calcium sulfite $CaSO_3 \cdot 2H_2O$.
Properties: White powder; loses water at 100°C. Soluble in sulfurous acid; slightly soluble in water.
Derivation: By the action of sulfurous acid on calcium carbonate.
Method of purification: Crystallization.
Grades: Technical; C.P.
Containers: 1-lb bottles; iron drums.
Uses: Textiles (antichlor); disinfectant in sugar industry, brewing; biological cleansing; food preservative and discoloration retarder; paper manufacture.

calcium sulfocarbolate. See calcium phenolsulfonate.

calcium sulfocyanate. See calcium thiocyanate.

calcium sulfophenate. See calcium phenolsulfonate.

calcium sulfophenylate. See calcium phenolsulfonate.

calcium superoxide. See calcium peroxide.

calcium tallate. See soaps, metallic.

calcium tannate.
Properties: Yellowish-gray powder. Soluble in dilute acids; slightly soluble in water.
Uses: Pharmaceuticals; adhesives.

calcium tartrate $CaC_4H_4O_6 \cdot 4H_2O$.
Properties: White, crystalline powder. Soluble in dilute acids; slightly soluble in water or alcohol.
Derivation: By the interaction of a calcium salt and crude cream of tartar.
Grades: Technical; C.P.
Containers: 1-lb bottles; wooden kegs.
Use: Tartaric acid.
Shipping regulations: None.*

calcium theobromine salicylate. See theobromine calcium salicylate.

calcium thiocyanate (calcium sulfocyanate, calcium rhodanate) $Ca(SCN)_2 \cdot 3H_2O$.
Properties: White hygroscopic crystals or powder. Soluble in water and alcohol.
Uses: Solvent for cellulose and polyacrylate; for parchmentizing; stiffening and swelling of textiles.

calcium thiosulfate (calcium hyposulfite) $CaS_2O_3 \cdot 6H_2O$.
Properties: White crystals; effloresces at 40°C; sp.gr. 1.872. Soluble in water; insoluble in alcohol.
Use: Medicine.

calcium tungstate (calcium orthotungstate; calcium wolframate; calcium wolframate, normal) $CaWO_4$.
Properties: White crystalline powder; sp.gr. 6.062. Soluble in ammonium chloride; insoluble in water; decomposed by hot acids.
Derivation: (a) By the interaction of calcium chloride and sodium tungstate. (b) Occurs in nature as scheelite (q.v.).
Method of purification: A slurry of powdered scheelite is treated with soda ash at about 370°C and 200 psi to form the soluble sodium tungstate. Insoluble impurities are filtered off and calcium tungstate is precipitated with lime.
Containers: Wooden barrels; fiber drums.
Uses: Luminous paints; fluorescent lamps; photography; x-ray pictures; medicine.
Shipping regulations: None.*
See next entry.

calcium tungstate, synthetic crystals $CaWO_4$.
Properties: Clear, colorless anisotropic tetragonal crystals, with good mechanical strength and excellent chemical stability; hardness 4.5-5; m.p. 1535°C; sp.gr. 5.9-6.1; refractive index approx. 1.93.
Uses: The strong fluorescence of the specially grown crystals on exposure to high-energy radiation recommends them as scintillation-counter crystals.
See previous entry.

*See "I.C.C. Shipping Regulations," page xiii.
Reference numbers refer to name of manufacturer. See "List of Manufacturers," page v.

calcium undecylenate $(CH_2CH(CH_2)_8COO)_2Ca$. A fine, white powder of limited solubility. M.p. 155°C.
Uses: Non-toxic bacteriostat and fungistat in cosmetics and pharmaceuticals.

calcium uranite. See autunite.

calcium wolframate. See calcium tungstate.

calcium wolframate, normal. See calcium tungstate.

calcium zirconate $CaZrO_3$.
Properties: Solid; m.p. 2345°C; sp.gr. 4.78. Soluble in nitric and other acids.
Uses: Chemical raw material, refractory.

calcium zirconium silicate $CaZrSiO_5$.
Properties: White solid; m.p. 2900°F; insoluble in water, alkalies; slightly soluble in acids; soluble in hydrofluoric acid.
Containers: 80-lb paper bags, 500-lb drums.
Use: Electrical resistor ceramics; glaze opacifier.

"Calcochrome" Dyes. [57] Trade name for a line of chrome colors applied in the dyeing of suiting, dress goods, overcoating, hats, blankets, outing shirts, and uniform cloth.

"Calcocid" Dyes. [57] Trade name for a line of acid dyestuffs used in the dyeing of wool and worsted goods, natural silk, jute, and in coloring diversified materials such as leather, soap, foodstuffs, feathers, artificial flowers, wood, paper, plastic materials, and in the preparation of color lakes.

"Calcodur" Dyes. [57] Trade name for a line of direct colors applied in the dyeing of cotton, rayon, and miscellaneous vegetable fibers.

"Calcofast" Wool Dyes. [57] Trade name for a line of metallized dyes containing chemically combined chromium used for dyeing wool in all forms including raw stock, tops, hand knitting, carpets and other yarns and piece goods. They can also be applied to leather, nylon, etc.

"Calcofluor" Dyes. [57] Trade name for a group of direct dyeing dyes which possess fluorescent properties. Used for dyeing cotton, linen, viscose, acetate, nylon, wool and certain synthetics. Used also in soaps as a brightener for textile use.

"Calcoform" Dyes. [57] Trade name for a selected group of direct cotton dyes which upon the aftertreatment of the dyeings in a solution of formaldehyde possess an increased fastness to water, perspiration, washing and common wet treatments. Used for the dyeing of ground shades on cotton and viscose rayon which are to be discharged, also for some sewing threads and hosiery.

"Calcogas." [57] Trademark for a series of dyes.

"Calcogene" Dyes. [57] Trade name for a line of sulfur colors used in the dyeing of cotton and other vegetable fibers such as linen.

"Calcoloid" Dyes. [57] Trade name for a line of colloidized vat dye pastes and powders used in the dyeing and printing of cotton and rayon. Processed to produce extremely fine dispersions, an important factor in the successful application of vat dyes.

"Calcomine" Dyes. [57] Trade name for a line of direct dyes for cotton and other vegetable fibers. Also applied to some extent to silk, wool, mohair, rayon, leather, and paper.

"Calcomites." [57] Trademark for sulfonamide preparations.

"Calco Mordant D." [57] Trademark. The sodium salt of a sulfur-phenol condensate used as a mordant for basic dyes.

"Calco" Naphthosol Dyes. [57] Trade name for a line of naphthols used as developers or preparers and a companion line of fast bases and salts to be used in combination therewith.

"Calconyl" Dyes. [57] Trade name for a line of coloring matters which are in substance stabilized combinations of a diazotized color base and a naphthol. They are used for dyeing or printing of fast shades on cotton and rayon.

"Calcophen" Dyes. [57] Trade name for a number of oil dyes that are of particular interest because of their non-subliming properties.
Uses: Varnish stains, foil coating, plastics.

"Calcosol" Dyes. [57] Trade name for a line of vat dyestuffs used in the dyeing and printing of cotton and rayon. Marketed in paste or powder form.

"Calco" Soluble Vat Dyes. [57] Trade name for a line of water-soluble sulfuric-acid esters of leuco vat dyes used for printing and dyeing many kinds of fabrics. They are extensively used in roller and screen printing of cotton fabrics in combination with stabilized azoic dyes ("Calconyls"), aniline black, etc. They are also used in the dyeing of cotton, viscose rayon, and wool. The dyeings possess vat-dye fastness.

"Calcosyn" Dyes. [57] Trade name for a line of direct dyeing dyes for the dyeing of certain synthetic fibers such as cellulose acetates and nylon.

"Calcotone" Pigment Pastes. [57] Trade name for a line of highly dispersed pigment pastes used whenever water suspensions of pigments are indicated. When used in combination with water-soluble binders, these pigment pastes can be used to print fabrics, both roller and screen, or to color sheers, laces, and nets. They also find use in coloring paper, both beater dyeing and coating.

"Calcozine" Dyes. [57] Trade name for a line of basic colors. Basic colors possess good affinity to mordanted cotton. Applied extensively also to leather and paper, and in pencils and color lakes.

"Calcozoic" Dyes. [57] Trade name for a line of stabilized azoic dyes developed for the printing of cotton, linen, and viscose fabrics.

calcspar. See calcite.

"Caldent." [172] Trade name for a dentifrice grade dicalcium phosphate dihydrate, $CaHPO_4 \cdot 2H_2O$, plus a proprietary additive.
Derivation: Lime and high quality phosphoric acid.
Containers: Bags.
Uses: As polishing agent in dentifrices.

"Caledon." [206] Brand name of proprietary line of vat dyestuffs derived from anthraquinone. Used for the dyeing and printing of cotton, rayon and silk and characterized by excellent fastness to washing, light, etc. Also used for coloring high grades of paper.

calendula (marigold; Mary-bud; goldbloom). Dried florets of Calendula officinalis.
Habitat: Southern Europe and Levant; cultivated everywhere.
Grades: Technical.
Containers: Boxes.
Uses: Coloring butter and margarine; adulterant for saffron; medicine.
Shipping regulations: None. *

calendulin. An amorphous, yellow, tasteless substance contained in marigold flowers and leaves (Calendula officinalis) which swells in water, though soluble in alcohol, alkalies, and acetic acid.
Use: See calendula.

"Calfalfa." [253] Brand name for a type of soil sulfur products.

"Calginate." [322] Trademark for calcium alginate (q. v.).

"Calgolac." [108] Trademark for a powdered, alkaline, sodium hexametaphosphate detergent.
Containers: 4-lb 5-oz packages (12/case); 100-lb drums.
Uses: Cleaning bars, fountains, and laboratory glassware and equipment.

"Calgon." [108] Trademark for a sodium phosphate glass commonly called "sodium hexametaphosphate." It has a molecular ratio of 1.1 Na_2O:1 P_2O_5 with a guaranteed minimum of 65% P_2O_5. Made from food-grade phosphoric acid and commercial soda ash by a thermal process. It is supplied in the form of powder, of agglomerated particles, and of broken glassy plates either pure or adjusted with mild alkalies. It is completely soluble in water in all proportions but is insoluble in organic solvents. It possesses sequestering, dispersing and deflocculating properties and precipitates proteins. In very low concentrations, it inhibits corrosion of steel and prevents the precipitation of slightly soluble, scale-forming compounds such as calcium carbonate and calcium sulfate. Used for softening water without precipitate formation as in dyeing, laundering, textile processing and washing operations; for precipitating proteins as in pretanning hides in the manufacture of leather; for dispersing clays and pigments; and as threshold treatment for scale and corrosion prevention in recirculating or once-through water systems.

"Calgon" Bouquet. [108] Trademark for a dry bead-like phosphate composition, rapidly soluble in water. Contains added perfume.
Containers: 1-lb boxes.
Uses: Water softener and conditioner for personal bathing.

"Calgon Composition T." [108] A complex glassy phosphate produced by a thermal process. White powder passing an 80 mesh sieve.
Containers: 100-lb bags.
Uses: Dispersing calcium carbonate pigments in the pulp and paper industry.

"Calgon Composition TG." [108] Colorless glass platelets of sodium zinc hexametaphosphate.
Containers: 100-lb bags.
Uses: Corrosion protection in recirculating cooling water and municipal water systems, and after mechanical cleaning of water mains where rapid film formation is important.

"Calgonite." [108] Trademark for an alkaline detergent composition containing "Calgon" as a principal ingredient. Generally recommended for spray-type mechanical washing operations where an alkaline cleaner is necessary and superior detergency with freedom from lime deposits is required. Laboratory "Calgonite" is a sudsing detergent composition containing "Calgon" as a principal ingredient to maintain glassware free from objectionable bacteria-retaining films. Effectively removes blood and culture media residues. Used for general hand washing of chemical and biological glassware and instruments.

"Calgreen." [147] Trademark for a non-separating combination of calcium arsenate and Air-Flo Green.
Containers: 4- and 50-lb bags.
Uses: For control of such insects as tomato hornworms, potato bugs, beet webworms, boll weevil, and bollworms.

caliche.
1. (Chile nitrate; Chile niter; Chile saltpeter; soda saltpeter). Crude sodium nitrate found in northern Chile, Bolivia, and United States (Nevada and California); used for fertilizer and dynamite mixtures. See also sodium nitrate.
2. Term used by geologists to refer to a hard subsurface soil layer in arid regions cemented by calcite and other minerals.

californite. A compact massive variety of vesuvianite (q. v.) found in California. Color, bright to yellowish-green. Takes a high polish and is used as a gem stone and an ornamental stone.

californium Cf. A synthetic radioactive element with atomic number 98. See actinide elements. The chemical properties of californium have been studied by tracer techniques and are similar to those of the other

*See "I.C.C. Shipping Regulations," page xiii.
Reference numbers refer to name of manufacturer. See "List of Manufacturers," page v.

transuranium elements. The trichloride, oxide and oxychloride have been prepared

calisaya bark. See cinchona bark, calisaya.

"Calktite." [326] Trade name for an acid- and alkali-proof caulking compound.
Properties: M.p. 15°F. A special grade has m.p. 55°F.
Containers: 35-, 70-, 175-, and 400-lbs.
Uses: Protective coating for masonry, acid tanks, and floors.

calmus. See calamus.

"Calnox." [89] Trademark for scale inhibitors, use in preventing the build-up of calcium salt scales from brines encountered in production of crude petroleum.

"Calo-Clor." [329] Trademark for a mercurial turf fungicide containing 73% mercury in chemical combination. It is a mixture of mercuric chloride (corrosive sublimate) with mercurous chloride (calomel) and inert ingredients.
Use: Prevention and control of brown patch and snow mold.

"Calocure." [329] Trademark for a mercurial turf fungicide especially compounded for positive control of brown patch. It contains 36.5% mercury in chemical combination.

"Calodorant B-1." [151] Trademark for mixed sulfide gas odorant, recommended for odorization of natural gas.

"Calodorant C." [151] Trademark for cyclic sulfide gas odorant, recommended for odorization of natural gas.

"Calodorant F." [151] Same as "Calodorant B-1."

"Calogreen." [329] Trademark for a turf fungicide composed of extremely finely divided form of mercurous chloride. Contains 85% mercury (insoluble in water). Long lasting protection against dollar spot and snow mold.

calomel. See mercurous chloride and calomel, native.

calomel, native (horn mercury; horn quicksilver) Hg_2Cl_2.
Properties: White, yellowish-gray, yellowish-white, gray or brown mineral, adamantine luster, pale yellowish-white streak. Consistency of horn; very sectile. Found as a coating in cavities with or near cinnabar. Contains 84.9% mercury. Soluble in aqua regia; insoluble in water. Sp. gr. 6.48; hardness 1-2.
Occurrence: United States (Texas), and Europe.
Use: Not an important source of mercury. See also mercurous chloride.

"Caloria." [51] Trademark for unique lubricants of both oil and grease consistency, which at elevated temperatures decompose, leaving no residue. Designed for bearing temperatures of 400°F and above, they are intended for those locations where ordinary lubricants leave harmful deposits. (See "Van Caloria").

calorific value. A measure of the quality of fuels, usually expressed as available Btu (q.v.).

calorizing. The process by which steel is coated with aluminum by heating it in aluminum powder. The aluminum forms an alloy with the steel surface and produces a thin, tightly adherent coating.

"Calsolene Oil HS." [206] Brand name for highly sulfonated oil, stable to hard water and acid and retaining wetting power in alkaline, neutral and acid conditions.

"Calsolene Oil HSA." [325] Sulfated ester; anionic wetting and dispersing agent. Used primarily for wet processing of hydrophobic fibers.

calx. See calcium oxide.

calx sulfurata. See lime, sulfurated.

"Cambar." [51] Trademark for high temperature block grease, suitable for plain bearings such as dryer rolls in paper mills. Light color of this product makes it especially suitable where lubricant stains are troublesome.

cambogia. See gamboge.

"Camoform" Hydrochloride. [330] Trademark for biallylamicol hydrochloride [6,6'-diallyl-alpha, alpha'-bis(diethylamino)-4,4'-bi-ortho-cresol dihydrochloride]. Used in medicine.

camomile oil. See chamomile oil, German; chamomile oil, Roman.

"Camoquin" Hydrochloride. [330] Trademark for amodiaquine hydrochloride [4-(7-chloro-4-quinolylamino)-alpha-diethyl-amino-ortho-cresol hydrochloride] Used in medicine.

2-camphanone. See camphor.

camphene $C_{10}H_{16}$.
Properties: Colorless crystals. Soluble in ether; slightly soluble in alcohol; insoluble in water. M.p. 48-51°C; b.p. 159-162°C.
Derivation: (a) By heating pinene hydrochloride with alkalies, aniline, or alkali salts, such as sodium acetate. (b) A constituent of certain essential oils.
Grades: Technical (46° m.p.)
Containers: Tins; drums; tanks.
Uses: Medicine; manufacture of synthetic camphor; camphor substitute.
Caution: When heated, gives off flammable vapors.
Shipping regulations: None.*

camphor (gum camphor; camphor, natural; 2-camphanone; camphor, synthetic) $C_{10}H_{16}O$. A ketone occurring naturally in the wood of the camphor tree (Cinnamomum camphora) native to Formosa and now cultivated in Florida and California.
Properties: Colorless or white crystals, granules or easily broken masses; characteristic odor; sp. gr. 0.99; m.p. 174-

*See "I.C.C. Shipping Regulations," page xiii.
Reference numbers refer to name of manufacturer. See "List of Manufacturers," page v.

179°C; slowly volatilizes at room temperature; insoluble in water; soluble in alcohol, ether, chloroform, carbon disulfide, and solvent naphtha; fixed and volatile oils.
Derivation: Steam distillation of the camphor-tree wood and crystallization. This product is called natural camphor and is dextrorotatory. Synthetic camphor, most of which is optically inactive, may be made from pinene, which is converted into camphene, (q. v.), which by treatment with acetic acid and nitrobenzene becomes camphor.
Containers: 1-lb tins to 250-lb barrels; drums.
Grades: Technical (synthetic, m. p. 163-168°C); U.S.P. XVI (m. p. 174-179°C).
Uses: Medicine (internal and external); plasticizer for "Celluloid," "Celluloid" film, cellulose nitrate, other explosives, and lacquers; insecticides and moth and mildew preventives; tooth powders; embalming; pyrotechnics; intermediate.
Caution: Flammable; gives off flammable vapors when heated which may form an explosive mixture with air.

camphor bromate (bromocamphor; brominated camphor; monobromated camphor) $C_{10}H_{15}BrO$.
Properties: Colorless crystals with slight camphor odor and taste. Also available as powder. Discolors in light and should be stored in cool, dark place. M. p. 76°C; b. p. 274°C; sp. gr. 1.449. Soluble in alcohol, ether, chloroform, and oils; insoluble in water.
Derivation: By heating camphor with bromine.
Method of purification: Crystallization.
Containers: 100-lb kegs; 25- and 100-lb drums.
Uses: Medicine; camphor derivatives.

camphor, bromated. See camphor bromate.

camphor, cantharides. See cantharidin.

camphoric acid $C_{10}H_{16}O_4$.
Properties: Colorless, odorless needles or scales. Soluble in alcohol, ether, fatty oils, and water; insoluble in chloroform. Sp. gr. 1.186 (20/4°C); m. p. 186-188°C.
Derivation: By oxidizing camphor with nitric acid.
Method of purification: Crystallization.
Grades: Technical.
Containers: 1-oz vials; ¼-, 1-, 5-lb bottles; 5-lb cans; fiber drums.
Uses: Celluloid; pharmaceuticals; medicine.
Shipping regulations: None.*

camphor, liquid. See camphor oil.

camphor, Malayan. See borneol.

camphor, natural. See camphor.

camphor oil (camphor, liquid).
Properties: Colorless natural oil with characteristic odor; sp. gr. varies from 0.870-1.040; refractive index 1.465-1.481 (20°C). Soluble in ether, chloroform; insoluble in alcohol.
Chief known constituents: Pinene, camphor, cineol, phellandrene, dipentene, safrol, and eugenol.
Derivation: By distilling the wood of the tree Cinnamomum camphora and separating the oil and solid camphor.
Grades: White, Japanese; Chinese; by-product (sassafrassy). Also sold as oil of camphor, rectified.
Containers: Drums.
Uses: Used as a substitute for turpentine oil in varnish manufacture and for cleaning type, electroplates, and cylinders in the printing industry; perfuming cheap soaps, especially soft soaps; shoe polish; hoof ointments; and hiding the odor of mineral oils; wagon greases and lubricating oils; medicine.
Shipping regulations: None.*

camphor, parsley. See apiol.

camphor, peppermint. See menthol.

camphor, pulsatilla. See anemonin.

camphor, synthetic. See camphor.

Canada balsam (Canada turpentine; turpentine, Canadian; balsam of fir).
Properties: Pale yellow, or greenish-yellow, transparent, viscous liquid. Agreeable aromatic pine-like odor. Feebly bitter, acrid taste. Slowly dries to a transparent varnish when exposed to the air; sp. gr. 0.983-0.997; refractive index (n 20/D) 1.52-1.54; acid number 84-87. Soluble in benzene, chloroform, and ether; insoluble in water.
Derivation: The oleoresin obtained from Abies balsamea.
Habitat: Canada and northern United States.
Uses: Medicine; cement for lenses; manufacture of fine lacquers; mounting in microscopy.
Not to be confused with balm of Gilead (q. v.).

Canada pitch. See hemlock pitch.

Canada snake root. See asarum.

Canada snake-root oil. See asarum canadense oil.

Canada turpentine. See Canada balsam.

Canadian asarabacca oil. See asarum canadense oil.

Canadian hemp. See apocynum.

Canamin clay. A colloidal clay from British Columbia. Particle size is very small and therefore has great adsorption capacity. Consists mainly of colloidal aluminum silicate.

cananga oil.
Properties: A yellowish essential oil, having a floral odor similar to that of oil of ylang ylang (q. v.). Sp. gr. 0.908-0.925; optical rotation −15° to −30°; refractive index, 1.495-1.503.
Grades: Regular (native); rectified (the latter being lighter in color, has better solubility in alcohol, and is more stable).

Derivation: By distillation from the flowers of the Javanese variety of Cananga odorata.
Containers: Cans.
Use: Perfumery, particularly for floral types.
Shipping regulations: None.*

canarium. See elemi gum.

canavanine $NH_2CNHNHOCH_2CH_2CHNH_2COOH$. An amino acid obtained from jackbean meal. It is found naturally in the L(+) form.
Properties: Crystals, from dilute alcohol; m.p. 184°C (dec); soluble in water; nearly insoluble in alcohol. Sulfate: crystals from dilute alcohol; m.p. 172°C (dec); soluble in water.
Use: Biochemical research.

candelilla wax.
Properties: Yellowish-brown, opaque to translucent solid wax. Soluble in chloroform, turpentine, carbon tetrachloride, trichlorethylene, toluene, hot petroleum ether and alkalies; insoluble in water. Sp. gr. 0.983; m.p. 67-68°C; saponification value 65; iodine number 37; refractive index 1.4555.
Derivation: Found as a greenish-gray coating on the entire surface of the wild candelilla plant of Mexico and Texas from which it is obtained by immersing the plant in boiling water, and skimming off the wax which rises to the surface.
Method of purification: Treatment with sulfuric acid or niter cake.
Impurities: Bark fragments, etc.
Grades: Crude; refined; powdered.
Containers: Bags; boxes.
Uses: Leather dressing; polishes; candle manufacture; cements; varnishes; substitute for carnauba wax and beeswax; electric insulating compositions; sealing wax; phonograph records; waterproofing and insect-proofing containers; paint removers; dentistry; paper sizes; rubber and rubber substitutes; stiffener for soft waxes.
Shipping regulations: None.*

candicidin. An antifungal antibiotic produced by Streptomyces griseus.

candidin. An antifungal antibiotic produced by Streptomyces viridoflavus.

candleberry. See myrica.

candle-nut oil. See lumbang oil.

candle pitch. See stearin and fatty-acid pitches.

candle-tar pitch. See stearin and fatty-acid pitches.

cane blackstrap. See molasses.

canella (white cinnamon; white wood bark; Bahama white wood; canilla).
Derivation: Bark of Canella winterana or C. alba.
Habitat: West Indies and Florida.
Grades: Technical.
Containers: Bales.
Uses: Medicine; condiment; addition to smoking tobacco.
Shipping regulations: None.*

canella oil.
Properties: Colorless; spicy odor and taste. Chief known constituents: eugenol; eucalyptol; caryophyllene. Sp. gr. 0.920-0.935; optical rotation +1°8'. Soluble in alcohol, ether, and chloroform.
Derivation: Distilled from the bark of Canella alba.
Method of purification: Rectification.
Grades: Technical.
Containers: Tins; glass bottles.
Uses: Medicine; perfume; condiment.
Shipping regulations: None.*

canescine. See deserpidine.

cane sugar ($C_{12}H_{22}O_{11}$). Sugar (sucrose) extracted from cane. Chemically identical with beet sugar.

canilla. See canella.

cannabin resin. See cannabis.

cannabis (cannabin resin; Indian hemp; Indian cannabis; hashish; guaza; marihuana). A habit-forming drug.
Derivation: Dried flowering tops of pistillate plants of Cannabis sativa.
Habitat: Persia; East India; cultivated in Europe, Asia and United States.
Grades: Technical.
Containers: Bags.
Uses: Medicine; corn removers; ointment and liniment.
Shipping regulations: None.*
Caution: Manufacture and transportation strictly regulated by law in U.S.A.

cannel coal. A variety of bituminous or subbituminous coal of uniform and compact fine-grained texture with a general absence of banded structure. It is dark gray to black in color, has a greasy luster, and is noticeably of conchoidal or shell-like fracture. It is noncaking; yields a high percentage of volatile matter; ignites easily; and burns with a luminous, smoky flame. (ASTM definition, ASTM D493-39.)

Cannizzaro reaction. The reaction of an aldehyde with alcoholic potassium hydroxide to produce an alcohol and the salt of an acid.

cantala (maguey). Hard, strong, light colored fibers obtained from the leaves of Agave cantala. Finer and more supple than sisal, but not as strong.
Sources: Philippines; India; Indonesia.
Uses: Twine and cordage.

"Cantaxin." [162] Trademark for ascorbic acid.

cantharides (blistering flies; cantharis; blistering beetle; Spanish flies; Russian flies). Dried insects. The species of most commercial importance is Cantharis vesicatoria.
Habitat: Southern and central Europe.
Grades: Chinese; Russian.
Use: Medicine and veterinary medicine.
Shipping regulations: None.*

cantharidin (cantharides camphor) $C_{10}H_{12}O_4$.
Properties: Colorless, crystalline scales; poisonous. Very slightly soluble in water and alcohol; slightly soluble in chloroform, acetone, acetic ether; soluble in fixed oils. M. p. 218°C; begins to sublime at about 90°C.
Derivation: From cantharides.
Method of purification: Crystallization.
Grades: Technical.
Containers: Glass bottles.
Use: Same as cantharides.
Shipping regulations: None. *

cantharis. See cantharides.

"Canthus." [51] Trademark for high viscosity, economical, straight mineral oils for use on gears and bearings with large clearances. Not recommended for low temperature service.

"Cantona." [51] Trademark for filtered steam cylinder oil of excellent quality. Atomizes readily and separates easily from condensate.

caoutchouc. See rubber, natural.

Cape gum. See arabic gum.

Cape ruby. See garnet.

capillarity. The attraction between molecules, both like and unlike, which results in the rise of a liquid in small tubes or fibers, or in the wetting of a solid by a liquid.

capillary rise. The rise or fall of a liquid along the sides of a narrow tube or in plant fibers.

capivi balsam. See copaiba resin.

"Capracyl." [28] Trademark for a group of neutral-dyeing acid colors that produce the highest possible degree of light fastness on nylon. Also suitable for dyeing wool, particularly in blends with cellulosic fibers.

capraldehyde. See decanal.

capric acid (decanoic acid; decoic acid; decylic acid) $CH_3(CH_2)_8COOH$. Occurs as a glyceride in natural oils.
Properties: White crystals. Unpleasant odor; soluble in most organic solvents, and dilute nitric acid; insoluble in water. Sp. gr. 0.8858 (40°C); b. p. 270°C (760 mm), 172.6°C (30 mm); m. p. 31.5°C; refractive index 1.4288 (40°C); acid number 308-315.
Derivation: Fractional distillation of coconut oil fatty acids.
Grades: Technical; 90%.
Containers: 1-gal bottles; 55-gal drums; tanks, carload lots.
Uses: Making esters for perfumes and fruit flavors; base for wetting agents; intermediates; plasticizer; resins.

capric aldehyde. See decanal.

caprilic acid. See caprylic acid.

caproic acid (hexanoic acid; hexylic acid; hexoic acid) $CH_3(CH_2)_4COOH$. Present in milk fats to extent of about 2%.
Properties: Oily, colorless or slightly yellow liquid; odor of limburger cheese. Soluble in alcohol and ether; slightly soluble in water. Sp. gr. 0.9276 (20/4°C); f. p. —4.0°C; b. p. 205°C (760 mm); 119°C (30 mm); refractive index 1.4168 (20°C); wt/gal 7.7 lbs; viscosity 0.031 poise (20°C).
Derivation: From crude fermentation of butyric acid; fractional distillation of natural fatty acids.
Method of purification: Rectification.
Grades: Technical; reagent to 99.8%.
Containers: 1-, 5-, 10-lb glass bottles; carboys; 31-, 55-gal drums; carload lots.
Uses: Analytical chemistry; flavors; manufacture of rubber chemicals, varnish driers, resins and pharmaceuticals.

epsilon-**caprolactam** (aminocaproic lactam) $\overline{CH_2(CH_2)_4NHCO}$.
Properties: White crystals; m. p. 68-69°C; sp. gr. 70% solution 1.05; refractive index (40°C) 1.4935, (31°C) 1.4965; soluble in water, chlorinated solvents, petroleum distillates and cyclohexene; heat of fusion 29 cal/g; heat of vaporization 116 cal/g; viscosity 9 cps at 78°C; vapor pressure 3 mm Hg at 100°C, 50 mm Hg at 180°C.
Containers: 80-lb paper bags; 300-lb fiber drums; tanks.
Uses: Manufacture of synthetic fibers (especially nylon 6), plastics, bristles, film, coatings, synthetic leather, plasticizers, and paint vehicles; cross-linking agent for curing polyurethanes. A very large use is for nylon tire cords.
Caution: Good ventilation is recommended to avoid any possible irritation from continuous breathing of its vapors or fine particles.

caproyl alcohol. See 1-hexanol.

caproyl chloride. A confusing name, since it might legitimately mean the chloride of either capric (C_{10}) or caproic (C_6) acids, although it seems usually to have meant the latter. For the C_{10} acid chloride, see decanoyl chloride; for the C_6, see hexanoyl chloride.

capryl alcohol. A term which should be replaced by more exact names wherever possible. This name has been used for both the 1-n-octanol and the 2-n-octanol, and it might also be assumed to mean a decyl alcohol. At the present time, the preponderance of usage favors its meaning the 2-n-octanol.

capryl aldehyde. See octanal.

capryl bromide. See octyl bromide.

caprylene. See octene.

caprylic acid (octanoic acid; octoic acid; octylic acid; caprilic acid) $CH_3(CH_2)_6COOH$.
Properties: Colorless oily liquid; slight unpleasant odor; burning rancid taste; sp. gr. 0.9105 (20/4°C); m. p. 16°C; b. p. 237.9°C (760 mm), 147.9°C (30 mm); refractive index 1.4278 (20°C). Very slightly soluble in water; soluble in alcohol

and ether.
Derivation: By saponification and subsequent distillation of coconut oil.
Method of purification: Crystallization or rectification.
Grades: Technical; 99%.
Containers: 1-gal jars; 55-gal drums; tanks; carload lots.
Uses: Synthesis of various dyes, drugs, perfumes, flavors; antiseptics and fungicides; ore separations; chemical raw material, and plasticizer.
Shipping regulations: None. *

caprylic alcohol. See 1-octanol.

caprylic aldehyde. See octanal.

caprylic bromide. See octyl bromide.

caprylic iodide. See octyl iodide.

capryl iodide. See octyl iodide.

capryloyl chloride. See octanoyl chloride.

caprylyl acetate. See n-octyl acetate.

caprylyl chloride. A confusing name, which might mean the alkyl chloride, octyl chloride, $C_8H_{17}Cl$, or the acid chloride, octanoyl chloride, $C_7H_{15}COCl$. See octanoyl chloride.

caprylyl peroxide. See octyl peroxide for shipping regulations.

capsicin (capsicum oleoresin). An ether extract of Capsicum frutescens. Reddish-brown, thick liquid with sharp peppery taste. Chief active constituent is capsaicin (an alkaloid with the formula $C_{18}H_{27}O_3N$). Insoluble in water; soluble in alcohol and ether. Poisonous when undiluted.
Grades: From domestic or African peppers.
Containers: Drums.
Uses: Medicine; in vinegar, pickles, etc.

capsicum (cayenne pepper; African pepper; chillies; bird pepper; red pepper).
Derivation: Dried ripe fruit of Capsicum frutescens; Capsicum annuum or the Louisiana sport pepper.
Habitat: Southern India; cultivated extensively in the tropics.
Grades: According to place of origin; N. F. XI.
Containers: Bags.
Uses: Medicine; condiment, as an adulterant for ginger, especially in ginger ale.
Shipping regulations: None. *

capsicum oleoresin. See capsicin.

"Captan." [183] Trademark for a gas odorant built on a mercaptan base. Clear liquid, lighter than water, similar in odor to butyl mercaptan.
Use: To add odor to odorless gases both for safety and for the detection of leaks.

captan (N-trichloromethylmercapto-4-cyclohexene-1,2-dicarboximide; N-trichloromethylthio-tetrahydrophthalimide) $C_9H_8O_2NSCl_3$.
Properties: White to cream powder; odor, pungent; m.p. 158-164°C; sp. gr. 1.5. Practically insoluble in water; partially soluble in acetone, benzene and toluene; slightly soluble in ethylene dichloride and chloroform.
Containers: 50-lb fiber drums; 5-lb bags.
Uses: Seed treatment; fungicide in paints, plastics, leather, and fabrics.
Caution: Avoid contamination of feed and foodstuffs. Avoid inhalation of dust or spray mist. Avoid prolonged or repeated contact with skin. (USDA Pesticides regulation; MCA warning label.)

"Captan 50-W." [1] Trade name for 50% wettable powder formulation containing captan (q. v.).

"Captax." [69] Trade name for a proprietary product, mercaptobenzothiazole. $\overline{C_6H_4NC(S)}\cdot SH$.
Properties: Yellow; odor distinct; sp. gr. 1.42; non-toxic; m.p. 170-175°C. Insoluble in water, gasoline; soluble in dilute caustic, alcohol, acetone, benzene, and chloroform.
Use: Accelerator of vulcanization in tire stocks, inner tubes, wire insulation, footwear, clothing, drug sundries,

caput mortuum. See iron oxide reds.

caramel (sugar coloring; burnt sugar).
Properties: Dark-brown, deliquescent powder or thick liquid; bitter taste; burnt sugar odor; sp. gr. (approx.) 1.35. Soluble in water and dilute alcohol.
Derivation: Obtained by carefully heating sugar, adding small quantities of sodium carbonate during the heating.
Grades: Technical; N. F. XI; also sold in solution.
Containers: 5-gal cans; barrels; drums.
Uses: Coloring foods, confectionery, sweetmeats, vinegar, liqueurs, malt beverages; tobacco flavoring; medicine.
Shipping regulations: None. *

caramiphen hydrochloride (1-phenylcyclopentanecarboxylic acid 2-diethylaminoethyl ester hydrochloride; diethylaminoethyl 1-phenylcyclopentane-1-carboxylate hydrochloride) $C_6H_5(C_5H_9)CO_2(CH_2)_2N(C_2H_5)_2\cdot HCl$.
Properties: Crystals; m.p. 145-146°C. Slightly soluble in water; soluble in alcohol.
Grade: N. N. D.
Use: Medicine.

carat. A unit of weight for jewels equal to 200 mg; also used to denote the proportion of gold in an alloy, e.g., 24 carat is pure gold while 18 carat is 75% gold.

caraway (carum; caraway seed).
Derivation: Dried ripe fruit of Carum carvi.
Habitat: Europe; central and western Asia; cultivated in England, Russia and United States.
Grades: Danish; Dutch; Polish; N. F. XI.
Containers: Boxes; bags.
Uses: Medicine; flavoring; liqueurs; condiment.
Shipping regulations: None. *

caraway oil (caraway-seed oil; carui oil).
Properties: Colorless or pale yellow, thin

*See "I.C.C. Shipping Regulations," page xiii.
Reference numbers refer to name of manufacturer. See "List of Manufacturers," page v.

liquid; characteristic odor; mild, spicy taste. Slightly soluble in water; soluble in alcohol. Sp. gr. 0.900-0.910 (25/25°C); b.p. 175-230°C; optical rotation +70° to +80°; refractive index 1.4840-1.4880 (20°C).
Chief known constituents: Carvone and dextro-limonene. Carvone content: 50 to 60%.
Derivation: Distilled from dried ripe fruit of Carum carvi.
Method of purification: Rectification.
Grades: Technical; N. F. XI.
Containers: Bottles; 25-lb tins; 10-, 55-gal drums.
Uses: Medicine; flavoring; manufacture of liqueurs; perfuming soaps.
Shipping regulations: None. *

caraway seed. See caraway.

caraway-seed oil. See caraway oil.

carbachol (carbamylcholine chloride; choline chloride carbamate) $NH_2COOCH_2CH_2 \cdot N(CH_3)_3Cl$.
Properties: White or faintly yellow crystals or as crystalline powder, odorless, or with faint amine odor, and hygroscopic. Soluble in water and alcohol; almost insoluble in chloroform and ether. M.p. 201-205°C (with decomposition).
Grade: U.S.P. XVI.
Use: Medicine.

"Carbamate." [55] Trade name for insecticides with the active ingredient ferbam (q.v.). Used on fruits, vegetables, tobacco and flowers.

carbamide. See urea.

carbamide peroxide. See urea peroxide.

carbamide phosphoric acid (urea phosphoric acid) $CO(NH_2)_2 \cdot H_3PO_4$.
Properties: White rhombic crystals. Very soluble in water and alcohol.
Containers: 350-lb drums.
Uses: Catalyst for acid-setting resins; flameproofing compositions; cleaning compounds; general acidulant.

carbamidine carbonate. See guanidine carbonate.

para-**carbamidobenzenearsonic acid.** See carbarsone.

carbamite. See sym-diethyldiphenylurea.

carbamylarsanilic acid. See carbarsone.

carbamylcholine chloride. See carbachol.

carbamylhydrazine hydrochloride. See semicarbazide hydrochloride.

carbamylmethylcholine chloride. See bethanechol chloride.

carbamylurea. See biuret.

carbanil. See phenyl isocyanate.

carbanilide. See diphenylurea.

carbanion. A negatively charged organic ion such as H_3C^-, R_3C^-, having one more electron than the corresponding free radical. Carbanions are short-lived but important intermediates in base-catalyzed polymerization and alkylation reactions. See carbonium ion.

"Carbansol." [243] Trademark for vat pigments.

"Carbanthrene." [243] Trademark for anthraquinone vat dyes.

carbapatite. See apatite.

Carba process. Method of making solid carbon dioxide by cooling the gas without added pressure being applied.

carbarsone (para-ureidobenzenearsonic acid; para-carbamidobenzenearsonic acid; carbamylarsanilic acid) $NH_2CONHC_6H_4AsO(OH)_2$.
Properties: White, almost odorless powder with slightly acid taste. M.p. 174°C. Slightly soluble in water and alcohol; very slightly soluble in chloroform and ether; soluble in solutions of alkali hydroxides and carbonates. Saturated solution is acid to litmus.
Derivation: By coupling sodium arsanilate with potassium cyanate.
Grade: U.S.P. XVI.
Use: Medicine.

"Carbasols." [223] Proprietary products consisting of natural resin esters condensed with modifying agents of the polyhydric alcohol-polybasic acid type.
Types: Available in wide variation as to physical properties such as color, softening point, and acid number.
Uses: Cellulose-ester lacquers; oleoresinous varnishes; printing ink vehicles.

carbazole (dibenzopyrrole; diphenylenimine) $(C_6H_4)_2NH$ (tricyclic).
Properties: White crystals with characteristic odor; m.p. 244-246°C; b.p. 352-354°C; soluble in alcohol and ether; insoluble in water.
Derivation: (a) From crude anthracene cake by selective solution of the phenanthrene with crude solvent naphtha, removal of the anthracene by conversion into a sulfonic derivative and extraction by means of water. (b) Synthetically, from ortho-aminodiphenyl.
Grades: Technical, 97%.
Containers: Barrels; fiber drums.
Uses: Manufacture of dyes; reagent; explosives; insecticides; lubricants; rubber antioxidants; as odor inhibitor in detergents.
Shipping regulations: None. *

carbazotic acid. See picric acid.

carbene. A positive or negative ion involving two-valent carbon and paired electrons, as in H_2C:, R_2C: or R_2C^{++} Different from methylene free radicals in that the latter have unpaired electrons. All these substances have only brief transitory existence and so are difficult or impossible to collect as such, but have important functions as intermediates in chemical reactions. See, for example, dichlorocarbene.

*See "I.C.C. Shipping Regulations," page xiii.
Reference numbers refer to name of manufacturer. See "List of Manufacturers," page v.

carbetapentane citrate [2-(diethylaminoethoxy) ethyl 1-phenylcyclopentyl-1-carboxylate citrate] $C_{20}H_{31}NO_3 \cdot C_6H_8O_7$.
Properties: White, odorless, crystalline powder. Melting range 90-95°C. Freely soluble in water, slightly soluble in alcohol; practically insoluble in ether.
Grade: N. F. XI.
Use: Medicine.

2-carbethoxycyclopentanone (ethyl cyclopentanone-2-carboxylate; ethyl 2-oxocyclopentanecarboxylate) $OC_5H_7COOC_2H_5$.
Properties: Colorless liquid with characteristic ester odor; b. p. 122-124°C (25 mm); flash point 191°F; refractive index 1.451 (25°C); density 1.0976 (0°C); soluble in equimolecular amounts of dilute alcohol.
Containers: Glass bottles; 5-gal cans; 50-gal drums.
Uses: Chemical intermediate; pharmaceutical intermediate.
Caution! Avoid inhalation of vapors and contact with skin.

beta-carbethoxyethyltriethoxysilane $C_2H_5OOC(CH_2)_2Si(OC_2H_5)_3$.
Properties: Colorless liquid. B. p. 246°C.
Use: Intermediate.

N-carbethoxypiperazine $C_7H_{14}N_2O_2$.
Properties: A colorless, somewhat viscous liquid; slight characteristic odor; b. p. 116-117°C (12 mm); 237°C (760 mm); refractive index 1.4756 (25°C). Miscible with water and common organic solvents.
Containers: 55-gal drums.
Use: Chemical intermediate.

beta-carbethoxypropylmethyldiethoxysilane $C_2H_5OOC(C_3H_6)CH_3Si(OC_2H_5)_2$.
Properties: Colorless liquid. B. p. 228°C.
Use: Intermediate.

carbide. See calcium carbide.

"Carbindone." [206] Brand name for a type of sulfur black dye used for textile printing.

carbinol.
1. A synomym for methyl alcohol, CH_3OH.
2. Hence, any compound of similar structure retaining the COH radical, and in which hydrocarbon radicals may be substituted for the hydrogens originally attached to the carbon. Thus isopropyl alcohol $(CH_3)_2CHOH$, and benzyl alcohol, $C_6H_5CH_2OH$, may be named dimethyl carbinol and phenyl carbinol respectively.

carbinoxamine maleate $ClC_6H_4CH(C_5H_4N)OCH_2CH_2N(CH_3)_2 \cdot C_4H_4O_4$. 2-[para-Chloro-alpha-(2-dimethyl-aminoethoxy)benzyl]pyridine maleate.
Properties: White, odorless, bitter crystalline powder; m. p. 116-121°C. Very soluble in water; freely soluble in alcohol and chloroform; very slightly soluble in ether; pH (1% solution) 4.6-5.1.
Grade: N. F. XI.
Use: Medicine.

"Carbitol." [214] Trademark for mono- and dialkyl ethers of diethylene glycol and their derivatives.

butyl "Carbitol"
See diethylene glycol monobutyl ether.
butyl "Carbitol" acetate
See diethylene glycol monobutyl ether acetate.
"Carbitol" acetate
See diethylene glycol monoethyl ether acetate.
"Carbitol" solvent
See diethylene glycol monoethyl ether.
dibutyl "Carbitol"
See diethylene glycol dibutyl ether.
diethyl "Carbitol"
See diethylene glycol diethyl ether.
n-hexyl "Carbitol"
See diethylene glycol monohexyl ether.
methyl "Carbitol"
See diethylene glycol monomethyl ether.

carbobenzoxychloride. See benzyl chlorocarbonate.

"Carbocell." [214] Brand name for a proprietary product. Amorphous carbon (99% carbon) (no foreign bond) with controlled pore size and permeability characteristics. Chemically inert except in strong oxidizing conditions. Free from thermal shock fracture. Available in tubes or rods up to 6 ¾ in. dia. x 36 in. long, and plates 12 in. x 12 in. Machinable; resistivity 0.007-0.008 ohm/cu in. Permeability 15-200 gals water/sq ft/minute at 5 lbs differential depending on grade.
Uses: Filtration; diffusion; steam sparging; anode; cathode.

carbodiimide. See cyanamide (1).

"Carbo-Dur." [184] Trademark for a hard, granular fast-wetting activated carbon.

"Carbofrax." [280] Trademark for refractory products composed principally of silicon carbide. Available as bonded refractories and cements.
Bonded Refractories
Properties: High refractoriness, high thermal conductivity, resistance to spalling, resistance to mechanical abrasion and resistance to oxidation. Contain 85% or more silicon carbide. The thermal conductivity is a straight line between 105.5 at 1100°F and 112.5 at 2900°F, the values being in Btu/sq ft/in. thickness/°F temperature difference/hr. The porosity is about 13% and permeability very low.
Uses: Bricks and special shapes for fuel-bed sections of boiler furnace walls, for retorts, piers, combustion chamber linings, combustion arches; tile used in hearths, muffle walls, retorts, still settings, etc., protection tubes for platinum couples and as "targets" in tube or block shapes for the newer types of optical and radiation pyrometers.
Refractory Cements
Vary in recommended temperature range for use and in texture.
Uses: Monolithic linings and patching; mortar for laying up "Carbofrax" brick; industrial applications such as forge and boiler furnaces, and crucible melting furnaces.

carbohydrase. An enzyme whose catalytic activity is directed toward the digestion or conversion of carbohydrates. Illustrations are amylase; invertase; maltase (q. v.).

carbohydrates. The class of compounds of carbon, hydrogen, and oxygen in which the latter elements are in the same proportions as in water, or in nearly these proportions. A more rigorous definition is that carbohydrates are aldehyde alcohols or ketone alcohols or compounds that on hydrolysis produce aldehyde or ketone alcohols. The best examples are the crystalline, soluble, sweet, relatively low molecular weight sugars, and the amorphous, tasteless and relatively insoluble high molecular weight starches and cellulose. Specific examples are sucrose $C_{12}H_{22}O_{11}$ and starch $(C_6H_{10}O_5)_x$.

"Carbo-Korez." [41] Trade name for a carbon-filled, synthetic-resin, acid-proof cement of the phenol-formaldehyde type used as a mortar cement where temperatures do not exceed 360°F. Especially good for high concentrations of sulfuric acid.

"Carbolac." [275] Trade name for series of channel carbon blacks for high grade enamels, lacquers and plastics. Available as:

"Carbolac 1." High color channel black.

"Carbolac 2." Standard high color channel black.

"Carbolac 46." High color channel black.

"Carbolan." [206] Brand name of proprietary line of acid dyestuffs giving results of extremely high resistance to milling and very good fastness to light. They are also used for printing on animal fibers.

carbolfuchsin (Ziehl's stain). A staining solution of fuchsin in alcohol and aqueous phenol used in the study of micro-organisms.

carbolic acid. See phenol.

carbolic oil (middle oil). Comprises the fraction having a boiling range of about 190-250°C obtained from distillation of coal tar, and containing naphthalene, phenol, and cresols.

2-carbomethoxy-1-methylvinyl dimethyl phosphate. See "Phosdrin."

carbomycin. An antibiotic isolated from products of Streptomyces halstedii when grown in suitable media by deep culture method. It inhibits growth of certain gram-positive bacteria such as staphylococci, pneumococci and hemolytic streptococci.

Properties: White, odorless, bitter powder; m.p. 195-220°C (dec). Freely soluble in chloroform; very slightly soluble in water; slightly soluble in alcohol and ether. Stable when protected from moisture; pH (saturated solution) 5.5-8.0.

Grade: N.N.D.

Use: Medicine.

Shipping regulations: None.*

carbon C. Non-metallic element (atomic weight 12; atomic number 6, group IV of periodic system) existing in two crystalline allotropic forms (diamond and graphite) and numerous amorphous forms of varying purity (charcoal, coke, carbon black, lampblack, and activated carbon). Isotopic carbon of atomic weight 13 and 14 also exists, and is of increasing importance as a tracer element (see carbon 14). Carbon is a constituent of all organic compounds and it also occurs in combined form in many inorganic substances (carbon dioxide of the air, limestone, etc.).

Properties: Black in varying degrees; other properties vary greatly with the form under consideration. Sp. gr. of amorphous form usually about 2, but may be less due to admixture with air or other materials. Graphite has sp. gr. 2.25, diamond 3.5. Carbon forms almost innumerable compounds; about half a million have been identified and very many more are undoubtedly possible. Most of these are made indirectly since carbon reacts very slowly with most materials, except for its tendency toward oxidation, which becomes noticeable in air at from 350 to 700°C depending on the form. Even at incandescence, oxidation in air is relatively slow. Carbon does not melt at ordinary temperatures. It sublimes above 3500°C, and the boiling point is about 4200°C. Thermal conductivity is high compared with other non-metals; graphite has thermal conductivity higher than many metals, and also has appreciable electrical conductivity. The thermal coefficient of expansion of carbon is 0.000001 to 0.000004 per °C. The specific heat varies from 0.16 at ordinary temperatures to 0.40 at 1500°C. In the form of ungraphitized structural or formed carbon, the tensile strength is 400 to 1000 psi, and the compressive strength usually about 2000 psi (but may be as high as 10,000 psi). Carbon is insoluble in common solvents, but soluble in some molten metals from which it crystallizes out as graphite.

For uses and further information, see the various specific forms of carbon.

carbon 14 (radiocarbon). Radioactive carbon of mass number 14.

Properties: Half-life, 5720 years; radiation, beta; radiotoxicity, moderately hazardous.

Derivation: The n, p reaction on N-14, usually by nuclear reactor irradiation of calcium nitrate.

Forms available: From the primary source as solid barium carbonate. Secondary suppliers market chemical compounds synthesized from the barium carbonate, including hydrocarbons, organic acids, alcohols, ketones, amino acids, sugars, steroids, and synthetic intermediates, labeled overall, or in specific positions in the molecule.

Uses: As a radiation source in thickness gauges used to measure thin plastic films;

*See "I.C.C. Shipping Regulations," page xiii.
Reference numbers refer to name of manufacturer. See "List of Manufacturers," page v.

C-14 is used extensively as a research tool and has aided in the study of such problems as catalytic petroleum processes; the role of carbon black in rubber; the function, distribution and elimination of drugs and pharmaceuticals in animals; immunization; the toxic effects of some common chemicals; photosynthesis; the mechanism of action of fumigants and insecticides; plant and animal nutrition and utilization of proteins, fats, etc; the mechanism of fermentation; thermal and photochemical exchange reactions; the diffusion of carbon in graphite; the mechanism of aging in steel; the diffusion of carbon in steel; flotation studies; etc.

Shipping regulations: Class D poison, radioactive material. Blue label. *

carbonado. See diamond.

Carbonate Remover. [28] Fine, white, free-flowing powder. Slightly soluble in water. Used to remove carbonates from cyanide plating solutions.

2, 3-carbonato-1-propanol. See glycerin carbonate.

carbon bisulfide. See carbon disulfide.

carbon black. Finely divided forms of carbon made by the incomplete combustion or thermal decomposition of natural gas or liquid hydrocarbons. The principal types, according to the method of production, are channel black, furnace black, and thermal black. See also under these names. Channel black is characterized by lower pH, higher volatile content, small particle size, and less chainlike structure between the particles; it is suitable for use in reinforcing natural rubber. Thermal black consists of relatively coarse particles and is used principally as a pigment. Furnace black produced from natural gas has an intermediate particle size while that produced from oil can be made in a wide range of controlled particle sizes and is particularly suitable for reinforcing synthetic rubber. Furnace blacks are now the most widely used. See also acetylene black and lamp black.

Properties: Black, amorphous powder. Sp. gr. 1.8-2.1; b.p. 4200°C. Insoluble in all solvents.

Containers: Multiwall paper bags; lined barrels; hopper cars.

Uses: Rubber and inks comprise the largest of many uses.

Shipping regulations: None. *

carbon, combined. As used in metallurgical discussions, carbon which has combined chemically with iron to form cementite, as distinct from graphitic carbon in iron or steel. See also pearlite; ferrite.

carbon cycle.

1. The progress of carbon from the air (carbon dioxide) to plants by photosynthesis (sugar and starches) then through the metabolism of animals to decomposition products which ultimately return it to the atmosphere in the form of carbon dioxide.

2. One of the processes by which the sun and other self-luminous astronomical bodies are thought to derive their energy. The net process is the combination of four hydrogen atoms to form helium. One mechanism, called the carbon cycle, involves successive additions of hydrogen atoms, followed by beta decay, to an initial carbon-12 atom, until a final step is reached in which the new nucleus breaks down to a helium atom and a carbon-12 is regenerated. The carbon thus functions as a catalyst for the process. At the temperatures prevailing in the sun all atoms are stripped of their electrons and the reaction is between the nuclei of the atoms, a thermonuclear reaction. Symbolically, the set of reactions is written

$$C^{12} + H^1 \rightarrow N^{13};\quad N^{13} \rightarrow C^{13} + e;$$
$$C^{13} + H^1 \rightarrow N^{14};\quad N^{14} + H^1 \rightarrow O^{15};$$
$$O^{15} \rightarrow N^{15} + e;\quad N^{15} + H^1 \rightarrow C^{12} + He^4.$$

carbon, decolorizing. Activated carbon, bone black, or other form of carbon having large surface area due to fine state of subdivision or porous character, so that it has capacity for removing colored, odoriferous and other substances from air, gas, or solution.

carbon, deodorizing. See carbon, decolorizing.

carbon dichloride. See perchloroethylene.

carbon dioxide (carbonic acid; carbonic anhydride) CO_2.

Properties: Colorless, odorless gas or heavy, volatile, colorless liquid, or white, snow-like solid. See dry ice. Gas: sp. gr. 1.53 (air = 1); liquid: sp. gr. (−37°C) 1.101; specific volume 8.76 cu ft/lb (70°F); solid: sp. gr. (−79°C) 1.56; m.p. (5.2 atm) −56.6°C; sublimes −78.5°C. Soluble in water.

Derivation: (a) Combustion of coal, coke or natural gas and decomposition of natural carbonates (lime and cement plants). Carbon dioxide is absorbed from the flue or kiln gases by sodium carbonate solutions or ethanolamines. (b) Fermentation of carbohydrates. (c) Action of acid on marble, limestone, or dolomite. (d) From natural springs or wells.

Grades: Technical; U.S.P. XVI; commercial, 99.5%; bone dry, 99.8%.

Containers: Solid: 50-lb blocks in insulated boxes. Liquid: steel cylinders; tank cars and trucks.

Uses (in approximate order of volume): Refrigeration of foods; carbonated beverages; industrial refrigeration; chemical intermediate (carbonates, urea, salicylic acid, etc.); fire extinguishers; inert atmospheres (welding); medicine. Recent uses are in wind tunnels, and to inject, as a saturated water solution, into the ground to force oil to the surface.

Warning! (solid) Extremely cold (109°F below zero). Causes severe burns. Liberates heavy gas which may cause suffocation. MCA warning label.

Shipping regulations: Gas and liquid: nonflammable gas; green gas label.*

carbon dioxide snow. See dry ice.

carbon dioxide, solidified. See dry ice.

carbon disulfide (carbon bisulfide) CS_2.
Properties: Clear, colorless or faintly yellow, flammable liquid; almost odorless when pure; the commercial article has a strong disagreeable odor; poisonous! Sp. gr. 1.260 at 25/25°C; b.p. 46.3°C; freezing point −111°C; wt/gal 10.48 lbs (25°C); refractive index 1.6232 (25°C); flash point −22°F; ignition point under some conditions as low as 100°C. Soluble in alcohol, benzene, and ether; slightly soluble in water.
Typical specifications: Carbon disulfide not less than 99.99%; boiling range within 1°C; free from objectionable foreign odor.
Derivation: (a) Reaction of natural gas or petroleum fractions with sulfur. (b) By heating sulfur and carbon (in the form of coal or coke) in either a direct-fired or an electric furnace and condensing the carbon disulfide vapors.
Method of purification: Treatment with lead acetate followed by lime water with subsequent distillation.
Impurities: Sulfur compounds.
Grades: 99.9%.
Containers: 5-, 10-, 55-gal drums; tank cars, 7,000-, 8,000-, 10,000-gals.
Uses: In viscose rayon; cellophane; manufacture of carbon tetrachloride and flotation agents; veterinary medicine; solvent for fats, resins, rubber, waxes, sulfur, and other chemical products; varnishes; lacquers; paint and varnish removers; rubber; textiles; fumigant; matches; preservative; pesticides.
Danger: Extremely flammable. Vapor harmful. Highly volatile. MCA warning label.
Shipping regulations: Flammable liquid. Red label. Cannot be shipped by express.*

carbon fluorides $(CF)_x$, C_4F. Solid nonconductive materials formed on the carbon anodes during electrolysis of molten potassium fluoride-hydrogen fluoride mixtures to yield elemental fluorine. C_4F is unstable at 60°C and higher temperatures. $(CF)_x$ forms only at higher temperatures.

carbon, gas (retort carbon; retort graphite; metallic carbon; glance coal).
Carbon, in a very dense form, found deposited in the upper parts of retorts used in coal-gas manufacture.
Uses: Manufacture of arc and battery carbons; dry cells.

carbon, graphitic. As used in discussions of metallurgy of iron and steel, practically pure carbon which forms in pig iron during the cooling process, because the absorbing power of iron for carbon decreases as its temperature falls. It exists in the iron in the form of tiny flakes which are distributed throughout the mass. Graphitic carbon gives pig iron the grayish-black appearance so often seen. The tendency of graphitic carbon is to weaken the metal, while combined carbon (q.v.) up to the limit of about 0.90% strengthens it. See also pearlite; cementite; ferrite.

carbon hexachloride. See hexachloroethane.

carbonic acid. Actually H_2CO_3, but customarily used as a synonym for the anhydride, carbon dioxide.

carbonic acid, diphenyl ester. See diphenyl carbonate.

carbonic anhydrase. An enzyme present in red blood cells which catalyzes the production of carbon dioxide and water from carbonic acid.
Use: Biochemical research

carbonic anhydride. See carbon dioxide.

carbonium ion. A positively charged organic ion such as H_3C^+, H_2RC^+, R_3C^+, $RC^+{=}O$, etc., having one less electron than the corresponding free radical and acting in subsequent chemical reactions as though the positive charge was localized on the carbon atom. Such ions can exist only when corresponding negative ions are also present, existence being demonstrated by electrical conductivity experiments. An electron-deficient carbon atom is extremely reactive and has only a transitory existence in most cases, but many organic rearrangement and replacement reactions are effectively explained in terms of a carbonium ion intermediate, including acid-catalyzed polymerization of propylene and other olefins. In this case propylene and hydrogen ion form a carbonium ion as follows:

$$H_3C\text{-}HC{=}CH_2 + H^+ \rightarrow H_3C\text{-}HC^+\text{-}CH_3$$

The latter then combines with another molecule of $H_3C{-}HC{=}CH_2$ to start chain growth.
The difference between a carbonium ion, a free radical and a carbanion may be illustrated as follows:

$$\left[\begin{array}{ccc} & & R \\ & & \cdot\cdot \\ R & : & C \\ & & \cdot\cdot \\ & & R \end{array}\right]^+ \quad \left[\begin{array}{ccc} & & R \\ & & \cdot\cdot \\ R & : & C\ \cdot \\ & & \cdot\cdot \\ & & R \end{array}\right]^0 \quad \left[\begin{array}{ccc} & & R \\ & & \cdot\cdot \\ R & : & C\ : \\ & & \cdot\cdot \\ & & R \end{array}\right]^-$$

carbonium ion free radical carbanion

carbonization of coal. Heating bituminous coal out of contact with air to obtain coke and the many valuable by-products, gaseous, liquid and solid. Low temperature carbonization (400-750°C) yields small quantities of gaseous products but large amounts of coal tar and liquids. Higher temperatures result in an increased ratio of gases to liquids. Some of the basic by-products are ammonia, fuel gas, light oil, and coal tar. These are the sources of many of the most important materials of the chemical industry.

carbonizing assistants. Materials for increasing the efficiency of dilute acid baths used for "carbonizing" wool. Carbonizing removes vegetable matter such as burrs from raw wool or reclaims wool fibers from mixed rags and waste by destroying cellulosic materials. Most wetting-out agents and emulsifiers unaffected by acids assist carbonization by increasing penetration.

carbon, metallic. See carbon, gas.

carbon, mineral. See graphite.

carbon monoxide CO.
Properties: Colorless gas, exceedingly faint metallic odor and taste; highly poisonous, inducing asphyxiation; 0.2% in air is poisonous and 0.43% will induce asphyxiation. Burns with a violet flame. Mixtures of carbon monoxide and air in certain proportions are flammable!
Constants: Sp. gr. 0.96716; b. p. −190°C; solidification point −199°C; specific volume 13.8 cu ft/lb (70°F).
Derivation: (a) Obtained almost pure by introducing a mixture of oxygen and carbon dioxide in contact with incandescent graphite, coke or anthracite. (b) Action of steam on hot coke or coal (water gas). (c) By-product in chemical reactions. (d) Combustion of organic compounds with limited amount of oxygen.
Absorbed by carbon, some metals and by a solution of cuprous chloride in hydrochloric acid or ammonia.
Grades: Commercial (98%); C. P. (99.5%).
Containers: Cylinders.
Uses: Chemical (methanol, ethylene, organic synthesis); fuels (gaseous); metallurgy (special steels, reducing oxides, nickel refining); zinc white pigments.
Shipping regulations: Flammable gas. Red gas label. *

"Carbonox." [236] Brand name for a proprietary product. An organic humic acid material used for treatment of drilling mud to reduce viscosity and gel strength. Also used to prepare emulsion muds characterized by low filtration rates, stability, and easy maintenance.
Containers: Multiwall paper bag containing 50 lbs.

carbon oxybromide. See bromophosgene.

carbon oxychloride. See phosgene.

carbon oxycyanide. See carbonyl cyanide.

carbon oxyfluoride. See carbonyl fluoride.

carbon, retort. See carbon, gas.

carbon tetrachloride (tetrachloromethane; perchloromethane) CCl_4.
Properties: Colorless liquid; peculiar odor, yielding heavy vapors; non-flammable; poisonous! Sp. gr. 1.585 (25/4°C); b. p. 76.74°C; freezing point −23.0°C; refractive index 1.4607 at 20°C; vapor pressure 91.3 mm (20°C); wt/gal 13.22 lbs (25°C); flash point, none; fire point, none. Miscible with alcohol, ether, chloroform, benzene, solvent naphtha, and most of the fixed and volatile oils; very slightly soluble in water.
Typical specifications: Carbon tetrachloride not less than 99.5%; volatility practically complete; residual odor, none; boiling range within 1°C.
Derivation: (a) By the interaction of carbon disulfide and chlorine in presence of a catalyzer (iron); (b) chlorination of methane or higher hydrocarbons at 250-400°C.
Method of purification: Treatment with caustic alkali solution to remove sulfur chloride, followed by rectification.
Grades: Technical; C. P.; N F. XI; electronic.
Containers: 5-, 10-, 55-gal drums; tank cars.
Uses (in approximate order of volume): Refrigerants and propellants, especially the chlorofluorohydrocarbons; metal degreasing; grain fumigants and insecticides; fire extinguishers; dry cleaning solvents; chlorinating organic compounds; general solvent; anthelmintic; in production of semiconductors.
Danger: Hazardous vapor and liquid; may be fatal if inhaled or swallowed. MCA warning label.
Shipping regulations: None. *

carbon trichloride. See hexachloroethane.

carbonyl. The divalent organic radical CO. Also used as a general term for metal compounds containing the CO group (i. e. $Ni(CO)_4$, nickel carbonyl).

N, N'-carbonylbis(4-methoxymetanilic acid) disodium salt $[C_6H_3(OCH_3)(SO_3Na)NH]_2CO$.
Properties: Gray paste; solids approx 70%.
Grades: Technical.
Use: Intermediate.

carbonyl bromide. See bromophosgene.

carbonyl chloride. See phosgene.

carbonyl cyanide (carbon oxycyanide) $CO(CN)_2$.
Properties: Colorless liquid. Unstable in the presence of water. Sp. gr. 1.124 at 20°C; b. p. 65.5°C (740 mm). Very poisonous.
Derivation: From diisonitrosoacetone.
Grades: Technical.
Uses: Organic synthesis; suggested military poison gas.

carbonyl fluoride (fluoroformyl fluoride; carbon oxyfluoride) COF_2.
Properties: Colorless gas. Unstable in the presence of water. B. p. −83°C; sp. gr. 1.139 (−114°C); m. p. −114°C. Very poisonous.
Derivation: Action of silver fluoride on carbon monoxide.
Grades:, Technical.
Uses: Organic synthesis; suggested military poison gas.

carbonyl iron. See iron pentacarbonyl.

carbonyl iron powder. Elementary iron in powdered form, 99.6-99.9% pure Fe. Particle size is usually 1 to 20 microns, according to type of powder supplied.
Derivation: By treating crude iron with carbon monoxide, under heat and pressure,

and subsequently heating the iron pentacarbonyl vapor obtained.
Containers: 1-, 5-lb tin; 25-lb pkg; 200-lb drum.
Uses: High frequency cores in electronics; magnetic fluids; alloying agents; catalysts; pharmaceuticals; powder metallurgy.

Carbortam. [337] A ferrotitanium containing also carbon, silicon, boron and aluminum; m.p. 2650°F; used to deoxidize and harden steel.

"Carborundum." [280] A trademark used on abrasives and refractories of silicon carbide, fused alumina and other materials.
Properties: For silicon carbide — Crystalline form ranges from small to massive crystals in the hexagonal system, the crystals varying from transparent to opaque with colors from pale green to deep blue or black; hardness of 9.17 in Mohs' scale; sp. gr. from 3.06-3.20; crushes into hard sharp abrasive granules; not affected by acids; slowly oxidizes at temperatures above 1000°C; good heat conductor; highly refractory. For fused alumina — See properties under the trademark "Aloxite."
Uses: Abrasive grains and powders for cutting, grinding and polishing; valve-grinding compounds; grinding wheels; stones; bones; rubbing bricks; coated abrasive products; tiles; antislip tiles and treads; refractory grains; tiles; bricks and blocks; grinding and polishing machines.

carbosand. Fine sand that has been treated with an organic solution and roasted in order to produce a material that can be sprayed onto oil slicks and aid in sinking such slicks, thereby destroying the fire hazard occasioned by the presence of oil on water.

"Carbose." [203] Trademark for a technical grade of sodium carboxymethylcellulose (Sodium CMC, NaCMC, CMC).
Properties: Free-flowing, white to pale cream, odorless powder; packed density, 42 lb/cu ft
Grades: "Carbose" D, "Carbose" I, "Carbose" IM, "Carbose" MX, and "Carbose" VL, differ chiefly in viscosity.
Containers: Bags and leverpak drums.
Uses: Detergency promoter; binder and adhesive; film former; drilling mud additive; textile and laundry sizing; surface and beater application in paper industry.
Hazards: Finely ground combustible material; do not allow accumulation of suspended and deposited dust.

"Carboseal." [214] Trademark for proprietary hygroscopic liquid compositions for joint sealing.
Properties: Clear, amber hygroscopic liquid; initial b.p. 280°F; water solubility (20°C) complete; sp. gr. 1.017 (60°F); wt/gal 9.19 lb (20°C).
Containers: 1-, 5-gal cans, 55-gal drums (8.5, 40, 490 lb net wt); tank cars.
Uses: Swelling and moistening agent for jute and other packing in cast-iron gas mains, to correct joint leakage and lay dust.

"Carbo-Sour." [244] A proprietary product consisting of highly soluble fluorine compounds.
Properties: White, dustless granules; highly soluble in water. Neutralizing value 30.2 oz sodium bicarbonate per lb.
Containers: 150-lb and 300-lb net fiber drums.
Uses: Laundry sour, especially when high solubility is desired.
Fire hazard: None.
Shipping regulations: None.*

"Carbo-Vitrobond." [41] Trade name for a carbon-filled, sulfur-based compound for use as a hot-pour, acid-proof cement where temperatures do not exceed 200°F. Has excellent resistance to hydrofluoric and nitric acids.

"Carbowax." [214] Trademark for polyethylene glycols and methoxypolyethylene glycols.
"Carbowax" polyethylene glycols are available in molecular weights ranging from 200 to 20,000.
Properties: Water white liquids to hard waxy solids; water soluble; good lubricity; heat stable; inert to many chemical agents; do not hydrolyze or deteriorate.
Uses: Water-soluble lubricants in rubber fabricating, textile processing, and metal forming; ointment bases for drugs and cosmetics; solvents for dyes, resins, proteins; plasticizers and dispersants for casein and gelatine compositions, glues, zein, and cork.
"Carbowax" methoxypolyethylene glycols are available in molecular weights of 350, 550, and 750. They are similar to polyethylene glycols of comparable weight and are designed for the manufacture of nonionic surface active agents through the preparation of their mono-fatty esters.

carbox fuel cells. See fuel cells.

"Carboxide." [214] Trademark for proprietary fumigant mixture of ethylene oxide and carbon dioxide.
Properties: Colorless gas. Composition not less than 90% carbon dioxide, not more than 10% ethylene oxide; residue not more than 0.1% by wt; vapor pressure not more than 725 psig (70°F).
Containers: 30-, 60-lb cylinders (net wt).
Uses: Fumigant and sterilizing agent to eliminate insects and bacteria. Particularly desirable as fumigants for foodstuffs, tobacco, cigars, furs, garments, rugs, furniture, mohair and woolen goods. Also exhibits valuable properties of ethylene oxide for controlling development of mold spores and thermophyllic bacteria.

carboxybenzene. See benzoic acid.

carboxyhemoglobin. Hemoglobin which has combined with carbon monoxide rather than oxygen, thus rendering it unable to transport oxygen. The affinity of hemoglobin for carbon monoxide is 200 times greater than that for oxygen, thus explaining the great toxicity of carbon monoxide.

carboxylase. A decarboxylase enzyme found in plant tissues which acts upon pyruvic acid, producing acetaldehyde and carbon dioxide.
Use: Biochemical research.

carboxylic. Term for the COOH group, the radical characteristic of all organic acids.

carboxymethylcellulose. See sodium carboxymethylcellulose.

carboxymethylmercaptosuccinic acid $HOOCCH_2SCH(COOH)CH_2(COOH)$. A heavy-metal chelate.
Properties: White powder; melting range 136-138°C. Water solubility, 147 g/100 g at 25°C; ethanol solubility, 76 g/100 g at 25°C.

carboxymethylpyridinium chloride hydrazide. See Girard's "P" reagent.

carboxymethyltrimethyl ammonium chloride hydrazide. See Girard's "T" reagent.

carboxypeptidase. A proteolytic enzyme found in the pancreas which catalyzes the hydrolysis of native food proteins. It acts upon polypeptides producing simpler peptides and amino acids.
Use: Biochemical research.

4-carboxyresorcinol. See beta-resorcylic acid.

6-carboxyuracil. See orotic acid.

carbromal. See bromodiethylacetylurea.

carburetted water gas. A city gas consisting of water gas in which the heating value (Btu) is increased and luminous quality (candle power) conferred by the addition of hydrocarbon gases and vapors obtained by high-temperature cracking of residual oils from petroleum refining operations. A typical composition: Illuminants, 13.3%; carbon monoxide, 30.4%; hydrogen, 37.7%; methane, 10.0%; ethane, 3.2%; carbon dioxide, 3.0%; oxygen, 0.4%; nitrogen, 2.1%; Btu/cu ft, 543.0; candle power, 22.1.

carbutamide. See 1-butyl-3-sulfanilylurea.

carcinogens. Substances which cause cancerous growths in living tissues. Among known carcinogens are 9,10-dimethyl-1,2-benzanthracene, 20-methylcholanthrene, benzo[a]pyrene (3,4-benzpyrene), found in coal, cigarette, and other tars. All are polynuclear hydrocarbons. Among other compounds thought to be carcinogens are urethane; the nitrogen mustards (these two types may both induce and inhibit tumors); some azo dyes (e.g., ortho-aminoazotoluene, 4-dimethylaminoazobenzene); beryllium, cobalt, selenium and nickel compounds; chromates; excess amounts of some hormones.

cardamom oil.
Properties: Colorless or pale-yellow essential oil; strongly aromatic, camphoraceous odor and taste. Sp. gr. 0.917-0.947 (25/25°C); refractive index 1.460; optical rotation +22° to +44° in 100 mm tube at 25°C. Insoluble in water, soluble in alcohol and ether. Keep well closed, cool, and protected from light.
Chief known constituents: Terpinene, dipentene, limonene, cineol, borneol.
Derivation: Distilled from the seeds of Elettaria cardamomum from Malabar and Ceylon.
Method of purification: Rectification.
Grades: Technical; N.F. XI.
Containers: Tins; glass bottles.
Uses: Flavoring; liqueurs; medicine.
Shipping regulations: None.*

cardamom seed.
Derivation: Dried, nearly-ripe fruit of Elettaria cardamomum.
Habitat: Malabar; cultivated in India and Ceylon.
Grades: Bleached; decorticated; green; N.F. XI.
Containers: Bags; cases.
Uses: Medicine; condiment; source of perfume extracts.
Shipping regulations: None.*

"Cardanol." [158] A mixture of 3-pentadecenyl and less saturated C_{15} phenols, as, $C_6H_4OH \cdot C_{15}H_{27}$.
Properties: Amber liquid, boiling range 180-230°C (1 mm); soluble in oils, waxes, and all organic solvents; insoluble in water, glycerine, and aqueous alkalies.
Uses: Aldehyde-reactive plasticizer for phenol-aldehyde resins, particularly for laminating purposes.

"Cardanol Bis-Phenol." [158] Stated to be 1,8-di(hydroxyphenyl) pentadecane $C_6H_4OH \cdot C_{15}H_{30} \cdot C_6H_4OH$.
Properties: Brown viscous liquid (10,000 cps at 25°C).
Uses: Base for epoxy resins; base for synthetic resins of phenol-formaldehyde type useful as wire enamel.

"Cardanol" Ethers. [158] A mixture of ethyl ethers of 3-pentadecenyl and less saturated C_{15} phenols, as, $C_6H_4OC_2H_5 \cdot C_{15}H_{27}$.
Properties: Light amber, low-viscosity liquid; boiling range 175-200°C (1 mm); pour point, −52°C.
Uses: Low-temperature plasticizer for GR-S rubber.

"Cardilate." [301] Trademark for erythrol tetranitrate (q.v.).

"Cardio-Green." [348] Trademark for indocyanine green, a diagnostic dye used in medicine.

"Cardolite" Brand Epoxy Resin Flexibilizer NC-513. [158] A fluid resin designed for use as a chemically-bound flexibilizer for epoxy resins.
Properties: Clear, deep amber liquid. Viscosity, 100 cps maximum at 25°C; epoxide equivalent of 475 to 575; specific gravity 0.960 to 0.975 at 25°C; open cup flash point 445°F.
Containers: Quart, 1-gal, 5-gal, and 55-gal drums.
Uses: It is non-extractable and is co-reactive

*See "I.C.C. Shipping Regulations," page xiii.
Reference numbers refer to name of manufacturer. See "List of Manufacturers," page v.

with epoxies. Reduces viscosity and improves solubility and resistance of epoxies to thermal and mechanical shock without adversely affecting strength, corrosion resistance, electrical or aging properties.

"Cardolite" Brand Friction Components. [158] Components based upon a phenolic-type liquid found in the fibrous outer shell of the cashew nut. (See cashew nutshell liquid.) These materials are produced in the form of friction-fortifying particles, and binding resins in liquid or powdered form.

Containers: 55-gal drums.

Uses: The resins are used in the formulation of brake linings, brake blocks, clutch facings and other friction units which have high heat resistance, and outstanding friction and wear characteristics.

"Cardolite" Brand Friction Fortifying Particles are used as additives to an asbestos-resin-filler matrix to stabilize and control the coefficient of friction, wear, noise and high temperature performance.

"Cardolite" Brand CNSL Liquid Binding Resins are heat-convertible polymers. Used to bond together the asbestos fillers and friction modifying particles, these resins provide a friction surface with relatively constant properties over a wide temperature range.

"Cardolite" Brand CNSL Powdered Binding Resins are heat convertible internally flexibilized resins for enhancing high temperature and friction characteristics of friction elements.

"Cardolite" Brand Polymeric Intermediates are used to prepare resins for coatings, saturating resins, binder resins, etc., where thermal stability, impact resistance and chemical resistance are needed.

"Cardosol" Brand Resin. [158] A water soluble ketone formaldehyde condensate which can be gelled and cured by alkali or heat.

Containers: 5-gal pails; 55-gal drums.

Use: Primarily used in conjunction with starch to provide water resistant adhesives for box board, coatings, for glass fibers and as a ceramic binder.

carica papaya. See papaya.

"Caricide." [57] Trademark for diethylcarbamazine.

caritol. See carotene.

carmania gum (Syrian gum). A tragacanth gum (q.v.) exuded from thorny shrubs of Astragalus in Syria.

"Carmethose." [305] Trademark for sodium carboxymethylcellulose, U.S.P. Used in medicine.

carmine (coccinellin). An aluminum lake of the pigment from cochineal (q.v.). Bright red pieces, easily powdered. Soluble in alkali solutions, borax; insoluble in dilute acids; slightly soluble in hot water.

Grades: Technical.

Containers: Tins.

Uses: Dyes, inks, indicator in chemical analysis; coloring food materials, medicines, etc.

Shipping regulations: None.*

carminic acid (cochinilin) $C_{22}H_{20}O_{13}$, a tricyclic compound. The essential constituent of carmine.

Properties: Dark, purplish-brown mass or bright-red powder. M.p., decomposes at 136°C; pH 4.8 yellow; pH 6.2 violet. Soluble in water, alcohol, alkali hydroxide solutions; insoluble in ether, benzene, chloroform.

Derivation: By extraction from the insects, Coccus cacti (cochineal).

Method of purification: Crystallization.

Grades: Technical.

Containers: Glass bottles.

Uses: Stain in microscopy; indicator in analytical chemistry; coloring proprietary medicines; pigment for fine oil colors; color photography; dyeing.

Shipping regulations: None.*

carnallite $KCl \cdot MgCl_2 \cdot 6H_2O$ or $KMgCl_3 \cdot 6H_2O$.

Properties: A natural hydrated double chloride of potassium and magnesium. White, brownish and reddish; streak, white; shining, greasy luster; strongly phosphorescent; bitter taste; deliquescent. Sp. gr. 1.62; hardness 1.

Occurrence: Germany; Alsace; New Mexico.

Use: A chief commercial source of manufactured potash salts.

"Carnation." [45] Trademark for white mineral oil, technical grade.

Properties: Sp. gr. 0.835-0.845 (60°F); Saybolt viscosity 65-75 (100°F); odorless and tasteless.

Uses: Cosmetic preparations; shell egg preservation; further organic synthesis.

carnauba wax (Brazil wax). One of the hardest and most expensive of the commercial waxes.

Properties: Hard, amorphous, light yellow to dark greenish brown brittle lumps; peculiar, agreeable odor. Sp. gr. 0.995 (15/15°C); m.p. 84-86°C; acid number 2-9; iodine number 13.5. Soluble in ether, boiling alcohol and alkalies; insoluble in water.

Derivation: An exudation from leaves of the wax palm, Copernica cerifera (Brazil).

Grades: No. 1 yellow; No. 2 N.C.; No. 2 regular; No. 3 N.C.; No. 3 chalky; powdered.

Containers: Bags; boxes.

Uses: Substitute for beeswax; shoe polishes; candles; leather finishes; varnishes; electric insulating compositions; furniture and floor polishes; phonograph records; carbon paper coating; waterproofing.

Shipping regulations: None.*

carnelian (cornelian). A clear pale to deep red chalcedony (q.v.).

carnosine (beta-alanylhistidine; ignotine) $C_9H_{14}N_4O_3$. An amino acid occurring in

*See "I.C.C. Shipping Regulations," page xiii.
Reference numbers refer to name of manufacturer. See "List of Manufacturers," page v.

muscle of many animals and man. It is found naturally in the L(+)-form.
Properties: M.p. 246-250°C (dec.); soluble in water. Nitrate: crystals; m.p. 222°C (dec.); soluble in water. Hydrochloride: crystals; m.p. 245°C (dec.); soluble in water. D(-)-carnosine: crystals; m.p. 260°.
Use: Biochemical research.

carnotite $K_2(UO_2)_2(VO_4)_2 \cdot 3H_2O$. A natural hydrated vanadate of uranium and potassium, usually found in sandstones and other sedimentary rocks.
Properties: Color bright yellow to lemon yellow, sometimes greenish yellow; luster dull or earthy, pearly or silky when coarsely crystalline. Soluble in acids. Radioactive. Usually occurs as a powder or in fine-grained aggregates.
Occurrence: Colorado, Utah, Arizona, New Mexico, South Dakota; Australia; Belgian Congo; U.S.S.R.
Use: Important ore of uranium; source of radium.
Shipping regulations: Poison; radioactive material. Red or blue label.*

Carnot's reagent. A reagent for the determination of potassium. It is an alcoholic solution of sodium bismuth thiosulfate, made from sodium thiosulfate and bismuth subnitrate.

"Carnube Wax." [354] A synthetic wax substitute for carnauba wax. It can be used to replace up to 75% of carnauba wax in bright-drying wax emulsions. M.p. 80-82°C; acid no. 78-80.

carob seed (carob bean, carob-tree bean, St. John's bread, algarroba bean, locust bean).
Derivation: From the tree Ceratonia siliqua. Seeds and pod contain a sweet pulp rich in sugar and gums.
Containers: Bales.
Use: As fodder; source of carob-seed gum.

carob-seed gum (locust-bean gum).
Properties: In powdered form, nearly pure white. Swells partially in cold water, but attains greater viscosity when heated. Insoluble in most organic solvents.
Typical specification: 12-14% moisture; acid insoluble ash 0.7-1.5%; protein content 5-6%.
Containers: Bags.
Derivation: Extracted from carob seeds.
Uses: As emulsifying agent in ice cream and cosmetic cream manufacture; in textile manufacturing as sizing and finishes, and as a substitute for tragacanth gum; pharmaceuticals; paint industry. See also locust kernel.

"Carolate." [415] Trade name for a self-emulsifying spermaceti-amide. A soft wax; bland odor and taste; pH of 1% solution 9.5.
Containers: 25-lb corrugated cartons.
Uses: Vehicle for cosmetics and pharmaceuticals.

"C-4" Aromatic Solvent. [11] Trade name for a clear alkylated aromatic hydrocarbon used as a medium-high-boiling (160-185°C) aromatic solvent for coating compositions and for processing purposes.

carony bark. See angostura.

Caro's acid (peroxysulfuric acid; persulfuric acid) H_2SO_5 or $HOSO_2OOH$.
Properties: White crystals; m.p. 45°C (decomposes). Is reported to have exploded spontaneously, and on addition of organic matter.
Derivation: Action of hydrogen peroxide on concentrated sulfuric acid; action of 40% sulfuric acid on potassium persulfate.
Use: See Caro's reagent.

Caro's reagent. A pasty mass of great oxidizing power (see Caro's acid). Used in testing aniline, pyridine, and alkaloids.

carotene (carotin, caritol, provitamin A) $C_{40}H_{56}$. A precursor of vitamin A occurring naturally in plants. It consists of 3 isomers; about 15% alpha, 85% beta, and 0.1% gamma. Carotene is a hydrocarbon member of a large class of pigments called carotenoids. It has the same basic molecular structure as vitamin A and is transformed to the vitamin in the animal liver.
Properties: Ruby-red crystals, easily oxidized on contact with air; m.p. (alpha) 188°C, (beta) 184°C, (gamma) 178°C; insoluble in water; slightly soluble in alcohol and ether; soluble in chloroform, benzene, and oils.
Source: Occurs as an orange-yellow pigment in plant and animal tissue, particularly in butter, eggs, sweet potatoes, alfalfa, barley, clover, rye, and wheat. The proportion is very small, only 10 to 40 mg of carotene per pound of fresh material.
Derivation: By extraction from carrots and palm oil concentrates; by a chromatographic process from alfalfa. beta-Carotene is also made by a microbial fermentation process from corn and soybean oil.
Grades: According to U.S.P. units of vitamin A. Sold as pure crystals or solutions in various oils.
Uses: Feed additive; pharmaceuticals; coloring and enriching margarine and dairy products.

carotenoids. A class of pigments which occurs in many vegetable oils and in some animal fats. They range in color mostly from yellow to deep red. They include the four hydrocarbons lycopene, alpha-, beta-, and gamma-carotene, and various derivatives.
Properties (general): Crystalline solids; soluble in fats and oils; insoluble in water, high melting; stable to alkali but unstable to acids and to oxidizing agents; color easily destroyed by hydrogenation or by oxidation; some are optically active.

carotin. See carotene.

carpaine $C_{14}H_{25}NO_2$.
Properties: White crystalline alkaloid.

*See "I.C.C. Shipping Regulations," page xiii.
Reference numbers refer to name of manufacturer. See "List of Manufacturers," page v.

Soluble in alcohol and ether; slightly soluble in water. M.p. 121°C; b.p. 215-235°C.
Derivation: By extraction from the leaves of Carica papaya.

carpaine hydrochloride $C_{14}H_{25}NO_2 \cdot HCl$.
Properties: White crystals; soluble in water, alcohol, and ether; m.p. 225°C (dec).

carrageen. See chondrus.

carrageenan (formerly carrageenin). The gelatinous extract of a seaweed called carrageen, Chondrus crispis (q.v.) or Irish moss. It is a complex carbohydrate made up of galactose, dextrose, and levulose residues and small quantities of pentosan or methylpentosan.
Properties: A water-soluble colloid, refined and dried to a free-flowing powder. Adsorbs water rapidly, dissolves readily in warm water, gels on cooling. A 3% solution forms a soft gel with m.p. 27-30°C, a 5% solution forms a firm gel with m.p. 40-41°C. There are two forms. One, extracted by cold water, gives a viscous solution in water. The other, extracted by hot water, gels on cooling.
Derivation: Extraction from the plants by hot water, filtering, concentrating the filtrate, and precipitating the carrageenan with alcohol.
Uses: In bacteriological cultures, in medicine; in foods as an emulsifier and stabilizer, and in general, as a protective colloid.

carrageenin. See carrageenan.

"Carrene 16." [54] Trademark for a solution of lithium bromide in water and used in absorption refrigeration machines.

"Carrene 500." [54] Trademark for an azeotropic mixture of 73.8% dichlorodifluoromethane and 26.2% unsymmetrical difluoroethane boiling at −28°F and used as a refrigerant.

carrier. As used in atomic tracer procedures, a substance which when added to a minute quantity (trace) of a like or similar substance will carry the trace with it through a chemical or physical process. Carriers make possible the study of the chemical behavior of radioactive substances formed in such small quantities that it would not be possible otherwise to observe their characteristic reactions.

carrot oil.
Properties: A light-yellow volatile oil having a spicy odor. Sp. gr. 0.870-0.944; optical rotation −8 to −37°; refractive index 1.482-1.491.
Chief known constituents: Carotene, pinene, limonene, palmitic acid, butyric or isobutyric acid.
Derivation: Distillation of the seeds of Daucus carota.
Uses: Liqueur and other flavors; to a small extent, perfumery.

carthamic acid. See carthamin.

carthamin (carthamic acid; safflor carmine; safflor red) $C_{21}H_{22}O_{11}$.
Properties: Dark-red powder with green luster. Slightly soluble in water; soluble in alcohol; insoluble in ether; solutions rapidly decompose.
Derivation: A glucoside coloring matter from Carthamus tinctorius.

carthamus (safflower; thistle saffron; American saffron; dyer's saffron; false saffron; bastard saffron). Florets of Carthamus tinctorius.
Properties: Red color, mixed with yellow; peculiar slightly aromatic odor.
Habitat: Levant and Orient; cultivated extensively in Europe and America.
Grades: Technical.
Containers: Bales.
Uses: Medicine; coloring cosmetics, liqueurs, butter, sweetmeats and various food products; dyeing artificial flowers; paints.
Shipping regulations: None.*

"Cartrax." [299] Trademark for a combination drug containing hydroxyzine hydrochloride and pentaerythritol tetranitrate. Used in medicine.

carui oil. See caraway oil.

carum. See caraway.

"Carum." [51] Trademark for grease-type lubricants prepared for use in chemical, food and similar processing industries where insolubility in the material being processed is essential. Intended for valves and pumps handling such materials.

carvacrol (isopropyl-ortho-cresol; 2-hydroxy-para-cymene) $(CH_3)_2CHC_6H_3(CH_3)(OH)$. An alcohol.
Properties: Thick, colorless oil; thymol odor; sp. gr. 0.976 (20/4°C); b.p. 237°C; m.p. 0°C; refractive index 1.523 (20°C). Insoluble in water; soluble in alcohol, ether, and alkalis.
Derivation: (a) From oil of origanum, from thyme, and summer savory; (b) also from para-cymene by sulfonation, followed by alkali fusion.
Uses: Perfumes; fungicides; disinfectant.
Shipping regulations: None.*

carvol. See carvone.

carvone (carvol) $C_{10}H_{14}O$. A quinone similar in general structure to carvacrol.
Properties: Pale-yellowish or colorless liquid; fine caraway-like odor. Sp. gr. 0.960; b.p. 230°C; refractive index (n 18/D) 1.4999. Soluble in alcohol, ether and chloroform; insoluble in water.
Derivation: From caraway, dill, spearmint oils.
Method of purification: Rectification.
Grades: Technical.
Containers: Tins; 1-, 5-lb glass bottles.
Uses: Medicine; flavoring; liqueurs; perfumery.
Shipping regulations: None.*

caryophyllin (oleanoic acid) $C_{30}H_{48}O_3$. Not to be confused with caryophyllene, a

sesquiterpene, $C_{15}H_{24}$.
Properties: White, odorless, silky needles; m.p. 310°C. Insoluble in water; soluble in alcohol and ether.
Derivation: From clove oil.

caryophyllus (clove). Dried flower buds of Eugenia caryophyllus having a strong fragrant, spicy odor.
Habitat: Malacca Islands; Zanzibar; Sumatra; South America; West Indies, etc.
Grades: Technical; N.F. XI; also as from Madagascar or Zanzibar.
Containers: Boxes; bags.
Uses: Medicine; manufacture of clove oil, eugenol, chocolate; in baking.
Shipping regulations: None.*

caryophyllus oil. See clove oil.

Casale system. One of the older processes for synthesis of ammonia from nitrogen and hydrogen gases with a promoted iron oxide catalyst. Characterized by use of high pressure (600-750 atmospheres), 500°C temperature, presence of ammonia (from recycle gas) in converter feed gas so as to slow down initial reaction rate and heat production. Conversion is about 40% per pass and the high pressure permits condensing the ammonia formed by passing the converted gas through water-cooled condensers.

cascade. A process or apparatus, usually in separation or purification, in which materials are passed through a multiplicity of identical or similar relatively simple operations, in order to multiply the separation or other effect that is achieved in a single simple operation. An outstanding example is the Oak Ridge diffusion plant for separating uranium isotopes by passing uranium fluoride mixtures through an extended series of diffusion cells, each of which causes a slight enrichment of the desired isotope. An ordinary bubble plate distillation tower is a much more frequently encountered example.

cascade particle. See fundamental particle.

"Cascamite." [65] Trademark for a powdered urea-formaldehyde resin glue; water-resistant, moldproof, stainfree.

cascara sagrada bark (sacred bark; chittem bark; chittim bark; Persian bark; bearberry bark; bearwood).
Derivation: Bark of Rhamnus purshiana.
Habitat: West coast of the United States.
Properties: The bark loses its emetic properties on being kept for one year. Odor distinct, taste bitter, and slightly acrid.
Grades: Technical; U.S.P. XVI.
Containers: Bales; multiwall paper sacks.
Use: Medicine.
Shipping regulations: None.*

cascarilla (eleuthera bark; sweet-wood bark; eluteria bark).
Derivation: Bark of Croton eluteria.
Habitat: West Indies.
Use: Medicine; sometimes added to smoking tobacco for flavor.

"Casco." [65] Trademark for a series of adhesives based on casein, seedmeal and soya proteins, furnished in dry powder form. For gluing wood and paper in a wide variety of applications.

"Cascola." [65] Trademark for a series of casein and dextrin type adhesives for carton sealing, paper laminating, tube winding and various packaging applications.

"Cascolac." [65] Trademark for a series of viscous, liquid, casein-based adhesives for general purpose paper gluing, packaging, label bonding, over-coating.

"Cascoloid." [65] Trademark for a series of casein base binders, emulsifiers, stabilizers, and thickeners for emulsion paints and coatings.

"Cascophen." [65] Trademark for a series of resorcinol, phenol-resorcinol, and phenolic resins in liquid and powder form. Used for waterproof wood adhesives, wet strength resins for paper, molding applications.

"Casco Resin." [65] Trademark for a series of liquid urea-formaldehyde resins, used in wood gluing operations, in wet-strength paper and in paper lamination to fortify starch compounds.

"Cascorez." [65] Trademark for a series of liquid and powdered polyvinyl adhesives for bonding wood, paper, fabric and other porous and semi-porous materials.

case hardening. A process of hardening a ferrous alloy so that the surface layer or "case" is made substantially harder than the interior or "core."

case-hardening compounds. Materials used to impart a hard surface to steel while the interior remains soft and tough. This is accomplished by heating the steel out of contact with air while packed in carbonaceous material, cooling it to black heat, reheating to a high temperature, and quenching. The compounds used are usually wood charcoal with sodium, potassium, or barium carbonates, cyanides, etc.
Shipping regulations: For mixtures containing cyanides, poison label.*

casein. The principal protein in milk (3% casein) and the main ingredient of cheese. It is a phosphoprotein (about 0.85% phosphorus and 0.76% sulfur) consisting of about 15 amino acids and has a molecular weight ranging from 75,000 to 375,000.
Properties (pure casein): White, tasteless, odorless, amorphous solid; sp. gr. 1.25-1.31; hygroscopic; stable when kept dry but deteriorates rapidly when damp. Soluble in dilute alkalies and concentrated acids; almost insoluble in water; precipitates from weak acid solutions.
Derivation: Acid casein: warm skim milk is acidified with dilute acid, the whey drawn off, the curd washed, pressed, ground and

dried. Rennet casein or paracasein: warm skim milk is treated with rennet extract. The curd contains combined calcium and calcium phosphate.
Grades: Acid precipitated (domestic edible, imported inedible); paracasein.
Uses: Paracasein, principally for plastics (see casein plastics); acid-precipitated, for paper coatings; glues; paints; adhesives; textile sizing; foods and feeds.
Shipping regulations: None.*

casein, acid-precipitated. See casein.

casein glue. See glue.

casein paints. Coatings in which casein replaces the ordinary drying oils, or is used as an emulsifying agent in emulsion and latex paints. Both types may be thinned with water.

casein plastic.
Properties: A very tough, nonflammable, thermoplastic or thermosetting material. It is readily colored, molded, and polished but has poor water resistance and dimensional stability.
Uses: Buttons, buckles, and novelty items which do not require dimensional stability.

casein, rennet. See casein.

casein-sodium (sodium caseinate).
Properties: White, coarse powder. Odorless, tasteless. Contains 65% proteins. Soluble in water (usually with turbidity).
Derivation: By dissolving casein in sodium hydroxide and evaporating.
Grades: Edible.
Uses: Medicine; foods; as emulsifier and stabilizer.

cashew gum (anacardium gum). The exudation from the bark of the cashew-nut tree, Anacardium occidentale. Hard, yellowish-brown gum, partly soluble in water. Used for inks, insecticides, pharmaceuticals, mucilage tanning agent, natural varnishes, bookbinders' gum.

cashew nut oil (acajou nut oil). The oil obtained from the edible kernel of Anacardium occidentale. Similar to almond or olive oil.

cashew nutshell liquid (cashew nutshell oil). The liquid or oil obtained from the spongy layer between the inner and outer shells of the nut of Anacardium occidentale. It is a by-product of the edible cashew-nut industry. A similar liquid is obtained from the semecarpus, or oriental cashew nut. The raw liquid contains about 90% anacardic acid, $C_{22}H_{32}O_3$, and a blistering compound containing sulfur. It is used as a vesicant and ant repellent. Most of the liquid used in commerce has been heated or treated with chemicals to make it safe to handle. The principal ingredient is then cardanol, a meta-phenol. The liquid is non-drying, but can be made drying by proper treatment. It polymerizes on heating and forms condensation products with aldehydes.
Containers: 10-lb tins; steel drums.
Uses: Varnishes and impregnating materials which are oil- and water-resistant, artificial rubber, plasticizers, germicides and insecticides, coloring materials and indelible inks, lubricants, and preservatives.

cassava starch.
Derivation: From cassava or manioc root, of the genus Manihot. By heating the damp starch in shallow pans, the granules burst and adhere, forming irregularly shaped, translucent kernels, known as tapioca.
Grades: Technical.
Containers: Burlap bags; wooden barrels.
Uses: Foodstuffs; laundry starches; adhesives; fuel alcohol; textile size.
Shipping regulations: None.*

Cassel brown. See Van Dyke brown.

Cassel earth. See Van Dyke brown.

Cassel green. See barium manganate.

Cassella's acid. See 2-naphthol-7-sulfonic acid; 2-naphthylamine-4,8-disulfonic acid.

Cassella's F acid. See 2-naphthylamine-7-sulfonic acid.

cassia bark. See cinnamon, cassia.

cassia buds. The dried unripe fruit of various species of Cinnamomum, with a cinnamon flavor and resembling small cloves. Not to be confused with cassia fistula.

cassia fistula (purging cassia; drumstick; Indian laburnum; pudding pipe; pudding stick; cassia pods).
Derivation: Dried fruits of Cathartocarpus fistula.
Habitat: Upper Egypt and East India; cultivated in tropical America and Africa.
Grades: Technical.
Containers: Bags.
Use: Medicine.

cassia oil (Chinese cinnamon oil; cinnamon, cassia oil; cinnamon oil, U.S.P. XVI).
Properties: Yellow or brownish limpid liquid; cinnamon-like odor; burning and intensively sweet taste; darkens and thickens on exposure to air; sp. gr. 1.045 to 1.063; optical rotation +1° to −1°; b.p. 240-260°C; refractive index 1.607-1.618; soluble in ether and chloroform; soluble in alcohol, acetic acid.
Chief known constituents: Cinnamic aldehyde (90-95%); cinnamyl acetate, methoxycinnamic aldehyde, phenyl propyl acetate, salicylic aldehyde, coumarin, and benzaldehyde.
Derivation: Distilled from leaves and twigs of Cinnamomum cassia.
Adulteration: Synthetic cinnamic aldehyde, kerosene, rosin, and benzyl acetate.
Grades: U.S.P. XVI; redistilled; technical; lead free.
Containers: Bottles; tins; cans.
Uses: Flavoring; perfumery; medicine; soaps; tanning leather.
Shipping regulations: None.*

cassia pods. See cassia fistula.

*See "I.C.C. Shipping Regulations," page xiii.
Reference numbers refer to name of manufacturer. See "List of Manufacturers," page v.

cassia pulp. Black viscous sweet mass composed of hydroxymethylanthraquinones, gum, tannin, albuminoids, and 60% sugar. Derived from cassia fistula pods. Used as medicine.

cassie oil. A floral absolute obtained by volatile solvent extraction. There are two types:
Cassie ancienne: Derived from Acacia farnesiana.
Cassie romaine: Derived from Acacia cavenia.
Use: Perfumery.

cassiterite (tinstone, wood tin, stream tin) SnO_2. Natural tin dioxide, usually found with igneous rocks.
Properties: Color brown, black, yellow, white; luster adamantine or dull submetallic; streak white; hardness 6-7; sp. gr. 6.8-7.1.
Occurrence: Malaya; Bolivia; Indonesia; Belgian Congo; Nigeria.
Use: Principal ore of tin.

"Castan." [51] Trademark for lubricants for open bearings or bearings with large clearances from which leakage or oil throwing should be prevented. Not suitable where moisture is present.

castanha oil (Brazil-nut oil).
Properties: Pale-yellow, odorless, fixed oil. Soluble in ether, carbon disulfide, and benzene; insoluble in water. Sp. gr. 0.9180; m.p. 0.4°C; saponification value 193; iodine number 106; refractive index (n 25/D) 1.4643.
Derivation: From the Brazil-nut Bertholletia excelsa.
Grades: Technical.
Containers: Tins; iron drums.
Uses: Soap; food; illumination.
Shipping regulations: None.*

cast iron. Any iron-carbon alloy that contains more than 1.7% carbon, and usually between 2 and 4.0%. Such iron usually contains 0.1 or 0.2% sulfur, 0.5-3% silicon, 0.5-1% manganese and up to 1% phosphorus. Cannot be shaped by hammering, rolling or pressing.

cast iron, alloy. Cast iron containing chromium, copper, molybdenum, nickel, or other steel-alloying elements in amounts from 0.1-5% for the purpose of improving strength and wear corrosion, or scaling resistance.

cast iron, gray. Cast iron with gray fracture and with its carbon largely in the uncombined state. The most common form of cast iron, easily melted and machined, relatively soft and tough. Properties depend upon composition, rate of cooling, and heat treatment.

cast iron, malleable. White cast iron that has been annealed after solidification in order to reduce carbon content and produce a product similar in many ways to mild steel.

cast iron, white. A cast iron with silvery surfaces where broken, low silicon content, and all its carbon chemically combined with iron, produced by sudden chilling of the molten iron. Very hard, brittle, and cannot be machined. Produced as an intermediate stage in making malleable cast iron and as a thin outer layer on the surface of gray cast iron.

Castner cell. (1) A mercury-cathode cell for the production of caustic soda and chlorine from brine. It consists of a rectangular concrete box separated into compartments by partitions that extend nearly to the bottom. A layer of mercury on the bottom serves as a seal between the compartments as well as the cathode. In operation, the cell is given a rocking motion, permitting the mercury to flow back and forth between the electrolysis compartment, where a sodium amalgam is formed, and the decomposition compartment, where the amalgam reacts with water to form sodium hydroxide. (2) An electrolytic cell for the production of sodium metal, hydrogen, and oxygen from fused sodium hydroxide. It has largely been superseded by the Downs cell for the production of sodium.

"Castolast H-W." [446] Trade name for a 93% high alumina castable cement bonded with low iron calcium aluminate cement. Shipped dry; with water addition, develops and maintains high strength, through 3200°F. Resistant to abrasion and impact.
Uses: Petrochemical unit liner; burner blocks and other high temperature applications. Can be cast, trowelled or applied with an air placement gun.

castor (castoreum).
Derivation: Dried preputial follicles with their secretions of the common beaver (Castor fiber). Solid unctuous masses contained in pairs of sacs, each about 2 in. in length. Characteristic irritating odor. Contains 40-70% resin.
Grades: Canadian or American; Russian.
Containers: Cans.
Uses: Medicine; perfume fixative.
A synthetic castor is also marketed.

castor bean. See ricinus; castor oil; castor bean oil meal.

castor bean oil meal (castor cake; castor pomace). The residue from extraction of oil from the castor bean (ricinus). The normal product contains 29.5% crude protein; 35.8% crude fiber; 13.2% nitrogen-free extract and 1.0% crude fat. The total digestible nutrients approximate 25%. The ash content of 7.5% is high in potash and phosphate.
Containers: Bag or bulk.
Uses: As animal feeds after removal of toxic ingredients or as fertilizer.
Shipping regulations: None.*

castor cake. See castor bean oil meal.

"Castordag." [46] Trademark for a concentrated colloidal dispersion of pure electric-furnace graphite in castor oil.

Properties: Liquid consistency; solids content 10%; average particle size 1 micron; maximum particle size 10 microns; sp. gr. 1.037; flash point 271°C; completely miscible with castor oil, alcohol.

Uses: Assembly lubricant for mechanical parts; formulation of hydraulic fluids.

castoreum. See castor.

castor meal. See castor bean oil meal.

castor oil (ricinus oil).

Properties: Pale-yellowish or almost colorless, transparent, viscid liquid, faint, mild odor and usually nauseating taste. It is a non-drying oil. Sp. gr. 0.960-0.970; saponification value 178; iodine value 85; solidifies at −10°C. Soluble in alcohol, ether, benzene, chloroform and carbon disulfide.

Derivation: From the seeds of the castor bean, Ricinus communis. They are cold pressed for the first grade of medicinal oil and hot pressed for the common qualities, about 40% of the oil content of the bean being obtained. Residual oil in the cake is obtained by solvent extraction.

Chief constituent: Ricinolein (glyceride of ricinoleic acid).

Grades: U.S.P. XVI; No. 1; No. 3; refined.

Containers: Drums; tanks.

Uses: Medicine; hydraulic fluids; high-grade lubricant; leather preservative; textiles (cotton dyeing, preparation of sulfonated oils, Turkey red oil); electric insulating compositions; toilet creams and hair dressings; special soaps; rubber compounding; plasticizer manufacture; production of sebacic acid, source of ricinoleate compounds.

See also castor oil, cracked; castor oil, dehydrated; blown oils.

Shipping regulations: None.*

castor oil, acetylated. See glyceryl triacetylricinoleate.

castor oil acid. See ricinoleic acid.

castor oil, blown. See blown oils.

castor oil, cracked. Product obtained on heating castor oil out of contact with air. Treatment of the cracked distillate forms a number of intermediates for the production of synthetic perfumes and flavors. Cracking products which have been identified are oenanthol (with benzaldehyde has a jasmin odor) and undecylenic acid (forms gamma undecalactone which gives a peach odor on further decomposition).

castor oil, dehydrated (DCO). A castor oil from which about 5% of the chemically combined water has been removed, and which, as a result, has drying properties similar to those of tung oil. Though it does not dry as rapidly as tung oil, its pale color and elasticity are desirable properties. It also increases the rate of drying of linseed oil. Dehydration is carried out commercially by heating the oil in the presence of catalysts, such as sulfuric and phosphoric acids, clays, and metallic oxides. The commercial product is offered in a wide range of viscosities and analytical constants. Used in paints and lacquers; alkyd resins.

castor oil, hydrogenated. Principally glyceryl tri-12-hydroxystearate (q.v.).

castor oil plant. See ricinus.

castor oil, soluble. See Turkey red oil.

castor oil, sulfonated. See Turkey red oil.

castor pomace. See castor bean oil meal.

"Castorwax." [202] Trademark for hydrogenated castor oil, the triglyceride of 12-hydroxystearic acid, obtained by controlled hydrogenation of pure castor oil.

Properties: A white, hard, brittle synthetic wax; m.p. 85°C; sp. gr. 0.990 (25°C); acid value 2; iodine value 3; saponification value 180; insoluble in most organic solvents at room temperature and compatible with ethyl cellulose, cellulose acetate butyrate, polyethylene (up to 25%), polymethacrylate, rosin, shellac, abietyl alcohol, natural and synthetic rubbers, insect and vegetable waxes.

Uses: In potting compounds, gasket and impregnating compositions, and wax blends where increase in grease and solvent resistance, hardness and melting point is desired; and as a blending agent and viscosity reducer in hot melts.

"Castrolite." [159] Proprietary product. Sulfonated castor oil, made by improved process of manufacture.

Grades: "Castrolite" 50%; "Castrolite" 75%.

Uses: As a softener and finishing oil for textiles; penetrant and leveling agent in dyeing. In cosmetics, shampoo oils, etc. (as a base oil); as a plasticizer; as an emulsifying agent for dispersing perfume and essential oils.

"Castung." [202] Trademark for dehydrated castor oil, a synthetic drying oil produced by chemical removal of the hydroxyl groups from castor oil to form additional double bonds. Available in several viscosities.

Uses: Non-yellowing drying oil for alkyds, paints and varnishes, for putty and calking compounds, for linoleum and oilcloth, and for modifying tall oil in coatings applications.

catalase. An oxidizing enzyme occurring in blood and tissues, which decomposes hydrogen peroxide. It can be isolated from animal tissue or molds and is used in food preservation (removing oxygen in packaged foods) and in decomposing residual hydrogen peroxide in bleaching and oxidizing processes.

"Catalin." [353] Brand name for a proprietary product. Phenol-formaldehyde resin.

Forms available: Rods, bars, tubes, sheets,

castings; is furnished in a wide range of colors; also colorless; nonflammable.
Uses: Buttons, buckles, toilet articles, radio cabinets; imitation jewelry; miscellaneous molded articles.

catalysis. A change in the rate of a chemical reaction caused by the presence of a small quantity of a substance (the catalyst) which remains unchanged in amount after the reaction is completed.

catalyst. A substance whose presence increases the rate of a chemical reaction. In some cases the catalyst functions by being consumed and regenerated, in other cases the catalyst seems not to enter the reaction and functions by virtue of surface characteristics of some kind. A negative catalyst (inhibitor, retarder) slows down a chemical reaction. Many common catalysts are powdered metals or other metallic compounds. Any kind of substance (solid, liquid, or gas) may be a catalyst, but only certain particular substances are catalysts for particular reactions. Thus finely powdered nickel is a catalyst to speed up the combination of hydrogen with liquid fats to produce solid fats. Gas masks for carbon monoxide atmospheres depend upon a metal oxide catalyst to speed up the oxidation of the monoxide to dioxide and so make the air safe to breathe. Vitamins and enzymes are organic catalysts of chemical processes in the body.

"Catalyst 1707." [55] Trade name for a metal oxide type dehydrogenation catalyst. Bulk density 67 lbs/cu ft (approx).
Containers: 100-lb bags; 400-lb drums.
Use: Catalyst for butadiene and styrene production.

catalytic reforming. Reforming (q.v.) in the presence of a catalyst.

"Catanac" SP. [57] Trademark for stearamidopropyldimethyl-beta-hydroxyethyl-ammonium dihydrogen phosphate. $[C_{17}H_{35}CONHCH_2CH_2CH_2N(CH_3)_2CH_2CH_2OH]-H_2PO_4$.
Properties: A tan waxy solid softening at 50°C. Has antistatic and surface active properties. Supplied as a 35% solution in an isopropyl alcohol water mixture. Light yellow liquid, not hygroscopic; flash point (open cup) less than 80°F.
Uses:
1. Antistatic for textiles, plastics, surface coatings, glass, wax polishes. Can be applied either by incorporation into molding compositions, or applied to surfaces.
2. Detergent in acid media, hard water, salt water. Also used as emulsifier, dispersing agent, rewetting and stripping agent, and mold lubricant.

cataphoresis. Usually identical with electrophoresis (q.v.). The term originally implied migration of a suspended particle in an electric field to the cathode only, but has received the more general meaning with increased knowledge of colloid chemistry.

catechol. See pyrocatechol.

catenane. A compound with interlocking rings, which are not chemically bonded, but which cannot be separated without breaking at least one valence bond. The model would resemble the links of a chain.

cathode. The negative terminal of an electrical cell, or vacuum tube, or other electrical device. In a vacuum tube, the cathode is the source of free electrons necessary for carrying the current. In electroplating baths, positively charged metallic ions travel toward the cathode.

cation. An ion having a positive charge. Cations in a liquid subjected to electric potential collect at the negative pole or cathode.

cation exchange. See ion exchange.

"Cation Exchange Resin Cleaner #1112." [210] Cleaner especially developed for use in case of cation exchange resin fouling.

cationic reagents (for flotation). Surface-active substances which have the active constituent in the positive ion. Used to flocculate and collect minerals that are not flocculated by the reagents such as oleic acid or soaps, in which the surface active ingredient is the negative ion. Reagents used are chiefly the quaternary ammonium compounds, e.g., cetyl trimethyl ammonium bromide.

catlinite (pipestone). A fine-grained silicate mineral related to pyrophyllite which is easily compressible, has high surface friction, and is used for gaskets in very high pressure work.

"Cato." [53] Trademark for a cationic derivative of starch, available in ungelatinized or gelatinized (cold water soluble) form. Used in manufacture of paper, warp sizing, etc.

cat's eye. A variety of natural silica or quartz (q.v.) used as a gem. Also a variety of chrysoberyl (q.v.).

caustic. (1) When used alone, the term usually alludes to caustic soda, sodium hydroxide (q.v.). (2) More generally, a strong base. (3) A class, or group, name given to certain chemicals and pharmaceutical products to indicate, or describe, their physiological effect. Caustics are products (such as sodium hydroxide, silver nitrate and the like) used for their corroding or disintegrating action on the skin and flesh. Their employment tends to cause a burning sensation and the destruction of living tissue.

caustic alcohol. See sodium ethylate.

caustic arsenic chloride. See arsenic trichloride.

caustic baryta. See barium hydroxide.

causticized ash. Combinations of soda ash and caustic soda in definite proportions marketed for purposes where an alkali is needed ranging in causticity between the two materials. Causticized ash is usually designated by its caustic soda content and the range of standard marketed products embraces 7%, 10%, 15%, 25%, 36%, 45%, and 67% of caustic soda.
Shipping regulations: None.*

caustic lime. See calcium hydroxide.

caustic, lunar. See silver nitrate, fused.

caustic potash. See potassium hydroxide.

caustic soda. See sodium hydroxide.

caustic Vienna (Vienna paste). A mixture of equal parts of potash and lime. Grayish-white, deliquescent powder or lumps.

caustic, white. See sodium hydroxide.

"CA V2B." Brand name for a hardenable corrosion-resisting non-galling alloy of the 19% chromium, 9% nickel type which contains in addition 2% copper, 3% molybdenum, 2.75% silicon and a small amount of beryllium. Machined in quenched annealed condition and precipitation hardened by 900°F heat treatment. Used for valve discs and plugs, bearing sleeves, gears and other bearing parts.

cave. See hot cell.

cavitation. The formation of a hole in a liquid just behind a rapidly moving propeller blade or other rapidly moving object.

cayenne pepper. See capsicum.

"Cazar." [51] Trademark for greenish black, lime-base greases of high quality, having high-viscosity mineral oil base and soft, smooth texture. Outstanding in adhesiveness and water resistance and well suited for use in cold weather.

Cb. Symbol for columbium, an obsolete name for niobium.

CBM. Abbreviation for chlorobromomethane (see bromochloromethane); also for constant boiling mixture. See azeotropic mixture.

CBW. Abbreviation for chemical biological weaponry.

cc. Abbreviation for cubic centimeter.

CC black. Abbreviation for conducting channel black. See channel black.

"CCC." [402] Trademark for a series of refined, pulverized high-calcium limestone products.
Uses: Whiting in calking compounds; filler or extenders in rubber, paints, ceramics, cements, etc.

"CCC-Diluent." [402] Trade name for a surface-treated pulverized limestone used as a conditioner and extender in pesticidal dust formulations.

"CCC" Trace Mineral Premix. [402] Proprietary product for livestock and poultry feed formulation.

Cd. Symbol for cadmium.

"CDB-59." [55] Trademark for potassium dichloroisocyanurate (q.v.).

"CDB-60." [55] Trademark for sodium dichloroisocyanurate q.v.).

"CDB-70." [55] Trademark for dichloroisocyanuric acid (q.v.).

"CDB-85." [55] Trademark for trichloroisocyanuric acid (q.v.).

CDP. Abbreviation for cytidine diphosphate. See cytidine phosphates.

Ce. Symbol for cerium.

"Cebicure." [123] Trademark for ascorbic acid for meat curing.

"Cebione." [123] Crystalline vitamin C, ascorbic acid (q.v.).

"Cebitate." [123] Trademark for sodium ascorbate for meat curing.

"Cedambrette." [342] Trade name for a mixture of natural and synthetics simulating cedarwood concentrates; used in perfumery for its woody note.

cedar camphor. See cedrol.

cedar gum.
Properties: Pale yellow to red or brownish tears. Swells greatly in water forming clear jellies.
Derivation: From Cedrela odorata or red cedar found in American tropics and West Indies.
Uses: Mucilage; cosmetics; pharmaceuticals.
Shipping regulations: None.*

cedar leaf oil. True cedar leaf oil is distilled from the leaves of Juniperus virginiana. The name has also been used as a synonym for thuja oil (q.v.) (from Thuja occidentalis). The properties of the two oils as described here are quite different.
Properties: Colorless liquid; sp. gr. 0.870-0.890; optical rotation +55° to 65° (20°C); soluble in alcohol and ether.
Chief constituents: Limonene, cadinene, borneol, and bornyl esters.
Containers: Cans; drums.
Uses: Medicine; microscopy; perfume.

cedarwood oil.
Properties: Volatile oil, colorless, pale yellow or greenish-yellow; mild, agreeable, persistent odor. Somewhat viscid and occasionally studded with cedar camphor crystals. Poisonous! Two principal varieties are known, designated here as (a) and (b).
Derivation: (1) Distilled from the wood of Juniperus virginiana. Probably represented by (a). (2) Collected as a by-product from drying kilns in lead-pencil manufacture. Probably represented by (b).
Chief constituents: Cedrol, cedrene.

Constants: (a) Sp. gr. 0.945-0.960 (15°C); optical rotation −27° to −45°; refractive index (n 20/D) 1.5020-1.5070; acid value up to about 1; ester value up to 6.5, after acetylation 26-42. (b) Sp. gr. 0.940-0.944 (15°C); optical rotation −40° to −46°22'; saponification value 2-4; ester value after acetylation 14-18.
Solubility in alcohol: (a) 1 vol in 10-20 vols of 90% alcohol, up to 6 vols of 95% alcohol, soluble in benzyl benzoate, fixed oils, mineral oil; insoluble in propylene glycol, glycerin; (b) in 5-6 vols of 95% alcohol.
Containers: Bottles; tins; drums.
Uses: Medicine; perfumery; perfuming soap; insectifuge; sanitary supplies; microscopy work.

cedarwood oil, Texas.
Properties: Colorless to yellow; slightly viscous liquid; fragrant odor; sp. gr. 0.950-0.960 (15°C); optical rotation −35° to −50°; refractive index (n 20/D) 1.5040-1.5070; soluble in all proportions of 95% alcohol, benzyl benzoate, mineral oil, fixed oils; insoluble in propylene glycol, glycerin.
Derivation: Steam distillation of chopped and ground wood of Juniperus Mexicana.
Method of purification: Rectification.
Containers: Glass, aluminum, or tin-lined vessels.
Uses: Soaps; sanitary supplies; polishes.

"Cedrene." [342] Trademark for terpenes from cedarwood oil.

"Cedrenol." [342] Trademark for crystalline alcohols from cedarwood oil.

"Cedrenone." [342] Trademark for ketone-like aromatics from cedarwood oil.

cedrol (cedar camphor) $C_{15}H_{26}O$, a tertiary terpene alcohol.
Properties: Crystalline substance having cedarwood odor.
Constants: M.p. 81°C min. Soluble in 11 parts of 95% alcohol.
Uses: Perfumery, for woody and spicy notes, and for the perfuming of disinfectants.

cedryl acetate $CH_3COOC_{15}H_{25}$.
Properties: Colorless liquid, having a light cedar odor. Sp. gr. 0.975-0.995; refractive index 1.496-1.510. Soluble in one volume of 90% alcohol.
Use: Perfumery.

"Cefro." [40] Trademark for rat, rabbit and mouse repellents containing ethyl-2,3,4,5-tetrachlorotetrahydro-2-furoate as the active ingredient. Available as a 5% emulsion in poly(vinyl acetate) and as a 40% granular formulation.
Properties (typical): Dark brown liquid with a fruity odor; insoluble in water; soluble in ethanol, acetone, benzene, hexane and ethyl acetate. Sp. gr. (20°C) 1.480; b.p. (0.1 mm) 83-85°C; refractive index (25°C) 1.493.
Derivation: Chlorination of ethyl 2-furoate.

"Celanese CL." [352] Trademark for a series of polyvinyl acetate emulsions. Available as:
"Celanese CL" 102: Fine particle size, water resistant homopolymer emulsion.
Uses: Paints, adhesive and paper coating specialties.
"Celanese CL" 202: Fine particle size, water resistant copolymer emulsion.
"Celanese CL" 203: Vinyl-acrylic copolymer emulsion.
Use: Improve durability of paints.
"Celanese CL" 204: Vinyl copolymer emulsion.
Uses: Improving scrub resistance and durability in paints.

"Celanese Solvent." [352] Trademark for a series of special solvents. Available as:
"Celanese Solvent" 203: Replacement for normal butyl alcohol in nitrocellulose lacquers, alkyd resin formulations and thinners; distillation range 115-120°C; flash point 100°F (open cup).
"Celanese Solvent" 601: Replacement for methyl ethyl ketone in vinyl and nitrocellulose applications; distillation range 74-84°C; flash point 10°F (open cup).
"Celanese Solvent" 901H: Replacement for butanol and methyl isobutyl carbinol in lacquers and brake fluids; distillation range 125-155°C; flash point 120°F (open cup).

"Celanthrene." [28] Trademark for a group of anthraquinone dyes designed especially for dyeing acetate and also suitable for application to nylon.

"Celcon." [421] Trademark for a highly crystalline acetal copolymer based on trioxane. See acetal resins.

"Celcure." [77] Trademark for acid cupric chromate, a wood preservative composed principally of copper sulfate, sodium dichromate and chromic or acetic acid.

celery fruits oil. See celery seed oil.

celery seed oil (celery fruits oil).
Properties: Limpid, greenish-yellow or colorless oil; characteristic odor; celery taste; sp. gr. 0.9236 (15°C); optical rotation +60° to +82°; refractive index 1.478-1.486; acid value up to 4; ester value 16-45, after acetylation 43-52; saponification value 178.1. Slightly soluble in water; soluble in alcohol, ether and chloroform.
Chief known constitutents: Limonene, phenols, sedanolide, sedanoic acid.
Derivation: Distilled from the seeds of Apium graveolens.
Containers: Bottles.
Uses: Flavoring; medicine.

celestial blue. Applied loosely to any of a number of iron blue pigments, usually containing considerable extender such as barytes.

celestite $SrSO_4$. Natural strontium sulfate, usually found in sedimentary rocks.
Properties: Colorless, white, pale blue, or red; luster vitreous to pearly. Resembles

barite (q.v.). Sp. gr. 3.95; hardness 3-3.5.
Occurrence: United States; Canada; Europe; Mexico.
Containers: Railroad cars.
Uses: Strontium chemicals; oil-well drilling mud; sugar refining; ceramics.

"Celite." [247] Trademark for diatomaceous earth and a line of products processed therefrom.
Properties: Color, white to pale brownish white, depending on grade and processing, calcined grades being pink to buff; sp. gr. 0.24-0.34; porous, capable of absorbing 300-400% water by weight; poor conductor of heat, sound, and electricity; resistant to acids except hydrofluoric; slowly soluble in hot alkali.
Typical analysis (ignited basis): Silica 92.7%; alumina 3.8%; ferric oxide 1.4%; lime and magnesia 1.0%; potash and soda 0.9%.
Occurrence: Lompoc, California.
Grades: Numerous types available depending on use.
Uses: Filtering; fillers; absorbent; abrasive in glass and metal polishing; catalyst carrier; ingredient in cements, flame-proofing agents, and other products.

"Celite" Filter Aids. [247] Trade name for products made from "Celite" diatomaceous earth.
Grades and Uses:
"Celite" Analytical Filter Aid: For rapid removal of gummy, gelatinous, flocculent or semi-collodial precipitates, and in purification of valuable chemicals and biologicals.
Filter-Cel: For clarity in liquids containing exceptionally small or colloidal suspended solids.
Filter-Cel Laboratory Standard: A calibrating filter aid used in making filtration comparisons.
Hyflo Super-Cel: Specially processed to give a flow rate five times faster than Filter-Cel.
Sorbo-Cel: For removing emulsified oil from contaminated waters in order to return an oil-free water for boiler feed or other process use.
Standard Super-Cel: Finely divided, heat-treated product for liquids containing moderately finely divided suspended matter.
No. 503: Fast filter aid for liquids with large amounts of coarse insoluble matter.
No. 505: Calcined, approximating Filter-Cel in performance, for special purposes.
No. 512: Flow rate about midway between Standard Super-Cel and Hyflo Super-Cel.
No. 521: For clarification of distilled spirits, liquors, cordials and tannin-bearing liquids in general, and where reaction may develop between iron and the liquid to be clarified.
No. 535: High flow rate and good clarity for filtering viscous and semi-viscous liquids.
No. 545: Highly porous with peak flow rate for clarifying viscous materials.

"Celite" Mineral Fillers. [247] Trade name for several grades of "Celite" powders.

Standard Grades	Average Particle Size in Microns
Snow Floss	1-2
Super Floss	2-4
Celite FC	4-6
Celite SSC	6-8
Celite HSC	7-9

Uses: In acetylene cylinders, adhesives, asphalts and pitches, battery boxes, catalyst carriers, cleansers and cleaners, crayons, decals, detergents, dyes, dynamite, coated fabrics, fertilizers, fuels, fumigants, gas purification, insecticides, insulating blocks, leather finishes, lens polishing, matches, mold wash, mold lubricant, paints and varnishes, papers, printing inks, plastics, polishes, polishing cloths, soaps, sound records, sponge rubber, seed coatings, and ultramarine blue.

"Celite" Preformed Catalyst Carriers. [247] Trade name for thermally stable catalyst carriers made of "Celite," available in different hardnesses and porosities, and in a complete line of shapes including aggregates in granular form, extruded pellets and spherular types.

"Celkate." [247] Trademark for finely divided hydrated synthetic magnesium silicates having high absorption properties; light tan in color; density 10-18 lb/cu ft; surface area 150-250 sq m/gram.
Grades: Available in various grades for purifying of petroleum base solvents, chemical and drug solutions, and for decolorizing of animal, fish, and vegetable oils.
Uses: As a filter agent to remove solids and selectively remove such solubles as color matter and free fatty acids; as an absorptive carrier of liquids; and as a conditioning agent to improve flow properties of dry powders.

cell. See electrolytic cell; dry cell; storage battery; fuel cells.

"Cellitazol." [307] Brand name of proprietary line of developed acetate dyestuffs. Used for the dyeing of acetate fibers. Characterized by good fastness to light, very good fastness to washing, etc.

"Celliton." [307] Brand name of proprietary line of disperse dyestuffs characterized by good fastness to light, washing, etc. Used for dyeing and printing acetate fibers.

cellobiose $C_{12}H_{22}O_{11}$. The product of the partial hydrolysis of cellulose, composed of two D-glucose molecules.
Properties: Colorless crystals; m.p. 225°C (dec); soluble in water; slightly soluble in alcohol; nearly insoluble in ether; insoluble in acetone.
Use: Bacteriology.

"Cellofax" WLD. [206] Brand name for a proprietary water-soluble cellulose derivative forming mucilages similar to starches and gums, but possessing the advantages that its solution does not ferment or develop a mold and that it has superior binding powers on pigment fillers.

Uses: In place of starch as a sizing agent for cotton and rayon; as a finishing agent for cotton, linen and rayon piece goods; in conjunction with the usual mucilages and gums for seasoning in the leather industry.

celloidin (celluidine; photoxylin). A form of pyroxylin (see nitrocellulose).

Properties: Slightly milky, white, transparent, tough gelatinous tablets, chips, or shreds. Soluble in a mixture of equal parts of alcohol and ether. Usually supplied immersed in water.

Derivation: A very pure nitrocellulose obtained by precipitation from an ether-alcohol solution of collodion cotton.

Grades: Technical.

Containers: Wooden kegs; bags.

Uses: Imbedding sections in microscopy; electrochemistry; photography; galvanoplastics; medicine.

Fire hazard: Dangerous.

Shipping regulations: Flammable solid. Yellow label.*

"Cellolyn." [266] Trademark for a series of synthetic resins especially designed for lacquers.

cellophane (see also acetate film). Film produced from wood pulp by the viscose process. It is transparent, strong, flexible, and highly resistant to grease, oil, and air. The base cellulose film is modified by softeners, flame-resisting materials and dyes; also by coating with other materials, to give a balanced combination of properties. Heat sealing and moistureproof grades, among others, are available.

Kinds: Basic film types include plain (nonmoistureproof) and moistureproof. Several modifications of the moistureproof variety are made to provide various degrees of moistureproofness, heat sealing properties and water resistant characteristics.

Available forms: Supplied in rolls and cut-to-size sheets.

Use: As a wrapper or protective package for fabricated articles and industrial applications.

"Cellosize." [214] Trademark for hydroxyethyl cellulose.

Properties: Snow-white, free-flowing powder; soluble in water. Upon drying solutions produce clear, colorless, odorless, and tasteless films which possess good heat and light stabilities, are readily soluble in water and insoluble in most organic solvents.

Viscosity types: WP-09 (5% aqueous soln), 70-110 cps; WP-3 (5% aqueous soln), 275-325 cps; WP-40 (5% aqueous soln), 4000 cps; WP-300 (5% aqueous soln), 30,000 cps; WP-4400 (2% aqueous soln); WP-15,000 (2% aqueous soln).

Containers: 4- and 8-lb Fiberpak containers; 40- and 125-lb Leverpak drums.

Uses: Protective colloid for emulsion polymerization; thickener for synthetic latices; warp size for cellulose acetate, rayon, and cotton; pigment suspending agent for liquid powders, leg make-up, shampoos, and creams; oil-impermeable coating for paper and fiber containers; carriers for pigments and colors in dyestuff pastes for textile printing; with added glyoxal for water-resistant films.

"Cellosolve." [214] Trademark for mono- and dialkyl ethers of ethylene glycol and their derivatives.

butyl "Cellosolve"
See ethylene glycol monobutyl ether.

butyl "Cellosolve" acetate
See ethylene glycol monobutyl ether acetate.

"Cellosolve" acetate
See ethylene glycol monoethyl ether acetate.

"Cellosolve" solvent
See ethylene glycol monoethyl ether.

dibutyl "Cellosolve"
See ethylene glycol dibutyl ether.

n-hexyl "Cellosolve"
See ethylene glycol monohexyl ether.

methyl "Cellosolve"
See ethylene glycol monomethyl ether.

methyl "Cellosolve" acetate
See ethylene glycol monomethyl ether acetate.

phenyl "Cellosolve"
See ethylene glycol monophenyl ether.

cell, standard. An electrolytic cell characterized by production of an electromotive force which is closely reproducible if suitable precautions are observed. The standard Weston cadmium-mercury cell produces an emf of 1.0183 volts at 20°C and is used as a primary international standard.

"Celluflex" 23. [352] Trademark for alkyl epoxystearate. Used as a low-temperature plasticizer and stabilizer for polyvinyl chloride and certain other polymers.

"Celluflex" 112. [352] Trademark for cresyldiphenyl phosphate (q.v.). Used for flame-resistance in vinyl formulations.

"Celluflex" 179. [352] Trademark for tricresyl phosphate, available as:

"Celluflex" 179A: Low specific gravity; low ortho content.
Uses: Flame-resistant plasticizer.

"Celluflex" 179C: General purpose grade.
Uses: Adhesive in air filters; coatings; films.

"Celluflex" 179 EG: Electrical grade.
Use: Insulation industry.

"Celluflex" CEF. [352] Trademark for tris (beta-chloroethyl) phosphate.

Properties: Clear, transparent liquid; sp. gr. 1.425 (20/20°C).

Use: Flame-retardant plasticizer.

"Celluflex" FR-2. [352] Trademark for tris-(dichloropropyl) phosphate.

Properties: Clear transparent liquid; sp. gr. 1.513 (20/20°C).

*See "I.C.C. Shipping Regulations," page xiii.
Reference numbers refer to name of manufacturer. See "List of Manufacturers," page v.

Use: Flame retardant for plastics and coatings.
Shipping regulations: None.*

"Celluflex" TPP. [352] Trademark for triphenyl phosphate (q.v.). Used as plasticizer in cellulose acetate and phenolic plastics.

"Celluguard." [352] Trademark for a water-glycol fire-resistant hydraulic fluid.

celluidine. See celloidin.

cellulase. A white and almost odorless enzyme capable of hydrolyzing and depolymerizing cellulosic polysaccharides of high molecular weight, including cellulose itself, into smaller fragments. It can be obtained from the fungus, Aspergillus niger.
Uses: Medicine, to aid in digestion of bulky materials; septic systems; brewing; extraction of essential oils.

"Celluloid." [352] Proprietary product consisting essentially of a solid solution of cellulose nitrate and camphor or other plasticizer with or without the presence of pigments and coloring matter. Available in form of sheets, 20 in. x 50 in. x 0.005 in. to over 1 in., cylindrical and profile rods in 60 in. lengths of various diameters, tubes in 60 in. lengths 0.020 in. minimum wall thickness, various diameters, films in continuous rolls in standard widths of 21 in. and 42 in. in thickness 0.003 in. to 0.010 in. Special sizes on request.
Properties: Clear and colored, transparent, translucent, and opaque in all shades, unlimited mottled and variegated effects; sp. gr. 1.35-1.60; tensile strength 5,000-10,000 psi; elongation 10-40%; Brinnell hardness 5-11 using 2.5 mm ball and 10 kg load; impact strength (Izod) 3.0-6.0 ft lbs/in of notch; ignition temperature 320-380°F; molding temperature 185-250°F; refractive index 1.5; dielectric strength for a $\frac{1}{8}$ in. sheet, 250-500 volts/mil at 60 cycles; dielectric constant 6.7-7.3 at 60 cycles. Soluble in organic solvents such as alcohols, ketones, and esters; insoluble in hydrocarbons, mineral oils, and mineral acids of low concentration at normal temperature; decomposed by alkalies and strong acids.
Workability: Fabricated by molding, dry and wet swedging, machine operations such as blanking, drilling, sawing, turning, milling, etc.
Containers: Wooden cases and cardboard containers.
Uses: Fabricated into innumerable articles such as toilet ware, fountain pens, toilet-seat covers, mathematical instruments, buttons, tool handles, advertising novelties, watch and clock crystals, toys, motion picture camera and x-ray films; bandages for surgery; rubber substitute in dentistry.
Fire hazard: Dangerous.
Shipping regulations: Flammable solid. Yellow label for express shipments.*

cellulose $(C_6H_{10}O_5)_n$. Cellulose is the preponderant and essential constituent of all vegetable tissues and fibers. (See anhydroglucose.) It is the basis of the textile and paper-making industries. Pure cellulose is most readily obtained from cotton by treatment with dilute alkalies and acids and thorough washing. The cellulose obtained in this manner is a white substance, sp. gr. about 1.45, retaining the form of the cotton fibers. Cellulose dissolves in Schweitzer's reagent (q.v.). When nitrated, it yields nitrocellulose (guncotton) used as such and in the manufacture of smokeless powders, collodion, "Celluloid," pyroxylin lacquers, and miscellaneous products. Cellulose with acetic anhydride and glacial acetic acid forms cellulose acetate (q.v.). With alkali and carbon disulfide, cellulose xanthate is formed and this is then converted to viscose rayon. Three forms of cellulose exist:
alpha-cellulose. alpha-Cellulose is taken as the fraction that can be filtered out of a mixture consisting of the fibrous material and sodium hydroxide solution (7.3%) of maximum dissolving power, after the fibers have previously been swelled with sodium hydroxide solution (17.5%). After separation, the alpha-cellulose is determined either by drying and weighing, or volumetrically by oxidation with potassium dichromate.
beta-cellulose. beta-Cellulose is taken as that fraction that precipitates at room temperature (15-35°C) after the filtrate has been acidified and is determined by the volumetric method.
gamma-cellulose. gamma-Cellulose is taken as the fraction that remains in solution after removing beta-cellulose, and is determined by the volumetric method.

The method of separating alpha-cellulose from the other two fractions is intended primarily for papers made from rags or chemical wood fibers. (ASTM definition, for alpha-, beta-, and gamma-cellulose; ASTM D-588-42.)

alpha-Cellulose has been suggested as a component of high energy rocket fuels.
See also wood pulp.

cellulose acetate (CA). A cellulose resin in which the cellulose is not completely esterified by acetic acid. See acetate process.
Properties: White flakes or powder; may be transparent, translucent, opaque, in sheet or film form. A thermoplastic resin, softening about 60-97°C and melting about 260°C. Sp. gr. 1.27-1.34; soluble in acetone, ethyl acetate, cyclohexanol, nitropropane, ethylene dichloride. As a plastic, it is notable for its toughness, high impact strength, low flammability, and ease of fabrication. It is subject to dimensional change due to cold flow, heat, or moisture absorption (1-7%).
Derivation: By acetylation and partial hydrolysis of cellulose. See acetate process.
Grades: Filtered and unfiltered. See "Tenite," "Plastacele," and "Lumarith" for

*See "I.C.C. Shipping Regulations," page xiii.
Reference numbers refer to name of manufacturer. See "List of Manufacturers," page v.

details regarding commercially available products. See also acetate (2); cellulose triacetate.
Containers: Fiber cartons or drums; multiwall paper sacks.
Uses: Manufacture of acetate fiber, lacquers, protective coating solutions, photographic film, transparent sheeting, thermoplastic molding composition, artificial leather.
Fire hazard: Relatively nonflammable compared with nitrocellulose which it has displaced in many applications for this specific reason.
Shipping regulations: None.*

cellulose acetate butyrate (CAB; cellulose acetobutyrate).
Properties: White flakes or granules, similar to cellulose acetate, and similarly convertible into plastic films, sheets, molded objects, etc.; sp. gr. 1.2; other properties may be varied at will according to proportions of acetate and butyrate, as well as by various conditions of manufacturing. Soluble in ketones, organic acetates, lactates, methylene, ethylene, and propylene chlorides and higher boiling solvents.
Derivation: By the reaction of purified cellulose with acetic and butyric anhydrides in the presence of sulfuric acid as catalyst and glacial acetic acid as solvent. The ratio of acetic and butyric components may be varied over a wide range.
Grades: According to butyryl content, as 17, 27, 38, 50%.
Containers: Fiber cartons and drums.
Uses: Manufacture of thermoplastic molding composition, photographic film, lacquers, protective coating solutions, protective strip coatings, etc.
Fire hazard: About the same as newsprint.
Shipping regulations: None.*

cellulose acetate propionate. Very similar to cellulose acetate butyrate but made with propionic anhydride instead of butyric anhydride.

cellulose-acetate rayon. Incorrect name for acetate. See acetate (2).

cellulose acetobutyrate. See cellulose acetate butyrate.

cellulose gum. A purified grade of sodium carboxymethylcellulose (q. v.).

cellulose, hydrated (hydrocellulose). Cellulose that has been caused to react with water (about 8-12%), forming a gelatinous mass.
Derivation: By mechanical pulverization and agitation with water, by the action of strong salt solutions, alkalies, or acids.
Use: In the manufacture of paper, vulcanized fiber, mercerized cotton, viscose rayon.

cellulose nitrate. See nitrocellulose.

cellulose nitrate sheeting.
Properties: Semirigid thermoplastic sheets in a variety of colors, transparent, translucent and opaque, including mottles, and shell and pearl effects. Thicknesses 0.005 in. and upward; sheet size 20 x 50 in. Several compositions are designed to meet requirements of different uses. Resistant to wear, hydrocarbons, dilute acids, dilute alkalies. Not resistant to ketones, esters, lower alcohols, glycol ethers, strong alkalies, strong oxidizing acids. Easily machined, cemented and finished. Discolored by prolonged exposure to sunlight. Degraded by temperatures above 240°F. Highly flammable.
Uses: Eyeglass frames; covering of toilet seats; hamper tops, etc.; covering of shoe heels; table-tennis balls; index tabs.
See also nitrocellulose.

cellulose, oxidized (cellulosic acid). Derivative of cotton cellulose produced by treatment with nitrogen dioxide. Is soluble in alkali but may be made to retain original form of the cellulose and much of its tensile strength. May also be powdered. The material is a copolymer of anhydroglucose and anhydroglucuronic acid, or on further oxidation may consist of polyanhydroglucuronic acid.
Properties: Slight charred odor; acid taste; soluble in aqueous organic bases, in dilute alkali, and in ammonium hydroxide, forming salts and esters. It is insoluble in water, acids, and common organic solvents. It slowly degrades at room temperatures and should be kept cool.
Grades: U.S.P. XVI; technical.
Containers: Glass bottles; fiber cans.
Uses: Surgery and medicine; ion-exchange medium; thickening agent.
Shipping regulations: None.*

cellulose propionate. Similar to cellulose acetate. See "Forticel."

cellulose sponge. A sponge of regenerated cellulose, highly absorbent, soft and resilient when wet, long-lasting. It will not scratch, can be sterilized by boiling and is not affected by ordinary cleaning compounds. The pores vary in size from coarse pore (the size of a pea) to fine pore (the size of a pinhead).
Use: Used commercially for many purposes, such as washing automobiles and trucks, walls and painted surfaces, windows, etc.; in the home for washing dishes, general cleaning, and in the bath; fine-pore sponge used in photographic laboratories.

cellulose sponge yarn. Cotton yarn core covered with cellulose sponge. Wound on ball multiple-end warps. Available in two diameters: A-31 (approximately $\frac{1}{4}$ in.) and B-20 (approximately $\frac{1}{8}$ in.), in buff and green.
Use: For making wet mops and weaving into cloths, pads, etc.

cellulose triacetate. A cellulose resin in which the cellulose is completely esterified by acetic acid. See acetate process.
Properties: White flakes; sp. gr. 1.2; soluble in chloroform, methylene chloride,

tetrachloroethane.

Derivation: By the reaction of purified cellulose with acetic anhydride in the presence of sulfuric acid as catalyst and glacial acetic acid as solvent, followed by very slight hydrolysis.

Grade: Flake.

Containers: Fiber cartons or drums.

Use: Protective coatings resistant to most solvents; textile fibers; base for magnetic tape.

Fire hazard: About the same as newsprint.

Shipping regulations: None.*

cellulose xanthate. A stage in the manufacture of viscose rayon. See viscose process. The xanthate has the composition ROCSSH, in which R represents the combining cellulose radical.

cellulosic acid. See cellulose, oxidized.

cellulosics. Resins made from cellulose. See preceding articles (cellulose acetate, etc.), cellophane, ethylcellulose, methylcellulose, nitrocellulose, sodium carboxymethylcellulose.

"Cellulube." [352] Trademark for a series of functional fluids (phosphate esters) combining fire-resistance and lubricating qualities. Available in controlled viscosities for industrial hydraulic and lubricant applications.

"Cellutherm." [352] Trade name for a series of synthetic lubricants based on trimethylolpropane esters. Available as:

"Cellutherm" 2505-A: High temperature lubricant for aircraft gas turbine engines operating at a bulk fluid temperature of 400°F.

"Cellutherm" 2712-B: Lubricating oil for aircraft gas turbine engines operating at bulk fluid temperature of 300°F.

"Celogen." [248] Trade name for para, para'-oxybis(benzenesulfonyl hydrazide), $H_2NNHSO_2C_6H_4OC_6H_4SO_2NHNH_2$, a nitrogen blowing agent for sponge rubber and expanded plastics.

Properties: Fine white crystalline powder; odorless; sp. gr. 1.52; decomposition range, 130-160°C; soluble in acetone; moderately soluble in ethanol and polyalkylene glycols; insoluble in benzene, ethylene dichloride, gasoline, and water; nonblooming; nondiscoloring; nonstaining; good storage stability under normal conditions of temperature; should be kept away from hot steam pipes, free flames, direct sunlight and similar sources of heat.

Uses: Used alone to produce a fine uniform cell structure in natural rubber, GR-S, neoprene, butyl and nitrile rubbers such as "Paracril." Also functions as an auxiliary with sodium bicarbonate to even out the irregularities in performance of the latter. Used alone for making expanded plastics especially expanded polyvinyl chloride from plastisols of "Marvinol" and similar products.

"CelogenAZ." [248] Trade name for azodicarbonamide, a nonstaining, nondiscoloring, odorless nitrogen blowing agent for sponge rubber and expanded plastics.

Properties: A fine yellow powder; sp. gr. 1.63; decomposition temperature 196°C in air, lower with use of promoters (see B-I-K); decomposition products are white; soluble in dimethylformamide, diethylene glycol (warm); insoluble in benzene, acetone, water and ethylene dichloride.

Uses: Produces a fine cellular structure in natural, SBR, neoprene and nitrile rubbers; produces blown vinyl products from plastisols of "Marvinol" and similar products, as well as blown products of polyethylene and polypropylene.

"Celontin." [330] Trademark for methsuximide. N-methyl-alpha-methyl-alpha-phenyl succinimide $C_{12}H_{13}NO_2$.

Properties: White to off-white crystalline powder. Soluble in alcohol and ether. Melting range 51-55°C.

Use: Medicine.

"Celoron." [281] Trademark for macerated canvas or paper-based industrial laminated or molded plastics.

Properties: Color, golden mottled brown or black; sp.gr. 1.35; high impact strength; unaffected by rapid temperature changes; resistant to heat, oil, water, and many chemicals; may be used continuously at 225-250°F.

Forms: Sheets; cut pieces; blanks; rings; molded parts.

Uses: Timing gears for automobile industry; electrical insulation; structural parts.

"Celotex." [351] Trademark for structural building and insulation board produced in large sheets. Made from bagasse (sugarcane fiber) and treated to be resistant to fungi, termites, and water penetration. Also available in tile and plank form, as plaster-base lath, sheathing and roof insulation. The name also covers roofing products, gypsum plasters, mineral wool, and hardboard.

"Cemad." [57] Trademark for a cement additive for the prevention of loss of fluid.

cement, aluminous (high alumina cement). A hydraulic-setting cement which contains at least 30 to 35% alumina (in contrast to Portland cement, which contains less than 5%). Aluminous cement attains its maximum strength more rapidly than Portland cement. It is also more resistant to solutions of sulfates. It exists in two modifications, sintered and fused.

Sintered types contain alumina up to 43% and a large proportion of iron oxide (up to 12.5%) which imparts a black color to the cements. They have a fusion point of about 2450°F and are represented by "Lumnite" and "Ciment Fondu." Fused types contain up to 52% alumina but only about 1% iron oxide, have a fusion point

*See "I.C.C. Shipping Regulations," page xiii.

Reference numbers refer to name of manufacturer. See "List of Manufacturers," page v.

up to about 2683°F and a higher total compression and tensile strength. They are represented by "Rolandshuette" (q. v.).

cementation. A process in which steel or iron objects are coated with another metal by immersing them in a powder of the second metal and heating to a temperature below the melting point of any of the metals concerned. Zinc, chromium, aluminum, and silicon are applied to iron or steel in this fashion.

cemented carbides (sintered carbides). Abrasive materials consisting of carbides of such metals as tungsten or tantalum bound together by a low-melting metal, usually cobalt. They are valuable for their hardness and durability, and are used in machining metals, plastics, porcelain, etc.

cement, chemical resisting. Portland cement that is somewhat more resistant to chemical action than the regular grade because of high tetracalcium aluminoferrate and low tricalcium aluminate content, and also because of additives such as water glass, calcium soaps, or other materials.

cement, H.E.S. See cement, high early strength.

cement, high alumina. See cement, aluminous.

cement, high early strength (cement, H.E.S.). A variety of Portland cement made from raw materials having a high lime-to-silica ratio. Contains a higher proportion of tricalcium silicate and hardens more quickly and with the evolution of more heat than regular Portland cement.

cement, hydraulic. See cement, aluminous and cement, Portland.

cementite Fe_3C. A carbide of iron formed in the manufacture of pig iron and steel. It is composed of 93.33% iron and 6.67% carbon. It is very hard and brittle and will scratch glass and feldspar, but not quartz. It is about two-thirds as magnetic as pure iron under an exciting current. It occurs in ordinary steels of more than 0.85% carbon and takes its name from cement steel, made by the cementation process, which contains a great deal of this carbide. See also ferrite; pearlite; carbon, combined; carbon, graphitic.

cement, low heat. A variety of Portland cement having higher tetracalcium aluminoferrate and dicalcium silicate content and less of tricalcium silicate and tricalcium aluminate than usual. The cement sets with the evolution of much less heat.

cement, Portland (cement, hydraulic). Very finely divided gray powder composed of compounds of lime, alumina, silica and iron oxide as tetracalcium aluminoferrate ($4CaO \cdot Al_2O_3 \cdot Fe_2O_3$), tricalcium aluminate ($3CaO \cdot Al_2O_3$), tricalcium silicate ($3CaO \cdot SiO_2$), and dicalcium silicate ($2CaO \cdot SiO_2$). These are abbreviated respectively as C_4AF, C_3A, C_3S and C_2S. Small amounts of magnesia, sodium, potassium, and sulfur are also present in combined form. The mixture has the property of hardening slowly when mixed into a paste with water. Hardening does not require air, and will occur under water. Portland cement is made by heating a powdered mixture of carefully chosen clay and limestone to incipient fusion, and then grinding the resulting clinker to a fine powder. A small portion of gypsum is usually added prior to or during the final grinding. There are five main types as regards usage:

Type I. In general construction where no special properties are required.

Type II. Moderate heat-of-hardening cement, for use in general construction exposed to moderate exposure to sulfate-bearing water or other chemicals or where moderate heat of hydration is desirable.

Type III. High early strength (H.E.S.) cement for use where high early strength is required.

Type IV. Low heat cement for use when a low heat of hydration is required.

Type V. Chemical resisting cements for use when high sulfate or chemical resistance is required.

See descriptions of Types III, IV, and V in preceding articles. See also cement, aluminous.

Containers: Multiwall paper sacks.

cement, pyroxylin. Mixtures containing cellulose nitrate, plasticizers and solvents which cause cellulose nitrate plastics to adhere to other bodies.

Shipping regulations: Flammable liquid. Red label may be required.*

cement rock. Argillaceous limestone used in the manufacture of Portland cement. Contains lime, silica, and alumina in varying proportions, and usually more or less magnesia.

cement, rubber. General term for solution of rubber in a hydrocarbon (such as gasoline or benzene).

Uses: As binder to hold materials in position until sewing or clamping is accomplished; as permanent bonds; as vulcanized seals; in shoe manufacturing; in automotive manufacture; as a sound deadener; adhesive for paper, and for repairing tires and tubes.

Shipping regulations: Flammable liquid. Red label may be required.*

"Centifoliol." [188] Trademark for a replacement for otto of rose and rose absolute.

Uses: Perfume and cosmetic compositions.

centigrade. The scale for measuring temperature in which 100° is the boiling point of water at standard atmospheric pressure, and 0° is the freezing point of water. A temperature given in centigrade degrees may be converted to the corresponding Fahrenheit temperature by the following

*See "I.C.C. Shipping Regulations," page xiii.
Reference numbers refer to name of manufacturer. See "List of Manufacturers," page v.

operation. Multiply the degrees centigrade by 1.8, and add 32 to the product. A temperature given in Fahrenheit degrees is converted to the corresponding centigrade temperature by subtracting 32 from the Fahrenheit temperature, and multiplying the remainder by 5/9. See absolute temperature for conversion of centigrade temperature to absolute or Kelvin scale.

centigrade heat unit. See chu.

centipoise. One one-hundredth of a poise. A poise is the unit of viscosity expressed as one dyne per second per square centimeter.

centistoke. One one-hundredth of a stoke. A stoke is equal to the viscosity in poises times the density of the fluid in grams per cubic centimeter.

"Century." [189] Trademark for a line of fatty acids, including stearic, oleic, polymerized, and mixed fatty acids.
"Century" 1210. Single pressed stearic acid.
"Century" 1220. Double pressed stearic acid.
"Century" 1230. Triple pressed stearic acid.
"Century" 1240. Supra (U.S.P.) stearic acid.
Uses: Buffing compounds; candles; cosmetics; crayons; emulsifiers; waxes; lubricating greases; shaving creams; metallic soaps; esters; paper coating; rubber compounding; pharmaceuticals; textile finishes.
"Century" 1005. Low titer redolene (oleic acid).
"Century" 1010. Redolene (oleic acid).
"Century" 1020. White (U.S.P.) oleic acid.
"Century" 1030. Low titer white (U.S.P.) oleic acid.
"Century" 1050. L.P. white (U.S.P.) oleic acid.
Uses: Textile soaps; esters; liquid soaps; waxes; emulsifiers; cosmetics; pharmaceuticals; polishes; plasticizers; lubricating oils; rubber compounding.
"Century" 1475. Mixed fatty acid; approximate composition C_{14} 10%, C_{16} 14%, C_{18} 24%, C_{20} 2%, polybasic acids 50%.
"Century" D-75. Polymerized fatty acid; a dicarboxylic acid, consisting of a mixture of dimer, trimer, and high molecular weight acids. Used in protective coatings, corrosion inhibitors; as a lube additive, and for bodying alkyds.
"Century" D-85. Polymerized fatty acid.
"Century" CD. Light colored mixed fatty acid containing short and long chain acids. Used in soaps, shampoos, dry cleaning compounds; alkyds; esters; lubricants.
"Century" 1480. Mixed fatty acid containing short and long chain acids.
"Century" 480. Mixed saturated fatty acids. Used in special esters, soaps, metallic stearates and lubricants.
"Century" 1405. Distilled tallow fatty acid, used in lubricants, esters, polishes, textile finishes, soap.

"Century" Hydrex. [189] Trade name for hydrogenated fatty acid products, including stearic acid, fish oil fatty acid, and tallow glyceride.
"Century" Hydrex 440. Hydrogenated stearic acid.
"Century" Hydrex 450. Hydrogenated stearic acid.
"Century" Hydrex 460. High quality hydrogenated stearic acid.
"Century" Hydrex 53. Hydrogenated fish oil fatty acid.
"Century" Hydrex 360. Hydrogenated tallow glyceride.
Uses: Buffing compounds; rubber compounding; candles; lubricating greases; crayons; emulsifiers; paper coatings; plasticizers; cosmetics; metallic stearates; textile finishes; polishes.

"Cenwax A." [189] Trade name for a hard, amorphous solid with practically no taste or odor, containing over 85% of 12-hydroxystearic acid.
Uses: Lubricants where good stability and water resistance are required, as in soaps, cosmetics, textile chemicals and insulation.

"Cenwax G." [189] Trade name for hydrogenated castor oil; principally the glyceride of 12-hydroxystearic acid.
Properties: Hard, wax-like solid; m.p. approximately 190°F; practically tasteless and odorless. Solid form: white to light-cream color; liquid form: colorless to light-straw color.
Uses: Lubricating grease where stability and water resistance are most important.

cephaelis. See ipecac.

cephalin. (kephalin). A group of phosphatides associated with lecithins found in brain tissue, nerve tissue, and egg yolk.
Properties: Yellowish, amorphous substance; characteristic odor and taste; insoluble in water and acetone; soluble in chloroform and ether; slightly soluble in alcohol.
Use: Medicine.

cera alba. See beeswax, bleached.

cera flava. See beeswax.

ceramics. Used in this dictionary to mean the ceramics industry, that is, the technology of producing fired clay and porcelain articles, their glazes, pigments, and modifiers. For the structural materials used in ceramics, see refractories. The two terms are often confused.

"Ceramol." [400] Trade name for a blend of cetyl and stearyl alcohols and higher alcohol sulfates. Melts from 50-60°C; acid number 1.0 max; saponification value 3 max; iodine number 5 max; acetyl value 185-195. Not alkaline to phenolphthalein.
Uses: Emulsifier for cosmetic creams, ointments, and lipsticks.

cerargyrite (chlorargyrite) AgCl. Natural silver chloride found in the oxidized zone of silver deposits.
Properties: Colorless, gray, yellowish, brownish; becomes violet brown or purple on exposure to light; luster resinous to adamantine; hornlike; can be cut with a

knife; sp. gr. 5.55; hardness 2.5.
Occurrence: Europe; South America; Colorado, Arizona, New Mexico, California.
Use: Ore of silver.
See also bromyrite.

"Ceratak." [128] Brand name for a grade of petroleum microcrystalline wax.
Properties: Color, amber; m.p. 155 or 165°F min.
Containers: 10-lb slabs, 8/carton or 168/pallet; 350-lb drums; tank cars.
Uses: Coating and laminating paper, foil, and board; impregnating and waterproofing fabrics.

"Ceraweld." [128] Brand name for a grade of petroleum microcrystalline wax.
Properties: Color, brown or amber; m.p. 155°F min.
Containers: 10-lb slabs, 8/carton or 168/pallet; 350-lb drums; tank cars.
Uses: Coating and laminating paper, foil, and board; impregnating and waterproofing fabrics.

"Cercor." [20] Trademark for honeycomb ceramic articles. A wide variety of ceramic materials can be utilized to produce articles having surface areas of 1000-2000 square feet per cubic foot of material and with a specific gravity of less than one.

cerebroside. A group of compounds found in brain and nervous tissue. Upon hydrolysis they yield a fatty acid, sphingosine, and a sugar, usually galactose.

"Ceres." [307] Trademark for a line of dyestuffs. Used for the coloring of wax, stearin, candles, etc.

"Ceresan" (2%). [28] Trademark for a seed disinfectant containing 2% ethyl mercuric chloride.
Containers: 1 ½-lb cans and 75-lb drums.
Use: For treatment of cotton, peanuts, and pea seeds to control seed-borne diseases and to reduce seed decay and check damping-off; as a short soak treatment for basal rot of narcissus bulbs.

"Ceresan" 75. [28] Trademark for liquid mercurial seed disinfectant. Active ingredients: ethyl mercuric 2,3-dihydroxypropylmercaptide and ethyl mercuric acetate. Colors seed red. "Ceresan" 100 and 200 are higher strength.
Containers: 1-, 5-, 30- and 55-gal drums.
Use: Control of certain seed-borne diseases, and to reduce losses from seed decay and damping-off of wheat, oats, barley, rye, sorghum, rice, flax and cotton.

"Ceresan" M. [28] Trademark for seed disinfectant containing ethyl mercuric paratoluene sulfonanilide.
Containers: 14-oz cans; 3-lb cans; 40- and 100-lb drums.
Use: As a dust or as a slurry treatment for the same disease control and crop seed uses as the "Ceresan" 75. Also sugar beet seed and control of snow mold of wheat.

"Ceresan" (New Improved). [28] Trademark for a seed disinfectant containing 5% ethyl mercuric phosphate.
Containers: 1-lb and 100-lb drums.
Use: Dip treatment for gladiolus corms. Also used as a dust treatment for the same purposes as "Ceresan" M.

ceresin wax (purified ozocerite; earth wax; mineral wax; cerosin; cerin).
Properties: White or yellow waxy cake; white is odorless; yellow has a slight odor. Sp. gr. 0.92-0.94; m.p. 68-72°C. Soluble in alcohol, benzene, chloroform, naphtha; insoluble in water.
Derivation: Purification of ozocerite by treatment with concentrated sulfuric acid and filtration through animal charcoal.
Grades: White; yellow.
Containers: Bags; cartons.
Uses: Candles; sizing; substitute for white wax; bottles for hydrofluoric acid; electric insulations; shoe and leather polishes; impregnating and preserving agent; lubricating compounds; wood fillers; floor polishes; antifouling paints; sizing and glossing paper; waxed papers; cosmetics; ointments and other pharmaceutical preparations; acid-proof coating for electrotypers' plates; matrix compositions; in admixture with rosin and sulfur for making printing forms; waterproofing textile fabrics.
Shipping regulations: None.*

"Ceresit." [76] Trademark for a line of water repellents for concrete, floor hardeners, masonry repair materials and allied protective products.

ceria. See ceric oxide; also rare earths.

ceric ammonium nitrate (cerium-ammonium nitrate; ammonium hexanitratocerate) $Ce(NO_3)_4 \cdot 2NH_4NO_3$.
Properties: Small prismatic, yellow crystals. Soluble in water and alcohol; almost insoluble in concentrated nitric acid; soluble in other concentrated acids.
Derivation: By electrolytic oxidation of cerous nitrate in nitric acid solution, and subsequently mixing solutions of cerium nitrate and ammonium nitrate, followed by crystallization.
Method of purification: Crystallization.
Grades: Technical.
Containers: Wooden kegs; tins.
Uses: Analytical chemistry; oxidant for organic compounds; scavenger in the manufacture of azides.
Fire hazard: Dangerous.
Shipping regulations: Oxidizing material. Yellow label.*

ceric ammonium sulfate $Ce(SO_4)_2 \cdot 3(NH_4)_2SO_4 \cdot 2H_2O$.
Properties: Yellow crystals, soluble in water and in acids.
Grades: Technical.
Containers: Bottles; fiber drums.

ceric hydroxide (ceric oxide, hydrated; cerium hydrate).
Properties: Whitish powder when pure.

*See "I.C.C. Shipping Regulations," page xiii.
Reference numbers refer to name of manufacturer. See "List of Manufacturers," page v.

The dry powder is best described as a hydrated oxide containing 85-90% ceric oxide. Soluble in concentrated mineral acids; insoluble in water.
Derivation: By treating a solution of a ceric salt with strong alkali. Reagent grade is prepared by adding a saturated solution of ceric ammonium nitrate to an excess of ammonium hydroxide.
Grades: Commercial; high purity.
Containers: Cans; bottles; fiber drums.
Uses: As a source of ceric sulfate; production of cerium salts; opacifier in glasses and enamels (imparts yellow color); production of shielding glass.

ceric oxide (cerium dioxide; cerium oxide; ceria) CeO_2.
Properties: Pale yellow, heavy powder (white when pure). Commercial article is brown. Sp.gr. 7.65; m.p. 2600°C. Soluble in sulfuric acid; insoluble in water and dilute acids; requires reducing agent with acid to dissolve the anhydrous oxide.
Derivation: By decomposing cerium nitrate by heat.
Impurities: Other rare-earth metal oxides.
Grades: Technical; high purity (99.8%).
Containers: Bottles; cans; fiber drums.
Uses: Ceramics; x-ray investigations; polishing glass; an optical sensitizer in photosensitive glass; a stabilizer to prevent browning of glass in radiation shields; opacifier in enamels.
Shipping regulations: None.*

ceric oxide, hydrated. See ceric hydroxide.

ceric sulfate (cerium sulfate) $Ce(SO_4)_2 \cdot 4H_2O$.
Properties: White or reddish-yellow crystals; sp.gr. 3.91; soluble in water (decomposes); soluble in dilute sulfuric acid.
Derivation: By the action of sulfuric acid on cerium carbonate.
Method of purification: Crystallization.
Grades: Technical; reagent.
Containers: Boxes; glass bottles.
Uses: Dyeing and printing textiles; analytical reagent; waterproofing; mildewproofing.
Shipping regulations: None.*

cerin. See ceresin wax.

cerinic acid. See cerotic acid.

cerite. One of the rare-earth minerals (q.v.). It is a hydrated silicate and contains from 60-70% of cerium and its allies, together with smaller amounts of iron, calcium and the yttrium earths. Color, between clove-brown and cherry-red to gray.
Constants: Sp.gr. 4.86; hardness 5.5.
Occurrence: Sweden.

cerium Ce. Rare-earth element; atomic number 58.
Properties: Gray, ductile, malleable metal; tarnishes in moist air; sp.gr. 6.78; m.p. 804°C; b.p. 2900°C; soluble in acids; decomposes water.
Occurrence: See rare-earth minerals.
Derivation: (a) Reduction of the chloride by calcium powder; (b) electrolysis of the fused chloride (see misch metal); (c) by-product from thorium nitrate production known as cerium residue is treated electrolytically with production of crude cerium metal containing also lanthanum and didymium.
Grades: Granules; ingots; rods (high purity).
Uses: Cerium salts; cerium-iron pyrophoric alloys; ignition devices; military signalling; illuminant in photography; reducing agent (scavenger in metallurgical work); catalyst in synthesis of ammonia; alloys for jet engines; solid state devices; rocket propellants.
See also misch metal.
Shipping regulations: None.*

cerium-ammonium nitrate. See ceric ammonium nitrate.

cerium carbonate. See cerous carbonate.

cerium chloride. See cerous chloride.

cerium dioxide. See ceric oxide.

cerium hydrate. See ceric hydroxide or cerous hydroxide.

cerium naphthenate.
Properties: Extremely rubbery material. Very difficult to dry. Very insoluble unless small quantities of organic stabilizers are used which make it easier to dissolve and form free-flowing solutions.
Derivation: By saponifying naphthenic acids and treating the sodium naphthenate formed with a suitable cerium salt.
Grades: The commercial product is not a pure one in the strictest sense of the term as the cerium salt used in making the naphthenate is a mixture of rare earths; e.g., lanthanum, neodymium, praseodymium, etc. However, these elements all have very similar properties and their presence has produced no noticeable detrimental effects in its commercial applications.
Uses: Inks (drier, improving water-resisting properties, preventing undue ink-absorption by paper, reducing emulsifying tendencies); waterproofing rope, sails and canvas products; paints and varnishes (drier, waterproofing agent); waterproofing agent for general purposes.

cerium nitrate. See cerous nitrate.

cerium oxalate. See cerous oxalate.

cerium oxide. See ceric oxide.

cerium sulfate. See ceric sulfate; cerous sulfate.

cermets. Refractory compositions made by bonding grains of ceramics, metal carbides, nitrides, etc. with metal. They combine the strength and toughness of the metal with the temperature resistance of the ceramic material and are intended for use in rocket motors, gas turbines, turbojet engines, and nuclear reactor mechanisms operating continuously at temperatures as high as 1800°F and, for short periods, as high as 4000°F. Niobium, tantalum, titanium, and zirconium, which wet both

*See "I.C.C. Shipping Regulations," page xiii.
Reference numbers refer to name of manufacturer. See "List of Manufacturers," page v.

ceramics and metals, are employed as brazing agents to bond the grains to the matrix. Cermets are made by powder metallurgy techniques in which a powdered mixture of refractory, metal, and brazing agent is molded to the desired form and subjected to high temperature and pressure. Typical cermets contain nickel with lead silicate; chromium with aluminum silicate; tungsten with beryllium and aluminum oxides; molybdenum with calcium and aluminum oxides.

"Cer-O-Cillin." [327] Trademark for penicillin "O," allylmercaptomethyl penicillin, an antibiotic obtained from cultures of Penicillium chrysogenum.

Use: Medicine; the antibacterial spectrum is similar to penicillin G, less likely to cause allergic reactions.

cerosin. See ceresin wax.

cerotic acid (hexacosanoic acid; cerinic acid) $CH_3(CH_2)_{24}COOH$. An acid obtained from beeswax, carnauba wax or Chinese wax.

Properties: White odorless crystals or powder; sp. gr. 0.8198 (100/4°C); m.p. 87.7°C; refractive index 1.4301 (100°C). Insoluble in water; soluble in alcohol, benzene, ether, acetone.

cerotin. See ceryl alcohol.

cerous carbonate (cerium carbonate) $Ce_2(CO_3)_3 \cdot 5H_2O$.

Properties: White powder; soluble in mineral acids (dilute); insoluble in water.

Derivation: By adding an alkali carbonate to a solution of a cerous salt.

cerous chloride (cerium chloride) $CeCl_3 \cdot xH_2O$.

Properties: White crystals; deliquescent; sp. gr. 3.88 (anhydrous); m.p. 848°C (anhydrous). Soluble in water, alcohol, and acids.

Derivation: By the action of hydrochloric acid on cerium carbonate or hydroxide.

Method of purification: Crystallization.

Grades: Technical and purified or reagent.

Containers: Kegs; fiber drums.

Uses: Incandescent gas mantles; spectrography; preparation of cerium metal.

Shipping regulations: None.*

cerous fluoride $CeF_3 \cdot xH_2O$.

Properties: Off-white powder, insoluble in water and acids.

Derivation: By treating cerous oxalate with hydrofluoric acid.

Grades: Technical.

Containers: Fiber drums.

Uses: In arc carbons to increase their brilliance; preparation of cerium metal.

Shipping regulations: None.*

cerous hydroxide (cerium hydrate). Approximate formula $Ce(OH)_3$.

Properties: White gelatinous precipitate; yellow, brown or pink when impurities are present. Soluble in acids; insoluble in water and alkali.

Derivation: Chief source is monazite sand.

Grades: Pure; crude.

Use: Pure form: To produce cerium salts, impart yellow color to glass and as an opacifying agent in glazes and enamels. Crude form: In the flaming arc lamp.

cerous nitrate (cerium nitrate) $Ce(NO_3)_3 \cdot 6H_2O$.

Properties: Colorless crystals; deliquescent. Soluble in water and alcohol.

Constants: M.p., loses $3H_2O$ at 150°C; b.p., decomposes at 200°C.

Derivation: By the action of nitric acid on cerous carbonate.

Method of purification: Crystallization.

Grades: Technical and purified.

Containers: Wooden barrels; fiber drums.

Uses: Incandescent gas mantles; medicine; reagent.

Shipping regulations: Oxidizing material. Yellow label.*

cerous oxalate (cerium oxalate) $Ce_2(C_2O_4)_3 \cdot 9H_2O$.

Properties: Yellowish-white, odorless, tasteless, crystalline powder; decomposes upon heating; soluble in dilute sulfuric and hydrochloric acids; very slightly soluble in water; insoluble in oxalic acid solution, alkalies, alcohol and ether.

Derivation: By extraction from monazite sand with oxalic acid, or with hydrochloric acid and conversion into the oxalate, followed by crystallization.

Method of purification: Recrystallization.

Grades: Pure; the commercial article consists of a complex mixture of oxalates of cerium, lanthanum and didymium, being more or less a by-product in the manufacture of thorium salts.

Containers: 5-, 25-lb boxes; 100-lb kegs; 200-lb barrels; multiwall paper sacks.

Uses: Medicine; isolation of the metals of the cerium group.

Shipping regulations: None.*

cerous sulfate (cerium sulfate) $Ce_2(SO_4)_3 \cdot xH_2O$.

Properties: White crystals or powder; soluble in water and in acids. M.p. 630°C (dehydrated); sp. gr. 2.886.

Derivation: Reagent grade is prepared by reducing a solution of ceric sulfate in sulfuric acid with hydrogen peroxide.

Grades: Technical and purified (reagent).

Uses: Developing agent for aniline black.

"Cerox." [88] Trademark for optical grade cerium oxide, approximately 90% purity. Used in glass polishing.

ceroxylin $C_{20}H_{32}O$. An amaroid obtained from the wax of Ceroxylon andicola, the wax-palm tree of South America.

Properties: White needles; m.p. greater than 100°C. Soluble in alcohol, ether, fatty oils.

"Cerrobase." [60] Trademark for the eutectic alloy of bismuth and lead. M.p. 255°F. Shrinks slightly after solidifying, later expands.

Uses: Proof casting forging dies; master patterns; mandrels for electroforming;

heat transfer medium in autoclaves; liquid seal in bright annealing and nitriding furnaces; molds for plastics; fusible foundry cores; filler for bending large diameter tubing, preventing wrinkles.

"Cerrobend." [60] Trademark for the eutectic alloy of bismuth, lead, tin and cadmium, m.p. 158°F. Expands during and after solidification.
Uses: As filler in thin walled tubing, prevents wrinkles in bending, melts out in hot water; assembly, checking, drilling, spotting fixtures in aircraft and automotive tooling; anchoring medium in precision machining jet engine components (buckets and blades).

"Cerrocast." [60] Trademark for a non-eutectic alloy of bismuth and tin. Melting range 281-338°F. Exhibits negligible volume change during and after solidification.
Uses: Soft metal dies for "lost wax" patterns; engraving machine patterns; split jaw chucks; molds for plastics; mandrels for electroforming.

"Cerrodent." [60] Trademark for a bismuth alloy, pouring about 200°F, for dental casting models.

"Cerrolows." [60] Trademark for bismuth alloys containing high percentages of indium. Ultra-low melting temperatures make them useful in prosthetic development, dental models, proof casting, anchorage and support of delicate work pieces during machining.

"Cerromatrix." [60] Trademark for a bismuth alloy with melting range 217-440°F, pouring at 250°F, expanding during and after solidification.
Uses: In die making to anchor punches; fastening bearings, bushings, and nonmoving parts in machinery; nests in drill jigs and dial feeding stations; sheet metal forming dies, etc.

"Cerrosafe." [60] Trademark for a bismuth, lead, tin, and cadmium alloy with melting range 158-190°F. Shrinks 15 minutes after solidification, then expands.
Uses: Accurate duplicate patterns; proof casting cavities, such as gun chambers, bullet molds; toy castings and hobby models; sprayed-on protective coating on wood patterns and core boxes.

"Cerroseal-35." [60] Trademark for an indium-tin alloy with m.p. near 250°F. Has characteristic of adhering to glass, mica, glazed ceramics, and quartz. Bonds to metals when used as regular solder.

"Cerrotru." [60] Trademark for a non-shrinking bismuth, tin eutectic alloy, m.p. 281°F.
Uses: Anchoring shafts in Alnico rotors, forming blocks in stretch presses; engraving machining models; special jaws in tangent tube bending; soft metal dies in "lost wax" process, mandrels in electroforming; molds by "dip" casting.

cerulean blue. A light blue pigment essentially cobaltous stannate $CoO \cdot n(SnO_2)$.

ceruse. See lead carbonate, basic.

cerussite $PbCO_3$. Natural lead carbonate, found in the upper zone of lead deposits.
Properties: Colorless, white, gray; luster adamantine; hardness 3-3.5; sp.gr. 6.55. Effervesces in nitric acid.
Occurrence: Colorado, Arizona, New Mexico, Idaho; Australia; Europe.
Uses: An ore of lead.

ceryl alcohol (cerotin) $C_{26}H_{54}O$. An alcohol obtained from Chinese wax (q.v.).
Properties: Colorless crystals; m.p. 79°C; insoluble in water; soluble in alcohol and ether.

ceryl cerotate $C_{52}H_{104}O_2$. The chief constituent of Chinese wax (q.v.). Colorless crystals with m.p. 84°C.

cesium Cs. One of the alkali group of elements, highest in the electromotive series (except the unstable francium); atomic number 55.
Properties: Silver-white, soft ductile metal; decomposes water, setting free hydrogen which ignites. Must be kept immersed in naphtha or kerosine. Sp.gr. 1.90; m.p. 28°C; b.p. 690°C. Soluble in acids and alcohol.
Derivation: By thermochemical reduction of cesium chloride with calcium.
Grades: Technical.
Containers: Glass bottles.
Uses: Photo-electric cells; radio tubes; as a "getter" in vacuum tubes. Has been suggested for: ion propulsion systems; plasma for thermoelectric conversion; heat transfer fluid in power generators; grain refining agent in metal.
Fire hazard: Dangerous.
Shipping regulations: Flammable solid. Yellow label.*

cesium 137. Radioactive cesium of mass number 137.
Properties: Half-life 30 years; radiation, beta; radiotoxicity, moderately hazardous. The beta decay of Cs-137 produces barium 137, which in turn is radioactive, emitting a 0.662 mev gamma ray with a 2.6 minute half-life. Most applications of Cs-137 depend on the fact that any Cs-137 preparation has an equivalent amount of the gamma-emitting Ba-137 daughter.
Shipping regulations: Class D poison, radioactive material. Red label.*

cesium alum. See cesium aluminum sulfate.

cesium aluminum sulfate (cesium alum) $CsAl(SO_4)_2 \cdot 12H_2O$.
Properties: Colorless crystals; sp.gr. 2.0215; m.p. 117°C; soluble in water; insoluble in alcohol.
Derivation: By adding a solution of cesium sulfate to a solution of potassium alum, concentrating and crystallizing.
Method of purification: Recrystallization.
Grades: Pure.

*See "I.C.C. Shipping Regulations," page xiii.
Reference numbers refer to name of manufacturer. See "List of Manufacturers," page v.

Containers: Glass bottles; tins.
Use: Mineral waters.
Shipping regulations: None.*

cesium ammonium bromide (ammonium-cesium bromide) $CsBr \cdot NH_4Br$.
Properties: White, crystalline powder; soluble in water.
Use: Medicine.

cesium antimonide. Used as a high-purity binary semiconductor.

cesium arsenide. Used as a high-purity binary semiconductor.

cesium bromide CsBr.
Properties: Colorless crystalline powder; sp. gr. 4.44; m.p. 636°C; b.p. 1300°C; soluble in water; slightly soluble in alcohol.
Grades: Technical; single pure crystals.
Use: Medicine; crystals for infrared spectroscopy; scintillation.

cesium carbonate Cs_2CO_3.
Properties: White, deliquescent, crystalline powder; b.p., decomposes at 610°C. Soluble in water, alcohol, and ether.
Derivation: By passing carbon dioxide into a solution of cesium oxide and subsequent crystallization.
Grades: Pure.
Containers: Glass bottles.
Uses: Brewing; manufacture of mineral waters.
Shipping regulations: None.*

cesium chloride CsCl.
Properties: Colorless crystals; sp. gr. 3.972; m.p. 646°C; sublimes at 1290°C. Soluble in water and alcohol.
Derivation: By the action of hydrochloric acid on cesium oxide and crystallization.
Method of purification: Recrystallization.
Grades: Pure.
Containers: Glass bottles; tins.
Uses: Medicine; brewing; manufacturing mineral waters; evacuation of radio tubes (positive ions supplied at surface of filament).
Shipping regulations: None.*

cesium dichromate $Cs_2Cr_2O_7$.
Properties: Reddish-white crystals. Soluble in water.

cesium dioxide Cs_2O_2.
Properties: Yellow needles; sp. gr. 4.25; m.p. 400°C; soluble in water and acids.
Grades: Technical; pure.
Use: Cesium salts.

cesium disulfate (acid cesium sulfate) $CsHSO_4$.
Properties: Colorless, rhombic prisms; sp. gr. 3.352; soluble in water.

cesium hydrate. See cesium hydroxide.

cesium hydroxide (cesium hydrate) CsOH.
Properties: Colorless or yellowish, fused, crystalline mass. Strong alkaline reaction. Hygroscopic. Caution! Keep well stoppered. Sp. gr. 3.675; m.p. 272.3°C. Soluble in alcohol and hot water.

cesium iodide CsI.
Properties: Colorless, crystalline powder; deliquescent. Sp. gr. 4.510; m.p. 621°C; b.p. 1280°C; soluble in alcohol and water.
Grades: Technical; single pure crystals.
Uses: Crystals for infrared spectroscopy; scintillation.

cesium nitrate $CsNO_3$.
Properties: Glittering, crystalline powder; saltpeter taste; sp. gr. 3.687; m.p. 414°C; b.p. (decomposes); soluble in water and acetone; slightly soluble in alcohol.
Derivation: By the action of nitric acid on cesium oxide and crystallization.
Method of purification: Recrystallization.
Grades: Pure.
Containers: Tins; glass bottles.
Use: Cesium salts.
Fire hazard: Dangerous.
Shipping regulations: Oxidizing material. Yellow label.*

cesium oxide Cs_2O.
Properties: Orange-red crystals; sp. gr. 4.36; m.p. decomposes 360-400°C; very soluble in water; soluble in acids.
Grades: Technical; pure.
Use: Cesium salts.

cesium oxides. See cesium oxide; cesium dioxide; cesium trioxide; cesium tetroxide.

cesium peroxide See cesium tetroxide.

cesium phosphide. Used as a high-purity binary semiconductor.

cesium-rubidium-ammonium bromide (rubidium-cesium-ammonium bromide; ammonium-cesium-rubidium bromide) $CsBr \cdot RbBr \cdot 6NH_4Br$.
Properties: White, crystalline powder; soluble in water.

cesium-rubidium bromide (rubidium-cesium bromide) $CsBr \cdot RbBr$.
Properties: White, crystalline powder; soluble in water; insoluble in alcohol.
Use: Medicine.

cesium-rubidium chloride (rubidium-cesium chloride) $CsCl \cdot RbCl$.
Properties: Colorless, crystalline powder; soluble in water; insoluble in alcohol.
Use: Medicine.

cesium silicate Cs_2SiO_3.
Properties: Yellow, crystalline powder; insoluble in water.
Derivation: By the interaction of a cesium salt and sodium silicate.

cesium sulfate Cs_2SO_4.
Properties: Colorless crystals; soluble in water; insoluble in alcohol. Sp. gr. 4.2434; m.p. 1010°C.
Derivation: By the action of sulfuric acid on cesium carbonate.
Method of purification: Crystallization.
Grades: Pure.
Containers: Kegs.
Uses: Brewing; mineral waters.
Shipping regulations: None.*

cesium sulfate, acid. See cesium disulfate.

cesium tetroxide (cesium peroxide) Cs_2O_4.
Properties: Yellow crystals; sp. gr. 3.77; m.p. 600°C; decomposes in water to CsOH; soluble in acids.
Grades: Technical; pure.
Use: Cesium salts.

cesium trioxide Cs_2O_3.
Properties: Chocolate-brown crystals; sp. gr. 4.25 (0°C); m.p. 400°C; decomposes in water; soluble in acids.
Grades: Technical; pure.
Use: Cesium salts.

"CET." [328] Brand name for a water-soluble cellulose-reactive modified urea-formaldehyde condensate used as a stabilization medium for cellulosic fabrics. It is an acid curing reactant, so dyed cloth should be neutral or slightly acid prior to finishing.

cetaceum. See spermaceti.

cetalkonium chloride (cetyldimethylbenzylammonium chloride) $C_6H_5CH_2N(CH_3)_2C_{16}H_{33}Cl$. A quaternary ammonium germicide.
Properties: Colorless, odorless, crystalline powder; m.p. 58-60°C. Soluble in water to form colorless, odorless solution having pH 7.2. Compatible with alkalies and antihistaminics. Soluble in alcohol, acetone, esters, carbon tetrachloride.
Containers: ¼-lb, 1-lb, 5-lb, 25-lb containers.
Uses: Medicine; germicide; fungicide; surface active agent.

cetane. See hexadecane.

cetane number. A rating for Diesel fuel comparable to the octane number rating for gasoline. It is the percentage of cetane ($C_{16}H_{34}$) which must be mixed with alpha-methyl naphthalene to give the same ignition performance, under standard conditions, as the fuel in question.

"Cetavlon." [207] Trademark for cetrimide, a mixture of dodecyl-, tetradecyl-, and hexadecyl-trimethylammonium bromides. A cationic detergent and bactericide used for skin sterilization and treatment of wounds and burns.

cetene. See 1-hexadecene.

cetin (cetyl palmitate; palmitic acid, cetyl ester) $C_{15}H_{31}COOC_{16}H_{33}$.
Properties: White crystalline wax-like substance. Chief constituent of commercial purified spermaceti. M.p. 50°C; b.p. 360°C; sp. gr. 0.832; refractive index (n 70/D) 1.4398. Soluble in alcohol and ether; insoluble in water.
Derivation: By solution from spermaceti.
Grades: Technical.
Containers: Wooden boxes.
Uses: Base for ointments, cerates, and emulsions; manufacture of candles, soaps, etc.
Shipping regulations: None.*

"Cetol." [430] Trade name for cetyl dimethyl benzyl ammonium chloride. F. D. A. approved for internal use; used in cough preparations.

"Cetone Alpha." [227] Trademark for a very highly refined alpha-isomethylionone.
Properties: Slightly yellow liquid; sp. gr. 0.925-0.929 (25/25°C); refractive index 1.500-1.5010 (20°C); flash point (Tag closed cup) 217°F; clearly soluble in 5 parts 70% alcohol.
Occurrence: Not found in nature.
Uses: Floral perfumes, particularly of a violet character.

cetraria (Iceland moss).
Properties: Gray fibrous bundles. Makes a gelatinous solution after being boiled in water.
Chief constituents: Cetraric acid, lichenostearic acid, fumaric acid and lichenine.
Derivation: The thallus of Cetraria islandica; habitat: North America; Europe; Japan.
Uses: Gelling and sizing agent in textiles, cosmetics, foods.

cetyl alcohol (alcohol C-16; cetylic alcohol; 1-hexadecanol; normal primary hexadecyl alcohol; palmityl alcohol) $C_{16}H_{33}OH$. A fatty alcohol.
Properties: White, odorless, tasteless crystals; sp. gr. 0.8176 (49.5°C); m.p. 49.3°C; b.p. 344°C; refractive index (n 79/D) 1.4283; soluble in alcohol and ether; insoluble in water.
Derivation: By saponifying spermaceti with caustic alkali; reduction of palmitic acid.
Method of purification: Crystallization; distillation.
Grades: Technical; cosmetic; N. F. XI.
Containers: Tins; cartons; drums; bags; steam-coiled tank cars.
Uses: Medicine; perfumery; emulsifier; cosmetics; base for making sulfonated fatty alcohols; to retard evaporation of water, when spread as a film on reservoirs, or sprayed on growing plants.
Shipping regulations: None.*

cetyl bromide $C_{16}H_{33}Br$.
Properties: Dark yellow liquid. Freezing point 15°C; b.p. 186-197°C (10 mm); sp. gr. 0.991 (25/25°C); lb/gal 8.25 (25°C); refractive index 1.460 (n 25/D); flash point 350°F. Soluble in ether; very slightly soluble in water, methanol.
Use: Synthesis.

cetyldimethylbenzylammonium chloride. See cetalkonium chloride.

cetyldimethylethylammonium bromide $C_{16}H_{33}(CH_3)_2C_2H_5NBr$. A quaternary ammonium salt (q. v.).
Properties: Paste.
Uses: Disinfectant; deodorant; germicide; fungicide; detergents.

cetyldimethylethylammonium chloride $C_{16}H_{33}(CH_3)_2C_2H_5NCl$. A quaternary ammonium salt (q. v.).

cetylic acid. See palmitic acid.

*See "I.C.C. Shipping Regulations," page xiii.
Reference numbers refer to name of manufacturer. See "List of Manufacturers," page v.

cetylic alcohol. See cetyl alcohol.

cetyl mercaptan (hexadecyl mercaptan) $C_{16}H_{33}SH$.
Properties: M.p. 18°C; b.p. 185-190°C (7 mm); sp. gr. 0.8474 (20/4°C); refractive index 1.4638 (n 20/D).
Grades: 95% (min.) purity.
Uses: Intermediates; synthetic rubber processing.

cetyl palmitate. See cetin.

cetyl pyridinium bromide $C_{16}H_{33}C_5H_5NBr$.
Properties: Cream colored waxy solid. Soluble in acetone, ethanol and chloroform.
Uses: Surface active agent; germicide.

cetylpyridinium chloride. The monohydrate of the quaternary salt of pyridine and cetyl chloride; $C_{16}H_{33}C_5H_5NCl \cdot H_2O$.
Properties: White powder with slight odor. M.p. 77-83°. Very soluble in alcohol, chloroform and water; very slightly soluble in benzene and ether; pH (1% soln) 6.0-7.0.
Grades: Technical; U.S.P. XVI.
Use: Medicine.

cetyltrimethylammonium bromide (hexadecyltrimethylammonium bromide) $C_{16}H_{33}(CH_3)_3NBr$. A quaternary ammonium salt.
Properties: White powder; soluble in water, alcohol and chloroform.
Grade: Technical.
Uses: Surface active agent; germicide.

cetyltrimethylammonium chloride $C_{16}H_{33}(CH_3)_3NCl$. A quaternary ammonium salt.

cevadilla. See sabadilla.

cevadilline. See veratrine.

cevadine. See veratrine.

"Cevalin." [100] Trademark for ascorbic acid, U.S.P.

cevine. See veratrine.

Ceylon gelatin. See agar-agar.

Ceylon isinglass. See agar-agar.

Cf. Symbol for californium.

CF. Abbreviation for citrovorum factor. See folinic acid.

CF black. Abbreviation for conducting furnace black. See furnace black.

CFE. Abbreviation for chlorotrifluoroethylene. Also used for polychlorotrifluoroethylene resins.

"CH-100." [244] Trademark for a product consisting of a caustic soda base combined with other alkalies, surface active agents, and sequestering agents.
Properties: Light cream-colored, flaked, mechanical mixtures; soluble in water; total Na_2O, 70.2%.
Uses: To prevent liming and rust staining of bottles and mechanical parts of soaker and "hydro" bottle washers in breweries and creameries. General deactivation of metal impurities.

chabazite $CaAl_2Si_4O_{12} \cdot 6H_2O$. Essentially a natural hydrous calcium aluminum silicate, usually containing some sodium and potassium. One of the zeolites (q.v.).
Properties: Color white, reddish, yellow, brown; luster vitreous; sp. gr. 2.1; hardness 4-5.
Occurrence: New Jersey, Colorado, Oregon; Europe.

chain reaction. A reaction between particles (atoms, molecules, or nuclei), in which one of the product particles of one reacting set is a reactant in the next set. The reaction between hydrogen and oxygen to produce water is a reaction of this kind, as are many polymerization reactions. More recently the term has had frequent use in reference to a type of nuclear reaction, fission, which is initiated by neutrons and is propagated by neutrons. When a U-235 nucleus absorbs a neutron, it splits or fissions with a great deal of energy, as well as emitting, on the average, about 2 ½ more neutrons, which may in turn be absorbed in other U-235 nuclei to propagate the reaction. In any given set of conditions of a fissioning system, neutrons may be lost through the walls, as well as captured by structural materials, or by the U-235 itself in a non-fission mode. When the conditions are such that exactly one neutron causes fission of a nucleus from each set of neutrons produced in the previous fission, the system is said to be critical, and the geometrical size of the system and the mass of fissionable material in the system are called, respectively, the critical size and critical mass. Control of either the neutron leakage or non-fission capture allows the adjustment of this ratio of the number of fissions in one generation of neutrons to the number in the previous generation. If the ratio is greater or less than one, the system is super- or subcritical and the rate of energy production rises or falls exponentially.

chalcanthite (blue vitriol, natural) $CuSO_4 \cdot 5H_2O$. Natural hydrous copper sulfate, occurring in the oxidized portions of some copper deposits.
Properties: Color deep azure blue; luster vitreous; soluble in water; sp. gr. 2.2; hardness 2.5.
Occurrence: Chile.
Use: Copper ore.

chalcedony. A microcrystalline form of native silica or quartz (q.v.).
Properties: Color variable; luster waxy; sp.gr. 2.6-2.65; hardness 6.5-7. Varieties include carnelian, chrysoprase, agate, bloodstone, and onyx.
Use: Ornamental material.

chalcocite (copper glance) Cu_2S. Natural cuprous sulfide, occurring with other copper minerals.
Properties: Color lead gray, tarnishing dull

black; luster metallic; sectile; sp.gr. 5.5-5.8; hardness 2.5-3.
Occurrence: Montana, Arizona, Utah, Nevada; Alaska; Chile; Mexico; Europe.
Use: Important ore of copper.

chalcogenides. See chalcogens.

chalcogens. The chemically related elements oxygen, sulfur, selenium, tellurium and polonium. The binary compounds are referred to as chalcogenides.

chalcopyrite (copper pyrites, yellow copper) $CuFeS_2$. Natural copper-iron sulfide, found in metallic veins and igneous rocks.
Properties: Color brass yellow, frequently tarnished bronze or iridescent; luster metallic; streak greenish black; sp.gr. 4.1-4.3; hardness 3.5-4. May carry gold or silver or mechanically intermixed pyrite.
Occurrence: Montana, Utah, Arizona, Tennessee; Europe; Chile; Canada.
Use: Important ore of copper.

chalk. (See also whiting.) A natural calcium carbonate composed of the calcareous remains of minute marine organisms. It varies in composition, properties and appearance ranging in color from snow-white through dull-white or grayish and from soft, incoherent and porous to hard and crystalline. It may contain up to 99% calcium carbonate in the form of calcite (q.v.) with silica, quartz, feldspar, zircon, rutile and other minerals as impurities. Phosphatic chalk contains up to 45% calcium phosphate. Glauconitic chalk is an admixture with grains of glauconite. Red chalk contains iron hydroxide.

chalk, cliffstone. One of the English chalks. It is much harder than ordinary chalks and is used to prepare a special grade of whiting, known as cliffstone Paris white.

chalk, drop. See chalk, prepared.

chalk, French. A variety of soapstone or steatite. See talc.

chalking. A natural process by which protective coatings develop a loose, powdery surface formed from the film. Chalking of paint films results from decomposition of the binder, principally through the action of ultraviolet rays. Although usually considered undesirable, chalking may be used advantageously in white exterior house paints as a means of shedding soil.

chalk, precipitated (calcium carbonate, precipitated). (See calcium carbonate for distinction from whiting and other chalks.)
Properties: Fine, white microcrystalline powder, the U.S.P. XVI grade containing not less than 98.0% calcium carbonate. Odorless, tasteless and stable in air. M.p., decomposes at 825°C with evolution of carbon dioxide. Density about 2.7. Soluble (with effervescence) in dilute acetic, hydrochloric, and nitric acids; practically insoluble in water; insoluble in alcohol.
Derivation: (a) By adding a boiling solution of calcium chloride to a boiling solution of sodium carbonate. (b) By passing carbon dioxide through milk of lime.
Grades: Technical; U.S.P. XVI.
Containers: Fiber cans; tins; glass bottles; multiwall paper sacks.
Uses: Medicine (antacid); dentifrices; baking powder; organic synthesis. For industrial uses see whiting.

chalk, prepared (drop chalk; calcium carbonate, prepared). (See calcium carbonate for distinction from whiting and other chalks.)
Properties: A very fine, white to grayish-white, amorphous powder, often formed in "conical drops." The N.F. XI grade contains not less than 97% calcium carbonate. Odorless, tasteless and stable in air. M.p., decomposes at 825°C with evolution of carbon dioxide. Soluble (with effervescence) in dilute acetic, hydrochloric and nitric acids; practically insoluble in water; insoluble in alcohol.
Derivation: By grinding some form of native calcium carbonate to a fine powder, agitating with water, allowing the coarser particles to settle, decanting the suspension and allowing the fine particles to settle slowly.
Grades: N.F. XI.
Containers: Fiber cans; tins; glass bottles; multiwall paper sacks.
Uses: Medicine (antacid); tooth powders; calcimine; polishing powders; silicate cements. For other uses, see whiting and chalk, precipitated.

chalybite. See siderite.

chamber acid. Sulfuric acid (q.v.) made by the chamber process.

chamber process. A process for manufacturing sulfuric acid from sulfur dioxide, air, and steam in the presence of nitrogen oxides as catalysts. The reaction takes place within large lead-lined chambers where the sulfuric acid-water mixture settles out as a fine mist.

The chamber process product is 50-55° Bé (60-70% H_2SO_4), which is suitable for manufacture of superphosphate fertiliqers. The product can be concentrated to 93% H_2SO_4.

chamomile oil, German (camomile oil; also called Hungarian chamomile oil.)
Properties: Viscous essential oil of deep-blue color, which by exposure to light and air turns into green and brown; characteristic odor; bitter, aromatic taste. Soluble in 95% alcohol, usually with separation of paraffin.
Constants: sp.gr. 0.922 to 0.956; acid value 9 to 50; ester value 3 to 33, after acetylation 117 to 155.
Adulterants: Cedar-wood oil, which reduces the congealing point. Unadulterated oils become viscous at 15°C and butyraceous at 0°C.
Derivation: Distilled from the flower heads of Matricaria chamomilla.
Uses: Flavoring; medicine.
Shipping regulations: None.*

*See "I.C.C. Shipping Regulations," page xiii.
Reference numbers refer to name of manufacturer. See "List of Manufacturers," page v.

chamomile oil, Hungarian. See chamomile oil, German.

chamomile oil, Roman. (camomile oil; anthemidis oil).
Properties: Soluble in 6 to 10 vols of 70% alcohol; in 1 to 2 vols of 80% alcohol (occasionally with separation of paraffin).
Constants: Sp.gr. 0.905 to 0.918 at 15°C; optical rotation −2° 30' to +3°; refractive index 1.442 to 1.457; acid value 1.5 to 14; ester value 214 to 317.
Chief known constituents: Esters of butyric, angelic and tiglic acids.
Derivation: Distilled from the flower heads of Anthemis nobilis.
Uses: Medicine; flavoring.
Shipping regulations: None.*

champaca oil. Fragrant fruity oil distilled from fresh flowers of Michelia champaca or M. congifolia, growing in the Philippines, Java and northern India. A leaf oil is also made. Small amounts are used in perfumery.

channel black. Carbon black made by impingement of a luminous natural-gas flame against an iron plate from which it is scraped at frequent intervals.
Properties vary widely but the material has an unusually fine state of subdivision and great surface area.
Grades: Conducting channel black (CC); hard processing (HPC); medium processing (MPC); easy processing (EPC).
Containers: Multiwall paper sacks.
Chief uses are as a reinforcing agent in rubber tires and as a pigment in printers' ink. Also used in other inks, and in paints, polishes, phonograph records, carbon paper, crayons, typewriter ribbons, pyrotechnic compositions, and insulating material. Also in case-hardening and crucible steel operations.
Shipping regulations: None.*

channel process. Method for making carbon black. See channel black.

charcoal. See also active carbon.
Derivation: A product of the destructive distillation of wood.
Grades: Technical, in lumps; powdered; briquettes.
Containers: Barrels; multiwall paper sacks.
Uses: Chemical (precipitant in the cyanide process, precipitant of iodine and lead salts from their solutions, catalyst, calcium carbide); decolorizing and filtering medium, absorbent in recovery of volatile solvents, gas absorbent; component of ordinary gunpowder and other explosives; fuel; poultry farming; arc light electrodes; decolorizing and purifying oils; artificial leather (solvent recovery); brewing (deodorant); metallurgy; heat insulating compositions; crayons; gasoline from casing-head gas; pharmaceutical preparations; plastics (solvent recovery); refrigeration (gas absorbent); sugar (decolorizing).
Fire hazard: Dangerous; ignites spontaneously when freshly calcined and exposed to air or when wet; hazardous when freshly ground and tightly packed.
Shipping regulations: Flammable solid. Yellow label.*

charcoal, animal. See animal black; blood charcoal; bone black; ivory black.

charcoal, bone. See bone black.

charcoal, mineral. See fusain.

charcoal, vegetable. See vegetable black; active carbon.

Charles' law. The volume of a sample of gas varies directly with the absolute or Kelvin temperature if the pressure remains constant.

Charlton white. See lithopone.

chars. See under specific charcoal or black.

chaulmoogra oil. (gynocardia oil; hydnocarpus oil).
Properties: Brownish-yellow oil or soft fat; characteristic odor; somewhat acrid taste. Soluble in ether, chloroform, benzene, solvent naphtha; sparingly soluble in cold alcohol; almost entirely soluble in hot alcohol, carbon disulfide.
Chief constituents: Glycerides of chaulmoogric and hydnocarpic acids.
Constants: Sp.gr. 0.940; saponification value 198 to 213; iodine value 96 to 104; acid value 21 to 27.
Derivation: Expressed from the seeds of Taraktogenos kurzii or Hydnocarpus anthelminthicus or wightianus.
Method of purification: Rectification.
Grades: Technical.
Containers: Tins; cases.
Use: Medicine (thought to have been useful in leprosy).
Shipping regulations: None.*

chaulmoogric acid (hydnocarpyl acetic acid) $CH_2CH_2CHCHCH(CH_2)_{12}COOH$. A cyclic fatty acid.
Properties: Colorless shiny leaflets; m.p. 68.5°C; soluble in ether, chloroform, and ethyl acetate.
Source: Chaulmoogra oil.
Use: Medicine; biochemical research.

chavicol methyl ether. See estragole.

checkerberry. See gaultheria.

"Cheelox." [307] Trademark for a series of organic chelating and sequestering agents, consisting of polycarboxylic acid derivatives of amines or polyamines or their salts, as, for example, ethylenediaminetetraacetic acid. Some types were formerly sold as "Nullapon."

chelate. See also sequestration. The type of compound or chemical union in which a central atom (most frequently a metal) is joined to others in the same molecule by both ordinary and coordinate valence forces. Thus in copper quinolinolate, a useful fungicide, the central copper atom is joined to two oxygen atoms by ordinary valence forces and to two nitrogen atoms by coordinate valences. Such linkages result in the formation of one or more heterocyclic rings in which the metal atom is part of the ring.

The metals in such compounds are usually not detectable by ordinary qualitative tests, and the formation of the properly chosen chelate compounds often serves as a sensitive test for metal ions in solution. Chelate formation is also applied practically in recovery or removal of traces of metal ions in industrial and biological processes.

chemical machining process.
1. The process of forming a three-dimensional photographic image within a piece of photosensitive glass, then dissolving it to leave a pattern whose shape is almost precisely complementary to that of the photographic image.
2. The phrase chemical machining is also applied to the use of etching or other types of chemical action to bring metal parts to desired dimensions with close tolerances.

chemical milling. See chemical machining process (2).

chemical red. A fine red pigment consisting, essentially, of red iron oxide. See also iron oxide reds and hematite, red.

chemical stoneware (brick, chemical). A clay pottery product which is widely employed to resist acids and alkalies. It is used for utensils, pipes, stopcocks, pumps, etc. Typical physical properties are as follows: Sp. gr. 2.2; hardness, scleroscope 100; ultimate tensile strength 2000 psi; ultimate compression strength 80,000 psi; modulus of rupture 5000 psi; modulus of elasticity 8×10^6 psi; sp. ht. 0.2; thermal conductivity 0.833 Btu/hr/sq. ft.; linear thermal expansion 2×10^{-6} per °F; H_2O absorption 0.4%.

Stoneware is made from special clays free from lime and iron, low in sand content, with low fire shrinkage; having the capacity of burning to a very dense body at low temperatures, and having sufficient plasticity to permit turning in a potter's wheel.

"Chemigum." [265] Trademark for a series of butadiene-acrylonitrile elastomers used for their exceptional resistance to oils and aromatic fuels. They are also known for their ease of processing and uniform high quality.

Properties: Characterized by low compression set, water absorption, and brittle point; high strength and elasticity when correctly compounded and vulcanized. Raw rubber has excellent bin aging characteristics and a mild, pleasant odor.

Uses: Gasoline hose; fuel cell interliners; oil resistant shoe soles; oil resistant automotive parts; hard packing compounds molded and extruded goods; abrasion resistant belt covers; friction compounds; gasket stocks; adhesives.

"Chemigum Latices." [265] Trademark for a series of butadiene-acrylonitrile latices possessing outstanding oil and solvent resistance characteristics. They are film forming and exhibit excellent adhesion and binding strength, mechanical stability, and ion tolerance.

Uses: Beater impregnation of paper; carpet backings; binders for nonwoven fabrics; adhesives; leather finishes; paper coatings.

chemiluminescence. The emission of absorbed energy (as light) due to a chemical reaction of the components of the system. It is normally considered to be an oxidation due to oxygen alone or to a compound which readily decomposes to form oxygen. An example is the oxidation of luminol in alcoholic solution. See luminescence.

"Chemipen." [412] Trademark for phenethicillin.

chemisorption. Chemisorption or chemical adsorption depends on chemical bond formation between the adsorbent and adsorbate, but in distinction from a chemical reaction, takes place only in a monolayer on the surface of the adsorbent as, for example, the adsorption of oxygen on a reactive iron catalyst. Chemisorption is thought to be essential in many catalytic reactions.

chemonite (copper arsenite, ammoniacal). A wood-preservative solution prescribed by Federal Specification TT-W-549 to contain copper hydroxide, $Cu(OH)_2$, 1.84%; arsenic trioxide, As_2O_3, 1.3%; ammonia, NH_3, 2.8%; acetic acid, 0.05%; water, as necessary to 100.0%.

"Chem-Rite A-22." [72] Trade name for chromate conversion coating for aluminum. Composed of hexavalent chromium compounds and inorganic activators. Produces light to iridescent gold color film on aluminum surface which increases corrosion resistance and provides base for paint and other organic coatings. Available for spray or brush coatings.

Containers: 1- and 5-gal packages; 55-gal drums.

"Chem-Rite C-55, Z-33." [72] Trade name for a single dip process for producing clear chromate coating on cadmium plated parts. Composed of hexavalent chromium compounds and inorganic activators. Z-33 is similar but for zinc plated parts.

"Chemtite." [333] Trade name for plating-rack enamels and stop-off coatings to prevent deposition of metal on the coated portion of articles to be plated.

chenopodium oil (wormseed oil, American; goosefoot oil).

Properties: Colorless or yellowish oil; characteristic penetrating, disagreeable odor; pungent, bitterish, burning taste. Soluble in 3 to 10 vols of 70% alcohol (inferior and adulterated oils do not yield a clear solution).

Chief known constituents: Ascaridole, $C_{10}H_{16}O_2$; para-cymene; ℓ-limonene.

Constants: Sp. gr. (good commercial oils) 0.965 to 0.990; (inferior oils), 0.93 to 0.965 (15°C); optical rotation −4° to −8° 50'; refractive index 1.4740-1.4790 (20°C).
Derivation: Distilled from the seeds and leaves of Chenopodium ambrosioides anthelminticum.
Containers: Cans.
Use: Medicine.
Shipping regulations: None.*

chenopodium oil, Levant. See wormseed oil, Levant.

cherry-bark oil, wild. An essential oil distilled from the bark of Prunus virginiana, Mill. It resembles oil of bitter almonds and consists largely of benzaldehyde and hydrocyanic acid.
Constants: Sp. gr. 1.045 to 1.050.
Uses: Medicine; aromatic beverages.
Shipping regulations: None.*

cherry bay. See cherry laurel leaves.

cherry juice. Liquid expressed from fresh ripe fruit of Prunus cerasus.
Properties: Clear liquid with aromatic, characteristic odor and sour taste. Affected by light. Color is red to reddish-orange. Sp. gr. 1.045-1.075; refractive index 1.3500; pH 3.0-4.0.
Grade: U.S.P. XVI.
Use: Flavoring agent for medicines.

cherry laurel leaves (English laurel; cherry bay). Fresh leaves of Prunus laurocerasus.
Habitat: Southeastern Europe; southwestern Asia.
Use: Flavoring agent similar to bitter almond, especially in the form of the water.

cherry-laurel oil.
Properties: Pale yellow liquid closely resembling oil of bitter almonds in its properties but distinguishable from it by its slightly different odor. Very poisonous! Hydrocyanic acid content 0.4 to 3.6% (rarely up to 8% and more).
Solubility in alcohol: In 2.5 to 4 vols of 60% alcohol (diminishing with age); in 1 to 2 vols of 70% alcohol.
Chief known constituents: Benzaldehyde, hydrocyanic acid, benzaldehyde cyanhydrin.
Constants: Sp. gr. 1.050-1.066 (occasionally) 1.0457 (15°C); optical rotation, usually inactive, but occasionally active +0°12' to −0°46'; refractive index 1.540 to 1.543; acid value 1.6-2.8.
Derivation: By macerating the leaves of Prunus laurocerasus, L. with water and then distilling with steam.
Contents: Bottles.
Uses: Flavoring; liqueurs.
Shipping regulations: None.*

chert. A fine-grained variety of native silica or quartz (q.v.). Similar to flint, but usually light in color. Fracture conchoidal, luster dull.

Chesney process. Method for producing magnesium metal and other magnesium products from sea water by precipitation with dolomitic lime as the means of separating the relatively small quantity of magnesium from the large volume of sea water.

chessylite. See azurite.

chi acid. See anthraquinone-1,8-disulfonic acid.

Chicago acid. See 8-amino-1-naphthol-5,7-disulfonic acid.

chicory root. Root of the plant Cichorium intybus.
Habitat: Cultivated in Europe.
Use: Substitute or adulterant for coffee; also used to impart flavor to coffee.

Chile niter. See caliche.

Chile nitrate. See caliche.

Chile saltpeter. See caliche.

chillies. See capsicum.

China bark. See quillaja.

China clay. See kaolin.

chinaldine. See quinaldine.

China oil. See Peru balsam.

China orange. See orange peel, sweet.

chinaphthol. See quinaphthol.

China-wood oil. See tung oil.

Chinese bean oil. See soybean oil.

Chinese blistering flies. See cantharides.

Chinese blue. See iron blues.

Chinese cinnamon. See cinnamon, cassia.

Chinese cinnamon oil. See cassia oil.

Chinese gelatin. See agar-agar.

Chinese isinglass. See agar-agar.

Chinese oil. See Peru balsam.

Chinese red. See chrome red.

Chinese rhubarb. See rhubarb.

Chinese tree wax. See Chinese wax.

Chinese vermilion. A name for red mercuric sulfide.

Chinese wax (insect wax; Chinese tree wax; vegetable spermaceti).
Properties: White to yellowish-white solid; nearly odorless and tasteless. Must not be confused with Japan wax (q.v.). Soluble in alcohol, chloroform, benzene, and naphtha. Insoluble in water.
Chief constituent: Ceryl cerotate.
Constants: Sp. gr. 0.970; m.p. 80-83°C; iodine number 1.4; saponification number 80-93.
Derivation: Secreted by an insect Coccus ceriferus. The wax is deposited on the branches of some trees and is removed by hand and then melted in boiling water to remove dirt, bark, etc.
Method of purification: Filtration.

Grades: Crude.
Containers: Burlap bags; wooden barrels; multiwall paper sacks.
Uses: Fine candles; coating interior and exterior of vegetable-tallow candles; medicine; paper size; furniture, leather, and shoe polishes; treating cotton fabrics.
Caution: Combustible.
Shipping regulations: None.*

Chinese white. See zinc oxide.

Chinese wood oil. See tung oil.

chinic acid. See quinic acid.

chinidine. See quinidine.

chiniofon. Mixture of 7-iodo-8-hydroxyquinoline-5-sulfonic acid ($C_9H_6NIO_4S$), its sodium salt, and sodium bicarbonate. Contains from 26.5-29.0% iodine.
Properties: Canary yellow powder with very slight odor; effervesces when moistened with water. Has bitter taste but leaves sweetish after-taste. Soluble in water; insoluble in alcohol, ether and chloroform.
Grade: N.F. XI.
Use: Medicine.

chinoidine. See quinoidine.

chinoline. See quinoline.

chinone. See quinone.

"Chip-Cal." [147] Trademark for low-lime calcium arsenate. Available in granular (48% tricalcium arsenate) and powder (85% tricalcium arsenate) form.
Containers: (granular) 18-, 36-lb bags; (powder) 50-lb bags.

"Chipcote." [147] Trademark for series of organic mercury seed treatments based on methyl mercury nitrile. Available as "Chipcote 25", a concentrate for use in slurry treaters and "Chipcote 75" for use undiluted in mist type and ready-mix treaters.
Containers: 5-, 30- and 55-gal drums.
Uses: Small grains, flax and cotton.

"Chip-Kil." [147] Brand name for a general purpose insecticide consisting of toxaphene, DDT and methyl parathion. "Chip-Kil S" contains strobane rather than toxaphene.
Containers: 5-, 30- and 55-gal drums.

chip sulfide. See sodium sulfide.

chitin. A glucosamine polysaccharide. The horny substance which is the principal constituent of the shells of crabs, lobsters, and beetles. It is also found in some fungi and bacteria.
Properties: White, amorphous, semitransparent mass; insoluble in the common solvents; soluble in concentrated hydrochloric, nitric, and sulfuric acids.
Use: Biological research.

chitinase.
Derivation: An enzyme from the puff ball, Calvatia gigantea.
Uses: Destroys fungi, such as black mold (Aspergillus niger), by breaking down chitin in cell walls.

chittem bark. See cascara sagrada bark.

chittim bark. See cascara sagrada bark.

chloanthite $(Ni, Co)As_2$. A natural nickel-cobalt arsenide, with nickel in excess of cobalt. Grades into smaltite (q.v.) and skutterudite, and has similar properties.

chlor-. See, preferably, chloro-.

chloracetic acid. See chloroacetic acid.

chloracetone. See chloroacetone.

chloracetophenone. See chloroacetophenone.

chloracetyl chloride. See chloroacetyl chloride.

chloral (trichloroacetic aldehyde) CCl_3CHO.
Properties: Colorless, mobile, oily liquid; penetrating odor.
Constants: Sp.gr. 1.5121; m.p. −57.5°C; b.p. 97.7°C; vapor pressure 35 mm (20°C); index of refraction (n 20/D) 1.4557; flash point 167°F; latent heat of vaporization 97.1 Btu/lb.
Soluble in alcohol, ether and chloroform; combines with water forming chloral hydrate.
Derivation: (a) By the chlorination of ethyl alcohol, addition of sulfuric acid, and subsequent distillation; (b) by the chlorination of acetaldehyde.
Method of purification: Rectification.
Grades: Technical.
Containers: 5-, 55-gal iron drums; glass bottles; tankcars.
Danger: Extremely hazardous liquid and vapor. Inhalation may cause delayed fatal lung injury. MCA warning label.
Uses: Organic synthesis; chloral hydrate; manufacture of DDT; and in liniments.
Shipping regulations: None.*

chloralamide. See chloral formamide.

chloral formamide (chloralamide; formamidated chloral) $CCl_3CHOHNHOCH$.
Properties: Colorless, lustrous crystals; odorless; slightly bitter taste. Soluble in water (hydrolyzes at 60°C), alcohol, ether and glycerol.
Constants: M.p. 114-115°C; decomposes at higher temperatures.
Derivation: Interaction of formamide and anhydrous chloral.
Containers: Amber-colored bottles.
Use: Medicine.
Shipping regulations: None.*

chloral hydrate ("knockout drops"; trichloroacetic aldehyde, hydrated) $CCl_3CH(OH)_2$.
Properties: Transparent, colorless crystals; aromatic, penetrating, slightly acrid odor and slightly bitter, caustic taste; poisonous! Slowly volatilizes when exposed to air. Soluble in water, alcohol, chloroform, and ether; also soluble in olive oil and turpentine oil.
Constants: Sp.gr. 1.901; m.p. 52°C; b.p. 97.5°C.
Derivation: By the action of $\frac{1}{5}$ of its volume

*See "I.C.C. Shipping Regulations," page xiii.
Reference numbers refer to name of manufacturer. See "List of Manufacturers," page v.

of water on chloral.
Method of purification: Crystallization.
Grades: Technical; U.S.P. XVI.
Containers: 20-, 40-lb bottles; jars.
Uses: Medicine; liniments.
Shipping regulations: None.*

chloral hydrate antipyrine (antipyrine chloral hydrate; hypnal) $C_{11}N_{12}N_2OCl_3CH(OH)_2$.
Properties: Colorless crystals; moderately soluble in water; soluble in alcohol; m.p. 67°C.
Derivation: By mixing antipyrine with hydrated chloral and crystallizing.
Use: Medicine.

chloralimide CCl_3CHNH.
Properties: Colorless, long crystalline needles; odorless. Not to be confused with chloral formamide (q.v.). Soluble in alcohol, ether, chloroform and oils; insoluble in water; m.p. 150-155°C.
Derivation: Heating chloral formamide on steam bath.
Use: Medicine.
Shipping regulations: None.*

chlorambucil (4-{para[bis(2-chloroethyl)-amino]phenyl} butyric acid) $(ClC_2H_4)_2NC_6H_4(CH_2)_3COOH$. A nitrogen mustard derivative.
Properties: Flattened needles; melts at 64-66°C. Soluble in ether.
Grade: N.N.D.
Use: Medicine.

chloramine
1. NH_2Cl. A colorless, unstable, pungent liquid; soluble in water; decomposes (slowly in dilute solution) to form nitrogen, hydrochloric acid, and ammonium chloride. Chloramine is an intermediate in the Raschig process for hydrazine (q.v.).
2. Also used as a synonym for chloramine-T.

chloramine-B. $C_6H_5SO_2NClNa$ (sodium benzenesulfonchloramine).
Properties: White powder with faint chlorine odor; soluble in water.

chloramine-T (sodium para-toluenesulfonchloramine) $CH_3C_6H_4SO_2NNaCl \cdot 3H_2O$. See also dichloramine-T.
Properties: White or slightly yellow crystals or crystalline powder. Contains not less than 11.5 nor more than 13% active chlorine. Slight odor of chlorine. Decomposes slowly in air, liberating chlorine. Not to be confused with NH_2Cl, which is also termed chloramine. Toxic if introduced in blood stream. Soluble in water; more soluble in boiling water; insoluble in benzene, chloroform, ether; decomposed by alcohol.
Derivation: By the action of ammonia on toluene-para-sulfonic chloride under pressure. The toluene para-sulfonamide produced is subjected to the action of sodium hypochlorite in the presence of an alkali and the chloramine produced by crystallization.
Purification: Crystallization.
Use: Medicine.
Shipping regulations: None.*

chloramphenicol
$NO_2C_6H_4CH(OH)CH(CH_2OH)NHCOCHCl_2$
D(-)Threo-1-(para-nitrophenyl)-2-dichloroacetamido-1,3-propandiol. An antibiotic derived from Streptomyces venezuelae or by organic synthesis. Effective against certain gram-negative organisms and rickettsia. It is now used mainly in treatment of typhoid and other diseases caused by organisms resistant to other antibiotics but controlled by chloramphenicol. It was the first substance of natural origin shown to contain aromatic nitro group.
Properties: Fine, white to grayish-white or yellowish-white, needlelike crystals or elongated plates. Bitter to taste, neutral to litmus, and reasonably stable in neutral or slightly acid solutions. M.p. 149-153°C; alcoholic solution is dextrorotatory while ethyl acetate solution is levorotatory. Very slightly soluble in water; freely soluble in alcohol, propylene glycol, acetone and ethyl acetate.
Derivation: Biological, by aerobic fermentation on wheat gluten medium. The filtrate is extracted with amyl acetate and purified by vacuum evaporation and crystallization. Synthetic, by a multistep batch process starting with para-nitrobromoacetophenone.
Grade: U.S.P. XVI.
Use: Medicine; feed supplement.

chloramphenicol palmitate
$C_{27}H_{42}Cl_2N_2O_6$. An antibiotic.
Properties: Fine, white, unctuous, crystalline powder; m.p. 86-92°C; faint odor; bland, mild taste. Insoluble in water; very slightly soluble in solvent hexane. Soluble in ether, acetone, chloroform; sparingly soluble in alcohol.
Grade: U.S.P. XVI.
Use: Medicine.

chloranil (tetrachloroquinone; tetrachloro-para-benzoquinone) $C_6Cl_4O_2$.
Properties: Yellow leaflets; m.p. 290°C; soluble in alcohol, ether, and benzene; insoluble in water.
Derivation: From phenol, para-chlorophenol, or para-phenylenediamine by treatment with potassium chlorate and hydrochloric acid.
Grades: Technical.
Containers: Iron drums.
Uses: Agricultural fungicide; dye intermediate; electrodes for pH measurements.
Caution: Avoid contact with skin. MCA warning label.
Shipping regulations: None.*

chloranthrene yellow. See flavanthrene.

chlorapatite. See apatite.

chlorargyrite. See cerargyrite.

"Chlorasol." [214] Trademark for a proprietary fumigant composition: 70.3% ethylene dichloride, 29.7% carbon tetrachloride

*See "I.C.C. Shipping Regulations," page xiii.
Reference numbers refer to name of manufacturer. See "List of Manufacturers," page v.

by weight.

Properties: Boiling range (760 mm) 75-78°C; sp.gr. 1.334-1.339 (20/20°C); 11.11 lb/gal (20°C); no flash.

Containers: 1-gal can; 5-, 55-gal drums; tankcars up to 10,000 gal.

Uses: A fumigant for meal, grain and clothes moths; grain weevils; grain, flour, and carpet beetles; the rice weevil; and book lice; as well as their larvae and eggs. Non-flammable under ordinary conditions of use; penetrates stored grain, rolled rugs, upholstered furniture, cartons, sacks, and stacked material.

chlorauric acid. See gold chloride.

"Chlorax." [147] Trademark for a weed and grass killer composed of sodium chlorate and sodium metaborate. Does not create fire hazard.

Containers: Dry: 4-lb cans; 50-lb bags; liquid: 5- and 55-gal drums.

Uses: To kill deep-rooted perennials and annual weeds and grasses.

"Chlorazol." [206] Brand name of line of direct dyes for cotton, silk, rayon, leather and paper.

chlorazotic acid. See aqua regia.

chlorbenside (para-chlorobenzyl para-chlorophenyl sulfide) $ClC_6H_4CH_2SC_6H_4Cl$. An agricultural toxicant used specifically as a miticide. Accepted as a generic name by Ent. Soc.

Properties: Crystals with almond-like odor (technical grades); m.p. 75-76°C; insoluble in water; soluble in most organic solvents; resistant to acid and alkaline hydrolysis.

Grade: Technical.

Use: Miticide for control of eggs and larvae of red spider mites.

Warning! May be irritating to skin.

chlorbenzene. See chlorobenzene.

chlorbenzol. See chlorobenzene.

chlor- compounds. See chloro- compounds.

chlorcyclizine hydrochloride $ClC_6H_4CH(C_6H_5)C_4H_8N_2CH_3 \cdot 2HCl$. 1-(para-Chloro-alpha-phenylbenzyl)-4-methylpiperazine dihydrochloride, an antihistaminic.

Properties: White, odorless, crystalline solid with bitter taste; m.p. 222-227°C; very soluble in water; soluble in chloroform and alcohol; practically insoluble in benzene and ether. Solutions acid to litmus; pH (1 in 100 solution) 4.8-5.5.

Grade: U.S.P. XVI.

Use: Medicine.

chlordan(e) $C_{10}H_6Cl_8$ (1,2,4,5,6,7,8,8-octachloro-4,7-methano-3a,4,7,7a-tetrahydroindane). Chlordan has been accepted as a generic name by Ent. Soc.

Typical specification: Colorless, odorless, viscous liquid; sp.gr. 1.57-1.67 (60/60°F); viscosity SSU 100 seconds (38°C); organic chlorine 64-67% by wt, purity 98%; b.p. 175°C (2 mm); refractive index (n 25/D) 1.56-1.57; soluble in many organic solvents; insoluble in water; miscible in deodorized kerosene; decomposes in weak alkalies.

Grades: Technical and pure.

Containers: Aluminum, aluminum-clad or in high-bake phenolic enamel-lined metal containers; 20-, 45-lb pails; 5-, 55-gal drums.

Uses: Insecticide; in oil emulsions, dusts and dispersible liquids; effective against aphids, squash bug, mosquito larvae, house fly, roaches.

Caution! Harmful if swallowed; absorbed through skin. MCA warning label.

chlordiazepoxide hydrochloride (7-chloro-2-methylamino-5-phenyl-3H-1,4-benzodiazepine 4-oxide hydrochloride) $C_{16}H_{14}ClN_3O \cdot HCl$.

Properties: Crystals; m.p. 212-218°C; soluble in water; sparingly soluble in alcohol; insoluble in ether and chloroform.

Use: Medicine.

"Chlorea." [147] Trademark for a non-selective weed and grass killer formulated with sodium chlorate and 3(para-chlorophenyl)-1,1-dimethylurea. Used for control of weeds and grass around buildings, telephone poles, parking areas, etc. Does not create fire hazard.

Containers: 50-lb bags.

chlorendic acid (hexachloroendomethylenetetrahydrophthalic acid) $C_9H_4Cl_6O_4$. 1,4,5,6,7,7-Hexachlorobicyclo-(2,2,1)-5-heptane-2,3-dicarboxylic acid.

Properties: Fine, white, free-flowing crystals. M.p. 208-210°C (sealed tube); loses water about 200°C. Forms a crystalline monohydrate. Very soluble in alcohol; slightly soluble in water, benzene, and carbon tetrachloride. Stable chlorine over 54% by weight.

Derivation: Hydrolysis of chlorendic anhydride.

Grades: Technical.

Containers: 50-, 250-lb fiber drums.

Uses: Fire-resistant polyester resins and paints; plasticizers; intermediate for dyes, fungicides, and insecticides.

chlorendic anhydride (hexachloroendomethylenetetrahydrophthalic anhydride) $C_9H_2Cl_6O_3$. 1,4,5,6,7,7-Hexachlorobicyclo-(2,2,1)-5-heptene-2,3-dicarboxylic anhydride.

Properties: Fine, white, free-flowing crystals. M.p. 239-240°C; sp.gr. 1.73. Readily soluble in acetone, benzene, toluene; slightly soluble in water, n-hexane, and carbon tetrachloride.

Derivation: By Diels-Alder reaction of maleic anhydride and hexachloropentadiene.

Grades: Technical.

Containers: 40-, 250-lb steel drums.

Uses: Flame-resistant polyester resins; hardening epoxy resins; chemical intermediate.

"Chloretone." [330] Trademark for chlorobutanol.

"Chlorex." [214] Trademark for 2,2'-dichloroethyl ether (q.v.).

"Chlorextol." [116] Trademark for a synthetic, non-flammable, non-sludging insulating liquid of high dielectric strength; can be used in place of oil if desired.

chlorhydrins. See chlorohydrins.

chlorhydrol. Aluminum chlorohydroxide complex $[Al_2(OH)_5Cl]$; pH 3.9-4.6. Freely soluble in water, solution stable.

chloride of lime. See chlorinated lime.

chloridizing. Heating in the presence of chlorine, as a step in the recovery of certain metals from their oxides or other compounds.

"Chlorimets." [47] A series of nickel-base cast alloys.

"Chlorimet" 2 contains 32% molybdenum, 3% iron max, 1% silicon and 0.10% carbon. Resistant to all concentrations of boiling hydrochloric acid, sulfuric acid concentrations between 30 and 93% at 80°C and phosphoric acid at very high temperatures. Shows resistance to other acids, bases and salts except those of a highly oxidizing nature.

"Chlorimet" 3 contains 18% molybdenum, 18% chromium, 3% iron max, 1.0% silicon and 0.07% carbon. Resistant to sulfuric acid concentrations up to 10% at boiling and all concentrations at moderate temperatures, hydrochloric acid and moist chlorine at intermediate temperatures and phosphoric acid at high temperatures. It also shows resistance to acetic, certain nitric acid solutions, hydrofluoric acid, bleach solutions, sodium and ammonium hydroxides and most other acids, bases and salts, particularly under oxidizing conditions.

chlorinated camphene. See toxaphene.

chlorinated isocyanuric acids. See dichloro- or trichloroisocyanuric acid, and potassium or sodium dichloroisocyanurate. Used as dry bleaches.

chlorinated lime (chloride of lime; bleaching powder); approximately $CaCl(ClO)\cdot 4H_2O$.

Properties: White powder; chlorine odor; m.p. (dec); decomposes in water, acids.

Derivation: By conducting chlorine into a box-like structure containing slaked lime spread upon perforated shelves.

Grades: 35-37% active chlorine; technical.

Containers: 10-lb cans; 58-, 100-, 130-, 300-, 325-, 453-, 800-lb steel drums; 415-lb wooden barrels.

Uses: Textile bleaching and numerous bleaching applications; organic synthesis; deodorizer; disinfectant.

Caution: Not combustible but evolves chlorine and at higher temperatures oxygen. With acids or moisture evolves chlorine freely at ordinary temperatures.

Shipping regulations: Dry, containing more than 8.80% available oxygen (39% available chlorine): oxidizing material. Yellow label.*

See also calcium hypochlorite.

chlorinated naphthalene. See chloronaphthalene.

chlorinated paraffin (chlorocosane). Light yellow to light amber liquid produced by chlorinating a paraffin oil. Typical average formula: $C_{24}H_{43.1}Cl_{6.9}$. Used as solvent for dichloramine-T, as a high pressure lubricant, and for fire-proofing textiles.

Paraffin wax which has been chlorinated is also called chlorinated paraffin.

Grades: 40%; 70%.

Containers: Drums.

chlorinated para red. A modification of para red that contains some chlorine. Much lighter than para or toluidine red and has excellent brilliancy but poorer heat resistance.

chlorinated polyether. See "Penton."

chlorinated polyolefins. See chlorinated rubber.

chlorinated polypropylene. A film-forming polymer used in coatings, inks, adhesives and paper coatings.

chlorinated rubbers (chlorinated polyolefins). Natural rubber, or more often, polyolefins, to which fairly large amounts of chlorine are added (up to 65% or more) in order to modify the properties of the elastomer. Some examples are: "Parlon," a chlorinated natural rubber, "Hypalon," a polyethylene, "Parlon P," an isotactic polypropylene, and "Butyl HT," a butyl rubber copolymer.

chlorinated soda. See Javelle water.

chlorinated tar camphor. See chloronaphthalenes.

chlorinated trisodium phosphate. See chlorinated TSP.

chlorinated TSP (chlorinated trisodium phosphate) $4(Na_3PO_4\cdot 11H_2O)\cdot NaOCl$. Active ingredients 3.25% min sodium hypochlorite and 91.75% min trisodium phosphate dodecahydrate. Inert ingredient less than 5% NaCl.

Properties: White crystalline water-soluble material stable under normal storage conditions. In solution has the properties of both trisodium phosphate and sodium hypochlorite.

Derivation: By reacting sodium phosphate, caustic soda and sodium hypochlorite.

Containers: Bags; drums; and bulk.

Uses: Cleaner and bactericide in dairies, food plants; dish washing compounds and scouring powders.

chlorine Cl. Element of atomic number 17; group VII of the periodic table.

Properties: Heavy, greenish-yellow gas, two and one-half times as heavy as air; or, clear amber liquid, one and one-half times as heavy as water; pungent, irritating odor. Caution! Poison! Soluble in water and alkalies.

Constants:

Gas: Liquefaction point 6 to 8 atm

*See "I.C.C. Shipping Regulations," page xiii.

Reference numbers refer to name of manufacturer. See "List of Manufacturers," page v.

(ordinary temperatures) or −34.1°C (760 mm); critical temperature 144°C; specific heat 0.115 cal/g/°C (15°C).
Liquid: Sp. gr. 1.4685 (0°C); b.p. −34.5°C; m.p. −101°C; vapor tension 6.62 atm (20°C); coefficient of expansion 0.00212 (20°C); latent heat of vaporization 68.8 cal/g (−34.1°C); liquid-gas ratio: 1 liter of liquid = 456.8 liters of gas (0°C and 760 mm); specific heat 0.226 cal/g/°C (0-24°C); heat of fusion 22.9 cal/g (−102°C); specific volume 5.4 cu ft/lb (70°F).
Derivation: Principally, by the electrolysis of sodium chloride brine in diaphragm cells or mercury cathode cells (q.v.). Chlorine is released at the positive electrode. Economically there is interest in the production of chlorine without the attendant sodium hydroxide. This is attained by the electrolysis of fused chlorides (sodium, potassium, lithium, or magnesium) and by the several chemical processes, the more important of which are: (a) the reaction of sodium chloride and nitric acid, (b) the oxidation of hydrogen chloride with air by means of ferric oxide-potassium chloride catalyst, and (c) the electrolysis of cupric chloride solution. In the latter process, the cupric chloride is regenerated with air and hydrochloric acid so that the over-all result is the oxidation of hydrochloric acid.
Grades: Technical (both gas and liquid); pure (99.5% min).
Containers: 100-, 150-lb steel cylinders; single unit tank cars of 30,000 lbs; multi-unit tank cars of 15 one-ton units (for liquid).
Uses (in approximate order of volume): Manufacture of chemicals which do not contain chlorine (ethylene glycol, tetraethyl lead, ethylene oxide); manufacture of chlorine-containing chemicals, broken down further as: (a) solvents (trichloroethylene, perchloroethylene, methylene chloride), (b) pesticides and herbicides (DDT, benzene hexachloride, toxaphene), (c) plastics and fibers (vinyl chloride, vinylidene chloride, "Kel-F"), (d) refrigerants and propellants ("Freons," "Genetrons," methyl chloride); pulp and paper; water and sewage treatment; textile bleaching; degassing of aluminum melts.
Danger: Hazardous liquid and gas under pressure; avoid breathing air containing gas; avoid contact with skin or eyes. MCA warning label.
Shipping regulations: Nonflammable gas. Green label.*

chlorine 36. Radioactive chlorine of mass number 36. Half-life about 440,000 years; radiation, beta; radiotoxicity, moderately hazardous.
Derivation: Separated from various isotopes produced during the pile irradiation of potassium chloride.
Forms available: As hydrochloric acid solution and as solid potassium chloride.
Uses: As a tracer in studying the salt water corrosion of metals, especially steel; the reaction mechanisms of chlorinated hydrocarbons; the location and flow of salt waters in porous media; etc.
Shipping regulations: Class D poison, radioactive material. Blue label.*

chlorine-bromide. See bromine-chloride.

chlorine dioxide ClO_2.
Properties: Dangerous! Explosive. Red-yellow gas; b.p. 10°C; very reactive, unstable. Soluble and decomposed in water. Dissolves in alkalies forming a mixture of chlorite and chlorate.
Derivation: Usually made at point of consumption from sodium chlorate, sulfuric acid and methanol, or from sodium chlorate and sulfur dioxide. Concentration of gas is limited to 10% to reduce explosion hazard.
Grades: Sold as hydrate, in frozen form.
Uses: Bleaching wood pulp, fats, oils and flour; bleaching and removing tastes and odors from water supplies and in swimming pools; odor control; maturing flour.
Shipping regulations (frozen hydrate): Oxidizing material. Yellow label. Not accepted for express.*

chlorine trifluoride ClF_3.
Properties: Nearly colorless gas or pale green liquid. B.p. 11.3°C; f.p. −83°C. Extremely reactive, comparable to fluorine. Reactions with organic compounds and with water take place with explosive violence.
Derivation: By reaction of chlorine and fluorine at 280°C and condensation of the product at −80°C. Obtained 99.0% pure.
Containers: Steel cylinders.
Uses: Incendiary; fluorinations; cutting oil well tubes; oxidizer in propellants.
Shipping regulations: Corrosive liquid. White label.*

chlorine water. A clear yellowish liquid, deteriorates on exposure to air and light. Made by saturating water with approximately 0.4% chlorine.
Use: Deodorizer, disinfectant, and also used medicinally.

chloriodized oil. Chlorinated and iodized vegetable oil. Contains 26.0-28.0% iodine in organic combination.
Properties: Pale yellow, viscous, oily liquid with faint, bland taste. Practically insoluble in water; slightly soluble in alcohol; freely soluble in benzene, chloroform and ether.
Derivation: Formed by chemical addition of iodine monochloride to oil.
Grade: U.S.P. XVI.
Use: Medicine.

chlorisondamine chloride $C_{14}H_{20}Cl_6N_2$.
4,5,6,7-Tetrachloro-2-(2-dimethylaminoethyl) isoindoline dimethylchloride. A quaternary ammonium compound.
Properties: Crystals; decompose 258-265°C. Soluble in water and alcohol.
Grade: N.N.D.
Use: Medicine.

"Chlor Kil." [55] Brand name for chlordane-based insecticides.

*See "I.C.C. Shipping Regulations," page xiii.
Reference numbers refer to name of manufacturer. See "List of Manufacturers," page v.

chlormerodrin ([3-chloromercuri)-2-methoxypropyl]urea) $ClHgCH_2CH(OCH_3)CH_2NHCONH_2$.
Properties: White, odorless powder with bitter metallic taste. Very soluble in sodium hydroxide; very slightly soluble in chloroform; slightly soluble in alcohol, methanol, and water. Stable to light and air; pH (0.5% solution) 4.3-5.0.
Grade: N.N.D.
Use: Medicine.

chlormethazanone $C_{11}H_{12}ClNO_3S$. 2-(4-Chlorophenyl)-3-methyl-4-meta-thiazanone-1-dioxide.
Properties: Crystals; m.p. 117°C; insoluble in water; slightly soluble in alcohol.
Use: Medicine.

chloroacetaldehyde $ClCH_2CHO$.
Properties (of 40% aqueous solution): Clear, colorless liquid with pungent odor. Boiling range 90-100°C; f.p. −16.3°C; sp.gr. 1.19 (25/25°C); refractive index 1.397 (25°C); wt/gal 9.9 lb (25°C). Soluble in water, acetone, methanol. At concentrations in water above 50% it forms an insoluble hemihydrate.
Uses: Chemical intermediate; fungicide; debarking logs or trees.

chloroacetaldehyde dimethyl acetal. See dimethyl chloroacetal.

chloroacetic acid (chloracetic acid; monochloroacetic acid) $CH_2ClCOOH$.
Properties: Colorless to light-brownish crystals, very deliquescent and caustic to the skin.
Constants: Sp.gr. 1.370 (70°C); crystallizing point, alpha form, 61.0 - 61.7°C; beta form, 55.5 to 56.5°C; gamma form, 50°C. The commercial material melts at 61 to 63°C; boiling range 186 to 191°C. Soluble in water, alcohol and ether.
Derivation: By the action of chlorine on acetic acid in the presence of acetic anhydride, phosphorus, or sulfur.
Grades: Technical; medicinal; reagent.
Containers: Drums.
Uses: Herbicide; production of carboxymethylcellulose, ethyl chloroacetate, glycine, synthetic caffeine, sarcosine, vitamins, EDTA, 2,4-D.
Shipping regulations: Corrosive liquid. White label. (Legal label name monochloroacetic acid.)*

chloroacetic anhydride (chloroethanoic anhydride) $(ClCH_2CO)_2O$.
Properties: Colorless to slightly yellow crystals with pungent odor; m.p. 51-55°C; soluble in chloroform and ether; hydrolyzes with water to chloroacetic acid.
Use: Intermediate.

ortho-**chloroacetoacetanilide** $CH_3COCH_2CONHC_6H_4Cl$.
Properties: White, crystalline solid. Resembles ethyl acetoacetate in chemical reactivity.
Constants: M.p. 107°C; vapor pressure 0.1 mm (20°C); flash point, none. Insoluble in water.
Typical specifications: Density 35 lbs/cu ft; m.p. 103 to 106°C.
Grades: Technical.
Containers: Fiber cartons. Net content 4, 20, 175 lbs.
Uses: Organic synthesis; dyestuffs.

chloroacetone (monochloroacetone; 1-chloro-2-propanone; chloracetone; chlorinated acetone) CH_3COCH_2Cl.
Properties: Colorless liquid; pungent irritating odor. Sp.gr. 1.162; b.p. 119°C; m.p. −44.5°C. Soluble in alcohol, ether and chloroform and water.
Derivation: By the chlorination of acetone.
Method of purification: Rectification.
Grades: Technical.
Containers: Iron drums.
Toxicity: Strong irritant.
Uses: Couplers for color photography; enzyme inactivator; insecticides; perfumes; antioxidant intermediate; medicine; organic synthesis.
Shipping regulations: Stabilized: poison; Class C; tear gas label; unstabilized: not accepted.*

chloroacetonitrile (chloroethane nitrile) $ClCH_2CN$.
Properties: Sp.gr. 1.2020-1.2035 (25/25°C); refractive index 1.4210-1.4240 (n 25/D); soluble in hydrocarbons, alcohols; insoluble in water.
Uses: Fumigant; intermediate.

chloroacetophenone (chloracetophenone; phenacylchloride; phenylchloromethylketone) $C_6H_5COCH_2Cl$. Two forms exist, designated as omega and para.
Properties: White crystals; m.p. (omega) 59°C; (para) 20°C; b.p. (omega) 245°C, (para) 236°C; soluble in alcohol and ether; insoluble in water.
Derivation: Action of chloroacetylchloride on benzene in presence of aluminum chloride.
Method of purification: Distillation and recrystallization.
Grades: C.P.; technical.
Containers: Glass bottles.
Use: War gas (lachrymator).
Shipping regulations: Class C poison; tear gas label. Legal label name: chloracetophenone.*

chloroacetyl chloride (chloracetyl chloride) $CH_2ClCOCl$.
Properties: Water-white liquid; pungent odor.
Constants: Sp.gr. 1.495 (0°C); b.p. 105 to 110°C; decomposes in water.
Derivation: (a) By the action of chlorine on acetyl chloride in sunlight. (b) By dropping phosphorus trichloride on chloroacetic acid.
Method of purification: Distillation.
Containers: Glass bottles; carboys.
Uses: Preparation of chloroacetaphenone (tear gas); chemical intermediate.
Shipping regulations: Corrosive liquid; white label.*

chloroacetylurethane $ClCH_2CONHCOOC_2H_5$.
Properties: Crystals; soluble in alcohol;

*See "I.C.C. Shipping Regulations," page xiii.
Reference numbers refer to name of manufacturer. See "List of Manufacturers," page v.

sparingly soluble in water. M.p. 129°C.
Derivation: By interaction of sodium urethane and ethyl chloroacetate.

chloroacrolein $H_2C:CClCHO$.
Properties: Colorless liquid. Caution! Very irritant!
Constants: Sp.gr. 1.205 at 15°C; b.p. 29 to 31°C (17 mm Hg).
Derivation: Chlorination of acrolein.

chloroamino- See aminochloro-.

para-**chloro**-ortho-**aminophenol.** See 2-amino-4-chlorophenol.

2-chloro-4-tert-**amylphenol** $C_5H_{11}C_6H_3ClOH$.
Properties: Sp.gr. 1.11 (20°C); boiling range 253-265°C; color, water-white; odor aromatic. Flash point 225°F.

meta-**chloroaniline** (meta-aminochlorobenzene) $ClC_6H_4NH_2$.
Properties: Colorless to light amber liquid; tends to darken during storage. Boiling range 228-231°C; f.p. −10.6°C.
Grades: Technical.
Containers: 525-lb drums; tankcars.
Uses: Intermediate for azo dyes and pigments; pharmaceuticals; insecticides; agricultural chemicals.
Caution! Toxic! Avoid breathing vapors or contact with skin.
Shipping regulations: Poison, class B. Poison label.*

ortho-**chloroaniline** (ortho-aminochlorobenzene) $ClC_6H_4NH_2$.
Properties: Amber liquid; amine odor; darkens on exposure to air. Distillation range 208-210°C; f.p. −2.3°C; sp.gr. 1.213 (20/4°C); refractive index 1.5896 (n 20/D). Miscible with alcohol and ether; insoluble in water.
Grades: Technical.
Containers: 550-lb drums; tankcars.
Use: Dye intermediate; standards for colorimetric apparatus; manufacture of petroleum solvents and fungicides.
Caution! Toxic! Avoid breathing vapor or contact with skin.
Shipping regulations: Poison, class B. Poison label.*

para-**chloroaniline** (para-aminochlorobenzene) $ClC_6H_4NH_2$.
Properties: White or pale yellow solid. M.p. 69.5°C; distilling range 229-233°C.
Grades: Technical.
Containers: 500-lb drums.
Use: Dye intermediate; pharmaceuticals; agricultural chemicals.
Caution! Toxic! Avoid breathing vapor or contact with skin.
Shipping regulations: Poison, class B. Poison label.*

4-chloroaniline-3-sulfonic acid $HSO_3C_6H_3ClNH_2$.
Properties: White to light-grey powder.
Containers: Fiber kegs; polythene-lined steel drums.
Use: Intermediate.

2-chloroanthraquinone $C_{14}H_7ClO_2$.
Properties: Resemble in general those of anthraquinone and 2-methylanthraquinone. M.p. 208-211°C. Insoluble in water; soluble in hot benzene.
Derivation: Prepared by condensing phthalic anhydride and chlorobenzene in the presence of anhydrous aluminum chloride to form para-chlorobenzoylbenzoic acid. Ring closure of the intermediate acid is brought about by heating in sulfuric acid solution.
Containers: Similar to those used for anthraquinone and 2-methylanthraquinone.
Use: Starting material for certain vat dyes.
Shipping regulations: None.*

chloroazodin $H_2NC(NCl)NNC(NCl)NH_2$.
alpha, alpha'-Azobis(chloroformamidine).
Properties: Bright yellow needles or flakes; faint odor of chlorine; slightly burning taste; sparingly soluble in alcohol; very slightly soluble in water; slightly soluble in glycerol and in glyceryl tri-acetate; very slightly soluble in chloroform. Solutions in glycerol and alcohol decompose rapidly on warming; all solutions decompose on exposure to light. Decomposes explosively at about 155°C; decomposition accelerated by contact with metals.
Use: Medicine.

chloroazotic acid. See aqua regia.

chlorobenzal. See benzyl dichloride.

ortho-**chlorobenzaldehyde** C_6H_4CHOCl.
Properties: Colorless to yellowish liquid; boiling range (typical) first drop 209°C, 50% 210.8°C, dry 215°C; f.p. 8.0°C (min); sp.gr. 1.240-1.245 (25/15°C). Soluble in alcohol, ether and acetone. Insoluble in water.
Containers: Glass carboys; 500-lb drums.
Uses: An intermediate in the preparation of triphenyl methane and related dyes, and for the synthesis of many organic chemicals such as pharmaceuticals and medicinals.

para-**chlorobenzaldehyde** C_6H_4CHOCl.
Properties: Colorless to yellowish powder; boiling range (typical) first drop 210°C, 50% 214°C; dry 220°C; f.p. 43°C (min). Soluble in alcohol, ether, and acetone; slightly soluble in toluene; insoluble in water and heptane.
Containers: 500-lb fiber drums.
Uses: An intermediate in the preparation of triphenyl methane and related dyes, and for the synthesis of organic chemicals such as pharmaceuticals and medicinals.

3-chloro-4-benzamido-6-methylaniline $ClC_6H_2NH_2CH_3(NHCOC_6H_5)$.
Properties: White solid; m.p. 198-199°C.
Uses: Azoic dyes; pigments.

chlorobenzanthrone $C_{17}H_9ClO$.
Properties: All isomers: yellow needles. Soluble in alcohol, benzene, toluene, acetic acid, etc.
Derivation: (a) From benzanthrone in acetic acid solution by treatment with chlorine.
(b) From benzanthrone in water suspension

by treatment with chlorine.
Method of purification: Crystallization from acetic acid.

chlorobenzene (chlorobenzol; chlorbenzene; chlorbenzol; phenyl chloride) C_6H_5Cl.
Properties: Clear, colorless, mobile, volatile, flammable liquid. Almond-like odor. Said to be mildly narcotic and is reputed to be less toxic than benzene. Sp. gr. 1.105 (25/25°C); b.p. 131.6°C; f.p. −45°C; wt/gal 9.19 lbs (25°C); refractive index 1.5216 (25°C); flash point 29°C; fire point 36°C; heat of vaporization 77.6 cal/g (b.p.); specific heat 0.30 cal/g/°C; dielectric constant 5.53 (1000 cycle); specific resistivity 7.8 x 10^9 ohms/cm. Miscible with most organic solvents; insoluble in water.
Typical specification: Boiling range within 1°C of b.p. 131.6°C (760 mm).
Derivation: (a) By passing dry chlorine into benzene to which a small aluminum-mercury couple is added as a carrier. (b) By passing chlorine into benzene in the presence of molybdenum chloride.
Grades: Technical.
Containers: 5-, 10-, 55-gal drums; tank cars.
Uses: Solvent for ethylcellulose, resins, paints, varnishes, lacquers; solvent mixtures; synthesis of phenol, DDT, aniline, picric acid, chloro- and nitrochlorobenzenes, intermediates, sulfur dyestuffs, military poison gases, drugs, perfumes; heat transfer medium; solvent in production of rubber, resins, drugs, perfumes and paints; carrier in dyeing synthetic fibers.
Warning! Vapor harmful. Keep away from heat and open flame. MCA warning label.
Shipping regulations: None.*

para-**chlorobenzenesulfonamide** $ClC_6H_4SO_2NH_2$.
Properties: White, odorless powder; m.p. 145-148°C; soluble in alcohol.
Grade: 98-99% purity.
Use: Intermediate for pharmaceuticals and resins.

para-**chlorobenzenesulfonic acid**-para-**chlorophenyl ester.** See para-chlorophenyl para-chlorobenzene sulfonate.

1-(para-**chlorobenzenesulfonyl**)**-3-propylurea.** See chlorpropamide.

para-**chlorobenzenethiol.** See para-chlorothiophenol.

para-**chlorobenzhydrol** (para-chlorobenzohydrol) $ClC_6H_4C(C_6H_5)HOH$.
Properties: White to off-white, crystalline powder; m.p. 57-61°C; insoluble in water; soluble in ether, alcohol, and benzene.
Grade: 95% pure.
Containers: Steel drums.
Use: Organic synthesis.

para-**chlorobenzhydryl chloride** $ClC_6H_4C(C_6H_5)HCl$.
Properties: A water-white to light straw colored liquid; refractive index 1.600 (20°C); b.p. 145°C (3 mm).

para-**chlorobenzohydrol.** See para-chlorobenzhydrol.

ortho-**chlorobenzoic acid** ClC_6H_4COOH.
Properties: Nearly white, coarse powder. Soluble in alcohol and ether; insoluble in water and toluene. M.p. 137°C.
Grades: Assay 97-98.5% (dry basis).
Containers: Wooden barrels (100 lbs); fiber drums.
Uses: Intermediate for the preparation of dyes, fungicides, pharmaceuticals and other organic chemicals.

para-**chlorobenzoic acid** ClC_6H_4COOH.
Properties: Nearly white coarse powder; m.p. 238°C; soluble in methanol, absolute alcohol, and ether; insoluble in water, 95% alcohol, and toluene.
Grade: Assay 97-98.5% (dry basis).
Containers: 100-lb wooden barrels; fiber drums.
Uses: Intermediate for the preparation of dyes, fungicides, pharmaceuticals and other organic chemicals.

chlorobenzol. See chlorobenzene.

para-**chlorobenzophenone** $ClC_6H_4COC_6H_5$.
Properties: White to off-white crystalline powder; m.p. 73-78°C; b.p. 332°C; soluble in acetone, benzene, carbon tetrachloride; ether and hot alcohol. Insoluble in water.
Grades: 93% pure.
Containers: Fiber drums.
Use: Intermediate.

ortho-**chlorobenzotrichloride** $ClC_6H_4CCl_3$.
Properties: Colorless liquid or solid. M.p. 29.37°C; b.p. 264.3°C; sp. gr. 1.5131 (25/4°C); refractive index (n 20/D) 1.5836. Soluble in alcohol, ether, and acetone; decomposes in water.
Containers: 125-lb glass carboys.
Uses: As an intermediate in the manufacture of pharmaceuticals, dyes, and other organic chemicals.

para-**chlorobenzotrichloride** $ClC_6H_4CCl_3$.
Properties: Water-white liquid; boiling range (typical) first drop 248°C, 50% 252°C; dry 257°C; f.p. (approx.) 3.8°C; sp. gr. 1.480-1.490 (25/25°C). Soluble in alcohol, ether, and acetone. Insoluble in water.
Containers: Glass carboys.
Uses: As an intermediate in the manufacture of pharmaceuticals, dyes, and other organic chemicals.

meta-**chlorobenzotrifluoride** $ClC_6H_4CF_3$.
Properties: Water-white aromatic liquid; b.p. 138°C (760 mm); freezing point −56°C; refractive index 1.446 (n 20/D); flash point 50°C (closed cup); sp. gr. 1.351 (15.5/15.5°C).
Containers: 55-gal steel drums.
Uses: Intermediate in dyes and pharmaceuticals; dielectrics; insecticides.

ortho-**chlorobenzotrifluoride** (ortho-chlorotrifluoromethylbenzene; ortho-chloro-

*See "I.C.C. Shipping Regulations," page xiii.

alpha, alpha, alpha-trifluorotoluene) $ClC_6H_4CF_3$.
Properties: Colorless liquid with aromatic odor; sp.gr. 1.379 (15.5/15.5°C); refractive index 1.456 (20°C); b.p. 152°C; f.p. −7.4°C; flash point (closed cup) 59°C; wt/gal 11.50 lb (15.5°C); vapor pressure 150 mm (100°C), 710 mm (150°C); viscosity 0.44 cps (210°F), 0.89 cps (100°F).
Containers: 55-gal steel drums.
Uses: Dye intermediate, chemical intermediate, solvent and dielectric fluid.

para-**chlorobenzotrifluoride** (para-chlorotrifluoromethylbenzene; para-chloro-alpha, alpha, alpha-trifluorotoluene) $ClC_6H_4CF_3$.
Typical specifications: Water-white liquid; aromatic odor; b.p. 139.3°C; f.p. −36°C; refractive index (n 20/D) 1.446; flash point (closed cup) 47°C; sp.gr. (15.5/15.5°C) 1.353; wt/gal 11.28 lbs (15.5°C); vapor pressure 29 mm (50°C), 220 mm (100°C); viscosity

°F	Centipoise	Centistoke
100	0.675	0.511
210	0.392	0.319

Containers: 55-gal steel drums.
Uses: Same as ortho-chlorobenzotrifluoride.

5-chloro-2-benzoxazolinone. See chlorzoxazone.

ortho-**chlorobenzoyl chloride** ClC_6H_4COCl.
Properties: Colorless liquid; boiling range (typical) first drop 227°C, 50% 232°C, dry 239°C; f.p. −4 to −6°C; sp.gr. 1.374 to 1.376 (25/15°C). Soluble in alcohol, ether, and acetone. Insoluble in water.
Grades: Technical.
Containers: Glass carboys.
Use: An intermediate in the manufacture of pharmaceuticals, dyes, and other organic chemicals.

para-**chlorobenzoyl chloride** ClC_6H_4COCl.
Properties: Colorless liquid; boiling range (typical) first drop 225°C, 50% 227°C, dry 233°C; f.p. 10-12°C; sp.gr. 1.364-1.367 (25/15°C). Soluble in alcohol, ether, and acetone. Insoluble in water.
Containers: Glass carboys.
Use: An intermediate in the manufacture of pharmaceuticals, dyes, and other organic chemicals.

para-**chlorobenzoyl peroxide.**
Shipping regulations: Oxidizing material. Yellow label.*

ortho-**chlorobenzyl chloride** $ClC_6H_4CH_2Cl$.
Properties: Colorless liquid; boiling range (typical) first drop 216°C, 50% 217.5°C, dry 222°C; f.p. below −30°C; sp.gr. 1.270-1.280 (25/15°C). Soluble in alcohol, ether, and acetone; insoluble in water.
Containers: Glass carboys.
Use: An intermediate for the preparation of organic chemicals, pharmaceuticals and dyes.

para-**chlorobenzyl chloride** $ClC_6H_4CH_2Cl$.
Properties: Colorless liquid or solid; f.p. 25-27°C; boiling range 218-230°C; sp.gr. 1.250-1.260; soluble in most organic solvents; insoluble in water.
Derivation: Chlorination of benzyl chloride, using iodine as the catalyst.
Containers: Glass carboys.
Grades: Technical.
Use: Intermediate for organic chemicals, pharmaceuticals and dyes.

para-**chlorobenzyl** para-**chlorophenyl sulfide.** See chlorbenside.

2-(para-**chlorobenzyl)pyridine** (2-(4-chlorobenzyl)pyridine) $ClC_6H_4CH_2C_5H_4N$.
Properties: Liquid; b.p. 310.5°C (760 mm); freezing point 8.4°C; sp.gr. 1.168 (25°C); refractive index 1.5865 (n 20/D); insoluble in water.
Use: Organic synthesis.

chlorobromo-. See bromochloro-.

2-chlorobutadiene-1,3. See chloroprene.

1-chlorobutane. See n-butyl chloride.

chlorobutanol (trichloro-tert-butyl alcohol; 1,1,1-trichloro-2-methyl-2-propanol; acetone chloroform) $Cl_3CC(CH_3)_2OH$.
Properties: Colorless to white crystals with characteristic odor and taste. Soluble in alcohol and glycerol; slightly soluble in water; readily soluble in ether, chloroform, and volatile oils. M.p. (anhydrous form) 97°C; m.p. (hemihydrate) 78°C; b.p. 167°C; sublimes easily.
Derivation: By action of potassium hydroxide on a solution of chloroform and acetone.
Grade: U.S.P. XVI.
Uses: Medicine; plasticizer for cellulose esters and ethers; preservative.

chlorocarbon. A compound of carbon and chlorine, or of carbon, hydrogen, and chlorine, such as carbon tetrachloride, chloroform, tetrachloroethylene, etc.

chlorochromic anhydride. See chromyl chloride.

chlorocosane. See chlorinated paraffin.

3-chlorocoumarin.
Properties: Slightly yellow crystalline solid. Freezing point, 118°C.
Grades: Technical.
Use: Tin plating solutions.

para-**chloro**-meta-**cresol.** Probably 4-chloro-3-methylphenol (OH group = 1), but also used by some authorities to refer to the 6-chloro-3-methylphenol. This is a prime example of poor nomenclature, since the exact compound referred to depends on whether the chlorine is considered to be para to the methyl or to the hydroxy group.

4-chloro-meta-**cresol.** See 4-chloro-3-methylphenol.

5-chloro-2-cyanoacetophenone.
Constants: M.p. 276-277°C.

*See "I.C.C. Shipping Regulations," page xiii.

Derivation: From a cyclic halogen compound by heating with copper cyanide in the presence of amines.

chlorodifluoroacetic acid $CClF_2 \cdot COOH$.
Properties: Colorless liquid; b.p. 122°C; f.p. 23°C. Completely soluble in water; miscible with most organic solvents. Very strong acid.
Uses: Catalyst, particularly for esterification and condensation reactions.

1,1,1-chlorodifluoroethane (1,1,1-difluorochloroethane) CH_3CClF_2.
Properties: Colorless, nearly odorless gas. B.p. -9.2°C; m.p. -130.8°C; sp. gr. 1.194 (-9°C). Insoluble in water.
Grades: Technical.
Containers: Cylinders.
Uses: Refrigerant; solvent.
Shipping regulations: Flammable gas. Red gas label.*

chlorodifluoromethane (difluorochloromethane; fluorocarbon 22) $CHClF_2$.
Properties: Colorless, nearly odorless gas. B.p. -40.8°C; freezing point -160°C; critical pressure 48.7 atm.
Grades: Technical.
Containers: Drums; cylinders.
Uses: Aerosol propellant; refrigerant; air conditioning; to make tetrafluoroethylene.
Shipping regulations: Nonflammable gas. Green label.*

2-chloro-1,4-dihydroxybenzene. See chlorohydroquinone.

2-chloro-10-(3-dimethylaminopropyl) phenothiazine. See chlorpromazine.

1-chloro-2,4-dinitrobenzene (dinitrochlorobenzene; dinitrochlorbenzol) $C_6H_3(NO_2)_2Cl$.
Properties: Pale yellow needles; soluble in alcohol; insoluble in water; sp.gr. 1.69; m.p. 27-53°C; b.p. 315°C.
Derivation: By the chlorination of dinitrobenzene.
Grades: Technical; fused.
Containers: Drums; tankcars.
Uses: Dyes; organic synthesis.
Shipping regulations: Poison, class B. Poison label. Legal label name is dinitrochlorbenzol.*

4-chlorodiphenyl sulfone. See para-chlorophenyl phenyl sulfone.

chloroethane. See ethyl chloride.

chloroethane nitrile. See chloroacetonitrile.

chloroethanoic anhydride. See chloroacetic anhydride.

chloroethene. See vinyl chloride.

beta-**chloroethyl acetate** $CH_2ClCH_2OOCCH_3$.
Properties: Colorless liquid; sp.gr. 1.152 (25/25°C); b.p. 145°C; freezing point -26°C. Soluble in alcohol, ether; slightly soluble in water.
Derivation: By subjecting vinyl acetate to the action of hydrogen chloride gas.

2-chloroethyl alcohol. See ethylene chlorohydrin.

beta-**chloroethylchloroformate** CH_2ClCH_2OOCCl.
Properties: Colorless liquid. Decomposed by alkaline solutions and hot water. Insoluble in cold water. Caution! Very irritant!
Constants: Sp.gr. 1.3825 (20°C); b.p. 152.5°C (752 mm).
Derivation: By bubbling gaseous phosgene into ethylene chlorhydrin at 0°C.

beta-**chloroethyl chlorosulfonate** $ClCH_2CH_2OSO_2Cl$.
Properties: Chloropicrin-like odor. Caution! Very irritant! Darkens on long storage and decomposes with evolution of hydrogen chloride. B.p. 101°C (23 mm).
Derivation: Interaction of sulfuryl chloride and glycol chlorohydrin. Also from action of sulfur trioxide on ethylene chloride below 45°C.
Grades: Technical.

chloroethylene. See vinyl chloride.

chloroform (trichloromethane) $CHCl_3$.
Properties: Clear, colorless, highly refractive, heavy, volatile liquid; characteristic odor; nonflammable; burning sweet taste. Keep from light. Miscible with alcohol, ether, benzene, solvent naphtha, fixed and volatile oils; slightly soluble in water.
Constants: Sp.gr. 1.485 (20/20°C); b.p. 61.2°C; freezing point -63.5°C; wt/gal 12.29 lbs (25°C); refractive index 1.4422 (25°C); flash point none.
Derivation: (a) By the reaction of chlorinated lime with acetone, acetaldehyde, or ethyl alcohol; (b) byproduct from the chlorination of methane.
Method of purification: Extraction with concentrated sulfuric acid and rectification.
Grades: Technical; C.P.; U.S.P. XVI.
Containers: Bottles; tins; drums; 8000- and 10,000-gal tank cars.
Uses (in approximate order of volume): Extraction and purification of penicillin and other antibiotics; intermediate for refrigerants and propellants and resins; dyes and drugs; anesthetic; general solvent; fumigant; insecticides.
Warning! Vapor harmful. Avoid prolonged or repeated breathing of vapor. MCA warning label.
Shipping regulations: None.*

chloroformoxime $ClHCNOH$.
Properties: Needles. Odor resembles that of hydrocyanic acid. Stable at 0°C; unstable at normal temperature. Small quantities volatilize. Large quantities decompose. Aqueous solutions slowly decompose. Soluble in water, alcohol, ether, benzene; slightly soluble in carbon disulfide.
Derivation: Interaction of hydrochloric acid and sodium fulminate (sodium isocyanate).
Use: Organic synthesis.

chloroformyl chloride. See phosgene.

chloroguanide hydrochloride $C_{11}H_{16}ClN_5 \cdot HCl$. 1-(para-Chlorophenyl)-5-isopropylbiguanide hydrochloride.
Properties: Crystals with bitter taste. M.p. 248-252°C. Soluble in alcohol; slightly soluble in water; insoluble in chloroform and ether; pH of saturated aqueous solution 5.8-6.3.
Use: Medicine.

chlorohydric acid. See hydrochloric acid.

chlorohydrin (alpha-chlorohydrin; 3-chloropropane-1,2-diol; glyceryl alpha-chlorohydrin) $CH_2OHCHOHCH_2Cl$.
Properties: Colorless liquid. Unstable; hygroscopic. The commercial grade is a mixture of the two isomers, alpha and beta, of which alpha is in a greater proportion.
Constants: Sp.gr. 1.326 (18°C); b.p. 213°C; boiling range 213-228°C (decomposes); wt/gal 11.012 lbs; freezing point −40°C; viscosity 2.388 poise (20°C). Miscible with some organic solvents and water; immiscible with oils.
Derivation: By passing hydrochloric acid gas into glycerol containing 2% acetic acid.
Grades: Technical.
Containers: Iron drums.
Uses: Solvent for acetylcellulose, glyceryl phthalate, resins; partial solvent for gums; solvent (60% water) for cellulose acetate; intermediate in organic synthesis; explosive.

chlorohydrins (chlorhydrins). See chlorohydrin; dichlorohydrin; ethylene chlorohydrin; epichlorohydrin; propylene chlorohydrin.

chlorohydroquinone (2-chloro-1,4-dihydroxybenzene; 2,5-dihydroxychlorobenzene [Cl=1]) $ClC_6H_3(OH)_2$.
Properties: White to light-tan fine crystals; m.p. 100°C; b.p. 263°C. Very soluble in water and alcohol; slightly soluble in ether.
Grades: Photographic; commercial.
Containers: Bottles; fiber drums.
Uses: Photographic developer; organic intermediate; dyestuffs; bactericide.
Shipping regulations: None.*

chlorohydroxybenzenes. See chlorophenols.

5-chloro-2-hydroxybenzophenone $C_6H_5COC_6H_3OHCl$.
Properties: Yellow crystals; nearly odorless; m.p. 93-95°C; soluble in alcohol, ethyl acetate, methyl ethyl ketone; insoluble in water.
Use: Light absorber, best at 320-380μ.

4-chloro-1-hydroxy-3,5-dimethylbenzene. Probably the so-called para-chloro-meta-xylenol (q.v.).

3-chloro-4-beta-hydroxyethylaminophenylmethyl sulfone.
Constants: M.p. 79-80°C.

2-chloro-4-(hydroxymercuri) phenol. See hydroxymercurichlorophenol.

4-chloro-1-hydroxy-3-methylbenzene. See 4-chloro-3-methylphenol.

6-chloro-3-hydroxytoluene. See 4-chloro-3-methylphenol.

5-chloro-7-iodo-8-quinolinol. See iodochlorohydroxyquinoline.

chloro-IPC (isopropyl-N-(3-chlorophenyl) carbamate; chloroisopropyl-N-phenylcarbamate; C-IPC) $C_6H_4ClNHCOOC_3H_7$.
Properties: Light tan powder; m.p. 41.4°C; vapor pressure (149°C) 2 mm; sp.gr. 1.18 (30°C); very slightly soluble in water.
Containers: 60-, 250-, 450-lb drums; for solution: 55-gal drums; tank cars.
Use: Pre-emergence herbicides; prevents sprouting of potatoes.
Caution! Harmful if swallowed. MCA warning label.

chloroisopropyl alcohol. See propylene chlorohydrin.

6-chloro-4-isopropyl-1-methyl-3-phenol. See chlorothymol.

chloroisopropyl-N-phenylcarbamate. See chloro-IPC.

chloromaleic anhydride C_4HClO_3.
Properties: Yellow liquid; sp.gr. 1.5; m.p. 10-15°C; b.p. 192°C.
Uses: Catalyst for epoxy resins; organic intermediate.

1-[3-(chloromercuri)-2-methoxypropyl] urea See chlormerodrin.

ortho-**chloromercuriphenol.** See ortho-hydroxyphenylmercuric chloride.

chloromethane. See methyl chloride.

chloromethapyrilene citrate. See chlorothen citrate.

3-chloro-4-methylaminophenylmethylsulfone.
Constants: M.p. 130°C.

3-chloro-2-methylaniline. See 2-amino-6-chlorotoluene.

3-chloro-4-methylaniline. 4-amino-2-chlorotoluene.

5-chloro-2-methylaniline. See 2-amino-4-chlorotoluene.

2-chloro-1-methylbenzene. See ortho-chlorotoluene.

3-chloro-1-methylbenzene. See meta-chlorotoluene.

4-chloro-1-methylbenzene. See para-chlorotoluene.

chloromethylchloroformate $ClCOOCH_2Cl$.
Properties: Mobile, colorless liquid. Penetrating, irritating odor. Hydrolyzed by hot and cold water. Decomposed by alkalies. Caution! Very irritant!
Constants: Sp.gr. 1.465 at 15°C; b.p. 106.5-107°C; vapor density 4.5 (air = 1); vapor tension 5.6 mm (20°C). Soluble in most organic solvents.

chloromethylchlorosulfonate $ClCH_2OClSO_2$.
Properties: Colorless liquid. Caution! Very irritant!
Constants: Sp. gr. 1.63; b.p. 49-50°C (14 mm).
Derivation: By longtime boiling of chlorosulfonic acid with chloromethylchloroformate; also from paraformaldehyde and chlorosulfonic acid.
Grades: Technical.

1-chloromethylethylbenzene. See ethylbenzyl chlorides.

alpha-**chloromethylnapthalene** (1-chloromethylnaphthalene) $C_{10}H_7CH_2Cl$.
Properties: Colorless to greenish-yellow liquid with sharp pungent odor; sp. gr. 1.182 (25/25°C); coagulation point 19.8°C; insoluble in water; soluble in usual organic solvents. Is very reactive.
Use: A lachrymator; intermediate.

4-chloro-3-methylphenol (4-chloro-1-hydroxy-3-methylbenzene; 6-chloro-3-hydroxytoluene; 4-chloro-meta-cresol; so-called para-chloro-meta-cresol) $C_6H_3CH_3OHCl$.
Properties: White or slightly pink crystals with phenolic odor; m.p. 64-66°C; volatile with steam; solubility 1:250 in water at 25°C; soluble in alkalies, organic solvents, fats and oils.
Uses: External germicide; preservative for glues, gums, paints, inks, textile and leather goods.

chloromethylphosphonic acid $ClCH_2PO(OH)_2$.
Properties: White hygroscopic solid; m.p. 85-95°C.
Uses: Intermediate for flameproofing agents, resins, lubricants, additives, plasticizers.

chloromethylphosphonic dichloride $ClCH_2POCl_2$.
Properties: Water-white to light straw liquid, highly reactive. Sp. gr. 1.638 (25°C); refractive index 1.4960-1.4970 (n 25/D).
Uses: Intermediate for flameproofing agents, resins, lubricants, additives, and plasticizers.
Caution! Highly reactive; vapor irritates eyes and lungs. Do not breathe vapor, dust, or spray mist.

"Chloromycetin." [330] Trademark for chloramphenicol.

chloronaphthalene (chlorinated tar camphor; chlorinated naphthalene). A group name for the products of the chlorination of naphthalene. According to the degree of chlorination the physical state varies from a thin, mobile liquid to a crystalline, amorphous wax. See following entries.
Caution! Avoid repeated contact with skin and inhalation of fumes or dust. MCA warning label.

chloronaphthalene oils.
Properties: Almost colorless, thin, mobile liquids. They leave no deposit when heated. These oils are: (1) free of moisture and will not absorb moisture, (2) neutral, and non-corrosive to metals, (3) high in dielectric strength, (4) poor supporters of combustion, (5) miscible with asphalt, wax, pitch, etc.
Constants: Sp. gr. 1.20-1.25 (68°F); liquid down to −25°F; congealing point −33°F; flash point (approx.) 350°F; volatile at 212°F (and slightly so at normal temperatures); b.p. 480-550°F; specific heat 0.282 (between 86 and 140°F).
Soluble in practically all organic solvent liquids and oils (the best are carbon tetrachloride and benzene); insoluble in caustic alkaline solutions and acid solutions except those which are powerful oxidizers.
Derivation: By chlorinating naphthalene.
Grades: Technical.
Containers: 55-gal steel drums; tank cars.
Uses: Plasticizers; carbon softener and remover; heat-transfer medium; solvent (for rubber, aniline and other dyes, mineral and vegetable oils, varnish gums and resins, waxes); fire-proofing agent.
Caution: Avoid repeated contact with skin. Avoid repeated inhalation of fumes and dust. MCA warning label.
Shipping regulations: None.*

chloronaphthalene waxes.
Properties: Crystalline, amorphous, synthetic wax. Produced in translucent, black, light, and varied colors. These waxes are: (1) free of moisture and will not absorb moisture, (2) neutral, and non-corrosive to metals, (3) high in dielectric strength and have an extraordinary specific inductive capacity, (4) non-supporters of combustion, (5) able to melt to a liquid of low viscosity.
Constants: Sp. gr. 1.40-1.7 (300°F); m.p. 190-265°F; b.p. 550-700°F.
Soluble in many organic solvent liquids and oils (when heated together); insoluble in caustic alkaline solutions and acid solutions except those which are powerful oxidizers.
Derivation: By chlorinating naphthalene.
Grades: Technical.
Containers: 130-lb drums.
Uses: Condenser impregnation; moisture-, flame-, acid-, insect-proofing of wood, fabric and other fibrous bodies; moisture- and flame-proofing covered wire and cable; solvent (for rubber, aniline and other dyes, mineral and vegetable oils, varnish gums and resins, and other waxes when mixed in the molten state).
Caution: Avoid repeated contact with skin. Avoid repeated inhalation of fumes and dust. MCA warning label.
Shipping regulations: None.*

5-chloro-2-beta-naphthylaminophenylmethyl sulfone.
Constants: M.p. 172-173°C.

alpha-**chloro**-meta-**nitroacetophenone** $NO_2C_6H_4COCH_2Cl$.
Properties: Off-white, free-flowing granules; approx. m.p. 95-100°C; soluble in chlorinated solvents; insoluble in water.
Uses: Bacteriostat and fungistat in cutting oils, water systems, paint, plastics, textiles; chemical intermediate.

*See "I.C.C. Shipping Regulations," page xiii.
Reference numbers refer to name of manufacturer. See "List of Manufacturers," page v.

4-chloro-2-nitro-5-aminobenzonitrile.
Constants: M.p. 213-214°C.
Derivation: From a cyclic halogen compound by heating with copper cyanide in the presence of amines.

2-chloro-4-nitroaniline (ortho-chloro-para-nitroaniline) $C_6H_3NO_2NH_2Cl$.
Properties: Yellow needles. Soluble in alcohol, benzene, ether; slightly soluble in water and strong acids. M.p. 107°C.
Derivation: (a) From 1,2-dichloro-4-nitrobenzene by heating with alcoholic ammonia. (b) From the chlorination of para-nitroaniline in acid solution.
Method of purification: Recrystallization.
Grades: Technical, as paste or powder.
Containers: Up to 225-lb barrels.
Use: As intermediate in the manufacture of dyes.
Shipping regulations: None.*

meta-**chloronitrobenzene** $C_6H_4ClNO_2$.
Properties: Yellowish crystals; sp.gr. 1.534; m.p. 44°C; b.p. 236°C; soluble in most organic solvents; insoluble in water.
Derivation: By chlorinating nitrobenzene in the presence of iodine and recrystallizing.
Grades: Technical.
Containers: Drums.
Use: Intermediate.
Caution: Do not inhale dust, or vapors when heated.
Shipping regulations: Poison, class B. Poison label.*

ortho-**chloronitrobenzene** $C_6H_4ClNO_2$.
Properties: Yellow liquid; sp.gr. 1.368; b.p. 245.5°C; m.p. 32.5°C; soluble in alcohol and benzene; insoluble in water.
Derivation: By nitrating chlorobenzene and purifying by rectification.
Grades: Technical.
Containers: Drums; tank cars.
Uses: Intermediate, especially for dyes.
Caution: Do not inhale vapors.
Shipping regulations: Poison, class B. Poison label.*

para-**chloronitrobenzene** $C_6H_4ClNO_2$.
Properties: Yellowish crystals; sp.gr. 1.520; m.p. 83°C; b.p. 242°C; soluble in organic solvents; insoluble in water.
Derivation: Nitration of chlorobenzene and recrystallization.
Grades: Technical.
Containers: Drums.
Use: Intermediate, especially for dyes.
Caution: Do not inhale dust, or vapors when heated.
Shipping regulations: Poison, class B. Poison label.*

2-chloro-5-nitrobenzene sulfonamide $ClNO_2C_6H_3SO_2NH_2$.
Properties: Greyish-white solid. Insoluble in water; soluble in benzene.
Uses: Dye and pharmaceutical intermediates.

6-chloro-3-nitrobenzenesulfonic acid, sodium salt $NaSO_3C_6H_3NO_2Cl$.
Properties: Off-white moist crystals.
Use: Intermediate.

4-chloro-3-nitrobenzoic acid $ClC_6H_3NO_2COOH$.
Properties: Light gray or white powder.
Use: Intermediate.

chloronitrobenzols. See chloronitrobenzenes.

4-chloro-3-nitrobenzotrifluoride (para-chloro-meta-nitrotrifluorotoluene) $C_6H_3CF_3NO_2Cl$.
Properties: Thin, oily liquid; sp.gr. 1.542 (15.5/15.5°C); m.p. -7.5°C; refractive index (n 20/D) 1.491; b.p. 222°C; soluble in organic solvents; insoluble in water.
Grade: 97.5%.
Use: Intermediate for dyestuffs; agricultural chemicals; pharmaceuticals.

4-chloro-2-nitrophenol $ClC_6H_3(NO_2)OH$.
Yellow powder; m.p. 86°C. Used as a dye intermediate.

4-chloro-2-nitrophenol, sodium salt $ClC_6H_3NO_2ONa$.
Properties: Red needle crystals, with one molecule of water of crystallization. Soluble in hot water.
Derivation: Nitration of para-dichlorobenzene followed by hydrolysis.
Method of purification: Recrystallization.
Grades: 90% anhydrous sodium salt, containing 80% base.
Uses: Dye intermediate; manufacture of 2-amino-4-chlorophenol.

4-chloro-2-nitrotoluene $ClNO_2C_6H_3CH_3$. Solid, m.p. 35-37°C; used as an intermediate.
Containers: Drums.

6-chloro-2-nitrotoluene $ClNO_2C_6H_3CH_3$. A solid; used as an intermediate.
Containers: Drums.

para-**chloro**-meta-**nitrotrifluorotoluene.** See 4-chloro-3-nitrobenzotrifluoride.

chloronitrous acid. See aqua regia.

chloropentafluoroethane.
Shipping regulations: Nonflammable gas. Green label.*

1-chloropentane. See n-amyl chloride.

chlorophenol. Strictly the term applies to the three isomeric monochlorophenols (see ortho-, meta-, and para-chlorophenol), but the term may also be used for any of the isomers of di-, tri-, tetra-, and up to pentachlorophenol. These are discussed under separate entries.

meta-**chlorophenol** (3-chloro-1-hydroxybenzene) C_6H_4OHCl.
Properties: White crystals with odor similar to phenol; discolors on exposure to air; sp.gr. 1.245; m.p. 33°C; b.p. 214°C.
Soluble in alcohol, ether and aqueous alkali; slightly soluble in water.
Derivation: From meta-chloroaniline through the diazonium salt.
Use: Intermediate in organic synthesis.
Danger! Rapidly absorbed through the skin. Causes severe burns. Avoid breathing vapor. Do not take internally. MCA warning label.

ortho-**chlorophenol** (2-chloro-1-hydroxybenzene) C_6H_4OHCl.
Properties: Colorless to yellow brown liquid with unpleasant penetrating odor. Slightly soluble in water; soluble in alcohol, ether, and aqueous sodium hydroxide. Volatile with steam. B.p. 175°C; freezing point 7°C; sp.gr. 1.265 (15.5°C).
Derivation: Chlorination of phenol.
Use: Organic synthesis such as manufacturing dyes.
Danger! Rapidly absorbed through skin. Causes severe burns. Avoid breathing vapor. Do not take internally. MCA warning label.

para-**chlorophenol** (4-chloro-1-hydroxybenzene) C_6H_4OHCl.
Properties: White crystals (yellow or pink when impure) with unpleasant penetrating odor. Very slightly soluble in water; soluble in benzene, alcohol, and ether. Volatile with steam; b.p. 217°C; m.p. 42-43°C; sp.gr. 1.306; refractive index (n 40/D) 1.5579. A 1% solution is acid to litmus.
Derivation: Direct chlorination of phenol; from chloroaniline through the diazonium salt.
Grades: N.F. XI; technical.
Containers: 55-gal drums.
Uses: Intermediate in synthesis of dyes and drugs; denaturant for alcohol; selective solvent in refining mineral oils; medicine (local).
Danger! Rapidly absorbed through skin. Causes severe burns. Avoid breathing vapor. Do not take internally. MCA warning label.

para-**chlorophenol sulfide.** White, almost odorless powder. Insoluble in water; soluble in alcohol, oils, and organic solvents. Used as mildewicide and preservative for paper, wood and cloth.

chlorophenothane. See DDT.

para-**chlorophenyl** para-**chlorobenzenesulfonate** (para-chlorobenzenesulfonic acid para-chlorophenyl ester) $(C_6H_4Cl)OSO_2(C_6H_4Cl)$.
Properties: Crystals from benzene. M.p. 86.5-86.8°C. Practically insoluble in water. Toxicity low to animals.
Use: Insecticide and acaricide.
Hazard: May cause skin irritations. MCA warning label.

3-(para-**chlorophenyl) -1, 1-dimethylurea** (monuron; CMU) $C_6H_4ClNHCON(CH_3)_2$.
Properties: White, crystalline, odorless solid. Very low solubility in water and hydrocarbon solvents. Stable toward oxidation and moisture. Solubility 230 ppm at 25°C in water and in No. 3 diesel oil. Decomposes at 185°C.
Use: Weed killer; pre-emergence herbicide.
Caution: Avoid breathing dust or mist. Avoid contact with skin, eyes and clothing. U.S.D.A. Pesticides regulation label.

3-(para-**chlorophenyl) -5-methylrhodanine** $C_{10}H_8ClNOS_2$. A fungicide and disinfectant used in seed treatment and nematode control.

chloro-ortho-**phenylphenol** (chloro-2-phenylphenol) $C_6H_3(OH)ClC_6H_5$.
Properties: Clear, colorless to straw colored, viscous liquid with faint characteristic odor; sp.gr. 1.228 (20/4°C); freezing point less than −20°C; boiling range 5-95% 146-158.7°C (5 mm); flash point 134°C; readily soluble in most organic solvents.
Composition: 80% 4-Chloro-2-phenylphenol; 20% 6-chloro-2-phenylphenol.
Use: Fungicide.

para-**chlorophenyl phenyl sulfone** (4-chlorodiphenyl sulfone) $ClC_6H_4SO_2C_6H_5$.
Properties: Dimorphic crystals; slight aromatic odor; tasteless; insoluble in water; soluble in most organic solvents. Relatively stable to acids and alkalies. Toxicity relatively low in animals.
Use: Acaricide (toxic to most grapes and pears).

chlorophenyltrichlorosilane $ClC_6H_4SiCl_3$. A mixture of isomers.
Properties: Colorless to pale yellow liquid. B.p. 230°C; sp.gr. 1.439 (25/25°C); refractive index (n 20/D) 1.5414; flash point (Cleveland open cup) 255°F. Readily hydrolyzed by moisture, with the liberation of hydrochloric acid.
Derivation: By Grignard reaction of silicon tetrachloride and chlorophenylmagnesium chloride.
Grades: Technical.
Use: Intermediate for silicones.

4-chlorophthalic acid $C_6H_3Cl(COOH)_2$.
Properties: Colorless crystals; m.p. 150°C; decomposes on further heating; soluble in alcohol and ether; insoluble in water.
Derivation: By the chlorination of phthalic acid.

chlorophyl. See chlorophyll.

chlorophyll (chlorophyl; leaf green; chromule). The green plant pigment that is involved in the process of photosynthesis; it is present in two forms, chlorophyll a and chlorophyll b, both of which are magnesium complex salts of phytol esters of porphin derivatives, and are related in structure to hemin, the red pigment of blood. Chlorphyll a has recently been synthesized by two different routes.
Properties:
Chlorophyll a: $C_{55}H_{72}MgN_4O_5$. Waxy blue-black microcrystals, usually aggregates of thin, lancet-like leaflets; m.p. 117-120°C. Freely soluble in ether, ethanol, acetone, chloroform, carbon disulfide, benzene; sparingly soluble in cold methanol; insoluble in petroleum ether. The alcoholic solution is blue-green with a deep-red fluorescence.
Chlorophyll b: $C_{55}H_{70}MgN_4O_6$. Waxy blue-black microcrystals. Sinters between 86 and 92°C. Sparingly soluble in petroleum ether, ligroin; freely soluble in absolute alcohol, ether. Ether solution has a

brilliant green color. Solutions with other organic solvents are usually green to yellow-green with red fluorescence.
Derivation: Alcoholic extraction of green plants; isolation by chromatography.
Grades: Aqueous, alcoholic or oil solutions; the water solutions are prepared by saponification of the oil-soluble chlorophyll.
Containers: Glass bottles.
Uses: Non-poisonous coloring agent for soaps, oils, fats, waxes, liquors, confectionery, preserves; cosmetics, perfumes; source of phytol; dye for leather; sensitizer for color film; antiknock agent; accelerator in vulcanizing of rubber; deodorizers; medicine (topical).

chlorophyllin. Product resulting from the controlled action of alcoholic potassium or sodium hydroxide on alcoholic leaf extracts. The methyl and phytyl groups are replaced by alkali but the magnesium is not replaced. Useful in food coloring, dyes, and medicine.

chloropicrin (chlorpicrin; nitrotrichloromethane; trichloronitromethane; nitrochloroform) CCl_3NO_2.
Properties: Pure product slightly oily, colorless, refractive liquid. Relatively stable; not decomposed by water or mineral acids. Caution! Very irritant! Protect eyes! Sp.gr. 1.692 (0°C); b.p. 112°C (760 mm); m.p. −69.2°C; coefficient of expansion 0.00102 (0°C); sp.ht. 0.235 (15-35°C); latent ht. of evaporation 59 calories; vapor density 5.69; vapor tension 16.91 mm Hg (20°C); volatility 184 mg/liter (20°C). Soluble in alcohol, benzene, carbon disulfide; slightly soluble in ether; insoluble in water.
Derivation: (a) Action of picric acid on calcium hypochlorite. (b) Nitrification of chlorinated hydrocarbons.
Grades: Commercial.
Containers: Metal drums.
Uses: Military poison gas; organic synthesis; dyestuffs (crystal violet); fumigant preparations; fungicides; insecticides; rat exterminator.
Danger! Poisonous if inhaled. MCA warning label.
Shipping regulations: Poison, class B; poison label. Legal label name chlorpicrin.*

6-chloropiperonyl ester chrysanthemummonocarboxylic acid. See barthrin.

chloroplatinic acid (platinic chloride) $H_2PtCl_6 \cdot 6H_2O$.
Properties: Red-brown crystals. Soluble in water, alcohol and ether.
Constants: Sp.gr. 2.431; m.p. 60°C.
Derivation: By solution of platinum in aqua regia, evaporation and crystallization.
Containers: Glass bottles.
Uses: Electroplating; platinizing pumice and the like for catalysts; etching zinc for printing; platinum mirrors; indelible ink; ceramics (producing fine color effects on high-grade porcelain); microscopy.

chloroprene (2-chloro-butadiene-1,3) $H_2C{:}CHCCl{:}CH_2$.
Properties: Colorless liquid; b.p. 59.4°C; sp.gr. 0.9583 (20/20°C); soluble in alcohol; slightly soluble in water.
Derivation: (a) By treatment of vinyl acetylene with cold hydrochloric acid; (b) from C_4 petroleum fractions.
Grades: Pure, 95% min.
Containers: Cylinders; tank trucks.
Use: The monomer from which neoprene synthetic rubber is made.

chloroprocaine hydrochloride (beta-diethylaminoethyl 2-chloro-4-aminobenzoate hydrochloride) $ClNH_2C_6H_3COOC_2H_4N(C_2H_5)_2 \cdot HCl$.
Grade: N.N.D.
Use: Medicine.

chloroprocaine penicillin O. See penicillin.

3-chloropropane-1,2-diol. See chlorohydrin.

1-chloro-2-propanol. See propylene chlorohydrin.

1-chloro-2-propanone. See chloroacetone.

3-chloropropene. See allyl chloride.

2-chloropropionic acid (alpha-chloropropionic acid) $CH_3CHClCOOH$.
Properties: Sp.gr. 1.260-1.268 (20°C); b.p. 183-187°C. Soluble in water.
Use: Intermediate in weed killers.

3-chloropropionic acid (beta-chloropropionic acid) CH_2ClCH_2COOH.
Properties: Crystals; m.p. 41°C; b.p. 200°C; soluble in water, alcohol, chloroform.
Use: Intermediate.

3-chloropropionitrile $ClCH_2CH_2CN$.
Properties: Colorless liquid; m.p. −51°C; flash point (closed cup) 168°F; refractive index (n 25/D) 1.4341; sp.gr. 1.1363 (25°C); b.p. 176°C (dec.). Miscible with acetone, benzene, carbon tetrachloride, alcohol, and ether.
Uses: Organic intermediate.

alpha-**chloropropylene.** See allyl chloride.

chloropropylene oxide. See epichlorohydrin.

3-chloro-1-propyne. See propargyl chloride.

6-chloroquinaldine $C_9H_5N(CH_3)Cl$.
Properties: Brownish-black oily crystalline mass.
Grade: Technical.
Use: Intermediate.

chloroquine $C_9H_6NClNHCH(CH_3)(CH_2)_3N(C_2H_5)_2$. 7-Chloro-4-(4-diethylamino-1-methylbutylamino)quinoline.
Properties: M.p. 87°C.
Derivation: From meta-chloroaniline and diethyl oxalacetate or diethyl malonate.
Use: Medicine. Usually dispensed as the phosphate (q.v.).

chloroquine diphosphate. See chloroquine phosphate.

*See "I.C.C. Shipping Regulations," page xiii.
Reference numbers refer to name of manufacturer. See "List of Manufacturers," page v.

chloroquine phosphate (chloroquine diphosphate) $C_{18}H_{26}ClN_3 \cdot 2H_3PO_4$. 7-Chloro-4-(4-diethylamino-1-methylbutylamino) quinoline diphosphate.
Properties: White, crystalline powder. Odorless; has bitter taste; affected by light. Freely soluble in water; almost insoluble in alcohol, chloroform and ether; solution has pH about 4.5. Exists in two forms: usual form melts 193-195°C; other form melts 215-218°C.
Grade: U.S.P. XVI.
Use: Medicine.

5-chlorosalicylanilide $ClC_6H_3OHCONHC_6H_5$.
Properties: White crystals; slightly soluble in water; soluble in alcohol, ether, chloroform and benzene.
Use: Intermediate for pharmaceuticals, dyes, pesticides.

5-chlorosalicylic acid $ClC_6H_3OHCOOH$.
Properties: White crystals; slightly soluble in water; soluble in alcohol, ether, chloroform and benzene.
Use: Intermediate for pharmaceuticals, dyes, pesticides.

N-chlorosuccinimide (NCS) $\overline{COCH_2CH_2CONCl}$.
Physical properties: White crystalline powder; m.p. 148-149°C; soluble in water; sparingly soluble in chloroform and carbon tetrachloride.
Uses: Chlorinating agent, disinfectant for swimming pools, and bactericide.

chlorosulfonic acid $ClSO_2OH$.
Properties: Colorless to light yellow, fuming, very corrosive liquid; pungent odor; sp.gr. 1.76-1.77 (20/20°C); m.p. −80°C; b.p. 151°C.
Derivation: By treating sulfur trioxide or fuming sulfuric acid with hydrochloric acid.
Grades: Technical.
Containers: 170-lb carboys; 1600-lb drums; 8000-gal tank cars.
Uses: Organic preparations, especially saccharin; military poison gas.
Danger: Causes severe burns. MCA warning label.
Shipping regulations: Corrosive liquid. White label.*

4-chlorosulfonylbenzoic acid $ClSO_2C_6H_4COOH$.
Properties: Light tan powder; soluble in benzene; slightly soluble in ether.
Use: Intermediate.

chlorosulfuric acid. See sulfuryl chloride.

chlorotetracycline. See chlortetracycline.

chlorotetrafluoroethane.
Shipping regulations: Nonflammable gas. Green label.*

chlorothen citrate (chloromethapyrilene citrate) $C_{14}H_{18}ClN_3S \cdot C_6H_8O_7$. N,N-Dimethyl-N'-(2-pyridyl)-N'-(5-chloro-2-thenyl)ethylenediamine citrate.
$ClC_4H_2SCH_2N(C_5H_4N)CH_2CH_2N(CH_3)_2 \cdot C_6H_7O_7$.
Properties: White, practically odorless crystalline powder. Slightly soluble in alcohol and water; practically insoluble in chloroform, ether and benzene. 1% solution is clear and colorless. pH (1% solution) 3.9-4.1. Melts at 112-116°; on further heating solidifies and remelts at 125-140° (dec).
Grade: N.F. XI.
Use: Medicine.

"Chlorothene." 233 Trademark for inhibited 1,1,1-trichloroethane (q.v.).
Typical specifications: Colorless liquid; b.p. 73-84°C; sp.gr. 1.319 (25/25°C); refractive index 1.435 (25°C); freezing point −50°C.
Use: Industrial solvent.

chlorothiazide (6-chloro-7-sulfamyl-1,2,4-benzothiadiazine-1,1-dioxide) $C_7H_6ClN_3O_4S_2$.
Properties: White or practically white odorless crystalline powder. Crystals decompose 342.5-343°C. Slightly soluble in water; soluble in alkaline aqueous solutions with decomposition upon standing or heating; slightly soluble in methanol and pyridine; insoluble in ether, benzene, chloroform.
Grade: U.S.P. XVI.
Uses: Medicine.

chlorothiazide sodium $C_7H_5ClN_3O_4S_2Na$.
Properties: Crystalline powder. Soluble in water.
Grade: N.N.D.
Use: Medicine.

para-**chlorothiophenol** (para-chlorobenzenethiol) ClC_6H_4SH.
Properties: Moist white to cream crystals; m.p. 52-55°C; b.p. 205-207°C. Soluble in most organic solvents.
Containers: Casks.
Uses: Oil additives; agricultural chemicals; plasticizers; rubber chemical; dyes; wetting agents and stabilizers.

chlorothymol (6-chloro-4-isopropyl-1-methyl-3-phenol) $CH_3C_6H_2(OH)(C_3H_7)Cl$.
Properties: White crystals or crystalline granular powder; characteristic odor; aromatic, pungent taste; becomes discolored with age; affected by light; m.p. 59-61°C. Soluble in benzene, chloroform, dilute caustic soda, alcohol; insoluble in water.
Derivation: Action of sulfuryl chloride on thymol in a solution of carbon tetrachloride.
Grades: N.F. XI.
Use: Bactericide.

alpha-**chlorotoluene.** See benzyl chloride.

meta-**chlorotoluene.** (3-chloro-1-methylbenzene) $CH_3C_6H_4Cl$.
Properties: Colorless liquid; sp.gr. 1.07218 (20/4°C); b.p. 161.6°C; f.p. −48.0°C.
Derivation: Diazotization of meta-toluidine followed by treating with cuprous chloride.
Caution! Vapor harmful. MCA warning label.
Uses: Solvent; intermediate.

ortho-**chlorotoluene** (2-chloro-1-methylbenzene) $CH_3C_6H_4Cl$.
Properties: Colorless liquid. B.p. 159.2°C; m.p. −35.1°C; sp.gr. 1.0776 (25/4°C);

refractive index (n 20/D) 1.5268. Miscible with alcohol, acetone, ether, benzene, carbon tetrachloride, and n-heptane; slightly soluble in water.
Derivation: By catalytic chlorination of toluene.
Containers: 45-, 100-, 115-lb carboys; 425-, 450-, 475-lb drums.
Caution! Vapor harmful. MCA warning label.
Uses: As a solvent and intermediate for organic chemicals and dyes.

para-**chlorotoluene** (4-chloro-1-methylbenzene) $CH_3C_6H_4Cl$.
Properties: Colorless liquid; boiling range (typical) first drop 162°C, 50% 162.4°C, dry 166°C; f.p. approx. 6.5°C; sp.gr. 1.065-1.067 (25/15°C); refractive index 1.5184 (22°C). Soluble in alcohol, ether, acetone, benzene, and chloroform. Insoluble in water.
Containers: Glass carboys or 425- or 450-lb iron drums.
Caution: Vapor harmful. MCA warning label.
Uses: As a solvent and intermediate for organic chemicals and dyes.

2-chlorotoluene-4-sulfonic acid (ortho-chlorotoluene-para-sulfonic acid) $CH_3C_6H_3(SO_3H)Cl$.
Properties: White glistening plates. Soluble in hot water.
Derivation: Chlorination of toluene-para-sulfonic acid.
Method of purification: Recrystallization from water.

2-chloro-para-**toluidine.** See 4-amino-2-chlorotoluene.

4-chloro-ortho-**toluidine.** See 2-amino-4-chlorotoluene.

6-chloro-ortho-**toluidine.** See 2-amino-6-chlorotoluene.

4-chloro-ortho-**toluidine hydrochloride.**
Shipping regulations: Poison, class B. Poison label. *

2-chloro-5-toluidine-4-sulfonic acid (6-chloro-meta-toluidine-4-sulfonic acid) $CH_3C_6H_2(NH_2)(SO_3H)Cl$.
Properties: Fine white crystals. Soluble in dilute caustic solution.
Derivation: From ortho-chlorotoluene-para-sulfonic acid by nitration and subsequent reduction.
Method of purification: Recrystallization of its sodium salt.
Containers: Fiber kegs; lined steel drums.
Use: Intermediate.

5-chloro-2-para-**toluidinephenylmethylsulfone.**
Constants: M.p. 136 to 137°C.

chlorotrianisene (tri-para-anisylchloroethylene) $(CH_3OC_6H_5)_2CCCl(C_6H_5OCH_3)$. A synthetic nonsteroid estrogen.
Properties: White, odorless, crystalline powder. M.p. 115-117°C. Freely soluble in acetone, benzene and chloroform; slightly soluble in ether; very slightly soluble in alcohol and water.
Grade: N.F. XI.
Use: Medicine.

chlorotrifluoroethylene (CFE; trifluorochloroethylene) $ClFC{:}CF_2$.
Properties: Colorless gas with faint ethereal odor; b.p. −27.9°C; m.p. −157.5°C; sp.gr. (liquid) (20°C) 1.305; critical temperature 107°C; critical pressure 39.0 atm; specific volume 3.3 cu ft/lb.
Derivation: From trichlorotrifluoroethane and zinc.
Grades: Technical; 99.0%.
Containers: Cylinders; tank cars; tank trucks. Shipped with inhibitor.
Uses: Polymerization to colorless oils, greases and waxes (see chlorotrifluoroethylene resins); also as a refrigerant (see "Freon 13").
Caution: Relatively toxic gas.
Shipping regulations: Flammable gas. Red label. Legal label name: trifluorochloroethylene. *

chlorotrifluoroethylene resins. Polymers of chlorotrifluoroethylene characterized by a high degree of chemical inertness. Virtually unaffected by inorganic acid, alkalies, oxidizing agents, and most organic solvents. For typical commercially available material see "Halon"; "Kel-F"; fluorothene.

chlorotrifluoromethane (trifluorochloromethane) $CClF_3$.
Properties: Colorless; non-toxic; nonflammable; non-corrosive liquid with an ethereal odor; b.p. −81.4°C.
Containers: Cylinders.
Uses: Refrigerant; hardening of metals; pharmaceutical processing.
Shipping regulations: Nonflammable gas; green label. *

ortho-**chlorotrifluoromethylbenzene.** See ortho-chlorobenzotrifluoride.

para-**chlorotrifluoromethylbenzene.** See para-chlorobenzotrifluoride.

ortho-**chloro**-alpha, alpha, alpha-**trifluorotoluene.** See ortho-chlorobenzotrifluoride.

para-**chloro**-alpha, alpha, alpha-**trifluorotoluene.** See para-chlorobenzotrifluoride.

chlorovinylarsinedichloride. See beta-chlorovinyldichloroarsine.

beta-**chlorovinyldichloroarsine** (dichloro(2-chlorovinyl)arsine; chlorovinylarsinedichloride; lewisite). Two isomers, probably cis and trans, are known. $ClCH{:}CHAsCl_2$.
Properties: Colorless liquid when pure. Impurities influence a color change ranging from violet to brown. Geranium-like odor. Decomposed by water and alkalies. Caution! Very irritant! Soluble in alcohol, benzene, and ether; slightly soluble in water. Sp.gr. 1.8855 (20°C); b.p. 190°C (decomposes), also 164.8°C; m.p. (given variously) −18.2 to + 0.1°C; sp. vol. 0.5302 (20°C); vapor density 7.2; vapor tension 0.394 mm (20°C); volatility 2300 mg/cu m (20°C); latent heat of vaporization 57.9; coefficient of thermal expansion

*See "I.C.C. Shipping Regulations," page xiii.
Reference numbers refer to name of manufacturer. See "List of Manufacturers," page v.

0.00094 (0 to 50°C). Inactivated by bleaching powder. Antidote is BAL.
Derivation: Condensation of arsenic trichloride with acetylene in the presence of aluminum or copper or mercury chloride. The mixed arsines are separated by fractionating.
Grades: Technical.
Use: Military poison gas; skin blistering agent.
Shipping regulations: Poison, class A by freight; not accepted by express! Poison gas label. Legal label name: lewisite.

beta-**chlorovinylmethylchloroarsine** $ClCH{:}CHAsClCH_3$.
Properties: Liquid; decomposed by water. Caution! Very irritant! B.p. 112-115°C (10 mm).
Derivation: Interaction of acetylene and methyldichloroarsine in the presence of aluminum chloride.

"Chlorowax." [244] Trademark for a series of liquid and resinous chlorinated paraffins containing from 40% to 70% chlorine by weight. They are odorless, nontoxic, nonflammable, and insoluble in water. Available in Grades LV, 40, 50, and 70. "Chlorowax" 70-S is a heat stabilized form of "Chlorowax" 70, a powdered resin.
Typical properties: Sp. gr. (25/25°C), color (Tag-Union Colorimeter), viscosity (Gardner tubes, poises at 25°C), respectively:
Grade LV: 1.13; 1.0; 5;
Grade 40: 1.15; 1.25; 25;
Grade 50: 1.26; 1.25; 125;
Grade 70: 1.65; 1.25; m.p. (Ball and Ring) 100°C.
Containers: "Chlorowax" LV, 40, and 50 in 50-lb and 525-lb steel drums. "Chlorowax" 70 and 70-S in 50-lb and 250-lb fiber drums.
Uses: The liquid grades, being of low volatility, are used extensively as plasticizers in paints, synthetic rubbers, and plastics. The resinous grades are used as modifiers to add moisture, chemical, and flame resistance to many paints, synthetic rubbers, and plastics.

para-**chloro**-meta-**xylenol** (probably 4-chloro-1-hydroxy-3,5-dimethyl benzene) $C_6H_2(CH_3)_2OHCl$. Crystals with phenolic odor.
Uses: Germicide; fungistat; preservative.
See also para-chloro-meta-cresol.

chlorpheniramine maleate (chlorprophenpyridamine maleate) $C_{16}H_{19}ClN_2 \cdot C_4H_4O_4$. 1-(para-Chlorophenyl)-1-(2-pyridyl)-3-dimethylaminopropane maleate.
Properties: White odorless crystals. M.p. 130-135°C. Slightly soluble in ether; soluble in alcohol, chloroform, and water. pH (1% solution) about 4.8.
Grade: U.S.P. XVI.
Use: Medicine.

chlorphenol red (dichlorosulfonphthalein) An acid-base indicator, showing color change from yellow to red over the pH range 4.8 to 6.4.

chlorphenoxamine
$CH_3(C_6H_5)C(C_6H_4Cl)OC_2H_4N(CH_3)_2 \cdot HCl$. beta-Dimethylaminoethyl(para-chloro-alpha-methylbenzhydryl) ether hydrochloride.
Properties: Crystals; m.p. 128°C; soluble in water.
Use: Medicine (antihistaminic).

chlorpicrin. See chloropicrin.

chlorpromazine (2-chloro-10-(3-dimethylaminopropyl)phenothiazine) $C_{17}H_{19}ClN_2S$.
Properties: Oily liquid; amine odor; alkaline reaction; b.p. 200-205°C, (0.8 mm).
Grade: N.N.D.
Use: Medicine.

chlorpromazine hydrochloride $C_{17}H_{19}ClN_2S \cdot HCl$.
Properties: White or slightly creamy white, odorless, crystalline powder. Darkens on prolonged exposure to light. Solutions acid to litmus. Soluble in water, alcohol, chloroform. Insoluble in ether and benzene. M.p. 195-198°C.
Grade: U.S.P. XVI.
Use: Medicine.

chlorpropamide 1-(para-chlorobenzenesulfonyl)-3-propylurea.
$C_3H_7NHCONHSO_2C_6H_4Cl$.
Properties: Crystals, m.p. 127-129°C; soluble in water, alcohol, chloroform.
Use: Medicine.

chlorprophenpyridamine maleate. See chlorpheniramine maleate.

chlorquinaldol (5,7-dichloro-8-hydroxyquinaldine; 5,7-dichloro-2-methyl-8-quinolinol) $CH_3C_9H_3N(OH)Cl_2$.
Properties: Yellow, crystalline, tasteless powder with a pleasant medicinal odor; m.p. 114°C; soluble in alcohol, chloroform; insoluble in water.
Grade: N.N.D.
Use: Medicine (bactericide and fungicide).

chlortetracycline (CTC; chlorotetracycline) $C_{22}H_{23}ClN_2O_8$. An antibiotic produced by the growth of Streptomyces aureofaciens in submerged cultures. It has a wide antimicrobial spectrum including many gram-positive and gram-negative bacteria, rickettsiae and several viruses. Its chemical structure is that of a modified naphthacene molecule. It is relatively non-toxic.
Properties: Golden-yellow crystals. M.p. 168-169°. Slightly soluble in water; very soluble in aqueous solutions above pH 8.5; freely soluble in the "Cellosolves," dioxane and "Carbitol"; slightly soluble in methanol, ethanol, butanol, acetone, ethyl acetate, and benzene; insoluble in ether and petroleum ether.
Derivation: By submerged aerobic fermentation, filtration, solvent extraction, and crystallization.
Use: Medicine (usually as hydrochloride); feed supplement; preservative for raw fish.

chlortetracycline hydrochloride
$C_{22}H_{23}ClN_2O_8 \cdot HCl$.
Properties: Odorless, yellow, crystalline

powder with bitter taste. Stable in air but affected by light. Soluble in solutions of alkali hydroxides and carbonates; practically insoluble in acetone, chloroform, dioxane and ether. pH (1 in 200 solution) 2.3-3.3.
Grade: N.F. XI.
Use: Medicine.

"Chlor-Trimeton Maleate." [321] Brand name of chlorpheniramine maleate.

chlorzoxazone (5-chloro-2-benzoxazolinone) $C_7H_4ClNO_2$. Derivative of zoxazolamine.
Grades: N.N.D.
Use: Medicine.

chocolate. The product formed by roasting and grinding fermented dried cacao beans (q.v.). Chocolate contains about 55% cocoa butter, some starch, traces of theobromine and tannin. Milk chocolate contains 30 to 35% cocoa butter, 12% milk solids and also added sugar.

chocolate fat. Cacao butter (q.v.) is the true chocolate fat and the highest-grade chocolates contain it exclusively. Substitutes are used, to a greater or less degree, in the lower-grade chocolates. These substitutes usually consist of vegetable fats, such as palm-nut or coconut stearin.

chocolate varnish. An alcoholic solution of gum benzoin and an edible resin. It may be used for hardening chocolate, thus preventing its turning white.

cholaic acid. See taurocholic acid.

cholecalciferol. See 7-dehydrocholesterol, activated.

choleic acids. A loose term applied to the complexes formed by deoxycholic acid (a bile acid) with fatty acids or other lipids, and with a variety of other compounds, including such aromatics as phenol and naphthalene. These complexes are similar to, and have been suggestive of, the complexes used in separation processes, such as the urea adducts, for large scale purification.

beta-**cholestanol.** See dihydrocholesterol.

5-cholesten-3 beta **-ol.** See cholesterol.

cholesterin. See cholesterol.

cholesterol (cholesterin; 5-cholesten-3 beta-ol) $C_{27}H_{45}OH$. The most common animal sterol, a monounsaturated, secondary alcohol of the cyclopentenophenanthrene system. Present in animals in part as the free sterol, and in part esterified with higher fatty acids. The primary precursor in biosynthesis appears to be acetic acid or sodium acetate. Cholesterol itself in the animal system is the precursor of bile acids, steroid hormones and provitamin D_3. Cholesterol is a universal tissue constituent and is the subject of widespread research, partly because of its role as a suspect in atherosclerosis.
Properties: White, or faintly yellow, almost odorless, pearly granules or crystals; affected by light; m.p. 148.5°C; b.p. 360°C (dec); sp.gr. 1.067 (20/4°C); levorotatory; specific rotation (25°C) −34 to −38°; insoluble in water; slightly soluble in alcohol; soluble in the usual fat solvents, vegetable oils and aqueous solutions of bile salts.
Source: Prepared from beef spinal cord by petroleum ether extraction of the nonsaponifiable matter; purification by repeated bromination.
Grades: Technical; U.S.P. XVI.
Containers: 1- and 5-lb glass bottles.
Uses: Medicine; emulsifying agent in cosmetic and pharmaceutical products; textile, leather and ink industries.

cholesterol pitch. See stearin and fatty-acid pitches.

cholic acid $C_{24}H_{40}O_5$. The most abundant bile acid. In bile it is conjugated with the amino acids glycine and taurine as glycocholic acid and taurocholic acid, respectively, and does not occur free.
Properties: The monohydrate crystallizes in plates from dilute acetic acid; bitter taste with sweetish aftertaste; anhydrous form, m.p. 198°C. Not precipitated by digitonin. Soluble in glacial acetic acid, acetone, and alcohol; slightly soluble in chloroform; practically insoluble in water and benzene.
Derivation: From glycocholic and taurocholic acids in bile; organic synthesis.
Containers: Drums.
Use: Medicine; biochemical research; pharmaceutical intermediate.

choline (beta-hydroxyethyl-trimethylammonium hydroxide) $(CH_3)_3N(OH)CH_2CH_2OH$. Has been called vitamin B_4. Known to be essential in the diet of rats, rabbits, chickens, and dogs. In man it is required for lecithin formation and can replace methionine in the diet. There is no evidence of disease in man due to choline deficiency. It acts as a dietary factor important in furnishing free methyl groups for trans-methylation; has a lipotrophic function.
Properties: Colorless viscous liquid; caustic, bitter taste; soluble in water, formaldehyde and in absolute methyl and ethyl alcohols; insoluble in ether, petroleum ether, benzene, carbon disulfide, carbon tetrachloride, and toluene; a strong base; breaks down into trimethylamine and glycol when heated; decomposes ammonium salts; stable to heat in acid solution; extremely hygroscopic.
Source: Food source: egg yolk, kidney, liver, heart, seeds, vegetables and legumes. Commercial source: synthetic preparation from trimethylamine and ethylene chlorohydrin or ethylene oxide.
Units: Amounts are expressed in milligrams of choline.
Uses: Medicine; nutrition; feed supplement; suggested for epoxy resins.
Shipping regulations: None.*
Usually used as choline chloride (q.v.).

*See "I.C.C. Shipping Regulations," page xiii.
Reference numbers refer to name of manufacturer. See "List of Manufacturers," page v.

choline bitartrate ($C_5H_{14}NO\cdot C_4H_5O_6$).
Properties: White crystalline powder; odorless or faint trimethylamine-like odor; acid taste; hygroscopic; soluble in water and alcohol. Insoluble in ether, chloroform and benzene.
Grades: N. F. XI.
Use: Medicine.

choline chloride $(CH_3)_3N(Cl)CH_2CH_2OH$. Same biological function as choline.
Properties: White crystals; salty, bitter taste; fishy odor; soluble in water and alcohol; insoluble in ether, petroleum ether, benzene, and carbon disulfide; stable to heat in acid solution; unstable in alkaline solution; extremely hygroscopic.
Source: Same as choline.
Units: Amounts are expressed in milligrams of choline.
Containers: Glass vials and bottles; 54-gal drums.
Grade: N. N. D.
Uses: Medicine; nutrition; animal feed supplement.
See also choline.

choline chloride carbamate. See carbachol.

choline dihydrogen citrate
$(CH_3)_3NC_2H_4OH\cdot H_2C_6H_5O_7$.
Properties: White, hygroscopic, crystalline, granular substance with acid taste; nearly odorless. M.p. 103-107.5°; freely soluble in water; very slightly soluble in alcohol; practically insoluble in benzene, chloroform, and ether. pH (25% solution) about 4.25.
Grade: N. F. XI.
Containers: Bottles; drums.
Use: Medicine.

choline gluconate
$HOCH_2CH_2N(CH_3)_3OOC(CHOH)_4CH_2OH$.
Properties: Straw colored, highly viscous mass. Amine-like odor and bitter taste. Soluble in water; sparingly soluble in alcohol; very slightly soluble in ether; practically insoluble in benzene and chloroform. pH (50% solution) 5.0-6.0.
Use: Medicine.

cholinesterase
1. (acetylcholinesterase). Enzyme specific for the hydrolysis of acetylcholine to acetic acid and choline in the body. It is found in the brain, nerve cells and red blood cells and is important in the mechanism of nerve action. See nerve gases.
Derivation: Prepared from bovine erythrocytes.
Uses: Biochemical experimentation; determination of phosphorus in insecticides and poisons.
2. "Pseudo" or nonspecific cholinesterase, prepared from horse serum. This esterase hydrolyzes other esters as well as choline esters. It is found in blood serum, pancreas and liver.
Both are commercially available.

choline succinate dichloride dihydrate. See succinylcholine chloride.

choline theophyllinate. See oxtriphylline.

"Cholografin Methylglucamine." [412] Trademark for methylglucamine iodipamide (q.v.).

"Cholografin Sodium." [412] Trademark for sodium iodipamide (q.v.).

cholylglycine. See glycocholic acid.

cholyltaurine. See taurocholic acid.

Chondodendron tomentosum extract. Aqueous preparation containing constituents of crude curare which are therapeutically effective. Curare activity is due almost entirely to presence of tubocurarine.
Properties: The purified extract is a clear, colorless aqueous solution stable to light and heat.
Derivation: A desiccated curare is obtained from a heavy syrup of bark and stems of Chondodendron tomentosum.
Grade: N. N. D.
Use: Medicine.

chondroitin sulfuric acid. A major constituent of the cartilaginous tissue in the body. It is a high molecular weight conjugated protein.
Use: Biochemical research.

chondrus (carrageen; Irish moss; pig-wrack; pearl moss; killeen; rock-salt moss). Dried plant of sea weed Chondrus crispus, edible when clean, and useful for its mucilaginous or gelatinizing qualities.
Properties: Yellowish-white when powdered. Forms a sol in hot water, a gel in the presence of salts. Insoluble in organic solvents.
Habitat: Irish coast; New England.
Grades: Sun bleached; chemically bleached; natural; N. F. XI.
Containers: Bales (220 lb).
Uses: Foods (thickening or suspending agent); clarifying agent; soaps; sizes; leather dressing; pharmaceuticals and medicine.

chop nut. See physostigma.

chorionic gonadotropin (HCG). Isolated from blood and urine of pregnant women; evidence indicates that it is secreted by the placenta. It is a glycoprotein containing about 11% galactose and having a molecular weight of about 100,000.
Properties: Rods or needle-like crystals. Soluble in water and glycols. Relatively unstable in aqueous solution; stable in dry form.
Units: One international unit equals the activity of 0.1 mg of a standard preparation.
Grade: N. N. D.
Use: Medicine; veterinary medicine.

"Chromacyl." [28] Trademark for a group of dyes that contain chromium in the dye molecule. Suitable for wool and nylon. These dyes are characterized by ease of application and good fastness.

chromated zinc chloride. See zinc chloride, chromated.

chromate of soda. See sodium chromate or sodium chromate, tetrahydrate.

chromate red. See chrome red.

chromatography. (See also gas chromatography.) A method of separation based on selective adsorption. A solution of the substance or substances desired is allowed to flow slowly through a column of adsorbent. Different substances will pass with different speeds down the column and will eventually be separated into zones. The column core can then be pushed out and the zones of material cut apart, or the zones can be eluted by passing more solvent down the column and collecting it in small fractions.

Partition chromatography involves the selective solution of the desired material between two solvents. The final solvent, usually water, is used to wet the solid material packed in the column, and the first solvent containing the desired material is poured into the column as above.

Paper chromatography is a micro method. A drop of the liquid to be investigated is placed near one end of a strip of paper. This end is immersed in solvent, which travels down the paper and distributes the materials present in the original drop selectively. Comparison with known substances makes identification possible.

"Chromax Castings." [350] Ferrous alloys containing 20-30% nickel and 15-20% chromium.

chrome alum. See chromium potassium sulfate.

chrome ammonium alum. See chromium ammonium sulfate.

chrome cake. A green form of salt cake (sodium sulfate) containing small amount of chromium. A by-product of sodium dichromate manufacture used in the paper industry.

chrome dye. A mordant dye, most frequently one in which sodium dichromate has been used as the mordant.

chrome glue. See glue.

chrome greens. Paint pigments which are a mixed precipitate of chrome yellow and iron blue. By varying the proportions of yellow and blue a wide range of hues is produced. Chrome greens have excellent lightfastness and good opacity and they are used extensively for almost all types of paints and enamels. They cannot be used where an alkaline condition is present either in paint or on the surface to which a finish is applied. This is due to the sensitivity of both the yellow and the blue to alkali. Chrome greens are produced in what are termed the C.P. and reduced grades. C.P. indicates commercially pure greens that are free from extenders such as barytes, clay, whiting, etc., while the reduced greens contain a base, generally a combination of barytes and clay, the clay being added to aid the suspending properties of the barytes. Chrome greens have the disadvantage of not being absolutely stable for color retention in the package when used in oil paints and enamels, due to the iron blue content. Iron blue is a highly oxidized iron ferrocyanide and in the presence of an oil that dries by oxidation the blue is partially deoxidized, which means that it loses some of its color or strength because of this chemical reaction. Thus, an aged chrome green paint, particularly a tint, may appear yellower in the package than when first made, but after it has been applied and allowed to dry, the shade will generally revert to nearly its original color. Chrome greens are widely used because of their brightness, opacity, lightfastness, excellent strength, and relatively low cost.

chrome-iron ore. See chromite.

"Chromekill 4A." [142] Trademark for a nearly neutral material composed of organic and inorganic reducing agents designed for the reduction of hexavalent chromium to trivalent chromium when it is present in small quantities in alkaline cleaning solutions and in electroplating baths.

"Chromel." [166] Trade name for a series of nickel-chromium alloys. Available as follows:

Grade A. Consists of 80% nickel, 20% chromium. Used for electric heating elements for temperature up to 2100°F.

Grade AA. A modified 80-20 type nickel-chromium alloy for use in controlled atmosphere furnaces requiring optimum resistance to corrosion, carburization and oxidation for temperatures up to 2200°F.

Grade C. Consists of 60% nickel, 16% chromium, balance mainly iron. For electric heating elements up to 1700°F; for rheostats and power resistor purposes.

Grade D. Consists of 35% nickel, 18.5% chromium, balance mainly iron. For heating elements up to 1400°F.

Grade R. A modified 80-20 type nickel-chromium alloy with high electrical resistivity and low temperature coefficient for use in precision resistors and potentiometers.

"Chromel-P." [166] Trademark for an alloy of approximately 90% nickel and 10% chromium, with carefully controlled minor ingredients. It is chiefly used as the positive element in "Chrome-Alumel" thermocouples and lead wire.

chrome-molybdenum steel. A basic open-hearth or electric furnace steel with both chromium and molybdenum. A typical low alloy steel may contain 0.3-1.1% chromium and 0.15-0.3% molybdenum.

Molybdenum high speed steels are often chromium-molybdenum types, with tungsten and vanadium additions.

See also molybdenum steels.

chrome-nickel steels. Usually basic open-hearth steel with chromium and nickel as the main alloying elements. Typical low alloy steel may contain nickel 0.45-1.75%. The 18.8 stainless steels (see steels, stainless) are the most common chrome-nickel steels.

chrome orange. See chrome red.

chrome oxide green (chromic oxide, hydrated). A pigment consisting of chromic oxide (q.v.) and not to be confused with chrome green. It is made by burning sodium dichromate with a reducing agent. The pure grade consists of 99% Cr_2O_3; sp. gr. 5.20. Chrome oxide green is one of the most permanent and indestructible pigments available and is fast to strong alkali and acids. It is weaker tinctorially and less brilliant than chrome greens; has good opacity and low oil absorption. The range of shades obtainable is comparatively small. It is useful in lime-proof paints and finishes that are to be applied to cement surfaces. The hydrated chromium oxide (blue-green in color) is also used as a paint pigment.

Containers: Bags, barrels; fiber drums.

chrome potash alum. See chromium-potassium sulfate.

chrome red (Chinese red; American vermilion; Austrian cinnabar; Derby red; Persian red; Victoria red; chromate red; chrome orange). A paint pigment consisting of basic lead chromate which corresponds to the formula $PbCrO_4 \cdot PbO$, with considerable variation in the proportions of the PbO and $PbCrO_4$.

Properties: Light orange to red powder; color varies depending on the alkalinity. Good body, working well in oil, but is darkened by sulfur and hydrogen sulfide. Good lightfastness. Soluble in acids and alkalies. Insoluble in water.

Derivation: (a) By digesting white lead with potassium dichromate and caustic soda. (b) By boiling chrome yellow (q.v.) with caustic soda or calcium hydroxide. (c) By the precipitation process from lead acetate or nitrate and sodium dichromate.

Grades: Technical, C.P.

Uses: Usually wherever an orange pigment is desired, unless it is necessary for some special purpose to employ a different type to obtain better brilliancy, alkali-fastness, non-settling properties or to avoid the use of a lead pigment. The darker shades find application in the manufacture of rust-inhibitive paints, and are generally referred to as basic lead chromates.

Note: Chrome red can be imitated by coloring white lead, orange lead, or barytes with certain of the coal-tar dyes, e.g., eosine.

chrome steel. See steels, stainless; iron, stainless; and chromium iron alloys.

chrome tanning. Process for the treatment of animal hides using a tanning solution of basic chromium sulfate, perhaps $Cr(OH)SO_4$, with concentration and acidity carefully controlled. Chrome tanning has almost completely replaced vegetable tanning in the production of upper-shoe leather. Ammonium salts are usually added to the tanning bath to control the pH of the skins.

chrome-vanadium steels. Steel made by basic open-hearth or electric furnace, containing up to 1.1% chromium and 0.15% vanadium. Used for high-quality springs.

chrome yellows (primrose chrome; primrose yellow; pale chrome; middle chrome; deep chrome; lemon chrome; lemon yellow; permanent yellow; yellow ultramarine). Yellow paint pigments of lead chromate, $PbCrO_4$.

Colors: A very light-greenish yellow to the lemon or light shade to a medium yellow. Medium yellow is about a normal lead chromate, containing 95% or more lead chromate, $PbCrO_4$. The lighter hues contain varying amounts of coprecipitated lead chromate and lead sulfate. Chrome yellows are very bright yellows and considerable progress has been made in recent years in improving their lightfastness, although even the best chrome yellows darken to some degree on exposure to light. The chrome yellows are considerably brighter and are available in a much greater range of shades than the iron oxide yellows. Chrome yellows are not alkali fast and they will not withstand extremely high baking temperatures without discoloration. Chrome yellows are used in paints and enamels, also kalsomine, but not in casein paints or finishes that are to be applied to surfaces that are alkaline, such as cement or stucco. Being lead pigments, chrome yellows will blacken in the presence of sulfides.

chromic acetate (chromium acetate) $Cr(C_2H_3O_2)_3 \cdot H_2O$.

Properties: Grayish-green powder or bluish-green, pasty mass. Soluble in water; insoluble in alcohol.

Derivation: By the action of acetic acid on chromium hydroxide. The solution is evaporated and crystallized.

Method of purification: Recrystallization.

Grades: Technical; paste; powder; C.P.; $7\frac{1}{2}$% solution.

Containers: 1-lb bottles; wooden barrels; solution in drums.

Uses: Textile mordant; tanning.

Shipping regulations: None.*

chromic acid (chromium trioxide; chromic anhydride) CrO_3. The name is in common use, although the true chromic acid, H_2CrO_4, exists only in solution.

Properties: Dark-purplish red crystals; soluble in water and ether. It is deliquescent and destructive to animal or vegetable tissues. Sp.gr. 2.67-2.82; m.p. 196°C.

Caution: Should not be brought into intimate contact with organic substances or other

*See "I.C.C. Shipping Regulations," page xiii.

reducing agents, as serious explosions are likely to result.
Derivation: (a) Sulfuric acid is added to a solution of sodium dichromate and the product is crystallized out; (b) chromite is fused with soda ash and limestone and then treated with sulfuric acid; (c) electrolytic process.
Method of purification: Crystallization.
Grades: Technical; C.P.
Containers: 100-, 400-lb drums; bottles; tins.
Uses: Chemicals (chromates; oxidizing agent, catalysts); chromium plating; intermediate; medicine (caustic); process engraving; anodizing; ceramic glazes; colored glass, metal cleaning; inks; tanning; paints; rubber pigment; textile mordant.
Danger! Strong oxidant. Contact with other material may cause fire. May cause burns or external ulcers. MCA warning label.
Shipping regulations: Oxidizing material. Yellow label.* Chromic acid solution: Corrosive liquid; white label.*

chromic anhydride. See chromic acid.

chromic bromide (chromium bromide) $CrBr_3 \cdot 6H_2O$.
Properties: Green crystals. Very hygroscopic. Soluble in alcohol and water. Sp.gr. 5.4.
Grades: Technical.

chromic chloride (chromium chloride; chromium sesquichloride) (a) $CrCl_3$ or (b) $CrCl_3 \cdot 6H_2O$.
Properties: (a) Violet crystals. Sp.gr. 2.76; sublimes about 1300°C. Occurs in both soluble and insoluble forms; the latter dissolves easily in water in the presence of a trace of chromous chloride or stannous chloride.
(b) Greenish-black or violet deliquescent crystals depending on whether or not chlorine is coordinated with the chromium. Sp.gr. 2.76; m.p. 83 or 95°C. Very soluble in water; soluble in alcohol; insoluble in ether.
Derivation: (a) By passing chlorine over a mixture of chromic oxide and carbon. (b) By the action of hydrochloric acid on chromium hydroxide.
Grades: Technical; C.P.
Containers: 1-, 5-lb bottles; 450-lb wooden barrels or fiber drums.
Uses: Chromium salts; intermediates; textile mordant; chromium plating; catalyst for polymerizing olefins.
Shipping regulations: None.*

chromic chloride (basic) $Cr_5(OH)_6Cl_9 \cdot 12H_2O$.
Properties: Sp.gr. 1.70. Readily soluble in water, methanol, ethanol and acetone.
Uses: Chromium compound intermediate; mordant in textile printing and dyeing; solvent tanning of leather.

chromic fluoride (chromium fluoride) $CrF_3 \cdot 4H_2O$ or $CrF_3 \cdot 9H_2O$.
Properties: Fine, green crystalline powder. Soluble in water and acids; insoluble in alcohol. Sp.gr. 3.78.
Derivation: By the interaction of chromium hydroxide and hydrofluoric acid.
Method of purification: Crystallization.
Grades: Technical; C.P.
Containers: 1-lb bottles; 400-lb wooden barrels.
Uses: Printing and dyeing woolens; mothproofing woolen fabrics.
Shipping regulations: None.*

chromic formate $Cr(OH)(HCOO)_2$.
Properties: Dark green liquid. Sp.gr. (20°C) 1.237; wt/gal 10.3 lb.
Grade: Technical.
Containers: Bottles.
Uses: Tanning agent; textile mordant; synthesis.

chromic hydrate. See chromic hydroxide.

chromic hydroxide (chromic hydrate; chromium hydroxide; chromium hydrate) $Cr(OH)_3$.
Properties: Green, gelatinous precipitate; decomposed to chromic oxide by heat. Insoluble in water; soluble in acids and strong alkalies.
Derivation: By adding a solution of ammonium hydroxide to the solution of a chromium salt.
Grades: Technical; C.P.
Containers: 1-lb bottles; wooden kegs; fiber containers.
Uses: Chromium salts and chromites; paint pigment.
Shipping regulations: None.*

chromic nitrate (chromium nitrate) $Cr(NO_3)_3 \cdot 9H_2O$.
Properties: Purple crystals; soluble in alcohol and water. B.p. 125.5°C (dec); m.p. 37°C.
Derivation: By the action of nitric acid on chromium hydroxide.
Shipping regulations: Oxidizing material. Yellow label.*

chromic oxide (chromium oxide; chromium sesquioxide; green cinnabar) Cr_2O_3.
Properties: Bright-green, crystalline powder; sp.gr. 5.04; m.p. 1990°C; insoluble in water, acids, and alkalies.
Derivation: (a) By heating chromium hydroxide; (b) by heating dry ammonium dichromate; (c) by heating sodium dichromate with sulfur and washing out the sodium sulfate.
Grades: Technical; C.P.; 99%.
Containers: 1-lb bottles; 100-lb kegs; 300-lb barrels.
Uses: Metallurgy; paint pigment; ceramics; catalyst in organic synthesis.
Shipping regulations: None.*
See also chrome oxide green.

chromic oxide, hydrated. See chrome oxide green.

chromic phosphate (chromium phosphate; Arnaudon's green; Plessy's green). (a) $CrPO_4 \cdot 6H_2O$; (b) $CrPO_4 \cdot 4H_2O$.
Properties: (a) Violet crystals; (b) green crystals. Soluble in acids; insoluble in water.

Derivation: (a) By the interaction of solutions of chromium chloride and sodium phosphate; (b) by mixing cold solutions of chrome alum and disodium hydrogen phosphate. Violet amorphous powder (not the hexahydrate) is formed which becomes crystalline on contact with water. On boiling it is converted into green crystalline hydrate.
Grades: Technical.
Containers: Wooden kegs.
Use: Paint pigment.
Shipping regulations: None.*

chromic sulfate (chromium sulfate) (a) $Cr_2(SO_4)_3$; (b) $Cr_2(SO_4)_3 \cdot 15H_2O$; (c) $Cr_2(SO_4)_3 \cdot 18H_2O$.
Properties: (a) Violet or red powder; (b) dark-green amorphous scales; (c) violet cubes. Sp.gr. (a) 3.012; (b) 1.867; (c) 1.70. (a) Insoluble in water and acids; (b) soluble in water; (c) soluble in water and alcohol.
Derivation: By the action of sulfuric acid on chromium hydroxide, with subsequent crystallization.
Method of purification: Recrystallization.
Grades: Technical; C.P.
Containers: 5-, 10-, 25-lb tins; 500-lb barrels.
Uses: Textile industries; green paints and varnishes; green ink; ceramics (glazes, green effects); tanning.
Shipping regulations: None.*

chromic sulfate, basic. Form of chromic sulfate used in tanning hides. Produced by reducing sodium dichromate with an inexpensive reducing agent.

"Chromindigen." [307] Trademark of mordant dyestuff. Used for the dyeing of wool and characterized by excellent fastness to light and very good fastness to fulling.

chromite (chrome iron ore) $FeCr_2O_4$. A natural oxide of ferrous iron and chromium, sometimes with magnesium and aluminum present. Usually found in magnesium- and iron-rich igneous rocks.
Properties: Color iron-black to brownish-black; streak dark brown; luster metallic to submetallic; sp.gr. 4.6; hardness 5.5.
Grades: Metallurgical; refractory; chemical.
Occurrence: California, Oregon; South Africa; Cuba; U.S.S.R.; Rhodesia; Philippines; Turkey.
Uses: Only commercial source of chromium and its compounds; metallurgy (chrome steel and ferrochrome); refractories; pigments.

chromium Cr. Element of atomic number 24, of group VI of the periodic system.
Properties: Hard, brittle, steel-gray metal. Does not tarnish in air. Resists very strong oxidizing agents due to passivity. Sp.gr. 7.1; m.p. 1900°C; b.p. 2200°C; soluble in acids, except nitric; soluble in strong alkalies; insoluble in water.
Derivation: The only important commercial source of chromium is chromite, $FeO \cdot Cr_2O_3$. American ore production is insignificant. Hence chromium is usually classed as "strategic" from the U.S. viewpoint. The metal is obtained by reducing the oxide by the thermite process using finely divided aluminum; reduction of chromite by carbon yields ferrochrome. Electrolytic chromium, made by electrolyzing chromium solutions, is a commercial product.
Forms available: Lumps, granules, powder; electroplates; high-carbon and low-carbon ferrochromium (about 70% Cr); high purity (99.97+%).
Uses: Stainless steels, chromium-plated ware, and many alloys characterized by high strength and corrosion resistance even at high temperatures; high purity grade used in nuclear energy and high temperature research.

chromium 51. Radioactive chromium of mass number 51.
Properties: Half-life 26.5 days; radiation, gamma and K.
Grade: N.N.D. (as radioactive $Na_2Cr^{51}O_4$).
Uses: Diagnosis of blood volume, blood cell life and cardiac output; etc.
Shipping regulations: Class D poison, radioactive material. Red label.*

chromium acetate. See chromic acetate.

chromium acetylacetonate $[CH_3COCHC(CH_3)O]_3Cr$.
Properties: Purple powder or red-violet crystals; m.p. 216°C; b.p. 340°C; insoluble in water; soluble in acetone and alcohol.
Derivation: Reaction of chromium chloride, acetylacetone and sodium carbonate.
Grades: Technical.
Use: Reduction of detonation of nitromethane.

chromium ammonium sulfate (ammonium chromium sulfate; chrome ammonium alum) $CrNH_4(SO_4)_2 \cdot 12H_2O$.
Properties: Green powder or deep violet crystals. Soluble in water; slightly soluble in alcohol. The aqueous solution is violet when cold; green when hot.
Grades: Technical.
Uses: Mordant; tanning.

chromium borides. At least three have been described: CrB, CrB_2, and Cr_3B_2. They have high melting points, are very hard and corrosion-resistant, and may be suitable for use in jet and rocket engines.
Properties: CrB, may be crystalline; sp.gr. 6.2; m.p. 1550°C; Mohs hardness 8.5; resistivity 67 μ-ohm cm (20°C). CrB_2, sp.gr. 5.15; m.p. 1850°C; hardness 2010 knoop; resists oxidation up to 1100°C. Cr_3B_2, may be crystalline; sp.gr. 6.1; Mohs hardness 9+; resistivity 116 μ-ohm cm (20°C).
Uses: Metallurgical additives; high temperature electrical conductors; cermets; refractories; coatings resistant to attack by molten metals.

chromium carbide Cr_3C_2.
Properties: Orthorhombic crystals; sp.gr.

6.65; microhardness, 2700 kg/sq mm (load 50 g); m.p. 1700°C; resistivity 95μ-ohm cm (room temperature). Has about best oxidation resistance at high temperatures of all metal carbides.
Uses: As gage blocks and hot extrusion dies; in powder form, as spray coating material; as components for pumps and valves which are chemically corrosion-resistant.

chromium carbonate. See chromous carbonate.

chromium chloride. See chromic chloride.

chromium copper. A copper-chromium alloy containing 8-11% chromium. Used in the manufacture of hard steels for increasing elasticity.

chromium fluoride. See chromic fluoride,

chromium hydrate. See chromic hydroxide.

chromium hydroxide. See chromic hydroxide.

chromium-iron alloys. Chromium in iron (1) increases resistance to oxidation and corrosion, (2) increases hardenability, (3) adds some strength at high temperatures, (4) resists abrasion and wear (with high carbon).

Chromium is used to form useful steels containing from 1-30% chromium. In low carbon ranges alloys are ductile; with high-carbon, very hard. Chromium is the essential component of stainless irons and steels.
See steel, stainless; ferrochromium.

chromium manganese antimonide. Brittle gray compound that exhibits magnetic properties only when above a definite temperature. If the composition of the compound is deliberately changed slightly, the temperature needed is shifted.

chromium naphthenate
Properties: Dark-green liquid or violet powder.
Derivation: By addition of chromium salts to solution of sodium naphthenate and recovery of the precipitate.
Grades: 6% chromium.
Containers: 400-lbs (net) standard packages.
Use: Paints (anti-chalking agent).

chromium nitrate. See chromic nitrate.

chromium oxide. See chromic oxide.

chromium oxychloride. See chromyl chloride.

chromium phosphate. See chromic phosphate.

chromium plating. Chromium plating is the process by which a thin, bright surface layer of metallic chromium is electrodeposited, usually from chromic acid-sulfuric acid baths, for protective and decorative purposes. "Hard-chromium" plating refers to the electrodeposition of thicker, very hard, chromium layers for engineering applications.

chromium potassium sulfate (chrome alum; chrome potash alum) $CrK(SO_4)_2 \cdot 12H_2O$.
Properties: Dark, violet-red crystals; efflorescent; sp.gr. 1.813. Soluble in water.
Derivation: By reducing potassium dichromate in dilute sulfuric acid with sulfurous acid.
Grades: Technical; C.P.
Containers: 1-, 5-lb bottles; 25-lb boxes; 100-lb kegs; 432- to 520-lb barrels; 500-lb casks.
Uses: Tanning industry (chrome-tan liquors); textile industry (mordant); photography (fixing bath); ceramics.
Shipping regulations: None.*

chromium sesquichloride. See chromic chloride.

chromium sesquioxide. See chromic oxide.

chromium steel. See steel, stainless; iron, stainless; and chromium-iron alloys.

chromium sulfate. See chromic sulfate.

chromium trioxide. See chromic acid.

chromogen. See chromophore.

"Chromogene." [307] Trademark for mordant dyestuffs used on wool and leather. Characterized by very good fastness to light, fulling, etc.

"Chromol." [244] Trademark for a series of raw oils emulsified with a non-ionic emulsifier and therefore stable to alum and salt. Used in the leather industry.
Containers: Non-returnable steel drums averaging 400-425 lbs net.

"Chromolan." [243] Trademark of metalized acid dyes.

chromophore. Chemical grouping which when present in an aromatic compound (the chromogen) gives color to the compound by causing a displacement of, or appearance of, absorbent bands in the visible spectrum. Dyes are sometimes classified on the basis of their chief chromophores, e.g.; —NO, nitroso dyes; $-NO_2$, nitro dyes; $-N=N-$, azo dyes, etc.

"Chromosorb." [247] Trademark for a line of closely graded (screened) calcined or flux-calcined diatomite aggregates. Available with or without acid washing. Calcined types are inherently pink color (Chromosorb P); flux-calcined types white colored (Chromosorb W). Hydrophobic (silicone treated) grades also available. Choice of grades available dependent on desired mesh or pore size.
Containers: 454 g. Chromosorb P, 300 g. Chromosorb W in glass bottles.
Use: In vapor-phase gas chromatography.

chromous bromide (chromium bromide) $CrBr_2$.
Properties: White crystals; changes to yellow on heating. Oxidizes in moist air but stable in dry air. Sp. gr. 4.356; soluble in water (blue color).

chromous carbonate (chromium carbonate) $CrCO_3$.
Properties: Grayish-blue amorphous mass; sp.gr. 2.75. Soluble in mineral acids;

*See "I.C.C. Shipping Regulations," page xiii.
Reference numbers refer to name of manufacturer. See "List of Manufacturers," page v.

slightly soluble in water containing carbon dioxide; insoluble in alcohol.

chromous chloride $CrCl_2$.
Properties: White, deliquescent needles; active reducing agent; very soluble in water.
Derivation: Reaction of the metal with anhydrous hydrogen chloride.
Grade: C.P.
Uses: Reducing agent; oxygen absorbent.

chromous oxalate $CrC_2O_4 \cdot H_2O$.
Properties: Yellow crystalline powder; soluble in water; active reducing agent.
Grades: C.P.

"Chromoxane." [307] Trademark of mordant dyestuffs. Used on wool; characterized by fairly good fastness to light, very good fastness to fulling, etc., and by relatively bright shade.

chromule. See chlorophyll.

chromyl chloride (chromium oxychloride; chlorochromic anhydride) CrO_2Cl_2.
Properties: Mobile, dark red liquid. B.p. 116°C; freezing point −96.5°C; sp.gr. 1.911. Fumes in air; reacts vigorously with water to form chromic acid, chromic chloride, hydrochloric acid, and chlorine. Miscible with carbon tetrachloride, tetrachloroethane, carbon disulfide.
Derivation: By heating sodium dichromate and sodium chloride with sulfuric acid.
Grades: Technical.
Containers: 400-g bottles; 85-, and 200-lb stainless steel drums.
Uses: Organic oxidations and chlorinations; solvent for chromic anhydride; chromium complexes and dyes.
Caution! Poison!
Shipping regulations: Corrosive liquid. White label.*

chrysamine G $C_{26}H_{16}N_4O_6Na_2$.
Properties: Yellowish-brown powder. Very sparingly soluble in water.
Derivation: By coupling diazotized benzidine with salicylic acid.

chrysamine R $C_{28}H_{20}N_4O_6Na_2$.
Properties: Yellowish-brown powder. Very sparingly soluble in water.
Derivation: By coupling diazotized tolidine with salicylic acid.

chrysanthemummonocarboxylic acid. Occurs in the pyrethrin group of natural and synthetic insecticides (cinerin I, pyrethrin I, allethrin, barthrin, cyclethrin, ethythrin, and furethrin). It is 2,2-dimethyl-3-(2-methylpropenyl)cyclopropanecarboxylic acid.

chrysarobin.
Properties: Microcrystalline orange-yellow powder; slight odor or odorless; tasteless. Soluble in chloroform and benzene; slightly soluble in ether and alcohol; very slightly soluble in water.
Derivation: Extraction from araroba.
Grades: Technical; U.S.P. XVI.
Containers: 1-oz vials; ¼-, 1-lb cartons; 5-, 10-, 50-lb cans.
Use: Medicine.
Caution: Causes dangerous inflammation if it enters the eye; irritating to mucous membranes.
Shipping regulations: None.*

chrysazin. See 1,8-dihydroxyanthraquinone.

chrysene (1,2-benzphenanthrene) $C_{18}H_{12}$ (a tetracyclic hydrocarbon).
Properties: Crystals; sp.gr. 1.274 (20/4°C); m.p. 254°C; b.p. 448°C; sublimes easily in a vacuum. Slightly soluble in alcohol, ether, glacial acetic acid. Insoluble in water.
Derivation: Product of distillation of coal tar.
Use: Organic synthesis.

chrysoberyl $BeAl_2O_4$ or $BeO \cdot Al_2O_3$.
A natural beryllium aluminate containing 19.8% beryllia. Color, various shades of green, yellow or sometimes raspberry or columbine red. Vitreous luster. Is not attacked by acids, but is decomposed by fused alkalies. Infusible by blowpipe.
Constants: Sp.gr. 3.5-3.84; hardness 8.5.
Occurrence: United States (Connecticut, New York, Maine); Brazil; Ceylon; Russia; Ireland.
Use: Gem stone.

chrysocale. An alloy composed of a large proportion of copper with some zinc and lead.

chrysocolla $CuSiO_3 \cdot 2H_2O$. A natural hydrous copper silicate, usually found in the oxidized portion of copper veins.
Properties: Color green, blue-green, brown, to black; streak white to pale blue; luster vitreous to earthy. Usually impure. Sp.gr. 2-2.4; hardness 2-4.
Occurrence: Arizona, New Mexico; Europe.
Uses: Minor ore of copper; gem stone.

chrysoidine (meta-diaminoazobenzene hydrochloride) $C_6H_5NNC_6H_3(NH_2)_2 \cdot HCl$.
Properties: Red-brown powder or large black shiny crystals with a green luster. Soluble in alcohol and water giving orange-brown solutions. Insoluble in ether; m.p. 117°C.
Uses: Chiefly for cotton and silk dyeing to obtain orange colors.

chrysolite. See olivine.

chrysophanic acid (1,8-dihydroxy-3-methylanthraquinone; chrysophanol) $C_{15}H_{10}O_4$.
Properties: Yellow crystals; m.p. 196°C; sublimes. Slightly soluble in water and cold alcohol; soluble in hot alcohol, chloroform and ether.
Source: Found in rhubarb root, cascara sagrada, senna leaves, goa powder.
Derivation: Oxidation of chrysarobin and other synthetic methods.
Use: Medicine.

chrysophanol. See chrysophanic acid.

chrysophenine. A bright yellow synthetic dye of the stilbene group, used in dyeing textiles and leather.

*See "I.C.C. Shipping Regulations," page xiii.
Reference numbers refer to name of manufacturer. See "List of Manufacturers," page v.

chrysoprase. An apple green variety of chalcedony (q. v.).

chrysotile. An important variety of light green to yellow-brown serpentine asbestos with delicate and easily separated fibers and a silky luster. See asbestos.

chu. Abbreviation for centigrade heat unit. It is the amount of heat required to raise the temperature of one pound of water one centigrade degree from 15°C to 16°C. It is sometimes called a pcu (pound centigrade unit).

chymosin. See rennin.

chymotrypsins. Enzymes found in the intestine which catalyze the hydrolysis of various proteins (especially casein) and protein digestion products to form polypeptides and amino acids.

chymotrypsinogen. A crystallizable enzyme occurring in the pancreas, which gives rise to chymotrypsin (q. v.).
Use: Biochemical research.

CI. Abbreviation for "Colour Index," the semiofficial description of dyes arranged by numbers assigned according to chemical classes. The more recent edition of "Colour Index" uses a different set of numbers from that of the first edition. See under dyes for the later numbers. The first edition numbers are all in four figures or less, whereas the later edition uses five-figure numbers.

"Cicoil." [90] Trade name of a raw, permanently liquid oiticica oil derived from the nuts of the Brazilian tree, Licania rigida Benth.
Properties: A drying oil replacement for tung oil. It is the glyceride of alpha-licanic acid (4-keto-9, 11, 13-octadecatrienoic acid). The keto groups in the acid radical give it unusual compatibility with lacquers, urea, melamine formol resins, vinyl resins and alkyds.
Constants: Viscosity (Gardner-Holdt) w-z (25°C); color (Gardner-Holdt), 9-11; heating test (ASTM), max. 17; sp. gr. 0.9770-0.9880 (20°C); refractive index 1.5090-1.5130 (25°C); acid value, max 4% F. F. A.; saponification value, 186-193; unsaponifiable matter, 1.5%.
Uses: Paints, oleoresinous varnishes, synthetic varnishes, linoleum, core oils, water proofing compounds and lacquers.

"Cicoil Standoil." [90] Trade name for "Cicoil" that has been heated to 230°C and held at that temperature for 15-20 min, then cooled. The viscosity is then about Z_4 (Gardner).

cigarette tar. The comparatively non-volatile residue from the burning of cigarette tobacco which appears in finely divided form in the smoke. Cigarette tar is known to contain minute traces of the highly aromatic ring compounds (especially benzo[a]-pyrene) found in coal tar which are supposed to cause cancer. See carcinogen.

"Ciment Fondu." An aluminous cement of the sintered type which originated in France. See cement, aluminous.

cimicifuga (black snake root; black cohosh; bugroot; bugbane; bugwort). Dried rhizome and roots of Cimicifuga racemosa.
Habitat: United States and Canada.
Grades: Technical.
Containers: Bales.
Use: Medicine.
Shipping regulations: None. *

cimicifugin (macrotin). An extract of resins and other bodies occurring in Cimicifuga. Yellowish-brown hygroscopic powder; soluble in alcohol.

cina. See santonica.

cincholepidine. See lepidine.

cinchona bark, calisaya. (Peruvian bark; calisaya bark; yellow cinchona bark; yellow calisaya bark; Jesuits' bark).
Derivation: Dried bark of Cinchona calisaya or other species of cinchona.
Habitat: South America; cultivated in Java, India, Jamaica, Ceylon, and West Africa.
Grades: Technical.
Containers: Bales; bags.
Use: Medicine.
Shipping regulations: None. *

cinchona bark, loxa. (loxa bark; crown bark; loja bark; huanuco bark; cuenca bark).
Derivation: Bark of Cinchona officinalis and other species of cinchona.
Habitat: Loxa and other parts of Ecuador; cultivated in India.

cinchona bark, succirubra (red cinchona; red Peruvian bark; red bark; St. Ann's bark).
Derivation: Dried bark of Cinchona succirubra or of its hybrids.
Habitat: South America; cultivated in Japan, Java, India, and Western Africa.
Containers: Bags.

cinchonidine $C_{19}H_{22}N_2O$, an alkaloid.
Properties: White prisms or powders. Odorless and bitter taste; protect from light. M. p. 210°C with decomposition; soluble in alcohol; slightly soluble in water and ether.
Derivation: By extraction of certain varieties of cinchona bark, and subsequent crystallization. The salts are formed by the action of the respective acid on the alkaloid.
Method of purification: Recrystallization.
Grades: Technical.
Containers: Vials; 5-, 25-, 50-, 100-oz tins.
Use: Medicine.
Shipping regulations: None. *

cinchonidine bisulfate $C_{19}H_{22}N_2O \cdot H_2SO_4 \cdot 5H_2O$.

cinchonidine hydrochloride $C_{19}H_{22} \cdot HCl \cdot 2H_2O$.

cinchonidine sulfate $(C_{19}H_{22}N_2O)_2 \cdot H_2SO_4 \cdot 3H_2O$.

cinchonine $C_{19}H_{22}N_2O$, an alkaloid.
Properties: White, shining prisms or needles;

*See "I. C. C. Shipping Regulations," page xiii.
Reference numbers refer to name of manufacturer. See "List of Manufacturers," page v.

m.p. 264.3°C; slightly soluble in water, alcohol, and ether. Aqueous solution is dextrorotatory.
Derivation: By extraction of the bark of various species of cinchona and subsequent crystallization. The salts are formed by the action of the respective acid on the alkaloids.
Method of purification: Crystallization.
Grades: Technical.
Containers: 1-oz vials; 5-, 25-, 100-oz tins.
Use: Medicine.
Shipping regulations: None.*

cinchonine bisulfate $C_{19}H_{22}N_2O \cdot H_2SO_4 \cdot 4H_2O$.

cinchonine hydrochloride $C_{19}H_{22}N_2O \cdot HCl \cdot 2H_2O$.

cinchonine nitrate $C_{19}H_{22}N_2O \cdot HNO_3 \cdot \frac{1}{2}H_2O$.

cinchonine sulfate $(C_{19}H_{22}N_2O)_2 \cdot H_2SO_4 \cdot 2H_2O$.

cinchophen (phenylcinchoninic acid; 2-phenylquinoline-4-carboxylic acid) $C_9H_5N(C_6H_5)(COOH)$.
Properties: Colorless needles or white to yellowish-white powder; slightly bitter taste; faint odor resembling benzoic acid; affected by light; m.p. 213-216°C. Slightly soluble in ether and alcohol; very slightly soluble in chloroform; practically insoluble in water.
Derivation: By heating alcoholic solution of aniline, pyruvic acid, and benzaldehyde.
Containers: 25-, 100-lb drums.
Use: Medicine.

cinchotine. An alkaloid of cinchona. Produced by the oxidation of cinchonine.

cinder. See slag.

cinene. See dipentene.

cineol. See eucalyptol.

cinerin I $C_{20}H_{28}O_3$. One of the four primary active insecticidal principles of pyrethrum flowers. It is the 3-(2-butenyl)-4-methyl-2-oxo-3-cyclopenten-1-yl ester of chrysanthemummonocarboxylic acid. See also cinerin II, pyrethrin I and II.
Properties: A viscous liquid, quickly oxidized in air; b.p. 200°C (0.1 mm). Insoluble in water; soluble in organic solvents; incompatible with alkalies.
Use: Insecticide.
Shipping regulations: None.*

cinerin II $C_{21}H_{28}O_5$. One of the four primary active insecticidal principles of pyrethrum flowers. It is the 3-(2-butenyl)-4-methyl-2-oxo-3-cyclopenten-1-yl ester of chrysanthemumdicarboxylic acid monomethyl ester. See also cinerin I, pyrethrin I and II.
Properties: A viscous liquid, quickly oxidized in air; b.p. 200°C (0.1 mm). Insoluble in water; soluble in organic solvents.
Use: Insecticide.
Shipping regulations: None.*

cinnabar (natural vermilion, liver ore) HgS. A natural mercuric sulfide, found in veins near recent volcanic rocks and hot springs.
Properties: Color cochineal red, scarlet, reddish brown to blackish; streak scarlet; luster adamantine to dull earthy when impure; sp.gr. 8.10; hardness 2.5. Soluble in aqua regia.
Occurrence: California, Nevada; Spain; Italy; Mexico; Yugoslavia.
Use: The only important ore of mercury.

cinnamaldehyde. See cinnamic aldehyde.

cinnamein. See benzyl cinnamate.

cinnamene. See styrene.

cinnamenol. See styrene.

cinnamic acid (beta-phenylacrylic acid; cinnamylic acid) $C_6H_5CH{:}CHCOOH$.
Properties: White, crystalline scales; soluble in benzene, ether, acetone, glacial acetic acid, carbon disulfide, oils; insoluble in water.
Constants: Congealing point 133°C (min); b.p. 300°C; soluble in alcohol and ether; slightly soluble in water.
Derivation: By heating benzaldehyde with sodium acetate in presence of a dehydrating agent (acetic anhydride) or by heating benzyl chloride with sodium acetate in an autoclave. Occurs naturally in Peru and Tolu balsams and styrax.
Grades: Technical; refined.
Containers: 1-, 5-lb glass bottles; 1-, 5-, 10-lb cans.
Uses: Medicine; perfumes.
Shipping regulations: None.*

cinnamic alcohol (cinnamyl alcohol; phenylallylic alcohol; styrylic alcohol; 3-phenyl-2-propen-1-ol; styryl alcohol) $C_6H_5CH{:}CHCH_2OH$.
Properties: White needles or crystals; hyacinth-like odor. Soluble in 3 volumes of 50% alcohol.
Constants: Congealing point 33°C (min) (pure), as low as 24°C (tech.); b.p. 257°C.
Derivation: (a) From oil of cassia or oil of cinnamon. Occurs as an ester. (b) Reduction of cinnamic aldehyde.
Method of purification: Recrystallization.
Grades: Technical.
Containers: 1-, 2-, 5-lb bottles; tin cans.
Uses: Perfumery, particularly for lilac and other floral scents.
Shipping regulations: None.*

cinnamic aldehyde (cinnamaldehyde; 3-phenylpropenal; cinnamyl aldehyde) $C_6H_5CH{:}CHCHO$.
Properties: Yellowish oil; cinnamon odor. Soluble in 5 volumes of 60% alcohol; very slightly soluble in water. Keep well stoppered.
Constants: Sp.gr. 1.048-1.052; refractive index 1.618-1.623; m.p. −8°C; b.p. 248°C.
Derivation: (a) From Ceylon and Chinese cinnamon oils. (b) By condensation of benzaldehyde and acetaldehyde.
Method of purification: Rectification.
Containers: 1-, 2-, 5-, 10-lb bottles; drums.
Uses: Flavors; spice perfumes.

cinnamic ether. Incorrect name for ethyl cinnamate.

cinnamol. See styrene.

cinnamon (cinnamon, Saigon).
Derivation: Bark of Cinnamomum loureirii.
Habitat: Anam (Cochin China); cultivated in Java, Sumatra and South America.
Grades: Technical; N. F. XI.
Containers: Boxes.
Uses: Medicine; source of cinnamon oil; flavoring; condiment.
Shipping regulations: None.*

cinnamon, artificial. Mixture of 10-25 parts cinnamic aldehyde, one part eugenol applied to a mixture of pecan shell powder and powdered toasted wheat bran, wood flour, bark flour, or similar materials. The oily mixture is dissolved in alcohol and the solution applied to the carrier powder and the solvent allowed to evaporate. Sometimes very small proportions of oil of cassia or oil of cinnamon are also added to the oily mixture.

cinnamon bark oil. See cinnamon oil.

cinnamon, cassia (cinnamon; cassia bark; Chinese cinnamon).
Derivation: Bark of Cinnamomum cassia.
Habitat: Southern China and Anam.
Uses: Medicine; source of cassia oil; flavoring; condiment.

cinnamon, cassia oil. See cassia oil.

cinnamon, Ceylon.
Derivation: Inner bark of the shoots of Cinnamomum zeylanicum.
Habitat: Ceylon, Sumatra, and Borneo; cultivated in tropical Africa, America, and Asia.
Grades: Technical; N. F. XI.
Containers: Boxes; bags.
Uses: Medicine; source of cinnamon oil; flavoring; condiment.
Shipping regulations: None.*

cinnamon leaf oil.
Properties: Pale yellow liquid, with pungent cinnamon odor.
Grades: (a) Ceylon; (b) Seychelles.
Chief known constituents: (a) Eugenol 70-90%; cinnamic aldehyde, up to 6%; cinnamic alcohol, safrole, linalool, benzyl benzoate; (b) same constituents; with eugenol content from 75-95%.
Constants: (a) Sp. gr. 1.03-1.06; optical rotation −1° 30' to +2° 20'; refractive index 1.525-1.540; (b) sp. gr. 1.02-1.06; optical rotation and refractive index, same as (a). Soluble in about 1.5 volumes of 70% alcohol; sometimes slightly opalescent.
Containers: Drums.
Derivation: Distilled from the leaves of Cinnamomum zeylanicum Nees.
Uses: Medicine; flavoring; perfumery.

cinnamon oil (cinnamon bark oil).
Properties: A light yellow volatile oil, having a spicy cinnamon odor, more delicate in the Ceylon variety.
Grades: (a) Ceylon; (b) Seychelles.
Chief known constituents: Cinnamic aldehyde (up to 70%); eugenol (up to 18%); phellandrene, benzaldehyde, linalool, and (for Seychelles variety) camphor.
Constants: Sp. gr. 1.014-1.040; refractive index 1.569-1.584; optical rotation, levo, up to −2°.
Derivation: Distilled from the bark of Cinnamomum zeylanicum Nees.
Containers: Bottles.
Uses: Medicine; flavoring; perfumery.
Shipping regulations: None.*

cinnamon oil, Chinese. See cassia oil.

cinnamon oil, U.S.P. XVI. See cassia oil.

cinnamon, Saigon. See cinnamon.

cinnamon, white. See canella.

cinnamon wood. See sassafras bark.

cinnamoyl chloride $C_6H_5CH{:}CHCOCl$.
Properties: Yellow crystals; m.p. 35°C; b.p. 170°C (58 mm); sp. gr. 1.1617 (45/4°C).
Use: Reagent for determination of water.

cinnamyl acetate $C_6H_5CH{:}CHCH_2OOCCH_3$.
Properties: Colorless liquid having a floral-spicy odor. Soluble in 4 volumes of 70% alcohol.
Constants: Sp. gr. 1.048-1.052; refractive index 1.539-1.542.
Use: Perfumery, as a fixative.

cinnamyl alcohol. See cinnamic alcohol.

cinnamyl aldehyde. See cinnamic aldehyde.

cinnamyl cinnamate (styracin) $C_9H_7O_2C_9H_9$.
Properties: Rectangular prismic crystals; m.p. 40°C (min). Soluble in alcohol, ether, benzene.
Derivation: Esterification of cinnamic acid with cinnamic alcohol.
Use: Perfumery.

cinnamylic acid. See cinnamic acid.

cinnamylic ether. Incorrect name for ethyl cinnamate.

"Cinnaryl." [233] Trademark for allyl cyclohexanepropionate used as food flavor and as a scent.

C-IPC. See chloro-IPC.

circa. Latin word for about or approximately, used here in describing properties. Abbreviation: ca.

"Cirkal." [244] Trademark for an alkaline detergent. Highly alkaline, rapidly soluble, non-sudsing powder.
Uses: In-place cleaning in dairies and food plants.
Hazards: Concentrated alkali; can cause burns.

"Cirrasol." [206] Brand name of proprietary textile softening agents and fiber processing assistants.

cirtine (false topaz). A form of native silica or quartz (q.v.), light yellow in color. The "Occidental topaz" or "Spanish topaz" of jewelers.

*See "I.C.C. Shipping Regulations," page xiii.
Reference numbers refer to name of manufacturer. See "List of Manufacturers," page v.

cis-. A prefix denoting that one of two geometrical isomers (q. v.) in which certain atoms or groups are on the same side of a plane. See also trans-. In this dictionary, these prefixes are disregarded in the alphabetizing.

"Cis-4." [303] Trademark for a cis-1,4-polybutadiene synthetic rubber.

citraconic acid (methylmaleic acid) $CH_3C(COOH):CH(COOH)$.

Properties: Hygroscopic colorless crystals. M.p. 91°C (dec); sp.gr. 1.62. Soluble in water, alcohol, ether; insoluble in benzene and petroleum ether.

Derivation: By carefully heating citric acid.

Grades: Technical; C.P.

citraconic anhydride (methylmaleic anhydride) $C_5H_4O_3$.

Properties: Colorless liquid. M.p. 7-8°C; b.p. 213-214°C; sp.gr. 1.25 (15/4°C). Soluble in ether.

Grades: Reagent.

Containers: Sealed tubules.

Uses: Reagent for alkalies, alcohols, and amines.

citral (geranial; geranialdehyde; 3,7-dimethyl-2,6-octadienal) $(CH_3)_2CCHCH_2CH_2C(CH_3)CHCHO$. As found in commerce, a mixture of alpha and beta isomers.

Properties: Mobile pale yellow liquid having strong lemon odor; sp.gr. 0.891-0.897 (15°C); refractive index 1.4860-1.4900 (20°C); not optically active. Soluble in 5 volumes of 60% alcohol; soluble in all proportions of benzyl benzoate, diethyl phthalate, glycerin, propylene glycol, mineral oil, fixed oils, and 95% alcohol; insoluble in water.

Derivation: Principal constituent of lemon grass oil and can be isolated by fractional distillation. Obtained synthetically by oxidation of geraniol, nerol, or linalool by chromic acid.

Grades: Technical; pure.

Containers: Glass bottles; tins; (synthetic) drums.

Uses: Perfumes; flavoring agent; intermediate for other aromatics.

citric acid (2-hydroxy-1,2,3-propane-tricarboxylic acid) $HOOCCH_2C(OH)(COOH)CH_2COOH \cdot H_2O$. Also available in anhydrous form.

Properties: Colorless translucent crystals, or as a white granular to fine crystalline powder; odorless; strongly acid taste; hydrated form is efflorescent in dry air. Sp.gr. 1.542; m.p. 153°C (anhydrous form). Very soluble in water and alcohol; soluble in ether.

Occurrence: Widely in living cells, both animal and plant. See TCA cycle.

Derivation: By mold fermentation of carbohydrates, including deep fermentation; from lemon, lime, pineapple juice, molasses.

Grades: Both hydrous (hydrated) and anhydrous; technical; C.P.; U.S.P. XVI.

Containers: Various, including bags, cartons, barrels, drums.

Uses: Preparation of citrates, flavoring extracts, confections, soft drinks, effervescent salts; medicines; antioxidant in foods; sequestering agent, including water conditioning; cleaning and polishing agent for stainless steel and other metals; alkyd resins; mordant.

citric acid cycle. See TCA cycle.

citrine ointment (mercuric nitrate ointment). A fatty ointment prepared from 7 parts mercury, 17 parts by weight of nitric acid and 76 parts of lard. When freshly made it has a greenish-yellow color but it becomes brown with age.

Containers: Jars and wooden tubs. Contact with all metal containers must be avoided.

Use: Pharmaceutical.

Shipping regulations: None.*

citrinin $C_{13}H_{14}O_5$. An antibiotic yellow pigment produced by Penicillium citrinin Thom and Aspergillus niveus; m.p. 170-171°C (decomposes). Yellow crystals; insoluble in water; soluble in alcohol, dioxane, dilute alkali solutions.

"Citroflex-2." [299] Trademark for triethyl citrate.

"Citroflex-4." [299] Trademark for tributyl citrate.

"Citroflex A-2." [299] Trademark for acetyl triethyl citrate.

"Citroflex A-4." [299] Trademark for acetyl tributyl citrate.

"Citronel 'B' and 'C'." [188] Brand name for lemon oil substitutes for technical use.

citronellal $C_9H_{17}CHO$. 3,7-Dimethyloct-6(or 7)-enal. Has both d- and ℓ-isomers. The d-isomer is described.

Properties: Colorless liquid having an intense lemon-like odor; sp.gr. 0.847-0.850; optical rotation +8° to +11°; b.p. 205°C; refractive index (n 20/D) 1.4566. Slightly soluble in water; soluble in alcohol and ether.

Derivation: From lemon, lemon grass, citronella oil and other oils.

Containers: Bottles; drums.

Uses: Soap perfumery; raw material for manufacture of hydroxycitronellal.

citronellal hydrate. See hydroxycitronellal.

citronella oil.

Properties: Light yellowish oil, with rather pungent, citrus-type odor. Soluble in 80% alcohol. Sp.gr. 0.887-0.906; refractive index 1.468-1.483; solutions are levorotatory.

Derivation: Steam-distilled from the grass of Cymbopogon nardus.

Grades: Ceylon; Java.

Containers: Tins; glass bottles.

Uses: Medicine; insect repellent; perfumes for soaps and disinfectants; source for manufacture of citronellal, geraniol, and products derived therefrom.

*See "I.C.C. Shipping Regulations," page xiii.
Reference numbers refer to name of manufacturer. See "List of Manufacturers," page v.

citronellol $C_9H_{17}CH_2OH$. A mixture of two structural isomers, 3,7-dimethyloct-6 (or 7)-enol.

Properties: Colorless liquid, having a somewhat rosy odor; sp.gr. 0.849-0.853; refractive index 1.456-1.458; optical rotation -1° 30' to +1° 30'. Soluble in two or more volumes of 70% alcohol.

Occurrence: In oils of citronella, geranium, rose, savin and other essential oils.

Derivation: From the oils above, or by reduction of citronellal or geraniol.

Containers: Bottles; drums.

Use: Perfumery, for floral odors, mainly of the rose types.

citronellyl acetate (citronellyl acetic ether) $C_{10}H_{19}OOCCH_3$.

Properties: Colorless liquid. Odor somewhat like that of bergamot oil.

Constants: Sp.gr. 0.884-0.891; b.p. 119-121°C (15 mm); optical rotation usually slightly dextro, up to +1°; refractive index 1.450-1.452. Soluble in 9 volumes of 70% alcohol.

Derivation: Action of acetic anhydride upon citronellol.

Grades: Technical.

Use: Perfumery.

citronellyl acetic ether. Incorrect name for citronellyl acetate.

citron yellow. See zinc yellow.

citrophen (para-phenetidine citrate) $C_3H_4OH(COOH)_3 \cdot C_6H_4(OC_2H_5)NH_2$.

Properties: White crystalline powder. M.p. 188°C. Slightly soluble in water.

citrovorum factor. See folinic acid.

citrulline $NH_2CONH(CH_2)_3CHNH_2COOH$. An arginine derivative. It is an amino acid found in watermelon juice, in the L(+) form.

Properties: Crystals from methanol—water mixture; m.p. 222°C; soluble in water; insoluble in methanol and ethanol.

Use: Biochemical research.

citrullus colocynthis. See colocynth.

citrus aurantium. See orange peel, sweet, and orange peel, bitter.

citrus flavonoid compounds. See bioflavonoids.

civet (zibeth).

Derivation: Unctuous secretion from the civet cat, Viverra civetta, Zibetha.

Habitat: Asia; Malakka Islands; Ethiopia; and East Indies.

Properties: Yellow to brown. Semi-solid. Soluble in hot alcohol and ether; insoluble in water.

Grades: Technical (crude, refined, artificial).

Containers: Horns of variable size ranging from 20-40 ozs each; 1-, 2-, 4-, 8-oz jars.

Use: Perfumery (fixative).

Shipping regulations: None.*

civettal. See "Tetraquinone."

"Civona." [28] Trademark for a hollow filament rayon fashion yarn. See rayon.

Cl. Symbol for chlorine.

clad metal. A special kind of combination in which two different metals are bonded by being rolled together with proper pressure, temperature and length of time. At the interface each metal diffuses sufficiently into the other to form an alloy and a permanent union.

clad steels. Composite materials consisting of a base of open-hearth steel that has a layer of pure nickel or nickel alloy bonded to it on one or on both sides. The plate is formed by pressure-welding in rolling mills at 2200°F. A solid solution of nickel and iron is formed at the junction. The physical properties of the cladding are very similar to those of the iron plate so that great stresses are not set up across the interface. Supplied with standard cladding, which may be 5, 10, 15, and 20% of the total metal thickness. Can be welded with arc or acetylene and can be obtained with different cladding on each side, if necessary.

clarain. One of the types of physical structure found in coal (see also fusain, durain, and vitrain). Clarain occurs as lustrous bands or striations, especially in bituminous coal. It is less friable than vitrain.

"Clarase." [212] Trademark for a product containing diastatic and proteolytic enzymes and, in addition, maltase, trypsin, rennet, erepsin, lipase, and hemolysin.

Properties: Dry, fine white powder, fully water-soluble, non-hazardous, non-flammable. Optimum pH for diastatic reaction 5.0-5.4, for protein solubilizing 7.0-7.5, for amino-acid liberating 6.2-6.6; optimum temperature 45°C.

Grade: For food products.

Containers: 1-, 5-, 10-, and 25-lb tin cans.

Uses: For starch conversions in the food industry; paste making; manufacturing of sizing materials; removal of colloidal starches from fruit juices.

"Clarite." [304] Trademark for a two part barium-cadmium organic vinyl stabilizer.

Properties: ("Clarite A") soft white powder; sp.gr. 2.34; refractive index 1.56 max; ("Clarite B") clear straw colored liquid; sp.gr. 0.91; refractive index 1.45.

Containers: ("Clarite A") fiberboard drums containing 75 lbs. ("Clarite B") metal drums containing 35 lbs net.

Uses: Heat and light stabilizers for vinyl extrusions, flooring and molded products.

clary sage oil.

Properties: Pale-yellow to light-yellow liquid, with warm odor suggestive of ambergris.

Derivation: Obtained by distillation of Salvia sclarea L. A floral absolute, solid, is obtained by volatile solvent extraction from the same plant.

Containers: Bottles.

Use: Perfumery.

*See "I.C.C. Shipping Regulations," page xiii.
Reference numbers refer to name of manufacturer. See "List of Manufacturers," page v.

clathrate compounds. See inclusion complexes.

Claude synthetic ammonia process. A European-developed process for synthesizing ammonia from its elements, characterized by high pressure and temperature with high conversion and no recirculation. Hydrogen is separated from coke-oven gas by its failure to liquefy, and nitrogen is obtained from liquid air.

Claude system. A process for the production of liquid air in which the compressed gas is made to perform work in an expansion engine and thus cool itself.

clay. General term for a great variety of aluminum silicate-bearing rocks of various compositions and degrees of purity. They are plastic when wet, and harden when fired. Typical minerals comprising the major proportion of any clay are the following:

kaolinite
$Al_2O_3 \cdot 2SiO_2 \cdot 2H_2O$
halloysite
$Al_2O_3 \cdot 3SiO_2 \cdot 2H_2O$
montmorillonite
$(Mg, Ca)O \cdot Al_2O_3 \cdot 5SiO_2 \cdot nH_2O$
illite
$K_2O, MgO, Al_2O_3, SiO_2, H_2O$;
all in variable amounts.

Clay usually contains other minerals including quartz, calcite, limonite, gypsum, and muscovite. Clay is usually formed by the weathering or alteration of aluminum-bearing rocks, such as granite or other igneous rocks containing feldspar.

Clays are used as ceramic raw materials for white ware, stoneware, pottery, glazing, tile, bricks, terra cotta, fire brick, crucibles, retorts, mortars, factory molds, drainage pipe; in the manufacture of cement; as a filler for paper, paint, rubber, linoleum, oil cloth; well-drilling mud; filtering; decolorizing oils and other liquids; in abrasives; in medicine; and in insecticide dusts.

The more important varieties of clays include kaolin or china clay, ball clay, fire clay, stoneware clay, bentonite, fuller's earth, activated clays, dusting clays, bleaching clays, calcined clays, colloidal clays, enamel clays, and filler clays. Numerous other special clays are named according to their use, as papermakers' clay, rubbermakers' clay, and sagger clay.

Containers: Bags.

"Clearate." [49] Tradename for a high grade soya lecithin. A surface-active substance and an effective anti-oxidant. Used to improve dispersion of ingredients in a mixture, as in manufacture of baked goods, confections, inks, cosmetics and paints.

cleavage. The ability of crystalline substances to split or break along definite planes. A crystal may cleave in one direction, as in mica; or in several directions, as in calcite, galena, or feldspar.

clemizole hydrochloride 1-para-Chlorobenzyl-2-pyrrolidylmethyl benzimidazole. Used in medicine as an antihistaminic.

"Clenesco." [428] Trademark for a line of cleaners for food plants and dairies. Contains one or more of the following: silicated alkali, complex phosphates, wetting agents, chromates, phosphated caustic, acids, solvents, chlorine, potassium soap, iodine, detergents, quaternary derivatives.

cleveite. A crystallized variety of uraninite (q. v.). It contains about 10% of the yttrium earths.

Cleve's acid. See 1-naphthylamine-6-sulfonic acid; or 1-naphthylamine-7-sulfonic acid; known also as Cleve's 1,6 acid and Cleve's 1,7 acid respectively.

cliffstone Paris white. A special grade of whiting (q. v.) made from a hard grade of English chalk.

"Climax Molybdenum." [67] Trademark for molybdenum concentrate and molybdenum sulfide.

"Climelt." [67] Trademark for molybdenum and tungsten metals and their respective alloys.

"Clipper Cleaner." [244] Trademark for a compound consisting of sodium carbonate as a base combined with complex phosphates and surface active agents.

Properties: Light cream, granular, dustless mechanical mix. Soluble in water; total Na_2O content 37.2-42.4%.

Containers: 40-lb (four 10-lb cans); 25-lb galvanized steel pails; 125-lb plywood drums; 325-lb wooden barrels.

Uses: Especially designed for creamery dairy barn cleaning; milking machines, etc; hand cleaning operations; tanks, vats.

"Clopane Hydrochloride." [100] Trade name for cyclopentamine hydrochloride (N, alpha-dimethylcyclopentaneethylamine hydrochloride; cyclopentyl-2-methylaminopropane hydrochloride).

Properties: A white, odorless, crystalline powder with a mild characteristic odor and a bitter taste; m. p. 113-116°C; soluble in water (1:1), alcohol (1:1.8), benzene (1:23.8), chloroform (1:1.3); slightly soluble in ether; pH (1% solution) about 6.2.

Use: Medicine.

"Clorafin." [266] Trademark for a series of chlorinated paraffins.

"Clorafin 40." Straw color; viscous liquid; 40-44% chlorine; viscosity 25-40 poises at 25°C.

"Clorafin 40V." Straw color; viscous liquid; 40-44% chlorine; viscosity 3-5 poises at 25°C.

"Clorafin 42." Light brown; viscous liquid; 40-44% chlorine; viscosity 25-40 poises at 25°C.

"Clorafin 42S." Straw color; viscous liquid; 40-44% chlorine; viscosity 25-40 poises at 25°C.

"Clorafin 50." Yellow; viscous liquid; 48-52% chlorine; viscosity 350-525 poises at 25°C.

Uses: Components of flame-proofing compositions for fabrics; flameproof paints, and flameproof adhesives; chemical resistant coatings, paints, and inks. Treated wood has a greatly reduced moisture absorption.

"Clorgran." [147] Brand name for free-flowing granular insecticides containing 5, 10 or 25% chlordan.
Containers: 5- and 50-lb bags.
Uses: For control of certain soil-inhabiting insects such as wireworms, and Japanese beetle grubs; crabgrass control.

closed-circuit grinding. A method of grinding or pulverizing in which the material that has not been sufficiently reduced in size is separated and returned for further grinding.

clove. See caryophyllus.

clove oil (caryophyllus oil).
Properties: A pale-yellow, thin liquid; darkens and thickens with age and exposure; strong aromatic odor; pungent and spicy taste. Soluble in ether and chloroform; soluble in 1 to 2 vols. and more of 70% alcohol; fresh, so-called extra-light oils in 2.5 to 3 vols. of 60% alcohol.
Chief known constituent: Eugenol (U.S.P. XVI specifies not less than 82%).
Constants: Sp.gr. 1.038-1.060; b.p. 250-260°C; refractive index 1.5270-1.5350; optical rotation to −1° 10'.
Derivation: Distilled from cloves, the unexpanded flowers of Eugenia aromatica (Eugenia caryophyllata).
Method of purification: Rectification.
Grades: Technical; U.S.P. XVI.
Containers: 1-, 6-lb bottles; drums.
Uses: Medicine (local); flavoring; dentistry; perfumery; confectionery; soaps.
Shipping regulations: None.*

club-moss. See lycopodium.

Cm. Symbol for curium.

"CM." [28] Trademark for a flame-retardant composition based on ammonium sulfamate and modified to prevent afterglow and to improve penetration.
Properties: Fine, white, granules; soluble in water; insoluble in dry cleaning solvents.
Containers: 50-lb fiber drums and 100-lb paper bags.
Use: Renewable type flame-retardant treatment for fabrics, paper, paper products, and other cellulosic materials.

CMC. See sodium carboxymethylcellulose.

CMP. Abbreviation for cytidine monophosphate. See cytidine phosphates.

CMU. Abbreviation for chlorophenyl-dimethylurea.

Co. Symbol for cobalt.

Co I. Abbreviation for coenzyme I. See nicotinamide adenine dinucleotide.

Co II. Abbreviation for coenzyme II. See nicotinamide adenine dinucleotide phosphate.

CoA. See coenzyme A.

coacervation. The salting out of a lyophilic sol into liquid droplets rather than into solid aggregates. The process is reversible and may be an intermediate stage in the coagulation of such sols.

"Coagulant Aid." [108] A series of effective combinations of polyelectrolytes and other materials.
Containers: Bags; drums.
Uses: Aid in the clarification of water for municipal and industrial uses.

coagulation (flocculation). The process of converting a finely divided or colloidally dispersed suspension of a solid into particles of such size that reasonably rapid settling occurs. This is usually accomplished by adding the salt of a di- or tri-valent metal. Thus alum, aluminum sulfate, and ferric sulfate are commonly added in clarifying water from suspended impurities.

Coahran process. Recovery of acetic acid from pyroligneous acid by extracting with ether. It is an improved version of the Brewster process (q.v.), but is basically the same. It is often referred to as the Brewster process.

coal. A natural solid combustible material consisting of amorphous elemental carbon with various amounts of hydrocarbons, complex organic compounds and inorganic materials. Coal was formed from prehistoric plant life and now occurs in layers with other sedimentary rocks.

Coal is classified into ranks according to its heating value, expressed in Btu/lb, and its fixed carbon content.

Rank	Fixed Carbon	Btu/lb
Anthracite	86-98%	13,500-15,600
Low; Medium Volatile Bituminous	69-86%	14,500-15,600
High Volatile Bituminous	46-69%	11,000-15,000
Subbituminous	46-60%	8,300-13,000
Lignite	46-60%	5,500- 8,300

Coal is most frequently specified in terms of its proximate analysis, giving the percentages of moisture, volatile combustible matter, fixed carbon, and ash. An ultimate analysis gives the percentages of the various elements present (C, H, O, N, and S). Coals are also described in terms of the petrographic constituents. Four distinct physical structures are recognized: clarain, durain, fusain, and vitrain (see also anthraxylon).

Coal has been surpassed by petroleum and natural gas as a source of energy in the U.S. It is an important source of chemical raw materials, which are obtained by destructive distillation, or coal liquefaction.

coal gas (bench gas; coke-oven gas). A mixture of gases produced by the destructive distillation of bituminous coal in highly heated fire-clay or silica retorts or in

*See "I.C.C. Shipping Regulations," page xiii.
Reference numbers refer to name of manufacturer. See "List of Manufacturers," page v.

by-product coke ovens. A typical gas analyzes 49.8% hydrogen, 29.5% methane, 4.0% illuminants, 8.5% carbon monoxide, 3.2% ethane, 1.6% carbon dioxide, 0.4% oxygen, 3.2% nitrogen. Candle power 16.6; Btu about 662 per cu ft.
Use: Domestic and industrial heating and lighting; source of coal tar, ammonia, benzene, toluene, xylene, and related materials.
Shipping regulations: Flammable gas; red gas label.*

coal, glance. See carbon, gas.

coal hydrogenation. See coal liquefaction.

coal liquefaction (coal hydrogenation). The conversion of coal into liquid hydrocarbons and related compounds by hydrogenation at elevated temperatures and pressures.

coal oil.
1. Kerosine made directly from crude petroleum (archaic).
2. The crude oil obtained by the destructive distillation of bituminous coal; or the distillate obtained from this crude oil, which may be used for illuminating purposes.

coal, soft. See bituminous coal.

coal tar.
Properties: A black, viscous liquid (or semi-solid), denser than water, with a characteristic naphthalene-like odor and a sharp burning taste; obtained in the destructive distillation of coal.
Soluble in ether, benzene, carbon disulfide, chloroform; partially soluble in alcohol, acetone, methanol, and benzene; only slightly soluble in water.
Grades: Crude; refined; U.S.P. XVI.
Containers: Tank-cars; barrels.
Uses: A major raw material for a great variety of dyes, drugs, and other organic chemicals. The most important single components, recovered by distillation and treatment with acids, alkalies and solvents, include benzene, toluene, xylene, phenol, naphthalene, anthracene, pyridine, carbazole, and phenanthrene. The crude or refined product or fractions thereof are also used for waterproofing, paints, pipecoating, roads, roofing, insulation, as pesticides, and in medicine.
Shipping regulations: May be classed as flammable liquid. Red label.*

coal-tar, colorless. See anthrasol.

coal-tar dyes. Dyes produced from the coal-tar hydrocarbons or their derivatives such as benzene, toluene, xylene, naphthalene, anthracene, aniline, etc.

coal-tar naphtha. See naphtha, solvent.

coal-tar pitch. A dark brown to black amorphous residue left after coal tar is redistilled. It amounts to 50-60% of the usual grades of coal-tar. A solid, melting at 150°F, it is used as a thermoplastic, in roofing, road surfacing; in the Hall electrolytic process for making aluminum.

coal-tar resins. See coumarone-indene resins.

coating clay. A high-grade, smooth, grit-free, white china clay for coating paper and textiles.

cobalamin. See vitamin B_{12}.

cobalamin concentrate (vitamin B_{12} activity concentrate). The dried partially purified product resulting from the growth of selected Streptomyces cultures or other cobalamin-producing microorganisms. The commercial product may contain harmless diluents or stabilizing agents.
Properties: Pink to brown granules or fine powder; may be hygroscopic; solutions affected by light; pH (1:200 solution) 4.0-8.0.
Grades: N.F. XI.
Use: Medicine; nutrition.

cobalt Co. Element of atomic number 27, group VIII of the periodic system.
Properties: Steel-gray, slightly pinkish, shining, hard, ductile, somewhat malleable metal; magnetic. Sp.gr. 8.9; m.p. 1493°C; b.p. 3550°C. Soluble (slowly) in dilute hydrochloric and sulfuric acid, (readily) in nitric acid. Forms useful coordination compounds.
Occurrence: Not found native. Principal ores are arsenides, sulfides, oxides, silicates, with nickel or iron (smaltite, cobaltite, chloanthite, linnaeite, niccolite, pentlandite). Principal sources of ores: Canada; Belgian Congo; Rhodesia; North Africa.
Derivation: From ore concentrates by roasting followed (a) by thermal reduction by aluminum, or (b) by electrolytic reduction of solutions of metal; (c) by leaching, with either ammonia or acid in an autoclave under elevated temperatures and pressures and subsequent reduction by hydrogen.
Forms available: Rondels (1 in. x 3/4 in.); shot; anodes; 150 and finer mesh powder.
Containers: 500-lb kegs; drums.
Uses: Chemical (cobalt salts, catalyst, oxidizing agent); electroplating; ceramics; lamp filaments; catalyst in hydrogenation of oils; coloring glass; inks; paints and varnishes; colored signs; cermets. Principal use is in alloys, especially cobalt steels for permanent supermagnets and cobalt-chromium high-speed tool steels; cemented carbides. Cobalt alloys are used at high temperatures, as in jet engines.
Shipping regulations: None.*

cobalt 60. Radioactive cobalt of mass number 60. One of the most common radioisotopes.
Properties: Half-life 5.3 years; radiation, beta and gamma; radiotoxicity, moderately hazardous. Cobalt 60 emits gamma rays which have about the same penetrating power as those from radium. However, radiocobalt is available in larger quantities and is cheaper than radium.
Derivation: Pile irradiation of cobalt oxide,

Co_2O_3, or of cobalt metal.
Forms available: Cobalt metal pellets or wire needles; cobaltous chloride in hydrochloric acid solution; solid cobaltic oxides; labeled compounds such as vitamin B_{12}.
Uses: Radiation therapy, such as the treatment of cancer; cobalt 60 is also used for radiographic testing of welds and castings; as a source of ions in gas-discharge devices; as the radiation source in liquid-level gages; for locating buried telephone and electrical conduits; as a source in portable radiation units; as a research aid in studying the permeability of porous media to flow of oil, the wearing quality of floor wax, the oil consumption in internal combustion engines, wool dyeing, etc.
Shipping regulations: Class D poison, radioactive material. Red label.*

cobalt acetate. See cobaltous acetate.

cobaltammines. Compounds containing the group $[Co(NH_3)_6]^{3+}$ or its derivatives in which some of the ammonia has been replaced by other groups or ions. The names hexammine, pentammine, etc., are used to indicate the number of ammonia groups present in any case.

Cobaltammines are prepared by adding excess ammonia to a cobaltous salt, exposing to air so that oxygen is absorbed and boiling so that oxidation of the cobalt occurs.

These compounds show none of the ordinary properties of cobalt. Different types of salts with various acid radicals are known. The ammonia in the ammines may be replaced, molecule for molecule, by other nitrogen compounds such as hydroxylamine or ethylene diamine, or by water, or by ions such as hydroxyl, chloride, nitrate, etc., or by groups such as nitro (NO_2). See coordination compounds.

cobalt-ammonium sulfate. See cobaltous-ammonium sulfate.

cobalt arsenate. See cobaltous arsenate.

cobalt black. See cobaltic oxide.

cobalt bloom. See erythrite.

cobalt blue (Thénard's blue; cobalt ultramarine; King's blue; Leyden blue; azure blue).
Properties: Blue to green pigment of variable composition, consisting essentially of mixtures of cobalt oxide and alumina, approximating cobaltous aluminate, $Co(AlO_2)_2$. Commercial samples show wide variation in cobalt content, the range being from 19-30%. Cobalt blue is said to be the most durable of all blue pigments, being completely unaltered by the atmosphere and only slightly subject to the action of chemical reagents.
Derivation: By heating alumina with any of the following: (a) cobaltous oxide, or a material yielding this oxide on calcination; (b) cobalt phosphate; (c) cobalt arsenate. Greenish shades may be made by incorporating zinc oxide in the batches.
Grades: Technical (called genuine, to distinguish it from the imitation, which is ultramarine blue).
Containers: 250-lb barrels.
Uses: Pigments in oil or water; cosmetics (eye-shadows, grease paints).

cobalt blue, imitation. See ultramarine blue.

cobalt bromide. See cobaltous bromide.

cobalt carbide Co_2C. A catalyst used in the Fischer-Tropsch process; made by passing carbon monoxide over finely divided cobalt metal under regulated conditions.

cobalt carbonate. See cobaltous carbonate, basic.

cobalt chloride. See cobaltous chloride.

cobalt chromate. See cobaltous chromate.

cobalt chromate, basic. See cobaltous chromate.

cobalt disalicylal ethylenediamine. See salcomine.

cobalt 2-ethylhexoate (cobalt octoate). Probably the cobaltous salt of 2-ethylhexoic acid, $C_4H_9CH(C_2H_5)COOH$.
Properties: Blue liquid, sp.gr. 1.013 (25°C).
Uses: Paint drier; whitener; catalyst.

cobalt ferrite. Developed as a new electromechanical switch, supposed to be 1000 times as fast as present switches. See ferrites.

cobalt glance. See cobaltite.

cobalt gluconate. Used as a dietary supplement.

cobalt hydrate. See cobaltic hydroxide and cobaltous hydroxide.

cobalt hydroxide. See cobaltic hydroxide and cobaltous hydroxide.

cobaltic acetylacetonate $Co(C_5H_7O_2)_3$.
Properties: Dark green or black crystals; sp.gr. 1.43; m.p. 241°C.
Derivation: Reaction of cobaltous carbonate with acetylacetone and peroxide.
Use: Vapor plating of cobalt.

cobaltic boride CoB.
Properties: Crystalline prisms; sp.gr. 7.25 (18°C); m.p. >1400°C. Decomposes in water; soluble in nitric acid.
Use: Ceramics.

cobaltic fluoride. See cobalt trifluoride.

cobaltic hydroxide (cobalt hydroxide; cobalt hydrate) $Co(OH)_3$, actually considered to be $Co_2O_3 \cdot 3H_2O$.
Properties: Dark-brown powder; soluble in cold concentrated acids; insoluble in water and alcohol.
Derivation: By the addition of sodium hydroxide to a solution of a cobaltic salt; by the action of chlorine on a suspension of cobaltous hydroxide; by the action of sodium hypochlorite on a cobaltous salt.
Grades: Technical.
Containers: Tins; kegs; drums.
Use: Cobalt salts.
Shipping regulations: None.*

cobaltic oxide (cobalt oxide; cobalt black) Co_2O_3. Sometimes incorrectly called cobalt peroxide.
Properties: Steel-gray or black powder. Soluble in concentrated acids; insoluble in water. Sp. gr. 4.81-5.60; m.p., decomposes at 895°C.
Derivation: By heating cobaltic hydroxide.
Grades: Technical; C.P.
Containers: 1-lb bottles; 100-lb drums; 350-lb kegs.
Uses: Pigment; coloring enamels; glazing pottery.
Shipping regulations: None.*

cobalt iodide. See cobaltous iodide.

cobaltite (cobalt glance) CoAsS. Silver-white to gray mineral; metallic luster. Contains 35.5% cobalt. Sp. gr. 6-6.3; hardness 5.5.
Occurrence: Canada; Belgian Congo; Sweden.
Use: An important cobalt ore; also used in ceramics.

cobalt linoleate. See cobaltous linoleate.

cobalt molybdate. A molybdenum catalyst (a gray-green powder) used in petroleum technology, in reforming and desulfurization.

cobalt monoxide. See cobaltous oxide.

cobalt naphthenate. See cobaltous naphthenate.

cobalt nitrate. See cobaltous nitrate.

cobalto-cobaltic oxide (tricobalt tetraoxide) Co_3O_4.
Properties: Steel-gray to black in anhydrous form. Insoluble in water, hydrochloric acid and nitric acid; soluble in sulfuric acid and fused sodium hydroxide; hygroscopic. Sp. gr. 6.07.
Derivation: By heating strongly other cobalt oxides in air. Thus, the commercial oxides contain a substantial quantity of Co_3O_4.
Uses: Ceramics; pigments; catalysts; preparation of cobalt metal; electronic chemicals.

cobalt octoate. See cobalt 2-ethylhexoate. See also soaps, metallic.

cobalt oleate. See cobaltous oleate.

cobalto-nickelous sulfate. See nickel-cobalt sulfate.

cobaltous acetate (cobalt acetate) $Co(C_2H_3O_2)_2 \cdot 4H_2O$.
Properties: Reddish-violet, deliquescent crystals. Soluble in water, acids, and alcohol; sp. gr. 1.7043; m.p. loses H_2O at 140°C.
Derivation: By the action of acetic acid on cobaltous hydroxide with subsequent crystallization.
Method of purification: Recrystallization.
Grades: Technical; pure crystalline; C.P.
Containers: Glass bottles; barrels.
Uses: Sympathetic inks; ingredient of varnishes used to color oilcloth; paint and varnish driers; catalyst; mineral supplement.
Shipping regulations: None.*

cobaltous aluminate. See cobalt blue.

cobaltous ammonium sulfate (cobalt ammonium sulfate; ammonium cobalt sulfate) $CoSO_4 \cdot (NH_4)_2SO_4 \cdot 6H_2O$.
Properties: Ruby-red crystals; soluble in water; insoluble in alcohol; sp. gr. 1.902.
Derivation: Crystallization of cobaltous sulfate with ammonium sulfate.
Method of purification: Recrystallization.
Grades: Technical; C.P.
Containers: Tins; glass bottles.
Uses: Ceramics; cobalt plating; catalyst.
Shipping regulations: None.*

cobaltous arsenate (cobalt arsenate) $Co_3(AsO_4)_2 \cdot 8H_2O$.
Properties: Violet-red powder. Soluble in acids; insoluble in water. Sp. gr. 2.948.
Derivation: By the interaction of solutions of sodium arsenate and of a cobalt salt.
Grades: Technical.
Containers: Wooden kegs; boxes.
Uses: Painting on glass and porcelain in light blue colors: coloring glass.
Shipping regulations: None.*
See erythrite for the mineral.

cobaltous bromide (cobalt bromide) $CoBr_2 \cdot 6H_2O$.
Properties: Red violet crystals. Soluble in water, alcohol, and ether. Anhydrous crystals are red. Sp. gr. 2.46; m.p., decomposes.
Derivation: By the action of bromine vapor or a mixture of bromine and hydrogen bromide vapors on heated cobalt. Purification by sublimation.
Grades: Technical; C.P.
Containers: Glass bottles.
Use: In hygrometers; catalyst.
Shipping regulations: None.*

cobaltous carbonate $CoCO_3$.
Properties: Red crystals; insoluble in water and ammonia; soluble in acids. Sp. gr. 4.13; m.p., decomposes. The cobalt carbonate of commerce is usually the basic salt (see following article).
Derivation: By heating cobaltous sulfate with a solution of sodium bicarbonate.
Uses: Ceramics; trace element added to soils and animal feed; temperature indicator, catalysts.

cobaltous carbonate, basic $2CoCO_3 \cdot 3Co(OH)_2 \cdot H_2O$. The cobalt carbonate of commerce.
Properties: Red violet crystals; soluble in acids; insoluble in cold water; decomposes in hot water. M.p., decomposes.
Derivation: By adding sodium carbonate to a solution of cobaltous acetate, followed by filtration and drying.
Grades: Technical; C.P.
Containers: 1-lb bottles; wooden barrels; bags.
Uses: Manufacturing cobaltous oxide; cobalt pigments; cobalt salts; intermediate.
Shipping regulations: None.*

cobaltous chloride (cobalt chloride) (a) $CoCl_2$; (b) $CoCl_2 \cdot 6H_2O$.

*See "I.C.C. Shipping Regulations," page xiii.
Reference numbers refer to name of manufacturer. See "List of Manufacturers," page v.

Properties: (a) blue, (b) ruby-red crystals. Soluble in water and alcohol. Sp. gr. (a) 3.348, (b) 1.924; m.p. (a) sublimes, (b) 86.75°C.
Derivation: By the action of hydrochloric acid on cobalt oxide with subsequent crystallization.
Method of purification: Recrystallization.
Grades: Technical; C.P.
Containers: Glass bottles; tins; drums.
Uses: Absorbent for ammonia; gas-masks; electroplating; sympathetic inks; hygrometers; in soils and animal feeds as needed trace element; vitamin B_{12}; flux for magnesium refining; solid lubricant; dye mordant; catalysts; barometers.

cobaltous chromate (basic cobalt chromate; cobalt chromate).
Properties: Brown or yellowish-brown powder. Variable composition. (Pure cobaltous chromate is $CoCrO_4$; gray-black crystals.) Soluble in mineral acids, in solution of chromium trioxide; insoluble in water.
Grades: Technical.
Use: Ceramics (tinting).

cobaltous citrate $Co_3(C_6H_5O_7)_2 \cdot 2H_2O$.
Properties: Rose-red crystals; m.p. 150°C ($-2H_2O$). Slightly soluble in water; soluble in dilute acids.
Uses: Vitamin preparation; therapeutic agents.

cobaltous cyanide (a) $Co(CN)_2 \cdot 2H_2O$, (b) $Co(CN)_2$.
Properties: (a) buff crystals, (b) blue-violet powder; sp. gr. (b) 1.872; m.p. (b) 280°C. Insoluble in water; soluble in potassium cyanide, hydrochloric acid, ammonium hydroxide.

cobaltous fluoride $CoF_2 \cdot 2H_2O$.
Properties: Rose-red crystals or powder; sp. gr. 4.46. Soluble in cold water, hydrofluoric acid. Decomposes in hot water.

cobaltous formate $Co(CHO_2)_2 \cdot 2H_2O$.
Properties: Red crystals; sp. gr. 2.129; m.p. 140°C ($-2H_2O$). Soluble in cold water.

cobaltous hydroxide (cobalt hydroxide; cobalt hydrate) $Co(OH)_2$.
Properties: Rose-red powder. Soluble in acids and ammonium salt solutions; insoluble in water and alkalies. Sp. gr. 3.597.
Derivation: By the addition of sodium hydroxide to a solution of a cobaltous salt.
Grades: Technical.
Containers: Tins; kegs; glass bottles.
Use: Cobalt salts; in preparation of paint and varnish driers; catalyst.
Shipping regulations: None.*

cobaltous iodide (cobalt iodide) $CoI_2 \cdot 6H_2O$.
Properties: Brownish-red crystals; loses iodine on exposure to air; sp. gr. 2.90. Soluble in water and alcohol. Anhydrous cobaltous iodide, CoI_2, is in form of black crystals, sp. gr. 5.68.
Derivation: Digestion of cobalt powder with iodine and water. Anhydrous cobaltous iodide is prepared by heating cobalt in iodine vapor.
Grades: Technical.
Containers: Glass bottles.
Use: In hygrometers.
Shipping regulations: None.*

cobaltous linoleate (cobalt linoleate) $Co(C_{18}H_{31}O_2)_2$.
Properties: Brown, amorphous powder. Soluble in alcohol, ether and acids; insoluble in water.
Derivation: By boiling a cobalt salt and sodium linoleate.
Grades: Technical.
Containers: 100-lb kegs; 350-lb barrels or fiber drums; multiwall paper sacks.
Use: Paint and varnish driers; especially enamels and white paints.
Shipping regulations: None.*

cobaltous naphthenate.
Properties: Brown, amorphous powder or bluish-red solid. Insoluble in water. Soluble in alcohol, ether, oils. Composition indefinite.
Derivation: By treating cobaltous hydroxide or cobaltous acetate with naphthenic acid.
Containers: Drums.
Uses: Paint and varnish driers.

cobaltous nitrate (cobalt nitrate) $Co(NO_3)_2 \cdot 6H_2O$.
Properties: Red crystals; deliquescent in moist air. Soluble in water and in acids. Sp. gr. 1.88; m.p. 56°C.
Derivation: By the action of nitric acid on cobalt hydroxide with subsequent crystallization.
Method of purification: Recrystallization.
Grades: Technical; C.P.
Containers: Glass bottles; wooden barrels.
Uses: Sympathetic inks; cobalt pigments; preparation of cobalt catalysts; additive to soils and animal feeds; vitamin preparations; hair dyes; porcelain decoration.
Caution: Fire hazard; dangerous, oxidizing material; in contact with organic or other readily oxidizable substances it will cause violent ignition or combustion.
Shipping regulations: Oxidizing material. Yellow label.*

cobaltous oleate (cobalt oleate) $Co(C_{18}H_{33}O_2)_2$.
Properties: Brown, amorphous powder. Soluble in alcohol and ether; insoluble in water. M.p. 235°C.
Derivation: By heating cobaltous chloride and sodium oleate, followed by filtration and drying.
Grades: Technical.
Containers: Wooden barrels; fiber drums.
Use: Paint and varnish driers.
Shipping regulations: None.*

cobaltous oxalate CoC_2O_4.
Properties: Reddish-white crystals; sp. gr. 3.021; insoluble in water; soluble in ammonium hydroxide.
Uses: Temperature indicator; preparation of catalysts (hydrated form).

cobaltous oxide (cobalt oxide; cobalt monoxide) CoO.
Properties: Grayish powder under most conditions; can form green-brown crystals. Soluble in acids and alkali hydroxides; insoluble in water. Sp. gr. 5.7-6.7; m.p. 1800°C (dec).
Derivation: By heating cobaltous carbonate in nitrogen or carbon monoxide; by heating cobalt in nitric oxide; by heating cobaltous sulfate in air.
Grades: Technical; ceramic.
Containers: 10-lb tins; 400-lb barrels and fiber drums.
Uses: Pigment in paints and ceramics; preparation of cobalt salts; catalyst; porcelain enamels; coloring glass; feed additive.
Shipping regulations: None.*

cobaltous perchlorate $Co(ClO_4)_2$.
Properties: Red needles; sp. gr. 3.327; soluble in water, alcohol.
Use: Chemical reagent.

cobaltous phosphate (cobalt phosphate) $Co_3(PO_4)_2 \cdot 8H_2O$.
Properties: Reddish powder; sp. gr. 2.769; slightly soluble in cold water; soluble in mineral acids; insoluble in alcohol.
Derivation: By the interaction of solutions of cobalt salts and sodium phosphate.
Grades: Technical.
Containers: Glass bottles; fiber drums.
Uses: Manufacturing cobalt pigments; coloring glass; painting on porcelain in light blue colors; animal feed supplement.
Shipping regulations: None.*

cobaltous resinate (cobalt resinate).
Properties: Brown-red powder; insoluble in water; soluble in oils.
Derivation: By cautiously heating a cobalt salt and rosin oil.
Grades: Technical; pure precipitated.
Containers: 300-lb barrels or fiber drums.
Use: Varnish drier.
Fire hazard: Dangerous.
Shipping regulations: Flammable solid. Yellow label. Legal label name: cobalt resinate, precipitated.*

cobaltous silicate Co_2SiO_4.
Properties: Violet crystals; sp. gr. 4.63; m.p. 1420°C; insoluble in water; soluble in dilute HCl.

cobaltous silicofluoride $CoSiF_6 \cdot 6H_2O$.
Properties: Pale red crystals; sp. gr. 2.087; soluble in water.
Use: Ceramics.

cobaltous succinate $Co(C_4H_4O_4) \cdot 4H_2O$.
Properties: Violet crystals; slightly soluble in cold water; soluble in alkalies; insoluble in alcohol.
Uses: Vitamin preparations; therapeutic agents.

cobaltous sulfamate $Co(NH_2SO_3) \cdot 3H_2O$. Soluble in water.

cobaltous sulfate (cobalt sulfate) (a) $CoSO_4$; (b) $CoSO_4 \cdot 7H_2O$.
Properties: Red powder; soluble in water. Sp. gr. (a) 3.472; (b) 1.918; m.p. (a) 989°C, (b) 96.8°C.
Derivation: By the action of sulfuric acid on cobaltous oxide.
Method of purification: Crystallization.
Grades: Technical; C.P.
Containers: 1-lb bottles; 100-lb kegs; 450-lb barrels.
Uses: Ceramics; pigments; glazes; in plating baths for cobalt; additive to soils and animal feeds; catalyst.
Shipping regulations: None.*

cobaltous tungstate (cobalt tungstate; cobalt wolframate) $CoWO_4$.
Properties: Reddish-orange powder; insoluble in water; soluble in hot concentrated acids; sp. gr. 8.42.
Derivation: By adding a sodium tungstate solution to a solution of a cobalt salt.
Grades: Technical.
Containers: Wooden barrels; tins; fiber drums.
Use: Pigment; drier for enamels, inks, paints; electronic field; antiknock agents.
Shipping regulations: None.*

cobalt oxide. See cobaltic oxide, cobaltous oxide, cobalto-cobaltic oxide. The commercial cobalt oxides are not usually definite chemical compounds but are mixtures of two or more cobalt oxides.

cobalt peroxide. Incorrect name for cobaltic oxide (q. v.).

cobalt phosphate. See cobaltous phosphate.

cobalt potassium cyanide $K_3Co(CN)_6$.
Properties: Yellow crystals; sp. gr. 1.906; m.p., decomposes; soluble in water; insoluble in alcohol.
Grades: Pure; electronic.
Uses (suggested): Microwave studies.

cobalt potassium nitrite (cobalt yellow; potassium cobaltinitrite; Fischer's salt; potassium hexanitrocobaltate III) $K_3Co(NO_2)_6$.
Properties: Yellow, microcrystalline powder. Slightly soluble in water; insoluble in alcohol; m.p., decomposes at 200°C.
Derivation: By adding potassium nitrite and acetic acid to a solution of a cobalt salt.
Method of purification: Crystallization.
Grades: Technical.
Containers: Glass bottles.
Uses: Medicine; yellow pigment; painting on glass or porcelain.
Shipping regulations: None.*

cobalt resinate. See cobaltous resinate.

cobalt selenite $CoSe_2O_3 \cdot 2H_2O$. A blue-red powder; insoluble in water.

cobalt silicide. A semiconductor material reported to have as much as 15% efficiency in converting heat to electricity in the temperature range 20-800°C.

cobalt soaps. See cobaltous linoleate, cobaltous naphthenate, cobaltous oleate, cobaltous resinate.

cobalt sulfate. See cobaltous sulfate.

cobalt tallate. Cobalt derivative of refined tall oil; of varying composition. Used as a drier in paints and varnishes. Available as liquid in 55-gal drums, solid in 430-lb drums.

cobalt tetracarbonyl $Co_2(CO)_8$.
Properties: Orange or dark brown crystals. Sp. gr. 1.78; m.p. 51°C, decomposing above this temperature. Insoluble in water; soluble in organic solvents as alcohol and ether.
Derivation: By the combination of finely divided cobalt with carbon monoxide under pressure.
Uses: High anti-knock gasoline; catalyst.

cobalt titanate Co_2TiO_4.
Properties: Greenish-black crystals; sp. gr. 5.07-5.12. Soluble in concentrated hydrochloric acid.

cobalt trifluoride (cobaltic fluoride) CoF_3.
Properties: Light brown, fine free-flowing powder; sp. gr. 3.88 (25°C); no odor, except HF odor developed in moist air; stable in sealed containers; reacts readily with moisture in the atmosphere to form a dark, almost black powder; reacts with water to form a black, finely divided precipitate (cobaltic hydroxide). As a fluorinating agent, yields one atom of fluorine and reverts to the difluoride.
Use: Fluorinating agent. The spent cobalt difluoride may be regenerated with elemental fluorine.

cobalt tungstate. See cobaltous tungstate.

cobalt ultramarine. See cobalt blue.

cobalt violet. See cobalt-ammonium phosphate.

cobalt wolframate. See cobaltous tungstate.

cobalt yellow. See cobalt-potassium nitrite.

"Cobenium." [155] Trademark for a heat-treatable, high cobalt alloy.
Typical analysis: Cobalt, 40%; chromium, 20%; nickel, 15%; molybdenum, 7%; manganese, 2%; beryllium, 0.04%; carbon, 0.15%; iron, balance.
Properties: Corrosion resistant; non-magnetic; resistant to set and fatigue; heat-treatable, high strength; high elasticity.
Uses: Spring material; general purpose corrosion resistant alloy.

"Coblac." [133] Trademark for a series of carbon black, nitrocellulose dispersions. A special process gives a high degree of dispersion of carbon black in nitrocellulose, thus making it possible to produce black lacquers without milling or grinding. Available in several types for pigmenting automotive lacquers, industrial lacquers, leather finishes, etc.

"Coblax." [51] Trademark for a series of inexpensive, dark oils used in "once through" lubrication of rough machinery. Good for outdoor, low-temperature use.

"Cobon." [169] Trademark for 2-nitroso-1-naphthol used for the colorimetric determination of cobalt. Sensitivity: 0.005 ppm cobalt.

C.O.C. Abbreviation for Cleveland open cup, a type of flash point test.

coca (erythroxylon; cuca; hayo; ipado).
Dried leaves of Erythroxylon coca, known commercially as Huanaco coca, or Truxillense rusby, known commercially as Truxillo coca. Contains a very small amount of cocaine.
Habitat: Bolivia; Chile; and Peru; cultivated in Java and British East and West Indies.

cocaine (methylbenzoylecgonine) $C_{17}H_{21}NO_4$.
An alkaloid.
Properties: Colorless to white crystals, or white crystalline powder; poisonous, habit-forming drug. Soluble in alcohol, chloroform, and ether; slightly soluble in water (solution is alkaline to litmus). The hydrochloric acid solution is levorotatory.
Constants: M.p. 98°C.
Derivation: By extraction of the leaves of Erythroxylon coca with sodium carbonate solution, treatment of the latter with dilute acid and extraction with ether, evaporation of the solvent, re-solution of the alkaloid and subsequent crystallization.
Method of purification: Recrystallization.
Grades: Technical; N.F. XI.
Containers: Vials; cans.
Use: Local anesthetic (medicine, dentistry). Sold subject to strict governmental supervision in most countries.
Shipping regulations: None.*

cocaine borate.
Properties: White, crystalline powder. Soluble in alcohol. Contains 68% (approx) cocaine.

cocaine hydrochloride $C_{17}H_{21}NO_4 \cdot HCl$.
Properties: Colorless crystals or white crystalline powder; poisonous, habit-forming drug. Soluble in water and alcohol; insoluble in ether. M.p. 183-195°C.
Derivation: By the action of hydrochloric acid on cocaine.
Method of purification: Crystallization.
Grades: Technical; U.S.P. XVI.
Containers: Vials; cans.
Use: Local anesthetic. Sold subject to strict governmental supervision in U.S.A. and most countries.

cocarboxylase (TPP; thiamine pyrophosphate chloride) $C_{12}H_{19}ClN_4O_7P_2S \cdot H_2O$. The coenzyme of the yeast enzyme carboxylase. It is the key substance in decarboxylation, an energy-producing reaction in the body.
Properties: Crystallizes from alcohol containing some hydrochloric acid; m.p. 240-244°C (dec); soluble in water; dry substance very stable.
Use: Biochemical research; medicine.

coccinellin. See carmine.

cocculin. See picrotoxin.

cocculus (fishberry; Malay fishberry).
Dried ripe fruit of Anamirta cocculus, a

*See "I.C.C. Shipping Regulations," page xiii.
Reference numbers refer to name of manufacturer. See "List of Manufacturers," page v.

woody, climbing plant. Poisonous!
Habitat: India.
Use: Medicine (vermifuge). Dangerous!
Shipping regulations: Poison, class B; poison label.*

coccus. See cochineal.

cochineal (coccus). A red coloring matter consisting of the dried bodies of the female insects of Coccus cacti, which live on cactus plants in Mexico, Central America, Algeria, and the East Indies. They are collected and killed by heat or by sulfur fumes. The coloring principle is carminic acid $C_{17}H_{18}O_{10}$.
Grades: Technical; N. F. XI; silver grain; black grain.
Containers: Boxes; fiber cans.
Uses: Coloring food, medicinal products, toilet preparations; manufacture of red and pink lakes and carmine; indicator in analytical chemistry; inks; dyeing.
Shipping regulations: None.*

cochineal, ammoniacal.
Properties: Brown tablets.
Derivation: Digesting cochineal with ammonia, adding gelatinous alumina, evaporating to dryness and cutting into cakes.
Use: Dyes a bluer shade than cochineal. To a limited extent used for blueing bleached cotton.

cochinilin. See carminic acid.

cocillana (guapi bark, cociliana, cocilliana). Dried bark of tree Guarea rusbyi.
Use: Medicine.

cocoa. See cacao.

cocoa butter. See cacao butter.

cocoanut cake. See coconut cake.

cocoanut oil. See coconut oil.

cocoa oil. See cacao butter.

"Cocoloid." [322] Trademark for an algin-carrageenin composition.
Properties: Tan-colored algin-carrageenin composition in granular form passing essentially through 40 mesh and having about 13% moisture. Soluble in milk at 160°F.
Grades: Refined.
Uses: A hydrophilic colloid especially prepared for use as a stabilizer for chocolate-milk products, sterilized cream, and other milk products.
Shipping regulations: None.*

coconut acid. Mixture of fatty acids derived from hydrolysis of coconut oil. Acid chain lengths vary from 6-18 carbons but are mostly 10, 12, and 14.
Grades: Distilled; double distilled.
Containers: Drums; tank cars.
Uses: Soaps; detergents; source of long-chain alkyl groups.

coconut butter. See coconut oil.

coconut cake (coconut palm cake; cocoanut cake; copra cake). The residual product from expression of oil from the seed of the coconut. See coconut oil meal.

coconut fiber. See coir.

coconut oil (coconut palm oil, cocoanut oil; coconut butter).
Properties: White, semi-solid, lard-like fat; characteristic odor. Chief constituent: the glyceride of lauric acid, but with appreciable amounts of the glycerides of capric, myristic, palmitic and oleic acids. Soluble in alcohol, ether, chloroform, and carbon disulfide.
Constants: Sp. gr. 0.92; saponification value 250-258; iodine value 8-9.5; m. p. 20-28°C.
Derivation: From the coconut (Cocos nucifera), the chief commercial supply coming from India, Ceylon and the South Sea Islands. The fresh meat of the nut is pressed, boiled in water or heated with solvents, and the oil extracted.
Method of purification: Filtration.
Grades: Crude; refined; Ceylon; Cochin; Manila.
Containers: 375-, 400-lb barrels; 1000-lb casks; 8000-gal tank cars.
Uses: Soaps; butter substitutes; foodstuffs; cosmetics; candles; emulsions; dyeing cotton; alkyd resins; lubricating greases; synthetic detergents; source of glycerin and fatty acids.
Shipping regulations: None.*
See also copra oil.

coconut oil meal. The dried and crushed form of coconut cake recovered from the hydraulic or expeller process of extraction of oil from the seed. The usual product of commerce contains 24.2% crude protein; 13.3% crude fiber; 35.7% nitrogen-free extract; 7.4% ether soluble (fat) and 6.0% ash. The total digestible nutrients approximate 72%.
Containers: Bulk or bags.
Uses: Animal feeds or as a fertilizer ingredient.
Shipping regulations: None.*

coconut palm cake. See coconut cake.

coconut palm oil. See coconut oil.

"C-O-C-S." [55] Trademark for insecticides containing copper oxychloride sulfate. Used on fruits and vegetables.

C. O. D. Abbreviation for chemical oxygen demand. See oxygen consumed; see also biochemical oxygen demand (B. O. D.) and dissolved oxygen (D. O.).

codecarboxylase (pyridoxal phosphate) $C_8H_8NO_2 \cdot H_2PO_4$. It is active as coenzyme in amino acid biosynthesis as well as in decarboxylation reactions.
Properties (characterized as the oxime): Nearly insoluble in water, alcohol, and ether; m.p. 229-230°C (dec.).
Source: Natural and synthetic. Synthesized (a) through the action of adenosine triphosphate, or phosphorus oxychloride, on pyridoxal, and (b) by phosphorylation of pyridoxamine followed by oxidation with

*See "I.C.C. Shipping Regulations," page xiii.
Reference numbers refer to name of manufacturer. See "List of Manufacturers," page v.

100% H_3PO_4. It occurs naturally in several decarboxylases (q. v.).
Use: Biochemical research.

codehydrogenase I. See nicotinamide adenine dinucleotide.

codehydrogenase II. See nicotinamide adenine dinucleotide phosphate.

codeine (methylmorphine) $C_{18}H_{21}NO_3 \cdot H_2O$. An alkaloid.
Properties: Colorless or white crystals or powder; poisonous! Effloresces slowly in dry air; affected by light; m.p. 154.9°C. Slightly soluble in water; soluble in alcohol and chloroform; levorotatory in acid and alcohol solutions.
Derivation: From opium by extraction and subsequent crystallization; also by the methylation of morphine.
Method of purification: Recrystallization.
Grades: Technical; N. F. XI.
Containers: Vials; cans.
Use: Medicine.
Shipping regulations: None.*

codeine hydrochloride $C_{18}H_{21}NO_3 \cdot HCl \cdot 2H_2O$. White crystalline powder, m.p. 264°C, used in medicine.

codeine phosphate $C_{18}H_{21}NO_3 \cdot H_3PO_4 \cdot 1\frac{1}{2}H_2O$.
Properties: White crystals or powder; odorless. Effloresces in dry air and is affected by light; m.p. 235°C. Soluble in water; slightly soluble in alcohol. Solutions are acid to litmus.
Grades: Technical; U.S.P. XVI.
Containers: Vials; cans.
Use: Medicine.

codeine sulfate $(C_{18}H_{21}NO_3)_2 \cdot H_2SO_4 \cdot 5H_2O$.
Properties: White crystals or powder; effloresces in dry air; affected by light; m.p. 278°C; specific rotation (1 in 50) −112.5 to −115°. Soluble in water; insoluble in alcohol, chloroform, and ether.
Grades: Technical; N. F. XI.
Containers: Vials; 25-, 100-oz cans.
Use: Medicine.
Shipping regulations: None.*

cod-liver oil (morrhua oil).
Properties: Pale yellow liquid, fixed, nondrying oil; characteristic odor, slightly fishy odor and taste. Soluble in ether, chloroform, ethyl acetate, petroleum ether, and carbon disulfide; slightly soluble in alcohol.
Chief constituents: Glycerides of palmitic, stearic acids; cholesterol, butyl alcohol esters, etc.
Constants: Sp.gr. 0.918-0.927; saponification value 180-192; iodine value 145-180; maumené test 102-113; acid value 204-207.
Derivation: From the livers of codfish (Gadus morrhua) and other species of Galidae. These are rendered by steam heat and the oil separated and chilled until the stearin solidifies, when it is pressed and the clear oil collected.
Method of purification: Filtration.
Grades: Pale; light-brown; dark-brown; N. F. XI.
Containers: 5-gal cans; 30-, 50-gal barrels; 8000-gal tank cars.
Uses: Medicine (for its vitamin A and D content); leather dressing; chamois-leather tanning.
Shipping regulations: None.*

codoil. See rosin oil.

"Codur." [333] Synthetic baking enamels having an alkyd and urea-formaldehyde resin base.

coenzyme. Comparatively low molecular weight organic substance which can attach itself to, and thus supplement, a specific protein to form an active enzyme system. The term coenzyme is considered to be synonymous with the term prosthetic group when used in connection with conjugated proteins which have enzyme activity.

coenzyme I. See nicotinamide adenine dinucleotide.

coenzyme II. See nicotinamide adenine dinucleotide phosphate.

coenzyme A (CoA). A coenzyme essential for the formation of acetylcholine and for acetylation reactions in the body. Pantothenic acid is found in the body as a constituent of coenzyme A. It has been completely synthesized, and is known to be built up from pantothenic acid, cysteamine, adenosine, and phosphoric acid.

coenzyme Q
$CH_3C_6(O)_2(OCH_3)_2[CH_2CH{:}C(CH_3)CH_2]_nH$. Found in animal organs and yeast. Is active in the citric acid cycle in carbohydrate metabolism. The n in the formula varies according to the source, as 10 if from beef heart, 6, 7, 8, or 9 if from microbial sources.

coesite. A mineral found in the sandstone lining the floor of Meteor Crater, Arizona, or made synthetically by subjecting quartz to very high pressure. It is very stable toward heating in hydrofluoric acid.

"Coethloblak." [133] Trademark for a black chip consisting of thoroughly dispersed carbon black in ethyl cellulose used in making jet black lacquers and tire paints. Available as:
"Coethloblak" CK-18. Composition: 31.25% medium high color impingement carbon black, 62.5% low viscosity ethyl-cellulose, 6.25% plasticizer.
Containers: 50- and 200-lb drums.

coffea. See coffee.

coffearine. See trigonelline.

coffee (coffea; caffea). Dried ripe seed of Coffea arabica or C. liberica.
Habitat: Ethiopia; Arabia; cultivated in Brazil, Colombia, Java, India, Hawaii, Central America.
Use: Roasted, for preparation of a beverage. Source of caffeine (q.v.); has been con-considered as raw material for production of plastics.

*See "I.C.C. Shipping Regulations," page xiii.
Reference numbers refer to name of manufacturer. See "List of Manufacturers," page v.

coffinite $U(SiO_4)_{1-x}(OH)_{4x}$ (or $USiO_4$ with appreciable $(OH)_4$ in place of some SiO_4). A naturally occurring uranium mineral. Color black; sp. gr. 5.1; luster adamantine; commonly fine grained and mixed with organic matter and other minerals.
Occurrence: Colorado, Utah, Wyoming, Arizona.
Use: An important ore of uranium in some mines on the Colorado plateau.

cognac ether. See ethyl cocoinate.

cognac oil. See ethyl oenanthate.

cogwheel ore. See bournonite.

cohosh, black. See cimicifuga.

cohune-nut oil. See cohune oil.

cohune oil (cohune-nut oil; corozo-nut oil).
Properties: Yellowish, fixed, semi-liquid fat. Soluble in ether and benzene; insoluble in water.
Constants: M.p. 18-20°C; saponification value 253.9; iodine number 12.9-13.6.
Derivation: From the cohune-nut of the palm, Attalea cohune, by expression.
Uses: Foods; soaps.

"Coilife." [308] Trademark for special epoxy resin encapsulation of random wound stators utilizing solventless epoxy resin formulations and rotational seasoning process.

coir. Fiber from the coconut shell. Used in ropes and mattings.

coke. Commonly used specifically to designate the residue from destructive distillation of coal. Also applied to the residues from destructive distillation of other types of carbonaceous materials, such as pitch, petroleum, etc.

coke-oven gas. See coal gas.

coke, petroleum. The solid residue remaining after destructive distillation of petroleum materials. The fixed or solid carbon content is 90-95%, and because of its purity such coke is used extensively in metallurgical processes; also for the Hall electrolytic process for aluminum.

coking coal. A coal suitable for making coke (q.v.). Different types of coals may be mixed to produce a good coking coal.

cola (kola; kola nuts; kola seeds; soudan coffee; guru).
Derivation: Seeds of Cola nitida or other species of Cola.
Habitat: West Africa; West Indies; India.
Containers: Bags.
Use: In soft drinks.

colamine. See ethanolamine.

Colburn process. A method of forming window glass. The molten glass is drawn up from the melt tank in a ribbon, rolled flat, annealed, and then sent to be cut into the desired size and shape. The rolling and annealing are done horizontally.

colchicine $C_{22}H_{25}NO_6$. An alkaloid.
Properties: Yellow crystals or amorphous powder; odorless or nearly so. Soluble in water, alcohol, and chloroform; moderately soluble in ether; affected by light; m.p. 135-150°C. Solutions are levorotatory.
Derivation: From Colchicum autumnale by extraction and subsequent crystallization. Has recently been synthesized.
Method of purification: Recrystallization.
Grades: Technical; U.S.P. XVI.
Containers: ⅛-, 1-oz vials; 5-, 10-, 15-grain vials; bottles.
Use: Medicine; to induce chromosome doubling in plants.
Caution: Very poisonous!
Shipping regulations: None.*

colchicine tannate.
Properties: Yellow powder; soluble in alcohol. Contains 38-40% colchicine.

colchicum (meadow saffron; autumn crocus; wild saffron; meadow crocus). Dried root and seed of Colchicum autumnale.
Habitat: Central and southern Europe and North Africa.
Grades: Technical.
Containers: Bales.
Uses: Extraction of colchicine; medicine.
Shipping regulations: None.*

cold area. A laboratory or plant area free from radioactive materials. See hot area.

cold light. See luminescence.

cold rubber. Synthetic rubber produced by polymerization at relatively low temperatures; specifically, SBR or butadiene-styrene type rubber produced by polymerization at about 40°F compared with the regularly used temperature of about 120°F. A special catalyst system is required, but the product has considerably improved strength and abrasion resistance compared with the rubber polymerized at the higher temperatures.

colemanite $Ca_2B_6O_{11}\cdot 5H_2O$. A natural hydrated calcium borate. Color, white or colorless; white streak; vitreous to dull luster. Contains 50.9% B_2O_3, 27.2% CaO, balance water.
Constants: Sp. gr. 2.26-2.48; hardness 4-4.5.
Occurrence: United States (California).
Use: One of the raw materials in the United States for boric acid, sodium borate, etc.

"Colfoam." [144] Trademark for urea-formaldehyde foams.
Properties: Snow-white, fluffy cubes or shreds of an interwoven crystalline nature containing 99% trapped air by volume. Weight 0.8 lb/cu ft. Thermal conductivity (K factor) at mean temperature of 75°F is 0.18. Odorless, mildewproof, chemically inert under low temperatures; able to withstand sustained temperatures up to 120°F; resistant to fire and corrosion.
Containers: 10-lb multiwall bags; cardboard cartons.

*See "I.C.C. Shipping Regulations," page xiii.
Reference numbers refer to name of manufacturer. See "List of Manufacturers," page v.

Uses: Low temperature insulation; reduce evaporation losses from crude oil tanks; reduce bulk density of plastics, ceramics, coatings, concrete and plaster.

collagen. Albuminoid comprising the major portion of the white fiber in connective tissues of the animal body, particularly in the skin, bones and tendons. Considered an aggregate of cyclopeptide micelles. Converted to gelatin by boiling with water.

"Collargol." [38] (colloidal silver). Trademark for a shiny, silver-gray solid containing approximately 78% silver in a soluble and stabilized form.
Properties: Dispersible in water.
Use: Medicine (antiseptic).

collaurin (collaurum; colloidal gold). An aqueous colloidal suspension of metallic gold free from other materials. Used in medicine.

collaurum. See collaurin.

2,4,6-collidine (2,4,6-trimethylpyridine) $(CH_3)_3C_5H_2N$.
Properties: Colorless liquid. B.p. 170.4°C; freezing point −44.5°C; sp.gr. 0.913 (20/20°C); refractive index (n 20/D) 1.4981. Soluble in alcohol; slightly soluble in water.
Grades: Technical (97.5% purity).
Containers: 1-, 5-, 55-gal drums.
Uses: Chemical intermediate; dehydrohalogenating agent.

collodion. See also pyroxylin.
Derivation: Solution of nitrated cellulose (mixture of trinitrocellulose and tetranitrocellulose) in ether and alcohol.
Properties: Pale yellow, syrupy liquid; very flammable; odor of ether. When exposed in thin layers, evaporation occurs to leave a tough colorless film.
Constants: Collodion U.S.P. XVI: Typical specifications: U.S.P. XVI formula pyroxylin 40 g, ether 750 cc, alcohol 250 cc (making a total of about 1000 cc); wt/gal 6.37 lbs (25°C)(approx); sp.gr. 0.765-0.775.
Collodion Flexible U.S.P. XVI: Collodion U.S.P. XVI plasticized with camphor and castor oil. Typical specifications: U.S.P. XVI formula camphor 20 g, castor oil 30 g, collodion U.S.P. XVI 950 g, (making a total of 1000 g); wt/gal 6.46 lbs (25°C)(approx).
Grades: Technical; U.S.P. XVI.
Containers: Glass bottles; 1-, 5-, 6-, 10-lb cans; 30-, 60-lb drums.
Uses: Photographic films; cementing; coating wounds and abrasions; patent and artificial leather; solvent for drugs; corn removers; process engraving and lithography; artificial pearls.
Fire hazard: Dangerous! Flammable! Flash point under 80°F; keep lights and fire away.
Shipping regulations: Flammable liquid. Red label.*

collodion cotton. See nitrocellulose.

collodion, flexible. See collodion.

colloid. Most frequently a special type of liquid mixture or suspension in which the particles of suspended liquid or solid are present in very finely divided form (i.e., particle size from about 1 to 500 millimicrons in diameter). The colloidal suspension of liquids in liquids is an emulsion.

Unlike ordinary suspensions, colloids do not exhibit the phenomenon of settling to a noticeable degree. Because of their exceedingly high ratio of surface area to volume, the rate of sedimentation is very slow, so that the slightest convection currents (as from small temperature differences) are sufficient to keep the particles in uniform distribution.

The suspended particles may contain from a few molecules to hundreds of molecules of small or average size. Such particles are actually too small to be seen by the ultramicroscope, but can be studied by means of reflected light, and can be resolved by the electron microscope. Suspensions in which the particles are very large molecules (proteins, polymers, etc.) may behave like colloids even though the particles are single molecules as large as half a micron in diameter. The realm of colloids also includes suspensions of finely divided liquids or solids in gases (fogs and smokes) and systems of thin films, bubbles or filaments whose thickness is of the above dimensions.

colloidal clay. See bentonite.

colloidal gold. See collaurin.

collophanite. See apatite.

"Colloresine." [307] Trademark for a series of thickening, binding, and finishing agents.
Composition: Sodium carboxymethyl cellulose; 99.5% active (dry basis).
Properties: White powder.
Uses: Primarily used as thickening, binding or finishing agent in paper, leather and textile printing trades. Thickening agent for latex dispersions. Used as builder in detergent formulations.
Grades: All "Colloresine" brands are chemically similar but differ in viscosity of aqueous solutions.
"Colloresine" HMS: Viscosity of 1% solution ranges from 100-300 cps.
"Colloresine" HV: Viscosity of 1% solution ranges from 1300-2200 cps.
"Colloresine" LV: Viscosity of 2% solution ranges from 25-50 cps.
"Colloresine" MV: Viscosity of 2% solution ranges from 300-600 cps.

colloxylin. See nitrocellulose.

colocynth (bitter apple; bitter cucumber; bitter gourd). Peeled, dried, unripe fruit of Citrullus colocynthis.
Habitat: Mediterranean region, Asia and Africa.
Grades: Technical; N.F. XI.
Containers: Boxes; barrels.

*See "I.C.C. Shipping Regulations," page xiii.
Reference numbers refer to name of manufacturer. See "List of Manufacturers," page v.

Use: Medicine.
Shipping regulations: None. *

Cologne brown. See Vandyke brown.

Cologne earth. See Vandyke brown.

Cologne spirits. A very pure grade of ethyl alcohol.
Shipping regulations: Flammable liquid. Red label. *

"Coloidex." [133] Trademark for a series of surface treated carbon blacks. Designed for cement and water systems. Sold in powder form. "Coloidex 3" is a regular color impingement carbon black.

colophony. See rosin.

"Colorex" Textile Stripper. [1] Trademark for titanium trichloride in aqueous solution with zinc chloride. Dark violet to black liquid.
Typical specifications: Grade No. 20: Titanium trichloride, 20%; zinc chloride, 16% approx; hydrochloric acid, 7-9%; water and stabilizer, 50-57%.
Grade No. 23: Titanium trichloride, 23%; zinc chloride, 19% approx; hydrochloric acid, 7-9%; water and stabilizer, 44-51%.
Containers: 5-, 13-gal carboys; 1-gal bottles (boxed).
Uses: Powerful reducing agent; dye stripper.
Shipping regulations: Corrosive liquid. White label. *

colorimetry. A means of analysis based on measuring the color intensity of a particular substance or a colored derivative of it.

color lake. See lake.

colorless dye. Synonym for optical bleach (q. v.).

colors. See dyes.

"Colorundum." [205] A balanced mixture of abrasive aggregates, mineral oxide color, and stearate for surfacing concrete floors, in a choice of colors.

Colour Index number. See dyes.

"Colton." [144] Trademark for polyvinyl homopolymer and copolymer resins, emulsions, and other polymer products.

"Columbia." [214] Trademark for activated carbons.
Properties: Hard, durable, inert pellets and granules. Low ash content. Unaffected by most chemical agents and conditions. Numerous grades available.
Constants: Real densities 1.75-2.10; apparent densities of individual particles 0.6-0.9; apparent densities of packed mass of particles 0.42-0.55; weights of packed mass of particles 26-34 lbs/cu ft; moisture contents 0.5-2.5%; ash contents 0.5-10%.
Containers: 1-, 5-lb friction top tin cans; 25-, 200-lb fiber drums.
Uses: Solvent recovery; gas purification; air conditioning; gas separation; gas masks; catalyst carriers.

columbite (tantalite, niobite) $(Fe,Mn)(Nb,Ta)_2O_6$. A natural oxide of niobium, tantalum, ferrous iron, and manganese, found in granites and pegmatites. Some tin or tungsten may be present in the mineral.
Properties: Color iron black to brownish black; streak dark red to black; luster submetallic; sp. gr. 5.2-7.9; hardness 6.
Occurrence: South Dakota, Colorado, New Mexico, Maine, North Carolina; Greenland; U.S.S.R.; Germany.
Use: Source of niobium and tantalum.

columbium Cb. Obsolete name for the element niobium. The latter name became official in 1949.

colza oil. See rape-seed oil.

"Comal." [51] Trademark for a grease for heavy industry. It is a mixed-base, oxidation-resistant grease formulated for use on plain and anti-friction bearings, and for application by grease gun or centralized system.

"Combiotic." [299] Trademark for a preparation containing penicillin and dihydrostreptomycin.

"Combistrep." [299] Trademark for a preparation containing streptomycin and dihydrostreptomycin.

common rosin. See rosin.

"Com-Plex." [101] Trademark for series of plasticizer extenders. Available as:
"Com-Plex" 4000: sp. gr. 0.880-0.895; boiling range 530-700°F; f. p. below −65°F; flash point 275°F.
"Com-Plex" 8200: sp. gr. 0.875-0.890; boiling range 545-700°F; f. p. below −65°F; flash point 275°F.
Uses: Vinyl processing industry.

complex compound. A compound formed by combination of substances that are themselves capable of independent existence. Examples are: double salts, hydrates, coordination compounds and inclusion complexes.
See also complex ion, and coordination compound.

complex ion. An ion formed by combination of a simpler ion with another ion or with an atom or a molecule. Thus copper ion Cu^{++} joins with ammonia molecules to form $Cu(NH_3)_4^{++}$, copper ammonia complex ion.

"Complexon." A German sequestrant of several varieties, similar to, and including ethylenediaminetetraacetic acid.

complexones. A group of aminopolycarboxylic acids basically derived from iminodiacetic acid and containing at least one $-N(CH_2COOH)_2$ group. These form stable complexes with many cations under varying conditions and these complexes are only very slightly ionized. The most commonly used of these is ethylenediaminetetraacetic acid (q. v.), which has been used in titrimetric procedures for the indirect

*See "I. C. C. Shipping Regulations," page xiii.
Reference numbers refer to name of manufacturer. See "List of Manufacturers," page v.

determination of metal ions and as a sequestering agent for separation and removal of metal contaminants. See also chelate; sequestration.

component. One of the minimum number of substances required to state the composition of all phases of a system. In the absence of chemical reaction, any one of the substances in a mixture.

composition metal leaf. See copper-aluminum alloy leaf.

compound. A substance composed of atoms or ions of two or more different elements. A compound has definite proportions by weight of the constituent elements and may thus be represented by a chemical formula. Each compound has its own characteristic properties, different from those of its elements, and from properties of other compounds.

Compound B. See corticosterone.

Compound E. See cortisone.

Compound F. See hydrocortisone.

compreg. A hardwood impregnated with a phenol-formaldehyde resin under heat and pressure.

"Compregnite." [65] Trademark for a phenol-formaldehyde liquid resin used for impregnation of wood to improve density and physical properties.

compressed petroleum gas. See liquefied petroleum gas.

"Compresto." [84] Trademark for an electrical conductor consisting of layers of shaped pure aluminum wires concentrically stranded about a single round core wire.

Compton effect. One of the principal processes by which high energy electromagnetic radiation, or gamma rays, interact with or are absorbed by matter. In the Compton process, the gamma ray frees an electron in matter as if the electron was unbound, dividing the momentum of the gamma ray between the ejected electron and a new gamma ray of lower energy going off in a new direction.

"Com-Sol" 176. [101] Trademark for a saturated highly aromatic hydrocarbon solvent.

Properties: Sp. gr. 0.829; refractive index 1.4657 (25°C); flash point 128°F; aromatic content 63%; naphthenic and paraffinic content 37%; boiling range 358-481°F.

Use: Replacement of dipentene or other terpene-type solvents in rubber solution formulations; solvent for chlorinated insecticides; preparation of resin solutions for paints and coatings.

"Conac S." [28] Trade name for N-cyclohexyl-2-benzothiazole sulfenamide, $C_6H_4SNCSNHCHC_4H_8CH_2$.

Properties: Cream colored powder; sp. gr. 1.27; m. p. 200-212°F.

Containers: 50-lb bags.

Use: To accelerate and improve the vulcanization of natural and synthetic rubber.

conarachin. A globulin derived with arachin from peanut meal. Differs from arachin largely in sulfur content.

"Concentals." [325] Brand name for a series of highly sulfated organic compounds. Used in cotton softening and finishing operations.

concentration.

1. The amount of a given substance in a stated unit of a mixture, solution, or ore. Common methods of stating concentration are per cent by weight or by volume, normality, weight per unit volume, as grams per cubic centimeter or pounds per gallon.

2. The process of increasing the amount of the given substance/unit of mixture, etc.

conchinine. See quinidine.

concrete.

1. A conglomerate of gravel, pebbles, broken stone, blast-furnace slag or cinders, termed the aggregate, embedded in a matrix of either mortar or cement, usually standard Portland cement in the U.S. Ancient and medieval concretes had lime-mortar matrices. Reinforced concrete, ferro-concrete, and armored concrete are concretes in which steel in various forms is used to strengthen the concrete. See cement, Portland.

2. (perfumery). A waxy solid which results from the solvent extraction (usually by refined ligroin) of a perfume source such as rose petals. When the concrete is dewaxed by use of a properly chosen second solvent, the desired essential oil remains. At this point it is called an absolute.

concrete, cellular. A light-weight concrete foam which may be made in several ways:

1. By the addition of aluminum powder to the concrete mix and applying heat, which sets hydrogen free to make the concrete cellular.

2. By whipping air into the mix containing an entraining agent.

3. By adding preformed foam to the mix. Such foams are made from a foaming agent such as dried blood, a stabilizer such as ferrous or aluminum sulfates, organic solvents, and a germicide such as chlorinated phenol or mercury salts.

condensation. A chemical reaction in which two or more molecules combine, with the separation of water or some other simple substance. If a polymer is formed, the process is called polycondensation (ASTM definition, ASTM D883-54T).

condensed asphalts. See blown asphalt.

"Con Det." [236] Trademark for a low foaming concentrated synthetic detergent for use in preparation of low-solids-content drilling muds for oil wells. Also provides lubricating and emulsifying properties and is

useful as a general purpose liquid detergent.
Containers: 5-gal cans and 55-gal drums.

condor vine. See condurango.

"Conductex." [133] Trademark for a group of carbon blacks designed to provide high electrical conductivity where required in rubber, plastics, coatings etc.

conduction (heat). The transfer of heat from one point to another within a body, or from one body to another when both bodies are in physical contact, in the absence of motion in the medium.

Conductive Silver Preparations. [28] Specially compounded materials containing silver powder in a suitable vehicle, with or without ceramic flux; can be coated on base materials such as titanate bodies, mica, glass, porcelain, steatite, plastics, wood, cloth and paper by stencil screening (squeegee), spraying, dipping, brushing, roller coating, banding wheel, or other suitable method. Fixed by air-drying, baking at low temperatures, or firing at elevated temperatures.
Uses: To produce capacitor electrodes, ceramic-to-metal solder seals, electrical shields, surfaces of high conductivity on nonconductive materials; as a base for electroplating on ceramic and nonceramic surfaces.

condurango (cundurango; eagle vine; mataperro; condor vine). Bark of Marsdenia cundurango.
Habitat: Ecuador and Peru.
Grades: Technical.
Containers: Bales.
Use: Medicine.
Shipping regulations: None.*

Condy's liquid. A dilute solution of potassium permanganate used as disinfectant.

cone flower. See echinacea.

conformation. The overall spatial arrangement of the atoms and groups in a polymer molecule, i.e., the general shape of a polymer molecule. See polymer, stereospecific. Also applied to spatial arrangement of atoms and groups in any complex molecule.

conglutin.
Properties: White powder; contains 18% nitrogen and 0.6% sulfur; used in medicine as a 6% solution. Soluble in weak alkaline solutions.
Derivation: A vegetable casein derived from almonds.

Congo copal. A natural fossil resin available in various commercial grades from water-white transparent to pale dust. The raw material is almost completely insoluble in all solvents but can be subjected to high temperatures and the Congo molecule cracked. Volatile compounds 20-25% of the original weight are given off during cracking at 650-675°F. The process is dangerous without special equipment due to flammability at that temperature. Can be purchased in processed form to avoid the fire hazard and to obtain a more uniform resin. The raw material has refractive index (n 20/D) 1.540-1.541; sp. gr. 1.05-1.07; direct acid number 92-115.
Containers: Bags.
Uses: Protective coatings. See also copal.

Congo red. Sodium diphenyl-bis-alpha-naphthylaminesulfonate $C_{32}H_{22}O_6N_6S_2Na_2$.
Properties: Brownish-red powder; soluble in water and alcohol; insoluble in ether; odorless; decomposes on exposure to acid fumes.
Derivation: Combination of tetraazotized benzidine and naphthionic acid.
Uses: Dye; medicine; indicator; biological stain.

conhydrine (oxyconiine; 2(alpha-hydroxypropyl) piperidine) $C_5H_{10}NCH(OH)C_2H_5$.
Properties: Colorless crystalline alkaloid; poisonous! Soluble in alcohol, ether, and chloroform; slightly soluble in water. M.p. 120.6°C; b.p. 220-226°C.
Derivation: By extraction of the seeds of Conium maculatum and subsequent crystallization.
Method of purification: Recrystallization.

coniferin $C_{16}H_{22}O_8$. A glucoside contained in pine bark and other conifers. When decomposed it yields coniferyl alcohol which can be oxidized to vanillin. Used as a raw material for manufacture of synthetic vanillin.

coniïne (propyl pyridine) $C_5H_{10}NC_3H_7$.
Properties: Colorless, oily liquid alkaloid; mousy odor; poisonous! Soluble in alcohol, ether, and oils; slightly soluble in water. M.p. −25°C; b.p. 166°C.
Derivation: By extraction of Conium maculatum and subsequent distillation.

coniïne hydrochloride $C_8H_{17}N \cdot HCl$.
Properties: White crystals; poisonous. Soluble in water and alcohol; insoluble in ether. M.p. 210-220°C.
Derivation: By the action of hydrochloric acid on coniïne.
Method of purification: Crystallization.

conium (poison hemlock; spotted hemlock; poison parsley; spotted cowbane). (The synonym hemlock is not officially recognized because of the confusion with the hemlock tree.) Poisonous!
Derivation: Full grown, but unripe carefully dried fruit of Conium maculatum.
Habitat: Europe; Asia; United States.
Grades: Technical.
Containers: Boxes.
Use: Source of coniïne.
Shipping regulations: None.*

conjugated double bonds. Two or more double bonds each separated from one another by a single bond, as in the structural formulas for butadiene-1,3 ($H_2C{=}CH{-}CH{=}CH_2$) or in maleic acid (the O=C−C=C−C=O skeleton).

conjugate layers. Two layers of a liquid system each composed of a different ternary mixture and in equilibrium with one another.

"Conpernik." [308] Trademark for an alloy of approximately equal proportions of iron and nickel having constant permeability over a range of low flux densities. It is used where constant inductance cores are required over a low range of inductions.

consistency. A term used in rheology to designate the property of a material or composition which is evidenced by its resistance to flow and is represented by a composite of its properties. For Newtonian liquids, consistency and viscosity are synonomous. For non-Newtonian liquids, it qualitatively represents plastic flow.

constant-boiling mixture. See azeotropic mixture.

cont. Abbreviation for containers or content.

contact acid. Sulfuric acid made by the contact process (q.v.).

contact process. A process for manufacture of sulfuric acid and oleum, in which the sulfur dioxide is oxidized to sulfur trioxide by contact with a platinum or vanadium pentoxide catalyst.

Air used for burning sulfur or pyrites is first dried by scrubbing with 95-98% sulfuric acid. After combustion, the burner gases are usually run through waste-heat boilers for the generation of steam. If pyrites is burned, it is necessary to remove dust by means of electrostatic precipitation, and to scrub out halogens in sulfuric-acid washtowers.

Prior to entry into the first catalytic converter, the burner gas temperature is adjusted to 575°C to obtain a high rate of conversion of most of the sulfur dioxide. Before entering the second converter, the temperature is adjusted to 450°C. The rate of conversion is relatively slow at this temperature but the equilibrium is such that 98% conversion can be realized. The sulfur trioxide is then cooled before going to the absorbers.

The converted gases pass through an oleum absorber, which is a packed tower. Sulfuric acid (98-99%) is trickled down the packing, and 20% oleum (100% sulfuric acid containing 20% free sulfur trioxide) leaves the tower. This may be further concentrated by distillation, the sulfur trioxide given off from the still being absorbed in 20% oleum in mechanically-stirred water-cooled vessels.

From the oleum absorber, the converted gases proceed to the final absorber, again a packed tower, where they are scrubbed with 98% sulfuric acid before being discharged into the atmosphere. The 98% acid is considerably concentrated in passing down this tower, so it is pumped to a diluting tank where water or dilute acid is added, and from which it is continuously circulated through the tower. Sulfuric acid (98%) to be sold is removed from the diluting vessel.

contact resins (impression resins, low-pressure resins). Synthetic thermosetting resins characterized by cure at relatively low pressure. The usual components are an unsaturated high molecular weight monomer such as an allyl ester, or a mixture of styrene or other vinyl monomer with an unsaturated polyester or alkyd. Cure requires heat and a catalyst as well as some pressure. The curing does not result in water formation as with phenol-formaldehyde resins.

contamination (radioactive). Radioactive materials which have been deposited anywhere that radioactivity is not desired. The removal of radioactive contamination is known as decontamination. Decontamination procedures vary with the type and intensity of the radiation and with the object involved. Many ordinary objects may be decontaminated by washing with water or with chemical solutions. Skin contamination is best treated by washing thoroughly with soap and water. Contaminated clothing should be appropriately stored until the extent of radioactivity is small enough to permit laundering. Objects and clothing still too radioactive to handle after preliminary treatment must be buried until the radiation level is safe, or they must be disposed of in a proper manner. No radioactive material should be burned in an open incinerator. Decontamination crews should wear protective clothing, footwear, gloves, masks, etc., depending upon the circumstances and the extent of contamination.

"Continental." [104] Trademark for a line of channel blacks used in natural and synthetic rubber, paints, inks and plastics.

Continex." [104] Trademark for furnace blacks used in rubber, plastics, paints, paper.

continuous distillation. Distillation in which a feed, usually of nearly constant composition, is supplied continuously to a fractionating column and the product is continuously withdrawn at the top, bottom, and sometimes at intermediate points.

Controlled Solubility Phosphates 15-J, 1-P, 1-R, 19-R. [108] Series of sodium-calcium phosphates in true glassy form with an active P_2O_5 content of 68%.

Containers: Available in pails, drums and cans.

Uses: Prevent the formation of "gyp" and scale in producing oil wells, heat treaters, salt water disposal systems, water floods.

control rod. See nuclear reactor.

convallaria (lily-of-the-valley; May lily; park lily; May blossom). Dried flowers, dried rhizome and roots of Convallaria majalis.

Habitat: United States; Europe and northern

Asia; cultivated in United States.
Use: Medicine.

convection (heat). The transfer of heat from one place to another by a moving gas or liquid. Natural convection results from differences in density that are caused by temperature differences. Thus warm air is less dense than cool air; the warm air rises relative to the cool air, and vice versa. Forced convection involves motion caused by pumps, blowers, or other mechanical devices.

"Convertit." [173] Trademark for an invertase preparation for production of stable soft cream centers in candies.

"Coolanol" 45.[58] Trademark for dielectric coolant for electronic equipment. A clear, amber liquid useful from −65° to 400°F.

"Coomassie." [206] Brand name for a proprietary line of acid dyestuffs.

coordination compound (Werner complex). A complex compound whose molecular structure contains a central atom bonded to other atoms by coordinate covalent bonds. These are bonds based on a shared pair of electrons both of which come from a single atom or ion. Examples of coordination compounds are cobalt III hexammine chloride $[Co(NH_3)_6]Cl_3$ and potassium chloroplatinate, $K_2[PtCl_6]$. In these cases the Co and Pt atoms are joined to the NH_3 and Cl respectively by coordinate covalent bonds.

Chelate compounds are a special kind of coordination compound (see chelate).

copaiba balsam. See copaiba resin.

copaiba oil.
Properties: Essential oil; colorless to yellowish or bluish; characteristic copaiba-balsam, pepperlike odor; bitter, grating, lingering taste.
Constants: Sp. gr. 0.88-0.91 (15°C); optical rotation −1 to −33°; refractive index 1.494-1.500; acid no. 0-1.9; insoluble in water; soluble in 5-6 vols 95% alcohol; soluble in ether, carbon disulfide.
Chief constituent: Caryophyllene, a sesquiterpene, $C_{15}H_{24}$.
Derivation: Distilled from copaiba balsam.
Containers: Bottles; 25-, 50-lb cans.
Use: Medicine.
Shipping regulations: None.*

copaiba resin (Jesuits' balsam; copaiba balsam; balsam capivi).
Properties: Transparent, viscous, light yellow to brownish-yellow liquid; peculiar odor. Soluble in alcohol, ether, chloroform, benzene and carbon disulfide; insoluble in water. Sp. gr. 0.940-0.990.
Derivation: The oleoresin from one or more South American species of copaiba.
Habitat: Brazil, Venzuela, and Colombia.
Grades: Technical. Copaiba resins are classified commercially according to port of export, and include Para, Maracaibo, Bahia, Marnaham, Cartagena and Maturin. The first two are the most important.
Containers: 1-, 5-lb bottles; 10-, 50- to 55-lb cans.
Uses: Medicine; varnishes; lacquers; brightening old paintings; tracing papers; tracing cloths; odor fixative.
Shipping regulations: None.*

copaivic acid $C_{20}H_{30}O_2$. A monobasic acid derived from the resin of copaiba.

copal. A class of natural resins, both recent and fossil. The principal recent, or soft, copals are Philippine, Manila, and pontianak; the principal fossil, or hard, copals are Congo and kauri.
Properties: Yellow to red, semitransparent, brittle lumps having a conchoidal fracture and vitreous luster. In general, the copals have higher acid numbers than the dammar resins. The soft copals are partly soluble in alcohol, chloroform, and turpentine. The hard copals are nearly insoluble in the usual solvents but, on strong heating, the resins lose 10-25% of their weight and become soluble in turpentine and linseed oil.
Habitat: East Indies; Philippines; Australia; Africa.
Grades: Technical; nubs, chips, seeds.
Containers: Bags.
Uses: Varnishes and lacquers.
Shipping regulations: None.*
See also animé; Zanzibar gum.

copal oils. Oil sometimes used in the preparation of oil varnishes. They are obtained by the dry distillation of copal.

"Copeel." [333] Trade name for a removable protective coating used during storage and shipment to protect painted and unpainted equipment.

"Copeenblak." [133] Trademark for a series of carbon dispersions in polyethylene for pigmenting cables and jackets, film and pipe stocks.

"Copel." [166] Trademark for 55-45 copper-nickel alloy used as a resistor material in the construction of electrical instruments where temperature coefficient of resistance must be very low.

copolymer. The substance produced by the polymerization or addition of two or more dissimilar monomers; as SBR synthetic rubber, from styrene and butadiene.

"Cop-o-zink." [93] Trade name for an intimate mixture of insoluble salts of copper and zinc in definite proportion.
Containers: 50-lb bags; 48-lb cases (12x4-lb bags).
Use: Fungicide with secondary benefit as source of nutritional trace elements. Synergistic action greatly enhances effectiveness. Compatible with DDT, BHC, organic and inorganic insecticides.

copper Cu. A metallic element of atomic number 29, of group Ib of the periodic system.

Properties: Distinctive reddish color; sp. gr. 8.96; m.p. 1083°C; b.p. 2325°C; ductile; good conductor of electricity. Dissolves readily in nitric and hot concentrated sulfuric acid; in hydrochloric and dilute sulfuric acid slowly but only when exposed to the atmosphere. More resistant to atmospheric corrosion than iron, forming a superficial layer of green basic carbonate. Readily attacked by alkalies.
Occurrence: Sometimes native and also in the following minerals: azurite, atacamite, azurmalachite, bornite, brochantite, chalcanthite, chalcocite, chalcopyrite (copper pyrites), chrysocolla, covellite, cuprite, enargite, malachite, stromeyerite, tennantite, tenorite, tetrahedrite. Leading producers are U.S.A.; Chile; Peru; Canada; Africa; U.S.S.R.
Derivation: Varies with the type of ore. With sulfide ores the steps may be (1) concentration (of low grade ores), (2) roasting, (3) formation of copper "matte" (40-50% Cu), (4) reduction of matte to "blister" copper (96-98%), (5) electrolytic refining to 99.9 + % copper.
Forms available: Ingots, sheet, rod, wire, tubing, shot, powder; high purity (impurities less than 10 ppm).
Uses: Electric wiring; switches, plumbing, heating, roofing material; chemical and pharmaceutical machinery; alloys (brass, bronze, monel metal, etc); electroplating; cooking utensils. Copper powder and massive copper used in making beryllium, beryllium-copper alloys, beryllium oxide. Flakes used as insulation for liquid fuels.

copper-8. Slang for copper-8-quinolinolate.

copper abietate (cupric abietate) $Cu(C_{20}H_{29}O_2)_2$.
Properties: Green scales; poisonous! Soluble in alcohol, and in oils, with fine green color; insoluble in water.
Derivation: By heating copper hydroxide with abietic acid.
Containers: Barrels.
Use: Preservative metal paint; fungicide.

copper acetate (cupric acetate; crystals of Venus; verdigris, crystallized) $Cu(C_2H_3O_2)_2 \cdot H_2O$.
Properties: Greenish-blue, fine powder; poisonous! Soluble in water, alcohol and ether.
Constants: Sp. gr. 1.9; m.p. 115°C; decomposes at 240°C.
Derivation: By the action of acetic acid on copper oxide and subsequent crystallization.
Use: Synthetic rubber.

copper acetate, basic (copper subacetate; verdigris; verdigris, blue; verdigris, green).
Properties: Masses of minute silky crystals either pale green or bright blue in color. Blue variety, approximate formula $(C_2H_3O_2)_2Cu_2O$. Green variety, approximate formula $CuO \cdot 2Cu(C_2H_3O_2)_2$. Coppery taste. Poisonous! The green rust with which uncleaned copper vessels become coated and which is commonly termed verdigris is a copper carbonate and must not be confused with true verdigris. Apart from its impurities, verdigris is a variable mixture of the basic copper acetates.
Soluble in acids; insoluble in alcohol; very slightly soluble in water.
Derivation: By the action of acetic acid on copper in the presence of air.
Method of purification: Crystallization.
Grades: Technical.
Containers: Wooden kegs; fiber drums.
Uses: Paint pigment; insecticide; fungicide; mildew preventive; mordant in dyeing and printing fabrics; copper acetoarsenite.
Shipping regulations: None.*

copper acetoarsenite (cupric acetoarsenite; Paris green; Schweinfurth green; imperial green; king's green; emerald green; new green; patgreen; moss green; mitis green; Vienna green; emperor green; parrot green; kaiser green; meadow green) $(CuO)_3As_2O_3 \cdot Cu(C_2H_3O_2)_2$.
Properties: Emerald-green powder; poisonous! Soluble in acids; insoluble in alcohol and water. Toxic to many plants.
Derivation: By reacting sodium arsenite with copper sulfate and acetic acid.
Grades: Technical; C.P.
Containers: 1-lb bottles; wooden kegs; fiber drums.
Uses: Pigment; insecticide; wood preservative preparations.
Warning! Poisonous if swallowed. MCA warning label.
Shipping regulations: Poison, class B. Poison label.*

copper acetylacetonate $Cu(C_5H_7O_2)_2$. Crystalline powder; slightly soluble in water. Resistant to hydrolysis. A chelating nonionizing compound.

"Copper-A Compound." [28] See tetracopper calcium oxychloride.

copperah oil. See copra oil.

copper albuminate (copper, albuminated).
Properties: Dark green scales. Soluble in dilute acids, alkalies.

copper albuminated. See copper albuminate.

copper alum. See copper, aluminated.

copper, aluminated (divine stone; copper alum; eye stone). A combination of copper and aluminum sulfates with potassium nitrate and some camphor.
Properties: Light green solid; soluble in water.
Use: As a caustic (medical).

copper-aluminum alloy leaf (composition metal leaf). Usually signifies the alloy containing equal parts of copper and aluminum, often in leaf or sheet form, and used as a master alloy.

copper amalgam.
Properties: Hard, brown leaflets. Contain 74% (approx) mercury and 24% (approx)

*See "I.C.C. Shipping Regulations," page xiii.
Reference numbers refer to name of manufacturer. See "List of Manufacturers," page v.

copper. Soluble in nitric acid.
Grades: Technical.
Use: Dental cement.

copper aminoacetate. See copper glycinate.

copper aminosulfate. See copper sulfate, ammoniated.

copper-ammonium chloride (ammonium-copper chloride; ammonium-cupric chloride) $CuCl_2 \cdot 2NH_4Cl \cdot 2H_2O$.
Properties: Blue or bluish-green crystals. Soluble in alcohol, water. Sp.gr. 1.98.

copper-ammonium rayon. See cuprammonium rayon.

copper arsenate.
Properties: Light blue, blue, or bluish-green powder. Variable composition. Contains 33% (approx) copper and 29% (approx) arsenic. Soluble in dilute acids, ammonium hydroxide; insoluble in alcohol, water.
Uses: Insecticide; fungicide.
Shipping regulations: Poison, class B. Poison label. *

copper arsenite (cupric arsenite; copper orthoarsenite; Scheele's green) $CuHAsO_3$; or, $Cu_3(AsO_3)_2 \cdot 3H_2O$; variable.
Properties: Fine, light-green powder; poisonous! Soluble in acids; insoluble in water and alcohol.
Constants: M.p., decomposes.
Derivation: By the interaction of copper sulfate and sodium arsenite.
Grades: Technical; C.P.
Containers: 1-lb bottles; wooden kegs; fiber drums.
Uses: Pigment (paints, wall paper, calico printing); insecticide.
Shipping regulations: Poison, class B. Poison label. *

copper arsenite, ammoniacal. See chemonite.

copperas. See ferrous sulfate.

copperas, blue. See copper sulfate.

copperas, green. See ferrous sulfate.

copperas, white. See zinc sulfate.

copper benzoate $(C_6H_5COO)_2Cu \cdot 2H_2O$.
Properties: Blue, crystalline, odorless powder; slightly soluble in cold water and alcohol.
Derivation: A salt of benzoic acid and a copper salt are reacted in water solutions.

copper-beryllium (beryllium-copper). Alloys having hardness of steel but which do not produce sparks when struck.
Uses: Fabrication of tools, apparatus and equipment for use where sparks would be hazardous, as in explosives operations or where explosive gas or gasoline mixtures are likely to be present.

copper bichromate. See copper dichromate.

copper, blister. See copper under "Derivation;" also see blister copper.

copper blue. See mountain blue and Bremen blue.

copper borate. See copper metaborate.

copper bromide (cupric bromide) $CuBr_2$.
Properties: Black crystalline powder or crystals; deliquescent. Soluble in acetone, alcohol, water. M.p. 498°C.
Grades: Technical; C.P.
Uses: Photography (intensifier); organic synthesis (brominating agent).

copper carbonate (cupric carbonate; copper carbonate, basic; artificial malachite; mineral green. For the native mineral see malachite) $Cu_2(OH)_2CO_3$.
Properties: Green powder; poisonous! Soluble in acids; insoluble in water.
Constants: Sp.gr. 3.7-4.0; decomposes 200°C.
Derivation: By adding sodium carbonate to a solution of copper sulfate, filtering and drying.
Grades: Technical; C.P.
Containers: 1-, 5-lb bottles; 25-lb boxes; 100-lb kegs; 300-, 400-lb barrels or fiber drums.
Uses: Pigments; pyrotechnics; insecticides; copper salts; coloring brass black; astringent in pomade preparations; antidote for phosphorus poisoning; smut preventive; fungicide.
Caution! Harmful if swallowed. MCA warning label.
Shipping regulations: None. *

copper carbonate, basic. See copper carbonate.

copper carbonate, basic, hydrated. See azurite

copper carbonate, blue. See azurite.

copper carbonate, green. See malachite.

copper, chessy. See azurite.

copper chloride (cupric chloride) (a) $CuCl_2$; (b) $CuCl_2 \cdot 2H_2O$.
Properties: (a) Brownish-yellow powder; hygroscopic; (b) green, deliquescent crystals. Poisonous! Soluble in water.
Constants: Sp.gr. (a) 3.054; (b) 2.39. M.p. (a) 498°C.
Derivation: (a) By the union of copper and chlorine. (b) Copper carbonate is treated with hydrochloric acid and the product crystallized.
Grades: Technical; C.P.; reagent.
Containers: 1-, 5-lb bottles; 25-lb boxes; bags; 100-lb kegs; 300-lb barrels or fiber drums.
Uses: Chemical (oxidizing agent; catalyst in Deacon chlorine process); mordant in dyeing and printing fabrics; sympathetic ink; disinfectant; pyrotechnics; wood preservation; metallurgy (refining copper, gold, silver, recovering mercury from its ores by the wet process, electroplating copper on aluminum); chrome brown; preservation of pulpwood and ground pulp; deodorizing and desulfurizing petroleum distillates; photography; water purifications; feed additive.
Shipping regulations: None. *
See also cuprous chloride.

*See "I.C.C. Shipping Regulations," page xiii.
Reference numbers refer to name of manufacturer. See "List of Manufacturers," page v.

copper chromate (basic cupric chromate) $CuCrO_4 \cdot 2CuO \cdot 2H_2O$.
Properties: Light chocolate-brown powder; poisonous! Loses water at 260°C. Soluble in nitric acid; insoluble in water.
Derivation: By the action of chromic acid on copper hydroxide.
Grades: Technical.
Containers: Wooden kegs; fiber drums.
Use: Dyeing (mordant).
Shipping regulations: None.*

copper compounds. For divalent (cupric) compounds, see under copper; for monovalent (cuprous), see under cuprous.

copper cyanide (cupric cyanide) $Cu(CN)_2$.
Properties: Green powder; exceedingly poisonous! Keep well stoppered! Soluble in acids and alkalies; insoluble in water.
Derivation: By the addition of potassium cyanide to a solution of copper sulfate, cupric cyanide is precipitated. This can be dried, but is not stable.
Grades: Technical.
Containers: Glass bottles; special drums.
Uses: Electroplating copper on iron; intermediates (introduction of the cyanide group in place of the amino radical in aromatic organic compounds).
Danger: Contact with acid liberates poisonous gas. MCA warning label.
Shipping regulations: None.*
See also cuprous cyanide.

copper, deoxidized. Copper metal specially treated to remove all or a part of the 0.05% oxygen normally present. It is more ductile than ordinary copper metal.

copper dichromate (copper bichromate; cupric dichromate) $CuCr_2O_7 \cdot 2H_2O$.
Properties: Brown-red crystals; soluble in water, alcohol, and ammonium hydroxide. Hygroscopic. Caution! Keep well stoppered! Sp. gr. 2.286.

copper dihydrazinium sulfate $CuSO_4(N_2H_4)_2 \cdot H_2SO_4$.
Properties: M.p. greater than 300°C, starts to decompose at 140°C; very slightly soluble in water, 250 ppm at 80°C.
Use: Foliage fungicide.
Caution! May cause skin and eye irritation. Harmful if swallowed or inhaled as mist or dust.

copper, electrolytic. Copper refined by electrolysis. The purest form of copper available commercially.

copper ethylacetoacetate $Cu(C_6H_9O_3)_2$.
Properties: Blue-green powder; m.p. 192-193°C. Insoluble in water; soluble in most organic solvents.
Use: Research and development.

copper ethylhexoate. See soaps, metallic.

copper ferrocyanide (cupric ferrocyanide) $Cu_2Fe(CN)_6 \cdot 7H_2O$.
Properties: Reddish-brown powder; very insoluble in water and acids; soluble in NH_4OH and KCN solutions. As a paint pigment, copper ferrocyanide retains desirable color, light-fastness, and chalking resistance and is compatible with high quality organic red and maroon pigments.
Uses: Paints and enamels; analytical test for traces of copper; inorganic osmotic membranes.

copper fluoride (cupric fluoride) $CuF_2 \cdot 2H_2O$.
Properties: Blue crystals; poisonous! Slightly soluble in water; soluble in acids.
Derivation: By decomposing copper carbonate with hydrofluoric acid and subsequent crystallization.
Method of purification: Recrystallization.
Grades: Technical.
Containers: Wooden kegs; fiber drums.
Uses: Ceramics; enamels.
Shipping regulations: None.*

copper fluosilicate (copper silicofluoride; cupric fluosilicate; cupric silicofluoride) $CuSiF_6 \cdot 4H_2O$.
Properties: Blue, hygroscopic crystals; poisonous! Soluble in water; slightly soluble in alcohol. Sp. gr. 2.158; decomposed by heat.
Derivation: By the interaction of copper hydroxide and hydrofluosilicic acid.
Method of purification: Crystallization.
Grades: Technical.
Containers: Wooden kegs.
Uses: Dyeing and hardening white marble; treating grape vines for "white disease."
Caution! Harmful if swallowed. MCA warning label.
Shipping regulations: None.*

copper glance. See chalcocite.

copper gluconate (cupric gluconate) $Cu(C_6H_{11}O_7)_2 \cdot H_2O$.
Properties: Odorless, light blue, fine, crystalline powder. Soluble in water; insoluble in acetone, alcohol, and ether.
Method of purification: Crystallization.
Grades: Pharmaceutical.
Containers: Cans; 25-lb fiber drums.
Use: Medicine; feed additive.
Shipping regulations: None.*

copper glycinate (copper aminoacetate; glycocoll-copper) $(NH_2CH_2COO)_2Cu$.
Properties: Blue triboluminescent crystals; m.p. 130°C. Slightly soluble in water and alcohol; insoluble in hydrocarbons, ethers and ketones.
Grades: Anhydrous; hydrated (one H_2O).
Uses: Catalyst for rapid biochemical assimilation of iron; electroplating baths.

copper green. See verte antique.

copper hemioxide. See copper oxide, red.

copper hydrate. See copper hydroxide.

copper hydroxide (cupric hydroxide; hydrated copper oxide; copper hydrate) $Cu(OH)_2$.
Properties: Blue powder; poisonous! Soluble in acids; insoluble in water.
Constants: Sp. gr. 3.368; m.p., decomposes.
Derivation: By the interaction of a solution of a copper salt with an alkali.
Grades: Technical.
Containers: Wooden kegs; fiber drums.

*See "I.C.C. Shipping Regulations," page xiii.
Reference numbers refer to name of manufacturer. See "List of Manufacturers," page v.

Uses: Copper salts; mordant; cuprammonium rayon; pigment; staining paper; feed additive.
Shipping regulations: None.*

copper 8-hydroxyquinoline. See copper 8-quinolinolate.

copper, indigo. See covellite.

"Copper Inhibitor 50." [28] Trade name for 50% disalicylalpropylenediamine, $HOC_6H_4CHNCH_2CH(CH_3)NCHC_6H_4OH$, and 50% aromatic solvent.
Properties: Amber colored liquid; sp. gr. 0.99.
Containers: 250-lb drums.
Use: To prevent catalytic action of copper on oxidation of natural and synthetic rubbers.

copper iodide. See cuprous iodide.

Copperized CZC Chromated Zinc Chloride. [28] Special wood preserving formulation based on zinc chloride, sodium dichromate, and copper chloride for severe service exposures. A companion product to CZC Chromated Zinc Chloride (q. v.).
Properties: Granular material with slight brownish color, or 50% water solution; sp. gr. of solution approximately 1.525; freezing point −35°F.
Containers: Solution: tank cars; dry: 575-lb drums.
Uses: For preserving wood by impregnation for use under severe service conditions against decay and termite attack; and for application as a fire retardant. Treated lumber is clean, odorless, paintable, and safe to handle.

copper lactate (cupric lactate) $Cu(C_3H_5O_3)_2 \cdot 2H_2O$.
Properties: Greenish-blue crystals or granular powder; soluble in water.
Uses: As a source of copper in copper plating; fungicides.
Caution! Harmful if swallowed. MCA warning label.

"Copper Lume." [72] Trade name for bright copper plating process. Prepared with copper cyanide, sodium cyanide and/or potassium cyanide and sodium and/or potassium hydroxide and organic and inorganic brightening agents.
Uses: Decorative plating applications in which smoothness and brightness of subsequent plated deposits are desirable.

copper matte. See copper; also matte.

copper-mercury iodide. See mercuric-cuprous iodide.

copper metaborate (copper borate; cupric borate) $Cu(BO_2)_2$.
Properties: Bluish-green, crystalline powder; insoluble in water; soluble in acids.
Derivation: By the interaction of copper hydroxide and boric acid.
Grades: Technical; C. P.
Containers: 1-lb bottles; kegs; tins.
Uses: Oil pigments; painting on porcelain; insecticides (especially wheat-rust compounds).
Caution! Harmful if swallowed. MCA warning label.
Shipping regulations: None.*

copper methane arsenate CH_3AsO_3Cu.
Properties: Greenish solid.
Derivation: Reaction of disodium methyl arsenate with copper salts.
Use: Algicide.

copper monoxide. See copper oxide, black.

copper naphthenate.
Properties: Green-blue solid. High germicidal power. Soluble in gasoline and mineral oil distillates. Dissolves in benzene to give green solution.
Derivation: Addition of solution of cupric sulfate to aqueous solution of sodium naphthenate.
Grades: 6, 8, 11 ½% copper.
Containers: 1- to 55-gal drums.
Uses: Wood, canvas, and rope preservative; insecticide; fungicide.

copper nitrate (cupric nitrate) (a) $Cu(NO_3)_2 \cdot 3H_2O$; (b) $Cu(NO_3)_2 \cdot 6H_2O$.
Properties: Blue, deliquescent crystals; poisonous! Soluble in water and alcohol.
Constants: Sp. gr. (a) 2.174; (b) 2.074. M. p. (a) 114.5°C; (b) 26.4°C; (a) decomposes 170°C.
Derivation: By treating copper or copper oxide with nitric acid. The solution is evaporated and product recovered by crystallization.
Method of purification: Recrystallization.
Grades: Technical; C. P.
Containers: 1-, 5-lb bottles; wooden barrels; drums.
Uses: Medicine; preparation of light-sensitive papers; analytical reagent; dyes; insecticide for vines; coloring copper black; electroplating; production of burnished effect on iron; paints, varnishes, enamels; pharmaceutical preparations; textiles; catalyst.
Fire hazard: Dangerous oxidizing material. In contact with organic or other readily oxidizable substances it will cause violent combustion or ignition.
Shipping regulations: Oxidizing material. Yellow label.*

copper nitrite. (basic copper nitrite; cupric nitrite) $Cu(NO_2)_2 \cdot 3Cu(OH)_2$; variable.
Properties: Green powder. Soluble (with decomposition) in dilute acids, ammonium hydroxide; slightly soluble in water.

copper nitrite, basic. See copper nitrite.

copper octoate. See soaps, metallic.

copper oleate (cupric oleate) $Cu(C_{18}H_{33}O_2)_2$.
Properties: Brown powder or greenish-blue mass; poisonous! Soluble in ether; insoluble in water.
Derivation: By the interaction of copper sulfate and sodium oleate.
Grades: Technical.
Containers: Tins.

Uses: Medicine; preserving fish nets and marine lines; fungicide; insecticide; ore flotation agent.
Shipping regulations: None.*

copper ore, gray. See fahlore.

copper ore, plush See cuprite.

copper ore, purple. See bornite.

copper ore, ruby. See cuprite.

copper ore, yellow. See chalcopyrite.

copper orthoarsenite. See copper arsenite.

copper oxide, black (cupric oxide; copper monoxide) CuO. For native black copper oxide see tenorite.
Properties: Brownish-black, amorphous or crystalline powder. Soluble in acids; insoluble in water. Sp. gr. 6.32; m.p. 1064°C.
Derivation: By the ignition of copper carbonate or copper nitrate.
Grades: Technical; C.P.
Containers: 1-, 5-lb bottles; 1-, 5-lb cans; 25-lb boxes; 100-lb kegs; 1,000-lb barrels or fiber drums.
Uses: Producing green or blue colors on glass, faïence, porcelain and stoneware; reagent in analytical chemistry; insecticide for potato plant; catalyst in the reduction of organic compounds; purification of hydrogen; batteries and electrodes; aromatic acids from cresols; electroplating; solvent for chromic iron ores; imitation precious stones; desulfurizing oils; rayon; medicine; paints.
Caution! Harmful if swallowed. MCA warning label.
Shipping regulations: None.*

copper oxide, hydrated. See copper hydroxide.

copper oxide, red. (cuprous oxide; copper protoxide; copper hemioxide; copper suboxide) Cu_2O. For the native ore see cuprite.
Properties: Reddish-brown crystalline powder. Soluble in acids; insoluble in water. Sp. gr. 5.75-6.09; m.p. 1210°C; b.p. 1800°C.
Derivation: (a) By the oxidation of finely divided copper. (b) By the addition of bases to cuprous chloride. (c) By the action of glucose on cupric hydroxide.
Grades: Technical; C.P.; 97% min (for pigments); also USN Type I (97%); USN Type II (90%).
Containers: 1-, 5-lb bottles; 25-lb cans; 25-lb boxes; 100-lb kegs; 1000-lb barrels.
Uses: Copper salts; ceramics; porcelain red glaze; red glass; electroplating; antifouling paints; fungicide.
Caution! Harmful if swallowed. MCA warning label.
Shipping regulations: None.*

copper oxychloride (cupric oxychloride) composition variable, possibly $3CuO \cdot CuCl_2 \cdot 3\frac{1}{2}H_2O$.
Properties: Bluish-green powder. Soluble in acids, ammonia; insoluble in water.
Caution! Harmful if swallowed. MCA warning label.
Use: Pigment; pesticide.

copper phenolsulfonate (copper sulfocarbolate) $[C_6H_4(OH)SO_3]_2Cu \cdot 6H_2O$.
Properties: Green prismatic crystals. Soluble in water and alcohol.
Derivation: Interaction of barium phenolsulfonate and copper sulfate.
Use: Medicine.
Shipping regulations: None.*

copper phosphate (cupric phosphate) $Cu_3(PO_4)_2 \cdot 3H_2O$.
Properties: Light-blue powder. Soluble in acids, ammonium hydroxide; insoluble in water.
Grades: Technical.
Caution! Harmful if swallowed. MCA warning label.
Uses: Analysis; medicine; fungicide.

copper phosphide (cupric phosphide) Cu_3P_2.
Properties: Grayish-black, metallic powder. Soluble in acid; insoluble in water. Sp. gr. 6.67.
Derivation: By heating copper and phosphorus.
Grades: Technical.
Containers: Wooden kegs; fiber drums.
Use: Manufacturing phosphor-bronze.
Shipping regulations: None.*

copper phthalate $C_8H_4O_4Cu$.
Properties: Fine blue powder; assay, minimum 95%; insoluble or very slightly soluble in common organic solvents or water.
Use: Fungicide.
Caution! Harmful if swallowed. MCA warning label.

copper phthalocyanine blue. See phthalocyanine pigments.

copper phthalocyanine green. See phthalocyanine pigments.

copper plating. The process for the production of a coating of substantially 100% pure copper on a metallic cathode. This process is carried out in an aqueous electrolyte, using copper anodes. Copper plating baths are of two types. The alkaline or cyanide bath in which copper is present as the cyanide is always used when copper is to be plated on die castings and on iron or steel, and is preferred for plating over brass or bronze. The acid bath, in which copper is present in the electrolyte as copper sulfate, is used for plating copper over nickel or over copper previously deposited on steel from the cyanide bath.

copper-potassium chloride (potassio-cupric chloride; potassium-copper chloride) $CuCl_2 \cdot 2KCl \cdot 2H_2O$.
Properties: Bluish-green crystals. Soluble in water.
Grades: Technical; C.P.; reagent.
Use: Analytical reagent.
Shipping regulations: None.*

copper-potassium cyanide. See potassium-copper cyanide.

*See "I.C.C. Shipping Regulations," page xiii.
Reference numbers refer to name of manufacturer. See "List of Manufacturers," page v.

copper potassium ferrocyanide (potassium copper ferrocyanide) $K_2CuFe(CN)_6 \cdot H_2O$.
Properties: Brownish-red powder. Insoluble in water. Used for pigment.

copper protoxide. See copper oxide, red.

copper pyrites. See chalcopyrite.

copper pyrophosphate. Used as a feed additive.

copper-8-quinolinolate (copper-8; copper 8-hydroxyquinoline) $Cu(C_9H_6ON)_2$.
Properties: Yellow-green nonhygroscopic, odorless powder. Solubility in neutral water about one part per million. Somewhat soluble in weak acids, soluble in strong acids. Insoluble in most organic solvents, but somewhat soluble in pyridine and quinoline. Solubilized copper-8 refers to the product formed by heating copper-8-quinolinolate with certain organic acids (naphthenic, lactic, stearic, etc.) or their salts. In such products the copper-8-quinolinolate does not settle out on standing, even after dilution with various solvents.
Derivation: From 8-quinolinol and copper salt such as copper acetate.
Grade: 10% active salt (1.8% Cu), solubilized.
Containers: Drums.
Uses: Fungicide and mildew-proofing of fabrics; in analysis for copper.

copper resinate (cupric resinate).
Properties: Green powder; poisonous! Soluble in ether and oils; insoluble in water.
Derivation: By heating copper sulfate and rosin oil and filtering and drying the precipitate.
Grades: Technical.
Containers: Wooden kegs or fiber drums.
Uses: Preservative metal paint, particularly for ships' bottoms; insecticide.
Shipping regulations: None.*

copper ricinoleate $Cu(C_{17}H_{32}OHCOO)_2$. A green plastic solid; soluble in water and aliphatic hydrocarbons; partially soluble in alcohols and glycols; soluble in ketones and aromatic hydrocarbons.
Warning! Poisonous!
Uses: Fungicides; insecticides.

copper salts. For divalent (cupric) salts, see under copper; for monovalent (cuprous), see under cuprous.

copper scale. A coating which forms on copper after heating. It is composed of a mixture of cupric and cuprous oxides.

copper selenate (cupric selenate) $CuSeO_4 \cdot 5H_2O$.
Properties: Light-blue crystals. Soluble in acids, ammonium hydroxide, water; insoluble in alcohol. Sp.gr. 2.559.

copper silicates. Complex mixtures precipitated by solutions of copper salts from sodium silicate solutions. They are used in pigments, catalysts and insecticides.

copper silicide. See silicon-copper.

copper silicofluoride. See copper fluosilicate.

copper-sodium chloride (sodio-cupric chloride; sodium copper chloride) $CuCl_2 \cdot 2NaCl \cdot 2H_2O$.
Properties: Light-green crystals. Soluble in water.

copper-sodium cyanide. See sodium-copper cyanide.

copper stearate (cupric stearate) $Cu(C_{18}H_{35}O_2)_2$.
Properties: Light blue, amorphous powder; poisonous! Soluble in ether, chloroform, benzene and turpentine; insoluble in water. M.p. about 170°C.
Derivation: By the interaction of copper sulfate and sodium stearate.
Grades: Technical.
Containers: Wooden kegs; tins; fiber drums.
Uses: Bronzing plaster statues; paint; see also soaps, metallic.
Shipping regulations: None.*

"Copper Steel." [251] Trademark for a copper-bearing steel with good corrosion resistance; available in hot-rolled and galvanized sheets.

copper steels. Copper (0.15-0.25%) in steel improves resistance to atmospheric corrosion and also the resistance to sulfuric acid. Similar proportions of copper (0.25-0.5%) also render steel more resistant to oxidation at higher temperature. Up to about 4%, copper increases the fluidity of the melt, improves tensile and yield strength with only minor loss in ductility. The copper may be alloyed with one-third nickel to prevent surface checking.
Used for structural steels where corrosion resistance is a factor.

copper subacetate. See copper acetate, basic.

copper suboxide. See copper oxide, red.

copper sulfate. (cupric sulfate; blue vitriol; blue stone; blue copperas) $CuSO_4 \cdot 5H_2O$.
Properties: Blue crystals or blue crystalline granules or powder, slowly efflorescing in air; white when dehydrated; nauseous metallic taste; poisonous! Found in nature as chalcanthite (q.v.). Soluble in water and slowly soluble in glycerin. Sp.gr. 2.284.
Derivation: (a) By the action of dilute sulfuric acid on copper or copper oxide in large quantities, with evaporation and crystallization.
Method of purification: Recrystallization.
Grades: Technical; C.P.; N.F. XI; also sold as monohydrate.
Containers: Multiwall paper sacks; drums.
Uses: The most important copper salt industrially; textile mordant; leather industry; germicides; insecticides; additive to some soils; pigments; electric batteries; electrolytic baths; copper salts; hair dyes; reagent in analytical chemistry; medicine; feed additive; improving casein glues; wood preservative; preservation of pulp wood and ground pulp; process engraving and lithography; ore flotation; destroying

algae and low forms of animal life in drinking water; in petroleum industry; synthetic rubber; steel manufacture; treatment of natural asphalts. The anhydrous salt is used as a dehydrating agent.
Caution! Harmful if swallowed. MCA warning label.
Shipping regulations: None.*

copper sulfate, ammoniated (cupric ammonia sulfate; ammonio-cupric sulfate; copper aminosulfate) $CuSO_4 \cdot 4NH_3 \cdot H_2O$.
Properties: Dark blue, crystalline powder; decomposes in air; soluble in water; insoluble in alcohol.
Derivation: By dissolving copper sulfate in ammonium hydroxide and precipitating with alcohol.
Method of purification: Crystallization.
Grades: Technical; C.P.
Containers: 1-lb bottles; wooden kegs; fiber drums.
Uses: Calico printing; manufacturing copper arsenate; insecticide; treating fiber products.
Caution! Harmful if swallowed. MCA warning label.
Shipping regulations: None.*

copper sulfide (cupric) CuS.
Properties: Black powder or lumps. Soluble in nitric acid; insoluble in water. Occurs as the mineral covellite. Sp.gr. 3.9-4.6; m.p. 1100°C.
Derivation: By passing hydrogen sulfide gas into a solution of a copper salt.
Grades: Technical; C.P.
Containers: 1-lb bottles; wooden barrels.
Uses: Copper metal; protective paint for vessels; dyeing with aniline black.
See also cuprous sulfide.
Shipping regulations: None.*

copper sulfocarbolate. See copper phenolsulfonate.

copper sulfocyanide. See cuprous thiocyanate.

copper tallate. See soaps, metallic.

copper tungstate (cupric tungstate; copper wolframate; normal copper tungstate) $CuWO_4 \cdot 2H_2O$.
Properties: Light-green powder; soluble in ammonium hydroxide; slightly soluble in acetic acid; insoluble in alcohol and water.

copper tungstate, normal. See copper tungstate.

copper undecylenate. Probably $[CH_2{:}CH(CH_2)_8COO]_2Cu$. Used as a fungicide.

copper uranite. See torbernite.

copper wolframate. See copper tungstate.

copper, yellow. See chalcopyrite.

"Coppralyte" 1085. [28] Trademark for a pale amber, mildly alkaline water solution of organic wetting agents; faint amine odor.
Use: As an antipitting agent for copper plating.

"Coppralyte" Plating Salt-Potassium Formulation. [28] A sodium formulation is also available. White powder, gives alkaline solution in water. Copper content (as Cu) min. 25.4%; free KCN min. 1.25%, max. 3.0%.
Use: For preparing copper plating baths for producing coatings on steel, zinc diecastings, and wire.

copra.
Derivation: The dried meat of coconut. Obtained from the South Sea Islands and the East Indies.
Grades: Cebu; Java; Macassar; South Sea; spot; sun-dried; Pacific coast; and Padang.
Containers: Bags.
Uses: For the extraction of coconut oil; confectionery; food.
Shipping regulations: None.*

copra cake. See coconut cake.

copra oil (copperah oil). The name applied to lower grades of coconut oil (q.v.).
Properties: White, wax-like, semi-solid; somewhat disagreeable odor; mild taste. Soluble in alcohol and ether; insoluble in water.
Chief constituents: Trimyristin and trilaurin.
Constants: Sp.gr. 0.910-0.926; m.p. 23-27°C; iodine number 8-9; saponification value 251-268; refractive index 1.441 (60°C).
Derivation: From the dried meat of the coconut Cocus nucifera by boiling and pressing.
Method of purification: Decolorizing with boneblack or fullers' earth.
Grades: Technical.
Containers: Boxes; wooden barrels; fiber drums; steel drums.
Uses: Soap; candles; food; medicine.
Shipping regulations: None.*

coprolites. Phosphatic nodules consisting of mixtures of calcium phosphate and calcium carbonate derived from the excrements of certain extinct fishes and reptiles. They were formerly used as phosphatic manures but are now rarely used due to the availability of sources of cheaper phosphates. They are mostly found in England and France, the best English grades containing up to 55 to 60% tricalcium phosphate and the French varieties containing a lower amount of phosphate.
Shipping regulations: None.*

coquina. A porous, coarse limestone composed of fragments of marine shells. Found in southern U.S. and used for road and building construction.

"Coraid." [108] Trademark for a powdered organic type corrosion inhibitor specifically for copper and copper alloys.
Containers: 65-lb drums.
Uses: Industrial cooling and other non-potable water systems.

"Co-Ral." [181] Trademark for O,O-diethyl O-3-chloro-4-methyl-2-oxo-2H-1-benzopyran-7-yl phosphorothioate (q.v.).

"Coral." [278] Trade name for a commercial grade of stereospecific polyisoprene rubber consisting essentially of cis-1,4-polyisoprene.
Properties: Similar to those of natural rubber, both unvulcanized and vulcanized.
Use: Replacement for natural rubber.
Hazards: None.
Shipping regulations: None.*

coral. Skeletons of the coral polyps found in the warmer oceans and consisting mainly of calcium carbonate colored with ferric oxide.

"Coramine." [305] Trademark for nikethamide, U.S.P. Used in medicine.

"Coray." [51] Trademark for general-purpose, low cold test, engine oils used for industrial machinery lubrication. Available in a wide range of viscosities.

"Corcast." [20] Trademark for petalite material for casting.
Properties: Cast articles have service temperatures up to 2500°F; expansion coefficient of about zero, and compressive strength of 2000-3000 psi; thermal conductivity of 0.723 Btu/ft/°F/hr. and density of 110 lbs/cu ft.
Containers: Packed in 100-lb cartons.
Uses: As tools, jigs, and fixtures for metal forming.

"Cordex." [74] Trademark for a cordage oil solution of copper napthenate containing 8% copper metal.
Use: Fungicide.

cordite. A smokeless powder which is a mixture of nitrocellulose and nitroglycerin with about 5% petrolatum added to thicken and stabilize the mixture. Materials are dissolved in acetone and mixed. Evaporation of the excess acetone leaves a gelatinous mass which is extruded into cords.

"Cordura." [28] Trademark for a textile viscose rayon yarn of relatively high tensile strength. See rayon.
Containers: Cones, tubes, cakes and beams.
Uses: In fabrics and some other tensile products where high strength is important.

"Coresinblak No. 3." [133] Trademark for a paste consisting of jet black impingement carbon thoroughly dispersed in an alkyd resin. Composition: 18% carbon black, 32% alkyd resin, and 50% solvent. Compatible with most medium and short oil alkyds.
Containers: 40- and 400-lb metal drums.
Use: For making high grade enamels, both air dry and baking.

"Corial Bottom." [307] Trademark for a series of leather finishing agents consisting of plasticized acrylic resin emulsion; 40-41% solids.
Properties: White liquids; can be diluted with water or alkaline aqueous solutions in all proportions without impairing emulsion stability. Compatible with materials commonly used in leather finishing, such as alkaline solubilized shellac, casein and latex. Can be mixed with either clear or pigment finishes.
Grades: The different types of "Corials" form films of varying flexibility, toughness, and transparency.
"Corial Bottom" E forms the hardest and least flexible film.
"Corial Bottom" EST produces the most flexible and tacky film.
"Corial Bottom" N similar to E, but more plastic and lower in tensile strength.

coriander (coriander seed). Dried, ripe fruit of Coriandrum sativum.
Habitat: Native to Italy; cultivated in India, Asia Minor, North Africa, United States.
Grades: Technical; N.F. XI.
Containers: Bags.
Uses: Medicine; condiment.
Shipping regulations: None.*

coriander oil.
Properties: Colorless or slightly yellowish liquid; aromatic odor; warm, spicy taste. Soluble in alcohol, ether, and chloroform.
Chief known constituents: Linalool; pinene.
Constants: Sp. gr. 0.863-0.878; refractive index 1.4665.
Derivation: Distilled from the fruit of Coriandrum sativum.
Method of purification: Rectification.
Grades: Technical; U.S.P. XVI.
Containers: Tins; glass bottles.
Use: Flavoring material.
Shipping regulations: None.*

coriander seed. See coriander.

coriandrol. Is d-linalool.

"Corilene." [206] Brand name for proprietary leather degreasing assistants.

cork. The light, porous, outer bark of the variety of oak tree known as cork-oak (Quercus suber).
Habitat: Southern Europe; northern Africa; now cultivated in southern United States.
Uses: Filler; stoppers; insulation; sound deadener; life preservers; gaskets; etc., linoleum manufacture.

cork black. A pigment obtained by charring cork.

corkboard. A mixture of ground cork and paper pulp formed into thick sheets for insulating purposes.

"Cormet A." [20] Trademark for sintered, porous articles of 99+% nickel.
Properties: Maximum yield strength of 10,000 psi; maximum operating temperature of 300°C. Available with pore diameters of 1 to 45 microns.
Uses: A non-contacting conveyor for glass of surface sensitive materials such as photographic film, adhesive materials, and plastic sheets.

corn. In United States and Canada: Indian corn or maize; in Great Britain and elsewhere: wheat, oats, or other grain.

cornelian. See carnelian.

"Corning." [20] Trademark for glass and glassware of various compositions and physical properties, and accessories used therewith.

"Corning Brand Glass No. 7280." [20] Trademark for substantially boron-free alkali-resistant glass.
Properties: Linear coefficient of expansion per °C 0.00000064 between 0-300°C; softening point 870°C; sp.gr. 2.61; very resistant to chemical attack.
Uses: Laboratory and pharmaceutical glassware; alkali-resistant tubing.

Cornish clay. Same as Cornish stone.

Cornish stone.
1. (Cornish clay). A partially weathered feldspar, used as flux and fusible ingredient in porcelain and tiles. Sometimes called china stone.
2. A quartz crystal from Cornwall.

corn oil (maize oil).
Properties: Pale yellow liquid; characteristic taste and odor. Insoluble in water; soluble in ether, chloroform, amyl acetate, benzene, and carbon disulfide and slightly soluble in alcohol. Sp.gr. 0.914-0.921; saponification value 188-193; iodine value 102-128.
Typical analysis: 98.6% triglycerides; 1.4% unsaponifiable; 1.0% sitosterols. Of the total fatty acids, 86% arc unsaturated (linoleic 56%, oleic 30%) and 13% saturated (palmitic 10%, stearic 3%).
Derivation: The germ of common corn (Indian corn, Zea mays) is removed from the grain and pressed.
Method of purification: Filtration.
Grades: Crude; refined; U.S.P. XVI; technical.
Containers: 375-lb barrels; 8000-gal tank cars.
Uses: Foodstuffs; soap; lubricants; leather dressing; rubber substitutes; lard substitutes; salad oil; hair dressings.
Shipping regulations: None.*

corn-oil-foots pitch. See stearin and fatty acid pitches.

corn-oil-pitch. See stearin and fatty acid pitches.

corn starch. See starch.

corn steep liquor. The dilute aqueous solution obtained by soaking corn kernels in warm 0.2% sulfur dioxide solution for 48 hours as the first step in the recovery of corn starch, corn oil, and gluten from corn. The solution contains mineral matter as well as soluble organic material extracted from the corn. It is used as a growth medium for penicillin and other antibiotics, and it is also concentrated and used as an ingredient of cattle feeds.

corn sugar. See dextrose.

corn syrup. See glucose.

"Corona." [188] Trademark for a certain grade of ylang ylang oil (q.v.).

coronizing. A process for producing an unusual corrosion-resistant coating on metals by electroplating a thin layer of nickel on the base metal.

corozo-nut oil. See cohune oil.

"Corpolin." [318] A proprietary product consisting of various organic materials in aqueous solution.
Properties: Clear, viscous liquid soluble in water and compatible with alkalies, gelatin, vegetable gums, and most other water-soluble materials; slightly alkaline reaction.
Containers: 120- and 600-lb lined drums.
Uses: Hygroscopic agent for use in vat-print paste and paper plasticizer; glycerin replacement.

corr. Abbreviation for corrected, usually applied with reference to a boiling point or other temperature.

corresponding states (reduced states). Two substances are in corresponding states when their pressures, volumes (or densities) and temperatures are proportional respectively to their critical pressures, volumes (or densities) and temperatures. If any two of these ratios are equal, the third must also be equal. This principle has been useful in the development of physical and thermal properties of substances.

corrinoids. Generic name for compounds of the vitamin B_{12} series containing the corrin nucleus.

corrosion. The conversion of iron, steel, and other alloys and metals into oxides, hydrated oxides, carbonates, or other compounds due to the action of air or water or both. The minor components present in the air or water are important factors in the rate of corrosion and the kind of corrosion products. Natural minor components such as carbon dioxide in air and water cause serious corrosion, but contaminants introduced by all types of air and water pollution usually accelerate corrosion. Salts, as in sea water, are serious causes of corrosion. Sulfur in fuels such as coal, oil, gasoline and natural gases and other fuels is also an important source of corrosion, so that removal treatments are common except for coal. Electric currents from power sources, or from differences in composition of materials in the ground or different parts of metal objects are also accelerators of corrosion.

"Corrosion Inhibitor NPA." [219] Tradename for nonylphenoxy acetic acid, $C_9H_{19}C_6H_4O$-CH_2COOH.
Properties: Light amber liquid. Miscible with organic solvents, insoluble in water; soluble in alkali. Flow point 0°C; viscosity (25°C) 6500 cps.
Uses: Corrosion inhibitor for turbine oils, lubricants, fuels; greases; hydraulic fluids, cutting oils and metal coolants.

corrosive sublimate. See mercuric chloride.

*See "I.C.C. Shipping Regulations," page xiii.
Reference numbers refer to name of manufacturer. See "List of Manufacturers," page v.

"Cortamp." [20] Trademark for zircon material for compaction forming.
Properties: Formed articles have service temperatures up to 4000°F; expansion coefficient of about 3×10^{-6} per °F; compressive strength of 10,000-12,000 psi; thermal conductivity of 1.08 Btu/ft/°F/hr., and density of 200 lb/cu ft.
Containers: 100-lb cartons.
Uses: As tools, jigs, and fixtures for metal forming.

"Cortate." [321] Brand name for deoxycorticosterone acetate.

"Cortef." [327] Trademark for hydrocortisone, U.S.P. grade.
Use: Medicine.

"Cortef" Cyclopentyl Propionate. [327] Trademark for hydrocortisone cyclopentyl propionate ($C_{29}H_{42}O_6$).
Properties: A white, odorless solid practically insoluble in water. Will dissolve in the following solvents to form 100 ml. of solution: 3.28 g. in 95% alcohol; 0.272 g. in ether; and 35.9 g. in chloroform.
Use: Medicine.

corticoid hormones. Adrenal cortical hormones produced or isolated from the cortex (external layer) of the adrenal gland. An animal dies if deprived of these hormones, and may be seriously ill if suffering from a lack or excess of them. These hormones have a strong effect on salt and water metabolism, carbohydrate and protein metabolism, and the ability of the animal to withstand various types of stress. The rate of synthesis and secretion of nearly all the adrenal steroids in the body is regulated by ACTH (q.v.), a hormone secreted by the pituitary gland.

Corticoid hormones now used in medicine include cortisone, hydrocortisone, deoxycortisone, fludrocortisone, prednisone, prednisolone, methyl prednisolone, triamcinolone, dexamethasone, and aldosterone. Some occur naturally in adrenal extract; others are modifications of the natural hormones. All are now made synthetically. They are derivatives of the cyclopentanophenanthrene nucleus.

corticosterone (Compound B) $C_{21}H_{30}O_4$. One of the less-active adrenal cortical steroid hormones.
Properties: Crystalline plates; m.p. 180-182°. Soluble in organic solvents; insoluble in water.
Derivation: Isolation from adrenal cortex extract; synthesis from deoxycholic acid
Use: Biochemical research.

corticotropin. See ACTH.

"Cortifan." [321] Brand name for hydrocortisone.

cortisol. See hydrocortisone.

cortisone (11-dehydro-17-hydroxycorticosterone; Compound E) $C_{21}H_{28}O_5$. One of the adrenal cortical steroid hormones. It has an effect on carbohydrate and protein metabolism and is used as an antiinflammatory agent. The total synthesis was first reported in 1951.
Properties: White crystalline solid; m.p. 220-224° (dec). Dextrorotatory in solutions. Slightly soluble in water; sparingly soluble in ether, benzene, and chloroform; fairly soluble in methanol, ethanol, and acetone.
Derivation: Isolation from adrenal gland extract (usually from cattle) (historical method); synthetically, from bile acids, from other steroids or sapogenins, by chemical means or microbiologically.
Use: Medicine (usually as acetate salt).

cortisone acetate (CA). The commonly used form of cortisone.
Properties: White or practically white, odorless, crystalline powder. Stable in air; m.p. about 240° (dec). Freely soluble in chloroform; soluble in dioxane and acetone; slightly soluble in alcohol; insoluble in water. Sensitive to light.
Derivation: See under cortisone.
Grade: U.S.P. XVI.
Use: Medicine.

"Cortogen Acetate." [321] Brand name for cortisone acetate.

"Cortone." [123] Trademark for cortisone.

"Cortril." [299] Trademark for hydrocortisone.

corundum Al_2O_3. Natural aluminum oxide, sometimes with small amounts of iron, magnesium, silica, etc. Found in metamorphic rocks, pegmatites and igneous rocks.
Properties: Color, variable; luster vitreous to adamantine; sp.gr. 4.02; hardness 9.
Varieties:
(a) Precious corundum. Vividly colored gem stones: true and oriental ruby (red); oriental topaz (yellow); oriental emerald (green); sapphire (blue); oriental amethyst (purple). The term sapphire is often used to cover all colors of corundum gem stones except red.
Occurrence: Burma; Siam; Ceylon; Montana, North Carolina; India; Australia.
Use: Gem stones.
(b) Ordinary Corundum. Light blue to gray, brown or black.
Occurrence: Eastern United States; Canada; South Africa; India.
Use: Abrasives.
(c) Emery. A mixture of corundum and magnetite or other iron oxide. Dull black, dark gray or bluish black color.
Occurrence: New York; Greece; Asia Minor.
Use: Various polishing and abrasive operations.
See also aluminum oxide; diaspore.

"Coryban." [299] Trademark for a combination drug containing purified hesperidin, ascorbic acid, salicylamide, acetophenetidin, caffeine, and prophenpyridamine maleate.

corynine. See yohimbine.

cosalite. A natural sulfide of lead and bismuth $2PbS \cdot Bi_2S_3$. Contains 42% bismuth. Found in United States (Washington); Sweden; Mexico.

cosaprin. See sodium acetylsulfanilate.

"Cosa-Signemycin." [299] Trademark for glucosamine-potentiated antibiotic combination containing tetracycline hydrochloride and triacetyloleandomycin.

"Cosa-Terramycin." [299] Trademark for glucosamine-potentiated oxytetracycline hydrochloride.

"Cosa-Tetracyn." [299] Trademark for glucosamine-potentiated tetracycline hydrochloride.

cosmic rays. Penetrating radiations appearing to come from a source in outer space. It is believed that the original rays are fairly simple. High altitude studies have shown that outside the earth's atmosphere most cosmic rays are protons with energies ranging from 1-100,000 billion electron volts, compared with a maximum of about 6 billion electron volts for particles produced in various man-made accelerators today (bevatron and cosmotron). The cosmic rays change when they pass through the atmosphere. They then consist mainly of mesons, protons, neutrons, electrons and photons. Almost all cosmic radiation absorbed at the earth's surface is of the latter type. Some cosmic rays have been detected under the surface of the earth in mines. Cosmic rays, like other high-energy atomic particles, are detected with cloud chambers. These contain a gas saturated with a vapor. A ray or particle passing through the chamber produces ions which serve as seeds to bring about condensation of the vapor to produce a fog track. This shows the path of the particles through the chamber. The range (distance traveled), the energy, and the nature of the particles causing the tracks can thus be determined.

"Cosmol." [221] Sperm oil esters with applications in textile chemicals, lubricants and cutting oils.

cosmotron. See cyclotron.

"Cosol." [21] Brand name for high boiling coal-tar solvents for use in alkyd resin enamels and synthetic lacquers.

"Costyreneblak." [133] Trademark for a series of carbon black polystyrene dispersions. Used for coloring polystyrene scrap. Available in chip form in several grades.

cotarnine chloride (cotarnine hydrochloride) $C_{12}H_{14}NO_3Cl$.

Properties: Yellow odorless crystalline powder; deliquescent, poisonous! Soluble in water and alcohol. M.p. 142-144°C.

Derivation: By the action of hydrochloric acid on an alkaloid prepared from narcotine by oxidation with nitric acid.

cotarnine hydrochloride. See cotarnine chloride.

cotarnine phthalate. $C_6H_4(COO)_2(C_{12}H_{14}O_3N)_2$.

Properties: Yellow powder. Soluble in water.

Use: Medicine.

cotton (gossypium). Staple fibers, usually 3/4-2 1/2 in. long, surrounding the seeds of various species of Gossypium. Cotton is the major textile fiber and is also an important source of cellulose, which constitutes 88-96% of the fiber.

Properties: Tensile strength 44,000-109,000 psi; elongation 3-7%; sp.gr. 1.54; moisture regain 7% (70°F, 65% relative humidity); yellows slowly at 250°F, decomposes about 300°F, burns readily. Decomposed by acids; swells in caustic but is undamaged. Soluble in cuprammonium hydroxide. Subject to mildew.

Sources: United States; Brazil; Egypt; India.

Uses: All types of apparel; industrial and household fabrics; automobile tires; upholstery.

See also cellulose; cellulose, oxidized; nitrocellulose.

cotton, acetylated. Cotton fibers, threads, or fabrics treated with acetic anhydride, acetic acid, and perchloric acid catalyst to improve the heat, rot, and mildew resistance by forming a surface coating of cellulose acetate.

cotton, aminized. A cotton fabric produced by reacting 2-aminoethylsulfuric acid with the cellulose of the fabric in a strongly alkaline solution. The treated cotton can take acid wool dyes and can be made rot-resistant and water-repellent.

cotton balls. See ulexite.

cotton, cyanoethylation of. Process for improving the properties of cotton by treatment with acrylonitrile. The cotton fabric is passed through a caustic bath, which induces mild swelling of the fiber and catalyzes the subsequent reaction with acrylonitrile. The fabric is then neutralized with acetic acid, washed and dried. The treatment leaves 3-5% nitrogen attached to the cellulose polymer. The cyanoethylated fiber is claimed to have permanent rot- and mildew-resistance, greater retention of strength after exposure to heat, improved receptiveness to dyes, and higher abrasion- and stretch-resistance than the original cotton.

cotton flock. Finely ground cotton rags used as a filler in plastics and as a finish on rubber fabrics.

cotton linters. See linters.

cotton oils (cotton spraying oils). Compounded oils sprayed (in the form of a fine mist) onto cotton to condition the fibers for yarn-making operations. Used to lubricate the fibers, to reduce static, "fly," and dust and generally improve the suppleness and strength of the fibers.

cottonseed cake (seed oil cake). The press cake derived from the extraction of cottonseed for its oil. When ground it is termed cottonseed meal.

cottonseed foots pitch. See stearin and fatty acid pitches.

cottonseed meal (cottonseed cake). The pulverized cottonseed press cake. Depending on the extractive process, varying percentages of protein will remain in the meal and it is normally sold with 36 to 45% protein content. The 42% product contains approximately 42% crude protein; 6% crude fiber; 25% nitrogen-free extract; 10% ether extract (fat) and 7% ash. The total digestible nutrient averages 79%. The ash is high in potash and phosphate.
Containers: Bulk or bags.
Uses: Animal feeds; fertilizer ingredient; filler for synthetic resin products.
Shipping regulations: None.*

cottonseed oil (seed oil).
Properties: Pale yellow or yellowish-brown to dark ruby-red or black-red, fixed, semi-drying oil depending on the nature and condition of the seed. The pure oil is odorless and has a bland taste. Soluble in ether, benzene, chloroform and carbon disulfide; slightly soluble in alcohol.
Constants: Sp.gr. 0.915-0.921; saponification value 190-198; iodine value 109-116.
Derivation: The seeds of the cotton plant (Gossypium herbaceum) are crushed in a mill, the meal is heated in iron kettles at 75-90°C and pressed in cloths, under 3,000-4,000 psi. An alternate process uses solvent extraction of the cottonseed.
Method of purification: Filtration.
Grades: Crude; refined; prime summer yellow; bleachable; U.S.P. XVI.
Containers: 375-lb barrels; 8000-gal tank cars.
Uses: Medicine; leather dressing; soap stock; lubricant; glycerol; base for cosmetic creams; food (oleomargarine and butter substitutes, in solid state as lard substitute, salad oil, cooking oil, packing sardines); waterproofing compositions; phonograph records.
Shipping regulations: None.*

cottonseed-oil pitch. See stearin and fatty acid pitches.

cotton solution. See nitrocellulose.

cotton spraying oils. See cotton oils.

cotton-stearin pitch. See stearin and fatty acid pitches.

Cottrell precipitator. An electrostatic device whereby negatively charged dust or fume particles are attracted to a wire electrode positively charged enclosed in a flue, the walls of which act as the other electrode. Widely used for treating sulfuric acid mist, cement mill dust, power-plant fly ash, metallurgical fumes, etc.

coulomb. The quantity of electricity which will deposit 0.001118 g of silver from a solution of silver nitrate.

coumarin (cumarin; benzopyrone; tonka bean camphor) $C_9H_6O_2$.
Properties: Colorless crystals, flakes or powder; fragrant odor similar to vanilla; bitter, aromatic burning taste; m.p. 69°C; b.p. 290°C. Soluble in 10 vols of 95% alcohol, in ether, chloroform, and fixed volatile oils; slightly soluble in water.
Derivation: (a) By heating salicylic aldehyde, sodium acetate and acetic anhydride; (b) fine grades are isolated from Tonka beans.
Method of purification: Crystallization.
Grades: Technical.
Containers: 1-, 5-, 20-lb tins; 25-, 200-lb fiber drums.
Uses: Deodorizing and odor enchancing agent; used in perfumes, soaps, tobacco, inks, rubber and other products where aromatic ingredients are required; flavoring; pharmaceutical preparations.
Shipping regulations: None.*

coumarone (cumarone; benzofuran) C_8H_6O, a bicyclic ring compound. Parent substance for the coumarone resins.
Properties: Colorless liquid; aromatic odor; sp.gr. 1.078; m.p. −18°C; b.p. 169°C. Insoluble in water; soluble in alcohol and ether.
Derivation: From the coal-tar naphtha fraction boiling between 150-200°C.
Use: See coumarone-indene resins.

coumarone-indene resins (coal tar resins). Resins obtained by heating mixtures of coumarone and indene (such as occur in the light-oil fraction from coal-tar refining) with sulfuric acid so as to cause polymerization to thermoplastic materials with softening points up to about 150°C.
Properties: Vary from fairly viscous liquids to hard resins; color pale yellow to nearly black; sp.gr. 1.05-1.10; soluble in hydrocarbon solvents, pyridine, acetone, carbon disulfide and carbon tetrachloride; insoluble in water and alcohol.
Uses: Components of aluminum paint, concrete curing compounds, pipe oils, rubber compounding, adhesives, chewing gum, printing inks, floor tile binding, phonograph records.
See "Nevindene" and "Paradene."

coumarouna bean. See tonka.

countercurrent. Applied to purification and other processes in which a liquid and a vapor stream, or two streams of immiscible liquid, or a liquid and a solid, are caused to flow in opposite directions and past or through one another with more or less intimate contact so that the individual substances present are more or less completely transferred to that stream in which they are more soluble or stable under the conditions existing. The streams leaving such a process are usually of higher purity than can be attained otherwise at equal cost. For example see extraction, liquid-liquid. Distillation with a fractioning column is also a typical countercurrent process, in which rising vapor is purified by contact with descending liquid. Leaching, washing and chemical reaction are frequently carried out in a countercurrent manner.

*See "I.C.C. Shipping Regulations," page xiii.
Reference numbers refer to name of manufacturer. See "List of Manufacturers," page v.

countercurrent extraction. See extraction, liquid-liquid.

counts. The external indication given by a radiation detector such as a Geiger counter of the amount of radioactivity to which the detector is exposed. The background counts are those which come from a source external to that being measured. Background counts may arise from cosmic rays, the presence of other radioactive materials in the location of the counter, electrical disturbances within the instrument, etc.

coupling. The combination of an amine or phenol with a diazonium compound to give an azo compound, the reaction by which azo dyes are prepared; thus meta-phenylenediamine $C_6H_4(NH_2)_2$ couples with benzene diazonium chloride $C_6H_5N_2Cl$ to produce the dye chrysoidine $C_6H_5N_2C_6H_3(NH_2)_2$.

"Covarnishblak." [133] Trademark for a series of dry powders consisting of carbon black dispersed in a hydrocarbon fossil resin. Used for pigmenting alkyd and oleoresinous enamels. Also useful in coloring rubber cements. Can be dissolved in the vehicle by simple stirring. Available in 50- and 200-lb drums.

covellite (indigo copper) CuS. Dark indigo-blue mineral, lead-gray to black streak, submetallic to resinous luster. Often shows fine purplish color when moistened. A mineral of secondary origin found in the enriched portions of copper sulfide veins. Contains 66.44% copper, 33.56% sulfur. Soluble in nitric acid; slowly soluble in hydrochloric acid.

Constants: Sp. gr. 4.59-4.64; hardness 1.5-2.

Occurrence: United States (Montana, Wyoming, California, Colorado, Utah, Alaska); Germany; Yugoslavia; Chile; Japan.

Use: Source of copper.

"Covinylblaks." [133] Series of dry chips consisting of carbon black thoroughly dispersed in vinyl resin. High grade vinyl coatings can be made from solution type "Covinylblaks" by dissolving chips in ketone solvents and compounding with other materials as required. Another series is available for coloring vinyl plastics such as wire jackets, vinyl sheets, etc.

Containers: 50- and 200-lb drums.

"Cowaxblak." [133] Trademark for a series of chips consisting of carbon black thoroughly dispersed in resinous chlorinated paraffin. Used for coloring vinyl, polystyrene, polyethylene and other plastics. Available in several grades.

Containers: 75- and 300-lb drums.

"Cowles Silicone Monomer M-5-124." [428] Trade name for a concentrated, stable solution of sodium propyl siliconate.

Properties: On dilution with water, the silicone becomes surface-active and is oriented toward surfaces that are siliceous or cellulosic in nature. On dehydration, it becomes insoluble and highly water repellent.

Uses: In paper pulp and cement slurries and asbestos fiber dispersions.

"Cowles SS-2DN." [428] Trade name for an aqueous, highly alkaline solution of sodium vinyl siliconate.

Properties: At the concentration and in the state supplied, the solution is quite stable. The critical pH of SS-2DN solution as supplied is 10.5. At and below this pH a rapid condensation to the polymer occurs. Above this pH the solution is quite stable and has a room temperature shelf life in excess of three years.

Uses: When acidified to a sol is used as a size for glass fibers.

Cox chart. A special semilogarithmic plot of vapor pressures vs. temperature especially useful for the petroleum hydrocarbons. The graph corresponding to each separate hydrocarbon is a straight line. All the lines appear to intersect at a point outside the chart.

cozymase. See nicotinamide adenine dinucleotide.

cp. Abbreviation for centipoise (q.v.).

C.P. Abbreviation for chemically pure. It signifies a minimum of impurities, but does not imply 100% pure.

"CP" 40. [306] Trademark for a chlorinated paraffin. Approximately $C_{24}H_{43}Cl_7$.

Properties: Light, honey-colored, viscous liquid. Its vapor pressure and evaporation rate are negligible at room temperature. Under prolonged storage in iron at elevated temperatures, "CP"-40 tends to develop acidity due to the liberation of hydrochloric acid, causing it to pick up iron and darken in color. B.p., decomposes; pour point 0°C; refractive index 1.508 (n 20/D); flash point (open cup) none; sp. gr. 1.185 (15.5/15.5°C); color, 1.5-2.5 (ASTM). Soluble in most organic solvents.

Uses: For waterproofing and fireproofing fabrics; as a plasticizer and extender for plastics, increases adhesiveness to metallic surfaces and adds to water and flame repellent qualities; ingredient in traffic paints and fire retardant paints.

"CPA-1800." [244] Trademark for a compound containing fluorine compounds and trivalent chromium.

Properties: Light green, free-flowing powder; soluble in water and chromic acid solutions.

Containers: 5-lb moisture-proof, fiber containers with tin-plated tops and bottoms; four 5-lb units per carton.

Uses: In the electroplating industry as an additive to the chromium plating solution; acts as catalyst to improve performance.

"C-P-B." [248] Trademark for dibutyl xanthogen disulfide.

Properties: Amber-colored, free-flowing liquid; sp. gr. 1.15; soluble in acetone,

*See "I.C.C. Shipping Regulations," page xiii.
Reference numbers refer to name of manufacturer. See "List of Manufacturers," page v.

benzol, gasoline, and ethylene dichloride; insoluble in water.

Use: Accelerator for proofing, pure gum hand-made druggist sundries and medical supplies, bathing shoes, bathing caps, novelties, and cold-cure cements.

CPR. Abbreviation for cyclonene-pyrethrin-rotenone. Applied to various insecticide formulations containing as active ingredients approximately 10 parts piperonyl cyclonene, 5 parts rotenone, and 1 part pyrethrin.

cps. Abbreviation for centipoise (q. v.).

Cr. Symbol for chromium.

cracking. The decomposition by heat, with or without catalysis, of petroleum or heavy petroleum fractions, with production of lower-boiling materials that are useful as motor fuels, domestic fuel oil, or other needed products. See also hydrocracking.

The term cracking is also applied more generally to thermal decomposition processes; thus ammonia (NH_3) may be cracked to nitrogen and hydrogen, and natural gas hydrocarbons such as methane (CH_4) are cracked into carbon and hydrogen, or into other hydrocarbons.

"Crag." [214] Trademark for agricultural chemicals including:

"Crag" Fly Repellent (active ingredient, butoxy polypropylene glycol). Colorless liquid, 100% active material.

Uses: Formulating dairy and livestock sprays for protection against biting and nuisance insects, such as horse flies, horn flies, stable flies, house flies, mosquitoes, and gnats.

"Crag" Fungicide 974 (active ingredient, 3,5-dimethyltetrahydro-1,3,5-2H-thiadiazine-2-thione). Wettable powder, 85% active material.

Uses: For the control of fungi and bacteria in the production and processing of paper, leather, glue, casein, starch, and other materials; for the control of dry rot of gladioli; for testing on other crops as a soil fungicide, nematocide and herbicide.

"Crag" Glyodin Solution (active ingredient, 2-heptadecyl glyoxaldine acetate) 34% active solution.

Uses: Control of scab, sooty blotch, Brooks spot, bitter rot, fly speck, and black rot of chrysanthemums; black spot of roses; rust of snapdragons.

"Crag" Herbicide-1 (SES) (active ingredient, sodium 2,4-dichlorophenoxyethyl sulfate). Water-soluble powder, 90% active material.

Uses: Preventing weeds in peanuts, strawberries, turf, asparagus, corn grown for seed, potatoes, perennial flowers, and nursery stock.

cramp bark. See viburnum opulus.

cranberry tree. See viburnum opulus.

cranes-bill. See geranium.

cream of tartar. See potassium bitartrate.

cream of tartar, borated. See potassium borotartrate.

cream of tartar, soluble. See potassium borotartrate.

creatine (N-methyl-N-guanylglycine; (alpha-methylguanido) acetic acid) $HN:C(NH_2)N(CH_3)CH_2COOH$. A nitrogenous acid found widely distributed in the muscular tissue of the body.

Properties: (Monohydrate) prisms from water; anhydrous at 100°C; decomposes 303°C; slightly soluble in water; insoluble in ether.

Source: Commercially isolated from meat extracts.

Grades: Technical; C.P.

Use: Biochemical research.

creatinine $C_4H_7N_3O$. The anhydride of creatine (q. v.); a metabolic waste product.

Properties: Leaflets from water; decomposes about 300°C; soluble in water; slightly soluble in alcohol; nearly insoluble in ether, acetone, and chloroform.

Use: Biochemical research.

creosote, beechwood. See creosote, wood-tar.

creosote carbonate.

Properties: Clear, colorless or yellowish, viscous liquid; slight creosote odor and taste, or odorless and tasteless. Soluble in alcohol; insoluble in water.

Derivation: Mixture of carbonates of various constituents of creosote.

Containers: 1-lb bottles; carboys.

Use: Medicine.

Shipping regulations: None.*

creosote, coal-tar (creosote oil; liquid pitch oil; tar oil).

Properties: Yellowish to dark green-brown, oily liquid; clear at 38°C or higher; characteristic odor; poisonous! Frequently contains substantial amounts of naphthalene and anthracene; distilling range 200-400°C; soluble in alcohol, benzene, and toluene.

Derivation: (a) Directly by the fractional distillation of coal-tar. (b) By redistillation of a coal-tar fraction. (c) A coal-tar from which the phenols and naphthalene have been partly extracted. (d) A mixture of two or more coal-tar fractions. Very similar materials are obtained from the tar accumulating from blast-furnace producer gas, and from similar gas-making operations.

Method of purification: Rectification.

Grades: Technical; crude; refined.

Containers: Iron drums; tank cars.

Uses: Wood preservative; disinfectants.

Caution: May cause skin irritation. MCA warning label.

Shipping regulations: None.*

creosote oil. See creosote, coal-tar.

creosote, wood-tar (creosote, beechwood).

Properties: Colorless or faintly yellow, oily liquid; characteristic smoky odor; caustic, burning taste. Miscible with

*See "I.C.C. Shipping Regulations," page xiii.
Reference numbers refer to name of manufacturer. See "List of Manufacturers," page v.

alcohol, ether, and fixed or volatile oils.
Constants: Sp. gr. 1.080; b.p. 203-220°C.
Derivation: A mixture of phenols and phenol derivatives obtained by the destructive distillation of wood-tar.
Grades: Technical; N.F. XI.
Containers: 44-lb carboys; 55-gal drums; tank cars.
Uses: Medicine; poison; ore-flotation agent.
Shipping regulations: None.*

cresidine (meta-amino-para-cresol, methyl ether) $CH_3C_6H_3(NH_2)OCH_3$.
Properties: White crystals; m.p. 51.5°C; b.p. 235°C. Soluble in alcohol and ether; sparingly soluble in hot water; volatile with steam.
Derivation: 2-Nitro-para-cresol, obtained by the action of nitrous and excess nitric acids upon para-toluidine, is methylated and reduced.
Method of purification: Distillation.
Grades: Technical.
Containers: Drums.
Use: Dyes.
Shipping regulations: None.*

"Creslan." [57] Trademark for an acrylic fiber.

cresol (methyl phenol; cresyl alcohol) $CH_3C_6H_4OH$. See also cresylic acid. Mixtures of the ortho-, meta-, and/or para-cresols are sold in a variety of grades, (see cresol, U.S.P. XVI for one grade) depending on color, distillation range, water, acid, or phenol content, use, and other criteria.
Containers: Drums; tank cars.
Uses: Phenolic resins; tricresyl phosphate; disinfectants; surfactants; flotation agents; scouring compounds; chemical intermediates; lube oil additives; enamel solvent.
Danger! Rapidly absorbed through skin. Causes severe burns. MCA warning label.

meta-**cresol** (meta-cresylic acid; meta-oxytoluene; 3-methylphenol) $CH_3C_6H_4OH$.
Properties: Colorless to yellowish liquid; phenol-like odor; poisonous! Soluble in alcohol, ether, and chloroform; slightly soluble in water.
Constants: Sp. gr. 1.034; m.p. 12°C; b.p. 203°C; wt/gal 8.66 lbs.
Derivation: By fractional distillation of crude cresol (from coal tar).
Method of purification: Rectification.
Grades: Technical (95-98%).
Containers: Drums.
Uses: Disinfectant; fumigating compositions; production of synthetic resins; photographic developer; nitrocresol explosives; ore flotation; intermediate; ink, paint and varnish remover; reclaiming rubber. See cresol.
Danger! Rapidly absorbed through skin; causes severe burns. MCA warning label.
Shipping regulations: None.*

ortho-**cresol** (ortho-cresylic acid; ortho-oxytoluene; 2-methylphenol) $CH_3C_6H_4OH$.
Properties: White crystals; phenol-like odor; poisonous! Soluble in alcohol, ether, and chloroform; slightly soluble in water.
Constants: Sp. gr. 1.0511; m.p. 30.9°C; b.p. 191°C; lbs/gal 8.61.
Derivation: By fractional distillation of crude cresol from coal tar.
Method of purification: Crystallization.
Grades: According to freezing point: 25°, 29°, 30°, 30.5°, etc.
Containers: 53-gal drums, tank trucks, tank cars.
Uses: Disinfectant; coumarin; organic intermediate. See cresol.
Danger! Rapidly absorbed through skin; causes severe burns. MCA warning label.
Shipping regulations: None.*

para-**cresol** (para-cresylic acid; para-oxytoluene; 4-methylphenol) $CH_3C_6H_4OH$.
Properties: Crystalline mass; phenol-like odor; poisonous! Soluble in alcohol, ether, and chloroform; slightly soluble in water; wt/gal 8.66 lb.
Constants: Sp. gr. 1.039; m.p. 35°C; b.p. 202°C.
Derivation: (a) By dehydrogenation of monocyclic terpenes to para-cymene, and subsequent oxidation. (b) By fractional distillation of crude cresol. (c) By fusing para-toluenesulfonic acid with caustic. (d) From benzene by the cumene process (see phenol).
Method of purification: Crystallization.
Grades: Technical; 98%.
Containers: Drums; tank trucks; tank cars.
Uses: Disinfectant; fumigating compositions; cresotinic acid; dyestuffs; organic intermediate. See cresol.
Danger! Rapidly absorbed through skin; causes severe burns. MCA warning label.
Shipping regulations: None.*

cresolphthalein. An acid-base indicator, changes from colorless to red between pH 8.2 and 9.8.

cresol purple. Meta-cresolsulfonphthalein, an acid-base indicator, showing color change from yellow to purple over the range pH 7.4-9.0.

cresol red. Ortho-cresolsulfonphthalein, an acid-base indicator, changes from orange to amber between pH 2 and 3 and from amber to red between pH 7.2 and 8.8.

cresol, U.S.P. XVI. A mixture of ortho-, meta-, and para-cresols.
Properties: Colorless, yellowish or pinkish liquid, highly refractive, becoming darker with age and exposure to light; phenol-like odor; sp. gr. 1.030-1.038; 90% by volume should distil between 195-205°C. Wt/gal 8.67 lbs. Slightly soluble in water, forming a cloudy solution; saturated solution is neutral or slightly acid to litmus; soluble in alcohol, ether, glycol and dilute alkalies.
Grade: U.S.P. XVI.
Derivation: From coal tar.
Containers: 53-gal drums; tank trucks; tank cars.
Uses: Disinfectant; antiseptic; textile soaps; wetting-out agents.
Danger! Rapidly absorbed through skin. Causes severe burns. MCA warning label.

"Cresophan." [19] Brand name for a proprietary product. Tertiary butyl derivative of cresols. High coefficient germicide and fungicide. Specific against fungoid skin diseases of the ringworm type (athlete's foot).

cresotic acid (cresotinic acid; hydroxytoluic acid) $C_6H_3COOHOHCH_3$. There are ten possible isomers, of which the most common is the 2-hydroxy-3-methylbenzoic acid, also known as ortho-cresotic acid or ortho-homosalicylic acid. The description which follows is of this isomer.
Properties: White crystals or powder; m.p. 166°C; b.p. about 250°C; insoluble in water; soluble in alcohol and ether.
Derivation: Treatment of ortho-cresol with caustic and carbon dioxide under pressure.
Containers: Fiber cans; drums.
Uses: Dye intermediate; research on plant growth inhibition.

cresotinic acid. See cresotic acid.

"Crestalkyd" Resins. [263] Proprietary product. Long, medium and short oil alkyd resins. Several grades: oil length varying from 30% to 80%, A. V. < 15.
Use: High quality paints, lacquers, enamels, and varnishes. Plasticizing resin for nitrocellulose and similar finishes.

"Crestamide" Resins. [263] Proprietary product. Pure and modified sulfonamide resins. Very pale or almost colorless; m.p. 70-90°C; A.V. < 20.
Uses: Cellulose lacquers; enamels; adhesives.

meta-**cresyl acetate** $CH_3C_6H_4OCOCH_3$.
Properties: Colorless oily liquid; odor similar to phenol; b.p. near 112°C; distils with steam; insoluble in water; soluble in common organic solvents.
Use: Medicine.

para-**cresyl acetate.** $CH_3C_6H_4OCOCH_3$.
Properties: Colorless liquid having a floral odor suggestive of narcissus. Soluble in 2.5 vols of 70% alcohol.
Constants: Sp.gr. 1.0532 (15°C); optical rotation (100 mm) 0°; refractive index 1.500 to 1.504; acid value 0.7; ester value 341.6.
Grades: Technical.
Use: Perfumery.

cresyl alcohol. See cresol.

cresyl diglycol carbonate (diethylene glycol bis(cresylcarbonate)) $C_{20}H_{22}O_7$.
Properties: Colorless liquid of low volatility. Sp.gr. 1.19 (20/4°C); boiling point approximately 250°C (2 mm); flash point 475°C; Saybolt viscosity 2170 centipoises; refractive index 1.523 (20°C); evaporation rate 0.006 mg/sq cm/hr(100°C). Insoluble in water (very stable to hydrolysis). Widely soluble in organic solvents. Compatible with many resins and plastics.

cresyl diphenyl phosphate ("Cellulflex 112"; "Santicizer" 140). $(CH_3C_6H_4)(C_6H_5)_2PO_4$.
Properties: Clear, transparent liquid; sp.gr. 1.200 (20/20°C).
Uses: Plasticizer with high degree of flame-resistance; good low temperature qualities; food packaging; gasoline additive.

ortho-**cresyl**-alpha-**glyceryl ether.** See mephenesin.

cresylic acid (See also cresol.) The trade designation for commercial mixtures of phenolic materials boiling above the cresol range. Cresylic acid consists of phenols, cresols, and xylenols and higher phenols in various proportions, according to its source and boiling range.
Derivation: Petroleum; coal tar. Imported cresylic acid (ADF, meaning American Duty-Free), is derived from coal tar.
Containers: 5-, 55-gal drums; tank trucks; tank cars.
Uses (in approximate order of volume):
(a) Petroleum-derived cresylic acid: tricresyl phosphate, disinfectants, metal cleaning compounds, phenolic resins, flotation agents, surfactants, chemical intermediates, oil additives. (b) Coal tar-derived cresylic acid: phenolic resins, wire enamel solvent, tricresyl phosphate, solvent refining of lubricating oils, disinfectants, scouring compounds, detergents; pesticides.

para-**cresyl methyl ether.** See methyl-para-cresol.

cresyl silicate $(CH_3C_6H_4O)_4Si$.
Properties: Colorless liquid; b.p. 450°C.
Derivation: Reaction of cresol and silicon tetrachloride.
Uses: Heat transfer fluid.

cresyl-para-**toluene sulfonate** (tolyl-para-toluene sulfonate) $CH_3C_6H_4SO_2C_6H_4CH_3$.
Properties: Brown, oily liquid; faint odor; sp.gr. 1.207; flash point 184°C; m.p. 69-70°.
Derivation: From reaction of para-toluenesulfonyl chloride with para-cresol.
Use: Has been used as a plasticizer.

"C.R.I." [239] A concentrated rust inhibiting germicide containing 12.5% (by weight) alkyl (mostly C_{12}, C_{14}, C_{16} but ranging from C_8 to C_{18}) dimethyl benzyl ammonium chloride as active ingredient. Inert ingredients are 40% 2-hydroxypropylamine nitrite, water (47.5%), and Fastusol Turquoise Blue.

critical. See chain reaction.

critical constants. A maximum or minimum value for a physical constant which is characteristic of the substance in question; e.g., the critical temperature is the temperature above which a gas cannot be liquefied by an increase in pressure.

critical humidity. The humidity value above which a solid salt will always become damp and below which it will always stay dry.

critical mass. See chain reaction.

critical size. See chain reaction.

critical solution temperature. The temperature above or below which two liquids are miscible in all proportions.

critical temperature. The temperature above which a gas cannot be liquefied by an increase in pressure.

croceic acid. See crocein acid.

crocein acid (croceic acid; Bayer's acid; 2-naphthol-8-sulfonic acid) $C_{10}H_6OHSO_3H$.
Derivation: Sulfonation of beta-naphthol at a low temperature and recrystallization from a salt.
Grade: Sold as the sodium salt in 100-lb fiber drums.
Use: Dye intermediate.

crocidolite (blue asbestos) $Na_2Fe_3^{+2}Fe_2^{+3}(Si_4O_{11})_2(OH)_2$. A natural sodium-iron silicate. One of the amphibole group of minerals. Lavender-blue or leek-green color; streak, same. One of the commercial forms of asbestos. Its chief use is in filters.

crocus (saffron; Spanish saffron; French saffron). Stigmas of Crocus sativus.
Habitat: Western Asia, Asia Minor, Egypt, France and Spain.
Containers: Boxes; tins.
Uses: Coloring; flavoring.
Shipping regulations: None.*

crocus martis. A name used for impure red ferric oxide pigments and polishing powders, usually produced by heating iron sulfate containing calcium sulfate, lime, or other inert filler. Also sometimes applied more generally to other impure oxides of red or yellow color.

crocus martis adstringens. Name given to red varieties of ferric oxide. See iron oxide reds.

crocus, polishing. A name given to red varieties of ferric oxide used for polishing or as pigments.

crocus, red. A red pigment based on ferric oxide but containing a large proportion of calcium sulfate or similar material. Similar to Venetian red.

"Cromophtals." [443] Trade name for organic pigments used for organic coatings, plastics and printing inks.

"Cronar." [28] Trademark for polyester photographic film base, a polyethylene terephthalate condensation polymer.

"Cronox." [89] Trademark for bactericides and corrosion inhibitors for metal surfaces in oil production and refining.

crossed discharge. Combined high and low frequency electric discharge stated to give high conversion of air to nitric oxide.

cross section (microscopic cross section; nuclear cross section). A term used particularly in nuclear physics and in nuclear reactor technology in regard to the reaction between nuclear particles. It is the effective target area that a particular nucleus exhibits in a nuclear process or reaction of a given kind. For example, the U-235 nucleus may react with neutrons to scatter them, capture them, or be fissioned by them, and is said to have a different cross section for each process. Cross sections are usually quoted in barns (q.v.). The rate of a nuclear reaction is the product of the flux of incident particles times the number of target particles times the cross section of the target particle. See also macroscopic cross section.

crotamiton (N-ethyl-ortho-crotonotoluide) $CH_3C_6H_4N(C_2H_5)COCH:CHCH_3$.
Properties: Light yellow, oily liquid with a faint fish odor. Very soluble in alcohol; practically insoluble in water; stable to light and air.
Grade: N. N. D.
Use: Medicine.

croton. See tiglium.

crotonaldehyde (2-butenal; crotonic aldehyde; beta-methyl acrolein; propylene aldehyde) $CH_3CH:CHCHO$. Commercial crotonaldehyde is the trans isomer.
Properties: Water-white mobile, flammable liquid; pungent, suffocating odor; turns to a pale yellow color in contact with light and air. An effective lachrymator. Slightly soluble in water; miscible in all proportions with alcohol, ether, benzene, toluene, kerosene, gasoline, solvent naphtha.
Constants: Sp. gr. 0.858 (15.6/4°C); b.p. 104-105°C; flash point 55°F; f.p. −74°C; vapor pressure 19.0 mm (20°C); specific heat 0.705; refractive index 1.4373 (20°C); wt/gal 7.11 lbs (20°C).
Derivation: Condensation of two molecules of acetaldehyde.
Grades: Technical; 87% water-wet form.
Containers: 5-, 10-, 55-, 110-gal drums.
Uses: Intermediate for n-butyl alcohol; solvent; preparation of rubber accelerators; purification of lubricating oils; insecticides; chemical warfare; fuel-gas warning agent; organic synthesis; leather tanning; alcohol denaturant.
Fire hazard: Dangerous.
Shipping regulations: Flammable liquid. Red label.*

crotonchloral. See butyl chloral.

crotonic acid (butenoic acid; beta-methacrylic acid) $CH_3CH:CHCOOH$. Exists in cis and trans isomeric forms, the latter being the stable isomer used commercially. The cis form melts at 15°C and is sometimes called isocrotonic acid.
Properties: White crystalline solid; sp. gr. 0.9730; m.p. 72°C; b.p. 185°C; soluble in water, ligroin, ethanol, toluene, acetone.
Derivation: Oxidation of crotonaldehyde.
Grade: 97%.
Containers: Glass bottles; fiber drums.
Uses: Synthesis of resins, polymers, plasticizers, drugs.

crotonic aldehyde. See crotonaldehyde.

croton oil (tiglium oil).
Properties: Brownish-yellow liquid; poisonous! Sp.gr. 0.935-0.950 (25°C); refractive index (n 40/D) 1.470-1.473. Soluble in ether, chloroform and fixed or volatile oils; slightly soluble in alcohol.
Chief constituents: Glycerides of stearic, palmitic, myristic, lauric and oleic acids and croton resin, a vesicant.
Derivation: By expression from the seeds of Croton tiglium.
Method or purification: Rectification.
Grades: Technical.
Containers: 25-lb tins; 1-, 5-, 10-lb glass bottles.
Use: Medicine.
Caution: Causes eruptions when applied to skin.
Shipping regulations: None.*

crotonolic acid. See tiglic acid.

crown bark. See cinchona bark, Loxa.

crown filler. A mineral filler, usually calcium sulfate or carbonate or a mixture thereof used in paper manufacture.

crown glass. A carefully made lime-soda-alumina glass with high refraction and low dispersion, used for optical glass.

crucible clays. Ball clays that are relatively refractory, for use in producing crucibles that will withstand high temperatures.

crucible steel. Steel which has been manufactured by the "crucible process" in which the steel is smelted in crucibles placed in a coke-fired furnace of special design. Very high grade steels are produced but are being largely replaced by electric furnace steels which permit more precision in forming alloys.

crude oil. See petroleum.

"Cryogel." [329] Trademark for an aluminum soap used as a low-temperature gelling agent for hydrocarbons.

cryogenics. The field of science dealing with the behavior of matter at very low temperatures and with low temperature techniques. The use of the liquefied gases oxygen, nitrogen and hydrogen, at temperatures to about −260°C, is now common industrial practice. Some useful electronic devices and specialized instruments, such as the cryogenic gyro, operate at liquid helium temperatures (about 4°K). Most cryogenic research centers around the anomalous behavior of substances at the liquid helium temperature or below.
See absolute temperature.

cryolite, natural (Greenland spar, icestone) Na_3AlF_6. A natural fluoride of sodium and aluminum. See also cryolite, synthetic.
Properties: Colorless to white; sometimes red, brown or black; luster vitreous to greasy; hardness 2.5; sp.gr. 2.95-3.0. A typical composition (exclusive of fluoride) is Al 13.23%, Na 32.71%, a little Mn_2O_3, MgO, and vanadic and phosphoric acids.
Occurrence: Greenland (only commercial source); Colorado; U.S.S.R.
Containers: 50-, 100-lb bags; 400-lb barrels.
Uses: As an electrolyte in the production and refining of aluminum from its oxide; in ceramics; insecticide; as a binder for abrasives; in electrical insulation; flux in open hearth process; explosives; polishes.

cryolite, synthetic (sodium fluoaluminate; sodium aluminum fluoride) Na_3AlF_6 or actually, perhaps, $2AlF_3 \cdot 6NaF \cdot 3CaF_2$. See also cryolite, natural.
Properties: A white, crystalline powder; m.p. 1000°C; refractive index 1.338. Insoluble in water; soluble in solutions of aluminum and ferric salts. Not compatible with alkaline materials.
Derivation: From fluorspar, sulfuric acid, hydrated alumina, and sodium carbonate.
Grades: Technical (low sulfate); insecticide; fluxing.
Containers and Uses: Same as cryolite, natural.

cryptohalite. See ammonium fluosilicate.

cryptopine $C_{21}H_{23}NO_5$.
Properties: White crystalline alkaloid; poisonous! Soluble in chloroform and boiling alcohol; insoluble in water and ether. M.p. 217-221°C.
Derivation: From opium, by extraction and crystallization.
Method of purification: Recrystallization.

cryptoxanthin (provitamin A; hydroxy-beta-carotene) $C_{40}H_{56}O$. A carotenoid pigment with vitamin A activity.
Properties: Garnet-red prisms with metallic luster from benzene and methanol; soluble in chloroform, benzene, and pyridine; slightly soluble in ligroin, petroleum ether, alcohol and methanol.
Occurrence: In many plants, egg yolk, butter, blood serum.
Uses: Nutrition; medicine.

crystal. A solid with characteristic shape and cleavage caused by arrangement of its atoms, ions, or molecules into a definite pattern or crystal lattice. Crystals have flat surfaces, sharp edges, and a definite angle between a given pair of surfaces. Real crystals of any size are never perfect in the sense that all their atoms or molecules are perfectly arranged on one perfect crystal lattice system. Any actual crystal is composed of a considerable number of small crystallites that do have perfect or nearly perfect lattice structure. At the boundaries of the crystallites there are irregularities or shifts in the lattice structure. The study of these crystal imperfections and of similar imperfections caused by atoms or molecules of foreign substances in the latter, is known as solid state physics. Most important properties and uses of crystals depend upon the crystal imperfections. Relatively large single crystals of a high degree of purity and perfection are used for optical and electrical purposes and for measurement of radioactivity.

crystal ammonia. See ammonium carbonate.

crystal carbonate. See sodium carbonate, monohydrate.

crystallite. A perfect portion of an ordinary crystal; i.e., a part of the crystal that has its atoms and molecules arranged in a perfect crystal lattice. Crystallites are usually quite small. An ordinary crystal is composed of an assemblage of a large number of such crystallites, some of which are almost perfectly aligned with one another while others are out of alignment to various extents.

crystallization. The process of forming crystals by cooling a molten material or a solution, or by evaporating a solution, or by condensing a vapor under suitable conditions. The crystals usually contain fewer impurities than the liquid (or vapor) from which they form, so that partial or fractional crystallization is much used for purification of materials. Large crystals usually are of higher purity than small ones, although not always. Formation of large crystals usually requires that the process proceed slowly. Large natural crystals or carefully grown synthetic crystals have important electrical uses. See crystal; crystallite.

crystallized verdigris. See copper acetate.

"Crystal O." [202] Trademark for an odorless and tasteless U.S.P. quality castor oil for cosmetic and pharmaceutical usage and for coating applications.

crystals of Venus. See copper acetate.

"Crystalustre." [333] Lacquers and varnishes having a clear cellulose base. Used in coating paper, cardboard, etc.

crystal violet. See methyl violet.

"Crystamet." [428] Trademark for sodium metasilicate pentahydrate.

"Crystarose" [188] $C_6H_5CH(CCl_3)OOCCH_3$. Brand name for a highly purified grade of trichloromethylphenylcarbinyl acetate. Relatively free of odor; enhances rose perfumes.

"Crystex." [1] Rubber-insoluble sulfur used as vulcanizing agent in natural and synthetic rubbers; 85% of the sulfur is insoluble at the usual milling temperatures. This metastable form is converted to the stable soluble sulfur at usual vulcanizing temperatures.
Containers: Multiwall paper bags, 50 lb net.
Use: In all rubber stocks which suffer from "bloom" in the uncured state, and in some latex compounds and naphtha cements.

"Crystic" Resins. [263] Proprietary products. Unsaturated polyester resins. Several grades. Maximum viscosity 40 poises (at 25°C).
Uses: Low pressure molding and laminating for glassfiber reinforced plastics; potting, casting, and embedding resins.

"Crystodigin." [100] Trademark for digitoxin, U.S.P.

"Crystolon." [249] Trademark for silicon carbide (q.v.).

Cs. Symbol for cesium.

C_2S. Abbreviation for dicalcium silicate, as used in cement.
See cement, Portland.

C_3S. Abbreviation for tricalcium silicate, as used in cement.
See cement, Portland.

"CS-137." [304] Trade name for a barium-sodium organic-complex vinyl stabilizer.
Properties: Creamy-white paste, sp.gr. 1.54, refractive index 1.48.
Containers: Supplied as a 70% solids paste in DOP. Metal drums containing 60 lbs net.
Uses: Used as a stabilizer for transparent organosols and solution coatings to impart superior light and weathering resistance.

CS Corrosion Inhibitor. [108] Granular, pale pink compound.
Containers: 20- and 100-lb drums.
Uses: Corrosion inhibitor for closed recirculating heating and cooling systems in air conditioning, low-pressure hot water heating, diesel engines and gas compressors.

C-stage resin (resite). The fully cured stage of phenol-formaldehyde resins (q.v.). Infusible and insoluble in all solvents. See A-stage resin.

CTC. Abbreviation for chlortetracycline.

CTP. Abbreviation for cytidine triphosphate. See cytidine phosphates.

Cu. Symbol for copper.

cubeb. See cubeba.

cubeba (cubeb; tailed pepper; Java pepper).
Derivation: Dried, unripe, but fully grown fruit of Piper cubeb.
Habitat: Southern Asia (Java, Borneo and Sumatra); cultivated in Ceylon and West Indies.
Grade: Technical.
Containers: Berries, in bags; powdered, in cases.
Use: Medicine.
Shipping regulations: None.*

cubebin $C_{20}H_{20}O_6$.
Properties: Inodorous white crystals; m.p. 131-132°C; soluble in chloroform and hot alcohol.
Derivation: A constituent of cubeba.

cubeb oil.
Properties: Colorless, pale greenish or yellowish liquid; characteristic odor of cubebs; warm camphoraceous taste. Soluble in alcohol, ether and chloroform.
Chief known constituents: Sesquiterpenes; cadinene; dipentene.
Constants: Sp.gr. 0.905-0.925; b.p. 175-280°C; refractive index 1.49-1.496; optical rotation -25 to -40°.

*See "I.C.C. Shipping Regulations," page xiii.
Reference numbers refer to name of manufacturer. See "List of Manufacturers," page v.

Derivation: Distilled from the unripe fruit of Piper cubeba.
Method of purification: Rectification.
Grades: Technical.
Containers: 10-lb tins; 1-, 5-lb glass bottles.
Use: Medicine.
Shipping regulations: None.*

cube root (barbasco root; timbo root).
Derivation: The dried roots of various species of Lonchocarpus.
Habitat: South and Central America.
Contains varying amounts of rotenone, depending on the grades, as, 5% rotenone.
Containers: Bags.
Use: Manufacture of insecticides and as a fish poison.

cubic centimeter (cc). The volume of a cube with sides one centimeter long. See also milliliter.

"Cubidow." [233] Trademark for compacted salt comprising either or both calcium and sodium chloride.

"Cubond." [294] Trademark for a copper brazing paste. Consists of metallic copper powder on cuprous oxide pigments of high purity in organic or petroleum vehicles which impart satisfactory suspension properties. Available in various grades, including additions of brazing fluxes, depending on specific end use.
Containers: Pints, gallons, and 5-gal pails.
Uses: Principally for joining steel parts together. In some instances has been used as a pyrometallurgical method for surface plating with copper.

"Cubor Dusts." [174] Trademark for insecticides containing 0.75 and 1.0% pure rotenone.
Containers: 4-, 50-lb bags.

cuca. See coca.

cucumber oil (gourd oil).
Properties: Greenish-yellow, fixed, drying oil; faint red fluorescence. Soluble in alcohol, ether and benzene; insoluble in water. Sp. gr. 0.923; m.p. −16°C; saponification value 188.7; iodine number 121.
Derivation: From the seeds of the cucumber, pumpkin, etc., by pressing.

cucurbita. See pepo.

cudbear (persio). A purplish-red powder prepared from certain lichens (Rocella de candolle, Locanora acharius).
Habitat: Norway, Sweden, European mountains, Mediterranean coast, Madeira, Mozambique, Madagascar, California.
Derivation: By macerating lichens with dilute ammonia and caustic soda, with fermentation.
Uses: For its tinctorial powers.
See also litmus and orchil.

cuenca bark. See cinchona bark, Loxa.

cuite. A term applied to degummed natural silk.

cullet. Broken glass added to the batch in glass manufacture.

"Culofix." [300] Trademark for dye fixatives for application to dyed textiles. Eliminates or minimizes color bleeding in water and/or in laundering.
"Culofix" L: Fatty cationic type to eliminate bleeding of direct dyes in water.
"Culofix" WFD: Cationic resin type which prevents color bleeding in laundering.

Culver's root. See leptandra.

cu. m. Abbreviation for cubic meter.

cumaldehyde. See cuminic aldehyde.

cumarin. See coumarin.

cumarone. See coumarone.

"Cumar" Resins. [175] Trademark for a series of neutral, stable, synthetic resins of the para-coumarone-indene type, manufactured from selected distillates of tar.
Properties: Wide compatibility permits use with a great variety of solvents, oils, resins, and waxes; mills readily into natural or synthetic rubber, acting as a softener and filler-wetting agent; imparts easy-flowing qualities to extruded stocks and gloss to molded goods; improves tensile elongation and tear resistance of GR-S compounds; available grades vary from semi-liquid to hard forms. Sp. gr. 1.08-1.14 (15.5/15.5°C); color, pale yellow to brown; softening range, 7-136°C; flash point, 160-278°C; iodine no., 41-44.
Grades: W (High Melting); V (Varnish); T3, T15 (Tile); MH, RH (Rubber); P10, P25 (Plastic); EX Dark (X Grades).
Containers: Flaked grades are sold in fiber drums; viscous grades, in steel barrels; solid grades, in metal cans.
Uses: In the manufacture of varnishes, floor tile, natural and synthetic rubber products, printing ink, adhesives, and waterproofing materials; for leather, electrical, radio, paper, and other industries.

"Cumate." [69] Trademark for proprietary preparation of copper dimethyldithiocarbamate $[(CH_3)_2NC(S)S]_2Cu$.
Properties: Dark brown powder, sp. gr. 175 ± .03; melts above 325°C; moderately soluble in acetone, benzene, chloroform; insoluble in water, alcohol, gasoline.
Uses: In SBR, primary accelerator; secondary accelerator with thiazoles. In butyl rubber, primary accelerator. For molded and extruded goods.
Also supplied in "rodform."

cumene (isopropylbenzene; isopropylbenzol; cumol) $C_6H_5CH(CH_3)_2$.
Properties: Colorless liquid. Soluble in alcohol, carbon tetrachloride, ether and benzene; insoluble in water.
Constants: Sp. gr. 0.8620; b.p. 152.7°C; wt/gal 7.19 lbs (25°C); freezing point -96°C; refractive index 1.489 (25°C); flash point 97°F; fire point 38°C; heat of vaporization 76.5 cal/g (b.p.); specific

*See "I.C.C. Shipping Regulations," page xiii.
Reference numbers refer to name of manufacturer. See "List of Manufacturers," page v.

heat 0.43 cal/g/°C; dielectric constant 2.3 (1,000 cycle); specific resistivity 8.6 x 10^{12} ohms/cm; viscosity 0.73 centipoises (25°C). Typical specifications: Boiling range < 3.0°C.

Derivation: (a) Distilled from coal tar naphtha fractions and from petroleum.

Method of purification: Rectification.

Grades: Technical; research; pure.

Containers: Drums; tank cars; tank trucks.

Uses: Additive to aviation gasoline; production of phenol, acetone, and alpha-methylstyrene; solvent.

Shipping regulations: None.*

cumene hydroperoxide (alpha-, alpha-dimethylbenzyl hydroperoxide) $C_6H_5C(CH_3)_2OOH$.

Properties: Colorless to pale yellow liquid, slightly soluble in water; readily soluble in alcohol, acetone, esters, hydrocarbons, chlorinated hydrocarbons.

Derivation: A solution or emulsion of cumene is oxidized with air at about 130°C.

Uses: Production of acetone and phenol; polymerization catalyst, particularly in redox systems, used for rapid polymerization.

Shipping regulations: Oxidizing material. Yellow label.*

cumenylamine. See cumidine.

cumic aldehyde. See cuminic aldehyde.

cumidine (para-isopropylaniline; cumenylamine) $(CH_3)_2CHC_6H_4NH_2$.

Properties: Colorless liquid; sp.gr. 0.957 (20/4°C); m.p. −63°C; b.p. 225°C. Insoluble in water; soluble in alcohol and ether.

See also pseudocumidine.

cumin. See cumin seed.

cuminic aldehyde (cumic aldehyde; cumaldehyde; para-isopropylbenzaldehyde) $(CH_3)_2CHC_6H_4CHO$.

Properties: Colorless to yellow liquid with a cumin odor; sp.gr. 0.976 (22°C); b.p. 235°C. Insoluble in water; soluble in alcohol and ether.

Use: Perfumery.

cumin oil.

Properties: Colorless or yellowish, limpid liquid; characteristic odor of cumin; sharp, spicy taste. Soluble in alcohol, ether and chloroform.

Chief known constituents: Cumene; cuminic aldehyde.

Constants: Sp.gr. 0.900-0.930; optical rotation +4° to +8°; refractive index 1.497-1.509 (20°C).

Derivation: Distilled from the fruit of Cuminum cyminum.

Method of purification: Rectification.

Grades: Technical.

Containers: 25-lb cans; 1-, 5-lb glass bottles.

Uses: Medicine; flavoring; perfumery.

Shipping regulations: None.*

cumin seed (cumin).

Derivation: Fruit of Cuminum cyminum.

Habitat: Mediterranean region and northern Africa.

Grades: Iranian; Moroccan; Turkish.

Containers: Bags.

Uses: Medicine; flavoring; curry powders. Source of an essential oil used in perfumery and from which certain chemicals can be made.

Shipping regulations: None.*

cumol. See cumene.

cumyl phenol.

Properties: White to tan crystals with characteristic phenol odor. Solidifying point 72.0°C; density 1.115 g/ml (25°C); distillation range 188.9-190.9°C (10 mm); flash point 180°C (COC); fire point 204°C (COC).

Containers: 200-lb fiber drums.

Uses: Intermediate for resins, insecticides, lubricants.

cundurango. See condurango.

"Cunife." [166] A ductile permanent magnet material composed of 60% copper, 20% nickel and 20% iron. Forms available: wire, strip, finished magnets.

"Cunilate." [8] Proprietary name for "solubilized copper-8-quinolinolate." There are many different types for various conditions of use. Properties vary chiefly with the type of agent used to promote the proper dispersion for each type. A typical "Cunilate" (#2174) contains 10% copper-8-quinolinolate with 2-ethylhexoic acid and an aromatic solvent as carrier, with 40% total solids.

Properties: (#2174) Sp.gr. 0.9542-0.9545 at 77°F; flash point 110°F; color, greenish yellow; pH 5.5-6.0.

Containers: Steel drums.

Uses: (#2174) For fungicidal treatment of fabrics, thread, cotton rope and other materials as well as in combination with textile treatments.

"Cunimene" 2243. [8] Trade name for cuprimagnesium dehydroabietyl amine 8-hydroxyquinolinium 2-ethylhexoate, a fungicidal product. A typical commercial product is light green in color, contains 94% solids, is dispersed in an aromatic solvent; has sp.gr. 1.09, and flash point 110°F. An emulsifiable variety is available.

"Cunimene" 2246. [8] Trade name for zinc-magnesium dehydroabietyl amine 8-hydroxy-quinolinium 2-ethylhexoate. A fungicide similar to "Cunimene" 2243.

"Cunisil-837." [324] Trade name for an alloy containing 97.50% copper, 1.90% nickel, and 0.60% silicon. This high-strength, corrosion-resistant alloy is available in round rod, with or without the final precipitation-hardening heat treatment. Used in electrical equipment.

cupellation process. A process for freeing silver, gold or other non-oxidizing metals from base metals which can be oxidized. The metallic mixture is placed in a cupel, which is a shallow, porous cup, and roasted in a blast of air. The base metal oxides are absorbed in the cupel, leaving the pure metal to be decanted.

cupferron (ammonium nitroso-beta-phenylhydroxylamine) $C_6H_5N(NO)ONH_4$.
Properties: Creamy-white crystals; m.p. 163-164°C. Soluble in water, alcohol and ether.
Derivation: By treating an ethereal solution of beta-phenylhydroxylamine with dry ammonia gas and amyl nitrite.
Method of purification: Recrystallization.
Grades: Pure; reagent.
Containers: Glass bottles.
Use: Reagent in analytical chemistry.
Shipping regulations: None.*

cuprammonium process. A process for making rayon by dissolving cellulose in an ammoniacal copper solution and spinning the solution into fibers, which are reconverted to cellulose by treatment with acid. Relatively pure alpha-cellulose (from cotton linters or treated wood pulp) is dissolved in a solution of ammonia and copper hydroxide or basic copper sulfate until the cellulose content reaches 7-10%. During an aging period, absorption of atmospheric oxygen causes degradation of the cellulose polymers and reduces the viscosity of the solution. When the proper viscosity has been attained, the solution is filtered, deaerated, and forced through minute openings of the spinneret into a bath of dilute sulfuric acid, regenerating the cellulose and stretching and thinning the filaments.

cuprammonium rayon (copper-ammonium rayon). Regenerated cellulose fibers made by the cuprammonium process. Available in continuous filament form.
Properties: Tensile strength (psi) 33,000-44,000; elongation 10-17%; sp.gr. 1.52-1.54; moisture regain 11-12.5% (70°F, 65% relative humidity); decomposes about 300°F; burns readily. Similar to cotton in chemical resistance, dyeing, and resistance to insects and mildew.
Uses: Knitted and woven material for wearing apparel such as dress fabrics.

cupreine (hydroxycinchonine; ultraquinine) $C_{19}H_{22}O_2N_2 \cdot 2H_2O$. One of the cinchona alkaloids.
Properties: Colorless crystals; m.p. (anhydrous) 198°C. Slightly soluble in water and alcohol; soluble in chloroform and ether.
Derivation: From cuprea bark Remijia pedunculata.

"Cuprex." [123] Trademark for a preparation for extermination of ectoparasites on men, animals, and plants.

cupric. Form of the word copper used in naming copper compounds in which the copper has a valence of two. See the corresponding compound under copper.

cupric chromate, acid. See "Celcure."

cupric chromate, basic. See copper chromate.

cupriethylene diamine.
Shipping regulations: Solution: Corrosive liquid. White label.*

"Cuprisote." [171] Trade name for a formula known as acid cupric chromate. When impregnated into the wood fiber by vacuum pressure process, it offers ultimate protection against decay and insect attack. Clean, non-toxic to humans, paintable after processing, non-leachable.

cuprite (ruby copper ore; red oxide of copper) Cu_2O. Crimson, scarlet, vermilion, deep or brownish-red, secondary mineral; adamantine or dull luster; brownish-red streak. Superior in hardness to cinnabar and proustite and differs from them in color or streak. Inferior in hardness to hematite. Contains 88.8% cuprous oxide, 11.2% oxygen. Soluble in nitric and concentrated hydrochloric acids.
Constants: Sp.gr. 5.85-6.15; hardness 3.5-4.
Occurrence: United States; England; Germany; France; Siberia; Australia; China; Peru; Bolivia.
Use: Source of copper.

cupro-magnesium. An alloy of copper and magnesium.

"Cupron." [155] Trademark for an alloy of 55% copper and 45% nickel (constantan).
Properties: Resistivity, 294 ohms per circular mil foot at 20°C; temp. coefficient of resistance ±0.00004/°C between 0-100°C; heat resistant below 1000°F; high thermal emf vs copper. Caution! Both terminals must be at same temperature.
Forms: Wire; ribbon; strip.
Uses: Rheostats and controls; resistors for electrical instruments.

cupro-tungsten. An alloy of copper and tungsten.

cuprous. Form of the word copper used in naming copper compounds in which the copper has a valence of one.

cuprous chloride $CuCl$ or Cu_2Cl_2. (Nantokite is the naturally occurring mineral.)
Properties: White cubical crystals; sp.gr. 3.53; m.p. 422°C; b.p. 1366°C; becomes greenish on exposure to air, and brown on exposure to light. Insoluble in water; soluble in acids or ammonia.
Derivation: Copper and cupric chloride solution or copper and hydrochloric acid in air.
Grades: Technical; C.P.; reagent.
Containers: Multiwall paper sacks; glass bottles; 100-lb drums.
Uses: Catalyst; preservative and fungicide; metallurgy; absorbent for carbon monoxide.
Caution! Harmful if swallowed. MCA warning label.

cuprous cyanide $Cu_2(CN)_2$ or $CuCN$. Copper content min. 70%; cyanide content min. 28.5%.
Properties: A creamy white powder; insoluble in water; soluble in sodium cyanide and

potassium cyanide; sp. gr. 2.9; m.p. 475°C. Poisonous!
Containers: 100-lb drums.
Use: In copper and brass cyanide electroplating solutions.
Danger! Contact with acid liberates poisonous gas. MCA warning label.
Shipping regulations: None.*

cuprous iodide CuI.
Properties: White to brownish-yellow powder. Soluble in ammonia and potassium iodide solutions; insoluble in water. Sp. gr. 5.653 at 15°C; m.p. 606°C.
Derivation: Interaction of solutions of potassium iodide and copper and iron sulfates.
Use: Medicine; feed additive.

cuprous oxide. See copper oxide, red.

cuprous sulfide. Cu_2S.
Properties: Black powder or lumps; soluble in nitric acid; insoluble in water. Occurs as the mineral chalcocite. Sp. gr. 5.52-5.82; m.p. about 1100°C.
Derivation: By heating cupric sulfide in a stream of hydrogen.
Grades: Technical.
Containers: Wooden barrels.
Uses: Copper metal; protective paint for vessels.
Shipping regulations: None.*

cuprous sulfite $Cu_2SO_3 \cdot H_2O$.
Properties: White crystalline powder; soluble in ammonium hydroxide, hydrochloric acid (with decomposition); insoluble in water. Sp. gr. 3.83.

cuprous thiocyanate (copper sulfocyanide) CuSCN.
Properties: Yellow-white amorphous powder. Insoluble in water; soluble in ammonia.
Uses: Manufacture of organic chemicals, antifouling paints; printing textiles.

Curacao aloe. A variety of aloe (q. v.).

"Curafos." [108] Trademark for food grade sodium hexametaphosphate and sodium tripolyphosphate used for curing, preventing undesirable color change, and loss of moisture in meat.
Containers: 100-lb bags and drums.

curare. The arrow poison of South American Indians. Intensely poisonous! Designated as tube, calabash, or pot type according to container in which it is packed and transported from point of origin.
Properties: Dark brown resinous extract, thick syrup or brittle amorphous solid; soluble in water and in aqueous alcohol.
Constituents: At least 40 alkaloids have been isolated, but of these d-tubocurarine (q. v.) is most useful in medicine and least toxic.
Source: From Chondodendron tomentosum and various species of Strychnos, in South America.
Use: In medicine, as a standardized extract.

curarine $C_{19}H_{26}N_2O$. An alkaloid obtained from curare. See also tubocurarine chloride.

"Curavis." [108] Trademark for a dry, pulverized, food grade, polymeric phosphate composition exhibiting high viscosity in water solution. For use in the meat packing industry.
Containers: 100-lb drums; 12-lb cartons of six 2-lb polyethylene bags.

curcuma (turmeric; curry; Indian saffron). Rhizome of Curcuma longa.
Habitat: China, East Indies and many tropical countries.
Grades: Technical.
Containers: Burlap bags.
Uses: Coloring foods; condiment (curry powder); textile dyeing; indicator in analytical chemistry.
Shipping regulations: None.*

curcuma oil (turmeric oil). An orange-yellow, slightly fluorescent liquid, which has slight odor of curcuma, produced by steam distillation of root of Curcuma longa. Sp. gr. about 0.94; b.p. 220-250°C (decomposed above 250°C). Soluble in one half volume of 90% alcohol. Consists of turmerol, valeric and caproic acids, and phellandrene.

curcumin (turmeric yellow) $(2\text{-}CH_3OC_6H_3\text{-}1\text{-}OH\text{-}4\text{-}CH{:}CHCO)_2CH_2$.
Properties: Orange-yellow needles, m.p. 177-179°C. Insoluble in water and ether; soluble in alcohol.
Derivation: The coloring principle from curcuma.
Uses: Dye; analytical reagent; food dye. As an acid-base indicator it is brownish-red with alkalies, yellow with acids; an indicator for boron.

"Cure-Set." [65] Trademark for a two-component adhesive (special latex and a catalyst vulcanizer) for bonding rubber and vinyl tile in on-grade installation where moisture may be a problem.

curie. The official unit of radioactivity, defined as exactly 3.70×10^{10} disintegrations per second. This decay rate is nearly equivalent to that exhibited by one gram of radium in equilibrium with its disintegration products. A millicurie (mc) is one thousandth of a curie. A microcurie (μc) is one millionth of a curie.

curine. See bebeerine.

curium Cm. A synthetic element with atomic number 96, first formed as the 242 isotope when plutonium was bombarded with helium ions in the cyclotron. See actinide elements. Curium is a silvery metal similar in properties to the rare earth, gadolinium. It is obtained by first preparing curium trifluoride, which is reduced with barium vapor at 1275°C. Curium forms compounds of the type CmF_3, Cm_2O_3, $Cm_2(C_2O_4)_3$, etc.

curled dock. See rumex.

"Curona." [173] Trademark for sodium isoascorbate as a curing aid in cured and comminuted meat products. It is used as an anti-oxidant in the meat packing industry for preserving the natural color of meat products, to shorten processing time and reduce shrinkage.

curry. See curcuma and also cumin seed.

cuscus oil. See vetiver oil.

cusparia bark. See angostura bark.

cutch extract A water-soluble tanning substance obtained by extracting the bark of several varieties of mangrove trees.

cut glass. Lead glass that has been carefully cut and polished.

cutting fluid. A liquid applied to a cutting tool to assist in the machining operation by washing away the chips or serving as a lubricant or coolant. Commonly used cutting fluids are: water; water solutions or emulsions of detergents and oils; mineral oils; fatty oils; chlorinated mineral oils; sulfurized mineral oils; mixtures of the foregoing oils.

cutting-tool lubricant. See cutting fluid.

cuttle fish bone. See sepia.

"Cyamite." [57] Trademark for an ammonia nitrate blasting agent which is not sensitive to the shock of an electric blasting cap, rifle slug or primacord. Bore hole must be primed with regular dynamite to shoot.

"Cyamon." [57] Trademark for an ammonium nitrate blasting agent which is designed for safe handling in the field. It is not sensitive to a blasting cap, rifle slug, primacord, flame or impact of heavy steel weights.

"Cyamon Primers" for Explosives. [57] Trademark for special primers sensitive to an electric blasting cap and primacord, used to detonate "Cyamon" blasting agents.

"Cyan." [57] Trade designation for a line of both blue and green phthalocyanine pigments.

"Cyana." [57] Trademark used in connection with the textile finishes obtained by applying "Aerotex" Resins and similar products.

"Cyanadur." [57] Trademark for a line of pigments.

"Cyanalube." [57] Trademark for a softener for use as a lubricant in sewing cloth.

"Cyanamer." [57] Trademark for an acrylic polymer.

cyanamide.

1. (cyanogenamide; carbodiimide) HN:C:NH or N⋮CNH$_2$.

Properties: Crystals; m.p. 42°C; sp. gr. 1.08. Very soluble in water, alcohol, ether.

Derivation: From calcium cyanamide and sulfuric acid.

2. Familiar term for calcium cyanamide (q.v.).

"Cyanatex." [57] Trademark for textile softener or finishing agent.

Cyanide-Chloride Mixtures. [28]
White, fused slabs, 3" x 1 3/4" x 1" (approx. 3 oz.) or white, crystalline powder.

30% Case Hardener. Sodium cyanide content min. 30%; sodium chloride min. 25%, max. 35%; f.p. approx. 1090°F (587°C); sp. gr. 2.12 at 77°F (25°C), 1.52 at 1562°F (850°C).

45% Cyanide-Chloride Mixture. Sodium cyanide content min. 45%, sodium chloride min. 45%, sodium carbonate max. 10%; f.p. approx. 1247°F (675°C); sp. gr. 1.91 at 77°F (25°C), 1.37 at 1580°F (860°C).

75% Cyanide-Chloride Mixture. Sodium cyanide content min. 75%, sodium chloride min. 12%, sodium carbonate max. 13%; f.p. approx. 1094°F (590°C); sp. gr. 1.74 at 77°F (25°C), 1.26 at 1580°F (860°C).

Containers: 100-lb cases; 100- and 200-lb drums.

Use: Case hardening of steel.

cyanide pulp. The mixture obtained by grinding crude gold and silver ore and dissolving the precious metal content in sodium cyanide solution.

"Cyanine Fast." [232] Brand name for a series of level-dyeing acid dyestuffs of good fastness properties.

cyanite (kyanite; disthene) Al_2OSiO_4. A natural silicate of aluminum, found in metamorphic rocks.

Properties: Color blue, green, white, gray; luster vitreous; sp. gr. 3.56-3.66; hardness 5 along length of crystal, 7 at right angles to this direction.

Occurrence: South Carolina, Virginia; India; Africa.

Use: Refractories.

See also andalusite, sillimanite.

cyanoacetamide (malonamide nitrile; propionamide nitrile) $CNCH_2CONH_2$.

Properties: White crystals; b.p., decomposes; m.p. 119°C; soluble in water and alcohol.

Derivation: Ammonolysis of cyanoacetic ester or by the dehydration of ammonium cyanoacetate.

Method of purification: Crystallization.

Containers: Fiber drums.

Uses: Organic pharmaceutical synthesis; plastic materials.

cyanoacetic acid (malonic nitrile) $CNCH_2COOH$.

Properties: White crystals; hygroscopic. Soluble in water, alcohol and ether. M.p. 66.1-66.4°C; decomposed at 160°C.

Derivation: By the interaction of sodium chloroacetate and potassium cyanide solution.

Method of purification: Crystallization.

Grades: Technical.

Containers: Tins.

Use: Organic synthesis.

Shipping regulations: None.*

"Cyanobrik." [28] Trademark for 97% sodium cyanide in 1-oz., pillow-shaped, briquette form.

cyanocarbons. A new class of compounds in which the cyanide radical (–CN) replaces hydrogen in organic compounds, as in tetracyanoethylene, $(CN)_2C:C(CN)_2$, which see. The compounds are quite reactive and form colored complexes with aromatic hydrocarbons.

"Cyanocel." [57] Trademark for chemically modified (cyanoethylated) cellulose.

cyanocobalamin. See vitamin B_{12}.

2-cyanoethyl acrylate $CH_2:CHCOOCH_2CH_2CN$.
Properties: Sp. gr. 1.0690; b.p., polymerizes when heated; f.p. –16.9°C; lbs/gal 8.9; flash point 145°F; soluble in water.
Uses: Forms polymers, copolymers for viscosity index improvers; adhesives; textile finishes and sizes.
Shipping regulations: None.*

cyanoethylation. Process for introducing the group, $-OCH_2CH_2CN$, into an organic molecule by reaction of acrylonitrile with a reactive hydrogen, such as that on a hydroxyl or amino group.
See also cotton, cyanoethylation of.

cyanoformic chloride CNCOCl.
Properties: Oily substance. Poison gas characteristics! B.p. 126-128°C (750 mm).
Derivation: Reaction between phthalyl chloride and the amide of ethyl oxalate.

"Cyanogas." [57] A proprietary product. Contains not less than 42% calcium cyanide; evolves hydrocyanic acid gas on exposure to atmospheric moisture.
Properties: Slate gray powder granules and flakes for different uses.
Containers: 4-oz, 1-, 5-, 25-, 100-lb metal containers.
Uses: For killing ants, rats, mice, moles, woodchucks, prairie dogs, and other pests in nests and burrows; for fumigating flour mills, warehouses, mushroom houses, and greenhouses; for fumigating citrus trees under tents to control scale insects; for fumigating grain storage.
Fire hazard: None.
Shipping regulations: Class B poison. Poison label.*

cyanogen C_2N_2.
Properties: Colorless gas; pungent penetrating odor; burns with a purple-tinged flame; extremely poisonous! Soluble in water, alcohol and ether. Sp. gr. 1.8064 (air = 1); liquefaction point –21°C; solidification point –34°C.
Derivation: Potassium cyanide solution is slowly dropped into copper sulfate solution; mercury cyanide is heated.
Grades: Technical; pure.
Containers: Liquefied cyanogen: iron cylinders.
Uses: Organic synthesis; poison gas in warfare.
Fire hazard: Dangerous.
Shipping regulations: Poison; class A. Poison gas label. Not accepted by express.*

cyanogenamide. See cyanamide (1).

cyanogen bromide (bromine cyanide) CNBr.
Properties: Prismatic, or acicular, transparent crystals. Penetrating odor. Slowly decomposed by cold water. Corrodes most metals. Caution! Very irritant and toxic! Soluble in alcohol, benzene, ether; sparingly soluble in water. Sp. gr. 2.02; b.p. 61.3°C (750 mm); m.p. 52°C; vapor density 3.6; vapor tension 63.3 (15°C); volatility 200,000 mg/cu m (20°C).
Derivation: (a) Action of bromine on potassium cyanide. (b) Interaction of sodium bromide, sodium cyanide, sodium chlorate, and sulfuric acid.
Grades: Technical.
Containers: Stoneware crocks; specially lined steel drums.
Uses: Organic synthesis; parasiticide; fumigating compositions; rat exterminants; cyaniding reagent in gold extraction processes; cellulose products treating agent; military poison gas.
Shipping regulations: Poison, class B. Poison label.*

cyanogen chloride CNCl.
Properties: Colorless liquid; poisonous! Vapors are highly irritant and very poisonous. Soluble in water, alcohol and ether. Sp. gr. 1.2; b.p. 12.5°C; m.p. –6.5°C; vapor density 2.1; vapor tension 682 mm (10°C); volatility 2,600,000 mg/cu m (15°C); heat of volatilization 109 cals (0°C); coefficient of thermal expansion 0.0015 (0°C).
Derivation: By the action of chlorine on moist sodium cyanide suspended in carbon tetrachloride and kept cooled to –3°C, followed by distillation.
Method of purification: Rectification.
Grades: Technical.
Containers: Iron cylinders.
Uses: Organic synthesis; manufacture of military poison gas.
Shipping regulations: Containing less than 0.9% water, poison liquid, class A. Poison gas label (by freight). Not accepted by express.*

cyanogen fluoride (fluorine cyanide) CNF.
Properties: Colorless gas. Forms a white, pulverulent mass if cooled strongly and sublimes at –72°C. Caution! Very irritant! Insoluble in water.
Derivation: Interaction of silver fluoride and cyanogen iodide.
Grades: Technical.
Containers: Glass containers (eventually, under the influence of light, it attacks the glass).
Uses: Organic synthesis; military poison gas (lachrymator).

cyanogen iodide (iodine cyanide) ICN.
Properties: Colorless needles; very pungent odor; acrid taste; violent poison! Soluble in water, alcohol, and ether; m.p. 146.5°C; sp. gr. 2.84.
Derivation: By heating a metal cyanide with iodine.
Method of purification: Crystallization.

*See "I.C.C. Shipping Regulations," page xiii.
Reference numbers refer to name of manufacturer. See "List of Manufacturers," page v.

Grades: Technical.
Containers: Glass bottles.
Use: Taxidermists' preservatives.
Shipping regulations: Poison, class B. Poison label. *

cyanogen mud. A mud formed in gas purifiers when coal gas is brought into contact with a saturated solution of ferrous sulfate for the removal of cyanogen. It contains about 30% Prussian blue.

"Cyanogran." [28] Trademark for a 97% sodium cyanide in granular form. White crystalline solid crushed to pass 100% through 10 mesh and retained on 50 mesh.
Containers: 100- and 200-lb. drums.

cyanoguanidine. See dicyandiamide.

"Cyanogum." [57] Trademark for a gelling agent.

cyano(methylmercuri)guanidine (methyl mercury dicyandiamide) $C_3H_6HgN_4$. Toxic! A disinfectant and fungicide used for treating seeds.
Danger! Poisonous by inhalation or swallowing. May cause skin irritation or delayed chemical burns. MCA warning label.

3-cyanopyridine C_5H_4NCN.
Properties: B.p. 206.2°C (760 mm); f.p. 49.6°C; soluble in water.

4-cyanopyridine C_5H_4NCN.
Properties: B.p. 195.4°C (760 mm); f.p. 78.5°C; partially soluble in water; soluble in most organic solvents.

cyanosilicones. Made from chlorosilanes (as trichlorosilane) and acrylonitrile.
Forms: Fluids, solutions, pastes, solids.
Uses: Adhesives; sealing and caulking; vehicle components.

cyanuramide. See melamine.

cyanuric acid (tricarbimide; tricyanic acid) $HO\underline{CNC(OH)NC(OH)N}\cdot 2H_2O$. See also isocyanuric acid, which is the ketone isomer.
Properties: White crystals; odorless; slight bitter taste. Soluble in water, alcohol and hot mineral acids. Sp.gr. 1.768.
Derivation: By heating urea or by the action of water on cyanuric chloride.
Method of purification: Crystallization.
Grades: Technical.
Containers: Glass bottles; tins.
Use: Organic synthesis.
Shipping regulations: None. *

cyanuric chloride (2,4,6 trichloro-1,3,5-triazine) $C_3N_3Cl_3$ (cyclic).
Properties: Crystals with pungent odor; sp. gr. 1.32; m.p. 146°C; b.p. 194°C (764 mm). Soluble in chloroform, carbon tetrachloride, hot ether, dioxane, ketones. Very slightly soluble in water (hydrolyzes in cold water).
Typical specifications: 97% pure.
Caution: Irritating to skin and eyes; handled commercially with proper precautions.
Containers: 50-lb steel drums; glass bottles.
Uses: Chemical synthesis; dyestuffs; pharmaceuticals; explosives; surfactants.

"Cyaqua." [57] Trademark for surface coating resins.

"Cyasorb." [57] An absorber for ultraviolet light.

"Cyclaine." [123] Trademark for an anesthetic consisting of hexylcaine hydrochloride.

cyclamate calcium. See calcium cyclamate.

cyclamates. Salts of cyclohexylsulfamic acid, $C_6H_{11}NHSO_3H$.

cyclamate sodium. See sodium cyclamate.

cyclamen alcohol. The alcohol corresponding to cyclamen aldehyde (q.v.), used as a stabilizer of cyclamen aldehyde.

cyclamen aldehyde (methyl para-isopropyl phenylpropyl aldehyde) $(CH_3)_2CHC_6H_4CH(CH_3)CH_2CHO$.
Properties: Colorless liquid having a heavy floral odor. Sp.gr. 0.949-0.959; refractive index 1.507-1.520. Soluble in 1 volume of 80% alcohol.
Use: Perfumery, for floral odors and soap perfumes.

cyclandelate (3,3,5-trimethylcyclohexyl mandelate) $C_6H_5CHOHCOOC_6H_8(CH_3)_3$.
Grade: N.N.D.
Use: Medicine.

cyclethrin 3-(2-cyclopentenyl)-2-methyl-4-oxo-2-cyclopentenyl ester of chrysanthemummonocarboxylic acid. As insecticide with applications similar to allethrin and other analogs. Accepted as generic name by Ent. Soc. See also furethrin, barthrin, ethythrin.
Properties: Viscous brown liquid, soluble in petroleum solvents and other common organic solvents. Formulated principally as liquid for spray applications corresponding to natural pyrethrins. Toxicity class of pyrethrin I and II.

cyclizine hydrochloride (1-diphenylmethyl-4-methylpiperazine hydrochloride) $(C_6H_5)_2CHC_4H_4N_2CH_3\cdot HCl$.
Properties: White crystalline powder or small colorless crystals. Odorless or nearly so; bitter taste; m.p. 285°C with decomposition; slightly soluble in water, alcohol, chloroform; insoluble in ether; pH (2% solution) 4.5-5.5.
Grade: U.S.P. XVI.
Use: Medicine.

cyclobarbital [5-(1-cyclohexenyl)-5-ethylbarbituric acid; tetrahydrophenobarbital] $C_{12}H_{16}N_2O_3$.
Properties: White crystals or crystalline powder; odorless; bitter taste; m.p. 170-174°C; soluble in alcohol or ether; very slightly soluble in cold water or benzene.
Derivation: Hydrogenation of phenobarbital with colloidal palladium in alcohol as a catalyst.
Use: Medicine.

cyclobarbital calcium (calcium cyclohexenylethylbarbiturate) $C_{24}H_{30}CaN_4O_6$.
Properties: White powder, somewhat soluble in water.
Grade: N.F. XI.
Use: Medicine.

cyclobutane (tetramethylene) C_4H_8.
Properties: Colorless gas. Sp.gr. 0.7083 (11°C); b.p. 13°C; m.p. −80°C.
Derivation: From petroleum.

cyclobutene (cyclobutylene) C_4H_6.
Properties: Gas; sp.gr. 0.733; b.p. 2.0°C.

cyclobutylene. See cyclobutene.

"Cyclocel." [99] Trademark for a reforming catalyst for gasolines and naphthas. Developed primarily for the Cyclo-version Process which incorporates catalyst reactivation facilities. Mesh grades 4/8, 4/10.
Typical analysis (volatile free basis): Al_2O_3 81.5%; Fe_2O_3 3.5%; TiO_2 3.5%; SiO_2 9.5%; Insolubles: 2.0%; Volatile Material: 2.0%.
Containers: 400-lb (net) steel drums.

cyclocitrylideneacetone. See ionone.

cyclocumarol $C_{20}H_{18}O_4$. A synthetic anticoagulant.
Properties: White, crystalline powder with slight odor. M.p. 164-168°. Insoluble in water; slightly soluble in alcohol.
Use: Medicine.

"Cyclodex." [74] Trademark for an emulsifiable cobalt catalyst to speed curing of latex paint films.

cycloform. See isobutyl-para-aminobenzoate.

cycloheptane (heptamethylene; suberane) C_7H_{14}.
Properties: Colorless liquid. Soluble in alcohol; insoluble in water. Sp.gr. 0.809; b.p. 118.1°C; m.p. −12°C; aniline equivalent −6.
Grades: Technical.
Use: Organic synthesis.

cycloheptanone (suberone) $C_7H_{12}O$.
Properties: Colorless, oil of peppermint odor; b.p. 179°C.
Uses: Research and intermediate.

cyclohexane (hexamethylene; hexanaphthene; hexahydrobenzene) C_6H_{12}.
Properties: Colorless, mobile liquid. Pungent odor. Somewhat similar to benzene but less toxic. Miscible with most lacquer solvents. Sp.gr. 0.779; b.p. 80.7°C; f.p. 6.3°C; refractive index 1.4263; aniline equiv. 7. Insoluble in water. Flash point (98% grade) −1°F.
Derivation: (a) By the catalytic hydrogenation of benzene. (b) Constituent of crude petroleum.
Grades: 85, 98, 99%.
Containers: Special metal cans and drums; tank cars.
Uses: Manufacture of nylon; solvent for cellulose ethers, fats, oils, waxes, bitumens; resins, crude rubber; extracting essential oils; chemical (organic synthesis, recrystallizing medium); paint and varnish remover; molding industry; glass substitutes.
Danger! Extremely flammable. Use with adequate ventilation. MCA warning label.
Shipping regulations: Flammable liquid; red label.*

cyclohexane carboxylic acid. See hexahydrobenzoic acid.

1,4-cyclohexanedimethanol $C_6H_{10}(CH_2OH)_2$. Cis and trans isomers are known.
Properties of pure (99.8% wt) substances: Liquid; b.p. 286.0°C (735 mm, cis-isomer), 283°C (735 mm, trans-isomer); m.p. 41-61°C; density (super-cooled) 1.0381 (25/4°C); flash point 330°F (COC); fire point 340°F (COC); refractive index (n 20/D) 1.4893. Soluble in water, ethyl alcohol.
Uses: Preparation of polyester films and protective coatings; reduction of reaction time in esterification.

cyclohexanol (hexahydrophenol) $C_6H_{11}OH$.
Properties: Colorless, oily liquid, camphor-like odor; hygroscopic. Sparingly soluble in water; miscible with most organic solvents and oils. Sp.gr. 0.937 (37/4°); m.p. 23°C; b.p. 160.9°C; wt/gal approximately 8 lbs; flash point 68°C; refractive index 1.465 (22°C).
Derivation: Phenol is reduced with hydrogen over active nickel at 160 to 170°F. The cyclohexanone is removed by condensing with benzaldehyde in the presence of alkali.
Grades: Technical.
Containers: 1-, 2-, 5-, 10-, 55-gal drums; tank cars.
Uses: Soap making to incorporate solvents and phenolic insecticides; manufacture of celluloid; source of adipic acid for nylon; textile finishing; solvent for rubber, nitrocellulose, resins, benzyl abietate, metallic soaps, dyes, vegetable, essential and mineral oils, cellulose esters and ethers; blending agent; lacquers; paints and varnishes; finish removers; dry cleaning; emulsified products; leather degreasing; polishes; solvent mixtures; plasticizers; plastics; germicides.
Caution: Vapor harmful. Use with adequate ventilation. MCA warning label.
Shipping regulations: None.*

cyclohexanol acetate (cyclohexanyl acetate) $CH_3COOC_6H_{11}$.
Properties: Colorless, nonflammable, nonexplosive, slightly toxic liquid. Has an odor resembling that of amyl acetate. Miscible with most lacquer solvents and diluents, and with halogenated and hydrogenated hydrocarbons. Soluble in alcohol; insoluble in water; sp.gr. 0.966; b.p. 177°C.
Uses: Solvent for nitrocellulose, cellulose ether, bitumens, metallic soaps, basic dyes, blown oils, crude rubber, many natural and synthetic resins and gums; lacquers.

cyclohexanone (pimelic ketone; ketohexamethylene) $C_6H_{10}O$.

*See "I.C.C. Shipping Regulations," page xiii.
Reference numbers refer to name of manufacturer. See "List of Manufacturers," page v.

Properties: Water-white to pale yellow liquid with acetone- and peppermint-like odor. Miscible with most solvents. B.p. 156.7°C; f.p. -47°C; sp.gr. 0.948; flash point (open cup) 129°F; refractive index (n 20/D) 1.4507; vapor pressure (212°F) 136 mm.
Typical specifications: Purity 97.7%; water 0.5% max; distillation range 2°C including 155.5°C; sp.gr. 0.9430-0.9460; refractive index 1.4460-1.4500; color, grams $K_2Cr_2O_7$/ liter, 0.005 max.
Derivation: By passing cyclohexanol over copper with air at 280°F; also by oxidation of cyclohexanol with chromic acid or oxide.
Containers: 5-gal. cans; 55-gal drums; tank cars.
Uses: Organic synthesis, particularly of adipic acid and caprolactam, polyvinyl chloride and its copolymers, and methacrylate ester polymers; solvent for DDT in aerosol bombs, also a solvent for basic dyes, fats, blown oils, waxes, crude rubber, resins, and various other materials; used in wood stains, paint and varnish removers, spot and stain removers, degreasing of metals, in polishes, as a leveling agent in dyeing and delustering silk.
Caution: Vapor harmful. MCA warning label.
Shipping regulations: None.*

cyclohexanone-delta (cyclohexanone-Δ) C_6H_8O. Volatile liquid (flash point 93°F) sometimes confused with cyclohexanone. There are two less hydrogen atoms per molecule in the so-called delta derivative, which is generally encountered only in the laboratory.

cyclohexanyl acetate. See cyclohexanol acetate.

cyclohexene (1,2,3,4-tetrahydrobenzene) C_6H_{10}.
Properties: Colorless liquid. Sp.gr. 0.8102; b.p. 83°C; m.p. −103.7°C; refractive index 1.445 (25°C); flash point 11°F; aniline equivalent 10; wt/gal 6.7 lbs (25°C). Soluble in alcohol; insoluble in water.
Grades: Technical; 95%; 99%; research.
Containers: Tank cars; barges.
Use: Organic synthesis; catalyst solvent; oil extraction.
Shipping regulations: Flammable liquid. Red label.*

3-cyclohexene-1-carboxaldehyde (1,2,3,6-tetrahydrobenzaldehyde) $CH_2CH{:}CHCH_2CH_2CHCHO$.
Properties: Liquid; sp.gr. 0.9721; b.p. 164.2°C; f.p. −100°C; lbs/gal 8.1 (20°C); flash point 135°F; slightly soluble in water.
Containers: Drums.
Uses: Intermediates; improved water resistance to textiles.
Shipping regulations: None.*

cyclohexenylethylbarbituric acid. See cyclobarbital.

cyclohexenylethylene. See vinyl cyclohexene.

cyclohexenyltrichlorosilane. $C_6H_9SiCl_3$.
Properties: Colorless liquid. B.p. 202°C; sp.gr. 1.263 (25/25°C); refractive index (n 25/D) 1.488; flash point (Cleveland open cup) 200°F. Readily hydrolyzed by moisture, with the liberation of hydrochloric acid.
Grades: Technical.
Use: Intermediate for silicones.
Shipping regulations: Corrosive liquid. White label.*

cyclohexylamine (hexahydroaniline; aminocyclohexane) $C_6H_{11}NH_2$.
Properties: Clear, nearly colorless liquid, with amine odor. Poisonous! Distillation range 132-137.5°C; sp.gr. 0.870-0.874 (15/15°C). Strong organic base; pH of 0.01% aqueous solution 10.5; forms an azeotrope with water, b.p. 96.4°C.
Grades: Technical (98%).
Containers: 375-lb drums; tank cars.
Uses: Corrosion inhibitor; boiler water treatment; petroleum additive; intermediate for dyes, insecticides, pharmaceuticals, synthetic sweetening agents.

cyclohexylbenzene. See phenylcyclohexane.

cyclohexyl bromide $C_6H_{11}Br$.
Properties: A liquid, not more than faintly yellow, having a penetrating odor. Sp.gr. 1.32-1.34 (25/25°C); refractive index 1.4926-1.4936 (25°C).

2-cyclohexylcyclohexanol $C_6H_{11}C_6H_{10}OH$.
Properties: Colorless liquid; freezing point 29°C; b.p. 271-277°C (760 mm); sp.gr. 0.977 (25/25°C); lbs/gal 8.13; refractive index 1.495 (25°C); flash point 255°F. Soluble in methanol and ether. Very slightly soluble in water.

cyclohexyl methacrylate $H_2C{:}C(CH_3)COOC_6H_{11}$.
Properties: Colorless monomeric liquid with pleasant odor. B.p. 210°C; refractive index (20°C) 1.4578; sp.gr. (20/20°C) 0.9626; viscosity (25°C) 5.0 centipoises; insoluble in water.
Uses: Optical lens systems; dental resins; potting resins for electronic assemblies.

1-cyclohexyl-2-methylaminopropane. See propylhexedrine.

para-**cyclohexylphenol** $C_6H_{11}C_6H_4OH$.
Properties: Crystals; m.p. 120°C (min).
Grades: Technical.
Use: Intermediate for resins and organic synthesis.

cyclohexylphenyl-1-piperidinepropanol hydrochloride. See trihexyphenidyl hydrochloride.

cyclohexyl stearate $C_6H_{11}OOCC_{17}H_{35}$.
Properties: Pale yellow color; sp.gr. 0.882 at 30/15.5°C; m.p. 26-28°C. Soluble in benzene, toluene and acetone; insoluble in water.
Uses: Plasticizer for natural and synthetic resins.

cyclohexylsulfamic acid. See "Hexamic Acid" and salts, calcium and sodium cyclamate.

cyclohexyl-para-toluenesulfonamide. See "Santicizer 1-H."

cyclohexyl trichlorosilane $C_6H_{11}SiCl_3$.
Properties: Colorless to pale yellow liquid. B.p. 206°C; sp.gr. 1.226 (25/25°C); refractive index (n 25/D) 1.4759; flash point (Cleveland open cup) 185°F. Readily hydrolyzed by moisture, with the liberation of hydrochloric acid.
Derivation: By Grignard reaction silicon tetrachloride and cyclohexylmagnesium chloride.
Grades: Technical.
Containers: 1-, 10-lb bottles; 100-lb drums.
Use: Intermediate for silicones.
Shipping regulations: Corrosive liquid. White label.*

"Cyclol." [293] Trademark for 2-hydroxymethyl-5-norbornene (q.v.).

cyclomethycaine 3-(2-Methylpiperidine)-propyl-para-cyclohexyloxybenzoate.
Properties: A white, odorless, crystalline powder. It is sparingly soluble in water, alcohol, and chloroform and very slightly soluble in acetone, ether and dilute acids.
Use: Medicine.

cyclomethycaine sulfate $C_{22}H_{34}ClNO_3 \cdot H_2SO_4$.
Properties: Crystals; somewhat soluble in water.
Grade: N.N.D.
Use: Medicine (surface anesthetic).

cyclonite (sym-trimethylene trinitramine; hexahydro-1,3,5-trinitro-sym-triazine; trinitrotrimethylenetriamine; RDX) $(CH_2)_3N_3(NO_2)_3$ (a cyclic molecule).
Properties: White crystalline compound; sp.gr. 1.82; m.p. 203.5°C. Soluble in acetone; insoluble in water, alcohol, carbon tetrachloride, and carbon disulfide; slightly soluble in methanol and ether.
Derivation: Ammonia and formaldehyde yield hexamethylene tetramine which is reacted with concentrated nitric acid.
Containers: Special lined drums.
Use: Very powerful explosive, 1.5 times as powerful as TNT.
Shipping regulations: Explosive, class A. High explosive label.*

"Cyclonol." [19] Trademark for powdered crystal form of 3,3,5 trimethylcyclohexanol.

cyclononane C_9H_{18}.
Properties: Colorless liquid. Sp.gr. 0.769; b.p. 170°C.

1,5-cyclooctadiene $HC{:}CH(CH_2)_2CH{:}CHCH_2CH_2$.
Properties: Liquid; freezing point −56.39°C; distillation range 301-303°F (technical); b.p. 149.34°C (pure); sp.gr. 0.88328 (20/4°C); lbs/gal 7.38; vapor pressure 0.50 psia (100°F); refractive index 1.4933 (20/D); flash point 100°F.
Grades: Technical; 95%; 99%.
Containers: Up to 1-gal bottles; 5-gal drums.
Shipping regulations: None.*

cyclooctane C_8H_{16}.
Properties: Colorless liquid. Sp.gr. 0.835; b.p. 148°C; m.p. 14°C.

cyclooctatetrene $HC{:}CHCH{:}CHCH{:}CHCH{:}CH$ or C_8H_8.
Properties: Colorless liquid; m.p. −27°C; b.p. 42.4°C (17 mm); sp.gr. 0.943 (0/4°C); refractive index 1.5394 (n 20/D). In spite of cyclic structure of molecule with alternate double bonds, the material does not resemble benzene. It behaves more like an aliphatic hydrocarbon, is relatively reactive and resinifies on standing in air.
Derivation: Polymerization of acetylene.

cycloparaffin. A hydrocarbon in which three or more of the carbon atoms in each molecule are united in a ring structure, and each of these ring carbon atoms is joined to two hydrogen atoms, or alkyl groups. The simplest members are cyclopropane (C_3H_6), cyclobutane (C_4H_8), cyclopentane (C_5H_{10}), cyclohexane (C_6H_{12}), and derivatives of these such as methylcyclohexane ($C_6H_{11}CH_3$).

1,3-cyclopentadiene $CH{:}CHCH{:}CHCH_2$. A colored liquid with sp.gr. 0.805 and b.p. 42.5°C; insoluble in water; soluble in alcohol, ether and benzene.
Derivation: From coal-tar.
Uses: Chemical intermediate; finishes.

cyclopentamine hydrochloride (1-cyclopentyl-2-methylaminopropane hydrochloride) $C_5H_9CH_2CH(CH_3)NHCH_3 \cdot HCl$.
Properties: White, crystalline powder. Mild characteristic odor and bitter taste. M.p. 113.0-116.0°C. Freely soluble in water, alcohol and chloroform; soluble in benzene; slightly soluble in ether. pH (1% solution) about 6.2.
Grade: N.N.D.
Use: Medicine.

cyclopentane (pentamethylene) C_5H_{10}.
Properties: Colorless liquid; soluble in alcohol; insoluble in water. Sp.gr. 0.7445 (20/4°C); b.p. 49.27°C; f.p. −94°C; refractive index 1.406 (20/D); flash point −37°C.
Derivation: From petroleum.
Grades: Technical; 95%, 99%; research.
Use: Solvent for cellulose ethers.
Containers: Steel drums; bottles.
Shipping regulations: Flammable liquid; red label.*

cyclopentanol (cyclopentyl alcohol) $CH_2CH_2CH_2CH_2CHOH$.
Properties: Colorless, viscous liquid; pleasant odor; sp.gr. 0.946 (20/4°C); refractive index 1.4575 (20°C); b.p. 139-140°C. Slightly soluble in water; soluble in alcohol.
Containers: Glass bottles; 5-gal cans; 55-gal drums.
Uses: Special perfume and pharmaceutical solvent; intermediate for dyes, pharmaceuticals and other organics.

cyclopentanone (adipic ketone) C_5H_8O.
Properties: A water-white, mobile liquid with a mild, distinctive ethereal odor;

*See "I.C.C. Shipping Regulations," page xiii.
Reference numbers refer to name of manufacturer. See "List of Manufacturers," page v.

b.p. 125-126°C (630 mm); sp.gr. 0.943; refractive index 1.437; flash point (closed cup) 87°F.
Containers: Glass bottles; 30-, 55-gal drums.
Use: Suggested intermediate for pharmaceuticals, biologicals, insecticides, and rubber chemicals.

cyclopentanone oxime C_5H_8NOH.
Nearly colorless and odorless crystalline material; m.p. 56°C; b.p. 196°C; soluble in water, alcohol. Used as intermediate in synthesis of amino acids, proline, ornithine, and citrulline.

cyclopentene $\overline{CH{:}CHCH_2CH_2CH_2}$.
Properties: Colorless liquid; sp.gr. 0.772; b.p. 44°C; f.p. −135.21°C; refractive index (20/D) 1.4225.
Grades: Technical; research.
Containers: Bottles.
Uses: Organic synthesis; plastics.
Shipping regulations: Flammable liquid. Red label.*

cyclopentenyl acetone [1-(1-cyclopentenyl)-2-propanone] $C_5H_7CH_2COCH_3$.
Properties: Clear, colorless liquid having a characteristic, ketone odor; b.p. 170°C; refractive index 1.4545-1.4550 at 25°C.

3-(2-cyclopentenyl)-2-methyl-4-oxo-2-cyclopentenyl ester of chrysanthemummonocarboxylic acid. See cyclethrin.

1-(1-cyclopentenyl)-2-propanone. See cyclopentenyl acetone.

cyclopentolate hydrochloride
$C_5H_8(OH)CH(C_6H_5)CO_2CH_2CH_2N(CH_3)_2 \cdot HCl$.
beta-Dimethylaminoethyl(1-hydroxy-cyclopentyl)-phenylacetate hydrochloride.
Properties: White, odorless, crystalline solid. M.p. 137-141°C; very soluble in water; freely soluble in alcohol; practically insoluble in ether; pH (1% solution) 5.0-5.4.
Grade: N.N.D.
Use: Medicine.

cyclopentylacetone (1-cyclopentyl-2-propanone) $C_5H_9CH_2COCH_3$.
Properties: Liquid; sp.gr. 0.893 (25/25°C); b.p. 180-184°C; refractive index 1.4420 (25°C).

cyclopentyl alcohol. See cyclopentanol.

cyclopentyl bromide (bromocyclopentane) C_5H_9Br.
Properties: Clear, mobile liquid with sweet aromatic odor. B.p. 137-138°C; sp.gr. 1.3866 (20/4°C); wt/gal 11.6 lb (20°C); refractive index (n 20/D) 1.4885; flash point (closed cup) 108°F. Insoluble in water.
Grades: Technical.
Containers: 5- and 13-gal carboys.
Uses: Organic synthesis (pharmaceuticals).
Shipping regulations: None.*

1-cyclopentyl-2-methylaminopropane hydrochloride. See cyclopentamine hydrochloride.

1-cyclopentyl-2-propanone. See cyclopentyl acetone.

cyclopentyl propionic acid $C_5H_9CH_2CH_2COOH$.
Properties: Liquid; b.p. 130-132°C (12 mm).
Containers: Glass bottles; 5-gal carboys.
Use: Intermediate.

cyclopentylpropionyl chloride (cyclopentylpropionic acid chloride) $C_5H_9CH_2CH_2COCl$.
Properties: Liquid; b.p. 81-82°C (10 mm).
Containers: Glass bottles; 5-gal carboys.
Use: Intermediate.
Shipping regulations: Corrosive liquid. White label.*

"Cyclophos." [172] Trade name for colorless crystalline, hydrated sodium polyphosphate. Cyclic compound.
Properties: Coagulates albumen and certain other proteins. Soluble in water.
Containers: 15-gal fiber drums.
Uses: Precipitation of proteins.

cyclophosphamide (endoxan) $C_7H_{15}Cl_2N_2O_2P$.
A chloroethyl phosphorus derivative, related chemically to the nitrogen mustards.
Use: Medicine.

cyclopropane (trimethylene) C_3H_6.
Properties: Colorless, flammable gas of characteristic odor resembling that of solvent naphtha and having a pungent taste. Sp.gr. 0.72-0.79; b.p. −32.9°C; m.p. −126.6°C. Soluble in alcohol.
Grades: Technical; U.S.P. XVI; 99.5% min.
Use: Organic synthesis; anesthetic.
Caution! Highly flammable gas. Forms flammable and explosive mixtures with air or oxygen.
Shipping regulations: Flammable gas. Red label.*

cyclopropanespirocyclopropane. See spiropentane.

cycloserine. See amino-3-isoxazolidone.

cyclosilane. See silanes.

"Cyclotene." [233] Trademark for a proprietary synthetic aromatic chemical. A white crystalline solid; soluble in propylene glycol, certain alcohols and water; odor and taste of maple-licorice with a definite suggestion of walnut.

cyclotron. A device usually used to accelerate positively charged particles such as protons, deuterons, and alpha particles to high energies (speeds). The cyclotron consists of a pair of flat, hollow, semi-circular electrodes (called Dees) enclosed in a vacuum chamber. The chamber is placed between the poles of a magnet. The ions which are formed by auxiliary equipment in the vacuum chamber travel within the Dees and are repeatedly accelerated by successive boosts when they cross the strong alternating electric field existing in the gap between the Dees. The paths of the ions are circular due to the influence of a steady magnetic field and increase in radius as the particles are accelerated to higher energies. Each time an ion passes through the gap between the Dees, it is again accelerated by the electric field. When the ion reaches the outer edge of the Dee, it is directed out of the chamber for use as a bombarding particle.

The synchrocyclotron (FM cyclotron) and the synchrotron are modified versions of the cyclotron. Both of the modifications produce particles of higher energies than the ordinary cyclotron since they compensate for the variation in the interval of time within which a particle traverses one Dee and crosses the electric field, once it has attained high speeds. The bevatron and cosmotron are synchrotrons and are capable of producing particles with energies in the billion electron volt range.

cycloversion. A process using bauxite as a catalyst for (1) desulfurization, (2) reforming, and (3) cracking of petroleum to form high octane gasoline.

"Cycopol" Resins. [57] Trademark for a series of styrene or substituted styrene-alkyd resins in organic solvent solution for air drying and baking type surface coatings.

cycrimine hydrochloride
$C_5H_9C(OH)(C_6H_5)CH_2CH_2C_5H_{10}N \cdot HCl$. Cyclopentylphenyl-3-(1-piperidyl)-1-propanol hydrochloride.
Properties: White, odorless, bitter solid. M.p. 241-244°C (dec). Slightly soluble in alcohol, chloroform, and water; practically insoluble in benzene and ether; pH (0.5% solution) 4.9-5.4.
Grade: N.F. XI.
Use: Medicine.

"Cydac." [57] Trademark for an accelerator for the vulcanization of rubber.

cydonia (quince seed).
Derivation: Seed of Cydonia vulgaris.
Habitat: Southern Asia and Europe; widely cultivated.
Containers: Bags.

"Cyfor Drylite." [57] Trademark for rosin size for use in paper manufacture.

"Cykelin." [64] Trademark for a series of linseed base dicyclopentadiene copolymers; 100% solids; produce varnish vehicles of quick, hard dry and excellent alkali and water resistance. Used in varnishes, enamels, aluminum paints.

"Cykelsoy." [64] Trademark for a soybean base dicyclopentadiene copolymer.
Properties and Uses similar to "Cykelin" (q.v.).

"Cylanto." [51] Trademark for unfiltered, steam-refined, steam-cylinder oils with good emulsibility and oiliness. Used under widely varying conditions of load, pressure and moisture.

"Cylesso." [51] Trademark for high quality steam-refined cylinder oils of high viscosity index and high flash point. Compounded and uncompounded grades available.

cylinder oil. A heavy lubricating oil used in engine or compressor cylinders.

cylinder stock. Viscous petroleum oils obtained as residues in the distillation of crude petroleum. Used for preparation of bright stock and steam cylinder oils.

"Cylo-Mar." [51] Trademark for marine cylinder oils, steam refined from selected, paraffinic crudes and designed to meet lubrication needs of all steam cylinders. Compounded and uncompounded grades available.

"Cymel." [57] Trademark for a line of synthetic resins based on melamine-formaldehyde filled with alpha-cellulose, cellulose, chopped fabric, glass fiber, asbestos fiber.

cymene (cymol; isopropyltoluene; isopropyltoluol; methylpropylbenzene) $CH_3C_6H_4CH(CH_3)_2$. The ortho-, meta-, and para-isomers are known.
Properties: Colorless, transparent liquids; aromatic odor. Sp.gr.: Ortho- 0.8748, meta- 0.862, para- 0.8551; m.p.: Ortho- −182°C, meta- −25°C, para- −73.5°C; b.p.: Ortho- 177°C, meta- 175.6°C, para- 176.5°C; refractive index, para- 1.489 (20°C). Soluble in alcohol, ether, and chloroform; insoluble in water.
Derivation: Mixed cymenes are produced from toluene by alkylation. Para-cymene occurs in several essential oils, and is made from monocyclic terpenes by dehydrogenation. These terpenes can be made from turpentine, or obtained as a by-product from the sulfite digestion of spruce pulp in paper manufacture.
Method of purification: Washing with sulfuric acid, water, and alkali.
Grades: Technical.
Containers: Drums; glass bottles.
Uses: Solvents; synthetic resin manufacture; metal polishes; organic synthesis (oxidation to hydroperoxides used as catalysts for synthetic rubber manufacture. Cymene alcohols are made by hydrogenating the hydroperoxides.) Pure para-cresol and carvacrol are made from para-cymene.

cymol. See cymene.

"Cynol" Agents. [57] Trademark for a series of rewetting, softening and defoaming agents used in the manufacture of paper. Available in liquid form.

"Cypan" Drilling Mud Conditioner. [57] Trademark for a synthetic organic chemical used to modify and control the properties of oil well drilling fluids.

"Cypel." [57] Trademark for a paper resin.

cypress oil.
Properties: Pale yellow liquid; characteristic odor. Soluble in alcohol, ether, and chloroform.
Chief known constituents: Pinene, cymene, camphene, furfural, terpineol, cypress camphor.
Constants: B.p. 160-250°C; optical rotation

+4° to +31°; sp. gr. 0.88-0.89.
Derivation: Distilled from the fresh leaves and tender shoots of the cypress tree Cupressus sempervirens.
Method of purification: Rectification.
Grades: Technical.
Containers: Tins; glass bottles.
Use: Medicine.
Shipping regulations: None.*

"Cyprex." [57] Trademark for dodine (a fungicide), N-dodecylguanidine acetate.

cypripedium (lady's slipper; American valerian; nerve root; Noah's ark; yellow moccasin flower). Dried rhizome and roots of Cypripedium bulbosum or other species of Cypripedium.
Habitat: Nova Scotia south to Alabama and west to Nebraska and Missouri.
Grades: Technical.
Containers: Bags.
Use: Medicine.
Shipping regulations: None.*

"Cyquest 40." [57] Trademark for a sequestrant.

"Cyrea." [57] Trademark for a urea feed compound.

"Cyron" Chemical Size. [57] Trademark for a synthetic sizing material for the paper industry. "Cyron" permits sizing of alkaline as well as acid paper and may be added either to the pulp or as a coating.

cysteamine (2-aminoethanethiol; mercaptamine; thioethanolamine) $HSCH_2CH_2NH_2$.
Properties: Crystals with unpleasant odor; oxidizes on contact with air; m.p. 97°C; soluble in water.
Uses: Medicine; in radiation sickness; believed to offer protection against radiation.

cysteine (alpha-amino-beta-thiolpropionic acid; amidothiolactic acid; beta-mercaptoalanine) $HSCH_2CH(NH_2)COOH$. An amino acid derived from cystine, occurring naturally in the L(+) form.
Properties: Colorless crystals; soluble in water, ammonium hydroxide, and acetic acid; insoluble in ether, acetone, benzene, carbon disulfide, and carbon tetrachloride.
Derivation: Hydrolysis of protein; degradation of cystine. Found in urinary calculi.
Uses: Biochemical and nutrition studies; believed to offer protection against radiation. Available commercially as L(+)-cysteine hydrochloride.

cysteine hydrochloride. The hydrochloride of L(+)-cysteine (q.v.).
Properties: Crystals; m.p. 175-178°C (dec); soluble in water, alcohol, and acetone.
Use: Biochemical research; medicine.
Commercially available.

cystine (beta, beta'-dithiobisalanine; di[alpha-amino-beta-thiolpropionic acid]) $HOOCCH(NH_2)CH_2SSCH_2CH(NH_2)COOH$. The chief sulfur-containing amino acid of protein.
Properties: White crystalline plates; soluble in water; insoluble in alcohol. Shows optical activity:
DL-cystine, m.p. 260°C
D(+)-cystine, m.p. 247-249°C
L(-)-cystine, m.p. 258-261°C with decomposition; the naturally occurring form.
Derivation: Hydrolysis of protein (keratin); organic synthesis. Found as small hexagonal crystals in urine.
Use: Biochemical and nutrition studies.

cytidine $C_9H_{13}N_3O_5$. The nucleoside consisting of D-ribose and cytosine.
Properties: White, crystalline powder; soluble in water, acid, alkali; insoluble in alcohol.
Derivation: From yeast ribonucleic acid. Also available as the hemisulfate, $(C_9H_{13}N_3O_5)_2 \cdot H_2SO_4$.

cytidine phosphates. Nucleotides used by the body in growth processes; important in biochemical and physiological research. Those isolated and commercially available (as sodium salts) are the monophosphate (CMP), the diphosphate (CDP) and the triphosphate (CTP).

cytidine-3-phosphoric acid. See cytidylic acid.

cytidylic acid (cytosylic acid; cytidine-3-phosphoric acid; cytosine nucleotide) $C_9H_{14}N_3O_8P$.
Properties: White crystalline powder; odorless, mild sour taste; m.p.: crystals from 50% alcohol, 230-233°C (with decomposition); crystals from water, 227°C (with decomposition). Soluble in water and dilute alkalies; slightly soluble in 50% alcohol; insoluble in alcohol and other organic solvents.
Derivation: From nucleic acid by hydrolysis.
Use: Experimental work.

cytisine (ulexine) $C_{11}H_{14}N_2O$. An alkaloid.
Properties: Colorless or yellowish-white crystals. Soluble in water and alcohol; insoluble in ether; m.p. 152-153°C.
Derivation: By extraction of the seeds of Cytisus laburnum, and many other Papilionaceae, and subsequent crystallization.

cytochemistry. The branch of biochemistry dealing with chemical compounds in and the chemical activity of animal and plant cells.

cytochrome c. The most abundant of the cytochromes. Molecular weight about 13,000.
Properties: Amorphous powder; stable to dilute acids, alkalies, and boiling.
Source: It is isolated from plasmolyzed yeast, ox and horse heart, algae, and wheat germ.
Use: Medicine; biochemical research.

cytochrome oxidase. An iron-porphyrin-containing protein which is an important enzyme in cell respiration. It catalyzes the oxidation of cytochrome c (q.v.) and is reduced itself in the reaction; it is then reoxidized by oxygen.

cytochromes. A class of iron-porphyrin proteins (see porphyrins) which are of great importance in cell metabolism. They are pigments which are found in the cells of nearly all animals and plants which use oxygen. There are several types of cytochromes which have been identified; cytochrome c (q.v.) is the most abundant and has been obtained in pure forms. The cytochromes and cytochrome oxidase have important functions in cell respiration.

cytosine $C_4H_5N_3O$. 2-Oxy-4-aminopyrimidine. A pyrimidine found in both ribonucleic and deoxyribonucleic acids, and certain coenzymes.

Properties: (monohydrate): Lustrous platelets; decomposes at 320-325°C. Slightly soluble in water and alcohol; insoluble in ether.

Derivation: Isolation following hydrolysis of nucleic acids; organic synthesis.

Use: Biochemical research.

cytosine nucleotide. See cytidylic acid.

cytosylic acid. See cytidylic acid.

"Cyuram." [57] Trademark for tetramethylthiuram disulfide pellets.

"Cyzac." [57] Trademark for surface coating resins.

"Cyzate." [57] Trademark for an accelerator for the vulcanization of rubber.

CZC Chromated Zinc Chloride. [28] Special formulation of zinc chloride and sodium dichromate.

Properties: Granular material with slight brownish color or 50% water solution; sp. gr. about 51.7°Bé; f.p. −35°F.

Containers: Solution: tank cars; dry material: 50- and 625-lb drums.

Uses: For preserving wood by impregnation against decay, termite attack; for application as a fire retardant. Treated lumber is clean, odorless, paintable, and safe to handle. See also Copperized CZC Chromated Zinc Chloride.

D

d-. Prefix meaning dextrorotatory. See D-.

D. Symbol for deuterium (q. v.).

D-. (or L-) Is a prefix signifying the stereoisomeric form of an organic substance, and means it has been correlated with the structure of D- or L- glyceraldehyde.. However, for amino acids, D- or L- refers to the configuration of the lowest numbered asymmetric center (alpha-carbon atom) while for carbohydrates it refers to the configuration of the highest numbered asymmetric center. Small d- and l- mean simply dextro- and levorotatory, but (+)- and (-)- are now preferred for these.
See optical isomerism.

2,4-D (2,4-dichlorophenoxyacetic acid) $C_6H_3(OCH_2COOH)Cl_2$.
Properties: White to yellow crystalline powder; not easily soluble in water or oils; soluble in alcohols. Stable; m. p. 138°C; b. p. 160°C (0.4 mm).
Derivation: By reaction of 2,4-dichlorophenol and chloroacetic acid in aqueous sodium hydroxide.
Forms available: Sodium salt (60-85% acid); amine salts (10-60% acid); esters (10-45% acid), such as the ethyl, isopropyl, or butyl esters. These forms are preferred to 2,4-D itself because they are dispersible in water and can be applied as sprays.
Grade: Technical.
Containers: Bags; 100- and 250-lb drums; (butyl, isopropyl esters) tank cars.
Use: Selective weed killer because of its growth-regulating action on plants. In general, plants of the grass family are more resistant than other groups.
Warning! Irritating to eyes, nose and throat. MCA warning label. Should be used with great caution around desirable plants, feed supplies, water supplies, etc.

"DA-1." [241] Trademark for a silica-alumina fluid catalyst (86.6% SiO_2, 13.2% Al_2O_3). Marketed in four grades (F-1, C-1, C-2, C-3). Available in hopper cars or drums for cracking petroleum gas oil fractions.

dachlaurin (D-L). Composed of 65% bromochloromethane and 35% carbon dioxide; used by Germans for extinguishing aircraft fires. Nitrogen sometimes added for use at high altitudes.

"Dacolyte." [28] Trademark for an addition agent for acid copper plating solution for heavy deposits. A dark brown, amorphous, fine powder; sharp odor.
Containers: 5-, 10-, 25-, and 50-lb fiber drums.

"Dacron." [28] Trademark for a polyester fiber made from polyethylene terephthalate. Available as filament yarn, staple, tow, and fiberfill. See also polyester fiber, dimethyl terephthalate, terephthalic acid, and polyethylene terephthalate.
Properties: Sp. gr. 1.38; tensile strength (psi) 53,000-141,000; break elongation 10-36%; moisture regain 0.4%; soluble in meta-cresol (hot), trifluoroacetic acid, and ortho-chlorophenol; m. p. 250°C.
Derivation: By reaction of dimethyl terephthalate and ethylene glycol. The resulting polymer is melt extruded through a spinneret and stretched.
Containers: Cores, tubes, bales, cartons and cases.
Uses: Apparel, curtains, ropes, belts, fire hose, filled products, and other textile and industrial applications.

dacthal (dimethyl tetrachloroterephthalate) $C_6Cl_4(COOCH_3)_2$.
Grades: 50% wettable powder, dusts.
Use: Pre-emergence treatment for crab grass control.

"dag" Dispersion No. 41. [46] Trademark for a concentrated colloidal dispersion of pure electric-furnace graphite in lacquer-diluent naphtha.
Properties: Liquid consistency; solids content 10%; average particle size 0.5 micron; maximum particle size 4 microns; sp. gr. 0.773; flash point 3°C; completely miscible with naphthas.
Uses: Dry-film metalworking lubricant for drawing, extruding, and stretch-forming, of aluminum and magnesium.
Shipping regulations: Flammable liquid. Red label.*

"dag" Dispersion No. 154. [46] Trademark for a concentrated colloidal dispersion of pure electric-furnace graphite in isopropanol.
Properties: Paste consistency; solids content 20%; average particle size 10 microns; sp. gr. 0.90; flash point 11°C; completely miscible with trichlorethylene, alcohols, ethers, esters, ketones, etc.
Uses: Electrically conductive coating for static bleeding, etc.; dry-film lubricant; opaque and transfer coating for graphic arts.
Shipping regulations: Flammable liquid. Red label.*

"dag" Dispersion No. 193. [46] Trademark for a concentrated dispersion of refined vermiculite in water.

Properties: Liquid consistency; solids content 32%; sp. gr. 1.26; b.p. 100°C; f.p. 0°C; completely miscible with water.

Use: Refractory coating for metal molding.

"dag" Dispersion No. 200. [46] Trademark for a concentrated colloidal dispersion of refined molybdenum disulfide in petroleum oil.

Properties: Liquid consistency; solids content 10%; average particle size 0.5 micron; sp. gr. 0.97; flash point 238°C; completely miscible with petroleum oil.

Uses: General industrial lubrication, especially where extreme-pressure characteristics are required; oil additive for internal-combustion engines, etc.; formulation of specialty lubricants.

"dag" Dispersion No. 2404. [46] Trademark for a concentrated colloidal dispersion of pure electric-furnace graphite in mineral spirits.

Properties: Liquid consistency; solids content 10%; average particle size 0.5 micron; maximum particle size 4 microns; sp. gr. 0.845; flash point 34°C; completely miscible with petroleum hydrocarbons.

Uses: General high-temperature lubricant for conveyor chains, bearings, etc.

"Dalamar." [28] Trademark for an azo yellow pigment.

Dalmatian insect powder. See pyrethrum flowers.

"Dalpac." [266] Trademark for several grades of butylated hydroxytoluene (di-tert-butyl-para-cresol). Supplied as flakes, powder, or liquid.

Uses: Antioxidant in food, animal feed, and nonfood industrial uses.

"Daltolac." [206] Brand name for proprietary alkyd resins for curing with isocyanates.

Dalton's law. The partial pressure of a gas or vapor in a perfect gaseous mixture is equal to its mole fraction in the mixture multiplied by the total pressure.

"Dalyde." [348] Trademark for dibromsalicylaldehyde, used in medicine.

damar. See dammar.

dammar (damar). A class of natural resins, principally of recent origin. Dammar resins are marketed under the classes Singapore, Siam, and East India. The latter class, which is of semi-fossil origin, is further graded as batu, pale, or black.

Properties: White to yellow, semitransparent lumps, having a conchoidal fracture. The dammar resins are all soluble in alcohol, benzene, turpentine, and oils; in general, they have lower acid numbers than the copals; other physical properties are highly variable.

Derivation: A resinous exudation from several species of trees in the East Indies and Malay.

Grades: Technical; bold, nubs, chips, and dust.

Containers: Bags and cases.

Uses: Spirit varnishes; impregnation of paper and fabric.

Shipping regulations: None.*

danaite. A cobalt-bearing iron arsenic sulfide, used as a source of arsenic.

dandelion. See taraxacum.

danthron. See 1,8-dihydroxyanthraquinone.

"Dantoin." [73] Trademark for 1,3-dichloro-5,5-dimethylhydantoin (q.v.).

DAP. Abbreviation for diallyl phthalate (q.v.).

"Dapon M." [55] Trademark for diallyl isophthalate prepolymer (q.v.).

"Daraprim." [301] Trademark for pyrimethamine, an antimalarial.

D'Arcet metal. See data under fusible alloys.

"Darco." [89] Trademark for activated carbon. Comes in various grades for use in sugar refining; removal of impurities from electroplating solutions; purification of dry-cleaning solvents; drug and chemical purification; and purification and decolorization of animal and vegetable oils, fats, and waxes.

"Daricon." [299] Trademark for oxyphencyclimine hydrochloride.

"Dariloid." [322] Trademark for a series of milk-soluble algin compositions, light tan or cream colored powders containing about 12% moisture. Hydrophilic colloids used as stabilizers in foods, especially milk and milk products.

Containers: 10-, 50-, 100-, 300-lb drums.

"Darlan." [119] Trademark for dinitrile fiber (copolymer of vinylidene dinitrile), a synthetic fiber notable for resilience, unusually high elastic recovery properties, resistance to storage at temperatures up to 330°F, dimensional stability to laundering and heat, resistance to the effects of outdoor exposure, and a soft, luxurious hand. This fiber is adaptable to the manufacture of many fabrics for sweaters, dresses, suits, coats, and can also be woven into a deep piled fabric with the softness, resilience, and appearance of natural fur.

"Daronyx." [328] Brand name for a dispersible organic solvent mixture that serves as dyeing assistant for "Dacron" polyester fiber.

"Dartal." [70] Trademark for a brand of thiopropazate hydrochloride, 1-(2-acetoxyethyl)-4-[3-(2-chloro-10-phenothiazinyl) propyl]piperazine dihydrochloride.

Use: Medicine.

"Darvan." [69] Trademark for a series of anionic surfactants which are used as dispersants, emulsion stabilizers, and latex stabilizers.

"Darvan No. 1". Sodium salts of polymerized alkyl naphthalene sulfonic acid; 77% active.

*See "I.C.C. Shipping Regulations," page xiii.
Reference numbers refer to name of manufacturer. See "List of Manufacturers," page v.

Properties: Buff powder; soluble in water to 40% .
Uses: Dispersant for latex compounding materials in water. Its wetting capacity is slight except towards certain specific materials. Stabilizer for SBR, natural and neoprene latices.
"Darvan No. 2." Sodium salts of polymerized substituted benzoid alkyl sulfonic acids; 84.5% active minimum.
Properties: Dark brown powder soluble in water to 25%.
Uses: Dispersant for high solids clay and filler slurries. Emulsifier for latex compounding ingredients.
"Darvan No. 3." Sodium salts of polymerized substituted benzoid alkyl sulfonic acids combined with an inert inorganic suspending agent; 84.5% active minimum.
Properties: Light gray-brown powder dispersible in water to 20%.
Uses: Dispersant for sulfur in water.
"Darvan No. 7." Aqueous solution of a polyelectrolyte; 35 ± 1% active.
Properties: Water white clear to slightly opalescent liquid.
Uses: Dispersant which causes minimum discoloration, is effective over a wide pH range, has low foaming tendency, and exhibits slight wetting capacity except towards certain specific materials. Stabilizer for emulsions and latices.

"Darvon." [100] Trademark for d-propoxyphene hydrochloride (q.v.).

DAS. Abbreviation for 4,4'-diamino-2,2'-stilbenedisulfonic acid.

dating. Any of a variety of methods used to determine the age of a naturally occurring substance or artifact. When a process is known to go on at a known rate in a material, leading to either the buildup or loss of a component, the age may be determined by an analysis that measures the amount of the buildup or loss. These processes may be chemical or nuclear, and the latter may be either spontaneous, or induced by cosmic rays. Examples of the former include the dating of ancient bones from the known rate of deposition of fluoride in bone when exposed to ground waters, and the dating of glass from the change in the vitrification. Nuclear methods have been uniquely valuable for dating since the rates of such processes are known not to be changed within the range of extremes of conditions found on earth. By measuring the amount of helium or of lead in uranium bearing minerals the age of the substance, or at least the length of time that the substance has existed as a solid deposit, may be calculated. By inference, this same measurement applied to igneous rocks gives this kind of an age of the earth. A similar measurement of the accumulation of the daughter products of thorium decay (q.v.) has also allowed the tentative dating of ocean sediments. With the discovery that radioactive carbon-14 is formed in the atmosphere from nitrogen and is thus incorporated in all living substances through the metabolism of carbon dioxide by plants, a process that ceases with death, and that the C-14 decays away after death at a known rate, the dating of many objects back through the first ice age has become possible. In general there is a most appropriate age range determinable by each of the above processes. The C-14 method applies to the interval from two to thirty thousand years of age, thorium in the millions of years, and uranium from hundreds of millions to billions of years.

daturine. See atropine.

daughter element. The element formed when another element undergoes radioactive decay. The latter is called the parent. The daughter may or may not be radioactive.

"Davco Granulated." [241] Brand name for mixed fertilizers.

"Davco Quality." [241] Brand name for mixed fertilizers.

"D-B-A." [248] Trade name for dibenzylamine $(C_6H_5CH_2)_2NH$.
Properties: Pale yellow to brown free-flowing liquid; sp. gr. 1.03; good storage stability; soluble in acetone, benzol, ethylene dichloride and gasoline. Insoluble in water.
Uses: Rubber activator; chemical intermediate.

DBM. Abbreviation for dibutyl maleate.

DBMC. Abbreviation for di-tert-butyl-meta-cresol.

"DB" Oil. [202] Trademark for a specially refined castor oil with minimum acidity and moisture content for dielectric and sonar applications, and for urethane polymers.

DBP. Abbreviation for dibutyl phthalate.

DBPC. Abbreviation for di-tert-butyl para-cresol.

"dbpc." [11] Trademark for di-tert-butyl-para-cresol, a solid alkylated phenol used as an antioxidant.
Properties: White to pale yellow colored solid; does not undergo reactions typical of phenols; insoluble in alkalies.
Grades: Technical; food.
Uses: Antioxidant for hydrocarbon materials; lubricating oil additive; stabilizer of gasoline for the prevention of gum formation; nonstaining rubber antioxidant; stabilizer for insecticidal preparations; antioxidant for industrial fats, greases, electrical insulating oils, turbine oils, and waxes. Preservative for edible oils and fats, fatty foods, baked goods, and animal and poultry feeds.

DBS. Abbreviation for dibutyl sebacate.

DCA. See deoxycorticosterone acetate.

DCB. Abbreviation for 1,4-dichlorobutane.

D & C dyes. Coal tar colors formerly certified by the Food and Drug Administration for use in drugs and cosmetics. They might contain no more than 20 ppm lead, 2 ppm arsenic, and 30 ppm other heavy metals. Similarly, dyes for foods also, FD & C dyes, were held to no more than 10 ppm lead, 1.4 ppm arsenic and a trace of other heavy metals, and were nontoxic in substantially larger quantities than would ever be used commercially. Examples are eosin (D & C Red No. 21); indigo (D & C Blue No. 6); uranine (D & C Yellow No. 8); amaranth (FD & C Red No. 2); indigo carmine (FD & C Blue No. 2); and yellow OB (FD & C Yellow No. 4).

DCHP. Abbreviation for dicyclohexyl phthalate (q.v.).

DCO. Abbreviation for dehydrated castor oil. See castor oil, dehydrated.

DCP. Abbreviation for dicapryl phthalate.

"D.C.P." [330] Trademark for calcium phosphate, dibasic, $CaHPO_4$.

DCPC. Abbreviation for dichlorophenyl methyl carbinol. See di(para-chlorophenyl)-ethanol.

"D-D." [125] Trademark for a soil fumigant which is 100% active and contains 1,3-dichloropropene, 1,2-dichloropropene, 3,3-dichloropropene, and related C_3 chlorinated hydrocarbons.
Properties: Pungent, dark brown liquid; approximate boiling range 50-115°C; slightly soluble in water; soluble in most common organic solvents; corrosive, especially to aluminum containers.
Containers: 30-gal drums, (300 lb net); 55-gal drums, (550-lb net).
Warning! May be fatal if swallowed, inhaled, or absorbed through skin. Hazardous vapor and liquid.
Shipping regulations: None.*

DDBSA. Abbreviation for dodecylbenzenesulfonic acid.

DDD. Abbreviation for dichlorodiphenyldichloroethane. See TDE.

DDDM. Abbreviation for 2,2'-dihydroxy-5,5'-dichlorodiphenylmethane. See dichlorophene.

DDH. Abbreviation for dichlorodimethylhydantoin.

DDM.
1. Abbreviation for 2,2'-dihydroxy-5,5'-dichlorodiphenylmethane. See dichlorophene.
2. Abbreviation for n-dodecyl mercaptan.

"DDM." [248] Proprietary name for a primary alkylmercaptan composed mainly of n-dodecyl mercaptan.
Properties: An off-white liquid; sp.gr. 0.84; soluble in most organic solvents; insoluble in water.
Uses: A polymerization modifier for SBR and nitrile rubbers.

DDP. Abbreviation for didecyl phthalate.

DDS. Abbreviation for diamino-diphenylsulfone. See sulfonyldianiline.

DDT (dichloro-diphenyl-trichloroethane; chlorophenothane; dicophane; 1,1,1-trichloro-2,2-bis(para-chlorophenyl)ethane) $(ClC_6H_4)_2CHCCl_3$. DDT has been accepted as a generic name by the Ent. Soc.
Properties: Colorless crystals or white to slightly off-white powder. Odorless or with slight aromatic odor. M.p. 108.5-109°C (pure); U.S.P. grade congeals at 89°C or above; b.p., decomposes. Insoluble in water; soluble in acetone, ether, benzene, carbon tetrachloride, kerosine, dioxane, and pyridine. Not compatible with alkaline materials.
Derivation: By condensing chloral or chloral hydrate with chlorobenzene in the presence of fuming sulfuric acid.
Grades: Technical; purified; aerosol; U.S.P. XVI.
Containers: Bottles; tins; bags; fiber drums.
Uses: Insecticide and miticide.
Caution! Harmful if swallowed; absorbed through skin when in solution. MCA warning label.
Shipping regulations: None.*

"DDT Gran." [147] Brand name for a free-flowing granular insecticide containing 5 or 10% DDT.
Containers: 50-lb bags.
Uses: Specially formulated for European corn borer and mosquito control.

DDVP (dimethyl dichlorovinyl phosphate) $(CH_3O)_2P(O)OCH{:}CCl_2$. Accepted as a generic name by the Ent. Soc. for a compound used as a systemic insecticide and as a fumigant in tobacco warehouses.
Containers: To 55-gal drums.

DE. Abbreviation for diatomaceous earth. See diatomite.

DEA. Abbreviation for diethanolamine.

"De-Acidite." [184] Trademark for a quaternary amine-type anion exchange resin.

Deacon's process. A process for the production of chlorine (q.v.) by passing hydrogen chloride and air through a heated tube. The original process was inefficient and has long been obsolete. The economic advantage of producing chlorine without attendant caustic soda as a byproduct has renewed interest in methods of oxidizing hydrogen chloride. One of the modifications of the Deacon's process now employed utilizes a catalyst of ferric oxide and potassium chloride.

dead-burned magnesia. See magnesite, dead-burned.

deadly nightshade. See belladonna.

dead soft steel. Normally a basic open-hearth steel, completely annealed, and with

*See "I.C.C. Shipping Regulations," page xiii.
Reference numbers refer to name of manufacturer. See "List of Manufacturers," page v.

carbon less than 0.1% and manganese from 0.2 to 0.5%.

deanol. See 2-dimethylaminoethanol.

"Dearborn Red." [141] Trade name for azo color pigments.
Properties: Good resistance to light and good outdoor durability, non-bleeding in water.
Grades: Light, medium and deep red shades.
Uses: Paints and enamels.

death's herb. See belladonna.

deblooming agents. Products added to mineral oils to mask fluorescence. Nitronaphthalene and yellow coal-tar dyes are among the products so used.

dec. Abbreviation for decomposes.

DEC. Abbreviation for beta-diethylaminoethyl chloride hydrochloride.

decaborane $B_{10}H_{14}$.
Properties: Crystals; stable indefinitely at room temperatures; decomposes slowly into boron and hydrogen at 300°C; density (25/4°C) 0.94; m.p. 99.7°C; sp.gr. 0.78 (100°C); b.p. 213°C (extrapolated). Very toxic! Slightly soluble in cold water; hydrolyzes in hot water; soluble in benzene, hexane, toluene. Forms shock-sensitive solutions with oxygenated and halogenated solvents.
Derivation: Obtained as a by-product of the pyrolysis of diborane or tetraborane.
Uses: Solid propellants; polymer synthesis; corrosion inhibitor; fuel additive; stabilizer; rayon delustrant; mothproofing agent; dye stripping agent; rubber vulcanization; reducing agent; fluxing agent; oxygen scavenger.
Shipping regulations: Flammable solid. Red label.*

decahydronaphthalene $C_{10}H_{18}$. Cis- and trans-forms are known.
Properties: Colorless liquid; aromatic odor. Insoluble in water; soluble in alcohol and ether.
Constants: Cis: Sp.gr. (20/4°C) 0.8927; m.p. −43.2°C; b.p. 194.6°C; refractive index (n 20/D) 1.48113. Trans: Sp.gr. (20/4°C) 0.8700; m.p. −31.5°C; b.p. 185.5°C; refractive index (n 20/D) 1.46968.
Derivation: By treatment of naphthalene in a fused state (above 100°C) with hydrogen in the presence of a catalyst such as finely divided copper or nickel oxide.
Grades: Technical.
Containers: 1-, 2-, 5-, 10-, 50-gal drums.
Uses: Solvent for oils, fats, waxes, resins, rubber, etc.; substitute for turpentine; cleaning machinery; stain-remover; shoe creams, floor waxes, etc.; cleaning fluids; lubricants; parallel uses to tetrahydronaphthalene (q.v.).
Shipping regulations: None.*

"Decalin." [28] Trademark for decahydronaphthalene ($C_{10}H_{18}$).

"Decalso." [184] Trademark for a precipitated gel-type sodium aluminosilicate cation exchanger.

decamethylenediamine $H_2N(CH_2)_{10}NH_2$.
Properties: Colorless liquid; b.p. 140°C (12 mm).

n-decanal (capraldehyde; capric aldehyde; n-decyl aldehyde; aldehyde C-10) $CH_3(CH_2)_8CHO$.
Properties: Colorless to light yellow liquid having a pronounced floral-fatty odor; sp.gr. 0.831-0.838 (15°C); refractive index (n 20/D) 1.427-1.431. Soluble in 80% alcohol, fixed oils, volatile oils, mineral oil; insoluble in water and glycerol.
Derivation: Occurs in lemongrass, citronella, orange, and many other oils. May be made synthetically by oxidation of the corresponding alcohol or reduction of the acid.
Use: Perfumery.

n-decane (decyl hydride) $CH_3(CH_2)_8CH_3$.
Properties: Colorless liquid; sp.gr. 0.7298; b.p. 174°C; m.p. −30°C; refractive index (n 20/D) 1.4114; flash point 44°C. Soluble in alcohol; insoluble in water.
Grades: Technical; 95%; 99%; research.
Containers: Bottles; drums.
Uses: Organic synthesis; solvent; standardized hydrocarbon.

decanedioic acid. See sebacic acid.

decanoic acid. See capric acid.

1-decanol (n-decyl alcohol; alcohol C-10) $CH_3(CH_2)_8CH_2OH$.
Properties: Colorless, water-white liquid. Sweet fat-like odor; sp.gr. 0.829; b.p. 232.9°C; m.p. 6°C; flash point (open cup) 180°F; refractive index (n 20/D) 1.4372. Insoluble in water (25°C); soluble in alcohol and ether.
Derivation: Reduction of coconut oil fatty acids.
Grades: Technical; high purity.
Containers: 55-gal drums; 10,000-gal tank cars.
Uses: Antifoam; perfume intermediate; making detergents,and esters; preparation of lube oil additives, plasticizers, adhesives, and metal polishes.

decanoyl chloride (sometimes called caproyl chloride) $CH_3(CH_2)_8COCl$. Available in 1-, 5-lb bottles; carboys and drums. Used as an intermediate.

decantation. Pouring or siphoning off the upper liquid from a precipitate or sediment or from a lower immiscible liquid as a partial means of separating the two phases.

decarboxylases. A group of enzymes in the living cell that remove carbon dioxide from various carboxylic acids without oxidation.

1-decene. See decylene.

"Deceresol." [57] Trademark for textile wetting and rewetting agents.

"Declomycin." [315] Trademark for demethylchlortetracycline (q.v.).

*See "I.C.C. Shipping Regulations," page xiii.
Reference numbers refer to name of manufacturer. See "List of Manufacturers," page v.

"Declostatin." [57] Trademark for demethylchlortetracycline hydrochloride and mystatin.

decoction. Pharmaceutical term for a liquid produced by boiling one or more drugs in water and filtering.

decoic acid. See capric acid.

decolorizing agents. Charcoals, blacks, clays, earths or other materials of highly adsorbent character which are used to remove undesirable color, as in the refining of sugar or oils. Also used in reference to bleaches involving a chemical reaction for removing color.

decontamination. See contamination.

"Decroline D." [307] Trademark for a stripping agent and discharge printing agent, consisting of normal zinc sulfoxylate formaldehyde; 90% active.
Properties: White, coarse, granular powder. Soluble in water.
Uses: Textile stripping agent; for best stripping results, "Decroline D" is applied in an acid liquor. Used as a discharge printing agent.

"Decroline X-4." [307] Trademark for a stripping agent consisting of basic zinc sulfoxylate formaldehyde; 88-90% active.
Properties: Fine, gray colored powder; soluble only in acid solutions.
Uses: Stripping agent primarily for wool and union yarn, piece goods and raw stock.

decyl acetate (acetate C-10) $CH_3(CH_2)_9OOCCH_3$.
Properties: Liquid with floral orange-rose odor; b.p. 187-190°C; sp. gr. 0.862-0.864; refractive index 1.426. Soluble in 80% alcohol, ether, benzene, glacial acetic acid; insoluble in water.
Derivation: By gently boiling together, for a long time, capric aldehyde and glacial acetic acid in the presence of zinc dust or powder; precipitating with water and distilling under reduced pressure.
Grades: Technical.
Use: Perfumery.

n-decyl alcohol. See 1-decanol.

decyl alcohol (mixed isomers). This is sold in drums and tank cars for many of the same uses as those of 1-decanol.

n-decyl aldehyde. See n-decanal.

decylamine $CH_3(CH_2)_9NH_2$.
Properties: Water-white; amine odor; boiling range 215-221°C; sp. gr. 0.797 (20/20°C); refractive index 1.437 (20°C); flash point 210°F.

decyl carbinol. See 1-undecanol.

decylene (1-decene) $C_{10}H_{20}$ or $H_2C{:}CH(CH_2)_7CH_3$.
Properties: Colorless liquid; sp. gr. 0.7396 (20/4°C); b.p. 172°C; m.p. −66.3°C; refractive index (n 20/D) 1.4220. Soluble in alcohol; insoluble in water.
Grades: Technical; high purity.
Use: Organic synthesis of flavors, perfumes, pharmaceuticals, dyes, oils, resins.

decyl hydride. See n-decane.

decylic acid. See capric acid.

decyl mercaptan $C_{10}H_{21}SH$.
Properties: Liquid; m.p. −26°C; b.p. 114°C (13 mm); sp. gr. 0.8410 (20/4°C); refractive index 1.4536 (n 20/D).
Grades: 95% (min.) purity.
Uses: Intermediate; synthetic rubber processing.

decyl-octyl methacrylate
$H_2C{:}C(CH_3)COO(CH_2)_nCH_3$. (n equals 7-9).
Containers: Drums.
Uses: Polymerizable monomer for plastics, molding powders, solvent coatings, adhesives, oil additives; emulsions for textile, leather, and paper finishing.

"Dednox." [423] Trademark for asphalt-gilsonite railroad protective coatings.

"DeeGee." [212] Trademark for a glucose oxidase-catalase enzyme system also called "DeeO" (q.v.).

"Deenate" 50W. [28] Trademark for agricultural and horticultural insecticide formulations based on DDT. Wettable powder containing 50% DDT Technical.
Containers: 3-, 4- and 50-lb bags.

"Deenax." [29] Trademark for di-tert-butyl-para-cresol (DBPC), an oxidation inhibitor used especially in waxes and natural fats.

"DeeO." [212] ("DeeGee"). Trademark for a glucose oxidase-catalase enzyme system, for removal of either oxygen or glucose.
Properties: Dry, fine white powder; water soluble; non-hazardous; nonflammable. Optimum pH range 4.5-8.0; optimum temperature 70-100°F.
Grade: For food products and analytical applications.
Containers: 1-oz to 1-lb bottles; 25- and 50-lb fiber drums.
Uses: In removing oxygen from canned or bottled beverages to prevent oxidation; in removing glucose from food products to increase shelf-life characteristics.

deep chrome. See chrome yellows.

deerberry. See gaultheria.

deer's tongue. See liatris.

"DEF." [181] Trademark for S,S,S-tributyl phosphorotrithioate (q.v.).

defecation. Purification; used specifically of the industrial clarification of sugar solutions.

deflagration. Sudden and sparkling combustion.

DeFlorez process. Process for the cokeless cracking of petroleum crudes in pipe stills. When the furnace pressure is 100 lbs, cracking takes place in the vapor phase; at 400 lbs pressure in the furnace, the unit operates as a liquid-phase cracking process.

defluorinated phosphate. Phosphate rock from which fluorine has been removed by high

temperature calcination.
Properties: Soluble in neutral ammonium citrate, 2% citric acid and 0.4% hydrochloric acid.
Grades: By % phosphorus, as 14, 18, 19% P.
Containers: Paper bags.
Use: Feed for stock and poultry.

"Defoamer PC-1244." [58] Trade name for an antifoam.
Properties: Clear light yellow liquid; sp. gr. 0.86 (60/60°C); viscosity, SUS at 100°F, 190; 7.72 lbs/gal. Soluble in benzene, toluene, kerosene, petroleum ether, carbon tetrachloride, isopropanol, tert-amyl alcohol, butyl "Cellosolve" and ethyl acetate. Insoluble in water, methanol, ethanol, and methyl "Cellosolve."
Containers: 55-gal nonreturnable black iron drums; 5-gal cans; tank cars.

defoaming agents (anti-foam agents). Products such as 2-octanol, sulfonated oils, or silicones, used for reducing foaming, which often interferes with processing operations. Such foams may be caused by casein, glue, gases, nitrogenous and other substances.

defoliant. A chemical agent which removes leaves from growing plants. The effects may be produced by a number of specific herbicides as well as pesticides and the result is a matter of degree of application. Minimal dosages of herbicides may remove leaves without damaging the plant structure or associated desirable portion of the plant. Preferable defoliants do not translocate largely beyond the application area. Examples of selective use are found in commercial leaf removal prior to harvest in the cotton and sugar beet industries. Defoliation has been also an undesirable associated effect with non-specific herbicides in areas adjacent to that where applied.

Examples are liquid or solid forms of commercial herbicides and/or pesticides of the group of the phenoxyacetic acids, carbamates, nitro compounds, of the organic types and arsenic acid, cyanides, cyanamides, cyanates, and thiocyanates, chlorates, of the inorganic type.

DEG. Abbreviation for diethylene glycol.

DEGN. Abbreviation for diethylene glycol dinitrate.

degras (sod oil; wool grease; tanning grease; leather grease).
Properties: Dark-brown unctuous fat; contains lanolin; disagreeable odor; sp. gr. 0.9322-0.9449; solidification point 38-40°C; saponification no. 84-127; iodine no. 15-21.5; acid no. 0.5-4.3; soluble in alcohol, ether, and benzene.
Derivation: Crude grease obtained by washing sheep's wool with trichloroethylene or by other washing processes.
Grades: American; English; common (15-17% free fatty acids); neutral (over 2% free fatty acids).
Containers: 400-lb drums.
Uses: Leather stuffing; belt-dressing compound; producing lanolin; printing inks; special soaps; varnishes.
Shipping regulations: None.*

degras, moellon.
Derivation: A by-product of the tannage of chamois leathers by impregnation with cod or menhaden fish oils. An oxidation of a part of the fatty acids of the oils takes place. When the tannage is complete, the excess of the oil contained in the skins is pressed out. This, when compounded, forms the moellon degras of commerce.
Grades: Anhydrous; 20% water; 30% water; 35% water.
Containers: Wooden barrels.
Uses: Stuffing leathers; belt dressing.
Shipping regulations: None.*

"D. E. H." [233] Trademark for epoxy resins.

de Haens salt. See antimony salt.

dehumidification. The removal of moisture (water vapor) from air. Also sometimes extended to analogous processes of removing a vapor from a gas mixture.

"Dehydratine No. 22." [205] Liquid silicone water repellent for coating masonry walls.

"Dehydratine No. 80." [205] Liquid admixture to accelerate hardening of concrete.

dehydration. The loss or removal of water from a substance or mixture either through ordinary drying or heating, or by absorption, adsorption, chemical reaction, condensation of the water vapor, or by centrifugal force or hydraulic pressure. The term dehydration is not usually applied to the loss of water from a water solution by evaporation or boiling.

"Dehydrite." [16] Trademark for anhydrous granular magnesium perchlorate.
Properties: Anhydrous salt, $Mg(ClO_4)_2$: Vapor pressure at ordinary temperature 0.000; stable to 250°C; absorbs over 30% of its weight of water by the formation of hydrates; will absorb a similar percentage of ammonia.
Spent "Dehydrite," $Mg(ClO_4)_2 \cdot 6H_2O$: M. p. 140-145°C; regenerated by heating gradually (10 hrs) to 250°C under vacuum (0.1 mm); channels do not form in the salt; deterioration is accompanied by contraction in volume, reducing tendency to clog.
Containers: 250-, 500-g bottles.
Uses: A dehydrating agent for use as a water absorbent in carbon combustions in steel analysis, in the ultimate analysis of organic substances, and in the drying of gases, including respiratory gases and carbon monoxide in air and blood.
Shipping regulations: May not be mailed.*

dehydro-. A prefix meaning removal of hydrogen, or sometimes removal of water, from a compound.

dehydroacetic acid (DHA)
$CH_3\overline{C:CHC(O)CH(COCH_3)C(O)O}$. 3-Acetyl-6-methyl-2(H)-pyran-2,4(3H)dione; 2-

acetyl-5-hydroxy-3-oxo-4-hexenoic acid delta-lactone.
Properties: Colorless, odorless, tasteless crystals. M.p. 108.5°C. Soluble in acetone, alcohol, and ether; insoluble in water.
Derivation: (a) By action of N-bromosuccinimide on ketene dimer; (b) by strong heating of acetoacetic ester.
Grades: Technical.
Uses: Fungicide and bactericide; plasticizer; chemical intermediate.

dehydroascorbic acid $\overline{OCOCOCOC}HCHOHCH_2OH$. The oxidized form of ascorbic acid (q.v.) with the same vitamin activity.
Properties: Needles; m.p. 225° (dec); soluble in water at 60°.
Derivation: Synthesized from ascorbic acid.
Use: Nutrition; medicine.

7-dehydrocholesterol (provitamin D_3) $C_{27}H_{44}O \cdot H_2O$. A sterol found in the skin of man and animals which forms vitamin D_3 upon ultraviolet irradiation.
Properties: Slender platelets from ether-methanol; m.p. 150°C; insoluble in H_2O; soluble in organic solvents.
Use: Nutrition; medicine.
See also following article.

7-dehydrocholesterol, activated (cholecalciferol; vitamin D_3) $C_{27}H_{44}O$. A free vitamin D_3, isolated in crystalline state from the 3, 5-dinitrobenzoate; produced by irradiation and equivalent in activity to the vitamin D_3 of tuna liver oil.
Properties: White colorless crystals. Unstable in light and air. Insoluble in water. Soluble in alcohol, chloroform and fatty oils. Melting range 84-88°C.
Grade: U.S.P. XVI.
Package: Hermetically sealed under nitrogen.
Use: Medicine, as antirachitic vitamin.

dehydrocholic acid $C_{24}H_{34}O_5$.
Properties: White, fluffy, odorless powder with bitter taste; m.p. 231-240°C. Almost insoluble in water; slightly soluble in ether and alcohol; soluble in chloroform, glacial acetic acid, and solutions of alkali hydroxides and carbonates.
Derivation: By oxidation of cholic acid.
Grade: U S.P. XVI.
Containers: Drums.
Use: Medicine; pharmaceutical intermediate.

dehydroepiandrosterone. See dehydroisoandrosterone.

dehydrogenase. An enzyme which catalyzes oxidation by the removal of hydrogen. See oxidase.

dehydrogenation. The process whereby hydrogen is removed from compounds by chemical means.

11-dehydro-17-hydroxycorticosterone. See cortisone.

dehydroisoandrosterone (dehydroepiandrosterone) $C_{19}H_{28}O_2$. An androgenic steroid; a metabolic product of the adrenal steroid hormones. It posses about one-third the androgenic activity of androsterone (q.v.).
Properties: Dimorphous: Needles with m.p. 140-141°; leaflets with m.p. 152-153°; precipitated by digitonin ; soluble in benzene, alcohol, and ether. Sparingly soluble in chloroform and petroleum ether.
Derivation: Isolated from male urine; synthesis from cholesterol or sitosterol.
Use: Medicine; biochemical research.
Also available as the acetate salt.

"Dehydrol." [141] Trade name for dehydrated castor oil used as a drying oil in the manufacture of varnishes and alkyd resins. It has excellent color retention, good drying properties, good water resistance, and good durability.

3-dehydroretinol. Vitamin A_2. See vitamin A.

dehydrothio-para-toluidine (2-(para-aminophenyl)-6-methylbenzothiazole) $CH_3\overline{C_6H_3SC(C_6H_4NH_2)N}$.
Properties: Long, yellowish, iridescent needles. Solutions have a violet-blue fluorescence. M.p. 191°C; b.p. 434°C. Soluble in alcohol; very slightly soluble in water.
Derivation: By heating para-toluidine and primuline base with sulfur and separation from the primuline base by distillation in vacuo.
Use: Dyestuff.

de-inkable inks. Special inks prepared so as to be readily removable thus facilitating re-use of the paper.

de-inking of paper. The removal of inks from paper by use of strong alkaline solutions such as soda-ash liquor, caustic soda or lime which dissolve varnish and free the ink carbon. Removal of the carbon is accomplished by use of colloidal agents such as talc or bentonite and by mechanical agitation with water.

deionized water. Water that has been purified of salts by passage in succession through a cation-exchange resin to replace metal ions such as calcium and iron by hydrogen ion, and then through an anion-exchange resin to remove both the hydrogen ions and the corresponding negative ions. When nonionic impurities are absent, such water is the equivalent of distilled water.

"Dekatyl." [28] Trademark for a line of cationic dyes especially suited for dyeing "Dacron" type 62 and 64 polyester fiber.

deKhotinsky cement. A thermoplastic cement, which is not attacked by water, sulfuric acid, nitric acid, hydrochloric acid, carbon disulfide, benzene, gasoline, or turpentine, and is very little affected by ether, chloroform, alkalies, but is readily dissolved by ethyl alcohol. Articles to be cemented are heated and cement is applied.
Derivation: Shellac and pine tar are heated together and stirred for several hours without overheating. Various degrees of hardness and variations in properties are

obtained by changing the proportions and the exact nature of the ingredients. Thus, the pine tar is sometimes replaced by a mixture of terpineol and wood creosote.

Grades: Hard, for cementing glass, metal, and porcelain; medium, for wood and for lathe and milling machine work.

"Delac-S." [248] Trade name for N-cyclohexyl-2-benzothiazole sulfenamide, a delayed action accelerator.

Properties: A cream colored powder; sp. gr. 1.27, melting range 95-100°C; soluble in acetone, ethylene dichloride and benzol; insoluble in water and gasoline.

Uses: An all-purpose accelerator for tires, footwear, soling and mechanical goods. Used in natural, SBR and nitrile rubbers.

"Delalutin." [412] Trademark for hydroxyprogesterone caproate (q. v.).

"Delamin." [266] Trademark for a technical grade of high molecular weight primary amine, derived from 18-carbon fatty acids of tall oil origin. Waxlike solid.

Uses: Beneficiation (flotation) of nonmetallic ores.

Delanium." An English product; an unusually hard form of carbon made from specially chosen coals. Mechanically strong, impermeable to liquids, easily fabricated. Used for heat exchangers, corrosion resistant linings and packings.

"Delatestryl." [412] Trademark for testosterone enanthate (q. v.).

"Delestrogen." [412] Trademark for estradiol valerate (q. v.).

delhi hard. A ferrous alloy (sp. gr. 7.75; m. p. 1500°C) containing in addition to iron 16.5 to 18% chromium, 1 to 1.1% carbon, 0.75 to 1% silicon, 0.35 to 0.5% manganese. It is resistant to cold ammonium hydroxide in all concentrations, and to mine and sea waters and moist sulfurous atmospheres.

"Delnav." [266] Trademark for technical grade of 2, 3-para-dioxanedithiol S, S-bis(O, O-diethyl phosphorodithioate) containing about 70% cis and trans isomers. Viscous brown liquid.

Uses: Miticide on cotton, citrus and ornamental plants; control of livestock pests.

"Delphicol." [57] Trademark for choline-methionine-inositol-folic acid-B.

delphinine $C_{34}H_{47}NO_9$.

Properties: White, crystalline alkaloid; poisonous! Soluble in alcohol and ether; insoluble in water.

Constants: Melts at 191-195°C with decomposition.

Derivation: By extraction from the seeds of Delphinium staphisagria.

"Delrin." [28] Trademark for acetal resin of composition $(\text{-OCH}_2\text{-})_n$ derived by polymerization of formaldehyde.

"Delsan." [28] Trademark for seed protectants containing insecticide-fungicide combinations, such as thiram plus dieldrin.

Uses: For treatment of beans, seed corn, and pea seed; for control of certain fungi and soil insects.

delta acid. See 2-naphthylamine-7-sulfonic acid.

"Delta-Cortef." [327] Trademark for prednisolone, a derivative of hydrocortisone, 11b, 17a, 21-trihydroxy-1, 4-pregnadiene-3, 20-dione ($C_{21}H_{28}O_5$).

Use: Medicine.

"Deltalin." [100] Trademark for calciferol, U. S. P.

"Deltasone." [327] Trademark for prednisone, a derivative of cortisone, 17a, 21-dihydroxy-1, 4-pregnadiene-3, 11, 20-trione ($C_{21}H_{26}O_5$).

Use: Medicine.

"Deltra." [123] Trademark for prednisone (q. v.).

"Deltyl." [227] Trademark for a mixture of isopropyl esters of lauric, myristic and palmitic acids. Colorless liquid; color No. 4 maximum; practically odorless. Available in two grades.

"Deltyl" Extra, predominantly isopropyl myristate.

Properties: Sp. gr. 0.847-0.854 (25/25°C); refractive index 1.433-1.436 (20°C); melting point max. 5°C.

"Deltyl" Prime, predominantly isopropyl palmitate.

Properties: Sp. gr. 0.850-0.855 (25/25°C); refractive index 1.435-1.429 (20°C); m. p. max. 14°C. Soluble in 4 parts of 90% alcohol; soluble in mineral, peanut, sesame, olive and almond oils; insoluble in water.

Uses: "Deltyl" Extra replaces vegetable or mineral oils in cosmetics.

delustrants. Chemical agents used to produce dull surfaces on synthetic fibers either before or after spinning so the resulting product more nearly resembles natural silk.

"Delvex." [100] Trademark for dithiazanine iodide (q. v.).

"Delvinal." [123] Trademark for sodium vinbarbital.

demecarium bromide $C_{32}H_{52}Br_2N_4O_4$. Decamethylenebis-(meta-dimethylaminophenyl-N-methylcarbamate) dimethobromide.

Properties: White, slightly hygroscopic powder; decomposes 162-167°C. Freely soluble in water, alcohol; sparingly soluble in acetone; insoluble in ether. Aqueous solutions are neutral, stable and may be sterilized by heat.

Use: Medicine.

"Demerol" Hydrochloride. [162] Trademark for meperidine hydrochloride.

11-demethoxyreserpine. See deserpidine.

demethylchlortetracyline $C_{21}H_{21}ClN_2O_8 \cdot 1.5\ H_2O$. (7-Chloro-6-demethyltetracycline.)

Properties: M.p. 174-178°C (decomposes).
Use: Medicine (antibiotic).

demeton. A commercial systemic insecticide; accepted as a generic name by the Ent. Soc. for O,O-diethyl-O(and S)-2-(ethylthio)ethyl phosphorothioate (q.v.).

DEMH. Abbreviation for 1,3-dibromo-5-ethyl-5-methylhydantoin (q.v.).

demulsifier. An agent which makes an oil resistant to emulsification.

"D.E.N." [233] Trademark for epoxy resins.

denaturants for alcohol. Chemicals or mixtures specified by the U.S. government (and other governments) for addition to ethyl alcohol to make it completely unsuitable for human consumption, and also to prevent recovery of the alcohol from the mixture. The denaturants do not prevent useful employment of the alcohol in industry and the arts.

The following are among the chemicals that are or have been used as denaturants: acetone, acetaldol, aldehydes, almond oil, ammonia water, animal oil, bay oil, benzoic acid, benzol, bergamot oil, brucine sulfate, butyl alcohol, camphor, cassia oil, cedar leaf oil, chloroform (crude), cinnamon oil, citronella oil, cinchonidine sulfate, clove oil, diethylphthalate, ethyl acetate, ethylamines, ethyl ether, ethyl propionate, emetine hydrochloride, eugenol, eucalyptol, gasoline, glycerine, guiacol, iodine, ipecac, kerosine, menthol, mercuric iodide, methanol (wood alcohol), nicotine, phenol, pine oil, pyridine bases, salicylic acid, soaps, sulfuric acid.

denatured alcohol. In the United States it is divided into two kinds as follows:
Completely denatured alcohol: Ethyl alcohol which has been rendered entirely unfit for beverage use by the addition of denaturants prescribed by the Federal government.
Specially denatured alcohol (SDA): Ethyl alcohol so treated with denaturants as to permit its use in a greater number of specialized arts and industries than completely denatured alcohol. The character of the denaturant or denaturants is such that it may be sold, possessed and used only pursuant to permit and bond except as otherwise provided in Government regulations.
Uses: Solvent or thinner (cellulose, resins, and similar products, for toilet products and pharmaceutical products used externally, for processing industrial, food and drug products, and for cleaning, preserving and flavoring materials); raw material for the synthesis of other chemicals; fluid such as anti-freeze or in brake fluids; fuel (airplane, motor, etc.); experimental studies; tobacco sprays and flavors (denatured rum used); manufacture of acetaldehyde.
Shipping regulations: Flammable liquid. Red label.*

dendritic. Pine-tree shaped; used of crystals or the particles in powders.

denier. The weight unit of a yarn: the weight in grams of 9000 meters of a yarn equals 1 denier. A 100-denier yarn is one of which 9000 meters weighs 100 grams.

DeNora cell. A mercury-cathode cell for the electrolytic production of chlorine and caustic soda. The electrolysis compartment consists of a long, narrow steel trough lined with chemically resistant stone, into which sodium chloride brine is pumped. The mercury cathode flows continuously along the bottom of the trough, becoming enriched in sodium, while chlorine is released from graphite anodes. The sodium amalgam is decomposed in a steel tower packed with graphite, where the sodium reacts with water to form hydrogen and sodium hydroxide. The mercury is pumped back to the electrolysis compartment.

"Densitol." [3] Trademark for brominated sesame oil.
Properties: Clear, reddish-brown, oily liquid with no taste or odor; sp. gr. 1.325-1.335 at 25°C.
Use: Weighting agent for citrus oils in the production of citrus emulsions for use in soft drinks.

density. Mass per unit volume, usually expressed in grams per cubic centimeter or in pounds per cubic foot or gallon. As a specific term it has been applied in electricity, photography and other areas with reference to length-force-time systems of units. See specific gravity.

density, apparent. The weight of a unit volume of powder, usually expressed as grams per cubic centimeter, determined by a specified method. (M.P.A. definition, M.P.A. Standard 9-50T). See density; specific gravity.

density, bulk. An alternate term for apparent density; used industrially to specify a simple dimensional measurement of powdered and granular materials in terms of weight of a unit volume. The measure may be carried out by the weight of material in a fixed volume container or by solvent displacement (usually water) of a known volume of material. See specific gravity.

dental plaster. See gypsum cements.

"Deo-Base." [45] Trade name for light petroleum distillate; superfine grade of kerosene without the objectionable odor of kerosene.

"Deodall # 1." [12] Trademark for a multi-purpose masking agent and deodorant. A mixture of synthetic aromatic chemicals and derivatives.
Uses: To deodorize organic solvents and petroleum products and mixtures containing them.

deodorants. Products used for destroying, masking, or modifying foul and unpleasant odors.

*See "I.C.C. Shipping Regulations," page xiii.
Reference numbers refer to name of manufacturer. See "List of Manufacturers," page v.

In physical form, deodorants may appear as cakes, blocks, balls, powders, ointments, pencils, creams, or liquids. They may or may not possess a strong, distinctive odor of their own and they may consist either of a single chemical, or a mixture. They find application in hospital procedures, in industry for various purposes, in housekeeping for treating atmospheric odors from cooking, refrigerators, garbage pails, water closets, drains, urinals, floors, dog kennels, and the like. The cosmetic industry markets products for masking perspiration odors and laundry and dry-cleaning establishments use one or several products in sequence for breaking down odors either from perspiration or from vomit and other stains.

deodorized oils. Oils which have been subjected to hydrogenation or other treatment to remove objectionable odors, in order to make them fit for human consumption.

deoxidizer. An agent which removes oxygen from a compound or from a molten metal.

deoxy-. Preferred prefix indicating replacement of hydroxyl by hydrogen in the parent compound. The meaning is identical to that of desoxy- and used interchangeably.

deoxyalizarin. See anthrarobin.

deoxycholic acid (desoxycholic acid) $C_{24}H_{40}O_4$. A bile acid; contains one less hydroxyl group than cholic acid.
Properties: Crystals; m.p. 172-73°. Not precipitated by digitonin. Practically insoluble in water and benzene; slightly soluble in chloroform and ether; soluble in acetone and solutions of alkali hydroxides and carbonates; freely soluble in alcohol.
Derivation: Isolation from bile; organic synthesis.
Containers: Drums.
Use: Medicine; precursor for organic synthesis of cortisone.
Also available as sodium salt.

11-deoxycorticosteroid. See deoxycorticosterone.

deoxycorticosterone (desoxycorticosterone; 4-pregnen-21-ol-3,20-dione; 11-deoxycorticosteroid) $C_{21}H_{30}O_3$. One of the adrenal cortical steroid hormones. Active in causing the retention of salt and water by the animal kidney.
Properties: Crystalline plates; m.p. 141-142°C. Freely soluble in alcohol and acetone.
Derivation: From adrenal cortex extract; synthesis from other steroids.
Use: Medicine (usually as acetate).

deoxycorticosterone acetate (desoxycorticosterone acetate; DCA; DOCA; desoxycortone acetate) $C_{23}H_{32}O_4$. The acetate salt of deoxycorticosterone.
Properties: White, or creamy white, crystalline powder; sensitive to light; m.p. 155-161°C; odorless; stable in air. Sparingly soluble in alcohol, acetone, and dioxane; slightly soluble in vegetable oils; practically insoluble in water.
Grade: U.S.P. XVI.
Use: Medicine.

deoxycorticosterone trimethylacetate (desoxycorticosterone trimethylacetate) $C_{26}H_{38}O_4$.
Properties: White, or creamy white, crystalline powder. Odorless; stable in air. Soluble in dioxane; sparingly soluble in acetone; slightly soluble in alcohol, methanol, ether; practically insoluble in water. M.p. 198-204°C.
Grade: U.S.P. XVI.
Use: Medicine.

deoxyephedrine hydrochloride. See methamphetamine hydrochloride.

deoxypentose nucleic acid. See deoxyribonucleic acid.

deoxyribonuclease. See pancreatic dornase.

deoxyribonucleic acid. (desoxyribonucleic acid; DNA; TNA; thymus nucleic acid; deoxypentose nucleic acid) $C_{39}H_{51}O_{25}N_{15}P_4$. Considered to be the prime genetic substance in the cell. Found in combination with protein in the chromosomes and genes. Constituents are adenine, guanine, cytosine, thymine, D-2-deoxyribose and phosphoric acid. It is now known that the molecule is a double helix in which the two strands of nucleotides are identical or at least probably genetically identical.
Properties: White or cream colored amorphous powder; odorless; slight sour taste; sp.gr. 1.700. Slightly soluble in water; readily soluble in alkali solutions; insoluble in alcohol and other organic solvents. Solutions are highly viscous.
Derivation: Can be extracted from the thymus gland; produced commercially from fish milt.
Uses: Biochemical and physiological research in the mechanisms of heredity, tumor growth and related processes.

D-deoxyribose (D-desoxyribose) $CH_2OHCHOHCHOHCH_2CHO$. A five-carbon-atom sugar that is unusual because there is no oxygen atom attached to the second carbon atom. It is found as a constituent of deoxyribonucleic acid (q.v.).

deoxyribose nucleic acid. See deoxyribonucleic acid.

"Depban." [233] Trademark for paraffin inhibitors for use in oil well equipment.

DEPC. Abbreviation for gamma-diethylaminopropyl chloride hydrochloride.

"Dependip." [200] Trademark for a petroleum solvent prepared by straight-run overhead distillation.
Properties: Water-white; sp.gr. 0.758 (60°F); wt/gal 6.31 lb (60°F); flash point, Tag closed cup, 52°F; considered nontoxic.
Caution: Flammable; keep fire and lights away.

Use: Solvent in rubber dipping cement.
Shipping regulations: Flammable liquid. Red label.*

dephlegmation. Partial condensation of vapor from distillation operation to produce liquid richer in higher boiling constituents than the original vapor. The residual vapor is of course richer in the lower boiling constituents.

depilatories. Products for removing hair from skin. Sulfides are largely used for this purpose. For example, the leather industry uses large amounts of sodium sulfide for unhairing hides. The cosmetic industry also markets various preparations for removing unsightly hair.

"Depilin." [57] A line of organic dehairing agents for the dehairing of hides.

"Depo"-Estradiol. [327] Trademark for estradiol 17-cyclopentylpropionate, ($C_{26}H_{36}O_3$).
Use: Medicine, in oil solution.

"Depo"-Testosterone. [327] Trademark for testosterone cyclopentylpropionate, ($C_{27}H_{40}O_3$); used in medicine.

dequalinium. Short for decamethylene-bis(4-aminoquinaldinium acetate). Used as an antibiotic oral antiseptic.

"D.E.R." [233] Trademark for epoxy resins.

"Deraspan." [233] Trademark for epoxy resins and curing agents.

derby red. See chrome red.

"Dergon." [300] Trademark for a series of liquid detergents used for textile scouring.
"Dergon" MF: Polyoxyethylene fatty ester composition.
"Dergon" OM: Alkanolamine fatty acid condensation product.
"Dergon" T: Highly sulfated fatty ester.

"Dergopal." [300] Trademark for a series of fluorescent whitening agents for natural and synthetic textile fibers. Different members applied to specific fibers. Includes both coumarin and stilbene types.

"Deriphat." [259] Trademark for a series of amphoteric surfactants. Salts or free acid forms of N-alkyl beta-aminopropionic acid or N-alkyl beta-iminodipropionic acid.
Derivation: Condensation of primary amines (from fatty acids) with acrylic monomers.
Grades: Flaked (sodium salts) and liquid (free acid or partial salt).
Containers: Flaked products in polyethylene-lined fiber drums. Liquid products in 55-gal steel drums and 5-gal pails.
Uses: Cosmetic and shampoo preparations; industrial and household cleaners; corrosion inhibitors; emulsion polymerization; emulsification; metal fabrication.

derris root. The root of the shrubs Derris elliptica and D. malaccensis. Toxic to lower animals and to insects but not to humans.
Constituents: Rotenone, toxicarol, tephrosin.
Use: Insecticides.

desacetyl lanatoside. See deslanoside.

desalting. Any process for making potable water from sea water or other saline waters. Distillation is the oldest method. Reuse of vapors through compressive distillation or multiple effect evaporation is practiced in order to limit heat consumption. Distillation with solar heat is expensive because the large areas required result in high equipment investments.

Electrodialysis is an inherently good method because the energy is used to remove the small proportion of salt from the relatively large amount of water instead of removing the water from the salt. Its practical use is restricted because of membrane deterioration, scale formation, and inefficient use of energy. Other proposed methods are freezing by direct contact of refrigerant with sea water, foam separation, liquid-liquid extraction, and various non-electric membrane processes, and ion exchange.

deserpidine (canescine; 11-demethoxyreserpine) $C_{32}H_{38}N_2O_8$.
Properties: Crystals; decomposes 230-234°C. An ester alkaloid.
Derivation: Isolated from Rauwolfia canescens Linn.
Grade: N.N.D.
Use: Medicine.

"Desiccite." [217] Brand name of adsorbent used for static dehumidification in protective packaging of metal equipment, food, and pharmaceuticals.

"Desicote." [274] Trademark for a mixture of hydrophobic monomers stabilized in chlorinated hydrocarbon and aromatic solvents.
Properties: Rapidly decomposes on contact with sorbed water of glass surfaces, leaving surface water-repellent.
Derivation: Mixed silanes.
Grades: Green label (approved for U.S. Parcel Post); yellow label (approved for Railway Express); red label (special precautions required for shipment).
Note: The Green label is the commercial stock in general use.
Containers: Two-ounce glass bottles with heavy molded cap and special cap liners packed in friction top steel can with absorbent filler to conform to safety requirements, approved by U.S. Post Office.
Uses: "Desicote" is a material which may be used safely on the most delicate glassware such as absorption cells of "Pyrex," "Vycor," and fused silica and which will not give any interference in the visible or ultraviolet range. May also be used for pH-sensitive glass electrodes and the like.
Hazards: Avoid contact with skin or mucous membrane; use adequate ventilation.
Shipping regulations: See grades above.*

desiodothyroxine. See thyronine.

deslanoside (desacetyl lanatoside) $C_{47}H_{74}O_{19}$.
Properties: Colorless or white crystals or

*See "I.C.C. Shipping Regulations," page xiii.
Reference numbers refer to name of manufacturer. See "List of Manufacturers," page v.

white, crystalline powder. Odorless; hygroscopic; m.p. 220-235°C. Insoluble in water; very slightly soluble in chloroform; slightly soluble in alcohol; soluble in methanol.
Grade: U.S.P. XVI.
Use: Medicine.
Caution: Very poisonous!

desmolase. An enzyme whose catalytic activity causes a split in or formation of a C-C bond without transfer of a radical or group within the substrate.

desorption. The process of removing an adsorbed material from the solid on which it is adsorbed. See adsorption. Desorption may be accomplished by heating or reduction of pressure or by the presence of another more strongly adsorbed substance or by a combination of these means.

desoxy-. Prefix identical in meaning to deoxy-, which is preferred.

desoxyanisoin $CH_3OC_6H_4COCH_2C_6H_4OCH_3$.
Properties: Off-white to buff, crystalline powder with a sweet, faint cinnamon-like odor; m.p. 110-112°C.

desoxycholic acid. See deoxycholic acid.

desoxycorticosterone. See deoxycorticosterone.

desoxycorticosterone acetate. See deoxycorticosterone acetate.

desoxycorticosterone trimethylacetate. See deoxycorticosterone trimethylacetate.

desoxycortone acetate. See deoxycorticosterone acetate.

desoxyephedrine hydrochloride. See methamphetamine hydrochloride.

"Desoxyn." [3] Trademark for methamphetamine hydrochloride (q.v.).

desoxypentose nucleic acid. See deoxyribonucleic acid.

desoxyribonucleic acid. See deoxyribonucleic acid.

D-desoxyribose. See D-deoxyribose.

destructive distillation. The decomposition of a material by heat and simultaneous distillation of the volatile products; e.g., the destructive distillation of coal to form coke, coal tar, and other liquid and gaseous products.

"Desyphed" Hydrochloride. [162] Trademark for methamphetamine hydrochloride.

"Detamide" 95. [428] Trade name for N,N-diethyl-meta-toluamide (q.v.).

"Detect-A-Leak." [414] Trademark for a leak-detecting fluid for pressure pipe lines.
Containers: 8 oz. cans; 1-, 5-, 55-gal. containers.

detergent. Formerly, a substance or mixture that has cleansing action due to a combination of properties including lowering of surface tension, wetting action, emulsifying and dispersing action and foam formation. Ordinary soap is the best known example. However, the word detergent is now coming to mean the synthetic variety, in distinction to soap, which is derived from natural fats and oils.
See also detergents, synthetic.

detergent alkylate. Generic term, particularly in the soap industry, applied to the reaction product of benzene or its homologues with a long-chain olefin (such as propylene trimer or tetramer) to produce an intermediate used in the manufacture of detergents. See, for example, dodecylbenzene.

"Detergent Alkylate No. 2." [136] Trade name for a mixture of alkyl benzenes produced by the alkylation of benzene with propylene polymer; sp.gr. 0.8708 (60/60°F). Used as raw material for producing alkyl aryl sulfonate type detergents or surface active materials.

detergents, synthetic (syndets). Materials which have a cleansing action like soap but are not derived directly from fats and oils. Synthetic detergents are surface active agents and have structurally unsymmetrical molecules containing both hydrophilic, or water-soluble, groups and hydrophobic, or oil-soluble hydrocarbon chains. There are three types, as follows:
(1) Anionic detergents form negatively charged ions containing the oil-soluble portion of the molecule. The ionizable group is the hydrophilic portion. Soap is an example of this class and the synthetic members are sodium salts of organic sulfonates or sulfates. Approximately 50% of all synthetic detergents are alkylaryl sulfonates (e.g., sulfonates of dodecylbenzene); about 20% of the total are sulfates of straight chain primary alcohols, either fatty alcohols or products of the Oxo process (e.g., sodium lauryl sulfate). Still another group are the lignin sulfonate derivatives.
(2) Cationic detergents, or invert soaps, ionize so that the oil-soluble portion is positively charged; the principal examples are quaternary ammonium halides. See, for example, benzethonium chloride; cetalkonium chloride. Outstanding germicidal activity overshadows the detergent applications of this type, which accounts for only about 1% of the total synthetic detergent production in the U.S.
(3) Nonionic detergents do not ionize but acquire hydrophilic character from an oxygenated sidechain, usually polyoxyethylene. The oil-soluble part of the molecule may come from fatty acids, alcohols, amides, or amines. By suitable choice of the starting materials and regulation of the length of the polyoxyethylene chain, the wetting, foaming, and detergent properties of nonionics may be varied greatly. Furthermore, compounds of this type, which comprises about 25% of the total synthetic detergent production, can be used in combination with either anionic or cationic detergents.

For examples, see "Ninol," and "Surfynol."

Synthetic detergents are now produced in greater volume than soaps, mainly because of their acceptance for household use. They are not only highly efficient cleansers, but are unaffected by hard water. Many of them can be used equally well in salt water or acid solutions. They are usually used in combination with inert diluents, such as sodium sulfate, or with builders, such as polyphosphates, polysilicates, or sodium carboxymethylcellulose.

Uses: Primarily, as household cleansers. Industrially, for textile scouring, bleaching, desizing, dyeing, printing, and finishing; metal cleaning and pickling; cleaning and sterilizing food processing equipment; in cosmetics; tanning and dyeing leather; ore flotation; fire fighting; emulsion polymerization of synthetic rubber; antibacterials (cationics).

determination. The ascertainment of the quantity or concentration of a specific substance in a sample.

"Dethdiet." [342] Trademark for red squill rodenticide concentrates.

"Dethmor." [342] Trademark for warfarin rodenticide concentrates.

deuterium D (heavy hydrogen). An isotope of hydrogen with atomic weight 2.016. M.p. 18.65°K; b.p. 23.6°K.

deuterium oxide. See heavy water.

deuteron (deuton). A nuclear particle having mass 2 and a positive charge of 1; identical with the nucleus of the deuterium atom.

deuton. See deuteron.

Devarda's alloy. See Devarda's metal.

Devarda's metal (Devarda's alloy).

Properties: Gray powder. Contains copper, aluminum, and zinc in the proportion of 50:45:5. Slightly soluble in hydrochloric acid.

Grades: Technical; reagent.

Use: Analysis (testing for nitrogen).

"Devcon." [445] Trade name for a plastic metal consisting of approximately 80% fine steel powders and 20% of an extremely strong plastic. Available in two types: "Devcon" A, a non-sagging putty-like material; and "Devcon" B, a semi-viscous liquid.

Containers: 1-, 4-lb and bulk containers.

Uses: For metal forming and repairing.

Hazards: Not to be taken internally or exposed to sensitive skin.

developed dyes. Those dyes that are produced by chemical reaction of the necessary materials after they have been absorbed by the fibers. Azoic dyes are the most important example.

developing agents. A term applied in the dyeing industry to certain organic compounds which, in combination with some other organic compound already deposited upon the fiber, will develop a colored compound, or if united with a dye already upon the fiber, will form a new coloring matter possessing a more desirable or a faster color. Also, a substance used in photography to produce a "latent" image by chemical reduction of a silver compound to metallic silver more rapidly in the portions exposed to light than in those not exposed.

devil's apple. See stramonium and podophyllum.

devil's dung. See asafetida.

"Devlex." [233] Trademark for oxazolidinone polymers.

dew point. Temperature at which air is saturated with moisture, or in general the temperature at which a gas is saturated with respect to a condensable component.

dexamethasone (9alpha-fluoro-16alpha-methylprednisolone) $C_{22}H_{29}FO_5$. A corticosteroid.

Properties: Crystals; m.p. 262-264°C. Soluble in water and organic solvents.

Use: Medicine and veterinary medicine.

dexbrompheniramine maleate

$C_{16}H_{19}BrN_2 \cdot C_4H_4O_4$. Dexbrompheniramine is the d-isomer of parabromdylamine, or 3-(para-bromophenyl)-3-(2-pyridyl)-N,N-dimethylpropylamine.

Properties: Crystals; soluble in water and alcohol; pH of 2% aqueous solution approx. 5.0.

Use: Medicine.

"Dexedrine." [71] Trademark for dextroamphetamine sulfate. See amphetamine sulfate.

"Dexet." [233] Trademark for aluminous cement, and related ingredients for use chiefly in oil wells.

"Dexon." [181] Trademark for para-dimethylaminobenzenediazo sodium sulfonate (q.v.).

dextran (macrose). Certain polymers of glucose which have chain-like structures and very high molecular weights (up to 200,000 or higher). It is produced from sucrose by Leuconostoc bacteria; occurs as slimes in sugar refineries, on fermenting vegetables or on dairy products. Clinical dextran is standardized to a low molecular weight (75,000); is made by partial hydrolysis and fractional precipitation of the high molecular weight particles.

Uses: As blood plasma substitute (clinical dextran); confections; lacquers; oil well drilling muds.

dextrin (British gum; amylin; gommeline; starch gum; artificial gum; vegetable gum; leiocom; sago dextrin; tapioca dextrin). An intermediate product formed by the hydrolysis of starches. Industrially it is made by treatment of various starches with dilute acids or by heating dry starch. The yellow or white powder or granules are soluble in water; insoluble in alcohol and ether. It is colloidal in properties and describes a class of substances, hence has no definite formula.

Uses: Adhesives; thickening agent; sizing paper and textiles; substitute for natural gums; food (polishing rice and cereal grains, polishing coffee, baking, etc.); decorating ceramics; matches; glass-silvering compositions, printing inks; felt manufacture; process engraving.

"Dextrinase." [212] Trademark for a fungal amylase which converts starches and dextrins to maltose and dextrose.

Properties: Dry, fine white powder, water-soluble, non-hazardous, non-flammable. Optimum pH 5.0, acetate buffer. Optimum temperature 40°C.

Grade: For food products and industrial applications.

Containers: 1-, 5-, 10-, and 25-lb drums.

Uses: For the production of syrups and other products high in reducing sugars.

dextrin, cassava.

Derivation: A dextrin prepared from cassava starch.

Grades: Technical.

Containers: Bags; barrels; fiber drums.

Uses: Adhesives; textile printing; confectionery; etc.

Shipping regulations: None.*

dextrin, crystallized. A brittle mass produced by decolorizing dextrin with animal charcoal and evaporating the product.

dextro-amphetamine phosphate. N.F. XI name for dextrorotatory amphetamine phosphate, monobasic.

dextro-amphetamine sulfate. U.S.P. XVI name for dextrorotatory amphetamine sulfate.

dextro-chlorpheniramine maleate. A dextrorotatory variety of chlorpheniramine maleate.

dextroglucose. See dextrose.

dextromethorphan hydrobromide

$C_{18}H_{25}NO \cdot HBr \cdot H_2O$. d-3-Methoxy-N-methylmorphinan hydrobromide.

Properties: Practically white crystals or crystalline powder possessing a faint odor; slightly soluble in water; freely soluble in alcohol and chloroform; insoluble in ether. Specific rotation 200 mg/10 ml solution +26 to +28°; pH (1 in 50 solution) 5.2-6.5.

Grade: N.F. XI.

Use: Medicine.

dextrorotatory. Having the property when in solution of rotating the plane of polarized light to the right or clockwise. Dextrorotatory compounds are given the prefix d- or (+) to distinguish them from their levorotatory, l- or (-) isomers.

See optical isomerism.

dextrose (dextroglucose; D(+)-glucose; baker's sugar; grape sugar; corn sugar)

$C_6H_{12}O_6 \cdot H_2O$. The sugar found in the blood of animals and occurring widely in plants.

Properties: Colorless crystals or white, crystalline or granular powder. Odorless with a sweet taste; sp.gr. 1.544; m.p. 146°C. Soluble in water; slightly soluble in alcohol.

Derivation: Hydrolysis of starch and starchy substances by action of hydrochloric acid; pure dextrose is obtained by complete conversion, under proper conditions; incomplete conversion gives glucose in a mixture of dextrose, dextrin and water. See also glucose.

Grades: Technical; U.S.P. XVI; C.P.; anhydrous; hydrated.

Containers: 1-, 5-lb bottles; 1-, 5-lb cans; 50- to 100-lb slabs; 100-lb bags; 400-lb barrels; fiber drums.

Uses: Confectionery; jams; various foods (especially infant foods); preparing tobacco; chrome-tanning liquors; logwood and dyewood liquors; brewing and wine manufacture; sizing, weighting and ingredient of printing pastes for textiles; solubilizer for natural gums; viscose and cuprammonium rayon precipitating baths; medicine; intermediate.

Shipping regulations: None.*

DFDT (difluorodiphenyltrichloroethane) $(FC_6H_4)_2CHCCl_3$. Fluorine analog of DDT, first developed in Germany.

Properties: A low-melting white solid; m.p. 45.5°C. A faint odor resembling ripe apples; less poisonous to warm-blooded animals and fish than DDT (according to U.S. Public Health Service), but unlike DDT it does not have a long residual value. It does not have broad killing power of DDT toward all insects but is more effective against flying insects, especially house flies.

Derivation: By condensing chloral and fluorobenzene in the presence of sulfuric acid or chlorosulfonic acid.

Uses: Poison; pesticide.

DFP. Abbreviation for diisopropyl fluophosphate.

"DHA." [233] Trademark for fungicides comprised of dehydroacetic acid and its salts.

DHS. Abbreviation for dihydrostreptomycin.

Di. Symbol for didymium.

di-. Prefix meaning two or twice. A compound not found under di- should be looked for under bi- or bis-, since bi-, bis-, and di- are nearly equivalent prefixes, assigned with slight differences in meaning for particular compounds, or according to customary usage.

diabase. A basic igneous rock, usually occurring in dikes or intrusive sheets, and composed essentially of plagioclase feldspar and augite with small quantities of magnetite and apatite. The plagioclase forms lath-shaped crystals lying in all directions among the dark irregular augite grains, giving rise to the peculiar diabasic or ophitic texture, which is a distinctive feature in the coarse-grained occurrences.

Occurrence: Canada, Connecticut, Maryland,

Massachusetts, Newfoundland, New Jersey, New York, Virginia, and Pennsylvania.
Uses: Building stone; cement aggregate; paving; railroad ballast.

"Diabinese." [299] Trademark for chlorpropamide.

diacetic acid. See acetoacetic acid.

diacetic ester. See ethyl acetoacetate.

diacetin (glyceryl diacetate) $CH_2O(OCCH_3)CHOHCH_2O(OCCH_3)$.
Properties: Hygroscopic liquid. It is a mixture of isomers. Sp. gr. 1.18; b.p. 259°C (approx.); refractive index 1.44. Miscible with water, benzene, and alcohol; the commercial mixture gels about −30°C.
Derivation: Heating one mole of glycerin with two moles of glacial acetic acid.
Grades: Technical.
Containers: Glass bottles; 5- and 10-gal cans.
Uses: Plasticizer and softening agent; solvent for cellulose derivatives, "Glyptal" resins, shellac.

diacetoacetyl ethylenediamine $(CH_2NHCOCH_2COCH_3)_2$.
Properties: White, crystalline solid. Slightly soluble in water. M.p. 164°C.

diacetone. See diacetone alcohol.

diacetone alcohol (diacetone; 4-hydroxy-4-methylpentanone-2; 4-hydroxy-2-keto-4-methylpentane) $CH_3COCH_2C(CH_3)_2OH$.
Properties: Colorless, pleasant-odored liquid. Flammable; sp. gr. 0.9406 at 20/20°C; b.p. (760 mm) 169.1°C; vapor pressure 1.1 mm (20°C); flash point 170°F; wt/gal 7.8 lbs (20°C); approximate change in wt. 0.0032 lb/gal/°F; viscosity 0.032 poise (20°C); f.p. −42.8°C; refractive index 1.42416 (20°C); coefficient of expansion 0.00097 (20°C).
Typical specifications: Acetone-free grade: Acidity not more than 0.05% (as acetic); sp. gr. 0.937-0.942 (20/20°C); boiling range (760 mm) below 135°C none, below 158°C not more than 5%, above 170°C none; dryness, miscible with 19 vol 60° Bé gasoline (20°C); color (500 mm tube) not more than 3 yellow Lovibond; average wt/gal 7.81 lbs (20°C). Technical grade: Acidity not more than 0.02% (as acetic); sp. gr. 0.915-0.920 (20/20°C); boiling range (760 mm) below 60°C none, below 160°C not more than 30%, above 170°C none; dryness, miscible with 19 vols 60° Bé gasoline (20°C); color not more than 3 yellow Lovibond (500 mm tube); average wt/gal 7.63 lbs (20°C).
Miscible with alcohols, aromatic and halogenated hydrocarbons, esters, and water. A constant boiling mixture with water has b.p. 99.6°C and contains approx. 13% diacetone alcohol.
Derivation: By the condensation of acetone.
Grades: Technical; acetone-free.
Containers: 1-, 5-gal cans; 55-gal (nonreturnable) drums; tank cars.
Uses: Solvent for nitrocellulose, cellulose acetate, various oils, resins, waxes, fats, dyes, tars; lacquers, dopes, coating compositions; wood preservatives; stains; rayon and artificial leather; imitation gold leaf; dyeing mixtures; celluloid cements; antifreeze mixtures; extraction of resins and waxes; preservative for animal tissue; metal-cleaning compounds; hydraulic compression fluids; stripping agent (textiles). The technical grade, containing acetone, has greater solvent power.
Shipping regulations: None.*

diacetonyl sulfide $(CH_3COCH_2)_2S$.
Properties: Crystals; b.p. 136-137°C (15 mm); m.p. 47°C.
Derivation: Interaction of chloroacetone and hydrogen sulfide gas.

diacetyl (biacetyl; butanedione; diketobutane; dimethyldiketone; dimethylglyoxal) $CH_3COCOCH_3$.
Properties: Yellow liquid; strong odor; extremely dilute solutions possess a distinct butter odor. Soluble in water, alcohol, and ether. Sp. gr. 0.990 (15/15°C); m.p. −3 to −4°C; b.p. 88-91°C; refractive index (n 18/D) 1.3933.
Grades: Technical; flavor grade.
Containers: Glass bottles; 5-lb aluminum containers.
Use: Aroma carrier of butter, vinegar, coffee, honey, etc.

diacetylaminoazotoluene (4-ortho-tolylazo-ortho-diacetotoluide) $[CH_3C_6H_4N:NC_6H_3(CH_3)N(CH_3CO)_2]$.
Properties: Crystalline powder. Color varies from yellowish-red, through rose to red. Acted upon by humidity in the atmosphere. Keep well stoppered! Soluble in alcohol, chloroform, and ether; as well as fats, oils, and greases; insoluble in water.
Constants: M.p. 74-76°C.
Use: Medicine (external).

3,5-diacetylamino-2,4,6-triiodobenzoic acid, methylglucamine salt. See methylglucamine diatrizoate.

diacetyldihydroxyphenylisatin (acetphenolisatin; endophenolphthalein) $C_{24}H_{19}O_5N$.
Properties: White, odorless, tasteless, crystalline powder; m.p. 241-242°C. Soluble in alcohol, ether, and benzene.
Use: Medicine.

diacetylenes. Unsaturated hydrocarbons containing two triple bonds, with the type formula C_nH_{2n-6}. The simplest member of the group is butadiyne or biacetylene, HC:CC:CH, a gas which boils at 10°C.

1,2-diacetylethane. See acetonyl acetone.

diacetylmethane. See acetylacetone.

diacetylmorphine (diamorphine; heroin) $C_{17}H_{17}NO(C_2H_3O_2)_2$.
Properties: White, odorless, bitter crystals or crystalline powder; poisonous; habit-forming drug! Soluble in alcohol.

M.p. 173°C.
Derivation: By the acetylization of morphine.
Method of purification: Crystallization.
Grades: Technical.
Containers: Tins; glass bottles.
Use: Prohibited in the United States because of its habit-forming character. Formerly used in medicine.
Shipping regulations: None.*

diacetylmorphine hydrochloride $C_{21}H_{23}O_5N \cdot HCl \cdot H_2O$.
Properties: White crystalline powder; m.p. about 230°C; soluble in water and alcohol; insoluble in ether.

diacetyl peroxide. See acetyl peroxide.

diacetyltannin. See acetyltannic acid.

"Diadem Chrome." [232] Brand name for a series of chrome dyestuffs suitable for application by the afterchrome method.

"Diademil." [412] Trademark for hydroflumethiazide (q.v.).

"Diagen." [28] Trademark for line of stabilized azoic colors that are used for printing cotton and rayon. They are tinctorially strong and have good fastness in full shades.

"Diagnex Blue." [412] Trademark for azuresin (q.v.).

"Dial." [305] Trademark for 5,5-diallybarbituric acid.
Use: Medicine.

dialdehyde resin tannage. A process for chemical tanning. Hide is treated with a dialdehyde followed by controlled polymerization with urea-formaldehyde, phenol-formaldehyde or similar resin-forming substances. Permanent chemical compounds involving the leather substances are formed and properties may be varied over a wide range.

dialdehyde starch. Starch in which the original anhydroglucose units have been partially oxidized to dialdehyde form by oxidation, for example, the product of the oxidation of cornstarch by periodic acid. See "Sumstar."
Uses: Thickening agent; tanning agent; binder for leaf tobacco and paper; adhesives.

dialkylchloroalkylamine hydrochlorides. Amine salts having the formula RCl·HCl, when R represents such groups as $(CH_3)_2NCH_2CH_2$- (beta-dimethylaminoethyl chloride hydrochloride); $(CH_3)_2NCH_2CH(CH_3)$- (beta-dimethylaminoisopropyl chloride hydrochloride), etc. Used in organic synthesis.

dialkylthioureas. Used as corrosion inhibitors. See dibutylthiourea; 1,3-diethylthiourea; diisopropylthiourea.

diallyl adipate $(CH_2CH_2COO)_2(CH_2CHCH_2)_2$.
Properties: Liquid; color-maximum #100 Pt-Co, characteristic odor; sp.gr. (20°C) 1.025.
Containers: 1-, 5-gal cans; 55-gal drums.
Use: Monomer.

diallylamine (di-2-propenylamine) $(CH_2{:}CHCH_2)_2NH$.
Properties: Liquid; sp.gr. (20°C) 0.7889; b.p. 112°C; f.p. −100°C; refractive index (n 20/D) 1.4404. Soluble in water.
Derivation: From allylamine or diallylcyanide.
Containers: Drums; tank cars.
Use: Intermediate.

diallylbarbituric acid (5,5-diallylbarbituric acid) $C_{10}H_{12}N_2O_3$.
Properties: White, odorless, crystals or crystalline powder; slightly bitter taste; soluble in alcohol or ether; slightly soluble in water; m.p. 171-173°C.
Use: Medicine.

diallyl cyanamide $(CH_2{:}CHCH_2)_2NCN$.
Properties: M.p. less than −70°C; b.p. 222°C; density 0.90.
Use: Organic intermediate; polymers.

diallyl diglycollate $(CH_2COO)_2O(CH_2CHCH_2)_2$.
Properties: Liquid; color-maximum #100 Pt-Co; characteristic odor; sp.gr. (20°C) 1.1113.
Containers: 1-, 5-gal cans; 55-gal drums.
Use: Monomer.

diallyl isophthalate $C_6H_4(COOCH_2CHCH_2)_2$.
Properties: Monomer is liquid; color-maximum #175 Pt-Co; mild characteristic odor; sp.gr. (20°C) 1.124. Prepolymer is solid; sp.gr. (25°C) 1.256.
Containers: Monomer in 1-, 5-gal cans, 55-gal drums; prepolymer in 37 ½-lb bags, 5-gal cans, 55-gal drums, tank cars.
Uses: Molding and laminating.

diallyl maleate $C_4H_2O_4(CH_2CHCH_2)_2$.
Properties: Colorless or straw-colored liquid. B.p. 109-110°C (3 mm); sp.gr. 1.077 (20°C); refractive index (n 20/D) 1.4699. Polymerizes readily when exposed to light or temperatures above about 50°C.
Grades: Technical.
Uses: Polymers and copolymers; insecticide formulations.

N,N-diallylmelamine $(C_3H_5)_2NCNC(NH_2)NC(NH_2)N$. A monomer, used for its polymers.

diallyl phthalate (DAP) $C_6H_4(COOCH_2CH{:}CH_2)_2$. The name is also used for the polymer.
Properties: Nearly colorless oily liquid; insoluble or limited solubility in gasoline, mineral oil, glycerin, glycols and certain amines. Soluble in most other organic liquids.
Typical specifications: Sp.gr. 1.120 (20/20°C); f.p. −70°C (viscous liquid); boiling range 158-165°C (4 mm); acidity (max) 0.03% (as acetic acid); odor mild, lachrymatory; flash point 330°F; fire point 359°F; vapor pressure 1.5 mm (150°C); refractive index 1.520 (25°C); viscosity 13 cps (20°C); surface tension 38 dynes per cm (20°C); thermal expansion 0.00076 from 10-40°C; wt/gal 9 lbs.
Containers: 1-, 5-gal cans; 55-gal drums; tank cars.
Uses: Primary plasticizer for most resins

which will polymerize if not inhibited; a polymerizable monomer which will polymerize with heat and catalyst into a clear, hard, insoluble polymer. It can be used to form low-pressure laminates with various fillers such as glass cloth, paper, etc. Used in electrical insulation.

diallyl sulfide. See allyl sulfide.

dialysis. Process in which smaller molecules are separated from larger ones in the same solution or mixture by use of a parchment or other semipermeable membrane that permits passage of the smaller but not the larger molecules. A common application is the removal of salts from solutions containing sugars or proteins. The process may also be used to separate large molecules in solutions from much larger colloidal particles that contain many molecules.

"Diam." [259] Trademark for N-alkyl 1,3-propylene diamines.
Derivation: The alkyl group is derived from coconut and tallow fatty acids.
Containers: 1-, 5-gal pails; 55-gal steel drums; tank trucks; tank cars.
Uses: Corrosion inhibitors; pigment treating; petroleum product additives; chemical intermediates.

"Diamid." [244] Trademark for a detergent formulation available in three grades: Regular, Concentrate, and Low-Foam.
Uses: Milk stone removal.

diamide hydrate. See hydrazine hydrate.

diamidogen sulfate. See hydrazine sulfate.

diamine. See hydrazine.

diamine sulfate. See hydrazine sulfate.

3,6-diaminoacridine. See acriflavine.

3,6-diaminoacridinium hydrogen sulfate. See proflavine sulfate.

meta-**diaminoazobenzene hydrochloride.** See chrysoidine.

diaminoazoxytoluene (azoxytoluidine) $C_6H_3(CH_3)(NH_2)N_2OC_6H_3(NH_2)(CH_3)$.
Properties: Yellow or orange crystals. M.p. 168°C; soluble in alcohol; insoluble in water.
Derivation: By alkaline reduction of para-nitro-ortho-toluidine.
Method of purification: Crystallization.
Grades: Technical.
Containers: Wooden kegs; fiber drums.
Use: Dye intermediate.
Shipping regulations: None.*

diaminobenzene. See phenylenediamine.

1,3-diaminobutane $NH_2CH_2CH_2CHNH_2CH_3$.
Properties: Water-white; amine odor; boiling range 143-150°C; sp. gr. 0.858 (20/20°C); refractive index 1.450 (20°C); flash point 125°F.

alpha, epsilon-**diaminocaproic acid.** See lysine.

diaminocarbanilide. See diaminodiphenylurea.

diaminodiethyl sulfide $S(CH_2CH_2NH_2)_2$.
Properties: Mobile, colorless liquid with amine-like odor. Miscible with water and benzene; insoluble in aliphatic hydrocarbons. B.p. 230-240°C; sp. gr. 1.054 (25°C).

diaminodihydroxyarsenobenzene dihydrochloride. See arsphenamine.

diaminodihydroxyarsenobenzene, sodium salt. See sodium arsphenamine.

di-para-**aminodimethoxydiphenyl.** See dianisidine.

diaminodiphenic acid (benzidinedicarboxylic acid) $C_6H_3(CO_2H)NH_2C_6H_3(CO_2H)NH_2$.
Properties: White crystals; soluble in alcohol and ether; insoluble in water.
Derivation: By boiling meta-nitrobenzaldehyde with caustic soda, reducing with zinc dust and acidifying.
Use: Dyestuff.

diaminodiphenyl. See benzidine.

diaminodiphenylamine (para-para'-diaminodiphenylamine) $HN(C_6H_4NH_2)_2$.
Properties: Yellowish crystals; soluble in alcohol and ether; insoluble in water. M.p. 158°C.

diaminodiphenylethylene. See para-diaminostilbene.

diaminodiphenylmethane $NH_2C_6H_4CH_2C_6H_4NH_2$.
Properties: Large silvery crystals; soluble in water, alcohol, ether, and benzene. M.p. 86°C.
Derivation: By heating formaldehyde anilide with aniline hydrochloride and aniline.
Uses: Dyes; source for diisocyanates.

diaminodiphenyl sulfate. See benzidine sulfate.

4,4'-diaminodiphenyl sulfone. See sulfonyldianiline.

diaminodiphenylthiourea (diaminothiocarbanilide) $(NH_2C_6H_4NH)_2CS$.
Properties: Colorless plates or crystalline solid; soluble in alcohol and ether; sparingly soluble in water. M.p. 195°C.
Derivation: By boiling para-phenylenediamine with carbon disulfide.

diaminodiphenylurea (diaminocarbanilide) $(NH_2C_6H_4NH)_2CO$.
Properties: Colorless plates; soluble in alcohol and hot water; sparingly soluble in cold water.

diaminodiphenylureadisulfonic acid $CO(NHC_6H_3NH_2SO_3H)_2$.
Properties: Colorless, needle-like crystals; slightly soluble in water.
Derivation: Action of phosgene upon either para-phenylenediaminesulfonic acid, or 4-nitraniline-3-sulfonic acid.

3,3'-diaminodipropylamine. See 3,3'-iminobispropylamine.

diaminoditolyl. See tolidine.

*See "I.C.C. Shipping Regulations," page xiii.
Reference numbers refer to name of manufacturer. See "List of Manufacturers," page v.

para, para'-**diaminoditolylmethane.** $NH_2C_7H_6CH_2C_7H_6NH_2$.
Properties: Glistening, crystalline plates; soluble in alcohol and ether; m.p. 149°C.
Derivation: By heating formaldehyde and ortho-toluidine.

1,2-diaminoethane. See ethylenediamine.

6,9-diamino-2-ethoxyacridine lactate monohydrate. See ethodin.

diaminoethoxyazobenzene hydrochloride. See ethoxazene.

di-para-**aminoethoxydiphenyl.** See ethoxybenzidine.

1,6-diaminohexane. See hexamethylenediamine.

3,6-diamino-10-methylacridinium chloride. See acriflavine.

3,6-diamino-10-methylacridinium chloride hydrochloride. See acriflavine hydrochloride.

diaminonaphthalene. See naphthylenediamine.

1,5-diaminopentane. See cadaverine.

1,2,4-diaminophenol $C_6H_3OH(NH_2)_2$.
Properties: Colorless crystals; m.p. 78-80°C with decomposition. Soluble in alcohol and ether.
Derivation: By the reduction of 1,2,4-dinitrophenol.
Method of purification: Crystallization.
Grades: Technical.
Containers: Wooden barrels; glass bottles; fiber drums.
Uses: Photographic developer; organic synthesis.
Shipping regulations: None.*

1,2,5-diaminophenol $C_6H_3OH(NH_2)_2$.
Properties: Colorless crystals; m.p. 68°C; soluble in water.
Derivation: By the reduction of 1,2,5-dinitrophenol.
Method of purification: Crystallization.
Grades: Technical.
Containers: Wooden barrels; glass bottles; fiber drums.
Use: Organic synthesis.
Shipping regulations: None.*

diaminophenol hydrochloride $C_6H_3(NH_2)_2OH \cdot 2HCl$.
Properties: Grayish-white crystals; soluble in water; slightly soluble in alcohol.
Derivation: By the interaction of dinitrophenol with iron and hydrochloric acid.
Method of purification: Crystallization.
Grades: Technical.
Containers: Kegs; boxes; glass bottles.
Uses: Photographic developer; dyeing furs and hair; analytical reagent.
Shipping regulations: None.*

2,4-diamino-6-phenyl-s-triazine. See benzoguanamine.

1,3-diaminopropane $NH_2CH_2CH_2CH_2NH_2$.
Properties: Water-white liquid; amine odor; boiling range 133-140°C; sp.gr. 0.886 (20/20°C); refractive index 1.459 (20°C); flash point 75°F.
Shipping regulations: Flammable liquid. Red label.*

2,6-diaminopyridine.
Properties: Crystals; f.p. 120.8°C; b.p. 285°C (760 mm); soluble in water.
Derivation: From 2-aminopyridine.

para-**diaminostilbene** (diaminodiphenylethylene) $C_6H_4(NH_2)CHCHC_6H_4(NH_2)$.
Properties: Colorless needles or plates; soluble in alcohol and ether; insoluble in water. M.p. 227°C.
Derivation: By the reduction of dinitrostilbene.

4,4'-diamino-2,2'-stilbenedisulfonic acid (DAS) $C_6H_3(NH_2)(SO_3H)CHCHC_6H_3(SO_3H)(NH_2)$.
Properties: Yellowish microscopic needles; soluble in alcohol and ether; insoluble in water.
Derivation: Boiling sodium salt of para-nitrotoluene-ortho-sulfonate in water with caustic soda and reduction with zinc dust.
Containers: Fiber kegs; polyethylene-lined steel drums.
Use: Dyestuffs.

diaminothiocarbanilide. See diaminodiphenylthiourea.

di-alpha-**amino-**beta-**thiolpropionic acid.** See cystine.

diaminotoluene. See toluene-2,4-diamine.

4,6-diamino-meta-**toluenesulfonic acid.** See meta-tolylenediamine sulfonic acid.

diammonium ethylenebisdithiocarbamate $NH_4S_2CNH(CH_2)_2NHCS_2NH_4$.
Properties: M.p. 72.5°C; very soluble in water.
Grades: 42% solution in water.
Containers: 5-, 30-gal drums.
Uses: Fungicide; intermediate; corrosion inhibitor.

diammonium hydrogen phosphate. See ammonium phosphate, dibasic.

diammonium phosphate. See ammonium phosphate, dibasic.

diamond. A mineral consisting essentially of carbon crystallized in the isometric system, usually in octahedral shape.
Properties: Colorless or white, sometimes blue, red, orange, green, black, yellow, brown; luster adamantine when cut, greasy when uncut; high refractive index and dispersion. Sp.gr. 3.51-3.53; hardness 10 (hardest substance known).
Varieties:
(a) Ordinary. Crystals ranging from stones of the first water (colorless and flawless) through various color tints, usually yellow, and full of flaws.
(b) Bort (boort). Stones too badly flawed or too off-color to be used in jewelry. Also applied to aggregates with a radial finely crystalline structure.
(c) Carbonado. A finely crystalline type, black in color, tough and compact.
(d) Synthetic. Have been made at high

temperatures and pressures (3000°F, 1.3 million psi).

Occurrence: South Africa (world's principal supply); Belgian Congo; Brazil; Arkansas; Venezuela; India; Borneo; Southwest Africa.

Uses: Gemstone; in glass cutters; diamond drill bits; metal cutting; wire dies; as a powder (diamond dust) for polishing gemstones and in abrasive wheels.

"Diamond Alkali Cell D-3." [244] Trade name for a high output electrolytic cell for the production of chlorine and caustic soda, having an operating range up to 34,000 amperes. At this load it is capable of producing 2270 pounds of chlorine per day, an output in excess of 21 pounds of chlorine gas per square foot of cellroom floor space. Operational features include high electrical efficiency and low hydrogen content in the chlorine cell gas. The cell is rectangular in shape (7'0" by 3'7") and consists of three major components. The bottom section is a cast iron anode base, from which vertical carbon anodes extend to distribute electrical current throughout the sodium chloride cell feed. Placed on the base is a cathode assembly, which has steel screens extending between the rows of anodes to receive the current. An asbestos fiber diaphragm is deposited on these screens to separate the products of electrolysis. The cell assembly is completed by placing a concrete head on top of the cathode.

"Diamond CR-80, 85." [244] Trade name for two slightly different vinyl chloride-vinyl acetate copolymers.

Properties: Fine, white powders, 100% through 40 mesh; specific viscosity 0.56 and 0.58; bulk density 38 and 35.5 lb/cu ft; volatiles less than 1 and 1.5%, respectively.

Containers: 50-lb paper bags.

Uses: In phonograph record stocks and in rigid calendered sheeting.

diamond dust. See diamond.

"Diamond FCR." [244] Trade name for a vinyl chloride-vinyl acetate copolymer. Used for vinyl asbestos flooring applications.

"Diamond PVC." [244] Trade name for a series of polyvinyl chloride resins. All are fine, white powders; 100% through 40 mesh; packed in 40- or 50-lb paper bags.

PVC-30: Specific viscosity 0.63; bulk density 0.57 g/cc; low molecular weight.

Uses: When blended with conventional resins, improves processability, reduces temperature requirements, and shortens production cycles.

PVC-35: Specific viscosity 0.85; bulk density 35 lb/cu ft; moisture 0.5% max. Low molecular weight; good flow properties.

Uses: Compounds based on this resin can be processed at the relatively low temperatures normally used for copolymers.

PVC-40: Average specific viscosity 1.10; bulk density 35 lb/cu ft; moisture 0.5% max. Intermediate molecular weight.

Uses: For calendering and molding; suited as a resin base for highly filled compounds.

PVC-62: Average specific viscosity 1.40; bulk density 20.6 lb/cu ft; moisture 0.5% max. High plasticizer absorption and rapid absorption rate.

Uses: Coated fabrics; calendered film, sheeting and tape stock; tubing and profile extrusion; elastomeric compounds for molded products. A cold dry blending resin.

PVC-70: Specific viscosity 1.80; bulk density 17 lb/cu ft; high molecular weight.

Uses: Designed for plastisols and organosols. Excellent flow or rheological properties exhibited by both fresh and aged compounds.

PVC-70F: Specific viscosity 1.80; bulk density 17 lb cu/ft; high molecular weight.

Uses: For plastisol and organosol dispersions.

PVC-450: Specific viscosity 1.35; bulk density 31 lb cu/ft; moisture 0.5% max.

Uses: Coated fabrics; heavy or clear calendered sheeting; calendered film-processing aid for high polymers; unplasticized and semi-rigid sheeting; dry blend extrusion compounds; elastomeric molding compounds.

PVC-500: Specific viscosity 1.55; bulk density 30 lb/cu ft; moisture 0.5% max; plasticizer adsorption, 32.

Uses: Dry blending, extrusion, calendering and molding.

"Diamond" Soda Crystals. [244] Brand name for sodium sesquicarbonate, $Na_2CO_3 \cdot NaHCO_3 \cdot 2H_2O$.

diamonds, synthetic. See diamond.

"Diamond Vulcanized Fibre." [281] Trade name for a line of products made from pure cotton rag paper chemically treated to form a permanently solid mass which is tough, hard, resilient, strong, of bonelike consistency, with excellent machining and forming qualities. Used for gears, machined or punched parts, electrical insulation, and many special uses.

diamorphine. See diacetylmorphine.

"Diamox." [57] Trade name for acetazolamide (2-acetylamino-1,3,4-thiadiazole-5-sulfonamide) $N_4S_2O_3C_4H_6$.

Properties: White crystalline powder; m.p. 252.3-252.6°C; very slightly soluble in water; slightly soluble in ethanol, acetone; soluble in dilute base.

Use: Medicine.

diamthazole dihydrochloride $C_{15}H_{23}N_3OS \cdot 2HCl$. 6-(2-Diethylaminoethoxy)-2-dimethylaminobenzothiazole dihydrochloride.

Properties: Crystals; decomposes 269°C. Soluble in water, ethanol and methanol.

Grade: N.N.D.

Use: Topical therapy (medicine); antifungal agent.

di-n-amylamine $(C_5H_{11})_2NH$.

Properties: Colorless liquid; b.p. 202-3°C (745 mm); very slightly soluble in water,

soluble in alcohol and ether. Sp. gr. 0.77-0.78 (20°C); diamylamine content at least 99%; acid-insoluble none; initial b.p. not below 175°C, not less than 95% boils below 202°C, final b.p. not above 218°C; wt/gal 6.45 lbs. Viscosity 0.01264 poise (20°C); refractive index 1.430 (20°C); surface tension 24.4 dynes/cm (13°C); specific heat 0.54 cal/g (20°C); heat of vaporization 83 cal/g; coefficient of expansion 0.00102 (20-60°C); vapor pressure 9 mm (26°C).
Derivation: From reaction of amyl chloride and ammonia.
Containers: 1-, 5-gal cans; 55-gal drums; tank cars.
Uses: Rubber vulcanization accelerators; flotation reagents; dyestuffs and corrosion inhibitors; solvent for oils, resins and some of the cellulose esters.
Shipping regulations: None.*

N, N-diamyl aniline (mixed isomers) $C_6H_5N(C_5H_{11})_2$.
Properties: Sp. gr. 0.898 (20°C); boiling range 276-292°C; color dark amber; odor faint aniline; flash point 260°F.

diamyl benzene $C_6H_4(C_5H_{11})_2$.
Properties: Sp. gr. 0.86 (20°C); b.p. 265°C; color water-white; odor aromatic; flash point 225°F.

diamyl chloronaphthalene $(C_5H_{11})_2C_{10}H_5Cl$.
Properties: Sp. gr. 1.06 (20°C); boiling range 352-377°C; color orange; odor naphthalenic; flash point 330°F.

diamylene $C_{10}H_{20}$.
Derivation: By polymerization of amylenes.

diamyl ether. See amyl ether.

2, 5-di(tert-amyl) hydroquinone [2, 5-di(tert-pentyl)hydroquinone] $(C_5H_{11})_2C_6H_2(OH)_2$.
Properties: Buff powder; m.p. 172°C (min.); sp. gr. 1.05 (25°C).
Uses: Protection of uncured rubber from oxidation; antioxidant for unsaturated resins and oils; in food packaging.

diamyl maleate $(CHCOOC_5H_{11})_2$.
Properties: Sp. gr. 0.981 (20°C); boiling range 263-300°C; color water-white; odor faintly alcoholic; flash point 270°F.

diamyl beta-naphthol $(C_5H_{11})_2C_{10}H_5OH$.
Properties: Sp. gr. 0.970 (20°C); boiling range 205-230°C (10 mm); color dark red; odor none; flash point 345°F.

diamyl nitrosamine $(C_5H_{11})_2NNO$.
Properties: Sp. gr. 0.891 (20°C); boiling range 120-130°C (10 mm); color straw; odor amine.

diamyl phenol (1-hydroxy-2, 4-diamylbenzene) $(C_5H_{11})_2C_6H_3OH$. Commercial form is a mixture of isomers including both secondary amyl and tertiary amyl groups mainly in 2, 4 positions.
Properties: Light straw-colored liquid with mild phenolic odor; miscible with both aliphatic and aromatic hydrocarbons; insoluble in water and 10% aqueous alkalies.
Typical specifications: Boiling range (ASTM 5-95%) 280-295°C; sp. gr. 0.930 (20°C); wt/gal 7.8 lbs (20°C); flash point (Tag open cup) 260°F.
Containers: 5-gal cans; 55-gal (non-returnable) black iron drums; 4000-, 6000-gal tank cars; tank trucks.
Handle with caution!
Uses: Synthetic resins; lubricating oil additives; rust preventives; plasticizers; surface active agents; synthetic detergents; antioxidants and antiskinning agents; rubber chemicals; and biologically active materials as effective germicides, fungicides, microbiocides, and insecticides.

di-tert-amylphenoxyethanol $(C_5H_{11})_2C_6H_3OCH_2CH_2OH$.
Properties: Clear to light-straw colored liquid; sp. gr. 0.960 (20°C); refractive index 1.5074; vapor pressure < 0.01 mm; boiling range 318-332°C; m.p. about −35°C. Insoluble in water.
Fire hazard: Flash point 149°C.

diamyl phthalate $C_6H_4(COOC_5H_{11})_2$.
Properties: Colorless, nearly odorless oily liquid. Sp. gr. (20°C) 1.022; wt/gal 8.52 lb (20°C); refractive index (25°C) 1.488; b.p. 342°C; m.p., less than −55°C; flash point (closed cup) 357°F.
Derivation: By esterification of phthalic anhydride with amyl alcohol in the presence of approximately 1% concentrated sulfuric acid as catalyst.
Method of purification: Distillation.
Grades: Technical.
Containers: Drums.
Use: Plasticizer.
Shipping regulations: None.*

diamyl sulfide (amyl sulfide) $(C_5H_{11})_2S$. A mixture of isomers.
Properties: Yellow liquid; sp. gr. 0.85-0.91 (20/20°C); distillation; initial b.p. (min) 170°C, not less than 95% above 180°C; average wt/gal 7.52; flash point (open cup) 185°F; refractive index 1.477 (19°C); surface tension 27.3 dynes/cm (13°C).
Containers: 1-gal can (7 lbs); 5-gal drum (35 lbs); 55-gal drum (380 lbs).
Uses: Used in preparing organic sulfur compounds, such as sulfones, sulfoxides, and other compounds through addition reactions; as a flotation agent in metallurgical processes; as a stench-producing agent (not as powerful as amyl mercaptan).

"Dianabol." [305] Trademark for methandrostenolone, N. N. D.
Use: Medicine.

dianhydrosorbitol. See sorbide.

"Dianil." [307] Direct colors of commercial fastness.

dianisidine (di-para-aminodi-meta-methoxydiphenyl; 3, 3'-dimethoxybenzidine) $[C_6H_3(OCH_3)NH_2]_2$.
Properties: White crystals; soluble in alcohol

*See "I.C.C. Shipping Regulations," page xiii.
Reference numbers refer to name of manufacturer. See "List of Manufacturers," page v.

and ether; insoluble in water. M.p. 137°C.
Derivation: The methyl ether of ortho-nitrophenol is reduced by zinc dust and caustic soda to the hydrazo compound, which is then rearranged with hydrochloric acid.
Method of purification: Crystallization.
Grades: Technical.
Containers: 350-lb wooden barrels; fiber drums; multiwall paper sacks.
Use: Dye intermediate.
Shipping regulations: None.*

di-para-anisyl-para-phenetylguanidine hydrochloride (acoin; guanicaine) $C_2H_5OC_6H_4NC(NHC_6H_4OCH_3)_2 \cdot HCl$.
Properties: White crystalline powder; incompatible with iodine and alkaline iodides. Soluble in water and alcohol. M.p. 176°C.
Use: Medicine (local).

diaphragm cell. A type of electrolytic cell for the production of caustic soda and chlorine from sodium chloride brine. The cell contains anode and cathode compartments separated by a porous diaphragm (usually asbestos fiber) to prevent mixing of the solutions. The brine is fed continuously to the anode compartment, where chlorine is released at the graphite anodes, and flows through the diaphragm to the iron cathode, where hydrogen is liberated and caustic soda accumulates in the liquid. The caustic solution, which is continuously drained from the cathode compartment, usually contains 9-12% sodium hydroxide, varying amounts of undecomposed sodium chloride, some dissolved chlorine, and traces of sodium chlorate and iron compounds. Although the mercury cathode cell (q.v.) produces a purer and more concentrated sodium hydroxide solution, it operates at a generally lower energy efficiency and is not as widely used in the U.S. Many types of diaphragm cells have been developed; for examples, see also Hooker, Nelson, Vorce cells, "Diamond Alkali Cell D-3."

"Di-Aqua." [244] Sodium alkyl aryl sulfonate in which the alkyl aryl portion is substantially all dodecylbenzene. Available in 40% and 80% active strengths, both in powder and flake form, also 40% beads.
Properties: Light tan powder (or flake); soluble in water.
Containers: Drums and bags.
Uses: Wetting agents; detergents; emulsifier.

diaspore $Al_2O_3 \cdot H_2O$. A natural hydrous aluminum oxide, occurring in bauxite (q.v.), and with corundum and dolomite.
Properties: White, gray, yellowish, and greenish; luster vitreous to pearly; sp.gr. 3.35-3.45; hardness 6.5-7.
Occurrence: Arkansas, Missouri, Pennsylvania; Switzerland; U.S.S.R.; Czechoslovakia.
Uses: As a refractory; abrasive; possible source of aluminum.

diastase, malt (amylase). An amylolytic enzyme.
Properties: Yellowish-white amorphous powder, or syrupy liquid. Soluble in water; almost insoluble in alcohol.
Derivation: The filtrate from the mash of malted grain is concentrated at low temperatures in vacuum. The sugar acts as preservative. The diastase converts insoluble starch and starch paste into soluble maltodextrins and maltose.
Containers: Wooden kegs; cans; fiber drums.
Uses: Desizing of textiles; making up colors for calico printing; finishing of textiles; medicine; bread-making; malted foods.
Shipping regulations: None.*

"Diatol." [192] Proprietary product. Said to be pure diethyl carbonate containing about 10% of unconverted anhydrous alcohol.
Constants: Flash point 82°F (approx); water solubility, 100 cc solvent dissolves 3.0 cc water (25°C); water-white color; distillation range, below 100°C not more than 15%, below 125°C not more than 50%, above 130°C none; mild odor, non-residual; purity, ester content as ethyl carbonate, not less than 90%; sp.gr. 0.954-0.957 (20/20°C).
Grades: Technical.
Containers: 1-gal cans; 5- to 36,000-lb (gross) drum cars; 6000- and 8000-gal tank cars.
Uses: A medium high-boiling nitrocellulose solvent, having the features of mild odor, stability and extremely low acidity. Because the acidity is due to the smallest trace of carbonic acid, it is considered as nearly a neutral solvent as it is possible to make. The extra ethyl alcohol content in "Diatol" contributes to solvent action, blending and flow. "Diatol" finds its chief use in lacquer formulation, and in the radio tube field.
Fire hazard: Combustible but not flammable; flash point over 80°F.

diatomaceous earth. See diatomite.

diatomite (diatomaceous earth; DE; kieselguhr; guhr; siliceous earth; tripolite; infusorial earth). A soft earthy rock composed of the siliceous skeletons of small aquatic plants called diatoms. As marketed it consists of light colored blocks, bricks, powder, or lumps resembling chalk or dried clay in appearance. Sp.gr. true 1.9-2.35, apparent 0.15-0.45; insoluble in acids except hydrofluoric; soluble in strong alkalies; able to absorb 1.5-4.0 times its own weight in water; a very poor conductor of sound, heat, and electricity.

Typical analysis:

silica (SiO_2)	86.89%
alumina (Al_2O_3)	2.32%
ferric oxide (Fe_2O_3)	1.28%
lime (CaO)	0.43%
potash (K_2O)	3.58%
water (H_2O)	4.89%

*See "I.C.C. Shipping Regulations," page xiii.
Reference numbers refer to name of manufacturer. See "List of Manufacturers," page v.

Occurrence: California, Oregon, Nevada, Washington, Utah, Idaho, Maryland, Virginia; Europe; Algeria; U.S.S.R.
Grades: Natural; chemical.
Containers: Bulk in cars, bags, multiwall paper sacks.
Uses: In filtration, clarifying, and decolorizing of liquids in the manufacture of sugar, beer, wine, liquors, fruit juices, water, dry cleaning fluid, oils and petroleum products, glue and adhesives, soap, shellacs and varnishes, etc.; as thermal insulating material in the form of blocks, brick, aggregates and cement; as filler in the manufacture of dynamite, sealing wax, match heads, textiles, plaster, plastics, paints, rubber, fertilizers; as an absorbent for liquids; as a mild abrasive; as a support and carrier for catalysts; in building material; in ceramics; in manufacture of water glass; ingredient of textile fireproofing agents.
Shipping regulations: None.*

diatrizoate methylglucamine. See methylglucamine diatrizoate.

diatrizoate sodium. See sodium diatrizoate.

1,4-diazabicyclo[2.2.2]octane
$CH_2CH_2NCH_2CH_2NCH_2CH_2$
Properties: Crystals, hygroscopic; m.p. 158°C; b.p. 174°C; forms crystalline hydrate; sublimes easily; soluble in water and organic solvents.
Uses: Suggested as catalyst for urethane foams and coatings; chemical intermediate.

"Diazanil." [307] Direct colors for diazotizing and developing on the fiber.

"Diazine." [243] Trademark of direct dyes, applied to cotton, diazotized and then coupled into phenols or amines.

1,3-diazine. See pyrimidine.

"Diazinon." (O,O-diethyl-O(2-isopropyl-6-methyl-4-pyrimidinyl) phosphorothioate) $[(CH_3)_2CHC_4N_2H(CH_3)O]PS(OC_2H_5)_2$.
Properties: Colorless liquid; b.p. 83-84°C (0.002 mm); slightly soluble in water, freely soluble in petroleum solvents, alcohol and ketones. More stable in alkaline than neutral or acid solutions.
Use: Agricultural insecticide having residual toxicity to flies. A cholinesterase inhibitor.
Toxicity: Similar to parathion (q.v.).

diazoaminobenzene (diazobenzeneanilide; benzeneazoanilide) $C_6H_5NNNHC_6H_5$.
Properties: Golden-yellow scales. Explodes on heating. Soluble in alcohol, ether, and benzene; insoluble in water.
Constants: M.p. 96°C; explodes when heated to 150°C.
Derivation: By the interaction of nitrous acid and an alcoholic solution of aniline.
Method of purification: Crystallization.
Grades: Technical.
Containers: Wooden kegs; fiber drums.
Uses: Organic synthesis; dyes.

4,4-diazoaminodibenzamidine
$NH_2C(NH)C_6H_4NNNHC_6H_4C(NH)NH_2$.
Properties: (trihydrate) yellow powder; m.p. 203°C (with decomposition); soluble in water at pH 6.1; solutions are unstable.
Grade: Veterinary.
Use: Veterinary medicine.

diazobenzeneanilide. See diazoaminobenzene.

para-**diazobenzenesulfonic acid** $C_6H_4SO_3N_2$.
Properties: White or slightly red crystals, or white paste. Soluble in water and ether; insoluble in alcohol.
Caution: May explode if heated.
Derivation: From sulfanilic acid, sodium nitrite and sulfuric acid.
Uses: Dyestuffs; reagent.

para-**diazodimethylaniline zinc chloride double salt** (para-dimethylaminobenzene diazonium chloride, zinc chloride double salt; para-diazotized aminodimethylaniline, zinc chloride double salt)
$(CH_3)_2NC_6H_4N_2Cl \cdot ZnCl_2$.
Properties: Yellow to orange (light sensitive) crystals.
Specifications: Moisture content 5-20%; zinc 17-23%; chloride 31-35%.
Containers: Bottles; fiber drums.
Uses: Rapid diazotype coupler, used in coatings for light-sensitive paper.
Shipping regulations: None.*

diazodinitrophenol.
Shipping regulations: Explosive, class A. Initiating explosive label. Not accepted by express.

para-**diazodiphenylamine sulfate.** Yellow-green solid with unpleasant odor. Sensitive to light; soluble in water. Used as a light sensitive diazo compound for coating on reproduction paper, giving direct positive prints of various colors with different developers or coupling agents.

1-diazo-2-naphthol-4-sulfonic acid
$C_{10}H_5N_2OSO_3H$.
Properties: Yellow needles in paste or dry form. Slightly soluble in water. Decomposed by heating above 100°C.
Derivation: Diazotization of 1-amino-2-naphthol-4-sulfonic acid and filtering of the diazo compound.
Method of purification: None.
Grades: Technical.
Containers: Wooden barrels; fiber drums.
Uses: Azo dyes; valuable chrome dyestuff component.
Fire hazard: The dry product can be ignited by sparks or a flame.

"Diazopon AN." [307] Trademark for a dispersing and stabilizing agent consisting of polyoxyethylated fatty alcohol; nonionic.
Properties: Clear, oily, yellow liquid; sp. gr. 1.03-1.04; soluble in water; stable to strong acids, alkalies and metallic ions.
Uses: Dispersing and stabilizing agent in the naphthol dyeing process to improve

fastness to rubbing. Used in dissolving fast color salts and has a stabilizing effect on the diazonium compound.

para-**diazotized aminodimethylaniline, zinc chloride double salt.** See para-diazodimethylaniline zinc chloride double salt.

diazotizing salts. See sodium nitrite.

"Diazyme." [212] Trademark for an amyloglucosidase which splits starch almost completely to glucose.

DIBA. See diisobutyl adipate.

dibasic dextro-amphetamine phosphate. N. F. XI name for dextrorotatory-amphetamine phosphate, dibasic.

ortho, ortho'-**dibenzamidodiphenyl disulfide** $C_{26}H_{20}O_2N_2S_2$.
Properties: Very light buff; practically odorless; sp. gr. 1.35 (approx); melting range 136-143°C.

1, 2, 7, 8-dibenzanthraquinone $C_{22}H_{12}O_2$.
Properties: Reddish-yellow needles. Soluble in acetic acid. M. p. 227°C (sublimes).

2, 3, 6, 7-dibenzanthraquinone (6, 13-pentacenequinone) $C_{22}H_{12}O_2$.
Properties: Yellow needles. Gives a blue color in concentrated sulfuric acid but changes to red when diluted. M. p. 388°C (sublimes). Soluble in sulfuric acid.

dibenzanthrone (violanthrone) $C_{34}H_{16}O_2$. Violet-blue vat dye.
Properties: Bluish-black powder. Soluble in nitrobenzene, conc. sulfuric acid.
Derivation: From benzanthrone.

3, 3'-dibenzanthronyl $C_{34}H_{18}O_2$.
Properties: Dark yellow needles. Soluble in conc. sulfuric acid. M. p. 412°C.

4, 4'-dibenzanthronyl $C_{34}H_{18}O_2$.
Properties: Yellow needles. Soluble in nitrobenzene; slightly soluble in benzene, alcohol, ether. M. p. 320°C.

1, 2, 5, 6-dibenzcarbazole $C_{20}H_{13}N$.
Properties: Needles.

dibenzofuran. See diphenylene oxide.

"Dibenzo G-M-F." [248] Trade name for dibenzoyl-para-quinonedioxime, $(C_6H_5COON)_2C_6H_4$.
Properties: Brownish grey powder; sp. gr. 1.37; starts to decompose above 200°C; good storage stability. Insoluble in acetone, benzol, gasoline, ethylene dichloride and water.
Uses: Non-sulfur vulcanizing agent for natural, SBR and butyl rubber, in tire-curing bags, gaskets and wire insulation to impart heat resistance.

dibenzopyrone. See xanthone.

dibenzopyrrole. See carbazole.

dibenzothiophene $\overline{C_6H_4C_6H_4S}$.
Properties: Colorless crystals; m. p. 97-98°C.
Uses: In cosmetics and pharmaceuticals; organic intermediate.

dibenzoyl. See benzil.

trans-**1, 2-dibenzoylethylene** $C_6H_5COCHCHCOC_6H_5$.
Properties: Yellow orange crystals; m. p. 111°C; soluble in glacial acetic acid, ethyl acetate, benzene and chloroform; sparingly soluble in alcohol; insoluble in water and petroleum ether.
Containers: Polyethylene-lined fiber drums.
Uses: Enzyme inhibitor, bactericide and chemical intermediate.

2, 4-dibenzoylresorcinol $C_6H_5COC_6H_2(OH)_2COC_6H_5$.
Properties: Light yellow crystals; nearly odorless; m. p. 125-128°C; soluble in alcohol, ethyl acetate, methyl ethyl ketone; insoluble in water.
Use: Light absorber, best at 280-370μ.

dibenzyl. See sym-diphenylethane.

N, N-dibenzylamine $HN(CH_2C_6H_5)_2$.
Properties: Colorless to light yellow liquid; sp. gr. 1.017 (20°C); refractive index 1.5730-1.5740 (n 25/D); distilling range 168-172°C (10 mm).
Containers: 200-lb, 400-lb steel drums.
Uses: Intermediate.

dibenzyl-para-aminophenol $(C_6H_5CH_2)_2NC_6H_4OH$.
Properties: Brown powder. Soluble in acetone, benzene, anhydrous methanol.
Constants: M. p. not lower than 110°C.

dibenzylaniline $C_6H_5N(CH_2C_6H_5)_2$.
Properties: Yellowish-white crystals. Soluble in alcohol and ether; insoluble in water.
Constants: M. p. 70°C; b. p. above 300°C.

dibenzyl ether $C_6H_5CH_2OCH_2C_6H_5$.
Properties: Unstable liquid. Faint, almond odor. Insoluble in water; soluble in most organic solvents.
Constants: Sp. gr. 1.035; b. p. 298-300°C; flash point 115-135°C.
Grades: Technical.
Uses: Plasticizer for nitrocellulose; in perfumery as a solvent for nitro-musks.

N, N'-dibenzylethylenediamine dipenicillin G. See benzathine penicillin G, under penicillin.

"Dibenzyline." [71] Trademark for phenoxybenzamine hydrochloride (N-phenoxyisopropyl-N-benzyl-beta-chloroethylamine hydrochloride), used in medicine.

N, N-dibenzylmethylamine $CH_3N(CH_2C_6H_5)_2$.
Properties: Colorless to light yellow liquid; Sp. gr. 0.99 (25°C); refractive index 1.5560-1.5590 (25°C); distilling range: 152-158°C (11 mm).
Containers: 200-lb, 400-lb steel drums.
Use: Intermediate.

dibenzyl sebacate $(C_6H_5CH_2OOC)_2(CH_2)_8$.
Properties: Light straw colored; b. p. 265°C (4 mm); sp. gr. 1.055 (30/20°C); complete non-volatility and excellent low-temperature

flexibility.
Containers: Drums; tank cars.
Use: Plasticizer, especially for plastic linings for containers.
Shipping regulations: None.*

dibenzyl succinate. See benzyl succinate.

2,5-dibiphenylyloxazole (BBO) $C_{27}H_{21}NO$.
Properties: Crystalline solid, m.p. 237-239°C.
Grade: Purified.
Use: Scintillation counter or as wave length shifter in solution scintillators.

"Dibistine." [305] Trademark for a compound containing tripelennamine hydrochloride U.S.P. and antazoline hydrochloride U.S.P.
Use: Medicine.

diborane (diboron hexahydride; boroethane) B_2H_6.
Properties: Colorless gas with repulsive odor. B.p. −92.5°C; m.p. −165°C; density 0.15 g/ml (17°C). Soluble in carbon disulfide; decomposes in water. Highly reactive; very flammable.
Derivation: (a) From boron trichloride or bromide and hydrogen; (b) by reaction of lithium aluminum hydride and boron trichloride in ether solution.
Containers: Gas cylinders.
Uses: Synthesis of organic boron compounds and metal borohydrides; polymerization catalyst for ethylene; fuel for air-breathing engines, and rockets; reducing agent.
Hazard: Very toxic.

diboron hexahydride. See diborane.

"Dibromantin." [109] Trade name for 1,3-dibromo-5,5-dimethylhydantoin.

dibromoacetylene BrC⋮CBr.
Properties: Heavy, colorless liquid. Disagreeable odor. Caution! Very toxic! Very unstable in the presence of oxygen! Decomposes with explosive violence if heated! Breaks down in damp air with production of powerful irritants! Soluble in most organic solvents.
Constants: Sp.gr. (approx) 2; b.p. 76-76.5°C.
Derivation: (a) Interaction of magnesium dibromoacetylene and an ethereal solution of cyanogen bromide. (b) Interaction of tribromoethylene and alcoholic potash.
Grades: Technical.
Use: Organic synthesis (halogenated ethylene).
Fire hazard: Dangerous. Flammable in air. Burns with a red flame.
Shipping regulations: Flammable liquid. Red label.*

9,10-dibromoanthracene $C_6H_4C_2Br_2C_6H_4$ (tricyclic).
Properties: Yellow crystals. Soluble in chloroform; slightly soluble in alcohol and ether; insoluble in water.
Constants: M.p. 221°C; b.p., sublimes.
Derivation: By the bromination of anthracene.
Method of purification: Crystallization.
Grades: Technical.
Containers: Wooden kegs; fiber drums.
Use: Organic synthesis.
Shipping regulations: None.*

ortho-**dibromobenzene** (benzene dibromide) $C_6H_4Br_2$.
Properties: Heavy liquid with pleasant, aromatic odor. B.p. 225.5°C; freezing point 7.13°C; sp.gr. 1.9767 (25/4°C); refractive index (n 20/D) 1.6155. Miscible with alcohol, acetone, ether, benzene, carbon tetrachloride, and n-heptane; insoluble in water.
Derivation: Interaction of benzene with an excess of bromine in presence of iron.
Grades: Technical.
Containers: Steel drums.
Uses: Solvent for oils; motor fuels; top-cylinder compounds; organic synthesis; ore flotation.
Shipping regulations: None.*

para-**dibromobenzene** (benzene dibromide) $C_6H_4Br_2$.
Properties: Colorless crystals. Soluble in alcohol and ether.
Constants: M.p. 89°C; b.p. 219°C; sp.gr. 2.261; refractive index (n 99/D) 1.5743.
Derivation: Obtained by the interaction of benzene with an excess of bromine in presence of a little iron.
Method of purification: Crystallization.
Impurities: Monobromobenzene.
Grades: Technical.
Containers: Tins; bags.
Use: Organic synthesis of dyestuffs and drugs; manufacture of intermediates.
Shipping regulations: None.*

N,N-dibromobenzenesulfonamide.
Properties: Solid; m.p. 109-111°C. Active bromine 50.4%.
Containers: Glass bottles; fiber drums.
Use: Halogenating agent.

dibromochloromethane $CHBr_2Cl$.
Properties: Clear, colorless heavy liquid. Sp.gr. 2.38; b.p. 116°C.
Containers: 5-gal carboys.
Uses: Organic synthesis.

1,2-dibromo-3-chloropropane $CH_2BrCHBrCH_2Cl$.
Properties: Amber to brown liquid; sp.gr. 2.05 (20°C); b.p. 195.5°C; freezing point 6.7°C; flash point (Tag open cup) 170°F.
Containers: 30-gal drums.
Uses: Soil fumigant for nematodes.
Shipping regulations: Poison, class B. Poison label.*

5,7-dibromo-2-chloro-3-pseudoindolone. See 5,7-dibromoisatin chloride.

dibromodiethyl sulfide $(CH_2CH_2Br)_2S$. The bromine analog of mustard gas.
Properties: White crystals. Hydrolyzed by water. Caution! Irritant! Soluble in alcohol, benzene, ether; insoluble in water.
Constants: Sp.gr. 2.05 (15°C); b.p. 240°C (decomposes); m.p. 31-34°C.

Derivation: Action of hydrobromic acid on an aqueous solution of thiodiglycol.
Grades: Technical.
Use: Organic synthesis.

dibromodiethyl sulfone $(CH_2CH_2Br)_2O_2S$.
Properties: Plates. Soluble in alcohol, benzene, ether. M.p. 111-112°C.
Derivation: Interaction of dibromodiethyl sulfide, chromic anhydride, and dilute sulfuric acid.

dibromodiethyl sulfoxide $(CH_2CH_2Br)_2OS$.
Properties: Glittering crystals. Soluble in alcohol, benzene, ether. M.p. 100-101°C.
Derivation: Interaction of benzoyl hydrogen peroxide and a hot solution of dibromodiethyl sulfide in chloroform.

dibromodifluoromethane CF_2Br_2.
Properties: Colorless heavy liquid; f.p. −141°C; b.p. 23.2°C; sp.gr. 2.288 (15/4°C); refractive index 1.399 (12°C). Insoluble in water; soluble in methanol and ether.
Grades: Pure (95.0% min).
Containers: Cylinders.
Uses: Synthesis of dyes, pharmaceuticals, quaternary ammonium compounds.

1,3-dibromo-5,5-dimethylhydantoin BrNCONBrCOC$(CH_3)_2$.
Properties: Free flowing cream colored powder with slight bromine odor. M.p. 187-191°C (decomposes); quite stable at 75°C. Soluble in benzene, chloroform, glacial acetic acid; slightly soluble in water and carbon tetrachloride; insoluble in hexane. Contains 55% active bromine, which is slowly released in aqueous solution.
Derivation: Bromination of dimethylhydantoin.
Grades: Technical.
Containers: Glass bottles; polyethylene-lined drums.
Uses: Controlled bromination and oxidation of organic compounds; water treatment; polymerization catalyst; potential germicide and sanitizer.

dibromoethane. See ethylene dibromide.

1,3-dibromo-5-ethyl-5-methylhydantoin (DEMH) BrNCONBrCOC$CH_3(C_2H_5)$.
Properties: Solid. Active bromine 52.5%.
Containers: Glass bottles; fiber drums.
Use: Halogenating agent.

2,4-dibromofluorobenzene $C_6H_3Br_2F$.
Properties: Colorless liquid; sp.gr. (20°C) 2.047; b.p. 214°C; refractive index (n 25/D) 1.5790. Insoluble in water; soluble in alcohol, acetone, ether, benzene, chloroform, ethyl acetate, and glacial acetic acid.
Uses: Intermediate for agricultural and pharmaceutical chemicals.
Caution: Avoid contact with eyes and skin.

dibromoformoxime CBr_2NOH.
Properties: Crystals. Not so toxic or irritant as the chloroformoximes.
Constants: M.p. 70-71°C. Distils between 75 and 85°C (3 mm).

dibromohydroxymercurifluorescein, disodium salt. See merbromin.

dibromoiodoethylene Br_2CCHI.
Properties: Liquid; sp.gr. 2.952 (24°C); b.p. 91°C (15 mm).
Derivation: Reaction of iodine and dibromoacetylene.

5,7-dibromoisatin chloride (5,7-dibromo-2-chloro-3-pseudoindolone) $C_8H_2Br_2ClNO$. Soluble in organic solvents such as benzene, chlorobenzene, etc.
Derivation: Isatin is gently warmed with bromine in concentrated sulfuric acid, giving 5,7-dibromoisatin which is then warmed with phosphorus pentachloride in an organic solvent.

1,2-dibromoisobutane. See isobutylene dibromide.

dibromomalonic acid HOOC·CBr_2COOH.
Properties: Light yellow needles or prisms; m.p. 147°C (decomposes).
Use: Intermediate for drugs and fine chemicals.

dibromomalonyl chloride ClOCCBr_2COCl.
Properties: Yellowish oily liquid; b.p. 75-77°C (15 mm).
Use: Chemical intermediate.

dibromomethane. See methylene bromide.

dibromomethyl ether $(CH_2Br)_2O$.
Properties: Colorless liquid. Decomposed by water. Caution! Very toxic! Soluble in acetone, benzene, ether; insoluble in water.
Constants: Sp.gr. 2.2; b.p. 154-155°C; m.p. −34°C.
Derivation: (a) The reaction product of paraformaldehyde and sulfuric acid is treated with ammonium bromide. (b) Interaction of hydrobromic acid and paraformaldehyde.

1,5-dibromopentane. See pentamethylene dibromide.

1,3-dibromopropane. See trimethylene bromide.

dibromopropanol (2,3-dibromo-1-propanol) $CH_2BrCHBrCH_2OH$.
Properties: Sp.gr. (20/4°C) 2.120; b.p. 219°C (760 mm), 90°C (8 mm). Soluble in acetone, alcohol, ether and benzene.
Uses: Intermediate in preparation of flame retardants, insecticides and pharmaceuticals.

2,6-dibromoquinone chlorimide $OC_6H_2Br_2NCl$.
Properties: Yellow crystals; m.p. 83°C; b.p. 121°C (decomposes).
Derivation: Reduction product of dibromonitrophenol is oxidized with chlorine in presence of alkali.

"Dibs." 57 Trademark for an accelerator for the vulcanization of rubber.

dibucaine $C_{20}H_{29}N_3O_2$, 2-n-Butoxy-N-(2-diethylaminoethyl)cinchoninamide.
Properties: Colorless or almost colorless powder; m.p. 62-65°C. Odorless; somewhat hydroscopic; affected by light. Soluble in hydrochloric acid and ether; slightly soluble in water.
Grade: U.S.P. XVI.
Use: Medicine (local anesthetic).

dibucaine hydrochloride $C_{20}H_{29}N_3O_2 \cdot HCl$. 2-Butoxy-N-(2-diethylaminoethyl)cinchoninamide hydrochloride.
Properties: Fine, white, lustrous crystals or as white powder. Odorless and very hygroscopic. Bitter, acrid taste with prolonged local anesthetic action. M.p. 95-100°C (dec). Sensitive to light. Soluble in water, alcohol, acetone, and chloroform; slightly soluble in cold benzene, ethyl acetate, and toluene. Solutions are acid to litmus; pH 5-6.
Grade: U.S.P. XVI.
Use: Medicine (local anesthetic).

dibutoline sulfate $(C_{15}H_{33}N_2O_2)_2SO_4$. Bis[dibutylcarbamate of ethyl-(2-hydroxyethyl)dimethylammonium] sulfate.
Properties: Hygroscopic powder; decomposes 166°C; soluble in water and benzene.
Grades: N.N.D.
Uses: Surface active agent in medicine.

2,5-dibutoxyaniline $C_6H_3(OC_4H_9)_2NH_2$.
Properties: M.p. 18°C; insoluble in water; soluble in organic solvents.
Containers: 250-lb drums.
Uses: Dyes; synthesis.

1,4-dibutoxy benzene. See hydroquinone di-n-butyl ether.

dibutoxyethyl adipate $(C_2H_4COOC_2H_4OC_4H_9)_2$.
Properties: Colorless, oily liquid. Sp.gr. (20/20°C) 0.997; f.p. −34°C; boiling range 205-215°C (4 mm); acidity, (max) 0.03% as acetic acid; mild, butyl type odor; flash point 370°F; fire point 440°F, vapor pressure <0.17 mm (150°C); viscosity, 12.5 cps (20°C); refractive index, 1.442 (25°C); surface tension 33 dynes/cm (20°C); thermal expansion 0.00078 (10-40°C); wt/gal 8 lbs. Insoluble or only slightly soluble in mineral oil, glycerine, glycols and some amines; soluble in most other organic liquids.
Containers: 5-gal cans (40 lbs net); 55-gal steel drums (450 lbs net).
Uses: Primary plasticizer for most resins, imparting flexibility at very low temperature, as well as stability to ultraviolet light, and flexibility.

dibutoxyethyl phthalate $C_6H_4(COOC_2H_4OC_4H_9)_2$.
Properties: Colorless liquid; f.p. −55°C; sp.gr. 1.06 (20°C); b.p. 220-230°C (4 mm); wt/gal 8.86 lbs; fast to light, water resistant.
Containers: 1-, 5-gal cans; 55-gal drums; tank cars.
Uses: Plasticizer for polyvinyl chloride, polyvinyl acetate and other resins.

dibutoxymethane $CH_2(OC_4H_9)_2$.
Properties: Colorless liquid. Wt/gal 6.97 lbs (20°C); refractive index 1.40615 (20°C); sp.gr. 0.838 (20/20°C); flash point 60°C (140°F); boiling range 164-186°C. Insoluble in water. Solubility of water in product 1.0 cc/100 cc (20°C).

dibutoxytetraglycol $(C_2H_4OC_2H_4OC_4H_9)_2O$.
Properties: Practically colorless liquid with characteristic odor. Slightly soluble in water (1.3% by wt); sp.gr. 0.9436 (20/20°C); lbs/gal 7.85 (20°C); b.p. 237°C (50 mm), 330°C (760 mm); vapor pressure less than 0.01 mm (20°C); freezing point −20°C; viscosity 5.7 centipoises (20°C); flash point 355°F; solubility of water in product 4.8% by wt (20°C); refractive index 1.4357.
Containers: 5-, 55-gal drums.
Use: Solvent; excellent solvent for DDT.

N,N-di-n-butyl acetamide $CH_3CON(C_4H_9)_2$.
Properties: Sp.gr. 0.890 (20°C); boiling range 245-250°C; color water-white; odor faint. Flash point 225°F.

di-n-butylamine $(C_4H_9)_2NH$.
Properties: Colorless liquid with amine odor. B.p. 159.6°C; freezing point −62°C; sp.gr. (20/20°C) 0.7613; wt/gal (20°C) 6.33 lbs; refractive index(n 20/D) 1.4175; flash point (open cup) 125°F. Insoluble in water; soluble in alcohol and ether; miscible with hydrocarbons.
Derivation: By reaction of butanol or butyl chloride with ammonia.
Grades: Technical.
Containers: 1-gal cans; 5-, 55-gal drums; tank cars.
Uses: Corrosion inhibitor; intermediate for emulsifiers, rubber accelerators, dyes, insecticides, and flotation agents.
Shipping regulations: None.*

di-sec-butylamine $(CH_3CHCH_2CH_3)_2NH$.
Properties: Water-white liquid; amine odor; boiling range 132-135°C; sp.gr. 0.754 (20/20°C); refractive index 1.412 (20°C); flash point 75°F.
Shipping regulations: Flammable liquid. Red label.*

N,N-di-n-butylaminoethanol $(C_4H_9)_2NCH_2CH_2OH$.
Properties: Sp.gr. 0.859 (20°C); boiling range 224-232°C; color water-white; odor faint, amine-like. Flash point 200°F.
Containers: 5-gal cans; 55-gal drums; tank cars.
Use: Synthesis.

3-dibutylaminopropyl-para-aminobenzoate sulfate. See butacaine sulfate.

N,N-di-n-butylaniline $C_6H_5N(C_4H_9)_2$.
Properties: Amber liquid; odor faint aniline. Sp.gr. 0.904 (20°C); boiling range 267-275°C; refractive index 1.519 (20°C). Soluble in alcohol and ether; insoluble in water. Flash point 230°F.

2,5-di-tert-butyl benzoquinone $[C(CH_3)_3]_2C_6H_2O_2$.
Properties: Yellow crystals; insoluble in water; soluble in ethyl acetate, acetone,

*See "I.C.C. Shipping Regulations," page xiii.
Reference numbers refer to name of manufacturer. See "List of Manufacturers," page v.

benzene; slightly soluble in ethyl alcohol.
Typical specifications: M.p. 149-151°C.
Containers: Fiber drums, 75 lbs net.
Uses: Oxidant; polymerization catalyst.

dibutyl butyl phosphonate $C_4H_9P(O)(OC_4H_9)_2$.
Properties: Colorless liquid, with mild odor. Stable. Insoluble in water, miscible with most common organic solvents; sp. gr. 0.948 (20/4°C); b.p. 127-128°C (2.5 mm); flash point 310°F (C.O.C.).
Containers: 5-gal, 55-gal steel drums.
Uses: Heavy metal extraction and solvent separation; gasoline additives; anti-foam agent; plasticizer; textile conditioner and antistatic agent.

dibutyl "Carbitol." [214] $C_4H_9O(C_2H_4O)_2C_4H_9$.
Trademark for diethylene glycol dibutyl ether (q.v.).

dibutyl "Cellosolve." [214] $C_4H_9OC_2H_4OC_4H_9$.
Trademark for ethylene glycol dibutyl ether (q.v.).

dibutyl chlorophosphate $(C_4H_9O)_2P(O)Cl$.
Properties: Water-white liquid; b.p. 103-106°C (1.5 mm); sp. gr. 1.0742 (25°C); refractive index 1.4289 (n 25/D). Soluble in common inert organic solvents; hydrolyzes slowly in water.
Uses: Intermediate in organic synthesis.

di-tert-butyl-meta-cresol (DBMC; 4,6-di-tert-butyl-3-methylphenol) $[C(CH_3)_3]_2CH_3C_6H_2OH$.
Properties: Crystalline solid; m.p. 62.1°C; b.p. 282°C; sp. gr. 0.912 (80/4°C); viscosity 9.9 centistokes (80°C), 1.42 centistokes (160°C); very soluble in ethanol, benzene, carbon tetrachloride, ethyl ether, and acetone; essentially insoluble in water, ethylene glycol, and 10% aqueous sodium hydroxide.
Typical specifications: Freezing point approx 48°C; bulk density, solidified material 56 lb/cu. ft.; flash point (open cup) 262°F.
Containers: 5-gal pails; 55-gal black iron drums.
Caution: Contact with eyes, skin, or clothing should be avoided.
Uses: Rubber reclaiming; rubber compounding; surface-active agents; resins and plasticizers; antioxidants and perfumes.

di-tert-butyl-para-cresol (DBPC; 2,6-di-tert-butyl-4-methylphenol; butylated hydroxytoluene; BHT) $[C(CH_3)_3]_2CH_3C_6H_2OH$.
Properties: White, crystalline solid; freezing point 70°C; b.p. 265°C; sp. gr. 1.048 (20/4°C); viscosity 3.47 centistokes (80°C), 1.54 centistokes (120°C); refractive index (n 75/D) 1.4859; soluble in methanol, ethanol, isopropanol, "Cellosolve" (12°C), petroleum ether, benzene, methyl ethyl ketone and linseed oil; insoluble in water and 10% sodium hydroxide.
Typical specifications: White to light-yellow color; free-flowing crystals; freezing point 68°C; flash point (open cup) 260°F; boiling range 257-266°C.
Grades: Technical; food; feed.
Containers: 5-gal pails; fiber drums containing 100 lbs net; tank cars.
Caution: Contact with the eyes, skin, or clothing should be avoided.
Uses: Antioxidant to stabilize hydrocarbons such as petroleum oils, gasoline, jet fuels, rubber, and vinyl monomers. Approved by the Food and Drug Administration for use as a food antioxidant; used in animal feeds, food packaging.

2,6-di-tert-butyl-alpha-dimethylamino-para-cresol ("Ethyl" Antioxidant 703) $(C_4H_9)_2C_6H_2OH[CH_2N(CH_3)_2]$.
Properties: Light yellow crystalline solid; m.p. 201°F; flash point 280°F (open cup). Insoluble in water and 10% sodium hydroxide; soluble in organic solvents.
Containers: 100-lb drums.
Use: Antioxidant in gasoline and oils.

dibutyl diphenyl tin $(CH_3CH_2CH_2CH_2)_2Sn(C_6H_5)_2$.
Properties: Clear, slightly greenish liquid; contains 30.7% Sn; b.p. 175°C (2 mm); refractive index 1.563 (17.5°C); sp. gr. 1.19.

di-tert-butyl disulfide $C(CH_3)_3SSC(CH_3)_3$.
Properties: Liquid; sp. gr. 0.9291 (60/60°F); boiling range 375-405°F; refractive index 1.491 (20°C); flash point 170°F (approx).
Containers: 1-, 5-, 54-gal drums.
Use: Intermediate.
Shipping regulations: None.*

n-dibutyl ether. See butyl ether.

dibutyl fumarate $(C_4H_9)OOCCH{:}CHCOO(C_4H_9)$.
Properties: Liquid; sp. gr. (20°C) 0.9873; b.p. 285.2°C; f.p. −15.6°C refractive index n (20°C) 1.4466; insoluble in water.
Containers: Drums; tank trucks; tank cars.
Use: Monomeric plasticizers; copolymers; intermediate.

dibutyl hexahydrophthalate.
Constants: Sp. gr. 1.005; flash point 152°C; boiling range 185-190°C.

2,5-di-tert-butyl hydroquinone $[C(CH_3)_3]_2C_6H_2(OH)_2$.
Properties: White powder; soluble in acetone, alcohol, benzene; insoluble in water, aqueous alkali.
Typical specifications: M.p. 210-212°C.
Grade: Technical.
Containers: Fiber drums, 75 and 150 lbs net.
Uses: Polymerization inhibitor; antioxidant; stabilizer against ultraviolet deterioration of rubber.

di-n-butyl itaconate
$CH_2{:}C(COOC_4H_9)CH_2(COOC_4H_9)$.
Properties: Clear, colorless liquid with slight odor. B.p. 145°C (10 mm); sp. gr. 0.9833 (22°C); refractive index (n 25/D) 1.442. Insoluble in water.
Uses: Resins; lube oil additives; plasticizers.

N,N-di-n-butyl lauramide $C_{11}H_{23}CON(C_4H_9)_2$.
Properties: Sp. gr. 0.861 (20°C); boiling range 200-230°C (3 mm); color straw; odor lauric acid.

*See "I.C.C. Shipping Regulations," page xiii.
Reference numbers refer to name of manufacturer. See "List of Manufacturers," page v.

dibutyl maleate
$C_4H_9OOCCH:CHCOOC_4H_9$.
Properties: Oily liquid. B.p. 280.6°C; freezing point, sets to a glass below −85°C; sp.gr. 0.9964 (20/20°C); wt/gal 8.3 lb (20°C); flash point (open cup) 285°F. Insoluble in water.
Grade: Technical.
Containers: 1-, 5-, 55-gal drums; tank cars.
Uses: Copolymers; plasticizers; intermediate.

2,6-di-tert-butyl-4-methylphenol. See di-tert-butyl-para-cresol.

4,6-di-tert-butyl-3-methylphenol. See di-tert-butyl-meta-cresol.

dibutyl oxalate $(COOC_4H_9)_2$.
Properties: Water-white, high-boiling liquid. Mild odor.
Constants: B.p. 240-250°C; refractive index 1.425; m.p. −30°C; wt/gal 8.24 lbs (approx)(20°C); coefficient of expansion/°C 0.00095; flash point 265°F (approx); viscosity (centipoises)(10% half sec. nitrocellulose solution) 800; water solubility, 100 cc solvent dissolves 0.5 cc water (25°C).
Typical specifications: Acidity, free acid as oxalic, not more than 0.05%; color water-white; distillation range below 240°C not more than 5%, below 248°C not less than 90%, above 255°C none; dryness, miscible without turbidity with 20 vol 60° Bé gasoline (20°C); non-volatile matter not more than 0.005 g/100 cc; odor mild; purity, ester content as dibutyl oxalate, not less than 99%; sp.gr. 0.989-0.993 (20/20°C). Miscible with most alcohols, ketones, esters, oils, hydrocarbons.
Derivation: By the standard esterification process using normal butyl alcohol and oxalic acid.
Grades: According to ester content 90%; 95%; 99-100%.
Containers: 1-gal (non-returnable) cans; 5-, 55-gal (non-returnable) steel drums; tank cars.
Uses: Organic synthesis; solvent.
Fire hazard: Combustible but not flammable.
Shipping regulations: None.*

di-tert-butyl peroxide $(CH_3)_3COOC(CH_3)_3$.
Typical specifications: Purity 97%; b.p. 111°C; sp.gr. 0.794 (20/4°C); refractive index (n 20/D) 1.389; flash point (Tag open cup) 65°F; stated to be stable when stored for long periods at temperatures up to 80°C; insoluble in water.
Containers: 1-, 6-, 30-lb polyethylene bottles; 98-lb steel drums.
Use: Polymerization catalyst for vinyl type monomers at temperatures above 100°C; intermediate.
Shipping regulations: Flammable liquid. Red label.*

2,4-di-tert-butylphenol $[(CH_3)_3C]_2C_6H_3OH$.
Properties: Tan crystalline solid; f.p. 48°C; b.p. 152-157°C (25 mm); sp.gr. 0.907 (60/4°C); lbs/gal 7.57 (60°C); flash point: 265°F. Soluble in methanol, ether; very slightly soluble in water.
Use: Intermediate.

2,6-di-tert-butylphenol $[(CH_3)_3C]_2C_6H_3OH$.
Properties: Light straw crystalline solid; m.p. 37°C; sp.gr. 0.914 (20°C); b.p. 253°C; flash point 245°F. Soluble in alcohol and benzene; insoluble in water.
Use: Intermediate.

dibutyl phosphite $(C_4H_9O)_2PHO$.
Properties: Water-white liquid; b.p. 95°C (1 mm); sp.gr. 0.9860 (25°C); refractive index 1.4228 (n 25/D); soluble in common organic solvents.
Containers: Carboys.
Uses: Solvent; antioxidant; intermediate.

dibutyl phthalate (DBP) $C_6H_4(COOC_4H_9)_2$.
Properties: A colorless, odorless, non-volatile, non-toxic, stable, oily liquid. Sp.gr. 1.0484 (20/20°C); f.p. −35°C; viscosity 0.203 poise (20°C); distillation range 227-235°C (37 mm Hg); flash point (Cleveland open cup) 340°F; wt/gal 8.72 lbs (68°F); approx. change in wt 0.0016 lb/gal/°F; refractive index 1.4915 (25°C); coefficient of expansion 0.00042/°F, 0.00078/°C; dilution ratio (nitrocellulose solution method) 2.7 with toluene, 1.7 with petroleum naphtha; b.p. 340.0°C; vapor pressure 1.1 mm (150°C). Miscible with the common organic solvents; very slightly soluble in water.
Derivation: By treating n-butyl alcohol with phthalic anhydride followed by purification, which results in a product unusually free from odor and color.
Grades: Technical, 99-100% dibutyl phthalate.
Containers: Crated cans (1-, 5-gals); 5-, 10-, 50-, 100-gal steel drums; tank cars and trucks.
Uses: Plasticizer; solvent for perfume oils; perfume fixative; textile lubricating agent; safety glass; leather dopes; insecticides; printing inks; resin solvent; paper coating; adhesives; as plasticizer in solid rocket propellants.
Shipping regulations: None.*

2,5-di-tert-butyl quinone $[C(CH_3)_3]_2C_6H_2O_2$.
Typical specifications: Yellow powder; m.p. 149-151°C; insoluble in water; soluble in alcohol, acetone, ethyl acetate, and benzene.
Use: Oxidizing agent.

dibutyl sebacate $C_4H_9OCO(CH_2)_8OCOC_4H_9$.
Properties: Clear, colorless, odorless liquid. B.p. 349°C (760 mm), 180°C (3 mm); f.p. −11°C; sp.gr. 0.936 (20/20°C); wt/gal 7.81 lb (20°C); refractive index 1.4395 (25°C); flash point 350°F. Insoluble in water.
Grade: Technical.
Containers: 1-, 5-, 55-gal drums; tank cars.
Uses: Plasticizer; rubber softener; component of cosmetics and perfumes; in sealing rings for food containers.
Shipping regulations: None.*

N,N-dibutyl stearamide $C_{17}H_{35}CON(C_4H_9)_2$.
Properties: Yellow; sp.gr. 0.860 (20/20°C);

*See "I.C.C. Shipping Regulations," page xiii.
Reference numbers refer to name of manufacturer. See "List of Manufacturers," page v.

boiling range 173-175°C (0.4 mm); flash point 420°F; fatty-acid odor.

di-tert-butyl sulfide $[(CH_3)_3C]_2S$.
Properties: Liquid; f.p. 12.3°F; boiling range 297-303°F; sp.gr. 0.8316 (60/60°F); wt/gal 6.93 lbs; refractive index 1.451 (20°C); flash point (approx) 100°F.
Containers: 1-, 5-gal drums.
Use: Intermediate.
Shipping regulations: None.*

dibutyl tartrate $C_4H_9OOCCHOHCHOHCOOC_4H_9$.
Properties: Liquid. M.p. 21°C; b.p. approx. 204°C(26 mm); refractive index 1.4463 (20°C); flash point 132.2°C (270°F); wt/gal 9.07 lbs (68°F). Typical specifications: Purity not less than 98% ester, by wt; sp.gr. 1.087-1.093 (20/20°C); acidity not more than 0.05%, calculated as tartaric acid; water, no turbidity when 1 vol is mixed with 19 vols of 60° Bé. gasoline at 20°C; color light straw. Miscible with the common organic solvents, oils, hydrocarbons.
Grade: Technical.
Containers: 1-gal cans; 5-, 55-gal steel drums. Net content 10, 45, 490 lbs.
Uses: Solvent for nitrocellulose, cellulose acetate; plasticizer for nitrocellulose, cellulose acetate, other cellulose esters and ethers, synthetic resins; lubricant; rubberized fabrics; lacquers; dopes; transfer inks.
Shipping regulations: None.*

dibutylthiourea $C_4H_9NHCSNHC_4H_9$.
Properties: White to light tan solid; m.p. 59-69°C; slightly soluble in water, soluble in methanol, ether, acetone, benzene, ethyl acetate; insoluble in gasoline.
Uses: Corrosion inhibitor; for pickling cast iron or carbon steel with hydrochloric acid; for pickling with sulfuric acid; for reducing corrosion of ferrous metals and aluminum alloys in brine; as intermediate.

dibutyltin diacetate $(C_4H_9)_2Sn(C_2H_3O_2)_2$.
Properties: Clear yellow liquid. B.p.130°C (2 mm); f.p. below 12°C. Soluble in water and most organic solvents.
Derivation: Reaction of acetic acid with dibutyltin oxide.
Uses: Stabilizer for chlorinated organics; catalyst for condensation reactions.

dibutyltin dichloride $(C_4H_9)_2SnCl_2$.
Properties: White crystalline solid; m.p. 43°C; b.p. 135°C (10 mm); sp.gr. 1.36 (50°C, liquid); refractive index 1.4991 (51°C).
Derivation: Reaction of butylmagnesium chloride with tin tetrachloride.
Use: Organo-tin intermediate.

dibutyltin di-2-ethylhexoate $(C_4H_9)_2Sn(O_2CC_7H_{15})_2$.
Properties: Waxy white solid. F.p. below 65°C. Insoluble in water; soluble in most organic solvents.
Derivation: Reaction of dibutyltin oxide with 2-ethylhexoic acid.
Uses: Catalyst for silicone curing; polyether foams.

dibutyltin dilaurate $(C_4H_9)_2Sn(OOCC_{11}H_{23})_2$.
Properties: Colorless crystals; sp.gr. 1.052 (20/20°C); f.p. 8°C; flash point 440°F; viscosity 45cps (20°C); insoluble in water.
Derivation: Reaction of lauric acid with dibutyltin oxide.
Containers: 1-, 5-, and 55-gal drums (8.5, 40 and 450 lbs).
Uses: Catalyst for polyurethane foams and resins; condensation catalyst; stabilizer for polyvinyl chloride resins; anthelmintic.

dibutyltin maleate $[(C_4H_9)_2Sn(OOCCH)_2]_x$.
Properties: White amorphous powder; m.p. 110°C. Insoluble in water; soluble in benzene and organic esters.
Derivation: Reaction of maleic acid with dibutyltin oxide.
Uses: Stabilizer for polyvinyl chloride resins; condensation catalyst.

dibutyltin oxide $[(C_4H_9)_2SnO-]_x$.
Properties: White powder; m.p. (decomposes). Insoluble in water.
Derivation: Hydrolysis of dibutyltin dichloride with caustic.
Uses: Condensation catalysts; intermediate for other organotins.

dibutyltin sulfide $[(C_4H_9)_2SnS]_3$.
Properties: Colorless oily liquid.
Derivation: Reaction of dibutyltin oxide with hydrogen sulfide.
Uses: Vinyl stabilizer; antioxidant; lubricating additive.

1,1-dibutylurea (N,N-dibutylurea) $NH_2CON(C_4H_9)_2$. Solidification point 22-25°C; boiling range 118-119°C (2-3 mm); soluble in alcohol and ether. Copolymerized with simple urea by the use of formaldehyde, yields modified resins that differ in nature from those made through the use of the mono-substituted ureas. These resins tend to be permanently thermoplastic.

DIC. Abbreviation for beta-diisopropylaminoethyl chloride hydrochloride.

dicalcium magnesium aconitate (calcium magnesium aconitate) $[C_3H_3(COO)_3]_2Ca_2Mg$.
Properties: White crystalline powder or lumps.
Derivation: By precipitation from cane molasses with lime.
Grades: Technical.
Containers: 100-lb paper bags.
Uses: Conversion to aconitic acid, tributyl aconitate and similar ester plasticizers.

dicalcium orthophosphate. See calcium phosphate, dibasic.

"Dicalcium Phosphate V-25." [172] Trade name for a dentifrice grade dicalcium phosphate dihydrate, $CaHPO_4 \cdot 2H_2O$, plus a proprietary additive.
Derivation: From lime and high quality phosphoric acid.
Containers: Bags and bulk boxes.
Use: Polishing agent in dentifrices.

*See "I.C.C. Shipping Regulations," page xiii.
Reference numbers refer to name of manufacturer. See "List of Manufacturers," page v.

"Dicalcium Phosphate Victor." [172] Trade name for a dentifrice grade dicalcium phosphate dihydrate, $CaHPO_4 \cdot 2H_2O$, plus a proprietary additive.
Derivation: Lime and high quality phosphoric acid.
Containers: Bags, drums, and bulk boxes.
Uses: Polishing agent in dentifrices.

dicalcium silicate $2\ CaO \cdot SiO_2$. One of the components of cement. See cement, Portland, and other cement articles. It is also obtained as a byproduct in electric furnace operation, and is used to neutralize acid soils.

"Dicalite." [218] Trade name for a large group of materials processed from diatomite (also called diatomaceous earth, diatomaceous silica, kieselguhr, and sometimes DE) and having a wide range of industrial uses. Crude diatomite averages about 90% silicon dioxide.
Properties: "Dicalite" materials are amorphous in form, comparatively soft and friable, free from gritty matter, and chemically inert. They consist of finely divided particles and are light in weight, varying in color according to grade from pure white, gray white, light cream to buff-pink. Gardner-Coleman oil absorption, 90 to 170; particle size, from 1 to 16% retained on a No. 325 mesh screen; maximum moisture content, 0.5 to 5%; average refractive index, 1.48.
Containers: 50-lb multi-wall paper bags.
Uses: "Dicalite" processed materials include filter aids for filtration of all types of liquids; fillers for paints, paper, asphalt products; etc; insulation for high temperature equipment; catalyst carriers, and conditioning agents for chemical fertilizers; general materials such as absorbents, extenders, concrete admixture, insecticide carriers, etc.

dicapryl adipate $C_8H_{17}OOC(CH_2)_4COOC_8H_{17}$. See note on capryl in next article.
Properties: Almost water-white liquid; b.p. (4 mm) 213-216.5°C. Good low-temperature qualities.
Use: Plasticizer used with vinyl resins and cellulose esters.

dicapryl phthalate (DCP; di-(2-octyl) phthalate) $(C_8H_{17}COO)_2C_6H_4$. The term capryl is common trade usage for the 2-octanol or sec-octyl derivative.
Properties: Nearly colorless, viscous liquid; b.p. (4.5 mm) 227-234°C; refractive index (20°C) 1.480; sp.gr. (25°C) 0.965; flash point 395°F. Insoluble in water; compatible with vinyl chloride resins and some cellulosic resins.
Grades: Technical.
Containers: Drums; tank cars.
Uses: Plasticizer for vinyl and cellulosic resins.

dicapryl sebacate $C_8H_{17}OOC(CH_2)_8COOC_8H_{17}$. See note on capryl above.
Properties: Light straw colored liquid; b.p. (4 mm) 231.5-239°C; non-volatile; excellent low-temperature flexibility.
Containers: Drums; tank cars.
Use: Plasticizer used with vinyl resins and acrylonitrile rubber.

dicapthan O-(2-chloro-4-nitrophenyl) O,O-dimethyl phosphorothioate. Used as insecticide with characteristics similar to parathion. Accepted as generic name by Ent. Soc.

dicetyl. See dotriacontane.

dicetyl ether (dihexadecyl ether).
Properties: Crystals; f.p. 54°C; b.p., decomposes at 300°C; sp.gr. 0.8117 (54/4°C).
Grade: 97% purity (min).
Uses: Electrical insulators; water repellents; lubricants in plastic molding and processing; antistatic substances; chemical intermediates.

dicetyl sulfide (dihexadecyl thioether; dihexadecyl sulfide) $(C_{16}H_{33})_2S$.
Properties: Solid; m.p. 57-58°C; b.p., decomposes; sp.gr. 0.8253 (60/4°C).
Grades: 95% (min) purity.
Uses: Organic synthesis (formation of sulfonium compounds).

dichlone (2,3-dichloro-1,4-naphthoquinone) $C_{10}H_4Cl_2O_2$.
Properties: Yellow needles; m.p. 193°C; soluble in xylene and ortho-dichlorobenzene; slightly soluble in ethyl alcohol, glacial acetic acid and carbon tetrachloride; almost insoluble in water.
Uses: Seed disinfectant; fungicide for foliage and textiles; insecticide.
Caution! Avoid contact with skin. MCA warning label.

dichlor-. See dichloro-.

dichloramine-T (para-toluenesulfondichloroamide) $CH_3C_6H_4SO_2NCl_2$. See also chloramine-T.
Properties: Pale yellow crystals or yellow crystalline powder, containing not less than 28% nor more than 30% active chlorine; chlorine odor; stable when pure; decomposed slowly by air, rapidly by impurities, petrolatums, kerosene, olive oil, and alcohol. Soluble in glacial acetic acid, chlorinated paraffin hydrocarbons, eucalyptol, benzene, chloroform and carbon tetrachloride; almost insoluble in water. M.p. 80°C.
Derivation: The product of a reaction between toluene-para-sulfonamine and calcium hypochlorite solution is acidified with acetic acid and subjected to extraction by chloroform. The chloroform solution is dried chemically, filtered and evaporated.
Use: Medicine.
Shipping regulations: None.*

"Dichloran." [430] Trade name for a high-alkyl dimethyl dichlorobenzyl ammonium chloride. Used as an antiseptic.

"Dichloricide." [123] Trademark for deodorizers, insecticides and insect repellents.

dichlorinated-. See dichloro-.

*See "I.C.C. Shipping Regulations," page xiii.
Reference numbers refer to name of manufacturer. See "List of Manufacturers," page v.

dichlorisone acetate 9alpha, 11beta-Dichloro-1,4-pregnadiene-17alpha, 21-diol-3,20-dione-21-acetate.
Properties: A chlorinated steroid. Available as liquid and cream.
Containers: Aerosol containers; bottles.
Uses: Topical therapy in medicine.

dichloroacetaldehyde $CHCl_2CHO$.
Properties: Colorless flammable liquid. with a penetrating pungent odor; density (25°C) 12.1 lbs/gal.
Containers: 55-gal drums; tank cars.
Use: Manufacture of insecticides.

dichloroacetic acid (bichloroacetic acid; Urner's liquid) $CHCl_2COOH$.
Properties: Colorless liquid; sp.gr. 1.5724 (13°C); m.p. −4°C; b.p. 193-194°C. Soluble in water, alcohol, and ether. Crystalline form, m.p. +9.3°C.
Derivation: By careful chlorination of acetic acid in the presence of iodine.
Containers: Bottles; 150-lb carboys.
Uses: Intermediate; pharmaceuticals.

2,5-dichloroacetoacetanilide $CH_3COCH_2CONHC_6H_3Cl_2$. White, crystalline solid. M.p. 96°C.

alpha,alpha-**dichloroacetophenone** $C_6H_5COCHCl_2$. Crystals; sp.gr. 1.34 (15°C); b.p. 247°C (dec); m.p. 20-21.5°C.

2,5-dichloroaniline $C_6H_3NH_2Cl_2$.
Properties: Light brown or amber-colored crystalline mass. Insoluble in water; soluble in alcohol, benzene, and dilute hydrochloric acid. M.p. 47-50°C; b.p. 251-252°C.
Derivation: Nitration of para-dichlorobenzene with subsequent reduction.
Method of purification: Steam or vacuum distillation.
Grade: Technical (96%).
Containers: 500-lb net barrels; kegs; fiber drums.
Use: Dye intermediate.
Shipping regulations: None.*

3,4-dichloroaniline $Cl_2C_6H_3NH_2$.
Properties: Crystals; m.p. 68-72°C; b.p. 272°C. Insoluble in water; soluble in most organic solvents.
Grade: Technical.
Containers: Up to tank cars.
Uses: Dye intermediate; intermediate for biologically active compounds.

2,5-dichloroaniline sulfate $Cl_2C_6H_3NH_2 \cdot \frac{1}{2}H_2SO_4$.
Properties: White powder.
Containers: Fiber kegs; polythene-lined steel drums.
Use: Intermediate.

2,4-dichlorobenzaldehyde $C_6H_3CHOCl_2$.
Properties: White crystalline solid; b.p. 233°C; m.p. 65-67°C. Soluble in methanol, absolute alcohol, ether and acetone; slightly soluble in ethanol; insoluble in water.
Containers: Fiber drums.
Use: As an intermediate in the manufacture of pharmaceuticals, dyes, and other organic chemicals.

2,5-dichlorobenzaldehyde $C_6H_3Cl_2CHO$.
Properties: White crystals; soluble in alcohol and ether.
Derivation: By the chlorination of benzaldehyde in presence of iodine or antimony.
Use: As an intermediate.

3,4-dichlorobenzaldehyde $C_6H_3Cl_2CHO$.
Properties: White crystalline solid; b.p. 230°C; m.p. 34-36°C; soluble in alcohol, ether, and acetone; slightly soluble in methanol and amyl ether; insoluble in water.
Containers: Fiber drums.
Uses: An an intermediate in the manufacture of pharmaceuticals, dyes, and other organic chemicals.

1,2-dichlorobenzene. See ortho-dichlorobenzene.

1,3-dichlorobenzene. See meta-dichlorobenzene.

1,4-dichlorobenzene. See para-dichlorobenzene.

meta-**dichlorobenzene** (1,3-dichlorobenzene) $C_6H_4Cl_2$.
Properties: Colorless liquid. Sp.gr. 1.288 (20/4°C); b.p. 172°C; m.p. −24°C; refractive index (20.9°C) 1.5457. Soluble in alcohol and ether; insoluble in water.
Derivation: By the further chlorination of mono-chlorobenzene.

ortho-**dichlorobenzene** (ortho-dichlorobenzol; 1,2-dichlorobenzene) $C_6H_4Cl_2$.
Properties: Colorless, limpid, volatile, stable, heavy liquid. Pleasant, aromatic odor; consists of a mixture composed principally of the ortho- compound which contains varying amounts of para- and meta-isomers. Considered nonflammable although it will burn when ignited.
Purified Grade: Sp.gr. 1.305 (25°C); wt/gal 10.9 lbs; chief impurities (para-dichlorobenzene, trichlorobenzene) not over 4%; boiling range 178-180.5°C; f.p. not below −19°C; flash point 79°C (Tag closed cup).
Technical Grade: Sp.gr. 1.284; wt/gal 10.7 lbs; chief impurities (para-dichlorobenzene, trichlorobenzene) not over 15%; boiling range 172-179°C; f.p. below −20°C. Miscible with most organic solvents; insoluble in water.
Derivation: By the further chlorination of mono-chlorobenzene.
Method of purification: Rectification.
Grades: Technical; purified.
Containers: 5-, 10-, 55-, 110-gal drums; tank cars.
Uses: Solvent for resins, tars, heavy greases, gums, hides, wool, waxes, sulfur, organic sulfur derivatives, acetylcellulose, and for oxides of nonferrous metals; paints, varnishes, lacquers; paint and varnish removers; metal polishes; polishing and cleaning compounds; spotting fluid for dry-cleaners; tar solvent valuable in removing tarry residues from stills and other equipment; organic synthesis;

*See "I.C.C. Shipping Regulations," page xiii.
Reference numbers refer to name of manufacturer. See "List of Manufacturers," page v.

termite exterminator; insecticide; solvent for repellents and preservatives in wood preservation; fumigant; removing sulfur from illuminating gases; heat-transfer fluid.
Shipping regulations: None.*

para-**dichlorobenzene** (1,4-dichlorobenzene; PDB) $C_6H_4Cl_2$. Widely used to protect clothing from moths.
Properties: White crystals; volatile (sublimes readily, leaving no residue); penetrating odor. Sp.gr. 1.458; b.p. 173.7°C; m.p. 53°C; flash point 67°C. Soluble in alcohol, benzene, and ether; insoluble in water.
Derivation: By the further chlorination of mono-chlorobenzene.
Grade: Technical.
Containers: Steel barrels; drums.
Uses: Insecticide; germicide; deodorant; dyes; intermediates; pharmacy; mothproofing compositions; agriculture (fumigating soil to control the peach-tree borer; destruction of the sugar-cane grub, etc.).

N,N-**dichlorobenzenesulfonamide** $C_6H_5SO_2NCl_2$.
Properties: Solid; fine crystals; white color; m.p. 68-71°C.
Containers: Glass bottles; fiber drums.
Uses: A source of positive chlorine.

3,3'-dichlorobenzidine
$C_6H_3ClNH_2C_6H_3ClNH_2$.
Properties: Crystalline solid; soluble in alcohol and ether; m.p. 133°C.
Derivation: By the chlorination of diacetylbenzidine and subsequent saponification.
Use: Intermediate for dyes and pigments.

4,4'-dichlorobenzilic acid ethyl ester. See ethyl 4,4'-dichlorobenzilate.

2,4-dichlorobenzoic acid $Cl_2C_6H_3CO_2H$.
Properties: White to slightly yellowish powder; m.p. 158-162°C. Soluble in alcohol, ether, acetone, 5% caustic. Insoluble in water and heptane.
Grades: Assay 98% min on dry basis, ash 0.20% max.
Containers: Wooden barrels; fiber drums.
Uses: Intermediate for the preparation of antimalarials, dyes, fungicides, pharmaceuticals and other organic chemicals.

3,4-dichlorobenzoic acid $Cl_2C_6H_3CO_2H$.
Properties: White to slightly yellowish powder; m.p. 202-204°C; soluble in alkali, alcohol, ether, and acetone; slightly soluble in diacetone; insoluble in water, ethylene dichloride, and toluene.
Grade: Technical.
Containers: 200-lb wooden barrels; fiber drums.
Uses: An an intermediate in the manufacture of pharmaceuticals, dyes, and other organic chemicals.

2,4-dichlorobenzotrichloride $Cl_2C_6H_3CCl_3$.
Properties: White crystalline solid; boiling range (typical) first drop 278°C, 50% 287°C, dry 292°C; f.p. (approx.) 45-48°C. Soluble in alcohol; insoluble in water.
Containers: Fiber drums.

3,4-dichlorobenzotrichloride $Cl_2C_6H_3CCl_3$.
Properties: Water-white liquid; boiling range (typical) first drop 276°C, 50% 281°C, dry 285°C; f.p. (approx.) 24.0°C; sp.gr. 1.585-1.590 (25/15°C). Soluble in alcohol, ether, and acetone; insoluble in water.
Grades: Technical.
Containers: Glass carboys.
Uses: As an intermediate in the manufacture of pharmaceuticals, dyes, and other organic chemicals.

2,4-dichlorobenzoyl chloride $Cl_2C_6H_3COCl$.
Properties: Colorless liquid; boiling range (typical) first drop 250°C, 50% 256°C, dry 260°C; f.p. 15-16°C; sp.gr. 1.500-1.510 (25/15°C). Soluble in alcohol, ether, and acetone; slightly soluble in heptane; insoluble in water.
Containers: Glass carboys.
Uses: As an intermediate in the manufacture of pharmaceuticals, dyes, and other organic chemicals.

3,4-dichlorobenzoyl chloride $Cl_2C_6H_3COCl$.
Properties: Colorless liquid; boiling range (typical) first drop 255°C, 50% 260°C, dry 270°C; f.p. 20-25°C; sp.gr. 1.508-1.513 (25/15°C). Soluble in alcohol, ether, and acetone; slightly soluble in heptane; insoluble in water.
Containers: Glass carboys.
Uses: As an intermediate in the manufacture of pharmaceuticals, dyes, and other organic chemicals.

2,4-dichlorobenzyl chloride $Cl_2C_6H_3CH_2Cl$.
Properties: Colorless liquid; boiling range (typical) first drop 245°C, 50% 248°C, dry 252°C; sp.gr. 1.415-1.420 (25/15°C). Soluble in alcohol, ether, and acetone; insoluble in water.
Containers: Glass carboys.
Uses: As an intermediate for the preparation of organic chemicals, pharmaceuticals, and dyes.

3,4-dichlorobenzyl chloride $Cl_2C_6H_3CH_2Cl$.
Properties: Colorless liquid; boiling range (typical) first drop 255°C, 50% 258°C, dry 260°C; sp.gr. 1.410-1.415 (25/15°C). Soluble in alcohol, ether, and acetone; insoluble in water.
Containers: Glass carboys.

1,1-dichloro-2,2-bis(para-ethylphenyl)ethane (diethyldiphenyldichloroethane; "Perthane"; 2,2-bis(para-ethylphenyl)-1,1-dichloroethane; di(para-ethylphenyl)dichloroethane) $CHCl_2CH(C_6H_4C_2H_5)_2$.
Properties: Crystals; m.p. 56-57°C. Soluble in acetone, kerosene, diesel fuel.
Caution: Similar to DDT as to toxicity.
Use: Insecticide, formulated as emulsifiable concentrate or wettable powder. Used especially in aerosols against insects, including moths.

1,4-dichlorobutane (tetramethylene dichloride; DCB) $ClCH_2(CH_2)_2CH_2Cl$.
Properties: Colorless, mobile liquid having a mild, pleasant odor. Sp.gr. 1.141

(20/4°C); boiling point 155°C (760 mm); flash point 104°F (Tag closed cup); refractive index (n 20/D) 1.4542. Insoluble in water; soluble in most common organic solvents.
Containers: 9-lb (1-gal) containers; 45-, 500-lb drums; 76,000-lb tank cars.
Use: Organic synthesis, including adiponitrile.

1,3-dichlorobutene-2 $ClH_2CCH:CClCH_3$.
Properties: Clear to straw-colored liquid; b.p. 125-130°C.

1,4-dichlorobutene-2 $ClH_2CCH:CHCH_2Cl$.
Properties: Colorless liquid, distinct odor. Miscible with benzene, alcohol, carbon tetrachloride. Immiscible with ethylene glycol, glycerine, and water.
Constants: B.p. 158°C (760 mm), 60°C (20 mm); m.p. 3.5°C; sp.gr. 1.1858 (25/4°C); refractive index (n 25/D) 1.4863.
Grades: Available as 95-98% trans-isomer, 2-5% cis-isomer. Above constants are for the pure trans-isomer.
Containers: Up to tank cars.
Use: Intermediate.
Caution! Contact with skin results in large blisters. Dilute vapor irritating to the eyes.

dichlorocarbene CCl_2. Exists only at low temperatures and pressures. M.p. −114°C; b.p. −20°C; decomposes on distillation at normal pressure to hexachloroethane and hexachlorobenzene. It reacts explosively with carbon; forms phosgene with oxygen.
Derivation: Reaction of carbon tetrachloride vapor with carbon at 1300°C and 10^{-3} mm Hg.
Use: Research.

2,4-dichloro-6-ortho-chloroanilino-s-triazine ("Dyrene") $(C_6H_5NCl)C_3N_3Cl_2$.
Properties: Tan crystalline solid; m.p. 159-160°C; insoluble in water.
Uses: Foliage fungicide.

dichloro(2-chlorovinyl)arsine. See chlorovinyldichloroarsine.

2,2'-dichlorodiethyl ether. See dichloroethyl ether.

dichlorodiethyl formal. See dichloroethyl formal.

dichlorodiethyl sulfide (mustard gas; dichloroethyl sulfide) $S(CH_2CH_2Cl)_2$.
Caution: Deadly vesicant war gas; causes conjunctivitis and blindness! Can be decontaminated by chloramines or bleaching powder.
Properties: Pure product: Colorless oily liquid. Technical product: Brown, oily liquid; pungent odor. Evaporates very slowly; absorbed by rubber; penetrates leather, textile fabrics. Decomposed by water. Sp.gr. 1.2741 (20°C); sp. volume 0.785 (20°C); b.p. 217.5°C (760 mm); m.p. 14.4°C (technical grade subject to impurities present); vapor density 5.4; vapor pressure 0.115 mm (20°C); volatility 0.625 mg/liter; latent ht. of fusion 25 cal; refractive index 1.53125; coefficient of thermal expansion 0.000881. Soluble in alcohol, benzene, ether, kerosene, chlorinated solvents, carbon disulfide, fats, and oils; sparingly soluble in water.
Derivation: Bubbling ethylene through sulfur chloride; also from thiodiglycol and hydrogen chloride.
Grades: Pure; technical (containing excess sulfur as a polysulfide).
Use: Organic synthesis; military poison gas.
Shipping regulations: Poison, class A. Poison gas label. Not accepted by express. Legal label name: mustard gas.*

dichlorodiethyl sulfone $(CH_2CH_2Cl)_2O_2S$.
Properties: Colorless crystals. Caution! Irritant! B.p. 179-181°C (14-15 mm); m.p. 52°C. Soluble in alcohol, chloroform, and ether; slightly soluble in water.

dichlorodifluoromethane (difluorodichloromethane; fluorocarbon-12) CCl_2F_2.
Properties: Colorless, odorless, noncorrosive gas. B.p. −29.8°C; m.p. −158°C; sp.gr. 1.486 (−30°C); critical pressure 43.2 atm. Slightly soluble in water; soluble in most organic solvents.
Derivation: (a) Reaction of carbon tetrachloride and hydrogen fluoride, in the presence of an antimony halide catalyst; (b) high temperature chlorination of vinylidene fluoride (vinylidene fluorides made by addition of hydrogen fluoride to acetylene).
Method of purification: Distillation.
Containers: Cylinders.
Uses: Refrigerant and air conditioner; aerosol propellant; plastics.
Shipping regulations: Nonflammable gas. Green label.*

1,3-dichloro-5,5-dimethylhydantoin (DDH) $ClNCONClCOC(CH_3)_2$.
Properties: White powder with mild chlorine odor. M.p. approximately 130°C; sublimes about 100°C without decomposition. Contains approximately 36% active chlorine. Slightly soluble in water with gradual liberation of hypochlorous acid; soluble in benzene, chloroform, ethylene dichloride, alcohol.
Derivation: By chlorination of dimethylhydantoin.
Grades: Technical.
Containers: Boxes; 200-lb drums.
Uses: Household laundry bleach; water treatment; mild chlorinating agent.

dichlorodinitromethane $Cl_2C(NO_2)_2$.
Properties: Liquid; b.p. 40°C (12 mm).
Derivation: Action of (fuming) nitric acid upon dichloroformoxime.

dichlorodiphenyldichloroethane. See TDE.

dichlorodiphenyltrichloroethane. See DDT.

1,1-dichloroethane. See ethylidene chloride.

1,2-dichloroethane. See ethylene dichloride.

sym-**dichloroethane.** See ethylene dichloride.

dichloroether. See dichloroethyl ether.

*See "I.C.C. Shipping Regulations," page xiii.
Reference numbers refer to name of manufacturer. See "List of Manufacturers," page v.

dichloroethoxymethane. See dichloroethylformal.

dichloroethyl acetate $CH_3COOCHClCH_2Cl$.
Properties: Water-white liquid. Sp. gr. 1.296 (20°C); boiling range: first drop 58°C, dry 65°C (13 mm); m. p. < −32°C; flash point 153°C; refractive index 1.444 (20°C); b. p., dec. Miscible with alcohol and ethyl ether. Immiscible with water.
Use: Organic synthesis.

dichloroethylarsine. See ethyldichloroarsine.

dichloroethylcarbonate $(ClH_2CCH_2O)_2CO$.
Properties: Colorless liquid. Slowly hydrolyzed by alkalies. Volatile in steam. Sp. gr. 1.3506 (20°C); b. p. 240°C (partial decomposition). Insoluble in water.
Derivation: By heating ethylene chlorohydrin and trichloromethylchloroformate together (under reflux).

sym-**dichloroethylene** (acetylene dichloride) ClHC:CHCl. Exists as cis and trans isomers.
Properties: Colorless, low-boiling, slightly toxic liquid. Pleasant odor. The hot vapors can be ignited but will not continue to burn unless heat is supplied. It decomposes slowly on exposure to air, light and moisture. Soluble in usual organic solvents; insoluble in water. Trans-isomer: sp. gr. 1.257; b. p. 47-49°C. Cis-isomer: sp. gr. 1.282; b. p. 58-60°C.
Derivation: Two stereoisomeric compounds made by the partial chlorination of acetylene.
Grades: Technical; as cis, trans, and mixture of both.
Containers: 300-, 550-lb drums.
Uses: Solvent for oils, gums, resins, waxes, rubber, shellac, and cellulose acetate; solvent mixtures for cellulose esters and ethers; dye extraction; perfumes; lacquers; thermoplastics; rubber; organic synthesis; medicine.
Warning: Flammable. MCA warning label.
Shipping regulations: Flammable liquid. Red label.*

dichloroethyl ether (dichloroether; dichloroethyl oxide; 2,2'-dichlorodiethyl ether; sym-dichloroethyl ether) $ClCH_2CH_2OCH_2CH_2Cl$.
Properties: Colorless, stable, non-corrosive liquid. Odor like that of ethylene dichloride. Vapor harmful! B. p. 178.5°C; sp. gr. 1.2220 (20/20°C); wt/gal 10.2 lbs (20°C); refractive index 1.457 (20°C); surface tension 41.8 dynes/cm (25°C); vapor pressure 0.7 mm (20°C); specific heat 0.369 cal (20-30°C); latent heat of evaporation 64.1 cal/g (178°C); flash point (closed cup) 55°C (131°F); apparent ignition temperature in air 396°C (745°F); viscosity 2.95 centipoise (20°C); f. p. −51.8°C. Miscible with most organic solvents; immiscible with the paraffin hydrocarbons; insoluble in water.
Derivation: Chlorination of ethyl ether.
Grades: Technical.
Containers: Glass bottles; iron drums; tank cars.
Uses: Solvent for fats, oils, waxes, gums, tars, pectins, resins, soaps, ethylcellulose; selective solvent for production of high grade lubricating oils; solvent mixtures for cellulose esters and ethers; textile scouring and cleansing; fulling compounds; wetting and penetrating compounds; organic synthesis; paints, varnishes, lacquers; finish removers; spotting and dry cleaning; soil fumigant.
Danger! Hazardous liquid and vapor. May be fatal if swallowed or inhaled. MCA warning label.
Shipping regulations: None.*

sym-**dichloroethyl ether.** See dichloroethyl ether.

dichloroethyl formal (dichlorodiethyl formal; bis(2-chloroethoxy)methane; dichloroethyoxymethane) $CH_2(OCH_2CH_2Cl)_2$.
Properties: Colorless liquid; b. p. 218.1°C; f. p. −32.8°C; sp. gr. 1.2339 (20/20°C); wt/gal 10.3 lb (20°C); flash point (open cup) 230°F. Slightly soluble in water; decomposed by mineral acids.
Grades: Technical.
Uses: Solvent; intermediate for polysulfide rubber.

dichloroethyl oxide. See dichloroethyl ether.

dichloroethyl sulfide. See dichlorodiethyl sulfide.

dichlorofluoromethane (fluorodichloromethane) $CHCl_2F$.
Properties: Colorless, nearly odorless gas. B. p. 8.9°C; f. p. −135°C; sp. gr. 1.426 (0°C); critical pressure 51.0 atm. Soluble in alcohol and ether; insoluble in water.
Grades: Technical.
Containers: Drums; cylinders.
Uses: Fire extinguishers; solvent; refrigerant; aerosol propellant.

dichloroformoxime CCl_2NOH.
Properties: Colorless, prismatic crystals. Disagreeable, penetrating odor. High vapor pressure. Slowly decomposes at normal temperatures, the rate depending on temperature and humidity.
Caution! Powerful irritant! B. p. 53-54°C (28 mm); m. p. 39-40°C. Soluble in water, alcohol, ether, and benzene.
Derivation: (a) Action of chlorine on fulminic acid, HONC. (b) Reduction of trichloronitrosomethane with either aluminum amalgam or hydrogen sulfide.

alpha-**dichlorohydrin** (alpha-propenyldichlorohydrin; glycerol dichlorohydrin; GDCH; dichloroisopropyl alcohol; 1,3-dichloro-2-propanol) $CH_2ClCHOHCH_2Cl$.
Properties: Colorless, slightly viscous, nonflammable, unstable liquid; faint chloroform-like odor. The commercial product is a mixture of the two isomers: 1,3-dichloro-2-hydroxypropane and 1,2-dichloro-3-hydroxypropane, of which the former is in a dominant amount. Miscible with most organic solvents, vegetable oils; slightly soluble in water. Sp. gr. 1.36-1.39;

*See "I. C. C. Shipping Regulations," page xiii.
Reference numbers refer to name of manufacturer. See "List of Manufacturers," page v.

m.p. −4°C; b.p. 174°C; refractive index 1.47-1.48; flash point 74°C; vapor pressure 7 mm.
Derivation: By the interaction of glycerol and dry hydrogen chloride gas and subsequent distillation.
Method of purification: Rectification.
Grades: Technical.
Containers: Drums.
Uses: Solvent for cellulose acetate, ethylcellulose, some types of cellulose nitrate, benzyl abietate, resins and gums, other products; solvent mixtures; intermediate in organic synthesis; paints, varnishes, lacquers; celluloid cements; water colors' binder; photographic lacquers.
Shipping regulations: None.*

5,7-dichloro-8-hydroxyquinaldine. See chlorquinaldol.

dichloroisocyanuric acid (dichloro-s-triazine-2,4,6-trione) $Cl_2H(NCO)_3$. A cyclic compound.
Properties: White, slightly hygroscopic, crystalline powder, granules. Density: (loose bulk, approx.) powder 34 lbs/cu ft; (granular) 53 lbs/cu ft. Active ingredient approximately 70% available chlorine.
Containers: 200-lb fiber drums.
Uses: Active ingredient in household dry bleaches, dishwashing compounds, scouring powders, and detergent sanitizers; replacement for calcium hypochlorite.

sym-**dichloroisopropyl alcohol.** See alpha-dichlorohydrin.

dichloroisopropyl ether $[ClCH_2C(CH_3)H]_2O$.
Properties: Colorless liquid; sp.gr. 1.1135 (20/20°C); b.p. 187.4°C (760 mm); vapor pressure 0.10 mm (20°C); flash point 185°F; wt/gal 9.3 lbs (20°C); coefficient of expansion 0.00096 (20°C); viscosity 0.0230 poise (20°C). Miscible with most oils and organic solvents; immiscible with water.
Grades: Technical.
Containers: 1-gal cans; 5- and 55-gal drums. Net content 9.0, 45, 500 lbs.
Uses: Solvent for fats, waxes, greases; extractant; paint and varnish removers; spotting agents; and cleaning solutions.
Shipping regulations: None.*

dichloromethane. See methylene chloride.

dichloromethylchloroformate $ClCOOCHCl_2$.
Properties: Colorless liquid. Decomposed by water and alkalies. Caution! Not so irritant as the mono- compound, but more toxic! Soluble in alcohol, benzene, and ether. Sp.gr. 1.56 (15°C); b.p. 110-111°C (760 mm); vapor density 5.7 (air = 1).
Derivation: (a) By chlorinating methyl formate; (b) by chlorinating methylchloroformate. In both methods the mixture of chloro-derivatives is then separated by fractionation.

sym-**dichloromethyl ether** (dichlorinated methyl oxide) $O(CH_2Cl)_2$.
Properties: Colorless, volatile liquid. Decomposed by heat and water; soluble in acetone, benzene, ethyl alcohol, and methyl alcohol; insoluble in water. Sp.gr. 1.315 (20°C); b.p. 105°C.
Derivation: (a) By the action of chlorine on methyl ether; (b) interaction of hydrochloric acid and formaldehyde, with subsequent dehydration of the chloromethyl alcohol formed.

alpha', beta-**dichloromethylethylketone** $ClCH_2COCH_2CH_2Cl$.
Properties: Liquid; slowly decomposes. Caution! Very irritant! B.p. 65°C (3 mm).
Derivation: Interaction of ethylene and chloroacetyl chloride in the presence of aluminum chloride.

5,7-dichloro-2-methyl-8-quinolinol. See chlorquinaldol.

dichloromethyl sulfate $(ClCH_2O)_2SO_2$.
Properties: Colorless, odorless liquid; soluble in alcohol, benzene, and ether. Sp. gr. 1.60 (20°C); b.p. 96-97°C (14 mm).
Derivation: (a) By bubbling sulfur trioxide through (cooled) dichloromethyl ether; (b) by heating chlorosulfonic acid with formaldehyde.

dichloronaphthalene. See chloronaphthalene.

2,3-dichloro-1,4-naphthoquinone. See dichlone.

2,5-dichloronitrobenzene (nitro-para-dichlorobenzene) $Cl_2C_6H_3NO_2$.
Properties: Pale yellow crystals.
Containers: Casks.
Use: Intermediate.

1,1-dichloro-1-nitroethane $H_3CC(Cl)_2NO_2$.
Properties: B.p. 124°C; sp.gr. 1.4153 (20/20°C); flash point 136°F.
Uses: Grain fumigant; solvent.

dichloropentane $C_5H_{10}Cl_2$. A mixture of the dichloro- derivatives of both normal and isopentane. About 40% are amylene dichlorides having two chlorine atoms attached to adjacent carbon atoms.
Specifications: Color clear and light yellow; sp.gr. 1.06-1.08 (20°C); acidity as HCl not over 0.025%; water content none; distillation 95% between 130-200°C; wt/gal 8.94 lbs.
Other properties: Surface tension 31.8 dynes/cm (25°C); viscosity 1.6 cps (25°C); specific heat 0.369 cal/g; heat of vaporization 68.5 cal/g (calc'd); water solubility negligible; water azeotrope at 80-97°C, 66% $C_5H_{10}Cl_2$ (approx.); kauributanol value 67 cc; evaporation rate at 109°F, 100% in 90 minutes.
Containers: 1-, 5-gal cans; 55-gal drums; 8000-gal tank cars.
Uses: Solvent for oils, greases, rubber, resins and bituminous materials; removal of tar; reclaiming rubber; paint and varnish removers; degreasing of metals; insecticide; soil fumigant; removal of wax deposits on sucker rods of oil-well equipment.
Fire hazard: Flash point (open cup) 105°F; fire point (open cup) 115°F.
Shipping regulations: None.*

*See "I.C.C. Shipping Regulations," page xiii.
Reference numbers refer to name of manufacturer. See "List of Manufacturers," page v.

dichlorophenarsine hydrochloride (3-amino-4-hydroxyphenyldichloroarsine hydrochloride) $C_6H_3(AsCl_2)(OH)NH_2 \cdot HCl$.
Properties: White, odorless powder; m.p. 200°C; soluble in water, solutions of alkali hydroxides and carbonates, and in dilute mineral acids.
Use: Medicine.

dichlorophene (2,2'-dihydroxy-5,5'-dichlorodiphenylmethane; 2,2'-methylene-bis-(4-chlorophenol); bis(5-chloro-2-hydroxyphenyl)methane; DDM; DDDM) $(C_6H_3ClOH)_2CH_2$.
Properties: Light tan, free-flowing powder; weakly phenolic odor; m.p. 177°C; vapor pressure 10^{-4} mm (100°C) and about 10^{-10} mm (25°C) (extrapolated value); mol wt 369.2. Soluble in acetone and alcohols; slightly soluble in benzene, toluene, carbon tetrachloride; insoluble in water.
Derivation: Condensation of para-chlorophenol with formaldehyde in the presence of sulfuric acid.
Grades: Pure and technical.
Containers: 1-, 5-, 25-, 50-, 100-lb cans; 150-lb drums.
Hazards: None, except those normally associated with inhalation of fine powders.
Uses: Fungicide and bactericide used to preserve textiles and prevent deterioration from molds, mildews, rots, mustiness, rusts and some types of rancidity; some dermatological and cosmetic applications; veterinary medicine.
Shipping regulations: None.*

2,4-dichlorophenol $Cl_2C_6H_3OH$.
Properties: White, low-melting solid; m.p. 45°C; b.p. 210°C. Soluble in alcohol and carbon tetrachloride; slightly soluble in water.
Derivation: By chlorination of phenol.
Grades: Technical.
Containers: Drums.
Uses: Organic synthesis.
Warning: Causes burns. MCA warning label.

2,4-dichlorophenoxyacetic acid. See 2,4-D.

2,4-dichlorophenoxyacetic acid, sodium salt. See 2,4-D.

3-(3,4-dichlorophenyl)-1,1-dimethylurea (diuron) $C_6H_3Cl_2NHCON(CH_3)_2$.
Properties: White, crystalline solid; m.p. 153-155°C; vapor pressure (30°C) 2×10^{-7} mm. Very low solubility in hydrocarbon solvents; approx. 42 ppm at 25°C in distilled water. Stable toward oxidation and moisture. Decomposes at 180°C.
Caution: Avoid breathing dust or mist. Avoid contact with skin, eyes, and clothing. U.S.D.A. Pesticides regulation label.
Use: Selective weed control in certain crops; or, at higher dosage rates, to control all vegetation for an extended period of time; pre-emergence herbicide.

di(para-chlorophenyl) ethanol (1,1-bis(para-chlorophenyl)ethanol; di(para-chlorophenyl)methyl carbinol; DMC; DCPC) $CH_3C(C_6H_4Cl)_2OH$.
Properties: Colorless crystals; m.p. 70°C; insoluble in water; soluble in common organic solvents.
Derivation: Reaction of 4,4'-dichlorobenzophenone with methyl magnesium bromide, followed by treatment with water.
Caution: Harmful if swallowed! Avoid contact with skin or eyes. Wash thoroughly after using. Avoid contamination of feed and foodstuffs. U.S.D.A. Pesticides Regulation label.
Use: Insecticide.

di(para-chlorophenyl) methyl carbinol. See di(para-chlorophenyl)ethanol.

dichlorophenyltrichlorosilane $Cl_2C_6H_3SiCl_3$. A mixture of isomers.
Properties: Straw-colored liquid; b.p. 260°C. Refractive index (n 20/D) 1.5638; flash point (Cleveland open cup) 286°F. Readily hydrolyzed by moisture, with the liberation of hydrochloric acid.
Derivation: Chlorination of phenyltrichlorosilane.
Grades: Technical.
Use: Intermediate for silicones.
Shipping regulations: Corrosive liquid. White label.*

3,6-dichlorophthalic acid $C_6H_2Cl_2(COOH)_2$.
Properties: Colorless, thick crystals; soluble in hot water.
Derivation: By oxidizing dichloronaphthalene tetrachloride with nitric acid.

1,2-dichloropropane. See propylene dichloride.

1,3-dichloro-2-propanol. See alpha-dichlorohydrin.

1,3-dichloropropene $CHCl{:}CHCH_2Cl$.
Properties: Exists in cis and trans isomeric forms, both colorless liquids. Sp. gr. 1.225 (20/4°); flash point (open cup) 95°F; insoluble in water; soluble in acetone, toluene, octane. Cis isomer b.p. 104°C, trans 112°C; refractive index (n 20/D) cis 1.469, trans 1.475.
Derivation: Chlorination of propylene.
Uses: Organic synthesis; soil fumigants.

alpha, alpha-**dichloropropionic acid, sodium salt.** See sodium 2,2-dichloropropionate.

2,6-dichlorostyrene $C_6H_3(CH{:}CH_2)Cl_2$.
Properties: B.p. 92-94°C (5 mm). Insoluble in water; soluble in most organic solvents. Will polymerize slowly on standing, unless inhibited.
Uses: Monomer and co-monomer in plastic research.

dichlorosulfonphthalein. See chlorphenol red.

dichlorotetrafluoroacetone $CClF_2COCClF_2$.
Properties: Colorless liquid; b.p. 45.2°C. Soluble in water and most organic solvents. Stable to acids but not alkalies.
Uses: Solvent in acidic media; complexing agent for active hydrogen compound separation.

sym-**dichlorotetrafluoroethane** (fluorocarbon-114; tetrafluorodichloroethane) $CClF_2CClF_2$.

*See "I.C.C. Shipping Regulations," page xiii.
Reference numbers refer to name of manufacturer. See "List of Manufacturers," page v.

Properties: Colorless, nearly odorless gas. B.p. 3.55°C; m.p. −94°C; critical pressure 32.3 atm.
Grades: Technical.
Containers: Cylinders up to 1 ton.
Uses: Solvent; fire extinguishers; refrigerant and air conditioner; aerosol propellants.

2,5-dichlorothiophene $C_4H_2Cl_2S$ (cyclic).
Properties: Colorless to light yellow liquid; b.p. 161°C.
Use: Intermediate.

2,4-dichlorotoluene $C_6H_3CH_3Cl_2$.
Properties: Colorless liquid; boiling range (typical) first drop 200°C, 50% 201°C, dry 202°C; f.p. (approx.) −13°C; sp.gr. 1.245-1.247 (25/15°C); refractive index 1.5480 (22°C). Soluble in alcohol, ether, and acetone; insoluble in water.
Containers: Glass carboys or drums.
Uses: As a high-boiling solvent and as an intermediate for organic synthesis.

3,4-dichlorotoluene $CH_3C_6H_3Cl_2$.
Properties: Colorless liquid. B.p. 208.9°C; m.p. −15.3°C; sp.gr. 1.2475 (25/4°C); refractive index (n 20/D) 1.5471. Miscible with alcohol, acetone, ether, benzene, carbon tetrachloride, and n-heptane; insoluble in water.
Containers: 100-lb carboys; 500-lb drums.
Uses: As a high boiling solvent and as an intermediate for organic syntheses.

dichloro-sym-triazine-2,4,6-trione. See dichloroisocyanuric acid.

beta,beta'-**dichlorovinylchloroarsine** $(ClCH{:}CH)_2AsCl$.
Properties: Yellow or yellowish-brown liquid when pure. Darker when impure. Decomposed by water. Caution! Very irritant! Soluble in alcohol, benzene, ether; insoluble in dilute acids.
Constants: Sp.gr. 1.702 (20°C); b.p. 230°C (decomposes); vapor density 8.1 (air = 1).
Derivation: Condensation of arsenic trichloride with acetylene in the presence of aluminum chloride. The mixed arsines are separated by fractionating.

beta,beta'- **dichlorovinylmethylarsine** $(ClCH{:}CH)_2AsCH_3$.
Properties: Liquid. B.p. 140-145°C (10 mm).
Derivation: Interaction of acetylene and methyldichloroarsine in the presence of aluminum chloride.

dichlorphenamide (1,3-disulfamyl-4,5-dichlorobenzene) $C_6H_2Cl_2(SO_2NH_2)_2$.
Properties: M.p. 239-241°C. Insoluble in water; soluble in alkaline solutions.
Use: Medicine.

dichromatic. Characterizing certain dyes and indicators for which different colors may be seen depending on the thickness of the solution viewed.

"**Dicodid.**" [9] Trademark for dihydrocodeinone; employed in medicine as bitartrate and hydrochloride salts.

dicophane. See DDT.

dicoumarol. See bishydroxycoumarin.

dicresyl glyceryl ether (glyceryl ditolyl ether). This may be a mixture of ortho-, meta-, and para-isomers.
$CH_3C_6H_4OCH_2CHOHCH_2OC_6H_4CH_3$.
Properties: Similar to cresyl glyceryl ether (see mephenesin) in properties. Sp.gr. 1.136; refractive index 1.549; boiling range 328-340°C.

dicresyl glyceryl ether acetate
$CH_3C_6H_4OCH_2CHOOCCH_3CH_2OC_6H_4CH_3$.
Properties: Fairly stable liquid.
Constants: Sp.gr. 1.115; b.p. 360°C; refractive index 1.53.

"**Dicumarol.**" [355] Trademark for bishydroxycoumarin (q.v.).

dicumyl peroxide $[C_6H_5C(CH_3)_2O]_2$. Used as a vulcanizing agent.
Shipping regulations: Solid and 50% solution: oxidizing material. Yellow label.*

"**Di-cup.**" [266] Trademark for a series of vulcanizing, polymerization, and cross-linking agents containing dicumyl peroxide (q.v.).

"**Dicurin**" **Procaine.** [100] Trademark for merethoxylline procaine (q.v.).

dicyandiamide (cyanoguanidine)
$NH_2C(NH)(NHCN)$.
Properties: Pure, white, crystals; sp.gr. 1.400 (25°C). Stable when dry. Melting range 207-209°C. Soluble in water and alcohol; sparingly soluble in ether.
Derivation: Polymerization of cyanamide in the presence of bases.
Method of purification: Crystallization.
Grades: 99% pure; technical.
Containers: 100-lb multiwall paper sacks.
Uses: Fertilizers; nitrocellulose stabilizer; organic synthesis, especially of melamine, barbituric acid and guanidine salts; pharmaceutical products; dyestuffs; explosives; retarding rancidity in fats and oils; fireproofing compounds; case-hardening preparations; cleaning compounds; soldering compounds; synthetic resins; vulcanizing accelerator; thinner for oil-well drilling muds; stabilizer in detergent compositions; modifier for starch products.
Shipping regulations: None.*

dicyclohexyl. See bicyclohexyl.

dicyclohexyl adipate $(CH_2CH_2CO_2C_6H_{11})_2$.
Properties: White, crystalline solid. Odorless. Compatible with most natural and synthetic resins. Soluble in most organic solvents; insoluble in water.
Constants: Sp.gr. 1.013 (40°C); b.p. 305-310°C; m.p. 37-38°C; density 45 lbs/cu. ft.; acidity (as adipic acid) less than 0.05%.

dicyclohexylamine $(C_6H_{11})_2NH$.
Properties: Clear, colorless liquid with faint amine odor; sp.gr. 0.913-0.919 (15/15°C); b.p. 256°C (760 mm); 135°C (25 mm); refractive index 1.4823 (n 25/D); flash point 110°C. Slightly soluble in water;

*See "I.C.C. Shipping Regulations," page xiii.
Reference numbers refer to name of manufacturer. See "List of Manufacturers," page v.

miscible with organic solvents. Toxic!
Containers: Drums; tank cars.
Uses: Intermediate; insecticides; plasticizer; corrosion inhibitors; antioxidants in rubber, lubricating oils, fuels; catalysts for paint, varnishes and inks; detergents; extractant.
Warning! Avoid breathing vapor or skin contact. Very toxic!

dicyclohexyl phthalate (DCHP). $C_6H_4(COOC_6H_{11})_2$.
Properties: White, granular solid; nonvolatile; mildly aromatic odor; sp. gr. (25/25°C) 1.20; m.p. 62-65°C.
Soluble in most of the organic solvents; practically insoluble in water; compatible with a large number of synthetic resins and plastics to which it imparts good electric properties, low volatile loss on oven-aging, and low water- and oil-absorption; high solvency for vinyl resins at normal processing temperatures.
Containers: Fiber drums of approximately 230-lb capacity.
Uses: Plasticizer for synthetic resins and plastics including nitrocellulose, ethyl cellulose, chlorinated rubber, polyvinyl acetate, polyvinyl chloride, and copolymers; when used to plasticize vinyl resins, produces tough, glossy compounds having a high modulus.

dicyclomine hydrochloride
$(C_6H_{11})(C_6H_{10})CO_2(CH_2)_2N(C_2H_5)_2 \cdot HCl$
2-Diethylaminoethyl bicyclohexyl-1-carboxylate hydrochloride.
Properties: Crystals; m.p. 164-166°C. Soluble in water.
Grade: N.N.D.
Uses: Medicine.

dicyclopentadiene $C_{10}H_{12}$.
Properties: Solid; sp.gr. 0.979 (20/20°C); b.p. 172°C; f.p. 33.6°C; lbs/gal 8.20 (60°F); refractive index 1.5073 (n 31/D); flash point 80-100°F (T.O.C.) Insoluble in water.
Grade: 96%.
Containers: Drums; tank cars.
Uses: Insecticides, resins, varnishes, ferrocene compounds, paints.

dicyclopentadiene dioxide $C_{10}H_{12}O_2$.
Properties: White crystalline powder; m.p. 180-184°C; sp.gr. 1.331 (25°C); slightly soluble in water; soluble in acetone and benzene.
Containers: Up to tank car lots.
Uses: Organic intermediate for epoxy resins, plasticizers, protective coatings.

dicyclopentadienyliron. See ferrocene.

dicyclopentadienyl metal compounds (ferrocenes). A class of stable hydrocarbon (cyclopentadienyl) complexes with metals. Most important member and first one to be discovered is the dicyclopentadienyl derivative of iron (see ferrocene). These compounds possess sandwich structures, that is, the two cyclopentadiene rings lie above and below the plane on which the metal atom is situated. They are mostly quite stable. For instance, they are inert to air, water, dilute acids and dilute bases. Most of the complexes of metals belonging to the first transition series on the periodic table, from titanium to nickel inclusive, have the same melting point, 173°C. They are also stable to high temperatures and to ultraviolet light.
Derivation: Interaction of cyclopentadiene sodium and the metal halide.
Uses: Suggested as coatings for missiles and satellites; jet fuels.

dicyclopentadienyltitanium chloride (titanium ferrocene) $(C_5H_5)_2TiCl_2$. A stereospecific catalyst having the typical ferrocene "sandwich" molecular structure (three-dimensionally).

DIDA. See diisodecyl adipate.

didecyl adipate $C_4H_8(COOC_{10}H_{21})_2$.
Properties: A light-colored liquid; sp. gr. 0.9181 (20/20°C); 7.7 lb/gal (20°C); b.p. 245°C (5 mm); vapor pressure 0.58 mm Hg (200°C); insoluble in water; viscosity 26.3 cps (20°C).
Use: Plasticizer.

didecylamine $[CH_3(CH_2)_9]_2NH$.
Properties: Light straw color; faint amine odor; boiling range 195-215°C (12 mm); sp. gr. 0.840 (20/20°C) (solid).

didecyl ether $(C_{10}H_{21})_2O$.
Properties: Liquid; m.p. 16°C; b.p. 170-180°C (6 mm); sp. gr. 0.819 (20/4°C); refractive index 1.4418 (n 20/D).
Grades: 95% (min) purity.
Uses: Electrical insulators; water repellent; lubricant in plastic molding and processing; antistatic substance; chemical intermediate.

didecyl phthalate (DDP) $C_6H_4(COOC_{10}H_{21})_2$.
Properties: A light-colored liquid; sp. gr. 0.9656 (20/20°C); 8.0 lb/gal (20°C); b.p. 261°C (5 mm); insoluble in water; vapor pressure 0.3 mm Hg (200°C); viscosity 113.2 cps (20°C).
Containers: Drums, tank cars; tank trucks.
Use: Plasticizer, especially for vinyl resins.

didecyl sulfide (didecyl thioether) $(C_{10}H_{21})_2S$.
Properties: m.p. 22°C; b.p. 205-206°C (4 mm); sp. gr. 0.831 (20/4°C); refractive index 1.4569 (n 33.5°C/D).
Grades: 95% (min.) purity.
Uses: Organic synthesis (formation of sulfonium compounds).

didecyl thioether. See didecyl sulfide.

didodecanyl thiodipropionate. See dilauryl thiodipropionate.

didodecylamine. See dilaurylamine.

didodecyl ether. See dilauryl ether.

didodecyl thioether. See dilauryl sulfide.

DIDP. See diisodecyl phthalate.

"Didrate." 342 Trademark for dihydrocodeine.

didymium.
1. The name applied to commercial

mixtures of rare earth elements obtained from monazite sand by extraction followed by the elimination of cerium and thorium from the mixture. The name is used like that of an element in naming mixed oxides and salts. The approximate composition of didymium from monazite, expressed as rare earth oxides is 46% lanthana, La_2O_3; 10% praseodymia, Pr_6O_{11}; 32% neodymia, Nd_2O_3; 5% samaria, Sm_2O_3; 0.4% yttrium earth oxides; 1% ceria, CeO_2; 3% gadolinia Gd_2O_3; 2% others. The mineral bastnaesite could also be a source of didymium mixtures. For uses, see didymium salts.
2. The name didymium has also been applied to mixtures of the elements praseodymium and neodymium since such mixtures were once thought to be an element and assigned the symbol Di.

didymium salts. Salts derived from commercial didymium mixtures (see didymium 1) are:
didymium acetate: $Di(C_2H_3O_2)_3 \cdot xH_2O$; pink crystals; soluble in water and acids.
didymium carbonate: $Di_2(CO_3)_3 \cdot xH_2O$; pink powder; insoluble in water; soluble in acids.
didymium chloride: $DiCl_3 \cdot 6H_2O$; lumps; soluble in water and acids.
didymium fluoride: DiF_3; nearly anhydrous pink powder; insoluble in water and acids.
didymium hydrate: Hydrated Di_2O_3; pink powder; insoluble in water; soluble in acids.
didymium nitrate: $Di(NO_3)_3 \cdot 6H_2O$; pink crystals; soluble in water and acids.
didymium oxide: Di_2O_3; brown powder; insoluble in water; soluble in acids.
Containers: Glass bottles; fiber and steel drums.
Uses: Coloring glass; decolorizing glass; in temperature-compensating capacitors for radio, television, and radar; in glass blowers' and welders' goggles (carbonate, oxide, or oxalate); in carbon arc cores (fluoride); in stainless steel (oxide); metallurgical research.

dieldrin (HEOD) $C_{12}H_{10}OCl_6$. The assigned common name for an insecticidal product containing not less than 85% of 1,2,3,4,10,10-hexachloro-6,7-epoxy-1,4,4a,5,6,7,8,8a-octahydro-1,4-endo, exo-5,8-dimethanonaphthalene. Obtained by oxidation of aldrin with peracids. See also endrin, a stereoisomer of dieldrin.
Properties: Light-tan, flaked solid, m.p. approx 150°C. Insoluble in water, methanol, aliphatic hydrocarbons; moderately soluble in most other common organic solvents. Compatible with most fertilizers, herbicides, fungicides and insecticides.
Grades: Emulsifiable concentrate; wettable powder; dust.
Containers: 175 lb fiber drums.
MCA warning labels: (dieldrin technical and formulations, 60% and over) Warning! Poisonous by skin contact, inhalation or swallowing; absorbed through skin. (Formulations 10 to 60%) Warning! Hazardous by skin contact, inhalation or swallowing. (Formulations less than 10%) Caution! Harmful by skin contact, inhalation or swallowing. (Similar labels required by U.S.D.A.)

dielectric. An electrical insulator, i. e., a material which will not conduct electric current. Thus different parts of a dielectric can have a difference in electric charges, and this will not be dissipated or only very slightly dissipated by conductivity. A dielectric is a material with electrical conductivity less than one millionth of a reciprocal ohm per centimeter. Materials with higher conductivity (10^{-6} to 10^{-3} mho/cm) are semiconductors, while ordinary electrical conductors such as metals have conductivity higher than 10^{-3} mho/cm.

dielectric constant. The dielectric constant of a material is the ratio of the electrical capacity of a condenser containing the material to the capacity of the same condenser with material replaced by a vacuum. It may be looked upon as a measure of ability of the material to maintain a difference in electrical charge over any specified distance.

dielectric liquids. Liquids of relatively high dielectric constant used in electrical capacitors, cables, switches, transformers, and circuit breakers to replace air and increase dielectric strength, and often also to improve heat dissipation. Commonly used materials are hydrocarbon oils, chlorinated hydrocarbons including chlorinated diphenyls, and silicone oils.

dielectric strength. The maximum electric field that an insulator or dielectric can withstand without breakdown, usually measured in kilovolts per centimeter. At breakdown a considerable current passes as an arc, usually with more or less decomposition of the material along the path of the current.

"Dielgran." [147] Brand name for a free-flowing granular insecticide containing 5 or 10% dieldrin.
Containers: 50-lb bags.
Uses: Control white grubs, wireworms, corn rootworms in corn and peanuts.

Diels-Alder reaction. A very general and important organic reaction for the synthesis of six-membered rings, discovered in 1928. It involves the addition of an ethylenic double bond to a conjugated diene, i. e., a compound containing two double bonds separated by one single bond, as in 1,3-butadiene ($CH_2=CH-CH=CH_2$) or cyclopentadiene. The ease of addition of the ethylenic compound is greatly enhanced by adjacent carbonyl groups, hence, maleic anhydride reacts quantitatively with hexachlorocyclopentadiene to form chlorendic anhydride.

dien. An abbreviation for diethylenetriamine, as used in formulas for coordination compounds. See also en; pn; py.

diene. An unsaturated hydrocarbon or diolefin having two double bonds, as 1, 3-butadiene, $CH_2=CH-CH=CH_2$. These double bonds are conjugated, whereas in 1, 4-pentadiene, $CH_2=CH-CH_2-CH=CH_2$, they are unconjugated, meaning separated by at least two single bonds.

"Diene." [278] Trade name for a commercial stereospecific polybutadiene rubber.
Properties: Unvulcanized: Narrow molecular weight distribution; soluble in aliphatic or aromatic hydrocarbons; free from gel; low "tackiness"; no elastic memory ("nerve").
Vulcanized: High resilience; excellent hysteretic properties; excellent resistance to abrasion and cold (brittle point <−125°F).
Containers: Boxes containing 75 lbs of "Diene" in polyethylene bags.
Uses: To enhance the low temperature and hysteretic properties of other elastomers.

dienestrol $HOC_6H_4C(CHCH_3)C(CHCH_3)C_6H_4OH$. 3, 4-Bis(para-hydroxyphenyl)-2, 4-hexadiene; a synthetic with estrogenic activity.
Properties: Colorless, odorless needles or powder; m.p. 227°C. Soluble in alcohol; practically insoluble in water. Sensitive to light.
Grade: U.S.P. XVI.
Use: Medicine.

Diesel Ignition Improver. A substance such as amyl nitrate, which is added to diesel fuels to improve fuel ignition and to raise the cetane number of the fuel. See "Ethyl."

diesel oil. Fuel for diesel engines obtained from distillation of petroleum. Its efficiency is measured by the so-called cetane number (q.v.). It is composed chiefly of aliphatic hydrocarbons. Its volatility is similar to that of gas oil. Also used in oil base drilling muds.

"Diesel-Treat." [108] Brand name for dry, granular, orange sodium dichromate, used as a corrosion inhibitor. Sold in 50-lb drums.
Uses: Closed cooling systems, particularly diesel engines; cooling tower systems.

diethanolamine (DEA; di(2-hydroxyethyl)-amine) $(HOCH_2CH_2)_2NH$.
Properties: Colorless crystals or liquid; active base; m.p. 28.0°C; b.p. 217°C (150 mm); sp.gr. 1.092 (30/20°C); flash point (open cup) 280°F; very soluble in water and alcohol; insoluble in ether, benzene.
Derivation: Ethylene oxide and ammonia.
Containers: Drums; tank cars.
Uses: Liquid detergents for emulsion paints, cutting oils, shampoos, cleaners and polishes; textile specialties; absorbent for acid gases; chemical intermediate for resins, plasticizers, etc.; solubilizing 2, 4-D.

diethanolamine lauryl sulfate. Used in synthetic detergents. Containers: Drums; tank cars.

2, 5-diethoxyaniline $NH_2C_6H_3(OC_2H_5)_2$.
Properties: Liquid; m.p. 83-85°C; insoluble in water; soluble in organic solvents.
Use: Intermediate.

1, 4-diethoxybenzene. See hydroquinone diethyl ether.

1, 1-diethoxyethane. See acetal.

diethoxyethyl phthalate. See diethyl glycol phthalate.

beta, beta'-**diethoxyethyl sulfide** $(CH_2CH_2OC_2H_5)_2S$.
Properties: Liquid.
Constants: Sp.gr. 0.9672 (20°C); b.p. 225°C (746 mm).

diethylacetal. See acetal.

diethyl acetaldehyde. See 2-ethylbutyraldehyde.

N, N-diethyl acetamide $CH_3CON(C_2H_5)_2$.
Properties: Sp.gr. 0.920 (20°C); boiling range 182-186°C; color water-white; odor faint. Flash point 170°F.

diethylacetic acid. See 2-ethylbutyric acid.

diethyl adipate $C_2H_5OCO(CH_2)_4OCOC_2H_5$.
Properties: Colorless liquid. Sp.gr. (25°C) 1.002; refractive index (25°C) 1.426; b.p. 245°C; m.p. −14°C. Insoluble in water.
Use: Plasticizer.

diethylaluminum chloride $(C_2H_5)_2AlCl$.
Properties: Colorless pyrophoric liquid. B.p. 208°C; f.p. −74°C. Flames instantly in contact with air; reacts violently in contact with water.
Derivation: Reaction of triethylaluminum with ethylaluminum sesquichloride.
Containers: Commercial quantities.
Uses: Polyolefin catalyst; intermediate in production of organometallics.

diethylaluminum hydride $(C_2H_5)_2AlH$. A pyrophoric mixture with triethylaluminum.
Derivation: Action of ethylene and hydrogen on aluminum.
Use: Catalyst reducing agent.

diethylamine $(C_2H_5)_2NH$.
Properties: Colorless liquid; ammoniacal odor; alkaline reaction. B.p. 55.5°C; freezing point −49.8°C; sp.gr. (20/20°C) 0.7062; wt/gal (20°C) 5.91 lbs; flash point (open cup) 5°F. Miscible with water, alcohol.
Derivation: From ethyl chloride and ammonia under heat and pressure.
Grades: Technical.
Containers: 1-, 5-gal cans; 55-gal drums; 8000-gal tank cars.
Uses: Rubber chemicals; textile specialties; selective solvent; dyes; flotation agents; resins; pesticides; polymerization inhibitors; pharmaceuticals; petroleum chemicals; electroplating; corrosion inhibitors.
Danger! Extremely flammable. May cause skin irritation. MCA warning label.
Shipping regulations: Flammable liquid. Red label.*

alpha-**diethylaminoaceto-2, 6-xylidide.** See lidocaine.

1-diethylamino-4-aminopentane. See 5-diethylamino-2-aminopentane.

*See "I.C.C. Shipping Regulations," page xiii.
Reference numbers refer to name of manufacturer. See "List of Manufacturers," page v.

5-diethylamino-2-aminopentane (1-diethylamino-4-aminopentane) $CH_3CH(NH_2)(CH_2)_2CH_2N(C_2H_5)_2$.
Properties: Liquid with an amine odor; sp. gr. 0.82; b.p. 142-144°C; soluble in water, alcohol and ether.
Use: Pharmaceuticals.

diethylaminoaniline. See para-aminodiethylaniline.

diethylaminoethanol $(C_2H_5)_2NCH_2CH_2OH$.
Properties: Colorless, hygroscopic liquid base combining the properties of amines and alcohols; b.p. 161°C; sp. gr. 0.88-0.89 (20/20°C); vapor pressure 21 mm (20°C); flash point 135°F; wt/gal 7.14 lbs (20°C).
Typical specifications A: Sp. gr. 0.880-0.890 (20/20°C); boiling range 158-165°C (760 mm).
Typical specifications B: Color clear and water-white; sp. gr. 0.880-0.890 (20/20°C); ; wt/gal 7.36 lbs; diethylaminoethanol between 99.5 and 101.0%; initial b.p. not below 158°C, final b.p. not above 164°C; freezing point −70°C; soluble in water, alcohol.
Grades: Technical.
Containers: 1-gal cans; 5-, 55-gal drums; tank cars.
Uses: Making water-soluble salts; fatty acid derivatives; textile softeners; pharmaceuticals; anti-rust compositions; emulsifying agents in acid media; derivatives containing tertiary amine groups.
Shipping regulations: None.*

diethylaminoethoxyethanol $(C_2H_5)_2NC_2H_4OC_2H_4OH$.
Properties: Sp. gr. 0.930-0.950 (20/20°C); boiling range 95% distils between 215.0-228.0°C.
Use: Intermediate.

beta-**diethylaminoethyl**-para-**aminobenzoate nitrate.** See procaine nitrate.

2-diethylaminoethyl benzilate hydrochloride. See benactyzine hydrochloride.

beta-**diethylaminoethyl chloride hydrochloride** (DEC) $(C_2H_5)_2NCH_2CH_2Cl\cdot HCl$. Used as intermediate in the manufacture of pharmaceuticals, and as an organic intermediate for attaching the diethylaminoethyl radical.

2-diethylaminoethyl-2-cyclopentyl-2-(2-thienyl)-hydroxyacetate methobromide. See penthienate bromide.

diethylaminoethyl diphenylacetate hydrochloride. See adiphenine hydrochloride.

beta-**diethylaminoethyl**-para-**ethoxybenzoate hydrochloride** (parethoxycaine hydrochloride) $C_2H_5OC_6H_4COOC_2H_4N(C_2H_5)_2\cdot HCl$. Used in medicine.

diethylaminoethyl 1-phenylcyclopentane-1-carboxylate hydrochloride. See caramiphen hydrochloride.

1-diethylamino-2-methylbenzene. See N,N-diethyl-ortho-toluidine.

5-diethylamino-2-pentanone $CH_3CO(CH_2)_3N(C_2H_5)_2$.
Properties: Liquid with an amine odor.
Use: Pharmaceuticals.

meta-**diethylaminophenol** $C_6H_4OHN(C_2H_5)_2$.
Properties: White, crystalline solid.
Constants: M.p. 78°C; b.p. 276-280°C. Soluble in alcohol, caustic soda, ether.
Derivation: Diethylaniline is sulfonated with oleum, and the resulting diethylaniline-meta-sulfonic acid fused with caustic soda.
Use: Dyes.

3-diethylaminopropylamine $(C_2H_5)_2NCH_2CH_2CH_2NH_2$.
Properties: Water-white; amine odor; b.p. 159°C; sp. gr. 0.82 (20/20°C); refractive index 1.442 (10°C); flash point 145°C (Tag C.C.).
Uses: Curing agent for epoxy resins; chemical intermediate.

gamma-**diethylaminopropyl chloride hydrochloride** (DEPC) $(C_2H_5)_2NCH_2CH_2CH_2Cl\cdot HCl$. Used in manufacture of pharmaceuticals. Organic chemical intermediate for attaching the diethylaminopropyl radical.

2,6-diethylaniline $(C_2H_5)_2C_6H_3NH_2$.
Properties: Brown liquid; f.p. 3°C; sp. gr. (20°C) 0.959; b.p. 243°C; flash point (open cup) 245°F. Soluble in toluene and alcohol; insoluble in water.
Uses: Chemical intermediate for pharmaceuticals, dyestuffs, pesticides, and other products.

N,N-**diethylaniline** $(C_2H_5)_2NC_6H_5$.
Properties: Colorless to yellow liquid.
Constants: Sp. gr. 0.9351; m.p. −38 to −39°C; b.p. 215-216°C; flash point 220°F. Soluble in alcohol and ether; slightly soluble in water.
Derivation: By heating aniline hydrochloride with alcohol at 180°C under pressure.
Method of purification: Rectification.
Grades: Technical.
Containers: Drums; tank cars.
Uses: Organic synthesis; dyestuff intermediate.
Shipping regulations: None.*

N,N-**diethylaniline**-meta-**sulfonic acid** $(C_2H_5)_2NC_6H_4SO_3H$.
Properties: White, crystalline solid; m.p. 270°C. Soluble in water.
Derivation: From diethylaniline by sulfonation with oleum.

diethylbarbituric acid. See barbital.

diethylbenzene $C_6H_4(C_2H_5)_2$. The commercial product is a mixture of isomers.
Properties: (1,2- or ortho-diethylbenzene): Colorless liquid; freezing point −31.4°C; b.p. 183.48°C (760 mm); refractive index (n 20/D) 1.5031; sp. gr. 0.8805 (20°C); soluble in alcohol, benzene, carbon tetrachloride, ether; insoluble in water.
(1,3- or meta-diethylbenzene): Colorless liquid; freezing point −83.92°C; b.p. 181.13°C (760 mm); refractive index (n 20/D) 1.4953; sp. gr. 0.8641 (20°C).

(1,4- or para-diethylbenzene): Colorless liquid; freezing point −43.2°C; b.p. 183.78°C (760 mm); refractive index 1.4949 (n 20/D); sp.gr. 0.8619 (20°C); soluble in alcohol, benzene, carbon tetrachloride, ether; insoluble in water.
Typical industrial specifications for mixture: B.p. range 179.8-184.8°C (760 mm); sp.gr. 0.865 (25/25°C); flash point 56°C; fire point 63°C; soluble in alcohol, benzene, carbon tetrachloride, ether; insoluble in water.
Containers: 55-gal drums, net weight 380 lbs; tank cars.
Uses: Intermediate; solvent.
Shipping regulations: None.*

diethylbromoacetamide (acetobromal) $C(C_2H_5)_2BrCONH_2$.
Properties: Crystalline powder; bitter, cooling taste; camphor-like odor. Decomposed by hot water.
Constants: M.p. 66-67°C. Slightly soluble in cold water; soluble in alcohol, benzene, ether, oils.
Use: Medicine.
Shipping regulations: None.*

di(2-ethylbutyl) azelate $C_6H_{13}OOC(CH_2)_7COOC_6H_{13}$. (2-Ethylbutyl = $(C_2H_5)_2CHCH_2- = C_6H_{13}-$.)
Properties: Very pale-yellow to water-white liquid; sp.gr. 0.9340 (20/20°C); viscosity 56 sec Saybolt (100°F); flash point 385°F; fire point 450°F; freezing point below −40°C; acid number below 1.0; faint odor; stable to heat, light, and hydrolysis.
Uses: As plasticizer for polyvinyl chloride and its copolymers as well as for cellulose esters.

di(2-ethylbutyl) phthalate (dihexyl phthalate) $C_6H_4(COOC_6H_{13})_2$.
Properties: Pale yellow liquid.
Typical specifications: Color, not more than 100 ppm Pt-Co std; sp.gr. 1.010-1.016 (20/20°C); b.p. 350°C (735 mm); acidity (as phthalic acid) 0.01% max; ester content 98% max.
Grades: Technical.
Containers: Steel drums, net 400 lbs.
Uses: Plasticizer for cellulose ester and vinyl plastics.

diethyl cadmium $(C_2H_5)_2Cd$.
Properties: Colorless pyrophoric liquid; b.p. 64°C (19 mm).
Derivation: Reaction of cadmium acetate with triethyl aluminum.
Uses: TEL production; synthesis of ketones from acid chlorides.

diethylcarbamazine citrate $C_{10}H_{21}N_3O \cdot C_6H_8O_7$. 1-Diethylcarbamyl-4-methylpiperazine dihydrogen citrate.
Properties: White, crystalline powder. Odorless or has slight odor; slightly hygroscopic. Very soluble in water; sparingly soluble in alcohol; practically insoluble in acetone, chloroform, and ether. M.p. 135-138°C.
Grade: U.S.P. XVI.
Use: Medicine.

N,N′-**diethylcarbanilide.** See sym-diethyldiphenylurea.

diethyl carbinol. See 3-pentanol.

diethyl "Carbitol." [214] Trademark for diethylene glycol diethyl ether.

diethyl carbonate (ethyl carbonate) $(C_2H_5)_2CO_3$.
Properties: Colorless liquid. Mild odor; stable.
Constants: Sp.gr. 0.975 (20/4°C); b.p. 126°C; m.p. −43°C; flash point 115°F (open cup). Miscible with alcohols, ketones, esters, aromatic hydrocarbons, some aliphatic solvents.
Derivation: Diethyl carbonate cannot be made by the usual esterification process, as carbonic acid is not reactive with ethyl alcohol. The successive steps in its manufacture are: (a) reacting chlorine and carbon monoxide to produce phosgene ($COCl_2$); (b) reacting phosgene with ethyl alcohol to make ethyl chlorocarbonate ($ClCO_2C_2H_5$); (c) reacting ethyl chlorocarbonate with anhydrous ethyl alcohol to produce diethyl carbonate. After the above steps, the crude diethyl carbonate is neutralized and redistilled.
Grades: Technical.
Containers: 1-gal (non-returnable) cans; 5-, 55-gal (non-returnable) steel drums; tank cars.
Uses: Solvent for nitrocellulose, cellulose ethers, many synthetic and natural resins; radio tube cathode fixing lacquers; organic synthesis.
Fire hazard: Combustible but not flammable; flash point over 80°F.
Shipping regulations: Requires no red caution label.*

O,O-diethyl O-3-chloro-4-methyl-2-oxo-2H-1-benzopyran-7-yl phosphorothioate ("Co-Ral") $C_{14}H_{16}ClO_5PS$.
Properties: Tan crystalline solid; m.p. 90-92°C; b.p. 20°C (10^{-7}mm); insoluble in water; soluble in aromatic solvents.
Uses: Insecticide, especially for control of pests attacking domestic animals.
Caution! May be harmful if swallowed, inhaled, or absorbed through the skin. Overexposure will result in cholinesterase depression.

diethyl chlorophosphate $(C_2H_5O)_2P(O)Cl$.
Properties: Water-white liquid; b.p. 60°C (2 mm); sp.gr. 1.1915 (25°C); refractive index 1.4153 (n 25/D). Soluble in common inert organic solvents; hydrolyzes in water.
Grades: Technical.
Use: Intermediate for organic synthesis.
Caution! Vapor intensely irritating to eye and lung tissue. Liquid can cause burns. Addition of water can cause a violent reaction.

diethylcyclohexane $(C_2H_5)_2C_6H_{10}$.
Properties: Liquid; sp.gr. 0.8037 (20/20°C); b.p. 174°C; f.p. −100°C; insoluble in water; flash point 125°F (open cup).

N,N-**diethylcyclohexylamine** $C_6H_{11}N(C_2H_5)_2$.
Properties: Clear, colorless liquid; b.p.

194.5°C. Soluble in ether and benzene; slightly soluble in water.
Grades: Technical.
Uses: Solvent; chemical intermediate.

diethyldichlorosilane $(C_2H_5)_2SiCl_2$.
Properties: Colorless liquid; b.p. 130.4°C; sp. gr. 1.053 (25/25°C); refractive index (n 25/D) 1.4309; flash point (Cleveland open cup) 77°F. Readily hydrolyzed by moisture, with the liberation of hydrochloric acid.
Derivation: By the reaction of powdered silicon and ethyl chloride at about 300°C, in the presence of copper powder.
Grades: Technical.
Containers: Bottles; 85-lb drums.
Use: Intermediate for silicones.
Shipping regulations: Corrosive liquid. White label.*

O, O-diethyl S-2-diethylaminoethyl phosphorothioate hydrogen oxalate $C_{10}H_{24}NO_3PS \cdot C_2H_2O_4$.
Properties: Crystals; m.p. 98-99°C.
Use: Insecticide.

diethyl diethylmalonate $(C_2H_5)_2C(COOC_2H_5)_2$. Colorless liquid; sweet odor; sp. gr. 0.984 (25/25°C). Used as an intermediate.

diethyldimethylmethane. See 3, 3-dimethylpentane.

diethyldiphenyldichloroethane. See 1, 1-dichloro-2, 2-bis(para-ethylphenyl)ethane.

sym-**diethyldiphenylurea** (N, N'-diethylcarbanilide; ethyl centralite; carbamite) $C_2H_5(C_6H_5)NCON(C_6H_5)C_2H_5$.
Properties: White crystalline solid; peppery odor; m.p. 79°C; b.p. 325-330°C; sp. gr. 1.12 (20°C); insoluble in water; soluble in organic solvents.
Uses: Stabilizer for nitrocellulose-based smokeless powder; age retarder in vulcanized rubber; stabilizer in solid rocket propellants.

diethylenediamine. See piperazine.

1, 4-diethylene dioxide. See 1, 4-dioxane.

diethylene disulfide. See 1, 4-dithiane.

diethylene ether. See 1, 4-dioxane.

diethylene glycol (dihydroxydiethyl ether) $CH_2OHCH_2OCH_2CH_2OH$.
Properties: Clear, colorless, practically odorless, syrupy liquid; extremely hygroscopic; non-corrosive; lowers freezing point of water. Miscible with water, ethyl alcohol, acetone, ether, ethylene glycol; immiscible with benzene, toluene, carbon tetrachloride.
Constants: B.p. 245.0°C; freezing point −8.0°C; sp. gr. 1.1184 (20/20°C); wt/gal 9.35 (15°C); refractive index 1.446 (25°C); surface tension 48.5 dynes/cm (25°C); viscosity 0.50 poise (15°C), 0.357 poise (20°C), 0.30 poise (25°C); vapor pressure 0.01 mm (30°C); specific heat 0.5509 (20°C); apparent ignition temperature in air 351°C (663°F); latent heat of evaporation 150 cal/g at b.p.; flash point 290°F; coefficient of expansion 0.00064 (20°C).
Typical specifications: Acidity not more than 0.02% (as acetic); color water-white; sp. gr. 1.117 to 1.120 (20/20°C); boiling range (760 mm) below 230°C none, below 240°C not more than 20%, below 250°C not less than 85%, below 270°C not less than 95%; water not more than 0.3%; average wt/gal 9.3 lbs (20°C).
Grades: Technical.
Containers: 5-, 10-, 55-, 110-gal drums; tank cars.
Uses: Textile lubricant, softening and finishing agent; conditioner and softener for casein, gelatin, vulcanized fibers, book-binding pastes, synthetic resins. Solvent for nitrocellulose, gums, resins, oils, organic compounds, dyes; moistening and softening agent for composition cork, glues, parchment, paper, tobacco, etc.; printing of textiles; manufacture of explosive diethylene glycol dinitrate; as a hygroscopic agent to remove moisture from natural gas; organic synthesis; as an antifreeze in sprinkler systems; cosmetics; straw products; herbicide.
Shipping regulations: None.*

diethylene glycol bis(n-butyl carbonate). See butyl diglycol carbonate.

diethylene glycol bis(chloroformate). See diglycol chloroformate.

diethylene glycol bis(cresyl carbonate). See cresyl diglycol carbonate.

diethylene glycol bis(phenyl carbonate). See phenyl diglycol carbonate.

diethylene glycol diacetate (diglycol acetate) $(CH_3COOCH_2CH_2)_2O$.
Properties: Colorless liquid. Miscible with water.
Constants: Sp. gr. 1.1159; b.p. 250°C; m.p. 19.1°C; flash point 275°F; vapor pressure 0.02 mm.
Grades: Technical.
Uses: Solvent for cellulose esters, printing inks, lacquers.

diethylene glycol dibutyl ether $C_4H_9O(C_2H_4O)_2C_4H_9$.
Properties: Practically colorless liquid with characteristic odor. Slightly soluble in water; sp. gr. 0.8853 (20/20°C); 7.4 lb/gal (20°C); b.p. 254.6°C (760 mm); vapor pressure 0.02 mm (20°C); freezing point −60.2°C; viscosity 2.39 cps (20°C).
Containers: 1-gal can; 5-, 55-gal drums; (7, 35, 400 lbs net wt).
Uses: High-boiling, inert solvent with application in extraction processes and in coatings and inks; diluent in vinyl chloride dispersions; extractant for uranium ores.

diethyleneglycol dicarbamate. See diglycol carbamate.

diethylene glycol diethyl ether. $(C_2H_5OC_2H_4)_2O$.
Properties: Colorless liquid; extremely stable; sp. gr. 0.9082 (20/20°C); b.p. 188.9°C (760 mm); flash point 180°F;

wt/gal 7.6 lb (20°C); freezing point −44.3°C. Soluble in hydrocarbons and water.
Grades: Technical.
Containers: 1-, 5-gal cans; 55-gal drums.
Uses: Solvent for nitrocellulose, resins, lacquers; high boiling medium and solvent for organic synthesis.

diethylene glycol dimethyl ether ("diglyme") $CH_3(OCH_2CH_2)_2OCH_3$.
Properties: Colorless liquid with mild odor. B.p. 162.0°C; m.p. −68.0°C; sp.gr. 0.9451 (20/20°C); flash point (open cup) 153°F; viscosity 1.089 cps (20°C). Miscible with water and hydrocarbons.
Grades: Technical.
Uses: Solvent; anhydrous reaction media for organo-metallic syntheses.

diethylene glycol dinitrate (DEGN; diglycol nitrate) $(O_2NOCH_2CH_2)_2O$.
Properties: Liquid; sp.gr. 1.377 (25/4°C); m.p. −11.3°C; b.p. 161°C; slightly soluble in water and alcohol; soluble in ether.
Uses: As a liquid rocket propellant, and also as an explosive plasticizer in solid rocket propellants.
Shipping regulations: Cannot be shipped by common carrier.*

diethylene glycol dipelargonate $(C_8H_{17}COOCH_2CH_2)_2O$. A simple ester of pelargonic acid primarily used as a secondary plasticizer for vinyls. Acid number 2.0; b.p. 229°C (5 mm); pour point 10°F, viscosity (S.U.V. at 110°C) 36 seconds.

diethylene glycol distearate. See diglycol stearate.

diethylene glycol monoacetate $HO(CH_2)_2O(CH_2)_2OOCCH_3$. Miscible with water and aromatic hydrocarbons. Solvent for nitrocellulose, cellulose acetate, camphor and rosin.

diethylene glycol monobutyl ether $C_4H_9OCH_2CH_2OCH_2CH_2OH$.
Properties: Colorless liquid; faint, characteristic butyl odor; b.p. 230.6°C; sp.gr. 0.9536 (20/20°C); wt/gal 8.0 lbs (20°C); refractive index 1.4316 (20°C); viscosity 0.0649 poise (20°C); vapor pressure 0.01 mm (20°C); specific heat 0.546 cal/g (20-25°C); flash point 240°F (open cup); coefficient of expansion 0.00088 (per °C) to 20°C; f.p. −68.1°C. Soluble in oils and water.
Grade: Technical.
Containers: 1- and 5-gal cans; 55-gal drums (net content 7.5, 35, 440 lbs); tank cars up to 10,000 gals.
Uses: Solvent for nitrocellulose, oils, dyes, gums, soaps, natural and synthetic resins; plasticizer intermediate.

diethylene glycol monobutyl ether acetate $CH_3CO(OC_2H_4)_2OC_4H_9$.
Properties: Colorless liquid, miscible with most organic liquids.
Constants: Sp.gr. 0.9810 (20/20°C); b.p. (760 mm) 246.8°C; vapor pressure $<$0.01 mm (20°C); flash point 240°F; wt/gal 8.2 lbs (20°C); nitrocellulose-xylene dilution ratio 1.8; coefficient of expansion 0.0010 (20°C); f.p. −32.3°C; viscosity 0.0356 poise (20°C).
Grades: Technical.
Containers: 1-gal cans; 5- and 55-gal drums (net content 8, 40, 450 lbs); tank cars up to 10,000 gals.
Uses: Solvent for oils, resins, gums, also for cellulose nitrate and synthetic resin coatings, where if substances are present which are good solvents for this ester, it will be retained by the film and act as a plasticizer; lacquers and coatings.

diethylene glycol monoethyl ether $CH_2OHCH_2OCH_2CH_2OC_2H_5$.
Properties: Colorless, hygroscopic liquid; mild, pleasant odor; slightly viscous; stable; b.p. 195-202°C; sp.gr. 1.0273 (20/20°C); refractive index 1.425 (n 25/D); wt/gal 8.6 lbs (20°C); miscible with water and the common organic solvents.
Grade: Technical.
Containers: 1-gal cans; 5- and 55-gal drums, net content 40, 80, and 450 lbs. Tank cars up to 10,000 gals.
Uses: Solvent for dyes, nitrocellulose, and resins; mutual solvent for mineral oil-soap and mineral oil-sulfonated oil mixtures; in the preparation of non-aqueous stains for wood; for setting the twist and conditioning yarns and cloth; textile printing; pyroxylin dope and plastics; textile soaps; lacquers; organic synthesis; brake fluid diluent.
Shipping regulations: None.*

diethylene glycol monoethyl ether acetate $CH_3COOCH_2CH_2OCH_2CH_2OC_2H_5$.
Properties: Colorless liquid; sp.gr. 1.0114 (20/20°C); b.p. (760 mm) 217.4°C; vapor pressure 0.05 mm (20°C); flash point 230°F; wt/gal 8.4 lbs (20°C); coefficient of expansion 0.00105 (20°C); f.p. −25°C; refractive index (30/D) 1.418; viscosity 0.0279 poise (20°C). Soluble in water; miscible with most organic solvents.
Grades: Technical.
Containers: 1-gal can; 5-, 55-gal drums; tank cars.
Uses: Solvent for cellulose esters, gums, resins; coatings and lacquers; printing inks.

diethylene glycol monohexyl ether $C_6H_{13}OC_2H_4OC_2H_4OH$.
Properties: Water-white liquid; sp.gr. 0.9346 (20/20°C); 7.8 lbs/gal (20°C); b.p. 259.1°C (760 mm); vapor pressure $<$0.01 mm (20°C); f.p. −33°C; viscosity 8.6 cps (20°C).
Containers: 1-gal cans; 5-, 55-gal drums (7.5, 35, 430 lbs net wt).
Use: High-boiling solvent.

diethylene glycol monolaurate. See diglycol laurate.

diethylene glycol monomethyl ether (2-(beta-methoxyethoxy) ethanol) $CH_3OCH_2CH_2OCH_2CH_2OH$.
Properties: Colorless liquid; refractive index (n 27/D) 1.4264; sp.gr. 1.0354

*See "I.C.C. Shipping Regulations," page xiii.
Reference numbers refer to name of manufacturer. See "List of Manufacturers," page v.

(20/4°C); b.p. 193°C; very soluble in water.
Containers: Drums; tank cars.
Uses: Solvent; brake fluid component; intermediate.

diethylene glycol monooleate. See diglycol oleate.

diethylene glycol monoricinoleate. See diglycol ricinoleate.

diethylene glycol monostearate. See diglycol monostearate.

diethylene glycol phthalate. See diglycol phthalate.

diethylene glycol stearate. See diglycol stearate.

diethylene oxide. See 1,4-dioxane.

diethylenetriamine $NH_2C_2H_4NHC_2H_4NH_2$.
Properties: Yellow liquid; ammoniacal odor. Strongly alkaline, hydroscopic, somewhat viscous liquid. Soluble in water, hydrocarbons. Liquid and vapor irritating to skin and eyes. Corrosive to copper and its alloys.
Constants: B.p. 206.7°C; sp.gr. 0.9542 (20/20°C); vapor pressure 0.37 mm (20°C); flash point 215°F; wt/gal 7.9 lbs (20°C); viscosity 0.0714 poise (20°C); coefficient of expansion 0.00088 (20°C).
Typical specifications: Sp.gr. 0.953-0.958 (20/20°C); boiling range 185-215°C (760 mm).
Grade: Technical.
Containers: 1-gal cans; 5-, 10-, 55-gal drums; tank cars.
Uses: Solvent for sulfur, acid gases, various resins, dyes; saponification agent for acidic materials; making derivatives.
Danger! Causes severe eye and skin burns. Avoid prolonged breathing of vapor. MCA warning label.
Shipping regulations: None.*

diethylenetriamine pentaacetic acid
$HOOCCH_2N[CH_2CH_2N(CH_2COOH)_2]_2$.
Properties: White, crystalline solid.
Grade: Technical.
Use: Chelating agent.

N,N-**diethylethanolamine** $(C_2H_5)_2NC_2H_4OH$.
Properties: Colorless liquid; refractive index 1.4400 (25°C); sp.gr. 0.8851 (20/20°C); 7.4 lbs/gal (20°C); b.p. 162.1°C (760 mm); vapor pressure 1.4 mm (20°C); viscosity 3.53 centipoise (20°C). Soluble in alcohol, ether, and benzene; completely soluble in water. Hygroscopic. Flash point 140°F. Solubility of water in compound complete at 20°C.
Uses: In the synthesis of antimalarials and of procaine; in corrosion inhibitors for protection of engine parts. Its fatty esters are of value as emulsifying agents for oils and waxes to be applied under acidic conditions.
Containers: Drums; tank cars.
Shipping regulations: None.*

diethyl ether. See ether.

diethyl ethoxymethylenemalonate
$C_2H_5OCHC(COOC_2H_5)_2$.
Properties: Liquid; sp.gr. 1.0855 (15/15°C); refractive index (n 20/D) 1.4625; b.p. 279-281°C with decomposition; insoluble in water.
Grade: 98% min (purity).
Containers: 55-gal drums.
Use: Synthesis.

uns-**diethylethylene.** See 2-ethyl-1-butene.

N,N-**diethylethylenediamine** $(C_2H_5)_2NC_2H_4NH_2$.
Properties: Colorless liquid; b.p. 145.2°C; sets to a glass below −100°C; sp.gr. 0.8211 (20/20°C); wt/gal 6.8 lb (20°C); flash point (open cup) 115°F. Miscible with water.
Grade: Technical.
Use: Intermediate.

diethyl ethylmalonate $C_2H_5CH(COOC_2H_5)_2$.
Properties: Clear colorless liquid; ester odor; sp.gr. 0.9994 (25/25°C).
Use: Intermediate.

diethyl ethylphosphonate $C_2H_5P(O)(OC_2H_5)_2$.
Properties: Colorless liquid with mild odor. Stable. Miscible with most common organic solvents. Soluble in water. Sp.gr. 1.025 (20/4°C); b.p. 82-83°C (11 mm); flash point 220°F.
Containers: 5-, 55-gal steel drums.
Uses: Heavy metal extraction and solvent separation; gasoline additives; antifoam agent; plasticizer; textile conditioner and antistatic agent.

O,O-diethyl S-2(ethylthio)ethyl phosphorodithioate ("Di-Syston") $C_8H_{19}O_2PS_3$.
Properties: Pale yellow liquid; b.p. 62°C (0.01 mm); low water solubility but soluble in most organic solvents.
Uses: Systemic insecticide for protection of crops against aphids and mites and other sucking insect pests.
Caution! Harmful if swallowed, inhaled or absorbed through the skin. Overexposure will result in cholinesterase depression.

O,O-diethyl O(and S)-2-(ethylthio)ethyl phosphorothioates (demeton; "Systox") $C_8H_{19}O_3PS_2$.
Properties: Colorless liquid; b.p. 134°C (2 mm); sp.gr. 1.118. Slightly soluble in water; soluble in most organic solvents.
Use: Systemic insecticide (absorbed by plant, which then becomes toxic to sucking and chewing insects).
Caution: May be harmful if swallowed, inhaled, or absorbed through the skin. Overexposure will result in cholinesterase depression.

diethylglycol phthalate (diethoxyethyl phthalate) $(C_2H_5OCH_2CH_2OOC)_2C_6H_4$.
Properties: Water-white to pale straw liquid; sp.gr. 1.115-1.120 (20/20°C); wt/gal 9.31 lbs.

di(2-ethylhexyl) adipate (DOA; dioctyl adipate) $C_4H_8[COOCH_2CH(C_2H_5)C_4H_9]_2$.
Properties: Light-colored oily liquid; sp.gr. 0.9268 (20/20°C); refractive index 1.4472; flash point 405°F; f.p. −70°C; b.p. 417°C

(760 mm), 214°C (5 mm); vapor pressure 2.60 mm (200°C); insoluble in water; viscosity 13.7 cps (20°C); 7.7 lb/gal (20°C).
Assay: 99% min.
Containers: 55-gal drums; tank cars and tank trucks.
Uses: Plasticizer, commonly blended with general purpose plasticizers, such as DOP and DIOP in processing polyvinyl and other synthetic resin compounds; solvent; aircraft lubes.

di(2-ethylhexyl) amine (dioctylamine) $[C_4H_9CH(C_2H_5)CH_2]_2NH$.
Properties: Water-white liquid with slightly ammoniacal odor; sp. gr. 0.8062 (20/20°C); 6.7 lbs/gal (20°C); b.p. 281.1°C (760 mm); vapor pressure 0.01 mm (20°C); viscosity 3.70 cps (20°C); flash point 270°F; high solubility in hydrocarbons and low solubility in water; solubility of water in 0.17% by wt; refractive index (n 20/D) 1.4420.
Uses: In the synthesis of dyestuffs, insecticides, emulsifying agents, and other organic compounds.

di(2-ethylhexyl) aminoethanol. See di(2-ethylhexyl)ethanolamine.

di(2-ethylhexyl) azelate (DOZ; dioctyl azelate) $(CH_2)_7[COOCH_2CH(C_2H_5)C_4H_9]_2$.
Properties: Odorless liquid; sp. gr. 0.919 (20/20°C); refractive index 1.4472; b.p. 376°C (760 mm); flash point 430°F (Cleveland open cup).
Uses: Plasticizer for vinyls; especially used as low-temperature plasticizer; base for synthetic lubricants.

di(2-ethylhexyl) ethanolamine (di(2-ethylhexyl)aminoethanol; dioctylaminoethanol) $[C_4H_9CH(C_2H_5)CH_2]_2N(CH_2)_2OH$.
Properties: Colorless liquid; insoluble in water; wt/gal 7.2 lb.
Grade: Technical.
Uses: Emulsifier; acid-stable wetting agent.

di(2-ethylhexyl) ether $[C_4H_9CH(C_2H_5)CH_2]_2O$.
Properties: Colorless, stable liquid with mild characteristic odor. Extremely low solubility in water (0.01% by wt); sp. gr. 0.8121 (20/20°C); 6.6 lbs/gal (20°C); b.p. 269.4°C (760 mm); vapor pressure <0.01 mm (20°C); sets to glass below −95°C; viscosity 2.89 cps (20°C). Solubility of water in 0.03% by wt (20°C); refractive index (n 20/D) 1.4325.
Uses: A high-boiling, inert reaction medium; also a component of certain foam breakers.

di(2-ethylhexyl) 2-ethylhexyl phosphonate $C_8H_{17}PO(OC_8H_{17})_2$.
Properties: Colorless liquid with a mild odor. Sp. gr. 0.908 (20/4°C); b.p. 160-161°C (0.26 mm); flash point 420°F. Insoluble in water; miscible with most common organic solvents.
Containers: 5-, 55-gal. drums.
Uses: Heavy metal extraction; solvent separation; gasoline additive; anti-foam agent; plasticizer; stabilizer; textile conditioner and antistatic agent.

di(2-ethylhexyl) hexahydrophthalate (dioctyl hexahydrophthalate) $C_6H_{10}[COOCH_2CH(C_2H_5)C_4H_9]_2$.
Properties: A light-colored liquid; sp. gr. 0.9586 (20/20°C); 8.0 lb/gal (20°C); b.p. 216°C (5 mm); vapor pressure 2.2 mm (200°C); insoluble in water; viscosity 42.1 cps (20°C).
Use: Plasticizer.

di(2-ethylhexyl) hydrogen phosphate (bis(2-ethylhexyl) hydrogen phosphate) $(C_8H_{17})_2HPO_4$.
Properties: Solid; sp. gr. 0.972 (20/4°C); flash point (Cleveland open cup) 340°F. Insoluble in water.
Use: Heavy metal extraction.

di(2-ethylhexyl) isophthalate (dioctyl isophthalate) $C_6H_4[COOCH_2CH(C_2H_5)C_4H_9]_2$.
Properties: B.p. 258°C at 10 mm; sp. gr. 0.984 (20/20°C); 8.2 lbs/gal; pour point +46°C; insoluble in water; viscosity 86.5 cps (20°C).
Use: Plasticizer.

di(2-ethylhexyl) maleate $C_8H_{17}OCOCHCHCOOC_8H_{17}$.
Properties: Liquid; b.p. 209°C (10 mm); f.p. sets to glass below −60°C; sp. gr. 0.9436 (20/20°C); wt/gal 7.9 lb (20°C); flash point (open cup) 365°F. Insoluble in water.
Grade: Technical.
Containers: 1-gal cans; 5-, 55-gal drums.
Uses: Copolymers; intermediate.

di(2-ethylhexyl) phosphite (bis(2-ethylhexyl) phosphite) $(C_8H_{17}O)_2PHO$. Mobile, colorless liquid with mild odor and a high degree of thermal stability. Insoluble in water (hydrolyzes very slowly); miscible with most common organic solvents. Sp. gr. 0.937 (20/4°C); b.p. 163-164°C (3 mm); flash point 330°F.
Containers: 5-gal, 55-gal drums.
Uses: Lubricant additive; intermediate; adhesive.

di(2-ethylhexyl) phosphoric acid (dioctyl phosphoric acid) $[C_4H_9CH(C_2H_5)CH_2]_2HPO_4$.
Properties: Liquid having strong acid properties; sp. gr. 0.973 (25/25°C); f.p. −60°C; refractive index 1.4420 (n 25/D); flash point 385°F; wt/gal 8.2 lbs. Insoluble in water; soluble in organic solvents.
Uses: Metal extraction and separation; intermediate for wetting agents and detergents.

di(2-ethylhexyl) phthalate (dioctyl phthalate; DOP) $C_6H_4[COOCH_2CH(C_2H_5)C_4H_9]_2$.
Properties: A light-colored odorless liquid; sp. gr. 0.9861 (20/20°C); m.p. −55°C; refractive index 1.4836; flash point 410°F; 8.2 lbs/gal (20°C); b.p. 231°C (5 mm); vapor pressure 1.3 mm Hg (200°C); viscosity 81.4 cps (20°C); insoluble in water; miscible with mineral oil.
Derivation: Reaction of 2-ethylhexyl alcohol and phthalic anhydride.
Containers: Drums; tank cars; tank trucks.
Uses: Plasticizer for many resins and synthetic rubbers.

di(2-ethylhexyl) sebacate (dioctyl sebacate) $(CH_2)_8(COOC_8H_{17})_2$.
Properties: Pale straw-colored liquid; sp. gr. 0.91 (25°C); refractive index 1.447 (28°C); b.p. 248°C (4 mm); m.p. −55°C; flash point (Cleveland open cup) 213°C; insoluble in water; partially compatible with cellulose acetate and cellulose acetate butyrate; compatible with ethyl cellulose, polystyrene, polyethylene, vinyl chloride, and vinyl chloride acetate.
Containers: Drums; tank cars.
Use: Plasticizer.

di(2-ethylhexyl) succinate (dioctyl succinate) $C_8H_{17}OCOCH_2CH_2COOC_8H_{17}$.
Properties: Liquid; b.p. 257°C (50 mm); f.p., sets to glass below −60°C; sp.gr. 0.9346 (20/20°C); wt/gal 7.8 lbs (20°C); flash point (open cup) 315°F; vapor pressure <0.01 mm (20°C); solubility in water <0.01% by wt (20°C).
Grade: Technical.
Containers: 1-gal cans; 5-, 55-gal drums.
Uses: Plasticizer; intermediate.
Shipping regulations: None.*

di(2-ethylhexyl) sodium sulfosuccinate. See dioctyl sodium sulfosuccinate.

diethyl isoamylethylmalonate $(C_2H_5)(C_5H_{11})C(COOC_2H_5)_2$. Colorless liquid; sweet odor; sp.gr. 0.950 (25/25°C). Used as an intermediate.

O,O-diethyl-O-(2-isopropyl-6-methyl-4-pyrimidinyl phosphorothioate. See "Diazinon."

diethylketone (metacetone; propione; 3-pentanone; ethyl propionyl) $C_2H_5COC_2H_5$.
Properties: Colorless, mobile, flammable liquid; acetone-like odor; soluble in alcohol and ether. B.p. 101°C; sp.gr. 0.816; m.p. −42°C; flash point (open cup) 55°F.
Derivation: By distilling sugar with an excess of lime.
Method of purification: Rectification.
Grade: Technical.
Containers: Iron drums; glass bottles.
Uses: Medicine; organic synthesis.
Fire hazard: Dangerous.
Shipping regulations: Flammable liquid. Red label.*

diethyl maleate $(HCCOOC_2H_5)_2$.
Properties: Water-white liquid; sp.gr. 1.0687; 8.92 lb/gal (20°C); refractive index (n 20/D) 1.4400; b.p. 225°C (760 mm); m.p. −11.5°C (approx.); viscosity 3.567 cps (20°C); flash point (Cleveland open cup) 250°F; dielectric constant 2.18 (calc)(25°C); surface tension 37.0 dynes/cm (20°C). Readily soluble in alcohol, diethyl ether, paraffinic hydrocarbons and most common organic solvents; sparingly soluble in water; readily hydrolyzed by alkaline solutions.
Derivation: By reacting maleic anhydride with ethyl alcohol in the presence of a catalyst.
Typical specifications: Purity 99-100% ester as diethyl maleate; sp.gr. 1.065-1.066 (25/25°C); boiling range not less than 90% distilling in a three-degree range including 129°C (40 mm); m.p. −10 to −12°C; color, water-white.
Containers: 1-gal cans; 5-, 55-gal drums; tank cars.
Uses: Organic synthesis.
Caution: Handle with care.
Shipping regulations: None.*

diethylmalonate. See ethylmalonate.

diethylmalonylurea. See barbital.

5,5-diethyl-1-methylbarbituric acid. See metharbital.

diethyl (1-methybutyl) malonate $[C_3H_7CH(CH_3)]CH(COOC_2H_5)_2$. Colorless liquid; ester odor; sp.gr. 0.969 (25/25°C). Used as an intermediate.

diethylmethylmethane. See 3-methylpentane.

N,N-diethylnicotinamide. See nikethamide.

diethyl para-nitrophenyl phosphate. See paraoxon (see under para).

O,O-diethyl-para-nitrophenyl thiophosphate. See parathion.

diethyl oxalate. See ethyl oxalate.

diethyl oxide. See ether.

di(para-ethylphenyl)dichloroethane. See 1,1-dichloro-2,2-bis(para-ethylphenyl)-ethane.

N,N-diethyl-para-phenylenediamine. See para-aminodiethylaniline.

diethyl phosphite $(C_2H_5O)_2HPO$.
Properties: Water-white liquid; b.p. 50-51°C (2 mm); sp.gr. 1.069 (25°C); refractive index 1.4061 (n 25/D); flash point 195°F (Cleveland open cup). Soluble in water, common organic solvents.
Containers: Carboys.
Uses: Paint solvent; lubricant additive; antioxidant; reducing agent; intermediate.

O,O-diethyl phosphorochloridothionate $(C_2H_5O)_2P(S)Cl$.
Properties: Colorless to light amber liquid; sp.gr. 1.196 (25/25°C); m.p. below −75°C; b.p. 49°C (<1 mm); refractive index 1.4705 (n 25/D); insoluble in water; soluble in most organic solvents. Stable at room temperature; slowly isomerizes at 100°C.
Uses: Intermediate for pesticides; oil and gasoline additives; flame retardants; flotation agents.
Caution: Exposure to vapor can cause irritation to eye and lung tissue.

diethyl phthalate (ethyl phthalate) $C_6H_4(CO_2C_2H_5)_2$.
Properties: Water-white, stable, odorless, nonflammable liquid; bitter taste; m.p. −40.5°C; refractive index 1.5002 (25°C); surface tension 37.5 dynes/cm (20°C); viscosity 31.3 centistokes (0°C); vapor pressure 14 mm (163°C), 30 mm (182°C), 734 mm (295°C); b.p. 298°C; coefficient of expansion per °F 0.00042, per °C 0.00076; dilution ratio with toluene 3.8, with petroleum naphtha 0.7; flash point

305°F; wt/gal 9.31 lb (approx.) (20°C); sp. gr. 1.120 (25/25°C). Miscible with alcohols, ketones, esters, aromatic hydrocarbons; partly miscible with aliphatic solvents.

Typical specifications: Free acid as phthalic, not more than 0.01%; color water-white; dryness, miscible without turbidity with 20 vols 60° Bé gasoline (20°C); odorless; purity ester content as diethyl phthalate 99-100%; sp. gr. 1.115-1.125 (20/20°C).

Derivation: By reacting phthalic anhydride with ethyl alcohol, followed by careful purification.

Grades: Technical.

Containers: 1-gal (nonreturnable) cans; 5-, 55-gal (nonreturnable) steel drums; tank cars.

Uses: Solvent for nitrocellulose, cellulose acetate; plasticizer; wetting agent; insecticidal sprays; camphor substitute; plastics; perfumery; as fixative and solvent; alcohol denaturant; mosquito repellents; plasticizer in solid rocket propellants.

Shipping regulations: None.*

2,2-diethyl-1,3-propanediol
$HOCH_2C(C_2H_5)_2CH_2OH$.

Properties: White crystals; m.p. 61.3°C; b.p. 160°C (50 mm); sp. gr. (at m.p.) 0.949; wt/gal 8.2 lb (60°C); flash point (open cup) 215°F. Soluble in water.

Grades: Technical; pharmaceutical.

Uses: Emulsifying agent; intermediate; medicine.

diethylpropion (1-phenyl-2-diethylamino-1-propanone hydrochloride)
$C_6H_5C(O)CH(CH_3)N(C_2H_5)_2 \cdot HCl$. Crystals; m.p. 168°C; used as a medicine.

diethylstilbestrol (stilbestrol)
$HOC_6H_4C(C_2H_5){:}C(C_2H_5)C_6H_4OH$.
3,4-Bis(para-hydroxyphenyl)-3-hexene. A non-steroid, synthetic estrogen, always in the trans form. It is the most active of the commonly-used stilbene-type compounds.

Properties: White, odorless crystalline powder; m.p. 169-172°C; almost insoluble in water; soluble in alcohol, chloroform, ether, fatty oils, and dilute alkali hydroxide.

Derivation: From anethole hydrobromide; from anisole; from anisoin or deoxyanisoin.

Containers: Bottles.

Grade: U.S.P. XVI.

Uses: Medicine; research; animal feeds.

diethylstilbestrol dimethyl ether
$(CH_3OC_6H_4CC_2H_5)_2$. White crystalline powder; m.p. 123-127°C.

diethylstilbestrol dipropionate
$C_2H_5CO_2C_6H_4C(C_2H_5)C(C_2H_5)C_6H_4CO_2C_2H_5$.
alpha, alpha-Diethyl-4,4'-stilbenediol dipropionate.

Properties: Odorless, tasteless, white, crystalline powder. M.p. 105-107°C. Readily soluble in acetone, hot alcohol, benzene, chloroform, ether and hot methanol; soluble in vegetable oils; very slightly soluble in water and dilute mineral acids; insoluble in aqueous alkalies. Suspension of 0.1 g. in 10 ml. diluted alcohol is neutral to litmus paper.

Derivation: From diethylstilbestrol by treatment with propionic anhydride in the presence of pyridine.

Grade: N.F. XI.

Use: Medicine.

diethyl succinate $(CH_2COOC_2H_5)_2$.

Properties: Clear, colorless liquid with faint pleasant odor. B.p. 216.2°C; m.p. −21°C; sp. gr. 1.0418 (20/20°C); wt/gal 8.7 lb (20°C); refractive index (n 20/D) 1.4201; flash point (open cup) 230°F. Miscible with alcohol and ether; slightly soluble in water.

Grades: Technical.

Uses: Plasticizer; intermediate.

diethyl sulfate (ethyl sulfate) $(C_2H_5)_2SO_4$.

Properties: Colorless liquid; faint, ethereal odor; irritating after-effect. Non-flammable. Noncorrosive; soluble in alcohol and ether; insoluble in water. Sp. gr. 1.1803; b.p. 208°C; vapor pressure 0.19 mm (20°C); flash point 250°F; wt/gal 9.8 lbs (20°C); coefficient of expansion 0.00091 (20°C); f.p. −24.4°C; viscosity 1.79 cps (20°C).

Typical specifications: Acidity not more than 0.03% (as sulfuric); purity not less than 98%; color, water-white; sp. gr. 1.177-1.182 (20/20°C); solubility, completely miscible with ethyl alcohol and ethyl ether; average wt/gal 9.82 lbs (20°C).

Derivation: By the action of fuming sulfuric acid on ethyl alcohol.

Method of purification: Rectification in vacuo.

Grades: Technical.

Containers: Drums; tank cars.

Use: As an ethylating agent in organic synthesis.

Shipping regulations: None.*

diethyl sulfide. See ethyl sulfide.

diethylsulfondimethylmethane. See sulfonmethane.

diethylsulfonmethylethylmethane. See sulfonethylmethane.

diethyl tartrate $C_4H_4O_6(C_2H_5)_2$.

Properties: Colorless thick, oily liquid; b.p. 280°C; m.p. 17°C; soluble in water and alcohol; sp. gr. 1.204 (20/4°C).

Uses: Plasticizer for automobile lacquers; solvent for nitrocellulose, gums, and resins.

diethylthioglycol $(CH_2CH_2OC_2H_5)_2S$.

Properties: Liquid; volatile in steam. Soluble in alcohol, benzene, and ether; slightly soluble in water. Sp. gr. 0.9672 (20°C); b.p. 225°C (746 mm).

1,3-diethylthiourea $C_2H_5NHCSNHC_2H_5$.

Properties: Buff solid; m.p. 68-71°C; slightly soluble in water; soluble in methanol, ether, acetone, benzene, and ethyl acetate; insoluble in gasoline.

Containers: Drums.

Uses: Corrosion inhibitor; for pickling iron or steel with hydrochloric acid or sulfuric acid; for reducing corrosion of ferrous metals and aluminum alloys in brine; as intermediate.

N, N-diethyl-meta-toluamide $CH_3C_6H_4CON(C_2H_5)_2$.
Properties: Colorless liquid; mild bland odor; b.p. 160°C (19 mm) and 111°C (1 mm); sp.gr. 0.996-1.002 (25/25°C); refractive index 1.5200-1.5235 (25°C); slightly soluble in water; soluble in alcohol, ether and benzene.
Grade: U.S.P. XVI.
Containers: 1- and 5-gal containers; 55-gal drums.
Uses: Insect repellents; resin solvent; film formers.
Warning: Avoid contact with eyes or lips.

N, N-diethyl-meta-toluidine $CH_3C_6H_4N(C_2H_5)_2$.
Properties: Crystals; b.p. 96-97°C (7 mm).
Grade: Technical.
Use: Dye intermediate.

N, N-diethyl-ortho-toluidine (1-diethylamino-2-methylbenzene) $CH_3C_6H_4N(C_2H_5)_2$.
Properties: Prisms from water; m.p. 72.3°C; b.p. 209°C; soluble in water, alcohol, and ether.
Derivation: From ortho-toluidine.

3, 9-diethyl-6-tridecanol (heptadecanol) $C_4H_9CH(C_2H_5)C_2H_4CH(OH)C_2H_4CH(C_2H_5)_2$.
Properties: Sp.gr. 0.8475; b.p. 309°C; flash point 310°F (open cup); refractive index 1.4531 (20°C). Insoluble in water.
Uses: Intermediate for synthetic lubricants, defoamers and surfactants.

1, 1-diethylurea $NH_2CON(C_2H_5)_2$. White solid; solidification point 41-75°C; when copolymerized with simple urea by the use of formaldehyde, it yields modified resins that differ in nature from those made through the use of mono-substituted ureas. These resins tend to be permanently thermoplastic.

diethyl valeramide. See valeryl diethylamide.

diffusion. The spontaneous mixing of one substance with another, due to the passage of the molecules of each substance through the empty spaces between molecules of the other substance. Best exemplified by gases, since any gas or mixture of gases will diffuse into others. Dissolved materials diffuse readily through liquids and very slow diffusion occurs within solids.

diffusion length. A property of materials of interest to nuclear engineers, especially of materials used in reactors for a moderator or reflector. It is a measure of the distance a thermal neutron diffuses after it is thermalized until it is captured. It is related to the density of the material and to the scattering and absorption cross sections.

difluophosphoric acid. See fluophosphoric acids.
Shipping regulations: (as difluorophosphoric acid, anhydrous) Corrosive liquid. White label.*

1, 1, 1-difluorochloroethane. See 1, 1, 1-chlorodifluoroethane.

difluorochloromethane. See chlorodifluoromethane.

difluorodichloromethane. See dichlorodifluoromethane.

4, 4'-difluorodiphenyl $FC_6H_4C_6H_4F$.
Properties: White, crystalline powder; aromatic odor. Soluble in alcohol, ether, chloroform, and oils; insoluble in water. Sp.gr. 1.04; m.p. 92-95°C; b.p. 254-255°C.
Derivation: By passing gaseous hydrofluoric acid into the product of the reaction of an excess of piperidine on diazotized benzidine.
Grades: Technical.
Containers: Glass bottles.
Use: Medicine.
Shipping regulations: None.*

difluorodiphenyltrichloroethane. See DFDT.

1, 1-difluoroethane (ethylidene fluoride) CH_3CHF_2.
Properties: Colorless, odorless gas. B.p. −24.7°C; m.p. −117°C; sp.gr. 1.004 (−25°C); index of refraction 1.255 (20°C). Insoluble in water.
Derivation: By addition of hydrogen fluoride to acetylene.
Grades: Technical.
Containers: Cylinders.
Uses: Refrigerant; aerosol propellant; intermediate.
Shipping regulations: Flammable gas. Red gas label.*

1, 1-difluoroethylene. See "Genetron" 1132A.

difoline. Powder containing cardioactive principles of digitalis.
Properties: Brown, amorphous powder soluble in water and alcohol; insoluble in ether.
Derivation: From digitalis leaves. Marketed as a solution or in tablets.

diformyl-meta-tolylenediamine.
Properties: Grayish-white; soluble in alcohol; melting range 173-175°C.
Derivation: From meta-tolylenediamine.

di-2-furfurylamine
Constants: B.p. 103-106°C (2-3 mm).
Derivation: Obtained in conjunction with tetrahydro-2-furfurylamine by hydrogenating nickel hydrofuramide in ethyl alcohol.

digallic acid. See tannic acid.

"Digifolin." [305] Trademark for purified digitalis glycosides.
Use: Medicine.

digitalin $C_{36}H_{56}O_{14}$. A digitalis glycoside.
Properties: Amorphous, white powder; m.p. 210-217°C; slightly soluble in water; soluble in alcohol.
Derivation: Solvent extraction of the seeds and leaves of Digitalis.
Containers: Bottles.
Use: Medicine.

digitalis (foxglove; purple foxglove; fairy gloves). Dried leaves of Digitalis purpurea.
Habitat: Southern and central Europe; cultivated in the United States.
Grades: Technical; U.S.P. XVI.
Containers: Boxes; drums.
Use: Medicine.
Shipping regulations: None.*

digitoxin $C_{41}H_{64}O_{13}$. Most active glycoside of Digitalis purpurea.
Properties: White, odorless, bitter leaflets or powder; m.p. 255-256°C; slightly soluble in water or ether; soluble in alcohol. Very poisonous!
Derivation: From digitalis leaves, usually Digitalis purpurea.
Containers: Bottles.
Grade: U.S.P. XVI.
Use: Medicine.

1,3-diglycidyloxybenzene. See resorcinol diglycidyl ether.

diglycocoll hydroiodide-iodine $[(NH_2CH_2COOH)_2 \cdot HI]+I_2$. Two moles of diglycocoll hydroiodide combined with two atomic weights of iodine. It contains 30.5 to 32.0 per cent of active iodine. Dark, almost black, lumpy powder with a strong odor of iodine. Freely soluble in water; only very slightly soluble in alcohol; however the iodine component is soluble. Solution in water (0.1%) is acidic (pH, about 3.0).
Use: Disinfection of drinking water.

diglycol acetate. See diethylene glycol acetate.

diglycol carbamate (diethylene glycol dicarbamate) $O(CH_2CH_2OCONH_2)_2$.
Properties: White crystalline substance; relatively stable to acid hydrolysis, but less stable to basic conditions.
Use: Manufacture of resins.

diglycol chlorohydrin. $ClCH_2CH_2OCH_2CH_2OH$.
Properties: Colorless liquid; miscible with water; sp.gr. 1.1698; b.p. 196.8°C; flash point 225°C; vapor pressure 0.17 mm.
Grades: Technical.

diglycol chloroformate [diethylene glycol bis(chloroformate)] $O(CH_2CH_2OCOCl)_2$
Properties: Liquid; b.p. 125-127°C (5 mm); soluble in acetone, alcohol, ether, chloroform, and benzene.
Use: In preparation of non-volatile plasticizers or modifying agents.

diglycolic acid $O(CH_2COOH)_2$.
Properties: White crystalline solid; m.p. 148°C; soluble in water and alcohol; pH of 10% aqueous solution 1.4. Forms a nonhygroscopic monohydrate at relative humidities above 72% at 25°C.
Containers: Multiwall paper bags (50 lbs net).
Uses: In the manufacture of resins and plasticizers; in organic syntheses.
Shipping regulations: None.*

diglycol laurate (diethylene glycol monolaurate) $C_{11}H_{23}COOC_2H_4OC_2H_4OH$.
Properties: Light straw-colored, oily liquid practically odorless, non-toxic, and edible. Sp.gr. 0.96. Dispersible in water; soluble in methanol, ethanol, toluene, naphtha, and mineral oil. Clearly miscible in certain proportions in cottonseed oil, acetone, and ethyl acetate.
Derivation: Lauric acid ester of diethylene glycol.
Containers: 1-gal cans (8 lbs); 5-gal cans (40 lbs); 55-gal drums (450 lbs).
Uses: Emulsifying agent for oils and hydrocarbon solvents in the manufacture of automobile, furniture, and shoe polishes; textile emulsions for lubrication, sizing, finishing, etc; sizing and coating emulsions for paper; lubricating and finishing emulsions for leather; fluid emulsions of oils for hand lotions, hair dressings, etc; cutting and spraying oils; dry-cleaning soap base; anti-foaming agent for casein, etc.
Shipping regulations: None.*

"Diglycol Laurate A, SE." [260] Proprietary brands of the modified ester of diethylene glycol and coconut fatty acids. Both are light yellow oily liquids, having a coconut odor. Acid value 2.0 max; iodine value 13.0 max; sp.gr. 0.974 (25°C); anionic. Dispersible in water.
Uses: Emulsifier and dispersing agent in cosmetic and other emulsion systems.

diglycol monostearate (diethylene glycol monostearate) $C_{17}H_{35}COOC_2H_4OC_2H_4OH$.
Properties: Small white flakes, available in regular or water-dispersible types.
Containers: Drums.
Uses: Emulsifier and thickener in cosmetics; mold release lubricant for die casting; temporary binder for ceramics and grinding wheels.

diglycol nitrate. See diethylene glycol dinitrate.

diglycol oleate (diethylene glycol mono-oleate;) $C_{17}H_{33}COOC_2H_4OC_2H_4OH$.
Properties: Light red, oily liquid; fatty odor. Soluble in ethanol, naphtha, ethyl acetate, methanol; partly soluble in cottonseed oil; insoluble in water. Sp.gr. 0.93; iodine value 65-75; titer below 0°C; pH (25°C) 7.7-8.2 (5% aqueous dispersion).
Derivation: Oleic acid ester of diethylene glycol.
Grades: Technical.
Containers: 1-gal cans (8 lbs); 5-gal cans (40 lbs); 55-gal drums (425 lbs).
Uses: Emulsifying agent for fluid water-in-oil emulsions for the manufacture of furniture polish, automobile polish; water-emulsion paints, and agricultural sprays.
Shipping regulations: None.*

"Diglycol Oleate 81, SE." [260] Proprietary brand of modified ester of diethylene glycol and oleic acid. Oily liquid; light yellow; fatty odor. Acid value 2.0 max; iodine value 78.0 max; sp.gr. 0.948 (25°C); anionic; dispersible in water. Flash point, open cup, 307°F.
Uses: Emulsifier and dispersing agent in cosmetic and other emulsion systems.

*See "I.C.C. Shipping Regulations," page xiii.
Reference numbers refer to name of manufacturer. See "List of Manufacturers," page v.

diglycol phthalate (diethylene glycol phthalate).
Properties: Pale yellow, liquid resin. Soluble in methanol, ethanol, acetone, ethyl acetate; partly soluble in toluol, naphtha, mineral oil, cottonseed oil; insoluble in water. Sp. gr. 1.29; saponification value 430-450; acid value 170-175.

diglycol ricinoleate (diethylene glycol monoricinoleate) $C_{17}H_{32}(OH)COOC_2H_4OC_2H_4OH$.
Properties: Light yellow liquid; m.p. below −60°C; sp. gr. 0.980 (25°C). Soluble in alcohol, acetone, and ethyl acetate; insoluble in water.
Grades: Technical.
Containers: 1-, 5-gal cans; 450-lb drums.
Uses: Plasticizer for synthetic resins and rubbers.

diglycol stearate (diethylene glycol distearate) $(C_{17}H_{35}COOC_2H_4)_2O$.
Properties: White, wax-like solid; faint fatty odor. Disperses in hot water; soluble (hot) in alcohol, oils, and hydrocarbons. M.p. 54-55°C; sp. gr. 0.9333 (20/4°C).
Derivation: Stearic acid ester of diethylene glycol.
Grades: Technical; cosmetic.
Containers: Drums.
Uses: Emulsifying agent for oils, solvents, and waxes; lubricating agent for paper and cardboard; suspending medium for titanium dioxide, carbon black, graphite, silica, etc. in the manufacture of polishes, cleaners, and textile delusterants; temporary binder for abrasive powders for the manufacture of abrasive and grinding wheels, also clays for ceramic insulation; protective coating for hygroscopic powders; thickening agent; pharmaceutical.
Shipping regulations: None.*

"Diglycol Stearate, SE." [260] Proprietary brand of modified ester of stearic acid and diethylene glycol. Waxy solid; white; mild fatty odor. Acid value 103 max; iodine value 7.0 max; m.p. 48-53°C; anionic; dispersible in water; flash point (open cup) 345°F.
Uses: Emulsifier in cosmetic and other type emulsion systems.

diglyme. Abbreviation for diethylene glycol dimethyl ether (q.v.).

digoxin. A cardiotonic digitalis glycoside. $C_{41}H_{64}O_{14}$.
Properties: Poisonous! Colorless to white crystals or white crystalline powder; odorless. Melts indistinctly at about 235°C (dec). Insoluble in water, chloroform, and ether; freely soluble in pyridine; soluble in dilute alcohol.
Derivation: From the leaves of Digitalis lanata.
Grade: U.S.P. XVI.
Use: Medicine.

"Di-Halo." [225] Trademark for bromochlorodimethylhydantoin (q.v.); active bromine 33% min., available bromine 66% min.; active chlorine 14% min., available chlorine 28% min. Used in production of disinfectants and special polymers and as reactive intermediate in synthesis.

dihexadecyl ether. See dicetyl ether.

dihexadecyl sulfide. See dicetyl sulfide.

dihexadecyl thioether. See dicetyl sulfide.

dihexyl. See n-dodecane.

di-n-hexyl adipate $(CH_2)_4(COOC_6H_{13})_2$.
Properties: Liquid; color, water-white to maximum 100 Pt-Co; sp. gr. (20°C) 0.939; refractive index (25°C) 1.438; surface tension (20°C) 32.7 dynes/cm; viscosity (20°C) 8.8 cps; b.p. (4 mm) 183-192°C (midpoint 191°C); water solubility (25°C) 0.1%; gasoline and oil solubility, complete.
Use: Rubber plasticizer in GR-S stocks at low temperatures.

di-n-hexylamine $[CH_3(CH_2)_5]_2NH$.
Properties: Water-white; b.p. 233-242°C; sp. gr. 0.788 (20/20°C); refractive index 1.434 (20°C); flash point 220°F.

di-n-hexyl maleate $C_6H_{13}OOCCH{:}CHCOOC_6H_{13}$.
Properties: Liquid; sp. gr. 0.9602 (20/20°C); b.p. 179°C (10 mm); refractive index 1.449 (20°C); vapor pressure, less than 0.01 mm (20°C); f.p. −70°C; viscosity 10.2 cps (20°C); solubility in water, less than 0.01% by wt (20°C).
Use: Preparation of resins.

dihexyl phthalate. See di(2-ethyl butyl) phthalate.

dihexyl sebacate $(CH_2)_8(COOC_6H_{13})_2$.
Properties: Light straw-colored liquid; b.p. (4 mm) 203°C.
Derivation: By reacting dodecyl alcohol with sebacic acid.
Containers: Drums; tank cars.
Use: Plasticizer for vinyl resins.
Shipping regulations: None.*

dihydrazine sulfate $(N_2H_4)_2 \cdot H_2SO_4$.
Properties: White, crystalline flakes; m.p. approx. 104°C; decomposes at about 180°C; very soluble in water; insoluble in most organic solvents.
Grades: 95% grade is available commercially.
Containers: Drums.
Use: Reducing agent.

dihydroabietyl alcohol (hydroabietyl alcohol) $C_{19}H_{31}CH_2OH$.
Properties: Solid; sp. gr. 1.007-1.008; refractive index 1.5280; vapor pressure 1.5×10^{-5} mm (25°C); m.p. 32-33°C; flash point 190°C; insoluble in water.
Containers: Drums.
Use: Plasticizer.

3,4-dihydrochlorothiazide. See hydrochlorothiazide.

dihydrocholesterol (beta-cholestanol; 3-beta-hydroxycholestane) $C_{27}H_{47}OH$. A sterol found in the feces. It differs from cholesterol in that it has no double bond.
Properties: White crystals; m.p. (monohydrate) 142°C; optical rotation α (25/D) = +23°. Soluble in fat solvents; insoluble

*See "I.C.C. Shipping Regulations," page xiii.
Reference numbers refer to name of manufacturer. See "List of Manufacturers," page v.

in water.
Derivation: By a series of oxidation and reduction reactions from cholesterol.
Use: Biochemical experimentation; pharmaceutical preparations.

dihydrocodeinone bitartrate
$C_{18}H_{21}NO_3 \cdot C_4H_6O_6 \cdot 2\frac{1}{2} H_2O$.
Properties: White, odorless, crystalline powder or crystals. Fairly soluble in water; slightly soluble in alcohol; insoluble in ether or chloroform. Affected by light.
Grade: N. F. XI.
Use: Medicine.

dihydrocoumarin. See benzodihydropyrone.

dihydrogen ferrous EDTA. See ethylenediaminetetraacetic acid salts.

2, 3-dihydroindene. See indan.

dihydromorphinone hydrochloride
$C_{17}H_{19}O_3N \cdot HCl$.
Properties: Fine, white, odorless, crystalline powder; soluble in water; sparingly soluble in alcohol; nearly insoluble in ether; affected by light.
Derivation: Reducing morphine in hydrochloric acid solution.
Grades: U. S. P. XVI.
Use: Medicine.

9, 20-dihydro-9-oxoanthracene. See anthrone.

2, 3-dihydro-4H-pyran C_5H_8O.
Properties: Colorless, mobile liquid; ether-like odor; b. p. 84.3°C (760 mm); freezing point −70°C; sp. gr. 0.927 (20/4°C); refractive index 1.4180 (25/D); flash point: 0°F (Tag closed cup). Solubility 1.6 g./100g. water (25°C); soluble in most organic solvents.
Containers: 1-, 5-, 55-gal drums; net weight: 7, 35 and 375 lbs, respectively.
Uses: Chemical intermediate.
Shipping regulations: Flammable liquid. Red label. *

1, 2-dihydro-3, 6-pyridazinedione. See maleic hydrazide.

dihydrostreptomycin (DHS) $C_{21}H_{41}N_7O_{12}$. A derivative of streptomycin (q. v.) in which the carbonyl group of the streptose (q. v.) portion has been reduced by the addition of two hydrogen atoms. It has antibiotic properties similar to streptomycin and is mainly used in the treatment of tuberculosis.
Derivation: By hydrogenation of streptomycin.
Use: Medicine (usually as sulfate salt).

dihydrostreptomycin sulfate
$(C_{21}H_{41}N_7O_{12})_2 \cdot 3H_2SO_4$. The most commonly used form of dihydrostreptomycin.
Properties: White or practically white powder; odorless or with very slight odor. Hygroscopic but stable toward light and air. Solutions are acid or nearly neutral to litmus and are levorotatory. Freely soluble in water; very slightly soluble in alcohol; practically insoluble in chloroform.
Grade: U. S. P. XVI.
Use: Medicine.

dihydrotachysterol $C_{28}H_{46}O$. A sterol.
Properties: Colorless or white crystals, or white crystalline powder. Odorless. Practically insoluble in water. Soluble in alcohol, ether, chloroform; sparingly soluble in vegetable oils; m. p. 123.5-129°C.
Grade: U. S. P. XVI.
Use: Medicine.

dihydroxyacetone (DHA; dihydroxypropanone) $HOCH_2COCH_2OH$.
Properties: Colorless, crystalline solid; m. p. 80°C; soluble in water and alcohol; nearly insoluble in ether; insoluble in petroleum ether; characteristic odor; sweet cooling taste.
Derivation: By the action of the sorbose bacterium on glycerol.
Containers: Drums.
Uses: Medicine; intermediate; emulsifier; humectant; plasticizers; fungicides; cosmetics (creates synthetic tan).

2, 4-dihydroxyacetophenone. See 4-acetylresorcinol.

dihydroxyaluminum aminoacetate (aluminum glycinate, basic; aluminum aminoacetate, basic; aluminum dihydroxyglycinate) $Al(OH)_2OOCCH_2NH_2$.
Properties: White odorless powder with faintly sweet taste; insoluble in water and organic solvents but dissolves in dilute mineral acids and solutions of fixed alkalies; forms fairly stable suspensions in water.
Derivation: Prepared by the addition of a solution of aluminum isopropoxide in isopropanol to an aqueous solution of glycine.
Grade: N. F. XI.
Use: Medicine.

1, 8-dihydroxyanthranol. See anthralin.

3, 4-dihydroxyanthranol. See anthrarobin.

1, 2-dihydroxyanthraquinone. See alizarin.

1, 5-dihydroxyanthraquinone (anthrarufin) $C_{14}H_6O_2(OH)_2$.
Properties: Yellow crystals; soluble in alcohol; very sparingly soluble in water. M. p. 280°C.
Derivation: By heating anthraquinone with boric acid and sulfuric anhydride.
Method of purification: Crystallization.
Impurities: 1, 8-dihydroxyanthraquinone.
Grades: Technical.
Containers: Wooden kegs; fiber drums.
Use: Dyes.
Shipping regulations: None. *

1, 8-dihydroxyanthraquinone (chrysazin; danthron) $C_{14}H_6O_2(OH)_2$.
Properties: Orange-colored powder or reddish-brown needles; m. p. 191°C; soluble in alcohol, sparingly soluble in water.
Derivation: From 1, 8-anthraquinone potassium disulfonate.
Grades: Technical; N. F. XI.

Containers: Kegs; fiber drums.
Uses: Dyes; medicine.
Shipping regulations: None.*

meta-**dihydroxybenzene.** See resorcinol.

ortho-**dihydroxybenzene.** See pyrocatechol.

para-**dihydroxybenzene.** See hydroquinone.

2,4-dihydroxybenzenecarboxylic acid. See beta-resorcylic acid.

4,5-dihydroxy-meta-**benzenedisulfonic acid disodium salt.** See disodium 1,2-dihydroxybenzene-3,5-disulfonate.

2,4-dihydroxybenzoic acid. See beta-resorcylic acid.

2,5-dihydroxybenzoic acid. See gentisic acid.

3,5-dihydroxybenzoic acid. See alpha-resorcylic acid.

2,5-dihydroxybenzoquinone $C_6H_2(OH)_2O_2$.
Properties: Yellow-orange solid, m.p. 216°C (dec.); soluble in concentrated sulfuric acid; slightly soluble in ethyl alcohol, acetone, water, benzene. Insoluble in petroleum ether.
Derivation: From hydroquinone.
Uses: Metal chelating; insecticides; polymerization inhibitor; tanning agent; dyestuff manufacture.
Toxic! Avoid skin and eye contact.

2,3-dihydroxybutane. See 2,3-butylene glycol.

2,5-dihydroxychlorobenzene. See chlorohydroquinone.

dihydroxydiaminomercurobenzene
$OHNH_2C_6H_3HgC_6H_3OHNH_2$. A mercury compound analogous to arsphenamine (q.v.), used in medicine as a source of mercury.
Shipping regulations: None.*

2,2'-dihydroxy-5,5'-dichlorodiphenylmethane. See dichlorophene.

dihydroxydiethyl ether. See diethylene glycol.

2,2'-dihydroxy-5,5'-difluorodiphenyl sulfide
$FC_6H_3(OH)S(OH)C_6H_3F$.
Properties: White amorphous solid; m.p. 119-121°C. Soluble in acetone, ether, chloroform, ethanol, ethyl acetate and glacial acetic acid; moderately soluble in benzene; insoluble in water, petroleum ether.
Uses: Fungicide (textile); agricultural chemical.

5,7-dihydroxydimethylcoumarin $C_{11}H_{10}O_4$.
Properties and uses closely resemble those of 5,7-dihydroxy-4-methylcoumarin.

dihydroxydiphenyl sulfone (sulfonyl diphenol)
$(C_6H_4OH)_2SO_2$. The commercial product is a mixture of the isomers, 4,4'-bishydroxyphenyl sulfone and 2,4'-bishydroxyphenyl sulfone.
Properties: White, free-flowing, odorless crystals. M.p. 217-227°C.
Grade: Technical.
Containers: 150-lb fiber drums.
Uses: Electroplating; phenolic resins; polyvinyl chloride resins; chemical intermediate.

5,5'-dihydroxy-7,7'-disulfonic-2,2'-dinaphthylurea (J acid urea; 6,6'-ureylenebis-1-naphthol-3-sulfonic acid).
Properties: (Crude) light gray paste. Soluble in water, very soluble in alkaline solution.
Derivation: Phosgenation of J acid.

para-**di-(2-hydroxyethoxy)benzene.** See hydroquinone, di(beta-hydroxyethyl) ether.

di(2-hydroxyethyl)amine. See diethanolamine.

N,N-dihydroxyethyl ethylenediamine
$(CH_2NHC_2H_4OH)_2$.
Properties: M.p. 98°C.
Use: Manufacture of textile-finishing assistants.

dihydroxyethyl sulfide. See thiodiglycol.

N,N-dihydroxyethyl-meta-**toluidine**
$CH_3C_6H_4N(C_2H_4OH)_2$.
Typical specifications: M.p. 62°C; color, light grey; dist. range 175-185°C (2 mm).

1,3-dihydroxy-4-hexylbenzene. See hexyl resorcinol.

3',4'-dihydroxy-2-isopropylaminoacetophenone hydrochloride $C_6H_3(OH)_2COCH_2NH(C_3H_7)\cdot HCl$.
A light colored crystalline powder with a faint odor. Used as an intermediate.

3,4-dihydroxy-alpha-**(methylaminomethyl)-benzyl alcohol.** See epinephrine.

1,8-dihydroxy-3-methylanthraquinone. See chrysophanic acid.

5,7-dihydroxy-4-methylcoumarin $C_{10}H_8O_4\cdot H_2O$.
Properties: Yellow to white; solid; fluoresces blue, absorbs ultraviolet light; melting range 270-285°C; insoluble in water; benzene, ether; soluble in alcohol and sodium hydroxide.
Derivation: From phloroglucinol.
Uses: In suntan oils as a sun screen; in clothes and wall paints as a whitening agent.

1,2-dihydroxynaphthalene $C_{10}H_6(OH)_2$.
Properties: Silvery plates. Soluble in alcohol and ether; sparingly soluble in water. M.p. 60°C.
Derivation: By reduction of beta-naphthoquinone with sulfurous acid.

1,3-dihydroxynaphthalene (naphthoresorcinol)
$C_{10}H_6(OH)_2$.
Properties: Transparent, crystalline plates; m.p. 124-125°C; soluble in alcohol, ether and water.
Derivation: By heating naphthalene-1,3-disulfonic acid with alkali at 230°C under pressure.
Grades: Technical; reagent.
Uses: Dyes; pharmaceuticals; analytical reagent for sugars, oils, glucuronic acid.
Shipping regulations: None.*

1,5-dihydroxynaphthalene $C_{10}H_6(OH)_2$.
Properties: White needles. Soluble in alcohol and ether; sparingly soluble in water. M.p. 260°C.

Derivation: By fusing naphthalene-1, 5-disulfonic acid with caustic soda.
Grade: Technical.
Containers: Wooden casks; fiber drums.
Use: Dyes.
Shipping regulations: None.*

1, 6-dihydroxynaphthalene $C_{10}H_6(OH)_2$.
Properties: White crystalline plates. Soluble in water. M.p. 136°C.
Derivation: By fusing naphthalene-1, 6-disulfonic acid with caustic soda.

1, 7-dihydroxynaphthalene $C_{10}H_6(OH)_2$.
Properties: Fine white needles. Soluble in alcohol and ether; sparingly soluble in water.
Constants: M.p. 158°C.
Derivation: By fusing naphthalene-1, 7-disulfonic acid with caustic soda.

1, 8-dihydroxynaphthalene $C_{10}H_6(OH)_2$.
Properties: White needles or plates. Soluble in alcohol and ether; sparingly soluble in water. M.p. 138°C.
Derivation: From naphthosulfone by fusion with caustic soda.

2, 3-dihydroxynaphthalene $C_{10}H_6(OH)_2$.
Properties: Colorless crystals. Soluble in alcohol and ether; sparingly soluble in water. M.p. 160°C.
Derivation: From 2-naphthol-3, 6-disulfonic acid by fusion with caustic soda.

2, 6-dihydroxynaphthalene $C_{10}H_6(OH)_2$.
Properties: White, crystalline plates. Soluble in alcohol and ether; sparingly soluble in water. M.p. 216°C.
Derivation: From 2-naphthol-6-sulfonic acid by fusion with caustic soda.

2, 7-dihydroxynaphthalene $C_{10}H_6(OH)_2$.
Properties: Long, white, crystalline needles or plates. Soluble in alcohol and ether; sparingly soluble in water. M.p. 186°C.
Derivation: From 2-naphthol-7-sulfonic acid by fusion with caustic soda.

2, 8-dihydroxy-3-naphthoic acid $C_{10}H_5(OH)_2COOH$.
Properties: Light green powder; slightly soluble in hot water; soluble in alcohol and acetone.
Use: Intermediate.

dihydroxyphenacyl chloride $C_6H_3(OH)_2COCH_2Cl$. Purple powder; used in medicine.

1-(3, 4-dihydroxyphenyl)-2-isopropylamino-ethanol esters. See isoproterenol hydrochloride and sulfate.

dihydroxypropanone. See dihydroxyacetone.

1, 2-dihydroxypropane. See propylene glycol.

4, 8-dihydroxyquinaldic acid. See xanthuremic acid.

8, 9-dihydroxystearic acid $C_{17}H_{33}(OH)_2COOH$.
Properties: White crystals, odorless, tasteless. Soluble in alcohol and ether; insoluble in water. M.p. 135°C.
Derivation: By heating dibromide of isooleic acid with silver oxide.

dihydroxysuccinic acid. See tartaric acid.

3, 5-dihydroxytoluene. See orcin.

dihyprylone (3, 3-diethyl-2, 4-dioxopiperidine) $C_9H_{15}NO_2$.
Properties: M.p. 102-107°C. Soluble in water, alcohol and chloroform.
Use: Medicine.

DII. Abbreviation for diesel ignition improver. See under "Ethyl."

diiodoacetylene IC⋮CI.
Properties: White crystals. Unpleasant odor. Light acts upon it, causing a gradual change in color to red and a separation of iodine. Caution! Very toxic and very volatile! Its vapors affect the eyes and mucous membranes! Soluble in alcohol, ether, benzene; insoluble in water. M.p. 78.5°C.
Derivation: By dissolving iodine in liquid ammonia and passing acetylene into the solution.
Grade: Technical.
Use: Organic synthesis.

diiodoaniline $C_6H_3I_2NH_2$.
Properties: Shining, brown crystals. Soluble in alcohol, ether, chloroform, ethyl acetate, and carbon disulfide; insoluble in water. M.p. 96°C; sp.gr. 2.75.
Derivation: By the action of iodine chloride on acetanilide, followed by saponification and distillation with steam.
Method of purification: Crystallization.
Grade: Technical.
Containers: Glass bottles; tins.
Use: Medicine.
Shipping regulations: None.*

diiodobrassidinic acid ethyl ester. See ethyl diiodobrassidate.

sym-**diiododibromoethylene** BrIC:CIBr.
Properties: Crystals; m.p. 95-96°C.
Derivation: Reaction of iodine and dibromoacetylene.
Use: Organic synthesis.

diiododiethyl sulfide $(CH_2CH_2I)_2S$.
Properties: Bright-yellow prisms. Slowly decomposes, the rate being accelerated by light and by heat. Hydrolyzed by alkali solutions. Caution! Very toxic! Soluble in alcohol, benzene, ether; insoluble in water. M.p. 62°C.
Derivation: Interaction of dichlorodiethyl sulfide with an acetic acid solution of sodium iodide.

diiodoform. See tetraiodoethylene.

diiodoformoxime CI_2NOH.
Properties: Crystals. Not so toxic or irritant as the chloroformoximes. M.p. 69°C.

diiodohydroxyquin (diiodohydroxyquinoline; 5, 7-diiodo-8-quinolinol) $C_9H_4NI_2OH$.
Properties: Colorless, or light yellowish to tan, microcrystalline powder; odorless or with faint odor. Stable in air. Melts with decomposition 200-215°C. Almost insoluble

*See "I.C.C. Shipping Regulations," page xiii.
Reference numbers refer to name of manufacturer. See "List of Manufacturers," page v.

in water; sparingly soluble in alcohol, acetone, and ether; soluble in hot pyridine and hot dioxane.
Derivation: By action of iodine monochloride on 8-hydroxyquinoline.
Grade: U.S.P. XVI.
Use: Medicine.

diiodohydroxyquinoline. See diiodohydroxyquin.

diiodomethane. See methylene iodide.

diiodo-4-phenolsulfonic acid. See sozoiodolic acid.

3,5-diiodo-4-pyridone-N-acetic acid, diethanolamine salt. See iodopyracet.

5,7-diiodo-8-quinolinol. See diiodohydroxyquin.

3,5-diiodosalicylic acid $C_7H_4O_3I_2$.
Properties: White to pale pink crystalline powder; slightly soluble in water.
Uses: Organic source of iodine for salt blocks and salt for animal nutrition.

diiodothyronine
$HOC_6H_4OC_6H_2I_2CH_2CH(NH_2)COOH$.
3,5-Diiodothyronine. A thyronine derivative which is an intermediate obtained in the manufacture of synthetic thyroxine; also, probably an intermediate in the synthesis of thyroxine by the thyroid gland.

diisobutyl adipate (DIBA)
$[C_2H_4COOCH_2CH(CH_3)_2]_2$.
Properties: Colorless, clear liquid. Odorless. Compatible with most natural and synthetic resins. Soluble in most organic solvents; insoluble in water.
Constants: Sp. gr. 0.950 (25°C); b.p. 278-280°C; m.p. −20°C; wt/gal 7.95 lbs; acidity (as adipic acid) less than 0.05%.
Containers: 1-, 5-gal cans; 55-gal drums; tank cars.
Use: Plasticizer.

diisobutyl aluminum chloride
$[(CH_3)_2CHCH_2]AlCl$.
Properties: Colorless liquid.
Derivation: Reaction of isobutylene and hydrogen on aluminum.
Use: Polyolefin catalyst.

diisobutyl aluminum hydride
$[(CH_3)_2CHCH_2]_2AlH$.
Properties: Colorless pyrophoric liquid; b.p. 105°C (0.2 mm).
Derivation: Reaction of isobutylene and hydrogen with aluminum.
Use: Reducing agent in manufacture of pharmaceuticals.

diisobutylamine $[(CH_3)_2CHCH_2]_2NH$.
Properties: Sp. gr. 0.745 (20°C); boiling range 136-140°C; color water-white; odor amine.
Fire hazard: Flash point 85°F.
Use: Intermediate.
Shipping regulations: None.*

diisobutylcarbinol. See 2,6-dimethyl-4-heptanol.

diisobutyl carbinyl acetate. See nonyl acetate.

diisobutylene. This term refers to a number of isomeric compounds of the formula C_8H_{16}, of which 2,4,4-trimethyl-pentene-1 and 2,4,4-trimethyl-pentene-2 are the most important since they are formed in appreciable amounts when isobutene (isobutylene) is polymerized.
Typical properties: Sp. gr. 0.7227 (60°F); boiling range 214-220°F.
Containers: Drums; tank cars.
Uses: Alkylation; intermediates; antioxidants; surfactants; lube additives; plasticizers; rubber chemicals.

alpha-**diisobutylene.** See 2,4,4-trimethyl-pentene-1.

beta-**diisobutylene.** See 2,4,4-trimethyl-pentene-2.

diisobutyl ketone (2,6-dimethyl-4-heptanone) $(CH_3)_2CHCH_2COCH_2CH(CH_3)_2$.
Properties: Colorless liquid. Stable. Miscible with most organic liquids.
Constants: Sp. gr. 0.8089 (20/20°C); b.p. 168.1°C (760 mm); vapor pressure 1.7 mm (20°C); flash point 140°F; wt/gal 6.7 lbs (20°C); freezing point −41.5°C; coefficient of expansion 0.00101 (20°C).
Typical specifications: Sp. gr. 0.808-0.813 (20/20°C) boiling range, 165-170°C (760 mm); acidity, not more than 0.2% (as acetic).
Grade: Technical.
Containers: 1-gal cans; 5-, 55-gal drums; tank cars.
Uses: Solvent for nitrocellulose, rubber, synthetic resins; lacquers; coating compositions; organic synthesis; roll-coating inks; stains; rubber.
Caution: Keep away from open flame and heat. Avoid breathing vapor. Avoid contact with skin. MCA warning label.
Shipping regulations: None.*

diisobutyl phenol. See octyl phenol.

diisobutyl phthalate $C_6H_4[COOCH_2CH(CH_3)_2]_2$.
Properties: Liquid; refractive index 1.4900 (n 25/D); sp. gr. 1.040 (20/20°C); flash point 345°F; b.p. 327°C.
Containers: 55-gal drums, tank trucks, tank cars.
Use: Plasticizer.

diisocyanates. Organic compounds containing two isocyanate groups (-NCO). They are formed by treating diamines (e.g., toluene-2,4-diamine, hexamethylenediamine, para, para'-diaminodiphenylmethane) with phosgene.
Uses: Production of polyurethane foams, resins, and rubber; incorporation in phenol-formaldehyde resins to improve water and alkali resistance; bonding rubber to rayon or nylon.

diisodecyl adipate (DIDA)
$C_{10}H_{21}OOC(CH_2)_4COOC_{10}H_{21}$.
Properties: Light colored, oily liquid; mild odor. Sp. gr. 0.918 (20/20°C); f.p. −71°C; boiling range 239-246°C; refractive index 1.450 (25°C); wt/gal 7.5 lbs. Insoluble or

limited solubility in glycerol, the glycols and some amines. Soluble in most other organic liquids.
Containers: 1-, 5-gal cans; 55-gal drums; tank cars; tank trucks.
Uses: Primary plasticizer for most resins; in sheeting, film and extrusions.

diisodecyl-4, 5-epoxy-tetrahydrophthalate. Plasticizer-stabilizer resistant to fungi; useful especially in vinyl plastics for outdoors. See "Flexol" PEP.

diisodecyl phthalate (DIDP) $C_6H_4(COOC_{10}H_{21})_2$.
Properties: Clear, volatile liquid with a mild odor; sp. gr. 0.966 (20/20°C); f. p. −50°C; b. p. 250-257°C (4 mm); refractive index n 25°C 1.483; viscosity 108 cps (20°C); wt/gal 8 lbs. Insoluble in glycerol, glycols and some amines; soluble in most other organic liquids.
Grade: Technical.
Containers: 55-gal drums; tank cars; tank trucks.
Use: Plasticizer for vinyl resins.

diisooctyl acid phosphate.
Shipping regulations: Corrosive liquid. White label.*

diisooctyl adipate (DIOA) $C_8H_{17}OOC(CH_2)_4COOC_8H_{17}$.
Properties: A light straw-colored liquid; mild odor; sp. gr. 0.924 (25°C); b. p. 214-226° (4 mm); f. p. −75°C.
Containers: 1-, 5-gal cans; 55-gal drums; tank cars.
Use: Plasticizer, especially for low temperatures.

diisooctyl azelate (DIOZ). A diester of azelaic acid used primarily as a plasticizer for vinyls and as a base for synthetic lubricant fluids. B. p. 237°C (5 mm); pour point −85°F, acid number 1.0; viscosity (S. U. V. at 110°C) 36 sec.
See also di-2-ethylhexyl azelate.

diisooctyl phthalate (DIOP) $(C_8H_{17}COO)_2C_6H_4$.
Isomeric esters obtained from phthalic anhydride and the mixed octyl alcohols made by the Oxo process (see isooctyl alcohol).
Properties: Nearly colorless, viscous liquid with a mild odor; b. p. 370°C; sp. gr. (20/20°C) 0.980-0.983; wt/gal (20°C) 8.20 lbs; flash point 410°F. Insoluble in water; compatible with vinyl chloride resins and some cellulosic resins.
Grade: Technical.
Containers: 1-, 5-, 55-gal drums; tank cars; tank trucks.
Uses: Plasticizer for vinyl, cellulosic, and acrylate resins and synthetic rubber.

diisooctyl sebacate (DIOS) $C_8H_{17}OOC(CH_2)_8COOC_8H_{17}$.
Properties: Liquid; sp. gr. 0.915 (25/25°C); flash point 440°F; pour point −40°C; viscosity 24 cps (20°C); 7.65 lb/gal.
Containers: Drums; tank cars.
Use: Plasticizer.
Shipping regulations: None.*

diisopropanolamine (DIPA) $(CH_3CHOHCH_2)_2NH$.
Properties: White crystalline solid; sp. gr. 0.9890 (45/20°C); b. p. 248.7°C; 8.2 lbs/gal (45°C); vapor pressure 0.02 mm (42°C); freezing point 42°C; viscosity 1.98 poise (45°C); miscible with water.
Grades: Technical.
Containers: Drums; tank cars.
Uses: Manufacture of emulsifying agents for polishes, textile specialties, leather compounds, insecticides, cutting oils, and water paints.

diisopropyl. See 2, 3-dimethylbutane.

diisopropylamine $[(CH_3)_2CH]_2NH$.
Properties: Colorless, volatile liquid with amine odor. B. p. 84.1°C; freezing point −96.3°C; sp. gr. (20/20°C) 0.7178; wt/gal (20°C) 6.0 lbs; flash point (open cup) 30°F. Miscible with water; soluble in most organic solvents.
Derivation: From isopropyl chloride and ammonia.
Grades: Technical.
Containers: 1-gal cans; 5-, 55-gal drums; tank cars.
Uses: Intermediate; catalyst.
Shipping regulations: Flammable liquid. Red label.*

diisopropylaminoethanol $[(CH_3)_2CH]_2NCH_2CH_2OH$.
Properties: Liquid; sp. gr. 0.873 (20/20°C); distillation range 188.0-192.0°C (760 mm); wt/gal 7.3 lbs; flash point 150°F.
Containers: 5-gal cans; 55-gal drums; tank cars.
Use: Intermediate.

beta-**diisopropylaminoethyl chloride hydrochloride** (DIC) $[(CH_3)_2CH]_2NCH_2CH_2Cl \cdot HCl$.
Used for organic synthesis, especially for introduction of the beta-diisopropylaminoethyl radical.

diisopropyl benzene $C_6H_4(CH_3CHCH_3)_2$.
Properties: Colorless liquid; f. p. below −50°C; boiling range 202-209°C (760 mm); sp. gr. 0.865 (25/25°C); wt/gal 7.20 lbs (25°C); refractive index 1.490 (n 25/D); flash point 170°F. Insoluble in water; soluble in methanol and ether.
Uses: Solvent; intermediate.

diisopropyl carbinol (2, 4-dimethylpentanol-3) $[(CH_3)_2CH]_2CHOH$.
Properties: Colorless liquid.
Constants: B. p. 140°C; m. p. below −70°C; wt/gal 6.9 lbs; flash point 49°C.
Grades: Technical.
Containers: 55-gal drums.
Uses: Solvent; organic synthesis (intermediate); denaturant.

diisopropyl cresol. Used as antioxidant or stabilizer in MYL (q. v.). See also isopropyl cresols.

diisopropyl dixanthogen $(C_3H_7OCS_2)_2$.
Typical specification: Yellow to greenish pellets; sp. gr. 1.28; m. p. 52°C (min); purity 98% (min.).

Insoluble in water; soluble in ethyl alcohol, acetone, benzene and gasoline.
Grades: Commercial.
Uses: Modifier in polymerization reactions; additive for lubricants; flotation reagent; fungicide or weed killer.
Handle with caution!

diisopropylene glycol salicylate $C_{13}H_{18}O_5$. The diisopropylene glycol mono ester of salicylic acid.
Properties: A clear light yellow liquid; sp.gr. 1.16 (25°C); refractive index 1.5150 (25°C); m.p. below −15°C; pH of 10% dispersion 5.3; soluble in aromatic paraffin and chlorinated solvents, in alcohols, esters, and vegetable oils. Totally absorbs ultraviolet radiation in range 2800-3200A.
Uses: As an ultraviolet light absorbent to protect human skin, plastics, paints, printing inks normally affected by the ultraviolet portion of the spectrum.

N, N-diisopropylethanolamine $[(CH_3)_2CH]_2NCH_2CH_2OH$.
Properties: A colorless liquid; sp.gr. 0.8742 (20°C); vapor pressure 0.08 mm (20°C); freezing point −39.3°C; flash point 175°F; slightly soluble in water.
Use: Synthesis.

diisopropyl fluophosphate (DFP; diisopropyl fluorophosphate; isoflurophate) $[(CH_3)_2CHO]_2POF$. Oily liquid; in the presence of moisture forms hydrogen fluoride.
Caution! Very dangerous in the slightest traces. Do not inhale vapors or allow contact with skin.
Properties: Sp.gr. 1.05; m.p. −82°C; b.p. 46°C (5 mm); refractive index (n 25/D) 1.3830; slightly soluble in water; soluble in alcohol and oils.
Use: Medicine (external) (very dilute concentrations).
One member of a series of compounds, fluophosphate alkyl esters, characterized by extremely high toxicity, marked miotic effects noted even in concentrations that are chemically indetectable. Some related less toxic members have been suggested as bactericides and insecticides.

diisopropyl fluorophosphate. Same as diisopropyl fluophosphate.

diisopropyl ketone (2,4-dimethylpentanone-3) $[(CH_3)_2CH]_2CO$.
Properties: Colorless, clear liquid.
Constants: B.p. 123.7°C; wt/gal 6.9 lbs.

diisopropylmethane. See 2,4-dimethylpentane.

2,6-diisopropylphenol $C_6H_3OH[CH(CH_3)_2]_2$.
Properties: Light straw-colored liquid; f.p. 18°C; sp.gr. 0.955 (20°C); b.p. 242°C; flash point 240°F (open cup). Soluble in toluene and alcohol; insoluble in water.
Use: Intermediate for synthetic resins, plasticizers, surface active agents.

diisopropylthiourea $(CH_3)_2CHNHCSNHCH(CH_3)_2$.
Properties: A grayish white solid; m.p. 138.5-142.5°C; slightly soluble in water; soluble in methanol, acetone, and ethyl acetate; insoluble in ether, benzene and gasoline.
Uses: Corrosion inhibitor; for pickling cast iron or carbon steel with hydrochloric acid; for pickling with sulfuric acid; for reducing corrosion of ferrous metals and aluminum alloys in brine; as intermediate.

diketene (acetyl ketene) $CH_3COCH{:}CO$.
Properties: Colorless, non-hygroscopic liquid; pungent odor; readily polymerizes on standing; sp.gr. 1.0897; m.p. −6.5°C; b.p. 127.4°C. Soluble in common organic solvents; insoluble in water.
Derivation: By spontaneous polymerization of ketene which is obtained by thermal decomposition of acetone, or from bromoacetylbromide and zinc.
Containers: Steel drums.
Uses: Source of acetoacetic esters, acetoacetanilide, phenylmethylpyrazolones, and benzoylacetone.
Shipping regulations: None.*

diketobutane. See diacetyl.

2,5-diketohexane. See acetonyl acetone.

2,5-diketopyrrolidone. See succinimide.

2,5-diketotetrahydrofurane. See succinic anhydride.

"Dilantin" Sodium. [330] Trademark for diphenylhydantoin sodium.

dilatancy. A term used in rheology to identify the flow property of certain suspensions in which the resistance to flow increases at a greater rate than the increase in the rate of flow. An example of this is quicksand or wet sea sand wherein the surface is rigid unless sustained pressure is applied, whereupon flow occurs.

"Dilaudid." [9] Trademark for dihydromorphinone; employed in medicine as the sulfate and hydrochloride salts.

dilaurylamine (didodecylamine) $(C_{12}H_{25})_2NH$.
Properties: M.p. 45°C; sp.gr. 0.89; almost insoluble in water.
Use: Chemical intermediate.

dilauryl ether (didodecyl ether) $(C_{12}H_{25})_2O$.
Properties: M.p. 33°C; b.p. 190-195°C (1 mm); sp.gr. 0.8147 (33/4°C).
Grade: 95% (min) purity.
Uses: Electrical insulators; water repellents; lubricants for plastic molding and processing; antistatic substances; chemical intermediates.

dilauryl sulfide (didodecyl thioether) $(C_{12}H_{25})_2S$.
Properties: M.p. 40-40.5°C; b.p. 260-263°C (4 mm); sp.gr. 0.8275 (40/4°C).
Grade: 95% (min) purity.
Uses: Organic synthesis (formation of sulfonium compounds).

dilauryl thiodipropionate (3, 3'-didodecyl thiodipropionate; didodecanyl thiodipropionate; thiodipropionic acid, dilauryl ester) $(C_{12}H_{25}OOCCH_2CH_2)_2S$.
Properties: Sp. gr. (solid, 25°C) 0.975; m.p. 37°C. Insoluble in water; soluble in most organic solvents. Extremely resistant to heat and hydrolysis.
Uses: Anti-oxidant; additive for high-pressure lubricants and greases; plasticizer and softening agent; preservative.

"Dilecto." [281] Trademark for a line of laminated plastics consisting of a variety of base materials and resins, available in sheets, tubes, or rods, of numerous grades for a variety of special mechanical and electrical insulation applications.

"Dileine." [78] Trademark for a series of gas-fading inhibitors used particularly in the dyeing of acetate fibers.

dilinoleic acid $C_{34}H_{62}(COOH)_2$.
Properties: Light yellow; sp. gr. 0.921 (100°C); refractive index 1.4851 (40°C); iodine value 80; N. E., 304. Heavy viscous liquid with slight odor.
Uses: As dibasic acid in alkyds as modifier; in polyamide resin; as polyester or metallic soap for petroleum additive; as emulsifying agent; in adhesives; as shellac substitute; to upgrade drying oils.

dilituric acid. See 5-nitrobarbituric acid.

dill. See anethum.

dill oil (anethum oil).
Properties: Pale yellow, volatile, essential oil; characteristic penetrating odor; sweetish taste, rapidly becoming sharp and burning. Soluble in alcohol, ether, benzene and chloroform.
Chief known constituents: Limonene and carvone.
Constants: Sp. gr. 0.895-0.915; optical rotation +70° to +80°.
Derivation: Distilled from the fruit of Anethum graveolens.
Method of purification: Rectification.
Grades: Technical.
Containers: Drums; glass bottles.
Uses: Flavoring agent; perfumery.
Shipping regulations: None. *

dill seed. See anethum.

"Diluex" and **"Diluex A."** [98] Proprietary names for two grades of Florida fuller's earth.
Properties: Finely divided grayish white powder; sp. gr. 2.2-2.4; low abrasiveness; compatible with complex organics, metallic inorganics, and plant derivative insecticides and fungicides; wets rapidly and disperses readily without forming lumps.
Containers: 50-lb multiwall paper bags.
Uses: Carrier for insecticides and fungicides; grinding or milling aid; conditioning agent in dusts and powders.

dimagnesium orthophosphate. See magnesium phosphate, dibasic.

dimagnesium phosphate. See magnesium phosphate, dibasic.

"Dimazine." [55] Trade name for uns-dimethylhydrazine (q. v.).

dimedone (1, 1-dimethyl-3, 5-diketocyclohexane; 5, 5-dimethyl-1, 3-cyclohexanedione) $(CH_3)_2C_6H_3O_2$.
Properties: Greenish-yellow needles, or prisms; m. p. 148-149°C; slightly soluble in cold water and petroleum ether; soluble in alcohol, chloroform, benzene.
Use: Reagent for the detection of ethyl alcohol and the identification of aldehydes.

dimefox (BFPO; bis(dimethylamino)fluorophosphate; tetramethyldiamidophosphoric fluoride) $[(CH_3)_2N]_2POF$. Dimefox is accepted as a generic name by the Ent. Soc.
Properties: Liquid; fishy odor; sp. gr. 1.1151 (20/4°C); b. p. 67°C (4.0 mm), 86°C (15 mm); refractive index 1.4267 (n 20/D). Soluble in water, ether, benzene; aqueous solutions are stable.
Derivation: Prepared by fluorination of bis(dimethylamido)phosphoryl chloride.
Use: A systemic pesticide, primarily for ornamental and non-food plants.

dimenhydrinate $C_{17}H_{22}NO \cdot C_7H_6ClN_4O_2$. 2-(Benzohydryloxy)-N, N-dimethylethylamine-8-chlorotheophyllinate.
Properties: Crystalline, white, odorless powder. Freely soluble in alcohol and chloroform; soluble in benzene; sparingly soluble in ether; slightly soluble in water. M. p. 102-107°; pH (saturated solution) 6.8-7.3.
Grade: U. S. P. XVI.
Use: Medicine.

dimension stone. A general term for stone sold in blocks and slabs, usually of specified sizes. Types of dimension stone include granite, limestone, marble, and sandstone (q. v.). Good dimension stone occurs in large uncracked blocks and has pleasing texture and color.
Use: Building stone, monuments, paving block, curbing and flagging.

dimer. A molecule formed by union of two identical simpler molecules. Also applied to the substances composed of such double molecules. Thus C_4H_8 is a dimer of C_2H_4. See polymer.

dimer acid. Coined name to describe a high molecular weight dibasic acid, which is liquid, stable, resistant to high temperatures, and which combines and polymerizes with alcohols and polyols to make a variety of products, such as plasticizers, lube oils, hydraulic fluids. Trimer acid, having three acid groups, is similar.

dimercaprol. See 2, 3-dimercaptopropanol.

2, 3-dimercaptopropanol (BAL; British Anti-Lewisite; dimercaprol; 1, 2-dithioglycerol) $CH_2(SH)CH(SH)CH_2OH$.
Properties: Colorless, oily, viscous liquid with strong, offensive odor of mercaptans. B. p. 80°C (1.9 mm), 140°C (40 mm);

m. p. 77°C; sp. gr. 1.2385 (25/4°C); refractive index (n 25/D) 1.5720. Soluble in vegetable oils; moderately soluble in water with decomposition; soluble in alcohol.
Derivation: (a) Bromination of allyl alcohol followed by reaction with sodium hydrosulfide; (b) hydrogenation of hydroxypropylene trisulfide.
Grades: U.S.P. XVI.
Uses: Medicine: antidote to Lewisite, organic arsenicals and heavy metals.

dimetan ($C_{11}H_{17}NO_3$). 5,5-Dimethyldihydroresorcinol dimethylcarbamate. An insecticide with some systemic properties. Accepted as a generic name by Ent. Soc.
Properties: Yellow crystals; m. p. 43-45°C; slightly soluble in water and oils but readily soluble in organic solvents. Mammalian toxicity reported low. A cholinesterase inhibitor.

1,4-dimethanesulfonoxybutane. See busulfan.

dimethicone $CH_3[Si(CH_3)_2O]_nSi(CH_3)_3$.
Properties: A colorless silicone oil consisting of dimethylsiloxane polymers (range in viscosities from 0.65 to 1,000,000 centistokes at room temperature). Viscosity grades above 50 centistokes are immiscible in water. Miscible with chloroform, ether.
Grade: N.N.D.
Uses: Ointments and topical drug ingredient; skin protectant.

dimethisoquin hydrochloride
$CH_3(CH_2)_2CH_2C_9H_5NOCH_2CH_2N(CH_3)_2 \cdot HCl$.
3-Butyl-1-(2-dimethylaminoethoxy)isoquinoline hydrochloride.
Properties: White powder, odorless, with bitter, numbing taste. M. p. 144-147°C. Freely soluble in alcohol; very slightly soluble in ether; soluble in water; pH (1% solution) 3.5-5.0.
Grade: N.F. XI.
Use: Medicine.

2,5-dimethoxyaniline $NH_2C_6H_3(OCH_3)_2$.
Typical specification: M. p. 79-81°C. Insoluble in water; soluble in organic solvents.
Grades: Technical.
Containers: Fiber drums.
Use: Intermediate for dyes, pharmaceuticals and insecticides.

2,5-dimethoxybenzaldehyde $(CH_3O)_2C_6H_3CHO$.
Properties: Flaked solid; m. p. 46-49°C; soluble in organic solvents; insoluble in water.
Use: Organic synthesis.

1,4-dimethoxybenzene. See hydroquinone dimethyl ether.

ortho-**dimethoxybenzene.** See veratrole.

dimethoxybenzidine. See dianisidine.

3,4-dimethoxybenzyl alcohol $C_6H_3(OCH_3)_2CH_2OH$.
Properties: Viscous, brown liquid or low melting solid.
Use: Organic synthesis.

1,2-dimethoxyethane. See ethylene glycol dimethyl ether.

dimethoxyethyl adipate
$CH_3OC_2H_4OOC(CH_2)_4COOC_2H_4OCH_3$.
Properties: Liquid; sp. gr. 1.075 (25°C); refractive index 1.439 (25°C); b. p. 185-190°C (11 mm); m. p. −16°C; slightly soluble in water.
Use: Plasticizer.

di(2-methoxyethyl) phthalate
$C_6H_4(COOCH_2CH_2OCH_3)_2$.
Properties: Oily liquid with mild odor; sp. gr. 1.172 (20/20°C); b. p. 340°C; f. p. −45°C; flash point 381°F (open cup).
Containers: 55-gal drums; tank cars; tank trucks.
Uses: Plasticizer, especially for cellulose acetate; solvent.

dimethoxymethane. See methylal.

3,4-dimethoxyphenethylamine. See homoveratrylamine.

3,4-dimethoxyphenylacetic acid. See homoveratric acid.

1-(3,4-dimethoxyphenyl)-2-nitro-1-propene $(CH_3O)_2C_6H_3CH{:}C(NO_2)CH_3$. Yellow crystals; m. p. 68-75°C; used as an intermediate.

dimethoxyphenyl penicillin sodium. See penicillin.

dimethoxystrychnine. See brucine.

dimethoxytetraglycol (tetraethylene glycol dimethyl ether) $CH_3(OCH_2CH_2)_4OCH_3$.
Properties: Water-white, practically odorless liquid. Stable; soluble in hydrocarbons, water.
Constants: Sp. gr. 1.0132 (20/20°C); b. p. 275.8°C (760 mm), 189°C (100 mm); vapor pressure < 0.01 mm (20°C); flash point 285°F; wt/gal 8.4 lbs (20°C); freezing point −29.7°C; viscosity 0.0405 poise (20°C); coefficient of expansion 0.00091 (20°C).
Typical specification: Sp. gr. 1.011-1.016 (20/20°C); boiling range 255-285°C (760 mm); acidity not more than 0.02% (as acetic).
Grades: Technical.
Containers: 1-gal cans; 5-, 55-gal drums. Net content 8, 40, 450 lbs.
Use: Solvent.
Shipping regulations: None.*

dimethyl. See ethane.

dimethylacetal (ethylidenedimethyl ether) $CH_3(OCH_3)_2CH$.
Properties: Colorless, flammable liquid; strongly aromatic odor. Soluble in water, alcohol, ether, and chloroform.
Constants: Sp. gr. 0.848 (25°C); b. p. 62-63°C.
Derivation: By heating acetaldehyde with methyl alcohol and glacial acetic acid, and distilling.
Method of purification: Rectification.
Grades: Technical.
Containers: Glass bottles; iron drums.
Uses: Medicine; organic synthesis.

Fire hazard: Dangerous.
Shipping regulations: Flammable liquid. Red label. *

dimethyl acetamide (DMAC) $CH_3CON(CH_3)_2$.
Properties: A colorless liquid; b.p. 166°C; sp. gr. 0.9366 (25°C); refractive index 1.4351 (25°C); miscible with water, aromatics, esters, ketones, and ethers.
Derivation: From dimethylamine.
Containers: 55-gal drums.
Uses: Solvent for plastics, resins and gums; intermediate; catalyst; in production of "Acrilan"; paint remover.

2,4-dimethyl acetophenone $CH_3COC_6H_3(CH_3)_2$.
Properties: Colorless liquid having a strong odor suggesting mimosa; sp. gr. 0.994-0.997; refractive index 1.532-1.534; soluble in four volumes of 60% alcohol.
Use: Perfumery.

dimethylamine (DMA) $(CH_3)_2NH$.
Properties: At ordinary temperatures dimethylamine is a gas with a strong ammoniacal odor; sp. gr. 0.6865 at -6°C; b.p. 6.88°C; m.p. -92.2°C; flash point of 25% solution (Tag open cup) 54°F; wt/gal (25% solution) approx 7.8 lbs (68°F); soluble in alcohol, ether, and water.
Derivation: By the interaction of methanol and ammonia over a catalyst at high temperatures. The mono-, di-, and trimethylamines are all produced. Yields are regulated by conditions.
Method of separation: Azeotropic distillation.
Grades: Technical (anhydrous, 25%, and 40% aqueous solutions); 99%.
Containers: Solution: 1-gal bottles; 5-, 55-gal drums; tank cars. Anhydrous: 25-, 50-, 100-, 1400-lb cylinders.
Uses: Acid gas absorbent; solvent; antioxidants; dyes; flotation agent; gasoline stabilizers; pharmaceuticals; textile chemicals; rubber accelerators; soaps and cleaning compounds; electroplating; dehairing agent; missile fuels; pesticide propellant.
Shipping regulations: Anhydrous: Flammable gas. Red gas label. Aqueous solution: Flammable liquid. Red label. *
Danger! Extremely flammable. Hazardous liquid and vapor under pressure. Liquid causes burns. Vapor extremely irritating. MCA warning label for anhydrous dimethylamine.

dimethylaminoaniline. See para-aminodimethylaniline.

dimethylaminoantipyrine. See aminopyrine.

dimethylaminoazobenzene (methyl yellow; butter yellow) $C_6H_5NNC_6H_4N(CH_3)_2$.
Properties: Yellow crystalline leaflets; m.p. 116°C; soluble in alcohol, ether, strong mineral acids, and oils; insoluble in water.
Derivation: Action of benzenediazonium chloride on dimethyl aniline.
Method of purification: Recrystallization.
Grades: Technical.
Containers: Glass bottles; tins.
Uses: Indicator in volumetric analysis (see indicators); also in test for peroxidized fats.
Shipping regulations: None. *

dimethylaminoazobenzene sulfonate (sulfobenzeneazodimethylaniline) $SO_3HC_6H_4N_2C_6H_4N(CH_3)_2$.
Properties: Violet crystals or powder; soluble in alcohol; slightly soluble in water.
Derivation: By the sulfonation of dimethylaminoazobenzene.
Method of purification: Crystallization.

para-**dimethylaminobenzaldehyde** $C_6H_4[N(CH_3)_2]CHO$.
Properties: Colorless crystalline plates. Soluble in hot water, alcohol and ether. M.p. 73°C; b.p. 176-177°C (17 mm).
Derivation: By mixing dimethylaniline, anhydrous chloral and phenol and allowing the mixture to stand. The phenol is removed by shaking with dilute caustic soda and the residue dissolved in water and hydrochloric acid and crystallized.
Method of purification: Recrystallization.
Grades: Technical; reagent.
Containers: Fiber drums.
Uses: Dyes; medicine.
Shipping regulations: None. *

para-**dimethylaminobenzene diazonium chloride, zinc chloride double salt.** See para-diazodimethylaniline, zinc chloride double salt.

para-**dimethylaminobenzenediazo sodium sulfonate** ("Dexon") $C_8H_{10}N_3O_3SNa$.
Properties: Solid; melts with decomposition above 200°C. Soluble in water.
Uses: Fungicide for protection of germinating seed and seedlings.

3-dimethylaminobenzoic acid $(CH_3)_2NC_6H_4COOH$.
Properties: Pale yellow crystals; m.p. 147-153°C.
Grades: Technical.
Use: Intermediate.

dl-**6-dimethylamino-4,4-diphenyl-3-heptanone hydrochloride.** See methadone hydrochloride.

4,4-dimethylaminodiphenylsulfone.
Constants: M.p. 179-180°C.

2-dimethylaminoethanol (deanol; dimethylethanolamine) $(CH_3)_2NCH_2CH_2OH$.
Properties: Colorless liquid with amine odor. B.p. 134.6°C; f.p. -59.0°C; sp. gr. (20/20°C) 0.8879; wt/gal (20°C) 7.4 lbs; refractive index (20°C) 1.4300; flash point (open cup) 105°F. Miscible with water, acetone, ether, and benzene.
Preparation: From ethylene oxide and dimethylamine.
Grades: Anhydrous and 70% aqueous soln.
Containers: 1-gal cans; 5-, 55-gal drums; tank cars.
Uses: Chemical intermediate in the synthesis of dyestuffs, textile auxiliaries, pharmaceuticals and corrosion inhibitors; medicine.
Shipping regulations: None. *

beta-**dimethylaminoethyl benzhydryl ether hydrochloride.** See diphenhydramine hydrochloride.

beta-**dimethylaminoethyl chloride hydrochloride** (DMC) $(CH_3)_2NCH_2CH_2Cl \cdot HCl$. Used in manufacture of antihistaminics and other pharmaceuticals. Organic intermediate for introduction of beta-dimethylaminoethyl radical.

(2-dimethylaminoethyl)-2-thenylaminopyridine. See methapyrilene.

(2-dimethylaminoethyl)-3-thenylaminopyridine. See thenyldiamine.

beta-**dimethylaminoisopropyl chloride hydrochloride** (DMIC) $(CH_3)_2NCH_2CHClCH_3 \cdot HCl$. Used in manufacture of analgesics and other pharmaceuticals.

ortho-**dimethylaminomethyl**-para-**butyl phenol** $C_6H_3(OH)[CH_2N(CH_3)_2](iso\text{-}C_4H_9)$.
Properties: Dark red liquid; odor phenolic, free of methylamine; sp. gr. 0.960 (25/25°C); refractive index 1.510 (25°C); distillation range 95-135°C (1 mm); m. p. 16-18°C; water content (Karl Fischer) 0.5%. Readily soluble in organic solvents; insoluble in water.

dimethylaminomethyl phenols
$C_6H_4OHCH_2N(CH_3)_2$. Exists as ortho-, meta and para- isomers, but the commercially available material is a mixture of ortho- and para-.
Properties: Dark red liquid; odor phenolic, free of methylamine; sp. gr. 1.020 (25/25°C); refractive index 1.530 (25°C); distillation range 80-130°C (2 mm); water content (Karl Fischer) 0.5%. Readily soluble in organic solvents; moderately soluble in water.

gamma-**dimethylamino**-beta-**methylpropyl chloride hydrochloride** (DMMPC) $(CH_3)_2NCH_2CH(CH_3)CH_2Cl \cdot HCl$. Used as intermediate for the preparation of tranquilizers, analgesics, anti-spasmodics, local anesthetics and hypotensive agents.

4-dimethylaminophenylmethylsulfone.
Constants: M. p. 167-168°C.
Grades: Technical.

1-dimethylamino-2-propanol
$(CH_3)_2NCH_2CHOHCH_3$.
Properties: Water-white; amine odor; b. p. 125.6°C; sp. gr. 0.850 (20/20°C); refractive index 1.421 (20°C); flash point 95°F; soluble in water and most organic solvents.
Use: Organic synthesis.

3-dimethylaminopropylamine
$(CH_3)_2NCH_2CH_2CH_2NH_2$.
Properties: Colorless liquid; b. p. 123°C (760 mm); sp. gr. 0.8100 (30°C); refractive index (n 25/D) 1.4328; flash point 95°F (Tag closed cup).
Uses: Curing agent for epoxy resins; organic intermediate.

1-dimethylamino-2-propyl chloride
$(CH_3)_2NCH_2CHClCH_3$. Yellow liquid which darkens with age; distillation range 113-120°C; refractive index 1.422-1.423 (25°C). Used as an intermediate.

gamma-**dimethylaminopropyl chloride hydrochloride** (DMPC) $(CH_3)_2NCH_2CH_2CH_2Cl \cdot HCl$. Used as intermediate for pharmaceutical and organic synthesis. Suggested as an intermediate for tranquilizers.

dimethylanilines. See xylidines.

N, N-dimethylaniline (aniline N, N-dimethyl) $C_6H_5N(CH_3)_2$.
Properties: Yellowish to brownish oily liquid. Soluble in alcohol and ether; insoluble in water. Sp. gr. 0.954; m. p. 2.5°C; b. p. 192.5-193.5°C; flash point 61°C; refractive index 1.5582.
Derivation: By heating a mixture of aniline, aniline hydrochloride and methyl alcohol (free from acetone) in an autoclave and distilling.
Method of purification: Rectification.
Grades: Technical; reagent.
Containers: Drums; tank cars.
Uses: Dyes; intermediates; solvent; manufacture of vanillin; stabilizer (acid acceptor).
Shipping regulations: None. *

dimethyl anthranilate (N-methyl methyl anthranilate) $CH_3COOC_6H_4NHCH_3$.
Properties: Colorless or pale yellow liquid with slight bluish fluoroescence, with grape-like odor; sp. gr. 1.132-1.138 (15°C); refractive index 1.578-1.581 (20°C); soluble in 3 volumes or more of 80% alcohol; soluble in benzyl benzoate, diethyl phthalate, fixed oils, mineral oils and volatile oils; insoluble in glycerin and somewhat soluble in propylene glycol; congealing point 18°C (4% methyl anthranilate impurity) to 10°C (20% methyl anthranilate impurity).
Derivation: Methylation of methyl anthranilate or esterification of N-methyl anthranilic acid.
Containers: Glass; aluminum or tin-lined cans.
Uses: Manufacture of perfumes, flavorings and drugs.
Shipping regulations: None. *

dimethylarsinic acid. See cacodylic acid.

1, 2-dimethyl benzene. See ortho-xylene.

1, 3-dimethyl benzene. See meta-xylene.

1, 4-dimethyl benzene. See para-xylene.

dimethylbenzidine. See tolidine.

dimethylbenzylcarbinol $C_6H_5CH_2C(CH_3)_2OH$.
Properties: Colorless or yellowish liquid; odor of hyacinth or lilac; sp. gr. 0.979 (16/4°C); m. p. 24°C; b. p. 228°C.
Use: Perfumery.
Shipping regulations: None. *

2, 5-dimethylbenzyl chloride
(2, 5-dimethyl-alpha-chlorotoluene) $C_6H_3(CH_3)_2CH_2Cl$.
Properties: B. p. 220-226°C. Insoluble in water; soluble in hydrocarbons, alcohols and ethers.
Grade: 98% min.

*See "I. C. C. Shipping Regulations," page xiii.
Reference numbers refer to name of manufacturer. See "List of Manufacturers," page v.

Uses: Intermediate for pharmaceuticals, dyes, perfumes, plasticizers, resins, wetting agents, germicides, etc.

alpha, alpha-**dimethylbenzyl hydroperoxide.** See cumene hydroperoxide.

2,2-dimethylbutane. See neohexane.

2,3-dimethylbutane (diisopropyl) $(CH_3)_2CHCH(CH_3)_2$.
Properties: Colorless liquid; b.p. 57.9°C; sp.gr. 0.66164 (20°C); f.p. −128.41°C; refractive index 1.37495 (20°C); flash point 29°C.
Derivation: Alkylation of ethylene with isobutane using aluminum chloride catalyst.
Grades: Technical; 95%; 99%.
Containers: Bottles; drums.
Uses: High octane fuel; organic synthesis.

2,2-dimethyl-1,3-butanediol $CH_3CH(OH)C(CH_3)_2CH_2OH$.
Properties: Liquid; sp.gr. 0.9700; b.p. 202.4°C; f.p. −12.8°C; wt/gal 8.1 lbs; very soluble in water.

3-dimethylcarbamoxyphenyltrimethylammonium bromide. See neostigmine bromide.

dimethyl carbate $C_{11}H_{14}O_4$. Bicyclo (2,2,1)-5-heptene-2,3-dicarboxylic acid dimethyl ester. Accepted as a generic name by the Ent. Soc.
Properties: A clear, oily liquid or crystalline solid; sp.gr. 1.165 (35/4°C); insoluble in water.
Derivation: By esterification of the Diels-Alder condensation product of maleic anhydride and cyclopentadiene.
Use: Insect repellent.

dimethylcarbinol. See isopropyl alcohol.

dimethyl carbonate. See methyl carbonate.

dimethyl chloroacetal (chloroacetaldehyde dimethyl acetal) $ClCH_2CH(OCH_3)_2$.
Properties: Colorless liquid with a pleasant odor; specifications for technical grade: boiling range 126-132°C; flash point 110°F; sp.gr. 1.082-1.092 (25/4°C); refractive index (n 25/D) 1.4110-1.4130; purity 97% (min); wt/gal 9.07 lbs.
Grades: Technical.
Containers: 1-gal glass bottles; 5-, 55-gal drums; tank cars.
Uses: Organic synthesis; pharmaceuticals; as a solvent.
Shipping regulations: None.*

2,5-dimethyl-alpha-**chlorotoluene.** See 2,5-dimethylbenzyl chloride.

dimethyl cyanamide $(CH_3)_2NCN$. An organic intermediate with unusual solvent properties; suggested for hydraulic fluids; corrosion inhibitors.

dimethylcyclohexane (hexahydroxylene). Mixture of ortho-, meta-, and para-isomers.
Properties: Water-white liquid of mild odor; sp.gr. 0.776 (15/15°C); boiling range 120°C; f.p. < −65°C; soluble in most common solvents; almost insoluble in water.
Use: Synthesis.

cis-**1,2-dimethylcyclohexane** (cis-ortho-dimethylcyclohexane) $C_6H_{10}(CH_3)_2$. The high-boiling 1,2-isomer.
Properties: Colorless liquid; sp.gr. (20/4°C) 0.7963; f.p. −50.1°C; b.p. 128.95°C; refractive index (20/D) 1.4359; flash point 16°C.
Grades: 99%; research.
Containers: Bottles.
Use: Organic synthesis.
Hazards: Flammable liquid.
Shipping regulations: Flammable liquid. Red label.*

trans-**1,2-dimethylcyclohexane** (trans-ortho-dimethylcyclohexane) $C_6H_{10}(CH_3)_2$. The low-boiling 1,2-isomer.
Properties: Colorless liquid; sp.gr. (20/4°C) 0.7761; f.p. −88.38°C; b.p. 122.77°C; refractive index (20/D) 1.42695; flash point 11°C.
Grades: 95%; 99%; research.
Containers: Bottles.
Use: Organic synthesis.
Hazards: Flammable liquid.
Shipping regulations: Flammable liquid. Red label.*

1,3-dimethylcyclohexane (meta-dimethylcyclohexane) $C_6H_{10}(CH_3)_2$.
Properties: Colorless liquid. Soluble in alcohol; insoluble in water. Sp.gr. 0.772; b.p. 121°C; m.p. −85°C; aniline equivalent 4.

1,4-dimethylcyclohexane(para) $C_6H_{10}(CH_3)_2$.
Properties: Colorless liquid; sp.gr. 0.767; b.p. 120.5°C; m.p. −86°C; aniline equivalent 6.

5,5-dimethyl-1,3-cyclohexanedione. See dimedone.

N-dimethylcyclohexaneëthylamine. See propylhexedrine.

dimethylcyclohexyl adipate $(CH_2CH_2COOC_6H_{10}CH_3)_2$.
Properties: Neutral, stable, colorless liquid.

1,2-dimethylcyclopentane $C_5H_8(CH_3)_2$.
Properties: Cis: b.p. 99.5°C; sp.gr. 0.772 (20°C). Trans: b.p. 91.8°C; sp.gr. 0.751 (20°C).
Grades: Technical.
Use: Organic synthesis.

dimethyl diaminophenazinechloride. See neutral red.

2,2-dimethyl-1,1-dianthraquinone. $C_{30}H_{18}O_4$.
Properties: Yellow crystals; soluble in hot nitrobenzene, aniline and chlorobenzene. M.p. 365-367°C.
Derivation: 1-Amino-2-methylanthraquinone is dissolved in sulfuric acid and sodium nitrite is added. The isolated and dried diazonium sulfate is stirred into acetic anhydride and copper powder added. Nitrogen is evolved and the combination takes place, forming the dianthraquinonyl derivative.
Method of purification: Crystallization from

solvents in which it is soluble.
Use: Intermediate for dyes.

2,5-dimethyl-2,5-di(tert-butylperoxy)hexane ("Lupersol 101")
Properties: Stable liquid; b.p. 50-52°C (0.1 mm Hg). Active oxygen 10.5% min.
Uses: Catalyst in polyethylene cross-linking, styrene polymerization; polyester resins.

dimethyldichlorosilane $(CH_3)_2SiCl_2$.
Properties: Colorless liquid; b.p. 70°C; m.p. -86°C; sp.gr. 1.062 (20°C); refractive index (n 25/D) 1.4023; flash point (Cleveland open cup) 16°F. Reacts with water to form complex mixture of dimethylsiloxanes, and liberates hydrochloric acid.
Derivation: Action of silicon on methyl chloride in presence of a copper catalyst, or by Grignard reaction from methyl chloride and silicon tetrachloride.
Grades: Technical.
Containers: Steel drums.
Use: Intermediate in production of dimethyl siloxane oils (silicone oils), silicone rubber, and silicone resins.
Shipping regulations: Flammable liquid. Red label.*

dimethyl dichlorovinyl phosphate. See DDVP.

5,5-dimethyldihydroresorcinol dimethylcarbamate. See dimetan.

1,1-dimethyl-3,5-diketocyclohexane. See dimedone.

dimethyldiketone. See diacetyl.

N,N'-dimethyl-N,N'-di-(1-methylpropyl)-para-phenylene diamine. A volatile, reddish-brown liquid. Forms a continuous protective film.
Use: Antiozonant in rubber.

dimethyl dioxane
$\overline{OCH(CH_3)CH_2OCH_2CH}(CH_3)$.
Properties: Water-white liquid. Soluble in water. Sp.gr. 0.9268; b.p. 117.5°C; flash point 75°F; vapor pressure 15.4 mm at 20°C.
Shipping regulations: Flammable liquid. Red label.*

dimethyldiphenylurea $(CH_3)_2(C_6H_5)_2N_2CO$.
Properties: White crystals. Soluble in alcohol, ether and benzene; insoluble in water. M.p. 120°C.
Derivation: By saturation of monomethylaniline with carbonyl chloride, removal of benzene by distillation, washing the residue with acid water and crystallizing from alcohol. The crystals are warmed with alcoholic ammonia, diluted with water to precipitate, washed with water, dissolved in alcohol and crystallized.
Method of purification: Recrystallization.
Grades: Technical.
Containers: Tins; glass bottles.
Use: Stabilizer for smokeless powder; explosives and nitro-compounds.
Shipping regulations: None.*

dimethylenemethane. See allene.

dimethylethanolamine. See dimethylaminoethanol.

dimethyl ether (methyl ether; methyl oxide; wood ether) CH_3OCH_3.
Properties: Colorless, flammable gas, or compressed liquid; soluble in water and alcohol. Sp.gr. 0.661; b.p. -24.5°C; f.p. -138°C.
Grades: Technical; 99.5%.
Containers: 25-, 50-, 100-, and 150-lb pressure cylinders.
Uses: Refrigerant; solvent; extraction agent; propellant for sprays; chemical (reaction medium); welding gas; various other uses.
Shipping regulations: Flammable gas. Red gas label.*

dimethyl ethyl carbinol. See tert-amyl alcohol.

dimethylethylene. See butene-2.

sym-**dimethylethylene glycol.** See 2,3-butylene glycol.

dimethylethyl-(3-hydroxyphenyl)ammonium chloride. See edrophonium chloride.

O,O-dimethyl-S-2-(ethylsulfinyl)ethyl phosphorothioate ("Meta-Systox R") $C_6H_{15}O_4PS_2$.
Properties: Amber liquid; b.p. 106°C (0.01 mm); sp.gr. 1.28 (20/4°C); soluble in water in all proportions.
Uses: Systemic insecticide.
Caution! May be harmful if swallowed, inhaled, or absorbed through the skin. Overexposure will result in cholinesterase depression.

N,N-dimethyl formamide (DMF) $HCON(CH_3)_2$.
Properties: Water-white liquid; non-corrosive; b.p. 152.8°C; m.p. -61°C; refractive index (n 25/D) 1.4269; sp.gr. 0.953-0.954 (15.6/15.6°C). Flash point (Tag open cup) 153°F. Miscible with water and most organic solvents, and many inorganic liquids.
Containers: 55-gal drums; tank cars; tank trucks.
Uses: Solvent for vinyl resins and acetylene, butadiene, acid gases, inorganic salts, some petroleum components; dyestuffs, and pharmaceuticals; used in making "Orlon."

dimethyl furan $\overline{OC(CH_3)CHCHC}(CH_3)$.
Properties: Colorless liquid; insoluble in water; sp.gr. 0.8900; b.p. 94°C; flash point 45°F.
Grades: Technical.
Shipping regulations: Flammable liquid. Red label.*

dimethyl glycol phthalate
$C_6H_4(COOCH_2CH_2OCH_3)_2$.
Properties: Colorless liquid; sp.gr. 1.17; b.p. 230°C.
Grades: Technical.
Containers: Glass bottles; tins; drums.
Uses: Solvent mixtures for cellulose esters; plasticizing mixtures for cellulose esters.
Shipping regulations: None.*

dimethylglyoxal. See diacetyl.

dimethylglyoxime (butane dioxime) $CH_3C(NOH)C(NOH)CH_3$.
Properties: White crystals or powder; m.p. 240-242°C; soluble in alcohol and ether; very slightly soluble in water.
Grades: Technical; C.P..
Containers: Tins; glass bottles.
Use: Analytical chemistry, especially as a reagent for nickel.

2,6-dimethyl-4-heptanol (diisobutylcarbinol) $[(CH_3)_2CHCH_2]_2CHOH$.
Properties: Colorless liquid; refractive index 1.423 (21°C); sp. gr. 0.8121 (20°C); b.p. 178°C (750 mm); insoluble in water; soluble in alcohol and ether.
Containers: Up to tank cars.
Uses: Surface active agents; lubricant additives; rubber chemicals; flotation agents; antifoam agent in textiles.

2,6-dimethyl-4-heptanone. See diisobutyl ketone.

2,6-dimethyl-5-hepten-1-al. See "Melonal."

2,6-dimethylheptene-3 $(CH_3)_2CHCH{:}CHCH_2CH(CH_3)_2$. Mixed cis and trans isomers.
Properties: Liquid; distillation range 128 to 129°C; sp. gr. (60/60°F) 0.722; refractive index (20/D) 1.412; flash point 70°F.
Grades: 95%.
Containers: Bottles; 5-gal drums.
Shipping regulations: Flammable liquid. Red label.*

2,5-dimethylhexadiene-1,5 $CH_2{:}C(CH_3)CH_2CH_2C(CH_3){:}CH_2$.
Properties: A water-white, flammable liquid with pleasant hydrocarbon odor; sp. gr. 0.740-0.760 (25/25°C); refractive index 1.426-1.429 (25°C); ASTM distillation, 90% distills between 114-123°C; soluble in hydrocarbons; insoluble in water.

2,5-dimethylhexadiene-2,4 $(CH_3)_2C{:}CHCH{:}C(CH_3)_2$.
Properties: Water-white, flammable liquid with pleasant hydrocarbon odor; sp. gr. 0.760-0.763 (25/25°C); refractive index 1.473-1.478 (25°C); ASTM distillation range, 95% distills between 131-138°C; insoluble in water; soluble in hydrocarbons.

2,5-dimethylhexane 2,5-dihydroperoxide $(CH_3)_2C(OOH)CH_2CH_2C(OOH)(CH_3)_2$.
Properties: Fine powder; 90% peroxide.
Use: High temperature catalyst for polyester premix compounds and silicone resins.
Shipping regulations: Oxidizing material. Yellow label.*

dimethylhexanediol (2,5-dimethylhexane-2,5-diol) $(CH_3)_2COH(CH_2)_2COH(CH_3)_2$.
Properties: White crystals; m.p. 88.5-89°C; b.p. 214-215°C; sp. gr. (20/20°C) 0.898. Soluble in water, acetone, and alcohol; insoluble in benzene, carbon tetrachloride, and kerosine.
Containers: Fiber drums.
Uses: Chemical intermediate.

dimethylhexynediol (2,5-dimethyl-3-hexyne-2,5-diol) $(CH_3)_2COHC{:}CCOH(CH_3)_2$.
Properties: White crystals; m.p. 94-95°C; b.p. 205-206°C; sp. gr. (20/20°C) 0.949. Soluble in water; slightly soluble in benzene, carbon tetrachloride, petroleum ether; very soluble in acetone, alcohol, and ethyl acetate.
Containers: 5-, 25-, 200-lb fiber drums.
Uses: Wire-drawing lubricant; antifoaming agent; coupling agent in resin coatings; chemical intermediate.

dimethyl hexynol $HC{:}CCOH(CH_3)CH_2CH(CH_3)_2$ (3,5-dimethyl-1-hexyne-3-ol).
Properties: Colorless liquid with camphor-like odor; b.p. 150-151°C; sets to a glass below −68°C; sp. gr. (20/20°C) 0.8545. Slightly soluble in water.
Containers: Drums.
Uses: Stabilizer for chlorinated organic compounds; surface active agent; intermediate.

dimethylhydantoin (DMH) $\overline{HNCONHCOC}(CH_3)_2$.
Properties: White, crystalline solid; m.p. 178°C; soluble in water, alcohol, and ether.
Derivation: (a) From acetone, urea, and ammonium carbonate; (b) from acetone, potassium cyanate and hydrocyanic acid.
Uses: Synthesis; preparation of water-soluble resins.

dimethylhydantoin-formaldehyde resin.
Properties: Light colored brittle resin; density 1.30 g/ml; dissolves readily in cold and hot water, methanol, ethyl acetate, methyl ethyl ketone, chloroform, methylene chloride, and hot glycerol; insoluble in benzene, xylene, petroleum ether, diethyl ether, trichloroethylene, and carbon tetrachloride.
Uses: Sizing; adhesives; blending agent.

uns-**dimethylhydrazine** (UDMH) $(CH_3)_2NNH_2$ 1,1-dimethylhydrazine.
Properties: Colorless, flammable liquid with ammonia-like odor; f.p. −58°C; b.p. 63°C; sp. gr. 0.782 (25°C); soluble in hydrocarbons.
Derivation: (a) Reaction of dimethylamine and chloramine; (b) reaction of a dimethylamine salt with sodium nitrite, followed by reduction of the product; (c) catalytic oxidation of dimethylamine and ammonia.
Containers: Drums; tank cars.
Uses: Component of jet and rocket fuels; chemical synthesis; stabilizer for organic peroxide fuel additives; absorbent for acid gases; photography.

dimethyl hydroquinone. See hydroquinone dimethyl ether.

dimethylhydroxybenzene. See xylenol.

dimethylhydroxyoctanal. See hydroxycitronellal.

N-3,4-dimethyl-5-isoazolylsulfanilamide. See sulfisoxazole.

dimethyl isophthalate $C_6H_4(COOCH_3)_2$.
Properties: Solid; set point (min) 66.5°C; b.p. 200°C (50 mm); flash point 280°F; soluble in most organic solvents; insoluble in water.
Containers: Polyethylene-lined drums.
Use: Plasticizer.

dimethylisopropanolamine $(CH_3)_2NCH_2CH(OH)CH_3$.
Properties: Sp. gr. 0.8645 (25/20°C); 7.4 lbs/gal (20°C); b.p. 125.8°C (760 mm); completely soluble in water; viscosity 1.51 cps (20°C); vapor pressure 9 mm (20°C); f.p. sets to glass below −85°C; refractive index 1.4189 (n 20/D); flash point 105°F; solubility of water in compound, complete at 20°C.
Uses: In the synthesis of methadone; other chemical synthesis. Combines the properties of tertiary amine and secondary alcohol.

dimethyl itaconate $CH_2{:}C(COOCH_3)CH_2(COOCH_3)$.
Properties: White crystals with slight odor; m.p. 36°C; b.p. 91.5°C (10 mm); sp. gr. 1.27 (24°C); refractive index (n 20/D) 1.441. Slightly soluble in water.
Grades: Technical.
Uses: Polymers and copolymers; plasticizers; intermediate.

dimethylketol. See acetylmethylcarbinol.

dimethylketone. See acetone.

dimethyl maleate $CH_3OOCCH{:}CHCOOCH_3$.
Properties: Sp. gr. 1.153; 9.62 lbs/gal; b.p. 200.4°C; flash point (Cleveland open cup) 235°F.

dimethylmethane. See propane.

O, O-dimethyl O-[4-(methylthio)-meta-tolyl] phosphorothioate ("Baytex") $C_{10}H_{15}O_3PS_2$.
Properties: Brown liquid; b.p. 105°C (0.01 mm); insoluble in water; soluble in most organic solvents.
Use: Insecticide.
Caution! May be harmful if swallowed, inhaled, or absorbed through the skin. Overexposure will result in cholinesterase depression.

2, 6-dimethylmorpholine
$\overline{OCH(CH_3)CH_2NHCH_2CH}(CH_3)$.
Properties: Liquid; sp. gr. 0.9346; b.p. 146.6°C; f.p. −85°C; very soluble in water; wt/gal 7.8 lbs; flash point 112°F.
Uses: Corrosion inhibitors; stabilizers for chlorinated solvents; manufacture of fast drying rubless floor polishes, rubber accelerators, germicides, and textile finishing agents.
Shipping regulations: None. *

dimethyl-alpha-naphthylamine $C_{10}H_7N(CH_3)_2$.
Properties: Colorless liquid; soluble in alcohol and ether; insoluble in water. Sp. gr. 1.045; b.p. 275°C.
Derivation: Action of methylsulfate on alpha-naphthylamine.
Method of purification: Fractional distillation.
Grades: C.P.; analytical.
Containers: Glass bottles.
Use: Determination of nitrites.
Shipping regulations: None. *

dimethyl-beta-naphthylamine $C_{10}H_7N(CH_3)_2$.
Properties: Crystalline solid; soluble in alcohol and ether; insoluble in water; sp. gr. 1.039 (70/70°C); m.p. 46°C; b.p. 305°C.
Derivation: By the interaction of dimethylamine and beta-naphthol.
Method of purification: Crystallization.

dimethylnitrobenzene. See nitroxylene.

O, O-dimethyl O-para-nitrophenyl phosphorothioate. See methyl parathion.

dimethyloctadienal. See citral.

dimethyloctadienol. See geraniol.

dimethyloctanediol (3, 6-dimethyloctane-3, 6-diol) $C_2H_5(CH_3)COH(CH_2)_2COH(CH_3)C_2H_5$.
Properties: White, waxy solid; m.p. 44°C; b.p. 241-242°C; sp. gr. (20/20°C) 0.919. Soluble in water, acetone, alcohol, benzene, carbon tetrachloride, and kerosene.
Containers: Fiber drums.
Uses: Non-foaming surface-active agent; chemical intermediate.

dimethyl octanol (3, 6-dimethyl-3-octanol) $(C_2H_5)CHCH_3(CH_2)_2COHCH_3(C_2H_5)$.
Properties: Colorless liquid; sweet rosy odor; sp. gr. 0.8366 (20/20°C); refractive index 1.4370 (n 20/D); b.p. 202-203°C; freezing point −67.5°C.
Uses: Perfumery, particularly for floral odors.

dimethyloctynediol
$C_2H_5(CH_3)COHC{\vdots}CCOH(CH_3)C_2H_5$
(3, 6-dimethyl-4-octyne-3, 6-diol).
Properties: White crystals; m.p. 55-56°C; b.p. 222°C; sp. gr. (solid, 20°C) 0.923, (liquid, 60°C) 0.908. Moderately soluble in water; slightly soluble in kerosine; very soluble in acetone, alcohol, benzene, and carbon tetrachloride.
Containers: Fiber drums.
Uses: Surface active agent; polymerization; intermediate.

dimethylol ethylene urea
$\overline{OCN}(CH_2OH)CH_2CH_2\overline{N}(CH_2OH)$. A cyclic urea, used in wrinkle-resistant textile finishes.

dimethylol ethyltriazone
$\overline{OCN}(CH_2OH)CH_2N(C_2H_5)CH_2\overline{N}(CH_2OH)$.
Used in wrinkle-resistant textile finishes.

dimethylol urea $CO(NHCH_2OH)_2$.
Properties: Colorless cyrstals; m.p. 126°C; soluble in water and methanol; insoluble in ether; capable of polymerization to synthetic resin.
Derivation: Combination of urea and formaldehyde in the presence of salts or alkaline catalysts.
Uses: The first stage in the formation of urea-formaldehyde resins; impregnating wood to increase hardness and fire resistance and to form self-binding laminations for plywood manufacture; in textiles for wrinkle resistance.

O, O-dimethyl S-4-oxo-1, 2, 3-benzotriazin-3(4H)-ylmethyl phosphorodithioate ("Guthion") $C_{10}H_{12}N_3O_3PS_2$.

*See "I.C.C. Shipping Regulations," page xiii.
Reference numbers refer to name of manufacturer. See "List of Manufacturers," page v.

Properties: Brown, waxy solid; m.p. 73-74°C. Very slightly soluble in water; soluble in most organic solvents.
Uses: Fruit insecticide.
Caution! May be harmful if swallowed, inhaled, or absorbed through the skin. Overexposure will result in cholinesterase depression.

2,2-dimethylpentane (trimethylpropylmethane) $(CH_3)_3CCH_2CH_2CH_3$.
Properties: Colorless liquid; sp. gr. 0.66956 (25°C); b.p. 79.205°C; m.p. −123.79°C; soluble in alcohol; insoluble in water.
Grades: Technical.
Use: Organic synthesis.

2,3-dimethylpentane $(CH_3)_2CHCH(CH_3)C_2H_5$.
Properties: Liquid; b.p. 89°C; sp. gr. (60/60°F) 0.699; refractive index, (20/D) 1.392; flash point 20°F.
Grades: 95%.
Containers: Bottles.
Shipping regulations: Flammable liquid. Red label. *

2,4-dimethylpentane (diisopropylmethane) $(CH_3)_2CHCH_2CH(CH_3)_2$.
Properties: Colorless liquid; sp. gr. 0.6684 (25°C); b.p. 80.5°C; refractive index 1.382 (n 20/D); flash point −12°C; m.p. −119°C; soluble in alcohol; insoluble in water.
Containers: Bottles and 5-gal drums.
Grades: 95%; 99%; research.
Shipping regulations: Flammable liquid. Red label. *

3,3-dimethylpentane (diethyldimethylmethane) $CH_3CH_2C(CH_3)_2CH_2CH_3$.
Properties: Colorless liquid; sp. gr. 0.68910 (25°C); b.p. 86.071°C; m.p. −134.46°C; soluble in alcohol; insoluble in water.

2,4-dimethylpentanol-3. See diisopropyl carbinol.

2,4-dimethylpentanone-3. See diisopropyl ketone.

dimethylphenol. See xylenol.

1,5-dimethyl-2-phenyl-4-aminopyrazolone. See 4-aminoantipyrine.

dimethyl-para**-phenylenediamine.** See para-aminodimethylaniline.

N,beta**-dimethylphenylethylamine.** See phenylpropylmethyl amine.

dimethyl phosphite $(CH_3O)_2P(O)H$.
Properties: Mobile, colorless liquid; mild odor; sp. gr. 1.200 (20/4°C); b.p. 72-73°C (25 mm); flash point 205°F. Soluble in water, and miscible with most common organic solvents.
Containers: 5-gal, 55-gal drums.
Uses: Lubricant additives; intermediate; adhesive.

dimethylphosphoramidocyanidic acid, ethyl ester. See tabun.

O,O-dimethyl phosphorochloridothioate $(CH_3O)_2P(S)Cl$.
Properties: Colorless to light amber liquid; b.p. 66-67°C (16 mm); sp. gr. 1.320 (25°C); refractive index 1.4795 (n 25/D). Soluble in alcohol, benzene, acetone, carbon tetrachloride, chloroform, ethyl acetate; slightly soluble in hexane; insoluble in water.
Stability: Very slight decomposition with storage; slowly isomerizes at 100°C.
Grades: 96-100% purity.
Caution: Exposure to vapor can cause irritation to eye and lung tissues.
Uses: Intermediates for insecticides, pesticides, fungicides; oil and gasoline additives; plasticizers; corrosion inhibitors; rubber accelerators; flame retardants; floatation agents.

dimethyl phthalate $C_6H_4(COOCH_3)_2$.
Properties: Colorless, odorless, lightfast, stable, non-toxic liquid; refractive index 1.5138 (25°C); heat of combustion 5769 cal/g; sp. gr. 1.189 (25/25°C); b.p. 282°C; flash point 300°F; wt/gal 9.93 lbs (68°F); solubility of water in dimethyl phthalate 1.8% by vol (25°C); coefficient of expansion 0.00042/°F, 0.00076/°C; dilution ratio (nitrocellulose solution method) 2.9 with toluene, not miscible with petroleum naphtha; vapor pressure < 0.1 mm (20°C).
Typical specifications: Acidity not more than 0.01% (as phthalic); color not more than 10 (Pt-Co scale); odor faint-not more than slightly aromatic; sp. gr. 1.189-1.191 (25/25°C); boiling range (760 mm) below 280°C none, above 285°C none, at least 95% within 2°C; purity not less than 99%; average wt/gal 9.93 lbs (20/20°C); refractive index 1.5145-1.5165 (20°C).
Miscible with common organic solvents; very slightly soluble in water.
Derivation: By means of the standard esterifying reaction between methyl alcohol and phthalic anhydride, followed by steps of isolation and purification.
Grades: Technical.
Containers: 55-gal nonreturnable drums; tank cars; tank trucks.
Uses: Plasticizer for nitrocellulose and cellulose acetate, resins, rubber and in solid rocket propellants; lacquers; plastics; rubber; coating agents; safety glass; molding powders; insect repellent; perfumes.
Shipping regulations: None. *

2,5-dimethylpiperazine (lupetazine) $(CH_3)_2C_4H_8N_2$.
Properties: Exists as both a cis- and a trans-isomer. The trans-isomer is a free-flowing non-hygroscopic white crystalline solid melting at 118°C. The cis-isomer is a colorless liquid freezing at 18°C; soluble in water and hydrocarbons. An available commercial mixture contains 75% of the trans- and 25% of the cis-isomer.
Uses: Pharmaceuticals; polyamide resins; fungicides; rubber accelerators; corrosion inhibitors; surface active agents; solvents.

*See "I.C.C. Shipping Regulations," page xiii.
Reference numbers refer to name of manufacturer. See "List of Manufacturers," page v.

N, N′-dimethylpiperazine (1, 4-dimethylpiperazine) $(CH_3)_2C_4H_8N_2$.
Properties: Sp. gr. (20/4°C) 0.8565; b.p. 131°C.
Uses: Curing agent for polyether urethane foams; intermediate for cationic surface-active agents.

dimethylpiperazine tartrate $(CH_3)_2C_4H_8N_2 \cdot C_4H_6O_6$.
Properties: White powder; pleasant acidulous taste; m.p. 250°C. Soluble in water.
Use: Medicine.
Shipping regulations: None. *

2, 6-dimethylpiperidine (2, 6-lupetidine) $(CH_3)_2C_5H_8NH$.
Properties: B.p. (760 mm) 127.9°C; sp. gr. (20/20°C) 0.8199; refractive index 1.4383 (n 20/D); soluble in water at 20°C in all proportions.
Use: Intermediate.

dimethylpolysiloxanes. See siloxanes.

2, 2-dimethylpropane. See neopentane.

2, 6-dimethylpyridine. See lutidine.

2, 7-dimethylquinoline $(CH_3)_2C_9H_5N$.
Typical specifications: M.p. −40°C (approx.); distillation range 140-150°C (20 mm); soluble in benzene and diethyl ether.
Uses: Organic synthesis; suggested as dye intermediate.

dimethyl sebacate $[(CH_2)_4COOCH_3]_2$. Plasticizer of ester type.
Properties: Liquid, water-white; sp. gr. 0.9896 (25/20°C); m.p. 24.5°C; flash point 145°C (293°F); b.p. approx. 294°C (760 mm); refractive index 1.4376 (20°C).
Grades: Technical.
Containers: 1-, 5- and 10-gal cans; tank cars.
Uses: Solvent or plasticizer for nitrocellulose, vinyl resins; intermediate.
Shipping regulations: None. *

dimethyl silicone. General term for a family of silicones of composition $[(CH_3)_2SiO]_x$, being the more volatile materials formed on hydrolysis of dimethyldichlorosilane. Colorless oils with b.p. ranging from 134°C (760 mm) (for x = 3) to 188°C (20 mm) (for x = 9), and presumably even higher boiling members exist.

dimethyl sulfate (methyl sulfate) $(CH_3)_2SO_4$.
Properties: Colorless liquid; vapors are very poisonous! Soluble in alcohol and ether; very slightly soluble in water. Sp. gr. 1.3516; m.p. −26.8°C; b.p. 188°C (dec).
Derivation: By adding fuming sulfuric acid to methyl alcohol and distilling in vacuo.
Method of purification: Rectification.
Grades: Technical.
Containers: Returnable drums; tank cars.
Uses: Methylating agent for amines and phenols.
Danger: Extremely hazardous liquid and vapor. Causes severe burns. MCA warning label.
Shipping regulations: Corrosive liquid. White label. *

dimethyl sulfide (methyl sulfide; methanethiomethane) $(CH_3)_2S$.
Properties: Colorless liquid; disagreeable odor. Soluble in alcohol and ether; insoluble in water. Sp. gr. 0.845 (20°C); m.p. −83°C; b.p. 37.5°C.
Derivation: (a) From kraft pulping black liquor, by heating it with inorganic sulfur compounds at high temperatures and pressures; (b) by interaction of a solution of potassium sulfide and methyl chloride in methanol.
Method of purification: Rectification.
Uses: Gas odorant; solvent for many inorganic substances; catalyst impregnator.
Shipping regulations: Flammable liquid. Red label. *

dimethylsulfonyloxybutane. See nitrogen mustards.

dimethyl sulfoxide (DMSO) $(CH_3)_2SO$.
Properties: Colorless liquid; b.p. 189°C; m.p. 18.45°C; sp. gr. 1.100 (20/20°C); specific heat 0.7; nearly odorless; slightly bitter taste; miscible with water.
Derivation: From dimethyl sulfide by a liquid phase oxidation process (a nitrogen oxide or nitric acid are reported to be used).
Containers: Drums; tank cars.
Uses: As a powerful solvent with low toxicity, for synthetic fibers, especially polyacrilonitrile fibers, industrial cleaners, pesticides, paint stripping; hydraulic fluids; treatment of wool felt.

dimethyl terephthalate (DMT) $C_6H_4(COOCH_3)_2$.
Properties: Colorless crystals; m.p. 140°C; sublimes above 300°C; insoluble in water; soluble in ether and hot alcohol.
Derivation: (a) Oxidation of para-xylene with nitric acid followed by esterification with methanol; (b) step-wise catalytic air oxidation of para-xylene with intermediate esterification of the para-toluic acid; (c) liquid phase oxidation of mixed xylenes followed by esterification.
Grade: Technical.
Uses: Polyester resins for film and fiber production.

dimethyl tetrachloroterephthalate. See dacthal.

N, N-dimethyl-N′-(alpha-thenyl)-N′-phenylethylenediamine hydrochloride. See methaphenilene hydrochloride.

N, N-dimethyl-N′-(3-thenyl)-N′-(2-pyridyl)-ethylenediamine hydrochloride. See thenyldiamine hydrochloride.

O, O-dimethyl 2, 2, 2-trichloro-1-hydroxyethylphosphonate ("Dipterex;" "Dylox;" "Neguvon") $(CH_3O)_2P(CHOHCCl_3)O$. White crystalline solid; m.p. 83-84°C; soluble in water. Used as an insecticide.
Caution! May be harmful if swallowed, inhaled or absorbed through the skin. Overexposure will result in cholinesterase depression.

(O, O-dimethyl O-(2, 4, 5 trichlorophenyl) phosphorothioate). See ronnel.

dimethyl tubocurarine chloride $C_{40}H_{48}Cl_2N_2O_6$.
Dimethyl ether of tubocurarine chloride.
Properties: White odorless, crystalline powder; when heated to 236°, decomposes with evolution of gas. Soluble in water and diluted sodium hydroxide; sparingly soluble in alcohol and diluted hydrochloric acid; very slightly soluble in chloroform; practically insoluble in benzene and ether.
Grade: N. N. D.
Use: Medicine.

dimethyl tubocurarine iodide $C_{40}H_{48}I_2N_2O_6$.
Dimethyl ether of d-tubocurarine iodide.
Properties: White to pale yellow, odorless, crystalline powder. When heated to about 257°, decomposes with evolution of gas. Slightly soluble in water, diluted hydrochloric acid and diluted sodium hydroxide; very slightly soluble in alcohol, benzene, chloroform, and ether.
Grade: N. F. XI.
Use: Medicine.

N, N'-dimethylurea (sym-dimethylurea; 1, 3-dimethylurea) $(CH_3NH)_2CO$.
Properties: Colorless prisms; sp. gr. 1. 14; m. p. 106°C; b. p. 270°C; soluble in water and alcohol; insoluble in ether.
Typical specifications: Setting point 104°C (min); 0. 05% amine; gray to white color; methylamine odor.
Use: Intermediate in synthesis of drugs.

1, 3-dimethylxanthine. See theophylline.

3, 7-dimethylxanthine. See theobromine.

dimolybdenum trioxide. See molybdenum sesquioxide.

dimyristyl amine (ditetradecylamine) $(C_{14}H_{29})_2NH$. Solid; m. p. 52°C; sp. gr. 0. 89. Almost insoluble in water. Used as an organic intermediate.

dimyristyl ether (ditetradecyl ether) $(C_{14}H_{29})_2O$.
Properties: M. p. 38-40°C; b. p. 238-248°C (4 mm); sp. gr. 0. 8127 (45/4°C).
Grade: 95% (min) purity.
Uses: Electrical insulators; water repellents; lubricants in plastic molding; antistatic substances; chemical intermediates.

dimyristyl sulfide (ditetradecyl sulfide; dimyristyl thioether) $(C_{14}H_{29})_2S$.
Properties: Solid; m. p. 49-50°C; b. p., decomposes; sp. gr. 0. 8258 (50/4°C).
Grades: 95% (min) purity.
Uses: Organic synthesis (formation of sulfonium compounds).

dimyristyl thioether. See dimyristyl sulfide.

N, N'-di-beta-**2-naphthyl**-meta-**phenylenediamine** $C_6H_4(NHC_{10}H_7)_2$.
Properties: Colorless needles; m. p. 191°C; sparingly soluble in alcohol; insoluble in water and ether.
Derivation: By heating meta-phenylenediamine with beta-naphthol and subsequent extraction with alcohol.
Method of purification: Crystallization.
Use: Organic synthesis.

N, N'-di-beta-**naphthyl**-para-**phenylenediamine** (DNPD) $C_6H_4(NHC_{10}H_7)_2$.
Typical specifications: Gray powder, set point 225°C (min); purity 98% (min); sp. gr. 1. 20; insoluble in water; slightly soluble in acetone and chlorobenzene.
Grade: Pure.
Uses: Antioxidant; stabilizer; polymerization inhibitor; intermediate in organic synthesis.

dinitraniline. Same as dinitroaniline.

dinitraniline orange. A pigment made from dinitraniline and beta-naphthol. It is a reddish shade orange that has excellent light-fastness. This pigment has better resistance to bleeding in oils and solvents than the ortho-nitraniline type and it is used for awning paints. It is suitable for trim paints and enamels and it matches the federal shade known as "international orange." It has comparatively poor heat resistance and therefore is not generally employed in finishes that are to be baked.

dinitroaminophenol. See picramic acid.

2, 4-dinitroaniline (2, 4-dinitraniline) $C_6H_3NH_2(NO_2)_2$.
Properties: Yellow crystals; slightly soluble in alcohol; insoluble in water. Sp. gr. 1. 615; m. p. 187. 5-188°C.
Derivation: By the nitration of para-nitroaniline with hot mixed acid.
Method of purification: Crystallization.
Grades: Technical; pure.
Containers: Fiber kegs; steel drums.
Use: Organic synthesis.
Shipping regulations: None. *

2, 4-dinitroanisole (2, 4-dinitrophenyl methyl ether) $CH_3OC_6H_3(NO)_2$. An effective ovicide.
Properties: Colorless to yellow monoclinic needles from water or alcohol; m. p. 88°C; sp. gr. 1. 341 (20/4°C), sublimes; slightly soluble in hot water; soluble in alcohol and ether.
Use: Effective against moths, furniture and carpet beetles, cockroaches, and body lice.

dinitrobenzene (binitrobenzene) $C_6H_4(NO_2)_2$.
Meta, ortho and para isomers.
Properties: Yellow crystals; soluble in alcohol; slightly soluble in water. Sp. gr.: meta 1. 546, ortho 1. 565, para 1. 6; m. p. meta 89. 9°C, ortho 117. 9°C, para 172-173°C; b. p. meta 302. 8°C, ortho 319°C, para 299°C.
Derivation: By nitration of nitrobenzene with hot mixed acid.
Method of purification: Crystallization.
Grades: Technical.
Containers: 5-, 25-, 50-, 100-lb drums.
Uses: Organic synthesis; dyes; camphor substitute in celluloid production.
Shipping regulations: Solid or liquid form, class B poison. Poison label. *

3, 5-dinitrobenzoyl chloride $(NO_2)_2C_6H_3COCl$.
Properties: Needles; m. p. 66-68°C; b. p. 196°C (12 mm); decomposed by water and alcohol.
Use: Reagent.

2, 4-dinitro-ortho-sec-**butylphenol** (2-sec-butyl-4, 6-dinitrophenol; DNBP) $CH_3(C_2H_5)CHC_6H_2(NO_2)_2OH$.
Properties: A reddish brown liquid, slightly soluble in water, soluble in alcohol and other organic solvents. Forms salts with metals and organic bases.
Uses: It is an excellent insecticide and ovicide, but due to its toxicity to plants must be used in the dormant growth season or as a salt form to reduce toxicity. It is also used as a herbicide for pre-emergence treatment.
Available commercially as the triethanolamine salt. See "DN-289."

dinitrochlorobenzene. See 1-chloro-2, 4-dinitrobenzene.

dinitrochlorbenzol. See 1-chloro-2, 4-dinitrobenzene.

4, 6-dinitro-ortho-**cresol** (DNOC; 4, 6-dinitro-2-methyl phenol) $CH_3C_6H_2(NO_2)_2OH$.
Properties: Yellow solid; m. p. 85. 8°C; very slightly soluble in water.
Use: Dormant ovicidal spray for fruit trees (highly phytotoxic and cannot be used successfully on actively growing plants).
Danger! Poisonous by swallowing or skin contact. Absorbed through skin. MCA warning label.

2, 4-dinitro-6-cyclohexylphenol. See dinitro-ortho-cyclohexylphenol.

dinitro-ortho-**cyclohexylphenol** (2, 4-dinitro-6-cyclohexylphenol; DNOCHP) $C_6H_{11}C_6H_2(NO_2)_2OH$. An insecticide which has partially replaced 4, 6-dinitro-ortho-cresol.
Use: Control of mites on citrus fruits.

dinitrodibenzyldisulfonic acid $(CH_2C_6H_3SO_3HNO_2)_2$.
Properties: Colorless plates or tablets; soluble in water, alcohol, and ether.
Derivation: By the oxidation of sodium para-nitrotoluene sulfonate with sodium hypochlorite in an excess of caustic soda.

3, 5-dinitro-2, 6-dimethyl-4-tert-**butylacetophenone.** See musk ketone.

2, 4-dinitro-4-hydroxydiphenylamine $(NO_2)_2C_6H_3NHC_6H_4OH$.
Properties: Yellow solid; m. p. 190°C; insoluble in water.
Derivation: Condensation of 2, 4-dinitro-1-chlorobenzene and para-aminophenol.

2, 6-dinitro-3-methoxy-4-tert-**butyltoluene.** See musk ambrette.

dinitronaphthalene $C_{10}H_6(NO_2)_2$. Isomers: (a) 1, 5-; (b) 1, 8-.
Properties: (a) Yellowish-white needles, (b) yellowish-white, thick, crystalline tablets. M. p. (a) 217°C; (b) 172°C. (a) Sparingly soluble in pyridine; (b) soluble in pyridine.
Derivation: By dissolving alpha-nitronaphthalene in sulfuric acid and adding nitric acid. The solution is heated to 80-90°C and cooled.
Method of purification: Crystallization.
Grades: Technical.
Containers: Wooden kegs; fiber drums.
Uses: Dyes, especially sulfur colors; intermediates.
Shipping regulations: None. *

2, 4-dinitro-1-naphthol-7-sulfonic acid (flavianic acid) $C_{10}H_6O_8N_2S$.
Properties: Yellow needles; m. p. 151°C; very soluble in water.
Uses: Intermediate; precipitant for organic bases; reagent for amino acids.

dinitrophenol $C_6H_3OH(NO_2)_2$. Commercial material is usually the mixture of 2, 3-, 2, 4-, and 2, 6-isomers that are formed by action of sulfuric-nitric acid mixtures on phenol.
Properties: Yellow crystals. (2, 3) sp. gr. 1. 681; m. p. 144°C. (2, 4) sp. gr. 1. 683; m. p. 114-115°C. (2, 6) m. p. 63°C. Soluble in alcohol and ether, also benzene and chloroform; slightly soluble in water (2, 3 most soluble).
Derivation: (a) By heating phenol with dilute sulfuric acid, cooling the product, and then nitrating, keeping the temperature below 50°C. (b) By nitration with mixed acid with very careful temperature control.
Method of purification: Crystallization.
Grades: Technical.
Containers: Galvanized or stainless steel drums; fiber cans; barrels.
Uses: Dyes, especially sulfur colors; picric acid; picramic acid; conservation of lumber timbers and poles; starting point in the manufacture of the photographic developer diaminophenol hydrochloride.
Fire hazard: Dangerous. None when wet.
Shipping regulations: Solutions, poison, class B. Poison label. *

2, 4-dinitrophenyl methyl ether. See 2, 4-dinitroanisole.

2, 4-dinitro-6-phenylphenol $(NO_2)_2(C_6H_5)C_6H_2OH$.
Use: Agricultural insecticide; chemical synthesis.

2, 4-dinitrosoresorcinol $C_6H_2(OH)_2(NO)_2 \cdot H_2O$.
Properties: Light brown powder; m. p. 162-163°C. Decomposes, sometimes violently. Soluble in water and most organic solvents.
Grade: Technical (13. 7% N).
Uses: Chelation of heavy metals; cross-linking agent.
Caution! Avoid contact with skin and breathing of dust.
Shipping regulations: Flammable solid. Yellow label. *

3, 5-dinitrosalicylic acid $C_6H_2(OH)(NO_2)_2COOH$.
Properties: Yellow crystals; slightly soluble in water; soluble in alcohol and benzene; m. p. 174°C.

*See "I. C. C. Shipping Regulations, " page xiii.
Reference numbers refer to name of manufacturer. See "List of Manufacturers, " page v.

Derivation: Nitration of salicylic acid.
Method of purification: Recrystallization.
Grades: C. P.
Containers: Glass bottles.
Uses: Determination of glucose.
Shipping regulations: None. *

dinitrostilbene disodium sulfonate (4, 4'-dinitrostilbene-2, 2'-disulfonic acid, disodium salt) $NaSO_3(NO_2)C_6H_3CH:CHC_6H_3(NO_2)NaSO_3$.
Properties: Yellow crystals; very soluble in hot water; moderately soluble in cold water; slightly soluble in alcohol and ether. M. p., decomposes.
Containers: Fiber kegs.

4, 4'-dinitrostilbene-2, 2'-disulfonic acid, disodium salt. See dinitrostilbene disodium sulfonate.

dinitrotoluene (dinitrotoluol; DNT) $C_6H_3CH_3(NO_2)_2$ (a) 2, 4-; (b) 3, 4-; (c) 3, 5-.
Properties: Yellow crystals; soluble in alcohol and ether; insoluble in water. Sp. gr. (a) 1. 3208, (b) 1. 32, (c) 1. 277; m. p. (a) 70. 5°C, (b) 61°C, (c) 92. 3°C.
Derivation: By nitration of nitrotoluene with hot nitrosulfuric acid.
Method of purification: Crystallization.
Grades: Technical.
Containers: Fiber drums.
Uses: Organic syntheses; toluidines; dyes; explosives.
Fire hazard: Dangerous. *

dinitrotoluol. See dinitrotoluene.

2, 6-dinitro-3, 4, 5-trimethyl-tert-butylbenzene. See "Musk Tibetene."

dinonyl adipate (DNA). Ester of nonyl alcohol (trimethyl hexanols as major component, dimethyl heptanols as minor component, as well as small amounts of other isomers).
Properties: B. p. 201-210°C (1 mm); sp. gr. 0. 926 (25°C); refractive index (n 20/D) 1. 4523; viscosity 14. 9 centistoke (100°F).
Derivation: By heating adipic acid and nonyl alcohol in the presence of a trace of an acidic catalyst and removing the water of reaction as an azeotrope with a solvent such as toluene or xylene.
Use: Plasticizer where special low-temperature properties are desired.

dinonyl carbonate $(C_9H_{19})_2CO_3$. Ester of nonyl alcohol; b. p. 135-140°C (0. 3 mm); sp. gr. 0. 894 (25°C); refractive index (n 20/D) 1. 4427.

dinonyl ether $C_9H_{19}OC_9H_{19}$.
Properties: B. p. 148-153°C (5 mm); sp. gr. 0. 817 (25°C); refractive index (n 20/D) 1. 4405. Dinonyl ether can be made from nonyl alcohol plus nonyl halide by the Williamson reaction.

dinonyl maleate $C_9H_{19}OOCCH:CHCOOC_9H_{19}$. Ester of nonyl alcohol; b. p. 157-167°C (0. 1 mm); sp. gr. 0. 941 (25°C); refractive index (n 20/D) 1. 4586; viscosity 6900 centistoke (−40°F), 17. 47 centistoke (100°F), 3. 50 centistoke (210°F).

dinonyl phenol $(C_9H_{19})_2C_6H_3OH$.
Properties: Insoluble in water, soluble in common organic solvents.
Use: Solvent.

dinonyl phthalate (DNP) $C_6H_4(COOC_9H_{19})$. Ester of nonyl alcohol.
Properties: B. p. 205-220°C (1 mm); sp. gr. 0. 979 (25°C); refractive index (n 20/D) 1. 4871; viscosity 55. 3 centistoke (100°F); flash point 420°F.
Derivation: By heating phthalic acid and nonyl alcohol in the presence of a trace of an acid catalyst and removing the water of reaction as an azeotrope with a solvent such as xylene or toluene.
Use: General purpose low-volatile plasticizer for vinyl resins; pure grade as stationary liquid phase in chromatography.

"Dinopol MOP" Plasticizer. [55] Brand name for mixed octyl phthalates, of the type of $C_6H_4(COOR)_2$.
Properties: Almost colorless, oily liquid; insoluble or only slightly soluble in glycerin, glycols, and certain amines. Soluble in most organic liquids.
Typical specifications: Sp. gr. (20/20°C) 0. 975±0. 003; f. p. −40°C; boiling range 232-267°C (4 mm); acidity, max. 0. 01% as acetic acid; flash point 435°F; fire point 505°F; vapor pressure 0. 01 mm (150°C); refractive index 1. 482 (25°C); viscosity 45 cps (20°C); surface tension 31 dynes/cm (20°C); thermal expansion 0. 00074 (10-40°C); wt/gal 8 lbs.
Containers: 1-, 5-gal cans; 55-gal drums; tank cars.
Uses: Primary plasticizer for most resins, imparting permanent flexibility, low-temperature flexibility, low water extraction, and heat stability.

DIOA. Abbreviation for diisooctyl adipate.

dioctadecylamine. See distearylamine.

dioctadecyl ether. See distearyl ether.

dioctadecyl sulfide. See distearyl sulfide.

3, 3'-dioctadecyl thiodipropionate. See distearyl thiodipropionate.

dioctyl adipate. Now more correctly named di(2-ethylhexyl) adipate (q. v.).

dioctylamine. See di-2-ethylhexylamine.

dioctylaminoethanol. See di(2-ethylhexyl) aminoethanol.

dioctyl azelate. See di(2-ethylhexyl) azelate.

dioctyl chlorophosphate (dioctyl phosphorochloridate) $(C_8H_{17}O)_2P(O)Cl$.
Properties: Water-white liquid; sp. gr. 0. 991 (25°C); refractive index 1. 445 (n 25/D); decomposes on distillation; soluble in common inert organic solvents; insoluble in water.
Uses: Intermediate in organic synthesis.

di-n-octyl, n-decyl adipate (DNODA).
Properties: Clear oily liquid; color, APHA 50 max; sp. gr. 0. 912-0. 920 (25/25°C);

*See "I. C. C. Shipping Regulations, " page xiii.
Reference numbers refer to name of manufacturer. See "List of Manufacturers, " page v.

refractive index 1.443-1.447 (25°C).
Containers: 5- and 55-gal black iron drums; tank cars.
Uses: Low temperature plasticizer.

di(n-octyl, n-decyl) phthalate (DNODP).
Properties: Clear oily liquid; color APHA 50 max; odor slight; acidity (as phthalic acid) 0.01% max; sp. gr. 0.968-0.977 (25/25°C).
Containers: 5-, 55-gal drums; tank cars.
Uses: Plasticizer for polyvinyl chloride.

dioctyl ether $(C_8H_{17})_2O$.
Properties: Liquid; m. p. −7°C; b. p. 291.7°C; sp. gr. 0.805 (17/4°C); refractive index 1.4329 (n 24/D).
Grades: 95% (min) purity.
Uses: Electrical insulator; water repellent; lubricant in plastic molding and processing; antistatic substance; chemical intermediate.

dioctyl hexahydrophthalate. See di(2-ethylhexyl) hexahydrophthalate.

dioctyl isophthalate. See di(2-ethylhexyl) isophthalate.

dioctyl phosphite (dioctyl phosphonate) $(C_8H_{17}O)_2P(O)H$.
Properties: Water-white liquid; b. p. 150-155°C (2-3 mm); sp. gr. 0.929 (25°C); refractive index 1.4418 (n 25/D); soluble in common organic solvents.
Containers: Carboys.
Uses: Solvent; antioxidant; intermediate.

dioctyl phosphonate. See dioctyl phosphite.

dioctyl phosphoric acid. See di(2-ethylhexyl) phosphoric acid.

dioctyl phosphorochloridate. See dioctyl chlorophosphate.

dioctyl phthalate. See di(2-ethylhexyl) phthalate.

di(2-octyl) phthalate. See dicapryl phthalate.

dioctyl sebacate. See di(2-ethylhexyl) sebacate.

dioctyl sodium sulfosuccinate (di(2-ethylhexyl) sodium sulfosuccinate) $C_{20}H_{37}NaO_7S$. An anionic surface-active agent.
Properties: White, wax-like, plastic solid with characteristic odor. Slowly soluble in water; freely soluble in alcohol and glycerin; very soluble in petroleum ether. Saponification value 240-253; stable in acid and neutral solutions; hydrolyzes in alkaline solutions.
Derivation: By esterification of maleic anhydride with 2-ethylhexyl alcohol followed by addition of sodium bisulfite.
Grade: N. F. XI.
Use: Medicine.

dioctyl succinate. See di(2-ethyl hexyl) succinate.

dioctyl sulfide (dioctyl thioether) $(C_8H_{17})_2S$.
Properties: Liquid; m. p. 0.5°C; b. p. 180°C (10 mm); sp. gr. 0.8419 (17/17°C); refractive index 1.4606 (n 20/D).
Grades: 95% (min) purity.
Uses: Organic synthesis (formation of sulfonium compounds).

dioctyl thioether. See dioctyl sulfide.

dioctyl thiopropionate. See 3,3'-(2-ethylhexyl) thiodipropionate.

diode. A two-electrode device (usually assembled from a semiconductor material and connecting wiring, etc.), having an anode and a cathode and which has marked unidirectional characteristics, as far as the behavior of an electrical current is concerned.

diode, crystal. A diode consisting of a semiconductor such as germanium or silicon, as one electrode, and a fine wire whisker resting on the semiconductor as the other electrode. Used as a rectifier or detector of microwave frequencies.

"Diodoquin." [70] Trademark for diiodohydroxyquin, U. S. P.

"Diodrast." [162] Trademark for iodopyracet.

"Diol." [51] Trademark for low carbon content and low pour point diesel engine lubricants for 2-stroke cycle, crankcase-scavenging engines and certain other diesels.

diolefin. See diene.

DIOP. Abbreviation for diisooctyl phthalate.

diopside, fused $CaMgSi_2O_6$. A synthetic diopside produced in the electric furnace and used as a refractory. Natural diopside is a mineral and is also sometimes used as a gem-stone.

diorite. A granitoid rock composed essentially of hornblende and feldspar. Quartz may be present in considerable amount, in which case the rock is called quartz diorite. Quarried for crushed rock in the District of Columbia, Virginia and many other states.

DIOS. Abbreviation for diisooctyl sebacate (q. v.).

diosma. See buchu.

1,4-dioxane (diethylene ether; 1,4-diethylene dioxide; diethylene oxide; dioxyethylene ether) $\underline{OCH_2CH_2OCH_2CH_2}$.
Properties: Flammable colorless liquid; faint pleasant ethereal odor; stable; miscible with water and most organic solvents. Vapor harmful.
Constants: B. p. 101.3°C; f. p. 11.8°C; sp. gr. 1.0356 (20/20°C); wt/gal 8.61 lbs (20°C); refractive index 1.4221 (20°C); surface tension 36.9 dynes/cm (25°C); vapor pressure 29.0 mm (20°C); viscosity 0.0131 poise (20°C); specific heat 0.420 cal/g (20°C); heat of fusion 33.8 cal/g; flash point 18°C (65°F)(ASTM open cup); latent heat of evaporation 98.6 cal/g at b. p.; heat of combustion 581 kg cal/mole; coefficient of expansion 0.00108 (20°C); electric conductivity $< 2 \times 10^{-8}$ recip. ohms

(25°C); miscible in all proportions with water.
Typical specifications: Acidity not more than 0.01% (as acetic); sp. gr. 1.030-1.038 (20/20°C); color water-white; odor mild, non-residual; f. p. not lower than 10°C; boiling range (760 mm) below 95°C none, above 103° none; average wt/gal 8.61 lbs (20°C).
Derivation: (a) Ethylene glycol by treatment with acid; (b) from beta, beta-dichloroethyl ether by treatment with alkali.
Method of purification: Distillation.
Grades: Reagent; technical.
Containers: 1-, 5-gal cans; 55-gal (non-returnable) drums; tank cars.
Uses: Solvent for cellulose acetate and other derivatives, fats, greases, natural and synthetic resins, oil-soluble dyes, mineral oils, vegetable oils, blown and heat-bodied oils; lacquers; paints; varnishes; paint and varnish removers; plastics; wetting and dispersing agent in textile processing, dye baths, stain and printing compositions; cleaning and detergent preparations; cements; cosmetics; deodorants; fumigants; emulsions; glues; polishing compositions; shoe creams; stabilizer for chlorinated solvents.
Warning! Flammable. Vapor harmful; tends to form explosive peroxides, especially when anhydrous. MCA warning label.
Shipping regulations: Flammable liquid. Red label. *

2,3-para-dioxanedithiol-S,S-bis-(O,O-diethylphosphorodithioate $C_{12}H_{26}O_6P_2S_4$.
Properties: Tan liquid. M. p. −20°C. Practically insoluble in water; partly soluble in hexane.
Use: Insecticide.

dioxolane $\underline{OCH_2CH_2OC}H_2$.
Properties: Water-white liquid; soluble in water; stable under neutral or slightly alkaline conditions. Sp. gr. 1.065; b. p. 74°C; flash point 35°F; vapor pressure 70 mm (20°C); wt/gal 8.2 lbs (20°C).
Grades: Technical.
Use: Suitable as a low-boiling solvent and extractant for oils, fats, waxes, dyes, and cellulose derivatives.
Shipping regulations: Flammable liquid. Red label. *

dioxolanes. Cyclic acetals resulting from action of an aldehyde and a glycol. Thus ethylene glycol and formaldehyde give the simplest compound of this type, dioxolane.

dioxolone-2. See ethylene carbonate.

dioxopurine. See xanthine.

dioxyanthraquinones. See dihydroxyanthraquinones.

dioxybenzenes. See dihydroxybenzenes.

dioxyethylene ether. See 1,4-dioxane.

dioxyline phosphate ("Paveril" Phosphate) $C_{22}H_{25}NO_4 \cdot H_3PO_4$. 6,7-Dimethoxy-1-(4'-ethoxy-3'-methoxy-benzyl)-3-methylisoquinoline phosphate.
Properties: White crystalline, odorless solid with a bitter taste; soluble in water; melts (with decomposition) 197-199°C.
Use: Medicine.

dioxynaphthalenes. See dihydroxynaphthalenes.

DIOZ. Abbreviation for diisooctyl azelate (q.v.).

DIPA. Abbreviation for diisopropanolamine (q.v.).

dipalmitylamine (dihexadecylamine) $(C_{16}H_{33})_2NH$. Solid; m. p. 65°C; sp. gr. 0.83; almost insoluble in water. Used as a chemical intermediate.

"Di-Paralene." [3] Trademark for chlorcyclizine hydrochloride (q.v.).

dipentaerythritol $(CH_2OH)_3CCH_2OCH_2C(CH_2OH)_3$. Found in technical pentaerythritol.
Containers: Up to carload or truckload quantities.
Uses: Paints and coatings.

"Dipentek." [138] Trade name for dipentaerythritol, technical.
Properties: An off-white, free-flowing powder. The molecule contains six primary hydroxyl groups, all of which are esterifiable.
Container: 50-lb multi-wall bags.
Uses: For high viscosity vehicles and fast drying alkyds.

dipentene (cinene; limonene, inactive; dℓ-paramentha-1,8-diene; cajeputene) $C_{10}H_{16}$. Commercial form is high in dipentene content, but also contains other terpenes and related compounds in varying amounts.
Properties: Colorless liquid; pleasant, lemon-like odor; sp. gr. (15.5/15.5°C) 0.847; b. p. 175-176°C (760 mm); flash point (closed cup) 43°C; wt/gal 7.15 lbs (15.5°C). Miscible with alcohol; insoluble in water.
Typical specifications: Sp. gr. 0.859-0.862 (15°C); moisture content, trace; refractive index 1.472-1.477 (25°C); kauri-butanol solvency test 88; flash point (closed cup) 114°F; acidity none; color water-white; wt/gal 7.20 lbs; distillation range 170°-200°C.
Derivation: (a) From various ethereal oils, particularly Levant wormseed oil, (b) By close fractionation of wood turpentine. (c) By-product in making synthetic camphor.
Grades: Steam distilled; destructively distilled.
Containers: 30-, 55-gal non-returnable galvanized drums; tank cars.
Uses: Solvent for oleoresinous products, rosin, ester gum, cumar and alkyd resins, waxes, metallic soap driers, rubber, etc.; rubber compounding and reclaiming of rubber; dispersing agent for oils, resins, resin-oil combinations, pigments and driers; paints, enamels, lacquers, and varnishes; general wetting and dispersing agent; printing inks; perfumes; substitute for turpentine and petroleum solvents in

floor waxes and furniture polishes; manufacture of synthetic resins, polyterpenes, chemicals.
Fire hazard: Flammable.
Shipping regulations: None. *

dipentene dioxide (limonene dioxide) $C_{10}H_{16}O_2$.
Properties: Liquid; sp. gr. 1.0287 (20°C); b. p. 242°C; f. p. −100°C; soluble in water.
Uses: Intermediate for plasticizers, epoxy resins; pharmaceuticals.

dipenteneglycol. See terpin hydrate.

dipentene monoxide (limonene monoxide) $C_{10}H_{16}O$.
Properties: Liquid; sp. gr. 0.929 (20°C).
Use: Organic intermediate; epoxy resins.

"Dipentene No. 122." [266] Trademark for technical grade dipentene (dℓ-limonene); colorless, mobile liquid; ASTM distillation range, 5-95%, 175-181°C.

"Dipentene No. 213." [266] Trade name for a mixture of monocyclic terpenes; colorless, mobile liquid; ASTM distillation range, 5-95%, 176-187°C.

"Dipentite." [306] Trademark for $C_6H_5OP(OCH_2)_2C(CH_2O)_2POC_6H_5$ (spiro).
Properties: White powder; b. p. 190°C (0.1 mm).
Containers: 12-gal pails, 25 lbs net.
Use: Stabilizer for resins.

2,5-di(tert-pentyl)hydroquinone. See 2,5-di(tert-amyl)hydroquinone.

diphemanil methyl sulfate (4-diphenylmethylene-1,1-dimethylpiperidinium methyl sulfate)
$(C_6H_5)_2CCCH_2CH_2N(CH_3)_2CH_2CH_2 \cdot CH_3SO_4$.
Properties: White or near white, bitter crystalline solid with faint characteristic odor. M. p. 189-196°C. Very slightly soluble in ether; slightly soluble in alcohol, chloroform, and water. Stable to heat and light; somewhat hygroscopic; pH (1% solution) 4.0-6.0.
Grade: N. F. XI.
Use: Medicine.

diphenadione (2-diphenylacetyl-1,3-indanedione) $C_{23}H_{16}O_3$.
Properties: Yellow, odorless crystals or crystalline powder. Practically insoluble in water, slightly soluble in acetone and in alcohol, soluble in benzene, ether, and in glacial acetic acid. Melting range: 144-150°C.
Grade: N. F. XI.
Use: Medicine.

diphenatrile. See diphenylacetonitrile.

diphenhydramine hydrochloride (2-benzhydryloxy)-N, N-dimethylethylamine hydrochloride; beta-dimethylaminoethyl benzhydryl ether hydrochloride)
$(C_6H_5)_2CHOCH_2CH_2N(CH_3)_2 \cdot HCl$.
Properties: White, odorless, crystalline powder. Darkens slowly on exposure to light; m. p. 166-170°C. Solutions practically neutral to litmus paper. Freely soluble in water, alcohol and chloroform; slightly soluble in acetone; very slightly soluble in benzene and ether.
Grades: U. S. P. XVI.
Use: Medicine.

diphenic acid (2,2'-biphenyldicarboxylic acid) $HOOCC_6H_4C_6H_4COOH$.
Properties: White needles; m. p. 228-229°C; soluble in hot water.
Use: Synthesis of dyes, detergents, pharmaceuticals.

"Diphenolic Acid." [424] (DPA). Trademark for 4,4-bis(4-hydroxyphenyl)pentanoic acid, $CH_3(C_6H_4OH)_2CCH_2CH_2COOH$.
Properties: White or light tan powder with a slight phenolic odor; m. p. 170-173°C. Soluble in hot water, alcohol and acetone; insoluble in benzene and cold water.
Uses: Paint formulations; coatings and finishes.

diphenyl (biphenyl) $C_6H_5C_6H_5$. Several crystalline forms are known.
Properties: White scales; pleasant odor. Soluble in alcohol and ether; insoluble in water. Sp. gr. approx 1; m. p. 70°C; b. p. 255°C.
Derivation: (a) By slowly passing benzene through a red hot iron tube. (b) By heating bromobenzene and sodium, with subsequent distillation.
Method of purification: Crystallization.
Grades: Technical.
Containers: Tins; steel drums; tank cars.
Uses: Organic synthesis; heat transfer agent; fungicides; dyeing assistant for polyesters.
Shipping regulations: None. *

diphenylacetic acid $(C_6H_5)_2CHCOOH$.
Properties: Colorless, odorless crystals; b. p. sublimes; m. p. 147.8-148.2°C; soluble in hot water, alcohol, ether, chloroform.

diphenylacetonitrile (diphenatrile) $(C_6H_5)_2CHCN$.
Properties: Yellow crystalline powder; m. p. 73-73.5°C; insoluble in water and very soluble in alcohol.
Uses: Preparation of diphenylacetic acid, para-diphenylethylamine and synthesis of anti-spasmodics; as an herbicide.

diphenylacetylene. See tolan.

2-diphenylacetyl-1,3-indanedione. See diphenadione.

diphenylamine (DPA; phenylaniline) $(C_6H_5)_2NH$.
Properties: Colorless to grayish crystals. Soluble in alcohol and ether; slightly soluble in water. Sp. gr. 1.159; m. p. 52.85°C; b. p. 302°C.
Derivation: By heating equal formula weights of aniline and aniline hydrochloride in an autoclave. The product is boiled with dilute hydrochloric acid to remove the unaltered aniline, and the residue is distilled.
Method of purification: Crystallization.
Grades: Technical; refined, flake and fused.
Containers: 350-lb barrels; fiber drums.
Uses: Antioxidant additives; stabilizers for plastics, including solid rocket propellants;

pesticides; explosives; dyes; pharmaceuticals.
Shipping regulations: None. *

diphenylaminechlorarsine. See phenarsazine chloride.

9,10-diphenylanthracene $C_{14}H_8(C_6H_5)_2$.
Properties: Crystals; m.p. 248-250°C; insoluble in water and alcohol; slightly soluble in toluene.
Grade: Purified.
Uses: As primary fluor or as wave length shifter in solution scintillators.

1,4-diphenylbenzene. See terphenyl.

diphenylbenzidine $C_6H_5HNC_6H_4C_6H_4NHC_6H_5$.
Properties: White powder. Insoluble in water; slightly soluble in alcohol and aromatic hydrocarbons. Sensitive to light. M.p. 242°C.
Derivation: Diphenylamine and fuming sulfuric acid.
Method of purification: Recrystallization.
Grades: C.P.; analytical.
Containers: Glass bottles.
Use: Determination of zinc and nitrites.
Shipping regulations: None. *

diphenylbromoarsine $(C_6H_5)_2AsBr$.
Properties: White crystals. Caution! Very irritant! M.p. 54-56°C.
Derivation: (a) Hydrobromic acid and diphenylarsenious oxide are heated together for about 4 hours at 115-120°C; (b) by action of arsenic tribromide on triphenyl arsine at 300-350°C.

1,3-diphenyl-2-buten-1-one. See dypnone.

diphenylcarbazide $(C_6H_5NHNH)_2CO$. Decomposes in light.
Properties: White crystals. Insoluble in water; soluble in alcohol and benzene. M.p. 173°C.
Derivation: Phenylhydrazine and urea.
Method of purification: Recrystallization.
Grade: C.P.
Containers: Glass bottles.
Use: Determination of copper and other metals.
Shipping regulations: None. *

diphenylcarbinol. See benzhydrol.

diphenyl carbonate (carbonic acid, diphenyl ester) $(C_6H_5O)_2CO$.
Properties: White, crystalline solid. White needles from alcohol. Can be halogenated and nitrated in characteristic manner. Readily undergoes hydrolysis and ammonolysis when treated respectively with inorganic bases, ammonia and amines. Soluble in acetone, hot alcohol, benzene, carbon tetrachloride, ether, glacial acetic acid and other organic solvents; insoluble in water. B.p. 302°C; m.p. 78°C; sp.gr. 1.1215 (87/4°C).
Grade: Technical.
Uses: Plasticizer and solvent; preparation of other carbonates.

diphenylchloroarsine $(C_6H_5)_2AsCl$.
Properties: Colorless crystals when pure. The technical product is a dark-brown liquid, which slowly changes into a semisolid, viscous mass. Decomposed by water (slowly). Insensitive to detonation. Caution! Very irritant! Soluble in carbon tetrachloride, chloropicrin, phenyldichloroarsine; practically insoluble in water.
Constants: Sp.gr. 1.363 (40°C) (solid), or 1.358 (45°C) (liquid); b.p. 333°C (in CO_2 atmosphere); m.p. 41°C; vapor pressure 0.0005 mm (20°C); volatility 0.68 mg/cu m (20°C); latent ht of volatilization 56.6 cal; sp. heat 0.217 cal; coefficient of thermal expansion 0.00075.
Derivation: Benzene and arsenic trichloride are heated together in the presence of aluminum chloride.
Grade: Technical.
Use: Military poison gas.
Shipping regulations: Solid, poison, class C. Tear gas label. *

diphenylcyanoarsine $(C_6H_5)_2AsCN$.
Properties: Colorless prisms. Characteristic odor resembling that of a mixture of bitter almonds and garlic. Slowly decomposed by water. Easily decomposed by alkali solutions. Caution! Very irritant! Soluble in alcohol, benzene, ether; slightly soluble in water.
Constants: Sp.gr. 1.45 (20°C); b.p. 213°C (21 mm); m.p. (given variously) 31-35°C; vapor pressure 0.0002 mm (20°C); volatility 0.1-0.15 mg/cu m (20°C).
Derivation: Interaction of hydrocyanic acid and diphenylarsenious oxide.

diphenyl decyl phosphite $(C_6H_5O)_2POC_{10}H_{21}$.
Properties: Nearly water-white liquid; sp.gr. 1.023 (25/15.5°C); m.p. 18°C; refractive index 1.5160 (n 25/D).
Containers: 55-gal containers.
Uses: Chemical intermediate; stabilizer for polyvinyl and polyolefin resins.

diphenyldichlorosilane $(C_6H_5)_2SiCl_2$.
Properties: Colorless liquid; b.p. 305°C; m.p. −22°C; sp.gr. 1.19 (20°C); refractive index (n 25/D) 1.5773; flash point (Cleveland open cup) 288°F. Readily hydrolyzed by moisture, with the liberation of hydrochloric acid.
Derivation: (a) By the reaction of powdered silicon and chlorobenzene in the presence of copper powder as catalyst; (b) by the reaction of phenylmagnesium chloride with silicon tetrachloride.
Grade: Technical.
Use: Intermediate for silicone lubricants.
Shipping regulations: Corrosive liquid. White label. *

diphenyldi-n-dodecylsilane $(C_6H_5)_2Si(C_{12}H_{25})_2$.
Properties: Colorless oil.
Derivation: Reaction of didodecyldichlorosilane with phenyl lithium.
Use: High-temperature lubricant.

diphenyldiethoxysilane $(C_6H_5)_2Si(OC_2H_5)_2$.
Properties: Colorless liquid; b.p. (12 mm) 164°C.
Derivation: Reaction of diphenyldichlorosilane with ethanol.

diphenyldiimide. See azobenzene.

2, 2-diphenyl-4-dimethylaminovaleronitrile $(C_6H_5)_2C(CN)CH_2CH(CH_3)N(CH_3)_2$. A granular, cream colored solid; m. p. 87°-92°C.
Precaution: Subject to Federal Narcotic Regulation.

diphenyleneimine. See carbazole.

alpha-**diphenylenemethane.** See fluorene.

diphenylenemethane oxide. See xanthene.

diphenylene oxide (dibenzofuran) $C_{12}H_8O$ (tricyclic).
Properties: Crystalline solid; m. p. 87°C; b. p. 288°C; insoluble in water; slightly soluble in alcohol, ether and benzene.
Derived from coal-tar.

1, 1-diphenylethane. See uns-diphenylethane.

1, 2-diphenylethane. See sym-diphenylethane.

uns-**diphenylethane** (1, 1-diphenylethane) $(C_6H_5)_2CHCH_3$.
Properties: Colorless liquid. Soluble in chloroform, ether, carbon disulfide. B. p. 286°C; sp. gr. 1.004 (20°C); m. p. —21.5°C.
Derivation: By the action of acetaldehyde upon benzene in the presence of concentrated sulfuric acid.
Grade: Technical.
Uses: Solvent for nitrocellulose; organic synthesis.

sym-**diphenylethane** (bibenzyl; dibenzyl; 1, 2-diphenylethane) $C_6H_5CH_2CH_2C_6H_5$.
Properties: White, crystalline needles or small plates. Soluble in alcohol, chloroform, ether, carbon disulfide; insoluble in water. Sp. gr. 0.9782; b. p. 284°C; m. p. 52°C.
Derivation: (a) By treating benzyl chloride with metallic sodium. (b) By the action of benzyl chloride on benzylmagnesium chloride.
Grade: Technical.
Use: Organic synthesis.

diphenyl ether. See diphenyl oxide.

diphenylethylene. See stilbene.

N, N-diphenylethylenediamine (ethylene diphenyldiamine) $C_6H_5NHCH_2CH_2NHC_6H_5$.
Properties: A cream-colored solid; sp. gr. 1.14; m. p. not definite, starts to soften at about 54°C while the main portion melts between 60-65°C; stable on storage; insoluble in water; soluble in acetone, ethylene dichloride, benzene, and gasoline; antioxidant.
Containers: 50-lb bags; 250-lb drums.
Use: As an antioxidant in rubber compounding.
Hazards: No health hazard when used in rubber or GR-S in amounts recommended.

diphenylguanidine (DPG; melaniline) $HN{:}C(NHC_6H_5)_2$. Isomeric symmetrical and unsymmetrical forms are known, the former being described here.
Properties: White crystalline powder; sp. gr. 1.13; m. p. 147°C; decomposes above 170°C; soluble in ethyl alcohol, carbon tetrachloride, chloroform, hot benzene and toluene; slightly soluble in water to give an alkaline solution.
Derivation: Treatment of aniline with cyanogen chloride.
Method of purification: Recrystallization.
Grade: Technical.
Containers: Bags; drums.
Uses: Acceleration of rubber vulcanization; primary standard for acids.
Shipping regulations: None. *

diphenylguanidine phthalate.
Properties: White to gray powder; m. p. 178°C; sp. gr. 1.20.
Containers: 50-lb bags; 150-lb drums.
Use: Rubber accelerator.

1, 6-diphenylhexatriene (DPH) $C_6H_5HC{:}CHCH{:}CHCH{:}CHC_6H_5$. Used as wave length shifter in solution scintillation counting.

diphenylhydantoin $\overline{HNCONHCOC}(C_6H_5)_2$.
Properties: White, odorless powder. Practically insoluble in water; soluble in hot alcohol; slightly soluble in cold alcohol, chloroform, ether; m. p. 292-299°C, with decomposition.
Grade: U. S. P. XVI.
Use: Medicine.

diphenylhydantoin sodium (phenytoin, soluble) $C_{15}H_{11}N_2O_2Na$. Sodium 5, 5-diphenylhydantoinate.
Properties: White, odorless powder; freely soluble in water; soluble in alcohol; practically insoluble in ether and in chloroform; somewhat hygroscopic; on exposure to air diphenylhydantoin is liberated.
Containers: Up to 100-lb drums.
Use: Medicine.

N, N'-diphenylhydrazine. See hydrazobenzene.

diphenylketone. See benzophenone.

diphenylmethane (benzylbenzene) $(C_6H_5)_2CH_2$.
Properties: Long colorless needles. Soluble in alcohol and ether; insoluble in water. Sp. gr. 1.0056; m. p. 26.5°C; b. p. 264.7°C.
Derivation: By condensation of benzyl chloride and benzene in presence of aluminum chloride.
Method of purification: Crystallization.
Grade: Technical.
Containers: Iron barrels.
Uses: Organic synthesis; dyes; perfumery.
Shipping regulations: None. *

diphenylmethane diisocyanate $OCNC_6H_4CH_2C_6H_4NCO$.
Derivation: para, para'-Diaminodiphenylmethane and phosgene.
Uses: Preparation of polyurethane resins; bonding rubber to rayon and nylon cord.

diphenylmethanol. See benzhydrol.

4-diphenylmethoxy-1-methylpiperidine hydrochloride. See diphenylpyraline hydrochloride.

3-diphenylmethoxytropane methanesulfonate. See benztropine methanesulfonate.

diphenylmethylchlorosilane $(C_6H_5)_2(CH_3)SiCl$.
Properties: Colorless liquid; b.p. 295°C.
Derivation: Grignard reaction of diphenyldichlorosilane with methylmagnesium chloride.
Uses: Intermediate; end stopper for silicone oils.

4-diphenylmethylene-1, 1-dimethylpiperidinium methyl sulfate. See diphemanil methyl sulfate.

1-diphenylmethyl-4-methylpiperazine hydrochloride. See cyclizine hydrochloride.

diphenylnaphthylenediamine $C_{10}H_6(NHC_6H_5)_2$.
Properties: Silvery, crystalline plates. Slightly soluble in alcohol; insoluble in water. M.p. 164°C.
Derivation: By heating 2, 7-dihydroxynaphthalene with aniline and aniline hydrochloride.
Method of purification: Crystallization.
Use: Organic synthesis.

diphenylnitrosamine. See N-nitrosodiphenylamine.

2, 5-diphenyloxazole (DPO; PPO) $C_{15}H_{11}NO$.
Properties: Solid; m.p. 70-72°C.
Grade: Scintillation.
Containers: Glass bottles.
Uses: Scintillation counter, or as wave length shifter in liquid scintillators..

diphenyl oxide (phenyl ether; diphenyl ether) $(C_6H_5)_2O$.
Properties: Colorless crystals, geranium odor. Soluble in alcohol and ether; insoluble in water. Sp. gr. 1.072-1.075; m.p. 27°C; b.p. 259°C.
Derivation: By the reaction of bromobenzene and sodium phenate heated under pressure.
Method of purification: Crystallization.
Grade: Technical.
Containers: Glass bottles; aluminum containers; demijohns; drums.
Uses: Organic synthesis; perfumery, particularly for soaps; heat transfer medium.
Shipping regulations: None. *

N, N'-diphenyl-meta-phenylenediamine. $C_6H_4(NHC_6H_5)_2$.
Properties: Flat crystalline needles. Soluble in hot alcohol; insoluble in water. M.p. 95°C.
Derivation: By heating resorcinol with aniline in presence of calcium chloride and zinc chloride at 210°C.
Method of purification: Crystallization.
Grade: Technical.
Containers: Wooden kegs; fiber drums.
Use: Organic synthesis.
Shipping regulations: None. *

N, N'-diphenyl-para-phenylenediamine (DPPD) $(C_6H_5NH)_2C_6H_4$.
Typical specifications: Gray powder; sp. gr. 1.20; m.p. 136°C min; ash-trace; purity 92% (min). Insoluble in water; soluble in acetone, benzene, monochlorobenzene and isopropyl acetate.
Grade: Commercial.
Uses: Flex-resistant antioxidant in natural and synthetic rubbers; stabilizer; polymerization inhibitor. Also as intermediate in manufacture of dyes, drugs, plastics, and detergents.

diphenyl phthalate $C_6H_4(COOC_6H_5)_2$.
Properties: White powder; m.p. approx 80°C; sp. gr. 1.28 (20°C); flash point 435°F; b.p. 405°C; wt/gal 10.68 lbs; refractive index 1.572 (74°C).
Typical specifications: Appearance white powder; melt clear and very light yellow color; crystallizing point 69°C min; acidity as phthalic 0.20% (max); moisture 0.5% (max); soluble in ketones, esters, and chlorinated hydrocarbons; insoluble in water.
Grade: Technical.
Containers: 100-lb fiber containers.
Uses: Plasticizer; plasticizing compositions for ethylcellulose, nitrocellulose, and various synthetic resins.

diphenyl-4-piperidinemethanol hydrochloride. See azacyclonol hydrochloride.

diphenyl-4-piperidylmethane $(C_6H_5)_2(C_5H_{10}N)CH$.
Properties: White to slightly off-white solid; difficultly soluble in water but readily soluble in dilute acids; moderately soluble in organic solvents such as alcohols, ketones, and aromatic hydrocarbons; f.p. 99.7°C min.
Use: Intermediate.

N, N'-diphenylpropylenediamine $C_6H_5NHCH_2CH(CH_3)NHC_6H_5$.
Properties: A clear, deep reddish-brown, semi-viscous liquid; sp. gr. 1.07; stable in storage; insoluble in water; soluble in acetone, ethylene dichloride, benzene, and gasoline; readily disperses.
Containers: 450-lbs.
Uses: Antioxidant for latex compounds of all kinds.
Hazards: No health hazards when used in rubber in the amount recommended.

diphenylpyraline hydrochloride (4-diphenylmethoxy-1-methylpiperidine hydrochloride) $(C_6H_5)_2CHOC_5H_9NCH_3 \cdot HCl$.
Properties: Crystals; m.p. 206°C. Soluble in water, alcohol, isopropanol; insoluble in ether and benzene.
Use: Medicine.

diphenyl-4-pyridyl carbinol $(C_6H_5)_2(C_5H_4N)COH$.
Properties: White solid; very weak base; slightly soluble in such solvents as methanol, ether, acetone, benzene; soluble in hot glacial acetic acid; m.p. 236-241°C.
Use: Intermediate.

diphenyl-4-pyridyl methane $(C_6H_5)_2(C_5H_4N)CH$.
Properties: White to pale yellow crystalline solid; moderately soluble in common organic solvents; b.p. 234°C (20 mm); f.p. 123°C min.
Use: Intermediate.

diphenylsilanediol $(C_6H_5)_2Si(OH)_2$.
Properties: White solid; m.p. 130-150°. Derived from hydrolysis of diphenyldichlorosilane. Used as a silicon chemical.

para, para'-**diphenylstilbene**
$C_6H_5C_6H_4CH:CHC_6H_4C_6H_5$.
Properties: Crystals; m.p. 308-310°C.
Uses: In purified form as fluor in plastic scintillators.

N,N'-diphenylthiourea. See thiocarbanilide.

diphenylurea (carbanilide)
$(NHC_6H_5)CO(NHC_6H_5)$.
Properties: Colorless prisms. Soluble in alcohol and ether; very slightly soluble in water. Sp. gr. 1.239; m.p. 235°C; b.p. 260°C.
Derivation: From aniline and phenylcyanate.
Method of purification: Crystallization.
Grades: Technical; reagent.
Containers: Wooden kegs; fiber drums.
Use: Organic synthesis.
Shipping regulations: None. *

diphenyl ortho-**xenyl phosphate**
$(C_6H_5O)_2(C_6H_5C_6H_4O)PO$.
Properties: Sp. gr. 1.20 (20°C); refractive index 1.582-1.590 (60°C); boiling range 250-285°C (5 mm); flash point 225°C; insoluble in water.
Use: Plasticizer.

diphosgene. See trichloromethylchloroformate.

diphosphopyridine nucleotide. See nicotinamide adenine dinucleotide.

dipicrylamine. See hexanitrodiphenylamine.

dip oils. See tar acids.

dipole. An assemblage of atoms or subatomic particles having equal electric charges of opposite sign separated by a finite distance; for instance the nucleus and orbital electron of a hydrogen atom, or the hydrogen and chlorine atoms of an HCl molecule.

dipole moment. In most molecules the atoms and their electrons and nuclei are so arranged that one part of the molecule has a positive electrical charge and other parts are therefore negatively charged with respect to the first mentioned part. The molecule therefore becomes a small magnet or dipole. When the molecule is subjected to changing electrical or magnetic fields, these interact with the dipole and the molecule is subjected to turning and twisting forces. The dipole moment (μ) is the distance in centimeters between the charges multiplied by the quantity of charge in electrostatic units.

dipotassium orthophosphate. See potassium phosphate, dibasic.

Dippel's oil. See bone oil.

dipping acid. See sulfuric acid.

dipropargyl (bipropargyl; 1,5-hexadiyne)
C_6H_6 or $HC:CCH_2CH_2C:CH$.
Properties: Colorless liquid. Soluble in alcohol; insoluble in water; sp. gr. 0.805; b.p. 85°C; m.p. −6.0°C.

dipropenyl. See 2,4-hexadiene.

di-2-propenylamine. See diallylamine.

di-n-propylamine $(C_3H_7)_2NH$.
Properties: Sp. gr. 0.741 (20°C); boiling range 105-109°C; color water-white; odor amine; wt/gal 6.2 lbs.
Containers: 5-gal cans; 55-gal drums; tank cars.
Fire hazard: Flash point 45°F.
Shipping regulations: Flammable liquid. Red label. *

dipropylene. See 2,4-hexadiene.

dipropylene glycol $(CH_3CHOHCH_2)_2O$.
Properties: Colorless, slightly viscous liquid. Soluble in toluene, water.
Constants: Sp. gr. 1.0252 (20/20°C); b.p. 231.8°C (760 mm); vapor pressure 0.01 mm (20°C); flash point 280°F; wt/gal 8.5 lbs (20°C); coefficient of expansion 0.00073 (20°C); viscosity 1.07 poise (20°C).
Typical specifications: Sp. gr. 1.034-1.039 (20/20°C); boiling range 215-240°C (760 mm); acidity not more than 0.01% (as acetic).
Grade: Technical.
Containers: 1-gal cans; 5-, 55-gal drums; tank cars.
Uses: Solvent for nitrocellulose, shellac; partial solvent for cellulose acetate; solvent mixtures; lacquers; coatings; printing inks.

dipropylene glycol dibenzoate
$C_6H_5CO_2CH_2CH(CH_3)OCH_2CH(CH_3)OCOC_6H_5$.
Properties: A light-colored liquid; sp. gr. 1.1271 (20/20°C); 9.4 lb/gal (20°C); b.p. 250°C (10 mm); vapor pressure 1.2 mm Hg (200°C); insoluble in water; viscosity 227 cps (20°C).
Use: Plasticizer.

dipropylene glycol monomethyl ether
$CH_3OC_3H_6OC_3H_6OH$.
Properties: Sp. gr. 0.950 (25/4°C); b.p. 189°C (760 mm), 74.5°C (10 mm); viscosity 3.5 cps (25°C); refractive index 1.419 (25°C); fire point 85°C; completely miscible with water, VM & P naphtha, acetone, ethanol, benzene, carbon tetrachloride, ether, methanol, monochlorobenzene and petroleum ether.
Containers: Drums; tank cars.
Uses: In many solvent applications; in hydraulic brake fluids.
Shipping regulations: None. *

dipropylene glycol monosalicylate (salicylic acid, dipropylene glycol monoester)
$C_3H_6(OOCC_6H_4OH)OC_3H_6OH$.
Properties: Light colored oil having faint characteristic fragrant odor; sp. gr. 1.16 (40°C); refractive index, about 1.52; soluble in alcohol; insoluble in water.
Uses: Ultraviolet light screening agents; protective coatings; plasticizers.

dipropylene triamine. See 3,3'-iminobispropylamine.

dipropyl ketone (butyrone; 4-heptanone)
$(CH_3CH_2CH_2)_2O$.
Properties: Stable, colorless liquid. Pleasant odor. Miscible with many of the organic solvents.
Constants: B.p. 143.7°C; m.p. −32.1°C;

*See "I.C.C. Shipping Regulations," page xiii.
Reference numbers refer to name of manufacturer. See "List of Manufacturers," page v.

sp. gr. 0.8162 (20/20°C); wt/gal 6.79 lbs (20°C); refractive index 1.4068 (20°C); surface tension 25.2 dynes/cm (25°C); viscosity 0.0074 poise (20°C); vapor pressure 5.2 mm Hg (20°C); flash point 49°C (120°F).
Grade: Technical.
Containers: 1- and 5-lb glass bottles; returnable steel drums.
Uses: Solvent for nitrocellulose, raw and blown oils, many natural and synthetic resins; lacquers; synthetic resin finishes.
Caution: Keep away from heat and open flame. Avoid breathing of vapor. Avoid contact with skin. MCA warning label.
Shipping regulations: None. *

dipropylmethane. See heptane.

dipropyl phthalate $C_6H_4(COOC_3H_7)_2$.
Properties: Sp. gr. 1.071 (25°C); refractive index 1.494 (25°C); b. p. 129-132°C (1 mm); solubility in water, 0.015% by weight.
Use: Plasticizer.

"Dipsanil." [206] Brand name for a textile waterproofing agent, consisting of an acid-stable aqueous emulsion of paraffin wax and aluminum acetate.

"Dipterex." [181] Trademark for O, O-dimethyl 2, 2, 2-trichloro-1-hydroxyethyl phosphonate.

dipteryx. See tonka.

alpha, alpha-**dipyridyl** $(C_5H_4N)_2$.
Properties: White crystals; m. p. 69-70°C; b. p. 272-273°C; slightly soluble in water; insoluble in alcohol, ether, benzene, chloroform, and petroleum ether.
Grade: Reagent.
Use: Reagent for iron determination.

2, 2'-dipyridylamine $(C_5H_4N)_2NH$.
Properties: Solid; f. p. 92.3°C (min); b. p. 222°C (50 mm); very slightly soluble in water.
Derivation: From 2-aminopyridine.
Use: Intermediate.

dipyridylethyl sulfide $[C_5H_4N(CH_2)_2]_2S$.
Properties: Sp. gr. 1.113 (25°C); refractive index 1.5841 (n 20/D); m. p. 1.5°C; soluble in water and all common organic solvents.
Grade: Technical (95% purity).
Uses: Synthesis of pharmaceuticals, dyestuffs, rubber chemicals, flotation agents, insecticides, fungicides, plasticizers, textile assistants, herbicides, oil additives, rust preventives, and pickling inhibitors.

dipyrone (1-phenyl-2, 3-dimethyl-5-pyrazolone-4-methylaminosulfonate sodium) $C_{13}H_{16}N_3O_4SNa \cdot H_2O$.
Properties: Almost white, water soluble, odorless, crystalline powder; faintly bitter taste; m. p. 172°C. Slightly soluble in ethyl alcohol; insoluble in ether and benzene.
Use: Medicine.

diquinine carbonate. See aristoquin.

direct dyes. Those soluble dyes that are taken up directly by fibers, presumably due to selective adsorption. Usually applied to cotton or union goods (cotton-wool mixtures). Dyeing assistants such as sodium chloride or sodium sulfate are used to obtain a higher concentration of dye on the fibers. These dyes are principally water-soluble sodium salts of sulfonic acids of azo dyes. Examples are Direct Blue 2B (C. I. 406), Direct Black EW (C. I. 581) and Direct Brown 3GO (C. I. 596).

diresorcinol (tetrahydroxydiphenyl) $(OH)_2C_6H_3C_6H_3(OH)_2$.
Properties: White to slight yellowish crystalline powder. Soluble in hot water and alcohol. M. p. 310°C.
Derivation: By fusing resorcinol and phenol with caustic soda.
Method of purification: Crystallization.
Grade: Technical.
Containers: Wooden kegs; fiber drums.
Use: Organic synthesis.
Shipping regulations: None. *

diresorcinolphthalein. See fluorescein.

N, N'-disalicylidene-1, 2-diaminopropane (N, N'-disalicylidene propylenediamine) $HOC_6H_4CH{:}NCH_2CH(CH_3)N{:}CHC_6H_4OH$. Used as a metal deactivator in motor fuels.

N, N'-disalicylidene propylenediamine. See N, N'-disalicylidene-1, 2-diaminopropane.

"Discaloy." [308] Trademark for an austenitic iron-base alloy containing nickel, chromium and relatively small proportions of molybdenum, titanium, silicon, and manganese. This alloy is precipitation-hardening and was developed primarily to meet the need for improved gas turbine discs, one of the most critical components of jet engines. It has exceptionally high creep strength combined with good ductility, resistance to notch sensitivity, and excellent oxidation resistance in the temperature range of 1000°F to 1350°F - the range in which gas turbine discs operate.

discharging agents. Substances capable of destroying a dye or mordant present within the fibers of a fabric. There are various methods of utilizing this property so that it is possible to produce a colorless figure upon a colored ground or a colored figure upon a differently colored ground.

"Discolite." [159] Trademark for sodium sulfoxylate formaldehyde, $NaHSO_2 \cdot CH_2O \cdot 2H_2O$.
White; available in powder, rice, pea, and chestnut sizes.
Uses: For discharge printing of textile fabrics.

disilanyl. See silanes.

disinfectants. Substances used on inanimate objects which destroy harmful microorganisms or inhibit their activity. They are not necessarily powerful enough to destroy spores. Some common disinfectants are phenol, cresol, guaiacol, thymol, mercurial or chlorinated phenols, various

*See "I. C. C. Shipping Regulations," page xiii.
Reference numbers refer to name of manufacturer. See "List of Manufacturers," page v.

coal-tar distillates, pine oil, formaldehyde, potassium permanganate, bleaching powder and the hypochlorites, hydrogen peroxide, and mercuric chloride. Their effectiveness is rated by the phenol coefficient test prescribed in U. S. Department of Agriculture Bulletin 198. See also antiseptics; sanitizers; phenol coefficient.

disintegration (radioactive or nuclear). A synonym for radioactive decay and for radioactivity. The transformation of one kind of nucleus into one or more different kinds of nuclei by bombardment with high energy particles such as neutrons, alpha particles, etc. is also known as disintegration. This process is called nuclear transformation.

disodium acetarsenate
$NaOOCCH_2As(OH)O(ONa)\cdot 2H_2O$.
Properties: White, crystalline powder. Soluble in water.
Derivation: By reacting sodium arsenite with sodium monochloracetate.

2, 7-disodium dibromo-4-hydroxymercurifluorescein. See merbromin.

disodium dibutyl-ortho**-phenylphenoldisulfonate.**
Properties: Light brown paste; insoluble in solvents that are immiscible with water; soluble in alcohol, acetone, dibutyl tartrate, ethylene glycol.
Containers: 185-lb steel drums.
Uses: Wetting, penetrating and spreading agent used in kier-boiling; scouring and dyeing textiles; industrial cleaners; deodorant preparations; insecticidal formulations; metal cleaning; stabilizer and wetting agent in latex used to treat cord or other fabrics.

disodium dihydrogen ethylenediaminetetraacetate. See ethylenediaminetetraacetic acid, disodium salt.

disodium 1, 2-dihydroxybenzene-3, 5-disulfonate (4, 5-dihydroxy-meta-benzenedisulfonic acid disodium salt; sodium catechol disulfonate) $C_6H_2(OH)_2(SO_3Na)_2$.
Properties: Non-hygroscopic crystals. Freely soluble in water; produces water-soluble, colored compounds with metal salts.
Use: Colorimetric reagent for iron, manganese, titanium, molybdenum.

disodium ethylenebisdithiocarbamate. See nabam.

disodium ethylenediaminetetraacetate. See ethylenediaminetetraacetic acid, disodium salt.

disodium methyl arsonate (DMA; methanearsonic acid, disodium salt) $CH_3AsO(ONa)_2$, sometimes with $6H_2O$.
Properties: Colorless crystalline solid; hygroscopic; m. p. > 355°C; soluble in water and methanol.
Derivation: Reaction of methyl chloride with sodium arsenate.
Grades: 55-65% powder concentrate; 31. 5% blend.
Containers: 50-, 100-lb fiber drums.
Uses: Pharmaceuticals; herbicide (crabgrass killer).
Caution: Avoid contact with skin or prolonged breathing of dust.

disodium orthophosphate. See sodium phosphate, dibasic.

disodium phenyl phosphate. See phenyl dihydrogen phosphate, disodium salt.

disodium phosphate. See sodium phosphate, dibasic.

disodium pyrophosphate. See sodium pyrophosphate, acid.

disodium sulfamylphenylazo-7-acetamido-1-hydroxynaphthalene-3, 6-disulfonate. See azosulfamide.

disodium [sulfonylbis-(para**-phenyleneimino)] dimethanesulfinate.** See sulfoxone sodium.

"Dispersall." [159] Trade name for ethylene oxide condensate.
Properties: Colorless, neutral, somewhat viscous liquid. Miscible with water in all proportions.
Uses: Dispersing agent; retardant and leveling assistant in the vat dyeing of cellulosic materials.

dispersing agents. Any materials added to a suspending medium to promote and maintain the separation of the individual, extremely fine particles of solids or liquids which are usually of colloidal size. Typical applications of dispersing agents include their use in the grinding of pigments for fine enamels and dispersing of certain water-insoluble dyes to secure uniform dyeing. The term is often used interchangeably with emulsifying agent or emulsifier (q.v.).

Dispersing Oil 10. [175] Brand name for a coal-tar oil distillate.
Properties: Clear, yellow-red liquid; sp. gr. 1. 020-1. 045 (15. 5/15. 5°C); distillation range, 220-300°C.
Containers: 55-gal steel drums.
Uses: A rubber softener, plasticizer, and reclaiming oil used similarly to "Bardol" B where light color is not a factor.

dispersion. A system of minute particles (solid, liquid, or gaseous) distinct and separate from one another and suspended in a liquid, gaseous, or solid medium. Usually applied as descriptive of colloidal particles (diameter 1-100 millimicrons) suspended in a suitable medium. Examples of dispersions: smog, homogenized milk, gels.

"Dispersite." [248] Trademark for water dispersions of natural, synthetic, and reclaimed rubbers and resins.
Uses: Adhesives for textiles, paper, shoes, leather, tapes; coatings for metal, paper, fabrics, carpets; protective (strippable) for saturating paper, felt, book covers, tape, jute pads; for dipping tire cords. Can be applied by spraying, spreading, impregnation, saturation.

"Dispersives Standard" 011, 012, 013. [108] Brand name for a series of dry, powdered, lignin-type dispersing agents.
Containers: 50-lb bags.
Uses: Prevent scale and sludge in steam boiler and preboiler equipment.

"Disperso." [104] Trademark for a wettable grade of zinc, calcium and other metallic stearates. Used where easy dispersion in water is desired.

"Dispersol." [206] Brand name of proprietary line of dispersed insoluble azo dyestuffs for application in dyeing and printing of acetate rayon.

displacement series. See electromotive series.

dissociation. The change of one substance into two or more new substances, usually under such circumstances that at least partial recombination occurs to form the original substance.

dissociation pressure. The pressure exerted by the gaseous product or products of decomposition of a liquid or solid substance. Usually refers to the equilibrium value of this pressure, but in some cases a non-equilibrium value may be implied.

dissolved oxygen (D. O.). One of the most important indicators of the condition of a water supply for biological, chemical and sanitary investigations. Adequate dissolved oxygen is necessary for the life of fish and other aquatic organisms and is an indicator of corrosivity of water, photosynthetic activity, septicity, etc. This measure is also a part of the biochemical oxygen demand (B. O. D.) evaluation.

distearylamine (dioctadecylamine) $(C_{18}H_{37})_2NH$.
Properties: Solid; sp. gr. 0.85; m. p. 69°C. Almost insoluble in water.
Use: Chemical intermediate.

distearyl ether (dioctadecyl ether) $(C_{18}H_{37})_2O$.
Properties: Solid; m. p. 58-60°C; b. p., decomposes.
Grade: 95% (min.) purity.
Uses: Electrical insulators; water repellents; lubricants in plastic molding and processing; antistatic substance; chemical intermediates.

distearyl sulfide (dioctadecyl sulfide; distearyl thioether) $(C_{18}H_{37})_2S$.
Properties: Solid; m. p. 68-69°C; b. p., decomposes; sp. gr. 0.8148 (70/4°C).
Grades: 95% (min.) purity.
Uses: Organic synthesis (formation of sulfonium compounds).

distearyl thiodipropionate (3, 3'-dioctadecyl thiodipropionate; thiodipropionic acid, distearyl ester) $(C_{18}H_{37}OOCCH_2CH_2)_2S$.
Properties: M. p. 55°C. Insoluble in water; very soluble in benzene and olefin polymers. Extremely resistant to heat and hydrolysis.
Toxicity: Low.
Uses: Anti-oxidant; additive; plasticizer and softening agent.

distearyl thioether. See distearyl sulfide.

disthene. See cyanite.

distillation. The process of separation consisting of vaporizing a liquid and collecting the vapor, which is usually condensed to a liquid. Pure substances may be distilled, but the process is almost always used to achieve separation of liquid mixtures because of the difference in composition between the liquid and the vapor formed from it. See destructive distillation, batch distillation, extractive distillation, rectification, dephlegmation, flash distillation, simple distillation, reflux, fractional distillation, azeotropic distillation, vacuum distillation, and molecular distillation.

1, 3-disulfamyl-4, 5-dichlorobenzene. See dichlorphenamide.

disulfiram. See tetraethylthiuram disulfide.

3, 5-disulfobenzoic acid $C_6H_3(HSO_3)_2COOH$.
Properties: White powder. Soluble in water.
Grade: C. P.
Uses: Intermediate for detergents, dyes and pharmaceuticals.

disulfuric acid. See sulfuric acid, fuming.

disulfuryl chloride. See pyrosulfuryl chloride.

"Disulphine." [206] Brand name of proprietary line of level dyeing acid dyestuffs for wool. They are also used for the dyeing of silk.

"Di-Syston." [181] Trademark for O, O-diethyl S-2-(ethylthio)ethyl phosphorodithioate (q. v.).

dita bark. See alstonia.

ditaine (echitamine) $C_{22}H_{28}N_2O_4 \cdot 4H_2O$.
Properties: White, thick, glistening, crystalline alkaloid; poisonous! Soluble in water, alcohol, ether, and chloroform. M. p. 206°C.
Derivation: By extraction from the bark of Alstonia scholaris.
Method of purification: Crystallization.
Grades: Technical.
Containers: Glass bottles.
Use: Medicine.
Shipping regulations: None. *

ditetradecylamine. See dimyristyl amine.

ditetradecyl ether. See dimyristyl ether.

ditetradecyl sulfide. See dimyristyl sulfide.

ditetrahydro-2-furfurylamine
Constants: B. p. 150-155°C (30 mm).
Derivation: By hydrogenating di-2-furfurylamine in ethyl alcohol.

"Dithane." [23] Trademark for agricultural fungicides based on salts of ethylene bisthiocarbamate. Supplied in zinc, manganese, and sodium forms as wettable powder or liquid concentrate.
Use: Control of fungus and mite diseases on plants.

1, 4-dithiane (diethylene disulfide) $\overline{SCH_2CH_2SCH_2CH_2}$.

*See "I. C. C. Shipping Regulations," page xiii.
Reference numbers refer to name of manufacturer. See "List of Manufacturers," page v.

Properties: White crystals. Volatile in steam. Soluble in alcohol, ether; slightly soluble in water.
Constants: B. p. 115.6°C (60 mm); m. p. 112°C.
Derivation: Interaction of dichloroethyl sulfide with sodium or potassium sulfide.
Use: Organic synthesis.

dithiane methiodide $C_4H_8S_2 \cdot CH_3I$.
Properties: Crystalline substance. Soluble in hot water; slightly soluble in alcohol; insoluble in ether. M. p. 174°C.
Derivation: Interaction of dichloroethyl sulfide and methyl iodide.

dithiazanine $C_{23}H_{23}IN_2S_2$. 3, 3'-Diethylthiadicarbocyanine iodide.
Properties: A blue cyanine dye; decomposes 248°C; slightly soluble in alcohol and methanol.
Grade: N. N. D.
Use: Medicine; photography.

beta, beta'-dithiobisalanine. See cystine.

dithiocarbamic acid (aminodithioformic acid) NH_2CS_2H.
Properties: Colorless needles, soluble in alcohol.
Uses: The metal salts of the acid are important as rubber accelerators, as are the thiuram disulfide derivatives. See thiuram, selenium diethyldithiocarbamate, zinc dibutyldithiocarbamate, zinc diethyldithiocarbamate, ziram. Also used as seed disinfectants.

2, 2'-dithiodibenzoic acid (dithiosalicylic acid) $(C_6H_4COOH)_2S_2$.
Properties: Tan to gray powder; m. p. 280°C (min.).
Containers: Drums.
Use: Intermediate for pharmaceuticals.

1, 2-dithioglycerol. See 2, 3-dimercaptopropanol.

dithione $C_{17}H_{21}O_5PS$. 7-Hydroxy-3, 4-tetramethylenecoumarin O, O-diethylthiophosphate.
Properties: Crystals; m. p. 88.0-88.5°C; Practically insoluble in water; limited solubility in organic solvents.
Use: Insecticide.

6, 8-dithioöctanoic acid. See d*l*-alpha-lipoic acid.

dithioöxamide (rubeanic acid) $SC(NH_2)C(NH_2)S$.
Properties: Stable orange-red powder; decomposes at 140°C; insoluble in water; soluble in acetone and chloroform. Forms highly colored stable complexes with many metal ions; can form a series of N, N'-derivatives.
Uses: Suggested for pigments; herbicides; metal deactivators; intermediates; plastics.

dithiosalicylic acid. See 2, 2'-dithiodibenzoic acid.

dithymol diiodide. See thymol iodide.

N, N-di-ortho-tolylethylenediamine $CH_3C_6H_4NHCH_2CH_2NHC_6H_4CH_3$.
Properties: Light brown to purple granular solid; sp. gr. 1.13; m. p. not definite, starts to soften at about 57°C with the main portion melting between 60-66°C; stable in storage; insoluble in water; soluble in acetone, ethylene dichloride, benzene and gasoline.
Containers: 50-lb bags; 250-lb drums.
Uses: Antioxidant for both natural and synthetic light colored rubber goods.
Hazards: No health hazard when used in rubber in the amounts recommended.

di-ortho-tolylguanidine (DOTG) $(CH_3C_6H_4NH)_2CNH$.
Properties: White powder; non-hygroscopic; non-toxic. Very slightly soluble in water; soluble in warm alcohol, from which it crystallizes on cooling. Sp. gr. 1.10; m. p. 179°C.
Derivation: Desulfurization of di-ortho-tolylthiourea with a lead compound in the presence of ammonia.
Containers: Drums (100-lbs net).
Use: To accelerate and improve vulcanization of natural and synthetic rubber compounds.
Shipping regulations: None.*

2, 7-di-para-tolylnaphthylenediamine $C_{10}H_6(NHC_6H_4CH_3)_2$.
Properties: Fine needles. Sparingly soluble in alcohol; insoluble in water. M. p. 237°C.
Derivation: By heating 2, 7-dihydroxynaphthalene with para-toluidine and para-toluidine hydrochloride.
Method of purification: Crystallization.

1, 3-di-para-tolylphenylenediamine $C_6H_4(NHC_6H_4CH_3)_2$.
Properties: Long needles. Soluble in alcohol and ether; insoluble in water. M. p. 137°C.
Derivation: By heating resorcinol and para-toluidine in presence of zinc chloride.
Method of purification: Crystallization.

di-ortho-tolylthiourea (DOTT) $SC(NHC_6H_4CH_3)_2$.
Properties: Colorless, crystalline leaflets; pungent odor; not hygroscopic. Soluble in alcohol, ether and benzene; insoluble in water. M. p. 144-148°C.
Derivation: By the interaction of ortho-toluidine and carbon disulfide.
Containers: Drums.
Use: Metal pickling inhibitor.

di-para-tolylthiourea $SC(NHC_6H_4CH_3)_2$.
Properties: White powder. Soluble in ether; slightly soluble in alcohol.
Derivation: From para-toluidine.
Method of purification: May be recrystallized from the above solvents.

ditridecyl phthalate (DTDP) $C_6H_4(COOC_{13}H_{27})_2$.
Properties: Sp. gr. 0.950 (25/25°C); b. p. 285 at 3.5 mm Hg; pour point 35°F; viscosity 190 cps (25°C).
Use: Plasticizer.

ditridecyl thiodipropionate (3, 3'-tetramethylnonyl thiodipropionate; thiodipropionic acid,

*See "I. C. C. Shipping Regulations," page xiii.
Reference numbers refer to name of manufacturer. See "List of Manufacturers," page v.

ditridecyl ester) $(C_{13}H_{27}OOCCH_2CH_2)_2S$.
Properties: Sp. gr. (25°C) 0.932. Insoluble in water; soluble in most organic solvents.
Uses: Stabilizer; plasticizer and softening agent for plastics; lubricant additive.

diuron. See 3-(3, 4-dichlorophenyl)-1, 1-dimethylurea.

divale. See belladonna.

divanadyl tetrachloride. See vanadyl chloride.

divi divi.
Derivation: The fruit of a West Indian tree Caesalpinia copiaria. Forms very thin pods about 3 inches in length.
Grades: Technical (pods; extract).
Containers: Pods: 100-, 200-lb and various size burlap bags; extract: 450- to 500-lb barrels.
Use: Tanning industry.
Shipping regulations: None. *

divine stone. See copper, aluminated.

divinyl acetylene $H_2C{:}CHC{\vdots}CCH{:}CH_2$. Trimer of acetylene, formed by passing it into a hydrochloric acid solution containing metallic catalysts.

divinyl B. See butadiene.

divinylbenzene (DVB; vinylstyrene) $C_6H_4(CH{:}CH_2)_2$, existing as ortho-, meta- and para-isomers. The commercial form contains the 3 isomeric forms together with ethylvinylbenzene and diethylbenzene.
Properties: (pure meta- isomer) Water-white liquid easily polymerized. B. p. 199.5°C; f. p. −66.90°C; sp. gr. 0.9289 (20°C); viscosity 1.09 centipoise (20°C); refractive index (n 20/D) 1.5772.
Divinylbenzene, 55%
Typical properties: Pale straw-colored liquid; f. p. −87°C; b. p. 195°C; sp. gr. 0.918 (25/25°C); insoluble in water; soluble in methanol and ether.
Grades: 50-60%; 20-25%.
Containers: Drums; tank cars.
Uses: Polymerization monomer for special synthetic rubbers, drying oils, ion-exchange resins, casting resins and other polymers.
Caution: Highly reactive compound. Polymerization, once started, may proceed with violence.
Shipping regulations: None. *

divinyl ether. See vinyl ether.

divinyl oxide. See vinyl ether.

3, 9-divinylspirobi-meta-**dioxane** (3, 9-divinyl-2, 4, 8, 10-tetraoxaspiroundecane.) $[CH_2{:}CHCH(OCH_2)_2]_2C$.
Properties: Liquid; m. p. 42°C; 120°C (2 mm); sp. gr. 1.251 (20/20°C); slightly soluble in water.
Uses: Intermediate and monomer.

divinyl sulfide $(CH_2{:}CH)_2S$.
Properties: Mobile liquid. Characteristic odor. Polymerizes readily. Sp. gr. 0.9174 (15°C); b. p. 85-86°C.
Derivation: Interaction of dichlorodiethyl sulfide and an alcoholic solution of potassium hydroxide.
Grades: Technical.

divinyl sulfone $CH_2{:}CHSO_2CH{:}CH_2$.
Properties: Liquid; sp. gr. 1.1788 (20/20°C); b. p. 234°C; f. p. −26°C; soluble in water; flash point 255°F.
Uses: As a monomer to make a series of polymers with diols, urea, and malonic esters; shrinkage control agent (textiles).

3, 9-divinyl-2, 4, 8, 10-tetraoxaspiroundecane. See 3, 9-divinylspirobi-meta-dioxane.

di-ortho-**xenyl phenyl phosphate** $(C_6H_5C_6H_4O)_2(C_6H_5O)PO$.
Properties: Sp. gr. 1.20 (60°C); refractive index 1.603-5 (60°C); boiling range 285-330°C (5 mm); flash point 250°C; insoluble in water.
Use: Plasticizer.

"Dixie." [110] Brand name for a line of dustless or uncompressed carbon black; used as a coloring agent for ink, paint, etc., and as a reinforcing agent for rubber.
Containers: Paper bags, cartons, bag units and bulk.
Closely related products include:
"Dixie BB": Medium color channel black for paints, plastics.
"Dixie EP": Ordinary grade channel black for news inks.
"Dixie R": Ordinary grade channel black.
"Dixie Perfecto": High color channel black for high gloss enamels and lacquers.
"Dixie 5": General purpose channel black for inks, paints.
"Dixie 15": Long flow channel black for lithographic ink, carbon paper.
"Dixie 20, 35, 40, 50, 60, 70, and 85": Furnace type blacks; classified respectively as SRF, GPF, HMF, FEF, HAF, ISAF and SAF, used principally in rubber.

"Dixie Clay." [69] Trademark for a hard clay proprietary product, a kaolin.
Properties: White to cream; sp. gr. 2.62 ± .03; fineness (through 325 mesh) 99.8%.
Uses: Filler for rubber.

"Dixsol." [319] Ammoniating solutions available in two general types, ammonia-ammonium nitrate solutions and ammonia-ammonium nitrate-urea solutions.

djenkolic acid $CH_2[SCH_2CH(NH_2)COOH]_2$. An amino acid isolated from the djenkol bean.
Properties: Rosettes of needles; slightly soluble in cold water; readily soluble in aqueous solutions of acids or alkalies; decomposes gradually 300-350°C.
Use: Biochemical research.

DKP. Abbreviation for dipotassium phosphate. See potassium phosphate, dibasic.

D-L. Abbreviation for dachlaurin.

DM. See phenarsazine chloride.

DMA. Abbreviation for dimethylamine or disodium methyl arsonate.

*See "I.C.C. Shipping Regulations," page xiii.
Reference numbers refer to name of manufacturer. See "List of Manufacturers," page v.

DMAC. Abbreviation for dimethylacetamide (q. v.).

DMC.
1. Abbreviation for dichlorophenyl methyl carbinol. See di(para-chlorophenyl) ethanol.
2. Abbreviation for beta-dimethylaminoethyl chloride hydrochloride.

DMDT. Abbreviation for dimethoxydiphenyltrichloroethane. See methoxychlor.

DMF. Abbreviation for dimethyl formamide.

DMF Antistall Additive. [28] Dimethyl formamide, used to prevent engine stalling during cool humid weather conditions.

DMH. Abbreviation for dimethyl hydantoin.

DMIC. Abbreviation for beta-dimethylaminoisopropyl chloride hydrochloride.

DMMPC. Abbreviation for gamma-dimethylamino-beta-methylpropyl chloride hydrochloride.

"DMP." [23] Trademark for dimethylaminomethyl-substituted phenols.
Use: Chemical intermediates, curing agents for epoxy resins, anti-oxidants.

DMPC. Abbreviation for gamma-dimethylaminopropyl chloride hydrochloride.

DMS. See dimethyl sulfide.

DMSO. Abbreviation for dimethyl sulfoxide.

"DMT." [266] Brand name for dimethyl terephthalate (q. v.).

DMU. See dimethylurea.

"DN-75." [233] Trademark for liquid agricultural fungicide containing dinitro-orthocyclohexylphenol (q. v.) for the control of psorosis.

"DN-111." [233] Brand name for a proprietary insecticide and miticide containing a salt of dinitro-ortho-cyclohexylphenol as active ingredient and used for controlling red spider and certain other mites on fruit and nut trees.

"DN-289." [233] Brand name for a proprietary insecticide and fungicide containing a salt of dinitro-sec-butylphenol and used as a dormant spray against certain insects and fungus diseases attacking fruit trees; also as a miticide and pesticide.

DNA. Abbreviation for deoxyribonucleic acid; also for dinonyl adipate.

DNBP. Abbreviation for dinitro-ortho-sec-butylphenol.

"DN-Dry Mix." [233] Brand name for a proprietary dormant fungicide and miticide. No. 1 contains dinitro-ortho-cyclohexylphenol as active ingredient. No. 2 contains dinitro-ortho-cresol.

"DN-Dust." [233] Brand name for a series of miticidal dusts containing dinitro-orthocyclohexylphenol and salts thereof, and used for controlling red spiders and certain other mites.

"D" Nickel. [283] Trademark for a wrought alloy containing approximately 4.5 per cent manganese. Used in the manufacture of spark plugs because of its improved resistance to attack by sulfur compounds at elevated temperatures; also for electronic components such as grid wires.

DNOC. Abbreviation for 4, 6-dinitro-orthocresol.

DNOCHP. Abbreviation for dinitro-orthocyclohexylphenol.

DNODA. Abbreviation for di(n-octyl, n-decyl) adipate.

DNODP. Abbreviation for di(n-octyl, n-decyl) phthalate.

"DN-Oil." [233] Trademark for solution of dinitrophenols in oil used as agricultural sprays.

DNP. Abbreviation for dinonyl phthalate.

DNPD. Abbreviation for N, N'-di-beta-naphthyl-para-phenylenediamine.

"DN-Sulfur Dust." [233] Proprietary insecticidal dust mixture containing 2, 4-dinitro-6-cyclohexylphenol and sulfur as active ingredients. Available in two forms: Regular and No. 10.

DNT. Abbreviation for dinitrotoluene.

D. O. Abbreviation for dissolved oxygen; see also biochemical oxygen demand (B. O. D.) and oxygen consumed (C. O. D.).

DOA. Abbreviation for dioctyl adipate. See the more exact name, di(2-ethylhexyl) adipate.

Dobbin's reagent. A reagent used as a test for caustic alkalies in soap.
Preparation: Mercuric chloride solution is added to potassium iodide solution until a permanent precipitate is obtained. The solution is filtered and 1 gram of ammonium chloride added. Dilute sodium hydroxide is then added until a precipitate is formed. The solution is filtered and made up to one liter.

D. O. C. Abbreviation for dichromate oxygen consumed. See oxygen consumed.

DOCA. Abbreviation for deoxycorticosterone acetate.

docosanoic acid. See behenic acid.

n-docosane $C_{22}H_{46}$.
Properties: Solid; m. p. 45.7°C; b. p. 230°C (15 mm); sp. gr. 0.778 (45/4°C); refractive index 1.4400 (n 45/D).
Grades: 95% (min) purity.
Uses: Organic synthesis; calibration; temperature-sensing devices.

1-docosanol. See behenyl alcohol.

cis-13-docosenoic acid. See erucic acid.

doctor solution. See doctor treatment.

doctor treatment. A method of improving or "sweetening" the odor of stocks of gasoline, petroleum solvents, or kerosine. A doctor solution consisting of sodium plumbite, Na_2PbO_2, is made by dissolving litharge in caustic soda solution, and the feed to be sweetened is passed through the doctor solution. The action of the sodium plumbite and the lead sulfide formed from it, in conjunction with free sulfur (either naturally present in the feed, or added) converts the disagreeable mercaptans to the pleasanter disulfides.

dodecafluoro-1-heptanol. See fluoroalcohols.

dodecanal. See lauryl aldehyde.

n-dodecane (bihexyl, dihexyl) $CH_3(CH_2)_{10}CH_3$.
Properties: Colorless liquid; sp. gr. 0.7504 (20/4°C); f.p. −10°C; b.p. 217°C; refractive index 1.4221 (20/D); flash point 71°C. Soluble in alcohol, acetone, ether; insoluble in water.
Grades: 95, 99%.
Containers: 1-, 5-, and 10-gal cans; drums.
Uses: Solvent; organic synthesis.

dodecanoic acid. See lauric acid.

n-dodecanol. See lauryl alcohol.

dodecanoyl peroxide. See lauroyl peroxide.

dodecene $C_{12}H_{24}$. Many possible isomers. See 1-dodecene and tetrapropylene.

1-dodecene (alpha-dodecylene) $H_2CCH(CH_2)_9CH_3$.
Properties: Colorless liquid; sp. gr. 0.7600 (20/4°C); m.p. −33.6°C; b.p. 213°C; refractive index (n 20/D) 1.4327; soluble in alcohol, acetone, ether, petroleum, coal tar solvents; insoluble in water.
Containers: 1-, 5-, 10-gal cans.
Uses: Flavors; perfumes; medicine; oils; dyes; resins.

dodecene oxide $CH_3(CH_2)_9\overline{CHCH_2O}$. A liquid; sp. gr. 0.836 (25°C); used as an organic intermediate.

dodecenylsuccinic acid $HOOCCH(C_{12}H_{23})CH_2COOH$.
Properties: Extremely viscous liquid; practically insoluble in water; completely soluble in oil.
Uses: Synthesis; corrosion inhibitor in oils; waterproofing.

dodecenylsuccinic anhydride $C_{12}H_{23}\overline{CHCOOOCCH_2}$. The normal and at least two branched chain dodecenyls are known commercially. The properties which follow are those of a branched chain compound.
Properties: Light yellow, clear, viscous oil; b.p. 180-182°C (5 mm); sp. gr. (25°C) 1.002; flash point 178°C (Cleveland open cup); viscosity (20°C) 400 centipoises, (70°C) 15.5 centipoises.
Containers: Drums.
Uses: Manufacture of alkyd and epoxy and other resins, anticorrosion agents, plasticizers, wetting agents for bituminous compounds, and vulcanizable products.

dodecyl acetate (acetate C-12; lauryl acetate) $C_{12}H_{25}OOCCH_3$.
Properties: Colorless liquid with a light fruity odor; sp. gr. 0.860-0.862; refractive index 1.430-1.433; b.p. 150.5-151.5°C; soluble in 3 volumes of 80% alcohol.
Derivation: By heating barium laurate and barium formate together in vacuo and reducing the crude lauryl aldehyde with zinc dust in the presence of glacial acetic acid.
Grades: Technical.
Containers: 1-, 5-, 10-gal cans.
Use: Perfumery.

n-dodecyl alcohol. See lauryl alcohol.

dodecyl aldehyde. See lauryl aldehyde.

dodecylaniline $C_{12}H_{25}C_6H_4NH_2$. (Probably the para-isomer).
Properties: Sp. gr. (25/25) 0.907-0.912; b.p. 340-350°C. Oil-soluble aromatic amine. Insoluble in water; soluble in most organic solvents.
Use: Intermediate.

dodecylbenzene (alkane; detergent alkylate). A commercial blend of isomeric, predominantly monoalkyl benzenes. The side chains are saturated, averaging twelve carbon atoms.
Derivation: Alkylation of benzene with isomeric dodecenes, obtained usually by polymerization of propylene. See tetrapropylene.
Grades: Technical.
Containers: 55-gal drums; tank cars.
Uses: Detergents of the alkyl aryl sulfonate type.

dodecylbenzenesulfonic acid (DDBSA) $C_{12}H_{25}C_6H_4SO_3H$. A liquid available in tank car and tank truck lots; used in making detergents.

n-dodecyl bromide. See lauryl bromide.

dodecyldimethyl(2-phenoxyethyl)ammonium bromide. See domiphen bromide.

alpha-**dodecylene.** See 1-dodecene.

N-dodecylguanidine acetate. See dodine.

n-dodecyl mercaptan $C_{12}H_{25}SH$. Many isomers of dodecyl mercaptan are possible and a variety of these occur in the technical material known as tert-dodecyl mercaptan or lauryl mercaptan or simply dodecyl mercaptan. See lauryl mercaptan.
Properties of n-isomer: Colorless liquid, boiling point 143°C; refractive index 1.4589; soluble in ether and alcohol; insoluble in water.

tert-**dodecyl mercaptan.** See lauryl mercaptan.

dodecylphenol $C_{12}H_{25}C_6H_4OH$. A mixture of isomers.
Typical properties: Light straw; sp. gr. 0.93 (20/20°C); flash point 250°F; boiling range 310-335°F; phenolic odor; soluble in

organic solvents; insoluble in water.
Containers: Drums; tank cars.
Uses: Solvent; intermediate for surface-active agents; oil additives; resins; fungicides; bactericides; dyes; pharmaceuticals; adhesives; rubber chemicals.

dodecyltrichlorosilane $C_{12}H_{25}SiCl_3$.
Properties: Colorless to yellow liquid; b. p. 288°C; sp. gr. 1.026 (25/25°C); refractive index (n 25/D) 1.4521. Readily hydrolyzed by moisture, with the liberation of hydrochloric acid.
Derivation: By Grignard reaction of silicon tetrachloride and dodecylmagnesium chloride.
Grades: Technical.
Containers: 1/2-, 1-, 8-lb bottles; 80-lb drums.
Use: Intermediate for silicones.
Shipping regulations: Corrosive liquid. White label. *

dodecyltrimethylammonium chloride $C_{12}H_{25}N(CH_3)_3Cl$.
Properties: White; surface tension, 0.1% in water, 33 dynes/cm (25°C); soluble in water and alcohol.
Uses: Germicides; moldicides; fungicides; textile fiber softeners; cationic emulsifiers; flotation reagents.

dodine (N-dodecylguanidine acetate) $C_{12}H_{25}NC(NH_2)_2 \cdot CH_3COOH$. Used as a fungicide.

dogbane. See apocynum.

dog button. See nux vomica.

dog-fish oil. See shark liver oil.

dog-tooth spar. See calcite.

dolomite. A mineral, and also a sedimentary rock composed mainly of the mineral.
Dolomite mineral. $CaMg(CO_3)_2$. A carbonate of calcium and magnesium. Color gray, pink, white, and various other colors; vitreous luster; sp. gr. 2.85; hardness 3.5-4; good cleavage in three directions; similar to calcite, but less soluble in acids (reacts with acid when powdered or with hot acid).
Dolomite rock. A rock similar to limestone, but composed principally of dolomite rather than calcite. Usually some calcite is present, along with argillaceous matter, iron and manganese carbonates, and quartz. Types intermediate between dolomite and limestone are common and are called dolomitic limestone or magnesian limestone.
Uses: Refractory for furnaces; manufacture of magnesium compounds, and magnesium metal; as building material; in fertilizers; stock feeds; paper making; ceramics; mineral wool.

dolomite, fused. A mixture of cubic crystals of calcium and magnesium oxides, fused in the electric furnace.

dolomitic limestone. See limestone; also dolomite.

dolomol. See magnesium stearate.

"Dolophine" Hydrochloride. [100] Trademark for methadone hydrochloride.

"Doloxide." [139] Trademark for a calcined dolomite, powdered; 91% through 200 mesh; used in chemical processes where a high magnesium burned lime is required.

dolphin oil. See porpoise oil.

domiphen bromide (dodecyldimethyl(2-phenoxyethyl)ammonium bromide) $C_6H_5OC_2H_4N(C_{12}H_{25})(CH_3)_2Br$. A quaternary ammonium salt.
Properties: Crystals; m. p. 112°C. Soluble in water and organic solvents.
Use: Medicine.

donaxine. See gramine.

Donovan's solution. Solution of arsenous and mercuric iodides.

"Donut Pyro." [172] Trade name for a grade of sodium acid pyrophosphate ($Na_2H_2P_2O_7$) possessing a controlled rate of reaction.
Derivation: Dehydration of monobasic sodium phosphate.
Uses: Leavening agent for doughnuts; for producing doughnuts in an automatic doughnut machine.

DOP. Abbreviation for dioctyl phthalate. See the more exact name, di(2-ethylhexyl) phthalate.

dopes. Sizing formulations consisting of solutions of nitrocellulose, cellulose acetate, or other cellulose derivatives. They are applied to crepe yarn to set the twist and assist creping, to airplane fabrics to make them taut, to balloon fabrics to make them less permeable to gases, and to leather to form a high-gloss finish.

"Doriden." [305] Trademark for glutethimide, N. N. D.
Use: Medicine.

"Dormison." [321] Brand name for methylparafynol.

Dorr strong-acid process. A process for making weak phosphoric acid by treating calcined phosphate rock with sulfuric acid (94%). The phosphoric acid produced is 35% P_2O_5 and is suitable for the manufacture of phosphate fertilizers.

"Dortan." [51] Trademark for light-colored, sulfurized cutting oils which permit work visibility. Grades available include both fatty oil compounded and extreme-pressure types.

dose (radiation). The amount of radiation delivered to a specified area or to the body of an individual. The permissible dose is the quantity of radiation which may be received by an individual over a given period with no detectable harmful effects. For x- or gamma ray exposure the permissible dose is 0.3 roentgen per week, measured in air. All persons working with radioactive materials are

*See "I.C.C. Shipping Regulations," page xiii.
Reference numbers refer to name of manufacturer. See "List of Manufacturers," page v.

expected to wear or carry some device for detecting or measuring the radiation received by them.

DOTG. Abbreviation for di-ortho-tolylguanidine.

dotriacontane (dicetyl) $CH_3(CH_2)_{30}CH_3$ or $C_{32}H_{66}$.
Properties: Crystals; sp. gr. 0.823; b. p. 310°C; m. p. 70°C.
Use: Research.

DOTT. Abbreviation for di-ortho-tolylthiourea (q. v.).

double bond. A type of union or structure in molecules of organic compounds, in which a pair of valence bonds joins a pair of carbon or other atoms, as in ethylene ($H_2C{=}CH_2$, or $H_2C{:}CH_2$).

"Double-Duty Sour." [244] A proprietary product consisting of boric acid, fluorine compounds and blue dye.
Properties: White powder; water soluble; neutralizing value, 24.0 oz sodium bicarbonate per lb.
Containers: 60-lb, 140-lb, 260-lb plywood drums.
Uses: Laundry sour and blue combination.

double nickel salt. See nickel ammonium sulfate.

double salts. See complex compound.

"Doubl" Rosin. [79] Trade name of an "FF" grade limed wood rosin.
Constants: M. p. (capillary tube) 70°C; m. p. (ball & ring) 93°C; acid number 109; color "FF"; 2.9% lime.
Containers: Non-returnable 18 gauge black-iron drums of about 500-lbs gross wt. Tare 14-16 lbs.
Uses: Battery wax; box toes; dry cleaning compounds; dry core binders; matches; pipe bending; printing ink; rock wool; roofing cement; shoe bottom fillers; smoking molds; waterproofing varnish-paper.

douglas-fir oil. See pine-needle oils.

"Dowanol." [233] Trademark applied to a series of alkyl and aryl mono ethers of ethylene glycol, propylene glycol, and various polyglycols; useful as solvents, plasticizers, hydraulic fluids, etc.

"Dowcarb." [233] Trademark for calcium carbonate slurry for use in the paper industry.

"Dowclene." [233] Trademark applied to a blended synthetic industrial solvent.

"Dow Corning 3, 4 and 5 Compounds." [149] Trade names for soft film silicone dielectric compounds for aircraft ignition systems, disconnect junctions, and for industrial electrical equipment. Nonmelting, water repellent and having low freezing points, they remain serviceable over wide temperature spans; retain dielectric properties in high humidity.

"Dow Corning 36 Emulsion." [149] Trade name for water dilutable mold lubricant for rubber and plastics. Effective at concentrations from 1 part emulsion to 35-200 parts water; does not decompose to form carbonaceous deposits; keeps mold maintenance to a minimum. Extremely fine particle size contributes to stability of the emulsion in storage and pumping, and to the better surface finish of molded products.
Uses: In molding tires, mechanical rubber goods, heels and soles, floor tiles, and hose.

"Dow Corning 200 Fluids." [149] Trade name for a variety of polydimethyl siloxanes having the general formula $(CH_3)_3Si[SiO(CH_3)_2]_nOSi(CH_3)_3$. With the value of n ranging from 0 to 2,000, these silicone fluids are available in a range of viscosity grades, from 0.65 to over 1,000,000 centistokes.
Properties: Clear, water-white liquids; low freezing points; high flash points; low volatility; resistance to breakdown due to repeated shear; and relatively flat viscosity-temperature slopes. Insoluble in water, incompatible with most organic polymers, heat-stable and oxidation resistant.
Uses: As damping media; dielectric fluids; release agents; lubricants for rubber or plastics and certain metal combinations; polishing agents; cosmetic bases; anti-floating additives for paints; water repellents for glass and ceramics.

"Dow Corning 550 and 710 Fluids." [149] Trade names for phenyl polysiloxane fluids which are characterized by unusual heat stability, resistance to oxidation and gumming, low volatility and high flash points.
Uses: As lubricants for clocks, oven timers and instruments; as heat transfer media; as liquid dielectrics.

"Dow Corning 33, 41 and 44 Greases." [149] Trade names for silicone greases serviceable from −100 to 450°F. Characterized by remarkable heat stability, resistance to oxidation and gumming, and serviceability over wide temperature span.
Uses: To lubricate motor and conveyor bearings; paper coating and flat glass drawing machines; timers; plastic gears; textile dryers and slashers; and as antiseize for hold down bolts and injection nozzles.

"Dow Corning Mold Release Fluid." [149] Trade name for heat-stable, 100% silicone oil used as a mold-release agent in molding rubber and plastics. Practically inert and non-volatile with a flash point in the range of 600°F. Insoluble in water, incompatible with most organic materials. Generally used as a 0.5 to 2% solution in chlorinated solvents, methyl ethyl ketone, naphtha, or white gasoline. Especially effective in the bead and parting-line release of heavy duty tires, deep-draw molding of plastics, and as a casing spray for green tires.

"Dow Corning Pan Glaze." [149] Trade name for a resinous silicone coating developed to replace pan grease in the baking of bread, buns, rolls, and other low sugar content bakery products. Properly applied, it provides easy release for 100 or more bakes.

"Dow Corning 802, 803, 804, 805, 806, and 840 Resins." [149] Trade names for silicone resins used in formulating heat-stable finishes and maintenance paints. Other available resins include silicone and modified silicones for bake-on and air-dry finishes, designed for service at temperatures to 1000°F, in corrosive atmospheres, or in severe weathering.

"Dow Corning 2103, 2104, 2105, 2106 Resins." [149] Trade names for silicone laminating resins used to bond glass cloth in the production of heat-stable, structural, and electrical insulating parts and components. Each resin is a Class H electrical insulating material rated for hottest spot temperatures of 180°C.

Uses: "Dow Corning 2103 and 2105 Resins" are designed for high pressure laminating techniques; "2104 and 2106 Resins," for low pressure laminating. Each is supplied as a solvent solution at a viscosity suitable for impregnating.

"Dow Corning 935, 994, 996 and 997 Varnishes." [149] Trade names for silicone electrical insulating varnishes, rated to withstand Class H hottest spot temperature service of 180°C. All are resistant to oxidation, and retain dielectric properties in high temperature service. All are supplied as solvent solutions at a viscosity most suitable for intended application. 935 and 994 Varnishes are designed for cloth coating and for bonding mica or inert mineral fillers. 996 and 997 Varnishes are dipping or impregnating materials for electrical equipment.

Dowell process. Process for increasing the production of oil wells by treatment with inhibited hydrochloric acid. The hydrochloric acid enlarges the drainage channels.

"Dowetch." [233] Trademark for magnesium photoengraving sheet, plate, and photoengraving chemicals.

"Dowex." [233] Trademark name for a series of synthetic ion-exchange resins made from styrene-divinylbenzene copolymers.

"Dowfax." [233] Trademark for a series of surface-active agents.

"Dowfax" 2A1: dodecyldiphenyl oxide disulfonic acid, sodium salt.

Uses: Liquid detergents; kier boiling of cotton; mercerizing cotton; oil well acidizing; water flood operations; sulfite; sulfate and semi-chemical processes in paper manufacturing.

"Dowfax" 9N9: nonyl phenol-ethylene oxide condensate having 9-10 moles of ethylene oxide.

Uses: Metal cleaning; formulation of household detergents; industrial cleaners, maintenance cleaners, agricultural toxicants; petroleum refining; manufacture of textiles, latex paints, paper and leather.

"Dowflake." [233] Brand name for calcium chloride, 77-80%, in a special flake form.

"Dowfume." [233] Trademark for a series of proprietary products used as fumigants, pesticides and insecticides.

"Dowicide." [233] Trademark applied to a series of phenolic compounds useful as disinfectants, fungicides, algicides and preservatives. Available in two types; water-soluble and oil-soluble.

"Dowlap." [233] Trademark for lamprey control products containing a chlorinated phenol as the active ingredient.

"Dowmetal." [233] Trademark applied to a series of magnesium alloys containing more than 85% of magnesium and characterized by their extreme lightness. Other properties such as strength, toughness, thermal conductivity, etc., vary according to the particular alloy. Available in all the usual fabricated forms, including sand castings, permanent mold castings, die castings, forgings, extruded shapes, plate, sheet and strip.

Downs cell. An electrolytic cell for the production of metallic sodium and chlorine from fused salt. The cell consists of a steel vessel lined with refractory and insulating brick. The anode is a graphite block projecting upward from the bottom of the cell and surrounded by a cylindrical steel or copper anode. A dome and collector ring above the electrodes keep the products separate. Specially purified and dried sodium chloride is used as the electrolyte and the sodium metal produced is 99.9% pure.

"Dowpac." [233] Trademark for a thermoplastic wetted walltype tower packing used in the aerobic biological oxidation of liquid wastes and in cooling towers.

"Dow-Per." [233] Trademark for perchloroethylene (q. v.), in a special drycleaning grade.

"Dowpon." [233] Trade name for a grass-killer, based on dalapon.

"Dowsol." [233] Trademark for alkyl phosphates for use in metal extraction.

"Dowtherm 209." [233] Trademark for an inhibited glycol ether that azeotropes with water. The azeotropic solution containing 47% water gives freeze protection to −45°F and has an azeotropic b. p. 209°F.

Use: Ebullient cooled engines.

"Dowtherm" A. [233] Trademark for a eutectic mixture of diphenyl and diphenyl oxide.

Properties: Colorless, non-corrosive liquid which is stable up to 725°F; sp. gr. 1.060 (25/25°C); b. p. 258°C; f. p. 12°C; specific heat 0.63 Btu/lb/°F at b. p.; soluble in all proportions in alcohol, carbon tetrachloride, and ether (25°C); insoluble in water.

Containers: 5-, 10-, 55-gal drums; 42, 84 and 475-lbs net.

Uses: As a heat-transfer medium.

*See "I. C. C. Shipping Regulations," page xiii.
Reference numbers refer to name of manufacturer. See "List of Manufacturers," page v.

"Dowtherm C." [233] Trademark for an isometric terphenyl mixture; m. p. 150°C; b. p. 340-390°C.
Uses: Heat transfer agent; moderator coolant in "organic moderated" nuclear reactors (it does not become radioactive).

"Dowtherm E." [233] Trademark for treated ortho-dichlorobenzene specially stabilized for heat transfer purposes in the range 300-500°F.
Properties: Clear, colorless liquid; sp. gr. 1.181 (212°F); b. p. 352°F; specific heat 0.31 Btu/lb/°F at b. p.; heat of vaporization, 118.9 Btu/lb at b. p.
Containers: 55-gal drums.

"Dowtherm SR-1." [233] Trademark for a specially inhibited ethylene glycol solution designed for low temperature operations, approximately −10 to +160°F.

doxylamine succinate $C_{17}H_{22}N_2O \cdot C_4H_6O_4$. 2-[alpha-(2-Dimethylaminoethoxy)-alpha-methylbenzyl] pyridine succinate; $CH_3C(C_6H_5)(C_5H_4N)OCH_2CH_2N(CH_3)_2 \cdot C_4H_6O_4$.
Properties: Cream to white powder with characteristic odor. M. p. 100-104°C. Very soluble in water; freely soluble in alcohol and chloroform; slightly soluble in ether and benzene; pH (1% solution) 4.9-5.1.
Grade: U. S. P. XVI.
Use: Medicine.

DOZ. Abbreviation for dioctyl azelate. See the more exact name, di(2-ethylhexyl) azelate.

DP. Abbreviation for dry point; used when a range of boiling temperatures is given. See also IBP.

DPA. Abbreviation for diphenylamine; also for "Diphenolic Acid."

DPG. Abbreviation for diphenylguanidine (q. v.).

DPH. Abbreviation for 1,6-diphenylhexatriene (q. v.).

DPN. Abbreviation for diphosphopyridine dinucleotide. See nicotinamide adenine dinucleotide.

DPO. Abbreviation for 2,5-diphenyloxazole (q. v.).

DPPD. Abbreviation for N,N'-diphenyl-para-phenylenediamine.

"DPS." [342] Trademark for bacitracin methylene disalicylate.

"D. P. Solution." [58] Trade name for a lacquer additive.
Properties: Pale straw-colored liquid; 8.2 lb/gal; 40% solution in equal parts of ethyl alcohol and acetone.
Containers: 5-, 55-gal drums.
Uses: Anticorrosion agent in lacquers and cotton solutions; light and heat stabilization of vinyl resins.

Dragendorff's solution. A solution of bismuth-potassium iodide used in analysis (testing for alkaloids). Caution! Keep away from light and keep well stoppered!

dragon's blood.
Properties: Deep red, amorphous lumps; m. p. 120°C. Soluble in alcohol, ether, and volatile and fixed oils. Insoluble in water.
Chief known constituents: Dracoalban, dracoresene, draconine, and esters.
Derivation: The resin from the surface of the fruit of several species of Daemonoraps; habitat: East Indies, Malay, Sumatra and Borneo.
Grades: According to source.
Containers: Tins; boxes.
Uses: Pigment; coloring plasters; lacquers; tooth-powders; coloring marble and stoneware; tanning extract; furniture polishes; colored papers; coloring toilet preparations; process engraving and lithography; pharmaceuticals; varnishes; paints.
Shipping regulations: None. *

"Drakeol." [25] Brand name for a series of proprietary, white mineral oils. Colorless, odorless and tasteless, they meet requirements in the appropriate U. S. P. XVI "Heavy Liquid Petrolatum" and "Light Liquid Petrolatum" classifications.

"Dramamine." [70] Trademark for U. S. P. dimenhydrinate.

"Drawcote." [204] Trademark for a class of compounds used as dry film lubricants in the cold working of metals. Commercially available in 300-lb fiber drums and wood barrels.

"Draw-ex." [51] Trademark for a group of drawing compounds suitable for hot and cold metal working. Oil-soluble and water-soluble grades are available, some being pigmented.

"DRC-14" Atlantic Reforming Catalyst. [241] Trademark for a platinum-on-silica-alumina-base catforming catalyst. For reforming of petroleum naphtha.

"Dresinate." [266] Trademark for salts (soaps) or rosins and modified rosins, used as emulsifiers, detergents and dispersants in soluble oils, cleaning compounds and other compositions.

"Dresinol." [266] Trademark for a series 40-45% solid dispersions in sodium hydroxide and ammonium hydroxide of modified rosins and special resin acids. They are compatible with animal glue, starch, dextrin, proteins, natural and synthetic rubber latices, polyvinyl acetate, polyvinyl chloride, and water dispersions of phenolic- and urea-formaldehyde-type resins. They are used with these materials to improve the wetting and penetration of, and adhesion to, polar surfaces. In specific cases improvements are also noted in bonding strength, speed in the development of tack, and other properties.

"Driacin." [416] Trademark for an ash-free organic salt of a hydrophobic, film-forming corrosion inhibitor.

*See "I.C.C. Shipping Regulations," page xiii.
Reference numbers refer to name of manufacturer. See "List of Manufacturers," page v.

Properties: Oily liquid; sp. gr. (60/60°F) 0.868; density, 7.2 lbs/gal; pour point, 45°F; flash point (Pensky-Martin), 120°F; viscosity (kinematic) 231 cs at 100°F, (universal) 1071 sec; insoluble in water; surface-active.
Uses: Sludge-dispersant additive for extending the storage life of cracked fuel oil and preventing filter or burner tip clogging; ingredient in rust-preventive formulations such as slushing oils.

"Dri-Clor." [204] Trademark for a powdered laundry bleach containing not less then 38% available chlorine.

"Dricoid." [322] A trademark for a series of algin-emulsifier compositions.
Properties: Light tan or cream colored algin-emulsifier compositions in dry powder form with about 12% moisture; soluble in water, milk, cream, and ice cream products.
Grades: Refined.
Containers: 10-, 50-, 100-, 300-lb drums.
Uses: A stabilizer-emulsifier composition for ice cream and pressurized whipped cream.
Shipping regulations: None.*

"Dri-Die." [311] Trademark for a thermally attrited hydrogel which produces an unshrunk amorphous silica particle (aerogel). The active ingredients are 95.3% amorphous silica gel (SiO_2, 90%) and 4.7% ammonium fluorosilicate. The material kills insects on contact through physical dehydration coupled with physical-chemical action.

"Drierite." [345] Proprietary product. A special form of anhydrous calcium sulfate having a highly porous granular structure and a high affinity for water. Absorbs water vapor both by hydration and capillary action.
Desiccating efficiency: Residual moisture in a gas amounts to 0.005 milligram per liter or 0.31 pound per million cubic feet at atmospheric pressure and 30°C. Corresponding vapor pressure of first hydrate, 0.005 mm.
Properties: Neutral, stable, constant in volume, inert except toward water. Insoluble in organic liquids, non-deliquescent, non-disintegrating, non-poisonous, non-corrosive. Color: white. Weight of granule forms: 60-62 lbs/cu ft.
Grades: Regular "Drierite," Indicating "Drierite" (turns blue to red in use) "Du-Cal Drierite" (for drying air and gases.)
Uses: All laboratory and industrial drying of solids, liquids, gases.

driers. Chemical additives used to accelerate the drying period of paints, varnishes, printing inks and the like by catalyzing the oxidation of drying oils or synthetic resin varnishes, such as alkyds, which dry by air curing. The usual driers are salts of metals with a valence of two or greater and unsaturated organic acids. The approximate order of effectiveness of the more common metals is cobalt, manganese, cerium, lead, chromium, iron, nickel, uranium and zinc. These are usually prepared as the linoleates, naphthenates and resinates of the metals. Paste driers are commonly the metal salts as acetates, borates, or oxalates dispersed in a drying oil.

"Dri-Film." [245] Trademark for a group of silicone resins designed to impart durable moisture and weather resistance to surfaces to which it is applied. Various types of the material are applied as a liquid and form a protective film over surfaces which protects from humidity, corrosive chemical atmospheres and dust.
Uses: Dri-Film 88 is used as a protective coating for electric motors, transformers, field coils, etc.; Dri-Film 144, a masonry water repellent; Dri-Film 432, used to impart durable water repellency and other properties including water-borne spot and stain resistance, improved hand and drape, increased flex abrasion resistance, and improved tear strength and wrinkle recovery.

drilling mud (oil well drilling mud). Mud used in drilling oil wells. It is sent down through the drilling pipe under high pressure and returns up through the annular space between the walls of the hole and the pipe. The mud helps control gas, oil and water pressures and helps maintain the walls of the hole. Its basic components are clay and water, but other materials are added to modify its properties. Some common additives are barite or celestite, to increase the specific weight, sodium tannate, to control the viscosity, lime or other caustic to increase the pH, gelatinized starches to prevent loss of water, and cellophane flakes to add bulk. Special clays such as bentonite are also used.

drilling-mud weighting materials. Class name given to materials which are added to drilling mud to control gas, oil, water, or formation pressures and to aid in maintaining the walls of the open hole.
See also bentonite.

"Drinalfa." [412] Trademark for methamphetamine hydrochloride (q. v.).

"Drinox." [401] Trade name for liquid insecticide seed treatment containing aldrin (30%).
Containers: 6-, 30-, 54-gal drums.
Use: As a seed treatment for small grains, sorghum, cotton, corn for protection against soil insects such as seed corn maggots, wireworms, etc.
Warning: Hazardous by skin contact, inhalation or swallowing. Keep away from heat and open fire; flash point above 115°F.

"Drinox H-34." [401] Trademark for liquid insecticide seed treatment containing heptachlor (24.5%).
Containers: 160 fl. oz. and 6-, 30-, 54-gal containers.
Use: As a seed treatment for small grains,

sorghum and corn for protection against soil insects such as seed corn maggots, wireworms, etc.
Warning: Hazardous by skin contact, inhalation or swallowing. Keep away from heat and open fires.

"Dri-N-Tite." 205 Cut-back asphaltic liquids for reimpregnating and coating built-up roofing systems.

"Driocel." 99 Trademark for a solid, granular desiccant used for the commercial scale drying of process liquids and gases. Manufactured from a selected grade of natural bauxite, reduced to the required particle size specification, and thermally activated to its maximum absorbing activity. Mesh grades 4/8, 4/10, 8/14.
Typical analysis (volatile free basis): Al_2O_3 81.5%; SiO_2 10.0%; TiO_2 3.5%; Fe_2O_3 3.0%; insolubles: 2.0%; volatile material: 6%.
Containers: 400-lb (net) steel drums.

"Driocel" S. 99 Trademark for a drying (dehydrating) medium for light liquid hydrocarbons. Developed specifically to overcome souring of light hydrocarbon liquids, such as LPG products in the final drying stage. Mesh grades 4/8, 8/14; tamped bulk density 55-57 lbs/cu ft.
Containers: 400-lb (net) steel drums.

"Dri-Pax." 241 Brand name for the small gram-size bags containing 1, 2, 3, and 5 grams of "Protek-Sorb" silica gel.

"Drisdol." 162 Trademark for crystalline vitamin D_2.

"Dri-Sol." 319 Trademark for a series of ammoniating solutions which are essentially anhydrous solutions of ammonia and ammonium nitrate. These solutions contain only 0.5% water.
Use: Fertilizer manufacture.

"Drisoy." 64 Trademark for a line of chemically treated soybean oils, for use as a general replacement of linseed oil.

"Dritomic." 50 Trademark for wettable sulfur used as fungicide on fruit and ornamentals.

"Dri-Tri." 334 Trademark for sodium phosphate, tribasic, Na_3PO_4.
Properties: White granules; soluble in hot water; pH (1% soln), 11.4; will cake heavily if exposed to moisture.
Grades: Technical; food processing.
Containers: 100-lb multiwall paper bags.
Uses: Detergents; dishwashing; water softening; all types of industrial and household cleaning.

"Driwall." 448 Trade name for transparent coatings (silicone); prevents water penetrating exterior walls of brick, stone, masonry.

drocarbil ($C_{16}H_{23}AsN_2O_7$). The acetarsone salt of arecoline.
Properties: Nearly white or slightly yellowish, odorless powder; freely soluble in water.
Grades: N. F. XI.
Use: Medicine.

drop black. A bone black or similar form of carbon which has been cast into tear or drop form by mixing with water and glue to give a paste-like mass.

drop chalk. See chalk, prepared.

drug and cosmetic dyes. See D & C dyes.

"Druid." 51 Trademark for heavy-bodied, dark green oils for journal bearing lubrication, where oil loss and wastage are too great to permit use of better oil.

drumstick. See cassia fistula.

dry battery. See dry cell.

dry-bone. See smithsonite.

dry cell (dry battery). An electrolytic cell (q. v.) in which the liquid electrolyte is soaked up in an absorbent powder to form a moist paste which functions satisfactorily in generating the electric current, but avoids the inconvenience of a liquid. Flashlight batteries are dry cells. Ammonium chloride is frequently used as the electrolyte.

dry chemical. Term applied to material for fire extinguishing systems; powdered chemical containing sodium bicarbonate with small percentages of added ingredients to render it free-flowing and water repellent. Used as a fire extinguisher on fires in electric equipment, oils, greases, gasoline, paints and flammable gases.

"Drycon." 173 Trademark for a concentrated resin base material for restoring sizing to dry cleaned garments.

"Dry-Flo." 53 Trademark for starch ester derivative containing hydrophobic groups. Especially free-flowing in dry state. Cannot be wetted with water, yet when moistened with a water-miscible solvent, it can be gelatinized and used to produce films with water repellent properties.
Uses: Dusting powder; no-offset spray for printing.

"Dryfol." 221 Trademark for a line of copolymer fish oils.

dry ice (carbon dioxide snow) CO_2. Solidified carbon dioxide.
Properties: White snow; sp. gr. 1.56 (−79°C); m. p. −78.5°C; sublimes; wt/cu ft 94-97.5 lb.
Derivation: High purity carbon dioxide gas is compressed and cooled to liquefaction, then expanded to vapor and snow in presses that compress the snow into cakes.
Containers: Blocks approximately 10" x 10" x 11" are wrapped in paper. Large shipments are made in heavily insulated refrigerator cars or trucks. Small shipments are made in specially insulated corrugated cardboard cartons or other special containers.
Uses: Refrigerant; carbonated beverages; producing inert atmospheres for mechanical and chemical applications; immobilization

for humane animal killing; machine tool coolant; grinding and tumbling aid; fire extinguishing medium; source of carbon dioxide for chemical reactions; preservative for fresh meat; cooling medium for treating metals; hardening of foundry molds and cores; shielding gas for welding.

Warning! Extremely cold. Causes severe burns. Liberates heavy gas which may cause suffocation. MCA warning label.

Shipping regulations: Nonflammable gas. Green label. *

drying oils. Oily, organic liquids which, when applied as a thin film, readily absorb oxygen from the air and "dry" to form a relatively tough, elastic substance. They are usually natural products such as linseed, tung, perilla, soybean, fish, and dehydrated castor oils but drying oils are also prepared by combination of the natural oils or their fatty acids with various synthetic resins (see alkyd resin and epoxy resin). The drying ability is due to the presence of unsaturated fatty acids, especially linoleic and linolenic acids, usually in the form of glycerides. The degree of unsaturation of an oil, and hence its drying ability, is expressed by its iodine number (ability to absorb iodine). The drying oils have the greatest capacity for iodine and the non-drying oils the least.

Uses: As binders in paints and varnishes.

"Drymet." [428] Trademark for sodium metasilicate, anhydrous (q. v.).

"Dryolene (VM & P)." [200] Trademark for a petroleum solvent prepared by straight-run overhead distillation.

Properties: Water-white; boiling range 205-287°F; sp. gr. 0.747 (60°F); flash point (Tag closed cup) 25°F; wt/gal 6.22 lbs (60°F).

Caution! Flammable; keep lights and fire away.

Uses: By paint and rubber cement manufacturers where fast setting and relatively slow final drying are preferred.

Shipping regulations: Flammable liquid. Red label. *

"Dryorth." [428] Trademark for anhydrous sodium orthosilicate.

"Dryseq." [428] Trademark for sodium sesquisilicate, anhydrous.

"DS-207." [304] Trademark for dibasic lead stearate, $2PbO \cdot Pb(C_{17}H_{35}COO)_2$, vinyl stabilizer.

Properties: Soft white unctuous powder, sp. gr. 2.02; refractive index 1.60.

Containers: Fiberboard drums containing 40 and 225 lbs.

Uses: Used as a stabilizer-lubricant for vinyl resins. Does not melt at processing temperatures.

D salt (2-naphthylamine-1,5-disulfonic acid, sodium salt) $NH_2C_{10}H_5(SO_3Na)_2$. Used to darken the mass tone of Lithol Reds.

DSP. Abbreviation for disodium phosphate. See sodium phosphate, dibasic.

DTDP. Abbreviation for ditridecyl phthalate (q. v.).

"Duart." [333] Clear varnishes and pigmented enamels formulated with China wood oil and adapted to produce wrinkle finishes.

"Duatok." [57] Trademark for sulfathiazole.

dubbin. See dubbing.

dubbing (dubbin). A mixture of cod oil and tallow used in the leather industry.

Dubb's asphalt. See sulfurized asphalts.

"Duclean" No. 1. [28] Trademark for inhibited sulfuric acid, technical 60° Bé. sulfuric acid containing an acid pickling inhibitor.

Properties: Clear, reddish-brown liquid; miscible with water; sp. gr. 1.706; f. p. below −10.8°C.

Containers: 185-lb glass carboys; 700-lb drums; tank cars; tank trucks.

Uses: In pickling iron and steel; for descaling and cleaning of steel and other metals under conditions where it is desirable to inhibit the acid corrosion of steel and other ferrous metals.

"Duclean" No. 2. [28] Trademark for hydrochloric acid solution inhibited; technical.

Properties: Clear, water-white liquid with pungent odor; miscible with water; sp. gr. 1.142; f. p. below −40°C.

Containers: 125-lb glass carboys; tank cars; tank trucks.

Uses: In pickling iron and steel; for removing deposits, including hard water scale, hydrated iron and rust from evaporators, condensers, water distributing lines, water well screens, etc., where it is important to inhibit acid corrosion of steel and other ferrous metals.

"Duco." [28] Trademark for a series of lacquers, enamels, undercoaters, sealers, rubbing and polishing compounds, household cement, paste wax for autos and furniture, and other compositions used by the automotive and furniture trades.

Dühring's rule. Relates the vapor pressures of similar substances at different temperatures. A straight or nearly straight line results if the temperatures at which a liquid exerts a particular pressure is plotted on a graph against the temperature at which some similar reference liquid exerts the same vapor pressure. Water is most frequently used as a reference liquid since its vapor pressure at various temperatures is well known.

"Dulac." [333] Trade name for lacquers and lacquer-enamels with a clear cellulose base which are used for various coating and finishing requirements.

dulcamara (bittersweet). Dried stem of climbing shrub Solanum dulcamara.

Habitat: Europe; North America.

*See "I.C.C. Shipping Regulations," page xiii.
Reference numbers refer to name of manufacturer. See "List of Manufacturers," page v.

dulcin (para-phenetolecarbamide; sucrol) $H_2NOCNHC_6H_4OC_2H_5$. This substance should not be confused with dulcitol (q. v.) which is sometimes called dulcin.

Properties: White, needle crystals or powder with taste about two hundred times as sweet as sugar. M. p. 173-174°C; soluble in alcohol and ether; moderately soluble in hot water.

Derivation: From para-aminophenol.

Uses: Substitute for sugar; improves flavor of castor and cod-liver oils.

Shipping regulations: None. *

dulcite. See dulcitol.

dulcitol (dulcite; melampyrit; dulcin; dulcose; euonymit) $C_6H_8(OH)_6$. A sugar.

Properties: White, crystalline powder; slightly sweet taste. Soluble in hot water; slightly soluble in cold water; very slightly soluble in alcohol. Sp. gr. 1.466; m. p. 188.5°C.

Derivation: By hydrogenation of lactose; occurs naturally in Melampyrum nemorosum.

Grades: Technical; reagent.

Containers: Glass bottles; fiber containers.

Uses: Bacteriology; medicine.

Shipping regulations: None. *

dulcose. See dulcitol.

"Dullatone." [328] Trademark for a series of delustering agents for textile fabrics, based largely on titanium pigments.

"Dullatone" 60. Cationic dispersion of titanium dioxide and other pigments.

"Dullatone" CA. Similar to above.

"Dullatone" CS. Self-precipitating insoluble inorganic salts.

"Dullatone" DG. Titanium dioxide dispersion in a cationic medium. A dark duller for use on blacks or navies without any frosting effects.

Dulong and Petit's law. The atomic heat capacity (atomic weight times specific heat) of elementary substances is a constant whose average value at room temperature is 6.2. A few elements, notably boron, carbon, and silicon, obey the law only at high temperatures.

"Dulux." [28] Trademark for a wide range of products for surface coating.

"Dulux" Foam Resin. [28] Trademark for polyester resin for use with isocyanates to produce flexible polyurethane foams.

dumortierite. Perhaps $8Al_2O_3 \cdot 6SiO_2 \cdot B_2O_3 \cdot H_2O$. A natural basic aluminum silicate. Bright, smalt-blue to greenish-blue in color. Vitreous luster. Sp. gr. 3.26-3.36; hardness 7.

Occurrence: United States (New York, Arizona, Nevada); France; Silesia; Norway.

Uses: Extensively used in spark-plug porcelain; special refractories.

"Dunkit." [33] Trade name for an immersion type stripping compound composed of cresylic and chlorinated compounds. Packaged in 55-gal drums. Used for removing organic coatings and carbon deposits in pipe lines, tanks, or on equipment. Non-flammable, but prolonged breathing and skin contact are hazardous.

dunnione. Probable formula $C_{15}H_{14}O_3$. An ortho-quinone.

Properties: Orange-red needles; m. p. 90°C; optical rotation +310°.

Derivation: A natural coloring matter of the naphthalene group found as a deposit on the leaves and flowers of Streptocarpus dunnii.

Uses: Paints; textiles.

"Duobel." [28] Trademark for high-velocity permissible explosives furnished in seven grades based upon velocity and cartridge count; poor water resistance.

Use: For mining coal where lump coal is not a factor.

"Duol." [28] Trademark for resinated forms of red azo pigments.

Uses: In printing inks, paint, and plastics.

"Duomeens." [15] These are N-alkyl trimethylene diamines. The alkyl group is derived from coconut, soya, and tallow fatty acids. Insoluble in water; soluble in organic solvents.

Containers: 55-gal openhead drums and tank cars.

Uses: Pigment dispersion; color flushing; pigment coatings; germicides; water proofing in textiles; wetting agents; extrusion lubricants; anti-static agents; lubricant improvers; water treatment.

Duo-Sol process. A commercial process for refining lubricating oils by extraction with a solvent consisting of liquid propane and a cresol base.

"Duponol." [28] Trademark for a line of surface active agents of the alcohol sulfate type. These have detergent emulsifying, dispersing, and wetting properties, and are used in the textile, paper, leather, cosmetic, shampoo, and electroplating industries.

"Duponol" 80. Sodium salt of technical octyl alcohol sulfate. Light yellow liquid. Wetting and penetrating agent effective in the presence of high concentrations of electrolytes.

"Duponol" C. White powder. Meets requirements of U. S. P. XIV for sodium lauryl sulfate. Used as emulsifying and dispersing agent for cosmetic, dental, and medical preparations.

"Duponol" D Paste. Sodium salt of unsaturated long chain alcohol sulfate. Yellow paste, used for scouring, emulsifying, dispersing and wetting agent and detergent.

"Duponol" EP. Alkylolamine salt of saturated long chain alcohol sulfate. Light yellow viscous liquid.

Use: As a detergent, wetting, dispersing and emulsifying agent having solubility, color stability, and foaming properties, particularly suited for shampoos.

"Duponol" G. Amine salt of saturated long chain alcohol sulfate. Oil-soluble light yellow paste.

Use: Powerful emulsifying agent for oils and solvents.
"Duponol" L-144-WDG. Sodium salt of a modified unsaturated long chain alcohol sulfate. Reddish-brown liquid.
Use: Emulsifying in dry cleaning and fungicidal sprays.
"Duponol" LS Paste. Sodium salt of unsaturated long chain alcohol sulfate. Light yellow fluid paste.
Use: Scouring and emulsifying agent in leather and textile industries.
"Duponol" ME Dry Powder. Sodium salt of technical lauryl alcohol sulfate containing minimum electrolyte. White powder.
Use: As emulsifying and dispersing agent in metal cleaning, electroplating, textile and leather processing.
"Duponol" OS. "Duponol" G plus long chain unsaturated alcohol. Clear yellow oil-soluble liquid.
Use: Very effective emulsifying agent for preparing oil-in-water emulsions.
"Duponol" QC. Sodium salt of technical lauryl alcohol sulfate. Fluid liquid. Carefully controlled and standardized to provide a raw material for shampoos which allows manufacture from a single product of types which vary from clear liquids to pastes.
"Duponol" SN. Sodium salt of saturated long chain alcohol sulfate. Clear light yellow liquid.
Use: As a detergent, wetting, dispersing and emulsifying agent with high foaming power designed for use in shampoos and hand cleaners.
"Duponol" ST. Alkylolamine salt of saturated long chain alcohol sulfate. Light yellow liquid.
Use: Preparation of liquid shampoos.
"Duponol" WA Dry Flakes. Sodium salt of technical lauryl alcohol sulfate. Light yellow flakes. Dry form of "Duponol" WA Paste.
"Duponol" WA Paste. Sodium salt of technical lauryl alcohol sulfate. Viscous white paste.
Use: As a detergent, wetting, dispersing and emulsifying agent in cosmetics, metal, leather, building materials, paper, pigment, and cleaning compound industries, and in shampoos.
"Duponol" WAQ. Sodium salt of technical lauryl alcohol sulfate. Fluid white paste.
Use: More fluid form of "Duponol" WA Paste designed primarily for shampoos.
"Duponol" WAQE. Sodium salt of technical lauryl alcohol sulfate. Fluid liquid.
Use: Carefully controlled and standardized for use as emulsifying agent in emulsion polymerization industry.
"Duponol" WAT. Alkylolamine salt of long chain alcohol sulfate. Clear yellow viscous liquid.
Use: As a detergent, wetting, dispersing and emulsifying agent of high solubility and good foaming, particularly suitable for shampoos.
"Duponol" WS. "Duponol" G plus long chain saturated alcohol. Light cream-colored waxy solid.
Use: A wax-soluble emulsifying agent for preparing water-in-oil emulsions.
"Duponol" XL. Modified long chain alcohol sulfate. Clear light yellow liquid.
Use: Shampoo base for high quality clear liquid shampoos.

"**Duracillin.**" [100] Trademark for procaine penicillin G, U. S. P.

"**Durad.**" [333] Synthetic resin varnishes, enamels, and undercoats having phenolic and alkyd synthetic-resin bases.

durain. One of the types of physical structure found in coal (see also clarain, fusain, and vitrain). Durain occurs as solid bands of varying thickness having a finely granular structure. It has a dull luster and a black to lead gray color. See also splint coal.

"**Duralon.**" [326] A furane thermosetting resin used in industrial and chemical applications. Available in various compounds suitable for casting, trowelling, molding, impregnating, and coating.
Uses: For industrial parts requiring resistance to acids, alkalies, and solvents.

duralumin. An aluminum alloy. It contains 4% copper, 0.5% magnesium, 0.25-1.0% manganese, and small amounts of iron and silicon. Resistant to corrosion by acids and sea water.
Uses: Aircraft parts, railroad cars, boats, machinery, etc.

"**Duramold C.**" [32] A water-hardening, cold hobbing die steel. Annealed to 90 Brinell, max. Composition: 0.06 C, 0.15 Mn, 0.10 Si.

"**Duramul.**" [333] Trade name for emulsion type undercoats and finishes.

"**Duranickel.**" [28] Trademark for a wrought, age-hardenable alloy containing approximately 94% of nickel. Has greater strength and hardness and the high resistance to corrosion which is characteristic of nickel.

"**Duranite.**" [448] Trade name of synthetic resin enamels for coating metal products.

"**Duranol.**" [206] Brand name of proprietary line of dispersed insoluble dyestuffs derived from anthraquinone for application in dyeing and printing of acetate rayon.

"**Duraplex.**" [23] Trademark for drying and non-drying oil-modified alkyd resins derived from phthalic anhydride, polyhydric alcohols, and vegetable oils. Air drying, baking, and non-drying grades, in solvent solution or viscous 100% resins. Produce tough, glossy, light-colored coatings with excellent durability.
Use: Primers, lacquers, and enamels; metal decorating; automotive coatings; furniture finishes; architectural enamels; inks.

"**Duraset.**" [248] Trademark for a series of products based on N-meta-tolyl phthalamic acid, $CH_3C_6H_4CONHC_6H_4COOH$.

Properties: M. p. 149-151°C (dec); nonvolatile; hydrolyzes rapidly in acid or alkaline media; only slightly soluble in water and benzene; soluble in acetone; soluble in ethyl, methyl, and isopropyl alcohols but decomposes rapidly.

Grades: 20W, a wettable powder.

Use: Flower and fruit setting hormone for research purposes.

Warning! Avoid inhalation and contamination of foodstuffs. Hazards, if any, not yet fully investigated.

"Durco." [47] A series of cast alloys which includes:

"Durco" D-10, a nickel base alloy containing 65% nickel, 23% chromium, 3.5% copper, 2% molybdenum and 1% manganese. It offers good resistance to acetic, sulfuric and phosphoric acids and basic solutions.

"Durco" 18-8-S, a stainless steel containing besides iron, 18-20% chromium, 8-10% nickel and 0.07% carbon (max). It offers good resistance to acetic, nitric and fatty acids and basic salts.

"Durco" 18-8-S-Mo, a stainless steel containing besides iron, 18-20% chromium, 8-10% nickel, 2-3% molybdenum and 0.07% carbon (max). It offers good resistance to phosphoric acid, hot acetic and fatty acids, sulfite liquors and sulfur dioxide.

"Durco" D-12, a chromic steel containing besides iron, 11.5-14% chromium, 1% (max) nickel and 0.15% carbon (max).

"Durcon." [47] A series of cast epoxy formulations for use in corrosion resisting equipment such as pumps, valves, fans, and others. This series of plastic materials is inert to hydrochloric acid, sulfuric acid to 80%, dilute nitric acid, alkalies and most organic acids and solvents. These materials are suitable for use to comparatively high temperatures.

durene (durol; sym-1,2,4,5-tetramethylbenzene) $C_6H_2(CH_3)_4$.

Properties: Colorless crystals; camphor-like odor. Soluble in alcohol, ether, and benzene; insoluble in water. Sublimes and is volatile with steam. Sp. gr. 0.838; m. p. 79-81°C; b. p. 189-191°C.

Derivation: By heating ortho-xylene and methyl chloride in presence of aluminum chloride. Occurs in coal tar.

Method of purification: Crystallization.

Grades: Technical.

Containers: Tins.

Use: Organic synthesis; plasticizers; polymers; fibers.

Shipping regulations: None. *

"Durez." [62] Trademark for a line of phenol-formaldehyde resins and molding compounds, and also including diallyl phthalate resins.

Properties: Specific gravities of molded objects range from 1.25 to 1.85 depending on filler and molding conditions; dielectric strengths up to 600 v/mil; tensile strengths 4000-7000 psi; flexural strengths 8000-12000 psi; impact 0.13 to 17.0 ft lb/in. (ASTM); moisture absorption 0.15 to 1.5% on 48 hours immersion (ASTM). Some of these properties require special grades as follows:

Grades:

Non-bleeding (for caps and closures of containers holding alcohol and other solvents, and for applications requiring sterilization up to 275°F).

Heat-resistant (containing mineral fillers) for service in the 300-500°F range (iron and utensil handles, toasters, coffee makers).

Impact resistant for gears, power transmission parts, bushings, castor wheels, tool parts, pulleys, instrument cases, machine housings.

Electrical and electronic grades, available in both phenolic and diallyl phthalate types. Used in ignition parts, resistor casings, coil forms, tube bases, distributor caps, commutator parts, circuit breakers.

Special purpose grades for resistance to water, soaps, detergents, corrosion, specific chemicals.

Bonding and impregnating resins such as for brake linings, abrasive wheels, mortars, laminated sheets and tubes. Coating resins for varnishes, enamels, inks, floor dressings.

"Durheat." [333] Trade name for clear cellulose-base lacquers, enamels, and varnishes which will withstand temperatures up to 250-300°F. Used largely on metals.

"Durichlor." [47] A ferrous cast alloy containing besides iron, 14.5% silicon, 0.85% carbon, 0.65% manganese, and 3.0% molybdenum. It offers the same high resistance to the conditions noted under "Duriron" and in addition gives excellent resistance to hot and cold hydrochloric acid and is superior to "Duriron" for most corrosive chloride salts. Not recommended for hydroxide, sulfites, and acidified ferric and cupric chlorides.

"Durimet-20." [47] A ferrous alloy in cast and wrought forms containing in addition to iron the following percentages of other elements: 29% nickel, 20% chromium, 3.5% copper (min), 2% molybdenum and 0.07% carbon (max). It offers good resistance to most sulfuric acid concentrations at 80°C and up to 10% at boiling, all concentrations of hot and cold acetic, fatty, mixed, phosphoric, sulfurous and nitric acids, solvents, sodium and ammonium hydroxides and many other salts, acids, and alkalies.

"Durindone." [206] Brand name of proprietary line of indigoid vat dyestuffs.

"Duriron." [47] A ferrous cast alloy containing besides iron, 14.5% silicon, 0.85% carbon and 0.65% manganese. It offers excellent resistance to all concentrations of hot and cold sulfuric, nitric, phosphoric and acetic acids, hypochlorite bleach solutions, chloroacetic acid and most metal-plating solutions. It is also resistant to all concentrations of cold hydrochloric acid and to dilute hydrochloric acid at moderate temperatures. It also shows good resistance to most other acids and salts except hydrofluoric acid, sulfurous acid, sulfites and highly oxidized metallic chlorides. It does not provide satisfactory resistance to most strongly basic solutions.

"Durisite." [326] A type of "Duralon" furane thermosetting resin.

"Duristone." [326] A type of "Duralon" furane thermosetting resin.

"Durite." [65] Trademark for a series of phenol-formaldehyde resins used in the manufacture of grinding wheels, brake linings, clutch facings, lamp-basing cements.

"Duro." [446] Trade name for an acid-proof brick made from materials of very low flux content. When fired stable minerals are formed which are insoluble in various acids and corrosive acidic solutions and resist alkaline solutions. Insoluble in mineral acids, except hydrofluoric.
Uses: Domes and dished bottoms of cylindrical tanks; capping tile; separators for electrolytic tanks; girders and supports for acid tower packing.

"Durobrite." [28] Trademark for cyanide zinc plating brightening agents. Amber liquids.

"Durofix." [188] Brand name for a proprietary product. A combined odorant and fixing agent.
Properties: Various odors available; faint amber color, will not discolor when exposed to air or sunlight or when incorporated in soaps; miscible with all essential oils, aromatic chemicals, and mixtures of same.
Uses: Soaps; bath salts; perfumes; toilet goods.

durol. See durene.

"Durol." [243] Trademark for milling blacks for wool and nylon.

dusting clays. Any finely divided pulverized clay that can serve as a diluent, carrier, or extender in the preparation of insecticide dusts, and which aids in adhesion of the insecticide to foliage.

dust-laying media. Calcium chloride, tars, liquid asphalt, heavy asphalt oils, solutions of petroleum asphalt in gas oils, crude oils, and emulsions of oils and water have all been used for road-dust laying.

Dutch liquid. See ethylene dichloride.

Dutch metal. A cheap imitation of gold leaf made of an alloy of copper and zinc.

Dutch oil. See ethylene dichloride.

Dutch pink. A yellow lake derived by absorbing quercitron, a yellow dye , on calcium carbonate or some similar inert material. Used as an artist's pigment, as wall paper coloring and to tone Brunswick Green. Sometimes referred to as Dutch yellow, English pink, or Italian pink.

Dutch process. Process for making white lead. Perforated lead disks are placed over, but not in contact with, 3% acetic acid (vinegar) and surrounded with tan-bark. The system is closed off for 100 days while the acid vapors, moisture, and CO_2 from the fermenting tanbark act on the lead to produce white lead (lead carbonate, basic). The white lead is broken away from unreacted lead, ground, floated in water and dried to give about 70% carbonate and 30% hydroxide.

Dutch yellow. Yellow dyestuff prepared from tetraazotized benzidine, salicylic acid, and sodium bisulfite. Used to obtain brownish-yellow shades on wool. The term has also been used synonymously with Dutch pink.

DVB. Abbreviation for divinylbenzene.

dwarf pine needle oil. See pine needle oil, Pinus mugo variety.

Dy. Symbol for dysprosium.

"Dyal." [141] Trade name for a series of resins and resin solutions of the alkyd type.

"Dyasist." [300] Trademark for dyeing assistants for use on synthetic fibers.
"Dyasist" 230: For dyeing 100% "Orlon" with cationic dyes.
"Dyasist" 281: Leveling agent for dyeing nylon tricot and other nylon filament fabrics, to eliminate or minimize Barré marks.

"Dybar." [28] Trademark for pellets containing 25% fenuron for control of oak, birch, locust, maple, pine, poplar and elm on fence rows, drainage ditches, utility and railroad rights-of-way, and spot treatment for control of bindweed.
Containers: 10-lb canisters; 50-lb bags.

dyclonine (4'-butoxy-3-piperidinopropiophenone hydrochloride) $C_{18}H_{27}NO_2 \cdot HCl$.
Properties: Crystals; m. p. 175-176°C; soluble in water, alcohol, acetone. Phenol coefficient 3. 6.
Grade: N. N. D.
Use: Medicine.

dyeing assistants. Any materials added to a dye bath to promote or control dyeing. The action of assistants differs with the classes of dyes, but in most cases they aid in level deposition of the dye, either by delaying its absorption, increasing its solubility, or assisting the dye solution to penetrate the material.

dye intermediates. See intermediate.

dye retarding agents. Materials added to dye baths to prevent, by decreasing absorption of the dyes, the rapid exhaustion of the bath.

dyer's alkanet. See alkanna.

dyer's saffron. See carthamus.

dyes. A dye is usually a colored organic compound or mixture, that can be applied to and give color to a second substance such as cloth, paper, plastic or leather, in a reasonably permanent fashion. Dyes characterized by satisfactory permanence in spite of exposure to light, air, and normal handling are referred to as "fast" dyes; others are "fugitive." Most dyes of commercial importance are synthesized from the aromatic hydrocarbons and related materials. Dyes are most frequently used to confer color on fabrics, but there are also a great number of other applications, including the following: paints and related materials; oil and gasoline; antifreeze mixtures or compounds; foodstuffs and fruit; inks and paper; rubber, resins, plastics; carbon paper and typewriter ribbons; soap, nail polish, cosmetics; polishes, waxes, and candles. See also pigments. Dyes are classified according to chemical composition, and also according to the way in which they behave during application. Vat, sulfur, direct, acid and azoic dyes have been used in greatest volume. Increases in usage have occurred for mordant, solvent, dispersed, basic, food, drug, cosmetic dyes and for fluorescent brighteners. The most important chemical types are the azo, anthraquinone, sulfur, indigoid and stilbene dyes. The chemical classes of coloring matters and their arrangement according to chemical structures have been designated numerically according to the "Colour Index" (1957 revision) as follows:

Chemical Class	Colour Index
Nitroso	No. 10000-10299
Nitro	10300-10999
Monazo	11000-19999
Disazo	20000-29999
Trisazo	30000-34999
Polyazo	35000-36999
Azoic	37000-39999
Stilbene	40000-40999
Diphenylmethane	41000-41999
Triarylmethane	42000-44999
Xanthene	45000-45999
Acridine	46000-46999
Quinoline	47000-47999
Methine	48000-48999
Thiazole	49000-49399
Indamine	49400-49699
Indophenol	49700-49999
Azine	50000-50999
Oxazine	51000-51999
Thiazine	52000-52999
Sulfur	53000-54999
Lactone	55000-55999
Aminoketone	56000-56999
Hydroxyketone	57000-57999
Anthraquinone	58000-72999
Indigoid	73000-73999
Phthalocyanine	74000-74999
Natural	75000-75999

dyes, disperse. These dyes fall mainly into three clearly defined chemical classes: (a) nitroarylamine, (b) azo and (c) anthraquinone, and almost all contain amino or substituted amino groups but no solubilizing sulfonic acid groups. They are water-insoluble dyes introduced as a dispersion or colloidal suspension in water and are absorbed by the fiber after which they may remain untreated or be aftertreated (diazotized) to produce the final color. Their use is primarily for cellulose acetate and nylon, polyester and other synthetic fibers and for thermoplastics.

dyes, leather. Natural dyes and pigments were originally and often still are used for dyeing leather. However, most types of synthetic dyestuffs were found to have an affinity for leather; particular basic and acid dyes for vegetable-tanned leather and direct and acid dyes for chrome-tanned leather.

dyes, natural. This group comprises all dyes obtained from animal or vegetable matter with little or no chemical processing. They are mainly mordant dyes but include some vat dyes, solvent dyes and others. Some illustrations are curcuma, saffron, carmine, litmus, indigo, chlorophyll, walnut extract. Since they are natural extracts, they are not expected to be chemically pure. The use is varied, including dyeing of natural fabrics, as artist's colors, spirit or water base inks, coloring of paper, as indicator colors.

"Dykast." [333] Trade name for lacquer enamels used especially on zinc and aluminum die castings.

"Dylan." [11] Trademark for soft polyethylene plastics having a low-polish, waxy appearance. Used for squeeze bottles, film packaging, other applications where flexibility is required.
See also "Super Dylan."

"Dylene." [11] Trademark for a light, rigid polystyrene used as a thermoplastic molding material. Available in a full range of colors and in both medium and high impact resistant types.

"Dylex." [11] Trademark for styrene-butadiene latices.

Properties: Stable, milky white latices with high solids content; fine particle size; low viscosity; high mechanical stability; good electrolyte and acid stability.

Grades: Styrene-butadiene ratios from 50-50 to 90-10.

Uses: Paper and textile coatings; films.

"Dylite." [11] Trademark for a free-flowing expandable polystyrene. Foam density controlled by application of heat. Used for low temperature thermal insulations, light weight core material in sandwich constructions, buoyant members, toys, novelties, and electronics.

"Dylox." [181] A selective insecticide containing O, O-dimethyl 2, 2, 2-trichloro-1-hydroxyethylphosphate (q. v.).

"Dymal." [141] Trade name for resins of the maleic anhydride or fumaric acid alkyd type.

"Dymerex." [266] Trademark for a light-colored, hard, thermoplastic resin; dimerized resin acids; softening point 152°C; acid number 143; color M; density 1.069 at 20°C.

"Dymetracin." [342] Trademark for sodium bacitracin methylene disalicylate.

"Dynahue." [204] Trademark for a complete laundry detergent designed for use on washable colored cottons.

dynamite. An industrial explosive which is detonated by blasting caps. The principal explosive ingredient is nitroglycerin or specially sensitized ammonium nitrate. Diethyleneglycol dinitrate, which is also explosive, is often added as a freezing point depressant. A dope such as wood pulp, and an antacid, as calcium carbonate, are also essential. See also blasting agents; blasting gelatin.
Shipping regulations: High explosive, Class A. High explosive label. *

"Dynawet." [233] Trademark for emulsifier and wetting agent for use in herbicides, fungicides, insecticides, etc.

"Dyne." [284] Trademark for a detergent-germicide recommended for dairy farm sanitation. Contains nonionic-iodine complexes. Claimed to be non toxic, non irritating, non staining when used as directed.

"Dynel." [214] Proprietary name for copolymer of vinyl chloride and acrylonitrile used as a textile fiber.
Properties: Irregular ribbon-shaped cross section filaments. Light cream color; easily dyed. Fire resistant and insect resistant. Unaffected by strong detergents and soaps and wide variety of inorganic acids, bases, and salts. Unaffected by hydrocarbons, dry cleaning solvents, and most other organic solvents. Acetone, cyclohexanone, and dimethyl formamide are solvents in varying degrees. Sp. gr. 1.30 (25°C); softening range 300-325°F.
Grades: Fiber sizes of 2, 3, 6, 12, and 24 denier and in staple lengths from 1 1/2 to 6 inches. Also available as 135,000 total denier tow.
Containers: Corrugated cartons, baled with steel tape containing 100 and 500 pounds net. Tow packed in cartons containing 150 pounds net.
Use: Textile fiber for use in pile, knitted and woven fabrics for apparel and industrial goods.

"Dyphene." [141] Trade name for pure phenolic resins characterized by extreme hardness, excellent chemical resistance and fast drying properties.

"Dyphenite." [141] Trade name for rosin-modified phenolic resins.

"Dyphos." [304] Trademark for dibasic lead phosphite ($2PbO \cdot PbHPO_3 \cdot 1/2H_2O$) vinyl stabilizer.
Properties: Fine white acicular crystals; sp. gr. 6.94; refractive index 2.25.
Containers: Fiberboard drums containing 40 and 300 pounds.
Uses: Heat, light and weathering stabilizer for vinyl and other chlorinated resins in paints and plastics. Special "XL" grade available for vinyl electrical insulation.
Caution! "Dyphos" should be stored in closed containers, away from open flame, and at temperatures not to exceed 400°F. Avoid exposure to sparks or static electricity by grounding equipment and using wooden scoops.

dyphylline [7-(2,3-dihydroxypropyl)theophylline] $C_{10}H_{14}N_4O_4$.
Properties: Crystals; extremely bitter taste; m.p. 158°C; freely soluble in water; soluble in alcohol and chloroform.
Grade: N.N.D.
Use: Medicine.

dypnone (phenyl alpha-methylstyryl ketone; 1,3-diphenyl-2-buten-1-one) $C_6H_5COCHC(CH_3)C_6H_5$. Stable, light colored liquid with a mild fruity odor.
Properties: B.p. (50 mm) 246°C; sp. gr. 1.093 (20/20°C); f.p., sets to a glass below −30°C; almost insoluble in water; flash point 350°F.
Uses: Softening agent, plasticizer, and perfume base. High absorption of ultraviolet light, low water solubility, low evaporation rate, and good solvent action make it suitable for use in coatings that must be stable to light.

"Dyrene." [181] Trademark for 2,4-dichloro-6-ortho-chloroaniline-s-triazine (q.v.).

dyscrasite. Ag_3Sb. A natural antimonide of silver. Color and streak silver-white; luster metallic, usually tarnished. Sp. gr. 9.74; hardness 3.5-4.
Occurrence: Germany; France; Canada.
Use: Ore of silver.

dysprosia. See dysprosium oxide.

dysprosium Dy. Rare earth element or lanthanide having atomic number 66.
Properties: Lustrous metal; m.p. 1465-1505°C; b.p. 2300° (approx). Reacts slowly with water; soluble in dilute acids.
Derivation: Reduction of the fluoride with calcium.
Source: See rare earth minerals.
Grades: High purity lumps or ingots.
See also rare earth metals.

dysprosium nitrate $Dy(NO_3)_3 \cdot 5H_2O$.
Properties: Yellow crystals; m.p. 88.6°C (in its water of crystallization). Soluble in water.
Derivation: Treatment of oxides, carbonates or hydroxide with nitric acid.
Grades: Up to 99.9% pure.
Containers: Glass bottles; fiber drums.
Shipping regulations: Oxidizing material. Yellow label. *

dysprosium oxide (dysprosia) Dy_2O_3.
Properties: White substance; many times more magnetic than ferric oxide; slightly hygroscopic, absorbing moisture and

*See "I.C.C. Shipping Regulations," page xiii.
Reference numbers refer to name of manufacturer. See "List of Manufacturers," page v.

carbon dioxide from the air. Sp. gr. 7.81 (27/4°C). Soluble in acids and alcohol.
See also rare earths.
Derivation: Ignition of hydroxides and oxyacids (carbonates, oxalates, sulfates, etc.).
Containers: Glass bottles; fiber drums.

dysprosium salts. Salts available commercially other than those listed here are the chloride, $DyCl_3 \cdot xH_2O$; fluoride, $DyF_3 \cdot 2H_2O$; arsenide, antimonide and phosphide. The last three are used as high purity binary semiconductors.

dysprosium sulfate $Dy_2(SO_4)_3 \cdot 8H_2O$.
Properties: Brilliant yellow crystals; stable at 110°C and completely dehydrated at 360°C. Soluble in water.
Derivation: Dissolving the hydroxide, carbonate or oxide in dilute sulfuric acid.
Use: For atomic weight determination.

"Dythal." [304] Trademark for dibasic lead phthalate vinyl stabilizer.
Properties: Fluffy white crystalline powder; sp. gr. 4.6; refractive index 1.99.
Containers: Fiberboard drums containing 40 and 250 lbs.
Uses: Heat and light stabilizer for general vinyl use. Excellent for high temperature vinyl insulation and opaque film and sheeting. Special "XL" grade available for vinyl electrical insulation.

"Dytol." [23] Trademark for aliphatic primary alcohols derived from natural fats and oils.
Uses: Additives for cosmetic creams, polymerization regulators for elastomers and plastics, detergents and viscosity index improvers for lubricating oils, finishing and softening agents for textiles, preparation of quaternary ammonium compounds, surfactants, water evaporation control and anti-foam.

"Dytrol." [58] Trademark for clear oily liquid resin plasticizer used as paint brush preservative, reconditioner; paint remover. Non-volatile; non-flammable; nontoxic; pleasant odor. Miscible with most common solvents, thinners and oils; softens paints, lacquers, varnishes; sp. gr. 1.088-1.091 (25/25°C); refractive index 1.507-1.510 (25°C); moisture content 0.1% maximum.
Containers: 5- and 55-gal drums.

E

E. A symbol first suggested for einsteinium. See actinide elements.

"E-2." [58] Trademark for ammonium nitrate fertilizer pellets, min. nitrogen content 33.5%, moisture 0.3%. No organic coating.
Shipping regulations: See ammonium nitrate. Consult manufacturer.

"E 1059." [181] See O,O-diethyl O(and S)-2-(ethylthio)ethyl phosphorothioates.

eagle vine. See condurango.

EAK. Abbreviation for ethyl amyl ketone.

earth. Originally a naturally occurring metal oxide, somewhat impure or diluted. Most frequently used for oxides of iron and other metals of similar chemical activity and occurrence. Many of these oxides are used as pigments (see earth, red; earths, green; iron oxide reds). The word earth is also used to describe classes of metal oxides, as, alkaline earths; rare earths. It is most frequently used in combination with another descriptive term as in the preceding examples, and is seldom used alone.

earth flax. See asbestos.

earth, infusorial. See diatomite.

earth-nut. See peanut.

earth-nut cake. See peanut cake.

earth oil. See petroleum.

earth pitch. See asphalt.

earth, red. A fine red pigment consisting essentially of red iron oxide. See also iron oxide reds.

earths, green (terre verte). Collective name for various pale bluish-green earths formed by disintegration of minerals, principally those of the hornblende type, and used as pigments. They are somewhat deficient in body and intensity of hue, and are now largely replaced by manufactured pigments.

earth wax. See ceresin wax.

East Indian copaiba balsam. See gurjun balsam.

East Indian copaiba balsam oil. See gurjun balsam oil.

East Indian geranium oil. See palmarosa oil.

eau de cologne. An alcoholic solution of a number of essential oils among which are bergamot, lemon, lime, lavender, rosemary, and neroli in varying amounts. Sold commercially as a perfume, toilet water, etc.

eau de javelle. See javelle water.

eau de Labarraque. See sodium hypochlorite; Labarraque's solution.

"EB-5." [233] Trademark for fumigant containing approximately 5% ethylene bromide (q.v.). Concentrations other than 5% are indicated by appropriate number.

ebm. See electron beam melted.

ebonite. Black, hard, vulcanized rubber used for valves, faucets, pipes, electric equipment, fountain pens, toilet articles, handles, etc., made from the cheaper grades of rubber, or from latex.

"Ebonol." [142] Trade name for various blackening processes including "Ebonol S" for blackening of steel, "Ebonol C" for blackening of copper and copper base alloys, and "Ebonol Z" for blackening of zinc. Materials are supplied in all cases as powders that are added to water at various temperatures to accomplish blackening.

ecgonidine. See anhydroecgonine.

echinacea (cone flower; black sampson; purple cone flower). Root of Brauneria pallida and B. augustifolia.
Habitat: North America.
Grade: Technical.
Containers: Bags; barrels.
Use: Medicine.

echitamine. See ditaine.

Eclipse. See benzopurpurin.

Eclipse Red. See benzopurpurin.

"Ecolid." [305] Trademark for chlorisondamine chloride N.N.D.
Use: Medicine.

economic poison. See pesticide.

"Econo-Sour." [244] A proprietary product consisting chiefly of fluorine compounds.
Properties: White powder; sparingly soluble in water; neutralizing value, 28.6 oz sodium bicarbonate per lb.
Containers: 150-lb and 300-lb fiber drums.
Uses: Laundry sour where economy is desired.
Shipping regulations: None.*

"Eco Spar." [333] Trade name for clear spar varnishes and enamels for use on exteriors.

ectylurea (2-ethyl-cis-crotonylurea) $H_2NCONHCOC(C_2H_5):CHCH_3$.
Properties: Crystalline solid; m. p. 198°C; slightly soluble in organic solvents.
Grade: N. N. D.
Use: Medicine.

edathamil calcium disodium. See calcium disodium EDTA.

EDB. Abbreviation for ethylene dibromide (q. v.).

"E-D-Bee." [88] Trade name for a concentrated fumigant for dormant treatment of bee hives and combs to control wax worms of bee-moths; contains ethylene dibromide.

Edeleanu process. A solvent extraction process using liquid sulfur dioxide for the removal of undesirable aromatics from heavy lubrication oils.

edestin. A vegetable globulin which can be obtained from hemp seed and certain other seeds.
Properties: White or light-brown crystals; soluble in water and dilute acids.
Use: Analytical determination of pepsin.

edible oils. Vegetable or animal oils, containing not over 0. 1% of free fatty acid, used for food purposes (cooking, salad-dressing and manufacture of oleomargarine or other butter substitutes, such as butter oils, deodorized oils, margarine oils, salad oils).

"Edicol." [206] Brand name for a line of dyestuffs used in coloring edible products, and specially prepared to conform to the food laws of the United Kingdom and other countries.

edrophonium chloride (dimethylethyl-(3-hydroxyphenyl)ammonium chloride) $HOC_6H_4N(CH_3)_2CH_2CH_3Cl$.
Properties: White, odorless, crystalline powder; m. p. 167-170°C (dec). Very soluble in water; freely soluble in alcohol; practically insoluble in ether. pH (1% solution) 4. 0-5. 0.
Grade: U. S. P. XVI.
Use: Medicine.

EDTA. Abbreviation for ethylenediamine-tetraacetic acid.

e.g. Abbreviation of "for example."

egg oil. See egg yolk.

egg yolk (egg oil). See also albumin, egg.
Properties: Yellow semi-solid mass; sp. gr. 0. 95; m. p. 22°C.
Derivation: From the eggs of hens, ducks, and geese. The yolks are separated from the whites and the yolks are then mixed with common salt; sometimes a small amount of borax is added as a preservative.
Grades: Technical; edible.
Containers: Wooden barrels; fiber drums.
Uses: Leather dressing; tanning; baking; dairy products; pharmaceuticals (ointments, emulsions, drug preparations); soap; perfumery.

eglantine. Name applied to isobutyl benzoate.

eglestonite. A natural mercury oxide-chloride, Hg_4Cl_2O. Found in Texas.

Egyptian asphalt. A glance pitch (q. v.) found in the Arabian desert between the Nile and the Red sea. It has sp. gr. 1. 10 (77°F), contains over 99% non-mineral content, is soluble in carbon disulfide and has a fusing point of 285°F (B & R).
Shipping regulations: None. *

Egyptianized clay. A clay to which tannin has been added in order to make it more plastic.

Egyptian privet. See henna.

Ehrlich 606. See arsphenamine.

Ei. A symbol used for einsteinium, but not sanctioned by the IUPAC.

eicosafluoro-1-undecanol. See fluoroalcohols.

eicosane $C_{20}H_{42}$. Most technical eicosane is a mixture of predominantly straight chain hydrocarbons averaging 20 carbon atoms to the molecule.
Properties (pure n-eicosane): White crystalline solid; f. p. 36. 7°C; b. p. 205°C (15 mm); refractive index 1. 4348 (n 20/D); sp. gr. 0. 778 (at melting point). Insoluble in water; soluble in ether. Can be readily chlorinated.
Specifications (technical eicosane): Melting range 35-36. 5°C; boiling range (10 mm) 10% 383°F; 90% 410°F.
Grades: Pure normal (99+ %); technical.
Uses: Cosmetics; lubricants; plasticizers; flameproofing fabrics.

eicosanoic acid. See arachidic acid.

1-eicosanol. See arachidyl alcohol.

eikonogen (1-amino-2-naphthol-6-sulfonic acid, sodium salt) $NH_2C_{10}H_5(OH)SO_3Na \cdot 2\frac{1}{2}H_2O$.
Properties: White powder which reduces silver salts. Soluble in water; insoluble in alcohol and ether.
Derivation: By the interaction of sodium carbonate and amino-2-naphthol-6-sulfonic acid.
Method of purification: Crystallization.
Grades: Technical.
Containers: Glass bottles.
Use: Photographic developer.
Shipping regulations: None. *

einsteinium (element 99) Es. A synthetic radioactive element with atomic number 99 first discovered in the debris from the 1952 hydrogen bomb explosion. Einsteinium has since been prepared in a cyclotron by bombarding uranium with accelerated nitrogen ions, in a nuclear reactor by irradiating plutonium or californium with neutrons, and by other nuclear reactions. The element is named for Albert Einstein. It has chemical properties similar to those of the rare earth holmium. Isotopes are known with mass numbers ranging from 246-253.
See also actinide elements.

ELA Elastomer Lubricating Agent. [28] Mixture of phosphate esters.

Properties: Pale yellow liquid.
Containers: Drums (425 lbs, net).
Use: As a lubricant for unvulcanized natural and synthetic rubber compositions.

elaidic acid (trans-9-octadecenoic acid) $CH_3[CH_2]_7CH:CH[CH_2]_7COOH$. The trans form of an unsaturated fatty acid of which the cis-form is oleic acid.
Properties: White solid; sp. gr. 0.8505 (79/4°C); m. p. 43.7°C; b. p. 288°C (100 mm); 234°C (15 mm); refractive index 1.4358 (79°C). Insoluble in water; soluble in alcohol, ether, benzene and chloroform.
Derivation: Synthetic product, from oleic acid by elaidinization.
Grade: Purified, 99+%.
Uses: In medical research and as reference standard in chromatography.

elaidinization. Originally, the reaction by which oleic acid is converted into elaidic acid but now used in the more general sense to indicate the conversion of any unsaturated fatty acid or related compound from the geometric cis to the corresponding trans form. Nitrous acid and selenium compounds are commonly used as catalysts for this reaction. The resulting trans acids are more stable to oxidative effects.

"Elaine." [242] Trade name for a brand of oleic acids.

"Elastex" 20-A Plasticizer. [175] Trademark for diisodecyl adipate, a plasticizer providing vinyl formulations with low-temperature performance.
Containers: 55-gal steel drums; tank trucks; tank cars.

"Elastex" 60-A Plasticizer. [175] Trademark for dioctyl adipate (q. v.), a plasticizer commonly used with general purpose plasticizers in processing polyvinyl and other synthetic resin compounds.
Containers: 55-gal steel drums; tank trucks; tank cars.

"Elastex" 50-B Plasticizer. [175] Trademark for butyl cyclohexyl phthalate (q. v.), used for processing natural and synthetic polymers and elastomers; also in nitrocellulose lacquers.
Containers: 55-gal steel drums; tank trucks; tank cars.

"Elastex" DCHP Plasticizer. [175] Trademark for dicyclohexyl phthalate (q. v.), used as a plasticizer in processing vinyls because of its excellent compatibility at elevated temperatures; also widely used in the formulation of heat-sealing paper coatings.
Grades: Cast solid; granular; or lump.
Containers: Cast, 55-gal destructible steel drums; granular or lump, fiber drums (250 lb net).

"Elastex" 10-P Plasticizer. [175] Trademark for diisooctyl phthalate (DIOP) (q. v.), a general purpose plasticizer for vinyl resins and synthetic rubbers.
Containers: 55-gal steel drums; tank trucks; tank cars.

"Elastex" 18-P Plasticizer. [175] Trademark for isooctyl isodecyl phthalate (q. v.), a plasticizer characterized by good retention of physical properties after heat-aging.
Containers: 55-gal steel drums; tank trucks; tank cars.

"Elastex" 28-P Plasticizer. [175] Trademark for dioctyl phthalate (DOP; di-2-ethylhexyl phthalate) (q. v.), a general purpose plasticizer used especially for vinyl resins and synthetic rubbers.
Containers: 55-gal steel drums; tank trucks; tank cars.

"Elastex" 40-P Plasticizer. [175] Trademark for butyl isodecyl phthalate (q. v.), a secondary plasticizer for PVC and PVC-PVAc copolymer resins; also used in the formulation of plastisols and organosols.
Containers: 55-gal steel drums; tank trucks; tank cars.

"Elastex" 48-P Plasticizer. [175] Trademark for butyl octyl phthalate, a primary plasticizer for vinyl resins.
Containers: 55-gal steel drums; tank trucks; tank cars.

"Elastex" 82-P Plasticizer. [175] Trademark for n-octyl n-decyl phthalate, a primary plasticizer for vinyl film, sheeting, hose, and wire and cable insulation.
Containers: 55-gal steel drums; tank trucks; tank cars.

"Elastex" 90-P Plasticizer. [175] Trademark for diisodecyl phthalate (DIDP) (q. v.), a low volatility plasticizer used to upgrade aging characteristics and also to stabilize viscosity of plastisols.
Containers: 55-gal steel drums; tank trucks; tank cars.

elastic glue. See glue.

elastin. An albuminoid which occurs in elastic tissue.
Properties: Yellow fibrous mass; insoluble in water, in dilute acids, alkalies and salt solutions, and in alcohol. Is partially digested by pepsin solution and wholly by trypsin.

elastomer. A synthetic polymer with rubber-like characteristics, but not necessarily in such a degree as to make the material practically useful. Examples of commercial products are butyl rubber, polyurethane rubber, and silicone rubber.

"Elastopar." [58] Trademark for N-methyl-N, 4-dinitrosoaniline (33⅓%) plus "Whitetex" Clay (66⅔%). Light colored, free flowing 30 mesh powder used for chemically modifying butyl rubber to give improved resilience; increased modulus; lower hardness; increased abrasion resistance; better low temperature flexibility; increased electrical resistance; reduced cold-flow and Mooney Viscosity.

elaterin $C_{20}H_{28}O_5$.
Properties: White, crystalline powder. Soluble in chloroform; slightly soluble in

alcohol, benzene, and ether. M.p. 216°C.
Derivation: From the juice of Echallium elaterium.
Method of purification: Crystallization.
Grade: Technical.
Containers: Glass bottles.
Use: Medicine.
Shipping regulations: None.*

elaterite (mineral caoutchouc). A variety of asphaltic pyrobitumen (q. v.).

elayl. See ethylene.

"Elbelan." [232] Brand name for a series of neutral-dyeing pre-metallized dyestuffs of outstanding fastness properties.

"Elchem" 1393. [28] Brand name for an addition agent, a grayish-white powder containing potassium cyanide and selenium; slightly hygroscopic, water-soluble, stable. Poisonous.
Containers: 25- and 100-lb fiber drums.
Uses: As brightening agent in cyanide copper plating baths.

"Elchem" 1396. [28] Brand name for an addition agent, a dark amber liquid; mildly alkaline.
Use: Brightener and antipitting agent for cyanide copper plating baths.

"Elchem" 1442. [28] Brand name for an addition agent, a mildly alkaline solution containing potassium cyanide and selenium. Poisonous.
Containers: 40-lb pails; 450-lb drums.
Uses: As a brightening agent and leveling agent in cyanide copper plating baths.

electrical oils. See insulating oils.

electric furnace refractories. See cermets.

electric steels. Steels which have been made in an electric furnace.

electrochemical equivalent. The number of grams of an element or group of elements liberated by the passage of one coulomb of electricity (one ampere for one second).

electrochemistry. Branch of chemistry which deals with the chemical change produced by an electric current and with the production of electricity from the energy released in a chemical reaction. Applications are in the ordinary dry cell, in lead plate storage batteries, electroplating, purification of copper, production of aluminum, fuel cells, and in corrosion of metals.

electrocortin. See aldosterone.

electrode, hydrogen. Platinum surface coated with platinum black, immersed in a solution and bathed with a stream of pure hydrogen gas. The potential developed depends on the equilibrium between the hydrogen gas and the hydrogen ions in solution. Used as the standard reference electrode.

electrodeposition. The precipitation of a material at an electrode as the result of the passage of an electric current through a solution or suspension of the material, e.g., copper from copper sulfate solution; rubber from latex.

electroforming. The reproduction of an object or pattern by electrodeposition. A mold of the object to be reproduced is made in a soft metal or in wax (by impression). The non-conductor mold surface is made conducting by coating with graphite. Some suitable metal is then deposited electrolytically on the mold surface. This mold is then (in most cases) a negative of the object to be produced. Most extensive use is in the phonograph record industry.

electroluminescence. The emission of light as a consequence of electrical discharges in gases. Typical examples are neon lights, mercury vapor lamps and lightning.

electrolysis. Decomposition by means of an electric current; the compound is split into positive and negative ions which migrate to and collect at the negative and positive electrodes.

electrolyte. A substance which dissociates into ions when in solution or a fused state and which will then conduct an electric current. Common examples are sodium chloride and sulfuric acid.

electrolyte acid. See battery acid.

electrolytic cell. A combination of a liquid or semi-liquid electrolyte (solution of a salt, acid, or base) and two solids (the electrodes) which generates an electric current when the electrodes are connected by an external wire. Flashlight batteries (see dry cell), storage batteries (q. v.), and fuel cells (q. v.) are special types of electrolytic cells. When electricity is generated in such a cell chemical changes occur at the electrodes so that either or both the electrodes and the electrolyte are gradually consumed.

The term cell or electrolytic cell is also loosely applied to somewhat similar arrangements in which an electric current from an external generator is passed through the cell, in order to cause chemical changes, as in the electrolysis of sodium hydroxide solution to produce hydrogen and oxygen gases.

"Electromanganese." [250] Trademark for an electrolytic manganese metal of high purity (99.9% min) used in the production of steel, aluminum and other metals and alloys.

electromotive series (displacement series; activity series). An arrangement of the metals in the order of their tendency to react with water and acids, so that each metal displaces from solution those below it in the series and is displaced by those above it. The arrangement of the more common metals is: potassium, sodium, magnesium, aluminum, zinc, iron, tin, lead, hydrogen, copper, mercury, silver, platinum, gold.

electron. A negatively charged particle with mass approximately 1/1860th that of the lightest atom (the hydrogen atom). Ordinary hydrogen atoms are composed of a single electron and a central nucleus consisting of a positively charged proton particle, the charge being equal to that on the electron but opposite in sign. Other atoms have two or more electrons and more complex nuclei. An electric current consists of a stream of electrons more or less freed from their association with a particular atom. Electrons sometimes behave like waves of energy instead of discrete particles. See also fundamental particle.

electron beam melted (ebm). Used of steels and alloys, which are melted in electron beam furnaces, a process competitive with vacuum arc melting.

Electronic Palladium. [28] Palladium metal with fluxes and vehicles for application on green ceramic bases that are fired at extreme temperatures and permit multiple laminations of electroded ceramic sheets used for capacitor manufacturer.
Containers: 1- to 32- oz jars.

Electronic Platinum. [28] Similar to Electronic Palladium.

electron volt (ev). A unit of energy which is defined as the energy acquired by any charged particle carrying a unit electric charge when it falls through a potential of one volt. It is equivalent to 1.60×10^{-12} ergs. One mev is equal to one million electron volts. One kev is equal to one thousand electron volts. One bev is one billion electron volts.

electrophoresis. The phenomenon of migration of suspended or colloidal particles in a liquid due to the effect of an e. m. f. or potential difference across immersed electrodes. Just as in ordinary electrolysis, the migration is toward electrodes of electric charge opposite to that of the particles. Most solids, being negatively charged, migrate to the anode, the exception being basic dyes, hydroxide sols, and colloids which have adsorbed positive ions, all of which are positively charged and migrate to the cathode. Migrated particles lose their charge at the electrode, and generally agglomerate around it. A small current is used to avoid agitation caused by eddy currents of electrolysis or by evolution of gas bubbles.

Electrophoresis is important in the study of proteins because the molecules of such materials act like colloidal particles and their charge is positive or negative according to whether the surrounding solution is acidic or basic. Thus, the acidity of the solution can be used to control the direction in which a protein moves upon electrophoresis.

electroplating. The deposition of a layer of metal (usually thin) on a base metal or conducting surface by electrolysis, i. e., the action of an electric current.

Electropolishing Solution. [28] A clear, dark amber corrosive liquid; glycolic acid-sulfuric acid formulation; acid content 85%. Freezing point below 32°F; nonflammable and nonexplosive; miscible with water.
Containers: 162-lb. carboys.
Uses: In electropolishing stainless steel.

electrowinning. Recovery of metals from ores by electrochemical processes.

"Electrunite." [251] Trademark for steel pipe and tubing, including both stainless and carbon steels.

element (elementary substance). A substance or kind of matter all of whose atoms are alike in the sense that they all have the same number of positive charges on the nucleus. The atoms however may have different masses, and in this case the element may have two or more isotopes. In a more conventional sense an element is a substance which can not be decomposed into or synthesized from other substances by ordinary chemical methods.

element 61. See promethium.

element 99. Now known as einsteinium. See also actinide elements.

element 100. Now known as fermium. See also actinide elements.

element 101. Now known as mendelevium. See also actinide elements.

element 102. Now known as nobelium (q. v.). See also actinide elements.

element 103. Tentatively named lawrencium (q. v.). See also actinide elements.

elementary particle. See fundamental particle.

elementary substance. See element.

elemi gum (canarium).
Derivation: A yellowish to brown resin from certain trees, Canarium commune in the Philippine Islands, Canarium maritanum in Mauritius and Amyris elemifera in Mexico and Brazil.
Soluble in alcohol, benzene, and turpentine.
Grades: Technical.
Containers: Bags.
Uses: Varnishes; lacquers; perfume fixative; ointments.
Shipping regulations: None. *

elemi oil.
Properties: An almost colorless, liquid oil; agreeable aromatic odor and taste. Soluble in alcohol, ether, chloroform, and carbon disulfide.
Chief known constituents: Limonene and phellandrene.
Constants: Sp. gr. 0.870-0.910.
Derivation: Distilled from the gum of the Manila elemi tree Canarium luzonicum.
Method of purification: Rectification.
Grades: Technical.
Containers: Tins; glass bottles.
Uses: Medicine; perfumes.
Shipping regulations: None. *

eleuthera bark. See cascarilla.

"Elf" Blacks. [275] Trade name for a series of standard color channel carbon blacks for

use in all types of printing inks and as a coloring agent in paints and plastics.
Available as:
"Elf" O. Specially treated for longer flow and low oil absorption.
"Elf" 1, 2 and 3. Darkest of the blacks in the Elf series.
"Elf" 4. Midway in color and strength in the Elf series.
"Elf" 5, 6, 7, 8. Largest particle size, lowest oil absorption and lowest strength of the Elf series.

"Elftex." [275] Trade name for a series of standard color furnace carbon blacks for use in inks and paints. Available as:
"Elftex" 5. Medium oil absorption.
"Elftex" 8. Lower oil absorption.

"Elipten." [305] Trademark for amino-glutethimide N. N. D.
Use: Medicine.

"Elite Fast." [232] Brand name for a series of neutral dyeing acid dyestuffs of excellent fastness to washing.

elixir of vitriol. See sulfuric acid, aromatic.

elixirs. Aromatic, sweetened, alcoholic liquids containing small quantities of active medicinal substances.

"Elkosin." [305] Trademark for sulfisomidine, N. N. D.
Use: Medicine.

ellagic acid. $C_{14}H_6O_8$. Found in galls.
Properties: Yellowish powder, odorless, tasteless. Soluble in alkaline solutions; insoluble in acid or neutral solutions.
Use: Medicine.

elm. See ulmus.

Elmer's. [65] Generic name applied to a series of products which include:
Elmer's Glue-All - fast-setting general-purpose, liquid polyvinyl acetate adhesive.
Elmer's Contact Cement - solvent-based cement for dry assemblage of plywood-phenolic laminates (Formica, Pionite) to wood or metal.
Elmer's Waterproof Glue - two-component resorcin-formaldehyde glue for use where water-proof or maximum durability wood bonds are required. Resists acids, alkalies, solvents, mildew, bacteria, cold or boiling water.
Uses: Boats, toys, outdoor furniture, sports equipment, etc.

"Eloma." [51] Trademark for transparent pipe coatings available in several grades for mill application.

"Elorine" Chloride. [100] Trademark for tricyclamol methochloride.

"Elprene." [41] Trade name for a synthetic-rubber coating of the neoprene type used as a general maintenance coating.

"El-Sixty." [58] Trademark for 1, 3-bis-(2-benzothiazolyl-mercaptomethyl)urea (q.v.).

eluteria bark. See cascarilla.

elutriation. A process of washing, decantation and settling which separates a suspension of a finely divided solid into parts according to their weight. It is especially useful for very fine particles below the usual screen sizes and is used for pigments, clay dressing, and ore flotation.

"Elvacet." [28] Trademark for thermoplastic, adhesive, and film-forming solutions and emulsions of polyvinyl acetate homo- and copolymers.
Polyvinyl Acetate Emulsion. Viscous, milk-white water dispersions which can be diluted with water. Available in various grades
Containers: 45-lb pails; 500-lb drums; 9, 000 to 37, 000-lb tank trucks; 37, 000-lb and 75, 000-lb tank cars.
Uses: Paint vehicles; adhesive base materials; binders; pigments and cement; textile finishes and sizes; paper coatings for grease-proofing and heat-sealing.
Polyvinyl Acetate Solution. Colorless, odorless resin available as a 60% solution in methanol.
Containers: 10- and 25-lb cartons; 50-lb multiwall bags; 100-lb drums.
Uses: Adhesive and binder; vehicle for metallic pigments; protective coatings on metal; heat-seal applications; grease-proofing; stiffening and permanent sizing.

"Elvanol." [28] Trademark for polyvinyl alcohol.
Properties: White to creamy-white powder; odorless; water-soluble synthetic resin.
Grades: Eight grades covering various degrees of hydrolysis in three viscosities (high, medium, low). Solutions are stable over long periods of time.
Containers: Cartons; multiwall bags; drums.
Uses: Sizes for textiles and papers; base for water-resistant laminating adhesives; adhesives and binders; molded products; emulsion stabilizer and thickener; photo-sensitive films.

"Elvasize." [28] Trade name for a two-step process for sizing paperboard. Borax solution, applied to the board in a pre-treatment, reacts with polyvinyl alcohol on the surface of the board to form a rigid, 3-dimensional gel, preventing excessive penetration of the size.

"Elvax." [28] Trademark for a series of high molecular weight vinyl resins. Wax-compatible, amorphous copolymers; translucent, white, 1/8" diameter, free-flowing pellets. Two basic varieties - one for hot-melt adhesive and coating formulation, the other for toughening waxes. Viscosities (at 30°C, 0. 25% in toluene) 0. 77 cp. and 0. 85 cp., respectively. Readily soluble in waxes, various low molecular weight resins and rosin esters; insoluble in aqueous mixtures and most polar solvents; density at 30°C 0. 95; extensibility 700% at break.
Containers: 50-lb bags.
Uses: As a wax additive, improves

toughness, reduces flaking, improves heat-seal bond strength. "Elvax"-wax coatings used for sheet-waxed cartonboard which retains low moisture vapor transmission, even after creasing. In laminating formulations for paper-to-paper and paper-to-cellophane, polyethylene, polystyrene, "Mylar" polyester film, or other plastic films, and to metal foils; as a base for hot-melt adhesives: as a low dielectric component in potting compounds.

emanon Em. A name at one time proposed for the element radon (q. v.).

embolite Ag(Cl,.Br). A natural chloride and bromide of silver. See cerargyrite and bromyrite.

embrittlement. A loss of toughness in a metal due to its becoming brittle. The causes are varied and frequently unknown. Embrittlement can lead to cracking and failure of the metal under stress.

"Emcol." [104] Brand name for a line of emulsifiers, germicides and detergents.

Emcol 61: Fatty acid amide for hair conditioning.

Emcol E-11: Alkyl benzyl triethyl ammonium chloride, 50% aqueous liquid, straw-colored. Used as germicide, textile assistant, sanitizing agent, deodorant.

Emcol E607: N(acyl colamino formylmethyl) pyridinium chloride (from lauric acid), white crystals; quaternary ammonium detergent germicide.

Emcol 4150: Fatty amide sulfonate (35% aqueous solution). Foaming agent, detergent, floating agent.

Emcol E607S: Stearoyl homolog of Emcol E607 used in hair rinses.

Emcol H Series: Agricultural pesticide emulsifiers and industrial solvent emulsion systems.

Emcol K-8300: Emulsifier-stabilizer for rubber latices and adhesives.

Emcol MST: Glycerol mono- and distearate; non-self-emulsifying; edible.

Emcol P1059: Amine dodecyl benzene sulfonate.

Emcol RDC-D: Diethylene glycol laurate, self-emulsifying.

Emcol RHT: Glycerol mono- and distearate, self-emulsifying. Used in medicine.

Emcol 5100 Series: Alkanol fatty acid condensates. Used as nonionic detergents, thickeners, cleaners, foam stabilizers, and emulsifiers.

emerald. See beryl.

emerald green. A pigment consisting of copper acetoarsenite (q. v.).

emerald, Oriental. See corundum.

"Emerox." [242] Trademark for azelaic acid.

"Emersol." [242] Trademark for a line of oleic acids, stearic acids, and special liquid vegetable fatty acids. The later are specifically for alkyd resins.

emery. See corundum.

"Emery." [242] A line of distilled animal and vegetable fatty acids and oleic and stearic acid esters.

emetin. See emetine.

emetine (emetin) $C_{29}H_{40}O_4N_2$. An alkaloid from ipecac.

Properties: White powder; m. p. 74°C; very bitter taste; darkens on exposure to light; poisonous! Soluble in alcohol and ether; slightly soluble in water.

Derivation: By extraction and crystallization from root of Cephalis ipecacuanha, or synthetically.

Use: Medicine.

Shipping regulations: None. *

emetine hydrochloride $C_{29}H_{40}O_4N_2 \cdot 2HCl$.

Properties: Odorless white crystalline powder; affected by light; m. p. 235-255°C; soluble in water and alcohol.

Grade: U. S. P. XVI.

Use: Medicine.

emf. Abbreviation for electromotive force.

"Emfac." [242] Trademark for pelargonic acid.

"Emlon." [194] Trademark applied to a series of resins for protective coatings and adhesives.

emodin (frangula emodin; frangulic acid) $C_{14}H_4O_2(OH)_3CH_3$. 1, 3, 8-Trihydroxy-6-methylanthraquinone. Found, either free or combined with a sugar in a glucoside, in Rhamnus, rhubarb, and rumex. A synthetic product is also available. Orange crystals; m. p. 256°C; soluble in alcohol; insoluble in water. Used in medicine.

"Emolein." [242] Trademark for a line of diesters for use in compounding synthetic lubricant fluids and greases.

emperor green. See copper acetoarsenite.

empirical formula. See formula, chemical.

"Empol." [242] Trademark for a polymerized fatty acid; a C_{36} dibasic acid made by the dimerization of polyunsaturated fatty acids. Very viscous liquid, acid value 180; saponification value 185; combining weight 305. Used in surface coatings, esters, lubricating greases, corrosion inhibitors, polyamides, polyurethane foams.

"Emralon" 310. [46] Trademark for a colloidal dispersion of polytetrafluoroethylene (PTFE) particles in a solution of phenolic resin.

Properties: 2 package liquid system; 50% PTFE, 50% resin; bakes for 1 hour at 300°F; flash point 46°F; friction coefficient 0. 05-0. 07.

Uses: As a clean, baked dry film where low coefficient of friction is needed under conditions of low load and low speed; small machine parts, rubber O-rings.

"Emralon" 320. [46] Trademark for a colloidal dispersion of polytetrafluoroethylene (PTFE) particles in a solution of a cellulosic resin.

*See "I. C. C. Shipping Regulations, " page xiii.

Properties: 2 package liquid system; 50% PTFE, 50% resin; air dry at room temperature; flash point 40°F; friction coefficient 0.05-0.07.
Uses: As a clean, air dry film type lubricant; very low friction coating of good release properties; small machine parts, wood parts, paper, cloth.

"Emtall." [242] Trade name for a line of fractionated tall fatty acids and distilled tall oils.

EMTS. Abbreviation for ethyl mercury para-toluene-sulfonanilide.

"Emulphogene AM-870." [307] Trademark for a nonionic surfactant, tallow alcohol-ethylene oxide reaction product; 100% active.
Properties: Soft wax; sp. gr. 1.04-1.05; soluble in water, acetone, alcohol, aromatic hydrocarbons and chlorinated hydrocarbons; insoluble in mineral oil and aliphatic hydrocarbons; stable to acid, alkali and metallic ions.
Uses: Emulsifier with good stability over a wide pH and temperature range; emulsifier for incorporation into natural and synthetic rubber latices to improve mechanical and chemical stability. In textile processing, used for acid degreasing of wool and to promote level dyeing. In the leather industry, emulsifying agent for mineral and raw neatsfoot oils; wetting agent for dry crusted leather; emulsifier in shoe waxes and polishes; surfactant for de-inking paper stock; dispersing and peptizing agent for pigments, clays, etc.

"Emulphor." [307] Trademark for a series of nonionic emulsifying agents and dispersants.
"Emulphor EL-620." Polyoxyethylated vegetable oil; 100% active.
Properties: Clear, yellow liquid; sp. gr. 1.04-1.05; soluble in water, hydrocarbon solvents; stable to metallic ions, alkalies and mild acids.
Uses: Emulsifying agent for oils, solvents and waxes; emulsifier for insecticide and herbicide formulations; emulsion stabilizer for polyester resins and polyvinyl acetate emulsion paints; in leather processing, used for degreasing, as a dyeing assistant and in fat liquor formulations; in paper processing; emulsifying and dispersing agent in manufacture of urethane foams and amino plastics.
"Emulphor EL-719." Polyoxyethylated vegetable oil; 96% min. activity.
Properties: Liquid; sp. gr. 1.06-1.07. Soluble in water.
Uses: Emulsifier for animal and vegetable fats; dispersing agent for pigments; promotes level dyeing in paper processing; degreasing, emulsifier and lubricant in leather processing; emulsion stabilizer for polyvinyl-acetate-emulsion paint; antistatic agent and lubricant for synthetic fibers.
"Emulphor ON-870." Polyoxyethylated fatty alcohol; 100% active.
Properties: Wax; density 1.03-1.04.
Uses: Emulsifier for mineral oils, fatty acids and waxes; stabilizer for natural or synthetic rubber latex emulsion; dyeing assistant and acid-degreasing agent for textiles; dispersant and degreasing agent in leather processing; emulsifier for aqueous dispersions of polyethylene.
"Emulphor VN-430." Polyoxyethylated fatty acid; 100% active.
Properties: Liquid; sp. gr. 0.90-1.00.
Uses: Emulsifier for mineral oils and liquid fatty acids; emulsifier for pesticides and cutting oils; used in paper, leather and textile processing; emulsifier in cosmetic industry.

"Emulsarin." [342] Trademark for blend of gums used for the preparation of pharmaceutical emulsions.

emulsifiable oil. See soluble oil.

emulsifier. See emulsion.

"Emulsifier STH." [307] Brand name for a corrosion inhibitor, sodium salt of an N(alkylsulfonyl)glycine; 88% active (min).
Properties: Dark brown, viscous liquid; density 1.00-1.05; stable to alkali; soluble in water, carbon tetrachloride, ethylene glycol, kerosene, mineral oil, and xylene.
Uses: Rust and galvanic corrosion inhibitor in polyvinyl chloride films; prevents corrosion in containers of some grades of carbon tetrachloride; provides good galvanic corrosion and rust protection of light metals, such as aluminum and magnesium.

emulsin (synaptase; amygdalase; beta-glucosidase). An enzyme catalyzing the production of glucose from beta-glucosides.
Properties: White powder; odorless and tasteless; capable of hydrolyzing glucosides such as amygdalin to glucose and the other component substances. Soluble in water; insoluble in ether and alcohol.
Source: Sweet almonds.
Derivation: By extracting an emulsion of almonds with ether, filtering the clear solution and precipitating the emulsin with alcohol.
Shipping regulations: None. *

emulsion. A substantially permanent heterogeneous liquid mixture of two or more liquids which do not normally dissolve in each other but which are held in suspension, one in the other, by mechanical agitation, or more frequently, by small amounts of additional substances known as emulsifiers. These modify the surface tension of the droplets to keep them from coalescing. Typical emulsions are milk, mayonnaise, and such pharmaceutical preparations as cod-liver oil emulsion or liquid petrolatum emulsion.

Typical emulsifiers are egg yolk, casein, and certain other proteins; soap; gums such as acacia, sea weed extracts, water-soluble cellulose derivatives; lignin, bentonite, and surface-active agents such as the

*See "I.C.C. Shipping Regulations," page xiii.
Reference numbers refer to name of manufacturer. See "List of Manufacturers," page v.

quaternary ammonium compounds, sulfonated oils, and polyhydric alcohol esters and ethers. Specific kinds of soaps include those from tallow, grease, fish oil, rosin acids. These are widely used in synthetic rubber manufacture. Mahogany soaps, i. e., the sodium salts of sulfonic acids from petroleum refining sludge, are used in synthetic resin production, as are sorbitan oleates and laurates, and polyoxyethylene esters. Stearic acid esters of glycerin, ethylene oxide, sorbitol, and glycols, and also lecithin, are used in food products.

Ammonium and amine fatty acid soaps are used in waxes and polishes. Various ones of these and others are widely used in stabilizing the emulsions of cutting oils, pharmaceuticals, drycleaning solvents, and mixtures used in the textile and leather industry.

emulsion paint. A paint, the vehicle of which is an emulsion of binder in water. The binder may be oil, oleoresinous varnish, resin, or other emulsifiable binder. (ASTM definition, ASTM D16-52). Latex is a common binder.

"Emulsol." [244] Trademark for a series of raw fish oils emulsified with soap.
Containers: Material packed in non-returnable steel drums averaging 400-425 lbs net.
Uses: In the leather industry.

en. An abbreviation for ethylenediamine as used in formulas for coordination compounds. An example is the cobalt complex $Co[en]_3(NO_3)_3$. See also dien; pn; py.

enamel.
1. A type of oil-base paint containing binders that form a film by oxidation or polymerization on exposure to air and having an outstanding ability to level off brush marks, etc., and form an especially smooth film. Enamels are usually intended for use as top coats and contain relatively less pigment than paint formulations for priming or surfacing. The term enamel is also applied to insulating varnishes for electrical equipment.
2. A hard, smooth ceramic coating on a metal, for purposes of decoration or corrosion resistance. Porcelain or other ceramic materials are applied to the surface as a pasty mixture and the metal object is heated to high temperature to melt the coating and thus form a continuous film.

enamel-brick clay. Similar to clays used for manufacture of buff face brick. See brick clay.

enamel clay. Ball clays which are capable of floating non-plastic enamel slips so that they will spray and dip more evenly. Enamel clays usually contain some alkali and must be as low as possible in carbon.

enamel oxides. Calcined oxide mixtures with varied compositions, used for coloring sheet steel, cast iron, or aluminum vitreous enamels.
Containers: 10-, 25-, 100-, 300- and 500-lb packages.

enamel, porcelain. See enamel; also Porcelain Enamels for Aluminum.

enanthic acid. See n-heptanoic acid.

"Enarax." [299] Trademark for a combination drug containing hydroxyzine hydrochloride and oxyphencyclimine hydrochloride.

enargite Cu_3AsS_4. A natural copper arsenic sulfide, found in metallic veins. May contain some antimony.
Properties: Color grayish-black to iron-black; streak grayish black, luster metallic; sp. gr. 4.45; hardness 3.
Occurrence: Montana, Utah, Colorado; Peru; Yugoslavia.
Use: Ore of copper and arsenic.

encapsulation. In general, any process in which a material or its individual pieces or particles are coated or covered with, imbedded in, or packaged in a plastic film, or sheath, or foam, or some similar containing arrangement. The meaning of the term includes processes in which a foam-forming resin is used to completely fill the spaces between an assembly of electrical and/or other components and their housings so these are completely imbedded in and supported by the foamed resin. The objective is to protect insulation, prevent movement and vibration of the several components and also to prevent resonance and other accoustical difficulties.
See also potting; microencapsulation.

endlichite. Mineral similar to vanadinite (q. v.) excepting that the vanadium is replaced by arsenic.

endo-. A prefix used in chemical names to indicate an inner position, specifically (a) in a ring rather than in a side chain, or (b) attached as a bridge within a ring.
See also exo-.

endophenolphthalein. See diacetyldihydroxyphenylisatin.

"Endor." [28] Trademark for a rubber peptizing agent containing activated zinc salt of pentachlorothiophenol $(C_6Cl_5S)_2Zn$, and 80% inert filler. Grayish green powder; sp. gr. 2.39.
Use: A peptizing agent for plasticizing natural rubber and all types of SBR (styrene butadiene rubber).
Containers: 100-lb drums.

endothermic. A process or change that takes place with absorption of heat.

"Endox." [142] Trade name for alkaline rust-removal and descaling products supplied in powder form. These products are added to water and the solutions are used both electrolytically and non-electrolytically for removal of rust and other oxides on iron and steel.

endoxan. See cyclophosphamide.

end point. In chemical analysis, the point during a titration at which a marked color change is observed, indicating that no more titrating solution is to be added.

endrin ($C_{12}H_{10}OCl_6$). The assigned common name for an insecticide 1, 2, 3, 4, 10, 10-hexachloro-6, 7-epoxy-1, 4, 4a, 5, 6, 7, 8, 8a-octahydro-1, 4-endo-endo-5, 8-dimethano-naphthalene. A stereoisomer of dieldrin which is the endo-exo isomer. See also aldrin.

Properties: White crystalline compound; m. p. approx. 200°C; insoluble in water, methanol. Moderately soluble in other common organic solvents. Compatible with non-acidic fertilizers, herbicides, fungicides and insecticides. Formulated as emulsifiable concentrates, wettable powders or dusts. Characteristics similar to dieldrin and aldrin. Not affected by alkalies; reacts with concentrated mineral acids, and is decomposed by acid catalysts.

MCA warning labels (formulations 2. 5% and over): Danger! Poisonous by skin contact, inhalation or swallowing; rapidly absorbed by skin. (Formulations less than 2. 5%): Warning! Hazardous by skin contact, inhalation or swallowing.

Shipping regulations: Liquid formulations containing 15% or more endrin and dry formulations of over 25%: Class B Poison. Poison label. *

"Enduro." [251] Trademark for a line of stainless and heat resisting steels. Available in the following groups and types:

Low Nickel Group (200 Series)

Types 201-202: Chromium 17, nickel 4, magnesium 6 and chromium 18, nickel 5, magnesium 8 respectively. Low nickel-content alternates for the 200 Series, having virtually the same properties as the 300 Series types.

Chromium-Nickel Group (300 Series)

Type 301: Chromium 16-18, nickel 6-8, carbon 0. 08-0. 20; used for applications requiring ductility and strength, such as railway cars, truck bodies, etc.

Type 302: Chromium 17-19, nickel 8-10, carbon > 0. 08-0. 20; especially suited to resist atmospheric corrosion and corrosive reagents; for dairy and chemical plant equipment; food and meat-processing machinery; high strength, light-weight structural members; and for resistance to oxidation at elevated temperatures.

Type 302B: Type 302 with 2-3 silicon; for resistance to oxidation in temperatures up to 1700°F; for annealing boxes, furnace parts, etc.

Type 303: A free-machining variation of Type 302 through addition of 0. 07 (min) selenium; machinability very good for chromium-nickel type—about 70% that of screw stock. Corrosion resistance same or little less than Type 302.

Type 304: Similar to Type 302 except carbon is kept to 0. 08 max. which permits its use in welded equipment subject to severe corrosion.

Type 305: A special modification of Type 302 to develop greater softness and less work-hardening; better adapted to successive drawing and spinning operations with less annealing than Type 302.

Type 309: Chromium 22-24, nickel 12-15; for resistance to oxidation up to 2000°F; fabricates, machines, and welds readily. High strength and creep at elevated temperatures. Not recommended for high sulfur conditions at high temperatures.

Type 309S: A variation of Type 309 with carbon 0. 08 (max.); for applications involving welding and corrosion resistance to eliminate carbide precipitation.

Type 310: Chromium 24-26, nickel 19-22, silicon 2 (max.); for maximum heat resistance. Best strength and creep at high temperatures, but may be attacked if sulfur is present in gases. Resistant to carburizing.

Type 316: A variation of Type 304 plus 2-3 molybdenum; resistant to acids encountered in paper and pulp processes, woolen dyeing and in chemical and pharmaceutical industries; recommended for severe corrosive conditions; good fabricating and welding properties.

Type 317: A modification of Type 316 with higher alloy content for applications requiring somewhat higher corrosion resistance than Type 316.

Type 321: A variation of Type 304 to which titanium has been added for eliminating intergranular corrosion at high temperatures; used for airplane collector rings, exhaust manifolds, and other high-temperature requirements.

Type 347: A variation of Type 304 plus columbium for applications similar to those for which Type 321 is recommended, but affording somewhat better corrosion resistance than titanium.

Straight-Chromium Group (400 Series)

Type 403: Chromium 11. 5-13, carbon 0. 15 (max.); used for applications where corrosion resistance and physical strength are needed at medium high temperatures.

Type 410: Chromium 11. 5-13. 5, carbon 0. 15 (max.); responds readily to heat treatment and is recommended where strength, toughness, and hardness are required; for pump shafts, valve seats, and stems, nuts and bolts, etc.

Type 414: A modification of Type 410 with addition of 1. 25-2. 5 nickel for somewhat better physical properties.

Type 416: Free-machining grade of Type 410 analysis. Machines nearly as well as Bessemer screw stock. Fairly resistant to the atmosphere, organic and fruit acids, etc. Can be hardened by heat treatment up to about 400 Brinell. Considerably more care and control required in forging operations than with Type 410.

Type 420: A straight chromium, high-carbon grade for heat treating for high hardness applications.

Type 420F: A high-carbon variation of

*See "I.C.C. Shipping Regulations," page xiii.

Reference numbers refer to name of manufacturer. See "List of Manufacturers," page v.

Type 416 having better mechanical properties.

Type 430: Chromium 14-18, carbon 0.12 (max); good corrosion resistance and heat resistance to 1500°F; general corrosion resistance, fabricating and welding properties inferior to Type 302; for bicycle fenders, oil-burner parts, etc.

Type 430F: A free machining modification of Type 430 with machinability about 85-90% of Bessemer screw stock.

Type 440: A variation of Type 430 with somewhat better mechanical properties.

Type 443: Chromium 18-23; high heat-resisting properties; good resistance to scaling, but strength and creep lower than chromium-nickel types; for furnace parts, etc.

Type 446: Chromium 23-27; heat resistance to 1900°F; not affected by sulfur gases; strength and creep at high temperatures not as good as the chromium-nickel steels.

Types 501 and 502: Chromium 4-6 with several carbon ranges to 0.25 and with or without addition of molybdenum or columbium, titanium, aluminum, and tungsten; additions of columbium, titanium, or aluminum practically eliminate air hardening on welding; corrosion and heat resistance considerably superior to that of carbon steels, and with fair strength at high temperature; for oil-refinery and furnace parts.

"Enduron." [3] Trademark for methyclothiazide (q. v.).

"Energex." [204] Trademark for a liquid detergent and emulsifying agent for dispersing moisture in dry cleaning solvents.

energy. Energy is defined as the capacity for doing work. Energy is work or anything that may be changed into work. Energy appears in various forms such as kinetic energy, potential energy, heat energy, electrical energy, chemical energy, and nuclear energy. The unit of energy is the erg which is a force of one dyne acting through one centimeter.

Kinetic energy is energy due to motion, and its numerical value in any case is one-half the mass of the body or particle that is moving multiplied by the square of the velocity, $E = 1/2\ mv^2$. Kinetic energy is always relative since the velocity of one body is always measured with respect to some point which itself may be moving.

Potential energy arises from the position of a body with respect to another body, and is equal to the work required to achieve this difference in position.

Heat energy is evidenced by the temperature of a body, and is due to molecular motion. Electrical energy arises from a difference in charge, and it along with chemical and nuclear energy are special forms of potential energy.

See also nuclear energy.

enfleurage. Method of extracting odoriferous components of flowers by means of fats or mixtures of fat and tallow, the process being carried out at room temperature to avoid decomposition of the desired perfumes. The latter are separated from the fat by washing with alcohol.

engine distillate. A petroleum distillate similar to naphtha but often of higher distillation range.

English bean. See tonka.

English laurel. See cherry laurel leaves.

English red. See iron oxide reds.

English vermilion. A precipitated mercury sulfide pigment, the color of which is a very light, bright vermilion shade. It is an opaque, non-bleeding red having a very low oil absorption and in paints and enamels it tends to darken on exposure to light. Because of its high cost, its use is quite restricted.

engraver's acid. See nitric acid.

"Engravoclor." [56] Brand name for ferric chloride solution specially prepared for photo-engraving process. Contains 43% ferric chloride. Sold only in 155-lb carboys.

"Enjay Butyl HT Series Rubber." [29] Brand name for isobutylene rubber containing chlorine. A new synthetic rubber where both chlorine and unsaturation are present in such a way that each may be utilized in vulcanization. The amount of each is small; however, they are sufficient to provide fast cure rates with conventional and new types of vulcanization systems. With its versatility of cure, it is found to be compatible with most other elastomers, such as butyl rubber, natural rubber, SBR, and neoprene. The outstanding property of this new polymer is its ability to resist heat up to 400°F. Other desirable properties are new compression set, good compression flexing, low permeability to gases, good tear strength, and good resistance to chemicals, oxidation and ozone.

"Enjay Butyl Latex." [29] Brand name for isobutylene type rubber in aqueous emulsion. The outstanding characteristics inherent in butyl are now available for application in the field of rubber latex. Superior gloss, brightness, softness and long lasting flexibility for paper coating, textile finishing and emulsion paints formulation are some of the attributes which butyl latex can bring to a wide variety of products. Butyl latex can be readily compounded with pigments, fillers, thickeners, and tackifiers, and is compatible with a wide range of resin and elastomer emulsions. This high-solids, odorless, water emulsion has freeze-thaw, mechanical and chemical stability.

"Enjay Butyl Rubber." [29] Brand name for isobutylene type rubber. A copolymer of isobutylene with a small proportion of isoprene to give a controlled, low degree of unsaturation. Processing is similar to that of natural rubber except that no premastication is required. Outstanding properties include: low permeability to gases; excellent resistance to ozone, weathering, heat, chemicals, tear and abrasion; and outstanding electrical properties. Used in inner tubes, curing bags and bladders, wire and cable insulation, molded goods for automobiles, industrial tank lining, hose and belting, adhesives, caulking and sealing compounds. Other desirable properties are low compression set, good compression flexing, and excellent color retention.

"Ennjay." [51] Trademark of several special grades of asphalt.

"Enovid." [70] Trademark for brand of norethynodrel with ethynylestradiol 3-methyl ether; 17 alpha-ethynyl-17-hydroxy-5(10)-estren-3-one with ethynyl-estradiol 3-methyl ether.
Use: Medicine.

"Enstrip." [142] Brand name of various solutions designed for stripping selectively one metal from another. All "Enstrips" with exception of "Enstrip L-88" are supplied as powders that are added to water for use.
"Enstrip S": For dissolving nickel, copper, zinc, cadmium and silver from steel.
"Enstrip 165S": For dissolving nickel from copper base alloys.
"Enstrip TL": For selective dissolving of tin and lead from steel and copper base alloys.
"Enstrip 103": For electrolytic dissolving of nickel, copper and other metals from steel.
"Enstrip L-88": A liquid for dissolving chromium, nickel and copper from the zinc base die castings.

"Entek." [142] Trade name for various organic compounds which are added in small amounts to rinse water to promote stain-free drying and produce a protective film which prevents subsequent tarnishing and corrosion of metals. Used after plating, pickling, cleaning or other wet processing of metals.

"Enth-Acid." [142] Trade name for a blend of acid salts, activators and surfactants which can be used as a replacement for liquid acids. Used for acid dipping of iron, steel, brass, copper, or zinc die castings prior to plating.

"Enthol." [142] Trademark for phosphoric acid-solvent mixtures designed for degreasing and oxide removal for such metals as steel, aluminum, and zinc. Materials are supplied in the liquid form.

"Enthonite." [142] Trademark for a polyvinyl plastisol coating used for coating metals; primarily for imparting resistance to acid, alkaline, and electroplating solutions. Material contains 100% solids and is applied by dipping.

"Enthox." [142] Trademark for salts that are added to water for producing chromate coatings on zinc and cadmium to withstand 100 or more hours in 20% salt spray.

"Entodon." [162] Trademark for propiodal.

entrainer. An additive for liquid mixtures that are difficult to separate by ordinary distillation. The entrainer usually forms an azeotrope with one of the components of the mixture and thereby aids in the separation of such a compound from the remainder of the mixture.

entrainment. The mist or fog of droplets of liquid carried off by the vapor of a boiling liquid, or more frequently from a liquid through which bubbles of gas or vapor are passing rapidly.

Ent. Soc. Abbreviation for Entomological Society of America.

enzymes. Definite chemical substances, protein in nature, which are formed in the living cells of plants and animals and which are necessary catalysts for the chemical reactions of biological processes. Enzymes are usually very specific in their catalytic behavior in that a given enzyme is effective for only one particular reaction. Enzymes are often classified by the kind of substance (substrate) that is consumed in the reactions that are catalyzed. For example, the term protease refers to enzymes that catalyze reactions that convert proteins to other substances. Carbohydrase (for carbohydrates) and lipase (for fats or glycerides) are corresponding terms. The manner of activity is also a basis for classification of all enzymes into four categories: (1) hydrolase, enzymes which catalyze the removal of water; (2) oxidase or dehydrogenase, which catalyze the transfer of electrons; (3) transferase, which cause a transference of a radical from one molecule to another; and (4) desmolase, which catalyze a split or form a C-C bond without group transfer. (It will be noted that enzymes named recently have the suffix -ase.) Corollary to the presence of enzymes are other substances, which have been termed coenzymes and antienzymes (q. v.).
Properties: Enzymes generally form colloidal suspensions in water and are insoluble in fat solvents. The range of molecular weights has been determined as between 20,000 and 483,000. The optimum temperature for enzyme activity occurs between 35° and 55°C. Inactivity occurs at 80°C, and enzymes are destroyed at 100°C. However, enzyme activity has been shown to be retained in seeds for over 100 years. The reaction velocity is governed by the quantity of enzyme present, the temperature, the pH of the media, the presence of certain metal ions and the accumulation of reaction products.
Uses: Enzymes cause hydrolysis (decompose fats, convert aldehydes into a mixture of acid and alcohol, invert sugars, digest

proteins); coagulate (clot blood, curdle milk); oxidize (convert alcohol into acetic acid). Most reactions activated by enzymes are reversible. Some common enzymes are: amylase, lipase, maltase, papain, ptyalin, pepsin, trypsin, urease, invertase, rennin, oxidase. For information on these see the specific enzymes.

eosin (eosine; bromeosin; tetrabromofluorescein) $C_{20}H_8Br_4O_5$.
Properties: Red, crystalline powder; soluble in alcohol and acetic acid; insoluble in water; the potassium and sodium salts are soluble in water.
Derivation: By the bromination of fluorescein.
Grades: Technical; pure.
Containers: Barrels; boxes; fiber drums.
Uses: Dyeing silk, cotton and wool; making red writing ink; coloring motor fuel.
Shipping regulations: None. *

eosin, soluble. See eosin, yellowish.

eosin, yellowish (eosin, soluble; eosin YS). Sodium or potassium salt of eosin; C. I. No. 768. $C_{20}H_6O_5Br_4Na_2$ or $C_{20}H_6O_5Br_4K_2$.
Properties: Brown to red crystals, yellow-red fluorescence in dilute aqueous solution, yellow-green fluorescence in dilute alcoholic solutions; soluble in water and alcohol.
Derivation: Prepared from brominated fluorescein.
Uses: Coloring ink, fabric, straw, paper; as a stain in microscopy; in preparing pink lakes; and in medicine.

eosin YS. See eosin, yellowish.

eosine. See eosin.

EP additives. See extreme pressure additives.

EPC black. Abbreviation for easy processing channel black. See channel black.

ephedrine (1-phenyl-2-methylaminopropanol; alpha-hydroxy-beta-methylaminopropylbenzene) $C_6H_5CH(OH)CH(NHCH_3)CH_3$. Optically active (levorotatory) form. See racephedrine for the inactive mixture of isomers.
Properties: White to colorless granules, pieces or crystals; unctuous to touch; hygroscopic; keep cool and protect from moisture; gradually decomposes on exposure to light. Soluble in water, alcohol, ether, chloroform, and oils. M. p. 33-40°C; b. p. 255°C.
Derivation: Isolation from stems or leaves of Ephedra, especially Ma huang (China and India).
Method of purification: Recrystallization.
Grades: Technical; N. F. XI.
Containers: Glass bottles.
Use: Medicine.
Shipping regulations: None. *

dℓ-ephedrine. See racephedrine.

ephedrine hydrochloride (1-phenyl-2-methylaminopropanol hydrochloride) $C_{10}H_{15}NO \cdot HCl$. Optically active (levorotatory) form. See racephedrine hydrochloride for the inactive mixture of isomers.
Properties: White, odorless powder or crystals. Soluble in water and alcohol; insoluble in ether. Affected by light. M. p. 217-220°C; specific rotation (1 in 20) between −33° and −33.5°.
Grade: N. F. XI.
Containers: Drums.
Use: Medicine (same as ephedrine).
Shipping regulations: None. *

ephedrine sulfate (1-phenyl-2-methylaminopropanol sulfate) $(C_{10}H_{15}NO)_2 \cdot H_2SO_4$. Optically active (levorotatory) form. See racephredine sulfate for the inactive mixture of isomers.
Properties: White, odorless powder or crystals. Soluble in water and alcohol. Affected by light. M. p. 245°C (with decomposition); specific rotation (1 in 20) −30° to −32°.
Derivation: By action of sulfuric acid on ephedrine.
Grades: U. S. P. XVI.
Containers: Glass bottles; drums.
Use: Medicine.
Shipping regulations: None. *

ephenamine (ℓ-N-methyl-1, 2-diphenyl-2-hydroxyethylamine hydrochloride) $C_6H_5CHOHCH(C_6H_5)NHCH_3 \cdot HCl$.
Use: To form a crystalline salt of penicillin G.

ephetonin. See racephedrine hydrochloride.

"Ephynal" Acetate. [190] Trademark for a brand of dℓ-alpha-tocopheryl acetate (q. v.).

epi.
1. A prefix denoting a bridge or intramolecular connection.
2. An abbreviation for epichlorohydrin.

"Epic." [51] Trademark for light yellow, liquid greases for textile mills and packaging machinery requiring a fluid lubricant that must not spatter and must "stay put." "Epic" lubricants have better oiliness and film strength than straight mineral oils.

epicarin (epicarine; 3-(2-hydroxy-1-naphthylmethyl)salicylic acid) $C_6H_3(OH)(COOH)(CH_2C_{10}H_6OH)$.
Properties: White to reddish powder; less toxic than naphthol. Soluble in alcohol, ether, and soaps; difficultly soluble in water. M. p. 199°C (with decomposition).
Preparation: From alpha-naphthol and orthocreosotic acid.
Use: Medicine.
Shipping regulations: None.*

epichlorohydrin (chloropropylene oxide; epi) CH_2OCHCH_2Cl.
Properties: Highly volatile, unstable, narcotic liquid. Chloroform-like odor; miscible with most organic solvents; immiscible with water, petroleum hydrocarbons.
Constants: Sp. gr. 1.1761 (20/20°C); b. p. 115.2°C; wt/gal 9.78 lbs; coefficient of expansion 0.00102 (20°C); vapor pressure

*See "I. C. C. Shipping Regulations," page xiii.
Reference numbers refer to name of manufacturer. See "List of Manufacturers," page v.

12.5 mm (20°C); f.p. −58.1°C; viscosity 1.12 cps (20°C); refractive index (n 25/D) 1.4358.
Grades: Technical.
Containers: Drums; tank cars.
Uses: Solvent for cellulose esters and ethers, gums, and resins; lacquers, paints, varnishes, coatings.
Shipping regulations: Poison, Class B. Poison label. *

epichlorohydrin resins. See epoxy resins.

epidosite. A rock consisting of epidote and quartz and sometimes containing gold. Banded epidosites are sometimes used for ornamental stones.

epidote $Ca_2(Al, Fe)_3(SiO_4)_3(OH)$. The principal member of a group of minerals for which the general formula is $R^{II}_2 R^{III}_3 (SiO_4)_3(OH)$ with R^{II} = Ca, Mn, Ce, and R^{III} = Al, Fe, Mn.
Properties: Color pistachio-green, yellow-green, greenish black to black; one good cleavage; sp. gr. 3.4; hardness 6-7.
Occurrence: Connecticut, California, New York, Colorado; Europe; Alaska.
Use: Sometimes as a gemstone.
See also epidosite, zoisite, allanite.

epinephrine (ℓ-methylaminoethanolcatechol; 3,4-dihydroxy-alpha-(methylaminomethyl-benzyl alcohol) $C_6H_3(OH)_2CHOHCH_2NHCH_3$. A hormone of the adrenal glands.
Properties: Light brown or nearly white, odorless crystalline powder; affected by light; m.p. 211-212°C; specific rotation (25°C) −50° to −53.5°; sparingly soluble in water; insoluble in alcohol, chloroform, ether, acetone, oils. Readily soluble in aqueous solutions of mineral acids, sodium hydroxide and potassium hydroxide.
Derivation: From the adrenal glands of sheep and cattle or synthetically from pyrocatechol.
Grades: U.S.P. XVI.
Containers: Glass bottles.
Uses: Medicine; shaving preparations.

epinephrine bitartrate $C_9H_{13}NO_3 \cdot C_4H_6O_6$.
Properties: White or gray, odorless, crystalline powder. Affected by light and air. Solutions are acid (pH about 3.5) to litmus. M.p. 147-152°C. Soluble in water; slightly soluble in alcohol; almost insoluble in chloroform and ether.
Grade: U.S.P. XVI.
Use: Medicine.

"Epiphen." [65] Trademark for an epoxy resin in liquid form. "Epiphen" ER-823 is used in adhesives for rubber, steel, aluminum or glass. Catalyst is supplied for specific end uses.

EPN (O-ethyl-O-para-nitrophenyl benzene-thionophosphonate) $C_6H_5P(C_2H_5O)(S)OC_6H_4NO_2$. The abbreviated name is commonly used and has been accepted as a generic name by the Ent. Soc.
Properties: Pure compound: light yellow crystals; m.p. 36°C; sp. gr. 1.5978 (30°C). Slightly soluble in water; freely soluble in most organic solvents. Decomposes in alkaline solutions.
Grades: Wettable powders and dusts.
Uses: As a pesticide.
Caution! Toxic! Is a cholinesterase inhibitor.

"EPN 300." [28] Insecticide containing 25% ethyl para-nitro-phenylthionobenzene phosphonate.
Uses: As a contact insecticide in agriculture to control certain mites, scales, and other insects. Also for use by qualified agencies for control of certain mosquito larvae.

"Epolene." [256] Trademark for a series of low-molecular-weight polyethylene resins. Available in both emulsifiable and non-emulsifiable types.
Properties: Rice-sized pellets; Ring and Ball softening point from 100°C to 114°C; Brookfield viscosity 340 cps to 16,000 cps (120°C). Color, Gardner scale, max. 2.
Containers: Paper bags (50 lbs net).
Uses: Floor polishes; textile finishes; rubber processing; paper coatings; inks; and as a modifying resin for use with plastic-grade polyethylene.

"Epon Curing Agent." [125] Trademark for curing agents C-111, D, U, Z, T, T-1, H_1, and H_2. By the addition of these curing agents, EPON resins can be hardened to form clear tough polymers with high physical strength, excellent chemical resistance, and good electrical properties.

"Eponite" 100. [125] Trademark for a water-dispersible liquid epoxy resin.
Properties: Colorless to pale yellow liquid; viscosity 90-150 cps (25°C); pH neutral; weight per gallon 10.2 lbs; slight characteristic odor.
Containers: 55-gal 18 gauge steel full removable head drum equipped with 2-inch, 3/4 inch Visegrip plated plugs in top head.
Uses: To impart durable crease resistance, shrinkage control embossing, glazing and minimum care effects on cotton, rayon and blended fabrics, especially where resistance to damage by chlorine is a requirement. Also used in combination with hydrolyzed or partially hydrolyzed polyvinyl acetate, starches, gums, cellulose ethers, selected resins or other chemical finishing agents to impart hand, stiffness, softness, dye-fixing or other fabric properties.
Caution! May polymerize rapidly, even violently, in presence of strong acids, strong bases and certain metallic salts. Store in closed, clean containers in cool place. Prevent all contact with skin or eyes.

"Epon" Resins. [125] Trademark for condensation products of epichlorohydrin and bisphenol-A having excellent adhesion, strength, chemical resistance and electrical properties when formulated into protective coatings, adhesives and structural plastics.
Grades and Uses: The solid "Epon" resins (1001, 1002, 1004, 1007 and 1009) are used primarily in solvent-applied coatings; "Epon" 1001 and 1002 are used in amine-cured

*See "I.C.C. Shipping Regulations," page xiii.
Reference numbers refer to name of manufacturer. See "List of Manufacturers," page v.

maintenance paints; "Epon" 1004 (after esterification with fatty acids) in air dried varnishes and enamels or blended with urea or melamine resins in baked films, and "Epon" 1007 and 1009 (unesterified) in chemically resistant baked finishes in combination with urea and phenolic resins. "Epon" resin films combine excellent hardness, durability and flexibility.

The lower molecular weight resins ("Epon" 815, 820, 826, 828, 830 and 836 and 1001) have been widely used in adhesives, castings, laminates, potting compounds and high solids or solvent free coatings. They have extremely low shrinkage which makes them uniquely suitable to structural applications.

"Eporal." [418] (Sulfonyldianiline).
Properties: Practically odorless, free-flowing yellowish powder; soluble in acetone, methanol, methyl ethyl ketone; insoluble in water.
Grade: Technical.
Containers: 10-lb tins; 100-lb drums.
Uses: As epoxy hardening agent.
Caution! Handle with care.

"Epoxol 7-4." [152] Trade name for a high purity epoxidized soybean oil.
Properties: Oxiarine oxygen, 7.0% min; iodine value, 2 max; Gardner color, 1 max; Gardner viscosity (25/25°C) 3.4-3.7 poises; acid no. o.2 mas; sp.gr. (25/25°C) 0.990-0.995.
Uses: As a stabilizing plasticizer for vinyl and other resins, or for polymerization alone or copolymerization with various other resins.
Containers: 5-, 55-gal drums; tank trucks or tank cars.

epoxy-. A prefix in organic nomenclature denoting an oxygen atom joined to each of two atoms which are already united in some other way, as -C-O-C-.

3,4-epoxycyclohexane carbonitrile $O(C_6H_9)CN$.
Properties: Liquid; sp. gr. 1.0929 (20/20°C); b.p. 244.5°C; f.p. −33°C; soluble in water.
Uses: Intermediate; stabilizer.

2,3-epoxy-2-ethylhexanol
$C_3H_7CHOC(C_2H_5)CH_2OH$.
Properties: Liquid; sp. gr. 0.9517 (20/20°C); b.p., decomposes; f.p. −65°C; slightly soluble in water.
Uses: Stabilizer; intermediate.

2,3-epoxy-1-propanol. See glycidol.

epoxy resins. The commercially available materials of this class are usually derived from bisphenol A and epichlorohydrin. Before curing they are viscous liquids or clear, brittle solids melting up to 155°C with molecular weights from 400 to 8000. Chemical nature is indicated by the formula, $H_2COCHCH_2O(RCHOHCH_2O)_nROCH_2CHOCH_2$ where R is $C_6H_4C(CH_3)_2C_6H_4$. Variations from this basic composition are produced by the ratio of the reactants, the conditions of reaction, and by the catalyst or agent and conditions used for final curing. Curing involves further reaction of the epoxy and hydroxy groups to cause chain growth and crosslinking. This is often accomplished by reaction with boron trifluoride amine catalysts or with polyamines, dibasic acid anhydrides or with phenolic or urea resins. Also, the epoxy and hydroxy groups may be esterified with unsaturated or saturated monobasic acids and curing then occurs as with alkyd resins. Heat may or may not be required for curing, depending on the various factors of composition, resin formation conditions, and curing agents used.

The epoxy resins are used for surface coatings, as adhesives and for laminating to produce plastic tanks, pipe, and aircraft parts, for casting plastic metal-forming tools and dies and for potting and encapsulation of electrical parts. There are also uses as plastic putty, solders and trowelling mixes. Epoxies show superior adhesion to metals and glass, and have limited shrinkage during cure.

Bisphenol A may be replaced by other diphenols, glycols or glycerine, but the resulting resins are of limited utility. The term epoxy plastic is also sometimes used for other types of high molecular weight compounds that contain the epoxy group, or are derived from such compounds. Thus "Carbowax" is sometimes designated as an epoxy.

EPR. Abbreviation for ethylene propylene rubber.

epsilon acid. See 1-naphthylamine-3,8-disulfonic acid.

"Epskol." [64] Trademark for a modified linseed oil used as a replacement for China wood oil in varnishes and enamels.

epsomite $MgSO_4 \cdot 7H_2O$. A natural hydrated magnesium sulfate found in salt deposits with gypsum, halite, etc. Also produced commercially by recrystallization. The name is sometimes incorrectly used to refer to the dissolved material in brines. See also kieserite and magnesium sulfate.
Properties: Colorless to white; luster vitreous to earthy; taste very bitter; sp. gr. 1.68; hardness 2-2.5.
Occurrence: Germany; Michigan, Wyoming, Utah, Washington.
Use: Raw material for commercial Epsom salts; used in the textile industry; tanning; fertilizer; paints and soaps; and medicine.

Epsom salts. See magnesium sulfate.

EPTC (ethyl N,N-di-n-propylthiolcarbamate) $C_2H_5COSN(C_3H_7)_2$. Used as a pre-emergence herbicide. Available in liquid and granular formulations.

equivalent weight. The equivalent weight of an element is the weight that will combine with or react with or can replace one atomic weight of hydrogen or one-half atomic weight of oxygen. The equivalent weight of an acid is the weight that contains one atomic weight of acidic hydrogen, i. e., the hydrogen that reacts during neutralization of acid with base. The equivalent weight of a base or hydroxide is the weight that will react with an equivalent weight of acid. Equivalent weights of other substances are defined in a similar manner, all these definitions being variants of the first sentence of this statement.

Er. Symbol for erbium.

"Erasol." [232] Brand name for a series of sulfoxylate reducing agents for use in the textile and paper industries.

"Eratrope." [232] Brand name for a series of discharge agents for removing vat dyes from cellulosic material.

"Eraydo." [135] A zinc-base alloy designed for use as a non-magnetic metal of higher annealing temperature and greater strength than commercial rolled zinc. Available in two types:

Type I. Pure zinc base with 0.7% to 1.25% copper. It is resistant to ordinary atmospheric conditions but is not stainless as it darkens in color upon exposure. It will not resist sea water and acid conditions. Used in operations requiring forming and drawing.

Type II. Pure zinc base with 0.7% to 1.25% copper and 0.01% magnesium. It is resistant to ordinary atmospheric conditions but is not stainless as it darkens in color upon exposure. It will not resist sea water and acid conditions. Used where greater hardness and rigidity are desired in flat plates requiring little, if any, bending, forming, or drawing.

Forms: Sheet and ribbon.

erbia. See erbium oxide.

erbium Er. Element with atomic number 68; one of the rare-earth elements of the yttrium subgroup.

Properties: Solid with metallic luster; insoluble in water; soluble in acids; salts are pink to red; sp. gr. 9.16 (15°C); m. p. 1400-1500°C; b. p. 2600°C (approx).

Derivation: Reduction of the fluoride with calcium.

Source: See rare earth minerals.

Grade: Lumps; ingots of high purity.

Uses: Nuclear control; special alloys.

erbium nitrate $Er(NO_3)_3 \cdot 5H_2O$.

Properties: Large reddish crystals; soluble in water, alcohol, ether, and acetone.

Derivation: Treatment of oxides, carbonates or hydroxides with nitric acid.

Grade: 99.9%.

Containers: Bottles; fiber drums.

Shipping regulations: Oxidizing material. Yellow label. *

erbium oxalate $Er_2(C_2O_4)_3 \cdot 10H_2O$.

Properties: Reddish microcrystalline powder; decomposes at 575°C. Very insoluble in water.

Use: Oxalates of the rare-earth metals are used to separate the latter from common metals.

erbium oxide (erbia) Er_2O_3.

Properties: Pink powder which readily absorbs moisture and carbon dioxide from the atmosphere. Sp. gr. 8.64; specific heat 0.065; infusible; insoluble in water; slightly soluble in mineral acids.

Derivation: By heating the oxalate or other oxy-acid salts.

Grades: 98-99%.

Containers: Bottles; fiber drums.

Uses: Phosphor activator; infrared-absorbing glass.

See also rare earths.

erbium salts. Besides those listed here, erbium chloride, $ErCl_3 \cdot xH_2O$, and erbium fluoride, $ErF_3 \cdot 2H_2O$, are also available commercially.

erbium sulfate $Er_2(SO_4)_3 \cdot 8H_2O$.

Properties: Pink monoclinic crystals; soluble in water; sp. gr. (given variously) about 3; dehydrated at 400°C.

Derivation: Dissolving hydroxides, carbonates, or oxides in dilute sulfuric acid.

Grades: 99.9%.

Containers: Bottles; fiber drums.

Use: To determine atomic weight of the rare-earth element.

Erdmann's reagent. A reagent used in testing for alkaloids. Made by the addition of concentrated sulfuric acid to a small quanitity of dilute nitric acid.

ergocalciferol (vitamin D_2). See vitamin D.

ergonovine $C_{19}H_{23}N_3O_2$. An alkaloid of importance in obstetrics.

Properties: Crystals with m. p. 162°C; freely soluble in lower alcohols, ethyl acetate, and acetone; slightly soluble in chloroform, more soluble in water than the other principal ergot alkaloids.

Derivation: From certain ergots; synthesized from lysergic acid and 2-amino-1-propanol.

Use: Medicine (as maleate salt).

ergonovine maleate $C_{19}H_{23}N_3O_2 \cdot C_4H_4O_4$.

Properties: White or faintly yellow microcrystalline powder; odorless; affected by light; soluble in water and alcohol; insoluble in ether and chloroform.

Grade: U. S. P. XVI.

Use: Medicine.

ergosterol (provitamin D_2) $C_{28}H_{44}O$. A plant sterol widely distributed in nature.

Properties: Colorless crystals; m. p. 166°C (with 1 ½ H_2O); b. p. 250°C (0.01 mm); sp. gr. 1.04; specific rotation −135° (in chloroform); insoluble in water; soluble in alcohol, benzene, ether. Affected by light and air and turns yellow.

Derivation: Synthesized by yeast from simple sugars; obtained from fungus ergot.

Use: Medicine; when irradiated with ultraviolet light, it has vitamin D activity.

ergot (secale cornutum; rye ergot). A fungus growth, Claviceps purpurea, on rye.

*See "I. C. C. Shipping Regulations," page xiii.

Reference numbers refer to name of manufacturer. See "List of Manufacturers," page v.

Habitat: Europe; cultivated in Spain and Russia.
Grades: Spanish; Russian; N. F. XI.
Containers: Bags of variable size; tin-lined drums.
Uses: Medicine; source of many important alkaloids.
Shipping regulations: None. *

ergotamine $C_{33}H_{35}N_5O_5$. An alkaloid obtained from certain ergots found in central Europe.
Properties: Hygroscopic crystals; affected by air, light, and heat; m. p. 212-214°C (dec). Freely soluble in chloroform; slightly soluble in benzene, acetone, and alcohol; almost insoluble in water and petroleum ether.
Use: Medicine (also used as tartrate salt).

ergotamine tartrate $(C_{33}H_{35}N_5O_5)_2 \cdot C_4H_6O_6$.
Properties: Colorless crystals or white crystalline powder; m. p. 177-184°C with decomposition; slightly soluble in water and alcohol.
Grades: U. S. P. XVI.
Use: Medicine.

ergotinine $C_{35}H_{39}O_5N_5$.
Properties: White crystalline alkaloid; poisonous! Soluble in alcohol and ether; insoluble in water. M. p. 205°C.
Derivation: By extraction of sclerotium of ergot Claviceps purpurea and crystallization.
Method of purification: Recrystallization.
Grades: Technical.
Containers: Glass bottles. (Protect from light.)
Use: Medicine.
Shipping regulations: None. *

"Ergotrate" Maleate. [100] Trademark for ergonovine maleate, U. S. P.

erigeron oil (fleabane oil; horseweed oil; butterweed oil).
Properties: Pale, yellow, limpid liquid, darkening and thickening with age and exposure; peculiar aromatic, persistent odor; aromatic, slightly pungent taste. Soluble in an equal vol. of 90% alcohol, but sometimes remains turbid after additions of several volumes of the solvent; also soluble in ether, chloroform, and carbon disulfide.
Chief known constituents: d-Limonene; terpineol; esters.
Constants: Sp. gr. 0.8565-0.868 (15°C); b. p. 175-180°C; optical rotation +52 to +83°; acid value 0; ester value 39 to 108, after acetylation 67 to 108.
Derivation: Distilled from the fresh, flowering herb of Erigeron canadensis.
Method of purification: Rectification.
Grades: Technical.
Containers: Glass bottles; cans.
Shipping regulations: None. *

eriodictyon (yerba santa).
Properties: Brownish fragments of leaves; slight, aromatic odor.
Derivation: Dried leaves of Eriodictyon californicum.
Habitat: California.
Grades: Technical; N. F. XI.
Containers: Boxes; fiber drums; bales.
Use: Medicine.

Erlanger blue. A name applied loosely to any of a number of the varieties of iron blue pigments. See iron blues.

"Ertrane." [309] Trademark for a surface active agent as additive for Portland and masonry cements to impart plasticity and air entraining properties thereto.

erucamide (erucyl amide) $C_{21}H_{41}CONH_2$.
Properties: Solid; sp. gr. 0.888; m. p. 75-80°C; iodine value 80-85; soluble in isopropanol; slightly soluble in alcohol and acetone.
Purity: 90% amide and 80% C_{22}.
Uses: Foam stabilizer; solvent for waxes and resins; emulsions.

erucic acid (cis-13-docosenoic acid) $C_8H_{19}CH{:}CH(CH_2)_{11}COOH$. A C_{22} fatty acid with one double bond. A homolog of oleic acid with four more carbons.
Properties: M. p. 33-34°C; b. p. 264°C (15 mm); iodine value 75.
Derivation: Fats and oils of mustard and rape seed.
Containers: 400-lb drums.
Uses: Preparation of dibasic acids and other chemicals.

erucyl alcohol $C_{22}H_{43}OH$. A C_{22} fatty alcohol having one double bond. It is a white, soft solid, almost odorless. Sp. gr. 0.8486; cloud point 81°F; boiling range 334-376°C; iodine value 82; flash point 395°F; soluble in alcohol and most organic solvents.
Impurities: Oleyl and linoleyl alcohols.
Derivation: Sodium reduction of erucic acid.
Uses: Lubricants; surfactants; petrochemicals; plastics; textiles; rubber.

erucyl amide. See erucamide.

"Erusticator." [204] Trademark for a rust remover which dissolves rust rapidly from fabrics.

"Erusto." [204] Trademark for a series of laundry and dry-cleaning products.

"Erusto-Cetic." [204] Trademark for a sour for wet cleaning. Fabric-safe; sets colors.

"Erustocide." [204] Trademark for a sour recommended especially for colored and white work.

"Erusto Filter Soap." [204] Trademark for free-flowing liquid dry-cleaning soap for use in petroleum solvent systems.

"Erustolin" B. [204] Trademark for a rust-removing sour where solubility is not of primary importance.

"Erusto Liqui-Blue No. 16." [204] Trademark for an all-purpose blue for use with or without sour.

"Erusto Oil, Paint & Grease Remover." [204] Trademark for an all-purpose spotter. Completely soluble in water or solvent.

"Erusto Salts Special." [204] Trademark for a quickly soluble, high-rust-removing sour.

"Erustosol." [204] Trademark for a liquid sour high in rust-removing qualities.

"Ervol." [45] Trademark for white mineral oil, N. F.
Properties: Sp. gr. 0.860-0.870 (60°F); Saybolt viscosity 125-135 (100°F); odorless and tasteless.
Uses: Pharmaceutical and cosmetic formulations; plasticizers; paper penetrants; foam depressants.

erythorbic acid (formerly called isoascorbic acid) $C_6H_8O_6$.
Properties: Shiny, granular crystals; decomp. 174°C. Soluble in water, alcohol, pyridine; moderately soluble in acetone; slightly soluble in glycerol.
Uses: Antioxidant (industrial and food) especially in brewing industry; reducing agent in photography, etc.

erythrene. See butadiene.

erythrite.
1. Synonym for erythritol.
2. (cobalt bloom) $Co_3(AsO_4)_2 \cdot 8H_2O$.
A natural hydrated cobalt arsenate found in the oxidized parts of cobalt and arsenic-bearing veins. Crimson, peach, red, pink or pearl gray in color, with adamantine or pearly luster. Contains 37.5% cobalt oxide. Soluble in hydrochloric acid. Sp. gr. 2.91-2.95; hardness 1.5-2.5.
Occurrence: United States (California, Colorado, Idaho, Nevada); Ontario.
Uses: Coloring glass and ceramics.

erythritol (tetrahydroxybutane; erythrite) $CH_2OHCHOHCHOHCH_2OH$. A tetrahydric alcohol found in Protocaccus vulgaris and other lichens of Rocella species.
Properties: White, sweet crystals; m.p. 121-122°C; b.p. 329-331°C; sp. gr. 1.45; soluble in water; slightly soluble in alcohol; insoluble in ether.

erythrityl tetranitrate (erythrol tetranitrate; tetranitrol; nitroerythrite) $CH_2ONO_2(CHONO_2)_2CH_2ONO_2$.
Properties: Crystals; m.p. 61°C. Explodes on percussion or heating! Soluble in alcohol, ether, and glycerol; insoluble in water.
Derivation: By nitration of erythritol.
Use: Medicine (diluted with lactose in non-explosive tablets).

"Erythrocin." [3] Trademark for erythromycin.

erythrol tetranitrate. See erythrityl tetranitrate.

erythromycin. $C_{37}H_{67}NO_{13}$. An antibiotic produced by growth of Streptomyces erythreus Waksman. It is effective against infections caused by gram-positive bacteria, including some beta-hemolytic streptococci, pneumococci, and staphylococci.
Properties: White or slightly yellow, odorless, bitter crystalline powder; m.p. 133-138°C. Freely soluble in alcohol, chloroform, and ether; very slightly soluble in water. Slightly hygroscopic. pH (saturated solution) 8-10.5. pH less than 4 is destructive. Alcoholic solution is levorotatory.
Grade: U.S.P. XVI.
Use: Medicine.

erythromycin ethyl carbonate $C_{40}H_{71}NO_{15}$. The ethyl carbonate ester of erythromycin.
Properties: White, crystalline powder, practically odorless and tasteless. Freely soluble in alcohol; slightly soluble in ether and water.
Grade: U.S.P. XVI.
Use: Medicine.

erythromycin glucoheptonate $C_{37}H_{67}NO_{13} \cdot C_7H_{14}O_8$. The glucoheptonate salt of erythromycin.
Properties: White crystalline, odorless powder. Freely soluble in water and alcohol; practically insoluble in ether. pH (2% solution) 6.0-7.5.
Grade: U.S.P. XVI.
Medicine.

erythromycin lactobionate $C_{37}H_{67}NO_{13} \cdot C_{12}H_{22}O_{12}$. The lactobionate salt of erythromycin.
Properties: White, practically odorless powder. Freely soluble in water and alcohol.
Grade: U.S.P. XVI.
Use: Medicine.

erythromycin propionate. Listed in the N.N.D.; used as medicine.

erythromycin propionate lauryl sulfate (propionyl erythromycin ester lauryl sulfate; "Ilosone").
Properties: A white substantially tasteless and odorless crystalline powder. It is almost insoluble in water and dilute acids, but is appreciably soluble in dilute alkalies. It melts with decomposition at 135-140°C.
Use: Medicinal.

erythromycin stearate $C_{37}H_{67}NO_{13} \cdot C_{18}H_{36}O_2$. Stearic acid salt of erythromycin, with an excess of stearic acid.
Properties: White or slightly yellow crystals or powder. Practically odorless or slight musty odor. Slight bitter taste. Saturated solution alkaline to litmus. Practically insoluble in water; soluble in alcohol, methanol, ether, chloroform.
Grade: U.S.P. XVI.
Use: Medicine.

erythrosine $C_{20}H_6I_4Na_2O_5$. Sodium (or potassium) salt of iodeosin.
Properties: Brown powder; forms cherry red solution in water; soluble in alcohol.
Use: Coloring.

erythroxylon. See coca.

Es. Symbol for einsteinium.

"Escalol 106." [10] A proprietary name for glyceryl para-aminobenzoate, a patented sunscreening compound.

"Escofos." [428] Trademark for complex phosphates and silicated alkali for soap regenerator and lime soap stripper action.

"Escon." [29] Trademark for a proprietary polypropylene resin thermoplastic.

esculin (aesculin; esculinic acid) $C_{15}H_{16}O_9 \cdot \frac{1}{2} H_2O$ (or $1\frac{1}{2} H_2O$). Glucoside derived from inner bark of Esculus hippocastanum L. (horse-chestnut).
Properties: White crystals, bitter taste; solutions show faint blue fluorescence. M. p. 160°C (dec); m. p. 205°C (anhydrous); b. p. 230°C (dec). Slightly soluble in cold water and alcohol; soluble in hot water and hot alcohol; very slightly soluble in ether; soluble in acetic acid, dilute alkalies, and hot chloroform.
Use: Sun tan lotions.

esculinic acid. See esculin.

eserine. See physostigmine.

"Esidrix." [305] Trademark for hydrochlorothiazide N. N. D.
Use: Medicine.

"Es-min-el." [93] Trade name for a granular mixture of water soluble salts of copper, manganese, zinc, iron, magnesium and boron.
Containers: 100-lb bags; 5-lb bags; bulk.
Uses: Trace element mineral mixture to alleviate plant nutritional deficiencies and maintain correct level. Applied directly to soil or mixed with NKP fertilizers.

"Es-min-el" Spray or Dust. [93] Trade name for a physical mixture of water-insoluble salts of copper, manganese, and zinc; a gray powder, similar to soil "Es-min-el," used for application to foliage as spray or dust.
Containers: 4-, 50-lb bags.

esparto (Spanish grass). A grass with a tough fiber, cultivated in Spain and North Africa, and used chiefly for cordage and especially in England for papermaking.

esparto wax. Hard vegetable wax extracted from esparto grass, having a m. p. of 75°C and the ability to blend well, emulsify easily, and impart smoothness to polishes and shoe-finishing preparations.

essential oils. Volatile oils derived from plants, and usually carrying the essential odor or flavor of the plant used. Chemically, essential oils are often principally terpenes (hydrocarbons), but many other classes of compounds are also found. They are to be distinguished from fixed oils such as linseed oil or coconut oil, in that the latter are glycerides of fatty acids and hence saponifiable. Essential oils (except for those containing esters) are unsaponifiable. Some essential oils are nearly pure single compounds, as oil of wintergreen, which is methyl salicylate. Others are mixtures, as spirits of turpentine (pinene, dipentene), and oil of bitter almond (benzaldehyde, hydrocyanic acid). Some contain resins in solution and are then called oleoresins or balsams (q. v.).
Properties: Volatile oils, of pungent taste, usually nearly colorless when fresh, but becoming darker and thick on exposure to the air; optically active; sp. gr. 0.850-1. 100. Soluble in alcohol, carbon disulfide, carbon tetrachloride, chloroform, petroleum ether and fatty oils; insoluble in water, except for individual constituents of some oils which may be partially water-soluble, resulting in a loss of these constituents during steam distillation.
Derivation: Formed and contained in flowers and plants to which they supply the characteristic odor commonly identified with the flower. Some can also be prepared synthetically.
Methods of extraction: (a) By steam distillation; (b) by pressing (fruit rinds); (c) by solvent extraction; (d) by maceration of the flowers and leaves in fat and treating the fat with a solvent; (e) by enfleurage (employed for those very delicate oils whose odors are destroyed by even moderate heat); i. e., exposing odorless fats to the exhalations of flowers until they become strongly charged with the perfume and then treating the fat with a solvent.
Uses: Perfumery; flavors; and thinning and extending precious metal preparations used in decorating ceramic ware.
See also essential oils, terpeneless. For additional data see under specific essential oil.

essential oils, terpeneless. Concentrated essential oils from which the terpenes and sesquiterpenes have been removed by fractional distillation under vacuum. Presence of terpenes in the oils causes deterioration of the odor or flavor on standing (oils become rancid).
Properties: More soluble in alcohol and more concentrated than ordinary essential oils. Essential oils sold as "concentrated" may not necessarily be terpeneless.

"Essex" (SRF.) [285] Proprietary brand name for semi-reinforcing furnace carbon black.
Properties: Sp. gr. 1. 77; free-flowing pellets, also available in fluffy, unpelleted form as "Essex-UC"; bulk density 35 lbs/cu ft; particle diameter 80 millimicrons; pH 9. 5; ash 0. 50% max; 99. 9% through 325 mesh screen; color (Nigro-meter) 99-100.
Containers: 50-lb paper bags or bulk.
Uses: As a reinforcing extender pigment for compounding in natural rubber and most synthetic rubbers. Contributes to good physical characteristics, at relatively low cost; as a black coloring agent in rubber, paper, plastics, paint, and ink.

"Esso-Journal." [51] Trademark for a hard, soda soap grease made with a cylinder stock base oil for service in the driving journal boxes of steam locomotives.

"Essolube." [51] Trademark for detergent motor oils for all types of gasoline and high-speed diesel engines. Three detergency levels are supplied for heavy and light duty service with various quality fuels.

"Esso-Mar." [51] Trademark for a high quality lucricating oil for circulating systems such as those of marine turbines. Two heavier grades are used primarily for diesel cylinder lubrication.

"Essorod." [51] Trademark for a grease lubricant for locomotive side rods. The lubricant has the correct melting point and consistency to provide adequate lubrication without leaking or throwing.

"Essotane." [51] Trademark for liquefied petroleum gases for domestic and industrial uses. Commonly called "bottled gas."

"Essotex." [51] Trademark for an emulsible fiber lubricant for processing on the woolen system. Prepared with a highly light-stable base, insuring storage stability. It emulsifies readily with hot or cold water.

"Essowax." [51] Trademark for fully refined paraffin wax available in slabs and in liquid form in wide range of melting points and hardness.

"Esstic." [51] Trademark for industrial oils, available in a number of grades suitable for lubricating many types of bearings. Recommended for use in enclosed units such as ring-oiled bearings of electric motors.

"Estan." [51] Trademark for light colored, general purpose, lime-base greases. Available in wide range of consistencies and suitable for all methods of application. Made with an oil having a minimum of internal friction and bearing drag.
See also "Van Estan."

"Estane." [119] Trademark for a thermoplastic polyurethane material; a poly (ester-urethane) elastomer which provides good physical and chemical properties without curing. Extremely tough and abrasion resistant with high tensile strength at high ultimate elongation, with solvent resistance—particularly to gasoline, low air permeability and exceptional low-temperature flexibility.
Uses: Wire and cable jacketing, fuel hose and tanks, belting, shoe heels.

est'd. Abbreviation for estimated.

ester gums (rosin esters). Hard synthetic resins produced by the esterification of natural resins (especially rosin) with polyhydric alcohols (principally glycerol, but also pentaerythritol).
Grades: By color.
Containers: 300-lb barrels and drums; multiwall paper sacks.
Uses: Paints, varnishes, and lacquers.

"Esteron." [233] Trademark for a series of weed and brush control products; they are formulated esters of 2, 4-D and 2, 4, 5-T.

esters. Organic compounds corresponding in structure to salts in inorganic chemistry. They may be considered as derived from the acids by the exchange of the replaceable hydrogen of the latter for an organic alkyl radical. Esters are not ionic compounds, but salts usually are.

acid	HNO_3, nitric acid
salt	KNO_3, potassium nitrate
ester	$C_2H_5NO_3$, ethyl nitrate

estersil. A fine, free-flowing, hydrophobic silica powder obtained by esterification of free silanol groups (-SiOH) on the surface of the silica particles with a monohydric alcohol. See "Valron."

"Estersol." [232] Brand name for a series of solubilized vat dyestuffs.

"Estinyl." [321] Brand name of ethinyl estradiol.

"Estonate." [88] Trademark for a DDT insecticide in both wettable powders and emulsifiable solutions.

"Estonmite." [88] Trademark for para-chlorophenyl para-chlorobenzene sulfonate; miticide; available as a dust base, wettable powder and emulsifiable solution; used as an ovicide, specific against the eggs of spider-mites.

"Estonox." [88] Trademark for toxaphene in a dust base, wettable powder and in a stabilized emulsifiable carrier; used for control of insects on cotton, seed alfalfa, sugar beets, beans and potatoes.

"Estor." [51] Trademark for naphthenic base crankcase oils containing detergent additives. Grades are available suitable for all degrees of service severity characterizing modern diesel operation.

estradiol $C_{18}H_{24}O_2$. One of the female sex hormones. It occurs in two isomeric forms, alpha and beta. Beta-estradiol has the greatest physiological activity of naturally occurring estrogens. The alpha form is relatively inactive. For commonly used preparations, see estradiol esters following, and ethinylestradiol.
Properties of beta form: White or slightly yellow; small crystals or crystalline powder; odorless; m. p. 173-179°C; almost insoluble in water; soluble in alcohol, acetone, dioxane, and in solutions of alkali hydroxides; sparingly soluble in vegetable oils.
Derivation: Isolation from human and mare pregnancy urine; commercial synthesis from cholesterol, ergosterol or diosgenin.
Grade: N. F. XI (beta form).
Use: Medicine.

estradiol benzoate $C_{18}H_{23}O \cdot C_7H_5O_2$. Benzoic ester of the beta isomer of estradiol (q.v.).
Properties: White or slightly yellow to brownish crystalline powder; m. p. 191-196°C; odorless; almost insoluble in water; soluble in alcohol, acetone, and dioxane; slightly soluble in ether; sparingly soluble in vegetable oils. Stable in air.
Grades: U. S. P. XVI.
Use: Medicine.

estradiol cyclopentyl propionate. The cyclopentylpropionic ester of the beta isomer of estradiol.
Properties: White, odorless, crystalline solid; m. p. 148-152°C; freely soluble in

chloroform and ether; practically insoluble in water and alkalies; slightly soluble in alcohol and methanol.
Grade: N. F. XI.
Use: Medicine.

estradiol dipropionate $C_{24}H_{32}O_4$. Dipropionic ester of the beta isomer of estradiol.
Properties: Small, white to off-white crystals or crystalline powder; m. p. 104-109°C. Almost insoluble in water; soluble in acetone, alcohol, and dioxane; sparingly soluble in vegetable oils. Solutions are dextrorotatory.
Grade: U. S. P. XVI.
Use: Medicine.

estradiol valerate (estradiol-17-valerate).
Properties: White crystalline solid; m. p. 143-150°C. Insoluble in water; soluble in sesame oil.
Grade: N. N. D.
Use: Medicine.

estragole (chavicol methyl ether; methyl chavicol; para-allylanisole) $C_6H_4(C_3H_5)(OCH_3)$.
Properties: Colorless liquid, having a characteristic anise odor; sp. gr. 0.965-0.975 (20/4°C); (n 17.5/D) 1.5230; b.p. 216°C. Soluble in alcohol and chloroform.
Occurrence: In estragon oil; basil oils; anise bark oil, and others.
Uses: Perfumes; flavors.

estragon oil (tarragon oil).
Properties: Colorless to yellowish-green essential oil; peculiar anise-like odor; aromatic but not sweet taste. Keep well stoppered. Solubility in alcohol: In 6 to 11 vols and more of 80% alcohol; in 1 vol and more of 90% alcohol.
Chief known constituent: Estragole.
Constants: Sp. gr. 0.900-0.945 (15°C); optical rotation + 2° to + 9°; refractive index 1.502 to 1.514; acid value up to 1; ester value 1 to 9, after acetylation 15.
Derivation: Distilled from the flowering herb of Artemisia dracunculus L.
Containers: Glass bottles; copper flasks.
Use: Flavoring.
Shipping regulations: None. *

"Estrex." [152] Trade name for a series of methyl, butyl, propyl and other esters of animal and vegetable fatty acids, primarily oleic and stearic acids.
Uses: As lubricants and lubricating oil additives in cosmetics, defoamers, leather tanning and penetrating oils.
Containers: 5-, 55-gallons; tank cars.

estriol $C_{18}H_{24}O_3$. Estriol is considered an excretory product and usually occurs in conjugation with glucuronic acid, forming estriol glucuronide, which has little biological activity.
Properties: White, odorless, microcrystalline powder; m. p. 282°C. Exhibits reddish fluorescence under filtered ultraviolet light. Undergoes phase change at 270-275°C. Practically insoluble in water; soluble in alcohol, dioxane, and oils.
Derivation: Isolation from human pregnancy urine; isolation from human placenta; organic synthesis.
Use: Medicine.

estrogens. A general term for female sex hormones. They are responsible for the development of the female secondary sex characteristics such as the deposition of fat and the development of the breasts. The naturally occurring estrogens, such as estradiol, estrone, and estriol, are steroids. Estrogens are produced by the ovary, and to a lesser degree, by the adrenal cortex and testis. Some synthetic nonsteroid compounds, such as diethylstilbestrol and hexestrol, possess estrogenic activity.
Use: Medicine.

"Estron." [115] Trademark for synthetic yarn and staple fiber, acetate tow for use in cigarette filter tips, tobacco smoke filters, and tobacco smoke filter tip rods.

estrone $C_{18}H_{22}O_2$. A steroid with some estrogenic activity. It is probably a metabolic product of alpha- and beta-estradiol. It has less estrogenic activity than beta-estradiol but more than estriol (q. v.) or alpha-estradiol.
Properties: Small, white crystals or white crystalline powder; m. p. 258-262°C; odorless; stable in air; insoluble in water; soluble in alcohol, acetone, dioxane, and in solutions of fixed alkali hydroxides.
Derivation: Isolation from human pregnancy urine; synthesis from ergosterol.
Grades: U. S. P. XVI.
Use: Medicine.

"Estynox." [202] Trademark for a group of epoxidized fatty oils and esters used as stabilizing plasticizers for nitrocellulose, ethylcellulose, polyvinyl chloride, natural and synthetic rubbers and other polymers.

"Etamon" Chloride. [330] Trademark for tetraethylammonium chloride.

"Etchalume." [142] Trademark for an alkaline detergent for aluminum designed for rapid etching and cleaning. Material contains ingredients to prevent caking or hardening of aluminum hydroxide formed during the etching process.

"Eternalure D-38." [328] A modified vinyl-type polymer emulsion for use as a textile size and finish of good durability and desirable hand for piece goods. It has high resistance to build up on rolls, cans, and clips during processing and drying and is an effective nylon hosiery finish and snag reducer.

"Ethafoam." [233] Trademark for polyethylene foam.

ethanal. See acetaldehyde.

ethanamide. See acetamide.

ethane (bimethyl; dimethyl; ethyl hydride; methyl-methane) C_2H_6.

Properties: Colorless gas; odorless; flammable; slightly denser than air; relatively inactive chemically; b. p. −88. 63°C; f. p. −183. 23°C (triple point); sp. gr. of liquid 0. 446 (0°C); sp. gr. of vapor (air = 1) 1. 04 (0°C, 760 mm); critical temperature 32. 1°C; critical pressure (absolute) 718 psi; specific heat at constant pressure 0. 897; specific heat at constant volume 0. 325; ratio of specific heats (cp/cv) 1. 224; explosive limits in air (per cent by volume) lower 3. 2, upper 12. 5; heat of combustion (approx.) 22, 300 Btu/lb or 1800 Btu/cu ft; flash point −135°C.
Derivation: Fractionation from natural gas.
Grades: 95%; 99%; research.
Containers: Steel cylinders.
Uses: Organic synthesis; refrigerant; fuel.
Shipping regulations: Flammable gas. Red gas label. *

ethanedioyl chloride. See oxalyl chloride.

ethane hydrate. See gas hydrates.

ethanethiol. See ethyl mercaptan.

ethanethiolic acid. See thioacetic acid.

ethanoic acid. See acetic acid.

ethanol. See ethyl alcohol.

ethanolamine (MEA; monoethanolamine; colamine; 2-aminoethanol; 2-hydroxyethylamine) $HOCH_2CH_2NH_2$.
Properties: Colorless, moderately viscous liquid. Ammoniacal odor. Strong base. Chemically active. Miscible with water; soluble in carbon tetrachloride, alcohol, chloroform.
Constants: Sp. gr. 1. 0179 (20/20°C); b. p. 170. 5°C, (760 mm); freezing point 10. 5°C; vapor pressure 0. 48 mm (20°C); flash point (open cup) 200°F; wt/gal 8. 5 lbs (20°C).
Typical specifications: Color water-white; sp. gr. 1. 017-1. 027 (20/20°C); boiling range (760 mm) not less than 90% over between 165 and 173°C; equivalent wt 61-63; av. wt/gal 8. 47 lbs (20°C).
Derivation: Reaction of ethylene oxide and ammonia gives a mixture of mono-, di-, and triethanolamines.
Grades: Technical; N. F. XI.
Containers: 1-, 5-gal cans; 55-gal drums; tank cars.
Uses: Scrubbing acid gases (H_2S, CO_2), especially in synthesis of ammonia, from gas streams; non-ionic detergents used in drycleaning, wool treatment, emulsion paints, polishes, agricultural sprays; chemical intermediates; pharmaceuticals; corrosion inhibitor.

ethanol formamide $HOCH_2CH_2NHOCH$.
Properties: Somewhat viscous liquid completely miscible with water, alcohol and glycerol; compatible with polyvinyl alcohol, many cellulosic and natural resins; b. p. 143°C at 2. 5 mm Hg; freezing point below −72°C; sp. gr. (25/4°C) 1. 170.

ethanol hydrazine. See beta-hydroxyethylhydrazine.

2-ethanolpyridine $C_5H_4NCH_2CH_2OH$.
Constants: B. p. 235°C with decomposition; f. p. −7. 8°C; sp. gr. 1. 091 (25°C); refractive index 1. 5366 (20°C); approx wt/gal 9 lbs.
Miscible with alcohol, water.

ethanolurea $NH_2CONHCH_2CH_2OH$. White; solidification point 71-74°C; its formaldehyde condensation products are permanently thermoplastic and water soluble. As increasing amounts of simple urea are mixed with ethanolurea, the condensation products gradually change from the pliable film-forming type of resin into the brittle types. This makes it possible to obtain almost any degree of water-solubility and flexibility that may be desired in the final resin. These modified resins formed with ethanolurea are compatible with polyvinyl alcohol, methyl cellulose, cooked starch and other water-dispersible materials.

"Ethasan." [58] Trademark for zinc diethyldithiocarbamate.

"Ethavan." [58] Trademark for ethyl vanillin.

"Ethazate." [248] Trademark for zinc diethyldithiocarbamate.
Properties: White powder; sp. gr. 1. 45; m. p. 173-178°C; moderately soluble in benzol and ethylene dichloride; slightly soluble in acetone; insoluble in water and gasoline.
Use: Accelerator for latex, dispersions, cements and proofing; insulated wire.
Also available as a 50% water dispersion, "Ethazate 50-D."

ethchlorvynol (1-chloro-3-ethyl-1-penten-4-yn-3-ol; beta-chlorovinyl ethyl ethynyl carbinol) $HC\vdots CCOH(C_2H_5)CH{:}CHCl$.
Properties: Colorless to yellow liquid with a pungent aromatic odor; darkens on exposure to light and to air; sp. gr. 1. 068-1. 071; refractive index, 1. 4765-1. 4800 (n 25/D). Immiscible with water; miscible with most organic solvents.
Grade: N. F. XI.
Use: Medicine.

ethene. See ethylene.

ethenol. See vinyl alcohol.

ether (ethyl ether; diethyl ether; sulfuric ether, anesthesia ether; ethyl oxide; diethyl oxide) $(C_2H_5)_2O$.
Properties: Very light, transparent, colorless, volatile, exceedingly flammable, mobile liquid; hygroscopic; pleasant aromatic, ethereal odor; burning and sweet taste. Caution! Stongly narcotic! Have no flames or sparking electric equipment anywhere that ether is being used. The vapor of ether, mixed with air, explodes when ignited.

Ethyl and isopropyl ethers have a tendency to form peroxides which are explosive. Consequently caution must be observed in distilling these ethers unless the peroxides present have first been destroyed by the addition of an easily oxidizable

*See "I. C. C. Shipping Regulations," page xiii.
Reference numbers refer to name of manufacturer. See "List of Manufacturers," page v.

material such as sodium sulfite in water solution.

The presence of peroxides may be determined by the addition of acidified sodium or potassium iodide solutions. A brown coloration developing within a few minutes by liberation of iodine from the iodide indicates that peroxides are present. Peroxides may be determined quantitatively by titrating the liberated iodine with standard sodium thiosulfate solution. Peroxide formation on storage may be inhibited by storing the ether in the presence of water or reducing agents.

B.p. 34.5°C; freezing point -116.2°C; sp. gr. 0.7147 (20/20°C); surface tension 17.0 dynes/cm (20°C); refractive index (n 20/D) 1.3526; viscosity 0.00233 poise (20°C); vapor pressure 442 mm (20°C); specific heat 0.5476 cal/g (30°C); flash point -40°C; latent heat of evaporation 83.96 cal/g at b.p.; electric conductivity 4×10^{-13} recip. ohms (25°C); explosive limits in air 2.34-6.15% (20°C); coefficient of expansion 0.00164 per °C; wt/gal 6 lbs (20°C).

Typical specifications: Anesthesia grade: Acidity not more than 0.002% (as acetic); sp. gr. 0.715-0.718 (20/20°C); color water-white; ethanol not more than 2% by weight; dryness, miscible without turbidity with an equal volume of carbon disulfide (20°C); average wt/gal 5.96 lbs (20°C).

Typical specifications: U.S.P. grade (Complies with specifications of U.S.P. XVI). Acidity, free acid not more than the equivalent of 1.6 cc N/50 sodium hydroxide per 100 cc; aldehydes none; color water-white; nonvolatile matter not more than 0.002 g per 100 cc; odor characteristic — no foreign residual odor; peroxides none; sp. gr. 0.713-0.716 (25/25°C).

Typical specifications: Absolute—A.C.S. (Conforms fully to U.S.P. XVI requirements for ether U.S.P. except for higher ether content and correspondingly lower sp. gr., which would be expected in the absolute or dehydrated product offered as absolute ether A.C.S.): acidity, free acid as acetic, not more than 0.0015%; aldehydes negative; color water-white; nonvolatile matter not more than 0.0015%; no foreign residual odor; peroxides negative; sp. gr. not over 0.7100 (25/25°C); substances darkened by sulfuric acid, negative.

Soluble in alcohol, chloroform, benzene, solvent naphtha, and oils; slightly soluble in water.

Derivation: By the action of sulfuric acid on ethyl alcohol, followed by distillation.

Method of purification: Rectification.

Grades: U.S.P. XVI (for anesthesia); A.C.S. Reagent; A.C.S. Absolute; C.P.; concentrated; U.S.P. 1880; washed; motor; electronic.

Containers: 30-, 100-lb drums; tank trucks.

Uses: Manufacture of smokeless powder; medicine; anesthetic; organic synthesis; analytical chemistry; priming gasoline engines; solvent for fats, oils, resins, waxes, gums and alkaloids; perfumery; pyroxylins; rayon; collodion; plastics; extractant in various processes; concentration of acetic acid; refrigerant; fumigant; dry-cleaning; motor fuels; alcohol denaturant.

Fire Hazard: Danger! Extremely flammable. Highly volatile. Tends to form explosive peroxides. MCA warning label.

Shipping regulations: Flammable liquid. Red label. *

ether (physics). The medium permeating all space postulated at one time by the wave theory of light. The usefulness of this concept is questioned by most modern physicists.

ethers. Chemically, ethers are compounds of neutral character derived from alcohols by elimination of water (one molecule of water from two molecules of alcohol). A better general characterization is that an ether is an organic compound in which an oxygen atom is interposed between two carbon atoms in the molecular structure. Only the lowest member of the series (methyl ether) is gaseous; most of them are liquid, and the highest are solid. The ethyl esters are often incorrectly called ethers, especially in earlier literature, i.e., ethyl acetate referred to as acetic ether. The word ether is also erroneously used instead of ethyl in such terms as ether benzoate. The term ether is regularly and correctly used as synonymous with the most common ether, i.e., diethylether. (See ether).

ethinamate (1-ethynylcyclohexyl carbamate) $C_6H_{10}(C\vdots CH)OOCNH_2$.

Properties: White essentially odorless powder. Very slightly soluble in water, freely soluble in alcohol, chloroform, and in ether; melting range: 95°C-98°C; pH of saturated solution 6.5-7.0.

Grade: N.F. XI.

Use: Medicine.

ethine. See acetylene.

ethinylestradiol (ethynylestradiol) $C_{20}H_{24}O_2$. An estrogen (female sex hormone).

Properties: Fine, white to creamy white, odorless, crystalline powder; sensitive to light. M.p. 142-146°C. Soluble in acetone, alcohol, chloroform, dioxane, ether and vegetable oils; practically insoluble in water; soluble in solutions of sodium or potassium hydroxide. Slightly dextrorotatory in dioxane solution.

Derivation: Preparation from estrone.

Grade: U.S.P. XVI.

Use: Medicine.

ethion (O,O,O',O'-tetraethyl-S,S-methylene-diphosphorodithioate; bis[S-(diethoxyphosphinothioyl)mercapto]methane) $C_9H_{22}O_4P_2S_4$. Accepted as generic name by Ent. Soc.

Properties: Liquid; sp. gr. 1.220 (20°C).

*See "I.C.C. Shipping Regulations," page xiii.

Reference numbers refer to name of manufacturer. See "List of Manufacturers," page v.

Slightly soluble in water; soluble in lactones, xylene, chloroform and methylated naphthalene. Toxicity considered moderate to mammals.
Use: Insecticide and miticide.

ethiops mineral. See mercuric sulfide, black.

ethisterone (pregneninolone; anhydrohydroxyprogesterone; ethynyltestosterone) $C_{21}H_{28}O_2$. A female sex hormone; a derivative of progesterone (q. v.) with similar activity. Effective when given by mouth.
Properties: White or slightly yellow crystals or as crystalline powder. Odorless; stable in air. Affected by light. M. p. 267-275° (dec.). Practically insoluble in water; slightly soluble in alcohol, chloroform, ether, and vegetable oils.
Derivation: From progesterone and other steroids.
Grade: U. S. P. XVI.
Use: Medicine.

ethocaine. See procaine hydrochloride.

"Ethocel." [233] Brand name for cellulose resin (ethylcellulose), thermoplastic molding granules. Available in colors and various forms for injection or extrusion molding, also in film and sheet.
Properties: Colorless (unless pigmented), odorless, tasteless and non-toxic, flexible and tough, weather- and shatter-resistant; widely soluble in organic solvents. Stable over a range from 200°F to -40°F.
Chemical properties: Not affected by water, strong or weak alkalies; resistant to weak acids, sunlight and age.
Uses: Decorative items; molded and extruded products; hot melt coatings; pigment dispersion; lacquers; varnishes; enamels; packaging; wire coating and wrapping, etc.

ethodin (6,9-diamino-2-ethoxyacridine lactate monohydrate) $C_{15}H_{15}N_3O \cdot C_3H_6O_3 \cdot H_2O$.
Properties: Pale yellow crystals; darken at 200°C. M. p. 235°C; slowly soluble in 15 parts water, soluble in 9 parts boiling water; soluble in 110 parts alcohol (22°C). Solutions are yellow, fluorescent and stable to boiling.
Purity: 97% (dry basis).
Containers: 1-oz, 1/4 lb, 1 lb.
Uses: Bactericide; surgical antisepsis; preparation of pure gamma globulin.

"Ethoduomeens." [15] Trade name for ethylene oxide adjuncts of the "Duomeens." They are used as emulsifiers and corrosion inhibitors in certain systems.

"Ethofats." [15] Trademark for a group of nonionic surface-active agents. Vary from low-melting solids to liquids in consistency. The solubility characteristics vary from oil to water miscibility and are identified according to an arbitrary solubility number. "Ethofats" with solubility numbers 20-30 are excellent sudsless detergents; others are useful dry-cleaning aids, emulsifying agents and emulsion breakers.
Containers: Standard 55-gal openhead drums.

ethoheptazine (ethyl heptazine) $C_{16}H_{23}NO_2$. 1-Methyl-4-carbethoxy-4-phenylhexamethyleneimine.
Properties: Liquid; sp. gr. 1.038 (26/4°C); b. p. 133-134°C (1.0 mm); refractive index 1.5210 (26°C).
Use: Medicine.

ethohexadiol. U. S. P. XVI name for 2-ethylhexanediol-1,3.

"Ethol." [232] Brand name for a series of spirit-soluble dyestuffs.

"Ethomeens." [15] Trademark for a family of tertiary aliphatic amines possessing one or two alkyl groups ranging from C_8-C_{18} in chain length and for certain natural-occurring mixtures of these. Can be obtained with varying degrees of cationic strengths from exceedingly strong to an almost non-ionic type. Good stability to strong alkalies and acids. The family members vary from solubility in oils to water miscibility.
Containers: Standard 55-gal openhead Quiklox type non-returnable drums of approx 440 lbs net weight.
Uses: As emulsifiers, wetting agents, herbicide and insecticide emulsifiers.

"Ethomids." [15] Trade name for mixtures of mono- and di-substituted amides ranging in form from dispersible solids to soluble waxes.
Derivation: By treating unsubstituted amides with ethylene oxide.
Containers: Standard 55-gal openhead drums.
Uses: Wetting agents; cotton detergents; emulsifiers for silicones used in textile finishes.

"Ethone." [227] Trademark for alpha-methyl anisal acetone, $CH_3OC_6H_4CHCHCOCH_2CH_3$ (1-para-methoxy-phenyl penten-1-one-3); min 99% pure.
Properties: White to pale yellow crystalline material; sharp dry odor, with a slight suggestion of butter; stable; causes discoloration; m. p., 60.0°C min.
Uses: In maple, berry and other flavors.

ethopropazine hydrochloride (10-(2-diethylaminopropyl) phenothiazine hydrochloride) $C_{19}H_{24}N_2S \cdot HCl$.
Properties: White or slightly off white, odorless, crystalline solid. Sparingly soluble in alcohol. Slightly soluble in water.
Grade: N. N. D.
Use: Medicine.

ethosuximide (alpha-ethyl-alpha-methylsuccinimide) $C_7H_{11}NO_2$.
Properties: White to off-white waxy powder or solid having a characteristic odor; soluble in water, alcohol and ether.
Use: Medicine.

ethovan. See ethyl vanillin.

ethoxazene (diaminoethoxyazobenzene hydrochloride) $C_{14}H_{16}N_4O \cdot HCl$. 4-(para-Ethoxyphenylazo)-meta-phenylenediamine hydrochloride. A reddish powder;

insoluble in water.
Use: Medicine.

para-**ethoxyacetanilide.** See acetophenetidin.

ethoxybenzidine (di-para-aminoethoxydiphenyl) $C_6H_4NH_2C_6H_3(OC_2H_5)NH_2$.
Properties: Glistening, flat needles. Soluble in alcohol; sparingly soluble in water. M.p. 135°C.
Derivation: By heating ethoxybenzidine sulfonic acid, obtained from benzeneazophenetolesulfonic acid, with water in an autoclave.

6-ethoxybenzothiazole-2-sulfonamide. See ethoxzolamide.

6-ethoxy-1,2-dihydro-2,2,4-trimethylquinoline ("Santoflex" AW.) $C_{14}H_{19}NO$.
Properties: Yellow liquid; refractive index 1.569-1.572 (25°C); sp. gr. 1.029-1.031 (25°C).
Containers: 55- and 5-gal drums.
Uses: A preservative to reduce oxidative loss of carotene in dehydrated alfalfa.

2-ethoxyethanol. See ethylene glycol monoethyl ether.

3-ethoxy-4-hydroxybenzaldehyde. See ethyl vanillin.

4-ethoxy-3-methoxybenzaldehyde $C_6H_3(OC_2H_5)(OCH_3)CHO$. White to light brown crystals having a slight vanillin odor; m.p. 62-64°C. Used as an intermediate.

4-ethoxy-3-methoxyphenylacetic acid $C_6H_3(OC_2H_5)(OCH_3)CH_2COOH$. An off-white powder; m.p. 119-122°C. Used as an intermediate.

1-ethoxy-2-methoxy-4-propenyl-benzene. See isoeugenol ethyl ether.

4-ethoxyphenol. See hydroquinone monoethyl ether.

ethoxyphenylurea. See dulcin.

ethoxytriglycol $C_2H_5O(C_2H_4O)_3H$.
Properties: Liquid; sp. gr. 1.0208 (20/20°C); 8.5 lbs/gal (20°C); b.p. 255.4°C (760 mm); vapor pressure less than 0.01 (20°C); freezing point −18.7°C; viscosity 7.80 cps (20°C); completely soluble in water.
Use: Synthesis of dyes.

ethoxzolamide (6-ethoxybenzothiazole-2-sulfonamide) $C_9H_{10}N_2O_3S_2$. Crystals; m.p. 188-190°C. Used in medicine.

"Ethron." 233 Trademark for polyethylene resins.

"Ethycol." 223 Proprietary products consisting of pigments dispersed in ethylcellulose and plasticizer.
Uses: Protective coatings and inks in which ethylcellulose is a major ingredient.

"Ethyl." 313 Trademark for a line of additives to hydrocarbon fuels and lubricants as well as other products.
Antiknock Compound-TEL-Motor Mix contains 61.5% tetraethyl lead, 17.86% ethylene dibromide, 18.81% ethylene dichloride, and 1.85% dye, kerosene and antioxidant. Freezing point is −28°F, flash point (open cup) greater than 245°F, density at 68°F 1.59 g/ml. Used to improve octane rating of motor fuels.
Antiknock Compound-TEL-Aviation Mix contains 61.5% tetraethyl lead, 35.68% ethylene dibromide, 2.91% dye, kerosene and antioxidant. Freezing point is +16°F, flash point (open cup) over 250°F, density 1.74 g/ml. Used to improve octane rating of aviation fuels.
Antiknock Compound-TEL-Motor 33 Mix (AK-33X) contains 57.5% tetraethyl lead, 7.0% methyl cyclopentadienyl mangenese tricarbonyl, 16.71% ethylene dibromide, 17.60% ethylene dichloride, 1.20% dye, kerosene, antioxidant. Freezing point is −28°F, flash point (open cup) greater than 233°F, density at 68°F 1.58 g/ml. The manganese compound promotes or extends the effect of tetraethyl lead in raising octane number, particularly of paraffinic fuels.
Antiknock Compound-TML-Motor Mix contains tetramethyl lead. Otherwise similar to corresponding TEL-Antiknock Compound.
Antiknock Compound-MLA-Motor Mix contains mixed methyl and ethyl lead derivatives. Otherwise similar to corresponding TEL and TML mixtures.
Ignition Control Compound 1 (ICC 1) is tris(chloroisopropyl)thionophosphate (q.v.).
Ignition Control Compound 2 is dimethylxylyl phosphate (q.v.).
Ignition Control Compound 3 is a mixture of methyl phenyl phosphates, containing 11% phosphorus. Also contains a corrosion inhibitor as a precaution against water contamination. Clear light straw colored liquid, sp. gr. 1.146 at 68°F, flash point (open cup) greater than 109°F, pour point −94°F; miscible with gasoline at 65°F; solubility in typical gasoline at 32°F is 18.5 volume per cent, in water at 86°F approximately 1 weight per cent; may hydrolyze on standing in presence of water. Used in motor fuels to control spark plug fouling, surface ignition, and motor rumble.
Caution: Do not take internally. Avoid contact with skin and eyes; flush immediately with much water if contact occurs.
Ignition Control Compound 4 is trimethyl phosphate (q.v.).
Multi-Purpose Additive (MPA) contains 52% mixed substituted oleamides, 37% isopropyl alcohol, 7% aromatic solvent, and 4% water. Clear amber liquid, density 0.888 g/ml at 68°F, flash point (closed cup) 75°F, pour point 50°F. Used to remove and prevent deposits on the throat walls of carburetors, to prevent carburetor icing, and as a corrosion preventative.
Metal Deactivator (MDA) contains 80% N,N'-disalicylidene-1,2-diaminopropane (N,N'-disalicylidene propylenediamine, formula $HOC_6H_4CH{:}NCH_2CH(CH_3)N{:}CHC_6H_4OH$) and 20% toluene solvent. Amber liquid, density 1.0672 g/ml at 68°F, flash point (open cup) 84°F, fire point 100°F. Soluble

in gasoline, insoluble in water. Used to neutralize the catalytic effect of copper in promoting fuel oxidation.

Warning: Avoid contact with skin and eyes; immediately flush with much water if contact occurs; contaminated clothing should be removed promptly and laundered by conventional methods before re-use. Do not store or handle near open flames.

Antioxidant 701: See 2, 6-di-tert-butylphenol.

Antioxidant 702: See 4, 4'-methylenebis(2, 6-di-tert-butylphenol).

Antioxidant 703: See 2, 6-di-tert-butyl-alpha-dimethylamino-para-cresol.

Antioxidant 733, a mixture of tert-butylphenols, contains a minimum of 75% 2,6-di-tert-butylphenol, 10-15% 2, 4, 6-tri-tert-butylphenol, and the remainder ortho-tert-butylphenol. Amber liquid, density 0.941 g/ml; freezing point 64°F, but supercools to 45°F; flash point (open cup) 230°F.

Used to inhibit formation of gum and peroxides in gasoline, and repress the formation of soluble and insoluble decomposition products of jet fuels during storage; as antioxidant for steam-turbine and industrial oils; also retards decomposition of antiknock compounds.

Antioxidant ZDP is dialkyl zinc dithiophosphate, available as straw colored oil or as solution in lubricating oil.

Used in lubricating oils to reduce wear in high output automobile engines. Also useful as an oxidation inhibitor and to protect bearing metals, particularly copper-lead and cadmium-silver alloys, from corrosive action of oil oxidation products.

Caution: Do not take internally. Avoid breathing of vapor. Avoid contact with skin and eyes. If skin contact occurs, immediately wash with soap and water; for eyes, flush with much water and get medical attention.

Diesel Ignition Improver (DII) is a mixture of primary amyl nitrates, consisting of 60% n-amyl nitrate, 5% isoamyl nitrate, and 35% 2-methyl-n-butyl nitrate. Light straw colored liquid with ethereal odor. Stable on storage. Density 0.998 g/ml; flash point (open cup) 120°F; pour point −190°F; insoluble in water. Used to raise cetane numbers of diesel fuels.

Handling: Not sensitive to thermal or mechanical shock, but keep containers closed and away from heat and flames.

Caution: Relatively non-toxic, but use with adequate ventilation and avoid breathing of vapor. Avoid contact with skin and eyes. In case of contact, wash skin with soap and water; for eyes, flush with much water.

Monopropellant 1 is n-propyl nitrate (NPN).

Monopropellant 2 is 60% ethyl nitrate, 40% n-propyl nitrate (60-40 EPN). White to straw colored liquid with ethereal odor; density 1.087 g/ml; flash point (open cup) 65°F; freezing point less than −100°C.

Caution: Highly flammable liquid. Shipped as a non-explosive, flammable liquid in steel drums. Store in well ventilated space.

Hazards: Avoid high concentrations of the vapor in air within enclosed spaces. Avoid undue and prolonged contact with the skin. In event of contact, the skin should be washed immediately with soap and water. Contaminated clothing should be laundered.

See monopropellant.

Oil Soluble Dye - Red. Methyl derivatives of azobenzene-4-azo-2-naphthol. (C.I. 248) Available in various powdered, granular, or bead forms for coloring gasoline.

Oil Soluble Dye - Orange. (C.I. 24) Benzene-azo-2-naphthol. Powder or flakes.

Oil Soluble Dye - Yellow. (C.I. 19) Para-dimethylaminoazobenzene. Flakes.

Oil Soluble Dye - Blue. 1-4-Diisopropylaminoanthraquinone. Available in powder and beaded forms.

Oil Soluble Dye - Bronze. Mixture of Red and Orange.

ethyl abietate (abietic acid ethyl ester) $C_{19}H_{29}COOC_2H_5$.

Properties: Amber-colored, viscous liquid which hardens upon oxidation. Soluble in ether, most varnish solvents; insoluble in water.

Constants: B.p. 350°C; flash point 178°C; m.p. 45°C; refractive index 1.4980; sp. gr. 1.02.

Derivation: (a) By heating together ethyl chloride and an alcoholic solution of rosin and caustic soda. (b) By reacting ethyl iodide with silver abietate.

"Ethylac." [204] Trademark for 2-benzothiazyl-N, N-diethylthiocarbamylsulfide.

Properties: Free-flowing, light yellow to tan powder; sp. gr. 1.27; m.p. 69°C (min).

Use: Rubber vulcanization accelerator.

ethylacetamide $CH_3CONHC_2H_5$.

Properties: Water-white; sp. gr. 0.920 (20/20°C); boiling range 206-208.5°C; flash point 230°F; faint odor.

ethyl acetamidocyanoacetate (acetamidocyanoacetic ester; ethyl N-acetyl-alpha-cyanoglycine) $NCCH(NHCOCH_3)COOC_2H_5$. Solid; m.p. 129°C. Used in amino acid and related compound synthesis.

ethyl acetanilide (ethyl phenylacetamide) $C_6H_5NC_2H_5COCH_3$.

Properties: White, crystalline solid. Faint odor. Soluble in most organic solvents. Insoluble in water.

Constants: Sp. gr. 0.994; b.p. 258°C; flash point 124°C; m.p. 54°C.

Grades: Technical.

Use: Substitute for camphor in the nitrocellulose industries.

ethyl acetate (acetic ether; acetic ester; vinegar naphtha) $CH_3COOC_2H_5$.

Properties: Colorless, fragrant, flammable liquid. Soluble in chloroform, alcohol and ether; slightly soluble in water.

Typical specifications: Commercial, 85-88%:

Acidity, free acid as acetic, not more than 0.02%; color water-white; distillation range 70-80°C; miscible without turbidity with 20 vols 60° Bé. gasoline (20°C); nonvolatile matter not more than 0.003 g per 100 cc; odor mild, non-residual; sp. gr. 0.883-0.888 (20/20°C).
95-98% Grade: Acidity, free acid as acetic, not more than 0.02%; color water-white; distillation range 73-80°C; miscible without turbidity with 20 vols 60° gasoline (20°C); non-volatile matter not more than 0.003 g per 100 cc; odor mild, non-residual; sp. gr. 0.895-0.899 (20°C).
99% Grade: Acidity, free acid as acetic, not more than 0.01%; color water-white; distillation range 75-80°C; miscible without turbidity with 20 vols 60° Bé. gasoline (20°C); non-volatile matter not more than 0.003 g per 100 cc; odor mild, non-residual; sp. gr. 0.900-0.904 (20/20°C); wt/gal 7.50 lbs; flash point 26°F (approx).
Derivation: By heating acetic acid and ethyl alcohol in presence of sulfuric acid, and distilling.
Method of purification: Rectification.
Grades: Commercial, 85-88%; 95-98%; 99%; technical; pure refined; N. F. XI (99%).
Containers: Drums; tank cars.
Uses: Lacquer and plastic solvent; general solvent; organic synthesis, flavoring; perfumery; smokeless powders; artificial fruit essences; bonbons and confections; artificial bristles and horsehair; pharmaceuticals; rayon. Sale is subject to strict government regulation.
Warning: Flammable. MCA warning label.
Shipping regulations: Flammable liquid. Red label. *

ethyl acetate, anhydrous. See ethyl acetate, grade 99%.

ethylacetic acid. See butyric acid.

ethyl acetoacetate (diacetic ester; acetoacetic ester) $CH_3COCH_2COOC_2H_5$.
Properties: Colorless liquid; fruity odor. Soluble in alcohol; slightly soluble in water.
Constants: Sp. gr. 1.0250 (20/4°C) m. p. -80°C; b. p. 180-181°C; wt/gal 8.5 lbs; vapor pressure 0.8 mm (20°C); flash point 185°F; coefficient of expansion 0.00101/°C.
Typical specifications: Acidity not more than 0.05% (as acetic); sp. gr. 1.023-1.028 (20/20°C); boiling range (50 mm) below 96°C not more than 5%, above 110°C none; purity not less than 97.5%; color not darker 8 mg potassium dichromate in 1 liter of water; solubility complete in alcohol, ether, and ethyl acetate; average wt/gal 8.54 lbs (20°C).
Derivation: By the action of metallic sodium on ethyl acetate, with subsequent distillation.
Method of purification: Redistillation.
Grades: Technical.
Containers: Drums; tank cars.
Uses: Organic synthesis; antipyrine; lacquers; dopes; plastics; manufacture of dyes, pharmaceuticals, antimalarials, and Vitamin B.
Fire hazard: Combustible but not flammable. Flash point over 80°F.
Shipping regulations: None. *

ethyl acetone. See methyl propyl ketone.

ethyl acetylene (1-butyne) $C_2H_5C{:}CH$.
Properties: Available as liquefied gas; b. p. 8.3°C; sp. gr. 0.669 (0/0°C); m. p. -130°C; flash point (Tag open cup) < 20°F; specific volume 7.2 cu. ft/lb. (70°F); insoluble in water.
Uses: Specialty fuel; chemical intermediate.
Shipping regulations: Flammable gas. Red gas label. *

ethyl N-acetyl-alpha-cyanoglycine. See ethyl acetamidocyanoacetate.

ethyl acetylsalicylate $C_2H_5OOCC_6H_4OOCCH_3$.
Properties: Colorless liquid. Insoluble in water; soluble in many organic solvents.
Constants: Sp. gr. 1.153; b. p. 272°C.
Use: Medicine.

ethyl acrylate $CH_2{:}CHCOOC_2H_5$.
Properties: Colorless liquid; b. p. 99.4°C; m. p. -72.0°C; sp. gr. 0.9230 (20/20°C); wt/gal 7.6 lbs (20°C); nearly insoluble in water; readily polymerized.
Derivation: (a) Ethylene cyanohydrin, ethyl alcohol, and dilute sulfuric acid; (b) Oxo reaction of acetylene, carbon monoxide, and ethyl alcohol in the presence of nickel or cobalt catalyst.
Grades: Technical (inhibited).
Containers: 1-gal cans; 5-, 55-gal drums; tank cars.
Uses: Polymers; acrylic paints; chemical intermediates.
See also acrylate resins.

ethylalbenzene. See phenylacetaldehyde.

ethyl alcohol (alcohol; grain alcohol; ethanol; fermentation alcohol; Cologne spirits; spirits of wine; ethyl hydroxide) C_2H_5OH.
Properties of pure 100% absolute alcohol (dehydrated alcohol): Colorless, limpid, volatile liquid; ethereal, vinous odor; pungent taste. Soluble in water, methyl alcohol, ether, and chloroform.
Properties of 95% ethyl alcohol: Refractive index 1.3651 (15°C); surface tension 22.8 dynes/cm (20°C); viscosity 0.0141 poise (20°C); vapor pressure 43 mm (20°C); specific heat 0.618 cal/g (23°C); flash point 14°C (57°F); sp. gr. (15.56°C) 0.816; b. p. 78°C.
Derivation: (a) From ethylene, either by direct catalytic hydration or by means of ethyl sulfate as an intermediate (phosphoric acid is one of catalysts used); (b) as a by-product of hydrocarbon synthesis from carbon monoxide and hydrogen, and as a by-product of methanol synthesis from these gases; (c) by fermentation of molasses, grains, or other carbohydrates, and sulfite pulp.
Grades: U. S. P. XVI (95% by vol); absolute; pure; denatured (see denatured alcohol); Cologne spirits; various proofs, as 190 proof. Note: One-half the proof number

gives the percent of ethyl alcohol by volume present in a mixture.
Containers: Drums, tank trucks, tank cars.
Uses: (including those for denatured alcohol) As a solvent and extraction medium; manufacture of intermediates, organic derivatives (especially acetaldehyde), dyes, synthetic drugs, synthetic rubber, detergents, cleaning solutions, surface coatings, cosmetics, pharmaceuticals, explosives; automobile radiator antifreeze; beverages; rocket fuel; and many specialized uses.
Fire hazard: Dangerous.
Shipping regulations: Flammable liquid. Red label. *

ethyl aldehyde. See acetaldehyde.

ethyl aluminum dichloride $C_2H_5AlCl_2$.
Properties: Clear, yellow pyrophoric liquid. B. p. (extrapolated) 194°C; f. p. 22°C; weight 10. 28 lbs/gal (25°C). Flames instantly with air; reacts violently with water.
Derivation: Reaction of aluminum chloride with ethyl aluminum sesquichloride.
Uses: Catalyst for olefin polymerization, aromatic hydrogenation; intermediate.
Shipping regulations: Flammable liquid. Red label. *

ethyl aluminum sesquichloride $(C_2H_5)_3Al_2Cl_3$.
Properties: Clear yellow liquid; b. p. 204°C; f. p. −20°C. Flames instantly with air; reacts violently with water.
Derivation: Reaction of ethyl chloride and aluminum.
Grades: Commercial.
Uses: Catalyst for olefin polymerization, aromatic hydrogenation; intermediate.
Shipping regulations: Flammable liquid. Red label. *

ethyl alpha-allylacetoacetate $CH_3COCH(CH_2CH{:}CH_2)COOC_2H_5$.
Properties: Water-white liquid; sp. gr. (20°C) 0. 989; wt/gal (20°C) 8. 24 lbs.
Containers: 30-gal aluminum drums (240 lbs net).
Uses: Intermediate for pharmaceuticals, perfumes, fungicides, insecticides, fine chemicals.

ethylamine (aminoethane) $CH_3CH_2NH_2$.
Properties: Colorless, volatile liquid or gas; ammonia odor; strong alkaline reaction. B. p. 16. 6°C; freezing point −81. 2°C; sp. gr. (liquid, 15/15°C) 0. 689; wt/gal (20°C) 5. 7 lbs; flash point (open cup) below 0°F. Miscible with water, alcohol, and ether.
Derivation: From ethyl chloride and alcoholic ammonia under heat and pressure.
Grades: Technical (anhydrous and 70% aqueous solution); pure, 98. 5% min.
Containers: 1-gal cans; 5-, 55-gal drums; tank cars.
Uses: Dye intermediates; solvent extraction; petroleum refining; stabilizer for rubber latex; detergents; organic synthesis.
Shipping regulations: Flammable liquid. Red label. *

ethylamine hydrobromide $C_2H_5NH_2 \cdot HBr$.
Properties: White, practically odorless granules; m. p. 158-161°C; very soluble in water.
Use: Intermediate where liquid ethylamine or liquid hydrobromic acid cannot be used.

ethyl-ortho-aminobenzoate. See ethyl anthranilate.

ethyl-para-aminobenzoate. (benzocaine) $C_6H_4NH_2CO_2C_2H_5$.
Properties: White, crystalline, odorless, tasteless powder; m. p. 88-92°C; exhibits local anesthetic properties when placed on the tongue. Soluble in dilute acids, less so in chloroform, ether, and alcohol, slightly soluble in almond or olive oil, very slightly soluble in water.
Derivation: By the ethylation of para-nitrobenzoic acid, followed by reduction.
Method of purification: Recrystallization.
Grades: Technical; pure; N. F. XI.
Containers: Glass bottles; drums.
Use: Medicine.
Shipping regulations: None. *

ethylaminoethanol. See ethylethanolamine. Mixed ethylaminoethanols (sold in up to tank car lots) may also contain diethylaminoethanol (q. v.).

ethyl amyl ketone (EAK; octanone-3) $CH_3CH_2CO(CH_2)_4CH_3$.
Properties: Colorless liquid, having a pungent odor. Soluble in 4 vols of 60% alcohol.
Constants: Sp. gr. 0. 819-0. 824; refractive index 1. 416.
Containers: Drums; tank cars.
Use: Perfumery; solvent.
Shipping regulations: None. *

N-ethylaniline $C_2H_5NHC_6H_5$.
Properties: Colorless liquid, becoming brown on exposure to light. Soluble in alcohol; insoluble in water.
Constants: Sp. gr. 0. 9631; m. p. −63. 5°C; b. p. 206°C; refractive index (n 20/D) 1. 5559.
Derivation: By heating aniline and ethyl alcohol in presence of sulfuric acid, with subsequent distillation.
Method of purification: Rectification.
Grades: Technical.
Containers: Iron drums; tank cars.
Use: Organic synthesis.
Shipping regulations: None. *

ortho-ethylaniline $C_6H_4(NH_2)C_2H_5$.
Properties: Brown liquid; f. p. −44°C; sp. gr. (20°C) 0. 982; b. p. 214°C; flash point (open cup) 208°F. Soluble in alcohol and toluene;;insoluble in water.
Uses: Chemical intermediate for pharmaceuticals, dyestuffs, pesticides, and other products.

ethyl anthranilate (ethyl-ortho-aminobenzoate) $C_6H_4(NH_2)COOCH_2CH_3$.
Properties: Colorless liquid, having an orange-flower and grape-type odor.
Constants: Sp. gr. 1. 117; refractive index

1.564; b.p. 260°C.
Uses: Perfumery and flavors, similar to methyl anthranilate (q. v.).

2-ethylanthraquinone $C_{14}H_8O_2C_2H_5$.
Properties: Buff to light yellow paste. M.p. 108.0°C.
Grades: Technical.
Use: Synthesis, especially of hydrogen peroxide.

"Ethyl" Antiknock Compound. [313] See "Ethyl."

ethylarsenious oxide C_2H_5AsO.
Properties: Colorless oil. Garlic-like, nauseating odor. Oxidizes in air and forms colorless crystals. Soluble in acetone, benzene, ether. Sp. gr. 1.802 (11°C); b.p. 158°C (10 mm).
Derivation: By hydrolysis of dichloroarsine.
Grade: Technical.
Use: Organic synthesis.

ethylbenzene (ethylbenzol; phenylethane) $C_6H_5C_2H_5$.
Properties: Colorless liquid; boiling point 136.187°C; refractive index 1.49594 (20°C); sp. gr. 0.86702 (20°C); f. p. 94.975°C; aniline equivalent 19; wt/gal 7.21 lbs (25°C); flash point 85°F; specific heat 0.41 cal/gm/°C; viscosity 0.64 centipoise (25°C).
Typical specifications (technical grade 95 mole % ethylbenzene): most probable impurities toluene and isopropylbenzene; distillation range 276.8-277.4°F; sp. gr. 0.872 (60/60°F); refractive index 1.496; soluble in alcohol, benzene, carbon tetrachloride, and ether; slightly soluble in water.
Derivation: (a) By heating benzene and ethylene in presence of aluminum chloride, with subsequent distillation; (b) by fractionation directly from the mixed xylene stream in petroleum refining.
Method of purification: Rectification.
Grades: Technical; pure; research.
Containers: Drums; tank cars.
Uses: Organic synthesis; solvent and diluent; intermediate in production of styrene.

ethyl benzoate (benzoic ether) $C_6H_5CO_2C_2H_5$.
Properties: Colorless, aromatic liquid; soluble in alcohol and ether; slightly soluble in hot water. Sp. gr. 1.043-1.046; m.p. −32.7°C; b.p. 212.9°C; refractive index 1.505.
Derivation: By heating ethyl alcohol and benzoic acid in presence of sulfuric acid.
Method of purification: Rectification.
Grade: Technical.
Containers: Iron drums; glass bottles.
Uses: Flavoring extracts; perfumery; solvent mixtures; lacquers; solvent for many cellulose derivatives and natural and synthetic resins.
Shipping regulations: None. *

ethylbenzol. See ethylbenzene.

ethyl benzoylacetate $C_6H_5COCH_2CO_2C_2H_5$.
Properties: Light yellow oil; boiling range 144-148°C (10 mm); sp. gr. 1.111-1.117 (20°C); flash point (open cup) 285°F. Soluble in most organic solvents; insoluble in water.
Derivation: Reaction of ethyl acetate and ethyl benzoate with metallic sodium.
Method of purification: Vacuum distillation.
Grade: 95% pure.
Uses: Dye and pharmaceutical intermediate.
Shipping regulations: None. *

ethyl ortho-benzoylbenzoate $C_6H_5COC_6H_4COOC_2H_5$.
Properties: Yellowish white solid; odorless; insoluble in water; soluble in alcohol, acetone, ethyl acetate, and benzol; m.p. not lower than 56-58°C; b.p. 325°C (760 mm).
Derivation: From benzoyl benzoic acid and alcohol.
Use: As a plasticizer for nitrocellulose, synthetic resins, etc.
Containers: 65-lb cans.
Shipping regulations: None. *

ethylbenzylaniline $C_6H_5N(C_2H_5)CH_2C_6H_5$.
Properties: Clear colorless oil; soluble in alcohol and ether; insoluble in water. Sp. gr. 1.034; b.p. 286°C.
Derivation: By heating ethyl aniline, benzyl chloride, and aqueous caustic soda, with subsequent distillation.
Method of purification: Redistillation.
Grade: Technical.
Containers: Iron drums.
Uses: Dyestuffs; organic synthesis.
Shipping regulations: None. *

ethylbenzyl chlorides (1-chloromethylethylbenzene) $ClCH_2C_6H_4C_2H_5$. Consists of 70% para- and 30% ortho-ethylbenzyl chloride.
Properties: Sp. gr. 1.0460-1.0475 (25/25°C); refractive index 1.5290-1.5305 (n 25/D); soluble in alcohols; insoluble in water.
Uses: Intermediate.

ethyl biscoumacetate (ethyl bis(4-hydroxycoumarinyl)acetate) $C_{22}H_{16}O_8$. A synthetic derivative of bishydroxycoumarin.
Properties: White, odorless, bitter, crystalline solid. M.p. 177-182°C. Another form melts 154-157°. Soluble in acetone and benzene; slightly soluble in alcohol and ether; insoluble in water.
Grade: N.N.D.
Use: Medicine.

ethyl bis (4-hydroxycoumarinyl) acetate. See ethyl biscoumacetate.

ethyl borate $(C_2H_5)_3BO_3$.
Properties: Colorless liquid. Mild odor. It is very stable to heat but hydrolyzes very rapidly in the presence of water, depositing boric acid in finely divided crystalline form.
Constants: B.p. 120°C; sp. gr. 0.863-0.864 (20/20°C); flash point 51.8°F; wt/gal 7.20 lbs (20°C); refractive index 1.37311 (20°C).
Typical specification: Boiling range 112 to 121°C, with not more than 5% distilling below 116°C; purity 98%; decomposes instantly with water.
Uses: Antiseptics; disinfectants; fireproofing of airplane fabrics; antiknock agent.

*See "I.C.C. Shipping Regulations," page xiii.
Reference numbers refer to name of manufacturer. See "List of Manufacturers," page v.

Fire hazard: Flammable.
Shipping regulations: Flammable liquid. Red label. *

ethyl bromide (bromoethane) C_2H_5Br.
Properties: Colorless, flammable, volatile liquid; soluble in alcohol and ether; sparingly soluble in water.
Constants: Sp. gr. 1.431 (20/4°C); b. p. 38.4°C; wt/gal 12-12.1 lbs; vapor pressure 386 mm (20°C); critical temperature 231°C; f. p. −119°C; surface tension (10°C) 25.48 dynes/cm, (20°C) 24.15, (40°C) 21.52; latent heat of vaporization 250.8 joules (34.4°C); heat of combustion 340 kg cal/mol (gas); dielectric constant, liquid 9.45 (20°C); constant minimum boiling mixture of ethyl alcohol 7 mole% and ethyl bromide 93 mole% 37.6°C.
Derivation: Red phosphorus is added to absolute ethyl alcohol, bromine is then slowly added to the mixture which is then distilled.
Method of purification: Rectification.
Grades: Technical (98%).
Containers: Drums; tank cars.
Keep tightly closed in a cool place protected from air and light.
Uses: Organic synthesis; medicine (anesthetic); refrigerant; solvent; grain and fruit fumigant.
Warning! Vapor harmful; MCA warning label.

ethyl bromoacetate $CH_2BrCOOC_2H_5$.
Properties: Clear, colorless liquid. Forms colorless needles when cooled by solid carbon dioxide-ether mixture. Partially decomposed by water. Soluble in alcohol, benzene, ether; insoluble in water. Has poison gas characteristics. Sp. gr. 1.53 (4°C); b. p. 168°C; m. p. −13.8°C; vapor density 5.8.
Derivation: Interaction of bromine and acetic acid in the presence of red phosphorus.
Grades: Technical.

ethyl butanoate. See ethyl butyrate.

2-ethylbutanol. See 2-ethylbutyl alcohol.

2-ethyl-1-butene (uns-diethylethylene) $CH_3CH_2(C_2H_5)C{:}CH_2$.
Properties: Colorless liquid; sp. gr. 0.6894 (20/4°C); b. p. 64.95°C; refractive index (n 20/D) 1.3969; soluble in alcohol, acetone, ether, petroleum, and coal-tar solvents; insoluble in water.
Typical specifications: Sp. gr. 0.6880 to 0.6920 (20/4°C); b. p. 64-66°C; refractive index (n 20/D) 1.3960 to 1.3995.
Grade: 95% pure.
Use: Organic synthesis of flavors, perfumes, medicines, dyes, resins.

3-(2-ethylbutoxy) propionic acid $CH_3CH_2CH(C_2H_5)CH_2OCH_2CH_2COOH$.
Properties: Water-white liquid; sp. gr. (20/20°C) 0.9600; b. p. (100 mm) 200°C; vapor pressure (20°C) < 0.1 mm; f. p., glass below −90°C; solubility in water (20°C) less than 0.01%.
Uses: Preparation of metallic salts for paint driers and gelling agents.

2-ethylbutyl acetate $C_2H_5CH(C_2H_5)CH_2OOCCH_3$.
Properties: Colorless liquid; mild odor. Sp. gr. 0.875-0.881 (20/20°C).
Typical specifications: Acidity not more than 0.10% as acetic; color water-white, sp. gr. 0.875 to 0.881 (20/20°C); boiling range, below 155°C none, above 164°C none (760 mm); purity not less than 90% ethylbutyl acetate; dryness, miscible with 19 vols. 60° Bé. gasoline at 20°C; average wt/gal 7.33 lbs (20°C).
Grade: Technical.
Use: Solvent for nitrocellulose, gums, resins, and lacquers.
Shipping regulations: None. *

2-ethylbutyl alcohol (2-ethylbutanol; hexyl alcohol, pseudo-) $CH_3CH_2CH(C_2H_5)CH_2OH$.
Properties: Colorless liquid; stable. Miscible with most organic solvents; slightly soluble in water. B. p. 148.9°C; sp. gr. 0.8328 (20/20°C); wt/gal 6.93 lbs (20°C); refractive index 1.4229 (20°C); surface tension 28.05 dynes/cm (28°C); viscosity 0.0563 poise (20°C); specific heat 0.586 cal/g at b. p.; flash point (ASTM open cup) 137°F; coefficient of expansion (per °C) 0.000892 to 20°C, 0.000921 to 55°C; vapor pressure 0.9 mm (20°C).
Typical specifications: Acidity not more than 0.02% (as acetic); color water-white; sp. gr. 0.830 to 0.835 (20/20°C); boiling range (760 mm) 140-155°C; dryness, miscible with 19 vol. of 60° Bé. gasoline (20°C); non-volatile matter not more than 0.005 g/100 cc; av wt/gal 6.93 lbs (20°C).
Grades: Technical.
Containers: 1-, 5-gal cans; 55-gal nonreturnable drums; tank cars.
Uses: Solvent for gums, oils, resins, waxes, dyes, other products; solvent; diluent; synthesis of perfumes, drugs; flavoring materials.

N-ethylbutylamine $C_2H_5NHCH_2CH_2CH_2CH_3$.
Properties: Water-white; amine odor; boiling range 110-113°C; sp. gr. 0.739 (20/20°C); refractive index 1.407 (20°C); flash point 65°F.
Shipping regulations: Flammable liquid. Red label. *

ethyl butyl carbonate $C_2H_5CO_3C_4H_9$.
Properties: Colorless liquid used as solvent for many natural and synthetic resins; in mixtures for nitrocellulose. Sp. gr. 0.92 to 0.93 (20°C); b. p. 135-175°C; flash point 50°C.

ethyl n-butyl ether (n-butyl ethyl ether) $C_2H_5OC_4H_9$.
Properties: Liquid; sp. gr. (20°C) 0.7528; m. p. −103°C; b. p. 92.2°C; flash point 30°F; vapor pressure (20°C) 43 mm; slightly soluble in water.
Containers: Cans; drums.
Uses: Extraction solvent; inert reaction medium.
Hazard! Flammable.
Shipping regulations: Flammable liquid. Red label. *

*See "I.C.C. Shipping Regulations," page xiii.
Reference numbers refer to name of manufacturer. See "List of Manufacturers," page v.

ethyl butyl ketone (3-heptanone) $CH_3CH_2CH_2CH_2COCH_2CH_3$.
Properties: Sp. gr. 0.8191 (20/20°C); boiling range 142.8 to 147.8°C; acidity as acetic 0.034%; 95% purity; wt/gal 6.8 lbs.
Containers: Drums; tank cars.
Uses: In solvent mixtures for air-dried and baked finishes; for polyvinyl and nitrocellulose resins.

2-ethyl-2-butylpropanediol-1, 3 (2-butyl-2-ethylpropanediol-1, 3) $HOCH_2C(C_2H_5)(C_4H_9)CH_2OH$.
Properties: White, crystalline solid; sp. gr. (50/20°C) 0.931; b. p. (50 mm) 178°C; f. p. 41.4°C; solubility in water (20°C) 0.8% by wt.
Use: Synthesis of lubricants, emulsifying agents; insect repellents; plastics.

2-ethylbutyl silicate $[CH_3C(C_2H_5)CH_2CH_2O]_4Si$.
Properties: Colorless liquid; b. p. (1 mm) 164°C.
Derivation: Reaction of silicon tetrachloride with 2-ethylbutanol.
Uses: Hydraulic fluid; heat transfer liquid.

2-ethylbutyraldehyde (diethyl acetaldehyde) $(C_2H_5)_2CHCHO$.
Properties: Colorless liquid; insoluble in water. Sp. gr. 0.8164 (20/20°C); b. p. (760 mm) 116.8°C; vapor pressure 13.7 mm (20°C); flash point 70°F; wt/gal 6.8 lbs (20°C); coefficient of expansion 0.00111 (20°C); f. p. −89°C; viscosity 0.60 cps (20°C).
Typical specifications: Sp. gr. 0.8170 to 0.823; boiling range 80 to 135°C (760 mm); acidity not more than 2.00% (as ethylbutyric).
Grades: Technical.
Containers: 1-gal cans; 5-, 55-gal drums.
Uses: Organic synthesis; pharmaceuticals; rubber accelerators; synthetic resins.
Shipping regulations: Flammable liquid. Red label. *

ethyl butyrate (ethyl butanoate) $C_3H_7CO_2C_2H_5$.
Properties: Colorless, non-toxic, volatile liquid; pineapple-like odor. Soluble in alcohol and ether; insoluble in water. Sp. gr. 0.8788; m. p. −93.3°C; b. p. 120.6°C; refractive index (n 20/D) 1.400.
Derivation: Ethyl alcohol and butyric acid are heated together in presence of sulfuric acid, with subsequent distillation.
Method of purification: Rectification.
Grades: Technical.
Containers: Iron drums; glass bottles.
Uses: Flavoring extracts; imparting a pineapple flavor; perfumery; solvent mixture for cellulose esters and ethers; many natural and synthetic resins; lacquers; safety glass.

2-ethylbutyric acid (diethyl acetic acid) $(C_2H_5)_2CHCOOH$.
Properties: Water-white liquid. Resembles butyric acid in most properties except that its odor is less pronounced and its water solubility limited. Sp. gr. 0.9225 (20/20°C); b. p. (760 mm) 190°C; vapor pressure 0.08 mm (20°C); flash point 210°F; wt/gal 7.7 lbs (20°C); coefficient of expansion 0.00093 (20°C); f. p. −9.4°C; viscosity 3.13 cps (20°C).
Grades: Technical.
Containers: 1-gal glass jugs; 5-gal carboys; 55-gal stainless steel drums.
Uses: Forms esters; intermediates used in making drugs, dyestuffs, chemicals.

ethyl caffeate $C_6H_3(OH)_2CH{:}CHCOOC_2H_5$.
Properties: Yellow to tan crystals; characteristic, aromatic odor; insoluble in water; very soluble in alcohol.
Typical specifications: Melting range 144-147°C; ash 0.1% (max). Moisture content 3% max.
Grade: C. P.
Containers: Bottles; fiber drums.
Uses: Food antioxidant.
Shipping regulations: None. *

ethyl caprate (ethyl decanoate; ethyl caprinate) $C_9H_{19}COOC_2H_5$.
Properties: Colorless liquid; fragrant odor. Soluble in alcohol and ether; insoluble in water; sp. gr. 0.862; b. p. 243°C.
Derivation: By heating capric acid, absolute alcohol and sulfuric acid, with subsequent distillation.
Method of purification: Redistillation.
Grades: Technical.
Containers: Iron drums; glass bottles.
Uses: Organic synthesis; manufacturing wine-bouquet and cognac essence.
Shipping regulations: None. *

ethyl caprinate. See ethyl caprate.

ethyl caproate (ethyl capronate; ethyl hexoate; ethyl hexanoate) $C_5H_{11}COOC_2H_5$.
Properties: Colorless to yellowish liquid; pleasant odor. Soluble in alcohol and ether; insoluble in water. Sp. gr. 0.873; b. p. 167°C.
Derivation: By heating absolute alcohol and n-caproic acid in presence of sulfuric acid, with subsequent distillation.
Method of purification: Rectification.
Grades: Technical.
Containers: Iron drums; glass bottles.
Uses: Organic synthesis; artificial fruit essences.
Shipping regulations: None. *

ethyl capronate. See ethyl caproate.

ethyl caprylate (ethyl octoate; ethyl octanoate) $CH_3(CH_2)_6COOC_2H_5$.
Properties: Colorless liquid; pineapple odor. Soluble in alcohol and ether; insoluble in water. Sp. gr. 0.873; m. p. −48°C; b. p. 207-209°C.
Derivation: By heating caprylic acid, alcohol, and sulfuric acid with subsequent distillation.
Method of purification: Rectification.
Grades: Technical.
Containers: Iron drums; glass bottles.
Uses: Flavoring; fruit essences.
Shipping regulations: None. *

ethyl carbamate. See urethane.

N-ethylcarbazole (9-ethylcarbazole).
Properties: Leaflets; soluble in ether and hot alcohol. M.p. 69-70°C; b.p. 175°C (5 mm).
Derivation: Action of ethyl chloride on potassium carbazolate.
Uses: Intermediate for dyes, pharmaceuticals, agricultural chemicals.

ethyl carbonate. See diethyl carbonate.

ethyl cellulose. An ethyl ether of cellulose.
Properties: White, granular, thermoplastic solid; variable properties depending on the degree to which hydroxyl radicals of cellulose have been replaced by ethoxy groups. The standard commercial grade has 47-48% ethoxy content (intermediate between compositions corresponding to two and three ethoxy groups per glucose unit of the cellulose). Greater ethoxy content increases solubility in organic solvents, lowers softening temperature, and decreases moisture absorption. The standard grade has sp. gr. 1.07-1.18; refractive index 1.47; high dielectric strength; softening point 100-130°C; and forms tough films, which retain flexibility to low temperatures. Soluble in most organic liquids, and compatible with resins, waxes, oils, and plasticizers; inert to alkalies and dilute acids.
Derivation: From alkali cellulose and ethyl chloride or sulfate; from cellulose and ethyl alcohol in presence of dehydrating agents.
Grades: Technical; N.F. XI. Sold in bags.
Uses: Adhesives; cable lacquers; extrusion wire insulation; injection plastics; protective coatings; hot-melt paper and cloth coating; pigment-grinding base; may be used as toughening agent for plastics, protective coatings, and textile finishing; printing inks; molding powders; proximity fuses; inhibitor tape for rockets; in vitamin preparations.

ethyl centralite. See sym-diethyldiphenylurea.

ethyl chloride (chloroethane) C_2H_5Cl.
Properties: Gas at ordinary temperature; compressed, a colorless, highly flammable, volatile liquid. Ether-like odor, burning taste. Stable and non-corrosive when dry but will hydrolyze in the presence of water or alkalies. Miscible with most of the commonly used solvents; slightly soluble in water. Sp. gr. 0.9214; m.p. −140.85°C; b.p. 12.5°C; critical point 187.2°C (52 atm; sp. gr. 0.33); latent heat of vaporization at 15.0°C 387 joules/g, at 25°C 385 joules/g; heat of combustion 316.7 kg cal/mole (gas); thermal conductivity gaseous state 0.873 (0°C); dielectric constant of liquid 6.29 (170°C) under its own pressure; vapor pressure 1000 mm (20°C); flash point (closed cup) −58°F.
Derivation: (a) From ethylene and hydrogen chloride; (b) by passing hydrogen chloride into a solution of zinc chloride and ethyl alcohol.
Method of purification: Distillation.
Grades: Technical; N.F. XI.
Containers: Cylinders; drums; tank cars.
Uses: Manufacture of tetraethyl lead and ethyl cellulose; anesthetic in medicine and dentistry; organic synthesis, alkylating agent; refrigeration; analytical reagent; solvent for phosphorus, sulfur, fats, oils, resins and waxes; insecticides.
Danger! Extremely flammable. MCA warning label.
Shipping regulations: Flammable liquid. Red label. *

ethyl chloroacetal $ClCH_2CH(OC_2H_5)_2$.
Properties: Water-white liquid with pleasant odor; sp. gr. 1.022 (20°C); boiling range: 54-61°C (20 mm); 149-153°C (760 mm); m.p. −32°C; flash point 117°F; refractive index 1.418 (20°C); soluble in alcohol and ethyl ether; insoluble in water.

ethyl chloroacetate $CH_2ClCO_2C_2H_5$.
Properties: Water-white, mobile liquid; pungent, fruity odor. Decomposed by hot water and alkalies. Soluble in alcohol, benzene, and ether; insoluble in water. Sp. gr. 1.1585 (20°C); b.p. 144.2°C; vapor density 4.23-4.46; flash point 54°C; refractive index (n 20/D) 1.4227.
Derivation: (a) By the action of chloroacetyl chloride on alcohol; (b) by treating chloroacetic acid with alcohol and sulfuric acid.
Method of purification: Distillation.
Grades: Technical.
Containers: Bottles; carboys; 55-gal drums.
Uses: Solvent; organic synthesis; military poison gas; vat dyestuffs.

ethyl chlorocarbonate (ethyl chloroformate) $ClCOOC_2H_5$.
Caution! Poisonous! Insoluble in water; soluble in alcohol, benzene, chloroform and ether. Wt/gal 9.46 lbs (approx) (20°C); coefficient of expansion per °F 0.00070, per °C, 0.00126; flash point 61°F (approx).
Typical specifications: Color water-white; dryness, miscible without turbidity with 20 vols 60° Bé gasoline at 20°C; free chlorine none; odor irritating, tear-producing; purity, ester content as ethyl chlorocarbonate, not less than 96%; sp. gr. (20/20°C) 1.135-1.139; b.p. 93-95°C; refractive index (n 20/D)1.3974.
Derivation: By reacting carbon monoxide with gaseous chlorine, producing phosgene ($COCl_2$) which is then reacted with anhydrous ethyl alcohol giving ethyl chlorocarbonate and splitting off hydrochloric acid. Owing to the poisonous and corrosive character of the main raw materials, all reactions take place in special acid-resistant equipment under constant technical control.
Grades: Technical.
Containers: 1-gal bottles; 5-gal carboys.
Use: Organic synthesis (intermediate in making diethyl carbonate and flotation agents).
Fire hazard: Flammable. Keep lights and fire away.

*See "I.C.C. Shipping Regulations," page xiii.
Reference numbers refer to name of manufacturer. See "List of Manufacturers," page v.

Shipping regulations: Corrosive liquid. White label. *

ethyl chloroformate. See ethyl chlorocarbonate.

ethylchlorosulfonate $C_2H_5OClSO_2$.
Properties: Colorless, oily liquid. Pungent odor; fumes in moist air. Decomposed by water. Attacks lead and tin, but copper mildly. Iron and steel not affected. Caution! Very irritant! Soluble in chloroform and ether; insoluble in water. Sp. gr. 1.379 (0°C); b. p. 152-153°C; vapor density 5 (air = 1); volatility 18,000 mg/cu m (20°C).
Derivation: (a) Action of fuming sulfuric acid on ethylchloroformate; (b) interaction of ethylene and chlorosulfonic acid.
Grades: Technical.
Uses: Organic synthesis; military poison gas.

ethyl cinnamate (ethyl phenylacrylate; cinnamic ether; cinnamylic ether) $C_6H_5CH{:}CHCOOC_2H_5$.
Properties: Limpid, oily liquid; strawberry-like odor. Soluble in alcohol and ether; insoluble in water. Sp. gr. 1.045-1.048; refractive index (n 20/D) 1.560; congealing point 7° (min); b. p. 271°C.
Derivation: By heating ethyl alcohol and cinnamic acid in presence of sulfuric acid.
Method of purification: Rectification.
Grades: Technical.
Containers: Bottles; cans; drums.
Uses: Perfumery; flavoring extracts.
Shipping regulations: None. *

ethyl citrate. See triethyl citrate.

ethyl cocoinate (cognac ether). Ethyl esters of mixed fatty acids.
Properties: Yellow, oily liquid; odor of russet apples. Soluble in alcohol and ether; insoluble in water. Sp. gr. 0.855.
Derivation: By the action of dry hydrochloric acid gas on an alcoholic solution of the fatty acids of coconut oil.
Method of purification: Rectification.
Grades: Technical.
Containers: Iron drums; glass bottles.
Use: Flavoring cognac.
Shipping regulations: None. *

ethyl crotonate $CH_3CH{:}CHCOOC_2H_5$.
Properties: Water-white liquid. Characteristic pungent persistent odor. Soluble in alcohol and ether; insoluble in water. Sp. gr. 0.9207 (20/20°C); b. p. 139°C; flash point 36°F; refractive index 1.4242 (20°C); wt/gal 7.65 lbs (20°C).
Grades: Technical.
Containers: Non-returnable. 1-, 5-gal cans; 55-gal drums.
Uses: Solvent and softening agent; lacquers; organic synthesis.
Shipping regulations: Flammable liquid. Red label. *

N-ethyl-ortho-crotonotoluide. See crotamiton.

2-ethyl-cis-crotonylurea. See ectylurea.

ethyl cyanide (propionitrile; propanenitrile) C_2H_5CN.
Properties: Mobile, colorless liquid; ethereal odor; poisonous! Soluble in alcohol and water. Sp. gr. 0.7829 (20/20°C); refractive index (n 20/D) 1.3664; b. p. 97.4°C; f. p. −92.9°C; flash point 61°F (open cup).
Derivation: By heating barium-ethyl sulfate and potassium cyanide, with subsequent distillation.
Method of purification: Rectification.
Grades: Technical.
Containers: Iron drums; glass bottles. Keep tightly closed.
Uses: Solvent; dielectric fluid; intermediate.
Shipping regulations: Flammable liquid. Red label. *

ethyl cyanoacetate (malonic ethyl ester nitrile) $CNCH_2COOC_2H_5$.
Properties: Colorless liquid; b. p. 206-208°C; m. p. −22.5°C; refractive index 1.41751 (20°C/D). Soluble in alcohol and ether; slightly soluble in water and alkaline solutions.
Derivation: Esterification of cyanoacetic acid with ethanol; reaction of an alkali cyanide and chloroacetic ethyl ester.
Method of purification: Vacuum distillation.
Grades: Reagent; technical.
Containers: Tin-lined steel drums.
Uses: Organic synthesis; pharmaceuticals; dyes.

ethylcyclohexane $C_2H_5C_6H_{11}$.
Properties: Colorless liquid; sp. gr. 0.787; boiling point 131.8°C; refractive index (n 20/D) 1.4330.
Grades: Technical.
Use: Organic synthesis.

ethylcyclopentane $C_2H_5C_5H_9$.
Constants: Sp. gr. 0.766; aniline equivalent 1; b. p. 103.5°C; refractive index (n 20/D) 1.4198.

ethyl cyclopentanone-2-carboxylate. See 2-carbethoxycyclopentanone.

ethyl decanoate. See ethyl caprate.

ethyldichloroarsine (dichloroethylarsine) $C_2H_5AsCl_2$.
Properties: Colorless, mobile liquid. Becomes yellowish under the action of light and air. Fruit-like odor (high dilution). Decomposed by water. Attacks brass, but not iron (dry). Caution! Very irritant! Soluble in alcohol, benzene, ether, and water.
Constants: Sp. gr. 1.742 (14°C); b. p. 156°C (decomposes); m. p. −65°C; coefficient of thermal expansion 0.0011; vapor density 6 (air = 1); volatility 20,000 mg/cu m (20°C); vapor pressure 2.29 mm (21.5°C).
Derivation: Chlorination of ethyl arsenious oxide.
Shipping regulations: Poison, class A. Poison gas label. * Not accepted by express.

ethyl 4,4'-dichlorobenzilate (4,4'-dichlorobenzilic acid ethyl ester; ethyl 2-hydroxy-2,2-bis(4-chlorophenyl) acetate) $(C_6H_4Cl)_2C(OH)COOC_2H_5$.
Properties: Viscous yellow liquid. Slightly soluble in water; soluble in most organic

*See "I.C.C. Shipping Regulations," page xiii.
Reference numbers refer to name of manufacturer. See "List of Manufacturers," page v.

solvents. Incompatible with alkaline materials or strong acids. Toxicity: Symptoms similar to those caused by DDT.
Use: Acaricide in spider-mite control; synergist for DDT.

ethyldichlorosilane $C_2H_5SiHCl_2$.
Properties: Colorless liquid; b. p. 75. 5°C; sp. gr. 1. 088 (25/25°C); flash point (Cleveland open cup) 30°F. Readily hydrolyzed by moisture, with the liberation of hydrogen and hydrochloric acid.
Derivation: By Grignard reaction of trichlorosilane and ethylmagnesium chloride.
Grade: Technical.
Use: Intermediate for silicones.
Shipping regulations: Flammable liquid. Red label. *

ethyldiethanolamine $C_2H_5N(CH_2CH_2OH)_2$.
Properties: Sp. gr. 1. 015 (20°C); boiling range 246-252°C; color water-white; odor amine.
Fire hazard: Flash point 255°F.
Uses: Solvent; detergents.

ethyl diiodobrassidate (diiodobrassidinic acid ethyl ester; iodobrassid) $CH_3(CH_2)_7CICI(CH_2)_{11}COOC_2H_5$ (trans).
Properties: White needles containing 41% iodine. Decomposes slowly in solution yielding free iodine; m. p. 37°C; soluble in alcohol, ether, chloroform, fatty oils; insoluble in water.
Use: Medicine.
Shipping regulations: None. *

ethyldimethylmethane. See isopentane.

ethyldipropylmethane. See 4-ethylheptane.

ethyl N, N-di-n-propylthiolcarbamate. See EPTC.

ethylene (olefiant gas; bicarburetted hydrogen; elayl; ethene) $H_2C:CH_2$.
Properties: Colorless gas with characteristic sweet odor and taste; flammable; forms an explosive mixture with air at a concentration of 3% (approx); m. p. −169°C; b. p. −102. 5°C; flash point −136°C; sp. gr. of liquid at 0°C 0. 610; vapor density (0°C, 760 mm) (air = 1) 0. 975; critical temperature 9. 5°C; critical pressure (absolute) 744 psi; explosive limits in air (per cent by volume) lower 3. 0, upper 34. 0. Has been found in auto exhaust fumes.
Typical specifications: Purity not less than 96% ethylene by gas volume, not more than 0.5% acetylene, not more than 4% methane and ethane; 13.4 cu ft/lb (15.6°C, 760 mm); slightly soluble in water, alcohol, and ethyl ether.
Derivation: Cracking of petroleum and natural gas. Ethylene is a major component of the refinery gas from cracking units, and is sometimes recovered therefrom by distillation or other means. In some cases the ethylene is used in further chemical reactions without purification. Some pure ethylene is produced by passing hot ethyl alcohol vapors over a suitable catalyst, such as activated alumina.
Grades: Technical (95% min); 99. 5% min; N. F. XI.
Containers: Cylinders and tube trailers.
Uses: The source of many tonnage ethyl and ethylene compounds (in approximate order of volume): ethylene oxide; polyethylene; synthetic ethyl alcohol; styrene; ethylene dichloride; ethyl chloride; ethylene dibromide, etc. It is also used in the coloring of fruit and blanching of vegetables; to increase growth rate of seedlings, vegetables and fruit trees; for oxyethylene welding and cutting of metals; in medicine (anesthetic).
Fire hazard: Dangerous.
Shipping regulations: Flammable gas. Red gas label. *

ethylene alcohol. See ethylene glycol.

ethyleneamine. See piperazine.

ethylenebis(iminodiacetic acid). See ethylenediaminetetraacetic acid.

ethylene bromide. See ethylene dibromide.

ethylene carbonate (glycol carbonate; dioxolone-2) $(CH_2O)_2CO$.
Properties: Colorless, odorless, low-melting solid. M. p. 36. 4°C; b. p. 248°C; sp. gr. (39/4°C) 1. 3218; refractive index (n 50/D) 1. 4158; flash point (open cup) 320°F. Miscible (40°) with water, alcohol, ethyl acetate, benzene, and chloroform. Soluble in ether, n-butanol, and carbon tetrachloride.
Derivation: Interaction of ethylene glycol and phosgene.
Grade: Technical.
Uses: Solvent for many polymers and resins; solvent extraction; organic synthesis of pharmaceuticals, rubber chemicals, textile finishing agents.

ethylene carboxylic acid. See acrylic acid.

ethylene chloride. See ethylene dichloride.

ethylene chlorobromide. See sym-bromochloroethane.

ethylene chlorohydrin (2-chlorethyl alcohol; glycol chlorohydrin) $ClCH_2CH_2OH$.
Properties: Colorless liquid; faint ethereal odor. Soluble in most organic liquids and completely miscible with water. Poisonous! Sp. gr. 1. 2045 (20/20°C); b. p. 128. 7°C (760 mm); refractive index (n 20/D) 1. 4419; vapor pressure 4. 9 mm (20°C); flash point 140°F; wt/gal 10. 0 lbs (20°C); coefficient of expansion 0. 00089 (20°C); f. p. −62. 6°C; viscosity 0. 0343 poise (20°C).
Typical specifications: Anhydrous grade: Acidity not more than 0. 02% (as hydrochloric); purity not less than 98. 0% ethylene chlorohydrin; color water-white (if shipped in glass); sp. gr. 1. 202-1. 208 (20/20°C); boiling range (760 mm) 122-135°C; solubility in water, completely miscible; average wt/gal 10. 01 lbs (20°C).
Typical specifications: 38% grade: Purity 38-42% ethylene chlorohydrin by weight; sp. gr. 1. 087-1. 097 (20/20°C); acidity not more than 0. 02% (as hydrochloric);

solubility in water, miscible with water without cloudiness; color water-white (if shipped in glass); odor mild, non-residual; average wt/gal 9.1 lbs (20°C).

Derivation: By action of hypochlorous acid on ethylene.

Method of purification: Rectification.

Grades: Refined; anhydrous.

Containers: (36 to 40% grade) bottles; jugs; carboys; (anhydrous) tank cars.

Uses: Solvent for cellulose acetate; solvent mixtures for ethylcellulose, cellulose acetate; synthesis of oil of rose, procaine, indigo, introduction of hydroxyethyl group in organic synthesis; to activate sprouting of dormant potatoes.

Danger! Extremely hazardous liquid and vapor. May be fatal if inhaled, swallowed or absorbed through skin. MCA warning label.

Shipping regulations: Poison, class B. Poison label. *

ethylene cyanide (ethylene dicyanide; succinonitrile) $C_2H_4(CN)_2$.

Properties: Colorless, waxy solid; soluble in alcohol, water, and chloroform. M.p. 57-57.5°C; b.p. 265.7°C (760 mm).

Derivation: By the interaction of ethylene dibromide and potassium cyanide in presence of alcohol.

Method of purification: Crystallization.

Use: Organic synthesis.

ethylene cyanohydrin (beta-hydroxy propionitrile) $HOCH_2CH_2CN$.

Properties: Poisonous; straw colored liquid; m.p. −46°C; b.p. 227-228°C (dec); sp. gr. 1.0404 (25/4°C); vapor pressure 0.08 mm (25°C), 20 mm (117°C), 760 mm (227-228°C). Miscible in all proportions with water, acetone, methyl ethyl ketone, ethanol, chloroform and diethyl ether. Insoluble in benzene, carbon tetrachloride, and naphtha.

Derivation: Ethylene oxide and hydrocyanic acid.

Grade: Technical.

Containers: Steel drums; tank cars.

Uses: Solvent for certain cellulose esters and inorganic salts. Organic intermediate.

ethylenediamine (1,2-diaminoethane) $NH_2CH_2CH_2NH_2$.

Properties: Volatile, colorless, alkaline liquid; ammonia odor. Strong base. Soluble in water, alcohol; slightly soluble in ether; insoluble in benzene.

Constants: Sp. gr. 0.8995 (20/20°C); wt/gal 7.50 lbs (20°C); 899 g/liter (20°C); b.p. 116-117°C; vapor pressure 10.7 mm (20°C); m.p. 8.5°C; viscosity 0.0154 poise (25°C); refractive index 1.4540 (26°C); dielectric constant 16.0 (18°C); ionization constant 7.1×10^{-5} (25°C); pH of 25% solution 11.9 (25°C); latent heat of vaporization 167 cal/g (calc); latent heat of fusion 77 cal/g (0°C); heat of solution 7.6 cal/mole (15°C); heat of combustion 452.6 cal/mole.

Typical specifications: Purity not less than 66% by wt; boiling range (760 mm) below 115°C none; above 122°C none; odor mildly ammoniacal; color water-white.

Derivation: By heating ethylene dichloride and ammonia, with subsequent distillation.

Method of purification: Redistillation.

Grades: Technical; U.S.P. XVI (67% solution); various strengths solutions.

Containers: 55-gal tin-lined drums; tank cars.

Uses: Solvent for albumin and fibrin; medicine; neutralizing acidity of oils; preparing casein and shellac solutions; stabilizing rubber latex; corrosion inhibitor in antifreeze solutions; textile lubricants; dyes; rubber accelerators; making of ethylenediamine nitrate, chlorate, and EDTA; dehairing fur skins; emulsifier; organic synthesis. To form water-insoluble polyamide resin adhesives from di- and trimerized unsaturated vegetable oil acids; heat sensitive adhesive in labelling and packaging.

Danger! Causes severe eye and skin burns. MCA warning label.

Shipping regulations: None. *

ethylenediamine mercury sulfate $HgSO_4 \cdot 2(CH_2NH_2)_2 \cdot 2H_2O$.

Properties: White powder. Contains approximately 43% mercury. Incompatible with sodium chloride. Do not expose to air. Soluble in water.

Use: Medicine.

Shipping regulations: None. *

ethylene diamine tartrate. Used to make piezoelectric crystals for control of electric frequencies, etc., as in televison.

ethylenediaminetetraacetic acid (EDTA; ethylenebisiminodiacetic acid; ethylenedinitrilotetraacetic acid) $(HOOCCH_2)_2NCH_2CH_2N(CH_2COOH)_2$. One of the most important organic chelating agents.

Properties: Colorless crystalline solid, decomposing at 240°C. Slightly soluble in water; insoluble in common organic solvents; neutralized by alkali metal hydroxides to form a series of water-soluble salts containing from one to four alkali metal cations.

Derivation: (a) By the addition of sodium cyanide and formaldehyde to a basic solution of ethylenediamine (forms the tetrasodium salt); (b) by heating tetrahydroxyethylethylenediamine with sodium or potassium hydroxide using a cadmium oxide catalyst (forms the tetrasodium salt).

Uses: In detergents, liquid soaps, shampoos, agricultural chemical sprays; for metal treatment such as cleaning and plating operations; for treatment of chlorosis; for decontamination of radioactive surfaces; as metal deactivator in vegetable oils, oil emulsions, pharmaceutical products, etc.; as an anticoagulant of blood; as an eluting agent in ion exchange; to remove insoluble deposits of calcium and magnesium soaps; in textiles to improve dyeing properties and scouring and detergent operations; in rubber and polymers; antioxidant; clarification of liquids; analytical chemistry; in heavy metal poisoning, to chelate lead, copper, etc.

ethylenediaminetetraacetic acid calcium disodium chelate. See calcium disodium EDTA.

ethylenediaminetetraacetic acid, disodium salt (disodium ethylenediaminetetraacetate; disodium dihydrogen ethylenediaminetetraacetate). Used as a dietary supplement or sequestrant in animal feeds; as a sequestrant in vinegar.

ethylenediaminetetraacetic acid salts (EDTA salts). A variety of these are available, with uses identical or similar to the acid. Among them are the tetrapotassium and tetrasodium salts, the disodium dihydrate, trisodium trihydrate or monohydrate, sodium ferric EDTA, dihydrogen ferrous EDTA, disodium calcium EDTA, and a similar range of disodium salts with magnesium, divalent cobalt, manganese, copper, zinc and nickel.

ethylene dibromide (EDB; 1,2-dibromoethane; ethylene bromide) CH_2BrCH_2Br.

Properties: Colorless, volatile, nonflammable liquid. Sweetish odor. Emulsifiable. Poisonous! Miscible with most solvents and thinners; slightly soluble in water. Sp. gr. 2.17-2.18 (20°C); wt/gal 18.1 lbs; b.p. 131°C; vapor pressure 17.4 mm (30°C); f.p. 9.10°C; surface tension 38.71 dynes/cm (20°C); latent heat of fusion at 9.55°C 56.62 joules/g, of vaporization at 130.8°C, 193.5 joules; dielectric constant 4.86 (18°C); refractive index 1.5357 (25°C); flash point none; fire point none; specific heat 0.18 cal/g/°C; specific resistivity 2.4×10^9 ohms/cm.

Typical specifications: Colorless; sp. gr. 2.17-2.19 (20/4°C); acidity less than 1 cc (0.01 N NaOH/100 cc oil); boiling range 90% within 2°C.

Derivation: By the action of bromine on ethylene gas.

Method of purification: Rectification.

Grade: Technical.

Containers: 55-gal drums; tank cars.

Uses: Scavenger for lead in gasoline; grain and fruit fumigant; solvent for fats, oils, waxes, gums, resins, other products; waterproofing preparations; celluloid; organic synthesis; wood insecticide; medicine.

Warning! Vapor harmful. Absorbed through skin. Avoid breathing vapor. MCA warning label.

Shipping regulations: None.*

ethylenedicarboxylic acid. See succinic acid.

ethylene dichloride (sym-dichloroethane; 1,2-dichloroethane; ethylene chloride; Dutch liquid; Dutch oil) CH_2ClCH_2Cl.

Properties: Colorless, oily liquid; chloroform-like odor; sweet taste. Stable in presence of water, alkalies, acids, or actively-reacting chemicals. Resistant to oxidation. Will not corrode metals. While it will burn, it does so with difficulty and the addition of 25% by volume of carbon tetrachloride is said to render the mixture safe under ordinary conditions. Miscible with most common solvents; slightly soluble in water.

Constants: B.p. 83.5°C; f.p. −35.5°C; sp. gr. 1.2554 (20/4°C); wt/gal 10.4 lbs; refractive index 1.444; flash point 70°F.

Derivation: By the action of chlorine on ethylene with subsequent distillation, with ethylene dibromide as catalyst.

Grade: Technical.

Containers: 1-, 5-gal cans; 55-gal nonreturnable drums; 6000- and 8000-gal tank cars.

Uses: Vinyl chloride; solvent for fats, oils, waxes, some alkaloids, camphor, rubber, "Bakelite," various resins, gums, other products; solvent mixtures for cellulose esters and ethers and other products; oil extraction; fumigants; dry-cleaning solvent mixtures and spotting; lacquers; paint, varnish and finish removers; metal degreasing; textile cleansing processes; soaps and scouring compounds; wetting and penetrating agents; organic synthesis; ore flotation; scavenger for TEL in gasoline.

Warning: Flammable. Vapor harmful. MCA warning label.

Shipping regulations: Flammable liquid. Red label.*

ethylene dicyanide. See ethylene cyanide.

ethylenedinitrilotetraacetic acid. See ethylenediaminetetraacetic acid.

ethylene diphenyldiamine. See N,N-diphenyl ethylenediamine.

ethylene glycol (ethylene alcohol; glycol) CH_2OHCH_2OH.

Properties: Clear, colorless, syrupy liquid; sweet taste; hygroscopic; lowers freezing point of water. Soluble in water, alcohol, and ether.

Constants: Sp. gr. 1.1155 (20°C); b.p. 197.2°C; f.p. −13.5°C; wt/gal 9.31 lbs (15/15°C); refractive index 1.430 (25°C); flash point 240.8°F; fire point 248°F.

Derivation: (a) By heating ethylene chlorohydrin with a solution of an alkali carbonate or bicarbonate; (b) oxidation of ethylene with air, followed by hydration of the ethylene oxide formed; (c) from formaldehyde, water and carbon monoxide, with hydrogenation of the resulting glycolic acid.

Grade: Technical.

Containers: 5-, 10-, 55-, 110-gal drums; tank cars. Net contents 45, 90, 465, 930 lbs; 8000 gals.

Uses: Coolant in motors; antifreeze in automobile radiators; brake fluids, glycol diacetate; polyester fibers; manufacture of low-freezing dynamite; dye solvent; extractant for various purposes; solvent mixtures for cellulose esters and ethers, especially cellophane; lacquers; resins; printing inks; wood stains; glue mixtures; leather dyeing; textile processing; tobacco; solvent for waxes, resins, organic chemicals, drugs, and other products.

Shipping regulations: None.*

ethylene glycol diacetate (glycol diacetate) $CH_3COOCH_2CH_2OOCCH_3$.

Properties: Colorless liquid; faint odor.

*See "I.C.C. Shipping Regulations," page xiii.
Reference numbers refer to name of manufacturer. See "List of Manufacturers," page v.

Soluble in alcohol, ether, benzene; slightly soluble in water (10%). Sp. gr. 1.1063 (20/20°C); b.p. 190.5°C (760 mm); vapor pressure 0.3 mm (20°C); flash point 220°F; wt/gal 9.2 lbs (20°C); f.p. −41.5°C; refractive index (n 20/D) 1.415.
Derivation: (a) Ethylene glycol and acetic acid; (b) ethylene dichloride and sodium acetate.
Containers: 1-gal cans; 5-, 55-gal drums.
Uses: Solvent for cellulose esters and ethers; resins; lacquers; printing inks; perfume fixative; non-discoloring plasticizer for ethyl and benzyl cellulose.
Shipping regulations: None.*

ethylene glycol dibutyl ether $C_4H_9OC_2H_4OC_4H_9$.
Properties: Practically colorless liquid with characteristic odor. Slightly soluble in water; sp. gr. 0.8374 (20/20°C); 7.0 lb/gal (20°C); b.p. 203.6°C (760 mm); vapor pressure 0.09 mm (20°C); freezing point −69.1°C; viscosity 1.34 cps (20°C).
Containers: 1-gal can, 5-, 55-gal drums; (6.5, 30, 380 lbs net wt).
Uses: High-boiling inert solvent, useful in specialized solvent and extraction applications.

ethylene glycol dibutyrate (glycol dibutyrate) $(CH_2OCOC_3H_7)_2$.
Properties: Liquid; sp. gr. (0°C) 1.024; refractive index (25°C), 1.424; b.p. 240°C; m.p. less than −80°C; solubility in water, 0.050% by weight.
Use: Plasticizer.

ethylene glycol diethyl ether
$C_2H_5OCH_2CH_2OC_2H_5$.
Properties: Colorless liquid; slight, ethereal odor; stable. Sp. gr. 0.8417 (20/20°C); b.p. 121.4°C (760 mm); vapor pressure 9.4 mm (20°C); flash point 95°F; wt/gal 7 lbs (20°C); f.p. −74°C; coefficient of expansion 0.00121 (20°C); viscosity 0.0065 poise (20°C). Partially miscible with water.
Grade: Technical.
Containers: 1-gal cans; 5-, 55-gal drums.
Uses: Organic synthesis (reaction medium); solvent, which added to colloidal systems such as detergents and wetting agents of limited water solubility, permits water dilution while reducing, without gelling, or clouding.

ethylene glycol diformate (glycol diformate) $HCOOCH_2CH_2OOCH$.
Properties: As supplied commercially is a mild odored, water-white liquid, soluble in water, alcohol and ether. Sp. gr. 1.2277 (20/20°C); 10.2 lbs/gal (20°C); b.p. 177.1°C (760 mm); vapor pressure 0.5 mm (20°C); f.p. −10°C.
Uses: In the presence of water, it hydrolyzes slowly, liberating formic acid; hence, it may be used where this characteristic is desirable. It is used in embalming fluids.

ethylene glycol dimethyl ether (GDME; 1,2-dimethoxyethane) $CH_3OCH_2CH_2OCH_3$.
Properties: Water-white liquid with a mild ether odor. Sp. gr. 0.8683 (20°C); b.p. 85.2°C; f.p. −69°C; refractive index 1.3792 (20/D); flash point 34°F (open cup); soluble in water and hydrocarbons; pH 8.2.
Use: Solvent.
Shipping regulations: Flammable liquid. Red label.*

ethylene glycol dipropionate (glycol propionate) $(CH_2OCOC_2H_5)_2$. Liquid; sp. gr. (15°C) 1.054; refractive index (25°C) 1.419; b.p. 211°C; m.p. less than −80°C; solubility in water, 0.16% by weight.
Use: Plasticizer.

ethylene glycol monoacetate (glycol monoacetate) $HOCH_2CH_2OOCCH_3$.
Properties: Colorless liquid; almost odorless; soluble in alcohol, ether, benzene, and toluene; partially soluble in water. B.p. 181-182°C; sp. gr. 1.108.
Derivation: (a) By heating ethylene glycol with acetic acid (glacial) or acetic anhydride; (b) by passing ethylene oxide into hot acetic acid containing sodium acetate or sulfuric acid.
Grade: Technical.
Use: Solvent for nitrocellulose, cellulose acetate, camphor.

ethylene glycol monobenzyl ether
$C_6H_5CH_2OC_2H_4OH$.
Properties: Water-white liquid; faint rose-like odor; sp. gr. 1.070 (20/20°C); b.p. (760 mm) 255.9°C; vapor pressure 0.02 mm (20°C); flash point 265°F; wt 8.9 lbs/gal (20°C).
Grade: Technical.
Containers: 1-gal glass jugs; 5- and 12-gal glass stoppered carboys; 8.5, 40 and 100 lb (net wt).
Uses: Solvent for cellulose acetate, dyes, inks, resins, perfume fixative; organic synthesis (selective hydroxyethylating agent); coating compositions for leather, paper, and cloth; lacquers.

ethylene glycol monobutyl ether (2-butoxyethanol) $HOCH_2CH_2OC_4H_9$.
Properties: Colorless liquid; mild odor; high dilution ratio with petroleum hydrocarbons; soluble in mineral oils and water.
Constants: B.p. 171.2°C; sp. gr. 0.9019 (20/20°C); wt/gal 7.5 lbs (20°C); refractive index 1.4190 (25°C); viscosity 0.0642 poise (20°C); specific heat 0.583; vapor pressure 0.76 mm (20°C); flash point 165°F; nitrocellulose-toluene dilution ratio 3.5; coefficient of expansion 0.00092 (20°C).
Grade: Technical.
Containers: 1-gal cans; 5- and 55-gal drums; (7.5, 35, 410 lbs); tank cars up to 10,000 gals.
Uses: Solvent for nitrocellulose resins; spray lacquers; brushing lacquers of four-hour type; varnishes; enamels; dry-cleaning compounds; varnish removers; textile (preventing spotting in printing or dyeing); mutual solvent for "soluble" mineral oils to hold soap in solution and to improve the emulsifying properties. In general, an inert solvent.
Shipping regulations: None.*

*See "I.C.C. Shipping Regulations," page xiii.
Reference numbers refer to name of manufacturer. See "List of Manufacturers," page v.

ethylene glycol monobutyl ether acetate
$C_4H_9OCH_2CH_2OOCCH_3$
Properties: Colorless liquid; fruity odor. Soluble in hydrocarbons and organic solvents; insoluble in water. B. p. 192. 2°C; sp. gr. 0. 9424 (20/20°C); f. p. −63. 5°C; flash point 190°F.
Grade: Technical.
Containers: 1-, 5-, 55-gal drums; tank cars.
Uses: High-boiling solvent for nitrocellulose lacquers, epoxy resins, multi-color lacquers; film coalescing aid for polyvinyl acetate latex.
Shipping regulations: None. *

ethylene glycol monobutyl ether laurate
$C_{11}H_{23}COO(CH_2)_2OC_4H_9$.
Properties: Liquid; sp. gr. (25°C) 0. 985; m. p. −10 to −15°C; insoluble in water.
Use: Plasticizer.

ethylene glycol monobutyl ether oleate
$C_{17}H_{33}COOCH_2CH_2OC_4H_9$.
Properties: Liquid; sp. gr. (25°C) 0. 892; m. p. less than −45°C; insoluble in water.
Use: Plasticizer.

ethylene glycol monoethyl ether (2-ethoxyethanol) $HOCH_2CH_2OC_2H_5$.
Properties: Colorless liquid, practically odorless; b. p. 135. 1°C (760 mm); sp. gr. 0. 9311 (20/20°C); wt/gal 7. 7 lbs (20°C); refractive index 1. 4060 (25°C); flash point 130°C; miscible with hydrocarbons and water.
Grades: Technical.
Containers: 1-gal cans; 5- and 55-gal drums; net contents 7. 5, 35, 420 lbs; tank cars up to 10, 000 gallons.
Uses: Solvent for nitrocellulose, natural and synthetic resins; mutual solvent for formulation of soluble oils; lacquers and lacquer thinners; dyeing and printing textiles; varnish removers; cleaning solutions; leather.
Shipping regulations: None. *

ethylene glycol monoethyl ether acetate
$CH_3COOCH_2CH_2OC_2H_5$.
Properties: Colorless liquid; mild, pleasant, ester-like odor; b. p. 156. 4°C; sp. gr. 0. 9748 (20/20°C); wt/gal 8. 1 lb (20°C); refractive index 1. 4030 (25°C); viscosity 1. 32 cps (20°C); flash point 135°F; f. p. −61. 7°C; coefficient of expansion 0. 00112 (20°C); vapor pressure 1. 2 mm (20°C). Miscible with aromatic hydrocarbons; slightly miscible with water.
Grade: Technical.
Containers: 1-gal cans; 5- and 55-gal drums; net contents 8, 40, 440 lbs; tank cars up to 10, 000 gals.
Uses: Solvent for nitrocellulose, oils and resins; retards "blushing" in lacquers; varnish removers; wood stains; textiles; leather.
Shipping regulations: None. *

ethylene glycol monoethyl ether laurate
$C_{11}H_{23}COO(CH_2)_2OC_2H_5$.
Properties: Liquid; sp. gr. (25°C) 0. 89; m. p. −7 to −11°C; insoluble in water.
Use: Plasticizer.

ethylene glycol monoethyl ether ricinoleate
$C_{17}H_{32}(OH)COO(CH_2)_2OC_2H_5$.
Properties: Liquid; sp. gr. (25°C) 0. 929; m. p. less than −10°C; insoluble in water.
Use: Plasticizer.

ethylene glycol monohexyl ether
$C_6H_{13}OCH_2CH_2OH$.
Properties: Water-white liquid; sp. gr. 0. 8894 (20/20°C); 7. 4 lbs/gal (20°C); b. p. 208. 1°C (760 mm); vapor pressure 0. 05 mm (20°C); f. p. −45. 1°C; viscosity 5. 15 cps (20°C).
Containers: 1-gal can; 5-, 55-gal drums. (7. 5, 35, 410 lbs net wt).
Use: High-boiling solvent.

ethylene glycol monomethyl ether (2-methoxyethanol) $CH_3OCH_2CH_2OH$.
Properties: Colorless liquid; mild, agreeable odor; stable; miscible with hydrocarbons, alcohols, ketones, glycols, water.
Constants: B. p. 124. 6°C (760 mm); sp. gr. 0. 9663 (20/20°C); wt/gal 8. 0 lbs (20°C); refractive index 1. 4021 (20°C); flash point 115°F; nitrocellulose-toluene dilution ratio 4. 0; f. p. −85. 1°C.
Grades: Technical.
Containers: 1-, 5-gal cans; 55-gal (nonreturnable) drums; net content 8, 40, 440 lbs; tank cars up to 10,000 gal.
Uses: Solvent for nitrocellulose, cellulose acetate, alcohol-soluble dyes, natural and synthetic resins; solvent mixtures; lacquers; enamels; varnishes; leather; perfume fixative; wood stains; sealing moisture-proof cellophane.
Caution! Vapor harmful. MCA warning label.
Shipping regulations: None. *

ethylene glycol monomethyl ether acetate
$CH_3COOCH_2CH_2OCH_3$.
Properties: Colorless liquid; pleasant, characteristic ester odor; stable; miscible with the common organic solvents; soluble in water.
Constants: Sp. gr. 1. 0067 (20/20°C); b. p. 145. 1°C (760 mm); vapor pressure 2. 0 mm (20°C); flash point 140°F; wt/gal 8. 4 lbs (20°C); toluene-nitrocellulose dilution ratio, 2. 3; f. p. −65. 1°C.
Grades: Technical.
Containers: 1-, 5-gal cans; 55-gal drums; net content 8, 40, 460-lbs. Tank cars up to 10,000 gal.
Uses: Solvent for nitrocellulose, cellulose acetate, various gums, resins, waxes, oils; textile printing; photographic film; lacquers; dopes.
Shipping regulations: None. *

ethylene glycol monomethyl ether acetyl ricinoleate
$C_{17}H_{32}(OCOCH_3)COOCH_2CH_2OCH_3$.
Properties: Liquid; sp. gr. 0. 966; refractive index 1. 460; boiling range 220-260°C; m. p. less than −60°C; flash point 218°C; insoluble in water.
Use: Plasticizer.

ethylene glycol monomethyl ether ricinoleate
$C_{17}H_{32}(OH)COOCH_2CH_2OCH_3$.
Properties: Liquid; sp. gr. (25°C) 0. 935;

m. p. less than −60°C; insoluble in water.
Use: Plasticizer.

ethylene glycol monomethyl ether stearate $C_{17}H_{35}COOCH_2CH_2OCH_3$.
Properties: Sp. gr. 0.890; m. p. 21°C; insoluble in water.
Use: Plasticizer.

ethylene glycol monooctyl ether $C_4H_9CHC_2H_5CH_2OCH_2CH_2OH$.
Properties: Colorless, odorless liquid; b. p. 228.3°C; sp. gr. 0.8859; flash point 230°F; vapor pressure 0.02 mm (20°C).
Uses: Solvent for cellulose esters, and as a plasticizer.

ethylene glycol monophenyl ether $C_6H_5OCH_2CH_2OH$.
Properties: Colorless liquid; faint aromatic odor; stable in presence of acids and alkalies; partially soluble in water. Sp. gr. 1.1094 (20/20°C); b. p. 240-248°C; vapor pressure 0.03 mm (20°C); flash point 250°F; phenol 0.3% (max); 9.2 lbs/gal.
Grades: Technical.
Containers: 1-, 5-gal cans; 55-gal drums; net content 9.0, 45 and 510 lbs.
Uses: Solvent for cellulose acetate, dyes, inks, resins; perfume and soap fixative; bactericidal agent; organic synthesis of plasticizers, germicides, perfume materials and pharmaceuticals.

ethylene glycol monoricinoleate $C_{17}H_{32}(OH)COO(CH_2)_2OH$.
Properties: Clear, moderately viscous, pale yellow liquid; mild odor; miscible with most organic solvents. Sp. gr. 0.965 (25/25°C); saponification value 170; hydroxyl value 270; solidifies at −20°C; insoluble in water.
Derivation: Castor oil and ethylene glycol.
Grade: Technical.
Containers: 5-gal cans; 55-gal drums.
Uses: Plasticizer; greases; urethane polymers.

"Ethylene Glycol Monostearate 40 and 70." 260
Proprietary brand of the stearic acid ester of ethylene glycol. 40 and 70 refer to monoester contents. Balance of composition is primarily diester with small amounts of free fatty acid and free glycol. Waxy solid, practically odorless, white. Iodine value 0.5; acid value 2.0 max; melting point 56-60°C (40 grade); 52-56°C (70 grade); insoluble in water.
Uses: Opacifier in cosmetics and hair preparations.

ethylene glycol silicate $(HOCH_2CH_2)_4SiO_4$.
Properties: Colorless liquid; slowly hydrolyzed by acids; miscible with water.
Uses: Non-volatile bonding agent for pigments; weather-proofing paints for protecting concrete, stone, brick, and plastic surfaces.

ethylene glycol stearate (glycol stearate) $C_{17}H_{35}COO(CH_2)_2OH$.
Properties: Yellow waxy solid; m. p. 57-60°C; sp. gr. 0.96 (25°C); soluble in alcohol, hot ether, and acetone; insoluble in water.
Containers: 150-lb cartons.

ethylene hydrate. See gas hydrates.

ethyleneimine (aziridine) $H_2\overline{CNHC}H_2$.
Properties: Clear colorless liquid; b. p. 56°C; sp. gr. (20/4°C) 0.832; miscible with water, organic solvents; highly corrosive but may be stored and handled in tin, lead, low carbon steel, polyethylene, "Kel-F" and "Teflon" plastics.
Derivation: From beta-chloroethylamine.
Caution: Polymerizes with explosive effect at times. Toxic!
Uses: Alkyl substituted forms, called alkyl aziranes, used as intermediates and for microbial control; aziridinyl compounds used also in polymers and as intermediates. Used in tonnage lots in Germany for textile and paper processing.
Shipping regulations: Inhibited form: Flammable liquid. Red label. *

ethylenenaphthalene. See acenaphthene.

ethylene oxide $\overline{CH_2CH_2O}$.
Properties: Colorless gas at ordinary temperatures. Mobile, colorless liquid at low temperatures. Soluble in the usual organic solvents and miscible with water in all proportions.
Constants: M. p. −111.3°C; b. p. 10.73°C; sp. gr. (20/20°C) 0.8711; wt/gal (20°C) 7.25 lbs; viscosity (0°C) 0.32 cps; flash point (Tag open cup) below −4°F.
Derivation: (a) By the action of caustic alkali on ethylene chlorohydrin; (b) oxidation of ethylene in the presence of a catalyst (e. g., silver).
Method of purification: Rectification.
Grades: Technical; pure 99.7%.
Containers: 175-lb cylinders; 400-lb drums; 4000- and 10,000-gal tank cars.
Uses (in approximate order of volume): Manufacture of ethylene glycol; di- and triethylene glycols; acrylonitrile; ethanolamines; polyglycols (solvents and lubricants); non-ionic detergents; petroleum demulsifier; miscellaneous uses, including sterilizing and fumigating; sweetening gasoline.
Danger: Extremely flammable. Vapor harmful. May cause burns. MCA warning label.
Shipping regulations: Flammable liquid. Red label. *

ethylene periodide. See tetraiodoethylene.

ethylene propylene rubber (EPR). A product of stereospecific copolymerization of ethylene and propylene, stated to show superior resistance to ozone, wear, and cracking.

ethylene tetraiodide. See tetraiodoethylene.

ethylene thiourea $\overline{NHCH_2CH_2NHC}S$.
2-Imidoazolidinethione.
Properties: White to pale green crystals, faint amine odor; m. p. 199-204°C; slightly soluble in cold water; very soluble in hot water; slightly soluble at room temperature in methanol, ethanol, acetic acid, naphtha,

*See "I.C.C. Shipping Regulations," page xiii.
Reference numbers refer to name of manufacturer. See "List of Manufacturers," page v.

but appreciably soluble in these at higher temperatures. Insoluble in ether, benzene, hexane, ethyl acetate, acetone, chloroform, dioxane, and butanol.
Containers: Fiber drums (150 lbs).
Uses: Suggested for use in electroplating baths, or as an intermediate for antioxidants, insecticides, fungicides, vulcanization accelerators, dyes, pharmaceuticals, synthetic resins.

ethylene urea resins. A type of amino resin (q. v.).

ethylethanolamine (ethylaminoethanol) $C_2H_5NHCH_2CH_2OH$.
Properties: Sp. gr. 0.914 (20°C); boiling range 167-169°C; color water-white; odor amine. Soluble in water, alcohol, and ether. Flash point 160°F.
Containers: 5-gal cans; 55-gal drums; tank cars.
Uses: Solvent; intermediate.
Shipping regulations: None. *

ethyl ether. See ether.

ethyl 3-ethoxypropionate $C_2H_5OCH_2CH_2COOC_2H_5$.
Properties: Liquid; sp. gr. (20°C) 0.9496; b. p. 170.1°C; vapor pressure (20°C) 0.9 mm; sets to glass at −100°C; slightly soluble in water.
Containers: Cans.
Uses: Intermediate for vitamin B_1; other chemicals.

ethylethylene. See butene-1.

ethylfluoroformate $FCOOC_2H_5$.
Properties: Liquid. Caution! Very irritant!
Constants: Sp. gr. 1.11 (33°C); b. p. 57°C.
Derivation: Interaction of ethylchloroformate and thallium fluoride.

ethylfluorosulfonate.
Properties: Liquid. Ethereal odor. Caution! Very irritant!
Grades: Technical.

ethyl formate $HCOOC_2H_5$.
Properties: Water-white, unstable liquid. Pleasant, aromatic odor; flammable. Miscible with benzene, ether, alcohol, water; gradual decomposition in water.
Constants: Sp. gr. 0.9236 (20/20°C); m. p. −80.5°C; b. p. 54.3°C; flash point −20°C; vapor pressure 200 mm (20.6°C), 300 mm (30.2°C); wt/gal 7.61 lb (68°F); refractive index 1.35975 (20°C); electric conductivity less than 1.45×10^{-9} reciprocal ohms (25°C).
Typical specifications: Purity 95% to 100% ester, by weight; acidity neutral to methyl orange (ethyl formate hydrolyzes in the presence of water); color water-white; odor pleasant, aromatic, non-residual; solubility in water 10% by volume (20°C); solubility of water in ethyl formate 17% by volume (20°C).
Derivation: By heating ethyl alcohol with formic acid in presence of sulfuric acid.
Grades: Technical.
Containers: 1-, 5-gal cans; 55-gal (tin-lined) steel drums. Net content 8, 38, 400 lbs.
Uses: Solvent for cellulose nitrate and acetate; acetone substitute; fumigant; larvicide; synthetic flavors; synthetic resins; medicine.
Danger: Extremely flammable. MCA warning label.
Shipping regulations: Flammable liquid. Red label. *

ethyl 3-formyl propionate $OCHC_2H_4COOC_2H_5$.
Properties: Liquid; sp. gr. 1.0625 (20/20°C); b. p. 190.9°C; f. p. less than −80°C; wt/gal 8.9 lbs; flash point 200°F. Somewhat soluble in water.
Uses: Solvents for lacquers; antibiotic extractions; acetic acid separations; coalescing acids for emulsion paints.

ethyl furoate $C_4H_3OCO_2C_2H_5$.
Properties: White leaflets or prisms. Insoluble in water; soluble in alcohol and ether.
Constants: Sp. gr. 1.1174 (20.8/4°C); m. p. 34°C.

ethyl glycine hydrochloride. See glycine ethyl ester hydrochloride.

ethyl glycocoll hydrochloride. See glycine ethyl ester hydrochloride.

4-ethylheptane (ethyldipropylmethane) $CH_3(CH_2)_2CHC_2H_5(CH_2)_2CH_3$.
Properties: Colorless liquid. Sp. gr. 0.730; b. p. 141.2°C; refractive index (n 20/D) 1.4109.
Grades: Technical.
Use: Organic synthesis.

ethyl heptanoate. See ethyl oenanthate.

ethyl heptazine. See ethoheptazine.

2-ethylhexaldehyde (butylethyl acetaldehyde; octyl aldehyde; 2-ethylhexanal) $C_4H_9CH(C_2H_5)CHO$.
Properties: Colorless, high-boiling liquid. Mild, characteristic odor. Miscible with most organic solvents; slightly soluble in water.
Constants: Sp. gr. 0.8205 (20°C); b. p. 163.4°C; vapor pressure 1.8 mm (20°C); flash point 125°F; wt/gal 6.8 lbs.
Typical specifications: Acidity not more than 2.0% as 2-ethylhexoic acid; color (500-mm tube) not more than 5 yellow Lovibond; sp. gr. 0.820-0.825 (20/20°C); boiling range, (760 mm) not more than 5% distills below 160°C, not less than 95% distills below 165°C; dryness, miscible with 19 vol 60° Bé. gasoline at 20°C; non-volatile matter not more than 0.002 g/100 cc; average wt/gal 6.85 lbs (20°C).
Grades: Technical.
Containers: 1-, 5-gal cans; 55-gal non-returnable drums.
Uses: Organic synthesis; perfumes.

2-ethyl hexanal. See 2-ethylhexaldehyde.

2-ethylhexanediol-1,3 (ethohexadiol) $C_3H_7CH(OH)CH(C_2H_5)CH_2OH$.

*See "I.C.C. Shipping Regulations," page xiii.
Reference numbers refer to name of manufacturer. See "List of Manufacturers," page v.

2-Ethyl-3-propyl-1, 3-propanediol.
Properties: Practically colorless, somewhat viscous, odorless liquid; hygroscopic; irritating to eyes. Sp. gr. 0.9422 (20/20°C); 7.8 lb/gal (20°C); b.p. 244°C (760 mm); vapor pressure less than 0.01 mm (20°C); freezing point below −40°C; refractive index 1.4465-1.4515; viscosity 323 cps (20°C); soluble in alcohol and ether; partially soluble in water.
Grades: U.S.P. XVI (as ethohexadiol); also industrial grade.
Uses: Insect repellent; cosmetic ingredient; vehicle and solvent in the formulation of printing inks; medicine.

ethyl hexanoate. See ethyl caproate.

2-ethylhexanol. See 2-ethylhexyl alcohol.

2-ethylhexenal. See 2-ethyl-3-propylacrolein.

2-ethyl-1-hexene $CH_3(CH_2)_3(C_2H_5)C{:}CH_2$.
Properties: Colorless liquid; sp. gr. 0.7270 (20/4°C); b.p. 120°C; refractive index (n 20/D) 1.4157; soluble in alcohol, acetone, ether, petroleum, coal-tar solvents; insoluble in water.
Grades: 95% min purity.
Use: Organic synthesis of flavors, perfumes, medicines, dyes, resins.

ethyl hexoate. See ethyl caproate.

2-ethylhexoic acid $C_4H_9CH(C_2H_5)COOH$.
Properties: Mild-odored liquid; slightly soluble in water; sp. gr. 0.9077 (20/20°C); 7.6 lb/gal (20°C); b.p. 226.9°C (760 mm); vapor pressure 0.03 mm (20°C); freezing point −118.4°F; viscosity 7.73 cps (20°C); acid number 370.
Containers: Drums; tank cars.
Uses: As a herbicide and pesticide; souring agent. Its metallic salts, particularly lead, manganese, cobalt and zinc, are used as high-quality paint and varnish driers. Ethylhexoates of light metals, such as lithium, magnesium, calcium, and aluminum, have the property of converting certain mineral oils to greases. Aluminum 2-ethylhexoate is an excellent gelling agent for liquid hydrocarbons such as gasoline and common petroleum fractions used in coating thinners. High molecular weight esters of this acid are especially useful as plasticizers.

2-ethylhexyl. An eight-carbon radical of the formula $CH_3(CH_2)_3CH(C_2H_5)CH_2$—. Many of its compounds were formerly called the general name octyl (q. v.).

2-ethylhexyl acetate (octyl acetate) $CH_3COOCH_2CHC_2H_5C_4H_9$.
Properties: Water-white, stable liquid; very slightly soluble in water; miscible with alcohol.
Constants: Sp. gr. 0.8733 (20°C); b.p. 198.6°C; m.p. about −80°C; vapor pressure 0.4 mm (20°C); flash point 180°F; wt/gal 7.3 lb (20°C); coefficient of expansion 0.00083 (20°C); viscosity 0.0154 poise (20°C).
Grades: Technical (about 95%).
Containers: 1-, 5-gal cans; 55-gal drums.
Use: Solvent for nitrocellulose, resins, lacquers, baking finishes.

2-ethylhexyl acrylate
$CH_2{:}CHCOOCH_2CH(C_2H_5)C_4H_9$.
Properties: Liquid, pleasant odor; sp. gr. 0.8869; b.p. (50 mm) 130°C; vapor pressure at 20°C 0.1 mm; sets to glass at −90°C; flash point 180°F; insoluble in water.
Containers: 400-lb drums; tank cars.
Use: Monomers for plastics, protective coatings, paper treatment. Used widely in water-based paints.

2-ethylhexyl alcohol (2-ethylhexanol; octyl alcohol) $CH_3(CH_2)_3CHC_2H_5CH_2OH$.
Properties: Colorless, slightly viscous liquid. Miscible with most organic solvents; slightly miscible with water.
Constants: Sp. gr. 0.83 (20°C); b.p. 183.5°C; vapor pressure 0.36 mm (20°C); refractive index 1.4300 (20°C); specific heat 0.564 cal/g (25°C); wt/gal 6.9 lb (20°C); flash point 81°C.
Derivation: Aldolization of acetaldehyde or of butyraldehyde, followed by hydrogenation.
Grades: Technical.
Containers: 1-, 5-gal cans; 55-gal drums; tank cars.
Uses: Defoaming agent; wetting agent; organic synthesis; solvent for: gums, resins, waxes, mineral, vegetable and animal oils, fats, dyestuffs, other products; solvent mixtures for nitrocellulose, paints, lacquers, baking finishes; penetrants for mercerizing cotton; textile finishing compounds; making plasticizers; inks; rubber; paper; lubricants; photography; clays; dry cleaning.

2-ethylhexylamine $C_4H_9CH(C_2H_5)CH_2NH_2$.
Properties: Sp. gr. 0.7894 (20/20°C); 6.56 lb/gal (20°C); b.p. 169.2°C (760 mm); vapor pressure 1.2 (20°C); viscosity 1.11 cps (20°C); flash point 140°F. Soluble in water; solubility of water in (20°C), 25.3%.
Use: Synthesis of detergents, rubber chemicals, oil additives and insecticides.

N-2-ethylhexyl aniline
$C_6H_5NHCH_2CH(C_2H_5)C_4H_9$.
Properties: Light yellow liquid with mild odor. Sp. gr. (20/20°C) 0.9119; b.p. (50 mm) 194°C; vapor pressure (20°C) <0.01 mm; freezing point, sets to a glass below −70°C; viscosity (20°C) 7.4 cps; solubility in water (20°C) <0.01%.
Uses: Solvent; synthesis.

2-ethylhexyl bromide $C_4H_9CH(C_2H_5)CH_2Br$.
Properties: Sweet, water-white liquid, insoluble in water.
Uses: Introduction of the 2-ethylhexyl group in organic synthesis. Preparation of disinfectants, pharmaceuticals.

2-ethylhexyl chloride $C_4H_9CH(C_2H_5)CH_2Cl$.
Properties: Colorless liquid. Sp. gr. 0.8833 (20°C); b.p. 172.9°C; refractive index 1.4310; 7.33 lbs/gal; freezing point −135°C. Insoluble in water.

Grades: Technical.
Use: In the synthesis of cellulose derivatives, dyestuffs, pharmaceuticals, textile auxiliaries, insecticides, resins.

2-ethylhexyl octylphenyl phosphite $(C_8H_{17}O)(C_8H_{17}C_6H_4O)_2P$, or $(C_8H_{17}O)_2(C_8H_{17}C_6H_4O)P$.
Properties: Colorless to light yellow liquid with characteristic odor. Sp. gr. (20/4°C) 0.935-0.950; flash point (Cleveland open cup) 385-390°F; insoluble in water.
Containers: 1-, 4-lb bottles; 5-, 55-gal drums.
Use: Antioxidant; plasticizer; flame retardant; lubricating oil additive.

3, 3'-(2-ethylhexyl) thiodipropionate (dioctyl thiopropionate; thiopropionic acid, dioctyl ester) $(C_8H_{17}OOCCH_2CH_2)_2S$.
Properties: Sp. gr. (25°C) 0.952. Insoluble in water; soluble in most organic solvents.
Uses: Antioxidant; stabilizer and lubricant.

ethyl hydride. See ethane.

ethyl hydroxide. See ethyl alcohol.

ethyl 2-hydroxy-2, 2-bis(4-chlorophenyl) acetate. See ethyl 4, 4'-dichlorobenzilate.

ethyl hydroxyethyl cellulose (cellulose ether).
Properties: White granular solid; available in extra low, low, and high-viscosity types. Soluble in mixtures of aliphatic hydrocarbons with small quantity of alcohols.
Uses: Film former in silk screen and gravure printing inks and protective coatings.

ethyl alpha-**hydroxyisobutyrate** $(CH_3)_2COHCOOC_2H_5$.
Properties: Water-white liquid; sp. gr. 0.978-0.986 (20°C); b. p. 149-150°C; soluble in water, alcohol, and ether.
Grades: Technical.
Uses: Solvent for nitrocellulose and cellulose acetate; solvent mixtures for cellulose ethers; organic synthesis; pharmaceuticals.

ethylidene acetobenzoate (ethylidene benzoacetate) $C_6H_5COOCH(COCH_3)CH_3$.
Derivation: Interaction of benzoic acid and vinyl acetate in presence of catalysts.
Use: Solvent for cellulose acetate, nitrocellulose, and natural and synthetic resins.

ethylidene aniline $C_6H_5NCHCH_3$.
Properties: Dark red-brown, viscous liquid. Soluble in gasoline and benzene; insoluble in water. B. p. 205°C.
Derivation: Action of acetaldehyde on aniline.
Method of purification: Distillation.
Grades: Technical.
Containers: 100-lb veneer-covered paint kits.
Use: Accelerator for rubber vulcanization.
Shipping regulations: None. *

ethylidene benzoacetate. See ethylidene acetobenzoate.

ethylidene chloride (1, 1-dichloroethane) CH_3CHCl_2.
Properties: Colorless, neutral, mobile liquid; aromatic ethereal odor; hot saccharin taste. Nonflammable. Soluble in alcohol, ether, fixed and volatile oils; sparingly in water.
Constants: Sp. gr. 1.174 (17°C); b. p. 57-59°C; freezing point −98°C; refractive index (n 20/D) 1.4166.
Use: Medicine; extraction solvent; fumigant.

ethylidenediethyl ether. See acetal.

ethylidenedimethyl ether. See dimethylacetal.

ethylidene fluoride. See 1, 1-difluoroethane.

ethyl iodide (iodoethane) C_2H_5I.
Properties: Clear, colorless liquid; turns brown on exposure to light. Soluble in alcohol and ether; slightly soluble in water.
Constants: Sp. gr. 1.90-1.93 (25/25°C); m. p. −108°C; b. p. 72°C; refractive index (n 15/D) 1.5168.
Derivation: By digesting red phosphorus with absolute ethyl alcohol, after which iodine is added. The mixture is heated under a reflux condenser and finally distilled.
Grades: Technical.
Containers: Amber glass bottles; 5-gal carboys.
Uses: Medicine; organic synthesis.

ethyl iodoacetate $CH_2ICOOC_2H_5$.
Properties: Dense, colorless liquid. Decomposed by light and air; also (very slowly) by alkaline solutions and water. Caution! Very irritant to the eyes!
Constants: Sp. gr. 1.8; b. p. 179°C; vapor density 7.4; vapor pressure 0.54 mm (20°C).
Derivation: Interaction of potassium iodide with either ethyl bromo- or chloroacetate.

ethyl iodophenylundecylate. See iophendylate.

5-ethyl-5-isoamylbarbituric acid. See amobarbital.

ethylisobutylmethane. See 2-methylhexane.

ethyl isobutyrate $(CH_3)_2CHCOOC_2H_5$.
Properties: Colorless, volatile liquid. Soluble in alcohol and ether; slightly soluble in water.
Constants: Sp. gr. 0.870; b. p. 110-111°C; m. p. −88°C; refractive index (n 20/D) 1.3903.
Derivation: By heating isobutyric acid and ethyl alcohol, with subsequent distillation.
Method of purification: Redistillation.
Grades: Technical.
Containers: Iron drums; glass bottles.
Uses: Organic synthesis; flavoring extracts.

ethyl isocyanate C_2H_5NCO. Liquid; sp. gr. 0.898; b. p. 60°C; soluble in chlorinated and aromatic hydrocarbons.
Use: Pharmaceutical intermediate.

ethyl isothiocyanate. See ethyl thiocarbimide.

ethyl isovalerate (ethyl valerate; valerianic ether) $(CH_3)_2CHCH_2COOC_2H_5$.
Properties: Colorless oily liquid with pleasant, fruity odor. B. p. 135°C; m. p. −99°C; sp. gr. 0.868 (20/20°C); refractive

index (n 20/D) 1.4009. Slightly soluble in water; miscible with alcohol, ether, and benzene.
Derivation: By heating sodium valerate and ethyl alcohol in presence of sulfuric acid or hydrochloric acid, with subsequent distillation.
Method of purification: Rectification.
Grades: Technical.
Containers: Iron drums; 1-, 5-, 10-lb bottles.
Uses: Essential oils; perfumery; artificial fruit essences.
Shipping regulations: None. *

ethyl lactate $CH_3CHOHCOOC_2H_5$.
Properties: Colorless liquid. Mild odor. Miscible with water, alcohols, ketones, esters, hydrocarbons, oils.
Constants: Sp. gr. 1.020-1.036 (20/20°C); b. p. 154°C; flash point 158°F (Tag open cup); wt/gal 8.55 lbs (approx) (20°C).
Derivation: (a) By the esterification of lactic acid with ethyl alcohol; (b) by combining acetaldehyde with hydrocyanic acid to form acetaldehyde cyanohydrin, which is converted into ethyl lactate by treatment with ethyl alcohol and an inorganic acid such as hydrochloric acid.
Grades: Technical (96%).
Containers: 5-, 55-gal (non-returnable) steel drums. Drum cars (min. of 36,000 lbs gross).
Uses: Solvent for nitrocellulose, cellulose acetate, many cellulose ethers, resins; lacquers; paints; enamels; varnishes; emollient; gelatinant; stencil sheets; safety glass.
Shipping regulations: None. *

ethyl levulinate $CH_3CO(CH_2)_2COOC_2H_5$.
Properties: Colorless liquid; sp. gr. 1.012; b. p. 205-206°C; soluble in water; miscible with alcohol; refractive index (n 20/D) 1.4229.
Grades: Technical.
Use: Solvent for cellulose acetate and starch ethers.

ethyl magnesium bromide C_2H_5MgBr, dissolved in ether.
Properties: Flammable liquid; sp. gr. 1.01.
Containers: Glass bottles; 5-, 55-gal drums.
Use: Grignard reagent.
Hazard: Flammable.
Shipping regulations: Flammable liquid. Red label. *

ethyl magnesium chloride C_2H_5MgCl, dissolved in ether.
Properties: Flammable liquid; sp. gr. 0.85.
Containers: Glass bottles; 5-, 55-gal drums.
Use: Grignard reagent.
Hazard: Flammable.
Shipping regulations: Flammable liquid. Red label. *

ethyl malonate (malonic ester; diethylmalonate) $CH_2(COOC_2H_5)_2$.
Properties: A colorless liquid; typical sweet ester odor. Virtually insoluble in water; soluble in alcohol, ether, chloroform, and benzene.
Constants: B. p. 198°C at ordinary pressure; m. p. −50°C; sp. gr. 1.055 (25/25°C).
Derivation: By passing hydrogen chloride into cyanoacetic acid dissolved in absolute alcohol, with subsequent distillation.
Method of purification: Distillation.
Grades: Technical; C. P.
Containers: Carboys; drums.
Use: Intermediate, for barbiturates and certain pigments.
Shipping regulations: None. *

ethylmalonic acid $C_2H_5CH(COOH)_2$.
Properties: Colorless crystals. Soluble in water, alcohol, and ether.
Constants: M. p. 111.5°C; decomposes at 160°C.
Derivation: From alpha-bromobutyric acid heated with potassium mercuric cyanide and decomposed with potassium hydroxide.
Method of purification: Crystallization.

ethyl mercaptan (ethyl sulfhydrate; ethanethiol) C_2H_5SH.
Properties: Colorless liquid. Flammable; volatile. Penetrating garlic-like odor. Slightly soluble in water; soluble in alcohol, ether, petroleum naphtha. Sp. gr. 0.83907 (20/4°C); b. p. 36°C; m. p. −121°C; refractive index (n 20/D) 1.4305; flash point, below 0°C.
Derivation: By saturating potassium hydroxide solution with hydrogen sulfide, mixing with calcium ethylsulfate solution and distilling on a water bath.
Containers: 55-gal drums; 25-lb cylinders.
Shipping regulations: Flammable liquid. Red label. *

ethylmercuric acetate $C_2H_5HgOOCCH_3$.
Properties: White crystalline powder; m. p. 178°C; slightly soluble in water; soluble in many organic solvents; may be steam distilled.
Uses: Seed fungicide as dust or slurry with water.
Danger! Poisonous by inhalation or swallowing. May cause skin irritation or delayed chemical burns. MCA warning label.

ethylmercuric chloride C_2H_5HgCl.
Properties: Crystals; sp. gr. 3.482; m. p. 193°C. Insoluble in water; slightly soluble in ether; soluble in hot alcohol. Sublimes readily.
Derivation: Reaction of zinc diethyl and mercuric chloride.
Uses: Fungicide for seed or bulb treatment either alone or with other organic mercury compounds.
Danger! Poisonous by inhalation or swallowing; may cause skin irritation or delayed chemical burns. MCA warning label.

ethylmercuric phosphate $(C_2H_5HgO)_3PO$.
Properties: White powder, soluble in water; garlic-like odor.
Process: Reaction of ethylmercuric acetate with phosphoric acid.
Uses: Seed fungicide; timber preservative.
Warning: Danger! Poisonous by inhalation or swallowing; may cause skin irritation or delayed chemical burns. MCA warning label.

para-**ethylmercurithiobenzenesulfonate, sodium salt** ("Sulfo-Merthiolate")
$C_2H_5HgSC_6H_4SO_3Na$.
Properties: A white to cream-colored powder with a characteristic odor. It is sensitive to light. It is soluble in water and alcohol and almost insoluble in benzene and ether.
Use: Medicine.

ethylmercury 2,3-dihydroxypropyl mercaptide
$C_2H_5HgSCH_2CHOHCH_2OH$. Organic mercurial compound used as a fungicidal dust or in slurry treatment for control of seed-borne diseases and to reduce losses from seed decay and damping-off of wheat, oats, rye, etc.

ethylmercury-para-**toluenesulfonanilide** (EMTS; "Ceresan" M)
$C_6H_5N(HgC_2H_5)SO_2C_6H_4CH_3$.
Properties: Crystals; pungent odor; nearly insoluble in water.
Uses: As a dust or slurry for control of seed-borne diseases and of fungi by treatment of seeds or bulbs.
Danger! Poisonous by inhalation or swallowing; may cause skin irritation or delayed chemical burns. MCA warning label.

ethyl metaphosphate $C_2H_5PO_3$.
Properties: A colorless, odorless, hygroscopic, viscous liquid. Soluble in water (decomposes); soluble in chloroform.

ethyl methacrylate $H_2C{:}CCH_3COOC_2H_5$.
Properties: Colorless liquid; b.p. 119°C; m.p. below -75°C; sp. gr. 0.911; flash point (open cup) 95°F; insoluble in water; readily polymerized.
Derivation: Reaction of methacrylic acid or methyl methacrylate with ethyl alcohol.
Grades: Technical (inhibited).
Containers: Drums; tank cars.
Uses: Polymers; chemical intermediates.
See also acrylate resins.

5-ethyl-5(1-methyl-1-butenyl) barbituric acid. See vinbarbital.

ethyl methyl ether $C_2H_5OCH_3$.
Properties: Colorless liquid; sp. gr. 0.725; b.p. 10.8°C. Soluble in water, miscible with alcohol and ether.
Use: Medicine.
Shipping regulations: Flammable liquid. Red label. *

3-ethyl-3-methylglutarimide. See bemegride.

ethyl methyl ketone. See methyl ethyl ketone.

5-ethyl-1-methyl-5-phenylbarbituric acid. See mephobarbital.

ethyl methyl phenyl glycidate (so-called aldehyde C-16; "strawberry aldehyde")
$CH_3(C_6H_5)\underline{COCHCOOC_2H_5}$.
Properties: Colorless to yellowish liquid, having a strong odor, suggestive of strawberry. Soluble in 3 vols of 60% alcohol.
Constants: Sp. gr. 1.104-1.123; refractive index 1.509-1.511.
Uses: Perfumery; flavors.

5-ethyl-3-methyl-5-phenylhydantoin
$C_{12}H_{14}N_2O_2$.
Properties: Crystals with m.p. 136-137°C; insoluble in water. Sodium salt is soluble in water giving an alkaline solution.
Derivation: Prepared from 5,5-ethyl phenylhydantoin by action of one mole of dimethyl sulfate; by treating phenylmethylureidoacetonitrile with hydrochloric acid.
Use: Medicine.

alpha-**ethyl**-alpha-**methylsuccinimide.** See ethosuximide.

7-ethyl-2-methyl-4-undecanol (tetradecanol)
$C_4H_9CH(C_2H_5)C_2H_4CH(OH)CH_2CH(CH_3)_2$.
Properties: Liquid; sp. gr. 0.8355 (20/20°C); b.p. 264°C; flash point 285°C; insoluble in water.
Uses: Intermediate for synthetic lubricants, defoamers and surfactants.

ethylmorphine hydrochloride
$C_{19}H_{23}NO_3 \cdot HCl \cdot 2H_2O$.
Properties: White crystalline powder; poisonous! Odorless; soluble in water and alcohol; slightly soluble in ether and chloroform.
Constants: M.p., about 123°C (dec).
Derivation: Action of hydrochloric acid on ethylmorphine which is made by action of ethyl iodide on morphine in alkaline solution.
Method of purification: Crystallization.
Grades: Technical; N.F. XI.
Containers: Glass bottles.
Use: Medicine.
Shipping regulations: None. *

N-ethyl morpholine
$\underline{CH_2CH_2OCH_2CH_2N}CH_2CH_3$.
Properties: Colorless liquid. Miscible with water.
Constants: Sp. gr. 0.916 (20/20°C); b.p. 138°C; wt/gal 7.6 lbs (20°C).
Grades: Technical.
Uses: Useful as intermediate in manufacture of dyestuffs, pharmaceuticals, rubber accelerators and emulsifying agents. Solvent for dyes, resins, oils. Catalyst in making polyurethane foams.

ethyl mustard oil. See ethyl thiocarbimide.

ethyl myristate (ethyl tetradecanoate)
$CH_3(CH_2)_{12}COOC_2H_5$.
Properties: Liquid; sp. gr. 0.856; m.p. 12°C; b.p. 295°C. Insoluble in water; soluble in alcohol; slightly soluble in ether.

N-ethyl-alpha-**naphthylamine** (N-ethyl-1-naphthylamine) $C_{10}H_7NHC_2H_5$.
Properties: Colored oil; b.p. 305°C. Insoluble in water; soluble in alcohol and ether.
Use: Intermediate.

ethyl nitrate $C_2H_5NO_3$.
Properties: Colorless, flammable liquid; pleasant odor; sweet taste. Soluble in alcohol and ether; insoluble in water.
Constants: Sp. gr. 1.116; m.p. -112°C; b.p. 87.6°C.
Derivation: By heating alcohol, urea nitrate and nitric acid, with subsequent distillation.

Method of purification: Rectification.
Grades: Technical.
Containers: Iron drums; glass bottles.
Uses: Organic synthesis; drugs; perfumes; dyes; liquid rocket propellant.
Fire hazard: Dangerous.
Shipping regulations: Flammable liquid. Red label. *

ethyl nitrite $C_2H_5NO_2$.
Properties: Yellowish, highly aromatic, ethereal, flammable, exceedingly volatile liquid. Soluble in alcohol and ether; insoluble in water.
Constants: Sp. gr. 0.900; b. p. 16.4°C.
Derivation: (a) By the action of ethyl alcohol on nitrous oxide gas. (b) By treating alcohol with alkali nitrites and sulfuric acid.
Grades: Technical.
Containers: Hermetically sealed tubes; 1-lb bottles.
Keep tightly closed and in a cool place, protected from light.
Uses: Organic preparations; medicine.
Fire hazard: Dangerous.
Shipping regulations: Flammable liquid. Red label. *

O-ethyl-O-para-nitrophenyl benzenethionophosphonate. See EPN.

ethyl nonanoate. See ethyl pelargonate.

ethyl octanoate. See ethyl caprylate.

ethyl octoate. See ethyl caprylate.

ethyl oenanthate (ethyl heptanoate; oenanthic ether; cognac oil) $CH_3(CH_2)_5COOC_2H_5$.
Properties: Clear, colorless oil with fruity odor and taste; sp. gr. 0.87; b. p. 187°C; soluble in alcohol, chloroform and ether; insoluble in water.
Derivation: By heating oenanthic acid and ethyl alcohol in presence of sulfuric acid, and subsequent recovery by distillation.
Grade: Technical.
Containers: Iron drums; glass bottles.
Uses: Artificial cognac flavor; flavor for liqueurs and fruity-type soft drinks.
Shipping regulations: None. *

ethyl oleate $C_{17}H_{33}COOC_2H_5$.
Properties: Light-colored, yellowish oleaginous liquid. Insoluble in water; soluble in alcohol and ether. Solubility of water in product 1.0 cc/100 cc (20°C).
Constants: Wt/gal 7.27 lbs (20°C); refractive index 1.45189 (20°C); m. p. −32°C (approx); sp. gr. 0.867; flash point 175.3°C (347.5°F).
Typical specifications: Boiling range 205-208°C with some decomposition; ester content 99% by wt.
Uses: Solvent; plasticizer; lubricant; water-resisting agent.

"Ethylose." [421] A graft cellulose, i. e., a water insoluble hydroxyethylcellulose.

ethyl oxalate (diethyl oxalate) $(COOC_2H_5)_2$.
Properties: Colorless, unstable, oily, aromatic liquid. Combustible but not flammable. Miscible in all proportions with alcohol, ether, ethyl acetate, and other common organic solvents; only very slightly soluble in water and gradually decomposed by it.
Constants: Sp. gr. 1.09 (20/20°C); b. p. 186°C; m. p. −40.6°C; wt/gal 8.96 lbs (20°C) (approx); coefficient of expansion per °F 0.00056, per °C 0.00101; flash point 168°F (approx).
Derivation: By standard esterification procedure using ethyl alcohol and oxalic acid. The final purification, however, calls for unusual technique and equipment. The last traces of water are most difficult to remove and this is accomplished by a special step in the rectification.
Method of purification: Distillation.
Containers: 1-gal cans; 5-, 55-gal steel drums; tank cars.
Uses: Solvent for cellulose esters and ethers, many natural and synthetic resins; radio tube cathode fixing lacquers; dye intermediate; pharmaceuticals; perfume preparations; organic synthesis.
Fire hazard: Combustible but not flammable; flash point over 80°F.
Shipping regulations: None. *

ethyl oxide. See ether.

ethyl 2-oxocyclopentanecarboxylate. See 2-carbethoxycyclopentanone.

ethyl pelargonate (ethyl nonanoate; wine ether) $CH_3(CH_2)_7COOC_2H_5$.
Properties: Colorless liquid; sp. gr. 0.866 (18/4°C); b. p. about 220°C; m. p. −44°C; insoluble in water; soluble in alcohol and ether.
Use: Flavoring material for alcoholic beverages.

3-ethylpentane (triethylmethane) $(C_2H_5)_3CH$.
Properties: Colorless liquid; b.p. 93.468°C; freezing point −118.593°C; sp.gr. 0.69818 (20°C); refractive index (n 20/D) 1.3934; soluble in alcohol; insoluble in water.
Grades: Technical.
Use: Organic synthesis.

meta-ethylphenol (3-ethylphenol) $HOC_6H_4C_2H_5$.
Properties: Colorless liquid; m. p. −4°C; b. p. 214°C; sp. gr. 1.001; very slightly soluble in water; miscible with alcohol and ether.

para-ethylphenol (4-ethylphenol) $HOC_6H_4C_2H_5$.
Properties: Colorless needles; m. p. 46°C; b. p. 219°C; soluble in alcohol or ether; slightly soluble in water.

ethyl phenylacetamide. See ethyl acetanilide.

ethyl phenylacetate $C_6H_5CH_2COOC_2H_5$.
Properties: Colorless liquid, with a honey type odor.
Constants: Sp. gr. 1.027-1.032; refractive index, 1.498; b. p. 276°C. Soluble in 8 parts of 60% alcohol.
Uses: Perfumery; flavors.
Shipping regulations: None. *

ethyl phenylacrylate. See ethyl cinnamate.

5-ethyl-5-phenylbarbituric acid. See phenobarbital.

*See "I.C.C. Shipping Regulations," page xiii.
Reference numbers refer to name of manufacturer. See "List of Manufacturers," page v.

ethyl phenylcarbamate (phenylurethane; ethyl phenylurethane) $C_6H_5NHCOOC_2H_5$.
Properties: White, crystalline product; aromatic odor, clove-like taste. M. p. 51°C. Soluble in alcohol, ether and boiling water; insoluble in cold water.
Derivation: By the action of ethyl alcohol on phenyl isocyanate.
Shipping regulations: None. *

N, N-ethyl phenyl ethanolamine $C_6H_5NC_2H_5CH_2CH_2OH$.
Constants: Sp. gr. 1.04 (20/20°C); b. p. (740 mm) 268°C; wt/gal 8.7 lbs (20°C).
Typical specifications: Boiling range 260-276°C (740 mm).
Grades: Technical.
Containers: 1-gal cans; 5-, 55-gal drums.
Uses: Organic synthesis; dyestuffs.

2-ethyl-2-phenylglutarimide (glutethimide) $C_{13}H_{15}NO_2$.
Properties: Occurs as white, crystalline powder. A saturated solution is slightly acid. Freely soluble in acetone, ethyl acetate, and chloroform; soluble in ethanol and methanol; practically insoluble in water. Melting range: 85°-87°C.
Grade: N. F. XI (as glutethimide).
Use: Medicine.

ethyl phenyl ketone. See propiophenone.

ethylphenylurethane. See ethyl phenylcarbamate.

ethylphosphoric acid $C_2H_5H_2PO_4$.
Properties: Pale straw-colored liquid; sp. gr. 1.33 (25°C); can be neutralized with alkalies or amines to give water-soluble salts.
Purity: 97%, with remainder being orthophosphoric acid and ethyl alcohol.
Uses: Catalyst; rust remover; soldering flux; intermediate.

ethyl phthalate. See diethyl phthalate.

ethyl phthalyl ethyl glycolate $C_2H_5OCOC_6H_4COOCH_2COOC_2H_5$.
Properties: Sp. gr. (25°C) 1.180; refractive index (25°C) 1.498; b. p. (5 mm) 190°C; flash point 193°C; solubility in water, 0.018% by wt.
Use: Plasticizer.

5-ethyl-2-picoline. See 2-methyl-5-ethylpyridine.

1-ethyl-1-propanol. See 3-pentanol.

ethyl propionate (propionic ester) $C_2H_5COOC_2H_5$.
Properties: Practically water-white liquid. Odor resembling pineapples; soluble in alcohol and ether; in water 2.5% at 15°C; sp. gr. 0.895 (15.5°C); b. p. 99°C; m. p. −73°C; flash point 12°C; refractive index (n 20/D) 1.3844.
Derivation: By treating ethyl alcohol with propionic acid.
Method of purification: Distillation.
Grades: Commercial, 85-90% ester content.
Containers: 50-, 100-gal drums.
Uses: Solvent for cellulose ethers and esters, various natural and synthetic resins; fruit syrups; cutting agent for pyroxylin.
Fire hazard: Flammable.
Shipping regulations: Flammable liquid. Red label. *

ethylpropionyl. See diethyl ketone.

2-ethyl-3-propylacrolein (2-ethylhexenal) $C_3H_7CH{:}C(C_2H_5)CHO$.
Properties: Yellow liquid. Powerful odor. Sp. gr. 0.8518 (20/20°C); b. p. 175.0°C (760 mm); vapor pressure 1.0 mm (20°C); flash point 155°F; wt/gal 7.1 lbs (20°C); coefficient of expansion 0.00098 (20°C); viscosity 0.113 poise (20°C).
Typical specifications: Sp. gr. 0.8470-0.8530 (20/20°C); boiling range 165-185°C (760 mm); acidity not more than 2.00% (as butyric).
Grades: Technical.
Containers: 1-gal cans; 5-, 55-gal (tin-lined) drums.
Uses: Insecticide; organic synthesis (intermediate); warning agents and leak detectors.

2-ethyl-3-propylacrylic acid $C_3H_7CH{:}C(C_2H_5)COOH$.
Properties: A liquid; sp. gr. (20°C) 0.9484; m. p. −7.8°C; b. p. 232.1°C; vapor pressure (20°C) less than 0.01 mm; flash point 330°F; insoluble in water.
Uses: Pharmaceuticals; resins and plastics; lubricants.

4-ethylpyridine $C_5H_4NC_2H_5$.
Properties: B. p. 168°C; sp. gr. 0.9460 (20°C); refractive index 1.5018 (n 20/D); soluble in water.

ethyl pyridylethylacrylate $CH_2{:}CHCOOC_2H_4C_5H_3NC_2H_5$.
Properties: Liquid; sp. gr. 1.0458 (20°C); b. p. 181 (50 mm); f. p. −75°C; very slightly soluble in water.
Uses: Manufacture of synthetic plastics and fibers; adhesives, textile finishes and sizes.

ethyl salicylate $C_6H_4(OH)COOC_2H_5$.
Properties: Colorless liquid, with a faint odor of methyl salicylate. Soluble in ether and alcohol; insoluble in water. Sp. gr. 1.127-1.130; refractive index 1.523; b. p. 231-234°C.
Uses: Perfumery; flavors.

"Ethyl Selenac." [69] Trademark for a proprietary preparation of selenium diethyldithiocarbamate $[(C_2H_5)_2NC(S)S]_4Se$.
Properties: Yellow powder; sp. gr. 1.32 ± .03; melting range 59-85°C; soluble in benzene, carbon disulfide, chloroform; insoluble in water, dilute caustic, gasoline.
Uses: Vulcanizing agent, also primary accelerator in natural and butyl rubber and SBR; secondary accelerator (with thiazoles) for natural rubber and SBR.

ethyl silicate (tetraethyl orthosilicate) $(C_2H_5)_4SiO_4$.
Properties: Flammable, colorless liquid; faint odor; soluble in alcohol; hydrolyzed by water to an adhesive form of silica.
Constants: Sp. gr. 0.9356 (20/20°C); b. p.

*See "I.C.C. Shipping Regulations," page xiii.
Reference numbers refer to name of manufacturer. See "List of Manufacturers," page v.

168.1°C (760 mm); vapor pressure 1.0 mm (20°C); flash point 125°F; wt/gal 7.8 lbs (20°C); coefficient of expansion 0.00112 (20°C); f.p. −77°C; viscosity 0.0179 poise (20°C).
Grades: 29% SiO_2; 40% SiO_2.
Containers: 1-gal cans; 5-, 55-gal (tin-lined) drums; tank cars.
Uses: Preservative for stone, brick, concrete, plaster; weatherproof and acid-proof mortar and cements; refractory bricks, other molded objects; heat-resistant paints; chemical-resistant paints; protective coatings for industrial buildings and castings; lacquers; bonding agent; intermediate.

ethyl silicate, condensed. Light-yellow liquid with mild odor consisting of 85% by wt tetraethyl orthosilicate and 15% polyethoxy siloxanes. On hydrolysis or ignition, yields high purity, refractory silica. Used as an intermediate for siloxane compounds; for precision casting of high-melting alloys; pigments binder for paints; surface hardener for sandstones.

ethyl sodium oxalacetate $C_2H_5OOCC(ONa):CHCOOC_2H_5$. Light yellow powder; 92% pure.
Derivation: Reaction of pure ethyl acetate and diethyl oxalate with metallic sodium.
Containers: 175-lb fiber drums.
Uses: Dyes; synthesis.

ethyl sulfate. See diethyl sulfate.

ethyl sulfhydrate. See ethyl mercaptan.

ethyl sulfide (diethyl sulfide) $(C_2H_5)_2S$.
Properties: Colorless, oily liquid; garlic-like odor; soluble in alcohol and ether; slightly soluble in water; sp. gr. 0.837; m.p. −102°C; b.p. 92-93°C; refractive index (n 20/D) 1.4423.
Derivation: By heating potassium ethyl sulfate and potassium sulfide, with subsequent distillation.
Method of purification: Rectification.
Grades: Technical.
Containers: Iron drums.
Uses: Organic synthesis; special solvent.
Shipping regulations: None. *

ethylsulfuric acid (acid ethylsulfate; sulfovinic acid) $C_2H_5HSO_4$.
Properties: Colorless, oily liquid; soluble in water, alcohol and ether; sp. gr. 1.316; b.p. 280°C.
Derivation: By the action of sulfuric acid on ethyl alcohol.
Method of purification: Distillation.
Grades: Technical.
Containers: Glass bottles; carboys.
Uses: Medicine; precipitant for casein; organic preparations.

ethylsulfurous acid (sulfovinous acid) $C_2H_5HSO_3$.
Properties: Crystalline, unstable mass; soluble in alcohol, ether and alkalies.
Derivation: By the action of thionyl chloride on ethyl alcohol.
Method of purification: Crystallization.
Use: Organic synthesis.
Shipping regulations: None. *

ethyl tetradecanoate. See ethyl myristate.

4-ethyl-1,4-thiazane $C_4H_8NSC_2H_5$.
Properties: Colorless, mobile oil; soluble in water; sp. gr. 0.9929 (15°C); b.p. 184°C (763 mm).
Derivation: Interaction of dichlorodiethyl sulfide and an aliphatic amine in the presence of alcohol and sodium carbonate.

ethyl thiocarbimide (ethyl mustard oil; ethyl isothiocyanate) C_2H_5NCS.
Properties: Colorless liquid; pungent odor; inflames the skin; soluble in alcohol; insoluble in water. Sp. gr. 1.004 (15/4°C); m.p. −5.9°C; b.p. 131-132°C.
Derivation: By the interaction of thiocyanic ether and phosphorus pentachloride, with subsequent distillation.
Method of purification: Rectification.

ethyl thioethanol $C_2H_5SC_2H_4OH$.
Properties: Pale straw liquid; sp. gr. 1.015-1.025 (20/20°C); distillation range: 180-184°C.
Grade: 95% min.
Containers: 5-, 55-gal drums.
Uses: Synthesis; intermediate.

ethylthiopyrophosphate. See tetraethyldithiopyrophosphate.

"Ethyl Thiurad." [58] Trade name for tetraethylthiuram disulfide.

ethyl-para-toluene sulfonate $CH_3C_6H_4SO_3C_2H_5$.
Properties: Toxic; unstable solid; m.p. 33°C; b.p. 221.3°C; density 1.17; soluble in many organic solvents; insoluble in water.
Grades: Technical.
Use: Plasticizer for cellulose acetate; accelerator; ethylating agent.

N-ethyl-ortho-toluidine $C_6H_4(CH_3)NHC_2H_5$.
Properties: Colorless to yellowish oil; soluble in alcohol, ether, and hydrochloric acid; insoluble in water. Sp. gr. 0.9534; b.p. 214°C.
Derivation: By heating ethyl alcohol with ortho-toluidine and hydrochloric acid.
Containers: Barrels.

N-ethyl-ortho-toluidine-para-sulfonic acid $CH_3C_6H_3(NHC_2H_5)SO_3H$.
Properties: White solid; soluble in alkaline solution; slightly soluble in water.
Derivation: Sulfonation of ethyl-ortho-toluidine.
Method of purification: Recrystallization.

alpha-(N-ethyl-meta-toluidino)-meta-toluenesulfonic acid $CH_3C_6H_3N(C_2H_5)CH_2C_6H_4SO_3H$.
Properties: Light tannish-gray paste. Solids, approximately 70%.
Grade: Technical.
Use: Intermediate.

ethyl triacetylgallate $C_2H_5OOCC_6H_2(OOCCH_3)_3$.
Properties: Colorless crystals, or white crystalline powder. Insipid taste; odorless; soluble in warm alcohol, and acetone;

*See "I.C.C. Shipping Regulations," page xiii.
Reference numbers refer to name of manufacturer. See "List of Manufacturers," page v.

slightly soluble in ether, and alcohol; insoluble in water. M. p. 134-136°C.
Derivation: Action of acetic anhydride, or acetyl chloride, upon ethyl gallate.

ethyltrichlorosilane $C_2H_5SiCl_3$.
Properties: Colorless liquid. B. p. 99.5°C; sp. gr. 1.236 (25/25°C); refractive index (n 25/D) 1.4257; flash point (Cleveland open cup) 57°F. Readily hydrolyzed by moisture, with the liberation of hydrochloric acid.
Derivation: By reaction of ethylene and trichlorosilane in the presence of a peroxide catalyst.
Grades: Technical.
Containers: Bottles; 100-, 500-lb drums.
Uses: Intermediate for silicones.
Shipping regulations: Flammable liquid. Red label. *

"Ethyl Tuads." [69] Trademark for tetraethylthiuram disulfide, $[(C_2H_5)_2NC(S)S]_2$.
Properties: White to cream powder (also supplied as white to cream rods and as blue rods); sp. gr. 1.42 ± .03; melting range 142-156°F; soluble in carbon disulfide, benzol, chloroform; insoluble in water, dilute caustic, gasoline.
Uses: As vulcanizing agent and as primary accelerator in natural, butyl, nitrile rubbers and SBR. As secondary accelerator (with thiazoles) in natural and nitrile rubbers and in SBR. In extruded and molded goods, sponge, tires and tubes, wire and cable applications.

"Ethyl Tuex." [248] Trademark for tetraethylthiuram disulfide.
Properties: Gray powder; sp. gr. 1.17; m. p. 60°C min; soluble in acetone, benzol, gasoline, and ethylene dichloride; insoluble in water.
Uses: Accelerator for natural rubber wire insulation; druggist sundries; mechanicals; proofing; footwear.

ethyl urethane. See urethane.

ethyl valerate. See ethyl isovalerate.

ethyl vanillin (ethovan; bourbonal; vanillal; vanirom; 3-ethoxy-4-hydroxybenzaldehyde) $OHC_6H_3(OC_2H_5)CHO$.
Properties: Fine, white crystalline material having an intense odor of vanillin; affected by light; m. p. 76.5°C; soluble in alcohol, chloroform, and ether; slightly soluble in water.
Grades: N. F. XI.
Containers: 100-lb fiber drums.
Use: Flavors, as a replacement or fortifier of vanillin.

ethyl vinyl ether. See vinyl ethyl ether.

"Ethyl Zimate." [69] Trademark for zinc diethyldithiocarbamate, $[(C_2H_5)_2NC(S)S]_2Zn$.
Properties: White powder; sp. gr. 1.48 ± .03; melting range 171-182.5°C; moderately soluble in dilute caustic, benzene, carbon disulfide, chloroform; insoluble in water, gasoline.
Uses: Primary accelerator, secondary accelerator (with thiazoles) for natural and butyl rubber and for SBR.

ethyne. See acetylene.

ethynylation. Condensation of acetylene with a reagent such as an aldehyde to yield an acetylenic derivative. The best example is the union of formaldehyde and acetylene to produce butynediol.

1-ethynylcyclohexanol $HC{:}CC_6H_{10}OH$.
Properties: Colorless, low-melting solid with sweet odor. M. p. 30-31°C; b. p. 180°C; sp. gr. (20/20°C) 0.967. Slightly soluble in water.
Containers: Cans; drums.
Uses: Stabilization of chlorinated organic compounds; intermediate; corrosion inhibitor for mineral acids.

1-ethynylcyclohexyl carbamate. See ethinamate.

ethynylestradiol. See ethinylestradiol.

beta-**ethynyl ethanol.** See 3-butyn-1-ol.

ethynyltestosterone. See ethisterone.

ethythrin. Ethyl analog of allethrin, used as insecticide with applications similar to allethrin (q. v.). See also barthrin, cyclethrin and furethrin.

"Eticylol." [305] Trademark for ethinyl estradiol U. S. P.
Use: Medicine.

"Etrolene." [233] Trademark for veterinary drug and insecticide containing organophosphorus compounds.

Eu. Symbol for europium.

eucaine hydrochloride (beta-eucaine hydrochloride; betacaine hydrochloride; benzamine hydrochloride) $C_{15}H_{21}NO_2 \cdot HCl$.
Properties: White, odorless crystalline powder. M. p. about 268°C (dec); soluble in water, alcohol, and chloroform.
Derivation: By the action of hydrochloric acid on beta-eucaine.
Method of purification: Crystallization.
Grades: Technical.
Containers: Glass bottles.
Use: Medicine.
Shipping regulations: None. *

eucaine lactate (benzamine lactate) $C_{15}H_{21}NO_2 \cdot C_3H_6O_3$.
Properties: White crystalline powder. M. p. about 152°C. Soluble in water and alcohol.
Use: Medicine.

eucalyptol (cineol; cajeputol) $C_{10}H_{18}O$.
Properties: Colorless oil, camphor-like odor and pungent, cooling, spicy taste. Slightly soluble in water; miscible with alcohol, chloroform, ether, glacial acetic acid and fixed or volatile oils. Sp. gr. 0.921-0.923 (25°C); b. p. 174-177°C; congealing point not below 0°C; refractive index 1.4550-1.4600 (20°C).
Derivation: By fractionally distilling eucalyptus oil, followed by freezing.
Method of purification: Rectification or crystallization by freezing.

*See "I. C. C. Shipping Regulations," page xiii.
Reference numbers refer to name of manufacturer. See "List of Manufacturers," page v.

Grades: Technical; N. F. XI.
Containers: 25-lb cans; 100-lb drums.
Uses: Pharmacy; flavoring; perfumery.
Shipping regulations: None.*

eucalyptus (blue gum tree; Australian fever tree). Dried leaves of tree Eucalyptus globulus.
Habitat: Australia; cultivated in sub-tropics, Europe; northern Africa; south-western United States.
Uses: Source of eucalyptus oil; medicine.
Shipping regulations: None.*

eucalyptus oil.
Properties: Colorless or faintly yellowish, mobile essential oil; aromatic odor; pungent, spicy and cooling taste. Soluble in alcohol; almost insoluble in water. (N. F. grade): Sp. gr. (25/25°C) 0.905-0.925; (n 20/D) 1.4580-1.4700; m.p. not lower than −15.4°C.
Chief constituents: Eucalyptol, 70-90%; aldehydes, terpenes.
Derivation: Distilled from fresh leaves of Eucalyptus globulus and a few other species of Eucalyptus.
Method of purification: Rectification.
Grades: N. F. XI; rectified.
Containers: Bottles; drums.
Uses: Medicine; ore flotation; perfumery, especially for soaps.
Shipping regulations: None.*

eucalyptus resin oil. See eucalyptus tar.

eucalyptus tar (eucalyptus resin oil). The residue obtained when, following the distillation of eucalyptus leaves, the oil is treated with caustic soda. It consists of a dark-brown syrup.
Uses: Disinfectant; perfuming soaps.
Shipping regulations: None.*

eucatropine hydrochloride (1, 2, 6, 6-tetramethyl-4-mandeloxypiperidine hydrochloride) $C_{17}H_{25}O_3N \cdot HCl$.
Properties: White, granular, odorless powder; m.p. 183-186°C; solutions are neutral to litmus. Very soluble in water; freely soluble in alcohol and chloroform; insoluble in ether.
Grade: U. S. P. XVI.
Use: Medicine.

euclase $Be_2Al_2(SiO_4)_2(OH)_2$. A natural basic aluminum-beryllium silicate.
Properties: Colorless, light blue, or green; vitreous luster; sp. gr. 3.05-3.10; hardness 7.5.
Occurrence: Austria; Russia; Brazil; Peru; Tasmania.
Use: Gem stones.

euflavine. See acriflavine.

eugallol. See pyrogallol acetate.

eugenic acid. See eugenol.

eugenol (4-allyl-2-methoxyphenol; eugenic acid) $C_3H_5C_6H_3(OH)OCH_3$.
Properties: Colorless or yellowish liquid; oily; becomes brown in the air; spicy odor and taste. Soluble in alcohol, chloroform, ether, and volatile oils; very slightly soluble in water. Sp. gr. 1.064-1.070; b.p. 253.5°C; refractive index 1.5400-1.5420 (20°C); optically inactive.
Derivation: By extraction of clove oil with aqueous potash, liberation with acid and rectification in a stream of carbon dioxide.
Method of purification: Redistillation.
Grades: Technical; U. S. P. XVI.
Containers: Tins; drums.
Uses: Perfumes; essential oils; medicine (as an active germicide); production of isoeugenol for the manufacture of vanillin.
Shipping regulations: None.*

eugenol acetate (acetyl eugenol) $C_3H_5C_6H_3(OCH_3)OOCCH_3$.
Properties: Solid crystals, melting to colorless liquid at warm room temperature; spicy odor of cloves. Souble in 4 parts of 70% alcohol. Sp. gr. 1.080-1.082; refractive index 1.520; m.p. 27°C.
Uses: Perfumery.

"Eukanol" Bottom A. [307] Trademark for a liquid consisting of a ricinoleic acid-amide ester derivative, used as a cationic leather-finishing agent for bottom coating.

"Eulan" CN. [307] Trademark for a permanent mothproofing agent, consisting of sodium pentachlorodihydroxytriphenyl methane sulfonate; anionic.
Properties: Fine, slightly granular powder. Compatible with acid or chrome dyes as well as with anionic and non-ionic surfactants. Incompatible with basic dyes, cationic surfactants and leveling agents based on albumin decomposition products.
Uses: Permanent mothproofing agent for wool and mohair which is applied in the dyebath. This product is resistant to washing, dry cleaning, and bleaching.

"Eulan" NKU Extra. [307] Trademark for a mothproofing agent, consisting of dichlorobenzyl triphenyl phosphonium chloride; 31% active.
Properties: Powder; dissolves in boiling water with stirring.
Uses: Mothproofing agent for wool and mohair. Applied in a neutral bath following dyeing or wet finishing.

"Eulysine" A. [307] Trademark for a dyeing assistant; an organic amine. Liquid; sp. gr. 1.08.
Uses: Prevents bronzing of sulfur colors; crabbing assistant for woolen and worsted fabrics; pasting agent for all dyes except basics.

"Eunaphtol" AS. [307] Trademark for a naphthol dyeing assistant, composed of a modified lignin-sulfonate.
Properties: Brown liquid; sp. gr. 1.075-1.095; miscible with water.
Uses: Naphthol dyeing assistant which improves fiber penetration and increases stability of the naphtholating liquors. Also recommended for pasting and dissolving naphthols.

euonymit. See dulcitol.

euonymus (wahoo; arrow wood; Indian arrow wood; bitter ash; burning bush; strawberry tree; spindle tree). Dried root bark of Euonymus atropurpureus.
Habitat: United States, east of the Mississippi.
Grade: Technical.
Containers: Boxes; bags.
Use: Medicine.
Shipping regulations: None.*

euphorbia (pill-bearing spurge; snake weed; cat's hair; Queensland asthma weed; flowery-head spurge). Whole plant Euphorbia pilulifera or hirta.
Habitat: Queensland; India.
Grade: Technical.
Containers: Bales.
Use: Medicine.
Shipping regulations: None.*

"Eureka" Soldering Flux Crystal. [28] Trademark for soldering flux crystal composition based on zinc chloride and ammonium chloride.
Properties: White crystalline powder.
Containers: 50-lb and 600-lb drums.
Use: As a flux in soft soldering operations and preparation of liquid soldering fluxes.

"Eureka" Soldering Flux Liquid. [28] Trademark for soldering flux liquids based on zinc chloride and other ingredients.
Properties: Clear, colorless water solutions; various gravities and compositions.
Containers: 605-lb drums; 145- and 165-lb carboys.
Uses: As soldering fluxes for soft soldering and as solder and lead-coating fluxes for use in the automotive, can, refrigeration, wire and other industries.

"Euresol." [9] Trade name for acetoresorcin (resorcinol monoacetate).

europia. See europium salts.

europium Eu. Atomic number 63; one of the lanthanide or rare-earth elements of the cerium subgroup.
Properties: Steel gray metal, difficult to prepare. Malleable; m.p. 826°C; b.p. 1489°C (approx); sp.gr. 5.24. Oxidizes rapidly in air; may burn spontaneously. Is the most reactive of the rare earth metals; liberates hydrogen from water.
Derivation: Reduction of the oxide with lanthanum or misch metal.
Source: See rare-earth minerals.
Grades: High purity (ingots; lumps).

europium salts
europium chloride $EuCl_3 \cdot xH_2O$. Colorless to pale pink crystals; soluble in water. Obtained by treating the oxide with hydrochloric acid.
europium fluoride $EuF_3 \cdot 0\text{-}2H_2O$.
europium nitrate $Eu(NO_3)_3 \cdot 6H_2O$. Colorless to pale pink crystals; soluble in water. Obtained by treating the oxide with nitric acid.
europium oxalate $Eu_2(C_2O_4)_3 \cdot xH_2O$. White powder; insoluble in water; slightly soluble in acids. Grades: 25-50% and 99.8% europium salt. Impure grades may be colored.
europium oxide Eu_2O_3. White powder; insoluble in water; soluble in acids to give the corresponding salt. Hygroscopic, absorbs carbon dioxide from the air. Obtained by ignition of the oxalate. Grades: 25-50% and 99.8% europium salt. Impure grades may be colored.
europium sulfate $Eu_2(SO_4)_3 \cdot 8H_2O$. Colorless to pale pink crystals; slightly soluble in hot water, more soluble in cold. Stable in air. Obtained by treating the oxide with sulfuric acid.
Uses: In red- and infrared-sensitive phosphors; the oxide in nuclear-reactor control rods.

eutactic. Same meaning as tactic (q.v.).

eutectic alloys. See table under fusible alloys.

euxenite (loranskite)
$(Y, Ca, Ce, U, Th)(Nb, Ta, Ti)_2O_6$. A rare earth mineral (q.v.). Color brownish black; luster brilliant to vitreous; sp.gr. 5-5.9; hardness 5-6.
Occurrence: Norway; Madagascar; Canada; Pennsylvania.
Uses: Source of uranium, niobium, tantalum.

ev. Abbreviation for electron volt.

"Evanacid 3CS." [312] Brand name for carboxymethylmercaptosuccinic acid (q.v.).

"Evanohm." [155] Trademark for an alloy of 75% nickel, 20% chromium, 2.5% aluminum, and 2.5% copper.
Properties: Resistivity 800 ohms per circular mil ft at 20°C; temperature coefficient of resistance ± 0.00002/°C; very low thermal emf vs copper; high tensile strength in fine sizes; corrosion resistant; nonmagnetic; heat resistant to 600°C.
Forms: Wire; insulated wire.
Use: Precision wound resistors.

Evans blue $C_{34}H_{24}N_6Na_4O_{14}S_4$. A diazo dye used in medicine to measure blood plasma volume.
Properties: Bluish green or brown iridescent powder. Odorless and hygroscopic. Very soluble in water; very slightly soluble in alcohol; practically insoluble in benzene, carbon tetrachloride, chloroform, and ether. pH (0.5% solution) 5.5-7.5.
Grade: U.S.P. XVI.
Use: Medicine.

evaporation. Change from liquid to vapor state, either at elevated or normal temperatures. For example sugar solutions and salt solutions are heated in large vessels to remove excess water by evaporation so that solid crystals will form. Also, volatile organic solvents are lost if allowed to stand in open containers even at ordinary temperature.

EVE. Abbreviation for ethyl vinyl ether. See vinyl ethyl ether.

"Evenglo." [11] Trademark for polystyrene plastics available in a wide range of color and various degrees of light transmission and diffusion for use in lighting applications.

*See "I.C.C. Shipping Regulations," page xiii.
Reference numbers refer to name of manufacturer. See "List of Manufacturers," page v.

"Everdur." [324] Trademark for a group of five copper-silicon alloys, with compositions adjusted to hot and cold working, hot forging, welding, free machining, and for ingots for remelting and casting. The most widely used alloy in this group is "Everdur-1010," with a nominal composition of copper 95.80%, silicon 3.10%, manganese 1.10%. In most environments, "Everdur" alloys are equivalent to copper in corrosion resistance. They are generally resistant to corrosion by marine and industrial atmospheres, fresh water, non-oxidizing mineral acids, alkalies, and salt solutions under many conditions of use.

"Evipal." [162] Trademark for hexobarbital.

"Evipal" Sodium. [162] Trademark for hexobarbital sodium.

exaltolide. Macrocyclic lactone of musk-like odor, intense and powerful, used in perfumery.

exaltone. Macrocyclic ketone having powerful musky odor, used in perfumery.

"Excelsior" Carbon Blacks. [133] Trademark for paint blacks comprising the ordinary color group; high hiding, high tinting strength. "Excelsior," a powdered form of impingement carbon black is used as an all purpose paint black. "Excelsior" Beads comprise a dustless free flowing form of regular color impingement carbon black; show marked advantage over powdered form for ball mill grinding.
Containers: 25-lb bags.

exchange reaction. A process whereby atoms of the same element in two different molecules or in two different positions in the same molecule transfer places. Exchange reactions are usually studied with the aid of a tracer or tagged atom.

"Exkin." [74] Trademark for a series of anti-skinning agents of the volatile oxime type.
Use: Paints.

exo-. A prefix used in chemical names to indicate attachment to a side chain rather than to a ring.
See also endo-.

"Exolvent." [141] Trade name for a solvent made from aliphatic hydrocarbons.

"Exon." [35] Trademark for a series of polyvinyl resins, compounds and latexes composed of polymers and copolymers based on vinyl chloride.
Containers: Resins and compounds: 50-lb bags. Latexes: 55-gal drums; tank trucks; tank cars.
Uses: Resins: Molding, sheeting film, strip coatings, protective coatings, extrusions, paints, ink, adhesives, plastisols, flooring, phonograph records.
Compounds: Wire coatings and insulation, extruded shapes and profiles, moldings, etc.
Latexes: Coating and impregnating cloth and paper to improve strength, appearance and resistance.
Hazards: Avoid freezing latex.

exothermic. Referring to a process which is accompanied by evolution of heat.

exotic fuels. See high energy fuels. The name also refers to those rocket fuels of a non-chemical nature (see rocket propellants).

expander. A mixture of lampblack, barium sulfate, and an organic material usually derived from the lignin fraction of wood that increases the capacity of storage batteries presumably by coating the anode and thus preventing the deposit of lead sulfate on the underlying lead metal.

"Expandex." [319] Trade name for a blood volume expander.

explosive, high. A detonating explosive composed either of an explosive compound or a mixture of compounds, which on detonating has a high disruptive effect. The detonation is accomplished intentionally by use of a blasting cap or blasting cap and booster, but most high explosives are also sensitive to shock and high temperature. The actual explosion is the result of an almost instantaneous chemical reaction causing the liberation of large amounts of gases.
Shipping regulations: Explosives, Class A. *

explosive, low. An explosive which deflagrates (burns over a relatively sustained period) rather than detonates. See explosive, high.
Shipping regulations: Explosives, Class A. *

explosive oil. See nitroglycerin.

explosives, permissible. Explosives approved by the Bureau of Mines for use in blasting in coal mines.

expression. Removal of a liquid from a solid by pressing, as in manufacture of vegetable oils from meal cakes.

"Exsize." [114] Trademark for a series of enzyme desizing agents containing starch-liquefying and proteolytic enzymes.
Derivation: Produced by growing pure cultures of micro-organisms on select media.
Properties: Liquid and dry powder of low specific gravity and viscosity. Predominantly alpha-amylase with small amount of protease.
Use: Desizing agent.
Shipping regulations: None. *

extraction, liquid-liquid (solvent extraction). A process in which one or more components are removed from a liquid mixture by intimate contact with a second liquid which is itself nearly insoluble in the first liquid and dissolves the impurities and not the substance that is to be purified. In other cases the second liquid may dissolve, i. e. extract, from the first liquid, the component that is to be purified, and leave associated impurities in the first liquid. Thus penicillin, which is produced along with many impurities in a dilute aqueous broth, can be recovered by extraction with amyl acetate, which dissolves penicillin but not water or the impurities in it. Lubricating oil is improved in quality by liquid-liquid extraction with any one of several less common liquids such as phenol or furfural.

Liquid-liquid extraction may be carried out by simply mixing the two liquids with agitation, and then allowing the two liquids to separate by standing quietly. It is however often desirable and economical to use countercurrent extraction in which the two immiscible liquids are caused to flow past or through one another in opposite directions with rather intimate contact. Thus fine droplets of heavier liquid can be caused to pass downward through a body of the lighter liquid in a vertical tube or tower which has entrance ports for the heavy and light liquids at top and bottom respectively, and similar exits at the bottom and top for these two streams.

extractive distillation. A variety of distillation that always involves the use of a fractionating column, and which is characterized by use of a purposely added substance whose presence modifies the vaporization characteristics of the materials undergoing separation, so as to make them easier to separate. The additive substance is often referred to as a solvent, and this solvent is usually chosen to be much less volatile than any of the substances being separated. This solvent is added to the downflowing liquid reflux stream near the top of the column, and is removed from the still pot or reboiler at the base of the column. The addition of furfural to mixtures of butadiene and butene hydrocarbons in order to separate the butadiene more easily is an example of extractive distillation.

extract of malt (maltine).

Properties: Light brown, sweet, viscous liquid; contains dextrin, maltose, a little glucose, and an amylolytic enzyme. It is capable of converting not less than five times its weight of starch into water-soluble sugars. Soluble in cold water, but more readily soluble in warm water. Sp. gr. not less than 1.350 and not more than 1.430 (25°C).

Derivation: By infusing malt with water at 60°C, concentrating the expressed liquid at a temperature not exceeding 60°C, and adding 10% by weight of glycerol.

Grade: N. F. XI.

Use: Medicine.

Shipping regulations: None.*

extreme pressure additives (EP additives).

1. Materials added to cutting oils to impart high film strength. They are mainly sulfur, chlorine, and occasionally phosphorus compounds. Actual conditions, amounts, etc. are trade secrets.

2. Lubricating oil and grease additives added to prevent metal-to-metal contact in highly loaded gears. (See "Aroclor"). In some cases this is accomplished by using additives which react with the metal gears to form a protective coating.

"Ex-Tri." [233] Trademark for trichloroethylene (q. v.).

"Exzyme." [114] Trademark for a proteolytic enzyme compound with a proteolytic digesting value of 15000 units/gram. Water soluble.

Uses: In the dry cleaning industry for the removal of soils and stains due to blood, glues, serum, egg, and others of albuminous origin.

Shipping regulations: None.*

eyestone. See copper, aluminated.

*See "I.C.C. Shipping Regulations," page xiii.
Reference numbers refer to name of manufacturer. See "List of Manufacturers," page v.

F

F. Symbol for fluorine.

°F. Abbreviation for Fahrenheit.

"FA." [224] Trademark for furfuryl alcohol (q. v.).

"Fabrikoid." [28, 56] Trademark for pyroxylin-coated fabrics used for a wide variety of purposes, including bookbinding, footwear, belts, bindings and welts, case covering, and for luggage covering, lining, and trim.

"Fabrilite." [28] Trademark for vinyl resin plastic coated fabrics and also for unsupported vinyl materials. The latter are used for handbags, pocketbooks, looseleaf binders and notebooks, and for lamination to metal for use as decorative trim. The coated fabrics are used for general and automotive upholstery and trim, seat covers, belts, bookbinding, footwear, folding doors and partitions, sports clothes, luggage, wall coverings, etc.

F acid. See 2-naphthol-7-sulfonic acid and 2-naphthylamine-7-sulfonic acid.

factice, rubber. Rubber-like products made by reaction of sulfur or sulfur chloride with vegetable oils. Used in some soft rubber products such as erasers.

"Factolac." [342] Trademark for a blend of gums and sugars used as a flavor emulsifying agent.

FAD. Abbreviation for flavin adenine nucleotide (q. v.).

fahlore (gray copper ore) $(Cu, Fe, Zn, Ag)_{12}(Sb, As)_4S_{13}$. A group of minerals consisting essentially of sulfantimonides or sulfarsenides of copper. Substitution of iron, zinc, silver, mercury, and lead is known for part of the copper, and of bismuth for arsenic and antimony. See tetrahedrite, tennantite, freibergite, and schwatzite.

Fahrenheit. The scale of temperature in which 212° is the boiling point of water at standard atmospheric pressure, and 32° is the freezing point of water. See centigrade for method of converting from Fahrenheit temperature scale to centigrade. See absolute temperature for converting Fahrenheit temperatures to absolute Rankine scale.

"Fairprene." [28] Trademark for a variety of products including:

Industrial cements made from synthetic rubber compounds for cementing rubber sheets and coated fabrics to leather, fabric, paper, metal and wood.

Protective and waterstop tape used as a seam sealant for shipbuilding and aircraft use in oil-, air-, and water-tight structural joints.

Synthetic rubber coated fabrics and sheet stock including silicone rubber and "Viton" A fluoroelastomer. Used for diaphragms, gaskets, washers, packings, seam sealants, curing bags and blankets.

fairy gloves. See digitalis.

fall-out. The deposition upon the earth of the radioactive particles contained in the cloud which forms as a result of a nuclear explosion.

"Falone." [248] Trademark for tris(2, 4-dichlorophenoxyethyl) phosphite, a pre-emergence herbicide.

Properties: A viscous amber liquid; sp. gr. 1.434; m. p. 70°C-72°C; soluble in benzene, xylene and aromatic hydrocarbons; insoluble in water. Available as an emulsifiable concentrate (Falone 44E) and a granular solid (Falone 10G).

Uses: A pre-emergence herbicide for the control of annual broad-leaf weeds and grasses on white potatoes, peanuts, strawberries and corn.

Hazards: Do not store near seeds or fertilizer. Avoid contact with skin or eyes. Do not store near heat or open flame.

false saffron. See carthamus.

false topaz. See cirtine.

false unicorn. See helonias.

"Fanal." [307] Trademark for phosphotungstic lakes. Used for printing inks. Characterized by brilliancy of shade and good fastness to light.

"Fanox." [51] Trademark for compounded oils available in several viscosity grades for metal working, and for quenching and cutting oils.

faraday. The quantity of electricity that can deposit (or dissolve) one gram-equivalent weight of a substance during electrolysis. It is about 96,500 coulombs.

farnesol $C_{15}H_{25}OH$. 3, 7, 11-Trimethyl-2, 6, 10(or 11)-dodecatrienol.

Properties: Colorless liquid having a delicate but rather faint floral odor. Soluble in 3 vols of 70% alcohol.

Constants: Sp. gr. 0.885 (15°C); b. p. 145-146°C (3 mm).

Derivation: Found in nature in many flowers

and essential oils, such as cassie, neroli, cananga, rose, balsams, ambrette seed.
Use: Perfumery.
Shipping regulations: None. *

"Fastel." [206] Brand name of proprietary line of pigments prepared from basic dyestuffs by combination with acids of the phosphomolybdic and tungsto-groups. Used in the production of printing inks.

"Fas-Tin-Flux." [72] Brand name for a flux for hot tinning consisting of zinc chloride and additives.

"Fastusol." [307] Trademark of direct colors used on cotton, rayon, paper, leather, and silk. Characterized by very good fastness to light.

fat clay. See ball clay.

fat dyes. Oil-soluble dyes for candles, wax, etc.

fatliquoring agents. Tanning operations tend to deprive hides of their natural oils and they therefore become hard and stiff. It is necessary to replace these natural oils in the leather in order to soften it. This is accomplished by placing the hides in a drum and tumbling them in emulsions of oil-in-water. Such oil-in-water emulsions are known as fat liquors and are usually made from raw oils such as cod, neatsfoot, and the like, rendered soluble by emulsifying with typical dispersing agents such as sulfonated oils.

fat pitch. See stearin and fatty-acid pitches.

fats. Solid glyceryl esters of higher fatty acids such as stearic and palmitic. Such esters and their mixtures are soft solids at ordinary temperatures. Liquid fats are known as animal and vegetable oils.

fatty-acid pitches. See stearin and fatty-acid pitches.

fatty acids. Monobasic organic acids derived from natural fats and oils. The term is also applied to all the monobasic acids of the general formula $C_nH_{2n+1}COOH$, but the following discussion applies only to the acids derived from natural fats and oils. The acids whose molecules have an even number of carbon atoms (usually 8 to 22) arranged in a straight chain are by far the most common and may be either saturated or unsaturated. The most abundant acids have 16 or 18 carbon atoms and these are commercially the most important. For more detail on these acids see palmitic, stearic, oleic, linoleic, and linolenic acids. The last two plus arachidonic acid are sometimes termed "essential fatty acids" by biochemists and nutritionists. The term fatty acid is also used sometimes to refer to the glyceride from which the acid can be derived. Thus an oil may be stated to contain a high percentage of oleic acid, when in fact it contains a high percentage of the corresponding glyceride.
Derivation: By hydrolytic splitting of fats and oils with glycerol as a by-product; oxidation of hydrocarbons.
Uses: (Usually in the form of mixtures of several acids) Soaps and synthetic detergents, lubricants, rubber products, cosmetics, waterproofing, nutrition, and research.

fatty alcohols. Primary alcohols from C_8 to C_{20}, usually straight-chain. The name once referred to the source of their manufacture, such as the natural fats and wax-containing substances (e. g. , sperm oil and spermaceti). More recently high molecular weight alcohols have been produced synthetically, in particular by the Oxo process. The more important methods of production are (a) reduction of vegetable seed oils and their fatty acids with sodium, (b) catalytic hydrogenation of such oils at elevated temperatures and pressures and (c) hydrolysis of spermaceti and sperm oil by saponification and vacuum fractional distillation. The more important commercial saturated alcohols are octyl, decyl, lauryl, myristyl, cetyl and stearyl. The commercially important unsaturated alcohols such as oleyl, linoleyl and linolenyl are also normally included in this group. The odor of the alcohols tends to disappear as the chain length increases. The alcohols are used as solvents for fats, waxes and gums; as resin solvents or co-solvents; as ingredients in pharmaceutical salves and lotions; as lube oil additives; as detergent ingredients; as textile antistatic and finishing agents.

fatty amines. Normal aliphatic amines which have been derived from fats and oils. They may be saturated or unsaturated, primary, secondary, or tertiary, but the alkyl groups are straight-chained and have an even number of carbons in each. The length of the alkyl groups varies from 8-22 carbon atoms.
Derivation: Fatty acids are treated with ammonia and heated to form fatty acid amides which are converted to nitriles and reduced to the amine.
Uses: As organic bases; synthesis; soaps; plasticizers; medicinals; rubber manufacture.

fatty esters. Fatty acids with the active hydrogen replaced by the alkyl group of a monohydric alcohol. The esterification of a fatty acid, RCOOH, by an alcohol, R'OH, yields the fatty ester RCOOR'. The most common alcohol used is methanol, yielding the methyl ester, $RCOOCH_3$. The methyl esters of fatty acids have higher vapor pressures than the corresponding acids and thus can be distilled more easily.

fatty nitriles RCN. Organic cyanides derived from fatty acids.
Derivation: Fatty acids are treated with ammonia and heated to form fatty acid amides which are converted to nitriles.
Uses: Intermediates for fatty amines; lube oil additives; plasticizers.

Fauser process for ammonia. Process for synthesis of ammonia at about 200 atmospheres pressure, 500°C, with promoted iron catalyst, and 10-20% conversion of the hydrogen-nitrogen mixture. The hydrogen is obtained by electrolysis, the nitrogen from air by using extra hydrogen for combustion to remove the oxygen.

Fauser process for nitric acid. A batch process more commonly used in Europe than in the U. S., to produce nearly 100% nitric acid. The mixture of oxides of nitrogen and water from an ammonia oxidation converter is cooled to condense the water so it may be removed. The remaining mixture is refrigerated to −10°C at 10 atmospheres pressure in a tower. Dilute nitric acid and liquid nitrogen dioxide are produced. Oxygen is then introduced in proper proportion and its temperature kept at 70°C with 50 atmospheres pressure for several hours.

"Faxam." [51] Trademark for a series of general purpose, paraffin-base pale engine oils for application on plain bearings. Also useful as hydraulic media and for process needs.

fayalite. See olivine.

FCC. Abbreviation for fluid-cracking catalyst, as used in the petroleum refining industry. Examples are powdered silica-alumina, in which alumina is impregnated in dry synthetic silica gel, and various natural clays impregnated with alumina.

F. D. A. Abbreviation for Food and Drug Administration.

FD & C dyes. Food, drug and cosmetic dyes. See D & C dyes.

Fe. Symbol for iron.

feather ore. See jamesonite.

"Febis." [51] Trademark for high quality paraffin base oils containing various compounding agents to suit them for process uses. They find applications as ceramic mold oils, cotton spray oils and on leather aprons of tape condensers.

"Fedrazil." [301] Trademark for a combination of pseudoephedrine hydrochloride and chlorcyclizine hydrochloride (q. v.).

FEF black. Abbreviation for fast extruding furnace black. See furnace black.

Fehling's solution. A reagent used in analytical work as a test for sugars. The Association of Official Agricultural Chemists and most of the laboratories abroad use Soxhlet's modification. The official method of preparing this solution, according to the Book of Methods of the Association of Official Agricultural Chemists, is as follows:

Prepare by mixing immediately before use equal volumes of (a) and (b).

(a) Copper sulfate solution: Dissolve 34.639 g of copper sulfate ($CuSO_4 \cdot 5H_2O$) in water, dilute to 500 cc and filter through prepared asbestos.

(b) Alkaline tartrate solution: Dissolve 173 g of Rochelle salts and 50 g of sodium hydroxide in water, dilute to 500 cc, allow to stand for two days, and filter through prepared asbestos.

Benedict's modification of Fehling's solution, a one-solution preparation that keeps well, is used for the same purpose as the two-liquid Fehling's solution and is sometimes called Fehling's solution.

feldspar (felspar; potassium aluminosilicate). Feldspar is the general name for a group of sodium, potassium, calcium, and barium aluminum-silicates. It is the common mineral in igneous and metamorphic rocks. Commercially, feldspar usually refers to the potassium feldspars, orthoclase and microcline, with the formula $KAlSi_3O_8$, usually with a little sodium.

Properties (orthoclase): Color white, gray, flesh-colored and various other shades; luster vitreous; hardness 6; two good cleavages; sp. gr. 2.54-2.57. Insoluble in water or acids.

Grades: (feldspar in general) Usually based on silicon dioxide content, potassium-sodium ratio, iron content, and fineness of grinding.

Containers: Bags; carload lots.

Occurrence: North Carolina, Colorado, New Hampshire, South Dakota, California, Arizona, Wyoming, Virginia, Texas.

Uses: Pottery, enamel, and ceramic ware; glass, soaps; abrasive; bond for abrasive wheels; artificial teeth; cements and concretes; insulating compositions; fertilizer; poultry grit; tarred roofing materials. Nepheline syenite, aplite, talc, and pyrophyllite are now used in place of feldspar in a number of applications.

felspar. See feldspar.

2-fenchanol. See fenchyl alcohol.

fenchol. See fenchyl alcohol.

fenchone $C_{10}H_{16}O$.

Properties: Oil with camphor-like odor.

Constants: Sp. gr. 0.9465 (19°C); b. p. 193°C. Soluble in ether; insoluble in water.

Derivation: A ketone found (a) as dextro-fenchone in oil of fennel; (b) as levo-fenchone in oil of thuja.

fenchyl alcohol (fenchol; 2-fenchanol; 1-hydroxy-fenchane) $C_{10}H_{18}O$.

Properties: Colorless, crystalline substance.

Constants: Sp. gr. 0.962 (approx); boiling range 198-204°C; freezing point 30-34°C; refractive index 1.4626; optical rotation −0.1°.

Derivation: From turpentine and pine oil. Also synthetically.

Grades: Technical.

Uses: Solvent; organic synthesis.

fennel (sweet fennel; foeniculum).

Derivation: Dried, ripe fruit of cultivated

varieties of Foeniculum vulgare.
Habitat: Southern Europe and western Asia; widely cultivated.
Grades: Technical; graded by country of origin.
Containers: Bags.
Use: Medicine.
Shipping regulations: None. *

fennel oil (fennel-seed oil).
Properties: Colorless or pale yellowish essential oil; characteristic aromatic odor; bitter, camphor-like taste changing to sweetish, mild and spicy. Soluble in 5-8 vols of 80% alcohol, in 1 vol of 90% alcohol, in ether; insoluble in water.
Chief known constituents: Anethole, fenchone, pinene, camphene, dipentene, and phellandrene.
Constants: Sp. gr. 0.953-0.973 (15°C); b.p. 160-220°C; optical rotation +12° to +20°; refractive index 1.528-1.538. Congealing temperature not below 3°C.
Grades and derivation: Fennel oil U.S.P. XVI is derived by distillation from Foeniculum vulgare; Roman fennel oil from F. dulce. The Roman fennel does not contain fenchone.
Adulteration: Elimination of all or part of anethole content; turpentine oil; alcohol.
Containers: 1-, 5-, 10-lb bottles; 25-, 50-lb tins.
Uses: Medicine; liqueurs; perfumery; flavors.
Shipping regulations: None. *

fennel-seed oil. See fennel oil.

"Fenso." [51] Trademark for a high-speed quenching oil, useful where metallurgical and production requirements dictate more rapid metal cooling than is possible with a straight mineral oil.

fenugreek (trigonella foenum graecum).
Derivation: Seeds of Trigonella.
Habitat: Egypt; Asia Minor; France and Germany.
Grades: Technical.
Containers: Bags.
Uses: Medicine; veterinary medicine; spices, including curry; flavors, especially maple.
Shipping regulations: None. *

fenuron. Coined name for 3-phenyl-1, 1-dimethylurea.

FEP. Abbreviation for fluorinated ethylene propylene (q.v.), used as in FEP fluorocarbons, FEP resins.

ferbam (ferric dimethyldithiocarbamate) $[(CH_3)_2NCSS]_3Fe$. A fungicide.
Properties: Black or dark colored, fluffy powder; decomposes above 180°C; usually readily dispersible but very slightly soluble in water; pH of saturated solution 5.0.
Derivation: By addition of carbon disulfide to an alcoholic solution of dimethylamine and precipitation with a ferric salt.
Grades: 76% wettable powder; 87% technical powder.
Containers: 3-lb and 50-lb multiwall paper bags.
Caution! May cause irritation of eyes, nose, throat and skin. MCA warning label.
Use: Fungicide.

ferberite. See wolframite.

"Ferberk." [81] Trademark for ferric dimethyldithiocarbamate. See ferbam.

"Fergon." [162] Trademark for ferrous gluconate.

fergusonite $(Y, Er)(Nb, Ta)O_4$. An oxide of yttrium, erbium, niobium, and tantalum, sometimes containing small amounts of other rare earths, and uranium, zirconium, thorium, calcium, iron and titanium. Found in pegmatites.
Properties: Color gray, brown, or black; luster dull to vitreous; streak brown or gray; hardness 5.5-6.5; sp. gr. 5.6-5.8.
Occurrence: North Carolina, South Carolina, Virginia, Texas; Norway; Sweden; Africa.
Use: A rare earth mineral.

"Fermate." [28] Trademark for agricultural and horticultural fungicide based on ferbam (ferric dimethyldithiocarbamate).
Containers: 3-, 50-lb bags.
Uses: As a fungicide for control of certain diseases of apples, pears, grapes, tobacco, and ornamentals.

"Fermentase." [114] Brand name for a baker's malt syrup of the diastatic variety.

fermentation alcohol. See ethyl alcohol.

"Fermex." [173] Trademark for a diastatic (protease - amylase) enzyme supplement for baking fermentation used to improve quality and uniformity of bread and other yeast-raised bakery products.

Fermi age. A term used in neutron moderation or slowing-down theory applied to nuclear reactors. Neutrons are slowed in discrete steps by scattering off of the nuclei of the moderator, and if this process is approximated as a smooth process, the Fermi age is related to the mean square path length for a neutron to be slowed to a given energy.

fermion. Also called Fermi particle, or Fermi-Dirac particle. Consideration of symmetry properties of the wave-mechanical description of systems of particles allow the classification of the particles into two kinds. Fermions are particles with half-integral spin, obey the Pauli exclusion principle and only one such particle may occupy a given quantum state. See also boson.

fermium (element 100) Fm. A synthetic radioactive element with atomic number 100 first discovered in the debris from the 1952 hydrogen bomb explosion. Fermium has since been prepared in a nuclear reactor by irradiating californium, plutonium, or einsteinium with neutrons, in a cyclotron by bombarding uranium with accelerated oxygen ions, and by other nuclear reactions. The element is named for Enrico Fermi. It has chemical properties similar to those of the rare earth erbium.

Isotopes are known with mass numbers 254, 255, and 256.
See also actinide elements.

"Ferocator" Salts. [204] Trademark. Laundry sour with good neutralizing value, moderate rust-removing qualities.

ferric acetate, basic (iron acetate, basic) $Fe(C_2H_3O_2)_2OH$.
Properties: Red powder. Soluble in alcohol and acids; insoluble in water.
Derivation: By the action of acetic acid on ferric hydroxide with subsequent crystallization.
Grades: Technical.
Containers: Wooden barrels; fiber drums.
Uses: Medicine; textile industries.
Shipping regulations: None. *

ferric acetylacetonate $Fe(C_5H_7O_2)_3$.
Properties: Crystalline powder; m. p. 179-182°C. Slightly soluble in water, soluble in most organic solvents. Resistant to hydrolysis. A chelating non-ionizing compound.
Uses: Moderating and combustion catalyst; solid fuel additive catalyst; bonding agent; curing accelerator; intermediate.

ferric alginate (alginoid iron; algiron; iron alginate) $C_{76}H_{77}N_2O_{22}Fe_3$.
Properties: Brown powder. Tasteless. Contains approx 11% iron. Soluble in ammonia; insoluble in water.
Derivation: Interaction of ferric chloride and sodium alginate.
Use: Medicine.

ferric ammonium alum. See ferric-ammonium sulfate.

ferric ammonium citrate (iron ammonium citrate). Protect from light.
Properties: Thin, transparent, garnet red scales or granules, or as a brownish yellow powder; odorless (or slight ammonia odor); saline, mildly ferruginous taste; deliquescent; affected by light. Soluble in water; insoluble in alcohol.
Derivation: By the addition of citric acid to ferric hydroxide, then adding ammonium hydroxide, followed by filtration.
Grades: Technical; N. F. XI (brown and green varieties, the latter containing a lower percentage of iron).
Containers: Bottles; 100-lb drums.
Uses: Medicine; blueprint photography; feed additive.
Shipping regulations: None. *

ferric-ammonium oxalate (iron-ammonium oxalate; ammonioferric oxalate) $(NH_4)_3Fe(C_2O_4)_3 \cdot 3H_2O$.
Properties: Green crystals. Soluble in water and alcohol; sensitive to light.
Derivation: By the interaction of ammonium binoxalate and ferric hydroxide.
Method of purification: Crystallization.
Grades: Technical; C. P.
Containers: 1-lb bottles; drums.
Use: Blue print photography.
Shipping regulations: None. *

ferric-ammonium sulfate (iron-ammonium sulfate; ferric ammonium alum; ammonioferric sulfate) $FeNH_4(SO_4)_2 \cdot 12H_2O$.
Properties: Lilac to violet; efflorescent crystals. Soluble in water; insoluble in alcohol.
Derivation: By mixing solutions of ferric sulfate and ammonium sulfate, followed by evaporation and crystallization.
Method of purification: Recrystallization.
Grades: Technical; C. P.
Containers: Bottles; jars; barrels.
Uses: Medicine; analytical chemistry; textile dyeing (mordant).

ferric-ammonium tartrate (ammonium-iron tartrate; iron-ammonium tartrate).
Properties: Reddish-brown scales. Transparent. Sweet taste. Soluble in water; insoluble in alcohol.
Use: Medicine.

ferric arsenate $FeAsO_4 \cdot 2H_2O$.
Properties: A green or brown powder. Insoluble in water; soluble in dilute mineral acids.
Use: Insecticide.
Shipping regulations: Poison, class B. Poison label. *

ferric arsenite. A basic salt of variable composition.
Properties: Brownish-yellow powder. Soluble in acids; insoluble in water.
Uses: Combined with ammonium citrate (ferric ammonium citrate) (q. v.) and used in medicine.
Shipping regulations: Poison, class B. Poison label. *

ferric benzoate (iron benzoate) $Fe_2(OOCC_6H_5)_6$.
Properties: Brown powder. Slightly soluble in water, alcohol, and ether.
Derivation: By the interaction of ferric hydroxide and benzoic acid.
Method of purification: Crystallization.
Grades: Technical.
Containers: Boxes; glass bottles.
Use: Medicine.
Shipping regulations: None. *

ferric bichromate. See ferric dichromate.

ferric bromide (ferric tribromide; ferric sesquibromide; iron bromide) $FeBr_3$.
Properties: Dark-red, deliquescent crystals. Soluble in water, alcohol, and ether. M. p., sublimes.
Derivation: By the action of bromine on iron filings.
Method of purification: Crystallization.
Grades: Technical.
Containers: Boxes; glass bottles. Keep cool, well closed, and protected from light.
Uses: Medicine; analytical chemistry; bromine salts.
Shipping regulations: None. *

ferric cacodylate $Fe[(CH_3)_2AsO_2]_3$.
Properties: Yellowish-brown powder; odorless; poisonous. Moderately soluble in cold water, more so in hot water, less so in alcohol.

Use: Medicine.
Shipping regulations: None. *

ferric chloride, anhydrous (ferric trichloride; ferric perchloride; ferric sesquichloride; iron chloride; iron sesquichloride; iron trichloride; flores martis; iron perchloride; molysite) $FeCl_3$.
Properties: Black-brown solid; sp. gr. 2.8; m. p. about 300°C; very soluble in water, alcohol and glycerol.
Derivation: Action of chlorine on ferrous sulfate or chloride.
Grades: Anhydrous 96%; 42° Bé solution, photographic and sewage grades.
Containers: (Solid) barrels; (solution) carboys; tank cars.
Uses: Coagulant for sewage and industrial wastes, also in glycerin manufacture, in etching copper in photoengraving, as a mordant, and to produce decorative surface effects on ceramics. Also an oxidizing, chlorinating and condensing agent, disinfectant, pigment, medicine; feed additive.

ferric chloride hydrate $FeCl_3 \cdot 6H_2O$.
Properties: Orange-yellow very deliquescent crystals. M. p. 37°C; b. p. 280°C; decomposes to yield hydrochloric acid if exposed to moist air or light.
Uses: See ferric chloride, anhydrous.

ferric chromate (iron chromate) $Fe_2(CrO_4)_3$.
Properties: Yellow powder; soluble in acids; insoluble in water and alcohol.
Derivation: By adding sodium chromate to a solution of a ferric salt.
Grades: Technical.
Containers: Wooden kegs; fiber drums.
Uses: Metallurgy; ceramics (color); paint pigment.
Shipping regulations: None. *

ferric citrate (iron citrate) $FeC_6H_5O_7 \cdot 3H_2O$.
Properties: Reddish-brown scales. Keep away from light. Soluble in water; insoluble in alcohol.
Derivation: By the action of citric acid on ferric hydroxide, and crystallization.
Containers: Amber-glass bottles; cans; drums.
Uses: Medicine; blueprint paper.
Shipping regulations: None. *

ferric dichromate (iron dichromate; ferric bichromate) $Fe_2(Cr_2O_7)_3$.
Properties: Reddish-brown granules. Soluble in water and acids.
Derivation: By heating aqueous chromic acid and moist ferric hydroxide.
Containers: Wooden kegs; fiber drums.
Grades: Technical.
Use: Preparation of pigments.

ferric dimethyldithiocarbamate. See ferbam.

ferric ethylhexoate. See soaps, metallic.

ferric ferrocyanide (iron ferrocyanide; Prussian Blue). Blue pigment described under iron blues (q. v.).

ferric fluoride (iron fluoride) FeF_3.
Properties: Green crystals; sp. gr. 3.18; soluble in acids; soluble in water.
Grades: Technical.
Use: Ceramics (porcelain, pottery).

ferric glycerophosphate (iron glycerophosphate) $Fe_2[C_3H_5(OH)_2PO_4]_3 \cdot x\ H_2O$.
Properties: Yellowish scales, odorless and nearly tasteless; soluble in water; insoluble in alcohol.
Grades: Technical.
Containers: Glass bottles; drums.
Use: Pharmaceutical.
Shipping regulations: None. *

ferric hydrate. See ferric hydroxide.

ferric hydroxide (ferric hydrate; iron hydroxide; iron hydrate; iron oxide, hydrated; ferric oxide, hydrated) $Fe(OH)_3$.
Properties: Brown flocculent precipitate which dries to the oxide; sp. gr. 3.4-3.9; m. p., loses water below 500°C; soluble in acids; insoluble in water, alcohol and ether.
Derivation: Addition of ferrous sulfate solution to ammonia solution.
Grades: Technical; C. P.
Containers: 1-lb bottles; wooden barrels.
Uses: Pharmaceutical preparations; water purification; manufacturing pigments; rubber pigment; antidote for arsenic poisoning.
Shipping regulations: None. *

ferric hypophosphite (iron hypophosphite) $Fe(H_2PO_2)_3$.
Properties: White or grayish-white powder; odorless; tasteless. Caution: Explosion may occur if triturated or heated with nitrates, chlorates or other oxidizing agents. Slightly soluble in water; more so in boiling water.
Containers: Bottles; drums.
Use: Medicine.
Shipping regulations: None. *

ferric malate (iron malate) $Fe_2(C_4H_4O_5)_3$.
Properties: Brown, hygroscopic crystals. Keep well stoppered. Soluble in water and alcohol.
Derivation: By the interaction of ferric hydroxide and malic acid.
Method of purification: Crystallization.
Grades: Technical.
Containers: Amber-glass bottles.
Use: Medicine.
Shipping regulations: None. *

ferric naphthenate
Properties: A metallic soap.
Derivation: Fusion method, by heating naphthenic acids with the metallic oxide.
Containers: Drums.
Uses: Conditioning and waterproofing agent, sludge preventative, fungicide and paint drier.

ferric nitrate (iron nitrate) $Fe(NO_3)_3 \cdot 9H_2O$.
Properties: Violet crystals; soluble in water, and alcohol; decomposed by heat.
Derivation: By the action of concentrated nitric acid on scrap iron or iron oxide, and crystallizing.
Grades: Technical; C. P.

Containers: 1-, 5-lb bottles.
Uses: Dyeing (mordant for buffs and blacks); medicine; tanning; analytical chemistry.
Fire hazard: Dangerous.
Shipping regulations: Oxidizing material. Yellow label. *

ferric octoate. See soaps, metallic.

ferric oleate (iron oleate) $Fe(C_{18}H_{33}O_2)_3$.
Properties: Brownish-red lumps. Soluble in alcohol, ether, and acids; insoluble in water. See also soaps, metallic.

ferric oxalate $Fe_2(C_2O_4)_3$.
Properties: Pale-yellow amorphous scale or powder; odorless; decomposes on heating to 100°C. Soluble in water and acids; insoluble in alkali.
Grades: Technical; C. P.
Containers: 50-lb drums.
Uses: Catalyst in making oxygen; silvertone photographic printing papers.

ferric oxide (ferric oxide, red; iron oxide; red iron trioxide; iron sesquioxide; ferric trioxide) Fe_2O_3. See also iron oxide reds.
Properties: Dense, dark-red powder or lumps; sp. gr. 5.12-5.24; m. p. 1565°C; soluble in acids; insoluble in water.
Derivation: (a) Found in nature as hematite ore (see hematite); (b) by calcining ferrous sulfate or oxalate; (c) by dehydrating ferric hydroxide; (d) by-product in some industries.
Grades: Technical; 99.5% pure.
Containers: 100-lb kegs; 300-, 350-, 450- and 500-lb barrels; fiber drums; 800-lb casks.
Uses: Metallurgy; gas purification; paint pigment; polishing compounds; pigment in rubber products; mordant in dyeing textiles; laboratory reagent; catalyst in chemical processes; in medicine; feed additive; magnetic tapes in electronics.
Shipping regulations: None. *

ferric oxide, hydrated. See ferric hydroxide.

ferric oxide, red. See ferric oxide.

ferric oxide, yellow. Impure ferric oxide (water and calcium sulfate are the usual impurities). See iron oxide yellows. Natural hydrated yellow iron oxide (limonite) is sometimes referred to as yellow ferric oxide.

ferric perchloride. See ferric chloride.

ferric phosphate (ferric phosphate, insoluble; iron phosphate) $FePO_4 \cdot 2H_2O$.
Properties: Yellowish-white powder. Insoluble in water; soluble in acids; sp. gr. 2.87.
Derivation: By adding a solution of sodium phosphate to a solution of ferric chloride. The product is filtered and then dried.
Grades: Technical; C. P.
Containers: Drums.
Uses: Medicine; fertilizers; feed additive.
Shipping regulations: None. *

ferric phosphate, soluble (ferric phosphate with sodium citrate).
Properties: Bright green crystals; odorless; stable in air; affected by light; soluble in water; insoluble in alcohol.
Derivation: By adding sodium phosphate to ferric citrate.
Grades: N. F. XI; technical; C. P.
Caution: Protect from light.
Use: Medicine.
Shipping regulations: None. *

ferric-potassium citrate (iron-potassium citrate).
Properties: Brown scales. Hygroscopic. Odorless. Contains 16% (approx.) iron. Soluble in water; insoluble in alcohol.
Caution: Keep well closed.
Use: Medicine.

ferric potassium oxalate $K_3Fe(C_2O_4)_3 \cdot 3H_2O$.
Properties: Emerald-green, monoclinic crystals, sensitive to light. Soluble in water; very slightly soluble in alcohol.
Containers: Drums.
Use: Medicine.

ferric-potassium sulfate (iron-potassium sulfate; iron alum) $FeK(SO_4)_2 \cdot 12H_2O$.
Properties: Pale violet crystals. Soluble in water; insoluble in alcohol; sp. gr. 1.806.
Derivation: By mixing solutions of potassium sulfate and ferric sulfate and crystallizing.

ferric potassium tartrate. See iron potassium tartrate.

ferric pyrophosphate (iron pyrophosphate) $Fe_4(P_2O_7)_3 \cdot xH_2O$.
Properties: Yellowish-white powder; insoluble in water, soluble in dilute acid. 24% Fe minimum. Not to be confused with ferric pyrophosphate soluble.
Containers: Drums.
Uses: Source of nutritional iron and for enrichment of foods not subject to rancidity.
Shipping regulations: None. *

ferric pyrophosphate, soluble.
Properties: Apple-green crystals; insoluble in water; insoluble in alcohol. Protect from light.
Derivation: By adding sodium pyrophosphate to ferric citrate.
Containers: 225-lb drums.
Use: Medicine; feed additive.

ferric resinate (iron resinate).
Properties: Reddish-brown powder; soluble in ligroin, carbon disulfide, ether, oil of turpentine; slightly soluble in alcohol; insoluble in water.
Containers: Drums.
Use: Drier (paints, varnish). See also soaps, metallic.

ferric salicylate (iron salicylate).
Properties: Violet-gray powder. Variable composition. Caution! Keep away from light! Slightly soluble in water.
Use: Medicine.

ferric sesquibromide. See ferric bromide.

ferric sesquichloride. See ferric chloride.

ferric sesquioxide. See ferric oxide.

ferric sesquisulfate. See ferric sulfate.

ferric silicate (iron silicate). Light-brown powder. Decomposed by hydrochloric acid. Insoluble in water.

ferric-sodium citrate (iron-sodium citrate).
Properties: Light-brown scales. Soluble in water.
Use: Medicine.

ferric sodium oxalate (iron-sodium oxalate) $Na_3Fe(C_2O_4)_3 \cdot 4\frac{1}{2}H_2O$.
Properties: Emerald-green crystals; decomposed by heat or light; soluble in water and alcohol.
Derivation: By the interaction of sodium acid oxalate and ferric hydroxide.
Method of purification: Crystallization.
Grades: Technical.
Containers: Glass bottles; drums.
Caution: Protect from light.
Uses: Photography; blueprinting.

ferric stearate (iron stearate) $Fe(C_{18}H_{35}O_2)_3$.
Properties: Light-brown powder; soluble in alcohol and ether; insoluble in water.
Derivation: By the interaction of solutions of ferric sulfate and sodium stearate.
Grades: Technical.
Containers: Wooden kegs; fiber drums.
Use: Varnish driers. See also soaps, metallic.
Shipping regulations: None. *

ferric subsulfate. See ferric sulfate, basic.

ferric succinate (iron succinate) $(C_4H_4O_4)_2Fe_2(OH)_2$.
Properties: Reddish-brown powder; insoluble in cold water; partly decomposed by hot, forming a more basic salt; soluble in dilute acids.
Derivation: By addition of ferric chloride to a succinate solution.
Use: Medicine.
Caution: Protect from light.
Shipping regulations: None. *

ferric sulfate (iron sulfate; ferric trisulfate; iron tersulfate; iron sesquisulfate; ferric sesquisulfate; iron persulfate)
(a) $Fe_2(SO_4)_3$; (b) $Fe_2(SO_4)_3 \cdot 9H_2O$.
Properties: Yellow crystals or grayish-white powder. Soluble in water. Sp. gr. (a) 3.097, (b) 2.0-2.1; m.p., decomposes.
Derivation: By adding sulfuric acid to ferric hydroxide.
Impurities: Ferrous sulfate; water; sulfuric acid.
Grades: Technical; C.P.; partly hydrated.
Containers: Bottles; wooden barrels; bags.
Caution: Keep well closed and protected from light.
Uses: Pigments; medicine; reagent in analytical chemistry; iron alum manufacture; etching aluminum; disinfectant; textiles (dyeing and calico printing); water purification; soil conditioner for alkaline soils.
Shipping regulations: None. *

ferric sulfate, basic (basic iron sulfate; ferric subsulfate; Monsel's salt; iron subsulfate) $Fe_4O(SO_4)_5$.
Properties: Yellow, hygroscopic powder.
Derivation: By adding ferrous sulfate to hot dilute sulfuric and nitric acids and boiling until all the nitric acid is driven off and filtering if necessary.
Grades: Technical.
Caution: Keep well closed.
Shipping regulations: None. *

ferric tallate. See soaps, metallic.

ferric tannate (iron tannate; iron gallotannate) $Fe_2(C_{14}H_7O_9)(OH)_3$.
Properties: Dark-brown or bluish black powder, variable composition. Soluble in alkalies; insoluble in water, alcohol and ether; soluble in dilute acids.
Derivation: By the interaction of ferric acetate and tannic acid solutions.
Grades: Technical.
Containers: Glass bottles.
Use: Medicine.
Shipping regulations: None. *

ferric tribromide. See ferric bromide.

ferric trichloride. See ferric chloride.

ferric trioxide. See ferric oxide.

ferric trisulfate. See ferric sulfate.

ferric valerate $Fe_2(C_5H_9O_2)_2(OH)_4$.
Properties: Brick-red powder with valeric odor; soluble in alcohol; decomposed by boiling water.
Derivation: Interaction of sodium valerate and ferric sulfate in solution.
Use: Medicine.
Caution: Keep tightly closed and protected from light.
Shipping regulations: None. *

ferric vanadate (iron metavanadate) $Fe(VO_3)_3$.
Properties: Grayish-brown powder; soluble in acids; insoluble in water and alcohol.
Derivation: By adding a solution of a ferric salt to the liquor obtained by leaching vanadium ores with caustic soda solution or by lixiviating the slags obtained when vanadium ores are fused with soda ash, etc.
Grades: Technical.
Containers: Wooden kegs; fiber drums.
Use: Metallurgy.
Shipping regulations: None. *

"Ferri-Floc." [93] Trade name for partially hydrated ferric sulfate.
Properties: Contains a minimum of 20.5% water soluble-ferric ion; granular, free-flowing and stable.
Containers: 100-lb polyethylene lined bags; 200-lb drums; bulk.
Uses: Coagulant and flocculating agent for water, sewage and waste disposal; pickling agent for copper and brass alloys; fertilizer nutritional trace elements.

ferrimolybdite (molybdic ocher; iron molybdate; molybdite) $Fe_2(MoO_4)_3 \cdot 8H_2O$.
Natural hydrated molybdate of iron.
Properties: Color yellow; luster silky to earthy, usually occurs as fibrous crusts.
Occurrence: New Mexico, Arizona, California, Nevada, Pennsylvania.

ferrite.

1. Iron which, in pig iron or steel, has not combined with carbon to form cementite (q. v.). It exists in alpha, beta, gamma, and delta forms, which vary in magnetism and ability to dissolve cementite. See also carbon, combined; carbon, graphitic; pearlite; cementite.
2. A compound of ferric oxide with a strong basic oxide, as sodium ferrite $NaFeO_2$. See also ferrites.
3. Name applied to magnetic iron oxides (ferromagnetic oxides) having a definite crystal structure (spinels) and the formula $M^{++}Fe_2^{+++}O_4$, in which the divalent metal may be any which fits into the crystal lattice, often iron, nickel, zinc, or manganese. Magnetite itself, $FeFe_2O_4$, is a ferrite. The magnetic properties vary according to the divalent atom present, and ferrites are now tailored for their desired effect, as nickel aluminum ferrite: $Ni_{0.86}^{++}Cu_{0.1}^{++}Mn_{0.02}^{++}Co_{0.02}^{++}Al_{0.3}^{+++}Fe_{1.7}^{+++}O_4$.

Uses: In electronics as rectifiers, on memory or record tapes; and for permanent magnets; in television, radio, radar, and in computer and missile guidance systems.

ferro-alloys. Alloys of iron with some element other than carbon used as a vehicle for introducing such an element into the manufacture of steel. The element may alloy with the steel by solution or as the carbide, neutralize the harmful impurities by combining with them and separating from the steel as flux or slag before solidification. See specific ferro-alloys.

ferroboron. A ferro-alloy averaging 16.2% boron used as hardening agent in special steels. It also is an efficient deoxidizer. Boron steel is used in controlling the operating rate of the uranium-graphite piles used to produce plutonium.

"Ferrocarbo." [280] Trademark for briquetted or granular silicon carbide.

Uses: As a cupola addition in the production of gray iron, or as a ladle addition to steel. It disintegrates into its component elements and acts as a powerful deoxidizer and graphitizer. Machinability and strength of the iron or steel are increased with no loss of hardness.

ferrocene (dicyclopentadienyliron) $(C_5H_5)_2Fe$. A coordination compound of ferrous iron and cyclopentadiene in which the organic portions have typically aromatic chemical properties. See also dicyclopentadienyl metal compounds.

Properties: Orange, crystalline solid; camphor-like odor; m.p. 173-174°C; resists pyrolysis at 400°C; resistant to ultraviolet light. Insoluble in water, slightly soluble in benzene, ether and petroleum ether.

Derivation: From ferrous chloride and cyclopentadiene sodium.

Uses: Additive to jet fuels and furnace oils to improve efficiency of combustion and eliminate smoke; antiknock additive for gasoline; suggested for coating for missiles and satellites; high temperature lubricant; curing agent for rubber and silicone resins; intermediate for high temperature polymers; ultraviolet absorber.

ferrocenes. See dicyclopentadienyl metal compounds.

ferrocerium. See misch metal.

ferrocholinate (iron choline citrate chelate) $C_{11}H_{20}FeNO_9$.

Properties: Greenish-brown to reddish-brown amorphous solid; soluble in water, yielding a stable solution; soluble in acid and alkaline media.

Derivation: Prepared by the reaction of choline dihydrogen citrate with ferric hydroxide.

Grade: N.N.D.

Use: Medicine.

ferrochrome. See ferrochromium.

ferrochromium (ferrochrome). An alloy composed principally of iron and chromium used as a means of adding chromium to steels. It is available in several classifications and grades, generally containing between 60-70% chromium.

Low-carbon ferrochromium: Used for low-carbon high-chromium steels such as stainless steels. Available in six grades of carbon content from 0.03-2.00%. Melting range from 2500-3100°F for 0.03% carbon to 2340-2460°F for 2% carbon.

High-carbon ferrochromium: Used for making medium and high-carbon steels of relatively low chromium content. Available in several grades having 65-70% chromium and 4-9% carbon. Melting range 2230-2790°F.

High-nitrogen ferrochromium: Ferrochromium with nitrogen content about 1%. Available in high and low-carbon grades.

Foundry grade ferrochromium: An alloy especially designed for the ladle addition of chromium to cast iron. May also be used for steel. Contains 6-9% silicon, and 1-7% carbon.

Special ferrochromiums: There are several alloys designed for ladle additions, such as the SM ferrochromiums, which contain 4-6% each of manganese and silicon. Available in low and high-carbon grades. These alloys are distinguished by the ease with which they dissolve in liquid steel.

The ferrochromiums are available in a variety of crushed sizes and lumps up to 75 pounds.

See chromium iron alloys; steel, stainless; and iron, stainless.

ferroconcrete. See concrete.

ferroelectric. A crystalline solid material such as barium titanate, monobasic potassium phosphate or potassium-sodium tartrate (Rochelle salts) that over certain limited temperature ranges has a natural or inherent deformation (polarization) of the electrical fields or electrons associated with the atoms and groups in the crystal lattice. This results in the development of positive and negative poles and a consequent "direction" of polarization, which can be reversed when the crystal is exposed to an external electric field. Ferroelectric crystals are internally strained and as a consequence show unusual piezoelectric and elastic properties that result in uses in capacitors and transducers where external physical forces are to be related to electrical phenomena.

ferroferric oxide. See iron oxide, black.

"Ferrolon." [232] Brand name for a series of organic sequestering agents used in textile applications.

ferromagnesite. An iron-bearing variety of magnesite (q. v.) used for refractories owing to its ability to bond under heat.

ferromanganese. An alloy consisting principally of iron and manganese used as a vehicle for adding manganese in the manufacture of steel. It contains between 78-90% manganese, depending on the grade. Available in three classifications:

Standard ferromanganese: Used on all grades of steel including Bessemer, open hearth and electric for forgings, rolled products, and castings. Melting range 1960-3210°F. Typical analysis: manganese 78-82%; iron 12-16%; carbon 6-8%; silicon 1% (max); phosphorus 0. 3% (max); sulfur 0. 05% (max).

Low-carbon ferromanganese: Used for adding manganese to extremely low-carbon steels and alloys, such as stainless steels in which the carbon often must be kept below 0. 1%. Available in several grades with carbon ranging from 0. 07% to 0. 75% (max) and upper limits on silicon and phosphorus of 1% and 0. 06 to 0. 2% respectively. Melting range 2190 to 2220°F.

Medium-carbon ferromanganese: Used for ordinary low-carbon steels. Sold in one grade. Manganese 80-85%; carbon 1. 5% (max); silicon 1. 5% (max).

Ferromanganese is available in ground, crushed and lump sizes ranging from 80 mesh to 75-lb lumps, suitable for ladle or furnace addition.

See manganese steel.

ferromolybdenum. An alloy composed largely of iron and molybdenum used as a means of adding molybdenum in the manufacture of steel. Engineering steels rarely contain more than 1% molybdenum, stainless steels may contain 3% and tool steels can contain as much as 10% molybdenum. Ferromolybdenum is available in several grades in which molybdenum ranges from 55% to 75% and the maximum carbon content is either 0. 10%, 0. 60%, or 2. 50%. Ferromolybdenum is generally added in the furnace, since it does not oxidize under steel-making conditions. M. p. 2965°F (approx). Available in crushed sizes up to one inch.

ferron. See loretin.

ferrophosphorus. Alloys of iron and phosphorus used in the steel industry for adjustment of phosphorus content of special steels; particularly useful in preventing thin sheets from sticking together when rolled and annealed in bundles. Produced in two grades: (a) 18% phosphorus; (b) 25% phosphorus. The 25% grade has advantages incident to one-third less alloy addition to the molten steel.

Containers: Bulk and in casks.

"Ferro-Sequels." [57] Trademark for iron preparation in sustained release dosage form.

ferrosilicon. An alloy of iron and silicon used to add silicon to steel and iron. Small quantities of silicon deoxidize the iron and larger amounts impart special properties. (See silicon steels.) It is also used in a process for producing metallic magnesium. (See Pidgeon process.) Ferrosilicon is available in six grades:

15% ferrosilicon: Available both for blast furnace and electric furnace. Used as a deoxidizer and cleanser, and for blocking heats. Generally added in the furnace. The material supplied for electric furnace use is very low in phosphorus. Several grades range from 14-18% silicon. Typical analysis: silicon 14-18%; carbon 1% (max); phosphorus 0. 05% (max); sulfur 0. 04% (max). Melting range 2220-2250°F. Available in 50- and 100-lb bags.

25% ferrosilicon: Same uses and grades as 15% ferrosilicon, but contains 25% to 30% silicon. Melting range 2200-2400°F. Available crushed and in lump form.

50% ferrosilicon: Most widely-used grade as a deoxidizer and in production of killed and semikilled steels and for making high-silicon steels. Contains 46-52% silicon. Melting range 2210-2230°F. Available in various lump and crushed sizes.

75% ferrosilicon: Preferred for production of high-silicon steels because less cold alloy has to be added to the melt, and this grade of ferrosilicon generates heat when added to the molten steel, obviating the necessity for additional heating. This is a great advantage in producing steels with more than 2% silicon. Contains 74-79% silicon. Melting range 2200-2400°F. Furnished in various crushed and lump sizes.

85% ferrosilicon: Same comments as 75% ferrosilicon. Furnished in two grades, 80-85% silicon, and 85-90% silicon. Melting range 2200-2460°F. Supplied in crushed and lump sizes.

90-95% ferrosilicon: Same comments as 75% ferrosilicon. Melting range 2200-2520°F. In crushed and lump sizes.

Shipping regulations: Ferrosilicon containing between 45-70% silicon will give some evolution of poisonous gases when moisture is present, due to impurities, and is subject to stringent restrictions when being shipped by water.*

ferrosilicon process for magnesium. See Pidgeon process.

ferroso-ferric oxide. See iron oxide, black.

"Ferro-Sour." [244] A proprietary product consisting of fluorine compounds.
Properties: White powder, sparingly soluble in water; neutralizing value 26.9 oz sodium bicarbonate per lb.
Containers: 150-lb and 300-lb net fiber drums.
Uses: Laundry sour, low cost iron-removing type.
Fire hazard: None.
Shipping regulations: None. *

ferrotitanium. An alloy composed principally of iron and titanium used to add titanium to steel. Three classifications are available.
Low-carbon ferrotitanium: Used in producing stabilized austenitic stainless steel. Reduces tendency for intergranular corrosion in areas adjacent to welds. Available with the following ranges of composition: titanium 20-45%; aluminum 3-7% (max); silicon 4-13% (max); carbon 0.1% (max).
High-carbon ferrotitanium and
Medium-carbon ferrotitanium: Used as deoxidizers and scavengers in steel in quantities such that almost no titanium remains in the steel. Ferrotitanium is a final deoxidizer in killed steels. It prevents segregation and helps control grain size. Titanium 15-20%; aluminum 1-2%; silicon 2-3%; carbon 6-8% (high carbon), 3-5% (medium carbon).
Ferrotitanium is furnished in various lump, crushed, and ground sizes. Special compositions for special applications are available.

"Ferrotone." [33] Trade name for a phosphoric acid-base rust remover. Packaged in 6-gal carboys and 55-gal drums. Used for removal of rust, rust stains and scale from equipment and decorative surfaces as a replacement for hydrochloric, nitric and other strong fuming acids.

ferrotungsten. An alloy of iron and tungsten used as a means of adding tungsten to steel. See tungsten steels. Contain 70-80% tungsten and no more than 0.6% carbon. It is also available in special grades. Melting range 3000-5000°F. Despite high melting temperature, ferrotungsten dissolves readily in molten steel and addition is not difficult. Furnished in ground and crushed sizes up to one inch.

ferrous acetate (iron acetate)
$Fe(C_2H_3O_2)_2 \cdot 4H_2O$.
Properties: Greenish crystals when pure and unexposed to air; usually partly brown from action of air; soluble in water and alcohol; oxidizes to basic ferric acetate in air.
Derivation: By the action of acetic acid or pyroligneous acid on iron, with subsequent crystallization.
Method of purification: Recrystallization.
Grades: Technical; C.P.
Containers: 1-lb bottles; wooden kegs.
Uses: Textile dyeing; medicine; dyeing leather black; wood preservative.
Shipping regulations: None. *

ferrous ammonium sulfate (Mohr's salt; iron-ammonium sulfate)
$Fe(SO_4) \cdot (NH_4)_2SO_4 \cdot 6H_2O$.
Properties: Light-green crystals. Soluble in water; insoluble in alcohol. Sp. gr. 1.865.
Derivation: By mixing solutions of ferrous sulfate and ammonium sulfate, followed by evaporation and subsequent crystallization.
Method of purification: Recrystallization.
Grades: Technical; C.P.
Containers: Wooden kegs; glass bottles.
Caution! Keep well closed and protected from light.
Uses: Medicine; analytical chemistry; metallurgy.
Shipping regulations: None. *

ferrous arsenate (iron arsenate)
$Fe_3(AsO_4)_2 \cdot 6H_2O$.
Properties: Green, amorphous powder. Insoluble in water; soluble in acids.
Derivation: By the interaction of solutions of sodium arsenate and ferrous sulfate.
Grades: Technical.
Containers: Boxes; glass bottles.
Uses: Medicine; insecticide.
Shipping regulations: Poison, class B. Poison label. *

ferrous bromide (iron bromide)
$FeBr_2 \cdot 6H_2O$.
Properties: Green crystalline powder; very deliquescent. Soluble in water and alcohol; sp. gr. 4.636; m.p. 27°C.
Derivation: By the action of bromine on iron filings.
Method of purification: Crystallization.
Containers: Glass bottles.
Caution! Keep tightly closed and protected from light.
Use: Medicine.
Shipping regulations: None. *

ferrous chloride (iron chloride; iron dichloride iron protochloride). (a) $FeCl_2$; (b) $FeCl_2 \cdot 4H_2O$.
Properties: Greenish-white crystals. Soluble in alcohol and water; sp. gr. (a) 2.988; (b) 1.93.
Derivation: By the action of hydrochloric acid on an excess of iron, with subsequent crystallization.
Grades: Technical; C.P.
Containers: 1-, 5-lb bottles; 100-lb kegs; 200-lb barrels; 500-lb drums.
Caution! Keep tightly closed and protected from light.
Uses: Mordant in dyeing; metallurgy; pharmaceutical preparations; starting point in manufacture of ferric chloride.
Shipping regulations: None. *

ferrous fluoride (iron fluoride) $FeF_2 \cdot 8H_2O$. (Also known: FeF_2 and $FeF_2 \cdot 4H_2O$.)

Properties: Green crystals. Soluble in acids; slightly soluble in water; insoluble in alcohol and ether; sp. gr. 4.09 (anhydrous).
Grades: Technical.
Containers: Boxes.
Use: Ceramics.
Shipping regulations: None. *

ferrous fumarate $FeC_4H_2O_4$. Anhydrous salt of a combination of ferrous iron and fumaric acid. A stable, odorless, substantially tasteless, reddish-brown anhydrous powder. Contains 33% iron by weight. Does not melt at temperatures up to 280°C.
Containers: Fiber drums, 25-lb and 100-lb; Leverpak drums, 400-lb.
Uses: Medicine; dietary supplement.
Shipping regulations: None. *

ferrous gluconate (iron gluconate) $Fe(C_6H_{11}O_7)_2 \cdot 2H_2O$.
Properties: Yellowish gray or pale greenish yellow, fine powder or granules with slight odor. Solution (1 in 20) is acid to litmus. Soluble in water and glycerin; insoluble in alcohol.
Method of purification: Crystallization.
Grades: Pharmaceutical, U.S.P. XVI.
Containers: Cans; fiber drums.
Uses: Medicine; feed additive.
Shipping regulations: None. *

ferrous hydroxide $Fe(OH)_2$.
Properties: White or green amorphous powder which turns brown in air from oxidation to $Fe(OH)_3$. Soluble in acids, soluble in presence of ammonium salts; insoluble in water.

ferrous iodide (iron iodide; iron protoidodide) $FeI_2 \cdot 4H_2O$.
Properties: Crystalline, grayish-black masses. Soluble in water and alcohol.
Constants: Sp. gr. 2.873; m. p. 177°C;
Derivation: By the action of iodine on iron filings.
Grades: Technical.
Containers: 1-lb bottles; wooden barrels.
Caution! Keep tightly closed and protected from light.
Uses: Manufacture of alkali metal iodides; pharmaceutical preparations.
Shipping regulations: None. *

ferrous lactate (iron lactate) $Fe(C_3H_5O_3)_2 \cdot 3H_2O$.
Properties: Greenish-white crystals, slight peculiar odor. Moderately soluble in water; slightly soluble in alcohol.
Derivation: By interaction of calcium lactate with ferrous sulfate or direct action of lactic acid on iron filings.
Containers: 1-lb bottles.
Caution! Keep well closed and protected from light.
Use: Medicine; dietary supplement.

ferrous magnesium sulfate (iron-magnesium sulfate) $FeSO_4 \cdot MgSO_4 \cdot 6H_2O$.
Properties: Greenish-white crystals. Soluble in water.

ferrous manganese sulfate (iron-manganese sulfate) $FeMn(SO_4)_2 \cdot 12H_2O$.
Properties: Yellowish-white powder. Soluble in hot water.

ferrous oxalate (iron oxalate) $FeC_2O_4 \cdot 2H_2O$.
Properties: Pale yellow, odorless crystalline powder. Soluble in acids; insoluble in water. Sp. gr. 2.28. Releases carbon monoxide when heated.
Derivation: By the interaction of solutions of ferrous sulfate and sodium oxalate.
Grades: Technical; C. P.
Containers: 1-lb bottles; wooden kegs; boxes.
Uses: Medicine; photographic developer.
Shipping regulations: None. *

ferrous oxide (iron monoxide) FeO.
Properties: Black powder; sp. gr. 5.7; m. p. 1420°C. Insoluble in water; soluble in acid.
Derivation: Prepared from the oxalate by heating but the product contains some ferric oxide. (Caution: Poisonous carbon monoxide is released when ferrous oxalate is heated.)

ferrous phosphide (iron phosphide) Fe_2P. A ferrophosphorus.
Properties: Bluish-gray powder. Soluble in mineral acids (decomposes); insoluble in water. Sp. gr. 6.56; m. p. 1290°C.
Grade: 24-26% phosphorus.
Containers: Barrels; bulk.
Uses: To increase iron and steel casting fluidity; prevent sheet steel sticking in rolling; give wire stiffness and better surface for drawing.

ferrous quinine citrate (iron-quinine citrate).
Properties: Greenish-yellow, thin scales; somewhat deliquescent in air. Slightly soluble in water.
Containers: Glass bottles.
Use: Medicine.

ferrous sulfate (iron sulfate; iron vitriol; green copperas; copperas; green vitriol; sal chalybis) $FeSO_4 \cdot 7H_2O$.
Properties: Greenish crystals or granules, often brownish-yellow in color from oxidation and efflorescence. Odorless; soluble in water, with saline taste; insoluble in alcohol.
Constants: Sp. gr. 1.89; m. p. 64°C; pH (10% solution) 3.7.
Derivation: (a) A by-product from the pickling of steel and in many chemical operations. (b) By the action of dilute sulfuric acid on iron. (c) Oxidation of pyrites in air, followed by leaching and treatment with scrap iron. (d) As by-product from ilmenite.
Method of purification: Recrystallization.
Grades: Technical; anhydrous; C. P.; U. S. P. XVI.
Containers: Bottles; bags; barrels; bulk.
Uses: Water purification; source of other iron salts and oxides; fertilizer; feed additive; writing inks; pigments; medicine; textile industry; leather industry; photography; deodorizer; disinfectant; reagent in analytical chemistry; weed exterminator;

indigo dyes; wood-preservative compositions; metallurgy (electrolytic iron; precipitating gold from its solutions; etching aluminum); process engraving and lithography; synthetic rubber.

Shipping regulations: None.*

ferrous sulfate, exsiccated (dried ferrous sulfate). Ferrous sulfate deprived of about six of its seven molecules of water of crystallization. Grayish-white powder, slowly soluble in water; insoluble in alcohol.

Used in medicine instead of the regular sulfate in making pills (U.S.P. XVI).

ferrous sulfide (iron sulfide; iron protosulfide; iron sulfuret; iron monosulfide) FeS.

Properties: Dark-brown or black metallic pieces, sticks or granules. Soluble in acids; insoluble in water.

Constants: Sp. gr. 4.75-5.40; m.p. 1179°C.

Derivation: By fusing iron and sulfur.

Impurities: Arsenic.

Grades: Technical; C.P.

Containers: 100-lb boxes; 500-lb kegs; 900-lb barrels.

Uses: The manufactured sulfide is used for generating hydrogen sulfide; ceramics; making other sulfides.

Shipping regulations: None.*

See also pyrite.

ferrovanadium. An iron-vanadium alloy used to add vanadium to steel. Vanadium appears in engineering steels to the extent of 0.1-0.25%, and in high speed steels to the extent of 1-2.5% or higher. See vanadium steels. Since vanadium oxidizes readily, it is usually added to the ladle for open-hearth steel, although it may be added to the furnace when the basic electric furnace is being used. Several grades are available.

Typical composition: Vanadium 35-55%; silicon 1.5-12%; carbon 0.2-3.5%. Melting range 2700-2770°F. Furnished in a variety of lump, crushed, and ground sizes.

ferrozirconium alloys. See zirconium-ferro alloys.

ferruginous. Containing iron.

ferrum. Latin name for iron; hence the symbol Fe.

ferrum reductum. See iron, reduced.

fertile material. In nuclear technology, name given to any substance which is not capable of sustaining a chain reaction but which can be converted into a fissionable material in a nuclear reactor. Uranium 238 (converted to plutonium 239) and thorium 232 (converted to uranium 233) are fertile materials.

fertilizers, synthetic. Usually refers to mixed fertilizers, containing nitrogen, phosphorus and potassium (see N-P-K mixtures). However, fertilizers consisting of only one of these essentials are also used in very high volume, as ammonium nitrate, anhydrous ammonia, superphosphate, potassium sulfate.

"Fesolv." [428] Trademark for combination of fluorides containing antichlor.

Properties: Pale green, homogeneous, non-caking powder.

Use: For removal of iron and rust stains plus neutralizing in laundering.

Fessler compound. A combination of salts with flocculating properties similar to casein. Stated to be useful in removing undesirable copper and iron from wines.

fever bark. See alstonia.

F.F.A. Abbreviation for free fatty acid. Used in describing specifications for fatty esters, glycerides, oils, etc.

FF black. Abbreviation for fine furnace black. See furnace black.

FGAN. Fertilizer grade ammonium nitrate, which see. It is used in blasting agents as well as fertilizers, because its coating of kieselguhr and its prilled form make it safer to handle than the usual grades.

"Fiba-Weld." [328] Trademark for a series of water-soluble or dispersible textile finishing agents.

Fiba-Weld A: Antislip finish with fullness.

Fiba-Weld B: Softener used with Fiba-Weld A.

Fiba-Weld N: Antislip finish, rustling finish for taffeta.

Fiba-Weld PJ: Antislip not unduly building up hand.

fiberfill. A fiber which is designed specifically for use as a filling material in such products as pillows, comforters, quilted linings, furniture battings.

"Fiberfrax." [280] Trademark for ceramic fiber made from alumina and silica. Available in bulk as blown, chopped and washed; long staple; paper; rope; roving; blocks.

Properties: Retains properties to 2300°F and under some conditions used to 3000°F; light weight; inert to most acids and unaffected by hydrogen atmosphere; resilient.

Uses: High temperature insulation of kilns and furnaces; packing expansion joints; heating elements, burner blocks; rolls for roller hearth furnaces and piping; fine filtration; insulating electrical wire and motors; insulating jet motors; sound deadening.

"Fiberglas." [191] Trademark for a variety of products made of or with glass fibers or glass flakes including:

Insulating Wools: In batts, blankets, rolls and processed forms for insulating buildings, equipment and vehicles at high and low temperatures.

Mats and Rovings: Used for reinforcing organic and inorganic materials such as polyester and epoxy resins, bitumens, etc., to form boats, roofing materials or for protecting underground pipelines against corrosion.

Coarse Fibers: In the form of packs used for filtering gases or liquids as in heating and ventilating systems and chemical processes.

Acoustical products: Sound absorbent

materials for acoustically treating residential, commercial and industrial buildings.

Yarns: Made by twisting and plying fine staple fibers or continuous filaments of glass which are fabricated on looms, braiders and other equipment standard in the textile industry. Also used as electrical insulation and as a chemically stable base for various coatings.

fibers, natural. Filaments, threads and bristles of animal or vegetable origin, used for paper, textiles, cordage, or brushes. The animal fibers, of which silk and wool are most important, consist essentially of protein chains and are relatively resistant to acids but easily degraded by alkalies or chlorine bleaches. The vegetable fibers (further classified below) consist essentially of cellulose chains and are relatively resistant to alkalies and chlorine bleaches but are degraded by acids. Vegetable fibers are classified according to source as follows:
1. Seed hairs: cotton, kapok.
2. Leaves (hard fibers): abaca, agave (cantala, henequen, sisal and istle).
3. Stems (soft or bast fibers): flax, hemp, jute, kenaf, ramie.

fibers, synthetic. Filaments, threads and bristles produced synthetically by converting a natural or synthetic raw material to liquid form, forcing this through small openings and immediately converting the resulting liquid thread to solid form. See acetate, rayon, glass fiber, nylon, saran, protein fibers, "Acrilan," "Dacron," "Dynel," "Orlon," and "Vinyon HH."

"Fibertex." [236] Brand name for a fibrous material used as an additive to rotary drilling mud to prevent or restore lost circulation in gravelly, fractured, or creviced formations. Effective in most cases of lost circulation, except those taking place in caverns or extremely large crevices.

Containers: Multiwall paper bags containing 40 lbs.

"Fibertuff." [11] Trademark for glass-fiber reinforced polystyrene, supplied in rod-shaped pellets. Used for injection molding and characterized by increased rigidity and shock resistance.

"Fibra Flo." [247] Trademark for a line of filter aids comprised of physical mixtures in different proportions of either asbestos fiber and diatomite, asbestos fiber and cellulose fiber, or asbestos fiber alone.

Uses: As a filter element pre-coating material making possible easier cleaning of elements after a filtration cycle; in special applications to obtain sharp or sparkling clarity.

"Fibrax." [51] Trademark for sodium-soap grease, a lubricant for plain and anti-friction bearings operating at moderate speeds under high temperatures or heavy loads or both. It seals bearings against foreign matter, and the high viscosity base oil imparts adhesive properties.

"Fibrex." [160] Trademark for a non-metallic, non-inductive wire especially designed for overhead power lines. It has extreme resistance to abrasion caused by swaying branches of trees.

fibrid. Generic name for a fibrous form of synthetic polymeric material, used for example as a binder material in the manufacture of textryl. See textryl.

fibrin, muscle. See syntonin.

fibrinogen. A sterile fraction of normal human plasma, dried from the frozen state. In solution it has the property of being converted into insoluble fibrin when thrombin is added.

Properties: White or grayish amorphous substance.

Grade: U. S. P. XVI.

Use: Medicine (blood plasma clotting factor).

fibroin. The fibrous material in silk, a scleroprotein containing glycine and alanine; light yellow silk-like mass. Insoluble in water; soluble in concentrated alkalies and concentrated acids.

fibrolite. See sillimanite.

fibrous silica. An extracted glass filament which has a very high silica content with traces of iron, calcium, and magnesium. The fibers can be produced in a batted form or spun into thread and woven into cloths, tapes, sleevings, and other textile materials. Thermal and chemical properties are similar to those of vitreous silica.

ficin. A proteolytic enzyme hydrolyzing casein, meat, hide powder, edestin, fibrin, liver, and other protein-like material.

Properties: Buff to cream-colored powder with an acrid odor; very hygroscopic; partially soluble in water; insoluble in the usual organic solvents.

Source: Fig latex or sap. Commercially prepared by filtering and drying the latex.

Use: Brewing, cheese, meat, leather, and textile industries.

"Filmcol." [125] Trademark for an authorized proprietary ethyl alcohol solvent. "Filmcol" is particularly useful in the film processing, photographic, textile, rubber latex and printing ink industries. The absence of hydrocarbon solvent in the denaturant formula reduces deterioration of rubber press rollers in printing and permits use of "Filmcol" where the presence of even small volumes of hydrocarbon solvents might prove detrimental. In chemical manufacture, "Filmcol" may also be used as a reaction medium for purifying and recrystallizing operations. Other uses include solvent for shellac and spirit varnishes, cleaning compounds, adhesives and latent solvent for nitrocellulose lacquers and other cellulosics.

Authorized composition: Specially denatured

alcohol No. 1 (190 proof) 100 parts by volume; isopropyl alcohol, 10 parts by volume; methyl isobutyl ketone, 1 part by volume.
Typical properties: Reid vapor pressure 2 psi (100°F); lbs/gal 6.777 (60°F); mild residual odor.
Containers: 55-gal non-returnable drums; 6000 to 10000-gal tank cars; tank trucks.
Warning! Flammable; poisonous if swallowed. Avoid prolonged or repeated contact with skin. Protect eyes against splashes and maintain vapor concentrations at comfort levels.
Shipping regulations: Flammable liquid. Red label.*

"Filmex." [192] Trademark for a proprietary denatured alcohol solvent, subject to government regulations.
Properties: Colorless liquid; sp. gr. 0.81345 (60°F); nonvolatile; infinitely soluble in water; boiling range 77.4-79.3°C; flash point (Tag open cup) 64°F; apparent proof 191.20 (60°F).
Containers: 5- and 55-gal drums; tank cars; tank wagons.
Uses: Cleaning rubber rolls, plates, and type in printing industry; formulating printing inks; drying photographic films.
Hazards: Flash point under 80°F.
Shipping regulations: Flammable liquid. Red label.*

"Filmfast." [50] Trademark for a spray compound for improving distribution of insecticidal and fungicidal sprays.

filter alum. See aluminum sulfate.

"Filter-Cel." See "Celite" Filter Aids.

filter sand. Sand used to filter sediment and suspended matter from water.

filtration. The process of separating suspended solids from their liquid by forcing the latter through a porous medium.

"Filtrol." [217] Brand name of acid activated clays used as decolorizing adsorbents and catalysts.
Grades: Various, depending upon use.
Particle size: 85-95% through 200-mesh.
Containers: 50-lb multi-ply paper bags.
Uses: Adsorption of color and other impurities from petroleum, animal, vegetable and marine oils, and fats and waxes; re-refining of used crankcase and other industrial oils; petroleum cracking catalyst; catalyst.

fines. The portion of a powder composed of particles which are smaller than a specified size (MPA definition, MPA Standard 9-50T).

finishing compounds. Substances used in the final or finishing stages of manufacture of a product, usually textiles and leather, to make them suitable and marketable for specific purposes. Such compounds contain materials that impart softness, flexibility, stiffness, color, water and fire resistance, etc.

fir, balsam. See Canada balsam.

fire-brick. See brick, fire-.

fire-clay. Clays containing only small amounts of fluxing impurities, but high in silica, alumina and water, and, therefore, capable of withstanding high temperatures. They are usually light in color, ranging from gray to yellowish-red and exhibit wide variation in both chemical and physical properties. Some authorities believe no clay should be classed as a fire-clay unless its fusion point exceeds 1600°C.

They are usually grouped as plastic or flint. There are, however, some having properties which may be considered as intermediate and these are usually referred to as either semi-plastic or semi-flint.

The plastic clays are usually distinguished by a lower ratio of alumina to silica than the flint clays and also contain a higher percentage of impurities together with a lower fusion point. They break down readily on exposure to weather and quickly develop good bonding power. The flint clays are hard and are better able to resist weathering. Plasticity can only be developed in them either by considerable grinding and tempering or by prolonged weathering. They are highly refractory. Semi-plastic and semi-flint clays often possess a high degree of refractoriness sometimes approaching that of the flint clays.
Occurrence: United States (Missouri, Pennsylvania, Maryland, Ohio, Kentucky, New Jersey, Colorado).
Uses: Fire-brick, retorts, furnace linings, crucibles, terra cotta, glass-factory pots and tanks.

fire foam. A blanket of foam composed of alumina and carbon dioxide; produced by interaction of a solution of alum and one of sodium carbonate or sodium silicate and glue.

"Firefrax." [280] Trademark for a group of refractory cements made from kaolin or fire-clay base materials for applications where aluminum silicate cements are best suited.
No. 1. Air-setting super-refractory cement. Application range is from 60-3000°F.
Containers: Shipped in plastic form in water-tight metal drums of 40-, 100-, and 350-lb capacities.
Uses: Laying and repairing fireclay and silica brick work; bond for crushed fire-brick or ganister for patching furnace linings and for making rammed-up or monolithic linings, patching material for by-product coke ovens.
No. 2. Heat-setting refractory cement. Application range is from 2000-2750°F.
Containers: Shipped dry in 100-lb bags.
Uses: Laying and patching fireclay and silica brick in applications subjected to moderate temperature conditions, and as a wash for small pouring ladles in the non-ferrous foundry.

"Firemaster" BP4A. [426] Trademark for tetrabromobisphenol-A, $C_{15}H_{12}Br_4O_2$.

Properties: An off-white crystalline solid; m. p. 180 to 184°C.
Grade: Technical.
Containers: Fiber drums.
Uses: Flame retardant for plastics.
Shipping regulations: None.*

"Firemaster" PB5. [426] Trademark for pentabromophenol, C_6Br_5OH.
Properties: A lavender crystalline solid; m. p. 225 to 226°C.
Grade: Technical.
Containers: Fiber drums.
Uses: Flame retardant for plastics.
Shipping regulations: None.*

"Firemaster" PHT4. [426] Trademark for tetrabromophthalic anhydride, $C_8Br_4O_3$.
Properties: A pale yellow crystalline solid; m. p. 279.5-280.5°C.
Grade: Technical.
Containers: Fiber drums.
Uses: Flame retardant for plastics.
Shipping regulations: None.*

"Firemaster" T23P. [426] Trademark for tris-(2,3-dibromopropyl) phosphate, $(CH_2BrCHBrCH_2O)_3PO$.
Properties: A viscous, pale yellow liquid; density, 18.5 lb/gal; refractive index, 20°C: 1.5772.
Grade: Technical.
Containers: Steel drums.
Uses: Flame retardant for plastics.
Shipping regulations: None.*

fire point. The lowest temperature at which a liquid evolves vapors fast enough to support continuous combustion. It is usually close to the flash point.

fire retardant finishes. See flameproofing finishes.

fire sand. See furnace sand.

"Firestone" Nylon. [35] Trade name for the polyamide produced by polymerizing caprolactam.
Properties: Crystalline thermoplastic granules (pellets).
Grades: Available in various viscosities; regular and heat stabilized; in natural, black and white and standard colors.
Containers: Vacuum packed 25-lb cans.
Uses: Wire coating; extruded shapes; profiles; filaments; fibers and film; molding.
Hazards: None.

"Fi-Retard." [300] Trademark for a group of flame retardants for application to cotton, rayon, nylon and paper.

fir oil, Douglas. See pine needle oils.

firwood oil. See pine-needle oils.

Fischer's reagent. A reagent used as a test for sugars.
Preparation: 3 parts of sodium acetate and 2 parts of phenylhydrazine hydrochloride in 20 parts of water.

Fischer reagent for water. See Karl Fischer reagent.

Fischer's salt. See cobalt potassium nitrite.

Fischer-Tropsch process. The term is used for any one of several processes originating in Germany for producing hydrocarbons or their oxygenated derivatives from water gas or other mixtures of carbon monoxide and hydrogen.
1. In the Fischer-Tropsch normal pressure process a mixture of hydrogen and carbon monoxide (synthesis gas) is passed over cobalt, nickel, iron, or other catalysts at atmospheric pressure and 150-250°C to produce a complex mixture containing appreciable quantities of higher hydrocarbons useful as liquid fuels.
2. The name Fischer-Tropsch is also sometimes applied to related processes such as that in which pressures of 5 to 20 atmospheres are used to obtain higher yields of hydrocarbons.
3. Finally, the name is also sometimes applied to the Synthol process proposed by Fischer and Tropsch for producing a mixture of alcohols, aldehydes, ketones, and fatty acids from synthesis gas by use of alkalized iron catalysts.

fisetin $C_{15}H_{10}O_6$. 3,7,3',4'-Tetrahydroxyflavone.
Properties: Yellow needles. The coloring principle of the wood of Rhus cotinus, or young fustic. Soluble in alcohol and alkaline solutions; m.p. 330°C.
Use: Natural dyestuff.

fish berry. See cocculus.

fish oil. See cod-liver, halibut, herring, menhaden, porpoise, salmon, sardine, shark, and tuna oils.

fissiochemistry. The process by which a chemical change or reaction is brought about by nuclear energy; for example, the production of anhydrous hydrazine from liquid ammonia in a nuclear reactor.

fission, nuclear. The nuclear change of an atom into two atoms of approximately equal weight, as in the element uranium in the explosion of the atomic bomb, and similar processes.

fission products. The nuclear species (i. e., kinds of atomic nuclei) produced by the splitting (fission) of heavy element isotopes such as uranium 235 and plutonium 239. Thirty-four different elements form from the fission of uranium 235 by slow neutrons. These fall into two groups with mass numbers ranging from 85-100 and from 130-150. The most abundant, in order, are zirconium, molybdenum, neodymium, barium, xenon, ruthenium, cerium and calcium. Gross fission products are of potential commercial use for activating phosphors, for cold sterilization of foods and drugs, and for the manufacture of static eliminators.

Fittig's synthesis. The preparation of aromatic hydrocarbons by condensation of aryl halides with alkyl halides in the presence of metallic sodium.

*See "I.C.C. Shipping Regulations," page xiii.
Reference numbers refer to name of manufacturer. See "List of Manufacturers," page v.

"Fixanol." [206] Proprietary products for use in the after-treatment of material dyed with direct cotton dyestuffs by which the material is given very good fastness to water, perspiration, and cross dyeing. A selected range of direct dyestuffs are also marketed under the name "Fixanol." Also dispersing agents for use in viscose manufacture.

fixatives, perfume. See fixing agents, perfume.

fixed nitrogen. See nitrogen fixation.

fixing agents, chemical.
1. Substances which are instrumental in the fixation of various mordants upon textile material by uniting chemically with them and holding them upon the fiber until the dyes may have an opportunity to unite with them.
2. Substances which cause the actual precipitation of the mordant on the fiber by double decomposition.

fixing agents, mechanical.
1. Substances (e.g., albumin) which are capable of holding pigments permanently upon textile fibers.
2. Certain gums and starches which are capable of holding dyes and other substances upon textile fibers for a sufficient length of time to permit some desirable reaction to take place.

fixing agents, perfume (fixatives, perfume). A substance which, when added to a perfume, prevents the volatilization of its various components too rapidly, and tends to equalize the rates of volatilization of these components. Thus, it will not only increase the odor life of the perfume, but will tend to make the odor character continuously unchanged during the period of evaporation. For many years, the leading fixatives were the animal products, ambergris, civet, musk, and castoreum. Oleoresins, such as oakmoss, benzoin, styrax, and others, are both fixatives and contributors to the odor value of the perfume. Essential oils, as vetivert, patchouli, and orris, can serve as fixing agents, and many synthetics, such as artificial musks, macrocyclic musks, and others, are fixing agents.

flag, sweet. See calamus.

flake lead. See lead carbonate, basic.

flake white. See bismuth subnitrate; bismuth oxychloride.

"Flamenol." [245] Trademark for electrical conductors insulated with a vinyl halide resin such as plasticized polyvinyl chloride.

"Flameproofing Agent 313." [73] Trade name product. Inorganic phosphates.

Properties: White, odorless, crystalline powder; non-toxic and non-irritating to the skin. Readily soluble in water in all proportions.

Containers: 8-, 40-lb cont.; 400-lb drums.

Uses: Particularly recommended for the flame-proofing of cotton goods, paper, etc., where the elimination of after-glow is an important factor. "Flameproofing Agent 313" does not tender the material nor cause reduction in tensile strength. It is non-hygroscopic so that the materials treated with it do not become damp or sticky in humid weather, or harsh and brittle in dry weather.

flameproofing finishes (fire retardant finishes). Materials used on cellulose products (wood, paper, cotton and other textiles) to slow down their combustion. Fire retardant finishes in longest use are various water-soluble, inorganic salts, including ammonium bromide, borax, boric acid, phosphates such as diammonium phosphate or sodium phosphate, sodium tungstate and ammonium sulfamate, and mixtures of these. Such solutions are effective when added in amounts of about 7% of the weight of the fabric.

Finishes widely used on military and outdoor fabrics, designated as FWWMR (fire, water, weather, mildew resistant) are based on mixtures of antimony oxide, or oxychloride, and chlorinated organics such as chlorinated paraffin, chlorinated rubber, polyvinyl chloride or polyvinylidene chloride.

Most recent flameproofing agents are phosphorus compounds, such as the polymer resin formed by tetrakis(hydroxymethyl)-phosphonium chloride and melamine. Brominated triallyl phosphate is a similar monomer. Some phosphorus compounds apparently unite chemically with the cellulose and are effective flameproofers but weaken the fabric. The phosphorus resins polymerized or cured right in the fabric seem the most promising of the finishes.

"Flamort" Fire Retardants. [414] Trademark for a series of fire retardants.

"Flamort" T: Fire retardant for velvet, sateen, silk, damask, viscose rayon.

"Flamort" TC: Fire retardant for cotton, canvas.

"Flamort" Protextile: For flameproofing and mothproofing.

"Flamort" U: Fire retardant for acetate rayon, nylon.

"Flamort" WC: Colorless, fire retardant surface impregnation for wood and paper.

"Flamort" Plastic Coat Clear: Fire retardant for foliage, palm fronds.

Containers: "Flamort" T, TC, Protextile and U: 5-, 10-, 20-, 25-, 50-, 100-lb containers in fiber drums. "Flamort" WC: 6-, 12-, 15-, 21-, 28-, 51-, 99-, 198-lbs in fiber drums. "Flamort" Plastic Coat Clear: quart cans, gallon cans, 5-gal pails, 55-gal drums.

flash distillation. Distillation in which an appreciable proportion of a liquid is converted to vapor in such a way that the final vapor is in equilibrium with the final liquid.

flash point. The lowest temperature at which a combustible liquid will give off a flammable vapor which will burn momentarily.

"Flatack Glue No. 3309." [170] A specially compounded cake form animal-glue base adhesive. Clear yellow, which becomes transparent when reduced with water and applied on material. Suitable for paperbox manufacturing, luggage manufacturing, lining display cases, laminating, bonding, and bindery operations. Available in cake form, weighing approximately 9 pounds each.

flatting agents. Flatting agents are substances ground into minute particles of irregular shape and used in paints and varnishes to cut down reflected light or glare from the finished film.

flavanthrene (indanthrene yellow; chloranthrene yellow) $C_{28}H_{12}O_2N_2$. A vat dye.
Properties: Brownish-yellow needles. Soluble in dilute alkaline solutions.
Derivation: By the action of antimony pentachloride on beta-aminoanthraquinone in boiling nitrobenzene.
Use: Dyeing.
Shipping regulations: None. *

"Flavaxin." [162] Trademark for riboflavin.

flavianic acid. See 2,4-dinitro-1-naphthol-7-sulfonic acid.

flavin.
1. Isoalloxazine, $C_{10}H_6N_4O_2$, the nucleus of various natural yellow pigments. See riboflavin and flavin enzymes.
2. Quercetin (q. v.), $C_{15}H_{10}O_7 \cdot 2H_2O$, a yellow dye derived from oak bark.

flavin adenine nucleotide (FAD) $C_{27}H_{33}N_9O_{15}P_2$. A dinucleotide containing riboflavin phosphate linked to adenine mononucleotide in the order riboflavin, pyrophosphate, ribose, and adenine. It functions as a coenzyme for many enzymes, since it has the ability to accept hydrogen atoms, thus oxidizing the subtrate.
Derivation: From yeast, liver, kidney, heart, muscle.
Use: Biochemical research.

flavine.
1. Acriflavine hydrochloride (q. v.), a bacteriostatic agent.
2. Flavin (q. v.).

flavin enzymes (flavoproteins). Enzymes composed of protein linked to coenzymes which are mono- or dinucleotides containing riboflavin. Because of their distinctive color they are also called "yellow enzymes." The flavin enzymes function in tissue respiration as dehydrogenases, the hydrogen atoms being taken up by the riboflavin group.

flavin mononucleotide. See riboflavin phosphate.

"Flavite-5." [296] Trade name for limed rosin possessing a capillary tube melting point of (typically) 116°C., acid value 85, color Mary/Kate (Rosin Standards). Used in printing ink, paint, and varnish.

"Flavophosphine" GDC. [307] Trademark of a basic dyestuff. Used chiefly for the dyeing of leather.

flavoproteins. See flavin enzymes.

flax. Bast fibers, approximately 20 in. long, obtained from the stems of the linseed plant, Linum usitatissimum. Finer, stronger and more durable than cotton.
Sources: Russia, Italy, Ireland, France, Egypt.
Uses: Apparel and household fabrics (linens); thread; twine.

"Flaxedil." [57] Trademark for gallamine triethiodide(tris-1,2,3-diethylaminoethoxybenzenetriethiodide) $C_{30}H_{60}I_3N_3O_3$.
Properties: White, crystalline powder; slightly hygroscopic; odorless; slightly bitter taste; m. p. about 250°C; an allotropic form exists with m. p. about 150°C. Soluble in water, alcohol, dilute acetone, methanol; very slightly soluble in acetone, ether, benzene, and chloroform. A 2% solution in water is clear, colorless and has a pH of 6.2-7.5.
Use: Medicine.

flax seed. See linseed.

flax-seed oil. See linseed oil.

fleabane oil. See erigeron oil.

flea seed. See psyllium.

flea wort. See psyllium.

"Flectol" H. [58] Trademark for polymerized 1,2-dihydro-2,2,4-trimethylquinoline.
Properties: Tan powder; m. p. 120°C min; sp. gr. 1.08; soluble in benzene, alcohol, and naphtha.
Uses: Rubber antioxidant.

"Flexac." [144] Trademark for polyvinyl acetate emulsions.
Properties: High molecular weight emulsion, capable of film formation at low temperatures. Forms clear, glossy film, resistant to water spotting; solids 55-57%; viscosity 800-1200 cps (77°/60 RPM); mean particle size 0.5 micron; pH 5.0-6.5.
Grade: FA-5.
Containers: 55-gal. lined steel or fiber drums.
Uses: Low-cost paints; semi-gloss and floor paints; adhesives; textile sizes, binders and finishes; coatings.

"Flexalyn." [266] Trademark for a series of synthetic resins of glycol esters of rosin. Acid numbers 8-12; lbs/gal 8.33-8.91; color (U. S. D. A. rosin scale) M to N; refractive index 1.536-1.543; softening point (Hercules drop method) 45-63°C; viscosity (80% solution, by wt, in xylene) 271-640 cps (25°C).
Uses: Plasticizing resins, as tackifiers, or as tougheners, in organic coatings, adhesives, waxes, industrial textile sizings, floor coverings, etc.

"Flexamine G." [248] Trademark for a mixture of N,N'-diphenyl-para-phenylene-diamine and

*See "I.C.C. Shipping Regulations," page xiii.
Reference numbers refer to name of manufacturer. See "List of Manufacturers," page v.

a complex diarylamine - ketone reaction product.
Properties: Brownish grey granules; sp. gr. 1.20; melting range 75-90°C; soluble in acetone, benzol and ethylene dichloride; insoluble in water and gasoline.
Uses: A superflexing antioxidant used in tires, camelback, wire insulation, neoprene belting and molded soles.

"Flexbond." [144] Trademark for polyvinyl acetate copolymer emulsions.
Properties: High molecular weight, internally plasticized copolymer and terpolymers. Sweet, slight odor; milky white color; viscosity varies according to grade; range 100 to 850 cps (60 RPM-RVO Brookfield); % solids varies according to grade; range 51-57%; lbs/gal 9.0-9.1. Emulsions form clear, colorless films with excellent water resistance.
Grades: 800, 811, 855, 306, 100.
Containers: 55-gal lined steel or fiber drums.
Uses: Vehicle for paint formulations; adhesive formulations; coatings and sizings.

"Flexichem." [152] A trade name for a series of metallic soap based compounds that enjoy wide usage in many industrial applications ranging in use from internal and external lubricants to tableting aids in the manufacture of certain types of pharmaceuticals.
Containers: 50- to 300-lb bags and drums.

"Flexichem 'B'." [152] Trade name for sodium stearate, a mono salt with the formula $NaC_{18}H_{35}O_2$, available in both Industrial and Food Grades. This product is a uniform, white powder; odor bland; bulk density 27 lbs/cu ft; titer of fatty acids 58°-60°C; iodine value 2.
Containers: 150-lb drums.

"Flexichem 'CS'." [152] Trade name for calcium stearate, a di salt with the formula $Ca(C_{18}H_{35}O_2)_2$; available in both Industrial and Food Grades. Color white; odor bland; bulk density 20 lbs/cu ft; average particle size ranges from 5 to 7 microns; titer 58°-60°C; iodine value 2.
Containers: 50-lb drums; paper bags.

"Fleximet." [152] Same as "Flexichem."

"Flexol." [214] Trademark for a line of plasticizers including:
380: di(2-ethylhexyl)isophthalate (q. v.).
426: a mixed alcohol phthalate.
Properties: A light-colored liquid; sp. gr. 0.9941 (20/20°C); 8.3 lb/gal (20°C); b.p. 224°C (5 mm); vapor pressure 1.8 mm (200°C); insoluble in water; viscosity 60.7 cps (20°C).
810: a higher alcohol phthalate.
Properties: A light-colored liquid; sp. gr. 0.9729 (20/20°C); 8.1 lb/gal (20°C); b.p. 245°C (5 mm); insoluble in water; viscosity 101 cps (20°C).
10-10: didecyl phthalate (q. v.).
13-13: di(tridecyl) phthalate (q. v.).
10-A: didecyl adipate (q. v.).
A-26: di(2-ethylhexyl) adipate (q. v.).
B-400: a polyalkylene glycol derivative.
Properties: A light-colored liquid; sp. gr. 0.995 (20/20°C); 8.3 lb/gal (20°C); vapor pressure 0.8 mm (200°C); solubility in water 0.1% by wt (20°C); viscosity 152 cps (20°C).
CC-55: di-2-ethylhexyl hexahydrophthalate (q. v.).
DOP: di-2-ethylhexyl phthalate (q. v.).
EP-8: epoxy ester plasticizer and stabilizer.
Properties: Mol. wt. 416; sp. gr. 0.9221 (20/20°C); 7.7 lbs/gal; f. p. −14.5°C; oxirane oxygen 5.1%; iodine no. 1.5%.
EPO: polymeric epoxy plasticizer and stabilizer.
Properties: Mol. wt. 1000; sp. gr. 0.9956 (20/20°C); 8.3 lbs/gal; pour point 25°F; oxirane oxygen, greater than 6.8%; iodine no. less than 2%.
3G20: triethylene glycol decanoate (q. v.).
3GH: triethylene glycol di(2-ethylbutyrate) (q. v.).
3GO: triethylene glycol di(2-ethylhexoate) (q. v.).
4GO: polyethylene glycol di(2-ethylhexoate) $C_7H_{15}OCOCH_2(CH_2OCH_2)_xCH_2OCOC_7H_{15}$.
Properties: A light-colored liquid; sp. gr. 0.9892 (20/20°C); 8.2 lb/gal (20°C); b.p. 250°C (5 mm); insoluble in water; viscosity 25.1 cps (20°C).
JPO: polymeric epoxy plasticizer and stabilizer.
Properties: Mol. wt. 1000; sp. gr. 0.990 (20/20°C); 8.3 lbs/gal; pour point 25°F; oxirane oxygen greater than 6.3%; iodine no. less than 3.0%.
8N8: 2,2'(2-ethylhexamido)diethyl di(2-ethylhexoate) $(C_7H_{15}OCOC_2H_4)_2NCOC_7H_{15}$.
Properties: A light-colored liquid; sp. gr. 0.9564 (20/20°C); 8.0 lb/gal (20°C); b.p. 256°C (5 mm); vapor pressure 0.60 mm Hg (200°C); insoluble in water; viscosity 139.2 cps (20°C).
PEP: di(isodecyl)-4,5-epoxytetrahydrophthalate. $OC_6H_8(COOC_{10}H_{21})_2$.
Properties: Sp. gr. 0.9867 (20/20°C); 8.2 lbs/gal; pour point +38°C; oxirane oxygen 3%; iodine no. 1%.
R-2H: a polyester.
Properties: A viscous, amber liquid; sp. gr. 1.055 (20/20°C); 8.8 lb/gal (20°C); insoluble in water; viscosity 16,520 cps (20°C).
TOF: tri(2-ethylhexyl)phosphate (q. v.).

"Flexo Wax C." [73] Brand name for proprietary product. Long-chain hydrocarbon.
Properties: Orange-colored amorphous wax. Insoluble in water and alcohols; soluble in hot naphtha, toluene, mineral spirits, mineral oil. A non-crystalline wax with high adhesive properties. Replaces ceresin wax, and beeswax. M. p. 63-68°C; sp. gr. 0.82 (25°C); flash point 257°C.
Containers: 1-gal cans (10-lb slabs); 5-gal cans (82 lbs); 55-gal drums (540 lbs).
Uses: Buffing compounds; polishes; engravings; lithography; adhesive compounds; leather dressings; textile sizes and finishes; modeling waxes; coatings; etc. Adhesive for "Cellophane," cellulose acetate and

other materials ordinarily resistant to the usual adhesives.

"Flexo Wax C Light." [73] Trade name product. Long chain hydrocarbon.
Properties: Cream, amorphous wax; m. p. 62-64°C; soluble in hot hydrocarbons, mineral oil and vegetable oils (gels on cooling); insoluble in water and alcohol.
Containers: 10-, 80-lb containers; 360-lb drums.
Uses: Same as "Flexo Wax C" but is particularly recommended where a light-colored product is necessary or desirable. Also as an adhesive in the screen-printing process to hold fabrics firmly onto the table during printing.

"Flexricin." [202] Trademark for a group of castor oil derivatives including (1) alkyl ricinoleates, which are compatible with cellulosic resins, polyvinyl butyral, polyamide, rosin and shellac and used as plasticizers; (2) acetyl ricinoleates, the alkyl esters of acetylated ricinoleic acid used as plasticizers for nitrocellulose, ethyl cellulose, polyvinyl chloride resins and copolymers, natural and synthetic rubbers, and other polymers and as textile lubricants; and (3) polyol ricinoleates used for synthesis, cosmetics, brake fluids, in waxes and greases, and as antifoam agents.

"Flextack Glue No. 3318." [170] Cake form, animal-base glue, containing glycerin. Pale yellow, drying transparent.
Use: Recommended for paper, cardboard, leather, coated fabrics, and similar materials.

"Flexzone 3-C." [248] Trademark for N-isopropyl-N'-phenyl-para-phenylene diamine.
Properties: Purple-gray flakes; sp. gr. 1. 14; m. p. 70°C minimum; soluble in benzol, ethylene dichloride, gasoline, ethyl acetate. Insoluble in water.
Uses: An all-purpose antioxidant - antiozonant combining protection against flexing, ozone, heat and oxygen in rubber tires, wire and cable, hose, footwear and mechanical goods; also an anti-copper agent.

"Flexzone 6H." [248] Trademark for N-phenyl-N'-cyclohexyl-para-phenylene diamine.
Properties: Grey-violet powder; sp. gr. 1. 16; m. p. 103-107°C; soluble in acetone, benzol, MEK, ethyl acetate, ethylene dichloride; insoluble in water.
Uses: Antioxidant - antiozonant for rubber products where outdoor weather resistance and fatigue cracking are problems, such as tires, weather stripping, insulated wire and cable, hose and footwear. It is used mainly in natural and SBR rubber and blends thereof.

flint. A finely crystalline form of native silica or quartz (q. v.), similar to chert, chalcedony, and jasper. A typical analysis gives 98% silica, 1. 5% water, and traces of iron, aluminum, calcium, and organic matter.
Properties: Color smoky gray, brownish, blackish, dull yellowish; luster waxy to greasy; hardness 6. 5-7; sp. gr. 2. 60-2. 65. More easily soluble in hot caustic alkali than crystallized quartz.
Occurrence: Europe; U. S.
Uses: Abrasive; balls for ball mills; paint extender; filler for fertilizer, insecticides, rubber, plastics and road asphalt; in ceramics; flint glass; chemical tower packing.

flint glass. A glass in which lead and potassium replace a considerable part of the lime and soda of ordinary glass. This gives a softer, more fusible, more lustrous and brilliant glass with high refraction and low dispersion and therefore of use as an optical glass.

flocculation. See coagulation.

"Flo-chilled." [203] Trademark for a specific grade of anhydrous caustic soda; crystalline form, exceptionally dustless, free-flowing, uniform in particle size.

"Flomax 25." [304] Trademark for a barium-cadmium organic vinyl stabilizer.
Properties: Clear yellow liquid; sp. gr. 1.03; refractive index 1.48.
Containers: Metal drums containing 40 and 450 lbs.
Uses: Excellent heat stabilizer for severe processing conditions. Specially effective in highly loaded systems, extrusions and in solution coatings.

"Florco." [98] Processed Florida fuller's earth for the absorption of oils, grease, and other liquids from floors, decks and similar surfaces.

"Florence" Zinc Oxides. [268] Brand name for a group of zinc oxides manufactured by the French or indirect process, the pigment being made from zinc metal.
Grades: "Florence" White Seal —7, "Florence" Green Seal —8; "Florence" Red Seal —9. Shipped in 50-lb bags.
Uses: Enamels, lacquers, printing inks, plastics, linoleum, insulated wire and cable, rubber, soap, etc.

flores martis. See ferric chloride, anhydrous.

"Florex." [98] Florida fuller's earth, the adsorption capacity of which has been increased by a special patented extrusion process to exceed that of any known naturally active clay. It is used chiefly in adsorption refining, including the decolorization, clarification, and neutralization of mineral, vegetable, and animal oils, fats, waxes, and for the filtration of other products. "Florex" is also widely used in the vapor or liquid-phase treatment of cracked or straight-run gasoline in the Gray, Osterstrom, Linde, and other processes. Standard meshes include 16/30, 30/60, 60/100, 100/up and 200/up.

Florida phosphate. Phosphate rock from Florida, usually fluorapatite

$(CaF)Ca_4(PO_4)_2$, encountered as land pebble, hard rock, soft rock, or powder.

"Florigel." [98] Florida fuller's earth, selected and processed for maximum gel-forming characteristics. Forms completely reversible gel that is unusually stable in the presence of brines or electrolytes. Emulsifier for oils and slow-break asphalts. Suspension builder for pesticides, polishing agents, waxes, etc. Binder for catalytic and adsorptive materials. Adhesive for applying pesticides to plant surfaces.

"Florigel H-Y." An improved grade of "Florigel;" a high yield salt water drilling clay.

"Florinef Acetate." [412] Trademark for fludrocortisone acetate (q. v.).

"Florisil." [98] A synthetic analytical adsorbent having an exceptionally high capacity. It is valuable for the selective adsorption of vitamins, hormones, dyes, and many other compounds, and is generally applicable to chromatographic adsorption technique. "Florisil" is widely used in the Connor and Straub method for the combined determination of thiamin and riboflavin. It will not break down in aqueous solutions and is free from dusting. Various meshes are available but the 30/60 and 60/100 classifications are usually employed for analysis by the percolation method.

"Florite" Desiccant Grade. [98] Granular activated bauxite drying agent for hydrocarbon gases and liquids, air, oxygen, nitrogen, etc. Moisture-adsorptive capacity from 5 to 20%; regenerated when heated to 350°F. Will not dissolve or disintegrate in water. Recognized for long service in natural-gas-drying plants. Available in 4/8, 8/20, and 8/16 mesh sizes.

"Florite" Refining Grade. [98] An activated bauxite adsorbent that is used primarily for the decolorization of petroleum oils. It is particularly effective for sweetening and decolorizing petrolatums and waxes. "Florite" is also used as a catalyst, one of the more common applications being in the vapor-phase treatment of gasoline to convert mercaptan sulfur to hydrogen sulfide. The most popular percolation grade is the 20/60 mesh but other mesh classifications are available.

"Floropryl." [123] Trademark for diisopropyl fluorophosphate.
Use: Medicine.

florspar. See fluorspar.

flos-ferri. See aragonite.

flotation. A process for separating minerals from waste rock or solids of different kinds from one another. This is done by agitating the pulverized mixture of solids with water, oil, and special flotation chemicals. When properly chosen, the latter cause preferential wetting of solid particles of certain types by the oil, while other kinds are not wet. The latter are carried to the surface by the air bubbles and thus floated away and separated from the wetted particles. Most frequently a frothing agent is also used to stabilize the bubbles in the form of a froth which can be easily separated from the body of the liquid. This is froth flotation.

Flotation Oil. [175] Special grades of coal-tar oils.
Grades: No. 4 (containing approx. 25% of selected tar acids) and No. 634.
Containers: 55-gal steel drums.
Uses: Separating valuable portion of finely ground sulfide ores, particularly copper, for mineral concentration operations.

"Flotox." [253] Brand name for flotation sulfur products.

"Flovis." [73] Trade name product. Modified polyoxyethylene fatty-acid ester.
Properties: Cream to tan solid; m. p. 39-42°C; sp. gr. 1.02 (25°C); pH 5% aqueous dispersion 3-5 (25°C).
Containers: 8-, 50-lb containers; 450-lb drums.
Uses: Used especially in the textile and adhesive industries for stabilizing starch solutions, fluid or heavy paste, against "setting-up." The starch gel remains considerably more fluid and free from lumps than those not so treated. Any "set" that occurs can be overcome relatively easily by simple stirring.

"Flowbrite." [142] Trade name for a formulation of oils used at elevated temperature for the bright flowing of electroplated tin.

flower of paradise. See henna.

flowers of antimony. See antimony trioxide.

flowers of sulfur. See sulfur.

flowers of tin. See stannic oxide.

flowers of zinc. See zinc oxide.

flowery head spurge. See euphorbia pilulifera.

"Flozene." [45] Trademark for a series of lubricant white mineral oils.
Typical properties (medium): Sp. gr. 0.860-0.870 (60°F); Saybolt viscosity 40 (210°F); odorless.
Uses: Industrial lubricants.

fludrocortisone acetate (9-alpha-fluorohydrocortisone acetate) $C_{23}H_{31}FO_6$. 9-Alpha-17-alpha-hydroxycorticosterone-21 acetate.
Properties: White, odorless polymorphic crystals; m. p. 233-234°C. Sparingly soluble in alcohol and very slightly soluble in water.
Grade: N. N. D.
Uses: Medicine; veterinary medicine.

"Fludrocortone." [123] Trademark for fludrocortisone (9-alpha-fluorohydrocortisone).

"Fluid Ball." [413] Trademark for a propellant casting powder consisting of fully colloided nitrocellulose having an average particle diameter of 50 microns or less. Composition can include liquid or solid modifiers.

Used as binder constituent of modified double base solid rocket propellants.
Containers: Fiber drums, 100-lb. net weight.
Fire hazard: Dangerous.
Shipping regulations: Explosives. Red label. *

fluid catalytic process. A continuous process used mainly for making high octane gasoline by cracking various petroleum fractions in the presence of finely divided alumina-silica gel catalyst. The catalyst is held in suspension by causing the petroleum fraction to flow upward so that the mixture acts like a fluid. The catalyst is withdrawn overhead, separated from the cracked vapors, and sent to another reactor where hot air is the fluidizing medium. In this reactor, called the regenerator, the carbon is burned off the catalyst and the hot catalyst is then sent back to the cracking reactor.

fluidize. In general, to convert to a fluid state; but in recent technology the term refers to processes in which a finely divided solid is caused to behave like a fluid by bringing it into suspension in a moving gas or liquid. The solids so treated are frequently catalysts; hence, the term fluid catalysis. In such a case the fluidized catalyst is brought into intimate contact and causes a desired reaction in the suspending liquid or gas mixture. Local overheating of the catalyst is greatly reduced, and portions of catalyst can be easily removed for regeneration without shutting down the unit. There are also non-catalytic applications in which the fluidized solid enters into direct reaction with the liquid or solid.

flumethiazide $C_8H_6F_3N_3O_4S_2$. 6-(Trifluoromethyl)-1, 4, 2-benzothiadiazine-7-sulfonamide-1, 1-dioxide.
Properties: White crystalline powder; m. p. 305-307°C (decomposes). Sparingly soluble in water; soluble in lower alcohols; insoluble in ethyl acetate, benzene, toluene; soluble in dilute alkali but unstable in alkaline solution.
Grade: Pharmaceutical.
Use: Medicine.

fluoboric acid HBF_4.
Properties: Colorless, clear, strongly acid liquid; b. p. 130° with decomposition. Miscible with water, alcohol.
Derivation: Action of boric and sulfuric acids on fluorspar.
Uses: Production of fluoborates; specially purified solution used in patented process for electrolytic brightening of aluminum. Also used to form stabilized diazo salts.

"Fluolite." [206] Brand name of proprietary fluorescent whitening agents.

fluophosphate alkyl esters. See diisopropyl fluophosphate.

fluophosphates. See fluophosphoric acids.

fluophosphoric acids. Term used to designate several acids containing fluorine and phosphorus. Mono (H_2PO_3F) and di (HPO_2F_2) fluophosphoric acids (also called phosphorofluoridic and phosphorodifluoridic acids) are available as clear, rather viscous nonvolatile anhydrous liquids. Hexafluophosphoric acid (HPF_6) is available as a 65% solution in water. The mono acid dissolves in water with little hydrolysis. The hexa acid is stable in neutral and alkaline solutions. These acids are suggested for use as metal cleansers, electrolytic or chemical polishing agents, for the formation of protective coatings for metal surfaces and as catalysts.

The sodium, potassium, and calcium salts of monofluophosphoric acid, and the potassium, ammonium, tetraethylammonium and pyridinium salts of hexafluophosphoric acid are water-soluble, dry, nonhygroscopic solids, stable towards heat and in water solutions at ordinary temperatures. The m. p. of the sodium mono salt is 625°C, that of the potassium mono salt 825°C, while the potassium hexa salt melts at 575°C, pyridinium hexa salt at 170°C, and tetraethylammonium hexa salt at 255°C. The hexa salts can be stored in solution without decomposition of the PF_6 ion. Suggested uses include maintenance of fluoride atmospheres and the preparation of bactericides and fungicides.
Shipping regulations: Corrosive liquid. White label. *

fluoranthene (idryl) $C_{16}H_{10}$. A tetracyclic hydrocarbon.
Properties: Colored needles; f. p. 107°C; b. p. 250°C (60 mm); insoluble in water; soluble in ether and benzene.
Derivation: From coal tar.

fluorapatite. See apatite.

"Fluorel" Brand 2141 Elastomer. [158] A fully-saturated fluorinated polymer containing more than 60 per cent fluorine by weight and is non-flammable.
Properties: Light colored gum, with raw properties including specific gravity, 1.85; Shore "A" hardness, 40; embrittlement temperature, −50°F.; excellent storage stability; soluble in esters and ketones. Optimum resistance to fuming nitric and fuming sulfuric acids, acetic acid and hydrochloric acid, bases and common solvents such as aromatic, aliphatic and chlorinated hydrocarbons, alcohols and ethers. Outstanding performance in hydraulic fluids and synthetic lubricants at elevated temperatures. Rated for continuous service at +400°F. and can withstand +600°F. temperatures at limited service. Gum can be air oven aged for 16 hours at 600°F. and retain 65 per cent of original tensile strength. Set as little as 10 per cent after 70 hours at 250°F.
Containers: 5, 10, 50 and 100-lb fiber drums.
Uses: O-rings, gaskets; hoses; wire and fabric coatings; diaphragms; fuel cells; expellant bladders; sealants; insulation; containers.

fluorene (alpha-diphenylenemethane) $\overline{C_6H_4CH_2C}_6H_4$.

Properties: Small, white, crystalline plates; fluorescent when impure. Soluble in alcohol, ether, benzene and carbon disulfide; insoluble in water. M. p. 116°C; b. p. 295°C (with decomposition).

Derivation: By reduction of diphenylene ketone with zinc; from coal tar.

Method of purification: Crystallization.

Grades: Technical; 98% pure.

Containers: 1-, 5-, 10-, 25-, 50-lb packages, 200-lb barrels.

Uses: Resinous products; insecticides; dyestuffs.

Shipping regulations: None. *

fluorescein (resorcinolphthalein; diresorcinolphthalein; see also uranine) $C_{20}H_{12}O_5$. C. I. No. 766.

Properties: Orange-red, crystalline powder; very dilute alkaline solutions exhibit a very intense, greenish-yellow fluorescence by reflected light, while the solution is reddish-orange by transmitted light; m. p. decomposes at 290°C; soluble in dilute alkalies, boiling alcohol, ether and dilute acids, glacial acetic acid; insoluble in water, benzene, chloroform.

Derivation: By heating phthalic anhydride and resorcinol.

Grades: The sodium salt (uranine) and potassium salt are marketed.

Containers: Wooden kegs; fiber drums.

Uses: Dyeing sea water for spotting purposes; tracer to locate impurities in wells; dyeing silk and wool; medicine; indicator and reagent for bromine.

Shipping regulations: None. *

fluorescein sodium. U. S. P. XVI name for uranine.

fluorescence. The production of visible light (white or colored) or other radiation by a substance as the result of exposure to and absorption of other radiations of different wave length, such as ultraviolet light, or electric discharges in a vacuum tube. Those substances having this property are known as phosphors, the term usually being restricted to those solids that absorb ultraviolet and emit visible light. In ordinary fluorescent lighting, the tube contains mercury vapor and argon, and the inside walls of the tube are coated with the fluorescent substance, often a zinc or cadmium compound. The passage of an electric current through the mercury vapor-argon mixture produces invisible ultraviolet light which is absorbed by the phosphor and re-emitted as visible light. The whole process occurs at a relatively low temperature (hence called a "cold light" process). Fluorescence is also used to identify or analyze certain minerals quickly such as scheelite and also certain organic derivatives.

fluorescin $C_{20}H_{14}O_5$. A reduction product of fluorescein.

Properties: A light-yellow powder; m. p. 125-127°C; soluble in alcohol or ether; insoluble in water.

Uses: Medicine; reagent.

fluoridation. Addition of about 1 part per million of fluorine (as sodium fluoride, sodium fluosilicate or a similar compound) to public drinking water supplies. The objective is to reduce tooth decay among the persons using the water.

fluorinated ethylene propylene resin (FEP resin). A copolymer of tetrafluoroethylene and hexafluoropropylene with properties similar to polytetrafluoroethylene resin, but which can be melt processed in conventional molding and extrusion equipment for use as wire insulation, cable jacketing, tubing, sheeting, shaped objects, rods, bottles. It can also be applied by fluidized bed coating, and used as a heat-sealable film for printed circuitry, flexible cable, capacitors, gaskets, and seals. Useful up to 200°C and is tough at low temperatures.

Containers: 5-, 50-, 100-lb drums.

fluorinated paraffin. A paraffin oil or wax which has been fluorinated (part of the hydrogen replaced with fluorine), usually by substitution of fluorine in a chlorinated paraffin.

Uses: Inert lubricant and sealant; heat transfer medium.

fluorinating agent. A compound used to introduce fluorine into some other compound. The higher fluorides of cobalt (CoF_3), manganese (MnF_4), silver (AgF_2), and of chlorine and bromine are used for this purpose. The usual procedure involves replacement of chlorine. Thus carbon tetrachloride (CCl_4) is treated with fluorinating agents to produce dichlorodifluoromethane (CCl_2F_2) and similar compounds.

fluorine F. Element in group VII of the periodic classification. Atomic number 9, atomic weight 19.00. The most reactive nonmetallic element.

Properties: Pale yellow gas; b. p. −188°C; m. p. −217.8°C; density of gas 1.695 (air = 1); sp. gr. of liquid (−188°C) 1.108; sp. vol. 10.2 cu ft/lb (70°F). Corrosive and poisonous. Reacts vigorously with most oxidizable substances at room temperature, frequently with ignition. Forms fluorides with all elements except the inert gases.

Occurrence: Widely distributed to the extent of 0.03% of the earth's crust. The chief minerals are fluorapatite, cryolite, and fluorspar.

Derivation: Electrolysis of molten anhydrous hydrofluoric acid - potassium fluoride melts with special copper-bearing carbon anodes, steel cathodes, containers and monel screens.

Containers: Special steel cylinders. Available both as a liquid and a compressed gas.

Uses: As an oxidizer in rocket fuels; production of metallic and other fluorides, particularly cobalt fluoride, antimony

fluorides, uranium hexafluorides, and sulfur hexafluorides. The first two are commonly used to introduce fluorine into organic compounds, i. e. , the fluorocarbons. Numerous fluorine compounds are made from hydrogen fluoride which is obtained directly from fluorspar, so that elemental fluorine is not needed in the process. See fluorocarbons, fluosilicic acid, sodium, ammonium, lead, barium, and calcium fluosilicates, various metallic fluorides.
Shipping regulations: Flammable gas. Red gas label. *

fluorine cyanide. See cyanogen fluoride.

fluorite. See fluorspar.

fluoroacetic acid CH_2FCOOH.
Properties: Colorless crystals; m. p. 33°C; b. p. 165°C; soluble in water, alcohol.
Caution: Poison!

fluoroacetone CH_2FCOCH_3.
Properties: Yellow liquid. Pungent odor; sp. gr. 0. 967 (20°C); b. p. 72. 5°C.
Derivation: Interaction of thallium fluoride and bromoacetone.

fluoroacetophenone (phenacyl fluoride; phenyl fluoromethylketone) $C_6H_5COCH_2F$.
Properties: Brown liquid. Pungent odor. Caution! Irritant. B. p. 98°C (8 mm).
Derivation: By Friedel-Craft synthesis.

fluoroalcohols. A group of fluorine - containing alcohols are available commercially of the general formula $H(CF_2CF_2)_nCH_2OH$ in which n = 1 to 5. They include the

C3, 1H, 1H, 3H-tetrafluoro-1-propanol (n = 1).
C5, 1H, 1H, 5H-octafluoro-1-pentanol (n = 2).
C7, 1H, 1H, 7H-dodecafluoro-1-heptanol (n = 3).
C9, 1H, 1H, 9H-hexadecafluoro-1-nonanol (n = 4).
C11, 1H, 1H, 11H-eicosafluoro-1-undecanol (n = 5).

Properties: Density of liquids 1. 48-1. 66; refractive index 1. 318-1. 320 (lowest of all known organic compounds); fluorine content 57. 5-71. 4%.
Derivation: By free radical telomerization of tetrafluoroethylene with methanol.
Grade: Technical.
Uses: Solvents; organic synthesis.

para-**fluoroaniline** $FC_6H_4NH_2$.
Properties: Liquid; sp. gr. (25°C) 1. 1524; b. p. 187. 4°C; m. p. 0. 82°C; refractive index (n 20/D) 1. 5395.
Uses: Dye intermediate; preparation of para-fluorophenol.

fluorobenzene C_6H_5F.
Properties: Colorless liquid with benzene odor; sp. gr. (20°C) 1. 0252; refractive index (n 25/D) 1. 4636; b. p. 84. 9°C; freezing point 41. 9°C. Insoluble in water; miscible with alcohol, ether.
Uses: Insecticide and larvicide intermediate; identification reagent for plastic or resin polymers.

fluorocarbon-11. See trichlorofluoromethane.

fluorocarbon-12. See dichlorodifluoromethane.

fluorocarbon-22. See chlorodifluoromethane.

fluorocarbon-113. See trichlorotrifluoroethane.

fluorocarbon-114. See dichlorotetrafluoroethane.

fluorocarbon resins. This term includes polytetrafluoroethylene, polymers of chlorotrifluoroethylene (fluorothene), vinylidene fluoride ($H_2C{=}CF_2$) hexafluoropropylene (C_3F_6) and similar compounds. These polymers are thermoplastic, inert to chemicals and oxidation, have high heat stability, retain their useful properties at both extremely low and high temperatures, have high electrical resistance, low coefficient of friction, and are nonwettable and resistant to moisture. The materials are available as resins, powders, and dispersions, and as films, sheets, tubes, rods, and tapes. Some materials are rubberlike. Commercially available varieties are "Kel-F," "Teflon," "Fluorel," "Aclar" and "Halon."

fluorocarbons. Compounds of carbon and fluorine with or without hydrogen; analogs of hydrocarbons in which all or nearly all the hydrogen has been replaced by fluorine. They are characterized by extreme chemical inertness, do not burn, and are thermally stable to 500°F or more. They are more volatile and more dense than corresponding hydrocarbons, have low refractive indices, low dielectric constants, low solubilities, low surface tensions, but their temperature coefficient of viscosity is high although viscosities are comparable with those of hydrocarbons.
General uses are as aerosol propellants, blowing agents, fire extinguishing agents, lubricants and hydraulic fluids, flotation and damping fluids, liquid dielectrics and coolants. Specialized uses are as impregnants for electrical insulation, spray-on wax coatings for alkali cleaning tanks, permanent lubricants for instruments and clocks.
See also fluorocarbon resins, and fluorochemicals.

"Fluorochemical FC-101." [158] Trade name for an inert fluorochemical fluid with unique solubility characteristics, that does not dissolve nor blend with hydrocarbon oils or the additives or detergents widely used in heavy duty lubricants.
Properties: Colorless, odorless, non-flammable liquid insoluble in water, oils and most organic solvents.
Uses: Specifically designed for use as sight glass fluids in force feed lubricators.

"Fluorochemical Liquid, Inert, FC-43." [158] A fully-fluorinated product composed primarily of an isomeric mixture of the perfluoro amine $(C_4F_9)_3N$.
Properties: Colorless, odorless liquid;

*See "I. C. C. Shipping Regulations," page xiii.
Reference numbers refer to name of manufacturer. See "List of Manufacturers," page v.

specific heat, 0.27 Btu/lb at 77°F.; heat of vaporization, 29.8 Btu/lb at b. p.; density (77°F) 1.88 ± 0.02; boiling point, 337-355°F ; electric strength (77°F) 35 KV minimum; dielectric constant (75°F) 1.90; viscosity, centistokes (77°F) 2.8 ± 0.5.
Containers: 4-lb quart bottle in fiber carton; 1- and 5-gal. cans.
Uses: High temperature heat transfer medium; dielectric fluid; pressure medium; liquid pump sealant; and used in metering devices.

"Fluorochemical Liquid, Inert, FC-75." [158] A fully fluorinated product composed of a mixture of compounds containing eight carbon atoms, principally perfluoro ethers.
Properties: Colorless, odorless liquid. Boiling point 210-225°F.; electric strength (77°F) 35 KV minimum; density (77°F) 1.77 ± 0.02; specific heat (Btu/lb/°F at 77°F) 0.248; viscosity, centistokes (77°F) 0.65 minimum; dielectric constant, liquid (75°F) 1.86.
Containers: 2-lb glass bottles in fiber carton; 1- and 5-lb cans.
Uses: Same as for "Fluorochemical Liquid, Inert, FC-43."

fluorochemicals. Organic compounds, not necessarily hydrocarbons, in which a large percentage of the hydrogen directly attached to carbon has been replaced by fluorine. The presence of two or more fluorine atoms on a carbon atom usually imparts great stability and inertness to the compound and fluorine usually increases the acidity of organic acids.
Derivation: (a) Electrolysis of solutions in hydrogen fluoride (Simons process); (b) replacement of chlorine or bromine by fluorine with hydrogen fluoride in the presence of a catalyst such as antimony trifluoride or pentafluoride; (c) addition of hydrogen fluoride to olefins or acetylene.
Uses: Refrigerants; lubricants; heat transfer media; aerosol propellants; fire extinguishers; rodenticides; inert plastics; chemical intermediates.

"Fluorochemical Surfactants." [158] Trade name for a line of surfactants (FC-95, FC-98, FC-128 and FC-134) that are available as anionic and cationic types and can therefore be used in most aqueous systems.
Properties: Most stable and surface-active of all known surfactants. Have unique property of being surface active in organic liquids as well.
Containers: 1-lb pint can in fiberboard carton; 5-lb gallon can in fiberboard carton; 4-gal fiberdrum; 5-gal pail and 7-gal pail.
Uses: Lowering surface tension in a water system to 16 to 18 dynes per centimeter; possessing chemical stability in such systems as 90 per cent hydrogen peroxide, hydrazine and fuming nitric acid; offering complete stability in extreme oxidizing conditions (such as chrome plating baths); and to promote leveling emulsion coatings. These surfactants have the ability to adsorb to many surfaces from solution and provide corrosion and stain resistance. Use level for "Fluorochemical Surfactants" is extremely low. Therefore they can offer surfactant properties without adverse side effects normally associated with use levels of ordinary surfactants.

fluorodichloromethane. See dichlorofluoromethane.

fluoroethylene. See vinyl fluoride.

"Fluoroflex." [14] Trademark for a line of products processed from fluorocarbon resin compositions.

fluoroform. See "Genetron" 23.

fluoroformyl fluoride. See carbonyl fluoride.

9-alpha-**fluorohydrocortisone acetate.** See fludrocortisone acetate.

9-alpha-**fluoro-16**-alpha-**hydroxyprednisolone.** See triamcinolone.

"Fluorolubes." [306] Trademark for addition polymers of trifluorovinyl chloride; essentially linear polymers built up of a recurring unit, $-CF_2CFCl-$.
Properties: Odorless, non-toxic; high density; non-flammable; free from hydrogen; stable in presence of concentrated mineral acids, alkalies, hydrogen peroxide and other strong oxidizing agents; thermally stable at temperatures up to 300°C; insoluble in petroleum base oils; slightly soluble in the lower alcohols; soluble in benzene, ketones and most chlorinated liquids.
Uses: Lubricant and sealant for plug cocks, valves and vacuum pumps; gasket and packing impregnant; fluid for hydraulic equipment, heat exchange and instrument damping; low temperature lubricant; high density fluids.

9-alpha-**fluoro-16**-alpha-**methylprednisolone.** See dexamethasone.

para-**fluorophenol** FC_6H_4OH.
Properties: White crystalline solid; density (56°C) 1.1889; m. p. 48.2°C (stable form), 28.5°C (unstable form); b. p. 185.6°C (760 mm), 78°C (15 mm). Soluble in water.
Use: Fungicide; intermediate for pharmaceuticals and fungicides.

fluorophosphoric acids. See fluophosphoric acids.

"Fluorosol." [243] Trade name for an optical bleach.

fluorothene. A plastic polymer of trifluorochloroethylene. C_2F_3Cl. Resistant to concentrated acids and alkalies up to 300°F. Swelled by some chlorinated hydrocarbons.

fluorotrichloromethane. See trichlorofluoromethane.

fluorspar (fluorite, florspar) CaF_2. Natural calcium fluoride occurring in veins either alone or with metallic ores. Color yellow, green, purple, other colors; luster vitreous; good cleavage in 4 directions; hardness 4;

sp. gr. 3. 2; m. p. 1350°C.
Grades: Metallurgical, ceramic, and acid grades containing greater than 85%, 95%, and 98% CaF_2 respectively.
Occurrence: Illinois, Kentucky, Tennessee, New Hampshire, Colorado, New Mexico, Arizona, Nevada, Utah, Montana, Texas, California, Washington; Canada; Europe; Mexico.
Containers: Bulk in railroad cars or barges; 125-lb bags; 500-lb barrels.
Use: Flux in open hearth steel furnaces and in gold, silver, copper, and lead smelting; manufacture of hydrofluoric acid; in ceramics; for synthetic cryolite; in carbon electrodes; emery wheels; electric arc welders; certain cements; dentifrices. Also as a phosphor; paint pigment; catalyst in wood preservatives. Clear fluorspar is used in optical equipment.

fluosilicates. Salts of fluosilicic acid H_2SiF_6. See sodium fluosilicate; ammonium fluosilicate. Magnesium fluosilicate is used as a concrete hardener and in magnesium casting. Zinc fluosilicate is used as a concrete hardener and for preventing decay of wood. The copper salt has a similar use and the barium salt is used in insecticides and ceramic operations. Lead fluosilicate is used in electrorefining and plating of lead. The zinc, magnesium, copper and lead fluosilicates are soluble, in contrast with the sodium and potassium salts.

fluosilicic acid (hydrofluosilicic acid; silicofluoric acid; hydrosilicofluoric acid; hydrofluorosilicic acid; sand acid) H_2SiF_6.
Properties: Transparent, colorless, fuming liquid. Corrosive! Poisonous. Attacks glass and stoneware. Keep in wax, wood, or special plastics. Soluble in water.
Derivation: By-product of the action of sulfuric acid on phosphate rock containing fluorides and silica or silicates. The hydrofluoric acid acts on the silica to produce silicon tetrafluoride, SiF_4, which acts with water to form fluosilicic acid, H_2SiF_6.
Grades: Technical; C. P.
Containers: 10-gal carboys; 50-gal barrels; tank cars.
Uses: Ceramics (to increase hardness); disinfecting copper and brass vessels in breweries; in building (hardening cement, plaster of Paris, concrete flooring, preserving masonry); general disinfectant; technical paints; wood preservative and impregnating compounds. Also in electroplating and in the manufacture of sodium, ammonium, magnesium, zinc, copper, barium, lead and other fluosilicates.
Shipping regulations: Corrosive liquid. White label. Legal label name, hydrofluosilicic acid. *

fluosulfonic acid (fluosulfuric acid) HSO_3F.
Properties: Colorless, fuming, highly corrosive liquid; b. p. 163°C with slight decomposition; soluble in nitrobenzene, soluble in water with partial decomposition.
Derivation: Combination of anhydrous hydrogen fluoride with sulfuric acid or with sulfuric acid anhydride.
Uses: Catalyst in organic synthesis; electropolishing.
Shipping regulations: Corrosive liquid. White label. *

fluosulfuric acid. See fluosulfonic acid.

fluoxymesterone $C_{20}H_{29}FO_3$. (9 alpha-Fluoro-11 beta, 17 beta-dihydroxy-17 alpha-methyl-4-androstene-3-one).
Properties: Crystals; m. p. 270°C (decomposes); optical rotation +109° (in ethanol).
Grade: N. N. D.
Use: Medicine.

fluphenazine dihydrochloride 1-(2-hydroxyethyl)-4-[3-2(trifluoromethyl-10-phenothiazinyl)-propyl]-piperazine dihydrochloride.
Properties: White crystalline odorless powder; m. p. 238. 5-239. 5°C; sensitive to light and air. Soluble in water.
Grade: Pharmaceutical.
Use: Medicine.

flush colors. Pigments dispersed in oil, varnish, etc., the transfer from the water phase to the oil phase having been effected without the usual drying and subsequent grinding of the dry pigment. It is claimed that the flush color as sold to the consumer is ready for use without grinding.

flux.
1. (chemistry and metallurgy) A substance that promotes the fusing of minerals or metals or prevents the formation of oxides. For example, in metal refining an addition of some mineral to the furnace charge is made for the purpose of absorbing mineral impurities in the metal. A slag (q. v.) is formed which floats on the top of the bath and is run off.
2. (soldering and brazing) A substance which is applied to the portions to be united and which, on the application of heat, aids in the ready flowing of the solder and prevents the formation of oxides while the solder unites with the two parts to form a tight joint.
3. (ceramics) Any readily fusible glass or enamel used as a base or ground.
4. (physics) The rate of flow or transfer of electricity, magnetism, water, heat, energy, etc., the term being used to denote the quantity that crosses a unit area of a given surface in a unit of time.

fluxing lime. See calcium oxide.

flux oil. An oil for blending with asphaltic or bituminous materials for the purpose of softening or reducing their consistency. See also residual oils.

flux, white. Sodium nitrate and sodium nitrite.
Properties: White powder; strong oxidizer.
Containers: Tin cans.
Uses: Metallurgy; welding.
Fire hazard: Dangerous.

Shipping regulations: Oxidizing material. Yellow label. *

fly ash. The very fine ash produced by combustion of powdered coal with forced draft, and often carried off with the flue gases from such processes. Special equipment is necessary for effective recovery.

Fm. Symbol for fermium. See element 100.

FM cyclotron. See cyclotron.

FMN. Abbreviation for flavin mononucleotide. See riboflavin phosphate.

"Foam" A. [148] Trademark for a barium sulfate product, ground from Missouri baryta ore.

Properties: White, inert; water-ground, water-floated, and silk-bolted; water-soluble salts, < 0.01%; acid free; sp. gr. 4.46; wt/solid gal, 37.15 lbs; 1 pound bulks 0.02692 gal; oil absorption 7.7; barium sulfate, 99.0%; 100% through 325 mesh.

Containers: Paper bags, 50-lb net.

Uses: Inert extender pigment for paints, colors, rubber compounds, adhesives, grinding compounds, linoleum, oilcloth, lubricating compounds, photographic films, x-rays films, photographic paper, printing plates, textile fabrics, glass, ceramics.

"Foamasol." [328] A dark brown free-flowing liquid consisting of a mixture containing the sodium salt of a proteinaceous material. It is used in concrete mixtures, where light weight is desired, to secure foaming and dispersing action. The protein type constituent effectively reduces segregation, improves workability, and aids the formation of smooth finished surfaces. Shipped in 55-gal containers.

"Foamex." [73] Trademark product. Mixture of aliphatic esters.

Properties: Very pale yellow liquid. Insoluble in water; sp. gr. 0.96-0.97.

Containers: 1-gal cans (8 lbs); 5-gal cans (40 lbs); 55-gal drums (440 lbs).

Uses: Suggested for the retardation and prevention of foam in the manufacture of water solutions of glue, casein, shellac, gelatin, etc. For use in the manufacture of paper coatings, leather finishes, latex compositions, textile sizes, finishes, adhesives.

"Foamite." [75] A proprietary trademark covering fire extinguishing foam-making equipment, including the chemical ingredients of which the foam itself is composed, such as the stabilizer.

Properties of the stabilizer: Dark brown, thick liquid or powder. Colloidal properties. Soluble in water and in alkalies. Insoluble in alcohol and ether. Resins precipitated by acids. 12° Bé. and 18° Bé. (120°F) according to concentration.

Derivation: By a secondary extraction of spent licorice root after the first or mild extraction. The primary extract contains the sugar, starch, and glycyrrhizin; while the secondary contains the resinous materials of the root.

Containers: 5-, and 55-gal drums.

Uses: For the production of foam for the extinguishment of class "A" and class "B" fires, especially large fires in oil-tank farms.

Shipping regulations: None. *

foeniculum. See fennel.

folacin. See folic acid.

"Folex." [40] Trademark for a 75% emulsifiable concentrate of tributylphosphorotrithioite (see "Merphos") for use as a cotton defoliant. Contains 6 lbs of "Merphos" per gal.

"Foliafume." [342] Trademark for pyrethrin-rotenone plant spray concentrates.

folic acid (pteroylglutamic acid; folacin; PGA) $C_{19}H_{19}N_7O_6$. Considered a member of the vitamin B complex. At least three substances with folic acid activity are found in nature, one of which, pteroylglutamic acid, is made synthetically.

Properties (pteroylglutamic acid): Orange-yellow needles or platelets; tasteless; odorless; slightly soluble in water; insoluble in lipid solvents; stable to heat in neutral and alkaline solution; destroyed by heating with acid; inactivated by light.

Sources: Food source: green plant tissue, fresh fruit, liver, and yeast.

Commercial Source: Synthetic pteroylglutamic acid made by the reaction of 2,3-dibromopropanol, 2,4,5-triamino-6-hydroxy pyrimidine, and para-aminobenzoyl glutamic acid.

Grade: 10% feed grade; U.S.P. XVI.

Containers: Fiber drums.

Uses: Medicine; nutrition.

"Folin Decalso." [184] Treated precipitated gel-type sodium aluminosilicate cation exchanger employed for ammonia determination.

folinic acid (5-formyl-5,6,7,8-tetrahydropteroyl-L-glutamic acid; citrovorum factor) $C_{20}H_{23}N_7O_7$. A member of the folic acid group of vitamins and a growth factor for the bacterium Leuconostoc citrovorum. Folinic acid is an important metabolite of folic acid and may be the active form in cellular metabolism. It is an effective hematopoietic factor. Ascorbic acid and vitamin B_{12} are essential for the conversion of folic acid to folinic acid.

Properties: $d\ell$-L-form: Crystals; decompose 240-250°C; sparingly soluble in water.

Derivation: (a) Prepared by catalytic reduction of folic acid; (b) produced microbially.

Sources: Same as those for folic acid.

Uses: Medicine; nutrition; biochemical research.

"Folione." [227] Trademark for methyl heptine carbonate $CH_3(CH_2)_4CCCOOCH_3$. (Heptyne carboxylic acid methyl ester).

Properties: Colorless to pale yellow liquid; strong and somewhat overpowering odor in pure state; when fully diluted, odor is

similar to violets; sp. gr. (25/25°C) 0.919-0.924; refractive index (20°C) 1.446-1.449; soluble in 6 parts of 70% alcohol.
Uses: For violet or violet-leaf effect; blends well with ionones and other violet notes.

follicle-stimulating hormone (FSH). One of the hormones secreted by the anterior lobe of the pituitary gland. It increases the formation of sperm cells by the testis and of graafian follicles by the ovary. Found in pituitary tissue of man, horse, sheep and pig. It is a protein with some carbohydrate.
Properties: White solid; soluble in water and 50% alcohol. Isoelectric point at pH 4.5; activity decreased by action of heat or pepsin. Molecular weight about 70,000.
Use: Medicine.

"Folrosia." [227] Trademark for a cyclic alcohol with a deep red rose fragrance.
Properties: Colorless liquid; sp. gr. 0.914-0.918 (25/25°C); refractive index (n 20/D) 1.4650-1.4680. Clearly soluble in 3 parts of 60% alcohol, 1 part of 70% alcohol. Flash point TCC 203°F.
Occurrence: Not found in nature.
Uses: In neroli, muguet, lilac and rose scents.
Hazard: Avoid aluminum containers.

"Folvite." [57] Trademark for folic acid; (pteroylglutamic acid) $C_{19}H_{19}N_7O_6$.
Properties: Yellow to yellow-orange crystals, odorless; chars when heated beyond 250°C. Soluble in acetic acid, phenol, pyridine, alkalies, hot dilute hydrochloric acid, hot dilute sulfuric acid; slightly soluble in methanol; very slightly soluble in cold water, 1% w/v in boiling water; insoluble in acetone, chloroform, ether, benzene, and alcohol; pH about 8.4.
Use: Medicine.

"Folvron." [57] Trademark for folic acid and iron.

"Fo-Made." [162] Trademark for a foam stabilizer for beer.

"Fomrez." [104] Trademark for a line of polyester and polyether resins used in the manufacture of urethane foams.

"Fonoline." [45] Trademark for petrolatum, U.S.P., of soft consistency and low melting point.
Properties: White or yellow in color; m.p. 112-120°F; Saybolt viscosity (white), 55-75 (210°F); (yellow), 60-85 (210°F); odorless.
Uses: Pharmaceutical and cosmetic preparations; many industrial applications.

food dyes. Dyes which are used to color food products. These dyes must meet rigorous standards for non-toxicity, and should have a high degree of solubility in water, ethanol or glycerin and edible oils, and be resistant to acid or alkali decomposition. Numerous changes are occurring with respect to toxicity tests and standards for acceptance by the federal Food and Drug Administration (FDA) and similar bodies in other countries. The majority of the synthetic dyes that have been used in food were azo dyes but the violet, blue and green shades have required the use of sulfonated dyes of the triarylmethane and other classes. The most common natural dye used is carotene, suitable for yellow coloring of edible oils and fats.

food of the gods. See asafetida.

fool's gold. See pyrite.

foots (soapstock). The mixture of soap, oil, and impurities that precipitates when natural fatty oils are refined by treatment with caustic soda or soda ash. Usually contains 30 to 50% of free and combined fatty acids, and is used for manufacture of relatively low grade soaps, and also as a source of free fatty acids. A related usage of the term is for the suspended solid matter in crude oils.

"Forasite." [36] A phenolic water-soluble adhesive used in bonding comminuted wood products. This resin is applicable to both wet and dry process.
Properties: Solids (PMMA) 40-85%.

"Foremul." [78] Trademark for a series of polyethylene esters of fatty acids used as emulsifiers in the fat liquoring of leather, degreasing agents, dispersing and stabilizing agents, and in formulation of insecticides and agricultural sprays.

forensic chemistry. The use of chemical knowledge and methods in connection with legal matters, most often those associated with evidence of a crime. Of particular importance are the use of chemical microscopy with or without polarized light to observe crystal characteristics, the use of spot and color reactions, and various spectrophotometer applications. Special methods are necessary for poisonous metals, volatile and gaseous poisons, and nonvolatile organic poisons. Examination of documents to establish the identity and age of ink, and to restore erasures is a special variety of forensic chemistry.

formal. See methylal.

formaldehyde (oxymethylene; formic aldehyde; methanal) HCHO.
Properties: Colorless gas; suffocating pungent odor; poisonous! M.p. −92°C; b.p. −21°C; soluble in water, alcohol and ether. Polymerizes easily. Is usually handled as an aqueous solution, with or without methanol, which acts as an inhibitor of the polymerization.
Properties of 37% solution (also called formaldehyde; formalin; formol): Clear colorless liquid; sp. gr. 1.075-1.081; b.p. 98°C (approx); pH 3.0; flash point 152°F (with 8% methanol).
Derivation: Oxidation of synthetic methanol or low-boiling petroleum gases such as methane, ethane, propane, and butane. Silver or an iron-molybdenum oxide are

the most common catalysts.
Grades: Aqueous solutions: 37%, 45%, 50% inhibited (with varying percentages of methanol) or uninhibited (methanol-free); also available in solution in n-butanol, methanol, or urea; U. S. P. XVI (37% aqueous solution containing methanol). See also paraformaldehyde.
Containers: Drums, carboys, tank cars and tank trucks.
Uses: Urea resins; phenolic resins; pentaerythritol; hexamethylenetetramine; melamine resins; polyformaldehyde resins; other chemicals; dyes, medicine (disinfectant, germicide); embalming fluids; preservative; hardening agent; reducing agent, as in recovery of gold and silver; corrosion inhibitor in oil wells.
Warning! Causes irritation of skin, eyes, nose and throat. MCA warning label.
Shipping regulations: None. *

formaldehyde acetamide. See formicin.

formaldehyde aniline (formaniline) $C_6H_5NCH_2$.
Properties: Colorless to yellowish crystals; initial m. p. 133°C; b. p. 271°C; sp. gr. 1. 14, but these vary somewhat from sample to sample. Soluble in water, ether, alcohol.
Derivation: By condensation of formaldehyde and aniline.
Uses: Rubber accelerator; intermediate in organic synthesis.

formaldehyde cyanohydrin. See glycolonitrile.

formaldehyde-gelatin. See glutol.

formaldehyde, para-. See paraformaldehyde.

formaldehyde, polymerized. See paraformaldehyde and trioxane.

formaldehyde-para-toluidine (anhydro-formaldehyde-para-toluidine; methylene-para-toluidine) $(CH_3C_6H_4NCH_2)_x$.
Properties: White powder with grayish-yellow cast. Aromatic odor. Not toxic to skin; soluble in acetone; sp. gr. 1. 11.
Derivation: Reaction between formaldehyde and para-toluidine.
Uses: Rubber (vulcanizing accelerator); dyes.

formalin. See formaldehyde.

formamidated chloral. See chloral formamide.

formamide (methanamide) $HCONH_2$.
Properties: Clear, colorless, hygroscopic oily liquid; sp. gr. 1. 146; b. p. 200-212°C with partial decomposition beginning about 180°C; m. p. 2. 5°C. Soluble in water and alcohol.
Derivation: By the interaction of ethyl formate and ammonia, with subsequent distillation.
Method of purification: Rectification.
Grades: Technical.
Containers: Glass bottles, 55-gal iron drums, tank cars.
Uses: Exceptionally good solvent, softener, intermediate in organic synthesis.
Shipping regulations: None. *

formanilide (phenylformamide) C_6H_5NHCHO.
Properties: Colorless to yellowish crystals; soluble in alcohol and water; m. p. 48-50°C; b. p. 271°C.
Derivation: By the reaction of aniline and formic acid.
Method of purification: Crystallization.
Grades: Technical.
Containers: Tins; glass bottles.
Use: Medicine.
Shipping regulations: None. *

formaniline. See formaldehyde aniline.

"Formaset." [42] Proprietary products. Water soluble or dispersible synthetic urea-formaldehyde type resins.
Properties: Colorless liquids and pastes. Soluble in water at room temperatures.
Containers: 55-gal lined steel drums.
Uses: Creaseproofing and shrinkproofing of textiles.

"Formcel." [352] Trademark for a series of water-free formaldehyde solutions in alcohols.
Uses: Production of alcoholated urea and melamine resins, coatings, laminating and textile-treating resins; embalming fluids.

"Formica." [13] High-pressure laminated plastic sheets of melamine and phenolic impregnated materials for decorative applications, and industrial sheets, rods, and tubes of various thermo-setting resins such as silicones, epoxies, melamines, phenolics, etc. combined with various supporting materials such as paper, linen, canvas, glass, etc. for electrical, chemical, and mechanical applications.

formic acid (hydrogen carboxylic acid) HCOOH.
Properties: Colorless, fuming liquid; pungent penetrating odor; dangerously caustic. Soluble in water, alcohol and ether; sp. gr. 1. 2201 (20/4°C); m. p. 8. 3°C; b. p. 100. 8°C; flash point 156°F; lbs/gal (20°C) 10. 16; refractive index 1. 3719 (20°C).
Derivation: (a) By treatment of sodium formate and sodium acid formate with sulfuric acid at low temperatures, and distilling in vacuo; (b) by acid hydrolysis of methyl formate; (c) as a by-product in the manufacture of acetaldehyde and formaldehyde in hydrocarbon oxidation.
Method of purification: Rectification.
Grades: Technical; 85%; 90%; C. P.
Containers: 150-, 195-lb carboys.
Uses: Chemical (formates, organic esters, oxalic acid, allyl alcohol); dyeing and finishing of textiles and paper; laundry sour; the manufacture of fumigants, insecticides, refrigerants, solvents for perfumes, lacquers; electroplating; use also suggested in tanning and in wine manufacture to aid fermentation; medicine; brewing (antiseptic); food preservative; silvering glass; leather; cellulose formate; rubber coagulant; ore flotation.
Warning! Causes burns; avoid breathing vapor. MCA warning label.

Shipping regulations: Corrosive liquid; white label. *

formic aldehyde. See formaldehyde.

formicin (formaldehyde acetamide) $CH_3CONHCH_2OH$.
Properties: Colorless, syrupy liquid. Soluble in water, alcohol and chloroform. Sp. gr. 1.14-1.18.
Derivation: By the interaction of acetamide and formaldehyde, with subsequent distillation.
Method of purification: Rectification.
Grades: Technical.
Containers: Iron drums; glass bottles.
Use: Disinfectant.
Shipping regulations: None. *

formol. See formaldehyde.

formonitrile. See hydrocyanic acid.

"Formopon." [23] Trademark for sodium formaldehyde hydrosulfite, a powerful, water-soluble reducing agent, used in synthetic rubber polymerization, dying and printing of textiles with vat colors, stripping of dyed fabrics, bleaching of soap. "Formopon" Extra is the basic zinc salt, a powerful, acid-soluble reducing agent, used in stripping colors from dyed fabrics and in the manufacture of liquid strippers.

"Formrez." [104] Trademark for a line of resins used to produce urethane elastomers (solid cast urethanes).

"Formula 40." [233] Trademark for alkanolamine salts of 2,4-D weed killers.

"Formula 602." [33] Trade name for a solvent cleaner composed of chlorinated hydrocarbons and aliphatic hydrocarbons.
Properties: Boiling range 250-360°F. Flash point 135° (C.O.C.); fire point 160° (C.O.C.).
Containers: 55-gal drums.
Uses: For cleaning electrical equipment and grease removal. It leaves no film and evaporates considerably faster than Stoddard solvent.

formula, chemical. There are several kinds that should be distinguished from one another.
A molecular formula uses the symbols of the elements to indicate the number and kind of atoms in a molecule of the compound, as H_2O_2 for hydrogen peroxide, C_2H_2 for acetylene, C_6H_6 for benzene.
An empirical formula is the simplest formula that can be used to express the relative number of atoms that unite to form a molecule of a compound. It does not necessarily indicate the total number of atoms in the molecule. Thus CH might be used as the empirical formula for both acetylene and benzene.
In a structural formula the symbols are written out so as to indicate the way in which the atoms are located relative to one another in the molecule. Such formulas usually give information on the kind of valence bonds that join the atoms. Examples of structural formulas are:

```
                       H H
                       | |
                       C=C
                      /   \
H-C≡C-H            H-C     C-H
                      \\   //
                       C-C
                       | |
                       H H
acetylene            benzene
```

formula weight. The sum of the atomic weights represented in a chemical formula. Thus, since the atomic weight of hydrogen is 1, and oxygen is 16, the formula weight of water (H_2O) is 18.

"Formvar." [61] Trademark for polyvinyl formal resins (see under polyvinyl acetal resins). Grades available in powder form

	(1)	(2)	(3)	(4)
7/70	21,000	6	45	9
12/85	30,000	6	24	20
7/95 "S"	18,000	8	11	18
15/95 "S"	34,000	8	11	65
15/95 "E"	34,000	5.5	11	50

(1) Molecular weight (weight average)
(2) Average hydroxyl content (as % polyvinyl alcohol)
(3) Average acetate content (as % polyvinyl acetate)
(4) Viscosity (cps) determined with 5 g. resin made to 100 ml. with ethylene chloride at 20°C.

Uses: Wire enamels; electrical insulations; coatings; adhesives; films and molded materials.

formyl fluoride HCOF.
Properties: Colorless, mobile liquid. Decomposes slowly with formation of hydrofluoric acid and carbon monoxide. Caution! Approximately three times more toxic than chloropicrin! Soluble in water (dec). B.p. -26°C.
Derivation: Interaction of benzoyl chloride and a formic acid solution of potassium fluoride.
Grades: Technical.
Uses: Organic synthesis; suggested military poison gas.

5-formyl-5,6,7,8-tetrahydropteroyl-L-glutamic acid. See folinic acid.

forsterite. See olivine.

"Forthane." [100] Trademark for methylhexaneamine (q.v.).

"Forticel." [352] Trademark for a thermoplastic product consisting essentially of cellulose propionate and plasticizers with or without pigments and coloring matter. Available in pellet form for injection and extrusion molding.
Properties: Sp. gr. 1.16-1.24; tensile strength 1,800-6,900 psi; refractive index 1.47-1.48; soluble or softened in some ketones and esters; affected by alcohol and chlorinated solvents; resistant to inks, aromatic hydrocarbons and mineral oils.
Containers: 25-, 50-, 100-, 200-, 250-lb

cylinderical fiber containers.
Uses: Fabricated into pen and pencil barrels, desk accessories, telephone bases and handsets, spectacle frames and protective goggles, tool handles and display trays.

"Fortiflex." [421] Trademark for a thermoplastic for use in injection molding, extrusion, blow molding, and calendering. "Fortiflex" A is a high density polyethylene consisting mainly of long molecules with occasional short side branches. "Fortiflex" B is a copolymer of ethylene and butene-1.
Properties: Milk-white translucent 1/8 inch pellets. Colors are also available. Density: "Fortiflex" A 0.960; "Fortiflex" B 0.950; melt index 0.2-8; tensile strength 3,100-3,700 (psi); highest use temperature 225°F; flammability 1 (in/min).

"Fortracin." [342] Trademark for animal feed supplements of bacitracin methylene disalicylate.

"Fortrel." [352] A polyester type synthetic fiber.

"Fosbond." [204] Trademark for a group of chemicals used to provide a corrosion-resistant bond between zinc or ferrous metals and a paint film.

"Foscoat." [204] Trademark for a class of chemicals designed to provide a phosphate coating prior to cold working.

"Fosfodril." [55] Brand name for a glassy phosphate of high molecular weight (sodium hexametaphosphate).
Properties: Powder or granular, odorless, glassy, hygroscopic salt.
Containers: 100-lb paper bags.
Uses: In petroleum industry, to impart the proper thixotropic properties to the muds used in oil-well drilling; water treatment in oil-well flooding operations.

"Fos-Fol." [147] Brand name for a product containing 73.8% tributyl phosphorotrithioite.
Containers: 5- and 30-gal drums.
Use: Cotton defoliant.

"Fosfo" Rosin. [79] Trademark of an "FF" grade of limed wood rosin containing approximately 4.5% lime.
Constants: M.p. (capillary tube) 88°C; m.p. (ball & ring) 110°C; acid number 86; color "FF".
Containers: Non-returnable 18-gauge black-iron drums of about 500 lbs gross wt. Tare 14-16 lbs.
Uses: Box toes; matches; printing ink; smoking molds; pipe bending.

"Foslube." [204] Trademark for a class of lubricants used to impregnate a phosphate coating prior to cold working.

"Fospray." [204] Trademark for a group of compounds used for the spray application of a phosphate coat prior to cold working.

"Fosrinse." [204] Trademark for a class of chemicals used to render insoluble the acid salts remaining after phosphate treatment of ferrous metals.

fossil resin. See amber.

fossil wax. See ozocerite.

"Fosterite." [308] Trademark of a family of resins. Largest application is as "solventless" varnishes for electric insulation, also as a photoelastic resin and as a bond for impregnating and laminating asbestos sheets. Rods made of this plastic will carry a beam of light without the dispersion which occurs in air, making it possible to bend a beam of light.

"Fotoceram." [20] Trademark for crystalline ceramic articles made by processing chemically sculptured glass. These products are utilized primarily for high temperature electronic components such as circuit boards.

"Fotocol." [319] Trademark for an alcohol-type solvent.
Properties: Water-white; mild, non-residual odor; sp. gr. (60/60°F), 0.812-0.814; acidity as acetic acid, 0.01% max; distillation range, 74-80°C; non-volatile matter, 0.005 g/100 ml max; flash point (Tag closed cup), 38°F; wt/gal at 60°F, 6.77 lb.
Containers: 1-gal can; 5-, and 54-gal steel drums.
Uses: Formulation of printing inks; cleaning printing plates and type; drying photographic films; for zinc etchings; shoe dyes. Distribution governed by Bureau of Internal Revenue.
Shipping regulations: Flammable liquid. Red label.*

"Fotoform." [20] Trademark for articles produced by chemically sculpturing any one of several, pre-treated glass compositions.

foul gas. Coke-oven gas or natural gas containing appreciable amounts of hydrogen sulfide and similar contaminants.

"Foundrez." [36] A group of water-soluble phenol-formaldehyde and urea-formaldehyde resins for foundry applications. Also two-stage powdered phenolic resins for use in the shell molding process.

foundry clays. See fire clays.

foundry sand (molding sand). Sand used in making molds for casting metal. Desirable properties vary according to the metal to be cast, but cohesiveness, refractoriness, texture, permeability, and durability are important.

Fourcault process. A method of forming window glass. The molten glass is drawn up from the melt tank in a ribbon, rolled flat, annealed, then cut to the desired size and shape. The rolling and annealing are done while the glass is in the vertical position.

fousel oil. See fusel oil.

Fowler's solution. A solution of potassium arsenite of definite strength, made by

*See "I.C.C. Shipping Regulations," page xiii.
Reference numbers refer to name of manufacturer. See "List of Manufacturers," page v.

boiling arsenic trioxide with potassium bicarbonate solution. Used in medicine as a means of administering arsenic in soluble form.

foxglove. See digitalis.

foxglove blue. A name applied loosely to any of a number of the varieties of iron-blue pigments. See iron blues.

f.p. Abbreviation for freezing point.

Fr. Symbol for francium.

fractional distillation. Distillation in which rectification is used to obtain product as nearly pure as possible. In such operation a part of the vapor is condensed and the resulting liquid contacted with more vapor, usually in a fractionating column with plates or packing. The term fractional distillation is also applied to any distillation in which the product is collected in a series of separate fractions.

"Fractol." [51] Trademark for a medium viscosity grade of white mineral oil fully meeting U. S. P. requirements for internal use.

francium Fr. The name francium has been adopted for element No. 87. It appears to exist only as radioactive isotopes. One isotope is actinium K. Other isotopes have been made artificially. Francium 223 is the longest lived isotope, having a half-life of 21 minutes. Francium belongs to the alkali-metal family. The name virginium has been suggested earlier.

frangula emodin. See emodin.

frangulic acid. See emodin.

Frankfort black. See vegetable black.

frankincense. See olibanum.

frankincense oil. See olibanum oil.

franklinite $(Fe, Mn, Zn)(FeMn)_2O_4$. Black mineral resembling magnetite (q. v.). Luster, metallic or dull. Only slightly magnetic. Frequently associated with red zincite and yellow to green willemite. Slowly soluble in hydrochloric acid; sp. gr. 5-5. 2; hardness 6-6. 5.

Occurrence: United States (New Jersey).

Uses: Zinc is recovered as zinc white and the residue is smelted for spiegeleisen. Also has been ground for dark paints.

Frary metal. A lead-base bearing metal containing 97-98% lead alloyed with 1-2% each of barium and calcium. Excellent for low-pressure bearings at moderate temperatures.

Frasch process.

A means by which some 90% of the world's sulfur is obtained. Developed just prior to 1900 by Herman Frasch, the process involves the melting of sulfur underground and forcing it to the surface while it remains molten.

A site is drilled with ordinary oil-well equipment, and three concentric pipes, the smaller within the larger, are lowered down the well casing. The outer (larger) carries hot water, the inner (smaller) compressed air to aid in lifting the column of molten sulfur, and the intermediate line carries the rising sulfur mixed with air, water and steam.

The temperature of the water when injected is about 100°F above the melting point of sulfur. In operation, the water permeates the sulfur-bearing rock formation and melts the sulfur which flows into a pool at the base of the shaft. The water pressure forces the sulfur into the exposed end of the intermediate-sized pipe, where it proceeds toward the surface. Air from the center pipe, entering the system at 500 psi rises with the sulfur forcing it to the surface.

Fraude's reagent. See perchloric acid.

"Free-Flo" Soda. [244] Trademark for sodium bicarbonate treated to increase flowability and non-caking properties.

Properties: U. S. P. powdered bicarbonate of soda ($NaHCO_3$) with approximately 0. 5% tricalcium phosphate (Ca_3PO_4) added. Na_2O content 37%; bulk density 60 lb/cu ft.; specific gravity 2. 2.

Containers: 100-lb Multiwall bags, fiber drums, kegs, and bulk shipments.

Uses: Manufacture and compounding of self-rising flours, cake mixes, and sponge rubber.

free radical. An atom or group of atoms such that there is at least one unpaired electron. A few free radicals have been isolated and samples collected (see triphenylmethyl). Most are short-lived intermediates, and others are convenient and useful theoretical or hypothetical concepts. Examples are benzoate radicals ($C_6H_5COO\cdot$) in the decomposition of benzoyl peroxide, and ethyl radicals ($C_2H_5\cdot$)in the pyrolysis of tetraethyl lead, $Pb(C_2H_5)_4$. Free radicals are always materials with high reactivity and high energy, and can be collected and stored only with special precautions such as collection in solution or at very low temperatures and in the absence of all but inert solvents or diluents. Some efforts have been made to devise means of collecting free radicals for subsequent use to generate power.

See also carbonium ion.

"Freezene." [45] Trademark for a series of refrigeration white mineral oils.

Typical properties: (medium) sp. gr. 0. 875-0. 885 (100°F); Saybolt viscosity 140-150 (210°F); odorless.

Uses: Low temperature lubrication.

freezing point. For a pure substance the freezing point or melting point is the temperature at which the liquid and solid are in equilibrium with one another, i. e., at a higher temperature the solid will gradually melt and at a lower temperature the liquid will solidify.

For a mixture the freezing point or

melting point varies with the composition, but is still the temperature at which the liquid mixture is in equilibrium with the solid material that separates from it on cooling. This latter is usually one pure component of the mixture but may be a solid solution, or a mixture.

freibergite. A variety of fahlore containing up to 18% silver. Usually steel-gray, sometimes iron-black; streak reddish; sp. gr. 5.05.

Occurrence: Idaho, Colorado, Nevada; Germany.

Fremy's salt. See potassium bifluoride.

French saffron. See crocus.

"Freon." [28] Trademark for a line of fluorinated hydrocarbons used as refrigerants, propellants, blowing agents, fire extinguishing agents, and solvents. They are available in amounts from 5 to 650 lbs in cylinders and drums. Most are available in tank car and tank truck quantities. Mixtures are sold for use as aerosol propellants.

"Freon-11" or "F-11". Trichloromonofluoromethane CCl_3F. Used as a refrigerant in industrial and commercial air conditioning systems; also, in industrial process water and brine cooling to −40°F (−40°C); as blowing agent for rigid and resilient plastic foams. Solutions of "Freon-11" and "Freon-12" are used as propellants for a wide variety of aerosol products. Can be tailored to individual pressure requirements from 1 to 69 lb/sq in gauge at 70°F (21.1°C).

"Freon-12" or "F-12". Dichlorodifluoromethane CCl_2F_2. Used as a refrigerant in a wide variety of refrigeration and air conditioning systems, including industrial, commercial, household and automotive systems. "Freon-12" is used as a propellant for high pressure aerosols such as those containing insecticides; also for surface-coating products such as metallic and pigmented paints and lacquers. Solutions of "Freon-12" are used as aerosol propellants with other "Freon" products; also as low temperature solvents and as blowing agent for rigid and resilient plastic foams. Pressure 70 lb/sq in gauge at 70°F (21.1°C).

"Freon-13" or "F-13". Monochlorotrifluoromethane $CClF_3$. Used as a refrigerant in both direct and indirect industrial low temperature cascade systems; aircraft environmental test chambers; shrink fit, toughening and hardening of metals; pharmaceutical processing. Pressure 459 lb/sq in gauge at 70°F (21.1°C).

"Freon-13B1" or "F-13B1". Monobromotrifluoromethane $CBrF_3$; b.p. −72°F (−57.8°C) at 760 mm. Used as a refrigerant in uses similar to those for "F-13"; also for shrink fit, toughening and hardening of metals; pharmaceutical processing, and as a fire extinguishing agent. Pressure 199 lb/sq in gauge at 70°F (21.1°C).

"Freon-14" or "F-14". Tetrafluoromethane CF_4; b.p. −198.4°F (−128.0°C) at 760 mm; critical temperature −49.9°F (−45.5°C); critical pressure 542 lb/sq in absolute. Used as a refrigerant in extremely low temperature cascade systems for environmental testing, metal conditioning; pharmaceutical processing; freezing and storage of biological products and other cryogenic applications; direct coolant for infrared detector cells; an inert propellant gas to operate satellite guidance and stabilization rockets.

"Freon-22" or "F-22". Monochlorodifluoromethane $CHClF_2$. Used as a refrigerant in industrial and commercial low temperature refrigerating systems. Solutions of "Freon-22" with "Freon-11" and "Freon-12" are used as propellants for aerosol products requiring special solvent and pressure qualities; as a chemical intermediate for plastics manufacture. Used as a low temperature solvent in investment casting process for preparation of ceramic molds. Pressure approximately 122 lb/sq in gauge at 70°F (21.1°C).

"Freon-113" or "F-113". Trichlorotrifluoroethane CCl_2FCClF_2. Used as a refrigerant in industrial and commercial air conditioning systems; in process water and brine cooling to 0°F (−17.8°C). "Freon-113" is used with other "Freon" compounds as a propellant for aerosol products; as blowing agent for rigid and resilient plastic foams; as a chemical intermediate for plastics manufacture.

"Freon-114" or "F-114". Dichlorotetrafluoroethane $CClF_2CClF_2$. Used as a refrigerant in fractional horsepower household refrigerating systems and drinking water coolers employing rotary vane type compressors; also, in indirect industrial and commercial air conditioning systems and in process water and brine cooling to −70°F (−56.7°C); as high stability heat transfer medium. Solutions of "Freon-114" and "Freon-12" are used as propellants for aerosol products where active ingredients require extreme stability - cosmetics, for example; also used as blowing agent for rigid and resilient plastic foams. Provide pressure from 12 to 69 lb/sq in gauge at 70°F (21.1°C).

"Freon" C-318 propellant and dielectric gas. Octafluorocyclobutane cyclic $(CF_2)_4$. B.p. 21.1°F (−6.0°C) at 760 mm. Use: Food grade as propellant for aerosol food products. Tasteless and odorless. Pressure 25 psig at 70°F (21.1°C). Mixture with nitrous oxide as propellant for whipped cream dessert topping, salad dressing and other pressurized food products. As propellant for flavoring extracts, spices and other food additives. As propellant for inhalation pharmaceutical products. Technical grade as dielectric coolant and heat transfer agent for transformers, cables, waveguides, and electronic devices. Dielectric strength 100 KV per inch at 760 mm. Service temperature at least 250°C.

"Freon" BF Solvent. Tetrachlorodifluoroethane CCl_2FCCl_2F. Used as a solvent. Also, in mixtures with "Freon" MF and "Freon" TF to provide specific boiling points and properties. As a flash point retarder for hydrocarbons and some higher boiling solvents.

"Freon" MF Solvent. Trichloromonofluoromethane CCl_3F. Used as a cleaner for hermetic motor compressors, oxygen storage tanks, etc. In blends as a cleaner for motion picture and television film, lithographic plates, typewriters, etc. As a flash point retarder for flammable solvents. Also used as an inert, noncorrosive brine and low temperature heat transfer fluid such as in the cold treatment of brazed stainless steels.

"Freon" TF Solvent. Trichlorotrifluoroethane CCl_2FCClF_2. Used as a solvent; for a cold immersion cleaning, as a vapor degreasing solvent, for ultrasonic or combination ultrasonic-vapor degreasing cleaning. Used as a cleaner for precision equipment such as gyroscopes, for mechanical and electrical controls, precision instruments and gauges, motors and generators, electrical and electronic equipment and assemblies, motion picture film, television and magnetic tapes, printed circuits, and as a flash point retarder in blends. Also used in extraction processes.

"Frianite." [118] A processed anhydrous alkali aluminum silicate with a pH of 5.4 to 6.5. A diluent used in compounding dusting insecticides and wettable concentrates.

friar's cowl. See aconite.

Friedel-Crafts reaction. Reaction originally defined as the condensation of alkyl or aryl halides with benzene and its homologs in the presence of anhydrous aluminum chloride. This definition has been widened to include analogous processes.

"Frigol" White Mineral Oil. [338] Brand name for a proprietary product. Available in five grades:

	Viscosity	Temp.
"Frigol" 350	340/50	100°F
"Frigol" 250	250/60	100°F
"Frigol" 150	150/60	100°F
"Frigol" 100	95/105	100°F
"Frigol" 185	180/190	100°F

These oils are made from special, low cold-test stocks.

Containers: 1- and 5-gal cans; 15-, 30-, and 55-gal drums; tank cars.

Uses: In the manufacture and servicing of refrigerators and air-conditioning machines.

"Frillon." [233] Trademark for a 2, 4, 5-T product.

frit. A term used in the ceramic industry and applied to a semi-fused mass, the constituents of which originally were soluble or insoluble, fusible or infusible. By "fritting" (i.e., preliminary fusing) the original properties of the constituents are changed; thus, the soluble materials become insoluble and the infusible materials fusible. Accordingly, the substances which could not otherwise be used in a glaze batch, but which are absolutely necessary for the best results, can be used. Other advantages offered by fritting are minimizing danger to health of workers when using lead salts, inducing better suspension of heavy products in the batch, more even distribution of constituents of the batch; ability to fire the glazed ware at a lower temperature. Most of the glazes used on dinnerware and sanitary ware contain frit. See also glazes.

froth flotation. See flotation.

"FR-S." [278] Trademark for general purpose synthetic rubbers and latexes, composed of copolymers of butadiene and styrene. "Hot" Rubbers, "Cold" Rubbers, Oil Extended Rubbers and Latexes are each available in several numbered grades.

Containers: Rubber: 75-lb polyethylene-wrapped bales in paper bags; disposable corrugated skid boxes of 30 bales (approx. 1 long ton). Latex: 5-gal drums to tank cars.

Uses: Ruber: Tires, hose, belting and packing; molded and extruded automotive and industrial products; soles and heels; hard rubber. Latex: Adhesives; formed rubber; textile and rug backing paper coating and impregnation; modification of plastics to produce high impact strength; asphalt additive.

Hazards: Avoid freezing latex.

Shipping regulations: None.*

fructose (fruit sugar; D(-)-fructose; levulose) $C_6H_{12}O_6$. A sugar occurring naturally in a large number of fruits and in honey. It is the sweetest of the common sugars.

Properties: Yellowish, white crystals; soluble in water, alcohol and ether; m.p. 103-105°C (dec).

Derivation: By the hydrolysis of inulin; hydrolysis of beet sugar followed by lime separation.

Grades: Technical; N.F. XI.

Containers: Wooden barrels; tins; fibers drums.

Uses: Foodstuffs; medicine; preservative.

fructose-1, 6-diphosphate (FDP; fructosediphosphoric acid) $H_2PO_4(C_6H_{10}O_4)H_2PO_4$. Can be prepared from fructose and certain other sugars by the use of yeasts. It is known to take part in cell metabolism. It is usually handled in the form of its barium or calcium salts, $C_6H_{10}O_{12}P_2Ba_2$ and $C_6H_{10}O_{12}P_2Ca_2 \cdot H_2O$. These are white amorphous powders, soluble in ice water and dilute acid solutions; insoluble in hot water and alcohol.

Uses: Organic synthesis; experimental work in cell metabolism.

fructosediphosphoric acid. See fructose-1, 6-diphosphate.

fructosin. See levulosin.

"Fruit-freeze." [123] Trade name for ascorbic acid for home freezing use.

"Fruit Fresh." [123] Trademark for an ascorbic acid preparation for preserving color and flavor of fruits.

fruit sugar. See fructose.

FSH. Abbreviation for follicle-stimulating hormone.

FT black. Abbreviation for fine thermal black. See thermal black.

"Fuadin." [162] Trademark for stibophen.

fuchsin (basic fuchsin; magenta). A synthetic rosaniline dyestuff, a mixture of rosaniline and pararosaniline hydrochlorides.
Properties: Dark green powder or greenish crystals with a bronze luster; faint odor. Soluble in water and alcohol.
Grades: N. F. XI.
Uses: Textile and leather industries; as a red dye; pharmaceutical.
Shipping regulations: None. *

fuel cells.
1. Rubberized tanks for the storage of liquid fuels.
2. Electrochemical devices for the continuous production of electricity by conversion of the chemical energy of continually supplied fuel and oxidant. Like primary cells and storage batteries, fuel cells supply low-voltage direct-current energy; unlike these more conventional sources, fuel cells are converters rather than storage devices and produce electrical energy as long as they are supplied with fuel and oxidant.

As industrial sources of power fuel cells give promise of low cost electricity through the highly efficient utilization of cheap, readily available fuels. For special applications they have the advantages of storage batteries without the necessity for recharging.

The principal types of fuel cells under development are:

a. Hydrox, utilizing hydrogen fuel and air or oxygen. The reaction product, water, must be continuously removed from the cell, usually by evaporation or condensation. This is the most highly developed type of cell. One form uses aqueous potassium hydroxide electrolyte while another version employs an ion exchange resin as electrolyte. Still another uses natural gas (methane) as the source of hydrogen and molten alkali carbonates as electrolyte.

b. Carbox, utilizing carbonaceous fuel and oxygen or air. In one example the electrolyte consists of fused carbonate salts at 500-800°C and the electrode reactions involve the interconversion of carbon dioxide and carbonate ions. Carbon dioxide together with oxygen (or air) is supplied to the cathode (positive plate) and the fuel, which may be hydrogen, carbon monoxide, or gaseous hydrocarbons, supplied to the anode (negative plate).

c. Redox, in which the electrode reaction involves relatively expensive reactants, which are regenerated externally and reused. In one version, stannous salts are oxidized by bromine in the cell to yield electrical power while the resulting stannic compounds are reduced outside the cell with coal or other carbonaceous fuel. The bromine is regenerated by air oxidation of hydrogen bromide from the cell.

d. Consumable electrode. For special applications a reactive metal, such as sodium amalgam, may be supplied to a cell as fuel to provide a controlled source of energy.

fuel element. See nuclear fuel.

fuel oil. Any liquid or liquefiable petroleum product used for generation of heat or power, exclusive of oils with a flash point below 100°F and oils burned in cotton or woolwick burners (such as kerosine). No. 1 and 2 fuel oils are liquid. The former is used in vaporizing or pot type burners and has a boiling range of about 400 to 625°F. No. 2 is less volatile and is used in domestic heaters not requiring No. 1. No. 4 oil is liquid at room temperature but is very viscous. It is usually used in industrial furnaces with no preheating facilities. No. 5 and 6 fuel oils are solids which must be liquefied by preheating before burning. They are sometimes referred to as bunker fuels, and are used in ships, locomotives, and industrial power plants.

fuel rod. See nuclear fuel.

fuller's earth. A variety of clay-like material which has high natural adsorptive powers. It is usually composed largely of the clay mineral attapulgite with some montmorillonite. Some varieties of bentonite have similar properties after activation. The natural material is treated by extrusion, drying and milling before use.
Occurrence: Florida, Georgia, Texas, Missouri, Illinois, Kentucky, Tennessee, Mississippi, Alabama, California, Utah.
Containers: Bulk; burlap bags; paper bags.
Uses: Decolorizing of oils and other liquids; oil-well drilling muds; insecticides; floor-sweeping compounds; cosmetics; rubber filler; catalyst; carrier for catalysts; filtering medium; pigments.
Shipping regulations: None. *

fuller's herb. See saponaria.

fulling agents. Soap solutions used in "fulling" wood, an operation by which wool fibers are interlocked to a dense, felty condition by means of friction, heat, and moisture.

fulling assistants. Sulfonated oils, wetting agents, sulfonated fatty-alcohol salts, etc., used to increase efficiency of fulling agents.

fulminates. Salts of fulminic acid (isocyanic acid), HN:C:O. The fulminates are extremely explosive, and are used as detonators.

fumagillin $C_{27}H_{36}O_7$. An antibiotic substance produced by Aspergillus fumigatus.
Properties: Light yellow crystals from dilute methanol; m. p. 189-194°C. Insoluble in water, dilute acids, saturated hydrocarbons. Soluble in most other organic solvents.
Grade: N. N. D.
Use: Medicine.

fumaric acid (boletic acid; lichenic acid; allomaleic acid) HOOCCH:CHCOOH. The trans-isomer of maleic acid.
Properties: Colorless, odorless crystals; stable in air; sp. gr. 1.625; sublimes 200°C; m. p. 287°C (sealed tube); solubility in water (25°C) 0.63 g/100 g; solubility in alcohol (30°C) 5.76 g/100 g; insoluble in chloroform and benzene.
Derivation: Fermentation of molasses; isomerization of maleic acid; catalytic oxidation of benzene.
Method of purification: Crystallization.
Grades: Technical; crystals.
Containers: 250-lb drums; bags.
Uses: (in approximate order of volume): Polyester resins; alkyd resins for paints, varnishes, molding powders, etc; rosin esters and adducts, for furniture lacquers and quick-setting printing inks; upgrading natural drying oils (especially tall oil) to improve drying characteristics; in foods, to replace citric and tartaric acids (in very small amounts); mordant; organic synethesis; modified phenolics.
Shipping regulations: None.*

fumaryl chloride ClCOCH:CHCOCl.
Properties: Clear, straw-colored liquid; b. p. 158-160°C (760 mm), 62-64°C (13 mm); sp. gr. (20°C) 1.408.
Containers: 150-lb, 13-gal carboys.
Uses: Chemical intermediate in the preparation of pharmaceuticals, dyestuffs, and insecticides.
Shipping regulations: Corrosive liquid. White label.*

"Fumetrol." [288] Trademark for chemical additives to electroplating baths.
Types: 101 and 102 - fume and spray suppressants for use in chromium plating solutions. 201 and 205 - permanently stable, highly effective, fluorinated wetting agents for all chromium plating baths.

fuming liquid arsenic. See arsenic trichloride.

fundamental particle (elementary particle). Any one of what are thought to be the basic building blocks in the structure of matter. At any stage in the development of physical theory, a particle is called fundamental if it cannot be convincingly interpreted as composed of simpler units. In an earlier era it was thought that the atoms themselves were of this character. Subsequent work has shown that they are in fact assembled from more fundamental particles, electrons and nucleons, and a variety of other fundamental particles of extremely transitory existence have since been found from experimental work on high energy nuclear reactions, and especially from cosmic ray investigations. The following table summarizes the particles presently known:

Kind of Particle	Number of Given Kind	Mass in Electron Mass Units	Charge	Class
photon	1	0	0	photon
neutrino	2	0	0	lepton
electron	2	1	−	lepton
muon (mu meson)	2	206	−	lepton
pion (pi meson)	3	263-273	−, 0	meson
kaon (k meson)	6	966	+, 0	meson
proton	2	1836	+	nucleon
neutron	2	1838	0	nucleon
lambda particle	2	2182	0	hyperon
sigma particle	6	2328-2341	+, 0, −	hyperon
xi particle (cascade particle)	4	2570	0, −	hyperon

Each particle has its antiparticle, and each may have more than one charge type. In some cases the particle is identical with its own antiparticle. Thus the pi zero is its own antiparticle, for pi minus there is an anti pi plus, and for the sigma particles, which occur with charges plus, zero, and minus, there are anti particles with charges minus, zero, and plus respectively. According to this method of classifying, there are thirty-two separately identifiable fundamental particles in the present state of physical theory.

"Fungchex." [81] Trademark for a mixture of mercuric chloride and mercurous chloride. Used as a turf fungicide.

fungicides. Chemicals which are used as a means of control of fungus growth. Two types of fungicides are generally recognized: those which protect against the growth of fungus and those which are designed to eradicate the fungus already present. Examples of the former are copper compounds (for fruit and vegetables), organic mercurial compounds (for seed treatments), phenolic compounds, and metal organic compounds (as wood preservatives). Examples of the eradicant fungicides are lime sulfur, organic mercurials, formaldehyde, dinitro compounds, certain antibiotics, and the quaternary ammonium derivatives. The term fungistat is sometimes used to refer specifically to those fungicides that inhibit fungus growth but do not kill or destroy the fungus.

fungicidin. See nystatin.

fungistats. Substances that stop or inhibit fungal growth but do not actually kill the fungi.

"Fungitrol." [74] Trademark for a series of industrial fungicides. Available as:
"Fungitrol" 25: solubilized copper cupferron containing 3.6% copper.
"Fungitrol" 50: contains 5% zinc.
"Fungitrol" 100 W.D.: water-dispersible quaternary ammonium naphthenate.
"Fungitrol" 617: coconut amine salt of tetrachlorophenol.

"Fungizone." [412] Trademark for amphotericin B (q. v.).

"Fungizymes." [78] Trademark for a blended mixture of enzymes, deliming and activating salts used in the dehairing and bating of all types of hides.

furacrylic acid. See furylacrylic acid.

"Furafil." [224] Trademark for lignocellulose produced by the pressure digestion of the acidified residue remaining after extraction of furfural from agricultural raw materials.
Properties: Dark brown free-flowing powder; burnt sugar odor; absorbs its own weight of water or oil; bulk density 30-35 lbs/cu ft; pH (water extract) 2-3.
Grades: "Furafil C," "M," "100," and "Fur-Ag" (q. v.).
Containers: 65- and 100-lb bags; bulk (40 ton min).
Uses: Additive for bulk, absorbency or conditioning action; extender for phenolic glues for plywood; foundry facings; oil well drilling muds; fertilizer.

"Fur-Ag." [224] Trademark for a sterilized grade of "Furafil" lignocellulose. Used as an anti-caking agent and organic conditioner in mixed fertilizers.

furaltadone (furmethonol). 5-Morpholino-3-(5-nitrofurfurylideneamino)-2-oxazolidinone. M. p. 206°C (dec). Used in medicine.

furamide. See furoamide.

furan (furfuran, tetrol) HC:CHCH:CHO. A hetrocyclic ring compound.
Properties: Colorless liquid turning brown on standing. This color change is retarded if a small amount of water is added to the furan. This material is somewhat toxic and in a room its fumes should not be allowed to reach a very high concentration; sp. gr. 0.9444 (15°C); b. p. 32°C (758 mm); refractive index 1.4216 (n 20/D); insoluble in water; soluble in alcohol and ether.
Derivation: (a) Dry distillation of furoic acid from furfural; (b) especially as an intermediate in the production of adiponitrile.
Grades: Refined.
Containers: 7-lb (1-gal) containers; 35- and 375-lb drums; 32,000-lb tank cars.
Uses: Organic synthesis.
Shipping regulations: None. *

furancarboxylic acid. See furoic acid.

2,5-furandione. See maleic anhydride.

furan resins. The term refers to monopolymers of furfuryl alcohol, and also to resins obtained by condensation of phenol with furfural or furfuryl alcohol, and to furfural-ketone polymers. The resins are always of a dark color.

The furfuryl alcohol polymers are usually low in viscisity and are used as penetrants into wood, sand, gypsum, and chemical stoneware. After curing by means of heat and acid catalysts, these resins have superior resistance to acids, alkalies and solvents. This type of furan resin is also used as a plasticizer for vinyl resins.

The phenol-furfural resins are usually used to modify the properties of phenol-formaldehyde resins in the direction of improved flow properties or increased resistance to chemical deterioration. The resins also have good adhesive properties and are used as binders in grinding wheels and foundry cores, and as bonding mortars for joining other materials. When used with asbestos or fiberglass filler they serve for molded products useful for their chemical resistance. Furan resins are also used as modifiers of other resins such as epoxies to obtain coatings, castings, laminates, cements, and sealants that have desired penetration, adhesion, or inert character.

furazolidone $C_8H_7N_3O_5$ (N-(5-nitro-2-furfurylidene-3-amino-2-oxazolidone).
Properties: Yellow powder; odorless; m. p. 255°C. Slightly soluble in polyethylene glycol; insoluble in water, alcohol, and peanut oil.
Derivation: Synthetically from furfural, hydroxyethylhydrazine, and diethyl carbonate.
Grade: N. F. XI.
Use: Medicine.

furethrin (2-furfuryl-4-hydroxy-3-methyl-2-cyclopenten-1-one ester of 2,2-dimethyl-3(2-methyl propanyl)cyclopropane carboxylic acid). A synthetic analog of allethrin substituting the 2-furfuryl for allyl in the side chain. Used as insecticide in manner similar to allethrin. Accepted as generic name by Ent. Soc. See also allethrin, barthrin, ethythrin, cyclethrin.

furfural (ant oil, artificial; pyromucic aldehyde; furfuraldehyde; has been called furfurol) C_4H_3OCHO or OCH:CHCH:CCHO.
Properties: Colorless mobile liquid when very pure; changes to reddish brown upon exposure to light and air. Penetrating odor somewhat similar to benzaldehyde. Furfural forms condensation products with many types of compounds, phenol, amines, urea, etc. Soluble in alcohol, ether, and benzene; 8.3% soluble in water at 20°C.
Constants: Sp. gr. 1.1598 (20/4°C); m. p. -36.5°C; b. p. 161.7°C (760 mm); heat of vaporization 107.5 cal; refractive index 1.5260 (n 20/D); flash point (Tag open cup) 150-160°F.
Derivation: From oat hulls, rice hulls,

corn cobs by steam-acid digestion.
Grades: Technical; refined.
Containers: 1-, 5-, 10-gal cans; 55-gal drums; tank cars.
Uses: Solvent refining of lubricating oils, butadiene, rosin and other organic materials; solvent for nitrocellulose, cellulose acetate, shoe dyes; wetting agent; preparation of synthetic resins; weed killer; fungicide; furfural derivatives; adipic acid and adiponitrile; bituminous or concrete road construction; production of lysine; refining of rare earths and metals.
Fire hazard: Slight. *
Shipping regulations: None. *

furfural acetic acid. See furylacrylic acid.

furfural acetone $(C_4H_3O)CH{:}CHCOCH_3$.
Properties: Light-yellow crystals turning dark reddish-brown on standing. Insoluble in alcohol, ether, and chloroform. M. p. 37-39°C; b. p. 112-115°C (10 mm).
Derivation: Condensation of furfural with acetone in alkali solution (some difurfural acetone is also formed).
Containers: Bottles.
Grades: Technical.
Shipping regulations: None. *

furfural acetophenone $(C_4H_3O)CH{:}CHCOC_6H_5$
Properties: Oil; forms polymorphous crystals. Sp. gr. 1. 1140 (20°C); b. p. 317°C; 187°C (11 mm).
Derivation: Condensation of furfural with acetophenone in alkali solution.
Grades: Refined.
Containers: Bottles.

furfuraldehyde. See furfural.

furfural diacetate $C_4H_3OCH(OOCCH_3)_2$.
Properties: Colorless crystals; m. p. 52-53°C; b. p. 220°C. Insoluble in water; soluble in ether and alcohol.
Derivation: From furfural and acetic anhydride in the presence of stannous chloride.
Grades: Technical.
Containers: Bottles.
Shipping regulations: None. *

furfuraldoxime (furfural oxime) $C_4H_3OCHNOH$.
Alpha form (sym-form)
Properties: Soluble in water, alcohol, ether, and benzene. M. p. 90-91°C; b. p. 201-208°C with decomposition.
Derivation: From furfural and hydroxylamine.
Beta form (anti-form)
Properties: Needles (from ligroin). M. p. 74-75°C. Difficulty soluble in cold water; soluble in alcohol, ether, carbon disulfide, benzene, and glacial acetic acid.
Derivation: From furfural and hydroxylamine.
Grades: Technical (probably mixture).
Containers: Bottles.
Shipping regulations: None. *

furfural oxime. See furfuraldoxime.

furfural phenylhydrazone $C_4H_3OCHNNHC_6H_5$.
Properties: Leaflets; insoluble in water; soluble in alcohol and ether. M. p. 97-98°C.
Derivation: Furfural and phenylhydrazine.
Shipping regulations: None. *

furfuramide. See hydrofuramide.

furfuran. See furan.

furfurin $C_{15}H_{12}N_2O_3$. An isomer of hydrofuramide.
Properties: Brown needles; insoluble in water; soluble in alcohol and ether. M. p. 116°C.
Derivation: Action of dilute alkali on hydrofuramide.
Shipping regulations: None. *

furfurol. A misnomer for furfural.

furfuryl acetate $C_4H_3OCH_2OOCCH_3$.
Properties: Colorless liquid turning brown upon exposure to light and air; pungent odor. Insoluble in water; soluble in alcohol and ether. Sp. gr. 1. 1175 (20/4°C); b. p. 175-177°C; refractive index 1. 4627 (D).
Derivation: By treatment of furfuryl alcohol with acetic anhydride.
Containers: Bottles.
Grades: Refined.
Use: Flavor.
Shipping regulations: None. *

furfuryl alcohol (furyl carbinol) $C_4H_3OCH_2OH$ or $\underline{OCH{:}CHCH{:}C}CH_2OH$.
Properties: Colorless, mobile liquid becoming brown to dark-red upon exposure to light and air. Poisonous! Reacts with explosive violence with mineral acids (even when dilute) and some strong organic acids to form a black, brittle, insoluble, infusible resin. Soluble in alcohol, ether, chloroform, benzene. When freshly prepared it is soluble in water in all proportions, but upon standing for some months or upon exposure to slightly acid conditions it becomes more or less water-insoluble.
Constants: Sp. gr. 1. 1285 (20/4°C); b. p. 170°C (750 mm); refractive index 1. 4850 (25°C/D); flash point (open cup) 167°F.
Derivation: Catalytic hydrogenation of furfural.
Grades: Technical; refined.
Containers: 1-, 5-, 10-, 55-gal drums; tank cars.
Uses: Wetting agent; synthetic resins; penetrant; solvent for dyes and resins; rocket fuels.
Shipping regulations: None. *

alpha-**furfuryl amine** $C_4H_3OCH_2NH_2$.
Properties: Colorless liquid; soluble in water, alcohol, and ether. Sp. gr. 1. 0550 (17°C); b. p. 145°C (757 mm); refractive index 1. 4900 (17°C).
Derivation: (a) Reduction of furfuraldoxime or hydrofuramide; (b) furfural and ammonia.
Shipping regulations: None. *

alpha-**furfuryl mercaptan.** An essential aromatic constituent of roasted coffee beans used as a basic ingredient for synthetic coffee compositions and fortifier for natural coffee blends and flavor adjunct.

furil $C_4H_3OCOCOC_4H_3O$.
Properties: Yellow needles; practically insoluble in water; difficulty soluble in cold alcohol and ether; soluble in methyl alcohol and chloroform. M. p. 165-166°C.
Derivation: Atmospheric oxidation of furoin.
Grades: Refined.
Containers: Bottles.
Shipping regulations: None. *

furmethonol. See furaltadone.

furnace black. A carbon black made by burning natural gas or vaporized aromatic hydrocarbon oil in a closed furnace with about 50% of the air required for complete combustion. The combustion products are cooled by a water spray and the finely divided carbon is separated from the gases. Furnace black produced from oil can be made in a wide range of closely controlled particle sizes and is particularly suitable for use as a filler and reinforcing agent for synthetic rubber.
Grades: Conducting furnace black (CF); fine (FF); high modulus (HMF); high elongation (HEF); reinforcing (RF); semi-reinforcing (SRF); high abrasion (HAF); super abrasion (SAF); fast extruding (FEF).

furnace oil. Usually means No. 1 fuel oil. See fuel oil.

furnace sand (fire sand). Sand used to line furnace bottoms or walls, particularly in open hearth steel furnaces.

"Furnane" (Red) and (Black). [41] Trade names for two synthetic resin cements of the vinyl and furan type respectively used in combinations with acid tile for food plant floor construction.

"Furnex." [133] Trademark for semi-reinforcing furnace carbon black (SRF). Used in rubber for compounds needing only medium reinforcement but high resilience, low heat buildup. Provides easy mixing, good calendering and extrusion properties. Available in several types:
Regular "Furnex" for tire carcasses, wire and cable, footwear, mechanical molded and extruded goods, V-belts, industrial tires, hose and tubing, mats.
"Furnex" H for same application but where slightly higher extrusion and better processing are desirable at same cost.
"Furnex" NS (non-staining) where such properties are desirable as in tire carcasses, refrigerator gaskets, auto window channel and similar goods. Available in 25- and 50-lb bags and hopper cars.

furnish. Term used by paper makers to refer to mixtures containing the constituents of paper as supplied to the papermaking machine.

furoamide (pyromucamide; furamide) $C_4H_3OCONH_2$.
Properties: Crystals; sublimes partly at 100°C. M. p. 142°C.
Derivation: Treatment of furoyl chloride with ammonia.
Grades: Refined.
Containers: Bottles.
Shipping regulations: None. *

furoic acid (pyromucic acid; furane carboxylic acid) C_4H_3OCOOH or OCH:CHCH:CCOOH (ring closed between O and C).
Properties: Colorless crystals; m. p. 133-134°C; sublimes at 130°C (50-60 mm). Slightly soluble in cold water; very soluble in hot water, alcohol and ether; insoluble in paraffin hydrocarbons.
Derivation: Cannizzaro reaction from furfural; oxidation of furfural.
Purification: Sublimation; fractional crystallization from hot water.
Grades: Technical.
Containers: 5-, 10-, 25-, 50-lb packages; 100-lb fiber drums.
Uses: Preservative; bactericide; furoates for perfume and flavoring; fumigant; textile processing.
Shipping regulations: None. *

furoin $C_4H_3OCH(OH)COC_4H_3O$.
Properties: Light-brown needles; insoluble in water; soluble in methyl alcohol and ethyl alcohol. M. p. 138-139°C (corr.).
Derivation: Action of potassium cyanide on furfural.
Grades: Refined.
Containers: Bottles.
Shipping regulations: None. *

furoyl chloride C_4H_3OCOCl.
Properties: Colorless liquid; powerful lachrymator. Must be handled with extreme care. Soluble in ether; decomposes in water. M.p. −2°C; b.p. 176°C, 66°C (10 mm).
Derivation: Treatment of furoic acid with phosphorus pentachloride.
Grades: Refined.
Containers: Bottles.
Use: Substitute for chloropicrin in disinfecting grain elevators.
Shipping regulations: None. *

furylacrylic acid (furfural acetic acid; furacrylic acid) $C_4H_3OCH{:}CHCOOH$.
Properties: White powder; slightly soluble in cold water; easily soluble in hot water; soluble in alcohol, ether, and glacial acetic acid. M. p. 141°C; b. p. 117°C (8 mm), 286°C (760 mm).
Derivation: From furfural by Perkins condensation.
Grades: Technical.
Containers: Bottles.
Uses: Derivatives used in perfumes.
Shipping regulations: None. *

furyl carbinol. See furfuryl alcohol.

fusain (mother of coal; mineral charcoal). One of the types of physical structure found in coal (see also clarain, durain, and vitrain). Fusain is a dull, brittle material resembling charcoal. It reduces the caking properties of the coal in which it occurs.

"Fused Salt B." [337] Trade name for potassium zirconium chloride.

fused salts. Salts (i. e., ionic compounds) in the molten state. High temperatures are usually involved in maintaining the molten state. Sodium chloride is the principal ingredient in many fused salts.

Uses: Production of alkali and other metals by electrolysis: aluminum, sodium, magnesium, titanium, zirconium, niobium, tantalum; as a base for circulating liquid fuels in nuclear reactors; fluxing and descaling metals; heat transfer agents.

fusel oil (amyl alcohol, fermentation; grain oil; potato spirit; potato oil). A volatile, poisonous, oily mixture consisting largely of amyl alcohols. Isoamyl alcohol (isobutyl carbinol) and active amyl alcohol (2-methyl-1-butanol) are chief constituents. Ethyl, propyl, butyl, hexyl and heptyl alcohols as well as other alcohols have been separated. Acids, esters, and aldehydes are also present. Normal primary amyl alcohol (1-pentanol) is not found in fusel oil. An appreciable percentage of fusel oil in a liquor has an adverse influence on its taste and physiological effects, but minute amounts are necessary for the characteristic flavor.

Properties: Clear, colorless liquid; disagreeable odor.

Crude: (Typical specifications) Sp. gr. 0.8315 (20/20°C); equivalent wt/gal 6.92 lbs (20°C); distillation range, distillate up to 122°C, 34%, between 122 and 138°C 61%, above 138°C 3.4%; 1.68% ethyl alcohol by volume.

Refined: (Typical specifications) Color water-white; distillation range below 110°C none; below 120°C not more than 15%; below 130°C not less than 60%; above 135°C none; sp. gr. 0.811-0.815 (20/20°C); wt/gal 6.76 lbs (approx) (20°C). Flash point 123°F (open cup); soluble in all proportions 60° gasoline without turbidity; soluble in water, alcohol, and ether.

Derivation: Obtained as a by-product in the alcoholic fermentation of starch- or sugar-containing materials such as potatoes, grapes, beetroots, grain, etc.

Method of purification: Rectification.

Grades: Crude; refined.

Containers: 1-, 5-, 10-gal cans; 5-, 10-, 53-, 104-gal drums; 6000-, 8000-gal tank cars.

Uses: Chemicals (amyl ether, amyl acetate, pure amyl alcohols, nitrous ether, various esters); identification of alkaloids in analytical chemistry; explosives (gelatinizing agent); solvent for fats and oils; artificial fruit syrups; ice cream and soda-water essences; intermediate; pharmaceuticals (preparation and purification of alkaloids, hypnotics, amyl nitrite); nitrocellulose plastics; synthetic rubber; varnishes; lacquers; solvent for resins and waxes; and perfumery.

Fire hazard: Combustible but not flammable; flash point over 80°F.

Shipping regulations: None. *

fusible alloys (low-melting alloys; fusible metals). The term generally means alloys melting below 450°F (233°C). These are usually the binary, ternary, quaternary, and quinary mixtures of bismuth, lead, tin, cadmium, indium, and less frequently other metals. Eutectic alloys are relatively few in number and are the particular compositions that have definite and minimum melting points as compared with other compositions of the same metals. The more important eutectic alloys are listed in Table I.

Table I
Eutectic Alloys
Percentage Composition

Melting Temp. °C	Bi	Pb	Sn	Cd	Other
248.0	-	82	-	18	-
221.0	-	-	96	-	Ag 4
199.0	-	-	91	-	Zn 9
183.0	-	38	62	-	-
144.0	60	-	-	40	-
143.0	-	31	51	18	-
138.56[1]	58	-	42	-	-
138.5	57	-	43	-	-
130.0	56	-	40	-	Zn 4
124.3[2]	55.5	44.5	-	-	-
102.5	54	-	26	20	-
95.0	52	37	16	-	-
91.5	52	40	-	8	-
78.8	57	17	-	26	-
70.0[3]	50	26.7	13.3	10	-
70.0[4]	50	27	13	10	-
58.0[5]	49	18	12	-	In 21
46.89[6]	44.7	22.6	8.3	5.3	In 19.1

[1] Cerrotru; [2] Cerrobase; [3] Cerrobend; [4] Lipowitz's metal; [5] Cerrolow-136; [6] Cerrolow-117.

There are hundreds of non-eutectic fusible alloys. They become liquid over a range of temperatures. Typical non-eutectic alloys are listed in Table II.

Table II
Typical Non-Eutectic Alloys

Yield Temp. °C	Melting Range °C	Percentage Bi	Pb	Sn	Other
159	145-176	12.6	47.5	39.9	-
154	143-163	14	43	43	-
145	130-173	20	50	30	-
142	120-152	21	42	37	-
135	132-129	5	32	45	18 Cd
127	124-130	56	2	40.9	0.7 Cd 0.4 In
116[1]	103-227	48	28.5	14.5	9.0 Sb
111	95-143	33.3	33.4	33.3	-
100	95-114	59.4	14.8	25.8	-
96	95-104	56	22	22	-
89	83-92	52	31.7	15.3	1.0 Cd
72.5[2]	70-90	42.5	37.7	11.3	8.5 Cd
64.0[3]	61-65	48	25.6	12.8	9.6 Cd 4.0 In

[1] Cerromatrix; [2] Cerrosafe; [3] Cerrolow-147.

The fusible alloys may have various habits of expansion, some behaving typically on solidification, i. e., contracting, others expanding, and others exhibiting considerable "growth" only after solidification, which may continue for 500 to

*See "I.C.C. Shipping Regulations," page xiii.
Reference numbers refer to name of manufacturer. See "List of Manufacturers," page v.

1000 hours. Total growth may be as much as 0.008 inches/inch. All fusible alloys creep under relatively light continuous loads. They are characterized by their "let go" or yield temperature as determined by a standard test procedure. Hardness of fusible alloys ranges from 5 to 22 Brinell, tensile strength between 3000 and 13,000 psi and elongation from 0 to 300%.

The best known fusible alloys are indicated in Table III.

Table III
Common Fusible Alloys

Name	Approx M.P. °C	Percentage Composition Bi	Pb	Sn	Cd
Cerrotru alloy	138.6	58	-	42	-
Cerrobase alloy	123.8	55.5	44.5	-	-
Rose's alloy	100	50	28	22	-
Newton's metal	95	50	31	19	-
D'Arcet metal	93	50	25	25	-
Wood's alloy	71	50	24	14	12
Wood's metal	71	50	25	12.5	12.5
Lipowitz's metal	70	50	27	13	10
Cerrobend alloy	70	50	26.7	13.3	10

The chief uses are in fusible automatic sprinkler links, in fusible vents in compressed gas tanks, for dental castings, for anchoring punches in metal working tools, anchoring bearings and bushings, for heat transfer liquids, for high temperature liquid seals, for glass to metal seals, as fusible cores in forming operations including tube bending, and for spray coatings of patterns for protection during handling and storage.

fusible metals. See fusible alloys.

fusion. Ordinarily a synonym for melting, e.g., of a crystalline substance. Since melted substances tend to mix readily, the word has assumed the meaning of "melt and blend." The so-called fusion of protons to form helium as utilized in the hydrogen bomb indicates a union of two or more protons to form an element with great liberation of energy.

fusion reactor. See nuclear fusion.

FWWMR. Fire, water, weather, mildew resistant, as applied to fabrics or textiles. See flameproofing finishes.

"Fybrene." [45] Trademark for petrolatum, U.S.P., of medium melting point and medium consistency.
Properties: M.p. 115-125°F; Saybolt viscosity 65-90 (210°F); odorless.
Use: Paper industry.

"Fyrex." [172] Brand name for a substantially neutral ammonium phosphate; fine crystals; soluble in water.
Grades: "Fyrex," "Flexible Fyrex," with added softening agent; "Special Flexible Fyrex," containing both a softening agent and a penetrating agent.
Uses: For flame proofing textiles, wood, and fibers; in the manufacture of matches to prevent afterglow.

G

g. Abbreviation for gram.

"G-4." [12] Trademark for a brand of dichlorophene (q. v.).

"G-11." [12] Trademark for hexachlorophene U. S. P.

"G-942." [28] Trademark for specialty tanning product based on the partial sodium salt of a polymeric carboxylic acid.
Properties: Viscous straw-colored liquid approximately 25% solids; sp. gr. 0.97.
Containers: 55-gal resin-lined steel drums, 475 lbs net.
Uses: As a plumping and tanning agent, primarily for light weight skins. Suitable for most types of leather such as suede, drawn and smooth grain, garment, etc.

Ga. Symbol for gallium.

GA_3. Symbol used for gibberellic acid (q. v.).

G acid. 2-Naphthol-6, 8-disulfonic acid.

gadolinite $YFeBe_2(SiO_4)_2O_2$. A natural silicate of beryllium, iron, and the yttrium and rare earth metals. Black, greenish black, or brown; luster vitreous to greasy; hardness 6.5-7.0; sp. gr. 4.0-4.5; thermoluminescent.
Occurrence: Texas, Arizona, Colorado; Norway; Greenland.

gadolinium Gd. Element having atomic number 64; group III of the periodic table; one of the rare-earth elements of the yttrium subgroup.
Properties: Lustrous metal; sp. gr. 7.87; m. p. 1350°C; b. p. 2700°C (approx). Reacts slowly with water; soluble in dilute acid. Exhibits a high degree of magnetism, especially at low temperatures. Salts are colorless. Only one valence (+3).
Source: See rare-earth minerals.
Derivation: Reduction of the fluoride with calcium.
Grades: Ingots, lumps, turnings, all of high purity.
Uses: Crystalline compounds of gadolinium, especially gadolinium sulfate octahydrate, are used in magnetic method of obtaining extremely low temperatures. Gadolinium has a high thermal neutron capture cross-section; is used as an alloy in stainless steel for nuclear control. Also used as a scavenger for oxygen and nitrogen in titanium and its alloys.

gadolinium chloride $GdCl_3 \cdot xH_2O$. Colorless crystals, soluble in water. Purities up to 99.9% gadolinium salt.
Containers: Glass bottles, fiber drums.

gadolinium fluoride $GdF_3 \cdot 2H_2O$. Available up to 99.9% purity.
Containers: Glass bottles, fiber drums.

gadolinium nitrate $Gd(NO_3)_3 \cdot xH_2O$. Colorless crystals, soluble in water. Purities up to 99.9% gadolinium salt.
Containers: Glass bottles, fiber drums.
Shipping regulations: Oxidizing material. Yellow label. *

gadolinium oxalate $Gd_2(C_2O_4)_3 \cdot 10H_2O$. White powder, insoluble in water, slightly soluble in acids. Purities up to 99.9% gadolinium salt.
Containers: Glass bottles, fiber drums.

gadolinium oxide Gd_2O_3. White to cream-colored powder; sp. gr. 7.41; m. p. 2330°C; insoluble in water, soluble in acids to form the corresponding salts. Hygroscopic absorbs carbon dioxide from the air. Purities up to 99.8% gadolinium oxide.
Containers: Glass bottles, fiber drums.
Uses: Nuclear reactor control rods; neutron shields; catalysts; dielectric ceramics; filament coatings; special glasses; phosphor activator.

gadolinium sulfate $Gd_2(SO_4)_3 \cdot 8H_2O$. Colorless crystals, slightly soluble in hot water, more soluble in cold. Purities up to 99.9% gadolinium salt.
Containers: Glass bottles, fiber drums.
Uses: For obtaining extremely low temperatures by the magnetic method.

"GAF" Carbonyl Iron Powders. [307] Trademark for microscopic, almost perfect spheres of extremely pure iron. They are produced in eleven carefully controlled grades, ranging in particle size from 3 to 20 microns in diameter. The iron content of some types is as high as 99.6-99.9%.
Uses: In high frequency cores for radio, telephone, television, short wave transmitters, radar receivers, direction finders Also used as alloying agents, catalysts, in powder metallurgy and in magnetic fluids.

gaize cement. See pozzolana cement.

gal. Abbreviation for gallon.

galactin. See luteotropin.

galactose $C_6H_{12}O_6$. A monosaccharide commonly occurring in milk sugar or lactose.
Properties: White crystals; soluble in water and alcohol; slightly soluble in glycerol;

*See "I. C. C. Shipping Regulations," page xiii.
Reference numbers refer to name of manufacturer. See "List of Manufacturers," page v.

m. p. 165-168°C.
Derivation: By acid hydrolysis of lactose.
Grades: Technical.
Containers: Bottles; drums; boxes.
Uses: Organic synthesis; medicine.

D(+)-galacturonic acid $C_6H_{10}O_7$ or $COOH(CHOH)_4CHO$. A compound found as a major constituent of plant pectins (q. v.). It exhibits mutarotation, having both an alpha and a beta form.
Properties: The alpha form melts with decomposition at 156-159°C. Soluble in water; slightly soluble in hot alcohol; insoluble in ether.
Derivation: Hydrolysis of pectins.
Use: Biochemical research.

galena (galenite; lead glance) PbS. Natural lead sulfide.
Properties: Color lead gray; streak lead gray; luster metallic; good cubic cleavage; sp. gr. 7. 4-7. 6; hardness 2. 5. Soluble in strong nitric acid; also in excess of hot hydrochloric acid.
Occurrence: Idaho, Utah, Arizona, Missouri, Colorado, Montana, Oklahoma; Canada; South America; Africa.
Use: Chief ore of lead. Frequently recovered for the silver it sometimes contains.

galenite. See galena.

"Galex." [122, 235] Trade name for a stable non-oxidizing rosin consisting principally of dehydroabietic acid.
Properties: Light amber; acid number 157; saponification number 163; softening point (B & R Method) 66°C; sp. gr. (20/4°C) 1. 082; flash point (Cleveland open cup) 210°C; fire point (Cleveland open cup) 240°C; gasoline insolubles 0. 014%; benzene insolubles 0. 008%. Soluble in ordinary organic solvents, paraffin wax, beeswax, carnauba wax; compatible with GR-S, Neoprene, natural and reclaim rubber and many natural and synthetic resins.
Uses: Rubber-based pressure sensitive adhesives and tapes; water insoluble adhesives; extender for natural and synthetic resins; in the manufacture of soldering fluxes, metal salts, waterproofing compounds, greases and lubricants, soaps, ceramic printing vehicles, electrical insulation, rubber cement; intermediate for the making of chemicals.

gallamine triethiodide
$C_6H_3[OCH_2CH_2N(C_2H_5)_3I]_3$. Tri (diethylaminoethoxy) benzene triethyl iodide.
Properties: White, fluffy, hygroscopic powder; m. p. 150°C. Freely soluble in water; soluble in alcohol; very slightly soluble in chloroform; insoluble in ether.
Grade: N. N. D.
Use: Medicine.

gallic acid (3, 4, 5-trihydroxybenzoic acid) $C_6H_2(OH)_3CO_2H \cdot H_2O$.
Properties: Colorless or slightly yellow, crystalline needles or prisms. Soluble in alcohol and glycerol; sparingly soluble in water and ether.
Constants: Sp. gr. 1. 694; m. p. 222-240°C.
Derivation: By the action of mold on solutions of tannin or by boiling the latter with strong acid or caustic soda.
Grades: Technical.
Containers: Barrels.
Uses: Photography; writing ink; dyeing; manufacture of pyrogallol; tanning agent and manufacture of tannins; paper manufacture; pharmaceuticals; process engraving and lithography; analytical reagent.
Shipping regulations: None. *

gallium Ga. Element of atomic number 31, of group III of the periodic system.
Properties: Silvery-white metal; m. p. 29. 7°C; b. p. 1600°C; liquid may be undercooled to almost 0°C without solidifying; sp. gr. 5. 9 (25°C), more dense as a liquid than as a solid; soluble in acid, alkali and slightly soluble in mercury. Gallium reacts with most metals at high temperatures.
Occurrence: Traces are present in a variety of ores and minerals. It is prepared commercially from bauxite, containing approximately one ounce of gallium per ton. Also prepared commercially from zinc ores.
Derivation: Extraction of gallium as gallium chloride by ethyl ether or isopropyl ether and subsequent electrodeposition from a sodium gallate solution.
Uses: Gallium has been suggested for use as a backing material for optical mirrors and as a possible heat exchange medium in nuclear power reactors.

gallium antimonide. See gallium arsenide.

gallium arsenide GaAs.
Properties: Crystals; m. p. 1238°C. Used as a high-purity binary semiconductor. Is sometimes alloyed with indium arsenide in semiconductors. Gallium antimonide, GaSb, is used similarly.

gallium oxides. The sesquioxide, Ga_2O_3, and suboxide, Ga_2O, are known. Both are stable at ordinary temperatures.

gallium phosphide GaP. Pale orange, transparent crystals or whiskers up to 2 cm long; made by vapor phase reaction, at relatively low temperatures, between phosphorus and gallium suboxide. These crystals are intermediate between normal semiconductors and insulators or phosphors. They operate over a temperature range of −55 to 500°C.

gallium salts. Many salts are known, although they seem to have found little use. Gallium is usually trivalent, but sometimes divalent, as in gallium chloride, $GaCl_3$, and gallium dichloride, $GaCl_2$.

gallocyanine. $C_{15}H_{12}N_2O_5$. A dye made from gallic acid. Used as a biological stain.

gallotannic acid. See tannic acid.

galls (nutgalls: Aleppo galls; Mecca galls; Turkey galls). Excrescences on various kinds of oak trees resulting from the

*See "I. C. C. Shipping Regulations," page xiii.
Reference numbers refer to name of manufacturer. See "List of Manufacturers," page v.

deposition of insect eggs.
Grades: The best grades (55-60% tannic acid) come from Iran, Syria, Turkey, and Tripoli. The poorer grades come from Italy, France, Germany, and Austria.
Colors range from black through green to white, owing to different degrees of maturity, the darker being more mature and containing a greater percentage of tannin.
Containers: Wooden barrels; bags.
Use: Source of gallic and gallotannic acids; tanning industry; ink manufacture; medicine; textile printing; pharmaceuticals.
Shipping regulations: None. *

galvanized iron (hot-dip process). Iron coated with zinc by dipping the metal into a bath of molten zinc held at a temperature somewhat above the m. p. (810-875°F).

"Galvene." [206] Brand name of proprietary line of chemical restrainers used in acid pickling of iron, steel, and ferrous alloys. Recommended for use in hydrochloric and hydrochloric-nitric acid mixtures.

"Galvoline." [233] Brand name for a proprietary product consisting of a cored magnesium ribbon used as a continuous anode for the cathodic protection of buried pipe lines and other metal structures.

"Galvomag." [233] Trademark for magnesium alloy composition for use in anodes in cathodic protection.

"Galvopak." [233] Brand name for a proprietary product consisting of a magnesium anode packed with back-fill material. Used in the cathodic protection of buried pipe lines and other metal sturctures.

"Galvorod." [233] Trademark for a cored magnesium rod used as an anode in the cathodic protection of water-heater tanks.

gamboge (cambogia). A gum-resin from Garcinia hanburyi.
Habitat: East Indies.
Chief constituents: Cambogic acid, a resin and a gum.
Grades: Technical; N. F. XI.
Containers: Wooden barrels, fiber drums.
Uses: Medicine; paints.
Shipping regulations: None. *

gametocide. A substance which can control pollinization of plants by selectively killing plant sex cells (gametes). Some suggested gametocides are maleic hydrazide; sodium alpha, beta-dichloroisobutyrate.

gamma acid. See 7-amino-1-naphthol-3-sulfonic acid.

gamma compounds. See explanation under alpha compounds. For specific gamma compounds see under name of compound.

"Gammacorten." [305] Trademark for dexamethasone.
Use: Medicine.

gamma-globulin. A fraction of serum globulin which has been separated by electrophoresis and which contains most of the antibodies.
Use: Medicine.

gamma rays. Electromagnetic radiation similar to X-rays except that gamma rays originate in the nucleus of an atom whereas x-rays originate in the extra-nuclear structure. Gamma rays usually have higher energies and correspondingly shorter wavelengths than x-rays.

"Gammexane." [206] Trademark for benzene hexachloride.

"Gamtox." [253] Brand name for a pesticide formulation containing benzene hexachloride.

"Ganaseg." [412] Trademark for 4, 4'-diazoaminodibenzamidine with two molecules of aceturic acid.

gangue. The minerals and rock material mined with a metallic ore but valueless in themselves or used only as a by-product. They are separated from the ore in the milling and extraction processes. Common gangue materials are quartz, calcite, limonite, feldspar, pyrite, and rock of various kinds.

ganister. A highly refractory siliceous sedimentary rock used for the manufacture of refractory brick. Typical analysis: 98. 20% SiO_2, 0. 30% Fe_2O_3, 0. 90% Al_2O_3, 0. 15% CaO, 0. 10% MgO.
Occurrence: Pennsylvania, Virginia, Wisconsin, Ohio; Great Britain.

"Gantrisin." [190] Trademark for a brand of sulfisoxazole (q. v.).

garbage pitch. See stearin and fatty-acid pitches.

garden angelica. See angelica.

garden lavender. See lavender.

"Gardenol." [227] Trademark for methyl phenyl carbinyl acetate, $CH_3COOCHCH_3C_6H_5$, (styralyl acetate; sec-phenyl ethyl acetate).
Properties: Colorless liquid; powerful greenleaf scent, suggestive of gardenia; stable, not known to cause discoloration; sp. gr. (25/25°C) 1. 023-1. 026; refractive index (20°C) 1. 492-1. 496; soluble in 2 parts of 70% alcohol.
Uses: In gardenia scents, as well as sparingly in tuberose, jasmin, and other florals.

garden rosemary. See rosemary.

garden sage. See salvia.

gardjan balsam oil. See gurjun balsam oil.

gargan balsam oil. See gurjun balsam oil.

garget. See phytolacca.

garlic. See allium.

garlic oil.
Properties: Pale yellowish liquid; characteristic, exceedingly penetrating odor. Soluble in alcohol, ether, and carbon disulfide.
Chief known constituents: Allylpropyl disulfide and diallyl disulfide.
Constants: Sp. gr. 1. 053.

Derivation: Distilled from the bulb and herb of Allium sativum.
Method of purification: Rectification.
Grades: Technical.
Containers: Glass bottles.
Uses: Flavoring; medicine (skin irritant). Has been suggested as a bactericide.
Shipping regulations: None. *

garnet. A group of silicate minerals with the general formula $R_3^{II}R_2^{III}(SiO_4)_3$. R^{II} = Ca, Mg, Fe, or Mn; R^{III} = Al, Fe, Ti, or Cr. Garnets in nature are usually composed of mixtures of various garnet subspecies (see below).
Properties: Color variable; luster vitreous to resinous; streak white; hardness 6.5-7.5; sp. gr. 3.5-4.3. Usually well crystallized.
Varieties:
Grossularite $Ca_3Al_2(SiO_4)_3$.
Pyrope $Mg_3Al_2(SiO_4)_3$. Deep red to nearly black. Often transparent, and then used as a gem. Sometimes known as Bohemian garnet, Cape ruby.
Almandite (almandine) $Fe_3Al_2(SiO_4)_3$.
Andradite $Ca_3Fe_2(SiO_4)_3$.
Spessartite $Mn_3Al_2(SiO_4)_3$.
Uvarovite $Ca_3Cr_2(SiO_4)_3$.
Common garnet. Includes mixtures of almandite, andradite, and grossularite.
Rhodolite. A mixture of pyrope and almandite.
Occurrence: Widespread in metamorphic and some igneous rocks. New York, New Hampshire, North Carolina; India; Ceylon; Brazil; Czechoslovakia; Africa; U. S. S. R.
Use: Gem stone; abrasive.

garnet lac. See shellac.

garnierite $(Ni, Mg)_6(OH)_6Si_4O_{11} \cdot H_2O$. A natural hydrous nickel-magnesium silicate, occurring as a natural alteration of magnesium silicate rocks.
Properties: Color apple green; luster dull to earthy; greasy feel; streak white to greenish; hardness 2-3; sp. gr. 2.2-2.8.
Occurrence: New Caledonia; North Carolina; Oregon; Africa.
Use: Ore of nickel.

garspar. A mixture of finely ground glass and quartz, produced in the grinding of plate glass.
Use: Substitute for feldspar in ceramics; filler in battery boxes and rubber.

gas. One of the three states of matter. A material in the gaseous state is characterized by very low density and viscosity (relative to liquids and solids); relatively great expansion and contraction with changes in pressure and temperature; ability to diffuse readily into other gases; and ability to distribute itself readily with almost complete uniformity throughout the whole of any container.

A "perfect" gas is one which closely conforms to the simple gas laws for expansion and contraction (Boyle's Law, Charles' Law).

Use of the word "gas" in the sense of gasoline, or any fuel or illuminant, or for the anesthetic nitrous oxide, is scientifically inaccurate.

gas black (carbon black, channel black, furnace black). Finely divided carbon made by incomplete combustion or thermal decomposition of natural gas. Used as reinforcing agent in tire treads and other rubber products. See carbon black.

"Gas-Chrom" A. [425] "Gas-Chrom" S that has been acid washed to achieve high temperature stability and efficiency in gas chromotography.

gas chromatography. The process in which the components of a mixture are separated from one another by volatilizing the sample into a carrier gas stream which is passing through and over a bed of packing consisting of a 20 to 200 mesh solid support. The surface of the latter is usually coated with a relatively nonvolatile liquid designated the stationary phase. This gives rise to the term gas-liquid chromatography. If the liquid is not present the process is gas-solid chromatography which is also widely useful for analysis. As in other types of chromatography, different components move through the bed of packing at different rates and so appear one after another at the effluent end, where they are detected and measured by thermal conductivity changes, density differences, or various types of ionization detectors.

Gas chromatography is advantageous as a means of analysis of minute quantities of complex mixtures from industrial, biological, and chemical sources, and is also of potential value in actually preparing moderate quantites of highly purified compounds otherwise difficult to separate from the mixtures in which they occur.

"Gas-Chrom" P. [425] "Gas-Chrom" S that has been both acid and alcoholic-base washed to achieve high temperature stability and efficiency, and to eliminate active sites.

"Gas-Chrom" S. [425] Trademark for flux-calcined diatomaceous earth, screened so as to be suitable for use as a support for gas chromatography phases and coatings. Available in closely controlled mesh sizes 60-80, 80-100, 100-120, 100-140, 120-140, 140-200.

gas hydrates. A number of gases form clathrate compounds with water and these are known as gas hydrates. The compounds are solids, and are insoluble in water. They usually form and exist only at relatively low temperatures and high pressures. The solids are formed directly by contact of gas and liquid water. Anywhere from 6 to 18 molecules of water may combine with each molecule of gas, depending upon the nature of the gas.

Interest in the gas hydrates for many years was generated mainly because of the

*See "I.C.C. Shipping Regulations," page xiii.
Reference numbers refer to name of manufacturer. See "List of Manufacturers," page v.

nuisance of such compound formation in gas pipelines. In recent years, the compounds have been proposed as a means of precipitating water from salt solution (or sea water), thus yielding potable water.

The best known gas hydrates are those of ethane, ethylene, propane and isobutane. Other known hydrate formers include: methane and 1-butene, most of the fluorochloro refrigerant gases, nitrous oxide, acetylene, vinyl chloride, carbon dioxide, methyl and ethyl chloride, methyl and ethyl bromide, cyclopropane, hydrogen sulfide, methyl mercaptan and sulfur dioxide.

gas-liquid chromatography. See gas chromatography.

gas liquor. See ammonia liquor.

gas oil. A liquid petroleum distillate with a viscosity and boiling range between kerosine and lubricating oil. The boiling range is about 450-800°F. Used for absorption oil, and for cracking in petroleum refineries.

gasoline (petrol, motor spirits). A mixture of volatile hydrocarbons suitable for operation of an internal combustion engine. The major components are usually hydrocarbons with boiling points ranging from 60-200°C. These include straight-chain and branched-chain paraffins, naphthenes, and aromatic hydrocarbons, such as n-heptane, isooctane, methyl cyclohexane, benzene, and toluene. The usual source of gasoline is by distillation of petroleum and cracking, polymerization and other chemical reactions by which the naturally occurring petroleum hydrocarbons are converted to those that have superior fuel properties. Such catalytic chemical conversion of hydrocarbons has become increasingly important.

A gasoline must have the proper mixture of low-boiling and high-boiling components so that it changes to vapor in the most efficient way when used in a motor. This characteristic is usually specified in terms of boiling range, specific gravity, and vapor pressure. Gasoline must not contain or form any non-volatile or gummy materials, and must have the proper combustion characteristics as measured by "knock rating" (octane number). Other common specifications relate to color, sulfur content, and corrosion.

Practically all commercial gasolines contain small amounts of various additives such as lead tetraethyl to improve octane rating, antioxidants, corrosion inhibitors. See "Ethyl." Special gasolines may contain alcohols or other nonhydrocarbons.

Uses: Fuel for internal combustion engines; solvent; paint mixing; rubber cements; illuminant.

Fire hazard: Dangerous.

Shipping regulations: Flammable liquid. Red label. *

gasoline, alkylate. Gasoline made by alkylation (q. v.).

Gasoline Antioxidant No. 5. [28] A solution of N-n-butyl-para-aminophenol in alcohols having the following weight composition: 50% N-n-butyl-para-aminophenol $HOC_6H_4NHC_4H_9$; 30% anhydrous isopropanol $(CH_3)_2CHOH$; 20% anhydrous methanol CH_3OH.

The solution is readily soluble in gasoline in normal use concentrations. The use of benzene or other aromatic solvent is recommended for preparation of concentrated solutions. Sp. gr. 0.90.

Containers: 55-gal (400 lbs) steel drums.

Use: For reducing the formation of gum and precipitation of lead in gasoline.

Gasoline Antioxidant No. 22. [28] N, N'-di(sec-butyl)-para-phenylenediamine, $C_6H_4[NHCH(CH_3)(C_2H_5)]_2$, containing no solvent. A mobile liquid readily soluble in gasoline in all proportions and at all operating temperatures; sp. gr. 0.94.

Containers: 55-gal (425-lb) steel drums.

Use: For retarding the formation of gum and precipitation of lead in gasoline. Concentration required 0.001 to 0.005% by weight.

gasoline, casinghead. See gasoline, natural.

gasoline, cracked. Gasoline produced by the thermal and/or catalytic decomposition of high-boiling components of petroleum. In general such gasolines have higher octane ratings than gasoline produced by fractional distillation of petroleum. The difference is due to the prevalence of unsaturated, aromatic and branched-chain hydrocarbons in the cracked gasoline. The actual properties vary widely with the nature of the starting material, and the temperature, time, pressure and catalyst used in the cracking process.

gasoline, ethyl. See gasoline, leaded.

gasoline, leaded (gasoline, ethyl). Gasoline to which tetraethyl lead has been added to increase its antiknock properties.

See octane number.

gasoline, natural (gasoline, casinghead). A volatile gasoline obtained by recovering the butane, pentane, and hexane hydrocarbons present in small proportion in certain natural gases. It is used in blending to produce a finished gasoline with adjusted volatility.

gasoline, polymer. A gasoline produced by polymerization of low molecular weight hydrocarbons such as ethylene, propene, and butenes. It is used in small amounts for blending with other gasolines to improve their octane number.

gasoline, reformed. A high octane gasoline obtained from low octane gasoline by heating the vapors to a high temperature or by passing the vapors through a suitable catalyst.

gasoline, straight run. Gasoline produced from petroleum by distillation, without use of cracking or other chemical conversion processes.

"Gastex." [275] Trade name for semi-reinforcing furnace carbon black for use in rubber goods. High loading capacity, good resilience and flex resistance.
Use: Tire carcass and bead insulation, mechanical goods, footwear and soling, wire jackets, belts, hose, packings.

Gattermann-Koch reaction. The formation of aromatic aldehydes from phenols by the use of anhydrous hydrogen cyanide, dry hydrogen chloride, and an aluminum chloride or zinc chloride catalyst.

gaultheria (checkerberry; wintergreen; deerberry; boxberry; teaberry). The leaves of a small evergreen plant Gaultheria procumbens.
Habitat: Canada and northeastern United States.
Chief constituents: Methyl salicylate, arbutin, ericolin, and urson.
Grades: Technical.
Containers: Bales; boxes.
Use: Source of oil.
Shipping regulations: None. *

gaultheria oil. See wintergreen oil.

gaultheria oil, synthetic. See methyl salicylate.

Gay-Lussac acid. The sulfuric acid-nitrogen oxides mixture which is the product of the Gay-Lussac tower in the chamber process for manufacture of sulfuric acid.

This acid has a sulfuric acid strength of 60° Bé, and a nitrogen oxides content of 1-2% calculated as N_2O_3.

gaylussite $Na_2Ca(CO_3)_2 \cdot 5H_2O$. Natural hydrated carbonate of sodium and calcium, found in dry lakes.
Properties: Colorless to yellowish white and white, luster vitreous; hardness 2.5-3; sp. gr. 1.99.
Occurrence: California, Nevada; Venezuela.

"G.B.S." [28] Trademark for globular sodium bisulfate.
Properties: White, opaque or translucent, globular shaped pellets; soluble in water; aqueous solutions strongly acid in reaction.
Containers: 100- and 400-lb fiber drums.
Uses: An easily handled solid acid which can be substituted for sulfuric acid in many uses; as an ingredient in cleaning compositions and in the preparation of glass and ceramic glazes.

Gd. Symbol for gadolinium.

GDCH. Abbreviation for glycerol dichlorohydrin. See alpha-dichlorohydrin.

GDME. Abbreviation for glycol dimethyl ether. See ethylene glycol dimethyl ether.

GDP. Abbreviation for guanosine diphosphate. See guanosine phosphates.

Ge. Symbol for germanium.

gear case oil. See transmission oil.

Geiger counter. See Geiger-Mueller counter.

Geiger-Mueller counter. A common form of a nuclear radiation detector. It consists usually of a tubular cathode with a coaxial center wire anode, filled with one of several possible mixtures of gases. When a high voltage is impressed across the electrodes, ionizing radiation traversing the tube gives rise to conductivity pulses which may be electrically amplified and registered. Each ionizing event gives rise to one pulse, and the counter tube with its associated electrical circuitry "counts" the number of individual ionizing radiations.

"Gelamite." [266] Trademark for a semigelatin high explosive of relatively high weight strength of 65%; very good water resistance.
Uses: Underground mining; quarrying, construction, and general blasting.

gelatin. A protein obtained from collagen by boiling skin, ligaments, tendons, bones, etc. with water. Its production differs from that of glue in that the raw materials are selected, cleaned and treated with special care so that the resulting product is cleaner, purer and generally clearer and lighter in color than glue.
Properties: Colorless or slightly yellow, transparent, brittle, practically odorless, tasteless sheets, flakes, shreds, pellets, or a coarse or fine powder; swells up and absorbs five to ten times its weight of water; soluble in hot water, glycerol and acetic acid; insoluble in alcohol, chloroform, and other organic solvents.
Grades: Edible; photographic; technical; U.S.P. XVI.
Containers: Barrels; bags; drums; boxes; cases.
Uses: Photographic film; lithography; sizing; plastic compounds; textile and paper work; foods; rubber substitutes; adhesives; cements; capsules for medicinals; artificial silk; matches; light filters; clarifying agent; hectographic masses; bacteriology; medicine.

gelatin, Bengal. See agar-agar.

gelatin, Ceylon. See agar-agar.

gelatin, Chinese. See agar-agar.

gelatin, explosive. A powerful explosive formed by mixing nitrocellulose with about nine times its weight of nitroglycerin, the product being a gelatinized mass. It is less sensitive to shock and friction than ordinary dynamite.
Shipping regulations: Explosive, class A. High explosive label. *

gelatin, Japanese. See agar-agar.

gelatin, nutrient. A culture medium for bacteria consisting of gelatin, beef extract and peptone in various concentrations.

gelbin. See calcium chromate.

*See "I.C.C. Shipping Regulations," page xiii.
Reference numbers refer to name of manufacturer. See "List of Manufacturers," page v.

"Gelcarin." [124] Trade name for a line of carrageenan extractives, hydrocolloids, which may be derived from a number of sea plants in the class of Rhodophyceae, order of Gigartinales. Used in the food, pharmaceutical and cosmetic industries, as gelling agents; stabilizers for emulsions, suspensions, etc.

"Gelex." [28] Trademark for semi-gelatin dynamites (Nos. 1 to 5 inclusive) having plasticity and water-resistance ratings between the ammonia dynamites and Special Gelatins. Have a very good fume rating.
Containers: Cartridges (1 1/4 x 8 in.) per 50-lb case range from 110 for No. 1 to 150 for No. 5.
Use: In mining metallic ores, gypsum, limestone; in quarrying medium hard rock; and in construction work.

"Gelfoam." [327] Trademark for an absorbable surgical sponge.
Derivation: Processed from a non-antigenic gelatin solution, which is processed into a sponge-like form.
Use: Medicine.

"Gel-Kote." [448] Trade name for pigmented polyester resin coatings for polyester products.

"Gelloid." [78] Trademark for a line of purified extracts of various types of Irish moss seaweed which are rich in mucilagenous content. The gel-producing ingredient is known as calcium carragheen sulfate. Products of varying gel strength and solubility are prepared by a specialized extraction and filtration procedure with subsequent solvent treatment to give dry, odorless, non-toxic, edible powders used in extensive food and pharmaceutical applications, especially as stabilizers, emulsifiers, moisture retentive agents, and emollient bases.

"Gelobel." [28] Trademark for gelatin-type permissible explosives; high-density and high water-resistance ratings.
Use: For coal mining where high strength, high velocity, concentration of charge, and water resistance are desired.

gel paint (thixotropic paint). A paint formulation which has a semi-solid or gel consistency when undisturbed but which flows readily under the brush or when stirred or shaken. After removal of the stress, it becomes stiff again and has little tendency to spill, drip, or run. The thixotropic quality is obtained by the carefully controlled reaction of a relatively small proportion of a polyamide resin with an alkyd resin vehicle.

gelsemine $C_{20}H_{22}O_2N_2$. An alkaloid.
Properties: White crystals; m.p. 178°C; poisonous! Soluble in alcohol, ether, and dilute acids.
Derivation: From the rhizome and root of Gelsemium sempervirens.
Grades: Technical.
Containers: Tins.
Use: Medicine.
Shipping regulations: None.*

gelsemium (yellow jasmin; wild woodbine).
Properties: Yellow masses.
Chief constituents: Gelsemine, gelseminine and gelsemic acid.
Derivation: The dried rhizome and roots of Gelsemium sempervirens.
Habitat: Southern United States.
Grades: Technical.
Containers: Bales.
Uses: Medicine; extraction of gelsemine.
Shipping regulations: None.*

"Gelva." [61] (See also "Gelva" [276]). Trademark for vinyl acetate polymers. Resins are available as solids or solutions in a variety of viscosities indicated by a "V-number" (viscosity of 8.6% in benzene as cps). Aqueous emulsions have outstanding stability and generally contain 55% solids (homopolymers or copolymers) with emulsion viscosities from 200 to several thousand cps. T S types (e.g. T S-30) are free filming (forming water-resistant films) with particle sizes generally 1 micron or less. S types (e.g. S-55) have heterogeneous particle sizes and possess exceptional quick tack as an adhesive. Alkali-soluble copolymers (C types) are available as solids or as an emulsion.
Uses: Adhesives; binders; chewing gum bases; coatings; floor polishes; hot melt adhesives; paints; paper treatment; permanent starches; slush molding; textile sizes and finishes; and thickeners.

"Gelva." [276] (See also "Gelva" [61]). Proprietary name for polymerized vinyl acetate resins manufactured outside U.S.A. Available in several standard viscosities: V 1.5, V 2.5, V 7, V 15, V 25, V 45, and V 60.
Properties: Colorless; soluble in lacquer solvents and aromatic hydrocarbons; insoluble in aliphatic hydrocarbons.
Uses: Lacquers; adhesives; coatings; impregnation; chewing-gum base.

"Gelvatex." [61] Trademark for aqueous emulsions of vinyl acetate polymers and compounded compositions thereof.

"Gelvatol." [61] Trademark for polyvinyl alcohol resins. Available in 12 grades from partially hydrolyzed to fully hydrolyzed polymers at various viscosities. Compatible with a wide variety of natural and synthetic resins as well as other materials. Usually water-soluble but oil and fat resistant, but readily converted to water resistant. Films are strong, gas and grease proof, orientable to polarized light, suitable for water-soluble packaging. Used as adhesive; coating; emulsifier; hydraulic cement additive; textile sizes; paper coating.

gem-. Prefix which is an abbreviation of geminate, meaning two identical groups attached to the same carbon atom.

geminate. See gem.

*See "I.C.C. Shipping Regulations," page xiii.
Reference numbers refer to name of manufacturer. See "List of Manufacturers," page v.

"Genacryl." [307] Proprietary name for basic dyes for acrylic fibers.

"Genamid" 250. [259] Trademark for a liquid coreactant for epoxy resins. This is a highly refined resinous amine adduct. By combining Genamid 250 with epoxy resins, a room temperature cured copolymer is produced which is clear and tough. This cross linked copolymer has excellent physical strength and outstanding adhesion. Used for coatings, castings, laminates, adhesives, and concrete topping and patching compounds.

"GenEpoxy." [259] Trademark for epoxy resins.
Derivation: Reaction product of bisphenol-A and epichlorohydrin.
Grades: Liquid resins with viscosities from 500 centipoises to 16,000 centipoises, epoxide equivalent from 172 to 191. Solid resins with epoxide equivalents from 500 to 1,900. Solutions of solid resins in various solvents also available.
Containers: Liquids in 55-gal steel drums; tank trucks; tank cars. Solids in 200-lb fiber drums; 50-lb multiwall bags. Solutions in 55-gal steel drums; tank cars.
Uses: Coatings; adhesives; laminates; castings; esters for coatings; potting and encapsulating.

"Genetron." [50] Trademark for a line of fluorinated hydrocarbons. Members of the series include:
"Genetron" 21, dichlorofluoromethane, $CHCl_2F$.
"Genetron" 23, fluoroform, CHF_3.
"Genetron" 142B, 1,1-difluoro-1-chloroethane, CH_3CClF_2.
"Genetron" 152A, 1,1-difluoroethane, CH_3CHF_2.
"Genetron" 1132A, 1,1-difluoroethylene, $CH_2{:}CF_2$.
Uses: As aerosol propellants; refrigerants; low temperature solvents; monomers; heavy-duty, fluorocarbon plastics.

"Gen-Flo Latices." [179] "Gen-Flo 67" is a styrene-butadiene emulsion polymer containing approximately 67% styrene and 33% butadiene; recommended for manufacturing interior latex paints and textile applications.
"Gen-Flo 62" contains approximately 60% styrene and 40% butadiene and is recommended for paint, paper, textile and rug backing applications.
"Gen-Flo" Latices have good film clarity, excellent mechanical stability and alkali resistance.
Uses: Latex paints, paper coatings, adhesives, textile printing inks, sizes, etc.

"Genicop." [50] Trademark for a DDT-copper formulation which acts as an insecticide and fungicide.

genin. The steroid portion which is linked to a sugar residue in certain glycosides. Important genins are found in the digitalis glycosides which are used in medicine as heart stimulants.

"Genite" 923. [50] Trademark for 2,4-dichlorophenyl ester benzenesulfonic acid; available as 50% emulsifiable or 50% wettable powder. Miticide specific for European red mite and clover mite; gives long lasting control of early mites.

"Genithion." [50] Trademark for a parathion insecticide.

"Genitol." [50] Trademark for a DDT insecticide.

"Genitox." [50] Trademark for a DDT insecticide.

"Gensol No. 6." [79] Trade name for terpene solvent.
Properties: Sp. gr. (15.5°/15.5°C) 0.842; refractive index (20°C) 1.465; flash point (open cup) 111°F; Engler distillation 5%, 166°C; 50%, 170°C; 95%, 178°C; Kauri butanol value 59 (basis, toluene = 105).
Containers: 55-gal drums; tank cars.
Uses: Odorant for masking other odors; solvent and softener in rubber reclaiming; solvent in printing ink manufacture.

"Gen-Tac Latex." [179] Trademark for a latex containing vinyl pyridine, butadiene, and styrene.
Uses: Used to gain adhesion between natural and synthetic rubber to cotton, rayon, nylon and "Dacron" in applications such as tires, mechanical goods, V-belts, conveyor belts or any application where rubber-to-fabric adhesion is needed.

"Genthane-S." [179] Trademark for a polyurethane elastomer. It is processed on conventional rubber processing equipment and has good abrasion resistance, high tensile strength, ozone resistance and hot dry temperature properties.
Uses: Mechanical goods, grommets, packings, and extrusions.

gentian (yellow gentian; bitter root). Dried rhizome and roots of Gentiana lutea.
Habitat: Mountainous regions of Europe and Asia Minor.
Grades: Technical; N.F. XI.
Containers: Bales; barrels.
Uses: Medicine; liqueurs.
Shipping regulations: None.*

gentian violet. See methyl violet.

gentisic acid (2,5-dihydroxybenzoic acid) $C_6H_3(OH)_2COOH$.
Properties: Crystals; m.p. 199-200°C; soluble in water, alcohol, and ether; insoluble in carbon disulfide, chloroform, and benzene.
Use: Medicine, as sodium gentisate (q.v.).

"Gentro." [179] Trademark for a series of staining and non-staining vulcanizable synthetic polymers containing butadiene and styrene manufactured by the cold process. They include staining, oil-extended staining, light-colored non-staining, and light-colored, oil-extended, non-staining polymers. Approximately 1.25% antioxidants are added to insure protection of the polymer during production and storage.

Properties: Contains approximately 23.5% styrene and 76.5% butadiene. Polymers mixed with reinforcing pigments give a variety of properties, abrasion resistance, flexibility, color, etc.

Uses: Tires, mechanical goods, proofed goods, extruded goods, shoe soles, heels, housewares, sponge, etc.

"Gentro-Jet." [179] Trademark for a series of carbon blacks co-precipitated with synthetic SBR polymer. They are produced by a steam-whipped, dispersant-free process. The masterbatches contain staining polymers only and are made with the HAF, ISAF and SAF carbon blacks. Antioxidants are added to insure protection of the polymer during production and storage.

Properties: Black content ranges from 52 parts black per 100 parts polymer to 75 parts black on 137.5 parts oil-extended polymer. Masterbatch mixed with other compounding materials gives high tensile strength and abrasion resistance.

Uses: Tires, tread rubber, mechanical goods, conveyor belts, V-belts, etc.

geochemistry. The study of the chemical composition of the earth in terms of the physicochemical and geological processes and principles that produce and modify the minerals and rocks of the earth. Of practical importance in discovering and establishing the limits of ore deposits, and in understanding geologic phenomena generally.

geometric isomerism. The existence of isomeric forms of a compound because of symmetry or lack of symmetry about the double bond in the molecule of an organic compound. Certain types of ring structure also result in this type of isomerism. See cis-.

"Geon." [119] Trade name for a group of polyvinyl chloride resins, plastics, and latices. Available in the following types:

General purpose resins for calendering, coating, extrusion, and injection molding.

Uses: Sheeting for upholstery, luggage, and wall covering. Film for shower curtains, draperies, tablecloths, and rainwear. Extruded garden hose, chemical tubing, belting, and welting. Injection molded grommets, gaskets, electrical plugs, toys, and automotive parts.

Rigid type resins for compounding without plasticizer for use in extrusion and molding.

Typical properties:

Flexible "Geon": Sp.gr. 1.2-1.55; tensile strength (psi) 1500-3500; elongation 200-450%; hardness (Shore Durometer) 50-100A; service temperature (max) 170-220°F; flexibility temperature (min) 0 to −70°F; volume resistivity (Ohm-cm at 25°C) $1\text{-}10 \times 10^{12}$ to $4\text{-}7 \times 10^{14}$ (max); dielectric constant at 60 cycles at 25°C 5.5-9.1.

Rigid "Geon": Sp.gr. 1.32-1.4; tensile strength (psi) 6000-9000; elongation 5-25%; hardness (Shore Durometer) 70-85D; service temperature (max) 150°F; volume resistivity (Ohm-cm at 25°C) exceeds 10^{16} (max); dielectric constant at 60 cycles at 25°C 3.0-3.2.

Uses: Transparent, translucent, or opaque sheet for embossing, press polishing, or laminating. Extruded chemical piping, rod, sheet, and profiles. Injection molded pipe fittings. Vacuum-formed relief maps, point of display signs, and containers. Compression molded phonograph records.

Insulation types for electrical industry. Supplied as cubical granules.

Typical properties: Sp.gr. 1.3-1.4; hardness (Durometer) 80-91A, 82C; tensile strength (psi) 1800-3900; elongation 200-340%; insulation resistance (K value) 350-10,000; dielectric strength (volts/mil) 500-800.

Compounded types for extrusion and injection molding with a range of properties depending upon composition.

Properties:

Extrusion compounds: Sp.gr. 1.25-1.48; hardness (Durometer A) 68-95; tensile strength (psi) 1200-3350; ultimate elongation 250-355%; brittle temperature −10 to −40°C; water absorption (24 hours at 100°C) 1.1-3% gain; stock extrusion temperature 350-365°F.

Injection molding compounds: Sp.gr. 1.2-1.4; hardness (Durometer A) 64-80; tensile strength (psi) 1400-2600; ultimate elongation 275-350%; brittle temperature −25 to −30°C; water absorption (24 hours at 100°C) 0.8-2.5% gain; injection molding temperature 340-360°F; injection molding pressure (psi) 16,000-22,000.

Uses: Extruded products including belting, welting, tubing, and gasketing. Injection molded products are vacuum sweeper parts, handles, dolls, luggage, hair curlers, refrigerator drain rails, and automobile lock knobs.

Latices: Water dispersions of polyvinyl chloride resins. Unplasticized, plasticized, and internally plasticized forms available. Total solids approximately 50-57%; pH ranges from 6-9; sp.gr. 1.083-1.210. Properties of value include low moisture vapor transmission, low gas permeability, resistance to flame and many chemicals.

Uses: Decorative, washable, and wear resistant coatings for paper and fabric. Non-woven fiber binders, heat sealable binders and coatings, fabric sizes, flameproof coatings and impregnations, food packaging, and leather finishes.

Paste resins: Polymers of controlled even particle size for easy dispersion in a resin-plasticizer system (plastisol), or in a resin-plasticizer-diluent system (organosol). Supplied as a fine white powder with a specific gravity of approximately 1.4. Can be formulated for dip coating, spread coating, foaming, molding, or casting. Products exhibit typical "Geon" resin properties.

Uses: Spread coated luggage, upholstery fabric, window shades, and decorative paper. Spray coated woods, and sheet metals. Dip coated gloves, chemical racks, and boots. Cast film, toys, dolls, and prosthetic devices. Slush molded thin walled shapes. Open and closed cell foam and sponge for insulation and shock absorption, and cushioning applications.

Polyblend: A colloidal blend of nitrile rubber and polyvinyl chloride in which the rubber is used as the plasticizer. The nitrile plasticizer is non-migrating, non-extractable, and non-volatile. Polyblend can be used as a thermoplastic vinyl, as a vulcanizate, or to modify conventional vinyls and rubbers.

Uses: Sheet and film for food packaging, upholstery, luggage, and footwear. Extruded straps, belting, and tubing. Coatings for fabrics, paper, wood, and metal.

Solution resins: Designed for high solids solutions, these high molecular weight polymers and copolymers are soluble in ketone systems. A modified vinyl-vinylidene chloride copolymer is specifically designed for direct solubility in toluene, xylene, hydrocarbons, and aromatic naphthas. They form coatings that have good adhesion to tin, aluminum, steel, fabric, paper, and other cellulosic materials. Coatings may be sprayed, brushed, or rolled.

Uses: Water and fire resistant fabrics; paint and lacquer formulations.

Hi-temp "Geon": Polyvinyl dichloride. Thermoplastic. Capable of withstanding temperatures 60°F higher than other vinyls. Tensile strength reaches above 2000 psi at 212°F, and deformation at 264 psi occurs only at temperatures above 215°F. Also has high modulus and impact strength. Can be extruded and molded in conventional equipment.

Uses: Chemical process piping, tanks, ducts, hot and cold water lines, valves.

geophysics. The science of physics as applied specifically to the earth. Of importance in locating oil deposits, oil-well drilling, mining and the study of earthquakes, volcanoes and other terrestrial phenomena.

"Gerallol." [227] Trademark for a mixture of geraniol and citronellol. Available in 3 grades.

"Gerallol" Extra: Minimum purity 90%; colorless liquid.

"Gerallol" Prime: Minimum purity 85%; slightly yellow liquid.

"Gerallol" HC: Minimum purity 85%; slightly yellow to yellow liquid.

Properties: Light rosy odor; sp. gr. 0.84-0.87; refractive index 1.44-1.46.

Uses: In light bouquets and rose florals.

geranial. See citral.

geranialdehyde. See citral.

geraniol $(CH_3)_2C{:}CH(CH_2)_2C(CH_3){:}CHCH_2OH$ (3, 7-dimethyl-2, 6-octadienol).

Properties: Colorless to pale yellow, liquid oil; pleasant geranium-like odor; sp. gr. 0.870-0.890 (15°C); m. p. −15°C; b. p. 230°C; refractive index (n 20/D) 1.4710-1.4780; optical rotation −2° to +2°; soluble in alcohol and ether, mineral oil, fixed oils; insoluble in water, and glycerol.

Derivation: From geranium oil; also from citronella and palmarosa oils by forming the double compound with calcium chloride.

Grades: Standard; soap; synthetic.

Containers: Cans; drums.

Use: Perfumery.

Shipping regulations: None.*

geraniol acetate. See geranyl acetate.

geraniol butyrate (geranyl butyrate) $C_3H_7COOC_{10}H_{17}$.

Properties: Colorless liquid; b. p. 151°C (18 mm); rose type odor; sp. gr. 0.9008 (17/4°C); insoluble in water; soluble in alcohol, ether.

Occurs in several essential oils.

Use: In perfumes and soaps.

geraniol formate (geranyl formate) $HCOOC_{10}H_{17}$.

Properties: Colorless liquid; b. p. 113°C (15 mm); sp. gr. 0.927 (20/4°C); rose type odor; insoluble in water; soluble in alcohol and ether.

Occurs in several essential oils.

Use: In perfumes and soaps.

geranium (cranes-bill; storksbill; alum root) Dried rhizome of Geranium maculatum.

Habitat: Canada and eastern United States; south to Georgia.

Containers: Bags.

Use: Medicine.

geranium, blood. See sanguinaria.

geranium oil.

Properties: Pale yellow or greenish liquid; exceedingly agreeable rose-like odor. Slightly soluble in water; soluble in alcohol and ether; sp. gr. 0.886-0.898; optical rotation −7° to −12°; refractive index (n 20/D) 1.4650-1.47.

Chief known constituents: Geraniol; citronellol.

Derivation: Distilled from the herb of several species of Pelargonium, especially P. graveolens, P. capitatum, P. odoratissimum, and P. roseum.

Method of purification: Rectification.

Grades: Algerian; Bourbon.

Containers: Cans.

Uses: Perfumery; manufacture of rhodinol.

Shipping regulations: None.*

geranium oil, East Indian. See palmarosa oil.

geranium oil, rose. A geranium oil prepared in southern France by adding rose petals to the pelargonium plants during distillation.

geranium oil, Turkish. See palmarosa oil.

geranyl acetate (geraniol acetate) $CH_3COOC_{10}H_{17}$.

Properties: Sweet, fragrant, clear, colorless liquid; odor of lavender; sp. gr. 0.907-0.918 (15°C); b. p. 128-129°C (16 mm); optical rotation −2° to +2°; refractive index

*See "I.C.C. Shipping Regulations," page xiii.

Reference numbers refer to name of manufacturer. See "List of Manufacturers," page v.

(n 20/D 1.4580-1.4640; soluble in alcohol and ether; insoluble in water and glycerol.
Derivation: (a) Constituent of several essential oils. (b) By heating geraniol and sodium acetate with acetic anhydride.
Grade: Technical.
Containers: Cans; drums.
Use: Perfumery.
Shipping regulations: None. *

geranyl butyrate. See geraniol butyrate.

geranyl formate. See geraniol formate.

German black. See vegetable black.

German fungus. See agaric.

germanic acid. See germanium oxide.

germanic oxide. See germanium oxide.

germanium (Ge). Element of atomic number 32, group IV of the periodic system.
Properties: Grayish-white metal; a semiconductor of electricity, whose conductivity depends largely on the impurities present.
Constants: Sp. gr. 5.46; m.p. 958°C; does not volatilize below 1350°C.
Derivation: Recovered from residues from refining of zinc and other sources, by heating in the presence of air and chlorine. Found also in two ores, argyrodite and germanite.
Method of purification: The chloride is distilled, and then hydrolyzed to the oxide, which is reduced by hydrogen to the metal. Zone-melting is used for final purification, and single semiconductor crystals are made by vaporization of germanium diiodide under conditions such that it dissociates and deposits pure germanium. The impurities in germanium are of controlling importance in its use in transistors.
Uses: Medicine (in form of dioxide); metal is used in vacuum tubes as a high-resistance element, and most importantly, in electronic devices such as transistors and diode rectifiers.

germanium dioxide. See germanium oxide.

germanium oxide (germanic acid; germanic oxide; germanium dioxide) GeO_2.
Properties: White powder; soluble in alkalies; slightly soluble in acids, water.
Use: Ingredient of special glass mixtures; medicine.

germanium-potassium fluoride (potassium-germanium fluoride) K_2GeF_6.
Properties: White crystals; soluble in water (hot); insoluble in alcohol.
Grade: Technical.

germanium telluride GeTe. An efficient semiconductor.

German silver. Obsolete name for nickel silver.

German-silver solder. Series of alloys with relatively low silver content (10-20%) and varying amounts of copper, zinc, cadmium and tin. Used for the usual purposes of silver solder.

Germantown black. See lampblack.

germicide. A product which kills bacteria, but its power to prevent further growth is (perhaps) only secondary. There are many proprietary preparations on the markets. See also disinfectants.

getter alloys. Alloys used in radio and television tubes and electric bulbs to absorb residual substances which impair efficient operation.

ghatti gum. The gummy exudation from the stem of Anogeissus latifolia.
Properties: Colorless to pale yellow tears, rounded or vermiform. Almost tasteless and odorless; partially soluble in water. Can be solubilized by autoclaving.
Uses: Similar to arabic gum.

giant granite. See pegmatite.

gibberellic acid (gibberellin X; GA_3) $C_{19}H_{22}O_6$. A plant-growth-promoting hormone, a metabolite extracted first from fungi and later recognized as occurring in many plants. It is said to be a tetracyclic dihydroxylactonic acid.
Properties: Crystals; m.p. 233-235°C; slightly soluble in water; soluble in methanol, ethanol, acetone. Soluble in aqueous solutions of sodium bicarbonate and sodium acetate.
Uses: Promotion of plant growth, especially seedlings; stem elongation; malting of barley with improved enzymatic characteristics; improving fruit setting of citrus; growth control in grapes.

gibberellins. A series of at least four plant hormones, similar in chemical structure, isolated from certain fungi and other plants. These are known as GA_1, GA_2, GA_3 and GA_4 and are structural variations of gibberellic acid (GA_3), which is best known and most effective. Since their discovery they have been identified in trace quantities in many plants and appear to be produced within the plant to regulate certain physiological processes. Application as sprays, dusts or as aerosols to fruit, vegetables or ornamental plants have shown many diversified effects: stem elongation in biennials; root elongation; increase in dry weight of leaf; fruit enlargement; increased photosynthetic rate; improved fruit setting. The gibberellins appear to be non-toxic and are accepted for use on certain food crops.
See also plant hormones; auxins.

gibberellin X. See gibberellic acid.

"Gibrel." [123] Trademark for compound for promoting plant growth; the potassium salt of gibberellic acid.

giga- Prefix meaning 10^9 units (symbol = G). 1 Gg = 1 gigagram = 10^9 grams.

gigily oil. See sesame oil.

gilder's whiting. See whiting.

gilsonite (uintaite; uintahite). An asphaltite (q.v.), or solidified hydrocarbons, found

*See "I.C.C. Shipping Regulations," page xiii.
Reference numbers refer to name of manufacturer. See "List of Manufacturers," page v.

only in the United States, in Utah and Colorado. One of the purest (99.9%) natural bitumens. It mixes well with the fatty acid pitches in all proportions.

Properties: Color in mass, black; conchoidal fracture; bright to fairly bright luster; brown streak. Sp. gr. 1.05-1.10 (77°F); hardness (Mohs' scale) 2; penetration 0 (77°F); fusing point (K & S) 250-350°F, (B & R) 270-370°F. Behavior on heating in flame, softens and flows; trace to 1% mineral matter. Soluble in all proportions of carbon disulfide. (Usually melted in a varnish kettle, but also soluble in a lukewarm bath of naphtha under mechanical agitation.)

Grades: "Selects" or "firsts," conchoidal and lustrous fracture; "run of mine" or "seconds," semi-conchoidal and semi-lustrous fracture.

Containers: Wooden barrels; multiwall paper sacks; bulk.

Uses: Acid, alkali and waterproof coatings; black varnishes, lacquers, baking enamels and japans; asphalt paints; wire-insulation compounds; printing inks; mineral wax; dielectric compounds; battery boxes; linoleum and floor tile; paving; insulation; water-proofing; fabric saturants and preservatives; also used as source of petrochemicals in some refineries.

Shipping regulations: None. *

gin. An alcoholic beverage made by distilling alcohol through a mixture of herbs and berries (juniper, coriander, etc.) and adjusting to 85 to 100 proof.

gingelly oil. See sesame oil.

ginger (zingiber).

Properties: Irregularly branched pieces; aromatic odor; aromatic, burning taste.

Chief constituents: Ginger oil (volatile), a resin and gingerol.

Derivation: The dried rhizome of Zingiber officinale.

Habitat: Southern Asia; West Indies; Africa; cultivated in all tropical countries.

Grades: N. F. XI; Cochin; Jamaican; Nigerian, Sierra Leone.

Containers: Tins; boxes; bags.

Uses: Medicine; confectionery; condiment; soft drinks.

Shipping regulations: None. *

ginger oil.

Properties: A pale yellow thick liquid; characteristic odor; aromatic, somewhat burning taste; sp. gr. 0.877-0.888 (15°C); optical rotation −28° to −45°; refractive index 1.488-1.494 (20°C); soluble in alcohol, ether, and chloroform, also in benzyl benzoate, diethyl phthalate, mineral oil, and most fixed oils; insoluble in glycerin, propylene glycol, and water.

Chief known constituents: Citral, borneol and phellandrene.

Derivation: Steam distilled from the dried rhizome of Zingiber officinale.

Grade: Technical.

Containers: Glass bottles; tin-lined or aluminum drums.

Use: Flavoring; preparation of liqueurs and soft drinks.

gingili oil. See sesame oil.

gingily oil. See sesame oil.

Girard's "P" reagent (carboxymethylpyridinium chloride hydrazide; acethydrazidepyridinium chloride) $C_5H_5NClCH_2CONHNH_2$.

Properties: White to faintly pinkish crystals with little or no odor; m. p. 190-200°C; soluble in water; insoluble in oils.

Containers: Bottles; fiber drums.

Uses: Separation of aldehydes and ketones from natural oily or fatty materials; extraction of hormones.

Girard's "T" reagent (carboxymethyltrimethyl ammonium chloride hydrazide; trimethylacethydrazide ammonium chloride) $(CH_3)_3NClCH_2CONHNH_2$.

Properties: White to faintly pinkish crystals, little or no odor; m. p. 182-192°C; soluble in water; insoluble in oils. Very hygroscopic.

Containers: Bottles; fiber drums.

Uses: Separation of aldehydes and ketones from natural oily or fatty materials; extraction of hormones.

Girbotol absorption process (amine absorption process). A process for the removal of hydrogen sulfide or carbon dioxide from a gaseous mixture. An organic amine such as ethanolamine or diethanolamine, which are basic, is allowed to flow down a tortuous path through a tower where it is contacted by and absorbs the hydrogen sulfide or carbon dioxide from the gas to be purified while it is moving up the tower. The amine which is contaminated with hydrogen sulfide or carbon dioxide is sent from the bottom of the tower to a steam stripper, where it flows countercurrent to steam, which strips the hydrogen sulfide or carbon dioxide from it. The amine is then returned to the top of the absorption tower.

The process employing diethanolamine is widely used in the petroleum industry for purifying refinery and natural gases and for recovery of hydrogen sulfide for sulfur manufacture. Removal of carbon dioxide from gases is usually done with monoethanolamine.

gitoxin $C_{41}H_{64}O_{14}$. A digitalis glycoside.

Properties: White needles; m. p. 266 to 269°C; slightly soluble in water, alcohol, and chloroform.

Derivation: From leaves of digitalis.

Use: Medicine.

"Giv-Tan" F. [12] (2-ethoxyethyl para-methoxycinnamate) $CH_3OC_6H_4CH{:}CHCOOC_2H_4OC_2H_5$.

Properties: Slightly yellow viscous liquid; practically odorless; sp. gr. 1.1000-1.1040 (25/25°C); refractive index 1.5660-1.5690 (20°C); f. p. below -25°C. Insoluble in water and glycerol; soluble in alcohol and mineral oil.

Grade: 98% minimum.

*See "I.C.C. Shipping Regulations," page xiii.
Reference numbers refer to name of manufacturer. See "List of Manufacturers," page v.

Containers: 1-, 5-, 10-, 25-, 50-lb tins; 130-, 270-, 450-lb lacquer lined drums.
Uses: Sun-tan lotions.

"GK Compound." [41] Trade name for a hot-pour, bituminous compound used to joint concrete and vitrified clay sewer pipe.

glacial acetic acid. See acetic acid.

glance pitch (manjak). An asphaltite (q. v.) (solidified hydrocarbons) resembling gilsonite in external appearance but having a decidedly black streak instead of a brown one. It also has a larger percentage of fixed carbon. The best known glance pitch comes from the Barbadoes, the material found on the surface of the veins being harder, more brittle and of a higher fusing point than that at the lower mine levels.
Properties: Black in mass; conchoidal to hackly fracture; bright to fairly bright luster; black streak (on porcelain). Soluble in carbon disulfide (usually greater than 95%).
Constants: Sp. gr. 1. 10-1. 15 (77°F); hardness (Mohs' scale) 2; penetration 0 (77°F); fusing point (K & S) 250-350°F, (B & R) 270-375°F; behavior on heating, softens and flows in flame; mineral matter less than 5.
Shipping regulations: None. *
See also Syrian asphalt.

glass. An amorphous transparent or translucent brittle material usually made by fusion of silica, soda ash, lime, and salt cake or similar materials. Other materials are used for various special glasses. The term glass is also applied more generally to amorphous solids, formed by cooling of liquids, of any composition. Though physically solid, glass is considered by physicists to be an under-cooled liquid, since its structure is not crystalline but amorphous, as is characteristic of liquids.

The composition of ordinary window glass (soda lime glass, lime glass) may be expressed as approximately 20% sodium oxide, 5% calcium oxide, 70-75% silica and small amounts of other components. The sp. gr. is about 2. 65. Plate glass is merely a thicker, more carefully made and finished variety of ordinary glass. A new float process for plate glass looks promising. Ordinary glass is very slowly dissolved by water, but more rapidly attacked by alkaline solutions. See flint glass, crown glass, optical glass. See also "Pyrex," Fourcalt process, Colburn process (for forming glass sheet, etc.).

glass colors. This term is applied to chemicals or mixtures used to confer special properties on glass. In addition to the color-producing compounds separately cited under various colored glasses the following are included as glass colors:
Acid resistant: For tumblers and tableware; substantially impervious to organic acids found in foods.
Alkali resistant: For milk and beverage-bottle labels having great resistance to alkaline sterilizing solutions.
Leadless type: For special applications.
Sulfide resistant: For illuminating ware; free from atmospheric discolorization.
Very soft nonresistant: For special applications.

glass enamels. A series of finely ground fluxes, basically lead borosilicate, intimately blended with colored ceramic pigments. Different grades give characteristics of acid resistance, alkali resistance, sulfide resistance, or low lead release to meet requirements for various uses. Firing range 1000°-1400°F (537. 8°-760°C).
Containers: Dry color, 25-, 50-, 100- and 200-lb fiber drums; paste colors, 100-lb (5-gal) pails; thermofluid enamels, 50-lb (5-gal) pails, 200-lb (20-gal) drums.
Uses: For fired-on labels and decorations on glassware, tumblers, milk bottles, beverage bottles, glass containers, illuminating ware, architectural glass, and signs.

glass fiber. Generic name for a manufactured fiber in which the fiber-forming substance is glass (Federal Trade Commission).
Properties: A continuous filament or staple fiber having unusual resistance to heat and chemicals. It is the strongest fiber known and is perfectly elastic up to its ultimate strength. It is attacked by hydrofluoric acid and alkalies; resistant to most other chemicals and solvents. Colored by resin-bonded pigments or by dyeing an applied protein film. Non-flammable.
Constants: Tensile strength (psi) 204,000-220,000 (yarns), individual fiber, approximately 1,000,000; elongation (yarns and fabric) 1. 7-3. 3%; sp. gr. 2. 54; moisture regain, none; loses strength above 600°F, softens about 1500°F.
Derivation: Molten glass is drawn at high speed through fine orifices.
Uses: Electrical insulation; plastic laminates; filter cloth and paper; surgical sutures; fireproof curtains and drapes. Not suitable for wearing apparel.

glassine. A thin transparent paper used for packaging and made with addition of a urea-formaldehyde resin to improve strength characteristics.

glass, liquid. See sodium silicate.

glass, "Pyrex." See "Pyrex."

glass, safety. A thin plastic sheet is cemented between two sheets of plate glass. The plastic is usually polyvinyl butyral, but various other materials have been used. If the glass is broken, the pieces do not fly but remain attached to the plastic. The term is also applied to highly tempered solid glass.

glass sand. A sand suitable for making glass. The principal component is quartz. A typical analysis is SiO_2 99. 41%, Al_2O_3 0. 21%; Fe_2O_3 0. 07%, CaO 0. 07%, MgO 0. 68%.
Occurrence: New Jersey, Pennsylvania, West Virginia, Missouri, Illinois, Maryland.

*See "I.C.C. Shipping Regulations," page xiii.
Reference numbers refer to name of manufacturer. See "List of Manufacturers," page v.

glass, soluble. See sodium silicate.

glass, volcanic. See obsidian.

glass, water. See sodium silicate.

glass wool. A fibrous wool-like material composed of fine filaments of glass intermingled like ordinary wool. Used in chemical laboratories; also in some producer-gas plants as a dust-filtering agent, and widely used in insulation and air filters.

glauberite. A natural sodium-calcium sulfate. $Na_2SO_4 \cdot CaSO_4$. Found in Arizona and New Mexico.

Glauber's salt. See sodium sulfate decahydrate.

glaucarubin $C_{25}H_{36}O_{10}$. Crystals; decomposing at 250-255°C; slightly soluble in water. Obtained from the meal from Simaruba glauca seeds. Used in medicine.

glauconite $K_2(Mg, Fe)_2Al_6(Si_4O_{10})_3(OH)_{12}$. A natural hydrous silicate of potassium, aluminum, iron, and magnesium, found in greensands, and other sedimentary rocks. Color green, luster earthy; sp. gr. 2. 3.
Occurrence: New Jersey, Virginia.
Uses: Water softener; in foundry molds; fertilizer.

"Glaurin." [73] Trademark for diglycol laurate.

glaze. A term used in the ceramic industry. According to the sense in which it is used it may mean: (1) a vitreous coating on finished pottery or enamelware, (2) the mixed and powdered dry materials of the batch to be used for producing the vitreous coating, (3) an emulsion of these materials suspended in water ("wet glaze"). Glazes may consist of common salt or feldspar but are more usually mixtures of native silicates such as feldspar, kaolin, or Cornish stone with flint, sand, cullet, chalk, borax, soda, white lead, red lead, or litharge. See also frit.

glaze stains. Finely ground calcined oxides of cobalt, copper, iron and manganese used for coloring ceramic glazes.

gliadin (prolamin). A group of simple vegetable proteins or globulins found in gluten, the protein of wheat, rye and other grains. Wheat gliadin has the composition: 52. 7% C, 17. 7% N, 21. 7% O, 6. 9% H, 1. 0% S. Insoluble in water, soluble in 70-90% alcohol, soluble in dilute acid and in alkali.
Use: In chemical synthesis of spinal anesthetics and in pharmaceutical preparations.

"Glidcol-Regular and WW." [296] Trade name for anti-skinning agents used predominantly for protective coatings.

"Glidfoam." [448] Trade name for polyurethane resins, basic resins for producing polyurethane foam resin.

"Glidpol." [448] Trade name for polyester resin, basic material for molding and laminating of reinforced plastic products.

"Glid-Rez." [448] Trade name for clear and pigmented coatings made from butoxy resins. Used in finishes for wide variety of products.

"Glid-Tile." [448] Trade name for polyester coatings for masonry block, concrete, wood, metal, plastic and wallboard.

"Globaline." [282] Trade name for tetraglycine hydroperiodide.
Properties: Active iodine 39. 5-42. 6%; ash 0. 5% max; total nitrogen 7. 35-7. 75%; meets government specification MIL-T-283C.
Containers: Drums.
Uses: Used in manufacture of water purification tablets.

"Globar." [280] Trademark for silicon carbide heating elements and resistors, and accessories used therewith.
Properties: Elements have a working temperature up to 2750°F and this range can be extended to 3000°F for short periods of time; low coefficient of expansion; structure of elements not affected by rapid heating and quick cooling; resistance of elements remains practically constant at temperatures above 900°F.
Uses: Electric resistors and heating elements; terminals and other accessories for electric heating elements; electric heating appliances; electric furnaces.

globin. The protein of hemoglobin and similar conjugate proteins.
Source: Hemoglobin.
Use: Medicine.

globin zinc insulin injection. See insulin.

globulin. Any of a group of simple proteins which is coagulated by heat, insoluble in water, soluble in dilute solutions of salts, strong acids and strong alkalies. Enzymes and acids cause hydrolysis of these simple proteins to produce amino acids as the only products. Examples are immune serum or gamma globulins in blood, edestin in hemp seed, myosin in muscle. The blood globulins are used in medicine, and are isolated by alcohol fractionation.

glonoin oil. See nitroglycerin.

"Gloria." [45] Trademark for white mineral oil, U. S. P.
Properties: Sp. gr. 0. 875-0. 885 (60°F); Saybolt viscosity 200-210 (100°F); pour point 15°F; odorless and tasteless.
Uses: Pharmaceutical and cosmetic formulations; plastics; tobacco; paper; animal husbandry.

GLPC. Abbreviation for gas-liquid partition chromatography. See gas chromatography.

glucagon (hyperglycemic-glycogenolytic factor; HG-factor; HGF). It is produced by the alpha cells of the islands of Langerhans and also, evidence indicates, by the gastric mucosa. It is opposite in effect to insulin, causing hyperglycemia by accelerating liver glycogenolysis but has no effect upon muscle glycogen. It has been isolated and crystallized and appears to be a straight-chain polypeptide with a molecular weight of

about 3500. Small amounts of glucagon have been detected in commercial insulin preparations.
Properties: Crystals; insoluble in water at pH 7.0 but readily soluble above pH 10.0. Soluble in acid solutions. Retains most of its activity after incubation with alkali or cystine.
Use: Medicine; biochemical research.

"Glucarine B." [73] Trademark for glycol carbohydrate complex.
Properties: Water-white, syrupy, clear fluid. Soluble in water, methyl alcohol, ethyl alcohol, glycerin, diethylene glycol, "Cellosolve," "Carbitol" and ethylene glycol. Insoluble in toluene, mineral spirits, mineral oil, and vegetable oil. Sp. gr. 1.32 (21°C).
Containers: 1-gal cans (11 lbs); 5-gal cans (55 lbs); 55-gal drums (630 lbs).
Uses: Replaces glycerin where a cheaper, colorless product is desired, especially for cosmetic and technical purposes.

glucase. See maltase.

glucinum. Obsolete name for beryllium.

glucinum compounds. See corresponding beryllium compounds.

gluconic acid (glyconic acid; dextronic acid; glycogenic acid) $CH_2OH(CHOH)_4COOH$.
Properties: Colorless or nearly colorless light brown syrupy liquid with mild acid taste. Soluble in water and alcohol.
Derivation: Bacterial, chemical, or electrochemical oxidation of dextrose.
Grades: Technical, 50% solution.
Containers: Barrels, kegs, glass carboys, tank cars.
Uses: Preparation of pharmaceutical and food products; in cleaning and pickling metals; sequestrant; cleansers for bottle washing, etc.
Shipping regulations: None.*

delta-**gluconolactone** $CH_2OH\overline{CH(CHOH)_3C(O)O}$.
Properties: White crystals.
Constants: M.p. 155°C; b.p. (decomposes); readily soluble in water, slightly soluble in alcohol.
Derivation: From gluconic acid.

D(+)-**glucosamine** $CH_2OH(CHOH)_3CHNH_2CHO$.
Properties: Colorless needles from alcohol; m.p. 110°C (dec); very soluble in water; very slightly soluble in methanol and ethanol; insoluble in ether and chloroform.
Use: Biochemical research.

D(+)-**glucose.** Identical with dextrose (q.v.). D(+)-glucose is the term preferred by biochemists.

glucose (liquid) (starch syrup; corn syrup). Thick, syrupy liquid, a mixture of dextrose, maltose, and dextrins with about 20% water.
Properties: Colorless to yellowish; soluble in water, glycerin; sparingly soluble in alcohol.
Derivation: By incomplete hydrolysis of starch and starchy substances by action of hydrochloric acid.
Grades: Technical; U.S.P. XVI.
Containers: 10-gal cans; 50-gal barrels.
Uses: Ingredient in confectionery, jelly, etc.; reducing agent; in alcoholic fermentations; adulterant in dyewood extracts; tanning; pharmaceuticals; treating tobacco.

glucose oxidase. Suggested as an oxygen scavenger, hence a preservative, for foods such as dried eggs.

glucose 6-phosphate (glucose 6-phosphoric acid; Robison ester) $C_6H_{11}O_5 \cdot H_2PO_4$. An intermediate in carbohydrate metabolism. Usually handled as the barium or dipotassium salts, which are water-soluble.

glucose 6-phosphoric acid. See glucose 6-phosphate.

alpha-**glucosidase.** See maltase.

beta-**glucosidase.** See emulsin.

glucosides. See glycosides.

glucosulfone sodium $C_{24}H_{34}N_2O_{18}S_3Na_2$. para, para'-Diaminodiphenylsulfone-N, N'-di(dextrose sodium sulfonate). Available only in a mixture containing about 88.5% of anhydrous para, para'-diaminodiphenylsulfone-N, N'-di(dextrose sodium sulfonate) and about 11.5% of dextrose.
Properties: White to faintly yellow, odorless, sweet tasting, amorphous solid. Freely soluble in water.
Grade: N.N.D.
Use: Medicine.

D(+)-**glucuronic acid** $COOH(CHOH)_4CHO$. A widely distributed substance in both plants and animals. It is usually found as part of a larger molecule as in various gums, or combined with phenols or alcohols.
Properties: Exhibits mutarotation. The beta form has m.p. 165°C. Soluble in water and alcohol.
Derivation: From gum acacia.
Use: Biochemical research; medicine.

glucuronolactone $C_6H_8O_6$. The gamma lactone of glucuronic acid. Found in plant gums and animal connective tissues.
Properties: Colorless, odorless, white powder. Sp. gr. 1.76 (30/4°C); m.p. 172-178°C; soluble in water.
Derivation: From glucuronic acid.
Grades: Purified; N.N.D.
Containers: 10-g, 4-, 8-oz vials.
Uses: Growth factor; medicine; pharmaceutical intermediate.

glue. An impure or degraded form of gelatin obtained by action of heat and water on protein animal tissues of bones, hides, horns. It absorbs cold water with much swelling and dissolves in hot water, the solution solidifying to jelly on cooling. Fish glue is obtained by heating with water the heads, fins, and tails of fish. It has weak jellying properties and is generally made into liquid glue. Liquid glue is made by treating fish or common glue with acetic, nitric or hydrochloric acids. The property of gelatinizing is lost by this operation but the adhesiveness is more or less unchanged. A partly decayed glutin obtained from flour in starch making is sometimes called albumin glue. Bone glue is an animal glue made wholly from bones. Casein glue is an adhesive made by dispersing casein in alkaline solutions. There are many formulas and many kinds of casein. It can be used cold. Chrome glue is an insoluble product prepared by mixing glue with ammonium or potassium dichromate or with chrome alum. It is used as a glass cement and for waterproofing materials. Glycerol (flexible; elastic) glue is glue dissolved in glycerol.

"Gluflex." [157] Trade name for a clear, balanced, non-crystallizing product consisting of sucrose, levulose, dextrose, sodium bisulfite and moisture, and having a solids content of 81.5%.
Properties: Hygroscopic; amber-colored; protected against aldehyde tanning of protein; soluble in water, glycerin, glycols and slightly soluble in alcohol.
Containers: Steel drums (approx. 640 lbs. net).
Uses: Adhesives containing animal protein, gummed tapes.

gluside. See saccharin.

gluside, soluble. See saccharin, sodium.

glutamic acid (alpha-aminoglutaric acid) $COOH(CH_2)_2CH(NH_2)COOH$. A nonessential amino acid present in all complete proteins. The naturally occurring form is L(+)-glutamic acid. The monosodium salt, L(+)-sodium glutamate, is important commercially as a flavoring intensifier.
Properties:
DL-glutamic acid: (synthetic racemic mixture) crystals; m.p. 225-227°C (dec); slightly soluble in ether, alcohol, and petroleum ether; sp. gr. 1.4601 (20/4°C).
D(−)-glutamic acid: leaflets from water; m.p. 247-249°C (dec); sp. gr. 1.538 (20/4°C).
L(+)-glutamic acid: crystals; sublimes at 200°C; decomposes at 247-249°C; nearly insoluble in ether, acetone, cold glacial acetic acid; insoluble in ethyl alcohol and and methanol; sp. gr. 1.538 (20/4°C).
Derivation: Hydrolysis of vegetable protein (e.g., beet sugar waste, wheat gluten); organic synthesis based on acrylonitrile.
Containers: Fiber drums.
Use: Medicine; biochemical research; source of sodium salt.
See also sodium glutamate; glutamic acid hydrochloride.
Available commercially as L(+)-glutamic acid.

glutamic acid hydrochloride (alpha-aminoglutaric acid hydrochloride) $COOH(CH_2)_2CH(NH_2)COOH \cdot HCl$.
Properties: White crystalline powder; sp. gr. 1.525; m.p. 202-213°C (dec); specific rotation (25°C) + 23.5° to + 25.5°. Very soluble in water, liberating hydrochloric acid; almost insoluble in alcohol and ether.
Derivation: Hydrolysis of gluten; organic synthesis.
Grades: N. F. XI.
Use: In medicine.
See also glutamic acid.

glutamine $H_2NC(O)(CH_2)_2CH(NH_2)COOH$. Both the L- and DL- forms are handled.
Properties: White, crystalline powder; soluble in water; insoluble in most organic solvents. Should be kept dry and refrigerated. M.p. (L-form) 184-185°C (dec.); (DL-form) 176°C.
Containers: Bottles.
Uses: Medicine; culture media; biochemical research.

gamma-**glutamylcysteinylglycine.** See glutathione.

"Glutan H-C-L." [57] Trademark for glutamic acid hydrochloride.

glutaraldehyde $OHC(CH_2)_3CHO$. Handled as 25% aqueous solution, sp. gr. 1.066 (20/20°C); b.p. 101°C; f.p. −5.8°C.
Uses: For cross-linking protein and polyhydroxy materials; imparting improved water resistance to textile sizes; tanning agent.

glutaric acid (n-pyrotartaric acid; pentanedioic acid) $COOH(CH_2)_3COOH$.
Properties: Colorless crystals; m.p. 97°C; refractive index (n 106/D) 1.419; b.p. 302-304°C; soluble in water, alcohol, and ether.
Derivation: From cyclopentanone.
Method of purification: Crystallization.
Grades: Technical.
Containers: Glass bottles; kegs.
Use: Organic synthesis.
Shipping regulations: None.*

glutaric anhydride (pentanedioic acid anhydride) $CH_2(CH_2CO)_2O$.
Properties: M.p. 56.5°C; b.p. 148.8°C (12 mm). Soluble in benzene and toluene; highly soluble in water on complete hydrolysis.
Uses: Plasticizers; resin; lubricant; adhesive synthesis; dyes and pharmaceuticals.

glutaronitrile (trimethylenedicyanide; pentanedinitrile) $NC(CH_2)_3CN$.
Properties: Colorless to straw-colored viscous liquid; b.p. 144-146°C (13 mm); sp. gr. 0.989; soluble in water and alcohol, insoluble in ether and carbon disulfide.
Use: Chemical intermediate.

glutathione (gamma-glutamylcysteinylglycine) $C_{10}H_{17}O_6N_3S$. A universal component of the living cell. Contains glutamic acid, cysteine, and glycine chemically bound. Hydrolysis sets free these three constituent amino acids.
Properties: White crystalline powder; odorless; m.p. 190-192°C; mild sour taste; soluble in water; insoluble in alcohol.
Use: Experimental work in nutrition and

metabolism, of especial interest as relates to radiation damage.

gluten. A mixture of proteins usually derived from corn or wheat, but applicable to similar material from other grains. Gluten is the protein present in flour and bread. Different proteins are present in gluten from different sources.

Properties: A yellowish to gray powder, or a gray brown sticky tough mass that is insoluble in water, soluble in alkali and in strong acetic acid. Crude commercial gluten contains up to 20% starch as a filler.

Derivation: Corn is steeped in water and the gluten and starch become suspended in the water and thus separated from other corn components. Settling or centrifugation permits separation of the starch from the gluten. A similar process can be used for recovering gluten from wheat flour or other sources.

Containers: Drums and cans.

Uses: In certain special breakfast foods and other cereals and foods; as gluten meal for cattle food; for adhesives; for production of certain amino acids.

glutethimide. See 2-ethyl-2-phenylglutarimide.

glutin. Amorphous, odorless, tasteless protein, having great adhesive strength. Soluble in hot water.

Derivation: A constituent of glue and gelatin.

glutoform. See glutol.

glutol (glutoform; formaldehyde gelatin).

Properties: A hard, clear transparent mass which may be pulverized, or white to yellow odorless powder; insoluble in cold water; soluble in hot water under pressure.

Derivation: By the action of a solution of formaldehyde upon gelatin.

"Glycamide." [123] Trademark for glycarbylamide (q. v.) for use as a coccidiostat.

glycarbylamide (4, 5-imidazoledicarboxamide) $C_3H_2N_2(CONH_2)_2$. White powder, melting above 360°C; insoluble in water. Used as a coccidiostat for chickens.

glyceraldehyde (glyceric aldehyde) $HOCH_2CHOHCHO$. Isomeric with dihydroxyacetone. It is produced by the oxidation of sugars in the body. As the simplest aldose, the conformation of d- and ℓ- glyceraldehydes has been designated the reference standard for D- and L- carbohydrates and derivatives.

Properties: (DL glyceraldehyde) Tasteless crystals from alcohol-ether mixture; m. p. 145°C; insoluble in benzene, petroleum ether, pentane.

Grades: 40% aqueous solution.

Uses: Biochemical research; intermediate; nutrition; preparation of polyesters, adhesives; cellulose modifier; leather tanning.

glyceric aldehyde. See glyceraldehyde.

glyceride. An ester of glycerol and fatty acids in which one or more of the hydroxyl groups of the glycerol have been replaced by acid radicals. The latter may be identical or different, so that the glyceride may contain up to three different acid groups. Glycerides may be made synthetically. The most common ones are based on fatty acids and occur naturally in oils and fats. Mono- and triglycerides (q. v.) are of commercial importance. See under glyceryl for examples.

glycerin. See glycerol.

glycerin carbonate (hydroxymethylethylene carbonate; 2, 3-carbonato-1-propanol) $CH_2O(CO)OCHCH_2OH$. 4-Hydroxymethyl-1, 3-dioxol-2-one.

Properties: Pale yellow, odorless, hygroscopic liquid. Boiling range 125-130°C (0. 1-0. 2 mm); freezing point, supercools to a glass; sp. gr. 1. 4000 (20/4°C); refractive index (n 20/D) 1.4580; flash point 415°F. Miscible with water, alcohol, ether; soluble in ethylene dichloride; insoluble in carbon tetrachloride, benzene, and aliphatic hydrocarbons.

Grades: Technical.

Uses: Solvent; intermediate.

glycerin, dynamite. A grade of glycerol (q. v.).

glycerinophosphoric acid. See glycerophosphoric acid.

glycerol (glycerin; glycyl alcohol) $C_3H_5(OH)_3$. The name glycerol is preferred over glycerin since the former indicates its alcohol structure.

Properties: Clear, colorless, or pale yellow, odorless, syrupy liquid; sweet, warm taste; hygroscopic; sp. gr. (anhydrous) 1. 2653, (U. S. P. XVI) greater than 1. 249 (25/25°C); (dynamite) 1. 2620; m. p. 18°C; b. p. 290°C; soluble in water and alcohol (aqueous solutions are neutral); insoluble in ether, benzene and chloroform. Flash point 177°C.

Derivation: (a) From the spent lye liquor from the saponification of fats and oils in the soap industry, by precipitation of salt, albuminoids and metallic soaps of the higher fatty acids by iron persulfate (crude) or aluminum sulfate and concentration with subsequent steam distillation; (b) from propylene or allyl alcohol by chlorination, and hydrolysis; (c) from acrolein and hydrogen peroxide followed by reduction of the glyceraldehyde.

Method of purification: Redistillation; ion exchange techniques.

Grades: U. S. P. XVI; C. P. (for pharmaceutical and commercial purposes where highest grade of glycerol is required); saponification, soap lye, crude yellow distilled (for commercial purposes where color and extreme purity are not factors); high gravity or dynamite (dehydrated to 99. 8-99. 9% purity); natural; synthetic; etc.

Containers: Drums; tank cars.

Uses: Alkyd resins; explosives; ester gums; pharmacy; perfumery; plasticizer for regenerated cellulose; cosmetics; foodstuffs (preservative, sweetening); conditioning tobacco; liqueurs; solvent; printer's ink rolls; emulsifying agent; rubber stamp and

copying inks; binder for cements and mixes; anti-freeze; paper coatings and finishes; special soaps; lubricant and softener; bacteriostat; penetrant; solvent; hydraulic fluid; humectant.
Shipping regulations: None. *

glycerol boriborate. Pale yellow liquid obtained by heating glycerin, sodium borate and boric acid. Composition varies. Soluble in cold water, absolute alcohol, other alcohols, glycerin. Used as adhesive, binder, fabric softener, fire retardant on fabrics.

glycerol dichlorohydrin. See alpha-dichlorohydrin.

glycerol glue. See glue.

glycerol mannitan laurate. A mixed polyalcohol fatty acid ester.
Properties: Red-brown oily liquid. Dispersible or insoluble in water and most solvents.
Containers: 1-, 5-gal, 480-lb drums.
Uses: Wetting agent; spreading agent in insecticides.

glycerol monolaurate (glyceryl monolaurate; lauryl glycerin) $C_{11}H_{23}COOCH_2CHOHCH_2OH$.
Properties: Cream-colored, semi-solid paste; very faint odor. Dispersible in water; soluble in methanol, ethanol, toluene, naphtha, mineral oil, cottonseed oil, ethyl acetate.
Typical specifications: M. p. 23-27°C; sp. gr. 0.98; F. F. A. less than 2.5%; iodine value 6-8; pH 8.0-8.6 (25°C) (5% aqueous dispersion).
Derivation: An ester of glycerol and lauric acid. See monoglycerides.
Containers: 1-, 5-gal cans; 220-, 400-lb drums.
Uses: Emulsifying and dispersing agent for the manufacture of food products. Emulsifying agent for oils, waxes and solvents; antifoaming agent; dry-cleaning soap base.

glycerol monooleate (glyceryl monooleate) $(C_{17}H_{33})COOCH_2CHOHCH_2OH$.
Properties: A yellow oil; sp. gr. (25°C) 0.94; m. p. −5°C; insoluble in water; soluble in alcohol and most organic solvents.
Derivation: See monoglycerides.
Containers: 1-, 5-gal cans, 55-gal drums.

glycerol monoricinoleate (glyceryl monoricinoleate) $C_{17}H_{33}OCOOC_3H_5(OH)_2$.
Properties: Orange-red, oily liquid; titer below 0°C; sp. gr. 1.02; iodine value 65-70; F. F. A. less than 2.5%; pH 9.2-9.5 (25°C) (5% aqueous dispersion); soluble in methanol, ethanol, toluene, cottonseed oil, ethyl acetate; disperses in water; insoluble in naphtha, mineral oil.
Derivation: See monoglycerides.
Grades: Technical.
Containers: 1-, 5-gal cans; 55-gal drums.
Uses: Non-drying emulsifying agent; solvent; plasticizer; in polishes, in cosmetics, in textile, paper, and leather processing.

Lowers surface tension. High lubricating value even at low temperatures.

glycerol monostearate (GMS; glyceryl monostearate; monostearin) $(C_{17}H_{35})COOCH_2CHOHCH_2OH$.
Properties: Pure white or cream-colored, wax-like solid with faint odor, and fatty, agreeable taste. Affected by light.
Constants: M. p. 58-59°C (capillary tube); sp. gr. 0.97; F. F. A. less than 5%; iodine value 3-4; pH 9.3-9.7 (25°C) (3%).
Dispersible in hot water. Soluble (hot) in alcohol, oils and hydrocarbons.
Derivation: Stearic acid ester of glycerol. See monoglycerides.
Grades: Edible; cosmetic; N. F. XI.
Containers: 1-, 10-, 50-, 150-, 500-lb drums, barrels.
Uses: Thickening and emulsifying agent for margarine, shortenings and other food products; emulsifying agent for oils, waxes and solvents; protective coating for hygroscopic powders; cosmetics; opacifier; detackifier.

"Glycerol Monostearate 860." [260] Proprietary brand of blend of mono and diglycerides. "Glycerol Monostearate 866" is the dispersible type. Mono content 45%. Balance diester with small amounts of triester, free fatty acid and free glycerol. Waxy solid; white; mild fatty odor. Acid value 3.0 max (860); 8.0 max (866). Iodine value 3.0 max; m. p. 57-60°C; insoluble in water (860), dispersible in water (866); 860 is nonionic; 866 is anionic; flash point (860) (open cup) 448°F; (866) 394°F.
Uses: Emulsifier, thickener, opacifier in cosmetic, food, pharmaceutical, textile and other emulsion systems.

glycerolphosphoric acid. See glycerophosphoric acid.

glycerol phthalate. See glyceryl phthalate.

glycerol tributyrate. See glyceryl tributyrate.

glycerol tripropionate. See glyceryl tripropionate.

glycerol tristearate. See stearin.

glycerophosphoric acid (glycerolphosphoric acid, glycerinophosphoric acid) $C_3H_5(OH)_2H_2PO_4$.
Properties: Colorless, odorless liquid. Soluble in water and alcohol.
Constants: Sp. gr. 1.60; m. p. −20°C.
Derivation: By the interaction of glycerol and phosphoric acid.
Method of purification: Distillation.
Grades: Technical.
Containers: 1-oz, 1-lb bottles; carboys.
Uses: Medicine; manufacture of glycerophosphates.
Shipping regulations: None. *

glycerophosphoryl inositol (GPI). A phospholipid extractable in the water-soluble fraction from alfalfa, algae, and other plant matter. Characterized by a high degree of surface-active properties.
Use: Biochemical research.

glyceryl abietate. An ester gum used as a food additive in citrus-flavored beverages.

glyceryl benzoate (tribenzoin) $C_3H_5(OOCC_6H_5)_3$.
Properties: Colorless, non-hygroscopic, crystalline solid. Soluble in most anhydrous lacquer solvents with the exception of petroleum hydrocarbons.
Constants: Sp. gr. 1.25; m. p. 71°C.

glyceryl alpha-chlorohydrin. See chlorohydrin.

glyceryl diacetate. See diacetin.

glyceryl ditolyl ether. See dicresyl glyceryl ether.

glyceryl heptadecanoate. See intarvin.

glyceryl margarate. See intarvin.

glyceryl monoacetate. See acetin.

glyceryl monolaurate. See glycerol monolaurate.

glyceryl monooleate. See glycerol monooleate.

glyceryl monoricinoleate. See glycerol monoricinoleate.

glyceryl monostearate. See glycerol monostearate.

glyceryl phthalate (glycerol phthalate).
Properties: Water-white, solid resin. Insoluble in water. Soluble (hot) in methanol, ethanol, acetone, ethyl acetate. Partly soluble in toluene, naphtha.
Constants: Sp. gr. 1.29; saponification value 605-615; acid value 300-315; softening point about 67°C.
Grades: Technical.
Containers: 1-gal cans (10 lbs); 5-gal cans (50 lbs); 55-gal drums (580 lbs).
Uses: Synthetic resin used in the manufacture of varnishes, lacquers, etc.

glyceryl ricinoleate. See glyceryl triricinoleate.

glyceryl triacetate. See triacetin.

glyceryl tri-(12-acetoxystearate) (castor oil, acetylated and hydrogenated) $C_3H_5(OOCC_{17}H_{34}OCOCH_3)_3$.
Properties: Clear, pale yellow, oily liquid; mild odor; soluble in most organic solvents; insoluble in water. Sp. gr. 0.955 (25/25°C); saponification value 298; iodine value 3; solidifies at 4°C.
Derivation: Hydrogenation of acetylated castor oil.
Grade: Technical.
Containers: 5-gal cans; 55-gal drums; tank cars.
Uses: Plasticizer for nitrocellulose, ethylcellulose and polyvinyl chloride; lubricants; protective coatings.

glyceryl tri-(12-acetylricinoleate) (castor oil, acetylated) $C_3H_5(OOCC_{17}H_{32}OCOCH_3)_3$.
Properties: A clear, pale yellow, oily liquid; mild odor; soluble in most organic liquids; insoluble in water. Sp. gr. 0.967 (25/25°C); saponification value 300; iodine value 76; solidifies at −40°C.
Grade: Technical.
Derivation: Acetylation of castor oil.
Containers: 5-gal cans; 55-gal drums; tank cars.
Uses: Plasticizer for nitrocellulose, ethylcellulose and polyvinyl chloride; lubricants; protective coatings.

glyceryl tributyrate (tributyrin; glycerol tributyrate) $C_3H_5(OCOC_3H_7)_3$.
Properties: Solid; sp. gr. (20°C) 1.035; refractive index (20°C) 1.4359; b. p. 315°C; m. p. less than −75°C; solubility in water 0.010%.
Containers: Drums.
Use: plasticizer.

glyceryl tricaprinate. See tricaprin.

glyceryl tri-[12-hydroxystearate] (castor oil, hydrogenated) $C_3H_5(OOCC_{17}H_{34}OH)_3$. Glyceryl triricinoleate which has had the double bond of each ricinoleic group saturated with hydrogen.
Properties: Hard, brittle wax-like solid, yellowish cream to milk white in color.
Typical specifications: M. p. 86-88°C; sp. gr. (100/25°C) 0.899.
Derivation: Hydrogenation of castor oil.
Impurities: Glyceryl stearate.
Uses: Lubricants; metallic soaps; waxes; plasticizers, cosmetics; chemical intermediate. The lithium compound is used in high temperature special greases.

glyceryl trinitrate. See nitroglycerin.

glyceryl trioleate. See olein.

glyceryl tripalmitate. See tripalmitin.

glyceryl tripropionate (glycerol tripropionate) $C_3H_5(OCOC_2H_5)_3$.
Properties: Solid; sp. gr. (20°C) 1.078; refractive index (20°C) 1.431; b. p. (20 mm) 177-182°C; m. p. less than −50°C; solubility in water, 0.313% of weight.
Use: Plasticizer.

glyceryl triricinoleate (glyceryl ricinoleate) $C_3H_5(OOCC_{17}H_{32}OH)_3$. The triglyceride of ricinoleic acid. It constitutes about 80% of castor oil.
Properties: A light amber oil.
Derivation: Castor oil is refined.
Uses: An emulsifying oil.

glyceryl tristearate. See stearin.

glycidol (2, 3-epoxy-1-propanol) CH_2OCHCH_2OH.
Properties: Colorless liquid; b. p. 162°C; soluble in water, alcohol and ether.
Derivation: Treatment of monochlorohydrins with bases.
Uses: Stabilizer for natural oils; demulsifier; dye-leveling agent; stabilizer for vinyl polymers.

glycine (amino acid) (aminoacetic acid, glycocoll) NH_2CH_2COOH. The principal amino acid in sugar cane.
Properties: White, very sweet odorless crystals; m. p. 232-236°C with decomposition; sp. gr. 1.1607; combines with

hydrochloric acid to form the hydrochloride; soluble in water; insoluble in alcohol and ether.
Derivation: By the action of concentrated ammonium hydroxide on monochloroacetic acid, or by alkaline hydrolysis of gelatin.
Grades: Technical; N. F. XI.
Containers: Glass bottles; barrels.
Uses: Organic synthesis; medicine; biochemical research; buffering agent.
Caution: Not to be confused with the photographic developer, para-hydroxyphenylaminoacetic acid, also known as glycine, which is poisonous, nor with the perfume.

glycine (perfume). The extreme dilution of methyl-para-tolyl ketone gives a perfume resembling the odor of the climbing plant glycine (Wistaria sinensis), native to China and cultivated elsewhere. The name is also given to bouquets made from violet, lilac and jasmin ottos.

Not to be confused with the glycine which is aminoacetic acid nor with para-hydroxyphenyl glycine, which is poisonous.

glycine (photographic). See para-hydroxyphenyl glycine. Note that this compound is poisonous!

glycine ethyl ester hydrochloride (ethyl glycine hydrochloride; ethyl glycocoll hydrochloride) $H_2NCH_2COOC_2H_5 \cdot HCl$.
Properties: White to pale yellow crystals; m. p. 140°C; soluble in water and alcohol; insoluble in hydrocarbons.
Grades: 97% min purity.
Uses: Synthesis; rocket propellants.

glycobiarsol (bismuth N-glycolylarsanilate; bismuth para-glycolylaminophenylarsonate) $HOCH_2CONHC_6H_4AsO_3BiO$.
Properties: Odorless, yellowish white or flesh-colored, amorphous powder. Decomposes when heated. Very slightly soluble in alcohol and water; insoluble in benzene, chloroform and ether. pH (saturated solution) 2.8-3.5.
Grade: U. S. P. XVI.
Use: Medicine.

glycocholic acid (cholylglycine) $C_{26}H_{43}NO_6$. Occurs as sodium salt in bile. It is formed by the combination of the amino acid glycine with cholic acid (q. v.). As the sodium salt, it aids in the digestion and absorption of fats.
Properties: Crystallizes from water containing 1.5 moles H_2O. Becomes anhydrous at 100°C. Anhydrous form decomposes at 154-155°C. Practically insoluble in water. The sodium salt is soluble in water and alcohol.
Derivation: Precipitation from bile.
Use: Biochemical research.

glycocoll. See glycine (amino acid).

glycocoll copper. See copper glycinate.

glycogen (animal starch; liver starch) $(C_6H_{10}O_5)_n$. A glycose polysaccharide. It is the storage carbohydrate of the animal organism, found especially in the liver and rested muscle.
Properties: White powder; forms a dextrorotatory colloidal solution; partially soluble in water; sweet tasting.
Derivation: Isolated from liver by treatment with 30% sodium hydroxide solution and precipitating glycogen from the solution with alcohol.
Use: Biochemical research.

glycogenic acid. See gluconic acid.

glycol. See ethylene glycol; it is also a general term for dihydric alcohols.

glycol carbonate. See ethylene carbonate.

glycol chlorohydrin. See ethylene chlorohydrin.

glycol diacetate. See ethylene glycol diacetate.

glycol dibutyrate. See ethylene glycol dibutyrate.

glycol diformate. See ethylene glycol diformate.

glycol dimercaptoacetate
$HSCH_2COOCH_2CH_2OOCCH_2SH$. B. p. 137-139°C (2 mm); insoluble in water; soluble in organic solvents. Purity 95% min.
Uses: Crosslinking agent for rubbers; accelerator in curing epoxy resins.

glycol dipropionate. See ethylene glycol dipropionate.

glycolic acid (glycollic acid; hydroxyacetic acid) $CH_2OHCOOH$.
Properties: Colorless crystals, deliquescent; m. p. 78-79°C; soluble in water, alcohol and ether.
Derivation: From chloroacetic acid by boiling with water or aqueous alkali; by oxidation of glycol.
Method of purification: Crystallization.
Containers: Kegs; special lined fiber drums; tank cars.
Uses: Pesticides; plasticizers; salts as catalysts; pharmaceuticals; dyeing.
See also hydroxyacetic acid, technical.

glycollic acid. See glycolic acid.

glycol monoacetate. See ethylene glycol monoacetate.

glycolonitrile (glyconitrile; formaldehyde cyanohydrin) $HOCH_2CN$.
Properties: Mobile, colorless, odorless oil. Supplied commercially as a 70% aqueous solution stabilized with a small amount of phosphoric acid. B. p. 183°C (759 mm) (slight decomposition); m. p., does not solidify when cooled to −72°C. Sp. gr. 1.1039 (19°C); refractive index (n 25/D) 1.4090; electrolytic dissociation constant $K = 0.843 \times 10^{-5}$ (25°C); pH of stabilized 50% aqueous solution = 1.042 (20°C).
Derivation: Formaldehyde and hydrocyanic acid.
Containers: (70% solution) Drums; tank cars.
Uses: Solvent and organic intermediate.

glycol phthalate, polymerized.
Properties: Light-brown, soft wax. Sp. gr. 1.19; acid value 10-15; free fatty acid 1-2%; sap. value 26.5-28; pH 2.65 (25°C)

(5%); m. p. 55°C. Soluble (hot) in water, methanol, ethanol, toluene, acetone, and ethyl acetate. Insoluble (hot) in naphtha, mineral oil, cottonseed oil.
Derivation: Phthalic acid ester of polymerized glycol.

glycol propionate. See ethylene glycol dipropionate.

glycol ricinoleate.
Properties: A light-amber self-emulsifying oil which forms milky emulsions with cold water. It is readily miscible with alcohols, hydrocarbons, solvents, and oils.

glycol salicylate (glycol monosalicylate) $C_6H_4OHCOO(CH_2CH_2OH)$.
Properties: Colorless, oily liquid; soluble in alcohol, ether, chloroform, benzene, olive oil; less soluble in water. B. p. 169-170°C (12 mm).
Use: Medicine.
Shipping regulations: None. *

glycol stearate. See ethylene glycol stearate.

glycolylurea. See hydantoin.

"**Glycomuls.**" [73] Trade name for a series of sorbitol fatty acid esters, ranging from liquids to relatively high-melting wax-like solids and with varying surface-active characteristics. Used in foods, cosmetics, pharmaceuticals, chemical specialties.

glyconic acid. See gluconic acid.

glyconitrile. See glycolonitrile.

glycosides. A group of organic compounds, of abundant occurrence in plants, which can be resolved by hydrolysis into sugars and other organic substances, known as aglycones. Specifically glycosides are acetals which are derived from a combination of various hydroxy compounds with various sugars. They are designated individually as glucosides, mannosides, galactosides, etc. , and as a group are called glycosides. For example, the action of methanol on glucose or mannose or galactose yields methyl glycosides. Glycosides were formerly all called glucosides, but the latter term is more correctly applied to any glycoside having glucose as its sugar constituent.

"**Glycosine.**" [19] Brand name for a proprietary product, para-phenetolecarbamide. Synthetic sweetener, 200 times sweeter than sugar.

"**Glycosperses.**" [73] Trade name for a series of sorbitan esters modified with varying chain lengths of ethylene oxide, as well as reaction products of fatty acids and fatty alcohols with ethylene oxide. They are hydrophilic surface-active agents, used like "Glycomuls."

"**Glyco Wax**" **S932.** [73] Trademark for a fatty acid ester.
Properties: Cream-colored, brittle wax. Somewhat harder than beeswax. High luster; insoluble in water; soluble (hot) in toluene, naphtha, mineral spirits, mineral oil, vegetable oil. M. p. 58-60°C; sp. gr. 0. 965 (25°C); F. F. A. 0. 1%; iodine value, max 10; flash point 328°C.
Containers: 1-gal cans (8-lb slab); 5-gal cans (50-lb slab); 500-lb drums (50-lb slab).

"**Glycox Emulsifiers.**" [73] Trade name product. Modified esters of polyhydric alcohol ethers.

The "Glycox Emulsifiers" are a series of non-ionic emulsifiers and wetting agents suggested for the manufacture of agricultural sprays and insect toxicant emulsions, especially when emulsions are to be prepared under the severe conditions of hard water and solutions of electrolytes and acids. They are water soluble or dispersible and are soluble in a wide range of organic solvents. Of particular interest is their compatibility with anionic and cationic surface-active agents giving increased emulsifying and penetrating properties.

glycyl alcohol. See glycerol.

glycyrrhiza (licorice).
Properties: Glossy black flattened, cylindrical rolls or in masses. Characteristic sweet taste. Yields a brown powder when pulverized. Insoluble in absolute alcohol and ether.
Habitat: United States, southern Europe to central Asia.
Derivation: The evaporated extract of Glycyrrhiza glabra or glandulifera.
Grades: Technical; U. S. P. XVI.
Containers: Boxes; cartons; barrels.
Uses: Medicine; flavoring tobacco and candy. Ammoniated licorice extract is used as a foam producer.
Shipping regulations: None. *

"**Glydag**" **B.** [46] Trademark for a concentrated colloidal dispersion of pure electric-furnace graphite in 1, 3-butylene glycol.
Properties: Liquid consistency; solids content 10%; sp. gr. 1. 07; completely miscible with glycols, water.
Uses: Formulation of lubricants for mechanical rubber parts and surfaces.

"**Glydote.**" [206] Brand name of proprietary textile printing solvent and assistant.

"**Glyecine**" **A.** [307] Trademark for a dyeing assistant, comprising thiodiethylene glycol; 100% active.
Properties: Clear, thin, yellow liquid; sp. gr. 1. 18; soluble in water; unaffected by alkalies, reducing agents, ordinary acids and other ingredients present in printing pastes.
Uses: Hygroscopic agent in textile printing solvent for basic colors.

glyodin (2-heptadecylglyoxalidine acetate; 2-heptadecyl 2-imidazoline acetate) $C_{17}H_{35}C_3H_5N_2 \cdot CH_3COOH$.
Properties: Light orange crystals; m. p.

*See "I. C. C. Shipping Regulations," page xiii.
Reference numbers refer to name of manufacturer. See "List of Manufacturers," page v.

94°C; insoluble in water.
Use: Fungicide.
Caution! May irritate eyes and skin. MCA warning label. *

glyoxal OHCCHO.
Properties: Yellow crystals or light yellow liquid; mild odor; m. p. 15°C; b. p. 51°C; sp. gr. 1.26 (20/20°C); 10.0 lbs/gal (20°C). Vapor has a green color and burns with a violet flame; refractive index (n 20/D) 1.3826; polymerizes on standing or in presence of a trace of water. An aqueous solution contains monomolecular glyoxal, and reacts weakly to acid. Undergoes many addition and condensation reactions with amines, amides, aldehydes, and hydroxyl-containing materials.
Derivation: Oxidation of acetaldehyde.
Grades: Available as a 30% aqueous solution consisting of various hydrated forms of glyoxal with small amounts of chemically related substances such as glycolic acid, formic acid, glycol, and formaldehyde.
Containers: (solid) Tins or fiber drums; (liquid) drums; tank cars.
Uses: Used as an insolubilizing agent for compounds containing polyhydroxyl groups such as polyvinyl alcohol, starch, "Cellosize" hydroxyethyl cellulose, and cellulosic materials. Used for the insolubilizing of proteins such as casein, gelatin and animal glue, in embalming fluids, in leather tanning, and for dimensionally stabilizing or "shrink-proofing" rayon. Used with "Cellosize" hydroxyethyl cellulose it imparts wet strength to paper stocks without loss of absorbency. Also used as a reducing agent in dyeing textiles.

glyoxaline. See imidazole.

glyoxyldiureid. See allantoin.

"Glyptal" Resins. [245] Trademark for a group of alkyd-type synthetic resins and plasticizers.

"G-M-F." [248] Trade name for para-quinonedioxime, $HONC_6H_4NOH$.
Properties: Dark brown powder. Sp. gr. 1.40; decomposes above 215°C. Good storage stability. Slightly soluble in acetone. Insoluble in water, gasoline, benzol and ethylene dichloride.
Use: Rubber vulcanizing agent, used in conjunction with red lead for fast-curing high modulus stock.

GMP. Abbreviation for guanosine monophosphate. See guanosine phosphates; also guanylic acid.

GMS. Abbreviation for glycerol monostearate.

"G.N.S." No. 5. [79] Trademark for pine oil.
Properties: Color, yellow; sp. gr. (15.5°C) 0.932; refractive index (20°C) 1.482; flash point (open cup) 167°F. Engler distillation: 5% at 200°, 50% at 212°, 95% at 220°C.
Containers: 55-gal drums; tank cars.
Uses: Mining-flotation; textile dyeing and cleaning; laundries; disinfectants; insecticides; deodorants; cleaning compounds; coated paper; paint and varnish; paint (casein); pharmaceuticals.

goa powder. See araroba.

goethite FeOOH. A natural hydroxy-oxide of ferric iron.
Properties: Color yellowish brown to dark brown; streak yellowish brown; luster dull to adamantine; hardness 5-5.5; sp. gr. 4.37. Occurs as radiating fibrous masses. Probably most of the material formerly called limonite is actually goethite with adsorbed water.
Occurrence: Widespread. Minnesota, Michigan, Alabama, Georgia, Texas.
Use: Ore of iron.

gold Au. Element of atomic number 79, of group Ib of the periodic system.
Properties: Yellow, ductile metal; does not corrode in the atmosphere but is attacked by chlorine and by cyanide solutions in the presence of oxygen. Soluble in aqua regia; insoluble in acids.
Constants: Sp. gr. 19.2; m. p. 1062°C; b. p. 2530°C.
Derivation: Generally found native enclosed in quartz with iron pyrite and other minerals, or as gold telluride and recovered by amalgamation with mercury or solution in cyanide (sodium or potassium), followed by precipitation and fusion. See also amalgam, calaverite, krennerite, petzite, sylvanite. Gold is found all over the world, but the great producing centers today are South Dakota, Utah, Alaska, California, South Africa, Australasia, Canada (Ontario, Quebec, Northwest Territory, Klondike region in Yukon Territory and British Columbia), India, China, Russia and Mexico.
Forms available: Sheet, wire, tubing, granulated, leaf, foil. Generally used in the form of alloys with copper or other metals; the gold content being expressed in carats, that is, the number of parts of gold in 24 parts of the alloy. Gold powder and sheet are available up to 99.999% pure.
Containers: Canvas sacks; wooden boxes.
Uses: Coins; jewelry; gold salts; dentistry; amalgams; gilding; decoration; gold leaf; gold plating; laboratory ware; anodes; alloys; solders; in electronics, as semiconductors, etc.
Shipping regulations: None. * Sale in U.S. and Canada subject to governmental regulations.

gold 198. Radioactive gold of mass number 198.
Properties: Half-life, 2.7 days; radiation, beta and gamma; radiotoxicity, moderately hazardous.
Derivation: Pile irradiation of gold metal.
Forms available: Gold metal, colloidal gold (see radiogold) and gold sodium thiosulfate.
Uses: Internal radiation therapy; to detect leaks in bacterial filters; to locate solidification boundary in continuously cast aluminum; to determine metallic silver

*See "I.C.C. Shipping Regulations," page xiii.
Reference numbers refer to name of manufacturer. See "List of Manufacturers," page v.

in photographic materials, etc. The decay product of Au-198 is stable mercury, Hg-198, which may be distilled from aged neutron irradiated gold for the fabrication of monoisotopic mercury arc light sources.
Shipping regulations: Poison, class D, radioactive material. Red label. *

gold alkyl mercaptides. Used for "thermal" gold plating of plastics, stainless steel, aluminum, magnesium and titanium. These mercaptides decompose at comparatively low temperatures (400-500°F) to form a continuous metallic film. The compounds used are the tert-butyl, tert-dodecyl, tert-octyl and tert-hexadecyl gold mercaptides.

gold, artificial. See stannic sulfide.

gold bloom. See calendula.

gold, blue. An alloy used by jewelers. It contains about 75% gold and 25% iron.

gold bromide. See gold monobromide; gold tribromide.

gold bronze powder. See aluminum bronze powder.

gold, burnish. Liquid mixtures containing gold or organic compounds of gold in solution with or without platinum and silver, which may be added to whiten or impart a green cast to the product.

gold, ceramic decorating. Gold in the form of powder, paste, or liquid for application on ceramic materials. Combined with suitable fluxes and vehicles for particular application.
Brown gold (gold powders). Finely divided, dry preparations containing gold and usually a flux. For dry application by dusting on a tacky surface to produce heavy gold deposits.
Uses: Decorating china, pottery, tile, etc., with fired-on gold designs; for increasing gold content of other gold preparations.
Burnish golds. Finely divided gold powder combined with suitable fluxes and incorporated into low-viscosity vehicles for brush application. Produce heavy, stable, long-wearing gold deposits. Fired to a matt surface; the true gold surface texture developed by burnishing.
Uses: Decorating high grade china and ceramicware with fired-on gold designs.
Electronic golds. Special preparations for conductive coating applications where resistance to strong acids and migration is required.
Uses: For applications on semiconductors (transistors, diodes, etc.), special capacitors, and printed circuits.
Liquid bright golds. Liquid compositions suitable for application by brushing, stamping, spraying or stenciling (squeegee). Fired to a bright mirror surface.
Uses: Decorating pottery, glass, tile, terra cotta, enameled metals.
Paste golds. Heavy compositions of powdered gold and flux; suitable for printing by tissue transfer method or can be diluted for brush application. Fired to matt surface which is burnished to develop true gold texture.
Uses: Decorating high grade chinaware.

gold chloride (a) $AuCl_3$ (auric chloride; gold trichloride); (b) $AuCl_3 \cdot 2H_2O$; (c) $AuCl_3 \cdot HCl \cdot 4H_2O$ or $HAuCl_4 \cdot 4H_2O$ (chlorauric acid; gold trichloride acid).
Properties: Yellow to red crystals; decomposed by heat; soluble in water, alcohol and ether.
Derivation: The action of aqua regia on gold.
Method of purification: Crystallization.
Grades: Technical; C. P., usually as chlorauric acid.
Containers: Glass bottles.
Uses: Photography; gold plating; special inks; medicine; ceramics (enamels, gilding and painting porcelain); glass (gilding, ruby glass); manufacture of finely divided gold; manufacture of purple of Cassius.
Shipping regulations: None. *

gold, colloidal. See collaurin.

golden seal. See hydrastis.

gold hydrate. See gold hydroxide.

gold hydroxide (auric hydroxide; gold hydrate) $Au(OH)_3$.
Properties: Brown powder. Sensitive to light. Caution! Keep in amber bottle! The hydroxide is probably a hydrated trioxide of gold Au_2O_3, and loses water easily. Soluble in hydrochloric acid, solutions of sodium cyanide and alkali hydroxides; insoluble in water.
Grades: Technical.
Uses: Daguerreotypes; gilding liquids; medicine; porcelain; rubber colorization (false teeth); gold plating.

gold iodide (aurous iodide) AuI.
Properties: Greenish-yellow powder. Decomposes slowly. Very slightly soluble in water. Sp. gr. 8.25.

gold leaf. Thin plates of gold that have been placed between layers of gold-beater's skin and hammered until they have spread to a state of extreme fineness. Gold leaf is used for gilding works of art, fabrics, books, etc. See also Dutch metal.

gold monobromide (aurous bromide; gold bromide) AuBr.
Properties: Yellowish-gray mass. Decomposes at 165°C (approx). Insoluble in water.
Grades: Technical.

gold orange. See methyl orange.

gold oxide (auric oxide; auric trioxide; gold trioxide) Au_2O_3.
Properties: Brownish-black powder; decomposed by heat. Caution! Keep in dark bottle. Soluble in hydrochloric acid; insoluble in water.
Grades: Technical.
Uses: Daguerreotypes; gilding liquids; medicine; porcelain; rubber colorization.

gold-potassium bromide (potassium auribromide) $AuBr_3 \cdot KBr \cdot 2H_2O$.
Properties: Violet crystals. Soluble in alcohol, water.

gold-potassium chloride (potassium aurichloride) $AuCl_3 \cdot KCl \cdot 2H_2O$.
Properties: Yellow crystals. Soluble in water, alcohol and ether.
Derivation: By neutralizing chlorauric acid with potassium carbonate.
Grades: Technical.
Containers: Glass bottles.
Uses: Photography; painting porcelain and glass; medicine.
Shipping regulations: None. *

gold potassium cyanide (potassium cyanaurite) $KAu(CN)_2$.
Properties: White, crystalline powder; poisonous! Soluble in water; slightly soluble in alcohol; insoluble in ether.
Derivation: By the action of hydrocyanic acid on potassium aurate.
Method of purification: Crystallization.
Grades: Technical.
Containers: Glass bottles, 1 and 10 oz., net.
Uses: Medicine; electrogilding.

gold-potassium iodide (potassium auric iodide) $AuI_3 \cdot KI$.
Properties: Black, lustrous crystals; m. p. 150°C, with decomposition; soluble in solution of potassium iodide (dilute), water (decomposes).
Grades: Technical.

"Gold Shield." [319] Trademark for a highly refined grain alcohol; high purity and freedom from foreign odor or flavor.
Uses: Industrial, pharmaceutical, and cosmetic purposes.

gold-silicon alloy (silicon-gold alloy). Formed in amorphous foils 10 microns thick by cooling molten gold and silicon almost instantaneously by spreading on a moving wheel. The atoms are "frozen" before crystals can form. Used in electronics.

gold size. A mixture of 2 parts copal varnish, 1 part yellow ocher, 4 parts turpentine, and 8 parts boiled oil; an adhesive composition used to prepare the surface before applying gold leaf.

gold-sodium bromide (sodium auribromide) $AuBr_3 \cdot NaBr \cdot 2H_2O$.
Properties: Brown-black crystals. Soluble in water.
Use: Medicine.

gold-sodium chloride (sodium-gold chloride; sodium aurichloride; sodium chloraurate; sodium chloroaurate) $NaAuCl_4 \cdot 2H_2O$.
Properties: Yellow crystals; soluble in water and alcohol.
Derivation: By neutralizing chloroauric acid with sodium carbonate.
Method of purification: Crystallization.
Containers: Glass bottles.
Grades: Technical.
Uses: Photography; staining fine glass; decorating porcelain; medicine.
Shipping regulations: None. *

gold-sodium cyanide $NaAu(CN)_2$.
Properties: A yellow powder; soluble in water; contains 46% gold (min).
Containers: Bottles (1, 10-ozs., net).
Use: For gold-plating radar and electric parts, band instruments, razor holders, lamps, clocks, jewelry and tableware.

gold sodium thiomalate
$NaOOCCH(SAu)CH_2COONa \cdot H_2O$.
Properties: Fine, white to yellowish-white, odorless powder with metallic taste; affected by light. Very soluble in water; practically insoluble in alcohol and ether. Aqueous solutions are colorless to pale yellow; pH (5% solution) 5.8-6.8.
Derivation: Reaction of sodium thiomalate with a gold halide.
Grade: N. F. XI.
Use: Medicine.

gold-sodium thiosulfate (aurous sodium thiosulfate; sodium aurothiosulfate) $Na_3Au(S_2O_3)_2 \cdot 2H_2O$.
Properties: White crystals. Odorless; contain 37% (approx) gold. Darkens on exposure to light. Soluble in water; insoluble in alcohol.
Grade: N. F. XI.
Use: Medicine.

gold solder. A solder usually composed of gold, silver, copper, zinc, or brass and used principally by jewelers.

gold thioglucose. See aurothioglucose.

gold-tin precipitate. See gold-tin purple.

gold-tin purple (purple of Cassius; gold-tin precipitate).
Properties: Brown powder. Insoluble in water; soluble in ammonia.
Derivation: By the reaction of a neutral solution of gold chloride with stannous and stannic chlorides, yielding a mixture of colloidal gold and tin oxide in varying proportions.
Grade: Technical.
Containers: Tins; glass bottles.
Uses: Manufacture of ruby glass; coloring enamels; painting porcelain.
Shipping regulations: None. *

gold tribromide (auric bromide; gold bromide) $AuBr_3$.
Properties: Brownish-black powder; m. p. 160°C, with decomposition. Soluble in alcohol, ether, water (reddish-brown).
Uses: Analysis (testing for alkaloids, spermatic fluid); medicine.

gold tribromide, acid. See bromoauric acid.

gold trichloride. See gold chloride.

gold trichloride, acid. See gold chloride.

gold trioxide. See gold oxide.

gold, white. A jeweler's alloy consisting of about 75-85% gold, 8-10% nickel and 2-9% zinc.

*See "I.C.C. Shipping Regulations," page xiii.
Reference numbers refer to name of manufacturer. See "List of Manufacturers," page v.

gommeline. See dextrin.

goober cake. See peanut cake.

"Good-rite 2007." [119] (Styrene-butadiene resin). Trade name for a reinforcing and stiffening agent compatible with crude and American rubbers. Imparts hardness, abrasion resistance, and flexural strength.

"Good-rite Antioxidants." [119] Trade name for a group of antioxidant materials used in rubber.

"Good-rite Carbopol." [119] (Synthetic hydrophilic polymer). Trade name for a water sensitive synthetic gum with gel forming properties. Useful in the manufacture of cosmetics and pharmaceuticals.

"Good-rite CB." [119] (An acrylic resin). Trade name for a sand binder for foundry applications with the advantages of negligible gas formation, excellent collapsibility and shakeout, and good dimensional stability.

"Good-rite K-702." [119] A solution of polyacrylic acid (25%) in water; clear and colorless; infinitely dilutable with water; compatible with a wide variety of aqueous dispersions; pH 2 to 3; sp. gr. 1.09; viscosity 500 to 1200 cps at 78°F; not subject to hydrolysis or bacterial degradation; contains not over 0.2% other salts.

"Good-rite K-705." [119] A solution of ammonium polyacrylate (25%) in water; clear to very slight haze with trace of yellow color; pH 7 to 9; sp. gr. 1.10; viscosity 1000 to 2000 cps at 78°F; other properties as for K-702.

"Good-rite K-708." [119] A solution of sodium polyacrylate (25%) in water; pH 8 to 10; sp. gr. 1.14; viscosity 1200 to 2400 cps at 78°F; contains 1.25% sodium bicarbonate; other properties as for K-705.

"Good-rite K720 and K721." [119] Trade name for water soluble synthetic resins effective as flocculating agents in sedimentation and filtration systems.

"Good-rite" Plasticizers. [119] Trade name for a group of primary plasticizers for vinyls, rubbers, and other high polymer resins. The specific plasticizers are the dioctyl, octyl-decyl, and didecyl phthalate, or adipate esters.

"Good-rite" GP261 (DOP) is a standard plasticizer exhibiting all around properties in calendering, extrusion, and plastisol applications. The higher molecular weight phthalates (GP-265 and GP-266) have low volatility properties and are useful in high temperature applications. The adipate plasticizers (GP-233, GP-235, and GP-236) have low temperature flexibility properties. The octyl-decyl and didecyl adipates (GP-235 and GP-236) have low volatility.

"Good-rite" plasticizers are insoluble, or have limited solubility, in water, glycerines, and glycols. They are soluble in most other organic liquids. They are designed to impart permanent flexibility; low water and soapy water plasticizer extraction; good electrical properties; plastisol viscosity stability.

Containers: 5-gal cans; 55-gal drums; tank trucks; tank cars.

Typical properties:

GP-261 (dioctyl phthalate): light colored oily liquid; sp. gr. 0.983 (25/25°C); color 50 max (APHA); viscosity 49.5 cps (25°C); acid number 0.075 max; flash point 400°F; fire point 470°F; refractive index 1.4844 (25°C).

GP-265 (octyl decyl phthalate): light colored oily liquid; sp. gr. 0.970 (25/25°C); color 75 max (APHA); viscosity 65.0 cps (25°C); acid number 0.075 max; flash point 415°F; fire point 485°F; refractive index 1.4840 (25°C).

GP-266 (didecyl phthalate): light colored oily liquid; sp. gr. 0.964 (25/25°C); color 75 max (APHA); viscosity 69.4 cps (25°C); acid number 0.075 max; flash point 435°F; fire point 510°F; refractive index 1.4835 (25°C).

GP-233 (dioctyl adipate): light colored oily liquid; sp. gr. 0.925 (25/25°C); color 50 max (APHA); viscosity 12.6 cps (25°C); acid number 0.075 max; flash point 380°F; fire point 450°F; refractive index 1.4470 (25°C).

GP-235 (octyl decyl adipate): light colored oily liquid; sp. gr. 0.918 (25/25°C); color 75 max (APHA); viscosity 17.0 cps (25°C); acid number 0.075 max; flash point 390°F; fire point 465°F; refractive index 1.440 (25°C).

GP-236 (didecyl adipate): light colored oily liquid; sp. gr. 0.915 (25/25°C); color 75 max (APHA); viscosity 19.2 cps (25°C); acid number 0.075 max; flash point 415°F; fire point 495°F; refractive index 1.4505 (25°C).

"Good-rite Vultrol." [119] Trade name for a retarding and anti-scorch agent used in the manufacture of rubber.

goosefoot oil. See chenopodium oil.

"Goremul A." [73] Trademark for condensed glycol ester of high molecular weight fatty acids.

Properties: Tan to brown wax-like material. Dispersible in hot water. Soluble in acetone and ethyl acetate. Soluble (hot) in methanol, ethanol, toluene and cottonseed oil. Partly soluble in mineral spirits and mineral oil.

Constants: Sp. gr. 1.05 (25°C); iodine value 31.5-34.5; pH (5% dispersion) 2.5-3.1; m.p. 44-53°C.

Containers: 1-gal cans (8 lbs); 5-gal cans (50 lbs); 55-gal drums (500 lbs).

Uses: For the manufacture of emulsions stable to acids, salts, and esters; for fluid and paste emulsions containing salt, aluminum chloride, hydrochloric acid, glacial acetic acid, oxyquinoline sulfate, and other chemicals which are difficult to incorporate in emulsions of mineral oil, pine oil, toluene, amyl acetate, etc.

goslarite (zinc vitriol) $ZnSO_4 \cdot 7H_2O$. White or yellowish earthy mineral formed by the oxidation of sphalerite in damp locations, especially in the presence of iron sulfides. See also zinc sulfate.

gossypium. See cotton.

gossypol 1, 1', 6, 6', 7, 7'-hexahydroxy-3, 3'-dimethyl-5, 5'-diisopropyl-2, 2'-binaphthyl-8, 8'-dialdehyde. A naturally occurring polyphenol in cottonseed known to be toxic and which reduces the usefulness of cotton seed meal as a poultry feed supplement.
Properties: Appears to have three crystalline modifications, with m. p. of 184°C, 199°C, and 214°C. Insoluble in water; soluble in methanol, ethanol, ether, chloroform and dimethyl formamide. Soluble with decomposition in dilute aqueous ammonia and sodium carbonate.
Uses: Stabilizer for vinyl polymers; rubber antioxidant.

gossypose. See raffinose.

gourd oil. See cucumber oil.

GPI. Abbreviation for glycerophosphoryl inositol.

gr. Abbreviation for gravity.

graft copolymer. See graft polymer.

graft polymer (graft copolymer). A polymer molecule in which the main backbone chain of atoms has attached to it at various points side chains containing different atoms or groups from those in the main chain. The main chain may be either a copolymer or may be derived from a single monomer.

grahamite. An asphaltite resembling albertite in its jet-black luster. It varies in its physical properties, some deposits being fairly pure while others contain as high as 50% mineral matter.
Properties: Black in mass; conchoidal to hackly fracture; very bright to dull luster; black streak. Soluble in carbon disulfide (45-100%).
Constants: Sp.gr., pure varieties containing less than 10% mineral matter 1.15-1.20; impure varieties containing more than 10% mineral matter, 1.175-1.50 (77°F); hardness (Mohs' scale) 2-3; penetration 0 (77°F); fusing point (K & S) 350-600°F, (B & R) 370-620°F; behavior on heating in flame, variety showing conchoidal fracture decrepitates violently, variety showing hackly fracture softens, splits, and burns; mineral matter variable up to 50%.
Occurrence: United States (Colorado, Oklahoma, West Virginia), Mexico; Trinidad; Cuba.
Shipping regulations: None. *

Graham's salt. See sodium hexametaphosphate.

grain alcohol. See ethyl alcohol.

grain oil. See fusel oil.

grains of kermes. See kermes.

gram. One one-thousandth of a kilogram (q. v.). It is approximately the weight of one milliliter of water at 4°C.

gramicidin. An antibiotic produced by the metabolic processes of the bacteria Bacillus brevis. It is a polypeptide which is active against most gram-positive pathogenic (disease-causing) bacteria. It is one of the two antibiotic components of tyrothricin (q. v.) but has been isolated and used alone.
Properties: White crystalline platelets; m. p. 229-230°C; soluble in the lower alcohols, acetic acid and pyridine; moderately soluble in dry acetone and dioxane; almost insoluble in water, ether, and hydrocarbons. Depresses surface tension; forms a fairly stable colloidal emulsion in distilled water.
Derivation: From tyrothricin by extraction with a mixture of equal volumes of acetone and ether, followed by concentration in vacuo and dissolving in hot acetone.
Grade: N. F. XI.
Use: Medicine.

gramine (donaxine; 3-[dimethylaminomethylindole]) $C_{11}H_{14}N_2$. An indole alkaloid from barley.
Properties: Crystals; m. p. 134°C; slightly soluble in water; insoluble in petroleum ether.
Use: Biochemical research.

grana tilli. See tiglium.

granatum (pomegranate bark).
Derivation: Bark of stem and root of Punica granatum.
Habitat: Mediterranean region and eastern, western and southern Asia; cultivated in semitropical countries.
Grades: Technical.
Containers: Bags.
Use: Medicine.
Shipping regulations: None. *

granite. A crystalline granular igneous rock composed essentially of quartz and feldspar. Granites differ widely in their appearance and character.
Uses: Dimension stone; ornamental stone.

granulation. The process of converting a substance into small grain-like particles.

grapefruit oil (oil of shaddock).
Properties: A volatile oil; pale yellow to orange-yellow liquid, with the characteristic citrus note of grapefruit; sp. gr. 0.854-0.860 (15°C); refractive index (n 20/D) 1.4750-1.4780; soluble in benzyl benzoate, fixed oils, mineral oil; slightly soluble in propylene glycol; insoluble in glycerin; unstable in presence of strong alkalies, strong acids.
Derivation: Expressed from the fresh peel of Citrus decumana.
Containers: Drums.
Uses: Flavors; eau de cologne perfumes.

grapefruit oil, terpeneless. A thick brownish-yellow oil, having a strength about 30 times that of oil of grapefruit, used in eau de cologne perfumes and flavors in which its better solubility is desirable.

grapefruit seed oil. Oil expressed from grapefruit seeds collected from citrus canning plants. After a short fermentation which loosens adhering pulp, the seeds are drained, steamed till the hulls crack and dried to a moisture content of 2.5 to 3.0%. The oil is expressed by an expeller and the filter cake used as fertilizer or cattle feed. The oil is refined by a caustic wash and decolorizing carbon. The unrefined oil has a reddish-brown color, a pleasant nutlike aroma, and an intensely bitter taste; sp. gr. 0.9179-9.9199; refractive index (n 25/D) 1.4688-1.4700. It becomes cloudy at about 15°C (59°F), pasty at 4°C (39.2°F), buttery at -4°C (24.8°F), and solid at -10°C (14°F).
Uses: Lubricant for leather and textile fibers.

grape-seed oil (grape-stone oil; winestones oil; raisin-seed oil).
Properties: Yellow, liquid, fixed oil; unpleasant odor; bitter taste; soluble in benzene, solvent naphtha, and carbon disulfide; sp. gr. 0.9202-0.9350; solidification point -10° to -15°C; saponification value 178-180; iodine number 94-96.5; Maumené number 52-54.
Derivation: From the dried, ground seeds of the grape Vitis vinifera by steeping in water, heating and pressing, or by extraction with a volatile solvent.
Method of purification: Decolorization with bone-black.
Grades: Technical.
Containers: Tins; iron drums.
Uses: Lubricant; fuel; illumination; food; soapmaking.
Shipping regulations: None. *

grape-stone oil. See grape-seed oil.

grape sugar. See dextrose.

"Graphallast." [82] Trade name for a group of graphite and hydrocarbon oilless materials. Used for low friction bushings, bearings, and seals in submerged applications in many corrosive chemicals at normal temperatures below 150°F. Resistant to scuffing or abrasion when submerged.

"Graphalloy." [82] Trade name for a series of oilless, self-lubricating, long-life, low-friction materials consisting of graphite and a metal or alloy such as Babbitt, bronze, cadmium, copper, gold, or silver. These "Graphalloy" materials are widely used for bearings, bushings, seals, electric brushes, brush assemblies, brush and slip-ring assemblies, and non-freezing electric contacts. Many grades are available to meet most chemical applications. Cryogenic applications to —450°F; or to +750°F in air.

"Graphicell." [214] A proprietary product, 99% graphitic carbon, but otherwise similar to "Carbocell."

"Graphic Red." [141] Trade name for lithol red pigments.
Composition: Sodium, barium, strontium, or calcium salts of diazotized Tobias acid coupled with beta-naphthol.
Properties: They have fair light resistance, fair resistance to heat and good resistance to acid and alkali. They are non-bleeding in water and organic vehicles.
Grades: "Graphic Red" Y: light yellow shade red (sodium salt); "Graphic Red" M: medium red (barium or strontium salt); "Graphic Red" R: dark bluish red (calcium salt); "Graphic Maroon:" light or dark maroon (calcium salt).
Uses: Printing inks, paints, rubber, floor coverings, crayons.

graphite (black lead, plumbago). The crystalline allotropic form of carbon characterized by a hexagonal arrangement of the atoms. Occurs naturally in Madagascar, Ceylon, Mexico and numerous other places in deposits of varying purity. Also produced synthetically by heating petroleum coke to about 3000°C in an electric resistance furnace.
See also graphite, pyrolytic.
Properties: Relatively soft, greasy feel, steel gray to black color with a metallic sheen; sp. gr. 2.0-2.25 depending upon origin; apparent specific gravity artificial graphite 1.5-1.8. In fabricated forms, electric resistivity 800-1300 micro-ohm/cm; specific heat 0.16 at room temperature, 0.40 at 1500°C; tensile strength 400 to 2000 psi; compressive strength usually about 1700-7500 psi. See also carbon.
Grades: Powdered; flake; crystals.
Containers: Bags; fiber drums.
Uses: Granular or flake forms: pencils, crucibles, retorts, foundry facings, molds, lubricants, paints and coatings, boiler compounds, stove polish, powder glazing, electrotyping.
Fabricated forms: Cores, molds, chills, crucibles, electrodes (furnace), electrodes (electrolytic), chemical equipment, electronic anodes, motor and generator brushes.

graphite, blue. A complex of carbon, fluorine and hydrogen fluoride formed when certain types of graphite are used as anodes during production of elemental fluorine by electrolysis of molten potassium fluoride-hydrogen fluoride mixtures. Also formed when fluorine is passed through a suspension of graphite in hydrogen fluoride.

graphite, pyrolytic (pyrographite). A dense graphite, stronger and more resistant to heat than ordinary graphite, and expected to find use in rocket nozzles, missiles in general, nuclear reactors. It is made by a recrystallization process from ordinary graphite. Forms of the latter are heated in a stream of hydrocarbon gas. The carbon in the gas is deposited on the original form with the carbon crystals in alignment along the flat planes of the form. Such a graphite has a high tensile strength even at 5000°F. Sheets only one mil thick are impervious to

*See "I.C.C. Shipping Regulations," page xiii.

liquids and gases. Its destruction temperature is about 6600°F.

"Graphlon." [82] Trade name for a group of widely used graphite and resin materials that exhibit extremely high chemical inertness. "Graphlon" bearings, bushings, and seals withstand corrosive chemicals which would destroy most other materials. Operating range from −450° to +450°F in air.

"Graphmetex." [82] Trade name for a group of special thermosetting resins employed as insulating spacers and holders for electrical components. Inert to most chemicals and solvents; fungus resistant, with excellent dielectric properties and high resistivity.

GRAS. Abbreviation for generally recognized as safe. Used of food additives.

"Gravidox." [57] Trademark for thiamine-pyridoxine.

"Gravinol." [3] Trademark for a mixture of brominated olive oil and brominated corn oil.

Properties: Clear, reddish-brown, oily liquid with no taste or odor; sp. gr. between 1.305 and 1.315 at 25°C.

Use: Weighting agent for citrus oils in the production of citrus emulsions for use in soft drinks.

gray acetate. See calcium acetate.

gray antimony. See stibnite.

gray copper ore. See tennantite; tetrahedrite; fahlore.

gray manganese ore. See manganite.

grease oil. See lard oil.

green. Used in the chemical industries to mean uncured or untreated material.

green broom. See scoparius.

green cinnabar. See chromic oxide.

green copperas. See ferrous sulfate.

green glass. A chromium compound is used with ordinary glass. Cupric oxide gives blue green.

"Green-Gold." [28] Trademark for yellow azo nickel pigment.

Uses: In paint, printing ink, plastics, and other applications where lightfastness is required.

green hellebore. See veratrum.

Greenland spar. See cryolite, natural.

greenockite CdS. A native cadmium sulfide containing 77.7% cadmium. Honey, citron or orange-yellow color; orange-yellow to brick-red streak; adamantine to resinous luster. Usually found with sphalerite (q.v.).

Constants: Sp. gr. 4.9-5.0; hardness 3-3.5.

Occurrence: United Stated (Pennsylvania, Missouri, Arkansas); Scotland; Bohemia.

green oil. A term given in the Scottish shale-oil industry to the once-run crude shale oil after chemical treatment. Also applied to anthracene (q.v.).

greensalt. A wood preservative containing chromated copper arsenate.

green salt. See uranium tetrafluoride.

greensand marl. A mixture of clay and calcite with glauconite. The principal source of glauconite (q.v.).

green soap. See soap, soft.

green verditer. A paint pigment consisting of the hydroxycarbonate of copper.

See malachite.

green vitriol. See ferrous sulfate.

"Greenz." [48] Trademark for an ammonium lignin sulfonate containing 4.5% iron, for use as an agricultural spray or soil additive in the treatment of iron chlorosis.

Griess reagent. A reagent used in analytical work for detecting nitrous acid. It consists of 0.1 gram pure white alpha-naphthylamine dissolved in 100 cc of water to which is added a solution of 5 cc glacial acetic acid and 1 gram sulfanilic acid in 100 cc of water.

"Griffco." [309] Trademark for a polyvinyl acetate emulsion useful as a base for adhesives and paint.

Griffiths' white. See lithopone.

Grignard reagents. A very important class of reagents of synthetic organic chemistry, made by union of metallic magnesium with an organic chloride, bromide, or iodide, usually in the presence of an ether, and in the complete absence of water. These reagents have the general formula RMgX where R is an alkyl or aryl or other organic group, and X represents a halogen. The value of the reagents lies in their ease of reaction with water, carbon dioxide, alcohols, aldehydes, ketones, amines, etc., to produce a great variety of organic compounds, usually with good yields. Examples of Grignard reagents are ethyl magnesium chloride (C_2H_5MgCl), methyl magnesium bromide (CH_3MgBr), etc.

Grillo (Grillo-Schroeder) process. Uses magnesium sulfate as a carrier for platinum catalyst in widely used platinum contact process system in the synthesis of sulfuric acid in the United States.

grindelia (gum plant; tar weed).

Derivation: Dried leaves and flowering tops of various species of Grindelia.

Habitat: California, southwestern United States.

Grades: Technical.

Containers: Bags; bales.

Use: Medicine.

griseofulvin. An antifungal antibiotic derived from Penicillium griseofulvum. It is said to be 7-chloro-4, 6-dimethoxycoumaran-3-one-2-spiro-1'-(2'-methoxy-6'-methyl-cyclohex-2'-ene-4'-one).

"Groco." [410] Trade name for a line of animal and vegetable fatty acids.
Groco 2, 4, 8, 18: Distilled red oil (oleic acid).
Groco 5L: Low linoleic white oleine.
Groco 6: White oleine.
Groco 20: Distilled linseed fatty acids.
Groco 24, 26: Distilled coconut fatty acids.
Groco 27, 28: Distilled soya bean fatty acids.
Groco 29: Distilled corn fatty acids.
Groco 30: Distilled cottonseed fatty acids.
Groco 40, 41: Distilled tallow fatty acids.
Groco 45, 46: Distilled palm oil fatty acids.
Groco 53, 54, 55, 55L, 65, 65C: Distilled stearic acid.
Groco 56: Hydrogenated stearic acid.
Groco 57, 58: Hydrogenated tallow fatty acids.

grog. A term applied in the ceramic industry to various crushed refractory materials which are added to the batch to reduce lamination in plastic clays and also to reduce shrinkage on drying. Such materials crushed for this purpose are pottery, fire brick, quartz, quartzite, burned ware, saggers, kiln and boiler clinkers.

grossularite. See garnet.

ground-nut. See peanut.

ground-nut cake. See peanut cake.

ground-nut oil. See peanut.

growth hormone. See somatotropic hormone.

growth regulators (plants). See plant hormones; auxins.

growth substances (plants). See plant hormones; auxins.

GR-S. Abbreviation for government rubber-styrene. Designation for the standard synthetic rubber made in U. S. Government plants in 1943 and following years. Produced by emulsion polymerization of 75 parts butadiene and 25 parts styrene, and also referred to as Buna-S. The more accepted designation now is SBR (styrene-butadiene rubber).

G salt. The sodium or potassium salt of 2-naphthol-6, 8-disulfonic acid (G acid).

g-strophanthin. See ouabain.

"G. T. O. No. 5." [188] Brand name for a general purpose industrial odorant.

GTP. Abbreviation for guanosine triphosphate. See guanosine phosphates.

guacetin. See guaiacetin.

guaethol. See ethyl guaiacol.

guaiac (guaiac gum; guaiac resin). A resin from certain Mexican and West Indian trees, especially Guaiacum santum and G. officinale.
Soluble in alcohol, ether, acetone, chloroform and caustic soda.
Grade: Technical.
Uses: Medicine; source of antioxidants for fats and foods.
Shipping regulations: None. *

guaiacetin (guacetin; sodium pyrocatechin acetate; sodium phenoneacetate) $C_6H_4(OH)OCH_2COONa$.
Properties: White powder; soluble in water.
Derivation: By the action of sodium carbonate on phenoneacetate.
Method of purification: Crystallization.
Grade: Technical.
Containers: Glass bottles. Keep solution in well filled bottles and protected from light.
Use: Medicine.
Shipping regulations: None. *

guaiac gum. See guaiac.

guaiacol (methylcatechol; pyrocatechol methyl ester; ortho-methoxyphenol; ortho-hydroxyanisole) $OHC_6H_4OCH_3$.
Properties: Faintly yellowish, limpid, oily liquid or yellow crystals; characteristic aromatic odor. Guaiacol constitutes 60-90% of beechwood cresote. Soluble in alcohol, ether, chloroform and glacial acetic acid; moderately soluble in water. Sp. gr. 1.1395; m. p. 27.9°C; b. p. 205°C.
Derivation: (a) By extracting beechwood creosote with alcoholic potash, washing with ether, crystallizing the potash compound from alcohol and decomposing it with dilute sulfuric acid. (b) Also from ortho-anisidine by diazotization and subsequent action of dilute sulfuric acid.
Method of purification: Recrystallization.
Grade: Technical.
Containers: (Crystals). Drums and tins; (liquid) drums and carboys.
Uses: Medicine; preparation of catechol and guaiacol compounds.
Shipping regulations: None. *

guaiacol benzoate $C_6H_5COOC_6H_4OCH_3$.
Properties: White odorless, almost tasteless powder; m. p. 57-58°C; slightly soluble in water, soluble in hot alcohol, ether and chloroform.
Use: Medicine.

guaiacol carbonate (neutral guaiacol carbonate) $(C_7H_7O)_2CO_3$.
Properties: Small colorless crystals or white crystalline powder. Either slight aromatic odor and taste or odorless and tasteless. Soluble in chloroform, ether; less soluble in alcohol; insoluble in water; m. p. 86-88°C.
Derivation: (a) Reaction between sodium guaiacolate and carbonyl chloride. (b) Reaction between guaiacol and methyl chloroformate.
Containers: 1-lb cartons; 50-lb drums.
Use: Medicine.

guaiacol cinnamate $C_6H_5CHCHCOO(C_6H_4OCH_3)$.
Properties: White needles; tasteless; odorless. Soluble in alcohol, acetone, benzene, chloroform; insoluble in water. Fusing point 130°C.
Use: Medicine.
Shipping regulations: None. *

*See "I. C. C. Shipping Regulations," page xiii.
Reference numbers refer to name of manufacturer. See "List of Manufacturers," page v.

guaiac resin. See guaiac.

guaiac-wood oil.
Properties: A very thick and viscid oil, becoming crystalline at ordinary temperatures; very agreeable violet and tea-like odor. Soluble in alcohol, ether and chloroform.
Chief known constituent: Guaiol (tiglic aldehyde).
Constants: Sp. gr. 0.965-0.975; optical rotation −6° to −7°.
Derivation: Probably from Bulnesia sarmienli.
Method of purification: Rectification.
Grade: Technical.
Containers: Cans; glass bottles.
Use: Perfumes.
Shipping regulations: None. *

"Guai-A-Phene." [296] Trademark for a phenolic type anti-skinning agent used for the prevention of gelling, skinning, and oxidation in paint, varnishes, printing inks, linoleum, and the like.

guanicaine. See di-para-anisyl-para-phenetylguanidine hydrochloride.

guanidine (carbamidine; iminourea) $NHC(NH_2)_2$.
Properties: Colorless crystals; m. p. 50°C; decomposes at 160°C. Soluble in water and alcohol.
Derivation: (a) By heating calcium cyanamide with ammonium iodide. (b) By treating urea with ammonia under pressure. (c) By the action of aqua regia on dicyanodiamine. (d) From guanidine carbonate.
Grade: Technical.
Containers: Tins.
Use: Organic synthesis.
Shipping regulations: None. *

guanidine-aminovaleric acid. See arginine.

guanidine carbonate (carbamidine carbonate) $(H_2NCNHNH_2)_2 \cdot H_2CO_3$.
Properties: White granules. Soluble in water, slightly in alcohol and acetone.
Constants: Decomposes at 197-199°C without melting; sp. gr. 1.25.
Derivation: From dicyandiamide.
Grade: Technical, over 95% pure.
Containers: Fiber drums, 200 lbs, net.
Uses: As a strong organic alkali; in organic synthesis; soap and cosmetic products.

guanidine hydrochloride $NHC(NH_2)_2 \cdot HCl$.
Properties: White powder; m. p. about 183°C. Soluble in water and alcohol; pH of aqueous solution 6.2 for 10% solution.
Grades: 88% and 95% pure.
Containers: Paper bags or fiber drums.
Uses: Exceptionally water-soluble source of guanidine for organic syntheses.

guanidine nitrate $H_2NC(NH)NH_2 \cdot HNO_3$.
Properties: White granules. Soluble in water and alcohol; slightly soluble in acetone. Melting range 206-212°C.
Derivation: From cyanamide or dicyandiamide.
Grade: Technical, over 95% pure.
Containers: Fiber drums, 200 lbs, net; multiwall paper sacks.
Uses: In manufacture of explosives, and as an ingredient of explosive mixtures (guanidine nitrate itself is not explosive); disinfectants; photographic chemicals.
Shipping regulations: Oxidizing material. Yellow label. *

dℓ-alpha-**guanidinopropionic acid.** See N-amidinoalanine.

guanine $C_5H_5N_5O$ (2-amino-6-oxypurine). A purine that is a constituent of ribonucleic acid and deoxyribonucleic acid. Usual sources are guano, sugar beets, yeast, clover seed, and fish scales.
Properties: Amorphous or small colorless rhombic crystals. M. p. 360°C (dec). Insoluble in water; sparingly soluble in alcohol and ether; freely soluble in ammonium hydroxide, alkali hydroxides, and dilute acids.
Derivation: Isolation following hydrolysis of nucleic acids (usually from yeast); organic synthesis.
Use: Biochemical research.
Available as hydrochloride or hemisulfate.

guanine riboside. See guanosine.

guano. A term originally limited to true guano, which consists of the dried excrements, feathers, and carcasses of sea fowl. It was formerly obtained exclusively from certain islands off the coast of Peru and Chile. Now obtained, in addition, from other sources. Typical analysis, Peru guano: Moisture 23%, ammonia 11%, phosphoric acid 12%.
The significance of the term has been broadened to include other varieties, as bat guano, fish guano, phosphatic guano (leached of its nitrogen content by rain), etc. Guano is also known by the geographical location from which it is obtained, as Ballestas guano, Cantores guano.
Containers: Bags; bulk.
Use: Fertilizer.

guanosine (guanine riboside) $C_{10}H_{13}N_5O_5$. The nucleoside containing guanine and D-ribose.
Properties: White, crystalline, odorless powder with mild, saline, or saline and bitter taste. M. p. 237-240°C (dec). Very slightly soluble in cold water; soluble in boiling water, dilute mineral acids, hot acetic acid, and dilute bases; insoluble in alcohol, ether, chloroform and benzene.
Derivation: Found in pancreas, clover, coffee plant, and pollen of pines; prepared from yeast nucleic acid.
Use: Biochemical research.

guanosine monophosphate. See guanylic acid.

guanosine phosphates. Nucleotides used by the body in growth processes; important in biochemical and physiological research. Those isolated are the monophosphate (GMP), the diphosphate (GDP) and the triphosphate (GTP).

*See "I.C.C. Shipping Regulations," page xiii.
Reference numbers refer to name of manufacturer. See "List of Manufacturers," page v.

"Guantal." [58] Trademark for diphenylguanidine phthalate (q. v.).

guanylic acid (GMP; guanosine monophosphate) $C_{10}H_{14}N_5O_8P$. The nucleotide consisting of guanine, D-ribose, and phosphoric acid; important in growth processes of the body.
Properties: Dihydrate, in form of long prisms, becomes anhydrous at 118°C. Decomposes at 180°C. Acid to litmus. Soluble in cold water; freely soluble in hot water.
Derivation: Isolation from nucleic acid of yeast or pancreas.
Use: Biochemical research.

guanyl nitrosaminoguanylidene hydrazine.
Shipping regulations: Explosive, class A. Initiating explosive label. Not accepted by express. *

guanyl nitrosaminoguanyl tetrazene. See tetrazene.

guanyl urea sulfate (carbamylguanidine sulfate; dicyanodiamidine sulfate) $(C_2H_6ON_4)_2 \cdot H_2SO_4 \cdot 2H_2O$.
Properties: White powder, over 97% pure. Soluble in water and alcohol.
Derivation: From cyanamide or dicyandiamide.
Uses: Analytical reagent for nickel; in manufacture of dyes and in organic synthesis.

guapi bark. See cocillana.

"Guardkote." [125] Trademark for fast-setting two-component liquid systems based on "Epon" resins and designed specifically for highway resurfacing and repair. They are used in combination with sharp aggregates to waterproof and deslick portland cement or bituminous concrete pavements.

guar flour. See guar gum.

guar gum (guar flour). The ground endosperms of Cyanopsis tetragonoloba, cultivated in Pakistan as livestock feed. The water-soluble portion of the flour (85%) is called guaran, and consists of 35% galactose, 63% mannose; 5-7% protein.
Properties: Light-gray powder. Dispersible in hot or cold water. Has 5-8 times the thickening power of starch.
Grades: Industrial; technical; edible.
Containers: Bags.
Uses: Paper manufacture; foods; cosmetics; pharmaceuticals; textiles; printing; polishing; atomic metal processing.

guayule. A shrub native to Mexico, but also planted in western United States during 1941-1945. It has a fairly high rubber content, but because of a high percentage of resins and for other reasons it is not competitive as a source of rubber.

guaza. See cannabis.

Guggenheim process. A process for the manufacture of sodium nitrate from the Chilean nitrate ore, caliche, in which heat is efficiently utilized and handling costs are kept to a minimum.

guhr. See diatomite.

Guignet's green.
Derivation: Chrome green made by fusing potassium chromate and boric acid. The mass is washed, ground and dried. The chromium borate formed is decomposed by water. The final product is insoluble in water.
Grade: Technical.
Containers: Wooden kegs; fiber drums.
Use: Paint pigment.
Shipping regulations: None. *

"Gumafoam S." [328] Trademark for a product consisting of plasticized vegetable gum solution and providing a bodying finish to textile fabrics.

gum, ammoniac. See ammoniac.

gum arabic. See arabic, gum.

gum artificial. See dextrin.

gumbo clay. A series of fine-grained highly plastic and tough clays which are chiefly used in the manufacture of railroad ballast. They cannot be used for brick making due to their high shrinkage on burning.
Occurrence: West-central states of the United States.

gum camphor. See camphor.

gum lac. See shellac.

gum plant. See grindelia.

gum rosin. See rosin.

gums.
1. In general, the dried exudations of secretions of plants. They vary in properties with the plant and almost defy classification. Their nomenclature is equally uncertain. Common usage leans to a rough grouping under water-soluble gums (see 2), oleoresins, resins, and rubber or rubber-like substances such as chicle. See oleoresins and various headings under resins and rubber.
2. Specifically, plant exudations which are soluble in water or swell in water. They consist largely of carbohydrates and are hydrophilic colloids. Examples are arabic, tragacanth, agar.

Many natural gums and resins are listed commercially under the name gum, as gum arabic, gum dammar, gum tragacanth, etc. These are entered in this dictionary under the specific name, as arabic, dammar, tragacanth, etc.

gum sugar. See arabinose.

gum thus. See turpentine or olibanum.

gum turpentine. See turpentine.

guncotton. See nitrocellulose.

"Gunk." [186] Trademark for a line of soaps and compounds consisting of degreasing and decarbonizing solvents, acid and alkaline powders and liquids.
Containers: Pint, quart, 1-, 5-, 15-, 30-, and 50-gal drums.

Uses: In automotive, airmotive, industrial and commercial fields for carbon digesting, paint stripping, derusting and descaling.

"Gunk" Carbon Met. [186] Trade name for a fast-drying, halogenated aromatic hydrocarbon solvent for liquid degreasing of metals.

"Gunk" C.C. [186] Trade name for a mild aromatic solvent specifically for cleaning carburetors and fuel pumps; a non-caustic, self-emulsifying and self-scouring liquid.

"Gunk" Compound I.S. [186] Trade name for an industrial shampoo; a self-emulsifying and self-scouring, water-soluble degreaser for typewriters and hand wiping of machinery.

"Gunk Compound Motor Fizik." [186] Trademark for a solvent oil used for dissolving carbonaceous tars and gums from engine head and valves, pistons, rings, and crankcases of internal combustion engines and as a pour point depressant in crankcase oil and gas line anti-freeze.

"Gunk Hydro-Seal." [186] Trademark for a diphase self-emulsifying and self-scouring carbon digester for decarbonizing and paint stripping engines and metal structures.

"Gunk Motor Purge." [186] Trademark for a self-emulsifying and self-scouring solvent for washing gums and carbonaceous tars from the inside of gasoline or diesel engines through solvent action coupled with soap and water detergency.

"Gunk" Neo Met. [186] Trade name for a non-flammable, fast-evaporating solvent for dry cleaning metal parts and electrical components, specifically replacing carbon tetrachloride.

"Gunk" Super Concentrate Degreaser. [186] Trade name for a concentrated self-emulsifying, self-scouring, oil-soluble, water-miscible liquid soap to be extended with petroleum distillate and rinsed with water, for removing grease and oil from engines, floors, etc.

gun metal. An alloy of copper with 10% tin.

gun powder (black powder). A mixture of sodium nitrate or saltpeter, sulfur, and charcoal in varying proportions. A typical formula is as follows: 70-75% saltpeter, 10-14% sulfur, 14-16% charcoal. It is designated according to size: Mealed; superfine grain or FFG; fine grain or FG; large or coarse grain or LG; large grain for rifles RLG; mammoth.

Containers: Kegs; special fiber containers; tin cans.

Shipping regulations: Explosive; class A by freight. Not accepted by express. *

gunpowder, white (white powder). A mixture of 2 parts potassium chlorate, 1 part potassium ferrocyanide and 1 part sugar.

Shipping regulations: High explosive. *

gurjun balsam (East Indian copaiba balsam). Oleoresinous secretion from a species of Dipterocarpus tree in India and China. Similar to copaiba.

Use: Source of oil.

Shipping regulations: None. *

gurjun balsam oil (gargan balsam oil; gardjan balsam oil; East Indian copaiba balsam oil; Indian wood oil).

Properties: Yellow, somewhat viscous essential oil. Solubility in alcohol: In 7-10 vols of 95% alcohol (solubility not unlimited).

Chief known constituents: Sesquiterpenes known as gurjunenes.

Constants: Sp. gr. 0.918-0.930; optical rotation −35° to −130°; refractive index 1.501-1.505; acid value up to 1; ester value up to 8, after acetylation 6-10.

Derivation: By distilling the balsam and oils obtained from various species of genus Dipterocarpus.

Use: Adulterant of other volatile oils.

Shipping regulations: None. *

guru. See cola.

"Guthion." [181] Trademark for O,O-dimethyl S-4-oxo-1,2,3-benzotriazin-3(4H)-ylmethyl phosphorodithioate (q.v.).

gutta-percha (isonandra gutta) $(C_{10}H_{16})_x$. See also polyisoprene. A rubber-like material obtained from the milky juice of leaves and bark of the geni Palaquium and Payena of tropical Asia, South America, and the Philippines. The material is skimmed from a water layer, purified, and in the process turns yellowish or gray. Melting point approximately 100°C; partly soluble in carbon disulfide, chloroform, solvent naphtha, and warm benzene.

Grades: Red Macassar (superior); gutta siak and gutta soh.

Uses: Insulating electric wires, cables and conduits; dentistry; making impressions of medals; waterproofing; fastening incandescent electric bulbs into their sockets; machinery belting; cutlery handles.

Shipping regulations: None. *

"Guyatt." [51] Trademark for a dark, fibrous lubricant used in plain or roller bearings where a semifluid or a stiff grease possessing good high-temperature characteristics is desired and no water contamination occurs.

gynesine. See trigonelline.

gynocardia oil. See chaulmoogra oil.

gynocardic acid $C_{17}H_{33}COOH$. A term used to denote the acids of the oil expressed from the seeds of Gynocardia odorata; also the acids from chaulmoogra oil.

Properties: White leaflets; m.p. 67.5°C; insoluble in water; soluble in most organic solvents.

Shipping regulations: None. *

gyplure. A synthetic product, 12-acetoxy-1-hydroxy-cis-9-octadecene, used as a sex attractant for the male gypsy moth. The natural product, found in the female moths, is said to be d-10-acetoxy-1-hydroxy-cis-7-hexadecene.

gypsum $CaSO_4 \cdot 2H_2O$. A natural hydrated calcium sulfate.

Properties: White or colorless, sometimes tinted grayish, reddish, yellowish, bluish, or brownish. White streak; pearly, silky or vitreous luster; can be easily scratched by finger nail; sp. gr. 2.31-2.33; hardness 1.5-2. Loses $1\frac{1}{2}$ H_2O at 128°C, and $2H_2O$ at 163°C. Insoluble in water; soluble in ammonium salts, acids and sodium chloride.

Varieties:

Alabaster: Fine grained and compact.

Satin spar: Fine translucent fibrous varieties, pearly, opalescent appearance. Not to be confused with the variety of calcite also called satin spar.

Selenite: Colorless, clear crystals or in broad folia.

Rock gypsum: Dull-colored rock often with clay, calcium carbonate, silica.

Occurrence: Gypsum is mined in many countries.

Grades: Crude; ground; anhydrous.

Containers: Bags; multiwall paper sacks; shipped in bulk in boxcars and ships.

Uses: About one fourth of the total is sold uncalcined, chiefly for Portland cement retarder and agricultural use. Large-scale uses for the calcined material are: wallboard, lath, sheathing board, tile, and plasters, including special plasters such as plate-glass, pottery and dental plasters. See also gypsum cements. Other uses are: in metallurgy, paper, paints, textiles, baking powders, phosphors; source of sulfuric acid; smelting zinc ores; drying agent; fertilizer ingredient.

gypsum cements (plaster of Paris; Keene's cement; Parian cement; Martin's cement; Mack's cement). A group of cements which consist essentially of calcium sulfate and are produced by the complete or partial dehydration of gypsum, $CaSO_4 \cdot 2H_2O$. They usually contain additions of various sorts, these additions causing the differentiation of various special names. For example, Keene's cement contains alum or aluminum sulfate, Mack's cement contains sodium or potassium sulfate, Martin's cement contains potassium carbonate and Parian cement contains borax.

gypsum plaster. See gypsum cements.

H. Symbol for hydrogen.

HA. Abbreviation for hydroxyanisole. See hydroquinone, monomethyl ether.

Haber process. Original process for synthesis of ammonia from nitrogen and hydrogen. See ammonia synthesis and Haber-Bosch process.

Haber-Bosch process. Early process for synthesis of ammonia, operating at 200 atmospheres pressure, 550°C, with a promoted iron catalyst. Only eight per cent conversion of the nitrogen-hydrogen mixture was obtained per pass through the reaction zone, but recirculations increased the overall yield. Hydrogen and nitrogen were obtained from water gas and producer gas. See ammonia synthesis.

H acid. See 1-amino-8-naphthol-3, 6-disulfonic acid.

haematin. See hematin.

haemoglobin. See hemoglobin.

HAF black. Abbreviation for high abrasion furnace black. See furnace black.

hafnium Hf. Element of atomic number 72, group IVb of the periodic table.
Properties: Metal resembling zirconium; sp. gr. 13.1; m.p. about 2000°C; b.p. above 5400°C. It differs from zirconium in that it has a high (115 barns) thermal neutron cross section.
Occurrence: In most zirconium ores.
Derivation: Very difficult to separate from zirconium. The techniques employed consist of fractional crystallization, fractional precipitation, fractional decomposition of certain compounds, fractional distillation, solvent extraction, ion exchange and adsorption. The metal is prepared by the thermal decomposition of the tetraiodide. (See zirconium.)
Uses: Manufacture of tungsten filaments; most hafnium has been sold to the Atomic Energy Commission, and is used primarily as a control element in nuclear reactors.
Shipping regulations: Powder and sponge, wet or dry; flammable solid. Yellow label. *

hafnium carbide HfC. Used for nuclear reactor control rods. Has high thermal neutron absorption cross section and very high melting point, 3890°C (7030°F).

"Hagacide" No. 101. [108] Trademark for an organic biocide based on quaternary ammonium compounds carefully selected for maximum effectiveness in industrial water systems.
Uses: Inhibits the growth of protozoa, algae, fungi and bacteria. Also inactivates enzymes for better control of odors in air conditioning systems. Most effective for sulfate-reducing bacteria. Not recommended for potable waters.

"Hagacide" No. 106. [108] Trademark for an organic biocide based on quaternaries different from "Hagacide" 101 and carefully selected for maximum effectiveness in industrial water systems particularly with respect to slime bacteria.
Containers: 40-lb cans; 435-lb steel drums.
Uses: Inhibits growth of protozoa, algae, fungi and bacteria. Also inactivates enzymes for better odor control in air conditioning systems. Most effective on gram negative slime bacteria. Not recommended for potable waters.

"Hagacide" No. 203. [108] Trademark for a liquid organo-metallic biocide made up in a biocidal solvent for industrial water systems. Readily dispersible in water. Is preferentially adsorbed on wood and metal surfaces providing reserve power.
Containers: 34-lb cans; 360-lb steel drums.
Uses: Inhibits growth of fungi, algae, and sulfate reducing bacteria. Especially suited for cooling tower systems because of unique property of forming a microscopic biocidal film, resistant to water leaching on wood surfaces. Not recommended for potable waters.

"Hagacide" No. 204. [108] Trademark for a blended organic biocide containing both quaternaries and organo-metallic compounds specially formulated to provide a high degree of toxicity to all microorganisms encountered in industrial water systems.
Containers: 40-lb cans; 435-lb drums.
Uses: Inhibits growth of all organisms encountered in industrial cooling water systems. Forms biocidal film, resistant to water leaching on wood surfaces in systems providing residual treatment. Not recommended for potable waters.

"Hagafilm." [108] Trademark for a protective film-forming amine added to boilers or steamlines to inhibit condensate corrosion caused by oxygen and low pH. Available in liquid and solid forms.
Containers: Liquid: 5-gal cans, 55-gal drums; solid: 55-lb drums.
See also "Hagamin" and "Hagevap."

"Hagamin." [108] Trademark for a liquid, volatile, alkaline amine of either 99% or 50% active material content used in steam-boilers for inhibiting condensate corrosion by raising the pH value.

"Haganox." [108] A tan water-soluble powder.
Containers: 100-lb bags.
Uses: To chemically remove dissolved oxygen left in the feed-water after mechanical deaeration.

"Hagatreat" 168. [108] Trademark for a blend of several corrosion inhibitors, one of which is chromate.
Containers: 125-lb lined steel drums.
Uses: Provides corrosion control in recirculating cooling water systems in industrial plants.

"Hagevap." [108] Trademark for a powdered organic dispersing agent and antifoam compound used for minimizing scale formation in boilers.
Containers: 75-lb drums.

Hahnemann's soluble mercury. See mercurous nitrate, ammoniated.

"Halane." [203] Trademark for 1,3-dichloro-5,5-dimethylhydantoin. It contains 66% minimum available chlorine and reacts with water to liberate hypochlorous acid at a slow and controlled rate. Major uses are in dry household laundry bleaches and in water treating.

halazone (para-N,N-dichloro-sulfamylbenzoic acid; para-sulfondichloraminobenzoic acid) $HOOCC_6H_4SO_2NCl_2$.
Properties: White crystalline powder; strong chlorine odor; affected by light. Soluble in glacial acetic acid, benzene; slightly soluble in water, chloroform; insoluble in petroleum ether.
Constants: M. p. 195° with decomposition.
Grades: N. F. XI.
Uses: In the form of tablets as a powerful water disinfectant.
Shipping regulations: None.*

half life.
1. The time required for the decomposition of half of a sample of a radioactive substance, and thus a measure of the rate of such processes. Half lives vary from fractions of a second for some of the radioactive elements produced in recent nuclear studies to thousands of years for relatively stable radioactive elements such as uranium.
2. The term is also applied to any process in which a single substance changes in some way.

halibut liver oil (haliver oil).
Properties: Pale yellow to dark red liquid; characteristic, slightly fishy but not rancid odor and fishy taste; soluble in alcohol, ether, chloroform, and carbon disulfide; insoluble in water.
Constants: Sp. gr. 0.920-0.930; saponification number 160-180; iodine number 120-136; refractive index about 1.47.
Derivation: By expressing and boiling halibut livers.
Method of purification: Filtration.
Grades: Crude, refined, N. F. XI.
Containers: Wooden barrels, steel drums.
Uses: Medicine; source of vitamins A and D; leather dressing.
Shipping regulations: None.*

halides. Binary compounds of the halogens (q. v.).

halite (rock salt) NaCl. A natural sodium chloride found in the earth in beds varying from a few feet to over three thousand feet in thickness. Color, white or colorless, but often yellow, brown, deep blue from impurities; streak, white; vitreous luster; taste salty.
Constants: Sp. gr. 2.1-2.6; hardness 2.5.
Occurrence: United States (Utah, California, New York, Texas, Pennsylvania, Virginia, Michigan, Ohio, Louisiana); Russia; Italy; Poland; Spain; Transylvania; Alsace; Roumania; Germany; Peru; Switzerland; England; Austria; India; Siberia; China; Canada.
Uses: See sodium chloride.

haliver oil. See halibut liver oil.

halloysite $Al_2O_3 \cdot 3SiO_2 \cdot 2H_2O$. A variety of clay (q. v.), used to some extent in refractories and as a petroleum cracking catalyst.

Hall process. The electrolytic process by means of which metallic aluminum is recovered from aluminum oxide (usually bauxite). The aluminum oxide, after being purified from iron and other substances in preliminary chemical operations, is dissolved in a bath of molten cryolite (sodium aluminum fluoride). Passage of a direct current through this molten mixture results in formation and liberation of carbon dioxide at the carbon anodes, and production and collection of molten aluminum in the bottom of the cell, whose carbon lining serves as the cathode. The molten aluminum is perodically tapped through a suitable opening.

halocarbon plastics. Plastics based on resins made by the polymerization of monomers composed only of carbon and a halogen or halogens (ASTM D883-54T). The halocarbon plastics are characterized by extreme chemical resistance, excellent electrical properties, and good resistance to heat. See also "Fluorothene," "Kel-F," "Teflon."

halogenation. Incorporation of one of the halogen elements, usually chlorine or bromine, into a chemical compound. Thus benzene (C_6H_6) is treated with chlorine to form chlorobenzene (C_6H_5Cl), and ethylene (C_2H_4) is treated with bromine to form ethylene dibromide ($C_2H_4Br_2$). Compounds of chlorine and bromine are sometimes used as the source of the halogen, phosphorous pentachloride being a good example.

halogens. The chemically related elements, fluorine, chlorine, bromine, and iodine, and also astatine.

"Halon." [175] Trademark for chlorotrifluoroethylene synthetic resin (q. v.).

Properties: Zero moisture absorption; non absorbent; good abrasion resistance, impact, tensile, and compressive strength; non-flammable; excellent machinability; high volume and surface resistivity at high and low temperatures; low dielectric constant and good power factor at high temperature and high frequency; moldable by compression, extrusion, injection, and transfer molding techniques.

Types: "Halon" VK: Serviceable to 350°F; sp. gr. 2. 1; hardness (Rockwell R) 75; elongation (% at 75°F) 150.

"Halon" TVS: Serviceable to 390°F; sp. gr. 2. 16; hardness (Rockwell R) 95; elongation (% at 75°F) 205.

Caution: At processing temperatures small quantities of harmful gaseous products are released. Proper ventilation is recommended to remove these volatile products. Waste should be buried, not incinerated.

"Halopont." [28] Trademark for a line of pigment colors used for tinting white paper.

halothane (2-bromo-2-chloro-1, 1, 1-trifluoroethane) $CF_3CHBrCl$.

Properties: Non-flammable, volatile liquid; sweetish odor. Sp. gr. (20/4°C) 1. 86; b. p. 50. 2°C (760 mm), 20°C (243 mm). Light sensitive, may be stabilized with 0. 01% thymol. Slightly soluble in water; miscible with many organic solvents.

Containers: Bottles.

Grade: N. N. D.

Use: Medicine (anesthetic).

hamamelis (witch-hazel; winter bloom; snapping hazel; striped alder; tobacco wood).

Derivation: Dried leaves and bark (also twigs) of Hamamelis virginiana.

Habitat: North America (New England to Minnesota, southward to Louisiana).

Grades: Technical.

Containers: Bags; bales.

Uses: Alcoholic extract used in medicine; pharmacy; toilet preparations.

Shipping regulations: None. *

"Hanane." An English product — a mixture of bis-(dimethylamino)-fluorophosphine oxide and bis-(dimethylamino)-phosphonous anhydride. Colorless liquid, faint odor, vapor pressure 0. 6 mm at 25°C. Soluble in water and most organic liquids. Dangerous to handle. Used as a systemic insecticide.

"Hansa." [307] Trademark of proprietary line of pigments. Used for wallpaper, lacquer, plastics, paints, rubber and coated paper. Characterized by very good fastness to light, etc.

Hansa Yellow. A class name for a group of organic azo pigments that have good brightness, light-fastness and alkali resistance. Their tinting strength is about four times that of a good chrome yellow of approximately the same shade. Hansa Yellow has comparatively poor opacity in enamels and poor flow and is seldom used except where a non-toxic pigment is required, prohibiting the use of a chrome or cadmium yellow. The heat resistance of Hansa Yellow is not particularly good, and therefore it is not well adapted for use in baking enamels.

Containers: Barrels.

Hansgirg process. The production of magnesium from magnesium oxide by carbon reduction. Magnesium oxide is fed into an electric arc furnace lined with carbon where it is vaporized at a temperature of 2100°C. The mixture of magnesium vapor and carbon monoxide is withdrawn from the furnace and cooled to 200°C by diluting with natural gas (so the carbon monoxide will not oxidize the magnesium). A fine dust, containing 65% magnesium mixed with the oxide and carbon, is collected with electrostatic precipitators and sublimed at 750°C in electric retorts using high vacuum. The product thus obtained is better than 99% pure magnesium.

"Harbide." [446] Trade name for a silicon carbide brick, formed by impact pressing, having low permeability, dense impervious surfaces, high resistance to oxidation.

Uses: Ceramic kiln furniture, boiler furnace settings, recuperator tubes, radiant tubes, retorts, and in applications which are exposed to mechanical abrasion.

"Harchem." [189] Brand name for proprietary grades of sebacic acid, capryl alcohol (a sec-octyl alcohol) and methyl hexyl ketone (q. v.).

"Harchemex." [189] Trade name for a compound of mainly C_{14} and C_{16} straight chain primary alcohols in the approximate ratio of 2 to 1.

Properties: M. p. 32-36°C; color (NPA) 1. 5 max; sp. gr. 0. 840 (approx) at 30/20°C.

Uses: Wetting agents; germicidal quaternary ammonium compounds; lubricating oil additive.

"Harcure A." [189] Trade name for a linear polymeric anhydride curing agent for epoxy resins. Imparts excellent flexibility and good resistance to thermal and mechanical shock. Good electrical properties and holds up well under high operating temperatures.

Properties: Light tan solid; m. p. 75-82°C; sp. gr. 1. 0-1. 05 at 85°C; bulk density 5 lbs/gal.

Uses: In fluidized bed powders, in electrical encapsulations, in coatings and in impregnated glass fiber laminates.

hard coal. See anthracite.

hardened oils. See hydrogenated oils.

"Hard Hydrocarbon." [69] Brand name for an air-blown petroleum residue mineral rubber.

Properties: Black, brittle solid (also supplied as granular powder); sp. gr. 1. 04 ± . 03; softening point 300-315°F.

Uses: Diluent, extender and processing aid in rubber compounding.

hard lead. See antimonial lead alloys.

*See "I. C. C. Shipping Regulations, " page xiii.

Reference numbers refer to name of manufacturer. See "List of Manufacturers, " page v.

hardness. See Mohs' scale.

hardwood ashes. See wood ashes.

"**Harflex.**" [189] Trademark for a line of plasticizers.

Harflex 10: Dimethyl sebacate. Used where extreme efficiency and low temperature properties are of prime importance.

Harflex 40: Dibutyl sebacate. Claimed to be the most efficient non-toxic plasticizer made. F. D. A. approval for use in food wrap.

Harflex 50: Dioctyl sebacate. (See di(2-ethylhexyl) sebacate. Used where extreme low temperature flexibility, high efficiency and low volatility are required. Exhibits good resistance to extraction by soap and water, and has good heat and light stability.

Harflex 90: Dibenzyl sebacate. Low volatility; used in products designed for high temperature service.

Harflex 110: Diisodecyl phthalate.

Harflex 120: Diisooctyl phthalate.

Harflex 130: Isooctyldecyl phthalate.

Harflex 140: Dibutyl phthalate.

Harflex 150: Dioctyl phthalate. See di(2-ethylhexyl) phthalate.

Harflex 180: Dicapryl phthalate.

Harflex 210: Diisodecyl adipate.

Harflex 250: Dioctyl adipate.

Harflex 300: A high molecular weight polymeric plasticizer with a fairly low viscosity and good permanence. Compatible with polyvinyl chloride and various synthetic elastomers.

Properties: Color (Gardner) 5 max; sp. gr. 1.096 (25/25°C); viscosity (210°F) 120 ± 5 cts.

Containers: 480-lb drums.

Uses: Cable jackets, textile coatings, upholstery materials, baby pants.

Harflex 305: Polymeric plasticizer with a relatively low viscosity.

Properties: Color (Gardner) 5 max; sp. gr. 1.087 (25/25°C); viscosity (210°F) 95 ± 5 cts.

Containers: 480-lb drums.

Uses: In all types of vinyl formulations.

Harflex 320: General purpose polymeric plasticizer. Combines good permanence with easy processing and a very low viscosity. Superior to monomeric plasticizers in resistance to extraction by oils and aromatic hydrocarbons.

Properties: Color (Gardner) 4.5; sp.gr. 1.085 (25/25°C); viscosity (210°F) 86 ± 5 cts.

Containers: 480-lb drums.

Uses: Shoe liners, coated fabrics, flooring.

Harflex 325: A non-migratory polymeric plasticizer that combines good processing characteristics with excellent permanence; fast fusing and medium viscosity.

Properties: Color (Gardner) 7 max; sp.gr. 1.100 (25/25°C); viscosity (210°F) 135 ± 5 cts.

Containers: 480-lb drums.

Uses: In compounding plastisols, organosols, refrigeration and freezer gaskets, food belting, upholstery, electrical cable insulation and wall covering.

Harflex 330: A non-migratory plasticizer of excellent permanence; efficient and has very low odor, and low volatility.

Properties: Color (Gardner) 1 max; sp.gr. 1.086 (25/25°C); viscosity (210°F) 185 ± 5 cts.

Containers: 480-lb drums.

Uses: Vinyl refrigerator gaskets, in auto crash pad film, electrical wire insulation, synthetic rubbers.

Harflex 340: A polar polymeric plasticizer of medium viscosity. Contributes good adhesion, low migration and quick adhesion to a polyvinyl acetate adhesive. Also used in aiding viscosity stability of a compound.

Harflex 370: A very permanent polymeric plasticizer of relatively high viscosity.

Properties: Color (Gardner) 6 max; sp. gr. 1.06 (25/25°C); viscosity (210°F) 550 ± 50 cts.

Containers: 480-lb drums.

Uses: Plasticizer for all types of vinyl formulations; also used in conjunction with isocyanates to make urethane foam systems.

Harflex 375: An extremely fine polymeric plasticizer of high viscosity. Excellent permanence to extraction and migration.

Properties: Color (Gardner) 7 max; sp.gr. 1.055 (25/25°C); viscosity (210°F) 3600 ± 100 cts.

Containers: 480-lb drums.

Uses: Electrical tapes and wire coatings; vinyl foams.

Hargreaves process. The manufacture of sodium sulfate (salt cake) from sodium chloride and sulfur dioxide. A mixture of sulfur dioxide and air is passed through sodium chloride brine in a countercurrent manner to produce sodium sulfate and hydrochloric acid.

Harris process. Process for the removal of arsenic, antimony, tin, and zinc from virgin or secondary lead by agitating the molten metal with molten caustic soda and salt. All undesirable metals are oxidized and the oxides dissolved in the caustic with exception of silver which is removed in a subsequent desilvering operation.

hartshorn. See ammonium carbonate.

hartshorn oil. See bone oil.

HAS. Abbreviation for hydroxylamine acid sulfate.

"**Hascrome.**" [214] Trademark for iron-base impact-resistant, hard-facing rod containing chromium and manganese. Used for welding to wearing parts, such as crusher jaws and dipper teeth.

hashish. See cannabis.

"**Hastelloy.**" [214] Trademark for a series of nickel-base alloys, having high resistance to corrosives, such as hot hydrochloric acid, hot sulfuric acid, wet chlorine, etc. as well as excellent physical and mechanical properties. Used for agitators, autoclaves, concentrators, heat exchangers,

exhausters, evaporators, condensers, dryers, heating and cooling coils, injectors, blowers, burner parts, chlorinating equipment, pickling equipment, pyrometer equipment, thermometer wells, pipe and fittings, pumps, valves, kettles, tanks and vessels for all kinds of chemical plant service.
Also applies to certain nickel base high temperature alloys.

hatchettine. Synonym for hatchettite (q. v.).

hatchettite (hatchettine; adipocerite). A soft variety of ozocerite (q. v.) or mineral wax. Color yellowish white, yellow, or greenish yellow; fuses at about 120°F. Sp. gr. 0.90-0.98 at 77°F. Found in bogs and coal beds in Great Britain.

"Haveg." [349] Trademark for a series of corrosion resistant compounds fabricated into chemical process equipment. Available as:
"Haveg 31" - Xylenol resin-leached asbestos combination.
"Haveg 41" - Phenol resin-leached asbestos combination.
"Haveg 61" - Furane resin-leached asbestos combination.
"Haveg 043" - Phenolic resin-graphic combination.
"Haveg 6720" - Glass reinforced furane resin.
"Haveg 7710" - Glass reinforced polyester resin.
"Haveg 7790" - Glass reinforced polyester resin.
"Haveg 9710" - Glass reinforced epoxy resin.

"Havelast." [349] Trademark for an elastomeric binder or impregnant for various reinforcing materials such as "Sil-Temp", as fabric or rovings, asbestos, glass or graphite. Used in the rocket and missile industry when resiliency is desired.

hayo. See coca.

"Haystellite." [214] Trademark for tungsten carbide possessing great hardness and toughness.
Forms: Inserts; composite rod; tube rod.
Uses: Application, by welding, to oil-well drilling tools and other parts subjected to extreme abrasion.

Hb. Symbol for hemoglobin.

"HB-20." [58] Trademark for a partially hydrogenated alkylaryl hydrocarbon.
Specifications: Light amber to dark-brown liquid; refractive index (25°C) approx. 1.54; sp. gr. (25°/15.5°C) approx. 0.96.
Containers: 450-lb drums; tank cars.
Use: Extender plasticizer for vinyls.

"HB-40." [58] Trademark for partially hydrogenated terphenyl.
Properties: Clear, mobile, high-boiling hydrocarbon; almost colorless, with faint, pleasant odor; 8.37 lb/gal; refractive index 1.5675 ± 0.0075 (25°C); insoluble in water; miscible in all proportions at room temperature with a number of solvents and oils.
Uses: Plasticizer for vinyl compounds, for styrene water dispersion paints, paper coatings, adhesives and styrene casting resins.

H-bomb. Abbreviation for hydrogen bomb. See nuclear fusion.

HCCH. Abbreviation for hexachlorocyclohexane (q. v.).

HCG. Abbreviation for human chorionic gonadotropin.
See chorionic gonadotropin.

HCH. Abbreviation for hexachlorocyclohexane (q. v.).

He. Symbol for helium.

health physics. The branch of physics which deals with the protection of personnel from the hazards of radiation. The chief duties of a health physicist are to set the standard for safe levels of exposure to various radiations, to aid in the detection of radiation in order to avoid overexposures, and to develop suitable methods for protection against radiation.

heating oil. See fuel oil.

heat of fusion. The quantity of heat required to convert unit weight of a solid to the liquid state. This varies somewhat with temperature, and to a much less extent with pressure.

heat transfer salt. See "Hitec" Heat Transfer Salt.

heavy chemicals. Chemicals produced in tonnage and carload quantities at low prices, such as sulfuric acid, ammonia, soda ash.

heavy hydrogen. See deuterium.

heavy oils. Oils distilled over from coal-tar between 230 and 330°C, the exact range not at all definite.
Shipping regulations: None. *

heavy oxygen. See oxygen 18.

heavy spar. See barite.

heavy spar, artificial. See barium sulfate.

heavy sulfur. See sulfur.

heavy water. Term that may be applied to any of the isotopic varieties of water whose molecules are composed of atoms of hydrogen and/or oxygen having atomic weights greater than 1 and 16 respectively. Also applied to ordinary water containing a higher proportion of these heavy water molecules than is normally present in natural water. The term could also mean H_2O^{18}, D_2O^{18}, but is more commonly applied to deuterium oxide (D_2O) which is composed of deuterium (heavy hydrogen, atomic weight 2) and ordinary oxygen. Deuterium oxide is present to the extent of about 1 part to 5,000 of ordinary natural water. Deuterium oxide melts at 3.8°C, boils at 101.4°C and has sp. gr. of 1.1056 at 25°C. Heavy water is used as a

*See "I.C.C. Shipping Regulations," page xiii.
Reference numbers refer to name of manufacturer. See "List of Manufacturers," page v.

moderator in nuclear reactors.
Derivation: (a) During World War II, made by the U. S. government by fractional distillation. This was much more expensive than (b) the Spevack process, which is a dual temperature isotopic exchange in which water and hydrogen sulfide exchange deuterium and hydrogen so that two water streams result, one richer in heavy water than the other. Process (c) uses liquid ammonia and ammonia synthesis gas, containing nitrogen and hydrogen. By temperature manipulation the deuterium present in the synthesis gas is concentrated in the liquid ammonia and is further concentrated by distillation. The deuterium is then transferred to water and can be concentrated in a special Kuhn still to about 99.8% D_2O.
See also tritium (hydrogen of atomic weight 3) which combines with oxygen to give another variety of heavy water.

hectorite $Mg_{2.67}Li_{0.33}Si_4O_{10}(OH)_2$. One of the montmorillonite group of minerals that are principal constituents of bentonite clays. The composition varies because magnesium and lithium are exchangeable, replaceable by calcium, sodium and other elements. Hectorite occurs in the Mohave Desert and is used as a source of a commercial mineral gel with uses as an adsorbent, flocculating agent, stabilizer for suspensions, emulsions, and a film forming material.

hedeoma oil (American pennyroyal oil; pulegium oil).
Properties: A pale yellowish limpid liquid, essential oil; characteristic, pungent, mint-like odor and taste. Sensitive to light. Soluble in two or more parts of 70% alcohol, ether and chloroform; slightly soluble in water; sp. gr. 0.925-0.940; refractive index 1.482.
Chief constituents: Pulegone, hedeomol.
Derivation: Distilled from the leaves and tops of Hedeoma pulegioides.
Adulteration: Mineral oil, turpentine, resin oil. Detected by their difficult solubility in 70% alcohol.
Method of purification: Rectification.
Grades: Technical.
Containers: 5-, 10-lb bottles; 25-, 50-lb tins.
Uses: Medicine; insectifuge; perfumery; manufacture of pulegone and its derivatives.
Shipping regulations: None. *

alpha-**hederin.** See helixin.

hedonal (methylpropylcarbinolurethane) $CH_3CH_2CH_2CH(CH_3)OCONH_2$.
Properties: White crystalline powder; feeble aromatic odor and taste; soluble in alcohol, ether, organic solvents; sparingly soluble in cold water; more soluble in hot water. Fusing point 76°C. B. p. 215°C.
Use: Medicine.
Shipping regulations: None. *

HEF black. Abbreviation for high elongation furnace black. See furnace black.

Hehner number. The percent by weight of water-insoluble fatty acids in oils and fats.

helcosol. See bismuth pyrogallate.

helenine. A nucleoprotein derived from the mold Penicillium funiculosum and used with some success as an antiviral drug.

Helianthine B. See methyl orange.

"Helindon." [307] Trademark of vat dyestuffs. Used for the dyeing of wool. Characterized by excellent fastness properties.

"Helio." [307] Trademark of organic pigment dyestuffs. Used for paints, lacquers, printing inks, wallpaper, coated paper, rubber and organic plastics. Characterized by very good general fastness properties.

heliodor. See beryl.

"Heliogen." [307] Trademark for phthalocyanine dyestuffs. Used for paints, lacquers, printing inks, wallpaper, coated paper, rubber, and organic plastics. Characterized by outstanding fastness to light as well as brilliancy of shade.

"Heliogen." [342] Trademark for brand of atomic iodine solutions for general sanitizing purposes.

"Heliophan." [19] Trademark for (homo)-menthyl salicylate (q. v.).

heliotropin (piperonal; piperonyl aldehyde) $C_6H_3(CH_2OO)CHO$(bicyclic).
Properties: White, shining crystals; turns red-brown on exposure to light; sweet floral odor, typical of heliotrope; m. p. 35.5-37°C; b. p. 263°C; soluble in alcohol and ether; insoluble in water, glycerol.
Derivation: By oxidation of isosafrole.
Method of purification: Crystallization.
Grades: Technical; recrystallized.
Containers: Bottles; tins; drums.
Uses: Medicine; perfumery; suntan preparations.
Shipping regulations: None. *

"Heliozone." [28] Trademark for a rubber chemical.
Properties: Greenish, waxy material.
Containers: Drums (150-lbs net).
Use: To retard sun-checking and cracking of rubber and synthetic rubber.

helium He. Element of atomic number 2, group 0 of periodic table.
Properties: Colorless, odorless, tasteless inert gas; does not combine chemically with any other substance; forms hydrates under certain special circumstances; b. p. −260.0°C; f. p. −272.2°C, the lowest of any substance; density of gas 0.1785 g/l at 0°C, or 0.138 compared with air as unity; very slightly soluble in water; insoluble in alcohol.
Source: Liquefaction of all other components of certain natural gases; also as a by-product of liquid air processing for oxygen production.
Grade: U. S. P. XVI; technical; 99.9% pure min.

*See "I.C.C. Shipping Regulations," page xiii.
Reference numbers refer to name of manufacturer. See "List of Manufacturers," page v.

Containers: Cylinders; tank cars.
Uses: Inflation of balloons and dirigibles; diluent for oxygen or anesthetic gases in medicine; component of "air" supplied to men working in tunnels digging under pressure; also for filling luminescent electric-light tubes; low-temperature research; leak detection; inert shield for arc welding; carrier gas in gas chromatography; inert atmosphere in making semiconductors and processing titanium and zirconium; heat transfer agent in nuclear reactors; low temperature work in electronics.
Shipping regulations: Non-flammable gas. Green gas label. *

helium 3. A stable isotopic form of helium, one-millionth as abundant in nature as ordinary helium. Useful in theoretical investigations of nuclear chemistry.

helixin (alpha-hederin) $C_{41}H_{64}O_{11}$.
Properties: Solid; m. p. 256-257°C; insoluble in water, petroleum ether; soluble in alcohol, acetone, glacial acetic acid, dilute aqueous basic solutions.
Derivation: From ivy leaves.

"Helix" Rosin. [79] Trademark of a specially processed pale wood rosin containing approximately 5. 5% of chemically combined lime; a product containing very little free lime.
Constants: M. p. (capillary tube) 113°C; m. p. (ball & ring) 133°C; acid number 73; color "WG-WW."
Containers: Non-returnable light-weight galvanized drums of about 500 lbs gross wt. Tare 14-16 lbs.
Uses: Linoleum print paint; varnish-air drying; varnish-baking; varnish-gloss oil; printing ink.

hellebore.
(a) Black; dried rhizome of perennial evergreen herb Helleborus niger.
(b) White; dried powdered rhizomes of genus Helleborus.
(c) Green; dried rhizomes and roots of Veratrum viride.
Uses: Medical; white variety is used as insecticide.

helleborein $C_{37}H_{56}O_{18}$.
Properties: Glucoside crystallizable in yellow prisms. Tastes both sweet and bitter. Poisonous! Soluble in water, weak alcohol; less soluble in ether and absolute alcohol; m. p. 270°C.
Derivation: From black and green hellebore.
Use: Medicine.

"Helmerco" Colors. [57] A class of lakes which are but very slightly soluble in water. For use on paper stock by beater dyeing through the use of alum and rosin; also for coating and wall paper printing.

helonias (false unicorn; starwort). Dried rhizome and roots of Chamaelirium luteum.
Habitat: United States.
Containers: Bales.
Use: Medicine.

helvite $Mn_4Be_3(SiO_4)_3S$. A natural silicate of beryllium and manganese. Color red to brown; hardness 6; sp. gr. 3. 3.
Occurrence: New Mexico, Virginia; Europe; U. S. S. R.
Use: Possible source of beryllium.

hem. See heme.

hematein $C_{16}H_{12}O_6$. An oxidation product of hematoxylin, the coloring principle of logwood. Not to be confused with hematin.
Properties: Green to reddish-brown crystals; m. p. 250°C with decomposition; insoluble in water; slightly soluble in alcohol and ether; soluble in dilute sodium hydroxide giving a bright red color; soluble in ammonia with brownish-violet color.
Derivation: By adding ammonia to logwood extract and exposing to air.
Uses: Indicator; biological stain.

hematin (haematin) $C_{34}H_{32}N_4O_4 \cdot FeOH$. Not to be confused with hematein. The hydroxide of heme (q. v.).
Properties: Blue to brown-black powder; decomposes at 200°C without melting; soluble in alkalies, hot alcohol or ammonia; slightly soluble in hot pyridine; insoluble in water, ether and chloroform.
Derivation: By dissolving hemin in dilute potassium hydroxide, precipitating with acetic acid and recrystallizing from pyridine.

hematite. Same as hematite, red.

hematite, brown. See limonite.

hematite, red (red iron ore; bloodstone). Iron oxide (Fe_2O_3), with impurities.
Properties: Brilliant black to blackish red or brick red mineral with brown to cherry red streak and metallic to dull luster. Sp. gr. 4. 9-5. 3; hardness about 6.
Occurrence: Michigan, Minnesota, Wisconsin, Alabama and other parts of U. S. Also in numerous other parts of the world.
Uses: The most important ore of iron. Also certain varieties are used as paint pigments and for rouge.
See also iron oxide reds and ferric oxide for other synonyms and uses.

hematoporphyrin $C_{34}H_{38}O_6N_4$. Deep red crystals; soluble in alcohol; sparingly soluble in ether; insoluble in water. Obtained from hemin or hematin by the action of strong acids. It is non-toxic and is reported to be preferentially absorbed by cancerous tissues, making them fluoresce under ultraviolet light.

hematoxylin (hydroxybrasilin; logwood crystals) $C_{16}H_{14}O_6 \cdot 3H_2O$. The coloring principle of logwood.
Properties: Colorless to yellowish crystals, redden on exposure to light; m. p. 140°C (anhydrous); soluble in alcohol and ether; slightly soluble in water.
Derivation: By concentrating logwood liquor in evaporating pans and crystallizing.
Grades: Technical, sold on basis of tinctorial value.

Containers: 350-, 400-lb wooden barrels.
Uses: Textile and leather dyeing; manufacture of ink and stain in microscopy; indicator in the titration of alkaloids; medicine.
Shipping regulations: None. *

hematoxylon. See logwood.

heme (hem) $C_{34}H_{32}FeN_4O_4$. The non-amino acid portion of hemoglobin consisting of reduced (ferrous) iron bound to protoporphyrin (see porphyrins).

hemicellulose. A type of natural substance more complex than a sugar and less complex than cellulose, and occurring in woody tissue along with cellulose. Obtained in pure form from corn grain hulls (corn fiber) by lime water extraction.

hemimorphite. See calamine (mineral).

hemin (Teichmann's crystals) $C_{34}H_{32}N_4O_4FeCl$. The chloride of heme.
Properties: Crystals which are brown by transmitted light and steel blue by reflected light. Sinters at 240°C. Freely soluble in ammonia water; soluble in strong organic bases; insoluble in carbonate solutions, dilute acid solutions; insoluble but stable in water.
Derivation: By heating hemoglobin with acetic acid and sodium chloride.
Use: The isolation of hemin is used for the identification of blood stains.

hemlock bark. Bark of the hemlock fir Tsuga canadensis.
Habitat: Northern and western parts of the United States and Canada.
Uses: Tanning industry; boiler compounds; pharmaceutical preparations.
Shipping regulations: None. *

hemlock gum. Incorrect name for hemlock pitch (q. v.).

hemlock needle oil. See spruce oil.

hemlock oil. See spruce oil.

hemlock pitch (Canada pitch). The resinous exudation from Tsuga canadensis or hemlock.

hemoglobin (Hb). Suggested empirical formula: $(C_{738}H_{1166}FeN_{208}S_2)_4$. The important respiratory protein of the red blood cells; it is necessary in the transfer of oxygen from the lungs to the tissues and of the carbon dioxide from the tissues to the lungs.

Hemoglobin is a conjugated protein consisting of approximately 94% globin (protein portion) and 6% heme (q. v.). Each molecule of hemoglobin can combine with one molecule of oxygen to form oxyhemoglobin (HbO_2). The iron (found in the heme portion) must be in the reduced (ferrous) state to enable the hemoglobin to combine with oxygen. See also carboxyhemoglobin.

Oxyhemoglobin is available commercially as a brownish red powder or crystals; soluble in water. It is used in medicine, and is usually called hemoglobin.

hemp. Soft, white fibers, 3-6 ft long, obtained from the stems of Cannabis sativa. It is coarser than flax but stronger, more glossy, and more durable than cotton.
Sources: Central Asia, Italy, Russia, U. S.
Uses: Blended with cotton or flax in toweling and heavy fabrics; twine; cordage; packing.

hemp, Canadian. See apocynum.

hemp oil. See hempseed oil.

hempseed oil (hemp oil).
Properties: Light green, fixed, non-drying liquid; becomes brownish-yellow on standing. Soluble in ether, benzene and carbon disulfide. Sp. gr. 0.9255-0.9280; saponification value 172-192; Maumené number 97; iodine number 148.
Derivation: From hempseed Cannabis sativa, by pressing or extraction.
Grades: Technical.
Containers: Tins; iron drums.
Uses: Soft soap; paints; varnishes.

henbane. See hyoscyamus.

hendecane. See undecane.

hendecanoic acid. See undecanoic acid.

1-hendecanol. See 1-undecanol.

2-hendecanol. See 2-undecanol.

10-hendecen-1-al. See undecylenic aldehyde.

10-hendecenyl acetate. See undecylenyl acetate.

heneicosane $C_{21}H_{44}$ or $CH_3(CH_2)_{19}CH_3$.
Properties: Crystals; sp. gr. 0.778 (40°C); b. p. 215°C (15 mm); m. p. 40°C.

n-heneicosanoic acid $CH_3(CH_2)_{19}COOH$. A saturated fatty acid not normally found in natural fats or waxes. White crystalline solid; m. p. 74.3°C. Synthetic product available for organic synthesis; 99% purity.

henequen. Hard, strong, reddish fibers obtained from the leaves of Agave fourcroydes. It is similar to sisal but coarser and stiffer.
Source: Mexico, Cuba.
Use: Binder twine; rope.

henna (Egyptian privet; lawsonia alba; flower of paradise).
Derivation: Leaves of Lawsonia alba.
Habitat: Orient; Mediterranean region; southern Asia and Australia.
Grades: Technical.
Containers: Bags.
Uses: Medicine; hair dye.
Shipping regulations: None. *

"Hennig Purifier." [177] Trademark for a preparation having a soda-ash base and other materials. Produced as walnut-sized briquettes. Packed in 100-lb paper bags. Used as a ladle addition to produce cleaner steel by aiding in removal of dissolved oxides and silicates and fluxing non-metallic inclusions to slag.

Henry's law. When a liquid and a gas are kept in contact with one another for a period of time the weight of the gas that dissolves in a given quantity of liquid is proportional to the pressure of the gas above the liquid. Thus if ordinary air is kept in contact with water at ordinary atmospheric pressure, each kilogram of water dissolves 0.017 grams of oxygen at 20°C., while if this pressure is halved by doing the experiment at high altitude where the pressure is only half an atmosphere the water dissolves only 0.0085 grams of oxygen. The law holds true only for equilibrium conditions, i.e., when enough time has elapsed so that the quantity of gas dissolved is no longer increasing or decreasing.

hentriacontane (n-hentriacontane) $C_{31}H_{64}$ or $CH_3(CH_2)_{29}CH_3$.
Properties: Crystals; sp. gr. 0.781 (68°C); b.p. 302°C (15 mm); m.p. 68°C.

1-hentriacontanol (formerly confused with 1-triacontanol as myricyl or melissyl alcohol) $CH_3(CH_2)_{29}CH_2OH$. Possibly one of the constituents of beeswax.

HEOD. Abbreviation for hexachloroepoxyoctahydrodimethanonaphthalene. See dieldrin.

hepar calcis. See calcium sulfide and lime, sulfurated.

heparin. See heparin, sodium.

heparin sodium (heparin). The sodium salt of a complex organic acid present in mammalian tissues and having the properties of prolonging the clotting time of blood. Heparin appears to be a dextrorotatory polysaccharide built up from hexosamine and hexuronic acid units containing sulfuric acid ester groups; it has the properties of a polymer.
Properties: White or pale-colored, amorphous powder; nearly odorless, hygroscopic; soluble in water; insoluble in alcohol, benzene, acetone, chloroform, and ether; pH of one in 100 solution 6.0 to 7.5.
Derivation: Animal livers or lungs.
Grade: U.S.P. XVI.
Use: Medicine.

hepar sulfuris. See potassium sulfide.

heptabarbital (5-[1-cyclohepten-1-yl]-5-ethylbarbituric acid; 5-ethyl-5-cycloheptanylbarbituric acid) $C_{13}H_{18}N_2O_3$.
Properties: White, odorless crystalline powder; slightly bitter taste. M.p. 174°C. Very sparingly soluble in water; slightly soluble in alcohol; soluble in alkaline solutions. Forms water-soluble sodium, magnesium and calcium salts.
Grade: N.N.D.
Use: Medicine.

heptachlor $C_{10}H_7Cl_7$. 1,4,5,6,7,8,8-Heptachloro-3a,4,7,7a-tetrahydro-4,7-methanoindene. An insecticide similar to chlordan (q.v.).
Properties: White to light tan, waxy solid; m.p. 95-96°C; sp. gr. 1.57-1.59. Insoluble in water; soluble in xylene.
Containers: Drums.
Warning! May be fatal if swallowed. Avoid contamination of feed and foodstuffs. MCA warning label.

heptacosane $C_{27}H_{56}$ or $CH_3(CH_2)_{25}CH_3$.
Properties: Crystals; soluble in alcohol; insoluble in water; sp. gr. 0.804; b.p. 270°C (15 mm); m.p. 59.5°C.

heptadecane $C_{17}H_{36}$ or $CH_3(CH_2)_{15}CH_3$.
Properties: Leaflets; soluble in alcohol; insoluble in water; sp. gr. 0.778; b.p. 303°C; m.p. 22.5°C.

n-heptadecanoic acid (margaric acid) $CH_3(CH_2)_{15}COOH$. A saturated fatty acid not normally found in natural fats or waxes.
Properties: Colorless crystals; m.p. 61°C; sp. gr. 0.8355 (90.6/4°C); b.p. 363.8°C (760 mm), 230.7°C (16 mm); refractive index 1.4324 (70°C). Soluble in alcohol and ether; insoluble in water. Available as a 99% pure synthetic product; used in organic synthesis.

heptadecanol. Any saturated C_{17} alcohol. See, for example, n-heptadecanol and 3,9-diethyl-6-tridecanol.

n-heptadecanol $C_{17}H_{35}OH$.
Properties: Colorless liquid. Slightly soluble in water; sp. gr. 0.8475 (20/20°C); b.p. 308.5°C (760 mm); vapor pressure < 0.01 mm (20°C); flash point 310°F; wt/gal 7.1 lbs (20°C).
Grades: Technical.
Containers: 1-gal cans; 5-, 55-gal drums. Net content 7.0, 35, 380 lbs.
Uses: Organic synthesis; plasticizer; intermediates; perfume fixatives. For soap and cosmetic preparations; base for manufacture of wetting agents and detergents.

2-heptadecylglyoxalidine (2-heptadecylimidazoline) $C_{20}H_{40}N_2$ or $C_{17}H_{35}C_3H_5N_2$.
Properties: Waxy solid; m.p. 85°C; b.p. 200°C (2 mm); slightly soluble in water; soluble in alcohol, benzene; hydrolyzes on standing to form N-2-aminoethyl stearamide.
Derivation: Prepared by reacting stearic acid with ethylene diamine.
Use: Fungicide.

2-heptadecylglyoxalidine acetate. See glyodin.

2-heptadecylimidazoline. See 2-heptadecylglyoxalidine.

2-heptadecyl 2-imidazoline acetate. See glyodin.

heptafluorobutyric acid (perfluorobutyric acid) C_3F_7COOH.
Properties: Colorless hygroscopic liquid with sharp odor. B.p. 120°C (735 mm); f.p. −17.5°C; sp. gr. 1.641 (25°C); refractive index (n 25/D) 1.290; surface tension 15.8 dynes/cm (30°C). Miscible with water, acetone, ether, and petroleum ether; soluble in benzene and carbon tetrachloride; insoluble in carbon disulfide and mineral oil.
Derivation: By electrolysis of a solution of butyric acid in hydrogen fluoride.

Grades: Technical.
Containers: 1-, 3-, 10-lb bottles.
Uses: Intermediate; surfactant; acidulant.

"Heptagran." [147] Brand name for a granular insecticide containing 2 1/2, 10, or 25% heptachlor.
Containers: 50-lb bags.
Use: In preparation of insecticide-fertilizer combination.

heptaldehyde. See heptanal.

heptalin acetate. See methylcyclohexanol acetate.

heptamethylene. See cycloheptane.

heptanal (heptaldehyde; oenanthic aldehyde; oenanthal; aldehyde C-7) $C_6H_{13}CHO$.
Properties: Oily, colorless liquid, with a penetrating fruity odor. Keep well stoppered. Soluble in 3 volumes of 60% alcohol; slightly soluble in water, soluble in ether. Sp. gr. 0.814-0.819; refractive index 1.42; m. p. 43°C; b. p. 153°C.
Derivation: Castor oil, from decomposition of the ricinoleic acid glyceride.
Method of purification: Distillation.
Grades: Technical.
Containers: Up to 55-gal drums.
Uses: Manufacture of 1-heptanol; organic synthesis; perfumery; pharmaceuticals; rubber products.

heptane (dipropylmethane) $CH_3(CH_2)_5CH_3$.
Properties: Volatile, colorless liquid; highly flammable; freezing point −90.595°C; b. p. 98.428°C; refractive index (n 20/D) 1.38764; sp. gr. 0.68368 (20°C); flash point −1°C. Soluble in alcohol, ether, chloroform; insoluble in water.
Typical specifications for commercial grade normal heptane: Distillation range 200-210°F; vapor pressure 2.0 psi absolute (100°F) (max); color, Saybolt +30 (min); maximum sulfur content 0.01 wt %; corrosion, passes ASTM D 130-30 test; doctor test, sweet; principal diluent is methylcyclohexane with small amounts of other naphthenes, isoheptanes, isooctanes, toluenes.
Derivation: Fractional distillation of petroleum. Purified by rectification. Also from the oleoresin of Pinus sabiniana.
Grades: Commercial; 99%; spectro; ASTM; research.
Containers: Bottles; drums; tank cars.
Uses: Standard for octane rating and determinations; anesthetic; solvent; organic synthesis; preparation of laboratory reagents.
Fire hazard: Dangerous.
Shipping regulations: Flammable liquid. Red label. *

1,7-heptane-dicarboxylic acid. See azelaic acid.

heptanedioic acid. See pimelic acid.

n-heptanoic acid (enanthic acid; oenanthic acid; n-heptylic acid; heptoic acid) $CH_3(CH_2)_5COOH$.
Properties: Clear, oily liquid; unpleasant odor. Soluble in alcohol and ether; insoluble in water. Sp. gr. 0.9181 (20/4°C); m. p. −7.0°C; b. p. 221.9°C (760 mm); refractive index 1.4229.
Derivation: By oxidizing heptaldehyde with potassium dichromate and sulfuric acid.
Method of purification: Distillation.
Grades: Technical.
Containers: Iron drums.
Use: Organic synthesis; production of special lubricants for aircraft and brake fluids.
Shipping regulations: None. *

1-heptanol (heptyl alcohol; alcohol C-7) $C_7H_{15}OH$.
Properties: Colorless fragrant liquid; m. p. −34.6°C; b. p. 175°C (765 mm); sp. gr. 0.824 (20/4°C); refractive index (n 20/D) 1.4233.
Typical specifications: Sp. gr. 0.822-0.825 (20/4°C); m. p. −34 to −36°C; b. p. 173-175°C (765 mm); refractive index (n 20/D) 1.4225-1.4250. Soluble in water, alcohol and ether.
Derivation: From heptaldehyde by reduction.
Containers: Glass bottles; 1-, 5- and 10-gal cans; 55-gal lined drums.
Uses: Organic intermediate, solvent, cosmetic formulations.
Shipping regulations: None. *

2-heptanol. See methyl amyl carbinol.

3-heptanol $CH_3CH_2CH(OH)C_4H_9$.
Properties: Liquid; sp. gr. (20°C) 0.8224; m. p. −70°C; b. p. 156.2°C; flash point 140°F; slightly soluble in water.
Containers: Cans, drums, tank cars.
Uses: Flotation frother; solvent and diluent in coatings; intermediates.

2-heptanone. See methyl n-amyl ketone.

3-heptanone. See ethyl butyl ketone.

4-heptanone. See dipropyl ketone.

"Hepteen Base." [248] Trademark for a heptaldehyde-aniline reaction product.
Properties: A dark-brown, free-flowing liquid; sp. gr. 0.93; soluble in acetone, benzol, and ethylene dichloride; moderately soluble in gasoline; insoluble in water.
Uses: Accelerator for pure gum, inner tube, white tire sidewall, air-cured footwear.

1-heptene (1-heptylene) C_7H_{14} or $CH_3(CH_2)_4CH{:}CH_2$.
Properties: Colorless liquid; sp. gr. 0.6968 (20/4°C); b. p. 93.3°C; m. p. −119.2°C; refractive index (n 20/D) 1.3994; soluble in alcohol, acetone, ether, petroleum, coal-tar solvents; insoluble in water.
Typical specifications: Sp. gr. 0.6955-0.6980 (20/4°C); b. p. 93-94°C; refractive index (n 20/D) 1.3987-1.4005.
Grades: 95% purity.
Use: Organic synthesis of flavors, perfumes, medicines, dyes, resins.

2-heptene (2-heptylene) $CH_3(CH_2)_3CH{:}CHCH_3$. Both cis and trans isomers are known.
Properties: Colorless liquid; sp. gr. cis 0.708, trans 0.704, commercial 0.7010-0.7050 (20/4°C); b. p. trans 98.0°C, cis 98.5°C, commercial 97-99°C; refractive

index (n 20/D), cis and trans 1.406; flash point −4°C; soluble in alcohol, acetone, ether, petroleum, coal-tar solvents; insoluble in water.
Grades: 95%.
Containers: Bottles.
Use: Organic synthesis.
Shipping regulations: Flammable liquid. Red label. *

3-heptene (3-heptylene) C_7H_{14} or $C_3H_7CH{:}CHC_2H_5$.
Properties (mixed cis and trans isomers): Colorless liquid; b. p. 95°C; sp. gr. (60/60°F) 0.705; refractive index (20/D) 1.405; flash point −6°C.
Grades: 95%.
Containers: Bottles.
Shipping regulations: Flammable liquid; red label. *

heptenes C_7H_{14}.
Properties: Liquid. Sp. gr. (15.56/15.56°C) 0.711; b. p. 189.5°C.
Derivation: Olefin fraction produced by catalytic polymerization of propylene and butylene.
Containers: Up to tank cars.
Uses: Lubricant additive; catalyst; surfactants.

heptoic acid. See heptanoic acid.

"Heptuna Plus." [299] Trademark for a preparation containing hematopoietic factors, vitamins, and minerals. Used in medicine.

heptyl acetate $C_7H_{15}OOCCH_3$. A liquid with fruity odor.
Use: Preparation of fruit essences.

heptyl alcohol. See 1-heptanol.

heptylamine $C_7H_{15}NH_2$.
Properties: Colorless liquid; sp. gr. 0.777 (20/4°C); m. p. −23°C; b. p. 155°C; slightly soluble in water; soluble in alcohol or ether.

n-heptyl carbinol. See 1-n-octanol.

heptylene. See heptene.

heptyl formate $HCOOC_7H_{15}$.
Properties: A colorless liquid with fruity odor; b. p. 176.7°C; sp. gr. 0.894 (0°C).
Use: Artificial fruit essences.

heptyl heptoate.
Properties: A colorless liquid with fruity odor; sp. gr. 0.865 (19°C); b. p. 273-274°C (754 mm).
Use: Fruit essences.

heptyne carboxylic acid methyl ester. See "Folione."

herbicides (weed-killers). A term used to describe a group of materials, chemical in nature, which are used to destroy plant life. These may be used in contact with the seed, stem or leaf of a plant and kill all plant life without regard to species (non-selective herbicides) or kill only certain species (selective herbicides). Since the plants normally destroyed by application of chemicals are weeds in relation to the crop desired, the term weed-killer has been commonly used. The non-selective herbicides have been used for many years, and are generally inorganic in nature. Sodium arsenite, sodium chlorate, ammonium thiocyanate and sodium chloride are typical examples. These are used in areas where barren or sterile soil is desired such as railways, canals and roadways. The selective herbicides are generally organic compounds which kill only selected species in relation to the general plant life in the area. Recent developments, especially with means of removal of broadleaf weeds from cereal crops has expanded the demand for the materials about the home and on the farm. The prime development came with the application of the aryloxy and related compounds; e. g., 2,4-dichlorophenoxyacetic acid; and developments of other selective herbicides such as the carbamates and urea derivatives; chlorinated acids and phenols; dinitro compounds and many others. The compounds generally function as plant hormones and disturb the physiological processes of the susceptible plants.

"Hercoflex." [266] Trademark for liquid plasticizers for vinyl resins and other film-formers.

"Hercolyn." [266] Trademark for a pale, viscous liquid; hydrogenated methyl ester of rosin; acid number 8 max; sp. gr. 1.03 at 25/25°C.
Uses: Plasticizer for resins and for lacquers; softener and plasticizer in adhesives, floor tile, and other plastic and asphaltic compositions; wax modifier; low-cost transparentizer of paper; pigment-grinding medium for inks.

"Heresite." [17] Trademark for a series of pure phenol-formaldehyde resinous coatings and related products of the thermosetting type, and also some synthetic rubber coatings. Applied by spraying, dipping, or brushing, followed by baking or air drying.
Uses: Anticorrosive lining for machinery and equipment for chemical, food, drug, and petroleum industries, tank cars and containers.

"Herkolite." [245] Trademark for electrical insulation comprising sheet material united by an adhesive binder.

heroin. See diacetylmorphine.

"Herpoco." [266] Trademark for a fertilizer grade of ammonium nitrate; technical grade of ammonium nitrate.

herring oil.
Properties: Pale yellow to dark red liquid. Soluble in ether, chloroform, solvent naphtha and carbon disulfide.
Constants: Sp. gr. 0.9202-0.932; saponification value 179-194; iodine value 130-142; refractive index 1.478.
Derivation: By boiling and pressing herring.
Method of purification: Filtration.
Grades: No. 1; No. 2; No. 2, brown; winter pressed.

*See "I.C.C. Shipping Regulations," page xiii.
Reference numbers refer to name of manufacturer. See "List of Manufacturers," page v.

Containers: Wooden barrels, steel drums.
Uses: Soap; leather dressing; currying and finishing; lubricating special machinery.
Shipping regulations: None. *

hesperidin $C_{28}H_{34}O_{15}$. A naturally occurring product, a bioflavonoid, related to rutin and the vitamin P group.
Properties: Fine needles. M. p. 258-262°C; soluble in dilute alkalies, pyridine; very slightly soluble in water, acetone, benzene, and chloroform.
Derivation: Extraction from orange peel or other citrus fruits.
Uses: Medicine; food supplement. Usually administered as the methyl chalcone. See next entry.

hesperidin methyl chalcone. A bioflavonoid. Produced by methylation of hesperidin in an alkaline solution.
Uses: Medicine; food supplements.

Hess's law. The heat evolved or absorbed in a chemical process is the same whether the process takes place in one or in several steps; also known as the law of constant heat summation.

"Het" Acid. [306] Trademark for chlorendic acid.

"Het" Diol. [306] Trademark for $C_6H_2Cl_4$(endo CCl_2)$(CH_2OH)_2$.
Properties: White, odorless solid; m. p. 204°C; soluble in most organic solvents.
Use: Organic intermediate.

hetero-. A prefix meaning other or different. For example, heterocyclic refers to compounds in which more than one kind of atom is joined in a ring.

heteroauxin. See 3-indoleacetic acid.

heterogeneous reactor. See nuclear reactor.

heterolysis. The unsymmetrical breakdown of a covalent electron bond.

$$A:B = A^+ + B^-:$$

See homolysis.

heteromolybdates (heteropolymolybdates). A large group of complex molybdenum salts and acids in which the anion contains oxygen atoms and from two to eighteen hexavalent molybdenum atoms as well as one or more other metal or nonmetal atoms such as phosphorus, arsenic, iron and tellurium. The latter are referred to as hetero atoms and any one of about 35 elements may be present in this manner. Example: $Na_3PMo_{12}O_{40}$, sodium phospho-12-molybdate.
Properties: The molecular weights of these compounds range up to 3000. The acids and most of the salts are very soluble in water, and the acids and some salts are soluble in organic solvents such as ethers, alcohols and ketones. Salts of metals of high atomic weight (cesium, silver, mercury, lead) are of low solubility in water, and a few alkali metal salts are insoluble. Amine and alkaloid salts are usually insoluble.

The acids and salts are commonly highly hydrated, and many are highly colored. Some compounds are strong oxidizing agents, their reduced forms being of an intense blue color. The acids are fairly strong (pK 10^{-1} to 10^{-3}). There are usually several replaceable hydrogens and acid salts are common. The addition of strong bases converts the heteropoly derivatives into simple molybdates and oxygenated forms of the hetero atom (thus phosphorus appears as PO_4 and iron as $Fe(OH)_3$).
Uses: Phosphomolybdates and phosphotungstates are used as precipitants for basic dyes to form lakes and toners. The phospho- and silicomolybdate groups are of key importance in the functioning of certain enzymes. There are many uses in analytical chemistry.

heteropolymolybdates. See heteromolybdates.

HETP.
1. Abbreviation for hexaethyl tetraphosphate.
2. Abbreviation for height equivalent to a theoretical plate. It is the height of a distillation or fractionating column which gives a separation equivalent to that of a theoretical plate in the physical separation process involved. A theoretical plate may be defined as the one which produces the same difference in composition as exists at equilibrium between two phases.

"Hetrazan." [57] Trade name for diethylcarbamazine(1-diethylcarbamyl-4-methylpiperazine dihydrogen citrate) $N_3OC_{10}H_{21}\cdot O_7C_6H_8$.
Properties: White, slightly hygroscopic powder; m. p. 135-138°C; very soluble in aqueous solution; soluble in alcohol; insoluble in organic solvents; pH (1% solution) 4. 0; indefinitely stable at room temperature.
Use: Medicine.

"Hetron." [62] Trademark for a line of polyester resins for use in laminates and molding. Usually compounded with fillers of glass or textile fibers or fabric, or paper. The base resins are available in various grades and forms such as:
Rigid, with high fire retardance, high heat distortion point, low viscosity, low shrinkage. Used for machine housings, electrical insulating board, radomes, structural panels, tanks and ducts.
Chemically resistant, also have high heat resistance, high distortion point.
Light stabilized, with superior weathering resistance, for skylights, explosion windows, industrial glazing where flame resistance is required.
Boat resins, low viscosity pre-accelerated thixotropic resin for hand layup. For all types of contact molding, boats, machine housing, deep draft parts.
Spray gun type, low viscosity thixotropic resin for use in resin/fiberglass 2-component spray-type guns. Used for boat hull construction, other large parts.
Semi-rigid, superior in tensile strength and elongation, gives laminates with resistance

to impact or rapid cure cracks. Structural panels and members, boat hulls, machine housing.
High impact at low temperatures, for gel coating by hand layup, piping, boat covering.

heulandite $H_4CaAl_2(SiO_3)_6 \cdot 3H_2O$. A mineral; one of the zeolites (q. v.).
Properties: White to red, gray or brown; white streak; vitreous or pearly luster; sp. gr. 2.18-2.22; hardness 3.5-4.
Occurrence: India, Europe, United States, Nova Scotia.

"Heviwater." [233] Trademark for a solution of calcium and zinc chlorides for use in oil wells.

"Hexa-Betalin." [100] Trademark for pyridoxine hydrochloride, U. S. P.

hexabromoethane C_2Br_6.
Properties: Yellowish-white, rhombic needles. Slightly soluble in water, alcohol.
Constants: M. p. 210-215°C. (decomposes with separation of bromine).
Derivation: Action of bromine on diiodoacetylene.
Grade: Technical.
Use: Organic synthesis.

hexacalcium phytate. See calcium phytate.

hexachlorobenzene (perchlorobenzene) C_6Cl_6.
Properties: White needles. Soluble in benzene and boiling alcohol; insoluble in water.
Constants: M. p. 229°C; b. p. 326°C.
Derivation: By heating hexyl iodide with iodine chloride.
Method of purification: Crystallization.
Grade: Technical.
Containers: Up to 150-lb drums.
Use: Organic synthesis; seed treatment.
Shipping regulations: None.*

hexachlorobutadiene C_4Cl_6, or $Cl_2C{:}CClCCl{:}CCl_2$.
Typical specifications: Clear colorless liquid with mild characteristic odor; melting range −19 to −22°C; boiling range 210-220°C; refractive index (n 20/D) 1.552 (±.001); flash point, none; sp. gr. 1.675 (15.5/15.5°C); 13.97 lbs/gal (15.5°C); purity 98% (min); vapor pressure 22 mm (100°C); 500 mm (200°C); viscosity (100°F) 2.447 cps; 1.479 centistokes; (210°F) 1.131 cps; 0.724 centistokes; insoluble in water; compatible with numerous resins and plastics; soluble in alcohol and ether.
Containers: Glass bottles; tins; 55-gal steel drums.
Uses: Solvent for natural rubber, synthetic rubber and other polymeric substances; high-boiling nonflammable solvent; nonflammable heat-transfer liquid; transformer fluid and hydraulic fluid; clarifying mash before centrifuging; wash liquor for removing C_4 and higher hydrocarbons.
Shipping regulations: None.*

1, 2, 3, 4, 5, 6-hexachlorocyclohexane (BHC; HCCH; HCH; TBH; benzene hexachloride) $C_6H_6Cl_6$. The technical material, which is formed by the chlorination of benzene in actinic light, is a mixture of at least five of the nine possible isomers. Some isomers are without appreciable activity. The gamma isomer (lindane) is most active biologically and technical material is therefore graded according to its gamma content.
Properties: White or yellowish powder or flakes; color, odor, melting point and other properties vary with the isomeric composition. Vapor pressure about 0.5 mm Hg (60°C); stable toward moderate heat but decomposed by alkaline substances. Melting points of the pure isomers are: (alpha-trans) 157-158°C; (beta-cis) 297°C (sublimes); (gamma) 112.5°C; (delta) 138-139°C; (epsilon) 217-219°C.
Method of purification: Fractional crystallization. The technical material may run 10-15% gamma isomer, but can be brought up to 99% (at this point it is usually called lindane (q.v.).
Containers: Bags, fiber drums.
Warning! Harmful if swallowed. May cause irritation of skin and eyes. May be absorbed through skin. MCA warning label.
Uses: As a component of insecticides toxic to flies, cockroaches, aphids, grasshoppers, wire worms, and boll weevils.

hexachlorocyclopentadiene (perchlorocyclopentadiene) C_5Cl_6.
Properties: Pale yellow liquid having a pungent odor. B. p. 239°C; freezing point 9.6°C; sp. gr. 1.717 (15/15°C); wt/gal 14.30 lb (15.5°C); refractive index (n 25/D) 1.563; flash point, none. Toxic.
Grade: Technical.
Containers: 55-gal drums; tank cars.
Uses: Intermediate for non-flammable resins, dyes, pesticides, fungicides, pharmaceuticals.
Shipping regulations: None.*

hexachlorodiphenyl oxide $C_{12}H_4Cl_6O$.
Properties: Light yellow, very viscous liquid. B. p. 230-260°C (8 mm); sp. gr. 1.60 (20/60°C); lbs/gal 13.12 at 25°C; refractive index 1.621 (25°C); flash point, none. Soluble in methanol, ether. Very slightly soluble in water.
Use: Solvent; intermediate.

hexachloroendomethylenetetrahydrophthalic acid. See chlorendic acid.

hexachloroendomethylenetetrahydrophthalic anhydride. See chlorendic anhydride.

1, 2, 3, 4, 10, 10-hexachloro-6, 7-epoxy-1, 4, 4a, 5, 6, 7, 8, 8a-octahydro-1, 4-endo, endo**-5, 8-dimethanonaphthalene.** See endrin and the endo, exo isomer dieldrin.

1, 2, 3, 4, 10, 10-hexachloro-6, 7-epoxy-1, 4, 4a, 5, 6, 7, 8, 8a-octahydro-1, 4-endo, exo**-5, 8-dimethanonaphthalene.** See dieldrin and the endo, endo isomer endrin.

hexachloroethane (perchloroethane; carbon trichloride; carbon hexachloride) Cl_3CCl_3.

*See "I. C. C. Shipping Regulations," page xiii.
Reference numbers refer to name of manufacturer. See "List of Manufacturers," page v.

Properties: Colorless crystals; camphor-like odor; sp. gr. 2.091; m.p. 185°C; b.p. sublimes at 185°C. Soluble in alcohol and ether; insoluble in water.
Method of purification: Crystallization.
Grades: Technical.
Containers: 100-, 250-, 500-lb barrels.
Uses: Organic synthesis; retarding agent in fermentation; camphor substitute in "Celluloid" manufacture; rubber accelerator; pyrotechnics and smoke devices; solvent; explosives; medicine.
Shipping regulations: None. *

1,2,3,4,10,10-hexachloro-1,4,4a,5,8,8a-hexahydro-1,4,5,8-endo, exo-**dimethano-naphthalene.** See aldrin.

hexachloromethylcarbonate (triphosgene) $(OCCl_3)_2CO$. A lachrymator.
Properties: White crystals. Odor similar to that of phosgene. Decomposed by hot water and alkali hydroxides. Only slowly acted upon by cold water. Soluble in alcohol, benzene, ether.
Constants: Sp. gr. 2 (approx); b.p. 205-206°C (partial decomposition); m.p. 78-79°C.
Derivation: (a) By-product in making trichloromethylchloroformate from impure methylchloroformate. (b) Chlorination of dimethyl carbonate exposed to direct sunlight.

hexachloromethyl ether $O(CCl_3)_2$.
Properties: Liquid. Phosgene-like odor. Caution! Very irritant!
Constants: Sp. gr. 1.538 (18°C); b.p. 98°C (partial decomposition).
Derivation: Chlorination of dichloromethyl ether.

hexachloronaphthalene. See chloronaphthalenes.

hexachlorophene (2,2'-methylene bis-(3,4,6-trichlorophenol); bis-(3,5,6,-trichloro-2-hydroxyphenyl)methane; 2,2'-dihydroxy-3,5,6,3',5',6'-hexachlorodiphenyl methane) $(C_6HCl_3OH)_2CH_2$.
Properties: White, free-flowing powder; essentially odorless; m.p. 160.5°C min; mol wt 406.92. Freely soluble in acetone, alcohol and ether; soluble in chloroform; insoluble in water.
Derivation: Condensation of 3,4,5-trichlorophenol with formaldehyde in the presence of sulfuric acid.
Grades: Pure; U.S.P. XVI.
Containers: Cans; drums.
Uses: Bactericidal and bacteriostatic agent; finds application in antiseptic (and surgical) soaps; deodorant products including soaps, various cosmetics, dermatologicals.
Hazards: None, except those associated with inhalation of fine powders.
Shipping regulations: None. *

hexachloropropene. See hexachloropropylene.

hexachloropropylene (hexachloropropene; perchloropropylene) $CCl_3CCl{:}CCl_2$.
Properties: Water white liquid; b.p. 210°C; insoluble in water; miscible with alcohol, ether, chlorinated compounds.
Uses: Solvent; plasticizer; hydraulic fluid.

hexacontane $C_{60}H_{122}$. High molecular weight hydrocarbon.
Properties: Solid. M.p. 101°C.

hexacosanoic acid. See cerotic acid.

hexadecafluoro-1-nonanol. See fluoroalcohols.

n-hexadecane (cetane) $C_{16}H_{34}$.
Properties: Colorless liquid; sp. gr. 0.77335 (20/4°C); b.p. 286.5°C; m.p. 18.14°C; refractive index (n 20/D) 1.43435. Soluble in alcohol, acetone, ether; insoluble in water.
Typical specifications: Sp. gr. 0.7720-0.7743 (20°C); b.p. 170-174°C (30 mm); m.p. 17-18.2°C; refractive index (n 20/D) 1.4340-1.4360.
Grades: Technical; ASTM.
Uses: Solvents; organic intermediates; standardized hydrocarbons.

hexadecanoic acid. See palmitic acid.

1-hexadecanol. See cetyl alcohol.

hexadecanoyl chloride. See palmitoyl chloride.

1-hexadecene (cetene; alpha-hexadecylene) $CH_3(CH_2)_{13}CH{:}CH_2$.
Properties: Colorless liquid; m.p. 4°C; b.p. 274°C; sp. gr. 0.784 (15/4°C); refractive index (n 20/D) 1.4441; insoluble in water; soluble in alcohol, ether, petroleum, coal-tar solvents.
Derivation: Treatment of cetyl alcohol with phosphorus pentoxide.
Grade: 95% purity.
Use: Organic synthesis in flavors, perfumes, medicines, dyes, resins.

cis-**9-hexadecenoic acid.** See palmitoleic acid.

hexadecen-6-olide. See ambrettolide.

n-hexadecyl alcohol, primary. See cetyl alcohol.

alpha-**hexadecylene.** See 1-hexadecene.

hexadecyl mercaptan. See cetyl mercaptan.

tert-**hexadecyl mercaptan** $C_{16}H_{33}SH$.
Properties: Colorless liquid; boiling range (5 mm) 121 to 149°C; sp. gr. (60/60°F) 0.874; refractive index (n 20/D) 1.474; flash point 135°C.
Containers: Drums and tank cars.
Uses: Polymer modification.
Shipping regulations: None. *

hexadecyltrichlorosilane $C_{16}H_{33}SiCl_3$.
Properties: Colorless to yellow liquid. B.p. 269°C; sp. gr. 0.996 (25/25°C); refractive index (n 25/D) 1.4568; flash point (Cleveland open cup) 295°F. Readily hydrolyzed by moisture, with the liberation of hydrochloric acid.
Derivation: By Grignard reaction of silicon tetrachloride and hexadecylmagnesium chloride.
Grades: Technical.
Uses: Intermediate for silicones.
Shipping regulations: Corrosive liquid. White label. *

hexadecyltrimethylammonium bromide. See cetyltrimethylammonium bromide.

2,4-hexadiene (bipropenyl; dipropenyl; dipropylene) C_6H_{10} or $CH_3CH{:}CHCH{:}CHCH_3$.
Properties: Colorless liquid. Insoluble in water.
Constants: Sp. gr. 0.718; b.p. 82°C; aniline equivalent 29.

2,4-hexadienoic acid. See sorbic acid.

1,5-hexadiyne. See dipropargyl.

"Hexadow." [233] Brand name for proprietary insecticide formulations containing benzene hexachloride. Available as a concentrate for preparing dust mixtures and as a wettable powder.

hexa-2-ethylbutoxydisiloxane $[(CH_3CH(C_2H_5)CH_2CH_2O)_3Si]_2O$.
Properties: Colorless oil; b.p. 195°C (0.2 mm).
Derivation: Reaction of silicon tetrachloride, 2-ethylbutanol and water.
Uses: Aircraft hydraulic fluid.

hexaethyl tetraphosphate (HETP) (so-called).
Properties: Very toxic! Yellow liquid. Sp. gr. 1.26-1.28 (25/4°C); m.p. −90°C; refractive index 1.427; decomposes at high temperatures; soluble or miscible in water and many organic solvents; not soluble in kerosene; hydrolyzes in low concentrations in water; hygroscopic.
Containers: 55-gal drums.
Uses: Contact insecticide for control of aphids, thrips, spider-mites, soft scale and various other insects.
Shipping regulations: Solid or liquid form: poison, class B. Poison label. *

hexafluophosphoric acid. See fluophosphoric acids.

hexafluoropropylene C_3F_6.
Shipping regulations: Nonflammable gas. Green label. *

hexaglycerine. See trimethylol propane.

"Hexagon." [319] Trademark for a specially fractionated grain-sprout, highly refined ethyl alcohol.

hexahydric alcohols. See mannitol, sorbitol, and dulcitol.

hexahydroaniline. See cyclohexylamine.

hexahydrobenzene. See cyclohexane.

hexahydrobenzoic acid (cyclohexane carboxylic acid) $C_6H_{11}COOH$. One of the naphthenic acids.
Properties: Colorless monoclinic prisms; m.p. 31°C; b.p. 233°C; sp.gr. 1.048 (15/4°C); refractive index 1.4561 (33.8°C); slightly soluble in water, soluble in alcohol and ether.
Uses: Manufacture of paint and varnish driers, dry-cleaning soaps, lubricating oils; stabilizer for rubber.

hexahydrocresol. See methylcyclohexanol.

hexahydromethylphenol. See methylcyclohexanol.

hexahydrophenol. See cyclohexanol.

hexahydrophthalic anhydride (1,2-cyclohexanedicarboxylic anhydride) $C_6H_{10}(CO)_2O$.
Properties: Clear, colorless, viscous liquid which freezes to a glossy solid; solidifying point 35-36°C; b.p. 158°C (17 mm); sp.gr. (40°C) 1.19; miscible with benzene, toluene, acetone, carbon tetrachloride, chloroform, ethanol and ethyl acetate; only slightly soluble in petroleum ether.
Containers: 5-gal tins; 55-gal drums.
Uses: Chemical intermediate for alkyds, plasticizers, insect repellents and rust inhibitors.

hexahydropyridine. See piperidine.

hexahydrothymol. See menthol.

hexahydrotoluene. See methylcyclohexane.

hexahydro-1,3,5-trinitro-sym-triazine. See cyclonite.

hexahydroxycyclohexane. See inositol.

hexahydroxylene. See dimethylcyclohexane.

n-hexaldehyde (caproic aldehyde) $CH_3(CH_2)_4CHO$.
Properties: Colorless liquid. Sharp, aldehyde odor.
Constants: Sp.gr. 0.8156 (20/20°C); b.p. 128.6°C (760 mm); vapor pressure 10.5 mm (20°C); flash point 90°F; wt/gal 6.9 lbs (20°C); f.p. −56.3°C.
Typical specifications: Sp.gr. 0.820-0.826 (20/20°C); boiling range 90-150°C; acidity not more than 2.00% (as butyric).
Grades: Technical.
Containers: 1-gal cans; 5-, 55-gal drums.
Uses: Organic synthesis; starting point in making plasticizers, rubber chemicals, dyes, synthetic resins, insecticides,

"Hexalin." [28] Trademark for cyclohexanol (usually shipped with 2.25% methanol as anti-freeze).
Containers: 55-gal drums; 8000- and 10000-gal tank cars; tank trucks.
Uses: Solvent in lacquers, shellacs and varnishes; homogenizer and stabilizer in soap, dry cleaning and textile industries; intermediate for chemicals, plasticizers, and lubricating oil additives.

hexamethonium bromide $(CH_3)_3NBr(CH_2)_6NBr(CH_3)_3$. Hexamethylenebis(trimethylammonium bromide).
Properties: White, tasteless, crystalline material with faintly aromatic odor. Freely soluble in methanol and water; soluble in alcohol; insoluble in ether.
Use: Medicine.

hexamethonium chloride $(CH_3)_3NCl(CH_2)_6NCl(CH_3)_3$. Hexamethylenebis(trimethylammonium chloride).
Properties: White, crystalline, hygroscopic powder with faint odor. M.p. 289-292°C (dec); very soluble in water; soluble in alcohol, methanol and n-propanol; insoluble in chloroform and ether. Available commercially as unhydrated form or as dihydrate.

*See "I.C.C. Shipping Regulations," page xiii.
Reference numbers refer to name of manufacturer. See "List of Manufacturers," page v.

Grade: N. N. D.
Use: Medicine.

hexamethylbenzene $C_{12}H_{18}$ or $C_6(CH_3)_6$.
Properties: Colorless plates; soluble in alcohol; insoluble in water.
Constants: B. p. 265°C; m. p. 165.5°C.

hexamethyldiaminoisopropanol diiodide. See propiodal.

hexamethylene. See cyclohexane.

hexamethyleneamine. Incorrect term for hexamethylenetetramine (q. v.).

hexamethylenediamine (1,6-diaminohexane) $H_2N(CH_2)_6NH_2$.
Properties: Colorless leaflets; m. p. 39-42°C; b. p. 205°C; somewhat soluble in water, ethyl alcohol, and ether.
Use: Formation of high polymers, especially nylon.
Shipping regulations: (solutions) Corrosive liquid. White label. *

hexamethylenetetramine (methenamine; ammonioformaldehyde; aminoform; hexamine; erroneously "hexamethyleneamine") $(CH_2)_6N_4$.
Properties: White crystalline powder, or colorless, lustrous crystals; practically odorless; sp. gr. 1.27 (25°C); has irritating action on the skin; soluble in water, alcohol, and chloroform; insoluble in ether; m. p. 280°C, but sublimes at 260-263°C, partly decomposing. Burns with smokeless flame. Flash point 482°F.
Derivation: By the action of ammonia on formaldehyde.
Method of purification: Recrystallization.
Grades: Technical; N. F. XI (as methenamine).
Containers: 25- to 150-lb drums; 100-lb bags; 200-lb barrels.
Uses: In resins and plastics, largely as a curing agent; pharmaceuticals; rubber accelerator; explosives; fungicide for citrus fruits; corrosion inhibitor; insolubilizing agent for proteins; shrink-proofing textiles and adding elasticity to cellulosic fibers. Many compounds of the substance are on the market.
Shipping regulations: None. *

hexamethylenetetramine acetylsalicylate $(CH_2)_6N_4 \cdot COOHC_6H_4OOCCH_3$.
Properties: White crystalline powder or colorless crystals. Insipid taste. Soluble in water, and alcohol. M. p. 118-119°C.
Derivation: By the action of salicylic acid upon hexamethylenetetramine in the presence of alcohol, and acetylating.

hexamethylenetetramine mandelate (methenamine mandelate) $C_6H_{12}N_4 \cdot C_6H_5CHOHCOOH$.
Properties: A white crystalline powder with sour taste; practically no odor; m. p. 127-130°C; very soluble in water; pH solutions 4; soluble in alcohol.
Grade: U. S. P. XVI (as methenamine mandelate).
Use: Medicine.

hexamethylenetetramine salicylate (methenamine salicylate) $(CH_2)_6N_4 \cdot C_6H_4OHCOOH$.
Properties: White crystalline powder; pleasant acidulous taste; soluble in alcohol and water.
Use: Medicine.
Shipping regulations: None. *

hexamethylenetetramine tannin $(CH_2)_6N_4(C_{14}H_{10}O_9)_3$.
Properties: Light brown, odorless, tasteless powder. Contains 87 parts tannic acid, 13 parts hexamethylenetetramine.

hexamethylethane (2,2,3,3-tetramethylbutane; tert-butyltrimethylmethane) C_8H_{18} or $CH_3C(CH_3)_2C(CH_3)_2CH_3$.
Constants: B. p. 106.8°C; m. p. 104°C. Insoluble in water.

hexamethylphosphoric triamide $[N(CH_3)_2]_3PO$. Pale, water-soluble liquid; used as an ultraviolet inhibitor in polyvinyl chloride.

hexamethyltetracosahexene. See squalene.

hexamethyltetracosane. See squalane.

"Hexamic Acid." [3] Trademark for cyclohexylsulfamic acid, $C_6H_{11}NHSO_3H$.
Properties: Odorless, white crystalline solid with a sweet-sour taste; m. p. 178-181°C. It is a strong, stable acid, soluble in water and alcohol; insoluble in oils.
Uses: As an acidulant and flavoring adjunct for pharmaceuticals.
See also its salts, known as calcium and sodium cyclamate.

hexamine. See hexamethylenetetramine.

hexanaphthene. See cyclohexane.

n-hexane C_6H_{14} or $CH_3(CH_2)_4CH_3$.
Properties: Colorless, volatile liquid; faint, peculiar odor; highly flammable; sp. gr. 0.65937 (20/4°C); b. p. 68.742°C; m. p. −95°C; refractive index (n 20/D) 1.37486; flash point −23°C. Soluble in alcohol, acetone and ether; insoluble in water.
Derivation: By fractional distillation from petroleum.
Containers: Bottles; drums; tank cars, according to grade.
Grades: 85%, 95%, 99%, spectro, and research.
Uses: Solvent, especially extraction solvent for vegetable oils; liquid in low temperature thermometers; calibrations.
Fire hazard: Dangerous.
Shipping regulations: Flammable liquid. Red label. *

hexanedioic acid. See adipic acid.

hexanedione-2,5. See acetonyl acetone.

1,2,6-hexanetriol $HOCH_2CH(OH)CH_2CH_2CH_2CH_2OH$.
Properties: A water-white liquid; sp. gr. 1.1063; sets to glass below −20°C; b. p. at 5 mm 178°C; flash point 380°F; miscible with water.
Containers: Cans.
Uses: Alkyd and polyester resin intermediate; softener, moistening agent, and solvent.

hexanitrodiphenyl amine (hexil, hexyl, hexite, dipicrylamine) $(NO_2)_3C_6H_2NHC_6H_2(NO_2)_3$.
Properties: Yellow explosive solid; m. p. 238-244°C; decomposes at higher temperatures; insoluble in water, ether, benzene, chloroform, and alcohol; slightly soluble in acetone, acetic acid; soluble in alkalies and warm acetic or nitric acid.
Derivation: Nitration of diphenylamine.
Uses: Explosives; also in analysis for potassium.

hexanitromannite. See mannitol hexanitrate.

hexanoic acid. See caproic acid. Hexanoic acid is the term preferred by the International Union of Chemistry.

1-hexanol (hexyl alcohol, caproyl alcohol, amyl carbinol) $CH_3(CH_2)_4CH_2OH$.
Properties: Colorless liquid; sp. gr. 0.8186; m. p. −51.6°C; b. p. 157.2°C; wt/gal 6.8 lbs (20°C); refractive index 1.4169 (25°C). Insoluble in water; soluble in alcohol and ether.
Derivation: (a) By reduction of ethyl caproate; (b) from olefins by the Oxo process.
Method of purification: Vacuum distillation of technical grade.
Grades: Technical (90-99%); purified (99.8%).
Typical specifications: Purity 98.9% ; sp. gr. 0.820 (20/20°C); distillation, initial 150°C, dry point 156°C; flash point 144°F (TOC).
Containers: 1-, 5-gal cans; 55-gal drums (non-returnable); tank cars.
Uses: Organic synthesis (introduction of the hexyl group into detergents, hypnotics, antiseptics, perfume esters and other pharmaceuticals); raw material for flotation agents, lubricant additives (zinc dihexyl dithiophosphate), solvents, brake fluids and agricultural chemicals.
Shipping regulations: None. *

hexanoyl chloride (also confusingly called caproyl chloride) $CH_3(CH_2)_4COCl$.
Properties: Colorless liquid, b. p. 151-153°C; refractive index 1.4867 (n 20/D); decomposed by water and alcohol; soluble in ether and chloroform.
Containers: Bottles, carboys, steel drums.
Use: Chemical intermediate.

hexaphenyldisilane $[(C_6H_5)_3Si]_2$.
Properties: White powder; m. p. 352°C.
Derivation: Sodium condensation of triphenylchlorosilane.
Use: High-temperature applications.

"Hexaphos." [55] Brand name for a glassy phosphate of high molecular weight having superior water-softening properties (sodium hexametaphosphate.)
Properties: Powder, granules, flakes or plates, odorless, glassy hygroscopic salt.
Sequestering Power: Ca 16.3; Mg 2.9; Fe 0.03 (g/100 g. "Hexaphos") determined with initial pH of 10.
Grades: Adjusted; unadjusted.
Containers: 100 lb paper bags.
Uses: Water-softening; boiler-scale control; component of cleansers; laundry mixes, dishwashing compounds, pitch control in pulp industry, prevention of lime soap deposits in textile operations.
Shipping regulations: None. *

hexatriacontane $C_{36}H_{74}$.
Properties: Crystals; sp. gr. 0.797.

1-hexene (hexylene) $CH_3CH_2CH_2CH_2CH{:}CH_2$.
Properties: Colorless liquid; sp. gr. 0.6734 (20/4°C); b. p. 63.55°C; m. p. −139.8°C; refractive index (n 20/D) 1.3876; flash point −26°C.
Grades: 95%, 99%; research.
Containers: Bottles.
Uses: Synthesis of flavors, perfumes, medicines, dyes, resins.
Shipping regulations: Flammable liquid. Red label. *

2-hexene $CH_3CH_2CH_2CH{:}CHCH_3$.
Properties (of mixed cis and trans isomers): Colorless liquid; b. p. 68°C; refractive index (20/D) 1.3948; sp. gr. (60/60°F) 0.686; flash point −20°C.
Grades: 95%.
Containers: Bottles.
Shipping regulations: Flammable liquid. Red label. *

hexenol $C_6H_{11}OH$. Naturally occurring green note found in grass, etc. Fine perfume ingredient for imparting natural green freshness in perfumes.

hexestrol $HOC_6H_4CH(C_2H_5)CH(C_2H_5)C_6H_4OH$. para, para'-(1, 2-Diethylethylene)diphenol. A non-steroid synthetic estrogen.
Properties: Odorless, white, crystalline powder; m.p. 185-188°C. Freely soluble in ether; soluble in acetone, alcohol, and methanol; slightly soluble in benzene and chloroform; practically insoluble in water and dilute mineral acids. Dissolves in vegetable oils and in dilute sodium or potassium hydroxide. Sensitive to light.
Derivation: From anethole; by reaction of diacetyl peroxide on para-methoxy-n-propylbenzene.
Grade: N. F. XI.
Use: Medicine.

hexetidine $C_{21}H_{45}N_3$. 5-amino-1, 3-bis (beta-ethylhexyl)-5-methylhexahydropyrimidine.
Grade: N. N. D.
Use: Medicine.

hexil. See hexanitrodiphenyl amine.

hexite. See hexanitrodiphenyl amine.

hexobarbital (N-methyl-5-cyclohexenyl-5-methylbarbituric acid) $C_{12}H_{16}N_2O_3$.
Properties: White crystals; m.p. 145-147°C.
Use: Medicine.

hexobarbital sodium $C_{12}H_{15}N_2NaO_3$. (Sodium 5-(1-cyclohexenyl)-1, 5-dimethylbarbiturate).
Properties: White, crystalline, odorless hygroscopic powder with slightly bitter taste, discolors on exposure to air. Very soluble in water; freely soluble in alcohol; practically insoluble in ether. Aqueous solutions are alkaline to litmus, and

decompose on standing; pH (10% solution) 11-12.
Grade: N. F. XI.
Use: Medicine.

hexocyclium methylsulfate $C_{20}H_{33}N_2O \cdot CH_3SO_4$. N-(beta-Cyclohexyl-beta-hydroxy-beta-phenylethyl)-N'-methylpiperazine methosulfate.
Properties: White, crystalline solid melting between 200° and 210°C. Freely soluble in water; slightly soluble in chloroform; insoluble in ether.
Grade: N. N. D.
Use: Medicine.

"Hexogen." [230] Trademark for a series of paint driers made with odorless solvents, essentially solutions of metallic salts of 2-ethyl hexoic acid. Supplied in a variety of high metal concentrations.

hexoic acid. See caproic acid.

hexone. See methyl isobutyl ketone.

hexokinase. An enzyme which catalyzes the formation of adenosine diphosphate and hexose-6-phosphate from adenosine triphosphate, and glucose or fructose.
Use: Biochemical research.

hexyl. The straight-chain radical C_6H_{13}.

hexyl. See hexanitrodiphenyl amine.

hexyl acetate $CH_3COOC_6H_{13}$.
Properties: Colorless liquid; sp. gr. 0.8902 (0/0°C); b. p. 169.2°C; insoluble in water; very soluble in alcohol or ether.
Derivation: Commercial product is made from sec-hexyl alcohol.
Use: Solvent for cellulose esters and resins.

hexyl alcohol. See 1-hexanol.

pseudo-**hexyl alcohol.** See 2-ethyl butyl alcohol.

n-hexylamine $CH_3(CH_2)_5NH_2$.
Properties: Water-white; amine odor; boiling range 126-132°C; sp. gr. 0.767 (20/20°C); refractive index 1.419 (20°C); flash point 105°F.

n-hexyl bromide (1-bromohexane) $CH_3(CH_2)_5Br$.
Properties: Colorless to slightly yellow liquid; sp. gr. 1.165 (20/20°C); b. p. 155.5°C; soluble in alcohol, esters, ethers; insoluble in water.
Grade: 96-98% pure.
Use: Intermediate, for introduction of hexyl group.

hexylcaine hydrochloride
$C_6H_5COOCH(CH_3)CH_2NH(C_6H_{11}) \cdot HCl$. 1-Cyclohexylamino-2-propyl benzoate hydrochloride.
Properties: White, bitter powder with slight aromatic odor. M. p. 182-184°C. Freely soluble in alcohol and chloroform; practically insoluble in ether; fairly soluble in water; pH (5% solution) 4.0-6.0.
Grade: N. F. XI.
Use: Medicine.
Shipping regulations: None. *

n-hexyl "Carbitol." [214] $C_6H_{13}OC_2H_4OC_2H_4OH$. Trademark for diethylene glycol monohexyl ether (q. v.).

n-hexyl "Cellosolve." [214] Trademark for ethylene glycol monohexyl ether $C_6H_{13}OCH_2CH_2OH$ (q. v.).

hexylene. See 1-hexene.

hexylene glycol (4-methyl-2,4-pentanediol) $(CH_3)_2COHCH_2CHOHCH_3$. Colorless, nearly odorless liquid; sp. gr. 0.9216 (20/4°C); b. p. 198.3°C; refractive index (n 20/D) 1.4276; flash point (open cup) 230°F; wt/gal 7.69 lbs. Miscible with water.
Containers: 5-, 55-gal drums; tank cars.
Uses: Hydraulic brake fluids; printing inks; coupling agent and penetrant for textiles; fuel and lubricant additive; emulsifying agent.

n-hexyl ether $C_6H_{13}OC_6H_{13}$.
Properties: Practically colorless liquid with characteristic odor. Very slightly soluble in water; sp. gr. 0.7942 (20/20°C); 6.6 lbs/gal (20°C); f. p. −43.0°C; viscosity 1.68 cps (20°C); flash point 170°F.
Uses: Extraction processes, and in the manufacture of collodion, photographic film, and smokeless powder.

hexylic acid. See caproic acid.

hexyl mercaptan $C_6H_{13}SH$.
Properties: B. p. 149-150°C (768 mm); sp. gr. 0.8450 (20/4°C); refractive index 1.4492 (n 20/D).
Grades: 95% min purity.
Uses: Intermediates; synthetic rubber processing.

hexyl methacrylate $C_6H_{13}OOCC(CH_3){:}CH_2$. Polymerizable monomer for plastics, molding powder, solvent coatings, adhesives, oil additives; emulsions for textile, leather, and paper finishing.
Containers: Drums.

para-tert-**hexylphenol** $C_6H_{13}C_6H_4OH$.
Properties: Sp. gr. 0.986 (20/20°C); boiling range 155-165°C; refractive index 1.520 (20°C); flash point 285°F; water-white; faint phenol odor.
Uses: For synthesis of other organic compounds and for the preparation of resinous condensation products.

hexylresorcinol (1,3-dihydroxy-4-hexylbenzene) $C_6H_{13}C_6H_3(OH)_2$.
Properties: White to yellowish-white needle-shaped crystals with a faint, fatty odor and sharp, astringent taste which produces a sensation of numbness when placed on the tongue; slightly soluble in water; freely soluble in alcohol, methanol, glycerin, ether, chloroform, benzene and vegetable oils; m. p. 62-67°C; b. p. 178-180°C (8 mm).
Grades: U.S.P. XVI.
Caution! Irritating to respiratory tract and skin. Alcohol solutions are vesicant.
Containers: Bottles; fiber drums.
Use: Medicine.
Shipping regulations: None.*

*See "I.C.C. Shipping Regulations," page xiii.
Reference numbers refer to name of manufacturer. See "List of Manufacturers," page v.

hexyl trichlorosilane $C_6H_{13}SiCl_3$.
Shipping regulations: Corrosive liquid. White label. *

1-hexyne (butyl acetylene) $C_4H_9C{\vdots}CH$.
Properties: Water-white with characteristic odor. Sp. gr. 0.7152 (20/4°C); refractive index 1.3990 (20°C); b. p. 71.4°C (760 mm); f. p. −132°C.

hexynol (1-hexyn-3-ol) $CH_3(CH_2)_2CHOHC{\vdots}CH$.
Properties: Light yellow liquid with slightly piercing odor; b. p. 142°C; sp. gr. (20/20°C) 0.882; slightly soluble in water, and miscible with most hydrocarbons, chlorinated solvents, ketones, alcohols and glycols.
Containers: 7-, 35- and 380-lb drums.
Uses: Corrosion inhibitor for mineral acids; high temperature oil well acidizing inhibitor.

Hf. Symbol for hafnium.

"HF Alkylation Process." [416] See "Alkylation Process, HF."

Hg. Symbol for mercury (Latin: hydrargyrum).

HGF. See glucagon.

HG-factor. See glucagon.

HHDN. Abbreviation for hexachlorohexahydrodimethanonaphthalene. See aldrin.

"Hibitane." [207] Trademark for chlorohexidine (bis-para-chlorophenyldiguanidohexane). The diacetate and dihydrochloride are used as antibacterial agents.

"Hiblak." [133] Brand name for a series of aqueous carbon black dispersions designed for darkening concrete and mortar. Available as:
"Hiblak." 25% regular color impingement carbon black, 75% water and DA.
"Hiblak" AE. 25% regular color impingement carbon black, 75% water and DA. Also contains air entraining agent.

"Hi Calcium Phosphate." [172] Trade name for a special, crystalline, monocalcium phosphate with a high calcium content.
Containers: 100-lb paper bags.
Use: Manufacture of calcium-enriched flour.

"Hicolor F." [37] An established grade of wood cellulose.
Properties: Cellulose-alpha 93%, beta 3.2%, gamma 3.8%. 10% potassium hydroxide solution solubility 16.7%.
Uses: Production of viscose rayon yarns of high quality; manufacture of plastics, vulcanized fiber; saturating papers, and allied products.

"Hicolor G." [37] A wood cellulose used extensively for the manufacture of continuous filament viscose rayon and staple fiber.

Hi-D. [319] Ammonium nitrate fertilizer, in granular, dense form.

hiddenite. See spodumene.

"Hi-fax." [266] Trademark for several grades of high-density polyethylene product in two different densities, 0.945 and 0.960.
Properties: Natural or colored molding powder pellets.
Uses: Flexible plastic pipe; tubing; sheet; industrial moldings; wire and cable coatings; blown structures; blow molding and injection molding.

high alumina brick. See brick, alumina.

high-alumina cement. See aluminous cement.

high boiling phenols. A mixture containing predominantly meta-substituted alkyl phenols.
Typical specifications: Average molecular weight 150; sp. gr. (20°C) 1.033; b. p. 238-288°C; vapor pressure (20°C) 0.01 mm; sets to glass below −30°C; flash point 250°C; slightly soluble in water.
Use: Phenolic resins; solvents; fuel oil sludge inhibitor; germicides; rubber chemicals.

high cranberry. See viburnum opulus.

high energy fuels. Fuels for jets and rockets which have a higher performance rate than the common hydrocarbon fuels (kerosine, JP-6). The term has also been used to refer specifically to those jet and rocket fuels based on boron hydrides. See rocket propellants, and jet fuels.

"High-K." [67] Trademark for potassium chloride.

"High Speed." [28] Trademark for tinning flux based on zinc chloride and ammonium chloride.
Properties: Colorless water solution; sp. gr. 1.526; f. p. below −50°F.
Containers: 165-lb carboys; 670-lb drums.
Use: As fluxing solutions in the manufacture of tin plate, terne plate, tin strips, dairy equipment, refrigerator shelves, automobile radiators, and wire goods.

high vacuum distillation. See molecular distillation.

"Hi-Level." [108] Trademark for a white, free-flowing granular dish-machine compound designed for high-speed, heavy soil installations using soft or moderately soft water.
Containers: 125- and 350-lb drums.

hindered isocyanate. See isocyanate generator.

Hi-N-Dri. [319] Ammonium nitrate fertilizer. Granular form, denser than any other ammonium nitrate, with 25% less bulk.

"Hiperco." [308] Trademark for an alloy of cobalt and iron with varying percentages of cobalt, 27%, 35%, 50%, with the highest magnetic saturation of any known commercial alloy. Permeability above 100 oersteds is far superior to iron and the electrical sheet steels. It is used as a core material for motors, generators, and transformers where minimum weight and size are prime requisites.

"Hipernik." [308] Trademark for an alloy of approximately equal proportions of iron and nickel having high initial and maximum

*See "I.C.C. Shipping Regulations," page xiii.
Reference numbers refer to name of manufacturer. See "List of Manufacturers," page v.

permeability and extremely low losses. Used in audio and instrument transformers, relays, magnetic bridges, shields for electronic tubes, etc. "Hipernik V" is a highly oriented alloy possessing a very high remanence accompanied by a low coercive force, and is especially suitable for saturable reactors and magnetic amplifiers.

"Hippuran." 329 Trademark for brand of iodohippurate sodium, a water-soluble, x-ray contrast medium.

hippuric acid. (benzaminoacetic acid; benzolaminoacetic acid, benzoylglycocoll; benzoylglycin) $C_6H_5CONHCH_2COOH$.
Properties: Colorless crystals; sp. gr. 1.371 (20°C); m.p. 188°C; decomposes on further heating; soluble in hot water, alcohol or ether.
Uses: Organic synthesis and medicine.

"Hi-Ratio Silicate." 244 Trademark for an anhydrous homogeneous combination of caustic alkali and sodium silicate. Uniform flake; soluble in water with evolution of heat.
Containers: 450-lb net non-returnable open head steel drums.
Uses: Highly alkaline silicate for metal cleaning, laundering, and other applications.

"Hi-Sil." 177 Trademark for a hydrated silica pigment, used in producing exceptional tear and abrasion resistance in natural rubber; in the production of high concentrate insecticide powders; and as an anti-caking and conditioning agent in powdered or granular materials.
"Hi-Sil" 101. Pigment of extremely fine particle size (0.025 micron). Packed in 50-lb paper bags.
"Hi-Sil" 233. Slightly finer and less alkaline than "Hi-Sil" 101, and also more reinforcing in all elastomers, particularly neoprene and nitriles. Packed in 50-lb paper bags.
"Hi-Sil" X303. Designed especially for silicone rubber to provide easy compounding, a high level of physical properties, and extremely low water absorption. Packed in 25-lb paper bags.

"Histadyl." 100 Trade name for thenylpyramine hydrochloride [N,N-dimethyl-N'-(2-thenyl)-N'-(2-pyridyl)-ethylenediamine hydrochloride]; also known as methapyrilene hydrochloride, U.S.P.
Properties: White crystalline powder; bitter taste; m.p. 159-161°C; nonhygroscopic; soluble in water; requires protection from freezing.
Use: Medicine.

"Histalog." 100 Trademark for betazole.

histaminase. An enzyme occurring in the animal digestive system, that converts histidine to histamine.

histamine (4-aminoethylglyoxaline; 4-imidazole ethylamine) $NH_2CH_2CH_2C_3H_3N_2$.
Properties: White crystals; m.p. 83-84°C; b.p. 209-210°C (18 mm); soluble in water; slightly soluble in alcohol.
A product of the degradation of histidine, histamine occurs in animal and human body tissues and is liberated by injury to the tissue. It is found whenever a protein is decomposed by putrefactive bacteria. In the form of histamine hydrochloride or phosphate, it is used in medicine.

histamine phosphate $C_5H_9N_3 \cdot 2H_3PO_4$.
Properties: Colorless, odorless, long prismatic crystals; m.p. 140°C; soluble in water. Stable in air but affected by light. Solutions acid to litmus paper.
Grades: U.S.P. XVI.
Use: Medicine.

histidine (alpha-amino-beta-imidazolepropionic acid) $HOOCCH(NH_2)CH_2C_3H_3N_2$. An amino acid essential for rats. It is found naturally in the L(-) form.
Properties: Colorless crystals; soluble in water; insoluble in alcohol and ether; shows optical activity.
DL-histidine, m.p. 285-6°C with decomposition.
D(+)-histidine, m p. 287-8°C.
L(-)-histidine, m.p. 277°C with decomposition.
Derivation: Hydrolysis of protein; organic synthesis.
Uses: Medicine; feed additive; biochemical studies.
Available commercially as L(+)-histidine hydrochloride, and as the free base.

L(+)-histidine monohydrochloride.
(a) $C_6H_9N_3O_2 \cdot HCl \cdot H_2O$;
(b) $C_6H_9N_3O_2 \cdot HCl$.
Properties: Small colorless crystals; nearly odorless; salty taste; m.p. (a) 80°C, (b) 140°C; fairly soluble in water; insoluble in alcohol or ether.
Use: Medicine; intermediate.

histochemistry. The chemistry of animal tissues and fluids.

"Hitec" Heat Transfer Salt. 28 Trademark for a eutectic mixture of inorganic salts consisting of 53% potassium nitrate, 40% sodium nitrite and 7% sodium nitrate.
Properties: F.p. 288°F; slow thermal decomposition 800-1000°F; rapid decomposition 1500°F; density 1.933 (400°F), 1.698 (1000°F).
Containers: 25-lb wooden boxes; 150-, 400-lb fiber drums.
Use: Heat transfer in industrial applications involving heating and/or cooling in the 800-1000°F temperature range.
Hazard: Oxidizing material.
Shipping regulations: Oxidizing material. Yellow label. *

"Hi-Test Alkali." 244 Trademark for a compound available in two grades of fusion products of caustic soda with modifying alkaline salts. Rapidly soluble highly alkaline flakes; free rinsing.
Uses: Bottle washing, cleaning of dairy and food plant equipment.

*See "I.C.C. Shipping Regulations," page xiii.
Reference numbers refer to name of manufacturer. See "List of Manufacturers," page v.

"Hi-Time." [108] Trademark for a free-flowing, white, dust-free granular dish-machine compound designed for high-speed, heavy soil installations in hard water areas.
Containers: 125- and 350-lb drums.

"Hi-Tri." [233] Trademark for trichloroethylene (q. v.).

"Hi-White" Clay. [285] Proprietary brand name for a group of hydrous aluminum silicates (sedimentary kaolins) from Georgia.
Properties: Sp. gr. 2.60; bulk density, aerated, 18-20 lbs/cu ft; packed, 35-40 lbs/cu ft; creamy white; pH 6.0-7.0; air-floated; particle size 62-68% minus 2 microns; color brightness (G. E.) 75-82.
Containers: 50-lb multiwall bags or bulk.
Uses: A low-cost high-brightness, minimum abrasion filler clay for paper; also used in rubber, paint, certain ceramic applications, and in roofing granules.

HMF black. Abbreviation for high modulus furnace black. See furnace black.

HNM. Abbreviation for hexanitromannite. See mannitol hexanitrate.

Ho. Symbol for holmium.

hoarhound. See marrubium.

Hofmann's reaction. Reaction used for preparation of a primary amine from an amide by treatment with a halogen (bromine, usually) and caustic soda. The resulting amine has one less carbon atom than the amide used.

hog's bean. See hyoscyamus.

hogweed. See scoparius.

hole. A vacant place where there should normally be an electron or an atom in the orderly arrangement of electrons or atoms in a crystal or solid. Such a hole arises from the presence of an impurity, or is due to an imperfection in the crystal formation. See semiconductors.

holmia. Holmium oxide. See rare earths.

holmium Ho. Atomic number 67; group III of the periodic table; one of the rare-earth elements of the yttrium subgroup.
Properties: A solid with metallic luster; sp. gr. 8.764; m. p. 1475-1525°C; b. p. approx 2300°C. Reacts slowly with water; soluble in dilute acids.
Derivation: Reduction of the fluoride by calcium.
Grade: Regular high purity (lumps; ingots).

holmium chloride $HoCl_3 \cdot xH_2O$. Available as 45% Ho_2O_3; up to 99.9% Ho salts.
Containers: Glass bottles; fiber drums.

holmium fluoride $HoF_3 \cdot 2H_2O$. Available as 77% Ho_2O_3; up to 99.9% Ho salts.
Containers: Glass bottles; fiber drums.

holmium nitrate $Ho(NO_3)_3 \cdot 6H_2O$. Available as 42% Ho_2O_3; up to 99.9% Ho salts.
Containers: Glass bottles; fiber drums.

holmium oxide (holmia) Ho_2O_3.
Properties: Light yellow solid; slightly hygroscopic.
Grades: 98-99%.
Containers: Glass bottles; fiber drums.
Uses: Refractories; special catalyst.

holmium sulfate $Ho_2(SO_4)_3 \cdot 8H_2O$. Available as 46% Ho_2O_3; up to 99.9% Ho salts.
Containers: Glass bottles; fiber drums.

"Holocain" Hydrochloride. [162] Trademark for phenacaine hydrochloride.

holystones. A variety of sandstone used in blocks for rubbing down rough surfaces, particularly on ships.

"Holzon." [205] A plasticized rubber-based paint for coating floors and swimming pools in color.

homatropine $C_{16}H_{21}NO_3$. An alkaloid.
Properties: White crystals; poisonous! Slightly soluble in water; m. p. 95.5°C.
Derivation: Condensation of tropine and mandelic acid.
Containers: Vials.
Use: Medicine (usually used in the form of its salts).
Shipping regulations: None.*

homatropine hydrobromide $C_{16}H_{21}NO_3 \cdot HBr$.
Properties: White crystals or white crystalline powder; poisonous! Affected by light. M. p. 212°C with partial decomposition. Soluble in water and alcohol and chloroform; insoluble in ether.
Derivation: By the action of hydrobromic acid on homatropine.
Method of purification: Crystallization.
Grades: Technical; U. S. P. XVI.
Containers: Bottles.
Use: Medicine.
Shipping regulations: None.*

homatropine methylbromide $C_{16}H_{21}NO_3 \cdot CH_3Br$.
Properties: Odorless, white, crystalline powder with bitter taste. M. p. 190-198°C; affected by light. Very soluble in water; freely soluble in alcohol; almost insoluble in ether and acetone; soluble in acetone containing 20% water; solutions practically neutral to litmus.
Grade: U. S. P. XVI.
Containers: Bottles.
Use: Medicine.

homo-. A prefix meaning the same or similar, usually designating a homolog of a compound, differing in formula from the latter by an increase of CH_2.

homogeneous reactor. See nuclear reactor.

homologous series. A series of organic compounds that are all exactly alike except that each successive member has one more CH_2 group in its molecule than the next preceding member. For instance CH_3OH methanol, C_2H_5OH ethanol, C_3H_7OH propanol, C_4H_9OH butanol, etc., form a homologous series.

homolysis. The symmetrical breakdown of a covalent electron bond,
A:B = A· + B·. See heterolysis.

homomorphs. Molecules similar in size and shape. They need have no other characteristics in common. Many properties of several homomorphs can be predicted by knowing properties of one homomorph.

homophthalic acid $C_6H_4(CH_2COOH)COOH$.
Light tan powder, used as an intermediate.

homoquinine $C_{20}H_{24}O_2N_2 \cdot C_{19}H_{22}O_2N_2 \cdot 1, 2$, or $4H_2O$. A molecular combination of quinine and cuprëine obtained from cuprea bark (Remijia perdunculata). It was at one time thought to be an entirely new alkaloid.
M. p. 177° (dry).

ortho-**homosalicylic acid.** See cresotic acid.

4-homosulfanilamide hydrochloride. See maphenide hydrochloride.

homoveratric acid (3,4-dimethoxyphenylacetic acid) $(CH_3O)_2C_6H_3CH_2COOH$. Crystals; m. p. 94-101°C; very slightly soluble in water; soluble in most organic solvents.

homoveratronitrile $(CH_3O)_2C_6H_3CH_2CN$. A white, crystalline powder; m. p. 62-67°C.

homoveratrylamine (3,4-dimethoxyphenylethylamine) $(CH_3O)_2C_6H_3(CH_2)_2NH_2$. A colorless to pale yellow liquid with slight vanilla odor; sp. gr. 1.09 (25/25°C); solidifies 15°C; b. p. 295°C (with decomposition); refractive index 1.5442-1.5452 (25°C).

hoof and horn meals. The hoofs and horns of animals not needed for more valuable purposes are slightly cooked until they become friable and are then ground to a fine powder for use as fertilizer. They run from 4 to 6% water and 17-18% ammonia.
Shipping regulations: None.*

hoof oil. See neats-foot oil.

Hooker cell. A diaphragm-type electrolytic cell (see diaphragm cell) for the production of chlorine and caustic soda from sodium chloride. It consists of three major parts: (1) a bottom section carrying the vertical graphite anode blades, (2) a middle section bearing the cathode fingers, which are formed of heavy wire screen and fit between the anodes when the cell is assembled, and (3) the top section, which has a brine inlet and chlorine outlet. The top and bottom sections are made of rather massive concrete for thermal insulation. An asbestos diaphragm is formed on the surface of the cathode fingers before assembly of the cell. The cell operates at 85-95°C and produces 6-8 lbs of chlorine per day per square foot of floor space.

Hooke's law. When a load is applied to any elastic body so that the body is deformed or strained, then the resulting stress (the tendency of the body to resume its normal condition) is proportional to the strain. Stress is measured in units of force per unit area, strain is the extent of the deformation. For example when a bar of metal is subjected to a stretching load, the extent of the increase in length of the bar is directly proportional to the force per unit area, i. e., to the stretching load or stress. In general Hooke's law applies only up to a certain stress called the yield strength.

Hoopes process. An electrolytic method of purifying commercial aluminum to 99.99% aluminum content. A bath of fused cryolite, aluminum fluoride and barium fluoride of definite density is used. A heavy layer of molten aluminum-copper alloy at the bottom of the cell acts as the anode. A layer of lighter pure aluminum metal floating on top of the bath serves as the cathode. Pure aluminum from the aluminum-copper alloy collects at the cathode on passage of current through the cell.

hopcalite. A mixture of oxides of copper, cobalt, manganese and silver, used in gas masks as a catalyzer converting carbon monoxide to carbon dioxide. Not safe when nitroparaffin vapors are present.

hops. See humulus.

hops oil.
Properties: A brownish-yellow essential oil; strong penetrating odor. Soluble in alcohol, ether, and chloroform. Insoluble in water.
Chief known constituents: Humulene, geraniol and terpenes.
Constants: Sp. gr. 0.855 to 0.880; refractive index (n 20/D) 1.4775.
Derivation: Distilled from the strobiles of Humulus lupulus.
Method of purification: Rectification.
Grades: Technical.
Containers: Glass bottles; drums.
Use: Aromatizing beer and tobacco.
Shipping regulations: None.*

hordenine (anhaline) $(CH_3)_2N(CH_2)_2C_6H_4OH$. para-Hydroxyphenylethyldimethylamine. An alkaloid from barley.
Properties: Colorless crystals; m. p. 117-118°C; sublimes 140-150°C; soluble in alcohol, chloroform, and ether; slightly soluble in water, benzene, toluene, and xylene.
Use: Medicine.

horehound. See marrubium.

"Horizo." [244] Trademark for a double salt of hydrated trisodium phosphate and sodium hypochlorite. Rapidly soluble alkaline powder yielding 3.25% min. available chlorine calculated as NaOCl.
Uses: Sanitizer; deodorant.

"Hormodin." [123] Trade name for a formulation of indolebutyric acid.

hormones. Complex organic compounds which are formed by one organ and which act in a specific manner on the function of another organ or organs. Hormones are produced by the internal secretion of the ductless or

*See "I.C.C. Shipping Regulations," page xiii.
Reference numbers refer to name of manufacturer. See "List of Manufacturers," page v.

endocrine glands and co-ordinate the functions of organs by circulation in the bloodstream. Many of the hormones are steroids (see androgen, corticoid hormones, estrogens, steroids). See, of the non-steroid type, epinephrine; thyroxine; insulin.

See also plant hormones.

"Hornglaze." [205] A liquid emulsion of carnauba wax formulated to produce a self-polishing tough, hard, water-resistant finish over linoleum, rubber tile, asphalt tile, or hardwood floors.

"Hornlux." [205] A liquid which, by means of its composition, allows the penetration of a phenolformaldehyde condensation product down into a newly completed terrazzo or colored concrete floor. Applied with a rag; then surface excess is wiped off. Used to seal the porosity, densify, harden, and develop the color of concrete or new terrazzo floors.

horn silver. See cerargyrite.

"Hornstone." [205] Magnesium fluosilicate and zinc fluosilicate in powder form. Applied to a concrete floor surface, it reacts to form new binding materials and new and harder compounds. Available in liquid as well as crystal form.

"Horse Head." [268] Trademark for a line of zinc metals and alloys, zinc and titanium pigments, metal powders.

"Horse Head" Cadmium Metal Sticks. A proprietary product cast in stick form containing 99.99% metallic Cd. Used for electroplating and alloys.

"Horse Head" C. P. Zinc Metal. Proprietary product supplied in sticks. Used for chemical applications and for precious metal alloys.

"Horse Head" Lehigh Slab Zinc. A proprietary prime western slab zinc. Shipped in 45-lb slabs. Used for galvanizing; for making French process zinc oxide for brass and other alloys.

"Horse Head" Rolled Zinc. A proprietary product used for boiler plates; spinning; drawing.

"Horse Head" Slab Zinc. A product of retort distillation which exceeds the A. S. T. M. specifications for high-grade slab zinc. Used for galvanizing; brass and other alloys; dry cells; and manufacture of rolled zinc products.

"Horse Head" Special Slab Zinc. 99.99+% pure zinc slabs produced by a patented pyrometallurgical process; meets A. S. T. M. specifications for special high-grade zinc. Used for diecastings, mold castings, zinc plating, and galvanizing.

"Horse Head" Spiegeleisen. A proprietary product cast in pigs. Used in steel, grey iron, malleable iron.

"Horse Head" Titanium Dioxide. Brand name for a group of anatase and rutile white hiding pigments. Shipped in 50-lb bags. Used in interior and exterior paints, rubber, paper coatings.

"Horse Head" Zinc Anodes. "Horse Head" Special Slab Zinc to which minimum aluminum has been added. Cast in special molds; used for electrogalvanizing and cathodic protection of steel.

"Horse Head" Zinc Oxide. Brand name for several grades of zinc oxide.

"Horse Head" Metal Powders. For the powder metallurgy process for making small metal parts. Available in brass, bronze, copper, nickel, silver. Also available as metallic zinc powder for metal spraying, hearing-aid batteries, chemical refining, brazing, and brake linings.

"Horse Head Zamak." [268] ("Zamak"). Proprietary zinc alloy for die casting consisting of high purity zinc, aluminum, magnesium, and other alloying ingredients. Supplied in 19-lb ingots.

Uses: For zinc alloy die castings with excellent resistance to corrosion, retention of dimensions, and impact strength.

"Horse Head Zilloy." [268] ("Zilloy"). A proprietary rolled zinc alloy. Stiffer, stronger, and with greater creep strength or resistance to plastic flow than ordinary commercial grades of rolled zinc, but possessing the same corrosion-resisting characteristics as rolled zinc and fabricated and finished by similar methods. Furnished in flat sheets, coils, plates.

Uses: Particularly suitable for screen frames, screen guides, splines, weather strips, corner beadings, moldings, and stampings. Corrugated form for light-weight, low cost, permanent roofing and siding.

horsemint oil (monarda oil).

Properties: A yellowish-red or brownish-red essential oil; strong thyme-like odor. Soluble in alcohol, ether, and chloroform. Sp. gr. 0.920-0.936.

Derivation: Distilled from the herb Monarda punctata.

Method of purification: Rectification.

Constituents: Thymol, cymene.

Grades: Technical.

Containers: Glass bottles.

Use: Preparation of liniments.

Shipping regulations: None. *

horseradish root. See armoraciae radix.

horseweed oil. See erigeron oil.

hot cell (cave) The name given to the space and facilities used for the manipulation of highly radioactive material. A hot cell is usually a room with a floor area of 30 to 100 square feet inside with concrete walls 3 to 6 feet thick, and with appropriately weighty ceiling and floor. Hot cell sizes are usually quoted as the number of curies or kilocuries of radioactive material they may contain and still allow operation by personnel on the outside without excessive exposure to radiation.

Such a cell is usually equipped with some form of remote handling device, or "mechanical hands", and with utilities such as electric power, compressed air, steam, etc., all under remote control, and all carried through the shielding in offset holes in the concrete so that no direct beam of radiation may reach the outside. Viewing of the internal operation of a cell may be done through periscopes, television, or through windows of high density glass or zinc bromide solution. Typical operations carried out in a hot cell are chemical processing, machining, inspection, and physical testing.

hot dip process. See galvanized iron.

"#20 Hot Galvanizing Flux." [72] Brand name for a flux for hot galvanizing consisting of zinc ammonium chloride and organic additions.

Houdry process. Catalytic cracking process in which oil vapor is passed through a fixed catalyst bed consisting of activated hydrosilicate of alumina in molded form. When catalyst is fouled, the vapor stream is diverted to a second catalyst chamber while the first is cleaned by burning off the deposited coke with air. Used for producing higher octane gasoline from heavy distillate.

HPC black. Abbreviation for hard processing channel black. See channel black.

HS. Abbreviation for hydroxylamine sulfate.

"HSM." [65] Trademark for hemisulfur mustard used in medicine.

ht. Abbreviation for heat.

"H.T." [58] Trademark for monocalcium phosphate. Free-flowing beads.

HT-1. A type of nylon from phenylenediamine and iso- or terephthalic acid, intended for high temperature uses.

"HT-44." [212] Trademark for product containing a standardized mixture of amylolytic and proteolytic enzymes.

Properties: Dry, white powder containing buffer salts; nonhazardous; nonflammable. Optimum pH 6.8-7.0; optimum temperature 158°F.

Grade: Food.

Containers: All quantities; bulk, 300-lb drums.

Uses: Desizing of textile fabrics prior to finishing. For starch modification at higher temperatures; in food and other industries.

"HTH." [84] Brand name of a proprietary high test calcium hypochlorite product commercially available as a stable, water-soluble material in both granular and tablet form, containing a minimum of 70% available chlorine as calcium hypochlorite.

Uses: Bleaching; sterilizing; oxidizing.

"HTH-15." [84] Trademark for an all purpose germicide, disinfectant and stain remover. "HTH-15" contains 15% of available chlorine and yields sodium hypochlorite solutions directly when added to water.

Uses: Dairy and poultry farm sanitation; for sterilizing glasses and food utensils, and for general sanitation. It has also proved invaluable as a china dip for removing stains from dishes in restaurants, cafeterias, etc.

HTU. Abbreviation for height of a transfer unit. It is the height of a distillation column or fractionating tower in which unit separation is achieved by transfer from liquid to vapor or vice versa, of the materials being separated. Unit separation is defined by the differential equation that takes into account the varying concentrations along the column. HTU is also applied to extraction and other countercurrent separation processes.

huanuco bark. See cinchona bark, loxa.

Huber's reagent. Used for detecting free mineral acid. An aqueous solution of ammonium molybdate and potassium ferrocyanide. With the exception of boric acid and arsenic trioxide, free mineral acids produce a reddish-brown precipitate, or a turbidity with the reagent.

Hubl's reagent.

(a) 50 grams iodine dissolved in 1 liter of 95% alcohol.

(b) 60 grams mercuric chloride dissolved in 1 liter alcohol.

(c) Make up an iodine monochloride solution from (a) and (b). Add in excess to a known weight of the fat or oil dissolved in chloroform. The excess of iodine chloride can be estimated by the potassium iodide and thiosulfate method. By running a blank test, the amount of iodide absorbed can be estimated.

Use: Determination of iodine value of oils and fats.

huebnerite. See wolframite.

"Humatin." [330] Trademark for paromomycin sulfate (q.v.).

humectant. A term denoting afinity for water, with stabilizing action on the water content of an article; thus, a humectant keeps within a narrow range the moisture content fluctuations caused by wide-range humidity fluctuations. Example: glycerol.

humic acid. An acid substance found in humus and obtained as a by-product in making montan wax from lignite.

Properties: Chocolate-brown, dust-like powder; slightly soluble in water; dissolves in hot concentrated nitric acid to give a dark red color.

Use: In drilling muds.

humidity, absolute. The pounds of water vapor per pound of dry air in an air-water vapor mixture.

humidity indicators. Certain cobalt salts (e.g. cobaltous chloride) that change color as the humidity of their environment changes. Cobaltous compounds are pink when hydrated and greenish-blue when anhydrous.

humidity, relative. The percentage relation that the actual amount of water vapor present in a given volume of air at a definite temperature bears to the maximum amount of water vapor that would be present if the air were saturated with water vapor at that temperature.

humin. Any black amorphous solid material obtained in acid digestions of organic materials.

humulin. See lupulin.

humulon. An antibiotic constituent of hops (see lupulin). $C_{21}H_{30}O_5$. A complex derivative of benzene.
Properties: Crystallizes from ether; m.p. 66°C; bitter taste; decomposes slowly in air; soluble in most organic solvents; slightly soluble in boiling water.

humulus (hops). Carefully dried strobiles of Humulus lupulus.
Habitat: Europe and North America.
Containers: Bags.
Uses: Medicine (aromatic bitter); brewing beer and beer substitutes.
Shipping regulations: None.*

humus. A black or brown substance formed by the decay of vegetable matter. Contains carbon, oxygen, and nitrogen in the soluble portion. A slight amount of acid known as humic acid is present also. Humus increases the soil's capacity for absorbing and retaining water, reduces its tenacity, and is the cause of a more rapid and thorough absorption of the sun's rays.

humus, sour. Humus which contains humic and related acids due to decomposition under excess of moisture and lack of air.

"HX" (HPC). [285] Proprietary brand name for hard processing channel carbon black.
Properties: Sp. gr. 1.77; free-flowing pellets, also available in fluffy, unpelletted form as "HX-U". Bulk density 25lbs/cu ft; particle diameter 25 millimicrons; pH 4.1-4.5; ash 0.05% max; 99.9% through 325 mesh screen; color (Nigrometer) 82.5-83.5.
Containers: 50-lb paper bags or bulk.
Uses: As a reinforcing ingredient for compounding in natural and most synthetic rubbers, contributing to abrasion resistance, good tensile and tear strength; as a black coloring agent in rubber, etc.

hyacinthin. See phenyl acetaldehyde.

hyaluronic acid. A polymer of acetylglucosamine and glucuronic acid, occurring as alternate units, with a molecular weight of 200,000 to 400,000. Found in vitreous humor, synovial fluid, pathologic joints, group A and C hemolytic streptococci and skin, which probably contains the largest store of this substance. Hyaluronic acid appears to bind water in the interstitial spaces, forming a gel-like substance which holds the cells together. Its solutions are highly viscous. It is probably present in the body as a salt with inorganic bases.

hyaluronidase. An enzyme that breaks down the polymeric structure of the gel-form of hyaluronic acid, thus dissolving the gel which acts as the cementing mucoid of connective tissure and increasing the permeability of tissue to the diffusion of substances accompanying the enzyme. The enzyme occurs in poison glands of some snakes, in insects, pathogenic bacteria and in sperm cells; it also is believed to have possibilities in combating sterility and the spread of disease within the body.
Properties: Hyaluronidase for injection is a white amorphous solid. Solutions are colorless and odorless.
Grades: U.S.P. XVI.
Uses: Medicine.

"Hyamine." [23] Trademark for quaternary-ammonium-type bactericides, algaecides, and fungicides, supplied as water-soluble crystals or aqueous solutions. Non-irritating, low toxicity compounds.
Use: Restaurant and dairy sanitizing, general bactericidal applications, mildew-proofing of textiles, swimming pool algaecide.

"Hycar." [119] Trade name for various types of synthetic rubber.
"Hycar" Nitrile Rubbers: Copolymers of butadiene and acrylonitrile, divided into groups on the basis of acrylonitrile content. High acrylonitriles are characterized by their superior resistance to oil and solvents. Polymers of lower acrylonitrile content have a relatively lower oil and solvent resistance, but better resilience and low temperature flexibility.
Properties:
(high acrylonitrile): Sp. gr. (Polymer) 1.00; tensile strength (psi) 1500-3000; elongation 100-800%; hardness (Durometer A) 10-100; flexibility (lowest temp.) −30 to −50°F; resilience 20-50% recovery (Lupke Rebound); maximum service temperature 250-300°F.
(medium acrylonitrile): Sp. gr. (Polymer) 0.98; tensile strength (psi) 1000-3000; elongation 100-700%; hardness (Durometer A) 10-100; flexibility (lowest temperature) −50 to −70° F; resilience 30-60% recovery (Lupke Rebound); maximum service temperature 250-300°F.
(low acrylonitrile): Sp. gr. (Polymer) 0.95; tensile strength (psi) 1000-2500; elongation 100-700%; hardness (Durometer A) 10-100; flexibility (lowest temperature) −70 to −85°F; resilience 40-70% recovery (Lupke Rebound); maximum service temperature 250-300°F.
Uses: (high acrylonitrile): Oil well parts, fuel cell liners, fuel hose, gaskets, packing, oil seals, and other applications requiring highest resistance to aromatic fuels, oils, and solvents.
(medium acrylonitrile): General purpose oil resistant applications, shoe soles, kitchen

mats, sink topping, and printing rolls.
(low acrylonitrile): Gaskets, grommets, and O-rings where flexibility at very low temperatures is critical.

"Hycar" Polyacrylic Rubbers: Polymers of an acrylic acid ester. Useful in applications where oil and solvent resistance at high temperatures (to 425°F) is required. Polyacrylics are not recommended for use where flexibility below −10°F or high resistance to water and steam are important, but they possess outstanding resistance to heat, ultraviolet light, and flexural failure.
Properties: Sp. gr. (Polymer) 1.09; tensile strength (psi) 1000-2000; elongation 100-300%; hardness (Durometer A) 40-90; flexibility (lowest temperature) −10 to −20°F; resilience 20-40% recovery (Lupke Rebound); maximum service temperature 300-425°F.
Uses: Oil and gasoline hose, packings, automatic transmission gaskets, and conveyer belts.

"Hycar" Brominated Butyl Rubber: A butyl rubber modified by the addition of bromine in the polymer chain. The typical butyl properties of exceptionally low gas permeability, and excellent resistance to the effects of ozone, heat, and light are retained, while most of the shortcomings of butyl rubber are eliminated. Brominated butyl cures at a considerably faster rate, with more consistent results, and has the ability to cure when blended or in contact with other rubbers. In addition, it exhibits excellent adhesion to other rubbers and metals.
Uses: Inner tubes, steam hose, curing bags, and as a modifying component in blends with GR-S and crude rubbers.

"Hycar" Styrene Rubber: Polymer of styrene and butadiene. Oil soluble, primarily a binder for abrasive wheels, with some application in the manufacture of adhesives.

"Hycar" Latices: Nitrile, acrylic, and styrene polymers are all available as water emulsion with total solids ranging from 38-52%. They are compatible with a wide range of compounding ingredients, and are characterized by adhesion to natural and synthetic fibers. Principal applications include saturation and impregnation of paper to improve strength, density, and resistance to oil, water, and solvents; leather finishing, in which the ability to bind pigments is important; textile processing for resistance to abrasion, improved seam strength, and color fastness; in the manufacture of adhesives where compatibility of the latices with phenolic resins, vinyls, and casein allow flexibility in compounding to meet various requirements.

"Hycryl." [22] Trade name for a series of thickening agents. Available as:
"Hycryl" A-1000 - modified ammonium polyacrylate (water solution).
Properties: Straw colored, jelly like material; pH 8.0-8.5; sp. gr. (25°C) 1.01; viscosity (25°) 18,000-30,000 cps; solids 6.2-6.8; water resistant when dry.
"Hycryl" A-2000 - Latex dispersion which can be converted to ammonium polyacrylate solution through the addition of ammonia.
Properties: Milky white emulsion, solids 35%, pH 3.5-5.5. Initial viscosity - water thin; converted viscosity over 200,000 cps.
Uses: Anti-settling stabilizer for water dispersions of pigments; compounding rubber latices; thickener for alcoholic nylon coating solutions.

"Hydan." [28] Trademark for methionine hydroxy analogue calcium 90%.
Use: A source of methionine (an essential amino acid) for poultry, dog and livestock feeds.
Containers: 50-lb bags.

hydantoin (glycolylurea) $\overline{\text{NHCONHCOC}}H_2$.
Properties: White, odorless solid crystallizing in needles. Slightly soluble in water, ether; soluble in alcohols and solutions of alkali hydroxides. M. p. 220°C.
Grades: Technical.
Uses: Intermediate in the synthesis of pharmaceuticals, synthetic resins; textile lubricants and softeners.

"Hydeal" Process. [416] Patented process for catalytic hydrodealkylation of alkylbenzenes to benzene of high purity. Reaction occurs in the presence of hydrogen over a bed of solid catalyst of undisclosed nature, the products being benzene, the hydrocarbon corresponding to the dealkylated alkyl group, and unreacted hydrogen. Benzene yield from demethylation of toluene is 90-95% of theoretical; from xylenes, approximately 85-90% of theoretical. Also may be applied to the dealkylation of alkylnaphthalenes.

hydnocarpic acid $\overline{CH_2CH_2CHCHC}H(CH_2)_{10}COOH$. A cyclic fatty acid.
Properties: Plates; m. p. 59-60°C; soluble in chloroform; slightly soluble in organic solvents.
Source: Chaulmoogra oil.
Use: Medicine.

hydnocarpus oil. See chaulmoogra oil.

hydnocarpyl acetic acid. See chaulmoogric acid.

hydrabamine penicillin V. See penicillin.

hydrabamine phenoxymethylpenicillin. See penicillin.

hydracetin. See acetylphenylhydrazine.

"Hydrafine" Clay. [285] Proprietary brand name for chemically treated hydrous aluminum silicate (sedimentary kaolin) from Georgia.
Properties: Sp. gr. 2.60; brightness 86-87 G. E.; pH 6.5-6.8; particle size 90-94% minus 2 microns (as contrasted with 70-80% range of intermediate coating grades — see "Hydratex," "Hydrasperse").
Containers: 50-lb multiwall bags or bulk.
Uses: A premium type, extra fine particle size waterwashed coating clay for specialty paper and boxboard applications, paint and ink.

"Hydralase." [78] Trademark for a series of fungal enzymes used principally in the food industries for conversion of starches or dextrins to dextrose and of proteins to amino acids.

hydralazine hydrochloride (1-hydrazinophthalazine hydrochloride) $C_8H_5N_2NHNH_2 \cdot HCl$.
Properties: White, odorless, crystalline powder. M. p. 270-280°C (dec); very slightly soluble in ether and alcohol; slightly soluble in water; pH (2% solution) 3.5-4.5.
Grade: N. F. XI.
Use: Medicine.

"Hydraphtal." [28] Trademark for a light yellow liquid combining solvent and scouring properties.
Use: For heavy duty degreasing of textiles and leather.

hydrargol (mercuric succinimide; mercury imidosuccinate) $Hg[(CH_2CO)_2N]_2$.
Properties: White crystalline powder. Moderately soluble in water.
Derivation: By heating together succinic acid, ammonia, carbon dioxide and mercuric oxide.
Containers: Glass bottles; fiber cans.

hydrargyrum. Latin name for mercury.

"Hydrar" Process. [416] Patented process for the catalytic hydrogenation of benzene to cyclohexane, or higher aromatics to their corresponding cycloparaffins. The hydrogenation of benzene is virtually stoichiometric. The purity of the cyclohexane product is a function of the purity of the benzene feed.

hydrase. See hydrolase.

"Hydrasperse" Clay. [285] Proprietary brand name for chemically treated hydrous aluminum silicate (sedimentary kaolin) from Georgia.
Properties: Sp. gr. 2.60; brightness 84.5-85.5 G. E.; pH 6.0-7.0; particle size 77-80% minus 2 microns; contains water-dispersing agent; water-washed, spray dried.
Containers: 50-lb multiwall bags or bulk.
Use: A spray-dried intermediate type of paper coating clay; low viscosity, high brightness; with dispersant already added. Saves "make-down" time and assures thorough dispersion in practically any equipment.

hydrastina. See hydrastine.

hydrastine (hydrastina) $C_{21}H_{21}NO_6$.
Properties: White pulverulent poisonous alkaloid. Slightly soluble in water, alcohol and ether. Soluble in acetone and benzene. M. p. 132°C.
Derivation: By extraction of the root of Hydrastis canadensis, with subsequent crystallization.
Use: Medicine (usually used in the form of the hydrochloride, sulfate, tartrate, etc.).
Shipping regulations: None. *

hydrastine hydrochloride $C_{21}H_{21}NO_6 \cdot HCl$.
Properties: White crystals; poisonous! Hygroscopic. Soluble in water and alcohol; slightly soluble in ether. M. p. 116°C.
Derivation: By the action of hydrochloric acid on hydrastine.
Use: Medicine.

hydrastinine $C_{11}H_{13}NO_3$.
Properties: White crystalline alkaloid; poisonous! Soluble in alcohol and ether; slightly soluble in warm water. M. p. 116-117°C.
Derivation: By extraction of the root of Hydrastis canadenis, with subsequent crystallization. The salts are obtained by the action of the respective acid on the alkaloid.
Grades: Technical.
Use: Medicine.

hydrastis (golden seal; orange root; yellow root; yellow puccoon; turmeric root; Indian turmeric). Dried rhizomes and roots of Hydrastis canadensis.
Chief constituents: Alkaloids, berberine, canadine, hydrastine and hydrastinine.
Habitat: North America.
Grades: Technical.
Containers: Bales.
Use: Medicine (source of its alkaloids).
Shipping regulations: None. *

hydrate. A compound formed by the combination of water with some other substance, in which the water supposedly retains its molecular state as H_2O. The water combines in a definite weight ratio and the hydrate may be represented by a chemical formula. Most hydrates are decomposed by gentle heating. See, for example, gas hydrates.
See also complex compound.

hydrated alumina. See alumina trihydrate.

hydrated aluminum oxide. See alumina trihydrate.

hydrated silica. See silicic acid.

"Hydratex" Clay. [285] Proprietary brand name for a group of hydrous aluminum silicates (sedimentary kaolins) from Georgia.
Properties: Sp. gr. 2.60; brightness 83.5-85.0 G. E.; pH 4.5-5.0; particle size 71-73% minus 2 microns; water-washed, lump or pulverized.
Containers: 50-lb multiwall bags or bulk.
Uses: An intermediate paper coating clay with excellent covering power and flow characteristics for certain coating methods. Also used in special applications as a filler for paper and paint.

hydration. The process of absorption or combination of water with another substance. The term may apply both to processes involving chemical reaction and to those involving mere absorption. It is not usually applied in cases where a liquid solution results. In the paper industry it refers to a prolonged treatment in the beater whereby a viscous pulp is produced that gives water-resistance and crackle to the finished paper.

hydraulic cement. A cement that hardens under water, Portland cement being the prime example.

hydraulic lime. See lime, hydraulic.

hydrazine (hydrazine base; hydrazine, anhydrous; diamine) H_2NNH_2.
Properties: Colorless, fuming, corrosive, hygroscopic liquid; m. p. 2.0°C; b. p. 113.5°C; sp. gr. 1.004 (25/4°C); wt/gal 8.38 lbs; flash point (open cup) 126°F. Miscible with water and alcohol; insoluble in chloroform and ether. Strong reducing agent and diacidic base.
Derivation: From hydrazine hydrate, either by (a) various dehydration processes, (b) solvent extraction with ethylene glycol; or by action of anhydrous ammonia on hydrazine salts. The nuclear fission of ammonia will yield hydrazine and may prove feasible.
Grades: To 99% pure.
Containers: Glass carboys; steel drums.
Uses: Jet and rocket fuels; intermediate for agricultural chemicals, antioxidants, textile chemicals, explosives, photographic developers, blowing agents; scavenger for chlorine in hydrogen chloride, anhydrous; corrosion inhibitor and scavenger for oxygen.
Caution: Vapor explosive and toxic; especially dangerous to the eyes.
Shipping regulations: Corrosive liquid. White label. *

hydrazine, anhydrous. See hydrazine.

hydrazine base. See hydrazine.

hydrazine hydrate (diamide hydrate). $H_2NNH_2 \cdot H_2O$.
Properties: Colorless, fuming liquid; m. p. −51.7°C; b. p. 119.4°C; sp. gr. 1.032; wt/gal 8.61 lbs; flash point (open cup) 163°F. Miscible with water and alcohol; insoluble in chloroform and ether. Strong reducing agent; weak base.
Derivation: (a) Sodium hydroxide, chlorine, and ammonia react in aqueous solution to form a dilute solution of hydrazine (Raschig process). Sodium chloride is a by-product. (b) Oxidation of urea by sodium hypochlorite, as in the Raschig process.
Method of purification: Fractional distillation; concentration by flash distillation, conversion to the slightly soluble sulfate, and treatment of the latter with concentrated sodium hydroxide solution.
Grades: 85%, 100% (based on hydrate); also weaker solutions.
Containers: Glass carboys; steel drums.
Uses: See hydrazine; scavenger for oxygen in boiler feed water.
Caution: Vapor explosive and toxic, especially dangerous to the eyes.
Shipping regulations (if containing 50% or less of water): Corrosive liquid. White label. *

hydrazine monobromide $N_2H_4 \cdot HBr$.
Properties: White, crystalline flakes; m. p. 81-87°C; decomposes at about 190°C; soluble in water and lower alcohols; insoluble in most organic solvents.
Grade: 95%.
Use: Soldering flux.

hydrazine monochloride $N_2H_4 \cdot HCl$.
Properties: White crystalline flakes; m. p. 87-92°C, decomposes at about 240°C; soluble in water (37g/100g H_2O at 20°C); somewhat soluble in lower alcohols; insoluble in most organic solvents.

hydrazine sulfate (diamine sulfate; diamidogen sulfate) $NH_2NH_2 \cdot H_2SO_4$.
Properties: White crystalline powder; very soluble in hot water; soluble 1 part in 33 cold water; insoluble in alcohol; stable in storage but contact with alkalies and oxidizing agents should be avoided.
Typical specifications: Purity 99.8-100%; sp. gr. 1.37; m. p. 254°C; impurities: no ferrous oxide, no chloride; non-volatile matter 0.05% (max); ash 0.01% (max).
Grades: C. P.; technical.
Containers: Bottles; 25- to 125-lb fiber drums; multiwall paper sacks.
Uses: Strong reducing agent; manufacture of chemicals; condensation reactions; catalyst in making acetate fibers. Also used in tests of blood; analysis of minerals, slags and fluxes; determination of arsenic in metals; separation of polonium from tellurium; as fungicide; germicide; in adhesives.
Shipping regulations: None. *

hydrazinophthalazine hydrochloride. See hydralazine hydrochloride.

hydrazobenzene (N, N'-diphenylhydrazine) $C_6H_5NHNHC_6H_5$.
Properties: M. p. (min) 126°C; soluble in alcohol; nearly insoluble in water.
Grade: 95%.
Use: Synthesis.

"Hydrholac." [23] Trademark for plasticized nitrocellulose lacquer emulsions, including clear finishes, binders and colors. Produce flexible, lacquer-type, cleanable leather finishes from aqueous systems.
Use: Finishes on glove, garment, handbag and shoe leather.

hydrindene. See indan.

hydriodic acid (hydrogen iodide) HI.
Properties: Clear colorless or pale yellow liquid, an aqueous solution of hydrogen iodide, which is a gas at ordinary temperatures. A constant boiling solution is formed of sp. gr. 1.7 containing 57% hydrogen iodide. Hydriodic acid is a strong acid and an active reducing agent, highly corrosive. For anhydrous hydrogen iodide: sp. gr. 4.3737; m. p. −51.3°C; b. p. −35.6°C.
Derivation: (a) In gaseous form by decomposition of phosphoric iodide with a minimum amount of water. (b) By passing

hydrogen with iodine vapor over warm platinum sponge which acts as a catalyzer, and absorption in water. (c) By the action of iodine on a solution of hydrogen sulfide.
Grades: Technical, 47%; N. F. XI, diluted, 10%.
Containers: 150-lb carboys.
Uses: Medicinal; preparation of iodine salts; organic preparations; analytical reagent; disinfectant; pharmaceuticals.
Shipping regulations: Corrosive liquid. White label. *

hydroabietyl alcohol. See dihydroabietyl alcohol.

ortho-**hydrobenzoic acid.** See salicylic acid.

"Hydrobreak." 244 Trademark for a compound consisting of a balanced blend of buffered alkalies and a surface active agent.
Properties: White, granular, dedusted mechanical mix; soluble in water.
Uses: A general laundry and dairy cleaner for use in areas of medium hard water.
Containers: 125-lb plywood drums; 325-lb wooden barrels.

hydrobromic acid (hydrogen bromide) HBr in aqueous solution. See also hydrogen bromide, anhydrous.
Properties: Clear, colorless or faintly yellow liquid consisting of an aqueous solution of hydrogen bromide, which is a gas at ordinary temperatures. A constant boiling solution is formed, of sp. gr. 1.49, containing 48% hydrogen bromide. Hydrobromic acid is a strong acid and highly corrosive; sensitive to light. Sp. gr. (hydrogen bromide gas) 2.71 referred to air; m. p. −86.13°C; b. p. −68.7°C. Sp. gr. (48% solution) 1.488 (20/4°C).
Derivation: Gas: by passing hydrogen with bromine vapor over warm platinum sponge which acts as a catalyzer. The solution: by dissolving the gas in water, or by distilling from a mixture of sodium bromide and 50% sulfuric acid.
Impurities: Sulfuric acid, heavy metals, hydrochloric acid, hydriodic acid.
Grades: Technical 40%; medicinal 48%; 62%.
Containers: Glass bottles; carboys.
Uses: Medicine; analytical chemistry; organic preparations.
Shipping regulations: Corrosive liquid. White label. *

"Hydrocarb." 236 Brand name for a proprietary product. A soluble sodium humate compound for thinning or emulsifying drilling muds.
Containers: Multiwall polyethylene lined paper bags containing 50 lbs.

hydrocarbon. A compound which consists solely of the elements carbon and hydrogen.

hydrocarbon, acetylene. A hydrocarbon which contains at least one pair of triple bonded carbon atoms in its structure. It satisfies the general formula C_nH_{2n-2}. Also known under the family name of alkyne.

hydrocarbon, aromatic. A hydrocarbon characterized by a molecular structure involving one or more six-carbon-atom rings, and having properties similar to those of benzene which is the simplest member of this group. Toluene, xylene, naphthalene, anthracene, and phenanthrene are other key members of this series.

hydrocarbon, branched chain. A non-aromatic hydrocarbon in which not all the carbon atoms of the molecule are in a single chain. The simplest is isobutane, $(CH_3)_2CHCH_3$.

hydrocarbon C-2, C-3, etc. A hydrocarbon containing 2, 3, etc. carbon atoms per molecule.

hydrocarbon, olefin. Any hydrocarbon which contains at least one pair of double bonded carbon atoms in its structure. With a single double bond, it satisfies the general formula C_nH_{2n}. Also known under the family name of alkene. Ethylene and propylene are typical examples.

hydrocarbon, paraffin. A hydrocarbon in which the proportion of hydrogen to carbon is such as to satisfy the general formula C_nH_{2n+2}, e. g. methane CH_4, ethane C_2H_6, octane C_8H_{18}, etc. These are also known as saturated hydrocarbons. Their family name is alkane.

hydrocarbon, saturated. See hydrocarbon, paraffin.

hydrocarbon, straight chain. A hydrocarbon in which all the carbon atoms of the molecule are in a single unbranched chain. Such hydrocarbons are also designated as normal hydrocarbons, e. g. n-hexane $(H_3CCH_2CH_2CH_2CH_2CH_3)$.

hydrocellulose. See cellulose, hydrated.

hydrochloric acid (muriatic acid; chlorohydric acid; hydrogen chloride) HCl in aqueous solution. See also hydrogen chloride, anhydrous.
Properties: Clear, colorless or slightly yellow, fuming, pungent liquid; poisonous! A constant-boiling acid containing 20% hydrogen chloride is formed. Hydrochloric acid is a strong, highly corrosive acid. Gas: M. p. −111°C; b. p. −83.1°C. One liter of gas weighs 1.6392 g at 0°C. The commercial "concentrated" or fuming acid contains 38% of hydrogen chloride and has a sp. gr. 1.19. Soluble in water, alcohol and ether.
Derivation: (a) By-product from the chlorination of benzene and other hydrocarbons; (b) by the action of sulfuric acid on common salt; (c) by burning hydrogen, methane, or water gas in an atmosphere of chlorine; (d) Hargreaves process (q. v.).
Method of purification: Rectification (to remove arsenic); sometimes fractional distillation over ferrous chloride.
Impurities: Iron and arsenic.
Grades: U. S. P. XVI (35-38%); N.F. XI diluted (10%); technical (usually 18°, 20°,

or 22° Bé.); white; C. P.
Containers: Glass bottles; carboys (5, 6. 5, 13 gals); rubber-lined steel drums; rubber-lined tank cars.
Uses (in approximate order of volume): Acidizing (activation) of petroleum wells; chemical intermediates; ore reduction (manganese, radium, vanadium, tantalum, tin, tungsten); food processing (corn syrup, sodium glutamate); pickling and metal cleaning; industrial acidizing and general cleaning.
Warning! Causes burns; avoid contact with skin or eyes. MCA warning label.
Shipping regulations: Corrosive liquid. White label. *

hydrochlorothiazide (3, 4-dihydrochlorothiazide) $C_7H_8ClN_3O_4S_2$. 6-Chloro-7-sulfamyl-3, 4-dihydro-1, 2, 4-benzothiadiazine-1, 1-dioxide.
Properties: Crystals; m. p. 273-275°C. Practically insoluble in water; soluble in dilute ammonia or sodium hydroxide, in methanol, ethanol and acetone.
Use: Medicine.

hydrocinnamic acid $C_6H_5CH_2CH_2COOH$.
Properties: Crystals with hyacinth-rose odor. M. p. 46°C.
Derivation: Reduction of cinnamic acid with sodium amalgam.
Use: Fixative for perfumes.
Shipping regulations: None. *

hydrocinnamic alcohol. See phenyl propyl alcohol.

hydrocinnamic aldehyde. See phenyl propyl aldehyde.

hydrocinnamyl acetate. See phenyl propyl acetate.

hydrocortisone (17-hydroxycorticosterone; cortisol; hydrocortisone alcohol; compound F) $C_{21}H_{30}O_5$. One of the adrenal cortical steroid hormones. More active than cortisone (q. v.) as an anti-inflammatory agent.
Properties: White, odorless, crystalline powder; sensitive to light; bitter taste; m. p. 217-220°C with some decomposition. Freely soluble in dioxane and methanol; insoluble in ether and water; slightly soluble in alcohol, chloroform and acetone.
Derivation: Isolation from extracts of adrenal glands; synthesis from other steroids.
Grade: U. S. P. XVI.
Use: Medicine (also used as the acetate salt).

hydrocortisone acetate $C_{23}H_{32}O_6$. The acetate salt of hydrocortisone.
Properties: White, odorless crystalline powder; m. p. 216-223°C (dec). Very slightly soluble in ether; practically insoluble in water; slightly soluble in alcohol and chloroform. Sensitive to light.
Grade: U. S. P. XVI.
Containers: Bottles.
Use: Medicine.

hydrocortisone alcohol. See hydrocortisone.

hydrocortisone sodium succinate $C_{25}H_{33}NaO_8$.
Properties: White, odorless, hygroscopic, amorphous solid. Very soluble in water and alcohol; insoluble in chloroform; very slightly soluble in acetone.
Grade: U. S. P. XVI.
Use: Medicine.

"Hydrocortone." [123] Trademark for hydrocortisone (q. v.).

hydrocotarnine $C_{12}H_{15}NO_3 \cdot \frac{1}{2}H_2O$.
Properties: White crystalline alkaloid; poisonous! Soluble in alcohol and ether. Insoluble in water.
Constants: M. p. 50-55°C; decomposes at 100°C.
Derivation: From opium.
Grades: Technical.
Containers: Glass bottles.
Use: Medicine.
Shipping regulations: None. *

hydrocracking. The cracking of petroleum or its products in the presence of hydrogen. Special catalysts are used as, for example, platinum on a solid base of mixed silica and alumina.

hydrocyanic acid (prussic acid; hydrogen cyanide; formonitrile) HCN.
Properties: Water-white liquid at temperatures below 26. 5°C; faint odor of bitter almonds; usual commercial material is 96-99. 5% pure; vapors intensely poisonous; sp. gr. (liquid) 0. 6970 (gas) 0. 9348; b. p. 25. 6°C; freezing point −13. 3°C; flash point 0°F; soluble in water. The solution is weakly acidic; sensitive to light. When not absolutely pure or stabilized, it polymerizes spontaneously with explosive violence. Miscible in all proportions with water, alcohol; soluble in ether.
Derivation: (a) By treating a cyanide with dilute sulfuric acid. (b) By catalytically reacting ammonia and air with methane or natural gas. (c) By recovery from coke oven gases. (d) Decomposition of formamide. (e) From ammonia and hydrocarbons by electrofluid reactor.
Grades: Technical (96-98%); 2, 5 and 10% solutions. All grades usually contain a stabilizer, usually 0. 05% phosphoric acid, to prevent explosive polymerization.
Containers: Bottles; steel cylinders; tank cars.
Uses: Manufacture of acrylonitrile, acrylates, adiponitrile, cyanide salts, dyes, fumigants; chelates; military poison gas; rubbers and plastics.
Fire Hazard: The liquid burns like alcohol; gas is not flammable in ordinary fumigation concentrations. Danger! Poison gas. Flammable. MCA warning label.
Shipping regulations: Hydrocyanic acid (prussic), liquid stabilized: poison, class A. Poison gas label by freight; not accepted by express. Hydrocyanic acid solutions: poison, class B. Poison label. Hydrocyanic acid, unstabilized: not accepted by common carrier. *

"**Hydrodarco**"[89] Trademark for activated carbons used in municipal and industrial water purification.

hydrodealkylation. Hydrodealkylation is a petroleum refining process by which hydrogen under pressure is used to convert hydrocarbons in heavy reformates, naphthenic crudes or catalytic cracking recycle stocks. The process also converts organic sulfur compounds into hydrogen sulfide. Catalysts used may be calcined alkalized chromia-alumina or cobalt molybdenum oxides, but some versions of the process do not use a catalyst.

hydrodisodium phosphate. Sodium phosphate, dibasic.

hydrofining. Process for high pressure catalytic hydrogenation of low-grade petroleum fractions to produce material having more desirable properties. Used for up-grading of charging stock such as heavy crudes and refinery residues, most frequently to produce improved lubricants.

"**Hydroflo.**"[413] Trademark for a trinitrotoluene base explosive with free flowing characteristics.

Uses: For seismic prospecting; open pit mining.

Containers: $12\frac{1}{2}$ lb and 25-lb multiwall paper bags in 50-lb. shipping cases.

Fire hazard: Dangerous.

Shipping regulations: Explosives. Red label. *

hydroflumethiazide (trifluoromethylhydrothiazide) $C_8H_8F_3N_3O_4S_2$. 3, 4-Dihydro-6-(trifluoromethyl)-2H-1, 2, 4-benzothiadiazine-7-sulfonamide-1, 1-dioxide.

Properties: White, crystalline, odorless solid; m. p. 260-275°C. Insoluble in water and acid; soluble in dilute alkali but unstable in alkaline solutions.

Grade: Pharmaceutical.

Use: Medicine.

hydrofluoric acid (hydrogen fluoride) HF in aqueous solution. See also hydrogen fluoride, anhydrous.

Properties: Colorless, fuming, mobile, corrosive liquid. Poisonous, dangerous; produces terrible sores when allowed to touch the skin. Only a moderately strong acid, but unlike other acids will attack glass and any silica-containing material.

Derivation: Hydrogen fluoride gas is distilled from a mixture of calcium fluoride (fluorspar) and sulfuric acid. The gas is absorbed in water.

Grades: C. P.; technical; various strengths to 70%.

Containers: Wax bottles; lead jars; lead carboys; steel drums; steel tank cars.

Uses: Polishing, etching and frosting of glass; pickling copper, brass, stainless and other alloy steels; electropolishing of metals; cleaning stone and brick, purification of filter paper and graphite; acidizing oil wells; control of fermentation; dissolving ores; laundry sour; cleaning castings.

See also under hydrogen fluoride, anhydrous.

Danger! Hazardous liquid and vapor. Causes severe burns which may not be immediately painful or visible. MCA warning label.

Shipping regulations: Corrosive liquid. White label. *

hydrofluorosilicic acid. See fluosilic acid.

hydrofluosilicic acid. See fluosilic acid.

"**Hydrofol.**"[221] Trademark for a line of hydrogenated fatty acids and glycerides. Available in a variety of grades for specific applications in industrial chemicals, soaps, textile sizings and softeners, packing compounds, adhesives, lubricants, polishing compounds, paper coatings, textile and leather chemicals, waxes, printing inks, plasticizers, cosmetics and electrical insulations.

hydroforming process. Process for dehydrogenation and conversion of paraffinic hydrocarbons into cyclic and aromatic hydrocarbons by the use of heat, pressure, and catalysts in the presence of hydrogen. Used in producing motor fuels of high octane rating from ordinary or low-grade products, such as straight-run gasolines or light naphthas. In some cases the processes are operated to recover technical benzene, toluene, and xylene from petroleum.

hydrofuramide (furfuramide) $OC_4H_3CH(NCHC_4H_3O)_2$.

Properties: Light brown to white powder.

Constants: M. p. 117°C; boils about 250°C with decomposition. Insoluble in cold water; soluble in alcohol and ether.

Derivation: Treatment of furfural with ammonia in the cold.

Containers: 25-, 50-, and 100-lb drums.

Uses: Accelerator; hardening agent for resins; in fungicides.

hydrogen H. Element of atomic number 1. The lightest element.

Properties: Colorless gas; highly flammable. Sp. gr. 0.0694 referred to air; sp. volume 193 cu ft/lb (70°F); m. p. -259°C; b. p. -252°C. Very slightly soluble in water; alcohol, ether.

Derivation: (a) By the action of steam on natural gas at high temperatures, and subsequent purification; (b) by treatment of water gas with steam and absorption of the carbon dioxide; (c) dissociation of ammonia; (d) thermal decomposition of hydrocarbons or catalytic reforming of petroleum; (e) by reaction of iron and steam; (f) catalytic reaction of methanol and steam; (g) by electrolysis of water.

Grades: Technical; pure, from an electrolytic grade of 99.8% to ultra pure, with less than 10 ppm impurities. See also para-hydrogen.

Containers: Steel cylinders; tank cars of cylinders.

Uses: Production of synthetic ammonia and synthetic methanol; hydrogenation of

*See "I. C. C. Shipping Regulations," page xiii.
Reference numbers refer to name of manufacturer. See "List of Manufacturers," page v.

organic materials such as naphthalene, phenol, oils; reducing agent for organic synthesis and reduction of metallic ores; reducing atmospheres to prevent oxidation; as oxyhydrogen flame for high temperatures; atomic-hydrogen welding; small balloons (no longer used for dirigibles or passenger-carrying balloons); making hydrochloric and hydrobromic acids; rocket fuel.

Shipping regulations: Gas: Flammable gas. Red gas label. Liquid: Not accepted by common carrier. *

ortho-**hydrogen.** See para-hydrogen.

para-**hydrogen.** Type of molecular hydrogen preferred for rocket fuels. Molecular hydrogen (H_2) exists in two varieties, ortho and para, named according to their nuclear spin types. Ortho-hydrogen molecules have a parallel spin; para- an antiparallel spin. By cooling to liquid air temperature and use of a catalyst, the normal equilibrium of 3 ortho- to 1 para- is displaced and para-hydrogen may be isolated. It is being produced with less than 5 ppm impurities.

hydrogen 2. See deuterium.

hydrogen 3. See tritium.

hydrogenated oils (hardened oils). Oils treated with hydrogen in presence of a catalyst, usually nickel, thereby converting all or part of the oleic acid or olein (unsaturated) into stearic acid or stearin (saturated). The oils thus treated are rendered suitable for the manufacture of hard soaps (where previously they could only be used for making soft soaps) and for making lubricants. Hydrogenated oils are also used for making lard substitutes for foods, tanner's greases, varnishes, etc.

hydrogenation. Combination of hydrogen with another substance, usually an unsaturated organic compound, and usually under the influence of temperature, pressure, and catalysts (usually nickel). Thus unsaturated components of cottonseed oil and other oils are hydrogenated to produce solid fats.

hydrogen, atomic. See atomic-hydrogen welding.

hydrogen bomb. See nuclear fusion.

hydrogen bromide. See hydrobromic acid; hydrogen bromide, anhydous.

hydrogen bromide, anhydrous HBr.

Properties and derivation: See hydrobromic acid. The gas is liquefied under a pressure of 350 psi at 25°C.

Grades: Up to 99.8% min purity.

Containers: 15- and 150-lb cylinders.

Uses: As an agent in pharmaceutical synthesis, makes (1) bromides by direct reaction with alcohols, (2) intermediates for barbiturate manufacture, (3) acts as intermediate in the manufacture of synthetic hormones. In the petroleum industry, hydrogen bromide is used as an alkylation catalyst.

Shipping regulations: Nonflammable gas. Green label. *

hydrogen carboxylic acid. See formic acid.

hydrogen chloride. See hydrochloric acid; hydrogen chloride, anhydrous.

hydrogen chloride, anhydrous HCl.

Properties and derivation: See hydrochloric acid. Pure HCl is a colorless gas, and on liquefaction gives a colorless liquid.

Method of Purification: Solvent extraction is one method.

Containers: Cylinders.

Uses: In reactions where aqueous hydrochloric acid is not suitable, such as production of vinyl chloride from acetylene and alkyl chlorides from olefins. Also in polymerization, isomerization, alkylation, and nitration reactions.

Danger: Hazardous liquid and gas under pressure. Causes burns. Extremely irritating. MCA warning label.

Shipping regulations: Nonflammable gas. Green label. *

hydrogen cyanide. See hydrocyanic acid.

hydrogen dioxide. See hydrogen peroxide.

hydrogen electrode. See electrode, hydrogen.

hydrogen fluoride. See hydrofluoric acid; hydrogen fluoride, anhydrous.

hydrogen fluoride, anhydrous (hydrogen fluoride) HF. See hydrofluoric acid.

Properties: Colorless, fuming, mobile, corrosive liquid, or colorless corrosive gas, soluble in water. Poisonous, dangerous, produces severe burns which do not heal easily. The liquid and gas consist of associated molecules; the vapor density corresponds to HF only at high temperatures. F.p. −83°C; b.p. 19.5°C; sp. gr. liquid 0.988.

Derivation: Distillation from the reaction of calcium fluoride and sulfuric acid.

Grade: To 99.9% min purity.

Containers: Cylinders; tank cars.

Uses: Catalyst in alkylation, isomerization, condensation, dehydration, and polymerization reactions; fluorinating agent in organic and inorganic reactions; production of fluorine and aluminum fluoride; preparation of aqueous hydrofluoric acid; production of synthetic cryolite, of fluorides, fluoborates, fluosilicates, fluorocarbons; additive in liquid rocket propellants; refining of uranium.

Danger! Extremely hazardous liquid and vapor. Causes severe burns which may not be immediately painful or visible. MCA warning label.

Shipping regulations: Corrosive liquid. White label. *

hydrogen iodide. See hydriodic acid.

hydrogen ion concentration. See pH.

*See "I.C.C. Shipping Regulations," page xiii.

Reference numbers refer to name of manufacturer. See "List of Manufacturers," page v.

hydrogenolysis. The cleavage of a bond in an organic compound with simultaneous addition of a hydrogen atom to each fragment.

hydrogen oxide. See water.

hydrogen peroxide (hydrogen dioxide; peroxide) H_2O_2.

Properties: A colorless, heavy liquid, usually sold in aqueous solution of various strengths. Anhydrous hydrogen peroxide has sp. gr. 1.46; m.p. −2°C; b.p. 158°C; is soluble in water and alcohol. It is fundamentally unstable; the decomposition is slow with pure material but catalyzed by many impurities, especially metallic impurities. The commercial solutions commonly contain a preservative, such as acetophenetidin or acetanilide. Contact with the more concentrated solutions should be avoided. Hydrogen peroxide is an active oxidizing agent. The concentrated material may react explosively with combustible materials.

Derivation: (a) By the electrolytic oxidation of sulfuric acid or a sulfate to persulfuric acid or a persalt with subsequent hydrolysis and distillation of the hydrogen peroxide formed; (b) by decomposition of barium peroxide with sulfuric or phosphoric acids; (c) hydrogen reduction of 2-ethylanthraquinone followed by oxidation with air to regenerate the quinone and produce hydrogen peroxide; (d) electrical discharge through a mixture of hydrogen, oxygen, and water vapor.

Grades: U.S.P. XVI (3%); technical (3%, 6%, 27.5%, 30%, 35%, 50%, and 90%). Most common commercial strengths are 27.5%, 35%, 50%, and 70%.

Containers: Amber glass bottles; carboys; aluminum drums; tank trucks and tank cars.

Uses: Bleaching of textiles, wood pulp, hair, fur, straw, glue, gelatin, waxes, soap, etc; source or organic and inorganic peroxides; rocket fuel oxidizer and torpedo propellant; production of foam rubber and other porous materials; antichlor; dyeing; electroplating; antiseptic; laboratory reagent; blowing agent; epoxidation; hydroxylation; oxidation and reduction; viscosity control for starch and cellulose derivatives.

Caution! Strong oxidant. Avoid contact with skin or eyes. Fire and explosion hazard with concentrations greater than 65%. MCA warning label.

Shipping regulations: Corrosive liquid. White label, for solutions containing over 8% hydrogen peroxide (H_2O_2) strength by weight.*

hydrogen phosphide. See phosphine.

hydrogen, phosphoretted. See phosphine.

hydrogen sulfate. See sulfuric acid.

hydrogen sulfide (sulfuretted hydrogen) H_2S.

Properties: Colorless, flammable gas; offensive odor; sweetish taste; dangerously poisonous! Soluble in water and alcohol; sp. gr. 1.1895; m.p. −83.8°C; b.p. −60.2°C.

Derivation: (a) By the action of dilute sulfuric acid on a sulfide, usually iron sulfide; (b) by direct union of hydrogen and sulfur vapor at a definite temperature and pressure; (c) by heating sulfur with paraffin wax; (d) as a by-product of petroleum refining.

Containers: Usually prepared as wanted; also shipped in steel cylinders.

Grades: Technical 98.5%; purified 99.5% min.

Uses: Purification of hydrochloric and sulfuric acids; precipitating sulfides of metals; reagent in analytical chemistry; manufacture of elementary sulfur.

Danger: Poison liquid and gas under pressure. Flammable. MCA warning label.

Shipping regulations: Flammable gas. Red gas label.*

hydrogen tellurate. See telluric acid.

"Hydrogum." [36] Ester gum made with hydrogenated rosin.

Properties: Color WW to WG (U.S. Department of Agriculture rosin standards); acid number 4 to 6; melting range 120 to 132°F (capillary tube method). Insoluble in alcohol; soluble cold in acetates, coal-tar solvents, turpentine, and drying oils. Imparts fair resistance to abrasion, water, and weather.

Uses: Varnishes; oleoresinous vehicles; adhesives; chewing gum; etc.

hydrohydrastinine $C_{11}H_{13}NO_2$.

Properties: White crystalline alkaloid; poisonous! Soluble in alcohol and water. M.p. 66°C; b.p. 303°C.

Derivation: By extraction of Hydrastis canadensis and subsequent crystallization.

Grades: Technical.

Containers: Glass bottles.

Use: Medicine.

Shipping regulations: None.*

hydrol. See tetramethyldiaminobenzhydrol.

hydrolase (hydrase). An enzyme which catalyzes the removal of water from the substrate.

"Hydrolin." [413] Trademark for an ammonium nitrate base blasting agent which requires specially constructed primers for detonation.

Containers: Tin cans, 4½", 5½" and 8" diameters; 16⅔-lb., 25-lb, 40-lb and 50-lb. weights; 4½" and 5½" sizes packed in 50-lb. shipping cases.

Use: For seismic prospecting at sea.

Fire Hazard: Dangerous.

Shipping regulations: Oxidizing material. Yellow label.*

hydrolubes. Water-glycol base, non-flammable hydraulic fluids.

"Hydrolux." [42] Proprietary product. Stabilized reducing agent.

Properties: Colorless, syrupy liquid. Completely soluble in water.

Containers: 55-gal steel drums.
Uses: Assistant in vat and sulfur color dyeing and printing. Used in yarn dyeing on same types of colors, as extender and stabilizer for reducing agents.

hydrolysis. A chemical reaction in which water acts upon another substance to form one or more entirely new substances. Examples are the conversion of starch to glucose by water in the presence of suitable catalysts; the conversion of sucrose (cane sugar) to glucose and fructose by reaction with water, again in the presence of an enzyme or acid catalyst; the conversion of natural fats into fatty acids and glycerin by reaction with water in one process of soap manufacture; and the reaction of the ions of a dissolved salt to form various products, such as acids, complex ions, etc.

"Hydromagma." [123] Trademark for magnesium hydroxide paste.

"Hydron." [307] Sulfide blues derived from carbazole.

hydrophilic. Having an affinity for water; capable of uniting with or dissolving in water.

hydrophobic. Having an antagonism to water; not capable of uniting or mixing with water.

"Hydropol." [303] Trade name for a partially hydrogenated polybutadiene. This rubbery polymer, when properly compounded with vulcanizing ingredients, has its major use in gaskets, O-rings, etc., for use in rockets. Such compounds have shown good resistance to unsymetrical dimethylhydrazine and hydrazine rocket fuels. Another application for Hydropol compounds is as a material for bonding polyethylene to rubber, brass, or brass-plated metals.

hydroponics. Cultivation of plants using solutions of inorganic salts instead of earth. See nutrient solution.

"Hydro-Pruf." [300] Trademark for a silicone water repellent for fabrics. Applied with a catalyst at high curing temperatures.

hydroquinine hydrochloride $C_{20}H_{26}O_2N_2 \cdot HCl \cdot 2H_2O$.
Properties: White, crystalline, alkaloid salt; m.p. 235°C (anhydrous). Soluble in alcohol and water.
Derivation: Extraction from Cinchona.
Uses: Medicine; raw material for manufacture of cinchona alkaloids.
Shipping regulations: None.*

hydroquinol. See hydroquinone.

hydroquinone (quinol; hydroquinol; para-dihydroxybenzene) $C_6H_4(OH)_2$.
Properties: White crystals; soluble in water, alcohol and ether. Sp. gr. 1.330; m.p. 170°C; b.p. 285°C.
Derivation: Aniline is oxidized to quinone by manganese dioxide and is then reduced to hydroquinone.
Method of purification: Crystallization.
Grades: Technical; photographic.
Containers: Glass bottles; multiwall paper sacks; fiber drums.
Uses: Photographic developer; dye intermediate; medicine; antioxidant, inhibitor; stabilizer in decorating ceramic ware, in coating compounds for rubber, stone, and textiles, and in paints and varnishes, motor fuels and oils; antioxidant for fats and oils; inhibitor of polymerization.
Shipping regulations: None.*

hydroquinone, benzyl ethers. Monobenzyl ether of hydroquinone ($C_6H_5CH_2OC_6H_4OH$); dibenzyl ether of hydroquinone ($C_6H_5CH_2OC_6H_4OCH_2C_6H_5$).
Typical specifications:
Monobenzyl ether: Tan powder; m.p. 110°C (min); purity 90% (min); sp. gr. 1.26; ash 0.25% (max).
Dibenzyl ether: Tan powder; m.p. 119°C (min); purity 90% (min).
Insoluble in water; soluble in acetone, benzene, and chlorobenzene.
Grade: Commercial.
Uses:
Monobenzyl ether: Stabilizer, antioxidant, polymerization inhibitor and in organic synthesis.
Dibenzyl ether: Solvent and in perfumes, soap, plastics and pharmaceuticals.
Handle with caution!

hydroquinone di-n-butyl ether (1,4-dibutoxybenzene) $C_6H_4[O(CH_2)_3CH_3]_2$.
Typical specifications: White flakes with no appreciable odor; melting point 45-46°C; b.p. 124°C (1.3 mm), 158°C (15.0 mm); insoluble in water; soluble in benzene, acetone, ethyl acetate, and alcohol.

hydroquinone diethyl ether (1,4-diethoxybenzene) $C_6H_4(OC_2H_5)_2$.
Properties: White, granular solid with anise-like odor; b.p. 246°C; neither boiling caustic nor acid solution cause any hydrolysis; ability to absorb ultraviolet light; insoluble in water; soluble in benzene, acetone, ethyl acetate, and alcohol.
Typical specifications: Melting point 70-71°C; b.p. 234°C.

hydroquinone, di(beta-hydroxyethyl) ether (para-di-[2-hydroxyethoxy]benzene) $C_6H_4(OC_2H_4OH)_2$.
Properties: White solid. M.p. 94-96°C, b.p. 185-200°C (0.3 mm). Slightly soluble in water and most organic solvents; miscible with water at 80°C.
Uses: Reactant in preparation of polyesters, polyolefins, polyurethanes; reactant in preparation of hard waxy resins; raw material in organic synthesis.

hydroquinone dimethyl ether (1,4-dimethoxybenzene; dimethyl hydroquinone) $C_6H_4(OCH_3)_2$.
Properties: White flakes with sweet clover odor; b.p. 213°C; m.p. 56°C; density 1.0293 g/ml (65°C); viscosity 1.04 cps (65°C); dielectric constant 2.8; absorbs

*See "I.C.C. Shipping Regulations," page xiii.
Reference numbers refer to name of manufacturer. See "List of Manufacturers," page v.

ultraviolet light in range 2800-3100 A. Insoluble in water; soluble in benzene and alcohol.
Containers: Glass bottles; fiber drums.
Uses: Weathering agent in paints and plastics; fixative in perfumes; dyes; resin intermediate; cosmetics, especially sun-tan preparations.

hydroquinone glucose. See arbutin.

hydroquinone mono-n-butyl ether $CH_3(CH_2)_3OC_6H_4OH$.
Typical specifications: White flakes; m. p. 64-65°C; b. p. 115°C (1.4 mm); insoluble in water; soluble in benzene, acetone, ethyl acetate, and alcohol.

hydroquinone monoethyl ether (4-ethoxyphenol) $C_2H_5OC_6H_4OH$.
Typical specifications: White solid; m. p. 63-65°C; b. p. 246-247°C; slightly soluble in water; soluble in benzene, acetone, ehtyl acetate, and alcohol.
Uses: See hydroquinone monomethyl ether.

hydroquinone monomethyl ether (4-methoxyphenol; para-hydroxyanisole) $CH_3OC_6H_4OH$.
Properties: White to tan flakes; m. p. 52.5°C; b. p. 243°C; sp. gr. (20/20°C) 1.55. Slightly soluble in water; readily soluble in benzene, acetone, ethyl alcohol, and ethyl acetate.
Containers: Glass bottles; fiber drums.
Uses: Manufacture of antioxidants, pharmaceuticals, plasticizers, dyestuffs; and inorganic synthesis; stabilizer for chlorinated hydrocarbons and ethyl cellulose; inhibitor for acrylic monomers and acrylonitriles; ultraviolet inhibitor.

hydrosilicofluoric acid. See fluosilicic acid.

"Hydrosol." [232] Brand name for a proprietary product of the hydrosulfite class for wool bleaching.

hydrosulfite. This term refers to a sodium hydrosulfite (q. v.).

hydrosulfite-formaldehyde compounds. This term usually refers to mixtures of sodium formaldehyde hydrosulfite and sodium formaldehyde bisulfite, used as discharges and stripping or reducing agents in dyeing and other textile operations. In some cases the zinc derivatives are used. Derivation is by the action of formaldehyde on aqueous sodium hydrosulfite, or from zinc, formaldehyde, sulfur dioxide and sodium hydroxide.

"Hydrotan." [236] Brand name for a proprietary product. A soluble caustic-tannin compound containing one part caustic soda and two parts tannin for the alkaline tannate treatment of drilling muds. Allows greater accuracy and economy in treatment and eliminates the hazards of handling raw caustic soda.

"Hydrotex." [354] A highly concentrated paraffin emulsion, stable to hard water, acids, and salts. Used for waterproofing textiles, cardboard, leather, etc. in one-bath processes.

"Hydro-T-Metal." [135] An alloy containing 0.2-0.7% copper, 0.08-0.160% titanium, 0.002-0.010% manganese, 0.003-0.020% chromium, the remainder being zinc.
Properties: M. p. 792°F; tensile strength 24,000 psi (min); specific heat 0.096 Btu/lb/°F; density 0.258 lb/cu. in.
Derivation: From high grade slab zinc.
Grades: Hydro-T-Metal 100 and 200.
Containers: Boxes, crates and skids.
Uses: Hydro-T-Metal 100: Flashings; roofing, coping covers, cavity walls, bay windows, gutters, gravel stops, roof aprons, water table, termite shields, sidings, shingles, corrugated sheet etc. Hydro-T-Metal 200: Industrial application where a non-rusting, easily soldered metal required; can be spun, drawn and formed.

hydrotropes (hydrotrophes). Chemicals which have the property of increasing the aqueous solubility of various slightly soluble organic chemicals. Used especially in the formulation of liquid detergents.

hydrotrophes. See hydrotropes.

hydrous. This term is commonly and loosely used of materials to indicate the presence of an indefinite amount of water. In the case of certain minerals and some other compounds, it is used to mean the presence of a definite proportion of combined water. Thus certain hydroxides regularly are referred to as hydrous oxides by mineralogists, and minerals containing water of hydration are referred to as hydrous forms. The term hydrous oxide is also used to refer to oxides of aluminum, iron, etc. as they are precipitated, with indefinite amounts of water, from their aqueous solutions. In any case the term hydrous should be used and interpreted with caution.

hydrox fuel cell. See fuel cells.

hydroxocobalamin. See vitamin B_{12}.

hydroxy-acetal. The complete chemical name is hydroxycitronellal dimethyl acetal. This is an aromatic chemical used in the perfume industry. It produces a lily-like odor and can also be used in lilac, orange flower and many other compositions for the toilet-goods trade. It has the advantage of holding up in the presence of alkalies and does not cause any stinging or smarting when used in cold creams, lipsticks, or other toilet preparations.

para-**hydroxyacetanilide.** See para-acetylaminophenol.

hydroxyacetic acid. See glycolic acid.

hydroxyacetic acid, technical (70%) $HOCH_2COOH$. For the pure acid, see glycolic acid.
Properties: Light straw-colored liquid containing approximately 70% hydroxyacetic

acid with traces of other organic acids. Odor like burnt sugar; sp. gr. 1.27; m.p. 10°C; soluble in water, methanol, acetone, acetic acid; insoluble in hydrocarbons, ethers, esters, higher ketones.
Containers: 50-gal wooden barrels; 6000- and 10,000-gal tank cars.
Uses: In leather dyeing and tanning; textile dyeing; manufacture of cleaning, polishing, and soldering compounds; copper pickling; adhesives; in electroplating; breaking of petroleum emulsions; preparation of adhesive emulsions; with citric acid as a chelating agent for iron.

hydroxyacetone. See acetol.

ortho-**hydroxyacetophenone** $C_6H_4(OH)COCH_3$.
Properties: Greenish-yellow liquid with minty odor; sp. gr. (20.8°C) 1.1307; b.p. (717 mm) 213°C; refractive index (n 20/D) 1.5580; slightly soluble in water.

2-hydroxyadipaldehyde
$OHCCH_2CH_2CH_2CHOHCHO$.
Properties: (25% aqueous solution): Sp. gr. (20°C) 1.066; b.p. 37°C (50 mm); vapor pressure (20°C) 17 mm; f.p. −3.5°C; pH approx 3.0.
Containers: 55-gal drums.
Uses: Intermediate; insolubilizing agent for proteins and polyhydroxy materials; crosslinking agent for polyvinyl compounds; shrinkage control agent (textiles).

beta-**hydroxyalanine.** See serine.

5-hydroxy-3-(beta-**aminoethyl) indole.** See serotonin.

hydroxyamphetamine [para-(2-aminopropyl) phenol] $HOC_6H_4CH_2CHNH_2CH_3$.
Properties: Crystals with m.p. 125-126°C. Soluble in water, alcohol, chloroform, and ethyl acetate.
Derivation: From para-nitrobenzyl chloride and a salt of nitroethane or from anisaldehyde and nitroethane.
Use: Medicine.

hydroxyamphetamine hydrobromide (para-(2-aminopropyl) phenol hydrobromide) $HOC_6H_4CH_2CH(CH_3)NH_2 \cdot HBr$.
Properties: White, crystalline solid with faint odor. M.p. 189-192°C. Very soluble in water; freely soluble in alcohol; practically insoluble in benzene and ether; pH (2% solution) 4.5-5.5.
Grade: U.S.P. XVI.
Use: Medicine.

hydroxyanilines. See aminophenols.

ortho-**hydroxyanisole.** See guaiacol.

para-**hydroxyanisole.** See hydroquinone monomethyl ether.

9-hydroxyanthracene. See anthranol.

hydroxyapatite. See apatite.

para-**hydroxyazobenzene**-para-**sulfonic acid.** $HOC_6N_4NNC_6H_4SO_3H$.
Properties: Orange red crystals; very soluble in water.
Uses: Analytical reagent; precipitant for numerous organic bases.

meta-**hydroxybenzaldehyde** HOC_6H_4CHO.
Properties: Orange pink crystals; m.p. 101.5°C; slightly soluble in cold water; very soluble in hot water and aromatic hydrocarbons.
Uses: Intermediate for dyes, plastics, pharmaceuticals and bactericides; color reagent for Schiff's reagent; sensitizing agent in photographic emulsions.

ortho-**hydroxybenzaldehyde.** See salicylaldehyde.

para-**hydroxybenzaldehyde** HOC_6H_4CHO.
Properties: Colorless needles; soluble in alcohol, ether, or water; sp. gr. 1.129; m.p. 116°C (sublimes).
Use: Pharmaceuticals.

ortho-**hydroxybenzamide.** See salicylamide.

hydroxybenzene. See phenol.

ortho-**hydroxybenzoic acid.** See salicylic acid.

meta-**hydroxybenzoic acid** $C_6H_4(OH)COOH$.
Properties: White powder; m.p. 200°C. Soluble in water and hot alcohol.
Uses: Intermediate for plasticizers; resins; light stabilizers; petroleum additives; pharmaceuticals.

para-**hydroxybenzoic acid** $C_6H_4(OH)COOH \cdot H_2O$.
Properties: Colorless crystals; soluble in alcohol, water, and in ether. M.p. 210°C.
Derivation: By the interaction of para-aminobenzoic acid and nitrous acid.
Method of purification: Crystallization.
Grades: Technical.
Containers: Drums.
Uses: Intermediates; synthetic drugs.
Shipping regulations: None. *

2-(4-hydroxybenzoyl) benzoic acid
$C_6H_4(COOH)COC_6H_4OH$. Off-white or slightly yellow crystalline powder; m.p. 209-216°C. Used as an intermediate.

ortho-**hydroxybenzyl alcohol.** See salicyl alcohol.

hydroxybrasilin. See hematoxylin.

3-hydroxy-2-butanone. See acetylmethylcarbinol.

beta-**hydroxybutyraldehyde.** See aldol.

beta-**hydroxybutyric acid**
$CH_3CH(OH)CH_2COOH$.
Properties: Viscid, yellow mass; m.p. 48-50°C; b.p. 130°C (12 mm); very soluble in water, alcohol, and ether.
Derivation: By the interaction of acetoacetic acid and sodium amalgam.
Grades: Technical; reagent.
Containers: Wooden kegs.
Use: Intermediates.
Shipping regulations: None. *

hydroxy-beta-**carotene.** See cryptoxanthin.

1-hydroxy-4-chloro-2-nitrobenzene. See 4-chloro-2-nitrophenol.

hydroxychloroquine sulfate $C_{18}H_{26}ClN_3O \cdot H_2SO_4$. 7-Chloro-4-{4-[ethyl(2-hydroxyethyl)-amino]-1-methylbutylamino} quinoline sulfate.
Properties: White, crystalline, odorless powder. Bitter taste; pH of solutions 4.5. Freely soluble in water. Insoluble in alcohol, chloroform, ether. Exists in two forms: m.p. usual form 240°C; m.p. other form 198°C.
Grade: U.S.P. XVI.
Use: Medicine.

3-β-hydroxycholestane. See dihydrocholesterol.

hydroxycinchonine. See cupreine.

hydroxycitronellal (citronellal hydrate; 3,7-dimethyl-7-hydroxyoctanal; synthetic muguet) $C_{10}H_{20}O_2$.
Properties: Viscous, colorless or faintly yellow liquid; sweet lily-type odor; sp. gr. 0.925-0.930 (15°C); refractive index (n 20/D) 1.448-1.450; optical rotation (Java type) +9 to +10.5°; (Eucalyptus citriodora type) +0.5 to −0.5°; boiling range 94-96°C (1 mm). Soluble in alcohol (50%), fixed oils; slightly soluble in water, glycerol, and mineral oil.
Derivation: Hydration of citronellal (Java citronella or Eucalyptus citriodora).
Containers: Cans.
Uses: Perfumery (fixative, muguet odor). See also "Laurine."

hydroxycitronellal dimethyl acetal. See hydroxy-acetal.

hydroxycitronellal methyl anthranilate Schiff base. See "Aurantiol."

17-hydroxycorticosterone. See hydrocortisone.

2-hydroxy-para-cymene. See carvacrol.

3-hydroxy-para-cymene. See thymol.

1-hydroxy-2,4-diamylbenzene. See diamyl phenol.

2-hydroxydibenzofuran $HOC_{12}H_7O$.
Properties: Powder; m.p. 128-134°C.
Grade: Technical.
Use: Synthesis.

2-hydroxydibenzofuran-3-carboxylic acid $HOC_{12}H_6OCOOH$.
Properties: Powder; m.p. 268°C.
Grade: Technical.
Use: Synthesis.

14-hydroxydihydromorphinone. See oxymorphone.

hydroxydimethylbenzene. See xylenol.

7-hydroxy-3,7-dimethyloctan-1-al. See "Laurine."

2′-hydroxy-5,9-dimethyl-2-(2-phenethyl)-6,7-benzomorphan hydrobromide. See phenazocine.

hydroxydione sodium (hydroxydione sodium succinate) $C_{25}H_{35}NaO_6$. Sodium 21-hydroxy-pregnane-3,20-dione succinate. A steroid compound; white powder; m.p. 193°C with decomposition. Soluble in slightly alkaline solutions; in acetone and chloroform.
Grade: N.N.D.
Use: Medicine.

hydroxydiphenyl. See phenylphenol.

para-**hydroxydiphenylamine** (anilinophenol) $C_6H_5NHC_6H_4OH$.
Typical specifications: Gray solid leaflets; m.p. 50°C (approx); purity 98% (min); distillation range 155-210°C (3 mm); insoluble in water; soluble in alcohol, ether, acetone, chloroform, alkali, and benzene.
Handle with caution!

hydroxydiphenylmethanes. See benzylphenols.

ortho-**(2-hydroxy-3,6-disulfo-1-naphthylazo)-benzenearsonic acid** (thorin) $HOC_{10}H_4(SO_3H)_2NNC_6H_4AsO_3H_2$. A reagent for the colorimetric determination of microgram quantities of thorium.

2-hydroxyethanesulfonic acid. See isethionic acid.

hydroxyethylacetamide. See N-acetyl ethanolamine.

2-hydroxyethylamine. See ethanolamine.

hydroxyethylcellulose. See also "Cellosize."
Properties: Nonionic, water soluble ether of cellulose. Stable in concentrated salt solutions; grease and oil resistant.
Containers: Fiber drums.
Uses: Thickening and suspending agent; stabilizer for vinyl polymerization.

hydroxyethylethylenediamine (aminoethyl-ethanol amine) $NH_2CH_2CH_2NHCH_2CH_2OH$.
Properties: Hygroscopic liquid. Mild, ammoniacal odor. Soluble in water. Sp. gr. 1.0304 (20/20°C); b.p. 243.7°C (760 mm); vapor pressure 0.01 mm (20°C); flash point 275°F; wt/gal 8.6 lbs (20°C).
Grades: Technical.
Containers: 1-gal glass jugs; 5-, 55-gal drums.
Use: Raw material for textile compounds, including antifuming agents, dyestuffs, resins, rubber products, insecticides, and certain medicinals.

hydroxyethylethylenediaminetriacetic acid $HOOCCH_2N(CH_2CH_2OH)CH_2CH_2N(CH_2COOH)_2$.
Properties: Soluble in water and methanol.
Grades: Technical.
Use: Chelating compound.

beta-**hydroxyethylhydrazine** (ethanolhydrazine) $HOCH_2CH_2NHNH_2$.
Properties: Colorless, slightly viscous liquid; sp. gr. (20°C) 1.11; m.p. −70°C; boiling range (25 mm Hg) 145-153°C; flash point 224°F; completely miscible with water; soluble in lower alcohols; slightly soluble in ether.
Grade: 70%.
Use: Intermediate.

N-hydroxyethylmorpholine. See N-morpholine ethanol.

N-hydroxyethyl piperazine $HOCH_2CH_2NCH_2CH_2NHCH_2CH_2$.
Properties: Liquid; sp. gr. 1.0614 (20/20°C); b.p. 246.3°C; f.p. −10°C; flash point 255°F. Miscible with water.
Uses: Intermediate for pharmaceuticals, anthelmintics, surface active agents, and synthetic fibers.

N-2-hydroxyethylpiperidine. See 2-piperidinoethanol.

hydroxyethyltrimethylammonium bicarbonate $(CH_3)_3NCH_2CH_2OH \cdot HCO_3$. A quaternary ammonium compound. Sp. gr. 1.0965 (25/4°C); 9.15 lbs/gal (68°F); b.p., decomposes. Miscible with water.
Uses: Catalyst; intermediate.

beta-**hydroxyethyltrimethylammonium hydroxide.** See choline.

1-hydroxyfenchane. See fenchyl alcohol.

alpha-**hydroxyisobutyronitrile.** See acetone cyanhydrin.

4-hydroxy-2-keto-4-methylpentane. See diacetone alcohol.

hydroxylamine (oxammonium) NH_2OH. The free base is unstable.
Properties: Colorless crystals; decomposes when heated and explodes at 130°C. Soluble in alcohol, acids and cold water. Sp. gr. 1.227; m.p. 33°C; b.p. 70°C.
Derivation: By decomposing hydroxylamine hydrochloride or sulfate with a base and distilling in vacuo.
Method of purification: Redistillation.
Containers: Lead-lined steel drums.
Uses: Reducing agent; organic synthesis.
Shipping regulations: None.*

hydroxylamine acid sulfate (hydroxylammonium acid sulfate; HAS) $NH_2OH \cdot H_2SO_4$.
Properties: White to brown crystalline solid; commercial grade is wet with free sulfuric acid; soluble in water and methanol; slightly soluble in alcohol. Wt/gal 15-16 lbs (20°C); m.p. indefinite; pH of 0.1 M aqueous solution 1.6.
Containers: 5-, 15-gal drums; 1-gal cans;
Uses: Unhairing agent; photographic developer; purification agent for aldehydes and ketones; synthesis of dyes, pharmaceuticals; rubber chemicals; analytical reagent.

hydroxylamine hydrochloride (hydroxylammonium chloride) $NH_2OH \cdot HCl$.
Properties: Colorless hygroscopic crystals; soluble in water, glycerol and alcohol; insoluble in ether. M.p. 155°C; b.p., decomposes; pH of 0.1 M aqueous solution 3.4.
Derivation: Reduction of ammonium chloride, frequently by electrolysis.
Method of purification: Crystallization.
Grades: Technical; C.P.
Containers: Glass bottles; 1-gal cans; 25-, 100-lb fiber drums.
Uses: Organic synthesis; photographic developer; medicine; controlled reduction reactions.

hydroxylamine sulfate (HS; hydroxylammonium sulfate) $(NH_2OH)_2 \cdot H_2SO_4$.
Properties: Colorless crystals; solution has a corrosive action on the skin; m.p. 172°C dec; soluble in water; slightly soluble in alcohol.
Method of purification: Crystallization.
Grades: Technical.
Containers: Glass bottles; 1-gal cans; 40-, 100-, 400-lb fiber drums.
Uses: Unhairing agent; photographic developer; purification agent for aldehydes and ketones; chemical synthesis; reducing agent; textile chemical; oxidation inhibitor for fatty acids; catalyst; biological and biochemical research; making oximes for paints and varnishes; aid in rustproofing.
Shipping regulations: None.*

hydroxylammonium acid sulfate. See hydroxylamine acid sulfate.

hydroxylammonium chloride. See hydroxylamine hydrochloride.

hydroxylammonium sulfate. See hydroxylamine sulfate.

hydroxymercurichlorophenol (2-chloro-4-(hydroxymercuri) phenol) $C_6H_3Cl(HgOH)(OH)$. Insoluble in water and common organic solvents. Soluble in solutions of acids and alkalies with the formation of salts.
Use: A seed disinfectant.
Warning: Poisonous if inhaled or swallowed. May cause skin irritation. MCA warning label.

hydroxymercuricresol. A pesticide.
Warning! Poisonous if inhaled or swallowed. May cause skin irritation. MCA warning label.

ortho-**[(3-hydroxymercuri-2-methoxypropyl) carbamyl]-phenoxyacetic acid, sodium salt.** See mersalyl.

hydroxymercurinitrophenol. A pesticide.
Warning! Poisonous if inhaled or swallowed. May cause skin irritation. MCA warning label.

alpha-**hydroxy**-beta-**methylaminopropylbenzene.** See ephedrine.

3-hydroxy-3-methylbutanone-2 $CH_3COH(CH_3)C(O)CH_3$. Colorless liquid; sp. gr. 0.95 (25/25°C). Used as an intermediate.

7-hydroxy-4-methylcoumarin. See beta-methyl umbelliferone.

hydroxymethylethylene carbonate. See glycerin carbonate.

dℓ-alpha-**hydroxy**-gamma-**methylmercaptobutyric acid, calcium salt** (2-hydroxy-4-methylthiobutyric acid, calcium salt) $(CH_3SCH_2CH_2CHOHCOO)_2Ca$.
Properties: A free flowing light tan powder; soluble in water; insoluble in common organic solvents.
Uses: Animal feed; synthesis of pharmaceuticals.

*See "I.C.C. Shipping Regulations," page xiii.
Reference numbers refer to name of manufacturer. See "List of Manufacturers," page v.

2-hydroxymethyl-5-norbornene "Cyclol"; bicyclo (2, 2, 1)-hept-5-ene-2-methylol. $C_8H_{12}O$.
Properties: Stable, colorless liquid; a high-boiling solvent; miscible with most common solvents.
Containers: 55-gal drums.
Uses: Modification of condensation and addition polymers for coatings.

4-hydroxy-4-methylpentanone-2. See diacetone alcohol.

2-hydroxy-4-methylthiobutyric acid, calcium salt. See dl-alpha-hydroxy-gamma-methylmercaptobutyric acid, calcium salt.

hydroxynaphthalene. See naphthol.

3-hydroxy-2-naphthoic acid (beta-hydroxynaphthoic acid; 3-naphthol-2-carboxylic acid; beta-oxynaphthoic acid) $C_{10}H_6OHCOOH$.
Properties: Yellow rhombic leaflets; soluble in alcohol and ether. M. p. 217.5-219°C.
Derivation: By treating sodium 2-naphtholate with carbon dioxide under pressure.
Containers: Fiber drums.
Use: Dyes.
Shipping regulations: None.*

beta-**hydroxynaphthoic acid.** See 3-hydroxy-2-naphthoic acid.

beta-**hydroxynaphthoic anilide** (naphthol AS) $C_{10}H_6OHCONHC_6H_5$.
Properties: Cream-colored crystals. Sodium salt is soluble in water; m. p. 246.0°C.
Derivation: Condensation of beta-hydroxynaphthoic acid and aniline.
Method of purification: Recrystallized through sodium salt.
Use: Dyes.

2-hydroxy-1, 4-naphthoquinone $C_{10}H_5O_2(OH)$.
Properties: Yellow to orange-yellow needles or powder; m. p. 192-195°C (dec.). Redox potential 0.362 volts; dissociation constant 1.05×10^{-4}; soluble in glacial acetic acid, alcohol, and ether; slightly soluble in cold water, benzene, carbon tetrachloride and petroleum ether.
Uses: Intermediate for pharmaceuticals, henna hair and wool dye, bactericides; seed disinfectant.

3-(2-hydroxy-1-naphthylmethyl) salicylic acid. See epicarin.

4-hydroxy-3-nitrobenzenearsonic acid $HOC_6H_3(NO_2)AsO(OH)_2$.
Properties: Pale-yellow crystals.
Derivation: Treating para-hydroxyphenyl arsonate with nitric and sulfuric acids.
Use: Growth stimulator for chicks.

cis-**12-hydroxyoctadec-9-enoic acid.** See ricinoleic acid.

12-hydroxyoleic acid. See ricinoleic acid.

15-hydroxypentadecanoic acid lactone. See "Thibetolide."

3-hydroxyphenol. See resorcinol.

beta-para-**hydroxyphenylalanine.** See tyrosine.

para-**hydroxyphenylglycine** (glycine[photographic]; photo-glycin) $HOC_6H_4NHCH_2COOH$.
Properties: White to buff crystals or amorphous powder; m. p. 240°C (with decomposition); slightly soluble in water; soluble in alkaline solutions.
Derivation: By condensation of para-aminophenol with chloracetic acid.
Method of purification: Recrystallization.
Grades: Technical; photographic.
Containers: Barrels; bottles.
Uses: Photographic developer; cellulose and nitrocellulose acetate lacquers and varnishes.
Shipping regulations: None.*

2-hydroxyphenylmercuric chloride (chloromercuriphenol) HOC_6H_4HgCl.
Properties: White to faint pink, fine crystals; 0.1 parts in 100 soluble in water (25°C); soluble in hot water, alkali, and alcohol. M. p. 152°C.
Containers: Fiber drums; bottles.
Uses: Antiseptic; fungicide.
Caution! Dust or strong solution causes blistering of the skin unless washed off immediately.

1-(hydroxyphenyl)-2-methylaminoethanol hydrochloride. See phenylephrine hydrochloride.

11-alpha-**hydroxyprogesterone** $C_{21}H_{30}O_3$.
Properties: White crystalline powder; m. p. 163°C approx; specific rotation +179°; insoluble in water; soluble in alcohol.
Derivation: From progesterone by microbiological oxidation.
Use: A steroid intermediate.

17-alpha-**hydroxyprogesterone acetate.**
Grade: N. N. D.
Use: Medicine.

hydroxyprogesterone caproate (17-alpha-hydroxyprogesterone hexanoate)
Properties: White powder; m. p. 118-120°C; insoluble in water. Optical rotation 25°C/D +59° (1% in chloroform).
Grade: N. N. D.
Use: Medicine.

17-alpha-**hydroxyprogesterone hexanoate.** See hydroxyprogesterone caproate.

hydroxyproline HOC_4H_7NCOOH. (gamma Hydroxy-alpha-pyrrolidine carboxylic acid; 4-hydroxy-2-pyrrolidine carboxylic acid.)
Properties: Colorless crystals; very soluble in water; slightly soluble in alcohol; insoluble in ether; optically active.
DL-hydroxyproline, m. p. 261-262°C with decomposition.
L-hydroxyproline, m. p. 270°C (naturally occurring).
D-hydroxyproline, m. p. 274°C.
Derivation: Hydrolysis of protein (gelatin);

*See "I.C.C. Shipping Regulations," page xiii.
Reference numbers refer to name of manufacturer. See "List of Manufacturers," page v.

organic synthesis.
Use: Biochemical and nutrition investigations. Available commercially as L-hydroxyproline.

2-hydroxy-1, 2, 3-propane-tricarboxylic acid. See citric acid.

alpha-**hydroxypropionic acid.** See lactic acid.

alpha-**hydroxypropionitrile.** See lactonitrile.

beta-**hydroxypropionitrile.** See ethylene cyanohydrin.

hydroxypropylglycerin.
Properties: Pale straw-colored liquid; sp. gr. 1.084 (25/25°C); refractive index 1.459 (25°C); flash point 380°F; pour point −23°C; soluble in water and methanol.
Uses: Intermediate to alkyd resins and polyesters; plasticizer for cellulosics, glue, starch and many resins.

hydroxypropyl methylcellulose (methyl cellulose, propylene glycol ether).
Properties: White fibrous or granular powder. Swells in water producing clear to opalescent, viscous, colloidal solution. Insoluble in anhydrous alcohol, in ether, and in chloroform.
Grade: N. F. XI.
Use: Medicine (suspending agent).

2-(alpha-**hydroxypropyl) piperidine.** See conhydrine.

N-beta-**hydroxypropyl**-ortho-**toluidine** $CH_3C_6H_4NHCH_2CH(OH)CH_3$.
Typical specifications: Color, amber; distillation range, 170-180°C (20 mm); sp. gr. 1.035-1.045 (20/20°C); refractive index 1.540-1.550 (20°C).
Grade: Technical.
Containers: Fiber drums, 250-lbs net.
Use: Dye intermediate.

2-hydroxypyridine-N-oxide. Bactericidal agent related to aspergillic acid; made from pyridine-N-oxide.

8-hydroxyquinoline (8-quinolinol; oxyquinoline; oxine) C_9H_6NOH.
Properties: White crystals or powder; darkens when exposed to light; technical material usually tan; almost insoluble in water; soluble in alcohol, acetone, chloroform, benzene, also in formic, acetic, hydrochloric and sulfuric acids, and in alkalies; phenolic odor; m. p. 73-75°C; b. p. 267°C.
Grades: C. P.; technical.
Uses: For precipitating and separating metals; preparation of fungicides.

8-hydroxyquinoline benzoate $C_6H_5COOC_9H_6N$. Fungicide effective against Dutch elm disease.

8-hydroxyquinoline sulfate (8-quinolinol sulfate; oxyquinoline sulfate) $(C_9H_7NO)_2 \cdot H_2SO_4$.
Properties: Pale yellow powder; slight saffron odor; burning taste. Soluble in water; slightly soluble in alcohol; insoluble in ether.
Use: Antiseptic, antiperspirant, deodorant.

8-hydroxyquinoline-5-sulfonic acid $C_9H_5N(OH)SO_3H$.
Properties: Pale yellow, needlelike crystals or powder; soluble in water; slightly soluble in organic solvents; m. p. 213°C with decomposition.

4-hydroxysalicylic acid. See beta-resorcylic acid.

12-hydroxystearic acid $CH_3(CH_2)_5(CHOH)(CH_2)_{10}COOH$. A C_{18} straight chain fatty acid with an -OH group attached to the carbon chain; m. p. 79-82°C. It is produced by hydrogenation of ricinoleic acid.
Containers: 100-lb bags.
Uses: Lithium greases; chemical intermediates.

1, 12-hydroxystearyl alcohol (1, 12-octadecanediol). A long-chain fatty alcohol made by reduction of 12-hydroxystearic acid by replacing the -COOH group with a $-CH_2OH$.
Typical specifications: M. p. 69°C; boiling range 315-335°C.
Impurities: Stearyl alcohol.
Derivation: Hydrogenated castor oil.
Uses: Chemical intermediate; synthetic fibers; organic synthesis; pharmaceuticals; surface-active agents; plastics and resins; protective coatings.

hydroxystilbamidine isethionate (2-hydroxy-4, 4'-stilbenedicarboxamidine diisethionate) $C_{20}H_{28}N_4O_9S_2$.
Properties: Fine, yellow, crystalline powder. Odorless; stable in air but affected by light; pH of 1 in 100 solution 3.3-5.3; m. p. 280°C; soluble in water; slightly soluble in alcohol; insoluble in ether.
Grade: U. S. P. XVI.
Use: Medicine.

2-hydroxy-4, 4'-stilbenedicarboxamidine diisethionate. See hydroxystilbamidine isethionate.

hydroxysuccinic acid. See malic acid.

hydroxytitanium stearate $HO[(C_{18}H_{35}OO)_2TiO]_xH$.
Properties: Waxy solid; m. p., decomposes.
Derivation: Reaction of titanium esters with stearic acid.
Uses: Dispersant; cross-linking agent; water repellent.

alpha-**hydroxytoluene.** See benzyl alcohol.

hydroxytoluic acid. See cresotic acid.

5-hydroxytryptamine. See serotonin.

hydroxyzine hydrochloride $C_{21}H_{27}ClN_2O_2 \cdot 2HCl$. 1-(para-Chlorobenzhydryl) 4-[2-(2-hydroxyethoxy)ethyl]-piperazine dihydrochloride.
Properties: Occurs as white, odorless powder. Very soluble in water; freely soluble in acetone and alcohol; fairly soluble in chloroform, and practically insoluble in ether; melting range 196-204°C (dec.).
Grade: N. F. XI.
Use: Medicine.

"Hydrozin." [78] Trademark for normal zinc formaldehyde sulfoxylate used for discharge printing on acetate grounds and for stripping wool, acetates and nylon. Also used as a catalyst for polymerization of vinyl monomers.

hydrozincite (zinc bloom) $Zn_5(OH)_6(CO_3)_2$. A natural basic carbonate of zinc, found in the upper zones of zinc deposits.
Properties: Color white to gray or yellowish; luster dull to silky; fluorescent in ultraviolet light; sp. gr. 3.5-4.0; hardness 2.0-2.5.
Occurrence: Missouri, Pennsylvania, Utah, California, Nevada; Europe.
Use: An ore of zinc.

"Hyfac." [242] Trademark for a line of products including hydrogenated fish and tallow fatty acids, hydrogenated fish oil, hydrogenated castor oil, hydrogenated tallow, and 12-hydroxystearic acid.

"Hyform" Emulsions. [57] Trademark for water emulsions of pure paraffin wax, microcrystalline wax, or a modification of one of these waxes.
Uses: In the ceramic field these emulsions are used as binders for pressed pieces, lubricants for die or mold release, and plasticizers during mold forming.

"Hylene." [28] Trademark for a line of organic isocyanates.
"Hylene" M. Methylene bis(4-phenyl isocyanate) $(C_6H_4NCO)_2CH_2$. Pale yellow lumps; sp.gr. 1.20.
Containers: 45- and 500-lb drums.
Use: For making urethane adhesives and polymers.
"Hylene" M-50. Methylene bis(4-phenyl isocyanate) $(C_6H_4NCO)_2CH_2$; 50% solution in monochlorobenzene; dark brown liquid; sp. gr. 1.16.
Containers: 45- and 500-lb drums.
Use: In elastomer based adhesives.
"Hylene" MP. Bis phenol adduct of methylene bis(4-phenyl isocyanate).
Properties: Grayish white powder; water stable.
Containers: 125-lb drums.
Use: A bonding agent and adhesive assistant for adhering "Dacron" polyester fiber to rubber compositions in water emulsion or latex systems.
"Hylene" T. Toluene-2,4-diisocyanate $CH_3C_6H_3(NCO)_2$.
Properties: Water white to pale yellow liquid; sp. gr. 1.22.
Containers: 45- and 500-lb drums; tank trailers and tank cars.
Uses: In manufacture of urethane foam, finishes, adhesives and rubbers.
"Hylene" TM. Mixture of 80% toluene-2,4-diisocyanate and 20% toluene-2,6-diisocyanate $CH_3C_6H_3(NCO)_2$.
Properties: Sp. gr. 1.22; water white to pale yellow liquid.
Containers: 45- and 500-lb drums; tank trailers; tank cars.
Uses: In manufacture of urethane foam, finishes, adhesives and rubbers.
"Hylene" TM-65. Mixtures of 65% toluene 2,4-diisocyanate, and 35% toluene 2,6-diisocyanate (see "Hylene" TM).
Properties: Sp. gr. 1.22; water white to pale yellow liquid.
Containers: 45- and 500-lb drums; tank trailers; tank cars.
Uses: In manufacture of urethane foam, finishes, adhesives and rubber.

"Hyonic." [309] Trademark for a line of liquid detergents, including various detergent bases, synergists, and wetting agents.

hyoscine (ℓ-scopolamine) $C_{17}H_{21}NO_4$.
Properties: Thick, colorless, syrupy liquid alkaloid; poisonous! Soluble in water, alcohol, and ether. M.p. 50-59°C.
Derivation: By extraction of various Solanaceae, hyoscyamus, belladonna, etc.
Grades: Technical.
Containers: Glass bottles.
Use: Medicine (in form of hydrobromide U.S.P. XVI, hydrochloride, hydroiodide, sulfate, etc.).

hyoscine hydrobromide (scopolamine hydrobromide) $C_{17}H_{21}NO_4 \cdot HBr \cdot 3H_2O$.
Properties: White crystals or white, granular powder; odorless; poisonous! Soluble in water and alcohol; insoluble in ether; slightly soluble in chloroform. M.p. 191°C.
Derivation: By the action of hydrobromic acid on hyoscine.
Method of purification: Crystallization.
Grades: Technical; U.S.P. XVI.
Containers: 5-, 15-grain vials; glass bottles.
Use: Medicine.

hyoscine sulfate $(C_{17}H_{21}NO_4)_2 \cdot H_2SO_4 \cdot 2H_2O$.
Properties: White crystals; poisonous! Soluble in water and alcohol.

hyoscyamine $C_{17}H_{23}O_3N$ (isomeric with atropine).
Properties: White crystalline alkaloid; poisonous! Slightly soluble in water; soluble in alcohol, ether, and dilute acids. M.p. 108.5°C.
Derivation: By extraction of belladonna or scopola roots, and subsequent crystallization.
Use: Medicine (in form of hydrobromide, hydrochloride or sulfate).

hyoscyamine hydrobromide $C_{17}H_{23}NO_3 \cdot HBr$.
Properties: White odorless crystals or crystalline powder; poisonous; affected by light. M.p. 149-152°C; specific rotation (25°C) not less than −24°. Soluble in water and alcohol.
Derivation: By the action of hydrobromic acid on hyoscyamine.
Method of purification: Crystallization.
Grades: Technical; N.F. XI.
Containers: Vials; bottles.
Use: Medicine.

hyoscyamine hydrochloride $C_{17}H_{23}NO_3 \cdot HCl$.
Properties: White crystals; poisonous!

Soluble in water and alcohol; m. p. 149-151°C.
Derivation: By the action of hydrochloric acid on hyoscyamine.
Containers: Vials.
Use: Medicine.
Shipping regulations: None. *

hyoscyamine sulfate $(C_{17}H_{23}NO_3)_2 \cdot H_2SO_4 \cdot 2H_2O$.
Properties: White, odorless crystals or crystalline powder; deliquescent; affected by light. Soluble in water and alcohol.
Caution! Hyoscyamine sulfate is extremely poisonous.
Constants: M. p. not less than 200°C; specific rotation (25°C) not less than −24°.
Grades: Technical; N. F. XI.
Containers: Vials; bottles.
Use: Medicine.

hyoscyamus (henbane; hog's bean; insane root; poison tobacco; black henbane). Dried leaves and flowering tops of Hyoscyamus niger.
Chief constituents: Alkaloids hyoscyamine and hyoscine.
Habitat: Europe; Asia; United States; cultivated in England.
Grades: Technical; N. F. XI.
Containers: Bags; bales.
Uses: Medicine (similar to belladonna); source of alkaloids.
Shipping regulations: None.*

"Hypalon" 20. [28] Trademark for chlorosulfonated polyethylene.
Properties: White chips; sp. gr. 1. 10.
Containers: 50-lb bags or boxes.
Use: An elastomer giving products which are oil resistant, unattacked by ozone, have excellent color stability and resistance to heat, weather and chemicals.

"Hypalon" 30. [28] Trademark for chlorosulfonated polyethylene.
Properties: White chips; sp. gr. 1. 28.
Containers: 50-lb boxes.
Uses: A solution type elastomer designed for protective and decorative coatings, a paint base. Coatings are resistant to weathering, ozone, many oils and chemicals.

"Hypalon" 40. [28] Trademark for a chlorosulfonated polyethylene.
Properties: White chips; sp. gr. 1. 18.
Use: An elastomer designed for ease in processing.

"Hypaque" Sodium. [162] Trademark for diatrizoate sodium.

hyperglycemic-glycogenolytic factor. See glucagon.

hypergolic fuels. Rocket fuels or propellants which consist of combinations of fuels and oxidizers, which ignite spontaneously on contact. Examples are: hydrazine hydrate, methanol and hydrogen peroxide, aniline and nitric acid.

hypernic extract. See brasilin.

hyperon. The name of a group of fundamental particles (q. v.). A hyperon is a fundamental particle with a mass greater than the mass of a nucleon.

hypersorption. Process in which activated carbon selectively adsorbs less volatile components from a gaseous mixture, while the more volatile components pass on unaffected. Particularly applicable to separations of low-boiling mixtures such as hydrogen and methane, ethane from natural gas, ethylene from refinery gas, etc.

hypertensin. See angiotensin.

hyphylline (dyphylline) $C_7H_7N_4O_2CH_2CH(OH)CH_2OH$. 7-(2, 3-Dihydroxypropyl) theophylline. A derivative of theophylline.
Properties: White, almost odorless, extremely bitter, amorphous solid. M. p. 155-160°C. Freely soluble in water; practically insoluble in ether; slightly soluble in alcohol and chloroform. pH (1% solution) 6. 5-7. 0.
Use: Medicine.

hypnal. See chloral hydrate antipyrine.

hypnone. See acetophenone.

hypo. See sodium thiosulfate.

hypochlorous acid HOCl. An unstable, weak acid, existing only in solution. It is formed, together with hydrochloric acid, when chlorine is dissolved in water, and is responsible for the bleaching action of such solutions. Its salts are bleaching agents. See calcium and sodium hypochlorite.

alpha-**hypophamine.** See oxytocin.

beta-**hypophamine.** See vasopressin.

hypophosphorous acid H_3PO_2.
Properties: Clear, colorless or slight yellow liquid; sour odor. Soluble in water. Sp. gr. 1. 439; m. p. 26. 5°C. A strong monobasic acid and strong reducing agent, sold in solution.
Derivation: Heating concentrated baryta water with white phosphorus and decomposing the barium hypophosphite with sulfuric acid, filtering the liquid and concentrating under reduced pressure.
Method of purification: Distillation.
Grades: Technical; N. F. XI (30-32% solution, sp. gr. 1. 13); 50% purified.
Containers: Bottles; carboys.
Uses: Preparation of hypophosphites.
Shipping regulations: None. *

"Hyporice." [329] Trademark for a highly improved form of sodium thiosulfate, having very uniform rice-sized crystals.

hypoxanthine $C_5H_4N_4O$. An intermediate in the metabolsim of animal purines; also widely distributed in the vegetable kingdon.
Properties: White to cream powder; decomposes at 150°C; almost insoluble in cold water; slightly soluble in boiling water; soluble in dilute acids and alkalis.
Derivation: Deamination of adenine; reduction of uric acid.
Use: Biochemical research; biological media.

hypoxanthine riboside. See inosine.

hypoxanthine riboside-5-phosphoric acid. See inosinic acid.

"Hyprin" GP30. [233] Trademark for hydroxypropylglycerin (q. v.).

"Hyprose" SP80. [233] Trademark for octakis-(2-hydroxypropyl)sucrose (q. v.).

"Hyros." [79] Trademark of a special "FF" wood rosin, containing no lime or other inorganic chemicals, which is distinguished by its high melting point, high viscosity and excellent solubility. M. p. (capillary tube) 64°C, (ball and ring) 85°C; acid number 121; unsaponifiable matter 20%; color "FF".

Containers: Non-returnable 18-gauge black-iron drums of about 500 lbs gross wt. Tare 14-16 lb.

Uses: Adhesive tape; artificial Burgundy pitch; battery wax; belt dressings; box toes; branding paint; brewers' pitch; core oil; dry core binders; fireworks; linoleum cement; matches; pitch; printing ink; rock wool; roofing cement; rubber cement; smoking molds; synthetic rosin oil; tree banding; Venice turpentine; wire-coating compounds.

hyssop (ysop, isop). An aromatic mint (Hyssopus officinalis) from Europe, handled as leaves or as the dried herb. Contains a volatile oil with camphor-like taste. Used in medicine.

hyssop oil.

Properties: A colorless, liquid, essential oil; sensitive to light; soluble in alcohol, ether, chloroform, and benzene. Insoluble in water. Sp. gr. 0.932.

Derivation: Distilled from the herb Hyssopus officinalis.

Method of purification: Rectification.

Grades: Technical.

Containers: Glass bottles.

Uses: Medicine; preparation of liqueurs.

Shipping regulations: None. *

hysteresis. A lag or delay between a changing force and its resulting effect, so that if the causing force is first increased and then decreased, the effect will continue to increase for a period of time after the causing force has started to decrease and vice versa. As a consequence the effect will continue to be noticeable for a period of time after the cause has disappeared. Another consequence is that the magnitude of the effect will be different for a particular value of the causing force, depending on whether the latter is increasing or decreasing and even depending on the rate of increase or decrease. Thus the exact nature of the effect in any specific case will depend upon the previous history of cause and effect for that case.

The most common example is with respect to the magnetization of iron by changing an electric current. Other examples arise in the viscosity or resistance to flow of certain colloids when subjected to deforming forces.

"Hytakerol." [162] Trademark for dihydrotachysterol.

"Hy-ten-sl." [111] Trade name for a type of bronze made in five grades of hardness and consisting of 60-68% copper, 20-24% zinc, 3-7% aluminum, 2.5-5% manganese, 2-4% iron.

Properties: Silvery yellow color; close homogenous grain; non-corrosive; good wear resistance; high tensile strength and hardness; resistant to high fluid pressures.

Forms available: Castings; rods and bars; forgings.

Uses: Heavy duty gears, bearings, nuts, valves, worm wheels, tracks and rollers.

"Hytrol" O. [28] Trademark for cyclohexanone ($C_6H_{10}O$). Colorless liquid; b. p. 156.7°C; f. p. −47°C.

Containers: 55-gal drums; 8,000- and 10,000-gal tank cars.

Uses: As a solvent in the textile, paint and varnish industries, especially for vinyls; as a chemical intermediate; lube oil additive.

"Hywax 122 and 123." [403] Trademark for self-emulsifiable sperm waxes. They are made up largely of fatty alcohol esters of fatty acids in the C_{14} to C_{20} range, plus a small percentage of chemically reacted nitrogen compound which acts as an emulsifier.

Properties:	"Hywax 122"	"Hywax 123"
m.p.	49-50°C.	49-50°C.
iodine number	5.0 max.	5.0 max.
saponification number	95-110	95-110
viscosity (210°F) SSU	56	51
flash point	440°F.	400°F.

Containers: Polyethylene-lined, fiber drums; tank cars.

Uses: Automobile polishes and cleaners; broaching oils; buffing and polishing compounds, floor and furniture waxes; leather stuffing; lubricating grease stabilizer; metal drawing compounds; anti-blocking agent for plastic films; plastic lubricants and stabilizers; rubber processing aid; rust inhibitors; shaving creams; textile lubricants and softeners, etc.

*See "I.C.C. Shipping Regulations," page xiii.
Reference numbers refer to name of manufacturer. See "List of Manufacturers," page v.

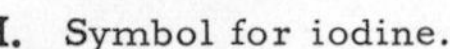

I. Symbol for iodine.

IA. Abbreivation for 3-indoleacetic acid.

IAA. Another abbreviation for 3-indoleacetic acid.

IBP. Abbreviation for initial boiling point; used when a range of boiling temperatures is given. See also DP.

ICC. Abbreviation for ignition control compound.

Iceland moss. See cetraria.

Iceland spar. See calcite.

ice stone. See cryolite, natural.

icthammol (ammonium ichthosulfonate; ammonium sulfoichthyolate)
Properties: Brownish-black, syrupy liquid, burning taste, characteristic empyreumatic odor. Incompatible with acids, alkaloids, carbonates, hydroxides, mercuric chloride. Soluble in water, alcohol-ether or alcohol-ether-water mixtures; partially soluble in alcohol and ether. Miscible with glycerol and fixed oils.
Derivation: An aqueous solution of sulfonated ammonium compounds derived from the action of sulfuric acid upon distillates from certain bituminous shales.
Grades: N. F. XI.
Containers: Drums.
Use: Medicine.
Shipping regulations: None. *

"Ichthymall." [329] Trademark for a brand of ichthammol, ammonium ichthosulfonate.

ichthyocolla. See isinglass.

ICSH. See interstitial-cell-stimulating hormone.

"Icyl." [206] Brand name of proprietary line of dyestuffs specially prepared for the dyeing of viscose rayon. They tend to cover the irregularities occasioned by varying batches of rayon.

I. D. Abbreviation for inside diameter. Used in describing apparatus.

ideal gas. A gas whose behavior can be predicted by Boyle's law, Charles' law, or the ideal gas equation through all ranges of temperature and pressure.

ideal solution. A solution which shows no change of internal energy on mixing, no attractive force between components, and follows Raoult's law over all ranges of temperature and concentration.

idocrase. See vesuvianite.

"Idonyx." [328] A free-flowing, yellow powder containing iodide-iodate salts, nonionic detergent, and complex phosphates. It provides 16% iodine when compounded with surface active agents and acid for the preparation of liquid and powdered type iodine germicides (iodophors).

idryl. See fluoranthene.

"Igenal." [307] Trademark of a line of dyestuffs for the coloring of chrome-tanned leather. Characterized by unusual tinctorial power on chrome leather.

"Igepal." [307] Trademark for a series of nonionic surfactants which are used as detergents, dispersants, emulsifiers and wetting agents. They are alkylphenoxypoly(ethyleneoxy)-ethanols, arising from the combination of an alkylphenol with ethylene oxide. The general formula is $RC_6H_4O(CH_2CH_2O)_nCH_2CH_2OH$, in which R may be C_8H_{17} or a high homolog.

"Igepon." [307] Trademark for a series of anionic surfactants used as detergents, wetting agents, emulsifiers, dispersants and foaming agents. The "Igepon" T and C types are sulfo-amides derived from N-methyltaurine or N-cyclohexyltaurine and fatty acids and have the general formula $RCON(R')CH_2CH_2SO_3Na$. "Igepon" A types are sulfo-esters derived from isethionic acid and a fatty acid and have the general formula $RCOOCH_2CH_2SO_3Na$. R and R' are alkyl groups.

Igewesky's reagent. An etching agent used in the microanalysis of carbon steels. It consists of a 5% solution of picric acid in absolute alcohol.

ignition control compound. A substance such as dimethylxylyl phosphate, methylphenyl phosphate, or trimethyl phosphate which is added to gasoline motor fuels to control spark plug fouling, surface ignition, and motor rumble. See "Ethyl."

ignition point. The minimum temperature at which ignition will occur and burning will continue without further heating or application of flame.

ignotine. See carnosine.

"Iletin." [100] Trademark for insulin injection U. S. P.

"Ilidar" Phosphate. [190] Trademark for a brand of azapetine phosphate (q. v.) used in medicine.

*See "I.C.C. Shipping Regulations," page xiii.
Reference numbers refer to name of manufacturer. See "List of Manufacturers," page v.

illinium. One of the names assigned to element of atomic number 61, based on spectroscopic evidence of the existence of the element reported in certain fractions from a complex fractional crystallization procedure. Element no. 61 is now promethium (q. v.).

illite. See clay.

"Illium." [314] Alloys composed of 56% nickel, 22.5% chromium, 6.5% iron, 6.5% copper, 6.4% molybdenum and small amounts of aluminum, manganese, silicon, and carbon to improve castability.
Properties: Tensile strength 60-73,000 lbs/sq. in (72°F), 24-33,000 lbs/sq in (1000°F); Brinell hardness 160-210; sp. gr. 8.31; m. p. 1300°C; resistant to heat and to corrosion by sulfuric, nitric, phosphoric and some mixed acids and salt solutions.
Use: Used in acid-pump parts, tanks, agitators, bearings, fittings, valves, heat-resistant thermometers, spray nozzles.

illuminants. Hydrocarbons other than methane present in carburetted water gas or similar gases. They include chiefly ethylene and the lower olefin and aromatic hydrocarbons, as well as ethane and higher paraffin hydrocarbons, and usually any acetylene, diolefins or other hydrocarbons soluble in fuming sulfuric acid as used in absorption gas-analysis procedures.

ilmenite (menaccanite; titanic iron ore) $FeO \cdot TiO_2$. Sometimes with some replacement of iron by magnesium or manganese. Iron-black mineral; black to brownish-red streak; submetallic luster. Resembles magnetite in appearance but is readily distinguished by feeble magnetic character.
Constants: Sp. gr. 4.5-5; hardness 5-6.
Occurrence: New York, Florida, North Carolina, Virginia, California, Wyoming, Arkansas; Canada; Sweden; U. S. S. R.; India. Also made synthetically.
Uses: Titanium paints and enamel; source of titanium metal; welding rods; titanium alloys; ceramics.

"Ilosone." [100] Trademark for propionyl erythromycin ester lauryl sulfate. See erythromycin propionate lauryl sulfate.

"Ilotycin." [100] Trademark for erythromycin, U. S. P.

"Ilotycin" Ethyl Carbonate. [100] Trademark for erythromycin ethyl carbonate, U. S. P.

"Ilotycin" Glucoheptonate. [100] Trademark for erythromycin for injection, U. S. P.

Imhoff sludge. See sewage sludge.

imidazole (glyoxaline) HNCHNCHCH. A dinitrogen ring compound. Colorless crystals, m. p. 90°C; b. p. 257°. Soluble in water, alcohol and ether. Base of many new compounds.

4,5-imidazoledicarboxamide. See glycarbylamide.

4-imidazole ethylamine. See histamine.

imidazo(4,5-d)pyrimidine. See purine.

3,3'-iminobispropylamine (dipropylene triamine; 3,3'-diaminodipropylamine) $H_2NC_3H_6NHC_3H_6NH_2$.
Properties: Liquid; sp. gr. 0.9307 (20/20°C); b. p. 240.6°C; f. p. −6.1°C; flash point 235°F; soluble in water.
Containers: 55-gal drums.
Uses: Intermediate for soaps, dyestuffs, rubber chemicals, emulsifying agents, petroleum specialties, insecticides, and pharmaceuticals.
Shipping regulations: Poison, class B. Poison label. *

iminourea. See guanidine.

imitation ultramarine blue. See ultramarine blue.

"Immedial." [307] Brand name of a line of sulfur dyestuffs. Used for the dyeing of cotton and rayon. Characterized by very good fastness to light and good fastness to washing and perspiration.

immune serum globulin. A sterile solution of globulins which contains those antibodies normally present in adult blood. Not less than 90% of the total protein is globulin. It is a transparent, nearly colorless, nearly odorless liquid. Must be kept refrigerated.
Derivation: From a plasma or serum pool of venous or placental blood from 1000 or more individuals.
Grade: U. S. P. XVI.
Use: Medicine.

"Impedex." [28] Trademark for sodium propionate.

imperial green. See copper acetoarsenite.

imperial red. See iron oxide reds.

"Impermex." [236] Brand name for a water-dispersible organic colloid, developed for the purpose of decreasing the water loss of drilling muds, even in muds highly contaminated with salt, salt water, cement, or any other water-soluble electrolyte.

"Implex." [23] Trademark for thermoplastic, high-impact acrylic molding powder, supplied in natural and colored forms. Maximum toughness, gloss, stain- and heat-resistant grades.
Use: Shoe heels, business-machine and musical instrument keys, housings, automotive parts, knobs, metallized parts, and other industrial and commercial products.

impramine $C_{19}H_{24}N_2 \cdot HCl$. 5-(3-Dimethylaminopropyl)-10,11-dihydro-5H-dibenz [b,f] azepine hydrochloride. Colorless crystals; m. p. 175°C; soluble in water. Used in medicine.

"Impranil." [422] Trade name for isocyanate-based adhesives used for textile finishes, including flock application.

"Impregnole." [42] Proprietary products. Aqueous dispersions of waxes and metallic salts.
Properties: Milk white emulsions. Disperse

*See "I. C. C. Shipping Regulations," page xiii.
Reference numbers refer to name of manufacturer. See "List of Manufacturers," page v.

readily in water above 50°C.
Containers: 50-gal steel drums.
Uses: Water repellent and spot proofing for textile fabrics.

impression resins. See contact resins.

"Impruvol." [11] Trademark for fuel and oil antioxidants.
"Impruvol" 33. An antioxidant for motor and aviation gasolines and jet fuels. Consists of 33 1/3 weight % solution of "dbpc" in toluene. Retards formation of gum, protects against induction system and engine deposits, does not impart color to gasoline.
"Impruvol" 20. An antioxidant for transformer and circuit breaker oils consisting of "dbpc," 20% by weight, dissolved in transformer oil.

impsonite. A variety of asphaltic pyrobitumen (q. v.) similar to albertite, black in color, with a black streak, sp. gr. 1.10-1.25, and fixed carbon 50-90%.
Occurrence: Oklahoma, Arkansas, Michigan, Nevada; South America.

In. Symbol for indium.

INAH. Abbreviation for isonicotinic acid hydrazide. See isoniazid.

"Inceloid." [176] Trade name for a line of waterproof adhesives for rubber and resins.

incendiary gels.
1. Mixtures of thermite (aluminum powder and iron oxide) suspended in oil set to a jelly with a small amount of soap, which undergoes spontaneous ignition on contact with air. Another type may contain magnesium in jellied oil.
2. Jellied gasoline combined with thickening agents such as "Napalm" or finely divided magnesium.

inclusion complexes. Crystalline mixtures, not true compounds, in which the molecules of one of the components are contained within the crystal lattice framework of the other component. The framework may be in the form of channels, cages or layers. The two compounds are present in constant but not stoichiometric proportions. The phenomenon depends upon the molecular dimensions of both components. Quite diverse substances can be combined, as argon in hydroquinone, or benzene in nickel cyanide-ammonia complex. The complexes are stable at ordinary temperatures. Melting or dissolving the crystals allows the entrapped component to escape. Inclusion complexes are also called adducts, or occlusion complexes.

Clathrate compounds are inclusion complexes in which molecules of one substance are completely caged within the other, as argon is caged within the hydroquinone crystals. Urea adducts are inclusion complexes of the channel type. In these, the complexing urea crystals wrap around the molecule of the other substance, usually a straight chain unbranched aliphatic hydrocarbon. Similar complexes are formed with thiourea.

The formation of inclusion complexes offers a means of separating chemically similar but physically different molecules. For example, the nickel cyanide-ammonia complex will remove benzene from hydrocarbon impurities and release it 99.992% pure. Similarly, normal alkanes or olefins may be separated from petroleum fractions by contact with saturated aqueous solutions of urea. The nature of the inclusion complexes also permits the isolation and handling of gases and liquids in a solid form.

Among materials which form such complexes are deoxycholic acid (choleic acids), dinitrobiphenyl, dextrin, zeolites and clay minerals, hydroquinone, and certain nitrogen compounds and basic organic zinc salts.

"Incoloy." [283] Trademark for an alloy containing approximately 32% nickel, 21% chromium and 46% iron. Used in high temperature applications because of its resistance to oxidation and carburization. Also useful in many corrosive environments. Made in both cast and wrought forms.

"Inconel." [283] Trademark for an alloy containing approximately 76% nickel, 16% chromium and 6% iron, Has high tarnish resistance, and is used extensively at high temperatures. Made in both cast and wrought forms.

"Inconel X." [283] Trademark for an age-hardenable wrought alloy containing approximately 73% nickel, 15% chromium, 0.8% aluminum, 2.5% titanium, and 0.85% columbium. Designed to have high stress-rupture strength and low creep rates under high stresses at temperatures up to 1500°F, as well as good resistance to corrosion and oxidation. Widely used in aircraft engines and structures and for high temperature springs.

"Incremin." [57] Trademark for lysine-vitamins.

"Indalone." [55] (n-butyl mesityl oxide oxalate). A proprietary product. A repellent applied directly on the skin principally for the biting stable and dog flies; also used for mosquitoes. A repellent used by the Armed Forces, 6-2-2 mixture, was a combination of 60% dimethyl phthalate, 20% 2-ethylhexanediol-1,3 and 20% "Indalone."

indan (hydrindene; 2,3-dihydroindene)
$CHCHCHCHCCCH_2CH_2CH_2$.
Properties: Colorless liquid; b. p. 176.5°C; m. p. −51.4°C; refractive index 1.5388 (16.4°C); sp. gr. 0.965 (20/4°C); insoluble in water; soluble in alcohol and ether.
Derivation: From coal tar.
Use: Base of interesting new derivatives.

indanthrene (indanthrone) $C_{28}H_{14}O_4N_2$. A blue vat dye or pigment, Colour Index No. 1106. The molecule consists of two anthraquinone nuclei linked through two NH groups.

Properties: Excellent durability and lightfastness, not decomposed by heating to 250°C. Soluble in dilute alkaline solutions.
Derivation: By fusion of beta-aminoanthraquinone with caustic potash in the presence of potassium nitrate.
Uses: Dyeing unmordanted cotton; as a pigment in quality paints and enamels.

"Indanthrene." [307] Brand name of vat dyestuffs. Used for the dyeing and printing of cotton, rayon, and silk. Characterized by excellent fastness to light, washing, chlorine, etc.

indanthrene yellow. See flavanthrene.

indanthrone. See indanthrene.

indene $\text{CHCHCHCHCCCHCHCH}_2$.

Properties: Colorless liquid, sp. gr. 1.006 (20/4°C), m.p. −3.5°C; b.p. 182°C; refractive index 1.5726 (n 25/D); flash point 173°F. Insoluble in water, soluble in most organic solvents, rapidly absorbs oxygen from the air, forms polymers by exposure to air and sunlight.
Derivation: Contained in the fraction of crude coal tar distillates which boils from 176 to 182°C.
Use: Preparation of synthetic resins (see coumarone-indene resins); interesting intermediate.

Indian apple. See podophyllum.

Indian arrow-wood. See euonymus.

Indian balsam. See Peru balsam.

Indian cannabis. See cannabis.

Indian grass oil. See palmarosa oil.

Indian gum. A series of gums obtained from trees found in the forests of India and Ceylon. It includes ghatti and karaya gums, as well as less well-known gums, gathered intermittently and not of uniform quality.

Indian hemp. See cannabis.

Indian laburnum. See cassia fistula.

Indian physic. See apocynum.

Indian pink. See spigelia.

Indian poke. See veratrum viride.

Indian red (iron saffron). A red (maroon) pigment formerly consisting of a variety of hematite imported from the East but now made artifically by calcining copperas to obtain the red ferric oxide pigment. There is no pigment, with possibly the exception of lithopone and artificial barium sulfate, which will approach Indian red in fineness of grain. It is also used for polishing gold, silver and other metals.
See also iron oxide reds.

Indian rhubarb. See rhubarb.

Indian saffron. See curcuma.

Indian squill. See squill.

Indian tobacco. See lobelia.

Indian tragacanth. See karaya gum.

Indian turmeric. See hydrastis.

Indian wood oil. See gurjun balsam oil.

Indian yellow.
1. (aureolin). A yellow pigment distinguished from other yellow pigments by being unaffected by hydrogen sulfide. It is durable, without action upon other pigments and is permanent in oils and water color. It consists of a double nitrite of cobalt and potassium and is prepared by adding excess of potassium nitrite solution to a solution of cobalt nitrate acidified with acetic acid. See cobalt potassium nitrite.
2. Also sometimes used for the yellow synthetic dye primuline.

India rubber. See rubber, natural.

indicator. A substance which by its color or other easily observable property indicates the presence or absence or concentration of some other substance, or the degree of reaction between two or more other substances.

The most common example is the use of acid-base indicators such as litmus, phenolphthalein, and methyl orange to indicate the presence or absence of acids and bases, or the approximate concentration of hydrogen ion in a solution. Typical examples are listed in the table.

Acid Base Indicators For Various pH Ranges

pH Range	Indicator
1.3- 3.0	Tropeoline OO A: red. B: yellow.
1.2- 2.8	Thymol blue A: red. B: yellow.
2.9- 4.0	Methyl yellow A: red. B: yellow.
3.1- 4.4	Methyl orange A: red. B: yellow-orange.
3.0- 4.6	Bromophenol blue A: yellow. B: purple.
3.8- 5.4	Bromcresol green A: yellow. B: blue.
4.2- 6.2	Methyl red A: red. B: yellow.
4.8- 6.4	Chlorphenol red A: yellow. B: red.
6.0- 7.6	Bromthymol blue A: yellow. B: blue.
6.4- 8.0	Phenol red A: yellow. B: red.
6.8- 8.0	Neutral red A: red. B: yellow-brown.
7.4- 9.0	Cresol purple A: yellow. B: purple.
8.0- 9.6	Thymol blue A: yellow. B: blue.
8.0- 9.8	Phenolphthalein A: colorless. B: red-violet.
9.3-10.5	Thymolphthalein A: colorless. B: blue.
10.11-12.0	Alizarine yellow R A: yellow. B: violet.

A = acid color, B = basic color.

Other common types of indicators are adsorption indicators, oxidation-reduction indicators and humidity indicators (see separate entries). Some highly colored substances serve as their own indicators (i.e., iodine, potassium permanganate, ceric sulfate, which change from their own characteristic colors in solution to colorless compounds). Other indicators are not easily classified (i.e., the blue color of starch and iodine and the red color formed by the action of ferric compounds on thiocyanates).

indicolite. A dark blue gem stone variety of tourmaline (q. v.).

indigo (indigotin; synthetic indigo blue) $C_{16}H_{10}N_2O_2$. C. I. No. 1177.
Properties: Dark blue, crystalline powder; bronze luster; sp. gr. 1.35; sublimes at 300°C (decomposes); soluble in aniline, nitrobenzene, chloroform, glacial acetic acid and concentrated sulfuric acid; insoluble in water, ether, and alcohol.
Derivation: Natural: by fermentation of the cut twigs and leaves of various species of Indigofera with water to decompose the glucoside, indican. The solution is oxidized to precipitate the indigo. This process is of historical interest only. Synthetic: By fusing phenyl glycine with alkali and sodium amide.
Grades: Technical; pure.
Containers: Tins; fiber drums.
Uses: Textile dyeing and printing inks; manufacture of indigo derivatives; ingredient of tobacco fertilizers; paints; laundering; analytical reagents.
Shipping regulations: None. *

indigo carmine (soluble indigo; indigo extract; sodium indigotindisulfonate; sodium coerulinsulfate) $C_{16}H_8N_2O_2(SO_3Na)_2$. C. I. No. 1180.
Properties: Blue powder or granules; slightly soluble in water (solutions are blue); very slightly soluble in alcohol.
Derivation: Indigotindisulfonic acid treated with soda.
Method of purification: Recrystallization.
Grades: Technical; U. S. P. XVI.
Containers: Wooden kegs; fiber drums.
Uses: Dyeing; medicine; analytical tests.
Shipping regulations: None. *

indigo extract. See indigo carmine.

indigoid dyes. Dyes whose molecular structure involves the indigo $C_{16}H_{10}O_2N_2$, or thioindigo, $C_{16}H_8S_2O_2$, groupings. Colour Index numbers are 73,000-73,999 (new edition). These are vat dyes and are applied principally on cotton. An example is indigo. This is used on cotton and rayon, and to a limited extent on silk. See also dyes.

"Indigolite." [78] Trademark for a combination of sodium formaldehyde sulfoxylate and a sulfonated quaternary base used to give a discharge on indigo-dyed grounds and discharge printing of vat dyestuffs.

indigo, soluble. See indigo carmine.

indigo, synthetic. See indigo.

indigotin. See indigo.

indirect dye. A mordant dye (q. v.).

indium In. Element of atomic number 49, of Group III of the periodic system.
Properties: Ductile, shiny, silver-white metal; softer than lead. Soluble in acids. Insoluble in alkalies. Sp. gr. 7.362; m. p. 156°C; b. p. 1450°C.
Occurrence: Not found native, but in a variety of zinc blende; the indium content is generally very low, sometimes to the extent of 0.1%. Sample ores of pegmatite in western Utah indicate this to be one of the largest indium-bearing ore deposits in the world.
Derivation: From certain zinc ores by chemical and electrolytic methods.
Forms available: Small ingots or bars, shot, pencils, wire, sheets, powder.
Purity: Technical; high purity (less than 10 ppm impurities).
Containers: Boxes; glass bottles.
Uses: In precious-metal alloys for jewelry and dental work; low-melting alloys; glass-sealing alloys; solders; lubricants; bearing metals; neutron indicator in atomic piles; semiconductors. With silver it forms a tarnish-resistant coating.
Shipping regulations: None. *

indium acetylacetonate $In(C_5H_7O_2)_3$. M.p. 186°C. Used as a catalyst.

indium chloride (indium trichloride) $InCl_3$.
Properties: White powder; hygroscopic; soluble in alcohol and water. Sp. gr. 4.
Derivation: Direct union of the elements or by the action of hydrochloric acid on the metal.
Grades: Technical.

indium compounds. Indium arsenide (InAs), antimonide (InSb), and phosphide (InP) are used as high purity binary semiconductors.

indium oxide (indium sesquioxide; indium trioxide) In_2O_3.
Properties: White to light yellow powder; soluble in acids (hot); insoluble in water; sp. gr. 7.179.
Derivation: By burning the metal in air or heating the hydroxide, nitrate, or carbonate.
Grades: Technical.

indium sesquioxide. See indium oxide.

indium sulfate $In_2(SO_4)_3$.
Properties: Grayish powder; deliquescent; soluble in water; sp. gr. 3.438; decomposed by heat.
Grades: Technical.

indium trichloride. See indium chloride.

indium trioxide. See indium oxide.

"Indo Carbon." [307] Trademark for sulfur dyestuffs. Used for the dyeing and printing of cotton and rayon. Characterized by very good fastness to light, washing, chlorine, etc; do not cause tendering of the fiber.

"Indofast." [438] Trademark for vat dyestuff pigments, including carbazole dioxazine violet. Used in paints, printing inks, and plastics.

indole (2,3-benzopyrrole) CHCHCHCHCCCHCHNH.

Properties: White to yellowish scales, turning red on exposure to light and air; odor unpleasant in high concentration but should not show a fecal quality. Is carcinogenic. Soluble in alcohol, ether, hot water, and fixed oils; insoluble in mineral oil and glycerol. M.p. 52°C; b.p. 254°C.
Derivation: By heating ortho-nitrocinnamic acid with potassium hydroxide and iron filings; from indigo, and by numerous syntheses. Also can be produced from 220 to 260° fraction from coal tar.
Method of purification: Recrystallization.
Grades: Technical; C.P.
Containers: Tins; glass bottles.
Uses: Chemical reagent; perfumery; medicine.
Shipping regulations: None.*

3-indoleacetic acid (IA; IAA; beta-indoleacetic acid; heteroauxin) $C_8H_6NCH_2COOH$. One of the plant hormones; see auxin.
Properties: Crystals; m.p. 168-170°C. The natural material is levorotatory; specific rotation 20/D is −3.8° in alcohol. Insoluble in water; soluble in alcohol and ether.
Use: To promote growth and rooting in plants.

beta-**indoleacetic acid.** See 3-indoleacetic acid.

indole-alpha-**aminopropionic acid.** See tryptophan.

3-indolebutyric acid $C_8H_6N(CH_2)_3COOH$.
Properties: White or off-white powder, essentially odorless. m.p. 123°C. Insoluble in water; soluble in alcohols and ketones.
Containers: Glass bottles; fiber drums.
Uses: Plant hormone, especially used in rooting plants.

"Indopol." [216] Trademark for synthetic monoolefin polymers of relatively high molecular weight used in caulking compounds, industrial sealants, adhesives, electrical insulation, surgical tapes, and also as a chemical intermediate.

"Indulin AT." [229] Trademark for an alkali lignin from the manufacture of paper pulp by the sulfate process.
Properties: Brown, free-flowing powder; insoluble in water and acids; soluble in aqueous alkali; sp. gr. 1.3; bulk density 25-30 lbs/cu ft; 2-6% moisutre; pH 3.5-5.5. Composition, dry basis: ash, 0.5% max; sulfur, 1.0-1.4%; methoxyl, 13.5-14.5%; total hydroxyl 8-10%; phenolic hydroxyl 2.0-2.1%. Unit combining weight with metals, 840. Reacts to form salts, ethers, esters, and other condensation products. Can be halogenated, nitrated, sulfited and sulfonated.
Containers: Packed in 50-lb net MWP bags.
Uses: Emulsion stabilization; storage battery plates; foam stabilization; protein precipitation; dyeing; electroplating; rubber reinforcing; rubber masterbatch production; ceramics deflocculation; ceramics binding; dispersing; drilling muds; and others.

"Indulin B" and "Indulin C." [229] Trademarks for sodium derivatives of Indulin AT.
Properties: Brown, free-flowing powders; soluble in water; will precipitate lignin in acids; sp. gr. 1.35, 1.46; bulk density 30-35 pounds/cu ft.; 2-8% moisture; pH 8-9, 9-10. Composition, dry basis: ash 8-12%, 18-22%; sulfur 1.2-1.6%, 1.4-1.8%; methoxyl, 12.5-13.5%, 10-12%. Same reactions as Indulin AT but have the difference of not requiring alkali for water solubility.
Containers: 50-lb MWP bags.
Uses: Same as Indulin AT.

"Indusoil." [228] Trademark for distilled or fractionated tall oils and tall oil products. Suffix used to further describe products. Typical examples are:
"Indusoil M-28."
Typical analysis: Acid number 190; rosin acids 28%; fatty acids 71%; unsaponifiables 2%; color (Gardner) 6.
"Indusoil L-3."
Typical analysis: Acid number 190; roson acids 3%; fatty acids 96%; unsaponifiables 2%; color (Gardner) 6.
"Indusoil JC-11." A product particularly designed for use as an emulsifier in the manufacture of synthetic rubber.
Uses: "Indusoil" products are used in many applications, such as adhesives, binders, cements, flotation, leather, paint and varnish, soaps, rubber and sanitary chemicals. Being a free fatty and/or rosin acid, with double bonds present, it will undergo saponfication, esterification, ethylene oxide condensation, decarboxylation, sulfonation, sulfurization, polymerization, and hydrogenation.

"Indusoil H-90." [228] Trademark for a tall oil rosin product.
Properties: A solid rosin product having similar properties to gum and wood rosin.
Uses: May be used in adhesives, binders, cements, ester gums, resins, soap, and rubber. Will undergo reactions common to rosin.

infrared. The region of the electromagnetic spectrum including wave lengths from 0.78 micron to about 300 microns (i.e., longer than visible light and shorter than microwave).

"Infrax." [280] Trademark for a refractory insulation, available only in brick form.
Properties: Suitable for use at temperatures up to 2700°F; high strength; thermal conductivity 2.5 Btu/sq ft/ inch thickness/°F temperature difference/hr; the weight of a standard 9 in. brick is 3 lbs.
Uses: Primary linings of electric furnaces

*See "I.C.C. Shipping Regulations," page xiii.
Reference numbers refer to name of manufacturer. See "List of Manufacturers," page v.

and kilns. Should be used as primary linings of fuel-fired furnaces only when protected by a cement facing.

infusions. Aqueous solutions obtained by treating drugs with hot or cold water, without boiling. Generally prepared by pouring boiling water upon the vegetable substance and macerating the mixture in a tightly closed vessel until the liquid cools. When not otherwise specified, they are of 5% strength, by weight.

infusorial earth. See diatomite.

ingot iron. Highly refined steel made by the basic open-hearth process with a maximum of 0.15% impurity. Due to high purity it has excellent ductility and resistance to rusting.

ingrain dyes. Insoluble dyes developed by impregnating any fabric with one or more of the intermediates and then producing the dye by reaction with another intermediate.

"Inhibisol." [33] Trade name for specially purified grade of "Penolene 643" (q. v.) to which a corrosion inhibitor has been added to make it safe for use on all metals, including aluminum.

inhibitor. General term for compounds or materials that have the effect of slowing down or stopping an undesired chemical change such as corrosion, oxidation or polymerization.

inks. See printing inks.

inosine (hypoxanthine riboside) $C_{10}H_{12}N_4O_5$. An important intermediate in animal purine metabolism. Also available as its barium salt.

Properties: (dihydrate) Crystallizes in needles from water; m. p. 90°C. Levorotatory in solution. Slightly soluble in water.

Derivation: By deamination of adenosine.

Use: Biochemical research.

inosinic acid (hypoxanthine riboside-5-phosphoric acid) $C_{10}H_{13}N_4O_8P$. An important intermediate in the synthesis and metabolism of animal purines.

Properties: Syrup with agreeable sour taste. Freely soluble in water and formic acid; very sparingly soluble in alcohol and ether.

Derivation: From meat extract or by enzymatic deamination of muscle adenylic acid.

Use: Biochemical research.

inositol (hexahydroxycyclohexane) $C_6H_6(OH)_6 \cdot 2H_2O$. A constituent of body tissue. There are 9 isomeric forms of inositol, of which i-inositol (myo-inositol or meso-inositol or specifically the cis-1, 2, 3, 5-trans-4, 6-hexahydroxycyclohexane) is the one having vitamin activity. It appears to prevent alopecia, promote growth, have lipotropic activity, influence gastro-intestinal motility, and to prevent the deposit of cholesterol in the liver.

Properties: White crystals; odorless; m. p. 224-227°C; sweet taste; soluble in water; insoluble in absolute alcohol and ether; stable to heat, strong acid and alkali; becomes anhydrous at 100°C; sp. gr. 1.524, 1.752 when anhydrous.

Source: Food source: Vegetables, citrus fruits, cereal grains, liver, kidney, heart and other meats. Commercial source: From corn steep liquor by precipitation and hydrolysis of crude phytate.

Units: Amounts are expressed in milligrams of inositol.

Grades: N. F. XI.

Containers: Bottles; cans; drums.

Uses: Medicine; nutrition; intermediate.

i-inositol. An optically inactive inositol isomer also referred to as myo-inositol in the literature.

inositol hexaphosphoric acid. See phytic acid.

INPC. Abbreviation for isopropyl N-phenylcarbamate. See IPC.

insane-root. See hyoscyamus.

insecticides. Chemical compounds which are used to control insects which are harmful, directly or indirectly, to man. The application of insecticides and subsequent control of the insect is achieved in several ways: (1) stomach poisons for insects which eat plant leaves or cloth fabric; (2) contact poisons which are applied directly to the body of the insect either by treating the material with which the insect comes in contact, or by introducing the toxic agent as a fumigant into the air the insect breathes, or (3) systemic insecticides in which the toxic agent is made a component of the plant itself. Examples of the stomach poisons are inorganic chemicals such as lead arsenate, sodium fluoride or sodium fluosilicates. Examples of contact poisons are natural plant extracts such pyrethrum, organic chlorinated compounds such as DDT, chlordane and the like, or many organic phosphates such as parathion. Examples of fumigants are hydrocyanic acid gas, para-dichlorobenzene and methyl bromide. Octamethylphosphoramide is a systemic insecticide, i. e., it is more toxic to insects after it has been absorbed into the leaves of the plants than it is as a direct poison.

"Insecti-sol." [25] Brand name for a proprietary highly purified odorless solvent exceeding CSMA minimum requirements for an insecticide base. This deodorized hydrocarbon distillate is water white and has a 170°F (min.) flash point, a 465-480°F distillation end point, and a 98% unsulfonatable residue. Also useful in food processing and cosmetic manufacture.

insect wax. See Chinese wax.

"Insidol." [300] Trademark for a wetting agent useful in textile processing where instantaneous penetration is required. A sulfodicarboxylic acid ester composition.

"Instant-Dri." [108] A solid form rinsing agent. Reduces surface tension of water.
Containers: 2 3/8 oz. bars; 1-oz Pak.

"Instantreat." [108] Trademark for a high purity specially processed powdered complex phosphate. Contains enough chlorinated ingredient to disinfect feed solution and feed equipment.
Containers: 6-lb cans; 100-lb drums.
Uses: Controls corrosion, lime scale and red water trouble in homes, hotels, restaurants and small industrial water systems.

insulating oils (electrical oils). Oils used as insulators and cooling medium in transformers, circuit-breakers, switches or other electric apparatus.

insulin. An important polypeptide hormone which originates normally in the beta cells of the isles of Langerhans situated in the pancreas. It is made up of 16 amino acids and small amounts of the heavy metals zinc, nickel, cobalt or cadmium. It regulates carbohydrate metabolism in the body by decreasing the blood glucose level. A systemic deficiency of insulin leads to diabetes, and the isolation of insulin has made it possible to control this disease.
Properties: White powder or hexagonal-shaped crystals; readily soluble in dilute acids and alkalies; soluble in water.
Derivation: Extraction of minced pancreas with acidified dilute alcohol, followed by precipitation with absolute alcohol.
Grades: U. S. P. XVI, in various solutions or suspensions, which include insulin injection; globin zinc insulin injection; isophane insulin suspension; protamine zinc insulin suspension; insulin zinc suspension.
Containers: Glass bottles, ampules, vials.
Use: Medicine.

insulin injection. See insulin.

insulin zinc suspension. See insulin.

"Insul-Mastic." [423] Trademark for a series of corrosion and insulation coatings based on asphalt-gilsonite.

"Insuloxide." [337] Trade name for zirconium oxide containing 95% ZrO_2 and 3-5% SiO_2. Cream colored powder with sp. gr. 5.4; average particle size 45 to 50 microns. Used as a thermal insulator and refractory where a prefused material is not required. Also used as a raw material in the production of super-refractories, and as a component in ceramic colors. See "Opax."

"Insurok." [63] A series of laminated and molding plastics characterized by durability, light-weight, resistance to many chemicals, and high dielectric strength. These plastics have use as a replacement for cast aluminum in airplane parts; and for electric uses as switches, distributors; commutators, etc.

"Intalox." [326] Trademark for particular shape of tower filling materials available in ceramic and carbon.

intarvin (glyceryl margarate; glyceryl heptadecanoate) $(C_{16}H_{33}COO)_3C_3H_5$. White lumpy masses; odor and taste of tallow; insoluble in water; freely soluble in chloroform or ether.

intermediate. A chemical produced because it is a necessary intermediate stage in the manufacture of one or usually more ultimate end-products such as dyes, drugs, etc.

interstitial-cell-stimulating hormone (ICSH) A gonadotropic hormone found in the pituitary gland. It is a glycoprotein and has been isolated from sheep pituitaries. ICSH stimulates the follicles of the ovary to develop into corpora lutea and stimulates the interstitial cells of the testis to secrete testosterone. It is commercially available in combination with follicle-stimulating hormone.
Properties: White powder; soluble in water. Destroyed by dilute acids and alkalies and by heating above 50°C.
Use: Medicine.

"Intocostrin." [412] Trademark for Chondodendron tomentosum extract (q. v.).

"Intracol." [83] Trade name for a series of long chain fatty acid amides and their acid salts; used as emulsifiers, dispersing agents and textile lubricants.

intrinsic angular momentum. See spin.

intumescence. The foaming and swelling of a plastic or other material when exposed to high surface temperatures or flames. Used with respect to polyurethane base coating materials for rocket reëntry application.

inulin (alant starch) $(C_6H_{10}O_5)_n$.
Properties: Horny, colorless, amorphous lumps or white powder; hygroscopic; sp. gr. 1.35. Soluble in hot water.
Derivation: A carbohydrate from the bulbs of Dahlia variabilis.
Grades: Technical.
Containers: Wooden barrels; fiber drums.
Uses: Diabetic bread; manufacture of fructose; diagnostic reagent.
Shipping regulations: None. *

in vacuo. Taking place in a vacuum; actually at a low pressure.

"Invermul." [236] Trademark for a basic emulsifier in the preparation of water in oil-type emulsion drilling muds for oil wells.
Containers: 50-lb multiwall paper bags.

invertase (sucrase; invertin). Enzyme produced by yeast and by the lining of the intestines. It is a white powder, soluble in water. It catalyzes the conversion of sucrose (ordinary sugar) to glucose and levulose (fructose) during fermentation of sugars.
Uses: Production of invert sugar for syrups and candy; analytical reagent for sucrose.

*See "I. C. C. Shipping Regulations," page xiii.
Reference numbers refer to name of manufacturer. See "List of Manufacturers," page v.

invertin. See invertase.

invert sugar. A mixture of 50% glucose and 50% fructose obtained by the hydrolysis of sucrose. It absorbs water readily, and is usually only handled as a syrup. Because of its fructose content, invert sugar is levorotatory in solution, and sweeter than sucrose. Invert sugar is often incorporated in products where loss of water must be avoided. Commercially it is obtained from the inversion of a 96% cane sugar solution.
Use: Food industry; brewing industry; medicine.

"Invin." [304] Trademark for a series of organic vinyl stabilizers: Available as:
"Invin 85." Barium-cadmium-zinc compound.
Properties: Clear light yellow liquid, sp. gr. 1.03, refractive index 1.48.
Containers: Metal drums containing 40 and 450 lbs.
Uses: Plastisol and organosol stabilizer. Provides excellent heat and light stability. Low viscosity. Excellent air release.
"Invin 91." Barium-cadmium compound.
Properties: Clear yellow liquid, sp. gr. 1.03, refractive index 1.48.
Containers: Metal drums containing 40 and 450 lbs.
Uses: Heat and light stabilizer for all types of clear and translucent film, sheeting and extrusions. Good resistance to yellowing and plate-out.

in vitro. A condition in which a reaction is carried out, or a process occurs, in a glass container, as a test tube or beaker.

in vivo. A condition in which a reaction is carried out, or a process occurs, in a living system, i.e., cells or tissues.

"Ioclide." [239] A water-soluble form of iodine, containing a nonionic detergent as a surface active agent. Contains 15.8% polyethoxopolypropoxypolyethoxyethanol-iodine complex, 15.2% nonylphenoxypolyethoxyethanol-iodine complex (provides 3.1% available iodine), and 0.2% hydrogen chloride, remainder water and other inert ingredients. Used as a germicide.

iodargyrite (iodyrite) AgI. A natural silver iodide. Color yellow or yellowish green, luster resinous to adamantine; streak shining yellow; hardness 1.5; sp. gr. 5.69.
Occurrence: Nevada, New Mexico, Arizona; Australia; Chile.
See also cerargyrite.

iodembolite. A variety of embolite (q.v.) containing iodine.

iodeosin (tetraiodofluorescein) $C_{20}H_8I_4O_5$.
Properties: Red powder; soluble in dilute alkalies; slightly soluble in alcohol and ether; insoluble in water.
Derivation: By the interaction of fluorescein and iodine in presence of iodic acid.
Grades: Technical.
Containers: Glass bottles.
Use: Indicator in analytical chemistry.
Shipping regulations: None. *

iodic acid HIO_3.
Properties: Colorless, rhombic crystals or white, crystalline powder. A moderately strong acid. Soluble in cold and hot water. Sp. gr. 4.629; m. p. 110°C (decomposes).
Derivation: By adding sulfuric acid to a solution of barium iodate and subsequent filtration and crystallization.
Method of purification: Crystallization.
Containers: 1-lb bottles; tins.
Grades: Technical.
Uses: Analytical chemistry; medicine.

iodic acid anhydride (iodic acid anhydrous; iodine pentoxide) I_2O_5.
Properties: White, crystalline powder. Soluble in dilute alcohol and water; insoluble in absolute alcohol, chloroform, ether, carbon disulfide. Sp. gr. 4.799; m. p. 300°C (decomposes).
Grades: Technical.
Uses: Oxidizing agent; organic synthesis.

iodic acid anhydrous. See iodic acid anhydride.

iodine I. Element of atomic number 53, group VII of the periodic table; the least reactive of the halogens.
Properties: Heavy, grayish-black plates or granules, having a metallic luster and characteristic odor. Readily sublimed, having a violet vapor. Poisonous; corrosive! Sp. gr. 4.98; m. p. 114.2°C; b. p. 184°C. Soluble in alcohol, carbon disulfide, chloroform, ether, carbon tetrachloride, glycerol, and alkaline iodide solutions; insoluble in water.
Derivation: (a) From oil-well brines, by oxidation with chlorine and absorption in sulfurous acid. (b) From the ashes of seaweeds or mother liquors of Chile saltpeter by the addition of sodium bisulfite solution.
Method of purification: Sublimation.
Grades: Technical; C. P.; U. S. P. XVI.
Containers: Bottles; 100- and 200-lb kegs; drums.
Uses: Medicine (germicide, antiseptic); organic compounds; dyes (aniline dyes, phthalein dyes); catalyst in intermediate manufacture; iodides; iodates; pharmaceuticals; leather manufacture; testing paper; process engraving and lithography; special soaps; analytical reagent; suggested for use in tungsten filament lamps.
Shipping regulations: None. *

iodine 131. Radioactive iodine of mass number 131.
Properties: Half-life 8 days; radiation, beta and gamma; radiotoxicity, moderately hazardous.
Derivation: By pile irradiation of tellurium and from the fission products of nuclear reactor fuels.
Forms available: As elemental iodine and in a weakly basic solution of sodium iodide in sodium sulfite; iodine 131 is also available in tagged compounds such

*See "I.C.C. Shipping Regulations," page xiii.
Reference numbers refer to name of manufacturer. See "List of Manufacturers," page v.

as dithymol diiodide, potassium iodate, diiodofluorescein, insulin, ACTH, etc.
Grade: N. N. D., as radio-iodinated serum albumin, and as sodium radio-iodide.
Uses: Diagnosis and treatment of goiter hyperthyroidism, and other thyroid disorders; iodine 131 is also used for internal radiation therapy; for locating brain tumors; in film gauges to measure film thicknesses of the order of one micron; for detecting leaks in water lines; as a source of radiation in oil field tests, as a tracer in chemical analysis; as a tracer in studying diet iodine for cattle, the functions of the thyroid gland, the efficiency of mixing pulp fibers, the thermal stability of potassium iodate in bread dough, chemical reaction mechanisms, etc.
Shipping regulations: Class D poison, radioactive material. Red label. *

iodine bisulfide. See sulfur iodine.

iodine bromide. See iodine monobromide.

iodine chloride. See iodine monochloride and iodine trichloride.

iodine cyanide. See cyanogen iodide.

iodine disulfide. See sulfur iodine.

iodine monobromide (bromine iodide) IBr.
Properties: Crystalline, purplish-black mass. Soluble in water with decomposition, alcohol, and ether. M. p. 42°C; b. p. 116°C (dec); sp. gr. 4.41.
Derivation: By the interaction of iodine and bromine.
Grades: Technical.
Containers: Glass bottles; metal boxes.
Use: Organic synthesis.
Shipping regulations: None. *

iodine monochloride ICl.
Properties: Reddish-brown, oily liquid; two solid forms, alpha and beta. Soluble in alcohol, water (with decomposition), and dilute hydrochloric acid.
Constants: M. p. (alpha) 27°C, (beta) 14°C; b. p. 101°C (dec); sp. gr. 3.78.
Derivation: By the action of dry chlorine on iodine.
Grades: Technical.
Containers: Glass bottles.
Uses: Analytical chemistry; organic synthesis.
Shipping regulations: Corrosive liquid. White label. *

iodine number (iodine value). The percentage of iodine that will be absorbed by a chemically unsaturated substance (vegetable oils, rubber, etc.) in a given time under arbitrary conditions. A measure of unsaturation.

iodine pentafluoride IF_5.
Properties: Fuming liquid; b. p. 100.5°C; m. p. 9.4°C; sp. gr. (liquid) 3.189 (25°C). Reacts violently with water; attacks glass.
Available in cylinders at 98.0% min purity.

iodine pentoxide. See iodic acid, anhydride.

iodine solution, strong. See Lugol's solution.

iodine tincture. A solution of iodine and potassium iodide or sodium iodide in alcohol; a reddish-brown liquid having the odors of iodine and alcohol; contains from 44-50% by volume of alcohol and 2 g of iodine per 100 cc.
Grades: U. S. P. XVI.

iodine trichloride ICl_3.
Properties: Orange-yellow, deliquescent, crystalline powder; pungent, irritating odor; poisonous! Soluble in water with decomposition, alcohol and benzene.
Constants: M. p. 33°C; sp. gr. 3.11.
Derivation: By interaction of iodine and chlorine.
Grades: Technical.
Containers: Amber glass bottles.
Use: Medicine; agent for introducing iodine and chlorine in organic synthesis.
Shipping regulations: None. *

iodine value. See iodine number.

iodipamide $(CH_2)_4(CONHC_6HI_3COOH)_2$. 3,3'-(Adipoyldiimino) bis(2,4,6-triiodobenzoic acid).
Properties: White, nearly odorless, crystalline powder, very slightly soluble in alcohol, chloroform, ether, and water; pH of saturated solution is between 3.5 and 3.9.
Grade: N.F. XI.
Use: Medicine.

iodipamide bis(N-methylglucamine) salt. See methylglucamine iodipamide.

iodisan. See propiodal.

iodival (2-iodoisovalerylurea) $(CH_3)_2HCCHICONHCONH_2$.
Properties: White powder containing 47% iodine. Soluble in alcohol, hot water; insoluble in cold water.
Constants: M. p. 180°C with decomposition.
Use: Medicine.
Shipping regulations: None. *

iodized oil. An iodine addition product of vegetable oil or oils containing 38-42% organically combined iodine.
Properties: Thick, viscous, oily liquid. Alliaceous odor. Oleaginous taste. Affected by air and light. Soluble in solvent naphtha.
Grade: U. S. P. XVI.
Use: Medicine (radiopaque medium).

alpha-**iodoacetophenone** $C_6H_5COCH_2I$.
Properties: Crystals. Soluble in alcohol, benzene, ether; insoluble in water.
Constants: B. p. 170°C (30 mm); m. p. 29.5 to 30°C.
Derivation: Interaction of chloroacetophenone and sodium iodide in presence of ethyl alcohol.

iodoalphionic acid
$C_6H_5CH(COOH)CH_2C_6H_2I_2OH$. beta-(4-Hydroxy-3,5-diiodophenyl)-alpha-phenylpropionic acid.
Properties: White crystals or as white or faintly yellowish powder, having a faint

*See "I.C.C. Shipping Regulations," page xiii.
Reference numbers refer to name of manufacturer. See "List of Manufacturers," page v.

characteristic odor and taste. M. p. 160-164°C (with decomposition). Stable in air but slightly discolored on prolonged exposure to light. Insoluble in water; readily soluble in alcohol and ether; slightly soluble in benzene and chloroform; soluble in both alkali carbonate and hydroxide solutions.
Grade: N. F. XI.
Use: Medicine.

iodobrassid. See ethyl diiodobrassidate.

iodochlorhydroxyquin. U. S. P. XVI name for iodochlorohydroxyquinoline.

iodochlorohydroxyquinoline (5-chloro-7-iodo-8-quinolinol; iodochlorhydroxyquin; iodochloroxyquinoline) C_9H_5ClINO.
Properties: Voluminous, spongy, brownish-yellow powder, with a slight, characteristic odor; m. p. 172°C; nearly insoluble in water and in alcohol; soluble in hot ethyl acetate and in hot glacial acetic acid.
Grade: U. S. P. XVI (as iodochlorhydroxyquin).
Containers: Drums.
Use: Medicine.

iodochloroxyquinoline. See iodochlorohydroxyquinoline.

iodoethane. See ethyl iodide.

iodoethylene. See tetraiodoethylene.

iodoform (triiodomethane) CHI_3.
Properties: Small, greenish yellow or lustrous crystals or powder; characteristic, penetrating odor. Soluble in alcohol, glycerol, chloroform, carbon disulfide and ether; insoluble in water.
Constants: Sp. gr. 4.08; m. p. 115°C.
Derivation: (a) By heating acetone or methyl alcohol with iodine in presence of an alkali or alkaline carbonate. (b) Electrolytically, by passing a current through a solution containing potassium iodide, alcohol and sodium carbonate.
Grades: Technical; N. F. XI.
Containers: Bottles; 100-, 300-lb drums.
Use: Medicine (external).
Shipping regulations: None. *

iodoformin (iodoformohexamethylenetetramine) $CHI_3 \cdot (CH_2)_6N_4$. Do not confuse with iodoformine, which is hexamethylenetetramine tetraiodide.
Properties: White crystalline powder, faint iodoform odor. Decomposes in wounds and boiling water liberating 75% free iodoform. Insoluble in water, alcohol, and ether. M. p. 178°C.
Derivation: Action of alcoholic iodoform solution on hexamethylenetetramine.
Use: Medicine.
Shipping regulations: None. *

iodoformohexamethylenetetramine. See iodoformin.

iodohippurate sodium $C_9H_7INaNO_3 \cdot 2H_2O$. Sodium ortho-iodohippurate dihydrate. Prepared as a white, crystalline powder which is very soluble in water. Aqueous solutions are clear and practically colorless. A 5% solution has pH 7.3-7.8. The free acid melts at 173-176°C.
Use: Medicinal.

7-iodo-8-hydroxyquinoline-5-sulfonic acid. See loretin.

2-iodoisovalerylurea. See iodival.

iodole (pyrrole tetraiodide; tetraiodopyrrole) C_4I_4NH.
Properties: Light grayish-brown or yellowish-gray crystalline powder; odorless, tasteless; decomposes 140-150°C; soluble in ether, alcohol, chloroform, fixed oils; difficultly soluble in water.
Derivation: Pyrrole (obtained from bone oil) is subjected to the action of a solution of iodine in potassium iodide. The precipitated iodole is dissolved in hot alcohol and precipitated by adding water.
Use: Medicine.
Shipping regulations: None. *

iodomethane. See methyl iodide.

iodopanoic acid. See iopanoic acid.

iodophor.
1. A complex of iodine with certain types of surface active agents that have detergent properties.
2. More generally, any carrier of iodine.

iodophosphonium. See phosphonium iodide.

iodophthalein (tetraiodophenolphthalein; nosophen) $C_{20}H_{10}I_4O_4$.
Properties: Pale yellow, odorless, tasteless powder. Decomposes about 200°C. Soluble in chloroform, ether, and solutions of alkalies; slightly soluble in alcohol; insoluble in water.
Use: Medicine.

iodophthalein sodium (TIPPS; tetraiodophenolphthalein sodium; nosophen sodium; soluble iodophthalein) $C_{20}H_8I_4O_4Na_2 \cdot 3H_2O$.
Properties: Pale blue to violet, odorless, hygroscopic crystals. Salty taste. Soluble in water; slightly soluble in alcohol. Gradually absorbs carbon dioxide from the air, becoming insoluble.
Grade: N.F.XI. (not less than 85% tetraiodophenolphthalein).
Use: Medicine.

2-iodopropane. See isopropyl iodide.

iodopyracet (3,5-diiodo-4-pyridone-N-acetic acid, diethanolamine salt) $C_5H_2I_2NOCH_2COONH_2(CH_2CH_2OH)_2$. A sterile water solution is listed in N. F. XI.
Use: Medicine (radio-opaque iodine compound).

iodosuccinimide (succiniodimide) $C_4H_4INO_2$.
Properties: Colorless crystals; m. p. 200-201°C; soluble in acetone, methanol; insoluble in carbon tetrachloride and ether. Decomposes in water.
Containers: Glass bottles.
Use: Iodinizing agent in synthetic organic chemistry.

*See "I. C. C. Shipping Regulations," page xiii.
Reference numbers refer to name of manufacturer. See "List of Manufacturers," page v.

iodothymol. See thymol iodide.

iodyrite. See iodargyrite.

"Ioflow." 329 Trademark for a potassium iodide mixture with 0.5% magnesium carbonate added to produce a free-flowing salt.

"Iokel." 244 Trade name for a phosphoric acid solution of a polyoxyethanol alkyl phenol condensate complex of elemental iodine.
Properties: Viscous, dark liquid with acidic reaction. Miscible in all proportions with water.
Uses: Detergent sanitizer; milk stone remover.

"Iomag." 329 Trademark for a potassium iodide mixture containing 90% potassium iodide and made free-flowing with 8% magnesium carbonate and 2% potassium hydroxide.

ion.
1. In solutions: an electrically charged atom or group of atoms.
2. In gases: electrically charged molecules. Ions may be either positively or negatively charged, indicating that an atom has either lost or gained one electron.

"Ionac." 210 Trademark for a line of cation and anion exchangers.
"Ionac" A-260. A quaternary amine type resin anion exchanger.
"Ionac" A-300. Tertiary amine, moderately basic type anion exchanger.
"Ionac" A-315. Weakly basic, polystyrene base, polyamine type resin anion exchanger.
"Ionac" A-540. Trimethylamine, strongly basic polystyrene quaternary amine type resin anion exchanger.
"Ionac" A-550. Strongly basic, dimethylethanolamine type resin anion exchanger.
"Ionac" A-580. Highly porous, medium basic, high capacity resin type anion exchanger.
"Ionac" A-590. Highly porous, medium basic, high capacity resin type anion exchanger.
"Ionac" C-50. Processed glauconite (naturally occurring greensand) cation exchanger.
"Ionac" C-100. Precipitated gel-type sodium aluminosilicate cation exchanger.
"Ionac" C-101. Treated precipitated gel-type sodium aluminosilicate cation exchanger employed for ammonia determination.
"Ionac" C-102. Treated precipitated potassium aluminosilicate cation exchanger.
"Ionac" C-150. Sulfonated coal type, acid resistant, cation exchanger.
"Ionac" C-240. High capacity, sulfonated, acid resistant, styrene type resin cation exchanger.
"Ionac" C-244. Cation exchanger in hydrogen form for use as a catalyst in place of strong acids.
"Ionac" C-245. High capacity ammonium cation exchanger.
"Ionac" C-250, C-255, C-260. Higher cross-linked forms of C-240.
"Ionac" C-270. Carboxylic cation exchanger containing only weakly acidic groups.
"Ionac" D-75 (ASMIT). Specially prepared decolorizing resin; prepared from bone char.
"Ionac" M-50. Permanganate-regenerated glauconite cation exchanger.
"Ionac" M-610. Mixed bed resin for demineralizing.
"Ionac" MA-3148, MA-3228. Anion exchange membranes for use in electrodialysis.
"Ionac" MC-3142, MC-3227. Cation exchange membranes for use in electrodialysis.
"Ionac" P-50. Hard, granular fast-wetting activated carbon.
"Ionac" R-50. A granular calcite employed for increasing the pH of water.

ion exchange. A reversible chemical reaction between a solid (ion exchanger) and a fluid mixture (usually an aqueous solution) by means of which ions may be interchanged. The superficial physical structure of the ion exchanger solid is not affected. The customary procedure is to pass the fluid through a bed of the solid, which is granular and porous, and has only a limited capacity for exchange. The process is essentially a batch type in which the ion exchanger, upon nearing exhaustion, is regenerated by the use of inexpensive brines, carbonate solutions, etc. (See zeolites).

The process was first used in water softening, but now has a much broader application. Ion exchange reactions are of four types: (1) base or cation exchange, as in the water-softening process, (2) hydrogen exchange, a special case of the first type, in which metal ions are replaced by hydrogen ion with the aid of a hydrogen zeolite, or resin, (3) anion exchange, as for instance the change of streptomycin sulfate to the chloride, and (4) acid removal, a special case of the third type, in which both hydrogen ion and the anion are removed by an amine resin (see ion exchange resins).
Some specific uses: Water softening; milk softening (substitution of sodium ions for calcium ions in milk); removal of iron from wine (substitution of hydrogen ions); recovery of chromate from plating solutions; uranium from acid solutions; streptomycin from broths; removal of formic acid from formaldehyde solutions; demineralization of sugar solutions; recovery of valuable metals from wastes; recovery of nicotine from tobacco-dryer gases; catalysis of reaction between butyl alcohol and fatty acids; recovery and separation of radioactive isotopes from atomic fission; analytical methods; desalting of water.

ion exchange, liquid. A misleading term for processes which involve chelation reactions rather than ion exchange. Liquid ion exchange is sometimes used to refer to the extraction of metals into a solution containing ethylenediaminetetraacetic acid (EDTA) or similar reagents that cause complex formation with metal ions.

ion exchange resins. Synthetic resins containing active groups (usually sulfonic, carboxylic, phenol, or substituted amino groups) that give the resin the property of combining with or exchanging ions between the resin and a solution. Thus a resin with active sulfonic groups can be converted to the sodium form and will then exchange its sodium ions with the calcium ions present in hard water. "Amberlite" resins are of this type. See also zeolites.

ion exclusion. The process in which a synthetic resin of the ion exchange type absorbs non-ionized solutes such as glycerine or sugar while it does not absorb ionized solutes that are also present in a solution in contact with the resin. Thus sodium chloride and glycerine can be separated by passage of their aqueous solution through a bed of particles of an ion exclusion resin.

ionic detergents. See detergents, synthetic.

"Ionite." [271] Trademark for a lignite-type material used for oil well drilling muds, as a low density filler in dark colored rubber, as an organic base filler for fertilizers, and as a source of humic acid.

ionization. The separation or dissociation of a molecule into atoms or groups of atoms that are of opposite electrical sign. This occurs spontaneously in many salts when they are dissolved in water, or when the salts are melted. Thus sodium chloride gives positive sodium ions and negative chloride ions. Molecules or atoms of gases are ionized by passage of an electric current through the gas. In this case electrons are removed from the molecule or atom, leaving it with a positive charge.

ionizing radiations. High speed or high energy particles or electromagnetic waves from radioactive sources, such as radium, or fission products, or from nuclear reactions of atomic or hydrogen bombs, nuclear piles, and accelerators or cyclotrons, and betatrons. The high speed high energy particles may be electrons, protons, neutrons, deuterons, or alpha particles with enough energy to ionize, i. e., remove electrons from ordinary atoms upon collision. The electromagnetic wave ionizing radiations are x-rays or gamma rays.

"Ionol." [125] Trademark for 2, 6-di-tert-butyl-4-methylphenol, a sterically hindered phenol with unusually good antioxidant properties. Unlike most phenols, "Ionol" is nonirritating to the skin and has a comparatively inert, non-acidic hydroxyl group.

Properties: M. p. 70°C; b. p. 265°C; sp. gr. (20/4°C) 1.048, sp. gr. (80/4°C) 0.899; flash point, Tag open cup, 260°C; soluble in methanol, ethanol, isopentane, benzene, toluene, methyl ketone, and linseed oil; insoluble in water and 10% sodium hydroxide.

Containers: 20-gal and 51-gal non-returnable fiber drums.

Uses: Antioxidant for such petroleum products as aviation gasoline, motor gasoline, transformer oils; turbine oils; hydraulic oils, refrigerator and similar industrial oils; for light colored natural or synthetic rubber compounds; stabilizer for neoprene and certain plastics.

Hazard: None.

Shipping regulations: None. *

"Ionol" CP. [125] Trademark for a highly purified 2, 6-di-tert-butyl-4-methylphenol (butylated hydroxytoluene) meeting rigid specifications as an antioxidant acceptable (subject to quantitative limitations) to the Food and Drug Administration and the Bureau of Animal Industry.

Properties: White crystalline solid soluble in fats and oils. Commercial product is 99 mol. % pure (min); f. p. 69.41°C min.

Containers: 20-gal non-returnable fiber drums with polyethylene liner.

Uses: To retard rancidification and extend shelf life of edible fats and oils from animal and vegetable sources, and of fat-containing foodstuffs in general.

Shipping regulations: None. *

ionone [alpha- or beta-cyclocitrylideneacetone; (alpha)4-(2, 6, 6-trimethyl-2(in beta-form is 1)-cyclohexenyl)buten-3-one-2] $C_{13}H_{20}O$. See also "Irisone."

Properties: Light yellow to colorless liquid, having violet-orris type odor. Soluble in alcohol; b. p. 126-128°C at 12 mm; sp. gr. 0.935 (25/25°C); refractive index 1.506 (20°C).

Grades: Alpha, beta, and mixtures of varying proportions.

Containers: Cans.

Derivation: Condensation of citronellal from lemon-grass oil with acetone.

Uses: Perfumery; synthesis of intermediates of vitamin A.

ion retardation. A process based on amphoteric (bifunctional) ion exchange resins containing both anion and cation adsorption sites. These sites will associate with mobile anions and cations in solution and thus remove both kinds of ions from solutions. These ions may be eluted by rinsing with water. Process can make clean separations of ionic-nonionic mixtures. Has also been suggested for demineralization of salt solutions.

iopanoic acid (iodopanoic acid; 3-(3-amino-2, 4, 6-triiodophenyl)-2-ethylpropanoic acid) $C_6HI_3NH_2CH_2CH(COOH)C_2H_5$.

Properties: Cream-colored, tasteless powder with faintly aromatic odor; m. p. 152-158°C (dec); darkens on exposure to light; soluble in acetone, ether, alcohol, chloroform and dilute alkalies; insoluble in water.

Grade: U. S. P. XVI.

Use: Medicine.

iophendylate

$IC_6H_4CH(CH_3)CH_2(CH_2)_7COOCH_2CH_3$.

A mixture of isomers of ethyl

iodophenylundecylate in uniform, but unknown proportions.
Properties: Colorless to pale yellow, odorless, viscous liquid; color darkens on long exposure to air; sp. gr. 1.248-1.257 (25/25°C); refractive index 1.5235-1.5255 (25°C); freely soluble in alcohol, benzene, chloroform, and ether; very slightly soluble in water.
Grade: U. S. P. XVI.
Use: Medicine.

iophenoxic acid $I_3(OH)C_6HCH_2CH(C_2H_5)COOH$. alpha-Ethyl-beta-(2,4,6-triiodo-3-hydroxyphenyl)propionic acid; alpha-ethyl-3-hydroxy-2,4,6-triiodohydrocinnamic acid.
Properties: White or creamy white crystalline powder with faint odor and characteristic taste. M. p. 146-149°C; very slightly soluble in water; soluble in alcohol, ether, alkaline solutions.
Grade: U. S. P. XVI (98-102%).
Use: Medicine (radiopaque medium).

"Iosan." [284] Trademark for a cleaner-sanitizer-disinfectant particularly formulated for use in the dairy field. Contains nonionic iodine complexes. Claimed to be non toxic, non irritating, non staining when used as directed.

"Iosol." [243] Trademark for solvent soluble dyes for plastics and lacquers.

iothiouracil sodium $CHNC(SNa)NC(OH)CI$. (sodium 5-iodo-2-thiouracil).
Properties: Odorless, white to light yellow crystalline powder; m. p. 235-240°C (dec); slightly soluble in alcohol and water; practically insoluble in acids. Usually obtained as dihydrate which is reasonably stable to moisture and sunlight at room temperature; pH 8.5-9.5 (2% solution).
Grade: N. N. D.
Use: Medicine.

IPA. Abbreviation for isopropyl alcohol.

ipado. See coca.

IPAE. See isopropylaminoethanols.

IPC (INPC; isopropyl N-phenylcarbamate) $C_6H_5NHCOOCH(CH_3)_2$.
Properties: White to gray crystalline needles, odorless when pure; m. p. 84°C (technical material); soluble in alcohol, acetone, isopropyl alcohol; slightly soluble in water.
Containers: Bottles; fiber drums.
Uses: Pre-emergence herbicide.
Caution! Harmful if swallowed. MCA warning label.

IPC, chloro-. See chloro-IPC.

ipecac (ipecacuanha; cephaelis). Dried root of Cephaelis ipecacuanha.
Habitat: Brazil and Bolivia; cultivated in India.
Grades: Technical; U. S. P. XVI.
Containers: Whole root: bags; powdered: boxes, barrels.
Uses: Medicine; tanning agent; manufacture of sweetmeats.
Shipping regulations: None. *

ipecacuanha. See ipecac.

ipomea (Mexican scammony root; jalap, orizaba). Dried root of Ipomoea orizabensis.
Grades: N. F. XI.
Containers: Bags.
Use: Medicine.
Shipping regulations: None. *

ipomea resin (Mexican scammony resin). A resin obtained by extraction with diluted alcohol from ipomea.
Properties: Brown or yellowish-orange translucent masses; odor characteristic; taste acrid; soluble in alcohol and chloroform.
Grade: N. F. XI.
Use: Medicine.

"Ipral." [412] Trademark for probarbital sodium (q. v.).

iproniazid phosphate (1-isonicotinyl-2-isopropylhydrazine phosphate) $C_9H_{13}N_3O \cdot H_3PO_4$.
Properties: M. p. 175.5-179°. Soluble in water; slightly soluble in alcohol; insoluble in chloroform and ether.
Grade: N. N. D.
Use: Medicine.

Ir. Symbol for iridium.

"Irgasan BS-200." [219] Trademark for 3,5,3'4'-tetrachlorosalicylanilide (TCSA).
Properties: White, free flowing non-hygroscopic powder; m. p. 161-3°C. Soluble in alkali and in most organic solvents; can be solubilized in soap and detergent formulations. Fluoresces when activated by ultraviolet light. Substantive to fabrics, for example cotton, nylon and wool.
Uses: Bacteriostat - fungistat, exhibiting activity against a broad spectrum of bacteria and fungi. Used in soaps, detergents, cosmetics, deodorants, cutting oils, plastics, cleaning compounds.

iridic bromide (iridium bromide; iridium tetrabromide) $IrBr_4$.
Properties: Hygroscopic powder. Soluble in alcohol, water.
Grades: Technical.

iridic chloride (iridium chloride; iridium tetrachloride) $IrCl_4$.
Properties: Brownish-black mass. Hygroscopic. Soluble in water, and alcohol.
Grades: Technical.
Uses: Analysis (testing for nitric acid in the presence of nitrous acid); in microscopic work.

iridium Ir. Element of atomic number 77, one of the platinum metals, group VIII of the periodic classification.
Properties: Silver-white metal. Limited ductility. Does not tarnish in air. On heating strongly a slightly volatile oxide is formed. Insoluble in acids, slowly soluble in aqua regia and in fused alkalies.
Constants: Sp. gr. 22.4; m. p. 2454°C; Brinell hardness (cast) 218.

Derivation: Occurs with platinum; remains insoluble when the crude platinum is treated with aqua regia; occurs as iridosmine.
Containers: Boxes.
Uses: As an alloy with platinum for jewelry; for electric contacts and thermocouples; commercial electrodes, and resistance wires; used for tipping pens when alloyed with osmium.

iridium 192. Radioactive iridium of mass number 192.
Properties: Half-life, 75 days; radiation, beta and gamma.
Derivation: Pile irradiation of iridium metal slugs.
Forms available: Iridium metal, or potassium or sodium chloroiridate in hydrochloric acid solution.
Uses: Radiography of light castings; treatment of cancer.
Shipping regulations: Poison, class D, radioactive material. Red label. *

iridium bromide. See iridic bromide.

iridium chloride. See iridic chloride.

iridium-potassium chloride (potassium chloroiridate; potassium-iridium chloride; potassium iridochloride) $IrCl_4 \cdot 2KCl$.
Properties: Dark-red crystals. Soluble in water (hot).
Grades: Technical.
Use: Black pigment (porcelain decoration).

iridium sesquioxide Ir_2O_3.
Properties: Black powder. Slightly soluble in hydrochloric acid (conc); insoluble in water.
Grades: Technical.
Use: Ceramics (porcelain decoration).

iridium-sodium chloride (sodium-iridium chloride) $IrCl_4 \cdot 2NaCl \cdot 6H_2O$.
Properties: Brownish-black crystals. Soluble in water.

iridium tetrabromide. See iridic bromide.

iridium tetrachloride. See iridic chloride.

iridosmine (osmiridium). A natural alloy of iridium and osmium containing some platinum, rhodium, ruthenium, iron, copper, palladium. Tin-white to light steel gray in color; streak, same; metallic luster. Composition is variable ranging from 10.0-77.2% iridium, 17.2-80.0% osmium, 0-10.1% platinum, 0-17.2% rhodium, 0-8.9% ruthenium, 0-1.5% iron, 0-0.9% copper, trace, palladium. Unattacked by aqua regia.
Constants: Sp. gr. 18.8-21.12; hardness 6-7.
Occurrence: United States (California, Oregon); Russia; Japan; Borneo; Australia; Tasmania; New Zealand; Canada; West Indies; Brazil; South Africa.
Uses: Very hard and resistant to corrosion. Fountain-pen point tips; surgical needles; watch pivots; compass bearings; hardening platinum (standard weights, jewelry); source of iridium and osmium.

iris. See orris.

Irish broom. See scoparius.

Irish moss. See chondrus.

"Irisol." [307] A fast alizarine direct violet.

"Irisone." [227] Trademark for a series of compounds consisting of alpha- and beta-ionones. Light yellow to colorless liquids, having floral odors, primarily violet.
Ionone Coeur is a refined mixture of alpha- and beta-ionone.
Ionone Pure, used in less expensive compounds, is also a mixture of alpha- and beta-ionone.
Ionone Beta Pure is a purified beta-ionone used in pharmaceuticals and for chemical synthesis.
Ionone Bis is used in soap compositions.

iron (ferrum) Fe. Element of atomic number 26, of group VIII of the periodic system.
Properties: Silvery-white, tenacious, lustrous, malleable, ductile metal, rarely found native except in meteorites. The only metal which can be tempered, i.e., hardened by heating and sudden cooling; heating and slow cooling make it very pliable. It is magnetic and can be magnetized, but soon loses its magnetism (steel retains it). It rapidly oxidizes (rusts) in damp or salty air and is corroded or dissolved by acids. It reacts with steam when hot to yield hydrogen and iron oxides. It is very brittle at very low tmperatures, softens at red-heat, and can be welded at white-heat.

Pure iron has a specific gravity of 7.85 and melts at 1530°C. It is difficult to purify and the pure metal is rarely encountered. Practically all commercial forms are impure in one way or another. The impurities are costly to remove or else are added to achieve desirable properties. Probably the purest form of iron available is powdered iron obtained by decomposition of iron pentacarbonyl (see carbonyl iron powder).
Major iron ores: Hematate; limonite; magnetite; siderite; and recently, taconite.
Derivation: By smelting the ores with limestone and coke in blast furnaces. The product is pig iron, which is treated further to produce the following common commercial forms.
Forms available: Pig iron, cast iron, wrought iron; steel; powdered iron; wire, filings, etc. (See iron, pig; iron, cast; iron, wrought; iron, ductile; steel; also various alloys of iron under ferro-.
Grades: N. F. XI (wire, filings, powder); powdered (electrolytic 99.9%, atomized, sponge, carbonyl, hydrogen reduced, electronic, milled pulverized).

iron 55. Radioactive iron of mass number 55.
Properties: Half-life 2.91 years; decays through K capture; radiotoxicity, very hazardous.

iron 59. Radioactive iron of mass number 59.
Properties: Half-life 46.3 days; radiation, beta and gamma; radiotoxicity, moderately

*See "I.C.C. Shipping Regulations," page xiii.
Reference numbers refer to name of manufacturer. See "List of Manufacturers," page v.

hazardous.

Derivation: Pile irradiation of iron metal, giving a product which contains iron 55 impurity. Both iron 55 and iron 59 are produced pure in the cyclotron. (See iron 55). Enriched samples of each are also available.

Forms available: Ferric chloride in hydrochloric acid solution; metallic iron; iron 59 is also available in tagged compounds such as hemoglobin, ferrous gluconate, ferrous ascorbate, etc., and in other forms by service irradiation.

Uses: For studies on the distribution of alloying elements in welds, the mechanism of corrosion by organic acids, engine friction wear, the lubricating qualities of engine oils, the chemistry of iron in sea water, mineral supplements in animal nutrition, liver functions and anemia, etc.

Shipping regulations: Poison, class D, radioactive material. Red label.*

iron acetate. See ferric acetate, basic; ferrous acetate.

iron acetate, basic. See ferric acetate, basic.

iron acetate liquor (iron liquor; black liquor; black mordant; iron pyrolignite).

Properties: Intensely black liquor, sometimes containing copperas or tannin. Absorbs oxygen from the air. Sp. gr. 1.09-1.115, containing 5-5.5% iron.

Derivation: (a) By the action of pyroligneous acid on iron filings; (b) double decomposition of ferrous sulfate with calcium pyrolignite.

Grades: According to specific gravity.

Containers: Wooden barrels.

Uses: Mordant, especially for alizarine and nitroso dyes, and for dyeing and printing logwood.

iron alginate. See ferric alginate.

iron alum. See ferric-potassium sulfate.

iron-ammonium citrate. See ferric-ammonium citrate.

iron-ammonium oxalate. See ferric-ammonium oxalate.

iron-ammonium sulfate. See ferric-ammonium sulfate and ferrous-ammonium sulfate.

iron-ammonium tartrate. See ferric-ammonium tartrate.

iron arsenate. See ferrous arsenate.

iron benzoate. See ferric benzoate.

iron black.

Properties: Fine black powder.

Derivation: By the action of zinc upon an acid solution of an antimony salt, a black antimony being precipitated as a fine powder.

Uses: Imparting the appearance of polished steel to papier maché and plaster of Paris.

Shipping regulations: None.*

iron blues. The iron blues are prepared by precipitating a ferrous ferrocyanide from a soluble ferrocyanide and ferrous sulfate. Subsequent oxidation produces a complex ferriferrocyanide whose shade and pigment properties are dependent upon the oxidizing agent, reactant concentrations, pH, temperature, size of batch and other conditions of manufacture. Common oxidants are nitric acid, sulfuric acid, potassium dichromate and sulfuric acid, perchlorates, and peroxides.

Properties: Semi-transparent pigment of powerful tinctorial strength. Insoluble in water, oils, alcohol, hot paraffin, organic solvents, and unaffected by dilute acids. Unstable to alkalies of all concentrations or reducing media. Resistance to light and ordinary baking temperatures allows it to be used for permanent industrial finishes and automobile finishes. Varying shades can be produced ranging from green to red tint with the mass tone from a reddish blue to a jet blue depending on oxidant and other conditions.

Containers: Barrels, carloads.

Uses: Paint and printing ink pigment, artist colors, laundry blue, paper dyeing, fertilizer ingredient.

Note: These blue pigments are known variously as Bronze blue, Milori blue, Chinese blue, Prussian blue, etc. Since the final shade is dependent on manufacturing conditions, purchasing is generally based on duplication of a shade rather than by a named pigment.

iron bromide. See ferric bromide and ferrous bromide.

iron buff (Nankin yellow). Ferric hydroxide dyed on cotton or cotton goods by steeping the latter in a solution of ferrous sulfate, basic ferric sulfate or ferric nitrate and precipitating the hydroxide on the fiber by means of calcium hydroxide solution, sodium hydroxide solution or soda ash.

iron carbonate, precipitated. See iron oxide, brown.

iron carbonyl. See iron pentacarbonyl.

iron, cast. Any iron-carbon alloy that contains more than 1.7% carbon, and usually between 2 and 4.0%. Such iron usually also contains 0.1 or 0.2% sulfur, 0.5 to 3% silicon, 0.5 to 1% manganese and up to 1% phosphorus. Cannot be shaped by hammering, rolling, or pressing.

cast iron, alloy. Cast iron containing chromium, copper, molybdenum, nickel, or other steel-alloying elements in amounts from 0.1 to 5% for the purpose of improving strength and wear, corrosion, or scaling resistance.

cast iron, gray. Cast iron with gray fracture and with its carbon largely in the uncombined state. The most common form of cast iron, easily melted and machined, relatively soft and tough. Properties depend upon composition, rate of cooling, and heat treatment.

cast iron, malleable. White cast iron that has been annealed after solidification in order to reduce carbon content and produce a product similar in many ways to mild steel.

cast iron, white. A cast iron with silvery surfaces where broken, low silicon content, and all its carbon chemically combined with iron, produced by sudden chilling of the molten iron. Very hard, brittle, and cannot be machined. Produced as an intermediate stage in making malleable cast iron and as a thin outer layer on the surface of gray cast iron.

iron chloride. See ferric chloride anhydrous and ferrous chloride.

iron choline citrate chelate. See ferrocholinate.

iron chromate. See ferric chromate.

iron citrate. See ferric citrate.

iron "cyanide." See ferric ferrocyanide.

iron dichloride. See ferrous chloride.

iron dichromate. See ferric dichromate.

iron, ductile. A malleable cast iron produced by the addition of sufficient magnesium and/or cerium to the melt to cause graphite to precipitate as spherulites rather than flakes. It has superior strength, ductility, toughness, machinability and corrosion resistance as compared to gray cast iron and has better castability, finish, and machinability than cast steel.

irone. See methylionone.

iron ethiops. See iron oxide, black.

iron ferrocyanide. See ferric ferrocyanide.

iron fluoride. See ferrous fluoride; ferric fluoride.

iron gallotannate. See ferric tannate.

iron, galvanized. See galvanized iron.

iron glance. Specular iron ore. See hematite, red.

iron gluconate. See ferrous gluconate.

iron glycerophosphate. See ferric glycerophosphate.

iron hydrate. See ferric hydroxide.

iron hydroxide. See ferric hydroxide.

iron hypophosphite. See ferric hypophosphite.

iron iodide. See ferrous iodide.

iron lactate. See ferrous lactate.

iron liquor. See iron acetate liquor.

iron-magnesium sulfate. See ferrous-magnesium sulfate.

iron malate. See ferric malate.

iron-manganese sulfate. See ferrous-manganese sulfate.

iron metavanadate. See ferric vanadate.

iron molybdate. See ferrimolybdite.

iron monosulfide. See ferrous sulfide.

iron monoxide. See ferrous oxide.

iron-nickel alloys (nickel-iron alloys).

Alloys of iron and nickel which are entirely austenitic in character at room temperature are properly called "iron-nickel alloys," otherwise they are considered nickel-steels. All alloys containing over 34% nickel are austenitic at all temperatures. With lower nickel content the heat treatment helps determine classification as an "alloy" or a steel.

The iron-nickel system has a peculiar range of properties of thermal expansion, thermoelasticity, and magnetic characteristics.

The alloy containing 36% nickel has a coefficient of thermal expansion of practically zero, with increases as the nickel content is raised or lowered. Within the 30-60% nickel range, practically any desired coefficient of expansion can be obtained and matched to any desired material so as to provide strainless junctions between the iron-nickel alloy and the second material at any temperature, or to permit controlled differential expansion as in bimetal thermostats.

Most metals lose stiffness with increasing temperature but iron-nickel alloys from 27-44% nickel have a positive thermal coefficient of elastic modulus, with a maximum coefficient at about 36% nickel. This property is useful in temperature-compensating devices, where the increased stiffness of the nickel counteracts the loss in stiffness of the other element. Thus springs can be constructed of steel and 36% nickel-iron alloy which maintain constant stiffness regardless of temperature. Substitution of 8-12% chromium for iron in the 36% nickel-iron alloy gives a metal with a thermal coefficient or "stiffness" of zero, that is, a constant modulus of elasticity, useful in watchsprings, tuning forks, bourdon tubes, etc.

Tremendous variations in magnetic properties are possible. Iron-nickel alloys (austenitic) of nickel content below 30% are non-magnetic at ordinary and at high temperatures. For the alloy with 30% nickel, room temperature is the magnetic transformation temperature, and alloys in this composition range show permeability varying with temperature, making possible temperature-compensating shunts in wattmeters and other electric instruments. From compositions of 35-90% nickel, the iron-nickel alloys show high permeability and low hysteresis loss. Small amounts of other alloying elements vary the magnetic properties greatly and permit "tailoring" of alloys to suit special requirements.

The iron-nickel alloys are not heat-hardenable, but addition of aluminum and titanium develops tempering characteristics.

See also nickel steels.

iron-nickel-chromium alloys. Ferrous alloys in which nickel exceeds chromium.

iron nitrate. See ferric nitrate.

iron oleate. See ferric oleate.

iron ore, bog. See limonite.

iron ore, brown. See limonite.

iron-ore cement. Cements in which ferric oxide replaces a large part of the alumina. There must be some alumina present, however. Iron-ore cement is rather slow setting and hardening, but is more resistant to sea water than is Portland cement. It is light to chocolate brown in color and has a specific gravity about 3.31, higher than Portland cement.

iron ore, chrome. See chromite.

iron ore, kidney. Red hematite (q. v.) exhibiting a fibrous or columnar structure and a nodular surface.

iron ore, magnetic. See magnetite.

iron ore, red. See hematite, red.

iron ore, spathic. See siderite.

iron ore, specular. See hematite, red.

iron ore, titanic. See ilmenite.

iron oxalate. See ferrous oxalate.

iron oxide. Ferrous oxide; ferric oxide; iron oxide, black. See also iron oxide pigments; iron oxide reds.

iron oxide, black (ferrosoferric oxide; ferroferric oxide; iron oxide, magnetic; iron ethiops; black rouge) $FeO \cdot Fe_2O_3$ or Fe_3O_4. See also the mineral form, magnetite.

Properties: Reddish or bluish black amorphous powder; sp. gr. 4.96; soluble in acids; insoluble in water, alcohol and ether.

Derivation: (a) Action of air, steam or carbon dioxide on iron. (b) Specially pure grade by precipitating hydrated ferric oxide from a solution of iron salts, dehydrating and reducing with hydrogen. (c) Occurs in nature as the mineral magnetite (q. v.).

Grades: Technical; pure (96% min.).

Containers: Wooden kegs; fiber drums; multiwall paper sacks, carloads.

Uses: Pigment; polishing compound; metallurgy; medicine; specially pure grades in magnetic inks and in ferrites for electronic industry.

iron oxide, brown (iron subcarbonate; iron carbonate, precipitated)

Properties: Reddish-brown powder, containing ferric carbonate with ferric hydroxide $Fe(OH)_3$, and ferrous hydroxide $Fe(OH)_2$ in varying quantities. Not a true oxide. Soluble in acids; insoluble in water and alcohol.

Derivation: By the interaction of solution of ferrous sulfate and sodium carbonate.

Grades: Technical.

Containers: Wooden barrels; fiber drums; multiwall paper sacks, carloads.

Use: Paint pigment.

Shipping regulations: None.*

iron oxide, hydrated. See ferric hydroxide.

iron oxide, magnetic. See iron oxide, black.

iron oxide pigments. The basic colors of these pigments are determined by chemical composition. Reds are ferric oxide, Fe_2O_3; yellows are hydrated ferric oxide, $Fe_2O_3 \cdot H_2O$, and blacks are ferroferric oxide, Fe_3O_4. The synthetic oxides are up to 96-99.5% pure. The natural oxides show a wide range of iron oxide content, and vary in color through red, yellow and brown. They include ochers, umbers, siennas, metallic browns, and red oxides. See iron oxide, black; iron oxide, brown; iron oxide reds; iron oxide yellows.

iron oxide process. A process for the removal of sulfides from a gas by passing the gas through a mixture of iron oxide, Fe_2O_3, and wood shavings. The iron oxide is converted to iron sulfide and can be regenerated by allowing the iron sulfide to contact air.

iron oxide reds (crocus martis adstringens; polishing crocus; purple oxide; red oxide; red stone; jeweler's rouge; rubigo; Indian red; red bole; bole; Armenian bole; caput mortuum; English red; angel red; chemical red; Pompeian red; Persian red; raddle; reddle; red rudd; red ochre; red chalk; red earth; Prussian red; Italian red; terra di Sienna; mineral rouge; blood red; pale oxide of iron; iron saffron; imperial red; Nuremberg red; scarlet red; Prague red Sinopis; Van Dyck red; Spanish oxide; Turkey red; Mars red; rouge de Mars; Pompey red; Venetian red). Pigments composed mainly of ferric oxide, Fe_2O_3. See also ferric oxide for a description of the pure material.

Some of these synonyms apply to relatively impure materials used as pigments; some apply to naturally occurring hematite (q. v.) of various degrees of purity, and before or after purification, heating or other treatment; others refer to synthetic materials prepared by various special methods. In most cases the terms are used loosely. When available, additional information is given under separate entries.

Grades: Sold usually on basis of iron oxide content and covering properties.

Containers: Fiber drums; multiwall paper bags (carloads).

Uses: Heavy-duty pigments, as in railway finishes, marine paints, metal primers; polishing compounds; pigment in rubber products; theatrical rouge; grease paints.

Shipping regulations: None.*

iron oxide, synthetic. See rouge.

iron oxide yellows. Hydrated ferric oxide, $Fe_2O_3 \cdot H_2O$. Precipitated pigments of much finer particle size and much greater tinctorial strength than the naturally occurring oxides such as ocher (q. v.). They have very low cost and are very useful for producing cream and buff-colored tints where the brightness of chrome yellows is not required. They have excellent light fastness

and resistance to alkali.
Containers: Fiber drums, multiwall paper bags.

iron pentacarbonyl (iron carbonyl) $Fe(CO)_5$.
Properties: Mobile yellow liquid. Decomposes on exposure to air or to light; poisonous! Soluble in nickel tetracarbonyl and most organic solvents; soluble with decomposition in acids and alkalies; insoluble in water. Sp. gr. 1.466 (18°C); b.p. 102.8°C (749 mm); decomposes at 200°C; m.p. −21°C.
Derivation: Finely divided iron is treated with carbon monoxide, in the presence of a catalyst such as ammonia.
Uses: Organic synthesis; anti-knock agent; source of pure iron powder. See carbonyl iron powder.
Shipping regulations: None. *

iron peptonate. See iron, peptonized.

iron peptonized (iron peptonate).
Properties: Dark brown granules or powder.
Derivation: Combination of iron oxide and peptone rendered soluble and sequestered (nonionic) by the presence of sodium citrate.

iron perchloride. See ferric chloride, anhydrous.

iron persulfate. See ferric sulfate.

iron phosphate. See ferric phosphate.

iron phosphide. See ferrous phosphide.

iron, pig. The impure iron produced by reduction of iron ore with limestone, coke and air in a blast furnace. This is the industrial source of all forms of iron and steel. See iron, cast, for approximate composition.

iron-potassium citrate. See ferric-potassium citrate.

iron-potassium sulfate. See ferric-potassium sulfate.

iron-potassium tartrate (tartrated iron; tartar, chalybeated; ferric-potassium tartrate) $Fe_2(C_4H_4O_6)_3 \cdot K_2C_4H_4O_6 \cdot H_2O$.
Properties: Thin, tansparent scales; deep garnet color; sweetish taste; astringent; arsenic limit 5 parts per million. Soluble in water, sparingly in alcohol.
Derivation: By submitting ferric oxide to the action of acid potassium tartrate for 24 hours, gently heating the product, adding distilled water, filtering, evaporating slowly to a syrup and drying.
Containers: Glass bottles.
Use: Medicine.
Shipping regulations: None. *

iron, powdered. See iron (under Grades); iron, reduced; carbonyl iron powder.

iron protochloride. See ferrous chloride.

iron protoiodide. See ferrous iodide.

iron protosulfide. See ferrous sulfide.

iron pyrites. See pyrite.

iron pyrites, white. See marcasite.

iron pyrolignite. See iron acetate liquor.

iron pyrophosphate. See ferric pyrophosphate.

iron-quinine citrate. See ferrous-quinine citrate.

iron red. A name given to red varieties of ferric oxide that are used as pigments. See iron oxide reds.

iron, reduced (ferrum reductum). Elementary iron obtained by chemical process in powdered form.
Properties: Grayish-black, amorphous, fine granular powder with no more than a slight luster. Stable in dry air.
Derivation: (a) By reducing ferric oxide, heated to a dull redness, in a stream of dry hydrogen; (b) by the decomposition of iron pentacarbonyl (see carbonyl iron powder) or (c) electrolytically.
Containers: 1-lb bottles; 1-, 5-, 25-, and 50-lb cans.
Uses: Medicine; organic synthesis; feed additive.

iron resinate. See ferric resinate.

iron saffron. See Indian red.

iron salicylate. See ferric salicylate.

iron salts. See ferric and ferrous salts.

iron sesquichloride. See ferric chloride, anhydrous.

iron sesquioxide. See ferric oxide.

iron sesquisulfate. See ferric sulfate.

iron, sherardized. See sherardizing.

iron silicate. See ferric silicate.

iron-sodium citrate. See ferric-sodium citrate.

iron-sodium oxalate. See ferric-sodium oxalate.

iron sponge. Finely divided porous form of iron made by reducing an iron oxide at such low temperatures that melting does not occur, usually by mixing iron oxide and coke and applying limited increase in temperature.
Uses: For precipitating copper or lead from solutions of their salts, removing sulfur compounds from coke-oven gas, and in electric-furnace steel operations.
Shipping regulations: When spent, or not properly oxidized: flammable solid. Yellow label by freight. Not accepted by express. *

irons, stainless. Alloys containing 3 to 28% chromium, with or without traces of nickel; essentially magnetic and ferritic in character. High chromium irons are brittle after welding. Most popular composition for fabrication is 15-18% chromium, 0.1% carbon (max.).
See ferritic stainless steel, under steel.

iron stearate. See ferric stearate.

ironstone clay. See argillaceous hematite and hematite, red.

ironstone clay, brown. See limonite.

iron subcarbonate. See iron oxide, brown.

iron subsulfate. See ferric sulfate, basic.

iron succinate. See ferric succinate.

iron sulfate. See ferric sulfate and ferrous sulfate.

iron sulfate, basic. See ferric sulfate, basic.

iron sulfide. See ferrous sulfide.

iron sulfuret. See ferrous sulfide.

iron tannate. See ferric tannate.

iron, tartrated. See iron-potassium tartrate.

iron tersulfate. See ferric sulfate.

iron tribromide. See ferric bromide.

iron trichloride. See ferric chloride, anhydrous.

iron trioxide. See ferric oxide.

iron vanadate. See ferric vanadate.

iron vitriol. See ferrous sulfate.

iron, wrought. Highly purified iron that has been uniformly admixed with a small proportion of slag, the mixing occurring while the iron is in a pasty stage somewhat below its melting point. This was the normal product of early iron forges because of the relatively low temperatures reached in such equipment. Wrought iron is now made from pig iron by special purification and mixing procedures referred to as "puddling" and "shotting."

A typical composition is carbon 0.1%, manganese 0.1%, phosphorus 0.08 to 0.16% (about half of this is in the slag), up to 0.035% sulfur, and 0.1 to 0.2% silicon (all in the slag, which itself is from 1 to 4% of the whole). Higher percentages of carbon and manganese generally indicate contamination with steel.

Wrought iron is relatively soft and malleable, has a fibrous structure due to the admixed slag, and shows great resistance to progressive corrosion.

"Irrathene." [245] Trademark for a thermosetting form of polyethylene (q. v.) formed by irradiation of polyethylene with high energy cathode rays (electrons). The product does not melt (up to 250°C) but oxidizes rapidly at elevated temperatures unless protected by an inhibitor. Its resistance to acids, alkalies, and solvents is superior to that of polyethylene and it has excellent electrical properties even at 200°C. It is available in films and tapes used for packaging and electrical insulation.

"Irron." [169] Trademark for 8-quinolinol-7-iodo-5-sulfonic acid used in the colorimetric determination of iron.

"IRS-2000 Latices." [52] Trademark for a series of latices containing emulsion copolymers of butadiene and styrene. Made by reacting a 50/50 ratio of monomers to a high degree of conversion in a hot emulsion recipe with a rosin soap as an emulsifier. Solids contents range from 40 to over 60%; viscosities range from 100 to 28,000 cps. Used in adhesives, chewing gum bases, mastic cements, carpet backing, fabric combining.

isanolic acid (erythrogenic acid). An 18-carbon fatty acid with acetylene triple bonds at the 9 and 11 positions and having also an ethylene bond but no hydroxyl substituent. Occurs in isano oil along with closely related bolekic acid which is a hydroxy acid. Similar acids are found in onguekoa seeds. Useful in modifying epoxy resin formulations.

isano oil. Fatty oil from an African tree of same name. When the oil is heated to 200°C, it polymerizes rapidly and may explode spontaneously. See isanolic acid.

isatic acid anhydride. See isatin.

isatic acid lactime. See isatin.

isatin (ortho-aminophenylglyoxalic lactime; ortho-aminobenzoylformic acid; isatic acid anhydride; isatic acid lactime) $C_6H_4COC(OH)N$(bicyclic).

Properties: Yellowish-red or orange crystals; bitter taste. Soluble in water, alcohol, and ether. M.p. 200-203°C.

Derivation: From indigo by oxidation.

Grades: Technical; reagent.

Uses: Dyestuffs; pharmaceuticals; analytical reagent.

Shipping regulations: None.*

isatoic anhydride $C_8NO_3H_4$. Tan powder.

Grade: Technical, 96% min.

Use: Intermediate for flavors and agricultural chemicals.

isethionic acid (2-hydroxyethanesulfonic acid) $HOCH_2CH_2SO_3H$.

Properties: Liquid; b.p. 100°C (dec); very soluble in water; insoluble in alcohol.

Uses: Detergents; surfactants; synthesis.

isinglass.

1. (Ichthyocolla). A pure white, odorless, tasteless gelatin prepared from the inner skins of the swimming bladders of fish, usually the sturgeon. Used as an adhesive and clarifying agent.
2. Mica.

isinglass, Bengal. See agar-agar.

isinglass, Ceylon. See agar-agar.

isinglass, Chinese. See agar-agar.

isinglass, Japanese. See agar-agar.

"Ismelin." [305] Trademark for guanethidine.

Use: Medicine.

iso-. A prefix denoting an isomer of a compound; specifically, denoting an isomer having a single, simple branching at the end of a straight chain. Iso- compounds will be found under I in this dictionary.

isoalloxazine (flavin) $C_{10}H_6N_4O_2$. An isomer of alloxazine. Derivatives of isoalloxazine

*See "I.C.C. Shipping Regulations," page xiii.
Reference numbers refer to name of manufacturer. See "List of Manufacturers," page v.

are widely distributed in plants and animals, usually as yellow pigments. See also riboflavin and flavin enzymes.

isoamyl acetate $CH_3COOCH_2CH_2CH(CH_3)_2$.
Properties: Colorless liquid; b. p. 142°C; m. p. −78.5°C; sp. gr. (15/4°C) 0.876; wt/gal (15°C) 7.30 lbs; flash point (closed cup) 92°F. Slightly soluble in water; miscible with alcohol and ether.
Derivation: Rectification of commercial amyl acetate.
Grades: Reagent; technical.
Uses: Flavoring; perfumes; solvent.
Caution! Avoid prolonged breathing of vapor, or prolonged or repeated contact with skin. Use with adequate ventilation. Keep away from heat and open flame. MCA warning label.
See also amyl acetate.

isoamyl alcohol, primary (3-methyl-1-butanol; isobutyl carbinol) $(CH_3)_2CHCH_2CH_2OH$.
See also fusel oil.
Properties: Colorless liquid; pungent taste; disagreeable odor. B. p. 132.0°C; m. p. −117.2°C; sp. gr. 0.813 (15/4°C); wt/gal 6.79 lb; refractive index (n 20/D) 1.4075; flash point (closed cup) 114°F. Slightly soluble in water; miscible with alcohol and ether.
Derivation: Distillation of fusel oil or the mixed alcohols resulting from the chlorination and hydrolysis of pentane.
Grades: Technical.
Containers: 1-, 5-, 55-gal drums; tank cars.
Uses: Photographic chemical; organic synthesis; pharmaceutical products; medicine; solvent; for determination of fat in milk; in microscopy.
Shipping regulations: None. *

isoamyl benzoate (amyl benzoate) $C_6H_5COOC_5H_{11}$.
Properties: Colorless liquid, with fruity odor. Sp. gr. 0.986 to 0.989; refractive index 1.493; b. p. 260°C. Soluble in 2 vols. of 80% alcohol; insoluble in water.
Use: Perfumery; flavors.

isoamyl benzyl ether (benzyl isoamyl ether) $C_5H_{11}OCH_2C_6H_5$.
Properties: Colorless liquid; fruity odor; sp. gr. 0.904-0.908; refractive index 1.481-1.485; soluble in 4 parts of 80% alcohol.
Grades: Technical.
Use: Soap perfumes.

isoamyl butyrate $C_5H_{11}OOCC_3H_7$.
Properties: Practically water white. Sp. gr. 0.866 to 0.868 (15.5°C); boiling range 150 to 180°C. Soluble in alcohol and ether; very slightly soluble in water.
Derivation: By treating isoamyl alcohol with butyric acid.
Method of purification: Distillation.
Grades: Commercial, 95 to 100% ester content.
Containers: Glass carboys; tin-lined drums.
Uses: Flavoring extracts; solvent and plasticizer for cellulose acetate.
Shipping regulations: None. *

isoamyl chloride $C_5H_{11}Cl$. Any of several compounds or mixtures thereof may be referred to by this name, since numerous isomers are possible, the most common of which is the following: 1-chloro-3-methylbutane $(CH_3)_2CH(CH_2)_2Cl$.
Properties: Colorless or slightly yellow liquid; b. p. 99.7°C (758 mm); sp. gr. 0.893; refractive index 1.410; insoluble in water; soluble in alcohol and ether.
Typical specifications: Assay 90%, sp. gr. 0.8725 to 0.8760 at 25/25°C.
Derivation: Isoamyl alcohol and hydrogen chloride, or chlorination of isopentane.
Containers: Glass bottles; carboys; tins.
Uses: (Mixtures, usually also containing normal amyl chloride) solvent (nitrocellulose, varnishes, lacquers, neoprene); in rotogravure inks, and for soil fumigation; also for organic compounds.
Shipping regulations: None. *

isoamyldichloroarsine $C_5H_{11}AsCl_2$.
Properties: Oily liquid. Somewhat agreeable odor. Decomposed by water. Caution! Very irritant! B. p. 88.5 to 91.5°C (15 mm).
Derivation: Interaction of phosphorus trichloride and isoamylarsenic acid.

alpha-**isoamylene.** See 3-methyl-1-butene.

beta-**isoamylene.** See 3-methyl-2-butene.

isoamylenes. Mixture containing high proportion of branched chain five-carbon olefins. Extracted from a low boiling gasoline fraction with sulfuric acid and recovered in high purity by distillation from the acid. Used largely for conversion to isoprene by dehydrogenation.

isoamyl furoate $C_4H_3OCO_2C_5H_{11}$.
Properties: Colorless liquid, becoming brown in light. Insoluble in water; soluble in alcohol and ether. Sp. gr. 1.0335 (20/4°C); b. p. 232 to 234°C, 135-137°C (25 mm); refractive index 1.4720.

isoamyl isovalerate. See isoamyl valerate.

isoamyl nitrite. See amyl nitrite.

isoamyl salicylate (amyl salicylate; orchidae) $C_6H_4OHCOOC_5H_{11}$.
Properties: A water-white liquid sometimes having a faint yellow tinge which should not be pink or red. Has a flowery orchid-like odor. Should not give a red ring when superimposed on a layer of sulfuric acid (indicating free amyl alcohol); sp. gr. 1.053-1.059 (15°C); refractive index (n 20/D) 1.5050-1.5080; optical rotation 0 to +2.30°; b. p. 280°C. Soluble in alcohol, ether; insoluble in water and glycerol.
Derivation: By esterifying salicylic acid with amyl alcohol. The ordinary article of commerce is the isoamyl ester.
Method of purification: Distillation.
Grades: A pure grade of at least 99% ester content which should not exceed 100% on analysis (indicating lower esters).
Containers: Carboys.
Uses: As an ingredient in perfumes and in

perfuming soap. Formerly used medicinally.
Shipping regulations: None. *

isoamyl valerate ("apple essence"; "apple oil"; isoamyl isovalerate; amyl valerianate; amyl valerate) $C_4H_9CO_2C_5H_{11}$.
Properties: Clear liquid; odor of apples when diluted with alcohol; sp. gr. 0.8812; b. p. 203.7°C; soluble in alcohol and ether; slightly soluble in water.
Derivation: By adding sulfuric acid to a mixture of amyl alcohol and valeric acid. Subsequent recovery by distillation.
Method of purification: Rectification.
Grades: Technical.
Containers: Iron drums; glass bottles.
Uses: Medicine; fruit essences.
Shipping regulations: None. *

isoascorbic acid. See erythorbic acid.

isobaric spin. See isotopic spin.

isobars.
1. Lines on a weather map connecting points with the same atmospheric pressure. Constant pressure lines on any type of graph.
2. Nuclides having the same mass number but different atomic numbers, in contrast to isotopes which have the same atomic number but different mass numbers. C-14 and N-14 are isobars.

isoborneol $C_{10}H_{17}OH$. A geometrical isomer of borneol.
Properties: White solid with camphor odor; m. p. 216°C (sublimes). More soluble in most solvents than borneol.
Derivation: By reduction of camphor.
Containers: Cans.
Uses: Perfumery; chemical esters.

isobornyl acetate $C_{10}H_{17}OOCCH_3$.
Properties: Colorless liquid. Pine-needle odor. Sp. gr. 0.978 ± 0.001 (20°C); b. p. 220 to 224°C.
Derivation: By heating camphene (50 to 60°C) with glacial acetic acid and sulfuric acid and separating by adding water.
Grades: Technical.
Containers: Cans.
Uses: Compounding pine-needle odors; toilet waters; bath preparations; antiseptics; theater sprays; soaps; making synthetic camphor.

isobornyl salicylate.
Properties: Viscous, colorless oil. Sweet odor. Ester content 96%.
Grades: Technical.
Uses: Perfumery (fixative); cosmetics (filter for suntan preparations).

isobornyl thiocyanoacetate $C_{13}H_{19}NO_2S$. The technical grade contains 82% or more of isobornyl thiocyanoacetate, also other terpenes and derivatives.
Properties: Yellow, oily liquid; terpene-like odor. Sp. gr. (25/4°) 1.1465; refractive index (25/D) 1.512; acid number 1.19. Very soluble in alcohol, benzene, chloroform, and ether; practically insoluble in water.
Derivation: By treating isoborneol with chloroacetyl chloride and potassium thiocyanate.
Uses: Insecticide; medicine.

isobornyl valerate $(CH_3)_2CHCH_2COOC_{10}H_{17}$.
Properties: Colorless, neutral liquid, oily taste, peculiar aromatic odor. Does not irritate the stomach. Soluble in alcohol, ether; sparingly soluble in water.
Constants: B. p. 132-138°C (12 mm); sp. gr. 0.954.
Use: Medicine.
Shipping regulations: None. *

isobutane (2-methylpropane; trimethylmethane) $(CH_3)_2CHCH_3$.
Properties: Colorless gas; characteristic natural gas odor; stable, does not react with water; has no corrosive action on metals; b. p. −11.73°C; f. p. −159.60°C; sp. gr. 0.5572 (20°C, at saturation pressure); flash point −83°C.
Typical specifications for a technical grade are as follows: Distillation range 9-15°F; sp. gr. of liquid 0.564 (60/60°F); API gravity 119.4 (60°F) (calculated); density of liquid (60°F) 4.70 lb/gal (calculated); vapor pressure 45.4 psi (70°F); purity not less than 95 mole % isobutane; principal impurity approximately 4% normal butane; sulfur content not to exceed 0.010 wt %, probably less than 0.005 wt %.
Derivation: An important component of natural gasoline, refinery gases, wet natural gas; also obtained by isomerization of butane.
Grades: Technical; also available as 99 mole % (pure grade), 99.96 mole % (research grade), and other high purity grades.
Containers: 16-gal pressure cylinders (net contents 68 lbs); 28-gal pressure cylinders; tank cars (10,000 gal).
Uses: Organic synthesis; refrigerant; fuel; starting material for liquid fuel synthesis, aerosol propellant.
Shipping regulations: Flammable gas. Red gas label. *

isobutane hydrate. See gas hydrates.

isobutanol. See isobutyl alcohol.

isobutanolamine. See 2-amino-2-methyl-1-propanol.

isobutene (2-methylpropene; isobutylene) $(CH_3)_2C{:}CH_2$.
Properties: Colorless very volatile liquid or easily liquefied gas; b. p. −6.9°C; m. p. −139°C; flash point −76°C; sp. gr. 0.6 (20°C); soluble in organic solvents. Polymerizes easily and also reacts easily with numerous materials.
Derivation: Gas mixtures containing considerable isobutene are obtain by fractionation of refinery gases resulting from cracking of petroleum.
Containers: Tank cars; cylinders.
Uses: Production of isoöctane, polymer gasoline, butyl rubber, polyisobutene resins, tert-butyl chloride, tert-butanol

and other derivatives, copolymer resins with butadiene, acrylonitrile, etc.
Shipping regulations: Flammable gas, Red gas label.*

isobutyl acetate $C_4H_9OOCCH_3$.
Properties: Colorless, neutral liquid; fruit-like odor. Soluble in alcohols, ether, and hydrocarbons; immiscible with water. B. p. 116-117°C; flash point 18°C; sp. gr. 0.8685 (15°C); refractive index 1.392 (approx.); wt/gal 7.23 lbs; m.p. -99°C.
Derivation: Treating isobutyl alcohol with acetic acid in the presence of catalysts.
Grades: Technical; solvent; perfume.
Containers: 55-gal drums; tank cars.
Use: Solvent for nitrocellulose and lacquers; perfumery.

isobutyl alcohol (isopropylcarbinol; isobutanol; 2-methylpropanol-1) $(CH_3)_2CHCH_2OH$.
Properties: Clear liquid; flammable. Soluble in water, alcohol, and ether. Sp. gr. 0.806 (15°C); b. p. 107°C; flash point 82°F; m. p. -108°C; refractive index 1.397 (15°C).
Derivation: Byproduct of synthetic methanol production, purified by rectification.
Containers: 55-gal drums, tank cars and trucks.
Uses: Manufacturing fruit essences; perfumes; organic synthesis; solvent; paint removers; fluorometric determinations; in liquid chromatography.
Precaution: Keep away from heat and open flames. Use adequate ventilation.

isobutyl aldehyde. See isobutyraldehyde.

isobutylamine $(CH_3)_2CHCH_2NH_2$.
Properties: Colorless liquid; amine odor; strongly caustic. Soluble in water, alcohol, ether, and hydrocarbons. Sp. gr. 0.731 (20°C); boiling range 66-69°C; flash point less than 20°F.
Containers: 55-gal drums; tank cars.
Uses: Organic synthesis; insecticides.
Shipping regulations: Flammable liquid. Red label.*

isobutyl-para-aminobenzoate (cycloform) $NH_2C_6H_4COOCH_2CH(CH_3)_2$.
Properties: White crystalline scales; m. p. 64-65°C. Almost insoluble in water; soluble in alcohol and vegetable oils.
Use: Medicine.

2-isobutylaminoethyl meta-aminobenzoate hydrochloride. See metabutethamine hydrochloride.

isobutyl benzoate $C_6H_5CO_2CH_2CH(CH_3)_2$.
Properties: Colorless liquid; characteristic odor; sp. gr. 1.002; b. p. 237°C; insoluble in water; miscible with alcohol and ether.
Uses: Perfumes; flavors.

isobutyl carbinol. See isoamyl alcohol, primary.

isobutyl cinnamate $C_{13}H_{16}O$. Colorless oil, amber fragrance. Sp. gr. 1.001 to 1.004; refractive index 1.541. Soluble in 2 vols. of 70% alcohol.
Use: Perfumery.

isobutylene. See isobutene.

isobutylene dibromide (1,2-dibromoisobutane) $(CH_3)_2CBrCH_2Br$.
Properties: Yellowish liquid; soluble in alcohol and ether; insoluble in water. Sp. gr. 1.798; b. p. 149°C.
Derivation: By the action of bromine on isobutene.

isobutyl furoate $C_4H_3OCO_2C_4H_9$.
Properties: Colorless liquid becoming brown in light. Insoluble in water; soluble in alcohol and ether. Sp. gr. 1.0383 (27.5/4°C); b. p. 221-223° (corr.); refractive index 1.4676 (27.5°C).

N-isobutylhendecenamide. See N-isobutylundecylenamide.

isobutyl isobutyrate $(CH_3)_2CHCOOCH_2CH(CH_3)_2$.
Properties: Colorless liquid; sp. gr. 0.875 (0/4°C); m. p. -80.7°C; b. p. 148.7° C; refractive index (n 20/D) 1.3999; insoluble in water; soluble in alcohol and ether.
Containers: Drums.

isobutyl lactate $CH_3CHOHCOOCH_2CH(CH_3)_2$.
Properties: Liquid; sp. gr. 0.964.

isobutyl mercaptan (2-methyl-1-propanethiol) $(CH_3)_2CHCH_2SH$.
Properties: Liquid, boiling 85-95°C; sp. gr. 0.8363 (60/60°F); flash point -10°C.
Grades: 95%.
Containers: Tank cars.
Shipping regulations: Flammable liquid. Red label.*

isobutyl propionate $CH_3CH_2COOCH_2CH(CH_3)_2$.
Properties: Water-white liquid; sp. gr. 0.86-0.8635 (20/20°C); b. p. 138°C; f. p. -71.4°C; refractive index 1.3975 (20°C); insoluble in water; very soluble in alcohol and ether.
Grades: Technical.
Containers: 55-gal drums; 10,000-gal tank cars.
Use: Paint, varnish, and lacquer solvent.

isobutyl salicylate $HOC_6H_4COOCH_2CH(CH_3)_2$.
Properties: Sp. gr. 1.064-1.065 (25°C); clear liquid, may have slightly yellowish tinge; b. p. 259°C. Soluble in alcohol; insoluble in water.
Use: Perfumery.

N-isobutylundecylenamide (N-isobutylhendecenamide) $CH_3(CH_2)_7CH{:}CHCONHC_4H_9$. A synergist for pyrethrum and used in insecticides.

isobutyl vinyl ether. See vinyl isobutyl ether.

isobutyraldehyde (isobutyl aldehyde) $(CH_3)_2CHCHO$.
Properties: Transparent, colorless, highly refractive liquid; pungent odor. Sp. gr. 0.794 (20/4°C); b. p. 64°C; m. p. -66°C; refractive index (n 20/D) 1.3730; flash point (closed cup) -40°F, (open cup) -11°F. Soluble in alcohol; insoluble in water.
Derivation: (a) Oxo process reaction of propylene with carbon monoxide and hydrogen; (b) dehydrogenation of isobutyl alcohol.

Method of Purification: Distillation.
Containers: 55-gal drums; tank cars.
Uses: Intermediate for rubber antioxidants and accelerators, for neopentyl glycol; organic synthesis.
Warning! Flammable. Use with adequate ventilation. Avoid prolonged or repeated contact with skin. MCA warning label.
Shipping regulations: Flammable liquid. Red label.*

isobutyric acid $(CH_3)_2CHCOOH$.
Properties: Colorless liquid; soluble in water, alcohol, and ether. Sp. gr. 0.946 to 0.950 (20/20°C); b. p. 154.4°C; f. p. −47°C; refractive index (n 20/D) 1.3930.
Grades: Technical.
Containers: Glass carboys; special steel and aluminum drums; tank cars and trucks.
Uses: In the manufacture of esters for solvents, flavors and perfume bases; as a disinfecting agent; in varnish; for deliming hides; as a tanning agent.

isobutyric anhydride $[(CH_3)_2CHCO]_2O$.
Liquid with boiling range 180 to 187°C; sp. gr. 0.951-0.956 (20/20°C).
Typical specification: Assay 98%.
Containers: 55-gal drums, tank trucks, tank cars.
Use: Intermediate.

isobutyronitrile (2-methylpropanenitrile; isopropyl cyanide) $(CH_3)_2CHCN$.
Properties: Colorless liquid; sp. gr. 0.773 (20/20°C); b. p. 107°C; m. p. −75°C; slightly soluble in water; very soluble in alcohol and ether.
Containers: Tank cars.
Uses: Chemical intermediate for insecticides and other applications.

isobutyroyl chloride (isobutyryl chloride; 2-methylpropanoyl chloride) $(CH_3)_2CHCOCl$.
Colorless liquid; refractive index 1.4079; density 1.017 (20/4°C); m. p. −90°C; b. p. 92°C; soluble in ether; reacts with water and alcohol.

isobutyryl chloride. See isobutyroyl chloride.

isocarboxazid $C_{12}H_{13}N_3O_2$. 1-Benzyl-2-(5-methyl-3-isoxazolylcarbonyl)hydrazine.
Properties: M. p. 105-107°C. Very slightly soluble in water; soluble in alcohol and chloroform; sparingly soluble in ether.
Use: Medicine.

isocetyl laurate $C_{11}H_{23}COOC_{16}H_{33}$.
Properties: Oily liquid with practically no odor; sp. gr. 0.858; f. p. below −65°C; viscosity 19.6 cp. at 25°C; insoluble in water; soluble in most organic solvents.
Uses: In cosmetics and pharmaceuticals as emollient, lubricant, fixative and solvent; plasticizer; mold release agent; textile softener.

isocetyl myristate $C_{13}H_{27}COOC_{16}H_{33}$.
Properties: Oily liquid with practically no odor; sp. gr. 0.857; f. p. −39°C; viscosity 25.6 cp at 25°C; insoluble in water; soluble in most organic solvents.
Uses: See isocetyl laurate.

isocetyl oleate $C_{17}H_{33}COOC_{16}H_{33}$.
Properties: Oily liquid with practically no odor; sp. gr. 0.862; f. p. −57°C; viscosity 29.0 cp at 25°C; insoluble in water; soluble in most organic solvents.
Uses: See isocetyl laurate.

isocetyl stearate $C_{17}H_{35}COOC_{16}H_{33}$.
Properties: Oily liquid with practically no odor; sp. gr. 0.857; f. p. 0°C; viscosity 32.0 cp at 25°C; insoluble in water; soluble in most organic solvents.
Uses: See isocetyl laurate.

isocholesterol. See lanosterol.

isocinchomeronic acid (2,5-pyridinedicarboxylic acid) $HOOC(C_5H_3N)COOH$.
Properties: Light tan powder, leaflets or prisms; no odor; m. p. 254°C, sublimes as nicotinic acid above this temperature; insoluble in cold water, alcohol, ether, benzene; slightly soluble in boiling water, boiling alcohol; soluble in hot dilute mineral acids.
Use: Intermediate for nicotinic acid, insecticides, polymers, dyestuffs.

isocrotonic acid. See crotonic acid.

isocyanate generator (hindered isocyanate)
An isocyanate derivative that decomposes to produce an isocyanate upon heating. For example in one type phenol is combined with an isocyanate, and the resulting urethane is stable at room temperature, but dissociates at 160° to produce the original phenol and isocyanate. These generators are used commercially in a mixture with a polyester which can be stored indefinitely, but which upon heating produces a polyurethane resin.

isocyanate resins. See polyurethane resins.

isocyanates. Compounds containing the isocyanate radical, -NCO. Monoisocyanates are in use, as in the treatment of cellulose to obtain a cellulose tricarbamate, but the term isocyanates usually refers to the diisocyanates (q.v.).

isocyanuric acid (s-triazine-2,4,6-trione) OCNHCONHCONH. The ketone isomer of cyanuric acid (q. v.), which is the triol. Derivatives of isocyanuric acid, as dichloro- and trichloroisocyanuric acid, and potassium and sodium dichloroisocyanurate, are used as bleaches and sanitizers.

isodecaldehydes $C_9H_{19}CHO$ (mixed isomers).
Properties: Sp. gr. 0.8290; b. p. 197.0°C; f. p. −80°C; insoluble in water; density 6.9 lb/gal; flash point 185°F.
Uses: Intermediate for pharmaceuticals, dyes, resins.

isodecane. See 2-methylnonane.

isodecanoic acid $C_{10}H_{20}O_2$ (mixture of branched chain acids, primarily trimethylheptanoic and dimethyloctanoic).
Properties: Liquid; b. p. 254°C (760 mm Hg), 137°C (10 mm Hg); sp. gr. 0.9019 (20/20°C);

*See "I. C. C. Shipping Regulations," page xiii.
Reference numbers refer to name of manufacturer. See "List of Manufacturers," page v.

f. p., glass below −60°C, very slightly soluble in water; viscosity 12.9 cps at 20°C; refractive index 1.4358 (n 20/D).
Uses: Intermediate for metal salts, ester type lubricants, plasticizers.

isodecanol $C_{10}H_{21}OH$ (mixed isomers).
Properties: Sp. gr. 0.8395; insoluble in water; flash point 220°F; b.p. 220°C.
Use: Antifoaming agent in textile processing.

isodecyl chloride $C_{10}H_{21}Cl$ (mixed isomers).
Properties: Sp. gr. 0.8767; b. p. 210.6°C; f. p. −180°C (sets to a glass); insoluble in water; flash point 200°F.
Uses: Solvent for oils, fats, greases, resins, gums; extractants, cleaning compounds; intermediates for insecticides, pharmaceuticals, plasticizers, polysulfide rubbers, resins, and cationic surfactants.

isodecyl octyl adipate.
Properties: Light colored, oily liquid; sp. gr. 0.924 (20/20°C); mid-b. p. 227°C (4 mm); refractive index 1.448 (25°C); viscosity 20 cps (20°C).
Use: Plasticizer.

isodurene (1, 2, 3, 5-tetramethylbenzene) $(CH_3)_4C_6H_2$.
Properties: Liquid. Soluble in alcohol and ether; insoluble in water. Sp. gr. 0.896; b. p. 197°C; m. p. −24°C.
Derivation: Occurs in coal tar.
Grades: Technical.
Use: Organic synthesis.

isoemodin. See aloe-emodin.

isoeugenol $(CH_3CHCH)C_6H_3OHOCH_3$. 1-Hydroxy-2-methoxy-4-propenylbenzene.
Properties: Pale yellow oil, spice-clove type odor. Soluble in alcohol, ether, and other organic solvents; slightly soluble in water. Sp. gr. 1.081-1.084; m. p. 19°C; b. p. 268°C; refractive index (n 19/D) 1.5739.
Derivation: From eugenol by isomerization with caustic potash.
Method of purification: Distillation.
Grades: Perfumers' grade.
Containers: 1- and 5-lb glass bottles; 1-, 5-, and 10-gal tins; drums.
Uses: Perfumes; vanillin.
Shipping regulations: None. *

isoeugenol acetate. See acetylisoeugenol.

isoeugenol ethyl ether (1-ethoxy-2-methoxy-4-propenylbenzene) $C_3H_5(CH_3O)C_6H_3OC_2H_5$.
Properties: Synthetic white crystalline powder; m. p. 64°C; insoluble in water; soluble in alcohol, ether, benzene.
Uses: As sweetening agent and odorant fixative.

isoflurophate. See diisopropyl fluophosphate.

isoheptanes C_7H_{16}. A mixture containing predominantly branched-chain heptane hydrocarbons.
Properties: Boiling range 82-93°C; sp. gr. 0.717 (60/60°F); refractive index 1.385 (n 20/D); aniline point 132°F; flash point −9°C.
Grades: Commercial.
Containers: Drums and tank cars; also 5- and 54-gal drums.
Uses: Solvent.
Shipping regulations: Flammable liquid. Red label. *

isohexadecyl derivatives. See isocetyl derivatives.

isohexane. See 2-methylpentane.

isohexanes C_6H_{14}. A mixture consisting primarily of branched-chain hexane hydrocarbons.
Properties: Boiling range 54 to 61°C; sp. gr. 0.671 (60/60°F); flash point −32°C.
Grade: Commercial.
Containers: 5-, 54-gal drums.
Uses: Solvent.
Shipping regulations: Flammable liquid. Red label. *

isolates, odoriferous. Pure chemical compounds derived from an essential oil or other natural perfume substance. Geranial (obtained from palmarosa oil or citronella oil); pinene (from turpentine); anethole (from anise oil).

isoleucine (alpha-amino-beta-methylvaleric acid; 2-amino-3-methylpentanoic acid). $CH_3CH_2CH(CH_3)CH(NH_2)COOH$. An essential amino acid, found naturally in the L(+) form.
Properties: Crystalline; slightly soluble in water; nearly insoluble in alcohol; insoluble in ether.
DL-isoleucine: (synthetic form) m. p. 292°C (dec).
L(+)-isoleucine: bitter taste; m. p. 283-284°C (dec).
D(-)-allo-isoleucine: m.p. 278°C (dec).
L(+)-allo-isoleucine: sweet taste; m. p. 280-281°C (dec).
Derivation: Hydrolysis of protein (zein, edestin); amination of the alpha-bromo-beta-methylvaleric acid.
Use: Medicine; nutrition; biochemical research.

"Isome." [342] Trademark for di-n-propyl 6, 7-methylene-dioxy-3-methyl-1, 2, 3, 4, -tetrahydronaphthalene-1, 2-dicarboxylate. Used in insecticides.

isomerization. Process for converting hydrocarbons or other organic compounds into compounds whose molecules have a different arrangement of atoms, but the same number and kinds of atoms. A very important example is the conversion of normal butane into isobutane, in connection with the production of isooctane and other high grade motor fuels.

isomers.
1. Molecules which contain the same number and kind of atoms but which differ in structure. This isomerism may be of several kinds. Butyl alcohol (C_4H_9OH or $C_4H_{10}O$) and ethyl ether ($C_2H_5OC_2H_5$ or $C_4H_{10}O$) have the same empirical formulas

but are entirely different kinds of chemical substances; normal ($CH_3CH_2CH_2CH_2OH$) and isobutyl alcohol ($[CH_3]_2CHCH_2OH$) differ in the shape of the molecules; sec-butyl alcohol ($CH_3CHOHCH_2CH_3$) exists in two forms that differ only as right and left hands differ. See also amyl alcohols; optical isomerism; geometric isomerism.
2. Nuclides (i.e., kinds of atomic nuclei) having the same atomic and mass numbers, but that exist in different energy states. One is always unstable with respect to the other, or both may be unstable with respect to a third. In the latter instance the energy of transformation in the two cases will differ.

isometheptene (2-methylamino-6-methyl-5-heptene) $(CH_3)_2C{:}CHCH_2CH_2CH(NHCH_3)CH_3$.
Properties: Colorless or slightly yellow, oily liquid with an amine-like odor. B.p. 175-177°C; sp. gr. 0.794-0.798 (25/25°C); refractive index 1.4428-1.4438 (n 25/D). Miscible with dilute mineral acids and most organic solvents; volatile with steam.
Use: Medicine.
Available also as the isometheptene tartrate, the hydrochloride N.N.D., and the mucate N.N.D.

alpha-**isomethylionone.** See "Cetone Alpha."

"Iso Mist Extra." [400] Trade name for isopropyl myristate, double distilled.
Properties: Saponification number 203-211; acidity as myristic acid 0.1% max; iodine number 1.0 (max); ester content 97.5% min.

isomorphism. The condition in which two or more entirely different substances have closely similar crystal structures, lattice dimensions, and chemical composition.

isonandra gutta. See gutta-percha.

isoniazid (N-isonicotinyl hydrazine; INAH; isonicotinic acid hydrazide) $C_5H_4NCONHNH_2$.
Properties: Colorless or white crystals or white crystalline powder; odorless; affected by air and light; m.p. 170-173°C; sparingly soluble in alcohol, slightly soluble in benzene and ether; freely soluble in water. Solutions practically neutral to litmus.
Grade: U.S.P. XVI.
Use: Medicine.

isonicotinic acid $\overline{CHCHNCHCHC}COOH$.
Properties: White, practically odorless powder; m.p. 314-317°C (sealed capillary); slightly soluble in water; pH of saturated aqueous solution at 20° 3.6.
Containers: Fiber drums.
Use: Synthesis of isoniazid and similar substances.

isonicotinic acid hydrazide. See isoniazid.

isonicotinoyl isopropylhydrazine phosphate. See iproniazid phosphate.

isonipecaine hydrochloride. See meperidine hydrochloride.

isooctane (2,2,4-trimethylpentane) $(CH_3)_2CHCH_2C(CH_3)_3$. See also 2-methylheptane.
Properties: Colorless liquid; sp. gr. 0.6919 (20/4°C); m.p. −107.4°C; b.p. 99.2°C (760 mm); refractive index (n 20/D) 1.3914. Insoluble in water, slightly soluble in alcohol and ether.
Grades: Technical; pure; research.
Containers: 5-, 54-gal drums.
Uses: Organic synthesis; motor fuel; used with normal heptane to prepare standard mixtures to determine anti-knock property of gasoline. See octane number.
Shipping regulations: Flammable liquid. Red label. *

isooctanes C_8H_{18}. A mixture of hydrocarbons, predominantly isomers of octane.
Properties: Liquid; flash point −11°C.
Containers: Drums; tank cars.
Shipping regulations: Flammable liquid. Red label. *

isooctanol. See isooctyl alcohol.

isooctene. Available commercial grade isooctene is a mixture whose principal components are branched chain heptenes and octenes.
Properties: Typical boiling range 190-200°F; bromine number 137; sp. gr. 0.726 (60/60°F).
Shipping regulations: Flammable liquid. Red label. *

isooctyl adipate. Plasticizer providing low temperature stability. Used in calendering film, sheeting, vinyl dispersions, extrusions.

isooctyl alcohol (isooctanol). General term that might be applied to any of the isomers of the formula $C_7H_{15}CH_2OH$ in which the eight carbon atoms form a branched chain structural arrangement. For practical purposes the term refers to a mixture of isomers made by the Oxo process. A selected C_7 hydrocarbon fraction is reacted with hydrogen and carbon monoxide gases in the presence of a special catalyst at pressures up to 3000 psi. The crude alcohol is recovered and purified.
Properties: Distillation range 182-195°C; wt/gal 6.95 lbs; sp. gr. (20/20°C) 0.832; flash point (Tag open cup) 180°F.
Containers: 55-gal drums; tank cars.
Uses: Combined with phthalic anhydride, maleic anhydride, adipic acid, sebacic acid, etc., to form plasticizers; intermediate for non-ionic detergents and surfactants; used in preparation of synthetic drying oils, cutting and lubricating oils, hydraulic fluids; resin solvent; emulsifier; antifoaming agent; and as an intermediate for introducing the isooctyl group into other compounds.
Shipping regulations: None. *

isooctyl isodecyl phthalate $C_8H_{17}OOCC_6H_4COOC_{10}H_{21}$.
Properties: Clear liquid; sp. gr. (20/20°C), 0.976; mild characteristic odor.

Grade: Technical.
Containers: Drums; tank trucks; tank cars.
Use: Plasticizer.

isooctyl palmitate $C_8H_{17}OOCC_{15}H_{31}$.
Properties: (typical specification) Clear liquid; sp. gr. 0.863 (20°C); acidity 0.2% max (palmitic); moisture 0.05% max; m.p. 6-9°C; b.p. 228°C (5 mm).
Uses: Secondary plasticizer for synthetic resins; extrusion aid and rubber plasticizer.

isooctylphenoxypolyoxyethylene ethanol (isooctylphenylpolyethylene glycol ether) $(CH_3)_3CCH_2C(CH_3)_2C_6H_4O(CH_2)_2O(C_2H_4O)_7C_2H_4OH$.
Properties: Slightly viscous pale amber-colored liquid; oily musty odor; m.p. 2-5°C; b.p. 150°C (initial) at 1 micron; density 1.06 g/ml (20°C); flash point 227°C.
Use: Surface active agent.

isooctylphenylpolyethylene glycol ether. See isooctylphenoxypolyoxyethylene ethanol.

isooctyl thioglycolate $HSCH_2COOCH_2C_7H_{15}$.
Properties: Water-white liquid with faint fruity odor; b.p. 125°C (17 mm); sp. gr. 0.9736 (25°C); refractive index 1.4606 (21°C); acid no., less than 1.
Grade: 99% (minimum purity).
Containers: Carboy, drum and ton lots.
Uses: Antioxidants, fungicides, oil additives, plasticizers, insecticides, stabilizers, polymerization modifiers.

isop. See hyssop.

isopentaldehyde C_4H_9CHO. A mixture of 5-carbon aldehydes consisting of valeraldehyde, 2-methyl butyraldehyde, 3-methyl butyraldehyde.
Properties: Water-white liquid with sharp odor; sp. gr. 0.8089 (20/20°C); b.p. 98.6°C; f.p. −95.4°C; water dissolves 0.85% aldehyde at 20°C; water soluble to 2.2% in the aldehyde.
Uses: Possible intermediate for bis-phenols, epoxy and polycarbonate resins, and modified formaldehyde resins.

isopentane (2-methylbutane; ethyldimethylmethane) $(CH_3)_2CHCH_2CH_3$.
Properties: Colorless, mobile, flammable liquid; pleasant odor; f.p. −159.890°C; b.p. 27.854°C; sp. gr. 0.61967 (20°C); soluble in hydrocarbons, oils, ether; very slightly soluble in alcohol; insoluble in water.
Typical specifications for a technical grade are as follows: Distillation range 81-85°F; sp. gr. of liquid 0.625 (60/60°F); API gravity 94.9 (60°C); refractive index (n 20/D) 1.354; contains not less than 95 mole% isopentane; principal impurity is normal pentane (4-5%); sulfur content not to exceed 0.005% by wt.
Typical specifications for a commercial grade: Distillation range 75-90°F; vapor pressure (100°F) 21 psia max; sulfur content 0.02 max wt %; doctor test negative; corrosion max 1; color, Saybolt min +30; non-volatile matter, gm/100 ml max 0.001.
Derivation: Fractional distillation from petroleum; purified by rectification.
Grades: Research (99.99%), pure (99%), technical (95%), commercial.
Containers: Pure: quart bottles; gallon bottles; 55-gal drums. Technical and commercial: 5-, 55-gal drums; tank cars (8000 gals).
Uses: Solvent; manufacture of chlorinated derivatives.
Fire hazard: Dangerous.
Shipping regulations: Flammable liquid. Red label. *

isopentanoic acid C_4H_9COOH. A mixture of 5-carbon acids consisting of approximately 55-65% 1-valeric acid, 35-45% 2-methylbutyric acid, and less than 5% 3-methylbutyric acid.
Properties: Water-white liquid with penetrating odor; sp. gr. 0.9388 (20/20°C); b.p. 183.2°C (760 mm Hg); vapor pressure 0.14 mm at 20°C; f.p. −44°C; water dissolves 3.24 wt% of acid at 20°C; acid dissolves 10.4% water at 20°C.
Uses: Possible intermediate for plasticizers, synthetic lubricants, pharmaceuticals, metallic salts, vinyl stabilizers; also possible extractant for mercaptans from hydrocarbons.

isophane insulin suspension. See insulin.

isophorone $\underline{COCHC(CH_3)CH_2C(CH_3)_2CH_2}$.
Properties: Practically water-white liquid; sp. gr. 0.9229 (20/20°C); 7.7 lbs/gal (20°C); b.p. 215.2°C (760 mm); vapor pressure 0.2mm (20°C); f.p. −8.1°C; viscosity 2.62 cps (20°C); flash point 205°F.
Possesses a high solvent power for vinyl resins and certain other synthetic resins, the cellulose esters and ethers, and many substances soluble with difficulty in other solvents; slightly soluble in water.
Containers: Drums; tank cars.
Use: Same as ethyl butyl ketone.

isophthalic acid (meta-phthalic acid) $C_6H_4(COOH)_2$.
Properties: Colorless crystals; m.p. 345-348°C; sublimes. Slightly soluble in water; soluble in alcohol, acetic acid; insoluble in benzene and petroleum ether.
Derivation: (a) Oxidation of meta-xylene with nitric acid in the presence of methanol and hydrolysis of the resulting ester; (b) oxidation of meta-xylene with sulfur in the presence of ammonia and water at 200-700°F and hydrolysis of the resulting mixture of amides and ammonium salts; (c) liquid phase oxidation of mixed xylenes; (d) direct oxidation of mixed alkyl aromatics with heavy metal salts and bromine as catalysts.
Containers: Drums; tank cars.
Grades: Technical.
Uses: Polyester and polyurethane resins; plasticizers.

isophthaloyl chloride (meta-phthalyl dichloride) $C_6H_4(COCl)_2$.

Properties: Crystalline solid; m.p. 41°C; b.p. 276°C; soluble in ether and other organic solvents; reactive with water and alcohol.
Uses: Intermediate; dyes; synthetic fibers; resins; films; protective coatings.

isopral (trichloroisopropyl alcohol) $CCl_3CH(CH_3)OH$.
Properties: Microscopic prisms, camphorous odor, aromatic taste. Absorbed through skin, subcutaneous tissue and the digestive tract. More toxic than chloral hydrate. Soluble in water, alcohol, ether. M.p. 49°C.
Use: Medicine.
Shipping regulations: None. *

isoprene (3-methyl-1,3-butadiene; 2-methyl-1,3-butadiene) $CH_2:C(CH_3)CH:CH_2$. The molecular unit of rubber. Building block of "natural" synthetic rubber (cis-1,4-polyisoprene).
Properties: Colorless, volatile liquid; f.p. −146°C; b.p. 34.08°C; refractive index 1.4216 (n 20/D); sp. gr. 0.6808 (20/4°C); flash point −48°C; insoluble in water; soluble in alcohol, ether; readily oxidizable and polymerizable.
Derivation: (a) From isoamylene by dehydrogenation; (b) from propylene, by dimerization and cracking; (c) refinery and coal gases and tars; (d) also produced in cracking of natural rubber and turpentine.
Containers: Bottles; drums; tank cars.
Uses: Manufacture of butyl rubber and synthetic natural rubber, also resins and chemicals.
Shipping regulations: Flammable liquid. Red label. *

isopropamide iodide (3-carbamoyl-3,3-diphenylpropyl diisopropyl methyl ammonium iodide) $C_{23}H_{33}N_2O$. Used in medicine.

isopropanol. See isopropyl alcohol.

isopropanolamine (MIPA) $CH_3CH(OH)CH_2NH_2$.
Properties: Liquid; slight ammonia odor; sp. gr. 0.9619; m.p. 1.4°C; b.p. 159.9°C; flash point 160°F; soluble in water. Emulsifying agent; forms soaps that are completely soluble in naphtha, gasoline, and mineral oil.
Containers: 1-gal cans; 5-, 55-gal drums.
Uses: Drycleaning soaps; soluble textile oils; wax removers; metal cutting oils; cosmetics; emulsion paints; plasticizers; insecticides.

isopropenyl acetate $CH_3COOC(CH_3):CH_2$.
Properties: Water white liquid; sp. gr. 0.9226; b.p. 97.4°C; f.p. −92.9°C; solubility in water 3.25% by weight (20°C); flash point 66°F.
Shipping regulations: Flammable liquid. Red label. *

isopropenylacetylene (2-methyl-1-buten-3-yne) $H_2C:C(CH_3)C\vdots CH$.
Properties: Colorless liquid; b.p. 33-34°C; freezing point −113°C; sp. gr. (20/20°C) 0.695; refractive index (n 20/D) 1.4168; flash point (Tag open cup) < 20°F; very slightly soluble in water and miscible with acetone, alcohol, benzene, carbon tetrachloride and kerosene.
Uses: Specialty fuel; chemical intermediate.

isopropenylchloroformate $ClCOOC(CH_3):CH_2$.
Properties: Liquid. Caution! Odor is very irritating! Sp. gr. 1.103 (20°C); b.p. 93°C (746 mm).
Derivation: Distillation of the reaction products of acetone and phosgene.

para-**isopropoxydiphenylamine** $C_6H_5NHC_6H_4OCH(CH_3)_2$.
Typical specifications: Dark gray flakes; sp. gr. 1.10; set point 78°C (min); purity 92% (min); ash 0.10% (max); insoluble in water; soluble in ethyl alcohol (2B), acetone, benzene and gasoline.

beta-**isopropoxyproprionitrile** (beta-alkoxypropionitrile) $(CH_3)_2CHO(CH_2)_2CN$.
Properties: Colorless to straw-colored liquid. Combines the chemical and physical properties of ethers and nitriles. M.p. −67°C; b.p. 82-86°C (25 mm Hg), 65-65.5°C (10 mm Hg). Sp. gr. 0.9058 (25°C). Slightly soluble in water; soluble in organic solvents.

isopropoxytitanium stearate. See titanium acylates.

isopropyl acetate $CH_3COOCH(CH_3)_2$.
Properties: Colorless, aromatic liquid. Stable; somewhat toxic; b.p. 89.4°C; sp. gr. 0.877 (15.5°C), 0.8690 (25/4°C); refractive index 1.378 (20°C); sp. ht. 0.46 cal/g; m.p. −73.4°C; heat of vaporization 135 Btu/lb; viscosity 0.49 cps (25°C); solubility in water 2.9 wt %; solubility of water 1.8 wt %; coefficient of expansion 0.00131/°C (approx); flash point 36°F; 7.17 lbs/gal (20°C). Miscible with most of the common organic solvents.
Specifications: Sp. gr. 0.860-0.862 (20/20°C); color water-white; acidity (acetic acid) 0.02% (max); moisture clear with 19 vols of naphtha; distillation initial 83°C (min); dry point 93°C (max); non-volatile matter 2 mg/100 cc; purity 85-88 (wt%); residual odor none.
Derivation: By reacting isopropyl alcohol with acetic acid in the presence of catalysts.
Grades: 95% grade; 85 to 88% grade.
Containers: 1-, 5-gal cans; 55-gal (nonreturnable) drums; tank cars 6000, 8000, and 10000 gals.
Uses: Solvent for nitrocellulose, fats, oils, waxes, gums, natural and synthetic resins; artificial leather; artificial silk; dopes; films; lacquers; plastics; synthetic perfumes; organic synthesis.
Warning! Flammable. MCA warning label.
Shipping regulations: Flammable liquid. Red label. *

isopropyl alcohol (IPA; dimethylcarbinol; sec-propyl alcohol; isopropanol; 2-propanol) $(CH_3)_2CHOH$.
Properties: Colorless, clear, mobile liquid; flammable.
Physical properties: B.p. 82.4°C; sp. gr.

*See "I.C.C. Shipping Regulations," page xiii.
Reference numbers refer to name of manufacturer. See "List of Manufacturers," page v.

0.7863 (20/20°C); refractive index 1.3756 (20°C); sp. ht. 0.65 cal/g; m. p. −88°C; critical temperature 235°C; critical pressure 53 atmospheres; vapor pressure 33 mm Hg at 20°C; flash point 72°F; heat of combustion 14,346 Btu/lb; heat of vaporization 288 Btu/lb; viscosity 2.1 cps (25°C); coefficient of expansion 0.00107/°C (approx). Soluble in water, alcohol and ether.

Specifications:

	91%	99%
Sp. gr. (20/20)	0.8175-0.8190	0.7900 (max)
Acidity (acetic acid) (max)	0.002%	0.002%
Distillation range	79.7-80.7°C	81.5-83°C
Non-volatile matter (mg/100 cc) (max)	2	2
Purity (volume %) (min)	91	99
Residual odor	none	none
Water dilution	clear	clear
Flash point, Tag open cup	80°F	72°F
Lbs/gal (20°C)	6.81	6.57

Derivation: By treatment of propylene with sulfuric acid and hydrolyzing.

Method of purification: Rectification.

Grades: 91%; 95%; 99%; N. F. XI (99%).

Containers: Tins; 55-gal drums; tank trucks up to 2000 gals; tank cars 6000 to 10,000 gals.

Uses: Manufacture of most of the national output of acetone and thus a source of acetic anhydride, diacetone alcohol, methyl isobutyl ketone, and other derivatives; solvent for essential and other oils, alkaloids, gums, resins, organic and inorganic compounds; latent solvent for cellulose derivatives; solvent mixtures; antistalling agent in liquid fuels; deicing agent for liquid fuels; pharmaceuticals; perfumes; lacquers; extraction processes; dehydrating agent; preservative; antifreeze; rocket fuel.

Fire hazard: Dangerous.

Shipping regulations: Flammable liquid. Red label. Legal label name isopropanol.*

isopropylamine (2-aminopropane) $(CH_3)_2CHNH_2$.

Properties: Colorless volatile liquid; amine odor; strong alkaline reaction. B. p. 32.4°C; f. p. −95.2°C; sp. gr. (20/20°C) 0.6881; wt/gal (20°C) 5.7 lbs; refractive index (n 15/D) 1.3770; flash point (open cup) below 0°F (values from −35°F to −9°F given by various authorities). Miscible with water, alcohol, and ether.

Derivation: From isopropyl chloride and ammonia under pressure.

Containers: 1-, 5-gal cans; 55-gal drums; tank cars.

Uses: Solvent; intermediate in synthesis of rubber accelerators, pharmaceuticals, dyes, insecticides, bactericides, textile specialties, and surface-active agents; dehairing agent; solubilizer for 2,4-D acid.

Danger! Extremely flammable. Vapor extremely hazardous. Liquid causes burns. MCA warning label.

Shipping regulations: Flammable liquid. Red label. *

para-**isopropylaminodiphenylamine.**

Use: Protecting rubber against ozone attack, flex cracking, and heat.

isopropylaminoethanols. A commercial mixture of approximately 60% isopropylethanolamine, $(CH_3)_2CHNHCH_2CH_2OH$, and 40% of isopropyldiethanolamine, $(CH_3)_2CHN(CH_2CH_2OH)_2$.

Properties: Amber to straw-colored liquid, distillation range 110-265°C; f. p. below −50°C; sp. gr. 0.91-0.94 (20/20°C); flash point 145-155°F (open cup).

Use: Synthesis of emulsifiers.

alpha-**(isopropylaminomethyl) protocatechuyl alcohol esters.** See isoproterenol hydrochloride and sulfate.

5-isopropylamino-1-pentanol $(CH_3)_2CHNH(CH_2)_4CH_2OH$.

Properties: Light-straw color; faint amine odor; boiling range 225-234°C; flash point 230°F.

para-**isopropylaniline.** See cumidene.

isopropyl antimonite $[(CH_3)_2CHO]_3Sb$.

Properties: Colorless liquid; b. p. 82°C at 7 mm Hg pressure.

Derivation: Reaction of antimony trichloride with isopropanol.

Uses: Cross-linking agent; flameproofing agent.

isopropylarterenol esters. See isoproterenol hydrochloride and sulfate.

para-**isopropylbenzaldehyde.** See cuminic aldehyde.

isopropylbenzene. See cumene.

isopropylbenzol. See cumene.

isopropyl bromide $CH_3CHBrCH_3$.

Properties: Colorless liquid; sp. gr. 1.304 (25/25°C); b. p. 58.5-60.5°C; f. p. −90°C; refractive index 1.422 (n 25/D); flash point none; slightly soluble in water; soluble in methanol, ether.

Uses: Synthesis of pharmaceuticals, dyes, other organics.

isopropyl butyrate $(CH_3)_2CHOOCC_3H_7$.

Properties: Colorless liquid; used in solvent mixtures for cellulose ethers; sp. gr. 0.8652 (13°C); b. p. 128°C.

Grades: Technical.

isopropylcarbinol. See isobutyl alcohol.

isopropyl chloride $CH_3CHClCH_3$.

Properties: Colorless liquid; sp. gr. 0.858 (25/25°C); b. p. 34.8°C; f. p. −117.6°C; refractive index 1.374 (n 25/D); flash point −45°F; slightly soluble in water; soluble in

methanol, ether.
Uses: Solvent; intermediate.
Shipping regulations: Flammable liquid. Red label. *

isopropyl N-(3-chlorophenyl)-carbamate. See chloro-IPC.

isopropyl-meta-cresol. See thymol.

isopropyl-ortho-cresol. See carvacrol.

isopropyl cresols. A mixture of di- and mono-isopropyl cresols used as an antioxidant. See MYL, thymol, and carvacrol.

isopropyl cyanide. See isobutyronitrile.

isopropyldiethanolamine. See isopropylamino-ethanols.

2-isopropyl-4 dimethylamino-5-methyl-phenyl-1-piperidine-carboxylate methyl chloride (Amo-1618). A plant tranquilizer or anti-gibberellin, which causes some plants to become dwarfs without affecting their growth or health otherwise.

isopropylethanolamine. See isopropylamino-ethanols.

isopropyl ether (diisopropyl ether) $(CH_3)_2CHOCH(CH_3)_2$.
Properties: Colorless liquid. Ethereal odor. Isopropyl ether is somewhat similar to ethyl ether in properties but does tend to form peroxides more readily than ethyl ether. (See ether). Consequently the presence or absence of peroxides should be determined and if present should be destroyed with sodium sulfite before distillation. B. p. 67.5°C; sp. gr. 0.723 (15.5/4°C); refractive index 1.368; m. p. −88°C; heat of combustion 16250 Btu/lb; heat of vaporization 124 Btu/lb; viscosity 0.32 cps (20°C); solubility in water 0.65% wt (25°C); solubility of water 0.025% wt (25°C); flash point 9°F; 6.05 lbs/gal (60°F). Miscible with most organic solvents and water.
Specifications: Sp. gr. 0.723-0.727 (20/20°C); water-white; acidity (acetic acid) 0.002% (max); moisture, clear with 19 vols of naphtha; distillation, initial b. p. 66°C (min), dry point 70°C (max); peroxides (active oxygen) 12 ppm (max); inhibitor, 5-12 mg/100 cc.
Grades: Technical.
Containers: 1-, 5-gal cans; 55-gal (non-returnable) drums; tank cars 6000, 8000, and 10,000 gal.
Uses: Solvent for animal, vegetable, mineral oils, waxes and resins; mixed with isopropanol may be used for de-waxing oils or de-oiling waxes; extraction of acetic acid from aqueous solutions; solvent for dyes in presence of small amount of alcohol; paint and varnish removers; spotting compositions; rubber cements.
Danger! Extremely flammable. Highly volatile. Tends to form explosive peroxides especially when anhydrous. MCA warning label.
Shipping regulations: Flammable liquid. Red label. *

isopropylethylene. See 3-methyl-1-butene.

isopropyl furoate $C_4H_3OCO_2C_3H_7$.
Properties: Colorless liquid, becoming brown in light. Insoluble in water; soluble in alcohol and ether; sp. gr. 1.0655 (23.7/4°C); b. p. 198.6°C (corr); refractive index 1.4682 (23.7°C).

isopropylideneacetone. See mesityl oxide.

para, para'-**isopropylidenediphenol.** See bisphenol A.

isopropyl iodide (2-iodopropane) CH_3CHICH_3.
Properties: Colorless liquid that discolors in air and light; miscible with chloroform, ether, alcohol, and benzene; slightly soluble in water; sp. gr. 1.703; m. p. −90°C; b. p. about 90°C; refractive index (n 20/D) 1.5026.
Grades: Pure; technical.
Uses: Organic synthesis; pharmaceuticals.

isopropyl meprobamate. See N-isopropyl-2-methyl-2-propyl-1,3-propanediol dicarbamate.

isopropyl mercaptan $(CH_3)_2CH(HS)$.
Properties: Extremely powerful unpleasant odor.
Derivation: Propylene and hydrogen sulfide.
Use: Standard for petroleum analysis.
Shipping regulations: Flammable liquid. Red label. *

2-isopropyl-5-methylbenzoquinone. See para-thymoquinone.

1-isopropyl-2-methylethylene. See 4-methyl-2-pentene.

N-isopropyl-2-methyl-2-propyl-1,3-propanediol dicarbamate (isopropyl meprobamate) $(CH_3)_2CHNHCOOCH_2C(CH_3)(C_3H_7)CH_2COO-NHCH(CH_3)_2$.
Properties: Crystals; m. p. 92-93°C. Sparingly soluble in water; insoluble in vegetable oils; soluble in many common organic solvents. Stable in dilute acids and alkalies.
Use: Medicine.

1-isopropyl-3-methyl-5-pyrazolyl dimethylcarbamate $C_{10}H_{17}N_3O_2$.
Derivation: By treating 1-isopropyl-3-methyl-5-pyrazolone with dimethylcarbamoyl chloride.
Use: Insecticide.

isopropyl myristate $CH_3(CH_2)_{12}CO_2CH(CH_3)_2$.
Properties: Colorless oil; practically odorless; sp. gr. 0.850-0.860; freezing point 3°C; refractive index (20°C) 1.435-1.438. Soluble in most organic solvents; insoluble in water.
Grades: Double-distilled.
Use: Cosmetics.

isopropyl percarbonate. See following entry.
Shipping regulations: Stabilized: Corrosive liquid. White label by freight. Not accepted by express. Unstabilized: Flammable solid. Red label by freight. Not accepted by express. *

isopropyl peroxydicarbonate. A catalyst which makes possible the production of medium density (0.95 sp. gr.) polyethylene by the conventional high pressure process. May also be used in polymerization of vinyl compounds. See preceding entry.
Properties: M. p. 46-50°F; decomposes 65°F; slightly soluble in water; soluble in many organic solvents.
Derivation: Reaction of isopropyl chloroformate with hydrogen peroxide in sodium hydroxide.

ortho-**isopropylphenol** $(CH_3)_2CHC_6H_4OH$.
Properties: Light yellow liquid, b. p. 214°C; f. p. 17°C; density 0.995 at 20°C; flash point (open cup) 95°C; insoluble in water; soluble in isopentane, toluene, ethyl alcohol, 10% sodium hydroxide.
Uses: Intermediate for synthetic resins, plasticizers, surface active agents, perfumes.

meta, para-**isopropylphenol** $(CH_3)_2CHC_6H_4OH$.
Properties: A solid mixture of the meta and para isomers, completely soluble in 10% sodium hydroxide; f. p. (meta) 25.9°C, (para) 63.2°C; b. p. (meta) 228.6°C, (para) 228.5°C.

isopropyl N-phenylcarbamate. See IPC.

4-isopropylpyridine $C_5NH_4C_3H_7$.
Properties: B. p. 182.2°C; density 0.9282 at 20°C; refractive index 1.4960 (n 20/D); solubility in 100 g of water at 20°C, 1.17 g; solubility of water in 100 g at 20°C, 19.4 g.

isopropyl titanate. See tetraisopropyl titanate.

isopropyltoluene. See cymene.

isopropyltoluol. See cymene.

isopropyltrimethylmethane. See 2,2,3-trimethylbutane.

isoproterenol hydrochloride (isopropylarterenol hydrochloride; alpha-(isopropylaminomethyl)protocatechuyl alcohol hydrochloride; 1-(3',4'-dihydroxyphenyl)-2-isopropylaminoethanol hydrochloride) $(HO)_2C_6H_3CH(OH)CH_2NHCH(CH_3)_2 \cdot HCl$.
Properties: White, odorless, slightly bitter, nonhygroscopic, crystalline solid; m. p. 167-172°C; affected by air and light. Freely soluble in water, soluble in alcohol, very slightly soluble in benzene; insoluble in chloroform and ether; 1% solution is clear and colorless and has pH 4.5-5.5. Aqueous solutions become pink upon standing.
Grade: U. S. P. XVI.
Use: Medicine.

isoproterenol sulfate (for synonyms, see isoproterenol hydrochloride) $(HO)_2C_6H_3CH(OH)CH_2NHCH(CH_2)_2 \cdot \frac{1}{2}H_2SO_4 \cdot 2H_2O$.
Properties: White, odorless, slightly bitter, hygroscopic, crystalline solid; m. p. 128° (dec); freely soluble in water, slightly soluble in alcohol, very slightly soluble in benzene and ether. 1% solution is clear and colorless and has pH 3.5-4.5; aqueous solutions become pink upon standing.
Grade: N. F. XI.
Use: Medicine.

isopulegol $C_{10}H_{17}OH$.
Properties: Water-white liquid. Mint-like odor. Sp. gr. 0.904-0.911; refractive index 1.471-1.474.
Grade: Technical.
Use: Perfumery (geranium and rose compounds).

isopulegol acetate $C_{10}H_{17}OOCCH_3$.
Properties: Water-white liquid. Mint-like odor. Sp. gr. 0.930-0.936.
Grade: Technical.
Use: Perfumery (geranium and rose compounds).

isopurpurin. See anthrapurpurin.

isoquinoline CHCHCHCHCCCHCHNCH.
Properties: Colorless plates or liquid; occurs in coal tar; also prepared synthetically; sp. gr. 1.099 (20°C); m. p. 23°C; b. p. 243°C; insoluble in water, soluble in dilute mineral acids and most organic solvents; refractive index (n 25/D) 1.6223.
Containers: Drums.
Grades: Technical (95% min).
Uses: Manufacture of pharmaceuticals (such as nicotinic acid), dyes, insecticides, rubber accelators, and in organic synthesis.

1,3-isoquinolinediol $C_9H_5N(OH)_2$.
Properties: Cream colored paste; solids, approx. 80%.
Grade: Technical.
Use: Intermediate.

isosafrole $C_{10}H_{10}O_2$.
Properties: Colorless, fragrant liquid; odor of anise; sp. gr. 1.117-1.120; refractive index 1.576; b. p. 253°C. Soluble in alcohol, ether, and benzene.
Derivation: Treatment of safrole with alcoholic potash.
Uses: Manufacture of heliotropin; perfumes; flavors.
Shipping regulations: None. *

isosorbide $OCH_2CHOHCHCHCHOHCH_2O$.
A polyol with a hydroxyl group attached to each of two cis-oriented saturated furan rings. Intermediate for pharmaceuticals.

isosorbide dinitrate (1,4,3,6-dianhydrosorbitol-2,5 dinitrate).
Use: Medicine.

isosterism. Similarity in physical properties of elements, ions, or compounds, due to similar or identical external electron arrangements.

isostilbene. See stilbene.

iso-syst. Proposed to designate a condition, experiment, or curve of constant composition.

isotactic. A type of polymer structure in which groups of atoms which are not part of the backbone structure are located either all above or all below the atoms in the backbone chain, when the latter are arranged so as to be all in one plane. See polymer, stereospecific.

"Isothan Q 15." 328 Trademark for a 20% solution of lauryl isoquinolinium bromide; a deep amber, water-soluble liquid with a pleasant, characteristic odor, used as a fungicide.

isotherm. Constant temperature line used on climatic maps or in graphs of thermodynamic relations, particularly the graph of pressure-volume relations at constant temperature.

isotones. Nuclides (i. e., kinds of atomic nuclei) which have the same excess of neutrons over protons.

isotopes. Varieties of a chemical element that differ in atomic weight, but are very nearly exactly alike in chemical properties. The difference arises because the atoms of the isotopic forms of an element differ in the number of neutrons in the nucleus. Ordinary chlorine is a mixture of isotopes having atomic weights 35 and 37, with the natural mixture having atomic weight about 35.46. Many of the elements similarly exist as mixtures of isotopes, and a great many new isotopes have been produced in the operation of nuclear devices such as the cyclotron.
Uses: Stable isotopes are used as tracers in all kinds of research, as calibration standards for mass spectrometers, as spectrochemical standards; and as standard sources for monochromatic light.
See tracer.

isotopic spin. A term that is introduced in one form of the theory of the atomic nucleus, a kind of quantum number similar to the ordinary spin quantum number, which considers the neutron and proton to be simply two different charge states of the same particle.

"Isotox." 253 Brand name for a type of insecticide product containing lindane.

isovaleral. See isovaleraldehyde.

isovaleraldehyde (isovaleral; isovaleric aldehyde) $(CH_3)_2CHCH_2CHO$. Occurs in orange, lemon, peppermint and other oils.
Properties: Colorless liquid; apple-like odor; sp. gr. 0.785; m. p. −51°C; b. p. 92°C; refractive index (n 20/D) 1.390. Soluble in alcohol and ether; slightly soluble in water.
Derivation: By the oxidation of isoamyl alcohol; also by Oxo process from petroleum.
Method of purification: Distillation.
Grade: Technical.
Containers: Iron drums; glass bottles.
Uses: Flavoring compounds; perfumes; pharmaceuticals; synthetic resins; rubber accelerators.

isovaleric acid $(CH_3)_2CHCH_2COOH$. Occurs in valerian, hop oil, tobacco and other plants.
Properties: Colorless liquid; disagreeable taste and odor; sp. gr. 0.931 (20/20°); b. p. 176°C; refractive index (n 20/D) 1.4043; m. p. −29°C ; slightly soluble in water; soluble in alcohol and ether.
Derivation: With other valeric acids, by distillation from valerian; by oxidation of isoamyl alcohol.
Uses: Medicine; flavors; perfumes.

isovaleric aldehyde. See isovaleraldehyde.

isovaleroyl chloride (3-methylbutanoyl chloride) $(CH_3)_2CHCH_2COCl$.
Properties: Colorless liquid; refractive index (n 24/D) 1.4136; density 0.9854 (24/4°C); b. p. 113°C; soluble in ether; reacts with water and alcohols.
Use: Intermediate in synthesis.

isovaleryl diethylamide. See valeryl diethylamide.

isovaleryl-para-phenetidine $C_2H_5OC_6H_4NHCOCH_2CH(CH_3)_2$.
Properties: White, glistening needles. Almost insoluble in water and ether; soluble in alcohol and chloroform.
Derivation: By heating isovaleric acid with para-phenetidine.

isoxsuprine hydrochloride (1-(para-hydroxyphenyl)-2-(1'-methyl-2'-phenoxyethylamino)propanol-1 hydrochloride) $HOC_6H_4CHOHCH(CH_3)NHCH(CH_3)CH_2$-$OC_6H_5 \cdot HCl$.
Properties: Bitter crystals; m. p. 201-208°C. Sparingly soluble in water; soluble in ethanol.
Use: Medicine.

istle. Short, hard fiber obtained from the leaves of various species of Agave of central Mexico. Jaumave istle is the best variety for brushes, yielding fibers which closely resemble animal bristles. Palma istle yields the lowest quality of fibers, used chiefly in twines and poor brushes, while tula istle is of intermediate quality.
Uses: Brushes; buffing wheels; upholstery; twine.

"Isuprel" Hydrochloride. 162 Trademark for isoproterenol hydrochloride (q. v.).

itaconic acid $CH_2:C(COOH)CH_2COOH$.
Properties: White, odorless crystals; m. p. 167-168°C; soluble in water, alcohols and acetone; sparingly soluble in other organic solvents.
Derivation: Submerged fermentation by mold of various carbohydrates.
Grades: Technical; refined.
Containers: Bags, carloads.
Uses: Copolymerizations; resins; plasticizers; lube oil additives; intermediate.

Italian red. See iron oxide reds.

"Itrumil." 305 Trademark for iothiouracil sodium (q. v.).

"I-Two." 342 Trademark for a brand of atomic iodine for industrial sanitizing.

IUPAC. Abbreviation for International Union of Pure and Applied Chemistry.

*See "I.C.C. Shipping Regulations," page xiii.
Reference numbers refer to name of manufacturer. See "List of Manufacturers," page v.

IVE. Abbreviation for isobutyl vinyl ether. See vinyl isobutyl ether.

"Ivo." [133] Trademark for series of bone blacks used in coatings for leather, paint, wallpaper, etc.
Containers: 250-lb drums.

ivory. A hard, white, close-grained substance which constitutes the greater part of the tusks of the elephant, mammoth, hippopotamus, narwhal and walrus. The best grades are those obtained from the elephant.

ivory, artificial. A substance resembling natural ivory made by mixing gypsum and stearic acid.

ivory black. An animal black produced from ivory. The term is sometimes erroneously applied to other animal blacks. Chief use is as a pigment.

J

jaborandi. See pilocarpus.

jaborandi oil (pilocarpus oil).
Properties: Bright yellow liquid; penetrating odor. Soluble in alcohol and ether.
Chief known constituents: Pilocarpine; ketones.
Constants: Sp. gr. 0.865-0.895; b.p. 180-200°C.
Derivation: Distilled from the leaves of Pilocarpus pennatifolius.
Method of purification: Rectification.
Containers: Glass bottles.
Uses: Medicine; hair tonics.
Shipping regulations: None.*

J acid. See 2-amino-5-naphthol-7-sulfonic acid.

J acid urea. See 5,5-dihydroxy-7,7-disulfonic-2,2-dinaphthylurea.

Jacquemart's reagent. A reagent used in analytical work as a test for ethyl alcohol. It consists of an aqueous solution of mercuric nitrate with nitric acid. On heating the liquid with the reagent, the mercury salt is partially reduced and if ethyl alcohol is present, yields a black precipitate on the addition of ammonia water. Methyl alcohol does not produce this reaction.

jade (jadeite; nephrite). A hard and extremely tough material of varying composition, greenish-white to deep green in color. Part of the so-called jade is jadeite, $NaAlSi_2O_6$, essentially a metasilicate of sodium and aluminum. Part is nephrite, essentially a metasilicate of iron, calcium and magnesium; and part is a variety of feldspar. Williamsite, a variety of serpentine, is sometimes mistaken for jade.
Occurrence: United States (Massachusetts), China, New Zealand, Philippines.
Use: Carved ornaments.

jadeite. See jade.

jalap.
Dried tuberous root of Exogonium purga.
Habitat: Mexico; cultivated in India.
Grades: Technical; N.F. XI.
Containers: Bags; barrels.
Use: Medicine.
Shipping regulations: None.*

jalap, orizaba. See ipomea.

jalap resin. Orange to reddish-brown fragments, or yellowish-gray to brown powder; slight odor; somewhat acrid taste; stable in air.
Derivation: Alcoholic extraction of jalap.
Grade: N.F. XI.
Use: Medicine.

Jamaica pepper. See pimenta.

jamesonite (feather ore) $Pb_4FeSb_6S_{14}$. A natural sulfantimonide of lead and iron, sometimes containing small amounts of copper and zinc.
Properties: Color and streak lead gray to gray black; luster metallic; hardness 2-3; sp. gr. 5.5-6.0.
Use: Minor ore of lead.

James' powder. See calcium phosphate, antimoniated.

Jamestown weed. See stramonium.

"Japalac." [448] Trade name for alkyd type decorative and interior-exterior protective coatings.

japan. A varnish yielding a hard, glossy, dark-colored film. Japans are usually dried by baking at relatively high temperatures. (ASTM definition, ASTM D16-52).

Japan agar. See agar-agar.

Japanese belladonna. See scopola.

Japanese gelatin. See agar-agar.

Japanese isinglass. See agar-agar.

Japan gelatin. See agar-agar.

Japan tallow. See Japan wax.

Japan wax. (Japan tallow; vegetable wax of Japan; sumac wax).
Derivation: From a species of Rhus by boiling the fruit in water.
Properties: A pale yellow solid wax; tallow-like rancid odor. Soluble in benzene and naphtha. Insoluble in water and in cold alcohol. Sp. gr. 0.970-0.980; m.p. 53°C.
Grade: Technical.
Containers: Cases.
Uses: Candles; wax matches; furniture polish; leather polishes; special soaps; substitute for beeswax; food packaging.
Shipping regulations: None.*

jasmine aldehyde. See alpha-amylcinnamic aldehyde.

jasmine oil.
Properties: Colorless, light yellow oil; characteristic odor. Soluble in alcohol, ether and chloroform. Sp. gr. 1.007-1.018; optical rotation +2.5° to +3.5°.
Chief known constituents: Benzyl acetate, linalyl acetate, linalol, indole, and jasmone.
Derivation: Alternate layers of the flowers of

Spanish jasmine, Jasmium grandiflorum, and cotton saturated with a fixed oil are exposed to the warmth of the sun in a closed vessel until the oil becomes impregnated, when it is expressed from the cotton. Also derived by extraction in petroleum ether or other volatile solvent.
Containers: Glass bottles; copper flasks.
Use: Perfumery.
Shipping regulations: None. *

jasmine, yellow. See gelsemium.

jasmone $C_{11}H_{16}O$. A ketone found in jasmine oil and other flower oils. Odor of jasmine; sp. gr. 0.944 (22°/0°C).

jasper. A finely crystalline form of quartz containing up to 20% impurities (iron oxide, iron hydroxide, clay, etc.). Usually red, yellow, dark green, grayish blue. Similar to flint and chert.
Use: Ornamental stone.

jaundice berry. See barberry.

jaune brilliant. See cadmium sulfide.

Java pepper. See cubeba.

Javelle water (eau de Javelle).
Derivation: A solution of potassium hypochlorite, prepared by adding potassium carbonate to a solution of calcium hypochlorite. (The term is also used for a solution of sodium hypochlorite made from soda ash and calcium hypochlorite).
Containers: Glass bottles; carboys.
Grades: Technical.
Uses: Bleaching agent; disinfectant.
Shipping regulations: None. *
See also Labarraque's solution.

"Javollal." [188] Trademark for an aromatic concentrate used as a substitute for oil of citronella.

"Jaysol." [29] Trademark for a proprietary ethyl alcohol composition containing an aliphatic solvent, methyl isobutyl ketone, and ethyl acetate.
Typical Properties: Apparent proof (60°F) 190; sp. gr. (20/20°C) 0.8125, (60/60°F) 0.8157; acidity (as acetic acid wt %) 0.002; color (Hazen) 20; non-volatile matter (mg/100 ml) 3; flash point (Tag closed cup) 58°F; approx. wt/gal 6.79 lb.
Containers: 55-gal drums; tank cars.
Shipping Regulations: Flammable liquid. Red label. *

"JC-60." [41] Trade name for a hot-pour, plastic-based compound with excellent adhesion and root-proof properties used to joint concrete and vitrified clay sewer pipe.

"Jellitac." [103] Trade name for a prepared dry wheat paste which is pre-gelatinized over hot rolls to make the starch water-soluble.

jelutong. See pontianak gum.

Jena glass. An early variety of glass of improved resistance to heat and shock, named from the location of its manufacture in Europe.

Jeppel's oil. See bone oil.

Jesuits' balsam. See copaiba resin.

Jesuits' bark. See cinchona bark, calisaya.

jet. A dense black type of coal which takes a good polish.
Occurrence: Colorado, England, Germany, France, Spain.
Use: Jewelry.

jet fuels. These fuels for jet engines are petroleum products similar to kerosene. A number of different types with somewhat different compositions and properties have been used. The important military jet fuels have been or are as follows:
JP-1 The earliest jet fuel. A naphthenic kerosene, obsolete in 1960.
JP-3 A gasoline-kerosene blend used in one-engine Navy aircraft. Superseded by other types.
JP-4 A widely used fuel consisting of approximately 65% gasoline and 35% light petroleum distillate, with rigidly specified properties.
JP-5 A specially refined kerosene having a flash point of 140°F and a freezing point of −40°C. Used by carrier-based aircraft because it can be stored aboard ship.
JP-6 A higher kerosene cut than JP-4, with fewer impurities, and used in advanced engines.
Commercial jet planes use ASTM Type A, A-1, or B. A and A-1 are kerosene types. The A-1 has lower freezing point and is used for long range flights; type A is the large volume fuel for short and medium range flights. Type B is a gasoline-kerosene type similar to JP-4.

"Jet-Milled." [344] Trademark for unusually finely ground mineral pigments.

"Jetron." [179] Trademark for a series of carbon blacks co-precipitated with synthetic SBR polymer, and dried in a Banbury dryer-mixer. Easier processing than the conventional oven-dried black masterbatches.

jeweler's rouge. See iron oxide reds.

Jews' pitch. See asphalt.

"Jiffix." [329] Trademark for an acid-hardening, ammonium thiosulfate, fixing bath. It is ready-mixed and rapid-acting.

jimson weed. See stramonium.

jonquil oil (narcissus oil). A colorless, very light, expensive oil from the flowers of Narcissus jonquilla which is used in perfumes.

Jordan almond. See almond, sweet.

josephinite. A natural alloy of nickel and iron containing 25.2% iron, 74.2% nickel, and 0.5% cobalt, and found in Oregon.

JP-1, JP-2, etc. These designate various types of military jet fuels. See jet fuels.

Judean pitch. See asphalt.

juniper (juniper berries). Berries, wood and tops of Juniperus communis.

*See "I.C.C. Shipping Regulations," page xiii.
Reference numbers refer to name of manufacturer. See "List of Manufacturers," page v.

Habitat: Northern Europe, Asia, and North America.
Grades: Technical.
Containers: Bags.
Uses: Medicine; gin; cordials; source of juniper oil.

juniper berries. See juniper.

juniper-berry oil (jupiter oil).
Properties: Essential oil. Colorless or faintly greenish-yellow; becomes darker and thicker with age and exposure to air; characteristic somewhat turpentine-like odor; bitter, burning taste.
Chief known constituents: Pinene, cadinene, juniper camphor.
Constants: Sp.gr. 0.865-0.882 (normal oil, 0.867-0.875); optical rotation up to −11°; refractive index 1.479-1.484; acid value to 3.0; ester value 2-8, after acetylation 18-23.
Solubility in alcohol: 1 vol in 5-10 vols of 90% alcohol (usually clear solutions obtained only with freshly distilled oils).
Derivation: Distilled from the fruit of Juniperus communis, L.
Grades: Technical; twice refined.
Containers: Bottles.
Uses: Medicine; veterinary practice; preparation of gin and liqueurs.
Shipping regulations: None.*

juniper gum. See sandarac gum.

juniper tar. See cade oil.

juniper tar oil. See cade oil.

juniper-wood oil. Oil obtained by steam distillation of juniper wood or twigs.
Constants: Sp.gr. 0.8692 (15°C); optical rotation −21°2'; refractive index 1.47111; acid value 0.9; ester value 6.7.
Solubility in alcohol: In 7 vols and more of 90% alcohol (with slight turbidity).
Containers: Glass bottles; cans.
Uses: Veterinary practice; medicine (external remedy).
Shipping regulations: None.*

jupiter oil. See juniper-berry oil.

"Jurnapak." [51] Trademark for a specialized wool yarn lubricant used primarily for packing armature journal boxes. Made with high quality, variegated-color wool yarn of correct length for proper journal box packing.

jute. Bast fibers, 4-10 ft long, obtained from the stems of several species of Corchorus, especially C. capularis. Contains a higher proportion of lignin and less cellulose than any other commercial vegetable fiber and has relatively poor strength and durability. The fibers are soft and lustrous but lose strength when wet. Among the vegetable fibers, jute is next to cotton in volume of consumption.
Sources: Bengal, Pakistan.
Grades: Technical.
Containers: Bales.
Uses: Burlap; sacking; linoleum; twine; packing; coarse paper.

"JZF." [248] Trademark for N, N'-diphenyl-para-phenylene diamine (q.v.).

*See "I.C.C. Shipping Regulations," page xiii.
Reference numbers refer to name of manufacturer. See "List of Manufacturers," page v.

K. Symbol for potassium.

K acid. See 1-amino-8-naphthol-4, 6-disulfonic acid.

kadaya gum. See karaya gum.

"Kadox" Zinc Oxides. [268] Brand name for a group of colloidal zinc oxides of extremely fine particle size, manufactured by the Palmerton process, the pigment being made by the combustion of metallic zinc.
Grades: Three grades developed especially for the rubber and paint industries. Shipped in 50-lb bags.
Uses: Extensively used in rubber for its extraordinary reinforcing effect and pronounced activating effect. For hardening slow-drying paints and to secure gloss and drying in alkyd vehicles.

kainite $MgSO_4 \cdot KCl \cdot 3H_2O$. A natural hydrated double salt of potassium and magnesium found in the European potash deposits. Color, white, gray, reddish or colorless; streak, colorless; vitreous luster. Contains 30% potassium chloride.
Constants: Sp. gr. 2.05-2.13; hardness 2.5-3.
Occurrence: Germany. One of the Stassfurt minerals. See potash.
Uses: Chemicals (potassium salts); fertilizer (as such).

kaiser green. See copper acetoarsenite.

"Kalite." [244] Brand name for calcium carbonate, surface coated.
Properties: Density as shipped, 60-65 lbs/cu ft; wt per solid gal, 22.07 lbs; color, light cream white; particle size, 1 micron approx.
Derivation: Precipitated calcium carbonate.
Containers: Multi-wall paper bags, 50-lbs net.
Uses: Rubber, plastics, drawing compounds.

kalium. Latin name for potassium.

kalsomine. See calcimine.

"Kam." [55] Brand name for a bottle-washing compound designed for dairy and beverage plants which are served by a water supply of normal or less hardness. Phosphated caustic soda. Available as white flakes (400-lb drums) or white solid (700-lb drums).
Shipping regulations: None.*

kanamycin sulfate $C_{18}H_{36}N_4O_{11} \cdot H_2SO_4$. Kanamycin is a wide-spectrum antibiotic. Its chemical name is 4(6?)-3-deoxy-3-amino-alpha-D-glucopyranosyl-6(4?)-6-deoxy-6-amino-alpha-D-glucopyranosyl-1, 2, 3-trideoxy-1, 3-diaminoscyllitol.
Properties: Yellowish crystals; decomposes over a wide range above 250°C; soluble in water; practically insoluble in methanol and ethanol.
Use: Medicine.

"Kaolex" Clay. [285] Proprietary brand name for a series of hydrous aluminum silicates (sedimentary kaolins) from Georgia and South Carolina for ceramics and refractories.
Properties: P.C.E. 33-35; sp. gr. 2.60; DMR to 350 psi for casting, jiggering, pressing and extruding. Air-floated or water-washed (lump or pulverized).
Containers: 50-lb multiwall bags or bulk.
"Kaolex" SC. Clean, white burning; low viscosity; moderate plasticity; low shrinkage; for casting slips and pressed bodies.
"Kaolex" D-6. An airfloated, plastic kaolin for jiggering, extrusion or pressing where high green strength and workability are essential. Clean, white burning, fine-grained, marked thixotropic tendencies, but used successfully in moderate amounts in casting slips.
"Kaolex" WW. A water-washed coarse-grained clay with fast casting rate, low viscosity in high gravity slips. Clean, white, open burning, low shrinkage, low reversible thermal expansion properties.
"Kaolex" BR. Airfloated, fine-grained; high suspension properties suitable for mill addition in porcelain enamels and glazes. With complete deflocculation yields slip of low viscosity and good stability.
"Kaolex" SH. Waterwashed, fine-grained, exceptionally white burning in vitrified compositions. Helps impart translucency in fine china; recommended for electrical porcelain.
"Kaolex" AX. Similar to SH, but treated with tetrasodium pyrophosphate; for refractory casting compositions using phosphate type deflocculants.
"Kaolex" 44. Similar to SH, but treated with barium carbonate; used as stabilizer in sanitary ware, in casting slips and some pressing operations to improve apparent plasticity without change in water content.

kaolin (China clay; white bole; bolus alba; argilla; porcelain clay; white clay; terra alba). A white-burning clay, which, due to its great purity, has a high fusion point and is the most refractory of all clays. Both English and domestic kaolins are used. The largest domestic producers

K

are Georgia, the Carolinas, Alabama and Florida.

Composition: Mainly kaolinite (40% alumina, 55% silica, plus impurities and water).

Properties: White to yellowish or grayish fine powder; sp. gr. 1.8-2.6. Insoluble in water and dilute acids. Has high lubricity (slipperiness) giving it good coating qualities.

Grades: Technical; N. F. XI; also graded on basis of color, and particle size.

Containers: Cartons; paper bags; drums, bulk.

Uses: Filler and coatings for paper; rubber; refractories; ceramics (porcelain, whiteware, stoneware, tile, electric insulation, slips and glazes); cements; fertilizers; chemicals (especially aluminum sulfate); insecticides; paint filler; linoleum.

Shipping regulations: None.*

See also kaolinite and aluminum silicate.

kaolinite $Al_2O_3 \cdot 2SiO_2 \cdot 2H_2O$. A clay mineral, rarely found pure, but it is the main constituent of kaolin and some other clays (q. v.).

Use: Suggested as an ion-exchange material.

kaon. See fundamental particle.

"Kao-Spheres." [99] Trade name for a spherical form of kaolin cracking catalysts.

Typical analysis (volatile free basis): SiO_2 52.9%; Al_2O_3 45.0%; Fe_2O_3 0.3%; TiO_2 1.7%; Na_2O and K_2O 0.05%. Tamped bulk density: 0.78 gm/cc.

kapoc oil. See kapok oil.

kapok. Cotton-like fibers obtained from the seed pods of various species of Ceiba and Bombax. The fibers are extremely light and resilient but are too brittle for spinning.

Sources: Indonesia, Philippines, Ecuador, West Africa.

Uses: Life jackets; insulation; pillows; upholstery.

kapok oil (kapoc oil).

Derivation: By pressing the seeds of Eriodendron anfractuosum and Bombax ceiba.

Habitat: The tropics.

Properties: Yellow-green oil; pleasant odor and taste. Soluble in alcohol, ether, and chloroform; sp. gr. 0.9235; saponification number 181-205; iodine value 117-129.

Grades: Technical.

Containers: Barrels.

Uses: Edible oil; ; soap stock.

Shipping regulations: None.*

"Kappadione." [100] Trademark for menadiol sodium diphosphate, U. S. P.

"Kapsol" Plasticizer. [55] Trademark for methoxyethyl oleate (q. v.).

"Karathane." [23] Trademark for an agricultural fungicide-miticide based on dinitro(1-methylheptyl) phenyl crotonate and supplied as a wettable powder or liquid concentrate. May be combined with most other insecticides and fungicides, except oil-based products.

Use: Controls powdery mildew and various species of mites on plants.

karaya gum (sterculia gum; Indian tragacanth, kadaya gum). A natural exudation from certain Indian trees, of the genus Sterculia. Color varies from large white tears to dark brown or black.

Properties: The chemical composition, viscosity and adhesive properties of the gum vary with the climate, elevation and soil in which the tree is grown. Properties also depend on freshness and time of storage. Viscosity greatly decreases over six months storage.

Typical specifications: Acid number varies from 13.4 to 21.3; moisture 11.6-15.3%, ash 6.1-7.0%; does not dissolve in water, but forms a translucent colloidal sol. Powdered gum swells in water.

Uses: Pharmaceuticals; textiles; foods; often as a substitute for tragacanth gum.

"Karbate." [214] (Impervious carbon and graphite). Trademark for carbon and graphite materials made impervious to fluids under pressure by impregnation with chemically resistant materials. Strength is increased by this impregnation but thermal conductivity is not lowered, nor are the other properties of carbon or graphite base modified to any extent. It is resistant to thermal shock and to attack by most non-oxidizing chemicals.

Properties: Sp. gr. 1.75-1.9; tensile strength 1700-2600 psi; compressive strength 9-11,000 psi; elastic modulus 20-30 psi x 10^5; electric resistivity 0.0003-0.0016 ohm inches; linear coefficient of thermal expansion, impervious carbon 0.0000054 per °C, impervious graphite, 0.0000043 per °C (100°C); thermal conductivity 3-86 Btu/hr/sq ft/°F/ft for carbon and graphite forms respectively. These materials are machinable.

Forms: Supplied as complete equipment items; also available in blocks, cylinders, tubes.

Uses: Pipe; fittings; valves; pumps; heat exchangers; towers and absorbers for chemical process equipment.

"Karkote." [423] Trademark for asphalt-gilsonite automotive protective coatings.

Karl Fischer reagent. A solution of iodine, sulfur dioxide and pyridine in methanol, or better, since the solution is more stable, in methyl "Cellosolve." It is used in the determination of water.

"Karmex." [28] Trademark for a wettable powder containing 80% diuron. Selective weed control in sugar cane, pineapple, caneberries, alfalfa, grapes, cotton and peppermint; also for same non-selective uses as "Telvar."

Containers: 2-lb canisters.

"Karmex" DL. [28] Trademark for suspension containing 28% diuron for pre-emergence weed control in cotton.

Containers: 1/2-gal cans.

"Kasil." [201] Trade name for various potassium silicate anhydrous glasses, anhydrous powders, and liquids.

"Katanol." [307] Trademark for a series of textile chemicals.

"Katanol" 0/50. Solubilized sulfur phenol condensate; 50% active.

Properties: Brown, viscous liquid; sp. gr. 1.2-1.3; soluble in water.

Uses: Mordant for basic dyes on cotton and rayon and for acid dyes on rayon.

"Katanol" W. Solubilized sulfur phenol condensate; 75% active.

Properties: Fine, tan powder; density 0.9-1.0; soluble in water.

Uses: Reserving agent for animal fibers in union dyebaths and in speck dyeing; wool reserving agent in dyeing of mixed fibers; mordant for basic dyes in paper processing.

"Katanol" WB Conc. Modified solubilized sulfur phenol condensate; 99% active.

Properties: Light gray, fine powder; density 0.6-0.7; soluble in water.

Uses: Wool reserving agent in union and speck dyeing.

"Katapol" PN-430. [307] Trade name for a cationic emulsifier and corrosion inhibitor, alkyl polyoxyethylene amine; 100% active.

Properties: Brown liquid; sp. gr. 0.94; soluble in water and hydrocarbons.

Containers: Drums; bulk.

Uses: Emulsifier for agricultural chemicals in hydrocarbon solvents; emulsifier for mineral oils; acid corrosion inhibitor for ferrous alloys; emulsifier in leather processing.

"Katapol" VP-532. [307] Trade name for a cationic surfactant, alkyl polyoxyethylene glycol amine; 20% active.

Properties: Cloudy, amber liquid; sp. gr. 1.00-1.02; soluble in water; stable to acid, alkali and metal ions; exhibits cationic properties in acid media and nonionic qualities in alkaline systems.

Containers: Drums.

Uses: Scouring agent for wool in acid or neutral solutions; scouring agent and dyeing assistant for shearlings; antistatic agent for polyester fibers and polyacrylonitrile fibers.

"Katapone" VV-328. [307] Trade name for a corrosion inhibitor, quaternary ammonium chloride in isopropanol; cationic; 85% active.

Properties: Deep amber, clear, viscous liquid; sp. gr. 0.99; soluble in water, 10% HCl solution, carbon tetrachloride, ethanol, ethylene glycol, xylene; stable to strong acids, mild alkalies, metallic ions.

Containers: 10-lb packages; 50-, 425-lb drums.

Uses: Acid corrosion inhibitor for steel, copper and aluminum; improved performance is obtained by admixture with an "Igepal"; corrosion inhibitor and bactericide in oil producing and processing equipment.

"Katigen." [307] Trademark for certain sulfur dyestuffs. Used for the dyeing of cotton and rayon. Characterized by good fastness to light, washing, etc. To a certain extent also used for the dyeing of paper.

kauri. A fossil (hard) copal resin (q. v.) derived from the kauri pine (Agathis australis) of New Zealand.

Uses: In varnishes and lacquers; to evaluate the solvent power of petroleum thinners for varnishes and paints.

kauri-butanol value. A measure of the solvent power of petroleum thinners used in paints and varhishes. The kauri-butanol value is the number of milliliters of the thinner required to cause cloudiness when added to 20 grams of a solution of kauri gum in butyl alcohol. The solution is prepared in the proportion of 100 grams of kauri in 500 grams of butyl alcohol.

"Kaurit." [440] Trademark for a series of urea resin adhesives for plywood, coreboard and particle board.

kava (kava-kava; ava-ava; kawa). Dried rhizome and roots of Piper methysticum of Polynesia; the fluid extract is used in medicine.

kava-kava. See kava.

kawa. See kava.

"Kaydol." [45] Trademark for white mineral oil, U. S. P.

Properties: Sp. gr. 0.880-0.895 (60°F); Saybolt viscosity 345-355 (100°F); odorless and tasteless.

Uses: Pharmaceutical and cosmetic formulations; plastics; tobacco; paper; animal husbandry.

"Kaylo." [191] Trademark for a hydrous calcium silicate high temperature insulation used on pipes, tanks, etc.

K-capture (K-radiation). A type of radioactive decay in which one of the electrons outside the nucleus of an atom is captured by the nucleus and immediately combines with a proton to form a neutron. The product of this radioactivity has the same mass number as the parent but the atomic number is one unit less. Thus Fe-55 with atomic number 26 decays by K-capture to form Mn-55, with atomic number 25. Terms synonomous with K-capture are K-electron capture and orbital electron capture.

Keene's cement. See gypsum cements.

"Kelacid." [322] A trademark for alginic acid.

Properties: A cream colored highly refined, fibrous powder passing essentially through 60 mesh and having a moisture content of about 8%; insoluble in water but swells rapidly.

Grades: Refined (complies with NF requirements).

Containers: 10-, 50-, and 200-lb drums.

Uses: Tablet disintegrant; hemostatic agent.

"Kelco-Gel." [322] Trademark for refined sodium alginate. Available as HV and LV grades.
Properties: Cream colored, semi-fibrous powder; viscosity 1% by weight about 400 cps (HV) and 50 cps (LV); about neutral pH.
Grade: Refined (N. F. XI grade).
Containers: 10-, 50-, 200-lb drums.
Uses: Thickening, suspending, stabilizing, binding and gelling agent for foods, pharmaceuticals, cosmetics, welding rods, ceramics, latex paints, industrial gels, pastes, coatings, films.

"Kelcoloid." [322] Trademark for the following propylene glycol alginate products.
Properties:
"Kelcoloid"LVF": A cream colored algin in powder form passing essentially through 80 mesh Tyler screen, soluble in hot or cold water to form a relatively low viscosity solution (1% by weight 120 cps) with pH about 4. Moisture content about 13%. Differs from sodium alginate in that it tolerates acid media in pH range 3 to 6 and is less reactive with heavy metal and alkaline earth ions.
"Kelcoloid HVF": Same as for "Kelcoloid LVF" except that water solutions have high viscosity (2% by weight about 7000 centipoises).
"Kelcoloid" O: Same as LVF except that water solutions have a lower viscosity (2% by weight about 115 cps).
Grades: Refined.
Containers: 10-, 50-, 200-lb drums.
Uses: As hydrophilic colloid – an emulsifier, thickener, stabilizer, suspending, foam stabilizing and whipping agent in aqueous media below pH 7.0. Suitable for food, pharmacuetical, cosmetic and industrial uses.
Shipping regulations: None. *

"Kelcosol." [322] Trademark for a sodium alginate product.
Properties:
"Kelcosol": Cream colored, highly refined fibrous powder passing essentially through 80 mesh and with moisture content about 11%; dissolves rapidly in hot or cold water to form clear, highly viscous solutions. Has about neutral pH value. (Viscosity 1% by weight about 1200 cps.) Forms firm gels in acid to neutral pH with alkaline earth salts such as calcium phosphates.
Grades: Refined. (Complies with N. F. requirements.)
Containers: 1/4-, 3/4-lb bottles; 10-, 50-, 200-lb drums.
Uses: See algin.
Shipping regulations: None. *

"Kelecin." [64] Trade name for a line of soybean lecithins; emulsifying and dispersing agents.
Grades: Plastic and fluid. Either grade single- or double-bleached.
Uses: Paints; printing inks; mastics; animal feeds; engine lubricants; rubber processing; plastics; cosmetics; and in margarine, animal and vegetable oils, pre-mixed bread formulations, chcolate candy.
Shipping regulations: None. *

K-electron capture. See K-capture.

"Kel-F." [158] Tradename for a line of fluorocarbon products, including polymers of trifluorochloroethylene and certain copolymers available as extrusion and molding powders, resins, dispersions, gums, oils, waxes and greases, that are characterized by high thermal stability, resistance to chemical corrosion, high dielectric strength, high impact, tensile and compressive strength.
"Kel-F" Plastics, a trifluorochloroethylene polymer: (C_2ClF_3). Colorless, nonflammable thermoplastic material; chemically inert; highly temperature resistant; exceptionally stable; high impact strength; resistant to thermal shock; impervious to corrosive chemicals and highly resistant to most organic solvents; high compression strength; zero moisture absorption; nonwettability; excellent electrical properties; wide temperature utility −320°F to +390°F (and under certain conditions as low as −460°F); excellent clarity. Note: "Kel-F" Plastics molding compounds may be plasticized with lower-molecular weight "Kel-F" Polymer Oil Fractions. The oil is a highly-fluorinated organic material with the same basic chemical structure as the high polymer. The plasticizer, where essential, renders the material softer, more pliable, tougher and gives an extended flex life to molded products. Plasticized molding powders with up to 25 per cent plasticizer added are available. "Kel-F" Brand Plastics can be molded by injection, extrusion, compression and transfer methods, producing parts that can be machined, engraved, cut, drilled, punched, sanded and polished. The powders can be converted into rods, tubes, sheets, laminates, film and wire coatings in addition to the molded shapes.
Uses: Valve diaphragms, tips and seats; gaskets; LOX lip seals; flow meters; O-rings; extruded tubing; electrical parts and encapsulating; self-locking nuts; vent seals; fuel bladders.
"Kel-F" Brand Dispersions provide protection for surfaces that do not lend themselves to the use of molded plastic because of size, design or construction. The dispersions are fine particles of "Kel-F" Plastic suspended in a volatile organic liquid and applied by spray, dip or spread coating techniques.
Uses: In pipe lines, trailer tanks, pumps, mixers, relays, transformers, storage tanks, flowmeters; insulating tape.
"Kel-F" Brand Elastomer gum is available in two grades, chemically similar, but slightly different in vulcanizate properties. The elastomer offers particular chemical inertness. Retains stability exposed to JP4 and JP5 fuels at temperatures to +400°F.
Uses: O-rings; gaskets; hose; fuel cells;

brake seals; diaphragms; wire coating; tubing; ducting; expellent bladders; shaft seals; flexi-liners; sealants.

"Kel-F" Oils, Waxes and Greases are exceptionally chemical and heat resistant and are unaffected by strong alkalis, acids and oxidants; hydrofluoric and fuming nitric acids, hydrogen peroxide and aqua regia. The materials remain stable at +500°F and will not carbonize or support combustion. They have a low thermal conductivity. The materials also offer high thermal stability and excellent load-bearing characteristics.

Uses: Compressor lubricants; hydraulic fluids; pump fluids; potting and sealing waxes; damping fluids for gyros; heat transfer media; lubricants for instruments; liquid dielectrics; sealants; lubricants for plug cocks and valves; synthetic lubricating blends.

"Kelgin." [322] A trademark for the following sodium alginate products.

Properties:

"Kelgin." An ivory-colored, highly refined, granular powder passing essentially through 40 mesh and having a moisture content of about 13%; dissolves in hot or cold water to give medium viscosity (1% by wt about 350 cps) of about neutral pH.

"Kelgin" F: Same as for "Kelgin" except finer mesh size (essentially through 80 mesh) and slightly lower viscosity (1% by wt about 300 cps).

"Kelgin" LV: Same as for "Kelgin" except gives a low viscosity solution (1% by wt about 50 cps).

"Kelgin" XL: Same as for "Kelgin" LV except gives a lower viscosity solution (1% by wt about 25 cps).

Grades: Refined (complies with N. F. requirements).

Containers: 10-, 100-, and 300-lb drums.

Uses: See algin.

Shipping regulations: None. *

"Kellin." [64] Trademark for a chemically treated linseed oil which polymerizes rapidly, dries fast, has excellent water resistance. Used in varnishes and enamels.

"Kelmar." [322] Trademark for a potassium alginate product.

Properties: A cream-colored, highly refined powder passing essentially through 150 mesh and having a moisture content of about 10%; dissolves in hot or cold water to give a medium viscosity solution of about neutral pH.

Grades: Refined.

Containers: 10-, 100-, 300-lb drums.

Uses: As a hydrophilic colloid especially prepared for use in dental impression gel compositions and other prosthetic uses. Such formulations have controlled setting time by chemical reaction at room temperature.

Shipping regulations: None. *

"Kelo Form." [19] Brand name for benzocaine U. S. P. XIII and other esters of para-amino-benzoic acid.

kelp. A large, coarse seaweed. Dried kelp contains from 2 to 4% ammonia, 1 to 2% phosphoric acid, 15 to 20% potash and traces of iodine.

Uses: Production of iodine, potash; sometimes as fertilizer.

Shipping regulations: None. *

"Kelsize." [322] Trademark for a sodium alginate product.

Properties: An ivory colored, refined granular powder passing essentially through 40 mesh and having a moisture content of about 13%; dissolves in hot or cold water to give medium viscosity solution (1% by weight about 200 cps) of about pH 9.

Grade: Technical; refined.

Containers: 10-, 100-, and 300-lb drums.

Uses: Paper sizing compositions and coatings; corrugated paperboard adhesives.

"Keltex." [322] Trademark for a sodium alginate product.

Properties: A tan-colored, refined, granular product passing essentially through 20 mesh and having a moisture content of about 13%; dissolves in hot or cold water to give a high viscosity solution (1% by weight about 250 cps) of slightly alkaline pH.

Grades: Technical; also available as "Keltex" S, passing through 10 mesh.

Containers: 10-, 100-, and 300-lb drums.

Uses: As a hydrophilic colloid; a suspending, thickening, emulsifying, stabilizing and filmforming agent in resin and casein emulsion paints, textile print pastes, textile sizing, adhesives, paper and board sizing, polishes, leather finishes, insecticides, boiler compounds, films and coatings, ceramics, etc.

Shipping regulations: None. *

"Kelthane." [23] Trademark for an agricultural miticide based on 1, 1-bis(chlorophenyl)-2, 2, 2-trichloroethanol and supplied as a wettable powder or emulsifiable concentrate.

Use: Miticide.

"Keltose." [322] Trademark for a gel-forming algin.

Properties: A cream-colored, refined powder passing essentially through 80 mesh and having a moisture content of about 10%; dissolves in hot or cold water to form a semi-gel.

Grades: Refined.

Containers: 10-, 50-, 100-, and 300-lb drums.

Uses: As a thickening, gel-forming and stabilizing agent for food applications such as bakery icings and meringues, confectioneries, fillings and toppings.

Shipping regulations: None. *

"Keltrol." [64] Trademark for a series of styrene-oil copolymers used in enamels, traffic paints, sealers, coatings, etc.

"Kelube." [322] Trademark for triisopropanolamine alginate.

Properties: Supplied in form of an aqueous paste containing 25% "Kelube." Soluble in water and certain organic solvents such as ethylene glycol, glycerine, propylene glycol, etc.

Uses: Industrial only. Used as thickener, suspending agent, plasticizer and emulsifier in processing of textiles, lubricants, inks, etc.

Kelvin. A scale of absolute temperature based on centigrade degrees. See absolute temperature.

"Kel-Vi'-Tol." [64] Trademark for linseed oil plus a polymerization catalyst used for blending with varnish oils to decrease kettle time.

"Kel-X-L." [64] Trademark for a high polyalcohol ester used as a general replacement for China wood oil in varnishes, enamels and calcicoaters. Polymerizes rapidly; dries fast, good water and alkali resistance.

"Kem." [141] Trade name for synthetic protective and decorative coatings.

"Kemadrin." [301] Trademark for procyclidine hydrochloride, used in medicine.

"Kem Cati-Coat." [141] Trade name for a synthetic catalyzed coating for building maintenance where exposed to severe corrosive conditions.

"Kemithal." [207] Trademark for thialbarbitone; 5-cyclo-hexenyl-5-allyl-2-thiobarbituric acid. The sodium salt is used in human and animal medicine. See "Anavenol."

"Kemprint." [141] Trade name for textile pigment emulsions used for printing on fabrics.

"Kenacort." [412] Trademark for triamcinolone (q. v.).

kenaf. Bast fibers obtained from the stems of Hibiscus sabdariffa, grown in Brazil and the West Indies. It is used as a substitute for jute in burlap coffee bags.

"Kenalog." [412] Trademark for triamcinolone acetonide (q. v.).

"Kendex." [167] Trade name for a group of resinous products. The resins vary in properties from hard friable substances to soft rubbery or waxy materials. They are used in coatings, sealers, laminants, adhesives, and other similar applications.

"Kenflex." [267] Trade name for a synthetic polymer of aromatic hydrocarbons.

Properties: Light brown solid or oily liquid; slight odor; non-volatile; compatible with most other polymers; low in electrical loss, high in dielectric strength; increases plasticity during processing without weakening product; soluble in common solvents except water and lower alcohols; high in wetting powder and an aid to dispersion; tacky, and a builder of tack in many elastomers.

Grades: "Kenflex" A (m. p. 160°F); "Kenflex" B (m. p. 80°F); "Kenflex" L (m. p. 30°F); "Kenflex" N (m. p. 35°F).

Uses: Processing and compounding aid for neoprene, "Hypalon," GR-S, vinyl compounds and other plastics; potting compounds; protective coatings; paper and textile coatings; insecticides; inks; and chemical synthesis.

"Kenmix." [267] Trade name for a line of dispersions of rubber compounding agents in a variety of vehicles, both wet and dry.

"Kennametal." [347] Trademark for a series of hard cemented carbides, made in three general classes: (a) Those having tungsten-titanium carbide $WTiC_2$, as their effective ingredient. Grades KM, K3H, K5H, K2S, ranging in Rockwell C hardness from 79. 0 to 82. 0, and transverse rupture strength from 200, 000 to 300, 000 psi. Used for steel cutting tools and wear-resistant parts. (b) Those consisting chiefly of tungsten carbide, WC. Grades K1, K6, K8, ranging in Rockwell C hardness from 76. 0 to 81. 0, and transverse rupture strength from 200, 000 to 325, 000 psi. Used for cutting tools on cast iron, non-ferrous, and non-metallic materials, and wear parts. Grades KE5 and KE7, Rockwell C hardness 76. 0 and 79.0, and transverse rupture strength 250, 000 and 325, 000 psi respectively. Special grades available only in extruded shapes, solid rounds, tubes, squares, and flats, for wear-resistant applications. (c) Special application grades for resisting abrasion and oxidation at high temperatures. Consist essentially of titanium carbide. Grade K138, Rockwell C hardness of 77. 0, transverse rupture strength of 175, 000 psi at room temperature. K138A has Rockwell C hardness of 75. 0 and transverse rupture strength of 150, 000 psi at room temperature. Has small percentage of other carbides, also. Adapted for particularly severe oxidizing conditions, or extended exposure to oxidizing temperatures. These grades retain a strength of 100, 000 psi at 1800°F.

"Kensol." [167] Trade name for a line of products of close-cut petroleum fractions.

Uses: Manufacture of heat set printing inks, base oils for aluminum foil roll, coolants for foil and sheet; high flash metal cleaners and other metal working applications where low viscosity and high flash are required.

"Kentanium." [347] Trademark for titanium-carbide alloy. Available in tubes, rods, bars, and highly complex shapes.

Properties: Young's modulus of elasticity (million psi), 50; tensile strength (psi) 135, 000; 0. 228 lbs/cu in; resists wear, oxidation, corrosion, and thermal shock. Retains high strength at temperatures where refractory alloys deform plastically.

Uses: Gas turbine blades, vane rings, impellers; diesel engine valve seats; hot rod mill guide inserts; bearings; bushings in contact with liquid metals; flame tubes; thermocouple protection tubes; anvils for spot welding; pressure sleeves; hot hardness tester balls.

kephalin. See cephalin.

"Kepone." [50] Trademark for decachloro-octahydro-1, 3, 4-metheno-2H-cyclobuta-[c, d]pentalen-2-one. Available as a 50% wettable powder for use on non-bearing citrus in Florida, specific against the citrus rust mite.

"KER." [11] Trademark for a series of epoxy resins.

"Keranol." [300] Trademark for a modified cationic softener for fabrics. Compatible with resin and other finishes. Also used on synthetic fibers as antistatic agent.

keratin. The protein of hair, hoofs, horns, etc.

"Keripon." [300] Trademark for a water soluble synthetic fatty ester with wetting and re-wetting properties for textile, leather and paper.

kerite.
1. A mixture of tar or asphaltum with sulfur and animal or vegetable oils. Used as a substitute for rubber.
2. One of the constituents of bitumen.

kermes (kermes berries; scarlet corns; grains of kermes; alkermes). The dried females of the shield louse Coccus ilicis. These insects contain a red dye which is the oldest coloring matter known. The coloring principle is considered to be kermesic acid $C_{18}H_{12}O_9$.

kermes berries. See kermes.

kernite $Na_2B_4O_7 \cdot 4H_2O$. A natural sodium borate, found in Kern County, California. Colorless to white; two good cleavages; luster vitreous to pearly; hardness 3; sp. gr. 1.95.
Use: Major source of borax and boron compounds.

kerocaine. See procaine hydrochloride.

kerogen. The organic oil-yielding matter present in oil shales. Not a definite compound but a complex mixture of various complex compounds that vary from one shale to another. Usually a soft brown powder, only slightly soluble in ordinary organic solvents. Contains small proportions of nitrogen and sulfur.

kerosine (kerosene). (The spelling kerosine is preferred by ASTM and ACS publications.) A distilled hydrocarbon from petroleum or shale oil, having a boiling range from about 150-300°C. Currently the most important uses are as fuel for jet engines, and as a heating fuel. Formerly used as an illuminant. Kerosine is also often used as a solvent for cleaning purposes; also in emulsions as an insecticide.

"Kessco." [260] Trade name for a series of fatty acid esters. Available as:
3354. Mono and diglyceride of liquid fatty acids.
3354-N. Mono and diglyceride of liquid fatty acids.
33D54. Diglyceride composition of liquid fatty acids.
8254. Polyoxyethylene fatty acid ester.
12254. Polyoxyethylene derivative of liquid fatty acid.
18254. Polyoxyethylene derivative of liquid fatty acid.
26254. Polyoxyethylene derivative of liquid fatty acid.
E-119. Complex polyhydroxy fatty acid ester.
E-122. Polyglycol ether fatty acid ester.
Properties: Yellow oily liquids; fatty odor; soluble in most organic solvents.
Uses: Emulsifiers; dispersants; wetting agents; lubricants; softeners.

"Kesscoflex." [260] Trade name for a series of plasticizers and solvents. Available as:
MCP. Dimethoxyethyl phthalate.
BCP. Dibutoxyethyl phthalate.
DOA. Di-2-ethylhexyl adipate.
DIOA. Di-isooctyl adipate.
BCO. Butoxyethyl oleate.
BS. Butyl stearate.
BO. Butyl oleate.
MCO. Methoxyethyl oleate.
MCS. Methoxyethyl stearate.
BCS. Butoxyethyl stearate.
BCL. Butoxyethyl laurate.
DBT. Dibutyl tartrate.
TRA. Glycerol triacetate.
DIA. Glycerol diacetate.

"Kesscomir." [260] Brand of isopropyl myristate with small amounts of palmitate and stearate esters.
Properties: Oily, colorless liquid; practically odorless; iodine number 1.0; sp. gr. 0.852 (25°C); f.p. −1°C; b.p. 164°C (4 mm); insoluble in water.
Uses: Emollient and solvent in cosmetics.

"Kesscowax" A21, A33. [260] Trade name for monostearate (acid stabilized). Mixtures of mono- and diglycerides plus an auxiliary emulsifier to render them acid stable and self-emulsifying.
Properties: White to cream, wax-like solids; slight fatty odor; iodine value, 3.0 max; m.p. 52.5-56.5°C (A21), 45-51°C (A33); partially soluble hot in aliphatic and aromatic solvents; water-dispersible.
Containers: 225-lb fiber drums.
Uses: Emulsifiers, opacifiers, and bodying agents for systems involving acids or acid reacting substances such as anti-perspirant, bleach, lemon and medicated creams and lotions.

"Kessco X-159." [260] Trade name for a fatty alkylolamide.
Properties: Waxy solid, light yellow, mild odor. M.p. 71.5°C; dispersible in hot water.
Uses: Thickener; opacifier; pearling agent.

"Kessco X-168." [260] Trade name for a substituted imidazoline.
Properties: Liquid; cationic; mild odor; dispersible in water. Soluble in most common organic solvents.
Uses: Wetting agent in acid solutions; emulsifier for oils and solvents.

"Kessco X-209." [260] Trade name for a fatty alkylolamide.
Properties: Light yellow liquid; anionic; mild odor. Soluble in water; limited solubility in most organic solvents.
Uses: Foam stabilizer for detergents; thickener for shampoos.

kesso oil. See valerian oil, Japanese.

"Ketac." [57] Trademark for ketone-formaldehyde resins.

ketene $H_2C:CO$.
Properties: Colorless, highly toxic gas; disagreeable taste; irritating to the lungs. Readily polymerizes; cannot be shipped or stored in a gaseous state.
Derivation: Pyrolysis of acetone or acetic acid by passing its vapor through a tube at 500-600°C.
Constants: M.p. −151°C; b.p. −56°C.
Containers: Steel bottles for intraplant transfer only.
Uses: Acetylating agent, generally reacting with compounds having an active hydrogen atom; reacts with ammonia to give acetamide. Starting point for making various commercially important products, especially acetic anhydride and acetate esters.

4-ketobenzotriazine (benzazimide; 4-keto-(3H)-1, 2, 3-benzotriazine) $C_7H_5N_3O$ (bicyclic).
Properties: Tan powder; m.p. 210°C (dec.); soluble in alkaline solutions and organic bases.
Uses: Organic synthesis.

alpha-**ketoglutaric acid** (2-oxopentanedioic acid) $HOOCCH_2CH_2COCOOH$. M.p. 113.5°C; soluble in water and alcohol. Important in cell metabolism.

beta-**ketoglutaric acid** (ADA; acetonedicarboxylic acid) $HO_2CCH_2COCH_2CO_2H$.
Properties: Colorless needles; m.p. 135°C (dec); soluble in water and alcohol; insoluble in benzene and chloroform.
Derivation: By heating dehydrated citric acid and concentrated sulfuric acid together.
Use: Organic synthesis.

ketohexamethylene. See cyclohexanone.

ketone, Michler's. See tetramethyldiaminobenzophenone.

ketonimine dyes. Dyes whose molecules contain the −NH=C= chromophore group. There are only two members in the class. These are auramine (Colour Index 655) and a closely related homolog, methyl aurin (Colour Index 656) in which a methyl group replaces one of the hydrogen atoms of aurin. These are basic dyes used on cotton with tannin or tartar emetic as mordants.

ketopropane. See acetone.

alpha-**ketopropionic acid.** See pyruvic acid.

gamma-**ketovaleric acid.** See levulinic acid.

kev. See electron volt.

Keyes process. A distillation process involving the addition of benzene to the constant-boiling 95% alcohol-water solution in order to obtain absolute (100%) alcohol. On distillation a ternary azeotrope mixture containing all three components leaves the top of the column while anhydrous alcohol leaves the column bottom. The azeotrope (which separates into two layers) is redistilled separately for recovery and reuse of the benzene and alcohol.

Kick's law. The amount of energy required to crush a given quantity of material to a specified fraction of its original size is the same no matter what the original size.

"Kierole TT." [42] Proprietary product. Solvent type soap.
Properties: Amber liquid. Disperses readily in water and alkaline kier liquors at 50°C up.
Containers: 55-gal steel drums.
Use: As assistant in obtaining uniform quality in cotton kier boiling process. Promotes uniform penetration of kier liquor.

kieselguhr. See diatomite.

kieserite $MgSO_4 \cdot H_2O$. A natural magnesium sulfate occurring in enormous quantities in the Stassfurt salt beds, Germany. Also found in Austria and India. See also epsomite and magnesium sulfate.

"killed" steels. Steel deoxidized by the addition of aluminum, ferrosilicon, etc., while the mixture is maintained at melting temperature until all bubbling ceases. The steel is quiet and begins to solidify at once without any evolution of gas when poured into the ingot molds.

killeen. See chondrus.

"Kilmag." [50] Trademark for a formulation of calcium arsenate for the control of certain fly maggots under poultry cages.

kilogram. The term has two meanings, as follows:
1. A mass identical with that of the international kilogram, which is carefully preserved at the International Bureau of Weights and Measures in France. This is approximately the mass of a liter of water just above the freezing point.
2. A force equal to the weight of one kilogram mass, measured at the surface of the earth at sea level.

kinetin $C_{10}O_9N_5O$. 6-Furfurylaminopurine.
Derivation: From desoxyribonucleic acid.
Use: Causes plant cells to divide; possibly can lead to discovery of an antikinetin which may be used in cancer treatment.

king's blue. See cobalt blue.

king's green. See copper acetoarsenite.

king's yellow. See orpiment (pigment).

kish.
1. Impure graphite which separates from

molten iron during the process of smelting in a blast furnace.
2. Dross on the surface of molten lead.

"Kjelgest." [16] Trademark for a pure grade of potassium sulfate with a nitrogen content not over 0.01%; containing less than 2% water and approximately 95% K_2SO_4.
Containers: 5- and 10-lb fiber drums.
Uses: Digestions in the determination of nitrogen by the official methods of the Association of Official Agricultural Chemists and of the American Oil Chemists' Society.
Shipping regulations: None. *

"Kleanrol." [28] Trademark for a soldering flux based on zinc chloride plus ammonium chloride.
Properties: White crystalline powder.
Containers: 50-lb and 500-lb drums.
Use: As a solder blanket flux in automatic soldering of sideseams in the manufacture of tinned containers.

"Klearol." [45] Trademark for a white mineral oil, technical grade.
Properties: Sp. gr. 0.828-0.838; Saybolt viscosity 55-65 (100°F); odorless and tasteless.
Uses: Cosmetic preparations; shell egg preservation; further organic synthesis.

"Kleenup." [253] Brand name for a type of insecticide and fungicide product containing petroleum oils.

Klein's reagent.
Derivation: A saturated solution of cadmium borotungstate, formula variously given, possibly $2CdO \cdot B_2O_3 \cdot 9WO_3 \cdot 18H_2O$.
Constants: Sp. gr. 3.28.
Use: For the separation of minerals by specific gravity.

"Kloben." [28] Trademark for an agricultural herbicide; a wettable powder containing 50% neburon, used as a selective weed control agent. Caution!
Containers: 5- and 25-lb fiber drums.

k meson. See fundamental particle.

"K Monel." [283] Trademark for a wrought age-hardenable alloy containing approximately 65% nickel, 30% copper and 3% aluminum; has high strength and hardness, good corrosion resistance, and is non-magnetic down to −150°F.

"knockout" drops. See chloral hydrate.

knock rating. See octane number.

"Knox-Out." [204] Trademark for a line of house hold and farm insecticides.

"Knox-Out Aerosol Insecticide." [204] Trademark for a mixture containing DDT, pyrethrins, and piperonyl butoxide for space spraying to control flying insects.

"Knox-Out Dual-Use Garden Dust." [204] Trademark for a mixture containing DDT, cryolite, and copper sulfate totaling 50.7% active ingredients. For control of most chewing garden insects. Packed in 2-lb cartons for application by duster or direct from the box.

"Knox-Out EQ-53 Emulsion Concentrate." [204] Trademark for a mixture containing 25% DDT. For dilution with water for moth-proofing washable woolens or for spraying or painting over large areas for residual control of crawling insects.

"Knox-Out Farm Insecticide." [204] Trademark for a powder containing 25% lindane. To be mixed with water for spraying or painting to control a wide range of pests and parasites of poultry, livestock and garden.

"Knox-Out Garden Dust." [204] Trademark for an insecticide containing 5% DDT, easily and readily applied with an ordinary dust gun or from the 1-lb sifter pack.

"Knox-Out Insecticide Powder." [204] Trademark for a 10% DDT insecticide powder.

"Knox-Out Insect Spray." [204] Trademark for a 5% DDT and pyrethrin oil-base double-use insecticide for space-spraying to kill flying insects and for spraying or painting on surfaces for residual control of crawling insects.

"Knox-Out Multi-Purpose Garden Dust." [204] Trademark for a combined insecticide and fungicide containing DDT, cryolite, and copper sulfate totaling 50.7% active ingredients. For control of most chewing garden insects.

"Knox-Out Roaches." [204] Trademark for a DDT-chlordane combination insecticide for spraying on crawling insects, or painting or spraying on surfaces for residual control.

"Ko-Blend IS." [179] A masterbatch of insoluble sulfur and non-staining SBR rubber. Masterbatch is made by co-precipitating the insoluble sulfur with the rubber in latex form. The non-staining antioxidant is added to insure protection of the masterbatch during production and storage. The masterbatch contains 50% insoluble sulfur.
Uses: To control sulfur bloom in applications such as light-colored mechanical goods, white sidewall tires, shoe soles, etc.

Koch's acid. See 1-naphthylamine-3,6,8-trisulfonic acid.

"Kodel." [115] A polyester type synthetic fiber.

kojic acid [5-hydroxy-2-(hydroxymethyl)-4-pyrone] $C_6H_6O_4$. An antibiotic substance.
Properties: Crystals; m.p. 152-154°C; soluble in water, acetone, alcohol; slightly soluble in ether; insoluble in benzene; mildly antibiotic.
Derivation: Fermentation of starches and sugars by certain molds.
Use: Chemical intermediate; metal chelates; insecticide; antifungal and antimicrobial agent.

kola. See cola.

kola nuts. See cola.

*See "I.C.C. Shipping Regulations," page xiii.
Reference numbers refer to name of manufacturer. See "List of Manufacturers," page v.

kola seeds. See cola.

Kolbe-Schmidt reaction. The preparation of salicylic acid or its derivatives from carbon dioxide and sodium or potassium phenolate.

"Kollidon." [440] Trademark for polyvinylpyrrolidone (q. v.). Available as:
"Kollidon" 17: K-value 17 ± 1.
"Kollidon" 25: K-value 25 ± 1.

"Kolo." [55] Trademark for series of insecticidal dusts and sprays containing sulfur.

"Koloc." [248] Trademark for a resin emulsion for application to cotton and wool. It stops the shrinking and felting of wool, increases tensile strength, improves wearing quality, reduces fiber loss in processing and service.

"Konakion." [190] Trademark for a brand of phytonadione, vitamin K_1(q. v.).

"Kontakt." [242] Trademark for a group of fat splitting reagents.

"Kopol." [36] Processed Congo copal gums in fused, esterified and modified (with ester gum) grades.
Properties: Color D to I (USDA rosin standards); acid number 10-85; melting point 165-200°F, capillary tube method. Soluble cold in acetates, coal-tar solvents, turpentine and drying oils. Imparts hardness, toughness, rubbing and polishing properties and tight adhesion.
Uses: Floor, rubbing and spar varnishes; high-bake enamels; floor paints; metal primers; etc.

"Kopoxite 159." [11] Trademark for resorcinol diglycidyl ether (q. v.).

koppite. A variety of pyrochlore (q. v.).

kordofan gum. See arabic gum.

"Koresin." [307] Brand name for a rubber tackifier; para-tert-butylphenol acetylene resin; 100% active.
Properties: Brittle, hard granules ranging in color from tan to brown; m.p. (capillary) 120-130°C; soluble in acetone, benzene, cyclohexane, petroleum ether, ethyl acetate, n-butanol; insoluble in water and ethanol.
Uses: Rubber tackifier for GR-S stocks and cement and for mixtures of natural rubber and GR-S; used as a resin in the manufacture of lacquers, inks, and varnishes.

"Korez." [41] Trade name for a silica-filled, synthetic-resin, acid-proof cement of the phenol-formaldehyde type especially good as a mortar cement for electro-refining work. Can be used up to 360°F.

"Korundal XD." [446] Trade name for a 90% alumina brick with high density. Resistant to corrosion and penetration by molten slags and fluxes, thermal shock, and can withstand load at temperatures up to 3000°F and higher.
Uses: Top checker courses of glass tank furnaces; aluminum alloy furnaces; electric furnaces producing alloys; piers; supporting arches and other constructions where heavy loads at high temperatures prevail.

"Kosmos." [110] Brand name for a line of carbon blacks similar to the "Dixie" line (q. v.).

Kourbatoff's reagents. Four etching agents used in the micro-analysis of carbon steels.
(a) A 4% solution of nitric acid in isoamyl alcohol.
(b) A 20% solution of hydrochloric acid in isoamyl alcohol. To this is added one-third of its volume of a saturated solution of nitroaniline or nitrophenol in alcohol.
(c) 1 part of a 4% solution of nitric acid in acetic anhydride to which is added 1 part each of methyl alcohol, ethyl alcohol and isoamyl alcohol.
(d) 3 parts of a saturated solution of nitrophenol added to 1 part of a 4% solution of nitric acid in ethyl alcohol.

kowrie gum. See copal.

"KP-23" Plasticizer. [55] Brand name for butoxyethyl stearate (q. v.).

"KP-90" Plasticizer. [55] Brand name for butyl epoxy stearate (q. v.).

"KP-140" Plasticizer. [55] Brand name for tributoxyethyl phosphate (q. v.).

"KP-201" Plasticizer. [55] Brand name for dicyclohexyl phthalate (q. v.).

Kr. Symbol for krypton.

K-radiation. See K-capture.

kraft paper. See wood pulp.

"Kralac A-EP." [248] Trademark for a high styrene-butadiene copolymer.
Properties: Hard, creamy white resin produced as relatively fine, friable granules which flow easily and are free from objectionable dusting; sp. gr. 1.04; softening range 185-200°F; soluble in the usual organic solvents for rubber.
Use: With natural and chemical rubbers, especially recommended for high-grade soles, tiling and molded parts. Main function is that of greatly increasing the hardness of elastomeric compounds.

"Kralastic." [248] Trademark for a series of ABS (acrylonitrile, butadiene, styrene) resins.
Properties: Granular rubber-plasticized resins; rigid and tough; dimensionally stable, light in weight; chemically resistant; good electrical properties.
Uses: Injection and extrusion applications; chemical pipe; cathode edge strips; cams, gears, cable floats, wheels, etc.

"Kralon." [248] Trademark for a series of rigid thermoplastic resin-rubber blends used for conduit and irrigation pipe, tool handles, automotive air ducts and other applications.

Krebs cycle. See TCA cycle.

Kremnitz white. Old name for white lead produced in Europe by the chamber process. See lead carbonate, basic.

krennerite $(Au, Ag)Te_2$. One of the gold-telluride group of minerals. Corresponds to the same general formula as sylvanite (q. v.), and calaverite (q. v.). Silver-white to pale yellow color.
Constants: Sp. gr. 8.35.
Occurrence: United States (Colorado), Rumania.

"KR Monel." [283] Trademark for a wrought age-hardenable alloy containing approximately 65% nickel, 30% copper and 3% aluminum which has had its machining qualities enhanced by a controlled carbon content.

Kroll process. The best known and widely used process for obtaining titanium metal. Titanium tetrachloride is reduced with magnesium metal at red heat and atmospheric pressure, in the presence of an inert gas blanket of helium or argon. Magnesium chloride and titanium metal are produced. Essentially the same process is also used for obtaining zirconium.

"Kromad." [329] Trademark for a broad-spectrum turf fungicide which effectively prevents and controls brown patch, dollar spot, copper spot and leaf-spot diseases.

"Krome Guard." [302] Trademark for pigmented and unpigmented paints and lacquers for use on metal surfaces.

"Kromfax" Solvent. [214] Trademark for thiodiethylene glycol (q. v.).

"Kromik." [141] Trade name for a multiple pigment rust-inhibiting primer for shop coatings or for first field coat on structural steel.

"Kromoid W." [244] Trademark for blend of vegetable, mineral, and animal waxes.
Containers: Non-returnable steel drums averaging 400-425 lbs net.
Use: To waterproof leather.

"Kromosperse." [74] Trademark for a pigment dispersant to produce tinting pastes compatible in many aqueous and oil solutions.

"Kronisol" Plasticizer. [55] Trademark for dibutoxyethyl phthalate (q. v.).

"Kronitex" AA Plasticizer. [55] Trademark for a grade of tricresyl phosphate.
Properties: Clear, nearly colorless oily liquid. Soluble in most other organic liquids; insoluble or limited solubility in certain amines, mineral oil, glycerin and glycols.
Typical specifications: Sp. gr. 1.165 (20/20°C); f. p. −33°C (stiff gel); boiling range 241-255°C (4 mm); acidity 0.01% (max) as acetic acid; phenols 0.03% (max); odor mild; flash point 500°F; fire point none up to 625°F; vapor pressure < 0.02 mm (150°C); refractive index 1.555 (25°C); viscosity 120 cps (20°C); surface tension 39 dynes/cm (20°C); coefficient of thermal expansion 0.00068 from 10-40°C; wt/gal 9.5 lbs.
Containers: 5-gal cans; 55-gal drums.
Uses: Primary plasticizer for most resins imparting flame retardance, low oil and water extraction, permanent flexibility, and stability at elevated temperatures. It is one of the least migratory plasticizers.

"Kronitex" K-3 Plasticizer. [55] Trademark for a grade of tricresyl phosphate.
Properties: Insoluble or limited solubility in certain amines, mineral oil, glycerine and glycols. Soluble in most other organic liquids.
Typical specifications: Sp. gr. 1.145 ± 0.005, (20/20°C); f. p. −23°C (stiff gel); boiling range, 250 to 266°C; acidity (max) 0.01% as acetic acid; odor, slight; flash point 505°F; fire point, none at 625°F; vapor pressure < 0.02 mm (150°C); refractive index 1.553 ± 0.002 (25°C); viscosity, 220-300 cps, (20°C); surface tension 37 dynes/cm (20°C); thermal expansion, 0.00067 per °C, (10° to 40°C); wt/gal 9 lbs.
Containers: 5-gal can (50 lbs net); 55-gal can (520 lbs net).
Uses: Primary plasticizers that impart flame retardance to plastics; lubricating oils; as an extreme pressure lubricant; a plasticizer for baking or regular lacquers; in hydraulic fluids; as a non-burning dust pick-up material for air filter systems, etc.

"Kronitex" MX Plasticizer. [55] Trademark for cresyl phenyl phosphate.
Properties: Insoluble or limited solubility in certain amines, mineral oil, glycerine, and glycols; soluble in most other organic liquids.
Typical specifications: Sp. gr. 1.195 ± 0.010 (20/20°C); freezing point −38°C (stiff gel); boiling range 235-255°C (4 mm); acidity (max) 0.01% as acetic acid; odor, very slight; flash point 460°F; fire point, none at 625°F; vapor pressure, < 0.02 mm Hg (150°C); refractive index 1.561 ± 0.002 (25°C); viscosity, 60-70 cps (20°C); surface tension, 42 dynes/cm (20°C); thermal expansion, 0.00067 per °C, (10°-40°C); wt/gal, 9.5 lbs.
Containers: 5-gal can (50lbs net); 55-gal can (545 lbs net).
Uses: As primary plasticizer imparting flame retardance to plastics; lubricating oils; as an extreme pressure lubricant; a plasticizer for baking or regular lacquers; in hydraulic fluids; as a non-burning dust pick-up material for air filter systems, etc.

"KRS-5." [134] Trade name for thallium bromide-iodide, synthetic optical crystal. Used as prisms and windows in infrared spectroscopy; infrared achromatic lenses; as lens components for microscope objectives for use in the infrared; military infrared optical instruments.

"Kryocide." [204] Trademark. A natural cryolite insecticide, as a dust, bait or spray for control of many chewing insect pests. "Kryocide-C" contains some copper and "Kryocide D-50" contains sulfur.

"Kryolith." [204] Trademark for sodium fluoaluminate (natural Greenland cryolite).

krypton Kr. Element of atomic number 36, zero group of periodic system. Colorless wholly inert gas—does not combine chemically with any element. Liquefies at −151.7°C.
Derivation: By fractional distillation of liquid air. Air contains 0.00005% of krypton.
Grades: Highest purity.
Containers: Hermetically sealed glass flasks.
Uses: Filling electric luminescent tubes.
Shipping regulations: Non-flammable gas. Green label.*

krypton 85. Radioactive krypton of mass number 85.
Properties: Half-life 10.3 years; radiations, beta, with a small component of gamma; radiotoxicity, relatively low.
Derivation: A fission product extracted from irradiated nuclear fuel.
Forms available: Gas of high chemical purity, but mixed with other isotopes of krypton, in sealed glass flasks.
Uses: Principally for self-luminous markers.
Shipping regulations: Poison, class D, radioactive material. Red or blue label.*

krypton 86. Isotope of krypton that is used in measurement of standard meter.
See meter.

"K-Stay." [69] Trademark for proprietary mixture of oil soluble sulfonic acid of high molecular weight in a petroleum base oil.
Properties: Amber liquid, sp. gr. 0.88-0.90; acid no. 1.0-1.1.
Uses: Processing aid for elastomers.

"KTPL" Resins. [11] Trademark for polystyrene resins, available as fine beads.
Properties: Clear, hard, thermoplastic, water-white resins. Lower molecular weight than molding grades of polystyrene. High chemical and light stability, resistant to acids and alkalies, have good solvent release.
Uses: Protective coatings; floor tiles.

KTPP. Abbreviation for potassium tripolyphosphate.

"Kubola." [51] Trademark for a brown, soda base, block grease of high melting point for use in open bearings. Grades are available which melt at various high temperatures and provide a uniform lubricating film on slow moving shafts.

kunzite. See spodumene.

"Kure-Blend MT." [179] A trademark for a masterbatch containing 50% tetramethyl thiuram disulfide and 50% non-staining SBR rubber. Masterbatch is made by co-precipitating the TMDT with the polymer in latex form and a non-staining antioxidant is added to protect the polymer during production and storage.
Uses: As an ultra accelerator in masterbatch form and applications such as mechanical goods, tubes, etc.

"Kuron." [233] Weed-killing composition containing 2-(2,4,5-trichlorophenoxy)propionic acid esters of mono-, di-, and tri-propylene glycol monobutyl ethers.

Kurrol's salt $NaPO_3(IV)$. A fibrous insoluble form of sodium metaphosphate obtained from the melt by undercooling to 550°C and seeding. This name has also been applied to material that was probably $NaPO_3$ II and/or III. See sodium meta-phosphate.

"Kutrol." [330] Trademark for uroenterone, an extract of pregnancy urine.

"Kutwell." [51] Trademark for emulsible, or soluble, cutting oils consisting of good quality base mineral oils to which emulsifiers are added. Water emulsions are used as coolants in metal-cutting operations.

kyanite. See cyanite.

"Kymene." [266] Trademark for a series of resins designed to contribute wet-strength to paper. Available as 30 and 40% aqueous solutions.

"Kynex." [315] Trademark for sulfamethoxypyridazine (q.v.).

kynurenic acid $C_{10}H_7NO_3$. A metabolic product of tryptophan.
Properties: Yellow needles; m.p. 282-283°C; soluble in hot alcohol; insoluble in ether.
Use: Nutrition studies.

"Kyrax" A. [144] Trademark for polyvinyl stearate.
Properties: White, waxy solid; m.p. 46-48°C; sp. gr. (20/20°C) 0.960-0.982; refractive index (n 55/D) 1.4550; iodine no. 3 max; acid no. 2 max; soluble in benzene, carbon tetrachloride, mineral spirits and halogenated propellants.
Containers: 22-, 120-, and 200-lb fiber drums.
Uses: Synthetic wax for polishes, coatings, mold release.

L

ℓ-. Prefix meaning levorotatory. See article under D-.

L-. Prefix referring to stereoisomeric form. See article under D-.

"L 13/59." See O,O-dimethyl-2,2,2-trichloro-1-hydroxyethylphosphonate.

"L-26." [304] Trade name for lead 2-ethyl hexoate, $[Pb(CH_3(CH_2)_3CH(C_2H_5)COO)_2]$.
Properties: Pale straw-colored viscous liquid; sp. gr. 1.52.
Containers: Metal drums (50 lbs net).
Uses: Hardener for silicone paint films and insulating varnishes, asphalt base removable coatings and reinforcing agent for lubricating oils and greases.

La. Symbol for lanthanum.

Labarraque's solution. Aqueous solution containing 4-6% sodium hypochlorite, an approximately equal amount of sodium chloride, and about 1% of sodium hydroxide or carbonate to stabilize the solution. Pale green color, chlorine odor. Keep in closed bottle, out of light. Used as a disinfectant and bleach.
See also Javelle water.

labdanum oil (ladanum oil).
Properties: A golden-yellow, essential oil; fine, ambergris odor. A crystalline body separates on standing. Soluble in alcohol, ether and chloroform. Sp. gr. 1.011.
Derivation: By distillation of the gum resin of Cistus ladaniferus.
Method of purification: Rectification.
Grade: Technical.
Containers: Glass bottles; tins.
Use: Perfumes.
Shipping regulations: None. *

labeled compound. See tracer.

lac (lacca). See shellac.

laccase. An oxidizing enzyme which oxidizes phenols to ortho- and para-quinones. It is found in the latex of the lac tree, in potatoes, sugar beets, apples, cabbages and other plants.

lac dye. A brilliant red dye obtained by maceration of crude lac.
See shellac.

lachrymators. Substances that function as tear gases.

L acid. See 1-naphthol-5-sulfonic acid and 1-naphthylamine-5-sulfonic acid.

lacmoid (resorcinol blue) $(HO)_2C_6H_3N[C_6H_2(OH)_3]_2$.
Properties: Lustrous, dark-violet, crystalline scales. Soluble in alcohol, ether, acetone, phenol, and glacial acetic acid; slightly soluble in water.
Derivation: From resorcinol by treatment with sodium nitrite.
Method of purification: Crystallization.
Grade: Technical.
Containers: Tins; glass bottles.
Use: Indicator in analytical chemistry.
Shipping regulations: None. *

lacmus. Chemically pure litmus (q. v.).

lacquer. A type of solvent-base paint that forms a film by evaporation of the solvent or by congealing from a molten state. The binders, or film-forming constituents, consist of cellulose esters or ethers, especially nitrocellulose, often in combination with alkyd resins. Examples of solvents are ethyl alcohol, methyl isobutyl ketone, butyl acetate, toluene and xylene. The term lacquer is also applied to the baking finish applied to the interior of food and beverage cans.
Grades: Dip, spray, brush.
Use: Coating metals, wood (especially furniture), etc.
Shipping regulations: May be flammable liquid. Red label. *

lac sulfur. See sulfur, lac.

lactalbumin. See albumin, milk.

lactams. Cyclic amides produced from amino acids by the removal of one molecule of water. An example is caprolactam (q. v.), $\overline{CH_2(CH_2)_4CONH}$, derived from epsilon-aminocaproic acid, $NH_2(CH_2)_5COOH$.

lactase. An enzyme present in intestinal juices and mucosa which catalyzes the production of glucose and galactose from lactose.
Use: Biochemical research.

lactic acid (alpha-hydroxypropionic acid; milk acid) $CH_3CHOHCOOH$.
Properties: Colorless or yellowish, odorless, hygroscopic syrupy liquid. B. p. (15 mm) 122°C; m. p. 18°C; sp. gr. 1.2. Miscible with water, alcohol, glycerin; soluble in ether; insoluble in chloroform, petroleum ether, carbon disulfide. Cannot be distilled at atmospheric pressure without decomposition; when concentrated above 50%, it is partially converted to lactic anhydride.
Derivation: (a) By fermenting starch, molasses, potatoes, etc. and neutralizing the acid as soon as formed with calcium or zinc carbonate. The solution of lactates is

*See "I.C.C. Shipping Regulations," page xiii.
Reference numbers refer to name of manufacturer. See "List of Manufacturers," page v.

concentrated and decomposed with sulfuric acid; (b) synthetically from sulfite pulp liquor.
Grades: Technical, 22% and 44%; edible, 50 to 80%; plastic, 50 to 80%; U. S. P. XVI (85-90%); C. P.
Containers: Barrels; carboys; 55-gal drums; tank trucks.
Uses (in approximate order of volume): Foods and beverages, as acidulant, flavoring, preservative; chemicals (salts, plasticizers, adhesives, pharmaceuticals); tanning; plastics and textiles.
Shipping regulations: None. *

lactic acid dehydrogenase. An enzyme found in animal tissues and yeast which acts upon lactic acid producing pyruvic acid.
Use: Biochemical research.

lactogenic hormone. See luteotropin.

lactonitrile (alpha-hydroxypropionitrile; acetaldehyde cyanohydrin) $CH_3CHOHCN$.
Properties: Straw-colored liquid, acid to methyl red; m. p. −40°C; b. p. 182-184°C (760 mm) (slight decomposition); sp. gr. 0.9919 (18.4°C); refractive index (18.4/D) 1.4058. Soluble in water and alcohol. Insoluble in petroleum ether and carbon disulfide.
Derivation: Acetaldehyde and hydrocyanic acid.
Grades: Technical; 95-97% purity.
Containers: Glass carboys.
Uses: Solvent; intermediate in production of ethyl lactate.

lactophenine (lactylphenetidine) $OC_2H_5C_6H_4NHCOCH(OH)CH_3$.
Properties: White, crystalline powder. Soluble in water; slightly soluble in alcohol. M. p. 118°C.
Derivation: By the action of lactic acid on phenetidine.
Method of purification: Crystallization.
Grade: Technical.
Containers: Tins; glass bottles.
Use: Medicine.
Shipping regulations: None. *

lactose (milk sugar; saccharum lactis) $C_{12}H_{22}O_{11}\cdot H_2O$.
Properties: White, hard, crystalline mass or white powder; sweet taste, odorless. Stable in air. Soluble in water; insoluble in ether and chloroform; very slightly soluble in alcohol.
Constants: Sp. gr. 1.525; m. p. decomposes at 203.5°C.
Derivation: From whey, by concentration and crystallization.
Method of purification: Recrystallization.
Grades: Crude; fermentation; spray dried; edible; U. S. P. XVI.
Containers: 100-lb bags; 100-, 225-lb drums; 175-, 200-lb barrels.
Uses: Pharmacy; infant foods; bacteriology; baking and confectionery; margarine and butter manufacture; pyrotechnics; medicine.
Shipping regulations: None. *

lactylphenetidine. See lactophenine.

lac wax. A wax obtained from lac consisting of myricyl and ceryl alcohols, free and combined with various fatty acids.
Shipping regulations: None. *

LAD. Abbreviation for lithium aluminum deuteride (q. v.).

ladanum oil. See labdanum oil.

"Ladex." [51] Trademark for an extreme-pressure grease used in heavy-duty roller bearings in steel mill equipment. Suitable for grease gun application and in pressure systems. Recommended for use at moderate speed and in the presence of water, but where excessive temperatures do not prevail.

lady's slipper. See cypripedium.

Lafon's reagent. Sulfuric acid solution of ammonium or sodium selenite, used as a test for codeine.

LAH. Abbreviation for lithium aluminum hydride.

lake. A special type of pigment consisting essentially of an organic soluble coloring matter combined more or less definitely with an inorganic base or carrier. It is characterized generally by a bright color and a more or less pronounced translucency when made into an oil paint.

Under this term are included two (and perhaps three) types of pigment: (1) the older, original type composed of hydrate of alumina dyed with a solution of the natural organic color, (2) the more modern and far more extensive type made by precipitating from solution various coal-tar colors by means of a metallic salt, tannin, or other suitable reagent, upon a base or carrier either previously prepared or coincidentally formed, and (3) a number combining both types in varying degree, which might be regarded as a third class. (ASTM definition, ASTM D 16-52).

Lakes are used extensively in the preparation of printing inks, lithographic inks, paints, and in the printing of wall paper and such materials.

lake dyes. Dyes used for the making of lakes by combination with or adsorption on salts of calcium, barium, chromium, aluminum, phosphotungstic acid, or phosphomolybdic acid.
See also mordant dyes.

Lake Red C. Red pigments made by coupling 2-chloro-5-aminotoluene-4-sulfonic acid with beta-naphthol and forming various metal salts.
Properties: Good resistance to bleeding; reasonable light resistance, good transparency; produces inks with good flow.
Grades: Resinated and non-resinated.
Uses: General purpose color for letterpress, gravure, flexographic, moisture set, heat set inks; specially for offset printing inks.

"Laktane." [51] Trademark for a solvent especially prepared for use as lacquer diluent

*See "I. C. C. Shipping Regulations," page xiii.
Reference numbers refer to name of manufacturer. See "List of Manufacturers," page v.

and in rotogravure printing inks. Its boiling range is typically 218-228°F.

lambda particle. See fundamental particle.

lamb mint. See peppermint.

"Laminac" Resins. [57] Trademark for a proprietary grade of polyester resin used mainly in the manufacture of reinforced plastics. Most widely used reinforcement is glass fiber. A wide variety of "Laminac" resins are available characterized by properties such as extremely high strength-to-weight ratio, water resistance, chemical resistance, and good electrical properties. Typical products fabricated include speedboats, radomes, tanks for water and chemical storage, and sports car bodies.

laminates. Products composed of thin layers or sheets united by an adhesive. Wood products such as plywood are made up of layers of veneer bonded with synthetic resin or casein glues. Safety glass may be made of layers of plastic hot-pressed together. Combinations of glass fabric, cloth, paper, plastics, etc., are in growing use for numerous purposes.

lamp black. A black or gray pigment made by burning low-grade heavy oils or similar carbonaceous materials with insufficient air, and in a closed system such that the soot can be collected in settling chambers. Properties are markedly different from carbon black. Used as black pigment for cements, ceramic ware, mortar, inks, linoleum, surface coatings, crayons, polishes, carbon paper, soap, etc.; ingredient of insulating compositions, liquid-air explosives, matches, fertilizers, furnace lutes, lubricating compositions; reagent in cementation of steel.
Containers: Bags.

"Lanalure." [328] Lanolin-containing emulsion for hosiery finish, easily compatible with resin and duller finishes. Is durable through ordinary washing. Does not build up on equipment, and serves as a release agent for hosiery from hosiery forms. Hosiery following a usual wash still exhibits a good hand.

"Lanamid." [243] Trademark for neutral-dyeing, pre-metallized dyes.

"Lanaset Resin." [57] Trademark. A melamine-formaldehyde resin applied to woven and knitted wool fabrics to control wool shrinkage and felting.

lanatoside C $C_{49}H_{76}O_{20}$. A glycoside obtained from the leaves of Digitalis lanata.
Properties: White crystals or powder; odorless; hygroscopic; extremely poisonous! M. p. about 250°C, with decomposition. Insoluble in water, soluble in dioxane and pyridine. Specific rotation (20°C) +32.0 to +34.5°.
Grades: N. F. XI.
Use: Medicine.

land pebble. A type of phosphate rock consisting of "pebbles" of phosphatic material in a clay and sand matrix. It sometimes contains a small amount of uranium, which is recovered as a by-product.
Occurrence: Florida.
Use: Source of phosphate for fertilizer and other uses.

langbeinite $K_2Mg_2(SO_4)_3$. A natural sulfate of potassium and magnesium, found in salt deposits.
Properties: Colorless, yellowish, reddish, greenish; luster vitreous; hardness 3.5-4; sp. gr. 2.83.
Occurrence: New Mexico, Germany, India.
Use: Source of potash.

"Lanitol." [300] Trademark for a group of alkylarylsulfonate type detergents.
"Lanitol" F: Flake; sodium salt.
"Lanitol" CW: Powder; same as flake, plus alkaline builders.
"Lanitol" KL: Liquid; triethanolamine salt.

"Lanoc." [206] Brand name of a range of mothproofing agents, for application to textiles.

"Lanole B." [42] Proprietary product. Blend of sodium oleate and solvents.
Properties: Clear amber colored liquid. Disperses readily in water at all temperatures.
Containers: 55-gal steel drums.
Uses: Scouring agent for the removal of tar, grease and paint from woolen textile fabrics primarily.

lanolin (wool fat, hydrous).
Properties: Yellowish-white, or nearly white, ointment-like mass, incorporated with not less than 25% and not more than 30% of water. Slight odor. Soluble in ether, chloroform; insoluble in water.
Derivation: A fat obtained from the wool of sheep.
Grades: Technical; U. S. P. XVI.
Containers: Drums.
Uses: Ointment base; cosmetics; leather-dressing, finishing and softening agent; rosin soaps; superfatting toilet soaps.
Shipping regulations: None. *

lanolin, anhydrous (wool fat; alapurin).
Properties: Brownish-yellow, tenacious, unctuous mass free of water and having not more than a slight odor. Soluble in benzene, ether, acetone, petroleum ether, and hot alcohol; sparingly soluble in cold alcohol; insoluble in water but can be mixed with about twice its weight of water without separation. M. p. 36-42°C.
Derivation: A fat obtained from the wool of sheep.
Grades: Technical; cosmetic; U. S. P. XVI.
Containers: 400-lb drums.
Uses: See lanolin.
Shipping regulations: None. *

lanosterol (isocholesterol) $C_{30}H_{50}O$. An unsaturated sterol closely related to cholesterol; m. p. 139-140°C.

"Lanoxin." [301] Trademark for preparations of digoxin (q. v.).

lanthana. See lanthanum oxide.

lanthanide series. Modern name given to the rare earth series of elements (see rare earth metals). The individual members of the series are called lanthanides or lanthanons.

lanthanons. See lanthanide series.

lanthanum La. Element of atomic number 57; group III of the periodic table; from the point of view of chemical behavior it is classed as one of the rare-earth elements of the cerium group.
Properties: White, malleable, ductile metal; oxidizes rapidly in air. Sp. gr. 6.18-6.19; m. p. 920°C; b. p. 4200°C; soluble in acids; decomposes water to lanthanum hydroxide and hydrogen.
Derivation: By electrolysis of lanthanum chloride and reduction of chloride or fluoride with calcium metal. For sources see rare-earth minerals.
Containers: Boxes.
Use: Lanthanum salts; electronic devices; pyrophoric alloys; rocket propellants, reducing agent.
Shipping regulations: None. *

lanthanum acetate $La(C_2H_3O_2)_3 \cdot xH_2O$.
Properties: White powder, soluble in water. Purities up to 99.9+%. Soluble in acids.
Containers: Glass bottles; fiber drums.

lanthanum ammonium nitrate $La(NO_3)_3 \cdot 2NH_4NO_3 \cdot 4H_2O$.
Properties: Colorless crystals; soluble in water.
Grades: Purities to 99.9+%.

lanthanum antimonide LaSb. Made in high purity for use as a binary semiconductor.

lanthanum arsenide LaAs. Made in high purity for use as a binary semiconductor.

lanthanum carbonate $La_2(CO_3)_3 \cdot H_2O$.
Properties: White powder; insoluble in water; soluble in acids.
Grades: Up to 99.9+% La salts.
Containers: Glass bottles, fiber drums.

lanthanum chloranilate $La_2(O:C_6Cl_2O_2:O)_3 \cdot nH_2O$. Used as a reagent for fluoride determination.

lanthanum chloride $LaCl_3 \cdot 7H_2O$.
Properties: White crystals; transparent; hygroscopic; (for anhydrous) sp. gr. 3.842 (25°C); m. p. 872°C. Soluble in alcohol, water; acids.
Derivation: Treatment of lanthanum carbonates or oxides with hydrochloric acid in an atmosphere of dry hydrogen chloride.
Grades: Purities to 99.9+%.
Containers: Glass bottles, fiber drums.
Uses: Anhydrous trichloride of rare-earth metal is often used to prepare the metal.

lanthanum fluoride LaF_3.
Properties: White powder, insoluble in water, acids.
Grades: Purities up to 99.9+%.
Containers: Glass bottles, fiber drums.

lanthanum nitrate $La(NO_3)_3 \cdot 6H_2O$.
Properties: White crystals; hygroscopic. Caution! Keep well stoppered. B. p. 126°C; m. p. 40°C. Soluble in alcohol, water, acids.
Grades: Purities to 99.9+%.
Containers: Glass bottles, fiber drums.
Uses: Antiseptic; gas mantles.
Shipping regulations: Oxidizing material. Yellow label. *

lanthanum oxalate $La_2(C_2O_4)_3 \cdot 9H_2O$.
Properties: White powder, insoluble in water, slightly soluble in acids.
Grades: Purities to 99.9+%.
Containers: Glass bottles, fiber drums.

lanthanum oxide (lanthana; lanthanum trioxide; lanthanum sesquioxide) La_2O_3.
Properties: White or buff, amorphous powder; sp. gr. 6.51 (15°C); m. p. 2315°C. Soluble in acids; insoluble in water; hisses in moist air like quicklime.
Derivation: By extraction from monazite sand; by ignition of hydroxide or oxyacid (oxalate, sulfate, nitrate, etc.); by direct combustion of free metal (burns with brilliant, white light).
Grades: Purities to 99.9+%.
Containers: Boxes, glass bottles, fiber drums.
Uses: Instead of lime in calcium lights; incandescent gas mantles; in optical glass; technical ceramics.
Shipping regulations: None. *

lanthanum phosphide LaP. Made in high purity for use as a binary semiconductor.

lanthanum sesquioxide. See lanthanum oxide.

lanthanum sulfate $La_2(SO_4)_3 \cdot 9H_2O$.
Properties: White crystals. Sp. gr. 2.821; refractive index (n 20/D) 1.564; soluble in alcohol; slightly soluble in water, acids.
Derivation: By dissolving hydroxide, carbonate or oxide in dilute sulfuric acid.
Grades: Purities to 99.9+%.
Containers: Glass bottles, fiber drums.
Uses: The sulfates of the rare-earth elements are often used for atomic weight determination of the element.

lanthanum trioxide. See lanthanum oxide.

lanthionine $S(CH_2CHNH_2COOH)_2$. A nonessential amino acid first obtained from deaminated wool.
Properties: Crystals; slightly soluble in water; insoluble in alcohol, ether, chloroform, and acetone.
DL-lanthionine: chars 240°; decomposes 286-292°C.
L(+)-lanthionine: darkens 245°C; decomposes 293-295°C.
D(−)-lanthionine: darkens 245°C; decomposes 293-295°C.
Use: Biochemical research.

"Lanum." [123] Trademark for purified wool fat prepared for medicinal and pharmaceutical use.

*See "I.C.C. Shipping Regulations," page xiii.
Reference numbers refer to name of manufacturer. See "List of Manufacturers," page v.

lapis lazuli (lazurite) $Na_{4-5}Al_3Si_3O_{12}S$. A natural sodium aluminum sulfosilicate, usually somewhat impure.
Properties: Color deep blue to greenish blue; luster vitreous; hardness 5-5.5; sp. gr. 2.4.
Occurrence: Afghanistan; U. S. S. R.; Chile; California.
Use: Ornamental stone; as coloring agent in cosmetics. Formerly a paint pigment (ultramarine) but now superseded by the artificial product.

larch agaric. See agaric.

larch turpentine. See turpentine, Venice.

lard (adeps). Purified internal fat of the hog.
Properties: Soft, white unctuous mass, faint odor, bland taste. Soluble in ether, chloroform, light petroleum hydrocarbons, carbon disulfide, insoluble in water. M. p. 36-42°C.
Chief constituents: Stearin, palmitin, olein.
Containers: Tins; steel drums.
Uses: Cooking; pharmacy (ointments, cerates); perfumery (pomades).

lard oil (grease oil).
Properties: Colorless or yellowish oil, with peculiar odor and bland taste. Soluble in benzene, ether, chloroform, and carbon disulfide; slightly soluble in alcohol.
Chief constituents: Olein, with a small percentage of the glycerides of solid fatty acids.
Constants: M. p. −2°C; refractive index (20/D) 1.470; sp. gr. 0.915; saponification value 195-196; iodine value 56-74.
Derivation: By cold pressing lard.
Adulterants: Cotton seed, petroleum oils.
Grades: Prime winter edible; prime winter inedible; off prime; extra no. 1; no. 1; no. 2.
Containers: Wooden barrels, steel drums; tank cars.
Uses: Lubricant; illuminant; metal cutting compounds; oiling wool; soap manufacture.
Shipping regulations: None. *

"Larex." [152] Trade name for a series of lard oils, composed of glycerine combined predominantly with unsaturated fatty acids; produced in nine grades to fit all kinds of uses, from high grade oils for such applications as precision lubrication to lower grade oils for uses in which quality is not critical, such as for compounding, fine drilling and machinery.
Containers: 5-, 55-gal; tank truck; tank car.

"Larothidol." [162] Trademark for a brand of bithionol.

"Larvacide." [401] Trade name for various products containing chloropicrin.

"Larvacide 15." [401] Trade name for a liquid fumigant containing chloropicrin (15%), carbon tetrachloride and carbon disulfide.
Containers: 30-, 50-gal drums.
Use: For control of insects and rodents in stored grain.
Caution: Poison label. May be fatal if inhaled or swallowed.

"Larvacide 100." [401] Trade name for commercially pure chloropicrin.
Containers: 1-lb bottles; 50-, 100-, 150-, 250-lb cylinders.
Uses: As a fumigant for control of storage insects, on stored grain; also used for soil fumigation to control fungi, weeds and soil insects.
Hazards: Toxic. Do not breathe vapors. Avoid contact with skin or clothing. Poison label.

"Larvacide 70 Aerosol." [401] Trade name for a fumigant containing chloropicrin (70%).
Containers: 50-, 100-, 150- and 250-lb cylinders.
Use: As a space fumigant for control of storage insects and rodents.
Warning: Poisonous vapor.

"Larvacide 85 Aerosol." [401] Trade name for a fumigant containing chloropicrin (85%) and methyl chloride (15%).
Containers: 50-, 100-, 150-, and 250-lb cylinders.
Use: For control of insects and rodents in grain storage buildings.
Warning: Poisonous vapor.

LATB. See lithium aluminum tri-tert-butoxyhydride.

"Latentacid E." [422] Trade name for ethyl para-toluenesulfonate; used as an accelerator in thermosetting resins such as urea-formaldehyde, melamine and furfural resins; also as an ethylating agent.

"Latentacid M." [422] Trade name for methyl para-toluenesulfonate; used as an accelerator in thermosetting resins such as urea-formaldehyde, melamine and furfural resins; also as a methylating agent.

latent heat. The quantity of heat absorbed or given off per unit weight of a material during a change of state such as ice to water, or water to steam.

latex. (Plural: latices or latexes; the former is preferred.) A milk-like fluid in which small globules or particles of natural or synthetic rubber or plastic are suspended in water. The milky sap from the rubber tree is the original example. In this material there is about 60% water, 35% rubber, and 5% other materials. Small amounts of impurities act as stabilizers to prevent settling. (Natural proteins serve this purpose in natural latex, and emulsifying agents are purposely added with synthetics.) Other impurities are also present (sugar-like components and salts in natural latex, polymerization catalyst residues in synthetics), which sometimes affect the products made from the latex. Ammonia is added to natural latex to prevent decomposition during shipment and storage. Latex is used in paints, in producing special papers, in adhesives, and to make foam and sponge rubber. The term latex is also applied to rubber products made directly from latex, as latex girdles, latex pillows, etc. Common synthetic latices include ordinary styrene-butadiene rubber, and polystyrene, polyvinyl chloride, polyvinylidene chloride, polyacrylate resins. Latex is also produced from reclaimed rubber.

latex paints. See emulsion paints.

"Lathanol LAL." [243] Brand name for a proprietary product, a highly refined sodium "lauryl" sulfoacetate; an organic detergent possessing wetting, scouring, emulsifying and dispersing properties; a prolific foamer.
Properties: White, dry powder; pH 6.9-7.1 in 0.25% water solution; stable to hard water; stable to acid and alkali in a pH range of 5.0-8.5; solubility in water solution 1% at 25°C, 25% at 100°C; surface tension (0.2% solution) 32 dynes/cm (25°C); hygroscopicity appreciable, keep containers sealed; sp. gr. 0.55; min active organic content 40%; odor sweet, pleasant; flavor practically tasteless.
Grades: Technical.
Containers: Non-returnable 55-gal fiber drums. (250-lbs net). Smaller packings if desired.
Uses: Tooth pastes; tooth powders; liquid dentifrices; foaming bath salts; shampoos; synthetic detergent bars.

latices. Plural of latex.

"Laticrete." [248] Trade name for a latex-based surfacing compound, a "flexible concrete" combining the resilient and long-wearing properties of rubber with the structural characteristics of concrete.
Properties: Light gray (can be pigmented); non-dusting; high chemical resistance; resiliency of wood flooring; flexible over wide temperature range; non-slipping; fire retardant and spark-proof; low noise factor.
Use: Flooring; tank lining; cement binder for brick, tile and glass; waterproofing tunnels, basements, channels; concrete repairing; highways; playgrounds, tennis courts, etc.

"Latyl." [28] Trademark for a group of finely dispersed dyes developed particularly for coloration of "Dacron" polyester fiber, on which they have exceptionally good light and wet fastness properties.

"Latyl" Carrier A. [28] Trademark designation for a white powder used to increase the dyeability of "Dacron" polyester fiber.

laudanidine (levo-laudanine; tritopine) $C_{20}H_{25}O_4N$. An alkaloid.
Properties: White crystals; m.p. 182-185°C; insoluble in water; soluble in alcohol, benzene, chloroform and slightly soluble in ether.
Derivation: From opium.

laudanine $C_{20}H_{25}NO_4$ (optically inactive form of laudanidine.)
Properties: Small prisms; poisonous! Soluble in benzene and chloroform; slightly soluble in alcohol and ether; m.p. 166°C.
Derivation: By extraction from opium.
Method of purification: Recrystallization.
Grades: Technical.
Containers: Glass bottles.
Use: Medicine.

laudanosine $C_{21}H_{27}NO_4$. An alkaloid.
Properties: White needles; poisonous! Soluble in alcohol, ether and benzene; insoluble in water. M.p. 89°C.
Derivation: From opium.
Method of purification: Recrystallization.
Grades: Technical.
Containers: Glass bottles.
Uses: Medicine.

laudanum (tincture of opium).
Properties: Brown liquid; poisonous! Soluble in alcohol and ether.
Derivation: Granulated opium dissolved in dilute alcohol and purified.
Grades: Technical; U.S.P. XVI.
Containers: Glass bottles.
Use: Medicine.
Shipping regulations: None. *

laughing gas. See nitrous oxide.

laundry blue. Materials used to color white cottons and linens with a blue tint in order to hide the yellow color produced by the alkali in the washing process. Usually a synthetic dye. Soluble Prussian blue is also used for this purpose.

lauraldehyde. See lauryl aldehyde.

laurel. See laurus.

laurel oil, volatile (sweet bay oil; laurel leaf oil).
Properties: A bright yellow liquid; aromatic odor; soluble in alcohol, ether, chloroform and benzene; sp. gr. 0.924.
Chief known constituents: Cineole; pinene.
Derivation: Distilled from the leaves or berries of Laurus nobilis.
Method of purification: Rectification.
Grades: Technical.
Containers: Glass bottles; cans; drums.
Uses: Medicine; flavors; perfumes.
Shipping regulations: None. *

Laurent's acid. See 1-naphthylamine-5-sulfonic acid.

Laurent's alpha acid. See 1-nitronaphthalene-5-sulfonic acid.

"Laurex." [248] Trademark for the zinc salts of a mixture of fatty acids in which lauric acid predominates.
Properties: Yellowish white granulated waxy powder; sp. gr. 1.15; m.p. 95-105°C; soluble in benzol; insoluble in acetone, gasoline, ethylene dichloride, and water.
Use: A fatty acid activator and plasticizer for use in all stocks as a processing aid.

lauric acid (dodecanoic acid) $CH_3(CH_2)_{10}COOH$. A fatty acid occurring in many vegetable fats as the glyceride, especially in coconut oil and laurel oil.

*See "I.C.C. Shipping Regulations," page xiii.
Reference numbers refer to name of manufacturer. See "List of Manufacturers," page v.

Properties: Colorless needles; sp. gr. 0.833; m.p. 44°C; b.p. 225°C (100 mm); refractive index 1.4323 (n 45/D); insoluble in water, soluble in alcohol and ether.
Derivation: Fractional distillation of mixed coconut or other acids.
Grades: 99.8% pure; technical.
Containers: 55-gal drums, tank cars.
Uses: Alkyd resins; wetting agents; soaps; detergents; cosmetics; insecticides; metallic soaps; chemical raw material.

lauric aldehyde. See lauryl aldehyde.

"Laurine." 227 Trademark for hydroxycitronellal (2,6-dimethyl-2-hydroxyoctanal-8) $(CH_3)_2C(OH)(CH_2)_3CH(CH_3)CH_2CHO$.
Properties: Colorless liquid; rather sweet and intense odor, characteristic of linden; stable; not likely to discolor; sp.gr., (25/25°C) 0.917-0.921; refractive index (20°C) 1.447-1.450; soluble in 1 part of 50% alcohol.
Uses: An extremely versatile perfume material, used extensively in such florals as lilac, hyacinth, jasmin, magnolia, narcissus, and the basis of many linden and lily-of-the-valley scents. In floral perfumes for soap, it shows good strength and tenacity.

lauroyl chloride $C_{11}H_{23}COCl$.
Properties: Water-white liquid; refractive index 1.445 (20°C); m.p. −17°C; b.p. 145°C (18 mm); decomposes in water and alcohol; soluble in ether.
Containers: 13-gal carboys; drums.
Use: Synthesis.

lauroyl peroxide (alperox C; dodecanoyl peroxide) $(C_{11}H_{23}CO)_2O_2$.
Properties: White, coarse powder; tasteless; faint odor; soluble in oils and in most organic solvents; slightly soluble in alcohols; insoluble in water; m.p. 53-55°C.
Grades: Technical (about 95%).
Containers: 1-lb (net) fiber containers; 100-lb polyethylene-lined drums.
Uses: Bleaching agent, intermediate and drying agent for fats, oils, and waxes; polymerization catalyst.
Shipping regulations: Oxidizing material. Yellow label. *

laurus (sweet bay; bay; noble laurel; laurel).
Derivation: Leaves and fruit of Laurus nobilis.
Habitat: Mediterranean region; cultivated in Mexico.
Grades: Technical.
Containers: Bags.
Uses: Medicine; source of laurel oil, volatile.
Shipping regulations: None. *

lauryl acetate. See dodecyl acetate.

lauryl alcohol (alcohol C-12; n-dodecanol) $CH_3(CH_2)_{10}CH_2OH$.
Properties: Colorless liquid, with floral odor; sp. gr. 0.830-0.836; refractive index 1.444; m.p. 24°C; b.p. 259°C. Soluble in 2 parts of 70% alcohol.
Derivation: Reduction of coconut oil fatty acids.
Containers: 55-gal drums, 8000-gal tank cars.
Uses: Synthetic detergents; lube additives; pharmaceuticals; rubber; textiles; perfumes.

lauryl aldehyde (lauric aldehyde; dodecyl aldehyde; aldehyde C-12 lauric; dodecanal; lauraldehyde) $CH_3(CH_2)_{10}CHO$.
Properties: Colorless liquid, becoming solid at cool temperatures, with a strong fatty floral odor; sp. gr. 0.828-0.836; refractive index 1.433-1.440; m.p. 44°C. Soluble in 90% alcohol; insoluble in water.
Use: Perfumery.

lauryl bromide (n-dodecyl bromide; 1-bromododecane) $C_{12}H_{25}Br$.
Properties: Clear, colorless to pale straw-colored mobile liquid with coconut odor and low volatility.
Typical properties: Sp. gr. 1.026 (25/25°C); boiling range (5-95% at 45 mm Hg) 151-208°C; f.p. −15.5°C.
Grade: Technical, approx 60% pure.
Derivation: Coconut oil.
Use: Possible intermediate for quaternary ammonium compounds.

lauryl chloride. Commercially, a mixture of n-alkyl chlorides, with $C_{12}H_{25}Cl$ dominant. A clear, water-white, oily liquid, with a faint fatty odor. Completely miscible with most organic solvents; slightly miscible with alcohol; immiscible with water.
Typical properties: Sp.gr. 0.863 (15.5/15.5°C); crystallization point −19°C; distillation range 112-160°C (5 mm); flash point 113°C; fire point 135°C.
Grades: Refined; technical.
Containers: 5-, 55-gal drums; tank cars.
Uses: Synthesis of esters, sulfides, lauryl mercaptan (used in styrene-butadiene polymerization), other organics.

lauryl glycerin. See glycerol monolaurate.

lauryl mercaptan (n-dodecyl mercaptan, tert-dodecyl mercaptan) $C_{12}H_{25}SH$ (approx.).
Properties (technical material, mixture of isomeric compounds): Water-white or pale-yellow liquid; mild characteristic odor; sp. gr. 0.85 (20/20°C); m.p. −7.5°C; distillation range 200-235°C at ordinary pressure, 100 to 200°C at 5 mm pressure; refractive index 1.45-1.47; insoluble in water; soluble in methanol, ether, acetone, benzene, gasoline, and ethyl acetate.
Grades: 95% min.
Containers: Steel drums; carboys; tank cars.
Uses: Manufacture of synthetic rubber and plastics, also in the synthesis of pharmaceuticals, and in insecticides and fungicides.

lauryl methacrylate $CH_2{:}C(CH_3)COO(CH_2)_{11}CH_3$. The commercial material is a mixture, containing also lower and higher fatty derivatives.
Containers: Drums.
Uses: Polymerizable monomer for plastics, molding powders, solvent coatings, adhesives, oil additives; emulsions for textile, leather, and paper finishing.

lauryl pyridinium chloride $C_5H_5NClC_{12}H_{25}$.
Properties: Mottled tan semisolid. Soluble in water and organic solvents. Flash point 165°C; fire point 175°C.
Grade: Technical, contains higher and lower fatty acid derivatives.
Containers: 100-, 375-lb polyethylene-lined drums.
Uses: Cationic detergent; dispersing and wetting agent; ingredient of fungicides and bactericides.

lautal. A hard aluminum alloy containing 4-5% copper, 1.5-2% silicon and fractional percentages of other metals such as iron, manganese, or magnesium.

"Lauxein." [58] Trademark for casein and soybean adhesives, dry powders good for low-temperature applications and glue bonding where water-resistance is desired.
Containers: Multiwall bags and fiber drums.
Uses: Bonding and cold setting glues used in the manufacture of plywood; furniture; case goods of all types.

"Lauxite." [58] Trademark for a series of urea, phenolic, melamine and resorcinol resins. Available in dry powders or liquids. Used for bonding, cold setting and impregnating adhesives and glues for furniture, plywood and aircraft; for hot and cold pressing; for radio frequency equipment; for molding of diversified components from granulated wood.

lava. A rock ejected from volcanoes. Lavas are composed mainly of silicates with a wide range in composition, the silica content varying from 40 to 80%, the balance generally consisting of oxides of aluminum, iron, calcium, magnesium, potassium and sodium, together with some water.

lavandin oil. Yellowish volatile oil having a lavender odor, although somewhat more camphoraceous. Its main component is linalool.
Derivation: Distillation of flowers of Lavandula latifolia fragrans. It is a cross of lavender and spike, native to the Alps. See also lavender oil.
Containers: Drums.
Use: Perfumery.
Shipping regulations: None.*

lavender (garden lavender; true lavender).
Properties: Grayish-lavender particles.
Chief constituent: Lavender-flower oil.
Derivation: The dried blossoms of Lavandula vera (officinalis).
Habitat: Mediterranean region.
Grades: Medium; ordinary; select.
Containers: Boxes; bales.
Uses: Medicine; insectifuge; perfumery; source of lavender oil.
Shipping regulations: None.*

lavender oil (lavender flower oil).
Properties: Essential oil; colorless, yellowish, or greenish-yellow; characteristic lavender odor; strongly aromatic; slightly bitter taste.
Chief known constituents: Linalool, linalyl acetate, geraniol, cumarin, furfurol, and borneol.
Constants: These vary considerably, especially with district from which oil is obtained. The following are U. S. P. XVI specifications: Sp. gr. 0.875-0.888; optical rotation −3° to −10°; refractive index 1.4590-1.4700; linalyl acetate content not less than 30%; angular rotation −3 to −10°; soluble in 4 vols. of 70% alcohol.
Derivation: Distilled from the fresh flowers of several species of the genus Lavandula.
Adulteration: Turpentine oil, cedar-wood oil, lavender-spike oil, terpinyl acetate, geranyl acetate, ethyl esters of the following acids: citric, oxalic, succinic, tartaric.
Containers: 1-, 5-, 10-lb bottles; cans.
Use: Perfumery.
Shipping regulations: None.*

lavender oil, terpeneless (See essential oils, terpeneless.)
Concentration: About 1.75-2 times that of the ordinary lavender oil.
Constants: Sp. gr. 0.893-0.898; optical rotation, about −5°.
Solubility in alcohol: 15 parts per 100 parts of 60% alcohol; 55 parts per 100 parts of 70% alcohol.
Shipping regulations: None.*

lavender-spike oil (spike oil; aspic oil; Spanish lavender oil; Spanish spike oil). See also lavandin oil.
Properties: Pale-yellow to yellow liquid; camphoraceous lavender-like odor; sp. gr. 0.900-0.915 (15°C); optical rotation −5° to +5°; refractive index (n 20/D) 1.4630-1.4680. Soluble in some dilution between one and three volumes of 70% alcohol, becomes hazy on further dilution; soluble in fixed oils; slightly soluble in glycerol; usually forms cloudy solutions with mineral oil.
Derivation: Steam distillation of flowers of Lavender latifolia; purified by rectification.
Uses: Soaps, bath preparations; masking odor in sprays and disinfectants.

"Lavenol 'A,' 'B,' and 'C'." [188] Brand names for a series of synthetic lavender oil substitutes of various types.

"Lavol." [188] Brand name for a substitute for linalyl acetate.

lawrencium. Name suggested for element 103, made by bombarding californium (no. 98) with boron 10 and boron 11 nuclei. Atomic weight believed to be about 257, and half life only about 8 seconds. See also actinide elements.

lawsonia alba. See henna.

Layor caranga. See agar-agar.

lazulite $MgAl_2(OH)_2(PO_4)_2$. A natural basic aluminum phosphate.
Properties: Color blue; luster vitreous; hardness 5-5.5; sp. gr. 3.0-3.1.
Occurrence: Europe; North Carolina, Georgia, California.
Use: Minor gem stone.

lazurite. See lapis lazuli.

lb. Abbreviation for pound.

L. C. L. Abbreviation for less than carload lots.

leaching. The process of extraction of a soluble component from a mixture with an insoluble component, by percolation of the mixture with a solvent, usually water, resulting in the solution and later separation of the soluble component. Synonymous with lixiviation. Examples: separation of tannin from barks, mineral salts from roasted ores.

lead (plumbum) Pb. An element of atomic number 82, of group IV of the periodic system. A heavy, malleable, ductile, gray, soft metal of low tensile strength rarely found native.

Properties: Sp. gr. 11.35 (20°C); m. p. 327.4°C; b. p. 1525-1620°C (760 mm). Soluble in dilute nitric acid; insoluble in water but slowly soluble in water containing weak acid; resists corrosion; relatively impenetrable by nuclear radiation.

Sources: Principally galena, cerussite, anglesite and pyromorphite; much lead is recovered from scrap lead at secondary smelters.

Derivation: By roasting and reduction of the ores.

Impurities: Silver, bismuth and copper.

Method of purification: Desilverizing, oxidation and electrolytic refining.

Types of lead: High purity (impurities less than 10 ppm), pure (99.9+%); pig lead; paste.

Types of pig lead:

Chemical lead. A trade term used to describe the undesilverized lead produced from the southeastern Missouri ores. Typical assay is 0.04-0.08% copper, 0.005-0.015% silver, and less than 0.005% bismuth.

Soft and desilverized leads. Used mainly in white lead, sheet, pipe, shot, and alloys.

Hard, or antimonial lead. Lead hardened or strengthened by up to 16% antimony. See antimonial lead alloys.

Acid and copper leads. Made by adding small proportions of copper to refined and desilverized lead. It is then equivalent to chemical lead.

Corroding lead. A lead refined until it is sufficiently pure for the manufacture of white lead by the corroding process.

Forms available: Ingots, sheet, pipe, shot, buckles or straps, grids, coils of pipe, tanks, valves, rod, wire, bars, cams, drums, traps, bends and lined or coated equipment; metallic paste; powder.

Uses (in approximate order of volume): Metal products (cable covering, solder, caulking lead, ammunition, bearing metals, sheet lead, pipes — the last two are especially important in the chemical industry – type metal, brass and bronze); storage batteries; chemicals (tetraethyl lead, misc.); pigments (red lead and litharge, white lead, others); miscellaneous (weights, annealing, galvanizing, as a shield in handling and shipping radioactive material, as sound-deadener or sound-proofer in plastics).

lead acetate (sugar of lead) $Pb(C_2H_3O_2)_2 \cdot 3H_2O$.

Properties: White crystals (commercial grades are frequently brown or gray lumps); poisonous! Absorbs carbon dioxide when exposed to air, becoming insoluble in water. Soluble in water; slightly soluble in alcohol; freely soluble in glycerol. Sp. gr. 2.50; m. p. loses $3H_2O$ at 75°C; b. p. 280°C, decomposes.

Derivation: By the action of acetic acid on litharge or thin lead plates.

Method of purification: Crystallization.

Impurities: Lead carbonate.

Grades: Powdered; granular; white crystals; C. P.

Containers: Barrels; multiwall paper sacks; drums.

Uses: Medicine; lead salts; textiles, with alum as a mordant in dyeing and printing cottons; weighting silk; indigo resist; waterproofing; manufacturing varnishes; lead driers; manufacture of chrome pigments; ingredient of hair dyes; analytical reagent.

Warning! Harmful dust. MCA warning label.

Shipping regulations: None. *

lead acetate, monobasic $Pb_2O(CH_3COO)_2$.

Properties: White powder; poisonous! Soluble in water, alcohol, and acids.

Derivation: By the interaction of lead oxide and acetic acid.

Method of purification: Recrystallization.

Grades: Technical; C. P.

Containers: Tins; glass bottles.

Uses: Lead salts; analytical chemistry; medicine.

lead alloys. See Tables under fusible alloys.

lead antimonate (Naples yellow; antimony yellow) $Pb_3(SbO_4)_2$.

Properties: Orange-yellow powder; very poisonous! Insoluble in water.

Derivation: By the interaction of solutions of lead nitrate and potassium antimonate, concentration and crystallization.

Method of purification: Recrystallization.

Grades: Technical.

Containers: Wooden kegs; fiber drums.

Uses: Paint pigment; staining glass, crockery and porcelain.

Shipping regulations: None. *

lead, antimonial. See antimonial lead alloys.

lead arsenate $Pb_3(AsO_4)_2$.

Properties: White crystals; very poisonous! Soluble in nitric acid; insoluble in water; sp. gr. 6 to 7.

Derivation: By the action of a soluble lead salt on a solution of sodium arsenate, concentration and crystallization.

Method of purification: Recrystallization.

Grades: Technical; C. P.

Containers: Powder: sacks; 100-lb barrels; paste: cases; tins; 100-lb kegs; 300- and

600-lb barrels; C. P.: multiwall paper sacks.
Use: Insecticide; herbicide.
Warning: Poisonous if swallowed. MCA warning label.
Shipping regulations: Solid: Class B poison. Poison label. *

lead arsenite $Pb(AsO_2)_2$.
Properties: White powder; soluble in nitric acid; insoluble in water.
Grades: Technical.
Use: Insecticide.
Shipping regulations: Solid: Class B poison. Poison label. *

lead azide $Pb(N_3)_2$.
Properties: Colorless needles; explodes at 350°C; a sensitive detonating agent. Should always be handled and shipped when submerged in water, to reduce sensitivity.
Derivation: The reaction of sodium azide with a lead salt.
Use: Primary detonating compound for high explosives.
Shipping regulations: Dextrinated type only. Explosive, Class A. Initiating explosive label. Not accepted by express. *

lead-base Babbitt.
1. A bearing metal with 10-15% antimony, 2-10% tin, up to 0.2% copper, with or without arsenic, and remainder lead. See Babbitt metal. Sometimes known as white-metal bearing alloys.
2. Another type of lead-base Babbitt contains alkaline-earth metals. Used in railway and diesel-engine bearings.

lead, black. See graphite.

lead, blue. A term applied to galena to distinguish it from white lead ore. It is also applied to lead sulfate, blue basic (q. v.).

lead borate $Pb(BO_2)_2 \cdot H_2O$.
Properties: White powder; poisonous! Soluble in dilute nitric acid; insoluble in water. Sp. gr. 5.598.
Derivation: By the interaction of solutions of lead hydroxide and boric acid, with subsequent crystallization.
Grades: Technical; C. P.
Containers: 1-, 5-lb bottles; wooden kegs.
Uses: Varnish and paint drier; waterproofed paints; lead glass; galvanoplastic work.
Shipping regulations: None. *

lead borosilicate. A constituent of optical glass, composed of a mixture of the borate and silicate of lead.

lead bromate $Pb(BrO_3)_2 \cdot H_2O$.
Properties: Colorless crystals. Poisonous! Soluble in hot water; sp. gr. 5.53; decomposes at about 180°C.
Grades: Technical.

lead bromide $PbBr_2$.
Properties: White powder. Slightly soluble in hot water; insoluble in alcohol. Sp. gr. 6.66; b. p. 916°C; m. p. 373°C.
Grades: Technical.

lead burning. Pieces of lead sheet and pipe are joined together by the use of solder or autogenous welding. Lead is a comparatively fusible metal and lead burning is accomplished as follows:

The parts to be joined are scraped clean and placed near together. Then a clean bar of the same metal is applied with an oxy-gas or oxy-hydrogen flame so that the three parts melt together locally. This is a simple process (yet it requires skill) and is sometimes used for lead roofing though principally used in making chemical equipment.

lead carbolate. See lead phenate.

lead carbonate $PbCO_3$. See also lead carbonate, basic.
Properties: White, powdery crystals; poisonous! Soluble in acids; insoluble in water and alcohol. Sp. gr. 6.43; decomposed by heat 315°C.
Derivation: By adding a solution of sodium bicarbonate to a solution of lead nitrate. Occurs in nature as cerussite.
Impurities: Basic lead carbonate.
Grades: Technical.
Containers: Wooden barrels; fiber drums.
Use: Paint pigment.
Shipping regulations: None. *

lead carbonate, basic (lead subcarbonate; white lead; BCWL; ceruse; lead flake) $2PbCO_3 \cdot Pb(OH)_2$.
Properties: White, amorphous powder; poisonous! Soluble in acids; insoluble in water; decomposes at 400°C; sp. gr. 6.14.
Derivation: (a) Dutch process. By the corrosion of lead buckles in pots by means of acetic acid and carbon dioxide generated by the fermentation of waste tan-bark. (b) Carter process. By treating very finely divided lead in revolving wooden cylinders with dilute acetic acid and carbon dioxide.
Grades: Dry; ground in oil; C. P.
Containers: Bottles; bags.
Uses: Paint pigment; putty; ceramic glazes.
Shipping regulations: None. *

lead chloride $PbCl_2$.
Properties: White crystals; poisonous! Slightly soluble in hot water; insoluble in alcohol and cold water. Sp. gr. 5.88; m. p. 498°C; b. p. 950°C.
Derivation: By the addition of hydrochloric acid or sodium chloride to a solution of a lead salt, with subsequent crystallization.
Method of purification: Recrystallization.
Grades: Technical; C. P.
Containers: 1-, 5-lb bottles; 400-lb fiber drums.
Uses: Preparation of lead salts; lead chromate pigments; analytical reagent.
Shipping regulations: None. *

lead, chocolate. A pigment prepared by first calcining lead oxide with about 30% copper oxide and then reducing the product of this operation to a uniform smoothness and homogeneity.

lead chromate $PbCrO_4$.
Properties: Yellow crystals; poisonous! Soluble in acids; insoluble in water; sp. gr. 6.123; m.p. 844°C.
Derivation: By interaction of solutions of sodium chromate and lead nitrate.
Method of purification: Washing.
Grades: Technical; C.P.
Containers: 1-lb bottles; tin cans; wooden barrels; fiber drums.
Use: Paint pigment.
Shipping regulations: None. *
See also chrome yellows.

lead chromate, basic. See chrome red.

lead-coating. Coatings of lead or lead-rich alloys are (1) deposited by dipping into the molten metal, after applying a layer of tin to secure good adhesion of the lead coating; (2) by electroplating from a fluosilicate or fluoborate bath, or (3) by spraying.
See metal spraying.

lead cyanide $Pb(CN)_2$.
Properties: White to yellowish powder; very poisonous! Slightly soluble in water; decomposes in acid.
Derivation: By the interaction of solutions of potassium cyanide and lead acetate.
Grades: Technical.
Containers: Wooden kegs; fiber drums.
Use: Metallurgy.
Shipping regulations: None. *

lead dioxide (lead oxide, brown; plumbic acid, anhydrous; lead peroxide; lead superoxide) PbO_2.
Properties: Brown, hexagonal crystals; will cause many materials to take fire if merely mixed with them owing to its strong oxidizing action; hence care needed in storing and shipping; poisonous! Soluble in glacial acetic acid; insoluble in water and alcohol; sp. gr. 8.91; m.p., decomposes.
Derivation: By adding bleaching powder to an alkaline solution of lead hydroxide.
Impurities: Lead chloride.
Grades: Technical; C.P.
Containers: 1-, 5-lb bottles; 1-, 5-lb cans; 200-lb fiber drums; barrels.
Uses: Oxidizing agent; medicine; oxidizing agent in dye and intermediate manufacture; electrodes; batteries; rubber substitutes; textiles (mordant, discharge in dyeing with indigo); matches; explosives; analytical reagent.
Shipping regulations: Oxidizing material. Yellow label. *

lead dust. Lead in very finely powdered form.

lead, electrolytic. Pure lead obtained by electrolytic deposition.

lead ethylhexoate. See soaps, metallic.

lead flake. See lead carbonate, basic.

lead fluosilicate (lead silicofluoride) $PbSiF_6 \cdot 2H_2O$.
Properties: Colorless crystals; soluble in water; decomposes when heated.
Grades: Technical.
Use: Solution for electrorefining lead.

lead formate $Pb(CHO_2)_2$.
Properties: Brownish-white, lustrous, very finely divided, crystalline substance; soluble in water; poisonous! Decomposes at 190°C.
Grades: Technical.
Containers: Multiwall paper sacks.
Use: Reagent in analytical determinations.

lead glance. See galena.

lead hydrate. See lead hydroxide.

lead hydroxide (lead hydrate; hydrated lead oxide) $Pb(OH)_2$ or $Pb_2O(OH)_2$.
Properties: White, bulky powder; poisonous! Soluble in alkalies; slightly soluble in water; soluble in nitric and acetic acid; sp. gr. 7.592; m.p., decomposes at 145°C; absorbs carbon dioxide from air.
Derivation: By the addition of sodium or ammonium hydroxide to a solution of a lead salt with subsequent filtration and drying.
Grades: Technical.
Containers: Cans.
Use: Lead salts.
Shipping regulations: None. *

lead hyposulfite. See lead thiosulfate.

lead iodide PbI_2.
Properties: Golden-yellow crystals or powder; odorless; poisonous! Soluble in potassium iodide and concentrated sodium acetate solutions; insoluble in water and alcohol; sp. gr. 6.12; m.p. 358°C; b.p. 872°C.
Derivation: By the interaction of lead acetate and potassium iodide.
Method of purification: Crystallization.
Grades: Technical; C.P.
Containers: Jars; tin boxes.
Uses: Bronzing; mosaic gold; printing; photography; medicine.
Shipping regulations: None. *

lead lining. Tanks are made with loose linings of sheet lead surrounded by a wooden, steel, or concrete exterior. Often the sheet lining is strapped, bolted or riveted with lead rivets to the reinforcing structure. Another method of joining lead to reinforcing metal is by means of a film of solder. The so-called homogeneous coating or lining consists in having the lead united to the reinforcing metal by means of a nonmetallic flux which forms a bond that holds its strength nearly to the melting point of lead.

lead linoleate (lead plaster) $Pb(C_{18}H_{31}O_2)_2$.
Properties: Yellowish-white paste; poisonous! Soluble in oils; insoluble in water.
Derivation: By heating a solution of lead nitrate with sodium linoleate.
Grades: Technical; fused (contains 26.5% Pb).
Containers: 125-, 400-lb drums.
Uses: Medicine; drier in paints and varnishes.
Shipping regulations: None. *

lead-manganese linoleate (manganese-lead linoleate).
Properties: Dark-brown, plaster-like mass; soluble in chloroform, hot linseed oil.
Grades: Technical.
Use: Drier in paints, varnishes, inks.

lead metasilicate. See lead silicate.

lead metavanadate. See lead vanadate.

lead molybdate $PbMoO_4$.
Properties: Yellow powder; poisonous! Soluble in acids; insoluble in water and alcohol.
Derivation: By adding a solution of lead nitrate to a solution of ammonium molybdate, concentration and crystallization.
Grades: Technical; C. P.
Containers: Bottles.
Use: Analytical chemistry; pigments (see molybdate oranges).
Shipping regulations: None. *
See also wulfenite.

lead (mono)nitroresorcinate.
Shipping regulations: Explosive, class A. Initiating explosive label. Not accepted by express. *

lead monoxide. See litharge.

lead beta-**naphthalenesulfonate** $Pb(C_{10}H_7SO_3)_2$.
Properties: White crystalline powder; poisonous! Soluble in alcohol; insoluble in water.
Derivation: By the action of lead acetate on beta-naphthalenesulfonic acid.
Method of purification: Crystallization.
Grades: Technical.
Containers: Kegs.
Use: Organic preparations.
Shipping regulations: None. *

lead naphthenate.
Properties: Soft, yellow, resinous semi-transparent material. Gives deposits in highly acid oils, but not when mixed with suitable quantities of cobalt or manganese. Soluble in alcohol; m. p. approx. 100°C.
Derivation: Addition of lead salt to aqueous sodium naphthenate solution.
Grades: Liquid: 16%, 24% Pb; solid: 37% Pb.
Containers: Steel drums; fiber drums.
Uses: Paint and varnish drier; wood preservative and insecticide; catalyst for reaction between unsaturated fatty acids and sulfates in the presence of air; lube oil additive to produce chatterless oils and extreme-pressure lubricants.
Shipping regulations: None. *

lead nitrate $Pb(NO_3)_2$.
Properties: White crystals; promotes combustion in contact with organic matter; poisonous! Soluble in water and alcohol; sp. gr. 4.53; decomposes between 205 and 223°C.
Derivation: By the action of nitric acid on lead.
Grades: Technical; C. P.
Containers: 1-, 5-lb bottles; 1-lb cartons; fiber drums.
Uses: Lead salts; medicine; mordant in dyeing and printing calico; matches; paint pigment; mordant for staining mother-of-pearl; oxidizer in the dye industry; sensitizer in photography; explosives; tanning; process engraving and lithography.
Fire hazard: Dangerous; oxidizing material. In contact with organic or other readily oxidizable substances it will cause violent combustion or ignition.
Shipping regulations: Oxidizing material. Yellow label. *

lead nitrite (basic lead nitrite; lead subnitrite).
Properties: Light-yellow powder; variable composition, essentially $3PbO \cdot N_2O_3 \cdot H_2O$. Soluble in dilute nitric acid. Easily decomposed.
Grades: Technical.

lead nitrite, basic. See lead nitrite.

lead ocher. See massicot (1).

lead octoate. See soaps, metallic.

lead oleate $[CH_3(CH_2)_7CH{:}CH(CH_2)_7COO]_2Pb$.
Properties: White powder or ointment-like granules or mass; poisonous! Soluble in alcohol, ether, turpentine, and benzene; insoluble in water.
Derivation: Reaction of oleic acid with lead hydrate or carbonate, or by the interaction of lead acetate and sodium oleate.
Grades: Technical.
Containers: Wooden kegs; 500-lb drums.
Uses: Varnishes; lacquers; paint drier; high-pressure lubricants.

lead, orange. See orange mineral.

lead orthophosphate, normal. See lead phosphate.

lead oxide, brown. See lead dioxide.

lead oxide, hydrated. See lead hydroxide.

lead oxide, red (red lead; minium; plumboplumbic oxide) Pb_3O_4.
Properties: Bright red powder; partly soluble in acids; insoluble in water. Sp. gr. reported variously 8.32-9.16; decomposes between 500 and 530°C.
Derivation: By carefully heating litharge in a furnace in a current of air.
Grades: Technical; 95%, 97%, 98%.
Containers: Bottles; 500-lb barrels; multiwall paper sacks.
Uses: Storage batteries; paints; glass; pottery and enameling; varnish; lead dioxide; purification of alcohol; packing pipe joints; rubber pigment; red pencils.
Shipping regulations: None. *

lead oxide, yellow. See litharge.

lead peroxide. See lead dioxide.

lead phenate (lead phenolate; lead carbolate) $Pb(OH)OC_6H_5$.
Properties: Yellowish to grayish-white powder; poisonous! Soluble in nitric acid; insoluble in water and alcohol.
Derivation: By boiling phenol with litharge.

lead phenolate. See lead phenate.

lead phenolsulfonate (lead sulfocarbolate) $Pb(C_6H_4OHSO_3)_2 \cdot 5H_2O$.
Properties: White crystals or powder. Soluble in water and alcohol.

lead phosphate (normal lead orthophosphate) $Pb_3(PO_4)_2$.
Properties: White powder; sp. gr. 6.9-7.3; m.p. 1014°C. Poisonous! Insoluble in water; soluble in acids and alkalies.
Grades: C.P.

lead phosphite, dibasic $2PbO \cdot PbHPO_3 \cdot \frac{1}{2}H_2O$.
Properties: Fine white acicular crystals; sp. gr. 6.94; refractive index 2.25. Insoluble in water.
Containers: 40-, 350-lb fiber drums.
Uses: Heat and light stabilizer for vinyl plastics and chlorinated paraffins. As an ultraviolet screening and anti-oxidizing stabilizer for vinyl and other chlorinated resins in paints and plastics.
Caution: Dibasic lead phosphite should be stored in closed containers, away from open flame, and at temperatures not to exceed 400°F. Avoid exposure to sparks or static electricity by grounding equipment and using wooden scoops.

lead plaster. See lead linoleate.

lead-potassium glass. See flint glass.

lead protoxide. See litharge.

lead, red. See lead oxide, red.

lead resinate.
Properties: Brown lustrous translucent lumps or yellow-white powder, or yellowish-white paste; poisonous! Insoluble in most solvents.
Derivation: By heating a solution of lead acetate and rosin oil.
Impurities: Lead oxide.
Grades: Precipitated, 23% Pb.
Containers: 50-lb kegs; 115-lb barrels.
Uses: Paint and varnish drier; textile waterproofing agent.
Shipping regulations: None. *

lead salicylate $Pb(C_6H_4OHCOO)_2 \cdot H_2O$.
Properties: White crystals; soluble in hot water and alcohol.
Containers: Drums.

lead sesquioxide Pb_2O_3.
Properties: Reddish-yellow powder. Soluble in alkalies and acids; insoluble in water.
Derivation: By gently heating metallic lead.
Grades: Technical.
Containers: Wooden kegs; fiber drums.
Uses: Medicine; ceramics; ceramic cements; metallurgy; varnishes; paint pigment.
Shipping regulations: None. *

lead-shot metal. An alloy of lead and arsenic. The arsenic content may range from 0.3 to 0.8% and may be added either in the form of white arsenic or arsenical dross. The arsenic imparts a greater fluidity to the metal and increases the tendency of the metal to assume a spherical form in passing through the air when dropped from the top of the lead-shot tower. With too little arsenic, the drops are pear-shaped, with too much they become double-convex. The addition of about 0.025% sodium sulfide to the water at the bottom of the tower prevents oxidation of the lead.

lead silicate (lead metasilicate) $PbSiO_3$. Formula and names of dubious accuracy.
Properties: White, crystalline powder; insoluble in most solvents.
Derivation: By the interaction of lead acetate and sodium silicate.
Grades: Technical.
Containers: Wooden kegs; fiber drums; multiwall paper sacks.
Uses: Ceramics; fireproofing fabrics.
Shipping regulations: None. *

lead silicate, basic (white lead silicate). A pigment made up of an adherent surface layer of basic lead silicate and basic lead sulfate cemented to silica.
Properties: Excellent film-forming properties with vegetable drying oils combined with low specific gravity.
Derivation: Fine silica is mixed with litharge and sulfuric acid. The mixture is then furnaced in a rotary kiln and ground to break up agglomerates.
Containers: Bags.
Use: As white lead pigment in exterior mixed pigment house paints.

lead silicates. Of various compositions. The anhydrous forms are made by roasting lead oxide with silica. Another means of preparation is by drying the reaction product of silica gel, litharge and acetic acid. Used in rubbers and films as fillers and to protect them against sunlight.

lead silicofluoride. See lead fluosilicate.

lead-silver Babbitt. A bearing metal with a small amount of copper (up to 0.2%), up to 5% tin, 10-15% antimony, from 2.5 to 5.1% silicon and remainder lead. See Babbitt metal.

lead-soap lubricants. Lead salts saponified with fats. These lubricants are hard at low temperatures, viscous at ordinary temperatures, but they become somewhat fluid on heating by friction. They are employed as "extreme-pressure lubricants." Due to their high melting point, they are not suited for high speed work.
Shipping regulations: None. *
See lead naphthenate; lead oleate; lead stearate; also metallic soaps.

lead-sodium hyposulfite. See lead-sodium thiosulfate.

lead-sodium thiosulfate (lead-sodium hyposulfite; sodium-lead hyposulfite; sodium-lead thiosulfate) $PbS_2O_3 \cdot 2Na_2S_2O_3$.
Properties: Heavy, small, white crystals. Soluble in solutions of thiosulfates.
Grades: Technical.
Use: Matches.

lead stannate $PbSnO_3 \cdot 2H_2O$.
Properties: Light-colored powder. Insoluble

in water. Approximate temperature of dehydration 170°C.
Uses: Additive in ceramic capacitors; pyrotechnics.

"Leadstar." [304] Trade name for normal lead stearate $[Pb(C_{17}H_{35}COO)_2]$ vinyl stabilizer.
Properties: Fine white unctuous powder; sp. gr. 1.41; refractive index 1.59.
Containers: Fiberboard drums containing 40 and 200 lbs.
Uses: Lubricating, stabilizing and water-proofing properties in vinyl plastics.

lead stearate $Pb(C_{18}H_{35}O_2)_2$. Poisonous!
Properties: White powder; m.p. 105°C; sp. gr. 1.323; soluble in ether and alcohol; insoluble in water.
Derivation: By heating a solution of lead acetate with sodium stearate.
Grades: Technical.
Containers: Tin cans; multiwall paper sacks.
Use: Varnish and lacquer drier; high-pressure lubricants; lubricant in extrusion processes; stabilizer for vinyl polymers; component of greases, waxes and paints.
Shipping regulations: None. *

lead styphnate. See lead trinitroresorcinate.

lead subcarbonate. See lead carbonate, basic.

lead subnitrite. See lead nitrite.

lead, sugar of. See lead acetate.

lead sulfate $PbSO_4$.
Properties: White, rhombic crystals; poisonous! Slightly soluble in hot water; insoluble in alcohol. Sp. gr. 6.12-6.39; m.p. 1170°C.
Derivation: By the interaction of solutions of lead nitrate and sodium sulfate.
Grades: Technical; C.P.
Containers: Bottles; boxes; kegs; barrels; multiwall paper sacks.
Use: Storage batteries; paint pigments.

lead sulfate, basic (white lead, sublimed; white lead sulfate). Approximate formula $PbSO_4 \cdot PbO$.
Properties: White monoclinic crystals; sp. gr. 6.92; m.p. 977°C; only slightly soluble in hot water or acids.
Grades: Vary from 72 to 85% lead sulfate and remainder lead oxide. Sold dry or ground in oil.
Derivation: Three methods are used:
(a) Lead sulfide ore (galena) is subjected to high temperatures in an oxidizing atmosphere.
(b) Molten lead is sprayed into a jet of ignited fuel gas and air in the presence of sulfur dioxide gas.
(c) Atomized metallic lead is mixed with water and sulfuric acid is added under controlled conditions.
Containers: Barrels; multiwall paper sacks.
Uses: Paints; ceramics; pigments in general; rubber industry.

lead sulfate, blue basic (sublimed blue lead, blue lead).
Composition: Lead sulfate (min) 45%, lead oxide (min) 30%, lead sulfide (max) 12%, lead sulfite (max) 5%, zinc oxide 5%, carbon and undetermined matter (max) 5%.
Properties: Blue-gray corrosion-inhibiting pigment; insoluble in water or alcohol.
Containers: Barrels.
Uses: Component of structural-metal priming coat paints; an excellent rust-inhibitor in paints; rarely used for color.
Derivation: By heating lead ores in special furnaces.

lead sulfide (plumbous sulfide) PbS.
Properties: Silvery, metallic crystals or black powder. Soluble in acids; insoluble in water and alkalies. Sp. gr. 7.13-7.7; m.p., decomposes.
Derivation: (a) Found in nature as the mineral galena (q.v.); (b) by passing hydrogen sulfide gas into an acid solution of lead nitrate.
Grades: Technical; C.P.
Containers: 1-lb bottles; wooden barrels; fiber drums.
Uses: Ceramics; metallic lead; infrared radiation detector.
Shipping regulations: None. *

lead sulfocarbolate. See lead phenolsulfonate.

lead sulfocyanide. See lead thiocyanate.

lead superoxide. See lead dioxide.

lead tallate. A lead derivative of tall oil.
Grades: Liquid, 16% Pb; solid, 24% Pb.
Containers: Drums.
Uses: See soaps, metallic.

lead tannate.
Properties: Amorphous, brownish-yellow powder; poisonous! Slightly soluble in alcohol and water.
Use: Medicine.

lead telluride PbTe. Single crystals used as a photoconductor and a semiconductor in thermocouples and the like.

lead, tellurium. An alloy containing 0.04-0.10% tellurium. An important alternative for hard lead. More resistant to corrosion by sulfuric acid than pure "chemical" lead as well as stronger and tougher. See types of lead, under lead.

lead tetraacetate $Pb(CH_3COO)_4$.
Properties: Colorless or faintly pink crystals, sometimes moist with glacial acetic acid. M.p. 175°C; density 2.228 (17°C); soluble in benzene, chloroform, nitrobenzene, hot glacial acetic acid.
Derivation: From red lead (Pb_3O_4) and glacial acetic acid in the presence of acetic anhydride.
Containers: Glass bottles; fiber drums.
Uses: Selective oxidizing agent in organic synthesis.
Caution! Poisonous! Avoid contact with skin. Provide adequate ventilation.
Shipping regulations: Poison, class B. Poison label. *

lead tetraethyl. See tetraethyl lead.

lead thiocyanate (lead sulfocyanate) $Pb(SCN)_2$.
Properties: A white or light-yellow, crystalline powder; soluble in potassium

*See "I.C.C. Shipping Regulations," page xiii.
Reference numbers refer to name of manufacturer. See "List of Manufacturers," page v.

thiocyanate, nitric acid and slightly soluble in cold water; decomposes in hot water. Sp. gr. about 3.8.
Containers: Drums (100 lbs net).
Uses: As an ingredient of priming mixture for small-arms cartridges; in safety matches; in dyeing.

lead thiosulfate (lead hyposulfite) PbS_2O_3.
Properties: White crystals; poisonous! Soluble in acids and sodium thiosulfate solution; insoluble in water. M. p., decomposes.
Derivation: By the interaction of solutions of lead nitrate and sodium thiosulfate, concentration and crystallization.
Shipping regulations: None. *

lead titanate $PbTiO_3$.
Properties: Pale-yellow solid; insoluble in water.
Derivation: Interaction of oxides of lead and titanium at a high temperature. Contains lead sulfate and lead oxide as impurities.
Use: Paint pigment.

lead trinitroresorcinate (lead styphnate) $C_6H(NO_2)_3(O_2Pb)$.
Properties: Explosive, exploding at 260-310°C; sp. gr. 3.1 for monohydrate and 2.9 for anhydrous. Monohydrate is monoclinic orange-yellow crystals; practically insoluble in water.
Shipping regulations: Explosive, class A. Initiating explosive label. Not accepted by express. *

lead tungstate (lead wolframate) $PbWO_4$.
Properties: Yellowish powder; poisonous! Soluble in acid; insoluble in water. Sp. gr. 8.235; m. p. 1130°C.
Derivation: By mixing solutions of lead nitrate and sodium tungstate, concentrating and crystallizing.
Method of purification: Recrystallization.
Grades: Technical.
Containers: Wooden kegs; fiber drums.
Use: Pigment.
Shipping regulations: None. *

lead vanadate (lead metavanadate; lead vanadinate) $Pb(VO_3)_2$.
Properties: Yellow powder; insoluble in water; decomposes in nitric acid.
Grades: Technical.
Containers: Glass bottles; fiber drums.
Uses: Preparation of other vanadium compounds; pigment.
Shipping regulations: None. *

lead vanadinate. See lead vanadate.

lead vitriol. See anglesite.

lead water. A 1% solution of basic lead acetate.

lead, white. See lead carbonate, basic; also lead silicate, basic; and lead sulfate, basic.

lead wolframate. See lead tungstate.

lead wool. Fine filaments or threads of metallic lead, prepared and used as a wooly mass for packing pipe joints.

lead yellow. See chrome yellows.

lead zirconate titanate (LZT) $PbTiZrO_3$. Forms piezoelectric crystals. Used as an element in hi-fi sets and as a transducer for ultrasonic cleaners.

leaf green. A very durable pale-green pigment obtained by igniting a mixture of chromic oxide and pure aluminum hydrate. A name also applied to chlorophyll (q. v.).

"Leafox" Agricultural Zinc Oxide. [268] Brand name for a commercially lead-free zinc oxide for agricultural purposes. It incorporates easily in water and makes a satisfactory spray, or can be used with other material for dusting.
Containers: 50-lb bags.
Uses: For the control of mottle leaf and little leaf in citrus plants.

leather, artificial. A material usually produced by coating a fabric with a dope consisting of a mixture of pyroxylin, castor or other oil, and pigments in an organic solvent. The solvent evaporates leaving a tough flexible coating. Ornamental effects can be produced by passing the finished leather through calender rolls or embossing presses.

leather, chamois. Made from the flesh layer of a split sheep skin by treating with fish oils, piling in contact with other similarly treated skins and allowing the fish oils to oxidize.

leather, chrome-tanned. See chrome tanning.

leather grease. See degras.

leatheroid. Tough fibrous material made in thin sheets or boards, similar to vulcanized fiber, and used for electric insulation.

Leblanc process. Obsolete process for manufacturing sodium carbonate from sodium sulfate and coke.

lecithin (lecithol; ovalecithin; phospholutein; phosphatidyl choline). Approximate formula: $CH_2(R)CH(R')CH_2OPO(OH)O-(CH_2)_2N(OH)(CH_3)_3$, R and R' being fatty acid groups. A group of phosphatides. They are mixtures of the diglyceride residues of stearic, palmitic, and oleic acids, linked to the choline ester of phosphoric acid.
Derivation: Egg yolk, other animal and vegetable sources, particularly soybeans and corn.
Grades: Edible; technical; bleached; purified; fluid; plastic.
Containers: Bottles; drums; carload lots.
Uses: Emulsifying agent, dispersant, wetting agent, penetrating agent, antioxidant; in vitamins for animal feeds; in margarine to improve consistency, in baked goods, pharmaceuticals, cosmetics; as a lubricant for textile fibers; in rubber and plastics fabrication; in inks, paints, polishes, sprays, pigment pastes and lubricating oils.

*See "I.C.C. Shipping Regulations," page xiii.
Reference numbers refer to name of manufacturer. See "List of Manufacturers," page v.

lecithol. See lecithin.

"Lecton." [28] Trademark for acrylic resin-coated glass fabrics and laminates used as electric insulating material because of thermal stability up to 130°C. Resistant to fluorinated hydrocarbons.

"Lectro." [304] Trademark for a series of vinyl stabilizers. Available as:

"Lectro 60." Lead silicate-complex vinyl stabilizer.

Properties: Fine white powder; sp. gr. 4.0; refractive index 2.1.

Containers: Multiwall paper bags (50 lbs net).

Uses: Economical stabilizer for vinyl electrical insulation and tape. Special "XL" grade available for vinyl electrical insulation.

"Lectro 77." Lead chlorophthalosilicate compound.

Properties: Fine white powder; sp. gr. 4.15.

Containers: Multiwall paper bags (50 lbs net).

Uses: Heat and light stabilizer for all types of electrical insulation in the 60°C through 80°C U.L. classes, including those requiring water immersion resistance.

"Lectro 78." Tetrabasic lead fumarate ($4PbO \cdot PbC_2H_2(COO)_2 \cdot 2H_2O$) compound.

Properties: Fine creamy-white powder; sp. gr. 6.54; refractive index 2.1.

Containers: Fiberboard drums containing 50 and 325 lbs.

Uses: Heat stabilizer for phonograph records, electrical insulation and electrical grade plastisols. Vulcanization agent for chlorosulfonated polyethylene.

"Ledate." [69] Trade name for lead dimethyldithiocarbamate $[(CH_3)_2NC(S)S]_2Pb$.

Properties: Gray; odor none; toxicity same as for any lead compound; m.p. above 258°C; sp. gr. 2.5. Insoluble in water, acetone, benzene, carbon disulfide, and gasoline.

Use: Vulcanization accelerator with litharge.

"Ledercillin." [57] Trademark for penicillin.

"Lederplex." [57] Trademark for Vitamin B complex.

"Ledinac." [57] Trademark for liver protein hydrolysate-amino acids.

lees black. Charcoal from wine lees.

"Lehigh" Leaded Zinc Oxides. [268] Brand name covering a range of leaded zinc oxides.

Grades: Several grades to meet requirements of individual uses. Materials with 35 and 50% lead calculated as $PbSO_4$ more generally in demand.

Containers: 50-lb bags.

Use: Extensively used in exterior paints and primers.

"Lemac." [65] Trademark for a series of polyvinyl acetates in bead form, at various molecular weights. Used in lacquers, adhesives, special coatings.

"Lemasize." [65] Trademark for alkali-soluble resin beads, which are dissolved in ammonia or soda ash solution, and applied in warpsizing of acetate and synthetic fibers.

Lemery's white precipitate. See mercury, ammoniated.

"Lemoflex." [65] Trademark for a series of internally-plasticized polyvinyl alcohols having excellent flexibility and cold-water solubility.

Uses: Adhesives; permanently plasticized films; cold-water-soluble films.

"Lemol." [65] Trademark for a series of polyvinyl alcohols in partially and fully hydrolyzed form at various molecular weights. Used in adhesives, emulsions, polymerization, film-coatings, polyester release agents, textile printing, finishing and sizing. Supplied as free-flowing powder.

lemon bioflavonoid complex. See bioflavonoids.

lemon chrome. See barium chromate.

lemon, essential salt of. See potassium binoxalate.

lemon-grass oil.

Properties: Two basic types are commercially available: (a) East Indian oil; dark yellow to light brown-red in color; pronounced heavy lemon-like odor; sp. gr. 0.900-0.910 (15/15°C); optical rotation −3° to +1°; refractive index (n 20/D) 1.4830-1.4890; soluble in alcohol; slightly soluble in glycerol. (b) "West Indian" oil; light yellow to light brown or orange; odor is lemon-like but of lighter character than East Indian oil; sp. gr. 0.875-0.900 (15/15°C); optical rotation −3° to +1°; refractive index (n 20/D) 1.4830-1.4890; soluble in fixed oils and alcohol; slightly soluble in glycerol.

Derivation: Steam distillation of Cymbopogon grasses: (a) East Indian oil from Cymbopogon flexuosus and Andropogon nardus var. flexuosus. (b) "West Indian" oil from Cymbopogon citratus and Andropogon nardus var. ceriferus.

Method of purification: Rectification.

Containers: Cans; drums.

Uses: Aromatic; isolates and ionones.

lemon oil (limonis oil).

Properties: Pale to deep-yellow or greenish yellow, limpid liquid; fragrant lemon-like odor; aromatic, mild, bitterish taste; sp. gr. 0.856-0.861 (sometimes 0.854); optical rotation +56° to +67°; refractive index 1.474-1.476; evaporation residue 2.1 to 6.6. Not perfectly soluble in 6 to 8 vols. of 90% alcohol; soluble in 0.5 to 1 vol. of 95% alcohol; soluble in carbon disulfide and glacial acetic acid.

Chief known constituents: Citral; limonene; phellandrene; levo- and dextro-pinene; geraniol; linalool; esters; aldehydes (nonylic, octylic and decylic). Citral

content 3.5 to 5%.
Derivation: From Citrus limon.
Grades: U. S. P. XVI; California; Messina.
Containers: Glass bottles; cans; drums.
Uses: Flavoring agent; soft drinks; perfumery; confectionery; polishes.

lemon oil, terpeneless. (See essential oils, terpeneless.)
Strength: 14 to 20 times that of lemon oil.
Constants: Sp. gr. 0.896-0.900; optical rotation 0° to −8°. Soluble 10 to 20 vols. in 100 vols. of 70% alcohol, and in all proportions in 90% alcohol.
Shipping regulations: None. *

lemon peel.
Derivation: The outer rind of the ripe fruit Citrus limon.
Habitat: Northern India, cultivated in Italy and West Indies, Spain, United States and other semi-tropical countries.
Grades: Technical.
Containers: Boxes.
Uses: Cooking; confectionery; lemon oil.
Shipping regulations: None. *

lemon yellow. See chrome yellows.

"Lenigallol." [9] Trade name for acetpyrogall (pyrogallol triacetate).

"Leonil" SA. [307] Trademark for a textile chemical specialty, a naphthalene sulfonic acid derivative; 77% active; anionic.
Properties: Fine, tan powder; soluble in water; stable to 10% H_2SO_4; density 0.55-0.65.
Uses: Leveling and penetrating agent in acid and chrome color dyeing of wool.

lepidine (gamma-methylquinoline; cincholepidine) $C_9H_6NCH_3$. An alkaloid.
Properties: An oily liquid; quinoline-like odor; turns red-brown on exposure to light. Sp. gr. 1.086; b.p. 266°C; solidifies about 0°C. Soluble in alcohol, ether, and benzene; slightly soluble in water.
Derivation: From cinchonine.
Method of purification: Rectification.
Grade: Technical.
Containers: Amber glass bottles.
Use: Organic preparations.
Shipping regulations: None. *

lepidolite (lithia mica) $K_2Li_3Al_4Si_7O_{21}(OH, F)_3$. A fluosilicate of potassium, lithium, and aluminum, found in pegmatites. Rubidium occurs as an impurity. A variety of mica (q. v.).
Properties: Color pink and lilac to gray; luster pearly; perfect micaceous cleavage; hardness 2.5-4; sp. gr. 2.8-3.0.
Occurrence: California, South Dakota, New Mexico, South Africa.
Use: Source of lithium and rubidium; flux in glass and ceramics production.

lepidomelane. See mica.

leptandra (Culver's root; black root).
Derivation: Dried rhizome and roots of Veronica virginica.
Habitat: Eastern United States.
Grade: Technical.
Containers: Bags.
Use: Medicine.

lepton. The name of a group of fundamental particles (q. v.). A lepton is a fundamental particle whose mass is equal to or less than that of a muon, or mu-meson. The group includes muons, electrons, neutrinos, and photons.

"Lethane." [23] Trademark for insecticide concentrates based on beta-butoxy-beta-thiocyanodiethyl ether. Supplied in petroleum distillate.
Use: Knockdown agent and toxicant in household, dairy and industrial insecticide sprays; mosquito larvicides.

lethargy. A term used in nuclear technology as a measure of the energy loss of neutrons in slowing down by multiple scattering in the moderator. It is the natural logarithm of the ratio of the initial energy to the energy of the state in question.

leucine (alpha-amino-gamma-methylvaleric acid; alpha-aminoisocaproic acid) $(CH_3)_2CHCH_2CH(NH_2)COOH$. An essential amino acid. Found naturally in the L(−) form.
Properties: White crystals; soluble in water; slightly soluble in alcohol; insoluble in ether; optically active (natural form). DL-leucine m.p. 332°C with decomposition. L(−)-leucine m.p. 295°C; sp. gr. 1.239 (18/4°C).
Derivation: Hydrolysis of protein (edestin, hemoglobin, zein); organic synthesis from the alpha-bromo acid.
Containers: Drums.
Use: Nutrition and biochemical studies.
Available commercially as DL-leucine.

leucite $KAl(SiO_3)_2$ or $K_2O{\cdot}4SiO_2$. A natural potassium-aluminum silicate found in lava. Color, white or gray; white streak; vitreous or greasy luster. Contains 21.5% potash.
Constants: Sp. gr. 2.45-2.50; hardness 5.5-6.
Occurrence: United States (Wyoming, Montana, Arkansas); Italy (most abundant source); Brazil; Sardinia; Bohemia; Asia Minor; Africa; Australia; Java; Borneo; Siberia.
Containers: Glass bottles; fiber cans.
Use: Possible source of potash.

leuco-alizarin. See anthrarobin.

leuco-compounds. See vat dyes.

leucogen. See sodium bisulfite.

leucoline. See quinoline.

"Leucosol." [28] Trademark for a line of vat colors especially prepared for textile printing.

"Leukanol." [23] Trademark for synthetic tanning assistants of the sulfonic-type, supplied in liquid and solid grades. Powerful dispersants for vegetable tannins and bleaches for chrome-tanned leather.
Use: Tanning and bleaching leather.

"Leukeran." [301] Trademark for chlorambucil (q. v.).

levallorphan tartrate (ℓ-N-allyl-3-hydroxymorphinan bitartrate) $C_{19}H_{25}NO \cdot C_4H_6O_6$.
Properties: White or practically white, odorless crystalline powder. M. p. 174-177°C. Soluble in water; sparingly soluble in alcohol; insoluble in ether and chloroform.
Grade: U. S. P. XVI.
Use: Medicine.

Levant wormseed. See santonica.

levarterenol (ℓ-norepinephrine; ℓ-arterenol; ℓ-alpha-(aminomethyl)-3, 4-dihydroxybenzyl alcohol) $C_6H_3(OH)_2CH(OH)CH_2NH_2$. A peripheral vasoconstrictor.
Properties: Microcrystals; occurs in adrenal glands. Decomposes at 217°C.
Use: Medicine.

levarterenol bitartrate (ℓ-norepinephrine bitartrate) $(HO)_2C_6H_3CHOHCH_2NH_2 \cdot C_4H_6O_6 \cdot H_2O$.
Properties: White or faintly gray crystalline, odorless powder; affected by air and light; m. p. 100-106°C; freely soluble in water; slightly soluble in alcohol; practically insoluble in ether and chloroform; pH (0. 1% solution) 3. 0-4. 0.
Grade: U. S. P. XVI.
Use: Medicine.

leveling agents. Compounds added to the dye bath in conjunction with certain dyes to assist in bringing about the level or even deposition of the latter.

"Leveller." [232] Brand name for a series of dyebath assistants and dispersing agents.

"Levelume." [72] Trade name for bright, high leveling nickel process. Prepared from nickel sulfate, nickel chloride, boric acid and organic addition agents. Solution is operable at high current densities (60 to 80 a. s. f.) and produces ductile, low stress deposits. Applications are in electrical appliance decorative plating, automotive trim, plumbing fixtures.

levisticum oil. See lovage oil.

"Levo-Dromoran" Tartrate. [190] Trademark for a brand of levorphanol tartrate, (q. v.), the levorotatory form of 3-hydroxy-N-methylmorphinan tartrate.

levonordefrin ℓ-1-(3', 4'-dihydroxyphenyl)-2-amino-1-propanol; ℓ-3, 4-dihydroxynorephedrine $C_6H_4(OH)_2(CHOHCHNH_2CH_3)$. White to buff-colored, odorless, crystalline solid. M. p. 205°-215°C. Practically insoluble in water; slightly soluble in acetone, chloroform, alcohol, and ether; freely soluble in aqueous solutions of mineral acids.
Grade: N. F. XI.
Use: Medicine.

"Levophed" Bitartrate. [162] Trademark for levarterenol bitartrate.

levorotatory. Having the property when in solution of rotating the plane of polarized light to the left or counterclockwise. Levorotatory compounds are given the prefix ℓ- to distinguish them from their dextrorotatory or d- isomers.
See also optical isomerism, and the prefix D-.

levorphanol tartrate (levo-3-hydroxy-N-methylmorphinan tartrate dihydrate) $C_{17}H_{23}NO \cdot C_4H_6O_6 \cdot 2H_2O$.
Properties: White, odorless, bitter crystalline powder; m. p. 114-116°C; insoluble in chloroform and ether; slightly soluble in alcohol and water; stable to light, air, heat and moisture; pH (1% solution) 3. 4-4. 0.
Grade: N. F. XI.
Use: Medicine.

levulic acid. See levulinic acid.

levulinic acid (gamma-ketovaleric acid; acetopropionic acid; levulic acid) $CH_3CO(CH_2)_2COOH$.
Properties: Crystals. B. p. 245-246°C (760 mm); m. p. 33-35°C; sp. gr. 1. 1447 (25/4°C); refractive index 1. 442 (16/D). Completely soluble in water, alcohols, esters, ethers, ketones, aromatic hydrocarbons. Insoluble in aliphatic hydrocarbons.
Containers: 5-, 55-gal containers.
Uses: Organic synthesis; pharmaceuticals; chrome plating; solder flux; stabilizer for calcium greases; control of lime deposits.
Shipping regulations: None. *

levulin, synthetic. See levulosin.

levulose. See fructose.

levulosin (levulin, synthetic; fructosin) $(C_6H_{10}O_5)_x$.
Properties: Deliquescent, amorphous, white solid; m. p. 140-145°C (dec); very soluble in water; slightly soluble in alcohol; insoluble in ether.
Derivation: Carbohydrate from the rhizomes of Helianthus tuberosus (Jerusalem artichoke); also produced by organic synthesis.

"Lewis." [204] Trademark designating different brands of household lye.

Lewis acids. See acid.

lewisite. See beta-chlorovinyldichloroarsine.

"Lewisol." [266] Trademark for a series of maleic alkyd-modified rosin-esters.
Uses: In lacquers and varnishes to give gloss, hardness, adhesion, and resistance to after-yellowing; also used in sanding sealers, enamels and printing inks.

Lewis process. A process for the production of carbon black from natural gas. The gas is burned in a limited supply of air. The black smoke is condensed with water, settled, filtered, dried, and powdered. The process produces a soft black useful in rubber compounding.

"Lexan." [245] Trademark for thermoplastic carbonate-linked polymers produced by reacting bisphenol A and phosgene. Used in molding applications and other industrial arts.
See also polycarbonate resins.

Leyden blue. See cobalt blue.

*See "I. C. C. Shipping Regulations," page xiii.
Reference numbers refer to name of manufacturer. See "List of Manufacturers," page v.

LH. Abbreviation for luteinizing hormone.

Li. Symbol for lithium.

liatris (deer's tongue; vanilla plant; wild vanilla).
Derivation: Leaves of Liatris odoratissima.
Habitat: United States (Virginia to Florida and Louisiana).
Grade: Technical.
Containers: Boxes.
Uses: Medicine; perfumery; flavoring tobacco.
Shipping regulations: None. *

"Librium" Hydrochloride. [190] Trademark for a brand of chlordiazepoxide hydrochloride (q. v.).

lichen blue. See litmus.

lichenic acid. See fumaric acid.

lichenin $C_6H_{10}O_5$.
Properties: Amorphous compound resembling and isomeric with starch. Soluble in water.
Derivation: From Iceland moss.
Shipping regulations: None. *

licorice. See glycyrrhiza.

lidocaine (alpha-diethylaminoaceto-2, 6-xylidide) $C_6H_3(CH_3)_2NHCOCH_2N(C_2H_5)_2$.
Properties: White or slightly yellow crystalline powder; characteristic odor; m. p. 66-69°C; b. p. 180-182°C (at 4 mm); soluble in alcohol, ether or chloroform; insoluble in water.
Derivation: By action of diethylamine on chloroacetylxylidide.
Grade: U. S. P. XVI.
Use: Medicine.

light oils. Fractional distillates from coal-tar, with b. p. range from 110-210°C, consisting of a mixture of benzene, pyridine, toluene, xylene, phenol and cresols.
Grade: Technical.
Containers: Tank cars; iron drums.
Uses: Source of benzene, solvent naphthas, toluene, xylene, phenol and cresols.
Fire hazard: Dangerous.
Shipping regulations: Flammable liquid. Red label. *
Note: The term is also sometimes used for oils of about the same b. p. range, but from other sources.

light red. A red pigment obtained by calcining yellow ocher.
See ochers.

lignaloe oil. See linaloe oil, Mexican.

"Lignasan." [28] Trademark for fungicide based on ethyl mercury phosphate.
Containers: 2-lb bags; 125-lb drums.
Use: For controlling sap stain or "blue stain" of lumber.

"Lignasan" X. [28] Trademark for product containing ethyl mercury phosphate.
Properties: Blue, free-flowing powder.
Containers: 125-lb drums; bulk; 4-oz envelopes packed in 50-lb drums.
Uses: Biological control of paper mill slime; preservation of starch and adhesives compositions and alum solutions.

lignin. The major non-carbohydrate constituent of wood and woody plants, comprising about one fourth of many such materials. It functions as a natural plastic binder for the cellulose fibers. Its chemical composition has been given as $(C_{10}H_{13}O_3)_x$ but this is undoubtedly oversimplified. These atoms are believed to be grouped structurally as $-(C_6H_4)(OCH_3)C_3H_6O-$ since there is evidence of benzene nuclei, methoxy groups, and 3-carbon side chains. There is also evidence of unsaturation.

Lignin is removed from wood by both the sulfate and soda paper pulp processes (see wood pulp), and limited amounts have been recovered from these sources and other wood waste.
Containers: 70-lb bags; 250-lb drums.
Uses: Stabilization of asphalt emulsions; rubber reinforcement; ceramic binder and deflocculant; dye leveler and dispersant; oil mud additive; precipitation of proteins; extender for phenolic plastics.

lignin sulfonates (lignosulfonates). Metallic sulfonate salts made from the lignin of sulfite pulp-mill liquors. See lignin. Molecular weights range from 1000 to 20, 000.
Properties: Light-tan to dark-brown powder; no pronounced odor; stable in dry form and relatively stable in aqueous solution; nonhygroscopic; no definite m. p. ; decompose above 200°C; sp. gr. about 1. 5. Generally give colloidal solutions or dispersions in water; practically insoluble in all organic solvents.
Uses: Dispersing agents in concrete and carbon black-rubber mixes; extenders for tanning agents; oil mud additives; ore flotation agents; production of vanillin, industrial cleaners, gypsum slurries, dyestuffs, pesticide formulations.
Commercially available as the salts of most metals and of ammonium.
See also sulfite waste liquor.

lignite. A low rank of coal between peat and subbituminous coal. The distinction of lignite from these materials is not sharp, as the transition from one to the other is gradual. Brown coal is a form of lignite closely related to peat. Lignites contain 20-45% moisture as mined and have heating values of 5500-8300 Btu/lb. They tend to disintegrate when exposed to weather and may ignite spontaneously. The principal U. S. deposits are in North Dakota, South Dakota, Montana, Texas, Louisiana, Mississippi, and Arkansas.

lignite wax. See montan wax.

lignocellulose. Plant tissue compounds containing cellulose, hemicellulose and lignin in a form of combination that is not well understood and probably varies with specific circumstances.

lignoceric acid (n-tetracosanoic acid) $CH_3(CH_2)_{22}COOH$. A long chain saturated

fatty acid found in minor quantities in most natural fats.
Properties: Crystals; m. p. 84.2°C; b. p. 272°C (10 mm); sp. gr. 0.8207 (100/4°C); refractive index 1.4287 (100°C); nearly insoluble in ethanol.
Source: Lignite and beechwood tar; peanut oil; sphingomyelin.
Grades: Technical; 99%.
Use: Biochemical research.

"Lignocol." [138] Trademark for an anti-skinning agent.
Properties: Clear, colorless to pale yellow liquid, miscible with alcohols, mineral spirits, toluene, vegetable oils and ester solvents; soluble in alkalies.
Containers: 40-, 450-lb drums.
Uses: An anti-skinning agent for paints, varnishes, enamels and other quick-drying finishes, particularly applicable in dipping tanks and industrial finishes where skinning is encountered; an anti-oxidant for printing inks and putties, and anti-gumming agent for hydrocarbon solvents.

lignosulfonates. See lignin sulfonates.

"Lignox." [236] Brand name for a proprietary soluble calcium lignosulfonate in dry powder form. Used for treatment of drilling mud containing calcium ions and in brine emulsion muds.
Containers: Asphalt-lined paper bags containing 50 lbs.

ligroin (petroleum ether; benzine). A saturated, volatile fraction of petroleum boiling in the range 20-135°C (58-275°F). It is used as a solvent, mostly in the laboratory.

The term ligroin should be used in place of benzine or petroleum ether. There is a special grade known as petroleum benzin (q. v.).
Fire hazard: Dangerous! Use with adequate ventilation; avoid prolonged breathing of vapor.
Shipping regulations: Flammable liquid. Red label. Legal label name: benzine or petroleum ether. *

"Lilial." [227] $C_{14}H_{20}O$. Trademark for para-tert-butyl-alpha-methylhydrocinnamaldehyde; [alpha-methyl-beta-(para-tert-butylphenyl) propionaldehyde].
Properties: Lilial is a clear, stable, slightly yellow liquid. Lilial Prime, a less highly refined grade, is slightly cloudy. Powerful odor recalling linden blossoms. Sp. gr. 0.942-0.949 (25/25°C); refractive index (n 20/D) 1.5030-1.5100; flash point TCC 204°C. Clearly soluble in 3 parts of 80% alcohol, 1 part of 90% alcohol.
Occurrence: Not found in nature.
Storage: Store in tightly-closed containers, preferably full. Avoid heat.
Uses: Building and bouquetting of lily, muguet, orange flower, lilac and other floral blends.

"Lily." [242] Trade name for a high grade of stearic acid. (Iodine value of 1.0 max.)

lily-of-the-valley. See convallaria.

Lima oil. See petroleum.

lime. In a narrow specific sense this refers to calcium oxide (q. v.). It is also used as a loose general term which refers to any of the various chemical and physical forms of quicklime, hydrated lime, and hydraulic lime used for any purpose. (Adapted from ASTM definition; ASTM C41-47.)

lime acetate. See calcium acetate.

lime, agricultural. Lime slaked with a minimum amount of water to form calcium hydroxide.

lime, air-slaked. Lime which has absorbed carbon dioxide and moisture from exposure to the atmosphere. It consists of a powder composed of calcium carbonate and calcium hydroxide.
Shipping regulations: None. *

lime-ammonium nitrogen. Ammonium nitrate with dolomite.

lime, chloride of. See chlorinated lime.

lime citrate. See calcium citrate.

lime, compounds of. See corresponding compounds of calcium.

lime, fat. A pure lime which combines readily with water to form a fine white powder, free from grit, and makes a smooth stiff paste with excess of water.
See also lime, lean.
Caution: Must not be loaded hot.

lime glass. See glass.

lime hydrate. Calcium hydroxide or hydrated lime.

lime, hydrated (slaked lime). A term usually applied to the commercial limes marketed as such after having been slaked with the correct quantity of water to yield a dry, fine powder. See also calcium hydroxide.
Containers: Multiwall paper sacks; wooden barrels.
Shipping regulations: None. *

lime, hydraulic. A variety of calcined limestone which, when pulverized, absorbs water without swelling or heating and gives a cement that hardens under water. The limestone burned for this purpose usually contains from 10-17% silica, alumina and iron and from 40-45% lime, with magnesia sometimes replacing lime to a considerable extent. Hydraulic limes slake more slowly than do ordinary limes and range all the way from those with feeble hydraulic properties to limes which harden well under water.
Caution: Must not be loaded hot.

lime hypophosphite. See calcium hypophosphite.

lime, lean. A lime which does not slake freely with water due to the fact that it has been prepared from limestone containing a high percentage of impurities, e. g., silica, iron, alumina, etc.

See also lime, fat.
Caution: Must not be loaded hot.

lime, liver of. See lime, sulfurated.

lime nitrate. See calcium nitrate.

lime-nitrogen. See calcium cyanamide.

lime oil (limette oil). An oil obtained from Citrus aurantifolia, which is grown in the West Indies and Mexico. Two different oils are produced, one by expression and the other by distillation of the peel, also as a by-product in the evaporation of lime juice used for the production of citric acid.
Expressed oil:
Properties: A golden-yellow oil hardly distinguishable from a high-grade lemon oil but having a more intense odor.
Chief known constituents: Citral, bisabolene, citropten. Citral content: 2.2-6.6% (Burgess).
Constants: Sp. gr. 0.878-0.901 (usually 0.880-0.884) (15°C); optical rotation +32° to +38°; refractive index 1.482-1.486; evaporation residue 10-18%; saponification value of the residue 160-181. Soluble in 4-10 vols of 90% alcohol (with turbidity, bluish fluorescence and separation of waxy constituents).
Distilled oil:
Properties: Unpleasant turpentine odor.
Chief known constituents: Terpineol, bisabolene. No citral.
Constants: Sp. gr. 0.860-0.870 (15°C); optical rotation +33° to +47°; refractive index 1.4702-1.4707.
Containers: Bottles; cans.
Uses: Extracts; flavoring; perfumery; toilet soaps; cosmetics.
Shipping regulations: None.*

lime oil, terpeneless. See essential oils, terpeneless.
Concentration: About 12-16 times that of the ordinary oil.
Constants: Sp. gr. (about) 0.92-0.93; optical rotation (about) -2°.
Shipping regulations: None.*

lime, quick. See calcium oxide.

lime saltpeter. See calcium nitrate.

lime slag cement. See slag cement.

lime, slaked. See lime, hydrated.

lime-soda process. The use of slaked lime (calcium hydroxide) and soda ash (sodium carbonate) to remove hardness in water. Hardness is caused by the presence of soluble calcium and magnesium salts (carbonates, chlorides, sulfates).

limestone $CaCO_3$. A rock composed mainly of calcium carbonate in the form of the mineral calcite (q.v.). Limestones are sometimes classed according to the impurities contained, e.g.,
Dolomitic limestone: Usually a limestone containing more than 5% magnesium carbonate.
See dolomite.
Magnesium limestone: A term used interchangeably with dolomitic limestone.
Argillaceous limestone: A limestone containing clay.
Siliceous limestone: A limestone containing sand or quartz.
Limestones are also named according to the formation in which they occur. See also marble.
Occurrence: Found in all parts of the United States, Canada, and most other countries.
Uses: Building stone; metallurgy (flux); manufacture of lime; source of carbon dioxide; agriculture; road ballast.

limestone, hydraulic. See lime, hydraulic.

limestone whiting. See whiting.

lime, sulfurated (calcic liver of sulfur; liver of lime; hepar calcis; calcium sulfide, crude; calx sulfurata). A mixture of calcium sulfide (q.v.) and calcium sulfate.
Properties: Yellowish-gray or grayish-white powder; odor of hydrogen sulfide. Soluble in acids; insoluble in water and alcohol.
Derivation: By roasting calcium sulfate with coke.
Grade: Technical.
Containers: Iron drums.
Uses: Medicine; depilatory; luminous paint.
Shipping regulations: None.*

lime-sulfur solution. A solution made by boiling together lime (50 lbs), sulfur (100 lbs) and water (100 gals) and diluting to one-tenth strength. Used extensively as a fungicidal spray on fruit trees and as a sheep dip.
Shipping regulations: None.*

limette oil. See lime oil.

lime-uranite. See autunite.

lime water.
Properties: Clear, colorless, odorless, alkaline aqueous solution of calcium hydroxide containing not less than 0.14 g of $Ca(OH)_2$ in each 100 cc at 25°C. (Note: The strength varies with the temperature at which the solution is stored.) Sp. gr. about 1.00 (25°C). Absorbs carbon dioxide from the air.
Grade: U.S.P. XVI.
Containers: Glass bottles; carboys.
Use: Medicine.
Shipping regulations: None.*

limonene dioxide. See dipentene dioxide.

limonene, inactive (or racemic or *dl*). See dipentene.

limonene monoxide. See dipentene monoxide.

limonis oil. See lemon oil.

limonite (brown hematite, brown ironstone clay, brown iron ore) $FeO(OH)\cdot nH_2O$. A natural iron oxide. Most limonite is now thought to be an amorphous form of goethite and other iron oxides with adsorbed and capillary water. Often mixed with small amounts of hematite, clay, sand, and manganese oxides.
Properties: Color dark brown to black, occasionally yellow brown; streak yellow

brown to brown; luster vitreous to dull; hardness 5-5.5; sp. gr. 3.6-4.0.
Varieties:
Bog iron. A variety of iron ore occurring as loose porous masses and found in marshy ground.
Yellow ocher (umber). Earthy material, mixed with clay. See also umber.
Brown clay (ironstone). Compact, often nodular masses, with clay.
Occurrence: Widespread.
Use: An ore of iron; yellow pigment.

linaloe oil. See linaloe oil, Mexican.

linaloe oil, Mexican (lignaloe oil).
Properties: Water-white or yellowish oil.
Chief known constituents: Linalool; geraniol; terpineol.
Constants: Sp. gr. 0.875-0.891 (15°C); optical rotation −3° to −14°; refractive index, 1.460-1.465; acid value up to 3.0; ester value 1-42.
Soluble in 1.5-2 vols of 70% alcohol; in 4-5 vols or more of 60% alcohol.
Derivation: By distilling the wood and fruit of several species of Bursera.
Habitat: Mexico.
Containers: 5-lb bottles; cans.
Use: Perfumery.
Shipping regulations: None.*

linaloe wood. The strongly fragrant wood of Cayenne and French Guiana which, upon distillation, yields linaloe oil (q. v.). When exported, it is in the form of cord wood from which the bark has not been removed. It is hard and heavy, cleaves easily, and the surface is reddish although yellow when freshly cut.
Shipping regulations: None.*

linalol. See linalool.

linalool (linalol)
$(CH_3)_2C{:}CHCH_2CH_2C(CH_3)OHCH{:}CH_2$.
3,7-Dimethyl-1,6-octadien-3-ol. Linalool is the ℓ-isomer; coriandrol is the d-isomer.
Properties: Colorless liquid; odor similar to that of bergamot oil and French lavender. Soluble in alcohol and ether.
Constants: Sp. gr. 0.858-0.868; b.p. 195-199°C.
Derivation: Derived from many essential oils, particularly from rosewood, petit grain, linaloe wood, bergamot, ho oil, and others.
Method of purification: Rectification.
Grades: Ex bois de rose; synthetic.
Containers: Glass bottles; drums.
Use: Perfumery.
Shipping regulations: None.*

linalyl acetate $C_{10}H_{17}C_2H_3O_2$.
Properties: Clear, colorless, oily liquid; odor of bergamot; b.p. 108-110°C; sp. gr. 0.908-0.920; refractive index (n 20/D) 1.450-1.458; soluble in alcohol, ether, diethyl phthalate, benzyl benzoate, mineral oil, fixed oils, alcohol; slightly soluble in propylene glycol; insoluble in water, glycerin.
Derivation: By the action of acetic anhydride on linalool in presence of sulfuric acid.
Method of purification: Rectification.
Grades: Technical (92%); 96-98%.
Containers: Glass bottles; drums.
Uses: Extracts; perfumery.
Shipping regulations: None.*

lindane (gamma-benzene hexachloride) $C_6H_6Cl_6$. The gamma isomer of 1,2,3,4,5,6-hexachlorocyclohexane.
Properties: White, crystalline powder with slight musty odor. M.p. 112.5°C. Freely soluble in acetone, benzene, and chloroform; soluble in alcohol; slightly soluble in ethylene glycol; practically insoluble in water.
Derivation: Chlorination of benzene in the presence of ultraviolet light. The mixture of stereoisomers, containing about 12% lindane, is separated by fractional crystallization.
Grade: U.S.P. XVI; technical.
Containers: Bags, drums.
Use: Insecticide; herbicide; medicine.
Hazard: (For dry formulations, 25% and over): Warning! May be fatal if swallowed. May be absorbed through skin. (Dry formulations less than 25%): Caution! Avoid breathing dust, vapor, or spray mist. Avoid contact with skin and eyes. MCA warning labels.

"Lindol." [352] Trademark product consisting of tricresyl phosphate.
Properties: Clear, transparent, oily liquid; color (Hazen Std.) 100; sp. gr. (20°C) 1.170; free phenols none; acidity not over 0.01%; ester value more than 99%; b.p. (10 mm) 260-275°C; pour point −15°C; flash point 225°C; surface tension 43 dynes/cm; specific heat 0.42 cal/g/°C; hydrolysis (16 hrs in boiling water) trace; solubility in water (25°C) 0.0004 ml/100 ml; refractive index (25°C) 1.553; viscosity 80 cps, 285 seconds, Saybolt Universal. Miscible in all proportions with well known lacquer solvents and diluents.
Containers: 1-, 5-gal cans; 55-gal drums.
Uses: Plasticizer in artificial leather, adhesives, lacquers and plastics of cellulose nitrate, ethylcellulose, vinyl resins, etc; blending agent for resins; grinding medium for pigments; fireretarding agent; lubricating agent.
Shipping regulations: None.*

"Lindsite." [88] Trademark for a rare earth oxide glass polishing agent. Contains cerium and other rare earth oxides in the proportion in which they occur naturally.

linear accelerator. A particle accelerator which accelerates electrons or ions through small voltages at a series of electrodes arranged in line, without the use of a magnet.

linnaeite Co_3S_4, or $(CoNi)_3S_4$. Steel-gray metallic mineral with reddish tarnish. Essentially cobalt sulfide but part of the cobalt is nearly always replaced by nickel and to a less extent by iron and copper.

*See "I.C.C. Shipping Regulations," page xiii.
Reference numbers refer to name of manufacturer. See "List of Manufacturers," page v.

Luster, metallic. Does not occur in large amounts.
Constants: Sp. gr. 4.8-5; hardness 5.5.
Occurrence: United States (Maryland, Missouri, Nevada), Germany.
Use: Source of cobalt and nickel.

"Linodoxine." [299] Trademark for a combination drug containing linoleic acid and pyridoxine hydrochloride (vitamin B-6).

"Linoil." [221] Trade name for a series of foundry core oils.

linoleic acid (linolic acid) $CH_3(CH_2)_4CH{:}CHCH_2CH{:}CH(CH_2)_7COOH$. An unsaturated fatty acid with two double bonds, widely distributed as a glyceride in the plant kingdom. It has been classed as the most important polyethenoid acid found in fats and oils. It is of commercial importance as a constituent of drying oils used in paints and varnishes. It is one of the so-called essential fatty acids in the human diet. Both the conjugated and nonconjugated acids are called linoleic acid (conjugated is 9,11-octodecadienoic, $CH_3(CH_2)_5CH{:}CHCH{:}CH(CH_2)_7COOH$; nonconjugated is 9,12-octadecadienoic, as in formula at beginning of article).
Properties: Colorless to straw-colored liquid; sp. gr. 0.905 (15/4°C); m.p. −5°C; b.p. 228°C (14 mm); refractive index 1.4710 (15°C). Insoluble in water; soluble in most organic solvents.
Commercial sources: Linseed oil; safflower oil; tall oil.
Grades: Technical, purified (99+%); edible grade free from chick edema.
Containers: Drums.
Uses: Soaps; special driers for protective coatings; emulsifying agents; medicine; foods; feeds; biochemical research.

linolein. A glyceride of linoleic acid. It is the constituent of linseed oil which induces the drying property.

linolenic acid (9,12,15-octadecatrienoic acid) $CH_3CH_2CH{:}CHCH_2CH{:}CHCH_2CH{:}CH(CH_2)_7COOH$. An unsaturated fatty acid containing three double bonds (a nonconjugated triene acid), which occurs as the glyceride in many seed fats, although often in small amounts. Linolenic acid is considered one of the essential fatty acids in the diet. It must also be present in drying oils in appreciable proportion to produce effective film formation and hence the natural oils such as linseed, perilla and hempseed possessing a linolenic acid content of 25 to 65% are in demand. Their principal use is in the manufacture of synthetic resins and of paints, varnishes, printing inks, etc.
Properties: Colorless liquid; soluble in most organic solvents, insoluble in water; sp. gr. 0.916 (20/4°C); m.p. −11°C; b.p. 230°C (17 mm).
Grade: Purified 99+%.
Uses: Medicine; biochemical research.

linolenyl alcohol $CH_3CH_2CH{:}CHCH_2CH{:}CHCH_2CH{:}CH(CH_2)_7CH_2OH$. The fatty alcohol derived from linolenic acid. It has a long straight carbon chain with three double bonds. Available commercially as a 50% product. Liquid at room temperature.
Typical specifications: Iodine value 190; cloud point 50.0°F; sp. gr. 0.864; color, white.
Derivation: Reduction of acid made from linseed oil.
Impurities: Oleyl and linoleyl alcohols with some saturated alcohols.
Uses: Protective coatings; flotation; lubricants; surface active agents; resins; synthetic fibers.

linoleum. Oxidized linseed oil, rosin and powdered cork are heated and the mass pressed hot onto canvas. Colors are obtained by mixing pigments in with the mass.

linoleyl alcohol $CH_3(CH_2)_4CH{:}CHCH_2CH{:}CH(CH_2)_7CH_2OH$. The fatty alcohol derived from linoleic acid. It has a long, straight carbon chain with two double bonds in it. Available commercially as a 50-60% pure alcohol. It is liquid at room temperature.
Typical specifications: Iodine value 137; sp. gr. 0.855; color, white; cloud point 59°F.
Derivation: Reduction of linoleic acid.
Impurities: Mostly oleyl alcohol with some linolenyl, and saturated alcohols.
Uses: Protective coatings; flotation; paper; surface active agents; resins; and leather.

linoleyltrimethylammonium bromide. An amorphous solid from yellow to very light brown in color. Very soluble in water, alcohol.
Uses: Germicide; deodorant; used as algicide and in slime control.

linolic acid. See linoleic acid.

linseed (flaxseed; linum).
Derivation: Ripe seeds of Linum usitatissimum (flax).
Habitat: Cultivated extensively.
Grades: Technical; N.F. XI.
Containers: Bags.
Uses: Source of linseed oil and cake; medicine as a demulcent and emollient.
Shipping regulations: None. *

linseed cake. The press cake formed when the seeds are crushed and the oil is extracted. See linseed oil meal.

linseed oil (flaxseed oil).
Properties: Golden-yellow, amber or brown oil with peculiar odor and bland taste; thickens and hardens on exposure to air, darkening and acquiring a pronounced taste. It is a typical drying oil. Soluble in ether, chloroform, carbon disulfide, petroleum ether, and turpentine; slightly soluble in alcohol.
Constants: These vary with source and treatment of the oil. Typical specifications

for the raw oil are: sp. gr. 0.931-0.936; iodine value, Wijs, (min) 177; saponification value 189-195; acid no (max) 4 (ASTM D 234-48); sp. gr. 0.925-0.935; iodine value, Hanus, (min) 170; saponification value 187-195.

Chief constituents: Glycerides of linolenic, oleic, linoleic and saturated fatty acids. The drying property is due to the linoleic and linolenic groups.

Derivation: From the seeds of the flax plant Linum usitatissimum by expression, or solvent extraction. Various refining and bleaching methods are used.

Grades: Raw (see linseed oil, raw); boiled (see linseed oil, boiled); double boiled; blown (see linseed oil, blown); varnish makers; refined (see linseed oil, refined).

Containers: Cans; drums; tank cars.

Uses: Paints; varnishes; linoleum and oilcloth; printing inks; core oils; linings and packings; synthetic resins; caulking; soap; pharmaceuticals.

linseed oil, blown. Linseed oil which is bodied, i.e., its viscosity is increased by having air bubbled through it while heated to 125°C. The reaction is mainly oxidation followed by polymerization of the oxidized molecules. The resulting product dries to a harder film than heat-bodied oils and is used largely in interior paints and enamels.

linseed oil, boiled. The term is a misnomer since the oil is not boiled. Small amounts of the oxides of manganese, lead, or cobalt, or their naphthenates, resinates or linoleates, are added to hot linseed oil. The substances so introduced are known as driers and serve to accelerate the drying of the oil. The "boiled oil" becomes thicker, more dense and assumes a darker hue.

linseed oil meal (linseed cake). The crushed and extracted residue from flaxseed (linseed). This is generally prepared by the "old process" consisting of crushing the seeds, cooking with steam, and hydraulic expression of the oil from the resulting cake. It is sold by the varying protein content. An example of a 33% protein meal has a typical analysis of 33.1% crude protein; 8.0% crude fiber; 34.5% nitrogen-free extract; 10.2% ether extract (fat) and contains approximately 74% total digestible nutrients.

Containers: Bags or bulk.

Uses: Animal feeds; fertilizer ingredient.

Shipping regulations: None. *

linseed oil, raw. Untreated raw oil from the flaxseed presses which is filtered through duck and flannel filter cloths in a plate and frame press. Yellow-brown or amber in color.

linseed oil, refined. Raw linseed oil which has been treated for the removal of solid fats, foreign matter and mucilaginous material. May be treated with acid or alkali, and fuller's earth used for bleaching and clarification of the final product.

"Linstyrol." [64] Trade name for styrenated linseed oil. Dries hard overnight and has good compatibility characteristics. Used for styrenated alkyds, vehicle reinforcement.

linters. Fleecy fibers from one-eighth to one-quarter of an inch in length, which adhere to cottonseed after it has been passed once through a cotton-gin. They are removed from the seed by a second ginning.

Uses: Rayon manufacture; pyroxylin; artificial leather; photographic films; plastics; explosives.

Shipping regulations: None. *

linum. See linseed.

lion's tooth. See taraxacum.

lipase. A class of enzymes which hydrolyze fats to glycerol and fatty acids. Lipase is abundant in the pancreas, but also occurs in gastric mucosa, in the small intestine, and in fatty tissue. It is found in milk, wheat germ, and various fungi. Commercial pancreatin and most trypsin preparations contain lipase. Commercial uses are in the manufacture of cheese and similar foods; for removal of fat spots in dry cleaning or grease accumulations.

lipid (lipide). A term used to define fats and fat-like materials. It includes all substances which are: (1) relatively insoluble in water but are soluble in the fat solvents (benzene, chloroform, acetone, ether, etc.), (2) related either actually or potentially to fatty acid esters, and (3) utilizable by the animal organism.

lipide. The form for lipid preferred by Chemical Abstracts.

dl-alpha-**lipoic acid** (6,8-dithiooctanoic acid; thioctic acid; POF)

$\underline{SSCH_2CH_2}CH(CH_2)_4COOH$. A pyruvate oxidation factor. Pyruvate is a normal intermediate in carbohydrate metabolism.

Properties: Crystals; m.p. 60-61°C; b.p. 160-165°C. Practically insoluble in water; soluble in fat solvents; forms a water-soluble sodium salt.

Sources: Food sources; yeast and liver.

Uses: Nutrition; biochemical research.

"Lipoiodine." [305] Trademark for ethyl diiodobrassidate.

Use: Medicine.

lipotropic agent. An agent which, because of its affinity for fats and oils, helps to regulate the metabolism of fat and cholesterol in the animal body. Inositol is an example.

Lipowitz's metal. See Tables I and III under fusible alloys.

lipoxidase. An enzyme which catalyzes the addition of oxygen to the double bonds of unsaturated fatty acids of plant origin.

Use: Biochemical research; whitening bread.

*See "I.C.C. Shipping Regulations," page xiii.
Reference numbers refer to name of manufacturer. See "List of Manufacturers," page v.

"Liqro." [228] Trademark for a whole tall oil obtained as a by-product of pulp and paper manufacture by the Kraft process.

Properties: Brown liquid; sp. gr. (60/60°F) 0.98-1.00; pour point, 45°F; viscosity 90-110 SSU at 210°F; moisture 0.4%; acid number 160-170; saponification number 165-175; rosin acids number 76-86; ash 0.4%.

Analysis: Fatty acids, 46-56%; rosin acids, 41-46%; sterols, high alcohols, etc., 6-9%.

Uses: Suitable for use as emulsifying agent for many applications such as asphalt emulsions, cutting oils, disinfectants, etc. Other uses include linoleum, core oils, soaps, and uses where a low cost mixture of rosin and fatty acids is required.

"Liquamast." [299] Trade name for veterinary product containing oxytetracycline hydrochloride and propylene glycol.

"Liquamycin." [299] Trademark for veterinary medicine containing oxytetracycline hydrochloride and propylene glycol.

liquation. The separation of two or more components of a mixture by heating to a temperature at which one component melts away leaving the others as solids. Used in the separation of an alloy constituent on heating or cooling.

liquefied petroleum gas (compressed petroleum gas; LPG). A compressed or liquefied gas obtained as a by-product in petroleum refining or natural gasoline manufacture. It usually consists of pure propane and butane, or a butane containing both normal and isobutanes. Its heating value is about 2600-3400 Btu/cu ft, depending on its composition.

Containers: Cylinders and drums; tank cars and tank trucks.

Uses: Domestic fuel, for small communities and isolated homes; industrial fuel; motor fuel; chemical synthesis; especially synthetic rubber.

Shipping regulations: Flammable gas. Red label. *

liquid air. Air which has been subjected to a series of compression, expansion and cooling operations until it liquefies. It may be a milky liquid due to the presence of carbon dioxide or transparent and somewhat blue in color if the carbon dioxide is removed.

Uses: Manufacture of oxygen and nitrogen; physical research.

liquid amber orientalis. See styrax.

liquid ammonia. See ammonia, liquid.

liquid bleach. See chlorine.

"Liquid Blue." [244] A proprietary blend of organic dyestuffs. Manufactured in red and green modifications.

Properties: Soluble in water in all proportions. Gives level tint, completely discharged by sodium hypochlorite bleach to eliminate build-up.

Containers: Quarts, 6 to a case; 2-oz bottles, 16 to a carton, 6 cartons to a case.

Uses: Laundry blue.

Liquid Bright Gold. See gold, ceramic decorating.

Liquid Bright Palladium. [28] Palladium metal with fluxes and vehicles, suitable for application to green ceramics by brushing, stamping, spraying or stenciling, followed by firing to a silvery mirror surface.

Containers: 100- and 500-g bottles.

Uses: Decorating pottery, glass, tile, terra cotta, enameled metals.

Liquid Bright Platinum. [28] See Liquid Bright Palladium.

liquid bronzes. See bronzing liquid.

liquid crystal. A liquid having the optical properties of a crystal.

liquid pitch oil. See creosote, coal tar.

liquid propellants. See rocket propellants.

"Liquid Resin SS." [328] Brand name for a water-soluble urea-formaldehyde resin used alone or with "NCF Paste" to procure a stiff finish on fabrics.

liquid rosin. See tall oil.

"Liqui-Moly." [289] Trademark for a series of extreme condition lubricants compounded with submicronized molybdenum disulfide (MoS_2) in a wide range of petroleum and synthetic oils and greases, volatile hydrocarbons and various resin vehicles. Because of its property of adhering tightly to metals, the molybdenum sulfide creates a load-bearing lubricative film between metal surfaces. The lubricants are suited for extreme pressure, high and low temperatures, and boundary lubrication problems. Molybdenum sulfide is thermally stable and retains its lubricity from −300° to 750°F in air and up to 2000°F in absence of air.

Containers: 1-pt, 1-gal cans; 5-gal steel pails.

liquor, ammonia. See ammonia liquor.

liquorice. See glycyrrhiza.

"Lissolamines" A and V. [206] Brand name for stripping assistants for the correction of faulty dyed and printed materials. The "A" brand is designed primarily for treating materials dyed with azoic colors and the "V" brand for materials dyed with vat dyestuffs.

liter. The volume of one kilogram of water at its temperature of maximum density (4°C) at standard atmospheric pressure. A liter is about 1.05 quarts, or 0.26 gallons.

"Lithafrax." [280] Trademark for a ceramic material made from beta-spodumene.

litharge (lead oxide, yellow; plumbous oxide; lead protoxide; lead monoxide) PbO. An oxide of lead made by heating metallic lead to 550°C. Exists in red and yellow

*See "I.C.C. Shipping Regulations," page xiii.
Reference numbers refer to name of manufacturer. See "List of Manufacturers," page v.

modifications. In one manufacturing process, litharge may be collected in cakes of from 1 to 1.5 tons in weight when it will cool very slowly. The inner part of the cake will swell up and form flakes of red litharge; the outer part, which is necessarily chilled more rapidly, solidifies in lumps of yellow oxide. The flake may be separated from the lump by sifting and marketed as such. The solid material that remains on the screens is ground wet, settled in water and dried. This product is known as levigated litharge. The colors of the commercial grades vary from canary yellow through lemon- to reddish-yellow or red, while a very pure product has the color of yellow ocher. Mechanical compression will turn the pure yellow varieties red. Litharge is soluble in alkalies and acids but insoluble in water. Sp. gr. 9.53; m.p. 888°C.
Grades: C.P.; fused; powdered.
Containers: Bottles; cartons; kegs; wooden and steel barrels; multiwall paper sacks.
Uses: (in approximate order of volume) Storage batteries; ceramic cements and fluxes, pottery and glazes, glass; chromium pigments; oil refining; varnishes, paints, enamels, ink, linoleum; insecticides; rubber; acid resisting compositions; match head compositions; leather tanning; other lead compounds; medicine.
Warning: Harmful dust. MCA warning label.
Shipping regulations: None.*
See also massicot.

litharge glass. A glass in which litharge (PbO) replaces part of the calcium oxide of ordinary lime-soda glass.

litharge-glycerin cement. Made by mixing glycerin with one sixth to one half portion of water and mixing with enough litharge to give a paste of desired consistency. Must be used as soon as mixed. Fillers retard the setting and avoid cracking. The product is somewhat resistant to acids.

"Lithcote." [145] Trademark for a line of protective coatings available as LC-19, LC-24, LC-25, LC-34, LC-73, LC-82D, baked modified phenolic or epoxy linings or as LC-610, a catalyzed epoxy coating.
Uses: To prevent corrosion and product contamination.

lithia mica. See lepidolite.

lithic acid. See uric acid.

lithium Li. Element of atomic number 3, group Ia of the periodic system; the lightest of the alkali elements, and the lightest solid element.
Properties: Soft silvery metal; must be kept under gasoline or kerosine or in inert gases. Surface changes from silver to gray on exposure to air. Sp. gr. 0.534; m.p. 179°C; b.p. 1317°C; hardness 0.6 Moh's scale; viscosity of liquid less than that of water; heat capacity equal to that of water. Good electrical conductor. Reacts with water, oxygen, and also with nitrogen gas at ordinary temperature.
Sources: Amblygonite, lepidolite, spodumene, petalite, Searles Lake brine (trona) containing dilithium-sodium phosphate.
Derivation: By electrolysis of a fused mixture of lithium chloride and potassium chloride.
Grades: 99.0%, available as ingots, rods, wire, ribbon, and shot.
Containers: Glass bottles containing kerosine; hermetically sealed copper or aluminum cartridges, 2.25 to 108 grams.
Uses: (in approximate order of volume) Thermonuclear weapons (H-bomb); reducing and hydrogenating agents (lithium hydride, lithium amide, lithium aluminum hydride); alloy hardeners; pharmaceuticals; lithium salts; Grignard agents. Scavenger and degasifier for both stainless and mild steels; deoxidizer in copper and copper alloys; catalyst in the production of synthetic "natural" rubber; selective reducing agent in organic synthesis; intermediate in vitamin A production; additive in rocket propellants; heat transfer medium.
Fire hazard: Dangerous.
Shipping regulations: Flammable solid. Yellow label.*

lithium acetate $LiOOCCH_3 \cdot 2H_2O$.
Properties: Deliquescent, colorless crystals. Pleasant taste leaving a salty aftereffect; m.p. 70°C. Soluble in alcohol and water.
Derivation: Reaction between lithium carbonate and acetic acid.
Use: Medicine.

lithium acetylsalicylate $LiOOCC_6H_4OOCCH_3$.
Properties: White, crystalline powder. Slightly hygroscopic. Soluble in water and alcohol; decomposes in moist air.
Derivation: By adding powdered, anhydrous lithium carbonate to an ethereal solution or suspension of acetylsalicylic acid and precipitating by means of ether.
Use: Medicine.

lithium alcoholates (lithium methylate, ethylate, n-propylate, isopropylate, n-butylate) $LiOCH_3$, $LiOC_2H_5$, $LiOC_3H_7$, $LiOC_4H_9$.
Properties: White powders which discolor slowly upon standing, changing from white to brown throughout the particle, even in closed bottles; decomposed by water. All soluble in their corresponding alcohols, in benzene, and in ether. Solubility increases with increasing molecular weight.
Grades: 98% alcoholate.
Uses: Condensation-type organic reactions of Claisen group; oxidation and reduction reactions; preparation of inorganic compounds of lithium where water would produce hydrolysis or hydrates, or where product would be too soluble in water.

lithium aluminate $LiAlO_2$.
Properties: White powder; m.p. above 1625°C; sp. gr. (25°C) 2.55. Insoluble in water.

Grades: Ceramic.
Containers: 100-, 250-lb fiber drums.
Uses: As a flux in high-refractory porcelain enamels.

lithium aluminum deuteride (LAD) $LiAlD_4$.
Properties: White to gray microcrystalline material, sp. gr. 1.02 g/cc. Stable in dry air at room temperature, but very sensitive to moisture. Decomposes above 140°C liberating deuterium. Soluble in diethyl ether, tetrahydrofuran. Slightly soluble in other low molecular weight ethers.
Preparation: By reacting aluminum chloride with lithium deuteride.
Containers: Glass bottles; vials.
Uses: Introduction of deuterium atom into molecule by reduction of same groups attacked by lithium aluminum hydride.
Hazards: Obtain detailed information on precautions before handling this material.
Shipping regulations: Flammable solid. Yellow label. *

lithium aluminum hydride (LAH) $LiAlH_4$.
Properties: Light porous white powder. Sp. gr. 0.917 g/ml. Sometimes turns gray on standing. Stable in dry air at room temperature, but very sensitive to moisture, even that in ordinary air. Decomposes to lithium hydride, aluminum metal and hydrogen above 130°C without melting. Soluble in diethyl ether, tetrahydrofuran, dimethyl "Cellosolve." Slightly soluble in dibutyl ether. Insoluble or very slightly soluble in hydrocarbons and dioxane.
Preparation: Reaction of aluminum chloride with lithium hydride.
Containers: Glass bottles; polythene bags placed in metal cans, up to 6-gallon capacity; steel drums, fiber cans.
Uses: Reducing agent for over 60 different functional groups, especially for pharmaceutical, perfume, and fine organic chemicals; source of hydrogen; propellant; catalyst in polymerizations.
Hazards: Caution! Obtain detailed information on precautions before opening containers of this material. May ignite spontaneously on grinding or rubbing, or from static sparks. Fires must be controlled by smothering with powdered limestone. All ordinary extinguishers must not be used.
Shipping regulations: Flammable solid. Yellow label. *

lithium aluminum hydride, ethereal $LiAlH_4$ plus ether.
Properties: Colorless solution in ether; very reactive to water.
Derivation: From lithium hydride and ether solution of aluminum chloride.
Use: See lithium aluminum hydride.
Shipping regulations: Flammable liquid. Red label. *

lithium aluminum tri-tert-butoxyhydride (LATB; lithium tri-tert-butoxyaluminohydride) $LiAl[OC(CH_3)_3]_3H$.
Properties: White powder; sp. gr. 1.03 g/cc. Stable in dry air but sensitive to moisture. Decomposes above 400°C with evolution of hydrogen. Soluble in the dimethyl ether of diethylene glycol, tetrahydrofuran, diethyl ether; slightly soluble in other ethers.
Containers: Glass bottles.
Uses: For stereospecific reductions of steroid ketones and for reductions of acid chlorides to aldehydes.
Hazards: Reacts with water to evolve hydrogen but does not usually ignite.
Shipping regulations: Flammable solid. Yellow label. *

lithium amide $LiNH_2$.
Properties: White crystalline solid; sp. gr. 1.18; melts 380-400°C; decomposes in water.
Derivation: Reaction of lithium hydride with ammonia.
Grades: 92-95% lithium amide.
Containers: 25-, 100-, 300-lb fiber drums.
Uses: Organic synthesis, including antihistamines and other pharmaceuticals.
Shipping regulations: Flammable solid. Yellow label. *

lithium arsenate $Li_3AsO_4 \cdot H_2O$.
Properties: White powder; sp. gr. 3.07 (15°C). Caution! Poisonous! Slightly soluble in water.

lithium benzoate $LiC_7H_5O_2$.
Properties: White crystals or powder; soluble in water and alcohol.
Derivation: Reaction of benzoic acid with lithium carbonate.
Containers: Drums.

lithium bichromate. See lithium dichromate.

lithium bitartrate $LiC_4H_5O_6 \cdot H_2O$. White crystals; soluble in water.

lithium borate. See lithium metaborate, lithium tetraborate.

lithium borohydride $LiBH_4$.
Properties: White to gray crystalline powder; decomposes in vacuum above 200°C; soluble in lower primary amines and ethers; sp. gr. 0.66; extremely hygroscopic.
Derivation: Reaction of sodium borohydride and lithium chloride.
Grades: Technical.
Containers: Glass bottles.
Uses: Source of hydrogen and other borohydrides; reducing agent for aldehydes, ketones and esters.
Shipping regulations: Flammable solid. Yellow label. *

lithium bromate $LiBrO_3$. White crystals; soluble in water.

lithium bromide LiBr.
Properties: White cubic deliquescent crystals, or as a white to pinkish white granular powder; odorless; sharp, bitter taste. Sp. gr. 3.464; m.p. 547°C; b.p. 1265°C; very soluble in water, alcohol, and ether. Slightly soluble in pyridine; soluble in methanol, acetone, glycol.

A hot concentrated solution dissolves cellulose. Forms addition compounds with ammonia and amines. Forms double salts with $CuBr_2$, $HgBr_2$, HgI_2, $Hg(CN)_2$, and $SrBr_2$.
Derivation: Reaction of hydrobromic acid with lithium carbonate.
Grades: N.F. XI; 53% (min) LiBr brine.
Containers: 5-, 50-lb glass jars; bags; brine in 400-, 700-lb steel drums.
Uses: Pharmaceuticals; air conditioning; humectant.

lithium butyl. See butyllithium..

lithium n-butylate. See lithium alcoholates.

lithium carbide Li_2C_2. A crystalline white powder; sp. gr. 1.65 (18°C); decomposes in water; soluble in acid, with evolution of acetylene.

lithium carbonate Li_2CO_3.
Properties: White powder; sp. gr. 2.111; m. p. 735°C; b. p., decomposes; more soluble in cold than in hot water; g/100 ml at 0°C 1.54, at 100°C 0.72; soluble in acids; insoluble in alcohol.
Derivation: Finely ground ore (spodumene) is roasted with sulfuric acid at 250°C. (Amblygonite, lepidolite, and dilithium sodium phosphate are also used as sources.) Lithium sulfate is leached from the mass and converted to the carbonate by precipitation with soda ash.
Grades: Technical; C. P.
Containers: 100-, 250-lb fiber drums.
Uses: Ceramics and glasses; pharmaceuticals; catalyst; lithium hydroxide and other lithium compounds; coating of arc-welding electrodes.

lithium chlorate $LiClO_3$.
Properties: Needlelike crystals, deliquescent; m. p. 128°C; decomposes at 270°C; more soluble in water than any other inorganic salt (313 g per 100 ml water at 18°C); very soluble in alcohol.
Uses: Air conditioning; inorganic and organic chemicals; propellants.

lithium chloride LiCl.
Properties: White deliquescent crystals; sp. gr. 2.068; m. p. 614°C; b. p. 1360°C; very soluble in water, alcohols, ether, pyridine, nitrobenzene.
Derivation: Reaction of lithium ores with chlorides.
Grades: Technical, 99% (min) assay; 35-40% brine, inhibited.
Containers: Crystals: 25-, 50-, 100-lb paper-lined drums; brine: 75-, 275-, and 500-lb steel drums.
Uses: Air conditioning; welding and soldering flux; dry batteries; heat-exchange media; salt baths; desiccant; humectant; general chemical; production of lithium metal; soft drinks and mineral water to reduce escape of carbon dioxide.
Caution! Apparently poisonous if taken internally in appreciable amounts, particularly if sodium chloride intake is low.

lithium chromate $LiCrO_4 \cdot 2H_2O$.
Properties: Yellow, crystalline deliquescent powder; soluble in water and forms a eutectic at -60°C. Soluble in alcohols.
Uses: Corrosion inhibitor; oxidizing agent for organic material, especially in the presence of light.

lithium citrate $Li_3C_6H_5O_7 \cdot 4H_2O$.
Properties: White powder or granules; m. p., decomposes; soluble in water; slightly soluble in alcohol.
Derivation: Reaction of citric acid with lithium carbonate.
Containers: 25-, 50-, 100-lb fiber drums.
Uses: Beverages; pharmaceuticals; dispersion stabilizer.

lithium cobaltite $LiCoO_2$. Dark blue powder; insoluble in water. The compound exhibits both the fluxing property of lithium oxide and the adherence-promoting property of cobalt oxide.
Use: Ceramics.

lithium deuteride LiD.
Properties: Gray crystals; sp. gr. 0.906 g/cc. Reacts slowly with moist air. Thermally stable to its melting point of 680°C. Insoluble in all inert organic materials.
Containers: Glass bottles.
Uses: Source of deuterium for reaction studies.
Shipping regulations: Flammable solid. Yellow label. *

lithium dichromate (lithium bichromate) $Li_2Cr_2O_7 \cdot 2H_2O$.
Properties: Yellowish-red, crystalline powder; deliquescent; soluble in water.
Caution! Keep well stoppered!
Constants: M. p. 130°C.

lithium ethylate. See lithium alcoholates.

lithium fluophosphate $LiF \cdot Li_3PO_4 \cdot H_2O$.
Properties: White crystals.
Derivation: By the interaction of lithium fluoride and lithium phosphate.
Grades: Technical.
Containers: Wooden kegs; fiber drums.
Use: Ceramics.
Shipping regulations: None. *

lithium fluoride LiF.
Properties: Fine white powder; sp. gr. 2.295; m. p. 870°C; b. p. 1670°C; slightly soluble in water; does not react with water at red heat; soluble in acids and hydrofluoric acid; insoluble in alcohol. Poisonous!
Derivation: Reaction of hydrofluoric acid with lithium carbonate.
Grades: Guaranteed 98% (min) LiF; C. P.; single pure crystals.
Containers: 25-, 50-, 100-lb fiber drums; barrels.
Uses: Welding and soldering flux; ceramics; heat exchange media; synthetic crystals in infrared and ultraviolet instruments; proposed for use in space components.

lithium greases. Greases using lithium soaps of the higher fatty acids as a base.

Water resistant; stable when heated above their melting point and cooled again. Used in aircraft and low temperature service. See lithium stearate; lubricating greases. Lithium hydroxystearate from hydrogenated castor oil is also widely used.

lithium hydride LiH.
Properties: White, translucent, crystalline mass, or powder. Commercial product is light bluish-gray due to minute amount of colloidally dispersed lithium. Sp. gr. (20°C) 0.82; m.p. 680°C; decomposition pressure nil at 25°C, 0.7 mm at 500°C, 760 mm at approx. 850°C. Decomposed by water, forming hydrogen and lithium hydroxide; insoluble in benzene and toluene; soluble in ether.
Derivation: Reaction of molten lithium with hydrogen.
Grades: 93-95%, based on hydrogen evolution.
Containers: Cans; cases; drums.
Uses: Desiccant; source of hydrogen; condensing agent in organic synthesis; preparation of lithium amide and double hydrides; nuclear shielding material.
Shipping regulations: Flammable solid. Yellow label. *

lithium hydroxide (a) LiOH (b) $LiOH \cdot H_2O$.
Properties: Colorless crystals; sp. gr. (a) 2.54 (b) 1.83; m.p. (a) 462°C; b.p. decomposes; soluble in water; slightly soluble in alcohol.
Derivation: Causticizing of lithium carbonate.
Grades: Technical; C.P.
Containers: 25-, 75-, 200-, 375-lb paper-lined steel drums.
Uses: Storage battery electrolyte; carbon dioxide absorbent; lubricating greases; general chemical; ceramics.

lithium hydroxystearate
$LiOOC(CH_2)_{10}CHOH(CH_2)_5CH_3$.
Properties: White powder; m.p. 205°C. Dissolves in hot petroleum oil to form greases.
Derivation: From hydrogenated castor oil.
Use: In greasemaking.

lithium hypochlorite LiOCl.
Use: Bleach.
Shipping regulations: Dry, containing more than 39% available chlorine, oxidizing material. Yellow label. *

lithium iodate $LiIO_3$.
Properties: White powder; sp. gr. 4.487 (25°C); m.p. 50-60°C (transition point from alpha to beta form); soluble in water; insoluble in alcohol.

lithium iodide (a) LiI; (b) $LiI \cdot 3H_2O$.
Properties: White crystals; soluble in water and in alcohol. Sp. gr. (a) 4.063, (b) 2.34 (25°C); m.p. (a) 446°C, (b) 72°C, loses water; b.p. 1171°C. Soluble in water and alcohol. Extremely hygroscopic.
Derivation: By the action of hydriodic acid on lithium hydroxide, with subsequent crystallization.
Method of purification: Recrystallization.
Containers: Glass bottles.
Uses: Medicine; mineral waters; photography.
Shipping regulations: None. *

lithium isopropylate. See lithium alcoholates.

lithium lactate $LiC_3H_5O_3$.
Properties: White odorless powder; very soluble in water with practically neutral reaction.
Use: Wherever a dry alkali lactate is required. Lithium lactate is non-hygroscopic and stable, whereas sodium and potassium lactates can be prepared in solution only.

lithium manganite Li_2MnO_3.
Properties: Reddish-brown powder. Insoluble in water. Extremely stable.
Containers: Drums.
Uses: Smelter addition in the manufacture of frit; as a mill addition; ceramic bonded grinding wheels.

lithium metaborate dihydrate $LiBO_2 \cdot 2H_2O$.
Properties: White crystalline powder; soluble in water; m.p. 840°C (anhydrous).
Grade: Ceramic.
Containers: 100-, 300-lb fiber drums.
Uses: Ceramics (flux in enamel cover coats; increases resistance to torsion).

lithium metasilicate (lithium silicate) Li_2SiO_3.
Properties: White powder; m.p. 1201°C; sp. gr. (25°C) 2.52; insoluble in water.
Grade: Ceramic.
Containers: 100-, 300-lb fiber drums; multiwall paper sacks.
Uses: Flux in glazes and ceramic enamels; welding rod coating.

lithium metavanadate. See lithium vanadate.

lithium methoxide. See lithium methylate.

lithium methylate (lithium methoxide) CH_3OLi.
Properties: White powder.
Derivation: Reaction of lithium with methanol.
See also lithium alcoholates.

lithium molybdate Li_2MoO_4. A white crystalline compound soluble in water. M.p. 705°C. Small amount, added as a mill addition, has given the necessary adherence for applying a white covercoat directly to steel. Also used as catalyst in petroleum cracking.

lithium nitrate $LiNO_3$.
Properties: Colorless powder; sp. gr. 2.38; m.p. 261°C; soluble in water and alcohol.
Derivation: Reaction of nitric acid with lithium carbonate.
Grades: Technical; commercially pure; reagent.
Containers: 25-, 50-, 100-lb steel drums.
Uses: Ceramics; pyrotechnics; salt baths; heat-exchange media; refrigeration systems; general chemical.
Shipping regulations: Oxidizing material. Yellow label. *

lithium nitride Li_3N.
Properties: Brownish-red crystals of hexagonal structure.
Chemical behavior: In dry air, no reaction at 25°C; at elevated temperatures, ignites

*See "I.C.C. Shipping Regulations," page xiii.
Reference numbers refer to name of manufacturer. See "List of Manufacturers," page v.

and burns intensely. Water vapor in moist air causes slow decomposition. Reacts with water giving lithium hydroxide and ammonia. Density approx. 1.3 g/cc (25°C); m. p. 845°C; decomposition pressure not measurable below 1250°C.

lithium orthophosphate. See lithium phosphate.

lithium perchlorate $LiClO_4$.
Properties: Colorless deliquescent crystals; sp. gr. 2.429; m. p. 236°C; decomposes at 290°C; has more available oxygen than liquid oxygen (on a volume basis). Reacts with $4NH_3$ to form an ammoniate. Forms $LiClO_4 \cdot 3H_2O$, colorless crystals with sp. gr. 1.84; m. p. 75°C. Soluble in water and alcohol.
Use: Suggested as an oxidizer in solid rocket propellants.

lithium peroxide Li_2O_2.
Properties: Fine white powder or a sandy yellow granular material; m. p. decomposes; density 2.14 g/cc (20°C). In closed container no detectable loss of available oxygen. Water solution at 20°C may be caused to decompose when catalyzed with manganese or iron salts. Soluble in water, 8% (20°C); anhydrous acetic acid, 5.6% (20°C); insoluble in absolute alcohol (20°C).
Uses: As supplier of active oxygen; commercial samples have 32.5% to 34% available oxygen content.
Shipping regulations: Oxidizing material. Yellow label. *

lithium phosphate (lithium orthophosphate) $2Li_3PO_4 \cdot H_2O$.
Properties: White, crystalline powder; soluble in acids; slightly soluble in water; sp. gr. 2.41.

lithium n-propylate. See lithium alcoholates.

lithium ricinoleate $LiOOCC_{17}H_{32}OH$.
Properties: A fine white powder. M. p. 174°C. Insoluble or very limited solubility in most organic solvents.
Derivation: Castor oil.
Uses: Alcoholysis and ester interchange catalyst.

lithium salicylate HOC_6H_4COOLi.
Properties: White or grayish-white crystals or granular powder; odorless; sweet taste; deliquescent in moist atmosphere. Soluble in water and alcohol.
Derivation: By heating a solution of salicylic acid and lithium carbonate until effervescence ceases.
Containers: Drums.
Use: Medicine.
Shipping regulations: None. *

lithium silicate. See lithium metasilicate.

lithium silicon.
Shipping regulations: Flammable solid. Yellow label. *

lithium stearate $LiC_{18}H_{35}O_2$.
Properties: White crystals; sp. gr. 1.025; m. p. 220°C; insoluble in cold and hot water, alcohol, and ethyl acetate; forms gels with mineral oils. Good lubricant.
Derivation: Reaction of stearic acid with lithium carbonate.
Grades: Grease; cosmetic.
Containers: 25-, 50-, 100-lb fiber drums; 50-lb asphalt-lined bags.
Uses: Cosmetics; plastics; waxes; greases; lubricant in powder metallurgy.

lithium sulfate $Li_2SO_4 \cdot H_2O$.
Properties: Colorless crystals; sp. gr. 2.06; m. p. 130°C; soluble in water, insoluble in 80% alcohol. Does not form alums.
Derivation: Reaction of sulfuric acid with lithium carbonate or with spodumene ore.
Grades: Technical and pharmaceutical.
Containers: 25-, 50-, 100-lb fiber drums.
Uses: Pharmaceutical products.

lithium tetraborate $Li_2B_4O_7 \cdot 5H_2O$.
Properties: White crystalline powder; m. p., loses water 200°C; very soluble in water, insoluble in alcohol.
Derivation: Reaction of boric acid with lithium carbonate.
Grades: Technical.
Containers: 25-, 50-, 100-lb fiber drums.
Use: Ceramics.

lithium titanate Li_2TiO_3. White powder. Insoluble in water. Exhibits strong fluxing properties when used in small percentages in titanium-bearing enamels. The insolubility permits its use as a mill addition in vitreous and semi-vitreous glazes.
Containers: Drums.

lithium tri-tert-butoxyaluminohydride. See lithium aluminum tri-tert-butoxyhydride.

lithium tungstate Li_2WO_4. White crystals; soluble in water.

lithium vanadate (lithium metavanadate) $LiVO_3 \cdot 2H_2O$.
Properties: Yellowish powder; soluble in water.
Use: Medicine.

lithium zirconate Li_2ZrO_3. White powder; insoluble in water. It has been found to be a very efficient flux in glasses containing zirconium dioxide. It is recommended as a flux in zirconium-opacified enamels, glazes and porcelains.

lithium-zirconium silicate $2Li_2O \cdot ZrO_2 \cdot SiO_2$. White powder. A strong flux in enamels, glazes and porcelains. It can be used in place of lithium zirconate.

lithocholic acid $C_{24}H_{40}O_3$. One of the bile acids; contains only one hydroxyl group.
Properties: Crystallizes in leaflets from alcohol; m. p. 184-186°C. Not precipitated by digitonin. Freely soluble in hot alcohol; soluble in ethyl acetate; slightly soluble in glacial acetic acid; insoluble in water, ligroin.
Derivation: From bile and gallstones; from deoxycholic acid or cholic acid.
Use: Biochemical research.

lithographic stone. A fine-grained limestone of uniform texture free from veins and spots.

*See "I.C.C. Shipping Regulations," page xiii.
Reference numbers refer to name of manufacturer. See "List of Manufacturers," page v.

It must be compact enough to take a good polish and yet sufficiently porous to absorb the grease of the draftsman's crayon.

lithol reds. Lithol reds are pigments made by combining the intermediates Tobias acid and beta-naphthol. This type of red is available as sodium, barium and calcium toners and lakes; the sodium is the lightest shade, the barium is what may be termed a medium shade and the calcium lithols are deep reds or maroons. Lithol reds are the lowest in cost of the organic reds and are widely used in enamels where maximum permanency is not required. They are non-bleeding in oil, but bleed slightly in lacquer solvents, and badly in soap solutions. They are used to quite an extent in exterior finishes, but they do not have the lightfastness of para reds or toluidine reds and therefore are not generally used in automotive finishes or bulletin paints. They find application in drum enamels, toy enamels, novelty finishes, etc. Resinated lithols contain a metallic resinate and they are generally brighter in mass tone and cleaner in tints, but lack the opacity of the non-resinated lithols and usually have poorer flow in enamels.

lithol rubine $OOCC_{10}H_5(OH)N{:}NC_6H_3(CH_3)SO_2OCa$.

An azo pigment used in paints, printing inks, plastics and cosmetics. Made by diazotizing para-toluidine-meta-sulfonic acid, and coupling with beta-oxynaphthoic acid. Pigment is used as the calcium salt.

lithophone. See lithopone.

lithopone (lithophone; Orr's white; Charlton white; Griffith's white; zinc baryta white; zinc sulfide white).

Properties: White powder, consisting of barium sulfate, zinc sulfide and zinc oxide.

Derivation: By mixing solutions of barium sulfide and zinc sulfate, filtering, washing and drying the precipitate. The latter is heated to redness, plunged into water while hot, ground with water, thoroughly washed and dried.

Grades: Technical.

Containers: 400-, 450-, 500-, 600-lb barrels; multiwall paper sacks.

Uses: Paint pigment; rubber industry; printers' white ink; pigment in filling white leathers, paper, linoleum, oilcloth and window-shade cloth; face powders; cosmetics, etc.

Shipping regulations: None. *

lithopone, cadmium. A lithopone in which cadmium replaces the zinc. Its uses are similar to lithopone (q. v.).

See also cadmium reds, in which selenium replaces some of the sulfur.

"Lithosol." [28] Trademark for a line of intermediates, dyes, and chemicals especially prepared for the manufacture of lakes and organic pigments. They also have some textile applications.

"Lithotone." [329] Trademark for a brand of paraformaldehyde-type, photomechanical film developer.

litmus (lacmus; lichen blue).

Properties: A blue, amorphous powder (frequently compressed into small cakes or sticks). Soluble in water; changes color with acidity of solution; red at pH 4.5, blue at pH 8.3.

Derivation: By treating various lichens (particularly Variolaria lecanora and V. rocella) with ammonia and potash and then fermenting the mass.

Grades: Technical.

Containers: Glass bottles; boxes.

Uses: Indicator in analytical chemistry.

Shipping regulations: None. *

litmus paper. White, unsized paper which has been dipped in an infusion of litmus in water. Used as an indicator in analytical chemistry.

liver of lime. See lime, sulfurated.

liver of sulfur. See potash, sulfurated.

liver ore. See cinnabar.

livers of antimony. Impure double sulfides of antimony, obtained by heating antimony pentasulfide with alkaline sulfides. Sodium sulfantimonate is the most important.

Properties: Brown powder; slightly soluble in water.

Shipping regulations: None. *

liver starch. See glycogen.

livingstonite $HgSb_4S_7$ or $HgS \cdot 2Sb_2S_3$.

Properties: Lead-gray mineral, metallic luster, red streak. Contains 24.8% mercury, 53.1% antimony, 22.1% sulfur. Resembles stibnite (q. v.) in form. Sp. gr. 4.1-4.8; hardness 2.

Occurrence: Mexico.

Use: Source of mercury.

lixiviation. See leaching.

"LNA." [65] Trademark for leucyl aminopeptidase subtrate used as diagnostic agent in medicine.

"Lo-Bax." [84] Trademark for a chlorine sterilizer designed especially for handlers of milk and other dairy and food products who require clear, fast-killing bactericidal solutions. It is a stable, quick-dissolving powder containing 50% available chlorine.

"Lobeite-5." [296] Trade name for limed rosin possessing a capillary tube melting point of typically 116°C, acid value 85, color Isaac/Harry (Rosin Standards). Used in printing ink, paint, and varnish.

lobelia (Indian or wild tobacco; asthma weed). Dried leaves and tops of Lobelia inflata.

Habitat: United States and Canada.

Grades: Technical.

Containers: Bags; bales.

Use: Medicine.

Shipping regulations: None. *

lobeline. Formula said to be $C_{22}H_{27}NO_2$.
Properties: Yellow honey-like liquid; poisonous! Crystallizes as needles from alcohol. Soluble in alcohol and chloroform; slightly soluble in ether.
Derivation: By extraction from the seeds of Lobelia inflata.
Method of purification: Rectification.
Grades: Technical.
Containers: Glass bottles.
Use: Medicine.
Shipping regulations: None. *

lobeline sulfate. The commercial material is a mixture of the sulfate salts of the total alkaloids of lobelia, with ℓ-lobeline predominating, and lesser amounts of lobelanine, lobelanidine, and lobelidine present.
Properties: Yellow crystalline solid; somewhat hygroscopic; soluble in water; slightly soluble in alcohol and chloroform.
Use: Medicine.

"Lockebrite." [245] Trademark for abrasive and cleaning material for porcelain insulators.

locust bean. See carob seed.

locust-bean gum. See carob-seed gum.

locust kernel. A purified carob-seed gum free from starch.
Use: Same as carob-seed gum.

lodestone. See magnetite.

loellingite (löllingite) $FeAs_2$. Natural iron arsenide, with some Co, Ni, Sb, and S.
Properties: Color silver white to steel gray; streak grayish black; hardness 5-5.5; sp. gr. 7.4.
Occurrence: Maine, New York, New Jersey, Colorado; Canada.
Use: Source of arsenic.

loess clay. See brick clay.

"Logo." [302] Trademark for solutions of organic polymeric substances used for surface treatment for plastics.

"Logofoil." [302] Trademark for a strippable protective coating composition.

"Logoquant." [302] Trademark for protective coating compositions for use on plastic articles.

"Logoset." [302] Trademark for synthetic enamels and varnishes.

logwood (hematoxylon).
Derivation: The heartwood of Hematoxylon campechianum. The raw logwood comes in the form of rough logs, 3 feet long, which are either ground or rasped into small chips. These chips, after being aged by being exposed to the atmosphere, are subjected to extraction.
Habitat: Central America; Mexico; and West Indies.
Uses: Textile and leather dyeing; medicine; ink-making.
Shipping regulations: None. *

logwood crystals. See hematoxylin.

loja bark. See cinchona bark, loxa.

löllingite. See loellingite.

"Lomar." [78] Trademark for a series of naphthalenesulfonic acid condensates used as powerful dispersing agents for dyes, pigments, graphite, clays, mica, carbon black, and other inert powders, especially in the cement, plaster, ceramic, paper, paint, and rubber industries.

"Lomax" Process. [416] Patented process of cracking in the presence of hydrogen to convert petroleum distillates ranging from kerosines to heavy vacuum gas oils into lower boiling products. Process is characterized by ability to make varying proportions of gasolines and light distillates from heavy vacuum gas oils, by production of greatly reduced quantities of light gases (when compared with conventional catalytic cracking), by the predominance of branched-chain isomers in the light hydrocarbons, and by the production of high quality middle oils.

"Lomotil." [70] Trademark for brand of diphenoxylate hydrochloride with atropine sulfate.

London purple, solid. An insecticide containing arsenic trioxide, aniline, lime, and ferrous oxide.
Warning! Poisonous if swallowed. MCA warning label.
Shipping regulations: Class B poison. Poison label. *

longwort. See pyrethrum root.

"Loosol." [256] Trade name for a light red liquid; sp. gr. 1.050 (15.6°C); boiling range 100-272°C. Used as an antioxidant in scrubbing oils and in gas manufacture.

loranskite. See euxenite.

loretin (ferron; 7-iodo-8-hydroxyquinoline-5-sulfonic acid) $C_9H_4N(I)(OH)(SO_3H)$.
Properties: Sulfur-yellow, almost odorless, tasteless, crystalline powder; m.p. 260-270°C, with some decomposition. Slightly soluble in water and alcohol; insoluble in ether.
Derivation: Obtained from the potassium salt of 8-hydroxyquinoline-5-sulfonic acid by the action of potassium iodide, bleaching powder and hydrochloric acid.
Uses: Medicine; colorimetric reagent for ferric iron.

"Lorexane." [207] Trademark for pure gamma-benzene hexachloride, the gamma isomer of hexachlorocyclohexane. A powerful insecticide for humans and animals.

"Lorfan" tartrate. [190] Trademark for a brand of levallorphan tartrate (q. v.).

"Lorite." [148] Trademark for a natural extender composed of predominantly acicular particles containing approximately 80% calcium carbonate and 20% diatomaceous earth. Sp. gr. 2.54; one pound bulks 0.04732 gal; oil absorption, 24; retained on No. 325 sieve, 1.0%.

Containers: 50-lb paper bags net.
Uses: "Lorite" has a flatting and bodying effect in paints and is recommended for sealers, flats, semi-glosses, and undercoaters, and for emulsion paints due to its low soluble salt content.

"Lorol." [28] Trademark for a series of technical grade straight chain (normal) even-numbered carbon alcohols ranging from C_8 (octyl) to C_{18} (stearyl), and including also mixtures of these.
Properties: Colorless liquids ("Lorol" 5 to 22) or white flakes ("Lorol" 24, 28). Sp. gr. about 0.83.
Containers: 55-gal drums or 8000-gal tank cars for liquids; 50- or 140-lb fiber containers for solids.
Composition data: For "Lorol" 5, 7, 9, approximate percentages by weight of alcohols are as follows:

	"Lorol" 5	"Lorol" 7	"Lorol" 9
decyl	2.6	2.5	1.0
lauryl	61.0	55.5	71.0
myristyl	23.0	21.0	26.0
cetyl	11.2	10.2	2.0
stearyl	2.2	10.8	–

"Lorol" 11 is approximately 80% lauryl, 20% myristyl.
"Lorol" 20 is essentially normal octyl alcohol.
"Lorol" 22 is essentially normal decyl alcohol.
"Lorol" 24 is cetyl alcohol meeting requirements of N. F. VIII.
"Lorol" 28 is technical grade stearyl alcohol. Also available as U. S. P. grade.
Uses: "Lorol" 5. Intermediate in manufacture of detergents, lubricating oil additives, quaternary ammonium compounds, mercaptans and textile finishing agents.
"Lorol" 7. Intermediate in manufacture of lubricating oil additives.
"Lorol" 9. Manufacture of lauryl chloride and its derivatives.
"Lorol" 11. Manufacture of lauryl chloride and its derivatives.
"Lorol" 20. Antifoam agent; intermediate in manufacture of plasticizers, lubricating oil additives and perfumes.
"Lorol" 22. Antifoam agent; intermediate in manufacture of quaternary ammonium compounds, lubricating oil additives, perfumes, and plasticizers.
"Lorol" 24. Emulsifying and softening agent in cosmetics, salves, and ointments.
"Lorol" 28. (Technical) Mold lubricant; intermediate in manufacture of lubricating oil additives; quaternary ammonium compounds and textile finishing agents.
"Lorol" 28 (U. S. P.). Ingredient in lipsticks, cosmetic creams, lotions, ointments, and salves.

"Lorridol." [354] A wetting and emulsifying agent.
Properties: Straw-colored organic liquid; sp. gr. 0.98; soluble in water; miscible with alcohol, glycerin, vegetable oils, carbon tetrachloride, cyclohexanol, and petroleum ether; pH of a 5% aqueous solution, 8.7.
Uses: As a wetting agent in textile processing; as an emulsifying agent in cosmetics, dry-cleaning fluids, leather polishes, shaving creams, insecticides, paints, etc.

"Lotol" Latex. [248] Trademark for a complete line of compounded latices, based on all types of synthetic and natural latices.

"Lotusate." [162] Trademark for talbutal.

"Loupole." [78] Trademark for a detergent used for boiling-off rayon piece goods at low and elevated temperatures; particularly recommended for graphite streak removal; has good compatibility with soap and other detergents.

lovage oil (levisticum oil).
Properties: A yellow-brown volatile oil, with a sharp odor. Soluble in alcohol, ether, chloroform, carbon disulfide, acetone, and benzene. Sp. gr.: Root oil, 1.00 to 1.049; fruit oil, 0.935; herb oil, 0.928.
Derivation: Distilled from the root, fruit or herb of Levisticum officinale.
Method of purification: Rectification.
Grade: Technical.
Containers: Glass bottles; copper flasks.
Uses: Flavors; and to a small extent, perfumes.
Shipping regulations: None. *

low-melting alloys. See fusible alloys.

low-pressure resins. See contact resins.

low-soda aluminas. Aluminum oxide (Al_2O_3) with less than 0.15% sodium oxide (Na_2O) content.
Use: High grade electric insulators and other ceramic bodies.

low-temperature carbonization. Destructive distillation of coal at relatively low temperatures (below 500°C) in order to produce the greatest possible yield of liquid products, and relatively small proportions of gases.

LOX. An abbreviation for liquid oxygen, especially when used as a rocket fuel.

loxa bark. See cinchona bark, loxa.

"Loxite." [277] Trademark for rubber cements, dispersions, and latices, consisting of natural and synthetic rubbers and reclaimed rubbers in latex form, dispersed in water, or dissolved in solvents.
Grades: Natural and synthetic latices and dispersions in water; reclaimed rubber dispersions; natural, synthetics, or reclaim in solvents.
Containers: 55-gal drums.
Uses: Impregnating and spreading paper and textiles; adhesives for rubber to metal, rubber to rubber, etc.
Hazards: Latices and dispersions: none. Keep from freezing. Solvent cements: same as solvent.

"LP Compound." [428] Trademark for soak cleaner for all metals except aluminum and bright zinc. Contains mild alkali, wetting agents and buffering agents.

LPG. Abbreviation for liquefied petroleum gas.

LSD. Abbreviation for lysergic acid diethylamide.
See lysergic acid.

"LTV" Silicone Compound. [245] Trade name for a group of low temperature vulcanizing silicone potting and embedding compounds that cure to a flexible, resilient solid at 70 to 80°C. Has a useful temperature range of −65 to 175°C with typical cure times on potted assemblies of 16 hours at 75°C or 72 hours at room temperature. LTV (low temperature vulcanizing) 602, the first grade of the material to become commercially available, is crystal clear. Has good electrical properties and other general properties typical of methyl silicone family.
Uses: Used to provide mechanical and dielectric protection for electronic components and assemblies including shock, vibration, moisture, ozone, corona and other environmental hazards of the space age.

Lu. Symbol for lutetium.

Lube Oil Additive 564. [28] Trade name for a solution of methacrylate polymers containing basic nitrogen in light neutral oil. Density 7.5 lbs/gal. Used to provide detergency for motor oils, particularly under low-temperature, low-power, stop-and-go driving conditions, and to improve viscosity index. Concentration required averages 2%.

lube oil additives. Chemicals added in small amounts to lubricating oils to impart special qualities, such as low pour point when chlorinated hydrocarbons are added. Other special properties imparted by typical additives are:

Property	Additive
low viscosity index	butene polymers
detergent and suspensoid properties	metallic stearate soaps
oxidation stability	calcium stearate
reduced foaming tendency	silicone compounds

lube oils. See lubricating oils.

lubricating greases. Lubricating greases are generally mixtures of a mineral oil or oils with one or more metallic soaps. The most common soaps are those of sodium, calcium, barium, aluminum, lead, lithium, potassium and zinc. Oils thickened with residuum, petrolatum or wax may be called greases. Some form of graphite may be added. Greases range in consistency from thin liquids to solid blocks and in color from transparent, stainless greases to heavy black residuum greases. The specifications for a grease are determined by the speed, load, temperature, environment, and metals in the desired application. Texture of grease may be smooth, buttery, ropy or stringy, fibrous, spongy, or rubbery. The texture does not necessarily indicate the viscosity of the grease, but is related to the formulation and methods of manufacture.
See also lubricating oils.

lubricating oils (lube oils). Selected fractions of refined mineral oils used for lubrication of moving surfaces. The term is also sometimes used to refer to transformer oils used for electrical insulating purposes and cooling. Lubricating oils usually have small amounts of additives to impart special properties such as viscosity index and detergency. They range in consistency from thin liquids to grease-like substances. In contrast to lubricating greases, lube oils do not contain solid or fibrous materials.

"Lubricin" N-1. [202] Trademark for a light viscosity derivative of castor oil that increases the oiliness and wetting power of mineral oils, decreases corrosion, and exhibits detergent action on tar, varnish, and carbon deposits. Used as as an additive for lubricating oils, motor fuels, and cutting oils.

"Lubricin" V-1. [202] Trademark for a lubricant which facilitates the processing of rigid vinyl plastics.

lubricity. The property of forming a lubricating film between moving surfaces, particularly when such surfaces are subjected to heavy loads and rapid movement. Lubricity depends partly upon the wetting ability of the film-forming material. Viscosity is somewhat related to lubricity but the relation is complicated. Oiliness is sometimes used as approximately equivalent to lubricity.

"Lubriseal." [16] Trademark for a lubricating, sealing, and rust inhibiting stopcock grease.
Properties: Tacky, smooth textured, odorless; stable; free from vegetable or animal oil; nearly acid and alkali proof and practically insoluble in water; vapor pressure less less than 10^{-5} mm (20°C).
Grades: Improved formula, m.p. approx. 40°C; high vacuum formula, m.p. approx. 50°C, for high vacuum (3×10^{-6} mm) work only.
Containers: 25-g collapsible tubes; 500-g cans.
Shipping regulations: None.*

"Lubrol." [206] Brand name of proprietary emulsifying and antistatic agents.

lucanthone hydrochloride (1-(2-diethylaminoethylamino)-4-methylthioxanthone hydrochloride).
Properties: Yellow crystals; m.p. 195-196°C. Freely soluble in water; slightly soluble in alcohol.
Containers: Bottles.
Use: Medicine.

"Lucidol." [154] Trademark for benzoyl peroxide and/or other solid organic peroxides.

luciferin. An albumin present in some animals which, under the influence of the enzyme luciferase, can be oxidized and exhibit bioluminescence, or "cold light."

"Lucite." [28] Trademark for line of methacrylate ester polymers available as granules, resins or syrup.
See acrylate resins.
"Lucite" 41, 42, 44, 45 and 46 are respectively the methyl, ethyl, n-butyl, isobutyl and mixed 50-50 n-butyl and isobutyl ester polymers.

"Ludox" AS. [28] Trademark for an aqueous colloidal sol containing approximately 30% silica and 0.25% ammonia; pH at 25°C 9.5-9.6; f.p. 32°F (0°C).
Containers: 55-gal drums.
Use: See "Ludox" HS and LS colloidal silicas. Suitable for applications requiring a volatile stabilizer which would be detrimentally affected by sodium ions present in other grades of "Ludox."

"Ludox" HS. [28] Trademark for an aqueous colloidal sol containing approximately 30% silica. Ratio, wt, $Na_2O:SiO_2$, 1:95; pH at 25°C 9.65-10.15; f.p. 32°F (0°C).
Containers: 55-gal drums; tank trucks; tank cars.
Uses: For antislip treatment of rayon, cotton, and wool staple for improved yarn strength; compounding and post-dip treatment of latex foams and films to improve modulus and effect rubber savings; modifying latex adhesives for improved bond strengths; for increased hardness and slip resistance in emulsion floor waxes; improved water and heat resistance of adhesive coatings and fibrous structures; for antisoil treatment of pile fabrics and miscellaneous substrates; frictionizing agent for paper; binder for inorganic materials; precoating photosensitized paper for color contrast.

"Ludox" LS. [28] Trademark for an aqueous colloidal sol containing approximately 30% silica. Ratio, wt, $Na_2O:SiO_2$, 1:285; pH at 25°C 8.2-8.7.
Use: See "Ludox" HS.

"Ludox" SM. [28] Trademark for an aqueous colloidal sol containing approximately 15% silica characterized by extremely small particle size averaging only 7 millimicrons. Ratio, wt, $Na_2O:SiO_2$, 1:159; pH at 25°C 8.1; f.p. 32°F (0°C).
Containers: 55-gal drums.
Uses: Binder for inorganic fibrous materials; antisoil applications to smooth surfaces; paper coatings to improve abrasion resistance, printability and other properties; latex adhesive compositions for improved adhesion.

"Lugatol." [440] Trademark for a series of hetero-polar azo dyestuffs with mainly anionic activity.
Uses: For leather of all types.

Lugol's solution (iodine solution, strong). An aqueous solution containing 5 g iodine and 10 g potassium iodide per 100 ml water.
Grade: U.S.P. XVI.
Use: Pharmaceutical.

"Lukens Carbon-Moly Steel Plate." [255] Molybdenum steel plate material made to ASTM A-204 specification.

"Lukens Chrome-Moly Steel Plate." [255] Chromium-molybdenum steel plate material made to ASTM A-387 specification.

"Lukens Clad Steels." [255] High alloy materials available in composite clad plate integrally and completely bonded to a variety of carbon or alloy backing steels by hot rolling. These include the stainless steels, types 405, 410, 430, 304, 304L, 309, 310, 316, 316Cb, 316L, 317, 317L, 321, 347, and nickel, low carbon nickel, Inconel, Monel, copper and cupro-nickel.

"Lukens Digester Steel." [255] A silicon-free rimmed steel which has been used for paper mill digesters.

"Lukens Manganese-Moly Steel Plate." [255] Manganese-molybdenum steel plate made to ASTM A-302 specification.

"Lukens Manganese-Vanadium Steel Plate." [255] High strength, low alloy steel plate made to ASTM A-242 specification.

"Lukens Nine Nickel Steel Plate." [255] Low carbon, high nickel steel plate made to ASTM A-353 specification.

"Lukens T-1 Steel Plate." [255] High strength proprietary steel plate for pressure vessel and structural applications and for abrasion resistance.

"Lumatex." [440] Trademark for a series of mineral and organic pigment dyestuffs which can be fixed on all types of textile fibers with suitable binders. Fixation by short curing at 100-130°C.
Uses: Screen and roller printing; dyeing.

lumbang oil (candle-nut oil).
Properties: A limpid, colorless or yellowish liquid; pleasant odor; bland taste. Soluble in alcohol, ether, chloroform, and carbon disulfide. Sp. gr. 0.920-0.927 (15°C); saponification value 190-193; iodine number 140-164; refractive index 1.4790 (15°C).
Derivation: From the candle-nut, the seed of Aleuritis moluccana, by expression.
Method of purification: Filtration.
Grades: Crude; refined.
Containers: Wooden barrels.
Uses: Illuminant; paints; caulking; soap manufacture; wood preservative.
Shipping regulations: None. *

"Luminal." [162] Trademark for phenobarbital.

luminescence. The emission of light (visible or invisible) without high temperature or incandescence; i.e., "cold light," as from a firefly. Both organic and inorganic substances may have the property of luminescence. A distinction is made between fluorescent and phosphorescent kinds of luminescence in that fluorescent substances

*See "I.C.C. Shipping Regulations," page xiii.
Reference numbers refer to name of manufacturer. See "List of Manufacturers," page v.

cease to shine when the exciting source is removed, while phosphorescent substances continue to radiate light for a characteristic period of time thereafter. See fluorescence; phosphorescence. The exciting source may be (1) photons, i.e., light or other radiation different in wave length from the light being emitted; (2) moving charged particles such as alpha, beta or gamma particles; (3) energy from chemical, biochemical or crystallographic changes. Special types of luminescence named according to the kind of substances involved or of the exciting source are bioluminescence, chemiluminescence, electroluminescence, photoluminescence and triboluminescence.

Common examples of luminescence in addition to the firefly are the light from neon lamps or from lightning, and the light from phosphor crystals of television tubes or fluorescent lighting tubes.

"Lumnite." An aluminous cement of the sintered type.
See cement, aluminous.

"Lump Coal." [28] Trademark for permissible dynamites (types C and CC) with very low velocity of detonation.
Use: For coal mining where maximum production of large-size coal is desired.

lunar caustic. See silver nitrate, fused.

lupanine $C_{15}H_{24}N_2O$. A poisonous alkaloid. Both dextro and levo forms are known. The latter is a colorless oil. The material has also been described as a yellow syrupy liquid with green fluorescence.
Properties: (d-form) White needles; m.p. 40°C; distils undecomposed at 220°C (10 mm); slightly soluble in water, with an alkaline reaction; soluble in alcohol, ether, and chloroform; refractive index (n 24/D) 1.5444.
Derivation: From the seeds of Lupinus albus and Lupinus angustifolius.
Method of purification: Crystallization.
Grade: Technical.
Containers: Glass bottles.
Use: Medicine.
Shipping regulations: None. *

"Luperco." [154] Trade name for compounds of benzoyl peroxide or some other organic peroxide finely dispersed with an organic or inorganic filler.

"Luperox." [154] Trademark for pastes of finely divided benzoyl peroxide dispersed in water or oils of various types.

"Lupersol." [154] Trade name for solutions of organic peroxides.

"Lupersol 101." [154] Trade name for 2,5-dimethyl-2,5-di(tert-butylperoxy)hexane (q.v.).

"Lupersol DDM." [154] Trade name for methyl ethyl ketone peroxide (q.v.).

lupetazine. See 2,5-dimethylpiperazine.

2,6-lupetidine. See 2,6-dimethylpiperidine.

lupinidine. See sparteine.

lupinine $C_{10}H_{19}ON$.
Properties: White, crystalline alkaloid; poisonous! Soluble in alcohol, ether, acetone, and chloroform; decomposed by water. M.p. 69-71°C; b.p. 255-257°C.
Derivation: By extraction from the seeds of Lupinus luteus and Lupinus niger.
Method of purification: Crystallization.
Grade: Technical.
Containers: Glass bottles.
Use: Medicine.
Shipping regulations: None. *

"Lupolen." [440] Trademark for series of polyethylenes of density 0.918-0.960 and melt index 0.2-20. Special grades contain polyisobutylene.
See "Oppanol B."

"Lupomin." [78] Trademark for a series of nitrogen compounds used for softening and finishing textiles.

"Luposec." [78] Trademark for a stable cationic emulsion of waxes with aluminum salts used in one-bath method for showerproofing or water-repelling all textile fabrics.

"Luposol." [78] Trademark for a prepared solution of locust bean gum used for finishing or as a printing paste.

lupulin (humulin). A bright, yellowish-brown granular powder of characteristic odor and hop taste, obtained from Humulus lupulus (hops). It contains lupulon and humulon.

lupulon. An antibiotic constituent of hops (see lupulin).
Properties: Crystallizes as prisms from methanol; m.p. 94°C; stable in absence of air, but turns yellow and develops odor on standing in air for several days. Soluble in alcohol; slightly soluble in water.

"Lurex." [233] Trademark for metallic yarns which are laminations of clear plastic films and aluminum foil or metallized film. Used in furnishing and apparel applications. The films generally used are cellulose-acetate-butyrate, or polyester film.

"Lusane." [28] Trademark for a special dye for cotton, with outstanding brillancy of shade and lightfastness.

"Lustan." [51] Trademark for a mold oil for glass machines. Comprises a low-carbon-base oil to which colloidal graphite has been added.

luster. The appearance of the surface of a substance in reflected light. The term is used particularly in describing minerals. Types of luster are: (a) metallic, like metals, or the mineral pyrite; (b) vitreous, like glass or quartz; (c) adamantine, exceedingly brilliant, like diamond; (d) resinous, like resin, or sphalerite; (e) dull, not bright or shiny; like chalk. Also called earthy luster. Other types are greasy, silky, and pearly.

Luster Colors. [28] Liquids containing resinates or similar organic compounds of various base or precious metals.
Containers: Precious metal lusters, 500-g (1-lb) bottles; base metal lusters, gal (7-lb) bottles.

"Lustrex." [58] Trademark for styrene resin powders or latices.

"Lustrone" Colors. [23] Trademark for plasticized nitrocellulose lacquers tinted with dyes.
Use: Transparent color effects on leather.

lutecium. Discontinued form for lutetium.

lutein. A yellow pigment isolated from the corpus luteum, and found in body fats and egg yolks. It is a carotenoid and is similar to, or identical with xanthophyll (q.v.).

luteinizing hormone. Obsolete term for interstitial-cell-stimulating hormone.

luteotropin (adenohypophyseal luteotropin; prolactin; galactin; lactogenic hormone). One of the hormones secreted by the anterior lobe of the pituitary gland. It aids in causing growth of the mammary gland and initiates milk secretion by the mammary gland; it also influences the activity of the corpus luteum, including the secretion of progesterone.
Properties: Crystalline protein; molecular weight about 33,300; almost insoluble in water; soluble in dilute acids and acidified methanol and ethanol.
Use: Medicine.

lutetia. See lutetium oxide. See also rare earths.

lutetium (formerly lutecium) Lu. Atomic number 71; group III of the periodic table; one of the rare-earth elements of the yttrium subgroup.
Properties: Metallic luster. Sp. gr. 9.849; b. p. 1900°C (approx); reacts slowly with water; soluble in dilute acids.
Derivation: Reduction of the fluoride with calcium.
Grades: Regular, high purity (ingots, lumps).
Uses: In nuclear technology.

lutetium chloride $LuCl_3 \cdot xH_2O$.
Purity: Up to 99.9% lutetium salts.
Containers: Glass bottles; fiber drums.

lutetium fluoride $LuF_3 \cdot 2H_2O$.
Purity: Up to 99.9% lutetium salts.
Containers: Glass bottles; fiber drums.

lutetium nitrate $Lu(NO_3)_3 \cdot 6H_2O$.
Purity: Up to 99.9% lutetium salts.
Containers: Glass bottles; fiber drums.
Shipping regulations: Oxidizing material. Yellow label. *

lutetium oxide (lutetia) Lu_2O_3.
Properties: White solid; slightly hygroscopic; absorbs water and carbon dioxide from the air.
Grades: Up to 99.9% purity.
Containers: Glass bottles; fiber drums.

lutetium sulfate $Lu_2(SO_4)_3 \cdot 8H_2O$.
Purity: Up to 99.9% lutetium salts.
Containers: Glass bottles; fiber drums.

2,3-lutidine (2,3-dimethylpyridine) $NC(CH_3)C(CH_3)CHCHCH$.
Properties: Liquid; b. p. 161.5°C; f. p. −15.5°C; density 0.949 (20°C); refractive index 1.5085 (n 20/D); solubility 13.3 g in 100 g water at 20°C; solubility of water in 100 g at 20°C, 131 g; soluble in water in all proportions below 16°C.
Purity: Commercial 95%.

2,4-lutidine (2,4-dimethylpyridine) $NC(CH_3)CHC(CH_3)CHCH$.
Properties: Liquid; b. p. 158.7°C; density 0.9325 (20°C); refractive index 1.5000 (n 20/D). Soluble in water in all proportions below 23°C.
Purity: Commercial 95%.

2,5-lutidine (2,5-dimethylpyridine) $NC(CH_3)CHCHC(CH_3)CH$.
Properties: Liquid; b. p. 157.3°C; density 0.9331 (20°C); refractive index 1.5005 (n 20/D); solubility 10.0 g in 100 g water at 20°C; solubility of water in 100 g at 20°C, 95 g; soluble in water in all proportions below 13°C.
Purity: Commercial 95%.

2,6-lutidine (2,6-dimethylpyridine) $NC(CH_3)CHCHCHCCH_3$.
Properties: Colorless oily liquid; peppermint odor; sp. gr. 0.932; b. p. 143°C; f. p. −6.6°C; refractive index 1.4973 (n 20/D). Derived from coal tar.
Typical specifications: 95% min. purity; f. p. −9.4°C min; distills (95%) within a 2°C range including 143.7°C.
Uses: Pharmaceuticals; resins; dyestuffs; rubber accelerators; insecticides.

3,4-lutidine (3,4-dimethylpyridine) $NCHC(CH_3)C(CH_3)CHCH$.
Properties: Liquid; f. p. −12°C.

3,5-lutidine (3,5-dimethylpyridine) $NCHC(CH_3)CHC(CH_3)CH$.
Properties: Liquid; b. p. 172.7°C; f. p. −6.6°C; density 0.944 (20°C); refractive index 1.5049 (n 20/D); solubility 3.3 g in 100 g water at 20°C; solubility of water in 100 g at 20°C, 55.4 g.
Purity: Commercial 95%.

2,6-lutidine-N-oxide $ONC(CH_3)CHCHCHCCH_3$.
Purity: 98% minimum.

"Lutocylol." [305] Trademark for ethisterone U. S. P.
Use: Medicine.

lututrin. A uterine-relaxing factor obtained from the corpus luteum of sow ovaries by a process of salting out followed by dialysis. It is a water-soluble protein or polypeptide.
Grade: N. N. D.

"Luxapole." [78] Trademark for compositions for treating textile fibers and particularly for finishing agents and dye penetrants.

"Luxol." [28] Trademark for a line of spirit and lacquer-soluble dyes. Used in lacquers, wood stains, spirit prints, inks.

"LX-685." [21] Designation for proprietary heat-reactive resin used in the manufacture of ready-mixed aluminum paints, grease and gasoline resistant coatings, floor and deck enamels and concrete curing compounds.

"Lyamine." [123] Trademark for lysine for use as an animal feed supplement.

"Lycedan." [91] Trademark for a brand of adenosine-5-phosphoric acid for medicinal use.

lycine. See betaine.

lycine hydrochloride. See betaine hydrochloride.

lycopodium (club-moss; vegetable sulfur).
Properties: Fine yellow powder.
Derivation: Spores of Lycopodium clavatum (club moss).
Habitat: North America, Asia and Europe.
Grades: Technical.
Containers: Multiwall paper sacks; cases.
Uses: Medicine; pyrotechnics.
Shipping regulations: None. *

lycorine $C_{16}H_{17}NO_4$. An alkaloid.
Properties: White crystals; poisonous! Slightly soluble in water, alcohol, and ether. M.p. 280°C.
Derivation: By extraction of Lycoris radiata.
Method of purification: Crystallization.
Grades: Technical.
Containers: Glass bottles.
Use: Medicine.
Shipping regulations: None. *

"Lycra." [28] Trademark for a spandex fiber. Available in yarn with denier sizes from 70 to 560.
Properties: Sp. gr. 1.0; tensile strength (psi) 9,000; break elongation 570%; moisture regain 0.3%; m.p. 250°C; tensile recovery 95% from 50% elongation; solution in dimethyl formamide (hot).
Uses: Foundation garments, swim wear, surgical hose, and other elastic products.
See also spandex.

Lydian stone (touchstone; basanite). A velvet-black form of quartz (q.v.) closely allied to or grading into chert, jasper or flint.
Use: Testing quality of precious metals (the metal tested is rubbed on the Lydian stone and the mark is checked against others made by alloys of predetermined composition).

lye. See sodium hydroxide and potassium hydroxide.

"Lykopon." [23] Trademark for sodium hydrosulfite. White crystalline powder, water-soluble, 94% min. $Na_2S_2O_4$. Powerful reducing agent for vat dyestuffs. Bleaching agent.
Use: Dyeing, printing, and stripping of textile fabrics; bleaching of soap, glue, etc.; oxygen scavenger in synthetic rubber manufacture.

lyophilic. Characterizing a material which readily goes into colloidal suspension; if into water, it may be termed hydrophilic. The colloid is supposedly stabilized by the formation of a protective layer of molecules of the dispersing medium about the suspended particles. Examples: glue, gelatin.

lyophilization. The freeze drying of biological materials. The material to be dried is first frozen and then placed in a high vacuum so that the water (ice) vaporizes into the vacuum without melting, and the non-water components are left behind in an undamaged state. Used for blood plasma, certain antibiotics and other heat-sensitive materials.

lyophobic. Characterizing a material which exists in the colloidal state without having any significant affinity for the medium. Such colloids are generally stabilized by the adsorption of ions and coagulate when the charge is neutralized. Examples: colloidal gold, colloidal arsenic sulfide.

lysergic acid $C_{16}H_{16}N_2O_2$. A product of ergot alkaloids which is used, as D-lysergic acid diethylamide (LSD), in medical research in the field of mental disorders. Lysergic acid has now been synthesized.
Properties: Crystallizes (with 1-2 molecules of water) in plates from water. M.p. 240°C (dec). It is amphoteric (behaves as both an acid and a base); moderately soluble in pyridine; slightly soluble in water and the usual, neutral organic solvents; soluble in alkaline and acid solutions.
Derivation: Alkaline hydrolysis of ergot alkaloids; organic synthesis beginning with indolepropionic acid.
Use: Medical research; synthesis of ergonovine. Derivatives of this acid may be of use as non-lethal incapacitating drugs in chemical warfare.

lysidine (methylglyoxalidine) $CH_3\overline{CNCH_2CH_2NH}$.
Properties: Colorless, hygroscopic crystals; mousy odor; m.p. 105°C; b.p. 195°C. Soluble in water, alcohol, and ether.
Derivation: From ethylenediamine hydrochloride and sodium acetate by dry distillation, decomposing the hydrochloride of the new base with concentrated potassium hydroxide and crystallizing.
Method of purification: Recrystallization.
Grades: Technical; 50% solution.
Containers: Glass bottles.
Use: Medicine.
Shipping regulations: None. *

lysine (alpha, epsilon-diaminocaproic acid) $NH_2(CH_2)_4CH(NH_2)COOH$. An essential amino acid.
Properties: Colorless crystals; soluble in water; slightly soluble in alcohol; insoluble in ether; optically active. D(−)-lysine, m.p. 224°C with decomposition.

L(+)-lysine, m. p. 224°C with decomposition.

Derivation: Fermentation or extraction of natural proteins; organic synthesis.

Containers: Glass bottles; fiber cans.

Uses: Biochemical and nutrition studies; pharmaceuticals; culture media; fortification of foods and feeds.

Commercially available as the DL- and L-lysine monohydrochloride.

"Lysol." [346] Trademark for a disinfectant containing soap and ortho-hydroxydiphenyl and xylenols as active ingredients.

Properties: Brown, oily liquid, cresylic odor, non-poison. Phenol coefficient 5; miscible with water and alcohol; sp. gr. 1.043.

Derivation: Saponification of a fatty oil and combination of resultant soap with xylenols and ortho-hydroxydiphenyl, followed by addition of propylene glycol.

Containers: Glass bottles; steel drums.

Uses: Disinfectant, germicide, fungicide, antiseptic.

lysozyme. An enzyme found in egg white which dissolves certain bacteria and hydrolyzes sugar linkages in glycoproteins.

"Lytron" Sand Conditioner. [58] Trademark for a finely divided, pale yellow powder, polyelectrolyte designed especially for foundry sand systems.

Containers: Nonreturnable fiber drums.

Use: Improves sand workability and packability for better metal casting molds.

LZT. Abbreviation for lead zirconate titanate.

M. Abbreviation for molar, used to characterize the concentration of a solution. A molar solution contains one gram-molecular weight of a substance in one liter of solution.

m-. Abbreviation for meta- (q. v.).

MAC. Abbreviation for methyl allyl chloride.

"Macaloid." [417] Trade name for beneficiated hectorite (q. v.) used to produce a mineral gel of the bentonite type.

macassar gum. See agar-agar.

mace. The coating (arillus) of nutmeg seeds (Myristic fragrans).
Habitat: Molucca Islands; cultivated in the tropics and East Indies, Ceylon, South America, India and Philippine Islands.
Grades: No. 1; No. 2; siftings.
Containers: Bales.
Use: Medicine.
Shipping regulations: None. *

mace oil.
Properties: Colorless or pale yellowish liquid; agreeable, aromatic odor, resembling oil of nutmeg. Soluble in alcohol, ether, and chloroform.
Chief known constituents: Pinene, dipentene, myristicol and myristicin.
Constants: Sp. gr. 0.91-0.93; optical rotation +10°.
Derivation: Distilled from the seed of Myristica fragrans.
Grades: Technical.
Containers: Cans; drums.
Uses: Flavoring; perfumery.
Shipping regulations: None. *

machine oil. A medium duty grade of lubricating oil used in lubrication of machine parts.

M acid. See 1-amino-5-naphthol-7-sulfonic acid.

Mack's cement. See gypsum cements.

Macquer's salt. See potassium arsenate.

macromole. See macromolecule.

macromolecule (macromole). A very large molecule containing hundreds or thousands of atoms, such as a polymer molecule. Most molecules contain relatively few atoms (two, three, etc., up to about ten). Colloidal particles are often referred to as macromolecules.

"Macroport" A. [249] Trademark for aluminum oxide catalyst carrier. Available as spheres, pellets and aggregate of fused aluminum oxide.

macroscopic cross section. The form of the cross section (q. v.) usually used in nuclear technology. It is the product of the cross section and the number of target particles per unit volume. It is also the reciprocal of the mean free path for the process in question.

macrose. See dextran.

macrotin. See cimicifugin.

madder.
Derivation: Pulverized root of Rubia tinctorum, a plant formerly cultivated in Europe and Asia Minor. The glucosides contained therein, when decomposed by fermentation, yield alizarin, now largely replaced by alizarin obtained from the anthracene oil of coal-tar.
See also madder lake.

madder lake. Madder lake is a pigment produced from the dyestuff known as alizarin red and takes the place of the coloring matter that was made many years ago from the madder plant. It is a bluish shade of red that is comparatively transparent, and its use in the paint industry is largely confined to stains where a permanent pigment is desired. It is used in making mahogany shades and is widely used in inks. It has very good permanency and is non-bleeding in oil and lacquer solvents. Madder lake is also used for artists' oil colors.

Maddrell's salt. $NaPO_3$ II, $NaPO_3$ III or mixtures thereof. These are insoluble crystalline forms of sodium metaphosphate produced, together with $NaPO_3$ I, by heating sodium phosphate, monobasic (NaH_2PO_4). At 300-475°C $NaPO_3$ II is produced and at lower temperatures, $NaPO_3$ III.
See sodium metaphosphate.

"Madribon." [190] Trademark for a brand of sulfadimethoxine (q. v.).

magenta. See fuchsin.

magic number. A name in common use to denote certain particularly stable numbers of neutrons or protons in nuclei. These were deduced from empirical observation of the number of stable isotopes or isotones of nuclei that had 2, 8, 20, 28, 50, 82, or 126 neutrons or protons or both. Modern theory, particularly the "shell" theory of the nucleus, has derived these numbers from the quantum numbers or

*See "I.C.C. Shipping Regulations," page xiii.
Reference numbers refer to name of manufacturer. See "List of Manufacturers," page v.

nuclei, based on an interpretation similar to the theory of the atom which has stable shells of electrons.

magister of bismuth. See bismuth subnitrate.

magister of lead. Old name for lead carbonate, basic.

magister of sulfur. Sulfur precipitated by acids from solutions of hyposulfites or polysulfides. It is in the amorphous form.

"Mag-Li-Kote." [139] Trademark for a burned dolomite, ground; used for coating pig casting machine molds, cinder ladles, stools and ingot molds.

"Maglite." [123] Trademark for light magnesium oxide.

magma.

1. In medicine, a class of preparations in which finely divided, freshly precipitated, insoluble, inorganic hydroxides are suspended in water to form a viscous, opaque mixture which may settle out on standing. Magmas of bismuth, magnesium and iron are used, commonly called milk of bismuth, milk of magnesia, etc.
2. In geology, a liquid, molten mass within the earth's crust (e.g., lava). The source of igneous rocks.

"Magmaster" Magnesite. [426] Trademark for dead burned magnesite.
Derivation: Synthetic.
Containers: Bulk and bags.
Uses: In the manufacture of refractories.

"Magnafloat." [250] Trademark for iron-oxide heavy media systems for the purification of coal, sand, gravel, and other similar materials.

magnalium. An alloy of aluminum and magnesium.

"Magnamycin." [299] Trademark for carbomycin.

magnesia. See magnesium oxide.

magnesia alba. See magnesium carbonate.

magnesia-alumina $MgO \cdot Al_2O_3$. A synthetic spinel.

magnesia, burnt. See magnesite, dead-burned.

magnesia, calcined. See magnesite, caustic-calcined.

magnesia, caustic-calcined. See magnesite, caustic-calcined.

magnesia-chromia $MgO \cdot Cr_2O_3$. A synthetic spinel.

magnesia, dead-burned. See magnesite, dead-burned.

magnesia, fused. Used as a refractory and to handle electricity at high temperatures. See "Magnorite."

magnesia, heavy. See magnesium oxide.

magnesia, light. See magnesium oxide.

magnesia magma (milk of magnesia). A white opaque, more or less viscous suspension of magnesium hydroxide in water from which varying proportions of water usually separate on standing.
Grade: U.S.P. XVI.
Use: Medicine.

magnesia mixture. One part each of ammonium chloride and magnesium sulfate dissolved in 8 parts of water to which 4 parts of ammonia water, of sp. gr. 0.96, are added and the whole filtered. Used in determination of phosphorus.

magnesia, refractory. See magnesite, dead-burned.

magnesite. The natural mineral, $MgCO_3$. The term magnesite is sometimes loosely used as a synonym for magnesia, as are also the terms caustic-calcined magnesite, dead-burned magnesite, and synthetic magnesite. See following articles and magnesium oxide.
Properties: White, yellowish, grayish-white, brown, mostly crystalline in form. Sp. gr. 3-3.12; hardness 3.5-4.5.
Occurrence: United States (California, Washington, Nevada); Austria; Greece.
Uses: To make the various grades of magnesium oxide; to produce carbon dioxide.
See also magnesium carbonate.

magnesite, burnt. See magnesite, dead-burned.

magnesite, calcined. See magnesite, caustic-calcined.

magnesite, caustic-calcined (caustic-calcined magnesia; calcined magnesite; calcined magnesia) Principally magnesia (magnesium oxide) MgO. The product obtained by firing magnesite, or other substances convertible to magnesia upon heating, at temperatures below 1560°C so that some carbon dioxide is retained (2-10%) and the magnesia displays adsorptive capacity or activity.
Grades: Technical; rubber.
Uses (in approximate order of volume): Magnesium oxychloride and oxysulfate cements; 85% magnesium oxide insulation; rubber; uranium processing; chemical processing (adsorption & catalysis); rayon; refractories; paper pulp; acid-neutralizing fertilizers; minor uses are welding rod coatings, fillers, glass constituents, abrasives.

magnesite, dead-burned (burnt magnesia; dead-burned magnesia; refractory magnesia; burnt magnesite) Magnesium oxide, MgO. The granular product obtained by burning (firing) magnesite, or other substances convertible to magnesia upon heating, above 1450°C long enough to form ... granules suitable for use as a refractory ... (ASTM definition). Synthetic magnesium hydroxide or chloride is sometimes used instead of magnesite as a source.
Grades: 85-87% (from magnesite ores); 97-99% (from sea water and brines).
Uses: Refractories, as grains or basic

brick, the latter especially in open hearth furnaces for steel, furnaces for nonferrous metal smelting, and in cement and other kilns.

"Magnesite H-W." [446] Trade name for an over 90% burned magnesia brick with minor added components to control properties. Resistant to molten metal and basic slags. Used in open hearth furnaces, electric steel furnaces, copper and nickel converters, refining industries, and other metallurgical industries.

magnesite, synthetic. Usually means magnesium oxide, MgO, as obtained from sea water, sea water bitterns, or well brines. The preliminary product is usually magnesium hydroxide or chloride, which is then heated, or sometimes treated with steam and heated in the case of the chloride, to obtain the oxide. Synthetic magnesite constitutes the purer grades of dead-burned magnesite.

magnesium Mg. Element of atomic number 12 of group II in the periodic table.

Properties: Silvery, malleable, moderately hard metal. Oxidizes and tarnishes in moist air but is stable in dry air. In finely divided form is easily ignited and burns with an intense white light. The solid form must be heated above its melting point before it will burn. Sp. gr. 1.69-1.75; m.p. 650°C; b.p. 1120°C. Soluble in acids; insoluble in water.

Derivation: (a) By electrolysis of fused magnesium chloride; (b) by reduction of magnesium oxide with ferrosilicon (Pidgeon process); (c) by reduction of magnesium oxide with carbon.

Method of purification: Distillation.

Forms available: Ingots; bars; 50-, 100- to 150-mesh powder; sheet and plate; rod; tubing; ribbon; castings. See also magnesium dust.

Uses (in approximate order of volume): Structural parts, including alloys (aircraft usage declining; use in missiles, automobiles is increasing); powder for pyrotechnics (including flares) and flash photography; production of iron, nickel, zinc, and other metals; magnesium compounds and organic synthesis; cathodic protection devices; reducing agent in production of titanium, zirconium, hafnium, uranium, beryllium, etc; optical mirrors; precision instruments; substitute for zinc in dry batteries.

Fire hazard: Dangerous! Combustible, particularly in form of powder, shavings or thin sheets.

Shipping regulations: Flammable solid. Yellow label, when the metal is in powder or scrap form. *

magnesium acetate (a) $Mg(OOCCH_3)_2$ or (b) $Mg(OOCCH_3)_2 \cdot 4H_2O$.

Properties: Colorless crystalline aggregate or monoclinic crystals. Acetic acid odor. (a) M.p. 323°C, sp. gr. 1.42, (b) m.p. 80°C, sp. gr. 1.45. Soluble in water and (dilute) alcohol.

Derivation: Interaction of magnesium carbonate and acetic acid.

Grades: Technical.

Uses: Calico printing (fixing aniline black); textiles (fixing eosins); medicine; deodorant, disinfectant and antiseptic.

magnesium acetylacetonate $Mg(C_5H_7O_2)_2$. Crystalline powder. Slightly soluble in water. Resistant to hydrolysis. A chelating non-ionizing compound.

magnesium acetylsalicylate $Mg(OOCC_6H_4OOCCH_3)_2$.

Properties: White, almost tasteless and odorless powder. Freely soluble in water; less soluble in alcohol.

Derivation: Interaction of acetylsalicylic acid and a magnesium salt; e.g., the carbonate, oxide, or hydroxide.

Use: Medicine.

magnesium ammonium phosphate (magnesium ammonium orthophosphate) $MgNH_4PO_4 \cdot 6H_2O$.

Properties: White powder; sp. gr. 1.71; m.p. decomposes to magnesium pyrophosphate, $Mg_2P_2O_7$; soluble in acids; insoluble in alcohol and water.

Derivation: By the interaction of solutions of a magnesium salt and ammonium phosphate.

Grades: Technical.

Containers: Glass bottles.

Uses: Medicine; fire retardant for fabrics; fertilizer.

Shipping regulations: None. *

magnesium ammonium orthophosphate. See magnesium ammonium phosphate.

magnesium ammonium sulfate (ammonium magnesium sulfate) $Mg(NH_4)_2(SO_4)_2 \cdot 6H_2O$.

Properties: Colorless crystals; sp. gr. 1.70; soluble in water.

Derivation: Solutions of magnesium and ammonium sulfates.

magnesium arsenate $Mg_3(AsO_4)_2 \cdot xH_2O$. White powder when pure, insoluble in water. Technical material is highly hydrated and made from magnesium carbonate and arsenic acid.

Containers: Fiber cans and drums; multiwall paper sacks.

Use: Insecticide.

Warning: Poisonous if swallowed. MCA warning label.

Shipping regulations: Poison, class B. Poison label. *

magnesium benzoate $Mg(C_7H_5O_2) \cdot 3H_2O$.

Properties: White crystalline powder; m.p. about 200°C; soluble in hot water and alcohol.

Use: Medicine.

Shipping regulations: None. *

magnesium biphosphate. See magnesium phosphate, monobasic.

magnesium borate $3MgO \cdot B_2O_3$ or $Mg(BO_2)_2 \cdot 8H_2O$.

Properties: Transparent, colorless crystals

or white powder. Soluble in alcohol, acetic acid, and inorganic acids; slightly soluble in water.
Derivation: By heating magnesium oxide, boric anhydride and potassium hydrogen fluoride.
Containers: Glass bottles; fiber cans.
Uses: Preservative; antiseptic; fungicide.
Shipping regulations: None. *

magnesium borocitrate $Mg(BO_2)_2 \cdot Mg_3(C_6H_5O_7)_2 \cdot 14H_2O$.
Properties: White powder or small, white, lustrous scales. Soluble in water.
Derivation: By mixing magnesium borate and magnesium citrate.
Grades: Technical.
Containers: Boxes; glass bottles.
Use: Medicine.
Shipping regulations: None. *

magnesium-boron fluoride.
Grade: Technical.
Containers: 145-lb steel drums.
Use: Metal flux.

magnesium bromate $Mg(BrO_3)_2 \cdot 6H_2O$.
Properties: White crystals or crystalline powder. Soluble in water; insoluble in alcohol.
Constants: Sp. gr. 2.29; m.p. loses $6H_2O$ (200°C); b.p. decomposes.
Derivation: By adding magnesium sulfate to a solution of barium bromate.
Method of purification: Recrystallization.
Grades: Pure; reagent.
Containers: Glass bottles; 25-lb tin boxes.
Use: Analytical reagent.
Fire hazard: Dangerous.
Shipping regulations: Oxidizing material. Yellow label. *

magnesium bromide $MgBr_2 \cdot 6H_2O$.
Properties: Colorless, very deliquescent crystals; bitter taste. Soluble in water; slightly soluble in alcohol.
Constants: M.p. about 165°C with loss of H_2O. Anhydrous m.p. 700°C.
Derivation: By the action of hydrobromic acid on magnesium oxide with subsequent crystallization.
Method of purification: Recrystallization.
Grades: Technical; C.P.
Containers: Glass bottles.
Uses: Medicine; organic syntheses.
Shipping regulations: None. *

magnesium-calcium chloride. See calcium-magnesium chloride.

magnesium carbonate (magnesium carbonate, precipitated; magnesia alba) $MgCO_3$. The term magnesium carbonate is generally reserved for the synthetic, pure variety. See also magnesite.
Properties: Very light odorless white powder; sp. gr. 3.04. Soluble in acids; insoluble in water.
Derivation: By mixing solutions of magnesium sulfate and sodium carbonate, filtering and drying.
Grades: Technical; U.S.P. XVI.
Containers: Bags or paper-lined barrels; 1-lb to 100-lb.
Uses: Magnesium salts; rubber pigments; inks; glass; pharmaceuticals, dentifrices and cosmetics; free-running table salts; mineral waters; filtering medium.

magnesium carbonate, precipitated. See magnesium carbonate.

magnesium chlorate $Mg(ClO_3)_2 \cdot 6H_2O$.
Properties: White powder. Bitter taste. Very hygroscopic. Caution! Keep well stoppered. Soluble in water; slightly soluble in alcohol.
Constants: Sp. gr. 1.8; m.p. 35°C; (decomposes at 120°C).
Containers: Glass bottles with ground stoppers.
Use: Medicine; defoliant; desiccant.
Shipping regulations: Oxidizing material. Yellow label. *

magnesium chloride (a) $MgCl_2$; (b) $MgCl_2 \cdot 6H_2O$.
Properties: Colorless or white crystals; deliquescent. Sp. gr. (a) 2.32, (b) 1.56; m.p. (a) 708°C, (b) loses $2H_2O$ at 100°C; b.p. (a) 1412°C, (b) decomposes to oxychloride. Soluble in water and alcohol.
Derivation: By the action of hydrochloric acid on magnesium oxide or hydroxide, especially the latter when precipitated from sea water or brines.
Method of purification: Recrystallization.
Grades: Technical (crystals, fused, flakes, granulated); C.P.
Containers: 100-lb bags; 350-, 575-, 700-lb drums.
Uses: For the electrolytic process of making magnesium metal; magnesium salts; disinfectants; fire extinguishers; fireproofing wood; magnesium oxychloride cement; refrigerating brines; ceramics; cooling drilling tools in drilling for saline deposits and to prevent the dissolution of salts; textiles (size, dressing and filling of cotton and woolen fabrics, thread lubricant; carbonization of wool); paper manufacture; road dustlaying compounds; floor sweeping compounds; flocculating agent and catalyst.
Shipping regulations: None. *

magnesium citrate, dibasic (acid magnesium citrate) $MgHC_6H_5O_7 \cdot 5H_2O$.
Properties: White or slightly yellow, odorless granules or powder. Soluble in water, insoluble in alcohol.
Derivation: Citric acid and magnesium hydroxide or carbonate.
Use: Medicine; dietary supplement.

magnesium cyclamate. A non-nutritive sweetener. See sodium cyclamate.

magnesium dioxide. See magnesium peroxide.

magnesium dust. Finely divided magnesium metal used in pyrotechnics, photographic flash-lights, and chemical preparations.
Shipping regulations: Flammable solid. Yellow label. *

magnesium fluoride (magnesium flux) MgF_2.
Properties: White crystals; exhibits

fluorescence by electric light. Soluble in nitric acid; insoluble in alcohol and water. Sp. gr. 3.0; m. p. 1396°C.
Derivation: By adding sodium fluoride or hydrofluoric acid to a solution of magnesium salt.
Grades: Technical; C. P.
Containers: 1-lb bottles; wooden barrels; multiwall paper sacks.
Uses: Ceramics; glass.
Shipping regulations: None. *

magnesium fluosilicate (magnesium silicofluoride) $MgSiF_6 \cdot 6H_2O$.
Properties: White, efflorescent crystalline powder; sp. gr. 1.788; decomposes on heating. Soluble in water.
Derivation: By treating magnesium hydroxide or carbonate with hydrofluosilicic acid.
Grades: Technical (crystals; solution).
Containers: Crystals: 400-lb barrels; drums. Solution: 400-lb barrels.
Uses: Ceramics; concrete hardeners; waterproofing; mothproofing.
Shipping regulations: None. *

magnesium flux. See magnesium fluoride.

magnesium formate $Mg(CHO_2)_2 \cdot 2H_2O$.
Properties: Colorless crystals; soluble in water; insoluble in alcohol and ether.
Derivation: By the action of formic acid on magnesium oxide.
Grades: Technical.
Containers: Boxes; glass bottles.
Uses: Analytical chemistry; medicine.
Shipping regulations: None. *

magnesium gluconate $Mg(C_6H_{11}O_7)_2 \cdot 2H_2O$.
Properties: Odorless, almost tasteless, white powder or fine needles. Soluble in water.
Derivation: Magnesia or magnesium carbonate dissolved in gluconic acid.
Method of purification: Crystallization.
Grades: Pharmaceutical.
Containers: Cans; fiber drums.
Uses: Medicine; has been proposed for use in dentifrices.
Shipping regulations: None. *

magnesium glycerinophosphate. See magnesium glycerophosphate.

magnesium glycerophosphate (magnesium glycerinophosphate) $MgPO_4 \cdot C_3H_5(OH)_2$.
Properties: Colorless powder; soluble in water; insoluble in alcohol.
Derivation: By the action of glycerophosphoric acid on magnesium hydroxide.
Method of purification: Crystallization.
Grades: Technical.
Containers: 1-, 5-lb bottles; 5-, 10-, 25-lb cans.
Use: Medicine; stabilizer for plastics.
Shipping regulations: None. *

magnesium hydrate. See magnesium hydroxide.

magnesium hydrogen phosphate. See magnesium phosphate, dibasic.

magnesium hydroxide (magnesium hydrate; in aqueous suspension: milk of magnesia; magnesia magma, q. v.) $Mg(OH)_2$.
Properties: White powder or milky liquid. Soluble in solutions of ammonium salts and dilute acids; almost insoluble in water and alcohol; sp. gr. 2.36; m. p. decomposes.
Derivation: By precipitation from a solution of a magnesium salt by sodium hydroxide. It occurs naturally as brucite (q. v.).
Grades: Technical; medicinal; N. F. XI (powder).
Containers: Wooden barrels or drums; glass bottles; carboys.
Uses: Sugar refining; medicine; dentifrices.
Shipping regulations: None. *

magnesium hypophosphite $Mg(H_2PO_2)_2 \cdot 6H_2O$.
Properties: White efflorescent crystals; sp. gr. 1.59; decomposes at 100°C to evolve phosphine. Soluble in water; slightly soluble in alcohol.
Derivation: By the action of hypophosphorous acid on magnesium oxide.
Method of purification: Crystallization.
Grades: Technical.
Containers: Glass bottles.
Use: Medicine.
Shipping regulations: None. *

magnesium hyposulfite. See magnesium thiosulfate.

magnesium iodide (a) MgI_2; (b) $MgI_2 \cdot 8H_2O$.
Properties: White, deliquescent, crystalline powder; discolors in air; soluble in water, alcohol, and ether. M. p. 632°C; sp. gr. (a) 4.48.
Derivation: By heating magnesium in iodine vapors.
Method of purification: Crystallization.
Grades: Technical.
Containers: Glass bottles.
Use: Medicine.
Shipping regulations: None. *

magnesium lactate $Mg(C_3H_5O_3)_2 \cdot 3H_2O$.
Properties: White crystals; very bitter taste; soluble in water; slightly soluble in alcohol.
Derivation: By the action of lactic acid on magnesium oxide, with subsequent crystallization.
Method of purification: Crystallization.
Grades: Technical.
Containers: Glass bottles.
Use: Medicine.
Shipping regulations: None. *

magnesium limestone. See limestone.

magnesium methoxide (magnesium methylate) $(CH_3O)_2Mg$.
Properties: Colorless crystalline solid; decomposes on warming.
Derivation: Reaction of magnesium and methanol.
Uses: Dielectric coating; cross-linking agent; to form stable gels; catalyst.

magnesium methylate. See magnesium methoxide.

magnesium mica. See phlogopite.

magnesium nitrate $Mg(NO_3)_2 \cdot 6H_2O$.
Properties: White crystals; soluble in water and alcohol. Deliquescent. Sp. gr. 1.464,

*See "I.C.C. Shipping Regulations," page xiii.
Reference numbers refer to name of manufacturer. See "List of Manufacturers," page v.

m.p. 95°C.
Derivation: By the action of nitric acid on magnesium oxide, with subsequent crystallization.
Method of purification: Recrystallization.
Grades: Technical; C.P.
Containers: 1, 5-lb bottles; drums.
Use: Pyrotechnics.
Fire hazard: Dangerous; oxidizing material. In contact with organic or other readily oxidizable substances it will cause violent combustion or ignition.
Shipping regulations: Oxidizing material. Yellow label. *

magnesium nitride Mg_3N_2.
Properties: Yellow-green crystals, insoluble in cold water; slightly soluble in alcohol or ether.

magnesium oleate $Mg(C_{18}H_{33}O_2)_2$.
Properties: Yellowish mass; soluble in linseed oil, hydrocarbons, alcohol, and ether; insoluble in water.
Derivation: By the interaction of magnesium chloride and sodium oleate.
Grades: Technical.
Containers: Wooden kegs; steel drums; fiber drums.
Uses: Varnish driers; in dry-cleaning solvents (to prevent spontaneous ignition); emulsifying agent; lubricant for plasticizers.
Shipping regulations: None. *

magnesium oxide (magnesia; periclase; magnesia, heavy; magnesia, light) MgO. The term magnesia is usually reserved for magnesium oxide that has been specially processed, and implies the purer varieties. See Derivation, below. The term periclase applies both to the natural mineral, MgO, and to the product obtained by calcining high purity magnesite ore or sea water magnesia above 1700°C.
Properties: White powder, either light or heavy depending upon whether it is prepared by heating magnesium carbonate or the basic magnesium carbonate; sp. gr. about 3.6 (varies); m.p. 2800°C; insoluble in water and alcohol; soluble in acids and ammonium salt solutions.
Derivation: (a) By calcining magnesium carbonate or basic magnesium carbonate; see also magnesite, caustic-calcined; magnesite, dead-burned; magnesite, synthetic; (b) by treating magnesium chloride with lime and heating, or by heating magnesium chloride in air.
Grades: Technical; C.P.; U.S.P. XVI; 99.5%; fused; low boron; rubber; semiconductor.
Containers: Fiber drums; multiwall paper sacks; tonnage lots.
Uses: High temperature refractories; electrical insulation; pharmaceuticals and cosmetics; rayon; filler, as in rubbers; oxychloride and oxysulfate cements; adsorption and catalysis; food packaging; semi-conductors; uranium processing; in space components, possibly for nosecones.

magnesium oxide, active. See magnesium oxide.

magnesium oxide, heavy. See magnesium oxide.

magnesium oxide, light. See magnesium oxide.

magnesium oxychloride cement (Sorel cement). A mixture of magnesium chloride and magnesium oxide that reacts with water to form a solid mass, presumed to be magnesium oxychloride. Fillers such as wood flour, sawdust, sand, powdered stone, talc, cork, and more recently, powdered metals, are usually present. A variety of proprietary mixtures are available.

magnesium palmitate $Mg(C_{16}H_{31}O_2)_2$.
Properties: Crystalline needles or white lumps; m.p. 121.5°C; insoluble in water and alcohol.
Use: Varnish drier; lubricant for plastics.

magnesium perborate $Mg(BO_3)_2 \cdot 7H_2O$.
Properties: White powder; sparingly soluble in water; decomposes with evolution of oxygen.
Derivation: By the action of peroxide or electrolytic oxidation of borate solutions.
Method of purification: Crystallization.
Containers: Glass bottles; metal cans.
Uses: Driers; bleaching; antiseptic.
Shipping regulations: Oxidizing material. Yellow label.

magnesium perchlorate (a) $Mg(ClO_4)_2$; (b) $Mg(ClO_4)_2 \cdot 6H_2O$.
Properties: White, crystalline, deliquescent; very soluble in water and alcohol; decomposes on heating; explosive when in contact with reducing materials!
Derivation: Magnesium hydroxide and perchloric acid.
Containers: Glass ground stoppered bottles; special non-returnable metal cans.
Use: (a) As a regenerable drying agent for gases.
Shipping regulations: Oxidizing material. Yellow label. *

magnesium permanganate $Mg(MnO_4)_2 \cdot 6H_2O$.
Properties: Bluish-black, friable, deliquescent crystals. Soluble in water.
Use: Medicine (strong antiseptic).

magnesium peroxide (magnesium dioxide) MgO_2.
Properties: White, tasteless, odorless powder. Decomposes above 100°C. Insoluble in water; soluble in dilute acids with formation of hydrogen peroxide. Available oxygen 28.4%. The technical grade is composed of 15% (min) magnesium peroxide and the rest is magnesium hydroxide.
Derivation: From sodium or barium peroxide with magnesium sulfate in a concentrated solution.
Grades: Technical; 15, 25, and 50%.
Containers: 25-, 100-, 200-lb drums.
Uses: Bleaching and oxidizing agent; medicine.
Fire hazard: Dangerous.

*See "I.C.C. Shipping Regulations," page xiii.
Reference numbers refer to name of manufacturer. See "List of Manufacturers," page v.

Shipping regulations: Oxidizing material. Yellow label. *

magnesium phosphate. See magnesium phosphate, dibasic; magnesium phosphate, monobasic; or magnesium phosphate, tribasic.

magnesium phosphate, dibasic (dimagnesium orthophosphate; dimagnesium phosphate; magnesium phosphate, secondary; magnesium hydrogen phosphate) $MgHPO_4 \cdot 3H_2O$.
Properties: White, crystalline powder; decomposes to pyrophosphate on heating. Soluble in dilute acids; slightly soluble in water. Sp. gr. 2.13.
Derivation: By the action of orthophosphoric acid on magnesium oxide.
Grades: Technical; C. P.
Containers: Glass bottles.
Use: Medicine; stabilizer for plastics; food additive.
Shipping regulations: None. *

magnesium phosphate, monobasic (magnesium biphosphate; acid magnesium phosphate; magnesium tetrahydrogen phosphate) $MgH_4(PO_4)_2 \cdot 2H_2O$.
Properties: White, hygroscopic; crystalline powder; decomposes to metaphosphate by heat. Soluble in water and acids; insoluble in alcohol.
Derivation: By the action of orthophosphoric acid on magnesium hydroxide.
Grades: Technical.
Containers: Boxes; barrels; kegs; bags.
Uses: Medicine; fireproofing wood; stabilizer for plastics.
Shipping regulations: None. *

magnesium phosphate, neutral. See magnesium phosphate, tribasic.

magnesium phosphate, secondary. See magnesium phosphate, dibasic.

magnesium phosphate, tribasic (magnesium phosphate, neutral; trimagnesium phosphate) $Mg_3(PO_4)_2 \cdot 8H_2O$ or $5H_2O$.
Properties: Fine, soft, bulky white powder; odorless and tasteless; sp. gr. 2.41; loses all water at 400°C. Soluble in acids; insoluble in water.
Derivation: Reaction of magnesium oxide and phosphoric acid at high temperatures.
Grades: Technical; reagent; N. F. XI.
Containers: Bags; barrels; drums.
Uses: Dentifrice polishing agent; pharmaceutical antacid, adsorbent, stabilizer for plastics; food additive.

magnesium phosphite $MgHPO_3 \cdot 3H_2O$.
Properties: White, crystalline powder; slightly soluble in water.

magnesium pyrophosphate $Mg_2P_2O_7 \cdot 3H_2O$.
Properties: White powder; soluble in acids; insoluble in alcohol and water. Sp. gr. 2.56; loses water at 100°C.

magnesium ricinoleate $Mg(OOCC_{17}H_{32}OH)_2$.
Properties: Coarse, yellow granules with faint fatty acid odor; m. p. 98°C; sp. gr. 1.03 (25/25°C). Used in cosmetics.

magnesium salicylate $Mg(C_7H_5O_3)_2 \cdot 4H_2O$.
Properties: Colorless, efflorescent crystalline powder. Soluble in water and alcohol.
Derivation: By the action of salicylic acid on magnesium hydroxide.
Method of purification: Crystallization.
Grades: Technical.
Containers: Wooden kegs; glass bottles.
Use: Medicine.
Shipping regulations: None. *

magnesium silicate $3MgSiO_3 \cdot 5H_2O$ (variable). See also magnesium trisilicate and serpentine.
Properties: Fine, white powder; sp. gr. 2.6-2.8; insoluble in water or alcohol.
Derivation: By the interaction of a magnesium salt and a soluble silicate.
Grades: Technical; C. P.
Containers: 1-lb bottles; wooden kegs; multiwall paper sacks.
Uses: Medicine; rubber industry; ceramics; glass; refractories; manufacture of permanently dry resins and resinous compositions; paints and varnishes (filler); animal and vegetable oils (bleaching agent); odor absorbent; filter medium; catalyst and catalyst carrier; anticaking agent in foods.
Shipping regulations: None. *

magnesium silicofluoride. See magnesium fluosilicate.

magnesium-sodium sulfate. See sodium-magnesium sulfate.

magnesium stannate $MgSnO_3 \cdot 3H_2O$.
Properties: White crystalline powder. Soluble in water. Approximate temperature of decomposition 340°C.
Use: Additive in ceramic capacitors.

magnesium stearate (dolomol) $Mg(C_{18}H_{35}O_2)_2$. Technical grade contains small amounts of the oleate and 7% magnesium oxide MgO.
Properties: Soft white light powder; sp. gr. 1.028; m. p. 88.5°C (pure), 132°C (technical); tasteless; odorless. Insoluble in water; soluble in hot alcohol.
Grades: Technical; U. S. P. XVI.
Containers: Fiber cans; multiwall paper sacks.
Use: Dusting powder; lubricant in making tablets; drier in paints and varnishes; flatting agent; in medicines; stabilizer and lubricant for plastics; emulsifying agent in cosmetics.

magnesium sulfate (a) $MgSO_4$; (b) (Epsom salts) $MgSO_4 \cdot 7H_2O$.
Properties: Colorless crystals, small usually needlelike; cooling, saline, bitter taste; neutral to litmus; sp. gr. (a) 2.65; (b) 1.678; soluble in glycerol; very soluble in water; sparingly soluble in alcohol.
Derivation: (a, b) By the action of sulfuric acid on magnesium oxide, hydroxide or carbonate; (b) mined in a high degree of purity in State of Washington, U.S.A.
Method of purification: Recrystallization.
Grades: Technical; C. P.; U. S. P. XVI.
Containers: Bottles; boxes; multiwall bags; kegs; barrels; bags; carloads.
Use: Medicinc; leather industry; fireproofing;

textiles (warp-sizing cotton goods, loading cotton goods, weighting silk, dyeing and calico printing); mineral waters; motion-picture snow; ceramics; explosives; matches; fertilizers; paper (sizing); cosmetic lotions; dietary supplement.
Shipping regulations: None. *

magnesium sulfite $MgSO_3 \cdot 6H_2O$.
Properties: White, crystalline powder; slightly soluble in water; insoluble in alcohol. M. p., loses $6H_2O$ at 200°C; b. p. decomposes.
Derivation: By the action of sulfurous acid on magnesium hydroxide.
Method of purification: Crystallization.
Grades: Technical.
Containers: Wooden barrels; glass bottles.
Uses: Medicine; paper pulp.
Shipping regulations: None. *

magnesium tetrahydrogen phosphate. See magnesium phosphate, monobasic.

magnesium thiosulfate (magnesium hyposulfite) $MgS_2O_3 \cdot 6H_2O$.
Properties: Colorless crystals; soluble in water; insoluble in alcohol. Sp. gr. 1.818.
Use: Medicine.

magnesium trisilicate. Approximately $Mg_2Si_3O_8 \cdot 5H_2O$.
Properties: Fine, white, odorless, tasteless powder; free from grittiness. Insoluble in water and alcohol; readily decomposed by mineral acids.
Derivation: Naturally occurring, as talc; by reaction of soluble magnesium salts with soluble silicates.
Grades: Technical; U. S. P. XVI.
Containers: Drums.
Uses: Industrial odor absorbent; decolorizing agent; antioxidant; medicine.

magnesium tungstate (magnesium wolframate) $MgWoO_4$.
Properties: White crystals; soluble in acids; insoluble in water and alcohol.
Derivation: By the interaction of solutions of magnesium sulfate and ammonium tungstate.
Grades: Technical.
Containers: Wooden kegs.
Uses: Fluorescent screens for x-rays; luminescent paint.

magnesium wolframate. See magnesium tungstate.

magnesium zirconium silicate $MgZrSiO_5$.
Properties: White solid; m. p. 3200°F; density 80 lbs/cu ft; insoluble in water, alkalies; slightly soluble in acids.
Containers: 80-lb paper bags; 500-lb drums.
Uses: Electrical resistor ceramics; glaze opacifier.

"Magnesol." [55] Brand name for a synthetic proprietary product, adsorptive magnesium silicate.
Properties: White solid; insoluble in water, mineral and vegetable oils, petroleum and chlorinated hydrocarbon solvents. Bulk density 24-28 g/100 cc.
Grades: Dry-cleaning; industrial.
Containers: 50-lb multiwall paper bags.
Uses: Solvent purification, clarification and recovery; oil refining; deodorizing and decolorizing of a large variety of oils and fats.
Shipping regulations: None. *

magnetic pyrites. See pyrrhotite.

magnetite (lodestone; iron ore, magnetic) Fe_3O_4, often with Ti, Mg.
See also iron oxide, black.
Properties: Black mineral; black streak; submetallic, or dull, to metallic luster. Contains 72.4% iron. Readily recognized by strong attraction by magnet. Soluble in powder form in hydrochloric acid. Decomposes at 1538°C to ferric oxide Fe_2O_3. Sp. gr. 4.9-5.2; hardness 5.5-6.5.
Occurrence: United States (New York, New Jersey, Pennsylvania, Arkansas, California, Washington, Utah); Cuba; Norway; Sweden; Germany; Siberia; Italy; Austria; Switzerland; Japan.
Use: Important ore of iron.

magnetohydrodynamic generator (MHD generator). A power generating device in which a very highly heated ionized gas containing free electrons (a plasma) is passed through a magnetic field in such a way that some of the electrons are caused to flow in an external circuit through electrodes that project into the generator. The plasma may consist of electrically heated air or argon, or hot combustion gases.

"Magnorite." [249] Trademark for fused magnesium oxide refractory products.
Properties: Sp. gr. 3.58; fusion point as high as 4750°F (2620°C) depending on purity; coefficient of thermal expansion 0.0000150; mean specific heat (30-1800°C) 0.29 cal/g/°C.
Grades and Uses: Available in grains of standard mesh sizes, cements, and shapes. Refractory grade for use in cements. Prefired shapes for lining metal melting furnaces and chemical reactor linings. Electrical grade as insulating medium in tubular type heating elements.

"Magon." [169] Trademark for 1-azo-2-hydroxy-3-(2,4-dimethylcarboxanilido)naphthalene-1'-(2-hydroxybenzene), used in the colorimetric determination of magnesium.

"Magron." [233] Vegetation maturant containing magnesium chlorate.

maguey. See cantala.

maize oil. See corn oil.

malacca nut. See semecarpus nut.

malachite (green carbonate of copper) $Cu_2(OH)_2CO_3$, $CuCO_3 \cdot Cu(OH)_2$ or $2CuO \cdot CO_2 \cdot H_2O$. Native hydrated basic copper carbonate.
Properties: Bright-emerald to grass-green color, sometimes nearly black. Pale green streak, silky, adamantine, or dull luster. 71.9% CuO, 19.9% CO_2, balance water.

*See "I.C.C. Shipping Regulations," page xiii.
Reference numbers refer to name of manufacturer. See "List of Manufacturers," page v.

Found with other copper ores often as an alteration product. Soluble in acids. Sp. gr. 3.9-4.03; hardness 3.5-4.

Occurrence: United States (New Jersey, Pennsylvania, Wisconsin, Arizona, New Mexico, Utah, Nevada); Russia; France; England; Germany; Cuba; Chile; Australia; South Africa; West Africa.

Uses: Source of copper; ornamental stoneware (table tops, vases, etc.); pigment (limited; malachite green usually refers to the organic dye).

malachite, artificial. See copper carbonate.

malachite green (benzaldehyde green; victoria green). A triphenylmethane dye, being the zinc double chloride, oxalate, or ferric double chloride of tetramethyl-para-aminotriphenylcarbinol. The term is sometimes applied to the mineral malachite.

Containers: Barrels.

Use: Dyeing.

malachite green toners. See phosphotungstic pigments.

malakin. See salicyl-para-phenetidine.

"Malaphos." [88] Trademark for anti-dusting, wettable powder and emulsifiable solution containing malathion.

malathion $C_{10}H_{19}O_6PS_2$. (O, O-dimethyl dithiophosphate of diethyl mercaptosuccinate). The common name (approved by the Ent. Soc.) of a phosphate insecticide with relatively low mammalian toxicity.

Properties: Yellow, high boiling liquid; (b.p. 156-157°C, under 0.7 mm with slight decomposition); m.p. 2.85°C; refractive index (n 25/D) 1.4985; sp.gr. 1.2315 (25°C); vapor pressure (20°C) approximately 0.00004 mm. Miscible with most polar organic solvents. Practically insoluble in water; stable to light but decomposes when heated to excessively high temperatures. Slow decomposition in the presence of iron, steel, tin plate, and copper. Hydrolyzed in the presence of alkaline materials.

Purity: Technical grade is 95+% pure.

Derivation: From diethyl maleate and dimethyl dithiophosphoric acid.

Containers: 500-lb lined metal drums.

Caution! Harmful if swallowed. Avoid prolonged breathing of dust or spray mist. Avoid prolonged contact with skin; wash thoroughly after using. Avoid contamination of feed and foodstuffs. (U.S. Pesticides Regulations label.) (MCA label similar.)

Uses: General insecticide including control of aphids, spider mites, scales, and house flies as well as a wide range of other sucking and chewing insects.

Malay fishberry. See cocculus.

male fern. See aspidium.

maleic acid (maleinic acid) HOOCCH:CHCOOH (cis isomer).

Properties: Colorless crystals, possessing a characteristic repulsive, astringent taste; faint odor. More toxic than the isomeric fumaric acid. Do not confuse with malic acid! Soluble in water, alcohol; very slightly soluble in benzene; sp. gr. 1.59; m.p. 130-131°C; at temperatures slightly above the m.p., is converted partly to fumaric acid.

Derivation: Catalytic oxidation of benzene; byproduct from manufacture of phthalic anhydride; catalytic oxidation of C_4 hydrocarbons.

Method of purification: Crystallization.

Grades: Technical; reagent.

Uses: Organic synthesis (malic, succinic, aspartic, tartaric, propionic, lactic, malonic, acrylic, hydracrylic acids); in the textile industry as such or in the form of various salts in the dyeing and finishing of cotton, wool and silk; in manufacture of synthetic resins; preservative for oils and fats.

Shipping regulations: None.* Poison label used and recommended by manufacturers.

maleic anhydride (2,5-furandione) HCC(O)OC(O)CH.

Properties: Colorless needles; sp. gr. 0.934 (20/4°C); m.p. 53°C; b.p. 200°C; soluble in acetone, hydrocarbons, ether, chloroform, petroleum ether.

Derivation: By passing a mixture of benzene vapor and air over a vanadium oxide catalyst at about 450°C. Phthalic anhydride or butenes are also used.

Grades: Technical; rods, flakes, lumps, briquettes, and molten.

Containers: 75-, 175-, 200-, 275-lb fiber drums (rod, flake, and lump); 575-lb iron drums (fused); tank cars (molten).

Uses: For polyester resins; alkyd coating resins; agricultural chemicals; paper sizing; drying oils; organic synthesis (malic, succinic, aspartic, tartaric, propionic, lactic, malonic acids); dyeing and finishing of textiles; preservative for oils and fats.

Warning! Causes burns. Avoid contact with eyes, skin or clothing. MCA warning label.

Shipping regulations: None.*

maleic hydrazide (1,2-dihydro-3,6-pyridazinedione) HCC(O)NHNHC(O)CH.

Properties: A solid, decomposing at 260°C; slightly soluble in hot alcohol; more soluble in hot water.

Uses: To kill crabgrass; treatment of tobacco plants; stops sprouting during storage of onions, potatoes, and carrots; as a growth inhibitor; sugar content stabilizer in sugar beets.

maleinic acid. See maleic acid.

maleo-pimaric acid. Reaction product of maleic anhydride and ℓ-pimaric acid; derived from pine gum.

Properties: Crystalline solid; m.p. about 225°C; soluble in most organic solvents; insoluble in water or aliphatic hydrocarbons.

malic acid (common malic acid; hydroxysuccinic acid; apple acid) $COOHCH_2CH(OH)COOH$. Do not confuse

with maleic acid, a highly poisonous derivative.
Properties: Colorless crystals; agreeable sour taste; sp. gr. (dℓ-form) 1.601, (d or ℓ form) 1.595 (20/4°C); m.p. (dℓ) 128°C, (d or ℓ) 100°C; b.p. (dℓ) 150°C (dec), (d or ℓ) 140°C (dec). Very soluble in water and alcohol; slightly soluble in ether.
Derivation: Occurs naturally in unripe apples, gooseberries, cherries, raspberries, tomatoes, mountain-ash berries, and syrup residues. Made synthetically by the catalytic oxidation of benzene to maleic acid which is converted to malic acid by heating with steam under pressure.
Grades: Technical, active and inactive. The natural material is levorotatory, but the synthetic material is inactive.
Containers: Glass jars; fiber cans; drums.
Uses: Medicine; manufacture of various esters and salts; in wine manufacture to age it by removing tartrates; food acidulant.
Shipping regulations: None.*

mallow, marsh. See althea.

malonamide nitrile. See cyanoacetamide.

malonic acid (methanedicarbonic acid) $CH_2(COOH)_2$.
Properties: White crystals; soluble in water, alcohol and ether. M.p. 132-134°C; b.p. decomposes; sp. gr. 1.63.
Derivation: From monochloroacetic acid by action with potassium cyanide, followed by hydrolysis.
Method of purification: Crystallization.
Grades: Technical.
Containers: Tins.
Use: Intermediate for barbiturates and pharmaceuticals.
Shipping regulations: None.*

malonic dinitrile $CH_2(CN)_2$ (propanedinitrile; methylene cyanide).
Properties: Colored crystals. M.p. 32.1°C; b.p. 220°C.
Use: Organic synthesis.

malonic ester. See ethyl malonate.

malonic ethyl ester nitrile. See ethyl cyanoacetate.

malonic methyl ester nitrile. See methyl cyanoacetate.

malonic nitrile. See cyanoacetic acid.

malonylurea. See barbituric acid.

malt. Yellowish or amber-colored grains of barley which have been partially germinated by artificial means. It contains dextrin, maltose and amylolytic enzymes and has an agreeable, characteristic odor and sweet taste. Black malt is grain which has been scorched in the drying process.
Uses: Amber: brewing; medicine (extract of malt). Black: coloring.
Shipping regulations: None.*

maltase (glucase; alpha-glucosidase). An enzyme which hydrolyzes maltose to glucose. Occurs in the small intestine, in yeast, molds and malt; usually associated with the enzyme amylase. Recovered in relatively pure form by treating yeast with toluene, chloroform or ethyl acetate, centrifuging to separate from the undissolved matter, and adding ammonia to avoid decomposition.

malt, extract of. See extract of malt.

maltine. See extract of malt.

maltine, French. Same as diastase, malt.

maltobiose. See maltose.

maltose (malt sugar; maltobiose) $C_{12}H_{22}O_{11} \cdot H_2O$. The most common reducing disaccharide; composed of two molecules of glucose. Found in starch and glycogen.
Properties: Colorless crystals; m.p. 102-103°C; soluble in water; soluble in alcohol; insoluble in ether.
Derivation: By the enzymatic action of diastase (usually obtained from malt extract) on starch.
Containers: Glass bottles; fiber cans; drums.
Uses: Nutrient; sweetener; culture media; stabilizer for polysulfides.
Shipping regulations: None.*

malt sugar. See maltose.

"Mam." [342] Trademark for dimethyl anthranilate. Used in perfumery.

mandarin oil (tangerine oil).
Properties: Essential oil; golden yellow (from mature fruit); olive green (from immature fruit). Characteristic refreshing odor.
Chief known constituents: Limonene; methyl esters of anthranilic and methylanthranilic acids.
Constants: Sp. gr. 0.854-0.859 (15°C); optical rotation +65 to +75°; refractive index 1.475-1.478; acid value up to 1.7; ester value 5 to 11; after acetylation 12.5. Soluble in 7 to 10 vols of 90% alcohol (with some turbidity).
Derivation: Expressed from the fresh peel of the mandarin orange Citrus nobilis, L.
Containers: Drums.
Uses: Flavoring; medicine.
Shipping regulations: None.*

mandelic acid (phenylglycolic acid; alpha-phenylhydroxyacetic acid; benzoglycolic acid; known also as amygdalic acid) $C_6H_5CHOHCOOH$. Exists in stereoisomeric forms. The properties are those of the dℓ-form.
Properties: Large white crystals or powder with a faint odor; sp. gr. 1.30; m.p. 117-119°C. Darkens on exposure to light. Soluble in ether; slightly soluble in water and alcohol.
Derivation: Hydrolysis of the cyanohydrin formed from benzaldehyde, sodium bisulfite, and sodium cyanide. Can be obtained from amygdalin, the glucoside found in almonds.
Containers: Glass bottles; 25-, 100-lb drums.
Use: Medicine; organic synthesis.

Mandelin's reagent. One gram of ammonium vanadate dissolved in 200 cc of conc. sulfuric acid. The reagent gives characteristic color tests with certain alkaloids.

mandelonitrile gentiobioside. See amygdalin.

mandelyl-para-phenetidine. See amygdophenine.

mandrake. See podophyllum.

maneb (MnEBD; manganese ethylenebisdithiocarbamate) $(\text{-SSCNHCH}_2\text{CH}_2\text{NHCSS-})\text{Mn}$. A fungicide obtained by treating a solution of nabam with manganous sulfate.
Properties: Very similar to zineb (q. v.). Brown powder; decomposes on heating.
Derivation: From disodium ethylene bisdithiocarbamate and a manganese salt.
Caution! May cause irritation of eyes, nose, throat, and skin. May be harmful if inhaled or swallowed. MCA warning label.
Use: Fungicide for foliage.

manganese Mn. Element with atomic number 25. A reddish-gray or silvery, brittle metallic element of group VII in the periodic table. Considered essential for plants and animals.
Properties: Softer than iron when pure; sp. gr. 7.2; m. p. 1245°C; b. p. 2097°C; decomposes water; slowly dissolves in dilute acids.
Occurrence: Never found native. Important manganese ores are pyrolusite, manganite, psilomelane, rhodochrosite, rhodonite, wad. Manganiferous iron, silver, and zinc ores are also important (franklinite). A large source is open hearth slags. Principal sources of ores: United States, Russia, India, Brazil, West Africa, Cuba.
Derivation: By reduction of the oxide with aluminum or carbon. Pure manganese is obtained electrolytically from sulfate or chloride solutions.
Grades: Technical; pure or electrolytic; powdered.
Containers: 500-, 900-lb barrels; drums; boxes.
Uses: Purifying and scavenging agent in the production of several metals; iron, copper, chrome-nickel, aluminum alloys, alloy steels; source of manganese chemicals.
Shipping regulations: None. *

manganese acetate $\text{Mn(C}_2\text{H}_3\text{O}_2)_2 \cdot 4\text{H}_2\text{O}$.
Properties: Pale red crystals; very soluble in water and alcohol; sp. gr. 1.59; m. p. 80°C.
Derivation: By the action of acetic acid on manganese hydroxide.
Method of purification: Crystallization.
Grades: Technical; C. P.
Containers: 1-lb bottles; wooden kegs; fiber cans and drums.
Uses: Textile dyeing; manufacturing bistre; catalyst in various chemical processes involving oxidation; leather tanning and finishing; paint and varnish (drier, boiled oil manufacture); fertilizers; food packaging.

manganese-ammonium sulfate (manganous ammonium sulfate) $\text{MnSO}_4 \cdot (\text{NH}_4)_2\text{SO}_4 \cdot 6\text{H}_2\text{O}$.
Properties: Light-red crystals; sp. gr. 1.83; soluble in water.

manganese arsenate. See manganous arsenate.

manganese, battery. See manganese dioxide.

manganese binoxide. See manganese dioxide.

manganese black. See manganese dioxide.

manganese, bog. See wad.

manganese borate MnB_4O_7.
Properties: Reddish-white powder; insoluble in water.
Derivation: By the action of boric acid on manganese hydroxide.
Grades: Technical; C. P.
Containers: 1-lb bottles; wooden kegs; fiber drums.
Use: Varnish and oil drier.
Shipping regulations: None. *

manganese-boron. An alloy of manganese and boron used in the making of brass, bronze and other alloys.

manganese bromide. See manganous bromide.

manganese-bronze. Alloy of 55 to 60% copper, 38 to 42% zinc, up to 3.5% manganese, with or without small amounts of iron, aluminum, tin or lead.

manganese carbonate (manganous carbonate) MnCO_3 (found as rhodocrosite).
Properties: Rose-colored crystals; almost white when precipitated. Soluble in dilute acids; insoluble in water. Sp. gr. 3.125; m. p., decomposes.
Derivation: A precipitate from the addition of sodium carbonate to a solution of a manganese salt followed by filtration, washing and drying.
Grades: Technical; C. P.
Containers: Bottles; bags; drums; carloads.
Uses: Manufacture of manganese salts; medicine; paint pigment; fertilizers; feed additive.
Shipping regulations: None. *

manganese chloride. See manganous chloride.

manganese chromate. See manganous chromate.

manganese citrate (manganous citrate) $\text{Mn}_3(\text{C}_6\text{H}_5\text{O}_7)_2$.
Properties: White powder; soluble in water in presence of sodium citrate.
Derivation: By the action of citric acid on manganese hydroxide.
Method of purification: Crystallization.
Grades: Technical.
Containers: Glass bottles.
Use: Medicine.
Shipping regulations: None. *

manganese dioxide (manganese binoxide; manganese black; battery manganese; manganese peroxide) MnO_2. The natural form is pyrolusite.
Properties: Black crystals or amorphous powder; soluble in hydrochloric acid; insoluble in water. Sp. gr. 5.026; m. p.

decomposes to Mn_2O_3 and oxygen at 535°C.
Derivation: (a) Found in nature as such. See pyrolusite. (b) By heating manganese oxide in a furnace in presence of oxygen. (c) Decomposition of manganous nitrate. (d) By certain electrolysis processes.
Grades: Technical; C. P.
Containers: Bottles; multiwall paper sacks; bags; drums; carloads.
Uses: In general, one of the prime oxidizing agents. For many uses, pyrolusite ore is interchangeable with the synthetic compound. Typical oxidizing uses are oxidation of leuco dye compounds, manufacture of purpurin from alizarin, of benzoic acid and benzaldehyde from toluene, production of hydroquinone. Also used in pyrotechnic mixtures, matches, and match box friction surfaces; dry cells and electrodes; for making other manganese compounds; in glass enamels and glazes as colorant, decolorizer and scavenger; as coloring agent in brick pigments and inks; for paint driers; fertilizer; feed additive; water treatment; rubber compounding; processing of uranium ore; in ferrites; in removal of hydrogen sulfide from gases; in manufacture of electrolytic zinc; in alloy steels, cast irons and wrought irons; medicine.
Shipping regulations: None. *

manganese dioxide, black. See pyrolusite.

manganese ethylenebisdithiocarbamate. See maneb.

manganese fluoride. See manganous fluoride.

manganese gluconate $Mn(C_6H_{11}O_7)_2 \cdot 2H_2O$.
Properties: Light pinkish powder or coarse pink granules. Soluble in water; insoluble in alcohol and benzene.
Method of purification: Crystallization.
Grades: Pharmaceutical.
Containers: Cans; fiber drums.
Uses: Medicinal; feed additive; dietary supplement.
Shipping regulations: None. *

manganese glycerinophosphate. See manganese glycerophosphate.

manganese glycerophosphate (manganese glycerinophosphate) $MnC_3H_7O_3 \cdot PO_3$.
Properties: Yellowish-white or pinkish powder; odorless; nearly tasteless. Soluble in water in presence of citric acid; slightly soluble in water; insoluble in alcohol.
Derivation: By the action of glycerophosphoric acid on manganese hydroxide.
Grades: Technical.
Containers: Glass bottles; boxes.
Use: Medicine (for its glycerophosphate content).
Shipping regulations: None. *

manganese green. See barium manganate.

manganese hydrate. Used as a synonym for manganese hydroxide. See manganic hydroxide; manganous hydroxide.

manganese hydrogen phosphate. See manganous phosphate, acid.

manganese hydroxides. See manganic hydroxide; manganous hydroxide.

manganese hypophosphite $Mn(H_2PO_2)_2 \cdot H_2O$.
Properties: Pink crystals or powder; odorless; tasteless.
Caution! An explosion may occur if manganese hypophosphite is heated or triturated with nitrates, chlorates, or other oxidants. Soluble in water; insoluble in alcohol.
Derivation: Interaction of manganese sulfate and calcium hypophosphite.
Method of purification: Crystallization.
Grades: Technical.
Containers: 1-lb bottles; 5-, 10-, 25-lb cans; 100-lb kegs; drums.
Use: Medicine; dietary supplement.
Shipping regulations: None. *

manganese iodide. See manganous iodide.

manganese lactate $Mn(C_3H_5O_3)_2 \cdot 3H_2O$.
Properties: Pale red crystals; soluble in water and alcohol.
Derivation: By the action of lactic acid on manganese hydroxide.
Grades: Technical.
Containers: Glass bottles.
Use: Medicine.
Shipping regulations: None. *

manganese-lead linoleate. See lead-manganese linoleate.

manganese linoleate $Mn(C_{18}H_{31}O_2)_2$.
Properties: Dark brown, plaster-like mass; soluble in linseed oil.
Derivation: By boiling a manganese salt, sodium linoleate and water.
Grades: Technical.
Containers: Wooden kegs; drums.
Uses: Paint and varnish drier; pharmaceutical preparations.
Shipping regulations: None. *

manganese monoxide. See manganous oxide.

manganese naphthenate.
Properties: Hard, brown, resinous mass. When precipitated in the cold it is a pale buff in color, but darkens immediately on solution. Soluble in mineral spirits, hardens on exposure to air. M. p. (approx) 130-140°C.
Derivation: Precipitation from mixture of soluble manganese salts and aqueous sodium naphthenate solution.
Containers: 50-lb steel drums.
Uses: Paint and varnish drier.
Shipping regulations: None. *

manganese nitrate. See manganous nitrate.

manganese oleate $Mn(C_{18}H_{33}O_2)_2$.
Properties: Brown, granular mass; soluble in oleic acid and ether; insoluble in water.
Derivation: By boiling manganese chloride, sodium oleate and water.
Grades: Technical.
Containers: Wooden kegs.

Uses: Medicine; varnish drier.
Shipping regulations: None. *

manganese oxalate $MnC_2O_4 \cdot 2H_2O$.
Properties: White crystalline powder; soluble in dilute acids; very slightly soluble in water. Sp. gr. 2.453; m. p., decomposes at 150°C.
Derivation: By adding sodium oxalate to manganese chloride.
Grades: Technical.
Containers: Wooden barrels.
Uses: Metallic manganese; paint and varnish drier.
Shipping regulations: None. *

manganese oxides. See manganous oxide; manganic oxide; manganese dioxide. Several other oxides of manganese are known, as manganic-manganous oxide Mn_3O_4; manganese trioxide MnO_3; and manganese heptoxide Mn_2O_7. These have not been found of importance commercially.

manganese peroxide. See manganese dioxide.

manganese phosphate. See manganous orthophosphate. See also manganous phosphate, acid.

manganese protoxide. See manganous oxide.

manganese pyrophosphate. See manganous pyrophosphate.

manganese resinate.
Properties: Dark, brownish-black mass or flesh colored powder. Soluble in hot linseed oil; insoluble in water.
Derivation: By boiling manganese hydroxide, resin oil and water.
Grades: Technical.
Containers: Wooden kegs; drums.
Use: Varnish and oil drier.
Fire hazard: Dangerous.

manganese sesquioxide. See manganic oxide.

manganese silicate. See manganous silicate.

manganese steel. See also ferromanganese.
1. Low-manganese steel. Manganese in steel counteracts brittleness from sulfur; inexpensively increases hardenability. In small quantities also influences grain size; used to make low-carbon, high-strength steels; increases tensile strength; reduces "hot shortness" or "red shortness" caused by sulfur, permitting metal to be hot worked; is also a deoxidizer.
2. Austenitic manganese steel (high-manganese steel; Hadfield's manganese steel). Composition 1.0 to 1.4% carbon, 10 to 14% manganese (usually 13-14%). An extremely tough non-magnetic alloy; high strength, high ductility, excellent resistance to wear; an outstanding material for resisting severe service combining abrasion and heavy impact. Available as castings up to several tons and as sheet, plate, and bar stock. Work hardens readily; has rather low yield strength at first, but deformation hardens it. Generally considered unmachinable, although it is possible to cut it with cemented carbide and cobalt high-speed steel tools. Most finishing is done with grinders; used in well-drill bits, crushers, power-shovel teeth, etc.

manganese sulfate. See manganous sulfate.

manganese sulfite. See manganous sulfite.

manganese tallate. Manganese salts of tall oil fatty acids. Used as a drier.

manganese-titanium.
Composition: Contains manganese, titanium, aluminum, iron, silicon.
Properties: (regular) M. p. 2650°F; (special) m. p. 2430°F.
Uses: (regular) Deoxidizer in high grade steel; (special) non-ferrous alloys deoxidizer.

manganic acetylacetonate $Mn[CH(COCH_3)_2]_3$.
Properties: Brown crystalline solid; m. p. 172°C.
Derivation: Reaction of a manganese salt with acetylacetone and sodium carbonate.

manganic fluoride MnF_3.
Properties: Red crystalline solid; sp. gr. 3.54; decomposed by water and by heat; poisonous!
Use: Fluorinating agent.

manganic hydroxide (manganese hydroxide; hydrated manganic oxide) $Mn(OH)_3$. Rapidly loses water to form $MnO(OH)$.
Properties: A brown powder; sp. gr. 3.258; m. p., decomposes; decomposes in acids; insoluble in water.
Derivation: By the action of oxygen on precipitated manganous hydroxide.
Grades: Technical.
Containers: Wooden barrels.
Uses: Pigment for fabrics; ceramics.
Shipping regulations: None. *

manganic oxide (manganese oxide; manganese sesquioxide) Mn_2O_3. In nature as manganite (q. v.).
Properties: Black, lustrous powder, sometimes tinged brown; very hard; sp. gr. 4.5; soluble in cold hydrochloric acid, hot nitric acid (decomposes), hot sulfuric acid; insoluble in water.

manganic oxide, hydrated. See manganic hydroxide.

"Manganin." [155] Trade name for an alloy of copper, manganese, and nickel.
Properties: Resistivity, 290 ohms per circular mil foot; low thermal emf vs. copper; temperature coefficient of resistance ±0.000015 between 15-35°C; when wound, subject to strains which must be relieved by artificially aging by baking at 250-280°F for 24-48 hours.
Forms: Wire; ribbon; shunt strip.
Uses: Resistors in Wheatstone bridges, decade boxes, potentiometers, etc.; shunts in DC ammeters.

manganite (gray manganese ore) $Mn_2O_3 \cdot H_2O$. Steel-gray to iron-black mineral, reddish-brown to black streak,

submetallic luster. Contains 62.4% manganese, 27.3% oxygen, 10.3% water. Formed in the same deposits as pyrolusite (q.v.), which is frequently an alteration product of manganite. Sp. gr. 4.2-4.4; hardness 4. Insoluble in water; soluble in hot sulfuric acid and hydrochloric acid.
Occurrence: United States (Michigan, Colorado); Germany; Sweden; England; Canada.
Use: An important ore of manganese.

manganosite. See manganous oxide.

manganous-ammonium sulfate. See manganese-ammonium sulfate.

manganous arsenate (manganese arsenate; manganous arsenate, acid) $MnHAsO_4$.
Properties: Reddish-white powder. Hygroscopic; poisonous! Soluble in acids; slightly soluble in water.

manganous arsenate, acid. See manganous arsenate.

manganous bromide (manganese bromide) $MnBr_2 \cdot 4H_2O$.
Properties: Red crystals; very soluble in water; deliquescent.
Derivation: Action of hydrobromic acid with manganese dioxide, manganous carbonate, or manganous hydroxide.

manganous carbonate. See manganese carbonate.

manganous chloride (manganese chloride) (a) $MnCl_2$; (b) $MnCl_2 \cdot 4H_2O$.
Properties: Rose-colored crystals; deliquescent. Sp. gr. (a) 2.478; (b) 1.913; m.p. (a) 650°C; (b) 87.5°C; b.p. (b) 106°C. Very soluble in water; slightly soluble in alcohol; insoluble in ether.
Derivation: By the action of hydrochloric acid on manganese dioxide, with subsequent crystallization.
Method of purification: Recrystallization.
Grades: Technical; C.P.
Containers: Drums.
Uses: Catalyst in the chlorination of organic compounds; paints (drier, manufacture of a brown pigment by reaction with a solution of dichromates); dyeing; pharmaceutical preparations; stimulant in fertilizer compositions; feed additive; dietary supplement.
Shipping regulations: None. *

manganous chromate (manganese chromate; manganous chromate, basic) $2MnO \cdot CrO_3 \cdot 2H_2O$.
Properties: Brown powder; slightly soluble in water with hydrolysis.

manganous chromate, basic. See manganous chromate.

manganous citrate. See manganese citrate.

manganous fluoride (manganese fluoride) MnF_2.
Properties: Reddish powder; soluble in acids; insoluble in water, alcohol, and ether. Sp. gr. 3.98; m.p. 856°C.
Derivation: By the action of hydrofluoric acid on manganous hydroxide.
Grades: Technical.

manganous hydroxide (manganese hydroxide) $Mn(OH)_2$. Occurs naturally as pyrochroite.
Properties: White-pink trigonal crystals; sp. gr. 3.258; hardness 2.5; decomposes with heat; insoluble in water and alkali; soluble in acids and ammonium salts.

manganous iodide (manganese iodide) (a) MnI_2; (b) $MnI_2 \cdot 4H_2O$.
Properties: (a) White deliquescent, crystalline mass; (b) rose crystals; sp. gr. (a) 5.01; m.p. (a) 638°C; b.p. (a) 1061°C; soluble in water with gradual decomposition.
Derivation: By the action of hydriodic acid on manganous hydroxide.

manganous nitrate (manganese nitrate) $Mn(NO_3)_2 \cdot 6H_2O$.
Properties: Pink crystals; very soluble in water, deliquescent; soluble in alcohol. Sp. gr. 1.82; b.p. 129°C; m.p. 26°C.
Grades: Technical; C.P.
Containers: Glass bottles; non-returnable t tins.
Uses: Ceramics; intermediates; catalyst; manganese dioxide.
Shipping regulations: Oxidizing material. Yellow label. *

manganous orthophosphate (manganese phosphate) $Mn_3(PO_4)_2 \cdot 7H_2O$.
Properties: Reddish-white powder; soluble in mineral acids; insoluble in water.
Derivation: By the action of orthophosphoric acid on manganous hydroxide.

manganous oxide (manganese protoxide; manganese monoxide; manganese oxide; manganosite) MnO.
Properties: Grass-green powder; soluble in acids; insoluble in water. Sp. gr. 5.09-5.18; m.p. 1650°C, but converted to Mn_3O_4 if heated in air.
Derivation: (a) By reduction of the dioxide in hydrogen. (b) By heating the carbonate with exclusion of air.
Grade: Technical.
Containers: Wooden barrels; iron drums; multiwall paper sacks.
Uses: Medicine; textile printing; analytical chemistry; catalyst in manufacture of allyl alcohol; ceramics; dry batteries; paints; colored glass; bleaching tallow.
Shipping regulations: None. *

manganous phosphate, acid (manganese hydrogen phosphate; manganese phosphate; manganous phosphate, secondary) $MnHPO_4 \cdot 3H_2O$.
Properties: Pink powder. Contains some tribasic phosphate. Soluble in acids; slightly soluble in water.
Use: Feed additive.

manganous phosphate, secondary. See manganous phosphate, acid.

manganous pyrophosphate (manganese pyrophosphate) (a) $Mn_2P_2O_7$; (b) $Mn_2P_2O_7 \cdot 3H_2O$.
Properties: White, amorphous powder;

*See "I.C.C. Shipping Regulations," page xiii.
Reference numbers refer to name of manufacturer. See "List of Manufacturers," page v.

sp. gr. (a) 3.71. Soluble in solutions of potassium or sodium pyrophosphate; insoluble in water.

manganous silicate (manganese silicate) $MnSiO_3$. Occurs naturally as rhodonite.
Properties: Red crystals or yellowish-red powder. Insoluble in water.
Constants: Sp. gr. 3.72; m.p. 1323°C.
Derivation: By the interaction of manganous salts with sodium silicate.

manganous sulfate (manganese sulfate) $MnSO_4 \cdot 4H_2O$.
Properties: Translucent, pale rose-red, efflorescent prisms. Soluble in water; insoluble in alcohol.
Constants: Sp. gr. 2.107; m.p. 30°C; anhydrous, m.p. 700°C; decomposes at 850°C.
Derivation: Byproduct of production of hydroquinone; or by the action of sulfuric acid on manganous hydroxide or carbonate.
Method of purification: Crystallization.
Grades: Technical; C.P.; fertilizer.
Containers: Bottles; boxes; kegs; 375-lb barrels; 600-lb casks; multiwall paper sacks.
Uses: Fertilizers; paints and varnishes; ceramics; paper; textile dyes; medicines; vitamins; fungicides; feed additive; dietary supplement; ore flotation; catalyst in viscose process; production of other manganese compounds.

manganous sulfite (manganese sulfite; manganous sulfite, normal) $MnSO_3$.
Properties: Grayish-black or brownish-red powder. Soluble in solution of sulfur dioxide; insoluble in water.

manganous sulfite, normal. See manganous sulfite.

mangle. See semecarpus nut.

mangrove.
Derivation: From Rhizophora mucronata.
Habitat: West Africa and Borneo.
Grades: Mangrove cutch: 30-35% tannin; liquid: 25% tannin.
Containers: Wooden barrels.
Use: Tanning industry.
Shipping regulations: None. *

Manila gum copal.
Properties: Generally pebble-like pieces of a pale brownish color. Soluble in ether, methyl alcohol and ethyl alcohol; partially soluble in amyl alcohol; insoluble in water.
Constants: Sp. gr. 1.062; m.p. 230-250°C.
Derivation: A copal resin imported from the Philippine Islands.
Grade: Technical.
Containers: Bags.
Uses: Spirits varnishes; enamel paints.
Shipping regulations: None. *
See also copal.

manila hemp. See abaca.

"Manilyl." [188] Brand name for a replacement for natural ylang ylang oil.

manioc root. See cassava starch.

manjak. See glance pitch.

manna. Solid sweetish exudation of Fraxinus ornus (manna ash). Small round lumps, yellow or grayish.
Constituents: Mannitol and derivatives.
Habitat: Mediterranean basin, Spain to Asia Minor.
Containers: Bags.
Use: Medicine.
Shipping regulations: None. *

manna sugar. See mannitol.

mannite. See mannitol.

mannitol (manna sugar; mannite) $C_6H_8(OH)_6$. Hexahydric alcohol.
Properties: White, crystalline powder, odorless, of faint, sweet taste; nonhygroscopic. Soluble in water; slightly soluble in lower alcohols and amines; almost insoluble in other organic solvents.
Constants: Sp. gr. 1.52; m.p. 165-167°C; specific rotation (20°C) between +23° and +24°; b.p. 290-295°C (at 3-3.5 mm).
Derivation: By hydrogenation of corn sugar or glucose; also by extraction from manna ash, or from seaweed.
Grades: Reagent; commercial; N.F. XI.
Containers: Bottles; fiber drums.
Uses: Base or excipient for tableting; ingredient in electrolytic condensers; basis of dietetic sweets; starting point for many derivatives.
Shipping regulations: None. *

mannitol hexanitrate(hexanitromannite; HNM; nitromannite; nitromannitol) $C_6H_8(ONO_2)_6$.
Properties: Colorless crystals, m.p. 112-113°. Explosive! Soluble in alcohol, acetone, ether, insoluble in water.
Derivation: By nitrating mannitol with mixed acid, purifying by precipitation from organic solvents, and stabilizing.
Grades: Technical; N.N.D.
Containers: Water-tight wooden barrels for wet shipment.
Uses: Explosive cap ingredient; medicine (admixed with a large proportion of carbohydrate).
Fire hazard: Dangerous.
Shipping regulations: Explosive, class A. Initiating explosive label. Not accepted by express. *

D(+)-**mannose** $C_6H_{12}O_6$. A carbohydrate found naturally in some plant polysaccharides.
Properties: Crystals from alcohol or acetic acid; sweet taste with bitter after-taste; m.p. 132°C (dec).
Derivation: By treating vegetable ivory with sulfuric acid; oxidation of mannitol.
Containers: Bottles; drums.
Use: Biochemical research.

"Mansulox." [250] Trademark for an oxygen-bearing manganese sulfur ladle addition used in making free machining steel.

"Man-Tan." [236] Brand name for a proprietary, economical tannin mud thinner containing selected ground mangrove bark.
Container: Multiwall paper bag containing 50 lbs.

*See "I.C.C. Shipping Regulations," page xiii.
Reference numbers refer to name of manufacturer. See "List of Manufacturers," page v.

manure salts. Potash salts containing high percentage of chloride and from 20-30% of potash (K_2O). Used in fertilizer.
See potash-magnesia double salt.

"Manzate." [28] Trademark for agricultural and horticultural fungicide based on maneb (manganese ethylenebisdithiocarbamate) for control of diseases on certain vegetables and fruits.
Containers: 3- and 4½-lb bags; 50-lb fiber drums.
Use: For spray or dust application of tomatoes, celery, potatoes, onions, carrots, and certain other fruits and vegetables.

"Mapharsen." [330] Trademark for oxophenarsine hydrochloride.

maphenide hydrochloride (para-aminomethylbenzenesulfonamide hydrochloride; 4-homosulfanilamide hydrochloride; marfanil) $(SO_2NH_2)C_6H_4CH_2NH_2 \cdot HCl$.
Properties: White crystals; freely soluble in water; m.p. 260-265°C.
Use: Medicine.

"Mapico Black." [133] Trademark for a synthetic magnetite (magnetic oxide of iron), fine and smooth, possessing remarkable coloring power for this type of pigment. Used in paint as a metal protective coating on steel surfaces; used as a tinting color in emulsion and other paints; widely used in cement products such as shingles and cement brick.

"Mapico Brown." [133] Trademark for proprietary iron oxides, much finer, smoother and stronger than natural earth browns; used in paint, cement, stucco, leather finishes, rubber, etc.

"Mapico Red." [133] Trademark for proprietary oxides of iron. Mapico Red #516 Medium and Dark represent the most readily dispersed oxides available. Mapico Reds #297, #347, #387, #477, and #567 developed especially for the rubber industry have excellent aging properties. These reds are also used in paints, lacquers, cement and stucco, leather finishes, roofing granules, asbestos shingles, electronic ferrites, etc.

"Mapico Tan." [133] Trademark for proprietary, heat-resistant tan colors used extensively in roofing granules and baking enamels. Also used in emulsion paints, enamels, leather finishes, and rubber.

"Mapico Yellow." [133] Trademark for proprietary ferric oxide hydrates, possessing from 4-6 times the tinting strength of natural earth yellows and in addition being finer and smoother. Used generally in paints, enamels, and lacquers, cement and stucco, leather finishes, etc.

maple sugar. Impure sucrose with flavor due to impurities characteristic of maple tree sap. Used as flavor.

"MAPO." [293] Trademark for tris[1-(2-methyl)-aziridinyl]phosphine oxide (q. v.).

"Maprofix" HM. [328] Trade name for sodium lauryl sulfate, 40% active powder. Used as a foaming detergent.

"Maprofix Paste." [328] Trademark for sodium lauryl sulfate. Its uses include: detergent, scouring agent, dyeing assistant, and dispersant in the textile industry.

"Maprofix TLS." [328] Trademark for triethanolamine lauryl sulfate. It is used as a detergent or as a foaming agent.

"Maprofix TLS-65." [328] Trade name for triethanolamine lauryl sulfate, 65% active; a paste.

"Maprofix WA; WAQ." [328] These specially refined sodium lauryl sulfates differ solely in their salt content. Both are widely used in liquid cream shampoos.

"Mapromol HSY." [328] Trademark for a product consisting of a blend of raw and sulfonated fatty alcohol, used as a textile softener.

"Maprotex." [328] Brand named product consisting of a sulfated fatty alcohol blend and applied in the textile industry as a continuous boil-off detergent for synthetics.

"MAPS." [293] Trademark for tris[1-(2-methyl)-aziridinyl]phosphine sulfide, $C_9H_{18}N_3PS$. Properties similar to "MAPO" but has lower water solubility.

"Marabond." [121] Trade name for a partially purified calcium lignosulfonate used in oil well cement retarders, foundry supplies, and some ceramic products.

"Maracarb" Chelating Agents. [121] Trade name for complex mixtures of the salts of lower molecular weight lignosulfonic acids and the salts of the alkaline reversion products of hexoses and pentoses which are produced from wood in the sulfite pulping process. Available in liquid and powder form. Used in fertilizers and agricultural chemicals.

"Maracell-E." [121] Brand name for a partially desulfonated sodium lignosulfate, developed for use as an organic agent for internal boiler treatments over a wide range of temperatures and pressure. Claimed to prevent or inhibit scale formation in boiler tubes, injectors, feed lines and economizers.

"Maraniol." [227] Trademark for 4-methyl-7-ethoxycoumarin $C_2H_5OC_6H_3OCOCHCCH_3$.
Properties: White crystals; walnut odor; stable; not known to cause discoloration; congealing point 113.0-114.0°; 1 gram is clearly soluble in 100 ml of 95% alcohol.
Occurence: Not found in nature.
Uses: In fougere and chypre compositions.

marany nut. See semecarpus nut.

"Marasperse." [121] Trademark for a line of lignosulfonates used as dispersants or emulsion stabilizing agents. The basic lignin monomer unit is a substituted phenyl propane. Available in various types for specific uses.

Properties: Brown powder; completely soluble in water, insoluble in oils and most organic solvents.
Containers: 50-lb multi-wall paper bags.
Uses: Dyestuffs; oil well drilling fluids; gypsum board; agricultural chemical formulations; industrial cleaners; carbon black dispersions; ceramics; concrete.

"Maratan." [121] Trade name for a highly purified sodium lignosulfonate retan agent for chrome retanned sole and insole leathers. Available in powder (50-lb paper bags) or liquid form (barrels; tank cars).

Marathon-Howard process. A treatment of waste sulfite liquor from sulfite pulp manufacture to recover chemicals and reduce stream pollution. The waste sulfite is treated with lime and precipitates (1) a calcium sulfite product for use in preparing fresh cooking acid for the sulfite pulp process and (2) a basic calcium salt of lignin sulfonic acid (see lignin sulfonates) which can be pressed and used as a fuel or can be used as raw material for vanillin, lignin plastics, and other chemicals. The remaining liquor, with its biological oxygen demand reduced 80 percent, is dumped into the stream as waste.

"Marax." [299] Trademark for a combination drug containing ephedrine sulfate, theophylline, and hydroxyzine hydrochloride.

marble. A coarse to fine granular or crystalline limestone, generally susceptible of a high polish. Marbles vary considerably in composition, structure and appearance. The term "marble" is rather loosely applied commercially and is sometimes used for ornamental stones which are not limestones.
Uses: Building and ornamental stone; (chips) artificial stone; source of carbon dioxide; neutralization of acids; (flour) abrasives for soaps.
See also whiting.

marcasite (white iron pyrites) FeS_2, as in pyrite. Pale brass-yellow mineral deepening in color after exposure. Metallic luster, streak nearly black. Resembles pyrite but can be distinguished from it by crystalline form, by chemical tests and by whiter color on fresh fracture. Contains 46.6% iron, 53.4% sulfur, sometimes with small amount of arsenic.
Constants: Sp. gr. 4.6-4.9; hardness 6-6.5.
Occurrence: United States (Illinois, Wisconsin, Missouri), Czechoslovakia, Germany, England.
Uses: Ore of iron; sulfuric acid manufacture; jewelry and costume decoration.

"Marcol." [51] Trademark for low viscosity N. F. white oils used in pharmaceutical and cosmetic products. Also used on food packaging machinery, cigar and cigarette machines and on candy and baking machinery where an innocuous lubricant is necessary.

"Marco" Resins. [263] Proprietary products. Unsaturated polyester resins. Several grades. Maximum viscosity 40 poises (at 25°C).
Uses: Low pressure molding and laminating for glass fiber reinforced plastics; potting, casting, and embedding resins.

"Marco" Resins. [421] Trademark for a series of styrenated polyester resins containing additives to suit them to various fabrication methods: hand layup, spray-up, matched die molding, and casting.

"Marcothix" Resins. [421] Trademark for a series of thixotropic forms of "Marco" resins. Their thixotropicity prevents excessive drainage during application to vertical surfaces.

"Marex." [322] Trade name for low viscosity ammonium alginate.
Properties: Cream colored powder; approximately 80 mesh; pH about 6; viscosity (1% by weight) about 85 cps.
Grade: Technical.
Containers: 10-, 50-, 100- and 300-lb drums.
Uses: Stabilizing, moisture controlling, suspending, plasticizing agent.

"Marezine." [301] Trademark for cyclizine hydrochloride and cyclizine lactate. Used in medicine.

marfanil. See maphenide hydrochloride.

margaric acid. See n-heptadecanoic acid.

margarine oils. Edible oils, used in the manufacture of oleomargarine, and containing not over 0.1% free fatty acid.

marialite. See wernerite.

marigold. See calendula.

marihuana. See cannabis.

"Marincate." [123] Trademark for magnesium trisilicate.

"Marinco." [123] Trademark for magnesium hydroxides, oxides and carbonates.

"Marinex." [51] Trademark for lubricating oil containing compounding suiting it for lubrication of the cylinders of multi-stage air compressors.

marjoram oil (calamintha oil).
Properties: Colorless, yellowish or greenish-yellow liquid; strong, penetrating odor. Soluble in alcohol, ether, and chloroform.
Chief known constituents: Terpineol; terpenes.
Constants: Sp. gr. 0.890-0.910; optical rotation +5° to +18°.
Derivation: Distilled from the flowering herb of Origanum marjorana L.
Method of purification: Rectification.
Grade: Technical.
Containers: Copper flasks; glass bottles.
Uses: Medicine; perfuming soaps; toilet preparations; flavors.
Shipping regulations: None.*

marking nut. See semecarpus nut.

*See "I.C.C. Shipping Regulations," page xiii.
Reference numbers refer to name of manufacturer. See "List of Manufacturers," page v.

marl. A natural mixture of clay and calcium or magnesium carbonate.

"Marlate." [28] Trademark for methoxychlor insecticides, supplied in 50% wettable powder and 24% emulsifiable oil formulations.
Use: For control of various insects on food and forage crops, on livestock, and in buildings, especially in dairy operations and elsewhere where a low toxicity agricultural insecticide is required.

"Marlex." [303] Trademark for a complete family of olefin polymers.
Low density polyethylene, used for coatings, films, blow molding.
"Marlex" 1478: Density 0.914 g/cc; melt index 7.8.
"Marlex" 1531: Density 0.915 g/cc; melt index 3.1.
"Marlex" 2030: Density 0.920 g/cc; melt index 3.0.
"Marlex" 2278: Density 0.922 g/cc; melt index 7.8.
"Marlex" 2331: Density 0.923 g/cc; melt index 3.1.
"Marlex" 2380: Density 0.923 g/cc; melt index 8.0.
"Marlex" 2420: Density 0.924 g/cc; melt index 2.0.
"Marlex" 2520: Density 0.925 g/cc; melt index 2.0.
Medium density polyethylene, used for films.
"Marlex" 2950: Density 0.929 g/cc; melt index 5.0.
"Marlex" 3328: Density 0.933 g/cc; melt index 2.8.
High density ethylene copolymer, used for sheets, blow molding, and injection molding.
"Marlex" 5003: Density 0.950 g/cc; melt index 0.3.
"Marlex" 5005: Density 0.950 g/cc; melt index 0.5.
"Marlex" 5012: Density 0.950 g/cc; melt index 1.2.
"Marlex" 5040: Density 0.950 g/cc; melt index 4.0.
"Marlex" 5065: Density 0.950 g/cc; melt index 6.5.
High density polyethylene, used for film, sheets, blow molding, and injection molding.
"Marlex" 6002: Density 0.960 g/cc; melt index 0.2.
"Marlex" 6009: Density 0.960 g/cc; melt index 0.9.
"Marlex" 6015: Density 0.960 g/cc; melt index 1.5.
"Marlex" 6035: Density 0.960 g/cc; melt index 3.5.
"Marlex" 6050: Density 0.960 g/cc; melt index 5.0.
High density tailored resin, used for heavy duty film, wire coatings, extruded pipes, injection molding.
"Marlex" TR-101: Density 0.940 g/cc; melt index 0.2.
"Marlex" TR-201: Density 0.950 g/cc; melt index 0.3.
"Marlex" TR-202: Density 0.950 g/cc; melt index 0.5.
"Marlex" TR-212: Density 0.960 g/cc; melt index 0.3.
"Marlex" TR-213: Density 0.970 g/cc; melt index 0.2.
"Marlex" TR-414: Density 0.955 g/cc; melt index 0.3.
Low density tailored resin, used for wire and cable coatings, injection molding.
"Marlex" TR-603: Density 0.916 g/cc; melt index 1.2.
"Marlex" TR-618: Density 0.915-0.920 g/cc; melt index 6.5-10.5.
"Marlex" TR-623: Density 0.921-0.925 g/cc; melt index 6.5-10.5.
"Marlex" TR-822: Density 0.923 g/cc; melt index 22.1.

"Marmax." [51] Trademark for specially compounded oils for use where wet conditions would wash off straight mineral oil. They emulsify readily with water and are used where excessive condensation exists or where water is unavoidably added.

Marme's reagent. A reagent used for testing for alkaloids. It is made by dissolving 20 parts of potassium iodide and 10 parts of cadmium iodide in 80 parts of water.

"Maroleum." [51] Trademark for a high quality, light colored, marine general purpose lime-base grease. Consistency suitable for application by hand or grease gun.

"Marplan." [190] Trademark for a brand of isocarboxazid (q. v.).

marrubium (horehound; hoarhound).
Derivation: Dried leaves and tops of Marrubium vulgare.
Habitat: Europe, central Asia and United States.
Grades: Technical.
Containers: Bales.
Uses: Medicine; confectionery.
Shipping regulations: None. *

Mars brown. See Mars pigments.

marsh gas. See methane.

marshmallow. See althea.

"Marsilid" Phosphate. [190] Trademark for a brand of iproniazid phosphate (q. v.).

Mars orange. See Mars pigments.

Mars pigments. Five pigments obtained by adding milk of lime to a solution of ferrous sulfate and calcining the precipitate formed. The different shades are obtained according to the temperature at which the calcination is conducted. These pigments are termed respectively Mars yellow, Mars orange, Mars brown, Mars red, and Mars violet. The Mars pigments are characterized by fine hues and great permanence.

Mars red. See Mars pigments. See also iron oxide reds.

Mars violet. See Mars pigments.

*See "I.C.C. Shipping Regulations," page xiii.
Reference numbers refer to name of manufacturer. See "List of Manufacturers," page v.

Mars yellow. See Mars pigments.

martensite. The chief constituent of hardened carbon tool steels. It is a solution of C or Fe_3C in beta-iron, or an exceedingly fine-grained alpha-iron with C or Fe_3C in atomic or molecular dispersion. Carbon content up to 1%. Easily obtained by quenching small bodies of hypereutectoid steel in cold water. More difficult to obtain in low-carbon steels.

Martin's cement. See gypsum cements.

"Marvibond Process." [248] A patented process for laminating decorative and protective "Marvinol" and other vinyls to metallic and non-metallic sheet by a high speed continuous method.

"Marvinol." [248] Trademark for polyvinyl chloride thermoplastic resins.

Properties: White, odorless, tasteless nonhygroscopic powders; resistant to water and corrosive elements; light and heat stable; properties vary with selection of stabilizer, plasticizer, pigment and filler.

Grades: Thirteen types for calendering, molding, extrusion, and coating including high, medium, and low molecular weight resins, plastisol resins and copolymer resins. Also includes a line of rigid, electrical, flexible extrusion and injection molding compounds.

Mary-bud. See calendula.

Maryland pink. See spigelia.

maser. The term is a condensation of the phrase "microwave amplification by stimulated emission of radiation." A maser emits radiation of very specific character as a consequence of resonance between the maser and the radiation it is absorbing, a response analogous to that of a tuning fork to sound waves of resonance frequency. In optical masers a suitable crystal (such as a synthetic ruby) is used to absorb and emit the radiation. The emitted light is of an extremely narrow frequency range, while the incident light may be of a range of frequencies. Radio masers consist of appropriate electronic circuitry and absorb and emit radio waves. Masers not only generate energy of narrow frequency range, but can serve to detect and greatly amplify minute quantities of incident frequencies if properly tuned.

mash. Mixture of malted barley (or other grain) and water used for preparing wort in brewing operations. Also mixture of grain, etc., prepared for fermentation in distilling, e.g., "sour mash whiskey."

"Masonite." [92] Trademark for composition hardboard made by defibrating wood chips into the fibrous state in the presence of steam at high pressure, refining the fibers, forming the fibers into (fibrous) mats, and compressing the mats into dense rigid panels in heated presses. The fiber is waterproofed with an emulsion having a paraffin base.

masonry cement. A group of special cements more workable than Portland cement and more plastic. Used as masonry mortars. Some are similar to waterproofed Portland cement while others are Portland cement mixed with hydrated lime, crushed limestone, diatomaceous earth, or granulated slag. Small additions of calcium stearate, petroleum and highly colloidal clays are sometimes made.

mass. The quantity of matter contained by a body, regardless of its location. Mass is constant, and is distinguished from weight, since the latter is affected by the distance of a body from the earth, i.e., by gravitation.

massecuite. A term applied in the sugar industry to the mixture of sugar and molasses prior to the removal of the molasses.

massicot.

1. Natural lead monoxide, PbO. Contains 92.8% lead. Found in United States (Colorado, Idaho, Nevada, and Virginia).
2. This term was formerly used in metallurgy to designate an oxide of lead corresponding to the same formula as litharge (PbO) but having a different physical state. It is formed by the oxidation of a bath of metallic lead at a temperature of about 345°C so that the oxide formed is not melted; sp. gr. 9.3, m.p. 600°C. If the oxide is melted, it is converted into litharge.

mass number. The number of neutrons and protons in the nucleus of an atom. Thus the mass number of ordinary helium is 4, that of ordinary carbon is 12, of ordinary oxygen 16, and ordinary uranium 238. A given nuclide (kind of nucleus) is characterized by its atomic number, equivalent to the number of protons which gives it its charge and thus determines the kind of element, and the number of neutrons which make up the remainder of its mass. Helium has two protons and two neutrons, mass number 4 and atomic number 2. Protons and neutrons each have very close to unit mass, and since the mass change associated with binding the particles together into the nucleus is also very small, the mass number is always within one-tenth unit of the atomic weight of the nuclide.

"Mastalone." [299] Trademark for an agricultural product containing oxytetracycline hydrochloride, oleandomycin, neomycin sulfate, and prednisolone. Used in veterinary medicine.

masterbatch. A quantity of rubber (usually synthetic) which has had non-rubber components added during the process of synthesis or other early states of the production of the rubber. Common non-rubber components are carbon black and various oils and similar extenders. Use of masterbatches achieves uniform mixtures without the cost of incorporation of additives by milling. Advantageous properties are sometimes achieved.

mastic. Same as mastic gum.

mastic gum (pistachia galls; mastiche; mastix; mastic). Solid resinous exudations of the tree Pistacia lentiscus.

Properties: Moderate yellow to pale greenish yellow, transparent tears with a dusty, glass-like luster; brittle; balsamic odor; turpentine taste. About 90% soluble in alcohol, 97% soluble in ether.
Habitat: Mediterranean islands.
Grades: N. F. XI; technical.
Containers: Bags.
Uses: Medicine; condiment; tooth cements; chewing gum; adhesive; lacquers; medical plasters; incense.
Shipping regulations: None. *

mastiche. See mastic gum.

mastic oil.
Properties: Colorless essential oil; characteristic, strongly balsamic odor. Soluble in 4-10 vols of 90% alcohol; in 0.2-2 vols of 95% alcohol.
Chief known constituent: Pinenes.
Constants: Sp. gr. 0.857-0.903 (15°C); b.p. 155-160°C; optical rotation +22 to +34°; refractive index 1.468-1.476; acid value up to 5; ester value 2.5-19, after acetylation 17-21.
Derivation: By distillation of mastic.
Use: Medicine.
Shipping regulations: None. *

mastix. See mastic gum.

"Mastolyn." [266] Trademark for a pale, hard, tough, phenolic-modified resin.

masurium Ma. Discarded name and symbol for technetium.

mata-perro. See condurango.

"Matawan." [72] Cleaners containing one or more of the usual sodium compounds such as the carbonate, silicate, phosphate, aluminate, borate, or caustic soda, as well as soaps or synthetic detergents and wetting agents.

matte. A product containing a metal sulfide, as obtained after roasting and fusion of sulfide minerals. Oxides or metals may also be present. Common examples are copper matte and nickel matte.

max. Abbreviation for maximum.

"Maxad." [123] Trademark for magnesium oxide for adsorption application.

"Maxade." [108] Trademark for a free-flowing, dust free, white granular dish-machine compound designed for high speed, heavy soil installations in hard or soft water areas.
Containers: 125-, 350-lb drums.

"Maxipen." [299] Trademark for potassium phenethicillin.

may apple. See podophyllum.

May blossom. See convallaria.

Mayer's reagent. See mercuric potassium iodide.

May lily. See convallaria.

"M-B-C" Fumigant. [88] Trademark for a methyl bromide fumigant with chloropicrin added.

MBMC. Abbreviation for monobutyl-meta-cresol. See tert-butyl-meta-cresol.

"3M Brand Fluorochemicals." [158] See "Fluorochemicals." [158]

"3M Brand Paper Chemical FC-805." [158] A paper sizing agent consisting of a chromium complex of a long chain fluorochemical in isopropyl alcohol solution.
Properties: Dark green liquid; sp. gr. 1.005; flash point, 70°F (isopropyl alcohol); soluble in water and short chain aliphatic alcohols, but on drying, becomes water resistant, insoluble and non-volatile.
Containers: 5-gal pail, 15-gal or 55-gal drums.
Uses: Imparts oil and water resistance to paper and paper board.

"3M Brand Perfluoro Carboxylic Acids." [158] See "Perfluoro Carboxylic Acids."

MBT. Abbreviation for mercaptobenzothiazole.

"MBTS." [28] Trade designation for (2-benzothiazolyl disulfide) $(\overline{C_6H_4SCN})_2S_2$.
Properties: Pale yellow powder or grains; sp. gr. 1.54; m.p. 155°C min; insoluble in benzene, ethylene dichloride, acetone, water and gasoline.
Containers: 50-lb bags.
Use: To accelerate and improve the vulcanization of natural and synthetic rubber and latex compounds.

"MC-3." [204] Trademark for a special blend of mild alkalies, complex phosphates, and wetting agents. Especially designed for dairy plant and farm equipment. Sequesters hardest water. Packaged in 5-lb cans, 125-, and 350-lb drums.

MCA. Abbreviation for Manufacturing Chemists' Association, Inc. This organization publishes a booklet of Warning Labels (Manual L-1), frequently revised, which are suggested for use in handling chemicals. The labels do not take the place of those required by law, but have achieved general recognition because of the care with which they have been prepared. The organization also publishes Chemical Safety Data Sheets, which give properties and information on specific chemicals for safe handling and use.

"Mc Namee Clay." [69] Trademark for a proprietary product, a soft kaolin clay.
Properties: White to cream; sp. gr. 2.62 ± .03; fineness (through 325 mesh) 99.7%.
Uses: Filler for rubber.

MCP (2-methyl-4-chlorophenoxyacetic acid) $CH_3ClC_6H_3OCH_2COOH$.
Properties: White crystalline solid;

*See "I.C.C. Shipping Regulations," page xiii.
Reference numbers refer to name of manufacturer. See "List of Manufacturers," page v.

m. p. 118-119°C. Free acid insoluble in water but sodium and amine salts are soluble.
Grades: Emulsifiable concentrates.
Use: Selective herbicide for control of weeds in cereal crops.
Warning! Irritating to eyes, nose, and throat. MCA warning label.

Md. Symbol for mendelevium (q. v.).

MDA. Abbreviation for metal deactivator. See "Ethyl."

"M. D. A." [138] Trade name for methylene disalicylic acid.
Properties: Non-hygroscopic light tan, coarse powder; stable in air (darkens in light); tends to decarboxylate at very high temperatures.
Containers: 100-, 250-lb fiber drums.
Uses: In alkyd resins and modified phenolic compositions for paints and varnishes; an intermediate for dyestuffs and the printing ink industry.

MDAC. Abbreviation for 4-methyl-7-diethylaminocoumarin (q. v.).

MEA. Abbreviation for monoethanolamine. See ethanolamine.

meadow crocus. See colchicum.

meadow green. See copper acetoarsenite.

meadow saffron. See colchicum.

"Mearlite." [270] Trademark for a synthetic nacreous pigment which contains no lead compounds. Mearlite G consists of a bismuth salt in the form of thin, plate-like crystals. Available both in the form of a suspension in conventional liquid vehicles and as a dry powder. See nacreous pigment.
Containers: 1-, 4-lb glass jars; 50-lb steel cans.

"Mearlmaid." [270] Trademark for natural pearl essence. See nacreous pigment.
Containers: 1-lb glass jars; 40-lb aluminum cans.

"Mebaral." [162] Trademark for mephobarbital.

mecamylamine hydrochloride (3-methylaminoisocamphane hydrochloride; N, 2, 3, 3-tetramethyl-2-norcamphanamine hydrochloride)

$CH_2CH_2CHCH_2CHC(CH_3)_2C(CH_3)(NHCH_3)\cdot HCl$.

Properties: White crystalline powder; almost odorless; m. p. 245°C with some decomposition; soluble in water, alcohol, chloroform; somewhat soluble in benzene, isopropyl alcohol; insoluble in ether.
Grade: USP XVI.
Use: Medicine.

Mecca balsam. See balm of Gilead.

Mecca galls. See galls.

mechanical pulp process. See wood pulp.

mechlorethamine hydrochloride [methyl-bis-(2-chloroethyl)amine hydrochloride] $CH_3N(CH_2CH_2Cl)_2\cdot HCl$. A nitrogen mustard.
Properties: White, crystalline, hygroscopic powder; poisonous, a nasal irritant and a vesicant; soluble in water; m. p. 108-111°C.
Grades: U. S. P. XVI.
Use: Medicine.

meclizine hydrochloride $C_{25}H_{27}ClN_2\cdot 2HCl\cdot H_2O$. (1-para-Chlorobenzhydryl-4-methylbenzylpiperazine dihydrochloride).
Properties: White or yellowish powder or crystals; slight odor; insoluble in water and ether; very soluble in chloroform, pyridine, and acid-alcohol-water mixtures; slightly soluble in dilute acids and alcohol.
Grade: U. S. P. XVI.
Use: Medicine.

meconic acid $HC_5O(O)(OH)(COOH)_2\cdot 3H_2O$.
Properties: White crystals; lose water of crystallization at 100°C; slightly soluble in water, alcohol and ether, acetone; gives reddish color with ferric chloride solution containing hydrochloric acid.
Derivation: From opium.
Method of purification: Crystallization.
Grades: Technical.
Containers: Glass bottles.
Use: Medicine.
Shipping regulations: None. *

meconin (opianyl) $C_{10}H_{10}O_4$. White crystals, soluble in hot water, alcohol, chloroform, essential oils. The lactone of meconinic acid; m. p. 102-103°C. A neutral principle of opium, occurring also in hydrastis; may be synthesized from guaiacol or by the oxidation of narcotine.
Shipping regulations: None. *

"Mecostrin." [412] Trademark for dimethyl tubocurarine chloride (q. v.).

MECSA. Abbreviation for (mono)octadecyl carboxymethylmercaptosuccinate.

"Medialan." [307] Trademark for surfactants.
"Medialan" LL-33.
Properties: Pale yellow liquid; soluble in water. Good foaming agent. Composition: 31% sodium N-lauroyl sarcosinate.
Uses: Surfactant with good lathering and cleansing properties suitable for a mild shampoo and other cosmetic preparations.
"Medialan" LL-99. Purified dry form of material described above; 94% active.

"Medi-Calgon." [108] Trademark for a gleaming white, powdered, sodium hexametaphosphate which has the ability to form a firm coagulum with tissue exudates.
Containers: 7-oz bottles; 5-lb cans.
Use: Medicine (topical applications).

Mediterranean squill. See squill.

medium yellow. See chrome yellows.

"Medo-Green." [58] Trademark for silage grade sodium metabisulfite $Na_2S_2O_5$.
Uses: Used in putting up grass silage to prevent bacterial activity, which causes fermentation and decay. Insures a sweet smelling silage of green color and high

*See "I. C. C. Shipping Regulations," page xiii.
Reference numbers refer to name of manufacturer. See "List of Manufacturers," page v.

carotene content and good palatability.
Containers: 100-lb bags.

medroxyprogesterone acetate (17 alpha-hydroxy-6 alpha-methyl progesterone acetate) $C_{24}H_{34}O_4$.
Properties: Crystals; m.p. 205-209°C.
Use: Medicine.

meerschaum (sepiolite) $H_4Mg_2Si_3O_{10}$, with partial replacement of Mg by Cu and Ni. A tough compact, natural hydrous magnesium silicate. Color, white, grayish-white, sometimes yellow or reddish.
Constants: Sp. gr. 2.0; hardness 2-2.5.
Occurrence: Turkey, Spain, Greece, Morocco, Czechoslovakia.
Uses: Tobacco pipes; used as a soap in Algeria and as a building stone in Spain.

mega-. Prefix meaning 10^6 units (symbol M). E.g., 1 megagram = 1,000,000 grams.

megass. See bagasse.

megilp. A linseed oil-mastic varnish mixture. Used as an ingredient of artists' oil paints.

meionite. See wernerite.

MEK. See methyl ethyl ketone.

"Mekomask." [188] Brand name for an industrial odorant especially effective when used with methyl ethyl ketone.

melaconite. See tenorite.

melamine (cyanuramide; 2,4,6-triamino-sym-triazine) $H_2N\underline{CNC(NH_2)NC(NH_2)N}$.
A cyclic trimer of cyanamide.
Properties: Pure white, monoclinic crystals. Sparingly soluble in water, glycol, glycerol, pyridine; very slightly soluble in ethanol; insoluble in ether, benzene, carbon tetrachloride. Sp. gr. 1.573 (250°C); m.p. 250°C.
Derivation: From cyanamide, dicyandiamide, or cyanuric chloride.
Method of purification: Recrystallization from water.
Grade: 99% minimum.
Containers: 80-lb bags; fiber drums.
Uses: Synthetic resins, organic syntheses.
Shipping regulations: None. *

melamine resins. Synthetic resins of the thermosetting type, made from melamine and formaldehyde. These resins are very versatile and widely used. The first step in resin formation is the production of trimethylol melamine, $C_3N_3(NHCH_2OH)_3$, the molecules of which contain a ring with 3 carbon and 3 nitrogen atoms, and have the $-NHCH_2OH$ groups attached to the carbon atoms of the ring. This molecule can combine further with others of the same kind through splitting-off of water from the hydrogen atom attached to nitrogen and the OH group of another molecule. If present, excess formaldehyde or excess melamine can also react with the trimethylol melamine or its polymers, so that there are endless possibilities of chain growth and cross-linking. The nature and degree of polymerization depends upon pH, but heat is always needed for curing. The lower molecular weight, uncured melamine resins are water soluble syrups. The higher molecular weight materials are less soluble or insoluble and are usually available as powders. These are often easily dispersible in water. The resins are very widely used as molding compounds with alpha-cellulose, wood flour or mineral powders as fillers, and with coloring materials also incorporated. Typical molding techniques are used to produce items such as utensils, containers, dishes, coffee makers, and items such as buttons, handles, lamp pedestals, fuse and switch boxes and various industrial and household objects. The resins are also used for laminating, for boilproof adhesives, for increasing wet strength of paper, and for textile treatment to achieve crease and wrinkle resistance, etc., and in leather processing.

Typical properties of molded objects are sp. gr. 1.5; tensile strength 7500 psi; compressive strength 45,000 psi; mold shrinkage 0.008; continuous heat resistance 210 to 230°F; heat distortion temperature 400°F; dielectric strength 320 volts per mil at 25°C; water absorption 0.1 to 0.6% in 24 hours. The finished resins do not discolor easily and are resistant to weathering, handling, scratching, and effects of ordinary water solutions.

Butylated melamine resins are formed by incorporating butyl or other alcohols during resin formation, whereupon the $NHCH_2OH$ groups become converted to $NHCH_2OC_4H_9$. These resins are soluble in paint and enamel solvents and lead to uses of melamines in surface coatings, often in combination with alkyds. These melamine surface coating resins give exceptional curing speed, hardness, wear resistance, and resistance to solvents, soaps, and foods.

Melamine-acrylic resins are water soluble and are used for formation of water-base industrial and automotive finishes.

Melamine Resins. [57] Typical examples of melamine resins:
No. 245-8: Solids 50%; solvent, butanol-xylol; color (Gardner) 1 max; viscosity (Gardner-Holdt) L-O at 25°C; hydrocarbon solvent tolerance 200; acid number, solid resin 1 max; approximately 8.3 lbs/gal.
No. 243-3: Solids 60%; solvent, petroleum spirits; color (Gardner) 1 max; viscosity (Gardner-Holdt) V-Y at 25°C; hydrocarbon solvent tolerance 300; acid number, solid resin 1 max; approximately 8.5 lbs/gal.
No. 248-8: Solids 55%; solvent, butanol-xylol; color (Gardner) 1 max; viscosity (Gardner-Holdt) N-Q at 25°C; hydrocarbon solvent tolerance 175; acid number, solid resin 1 max; approximately 8.4 lbs/gal.
No. 247-10: Solids 60%; solvent, butanol; color (Gardner) 1 max; viscosity

(Gardner-Holdt) T-W at 25°C; hydrocarbon solvent tolerance 1000; acid number, solid resin 1 max; approximately 8.4 lbs/gal.
Uses: See melamine resins.

melampyrite. See dulcitol.

melaniline. See diphenylguanidine.

melanin. A brownish-black pigment that occurs normally in the retina, skin, and hair of higher animals with the exception of albinos. Formed from tyrosine by the action of tyrosinase.

"Melaqua." [57] Trademark for water-soluble melamine acrylic coating resins.

"Meleine." [78] Trademark for a plasticized aliphatic triamide with surface tension reducer used for inhibiting against fume fading on acetate dyed fabrics.

meletin. See quercetin.

"Melhi." [266] Trademark for a dark, brittle, polymerized resin supplied in three types:
Type 2, softening point (Hercules drop method) 115-120°C.
Type 3, softening point (Hercules drop method) 115-120°C.
Type 4, softening point (Hercules drop method) 130°C.
Uses: Printing inks; adhesives; core oils; box toes; coal-tar emulsions; rubber compounding; varnishes.

"Melilotin." [19] Trademark for benzodihydropyrone (q.v.).

melin. See rutin.

melissic acid (triacontanoic acid)
$CH_3(CH_2)_{28}COOH$. A long-chain fatty acid.
Properties: Hard crystalline solid; m.p. 94°C; soluble in benzene and hot alcohol; insoluble in water.
Derivation: By oxidation of 1-triacontanol; occurs in minor amounts in many plant and insect waxes and in montan wax.
Use: Biochemical research.

melissyl alcohol. See 1-triacontanol and 1-hentriacontanol. The term melissyl alcohol, which has been used for both by various authorities, should be dropped.

melitose. See raffinose.

melitriose. See raffinose.

"Mellene." [188] $CH_3C_5H_4(O)(OH)$. Trademark for 2-hydroxy-3 methyl-2-cyclopentene-1-one. Maple licorice flavor. Used in flavors and perfumes to lend a soft character.

"Mellotone." [188] $CH_3C_5H_2O(O)(OH)$. Trademark for 3-hydroxy-2-methyl-1,4-pyrone. Caramel-butterscotch flavor; used in perfumes and flavors for a soft, malt-like effect.

"Melmac." [57] Trademark for certain products molded from melamine-formaldehyde resins.

"Melonal." [227] Trademark for 2,6-dimethyl hepten-2-al-7, $(CH_3)_2CCH(CH_2)_2CHCH_3CHO$; minimum 85% pure.
Properties: Yellow liquid; moderately stable, but not likely to cause discoloration. Sp. gr. (25/25°C) 0.845-0.855; refractive index (20°C) 1.441 to 1.447. Clearly soluble in 2 parts of 70% alcohol.
Used for its melon odor.

"Melostrength" Resin. [57] Trademark for a melamine-formaldehyde paper resin designed to improve wet and dry strength properties of paper. It is a pulp additive.

"Melsan." [28] Trademark for fungicide based on ethyl mercury phosphate and sodium pentachlorophenate.
Containers: 150-lb drums.
Uses: For prevention of surface molds and blue-stain organisms on freshly sawn lumber.

melting point. The temperature at which solid and liquid forms of a substance are in equilibrium. In common usage the melting point is taken as the temperature at which liquid forms as a small sample has its temperature increased gradually. A pure substance will melt sharply and completely over a narrow temperature range, while a mixture will melt gradually over a wide temperature range. Various special methods are required to obtain melting points for special purposes.

"Meltopax." [337] Trade name for a proprietary sodium zirconium silicate with approximately 54% ZrO_2, 26% SiO_2, 14% Na_2O. A white-cream powder with sp. gr. 3.9; bulk density 72 lb/cu ft; m.p. 2600°F; average particle size 44 microns max; insoluble in water and alkalies; slightly soluble in dilute mineral acids and hot concentrated sulfuric acid; soluble in hydrofluoric acid. Used in enamel frits, ceramic colors, special glasses.
Containers: 80-lb paper bags; 550-lb barrels; 30,000-lb carloads.

"Melurac" Resins. [57] Trademark for urea-melamine-formaldehyde condensation products used mainly as adhesives for bonding of veneers for the production of exterior grade plywood or for the assembly of wooden structures for outdoor use.

"Mema." [150] Trademark for a liquid seed disinfectant which contains 11.4% methoxyethylmercury acetate.
Containers: 1-gal jugs.
Uses: Controls seed-borne diseases of wheat, barley, and oats, cotton, flax and sorghum. Protects against seed rot and seedling blight.

MENA. Abbreviation for the methyl ester of naphthaleneacetic acid. See alpha-naphthaleneacetic acid, methyl ester.

menaccanite. See ilmenite.

menadiol sodium diphosphate (tetrasodium 2-methyl-1,4-naphthalenediol diphosphate)

$C_{10}H_5CH_3(OPO_3Na_2)_2 \cdot 6H_2O$.
Properties: White to pink powder with characteristic odor; hygroscopic; solutions neutral or slightly alkaline to litmus. Very soluble in water; insoluble in alcohol.
Grade: USP XVI.
Use: Medicine.

menadione (2-methyl-1, 4-naphthoquinone; menaphthone; vitamin K_3) $C_{10}H_5CH_3O_2$.
Properties: Yellow, crystalline powder; nearly odorless; m. p. 105-107°C; affected by sunlight. Soluble in alcohol, benzene, and vegetable oils; moderately soluble in chloroform and carbon tetrachloride; insoluble in water.
Derivation: Oxidation of beta-methylnaphthalene.
Grades: U. S. P. XVI.
Containers: Dark glass vials or bottles; fiber cans.
Use: Medicine; fungicides.
Caution: Menadione powder is irritating to the respiratory tract and to the skin. An alcohol solution has vesicant properties.
Shipping regulations: None. *

menadione sodium bisulfite
$C_{11}H_8O_2 \cdot NaHSO_3 \cdot 3H_2O$. A vitamin K derivative.
Properties: White, crystalline, odorless, hygroscopic powder; sensitve to light; soluble in water; slightly soluble in alcohol; almost insoluble in ether and benzene.
Derivation: Reaction of sodium bisulfite with menadione.
Grade: U. S. P. XVI; technical.
Use: Medicine.
Shipping regulations: None. *

menaphthone. See menadione.

mendelevium (element 101) Md. Synthetic radioactive element produced in a cyclotron by bombarding einsteinium with alpha particles. The element is named for D. I. Mendeleev. At the time of discovery, only seventeen atoms were prepared and identified. Mendelevium decays by spontaneous fission with a half-life of about a half hour. It is believed to have chemical properties similar to those of the rare earth thulium.

menhaden oil (pogy oil; mossbunker oil).
Properties: A yellowish-brown or reddish-brown liquid; characteristic odor. Soluble in ether, benzene, naphtha, and carbon disulfide. Sp. gr. 0.927-0.933; saponification value 191-196; iodine value 139-180; refractive index 1.480.
Derivation: By cooking or pressing the body of the menhaden (mossbunker) fish. Winter oils are made by chilling which separates stearin.
Method of purification: Filtration and bleaching with fuller's earth.
Grades: Prime crude; brown strained; strained; bleached; winter oil; bleached winter white oil. Also sometimes graded: A, extra pale; B, pale; C, brown; D, dark brown.
Containers: Drums; tank cars.
Uses: Leather dressing; substitute for linseed oil in making patent leather; chamois tanning; making fats by hydrogenation; soap making, after hydrogenation; tempering steel; adulterating codliver oil; printing and lithographic inks (linseed oil substitute in paints and linoleum).
Shipping regulations: None. *

dℓ-para-**mentha-1, 8-diene.** See dipentene.

menthanediamine
$(CH_3)_2C(NH_2)\underline{CHCH_2CH_2C(CH_3)(NH_2)CH_2}CH_2$.
A primary alicyclic diamine; also a tert-alkylamine; a low viscosity liquid.
Containers: Drums; tank cars.
Uses: Curing agent for epoxy resins; chemical intermediate.

para-**menthane hydroperoxide.**
Properties: A clear, pale yellow liquid; sp. gr. 0.910-0.925 (15.5/4°C); refractive index 1.460-1.475 (20°C).
Use: Catalyst for rubber and polymerization reactions; coatings.
Shipping regulations: Oxidizing material. Yellow label. *

menthol (hexahydrothymol; methylhydroxyisopropylcyclohexane; peppermint camphor) $CH_3C_6H_9(C_3H_7)OH$.
Properties: White crystals with strong minty-cooling odor and taste; constants will vary according to grade; m. p. from 32.5 to 43°C; congealing temperatures from 27 to 41°C; soluble in alcohol, ether, chloroform, light petroleum solvents, glacial acetic acid, liquid petrolatum and fixed or volatile oils; slightly soluble in water.
Derivation: By freezing from peppermint oil; by synthesis from citronellal and by other syntheses.
Grades: Technical; U. S. P. XVI; Brazilian, Japanese (levo, from peppermint oil; levo or racemic, synthetically).
Containers: Glass bottles; tins; lined fiber cans, cases.
Uses: Medicine; perfumery; confectionery; liqueurs.
Shipping regulations: None. *

menthol acetic ester. See menthyl acetate.

menthol valerate (menthyl isovalerate) $(CH_3)_2CHCH_2COOC_{10}H_{19}$.
Properties: Colorless liquid; mild pleasant odor; cooling, faintly bitter taste; sp. gr. 0.907 (15/4°C); insoluble in water; soluble in alcohol, chloroform, ether, and oils.
Derivation: By the action of valeric acid on menthol.
Grades: Technical.
Containers: Glass bottles.
Use: Medicine.

menthone $C_{10}H_{18}O$.
Properties: Colorless, oily, mobile liquid; slight peppermint odor; slightly soluble in water; soluble in organic solvents. Sp. gr. 0.897 (15°C); b. p. 207°C.
Derivation: A ketone found in oil of peppermint.
Containers: Glass bottles; tins.
Shipping regulations: None. *

menthyl acetate (menthol acetic ester) $C_{10}H_{19}OOCCH_3$.
Properties: Colorless liquid. Menthol-like odor. Slightly soluble in water; miscible with alcohol and ether. B. p. 227 to 228°C; sp. gr. 0.922-0.927; optical rotation −72° 47' to −73° 18'; refractive index 1.447.
Derivation: (a) By boiling menthol with acetic anhydride in the presence of sodium acetate. (b) Found in peppermint oil.
Grades: Technical.
Containers: Glass bottles.
Use: Perfumery; flavoring.

menthyl ethoxyacetate. See menthyl ethylglycolate.

menthyl ethylglycolate (menthyl ethoxyacetate) $C_{10}H_{19}OCOCH_2OC_2H_5$.
Properties: Colorless oily liquid, odorless. Decomposed by alkalies with release of menthol. Less irritating to mucous membranes than menthol. Soluble in alcohol and ether; slightly soluble in water.
Containers: Glass bottles.
Use: Medicine.
Shipping regulations: None. *

menthyl isovalerate. See menthol valerate.

menthyl salicylate $C_6H_4(OH)COOC_{10}H_{19}$.
Properties: Colorless liquid; miscible with alcohol, ether, chloroform, and fatty oils in all proportions. Insoluble in water; soluble in organic solvents.
Containers: Glass bottles.
Use: Medicine; sunscreen preparations.

homo-**menthyl salicylate.** A homolog of menthyl salicylate.
Properties: Light yellow almost odorless oil, neutral and non-irritating to the skin. Absorbs to a great extent the rays in sunlight causing skin burning (about 2940 to 3200 A). Insoluble in water; soluble in alcohol, chloroform and ether.
Containers: Glass bottles; iron drums.
Uses: Ultraviolet filter for anti-sunburn creams and oils; analgesic properties for relieving sunburn.
Shipping regulations: None. *

"Mentor." [51] Trademark for non-viscous oils of the mineral seal type, used mainly as absorption media in vapor recovery systems, such as those recovering coal-tar solvents and casing-head gasoline.

"Mentor" Beads. [86] Trademark for an alkylarylsulfonate detergent.
Uses: Bleaching assistant; detergent; dyeing and carbonizing assistant; emulsifying; kier-boiling; wetting; dispensant; wool scouring; acid fulling.

MEP. Abbreviation for methyl ethyl pyridine.

mepazine acetate (10[(1-methyl-3-piperidyl)-methyl]phenothiazine acetate) $C_{19}H_{22}N_2S \cdot CH_3COOH$.
Grade: N. N. D.
Use: Medicine.

mepazine hydrochloride (10-[(1-methyl-3-piperidyl)methyl]phenothiazine hydrochloride) $C_{19}H_{22}N_2S \cdot HCl \cdot H_2O$.
Properties: Crystals; photosensitive; slightly bitter taste. Very slightly soluble in water; soluble in absolute ethanol, chloroform; practically insoluble in ether, benzene.
Grade: N. N. D.
Use: Medicine.

mepenzolate methylbromide $(C_6H_5)_2C(OH)COOC_5H_9NCH_3 \cdot CH_3Br$. 1-Methyl-3-piperidyl benzilate methylbromide.
Grades: N. N. D.
Use: Medicine.

meperidine hydrochloride (isonipecaine hydrochloride; pethidine hydrochloride) $C_{15}H_{21}O_2N \cdot HCl$. Ethyl-1-methyl-4-phenylpiperidine-4-carboxylate hydrochloride.
Properties: Fine, white, odorless, crystalline powder; m. p. 186-189°C; stable in air; very soluble in water; soluble in alcohol; sparingly soluble in ether; aqueous solutions acid to litmus.
Grade: U. S. P. XVI.
Use: Medicine.

mephenesin (3-ortho-toloxy-1,2-propanediol; ortho-cresyl-alpha-glyceryl ether) $CH_3C_6H_4OCH_2CH(OH)CH_2OH$.
Properties: Crystalline white powder with faint odor and bitter taste; m. p. 70-73.5°C. Freely soluble in alcohol, chloroform and ether; sparingly soluble in benzene and water. pH (saturated solution) about 6.
Grade: N. F. XI.
Use: Medicine.

mephenesin carbamate (3-ortho-toloxy-1,2-propanediol-1-carbamate) $CH_3C_6H_4OCH_2CH(OH)CH_2OCONH_2$.
Properties: White crystals; m. p. 93°C; hemihydrate m.p. 80-84°C; sparingly soluble in water; slightly soluble in chloroform; freely soluble in alcohol.
Grade: N. N. D.
Use: Medicine.

mephenoxalone ("Trepidone"; 5-(ortho-methoxyphenoxymethyl)-2-oxazolidinone) $C_{11}H_{13}NO_4$.
Properties: M. p. 139-141°; insoluble in water.
Use: Medicine.

mephentermine (N, alpha, alpha-trimethylphenethylamine) $C_6H_5CH_2C(CH_3)_2NHCH_3$.
Properties: Clear, colorless to pale yellow liquid with a fishy odor; very soluble in alcohol; practically insoluble in water.
Grade: N. N. D.
Use: Medicine.

mephentermine sulfate (N-alpha, alpha-trimethylphenethylamine sulfate) $(C_{11}H_{17}N)_2 \cdot H_2SO_4 \cdot 2H_2O$.
Properties: White, odorless crystals or crystalline powder; solutions are acid to litmus. Soluble in water; slightly soluble in alcohol; practically insoluble in chloroform.
Grade: U. S. P. XVI.
Containers: Bottles.
Use: Medicine.

mephobarbital (5-ethyl-1-methyl-5-phenylbarbituric acid) $C_{13}N_{14}N_2O_3$.
Properties: White, tasteless, odorless crystals; m.p. 177-181°C. Soluble in chloroform; slightly soluble in alcohol and ether; very slightly soluble in water. Dissolves in fixed alkali hydroxides and carbonates.
Grades: U.S.P. XVI.
Use: Medicine.

meprobamate (2-methyl-2-n-propyl-1,3-propanediol dicarbamate) $H_2NCOOCH_2C(CH_3)(C_3H_7)CH_2OOCNH_2$.
Properties: White powder; m.p. 103-107°C; characteristic odor and bitter taste; slightly soluble in water and ether; soluble in alcohol and acetone.
Grade: U.S.P. XVI.
Use: Medicine.

meprylcaine hydrochloride (2-methyl-2-propylaminopropyl benzoate hydrochloride) $C_6H_5COOCH_2C(CH_3)_2NHCH_2CH_2CH_3 \cdot HCl$.
Properties: White, odorless, crystals; m.p. 150-152°C; soluble in water, alcohol, chloroform; slightly soluble in acetone.
Grade: N.F. XI.
Use: Medicine.

"MER/29." Chemical name: 1-[para-(beta-diethylaminoethoxy)phenyl]-1-(para-tolyl)-2-(para-chlorophenyl)ethanol. Stated to be a cholesterol inhibitor.

"Merac." 204 Trademark for a rubber latex accelerator.
Properties: Dark brown liquid; sp. gr. (20/20°C) 1.025-1.035; viscosity, 9.78 cps at 25°C; flash point 185°F; freezing point −25°C (max); soluble in water, methanol, acetone, and ethyl acetate; insoluble in ethyl ether, benzene, and hexane.
Use: Water-soluble rubber latex accelerator.

meralluride. Methoxyhydroxymercuripropylsuccinylurea ($C_9H_{16}HgN_2O_6$) and theophylline ($C_7H_8N_4O_2 \cdot H_2O$) in approximately molecular proportions.
Properties: White to slightly yellow powder; affected slowly by light; soluble in glacial acetic acid and solutions of alkali hydroxides; slightly soluble in water; moderately soluble in hot water. Saturated solution is acid to litmus.
Grade: U.S.P. XVI.
Use: Medicine.

"Merantine." 232 Brand name for a series of Brilliant acid dyestuffs of good fastness to washing and moderate fastness to light.

merbromin (dibromohydroxymercurifluorescein disodium salt; 2,7-disodiumdibromo-4-hydroxymercurifluoroscein) $C_{20}H_8Br_2HgNa_2O_6$.
Properties: Iridescent, green scales or granules; odorless; soluble in water; insoluble in alcohol, acetone, chloroform, or ether.
Derivation: From dibromofluorescein and mercuric acetate.
Grades: Technical; N.F. XI.
Containers: 25-, 100-lb drums.
Use: Medicine.

"Mercadium." 266 Trademark for group of insoluble cadmium, mercury sulfide pigments. Brilliant, light-fast, chemically stable colors ranging in shade through orange, red, and maroon.
Uses: Plastics; rubber; protective coatings; and printing inks.

mercaptamine. See cysteamine.

mercaptans. A group of organic compounds resembling alcohols but having the oxygen of the hydroxyl group replaced by sulfur, as, ethyl mercaptan, C_2H_5SH. They have a particularly strong disagreeable odor.

mercaptoacetic acid. See thioglycolic acid.

beta-**mercaptoalanine.** See cystein.

2-mercaptobenzoic acid. See thiosalicylic acid.

mercaptobenzothiazole (MBT)
CHCHCHCHCCSC(SH)N.
Properties: Yellowish powder. Slight odor (depends on degree of purification). Nontoxic. Soluble in dilute caustic, alcohol, acetone, benzene, chloroform; insoluble in water and gasoline. Sp. gr. 1.42; m.p. 179°C.
Grades: Technical.
Containers: 1-, 5- and 10-lb fiber cans; 55-lb drums; multiwall paper sacks.
Uses: One of the most widely used accelerators for rubber vulcanization; improves resistance of rubber to oxidation and abrasion; generally used in tire treads and carcasses, mechanical specialties, etc. Also used as a fungicide and as a corrosion inhibitor.

mercaptoethanol $HSCH_2CH_2OH$.
Properties: Water-white, mobile liquid with characteristic odor; sp. gr. 1.1168 (20/20°C); 9.29 lbs/gal (20°C); b.p. 157.1°C (760 mm); vapor pressure 1.0 mm (20°C); viscosity 3.43 cps (20°C); completely soluble in water, benzene, ether, and most organic solvents; flash point 165°F; possesses a sulfhydryl group and a hydroxyl group, thus sharing the chemical reactivity of a mercaptan and an alcohol. Has the ability to add certain types of unsaturated compounds to form stable hydroxyethyl sulfides. F.p., sets to a glass below −100°C; refractive index (n 20/D) 1.5011.
Containers: 1-, 5-, and 10-lb fiber cans; 55-lb drums; multiwall paper sacks.
Uses: Used as a solvent for dyestuffs; as an intermediate for producing dyestuffs, pharmaceuticals, rubber chemicals, flotation agents, insecticides, plasticizers, textile assistants and other compounds; as a water-soluble reducing agent; as a non-nitrogeneous sulfhydryl reagent in the investigation of proteins.
Shipping regulations: None.*

beta-**mercaptoethylamine hydrochloride** $HS(CH_2)_2NH_2 \cdot HCl$.

Properties: Hygroscopic white powder; m. p. 69°C; very soluble in water and ethyl alcohol.

2-mercapto-4-hydroxypyrimidine. See thiouracil.

mercaptomerin sodium $C_{16}H_{25}HgNNa_2O_6S$. Disodium N-[3-(carboxymethylmercaptomercuri)-2-methoxypropyl]-alpha-camphoramate.

Properties: Hygroscopic white powder or amorphous solid; freely soluble in water; soluble in alcohol; slightly soluble in ether and chloroform; practically insoluble in benzene.
Grade: U. S. P. XVI.
Use: Medicine.

beta-**mercaptopropionic acid** $HSCH_2CH_2CO_2H$.
Properties: M. p. 16.8°C; b. p. 110.5-111.5°C (15 mm).

6-mercaptopurine (6-MP) $C_5H_4N_4S$. A sulfur-containing purine base not found in animal nucleoproteins.
Properties: Yellow, crystalline powder; m. p. 308°C (decomposes); nearly odorless; insoluble in water, acetone, ether; soluble in hot alcohol and dilute alkali solutions; slightly soluble in dilute H_2SO_4.
Grade: U. S. P. XVI.
Use: Medicine.

mercaptosuccinic acid. See thiomalic acid.

2-mercaptothiazoline C_3H_4NSSH. Creamy-white crystals.
Use: Synthesis of pharmaceuticals.

mercerized cotton. Cotton which has been passed through a cold bath of sodium hydroxide and afterwards washed with hot and cold water. The process causes a shrinking of the fiber with an increased attraction for coloring matter and imparts a luster to the fiber.

mercerizing assistants. Compounds used to increase the penetration of mercerizing baths. Cresylic acid compounds and derivatives, special sulfonated oils and other wetting agents are typical materials used.

"Mercote." [123] Trademark for vitamin products.

mercuric acetate (mercury acetate) $Hg(C_2H_3O_2)_2$.
Properties: White, crystalline powder; poisonous! Soluble in alcohol and water; sensitive to light. Sp. gr. 3.2544.
Method of purification: Crystallization.
Grades: Technical; C. P.
Containers: Glass bottles; 25-lb jars.
Uses: Medicine; catalyst in organic synthesis.
Shipping regulations: Poison, class B. Poison label. *

mercuric-ammonium chloride (mercury-ammonium chloride) $HgCl_2 \cdot 2NH_4Cl \cdot 2H_2O$.
Properties: White powder; soluble in water; slightly soluble in alcohol. Poisonous!
Use: Medicine.
Shipping regulations: Poison, class B. Poison label. *
See also mercury, ammoniated.

mercuric arsanilate. See mercury atoxylate.

mercuric arsenate (mercury arsenate) $HgHAsO_4$.
Properties: Yellow powder; poisonous! Soluble in hydrochloric acid; slightly soluble in nitric acid; insoluble in water.
Grades: Technical.
Uses: Medicine; waterproof paints; antifouling paints.
Shipping regulations: Poison, class B. Poison label. *

mercuric-barium bromide (barium-mercury bromide; mercury-barium bromide) $HgBr_2 \cdot BaBr_2$.
Properties: Colorless, crystalline mass. Very hygroscopic. Poisonous! Soluble in water.
Shipping regulations: Poison, class B. Poison label. *

mercuric-barium iodide (barium-mercury iodide; mercury-barium iodide) $HgI_2 \cdot BaI_2 \cdot 5H_2O$.
Properties: Reddish or yellow, crystalline mass; unstable; deliquescent. Poisonous! Soluble in alcohol and water.
Grades: Technical.
Uses: Micro-analysis (testing for alkaloids); preparing Rohrbach's solution.
Shipping regulations: Poison, class B. Poison label. *

mercuric benzoate (mercury benzoate) $Hg(C_7H_5O_2)_2 \cdot H_2O$.
Properties: White crystals; poisonous; m. p. 165°C; sensitive to light. Soluble in solutions of sodium chloride and ammonium benzoate; slightly soluble in alcohol and water.
Derivation: By the interaction of a mercuric salt and sodium benzoate.
Grades: Technical.
Containers: Glass bottles.
Use: Medicine.
Shipping regulations: Class B poison. Poison label. *

mercuric biniodide. See mercuric iodide.

mercuric bromide (mercury bromide) $HgBr_2$.
Properties: White, rhombic crystals; poisonous! Sensitive to light. Soluble in alcohol and ether; sparingly soluble in water. Sp. gr. 5.74; m.p. 235°C; b.p. 322°C.
Derivation: By adding potassium bromide to a solution of a mercuric salt and crystallizing.
Method of purification: Recrystallization.
Grades: Technical; C. P.
Containers: Glass bottles.
Use: Medicine.
Shipping regulations: Class B poison. Poison label. *

mercuric chloride (corrosive sublimate; mercury bichloride; mercury chloride, corrosive) $HgCl_2$.
Properties: White crystals or powder;

*See "I. C. C. Shipping Regulations," page xiii.
Reference numbers refer to name of manufacturer. See "List of Manufacturers," page v.

very poisonous! May be fatal if swallowed. Do not breathe dust. Soluble in water, alcohol, ether, pyridine, glycerine and acetic acid ester. Sp. gr. 5.32; m. p. 265°C; b. p. 303°C.
Derivation: (a) Direct combination of chlorine with mercury heated to volatilizing point; (b) by subliming mercuric sulfate with common salt.
Method of purification: Recrystallization and sublimation.
Impurities: Mercurous chloride.
Grades: Technical; lump; crystals; granular; powder; C. P.; N. F. XI.
Containers: Glass bottles; 25-lb boxes; 200- to 250-lb kegs; multiwall paper sacks, drums.
Uses: Manufacture of calomel and other mercury compounds; organic synthesis (catalyst in polyvinyl chloride production and in brominating); analytical reagent; medicine; metallurgy (metal coating, electroplating aluminum, manufacture of tin and zinc alloys of fine structure, bronzing steel); tanning; fungicide, insecticide and wood preservative; embalming fluids; textile printing; dry batteries; photography; process engraving and lithography.
Danger: May be fatal if swallowed. MCA warning label.
Shipping regulations: Class B poison. Poison label. *

mercuric chloride, ammoniated. See mercury, ammoniated.

mercuric-cuprous iodide (copper-mercury iodide; mercury-copper iodide) $HgI_2 \cdot 2CuI$.
Properties: Dark red, crystalline powder; sp. gr. 6.12; poisonous! Insoluble in alcohol and water.
Grades: Technical.
Use: Thermoscopy (detecting overheating of machine bearings) by reversible color change.
Shipping regulations: Poison, class B. Poison label. *

mercuric cyanate. See mercury fulminate.

mercuric cyanide (mercury cyanide) $Hg(CN)_2$.
Properties: Colorless, transparent prisms, darkened by light; poisonous! Soluble in water and alcohol. Sp. gr. 4.018; m. p. decomposes.
Derivation: By the interaction of mercuric oxide and an aqueous solution of hydrocyanic acid.
Method of purification: Crystallization.
Grades: Technical; C. P.
Containers: Amber glass bottles; wooden kegs; fiber drums.
Uses: Medicine; germicidal soaps; manufacturing cyanogen gas; photography.
Shipping regulations: Class B poison. Poison label. *

mercuric dichromate. See mercury dichromate.

mercuric dioxysulfate. See mercuric subsulfate.

mercuric fluoride (mercury fluoride) HgF_2.
Properties: Transparent crystals; sp. gr. 4.00; m. p., decomposes; moderately soluble in water and alcohol; poisonous!
Derivation: Mercuric oxide and hydrofluoric acid.
Uses: Synthesis of organic fluorine compounds.
Shipping regulations: Poison, class B. Poison label. *

mercuric iodate (mercury iodate) $Hg(IO_3)_2$.
Properties: Amorphous, white powder; poisonous! Soluble in hydrochloric acid, hydrobromic acid, hydriodic acid, water (containing sodium chloride or potassium iodide); insoluble in alcohol and water.
Containers: Glass bottles; fiber cans.
Use: Medicine.
Shipping regulations: Poison, class B. Poison label. *

mercuric iodide (mercuric biniodide; mercury biniodide; mercury iodide, red; mercury iodide, yellow) HgI_2. Poisonous!
Properties: (a) Red, tetragonal crystals; turn yellow when heated to 150°C, returning to red on cooling; (b) Yellow, rhombic crystals. Soluble in boiling alcohol, and in solutions of sodium thiosulfate or potassium iodide or other hot alkali chloride solutions; almost completely insoluble in water. Sp. gr. (a) 6.2-6.35, (b) 5.91-6.06; m. p. (a) 241-257°C, (b) 241°C; b. p. (a) 349°C, (b) 349°C.
Derivation: (a) By the direct union of mercury and iodine. (b) As a precipitate by adding potassium iodide to a solution of a mercuric salt. Yellow form precipitates from alcoholic solutions.
Grades: Technical; reagent; N. F. XI.
Containers: Bottles, drums.
Use: Medicine; analytical reagents (Nessler's reagent; Mayer's reagent).
Shipping regulations: Solid and solution: poison, Class B. Poison label. *
See also mercuric potassium iodide.

mercuric lactate $Hg(C_3H_5O_3)_2$.
Properties: White crystalline powder; soluble in water; decomposed by heat.
Use: Medicine.
Shipping regulations: Poison, Class B. Poison label. *

mercuric nitrate (mercury nitrate; mercury pernitrate) $Hg(NO_3)_2 \cdot H_2O$.
Properties: Colorless crystals or white deliquescent powder; poisonous! Sp. gr. 4.3; m. p. 79°C; decomposed by heat. Soluble in water and nitric acid; insoluble in alcohol.
Derivation: By the action of hot nitric acid on mercury.
Method of purification: Crystallization.
Impurities: Mercurous nitrate.
Grades: Technical; C. P.
Containers: 1-, 5-lb bottles; 10-, 25-lb jars; 100-lb kegs.
Uses: Nitration of aromatic organic compounds; medicine; felt manufacture; manufacture of mercury fulminate.

Fire hazard: Dangerous.
Shipping regulations: Oxidizing material. Yellow label and poison label. *

mercuric nitrate ointment. See citrine ointment.

mercuric oleate (mercury oleate)
Properties: Yellowish to red liquid, semisolid or solid mass; poisonous! Slightly soluble in fixed oils; insoluble in water.
Derivation: By mixing yellow mercuric oxide with oleic acid.
Grades: Technical; N. F. XI.
Containers: Glass bottles.
Uses: Medicine; antiseptic; antifouling paints.
Shipping regulations: Poison, Class B. Poison label. *

mercuric oxide, red (red precipitate; mercury oxide, red) HgO.
Properties: Heavy, bright, orange-red powder; very poisonous! Soluble in acids; soluble in dilute hydrochloric and nitric acids; slightly soluble in water, more so after boiling; insoluble in alcohol and ether. Sp. gr. 11.00-11.29 (11.21 for finely ground); m. p. decomposes.
Derivation: By heating mercurous nitrate.
Grades: Technical; C. P.
Containers: 1-lb bottles; 1-, 5-, 25-lb boxes; 200-lb kegs, 50-lb drums.
Uses: Chemicals (oxidizing agent, mercury salts, desulfurization of organic compounds); paint pigment; perfumery and cosmetics; pharmaceuticals; ceramics (pigment); batteries; polishing compounds; analytical reagent; antifouling paints.
Shipping regulations: Class B poison. Poison label.*

mercuric oxide, yellow (mercury oxide, yellow; yellow precipitate) HgO. Differs from red mercuric oxide in the fineness of its particles.
Properties: Light, amorphous, orange-yellow powder; odorless; stable in air but turns dark on exposure to light; finer powder than the red form; sp. gr. 11.03 (27.5°C); m. p., decomposes. Slightly soluble in cold water, more so after boiling; soluble in dilute hydrochloric and nitric acids, potassium iodide solution, concentrated solutions of alkaline-earth chloride, magnesium chloride; insoluble in alcohol.
Derivation: (a) By the action of either potassium hydroxide or sodium hydroxide on mercuric chloride. (b) By the action of sodium carbonate upon mercuric nitrate solution.
Grades: C. P.; technical; N. F XI.
Containers: Bottles; 25-, 50-, 100-lb drums.
Use: Medicine; see mercuric oxide, red.
Shipping regulations: Poison, Class B. Poison label.*

mercuric oxycyanide $HgO \cdot Hg(CN)_2$.
Properties: White crystalline powder; sp. gr. 4.44; explodes on heating; poisonous! Moderately soluble in water.
Use: Medicine.
Shipping regulations: Class B poison. Poison label. *

mercuric phosphate (normal mercuric phosphate; neutral mercuric phosphate; trimercuric orthophosphate; mercuric phosphate, tertiary; mercury phosphate) $Hg_3(PO_4)_2$.
Properties: Heavy, white or yellowish powder, poisonous. Soluble in acids; insoluble in alcohol, water.
Use: Medicine.
Shipping regulations: Poison, Class B. Poison label. *

mercuric phosphate, neutral. See mercuric phosphate.

mercuric phosphate, normal. See mercuric phosphate.

mercuric phosphate, tertiary. See mercuric phosphate.

mercuric-potassium cyanide (mercury-potassium cyanide) $Hg(CN)_2 \cdot 2KCN$.
Properties: Colorless crystals; very poisonous! Soluble in water and alcohol.
Derivation: By mixing mercuric and potassium cyanides and crystallizing.
Method of purification: Recrystallization.
Grades: Technical.
Containers: Wooden kegs; fiber drums.
Use: Silvering glass in mirror manufacture.
Shipping regulations: Class B poison. Poison label. *

mercuric-potassium iodide (Mayer's reagent; potassium mercuric iodide) See also Nessler's reagent. K_2HgI_4 or $2KI \cdot HgI_2$.
Properties: Odorless, yellow crystals, deliquescent in air. Crystallizes with either 1, 2, or 3 molecules of water. The commercial product is the anhydrous form containing about 25.5% mercury. Sp. gr. 4.29. Neutral or alkaline to litmus. Very soluble in water; soluble in alcohol, ether and acetone.
Derivation: (a) By evaporating Nessler's reagent. (b) By the action of hydrochloric acid and potassium iodide on mercuric cyanide or mercuric chloride. (c) By the action of potassium iodide on mercuric oxide.
Grades: C. P.
Containers: 1-lb and 5-lb bottles.
Uses: Medicine; chemical analysis.
Shipping regulations: Class B poison. Poison label. *

mercuric salicylate (salicylated mercury).
Properties: White powder, yellow or pink tinge. A compound of mercury and salicylic acid of somewhat varying composition, mercury replacing both phenolic and carboxylic hydrogen. Contains not less than 54% nor more than 59.5% mercury. Odorless; tasteless; poisonous! Soluble in solutions of the fixed alkalies or their carbonates, and in warm solutions of the alkali halides; almost insoluble in water and alcohol.
Derivation: By gently heating freshly precipitated yellow mercuric oxide and

salicylic acid in presence of water.
Containers: Ampules, capsules; 1-lb bottles.
Use: Medicine.
Shipping regulations: Class B poison. Poison label. *

mercuric-silver iodide (mercury-silver iodide; silver-mercury iodide) $HgI_2 \cdot 2AgI$.
Properties: Yellow powder; becomes red at 40-50°C. Sp. gr. 6.08; soluble in solutions of potassium cyanide or potassium iodide; insoluble in acids (dilute), and water.
Grades: Technical.
Containers: Glass bottles; fiber cans.
Use: Thermoscopy (detecting overheating in journal bearings).
Shipping regulations: Poison, class B. Poison label. *

mercuric stearate (mercury stearate) $(C_{17}H_{35}CO_2)_2Hg$.
Properties: Yellow, granular powder; soluble in fatty oils; slightly soluble in alcohol.
Uses: Medicine; germicide.
Shipping regulations: Poison, class B. Poison label. *

mercuric subsulfate (basic mercuric sulfate; mercuric dioxysulfate; turpeth mineral; turbith mineral) $Hg(HgO)_2SO_4$.
Properties: Heavy, lemon-yellow powder or bright-yellow scales. Poisonous! Turns red and brown on heating, yellow on cooling. Volatile at red heat, decomposing into mercury, mercurous sulfate, oxygen and sulfur dioxide. Soluble in sulfuric acid, dilute hydrochloric acid, dilute nitric acid, acetic acid; very slightly soluble in water, more so in hot water. Sp. gr. 6.444.
Derivation: Addition of water to normal mercuric sulfate.
Containers: Glass bottles; fiber cans.
Use: Medicine.
Shipping regulations: Class B poison. Poison label. *

mercuric succinimide. See hydrargol.

mercuric sulfate (mercury persulfate; mercury sulfate) $HgSO_4$.
Properties: White, crystalline powder; poisonous! Soluble in acids; insoluble in alcohol. Sp. gr. 6.466; m. p., decomposes at red heat.
Derivation: By the action of sulfuric acid on mercury, with subsequent crystallization.
Method of purification: Recrystallization.
Grades: Technical; C. P.
Containers: 1-, 5-lb bottles; 25-lb jars; 100-lb kegs; fiber drums.
Uses: Medicine; producing calomel and corrosive sublimate; catalyst in the conversion of acetylene to acetaldehyde; extracting gold and silver from roasted pyrites; galvanic batteries.
Shipping regulations: Class B poison. Poison label. *

mercuric sulfate, basic. See mercuric subsulfate.

mercuric sulfide, black (ethiops mineral; mercury sulfide, black) HgS.
Properties: Black powder; poisonous! Soluble in sodium sulfide solution; insoluble in water, alcohol, and nitric acid. Sp. gr. 7.55-7.70; m. p., sublimes at 446°C.
Derivation: By passing hydrogen sulfide gas into a solution of a mercury salt or the reaction of mercury with sulfur.
Grades: Technical; C. P.
Containers: 1-lb bottles; wooden kegs; fiber drums.
Use: Pigment for coloring horn.
Shipping regulations: Poison, class B. Poison label. *

mercuric sulfide, red (vermilion; quicksilver vermilion; Chinese vermilion; red mercury sulfide; artificial cinnabar; red mercury sulfuret) HgS.
Properties: Fine, bright scarlet powder; poisonous! Insoluble in water and alcohol. Sp. gr. 8.06-8.12; m. p., sublimes at 446°C.
Derivation: By heating mercury and sulfur, with subsequent recovery by sublimation. A precipitated form is known as English vermilion (q. v.).
Method of purification: Resublimation.,
Grades: Technical; C. P.
Containers: Wooden kegs; glass bottles; fiber drums.
Uses: Medicine; paint pigment; rubber pigment; plastics (pigment); coloring sealing wax.
Shipping regulations: Poison, class B. Poison label. *

mercuric sulfocyanate. See mercuric thiocyanate.

mercuric sulfocyanide. See mercuric thiocyanate.

mercuric thiocyanate (mercuric sulfocyanate; mercuric sulfocyanide; mercury sulfocyanate; mercury thiocyanate) $Hg(SCN)_2$.
Properties: White powder; poisonous! Soluble in alcohol; slightly soluble in water. M. p. decomposes.
Derivation: By precipitation of mercuric nitrate with ammonium sulfocyanate and subsequent solution in a large amount of hot water and crystallizing.
Method of purification: Recrystallization.
Grades: Technical.
Containers: Glass bottles.
Uses: Photography; producing "Pharoah's serpents."
Shipping regulations: Class B poison. Poison label. *

"Mercurochrome." 348 Trademark for merbromin (q. v.).
Use: As a general antiseptic, most commonly in the form of a 2% aqueous solution. It is relatively non-irritating and non-toxic.

mercurol (mercury nucleate).
Composition: Contains 20% mercury. Poison!
Properties: Brown powder. Soluble in water; insoluble in alcohol.
Use: Medicine.

*See "I.C.C. Shipping Regulations," page xiii.
Reference numbers refer to name of manufacturer. See "List of Manufacturers," page v.

Shipping regulations: Poison, class B. Poison label. *

mercurophylline. Mixture of sodium salt of N-[3-(hydroxymercuri)-2-methoxypropyl]-camphoramic acid and theophylline in approximately molecular proportions.
Properties: White or slightly yellow, odorless powder; somewhat hygroscopic; slowly darkens in light; solutions are alkaline to litmus; soluble in water and in alcohol; insoluble in ether and mineral oils.
Grade: N. F. XI.
Use: Medicine.

mercurous acetate (mercury proto-acetate; mercury acetate) $HgC_2H_3O_2$.
Properties: Colorless scales or plates. Decomposed by boiling water and by light into mercury and mercuric acetate. Poisonous! Slightly soluble in water; insoluble in alcohol and ether; soluble in dilute nitric acid.
Derivation: Reaction of sodium acetate with mercurous nitrate solution acidified with nitric acid.
Grades: Technical.
Shipping regulations: Poison, class B. Poison label. *

mercurous arsenite (mercury arsenite).
Properties: Brown powder; variable composition. Unstable. Soluble in nitric acid; insoluble in water. Poisonous!

mercurous bromide (mercury bromide) HgBr or Hg_2Br_2.
Properties: White powder or colorless crystals. Odorless, tasteless. Becomes yellow on heating, returning to white on cooling. Darkens on exposure to light. Soluble in fuming nitric acid (prolonged heating), hot concentrated sulfuric acid, hot ammonium carbonate or ammonium succinate solutions; sparingly soluble in water; insoluble in alcohol and ether.
Constants: Sp. gr. 7.307; sublimes at 340-350°C; m. p. 405°C.
Derivation: (a) Action of potassium bromide on solution of mercurous nitrate in dilute nitric acid. (b) Sublimation from mixture of mercury and mercuric bromide. (c) Cooling a hot solution of the salt in mercurous nitrate.
Containers: Glass bottles; fiber cans.
Use: Medicine.
Shipping regulations: Class B poison. Poison label. *

mercurous chlorate (mercury chlorate) $Hg_2(ClO_3)_2$.
Properties: White crystals. Explodes with combustible substances. Caution! Keep away from light! Sp. gr. 6.409; m. p. 250°C (decomposes). Soluble in alcohol and water.
Shipping regulations: Oxidizing material. Yellow label. Poison, class B. Poison label. *

mercurous chloride (mercury monochloride; mercury protochloride; mercury chloride, mild; calomel) Hg_2Cl_2.
Properties: White, rhombic crystals or crystalline powder; non-poisonous. Odorless, stable in air, but darkens on exposure to light. For the natural product, see calomel, native. Insoluble in water, ether, alcohol and cold dilute acids. Sp. gr. 6.993; m. p. 302°C; b. p. 384°C. Decomposed by alkalies. Poisonous if taken internally in quantity.
Derivation: By heating mercuric chloride and mercury, with subsequent sublimation.
Method of purification: Sublimation.
Impurities: Mercuric chloride.
Grades: Technical; C. P.; N. F. XI.
Containers: Bottles; 25-, 50-, 100-lb drums.
Uses: Medicine; pyrotechnics; fungicide.
Shipping regulations: None. *

mercurous chromate (mercury chromate) Hg_2CrO_4.
Properties: Brick-red powder. Variable composition. Decomposes on heating. Soluble in nitric acid (conc.); insoluble in alcohol and water.
Grades: Technical.
Use: Ceramics (coloring green).

mercurous gluconate.
Shipping regulations: Poison, class B. Poison label. *

mercurous iodide (mercury protoiodide) HgI or Hg_2I_2.
Properties: Bright yellow, amorphous powder, becoming greenish on exposure to light due to decomposition into metallic mercury and mercuric iodide. Becomes dark yellow, orange and orange-red on heating. Undergoes same color change in opposite order on cooling. Odorless and tasteless. Soluble in castor oil, liquid ammonia, aqua ammonia; insoluble in water, alcohol, and ether. Sp. gr. 7.6445-7.75; sublimes at 110-120°C; m. p. 290°C (with partial decomposition).
Derivation: (a) Action of potassium iodide on a mercurous salt. (b) Boiling a solution of mercurous nitrate containing nitric acid with excess of iodine.
Grades: Technical.
Containers: Bottles; jars; kegs; drums.
Use: Medicine (external).
Shipping regulations: Class B poison. Poison label. *

mercurous nitrate, ammoniated (Hahnemann's soluble mercury; black precipitate; ammoniated mercury nitrate) Composition uncertain.
Properties: Black to grayish-black powder; poisonous! Sensitive to light. Soluble in acids; insoluble in water and alcohol.
Derivation: By adding ammonium hydroxide to a solution of mercuric nitrate.
Grades: Technical.
Containers: Dark amber glass bottles.
Use: Medicine.
Shipping regulations: Class B poison. Poison label. *

mercurous nitrate, hydrated $HgNO_3 \cdot H_2O$.
Properties: Short prismatic crystals;

effloresces and becomes anhydrous in dry air. Sensitive to light. Soluble in small quantities of warm water (hydrolyzes in larger quantities), water acidified with nitric acid, boiling carbon disulfide, methylamine; slightly soluble in benzonitrile; insoluble in liquid ammonia. Sp. gr. 4.785 (3.9°C); m.p. 70°C, decomposes.
Derivation: By the action of cold dilute nitric acid upon an excess of mercury and warming slightly.
Grades: Technical; C.P.
Containers: Bottles; jars; kegs; fiber drums.
Uses: Medicine; cosmetics; analytical agent.
Shipping regulations: Class B poison. Poison label. *

mercurous oxide Hg_2O.
Properties: Black powder; sp. gr. 9.8; decomposes at 100°C; soluble in acids; insoluble in water.
Derivation: Action of sodium hydroxide on mercurous nitrate.
Shipping regulations: Poison, class B. Poison label. *

mercurous phosphate (mercurous phosphate, neutral; mercurous phosphate, normal; mercury phosphate; mercurous phosphate tertiary; trimercurous orthophosphate) Hg_3PO_4.
Properties: Heavy, white powder; variable composition. Sensitive to light. Soluble in nitric acid; insoluble in alcohol, phosphoric acid, and water.
Use: Medicine.

mercurous phosphate, neutral. See mercurous phosphate.

mercurous phosphate, normal. See mercurous phosphate.

mercurous phosphate, tertiary. See mercurous phosphate.

mercurous sulfate Hg_2SO_4.
Properties: White to yellow crystalline powder; soluble in hot sulfuric acid, dilute nitric acid; almost insoluble in water. Sp. gr. 7.56.
Derivation: (a) Dissolving mercury in sulfuric acid (2:3) and heating gently. (b) Adding sulfuric acid to mercurous nitrate solution.
Grades: Technical; C.P.
Containers: 1-, 5-lb bottles.
Uses: Chemical (admixed with sulfuric acid as a catalyst, in oxidation of naphthalene to phthalic acid); laboratory batteries (Clark cell, Weston cell).
Shipping regulations: Class B poison. Poison label. *

mercury (quicksilver; hydrargyrum) Hg. Element with atomic number 80, group IIb of the periodic table.
Properties: A silvery, liquid, metallic element, sometimes found native; poisonous! Insoluble in hydrochloric acid; soluble in sulfuric acid upon boiling; readily and completely soluble in nitric acid. Insoluble in water, alcohol, and ether. Sp. gr. 13.59; m.p. −38.85°C; b.p. 357.33°C.
Derivation: By heating cinnabar in air, or with lime.
Method of purification: Distillation. An important proportion of used mercury is recovered by redistillation.
Grades: Technical; N.F. XI; virgin; redistilled.
Containers: 1-, 5-, 10-lb bottles or jugs; larger amounts in 76-lb flasks only.
Uses: Mercury cells and other electrical apparatus; mercury vapor boilers; mercury vapor lamps; barometers; thermometers; mercurials for medicine and pesticides; dental preparations; amalgams; antifouling paint; catalyst.
Shipping regulations: None. *

mercury acetate. See mercuric acetate; mercurous acetate.

mercury para-aminophenyl arsenate. See mercury atoxylate.

mercury, ammoniated (mercuric chloride, ammoniated; ammonobasic mercuric chloride; ammoniated mercury chloride; white precipitate; white precipitate, fusible; aminomercuric chloride; mercury cosmetic; Lemery's white precipitate) $HgNH_2Cl$.
Properties: White, pulverulent lumps or powder; earthy, metallic taste; odorless; stable in air; darkens on exposure to light; poisonous! Soluble in ammonium carbonate and sodium thiosulfate solutions and in warm acids; insoluble in water and alcohol.
Derivation: By precipitating mercuric chloride with ammonium hydroxide in excess.
Grades: U.S.P. XVI; technical.
Containers: 1-, 5-, 25-, 50-lb boxes; 100-lb kegs; 200-lb barrels or fiber drums.
Use: Medicine.
Shipping regulations: Class B poison. Poison label. *
See also mercuric ammonium chloride.

mercury-ammonium chloride. See mercuric-ammonium chloride.

mercury arsenate. See mercuric arsenate.

mercury arsenite. See mercurous arsenite.

mercury atoxylate (mercury para-aminophenylarsenate; mercuric arsanilate) $Hg[OOAs(OH)C_6H_4NH_2]_2$.
Properties: White powder containing 31.8% mercury; very slightly soluble in water. Used as a 10% suspension in olive oil, or as a 5% ointment.
Use: Medicine.
Shipping regulations: Poison, Class B. Poison label. *

mercury-barium bromide. See mercuric-barium bromide.

mercury-barium iodide. See mercuric-barium iodide.

mercury benzoate. See mercuric benzoate.

mercury bichloride. See mercuric chloride.

mercury bichromate. See mercury dichromate.

*See "I.C.C. Shipping Regulations," page xiii.
Reference numbers refer to name of manufacturer. See "List of Manufacturers," page v.

mercury biniodide. See mercuric iodide.

mercury bromide. See mercuric bromide and mercurous bromide.

mercury-cathode cell. An electrolytic cell for the production of caustic soda and chlorine from sodium chloride brine. Continuously fed brine is decomposed in one compartment between graphite anodes, where chlorine is liberated, and a mercury cathode, where a sodium amalgam is formed. The amalgam flows continuously or intermittently to a second compartment where it is decomposed with water, forming a caustic solution. The decomposition is usually performed electrolytically by making the amalgam anodic with respect to an iron or graphite cathode. Pure water is supplied to the decomposition compartment at such a rate as to maintain a constant concentration of caustic in the product. With respect to the diaphragm cell, the mercury cathode cell has the advantages of producing a very pure caustic and, generally, a more concentrated solution (50-70%); it has the disadvantages of a higher operating voltage and lower efficiency (52-55%), and a high capital investment in mercury. For examples of mercuty-cathode cells, see Castner cell and DeNora cell.

mercury chlorate. See mercurous chlorate.

mercury chloride. See mercuric chloride and mercurous chloride.

mercury chloride, ammoniated. See mercury, ammoniated.

mercury chloride, corrosive. See mercuric chloride.

mercury chloride, mild. See mercurous chloride.

mercury chromate. See mercurous chromate.

mercury, colloidal.
Properties: Clear, dark-brown liquid. Faint alkaline reaction. Contains 1% colloidal mercury. Caution! Do not subject to bright light or to wide temperature variations.
Use: Medicine.

mercury-copper iodide. See mercuric-cuprous iodide.

mercury, cosmetic. See mercury, ammoniated.

mercury cyanide. See mercuric cyanide.

mercury dichromate (mercuric dichromate; mercury bichromate) $HgCr_2O_7$.
Properties: Heavy, red, crystalline powder; soluble in acids; insoluble in water.

mercury fluoride. See mercuric fluoride.

mercury fulminate (mercuric cyanate) $Hg(CNO)_2$.
Properties: Gray crystalline powder; explodes when dry under the slightest friction or shock; must be kept moist until used. Soluble in alcohol, ammonium hydroxide and hot water; slightly soluble in cold water. Sp. gr. 4.42; m.p., explodes.
Derivation: By treating mercury with strong nitric acid and alcohol.
Grades: Technical.
Containers: Canvas bags in stone crocks filled with water; in lots of five pounds in glass bottles.
Uses: Manufacture of caps and detonators for producing explosions for military, industrial and sporting purposes.
Fire hazard: Dangerous; high explosive.
Shipping regulations: In dry form: Forbidden explosive. Not accepted by common carrier. Wet: Explosive, class A. Initiating explosive label. Not accepted by express.*

mercury, horn. See calomel, native.

mercury imidosuccinate. See hydrargol.

mercury iodate. See mercuric iodate.

mercury iodide, red. See mercuric iodide.

mercury iodide, yellow. See mercuric iodide.

mercury monochloride. See mercurous chloride.

mercury naphthenate.
Properties: A dark amber liquid; soluble in lubricating oils and mineral spirits. Wt/gal 10.4 lbs.
Grades: 25% mercury.
Containers: 80-lb drums.
Uses: Mildew-resistance promoter in paints; antiknock compound.
Shipping regulations: Poison, class B. Poison label.*

mercury nitrate. See mercuric nitrate.

mercury nitrate, ammoniated. See mercurous nitrate, ammoniated.

mercury nucleate. See mercurol.

mercury oleate. See mercuric oleate.

mercury oxide, red. See mercuric oxide, red.

mercury oxide, yellow. See mercuric oxide, yellow.

mercury oxycyanide. See mercuric oxycyanide.

mercury pernitrate. See mercuric nitrate.

mercury persulfate. See mercuric sulfate.

mercury phosphate. See mercurous phosphate and mercuric phosphate.

mercury-potassium cyanide. See mercuric-potassium cyanide.

mercury protoacetate. See mercurous acetate.

mercury protochloride. See mercurous chloride.

mercury protoiodide. See mercurous iodide.

mercury, salicylated. See mercuric salicylate.

mercury-silver iodide. See mercuric-silver iodide.

mercury stearate. See mercuric stearate.

mercury sulfate. See mercuric sulfate.

mercury sulfide, black. See mercuric sulfide, black.

mercury sulfide, red. See mercuric sufide, red.

mercury sulfocyanate. See mercuric thiocyanate.

mercury sulfuret, red. See mercuric sulfide, red.

mercury thiocyanate. See mercuric thiocyanate.

merethoxylline procaine. Consists of a mixture of one mole of the procaine salt of merethyoxylline and 1.4 moles of theophylline. It is prepared by dissolving the merethoxyline and procaine in water and adding the theophylline. It is not isolated from solution.

Merethoxylline is a mercuri compound, $C_{15}H_{19}HgNO_6$, or ortho-(N-gamma-hydroxymercuri-beta-hydroxyethoxypropylcarbamido)phenoxyacetic acid. It is a white powder, m.p. 138-141°C; practically insoluble in water; soluble in alkaline solutions.

Grade: N.N.D.

Use: Medicine.

"Merez" Metal Resinates. [296] Trademark for zinc calcium metal resinates in grades "Merez" A, B, C, and D having typical melting points (capillary tube) 167°C, 157°C, 147°C, and 136°C respectively. Used in paint, varnish, and printing ink.

"Mergamma." [150] Trademark for wireworm killer and seed disinfectant containing a combination of 40% gamma isomer of benzene hexachloride and 1.93% phenylmercury urea.

Containers: 12-oz or 3-lb tins; 40-lb drums.

"Merlon." [58] Trademark for a line of textile treating resins based on vinyl acetate, styrene, or vinyl butyral, or other synthetic polymers.

"Merol-S." [296] Trade name for a zinc resinate in mineral spirits solution. Used in paint, varnish, and printing ink.

"Merox" Process. [416] Patented process for extracting easily removed mercaptans and converting the remaining mercaptans in naphthas to disulfides, thereby yielding a product sweet by the doctor test. The combination of extraction and sweetening is applicable to gasoline and lighter hydrocarbons, and the sweetening step to many jet fuels and kerosines.

Process employs caustic soda which is regenerated by blowing with air in the presence of a metal chelate catalyst, to oxidize the mercaptans to disulfides. Disulfides are decanted from the caustic in the extraction application, while in the sweetening application they remain in the hydrocarbon.

"Merpentine." [28] Trademark for a surface-active agent, a red-brown liquid with good wetting and solvent properties.

Use: As a wetting and scouring agent in the yarn-dyeing and leather industries.

"Merphos." [40] Proprietary name for tributylphosphorotrithioite $(C_4H_9S)_3P$.

Properties: Technical grade is a nearly colorless liquid; insoluble in water; soluble in a variety of organic solvents. B.p. (0.08 mm) 115-134°C; sp. gr. (20°C) 1.02; refractive index (25°C) 1.542.

Derivation: Reaction of butyl mercaptan with phosphorus trichloride.

Purity: 95%.

Use: Active ingredient of cotton defoliant "Folex."

"Merpol." [28] Trademark for a line of wetting, scouring, and dyeing assistants used principally in the textile, paper, and leather industries.

mersalyl (ortho-[(3-hydroxymercuri-2-methoxypropyl)carbamyl]phenoxyacetic acid, sodium salt; mersalyl sodium) $NaOOCCH_2OC_6H_4CONHCH_2CH(OCH_3)CH_2$-HgOH.

Properties: White or almost white, somewhat deliquescent, crystalline powder; odorless with a bitter taste; decomposed gradually by light; soluble in water and alcohol; insoluble in chloroform and ether.

Grade: N.F.XI.

Use: Medicine.

mersalyl sodium. See mersalyl.

"Mersize." [58] Trademark for a paper sizing agent in paste and dry form.

Containers: Liquid form: 6000-, 8000-, and 10,000-gal tank cars; 500-lb drums. Dry form: 50-lb bags.

Use: In manufacture of water-resistant paper.

"Mersolite-1." [81] Trademark for phenylmercuric hydroxide (q.v.).

"Mersolite-2." [81] Trademark for phenylmercuric chloride (q.v.).

"Mersolite-7." [81] Trademark for phenylmercuric nitrate (q.v.).

"Mersolite-8, -88, -94." [81] Trademark for phenylmercuric acetate (q.v.).

"Mersolite-90." [81] Trademark for phenylmercuric borate (q.v.).

"Mertax." [58] Trade name for purified 2-mercaptobenzothiazole.

Properties: Light yellow powder; m.p. 175°C min; mercaptobenzothiazole content 97% min; sp. gr. 1.42.

Uses: Rubber accelerator.

"Merthiolate." [100] Trademark for thimerosal, N.F.

Use: Medicine.

"Mertone." [58] Trademark for a colloidal silica precoat for blueprint and other reproduction papers.

Properties: Liquid; 15% and 30% colloidal dispersion of silica in water; odorless.

Containers: 55-gal lined steel drums.

mescaline (3,4,5-trimethoxyphenethylamine) $(CH_3O)_3C_6H_2CH_2CH_2NH_2$. A poisonous alkaloid from mescal buttons (the flowering heads of certain types of cactus).
Properties: Crystals; m.p. 35-36°C; b.p. 180°C; soluble in water, alcohol, chloroform, benzene; nearly insoluble in ether; takes up carbon dioxide from the air.
Use: Biochemical and medical research; non-lethal incapacitating drug.

mesitylene (1,3,5-trimethylbenzene; sym-trimethylbenzene) $C_6H_3(CH_3)_3$. A colored liquid; sp.gr. 0.863; m.p. −52.7°C; b.p. 164.6°C; insoluble in water; soluble in alcohol and ether. Derived from coal tar.
Use: Intermediate.

mesityl oxide (isopropylideneacetone; methyl isobutenylketone; 4-methyl-3-penten-2-one) $(CH_3)_2C:CHCOCH_3$.
Properties: Oily, colorless liquid. Honey-like odor. Partially soluble in water; miscible with alcohols, ethers.
Constants: Sp. gr. 0.8569 (20/20°C); b.p. 130-131°C (760 mm); vapor pressure 8.7 mm (20°C); toluene dilution ratio 3.8; flash point 90°F; wt/gal 7.1 lbs (20°C); coefficient of expansion 0.00107 (20°C); freezing point −46.4°C; viscosity 0.0060 poise (20°C).
Typical specifications: Sp. gr. 0.852-0.856 (20/20°C); boiling range 123-132°C (760 mm); acidity not more than 0.01% (as acetic).
Derivation: By the dehydration of acetone or diacetone alcohol.
Grades: Technical.
Containers: 1-gal cans; 5-, 55-gal drums; carloads; tank cars.
Caution: Keep away from heat and open flame. Avoid breathing of vapor. Avoid contact with skin. MCA warning label.
Uses: Solvent for cellulose esters and ethers, oils, gums, resins, lacquers, roll-coating inks, stains; ore flotation; paint and varnish-removers; insect repellent.
Shipping regulations: None.*

meso-. A prefix meaning middle or intermediate; specifically:
1. An optical isomer which is inactive as a result of internal compensation.
2. An intermediate hydrated form of an inorganic acid.
3. Designating a middle position in certain cyclic organic compounds, or
4. A ring system characterized by a middle position of certain rings.

meso-inositol. See inositol.

meson. The name given to any of the group of fundamental particles (q.v.) whose masses are intermediate between the mass of an electron and the mass of a nucleon.

mesoxalylurea. See alloxan.

mesquite gum.
Properties: Irregular, colorless or dark amber-brown round pieces. Resembles gum arabic in solubility.
Derivation: From Prosopis juliflora DC., found in New Mexico, Texas and south-western United States.
Uses: Cattle food; source of tannin.
Shipping regulations: None.*

"Mestinon" Bromide. [190] Trademark for a brand of pyridostigmin bromide (q.v.).

mesyl chloride. See methanesulfonyl chloride.

meta-. A prefix. For definition of meta- compounds see under ortho-. In this dictionary, meta- is disregarded in the alphabetizing; e.g., for meta-cresol, see cresol.

metabolite. An intermediate material produced and used in the processes of a living cell or organism. Metabolites are used for replacement and growth in living tissue, and are also broken down to be a source of energy in the body. Examples are nucleic acids, enzymes, glucose, cholesterol, and many similar substances.

metabutethamine hydrochloride (2-isobutyl-aminoethyl meta-aminobenzoate hydrochloride)
$H_2NC_6H_4COOCH_2CH_2NHCH_2CH(CH_3)_2 \cdot HCl$.
Properties: White, odorless, crystalline solid; m.p. 181-184°C; soluble in water; slightly soluble in alcohol, acetone, chloroform.
Grade: N.F. XI.
Use: Medicine.

metabutoxycaine hydrochloride (2'-diethyl-aminoethyl 3-amino-2-butoxybenzoate hydrochloride)
$C_6H_3(NH_2)(OC_4H_9)(COOCH_2CH_2N(C_2H_5)_2 \cdot HCl$.
Properties: White, odorless, crystals; m.p. 117-120°C; very soluble in water, alcohol; slightly soluble in acetone, chloroform; very slightly soluble in ether.
Grade: N.F. XI.
Use: Medicine.

metacetone. See diethylketone.

metacinnabarite. A mineral of the same composition as cinnabar (q.v.) but black in color. Occurs with cinnabar. A variety, guadalcazarite, contains some selenium and zinc. Used as a source of mercury.

"Metadelphene." [266] Trademark for N,N-diethyl-meta-toluamide (q.v.).

metaformaldehyde. See sym-trioxane.

"Metafos." [164] Trademark for a sodium phosphate glass commonly called sodium hexametaphosphate (q.v.), a linear polymer containing a minimum of 67% P_2O_5. Soluble in water; insoluble in organic solvents.
Grades: Beads, granular, powder.
Containers: Bags; fiber drums.
Uses: In general, as a sequestering, dispersing and deflocculating agent; specifically, to sequester alkaline earth and heavy metal ions; to prevent scaling and corrosion of pipes; to soften water; to disperse pigments and clays in paper making and oil well drilling.

"Metalate." [244] Trade name for a white, crystalline material containing 29.8%

*See "I.C.C. Shipping Regulations," page xiii.
Reference numbers refer to name of manufacturer. See "List of Manufacturers," page v.

total Na_2O.
Containers: 125-lb and 325-lb net drums; 325-lb net barrels.
Uses: Laundry builder and break compound; also used for miscellaneous cleaning.

metal deactivator. Substance that is added to gasoline motor fuels to neutralize the catalytic effect of copper in promoting fuel oxidation. See following article for a typical commercial material.

Metal Deactivator. [28] A 78% solution of N, N'-disalicylidene 1, 2-diaminopropane and related compounds, in an aromatic solvent.
Properties: Brown liquid at temperatures above −15°C; sp. gr. 1.076; miscible in gasoline in all proportions above 20°C and greater than 1% at −30°C. Also miscible in most gasoline antioxidants in all proportions above 0°C.
Containers: 30-gal steel drums.
Use: For improving the storage stability of petroleum distillates containing dissolved copper. Concentration required: 0.0005-0.002% by weight.

metaldehyde $(CH_3CHO)_n$. A polymer of acetaldehyde, in which n usually is 4 or 6.
Properties: White prisms, partial regeneration of acetaldehyde when heated above 80°C. Soluble in benzene, chloroform; slightly soluble in alcohol, ether; insoluble in water.
Constants: Sublimes 112-115°C; m. p. 246°C.
Use: As a fuel to replace alcohol; to destroy snails.

metal dyes. Aluminum may be dyed by first producing on its surface an oxide film, which then adsorbs dyestuff in a subsequent operation. Steel may be similarly treated. Alizarin Cyanin RR, Alizarin Green S, Nigrosine 2Y and Naphthalene Blue RS have been mentioned as suitable for these purposes.

"Metalflake." [233] Trademark for precision-cut particles of metallic foil used in paint for automotive finishes, or incorporated in paper, plastics, leathers and fabrics.

"Metallac." [309] Trademark for zinc stearate (q. v.).

metallic fiber. Generic name for a manufactured fiber composed of metal, plastic-coated metal, metal-coated plastic, or a core completely covered by metal (Federal Trade Commission). A common type is aluminum filament covered with cellulose acetate butyrate. It may be colored by a dye in the plastic or by gold or silver glitter. Used as a decorative yarn.

metallized dyes. Soluble dyes including any one of a variety of metals chemically combined, applied to wool in an acid bath, by use of sodium chloride to salt out the dye onto the fiber.

metallography. Study of the structure of metals and alloys, principally by use of the microscope.

metal protection. See cadmium plating, calorizing, chromium plating, clad metal; copper plating, galvanized iron, lead coating, lead lining, lead burning, metal spraying, nickel plating, rubber-plated metal, sherardizing.

metals, fusible. See fusible alloys.

metals, powdered. Metals are produced in powdered form for a variety of uses in several industries. In this form they are the raw materials for the broad processing field known as powder metallurgy. Metal powders range in size from −325 mesh (0.045-0.060 mm "diameter") to +100 mesh, and are available in practically all industrial metals. They are produced by machining, milling, shotting, granulation, atomizing, condensation, reduction, chemical precipitation or electrodeposition. The properties and purities of the powders vary with the method of preparation.
Uses: In the electric, automotive, machinery, tool and refractory-metal industries; metallic paint pigments; flares and incendiary bombs; brazing materials; calorizing, metal-spraying, metallurgical agents; heat generating agents; catalysts, etc.
Caution: When finely divided as in a powder, even ordinary metals are often subject to spontaneous heating and combustion, and their dust may constitute an explosive hazard. Appropriate information and precautions should be used to suit each specific case.
See also powder metallurgy.

metal spraying. A method of spraying metal through a gun for the purpose of coating different objects. The gun commonly used is the gas gun. In this gun the metal to be sprayed is introduced in the form of wire through a central tube. Through an annular space surrounding this tube a gaseous mixture, either hydrogen or acetylene with oxygen or air, is passed, which upon burning at the orifice melts the wire in the inner part of the conical flame. Compressed air or other gas is passed through an outer annular space for the purpose of atomizing and spraying the metal as it is melted.

"Metalyn." [266] Trademark for the distilled methyl ester of tall oil; amber liquid; acid number 5 max; viscosity 70-100 seconds (Saybolt Universal, at 100°F).

"Metandren." [305] Trademark for methyltestosterone U. S. P.
Use: Medicine.

metanilic acid (meta-sulfanilic acid; meta-aminobenzenesulfonic acid) $C_6H_4(NH_2)SO_3H$.
Properties: Small colorless needles. Soluble in water, alcohol, and ether.
Derivation: By the reduction of meta-nitrobenzenesulfonic acid. Nitrobenzene is sulfonated until the product is soluble in water. The mixture is then poured into water and reduced with iron, made alkaline with lime and the lime salt dissociated with sodium carbonate.

Grades: Technical.
Containers: 250-, 300-lb barrels; drums.
Uses: Dyes; medicine.
Shipping regulations: None. *

"Metaphen." [3] Trademark for nitromersol (q. v.).

metaraminol bitartrate (ℓ-meta-hydroxy-norephedrine bitartrate) $HOC_6H_4CH(OH)CH(CH_3)NH_2 \cdot C_4H_6O_6$.
Properties: White, practically odorless, crystalline powder; melting range 170-176°C; soluble in water; somewhat soluble in alcohol; insoluble in ether and in chloroform; pH of 5% solution 3.2 and 3.5.
Grade: N. F. XI.
Use: Medicine.

"Metasap." [309] Trademark for a series of polyvalent metal and lithium soaps of higher fatty acids having a general use in the industrial arts.

"Metasol." [452] A line of mercury derivatives used as bactericides and fungicides. They are diphenylmercuric ammonium propionates.

"Meta-Systox R." [181] Trademark for O,O-dimethyl-S-2-(ethylsulfinyl)ethyl phosphorothioate (q. v.).

"Metavis." [309] Trademark for a metallic soap composition (as, aluminum tristearate) for thickening oils and for producing lubricating greases.

meter. The basic unit of length of the metric system. It is 39.37 inches long. It was originally defined as exactly one ten-millionth of the distance from the equator to the North Pole. The practical standard was and is a carefully preserved platinum-iridium meter bar kept at Paris under terms of an international treaty. A meter is now defined as 1,650,763.73 wave lengths of the orange-red line of the isotope krypton 86. This standard based on recent scientific advances, is practical, exact, and invariable.

"Methac." [65] Trademark for a series of blends of methyl acetate with methanol in varying proportions. Used in lacquer solvents, paint removers, organic synthesis.

methacetin (acetanisidine; para-methoxy-acetanilide) $CH_3OC_6H_4NHCOCH_3$.
Properties: White, crystalline powder; feebly bitter taste. Soluble in alcohol, acetone and dilute acids; insoluble in a water. M. p. 127.1°C.
Derivation: By the acetylation of para-anisidine.
Grades: Technical; pure.
Containers: Tins; glass bottles.
Use: Medicine.
Shipping regulations: None. *

methacholine bromide (acetyl-beta-methyl-choline bromide) $CH_3COOCH(CH_3)CH_2N(CH_3)_3Br$.
Properties: White, crystalline, very hygroscopic powder with slight fishy odor. M. p. 147-150°C. Readily soluble in alcohol and water; insoluble in benzene and ether. pH (5% solution) about 5.
Grade: N. F. XI.
Use: Medicine.

methacholine chloride (acetyl-beta-methyl-choline chloride) $CH_3COOCH(CH_3)CH_2N(CH_3)_3Cl$.
Properties: Colorless or white crystals or white crystalline powder; odorless or very slight odor; very deliquescent; very soluble in water; freely soluble in alcohol and chloroform; insoluble in benzene and ether. Solutions neutral to litmus paper; m. p. 170-173°C.
Grade: N. F. XI.
Use: Medicine.

"Methacrol." [28] Trademark for a line of thermoplastic resin dispersions that are used as finishing agents in the textile and leather industries.

methacrolein (methacrylaldehyde) $CH_2{:}C(CH_3)CHO$.
Properties: Liquid; sp. gr. 0.8474 (20/20°C); b. p. 68.0°C (760 mm); flash point 5°F (Cleveland open cup); solubility in water at 20°C 5.9% by weight. Shipped with 0.1% hydroquinone as polymerization inhibitor.
Uses: Copolymers; resins.
Shipping regulations: Flammable liquid. Red label. *

methacrylaldehyde. See methacrolein.

methacrylate esters. Esters of methacrylic acid having the formula $CH_2{:}C(CH_3)COOR$, where R is usually methyl, ethyl, isobutyl, or n-butyl-isobutyl (50-50). They are supplied commercially as the polymers. See acrylate resins.

methacrylate resins. See acrylate resins.

methacrylatochromic chloride.
$H_2C{:}C(CH_3)\underline{C{:}OCrCl_2OHCrCl_2O}$.
Properties: Water-soluble solid.
Process: Reaction of methacrylic acid with basic chromic chloride.
Uses: Water repellent; nonadhesive; insolubilizer for vinyl polymers.

methacrylic acid (monomer) (alpha-methacrylic acid) $CH_2{:}C(CH_3)COOH$.
Properties: Colorless liquid; m. p. 15-16°C; b. p. 161-162°C; sp. gr. 1.015 (20°C). Soluble in water, alcohol, ether, most organic solvents. Undergoes polymerization readily to give water-soluble polymers.
Derivation: Reaction of acetone cyanohydrin and dilute sulfuric acid; oxidation of isobutylene.
Grades: 40% Aqueous solution, b. p. 76-78°C (25 mm); crude monomer 85% pure; glacial (98%).
Containers: 13 ½-gal carboys; 55-gal drums; tank cars.
Uses: Monomer for large-volume resins and polymers; organic synthesis. Many of the polymers are based on esters of the acid, as the methyl, butyl, or isobutyl esters.
See acrylate resins.

beta-**methacrylic acid.** Name sometimes used for trans form of crotonic acid (q. v.).

"Methadone." [3] Trademark for dℓ-6-dimethylamino-4, 4-diphenyl-3-heptanone.
Derivation: Reaction of diphenylacetonitrile with ℓ-di-methylamino-2-chloro-propane.
Use: Medicine.

methadone hydrochloride (dℓ-6-dimethylamino-4, 4-diphenyl-3-heptanone hydrochloride)
$(C_6H_5)_2C(COC_2H_5)CH_2CH(CH_3)N(CH_3)_2 \cdot HCl$.
Properties: A crystalline substance with bitter taste; no odor; m. p. 232-235°C; soluble in water and alcohol; slightly soluble in isopropanol, and chloroform; practically insoluble in ether and glycerol. pH (1% aqueous solution) 4. 5-6. 5.
Grade: U. S. P. XVI.
Use: Medicine.

methadrine. A non-lethal incapacitating drug.

"Methalate 'C'." [188] Brand name for a proprietary product. Possesses an odor similar to methyl salicylate for which it is an effective substitute.
Uses: Fly sprays; also can be used as a low-priced aromatic for covering objectionable odors present in mixtures of fixed oils, animal or vegetable, especially such oils as are used in the paint industry and in low-grade soaps.

methallenestril (beta-ethyl-6-methoxy-alpha, alpha-dimethyl-2-napthalenepropionic acid) $CH_3OC_{10}H_6CH(C_2H_5)C(CH_3)_2COOH$.
Properties: Crystals; m. p. 132. 5°C. Soluble in ether, vegetable oils.
Grade: N. N. D.
Use: Medicine.

methallyl acetate. See methylallyl acetate.

methallyl alcohol. See 2-methyl-2-propen-1-ol.

methallyl chloride. See methylallyl chloride.

methaminodiazepoxide. See chlordiazepoxide hydrochloride.

methamphetamine hydrochloride (deoxyephedrine hydrochloride; 1-phenyl-2-methylaminopropane hydrochloride)
$C_6H_5CH_2CH(CH_3)NHCH_3 \cdot HCl$.
Properties: White crystals or white crystalline powder; m. p. 171-175°C; odorless; aqueous solution acid to litmus paper; soluble in water, alcohol, and chloroform; very slightly soluble in absolute ether.
Grade: N. F. XI.
Use: Medicine.

methanal. See formaldehyde.

methanamide. See formamide.

methandrostenolone (17 alpha-methyl-17 beta-hydroxyandrosta-1, 4-dien-3-one). An anabolic hormone; anabolic steroid.
Use: Medicine (deposition, synthesis, utilization of protein).

methane (marsh gas, methyl hydride) CH_4. The first member of the aliphatic hydrocarbons series.
Properties: Colorless, odorless, tasteless, flammable gas; lighter than air, practically inert toward sulfuric acid, nitric acid, alkalies, and salts but reacts with chlorine and bromine in light (explosively in direct sunlight). Forms explosive mixture with air or oxygen; b. p. −161. 6°C; m. p. −182. 5°C; density of vapor 0. 554 (0°C) (760 mm, air = 1); critical temperature −82. 1°C; critical pressure 672 psia; explosive limits in air 5. 3-14. 0 per cent by volume; heating value 1009 Btu/cu ft. Soluble in alcohol, ether; slightly soluble in water.
Occurence: Chief component of most natural gas and of marsh gas; a major component of coal gas; present to some extent in air in coal mines.
Derivation: From natural gas by absorption or adsorption methods; specially pure methane is obtained by supercooling and distillation.
Containers: High pressure pipe lines; high pressure cylinders.
Grades: Research, 99. 65%; pure, 99%; technical, 95%; Btu grade must have heating value of 1000 ±3 Btu/cu. ft. at 60°F and 30 inches Hg pressure.
Uses: A source of petrochemicals by its conversion to hydrogen and carbon monoxide by steam cracking or partial oxidation. Important products are methanol, acetylene, hydrogen cyanide. Chlorination produces carbon tetrachloride, chloroform, methylene chloride, and methyl chloride. In the form of natural gas, methane is used as a fuel, and is also the source of carbon black.
Shipping regulations: Flammable gas. Red gas label. *

methanearsonic acid, disodium salt. See disodium methylarsonate.

methanecarboxylic acid. See acetic acid.

methanedicarbonic acid. See malonic acid.

methanesulfonyl chloride (mesyl chloride) CH_3SO_2Cl.
Properties: Pale yellow liquid; b. p. 164°C; f. p. −32°C; soluble in most organic solvents; insoluble in water (slowly hydrolyzes).
Containers: 5-gal, 55-gal drums.
Grades: 98%; 99+%.
Uses: Intermediate.

methanethiol. See methyl mercaptan.

methanethiomethane. See dimethyl sulfide.

methanoic acid. See formic acid.

methanol (methyl alcohol, wood alcohol) CH_3OH.
Properties: Clear, colorless, mobile, volatile, flammable liquid; poisonous! Soluble in water, alcohol, and ether; sp. gr. 0. 7924; m. p. −97. 8°C; b. p. 64. 5°C; wt/gal 6. 59 lbs (20°C); refractive index 1. 329 (20°C); surface tension 22. 6 dynes/cm (20°C); viscosity 0. 00593 poise (20°C);

*See "I. C. C. Shipping Regulations," page xiii.
Reference numbers refer to name of manufacturer. See "List of Manufacturers," page v.

vapor pressure 92 mm (20°C); specific heat (vapor) 0.39 cal/g (77°C); latent heat of evaporation 262.8 cals/g at b.p.; heat of fusion 23.6 cals/g; heat of combustion 170.9 kg cals/mol; flash point (open cup) 60°F; dielectric constant 31.2 (20°C); critical pressure 78.7 atm; critical temperature 240°C; average coefficient of cubical expansion 0.0012 from 0-61°C.

Derivation: (a) By high pressure catalytic synthesis from carbon monoxide and hydrogen. (b) Through partial oxidation of natural gas hydrocarbons. (c) By purification of the pyroligneous acid resulting from destructive distillation of wood.

Method of purification: Rectification.

Containers: Various, up to 8000-gal tank cars.

Grades: Technical; C. P. Electronic (used to cleanse and dry components).

Uses: (in approximate order of volume): Production of formaldehyde; automotive antifreeze; chemical synthesis (methyl amines, methyl chloride, methyl methacrylate, etc.); general solvent (including surface coatings, paint removers, inks, and adhesives); aviation fuel (for water injection); denaturant for ethyl alcohol; rocket fuel; dehydrator for natural gas.

Danger: Flammable; vapor harmful. May be fatal (or cause blindness) if swallowed. Cannot be made non-poisonous. MCA warning label.

Shipping regulations: Flammable liquid. Red label. *

methantheline bromide $C_{21}H_{26}BrNO_3$. beta-Diethylaminoethyl-9-xanthene carboxylate methobromide.

Properties: White or nearly white, practically odorless, microcrystalline powder with very bitter taste; m.p. 171-177°C; very soluble in water; freely soluble in alcohol and chloroform; practically insoluble in ether. Water solution decomposes on standing.

Grade: N. F. XI.

Use: Medicine.

methaphenilene hydrochloride (N,N-dimethyl-N'-(alpha-thenyl)-N'-phenylethylenediamine hydrochloride) $C_{15}H_{20}N_2S \cdot HCl$.

Properties: White to pale yellow, crystalline powder with faint odor; m.p. 184-189°C. Soluble in water; sparingly soluble in alcohol and chloroform; practically insoluble in ether. pH (2% solution) 5-7.

Use: Medicine.

methapyrilene (2-[(2-dimethylaminoethyl)-2-thenylamino]pyridine) $C_4H_3SCH_2N(C_5H_4N)CH_2CH_2N(CH_3)_2$.

Properties: Liquid; b.p. (0.45 mm) 125-135°; b.p. (3 mm) 173-175°.

Derivation: Prepared by condensing 2-(2-thienylmethyl)aminopyridine with dimethylaminoethyl chloride in presence of sodamide, or by condensing N,N-dimethylaminoethyl-alpha-aminopyridine with 2-thienylmethyl chloride.

Use: Medicine (more widely used in combination with other drugs and as various salts, especially the hydrochloride).

methapyrilene fumarate $C_{14}H_{19}N_3S \cdot 1\frac{1}{2}C_4H_4O_4$.

Properties: White, crystalline powder with a faint aromatic odor. M.p. 133-136°C. Soluble in water, slightly soluble in alcohol, and insoluble in ether. An aqueous solution is acidic.

Containers: 25-lb, 100-lb fiber drums.

Use: Medicine.

methapyrilene hydrochloride $C_4H_3SCH_2N(C_5H_4N)CH_2CH_2N(CH_3)_2 \cdot HCl$, or $C_{14}H_{19}N_3S \cdot HCl$.

Properties: White, crystalline powder with faint odor. M.p. 159-163°C. Very soluble in water; freely soluble in alcohol and chloroform; practically insoluble in benzene and ether. pH (5% solution) 5-6.

Containers: Drums.

Grades: N. F. XI.

Use: Medicine.

"Methar." [49] Trade name for a disodium methyl arsenate selective crabgrass killer.

metharbital (5,5-diethyl-1-methylbarbituric acid) $(C_2H_5)_2\underline{CC(O)N(CH_3)C(O)NHC}(O)$.

Properties: A white, crystalline powder with faint aromatic odor; m.p. 151-155°C; slightly soluble in alcohol and ether; very slightly soluble in water.

Grade: N. F. XI.

Use: Medicine.

"Methasan." [58] Trademark for zinc dimethyldithiocarbamate.

"Methasol." [206] Brand name of proprietary line of spirits-soluble coloring matters.

"Methazate." [248] Trademark for zinc dimethyldithiocarbamate.

Properties: White powder; sp. gr. 1.68; m.p. 240-255°C; only slightly soluble in acetone, benzol, ethylene dichloride, gasoline, and water.

Uses: For wire insulation; proofing; footwear; latex; dispersions; cements and sundries.

methazolamide $CH_3CON{:}\overline{CSC(SO_2NH_2)NNCH_3}$ (5-acetylimino-4-methyl-delta-1,3,4 thiadiazoline 2-sulfonamide).

Properties: Solid; m.p. 213-4°C; can be recrystallized from water.

Use: Medicine.

"Methedrine." [301] Trademark for methamphetamine hydrochloride, used in medicine.

methenamine. U.S.P. name for hexamethylenetetramine (q.v.).

methenamine mandelate. U.S.P. name for hexamethylenetetramine mandelate (q.v.).

methenamine salicylate. U.S.P. name for hexamethylenetetramine salicylate.

methenyl tribromide. See bromoform.

methetharimide. See bemegride.

methimazole (1-methyl-2-mercaptoimidazole) $\underline{N(CH_3)C(SH)NCHC}H$.

*See "I.C.C. Shipping Regulations," page xiii.

Reference numbers refer to name of manufacturer. See "List of Manufacturers," page v.

Properties: White to pale buff, crystalline powder; almost no taste and a very faint odor; sensitive to light; m. p. 144-147°C; soluble in water, alcohol, and chloroform; slightly soluble in ether. Aqueous solution neutral to litmus.
Grade: U. S. P. XVI.
Use: Medicine.

methiodal sodium (sodium iodomethanesulfonate) ICH_2SO_3Na.
Properties: A white crystalline powder; odorless with slight salty taste followed by sweetish after-taste. Decomposes on exposure to light; solutions are neutral to litmus; soluble in water; very soluble in methanol; slightly soluble in alcohol; practically insoluble in acetone, ether and benzene.
Derivation: From sodium sulfite and methylene iodide.
Grade: N. F. XI.
Use: Medicine (as radiopaque contrast medium).

methionine $CH_3SCH_2CH_2CH(NH_2)COOH$. An optically active essential sulfur-containing amino acid important in biological trans-methylation processes. The levo form is the biologically active form.
Properties (of DL racemic mixtures): White, crystalline platelets or as powder. Faint odor. Soluble in water, dilute acids, and alkalies; very slightly soluble in alcohol; practically insoluble in ether. pH (1% aqueous solution) 5. 6-6. 1.
Derivation: Hydrolysis of protein; organic synthesis.
Containers: Glass vials, 1-, 5-, 10-lb glass bottles; 5-, 10-lb fiber cans; 55-lb fiber drums.
Grade: N. F. XI; feed, 98%.
Use: Pharmaceuticals; cosmetics; nutrition and biochemical studies; food and feed supplement.
Shipping regulations: None. *

methionine hydroxy-analog calcium $(CH_3SCH_2CH_2CHOHCOO)_2Ca$. Calcium salt of DL-alpha-hydroxy-gamma-methylmercaptobutyric acid. Free methionine hydroxy analog is a metabolite in methionine utilization.
Use: Feed supplement.

methocarbamol $C_{11}H_{15}NO_5$. Chemical name is 3(-ortho-methoxyphenoxy)-1, 2-propanediol 1-carbamate.
Properties: M. p. 92-94°C; soluble in water, alcohol and propylene glycol.
Use: Medicine.

"Methocel." [233] Trademark for methylcellulose (q. v.). Available for use as a thickening, emulsifying and dispersing agent. Also as a grease-proofing and gloss ink improvement agent.

methohexital (alpha-dl-5-allyl-1-methyl-5-(1-methyl-2-pentynyl)barbituric acid) $C_{14}H_{18}N_2O_3$.
Properties: White crystalline powder; essentially odorless; m. p. 91. 5°-96°C. Soluble in alcohol, chloroform and dilute alkalies; insoluble in water.
Containers: Fiber drums with polyethylene bags.
Use: Medicine.

methotrexate (amethopterin; 4-amino-10-methylfolic acid) $C_6N_4H(NH_2)_2CH_2N(CH_3)C_6H_4CONHCH-(COOH)CH_2CH_2COOH$, or $C_{20}H_{22}N_8O_5$.
Properties: Orange-brown crystalline powder. Insoluble in water, alcohol, chloroform, ether. Slightly soluble in dilute hydrochloric acid; soluble in dilute solutions of alkali hydroxides and carbonates. Folic acid antagonist.
Grade: U. S. P. XVI.
Caution! Extremely poisonous.
Use: Medicine.

methoxamine hydrochloride $(CH_3O)_2C_6H_3CH(OH)CH(CH_3)NH_2 \cdot HCl$. beta-Hydroxy-beta-(2, 5-dimethoxyphenyl) isopropylamine hydrochloride.
Properties: Colorless or white, plate-like crystals or white, bitter, crystalline powder; odorless or very slight odor; m. p. 212-216°C; freely soluble in water and alcohol; very slightly soluble in ethyl acetate; almost insoluble in chloroform and ether.
Grade: U.S.P. XVI.
Use: Medicine.

"Methoxone." [150] Trademark for a selective weed killer containing 23% sodium salt of 2-methyl-4-chlorophenoxyacetic acid. Used for control of broad-leaved weeds in rice, oats, etc.
Containers: 5-, 30- and 55-gal drums.

"Methox" Plasticizer. [55] Trademark for dimethoxyethyl phthalate.
Properties: Nearly water-white oily liquid; soluble in most other organic liquids; insoluble or limited solubility in glycerin, glycols, and certain amines.
Typical specifications: Sp. gr. 1. 171 ± 0. 003 (20/20°C); f. p. −45°C (stiff gel); boiling range 190-203°C (4 mm); acidity 0. 03% (max) as acetic acid; odor mild; flash point 380°F; fire point 440°F; vapor pressure < 0. 25 mm (150°C); refractive index 1. 500 (25°C); viscosity 53 cps (20°C); surface tension 3 dynes per cm (20°C); thermal expansion 0. 00076 from 10-40°C; 9. 5 lbs/gal.
Containers: 5-gal cans (50-lbs net); 55-gal steel drums (530 lbs net).
Uses: Primary plasticizer, especially for cellulose acetate, imparts stability to ultra-violet light, low oil extraction, and permanent flexibility.

methoxsalen (8-methoxypsoralen)
OCH:CHC:CC(OCH₃):CC:(CH)CH:CHC(O)O.
Properties: White to cream colored, odorless, crystalline solid. Slightly soluble in alcohol, practically insoluble in water.
Grades: N. N. D.
Use: Medicine (suntan accelerator, sunburn protector).

*See "I. C. C. Shipping Regulations," page xiii.
Reference numbers refer to name of manufacturer. See "List of Manufacturers," page v.

para-**methoxyacetanilide.** See methacetin.

methoxyacetic acid CH_3OCH_2COOH.
Properties: Liquid, having a m. p. (min) 7.7°; boiling range (733) (min) 197-198°C; sp. gr. (25/4°C) 1.1738; refractive index (25°C) 1.415; flash point 260°F; fire point 260°F; acid number (min) 612.
Use: Synthesis.

para-**methoxyacetophenone** (para-acetoanisole; para-acetylanisole) $CH_3OC_6H_4COCH_3$.
Properties: Crystalline tablets; b. p. 258°C; congealing point 36.5°C. Soluble in alcohol and ether.
Derivation: Interaction of anisole and acetyl chloride in the presence of aluminum chloride and carbon disulfide.
Grades: Technical.
Use: Perfumery, for floral odors.

methoxyacetyl-para-phenetidine $CH_3OCH_2CONHC_6H_4OC_2H_5$.
Properties: White needles; tasteless but becoming bitter on chewing. Soluble in alcohol, ether, chloroform, volatile oils, and boiling water; much less so in cold water. M. p. 98°C.
Derivation: By heating methoxyacetic acid and para-phenetidine.
Use: Medicine.
Shipping regulations: None. *

ortho-**methoxyaniline.** See ortho-anisidine.

para-**methoxyaniline.** See para-anisidine.

para-**methoxybenzaldehyde.** See anisaldehyde.

methoxybenzene. See anisole.

para-**methoxybenzoic acid.** See anisic acid.

para-**methoxybenzyl acetate.** See anisyl acetate.

para-**methoxybenzyl alcohol.** See anisic alcohol.

para-**methoxybenzyl formate.** See anisyl formate.

3-methoxybutanol $CH_3CH(OCH_3)CH_2CH_2OH$.
Properties: Liquid; sp. gr. 0.9229; b. p. 161.1°C; vapor pressure (20°C) 0.9 mm; sets to glass at −85°C; soluble in water.
Uses: High-boiling lacquer solvent; coupling agent for brake fluids; intermediate for plasticizers, herbicides; film-forming additive in PVA emulsions; solvent for pharmaceuticals.

methoxychlor (methoxy DDT; DMDT) $Cl_3CCH(C_6H_4OCH_3)_2$. 2,2-Bis(para-methoxyphenyl)-1,1,1-trichloroethane; dimethoxydiphenyl trichloroethane.
Properties: White, crystalline solid; m. p. 89°C; insoluble in water. Less toxic than DDT to higher animals. Not compatible with alkaline materials.
Derivation: Reaction of methyl phenyl ether and chloral hydrate.
Containers: Drums.
Grades: Technical.
Use: Insecticide, effective against mosquito larvae and house flies. Especially recommended for use around dairy barns.

methoxy DDT. See methoxychlor.

2-methoxyethanol. See ethylene glycol monomethyl ether.

2-(beta-methoxyethoxyl) ethanol. See diethylene glycol monomethyl ether.

2-methoxyethylmercury acetate $CH_3OCH_2CH_2HgOOCCH_3$. Toxic! A fungicide and disinfectant used in treating seeds.

methoxyethyl oleate
Properties: Oily liquid, mild odor. F. p. below −18°C; sp. gr. 0.898 (25°C); boiling range 180-206°C (4 mm Hg); flash point (open cup) 385°C; viscosity, 8 cp (25°C).
Uses: Plasticizer and solvent.

methoxyethyl stearate
Properties: Oily liquid, mild odor; f. p. 19 to 24°C; boiling range 186-205°C (4 mm Hg); flash point (open cup) 378°C; viscosity 9 cp at 25°C.
Uses: Plasticizer and solvent.

3-methoxy-4-hydroxybenzaldehyde. See vanillin.

methoxyhydroxymercuripropylsuccinyl urea $C_9H_{16}HgN_2O_6$.
Properties: Bitter crystals, m. p. 198.5°C.
Derivation: Made by the mercuration of allylsuccinylurea. See merallurIde.

4'-methoxy-2-(para-methoxyphenyl) acetophenone. See desoxyanisoin.

2-methoxy-5-methylaniline. See cresidine.

4-methoxy-4-methylpentanol-2 ("Pent-Oxol") $CH_3C(CH_3)(OCH_3)CH_2CHOHCH_3$.
Properties: Liquid. Boiling range 163.8-167°C.
Use: Solvent for resin coating formulation.

4-methoxy-4-methylpentanone-2 ("Pent-Oxone") $CH_3C(CH_3)(OCH_3)CH_2COCH_3$.
Properties: Water-white liquid; boiling range 147-163°C; flash point 141°F.
Derivation: Diacetone alcohol.
Use: Solvent for a variety of resin coatings.

methoxymethyl salicylate. See salicylic acid methoxymethyl ester.

2-methoxynaphthalene. See beta-naphthyl methyl ether.

1-methoxy-4-nitrobenzene. See para-nitroanisole.

methoxyphenamine hydrochloride $CH_3OC_6H_4CH_2CH(CH_3)NHCH_3 \cdot HCl$. 2-(ortho-Methoxyphenyl) isopropylmethylamine hydrochloride.
Properties: Crystalline, white powder which is odorless and bitter; m. p. 124-128°C; freely soluble in alcohol, chloroform, and water; slightly soluble in ether and benzene; pH (5% solution) 5.3-5.7.
Grade: N. N. D.
Use: Medicine.

4-methoxyphenol (para-methoxyphenol). See hydroquinone monomethyl ether.

ortho-**methoxyphenol.** See guaiacol.

*See "I.C.C. Shipping Regulations," page xiii.
Reference numbers refer to name of manufacturer. See "List of Manufacturers," page v.

3-(ortho-**methoxyphenoxy)-1, 2-propanediol 1-carbamate.** See methocarbamol.

1-(para-**methoxyphenyl)-1-penten-3-one.** See "Ethone."

methoxypolyethylene glycols. A series of compounds with properties similar to the polyethylene glycols of comparable molecular weight. Slightly viscous liquids to soft wax-like solids. Used for manufacture of detergents and emulsifying and dispersing agents through the preparation of the mono-fatty-acid derivatives.

methoxypromazine maleate $C_{18}H_{22}N_2OS \cdot C_4H_4O_4$. [2-Methoxy-10-(3'-dimethylaminopropyl)-phenothiazine maleate].
Properties: M. p. 141-145°C; slightly soluble in water, methanol; soluble in chloroform, dimethylformamide.
Uses: Medicine.

para-**methoxypropenylbenzene.** See anethole.

para-**methoxypropiophenone** $C_2H_5COC_6H_4OCH_3$.
Properties: Clear, colorless liquid; distillation range 110-140°C at 3 mm Hg pressure; refractive index, 1.543 to 1.545 (25°C).

3-methoxypropylamine $CH_3OCH_2CH_2CH_2NH_2$.
Properties: Colorless liquid; b. p. 116°C (760 mm); sp. gr. 0.8615 (30°C); refractive index (n 25/D) 1.4153; flash point 90°F Tag open cup; miscible with water, ethanol, toluene, acetone, carbon tetrachloride, hexane, and ether.
Use: Organic intermediate.

8-methoxypsoralen. See methoxsalen.

methoxytriethylene glycol acetate. See methoxytriglycol acetate.

methoxytriglycol $CH_3O[C_2H_4O]_3H$.
Properties: Sp. gr. 1.0494; b. p. 193.6°C; flash point 245°F; infinitely soluble in water.

methoxytriglycol acetate (methoxytriethylene-glycol acetate) $CH_3COO(C_2H_4O)_3CH_3$.
Properties: Practically colorless liquid with pleasant fruity odor; sp. gr. 1.0940 (20/20°C); 9.2 lbs/gal (20°C); b. p. 244.0°C (760 mm).
Uses: As an "anti-dusting agent" for finely powdered materials, especially as an "anti-sneeze" for certain dyestuffs. Low volatility suggests its use as water-soluble plasticizer for resins, casein, etc.

methscopolamine bromide (scopolamine methylbromide; epoxytropine tropate methylbromide) $C_{17}H_{21}NO_4 \cdot CH_3Br$.
Properties: White, odorless, bitter, crystalline powder; sparingly soluble in alcohol, freely soluble in water; pH of 1% water solution is 7.2-7.8.
Grade: N. N. D.
Use: Medicine.

methscopolamine nitrate (scopolamine methylnitrate; epoxytropine tropate methyl nitrate) $C_{18}H_{24}N_2O_4$, or $C_7OH_9N^+(CH_3)_2OCOCH(CH_2OH)C_6H_4 \cdot NO_3^-$.
Properties: White, odorless, tasteless, crystalline powder; freely soluble in alcohol and in water; pH (0.05% in water) is 5.0-5.4.
Grades: N. N. D.
Use: Medicine.

methyclothiazide. 6-Chloro-3-chloromethyl-2-methyl-7-sulfamyl-3, 4-dihydro-1, 2, 4-benzothiadiazine-1, 1-dioxide.
Properties: White crystalline solid; m. p. 220-222°C; readily soluble in sodium hydroxide solution.
Use: Medicine.

methyl abietate $C_{19}H_{29}COOCH_3$.
Properties: Colorless to yellow liquid; sp. gr. 1.033-1.043 (20°C); refractive index 1.525-1.535; flash point 180-220°C; b. p. 365°C. Miscible with most organic solvents.
Containers: Drums; carloads.
Grades: Technical.
Uses: Solvent and plasticizer; lacquers; varnishes; linoleum; coating compositions.

N-**methylacetanilide** $C_6H_5N(CH_3)COCH_3$.
Properties: Needles or long tablet-like crystals; soluble in hot water and dilute alcohol. B. p. 240-250°C; m. p. 101°C.
Derivation: By heating acetylchloride and methylaniline.
Use: Medicine.
Shipping regulations: None. *

methyl acetate $CH_3CO_2CH_3$.
Properties: Colorless, volatile, flammable liquid; fragrant odor. Miscible with the common hydrocarbon solvents; partially soluble in water. Sp. gr. 0.92438; m. p. −98.05°C; b. p. 54.05°C; flash point −16°C; refractive index 1.3619 (20°C); wt/gal 7.76 lbs (20°C).
Typical specifications, technical grade: Methyl acetate 82-85%; acidity as acetic 0.01% (max); boiling range 52-58°C; water: substantially dry; color water-white; chlorides none; iron none; dryness test 10 volume; sp. gr. 0.904-0.914 (20°C); wt/gal 7.54 lbs (20°C).
Typical specifications, C. P. grade: Methyl acetate 97% min; acidity as acetic 0.005% (max); boiling range 55-58°C; water: substantially dry; color water-white; chlorides none; iron none; dryness test 20 volume; sp. gr. 0.930-0.940 (20°C); wt/gal 7.76 lbs (20°C).
Derivation: By heating methyl alcohol and acetic acid in presence of sulfuric acid and distilling.
Method of purification: Rectification.
Grades: Technical; C. P.
Containers: 5-gal cans; 55-gal drums; tank cars.
Uses: Solvent; extracts; perfumery; artificial leather; plastics; solvent for nitrocellulose and acetyl cellulose, cellulose esters; paints, varnishes and lacquers.
Fire hazard: Dangerous.
Shipping regulations: Flammable liquid. Red label. *

methylacetic acid. See propionic acid.

methyl acetoacetate $CH_3COCH_2CO_2CH_3$.
Properties: Colorless liquid; soluble in alcohol; sp. gr. 1.0785 (20/20°C); b.p. 171.7°C; vapor pressure 0.7 mm (20°C); flash point 180°F; wt/gal 9.0 lbs (20°C); f.p. −31.9°C.
Typical specifications: Acidity not more than 0.03% (as acetic); sp. gr. 1.074-1.079 (20/20°C); boiling range, below 91°C (50 mm) not more than 10%, from 90-95°C not less than 85%; purity not less than 95%; color not darker than 8 mg $K_2Cr_2O_7$ in 1 liter of water; solubility complete in alcohol, ether, and ethyl acetate; average wt/gal 8.96 lbs (20°C).
Grades: Technical.
Containers: 1-gal glass jugs; 5-, 12-gal glass carboys; 55-gal drums.
Uses: Solvent for cellulose ethers; ingredient of solvent mixtures for cellulose esters; organic synthesis.

methyl acetone (not a true chemical compound).
Properties: A water-white, anhydrous liquid, consisting of various mixtures of acetone, methyl acetate, and methanol. Miscible with hydrocarbons, oils, and water.
Derivation: A byproduct in the wood-distillation industry.
Grades: Technical (natural and synthetic).
Containers: 1-, 5-gal cans; 55-gal drums; tank cars.
Uses: Solvent for nitrocellulose, cellulose acetate, rubber, gums, resins; lacquers; paint and varnish removers; rubber goods; plastics; cements; artificial leather; gas mantles; extracts; extracting perfumes; dewaxing natural gums.
Shipping regulations: Flammable liquid. Red label. *

methylacetophenone (methyl tolyl ketone).
Properties: Colorless or pale-yellow liquid; fragrant, coumarin odor. Soluble in 7 parts of 50% alcohol. Sp. gr. 1.001-1.004; refractive index 1.533-1.535.
Derivation: Action of acetic anhydride on toluene.
Use: Perfumery.
Shipping regulations: None. *

methylacetylene (propyne) $CH_3C\vdots CH$.
Properties: Available as liquefied gas; b.p. −23.1°C; freezing point −101.5°C; specific volume 9.7 cu ft/lb (70°F).
Uses: Specialty fuel; chemical intermediate.
Shipping regulations: (15-20% propadiene mixture) Flammable gas. Red gas label. *

methyl acetylricinoleate
$C_{17}H_{32}(OCOCH_3)COOCH_3$.
Properties: A clear, pale yellow, low viscosity, oily liquid; mild odor; sp. gr. 0.938 (25/25°C); saponification value 301; iodine value 75; solidifies at −26°C. Soluble in most organic liquids; insoluble in water.
Derivation: Castor oil, methyl alcohol and acetic anhydride.
Grade: Technical.
Containers: 5-gal cans; 55-gal drums; tank cars.
Uses: Plasticizer; lubricant; protective coatings; synthetic rubbers; vinyl compounds.

methyl acetylsalicylate $CH_3COOC_6H_4COOCH_3$.
Properties: White crystals; m.p. 52°C; b.p. 134-136°C (9 mm).
Derivation: By heating methyl salicylate with a slight excess of acetic anhydride, adding alcohol, then water, and separating the precipitate.
Use: Medicine; synthesis.

beta-**methylacrolein.** See crotonaldehyde.

methyl acrylate $CH_2{:}CHCOOCH_3$.
Properties: Colorless, mobile, volatile liquid; b.p. 80.5°C; m.p. −76.5°C; vapor pressure (20°C) 65 mm; sp. gr. (20/20°C) 0.9574; wt/gal 8.0 lbs; slightly soluble in water; readily polymerized.
Derivation: (a) Ethylene cyanohydrin, methanol, and dilute sulfuric acid; (b) Oxo reaction of acetylene, carbon monoxide, and methanol in the presence of nickel or cobalt catalyst.
Grades: Technical (inhibited).
Containers: 1-gal cans; 5-, 55-gal drums; tank cars.
Uses: Polymers; amphoteric surfactants; vitamin B_1; chemical intermediate.
See also acrylate resins.

beta-**methylacrylic acid.** See crotonic acid.

methylal (dimethoxymethane; formal)
$CH_3OCH_2OCH_3$.
Properties: Colorless, volatile, flammable liquid; chloroform-like odor; pungent taste; m.p. −104.8°C; sp. gr. 0.86 (20/4°C); b.p. 42.3°C; soluble in water at 20°C to extent of 32 wt %; completely soluble in alcohol, ether, and hydrocarbons; flash point (open cup) 0°F (approx.).
Typical specifications: Methylal content 97% (min); boiling range 42.0-43.5°C; refractive index (n 20/D) 1.3525-1.3545; aldehydes, % by weight 0.10% (max); water content, % by weight 1.50% (max); acidity as acetic, % by weight 0.10% (max); sp. gr. 0.860-0.863 (20/4°C).
Containers: Glass bottles; steel drums.
Uses: Solvent; starting material for organic synthesis; for perfumes, adhesives and protective coatings; as a special fuel.
Fire hazard: Dangerous.
Shipping regulations: Flammable liquid. Red label. *

methyl alcohol. See methanol.

methylallyl acetate $CH_2{:}C(CH_3)CH_2OOCCH_3$.
Properties: Colorless liquid; sp. gr. 0.9162 (20°C); wt/gal 7.6 lbs.
Uses: Monomer; preparation of methallyl derivatives.

methylallyl chloride (methallyl chloride; MAC)
$CH_2{:}C(CH_3)CH_2Cl$.
Properties: A colorless volatile liquid; disagreeable odor; b.p. 72°C; flash point 14°F; m.p. below −80°C; sp. gr. 0.926.
Uses: Insecticide; fumigant; synthesis.
Containers: 1 gal and 30-gal cans; 5-, and

*See "I.C.C. Shipping Regulations," page xiii.
Reference numbers refer to name of manufacturer. See "List of Manufacturers," page v.

55-gal drums; tank cars.
Shipping regulations: Flammable liquid. Red label. *

methylaluminum sesquibromide $(CH_3)_3Al_2Br_3$.
Properties: Cloudy yellow liquid at 25°C; f.p. −4°C; b.p. (extrapolated, at 760 mm Hg) 166°C; density 1.514 g/ml (25°C).
Uses: Catalyst for polymerization of olefins; catalyst for hydrogenation of aromatics; chemical intermediate.
Caution: Flames instantly on contact with air; reacts violently in contact with water.
Shipping regulations: Flammable liquid. Red label. *

methylaluminum sesquichloride $(CH_3)_3Al_2Cl_3$.
Properties: Clear, colorless liquid at 25°C; f.p. 22.8°C; b.p. (extrapolated, at 760 mm Hg) 143.7°C; density 1.1629 g/ml (25°C), 9.705 lb/gal (25°C).
Uses: Catalyst for polymerization of olefins; catalyst for hydrogenation of aromatics.
Caution: Flames instantly on contact with air; reacts violently on contact with water.
Shipping regulations: Flammable liquid. Red label. *

methylamine (monomethylamine) CH_3NH_2.
Properties: Flammable gas. Strong, ammoniacal odor; b.p. −6.79°C; m.p. −92.5°C; flash point of 30% solution (Tag open cup) 34°F; soluble in water, alcohol ether.
Derivation: Interaction of methanol and ammonia over a catalyst at high temperature. The mono-, di-, and trimethylamines are all produced, and yields are regulated by conditions.
Grades: Technical (anhydrous; 30-40% aqueous solutions).
Containers: 1-gal bottles; 55-gal steel drums; tank cars; cylinders (anhydrous).
Uses: Intermediate for vulcanization accelerators, dyes, pharmaceuticals, insecticides, fungicides, surface active agents; tanning; dyeing of acetate textiles; fuel additive; polymerization inhibitor; component of paint removers; solvent; photographic developer; rocket propellant.
Shipping regulations: Gas: flammable gas. Red gas label. Solution: flammable liquid. Red label. *

methylaminoacetic acid. See sarcosine.

methylaminobenzoate. See methyl anthranilate.

methyl ortho-**aminobenzoate.** See methyl anthranilate.

methylaminodimethylacetal $(CH_3O)_2CHCH_2NHCH_3$.
Properties: Water-white to slightly yellow, clear liquid having a sharp ammoniacal odor; refractive index 1.406-1.409 (n 20/D); sp.gr. 0.924 (25°/25°C).

ℓ-methylaminoethanolcatechol. See epinephrine.

para-**methylaminoethanolphenol tartrate.** See phenylephrine tartrate.

2-(methylamino)glucose. See N-methylglucosamine.

methyl-meta-**amino**-para-**hydroxybenzoate** (orthoform; orthoform-new) $CH_3OOCC_6H_3(NH_2)(OH)$.
Properties: White powder, odorless, tasteless; soluble in alcohol; almost insoluble in water; m.p. 141-143°C.
Use: Medicine.

3-methylaminoisocamphane hydrochloride. See mecamylamine hydrochloride.

2-methylamino-6-methyl-5-heptene. See isometheptene.

N-methyl-para-**aminophenol** $CH_3NHC_6H_4OH$.
Properties: Colorless needles; poisonous! Solutions have irritating effect on the skin. Soluble in water, alcohol and ether. M.p. 87°C.
Derivation: (a) By the interaction of hydroquinone and methylamine. (b) By the methylation of para-aminophenol hydrochloride.
Method of purification: Recrystallization.
Grades: Technical.
Containers: Wooden barrels; glass bottles; fiber drums; multiwall paper sacks.
Uses: Organic synthesis; photographic developer.
Shipping regulations: None. *

N-methyl-para-**aminophenol sulfate** $HOC_6H_4NHCH_3 \cdot \frac{1}{2}H_2SO_4$.
Properties: Colorless needles; m.p. 250-260°C with decomposition; soluble in water and alcohol; insoluble in ether. Discolors in air.
Derivation: By methylation of para-aminophenol and conversion of the resulting methylated base by neutralization with sulfuric acid.
Method of purification: Recrystallization from water.
Grades: C.P.; photographic.
Containers: Kegs; bottles; barrels; fiber cans.
Use: Photographic developer.
Shipping regulations: None. *

methylamyl acetate (methylisobutyl carbinol acetate) $CH_3COOCH(CH_3)CH_2CH(CH_3)_2$.
Properties: Colorless liquid. Mild, agreeable odor. Sp. gr. 0.8595 (20/20°C); b.p. 146.3°C; vapor pressure 3 mm (20°C); flash point 105°F; wt/gal 7.1 lbs (20°C); nitrocellulose-toluene dilution ratio, 1.7; f.p. −63.8°C.
Typical specifications: Acidity not more than 0.02% (as acetic); purity not less than 95% methyl amyl acetate; sp.gr. 0.855-0.860 (20/20°C); boiling range (760 mm) below 140°C none, above 150°C none, not more than 5% distils below 143°C, not less than 95% distills below 148°C; color water-white; dryness, miscible with 19 vol 60° Be. gasoline (20°C); average wt/gal 7.14 lbs (20°C).
Grades: Technical.
Containers: 1-, 5-gal cans; 55-gal (non-returnable) drums; tank cars.
Uses: Solvent for nitrocellulose lacquers.
Caution! Keep away from heat and open

flame. Avoid breathing vapor. Avoid contact with skin. MCA warning label.
Shipping regulations: None. *

methylamyl alcohol (methylisobutyl carbinol; MIBC; 4-methylpentanol-2) $(CH_3)_2CHCH_2CH(CH_3)OH$.
Properties: Colorless, stable liquid. Miscible with most common organic solvents, water. B.p. 131.8°C; sp.gr. 0.8079 (20/20°C); wt/gal 6.72 lbs (20°C); refractive index 1.4089 (25°C); vapor pressure 3.8 mm Hg (20°C); flash point 120°F.
Typical specifications: Acidity not more than 0.01% (as acetic); color water-white; sp.gr. 0.806-0.811 (20/20°C); boiling range (760 mm) below 125°C none, above 135°C none; dryness, miscible with 19 vol 60° Bé. gasoline (20°C); average wt/gal 6.72 lbs (20°C)
Grades: Technical.
Containers: 1-gal cans; 5-gal cans; 55-gal drums; tank cars.
Uses: Solvent for various dyestuffs, oils, gums, resins, waxes; lacquers; organic synthesis.
Caution! Keep away from heat and open flame. Avoid breathing vapor. Avoid contact with skin. MCA warning label.
Shipping regulations: None. *

methyl-n-amyl carbinol (heptanol-2; heptyl alcohol) $CH_3(CH_2)_4CHOHCH_3$.
Properties: Stable colorless liquid; mild odor; miscible with common organic liquids. Sp.gr. 0.8187 (20/20°C); b.p. 160.4°C (760 mm); vapor pressure 1.0 mm (20°C); flash point 160°F; wt/gal 6.8 lbs (20°C).
Typical specifications: Sp.gr. 0.816-0.821 (20/20°C); boiling range (760 mm) 155 to 165°C; acidity not more than 0.03% (as acetic).
Grades: Technical.
Containers: 1-gal cans; 5-gal drums; 55-gal drums; tank cars.
Uses: Solvent for synthetic resins; frothing agent in ore flotation.
Shipping regulations: None. *

methyl-n-amyl-ketone (2-heptanone) $CH_3CH_2CH_2CH_2CH_2COCH_3$.
Properties: Water-white liquid. Stable. Slightly soluble in water; miscible with the usual organic lacquer solvents.
Constants: Sp.gr. 0.8166 (20/20°C); b.p. 150.6°C (760 mm); vapor pressure 2.6 mm (20°C); flash point 120°F; refractive index 1.4110 (20°C); wt/gal 6.8 lbs (20°C) nitrocellulose-toluene dilution ratio 3.9; f.p. −26.9°C.
Typical specifications: Acidity not more than 0.05% (as acetic); purity not less than 95%; color water-white; odor mild, non-residual; sp.gr. 0.816-0.821 (20/20°C); boiling range (760 mm) below 147°C none, above 154°C none, below 149°C not more than 5%, below 152°C not less than 95%; dryness, miscible with 19 vol 60° Bé. gasoline (20°C); non-volatile matter not more than 0.005 g/100 cc; average wt/gal 6.81 lbs (20°C).
Grades: Technical.
Containers: 1-, 5-gal cans; 55-gal drums; carloads.
Caution: Keep away from heat and open flame. Avoid prolonged breathing of vapor, or prolonged or repeated contact with skin. MCA warning label.
Uses: Solvent for nitrocellulose; lacquers; inert reaction medium.
Shipping regulations: None. *

N-methylaniline $C_6H_5NH(CH_3)$.
Properties: Colorless to reddish-brown, oily liquid; discolors on standing. Soluble in alcohol, ether, water and chloroform. Sp.gr. 0.991; m.p. −57°C; b.p. 190-191°C.
Derivation: By heating methyl iodide with aniline and subsequent distillation.
Method of purification: Rectification.
Grades: Technical.
Containers: Iron drums; tank cars.
Use: Organic synthesis; solvent; acid acceptor.
Shipping regulations: None. *

alpha-**methylanisalacetone.** See "Ethone."

5-methyl-ortho-**anisidine** (methylcresidine) $C_6H_3(NH_2)(OCH_3)(CH_3)$.
Properties: Light amber to pink egg-sized lumps, with unpleasant aromatic odor; f.p. 50.5-51.8°C; darkens somewhat on storage.
Grade: 99% min.
Containers: 270-lb polyethylene-lined fiber drums.
Use: Intermediate.

methylanisole. See methyl-para-cresol.

1-methylanthracene. See alpha-methylanthracene.

alpha-**methylanthracene** (1-methylanthracene) $C_{15}H_{12}$ or $C_6H_4(CH)_2C_6H_3CH_3$ (a tricyclic aromatic).
Properties: Colorless leaflets. Soluble in alcohol; insoluble in water. Sp.gr. 1.101; b.p. 200°C; m.p. 86°C.
Grades: Technical.
Use: Organic synthesis.

methyl anthranilate (methyl ortho-aminobenzoate) $H_2NC_6H_4CO_2CH_3$.
Properties: Colorless to pale-yellow liquid with bluish fluorescence; grape-type odor; sp.gr. 1.167-1.175 (15°C); refractive index (n 20/D) 1.5820-1.5840; b.p. 135°C; congealing point 23.8°C (min); soluble in 5 volumes or more of 60% alcohol; soluble in ether, benzyl benzoate, diethyl phthalate, fixed oils, propylene glycol, volatile oils; slightly soluble in water, mineral oil; insoluble in glycerin.
Derivation: By heating anthranilic acid and methyl alcohol in presence of sulfuric acid, with subsequent distillation. Occurs naturally in many flower oils.
Method of purification: Recrystallization.
Grades: Technical.
Containers: 1-, 5-, 10-lb bottles; 25-lb tins; 50-lb cases; 100-lb drums.
Uses: Flavoring; perfume in cosmetics and pomades.
Shipping regulations: None. *

*See "I.C.C. Shipping Regulations," page xiii.
Reference numbers refer to name of manufacturer. See "List of Manufacturers," page v.

methylanthraquinone $CH_3C_6H_3(CO)_2C_6H_4$ (tricyclic).
Properties: White needles; soluble in ether and benzene; very slightly soluble in alcohol; insoluble in water. M. p. 177°C; b. p. sublimes.
Derivation: By heating anthraquinone and methyl alcohol in presence of sulfuric acid.
Method of purification: Crystallization.
Grades: Technical.
Containers: Wooden barrels.
Use: Organic synthesis.
Shipping regulations: None. *

methyl arachidate (methyl eicosanoate) $CH_3(CH_2)_{18}COOCH_3$. The methyl ester of arachidic acid.
Properties: Waxlike solid; m. p. 45. 8°C; b. p. 284°C (100 mm), 216°C (10 mm); refractive index 1. 4352 (50°C). Insoluble in water, soluble in alcohol and ether.
Derivation: Esterification of arachidic acid with methanol and vacuum distillation.
Grades: Purified (99. 8%+).
Use: Special synthesis; intermediate for pure arachidic acid; reference standard for gas chromatography; medical research.

methyl arecaidinate. See arecoline.

methylated spirits. A variety of denatured alcohol (q. v.), containing 5 gallons methanol per 100 gallons 190 proof ethyl alcohol.

methylated trimethylolmelamine $C_3N_3(NHCH_2OH)_2(NHCH_2OCH_3)$(cyclic). A melamine derivative proposed for use on cotton fabrics to increase wrinkle resistance. May be used in partly polymerized form, and as di- and trimethylated as well as in the monomethylated form shown here. Chlorine bleach reacts with the hydrogen atoms attached to nitrogen.

methyl behenate (methyl docosanoate) $CH_3(CH_2)_{20}COOCH_3$. The methyl ester of behenic acid.
Properties: Waxlike solid; m. p. 53. 2°C; b. p. 215. 5°C (3. 75 mm); refractive index 1. 4262 (80°C). Insoluble in water, soluble in alcohol and ether.
Derivation: Esterification of behenic acid with methanol followed by fractional distillation.
Grades: Purified (99. 8%+).
Use: Special synthesis; intermediate for pure behenic acid; biochemical and medical research; reference standard in gas chromatography.

methylbenzaldehydes. See tolyl aldehydes.

methylbenzene. See toluene.

methylbenzethonium chloride $(CH_3)_3CCH_2C(CH_3)_2C_6H_3(CH_3O)(CH_2)_2O$-$(CH_2)_2N(CH_3)_2(CH_2C_6H_5)Cl \cdot H_2O$. Benzyldimethyl(2-[2-(para-1, 1, 3, 3-tetramethylbutylcresoxy) ethoxy] ethyl) ammonium chloride.
Properties: Colorless, odorless crystals with bitter taste; m. p. 161-163°C; readily soluble in alcohol, hot benzene, "Cellosolve," chloroform, and water; insoluble in carbon tetrachloride and ether.
Grade: U. S. P. XVI.
Use: Medicine (bactericide).

methyl benzoate (niobe oil) $C_6H_5CO_2CH_3$.
Properties: Liquid of fragrant odor. Colorless. Oily. Sp. gr. 1. 085-1. 088; refractive index 1. 514; m. p. -12. 3°C; b. p. 198. 6°C. Soluble in 3 parts of 60% alcohol, soluble in ether; very slightly soluble in water.
Derivation: (a) By heating methyl alcohol and benzoic acid in presence of sulfuric acid. (b) Passing dry hydrogen chloride through a solution of benzoic acid in methanol. (c) Occurs naturally in oils of clove, ylang ylang, tuberose.
Method of purification: Rectification.
Grades: Technical.
Containers: 5-, 10-lb bottles; 40-lb tins; drums.
Uses: Perfumery; solvent for cellulose esters and ethers, resins, rubber.
Shipping regulations: None. *

methylbenzoic acid. See ortho-, meta-, and para-toluic acid.

methylbenzol. See toluene.

methyl ortho-**benzoylbenzoate** $C_6H_5COC_6H_4COOCH_3$.
Properties: Sp. gr. 1. 190 (25°C); refractive index 1. 587 (25°C); vapor pressure 4. 0 mm (175°C); b. p. 351°C; m. p. 40°C; flash point, 175°C; very slightly soluble in water.
Use: Plasticizer.

methylbenzoylecgonine. See cocaine.

methyl benzoylsalicylate (benzoylsalicylic acid, methyl ester) $C_6H_5COOC_6H_4COOCH_3$.
Properties: White, acicular crystals; m. p. 85°C; incompatible with acids or highly acid salts. Soluble in alcohol and chloroform; slightly soluble in ether; insoluble in water.
Use: Medicine.

alpha-**methylbenzyl alcohol.** See styralyl alcohol.

alpha-**methylbenzylamine** $C_6H_5CH(CH_3)NH_2$.
Properties: Water-white liquid, mild ammoniacal odor; sp. gr. 0. 9535 (20/20°C); refractive index 1. 5366 (20°C); b. p. 188.5°C (760 mm); vapor pressure 0. 5 mm (20°C); f. p., sets to a glass below -65°C; flash point, 175°F (Cleveland open cup); soluble in most organic solvents and hydrocarbons; somewhat soluble in water.
Uses: Synthesis; emulsifying agent.

alpha-**methylbenzyldiethanolamine** $C_6H_5CH(CH_3)N(C_2H_4OH)_2$.
Properties: Dark amber liquid, ammonia-like odor; sp. gr. 1. 0812 (20°C); b. p. 244°C (50 mm); vapor pressure less than 0. 01 mm (20°C); sets to glass at -7°C; moderately soluble in water.
Grade: Technical.
Uses: Emulsifying agents, textile specialties, quaternaries.
Shipping Regulations: None. *

alpha-**methylbenzyldimethylamine** $C_6H_5CH(CH_3)N(CH_3)_2$.
Properties: Sp. gr. 0.9044 (20/20°C); b.p. 195.6°C (760 mm); vapor pressure 0.6 mm (20°C); f.p., sets to a glass below −70°C; refractive index 1.5024 (20°C); viscosity 1.85 cps (20°C); slight solubility in water.
Use: Polymerization catalyst.

alpha-**methylbenzyl ether** $C_6H_5CH(CH_3)OCH(CH_3)C_6H_5$.
Properties: Straw yellow, mobile liquid with faint odor; sp. gr. 1.0017 (20/20°C); b.p. 286.3°C (760 mm); vapor pressure less than 0.01 mm (20°C); f.p., sets to a glass below −30°C; very slightly soluble in water; flash point 275°F (Cleveland open cup); soluble in most organic solvents.
Uses: Solvent; styrenating agent; softener for synthetic rubbers.

methyl-bis-(2-chloroethyl)amine hydrochloride. See mechlorethamine hydrochloride.

methyl blue. Sodium triphenyl-para-rosaniline sulfonate, a dark blue powder or dye used in medicine as an antiseptic and in biological and bacteriological stains.

methyl borate. See trimethyl borate.

methyl bromide (bromomethane) CH_3Br.
Properties: Colorless, transparent, easily liquefied gas, or volatile liquid; burning taste; chloroform-like odor. Nonflammable. Poisonous! Miscible with most organic solvents; forms a voluminous crystalline hydrate with cold water. Sp. gr. 1.732 (0°C); b.p. 4.6°C; vapor pressure 1250 mm (20°C); m.p. −84°C.
Derivation: By the action of bromine on methyl alcohol in presence of phosphorus, with subsequent distillation.
Method of purification: Rectification.
Grades: Technical; pure (99.5% min).
Containers: Steel cylinders.
Uses: Organic synthesis; low-boiling solvent; refrigerant; fire-extinguishing agent; soil and grain fumigant.
Danger: Causes burns; vapor extremely hazardous; highly volatile. MCA warning label.
Shipping regulations: Class B poison. Poison label. *

methyl bromoacetate $BrCH_2COOCH_3$.
Properties: Colorless to straw-colored liquid; f.p. below −50°C; b.p. 145.0-146.7°C; sp. gr. 1.655 (25/25°C); refractive index 1.456 (25°C); very slightly soluble in water; soluble in methanol, ether.
Uses: Synthesis of weed killers, dyes, vitamins, pharmaceuticals; lachrymator.

2-methyl-1,3-butadiene. See isoprene.

3-methyl-1,3-butadiene. See isoprene.

2-methylbutane. See isopentane.

2-methyl-2-butanethiol. See tert-amyl mercaptan.

2-methyl-1-butanol (amyl alcohol, primary, active; sec-butyl carbinol) $CH_3CH_2CH(CH_3)CH_2OH$. The active alcohol from fusel oil. The synthetic product is a racemic mixture of both dextro- and levorotatory compounds and, therefore, not optically active.
Properties: Colorless liquid; sp. gr. 0.81-0.82 (20°C); f.p., less than −70°C; b.p. 128°C; refractive index (20°C) 1.41. Slightly soluble in water; miscible with alcohol and ether.
Derivation: Occurs in fusel oil (q.v.); is made synthetically by fractional distillation of the mixed alcohols resulting from the chlorination and alkaline hydrolysis of pentane.
Containers: 1- and 5-gal cans; 55-gal drums.
Uses: Solvent; organic synthesis (introduction of active amyl group); lubricants; plasticizers; additives for oils and paints.
Fire hazard: Flash point (open cup) 120°F.
Shipping regulations: None. *

2-methyl-2-butanol. See tert-amyl alcohol.

3-methyl-1-butanol. See isoamyl alcohol, primary.

3-methyl-2-butanone. See methyl isopropyl ketone.

methylbutanoyl chloride. See isovaleroyl chloride.

2-methyl-1-butene C_5H_{10} or $H_2C{:}C(CH_3)CH_2CH_3$.
Properties: Colorless, very volatile flammable liquid; disagreeable odor; b.p. 31.11°C; refractive index 1.378 (n 20/D); sp. gr. 0.650 (20/20°C); f.p. −137.52°C; flash point −48°C; soluble in alcohol; insoluble in water.
Derivation: Refinery gas.
Grades: 95%, 99%, and research.
Containers: Bottles, drums.
Shipping regulations: Flammable liquid. Red label. *

2-methyl-2-butene. See 3-methyl-2-butene.

3-methyl-1-butene (isopropylethylene; alpha-isoamylene) C_5H_{10} or $H_2C{:}CHCH(CH_3)_2$.
Properties: Colorless extremely volatile liquid; disagreeable odor; b.p. 20.1°C; refractive index 1.3643 (n 20/D); sp. gr. 0.6272 (20°C); f.p. −168.5°C; flash point −57°C; soluble in alcohol; insoluble in water.
Derivation: Cracking of petroleum; a component of refinery gas.
Grades: Research, 99% min.; technical 95% min.
Containers: Cylinders under pressure.
Uses: Organic synthesis; high-octane fuel manufacture.
Shipping regulations: Flammable liquid. Red label. *

3-methyl-2-butene (2-methyl-2-butene; trimethylethylene; beta-isoamylene) C_5H_{10} or $H_3CCH{:}C(CH_3)_2$.
Properties: Colorless volatile flammable liquid, disagreeable odor; b.p. 38.51°C; refractive index 1.387 (n 20/D); sp. gr. 0.6623 (20/4°C); f.p. −133.83°C; flash point −46°C; soluble in alcohol; insoluble in water.

Derivation: Cracking of petroleum; a component of refinery gas.
Grades: 90%, 95% (technical), 99% (pure), and research.
Containers: Bottles; drums; tank cars.
Uses: Organic synthesis; dental and surgical anesthetic; high octane fuel manufacture.
Shipping regulations: Flammable liquid. Red label. *

2-methyl-2-butenoic acid. See angelic acid.

methylbutenol (2-methyl-3-buten-2-ol) $(CH_3)_2COHCH{:}CH_2$.
Properties: Clear, colorless liquid; b. p. 96-97.5°C; freezing point −30.5°C; sp. gr. (20/20°C) 0.8249; refractive index (n 20/D) 1.4163; soluble in water and miscible with acetone, benzene, carbon tetrachloride, and kerosene.
Containers: 7-, 35- and 380-lb drums.
Uses: Chemical intermediate, for perfumes, pharmaceuticals.

2-methyl-1-buten-3-yne. See isopropenylacetylene.

5-(1-methylbutyl) barbituric acid $C_3H_7CH(CH_3)C_4N_2O_3H_3$.
Properties: A crystalline powder, usually white (sometimes pale pink); m. p. 165.0°-168.0°C.

1-methyl-4-tert-butylbenzene (para-tert-butyltoluene) $(CH_3)_3CC_6H_4CH_3$.
Properties: Sp. gr. (20/20°C), 0.857-0.863; b. p. 192.8°C; insoluble in water.
Grades: Technical.
Containers: Tank cars; drums.
Uses: Solvent; intermediate.

methyl butyl ketone (propylacetone) $CH_3COC_4H_9$.
Properties: B. p. 127.2°C (760 mm); sp. gr. 0.830 (20/20°C); refractive index 1.4024 (20°C); vapor pressure 10 mm (20°C); soluble in water, alcohol, ether.
Grades: Technical.
Use: Solvent.
Shipping regulations: None. *

2-methyl-6-tert-butylphenol $C_6H_3(OH)(CH_3)(tert\text{-}C_4H_9)$.
Properties (typical): Crystalline solid; light straw color; m. p. 28°C; density 0.9618 (30°C); b. p. 230°C; flash point (open cup) 107°C; soluble in methyl ethyl ketone, ethyl alcohol, benzene, isooctane; insoluble in water.

alpha-**methyl**-beta-(para-tert-**butylphenyl) propionaldehyde.** See "Lilial."

methylbutynol $HC{:}CCOH(CH_3)_2$. 2-Methyl-3-butyn-2-ol.
Properties: Colorless liquid. B. p. 104-105°C; m. p. 2.6°C; sp. gr. (20/20°C) 0.8672; refractive index (n 20/D) 1.4211; flash point (Tag open cup) 77°F. Miscible with water; soluble in most organic solvents.
Containers: 7-, 35-, 385-lb drums; tank trucks and tank cars.
Uses: Stabilizer in chlorinated solvents; viscosity reducer and stabilizer; electroplating brightener; intermediate in synthesis of hypnotics and isoprenoid chemicals such as vitamin A, ionone and perfume alcohols; solvent for alcohol-soluble nylon and polyamide resins.
Shipping regulations: Flammable liquid. Red label. *

methyl butyrate $CH_3CH_2CH_2COOCH_3$.
Constants: Sp. gr. 0.898 (20°C); b. p. 102°C.
Grades: Technical.
Uses: Solvent for ethylcellulose; solvent mixture for nitrocellulose; "Celluloid."

methyl caprate (methyl decanoate) $CH_3(CH_2)_8COOCH_3$.
Properties: Colorless liquid; sp. gr. 0.8733 (20/4°C); m. p. −13.3°C; b. p. 224°C (760 mm), 130.6°C (30 mm); refractive index 1.4237 (25°C). Insoluble in water, soluble in alcohol and ether.
Derivation: Esterification of capric acid with methanol or alcoholysis of coconut oil; purified by fractional vacuum distillation.
Grades: Technical; 99.8% pure.
Use: Intermediate for detergents, emulsifiers, wetting agents, stabilizers, resins, lubricants, plasticizers, textiles.

methyl caproate (methyl hexanoate) $CH_3(CH_2)_4COOCH_3$. The methyl ester of caproic acid.
Properties: Colorless liquid; sp. gr. 0.8850 (20/4°C); m. p. −71°C; b. p. 151.2°C (760 mm), 63.0°C (30 mm); refractive index 1.4049 (20°C); insoluble in water; soluble in alcohol and ether.
Derivation: Esterification of caproic acid with methanol or alcoholysis of coconut oil.
Method of purification: Vacuum fractional distillation.
Grades: Technical; 99.8+%.
Uses: Intermediate for caproic acid, detergents, emulsifiers, wetting agents, stabilizers, resins, lubricants, plasticizers, textiles.

methyl caprylate (methyl octanoate) $CH_3(CH_2)_6COOCH_3$. The methyl ester of caprylic acid.
Properties: Colorless liquid; sp. gr. 0.8784 (20/4°C); m. p. −37.3°C; b. p. 192°C (759 mm), 98.3 (30 mm); refractive index 1.4152 (25°C). Insoluble in water; soluble in alcohol and ether.
Derivation: (a) Esterification of caprylic acid with methanol, (b) alcoholysis of coconut oil.
Method of purification: Vacuum fractional distillation.
Grades: Technical; 99.8%.
Containers: 55-gal drums.
Uses: Intermediate for caprylic acid, detergents, emulsifiers, wetting agents, stabilizers, resins, lubricants, plasticizers, textiles.

methyl "Carbitol." [214] Trademark for diethylene glycol monomethyl ether $CH_3OCH_2CH_2OCH_2CH_2OH$ (q. v.)

methyl carbonate (dimethyl carbonate) $CO(OCH_3)_2$.

*See "I.C.C. Shipping Regulations," page xiii.
Reference numbers refer to name of manufacturer. See "List of Manufacturers," page v.

Properties: Colorless liquid. Pleasant odor. Miscible with acids and alkalies. Stable in the presence of water. Soluble in most organic solvents; insoluble in water. Sp. gr. 1.0718 (20°C); b.p. 90.6°C; m.p. −0.5°C.
Derivation: Interaction of phosgene and methyl alcohol.
Grades: Technical.
Use: Organic synthesis.
Shipping regulations: None. *

methylcatechol. See guaiacol.

methyl "Cellosolve." [214] Trademark for ethylene glycol monomethyl ether, $CH_3OCH_2CH_2OH$ (q.v.).

methyl "Cellosolve" acetate. [214] Trademark for ethylene glycol monomethyl ether acetate $CH_3COOCH_2CH_2OCH_3$ (q.v.).

methylcellulose. A methyl ether of cellulose.
Properties: Grayish white, fibrous powder; aqueous suspensions neutral to litmus. Swells in water and produces a clear to opalescent, viscous, colloidal solution. Insoluble in alcohol, ether and chloroform; soluble in glacial acetic acid; unaffected by oils and greases; stable up to about 300°C; flammable when ignited; stable to light.
Derivation: From cellulose by conversion to alkali cellulose and then reacting this with methyl chloride, dimethyl sulfate, or methyl alcohol, and dehydrating agents. The proportions of the reacting materials are varied to control the properties of the product, such as water solubility and viscosity of the water solutions.
Grades: U.S.P. XVI; technical.
Containers: 45-, 50-lb cartons, bags, car loads.
Uses: Medicine; as dispersing, thickening, emulsifying and sizing agent, and as an adhesive.

methylcellulose, propylene glycol ether. See hydroxypropylmethylcellulose.

methyl cerotate (methyl hexacosanoate) $CH_3(CH_2)_{24}COOCH_3$. The methyl ester of cerotic acid.
Properties: Waxlike solid. Insoluble in water, soluble in alcohol and ether; m.p. 62.9°C; b.p. 237°C (1.95 mm); refractive index 1.4301 (80°C).
Derivation: Esterification of cerotic acid with methanol.
Grades: Purified (99+%).
Uses: Intermediate in special synthesis; medical research; reference standard for gas chromatography.

methyl chavicol. See estragole.

methyl chloride (chloromethane) CH_3Cl.
Properties: Colorless, non-corrosive, liquefiable gas which is transparent in both the gaseous and the liquid state. Faintly sweet, ethereal odor; non-irritant but poisonous; sp.gr. 0.92 (20°C); b.p. −23.7°C (760 mm); m.p. −97.6°C; flash point below 32°C; refractive index 1.3712 (−23.7°C); critical temperature, 143°C; critical pressure 970 psi absolute; specific heat (C_p = 0.24, C_v = 0.20) C_p/C_v = 1.20; wt/gal 7.68 lbs (20°C); soluble in water, alcohol, chloroform, benzene, carbon tetrachloride, glacial acetic acid.
Derivation: Chlorination of methane; action of hydrochloric acid on methyl alcohol in presence of sulfuric acid.
Grades: Pure (99.5% min), technical, and 2 refrigerator grades.
Containers: 60-, 70-, 100-, 145, 300-lb cylinders, and in tank cars (10,000 gal).
Uses: Catalyst carrier in low temperature polymerization as of synthetic rubber, silicones; refrigerant; medicine; as a fluid for thermometric and thermostatic equipment, as a methylating agent in organic synthesis, such as methylcellulose; and as an extractant and a low-temperature solvent; propellant in high pressure aerosols.
Explosive hazard: Moderate; explosions stated to be practically impossible in household refrigeration installations.
Warning! Flammable liquid and gas under pressure. Vapor harmful. MCA warning label.
Shipping regulations: Flammable gas. Red gas label. *

methyl chloroacetate $ClCH_2COOCH_3$.
Properties: Colorless liquid; sp.gr. 1.236 (20/4°C); m.p. −32.7°C; b.p. 131°C; slightly soluble in water; miscible with alcohol and ether.
Use: Solvent.

methylchloroform. See 1,1,1-trichloroethane.

methyl chloroformate $ClCOOCH_3$.
Properties: Colorless liquid. Decomposed by hot water. Stable to cold water. Caution! Very irritant! Soluble in methyl alcohol, ether, and benzene. Sp. gr. 1.23 (15°C); b.p. 71.4°C; vapor density 3.9 (air = 1).
Derivation: Reaction between methyl alcohol and carbonyl chloride.
Grades: Technical (95% min).
Uses: Military poison gas (lachrymator); organic synthesis; insecticides.
Shipping regulations: Corrosive liquid. White label. *

methylchloromethyl ether $ClCH_2OCH_3$.
Properties: Liquid; sp.gr. 1.0625 (10/4°C); m.p. −103.5°C; b.p. 59.5°C; decomposes in water; soluble in alcohol and ether.
Shipping regulations: Flammable liquid. Red label. Not accepted by express. *

2-methyl-4-chlorophenoxyacetic acid. See MCP.

2-methyl-4-chlorophenoxypropionic acid $CH_3C_6H_3ClOCH_2CH_2COOH$. Similar to 2,4-D.
Use: Weed killer.

methyl chlorosilanes. These compounds are used as intermediates in the formation of silicones or siloxanes. They also have the property of reacting with hydroxyl

*See "I.C.C. Shipping Regulations," page xiii.
Reference numbers refer to name of manufacturer. See "List of Manufacturers," page v.

groups on many types of surfaces to produce a permanent thin surface film of silicone that confers water-repellent properties on the surface. The common examples are methyltrichlorosilane, dimethyldichlorosilane, and trimethylchlorosilane (q. v.).

methyl chlorosulfonate CH_3OSO_2Cl.
Properties: Colorless liquid. Pungent odor. Decomposed by water. Caution! Very irritant! Soluble in alcohol, carbon tetrachloride, chloroform; insoluble in water. Sp. gr. 1.492 (10°C); b. p. 133-135°C (dec); m. p. -70°C; vapor density 4.5 (air = 1).
Derivation: Interaction of sulfuryl chloride and methyl alcohol.
Containers: Steel bottles.
Grades: Technical.
Uses: Organic synthesis; military poison gas.

methyl cinnamate $C_6H_5CH{:}CHCOOCH_3$.
Properties: White crystals; strawberry-like odor. Soluble in alcohol and ether; insoluble in water. Sp. gr. 1.0415; m. p. 34°C; b. p. 259.6°C.
Derivation: By heating methyl alcohol, cinnamic acid and sulfuric acid, with subsequent distillation.
Method of purification: Recrystallization.
Grades: Technical.
Containers: 1-, 5-, 10-lb bottles; 25-lb tins.
Uses: Perfumes; flavoring confectionery.
Shipping regulations: None. *

methyl cinnamyl ketone. Benzylideneacetone.

methylcoumarin $C_{10}H_8O_2$.
Properties: White crystals with a vanilla flavor; exists as alpha and beta forms; m. p. (alpha) 90°C, (beta) 82°C; both forms are soluble in alcohol.
Uses: Perfumes; flavoring.

methyl-para-cresol (methylanisole; para-cresyl methyl ether) $CH_3C_6H_4OCH_3$. (The synonyms illustrate better nomenclature.)
Properties: Colorless liquid with a strong floral odor. Sp. gr. 0.967-0.969; refractive index 1.512-1.514. Soluble in 7 parts of 70% alcohol.
Containers: Glass bottles.
Use: Perfumery.
Shipping regulations: None. *

cis-alpha-**methylcrotonic acid.** See angelic acid.

trans-alpha-**methylcrotonic acid.** See tiglic acid.

methyl cyanide. See acetonitrile.

methyl cyanoacetate (malonic methyl ester nitrile) $CNCH_2COOCH_3$.
Properties: Colorless liquid; b. p. 203°C (115°C at 16 mm); m. p. -22.5°C; sp. gr. 1.1225 (15/4°C); soluble in water, alcohol and ether.
Derivation: Esterification of cyanoacetic acid with methanol; reaction of an alkali cyanide and chloracetic methyl ester.
Method of purification: Vacuum distillation.
Containers: 55-gal tinlined steel drums, 15-gal tinlined steel drums; 50-lb boxed tins; 25-, 5-lb bottles.
Uses: Organic synthesis; pharmaceuticals; dyes.
Shipping regulations: None. *

methyl cyanoformate $CNCOOCH_3$.
Properties: Colorless liquid. Ethereal odor. Decomposed by alkalies and water. Soluble in alcohol, benzene, ether. Sp. gr. (approx) 1.00 (20°C); b. p. 100°C.
Derivation: Methylchloroformate is dissolved in methanol and subjected to the action of (hot) sodium or potassium cyanide.
Containers: Glass bottles.
Use: Organic synthesis.
Shipping regulations: None. *

methylcyclohexane (hexahydrotoluene) C_7H_{14}.
Properties: Colorless liquid; sp. gr. 0.769; b. p. 100.8°C; m. p. -126.9°C; refractive index 1.42312; flash point -6°C.
Source: Petroleum.
Grades: Technical (95%); 99%, and research.
Containers: Glass bottles; drums.
Uses: Solvent for cellulose ethers; organic synthesis.
Shipping regulations: Flammable liquid. Red label. *

methylcyclohexanol (hexahydromethyl phenol; hexahydrocresol) $CH_3C_6H_{10}OH$ or $C_7H_{14}O$.
Properties: Colorless, viscous liquid; aromatic, menthol-like odor; non-explosive; slightly toxic.
Derivation: (a) A mixture of three isomeric (ortho, meta and para) cyclic secondary alcohols made by the hydrogenation of cresol; (b) catalytic oxidation of methylcyclohexane.
Grades: Technical.
Containers: 1-, 5-gal cans; 55-gal drums.
Uses: Solvent for cellulose esters and ethers for lacquers; antioxidant for lubricants; blending agent for special textile soaps and detergents.
Caution: Vapor harmful. MCA warning label.
Shipping regulations: None. *

methylcyclohexanol acetate (heptalin acetate) $C_7H_{13}OOCCH_3$.
Properties: Colorless, non-explosive, non-flammable, non-toxic, non-corrosive liquid. Has an ester-like odor. Has a slower rate of evaporation than amyl acetate. B. p. 176-193°C; sp. gr. 0.941; flash point 65°C; toluene dilution ratio 2.5.
Derivation: Catalytic hydrogenation and esterification of cresols by means of acetic acid.

methylcyclohexanone $CH_3C_5H_9CO$.
Properties: Water-white to pale-yellow liquid. Acetone-like odor. It is a mixture of cyclic ketones. Closely resembles cyclohexanone in physical properties, miscibility, tolerance for non-solvents, and solvent action.
Constants: B. p. 160-170°C; sp. gr. 0.925; flash point 130°C; benzene dilution ratio 5.5; toluene dilution ratio 5.5; xylene

dilution ratio 7.0.
Derivation: By high-temperature, catalytic hydrogenation of cresols or by the dehydrogenation of methylcyclohexanol.
Grades: Technical.
Containers: Glass bottles; 1- and 5-gal cans; 55-gal drums.
Uses: Solvent; lacquers.
Caution: Vapor harmful. MCA warning label.
Shipping regulations: None. *

methylcyclohexanone glyceryl acetal $CH_3C_6H_9O_2C_3H_5OH$(spiro rings).
Properties: Sp. gr. 1.074 (20°C); refractive index 1.474 (20°C); b.p. 130-140°C (20 mm); flash point, 113°C; insoluble in water.
Use: Plasticizer.

methylcyclohexanyl oxalate $(CH_3C_6H_{10}OOC)_2$.
Properties: Colorless, odorless, neutral, anhydrous, stable liquid comprising a mixture of isomers. Miscible with most lacquer solvents and diluents.

4-methylcyclohexene-1 $C_6H_9CH_3$.
Properties: Sp. gr. 0.804 (60/60°F); initial b.p. (of available commercial grade) 216.7°F; vapor pressure 0.20 psia (100°F); flash point approx. 30°F.
Containers: Bottles; 5-gal drums.
Grade: Technical.
Shipping regulations: Flammable liquid. Red label. *

6-methyl-3-cyclohexene carboxaldehyde $CH_3CHCH_2CH{:}CHCH_2CHCHO$.
Properties: Sp. gr. 0.9484; b.p. 176.4°C; f.p. −39.0°C; solubility in water, 0.3% by weight at 20°C.
Use: Intermediate for chemical synthesis.

N-methyl-5-cyclohexenyl-5-methylbarbituric acid. See hexobarbital.

N-methylcyclohexylamine $C_6H_{11}NHCH_3$.
Properties: Water white liquid; sp. gr. 0.86 (20°C); soluble in alcohol and ether; slightly soluble in water.
Typical specification: Purity 99%; distillation range, 5-95 cc within 2°C, including 149°C, corrected to 760 mm.
Uses: Intermediate; solvent; acid acceptor.

methylcyclopentadiene dimer (methyl-1,3-cyclopentadiene) $C_{12}H_{16}$.
Properties: Sp. gr. 0.9341 (20/4°C); b.p. 78-183°C. Insoluble in water; very soluble in alcohol, benzene and ether.
Uses: High energy fuels; plasticizers; resins; surface coatings; perfumes; pharmaceuticals; stabilizers; dyes; additives.

methylcyclopentadienyl manganese tricarbonyl $CH_3C_5H_4Mn(CO)_3$.
Derivation: Methylcyclopentadiene with manganese carbonyl.
Use: Gasoline additive; a promotor for the antiknock action of tetraethyl lead. See "Ethyl" Motor 33 Mix, under "Ethyl."

methylcyclopentane $C_5H_9CH_3$.
Properties: Colorless liquid. Sp. gr. 0.750; m.p. −142.5°C; aniline equivalent 4; refractive index 1.40983 (20°C); b.p. 72°C (742 mm); C.S.T. aniline 34.0°C; flash point −27°C.
Grades: Technical (95%), 99%, and Research.
Containers: Glass bottles; cans; steel drums.
Use: Organic synthesis.
Shipping regulations: Flammable liquid. Red label. *

methylcytosine $C_5H_7N_3O$. 5-Methyl-2-oxy-4-aminopyrimidine. A pyrimidine found in deoxyribonucleic acids, nucleotides, and nucleosides.
Properties: Crystallizes in prisms from water (may contain ½ molecule H_2O). M.p. 270° (dec); soluble in water.
Derivation: Isolation following hydrolysis of certain deoxyribonucleic acids; also synthetically.
Use: Biochemical research.

methyl decanoate. See methyl caprate.

methyl dichloroacetate $Cl_2CHCOOCH_3$.
Properties: Liquid; sp. gr. 1.3759-1.3839 (20/20°C); refractive index (n 20/D) 1.4374-1.4474.
Grades: 99.0% pure.
Containers: 5-, 13-gal carboys; 55-gal drums.
Use: Organic intermediate.

methyldichloroarsine CH_3AsCl_2.
Shipping regulations: Poison, class A. Poison gas label by freight. Not accepted by express. *

methyldichlorosilane CH_3SiHCl_2.
Properties: Colorless liquid; b.p. 41°C; sp. gr. 1.10 (27°C).
Derivation: Byproduct from reaction of methyl chloride with silicon and copper.
Uses: Manufacture of siloxanes.
Shipping regulations: Flammable liquid, red label. *

methyl dichlorostearate $C_{17}H_{33}Cl_2COOCH_3$ (approx).
Properties: Light yellow, oily liquid with a slight fatty odor. Completely miscible with most organic solvents; freezing range +7 to −5°C; b.p. decomposes 250°C; sp. gr. 0.997 (15.5/15.5°C); refractive index 1.4599 (n 25/D); flash point 181°C; fire point 210°C.
Containers: 55-gal steel drums.
Uses: Intermediate; plasticizer extender.

methyldiethanolamine $CH_3N(C_2H_4OH)_2$.
Properties: Colorless liquid with amine-like odor. Miscible with benzene, water. Sp. gr. 1.0418 (20°C); b.p. 247.2°C (760 mm); wt/gal 8.7 lb; coefficient of expansion 0.00073; vapor pressure <0.01 mm (20°C); f.p. −21.0°C; viscosity 1.01 poise (20°C); refractive index 1.4699.
Grades: Technical.
Containers: Glass bottles; 1-, 5- and 10-gal cans; 55-gal non-returnable drums.
Use: Intermediate in manufacture of textile auxiliaries, resins, dyestuffs, insecticides, emulsifying agents, corrosion inhibitors and for the absorption of

acidic gases.
Shipping regulations: None. *

4-methyl-7-diethylaminocoumarin $CH_3C_9H_4O(O)N(C_2H_5)_2$.
Properties: Granular; light tan color; m.p. 68-72°C; gives a bright blue-white fluorescence in very dilute solutions in either daylight or ultraviolet light; soluble in aqueous acid solutions, resins, varnishes, vinyls and nearly all common organic solvents such as alcohols, esters, ketones, ethers, glycols, etc; slightly soluble in aliphatic hydrocarbons.
Uses: Optical bleach in textile industry; in overprint and clay casein coatings for paper, labels, book covers, etc; to lighten plastics, resins, varnishes and lacquers; as an invisible marking agent.

6-methyldihydromorphinone hydrochloride. See metopon hydrochloride.

2-methyl-3-dimethylaminopropiophenone hydrochloride $C_6H_5COCH(CH_3)CHN(CH_3)_2 \cdot HCl$. An off-white crystalline powder; m.p. not less than 150.0°C.

methyl N-3, 7-dimethyl-7-hydroxyoctylidene-anthranilate. See "Aurantiol."

methyldioxolane (2-methyl-1, 3-dioxolane) $OCH_2CH_2OCH(CH_3)$.
Properties: Water-white liquid. Soluble in water. Sp.gr. 0.982 (20/20°C); b.p. 81°C.
Grades: Technical.
Uses: As extractant and solvent for oils, fats, waxes, dyestuffs and cellulose derivatives, especially cellulose acetate.

ℓ-N-methyl-1, 2-diphenyl-2-hydroxyethylamine hydrochloride. See ephenamine.

methyldipropylmethane. See 4-methylheptane.

methyl docosanoate. See methyl behenate.

methyl dodecanoate. See methyl laurate.

methyl eicosanoate. See methyl arachidate.

methyl elaidate
$CH_3(CH_2)_7CH{:}CH(CH_2)_7COOCH_3$. The methyl ester of elaidic acid (trans-octadec-9-enoic acid).
Properties: Colorless liquid, insoluble in water, soluble in most organic solvents. Sp.gr. 0.8702 (25°C); m.p. <15°C; b.p. 213.5°C (15 mm); refractive index 1.4462 (25°C).
Derivation: Prepared by elaidinization and esterification.
Use: Pure grade (99+%) used in biochemical research.

N, N'-methylenebisacrylamide
$CH_2(NHCOCH{:}CH_2)_2$.
Properties: Colorless, crystalline solid; m.p. 185°C; sp.gr. 1.235 (30°C).
Uses: Organic intermediate, cross linking agent.

4, 4'-methylenebis(2-chloroaniline). See "Moca."

2, 2'-methylenebis(4-chlorophenol). See dichlorophene.

4, 4'-methylenebis(2, 6-di-tert butylphenol)
$[(C_4H_9)_2C_6H_2(OH)]_2CH_2$.
Properties: Light yellow powder; m.p. 309°F; b.p. 553°F (40 mm); flash point (open cup) 424°F. Insoluble in water and 1.0 N sodium hydroxide.
Uses: Oxidation inhibitor and antiwear agent for motor oils, aviation piston engine oils, and industrial oils.

methylenebis(phenylene isocyanate). See diphenylmethane diisocyanate.

2, 2'-methylenebis(3, 4, 6-trichlorophenol). See hexachlorophene.

methylene blue (methylthionine chloride) $C_{16}H_{18}N_3SCl \cdot 3H_2O$. (medicinal); $(C_{16}H_{18}N_3SCl)_2 \cdot ZnCl_2 \cdot H_2O$ (dye).
Properties: Dark green crystals or powder with bronze-like luster; odorless or slight odor; stable in air. Soluble in water, alcohol, chloroform. Water solutions are deep blue.
Derivation: By oxidation of para-aminodimethylaniline with ferric chloride in the presence of hydrogen sulfide. The dye is the zinc chloride double salt of the chloride; the medicinal product is the hydrochloride of the base.
Grades: U.S.P. XVI; technical.
Containers: Bottles; 25-, 100-lb drums.
Uses: Medicine; dyeing cotton and wool; biological and bacteriological stains; reagent in oxidation-reduction titrations in volumetric analysis; indicator.
Shipping regulations: None. *

methylene bromide (dibromomethane) CH_2Br_2.
Properties: A clear, colorless liquid; sp.gr. 2.47; solidifies −52°C; b.p. 97°C; slightly soluble in water; miscible with alcohol, ether, chloroform and acetone.
Containers: 5-gal carboys.
Uses: Organic synthesis; solvent; heavy gauge liquid.

methylene chloride (methylene dichloride; dichloromethane) CH_2Cl_2.
Properties: Colorless, volatile liquid; penetrating ether-like odor; poisonous when inhaled! Soluble in alcohol and ether; insoluble in water. Sp.gr. 1.335 (15/4°C); m.p. −97°C; b.p. 40.1°C; 11.07 lbs/gal (20°C); refractive index (n 20/D) 1.4244; viscosity (20°C) 0.430 cps.
Derivation: By the chlorination of methyl chloride and subsequent distillation.
Method of purification: Rectification.
Grades: Technical; paint remover.
Containers: Glass bottles; 55-gal drums; 8,000 and 10,000-gal tank cars; tank trucks.
Uses: Component of paint removers; in manufacture of special photographic film; fumigant; solvent for alkaloids, bitumens, crude rubber, oils, resins, waxes, and many organic compounds; solvent mixtures; solvent mixtures for cellulose esters and ethers; textile and leather coatings; lacquers; fire-extinguishing compositions; refrigeration; local anesthetic in dentistry; extraction of oils, fats, perfumes, flavors, and drugs; spotting agent; degreasing;

dewaxing; chemical (organic synthesis); as a propellant for aerosols.
Caution: Avoid prolonged or repeated contact with skin, breathing of vapor. MCA warning label.
Shipping regulations: None. *

methylene chlorobromide. See bromochloromethane.

methylenecitrosalicylic acid (salicitrin) $C_{21}H_{16}O_{11}$.
Properties: White, crystalline powder. Soluble in alkaline solutions and alcohol; almost insoluble in water. M. p. 150-152°C.
Use: Medicine.

methylene cyanide. See malonic dinitrile.

4,4'-methylenedianiline $NH_2C_6H_4CH_2C_6H_4NH_2$.
Properties: Yellow to light brown crystalline solid with faint amine-like odor; f. p. 89.0°C (min); almost insoluble in water and carbon tetrachloride; soluble in acetone and methanol.
Use: Intermediate for adhesives, including 4,4'-methylenediphenyl diisocyanate.

methylene dichloride. See methylene chloride.

3,4-methylenedioxypropylbenzene $C_3H_7C_6H_3O_2CH_2$.
Properties: Sp. gr. 1.065 (25/25°C); somewhat soluble in alcohol.
Uses: Essential oil compositions.

methylene ditannin. See tannoform.

methylene iodide (diiodomethane) CH_2I_2.
Properties: Yellow liquid. Soluble in alcohol and ether; insoluble in water. Sp. gr. 3.33; m. p. 6°C; b. p. 180°C.
Derivation: By heating iodoform with an alcoholic solution of sodium acetate and subsequent distillation.
Method of purification: Rectification.
Grades: Technical; C. P.
Containers: Steel drums; glass bottles.
Uses: Separating mixtures of minerals; organic synthesis.
Shipping regulations: None. *

methylene-para-toluidine. See formaldehyde-para-toluidine.

methylergonovine maleate $C_{14}H_{11}NHNCH_3CONHCH(C_2H_5)CH_2OH \cdot H_2C_4H_2O_4$.
Properties: White to pinkish tan powder; odorless; bitter taste; unstable to light and to heat; soluble in water, alcohol; very slightly soluble in chloroform, ether; pH of solution (1 in 5000) is between 4.4 and 5.2.
Grade: N. F. XI.
Use: Medicine.

N-methylethanolamine $CH_3NHC_2H_4OH$.
Properties: Liquid; sp. gr. 0.9414; b. p. 159.5°C; vapor pressure 0.7 mm (20°C); f. p. −4.5°C; flash point 165°F; soluble in water.
Containers: Cans; drums.
Uses: Textile chemicals; pharmaceuticals.

methyl ether. See dimethyl ether.

4-methyl-7-ethoxycoumarin. See "Maraniol."

methylethylcarbinol. See sec-butyl alcohol.

methyl ethyl diketone. See acetyl propionyl.

2-methyl-2-ethyl-1,3-dioxolane $(CH_3)(C_2H_5)\underbrace{COCH_2CH_2O}$.
Properties: Sp. gr. 0.9392; b. p. 117.6°C; f. p. −81.96°C; flash point 74°F; solubility in water 2.2% by weight.
Shipping regulations: Flammable liquid. Red label. *

methylethylene glycol. See propylene glycol. However, in Europe this term, written methyl ethylene glycol, is regarded as synonymous with ethylene glycol monomethyl ether.

sym-**methylethylethylene.** See n-beta-amylene.

3,3-methylethylglutarimide. See bemegride.

methylethylglyoxal. See acetyl propionyl.

methyl ethyl ketone (ethyl methyl ketone; 2-butanone; MEK) $CH_3COC_2H_5$.
Properties: Colorless liquid; acetone-like odor; flammable; b. p. 79.6°C; sp. gr. 0.8255 (0/4°C), 0.805 (20/4°C), and 0.7997 (25/4°C); refractive index 1.379 (20°C); sp. heat 0.549 cal/g; m. p. −86.4°C; heat of combustion 14520 Btu/lb; heat of vaporization 191 Btu/lb, viscosity 0.40 cps (25°C); solubility in water 22.6 wt %; solubility of water 9.9 wt %; coefficient of expansion 0.0013/°C (approx); flash point 24°F; 6.71 lbs/gal (20°C). Soluble in water, alcohol and ether; miscible with oils.
Typical specifications: Specific gravity 0.805–0.807 (20/20°C); color water-white; acidity (acetic acid) 0.002% (max); moisture — clear with 19 vols of naphtha; distillation — initial 79°C (min), dry point 81.5°C (max); non-volatile matter 2 mg/100 cc; purity 98.0% wt (min); residual odor — none.
Derivation: From mixed n-butylenes and sulfuric acid to cause hydrolysis, followed by distillation to separate sec-butyl alcohol, which is dehydrogenated; by controlled oxidation of butane.
Method of purification: Rectification.
Grades: Technical.
Containers: Cans, 1 and 5 gal; drums, 55 gal; tank wagons up to 2000 gal; tank cars 6000 or 8000 gal.
Uses: Lacquers; dewaxing of lubricating oil; paint removers; cements and adhesives; celluloid; dopes; organic synthesis; manufacture of smokeless powder; solvent; artificial leather dressings; dyes; cleaning fluids; shoe manufacture; printing.
Warning! Flammable. MCA warning label.
Shipping regulations: Red label. *

methyl ethyl ketone peroxide. See "Lupersol DDM."

2-methyl-5-ethylpyridine (MEP; aldehydine; aldehyde collidine; 5-ethyl-2-picoline) $CH_3C_5H_3NC_2H_5$.

Properties: Colorless liquid; sharp penetrating odor; sp. gr. 0.921 (20/20°); b. p. 178.3°C (760 mm); f. p. −70.3°C; flash point 165°F (Cleveland open cup); refractive index (n 20/D) 1.4970; slightly soluble in water.
Derivation: Paraldehyde is treated with ammonia under high pressure and in the presence of ammonium acetate as a catalyst. Picolines and other substituted pyridines are byproducts.
Method of purification: Fractional distillation.
Grades: Technical.
Containers: Drums; tank cars.
Uses: Nicotinic acid and nicotinamide; vinyl pyridines for copolymers; intermediates for germicides and textile finishes; corrosion inhibitor for chlorinated solvents.
Shipping regulations: None. *

N-methylformanilide $C_6H_5N(CH_3)CHO$.
Properties: Colorless to light yellow liquid; refractive index 1.5570-1.5600 (n 25/D); distillation range 127-131°C (16 mm).
Grade: 95% min.
Containers: 30-, 55-gal steel drums.
Use: Organic synthesis.

methyl formate $HCOOCH_3$.
Properties: Colorless, flammable liquid; agreeable odor. Saponified by water or alkaline solutions. Soluble in water, alcohol and ether.
Constants: Sp. gr. 0.950-0.980 (20/20°C); m. p. −99.8°C; b. p. 31.8°C; flash point −32°C (−25.6°F); vapor pressure 0°C 195.0 mm, 10°C 309.4 mm, 20°C 476.4 mm, 30°C 707.9 mm; wt/gal 8.03 lbs (68°F); electric conductivity 3.6×10^{-5} reciprocal ohm (25°C); refractive index 1.3431 (20°C).
Derivation: By heating methyl alcohol with sodium formate and hydrochloric acid, with subsequent distillation.
Method of purification: Rectification.
Grades: Technical; refined.
Containers: Iron drums; tank cars.
Uses: Organic synthesis; cellulose acetate solvent; making military poison gases; fumigant; larvicides.
Fire hazard: Dangerous.
Shipping regulations: Flammable liquid. Red label. *

2-methylfuran $C_4H_3OCH_3$.
Properties: Colorless, mobile liquid; ether-like odor; m. p. −88.68°C; b. p. 63.2-65.6°C (760 mm); sp. gr. 0.913 (20/4°C); refractive index: 1.4320 (20/D); flash point −16°F (Tag open cup). Practically insoluble in water, 0.3 g/100 g water. Infinitely miscible with most organic solvents. Forms a binary azeotrope with methanol, a ternary azeotrope with methanol-water.
Containers: 1-, 5-, 55-gal containers; net weight: 7, 35, and 375 lbs, respectively.
Use: Chemical intermediate.
Shipping regulations: Flammable liquid. Red label. *

methyl furoate $C_4H_3OCO_2CH_3$.
Properties: Colorless liquid, turning yellow in light. Pleasant odor. Practically insoluble in water; soluble in alcohol and ether. Sp. gr. 1.1739 (15/15°C); b. p. 181.3°C (corr); refractive index 1.4860 (20°C/D).
Derivation: By the usual esterification methods for furoic acid.
Containers: Amber-glass bottles; 1-, 5- and 10-gal tins.
Uses: Solvent; organic synthesis.
Shipping regulations: None. *

N-methylglucamine $CH_2OH(CHOH)_4CH_2NHCH_3$.
Properties: White crystals; m. p. 128°C. Soluble in water; slightly soluble in alcohol.
Preparation: From glucose and methylamine.
Uses: Detergents, pharmaceuticals, dyes.

methylglucamine diatrizoate (diatrizoate methylglucamine; 3,5-diacetylamino-2,4,6-triiodobenzoic acid, methylglucamine salt) $(CH_3CONH)_2C_6I_3COOH \cdot CH_3NHCH_2$-$(CHOH)_4CH_2OH$.
Properties: Available in solution for injection as a clear, colorless to pale yellow, slightly viscous liquid with a pH between 6.0-7.6.
Grade: U. S. P. XVI.
Use: Medicine (radiopaque medium).

methylglucamine iodipamide (iodipamide bis(N-methylglucamine) salt) $C_{20}H_{14}I_6N_2O_6 \cdot 2C_7H_{17}NO_5$. 3,3'-(Adipoyldiimino)bis-2,4,6-triiodobenzoic acid, methylglucamine salt.
Properties: Crystals; soluble in water. Solutions are radio-opaque.
Grades: U. S. P. XVI (solution for injection).
Use: Roentgenographic contrast medium.

methyl-alpha-D-glucopyranoside. See methyl glucoside.

N-methyl-L-glucosamine (2-methylamino glucose) $CH_2OH(CHOH)_3CH(NHCH_3)CHO$ (this formula only approximate). A hexosamine found in streptomycin.
Properties (of hydrochloride): Crystals with m. p. 160-163°. Freely soluble in water. Exhibits mutarotation in solution.
Derivation: Hydrolysis of streptomycin.

methyl glucoside (methyl alpha-D-glycopyranoside) $CH_2OH\underline{CH(CHOH)_3CH}OOCH_3$.
Properties: Odorless, white crystals. M. p. 168°C; b. p. 200°C (0.2 mm); specific optical rotation (aqueous solution) +158.9° (20°C); sp. gr. (30/4°C) 1.46. Soluble in water and 80% alcohol; slightly soluble in methanol; insoluble in ether.
Derivation: (a) By treating dextrose with methanol in the presence of hydrochloric acid or cation exchange resin; (b) enzymatic synthesis from yeast.
Grades: Technical.
Containers: 100-lb multiwall paper bags.
Uses: Alkyd resins; drying oils; plasticizer for phenolic, amine, and alkyd resins; nonionic surfactants.

methyl glycocoll. See sarcosine.

methyl glycol. See propylene glycol. However, in Europe this term is regarded as synonymous with ethylene glycol monomethyl ether.

methylglyoxalidine. See lysidine.

(alpha-**methylguanido) acetic acid.** See creatine.

N-methyl-N-guanylglycine. See creatine.

methyl heneicosanoate $CH_3(CH_2)_{19}COOCH_3$. The methyl ester of heneicosanoic acid.
Properties: White waxlike solid. Insoluble in water, soluble in alcohol and ether. M. p. 48-9°C; b. p. 207°C (3.75 mm).
Grades: Purified 96%, 99.5%.
Uses: Intermediate in organic synthesis; medical research.

methyl heptadecanoate (methyl margarate) $CH_3(CH_2)_{15}COOCH_3$. The methyl ester of heptadecanoic acid (margaric acid).
Properties: White waxlike solid. Insoluble in water, soluble in alcohol and ether. M. p. 29°C; b. p. 184-7°C (760 mm); 130°C (1 mm).
Grades: Purified 96%, 99.5%.
Uses: Intermediate in organic synthesis; medical research.

2-methylheptane $(CH_3)_2CH(CH_2)_4CH_3$. Also known as isooctane (q. v.).
Properties: Colorless liquid; sp. gr. 0.6979 (20°C); f. p. −109°C; b. p. 117.6°C (760 mm); refractive index (20/D) 1.3949. Insoluble in water, soluble in alcohol and ether.
Use: Organic synthesis.

3-methylheptane $C_2H_5CH(CH_3)(CH_2)_3CH_3$.
Properties: Colorless liquid; m. p. −120.5°C; b. p. 118.927°C; sp. gr. 0.70582 (20/4°C); refractive index 1.39849 (n 20/D).
Grades: 99%, 95%.
Use: Calibration; organic synthesis.

4-methylheptane (methyldipropylmethane) C_8H_{18} or $CH_3(CH_2)_2CHCH_3(CH_2)_2CH_3$.
Properties: Colorless liquid. Soluble in alcohol and ether; insoluble in water. Sp. gr. 0.7161; b. p. 122.2°C.
Grades: Technical.
Use: Organic synthesis.

methylheptenone (6-methyl-5-hepten-2-one) $(CH_3)_2C{:}CH(CH_2)_2COCH_3$. Constituent of lemon grass oil and many other essential oils.
Properties: Colorless liquid; insoluble in water but miscible with alcohol or ether. Sp. gr. 0.860 (20°C); m. p. −67.1°C; b. p. 173-174°C.
Derivation: From oil of lemon grass or by controlled oxidation of corresponding secondary alcohol.
Containers: Glass bottles; tins.
Uses: Organic synthesis; inexpensive perfumes.
Shipping regulations: None. *

methyl heptine carbonate $CH_3(CH_2)_4C{:}CCOOCH_3$. See also "Folione."
Properties: Colorless liquid, having an extremely strong violet-type odor. Sp. gr. 0.919-0.923; refractive index 1.446-1.450. Soluble in 4 parts of 70% alcohol.
Derivation: From heptaldehyde.
Containers: Glass bottles; tins.
Use: Perfumery, particularly for violet-type odors.
Shipping regulations: None. *

methyl hexacosanoate. See methyl cerotate.

methyl hexadecanoate. See methyl palmitate.

methylhexamine. See 4-methyl-2-hexylamine.

2-methylhexane (ethylisobutylmethane) C_7H_{16} or $(CH_3)_2CH(CH_2)_2CH_2CH_3$.
Properties: Colorless liquid. Soluble in alcohol; insoluble in water. Sp. gr. 0.6789; b.p. 90.0°C; f. p. −118.5°C; refractive index 1.38498 (20°C); C. S. T. aniline 73.6°C.
Grades: Technical.
Containers: Glass bottles; 1-, 5-, and 10-gal tins; 55-gal drums.
Use: Organic synthesis.

3-methylhexane $H_3CCH_2CH(CH_3)CH_2CH_2CH_3$.
Properties: Colorless liquid; b. p. 92°C; sp. gr. 0.692 (60/60°F); refractive index 1.388 (n 20/D); flash point −4°C.
Grades: Technical 95%.
Containers: Bottles and 5-gal drums.
Shipping regulations: Flammable liquid. Red label. *

methylhexaneamine. See 4-methyl-2-hexylamine.

methyl hexanoate. See methyl caproate.

5-methyl-2-hexanone. See methyl isoamyl ketone.

1-methyl-2-hexylamine. See 2-aminoheptane.

4-methyl-2-hexylamine. (2-amino-4-methylhexane; methylhexaneamine; methylhexamine) $CH_3CH_2CH(CH_3)CH_2CH(NH_2)CH_3$.
Properties: Colorless to pale yellow liquid with amine odor; b. p. 130-135°C; density 0.762 to 0.765; refractive index 1.4150 to 1.4175 (n 25/D). Very slightly soluble in water; soluble in alcohol, chloroform, ether, dilute acids.
Grade: N. N. D.
Use: Medicine.

methylhexylcarbinol. See 2-n-octanol.

methyl hexyl ketone (2-octanone) $CH_3COC_6H_{13}$.
Properties: Colorless liquid with pleasant odor; camphor taste; sp. gr. 0.82 (20/4°C); m. p. −20.9°C; b. p. 173.5°C; distillation range 166-173°C; refractive index 1.416 (20°C); insoluble in water; soluble in alcohol, hydrocarbons, ether, esters, etc.
Preparation: By distilling sodium ricinoleate with caustic soda.
Containers: Glass bottles; 1-, 5- and 10-gal tins; 55-gal drums.
Uses: Perfumes; high-boiling solvent, especially for epoxy resin coatings; a major constituent in leather finishes; odorant; as an anti-blushing agent for

nitrocellulose lacquers.
Shipping regulations: None. *

methylhydrazine CH_3NHNH_2.
Properties: Colorless hygroscopic liquid; sp. gr. 0.874 (25°); m.p. −52.4°C; b.p. 87.5°C. Slightly soluble in water; soluble in alcohol and ether.
Uses: Missile propellant; intermediate; solvent.
Shipping regulations: Flammable liquid. Red label. *

methyl hydride. See methane.

methyl hydrogen sulfate. See methylsulfuric acid.

17-alpha-methyl-17-beta-hydroxyandrosta-1, 4-dien-3-one. See methandrostenolone.

methyl-para-hydroxybenzoate. See methylparaben.

methylhydroxybutanone (3-methyl-3-hydroxy-2-butanone) $(CH_3)_2COHCOCH_3$.
Properties: Clear, colorless liquid with sweet, camphor-like odor; b.p. 140.3°C; freezing point −86.5°C; sp. gr. (20/20°C) 0.9553; refractive index (n 20/D) 1.4153; miscible with water, acetone, benzene, mineral spirits.
Containers: 7-, 40- and 440-lb drums.
Uses: Specialty solvent; chemical intermediate; flavor formulations.

1-methyl-2-hydroxy-4-isopropyl benzene $C_6H_3(CH_3)(OH)(isoC_3H_7)$. Used in food packaging. See isopropylcresols.

methylhydroxyisopropylcyclohexane. See menthol.

methyl 12-hydroxystearate. $C_{17}H_{34}OHCOOCH_3$.
Properties: White, waxy solid in the form of short flat rods; m.p. 48°C; acid value 4; saponification value 177; iodine value 5; insoluble in water, limited solubility in organic solvents.
Uses: Adhesives, inks; cosmetics; greases.

3-methylindole. See skatole.

beta-methylindole. See skatole.

methyl iodide (iodomethane) CH_3I.
Properties: Colorless liquid; turns brown on exposure to light; sp. gr. 2.24-2.27 (25/25°C); m.p. −66.1°C; b.p. 42°C; refractive index 1.526-1.527 (25°C). Soluble in alcohol and ether; insoluble in water.
Derivation: By the interaction of methyl alcohol, sodium iodide and sulfuric acid, with subsequent distillation.
Method of purification: Rectification.
Grades: Technical.
Containers: Amber-glass bottles and carboys, 90 lbs net, 120 lbs gross.
Caution: Burns skin; avoid contact.
Uses: Medicine; organic synthesis; in microscopy and in testing for pyridine.

methylionone (irone) $C_{14}H_{22}O$.
Properties: Colorless to amber-yellow liquid, with soft violet odors.
Grades: Several isomers are available, as alpha, beta, delta, gamma, and mixtures of such. The constants vary according to the content of these isomers, but fall approximately within the following limits: Sp. gr. 0.926-0.939; refractive index 1.501-1.504; b.p. 144 (16 mm); soluble in alcohol; insoluble in water.
Derivation: Oil of orris.
Containers: Glass bottles; cans; drums.
Use: Perfumery.
Shipping regulations: None. *

gamma-methylionone. See "Cetone Alpha."

methyl isoamyl ketone (5-methyl-2-hexanone) $CH_3COC_2H_4CH(CH_3)_2$.
Properties: Colorless, stable liquid; pleasant odor. Sp. gr. 0.8132 (20/20°C); b.p. 144°C; f.p. −73.9°C; wt/gal 6.77 lbs; flash pt. 110°F (open cup). Slightly soluble in water; miscible with most organic solvents.
Containers: Drums; tank cars.
Uses: Solvent for nitrocellulose, cellulose, acetate butyrate, acrylics, and vinyl copolymers.

methylisobutenylketone. See mesityl oxide.

methylisobutyl carbinol. (MIBC) See methylamyl alcohol.

methylisobutyl carbinol acetate. See methylamyl acetate.

methyl isobutyl ketone (hexone; 2-methyl-4-pentanone) $(CH_3)_2CHCH_2COCH_3$.
Properties: Colorless, stable liquid. Pleasant odor. Slightly soluble in water; miscible with most organic solvents. Sp. gr. 0.8042 (20/20°C); b.p. 115.8°C; f.p. −80.4°C; wt/gal 6.68 lbs (20°C); vapor pressure 15.7 mm (20°C); refractive index 1.3959 (20°C); surface tension 25.4 dynes/cm (25°C); viscosity 0.0059 poise (20°C); specific heat 0.496 cal/g (25°C); latent heat of evaporation 86.0 cals/g at b.p.; flash point 23°C (74°F); coefficient of expansion 0.00114 (per °C) to 20°C, 0.001170 (per °C) to 55°C; explosive limits % by volume in air: lower 1.4%, upper 7.5%.
Typical specifications: Acidity not more than 0.02% (as acetic); purity not less than 95% by weight; sp. gr. 0.799-0.804 (20/20°C); color water-white; boiling range (760 mm) below 111°C none, above 117°C none; average wt/gal 6.71 lbs (60°F).
Derivation: Mild hydrogenation of mesityl oxide.
Grades: Technical, 98.5%.
Containers: 1-, 5-gal cans; 55-gal drums; tank cars.
Uses: Solvent for nitrocellulose, certain types of cellulose ethers, camphor, oils, fats, waxes, and various natural and synthetic gums and resins; solvent mixtures for cellulose acetate, lacquers, finishes; lacquers; extraction processes; organic synthesis.
Warning: Flammable. MCA warning label.
Shipping regulations: Flammable liquid. Red label. *

methylisoeugenol (propenyl guaiacol) $C_3H_5C_6H_3(OCH_3)_2$.

Properties: Colorless to light-yellowish liquid, with spicy odor. Sp. gr. 1.050-1.053; b.p. 262-264°C; refractive index 1.566-1.569. Soluble in 2 parts of 70% alcohol.
Containers: Glass bottles; drums.
Use: Perfumery, for carnation and other types.

methyl isonicotinate $C_5NH_4COOCH_3$.
Properties: Clear amber to red liquid; mild odor, f.p. 145°C; sp. gr. 1.15 (20/20°C).
Use: Intermediate for synthesis of isonicotinic acid hydrazide.

methyl isopropenyl ketone $CH_3COC(CH_3){:}CH_2$.
Properties: Liquid; sp. gr. 0.854 (20°C); undergoes polymerization and copolymerization readily.
Use: Plastics.
Shipping regulations: Flammable liquid. Red label. *

methylisopropoxyfluorophosphine oxide. See Sarin.

methyl isopropyl ketone (3-methyl-2-butanone) $CH_3COCH(CH_3)_2$.
Properties: A colorless liquid; b.p. 93°C; m.p. -92°C; refractive index (n 16/D) 1.38788; density 0.815 (15/4°C); very slightly soluble in water; soluble in alcohol, ether.
Derivation: Synthetic, and also a fermentation by-product.
Warning: Flammable. MCA warning label.

5-methyl-2-isopropylphenol. See thymol.

methyl-para-isopropylphenyl propyl aldehyde. See cyclamen aldehyde.

methyl lactate $CH_3CHOHCOOCH_3$.
Properties: Liquid. Miscible with most organic liquids, water. B.p. 144.8°C; m.p. (approx) -66°C; refractive index 1.4156 (20°C); flash point 51.7°C (125°F); wt/gal 9 lb (68°F).
Typical specifications: Purity not less than 95% ester, by weight; sp. gr. 1.087-1.097 (20/20°C); acidity not more than 0.15%, calculated as lactic acid; water no turbidity when one volume is mixed with 19 volumes of 60° Bé. gasoline (20°C); color water-white; distillation range below 115°C none, between 141°C and 145°C not less than 60%, above 155°C none.
Grades: Technical.
Containers: 1-gal cans; 5-, 55-gal steel drums.
Uses: Solvent for cellulose acetate, nitrocellulose, cellulose acetobutyrate, cellulose acetopropionate; lacquers; stains.
Shipping regulations: None. *

methyllactonitrile. See acetone cyanohydrin.

methyl laurate (methyl dodecanoate) $CH_3(CH_2)_{10}COOCH_3$. The methyl ester of lauric acid.
Properties: Water-white liquid; sp. gr. 0.8702 (20/4°C); m.p. 4.8°C; b.p. 262°C (766 mm), 160°C (30 mm); refractive index 1.4301 (25°C). Insoluble in water; non-corrosive.
Typical specifications: 95% methyl laurate, remainder methyl caprate and methyl myristate; acid value less than 1; saponification number 260.
Derivation: From coconut oil.
Method of purification: Vacuum fractional distillation.
Grades: 69, 74, 90, 96, 99.8%.
Containers: Cans; drums; tank cars.
Uses: Intermediate for detergents, emulsifiers, wetting agents, stabilizers, resins, lubricants, plasticizers, textiles.

methyl lauroleate $CH_3CH_2CH{:}CH(CH_2)_7COOCH_3$. The methyl ester of lauroleic acid.
Properties: Colorless liquid; insoluble in water; soluble in common organic solvents.
Grades: Purified product, 99.5%.
Uses: In medical research, organic synthesis and as a reference standard for gas chromatography.

methyl lignocerate (methyl tetracosanoate) $CH_3(CH_2)_{22}COOCH_3$. The methyl ester of lignoceric acid.
Properties: Waxlike solid. Insoluble in water, soluble in alcohol and ether; m.p. 57.8°C; b.p. 232°C (3.75 mm); refractive index 1.4283 (80°C).
Derivation: Esterification of lignoceric acid with methanol followed by vacuum distillation.
Grades: Purified (99.8%+).
Uses: Intermediate in special synthesis; medical research; reference standard in gas chromatography.

methyl linoleate
$CH_3(CH_2)_4CH{:}CHCH_2CH{:}CH(CH_2)_7COOCH_3$. The methyl ester of linoleic acid (cis, cis-octadec-9,12-dienoic acid).
Properties: Colorless oil, insoluble in water; soluble in alcohol and ether. Sp. gr. 0.8886 (18/4°C); m.p. -35°C; b.p. 212°C (16 mm); refractive index 1.4593 (25°C).
Derivation: Urea fractionation and vacuum distillation of methyl esters of safflower oil.
Grades: Technical; purified (99+%).
Uses: Intermediate for detergents, emulsifiers, wetting agents, stabilizers, resins, lubricants, plasticizers, textiles.

methyl linolenate
$CH_3CH_2CH{:}CHCH_2CH{:}CHCH_2CH{:}CH(CH_2)_7$-$COOCH_3$. The methyl ester of linolenic acid (cis, cis, cis-octadec-9,12,15-trienoic acid).
Properties: Colorless liquid; insoluble in water; soluble in alcohol and ether. Sp. gr. 0.892 (20/4°C); m.p. less than 15°C; b.p. 207°C (14 mm); refractive index 1.4632 (40°C).
Derivation: Esterification and vacuum fractional distillation.
Uses: Organic synthesis and biochemical and medical research.

methylmagnesium bromide CH_3MgBr.
Properties: Solutions in ether. Flammable liquid.
Derivation: Reaction of magnesium and methyl bromide.

Containers: Glass bottles; 5- and 55-gal drums.
Uses: Alkylating agent in organic synthesis; Grignard reagent.
Warning! Flammable!
Shipping regulations: Flammable liquid. Red label. *

methylmagnesium iodide CH_3MgI.
Properties: Solutions in ether. Flammable liquid.
Derivation: Reaction of magnesium and methyl iodide.
Containers: Glass bottles; 5- and 55-gal drums.
Uses: Alkylating agent in organic synthesis; Grignard reagent.
Warning! Flammable!
Shipping regulations: Flammable liquid. Red label. *

methylmaleic acid. See citraconic acid.

methylmaleic anhydride. See citraconic anhydride.

methyl margarate. See methyl heptadecanoate.

methyl mercaptan (methanethiol) CH_3SH.
Properties: Water white liquid when below boiling point, or colorless gas; powerful unpleasant odor; highly flammable; toxic. M. p. −121°C; density 0.87 (20°C); flash point below 0°C; b. p. 7.6°C; insoluble in water; soluble in alcohol, ether, petroleum naphtha.
Grades: 98.0% purity.
Containers: 180-lb cylinders; tank cars.
Use: Synthesis.
Shipping regulations: Flammable gas. Red gas label. *

1-methyl-2-mercaptoimidazole. See methimazole.

methylmercury dicyandiamide. See cyano-(methylmercuri)guanidine.

methylmercury oxyquinolinolinate $C_9NH_{14}OHgCH_3$.
Properties: Brown solution. Pure compound not isolated.
Use: Seed fungicide.

methyl metaborate CH_3BO_2. Used in fire extinguishers.

methyl methacrylate $CH_2{:}C(CH_3)COOCH_3$. An important monomer.
Properties: Colorless, mobile, volatile liquid; b. p. 101°C; m. p. −48.2°C; sp. gr. (25/25°C) 0.940; flash point (open cup) 85°F; slightly soluble in water; readily polymerized.
Derivation: Acetone cyanohydrin, methanol, and dilute sulfuric acid.
Grades: Technical (inhibited).
Containers: 440-lb drums; tank cars.
Uses: Polymers and copolymers. See also acrylate resins.
Warning! Flammable; may cause skin irritation. MCA warning label.
Shipping regulations: Flammable liquid. Red label. *

methyl methacrylate resins. See acrylate resins.

methylmethane. See ethane.

N-methyl methyl anthranilate. See dimethyl anthranilate.

methyl beta-methylthiopropionate. Material occurring in pineapple. Essential in reconstituting true pineapple flavor.

methylmorphine. See codeine.

N-methyl morpholine $\underbrace{CH_2CH_2OCH_2CH_2N}CH_3$.
Properties: Water-white liquid. Forms constant boiling mixture with water containing 25% water and boiling at 97°C. Miscible with benzene, water. Sp. gr. 0.921 (20/20°C); b. p. 115.4°C.
Grades: Technical.
Containers: Glass bottles; tins; drums.
Uses: Catalyst in polyurethane foams; extraction solvent; stabilizing agent for chlorinated hydrocarbons; self-polishing waxes; oil emulsions; corrosion inhibitors; pharmaceuticals.

methyl myristate (methyl tetradecanoate) $CH_3(CH_2)_{12}COOCH_3$. The methyl ester of myristic acid.
Properties: Colorless liquid at room temperature; m. p. 17.8°C; b. p. 186.8°C (30 mm), 157.5°C (1 mm); refractive index 1.4351 (25°C). Insoluble in water.
Derivation: (a) Esterification of myristic acid with methanol, (b) alcoholysis of coconut oil with methanol.
Method of purification: Vacuum fractional distillation.
Grades: Technical (93%); purified (99.8+%).
Containers: 55-gal drums.
Uses: Intermediate for myristic acid, detergents, emulsifiers, wetting agents, stabilizers, resins, lubricants, plasticizers, textiles, animal feeds.

methyl myristoleate $CH_3(CH_2)_3CH{:}CH(CH_2)_7COOCH_3$. The methyl ester of myristoleic acid (cis-tetradec-9-enoic acid).
Properties: Colorless liquid; insoluble in water; soluble in alcohol and ether. B. p. 108.9°C (1 mm).
Uses: Purified product used in medical research and organic synthesis.

alpha-**methylnaphthalene** $C_{10}H_7CH_3$. Colorless liquid derived from coal tar.
Properties: Sp. gr. 1.025; m. p. −22°C; b. p. 240-243°C; insoluble in water; soluble in alcohol and ether.
Uses: Primary reference fuel in standardization of Diesel engine fuels; organic synthesis.

beta-**methylnaphthalene** $C_{10}H_7CH_3$.
Properties: Solid; sp. gr. 0.994 (40/4°C); b. p. 241-242°C; m. p. 35.1°C; insoluble in water; soluble in alcohol and ether.
Derivation: From coal tar.
Grades: Technical, 95% min.
Containers: 1-, 5-, 55-gal drums.

Use: Organic synthesis; insecticides.
Shipping regulations: None. *

methyl-beta-naphtholate. See beta-naphthyl methyl ether.

2-methylnaphthoquinone. See menadione.

methyl naphthyl ether. See beta-naphthyl methyl ether.

methyl naphthyl ketone $C_{10}H_7COCH_3$.
Properties: White crystalline material, with a sweet orange-blossom odor. Soluble in 5 parts of 95% alcohol.
Constant: Congealing point 53°C.
Containers: Glass bottles; fiber cans.
Use: Perfumery.
Shipping regulations: None. *

methyl nicotinate $C_5H_4NCOOCH_3$.
Properties: White to straw-colored crystals, darkening and becoming reddish on standing; f. p. 37.5°C min; mild pleasant odor.
Purity: 95% minimum.

methyl nitrate CH_3NO_3.
Properties: Explosive liquid. B. p. 66°C; sp. gr. 1.217 (15°C); slightly soluble in water; soluble in alcohol and ether.
Derivation: By reaction of nitric acid and methanol in the presence of urea.
Use: Rocket propellant.

methylnitrobenzene. See nitrotoluene.

methylnitrobenzol. See nitrotoluene.

methyl nonadecanoate $CH_3(CH_2)_{17}COOCH_3$. The methyl ester of nonadecanoic acid.
Properties: White waxy solid; m. p. 39.5°C; b. p. 190.5°C (3.75 mm); insoluble in water; soluble in alcohol and ether.
Grades: Purified 96%, 99.5%.
Uses: Intermediate in organic synthesis; medical research.

2-methylnonane (isodecane) $(CH_3)_2CH(CH_2)_6CH_3$.
Properties: Colorless liquid; sp. gr. 0.728; m. p. −74.7°C; b. p. 167°C.

methyl nonanoate (methyl pelargonate) $CH_3(CH_2)_7COOCH_3$. The methyl ester of pelargonic acid.
Properties: Colorless liquid; m. p. −35°C; b. p. 213.5°C (760 mm), 82-84°C (11 mm); sp. gr. 0.877 (18°C); refractive index 1.4302 (25°C). Insoluble in water; soluble in alcohol and ether.
Derivation: Esterification of nonanoic (pelargonic) acid with methanol followed by fractional distillation.
Grades: Purified (96+%).
Uses: Reference standard for gas chromatography; intermediate in organic synthesis; medical research.

methyl 2-nonenoate. See "Neofolione."

methylnonylacetaldehyde (aldehyde C-12 MNA) $CH_3(CH_2)_8CH(CH_3)CHO$.
Properties: Colorless liquid, with strong odor of fatty-orange character. Soluble in 3 volumes of 80% alcohol.
Constants: Sp. gr. 0.824-0.828; refractive index 1.432-1.435.
Containers: Glass bottles.
Use: Perfumery.
Shipping regulations: None. *

methyl nonyl ketone (2-undecanone) $CH_3COC_9H_{19}$.
Properties: Oily liquid, strong odor. Soluble in 2 parts of 70% alcohol.
Constants: Sp. gr. 0.822-0.826; b. p. 225°C; refractive index 1.429-1.433.
Derivation: A ketone found in oil of rue; also made synthetically.
Containers: Glass bottles.
Use: Perfumery.
Shipping regulations: None. *

methyl octadecanoate. See methyl stearate.

methyl octanoate. See methyl caprylate.

methyl 2-octynoate. See "Folione."

methylol dimethylhydantoin $(CH_3)_2CN(CH_2OH)CONHCO$ (ring linking C and CO).
Properties: A white, odorless crystalline solid; m. p. 99-103°C; soluble in water, methanol, acetone; slightly soluble in ethyl acetate; insoluble in hydrocarbons, trichloroethylene, carbon tetrachloride and diethyl ether.
Uses: Textile and paper finishing; neutral source of formaldehyde.

methyl oleate $CH_3(CH_2)_7CH{:}CH(CH_2)_7COOCH_3$. The methyl ester of oleic acid (cis-octadec-9-enoic acid).
Properties: Clear to amber liquid. Faint fatty odor. Soluble in alcohols and most organic solvents; insoluble in water; sp. gr. 0.8739 (20°C); m. p. −19.9°C; b. p. 218.5°C (20 mm); refractive index 1.4505.
Derivation: Esterification of oleic acid; vacuum fractional distillation; solvent crystallization.
Grades: Technical; purified 99+%.
Uses: Technical grades as intermediate for detergents, emulsifiers, wetting agents, stabilizers, textile treatment; plasticizers for duplicating inks, rubber, waxes, etc. Purified grade in biochemical research; chromatographic reference standard.
Shipping regulations: None. *

methylol riboflavin. A mixture of methylol derivatives of riboflavin (q. v.) exhibiting the same activity.
Properties: Orange to yellow hygroscopic powder; nearly odorless; soluble in water; nearly insoluble in alcohol, benzene, chloroform, and ether. It is dextrorotatory. The dry powder is unstable and upon standing loses biological activity due to the liberation of formaldehyde.
Derivation: Formed by the action of formaldehyde on riboflavin in weakly alkaline solutions.
Grade: N. N. D.
Use: Nutrition; medicine.

methylol urea $H_2NCONHCH_2OH$.
Properties: Colorless crystals; m. p. 111°C; soluble in water and methanol; insoluble in ether; capable of polymerization to synthetic resin.
Derivation: Combination of urea and form-

aldehyde, in the presence of salts or alkaline catalysts.
Containers: Fiber cans and drums.
Uses: The first stage in the formation of urea-formaldehyde resins; in molding adhesives; and in treating textiles and wood.
Shipping regulations: None. *
See also urea-formaldehyde resins.

methylol ureas. See methylol urea and dimethylol urea.

"Methylon." 245 Trademark for resinous compositions made from condensation products of substitued aromatic hydrocarbons and aldehydes used in the surface coating and protective arts.

methyl orange (para-(para-dimethylamino phenylazo)-benzene sulfonate of sodium; Helianthine B; orange III, gold orange; tropeolin D) $(CH_3)_2NC_6H_4NNC_6H_4SO_3Na$.
Properties: Orange-yellow powder; soluble in water, insoluble in alcohol.
Use: Acid-base indicator, red in acid, yellow-orange in alkaline, pH range 3.1-4.4.
See indicators.

methyl-orthophosphoric acid. See methylphosphoric acid.

methyl oxide. See dimethyl ether.

methyl palmitate (methyl hexadecanoate) $CH_3(CH_2)_{14}COOCH_3$. The methyl ester of palmitic acid.
Properties: Colorless liquid; m. p. 29.5°C; b. p. 211.5°C (30 mm), 180.5°C (10 mm); refractive index 1.4310 (45°C). Insoluble in water, soluble in alcohol and ether.
Derivation: Esterification of palmitic acid with methanol or alcoholysis of palm oil. Vacuum distillation.
Grades: 80%; pure (99.8%).
Uses: Intermediate for detergents, emulsifiers, wetting agents, stabilizers, resins, lubricants, plasticizers, textiles, animal feeds, medical research.

methyl palmitoleate $CH_3(CH_2)_5CH{:}CH(CH_2)_7COOCH_3$. The methyl ester of palmitoleic acid (cis-hexadec-9-enoic acid). Colorless liquid; m.p. <15°C; b.p. 140-141°C (5 mm). Soluble in alcohol and ether, insoluble in water. Purified product used in organic synthesis, medical and biochemical research. Prepared by crystallization and vacuum fractional distillation.

methylparaben (methyl para-hydroxybenzoate) $C_8H_8O_3$.
Properties: Colorless crystals, or white crystalline powder; m. p. 125-128°C; odorless or faint characteristic odor; slight burning taste; soluble in water, alcohol, ether; slightly soluble in benzene and in carbon tetrachloride.
Grades: U. S. P. XVI.
Use: Medicine; food additive (preservative).

methyl parathion (O, O-dimethyl O-para-nitrophenylphosphorothioate) $(CH_3O)_2P(S)OC_6H_4NO_2$. The methyl homolog of parathion and accepted as a generic name by the Ent. Soc.
Properties: White crystalline solid; m. p. 35-36°C; sp. gr. 1.358 (20/4°C); refractive index 1.5515 (35°C). Slightly soluble in water; miscible in all proportions with acids and alcohols, esters and ketones. Slowly decomposed by acid solutions, rapidly in dilute alkalies. Toxicity similar though less than parathion.
Grades: Emulsifiable concentrates, wettable powders and dusts.
Containers: Drums.
Uses: Control of insects under approved conditions.
Danger! MCA warning label. Poisonous by skin contact, inhalation or swallowing; rapidly absorbed through skin; repeated exposure may, without symptoms, be increasingly hazardous.
Shipping regulations (dry or liquid): Poison, class B. Poison label. *

methyl pelargonate. See methyl nonanoate.

methyl pentadecanoate $CH_3(CH_2)_{13}COOCH_3$. The methyl ester of pentadecanoic acid.
Properties: Colorless liquid. Insoluble in water; soluble in alcohol and ether. Sp. gr. 0.8618 (25/4°C); m. p. 18.5°C; b. p. 199°C (30 mm); index of refraction 1.4374 (25°C).
Grades: Reagent 96% and 99.5%.
Uses: Intermediate in organic synthesis; medical research.

methylpentadiene C_6H_{10}. Numerous isomers are possible. Commercially available mixture contains 2- and 4-methyl-1, 3-pentadiene.
Properties: Sp. gr. 0.7184 (20/4°C); b. p. 75-77°C; flash point −30°F; reactive with halogens, hydrohalogens, sulfur dioxide and maleic anhydride. Caution! Highly flammable.
Containers: Glass bottles.
Uses: Organic synthesis; alkyd resins and other polymers.
Shipping regulations: Flammable liquid. Red label. *

2-methylpentaldehyde $C_3H_7CH(CH_3)CHO$.
Properties: Sp. gr. 0.8092; b. p. 118.3°C; f. p. −100°C; solubility in water 0.42% by weight; flash point 68°F.
Uses: Intermediates for dyes, resins, pharmaceuticals.
Shipping regulations: Flammable liquid. Red label. *

2-methylpentane (isohexane) $CH_3(CH_2)_2CH(CH_3)_2$.
Typical specifications: F. p. −244.62°F; b. p. 140.1°F; refractive index 1.372 (20°C); sp. gr. 0.658 (60/60°F); flash point −10°F.
Grades: 95%, 99% and research.
Containers: 1-gal bottles; 5-gal drums.
Shipping regulations: Flammable liquid. Red label. *
See also isohexanes.

3-methylpentane (diethylmethylmethane) C_6H_{14} or $CH_3CH_2CHCH_3CH_2CH_3$.
Properties: Colorless liquid. Soluble in alcohol; insoluble in water; slightly soluble

in ether.
Constants: Sp. gr. 0.676 (20/4°C); b.p. 64.0°C; refractive index 1.37662 (20°C); C.S.T. aniline 69.3°C.
Grades: Technical (95%); 99%; and research.
Containers: Glass bottles; drums.
Use: Organic synthesis.
Shipping regulations: Flammable liquid. Red label.*

2-methyl-1,3-pentanediol
$C_2H_5CH(OH)CH(CH_3)CH_2OH$.
Properties: Sp. gr. 0.9745; b.p. 220.3°C; f.p. −30°C; infinitely soluble in water.
Uses: Solvent; coupling agent.
See also hexylene glycol, which is a mixture of isomers.

2-methyl-2,4-pentanediol
$CH_3CHOHCH_2COH(CH_3)CH_3$.
Properties: Colorless liquid. Completely miscible with water and most organic solvents including lower aliphatic hydrocarbons.
Constants: Sp. gr. 0.9235 (20/20°C); b.p. 197.1°C; viscosity 34 cps (20°C); vapor pressure 10.8 mm (95.2°C); 334 mm (169.7°C); flash point 94°C (open cup); wt/gal 7.59 lbs (20°C).
Uses: Coupling agent; chemical synthesis.
See also hexylene glycol, a mixture of isomers.

4-methyl-2,4-pentanediol. See hexylene glycol, a mixture of isomers.

2-methylpentanoic acid.
$CH_3CH_2CH_2CH(CH_3)COOH$.
Properties: Water white liquid; sp. gr. 0.9242 (20/20°C); b.p. 196.4°C; vapor pressure 0.02 mm (20°C); f.p. sets to glass below −85°C. Solubility in water 1.3% by wt (20°C); solubility of water in, 2.9% by wt (20°C).
Uses: Suggested for synthetic lubricants; plasticizers, vinyl stabilizers; metallic salts; alkyd resins.

2-methyl-1-pentanol $C_3H_7CH(CH_3)CH_2OH$.
Properties: Sp. gr. 0.8252; b.p. 148.0°C; vapor pressure 1.1 mm Hg (20°C); solubility in water 0.31% by weight; flash point 135°F.

4-methyl-2-pentanol. See methylamyl alcohol.

2-methyl-4-pentanone. See methyl isobutyl ketone.

2-methyl-1-pentene (1-methyl-1-propylethylene) C_6H_{12} or $CH_2{:}C(CH_3)CH_2CH_2CH_3$.
Properties: Colorless liquid; sp. gr. 0.6820 (20/4°C); b.p. 62.2°C; f.p. −191.39°C; refractive index 1.3925 (n 20/D); soluble in alcohol, acetone, ether, petroleum, coal-tar solvents; insoluble in water.
Typical specifications: Sp. gr. 0.6805-0.6835 (20/4°C); b.p. 61-63°C; refractive index 1.3915-1.3940 (n 20/D).
Grades: 95%, 99%; research.
Containers: Bottles.
Uses: Organic synthesis; flavors; perfumes; medicines; dyes; oils; resins; plastics.

4-methyl-1-pentene $H_2C{:}CHCH_2CH(CH_3)CH_3$.
Properties: Liquid; f.p. −244.7°F; b.p. 128.4°F; sp. gr. 0.6640 (20/4°C); refractive index 1.38265 (n 20/D); flash point −25°F.
Grades: 95%, 99%; research.
Containers: Up to one gallon bottles.
Shipping regulations: Flammable liquid. Red label.*

4-methyl-2-pentene cis, trans mixture (1-isopropyl-2-methylethylene) C_6H_{12} or $CH_3CH{:}CHCH(CH_3)_2$.
Properties: Colorless liquid; sp. gr. 0.670 (20/4°C); b.p. 55°C; refractive index 1.388 (n 20/D); soluble in alcohol, acetone, ether, petroleum, coal-tar solvents; insoluble in water.
Typical specifications: Sp. gr. 0.668-0.672 (20/4°C); b.p. 54-56°C; refractive index 1.388 (n 20/D).
Use: Organic synthesis.
Shipping regulations: Flammable liquid. Red label.*

cis-**4-methyl-2-pentene** (low boiling 4-methyl-2-pentene) $H_3CCH{:}CHCH(CH_3)_2$.
Properties: Liquid; f.p. −211.7°F; b.p. 133.2°F; sp. gr. 0.674 (60/60°F); flash point approximately −25°F.
Containers: Up to one gallon bottles.
Grades: 95%, 99%; research.
Shipping regulations: Flammable liquid. Red label.*

trans-**4-methyl-2-pentene** (high boiling 4-methyl-2-pentene) $H_3CCH{:}CHCH(CH_3)_2$.
Properties: Liquid; f.p. −221.6°F; b.p. 137.1°F; sp. gr. pure compound 0.6688 (20/4°C); sp. gr. technical grade 0.674 (60/60°F); vapor pressure 7.1 psia; flash point −30°F.
Grades: 95%, 99%; research.
Containers: Up to one gallon bottles.
Shipping regulations: Flammable liquid. Red label.*

methylpentynol $HC{:}CCOH(CH_3)CH_2CH_3$.
3-Methyl-1-pentyn-3-ol.
Properties: Colorless liquid; b.p. 121-122°C; m.p. −30.6°C; sp. gr. (20/20°C) 0.8721; refractive index (n 20/D) 1.4318; flash point (Tag open cup) 101°F. Moderately soluble in water; miscible with acetone, benzene, carbon tetrachloride, kerosine.
Containers: 7-, 35-, 385-lb drums.
Uses: Stabilizer in chlorinated solvents; viscosity reducer and stabilizer; electroplating brightener; intermediate in synthesis of hypnotics and isoprenoid chemicals such as vitamin A, ionone and perfume alcohols; solvent for alcohol-soluble nylon and polyamide resins.

methylphenethylamine. See amphetamine.

methylphenidate hydrochloride (methyl-alpha-phenyl-2-piperidineacetate hydrochloride) $C_{14}H_{13}NO_2 \cdot HCl$.
Properties: Crystals; decompose 195°C. Soluble in water.
Grade: N.N.D.
Use: Medicine.

*See "I.C.C. Shipping Regulations," page xiii.
Reference numbers refer to name of manufacturer. See "List of Manufacturers," page v.

methylphenol, 2-, 3-, or 4-. See ortho-, meta- or para-cresol respectively.

methyl phenylacetate $C_6H_5CH_2COOCH_3$.
Properties: A colorless liquid having a fine honey-like odor. Soluble in 5 parts of 60% alcohol.
Constants: Sp. gr. 1.062-1.066; refractive index 1.506-1.509.
Containers: Glass bottles; 1-, 5- and 10-gal tins; 55-gal steel drums.
Uses: Perfumery; flavors for tobacco.
Shipping regulations: None. *

methyl phenyl carbinol. See styralyl alcohol.

methyl phenyl carbinyl acetate. See styralyl acetate; see also "Gardenol."

methylphenyldichlorosilane $CH_3(C_6H_5)SiCl_2$.
Properties: Colorless liquid; b.p. 82°C (13 mm); sp. gr. 1.19.
Derivation: From chlorobenzene Grignard reagent and methyltrichlorosilane, or from benzene and methyldichlorosilane.
Uses: Manufacture of silicones.

methyl phenyl ether. See anisole.

3-methyl-3-phenylglycidic acid ethyl ester. See ethyl methylphenyl glycidate.

2-methyl-2-phenylpropane. See tert-butylbenzene.

3-methyl-1-phenyl-5-pyrazolone. See 1-phenyl-3-methyl-5-pyrazolone.

6-methyl-2-phenylquinoline-4-carboxylic ethyl ester. See neocincophen.

N-methyl-2-phenylsuccinimide. See phensuximide.

2-methyl-9-phenyl tetrahydro-1-pyridindene bitartrate. See phenindamine tartrate.

methylphloroglucinol (2,4,6-trihydroxytoluene) $C_6H_2(OH)_3CH_3$.
Properties: Cream to light tan, fine crystals; odorless; m.p. 210-214°C; soluble in water, alcohol, and ether; insoluble in benzene.
Containers: Bottles; fiber drums.
Uses: A very reactive coupling agent; potential dye and plastic intermediate.
Shipping regulations: None. *

methylphosphoric acid. (methyl orthophosphoric acid) $CH_3H_2PO_4$.
Properties: Pale straw-colored liquid; sp. gr. 1.42 (25°C); can be neutralized with alkalies or amines to give water-soluble salts.
Purity: 97% with remainder being orthophosphoric acid and methyl alcohol.
Containers: Glass bottles; 1-, 5- and 10-gal tins; 55-gal steel drums.
Uses: Textile and paper processing compounds; catalysts in urea-resin formation; polymerizing agents for resin and oils; rust remover; soldering flux; chemical intermediate.
Shipping regulations: None. *

N-methyl piperazine $CH_3\underline{NCH_2CH_2NHCH_2CH_2}$.
Properties: Liquid; sp. gr. 0.9038; b.p. 138.0°C; f.p. -6.4°C; infinitely soluble in water; flash point 108°F.
Uses: Intermediate for pharmaceuticals, surface agents, synthetic fibers.

2-methylpiperidine (2-pipecoline) $C_5NH_{10}CH_3$.
Properties: Liquid; b.p. 118.2°C; f.p. -4.2°C; sp. gr. 0.8401 (20/20°C); refractive index 1.4457 (n 20/D); soluble in water in all proportions at 20°C.

6-alpha-**methylprednisolone-21-acetate.**
Properties: M.p. 205-208°C.
Use: Medicine (steroid).

methylprednisolone sodium succinate. A corticosteroid.

2-methylpropane. See isobutane.

2-methylpropanenitrile. See isobutyronitrile.

2-methyl-1-propanethiol. See isobutyl mercaptan.

2-methyl-2-propanethiol. See tert-butyl mercaptan.

2-methyl-1-propanol. See isobutyl alcohol.

2-methyl-2-propanol. See tert-butyl alcohol.

2-methylpropanoyl chloride. See isobutyroyl chloride.

2-methylpropene. See isobutene.

2-methyl-2-propen-1-ol (methallyl alcohol) $H_2C{:}C(CH_3)CH_2OH$.
Properties: Colorless liquid; sp. gr. 0.8515 (20/4°C); b.p. 110-116°C; soluble in water, alcohols, ethers.
Uses: Suggested as intermediate for pharmaceutical, insecticide, dyestuff; perfume, flavor, resin, plastics, and rubber products.

methyl propionate $CH_3CH_2COOCH_3$.
Properties: Clear, colorless liquid. Soluble in most organic solvents; somewhat soluble in water.
Constants: Sp. gr. 0.937 (4°C); boiling range 78.0-79.5°C; flash point -2°C; wt/gal 7.58 lbs.
Grades: Technical.
Containers: 55-gal drums; tank cars.
Uses: Solvent for cellulose nitrate; solvent mixtures for cellulose derivatives; lacquers, paints, varnishes; coating compositions.
Shipping regulations: Flammable liquid. Red label. *

2-methyl-2-propylaminopropyl benzoate hydrochloride. See meprylcaine hydrochloride.

methylpropylbenzene. See cymene.

methyl propyl carbinol. See 2-pentanol.

methyl propyl carbinol urethane. See hedonal.

methyl n-**propyl ether** $CH_3OCH_2C_2H_5$.
Properties: Colorless liquid; b.p. 37°C; sp.gr. 0.738 (20°C); soluble in alcohol and ethyl ether. Slightly soluble in water.
Use: Of possible use as an anesthetic.

1-methyl-1-propylethylene. See 2-methyl-1-pentene.

*See "I.C.C. Shipping Regulations," page xiii.
Reference numbers refer to name of manufacturer. See "List of Manufacturers," page v.

methyl propyl ketone (ethyl acetone, 2-pentanone; MPK) $CH_3COC_3H_7$.
Properties: Water-white liquid. The commercial material consists of a mixture of methyl propyl and diethyl ketones in the approximate ratio of 3 to 1 and contains at least 97% of these ketones, the balance being sec-amyl alcohol. Soluble in alcohol and ether; insoluble in water.
Constants: Sp. gr. 0.809 (20/20°C); b.p. 101.7°C; refractive index 1.3895 (20°C); viscosity 0.473 centipoise (25°C); flash point 45°F (7.2°C) (closed cup).
Typical specifications: Color water-white; water-miscible without turbidity with 19 vols of 60°Bé gasoline (20°C); acidity less than 0.003% (as acetic acid); distillation range more than 90% distils over between 100 and 103°C.
Grades: Technical.
Containers: Glass bottles; 1-, 5- and 10-gal tins; 55-gal steel drums.
Uses: Solvent; substitute for diethyl ketone.
Shipping regulations: Flammable liquid. Red label. *

2-methyl-2-n-propyl-1,3-propanediol dicarbamate. See meprobamate.

2-methylpyridine. See alpha-picoline.

3-methylpyridine. See beta-picoline.

4-methylpyridine. See gamma-picoline.

N-methylpyrrole $C_4NH_4CH_3$.
Properties: Liquid; b.p. 112°C; f.p. −57°C; density 0.914 (20°C); refractive index 1.4898 (n 17/D); flash point 61°F; slightly soluble in water.
Grade: 98% min. purity.
Containers: 55-gal steel drums; 1-gal, 5-gal cans.
Use: Organic synthesis.
Shipping regulations: Flammable liquid. Red label. *

N-methylpyrrolidine $CH_3\overline{NCH_2CH_2CH_2CH_2}$.
Properties: Colorless liquid, ammonia-like odor; refractive index 1.4200 to 1.4230 (25°C).

N-methyl-2-pyrrolidone $CH_3\overline{NCH_2CH_2CH_2CO}$.
Properties: M.p. −24°C; b.p. 202°C; flash point 204°F; miscible in all proportions with water, various organic solvents, castor oil.
Derivation: High pressure synthesis from acetylene and formaldehyde.
Uses: Solvent for resins, acetylene; petroleum processing; spinning agent for polyvinyl chloride; intermediate.

alpha-**methylquinoline.** See quinaldine.

gamma-**methylquinoline.** See lepidine.

methyl red $(CH_3)_2NC_6H_4NNC_6H_4COOH$.
para-Dimethylaminoazobenzenecarboxylic acid.
Properties: Dark-red powder or violet crystals; m.p. 180°C; insoluble in water; soluble in alcohol, ether, glacial acetic acid.
Use: Acid-base indicator in the range pH 4.2-6.2 (red to yellow).
See indicators.

methylresorcinol. See orcin.

methyl ricinoleate
$CH_3(CH_2)_5CH(OH)CH_2CH:CH(CH_2)_7COOCH_3$.
The methyl ester of ricinoleic acid.
Properties: Colorless liquid, insoluble in water; soluble in alcohol and ether. Sp. gr. 0.9236 (22/4°C); m.p. −4.5°C; b.p. 245°C (10 mm); refractive index 1.4628.
Derivation: Esterification of ricinoleic acid or alcoholysis of castor oil; purification by vacuum distillation.
Grades: Technical; purified (99+%).
Containers: 5-gal cans; 55-gal drums.
Uses: Plasticizer; lubricant; cutting oil additive; wetting agent.

methylrosaniline chloride. U.S.P. XVI name for methyl violet (q.v.).

methyl salicylate. Methyl salicylate U.S.P. XVI may be derived either synthetically or naturally, as below, and official synonyms include gaultheria oil, wintergreen oil, betula oil, and sweet-birch oil. $C_6H_4OHCOOCH_3$.
Properties: Colorless, yellowish, or reddish liquid oil; odor of wintergreen; refractive index 1.535-1.538; sp. gr. 1.180-1.185; m.p. −8.3°C; b.p. 222.2°C; soluble in 7 parts of 70% alcohol; soluble in ether and in glacial acetic acid; sparingly soluble in water.
Derivation: By heating methanol and salicylic acid in presence of sulfuric acid, or by distillation from leaves of Gaultheria procumbens or bark of Betula lenta.
Method of purification: Rectification.
Grades: Technical; U.S.P. XVI.
Containers: 1-, 5-lb bottles; 10-, 55-gal tin-lined drums.
Uses: Medicine; flavoring; perfumery; solvent for cellulose derivatives; insecticides; polishes; printing and copying inks.
Shipping regulations: None. *

"Methyl Selenac." [69] Trademark for proprietary product, selenium dimethyldithiocarbamate $[(CH_3)_2NC(S)S]_4Se$.
Properties: Yellow powder (also supplied in "rodform"); sp. gr. 1.58±.03; melting range 140-172°C; slightly soluble in carbon disulfide, benzene, chloroform; insoluble in water, dilute caustic, gasoline.
Uses: Vulcanizing agent, for natural and butyl rubbers and SBR. Primary accelerator in natural and butyl rubber and SBR. Secondary accelerator (with thiazoles) in natural rubber and SBR.

methyl silicone. General term for the most common and important variety of silicones, having composition $[(CH_3)_2SiO]_x$, $[(CH_3)_2SiO_3]_y$, etc., and having properties of oils, resins, or rubber according to the molecular size and arrangement.
See silicones and siloxanes.

methyl stearate (methyl octadecanoate) $CH_3(CH_2)_{16}COOCH_3$. The methyl ester of

stearic acid.
Properties: Colorless crystals; m. p. 37.8°C; b. p. 234.5°C (30 mm), 204.5°C (10 mm); refractive index 1.4328 (50°C). Insoluble in water; soluble in ether and alcohol.
Derivation: Esterification of stearic acid with methanol or alcoholysis of stearin with methanol.
Method of purification: Vacuum fraction distillation.
Impurities: Most technical methyl stearate is 55% stearate and 45% methyl palmitate.
Grades: Distilled; pressed; technical; pure 99.8+%.
Containers: 1-, 7-, 35-lb cans; 210-, 380-, 400-lb drums.
Uses: Intermediate for stearic acid, detergents, emulsifiers, wetting agents, stabilizers, resins, lubricants, plasticizers and textiles.

alpha-**methylstyrene** $C_6H_5C(CH_3){:}CH_2$.
Properties: Colorless liquid, subject to polymerization by heat or catalysts; b. p. 165.38°C (760 mm); m. p. −23.21°C; sp. gr. 0.9062 (25/25°C); viscosity 0.940 cps (20°C); flash point 136°F; fire point (COC) 136°F; refractive index 1.5359 (25/25°C); slightly soluble in water; lower explosive limit, vol 0.90%.
Typical commercial sample: alpha-methyl styrene 98.5-99.5%. A polymerization inhibitor such as tert-butyl catechol is usually also present.
Derivation: From benzene and propylene by use of aluminum chloride and hydrogen chloride to yield cumene which is then dehydrogenated.
Hazard: Contact with skin and breathing of vapors must be avoided.
Containers: Glass bottles; steel drums.
Use: Polymerization monomer.
Shipping regulations: None. *

methyl styryl ketone. See benzylidene acetone.

methylsuccinic acid. See pyrotartaric acid.

methyl sulfate. See dimethyl sulfate.

methyl sulfide. See dimethyl sulfide.

methylsulfonal. See sulfonethylmethane.

methylsulfuric acid (acid methyl sulfate; methyl hydrogen sulfate) CH_3OSO_2OH or CH_3HSO_4.
Properties: Oily liquid. Soluble in anhydrous ether; slightly soluble in alcohol, water.
Constants: B. p. 188°C; sp. gr. 1.352; f. p. −27°C.
Derivation: Interaction of methyl alcohol and chlorosulfonic acid.
Shipping regulations: None. *

N-methyltaurine. Available in commercial quantities as an aqueous solution of the sodium salt, sodium N-methyltaurate, or $CH_3NHCH_2CH_2SO_3Na$.
Properties (of solution): Clear, light-colored liquid, about 34-36% sodium salt, sp. gr. 1.21 (25/4°C). At freezing point (−28°C average) becomes a suspension of white crystals.
Uses: Intermediate for detergents, dyestuffs, pharmaceuticals, and other organics.

methyltestosterone $C_{20}H_{30}O_2$. 17-Methyltestosterone. A synthetic androgenic steroid.
Properties: White or creamy white crystals or crystalline powder; odorless; stable in air; slightly hygroscopic; affected by light; m. p. 163-168°. Soluble in alcohol, methanol, ether, and other organic solvents; sparingly soluble in vegetable oils; insoluble in water.
Derivation: By organic synthesis.
Grade: U. S. P. XVI.
Containers: 100-g bottles.
Use: Medicine.

methyl tetracosanoate. See methyl lignocerate.

methyl tetradecanoate. See methyl myristate.

2-methyltetrahydrofuran $C_4H_7OCH_3$.
Properties: Colorless, mobile liquid; ether-like odor; b. p. 80.2°C (760 mm); f. p. −136°C; sp. gr. 0.854 (20/4°C); refractive index 1.4025 (25/D); flash point 12°F (Tag closed cup). Solubility in water 15.1 g/100 g water (25°C). Solubility in water increases with a decrease in temperature. Infinitely soluble in most organic solvents.
Containers: 1-, 5-, and 55-gal drums weighing 7, 35, and 375 lbs, respectively.
Uses: Chemical intermediate; reaction solvent.
Shipping regulations: Flammable liquid. Red label. *

methyltheobromine. See caffeine.

methyl 2-thienyl ketone. See 2-acetylthiophene.

meta-**(methylthio)aniline.** See meta-methylthioniline.

meta-**methylthioniline** (meta-(methylthio)aniline) $H_2NC_6H_4SCH_3$.
Properties: Pale yellow oil. Sp. gr. 1.140 (25°C); b. p. 163-165°C (16 mm); f. p. −3.0°C. Insoluble in water; soluble in alcohol, benzene, acetic acid.
Use: Pharmaceutical intermediate.

methylthionine chloride. See methylene blue.

6-methyl-2-thio-4-oxypyrimidine. See 6-methylthiouracil.

6-methyl-2-thiouracil (6MT; 6-methyl-2-thio-4-oxypyrimidine) HNC(S)NHC(O)CHCCH$_3$.
Properties: White, odorless crystalline powder with pronounced bitter taste; m. p. 326-331°C (dec); sublimes readily when heated in platinum dish; very slightly soluble in water; sparingly soluble in alcohol; slightly soluble in ether and chloroform; practically insoluble in benzene; freely soluble in ammonia and solutions of alkali hydroxides. Stable on heating in alkaline solution.
Grade: U. S. P. XVI.
Use: Medicine.

methyl para-**toluate** $CH_3C_6H_4COOCH_3$.
Properties: White crystalline solid.
Use: Organic synthesis.

*See "I. C. C. Shipping Regulations," page xiii.
Reference numbers refer to name of manufacturer. See "List of Manufacturers," page v.

methyl para-**toluenesulfonate** $CH_3C_6H_4SO_3CH_3$.
Properties: Solidification point 24°C; b. p. 157°C (8 mm); decomposes 262°C (760 mm).
Grade: 96% min.
Uses: Accelerator; methylating agent.
Caution: May cause skin irritations when handled.

methyl tolyl ketone. See methyl acetophenone.

methyltrichlorosilane CH_3SiCl_3.
Properties: Colorless liquid. B. p. 66.4°C; sp. gr. 1.270 (25/25°C); refractive index (n 25/D) 1.4085; flash point (Cleveland open cup) 47°F. Readily hydrolyzed by moisture, with the liberation of hydrochloric acid.
Derivation: By Grignard reaction of silicon tetrachloride and methylmagnesium chloride.
Grade: Technical.
Use: Intermediate for silicones.
Shipping regulations: Flammable liquid. Red label. *

methyl tricosanoate $CH_3(CH_2)_{21}COOCH_3$. The methyl ester of tricosanoic acid.
Properties: White waxlike solid. Insoluble in water, soluble in alcohol and ether. M. p. 55-56°C.
Grades: Purified 96% and 99.5%.
Uses: Intermediate in organic synthesis; medical research.

methyl tridecanoate $CH_3(CH_2)_{11}COOCH_3$. The methyl ester of tridecanoic acid.
Properties: Colorless liquid. Insoluble in water, soluble in alcohol and ether; m. p. 5.5°C; b.p. 130-132°C (4 mm); refractive index 1.4327 (25°C).
Derivation: Esterification of tridecanoic acid with methanol followed by fractional distillation.
Grades: Purified 96% and 99.5%.
Uses: Intermediate in organic synthesis; medical research; reference standard in gas chromatography.

methyl trimethylolmethane. See trimethylol ethane.

methyl-1, 2, 2-trimethylpropoxyfluorophosphine oxide. See Soman.

methyltrinitrobenzene. See trinitrotoluene.

"Methyl Tuads." [69] Trademark for tetramethylthiuram disulfide, $[(CH_3)_2NC(S)S]_2$.
Properties: Buff to light gray solid (also supplied in "rodform"); sp. gr. 1.27 ± .03; melting range 142 -156°C; soluble in benzene, carbon disulfide, chloroform; insoluble in water, dilute caustic, gasoline.
Uses: Vulcanizing agent and primary accelerator in natural, nitrile and butyl rubbers and in SBR. As secondary accelerator in natural and nitrile rubbers and SBR. Use in coated fabrics, extruded and molded goods, inner tubes, wire and cable.

beta-**methylumbelliferone** (7-hydroxy-4-methylcoumarin; BMU) $C_{10}H_8O_3$.
Properties: White to light tan powder; m. p. 186-188°C; soluble in concentrated sulfuric acid; partly soluble in ethanol, isopropanol, 5% aqueous sodium carbonate solution; very slightly soluble in water and white mineral oil; very dilute aqueous alkaline solutions give a bright blue-white fluorescence in daylight or ultraviolet light.
Grade: Technical.
Containers: 100 lb fiber drums.
Uses: As an optical bleach in soaps, starches, and laundry products; effective sunscreen in suntan lotions.

methyl undecanoate $CH_3(CH_2)_9COOCH_3$. The methyl ester of undecanoic acid.
Properties: Colorless liquid. Insoluble in water, soluble in alcohol and ether; b. p. 123°C (10 mm); refractive index 1.4270 (25/4°C).
Derivation: Esterification of undecanoic acid with methanol followed by fractional distillation.
Grades: Purified 96%; 99.5%.
Uses: Organic intermediate for synthesis; medical research.

5-methyluracil. See thymine.

methylvinyldichlorosilane $(CH_3)(C_2H_3)SiCl_2$.
Properties: Liquid; b. p. 92°C; sp. gr. 1.08 (25°C).
Derivation: From methyldichlorosilane and acetylene or vinyl chloride.
Use: Manufacture of silicones.

methyl vinyl ether. See vinyl methyl ether.

2-methyl-5-vinylpyridine $CH_3C_5NH_3CH{:}CH_2$.
Properties: Clear to faintly opalescent colorless liquid; sp. gr. 0.978-0.982 (20/20°C); b. p. 181°C (760 mm); refractive index 1.5400-1.5454 (20°C).
Containers: Drum, tank trucks; tank cars.
Use: Synthesis.

methyl violet (gentian violet; methylrosaniline chloride; crystal violet). This term is most frequently applied to mixtures containing hexa- and pentamethylpara-rosaniline hydrochloride. Sometimes refers specifically to the hexamethyl derivative ($C_{25}H_{30}ClN_3$).
Properties: Dark green powder or crystals with metallic luster. Soluble in water, alcohol, glycerin, chloroform; insoluble in ether.
Derivation: Dimethyl aniline and phosgene, or dimethyl aniline, cupric chloride and phenol.
Grades: U. S. P. XVI; as methylrosaniline chloride.
Uses: An acid-base indicator for the range pH 2-3.1, yellow in acid, violet in alkali; in medicine; textile dye; pigment in ink, carbon paper, typewriter ribbons.

methyl yellow. See dimethylaminoazobenzene.

"Methyl Zimate." [69] Trademark for zinc dimethyldithiocarbamate, $[(CH_3)_2NC(S)S]_2Zn$.
Properties: White powder (also supplied as white rods and as pink rods); sp. gr. 1.71 ± 0.3; melting range 242-257°C; moderately soluble in dilute caustic, benzene, carbon disulfide, chloroform; insoluble in water, gasoline.
Uses: Primary accelerator; secondary

*See "I.C.C. Shipping Regulations," page xiii.
Reference numbers refer to name of manufacturer. See "List of Manufacturers," page v.

accelerator (with thiazoles) in natural and butyl rubber and in SBR. For extruded and molded goods, wire and cable, footwear and general mechanical goods.

methyprylon (3,3-diethyl-5-methyl-2,4-piperidinedione) $C_5NH_4(O)_2(C_2H_5)_2(CH_3)$.
Properties: Nearly white, crystalline powder; slight characteristic odor; bitter taste; melting range 74°-77°C; soluble in water; very soluble in alcohol, in chloroform, in ether, and benzene.
Grade: N. F. XI.
Use: Medicine.

"Meticortelone." [321] Brand name for prednisolone (q. v.).

"Meticorten." [321] Brand name for prednisone (q. v.).

"Metol." [134] Trademark for methyl-para-aminophenol sulfate (q. v.).

metopon hydrochloride (6-methyldihydromorphinone hydrochloride) $C_{18}H_{21}O_3N \cdot HCl$. A morphine derivative.
Properties: White, odorless, crystalline powder; very soluble in water; sparingly soluble in alcohol; slightly soluble in chloroform; very slightly soluble in ether; insoluble in benzene.
Use: Medicine.

"Metrazol." [9] Trademark for pentylenetetrazol (pentamethylenetetrazol).

"Metron." [88] Trademark for an emulsible solution containing methyl parathion.

"Metso." [201] Trademark for sodium metasilicate, sesquisilicate, orthosilicate and detergent mixtures based on these.

"Metubine Iodide." [100] Trade name for dimethyl-tubocurarine iodide (O-methyl-d-tubocurarine iodide).
Properties: A white to pale yellow, odorless, crystalline powder; decomposes at 257°C; slightly soluble in water, dilute hydrochloric acid and dilute NaOH; very slightly soluble in alcohol; practically insoluble in benzene, chloroform, and ether.
Use: Medicine.

"Metycaine." [100] Trademark for piperocaine hydrochloride, U. S. P.

mev. See electron volt.

Mexican scammony resin. See ipomea resin.

Mexican scammony root. See ipomea.

mexico seed. See ricinus.

Mg. Symbol for magnesium.

mg. Abbreviation for milligrams.

MH. See maleic hydrazide.

"MH-30." [248] Trademark for a 30% solution of maleic hydrazide (q. v.).
Properties: Water-soluble liquid containing 30% by weight of maleic hydrazide as the diethanolamine salt. Contains 3 lbs of maleic hydrazide per gallon.
Uses: To prevent sucker development on tobacco; to control wild onions and garlic; to temporarily inhibit growth of grasses along highways; weed control on cranberries; to reduce freeze injury to citrus trees; chemical pruning of citrus trees; chemical thinning of peaches; as a preharvest spray to prevent potato sprouting in storage.

"MHA." [58] Trademark for methionine hydroxy analogue, calcium salt (q. v.).

MHD generator. See magnetohydrodynamic generator.

MHM. Abbreviation for monohydroxymethane. See methanol.

miazine. See pyrimidine.

MIBC. Abbreviation for methylisobutyl carbinol. See methylamyl alcohol.

mica. A group of silicates of varying chemical composition, but with similar physical properties and atomic structure. They all have an excellent cleavage, and can be split into very thin flexible elastic sheets. All contain hydroxyl, an aluminum silicate group, and an alkali. They are common in igneous and metamorphic rocks, and in some sedimentary rocks.
Commercially, mica usually refers to muscovite or phlogopite (q. v.).
See also lepidolite and vermiculite.

"Micabond." [281] Trademark for an electrical insulation material consisting primarily of mica with electrical insulating binders.
Forms: Tape; tubing; segments; plate; fabricated parts.
Uses: Motors; insulation against heat.

"Micarta." [308] Trademark. A group of laminated plastics used as sheets, rods, tubes, and special molded shapes.
Properties: Various colors—usually black or brown; sp. gr. 1.25-1.80; tensile strength 8000-20,000 psi; compression strength 25,000-70,000 psi; dielectric strength up to 700 volts/mil. Insoluble and resistant to water, organic solvents, dilute alkalies, and non-oxidizing acids.
Composition: Paper or fabric of cellulose, glass, asbestos, or synthetic fibers bonded with phenolic or melamine resins and cured at elevated temperature and pressure.
Uses: Plating barrels; rayon-manufacturing equipment; pickling tanks; electric and thermal insulation; oil-handling equipment; steel rolling-mill bearings; chemical-handling valve bodies; paper-mill suction box covers and equipment.

mica schist. A variety of laminated metamorphic rock composed of mica with quartz, feldspar, and other silicate minerals. Sometimes used as a refractory material or to make ground mica for roofing.

mica, synthetic. Usually fluorophlogopite (a fluorine derivative of phlogopite), made by (a) treating potassium fluosilicate with alumina, under pressure and heat, or (b) melting basic oxides, fluorides and feldspar together. The product has higher

temperature stability than natural mica, and its dielectric properties and machinability are about the same.
Uses: Electrical-electronic field.

"Micatex." [236] Brand name for mica prepared for addition to drilling fluids to reduce water loss to the formation and for overcoming mild losses of circulation. An effective seal is formed over mildly permeable formations when the mud in which it is entrained forces the material against the formation. Will not disintegrate appreciably, nor will it corrode or abrade slush-pump liners or other metal or moving parts of the mud system.

Michler's hydrol. See tetramethyldiaminobenzhydrol.

Michler's ketone. See tetramethyldiaminobenzophenone.

"Micratized." [309] Trademark for a vitamin-containing product for fortifying foods and feeds with vitamins.

"Micris." [101] Trademark for petroleum hydrocarbon waxes that are selectively processed from residual petroleum stocks to produce amorphous type microcrystalline waxes.
Uses: Waterproofing, insulating, sealing, and preserving coatings in the paper, textile, electrical and packaging trades; components of paint, ink, and polish.

micro-. Prefix meaning 10^{-6} units (symbol μ). E.g., 1 microgram = 0.000001 grams.

microballoons. Tiny vinyl plastic spheres which are used to form a protecting layer over liquid surfaces, such as oils in big tanks, to reduce evaporation.

"Micro-Cel." [247] Trademark for line of finely divided hydrated synthetic calcium silicates.
Properties: White to light gray color range dependent on grade. Density (apparent) 5-10 lbs/cu ft; pH 7-10; absorption (water) 300-600%.
Uses: Inert extenders; absorbents; bulking agents; pesticide-carrier inerts.

microchemistry. The branch of chemistry that deals with procedures that require the handling of very small quantities of materials. Various common chemical operations such as weighing, preparation and purification, analysis, testing, are carried out on a scale ten to thousands of times smaller than is possible by ordinary laboratory procedures.

microcidine. See sodium beta-naphtholate.

microcosmic salt. See sodium-ammonium phosphate.

microcrystalline waxes. See waxes, microcrystalline.

microcurie. See curie.

"Micro-dritomic." [50] Trademark for a micron-fine wettable sulfur agricultural fungicide.

microencapsulation. The production of a material in very small capsules, about 20 to 150 microns in diameter. Gelatin is widely used as the encapsulating agent. The process is used for adhesives, carbon for carbon paper, and many volatile, toxic, or odorous substances. The advantage is that the capsules remain stable and inert until broken down by heat or pressure.

microfractor. A type of multistage molecular fractionation apparatus.

microlite $(Na,Ca)_2Ta_2O_6(O,OH,F)$. A natural hydrous oxide of sodium, calcium and tantalum, found in pegmatites. Commonly contains some niobium also.
Properties: Color yellow, brown, red, green; streak yellowish or brownish; luster vitreous; hardness 5-5.5; sp. gr. 4.2-4.4. Isomorphous with pyrochlore.
Occurrence: New Mexico, California, Connecticut, Virginia; Europe.
Use: Ore of tantalum.

"Microliths." [443] Trade name for pigment dispersions for organic coatings, inks and plastics.

"Micromet." [108] Trademark for a specially formulated phosphate glass, slowly soluble in water, and used to inhibit scale, corrosion, and red water in water systems and air conditioning systems.

micron. Short unit of length in the metric system. One millionth of a meter; 10^{-4} centimeter; 10^{-3} millimeter (a meter is 39.37 inches).

"Micronex." [133] Trademark for series of impingement carbon blacks produced from natural gas. Used to obtain good tensile strength and abrasion resistance, cracking and tear resistance and safe processing. Used primarily in natural rubber truck treads and carcasses, mining cable covers and wire jacket compounds, camel back and heavy duty footwear. Grades available are Standard Micronex (medium processing channel); Micronex W-6 (easy processing channel). Available in 25- and 50-lb bags and hopper cars.

microscopic cross section. See cross section; see also macroscopic cross section.

"Microsols." [443] Trade name for pigment dispersions for aqueous applications.

"Microthene." [192] Trademark for a series of finely divided polyethylene resins. Used for coating and molding.

"Microthion." [412] Trademark for thiostrepton (q.v.).

Microtraps. [241] Tailored zeolites. A family of crystalline alumino-silicates with a three dimensional network structure of silica and alumina tetrahedra, characterized by a repeating three dimensional network of large, open alumino-silicate "cages" interconnected by smaller uniform sized pores. This structure makes the Microtraps excellent desiccants, with capacities less sensitive to relative saturation and elevated temperatures than most common adsorbents.

This structure also provides the unique ability to adsorb small molecules within the "cages" while excluding larger molecules that will not pass through the pores. They can effect sharp, selective separation of molecules on the basis of size or shape.

Three types, with pores of 4A, 5A, and 13A are available in various forms: 4-8 and 8-12 mesh beads, 8-14 and 14-30 mesh granules, and fine powder.

Containers: 5-, 25-, 50-, 100-lb air tight cans.

middle chrome. See chrome yellows.

middle oil. See carbolic oil.

middlings. The granular part of the interior of the wheat berry obtained in the process of milling. This product, when reduced by grinding to the desired fineness, produces the finest quality of flour. See also sharps.

"Midicel." [330] Trademark for sulfamethoxypyridazine (q. v.).

Grade: U. S. P.

Use: Medicine.

"Migral." [301] Trademark for a combination of ergotamine tartrate chlorcyclizine hydrochloride and caffeine (q. v.).

migration area. A term used in nuclear technology as a measure of the moderation or slowing down of neutrons. It is one sixth of the mean square distance a neutron travels before thermal capture.

"Mike" Sulfur. [233] Brand name for agricultural sulfur with average particle size 3 to 4 microns.

mil. One thousandth (0. 001) of an inch.

mildew preventives. Compounds used to prevent the growth of parasitic fungi, usually stain-producing, on such organic materials as textiles, leather, paper, farinaceous products, etc. Compounds most widely employed include cresols, phenols, benzoic acid, formaldehyde and organic derivatives or salts of copper, zinc and mercury.

milfoil. See achillea.

"Milibis." [162] Trademark for glycobiarsol.

milk acid. See lactic acid.

milk glass. Translucent or nearly opaque milk-colored glass produced by adding calcium fluoride and alumina to an ordinary glass.

milk of bismuth. See magma.

milk of iron. See magma.

milk of lime. Calcium hydroxide suspended in water. See magma.

milk of magnesia. See magnesia magma; magnesium hydroxide.

milk of sulfur. See sulfur, lac.

milk sugar. See lactose.

"Millcot." [51] Trademark for lubricating oils primarily designed for packaging machinery and textile mill equipment requiring an economical, non-spattering, non-creeping product. Several grades contain special film strength additive.

"Millical." [244] Brand name for calcium carbonate.

Properties: Oil absorption, 58-62; density as shipped, 38-42 lbs/cu ft; wt/solid gal, 22. 07 lbs; color, light cream white; particle size, 1 micron approx.

Derivation: Precipitated calcium carbonate.

Containers: Multi-wall paper bags, 50 lbs net.

Uses: Rubber, plastics, drawing compounds.

millicurie. See curie.

milliliter. A thousandth of a liter, which is the volume occupied by one kilogram of pure water at 4°C and 760 mm pressure. One milliliter (ml) equals 1. 000027 cubic centimeters (cc).

millimicron ($m\mu$). One-thousandth of a micron, or 10 angstrom units.

Millon's reagent. A reagent used in analytical work as a test for albumin. It is prepared by dissolving mercury in an equal weight of nitric acid of sp. gr. 1. 41, diluting the solution to twice its volume, allowing to stand and then decanting the liquid from the precipitate.

millstone. See buhrstone.

"Milontin." [330] Trademark for phensuximide (q. v.).

milorganite. An activated sludge marketed in dry granular form by the Milwaukee sewage disposal plant. Contains 5-10% moisture, 6. 5-7. 5% ammonia, 2. 5-3. 5% available phosphoric acid, 3-4% total phosphoric acid.

Use: Fertilizer.

Milori blue. A name applied loosely to any of a number of the varieties of iron blue pigments. See iron blues.

mimetite $Pb_5(AsO_4)_3Cl$. A natural chloride and arsenate of lead.

Properties: Color yellow to yellowish brown; streak white; luster resinous to subadamantine; sp. gr. 7. 24; hardness 3. 5-4. Forms a continuous series with pyromorphite.

Occurrence: California, Arizona, Nevada; Europe.

Use: A minor ore of arsenic.

mimosa bark. See wattle bark.

min. Abbreviation for minimum.

"Minecoat." [323] Trademark for a coal tar coating for mine interiors.

"Mine Gel." [413] Brand name applied to a series of semi-gelatin dynamites.

Containers: Packaged in cartridges of 7/8" diameter and up, in 50-lb. shipping

cases.
Uses: Underground mining; quarrying; construction and general blasting.
Fire hazard: Dangerous.
Shipping regulations: Explosives. Red label. *

mineral black. Black pigments made by grinding and/or heating black slate, shale, slaty coal, coke and coal. Pigment for various inks, coatings, surface coatings, leather finishes, plastics, etc.

mineral blue. Applied loosely to any of a number of varieties of iron blue pigments, usually containing considerable extender such as alumina.

mineral butter. See antimony trichloride.

mineral caoutchouc. See elaterite.

mineral charcoal. See fusain.

mineral cotton. See mineral wool.

"Mineralead." [41] Trade name for a hot-pour, sulfur-based compound used to joint cast iron bell and spigot pipe.

mineral fat. See petrolatum.

mineral graphite. See talc.

mineral green. See copper carbonate.

mineral jelly. See petrolatum.

mineral oil. Any liquid product of petroleum within the viscosity range of products commonly called oils.

mineral oil, white. Synonym for petrolatum, liquid (q. v.).

mineral pitch. See asphalt.

mineral rouge. See rouge (1) and iron oxide reds.

mineral rubber.
1. A term applied to asphaltites such as gilsonite (q. v.) and grahamite (q. v.).
2. Blown asphalts (q. v.).
Constants: Sp. gr. 1.00; m. p. about 100°F.
Uses: Rubber compounding; flux for asphaltites; ingredient of protective coatings; paints.

mineral seal-oil. A distilled and refined oil, having a boiling range higher than kerosine, but lower than gas oil.

mineral spirits. See naphtha, painters'.

"Mineral Spirits No. 10." [200] Trade name for a petroleum solvent.
Properties: Water-white color; boiling range 310-377°F; sp. gr. 0.779 (60°F); wt/gal 6.49 lbs (60°F); flash point 103°F.
Containers: Drums, tank wagon, tank car.
Uses: Paint, varnish and enamel thinner, metal cleaning, degreasing, herbicidal spray.

mineral superphosphate. See superphosphate.

mineral thinner. See naphtha, painters'.

mineral turpentine. See naphtha, painters'.

mineral wax. See ozocerite and ceresin wax.

mineral wool (mineral cotton; silicate cotton; slag wool; rock wool). A mass of fine intertwined fibers formed by blowing air or steam through molten rock or slag. Poor conductor of heat and sound; fire- and insect-proof. Used for insulation, as a binder and filler for synthetic resin-bonded panels; used for special structural and insulating purposes, filtering medium, fireproofing material.

minium Pb_3O_4. Natural red oxide of lead. Found in Colorado, Idaho, Utah, Wisconsin. See also lead oxide, red.

minium, iron. A name sometimes given to hematite or ferric oxide.

"Min-U-Sil." [436] Trade name for micron-sized silica.
Properties: High silica, low iron content (99.9% pure SiO_2); closely controlled particle size distribution.
Grades: Available in four uniform grades; 5, 10, 15 and 30 micron.
Containers: 50-lb multiwall paper bags.
Uses: Semi-reinforcing filler in silicone rubber; improved ceramic raw material and filler in plastics, paints, wood fillers, etc.

"Miokon Sodium." [329] Trademark for sodium diprotrizoate (q. v.), a water-soluble x-ray contrast medium.

MIPA. Abbreviation for monoisopropanolamine. See isopropanolamine.

mirabilite $Na_2SO_4 \cdot 10H_2O$. A natural hydrated sodium sulfate corresponding to the crystallized sulfate sold commercially as Glauber's salt. White or faintly greenish color, vitreous luster. Contains 19.3% Na_2O, 24.8% SO_3, balance water.
Constants: Sp. gr. 1.48; hardness 1.5-2.
Occurrence: United States (Indiana, Utah), Russia, Spain, Sicily, Chile, England, Italy.
Uses: See Glauber's salt.

"Mirasol Resins." [223] Proprietary products consisting of alkyd type resins. Epoxy resin esters are also marketed under this name.
Types: Available in all modifications including drying oils, semi-drying oils, non-drying oils, natural and phenolic resins.
Uses: Air-drying and baking finishes including architectural, lacquer, wrinkle, hammer and other industrial enamels; also printing inks and textile finishes.

"Miravar." [223] Proprietary products consisting of oleoresinous varnishes.
Uses: Industrial finishes and wrinkle enamels.

mirbane essence. See nitrobenzene.

mirbane oil. See nitrobenzene.

misch metal. The primary commercial form of mixed rare earth metal, prepared by the electrolysis of fused rare earth chloride mixtures. Misch metal contains 94-99% rare earth metals plus traces of calcium, carbon, aluminum, silicon and iron. A typical composition is 52% cerium, 18% neodymium, 5% praseodymium, 1% samarium, 24% others, including lanthanum. Lanthanum-enriched misch metal contains about 27% lanthanum. Some grades are nearly free of cerium. Ferrocerium is an alloy of misch metal and iron.

Properties: Sp. gr. about 6.67; m.p. about 1200°F.

Form: Waffle-like plates weighing 40 to 60 lbs packed in oiled paper, immersed in oil, or painted with vinyl paint.

Uses: Lighter flints; in ferrous and nonferrous alloys, imparting good low temperature impact resistance, reducing forge cracking and improving rolling properties when added to steels; improving fluidity, hot workability, oxidation resistance and strength when added to cast iron; increasing high temperature strength when added to aluminum and magnesium alloys; improving high temperature oxidation resistance when added to nickel alloys; and acting as a deoxidizer when used in copper and copper alloys, as a getter alloy.

"Misco 18-8." [209] Trade name for iron-based alloys with principal elements: nickel 10%, chromium 20%, silicon 1.25%, manganese 0.6%. Available in 5 grades. Carbon contents vary with the grades and for two of the grades an addition of 2.75% molybdenum is made.

"Misco CF 3": 0.03% max carbon, 0.50% max molybdenum.

"Misco CF 8": 0.08% max carbon, 0.50% max molybdenum.

"Misco CF 3M": 0.03% max carbon, 2.75% molybdenum.

"Misco CF 8M": 0.08% max carbon, 2.75% molybdenum.

"Misco CF 8C": 0.08% max carbon, columbium 10 × carbon %.

These alloys are used for castings in corrosion resistant service at sub zero, normal, and elevated temperatures. Each has its own particular advantages. CF3 is a low carbon grade (suitable for welding without heat treatment) with good corrosion resistance to strongly oxidizing corrosive media such as nitric acid. CF8 is commonly used for service at temperatures below −400°F, where it retains high impact resistance. CF3M and CF8M have good resistance to reducing corrosive media and enhanced resistance to sea water corrosion and pitting. CF3M and 8M are more resistant to weakly oxidizing media than the alloy without molybdenum. CF8C is used where welding in the field is necessary or for service at prolonged elevated temperatures in the range 800-1600°F, where columbium acts as a stabilizer for the carbon present in the alloy.

"Misco 20." [209] Trade name for a complex iron base alloy which contains the following alloying elements: 20% chromium, 29% nickel, 2.25% molybdenum, 3% copper. It has corrosion resistance to sulfuric acid, in all concentrations, and to strongly reducing chemicals, superior to the 18-8 grades. It is also superior in hot chloride salt solutions and weak acids. Its resistance to nitric acid is similar to the 18-8 grades.

"Misco CA15." [209] Trade name for an iron base alloy with 12% chromium as the alloying element, with 0.15% carbon, 0.60% manganese, and 10% silicon as minor constituents.

The alloy has a ferritic structure and is hardenable to 350° Brinell and has good resistance to many organic media in relatively mild service. Varying carbon ranges are available and the alloy has some heat resistant applications.

"Misco CE30." [209] Trade name for an iron base alloy with 29% chromium, 9% nickel as alloys, 1% silicon, and 0.25% carbon as minor constituents. Good corrosion resistant alloy to such media as nitric acid, sulfurous acid, and most oxidizing acids. It is a two phase alloy as cast and can be used in the range 800°-1400°F without loss of corrosion resistance exhibited by 18-8 at these temperatures. Particularly adaptable to paper mill corrosion resistant castings.

"Misco HH." [209] Trade name for an iron base alloy with 25% chromium, 12% nickel as alloys, and 0.40% carbon, 0.60% manganese, and 1.5% silicon as minor constituents. Used primarily for high temperature applications up to 2000°F for its resistance to sulfur compounds in furnace gases originating from high sulfur fuels. Available in two grades:

Type 1, a partially ferritic grade, has good ductility at 1800°F;

Type 2, a fully austenitic alloy, has greater strength at high temperature but lower ductility.

"Misco HT." [209] Trade name for an iron base alloy with 35% nickel, 15% chromium as alloys, and 0.60% carbon, 0.60% manganese, and 1.75% silicon as minor constituents. Used in high temperature applications for temperatures up to 2050°F where strength and resistance to oxidation at those temperatures are required. Its use is restricted to nonhigh sulfur compound atmospheres.

"Misco HUC." [209] Trade name for an iron base alloy with 38% nickel, 18% chromium, and 1.5% columbium as alloys, and 0.50% carbon, 0.60% manganese, 1.50% silicon as minor constituents. It is used for greater strength applications than the HT grade with greater resistance to oxidation. Other elements may be added to increase the hot strength and the usual limit in temperature is 2100°F.

"Misco HW." [209] Trade name for a nickel base alloy with 60% nickel, 12% chromium, with 0.50% carbon, 0.60% manganese, and 20% silicon as minor constituents. The iron content is approx 25%. It is used at high temperature and has good oxidation resistance up to 2050°. It is superior to HT alloy in its resistance to thermal shock. The alloy is magnetic. It is not recommended for high sulfur gas media.

"Misco HX." [209] Trade name for a nickel base alloy with 66% nickel, and 17% chromium, with 0.50% carbon, 0.60% manganese and 2.0% silicon as minor constituents. It is used in high temperature applications and has good oxidation resistance up to 2100°F. It is superior to HT alloys in its resistance to thermal shock. The alloy is nonmagnetic and is not recommended for high sulfur applications.

"Misco K." [209] An iron-base alloy which in addition to iron contains the following principal elements: nickel 20%; chromium 25%; silicon 1.5%; carbon 0.30%. This alloy finds usefulness as a general heat-resisting material showing excellent oxidation resistance at temperatures up to 2000°F together with excellent strength.

"Misco XM." [209] Trade name for an iron base alloy with 18% nickel, 16% cobalt, 22.5% chromium, 2.5% molybdenum, 2% tungsten, 0.5% columbium as alloys, and 0.6% manganese, 1.0% silicon, and 0.20% carbon as minor elements. Used for high strength requirements at temperatures up to 2200°F; is not so susceptible to sulfur compound attack as the nickel base alloys. It has approximately twice the strength of HT alloy at elevated temperatures.

"Miscrome 4." [209] An iron-chromium alloy, containing in addition to iron the following principal elements: chromium 12.5%; nickel under 0.80%; silicon 1.4%; manganese 1.0%; carbon 0.12%. This alloy exhibits good corrosion resistance to the milder types of environment such as atmospheric corrosion, oxidation resistance up to 1200°F, mild oxidizing acids, etc. It is useful as a material of construction where high room temperature strength and hardness is required in addition to corrosion resisting properties.

mispickel. See arsenopyrite.

"Mistron." [38] Trademark for ultra-fine particle size magnesium silicates available in a variety of grades ranging in both chemical purity and particle size. Used in paints as an extender pigment for viscosity and gloss control. Finer grades used as a partial replacement for titanium dioxide in both paints and paper coatings. Excellent brightness, nonabrasiveness and ease of dispersion. Compatible in both oleoresinous and latex emulsion coatings.

Uses: White reinforcing pigment in natural and synthetic rubbers; in blends with carbon black in rubber compositions.

miticide. A substance which kills mites, small animals of the spider class, among them the European red mite and the common red spider which infest fruit trees.

mitis green. See copper acetoarsenite.

"Mitox." [253] Brand name for para-chlorobenzyl para-chlorophenyl sulfide. Used as a specific miticide or acaracide product.

mixed acid (nitrating acid). Any mixture of sulfuric and nitric acids used for nitrating, e.g., in the manufacture of explosives, plastics, etc. Standard mixed acid consists of 36% nitric acid and 61% sulfuric acid.

Danger: Causes severe burns; vapor extremely hazardous; may cause nitrous gas poisoning; spillage may cause fire or liberate dangerous gas. MCA warning label.

Shipping regulations: Corrosive liquid. White label.*

mixed lead alkyls. Mixtures containing various methyl and ethyl derivatives of tetraethyl lead and tetramethyl lead. Thus methyl triethyl lead, dimethyl diethyl lead and ethyl trimethyl lead may all be present with or without tetraethyl and tetramethyl lead. Used as antiknock agents in motor fuels.

mixture. A kind or sample of matter containing two or more substances that are not chemically united, and can therefore be separated by taking advantage of differences in their physical properties, such as solubility in a solvent, difference in boiling point or freezing point, etc.

MKP. Abbreviation for monopotassium phosphate. See potassium phosphate, monobasic.

ml. Abbreviation for milliliter.

MLA. Abbreviation for mixed lead alkyls (q.v.).

mm. Abbreviation for millimeter.

Mn. Symbol for manganese.

MnEBD. Abbreviation for manganese ethylenebisdithiocarbamate. See maneb.

Mo. Symbol for molybdenum.

"Mobilcer." [331] Brand name for a line of wax emulsions.

Uses: Sizes for paper, paperboard, particle board, textiles and cordage; binders in ceramics; carriers for pesticides; coating of fruits and vegetables; plasticizing of laundry starch; treatment of nursery stock; end-checking treatments for timber and lumber; in the manufacture of foam rubber and polymeric latex products; curing of concrete.

"Mobil-Kote." [331] Brand name for a light oil used as a temporary corrosion-resistant film. Not suitable for outdoor storage or heavy-duty service.

"Mobilpar." [331] Brand name for a number of emulsified, or emulsifiable petroleum products.

Uses: In the lubrication of textile fibers; as plasticizers for starch formulas in

*See "I.C.C. Shipping Regulations," page xiii.
Reference numbers refer to name of manufacturer. See "List of Manufacturers," page v.

the textile industry; as softeners and wetting-out agents in the paper and textile industries; and as foam control agents in aqueous processes.

"Mobil Sorbead Desiccants." [331] Brand name for a line of bead desiccants.

Uses: For the removal of moisture from gases in static and dynamic systems; for the recovery of hydrocarbons from natural gas streams.

"Mobilwax." [331] Brand name for petroleum waxes of both the paraffinic and microcrystalline types.

"Moca." [28] Trademark for methylene-bis-ortho-chloroaniline, $CH_2(C_6H_4ClNH_2)_2$.

Properties: Tan colored, coarsely-ground lumps; sp. gr. 1.39; m. p. 100-105°C; soluble in hot methyl ethyl ketone, acetone.

Containers: 50-lb drums.

Use: Curing agent for "Adiprene" L urethane rubber, other urethane rubbers, and epoxy resins.

modacrylic fiber. Generic name for a manufactured fiber in which the fiber-forming substance is any long chain synthetic polymer composed of less than 85% but at least 35% by weight of acrylonitrile units, $-CH_2CH(CN)-$ (Federal Trade Commission). Other chemicals such as vinyl chloride are incorporated as modifiers.

Uses: Deep pile and fleece fabrics; industrial filters; carpets; underwear; blends with other fibers. See "Dynel."

"Mod-Epox." [58] Trademark for modifier for epoxy resins. Reduces viscosity of liquid epoxy resins, accelerates cure; improves strength, electrical and adhesive characteristics. Used in tool and die manufacture; electric potting compounds; encapsulations; adhesives; surface coatings and body solders.

Containers: 5- and 55- gal drums.

moderator. A substance of low atomic weight such as beryllium, carbon (graphite) or deuterium (in heavy water) which is capable of reducing the speed of neutrons but which has little tendency toward neutron absorption. The neutrons lose speed when they collide with the atomic nuclei of the moderator. Moderators are used to adjust the speed of neutrons in nuclear fission reactors since slow neutrons are most likely to produce fission. A typical graphite-moderated reactor may contain 50 tons of uranium for 472 tons of graphite.

"Modicol." [309] Trademark for stabilizers for natural and synthetic latex, including:

"Modicol N." A nonionic, liquid fatty amido condensate used to improve chemical resistance and viscosity uniformity.

"Modicol S." An anionic, liquid sulfonated fatty product used to improve mechanical stability and resistance to acids.

modified sodas. See sodas, modified.

"Modiphats." [259] Trade name for the polyhydric alcohol esters of a series of fatty acids.

Uses: Textile processing; plasticizers.

"Modulex" (HMF). [285] Proprietary brand name for high modulus furnace carbon black.

Properties: Sp. gr. 1.77; free-flowing pellets, (also available in fluffy, unpelleted form as "Modulex-UC"); bulk density 35 lbs/cu ft; particle diameter 65 millimicrons; pH 10.0; ash 0.75% max; 99.9% thru 325 mesh screen; color (Nigrometer) 95-96.

Containers: 50-lb paper bags or bulk.

Uses: As a reinforcing ingredient for compounding in natural and most synthetic rubbers, contributing to abrasion resistance, good tensile and tear strength; as a black coloring agent in rubber, paper, plastics, paint and ink.

"MODX." [94] A proprietary mixture of inorganic and organic acetates containing 25% diphenylethylenediamine.

Properties: Cream colored granules; odorless; sp. gr. 1.34; m. p. indefinite - fluxes at milling temperature; ash 31-37%. Activates aldehyde amines, thiazoles and dithiocarbamates; imparts high modulus; stains very light-colored stock to slight extent; improves aging.

Containers: 300-lb drums.

Uses: As an activator and age resistor in compounding mechanicals, tire, tubes, heels, and soles; vulcanization leveler and activator with age-resisting properties for natural rubber, GR-S and Buna-N.

moellon degras. See degras, moellon.

"Mogul." [275] Trade name for a series of channel long-flow carbon blacks for use in inks. Available as:

"Mogul." Blackest mass-tone and high jet strength.

"Mogul Special." Longer flow and higher loading capacity than "Mogul."

"Mogul A." Longest flow black.

Mohr's salt. See ferrous-ammonium sulfate.

Mohs' scale. A scale of hardness of minerals, running from one to ten, with talc as the softest and diamond as the hardest. The hardness of a mineral is determined by which minerals of the scale will scratch it, and vice versa.

1.	talc	6.	orthoclase
2.	gypsum	7.	quartz
3.	calcite	8.	topaz
4.	fluorite	9.	corundum
5.	apatite	10.	diamond

Other useful hardnesses are: fingernail, a little over 2; penny, about 3; pocket knife, a little over 5; window glass 5.5; and a steel file 6.5.

The difference in hardness between corundum, H = 9, and diamond, H = 10, is greater than the difference between talc, H = 1, and corundum. A modified scale has been proposed in which quartz is 8, topaz 9, garnet 10, fused alumina 12, silicon carbide 13, boron carbide 14, and diamond 15.

mol. Abbreviation for molecular.

"Molacco." [133] Trademark for furnace carbon black characterized by blue tone and high loading capacity. Available in powdered and bead forms. Used in protective and decorative coatings and printing inks. Containers: In powdered form, 25-lb bags; bead form, 25- and 50-lb bags.

molar.
1. A molar solution is one that contains one molecular weight in grams (one mole) of its dissolved substance in one liter of solution.
2. Molar quantities are quantities proportional to the molecular weights of the substances concerned.

molasses. The definition of molasses varies in different countries, and in the United States it is not the same in the cane-sugar industry as in the beet-sugar industry.

In the raw cane-sugar industry in the United States molasses is defined as the syrupy mother liquor which is left after sucrose has been removed from sugar-cane juice by concentration, crystallization, and separation of the sugar crystals (usually in centrifugals). If only one crop of crystals has been removed, the mother liquor is termed "first molasses"; if a second crop has been removed after reconcentration of the first mother liquor, the resulting product is termed "second molasses," and so on. The final mother liquor from which no more sugar can be extracted in factory practice by the above process is termed "final molasses," "blackstrap molasses," or briefly "blackstrap."

The final mother liquor obtained in cane-sugar refineries is not termed molasses, but "refiner's syrup," or "barrel syrup."

In the United States beet-sugar industry only the final mother liquor, obtained after concentration, crystallization, and centrifugation of beet juice is termed "molasses." If the Steffen process of desugarization is practiced, the final mother liquor is known as "Steffen molasses."

There are large variations in the composition of different kinds of molasses or final mother liquors. Typical analyses, in round figures, are about as follows.

Per Cent	Cane Blackstrap	Beet Molasses	Barrel Syrup
Sucrose	30	50	35
Reducing sugars	20	trace	25
Ash	10	10	6
Organic non-sugars	20	20	14
Water	20	20	20

Uses: Food; feed; raw material for acetone and butanol, for citric acid, and especially for ethyl alcohol.

molasses, beet. See molasses.

molasses, lactose. Molasses obtained from the preparation of milk sugar.

molding sand. See foundry sand.

mole. A unit quantity in chemistry. An amount of a substance in grams (gram mole) or pounds (pound moles) which corresponds to the sum of the atomic weights of all the atoms appearing in the molecule. The spelling mol is also used but does not have official sanction.

molecular distillation (high vacuum distillation). Distillation at low pressures of the order of 0.001 mm. A molecular distillation is distinguished by the fact that the distance from the surface of the liquid being vaporized to the condenser is less than the mean free path (the average distance traveled by a molecule between collisions) of the vapor at the operating pressure and temperature. This distance is usually of the order of magnitude of a few inches. This process is useful in separation of extremely high boiling and heat-sensitive materials such as glycerides and some vitamins.

molecular formula. See formula, chemical.

molecular sandwich. A type of molecular structure in which a transition metal (one from the central part of the periodic table) lies between two aromatic rings such as C_5H_5-, as in ferrocene.

See dicyclopentadienyl compounds.

molecular sieves. Zeolites or similar materials whose atoms are arranged in a crystal lattice in such a way that there are a large number of small cavities interconnected by smaller openings or pores of precisely uniform size. Normally these cavities contain water molecules, but upon heating, this water is driven off without any change in the remaining crystal lattice. The network of cavities and pores may occupy 50% of the total volume of the crystals.

Molecular sieves have a strong tendency to readsorb water. In the absence of water they will adsorb other molecules that are small enough to pass through the pores. These small molecules may thus be separated from a mixture with larger molecules.

A few natural zeolites exhibit molecular sieve characteristics to a limited degree. Synthetic zeolites are available in two sizes (pore openings 4 and 5 angstrom units in diameter) with high capacity for adsorption and regeneration even when used at elevated temperatures.

Uses: Drying gases such as air, hydrogen, natural gas, refinery gas, ethylene; drying liquids such as benzene, alcohols, hydrocarbons, fluorocarbons; separation of ethylene from carbon dioxide, carbon dioxide from annealing gas, hydrogen sulfide from natural gas; removal of normal paraffins from light naphthas. Also as a carrier for volatile, toxic, odoriferous, or reactive compounds, which can then be released by heat or displacement at the desired time and place. Materials that have been so carried include organic and metallo-organic compounds, halogen elements, acid gases, water, perfume, catalysts, pesticides, fumigants, ripening agents, radioactive isotopes, blowing agents, antioxidants in rubbers and plastics.

molecular weight. The sum of the atomic weights of the atoms in a molecule. Thus the molecular weight of methane gas CH_4 is 16, the atomic weights being C=12, H=1. The chemical formula used in such a calculation must be the true molecular formula of the substance designated. For example, the molecular formula of ordinary oxygen is O_2, and the molecular weight is 32 (atomic weight of O=16). For ozone the proper molecular formula is O_3, and the molecular weight is 48. The true molecular weight of a gas or vapor is found in the case of a brand new compound by measuring the volume of a given weight and then calculating the weight of 22.4 liters at 0°C and 760 mm. For liquids and solids more complicated means must be used. Once the molecular weight is known, the correct molecular formula can be written.

"Molex" Process. [416] Patented process employing molecular sieves to separate normal paraffins from mixtures with isoparaffins and other types of hydrocarbons. Products consist of a normal paraffin stream of high purity and a second stream containing the remaining hydrocarbons in the original mixture. The normal paraffins are excellent components of jet fuels or as raw materials for further chemical synthesis. The denormalized stream, if within the gasoline boiling range, will have higher antiknock quality than the original mixture containing the normal paraffins.

"Mollescal CA." [307] Brand name for a leather chemical; a combination of mild organic bases, wetting and antiseptic agents.
Properties: Clear, thin, brown liquid; sp. gr. 1.13-1.15; readily soluble in water.
Uses: Used in the leather trade to facilitate the soaking of dry hides and skins. This product retards putrefaction, aids in the emulsification and removal of fat, results in leather of smoother grain and firmer flanks.

"Mollisan AS." [328] A multipurpose lubricant and softener for textile finishing providing improved tear strength and abrasion resistance when used as an additive or topping treatment for urea or melamine resin finishes. It is effective as an anti-static agent for synthetics and enhances the sewability of resin-finished fabrics. Also a non-yellowing full-bodied softener for cottons; a yarn lubricant and anti-static; a napping assistant.

molucca grains. See tiglium.

"Moly." [67] Trademark for a series of molybdenum-containing compounds used for seed treatment as a foliar spray, fertilizer additive and for similar related uses.

molybdate chrome orange. See molybdate orange.

molybdate orange (molybdenum orange; molybdate chrome orange). An inorganic pigment which is a solid solution of lead chromate, lead molybdate, and lead sulfate.
Properties: Fine dark orange or light red powder.
Derivation: By adding solutions of sodium chromate, sodium molybdate and sodium sulfate to a lead nitrate solution under carefully controlled conditions and filtering off the precipitate.
Containers: Barrels.
Uses: Printing inks; paints; plastics.

molybdenite (molybdenum glance) MoS_2. Natural molybdenum sulfide found in igneous rocks and metallic veins.
Properties: Color, bluish-lead gray; streak gray-black; luster metallic; one perfect cleavage; greasy feel; hardness 1-1.5; sp. gr. 4.6-4.8. Similar in appearance to graphite. Soluble in sulfuric and strong nitric acids.
Occurrence: Colorado, Utah, New Mexico; Canada; Europe; Australia; Mexico.
Use: Principal ore of molybdenum.

molybdenite concentrate. Commercial molybdenite ore after the first processing operations. Contains about 90% molybdenum disulfide along with quartz, feldspar, water, and processing oil.

molybdenum Mo. Metallic element of atomic number 42, in group VI of the periodic table.
Properties: Gray metal or black powder; of wide but not abundant distribution. It is a necessary trace element for some crops. See molybdenite and wulfenite. Insoluble in hydrochloric or hydrofluoric acid, ammonia, sodium hydroxide, or dilute sulfuric acid; soluble in hot concentrated sulfuric or nitric acids; insoluble in water. Sp. gr. 10.2; m. p. 2620°C; high strength at very high temperatures.
Derivation: By aluminothermic, hydrogen, or electric furnace reduction of molybdic anhydride.
Forms available: Rods, wire, powder; ingots (from powder); high ductility sheets; concentrates; also as large single crystals.
Purity: Rods and wire 99.9%; powder 95%.
Containers: Wooden barrels; cartons.
Uses: Metallurgy (alloy steels); as wire (windings for electric resistance furnaces, construction of spider which supports tungsten filaments in some incandescent lamps, welded to "Pyrex" glass in construction of plate standards, grids and filament supports in radiotrons); as sheet in manufacture of some types of radiotrons; substitute for platinum in contact-making and breaking devices; points for spark plugs; certain parts of x-ray tubes and equipment; in molybdenum compounds, for plating other metals, and especially as

lubricants and catalysts, but also in fertilizers, pigments and dyes, metal finishing compounds, ceramic enamels and glazes, drugs, and reagents.
See also heteromolybdates.
Shipping regulations: None. *

molybdenum acetylacetonate $Mo(C_5H_7O_2)_n$. Crystalline powder. Slightly soluble in water. Resistant to hydrolysis. A chelating non-ionizing compound.

molybdenum aluminide. A cermet which can be flame-sprayed.

molybdenum anhydride. See molybdenum trioxide.

molybdenum boride MoB. Powder used extensively by electronics industry for metal brazing.

molybdenum dioxide MoO_2.
Properties: Lead-gray, non-volatile powder; insoluble in hydrochloric and hydrofluoric acids; sparingly soluble in sulfuric acid and in bases; sp. gr. about 6.4.
Derivation: Reduction of molybdenum trioxide or molybdates by hydrogen; partial oxidation of metallic molybdenum.

molybdenum disilicide $MoSi_2$. A cermet. Powder, not affected by air up to 3000°F, not attacked by most inorganic acids, including aqua regia. Has high stress-rupture strength. Available as cylinders, lumps, granules and powder; may be coated on materials by vapor deposition and by flame spraying. Forms oxidation-resistant coatings. Used in electrical resistors.

molybdenum disulfide (molybdic sulfide, molybdenum sulfide) MoS_2. See also molybdenite.
Properties: Black, lustrous powder; sp. gr. 4.80; m. p. 1185°C; soluble in aqua regia, sulfuric acid (conc); insoluble in water.
Derivation: Purification of molybdenite; by the reaction of sulfur or hydrogen sulfide on molybdenum trioxide.
Uses: Lubricants in greases, oil dispersions, resin-bonded films, dry powders, etc., especially for use at extreme pressures and high vacua; also used as a hydrogenation catalyst.

molybdenum glance. See molybdenite.

molybdenum hexacarbonyl $Mo(CO)_6$.
Properties: White shiny crystals; decomposing at 150°C without melting; sp. gr. 1.96; b. p. about 155°C; vapor pressure at 20°C about 0.1 mm Hg, at 101°C about 43 mm Hg; insoluble in water; soluble in ceresin, paraffin oil, benzene, aminoanthraquinone; slightly soluble in ether and other organic solvents.
Derivation: From molybdenum pentachloride by reaction with zinc dust and carbon monoxide in ether at high pressures.
Uses: Plating molybdenum, i.e., molybdenum mirrors.
Shipping regulations: None.*

molybdenum hexafluoride MoF_6.
Properties: White crystalline compound; m. p. 17.5°C; b. p. 35°C; sp. gr. (liq.) about 2.5. Reacts readily with water.
Derivation: Direct action of fluorine on molybdenum metal.
Uses: Important in the separation of molybdenum isotopes.

molybdenum lakes. See phosphomolybdic pigments.

molybdenum orange. See molybdate orange.

molybdenum III oxide. See molybdenum sesquioxide.

molybdenum oxides. See molybdenum sesquioxide; molybdenum dioxide; molybdenum trioxide.

molybdenum pentachloride $MoCl_5$.
Properties: Green-black solid; dark red as liquid or vapor; m.p. 194°C; b.p. 268°C; sp.gr. 2.9. Hygroscopic, reacting with water and air; soluble in dry ether, dry alcohol, and other anhydrous organic solvents.
Derivation: Direct action of chlorine on finely divided molybdenum metal.
Uses: Intermediate in preparation of molybdenum hexacarbonyl which in turn is used for making molybdenum mirrors; as general intermediate; as catalyst; for spraying molybdenum coatings.

molybdenum sesquioxide (dimolybdenum trioxide, molybdenum III oxide) Mo_2O_3. Known only in the hydrated form, $Mo(OH)_3$, although commonly assigned the formula Mo_2O_3. A compound formed by a dry reaction of molybdenum and oxygen which approximates the composition of the sesquioxide is probably a mixture of molybdenum and molybdenum dioxide.
Properties: Gray-black powder; slightly soluble in acids; insoluble in alkalies and water.
Derivation: Zinc reduction of acid solutions of molybdic acids and molybdates; electrolytic deposition from acid solutions of molybdates.
Uses: Catalyst in organic synthesis; decoration and protection for metal articles; feed additive.

molybdenum silicide. Alloy of 60% molybdenum, 30% silicon, and 10% iron, used as means of introducing molybdenum into steel.

molybdenum steels. Molybdenum in steel has the following effects:
1. Raises grain-coarsening temperature of austenite (refines grain).
2. Deepens hardening.
3. Counteracts tendency toward temper brittleness.
4. Raises hot and creep strength, red hardness.
5. Enhances corrosion resistance in stainless steel.
6. Forms abrasion-resisting particles.

Iron-molybdenum alloys containing 6 to 30% molybdenum can be age-hardened.

Minimizes temper embrittlement in low-alloy steel. The most effective metal that can be added to increase strength at elevated temperatures; increases creep resistance. Used as substitute for tungsten in high-speed steels on the basis of 1 part molybdenum to 2 parts tungsten. Molybdenum high-speed steels have greater tendency to surface decarburization when heated in oxidizing atmospheres, which has restricted their application.

In engineering steels, molybdenum rarely exceeds one per cent, in tool steels may be 10%, stainless steels 3%.

See also ferromolybdenum.

molybdenum sulfide. See molybdenum disulfide.

molybdenum trioxide (molybdenum anhydride; molybdic oxide; molybdic acid anhydride) MoO_3.

Properties: White at ordinary temperatures, yellow at elevated temperatures; sp. gr. 4.69; m.p. 795°C; b.p. 1264°C. Sparingly soluble in water; very soluble in excess alkali with formation of molybdates; soluble in concentrated mixtures of nitric and hydrochloric acids or nitric and sulfuric acids. Two hydrates are known: $MoO_3 \cdot H_2O$ and $MoO_3 \cdot 2H_2O$. Readily combines with acids and bases to form a series of polymeric compounds.

Derivation: Roasting of molybdenite; by ignition of the metal, the sulfides, the lower oxides, and of molybdic acids.

Purification: Sublimation.

Grades: Technical; pure; reagent, A.C.S.

Containers: Bottles; boxes; kegs; drums; carload lots.

Uses: Source material for preparation of molybdenum compounds; agriculture (as source of needed molybdenum in soil); analytical chemistry; manufacture of metallic molybdenum; introduction of molybdenum in alloys; corrosion inhibitor; ceramic glazes; enamels; pigments; catalyst in petroleum industry; medicine.

Shipping regulations: None. *

molybdic acid. Molybdic acid of commerce is either an ammonium molybdate (molybdic acid 85%) or molybdenum trioxide. The use of the term interchangeably for either of these compounds has caused confusion.

Solutions of molybdic acid are very complex chemically since they show a great tendency to polymerize into di-, tri-, tetra-, poly-, etc., molybdates.

Containers: Drums.

molybdic acid, anhydride. See molybdenum trioxide.

molybdic ocher. See ferrimolybdite.

molybdic oxide. See molybdenum trioxide.

molybdic sulfide. See molybdenum disulfide.

molybdite. See ferrimolybdite.

moly-blacks. Black, lustrous decorative coatings consisting mainly of molybdenum. The coatings are usually applied electrolytically from a bath containing soluble molybdates. However in certain cases some may be applied by immersion. The coatings are used commercially for blackening zinc or zinc-base alloys.

"Moly-Gro." [433] Trademark for molybdenum compounds used to correct molybdenum deficiency in fertilizers, soils and crops.

"Molykote." [199] Trademark for lubricants containing a highly purified molybdenum disulfide powder which resembles graphite in appearance, but contains none. Used in friction problems which involve galling, welding and seizing; recommended for friction problems at high or low temperatures, in dusty atmospheres, radiation, vacuum, and liquid oxygen environment. Also used for similar mating metals; fretting; high starting friction; and rubber on metal. Available as powder, greases, dispersions, sticks and bonded lubricant coatings.

"Molynamel." [289] Trademark for resin-bonded molybdenum disulfide lubricating enamels. A variety of "Liquid-Moly."

molysite. See ferric chloride, anhydrous.

"Moly-Sulfide." [67] Trademark for molybdenum disulfide.

Uses: Lubricant; lubricant additive; filler.

"Monacide." [405] Brand name for a series of insecticides, especially with designation "5% DDVP" a spray for tobacco warehouses. DDVP is an O-dimethyl-2,2-dichlorovinyl phosphate.

"Monad." [86] Trademark for series of detergents available as:

"Monad" G: Neutral sulfated monoglyceride.

"Monad" SF, High A.I.: Neutral sulfated monoglyceride, salt free.

Uses: Wetting; emulsifier; lime soap dispersant; softening agent.

"Monalit." [405] Trade name for a series of yarn conditioning compounds, used in spinning and weaving mills, to set the twist of the yarn after spinning or twisting to facilitate more perfect weaving.

"Monamid." [405] Trademark for a series of surface active agents known as "Super" amides, having an amide content up to 93%. Made by condensing fatty acids and esters and ethanolamines to form alkylolamides.

Properties: Most grades are practically odorless; light color. Foam boosters, stabilizers, and viscosity builders. Several grades are:

150-L: Lauric acid and diethanolamine.

150-AD: Coconut fatty acid and diethanolamine.

150-M: Myristic acid and diethanolamine.

Uses: Shampoos, bubble baths; liquid industrial and household cleaning and chemical specialty compounds.

"Monamine." [405] Trademark for a series of surface active agents known as fatty acid amine condensates.

Properties: Commonly amber color with slight odor. Good foamers, detergents,

*See "I.C.C. Shipping Regulations," page xiii.

Reference numbers refer to name of manufacturer. See "List of Manufacturers," page v.

and emulsifiers. Soluble in water; soluble or dispersible in aromatic and chlorinated aliphatic solvents. Several grades are:
AD-100: Coconut fatty acid, 98-99% active material.
AA-100: Distilled coconut fatty acid, 96-98% active material.
ACO-100: Lauric acid, 98-99% active material.
Uses: Dish and glass cleaners; car wash; floor cleaner; liquid hand soap; paper and textile softeners.

"Monamulse." [405] Trademark for a series of compounded emulsifiers for paints, plastics, graphic arts, agricultural sprays and other fields.

"Monaquest." [405] Trademark for a series of organic amino polycarboxylic acid sequestering agents useful in complexing, chelating or sequestering of hard water salts and other polyvalent metals.

"Monarch." [266] Brand name for group of insoluble phthalocyanine pigments producing bright shades of blue and green. Lightfast, high strength.
Uses: Protective coatings; paper; textiles; inks.

"Monarch." [275] Trade name for series of channel carbon blacks for use in paints, plastics and paper. Available as:
"Monarch 71." High color channel black.
"Monarch 74." Medium color channel black.
"Monarch 81." Regular color all-purpose channel black.

monarda oil. See horsemint oil.

"Monastral." [28] Trademark for a line of insoluble phthalocyanine pigments producing extremely bright shades of blue and green with excellent fastness properties and high tinctorial strength. Used in the lake, wallpaper, paper, textile, and other industries.

"Monastrip." [405] Trade name for a solvent-stripper for uncured and cured epoxy, polyester, and silicone rubber potting, casting, and encapsulating compounds.

"Monawet." [405] Brand name for sodium dioctyl, sodium di-hexyl and sodium di-isobutyl sulfosuccinates, a group of surfactants with unusually fast wetting characteristics. 5/100th of 1% - 1 second Draves sinking time. Used for fast wetting, spreading and penetration, also for polymerization emulsions.

monazite Ce, La, $Th(PO_4)$. A natural phosphate of the rare earth metals, principally the cerium and lanthanum metals, usually with some thorium. Yttrium, calcium, iron, and silica are frequently present. Monazite sand is the crude natural material and is usually purified from other minerals before entering commerce.
Properties: Color yellowish to reddish brown; luster vitreous to resinous; streak white; hardness 5-5.5; sp. gr. 4.9-5.3.
Occurrence: North Carolina, South Carolina, Idaho, Colorado, Montana, Florida; Brazil; India; Australia; Canada.
Uses: Important source of thorium, cerium, and other rare earth metals and compounds.

Mond process for nickel. Mixed metallic ores obtained from roasting of crude ores are heated from 50-80°C in a stream of producer gas. Oxides other than nickel are reduced to the metallic state while nickel forms nickel carbonyl $[Ni(CO)_4]$ which passes off as a vapor. The vapor is subsequently resolved into gaseous carbon monoxide and free nickel which is deposited in a lustrous, mirror-like form.

"Monel." [283] Trademark for an alloy containing approximately 66% nickel and 31% copper. Has good resistance to many corrosive environments and is tough and strong. Made in both cast and wrought forms.
See also "K," "KR," "R" and "S" "Monel."

"Monex." [248] Trademark for tetramethylthiuram monosulfide.
Properties: Yellow powder; sp. gr. 1.40; m. p. 104-107°C; soluble in acetone, benzol, and ethylene dichloride; insoluble in water and gasoline.
Uses: Accelerator for natural rubber wire insulation; druggist sundries; mechanicals; proofing; footwear; sponge rubber and transparent pure gum stocks and for all types of GR-S.

monitoring. Periodic or continous examination by means of suitable instruments to determine the amount of radiation or radioactive contamination present in a particular location or on an individual.

monkshood. See aconite.

mono-. Prefix denoting single radical; see under specific compound; e. g., monoacetin, see acetin.

mono acid F. See 2-naphthol-7-sulfonic acid.

monoanhydrosorbitol. See sorbitan.

monoazo dyes. See azo dyes.

monobasic. Acids with one, two, and three displaceable hydrogen atoms per molecule are termed mono, di, and tribasic acids respectively. Monobasic, dibasic and tribasic salts are salts which are formed with displacement of one, two, and three hydrogen atoms respectively from the acid. These terms are commonly applied only to salts of tribasic acids, e.g. the orthophosphates.

"Monobed." [23] Trademark for intimate mixtures of "Amberlite" cation and anion exchange resins.
Use: Complete removal of ionizable impurities from water and other solutions in a one-step treatment.

"Monobel." [28] Trademark for low-velocity permissible dynamites furnished in six grades based upon velocity and cartridge count. Fair to poor water resistance.
Use: For mining coal where lump coal is a factor.

monobenzone (para-benzyloxyphenol; monobenzyl ether of hydroquinone) $C_6H_5CH_2OC_6H_4OH$.
Properties: White, odorless, crystalline powder; little taste; melting range 115-118°C; soluble in alcohol and acetone; insoluble in water.
Grade: N. F. XI.
Use: Medicine.

"Monobromantin." [109] Trade name for 3-bromo-5, 5-dimethylhydantoin.

monobromated camphor. See camphor bromate.

monocalcium phosphate. See calcium phosphate, monobasic.

"Monochrome." [232] Brand name for a series of chrome dyestuffs suitable for application to wool by the metachrome process.

"Monochrome." [307] (U. S. A.). Brand name of a series of mordant dyestuffs. Used for the dyeing of wool. Characterized by very good fastness properties.

"Monocite" H-100. [28] Trademark for a special grade of methyl methacrylate monomer used in manufacture of cast acrylic sheets.
Containers: Insulated tank trucks.

"Monodral" Bromide. [162] Trademark for penthienate bromide.

"Monofrax." [442] Trademark for fused cast refractory articles.
Properties: Extremely low permeability, high resistance to fluxes, slags, and glasses.
Uses: Blocks, bricks and shapes for very pure and highly refractory glass, chemical and metallurgical applications.

monoglycerides. Glycerol esters of fatty acids in which only one acid group is attached to the glycerol group. A typical formula is $RCOOCH_2CHOHCH_2OH$. Small amounts of monoglycerides occur naturally.
Derivation: Produced synthetically by the alcoholysis of fats with glycerol, yielding a mixture of mono-, di-, and tri-glycerides which is predominantly mono-glycerides.
Uses: Emulsifiers; cosmetics; lubricants.
See glycerol monostearate, glycerol monolaurate, etc.

"Monolene." [203] Trademark for N-(2-hydroxypropyl)ethylenediamine. This product is a colorless to faint yellow liquid with a boiling point of 130°C at 22 mm. It is used as a chemical intermediate since it will undergo reaction with a wide range of functional organic compounds.

"Monolite." [206] Proprietary name of line of insoluble lake colors. Used for the manufacture of paint and printing inks, and for the printing of wallpapers and the preparation of coated papers.

monomer. A molecule or compound usually containing carbon and of relatively low molecular weight and simple structure, which is capable of conversion to polymers, synthetic resins or elastomers by combination with itself or other similar molecules or compounds. Thus, styrene is the monomer from which polystyrene resins are produced; vinyl chloride and vinyl acetate are the monomers from which "Vinylite" resins are obtained. Other common monomers are methyl methacrylate for "Lucite" or "Plexiglas"; adipic acid and hexamethylenediamine for nylon; and styrene and butadiene for SBR synthetic rubber.

"Monopentek." [138] Trade name for monopentaerythritol.
Containers: 50-lb multiwall bags.
Uses: For certain alkyds and other resins of low or special viscosity requirements.

"Monoplex." [23] Trademark for monomeric, liquid plasticizers for polyvinyl chloride and other high polymers. Primarily esters, but also some epoxides which impart heat and light stability.
Use: Plasticizers, stabilizers, processing aids.

"Monopole Oil." [78] Trademark for a double sulfonated castor oil used in the textile trade in bleaching, dyeing, finishing; as a dispersant for dye solutions in printing; also used for its solvent and detergent properties in hand cleaners, polishes, household preparations, bath oils, detergents, etc.

monopropellant. A propellant which combines fuel and oxidizer in one compound or mixture. Gunpowder is an example of a solid monopropellant. Liquid monopropellants, for rockets, include: methyl nitrate; nitromethane; a mixture of hydrocarbons with tetranitromethane; a mixture of methyl nitrate and methanol.
Warning! Very dangerous materials.
See also rocket propellants.

monosodium glutamate. See sodium glutamate.

"Monosol." [206] Brand name of proprietary line of soluble colors especially adapted for converting to insoluble pigments, which are used in the manufacture of printing inks and paints.

monostearin. See glycerol monostearate.

monosulfonic acid F. See 2-naphthol-7-sulfonic acid.

"Monosulph." [309] Trademark for an anionic penetrant and emulsifying agent, 68% highly sulfonated castor oil.
Uses: Textile dyeing assistant; fatliquor for suede leather; paper coating evener; plasticizer for starch, glues; emulsifier for latex.

"Monotan." [78] Trademark for a resinous type of synthetic tanning material used for all types of leather tannage, either alone or in combination with mineral, vegetable and other synthetic tanning compounds.

"Mono Thiurad." [58] Trademark for tetramethylthiuram monosulfide.

"Monsanto Detergent MXP." [58] Trade name for a non-dusty, built detergent based on a non-ionic synthetic wetting agent.
Use: Detergent with controlled sudsing action.
Containers: 100-lb bags; 100-, 300-lb drums.

"Monsanto Fire Retardant A." [58] Trade name for a substantially neutral mixture of ammonium phosphates containing a wetting agent to improve penetration.
Use: Flame-retardant treatment of wood and paper.
Containers: 100-lb bags; 100-, 200-lb fiber drums.

"Monsanto Penta." [58] Trademark for pentachlorophenol, technical.

Monsel's salt. See ferric sulfate, basic.

Monsel's solution. See ferric sulfate, basic.

montan wax (lignite wax).
Properties: White, hard wax; crude product, dark brown; m. p. 80-90°C. Soluble in carbon tetrachloride, benzene, and chloroform; insoluble in water.
Derivation: By countercurrent extraction of lignite. American and German lignite are usual sources.
Method of purification: Distillation with superheated steam.
Grades: Crude; refined.
Containers: Bags.
Uses: Substitute for carnauba and beeswax; increasing hardness of wax compositions; shoe polishes; furniture polishes; phonograph records; roofing paints; rendering paints waterproof; adhesive pastes; candles; hardener for fat compositions; electric insulating compositions; paper-sizing compositions; carbon papers; wire coating; wax sprays.

"Montar." [58] Trademark for a series of synthetic coal tar resins, available in several grades ranging in softening point, 120-250°C.

Mont Cenis process. Ammonia synthesis process using relatively low pressures and temperatures, an iron cyanide catalyst, and obtaining relatively low yields.

monticellite. See olivine.

montmorillonite $Al_2O_3 \cdot 4SiO_2 \cdot H_2O$, with many variations. Generally about one-sixth of the aluminum atoms are replaced by magnesium atoms, and varying amounts of hydrogen, sodium, potassium, calcium, and magnesium are loosely combined. It is commonly found in bentonite.
Properties: Soft mineral that becomes claylike or powdered after being wet.
Uses: Carrier (i. e., it absorbs various materials without swelling); diluent and extender for powdered materials; coagulation aid; coating agent; aid in bleaching or decolorizing oils and chemicals; filtering agent.

monuron. Coined name for 3-(para-chlorophenyl)-1, 1-dimethylurea (q. v.).

moonstone. A gem stone which is a variety of orthoclase or albite or intermediary mixtures.
See feldspar.
Occurrence: United States (California, Virginia, Pennsylvania); Ceylon; Switzerland; Brazil; Australia; Canada.

"Morcowet." [258] Trademark for wetting agents for textile and general usage.

mordant dye. One that requires the use of a third substance (see mordants) to affix or bind the dye to the fiber. The mordant is usually a three-valent chromium derivative formed from chromate or dichromate, but metallic hydroxides, tannic acid and other substances are sometimes used. The mordant forms a lake with the dye and the color is altered according to the metal of the mordant.

"Mordantine." [165] Trade name for liquid antimony lactate containing 11% available antimony oxide. Completely soluble in cold water. Recommended as a replacement for technical tartar emetic.
See also "Antilac."

mordanting assistants. Chemicals such as lactic, oxalic and sulfuric acids, tartrates, etc., used in conjunction with mordants to bring about a gradual decomposition of the latter, and to assist in producing a uniform deposition of the actual mordant upon and within textile materials.

mordant rouge (red liquor; red acetate). A solution of aluminum acetate in acetic acid used in dyeing and calico printing. It is made by dissolving aluminum hydroxide in acetic acid, or by decomposing lead or calcium acetates with aluminum sulfate or alum. Calcium acetate yields the best red liquor; that made from lead acetate is not entirely free from lead, which dulls the shade of delicate colors. It contains sulfate of the alkali metal when made from alum and decomposes more readily than when made from aluminum sulfate.
See also aluminum acetate.

mordants. Substances capable of uniting with both dyes and textile fibers so as to improve the bond between dye and textile and give improved textures.

The most important mordants are sodium dichromate or chromium complexes of various kinds. Copper compounds are sometimes used, and various other metals have the effect of aiding the action of the mordant. The mordant treatment may be applied before, along with, or subsequent to the application of the dye to its textile. Mordants are used with acid dyes, basic dyes, direct dyes and sulfur dyes, although the term metallizing may be used in some cases. Premetallized dyes contain chromium in the dye molecule.

"Morecrop." [139] Trademark for a dolomitic, hydrated lime having a neutralizing value reported as 166% in terms of calcium carbonate; used for adjusting soil pH and furnishing the elements magnesium and calcium.

"Morester." [258] Trademark for a line of saturated polyester resins for use in the synthesis of polyurethanes and rubbers.

"Morflex." [258] Trademark for a series of plasticizers for vinyl, other synthetic resins, and rubbers. These include adipates, phthalates, sebacates, azelates, polymerics and other special products.
Containers: Drums; tank cars.

morganite. See beryl.

morin $C_{15}H_{10}O_7 \cdot 2H_2O$.
Properties: Colorless needles. One of the two coloring principles of old fustic (yellow Brazil-wood). Soluble in boiling alcohol, alkaline solutions; slightly soluble in water.
Derivation: By precipitation from an extract of old fustic.
Uses: Mordant dye; analytical reagent.
Shipping regulations: None. *

"Morlex." [214] Trademark for corrosion inhibitors, for example:
Corrosion Inhibitor A (a mixture of 91% morpholine in water).
Properties: Water-white liquid. Ammoniacal odor. Sp. gr. 1.0185-1.0235 (20/20°C); wt/gal 8.5 lb (20°C); flash point 100°F.
Containers: 1-gal can; 5-, 55-gal drums (8.5, 40, and 450 lb).
Uses: Corrosion inhibitor for steam boilers and steam heating systems.

"Mornidine." [70] Trademark for brand of pipamazine, 10-[3-(4-carbamoylpiperidino)propyl]-2-chlorphenothiazine.
Use: Medicine.

"Mornop." [51] Trademark for oil-soluble cutting oil base of sulfurized type. It is blended with regular cutting fluids to permit difficult machining operations on units normally working on low, free cutting or medium carbon steels.

"Moroc." [244] Trademark for a silicate binder, containing $Na_2O \cdot 4SiO_2 \cdot 2H_2O$ + organic compounds.
Properties: Liquid; gravity 48-52°Bé; sp. gr. 1.50-1.60.
Grades: No. 1, 2, 3 and 4.
Containers: 5-gal pails; 55-gal drums.
Uses: Sand core and mold binder for foundries, cured with carbon dioxide.

Morocco gum. See arabic gum.

"Moroc" Core Paste. [244] Trademark for an accelerated air-setting adhesive.
Properties: Black, viscous liquid. Sp. gr. 1.62.
Containers: 5 gal pails; 55 gal drums.
Uses: Adhesives for foundry cores and molds.

"Morpasol." [170] Trademark for a non-aqueous dispersion of polyvinyl chloride resins in plasticizers plus heat-and-light stabilizers, fillers, colors and other special-purpose chemicals. Resin particles are solvated by the plasticizer when heated to 350°F and fuse into a homogeneous solid plastic without loss of weight. The cured plastisol is abrasion resistant, fire-proof, grease-proof, water-proof, freeze resistant and electrically insulating. Viscosity is adjusted for use in rotary casting and slush molding, reverse roller coating, knife coating, cold dipping, flow coating, hot dipping, etc.

"Morpel." [258] Trademark for a series of synthesized petroleum sulfonates for lubricating oil additives and other uses.

morphine $C_{17}H_{19}NO_3 \cdot H_2O$.
Properties: White crystalline alkaloid; poisonous! Slightly soluble in water, alcohol and ether. M. p. 254°C.
Derivation: From opium by extraction and crystallization. Opium contains about 10% morphine.
Method of purification: Recrystallization.
Containers: 1/8-, 1-oz bottles; 5-oz tins; 100-oz cans.
Uses: Medicine (in form of acetate, hydrochloride, tartrate, and other soluble salts).

para-**morphine.** See thebaine.

morphine acetate $C_{17}H_{19}NO_3 \cdot C_2H_3O_2 \cdot 3H_2O$.
Properties: White, crystalline or amorphous powder; poisonous! Soluble in water and alcohol; insoluble in ether. M. p. 200°C.
Derivation: By heating morphine and acetic acid in presence of sulfuric acid.
Method of purification: Crystallization.
Grade: Technical.
Containers: 1/8-, 1-oz bottles; 5-oz tins.
Use: Medicine.

morphine benzyl ether hydrochloride. See peronine.

morphine bimeconate. See morphine meconate.

morphine hydrobromide $C_{17}H_{19}NO_3 \cdot HBr \cdot 2H_2O$.
Properties: Orthorhombic needles; light sensitive; soluble in water and alcohol.
Containers: To 100-oz cans.
Use: Medicine.

morphine hydrochloride $C_{17}H_{19}NO_3 \cdot HCl \cdot 3H_2O$.
Properties: Poisonous! White, needle-like crystals, white crystalline powder, or as cubical masses. Odorless; affected by light and air; solutions are acid to litmus; soluble in water, alcohol, and glycerin; insoluble in chloroform and ether.
Derivation: By the action of hydrochloric acid on morphine.
Method of purification: Crystallization.
Grade: N. F. XI.
Containers: 1/8-, 1-oz bottles; 5-oz tins; 100-oz cans.
Use: Medicine.

morphine meconate (morphine bimeconate) $(C_{17}H_{19}NO_3)_2 \cdot C_7H_4O_7 \cdot 5H_2O$.
Properties: Yellowish-white, crystalline powder; poisonous! Soluble in water and alcohol.

Derivation: From opium by extraction.
Method of purification: Crystallization.
Grade: Technical.
Containers: ⅛-, 1-oz bottles; 5-oz tins.
Use: Medicine.

morphine methyl bromide. See morphosan.

morphine nitrate $C_{17}H_{19}NO_3 \cdot HNO_3$.
Properties: White powder; darkens when exposed to light; poisonous! Soluble in water.
Derivation: By the action of nitric acid on morphine.
Method of purification: Crystallization.
Grade: Technical.
Containers: ⅛-, 1-oz bottles; 5-oz tins.
Use: Medicine.

morphine sulfate $(C_{17}H_{19}NO_3)_2 \cdot H_2SO_4 \cdot 5H_2O$.
Properties: White, feathery, silky crystals; cubical masses of crystals, or as white crystalline powder; odorless; poisonous! M.p. 250°C, decomposes; soluble in water; slightly soluble in alcohol; insoluble in ether and chloroform. Affected by air and light.
Derivation: By the action of sulfuric acid on morphine.
Method of purification: Crystallization.
Grade: U.S.P. XVI.
Containers: ⅛-, 1-oz bottles; 5-oz and 100-oz tins; cubes in 1-oz bottles only.
Use: Medicine.

morpholine (tetrahydro-1, 4-oxazine) $\overline{OCH_2CH_2NHCH_2CH_2}$, or C_4H_8ONH.
Properties: Colorless, mobile, hygroscopic liquid with a characteristic amine-like odor. A mild base. Miscible with water; soluble in alcohol and ether. B.p. 128.9°C; m.p. −4.9°C; sp. gr. (20/20°C) 1.002; wt/gal (20°C) 8.34 lbs; vapor pressure (20°C) 6.6 mm; viscosity (20°C) 2.23 cps; flash point (open cup) 100°F.
Derivation: By reaction of ethylene oxide and ammonia.
Grades: Technical; 98%.
Containers: 1-gal cans; 5-, 55-gal drums; tank cars.
Uses: Solvent for dyes, resins, waxes; making emulsifying agents; emulsifying agent; organic synthesis; additive to boiler water; water-resistant adhesives and polishes; corrosion inhibitor.
Warning! May cause eye burns and skin irritation. Absorbed through skin. MCA warning label.
Shipping regulations: None.*

morpholine ethanol (N-hydroxyethylmorpholine) $C_4H_8ONCH_2CH_2OH$.
Properties: Colorless liquid; miscible with water; sp. gr. 1.0724; b.p. 225.5°C; flash point 210°F.

6-(N-morpholino)-4, 4-diphenyl-3-heptanone hydrochloride (heptalgin)
$C_4H_8ONCH(CH_3)CH_2C(C_6H_5)_2COC_2H_5 \cdot HCl$.
Properties: M.p. 225°C (dec); (free base melts at 76°C); soluble in water, alcohol, methanol, chloroform; insoluble in acetone, benzene, ethyl acetate.
Use: Medicine.

morphosan (morphine methyl bromide) $C_{17}H_{19}NO_3 \cdot CH_3Br$.
Properties: Needles; m.p. 265°C. Soluble in water.
Use: Medicine.

morrhua oil. See cod-liver oil.

"Morsperse." [258] Trademark for dispersing agents for pigments and similar uses.

"Morton Soil Drench C." [401] Trade name for a concentrate containing methylmercury dicyandiamide (2.2%).
Uses: As a soil drench for control of damping-off organisms; as a foliage spray and dip for cuttings, bulbs and corms to control fungal diseases.
Warning! This liquid is poisonous when inhaled, swallowed or absorbed through the skin. Do not breathe vapors. Do not get in eyes, on skin or on clothing. Handle carefully.

mosaic gold. See stannic sulfide.

moschus. See musk.

"Moskene." [227] Trademark for 1, 1, 3, 3, 5-pentamethyl-4, 6-dinitroindane, $C_{14}H_{18}N_2O_4$.
Properties: Pale yellow crystals; odor resembles musk ketone, having power and depth; m.p. 132°C min; rather stable: may cause discoloration. Congealing point, min. 132.0°C. Soluble in 100 g of solvent, as follows: 4.19 g in methyl "Carbitol"; 16.7 g in benzyl benzoate; 24.2 g in diethyl phthalate; 25.9 g in dimethyl phthalate; 1.05 g in 95% alcohol.
Uses: Blends well with perfumes of a musk character, particularly in fougeres and lavenders. Odor value lies between musk ketone and musk ambrette, but it has a more pronounced ambrette-seed character than either of these.

mosoi flower oil. See ylang ylang oil.

mossbunker oil. See menhaden oil.

moss green. See copper acetoarsenite.

mother of coal. See fusain.

"Moth-Snub." [300] Trademark for dieldrin (q.v.).

motor spirits. See gasoline.

mountain blue (copper blue).
Derivation: The mineral azurite, in ground form.
Grades: Technical.
Containers: Kegs; boxes; fiber drums.
Use: Paint pigment.
Shipping regulations: None.*

mountain cork. See asbestos.

mountain paper. A variety of asbestos occurring in thin, flexible, tough sheets.

mountain tobacco. See arnica flowers.

mousse de chene. See oakmoss resin.

mowrah fat. A vegetable butter.
Properties: Yellow, semi-liquid fat; relatively unsaturated; bitter, aromatic taste;

*See "I.C.C. Shipping Regulations," page xiii.
Reference numbers refer to name of manufacturer. See "List of Manufacturers," page v.

characteristic odor similar to that of cacao beans; soluble in ether, chloroform, light petroleum hydrocarbons and carbon disulfide; sp. gr. 0.894-0.898; m.p. 23-29°C; saponification value 188-194; iodine value 58-67.
Derivation: From the seeds of Bassia latifolia.
Habitat: India (northern provinces).
Grades: Crude; refined.
Containers: Wooden barrels.
Use: Soaps.
Shipping regulations: None.*

6-MP. Abbreviation for 6-mercaptopurine.

M.P.A. Abbreviation for Metal Powder Association.

MPA. Abbreviation for multipurpose additive. See "Ethyl."

"M-P-A." [202] Trademark for a gelling agent used in paints and lacquers to prevent pigment settling and excess film flow. Paste form; 24% to 40% solids in choice of mineral spirits, xylene or toluene.
Containers: 5-gal pails; 55-gal drums.

MPC black. Abbreviation for medium processing channel black. See channel black.

MPK. Abbreviation for methyl propyl ketone.

"MPS-500." [306] Trademark for a plasticizer consisting of a stabilized chlorinated fatty acid ester.
Properties: F.p. −39°C; fire point 252°C; sp. gr. 1.19 (25/15.5°C); flash point (open cup) 164°C; soluble in chlorinated hydrocarbons, aliphatics, aromatics, alcohols, esters, ketones; insoluble in water; not compatible with cellulose acetate, nitrocellulose, polyvinyl butyral. Imparts excellent electrical properties, high strength, flame retardance, and low-temperature flexibility.
Containers: 55-gal lacquer-lined steel drums; tank cars.
Uses: Plasticizer, sole or secondary, for resins and rubbers, especially for sheeting, wire covering, shoe soles, electrical extrusion compounds.

mrep. Abbreviation for milliroentgen equivalent physical. See roentgen.

"M-S." [241] Trademark for a silica-alumina petroleum cracking catalyst (86.6-74.7% silica, 13.2-25% alumina). Marketed in five grades differing in particle size: F-3, F-2, F-1, C-1, C-2; sold in hopper cars or drums for use in cracking petroleum gas oil fractions.

MSG. Abbreviation for monosodium glutamate. See sodium glutamate.

MSP. Abbreviation for monosodium phosphate. See sodium phosphate, monobasic.

6 MT. Abbreviation for 6-methylthiouracil.

MT black. Abbreviation for medium thermal black. See thermal black.

"M & T Catalyst T-8." [288] Trade name for dibutyltin di-2-ethylhexoate.
Properties: Waxy white solid; sp. gr. 1.070 (25°C); m.p. 54-60°C.
Containers: 50- and 150-lb fiber drums with polyethylene liners.
Use: In the manufacture of polyether-based urethane foams via the "one-shot" method and in a silicone emulsion system for the treatment of textiles for water and stain repellency.

"M & T Catalyst T-9." [288] Trade name for a "stannous" type catalyst useful in "one-shot" polyether urethane foams.
Properties: Pale yellow liquid; sp. gr. 1.26 (25°C); lbs/gal 10.4; soluble in most organic solvents.
Containers: 55-gal steel drums; 6-gal steel containers; 1-gal metal containers.

"M & T Catalyst T-12." [288] Trade name for an organotin catalyst.
Properties: Oily liquid; solid below room temperature; yellow color; sp. gr. 1.05 (25°C); setting pt. 17 to 20°C.
Containers: 50- and 450-lb phenolic lined steel drums.
Uses: Extremely effective in the catalyst system for the production of "one-shot" polyether urethane foams.

"M & T Catalyst T-18." [288] Trade name for a liquid stannous oleate catalyst especially useful in "one-shot" polyurethane foams.
Properties: Light to medium yellow liquid; sp. gr. 1.03 (25°C); lbs/gal 8.6.
Containers: 55-gal steel drums; 6-gal steel containers; 5-qt metal containers.

MTD. Abbreviation for meta-tolylenediamine. See toluene-2,4-diamine.

"M & T Flame Retarder." [288] Trade name for an antimony-containing compound with a high degree of flame retardancy.
Properties: Fine white powder; sp. gr. 4.66 (25°C); refractive index 1.75 (25°C); insoluble in organic solvents; very slightly soluble in water.
Uses: In vinyls and other plastics.

"MTM." [303] Trademark for a mixture of tertiary C_{12}, C_{14}, and C_{16} aliphatic mercaptans.
Properties: Boiling range 88-142°C (5 mm); sp. gr. 0.866 (60/60°F); refractive index 1.468 (n 20/D); flash point 96°C.
Containers: Drums and tank cars.
Uses: Polymer modifications.
Hazards: Flammable liquid.
Shipping regulations: None.*

mucic acid (saccharolactic acid; tetrahydroxyadipic acid) $HOOC(CHOH)_4COOH$.
Properties: White crystalline powder; m.p. about 210°C (dec); soluble in water; insoluble in alcohol.
Derivation: By the oxidation of lactose or similar carbohydrates with nitric acid.
Method of purification: Crystallization.
Grades: Technical.
Containers: Tins.

Uses: Organic synthesis; substitute for tartaric acid.
Shipping regulations: None.*

mucilage. Originally a decoction of linseed, foenugreek seeds, and marshmallow root, but more generally applied to any adhesive paste.

mucins. Complex naturally occurring compounds consisting of amino sugars (such as glycosamine and galactosamine), glucuronic acid and sulfuric acid.
Use: Medicine.

muguet, synthetic. See hydroxycitronellal.

"Mullfrax." [280] Trademark for refractory products made from mullite produced in electric furnaces. Available as bonded refractories and refractory cements.
Properties: High refractoriness, chemical stability and strength at high temperatures, resistance to spalling.
Bonded refractories:
"Mullfrax." Electric furnace mullite. The heat transmission coefficient averages 15 Btu/sq ft/inch thickness/°F/hr; porosity, 22%; permeability, low.
Uses: Brick for piers, support arches and other load-bearing constructions in furnaces and kilns, glass-tank ports and glass-tank superstructures, linings for indirect-arc electric furnaces, shapes to fit special furnace and kiln requirements, burner blocks in powdered coal, oil and gas-fired furnaces and kilns.
"Mullfrax" C. The coefficient of thermal conductivity averages 12 to 14 Btu/sq ft/inch thickness/°F/hr.
Uses: Brick for linings of indirect-arc electric as well as fuel-fired direct and crucible type melting furnaces for non-ferrous metals and alloys; tuck stones; checker brick and superstructures in glass tanks; burner blocks for powdered coal, oil, or gas-fired furnaces; for load-bearing in kilns; supporting muffle walls and floors, and electric furnace roofs.
Refractory cements:
"Mullfrax." Electric furnace mullite refractory cements. Used in patching and monolithic linings, and in laying "Mullfrax" brick and other less refractory materials.
"Mullfrax" S. Converted kyanite refractory cements. Used for patching and monolithic linings of fuel-fired and electric furnaces, and in laying "Mullfrax" S brick.

mullite $3Al_2O_3 \cdot 2SiO_2$. An aluminum silicate formed by heating other aluminum silicates (such as cyanite, sillimanite, and andalusite) to high temperatures, and the only stable member of the group. Also found in nature, but rare. It was at one time stockpiled by the government.
Properties: Colorless crystals; sp. gr. 3.15; m.p. 1810°C; insoluble in water.
Containers: 50-lb bags.
Uses: Refractories; glass. See also aluminum silicates.

"Mulsor." [83] Brand name for a line of emulsifiers; chiefly long chain fatty acid esters of the glycols and related compounds.

"Multifex MM." [244] Brand name for ultra fine calcium carbonate:
Properties: Wt/solid gal 22.07 lbs; color, white; particle size, 0.04-0.06 microns.
Derivation: Precipitated calcium carbonate.
Containers: Multi-wall paper bags, 50 lbs net.
Uses: Paint, plastics, rubber, inks.

"Multimet." [214] Trademark for an iron-base alloy. The wrought alloy is suitable for service up to 1500°F without the necessity of age-hardening; it can be drawn, spun, rolled, flanged, and dished cold. Used for rotors, turbine blading, bolts for high-temperature applications. The cast alloy (low carbon) is used for investment castings, sand castings, turbine blading, nozzles, and mandrels.

"Multi-Sperse." [141] Trade name for stir-in pulps compatible with all types of latex paints.
Properties: Disperse easily with simple stir-in techniques. Excellent compatibility; maximum stability; good alkali stability; good light resistance; contain non-ionic surfactants.
Grades: Yellows, oranges, reds, blues, greens, violets and black.
Uses: For all types of latex paints, including acrylic, butadiene-styrene types and polyvinyl acetate.

mu meson. See fundamental particles.

muon. A mu meson, one of the fundamental particles (q.v.).

"Murano." [270] Trademark for a synthetic nacreous pigment which has twin inherent colors, one being observed by reflected light and the second by transmitted light, the two colors being complementary. The pigment particles consist of a lead compound in the form of thin, plate-like crystals. The twin colors are usually seen simultaneously, since part of the light reaches the eye by direct reflection from the surface of the crystals and part of the light is reflected from within the crystal layer and is transmitted through the intervening crystals. (See nacreous pigment.)
Containers: 1-, 4-lb glass jars; 50-lb steel cans.
Uses: To produce a play of colors when coated on surfaces or incorporated in transparent plastics or plastic film.

muriatic acid. Old term for hydrochloric acid.

murillo bark. See quillaja.

muscle adenylic acid. See adenylic acid.

muscle fibrin. See syntonin.

muscone. See musk.

*See "I.C.C. Shipping Regulations," page xiii.
Reference numbers refer to name of manufacturer. See "List of Manufacturers," page v.

muscovite (white mica, potassium mica, isinglass) $KAl_2AlSi_3O_{10}(OH)_2$. A natural hydrous potassium aluminum silicate of the mica group (q. v.).
Properties: Colorless, yellowish, brownish or reddish; luster vitreous to silky or pearly; hardness 2-2.5; sp. gr. 2.76-3.1; one perfect cleavage. Grades are based on color, size, and other physical properties. Muscovite is used as sheets and in ground form.
Occurrence: India, Brazil, North Carolina, New Hampshire, South Dakota, Georgia, Connecticut.
Uses: As a dielectric in electrical equipment in capacitors, coils, spark plugs, and radio tubes; in special gaskets; in metallurgical equipment as windows of furnaces and other items; in toasters and similar electrical appliances; as diaphragms in acoustic equipment; in heat-resistant applications in the form of sheets, tape and cloth; as ground mica in paints; filler in rubber; in roofing material; in oil-well muds; in coated fabrics; as a lubricant.

musk (moschus).
Properties: An unctuous brownish semi-liquid when fresh; dried, in grains or lumps with color resembling dried blood. The odorbearing constituent is muscone or muskone, $CH_3C_{15}H_{27}O$, a 15-carbon ring with ketone oxygen. See also musk, synthetic.
Derivation: Secretion from preputial follicles of the musk deer, Moschus moschiferous.
Habitat: Northern Asia, Tonquin, and Tibet.
Grades: Tonquin, Cabardine, Yeman, Assam or Nepal.
Containers: Metal boxes; fiber drums.
Uses: Medicine; perfumery; flavoring sweetmeats; toilet soaps; mothproofing agent. It is a powerful, but expensive, fixative in perfumery.
Shipping regulations: None.*

musk ambrette (2,6-dinitro-3-methoxy-4-tert-butyltoluene) $CH_3C_6H(NO_2)_2(OCH_3)C(CH_3)_3$.
Properties: White to yellow powder, with heavy musky odor of ambrette seed. Soluble in benzyl benzoate, diethyl phthalate, dimethyl phthalate, fixed oils, volatile oils, mineral oil and methyl "Carbitol"; sparingly soluble in alcohol. Insoluble in water and glycerin.
Constants: Congealing point 83.4°C.
Containers: Tin cans; fiberboard containers; barrels.
Use: Perfumery (fixative).
Shipping regulations: None.*

musk ketone (3,5-dinitro-2,6-dimethyl-4-tert-butylacetophenone) $CH_3OCC_6(NO_2)_2(CH_3)_2C(CH_3)_3$.
Properties: White to yellow crystals, having a sweet musk odor. Soluble in benzyl benzoate, diethyl phthalate, dimethyl phthalate, fixed oils, volatile oils; somewhat soluble in methyl "Carbitol"; sparingly soluble in alcohol; insoluble in water, glycerin, propylene glycol.
Containers: Tin cans; fiberboard containers; barrels; drums.
Use: Perfumery (fixative).
Shipping regulations: None.*

muskmallow. See ambrette seed.

muskone. See musk.

musk-root. See sumbul.

musk seed. See ambrette seed.

musk, synthetic. A number of compounds are so called. They fall into two categories:
(a) Macrocyclic musks. These are ketones and lactones with large cycles (15- or 16-carbon rings), structurally resembling the odoriferous principles of natural musk, civet, and musky-type plants. Among these are ambrettolide, civetone, muskone, exaltolide.
(b) Nitrated compounds, usually nitrated tert-butyl-toluenes or xylenes or related compounds. The three most commonly used in perfumery are musk ambrette, musk ketone, and musk xylol or musk xylene.

"Musk Tibetene." [227] Trademark for 2,6-dinitro-3,4,5-trimethyl-tert-butylbenzene, $(CH_3)_3(NO_2)_2C_6C(CH_3)_3$, 5-tert-butyl-4,6-dinitrohemimellitene.
Properties: Pale yellow crystals; resembles the odor of musk ketone, but has a slightly heavier and sweeter character; stable; may cause discoloration. Congealing point, min 135.0°C. Soluble in 100 g solvent, as follows: 5.4 g in methyl "Carbitol"; 27.4 g in benzyl benzoate; 13.7 g in diethyl phthalate; 15.6 g in dimethyl phthalate; 1.39 g in 95% alcohol.
Uses: Used to impart a desired animal note to any perfume for lotions, powders, creams, or soaps. Is stable to light. Eliminates discoloration in soap, perfume, and powder containing stearates, or a bath salt containing sesquicarbonates.

musk xylol (musk xylene; 2,4,6-trinitro-1,3-dimethyl-5-tert-butylbenzene) $(NO_2)_3C_6(CH_3)_2C(CH_3)_3$.
Properties: White to yellow crystals, with powerful odor of musk.
Constants: Congealing point 111.7°C or 104-106°C. May have two congealing points.
Soluble in benzyl benzoate, diethyl phthalate and dimethyl phthalate, fixed oils and volatile oils; sparingly soluble in methyl "Carbitol"; very slightly soluble in alcohol. Insoluble in water, glycerin, propylene glycol.
Containers: Tin cans, fiberboard containers, barrels, drums.
Uses: Perfumery (fixative).
Shipping regulations: None.*

mustard, black (sinapis nigra; mustard, brown).
Derivation: Seed of Brassica nigra or Brassica juncea.
Occurrence: United States, Europe, and Asia.

Grades: Technical; N. F. XI.
Containers: Tins.
Uses: Medicine; condiment; mustard oil.
Shipping regulations: None. *

mustard, brown. See mustard, black.

mustard gas. See dichlorodiethyl sulfide.

mustard oil, artificial. See allyl isothiocyanate.

mustard oil, black. See mustard oil, volatile.

mustard oils. Organic compounds having the formula R-N=C=S, in which R is an alkyl or aryl radical, -NCS an isothiocyanate group. The name is derived from its best known member, allyl isothiocyanate, (q. v.) which is the characteristic ingredient of mustard oil.

mustard oil, volatile (black mustard oil; sinapis oil).
Properties: Colorless to pale yellow limpid liquid slowly changing to reddish-brown on exposure to light. Pungent, acrid odor and taste.
Chief known constituents: Allyl isothiocyanate, carbon disulfide, allyl cyanide.
Soluble in 7 to 10 vols of 70% alcohol; in 2.5 to 3 vols of 80% alcohol; clear solutions in all proportions with 90% alcohol.
Constants: Sp. gr. 1.016-1.022 (rarely 1.030); refractive index 1.52681-1.52804; b. p. 148-154°C (760 mm).
Derivation: From the seeds of Sinapis nigra, L., and Sinapis juncea, L. These are ground, rendered free of their fatty oil content by hydraulic pressure, mixed with warm water, allowed to ferment and then distilled.
Containers: Bottles.
Use: Medicine; proprietary liniments.
Shipping regulations: None. *

mustard, white (sinapis alba; mustard, yellow).
Derivation: Seed of Sinapis alba.
Occurrence: United States, Europe, and Asia.
Grades: Technical.
Containers: Tins; bags.
Uses: Condiment.

mustard, yellow. See mustard, white.

Muthmann's liquid. See acetylene tetrabromide.

Mv. Symbol used for mendelevium, but not sanctioned by IUPAC.

MVE. Abbreviation for methyl vinyl ether. See vinyl methyl ether.

"MX." [280] Trademark for fiber-bonded abrasives.
Properties: High tensile strength and resistance to impact and heat shock; unusually resilient.
Uses: For finishing and polishing flutes of taps, drill end mills, reamers, etc.; removing burrs from milling and drilling operations; breaking edges of cast aluminum parts, etc.; cleaning cast iron molds; removing flash from molded plastics.

MXD6. A type of nylon from meta-xylylenediamine and adipic acid.

"My-B-Den." [272] (Sodium salt of adenosine-5-monophosphoric acid) $C_{10}H_{14}N_5O_7P$. Trade name for a biochemical substance prepared by enzymatic phosphorylation of adenosine from muscle sources and purified to give the crystalline free acid and its sodium salt. A5MP is concerned with numerous basic biochemical reactions, and is used in medicine and biochemical research.

"Mycoban." [28] Trademark for sodium and calcium propionates. These salts inhibit the growth of many fungi and of some microorganisms, particularly bacillus mesentericus, for commercially significant periods of time. Because of this property they find application in many industries, particularly to inhibit mold and rope in bread, rolls (including "brown and serve"), pie crust, etc. However, the presence of the calcium ion in some cases restricts the use of "Mycoban" calcium propionate.
Containers: 250-lb fiber drums; cartons of six 10-lb paper bags (sodium type only).

"Mycostatin." [412] Trademark for nystatin (q. v.).

MYL. A powder used during World War II for the control of body lice which carry typhus fever. Its composition included:

Pyrethrins (20% pyrethrum extract)	0.2%
N-isobutylundecylenamide (synergist)	2.0
2,4-dinitroanisole (ovicide)	2.0
isopropyl and diisopropyl cresols (anti-oxidants)	0.25
pyrophyllite (diluent) - to make 100%.	

"Mylar." [28] Trademark for film of polyethylene terephthalate.

"Mylase." [173] Trademark for a fungal amylase in powder form for sirup and dextrin conversions.

"Myleran." [301] Trademark for 1,4-dimethanesulfonoxybutane, (GT 41), used in medicine.

mylone. 3,5-Dimethyl-2$\underline{H}$-1,3,5-tetrahydrothiadiazine-2-thione. A soil fumigant.

myo-inositol. See inositol.

myrbane, essence of. See nitrobenzene.

"Myrcene-85." [296] Brand name for a special grade of the triply unsaturated aliphatic hydrocarbon, $C_{10}H_{16}$, 7-methyl-3-methylene-1,6-octadiene. Used in preparation of flavor and odor chemicals.

myrcia oil (bay oil; bayleaf oil).
Properties: Essential oil; yellow color, becoming brown on exposure to air; pleasant clove-like odor; pungent, spicy taste; phenol content 50-65%. Sp. gr. 0.950-0.990 (25/25°C), going as low as 0.939 for poor-quality oils; optical rotation 0 to −3°; refractive index 1.507-1.516 (20°C). Soluble in 1-2 vols. of 70% alcohol (if

freshly distilled; solubility decreases rapidly with age).
Derivation: By distillation of the leaves of Pimenta racemosa (Pimenta acris). Note: Many of the species of the genera Pimenta and Myrica closely resemble one another. Thus, often, a mixture of the leaves is distilled unless great care is exercised in the gathering.
Adulteration: Kerosene; alcohol; clovestem oil.
Grades: N. F. XI.
Containers: Tins; glass bottles; drums.
Uses: Flavors; perfumes; bay rum.
Shipping regulations: None.*

myrica (candleberry; bayberry; wax myrtle; wax berry; tallow shrub). Bark of Myrica cerifera or carolinensis.
Habitat: Maryland to Florida, west to Texas and Arkansas.
Grades: Technical.
Containers: Bales; multi-wall paper sacks.
Uses: Medicine; source of bayberry wax.
Shipping regulations: None.*

myricyl alcohol. See 1-triacontanol and 1-hentriacontanol. The term myricyl alcohol, which has been used for both by various authorities, should be dropped.

myristica (nutmeg). Kernel of the ripe seed of Myristica fragrans.
Habitat: Southern Asia and Moluccas; cultivated in many tropical countries.
Grades: Technical; N. F. XI.
Containers: Bags.
Uses: Medicine; condiment; damaged seeds used as source of nutmeg oil.
Shipping regulations: None.*

myristic acid (tetradecanoic acid) $CH_3(CH_2)_{12}COOH$.
Properties: Oily, white crystalline solid. Soluble in alcohol and ether; insoluble in water.
Constants: Sp. gr. 0.8739 (80°C); b. p. 326.2°C (760 mm); 204.3°C (20 mm); m. p. 54.4°C; refractive index 1.4310 (n 60/D).
Derivation: Fractional distillation of coconut acid.
Containers: Cans; bags; 55-gal drums; tank cars.
Grades: Technical; 99.8%.
Uses: Soaps, cosmetics; synthesis of esters for flavors and perfumes.

myristica oil. U. S. P. XVI name for nutmeg oil (q. v.).

myristoyl peroxide $(C_{13}H_{27}CO)_2O_2$.
Properties: Soft granules; 90% peroxide.
Use: Catalyst for vinyl type monomers.

myristyl alcohol (1-tetradecanol) $C_{14}H_{29}OH$.
Properties: Sp. gr. 0.8355 at 20/20°C; b. p. (760 mm) 264.1°C, (20 mm) 177.1°C; m. p. 38°C; vapor pressure 0.01 mm (20°C); flash point 285°F; wt 7.0 lbs/gal (20°C); coefficient of expansion 0.00083 (20°C); viscosity 0.366 poise (20°C). Insoluble in water; soluble in alcohol and ether.
Grades: Technical.
Containers: 1-gal cans; 5-, 55-gal drums.
Uses: Organic synthesis; plasticizer; antifoam agent; intermediate; perfume fixative for soaps and cosmetics; base for the manufacture of wetting agents and detergents; ointments and suppositories.

myristyl chloride. See tetradecyl chloride.

myristyl mercaptan. See tetradecyl mercaptan.

"Myrj." [89] Trademark for each of a series of nonionic, low-melting point, waxy surface-active agents, ranging from water-dispersible to water-soluble and varying in oil solubility. Essentially neutral polyoxyethylene derivatives of fat-forming fatty acids.

myrrh gum. The gum resin of Commiphora myrrha or other species of myrrh.
Habitat: Nubia, Somaliland and Arabia. Partially soluble in water, alcohol and ether.
Grades: Technical; N. F. XI.
Containers: Bags; cases.
Uses: Dentifrices; perfumery; protective in pharmaceuticals.
Shipping regulations: None.*

myrrh oil.
Properties: Yellowish, rather viscid liquid; strong odor. Soluble in 80% alcohol.
Chief constituents: Cuminic aldehyde, eugenol, meta-cresol, pinene and dipentene.
Constants: Sp. gr. 0.988-1.007; b. p. 220-235°C.
Derivation: Distilled from the gum-resin, myrrh.
Grades: Technical.
Containers: Copper flasks; glass bottles.
Use: Medicine; perfumery.
Shipping regulations: None.*

myrtle oil.
Properties: Light, yellow liquid; agreeable aromatic odor. Soluble in 80% alcohol.
Chief known constituents: Cineole, borneol, d-pinene, dipentene.
Constants: Sp. gr. 0.89-0.92; optical rotation +10° to +30°.
Derivation: Distilled from the leaves of Myrtus communis.
Method of purification: Rectification.
Grades: Technical.
Containers: Iron drums; glass bottles.
Uses: Medicine; perfumery (fixative); flavors.
Shipping regulations: None.*

myrtle wax. See bayberry wax.

"Mysoline." [207] Trademark for primidone, 5-ethyl-5-phenyl-hexahydropyrimidine-4,6-dione. Used in medicine.

*See "I.C.C. Shipping Regulations," page xiii.
Reference numbers refer to name of manufacturer. See "List of Manufacturers," page v.

N

N. Symbol for the element nitrogen (q. v.). The names of certain compounds (such as N,N-dibutyl urea) contain this symbol as an indication that the group or groups appearing next in the name (i. e., the butyl groups in the example cited) are joined to the nitrogen atoms in the molecule of the compound under discussion. In this book the alphabetical arrangement ignores such N symbols appearing in the name of a compound. Thus N, N-dibutyl urea is found under D.

N. Abbreviation for normal, as used in analytical chemistry. See normal (2).

n-. Abbreviation for normal. See normal (1). In this book the prefix n- is ignored in the alphabetical arrangement.

Na. Symbol for sodium.

"NA-22." [28] Trade name for (2-mercaptoimidazoline) $\underline{CH_2CH_2NC(SH)NH}$.
Properties: A white powder; sp. gr. 1. 42; m. p. not lower than 195°C.
Containers: 125-lb drums.
Use: To accelerate and improve the vulcanization of neoprene.

nabam (disodium ethylenebisdithiocarbamate) $NaSSCNHCH_2CH_2NHCSSNa$.
Properties: Colorless crystals when pure; easily soluble in water.
Derivation: Addition of carbon disulfide to an alcoholic solution of ethylenediamine followed by neutralization with sodium hydroxide; or by reaction of ethylenediamine with carbon disulfide in aqueous sodium hydroxide.
Grades: Usual commercial form is a 19% aqueous solution.
Caution! (For 19% solutions and over). May cause skin irritation. Avoid contact with skin and eyes. In case of contact, flush with plenty of water; for eyes get medical attention. MCA warning label.
Uses: Plant fungicide, and starting material for formation of derivatives that are also plant pesticides.

"Nabor." [243] Trademark for basic dyes intended especially for dyeing acrylic fibers.

"Nacan." [243] Brand name for sodium metanitrobenzenesulfonate. A protective antireduction agent.

"Naccogene." [243] Trademark for azoic compositions.

"Nacconate." [243] Trade name for diisocyanates.

"Nacconate 65." [243] (TDI mixed isomers). A mixture of 65% 2, 4-tolylene diisocyanate and 35% 2, 6-tolylene diisocyanate.
Properties: Clear, faintly yellow liquid containing 99. 5% tolylene diisocyanate isomers; b. p. 118-120°C (10 mm); sp. gr. at 20°, 1. 22; soluble in acetone, benzene, carbon tetrachloride, chlorobenzene, kerosine and nitrobenzene.
Containers: 55- and 5-gal special lined drums.
Uses: In the manufacture of foams, coatings for wire and other polyurethane products.
Handling: Use protective clothing including goggles and gloves. In case of contact with skin wash with soap and water or alcohol followed by soap and water. Do not put water, solvent or other material into the container.
Shipping regulations: Poison, class B. Poison label.*

"Nacconate 80." [243] A mixture of 80% 2, 4-tolylene diisocyanate and 20% 2, 6-tolylene diisocyanate. For properties, etc., see "Nacconate 65."

"Nacconate 100." [243] (2, 4-tolylene diisocyanate, TDI), $CH_3C_6H_3(NCO)_2$.
Properties: Clear, faintly yellow liquid, solidifying at 20°C; b. p. 118-120°C (10mm); sp. gr. (20°C) 1. 22; purity 99. 5%.
Containers: 55- and 5-gal special lined drums.
Uses: Specialty casting resins.
Handling: See "Nacconate 65."
Shipping regulations: Poison, class B. Poison label.*

"Nacconate 200." [243] (3, 3'-bitolylene-4, 4'-diisocyanate; 3, 3'-dimethylbiphenylene-4, 4'-diisocyanate) $OCN(CH_3)C_6H_3C_6H_3(CH_3)NCO$.
Properties: Small, very pale yellowish flakes of 99. 5% purity; solidification point 69. 6°C; sp. gr. (80°C) 1. 197; soluble in acetone, benzene, carbon tetrachloride, chlorobenzene, kerosine and nitrobenzene.
Containers: 275-lb fiber drums.
Uses: Urethane elastomers.
Handling: Use protective clothing including goggles and gloves with good ventilation. In case of contact with skin, wash with alcohol followed by soap and water. Do not put water, solvent or other material in container.
Shipping regulations: Poison, class B. Poison label.*

"Nacconate 300." [243] (diphenylmethane-4, 4'-diisocyanate; methylene-di-para-phenylene isocyanate; 4, 4'-diphenylene-methane diisocyanate; MDI) $CH_2(C_6H_4NCO)_2$.

*See "I.C.C. Shipping Regulations," page xiii.
Reference numbers refer to name of manufacturer. See "List of Manufacturers," page v.

Properties: Fused light yellow solid; solidification point 37.2°C; sp. gr. (70°C) 1.197; soluble in acetone, benzene, carbon tetrachloride, chlorobenzene, kerosine and nitrobenzene.
Containers: 5-gal special lined drums.
Uses: Production of urethane elastomers, coatings, spandex-type fibers.
Handling: Use protective clothing, including goggles and gloves, with good ventilation. Wash with soap and water or alcohol followed by soap and water in case of contact. Do not put water, solvent or other material in container.
Shipping regulations: Poison, class B. Poison label. *

"Nacconol." [243] Trade name for a series of alkyl aryl sodium sulfonate detergents with wetting, emulsifying and dispersing properties. All available in technical grade in bags, fiber drums, steel drums.
Properties: (Typical of solid forms). Solubility in water 9.0% at 25°C; 21% at 100°C; in 1% acetic acid 11.0% at 25°C, 23% at 100°C; in alkaline solution of pH 10.5, 7% at 25°C, 23% at 100°C; in common organic solvents 0.3 to 10%; barium and lead salts sparingly soluble; other metallic salts 0.5 to 10.0%; pH approximately 7.0 in 1% water solution. Stable to hard water, 15% boiling sulfuric acid, 10% boiling sodium hydroxide and moderate concentrations of oxidizing and reducing agents. Surface tension of 0.10% solution, 32.8 dynes/cm at 30°C. Slightly to highly hygroscopic; faint odor; bitter flavor.
Uses: Washing, scouring, wetting agents and dyeing assistants for textile fibers and fabrics, leather and paper. Washing and scouring agents for power laundering, rugs and upholstery, automobiles, refrigerators, glass, painted surfaces, city streets, etc. Used in conjunction with alkalies for metal cleaning, bottle washing in dairies, breweries, and carbonated beverage plants, commercial dishwashing, scouring powders, etc. Also used in adhesives, dyestuffs, buffing compounds, etc.

The solid forms described above include "Nacconol" Beads, DB, HG, NR, NRSF, SW, Z. Liquid forms include "Nacconol" 60S, SL, SZA. "Nacconol" 60S and SL are straw-colored, free-flowing, viscous liquids, used in detergents; SZA is an opaque acid type used as a metal cleaner.
Shipping regulations ("Nacconol" SZA): Corrosive liquid. White label.*

"Naccosol" A. [243] Brand name for a proprietary sulfonated alkylated aromatic compound, primarily a wetting agent.
Properties: Light tan, dry powder; pH approx. 7 in 1% water solution; stable to hard water, 10% boiling sulfuric acid, 10% boiling sodium hydroxide, normally used concentrations of oxidizing and reducing agents; very soluble in water at 25°C; surface tension (1% solution) 35.5 dynes/cm (25°C); hygroscopicity very high, keep containers covered; sp. gr. 0.51; odor faint.
Grades: Technical.
Containers: Non-returnable 55-gal fiber drums, 250 lbs net.
Uses: A general surface-active agent especially useful as a solubilizing agent and as a wetting agent over a wide pH range.
Fire hazard: None.
Shipping regulations: None. *

"Naccotan" A. [243] Brand name for a proprietary alkyl aryl sodium sulfonate, an organic retanning agent.
Properties: Light brown, dry fine flakes; pH approx. 7.0 in 1% water solution; stable to hard water, 10% boiling sulfuric acid, and 10% boiling sodium hydroxide; very soluble in water at 25°C; hygroscopicity slight, keep containers covered; sp. gr. 0.36; odorless.
Grades: Technical.
Containers: Paper bags only, 75 lbs net.
Uses: Used in the retanning of chrome-tanned leather; useful in the dispersion of thick slurries.
Fire hazard: None.
Shipping regulations: None. *

"Naccufix." [243] Copper organo-complex for fixing direct dyes on cellulose fibers.

"Nacelan." [243] Trademark of water-insoluble dyes dispersed in water and forming solid solutions in synthetic fibers.

nack. See NaK.

"Naco." [241] Brand name for mixed fertilizers.

nacreous pigment. Nacreous or pearlescent pigments are substances which produce a pearly luster. They may be applied as surface coatings, as in simulated pearls, or may be incorporated in plastics, as in plastic simulated mother-of-pearl shirt buttons. The pigment particle is generally a very thin crystalline platelet of high index of refraction. The crystals are readily oriented into parallel layers because of their shape. Being transparent, each crystal reflects only part of the incident light reaching it, and transmits the remainder to the crystals below. The nacreous effect is obtained from the simultaneous reflection of light from the many parallel microscopic layers.

Pearl essence, also known as natural pearl essence, is the original nacreous pigment in which the pigment particles are the guanine crystals obtained from fish scales and fish skin. The crystals are extremely thin plates in the form of elongated hexagons. Synthetic nacreous pigments, also known as synthetic pearl essence, are generally inorganic substances which are crystallized in the form of thin plates. These are generally lead or bismuth compounds.

Nacreous pigments are usually available in the form of suspensions in a liquid vehicle, which may be water, a plasticizer, resin, or lacquer, depending on the ultimate application in which the material is to be used.

"Nacromer." [270] Trademark for synthetic nacreous pigments, consisting of lead salts crystallized in the form of extremely thin plates and suspended in a suitable liquid vehicle. Various grades, which range from a brilliance surpassing that of the natural pearl to a soft, velvety luster.

See nacreous pigment.

Containers: 1-, 4-lb glass jars; 50-lb steel cans.

NAD. See nicotinamide adenine dinucleotide.

"Nadic Anhydride." [243] Trademark for endo-cis-bicyclo[2.2.1]-5-heptene-2,3-dicarboxylic anhydride (4-endomethylenetetrahydrophthalic anhydride) $C_7H_8(CO)_2O$.

Properties: White crystals; m.p. 164-165°C; soluble in benzene, toluene, acetone, carbon tetrachloride, chloroform, ethanol, and ethyl acetate; slightly soluble in petroleum ether.

Containers: 250-lb fiber drums (35-gal).

Uses: Chemical intermediate in the manufacture of alkyd and polyester ester resins; curing agent for epoxies; modifying agent for drying oils for emulsion paints; copolymerizing agent with vinyl and vinylidene chlorides; intermediate for softening agents for rubber, plasticizers, wire enamel resins, surfactants, insectifuges, high-boiling solvents, synthetic lubricants, textile assistants and penetrants.

"Nadic Methyl Anhydride." [243] $C_{10}H_{10}O_3$. Trademark for methylbicyclo[2.2.1]heptene-2,3-dicarboxylic anhydride isomers.

Properties: Clear, colorless to light-yellow viscous liquid; viscosity 175-225 cps (25°C); refractive index 1.500-1.506 (n 20/D); sp. gr. 1.200-1.250 (20/20°C). Miscible in all proportions with acetone, benzene, naphtha, and xylene.

Grades: Technical.

Containers: 15-, 55-gal metal drums.

Uses: Curing agent for epoxy resins; intermediate for polyesters, alkyd resins, and plasticizers.

"Nadone." [243] Trade name for cyclohexanone.

NADP. Abbreviation for nicotinamide adenine dinucleotide phosphate.

naepaine hydrochloride $H_2NC_6H_4COOCH_2CH_2NHCH_2(CH_2)_3CH_3 \cdot HCl$. 2-Amylaminoethyl-para-aminobenzoate hydrochloride.

Properties: Fine white odorless powder which, when applied to the tongue, gives bitter taste followed by sensation of numbness. Soluble in water; sparingly soluble in alcohol; insoluble in benzene, chloroform, and ether. Free base (m.p. 65°C) separates as a solid from solution of the hydrochloride on the addition of sodium hydroxide or carbonate solution. The hydrochloride is dimorphic. The form which crystallizes from water melts at 153.5° while that crystallized from amyl alcohol melts at 176°. Aqueous solutions are acid to litmus.

Grade: N.F. XI (m.p. 175° to 177°).

Use: Medicine (local).

NaK (sodium-potassium alloy; nack).

Properties: Soft silvery solid or liquid; must be kept away from air and moisture. Very reactive!

Grades: (a) 78% K, 22% Na; m.p. −11°C; b.p. 784°C; density 0.847 g/cc (100°C); (b) 56% K, 44% Na; m.p. 19°C; b.p. 825°C; density 0.886 g/cc (100°C).

Containers: Stainless steel cans (3, 10, 25, 200 lbs); small glass ampules.

Uses: Heat exchange fluid; electric conductor; for organic synthesis and catalysis.

Shipping regulations: Flammable solid Yellow label. *

"Nakta." [51] Trademark for lime-base greases made with highly refined mineral oil and suitable for low temperature dispensing and use. Available in a wide range of consistencies for all types of mine car plain and roller bearings.

"Nalan" RF Durable Water Repellent. [28] Light cream-colored viscous paste dispersion.

Uses: Along with thermosetting resins as a softener modifier to obtain improved crease resistance and wash-and-wear properties on cotton fabric.

"Nalcite HCR." [182] Cation exchanger; sulfonated hydrocarbon of the styrene-base type.

"Nalcoag." [182] Colloidal silica sol for use as a bonding agent in ceramics.

"Nalline." [123] Trademark for nalorphine.

nalorphine hydrochloride (N-allylnormorphine hydrochloride) $C_{19}H_{21}NO_3 \cdot HCl$, a derivative of morphine.

Properties: White or practically white, odorless, crystalline powder; m.p. 260-263°C; affected by light and air; soluble in water; slightly soluble in alcohol; insoluble in chloroform and ether; soluble in diluted alkali hydroxide solutions; pH (0.5% solution) 4.4-5.5.

Grade: U.S.P. XVI.

Use: Medicine.

"Nalzin N." [304] Trademark for a zinc organic vinyl stabilizer.

Properties: Clear straw-colored liquid; sp. gr. 0.95; refractive index 1.51.

Containers: 40- and 400-lb metal drums.

Uses: Adjunct for barium-cadmium stabilizers to impart sulfide stain resistance.

nandrolone phenpropionate $C_{27}H_{34}O_3$. 19-Nor-delta^4-androstene-17 beta-ol-3-one-beta-phenyl propionate.

Use: Medicine.

Nankin yellow. See iron buff.

"Nankor." [233] Trademark for organophosphorus compounds.

nano-. Prefix meaning 10^{-9} units (symbol n). 1 ng = 1 nanogram = 0.000000001 gram.

nantokite. See cuprous chloride.

napalm. An aluminum soap of a mixture of oleic, naphthenic and coconut fatty acids.
Properties: Becomes viscous when shaken, makes gasoline thicken or jell.
Uses: In flame throwers and fire bombs; future possibility of use in preparing an almost solid gasoline for convenient storage, and of increasing the production of oil and gas wells by pumping in a mixture of crude oil, napalm and sand instead of using explosives and acids.

naphazoline hydrochloride $C_{14}H_{14}N_2 \cdot HCl$. 2-(1-Naphthylmethyl)imidazoline hydrochloride.
Properties: White, crystalline powder. Odorless, with bitter taste; m. p. 253-258°C. Solutions are neutral to litmus paper. Freely soluble in water and alcohol; very slightly soluble in chloroform and practically insoluble in ether.
Grade: N. F. XI.
Use: Medicine.

naphtha.
1. The word is usually applied to a narrow-boiling-range fraction of petroleum with volatility somewhere between that of gasoline and kerosine. For specific naphthas derived from petroleum see naphtha, petroleum; naphtha, blending; naphtha, painters; rubber solvent.
2. Solvent naphtha, and heavy naphtha are terms applied to aromatic solvents derived from coal tar. See under the corresponding naphtha.

naphtha, blending. A petroleum fraction with volatility similar to the higher boiling fractions of gasoline. It is used primarily in blending with natural gasoline to produce a finished gasoline of proper volatility.

naphthacene (tetracene; 2,3-benzanthracene; rubene; chrysogen) $C_{18}H_{12}$. The molecule consists of four benzene rings fused together.
Properties: Orange solid; density 1.35; m. p. about 350°C; not easily soluble in any ordinary solvent; slight green fluorescence in daylight.
Occurrence: In commercial anthracene.

naphtha, cleaners'. A dry-cleaning fluid derived from petroleum and similar to Stoddard solvent (q. v.), but not necessarily meeting all its specifications.

naphtha, heavy (crude heavy solvent naphtha).
Properties: Deep amber to dark red liquid; a mixture of xylene and higher homologs; flammable. Sp. gr. 0.885-0.970; b. p. 160-220°C (about 90% at 200°C); flash point not above 100°F.; evaporation 303 minutes.
Derivation: (a) From coal-tar by fractional distillation. (b) From illuminating gas by scrubbing and distilling the resulting oil.
Grades: Technical.
Containers: Drums; tank cars.
Uses: Coumarone resins; solvent for dark-colored paints and enamels; solvent for asphalts, road tars, pitches, etc.; in saturating asbestos-board, brake linings, and the like; cleansing compositions; illuminant; process engraving and lithography; rubber cements (solvent); naphtha soaps.

naphthalene (tar camphor; naphthalin) $C_{10}H_8$ or CHCHCHCHCCCHCHCHCH.
Properties: White crystalline, volatile flakes; strong coal-tar odor. Soluble in benzene, absolute alcohol and ether; insoluble in water. Sp. gr. 1.152; m. p. 80.05°C; b. p. 217.96°C.
Derivation: (a) From the coal-tar oils boiling between 170 and 230°C by crystallization and distillation. (b) From petroleum fractions after various catalytic processing operations.
Purification: Distillation, crystallization, hydrogenation, sodium treatment.
Grades: Flakes; balls; blocks; cubes; liquid; grains (rice); powder crushed; tablets; crude; C. P.
Containers: 1-lb bottles; 1-lb cans; 175-, 250-lb barrels; crude: 200-lb bags; tank trucks; cars.
Uses: Intermediate (phthalic anhydride, naphthal, "Tetralin," "Decalin," chlorinated naphthalenes, naphthyl and naphthol derivatives, dyes); moth repellent; fungicide; explosives; cutting fluid; lubricant; synthetic resins; synthetic tanning; preservative; solvent; textile chemicals; emulsion breakers; scintillation counters.
Shipping regulations: None. *

alpha-**naphthaleneacetic acid** (1-naphthylacetic acid) $C_{10}H_7CH_2COOH$.
Properties: White crystals, odorless; m. p. 132-135°C. Soluble in acetone, ether, and chloroform; slightly soluble in water and alcohol.
Grades: Usually supplied in dilute form, either as a powder or liquid solution ready for use.
Containers: Powder: fiber cans or multiwall paper sacks; solution: glass bottles and carboys.
Uses: Used as a solution for inducing the rooting of plant cuttings; for spraying apple trees for prevention of early drop.
Fire hazard: None, unless a flammable carrier is used.
Shipping regulations: None. *

alpha-**napthaleneacetic acid methyl ester** (MENA) $C_{10}H_7CH_2COOCH_3$. A plant growth regulator or hormone, used for delaying sprouting of potatoes; weed control, thinning of peaches, etc.

naphthalene, chlorinated. See chloronaphthalene.

naphthalenediamine. See naphthylenediamine.

naphthalene-1,5-disulfonic acid (Armstrong's acid) $C_{10}H_6(SO_3H)_2$.
Properties: White crystalline solid; soluble in water.
Derivation: Sulfonation of naphthalene with oleum at low temperature and separation from the 1,6 isomer.
Method of purification: Recrystallization from water.

*See "I.C.C. Shipping Regulations," page xiii.
Reference numbers refer to name of manufacturer. See "List of Manufacturers," page v.

Grades: Technical.
Containers: Oak barrels.
Use: Intermediate for dyes.
Shipping regulations: None. *

naphthalene-2, 7-disulfonic acid $C_{10}H_6(SO_3H)_2$.
Properties: White crystalline solid; soluble in water.
Derivation: Sulfonation of naphthalene at high temperature and separation from 2, 6-isomer.
Method of purification: Recrystallization.
Grades: Technical.
Containers: Oak barrels; fiber drums.
Use: Intermediate for dyes.
Shipping regulations: None. *

alpha-**naphthalenesulfonic acid** $C_{10}H_7SO_3H \cdot H_2O$.
Properties: Deliquescent crystals; soluble in water, alcohol, and ether. M. p. 90°C.
Derivation: By the interaction of naphthalene and sulfuric acid.
Method of purification: Crystallization.
Containers: Wooden barrels; kegs.
Grades: Technical.
Uses: Starting point in the manufacture of alpha-naphthol, alpha-naphtholsulfonic acid, alpha-naphthylaminesulfonic acid; solvent (sodium salt) for phenol in the manufacture of disinfectant soaps.
Shipping regulations: None. *

beta-**naphthalenesulfonic acid** $C_{10}H_7SO_3H$ or $C_{10}H_7SO_3H \cdot H_2O$.
Properties: Non-deliquescent, white plates; m. p. 124-125°C. Soluble in water, alcohol, and ether.
Derivation: By the sulfonation of naphthalene.
Method of purification: Crystallization.
Grades: Technical.
Containers: Wooden barrels; fiber drums.
Uses: Starting point in the manufacture of beta-naphthol, beta-naphtholsulfonic acid, beta-naphthylaminesulfonic acid; etc.
Shipping regulations: None. *

1, 3, 6-naphthalenetrisulfonic acid, trisodium salt $C_{10}H_5(SO_3Na)_3$.
Properties: Fine buff crystals.
Grades: Technical.
Containers: Fiber drums, 200 lbs net, 220 lbs gross; bottles.
Use: Diazo type stabilizer.
Shipping regulations: None. *

naphthalic acid. See phthalic acid.

naphthalin. See naphthalene.

"Naphthanil." [28] Trademark for a series of dye bases; products used in combination with the various "Naphthanil" prepares to form insoluble azo dyes. Prior to coupling the bases must first be diazotized to form the diazo salt. Widely used on cotton and rayon textiles.

naphtha, painters' (naphtha, V. M.&P.; varnish makers' naphtha; petroleum spirits; petroleum thinner; mineral spirits; turpentine substitute; mineral thinner; mineral turpentine.)
Any of a number of narrow-boiling-range fractions of petroleum with boiling points of about 200 to 300°F according to the specific use. Used as thinners in paints and varnish.
Note: The term "turpentine substitute" is misleading and should not be used.
Fire hazard: Flammable. MCA warning label.
Shipping regulations: Flammable liquid. Red label. Legal label name: "turpentine substitute."*

naphtha, petroleum. A generic term applied to refined, partly refined, or unrefined petroleum products and liquid products of natural gas, not less than 10 percent of which distil below 347°F (175°C), and not less than 95 percent of which distil below 464°F (240°C) when subjected to distillation in accordance with the Standard Method of Test for Distillation of Gasoline, Naphtha, Kerosine, and Similar Petroleum Products (ASTM D 86).
Note: The naphthas used for specific purposes such as cleaning, manufacture of rubber, manufacture of paints and varnishes, etc., are made to conform to specifications which may require products of considerably greater volatility than that set by the limits of this generic definition.
Shipping regulations: Flammable liquid. Red label may be required. *

naphtha, solvent (160° benzol; coal-tar naphtha).
Properties: A mixture of small quantities of benzene and toluene with xylene and higher homologs, from coal-tar. (a) Crude: dark straw-colored liquid; flammable. (b) Refined: water-white liquid; flammable.
Constants: Sp. gr. (a) 0. 862-0. 892, (b) 0. 862-0. 872; b. p. (a) about 160°C (80%), (b) about 160°C (90%); flash point (a) and (b) about 25. 6°C.
Derivation: From coal-tar by fractional distillation; from illuminating gas by scrubbing and distilling the resulting oil.
Grades: Dark straw; water-white.
Containers: 55-, 110-gal iron drums; 8000-gal tank cars.
Uses: Solvent; xylene; cumene; nitrated solvent naphtha for incorporation in dynamite compositions.
Fire hazard: Dangerous.
Shipping regulations: Flammable liquid. Red label. *

naphtha, solvent, crude heavy. See naphtha, heavy.

naphtha, V. M. &P. See naphtha, painters'. The initials mean Varnish Makers' and Painters'.

naphthenes. Cycloparaffin hydrocarbons, generally derivatives of cyclopentane (C_5H_{10}) or cyclohexane (C_6H_{12}) occurring in mixture of petroleums of various origins.

naphthenic acids. A class of acids derived from petroleum, particularly that of a non-paraffinic character.
It is stated that these acids are produced in part by oxidation of certain readily oxidized cycloparaffins (naphthenes) during

distillation or other petroleum-refining operations. The acids dissolve in the aqueous caustic solutions used in refining, and are recovered by reacidification and purification from admixed oil, which is usually reduced to 1% or less.

Individual naphthenic acids are isolated with difficulty. The known members of this group are apparently colorless or faintly colored oily liquids, volatile with steam, and boil without appreciable decomposition in the range 200 to 300°C. The most common are derivatives of cyclopentane such as C_5H_9COOH. Similar derivatives of cyclohexane and cycloheptane are also common, and a great variety of homologs and higher molecular weight analogs has been noted. The materials are slightly soluble in water, but freely soluble in hydrocarbons and organic solvents.

Commercial naphthenic acid is a mixture, usually dark colored, and corrosive to metals; malodorous. The acids present have molecular weights in the range 180 to 350.

Uses: Production of sodium, calcium, aluminum, chromium, cobalt, copper, manganese, zinc and lead naphthenates. Their principal uses are as paint driers, wood preservatives; limited use in extreme pressure lubricants. See under separate entries. The free naphthenic acids have been proposed as suitable solvents for vulcanized rubber, various resins and gums such as copal, dammar, sandarac, and mastic and for aniline dyes; for reduction of viscosity of colloidal solutions, clarifying agents for mineral oils; production of detergents and wetting agents; insecticides; additive to wood oil to permit drying without cracking.

naphthionic acid (1-aminonaphthalene-4-sulfonic acid; 1-naphthylamine-4-sulfonic acid; 4-amino-1-naphthalenesulfonic acid) $C_{10}H_6(NH_2)SO_3H$.

Properties: White crystals or powder. Soluble in alcohol and ether.

Derivation: By baking a mixture of equal molecules of alpha-naphthylamine and sulfuric acid with which about 3% of oxalic acid is incorporated.

Method of purification: Crystallization.

Grade: Technical.

Containers: Wooden barrels.

Use: Dye intermediate.

Shipping regulations: None. *

1-naphthol. See alpha-naphthol.

2-naphthol. See beta-naphthol.

alpha-**naphthol** (1-naphthol; 1-hydroxynaphthalene) $C_{10}H_7OH$.

Properties: Colorless or yellow prisms or powder; disagreeable taste. Soluble in benzene, alcohol and ether; slightly soluble in water. Sp. gr. 1.224 (4°C), 1.0954 (95/4°C); m.p. 96°C; b.p. 278°C; volatile in steam; sublimes; refractive index 1.6206 (98.7°C).

Derivation: By fusing sodium alpha-naphthalene sulfonate and caustic soda. The melt is decomposed with hydrochloric acid and distilled.

Method of purification: Redistillation.

Impurities: Beta-naphthol.

Grades: Technical; pure.

Containers: 55-gal drums.

Uses: Dyes; organic synthesis; synthetic perfumes.

Shipping regulations: None. *

beta-**naphthol** (2-naphthol; 2-hydroxynaphthalene) $C_{10}H_7OH$.

Properties: White, lustrous, bulky leaflets, or white powder; darkens with age; faint phenol-like odor; stable in air but darkens on exposure to sunlight; sp. gr. 1.217; m.p. 121.6°C; b.p. 285°C; flash point (closed cup) 310°F. Soluble in alcohol, ether, chloroform, glycerol, oils and alkaline solutions; slightly soluble in water.

Derivation: By fusing sodium beta-naphthalene sulfonate with caustic soda. The product is distilled in vacuo.

Method of purification: Sublimation.

Grades: Technical; sublimed; resublimed; N.F. XI.

Containers: 1-lb cartons; 250-, 300-, 350-lb wooded barrels or fiber drums.

Uses: Dyes; pigments, anti-oxidants for rubber, fats, oils, insecticide; synthesis of fungicides, pharmaceuticals, perfumes.

Shipping regulations: None. *

naphthol AS. See beta-hydroxynaphthoic anilide.

beta-**naphthol benzoate.** See benzonaphthol.

3-naphthol-2-carboxylic acid. See 3-hydroxy-2-naphthoic acid.

1-naphthol-3,6-disulfonic acid $C_{10}H_5(SO_3H)_2OH$.

Derivation: Fusion of sodium naphthalene-1,3,6-trisulfonate with caustic soda, or by diazotization of 1-naphthylamine-3,6-disulfonic acid and treatment with sulfuric acid.

Use: Dye intermediate.

1-naphthol-4,8-disulfonic acid (Schoelkopf's acid) $C_{10}H_5OH(SO_3H)_2$.

Properties: Colorless crystals.

Derivation: Decomposition of 1-naphthylamine-4,8-disulfonic acid by diazotization and acidifying with heat.

Use: Dye intermediate.

2-naphthol-3,6-disulfonic acid (R acid; beta-naphtholdisulfonic acid) $C_{10}H_5(OH)(SO_3H)_2$.

Properties: Deliquescent, colorless, silky needles; soluble in water, alcohol and ether.

Derivation: By heating beta-naphthol with sulfuric acid (98%), dissolving the melt in water and adding salt.

Method of purification: Crystallization or through the hard, soluble calcium-sodium salt.

Grades: Technical, with varying amounts of 2-naphthol-6-sulfonic acid. The common article of trade is the disodium salt, an odorless, gray, dry, ground powder.

Containers: Wooden barrels; multiwall paper sacks.

Use: Manufacturing azo dyes for textiles, lakes and foods.

2-naphthol-6,8-disulfonic acid, potassium salt (G salt) $C_{10}H_5OH(SO_3K)_2$.
Properties: Very small white needles, dry or in paste. Very soluble in water.
Derivation: Sulfonating beta-naphthol in oleum, diluting in water and salting out with potassium sulfate.
Method of purification: Recrystallization.
Grades: Technical; not more than 1.0% R and Schaeffer salts.
Containers: Wooden barrels; fiber drums.
Use: Azo dyes for textiles and lakes.
Shipping regulations: None.*

beta-**naphthol methyl ether.** See beta-naphthyl methyl ether.

beta-**naphthol sodium.** See sodium beta-naphtholate.

1-naphthol-4-sulfonic acid (N W acid; Neville and Winther's acid; alpha-naphtholsulfonic acid) $C_{10}H_6(OH)SO_3H$.
Properties: Transparent plates. Soluble in water. M.p. 170°C.
Derivation: From the sodium salt of naphthionic acid by hydrolyzing the amino group.
Grades: Technical. Also commonly handled as the sodium salt, an odorless cream to light tan, dry ground powder.
Containers: Wooden barrels; fiber drums.
Use: Dyes.
Shipping regulations: None.*

1-naphthol-5-sulfonic acid (L acid; alpha-naphtholsulfonic acid) $C_{10}H_6(OH)SO_3H$.
Properties: White solid. Soluble in water.
Derivation: (a) From naphthalene-1,5-disulfonic acid by fusion with caustic soda. (b) From 1-naphthylamine-5-sulfonic acid by diazotizing and boiling the diazo solution with dilute sulfuric acid.
Method of purification: Recrystallization from hot water.
Grades: Technical. See also 1-naphthol-5-sulfonic acid, sodium salt.
Containers: Wooden barrels; fiber drums.
Use: Dyes.
Shipping regulations: None.*

2-naphthol-1-sulfonic acid $C_{10}H_6(OH)SO_3H$.
Properties: White crystalline solid; soluble in water.
Derivation: By sulfonating beta-naphthol with 2 to 2.5 parts of 90 to 92% sulfuric acid at about 40°C.
Method of purification: Recrystallization from water.
Grade: Technical.
Containers: Wooden barrels; fiber drums.
Use: Intermediate for Tobias acid.
Shipping regulations: None.*

2-naphthol-6-sulfonic acid (Schaeffer's acid; beta-naphtholsulfonic acid) $C_{10}H_6(OH)SO_3H$.
Properties: White leaflets; soluble in water and alcohol. M.p. 122°C.
Derivation: By sulfonation of beta-naphthol and separation from the croceine acid formed simultaneously.
Method of purification: Recrystallization from water.
Grades: Technical. Also commonly sold as the sodium salt, a cream to light tan, dry ground powder with faint naphthol odor.
Containers: Wooden barrels; fiber drums.
Use: Dyes.
Shipping regulations: None.*

2-naphthol-7-sulfonic acid (Cassella's acid; mono-sulfonic acid F; F acid; mono acid F; beta-naphtholsulfonic acid) $C_{10}H_6(OH)(SO_3H)$.
Properties: White crystals; soluble in water and alcohol; m.p. 89°C.
Derivation: By fusion of naphthalene-2,7-disulfonic acid with caustic soda or by heating the acid with an aqueous solution of caustic soda in an autoclave.
Grade: Technical.
Containers: Wooden barrels; fiber drums.
Use: Dyes.
Shipping regulations: None.*

2-naphthol-8-sulfonic acid. See croceine acid.

alpha-**naphtholsulfonic acids.** See 1-naphthol-4-sulfonic acid and 1-naphthol-5-sulfonic acid.

beta-**naphtholsulfonic acids.** See 2-naphthol-6-sulfonic acid and 2-naphthol-7-sulfonic acid.

1 naphthol-5-sulfonic acid, sodium salt (L-acid salt) $C_{10}H_6(OH)(SO_3Na)$.
Properties: Gray to light tan dry powder. Odorless.
Purity: 70% min; inorganic content 23% max.
Containers: 175 lb fiber drums.

1,4-naphthoquinone (alpha-naphthoquinone) CHCHCHCHCCCOCHCHCO.
Properties: Greenish yellow powder; odor like benzoquinone; m.p. 123-126°C; begins to sublime below 100°C; very slightly soluble in water; soluble in ethyl alcohol, ethyl ether, chloroform, benzene, and acetic acid.
Uses: Polymerization regulator for rubber and polyester resins; synthesis of dyes and pharmaceuticals; fungicide; algaecide.

alpha-**naphthoquinone.** See 1,4-naphthoquinone.

naphthoquinoneoxime. See nitrosonaphthol.

1,2-naphthoquinone-4-sulfonic acid (beta-naphthoquinone-4-sulfonic acid) $C_{10}H_5(O)_2SO_3H$.
Derivation: Oxidation with nitric acid of 2-amino-1-naphthol-4-sulfonic acid or 1-amino-2-naphthol-4-sulfonic acid.
Uses: Dye intermediate; identification of sulfonamide derivatives.

beta-**naphthoquinone-4-sulfonic acid.** See 1,2-naphthoquinone-4-sulfonic acid.

naphthoresorcinol. See 1,3-dihydroxynaphthalene.

"Naphthosol." [57] Trademark for a series of dyestuffs and intermediates.

1,8-naphthosultam-2,4-disulfonic acid (sultam acid) $C_{10}H_4(SO_3H)_2NHSO_2$. (The $-NHSO_2-$ group forms a third ring by attachment at the 1,8 positions.)
Properties: White solid; soluble in water;

slightly soluble in alcohol.
Derivation: Oleum sulfonation of 1-naphthylamine-8-sulfonic acid or 1-naphthylamine-4, 8-disulfonic acid.
Grade: Technical.
Containers: Not sold as such.
Use: Intermediate for Chicago acid.
Shipping regulations: None. *

beta-**naphthoxyacetic acid** (2-naphthoxyacetic acid) $C_{10}H_7OCH_2COOH$. A plant hormone.
Properties: Crystals; m. p. 156°C. Soluble in water, alcohol, acetic acid.

1-naphthylacetic acid. See naphthaleneacetic acid.

alpha-**naphthylamine** $C_{10}H_7NH_2$.
Properties: White crystals becoming red on exposure to air. Soluble in alcohol and ether; slightly soluble in water. Sp. gr. 1. 13; m. p. 50°C; b. p. 301°C.
Derivation: By the reduction of alpha-nitronaphthalene with iron and hydrochloric acid. The mass is then mixed with milk of lime and distilled.
Method of purification: Crystallization.
Grade: Technical.
Containers: 225-, 300-, 350-lb wooden barrels; multiwall paper sacks.
Warning: Harmful dust and vapor. Repeated absorption may result in bladder tumors. MCA warning label.
Shipping regulations: None. *

beta-**naphthylamine** $C_{10}H_7NH_2$.
Properties: White to reddish lustrous leaflets. Soluble in hot water, alcohol, ether, and benzene. Commercial: f. p. 109. 5°C; sp. gr. 1. 061 (98/4°C); b. p. 306°C.
Derivation: From beta-naphthol by heating in an autoclave with ammonium sulfite and ammonia.
Method of purification: Distillation.
Grade: Technical.
Containers: Wooden barrels; fiber drums; multiwall paper sacks.
Use: Dyes.
Warning: Harmful dust and vapor. Repeated absorption may result in bladder tumors. MCA warning label.
Shipping regulations: None. *

1-naphthylamine-3, 8-disulfonic acid (epsilon acid) $C_{10}H_5(NH_2)(SO_3H)_2$.
Properties: White crystalline scales; soluble in hot water.
Derivation: Naphthalene-1, 5- and 1, 6-disulfonic acids are nitrated and reduced, resulting in 1-naphthylamine-3, 8- and 4, 8-disulfonic acids. The separation is effected by crystallizing out the acid sodium salt of 1-naphthylamine-3, 8-disulfonic acid.
Grade: Technical.
Containers: Wooden barrels; fiber drums.
Use: Dyes.
Shipping regulations: None. *

2-naphthylamine-3, 6-disulfonic acid (amino R acid) $C_{10}H_5(NH_2)(SO_3H)_2$.
Properties: Light brown to tan granular paste. Characteristic odor.
Purity: 50% min.

2-naphthylamine-4, 8-disulfonic acid (C acid; Cassella acid; 3-amino-1, 5-naphthalenedisulfonic acid) $C_{10}H_5(NH_2)(SO_3H)_2$.
Properties: White crystalline solid; soluble in water (slightly).
Derivation: Reduction of 2-nitronaphthylamine-4, 8-disulfonic acid.
Method of purification: Recrystallization of sodium salt from water.
Use: Dyestuffs.

2-naphthylamine-5, 7-disulfonic acid (amino-J acid; 6-amino-1, 3-naphthalenedisulfonic acid) $C_{10}H_5(NH_2)(SO_3H)_2$.
Properties: Crystallizes in white lustrous leaflets from water and in long needles from hydrochloric acid solutions.
Derivation: By sulfonation of either 2-naphthylamine-5-sulfonic acid or 2-naphthylamine-7-sulfonic acid.
Grade: Technical.
Containers: Wooden barrels; fiber drums.
Use: Dyes.
Shipping regulations: None. *

2-naphthylamine-6, 8-disulfonic acid (amino-G acid; 7-amino-1, 3-naphthalenedisulfonic acid) $C_{10}H_5(NH_2)(SO_3H)_2$.
Properties: White crystalline solid; soluble in water.
Derivation: (a) From G acid by heating sodium salt with ammonia and sodium bisulfite solution in an autoclave under pressure. (b) Sulfonation of beta-naphthylamine.
Method of purification: Recrystallization from water.
Grade: Technical.
Containers: Wooden barrels; fiber drums.
Use: Dyes.
Shipping regulations: None. *

2-naphthylamine-1, 5-disulfonic acid, sodium salt. See D salt.

alpha-**naphthylamine hydrochloride** $C_{10}H_7NH_2 \cdot HCl$.
Properties: White to gray, crystalline powder; soluble in water, alcohol, and ether.
Derivation: By the action of hydrochloric acid on alpha-naphthylamine.
Method of purification: Crystallization.
Grades: Technical; C. P.
Containers: 1-lb bottles; wooden kegs; fiber drums.
Uses: Dyes; organic synthesis.
Shipping regulations: None. *

1-naphthylamine-4-sulfonic acid. See naphthionic acid.

1-naphthylamine-5-sulfonic acid (Laurent's acid; L acid; 5-amino-1-naphthalenesulfonic acid) $C_{10}H_6NH_2(SO_3H)$.
Properties: Anhydrous, white or pinkish crystalline needles; greenish fluorescence in dilute aqueous solution. Soluble in water.
Derivation: (a) From alpha-naphthylamine by sulfonation with oleum. (b) From alpha-naphthalenesulfonic acid by nitration, reduction and separation from 1-naphthylamine-8-sulfonic acid also formed.
Method of purification: Crystallization.
Grades: Technical (not more than 2% 1, 8 acid).

Containers: Wooden barrels; fiber drums.
Use: Azo dyes.
Shipping regulations: None. *

1-naphthylamine-6 and 7-sulfonic acid (Cleve's acid; 5- and 8-amino-2-naphthalenesulfonic acids) $C_{10}H_6NH_2SO_3H$.
Properties: Colorless needles; not very soluble in water.
Derivation: Sulfonation, nitration, reduction and either separation of the combined 1, 6 and 1, 7 acids or separating by different concentrations and salting out.
Method of purification: Recrystallization.
Grades: Technical; either mixture of 1, 6 plus 1, 7 acids or each separate.
Containers: Wooden barrels; fiber drums.
Uses: Azo and diazo dyes.
Shipping regulations: None. *

1-naphthylamine-8-sulfonic acid (peri acid; S acid; Schoelkopf's acid; 8-amino-1-naphthalenesulfonic acid) $C_{10}H_6NH_2SO_3H$.
Properties: White needles; slightly soluble in water.
Derivation: Together with 1-naphthylamine-5-sulfonic acid by sulfonating naphthalene, nitrating, reducing and separating the combined precipitated acids with soda ash. The insoluble 1, 8 sodium salt is filtered off from the 1, 5 sodium salt solution and transposed into free acid with muriatic acid.
Grades: Technical, with less than 1% 1, 5 acid.
Containers: Wooden barrels; fiber drums.
Use: Mostly as phenyl-1-naphthylamine-8-sulfonic acid for azo dyes.
Shipping regulations: None. *

2-naphthylamine-1-sulfonic acid (Tobias acid; 2-amino-1-naphthalenesulfonic acid) $C_{10}H_6(NH_2)(SO_3H)$.
Properties: Crystallizes in white needles. Soluble in hot water.
Derivation: Sodium 2-naphthol-1-sulfonate (from beta-naphthol and sulfuric acid at 40°C) is heated with ammonium hydrogen sulfite and ammonia in an autoclave at from 100 to 150°C.
Method of purification: Precipitation from dilute solution of sodium salt.
Grade: Technical.
Containers: Wooden barrels; fiber drums.
Use: Dyes.
Shipping regulations: None. *

2-naphthylamine-6-sulfonic acid (Broenner's acid; 6-amino-2-naphthalenesulfonic acid) $C_{10}H_6(NH_2)SO_3H$.
Properties: Colorless needles; soluble in boiling water.
Derivation: By heating sodium 2-naphthol-6-sulfonate with concentrated ammonia in an autoclave at 180°C.
Grades: Technical. Also commonly sold in the form of the sodium salt, an odorless, light gray to pink powder.
Containers: Barrels; kegs; fiber drums.
Use: Dyes.
Shipping regulations: None. *

2-naphthylamine-7-sulfonic acid (Cassella's acid F; Bayer's acid; F acid; delta acid) $C_{10}H_6(NH_2)SO_3H$.
Properties: Colorless crystals; soluble in water, alcohol, and ether.
Derivation: (Cassella's acid F.) By heating sodium 2-naphthol-7-sulfonate with aqueous ammonia and ammonium acid sulfite in an autoclave.
Grade: Technical.
Containers: Wooden barrels; kegs; fiber drums.
Use: Dyes.
Shipping regulations: None. *

2-naphthylamine-8-sulfonic acid (Badische acid). Similar to other naphthylamine-sulfonic acids.

1-naphthylamine-3, 6, 8-trisulfonic acid (Koch's acid; 8-amino-1, 3, 6-naphthalenetrisulfonic acid) $C_{10}H_4(NH_2)(SO_3H)_3$.
Properties: White solid; soluble in water (slightly).
Derivation: Naphthalene is sulfonated to naphthalene-1, 3, 6-trisulfonic acid using oleum; and this trisulfonic acid is nitrated cold and then reduced with iron.
Method of purification: Recrystallization from water.
Grades: Technical.
Containers: Wooden barrels; fiber drums.
Use: Dyes.
Shipping regulations: None. *

naphthyl benzoate. See benzonaphthol.

naphthylenediamine (diaminonaphthalene; naphthalenediamine) $C_{10}H_6(NH_2)_2$. There are eight isomers. The following properties are those of the 1, 5 isomer.
Properties: Colorless crystals; m.p. 190°C; b.p., sublimes. Soluble in alcohol and hot water; very sparingly soluble in cold water.
Derivation: (a) By the reduction of alpha-dinitronaphthalene. (b) By heating dihydroxynaphthalene with aqueous ammonia.
Method of purification: Crystallization.
Grades: Technical.
Containers: Wooden kegs; fiber drums.
Use: Organic synthesis.
Shipping regulations: None. *

naphthyleneëthylene. See acenaphthene.

N-alpha-**naphthylethylenediamine dihydrochloride** $C_{10}H_7NHCH_2CH_2NH_2 \cdot 2HCl$. Colorless crystals; soluble in water. Used as reagent for the quantitative determination of sulfa-drugs.

beta-**naphthyl ethyl ether** (bromelia; nerolin II; nerolin; 2-ethoxynaphthalene) $C_{10}H_7OC_2H_5$.
Properties: White crystals; orange-blossom odor; congealing pt. 35°C; soluble in 5 parts of 95% alcohol.
Derivation: By the interaction of beta-naphthol and ethyl alcohol in presence of sulfuric acid.
Method of purification: Crystallization.
Grades: Technical.
Containers: Tins.
Uses: Perfumes; soaps.
Shipping regulations: None. *

1-naphthyl N-methylcarbamate $C_{10}H_7OOCNHCH_3$.
Derivation: Synthesized directly from 1-naphthol and methyl isocyanate or from naphthyl chloroformate (1-naphthol and phosgene) plus methylamine.
Use: Insecticide.

beta-**naphthyl methyl ether** (yara-yara; beta-naphthol methyl ether; 2-methoxynaphthalene; methyl naphthyl ether; methyl beta-naphtholate) $C_{10}H_7OCH_3$.
Properties: White, crystalline scales; soluble in alcohol and ether; insoluble in water. M. p. 72°C; b. p. 274°C.
Derivation: (a) By heating beta-naphthol and methyl alcohol in presence of sulfuric acid. (b) By methylating beta-naphthol with dimethyl sulfate.
Method of purification: Crystallization.
Grades: Technical.
Containers: Tins.
Use: Perfumery (soaps).
Shipping regulations: None. *

alpha-**naphthylphenyloxazole** (NPO; ANPO; 2-(1-naphthyl)-5-phenyloxazole) $C_{19}H_{15}NO$.
Properties: Solid; m. p. 104-106°C.
Grade: Scintillation.
Containers: 100-g bottles.
Use: Scintillation counter or wave length shifter in solution scintillators.

alpha-**naphthyl salicylate** $HOC_6H_4COOC_{10}H_7$.
Properties: White, crystalline powder, incompatible with alkalies and ferric compounds. Soluble in alcohol, ether, and fixed oils; insoluble in water. M. p. 83°C.
Derivation: By the action of salicylic acid upon alpha-naphthol.
Use: Medicine.
Shipping regulations: None. *

beta-**naphthyl salicylate** (salicylic naphthyl ester) $C_6H_4OHCOO(C_{10}H_7)$.
Properties: White, shining crystals; odorless, tasteless; decomposed by the alkaline pancreatic fluid; undecomposed by cold acids or alkalies. Soluble in hot alcohol, ether, benzene; soluble with difficulty in cold alcohol; insoluble in water. M. p. 95°C.
Derivation: By the action of salicylic acid on beta-naphthol.
Containers: Glass bottles; fiber cans.
Use: Medicine.
Shipping regulations: None. *

alpha-**naphthylthiourea** (ANTU) $C_{10}H_7NHCSNH_2$.
Properties: Odorless gray powder; m. p. 198°C. Insoluble in water and only very slightly soluble in most organic solvents.
Derivation: From alpha-naphthylthiocarbamide and alkali or ammonium thiocyanate.
Containers: Fiber cans.
Use: Rodenticide.
Warning: Poisonous if inhaled or swallowed. MCA warning label.

Naples yellow. See lead antimonate.

narceine $C_{23}H_{27}O_8N \cdot 3H_2O$.
Properties: White, silky crystals; bitter taste; odorless. Crystallizes from water at 60°C with 2 molecules of water, loses these at 100°C and a further molecule at 140°C. The fused residue is a mixture of bases. Soluble in alcohol and boiling water; less soluble in cold water; insoluble in ether. M. p. 150-160°C (commercial), 170-171°C (pure base).
Derivation: Occurs in opium. May be obtained from narcotine.
Use: Medicine.
Shipping regulations: None. *

narcissus oil. See jonquil oil.

narcosine. See noscapine.

ℓ-alpha-**narcotine.** See noscapine.

naringenin-7-rhamnoglucoside. See naringin.

naringenin-7-rutinoside. See naringin.

naringin (naringenin-7-rhamnoglucoside; naringenin-7-rutinoside; aurantiin) $C_{27}H_{32}O_{14}$.
Properties: A bioflavonoid. Crystals; m. p. 171°C; bitter taste. Soluble in acetone, alcohol, warm acetic acid.
Source: Extracted from flowers and rind of grapefruit and immature fruit.
Uses: Beverages; medicine; food supplement.

"Narlene." [233] Trademark for organophosphorus compounds.

native paraffin. See ozocerite.

"Natox." [172] Brand name for sodium oxalate. $Na_2C_2O_4$.
Properties: Grayish, crystalline powder. Minimum of 88% $Na_2C_2O_4$.
Containers: 100-lb paper bags.
Uses: Insolubilizer in the manufacture of wallboard cement and insulating materials; tanning of kid skin; fireworks.

natrium. The Latin name for sodium; hence the symbol Na in chemical nomenclature.

natroalunite. See alunite.

natrolite $Na_2Al_2Si_3O_{10} \cdot 2H_2O$. A mineral of the zeolite group. See zeolites.
Properties: Colorless or white to gray, yellow, greenish or red; sp. gr. 2.2-2.25; hardness 5-5.5.

"Natrosol 250." [266] Trademark for a nonionic, water-soluble cellulose ether.
Properties: White powder; readily soluble in hot or cold water.
Uses: Thickener; binder; protective colloid; suspending agent in latex paints and emulsion polymerization.

"Natsyn." [265] Trademark for a series of cis-1,4-polyisoprene synthetic rubbers essentially duplicating the chemical structure of natural rubber.

Natta catalysts. A particular type of stereospecific catalyst (q. v.) made from titanium chloride and aluminum alkyl or similar materials by a special process which includes grinding the materials together to produce an active catalyst surface. Also used to designate stereospecific catalysts in general.

natural cement. A hydraulic cement produced by pulverizing and then heating naturally occurring rock (cement rock) containing appropriate proportions of limestone, clay, magnesia and iron. Ignition temperatures are usually lower than for Portland cement. Final pulverizing is necessary as with Portland cement.

natural gas. A mixture of the low molecular weight paraffin series hydrocarbons methane, ethane, propane and butane with small amounts of higher hydrocarbons, also frequently containing small or large proportions of nitrogen, carbon dioxide, hydrogen sulfide and occasionally small proportions of helium. Methane is almost always the major constituent. Natural gas accompanying petroleum always contains appreciable quantities of ethane, propane, butane, as well as some pentane and hexane vapors and is known as "wet" gas. "Dry" gas contains little of these higher hydrocarbons. See also sour gas.

The exact composition of natural gas varies with locality. The heating value of natural gas is usually over 1000 Btu/cu ft unless nitrogen or carbon dioxide are important components of the gas.

Uses: Natural gas is used directly as a fuel and the higher hydrocarbons in it are also recovered for blending in motor fuel, and for use as liquefied fuel gases. Manufacture of carbon black is a major use, as is the use as a starting material for chemical synthesis of ammonia, acetylene, vinyl chloride, methanol, formaldehyde, ethanol, acetaldehyde. The most recently developed use is for production of synthesis gas from which liquid fuels and oxygenated aliphatic organic compounds are produced.

Containers: Limited amounts are shipped in high pressure cylinders.

Shipping regulations: Flammable compressed gas. Red gas label.*

"Naturetin." [412] Trademark for bendroflumethiazide (q. v.).

"Naugapol." [248] Trademark for a series of butadiene-styrene copolymers which have received special processing for minimum water soluble salts and low ash content. Included are "Naugapol K" series which are masterbatches of high styrene resin ("Kralac" A - EP) and "Naugapol" elastomers. Uses for "Naugapol" include wire and cable insulation, mechanical goods, adhesives and cements. The "K" series are used for shoe soles, floor tile and wire and cable insulation.

"Naugatex." [248] Trademark for a series of synthetic latices. The 2000 and 2100 types are hot and cold polymerized SBR types respectively. The 2600 series are nitrile latices. The 2700 series are high styrene or resin latices. These latices have a wide range of properties and uses, such as textile applications, paper saturation, and beater addition; tire cord; rug backing; paints; chewing gum and foam sponge.

"Naugawhite." [248] Trademark for an alkylated phenol antioxidant.

Properties: A slightly viscous clear amber liquid; sp. gr. 0.96; soluble in acetone, ethylene dichloride, benzol and gasoline; insoluble in water. Also available as "Naugawhite Powder", sp. gr. 1.19.

Uses: A nondiscoloring, nonstaining, general purpose antioxidant for rubber and latex in foam sponge, tire carcass, refrigerator gaskets, footwear, proofing, wire insulation and sundries. Used in all types of natural and synthetic rubbers.

naval stores. Historically, the pitch and rosin used on wooden ships. The term now includes all the modern products from pine wood and stumps, including rosin, turpentine and pine oils, from either gum or wood, and also tall oil and its derivatives.

"Navee '42." [33] Trade name for an emulsifiable degreasing compound consisting of high kauri-butanol solvents and emulsifying agents. Flash point 150°F min. Used for marine and industrial degreasing and light duty carbon removal. Also available as "Navee 427," a variation non-corrosive to aluminum, especially designed for the aircraft industry.

"Naxol." [243] Trade name for cyclohexanol.

Nb. Symbol for niobium.

NBA. Abbreviation for N-bromoacetamide.

NBC. [28] Sun or ozone cracking inhibitor. Nickel dibutyldithiocarbamate. $[(C_4H_9)_2NC(S)]_2Ni$.

Properties: Green powder or flakes; sp. gr. 1.26; m. p. 86°C or higher.

Containers: 250-lb drums.

Uses: To prevent cracking due to ozone of SBR (styrene butadiene rubber) stocks; to improve heat resistance and retard discoloration of neoprene compounds.

NBR. Nitrile-butadiene rubber. See acrylonitrile rubber.

NBS. Abbreviation for N-bromosuccinimide.

NC. Abbreviation for nitrocellulose.

"NCF Paste." [328] Brand name for a water-soluble urea-formaldehyde resin, used in the textile industry to produce a durable creaseproof, dimensional stabilizing finish. It is snow-white, with a faint odor of formaldehyde.

NCS. Abbreviation for N-chlorosuccinimide.

Nd. Symbol for neodymium.

NDGA. Abbreviation for nordihydroguaiaretic acid.

Ne. Symbol for neon.

"Neantine." [227] Trademark for diethyl phthalate (phthalol; ethyl phthalate) $C_6H_4(COOC_2H_5)_2$, 99% pure.

Properties: Colorless liquid; when purified, is completely odorless; otherwise develops a very faint and slightly sweet odor; stable;

*See "I.C.C. Shipping Regulations," page xiii.

will not discolor. Sp. gr. (25/25°C) 1.115-1.118; refractive index 1.499-1.502. Clearly soluble in 5 parts of 60% alcohol.
Uses: Employed as solvent for synthetic musks and other solid aromatic substances; also a denaturant for perfumery alcohol.

"Neatex." [152] Trade name for neat's foot oil; produced in seven grades of different cold tests for various uses such as belt and leather dressings, lubricating oils and greases, sulfonated oils, silk soaking, tanning oils, textile tapes, veterinarian supplies.

neatsfoot oil (bubulum oil; hoof oil).
Properties: A fixed, pale yellow oil with a peculiar odor. Soluble in alcohol, ether, chloroform, and kerosene. Sp. gr. 0.916; saponification value 194-199; iodine value 70.
Derivation: By boiling in water the shinbones and feet (deprived of hoofs) of cattle and separating the oil from the fat obtained.
Adulterants: Rape, cotton-seed, fish and mineral oils.
Grades: 15°; 20°; 30°; 40°F cold test; being the temperature in degrees F at which stearin separates.
Containers: Tins; drums.
Uses: Leather industry for "fat-liquoring"; waterproofing and softening leather; lubricant; oiling wool.
Shipping regulations: None. *

"Nebony." [21] Brand name for hydrocarbon, dark-colored thermoplastic resins of good odor available in grades from tacky medium-hard to glossy, brittle solid. Soluble in ketones, esters, terpenes, naphthenes, aromatic and chlorinated solvents and partially soluble in aliphatic and ether-alcohol solvents. Used in phonograph records, sound deadening compounds, rubber compounding, electrical insulating compounds, plastics, floor tile, adhesives, wax compounds, pipe coating oils, and paper impregnants.

"Nebula." [51] Trademark for multi-purpose grease having ability to serve under heavy loads at high speeds and under extremes of temperatures.

neburon. See 1-n-butyl-3-(3,4-dichlorophenyl)-1-methylurea.

"Nectadon." [123] Trademark for noscapine.

"Neelium." Quaternary alloy of bismuth, tellurium, selenium, antimony. Possible semiconductor for thermoelectric cooling.

"Neelium." [41] Trade name for a synthetic rubber coating of the neoprene type which can be applied in films up to 20 mils in thickness.

"Neetol." [244] Trademark for a series of alkaline oils, based primarily on neatsfoot oil.
Containers: Non-returnable steel drums averaging 400-425 lbs net.
Uses: Used by the leather industry and referred to as mayonnaise type fat-liquor.

"Negamine 142A." [83] Trademark for a textile finishing agent consisting of the amino esters of long chain fatty acids.

"Negatan." [100] Trademark for negatol (q. v.).

negatol ("Negatan"). A condensation product of meta-cresol-sulfonic acid with formaldehyde. A polymerized dihydroxydimethyldiphenylmethane-disulfonic acid. It is dispersible in water forming colloidal solutions which are very acidic. The pH of a 5% dispersion is about 1.0.
Use: Medicine.

"Neguvon." [181] Parasiticide and anthelmintic for domestic animals, containing O,O-dimethyl 2,2,2-trichloro-1-hydroxyethyl phosphonate (q. v.).

"Nekal." [307] Trademark for a series of wetting and dispersing agents.
"Nekal" BA-75: Sodium alkylnaphthalenesulfonate; 65% active; anionic.
Properties: Powder; density 0.73.
Uses: Dyeing and leveling agent in leather processing; wetting, dispersing and penetrating agent in textile processing; wetting agent in agricultural chemicals; dispersing agent in plastics and synthetic latices; stabilizer in latex formulations.
"Nekal" BX-78: Sodium alkylnaphthalenesulfonate; approximately 20% sodium sulfate.
Properties: Powder; density 0.55-0.70.
Uses: Wetting, penetrating, pasting agent in textile processing; wetting and dispersing agent in leather processing; wetting agent in paper manufacture; wetting agent for dry-color pigments; extender for paper-coating formulations; surfactant for latex polymerization and emulsification.
"Nekal" NF: Sodium alkylnaphthalenesulfonate; anionic.
Properties: Liquid; water-soluble and stable to acids, alkalies and hard water; sp. gr. 1.04.
Uses: Dispersant for solids in oils; wetting and penetrating agent for padding and long-liquor drying with vat, naphthol, sulfur and direct colors.
"Nekal" NS: Sulfonated aliphatic polyester; approximately 22% active.
Properties: Liquid; sp. gr. 1.04; low foaming.
Uses: Wetting and penetrating agent in textile processing; rewetting agent for "Sanforizing."
"Nekal" WS-21: Sulfonated aliphatic polyester; 18-20% active.
Properties: Clear, colorless liquid; soluble in water, stable in hard and soft water; moderately stable in acid and alkaline liquors.
Uses: Wetting and rewetting agent for use in both hot and cold solutions; wetting and penetrating agent in various phases of textile processing; surfactant in polyvinyl acetate emulsion paint formuations.
"Nekal" WS-25. Sulfonated aliphatic polyester; 55-60% active.
Properties: More concentrated solution of "Nekal WS-21."
Uses: See "Nekal WS-21."

*See "I.C.C. Shipping Regulations," page xiii.
Reference numbers refer to name of manufacturer. See "List of Manufacturers," page v.

"Nelio" Dipentene. [296] Brand name for commercial cut of terpene hydrocarbons possessing a boiling range similar to that of the chemical compound dℓ-limonene. Used in paints for wetting and dispersing properties; in rubber reclaiming as a swelling and penetrating agent; in resin manufacturing.

"Nelio" Gum Rosin. [296] Brand name for a proprietary product. A relatively new type of rosin which is absolutely clean, being clear and free from dirt and sand. Refined in aluminum stills under chemical control, and embodying absolute uniformity. Available in grades WW to H. "Nelio" gum rosin is processed to the following specifications: Acid number 161-168 (mg KOH); saponification number 170-180 (mg KOH); softening point by ring and ball method; (Grades X, WW, WG) 168-176°F, (Grades N and below) 171-176°F; ash max. 0.04%; unsaponifiable 5-9%; turpentine content 0.2-0.6%.

"Nelio" Gum Spirits Turpentine. [296] Brand name for a proprietary product, a pure gum spirits of turpentine.

"Nelio" Alpha-Pinene. [296] Brand name for a proprietary product. Typical purity of 98% as alpha-pinene.

"Nelio" Refined Sulfate Wood Turpentine. [296] Brand name for a proprietary product, the by-product of the sulfate pulping operations located predominately in southern U. S.; similar to gum turpentine.

"Nelio" Resin. [296] Brand name for a proprietary product, a liquid rosin for use in the manufacture of paints, varnishes, and enamels, available in pale and medium grades containing 80% "Nelio" Gum Rosin and 20% turpentine.

"Nelio" Terpineol. [296] Brand name for a proprietary product, a water-white viscous liquid with a characteristic woody lilac odor. Contains a minimum alcohol content as terpineol of 97% consisting principally of the alpha isomer. Used primarily in the preparation of essential oils and industrial and soap perfumes.

Nelson cell. A diaphragm-type electrolytic cell (see diaphragm cell) for the production of chlorine and caustic soda. The Nelson cell, once widely used but now largely replaced by Vorce and Hooker cells, is of rectangular design. A steel outer tank collects the caustic and hydrogen. A row of carbon anodes is suspended from the lid and a U-shaped cathode of perforated sheet steel is located between the anodes and the tank. The cathode is lined with a diaphragm of asbestos paper, which separates the brine in the anode compartment from the caustic in the outer cathode chamber.

"Nemafume." [88] Trademark for 1,2-dibromo-3-chloropropane; 99% purity.

"Nemagon." [125] Trademark for a soil fumigant containing 97% of 1,2-dibromo-3-chloropropane and 3% of other active compounds.
Properties: Pungent, brown liquid; b.p. approx. 195°C; setting point 7°C; slightly soluble in water; soluble in most common organic solvents.
Containers: 30-gal unlined metal drums (505 lb net).
Warning! (technical product and formulations 20% and over): Harmful liquid and vapor; (liquid) combustible, keep away from heat and open flame.
Caution! (formulations less than 20%): May cause irritation of skin, eyes, nose and throat.

nematocide. An agent which is destructive to nematodes (roundworms or threadworms).

"Nembutal." [3] Trademark for pentobarbital sodium.

"Nemex." [401] Trade name for a mixture of chloropicrin (50%) and chlorinated C_3 hydrocarbons, including 1,3-dichloropropene, 1,2-dichloropropane.
Uses: Soil fumigant for nematodes and fungi.
Warning: Poisonous liquid and vapor. Do not breathe vapor. Do not get in eyes, on skin or clothes.

neo-.
1. A prefix meaning new and designating a compound related in some way to an older one.
2. A prefix indicating a hydrocarbon in which at least one carbon atom is connected directly to four other carbon atoms; as, neopentane.

"Neo-Antergan." [123] Trademark for pyrilamine.

"Neo-Aristocort." [57] Trademark for neomycin-triamcinolone acetonide.

neoarsphenamine. Consists chiefly of sodium 3,3'-diamino-4,4'-dihydroxyarsenobenzene-N-methanol sulfoxylate: $NH_2OHC_6H_3As{:}AsC_6H_3OHNH(CH_2O)SONa$.
Properties: Yellow powder, containing not less than 19% arsenic. Odorless or slight odor. Poisonous! Soluble in water and glycerol; slightly soluble in alcohol; almost insoluble in acetone, chloroform, ether.
Containers: Ampules.
Use: Medicine (the same as arsphenamine).
Shipping regulations: None.*

"Neobiotic." [299] Trademark for neomycin sulfate.

"Neobon." [299] Trademark for a preparation containing vitamins, minerals, and other factors. Used in medicine.

"Neobon." [41] Trade name for a synthetic rubber membrane and coating of the neoprene type for protecting concrete and steel tanks from corrosives.

"Neochel." [288] Trademark for chemical additives to electroplating baths. Liquid formulation for use in all cyanide copper and bronze plating solutions which replaces Rochelle salt or proprietary materials in these baths.

neocinchophen (6-methyl-2-phenylquinoline-4-carboxylic ethyl ester) $C_{19}H_{17}NO_2$.
Properties: White to pale-yellow, crystalline powder; odorless and tasteless; permanent in air; affected by light. M. p. 75-76°C. Nearly insoluble in water; soluble in hot alcohol and strong acids; very soluble in chloroform and ether.
Derivation: Synthetic.
Method of purification: Crystallization.
Grades: N. F. XI.
Containers: Glass bottles; tin cans; paper-lined barrels; well closed, light-resistant containers.
Use: Medicine.
Shipping regulations: None. *

"NeoCoat." [204] Trademark for a self curing liquid neoprene maintenance coating for steel, concrete, and wood surfaces. Resistant to corrosive atmospheric conditions, splash and spill of chemicals, and to abrasion. Shipped in 1-gal and 5-gal cans.

neodymia. Neodymium oxide. See neodymium salts. See also rare earths.

neodymium Nd. Element having atomic number 60; group III of the periodic table; one of the rare-earth elements of the cerium subgroup.
Properties: Yellowish metal; tarnishes easily; sp. gr. 7.004; m. p. 1024°C; b. p. about 3150°C; ignites to oxide (200-400°C); liberates hydrogen from water; soluble in dilute acids. Color of salts rose-red.
Derivation: For source and isolation see rare earth minerals; metal produced by reduction of the chloride or fluoride with calcium powder.
Containers: Boxes.
Uses: Neodymium salts; electronics; alloys; in colored glass; to increase heat resistance of magnesium in aircraft and missiles.
Shipping regulations: None. *
See also didymium.

neodymium salts.
neodymium acetate $Nd(C_2H_3O_2)_3 \cdot xH_2O$. Pink powder, soluble in water.
neodymium ammonium nitrate $Nd(NO_3)_3 \cdot 2NH_4NO_3 \cdot 4H_2O$. Pink crystals, soluble in water. Technical grade contains 75% neodymium salt. Principal impurities praseodymium and samarium.
neodymium carbonate $Nd(CO_3)_3 \cdot xH_2O$. Pink powder, insoluble in water, soluble in acids. Grades 75%, 95%, and 99% neodymium salt.
neodymium chloride $NdCl_3 \cdot xH_2O$. Pink lumps, soluble in water and acids. Grades 75%, 95%, 99%, and 99.9% neodymium salt. Used as source of anhydrous chloride for preparation of the metal.
neodymium fluoride. Pink powder, insoluble in water. Grades 65%, 75%, 99%, and 99.9% neodymium salt.
neodymium nitrate $Nd(NO_3)_3 \cdot 6H_2O$. Pink crystals, very soluble in water. Grades 75%, 95%, 99%, and 99.9% neodymium salt.
Shipping regulations: Oxidizing material. Yellow label. *
neodymium oxalate $Nd_2(C_2O_4)_3 \cdot xH_2O$. Pink powder insoluble in water, slightly soluble in acids. Grades 75%, 95%, and 99% neodymium salt.
neodymium oxide (neodymia) Nd_2O_3. Pure product a blue-gray powder. Technical grade a brown powder. Grades 65%, 75%, 85%, 95%, 99% and 99.9% oxide. Insoluble in water, soluble in acids. Hygroscopic, absorbs carbon dioxide from the air. Used in ceramic capacitors, coloring glass; catalysts.
neodymium sulfate $Nd_2(SO_4)_3 \cdot 8H_2O$. Pink crystals, soluble in cold water; sparingly soluble in hot water. Grades 75%, 99%, and 99.9% neodymium salt.
Containers: Bottles, fiber and steel drums.
Uses: Decolorizing glass; coloring glass used in glass blowers' and welders' goggles, tableware, etc.

"Neo-Fat" Products. [15] Trademark for a series of fatty acids produced by fractional crystallization and distillation processes.
Containers: Lined bung drums; lined open head drums; or 50-lb paper bags; all also in aluminum tank cars.
Various products are available:
"Neo-Fat" 8. Proprietary name for caprylic acid.
Properties: Acid value 390; iodine value 0.8; titer 15°C; color Lovibond (5 ¼) 1.0R-20Y; unsap. 0.2%; moisture 0.2% approx. 7.5 lbs/gal.
Uses: Chemical raw material; fatty alcohols; metal salts; detergents; flotation; paper coating.
"Neo-Fat" 10. Proprietary name for capric acid.
Properties: Acid value 326; iodine value 0.8; titer 30°C; color Lovibond (5 ¼) 3.0R-10Y; color stab. Lovibond (5 ¼) 3.0R-35Y; unsap. 0.2%; moisture 0.2%; approx. 7.5 lbs/gal.
Uses: Chemical raw material; perfumes; flavors; alkylolamides; flotation; paper coating.
"Neo-Fat" 12. Proprietary name for lauric acid.
Properties: Acid value 280; iodine value 0.8; titer 42.5°C; color Lovibond (5 ¼) 0.8R-8Y; color stab. Lovibond (5 ¼) 2.0R-20Y; unsap. 0.1%; moisture 0.2%; approx 7.5 lbs/gal.
Uses: Alkyd resins; wetting agents; soaps; cosmetics; insecticides; metallic soaps; chemical raw material.
"Neo-Fat" 14. Proprietary name for myristic acid.
Properties: Acid value 246; iodine value 0.8; titer 50°C; color Lovibond (5 ¼) 0.8R-8Y; color stab. Lovibond (5 ¼) 2.5R-25Y; unsap. 0.1%; moisture 0.2%; approx 7.5 lbs/gal; odor, bland and characteristic.
Uses: Alkyd resins; wetting agents; soaps; cosmetics; insecticides; metallic soaps; chemical raw material; specialty lubricating greases.
"Neo-Fat" 16. Proprietary name for palmitic acid.

Properties: Acid value 218; iodine value 1.0; titer 57.5°C; color Lovibond (5 1/4) 0.8R-8Y; color stab. Lovibond (5 1/4) 2.0R-20Y; unsap. 0.2%; moisture 0.2%; approx 7.5 lbs/gal.
Uses: Chemical raw material; esters; plasticizers; metallic soaps; greases; detergents; cosmetics; shaving cream; fatty alcohols; sulfonic acids; paper coating.

"Neo-Fat" 16-54. Proprietary name for eutectic palmitic-stearic acid.
Properties: Acid value 212; iodine value 1.0; titer 54°C; color Lovibond (5 1/4) 1.5R-15Y; unsap. 0.3%; moisture 0.3%; approx 7.5 lbs/gal; appearance, hard waxy amorphous.
Uses: Chemical raw material; esters; plasticizers; metallic soaps; greases; detergents; cosmetics; shaving cream; fatty alcohols; sulfonic acids; paper coating.

"Neo-Fat" 18. Proprietary name for stearic acid.
Properties: Acid value 179; iodine value 1.5; titer 67.5°C; color Lovibond (5 1/4) 1.0R-10Y; unsap. 0.3%; moisture 0.3%; approx 7.5 lbs/gal.
Uses: Chemical raw material; esters; plasticizers; metallic soaps; greases; detergents; cosmetics; shaving cream; fatty alcohols; sulfonic acid.

"Neo-Fat" 18-53. Proprietary name for single pressed stearic acid.
Properties: Acid value 209; iodine value 12; titer 53.0°C; color Lovibond (1") 1.5R-15Y; unsap. 0.4%; moisture 0.2%; approx 7.5 lbs/gal.
Uses: Buffing compounds; polishes; rubber compounding.

"Neo-Fat" 18-54. Proprietary name for double pressed stearic acid.
Properties: Acid value 209; iodine value 6; titer 54.1°C; color Lovibond (5 1/4) 0.5R-2Y; Gardner color after heating 1 hr at 150°C, 2; unsap. 0.6%; moisture 0.2%; approx 7.5 lbs/gal; peroxide index 0.1; buffing stick test, must paste; odor bland.
Uses: Paper coating; recording cylinders; cosmetics; candles; buffing compounds; esters; plasticizers.

"Neo-Fat" 18-55. Proprietary name for triple pressed stearic acid.
Properties: Acid value 207.5; iodine value 1.5; titer 55.0°C; color Lovibond (5 1/4) 0.5R-2Y; unsap. 0.4%; moisture 0.2%; approx 7.5 lbs/gal; odor bland.
Uses: Paper coating; recording cylinders; cosmetics; candles; buffing compounds; esters; plasticizers; shaving cream.

"Neo-Fat" 18-59. Proprietary name for rubber grade stearic acid.
Properties: Acid value 198; iodine value 8; titer 56°C; color Lovibond (1") 4.0R-20Y; unsap. 1.5%; moisture 0.3%; approx 7.5 lbs/gal.
Uses: Rubber compounding; buffing compounds; recording cylinders; water proofing cement.

"Neo-Fat" 18-61. Proprietary name for mixture of stearic - palmitic acids.
Properties: Acid value 202; iodine value 2; titer 62°C; color Lovibond (5 1/4) 1.0R-10Y; unsap. 0.4%; moisture 0.2%; approx 7.5 lbs/gal.
Uses: Chemical raw material; polishes; buffing compounds; water proofing cement.

"Neo-Fat" 55. Proprietary name for distilled palm fatty acid.
Properties: Acid value 207; iodine value 50; titer 45°C; color Lovibond (5 1/4) 1.0R-10Y; unsap. 0.5%; moisture 0.5%; approx 7.5 lbs/gal.
Uses: Chemical raw material; chemical specialties; soaps; polishes.

"Neo-Fat" 65. Proprietary name for distilled animal fatty acids.
Properties: Acid value 202; iodine value 60; titer 41°C; color Lovibond (1") 2R-20Y; unsap. 1.5%; moisture 0.5%; approx 7.5 lbs/gal.
Uses: Chemical raw material; lubricating greases; soaps; chemical specialties.

"Neo-Fat" 92-04. Proprietary name for low titer crystallized distilled white oleic acid.
Properties: Acid value 199; iodine value 93; titer 3°C; color Lovibond (5 1/4) 1R-10Y; unsap. 1.0%; moisture 0.3%; approx 7.5 lbs/gal.
Uses: Chemical raw material; self-polishing wax compounds; esters; plasticizers; soaps.

"Neo-Fat" 94-04. Proprietary name for low titer crystallized red oil.
Properties: Acid value 199; iodine value 93; titer 9°C; color Lovibond (1") 2R-15Y; unsap. 1.0%; moisture 0.3%; approx 7.5 lbs/gal.
Uses: Chemical raw material; self-polishing wax compounds; esters; plasticizers; ore flotation; soaps.

"Neo-Fat" 94-10. Proprietary name for 8-11 titer crystallized red oil.
Properties: Acid value 198; iodine value 93; titer 9°C; color Lovibond (1") 2R-15Y; unsap. 1.0%; moisture 0.3%; approx 7.5 lbs/gal.
Uses: Chemical raw materials; self-polishing wax compounds; esters; plasticizers; ore flotation; soaps.

"Neo-Fat" 255. Proprietary name for stripped coco fatty acids.
Properties: Acid value 255; iodine value 12; titer 28°C; color Lovibond (5 1/4) 1.5R-10Y; unsap. 1.0%; moisture 0.5%; approx 7.5 lbs/gal.
Uses: Soaps; shampoos; cosmetics; esters; plasticizers.

"Neo-Fat" 263. Proprietary name for mixture of lauric and myristic acids, 50-50 blend.
Properties: Acid value 262; iodine value 1.0; titer 35°C; color Lovibond (5 1/4) 1R-10Y; unsap. 0.2%; moisture 0.2%; approx 7.5 lbs/gal.
Uses: Alkyd resins; wetting agents; soaps; cosmetics; insecticides; metallic soaps; chemical raw material; specialty lubricating greases.

"Neo-Fat" 265. Proprietary name for distilled coco fatty acids.
Properties: Acid value 265; iodine value 11; titer 24°C; color Lovibond (5 1/4) 2.0R-15Y; unsap. 0.5%; moisture 0.3%; approx 7.5 lbs/gal.

Uses: Alkyd resins; soaps; shampoos; esters.

"Neo-Fat" 270. Proprietary name for a mixture of lauric and myristic acids, 70-30 blend.

Properties: Acid value 270; iodine value 1.0; titer 33°C; color Lovibond (5¼) 1R-5Y; color stab. Lovibond (5¼) 3.5R-35Y; unsap. 0.2%; moisture 0.2%; approx 7.5 lbs/gal.

Uses: Alkyd resins; wetting agents; soaps; cosmetics; insecticides; metallic soaps; chemical raw material; specialty lubricating greases.

"Neo-Fat" 280. Proprietary name for a mixture of capric-myristic fatty acids.

Properties: Acid value 278; iodine value 1.0; titer 35°C; color Lovibond (5¼) 1.0R-10Y; color stab. Lovibond (5¼) 3.5R-35Y; unsap. 0.2%; moisture 0.2%; approx. 7.5 lbs/gal; Gardner 1.

Uses: Alkyd resins; wetting agents; cosmetics; insecticides; metallic soaps; chemical raw material; specialty lubricating greases.

"Neo-Fat" 360. Proprietary name for mixture of caprylic-capric acids.

Properties: Acid value 361; iodine value 1; titer 5°C; color Lovibond (5¼) 2R-20Y; unsap. 0.5%; moisture 0.5%; approx 7.5 lbs/gal.

Uses: Chemical raw material; perfumes; flavors; alkylolamides; flotation; paper coating; metal salts; detergents; fatty alcohols.

"Neofinish." [159] Trade name for a non-ionic compound dispersible in hot water. Used as a softener for all textile fibers.

"Neofolione." [227] Trademark for methyl 2-nonenoate. $CH_3(CH_2)_5CH{:}CHCOOCH_3$, 98.5% pure.

Properties: Colorless to slightly yellow liquid; stable; will not discolor. Sp. gr. (25/25°C) 0.893-0.898; refractive index (20°C) 1.440-1.444; clearly soluble in 4 parts of 70% alcohol.

Uses: In high class perfumes, in small quantities, where it will create a violet-leaf character.

"Neo-Germ-I-Tol." [430] Trade name for a high-alkyl dimethyl benzyl ammonium chloride. Used in the preparation of sanitary maintenance products; effective in hard water.

neohexane (2,2-dimethylbutane) C_6H_{14} or $C_2H_5C(CH_3)_3$.

Properties: Colorless volatile liquid; b.p. 49.7°C; refractive index 1.3659 (25°C); sp. gr. 0.6570 (25°C); freezing point −99.7°C; flash point −32°C; characterized by a very high octane rating in internal-combustion engines.

Derivation: By the thermal or catalytic union (alkylation) of ethylene and isobutane, each of which is recovered from refinery gases resulting from the cracking of petroleum.

Grades: 95%, 99%, research.

Containers: Bottles; drums.

Use: As a component of motor and aviation fuels with very high octane ratings.

Shipping regulations: Flammable liquid. Red label. *

"Neo-Iopax." [321] Brand name for sodium iodomethamate.

"NeoLine." [204] Trademark for a liquid neoprene, externally catalyzed, for coating immersed surfaces of steel and concrete. Applied by brush, spray or roller for heat curing or air curing. Resists alkalies, moderate concentrations of acids, except oxidizing agents, and many solvents up to 220°F. Shipped in 1-gal and 5-gal cans.

"Neolith." [434] Brand name for litharge in a ceramic grade. Reddish-brown pellets; relatively dust-free; free flowing; 5% max on 20 mesh, 5% max through 250 mesh. Poison!

Containers: 50-, 100-lb bags; 700-lb drums; railroad hopper cars.

Uses: Ceramics; glass; allied industries.

"Neolyn." [266] Trademark for a series of soft or medium hard modifying resins. For use in adhesives, lacquers, organo-sols, plastisols, and floor tile.

"Neomerpin." [28] Trademark for a line of surface active agents based on aromatic sulfonic acid.

Properties: Yellow, translucent liquid.

Use: As a wetting, scouring, and emulsifying assistant for the textile and leather industries.

neomycin. An antibiotic; a metabolic product of Streptomyces fradiae. It consists of a family of at least four related compounds, neomycins A, B, and C, and fradicin. However the marketed form (as the sulfate) appears to be a single compound. The chemical structure is believed to be similar to that of streptomycin. Neomycin is active against gram-positive, acid-fast and gram-negative bacteria. It is most widely used for treating certain types of skin infections. It is stable and active in alkaline solution.

Use: Medicine (as the sulfate salt).

neomycin sulfate. The sulfate salt of neomycin (q.v.).

Properties: White to slightly yellow crystals or powder; odorless or practically odorless; hygroscopic. Solutions are dextrorotatory. Very soluble in water, very slightly soluble in alcohol and insoluble in acetone, chloroform and ether.

Grades: U.S.P. XVI; commercial.

Use: Medicine (antibiotic); cosmetic, textile, paper industries.

neon Ne.

Properties: Colorless wholly inert gas; does not combine chemically with any element. An element of atomic number 10, group 0 of the periodic system. Liquefies at −245.92°C.

Derivation: By fractional distillation of liquid air. It constitutes 0.0012% of normal air.

Grades: Technical; highest purity.
Containers: Technical, steel cylinders; H. P., hermetically sealed glass flasks.
Use: Filling luminescent electric tubes and photoelectric bulbs.
Shipping regulations: Nonflammable gas. Green label. *

neonicotine. See anabasine.

neopentane (2,2-dimethylpropane; tetramethylmethane) C_5H_{12} or $C(CH_3)_4$. Present in small amounts in natural gas.
Properties: Colorless gas or very volatile liquid; b. p. 9.5°C; sp. gr. 0.613 (0/0°C); m. p. −20°C; soluble in alcohol; insoluble in water.
Grades: Technical 95%, pure 99%, research 99.9%.
Containers: Cylinders under low pressure.
Shipping regulations: Flammable liquid. Red label. *

neopentyl glycol $HOCH_2C(CH_3)_2CH_2OH$.
Properties: White, crystalline solid; boiling range, 95% between 204-208°C (760 mm); m. p. 120-130°C.
Containers: Fiber drums.
Use: Polyester foams.

"Neopones." [449] Trademark for a series of nonionic surface active agents useful as wetting agents, foaming agents, dye retardants and detergents for automatic laundry machines.

neoprene. A type of elastomers based on polymers of 2-chlorobutadiene-1,3.
Properties: Creamy white or amber chips, or dark brown putty-like solids; sp. gr. 1.23.
Containers: 50-lb bags.
Uses: In oil-, solvent-, heat-, and weather-resistant resilient products; for quick-setting, high-strength adhesive cements; paints and putties, for lining tanks and chemical equipment; crepe soles for shoes; binder for rocket fuels.

neoprene latices. Water emulsions of polymerized 2-chlorobutadiene-1,3. White milky liquids, with solid content running from 34 to 60%. Sp. gr. 1.06-1.15.
Containers: Drums; tank cars; tank trucks.
Uses: See under neoprene.

"NeoPrime A, B, C." [204] Trademark for chlorinated rubber base primers, used with liquid neoprene coatings to bond the neoprene coating to the underlying surface.

"Neoprontosil." [162] Trademark for azosulfamide.

neopyrithiamine. See pyrithiamine.

"Neo-Silvol." [330] Trademark for colloidal silver iodide.

"Neosol" Solvent. [125] Trademark for ethyl alcohol proprietary solvent based on a formulation approved by the Bureau of Internal Revenue.
Authorized composition: Specially denatured alcohol No. 1 (190 Proof), 100 parts by volume; methyl isobutyl ketone, 1 part by volume; ethyl acetate, 1 part by volume; aviation gasoline, 1 part by volume.
Typical properties: Reid vapor pressure (psi at 100°F), 1.7; residual odor, none; lbs/U. S. gal at 60°F, 6.78; flash point 45°F.
Containers: 55-gal non-returnable drums; 6000-, 10,000-gal tank cars; tank trucks.
Uses: Solvent for shellac and other spirit varnishes; latent solvent in nitrocellulose lacquers; solvent for adhesives and coating compounds, inks, and spot remover formulations.
Warning! Flammable. Poisonous if swallowed. Avoid contact with skin. Protect eyes against splashes and maintain vapor concentrations at comfort levels.
Shipping regulations: Flammable liquid. Red label. *

"Neo Spectra." [133] Trademark for a series of jet impingement carbon blacks for automotive enamels and all types of applications requiring high jetness. Available as:
"Neo Spectra Mark I." For specialty applications requiring very high jetness in powder form only.
Container: 25-lb bags.
"Neo Spectra Mark II." Standard black for top quality enamels and lacquers. High gloss and blackness, fast dispersion vehicle seeking.
Containers: 25-lb bags (powder form) and 5-lb bags (bead form).
"Neo Spectra III." Medium high color, combines excellent quality and economy for enamels, lacquers, synthetic fibers and plastics.
Container: 10-lb bag (powder form only).

neostigmine bromide (3-dimethylcarbamoxyphenyltrimethylammonium bromide) $(CH_3)_2NCOOC_6H_4N(CH_3)_3Br$.
Properties: White, crystalline powder; odorless and of bitter taste; m. p. 167°C (dec); very soluble in water; soluble in alcohol; practically insoluble in ether.
Grades: U. S. P. XVI.
Use: Medicine.

neostigmine methylsulfate $(CH_3)_2NCOOC_6H_4N(CH_3)_3 \cdot SO_4CH_3$.
Properties: White crystalline powder; odorless; bitter taste; soluble in water and less in alcohol; m. p. 142-145°C.
Grades: U. S. P. XVI.
Use: Medicine.

"Neo-Synephrine" Hydrochloride. [162] Trademark for phenylephrine hydrochloride.

"Neotex." [133] Trademark for furnace carbon blacks used in rubber, printing inks and protective coatings. Characterized by low structure and low oil absorption. Available in three grades, "Neotex" 100, "Neotex" 130, and "Neotex" 150.
Containers: 25- and 50-lb bags; hopper cars.

"Neothane." [265] Trademark for a series of solid polyurethane elastomers.
Uses: Solid tires; shoe soles and heels;

industrial rolls, gears, bearings, belts, bumpers, gaskets, seals, mounts, pedals, etc.

"Neo-Tone Sour." [244] Proprietary product consisting chiefly of fluorine compounds and complex phosphates.
Properties: White granular material; soluble in water; neutralizing value, 21.3 oz sodium bicarbonate per lb.
Containers: 150-lb and 300-lb net fiber drums.
Uses: Laundry sour of the iron-removing type.

"Neotran." [233] Trademark for insecticidal preparations with bis(p-chlorophenoxy)-methane as active ingredient.

"Neowet." [159] Trade name for a slightly yellowish somewhat viscous liquid surface active agent having a faint color and containing 33.3% active ingredients. It is a complex polyethylene ether and is non-ionic.
Uses: For use with enzymatic desizing agents and for general wetting purposes.

"Neowet X." [159] Trade name for organic ether sulfonate.
Properties: Water-white, slightly viscous liquid with slight ethereal odor. Water soluble and contains 20% active ingredients.
Uses: Detergent; surface active agent.

"Neozone." [28] Trademark for a line of rubber antioxidants.
"Neozone" A. Phenyl-alpha-naphthylamine, $C_{10}H_7NHC_6H_5$. Yellow pellets which turn dark purple on exposure to air or light; sp.gr. 1.22; f.p. not lower than 50°C.
"Neozone" D. Phenyl-beta-naphthylamine, $C_{10}H_7NHC_6H_5$. Fine light gray powder; sp.gr. 1.24; f.p. not lower than 106°C.
Containers: (A) 250-lb, (D) 160-lb drums.
Uses: For natural and synthetic rubbers; (D) stabilizer in manufacture of SBR.

"Neozyme." [159] Trademark for a combination of proteolytic and amylolytic enzymes, together with a small proportion of a fat-splitting enzyme.
Uses: Desizing fabrics sized with starch or gelatin or a combination of both.

nepheline. See nephelite.

nepheline syenite. An igneous rock composed mainly of feldspar and nephelite, being high in alumina and quartz-free. Magnetite and mica are sometimes present.
Occurrence: Ontario; less important amounts in Arkansas; New Jersey.
Uses: Extensively used in glass manufacture; pottery; porcelain; roofing; enamels.

nephelite (nepheline) $(Na, K)(Al, Si)_2O_4$. Essentially a silicate of sodium, found in silica-poor igneous rocks.
Properties: Colorless, white, yellowish; luster vitreous to greasy; hardness 5.5-6; sp. gr. 2.55-2.65. Eleolite is a massive variety with a greasy luster.
Occurrence: U.S.S.R.; Ontario; Norway; South Africa; Maine, Arkansas, New Jersey.
Uses: Ceramic and glass manufacture; enamels; source of potash.

nephrite. See jade.

"Nepoxide." [41] Trade name for a synthetic resin coating of the epoxy type which exhibits excellent adhesive properties and resistance to general chemicals and solvents. Can be deposited in high film thickness.

"Neptazane." [315] Trademark for methazolamide (q. v.).

neptunium Np. A synthetic element, having atomic number 93, first formed by bombarding uranium with high-speed deuterons. Several isotopes have since been prepared ranging in half-lives from 7.3 minutes to about 2 million years and in mass numbers from 231 to 241. Neptunium 237, the longest-lived isotope, has been found naturally in extremely small amounts in uranium ores. It is not believed to be primeval but to be formed by the action of stray neutrons on uranium. Neptunium 237 is produced in weighable amounts as a by-product in the production of plutonium 239.

Metallic neptunium is obtained by first preparing neptunium trifluoride, which is reduced with barium vapor at 1200°C. It is a silvery white metal; m.p. 640°C; sp. gr. 17.7. Neptunium is similar chemically to uranium, forming analogous compounds such as NpF_3, NpF_6, NpF_4, NpO_2, Np_3O_8, etc.

neptunium decay series. The series of short lived and little known elements produced as successive intermediate products when the elements uranium 237 and plutonium 241 undergo radioactive disintegration through neptunium 237 and finally into bismuth 209. The latter is not radioactive and thus is the end element of the series.

"Neran." [206] Brand name for a peptized casein product used for seasoning light leathers.

nerol $C_{10}H_{17}OH$. The trans isomer of geraniol.
Properties: Colorless liquid; rose-neroli odor.
Derivation: Iodization of geraniol with hydriodic acid, followed by treatment with alcoholic soda.
Containers: Glass bottles; drums.
Use: Perfumery.
Shipping regulations: None.*

nerolidol $C_{15}H_{26}O$. A sesquiterpene alcohol.
Properties: Straw-colored liquid with an odor similar to rose and apple. Sp. gr. 0.878; refractive index 1.480-1.482; stable in air; soluble in alcohol.
Occurrence: Found in cabreuva oil, balsam Peru, and oils of orange flower, neroli, sweet orange, and ylang ylang.
Use: Perfumery.

nerolin. See beta-naphthyl ethyl ether.

nerolin II. See beta-naphthyl ethyl ether.

neroli oil. See orange flower oil.

*See "I.C.C. Shipping Regulations," page xiii.
Reference numbers refer to name of manufacturer. See "List of Manufacturers," page v.

"Nerolon." [19] Brand name for beta-naphthyl methyl ketone, an aromatic chemical used primarily in soap.

"Nerone." [227] Trademark for a synthetic neroli ketone.
Properties: Yellow liquid, having a fresh leafy odor, recalling petitgrain oil; sp. gr. 0.910-0.915 (25/25°C); refractive index (n 20/D) 1.4650-1.4750; flash point TCC above 212°F. Soluble in 9 parts of 70% alcohol, 2 parts of 80% alcohol.
Occurrence: Not found in nature.
Uses: In formulating mimosa, new mown hay, and similar types; also in intensifying castoreum and oakmoss effects.

"Nerosol." [342] Trademark for a blend of sesquiterpenic alcohols used in perfumery for peppery note.

nerve gases (nerve poisons). Highly toxic chemical warfare agents developed in Germany during World War II. Structurally, they are organic derivatives of phosphoric acid (principally, alkyl phosphates, fluorophosphates, and thiophosphates). Like the insecticides diisopropyl fluorophosphate, OMPA, parathion, and tetraethyl pyrophosphate, they inhibit the enzyme cholinesterase and cause acetylcholine poisoning and the resulting cessation of nerve transmission. The nerve gases are colorless, odorless, tasteless liquids of low volatility. They are absorbed rapidly through the eyes, lungs, or skin and are approximately 10 to 100 times as toxic to man as hydrogen cyanide. Atropine sulfate is used in the treatment of nerve gas poisoning. The principal German nerve gases were Sarin, Soman, and Tabun (q.v.). Many recent pesticides have the same general structure.

nerve poisons. See nerve gases.

nerve root. See cypripedium.

Nessler's reagent. Solution of mercuric iodide in potassium iodide, used in detecting the presence of ammonia, particularly in very small amounts.

Neuberg blue. A mixture of copper blue (powdered azurite) and an iron blue (Prussian blue). It can be more easily ground in oil than pure copper blue.

"Neubrite." [428] Trademark for ammonium and zinc silicofluorides containing optical brightener and an antichlor.
Properties: Yellow, free-flowing, dustless granules.
Uses: As neutralizer and whitener in laundering.

"Neufume DMX." [328] Brand name for a cationic dispersion of an aromatic amine, used in the textile industry as an acetate fume-proofing agent; substantive and durable in liquid or paste form.

neurine $CH_2{:}CHN(CH_3)_3OH$ (trimethylvinylammonium hydroxide). A poisonous ptomaine formed during putrefaction by the dehydration of choline.
Properties: Syrupy liquid; fishy odor; absorbs carbon dioxide from the air; soluble in water and alcohol; very poisonous!
Use: Biochemical research.

"Neusol." [428] Trademark for ammonium and zinc silicofluorides, containing antichlor.
Properties: Light blue, granular, dustless and non-caking.
Uses: To mothproof and neutralize fabrics in laundering.

"Neutral 50." [244] A proprietary product consisting of synthetic detergents, complex phosphates, and silicates.
Properties: A light buff powder; soluble in water. Total alkali as Na_2O, 13.6%; active Na_2O, 3.1%.
Containers: 275-lb fiber drums; 100-lb fiber drums; 5-lb cans, 6 to a case; 20-lb galv. steel pail.
Uses: For dairy, dishes, and general hand cleaning where a mildly alkaline non-abrasive cleaner is used.
Fire hazard: None.
Shipping regulations: None.*

"Neutralite." [184] A granular calcite employed for increasing the pH of water.

neutralization. In everyday language the chemical reaction between an acid and a base in which they are changed into a salt and water. In farming, soil acids are treated with hydrated lime (calcium hydroxide, a base). The formic acid of certain insect bites is neutralized with ammonia, a base.

A more precise definition of neutralization is that it is the reaction between hydrogen ion from an acid and hydroxyl ion from a base to produce water.
See acid; base.

neutral oils. Lubricating oil of medium or low viscosity obtained by distillation and dewaxing of crude petroleum or its cracking products.

neutral red (toluylene red) $(CH_3)_2NC_6H_3N_2C_6H_2CH_3NH_2 \cdot HCl$ (tricyclic).
Properties: Green powder; dissolves in water or alcohol to give red color.
Use: Acid-base indicator in the range pH 6.8-8.0 (red in acid, yellow brown in alkali). See indicators.

neutral soap. See soap, hard.

neutral sodas. See sodas, modified.

neutral spirits. A name for ethyl alcohol.

"Neu-Tri." [233] Trademark for trichloroethylene (q.v.).

"Neutrigan." [307] 100% proprietary compound.
Properties: Fine, nearly white powder; density 1.25; soluble in water.
Uses: Neutralizing agent for chrome tanned leather. Permits fuller yields with acid and direct colors and better absorption of fat liquor by the leather.

neutrino. A subnuclear particle whose existence was first hypothecated to satisfy the laws of

*See "I.C.C. Shipping Regulations," page xiii.
Reference numbers refer to name of manufacturer. See "List of Manufacturers," page v.

conservation of energy and momentum in nuclear transformations involving electrons. Its existence has since been proven by direct observation. The neutrino has no charge and probably no rest mass and thus shows almost no interaction with matter, making any direct experimental observation of the particle extremely difficult.

See fundamental particle.

"Neutrol." [217] Brand name of acid activated clay used as decolorizing adsorbent for vegetable and animal fats and oils.

See also "Filtrol."

"Neutrolene." [244] Trademark for a series of non-ionic, moisture free, fatty oils.

Containers: Non-returnable steel drums averaging 400-425 lbs net.

Uses: As a fatliquor in the leather industry. Also may be used as an emulsifier for raw oils.

"Neutroleum." [188] Trademark for an all-purpose deodorizing agent. Claim is made that if properly used it will completely and permanently deodorize liquid insecticides, waxes, polishes, glues, linoleums, ink, paints, varnishes, cleaner's naphtha, para blocks, naphthalene, turpentine substitutes, solvents, and diluents, including lacquer diluents, petroleum, and solvents.

"Neutrolox." [204] Trademark for a very high grade ammonium chloride prepared for use in textile finishing plants.

Properties: Fine white powdery material; extremely soluble in water and solutions containing customary concentrations; pH 6-7.

Use: To neutralize textiles containing caustic soda from mercerizing, scouring, or bleaching operations, giving the textiles so treated a pH of 6-7 after drying.

neutron. A subnuclear particle having very nearly the same mass and size as the proton, but without an electric charge. Free neutrons are unstable and undergo radioactive decay to protons. Neutrons appear to be stable, however, when bound in atomic nuclei, and the latter are now thought to be composed exclusively of protons and neutrons, stable assemblages when the total number of particles and the neutron/proton ratio lies within certain limits. Neutrons interact with almost all nuclei and are absorbed. Since the neutron has no charge and therefore experiences no repulsive forces by atomic or nuclear charges it is readily transmitted through and absorbed by matter. A nuclear reactor is a copious source of neutrons. Depending on the kinetic energy that they may have they are classed as fast, epithermal, and thermal, or slow.

See fundamental particle.

"Neutronyx." [328] A series of compounds which are proprietary surface-active non-ionic agents, used as emulsifying, wetting, dispersing agents and detergents. The 300 series are fatty acid esters of polyethylene glycol and the 600 series are polyoxyethylated alkyl phenols.

"Neutronyx" 834: Oil-soluble emulsifying agent.

"Neutronyx" 330: Water-soluble, coupling and auxiliary emulsifier.

"Neutronyx" 331: Water-soluble, coupling and auxiliary emulsifer.

"Neutronyx" 332: Water-dispersible, thickening agent.

"Neutronyx" 333: Water-soluble, foaming agent.

"Neutronyx" 600: Oil- and water-soluble emulsifer; detergent.

"Neutronyx" 611: Water-soluble, low foam; low cloud; detergent.

"Neutronyx" 640: Water-soluble; extremely high cloud point.

"Neutroscents." [188] Trademark for a series of perfumes designed particularly to cover objectionable odors; available in water-soluble form for sprays, air-conditioning apparatus, and other dispersion devices. Also available in a highly concentrated form for incorporation into technical products.

"Nevillac." [21] Brand name for a series of (alkyl) hydroxy resins. Used in adhesives, lacquers, paper coatings, special inks and varnishes.

Neville and Winther's acid. See 1-naphthol-4-sulfonic acid.

"Nevillite." [21] Trademark for water-white cycloparaffin resin.

Uses: Pressure-sensitive adhesives, white rubber products, special paper coatings, polyethylene modifier.

"Nevindene." [21] Brand name for high melting coumarone-indene resins of extreme hardness used for dental compounds, fast-drying varnishes, rotogravure inks, aluminum paints and insulating compounds.

"Nevinol." [21] Brand name for a plasticizing and solvent oil used as a stable plasticizer for resins and gums; also used in fly paper, adhesives, inks, aluminum pastes, waterproofing compounds, and rubber-resin finishes.

new green. See copper acetoarsenite.

Newport "S." [79] Trade name for a pale terpene resin.

Properties: Color X-WG; m. p. (capillary tube) 100°C; m. p. (ball and ring) 118°C; sp. gr. (25°C) 1.03. Petroleum soluble type.

Containers: Non-returnable, light gauge metal drum containing approx 500 lbs net.

Uses: Adhesives, stiffening agent for textiles.

Newport "V40." [79] Trade name for a pale terpene resin.

Properties: Color X-WG; m. p. (capillary tube) 100°C; m. p. (ball and ring) 118°C; sp. gr. (25°C) 1.05. Alcohol soluble type.

Containers: Non-returnable, light gauge metal drums containing approx 500 lbs net.
Uses: Paper coatings; floor coatings; in half-second butyrate coatings; etc.

"Newport White" Pine Oil. [79]
Properties: Color, white; sp. gr. (15.5°C) 0.934; refractive index (20°C) 1.4820; polymerized residue 0.4%; flash point (open cup) 173°F; Engler distillation 5%, 204°C; 50%, 213°C;
Containers: 55-gal drums; tank cars.
Uses: Mining-flotation; textile dyeing and cleaning; laundries; disinfectants; insecticides; deodorants; cleaning compounds; coated paper; paint and varnish; paint (casein); pharmaceuticals.

new silver. See German silver.

Newtonian flow. A term used in rheology to describe a type of flow occurring in a liquid system where the rate of shear is directly proportional to the shearing force. It can occur, ideally, under the influence of an infinitesimally small force. Mineral oils, at low rates of shear, exhibit Newtonian flow. When rate of shear is not directly proportional to the shearing force, flow is described as non-Newtonian. In general, Newtonian flow is exhibited by relatively stiff, plastic-like materials.

Newton's alloy. See table under fusible alloys.

N. F. Abbreviation for National Formulary, an official list of drugs published by the American Pharmaceutical Association. The latest edition at the time this dictionary was written was the 11th, noted as N. F. XI.

"N-Glo-5," "N-Glo-5-Y." [79] Trade names for two gloss oils, i. e., solutions of limed rosin in mineral spirits.
Properties:

	"N-Glo-5"	"N-Glo-5-Y"
Acid value (on solution)	44	46
Concentration (total solids)	61%	64%
Viscosity (Gardner-Holdt)	K	Y
Color (Hellige)	8-9	8-9
Per cent lime	5%	5%

Containers: 55-gal drums; tank cars.
Uses: Paint and varnish; sizing varnish.

Ni. Symbol for nickel.

"Niacet." [214] Trademark for vinyl acetate and various metallic acetate salts, including aluminum formoacetate, copper acetate, potassium acetate, sodium acetate, sodium diacetate, zinc acetate and "Niaproof" aluminum acetate.

"Niacide." [55] Trademark for fungicidal products containing dimethyl dithiocarbamates used mainly for scab control.

niacin. See nicotinic acid.

niacinamide. See nicotinamide.

"Niagaramite." [55] Trademark for miticide containing 15% aramite (q. v.).

nialamide $C_5H_4NCO(NH)_2(CH_2)_2CONHCH_2C_6H_5$. Designated chemically as 1-(2-(benzylcarbamyl) ethyl)-2-isonicotinoylhydrazine. It is a white, crystalline powder of low solubility in water and good solubility in slightly acid solutions. It is stable in crystalline form, suspension and solution. Used in medicine as an amine oxidase inhibitor.

"Niamid." [299] Trademark for nialamide.

"Niaproof." [214] Trademark for a water-repellent compound. Substantially a soluble basic aluminum acetate salt.
Properties: Fine white powder readily soluble in water. Aluminum oxide (Al_2O_3) value 35.5-37%; sulfates, chlorides, trace; insoluble 0.2% (max); pH in water solutions 4.7-4.8.
Grades: Technical.
Containers: 25-lb non-returnable Fiberpaks; 200-lb non-returnable Leverpaks.
Uses: Source of aluminum ion for water-repellent finishes for textile, paper, and leather products, particularly in processes using wax or soap emulsions.

"Niatex." [214] Trademark for antistatic agents. (Antistatic Agent AG-2).
Properties: Viscous, light-colored liquid; 20% active aqueous solution; gives a durable antistatic finish to synthetic fibers and fabrics.

"Niatox." [55] Trademark for a line of DDT sprays and dusts.

"Niax." [214] Trademark for a series of polymeric propylene oxide polyols having molecular weights from 300 to 6000 and hydroxyl groups varying from 2 to 6 and for a series of catalysts for polyurethane foams and resins.
"Niax" PPG diols 425, 1025, 2025, 3025, 4025. Adducts of propylene oxide to dipropylene glycol; colorless liquids; 425 is water soluble; 1025 through 4025 are water insoluble. The numbers are the average molecular weights.
"Niax" triols LHT-240, LHT-112, LHT-67, LHT-42, LHT-34, LHT-28. Adducts of propylene oxide to 1,2,6-hexanetriol. Colorless liquids, water insoluble. The numbers are the average hydroxyl number.
"Niax" triols LG-168, LG-56, LM-52. Adducts of propylene oxide to glycerine. Colorless liquids, water insoluble; numbers are the average hydroxyl number.
Containers: 1-, 5-, 55-gal drums (8, 40 and 460 lbs); tank cars up to 10,000 gal.
Uses: Intermediates for flexible, semi-rigid and rigid polyurethane foams; polyurethane elastomers, and coating resins.
"Niax" catalyst D-22. Dibutyl tin dilaurate (q. v.).

"Ni-Bral." [283] Trademark for a nickel aluminum bronze containing approximately 5% nickel, 9.5% aluminum, 3-4% iron, 1.5-3.0% manganese, balance copper.

This alloy provides tensile strengths over 90,000 psi and yield strengths over 45,000 psi and about 18% elongation. Because of its superior resistance to corrosion-erosion in sea water it is preferred for heavy duty ship propellers and sea water pumps, valves, etc.

"Nibrite." [134] Trademark for brightener for nickel plating solutions.

"Nicarb." [123] Trademark for nicarbazin for use as a coccidiostat. Nicarbazin is an equimolar complex of 4,4'-dinitrocarbanilide and 2-hydroxy-4,6-dimethylpyrimidine. It forms crystals, decomposes at 265-275°C; insoluble in water.

nicarbazin. See "Nicarb."

niccolite (arsenical nickel) NiAs. Arsenic replaced to some extent by antimony or sulfur, and nickel by iron or cobalt.
Properties: Pale copper-red mineral with dark tarnish, metallic luster. Contains 43.9% nickel; soluble in concentrated nitric acid. Sp. gr. 7.3-7.67; hardness 5-5.5.
Occurrence: United States (Nevada, Connecticut, Michigan); Canada (Ontario, Newfoundland); Argentina; Germany; Portugal.
Use: Nickel ore.

"Nichrome." [350] Trademark for an alloy containing 60% nickel, 24% iron, 16% chromium, 0.1% carbon. It is used principally for electric resistance purposes. "Nichrome" castings, which contain 60% nickel, 25% iron, 15% chromium and 0.7% carbon, are resistant to cold sulfuric acid in all concentrations and to hot (not boiling) sulfuric acid in all strengths except concentrated (95%). It also offers good resistance to mine and sea waters and moist sulfurous atmospheres.

nickel Ni. An element of atomic number 28 in group VIII of the periodic system.
Properties: Hard silvery metal; takes a high polish; sp. gr. 8.908; m.p. 1455°C; good resistance to corrosion; attacked only very slowly by hydrochloric and sulfuric acids, more readily by nitric acid; very resistant to strong alkalies.
Occurrence: Chiefly as pentlandite at Sudbury, Ontario and garnierite in New Caledonia; Cuba; Norway; recent discoveries in U.S.S.R. and in the Northwest Territory of Canada.
Derivation: The sulfide ore is refined by flotation and roasting to sintered nickel oxide, and either marketed as such or reduced to metal which is cast into anodes and refined electrolytically or by the carbonyl process. (See Mond process.) Nickel is extracted from the Cuban iron ores by ammonia leaching.
Grades: Electrolytic; ingot; pellets; shot; sponge; powder.
Uses: Construction material, mainly in the form of alloys such as wrought and cast low-alloy steels, stainless steels, "Monel," and "Inconel" because of its strength and resistance to corrosion; magnetic alloys; electroplating; used in the alkaline (Edison) storage battery; finely divided nickel as a catalyst in organic syntheses.
Shipping regulations: Catalyst, spent, activated forms: Flammable solid. Yellow label. *

"330" Nickel. [283] Trademark for a wrought alloy containing at least 99% nickel.
Use: An anode material for vacuum tubes and other electronic components.

nickel acetate $Ni(OOCCH_3)_2 \cdot 4H_2O$.
Properties: Green, monoclinic crystals. Effloresces somewhat in air. Sp. gr. 1.74; decomposes on heating. Soluble in water and alcohol.
Derivation: (a) By heating nickel hydroxide with acetic acid in the presence of metallic nickel. (b) Interaction of nickel sulfate and lead acetate.
Grade: Technical.
Containers: Glass bottles; fiber cans; barrels.
Uses: Textiles (mordant); nickel plating.

nickel acetylacetonate $Ni[OC(CH_3){:}CHCOCH_3]_3$.
Properties: Green crystals; m.p. 228°C.
Derivation: Reaction of nickel chloride with acetylacetone and ammonia.

nickel alloys See iron-nickel alloys, also "Alumel," "Balco," "Chlorimets," "Chromel," "Chromel P," "Copel," "Cupron," "D" Nickel, "Duranickel," "Durco," "Evanohm," "Hastelloy," "Illium," "Inconel," "Inconel X," "K Monel," Lukens Monel-Clad Steel, Lukens Inconel-Clad Steel, Lukens Nickel-Clad Steel, "Monel," "Nichrome," "Ni-Span," "R Monel," "S Monel," and "Tophet A, C, and D."

nickel aluminide A cermet which can be flame-sprayed.

nickel-ammonium chloride (ammonium-nickel chloride) (a) $NiCl_2 \cdot NH_4Cl$; (b) $NiCl_2 \cdot NH_4Cl \cdot 6H_2O$.
Properties: (a) Yellow powder; (b) green crystals; sp. gr. 1.65. Soluble in water; deliquescent.
Grade: Technical.
Containers: Glass bottles; fiber cans.
Uses: Electroplating; dyeing (mordant).

nickel-ammonium sulfate (nickel salts, double; ammonium-nickel sulfate) $NiSO_4 \cdot (NH_4)_2SO_4 \cdot 6H_2O$.
Properties: Green crystals; decomposed by heat. Soluble in water; less in ammonium sulfate solution; insoluble in alcohol. Sp. gr. 1.929.
Derivation: An aqueous solution of nickel sulfate is acidified with sulfuric acid; then an aqueous solution of ammonium sulfate is added. On concentrating, crystals of the double sulfate separate out.
Method of purification: Recrystallization.
Grades: Technical; C.P.
Containers: Barrels; kegs; boxes; fiber drums.
Use: Nickel electrolyte for electroplating.
Shipping regulations: None. *

nickel arsenate (nickelous arsenate) $Ni_3(AsO_4)_2$.
Properties: Yellow-green powder; soluble in acids; insoluble in water. Sp. gr. 4.98.
Grade: Technical.
Containers: Fiber cans.
Use: Catalyst (hardening fats used in preparing soap).
Shipping regulations: Poison, class B. Poison label. *

nickel black. See nickelic oxide.

nickel bromide (nickelous bromide) (a) $NiBr_2$ (b) $NiBr_2 \cdot 3H_2O$.
Properties: (a) Brownish-yellow solid or yellow, lustrous scales. (b) Deliquescent, greenish scales. Soluble in water, alcohol, ether, and ammonium hydroxide. Sp. gr. (a) 4.64; m.p. (b) loses water of crystallization at about 200°C.
Derivation: By the action of hydrobromic acid on nickel oxide.

nickel carbonate $NiCO_3$.
Properties: Light-green crystals; insoluble in water; soluble in acids; m.p., decomposes.
Derivation: By the addition of sodium carbonate to a solution of nickel sulfate.
Containers: Barrels.

nickel carbonate, basic. Uncertain composition, represented variously as $NiCO_3 \cdot 2Ni(OH)_2 \cdot 4H_2O$ or $2NiCO_3 \cdot 3Ni(OH)_2 \cdot 4H_2O$.
Properties: Light green crystals or brown powder. Sp. gr. 2.6. Insoluble in water, soluble in ammonia and dilute acids.
Derivation: By the addition of soda ash to a solution of nickel sulfate.
Grade: Technical.
Containers: 100-lb drums; 400-lb barrels.
Uses: Electroplating; preparation of nickel catalysts; ingredient of ceramic colors and glazes.
Shipping regulations: None. *

nickel carbonyl (nickel tetracarbonyl) $Ni(CO)_4$.
Properties: Colorless, or yellow volatile liquid; flammable; poisonous! Dangerous to inhale! Soluble in alcohol and concentrated nitric acid; insoluble in water. Sp. gr. 1.3185; m.p. −25°C; b.p. 43°C; vapor decomposes at 60°C.
Derivation: By passing carbon monoxide gas over finely divided nickel.
Grades: Technical.
Containers: Iron cylinders.
Use: For production of metallic nickel by Mond process; gas plating.
Fire hazard: Dangerous!
Shipping regulations: Flammable liquid. Red label (by freight; not accepted by express). *

nickel chloride (a) $NiCl_2$ (b) $NiCl_2 \cdot 6H_2O$.
Properties: (a) Brown scales; deliquescent in moist air. (b) Green scales; deliquescent in moist air. Soluble in water and ammonium hydroxide. Sp. gr. (a) 2.56; m.p., sublimes.
Derivation: By the action of hydrochloric acid on nickel oxides.
Method of purification: Crystallization.
Grades: Technical; C. P.
Containers: 1-, 5-lb bottles; 275-lb barrels; 100-, 50-lb kegs; 25-lb boxes; fiber drums; multiwall paper sacks.
Uses: Nickel-plating cast zinc; manufacture of sympathetic ink; antiseptic; absorbent for ammonia gas in military and industrial gas masks.
Shipping regulations: None. *

nickel-chromium steels. A series of low-alloy and high-alloy steels, containing both nickel and chromium, characterized by hardness and toughness.
Typical low-alloy compositions: Ni 1.10-1.40, Cr 0.55-0.90%. Nickel is usually about twice the chromium content. High-alloy members are usually stainless steels.
See under steel.

nickel-cobalt sulfate (cobalto-nickelous sulfate).
Properties: Reddish-brown, crystalline mass. Soluble in water.
Grades: Technical.
Containers: Glass bottles; fiber cans.
Uses: Blackening brass, zinc; dyeing, printing (mordant).
Shipping regulations: None. *

nickel cyanide $Ni(CN)_2 \cdot 4H_2O$.
Properties: Apple-green plates or powder; poisonous! Soluble in ammonium hydroxide and potassium cyanide solution; insoluble in water and acids. M.p., loses $4H_2O$ at 200°C; b.p., decomposes.
Derivation: By adding potassium cyanide to a solution of a nickel salt.
Grade: Technical.
Containers: Wooden kegs; glass bottles; fiber drums; multiwall paper sacks.
Uses: Metallurgy; electroplating; galvanoplastic work.
Shipping regulations: Class B poison. Poison label. *

nickel dibutyldithiocarbamate $Ni[SC(S)N(C_4H_9)_2]_2$.
Use: Rubber compounding.
See zinc dibutyldithiocarbamate.

nickel formate $(HCOO)_2Ni \cdot 2H_2O$.
Properties: Green crystals; sp. gr. 2.15; soluble in water.
Grades: Technical.
Containers: Barrels; kegs; fiber drums.
Use: Production of nickel catalyst for hydrogenation.

nickel hydroxide (a) Nickelous $4Ni(OH)_2 \cdot H_2O$ (b) Nickelic $Ni(OH)_3$.
Properties: (a) Pale green powder; (b) black powder; (a) soluble in acids, ammonium hydroxide; insoluble in water and alkalies. Sp. gr. (a) 4.36; m.p. (a) decomposes; (b) decomposes.
Derivation: (a) By adding caustic soda to a solution of nickelous salt. (b) By adding a hypochlorite to a solution of a nickel salt.
Grades: Technical; C. P.
Containers: Wooden kegs; glass bottles; fiber drums.
Use: Nickel salts.

nickelic oxide (nickel peroxide; nickel sesquioxide; black nickel oxide; black nickel) Ni_2O_3.
Properties: Gray-black powder; soluble in acids; insoluble in water. Sp. gr. 4.84; m.p., is reduced to NiO at 600°C.
Derivation: By gentle heating of the nitrate or chlorate.
Grades: Technical; C.P.
Containers: 1-lb bottles; tins; barrels.
Use: Storage batteries.

nickel iodide (nickelous iodide) NiI_2 or $NiI_2 \cdot 6H_2O$ (loses water at 43°C).
Properties: Black, crystalline powder or blue-green crystals. Hygroscopic. Sublimes at 797°C without melting. Soluble in alcohol, water; sp. gr. 5.834.

nickel-iron alloys. See iron-nickel alloys.

"Nickel-Lume." [72] Trade name for organic bright nickel process; prepared from nickel sulfate, nickel chloride, boric acid and organic addition agents.

nickel matte. See matte.

nickel nitrate $Ni(NO_3)_2 \cdot 6H_2O$.
Properties: Green, deliquescent crystals. Keep well stoppered. Soluble in water and alcohol.
Constants: Sp. gr. 2.065; m.p. 56.7°C; b.p. 136.7°C.
Derivation: By the action of nitric acid on nickel oxide.
Method of purification: Crystallization.
Grades: Technical; reagent.
Containers: Glass bottles; drums.
Uses: Nickel plating; preparation of nickel catalysts; manufacture of brown ceramic colors.
Fire hazard: Dangerous.
Shipping regulations: Oxidizing material. Yellow label. *

nickel nitrate ammoniated (nickel nitrate tetrammine) $Ni(NO_3)_2 \cdot 4NH_3 \cdot 2H_2O$.
Properties: Green crystals. Soluble in water; insoluble in alcohol; decomposes in air.
Derivation: By adding ammonium hydroxide to a nitric acid solution of nickel nitrate, with subsequent crystallization.
Method of purification: Recrystallization.
Grade: Technical.
Containers: Glass bottles; wooden kegs; fiber drums.
Use: Nickel plating.
Fire hazard: Dangerous.
Shipping regulations: Oxidizing material. Yellow label. *

nickel nitrate tetrammine. See nickel nitrate ammoniated.

nickelous arsenate. See nickel arsenate.

nickelous bromide. See nickel bromide.

nickelous chloride. See nickel chloride.

nickelous iodide. See nickel iodide.

nickelous oxide. See nickel oxide.

nickelous phosphate. See nickel phosphate.

nickelous phosphate, tertiary. See nickel phosphate.

nickel oxide (nickelous oxide; nickel protoxide; green nickel oxide) NiO.
Properties: Green powder, becoming yellow; is found in nature as the mineral bunsenite. Soluble in acids and ammonium hydroxide; insoluble in water.
Constants: Sp. gr. 6.6-6.8; absorbs oxygen at 400°C forming Ni_2O_3 which is reduced to NiO at 600°C.
Derivation: By heating nickel hydroxide or nitrate.
Grades: Technical; C.P.
Containers: Wooden kegs; glass bottles; fiber drums; barrels.
Uses: Nickel salts; porcelain painting.
Shipping regulations: None. *

nickel oxide, black. See nickelic oxide.

nickel oxide, green. See nickel oxide.

nickel peroxide. See nickelic oxide.

nickel phosphate (nickelous phosphate; trinickelous orthophosphate; tertiary nickelous phosphate) $Ni_3(PO_4)_2 \cdot 7H_2O$.
Properties: Light-green powder. Soluble in acids, ammonium hydroxide; insoluble in water.
Grade: Technical.
Uses: Electroplating; making "nickel yellow."

nickel plating. An electrolysis process in which nickel is deposited onto another metal or other material by electrolysis of a nickel salt solution. Usually the solution contains both the chloride and sulfate. The plate is soft or hard, dull or shiny, etc., according to the pH and other conditions and according to what additives are present. Common additives are a wetting agent such as lauryl sulfate, which prevents pitting, and brightening agents which may be sulfonates or other organic compounds. Impure nickel ingots are also sometimes converted to pure nickel by a similar process during which nickel is dissolved from the ingot (which is made the anode), while deposition of pure metal takes place on a cathode.

It is estimated that over one-half billion square feet of surface are nickel plated per year. Chromium plating is almost always preceded by a layer of nickel plate. The layer of nickel ranges from 0.0001 to 0.06 inches depending upon the application.

Applications are for automobiles, appliances, furniture, tools, machine parts, hardware, plumbing fixtures, bicycles, scales, cameras, wire products, business machines, ornaments, utensils, musical instruments, radios, all types of electrical and electronic equipment, electrotype, jewelry, clocks, stoves. Electroforming is also used. Sometimes nickel alloys with cobalt or tungsten are plated.

nickel-potassium sulfate (potassium-nickel sulfate) $NiSO_4 \cdot K_2SO_4 \cdot 6H_2O$.
Properties: Blue-green crystals. Soluble in water. Sp. gr. 2.124.

nickel protoxide. See nickel oxide.

nickel-rhodium. Alloys containing nickel and from 25-80% of rhodium; but sometimes also some platinum, iridium, palladium, molybdenum, tungsten, copper, iron, or cobalt.
Uses: Electrodes; chemical apparatus; reflectors; pen points.

nickel salts, double. See nickel ammonium sulfate.

nickel salts, single. See nickel sulfate.

nickel sesquioxide. See nickelic oxide.

nickel silvers. Non-ferrous alloys of the following compositions: (a) Nickel silver 18% A contains 65% copper, 18% nickel and 17% zinc. It offers good resistance to cold dilute sulfuric and hydrochloric acids and to hot dilute sulfuric acids under certain conditions of operation. It is resistant to cold acetic acid in all concentrations and to hot (not boiling) acetic acid up to 10%, to sodium hydroxide under all conditions and to sea water and moist sulfurous atmospheres. (b) Nickel silver 18% B contains 55% copper, 18% nickel and 27% zinc. It offers approximately the same resistance to corrosion as the 18% A alloy.

nickel stannate $NiSnO_3 \cdot 2H_2O$.
Properties: Light colored crystalline powder; approx. temperature of dehydration 120°C.
Uses: Additive in ceramic capacitors.

nickel steel. When nickel is introduced in amounts up to approximately 5% the effect is to increase strength and hardness without a comparative decrease in ductility. Nickel steels are particularly suitable for case-hardening. Nickel markedly improves corrosion resistance. Along with aluminum it imparts age-hardening characteristics to iron. (1) Nickel strengthens unquenched or annealed steels, (2) toughens pearlite-ferritic (medium carbon) steels, especially at low temperatures, (3) renders high chromium-iron alloys austenitic.
See also iron-nickel alloys; steel.

nickel sulfate (nickel salts, single; blue salt) (a) $NiSO_4$; (b) $NiSO_4 \cdot 6H_2O$; (c) $NiSO_4 \cdot 7H_2O$.
Properties: (a) Yellow-green crystals; (b) blue or emerald green crystals; (c) green crystals. All the sulfates are soluble in water; (b) and (c) are soluble in alcohol; (a) is insoluble in alcohol and ether.
Constants: Sp. gr. (a) 3.418, (b) 2.031, (c) 1.98; m. p. (a) loses SO_3 at 840°C, (b) loses $6H_2O$ at 280°C, (c) 98-100°C.
Derivation: By the action of sulfuric acid on nickel oxide.
Grades: Technical; C. P.
Containers: Bags; 300-, 400-lb barrels; 100-lb kegs; 25-lb boxes; fiber drums; carloads.
Uses: Manufacture of nickel-ammonium sulfate; nickel catalysts; nickel plating; mordant in dyeing and printing textiles; blackening zinc and brass; paints and varnishes; ceramics.
Shipping regulations: None. *

nickel tetracarbonyl. See nickel carbonyl.

nickel-titanium.
Composition: Nickel, titanium, àluminum, iron, silicon.
Properties: M. p. 2700°F.
Uses: Titanium source for nickel base alloys.

nicometh. See methyl nicotinate.

"Nicon." [169] Trademark for diethyldithiocarbamate used in the colorimetric determination of nickel.

"Niconyl." [330] Trademark for isonicotinic acid hydrazide (isoniazid U. S. P.).

nicotinamide (niacinamide; nicotinic acid amide) $C_5H_4NCONH_2$. Same biological function as nicontinic acid (q. v.).
Properties: Colorless needles; m. p. 129°C; stability as for nicotinic acid; soluble in ethyl alcohol and water; bitter taste.
Sources: Synthetic made by conversion of nicotinic acid to the amide.
Grades: U. S. P. XVI.
Containers: Bottles; fiber drums.
Uses: Medicine; nutrition.
Also commercially available as nicotinamide hydrochloride.

nicotinamide adenine dinucleotide. Name recommended by the International Union of Biochemistry and IUPAC. (NAD; diphosphopyridine nucleotide; DPN; cozymase; coenzyme I; Co I; codehydrogenase I). $C_6H_6N_2O \cdot C_5H_8O_3 \cdot PO_3 \cdot O \cdot HPO_3 \cdot C_5H_8O_3 \cdot C_5H_4N_5$. A co-enzyme necessary for the alcoholic fermentation of glucose, and the oxidative dehydrogenation of other substrates.
Properties: A white hygroscopic powder; soluble in water; stable for about a week in aqueous solutions.
Source: Yeast. Commercially, is isolated from yeast and purified by ion-exchange chromatography.
Grades: 75% and 85% level of purity.
Use: Biochemical research; chromatography.

nicotinamide adenine dinucleotide phosphate. Name recommended by the International Union of Biochemistry and IUPAC. (NADP; triphosphopyridine nucleotide; TPN; phosphocozymase; codehydrogenase II; coenzyme II; Co II) $C_{21}H_{29}N_7O_{17}P_3$. The coenzyme of apozymase, necessary for the alcoholic fermentation of glucose, and the oxidative dehydrogenation of other substrates.
Derivation: From yeast. NADP is prepared by enzymes from NAD and purified by ion exchange chromatography.
Use: Biochemical research.
Commercially available as the sodium salt of the oxidized form.

nicotine $C_{10}H_{14}N_2$ or $C_5H_4NC_4H_7NCH_3$. (beta-Pyridyl-alpha-N-methylpyrrolidine).
Properties: Alkaloid from tobacco; thick

*See "I. C. C. Shipping Regulations," page xiii.
Reference numbers refer to name of manufacturer. See "List of Manufacturers," page v.

water-white levorotatory oil, turning brown on exposure to the air; poisonous! Hygroscopic; soluble in alcohol, chloroform, ether, kerosene, water, and oils. B.p. 247°C (dec); sp. gr. 1.00924.

Derivation: By distilling tobacco with milk of lime and extracting with ether.

Grades: Technical.

Containers: Glass bottles; tins.

Uses: Medicine; insecticide (horticultural purposes); tanning.

Shipping regulations: Poison, class B. Poison label. *

nicotine dusts.

Caution: Harmful if inhaled or swallowed. Avoid excessive exposure to dust; avoid contact with skin and eyes; wash thoroughly after handling. Store away from feed and foodstuffs. The percentage and form of nicotine present may call for poison labeling under various state laws. MCA warning label.

nicotine salts.

(a) Hydrochloride: $C_{10}H_{14}N_2 \cdot 2HCl$.

(b) Salicylate: $C_{10}H_{14}N_2 \cdot C_7H_6O_3$.

(c) Sulfate: $(C_{10}H_{14}N_2)_2 \cdot H_2SO_4$.

(d) Tartrate: $C_{10}H_{14}N_2 \cdot 2C_4O_6H_6 \cdot 2H_2O$.

Properties: (a) Colorless oil; poisonous! (b) White crystals; poisonous! (c) White crystals; poisonous! (d) White plates; poisonous! All the salts are soluble in water, alcohol and ether. M.p. (b) 117.5°C.

Derivation: By the action of the respective acid on the alkaloid.

Method of purification: Crystallization.

Grades: Technical (sulfate, 40% grade).

Containers: Glass bottles; for crystals fiber cans; for liquid non-returnable tins.

Uses: Medicine; insecticide (horticultural and general purposes).

Danger: Poisonous if swallowed; absorbed through skin. MCA warning label.

Shipping regulations: Poison, class B. Poison label. *

nicotinic acid (niacin; pyridine-3-carboxylic acid) C_5H_4NCOOH. The antipellagra vitamin, essential to many animals for growth and health. In man, nicotinic acid is believed necessary along with other vitamins for the prevention and cure of pellagra. It functions in protein and carbohydrate metabolism. As a component of two important enzymes, coenzyme I and coenzyme II, it functions in glycolysis and tissue respiration.

Properties: Colorless needles; odorless; m.p. 236°C; sublimes above melting point; sour taste; soluble in water and alcohol; insoluble in most lipid solvents; quite stable to heat and oxidation.

Units: Amounts of nicotinic acid are expressed in milligrams.

Sources: Food sources: meat, fish, milk, whole grains, yeast. Commercial sources: synthetic nicotinic acid is made by oxidation of nicotine, quinoline, or 2-methyl-5-ethylpyridine (from ammonia and formaldehyde or acetaldehyde).

Containers: Glass vials, bottles, fiber cans and drums.

Grades: U.S.P. XVI; 50 and 80% USP, blended with soy flour (animal feeds).

Uses: Medicine; nutrition; feeds; enriched flours.

See also nicotinamide.

nicotinic acid amide. See nicotinamide.

niello. The black metallic-like mixture of the sulfides of copper, silver and lead that has been used since the 11th century to inlay ornamental designs engraved in metal, usually silver. In earlier periods silver sulfide seems to have been used as the niello material.

nifuroxime (anti-5-nitro-2-furaldehyde)oxime $C_4OH_2NO_2CHNOH$.

Properties: White to pale yellow powder; becomes tan on standing; slightly soluble in water; fairly soluble in alcohol; very soluble in dimethylformamide.

Grade: N.F. XI.

Use: Medicine.

nigre. The dark-colored layer, containing some soap as well as salts and impurities, formed in soap manufacture as an intermediate layer between the layers of soap proper and lye.

nigrosine. A class of blue or black dyes, some soluble in water, some in alcohol and some in oil, used in manufacture of ink and shoe-polish and in dyeing leather, wood, textiles, etc.

"Ni-Hard." [283] Trademark for abrasion-resistant martensitic white cast irons containing approximately 4.5% nickel and 1.5% chromium. Hardnesses of 550 to 700 BHN are available. Used in service wherever abrasion resistance under mild impact is desired.

nikethamide (N,N-diethylnicotinamide; pyridine-3-carboxylic acid, diethylamide) $C_5H_4NCON(C_2H_5)_2$.

Properties: Clear, colorless to pale yellowish, somewhat viscous liquid, which crystallizes on standing in the cold and melts again as the temperature rises; faint, characteristic, aromatic odor and peculiar, bitter taste. Solutions are clear and nearly colorless and have no more than a faint odor of diethylamine. Miscible with water, alcohol, and ether; sp. gr. 1.058-1.066; congealing range, 22-24°C; refractive index (25°C) 1.522-1.524; pH (1 in 4 solution) 6.5-7.5.

Grades: N.F. XI.; technical.

Containers: Carboys.

Use: Medicine.

nil alba. See zinc oxide.

"Nile." [307] Bright basic blue for discharge printing.

"Nilevar." [70] Trademark for brand of norethandrolone, 17 alpha-ethyl-17-hydroxynorandrostenone. Used in medicine.

"Nilite." [28] Trademark for a series of nitrocarbonitrate blasting agents.

"Nilofoam." [300] Trade name for a silicone defoaming agent.

"Nilstain." [155] Trademark for a line of stainless steel alloys.

niobe oil. See methyl benzoate.

niobic acid $Nb_2O_5 \cdot nH_2O$. The term includes all hydrated forms of Nb_2O_5. Niobic acid forms as a white insoluble precipitate when a potassium hydrogen sulfate fusion of a niobium compound is leached with hot water or when niobium fluoride solutions are treated with ammonium hydroxide. It is soluble in concentrated sulfuric acid, concentrated hydrochloric acid, hydrofluoric acid, and in bases. The formation of niobic acid is important in the analytical determination of niobium.

niobite. See columbite.

niobium Nb (formerly columbium, Cb). Element of atomic number 41, group V of the periodic system. It is a comparatively plentiful metal.

Properties: Gray or silver-white, hard metal, not readily tarnished; sp. gr. 8.57; m.p. 2415 ± 15°C. Reacts with oxygen and the halogens only when heated. Insoluble in acids except mixed nitric and hydrofluoric; attacked by fused alkalies.

Occurrence: In columbite.

Derivation: (a) Reduction of the complex alkali fluoride with sodium, or of the oxide with calcium, aluminum, or hydrogen. (b) Heating niobium oxide and niobium carbide together in a vacuum to produce the pure metal. Large single crystals are produced by arc-fusion. High purity alloys are prepared by electron beam melting processes.

Uses: In chromium steel; "getter" in vacuum techniques; nuclear energy equipment; cermets (carbide); alloys in jet engines, missiles, rockets; as rotor at liquid helium temperatures in gyroscopes.

niobium carbide NbC.

Properties: Lavender-gray powder; insoluble in water and in all acids except a mixture of nitric and hydrofluoric acids. M.p. about 3500°C; hardness 2400 kg/sq. mm; sp. gr. 7.82.

Derivation: By direct combination of niobium with carbon or by the reduction of niobium oxide with lampblack.

Uses: Cemented carbide tipped tools; certain special steels; preparation of niobium metal.

niobium chloride (niobium pentachloride) $NbCl_5$.

Properties: Yellow crystalline solid; soluble in alcohol, ether, carbon tetrachloride, hydrochloric acid, conc. sulfuric acid. M.p. 194°C; b.p. 241°C; sp. gr. 2.75. Deliquescent; decomposes in moist air with evolution of hydrogen chloride fumes. Caution! Keep well stoppered.

Derivation: Direct combination of niobium and chlorine; chlorination of niobium oxide in the presence of carbon.

Containers: Available in commercial quantities.

Uses: Preparation of pure niobium; intermediate.

niobium oxide (niobium pentoxide) Nb_2O_5.

Properties: White powder; insoluble in acids except hydrofluoric; soluble in fused potassium hydrogen sulfate, or carbonates or hydroxides of the alkali metals. Sp. gr. 4.5-5.0.

Derivation: Strong ignition of niobic acid.

Uses: Intermediate; in electronics.

niobium pentachloride. See niobium chloride.

niobium pentoxide. See niobium oxide.

niobium-potassium oxyfluoride (potassium-niobium oxyfluoride; potassium oxyfluoniobate) $K_2NbOF_5 \cdot H_2O$.

Properties: White lustrous leaflets. Greasy to touch. Soluble in water.

Uses: Separation of niobium from tantalum; electrolytic preparation of niobium metal.

niobium-tin Nb_3Sn. Used for special wire for superconducting magnets to obtain high magnetic fields for use in communication, and containment of thermonuclear fusion plasmas.

niobium-uranium alloys. Niobium alloyed with 20% uranium yields a nuclear fuel which maintains tensile strength and hardness at 1600°F. The fuel is used in nuclear reactors such as gas-cooled units.

"Ni-O-Nel." [283] Trademark for an alloy containing approximately 42% nickel, 22% chromium, 30% iron and 3% molybdenum. Has outstanding corrosion resistance, particularly toward sulfuric and phosphoric acids.

"Ni-Plex." [72] Trade name for a blend of mildly alkaline organic chemicals used to remove nickel deposits from any base metal.

"Niran" Insecticide. [58] Trademark for parathion (q.v.).

"Ni-Resist." [283] Trademark for a series of corrosion resisting austenitic nickel cast irons containing from 17 to 35% nickel. Most types contain chromium between 2 and 4%. These irons are used for resistance to heat and corrosion and for selected thermal expansivities.

"Nirus AOL." [83] Trademark for a cationic, oil soluble, neutral, single compound used as a corrosion inhibitor in the petroleum industry. Low viscosity; tan fluid.

Containers: 400-lb steel drums.

"Nisentil" Hydrochloride. [190] Trademark for a brand of alphaprodine hydrochloride (q.v.).

"Ni-Span-C." [283] Trademark for a wrought age-hardenable nickel-iron alloy containing about 42% nickel, which has high tensile strength and a substantially constant modulus of elasticity over the temperature range of minus 50°F to plus 150°F. Widely used for Bourdon tubes and hair springs.

niter (saltpeter) KNO_3. A natural potassium nitrate. Color, white, gray, or colorless; streak, white; vitreous luster. Contains 46.5% K_2O, 53.5% N_2O_5. Found as white crusts, needle-like crystals and silky tufts in limestone caverns or as incrustations upon the earth's surface or on rocks. Also found in hot, dry countries as an efflorescence in soils containing human or animal excrement. In such countries advantage has been taken of this and soil, plant ashes and decomposing organic matter (manure) are built into mounds, moistened periodically and the niter finally extracted with water.
Constants: Sp. gr. 2.09-2.27; hardness 2.
Occurrence: United States (Kentucky, Wyoming), India, Egypt, Algeria, Iran, Spain, France, Germany.
Uses: See potassium nitrate.
Shipping regulations: Oxidizing material. Yellow label. *

niter cake. See sodium bisulfate.

niter, Chile. See caliche.

niton. See radon.

nitralloy. See nitriding steel.

"Nitramex" No. 2-H. [28] Trademark for a high density blasting agent of "Nitramon" type.

nitramine. See tetryl.

"Nitramite." [28] Trademark for a low density blasting agent of "Nitramon" type.

"Nitramon." [28] Trademark for an ammonium nitrate base blasting agent which is not detonated by impacts from rifle bullets, sledge hammers, or even by the heat from blow torches. Is detonated by specially constructed primers.
Use: Various modifications serve for large and small drill hole blasting, and for seismic prospecting on land and at sea.

nitranilic acid (2,5-dihydroxy-3,6-dinitroquinone) $C_6O_2(NO_2)_2(OH)_2$.
Properties: Flat yellow crystals; loses water at 100°C; decomposes explosively at 170°C; soluble in water and alcohol; insoluble in ether.

ortho-**nitraniline orange.** Ortho-nitraniline orange is produced from the intermediates ortho-nitraniline and beta-naphthol. It is a very bright, light shade organic orange having good light fastness when used by itself in enamels. It is used for awning paints, and since it has comparatively good alkali fastness it is used in the manufacture of casein paints. One disadvantage is that ortho-nitraniline orange bleeds badly in oils and lacquer solvents.

nitranilines. See nitroanilines.

nitrating acid. See mixed acid.

nitre. See niter (saltpeter).

nitre cake. See sodium bisulfate.

nitre, Chile. See caliche.

"Nitrelmang." [250] Trademark for a high purity nitrided grade of "Electromanganese" (q. v.).

"Nitretamin Phosphate." [412] Trademark for trolnitrate phosphate (q. v.).

"Nitrex." [248] Trademark for synthetic rubber latices of the butadiene-acrylonitrile type; used for paper saturation, leather finishing, and as a plasticizer for resin latices.

nitric acid (aqua fortis; engraver's acid; azotic acid) HNO_3.
Properties: Transparent, colorless or yellowish, fuming, suffocating, caustic and corrosive liquid. Miscible with water.
Constants: B. p. (decomposes) 83°C; m. p. −41.59°C; sp. gr. 1.504 (25/4°C); vapor pressure 62 mm (25°C); refractive index 1.3970 (n 24/D); viscosity 0.761 cp (25°C).
Derivation: (a) Oxidation of ammonia by air in the presence of a platinum catalyst. The product is approximately 60% nitric acid, 40% water. Concentration is achieved by distillation with sulfuric acid, or by extractive distillation with magnesium nitrate. See also the Fauser process. (b) Formerly from Chile saltpeter by action of sulfuric acid.
Grades: Technical; pure. See also nitric acid, fuming.
Strength of solutions: 36, 38, 40, 42°Bé; 58-63.5%; 95%.
Containers: Bottles; carboys (carloads); tank cars.
Uses: Primary use is manufacture of ammonium nitrate for fertilizer and explosives. Lesser but important uses are organic synthesis (dyes, drugs, explosives, cellulose nitrate, nitrate salts); also in metallurgy, photoengraving; etching steel; ore flotation; medicine.
Danger! Causes severe burns; vapor extremely hazardous. May cause nitrous gas poisoning. Spillage may cause fire or liberate dangerous gas. MCA warning label.
Shipping regulations: Corrosive liquid. White label. *

nitric acid, fuming.

1. White fuming nitric acid (WFNA) contains more than 97.5% nitric acid, less than 2% water, and less than 0.5% oxides of nitrogen. It is a colorless or pale yellow liquid which fumes strongly. It is decomposed by light or elevated temperatures, becoming red in color from nitrogen tetroxide.
2. Red fuming nitric acid (RFNA) contains more than 86% nitric acid, approximately 6-15% oxides of nitrogen (as nitrogen tetroxide) and less than 5% water.

Derivation: From dilute nitric acid, nitrogen tetroxide, and oxygen by the Fauser process.
Grades: Technical.
Containers: Stainless steel or aluminum drums.
Uses: Preparation of nitro- compounds; oxidizer in liquid rocket propellants.
Danger! Causes severe burns; vapor extremely hazardous; may cause nitrous gas poisoning. Spillage may cause fire or liberate

dangerous gas. MCA warning label.
Shipping regulations: Corrosive liquid. White label.*

nitric oxide NO. See also nitrogen dioxide.
Properties: A colorless gas (readily reacts with oxygen at room temperature to form nitrogen dioxide, NO_2, a reddish-brown gas). B.p. −150.0°C; m.p. −161.0°C; sp.gr. at b.p. 1.27; slightly soluble in water.
Grades: Pure (99%).
Containers: Cylinders.
Uses: An important intermediate stage in the manufacture of nitric acid from ammonia. Also formed from atmospheric oxygen and nitrogen in the electric-arc process for fixation of nitrogen.
Shipping regulations: Poison, class A. Poison gas label. Not accepted by express.

nitriding. A process of case hardening in which a ferrous alloy, usually of special composition, is heated in an atmosphere of ammonia or in contact with nitrogenous material to produce surface hardening by absorption of the nitrogen without quenching.

nitriding steel (nitralloy). The alloys used for nitriding are known as nitralloy. They contain aluminum as an alloying constituent. Several types of nitralloy are available with ranges of composition as follows: aluminum 0.85-1.2%; carbon 0.20-0.45%; chromium none to 1.8%; molybdenum 0.15-1.00%; manganese 0.4-0.7%; silicon 0.2-0.4%.

Nickel nitriding steels are also sometimes used, and stainless steel and others not containing aluminum are also nitrided, although the case is not nearly as hard as nitralloys.

nitrile. An organic compound containing the −CN grouping, for example, acrylonitrile $CH_2{:}CHCN$.

nitrile-butadiene rubber. See acrylonitrile rubber.

nitrile rubber. See acrylonitrile rubber.

nitrile-silicone rubber (NSR). Combines the characteristic properties of silicones with the oil resistance of nitrile rubber. Resistant to jet fuels, solvents and hot oils.

nitrilomalonamide. See cyanoacetamide.

nitrilotriacetic acid. See triglycollamic acid.

para-**nitroacetanilide** $NO_2C_6H_4NHCOCH_3$.
Properties: White crystals; soluble in alcohol and ether; very slightly soluble in cold water. Soluble in hot water, in potassium hydroxide solution. M.p. 214-216°C.
Derivation: By acetylating aniline, then nitrating.
Method of purification: Crystallization.
Grades: Technical.
Containers: 300-, 325-, 375-lb barrels.
Use: Manufacture of para-nitraniline.

meta-**nitroacetophenone** $C_6H_4NO_2COCH_3$. Light yellow amorphous solid.

para-**nitro**-ortho-**aminophenol** $C_6H_3OHNH_2NO_2$.
Properties: Yellow-brown leaflets containing water of crystallization melting at 80 to 90°C; anhydrous melts at 154°C. Soluble in acid.
Derivation: From dinitrophenol.
Method of purification: Recrystallization from hydrochloric acid.
Grades: Technical.
Containers: Wooden barrels; fiber drums.
Use: Dyes.
Shipping regulations: None.*

meta-**nitroaniline** (meta-nitraniline) $NO_2C_6H_4NH_2$.
Properties: Yellow needles; sp.gr. 1.43; m.p. 111.8°C; b.p. 285°C; soluble in alcohol and ether; very slightly soluble in water.
Derivation: From aniline by nitration after acetylation, with subsequent removal of the acetyl group by hydrolysis.
Method of purification: Crystallization.
Grades: Technical (crystals; paste).
Containers: Drums.
Uses: Color test for pine wood; dye intermediate.
Danger! Hazardous solid and vapor. Rapidly absorbed through skin. MCA warning label.

ortho-**nitroaniline** (ortho-nitraniline) $NO_2C_6H_4NH_2$.
Properties: Orange-red needles; sp.gr. 1.443; m.p. 69.7°C. Soluble in alcohol and ether; very slightly soluble in water.
Derivation: From aniline by nitration after acetylation, with subsequent removal of the acetyl group by hydrolysis.
Method of purification: Crystallization.
Grades: Technical (flaked).
Containers: Drums.
Uses: Dye intermediate.
Danger! Hazardous solid and vapor. Rapidly absorbed through skin. MCA warning label.

para-**nitroaniline** (para-nitraniline) $NO_2C_6H_4NH_2$.
Properties: Yellow needles; sp.gr. 1.437; m.p. 148°C; soluble in alcohol and ether; very slightly soluble in water.
Derivation: From aniline by nitration after acetylation, with subsequent removal of the acetyl group by hydrolysis.
Method of purification: Crystallization.
Grades: Technical.
Containers: Drums.
Uses: Dye intermediate, especially para-nitraniline red.
Danger: Hazardous solid and vapor. Rapidly absorbed through skin. MCA warning label.
Shipping regulations: Poison, class B. Poison label.*

ortho-**nitroanisole** $C_6H_4OCH_3NO_2$.
Properties: Light reddish or amber liquid. Soluble in alcohol and ether; insoluble in water; sp.gr. 1.255 (20/20°C); crystallizing point 9.6°C; boiling range 268-271°C.
Derivation: From ortho-nitrophenol by methylation or from ortho-nitrochlorobenzene by action of methanol (methyl alcohol) and caustic soda.
Method of purification: Distillation.

Grades: Technical.
Containers: Galvanized drums; tank cars.
Uses: Organic synthesis; manufacture of intermediates for dyes and pharmaceuticals.
Shipping regulations: None. *

para-**nitroanisole** (1-methoxy-4-nitrobenzene) $NO_2C_6H_4OCH_3$.
Properties: Colored monoclinic crystals; m.p. 54°C; b.p. 260°C. Insoluble in water; soluble in alcohol, ether.
Use: Intermediate.

5-nitrobarbituric acid (dilituric acid) $O_2NHCCONHCONHCO$.
Properties: Prisms and leaflets from water; m.p. 176°C (dec); slightly soluble in water; soluble in alcohol and sodium hydroxide solution; insoluble in ether.
Grades: Reagent; technical.
Use: Microreagent for potassium.

3-nitrobenzaldehyde (meta-nitrobenzaldehyde) $NO_2C_6H_4CHO$.
Properties: Yellowish, crystalline powder; m.p. 58°C; b.p. (23 mm) 164°C; almost insoluble in water; soluble in alcohol, chloroform, ether.
Grades: Technical.
Uses: Synthesis of dyes, pharmaceuticals, surface active agents; vapor phase corrosion inhibitor; antioxidant for chlorophyll; mosquito repellent.

nitrobenzene (oil of mirbane; essence of mirbane; essence of myrbane) $C_6H_5NO_2$.
Properties: Bright yellow crystals or yellow, oily liquid; poisonous! Soluble in alcohol, benzene, and ether; very slightly soluble in water. Sp. gr. 1.19867; m.p. 5.70°C; b.p. 210.85°C.
Derivation: From benzene by nitrating with nitric acid.
Method of purification: By washing and distilling with steam, then redistilling.
Impurities: Unconverted benzene.
Grades: Technical.
Containers: Glass bottles; various tins; 500-, 1000-lb iron drums. Carloads, tank cars.
Uses: Major use is manufacture of aniline; also used to produce pyroxylin insulating compounds; solvent for cellulose ethers; modifying esterification of cellulose acetate; ingredient of metal polishes and shoe polishes; raw material for manufacture of benzidine, quinoline, azobenzene, etc.
Danger: Hazardous liquid and vapor, rapidly absorbed through skin. MCA warning label.
Shipping regulations: Class B poison. Poison label. *

para-**nitrobenzeneazoresorcinol.** $NO_2C_6H_4N_2C_6H_3(OH)_2$.
Properties: Red crystals. Slightly soluble in water; soluble in nitrobenzene. M.p. 198°C.
Derivation: Diazotized para-nitroaniline is coupled with resorcinol.
Method of purification: Recrystallization.
Grades: Analytical.
Containers: Glass bottles.
Use: Determination of magnesium.
Shipping regulations: None. *

para-**nitrobenzene azosalicylate sodium salt.** See alizarin yellow R.

nitrobenzoic acid $C_6H_4(NO_2)COOH$. (a) meta-, (b) ortho-, (c) para-(nitrodracylic acid).
Properties: Yellowish-white crystals. (a) Soluble in alcohol and ether; slightly soluble in water; (b) Soluble in water, alcohol and ether; (c) Soluble in alcohol; sparingly soluble in water.
Constants: (a) Sp. gr. 1.494; m.p. 140-141°C; (b) Sp. gr. 1.575; m.p. 147.7°C; (c) Sp. gr. 1.5497; m.p. 238°C.
Derivation: (a and b) By the nitration of benzoic acid; (c) By the oxidation of para-nitrotoluene by hot chromic acid mixture.
Method of purification: Crystallization.
Grades: Technical.
Containers: Fiber drums; car loads.
Uses: (a and b) Organic synthesis; (c) preparation of anesthetics and as intermediate in the manufacture of dyes and sun screening agents.
Shipping regulations: None. *

meta-**nitrobenzotrifluoride** (3-nitrobenzotrifluoride; meta-nitrotrifluoromethylbenzene; meta-nitro-alpha, alpha, alpha-trifluorotoluene) $NO_2C_6H_4CF_3$.
Properties: Pale straw, thin oily liquid, aromatic odor; distillation range 200.5-208.5°C; f.p. −5.0°C; sp. gr. (15.5°C) 1.437; b.p. 203°C; flash point (open cup) 101°C; fire point (open cup) 102°C; wt/gal 11.98 lbs (15.5°C); viscosity 2.35 cps (100°F); soluble in organic solvents; insoluble in water.

meta-**nitrobenzoyl chloride** $NO_2C_6H_4COCl$.
Properties: Yellow to brown liquid; partially crystallized at room temperature; m.p. 34°C (approx); b.p. 278°C; soluble in ether; decomposes in water and alcohol.
Use: Manufacture of dyes for fabrics and color photography; intermediate in preparation of pharmaceuticals.

para-**nitrobenzoyl chloride** $NO_2C_6H_4COCl$.
Properties: Yellow crystalline solid; m.p. 72°C; b.p. 154°C (15 mm); decomposes in water and alcohol; soluble in ether.
Use: Intermediate for procaine hydrochloride, dyestuffs.

para-**nitrobenzyl cyanide** (para-nitro-alpha-tolunitrile) $NO_2C_6H_4CH_2CN$.
Properties: Crystals; m.p. 116-118°C. Insoluble in water; soluble in alcohol and ether.
Derivation: Action of concentrated nitric acid on benzyl cyanide.
Uses: Intermediate for dyestuffs, pharmaceuticals, penicillin precursors, local anesthetics; preparation of para-nitrophenylacetic acid.

ortho-**nitrobiphenyl** (ONB; ortho-nitrodiphenyl) $C_6H_5C_6H_4NO_2$.
Properties: Light-yellow to reddish

crystalline solid or liquid; sp. gr. 1.203 (25/25°C); 10 lbs/gal; crystallizing pt. 34.5°C (min); refractive index 1.613 approx (25°C); b.p. 330°C approx. Soluble in carbon tetrachloride, mineral spirits, pine oil, turpentine, benzene, acetone, glacial acetic acid, and perchlorethylene.
Derivation: By controlled nitration of biphenyl.
Uses: Plasticizer; imparts fungicidal properties to textiles; dye intermediate.
Warning! This material can cause serious damage to respiratory passages and liver.

nitrobromoform. See bromopicrin.

"Nitro BT." [65] Trademark for tetrazolium salt (q.v.).

2-nitro-1-butanol $CH_3CH_2CHNO_2CH_2OH$.
Properties: Colorless liquid; solubility in water 20 g/100 cc (20°C); sp. gr. 1.133 (20/20°C); b.p. 105°C (10 mm); m.p. −48 to −47°C; wt/gal 9.44 lbs (20°C); refractive index 1.4390 (20°C); pH of 0.1 M solution 4.51.
Containers: 5- and 55-gal drums; 1-gal cans.

nitrocalcite. See calcium nitrate.

nitrocarbonitrate ("Nilite"). A blasting agent consisting of ammonium nitrate sensitized with diesel oil.
Shipping regulations: Oxidizing material. Yellow label. *

nitrocellulose (cellulose nitrate; cotton solution; nitrocotton; guncotton; collodion cotton). A cotton-like or pulp-like material of variable composition obtained by treating cellulose (in the form of linters, cotton waste, cotton wool, tissue paper or wood pulp) with a mixture of concentrated nitric and sulfuric acids, the excess of which is removed by washing, digesting and boiling procedures. The name cellulose nitrate is technically correct although nitrocellulose is more commonly used. By varying the strength of the acids, the temperature, the time of nitration and the proportion of acids to cellulose, products of widely different properties are obtained. These are classified according to percentage of nitrogen, which usually ranges from 10 to 14. The latter corresponds approximately to the empirical formula $C_6H_7O_5(NO_2)_3$. Nitrocellulose of approximately 12.5 to 13.5% nitrogen is used for explosives such as guncotton and some dynamites. This high-nitrogen form is soluble in acetone, but insoluble in ether-alcohol mixtures. When nitrogen content is in the range 10.5 to 12.2%, the material is referred to as soluble nitrocellulose because it is soluble in ether-alcohol mixtures and is used for preparation of collodion, "Celluloid," plastics, and fastdrying lacquers. The name pyroxylin (q.v.) has sometimes been applied to cellulose nitrate, its solutions, and products. It is an official U.S.P. XVI grade. The most recently developed major use is as the basic material in solid mono-propellants, including those for rockets.
Fire hazard: Dangerous.
Shipping regulations: Various, according to whether dry or wet, and to solvent used. High explosive, red or yellow labels. *

nitrochloro derivatives. See under the corresponding chloronitro derivative.

nitrochloroform. See chloropicrin.

nitrocobalamin. Vitamin B_{12C}. See vitamin B_{12}.

"Nitrocols." [223] Proprietary products consisting of pigments dispersed in nitrocellulose and plasticizer. Available in two forms, chip and paste.
Grades and Uses:
Super 1A. A high-color carbon black dispersed in ½ sec. nitrocellulose and dibutyl phthalate. Used in high-grade jet black lacquer finishes.
No. 250. A medium-color carbon black dispersed in ½ sec. nitrocellulose and dibutyl phthalate, with a high pigment concentration. Used in industrial-type lacquers.
No. 9. Titanium dioxide (anatase) dispersed in ½ sec. nitrocellulose and dibutyl phthalate. Used in high-gloss white lacquer enamels.
No. 73. Titanium dioxide dispersed in 5-6 sec. nitrocellulose and dibutyl phthalate. Used in high-gloss white lacquer enamels.
No. 90. Titanium dioxide (rutile) dispersed in ½ sec. nitrocellulose and dibutyl phthalate. Used in high-gloss white lacquer enamels.

nitrocotton. See nitrocellulose.

2-nitro-para-cresol (4-methyl-2-nitrophenol) $NO_2(CH_3)C_6H_3OH$.
Properties: Yellow crystals; density 1.24 g/ml (38/4°C); m.p. about 35°C; b.p. 125°C (25 mm); slightly soluble in water; soluble in alcohol, ether.
Use: Intermediate.

nitrodichloro derivatives. See the corresponding dichloronitro derivative.

ortho-**nitrodiphenyl.** See ortho-nitrobiphenyl.

ortho-**nitrodiphenylamine** $C_6H_5NHC_6H_4NO_2$.
Properties: Red-brown crystalline powder; m.p. 75-76°C.
Containers: Fiber kegs.
Use: Intermediate.

nitrodracylic acid. See para-nitrobenzoic acid.

nitro dyes. Dyes whose molecules contain the NO_2 chromophore group in their structure, and whose Colour Index is from 10,300-10,999.

nitroerythrite. See erythrityl tetranitrate.

nitroethane $CH_3CH_2NO_2$.
Properties: Colorless liquid. Solubility in water 4.5 cc/100 cc (20°C); solubility of water in nitroethane 0.9 cc/100 cc (20°C).
Constants: Sp. gr. 1.052 (20/20°C); freezing point −90°C; b.p. 114°C (760 mm); vapor pressure (15.6 mm) (20°C); flash point 106°F (Tag open cup); wt/gal 8.75 lbs (68°F); refractive index 1.3917 (20°C).

Derivation: By reaction of ethane with oxides of nitrogen or nitric acid under pressure.
Containers: 5- and 55-gal drums; tank cars.
Uses: Solvent for nitrocellulose, cellulose acetate, cellulose acetopropionate, cellulose acetobutyrate, vinyl, alkyd, and many other resins, waxes, fats and dyestuffs; chemical synthesis.
Shipping regulations: None. *

2-nitro-2-ethyl-1,3-propanediol $CH_2OHC(C_2H_5)NO_2CH_2OH$.
Properties: White, crystalline solid; m. p. 56-57°C; b. p. decomposes (10 mm); pH 0.1 M aqueous solution 5.48; soluble in organic solvents; very soluble in water.
Containers: Fiberpak boxes.
Shipping regulations: None. *

5-nitro-2-furaldehyde semicarbazone. See nitrofurazone.

nitrofurantoin $C_8H_6N_4O_5$. N-(5-nitro-2-furfurylidene)-1-aminohydantoin.
Properties: Yellow, bitter powder with slight odor. M. p. (dec) 270-272°C. Very slightly soluble in alcohol and practically insoluble in ether and water.
Grade: U.S.P. XVI.
Use: Medicine (antibacterial agent).

nitrofurazone (5-nitro-2-furaldehyde semicarbazone) $C_6H_6N_4O_4$.
Properties: Odorless, lemon-yellow, crystalline powder which turns brownish black on heating and decomposes between 236 and 240°. Nearly tasteless but develops bitter aftertaste. Fairly soluble in alcohol and propylene glycol. Slightly soluble in water and polyethylene glycol mixtures. Practically insoluble in chloroform or ether.
Derivation: By simultaneous acetylation and nitration of furfural, followed by reaction with semicarbazide and sulfuric acid.
Grades: N.F. XI.
Use: Medicine (antibacterial agent).

"Nitrogation." [125] Trademark for anhydrous ammonia specifically intended for direct injection into the irrigation stream for purpose of soil fertilization.

nitrogen N. Element of atomic number 7, of group V of the periodic system.
Properties: Colorless, odorless, tasteless diatomic gas constituting about four-fifths of the air; colorless liquid, chemically rather inert; sp. gr. (gas) 0.96737, referred to air; (liquid) 0.804; (solid) 1.0265; m. p. −210°C; b. p. −195.5°C. Soluble in water; slightly soluble in alcohol.
Derivation: From liquid air by fractional distillation.
Impurities: Argon and other "rare gases"; oxygen.
Grades: U.S.P. XVI; prepurified 99.966% min; extra dry 99.7% min; water pumped 99.6% min.
Containers: Steel cylinders.
Uses: Production of ammonia, acrylonitrile, cyanamide, cyanides, nitrides; inert gas for purging, blanketing, and exerting pressure; for freezing out gaseous impurities in electronic manufacturing operations; for chilling in aluminum foundries; for bright annealing of steel; for cooling agent in low temperature processes; as an inert pressuring and blanketing gas in missiles.

nitrogenase. Enzyme which fixes nitrogen and can be isolated from soil bacteria, namely, Clostridium pasteurianum. It is possible to synthesize ammonia from nitrogen and hydrogen without high temperatures and pressures by means of nitrogenase. Pyruvic acid is an adjunct of the reaction that occurs.

nitrogen chloride. See nitrogen trichloride.

nitrogen dioxide (nitrogen peroxide) NO_2.
Properties: A red-brown gas which exists in varying equilibrium with other oxides of nitrogen as the temperature is varied. When heated, it becomes less brown, then yellow, and finally colorless at 150°C because the NO_2 gradually dissociates into colorless NO (nitric oxide) and oxygen gas. When cooled, nitrogen dioxide condenses at 21.15°C to a yellow-brown liquid (sp. gr. 1.45 at 20°C), and freezes at −11.2°C to a colorless solid. This color change marks the gradual conversion of the brown NO_2 gas to colorless N_2O_4 (nitrogen tetroxide). Under most circumstances samples of nitrogen dioxide consist of an equilibrium mixture of NO_2 and N_2O_4. Either of these oxides will dissolve in water to form nitric acid.
Containers: 125-, 150-, 2000-lb cylinders; tank cars.
Grades: Pure, 99.5% min.
Uses: Intermediate in production of nitric acid; nitrating agent; oxidizing agent; catalyst; oxidizer for rocket fuels.
Danger! Extremely dangerous liquid and gas. Inhalation may cause fatal lung injury. MCA warning label.
Shipping regulations: Poison, class A; poison gas label by freight; not accepted by express. Nitrogen dioxide is the legal label name for shipping regulation purposes. *

nitrogen fixation. The conversion of the nitrogen of the air into a useful compound. This is accomplished naturally through the action of the bacteria that exist on the roots of leguminous plants (recently duplicated in the laboratory) and also by the action of lightning in causing combination of nitrogen and oxygen of the air.

The direct combination of nitrogen and hydrogen (see ammonia synthesis) is by far the most important and widely used industrial procedure. The union of calcium carbide and nitrogen to form calcium cyanamide is a little used process. The direct combination of nitrogen and oxygen (arc process) has been used but has not been continued. Various other reactions can be caused to take place but have not been used successfully on a large scale.

nitrogen fluorine compounds. Among these are nitrogen trifluoride (q. v.), dinitrogen tetrafluoride, dinitrogen difluoride, and difluoramine. These materials are powerful oxidizers for both liquid and solid propellants for rocket propulsion.

nitrogen monoxide. See nitrous oxide.

nitrogen mustards. A class of compounds with fishy odor and lachrymatory properties, and of importance in cancer treatment and research. They are named from their similarity in structure to mustard gas (dichlorodiethyl sulfide). The sulfur of the mustard gas is replaced by an amino nitrogen, so that the typical nitrogen mustards are halogenated alkylamines, such as methyl bis(2-chloroethyl)amine, $(CH_2ClCH_2)_2NCH_3$ (see mechlorethamine hydrochloride). Other examples are triethylene melamine, and triethylene thiophosphoramide (q. v.), and triethylene phosphoramide. The drug "Myleran" or 1, 4-dimethylsulfonyloxybutane, $(-CH_2CH_2OSO_2CH_3)_2$, is somewhat similar in structure.

nitrogen oxides. See nitrous oxide, nitric oxide, and nitrogen dioxide. Nitrogen tetroxide is described under nitrogen dioxide.

nitrogen peroxide. See nitrogen dioxide.

nitrogen solution. Mixture of 60 parts of ammonium nitrate and 40 parts of 50% aqua ammonia for neutralizing superphosphate in making fertilizer.
Containers: Tank cars.

nitrogen tetroxide N_2O_4. See nitrogen dioxide.

nitrogen trichloride (nitrogen chloride) NCl_3.
Properties: Yellow oil or rhombic crystals; sp. gr. 1. 653; explodes at 95°C; b. p. less than 71°C; m. p. less than -40°C; insoluble in cold water, decomposes in hot water; soluble in chloroform, phosphorus trichloride, and carbon disulfide; formerly used in bleaching and aging of flour. Poisonous.

nitrogen trifluoride NF_3. Colorless gas, colorless mobile liquid. Useful as oxidizer for high energy fuels; also used in synthesis.

nitroglycerin (explosive oil; glyceryl trinitrate; glonoin oil; trinitroglycerin) $CH_2NO_3CHNO_3CH_2NO_3$.
Properties: Pale yellow, thick, flammable, explosive liquid. Soluble in alcohol and ether; slightly soluble in water.
Constants: Sp. gr. 1. 6009; freezing point 13. 1°C; explosion point 260°C.
Derivation: By dropping glycerol through cooled, mixed acid and stirring, followed by repeated washing with water.
Grades: Technical.
Containers: Tin cans.
Uses: Explosive; production of dynamite and other explosives; medicine; explosive plasticizer in solid rocket propellants; possible liquid rocket propellant.
Fire hazard: Dangerous.
Shipping regulations: Cannot be shipped by common carrier. *

nitroguanidine $H_2NC(NH)NHNO_2$.
Properties: Yellow solid; m. p. 246°C; soluble in alcohol.
Derivation: Made by dissolving guanidine nitrate in concentrated sulfuric acid and then diluting.
Uses: Explosives and smokeless powders; anti-muzzle-flash agent for solid propellants.
Containers: Fiber drums.
Shipping regulations: Class A explosive. High explosive label. Not accepted by express. *

nitrohydrochloric acid. See aqua regia.

3-nitro-2-hydroxybenzoic acid. See meta-nitrosalicylic acid.

4-nitro-3-hydroxymercuri-ortho-cresol anhydride. See nitromersol.

"Nitrojection Ammonia." [125] Trademark for anhydrous ammonia specifically intended for direct injection into the soil by tractor equipment for purposes of soil fertilization.

nitromannite. See mannitol hexanitrate.

nitromannitol. See mannitol hexanitrate.

nitromersol (4-nitro-3-hydroxymercuri-ortho-cresol anhydride) $C_6H_2(CH_3)(NO_2)(OHg)$.
Properties: Brownish yellow or yellow granules or powder. Odorless and tasteless. Insoluble in water; insoluble in alcohol, acetone, ether; soluble in solutions of alkalies, ammonia, by opening the anhydride ring and salt formation.
Grade: N. F. XI.
Use: Medicine.

nitromethane CH_3NO_2.
Properties: Colorless liquid. Solubility in water 9. 5 cc/100 cc (20°C); solubility of water in nitromethane 2. 2 cc/100 cc (20°C).
Constants: Sp. gr. 1. 139 (20/20°C); b. p. 101°C (760 mm); vapor pressure 27. 8 mm (20°C); flash point 112°F (Tag open cup); wt/gal 9. 5 lbs (68°F); refractive index 1. 3817 (20°C); freezing point -29°C.
Derivation: By reaction of methane with oxides of nitrogen or nitric acid under pressure.
Containers: 5-, and 55-gal drums and 1-gal cans.
Uses: Solvent for nitrocellulose, cellulose acetate, cellulose acetopropionate; cellulose acetobutyrate, vinyl, alkyd, and many other resins, waxes, fats and dyestuffs; chemical synthesis; rocket fuel.
Shipping regulations: None. *

2-nitro-2-methyl-1, 3-propanediol $CH_2OHC(CH_3)NO_2CH_2OH$.
Properties: White, crystalline solid. Solubility in water 80 g/100 cc (20°C).
Constants: M. p. 147-149°C; b. p. decomposes (10 mm); pH 0. 1 M solution 5. 42.
Containers: Fiberpak boxes.

nitromuriatic acid. See aqua regia.

*See "I. C. C. Shipping Regulations, " page xiii.
Reference numbers refer to name of manufacturer. See "List of Manufacturers, " page v.

nitron (1,4-diphenyl-3,5-endo-anilino-4,5-dihydro-1,2,4-triazole) $C_{20}H_{16}N_4$.
Properties: Lemon-yellow, fine crystalline needles. Soluble in chloroform, acetone, and acetic acid ester; slightly soluble in ether and alcohol.
Derivation: Triphenylaminoguanidine (prepared from thiocarbanilide and phenylhydrazine), is heated with formic acid, the product diluted with much water, filtered, and precipitated with ammonium hydroxide. The product is dissolved in chloroform, the solution concentrated and allowed to crystallize.
Grades: Reagent; technical.
Containers: Glass bottles.
Use: Reagent for the detection of the nitrate ion (NO_3^-) in very dilute solutions.
Shipping regulations: None. *

"Nitron." [58] Trademark for cellulose nitrate sheeting.

alpha-**nitronaphthalene** $C_{10}H_7NO_2$.
Properties: Yellow crystals. Soluble in alcohol and ether; insoluble in water.
Constants: Sp. gr. 1.331; m. p. 61°C; b. p. 304°C.
Derivation: By the action of a mixture of nitric and sulfuric acids on finely ground naphthalene.
Method of purification: Crystallization.
Grades: Technical.
Containers: Wooden barrels; kegs; fiber drums.
Uses: Dyes; naphthylamine; is added to mineral oils to mask their fluorescence.
Shipping regulations: None. *

1-nitronaphthalene-5-sulfonic acid. (Laurent's alpha acid) $C_{10}H_6(NO_2)(SO_3H)$.
Properties: Pale yellow needles. Soluble in water, alcohol, and ether.
Derivation: By sulfonating nitronaphthalene with a mixture of chlorohydrin and sulfuric acid.
Method of purification: Crystallization.
Grades: Technical.
Containers: Wooden barrels; fiber drums.
Uses: Dyes.
Shipping regulations: None. *

nitronium perchlorate NO_2ClO_4.
Properties: White crystalline solid; m. p. 120-140°C; hygroscopic; noncorrosive; soluble in water to form nitric and perchloric acids; highly reactive.
Derivation: From ozone, nitrogen dioxide, chlorine dioxide.
Use: Suggested as propellant oxidizer.

nitroparaffins. Organic compounds derived from paraffin hydrocarbons by replacement of one or more hydrogen atoms by a nitro (NO_2) group. Examples are nitromethane (CH_3NO_2) and nitroethane ($C_2H_5NO_2$). Useful as solvents and raw materials for synthesis.

para-**nitrophenetole** $NO_2C_6H_4OC_2H_5$.
Properties: Crystallizes in prisms. Soluble in alcohol and ether.
Constants: M. p. 58°C; b. p. 283°C.
Derivation: This is prepared by ethylation of para-nitrophenol with ethyl chloride.
Method of purification: May be recrystallized from alcohol.
Grades: Technical.
Containers: Drums.
Uses: Dyes and other intermediates.
Shipping regulations: None. *

nitrophenide [bis(3-nitrophenyl) disulfide] $(NO_2C_6H_4)_2S$.
Properties: Yellow, rhomboid crystals; m. p. 83°C; insoluble in water; soluble in ether; slightly soluble in alcohol.
Derivation: Reduction of meta-nitrobenzenesulfonyl chloride with hydriodic acid.
Use: Veterinary medicine and pharmaceutical intermediate.

meta-**nitrophenol** $NO_2C_6H_4OH$.
Properties: Pale yellow crystals; sp. gr. 1.485 (20°C); m. p. 96-97°C; b. p. 194°C (70 mm).
Derivation: Diazotized meta-nitroaniline is boiled with water and sulfuric acid.
Use: As an indicator.

ortho-**nitrophenol** $NO_2C_6H_4OH$.
Properties: Yellow crystals; sp. gr. 1.295 (45°C), 1.657 (20°C); m. p. 44-45°C; b. p. 214°C; soluble in hot water, alcohol, ether.
Derivation: Action of dilute nitric acid on phenol at low temperature; para-nitrophenol formed at same time. They are separated by steam distillation.
Containers: Glass bottles; fiber cans; drums.
Uses: Intermediate in organic synthesis; indicator.
Warning! Hazardous solid. Absorbed through skin. MCA warning label.
Shipping regulations: None. *

para-**nitrophenol** $NO_2C_6H_4OH$.
Properties: Yellowish monoclinic prismatic crystals; sp. gr. 1.479-1.495 (20°C); m. p. 111.4-114°C (sublimes); b. p. 273°C (dec); soluble in hot water, alcohol, ether.
Derivation: See ortho-nitrophenol.
Containers: Glass bottles; fiber cans; drums.
Uses: Intermediate in organic synthesis; indicator.
Warning! Hazardous solid. Absorbed through skin. MCA warning label.
Shipping regulations: None. *

para-**nitrophenylacetic acid** (para-nitro-alpha-toluic acid) $NO_2C_6H_4CH_2COOH$.
Properties: Colored needles. M. p. 152-3°C. Slightly soluble in cold water; soluble in alcohol and chloroform.
Derivation: Hydrolysis of para-nitrobenzyl cyanide with 50% sulfuric acid.
Uses: Intermediate for dyestuffs, pharmaceuticals, penicillin precursors, local anesthetics.

4-nitrophenylarsonic acid $NO_2C_6H_4AsO(OH)_2$.
Properties: Crystalline solid.
Derivation: Nitration of phenylarsonic acid.
Uses: Veterinary medicine.

"Nitrophoska." [440] Trademark for a series of highly concentrated fertilizers. Available

as (N:P:K ratios shown):
"Nitrophoska" Red 13:13:20;
"Nitrophoska" Green 15:15:15;
"Nitrophoska" Grey 10:8:18;
"Nitrophoska" Blue 12:12:19;
"Nitrophoska" Yellow 15:15:6 + 4% magnesium oxide;
"Nitrophoska" Blue Special 12:12:17 + 2% magnesium oxide plus (in 100 kg) about 880 g borax, 270 g manganese sulfate, 150 g copper sulfate, 2.5 g cobalt sulfate, 90 g zinc sulfate;
Bor-"Nitrophoska" Red 13:13:20 + 2% borax.

The nitrogen in these fertilizers is about half in the form of quick-acting nitrate nitrogen and the remainder in the form of long-lasting ammonia nitrogen. The phosphoric acid is all available to the plants; one third as quick-acting (water-soluble) and two thirds as long-lasting (citrate-soluble) phosphoric acid. The potash is completely water-soluble. In "Nitrophoska" Red, Green, and Grey the potash originates from potassium chloride, whereas in "Nitrophoska" Blue and Yellow the potash originates from potassium sulfate. The magnesia in "Nitrophoska" Yellow originates from magnesium sulfate.

nitrophosphate. A nitrogen-phosphorus fertilizer produced by the action of nitric acid or a mixture of nitric and sulfuric or phosphoric acids on phosphate rock. Potassium salts usually are added to produce complete fertilizers.
Typical analysis: Available nitrogen 15%, available P_2O_5 15%, available K_2O 15%.
Containers: Bags, bulk, carloads.
Use: Fertilizer.
See also superphosphate; triple superphosphate.

1-nitropropane $CH_3CH_2CH_2NO_2$.
Properties: Colorless liquid. Solubility in water 9.5 cc/100 cc (20°C); solubility of water in 1-nitropropane 0.5 cc/100 cc (20°C).
Constants: Sp. gr. 1.003 (20/20°C); b.p. 132°C (760 mm); vapor pressure 7.5 mm (20°C); flash point 120°F (Tag open cup); wt/gal 8.4 lbs (68°F); refractive index 1.4015 (20°C); freezing point −108°C.
Derivation: By reaction of propane with nitrogen oxides or nitric acid under pressure.
Containers: 5-, 55-gal drums and 1-gal cans; tank cars.
Uses: Solvent for many resins, waxes, fats and dyestuffs; chemical synthesis.
Shipping regulations: None. *

2-nitropropane $CH_3CHNO_2CH_3$.
Properties: Colorless liquid. Solubility in water 1.7 cc/100 cc (20°C); solubility of water in 2-nitropropane 0.6 cc/100 cc (20°C).
Constants: Sp. gr. 0.992 (20/20°C); b.p. 120°C (760 mm); vapor pressure 12.9mm (20°C); flash point 103°F (Tag open cup); wt/gal 8.3 lbs (68°F); refractive index 1.3941 (20°C); freezing point −93°C.
Derivation: By reaction of propane with nitrogen oxides or nitric acid under pressure.
Containers: 5-, 55-gal drums; 1-gal cans; tank cars.
Uses: Solvent for many resins, waxes, fats and dyestuffs; chemical synthesis; used as wetting agent in grinding pigments.
Shipping regulations: None. *

meta-**nitrosalicylic acid** (3-nitro-2-hydroxybenzoic acid) $C_6H_3COOH(OH)NO_2$.
Properties: Yellowish crystals. Soluble in water and in alcohol. M.p. 144°C.
Derivation: By the nitration of salicylic acid.
Method of purification: Crystallization.
Grades: Technical.
Containers: Tins; kegs; fiber drums.
Uses: Intermediate; azo dyes.
Shipping regulations: None. *

meta-**nitrosalicylic acid, methyl ester** $C_6H_3(OH)(NO_2)COOCH_3$.
Properties: M.p. 130°C; insoluble in water; soluble in organic solvents.

"Nitrosan." [28] Trademark for a nitrogen-releasing chemical blowing agent containing 70% N,N'-dimethyl-N,N'-dinitrosoterephthalamide (NTA) and 30% white mineral oil by weight.
Properties: Nonhygroscopic crystalline powder; sp. gr. 1.2 g/cc; slightly to moderately soluble in common organic solvents. A smooth and rapid evolution of nitrogen gas is obtained when heated at 100°C. Medium and strong bases cause it to react violently and give off toxic diazomethane gas. Exposure to sunlight causes color change from yellow to green.
Containers: Nine 5-lb fiberpaks per cardboard carton.
Uses: For preparing open- and closed-cell chloride foam. Also used in expanding elastomers where milling temperatures do not exceed 50°C.
Hazards: Flammable solid. Keep away from open flame.
Shipping regulations: Flammable solid. Yellow label. *

para-**nitrosodimethylaniline** $NOC_6H_4N(CH_3)_2$.
Properties: Green leaflets. Soluble in alcohol and ether; insoluble in water; m.p. 93°C.
Derivation: By action of nitrous acid on N-dimethylaniline.
Method of purification: Crystallization.
Grades: Technical.
Containers: 120-lb barrels; fiber drums.
Uses: Production of methylene blue; vulcanization accelerator.
Shipping regulations: None. *

N-nitrosodiphenylamine (diphenylnitrosamine; nitrous diphenylamide) $(C_6H_5)_2NNO$.
Typical specifications: Yellow to brown or orange powder or flakes; m.p. 62 to 67°C; sp. gr. 1.23; insoluble in water; soluble in alcohol, acetone, benzene, and ethylene dichloride. Somewhat soluble in gasoline.
Containers: Fiber drums.
Uses: Retarder of vulcanization of rubber;

*See "I.C.C. Shipping Regulations," page xiii.
Reference numbers refer to name of manufacturer. See "List of Manufacturers," page v.

pesticide.
Shipping regulations: None. *

nitroso dyes (quinone oxime dyes). Dyes whose molecules contain the -NO or =NOH chromophore group in their structure and whose Colour Index is from 10,000-10,299 (2nd ed.)

nitrosoguanidine ONNHC(NH)(NH_2). A yellow powder used as fuel in percussion primers.
Shipping regulations: Explosive, class A. Initiating explosive label. Not accepted by express. *

nitrosonaphthol (naphthoquinoneoxime) $NOC_{10}H_6OH$. Several isomers are available; the following description is for alpha-nitroso-beta-naphthol, or 1-nitroso-2-naphthol.
Properties: Yellow needles. Soluble in alcohol and ether; insoluble in water. M. p. 110°C.
Derivation: By the action of nitrous acid on beta-naphthol.
Grades: Technical.
Containers: Wooden kegs; fiber drums.
Uses: Organic synthesis; prevention of gum formation in gasoline; analytical reagent.

para-**nitrosophenol** C_6H_4OHNO.
Properties: Crystallizes in light brown leaflets which decompose at 140°C. Soluble in alcohol, ether and acetone; moderately soluble in water. M. p. 140°C.
Derivation: From phenol by action of nitrous acid in the cold.
Grades: Technical.
Containers: Steel barrels kept tightly covered.
Use: Dyes.
Fire hazard: Ignites instantly when small amounts of acid or alkali are dropped into it; if impure it sometimes explodes by self-ignition; if ignited it burns explosively.
Shipping regulations: None shipped; for intra-plant transfer must be handled in tightly covered steel barrels.

nitroso rubber. Copolymer of tetrafluoroethylene and trifluoronitrosomethane, with unit structure $(-N(CF_3)OCF_2CF_2-)_n$.
Properties: Clear transparent gum, nonflammable, resistant to chemicals and solvents except the "Freons"; flexible at −40°F.

nitrostarch (starch nitrate) $C_{12}H_{12}(NO_2)_8O_{10}$.
Properties: Orange-colored powder; contains 16.5% nitrogen; highly explosive; soluble in ether-alcohol.
Containers: Steel drums or earthenware pots containing water.
Use: Explosives.
Shipping regulations: Various, according to whether dry or wet, and to solvent used. High explosive, red, or yellow labels. *

beta-**nitrostyrene** $C_6H_5CH{:}CHNO_2$.
Properties: M. p. 58°C. Available as a 30% solution in styrene.
Use: Chain stopper in styrene type polymerization.

para-**nitrosulfathiazole** $NO_2C_6H_4SO_2NHC_3H_2NS$. A sulfathiazole derivative.
Properties: Pale yellow powder; odorless; slightly bitter taste; m. p. 258-266°. Slightly soluble in alcohol, very slightly soluble in chloroform, ether and water and practically insoluble in benzene. Freely soluble in solutions of fixed alkali hydroxides.
Grades: N. F. XI.
Use: Medicine.

nitrosyl chloride NOCl. One of the oxidizing agents in aqua regia.
Properties: Yellow-red liquid; b. p. −5.5°C; yellow gas, poisonous; corrosive; decomposed by water.
Grades: Pure, 93% min.
Containers: Steel cylinders.
Uses: Synthetic detergents; catalyst; intermediate.
Shipping regulations: Non-flammable compressed gas. Green label. *

nitrosylsulfuric acid $ONOSO_3H$ with H_2SO_4.
Properties: Clear, straw-colored, oily liquid. Approximately 40% nitrosylsulfuric acid and 54% sulfuric acid. Pure $ONOSO_3H$: crystalline, m. p. 73°C.
Grades: Technical.
Containers: Carboys; drums.
Use: In the manufacture of dyes and intermediates.
Shipping regulations: Corrosive liquid. White label. *

nitrotoluene $NO_2C_6H_4CH_3$ (methylnitrobenzene; methylnitrobenzol).
Properties: Soluble in alcohol, ether, and benzene; insoluble in water.
(a) meta-: Yellow crystals; sp. gr. 1.1570; m. p. 16°C; b. p. 230-231°C.
(b) ortho-: Yellow liquid; sp. gr. 1.163 (20/4°C); m. p. −9.55°C; b. p. 222.3°C.
(c) para-: Yellow crystals; sp. gr. 1.2856; m. p. 51.4°C; b. p. 237.7°C.
Derivation: (a) From meta-nitro-para-toluidine. (b) and (c) From toluene by nitration and separation by fractional distillation.
Grades: Technical.
Containers: Fiber drums; multiwall paper sacks; tank cars.
Uses: (a) Organic synthesis. (b) and (c) For production of toluidine, tolidine, fuchsine and various synthetic dyes.
Shipping regulations: None. *

para-**nitrotoluene**-ortho-**sulfonic acid** (4-nitrotoluene-2-sulfonic acid) $NO_2C_6H_3(CH_3)SO_3H$.
Properties: Crystallizes from water in pale yellow prisms. Soluble in alcohol, ether and chloroform. M. p. 133.5°C.
Derivation: From para-nitrotoluene by sulfonation with oleum.

para-**nitro**-alpha-**toluic acid.** See para-nitrophenylacetic acid.

meta-**nitro**-para-**toluidine** (3-nitro-4-toluidine) $NO_2C_6H_3(CH_3)NH_2$.

Properties: Orange-red crystals; soluble in alcohol and concentrated sulfuric acid. M. p. 117°C.
Derivation: From acetyl-para-toluidine by nitration.
Method of purification: Recrystallization.
Grades: Technical.
Containers: Fiber drums; multiwall paper sacks.
Use: Dyes.
Shipping regulations: None. *

para-**nitro**-ortho-**toluidine** (4-nitro-2-toluidine) $NO_2C_6H_3(CH_3)NH_2$.
Properties: Yellow crystalline solid. Soluble in alcohol and ether. M. p. 104°C.
Derivation: From ortho-toluidine by nitration.
Method of purification: Crystallization.
Grades: Technical.
Containers: Barrels; fiber drums; multiwall paper sacks.
Use: Dyes.
Shipping regulations: None. *

para-**nitro**-alpha-**tolunitrile.** See para-nitrobenzyl cyanide.

nitrotrichloromethane. See chloropicrin.

meta-**nitrotrifluoromethylbenzene.** See meta-nitrobenzotrifluoride.

2-nitro-4-trifluoromethylbenzonitrile.
Constants: B. p. 156-158°C (18-19 mm); m. p. 47-48°C.
Derivation: From a cyclic halogen compound by heating with copper cyanide in the presence of amines.
Grades: Technical.
Use: Dyes.

meta-**nitro**-alpha, alpha, alpha-**trifluorotoluene.** See meta-nitrobenzotrifluoride.

nitrourea $NH_2CONHNO_2$.
Properties: White crystalline powder; slightly soluble in water; soluble in alcohol or ether; m. p. 158-159°C.
Derivation: Urea nitrate and concentrated sulfuric acid.
Containers: Steel drums.
Caution! Decomposed by heat. Explosive.
Shipping regulations: Explosive, class A; high explosive label. *

nitrous diphenylamide. See N-nitrosodiphenylamine.

nitrous oxide (nitrogen monoxide; laughing gas) N_2O.
Properties: Colorless, sweet-tasting gas; nonflammable; condensable into a colorless liquid; sp. gr.: gas, 1.52 referred to air; liquid, 1.22 (−89°C); m. p. liquid, −102°C; b. p. −89.8°C; soluble in alcohol and concentrated sulfuric acid; slightly soluble in water.
Grades: Pure, 98.0% min; U. S. P. XVI.
Containers: Steel cylinders.
Uses: Anesthetic in dentistry and surgery; food aerosols.
Shipping regulations: Nonflammable gas. Green label. *

"Nitrox." [292] See "Solvay Nitrox."

nitroxanthic acid. See picric acid.

nitroxylene (dimethylnitrobenzene) $C_6H_3(CH_3)_2NO_2$. (a) 4-nitro-ortho-xylene; (b) 4-nitro-meta-xylene; (c) 2-nitro-para-xylene.
Properties: (a) Pale yellow, crystalline needles. (b) Yellow liquid becoming red-brown on exposure. (c) Pale yellow liquid becoming red-brown on exposure. Soluble in alcohol and ether; insoluble in water. Sp. gr. (a) 1.139, (b) 1.135, (c) 1.132; m. p. (a) 29°C, (b) 2°C; b. p. (a) 258°C, (b) 246°C, (c) 240°C.
Derivation: By nitrating xylene, resulting in a mixture of the three nitroxylenes, consisting largely of the 4-nitro-meta-xylene.
Method of purification: Rectification.
Grades: Technical.
Containers: Iron drums.
Uses: Organic synthesis; gelatinizing accelerators for pyroxylin.
Shipping regulations: Poison, class B. Poison label. *

"Ni-Vee." [283] Trademark for nickel tin bronzes containing 5% nickel, 5% tin, 2% zinc and 0 to 20% lead, balance copper. Simple heat treatments applied to these compositions improve the cast strength. Useful in most engineering applications requiring strengths up to 85,000 psi tensile strength.

nivenite. A variety of uraninite (q. v.) containing rare earth metals.

N. N. D. Abbreviation for New and Nonofficial Drugs, an annual compilation published by the Council on Drugs of the American Medical Association. In general it lists the more recent drugs not included in the U. S. P. or N. F. This dictionary has descriptions of most of the drugs (those which are distinct chemical substances) in the 1960 N. N. D., and where the grade N. N. D. is cited, it means that material is in the 1960 edition.

"NNO." [89] Trademark for glycerol mannitan laurate, used as a wetting agent and spreader for contact insecticides, insecticidal stomach poisons, and hormone sprays.

"NNOR." [89] Trademark for a combination of "NNO" and rotenone, used for control of common greenhouse and garden insects.

No. Symbol for the element nobelium (q. v.). See also actinide elements.

Noah's ark. See cypripedium.

nobelium No. Synthetic radioactive element number 102; one of the actinide series of elements; produced in a cyclotron by bombarding curium 244 with nuclei of carbon 13 accelerated to high energies. The original discovery was at the Nobel Institute of Physics in Stockholm, Sweden. The element has a half-life of only about 10 minutes.

Like the other actinide elements of atomic numbers near 100, the element probably has properties similar to those in the rare earth metals.

noble laurel. See laurus.

noble metals. Gold, silver, mercury, platinum, palladium, iridium, rhodium, ruthenium and osmium.

"NOBS" No. 1 Accelerator. [57] Trademark for a selected blend of N-oxydiethylene benzothiazole-2-sulfenamide and benzothiazyl disulfide. A delayed-action accelerator for use in furnace black-rubber stocks where processing safety is important.

"NOBS" Special Accelerator. [57] Trademark for N-oxydiethylene benzothiazole-2-sulfenamide. A delayed-action accelerator for use in furnace black-rubber stocks where processing safety is important. Imparts greater scorch protection than "NOBS" No. 1.

"No Bunt." [147] Brand name for a wettable powder seed disinfectant containing hexachlorobenzene.
Containers: 5-lb bags; 40- and 75-lb drums; liquid 1-gal jugs and 12 ½- and 25-gal drums.
Uses: Control of covered smut of wheat and loose and covered smut of sorghum.

"Noctec." [412] Trademark for chloral hydrate (q. v.).

"No D-K." [256] Trade name for a wood preservative.

"Nogas." [28] Trademark for metal pickling agent.
Properties: Amorphous brown powder, soluble in water and acids. Will give foam-producing solutions when dissolved in water or acids.
Containers: 65-lb fiber drums; 325-lb bbls.
Use: As an addition agent in acid pickling of iron and steel products to produce a foam blanket and to improve the atmosphere around pickling operations by elimination of acid mist particles usually given off by the pickling operation.

"Nolane." [342] Trademark for 2, 6-allyl-dichlorosilane-resorcinol, used for priming glass fibers, i. e., prior to laminating.

"Noludar." [190] Trademark for a brand of methyprylon (q. v.).

nonadecane $C_{19}H_{40}$ or $CH_3(CH_2)_{17}CH_3$.
Properties: Leaflets; soluble in alcohol; insoluble in water; soluble in ether; sp. gr. 0.777; m. p. 32°C; b. p. 330°C.
Grades: Technical.
Use: Organic synthesis.

n-nonadecanoic acid $CH_3(CH_2)_{17}COOH$. A saturated fatty acid normally not found in natural vegetable fats or waxes.
Properties: Colorless crystals; m.p. 68.7°C; b. p. 297°C (100 mm); soluble in alcohol and ether; insoluble in water. Synthetic product available 99% pure for organic synthesis.

gamma-**nonalactone.** See gamma-nonyl lactone.

nonanal (pelargonic aldehyde; n-nonyl aldehyde; aldehyde C-9) $C_8H_{17}CHO$.
Properties: Colorless liquid with an orange-rose odor; sp. gr. 0.822-0.830; refractive index 1.424-1.429. Soluble in 3 volumes of 70% alcohol.
Containers: Glass bottles; 1-, 5- and 10-gal tins.
Use: Perfumery.
Shipping regulations: None. *

nonane (nonyl hydride) C_9H_{20} or $CH_3(CH_2)_7CH_3$.
Properties: Colorless liquid. Soluble in alcohol; insoluble in water. Sp. gr. 0.722; b. p. 150.7°C; m. p. −51°C; refractive index 1.40561 (20°C); C. S. T. aniline 74.9°C; flash point 44°C.
Grades: Technical (95%); 99%; research.
Containers: Glass bottles; drums.
Use: Organic synthesis.
Hazard: Flammable liquid.
Shipping regulations: Red label not required. *

nonanedioic acid. See azelaic acid.

nonanoic acid. See pelargonic acid.

nonanol. See nonyl alcohol.

nonanoyl chloride. See pelargonyl chloride.

nondrying oils. See drying oils.

nonene. See nonylene.

"Non-Fer-Al." [244] Brand name for a high purity precipitated calcium carbonate.
Properties: Oil absorption, 30-35; density as shipped, 55-60 lbs/cu ft; wt/solid gal, 22.07 lbs; color, pure white; particle size, 5-10 microns.
Derivation: Precipitated calcium carbonate.
Containers: Multiwall paper bags, 50-lbs net.
Uses: Paint, plastics, rubber.

non-geminate. See gem-.

"Nonic." [204] Trademark for a line of non-ionic surface-active agents based on polyethylene glycol tert-dodecylthioether or related compounds.
Uses: Detergents; paper manufacture; grease removal; wetting agent; emulsifier; clouding agent.

nonionic detergents. See detergents, synthetic.

"Nonisol." [219] Trademark for a series of nonionic surface-active fatty acid esters of higher polyglycols.
Properties: Light color; bland odor. Free fatty acid content is maintained at less than 3% and ash content less than 0.1%. Comparatively low melting points, high boiling points and low vapor pressure. Either soluble or readily dispersible in cold water; characteristically insoluble in hot water. Solubility in presence of electrolyte is good. All soluble in polar and semi-polar solvents with the exception of glycerine and glycols; solubility in aliphatic hydrocarbons varies inversely with water solubility. Stable to

heating at 200°C or autoclaving.
Uses: Cosmetics; solvent emulsions; polishes; rust preventative oils; insecticide and agricultural sprays.

n-nonoic acid. See pelargonic acid.

nonoic acids. Acids of the formula $C_8H_{17}COOH$, of which there are many possible isomers. Pelargonic acid (q. v.) is the normal or straight-chain acid. Various mixtures of branched chain nonoic acids are recovered from the products of the Fisher-Tropsch process.

"Nonox." [206] Brand name for rubber antioxidant.

"Nonoxol." [206] Brand name for antioxidants for oils, fats and oxygenated soap powders.

non-viscous neutral oil. A neutral oil of viscosity lower than 135 S. U. S. at 100°F.

nonwoven fabrics. Staple lengths of any of the common textile fibers mechanically positioned into a random pattern, and then bonded with suitable resins or plastics to form sheets of the nonwoven fabric.

nonyl acetate $C_9H_{19}OOCCH_3$. Exists in numerous isomeric forms so that any one of several compounds may be indicated by this name and formula, particularly since the Oxo and Fisher-Tropsch synthesis have been developed. See following:

n-nonyl acetate (acetate C-9) $CH_3COO(CH_2)_8CH_3$.
Properties: Colorless liquid; strong and pungent odor; sp. gr. 0.864-0.868; refractive index 1.422-1.426. Soluble in 4 volumes of 70% alcohol.
Containers: Glass bottles; 1-, 5- and 10-gal tins.
Use: Perfumery.
Shipping regulations: None. *

diisobutyl carbinyl acetate $(C_4H_9)_2CHOOCCH_3$.
Properties: Colorless liquid; sp. gr. 0.8530 (20/20°C); b. p. 192.4°C; f. p. −48.1°C; refractive index 1.4152 (20°C); slightly soluble in water (0.02 wt% at 20°C).

nonyl alcohol $C_9H_{19}OH$. Exists in numerous isomeric forms, any one of which may be referred to by the above name and formula. See following:

n-nonyl alcohol (nonanol; alcohol C-9; octyl carbinol; pelargonic alcohol) $CH_3(CH_2)_7CH_2OH$.
Properties: Colorless liquid, with rosy odor; sp. gr. 0.826-0.829; refractive index 1.431-1.435; m. p. −5°C; b. p. 215°C. Soluble in 7 volumes of 50% alcohol; insoluble in water.
Containers: Glass bottles; 1-, 5- and 10-gal tins.
Uses: Perfumery; flavors.
Shipping regulations: None. *

nonyl aldehyde $C_8H_{17}CHO$. There are available mixtures of various isomeric nonyl aldehydes. These are produced by the Fisher-Tropsch or related process, and properties vary with the particular source. For the n-nonyl aldehyde, see nonanal.

n-nonylamine $C_9H_{19}NH$.
Properties: B. p. 75-85°C (20 mm); sp. gr. 0.798 (25°C); refractive index 1.4366 (n 20/D); may be prepared by the catalytic reduction of nonanonitrile and from nonyl halides by conventional techniques.

tert-nonylamine. Principally tert-$C_9H_{19}NH_2$ and tert-$C_{10}H_{21}NH_2$.
Properties: Boiling range 160-174°C; sp. gr. 0.789 (25°C); refractive index 1.428 (25°C); flash point 120°F; insoluble in water; soluble in common organic solvents, especially in petroleum hydrocarbons.
Containers: Drums; tank cars.
Caution! Avoid prolonged contact with skin; prolonged inhalation.
Uses: Intermediate for rubber accelerators, insecticides, fungicides, dyestuffs, pharmaceuticals.

nonylbenzene $C_9H_{19}C_6H_5$.
Properties: Light-straw; faint aromatic odor; boiling range 245-252°C; sp. gr. 0.864 (20/20°C); refractive index 1.488 (20°C); viscosity 41.9 cps (20°C); flash point 210°F.
Uses: Raw material for the manufacture of surface-active agents.

nonyl benzoate $C_9H_{19}OOCC_6H_5$.
Properties: B. p. 140-145°C (5 mm); sp. gr. 0.961 (25°C); refractive index 1.4911 (n 20/D); viscosity 5.35 centistokes (100°F), 1.68 centistokes (210°F).

nonyl bromide $C_9H_{19}Br$.
Properties: B. p. 81-85°C (10 mm); sp. gr. 1.101 (25°C); refractive index 1.4583 (n 20/D).
Derivation: High yields of nonyl bromide are obtained by passing hydrogen bromide into the alcohol while heating or by refluxing with aqueous hydrobromic acid in the presence of an acid catalyst.

nonyl chloride $C_9H_{19}Cl$.
Properties: B. p. 58-63°C (8 mm); sp. gr. 0.878 (25°C); refractive index 1.4379 (n 20/D).
Derivation: Nonyl alcohol reacts with hydrogen chloride at elevated temperatures and pressure to give nonyl chloride. It can also be readily made by refluxing a mixture of concentrated hydrochloric acid with the alcohol in the presence of zinc chloride.

1-nonylene (1-nonene) C_9H_{18} or $CH_3(CH_2)_6CH:CH_2$.
Properties: Colorless liquid; soluble in alcohol; insoluble in water. Sp. gr. 0.7433; b. p. 149.9°C.
Grades: Technical.
Use: Organic synthesis; wetting agent; lube oil additive.

nonylenes (nonenes). A mixture of hydrocarbons boiling about 279-296°F; sp. gr. 0.7440 (60°F); used for alkylation and as a chemical intermediate for surfactants and Oxo alcohols.

nonyl hydride. See nonane.

*See "I.C.C. Shipping Regulations," page xiii.
Reference numbers refer to name of manufacturer. See "List of Manufacturers," page v.

n-nonylic acid. See pelargonic acid.

gamma-**nonyl lactone** (gamma-nonalactone; aldehyde C-18; prunolide; nonanolide-1, 4; coconut aldehyde; gamma-n-amylbutyrolactone) $C_9H_{16}O_2$.
Properties: Yellowish to almost colorless liquid; coconut-like odor; sp. gr. 0.956-0.963; refractive index 1.447; soluble in 5 volumes of 50% alcohol.
Uses: Perfumery; flavors.

nonyl methacrylate. Ester of nonyl alcohol.
Properties: B. p. 76-78°C (1 mm); sp. gr. 0.880 (25°C); refractive index 1.4422 (n 20/D).

nonyl nonanoate. Ester of nonyl alcohol.
Properties: B. p. 148-166°C (10 mm); sp. gr. 0.863 (25°C); refractive index 1.4419 (n 20/D).

nonyl phenol $C_9H_{19}C_6H_4OH$. Trade designation for a mixture of isomeric monoalkyl phenols, predominantly para-substituted.
Properties: Pale yellow, viscous liquid with a slight phenolic odor. Insoluble in water; soluble in most organic solvents.
Typical specifications: Sp. gr. (20/20°C) 0.94-0.95; boiling range (95%) 283-302°C; color (APHA) 200; hydroxyl number 240-255.
Grades: Technical.
Containers: 5-, 55-gal drums; tank cars.
Uses: Intermediate for surface active agents, lube oil additives; wetting agents, stabilizers, petroleum demulsifiers, fungicides; bactericides; dyes; drugs; adhesives; rubber chemicals; phenolic resins and plasticizers.

nonyl thiocyanate $C_9H_{19}SCN$.
Properties: B. p. 84-86.5°C (1 mm); sp. gr. 0.919 (25°C); refractive index 1.4696 (n 20/D); nonyl thiocyanate can be made from nonyl chloride by refluxing with alcoholic sodium thiocyanate solution using conventional technique.

nonyl trichloroarsine
Shipping regulations: Corrosive liquid. White label. *

"No-Odorol." [57] A sulfonated oil used for textile finishing and softening. Recommended for cotton, silk and rayon fabrics, whether dyed, printed or bleached. Fabrics finished with "No-Odorol" possess a full soft, drapy hand, free from development of discoloration or odor.

"Nopalcol." [309] Trademark for a series of polyoxyethylene fatty acid derivatives (esters, amido amines, amides) used as detergents, penetrants, emulsifying or wetting agents.

"Nop-cap." [309] Trademark for calcium pantothenate containing products for the fortification of foods and feeds.

"Nopcay." [309] Trademark for a series of dry vitamin-containing products for fortifying animal and poultry foods and feeds.

"Nopcocell." [309] Trademark for polyurethane foamed plastic for general uses.

"Nopcofoam." [309] Trademark for a series of polyurethane-foamed plastics in sheet, slab, block, and like form.

"Nopcogen." [309] Trademark for a series of fatty acid derivatives, including alkylolamides, polyamine condensates, alkylolamine condensates, and sulfonated amine condensates, used as softeners, detergents, wetting, or emulsifying agents.

"Nopcolene." [309] Trademark for a series of fat-liquoring and finishing compositions.

"Nopcosant." [309] Trademark for a series of naphthalene sulfonated derivatives for general use in the industrial arts.

"Nopcoset." [309] Trademark for a series of polyvinyl acetate emulsions for general use in the industrial arts, particularly for use in the paper, textile, paint, leather and wood-working industries.

"Nopcosulf." [309] Trademark for a series of sulfated oils for general use in the industrial arts.

"Nop-Dry." [309] Trademark for high potency vitamin-containing solids for fortifying foods and feeds with vitamins.

"Nopol." [296] Trade name for a synthetic bicyclic primary alcohol derived from beta-pinene, a major constituent of turpentine. Used as a chemical raw material for perfumery, wetting agents.

NOPON. See para-bis[2-(5-alpha-naphthyloxazolyl)benzene].

"Nop-Sol." [309] Trademark for water-dispersible vitamin concentrates.

nor-. A prefix signifying normal and indicating the parent from which another compound may theoretically be derived, usually by removal of one or more carbon atoms (with attached hydrogen). It is used rather ambiguously in terpene and steroid chemistry, and differs completely from the meanin of normal which is shown by the abbreviation n-. See normal.

"Norad." [121] Trade name for a water-soluble adhesive used to prevent slippage of palletized multiwall bags. Shipped in 55-gal steel drums.

"Norane" R. [42] Proprietary compound. A quaternary type of compound with thermosetting characteristics.
Properties: Tan colored paste. Disperses in water at 70°C for application.
Containers: 30-35 gal fiber containers.
Uses: Applied to cotton fabrics to produce durable water-repellent finish.

"Norane" Silicone. [42] Proprietary product. A 30% silicone resin emulsion.
Properties: White emulsion. Disperses readily in water and resin mixtures of neutral pH at 30°C.
Containers: 55-gal lined steel drums.

Use: Used alone or in combination resins with suitable catalysts on cotton, rayon, acetate, nylon and blends for durable water repellent finish.

"Norane" 4 Star A. [42] Proprietary product. A hydrophobic thermosetting resin composition.
Properties: Cream colored paste. Disperses readily in water at 40°C.
Containers: 55-gal steel drums.
Use: Durable type of water repellent treatment for cotton, rayon and their blends.

"Norane" 4 Star GG-2. [42] Proprietary compound. A special fatty acid amide condensation suitable for use with textile resins.
Properties: A viscous white emulsion, disperses in water at 50°C.
Containers: 55-gal steel drums.
Uses: Durable water repellent for cellulosic textiles. When combined with thermosetting resins, shrinkproofing and creaseproofing effects can also be obtained simultaneously.

"Norane" W1 and W2. [42] Proprietary compounds. Wax emulsion and metallic salt combination; when combined, make water-repellent.
Properties: Forms white dispersion in water at 85°C.
Containers: 55-gal steel drums for each.
Use: Two-product fabric water-repellent combination applied from single bath for woolen or nylon textile fabrics especially where durable repellency is desired.

"Norbide." [249] Trademark for boron carbide (q. v.).

nordefrin hydrochloride
$C_6H_3(OH)_2CH(OH)CH(NH_2)(CH_3) \cdot HCl$. dℓ-1-(3',4'-Dihydroxyphenyl)-2-amino-1-propanol hydrochloride; dℓ-3,4-dihydroxy-norephedrine hydrochloride.
Properties: White crystals; melting range 175-180°C; soluble in water, alcohol; insoluble in ether.
Grade: N. F. XI.
Use: Medicine.

nordhausen acid. Fuming sulfuric acid of sp. gr. 1.86-1.90.

nordihydroguaiaretic acid (NDGA) $[C_6H_3(OH)_2CH_2CH(CH_3)]_2$. 4,4'-(2,3-Dimethyltetramethylene)dipyrocatechol).
Properties: Crystals from acetic acid, m. p. 184-185°C. Soluble in methanol, ethyl alcohol, ether; slightly soluble in hot water, chloroform; nearly insoluble in benzene, petroleum ether.
Derivation: Extraction from guaiacum; also synthetically.
Uses: Food grade antioxidant to retard rancidity of fats and oils.

dℓ-norephedrine hydrochloride. See phenylpropanolamine hydrochloride.

ℓ-norepinephrine. See levarterenol.

ℓ-norepinephrine bitartrate. See levarterenol bitartrate.

norethandrolone (17-alpha-ethyl-19-nortestosterone; 17-alpha-ethyl-17-hydroxy-19-nor-4-androsten-3-one) $C_{20}H_{30}O_2$.
Properties: Crystals; m. p. 140-141°C. Insoluble in water; soluble in alcohol, benzene, ether, ethyl acetate.
Derivation: Prepared by catalytic hydrogenation of 17-alpha-ethynyl-19-nortestosterone.
Grade: N. N. D.
Use: Medicine.

norethindrone (19-nor-17-alpha-ethynyltestosterone; 17-alpha-ethinyl-19-nortestosterone) $C_{20}H_{26}O_2$.
Properties: White to creamy white, odorless; crystalline powder. Melting range 202-208°C.
Derivation: Prepared from 19-nor-4-androstene-3,17-dione.
Grade: N. D.
Use: Medicine.

19-nor-17-alpha-**ethynyltestosterone.** See norethindrone.

Norge niter. See calcium nitrate.

"Norisodrine Sulfate." [3] Trademark for isoproterenol sulfate (q. v.).

"Norit." [107] Trade name for activated adsorption carbons of vegetable origin.
Grades: Powder and several granular sizes in different qualities for different applications.
Containers: Powder form: 50-lb bags; 100-, 150-lb drums. Granular form: 100-, 200-lb drums.
Uses: Decolorizing; adsorption of impurities, bad odors and tastes; catalyst carrier; solvent recovery; gas mask carbon.

norleucine (alpha-aminocaproic acid) $CH_3(CH_2)_3CH(NH_2)COOH$. A nonessential amino acid found naturally in the L(+) form.
Properties: Leaflets, crystallized from water.
DL-norleucine: Soluble in water; slightly soluble in alcohol; soluble in acids; decomposes 327°C. Available commercially.
L(+)-norleucine: Slightly sweet; sublimes 275-280°C; m. p. 301°C (dec).
D(-)-norleucine: Bitter; partially sublimes 275-280°C; m. p. 301°C (dec).
Derivation: Found in traces in proteins; organic synthesis.
Use: Biochemical research.

"Norlig" Binders and Dispersants. [121] Trade name for a line of unmodified or partially modified lignosulfonates derived from spent sulfite liquor. Available in both powder and liquid forms. Liquids are darker brown in color and more sticky as solids concentration is increased.
Uses: Road binder; linoleum paste; foundry products; leather tanning; gypsum board.

"Norlutate." [330] Trade name for norethindrone acetate, 17-alpha-ethinyl, 19 nortestosterone acetate, $C_{22}H_{28}O_3$.
Properties: White to creamy white crystalline powder. Soluble in dioxane, insoluble in water. Melting range: 157-163°C.
Use: Medicine.

"Norlutin." [330] Trademark for norethindrone, (q. v.).

normal.
1. A term, abbreviated as the prefix n-, used to designate those hydrocarbons or hydrocarbon radicals whose molecules contain a single unbranched chain of carbon atoms. Thus normal butane or n-butane is the compound whose molecular structure is indicated by the formula $HCH_2CH_2CH_2CH_3$.

See also amyl alcohol for examples.

2. A term used in analytical chemistry and abbreviated as N, to describe a solution containing one gram equivalent weight of reactive material per liter of solution.

See also nor-.

"Normasal." [304] Trademark for normal lead salicylate vinyl stabilizer.
Properties: Soft creamy-white crystalline powder; sp. gr. 2.36; refractive index 1.78.
Containers: Fiberboard drums containing 25 and 200 lbs.
Uses: Stabilizer or costabilizer for vinyl flooring and other vinyl compounds; in paints to impart durability, retard chalking and prolong color fastness.

"Normocytin." [57] Trademark for vitamin B_{12}.

norphytane. See "Pristane."

"Norva." [51] Trademark for non-soap grease for specialized applications as, for instance, where high temperature prevails. Has water resistance and good adhesiveness.

Norway saltpeter. See ammonium nitrate.

Norwegian saltpeter. See calcium nitrate.

noscapine (ℓ-alpha-narcotine; narcosine). An isoquinoline alkaloid of opium. $C_{22}H_{23}NO_7$.
Properties: Fine white crystalline powder. M.p. 176°C; sublimes at 150-160°C. Insoluble in water; practically insoluble in vegetable oils; slightly soluble in hot solutions of potassium and sodium hydroxide; soluble in most organic solvents. Salts formed with acids are dextrorotatory and unstable to water.
Derivation: Prepared from the seed capsules of Papaver somniferum.
Grade: N.F. XI.
Use: Medicine.

noscapine hydrochloride.
Properties: Fine, white powder; odorless; soluble in water, chloroform, methanol, ethanol; slightly soluble in acetone; insoluble in ether; pH of solution (5%) is between 2.5 and 3.5.
Use: Medicine.

nosophen. See iodophthalein.

nosophen sodium. See iodophthalein sodium.

novaculite. An exceedingly fine-grained quartzose rock, used as an abrasive. Occurs in Arkansas, Georgia, Massachusetts, North Carolina, Oklahoma and Tennessee.

"Novadelox." [282] Proprietary name for a mixture of benzoyl peroxide and an inert filler used to bleach flour.

"Novaldin." [162] Trademark for dipyrone.

"Novalgin." [162] Trademark for dipyrone.

"Novatone." [19] Brand name for a proprietary product. Anisic ketone; aromatic chemical. Fused crystals; m.p. 35°C; used in soaps and perfume compounds. Aubepine odor.

"Novege." [233] Trademark for 2,4,5-T compositions.

novobiocin calcium $C_{62}H_{70}CaN_4O_{22} \cdot 2H_2O$. Calcium salt of an antibacterial substance produced by the growth of Streptomyces niveus, or S. spheroides.
Properties: White or practically white crystalline powder. Odorless; soluble in alcohol; slightly soluble in water, ether; very slightly soluble in chloroform; pH of saturated solution 6.5-8.5.
Grade: U.S.P. XVI.
Use: Medicine (antibiotic).

novobiocin sodium $C_{31}H_{35}N_2NaO_{11}$. Sodium salt of an antibacterial substance produced by the growth of Streptomyces niveus or S. spheroides.
Properties: White or practically white crystalline powder; odorless; hygroscopic. Very soluble in water; soluble in alcohol and glycerol; pH of saturated solution 6.5-8.5.
Grade: U.S.P. XVI.
Use: Medicine (antibiotic).

"Novocaine." [162] Trademark for a brand of procaine hydrochloride.

novolaks. Thermoplastic, soluble phenol-formaldehyde resins (q. v.) obtained by the use of acid catalysts or of excess phenol. They can be cured to the thermosetting, insoluble form with hexamethylenetetramine. Used principally in varnishes.

"Novoviol." [188] Brand name used to identify a series of ionones.

"Noxfish." [342] Trademark for rotenone fish-toxicant compositions.

Np. Symbol for neptunium.

N-P-K mixtures. Fertilizers containing nitrogen, phosphorus and potassium. These are usually characterized by numbers such as 5-10-10, meaning 5% nitrogen, 10% phosphorus as P_2O_5, and 10% potassium as K_2O. The percentages show the amount of available N, P or K, rather than the total amount present.

NPN. Abbreviation for n-propyl nitrate.

NPO. See alpha-naphthylphenyloxazole.

"NSAE." [328] Brand name of an alkyl aromatic sulfonate that is a highly effective anionic wetting agent in concentrated solutions. In the textile processing industry it is used as an assistant in detergent and dispersing operations, with moderate foaming action.

It is also used in fruit and vegetable washing, and in the manufacture of insecticides and fungicides; as a conditioner for paper mill felts; and a pigment dispersant in printing.

NSR. Abbreviation for nitrile silicone rubber.

NTP. Abbreviation for normal temperature and pressure, i.e. 0°C and 760 mm pressure.

n-type crystal. See transistor.

"Nuact Paste." [74] Trademark for a lead "feeder" drier to counteract loss of drying properties in paints during storage.

"Nuade." [74] Trademark for rubber base product which increases tack of roller mill pastes resulting in faster, lower cost grinding.

"Nubrite." [134] Trademark for brightener for nickel plating solutions.

"Nuchar Activated Carbon." [228] Trade name for carbon of vegetable origin.

Grades: Industrial and water purification.

Containers: Multiwall paper bags; bulk hopper cars.

Uses: Deodorization and decolorization.

nuclear chemistry. A term used to denote the chemical aspects of the study of the atomic nucleus, the central, positively charged portion of the atom. Nuclear chemistry includes the investigation of the fission or disruption of nuclei, their joining or fusion, and also the characterization of the properties of the reaction products.

nuclear cross section. See cross section.

nuclear energy. Energy resulting from the rearrangement of the nuclei of atoms, either the fragmentation of heavy nuclei as in the fission of U-235 or plutonium into two approximately equal parts, or the formation of heavier nuclei from light ones as in the fusion of hydrogen into helium. The energy is equivalent to a slight loss of mass in the system undergoing such a transformation according to the Einstein mass-energy relationship ($E = mc^2$).

The energy manifests itself predominantly as the kinetic energy of the fragments, especially the energy of fission as released in a nuclear reactor. A smaller portion of the energy is emitted as gamma rays from the fission process, or as beta and gamma rays from the fission products. The fission products may be separated from the nuclear fuel and used as a separate source of nuclear energy. Power plants for satellites have been designed using the nuclear energy of fission products to generate electrical energy for radio communication. Radioactive isotopes prepared by irradiation of materials in nuclear reactors have also been used as a source of nuclear energy. In one or another of these forms nuclear energy has been applied to chemical processes such as the synthesis of nitric acid from air, the synthesis of hydrazine, the cross-linking (hardening) of polymers, and the sterilization of pharmaceuticals.

nuclear fuel. Any material which may undergo the appropriate reaction and be the source of energy in a fusion or fission nuclear reactor. At the present state of technology the term usually means uranium, thorium, or plutonium, either as natural materials, or enriched or synthesized isotopes of these elements in some chemical form and physical state suitable for the reactor in question. They are used as solutions, or as shaped metals, oxides, or carbides, giving rise to the terms fuel element, fuel plate, or fuel rod.

nuclear fusion. A nuclear reaction in which the nuclei of light atoms fuse together to form heavier nuclei, and since the heavier ones are more stable, energy is released in the process. These energy releases are large, and much research is being devoted to developing the process as a practical power source. Very high temperatures are necessary in order that the nuclei will have kinetic energies high enough to overcome the repulsive forces between the electric charges on the nuclei. The source of energy in the sun is a reaction of this kind in which the nuclei of hydrogen atoms are combined to form helium at the temperature of the sun, and when high temperature is the driving force of the reaction it is called a thermonuclear reaction. Uncontrolled release of fusion energy has been achieved on earth in the hydrogen bomb in which a fission reaction acts as the trigger and supplies the initial high temperature to start the fusion reaction. If controlled fusion can be achieved, it will have the advantages of a cheap and practically unlimited source of energy (deuterium) and no radioactive by-products such as are produced in a fission reactor, though an installation will probably necessarily be quite large with a high capital cost.

nuclear magnetic resonance. An electromagnetic field method of determining whether hydrogen atoms in a molecule are part of a CH_3- group, a $-CH_2-$ group, an -OH group, or other characteristic grouping. The method also indicates the number of hydrogen atoms in each category.

The method is carried out by placing the sample in a strong constant magnetic field, and then applying a perpendicular radio frequency alternating magnetic field. At certain frequencies of the latter field the hydrogen atom nucleus will absorb and emit energy, and the frequency and amount of energy depend on the characteristic grouping in which the hydrogen atom is located. The method is particularly useful in complex substances with many hydrogen atoms.

nuclear reaction. Any process or change that involves the nucleus of an atom.

Spontaneous radioactivity is a nuclear reaction for which only a single nucleus is required. Nuclear reactions also occur when a nucleus is bombarded with particles such as neutrons, protons, positrons, etc. Fission of uranium 235, and fusion of hydrogen into helium are other types of nuclear reactions. Nuclear reactions are usually accompanied by large energy changes, that are millions of times greater than those of ordinary chemical reactions such as combustion.

nuclear reactor. A device for the controlled liberation of nuclear energy, particularly from the fission of uranium. The essential parts of a reactor are (1) the nuclear fuel (q. v.); (2) the moderator, which serves to slow down the energetic neutrons produced in fission, because low energy neutrons are more effective at inducing fission than fast ones. Graphite, beryllium, heavy water, or ordinary water are the usual moderators; (3) the coolant - water, mercury, molten sodium or molten salt, or a gas - which is circulated through the reactor core to remove the heat energy of fission; (4) the shield, usually of ordinary or high density concrete which surrounds the reactor to absorb the nuclear radiations associated with the fission process for the protection of personnel; (5) the control system - radiation detectors which monitor the radiation level, especially the neutron flux level in the reactor core, and through electronic and hydraulic mechanisms determine the degree of insertion of control rods. These are rods of neutron-absorbing material that soak up excess neutrons and maintain the level of operation at a given value.

The term reactor may be modified for several purposes. The terms fast, intermediate or epithermal, and thermal refer to the energy of the neutrons propagating the chain reaction. A fast reactor will have no moderator. In a heterogeneous reactor the lumps or rods of fissionable material are imbedded at regular intervals in a mass of the moderator. In a homogeneous reactor the fuel is dissolved in a liquid moderator or carrier and the entire solution is circulated as coolant. Breeder reactors use surplus neutrons to produce more fissionable material; thus in one type uranium 238 is converted to U-235.

Nuclear reactors are generally classified into types depending on the purposes for which they are intended. Power reactors are associated with electrical generating equipment and use the heat from the reactor to produce electricity. Test reactors are designed to have regions of very high radiation density for the exposure of materials for engineering test. Production reactors are used for the production of fissionable material like plutonium or to make radioactive isotopes. Research reactors are designed for flexibility of application, for the study of reactor behavior or as a source of nuclear radiations for studies in the physical and biological sciences.

Sizes of reactors are usually quoted as gross heat energy output, but in electrical units as so many thermal kilowatts. Power reactor sizes however are sometimes quoted in net electric kilowatts of electric power produced. The size range from small research reactors to large power reactors is from a few watts to several hundred thousand kilowatts.

nuclear transformation. See disintegration.

nucleic acid, metallic salts. Water-soluble salts of variable composition, depending on conditions of formation. The percentage of metal varies within definite limits.

nucleic acids. Complex compounds found in all living cells. They are usually found chemically bound to proteins to form nucleoproteins. Nucleic acids are of high molecular weight and are easily changed by many mild chemical reagents. They contain carbon, hydrogen, oxygen, nitrogen (15-16%), and phosphorus (9-10%).

The fundamental units of nucleic acid are nucleotides (q. v.); nucleic acids are polynucleotides in which the nucleotides are linked by phosphate bridges. Upon extensive heating in the presence of water (hydrolysis), nucleic acids yield a mixture of purines and pyrimidines, D-ribose or D-deoxyribose, and phosphoric acid. Nucleic acids are subdivided into two types: ribonucleic acid (RNA), containing the sugar D-ribose; and deoxyribose nucleic acid (DNA) containing the sugar D-deoxyribose. Good sources of nucleic acids are salmon, thymus, yeast, and wheat kernel embryo.

Uses: Biochemical and medical research in heredity, virus diseases and cancer.

nucleon. General name applied to neutrons and protons, the essential constituents of atomic nuclei, and also used as a class name for fundamental particles of that mass. See fundamental particle.

nucleonics. A general term, not defined by other than common usage, but which is usually taken to encompass the broad areas of science and technology involving nuclear phenomena, with emphasis on the applied aspects.

nucleoprotein. A type of protein universally present in the nuclei and the surrounding cytoplasm of living cells. A nucleoprotein is composed of a protein, which is rich in basic amino acids, and a nucleic acid (q. v.). The nucleic acid portion is isolated and used in medical and biochemical research.

nucleosides. Compounds of importance in physiological and medical research, obtained during partial decomposition (hydrolysis) of nucleic acids (q. v.), and containing a purine or pyrimidine base linked to either D-ribose, forming ribosides, or D-deoxyribose, forming deoxyribosides. For specific nucleosides see adenosine, cytidine, guanosine and uridine.

nucleotides. The fundamental units of nucleic acids (q.v.); some are important coenzymes. The nucleotides found in nucleic acids are phosphate monoesters of nucleosides (q.v.); examples are adenylic acid, guanylic acid, and uridylic acid.

The term nucleotide is also applied to compounds not found in nucleic acids and which contain substances other than the usual purines and pyrimidines. Nucleotides of the latter type are modified vitamins and function as coenzymes; examples are riboflavin phosphate (flavin mononucleotide) flavin adenine nucleotide, diphosphopyridine nucleotide, triphosphopyridine nucleotide, and coenzyme A.

nucleotinphosphoric acid. See thyminic acid.

nucleus.

1. The dense positively charged central portion of an atom. This contains essentially all of the mass of the atom and determines its kind. Its linear dimension is only about one ten-thousandth of the whole atom. The remainder of the volume of the atom is the electron cloud which neutralizes the nuclear charge.
2. The central portion of a living cell.

nuclide. A particular kind of atom, characterized by the mass, the charge (number of protons), and the energy content of its nucleus. A radionuclide is a radioactive nuclide.

"NuGreen." [28] Trademark for a fertilizer compound containing 45% urea nitrogen.
Properties: Tiny round green pellets, much like buckshot in size and shape, specially treated to maintain free-flowing, non-caking characteristics.
Containers: 80-lb paper bags.
Use: For spray application to the foliage of plants or for soil application either direct or in mixed fertilizers.

"NuGreen" L-B. [28] Trademark for a fertilizer compound containing 45% urea nitrogen, less than 0.2% biuret.
Properties: Light tan granular solid specially treated to maintain relatively free-flowing, non-caking characteristics.
Containers: 80-lb paper bags.
Use: As a low-biuret nitrogen fertilizer for spray application to citrus crops.

"Nu-Iron." [93] Trade name for an iron supplement for plants.
Properties: Yellow powder; fine particle size; water insoluble; non-hygroscopic; iron in neutral form, 30%; stable in storage; forms good water suspension.
Containers: 50-lb bags.
Uses: Corrects iron deficiencies in plants, particularly ornamentals, grasses and vegetables. Most effective as foliar spray or dust.

"Nulix 15." [79] Trade name for limed polymerized wood rosin. Approx. 6.5% lime added per 100 lbs. of rosin.
Constants: M.p. (capillary tube) 150°C; m.p. (ball and ring) 168°C; acid number 55; color M.
Containers: Non-returnable, light gauge galvanized drums of about 500 lbs net. Tare 14-16 lbs.
Uses: Linoleum print paint, paint and varnishes, gloss oils, printing inks.

"Nullapon." [307] Now carried under "Cheelox" (q.v.).

"Nulocrystal." [157] Brand name for a partial invert sugar, proprietary pure-food non-crystallizing product, consisting of sucrose, levulose, dextrose and moisture.
Properties: Practically colorless; solids content, 77%; soluble in water, glycerin, glycols, and slightly soluble in alcohol.
Containers: Drums and pails.
Uses: Confectionery; baking; beverages; syrups, pharmaceuticals; animal and vegetable adhesives; paper (as a plasticizer); and varied other uses where glycerin can be used.

"Nulomoline." [157] Brand name for invert sugar; proprietary, pure-food product consisting of levulose, dextrose and moisture.
Properties: Practically colorless; solids content, 77%; soluble in water, glycerin, glycols, and slightly soluble in alcohol.
Containers: Drums and pails.
Uses: Confectionery; baking; beverages; syrups; pharmaceuticals; animal and vegetable adhesives; paper (as a plasticizer) and various other uses, where glycerin can be applied.

"Nu-Manese." [93] Trade name for a manganese supplement for plants.
Properties: Brownish-black powder; water insoluble; acid soluble; essentially MnO.
Containers: 100-lb bags and bulk.
Uses: Source of manganese as a nutritional trace element. Suitable for direct soil application or spraying or dusting.

"Number 20 Flux." [56] Brand name for a specially prepared galvanizing flux. Very water soluble, crystalline material consisting mainly of zinc chloride ($ZnCl_2$) and ammonium chloride (NH_4Cl) with foaming and wetting agents added. Sold in 400-lb barrels.

"Nu-Met." [244] Trademark for an alkaline detergent. Moderately alkaline sudsing detergent yielding available chlorine in solution.
Uses: Manual cleaning in dairy farms, milk plants, food processing plants.

"Numet." [439] Trade name for depleted uranium.
Properties: Uranium metal wherein U-235 isotopic content has been reduced below 0.7 of 1% as found in normal uranium. Has high structural strength coupled with high density of 18.9 g/cc.
Derivation: Processed from uranium hexafluoride as tailing from the gaseous diffusion plants.
Uses: Balance weights for aircraft, high speed rotors in gyro-compasses, gamma

radiation shielding, radioisotope transportation casks and fuel element transfer casks; in general as structural material applicable to radiation shielding.
Hazards: Has slight radioactivity; use must be licensed with Atomic Energy Commission.

"Nuodex 100 V.T." [74] Trademark for quaternary ammonium naphthenate.
Use: Fungicide for vinyl formations.
Also available as "Nuodex 100 S.S.," solvent soluble formulation.

"Nuogel." [74] Trademark for a series of aluminum soaps, used to produce stable gels in aliphatic hydrocarbons.

"Nuogel" A.O. [74] Trademark for aluminum octoate, used to produce stable gels in many hydrocarbons.

"Nuolates." [74] Trade name for driers based on metallic salts of tall oil acids.

"Nuophene." [74] Trademark for a technical grade of dihydroxydichlorodiphenylmethane.
Use: Industrial fungicide.

"Nuosperse 657." [74] Trademark for a combination of surface active agents which react synergistically in many systems to develop effective wetting, dispersing, and anti-settling properties; also has film forming properties.

"Nuostabe." [74] Trademark for a series of vinyl stabilizers and fungicides. Many of them are metallic soaps or metal-organic complexes.

"Nupercainal." [305] Trademark for dibucaine, a local anesthetic.

"Nupercaine." [305] Trademark for dibucaine hydrochloride, an anesthetic.

"Nu-Pon." [448] Trade name for epoxy resin primers and enamels for household appliances, metal products, and corrosion resistant applications.

"Nuray." [51] Trademark for highly refined straight mineral, lubricating oils of paraffinic type, which serve general plant use as lubricants and hydraulic mediums.

Nuremberg red. See iron oxide reds.

"NuRexform." [28] Trademark for arsenate of lead powder, colored pink to comply with certain state laws.

"Nurish." [241] Trade name for a water soluble fertilizer (20-20-20) derived from diammonium phosphate, urea and muriate of potash. Sold in 1-, 50-lb polyethylene bags.

"Nusope" 33-A. [74] Trademark for liquid sodium naphthenate, used as pigment wetting and dispersing agent in aqueous systems and as an emulsifier.

Nusselt number. A number used in heat transfer studies and calculations to compare heat losses by conduction from various shaped objects under various conditions. It combines into a single number the actual heat loss (Q), the temperature difference (ΔT) between the body and its surroundings, the size (d) and shape of the body and the thermal conductivity (k) of the fluid surrounding the object, in the equation

$$Nu = Qd/\Delta Tk.$$

nutgalls. See galls.

nutmeg. See myristica.

nutmeg oil (myristica oil).
Properties: Thin, colorless or pale-yellow liquid, volatile oil; strong nutmeg odor; warm, spicy taste; sp. gr. 0.880-0.910 for East Indian oil, 0.854-0.880 for West Indian oil; optical rotation +8 to +30° for East Indian oil, +25 to +45° for West Indian oil; refractive index 1.4740-1.4880 for East Indian oil, 1.4690-1.4760 for West Indian oil (both at 20°C); soluble in 90% alcohol, carbon disulfide and glacial acetic acid.
Chief known constituents: Myristicin; pinene; dipentene.
Derivation: By distillation from nutmegs, Myristica fragrans.
Grades: Technical; U.S.P. XVI.
Containers: Glass bottles; cans.
Uses: Medicine; flavoring; perfumery.

"Nuto." [51] Trademark for lubricating oils of good color and high resistance to oxidation; recommended for circulating and hydraulic systems.

"Nutralac." [244] Proprietary name for a hydrated compound consisting of sodium carbonate and sodium bicarbonate.
Properties: White, needle-like crystals, free-flowing; soluble in water; total Na_2O content, 41.45%; density, 40-50 lbs/cu ft.
Containers: 80-lb wooden kegs; 100-lb multiwall paper bags; 280-lb wooden barrels.
Uses: Dairy and food industry for neutralizing acidity in cream and related foods; compounding both salts and hand dishwashing preparations; leather tanning; and textile processing.

nutrient solution. A water solution of minerals necessary for plant growth which is used instead of soil, the plants being supported by mechanical means. Such solutions contain various proportions of potassium, phosphorus, calcium, sulfur, and magnesium, together with traces of iron, boron, zinc, and copper. They are extensively used for commercial growing of flowers and vegetables, and also to some extent for house plants.
See hydroponics.

"Nuvis." [74] Trademark for a series of bodying agents for viscosity control and prevention of sag and pigment settling in paint systems.

nux vomica (quaker buttons; poison nut; dog button; vomit nut). Dried ripe seed of Strychnos nux vomica. Odorless; pale brown to olive color.
Habitat: Southern Asia and Northern Australia.
Grades: Technical; N.F. XI.

Containers: 70-lb bags; powder, 200- to 250-lb barrels.
Uses: Medicine; source of alkaloids strychnine and brucine.
Shipping regulations: None. *

"Nu-Z." [93] Trade name for a nutritional zinc compound, 52-55% Zn; white powder; water insoluble; stable in storage.
Containers: 50-lb bags.
Uses: Corrects zinc nutritional deficiency in plants through foliar absorption. An additive to animal feed mixtures and salt cakes.

N.W. acid. Abbreviation for Neville and Winther's acid. See 1-naphthol-4-sulfonic acid.

"Nydrazid." [412] Trademark for isoniazid (q.v.).

nylidrin hydrochloride $HOC_6H_4CH(OH)$-$CH(CH_3)NHCH(CH_3)(CH_2)_2C_6H_5 \cdot HCl$.
para-Hydroxy-alpha-[1-(1-methyl-3-phenylpropyl-amino)-ethyl] benzyl alcohol hydrochloride.
Properties: White, odorless, tasteless, crystals or powder; slightly soluble in water, alcohol; very slightly soluble in chloroform, ether; pH of 1% solution is between 4.5 and 6.5.
Grade: N.F. XI.
Use: Medicine.

nylon. This word is a generic term for any long-chain polymeric amide which has recurring amide groups -CONH- as an integral part of the main polymer chain. The term does not refer to a particular product but rather to a family of chemically related materials which may be fabricated and used in many different physical forms. The chemical forms are described in separate articles following this one, under nylon 4, nylon 6, etc. The physical forms are described as follows within this article.

nylon bristles. See nylon monofilaments.

nylon fiber. Generic name for a manufactured fiber in which the fiber-forming substance is any long-chain synthetic polyamide having recurring amide groups (-CONH-) as an integral part of the polymer chain (Federal Trade Commission).

nylon molding powders. The descriptive material below applies to one particular nylon resin molding powder, and may be inapplicable to molding powders made from nylons of chemically different types or containing different auxiliary ingredients.
Properties: In molded or extruded form, the natural color is translucent cream white. Colored material is available in some types. Outstanding characteristics are toughness over a wide range of temperatures; strength in thin sections, along with ability to be molded in thin sections; resilience; abrasion-resistance; good bearing characteristics; dimensional stability at temperatures as high as 275°F, and hence ability to be steam-sterilized; low specific gravity; generally good resistance to chemicals and solvents; good dielectric properties; self-extinguishing character.
Uses: Molding powder for coilforms, sheathing of insulated wire, electrical insulation, sterilizable utensils, brush-backs, combs, gear and bearings, slide fasteners, impact tools and machine parts.

nylon monofilaments. Nylon in the form of relatively coarse, flexible monofilaments, available in a rather wide range of diameters. Characterized by a high degree of toughness, strength, and durability; resistant to chemicals and heat. Commercial products made from nylon monofilaments include such items as fishing leaders, snells, and lines; level bristles for toothbrushes, hair-brushes, and industrial brushes; racket strings; surgical sutures; and tapered paintbrush bristles.

nylon plastic. A thermoplastic material, similar in chemical structure to nylon fiber-forming polymeric amides and derivable from the same basic substance.
Properties: Nylon plastic is available in a range of properties depending upon the chemical type and auxiliary ingredients, as follows: tensile strength (73°F) 5000-9600 psi; elongation (73°F) 35-300%; modulus of elasticity (73°F) 38,000-285,000 psi; dielectric constant (60 cycles) 4.1-10.7, (10^6 cycles) 3.4-4.5; power factor (60 cycles) 0.014-0.19, (10^6 cycles) 0.03-0.14; mold shrinkage 0.010-0.015 in. per in.; compression ratio 2.1-2.5.

nylon resin, soluble. The descriptive material below applies to one particular nylon resin, and may be inapplicable to other soluble polyamide resins.
Properties: Resistant to aqueous alkali solutions (hot or cold), to oxygen and to ozone. At temperatures below 25°C it is little affected, chemically, by nonoxidizing acids. Soluble in phenols and in lower aliphatic alcohols, particularly methanol and ethanol, with aid of heat. Insoluble in most other solvents, including aliphatic and aromatic hydrocarbons, halogenated hydrocarbons, ketones, esters, carbon disulfide, water.
Containers: Fiber drums; multiwall paper bags.
Uses: Adhesives for wood, textiles, metals, glass; hydrocarbon barrier in fuel cells; coatings for textiles and paper; protective coating for thread; finishes; stiffeners and binders for textiles.

nylon staple and tow. Crimped nylon fibers in a variety of lengths and deniers. Nylon staple is converted on standard textile equipment into 100% nylon spun yarns, or blended with natural or other synthetic fibers.
Containers: Bales or cartons.
Use: In all branches of textile industry, as in yarns for sweaters and men's hosiery; woven suiting fabrics.

nylon yarn. Continuous single and multifilament types for use in all branches of the textile industry.
Containers: Bobbins, tubes, and beams.

nylon 4. A type of nylon made from pyrrolidone.

nylon 6. A nylon obtained by polycondensation of caprolactam.
Properties of fiber: Tensile strength (lbs/sq in.) 73,000-120,000; elongation 16 to 42%; sp. gr. 1.14; moisture absorbency 8% at 95% relative humidity; m.p. 215°C.
Use: Fiber and molding resins; used in tires.

nylon 66 (nylon 6,6; nylon 6/6). A nylon obtained by the condensation of hexamethylenediamine with adipic acid (polyhexamethylene adipamide).
Properties: Sp. gr. 1.14; tensile strength (psi) 58,000-134,000; break elongation 16-42%; moisture regain 4.2% (70°F, 65% R.H.); m.p. 250°C; soluble in 90% formic acid, meta-cresol.
Containers: Bobbins, tubes, beams, bales, and cartons.
Uses: Fiber for apparel and home furnishings, tires, tarpaulins, etc.

nylon 610 (nylon 6,10; nylon 6/10). A nylon obtained by the condensation of hexamethylenediamine with sebacic acid. Used for brush bristles and monofilaments.

nylon 7. A comparatively new nylon which is a polymer of ethyl aminoheptanoate, a 7-carbon acid ester. It has a higher softening temperature (430°F) than the older nylons and is especially suitable for tire cords.

nylon 8. A type of nylon made from caprolactam.

nylon 9. A type of nylon made from 9-aminononanoic acid.

nylon 11. A type of nylon made from 11-aminoundecanoic acid, and used commercially for fiber and molding purposes.

nylon, elastic. A modification of nylon 610 in which sebacic acid is condensed with hexamethylenediamine and a relatively small amount of an alkyl-substituted hexamethylenediamine.

nystatin (fungicidin) $C_{46}H_{77}NO_{19}$. An antifungal agent.
Properties: Yellow to light tan powder; odor suggestive of cereals; hygroscopic; affected by light, heat, air and moisture. Sparingly soluble in methanol, ethanol; insoluble in water, chloroform, ether and benzene. In solution is rapidly inactivated by acids and bases.
Derivation: Produced by fermentation with Streptomyces noursei and aureus.
Grade: U.S.P. XVI.
Use: Medicine.

"Nytal." [69] Trademark for a proprietary talc, or magnesium-calcium silicate.
"200": White; sp. gr. 2.85±.03; fineness (through 325 mesh) 98%.
"300," "400": Finer size particles than "200" but similar in composition and properties.
Uses: Dusting uncured rubber; filler in specialized applications.

nytril. Generic name for a manufactured fiber containing at least 85% of a long-chain polymer of vinylidene dinitrile, $-CH_2C(CN)_2-$, where the vinylidene dinitrile content is no less than every other unit in the polymer chain (Federal Trade Commission).
Properties: Soft, resilient fabric is obtained; is easy to clean; does not pill; resists wrinkling, and retains shape after pressing.
Uses: Fur-like pile fabrics; sweaters; yarns; blended fabrics for coats and suits.

O. Symbol for oxygen.

o-. Abbreviation for ortho- (q. v.).

oakmoss resin (mousse de chêne). Concrete oleoresin.
Derivation: Extracted from Evernia prunastri and E. furfuracea, lichens growing on oak, spruce and fruit trees; collected principally on French and Italian mountains bordering the Mediterranean.
Use: Perfumery (important fixative).
Shipping regulations: None. *

Oberphos process. A process for the granulation of superphosphate fertilizers by heating to 300°F at a pressure of 90 psi in an autoclave.

"Obra." [51] Trademark for stiff consistency, soda soap greases, dark green in color for specialized open bearing uses.

"Obron." [299] Trademark for a vitamin and mineral preparation. Used in medicine.

obsidian (volcanic glass). An igneous rock composed largely of silica which has been fused and cooled to a glassy rather than a crystalline condition. Usually dark in color, with a vitreous luster and fracture. Used in acid-concentrating plants on account of its acid-resisting properties.

O.C. Abbreviation for oxygen consumed (q. v.).

occlusion compounds. See inclusion complexes.

"Ocenol." [28] Trademark for a mixture of fatty alcohols, principally oleyl alcohol. Used as a chemical raw material, plasticizer, antifoam agent, and lubricant.

ocher, burnt. See ochers.

ochers. A name given to various native earthy materials used as pigments and consisting, essentially, of hydrated ferric oxide, admixed with clay and sand in varying amounts and in impalpable subdivision. When carrying much manganese they grade into umbers (q. v.).

Ochers are either yellow, brown, or red. Depending on the impurities present some ochers can be marketed after such simple preliminary treatment as drying, grinding and bolting; others, however, require more elaborate treatment, including calcining, in order to obtain the desired shade. The best reds are sometimes obtained by calcining the yellow varieties. They are called burnt ochers. Others, again, are obtained by calcining copperas or as a residue from the roasting of iron pyrites.

In general, the native yellows and browns are varieties of limonite (q. v.) and the native reds varieties of hematite (q. v.). One variety of the red ocher is known as scarlet ocher. Their value as pigments depends not only on depth of color but also on the amount of oil required as a vehicle.

ocher scarlet. See ochers.

ocotea oil. A volatile oil derived from Ocotea cymbarum, and used for its saffrole content for the manufacture of heliotropin.
Containers: Drums.

"OCPN." [233] Trademark for a chloronitroaniline, $C_6H_3ClNO_2NH_2$, a dye and pigment intermediate.
Properties: Yellow, crystalline powder; m. p. 108.4°C; insoluble in water; soluble in methanol and ether.

9,11-octadecadienoic acid. See linoleic acid.

9,12-octadecadienoic acid. See linoleic acid.

octadecadienol. See linoleyl alcohol.

n-octadecane $C_{18}H_{38}$ or $CH_3(CH_2)_{16}CH_3$
Properties: Colorless liquid; sp. gr. 0.7767 (28/4°C); b. p. 318°C; m. p. 28.0°C; refractive index 1.4369 (n 28/D). Soluble in alcohol, acetone, ether, petroleum, coal-tar hydrocarbons; insoluble in water.
Typical specifications: Sp. gr. 0.7750-0.7769 (28/4°C); b. p. 190-193°C (30 mm); m. p. 26.5-28.0°C; refractive index (n 28/D) 1.436-1.470.
Containers: Glass bottles.
Uses: Solvents; organic synthesis; calibration.
Shipping regulations: None. *

1,12-octadecanediol. See 1,12-hydroxystearyl alcohol.

n-octadecanoic acid. See stearic acid.

1-octadecanol. See stearyl alcohol.

n-octadecanoyl chloride. See stearoyl chloride.

9,12,15-octadecatrienoic acid. See linolenic acid.

octadecatrienol. See linolenyl alcohol.

9-octadecen-1,12-diol. See ricinoleyl alcohol.

1-octadecene $C_{18}H_{36}$ or $CH_3(CH_2)_{15}CH{:}CH_2$.
Properties: Colorless liquid; sp. gr. 0.7884 (20/4°C); refractive index (n 20/D) 1.4456; b. p. 180°C (15 mm); soluble in alcohol, acetone, ether, petroleum, coal-tar solvents; insoluble in water.
Typical specifications: Sp. gr. 0.788-0.790 (20/4°C); m. p. 16-18°C; b. p. 177-181°C, (15 mm); refractive index 1.4450-1.4475

*See "I.C.C. Shipping Regulations," page xiii.
Reference numbers refer to name of manufacturer. See "List of Manufacturers," page v.

(n 20/D).
Grades: 95% min. purity.
Containers: Glass bottles.
Uses: Organic synthesis of flavors, perfumes, medicines, dyes, oils, resins, plastics.
Shipping regulations: None. *

octadecene-octadecadieneamine. See oleyllinoleylamine.

cis-**9-octadecenoic acid.** See oleic acid.

trans-**9-octadecenoic acid.** See elaidic acid.

octadecenol. See oleyl alcohol.

cis-**9-octadecenoyl chloride.** See oleyl chloride.

octadecenyl aldehyde $C_{17}H_{35}CHO$.
Properties: B.p. 167°C (20 mm); refractive index 1.4620 (25°C); sp. gr. 0.847 (25°C).
Uses: Preparation of octadecyl aldehyde and octadecylamine. Intermediate for other products for use as vulcanization accelerators, rubber antioxidants, synthetic drying oils and pesticides.
Shipping regulations: None. *

octadecyl alcohol. See stearyl alcohol.

octadecyl carboxymethylmercaptosuccinate (MECSA). Mentioned as a preservative for flavor of cooking oils, functioning by chelation.

octadecyldimethylbenzylammonium chloride $C_{18}H_{37}(CH_3)_2(C_6H_5CH_2)NCl$. White, crystalline powder. Very soluble in water, chloroform. Soluble in benzene, acetone, xylol. A typical quaternary ammonium salt.

octadecyl mercaptan. See stearyl mercaptan.

octadecyl trichlorosilane.
Shipping regulations: Corrosive liquid. White label. *

octafluorocyclobutane ("Freon-C318") C_4F_8. Chemically inert dry gas; high electric strength; used for electrical equipment.

octafluoro-1-pentanol. See fluoroalcohols.

octafluoropropane. See perfluoropropane.

octahedrite. See anatase.

octakis(2-hydroxypropyl) sucrose.
Properties: Very viscous straw-colored liquid; sp. gr. 1.170 (70/20°C); refractive index 1.485 (25°C); pour point 38°C; flash point 485°F; soluble in water, methanol and ether.
Uses: Crosslinking agent for urethane foams; plasticizer for cellulosics; glue, starch and many resins.

octamethyl pyrophosphoramide (OMPA; schradan) $OP[N(CH_3)_2]_2OP[N(CH_3)_2]_2O$. Bis(bisdimethylaminophosphonous) anhydride. Caution! Very toxic.
Properties: Viscous liquid; sp. gr. 1.137; b.p. 120-125°C (0.5 mm); refractive index (n 25/D) 1.462. Miscible with water; soluble in most organic solvents; hydrolyzed in the presence of acids, but not by alkalies or water alone.
Use: A major commercial systemic insecticide for plants; is absorbed by the plant, which then becomes toxic to sucking and chewing insects.

"Octamine." 248 Trade name for a reaction product of diphenylamine and diisobutylene.
Properties: Light brown granular waxy solid; sp. gr. 0.99; m.p. 75-85°C; good storage stability. Soluble in gasoline, benzol, ethylene dichloride and acetone. Insoluble in water.
Uses: Antioxidant in natural, SBR, neoprene and nitrile rubbers. Used in tire carcass, footwear, molded heels and soles, proofing, sponge, automotive rubber, wire insulation and tiling.

octanal (n-octyl aldehyde; aldehyde C-8; caprylic aldehyde) $CH_3(CH_2)_6CHO$.
Properties: Colorless liquid with strong fruity odor; sp. gr. 0.820-0.830; refractive index 1.418-1.425; b.p. 163°C. Soluble in 70% alcohol.
Containers: Glass bottles.
Uses: Perfumery; flavors.

n-octane C_8H_{18} or $CH_3(CH_2)_6CH_3$.
Properties: Colorless liquid; sp. gr. 0.7026 (20/4°C); refractive index (n 20/D) 1.39745; b.p. 125.667°C; m.p. -56.798°C; flash point 22°C; soluble in alcohol, acetone, ether; insoluble in water.
Typical specifications: Sp. gr. 0.702-0.7045; b.p. 125-126°C; refractive index 1.3970-1.4000.
Grades: 95%; 99%; research.
Containers: Bottles; 5-gal drums; tank cars.
Uses: Solvent; organic synthesis; calibrations.
Shipping regulations: Flammable liquid. Red label. *

1,8-octanedicarboxylic acid. See sebacic acid.

octanedioic acid. See suberic acid.

octane number. A number indicating the degree of knocking of a fuel mixture under standard test conditions. Pure normal heptane is arbitrarily assigned zero octane number, while isooctane is assigned 100. A rating of 75 for a given fuel indicates that its degree of knocking in the standard test is equal to that of a mixture of 75 parts isooctane and 25 parts n-heptane.

octanoic acid. See caprylic acid.

1-n-octanol (n-octyl alcohol, primary; alcohol C-8; heptyl carbinol) $CH_3(CH_2)_6CH_2OH$. In industrial practice, the term octyl alcohol has been used for both 1-octanol and 2-ethylhexanol, which latter is also sometimes called isooctanol. The term capryl alcohol has been used for both 1-octanol and 2-octanol. It therefore seems preferable to distinguish the normal primary alcohol as 1-n-octanol.
Properties: Colorless liquid with penetrating, characteristic, aromatic odor; sp. gr. 0.826 (20°C); b.p. 194-195°C (760 mm), 108.7°C (30 mm); refractive index 1.430 (20°C); m.p. -16°C. Miscible with alcohol, chloroform, ether; immiscible with water.
Derivation: By reduction of caprylic acid.

Grades: Technical; C. P.; pure; perfume.
Containers: Glass bottles; 1-, 5-, and 10-gal tins; drums; tank cars.
Uses: Perfumery; cosmetics; organic synthesis; solvent; manufacture of high boiling esters; as an anti-foaming agent.

2-n-octanol (sec-n-octyl alcohol; methyl hexyl carbinol) $CH_3(CH_2)_5CHOHCH_3$. Frequently and confusingly called capryl alcohol (q. v.).
Properties: Colorless, oily liquid; refractive; characteristic, disagreeable, but aromatic odor. Miscible with alcohol, ether; immiscible with water. Sp. gr. 0.825 (15°C); b. p. 178-179°C; m. p. −38°C; refractive index 1.437 (20°C).
Derivation: By distilling sodium ricinoleate with an excess of sodium hydroxide.
Grades: Technical; pure.
Containers: Bottles; drums; tank cars.
Uses: Solvent; manufacture of plasticizers, wetting agents, foam control agents, hydraulic oils, petroleum additives, perfume intermediates.
Shipping regulations: None. *

2-octanone. See methyl hexyl ketone.

3-octanone. See ethyl amyl ketone.

octanoyl chloride (capryloyl chloride; sometimes called caprylyl chloride) $CH_3(CH_2)_6COCl$.
Properties: Clear, water-white to straw-colored liquid with characteristic pungent odor. Miscible with most common solvents; reacts with alcohol and water. Fairly stable below the boiling point. Freezing point below −70°C; distillation range 183-212°C; sp. gr. 0.9576 (15.5/15.5°C); refractive index 1.4357 (n 20/D); flash point 82°C; fire point 87°C.
Containers: 5-, 13-gal carboys.
Use: Synthesis.

"Octasol." [134] Trademark for metal salts of 2-ethylhexoic acid, in liquid form. Soluble in most hydrocarbons, vegetable oils, paint vehicles, etc. Used as paint driers, wetting agents, catalysts, etc.

1-octene (1-octylene; 1-caprylene) C_8H_{16} or $CH_3(CH_2)_5CH{:}CH_2$.
Properties: Colorless liquid; sp. gr. 0.7160 (20/4°C); b. p. 121.27°C; m. p. −102.4°C; refractive index (n 20/D) 1.4088; flash point 21°C; soluble in alcohol, acetone, ether, petroleum, coal-tar solvents; insoluble in water.
Typical specifications: Sp. gr. 0.7150-0.7175 (20/4°C); b. p. 121-123°C; refractive index (n 20/D) 1.4070-1.4105.
Grades: 95%, 99%, research.
Containers: Glass bottles; drums.
Use: Organic synthesis.
Shipping regulations: Flammable liquid. Red label. *

2-octene C_8H_{16} or $CH_3(CH_2)_4CH{:}CHCH_3$. Cis and trans forms exist.
Properties: Colorless liquid; sp. gr. cis 0.7243, trans 0.7199, commercial 0.7185-0.7200 (20/4°C); b. p. cis 125.6°C, trans 125.0°C, commercial 124.5-127°C; m. p. −94.04°C; refractive index cis 1.4150, trans 1.4132, commercial 1.4120-1.4145 (n 20/D); flash point (mixed isomers) 21°C; soluble in alcohol, acetone, ether, petroleum, coal-tar solvents; insoluble in water.
Containers: Glass bottles; drums.
Grade: 95%.
Use: Organic synthesis.
Shipping regulations: Flammable liquid. Red label. *

"Octin." [9] Trademark for isometheptene (2-methyl-amino-6-methyl-heptene-5) employed as the mucate and hydrochloride salts.

"Octin" Mucate. [9] Trademark for the mucic acid salt "Octin" (isometheptene). The mucate salt is a white crystalline, bitter powder; m. p. 144-151°C (with decomposition). It is very soluble in water, and very slightly soluble in ether and in chloroform.
Use: Medicine.

octoic acid. See caprylic acid.

octyl. The general name describing all eight-carbon radicals having the formula C_8H_{17}-. It has often been used interchangeably for the 2-ethylhexyl isomer (q. v.).

octyl acetate. See n-octyl acetate and 2-ethyl hexyl acetate.

n-octyl acetate (acetate C-8; caprylyl acetate) $CH_3COO(CH_2)_7CH_3$.
Properties: Colorless liquid with strong floral-fruity odor. Slightly soluble in water; soluble in alcohol and most other organic liquids. Sp. gr. 0.865-0.869; refractive index 1.419-1.422; b. p. 199°C.
Containers: Glass bottles.
Uses: Perfumery; flavors.

octyl alcohol. See 1-octanol; 2-octanol; 2-ethylhexyl alcohol.

n-octyl alcohol, primary. See 1-n-octanol.

n-octyl alcohol, secondary. See 2-n-octanol.

octyl aldehyde. See octanal and 2-ethylhexaldehyde.

n-octylamine $CH_3(CH_2)_7NH_2$.
Properties: Water-white; amine odor; boiling range 170-179°C; sp. gr. 0.779 (20/20°C); refractive index 1.431 (20°C); flash point 140°F.

tert-**octylamine** $(CH_3)_3CCH_2C(CH_3)_2NH_2$.
Properties: Liquid; b. p. 137-143°C; sp. gr. 0.771 (25°C); refractive index 1.423 (25°C); flash point 92°F (o. c.); insoluble in water; soluble in common organic solvents, especially petroleum hydrocarbons.
Containers: Drums; tank cars.
Caution! Avoid prolonged contact with skin, prolonged inhalation.
Uses: Intermediate for rubber accelerators, insecticides, fungicides, dyestuffs, pharmaceuticals.

N-octylbicycloheptene dicarboximide $C_8H_{17}NC_9H_8O_2$.
Properties: Liquid; b. p. 158°C (2 mm); sp. gr. 1.05 (18°C); refractive index 1.505

(n 20/D). Miscible with most organic solvents and oils.
Derivation: From maleic anhydride, cyclopentadiene and 2-ethylhexylamine.
Use: Insecticide and pesticide synergist.

octyl bromide (capryl bromide; caprylic bromide) $CH_3(CH_2)_6CH_2Br$.
Properties: Colorless liquid. Miscible with alcohol, ether; immiscible with water. Sp. gr. 1.118 (15°C); b.p. 202°C; m.p. -55°C; refractive index 1.4503 (25°C).
Grades: Technical.
Use: Organic synthesis.

octyl carbinol. See nonyl alcohol.

n-octyl n-decyl adipate.
Properties: Liquid, mild characteristic odor; sp. gr. 0.92-0.98 (20/20°C).
Containers: 1-gal and 5-gal cans; 55-gal drums; tank cars.
Use: Low temperature plasticizer.

n-octyl n-decyl phthalate.
Properties: Clear liquid; mild characteristic odor. Sp. gr. 0.972-0.976 (20/20°C).
Containers: 55-gal drums; tank trucks; tank cars.
Uses: Plasticizer for vinyl resins.

octylene. See octene.

octylene glycol titanate.
Properties: Light-yellow solid.
Derivation: Reaction of butyl titanate with octylene glycol.
Uses: Cross-linking agent; surface active agent.
See also titanium chelates.

octylene oxide Mixed $CH_3(CH_2)_5\overline{CHCH_2O}$ and $CH_3(CH_2)_4\overline{CHCH(CH_3)O}$.
Use: Organic intermediate; additive to epoxy resins. Sp.gr. of liq. 0.830 (25°C).

octyl hydride. See octane.

octylic acid. See caprylic acid.

octyl iodide (caprylic iodide; secondary normal capryl iodide) $CH_3(CH_2)_5CHICH_3$.
Properties: Oily liquid. Caution! Keep away from light and air! Sp. gr. 1.318 (18°C); b.p. (approx) 210°C (dec).
Grades: Technical.
Containers: Amber glass ground stoppered bottles.
Use: Organic synthesis.
Shipping regulations: None.*

n-octyl mercaptan $C_8H_{17}SH$.
Properties: A water-white liquid with mild odor; b.p. 199°C; sp. gr. 0.8395 (25/4°C); refractive index 1.4497 (25°C).
Grades: Technical (95% min mercaptan).
Uses: Polymerization conditioner; synthesis.

tert-octyl mercaptan $C_8H_{17}SH$.
Properties: Boiling range 154-166°C; sp. gr. 0.848 (60/60°F); refractive index 1.454 (n 20/D); flash point 41°C.
Grades: 95%.
Containers: Drums; tank cars.
Uses: Polymer modification.
Shipping regulations: None.*

n-octyl methacrylate $H_2C:C(CH_3)COOC_8H_{17}$.
Properties: A water-insoluble colorless liquid; polymerizes to a colorless resin if a stabilizer is not added.

octyl peroxide (caprylyl peroxide).
Shipping regulations: Solution: oxidizing material. Yellow label.* Legal label name is caprylyl peroxide.

octyl phenol (diisobutyl phenol) $C_8H_{17}C_6H_4OH$.
Properties: White flakes; congeals 72-74°C; sp. gr. 0.89 (90°C); b.p. 280-302°C; hydroxyl coefficient 259-275. Insoluble in hot and cold water. Limited solubility in alkalies. Soluble in 1-1 mixture of methanol and 50% aqueous potassium hydroxide, also in alcohol, acetone, benzene, ether, chloroform, carbon tetrachloride.
Containers: Bags; drums; carloads; tank cars.
Uses: Non-ionic surfactants; plasticizers; antioxidants; fuel oil stabilizer; intermediate for resins, fungicides, bactericides, dyestuffs, adhesives, rubber chemicals.
Shipping regulations: None.*

octylphenoxy polyethoxyethanol
$(CH_3)_3CCH_2C(CH_3)_2C_6H_4O(CH_2CH_2O)_xH$. Anhydrous liquid mixture of mono-para-(1,1,3,3,-tetramethylbutyl)phenyl esters of polyethylene glycols in which x varies from 5 to 15.
Properties: Clear, yellow, viscous liquid; faint odor; bitter taste; sp. gr. 1.060; refractive index 1.489 (25°C); soluble in water, alcohol, acetone, benzene, toluene; insoluble in hexane; pH is between 7 and 9.
Grade: N. F. XI.
Use: Medicine; food packaging; probably as a plasticizer for films.

para-octylphenyl salicylate
$C_6H_4OHCOOC_6H_4C_8H_{17}$. Prevents photo-oxidation in polyethylene and polypropylene.

octyl phosphate. See trioctyl phosphate.

octyl trichloroarsine.
Shipping regulations: Corrosive liquid. White label.*

"Odamask." [342] Trademark for deodorizing and reodorizing aromatics for industrial use.

"Odrenes." [12] Trademark for a series of economical industrial perfumes possessing masking and reodorizing properties. Versatile compositions containing blends of natural and synthetic chemicals resulting in many different odor types.

"Odrex." [188] Brand name for an industrial odorant.

oenanthal. See heptanal.

oenanthic acid. See n-heptanoic acid.

oenanthic aldehyde. See heptanal.

oenanthic ether. See ethyl oenanthate.

oenanthylic compounds. Same as oenanthic compounds.

"Oenethyl." [9] Trademark for 2-methyl-aminoheptane; employed as the hydrochloride salt.

*See "I.C.C. Shipping Regulations," page xiii.
Reference numbers refer to name of manufacturer. See "List of Manufacturers," page v.

oenology. The study of wines.

"Ohopex" Q-10 Plasticizer. [55] Trademark for fatty phthalic acid esters.
Properties: Slightly yellow, oily liquid.
Typical specifications: Sp. gr. 0.957 ± 0.005 (20/20°C); f.p. -50°C (stiff gel); boiling range 215-235°C (4 mm); acidity 0.1% (max) as acetic acid; flash point 415°F; fire point 480°F; refractive index 1.477 (25°C); viscosity 47 cps (20°C); vapor pressure < 0.06 mm Hg (150°C); surface tension 32 dynes/cm (20°C); wt/gal 8 lbs. Insoluble or limited solubility in glycerin, glycols and certain amines. Soluble in most other organic liquids.
Containers: 1-, 5-gal cans; 55-gal drums; tank cars.
Uses: Primary plasticizer for most resins imparting permanent flexibility, low water extraction, excellent hand and drape, low temperature flexibility, high gloss, and heat stability. An excellent plasticizer for extrusions or polishing stocks.

"Ohopex" R-9 Plasticizer. [55] Trade name for mixed octyl fatty acid esters.
Typical specifications: Sp. gr. 0.864±0.005 (20/20°C); freezing point -10°C (approx); boiling range 210-239°C; acidity (max) 0.3% as acetic acid; odor mild, characteristic; flash point 410°F; fire point 475°F; vapor pressure < 0.06 mm Hg (150°C); refractive index 1.452±0.002; viscosity 14.2 cps (20°C); surface tension 29 dynes/cm (20°C); thermal expansion 0.00083/°C (10-40°C); wt/gal 7 lbs. Limited solubility in many compounds containing OH group such as alcohols, "Cellosolves," glycols, glycerine, etc. Soluble in most other organic liquids.
Containers: 1-, 5-gal cans; 55-gal drums; tank cars.
Uses: Primary plasticizer for chloroprene and many other types of synthetic rubber, ethyl cellulose and chlorinated rubber.

"1320 Oil." [175] Brand name for a semi-refined coal-tar solvent.
Properties: Light amber color; not over 5% distills to 135°C; not less than 95% distills to 200°C; sp. gr. 0.860 (15.5/15.5°C); approx. 7.46 lbs/gal.
Containers: 55-gal steel drums.
Uses: A medium-boiling paint and enamel solvent; industrial solvent where slight color is permissible.

oil blue. A violet-blue pigment consisting of copper sulfide. It is not very durable but when used in varnish, which protects it from the action of air, it is fairly permanent.

"Oil Blue A." [28] Trade name for proprietary grade of 1,4-di(isopropylamino)anthraquinone, a petroleum dye used to color gasoline, etc.

oil bois de rose Brazilian (rosewood oil; bois de rose oil).
Properties: Pale-yellow to yellow liquid; fragrant odor; sp. gr. 0.8750-0.8950 (15°C); optical rotation -4° to +5°; refractive index 1.4620-1.4685 (n 20/D); total alcohols (as $C_{10}H_{18}O$) 84-92%; soluble in 6 volumes of 60% alcohol and in 2 volumes of 70% alcohol; soluble in fixed oils, propylene glycol, benzyl benzoate; slightly soluble in glycerin.
Derivation: Steam distillation of the so-called Brazilian linaloe tree (the botanical derivation for the tree is obscure).
See linaloe oil, Mexican.
Containers: Glass bottles; copper flasks.
Use: Perfumery.
Shipping regulations: None. *

"Oil Bronze." [28] Trade name for blend of Oil Red and Oil Orange, petroleum dyes used to color gasoline, etc.

oil cakes. The residue obtained after the expression of vegetable oils from the oil-bearing seeds. These cakes are valuable both as cattle feed and fertilizer. When ground they are known as meal. For further data see under specific headings, such as cottonseed cake and meal; peanut oil meal, etc.

"Oildag." [46] Trademark for a concentrated colloidal dispersion of pure electric-furnace graphite in petroleum oil.
Properties: Liquid consistency; solids content 10%; average particle size 0.5 micron, max. particle size 4 microns; sp. gr. 0.977; flash point 196°C; completely miscible with petroleum oils.
Uses: General industrial lubrication; oil additive for internal-combustion engines; formulation of glass-mold oils and specialty lubricants; meets U. S. Military Specification MIL-L-3572, Grade B.

"Oilfos." [58] Trademark for glassy sodium phosphate, a dry powder used exclusively for controlling viscosity of oil well drilling muds.
Containers: 100-lb net paper bags; triotex bags.

oil gas. A gas made by the interaction of oil vapors and steam at high temperatures by methods similar to those used for water gas production. Heating value about 554 Btu/cu ft. A typical analysis is illuminants 4.2%, carbon monoxide 10.4%, hydrogen 47.6%, methane 27.0%, carbon dioxide 4.6% oxygen 0.4%, nitrogen 5.8%.

"Oil Orange." [28] Trade name for proprietary grade of phenylazo-2-naphthol, a petroleum dye used to color gasoline, etc.

oil plant. See ricinus.

"Oil Red." [28] Trade name for proprietary grade of methyl derivatives of azobenzene-4-azo-2-naphthol, a petroleum dye used to color gasoline, etc.

oils. Any liquid of relatively high viscosity and slippery feel is likely to be called an oil. The term is applied to a great many different kinds of substances, many of which are described under specific entries such as coconut oil, olive oil, peanut oil, shale oil, myrrh oil.

There are three major categories of oils: (a) petroleum or mineral or hydrocarbon oils derived from crude petroleum; (b) fatty oils which are glycerol esters and derived from vegetable or animal fats or similar materials; (c) essential oils (q. v.), derived from plants, usually possessing a characteristic odor or flavor, usually not esters but more often terpene hydrocarbons.

The term oil is also applied to other oily substances that do not come under the preceding headings. Examples of these are oil of vitriol and oil of mirbane.

oil shale. A sedimentary rock with relatively high organic content (30 to 60% volatile matter and fixed carbon) that yields an oil when heated in the absence of air, but does not yield oil when extracted with ordinary solvents. Typical shales yield from 20 to 50 gallons of crude oil per ton, this oil being of a relatively unsaturated or olefinic character compared with petroleum.

See also kerogen.

oil varnish. See varnish.

oil well drilling mud. See drilling mud.

oil white. Usually mixtures of lithopone and white lead or zinc white (q. v.). It may also contain gypsum, magnesia, whiting, or silica. Used as a white-lead substitute.

Shipping regulations: None. *

"Oil Yellow N." [28] Trade name for proprietary grade of para-dimethylaminoazobenzene, a petroleum dye used to color gasoline, etc.

oiticica oil.

Properties: Light yellow drying oil which slowly solidifies to a buttery consistency unless first heated for about 30 minutes at 210-220°C. Hence properties vary according to whether the oil is raw or heat-treated (semi-polymerized).

Typical specifications:

(raw) A semi-solid white or cream-colored mass; sp. gr. 0.965 (min); acid value 5.0 (max); saponification value 185 (max); refractive index 1.514 (25°C) (min); moisture and volatile matter 0.9% (max); polymerization, by the Browne heat test, 22 min (max); Gardner-Holdt viscosity O to P; color, according to the 1933 Gardner scale, 10 (max).

(semi-polymerized): An amber-colored viscous liquid. Sp. gr. 0.975 (min); acid value 7.5 (max); refractive index 1.507 (25°C) (min); moisture and volatile matter 0.5% (max); Gardner-Holdt viscosity Z-2 to Z-3; color, by the 1933 Gardner scale, 12 (max).

Derivation: By expression from the seeds of the Brazilian oiticica tree, *Licania rigida*.

Chief constituents: Glycerides of alpha-licanic acid (4-keto-9, 11, 13-octadecatrienoic acid).

Containers: Steel drums; tank trucks.

Uses: Drying oil in paints, varnishes, etc.; substitute and adjunct for tung oil.

"Oitioil." [90] Trade name for a raw oiticica oil.

Properties: Viscosity, M to O (Gardner-Holdt 25°C); color, max 12 (Gardner); heating test (ASTM), max 22 min; sp. gr. 0.960-0.980 (20°C); refractive index 1.5150-1.5190 (25°C); acid value, max 4% F. F. A.; saponification value 186-193; unsaponifiable matter, 1.5%.

Uses: Drying oil for paints and varnish; substitute for tung oil.

"OKO." [64] Trade name for a line of polymerized linseed oils produced by vacuum polymerization to remove all traces of decomposed matter. Low acid, light color, minimum after-yellowing.

Uses: Paints, varnishes, enamels, flats, printing inks, brake linings.

okonite. A mixture of raw rubber and the black residue obtained in the purification of ozocerite which is vulcanized to give a flexible, tough, waterproof insulator.

"Olancha Clay." [38] Trade name for a natural montmorillonite used in the preparation of insecticide concentrates. Unusually high absorptive capacity for DDT, toxaphene and other types of insecticides. Supplied in several different mesh sizes and in both powdered and granular form.

old yellow enzymes. See Warburg's yellow enzyme.

oleamide cis-$CH_3(CH_2)_7CH{:}CH(CH_2)_7CONH_2$.

Properties: Ivory-colored refined powder; approximate melting point 72°C; sp. gr. 0.94.

Grades: Refined.

Uses: Slip-agent for extrusion of polyethylene; wax additive; ink additive.

oleandomycin phosphate $C_{35}H_{61}NO_2 \cdot H_3PO_4$.

Properties: White, odorless, crystals or powder; soluble in water, alcohol; slightly soluble in ether.

Derivation: Streptomyces antibioticus.

Grade: N. F. XI.

Use: Medicine.

oleanolic acid. See caryophyllin.

"Olefane." [429] Trademark for polypropylene film.

olefiant gas. See ethylene.

olefin fiber. Generic name for a manufactured fiber in which the fiber-forming substance is any long chain synthetic polymer composed of at least 85% by weight of ethylene, propylene, or other olefin units (Federal Trade Commission).

Properties: Extreme light weight; good abrasion resistance; easily cleaned.

Uses: Seat covers for automobiles; outdoor furniture; marine ropes; shoe fabrics; belts; handbags.

olefins. A class of unsaturated hydrocarbons of the general formula C_nH_{2n} and named

*See "I.C.C. Shipping Regulations," page xiii.

Reference numbers refer to name of manufacturer. See "List of Manufacturers," page v.

after the corresponding paraffins by adding "ene" or "ylene" to the stem. Characterized by relatively great chemical activity. See ethylene, propylene, and butenes for typical examples.

oleic acid (cis-9-octadecenoic acid; red oil) $CH_3(CH_2)_7CH:CH(CH_2)_7COOH$. A monounsaturated fatty acid; a common component of almost all naturally occurring fats as well as tall oil. Most commercial oleic acid is derived from animal tallow or natural vegetable oils.
Properties: Commercial grades: Yellow to red oily liquid; lardlike odor; darkens on exposure to air. Insoluble in water; soluble in alcohol, ether and most organic solvents. It is itself a good solvent for other oils, fatty acids and oil-soluble materials.
Purified grades: Water-white liquid; sp. gr. 0.8905 (20/4°C); m. p. 13.2°C; b. p. 286°C (100 mm), 225°C (10 mm); refractive index 1.4599 (20°C).
Derivation: The free fatty acid is obtained from the glyceride by hydrolysis, steam distillation and separation by crystallization or solvent extraction. Filtration from the press cake results in the commercial oleic acid of commerce (red oil) which is repurified and bleached for specific uses. Several grades are available containing varying proportions of other acids as impurities, such as linoleic, linolenic, myristic, palmitic, and stearic acids.
Grades: Variety of technical grades; grade free from chick edema factor; U. S. P. XVI; purified 99+%.
Containers: Bottles; barrels; tank trucks; tank cars.
Uses: Soap base; manufacture of oleates; ointments; cosmetics; polishing compounds; lubricants; ore flotation; organic synthetic intermediate; surface coatings.
Shipping regulations: None. *

olein (triolein, glyceryl trioleate) $(C_{17}H_{33}COO)_3C_3H_5$. The triglyceride of oleic acid, occurring naturally in most fats and oils. It constitutes about 70-80% of olive oil.
Properties: Yellow, oily liquid; sp. gr. 0.915; m. p. −4 to −5°C; soluble in chloroform, ether, carbon tetrachloride; slightly soluble in alcohol.
Impurities: Stearin, linolein.
Derivation: Refined natural oils.
Uses: Textile lubricants.

oleoresin black pepper. See oleoresin pepper.

oleoresin capsicum.
Derivation: From the fruit of Capsicum frutescens (cayenne or African pepper) by ether or acetone extraction.
Grade: N. F. XI. Same as capsicin (q. v.).

oleoresin cubeb.
Derivation: From the fruit of Piper cubeba by alcohol extraction. Soluble in alcohol and ether.
Grade: Technical.
Containers: Glass bottles; boxes.
Use: Medicine.
Shipping regulations: None. *

oleoresin ginger.
Derivation: From rhizome of Zingiber officinale, by acetone, alcohol, or ether extraction. Soluble in alcohol and ether.
Grades: Technical; N. F. XI.
Containers: Glass bottles; boxes.
Use: Medicine.
Shipping regulations: None. *

oleoresinous varnish. An oil varnish. See varnish.

oleoresin pepper (oleoresin black pepper).
Derivation: By acetone extraction of the fruit of Piper nigrum.

oleoresins.
Derivation: Semi-solid mixtures of the resin and the essential oil of the plant from which they exude. They have a pungent taste and a peculiar odor and are sometimes referred to as balsams.
See also benzoin gums; Peru, tolu and styrax balsams.

oleovitamin A. Natural vitamin A in oil. It may be (1) fish liver oil, or (2) fish liver oil diluted with an edible vegetable oil, or (3) a solution of vitamin A concentrate in fish-liver oil or in an edible vegetable oil. The vitamin A concentrate may be derived from natural (animal) sources, from synthetic vitamin A, or from its fatty acid esters. It is a thin oily liquid with a fishy but not rancid odor and taste. Unstable to air and light; insoluble in water; soluble in alcohol, ether and vegetable oils.
Grade: U. S. P. XVI.
Use: Nutrition.

oleovitamin D synthetic. A form of vitamin D (q. v.). It is a solution of calciferol or activated 7-dehydrocholesterol in an edible vegetable oil, or a solution in an edible vegetable oil of the products of activated ergosterol or 7-dehydrocholesterol.
Properties: A clear, colorless to light yellow, oily liquid; nearly odorless; bland taste; slightly soluble in alcohol; miscible with ether and with chloroform.
Units: One U. S. P. unit of vitamin D activity is the specific biologic activity of 0.025 microgram of vitamin D_3.
Grades: U. S. P. XVI.
Uses: Medicine; nutrition.

oleoyl chloride (cis-9-octadecenoyl chloride) $CH_3(CH_2)_7CH:CH(CH_2)_7COCl$.
Properties: Liquid; b. p. 175-180°C (3 mm); soluble in hydrocarbons and ethers; reacts slowly with water.
Containers: Bottles; carboys; 55-gal drums.
Uses: Chemical intermediate.

oleum. The Latin name for oil. Also applied to fuming sulfuric acid (q. v.).

oleyl alcohol (octadecenol) $CH_3(CH_2)_7CH:CH(CH_2)_7CH_2OH$.
The fatty alcohol derived from oleic acid. It has a long straight carbon chain with one double bond in it. Available commercially

as 80-90% pure. Clear liquid at room temperature.
Typical specifications: Iodine value 88, cloud point 20°F, boiling range 282-349°C.
Impurities: Linoleyl, myristyl and cetyl alcohols.
Derivation: Reduction of oleic acid.
Uses: Chemical synthesis, resins, petroleum additives, surface active agents, polymers.

oleylhydroxamic acid $C_{17}H_{33}CONHOH$.
Properties: Waxy solid; off-white color; sp. gr. 0.897 (70/25°C); insoluble in water; soluble in aqueous potassium hydroxide and organic solvents.

oleyl-linoleylamine (octadecene-octadecadiene-amine).
Properties: Highly unsaturated primary amine; soluble in many organic solvents; insoluble in water.
Constants: Sp. gr. 0.83; m. p. 19°C; b. p. 198-209°C; amine no. 200-210; iodine value 90 min.
Use: Organic intermediate.

oleyl methyl tauride. See sodium N-methyl-N-oleoyl taurate.

olibanum (frankincense; gum thus).
Distilled from the dried exudation of Boswellia species.
Habitat: Nubia; Egypt and Somaliland.
Grades: Technical.
Containers: Kegs.
Uses: Pharmacy; incense; fumigating preparations; perfumery (fixative).
Shipping reguations: None. *

olibanum oil (frankincense oil).
Properties: Colorless or yellowish oil having an agreeably balsamic and faintly lemon-like odor. Soluble in 4 to 6 vols. of 90% alcohol, occasionally with slight turbidity. Soluble in ether, chloroform, and carbon disulfide.
Chief known constituents: Pinene; phellandrene; dipentene.
Constants: Sp. gr. 0.876-0.892 (15°C).
Derivation: By distilling gum thus.
Containers: Cases; bottles.
Use: Medicine.
Shipping regulations: None. *

oligomer. Name suggested for a polymer compound molecule consisting of only a few monomer units, as 2, 3, 4.

olivenite $Cu_2(AsO_4)(OH)$. A natural basic arsenate of copper. Color various shades of green, brown, and gray; luster adamantine to vitreous, streak olive green to brown; sp. gr. 4.4; hardness 3.
Occurrence: England; Chile; Utah.

olive oil (sweet oil).
Properties: Pale yellow or greenish-yellow, fixed, liquid oil; slight characteristic odor and taste; faintly acrid aftertaste. Soluble in ether, chloroform and carbon disulfide; sparingly soluble in alcohol.
Chief known constituents: Olein; palmitin.
Constants: Sp. gr. 0.910-0.918; saponification value 188-196; iodine value 77-88.
Derivation: By expressing the pulp of the fruit of the olive tree, Olea europea. The best oil comes from fruit not quite ripe. The crude oil is washed and filtered. The cake is subjected to further pressings and finally solvent extraction, a lower grade of oil being produced each time.
Impurities: Free fatty acids, sediment, water and adulterants.
Adulterants: Cottonseed, peanut, sesame and poppy oils.
Grades: U. S. P. XVI; edible; commercial; sulfur oil (olive oil foots). The edible oil and also the commercial oil are obtained by expression, and the last grade by extraction usually with carbon disulfide.
Containers: Drums.
Uses: As food (substitute for butter in Italy and other countries); in ointment, liniments, etc.; for manufacture of Castile soap; special textile soaps; lubricant; wool oil; tanning.

olive oil foots. See olive oil (grades).

olivine (chrysolite) $(Mg, Fe)_2SiO_4$. Natural magnesium-iron silicate, found in igneous and metamorphic rocks, meteorites, and blast furnace slags. A complete series exists from Fe_2SiO_4 to Mg_2SiO_4. Some of the more important varieties are:
Forsterite Mg_2SiO_4. Color green; luster vitreous; hardness 6.5-5-7; sp. gr. 3.2-3.3.
Fayalite Fe_2SO_4. Color yellow to black; luster vitreous; hardness 6.5-7; sp. gr. 3.6.
Peridot is a transparent green gem variety.
Other minerals often grouped with olivine are monticellite $CaMgSiO_4$, and tephroite Mn_2SiO_4.
Occurrence: North Carolina, Washington; New Zealand; Europe.
Grades: Crude, 20 mesh, 100 mesh.
Use: Refractories; cements; possible source of magnesium metal.

"OLOA." [151] Brand name for mixtures of metal-organic and/or organic compounds in a lubricating oil carrier. Used to fortify well-refined base stocks to yield motor oils that minimize engine deposits, engine wear, bearing corrosion, engine rusting, and friction between rubbing surfaces.

"Omamids." [84] Trademark for a series of thermoplastic resins of the polyamide type.
Properties: "Omamid" S and "Omamid" C are solid, water insoluble, amber colored, thermoplastic resins, readily dispersible on heating in many organic systems.
Uses: In surface coatings, inks, and adhesives, as thixotropic viscosity modifiers and gelling agents. Impart "built in" thixotropy to solvent or oil based systems.

"Omazene." [84] Trade name for copper dihydrazinium sulfate (q. v.). Available as 50% wettable powder.
Use: Foliage fungicide.
Caution! Avoid skin exposure and inhaling dust or mist.

*See "I.C.C. Shipping Regulations," page xiii.
Reference numbers refer to name of manufacturer. See "List of Manufacturers," page v.

OMC. Abbreviation for oxidized microcrystalline waxes (q. v.).

"Omnadin." [162] Trademark for prolipin.

OMPA. Abbreviation for octamethyl pyrophosphoramide.

"O-M Special." [188] Brand name for an odor modifier for isopropyl alcohol.

ONB. Abbreviation for ortho-nitrobiphenyl.

"Oncor." [304] Trademark for a series of pigments. Available as:

"Oncor 23A." Antimony silico-oxide pigment.
Properties: White powder; sp. gr. 3.6. Tinting strength in oil 160.
Containers: Multiwall paper bags (50 lbs net).
Uses: Imparts flame resistance to halogenated plastic and paint compositions.

"Oncor 45X." Basic silicate white lead pigment.
Properties: White powder; sp. gr. 4.0. Tinting strength in oil 70.
Containers: Multiwall paper bags (50 lbs net).
Uses: Imparts film durability and good performance in house paints.

"Oncor M50." Basic lead silico-chromate pigment.
Properties: Orange colored powder; sp. gr. 4.1. Resistance to exposure excellent.
Containers: Multiwall paper bags (50 lbs net).
Uses: Imparts corrosion resistance, exposure durability and color retention to anti-corrosive paints.

"Oncor T15." Lead silico-titanate pigment.
Properties: Off-white colored powder; sp. gr. 4.05. Tinting strength in oil 200.
Containers: Multiwall paper bags (50 lbs net).
Uses: Imparts color retention to tinted outside house paints.

"Ondal" A Oxidizing Agent. [28] White powder.
Use: For one-bath oxidizing and soaping of vat colors in the textile industry.

"Ondelette." [28] Trademark for a random slubbed rayon fashion yarn. See also rayon.

onion oil.
Properties: Yellowish liquid; penetrating odor. Soluble in ether, chloroform, and carbon disulfide.
Chief known constituent: Allylpropyl bisulfide.
Constants: Sp. gr. 1.035-1.045.
Method of purification: Rectification.
Grades: Technical.
Containers: Glass bottles.
Use: Flavoring.

onyx. A form of silica or quartz (q. v.), essentially a chalcedonic silica. Contains colored bands (black and white, red and white) like agate (q. v.) but the bands are straight and the layers in even planes. See also sardonyx. Used for cameos. Most of the commercial onyx has been artificially colored. Onyx marble is a form of limestone.

"Onyxide 75%." [328] Brand name for a 75% concentration of alkenyl dimethyl ethyl ammonium bromide in which the alkenyl radical contains predominantly 18 carbon atoms. It is a cationic surface-active compound supplied as a tan paste of pleasant characteristic odor. Use dilutions are virtually odorless, colorless and slightly bitter in taste.
Containers: 15-, 30-, 50-gal specially lined steel drums.
Suggested uses: For the control of algae and slime in swimming pools, cooling towers, and air conditioning units.

"Onyxol 336." [328] A brand-named liquid detergent, wetting and dispersing agent, and thickener. It is a lauric acid alkanolamine condensate, employed as liquid household detergent ingredient, carpet and rug wet cleaner, for wallpaper removal, and as a dispersant.

"Onyxol 9162." [328] A brand-named, amber-colored liquid 96% active that can be used in all types of wool scouring machines, also as an auxiliary emulsifying agent, and an automobile cleaner. It is a coconut fatty acid alkanolamine condensate.

"Onyxsan HSB; -S; -S-50." [328] Cation-active permanent softeners, particularly effective in the softening and finishing of rayons and cottons. "Onyxsan"-treated fabrics do not become rancid in storage, do not mark-off, and are compatible with many other special finishing products.

opal. A form of native amorphous hydrated silica containing 3-13% water. It may be transparent or opaque; vitreous, dull, or pearly luster; and varied in color, or colorless with varied internal reflections. The play in color is thought to be due to thin curved layers with a different refractive index from the material around them, thus breaking the light up into various prismatic colors. Sp. gr. 1.9-2.3; hardness 5.5-6.5.
Use: Gem stone.

"Opalon." [58] Trademark for vinyl chloride.

"Opalwax." [202] Trademark for hydrogenated castor oil. See "Castorwax" and glyceryl tri-12-hydroxystearate.

"Opax." [337] Trade name for 90% ZrO_2 (zirconium oxide) containing 7% SiO_2, 0.6% Al_2O_3, 0.8% Na_2O. White cream powder; sp. gr. 5.4; m. p. 4500°F; average particle size 15 microns. Used as opacifier for glazes, enamels and ceramic pigments, in dinner ware, art ware, wall tile and sheet steel enamels. Promotes craze resistance and color stability.
Containers: 80-lb paper bags; 500-lb barrels; 36,000-lb carloads.

"Opax S." [337] Similar to "Opax" but contains 88% ZrO_2, 7% SiO_2; sp.gr. 5.2; and has average particle size 5 microns. The finer particle size results in greater opacity in enamels or glazes.

"Opex." [141] Trade name for nitrocellulose lacquer for automotive and industrial finishes.

OPG. Abbreviation for oxypolygelatin (q. v.).

"Ophthaine." 412 Trademark for proparacaine hydrochloride (q. v.).

opianyl. See meconin.

opium.

Derivation: The air-dried, milky exudation obtained by incising the unripe capsules of Papaver somniferum L. or its variety album De Candolle (Fam. Papaveraceae).

Properties: Gum opium: more or less rounded, somewhat flattened masses; externally dark brown; internally dark brown, interspersed with lighter areas; strong, peculiar odor; bitter, somewhat acrid taste; somewhat plastic when fresh, becoming hard and brittle or tough on keeping; sp. gr. 1.336. Official opium, for purposes other than the preparation of alkaloids or their salts, must yield, when dried and powdered, not less than 9.5% and not more than 10.5% anhydrous morphine. When used as a source for tincture or extract, it must contain, when dry, not less than 7.5%.

Granulated opium: A U. S. P. XVI product produced by drying gum opium below 70° followed by granulation and screening so that no more than 10% passes a 60-mesh screen and all passes a 10-mesh screen.

Powdered opium: U. S. P. XVI grade consisting of a fine powder with 10-10.5% anhydrous morphine.

Gum opium, granulated opium and powdered opium are listed in the U. S. P. XVI.

Denarcotinized or deodorized opium: Powdered opium from which has been removed its odor and nauseating substances.

Commercial varieties: (1) Asia Minor, Smyrna, Constantinople; (2) Macedonian; (3) Bulgarian; (4) Persian; (5) Indian; (6) Chinese; (7) Egyptian.

Containers: Tins.

Use: Medicine.

Shipping regulations: None. *

opium gum. Crude opium.

opium, tincture of. See laudanum.

"Oppanol B." 440 Trademark for isobutene (q. v.) in various stages of polymerization. Varies from oily liquid through highly viscous material to rubberlike solid according to degree of polymerization.

Uses: Insulation and lining material, and production of adhesives and sealing compounds in the rubber industry.

optical bleach. See optical brightener.

optical brightener (optical bleach). A dyestuff which will absorb ultraviolet energy and emit the energy in the visible range. By selection of the color of the emitted light the sum of the total visible light reflection can enchance the surface reflection as desired or retard undesirable shades. Optical brighteners are used in soaps, detergents, paper, textiles and plastics. Typical examples are stilbene-triazine derivatives. See also white dye.

optical crystals. Crystals, either naturally occurring, or, more frequently, synthetic, which are used for infrared and ultraviolet optics, piezoelectric effects, and short wave radiation detection. Examples are sodium chloride, potassium iodide, silver chloride, calcium fluoride, and, for scintillation counters, such organic materials as anthracene, naphthalene, stilbene, and terphenyl.

optical glass. Carefully made glass of great uniformity and usually special composition to give desired transmission, refraction, and dispersion of light.

optical isomerism. The existence of two stereoisomeric forms of a compound which differ in properties only with respect to the direction in which they rotate the plane of polarized light, the angles of their crystal faces, and some related properties. Mixtures of such compounds with one another are not separable by ordinary means. The situation arises with compounds having an asymmetric carbon atom in their molecules. The classical examples are the dextrorotatory and levorotatory tartaric acids studied by Pasteur, but hundreds of other cases are known. A common example is the active amyl alcohol from fusel oil (2-methyl-1-butanol). This has the formula $CH_3CH_2C^*H(CH_3)CH_2OH$ in which the starred carbon is asymmetric since it is joined to CH_3CH_2, H, CH_3, and CH_2OH.

optical rotation. The property of some substances of rotating the plane of vibration of polarized light through an arc to an extent characteristic of the substance.

"Ora-Lutin." 330 Trademark for anhydrohydroxyprogesterone.

orange III. See methyl orange.

orange cadmium. See cadmium sulfide.

orange flower oil (neroli oil).

Properties: Essential oil; pale yellow, fluorescent liquid becoming brownish red on exposure to light; pleasant orange blossom odor; bitter aromatic taste; sp. gr. 0.863-0.880 (25/25°C); optical rotation +1.5° to +9.1° (25°C). Soluble in an equal volume of alcohol.

Chief known constituents: Limonene, linalool, methyl anthranilate, geraniol, linalyl acetate.

Derivation: Distilled from the flower of Citrus aurantium L.

Grades: N. F. XI.

Containers: Glass bottles.

Uses: Perfumery; flavoring.

Shipping regulations: None. *

orange lakes. Pigments made by precipitating an orange dyestuff on a base, usually of aluminum hydrate. They are transparent and vary from poor to fair with respect to lightfastness, depending upon the particular dyestuff used. Their principal use is in the production of transparent metal

*See "I. C. C. Shipping Regulations," page xiii.
Reference numbers refer to name of manufacturer. See "List of Manufacturers," page v.

coatings for cans, bottle caps, etc. They are non-bleeding in oil and withstand the high temperatures at which finishes of this type are usually baked. They are not sufficiently lightfast for sign coatings that are to be subjected to exterior exposure. When used in combination with the transparent yellow lakes a "gold" effect is produced.

orange mineral. A lead oxide pigment made in a furnace by roasting lead carbonate or sublimed litharge; it is a very bright orange, but has low tinting strength. It is not all that is to be desired from the standpoint of permanency as on weathering a chemical change occurs and a white chalking or scum develops on the surface, probably due to the formation of some carbonate. It has very good opacity, very low oil absorption and is sometimes employed as a base for vermilion, permanent (q. v.). "Orange mineral" is not used in pigmented enamels to any very great extent, but is employed in primers for metal surfaces.

orange oils. See orange peel oil, sweet; orange peel oil, bitter; orange flower oil; orange oil, terpeneless.

orange oil, sweet. See orange peel oil, sweet.

orange oil, terpeneless.
Concentration: About 35 to 50 times that of the ordinary orange oil.
Constants: Sp. gr. about 0.894; optical rotation, varies between wide limits according to the degree to which the terpenes have been extracted. Soluble in 50 parts per 100 parts of 70% alcohol.
Shipping regulations: None. *

orange oil, U.S.P. XVI. See orange peel oil, sweet.

orange peel, bitter.
Derivation: Dried rind of unripe but fully grown fruit of Citrus aurantium, L.
Grades: Technical; N. F. XI.
Containers: Bags; bales.
Use: Flavoring; medicine.
Shipping regulations: None. *

orange peel oil, bitter. A volatile oil.
Properties: Similar to orange peel oil, sweet (q. v.) excepting that the taste is bitter.
Constants: Sp. gr. 0.845-0.851 (25/25°C); optical rotation +88° to +98° (25°C); refractive index 1.4725-1.4755 (20°C).
Soluble in 4 volumes alcohol, in dehydrated alcohol and glacial acetic acid.
Derivation: Expressed from the fresh peel of Citrus aurantium, L.
Containers: Bottles; tins; drums.
Uses: Flavoring; perfumery; medicine.
Shipping regulations: None. *

orange peel oil, sweet (orange oil, sweet; orange oil, U.S.P. XVI.)
Properties: Yellow to yellowish-brown essential oil; mild, aromatic, not bitter taste; characteristic orange odor.
Chief known constituents: Limonene.
Constants: Sp. gr. 0.848-0.853 (15°C); optical rotation +95° 30' to +98°; refractive index 1.473-1.475.
Soluble in 7 to 8 vols of 90% alcohol (usually not clear) and in glacial acetic acid.
Derivation: Expressed from the peel of Citrus aurantium, L., subspecies sinensis.
Grade: U.S.P. XVI.
Containers: Bottles; tins; copper flasks.
Uses: Flavoring; perfumery; medicine.
Shipping regulations: None. *

orange peel, sweet (sweet orange; Portugal orange; China orange).
Derivation: Rind of the fresh fruit of Citrus aurantium.
Habitat: Northern India, Spain and West Indies; cultivated near the Mediterranean, Florida, California, etc.
Grades: Technical.
Containers: Bags.
Uses: Flavoring; medicine.
Shipping regulations: None. *

orange root. See hydrastis.

orange toners. Dyestuffs consisting of diazo compounds coupled to diacetoacetic acid arylides and containing no sulfonic or carboxylic groups. Used in printing inks.

"Orasols." [443] Trade name for dyes, soluble in organic solvents, used for organic coatings, inks and plastics.

"Ora-Testryl." [412] Trademark for fluoxymesterone (q. v.).

"Oratol." [78] Trademark for a neutralized sulfonated ester of a higher alkanolamide, containing added solvents for increased scouring efficiency. Used for scouring rayons, acetates, wool, and mixed fabrics previous to dyeing, to remove sizings, lubricants, and weaving stains.

orbital angular momentum. See spin.

orbital electron capture. See K-capture.

orcanette. See alkanna.

orchidaé. See isoamyl salicylate.

orchil (archil; orseille).
Properties: Dark brown-red paste or aqueous extract; a coloring matter obtained from various species of lichens, roccella, variolaria, lecanora, etc.
Habitat: Azores, Canary Islands, Mediterranean region.
Chief constituents: Orcin and orcein.
Derivation: By macerating lichens with dilute ammonia and caustic soda, allowing to ferment, and adding sulfuric acid and salt.
Grades: Paste; extract.
Containers: Glass bottles.
Uses: Dyeing, particularly carpet yarns or to modify the effect of other dyes.
Shipping regulations: None. *

orcin (dihydroxytoluene; methylresorcinol; orcinol) $CH_3C_6H_3(OH)_2 \cdot H_2O$. 1-Methyl-3,5-dihydroxy-benzene.
Properties: White, crystalline prisms, becoming red in air; intensely sweet,

*See "I.C.C. Shipping Regulations," page xiii.
Reference numbers refer to name of manufacturer. See "List of Manufacturers," page v.

unpleasant taste. Soluble in water, alcohol, and ether. Sp. gr. 1.2895; m. p. (anhydrous) 107°C, (hydrated) 56°C; b. p. 287-290°C.
Derivation: By fermentation of various species of lichens (roccella), and extraction.
Grades: Technical.
Containers: Glass bottles.
Use: Medicine; reagent for certain carbohydrates.
Shipping regulations: None. *

orcinol. See orcin.

ordeal bean. See physostigma.

ore. An aggregation of valuable minerals and gangue (q. v.) from which one or more metals can be extracted at a profit.

ore flotation. See flotation.

"Orefraction." [288] Trademark for domestic zircon. Used for ceramic and foundry purposes.

"Oreton." [321] Brand name for testosterone propionate.

"Oreton Methyl." [321] Brand name for methyltestosterone.

organometallic compounds. Compounds containing carbon and a metal. Ordinary metallic carbonates (calcium carbonate, etc.) are excluded and also metallic salts of common organic acids. Examples of organometallic compounds are Grignard compounds such as CH_3MgI, metallic alkyls such as butyllithium (C_4H_9Li), tetraethyl lead, triethyl aluminum; or tetrabutyl titanate, sodium methylate, dibutyl tin dilaurate, copper phthalocyanine, zineb, ethylmercuric acetate, various arsonic acids, stearato chromic chloride, ferrocene, nickel carbonyl.

organo-phosphates. Fertilizer ingredients consisting of phosphoric esters of glycerol, glycol, sorbitol, glucose, etc., which retain their solubility when in contact with the soil long enough to penetrate into the deeper soil layer and thus supply phosphorus to the deeper part of the plant root system.

organosol. Colloidal dispersion of any insoluble material in an organic liquid, but more specifically the finely divided or colloidal dispersion of a synthetic resin in plasticizer in which dispersion the volatile content exceeds 5% of the total weight. See plastisol.

Oriental cashew nut. See semecarpus nut.

Oriental sweet gum. See styrax.

Orient yellow. See cadmium sulfide.

origanum oils. Essential oils obtained from various species of the genus Origanum indigenous to the Mediterranean countries. The botanical origin of the various oils cannot always be determined due to the large number of species and also, because commercial oils are not always distilled exclusively from a single species. All origanum oils contain carvacrol, together with cymene and sometimes linalool.
Physical characteristics of typical oil from Cyprus: Light-yellow color but becomes darker on exposure to the air; sp. gr. 0.962-0.967 (15°C); optical rotation inactive or slightly dextrogyrate up to +0° 20'; soluble in 2 to 3 vols. and more of 70% alcohol.
Containers: Bottles; tins.
Uses: Flavoring; pharmaceutical.
Shipping regulations: None. *

"Orion Red." [141] Trade name for azo red pigments derived from beta-naphthol.
Properties: Good light resistance, good heat resistance, non-bleeding in water and organic solvents. Yellowish-red, clean in masstone and tint.
Uses: Printing inks, rubber, plastics.

"Orizon." [58] Trademark for polyethylene.

"Orlon." [28] Trademark for an acrylic fiber. Available in various types of staple and tow. See also acrylic fiber, "Orlon Cantrece," "Orlon Sayelle."
Properties: Sp. gr. 1.14-1.17; tensile strength (psi) 32,000-39,000; break elongation 20-28%; moisture regain 1.5% (70°F, 65% R. H.); softens at 455°F; soluble in butyrolactone (hot), dimethyl formamide (hot), ethylene carbonate (hot). Resistant to mineral acids; fair to good resistance to weak alkalies. Insoluble in alcohol, acetone, benzene, carbon tetrachloride, and petroleum ether; soluble in dimethyl sulfoxide, maleic anhydride, ethylene carbonate, nitriles, and nitrophenols.
Derivation: A solution of polymerized acrylonitrile is forced through minute holes of a spinneret, the solvent is removed and the resulting fiber is stretched.
Containers: Bales and cartons.
Uses: In apparel, home furnishings, and industrial applications.

"Orlon Cantrece." [28] Trademark for "Orlon" as a filament yarn.

"Orlon Sayelle." [28] Trademark for a bi-component acrylic fiber available as staple and tow only.

ornithine (2,5-diaminovaleric acid) $NH_2(CH_2)_3CH(NH_2)COOH$. A nonessential amino acid produced by the body and important in protein metabolism.
Properties: L(+)-ornithine: Crystals from alcohol-ether; m. p. 140°C; soluble in water and alcohol.
DL-ornithine: Crystals from water; slightly soluble in alcohol.
Derivation: Isolated from proteins after hydrolysis with alkali.
Use: Biochemical research.

"Oronite." [151] First word of each of a series of trademarks or trade names for products derived from petroleum. Included are:
"Oronite Alkane". An alkyl aromatic hydrocarbon used as raw material for synthetic

detergents.
"Oronite SA88." Sulfonic acid from "Alkane."
"Oronite Detergent Slurry." Neutralized sulfonic acid from "Alkane."
"Oronite Wetting Agent S." Paste type sulfonate from "Alkane."
"Oronite Dispersants NI-E, NI-O, NI-W." Non-ionic surfactants of the alkyl phenol ethylene oxide type, for use in low-foaming detergents and emulsifiers in retail and industrial syndets.
"Oronite ADE-50." Quaternary ammonium compound.
"Oronite ABC." Alkyl benzyl chloride used for manufacture of quaternary ammonium compounds.
"Oronite Aromatics ABH." High molecular weight liquid alkyl aromatic hydrocarbons.
"Oronite LPG." Mercaptan type gas odorant for odorization of liquefied petroleum gas and natural gas.

"Oropon." [23] Trademark for proteolytic enzyme concentrates, formulated with or without deliming salts. Various grades include enzymes of pancreatic, fungal or bacterial origin, to remove undesirable protein matter and excess lime from dehaired animal skins.
Use: Bating or puering of skins in preparation for leather tanning.

"Orotan" TV. [23] Trademark for a synthetic tanning agent with attributes of vegetable tannins. Dark-red viscous solution, 31% tannin. Imparts high degree of tannage, strength, fullness and solidity to leather. Also solubilizing, penetrating and bleaching agent.
Use: Tanning of leather.

orotic acid (uracil-6-carboxylic acid; 6-carboxyuracil) $C_4N_2H_3(O)_2COOH$. It is found in cow's milk and has also been isolated from certain strains of molds (Neurospora). Orotic acid is a growth factor for certain micro-organisms.
Properties: Crystals with m. p. 345-346°.
Containers: 10-, 25-, 100-, and 1000-g.
Use: Biochemical research, especially the biosynthesis of nucleic acids.

orphenadrine hydrochloride (N, N-dimethyl-2-ortho-methyl-alpha-phenylbenzyloxyethylamine hydrochloride) $C_6H_5CH(CH_3C_6H_4)O(CH_2)_2N(CH_3)_2 \cdot HCl$.
Properties: Crystals; m. p. 156-157°C; soluble in water, alcohol, chloroform. Slightly soluble in acetone, benzene; insoluble in ether.
Grade: N. N. D.
Use: Medicine.

orpiment (arsenic yellow, auripigment, king's yellow, royal yellow, yellow arsenic sulfide) As_2S_3. A term applied to arsenic trisulfide (q. v.) whether obtained native as the mineral or artificially by chemical reaction.
Properties of the mineral: Color lemon yellow; luster resinous to pearly; good micaceous cleavage; hardness 1. 5-2; sp. gr. 3. 5.
Occurrence: Utah, Nevada, Wyoming, Peru, Rumania.
Use: Dye; tanning. Now displaced by the artificial product.
Shipping regulations: Class B poison. Poison label. *

orris (orris root; white flag; iris).
Rhizome of Iris florentina, I. germanica, I. pallida.
Habitat: Central and southern Europe.
Grades: Florentine, Verona. Whole, powdered, fingers.
Containers: Whole: bales; powdered: barrels, boxes.
Uses: Medicine; tooth powders; perfumery.
Shipping regulations: None. *

orris oil (orris root oil).
Properties: Semi-solid, yellowish, fatty, volatile oil; slightly dextrogyrate.
Chief known constituents: Myristic acid, oleic acid, irone, and their methyl esters.
Constants: M. p. 44-50°C; acid value 213-222. Soluble in alcohol, ether and chloroform.
Derivation: Distilled from the rhizome of Iris florentina.
Method of purification: Rectification.
Grades: Technical.
Containers: Copper flasks; glass bottles.
Use: Perfumes.
Shipping regulations: None. *

orris root. See orris.

Orr's white. See lithopone.

orseille. See orchil.

orthamine. See ortho-phenylenediamine.

orthite. See allanite.

ortho-. A prefix. Ortho-, meta- and para-compounds are di- substitution products derived from benzene in which the substituting radicals or groups are structurally placed in certain definite positions in the benzene nucleus. When two substituting groups (A and B) are in such position that they are attached to adjoining carbon atoms of the benzene nucleus (which contains six carbon atoms arranged in the form of a hexagon), B is in the ortho position with respect to A, and the compound is an ortho-compound. If A and B are so attached that they have a third carbon atom of the nucleus between them, a meta-compound results. If A and B are attached to opposite atoms in the nucleus (two other carbon atoms between them) a para-compound results. The ortho-compounds will be found under the name of the compound, as: ortho-cresol, see cresol. In other words, for organic compounds in this dictionary, ortho-, meta-, and para- are not used in alphabetizing.

In inorganic chemistry the prefix ortho- designates the most highly hydrated acid, or its salt, to contrast with the meta- or less hydrated acid or salt. For example $H_3PO_4(P_2O_5 \cdot 3H_2O)$ is orthophosphoric acid and $HPO_3(P_2O_5 \cdot H_2O)$ is metaphosphoric acid. For inorganic compounds in this dictionary, ortho- and meta- prefixes are used in alphabetizing.

orthoarsenic acid. See arsenic acid.

orthoboric acid. See boric acid.

"Orthobrite." [244] Trademark for a laundry detergent.
Properties: Rapidly soluble yellow flake with a strongly alkaline reaction.
Uses: Soap builder and break compound for commercial laundries.

"Orthochrom." [23] Trademark for pigmented plasticized nitrocellulose lacquers and thinners. Produce durable, washable, flexible, colored lacquer finishes of good light fastness.
Use: Finishing of belt, garment, upholstery and other leathers.

"Orthocide." [253] Brand name for a type of fungicide products containing captan.

orthoclase. See feldspar.

"Orthoclear." [23] Trademark for permanently plasticized nitrocellulose binders and lacquers in various solvents. Produce clear, durable, flexible finishes.
Use: Topcoat finishes for glazing or high gloss leather coatings.

orthocoll. See potassium guaiacol sulfonate.

"Orthodull." [23] Trademark for a dulling agent dispersion.
Uses: Additive for "Orthochrom" and "Ortholite" finishes.

orthoform. See methyl-meta-amino-para-hydroxybenzoate.

orthoformic ester. See triethyl orthoformate.

orthoform, new. See methyl-meta-amino-para-hydroxybenzoate.

"Ortho-Klor." [253] Brand name for insecticide products containing chlordan.

"Ortholate." [244] Trade name for a laundry alkali. White, rapidly soluble flake with sequestering and dispersive properties. Strongly alkaline.
Uses: Soap builder and break compound for commercial laundries.

"Ortholeum." [28] Trademark for line of lubricant additives.
"Ortholeum" 162. A lubricant assistant consisting of mono- and di-alkyl phosphates RH_2PO_4 and R_2HPO_4.
Properties: Light brown or pale amber viscous liquid; f. p. about 15°C; sp. gr. 0.99; miscible with oil in all proportions.
Containers: 55-gal steel drums (400 lbs net).
Use: In oils to improve film strength and to impart rust preventive properties. Concentrations vary with nature of applications.
"Ortholeum" 300. A grease stabilizer comprising a mixture of aromatic amines providing a combination of an antioxidant and a metal deactivator. A tan colored flaked solid.
Containers: 51-gal leverpak (175 lbs net).
Use: As a stabilizer for greases. Concentration required is usually 0.1 to 1.0% by weight.

"Ortholite." [23] Trademark for clear and pigmented vinyl lacquers, binders, and solvents. Produce finishes of outstanding abrasion resistance and low temperature flexibility.
Use: Finishes on upholstery, automotive, luggage, and case leathers.

"Orthol-K." [253] Brand name for pesticide products containing phytonomic or similar type petroleum oils.

"Orthophen." [204] Trademark for a special blend of mixed amyl phenols for the paint industry.
Properties: Straw colored liquid; sp. gr. 0.95-0.97 (30°C); distillation, 95% between 235-270°C; flash point 200°F; solidification point < -10°C.
Use: Anti-skinning agent.

"Orthophos." [253] Brand name for a type of insecticide containing parathion.

orthophosphoric acid. See phosphoric acid.

"Orthorix." [253] Brand name for a type of parasiticide containing lime sulfur.

"Orthosil." [204] Trademark. Also called "Pennsalt" Cleaner Number 30. A quick-acting detergent used in heavy-duty metal cleaning. Removes drawing, cutting, and other fabricating oils and greases from steel and brass before plating, enameling, lacquering, tinning, bonderizing, and other final finishes. For use in power washers and electrolytic cleaning tanks. "Orthosil" is a practically anhydrous, water-soluble, sodium orthosilicate in granular form. It combines high alkaline concentration and electric conductivity with excellent buffer action. Special "Orthosils" made for still tank cleaning have suitable wetting and water softening additions. Packed in 400-lb steel drums.

"Orthosolv." [292] Trademark for a product that combines the properties of straight ortho-dichlorobenzene with the advantages of mixing readily with water, thus making possible the preparation of solutions of different strengths. Used to control objectionable odors of sewage plants, refuse cans and trucks, etc.

orthotungstic acid. See tungstic acid.

"Orthoxine" Hydrochloride. [327] ($C_{11}H_{18}ClNO$). Trademark for methoxyphenamine hydrochloride; beta-(o-methoxyphenyl)-isopropyl-N-methylamine hydrochloride.
Properties: Colorless crystals; m. p. 125-128°C; very soluble in water, alcohol; insoluble in ether and benzene.
Derivation: Synthetic.
Use: Medicine.

"Ortolan." [440] Trademark for a series of metal complex dyestuffs for dyeing and printing on wool, silk and polyamide fibers.

"Orzan." [48] Trade name for a group of spent sulfite liquor products.

"Orzan" A: ammonium lignosulfonate plus wood sugars, available as dark brown 50% solids solution in 500-lb drums, tank cars, or brown free-flowing powder in 50-lb bags.

"Orzan" AH: partially polymerized ammonium lignosulfonate plus wood sugars, a dark brown free flowing powder in 25-lb bags.

"Orzan" S: sodium lignosulfonate plus wood sugars, available as dark brown 50% solids solution in 500-lb drums, tank cars, or light brown free-flowing powder in 50-lb bags.

Uses: Dispersant; flotation reagent; chelating agent; emulsion stabilizer; binder; tannin extender; adhesive ingredient.

Os. Symbol for osmium.

"OS-45" Type III. [58] Trademark for hydraulic fluid. Clear amber liquid; useful from -65° to 400°F.

"OS-45" Type IV. [58] Trademark for hydraulic fluid. Clear amber liquid; useful from -65° to 550°F.

oscine. See scopoline.

osmic acid anhydride (osmine tetroxide; perosmic acid anhydride; perosmic oxide) OsO_4.

Properties: Yellowish crystals; very pungent, disagreeable odor; vapor irritating when breathed; highly poisonous! Soluble in water, alcohol, and ether. Sp. gr. 4.90; m. p. 40°C; b. p. 130°C.

Derivation: By heating powdered osmium in air, or by treating it with nitric acid, aqua regia, or chlorine.

Method of purification: Crystallization.

Grades: Technical.

Containers: Glass bottles.

Uses: Medicine; microscopic reagent; photography; incandescent gas mantles; catalyst in organic synthesis.

osmic-ammonium chloride. See osmium-ammonium chloride.

osmiridium. See iridosmine.

osmium Os. Element having atomic number 76 and in group VIII of the periodic system.

Properties: Hard white metal of the platinum group. Has a bluish cast. On heating in air gives off a pungent poisonous fume of osmium tetroxide. Insoluble in acids and aqua regia; attacked by fused alkalies.

Constants: Sp. gr. 22.5; m. p. 2700°C.

Derivation: Occurs with platinum from which it is recovered during the purification process. Also occurs with iridium as a natural alloy called iridosmine.

Uses: Hardener for platinum; pen points; catalysts.

Shipping regulations: None. *

osmium-ammonium chloride (osmic-ammonium chloride) $(NH_4)_2OsCl_6$.

Properties: Red powder. Contains 43.5% osmium. Soluble in alcohol, water.

osmium chloride (osmium dichloride; osmous chloride) $OsCl_2$.

Properties: Dark green needles. Hygroscopic. Caution! Keep away from air! Soluble in alcohol, ether, water.

osmium dichloride. See osmium chloride.

osmium-potassium chloride (potassium-osmic chloride) K_2OsCl_6.

Properties: Dark red, octahedral crystals. Nearly black. Contains 39.6% osmium. Soluble in alcohol, water.

osmium-sodium chloride (sodium-osmic chloride) Na_2OsCl_6.

Properties: Orange, rhombic prisms. Contains 40.3% osmium. Unstable. Soluble in alcohol, water.

Grades: Technical.

Use: Catalyst (organic oxidation).

osmium tetroxide. See osmic acid anhydride.

osmocene. Dicyclopentadienylosmium. Like ferrocene (q. v.).

"Osmon." [169] Trademark for 1-naphthylamine-4,6,8-trisulfonic acid used in the colorimetric determination of osmium.

osmosis. Passage of solvent from pure solvent into a solution, or from a less to a more concentrated solution, through a membrane which is permeable to the solvent but not to the solute. Important in biological processes.

osmotic pressure. The excess pressure which when applied to a solution will just prevent osmosis (q. v.). In ideal very dilute solutions the osmotic pressure is equal to the pressure which the solute would exert if it were an ideal gas at the same temperature and in the same volume. The molecular weights of solutes may be calculated from measured osmotic pressures. The method is practical only with large molecules such as proteins and high polymers.

osmous chloride. See osmium chloride.

os sepiae. See sepia (2).

"Ostensin." [24] Trademark for trimethidinium methosulfate (q. v.).

Othmer process. Production of acetic acid from pyroligneous acid by azeotropic distillation. The pyroligneous acid is first stripped of methanol and then distilled in the azeotropic dehydrating column using an entrainer such as butyl acetate to separate the acetic acid from the water. The acetic acid is then further refined to separate the tar.

"Otrivin." [305] Trademark for xylometazoline. Used in medicine.

otto of rose oil. See rose oil.

ouabain (G-strophanthin; strophanthin thoms) $C_{29}H_{44}O_{12} \cdot 8H_2O$. A glucoside.

Properties: White odorless crystals; extremely poisonous; soluble in water and alcohol; specific rotation (alpha) -31 to -32.5° 20/D (anhydrous). M. p. 190° (dec).

Grades: U.S.P. XVI.
Caution: Extremely poisonous!
Containers: Tight, light-resistant bottles.
Use: Medicine.

ouricury wax. A vegetable wax exuded by the leaves of Cocos coronapa (South America).
Properties: Brown; acid value 10; saponification value 80; sp. gr. 0.970 (15°C); m.p. 83°C; foreign matter (dirt, etc.) sometimes 18%.
Containers: Bags.
Use: A substitute for carnauba wax.

outgassing of metals. The removal of gas from a metal by heating at a temperature somewhat below melting, while maintaining a vacuum in the space around the metal. Usually done before melting but may be done afterward.

ovalbumin. See albumin, egg.

ovalecithin. See lecithin.

Overglaze Colors. [28] Finely ground mixtures of pigments and low-melting glasses suitable for use over standard ceramic glazes. Temperature range 1300°-1500°F; cones 018-014.
Uses: For bands, decalcomania, and similar decorative designs fired on china, pottery, terra cotta, tile, and other glazed ceramic surfaces.

"Ovocylin." [305] Trademark for estradiol.
Use: Medicine.

"O & W Compound." [244] Trademark for a compound consisting of a blended clay-alkali mix.
Properties: Light grey mechanical mixture; forms colloidal alkaline solution in water with surface active and soil suspending properties.
Containers: 100-lb multi-wall paper bags; 350-lb fiber drums.
Uses: Commercial laundries; for cleaning heavily soiled work clothing and wiping cloths; low temperature washing; dark fabrics.

"OXAF." [248] Trade name for the zinc salt of 2-mercaptobenzothiazole. $Zn(SCNSC_6H_4)_2$.
Properties: White to pale yellow powder. Sp. gr. 1.63; melting range, decomposes without melting when heated to 200°C or over. Excellent storage stability. Slightly soluble in ethylene dichloride and acetone. Insoluble in water, benzol, and gasoline.
Handling precautions: None. Approved for use in rubber stock in contact with the skin.
Use: Rubber accelerator; all types generally but more especially in latex, foam sponge, wire insulation, air cured footwear, druggist sundries and specialties.

oxalic acid $HOOCCOOH \cdot 2H_2O$.
Properties: Transparent, colorless crystals; poisonous! Soluble in water, alcohol and ether.
Constants: Sp. gr. 1.653; m.p. 187°C of anhydrous form, 101.5°C for dihydrate.
Derivation: (a) Carbon monoxide and hot sodium hydroxide under pressure react to give sodium formate. This is converted to sodium oxalate in an autoclave at 400°C. Calcium oxalate is formed by reaction with calcium hydroxide, and sulfuric acid then used to produce oxalic acid. (b) Oxidation of carbohydrates such as sawdust with nitric acid, or other reactions of carbohydrates with dilute acids or alkalies.
Method of purification: Crystallization.
Grades: Technical (crystals and powder); C.P.
Containers: 100-lb bags; kegs; 300-lb barrels; 100-, 375-lb drums.
Uses: Automobile radiator cleanser; acid rinse in laundries; leather tanning and processing agent; purifying agent in manufacture of glycerol, glycollic acid; formic acid and esters, dextrin from starch, purification of tartaric acid and cream of tartar; bleaching agent; photography; medicinals; dyes and inks; purifying stearin; component of metal polishes; textile treating baths; ink and rust removers; cleansing agent in breweries; precipitating agent for rare earths; wood cleansing compositions; engraving and lithography; catalyst for some organic reactions.
Warning: Harmful if swallowed; causes skin irritation. MCA warning label.
Shipping regulations: None.*

oxalyl chloride (ethanedioyl chloride) $(COCl)_2$.
Properties: Colorless liquid. If cooled to −12°C, solidifies to a white, crystalline mass. Gives off carbon monoxide on heating. Decomposed by water and alkaline solutions. Caution! Very toxic! Soluble in ether, benzene, chloroform.
Constants: B.p. 64°C; m.p. −12°C; sp. gr. 1.43.
Derivation: Interaction of oxalic acid and phosphorus pentachloride.
Grades: Technical.
Containers: Steel drums.
Uses: Military poison gas; chemical (chlorinating agent in organic synthesis).

oxamide $NH_2COCONH_2$.
Properties: A white, odorless powder; m.p. 419° (dec) (probably the highest melting organic compound); insoluble in water; very slightly soluble in alcohol and ether.
Containers: Carload lots.
Use: Stabilizer for nitrocellulose preparations.

oxammonium. See hydroxylamine.

oxamycin. See amino-3-isoxazolidone.

oxanamide (2-ethyl-3-propylglycidamide) $\underline{OCH(C_3H_7)C}(C_2H_5)CONH_2$.
Use: Medicine.

oxetane (trimethylene oxide), $\overline{CH_2OCH_2CH_2}$. An oxetane group would be $=\underline{COCH_2C}=$ and is one kind of an epoxy group.

oxetane resins. See "Penton."

oxidase. An enzyme whose activity results in the transfer of electrons on the substrate; an oxidizing enzyme.

*See "I.C.C. Shipping Regulations," page xiii.
Reference numbers refer to name of manufacturer. See "List of Manufacturers," page v.

oxidation. Originally, the combination of oxygen with some substance. Now, any chemical change in which the oxidation state (positive valence) of an element is increased. According to atomic theory, a change in which an atom loses one or more electrons. Since some other atom gains the electrons, the latter atom undergoes the opposite change, reduction. Hence an oxidation is always accompanied by a reduction.

oxidation-reduction indicators. Substances that have a color in the oxidized form different from that of the reduced form, and that can be reversibly oxidized and reduced. Thus if diphenylamine is present in a ferrous sulfate solution to which potassium dichromate is being added, a beautiful violet color is formed with the first drop of excess dichromate.

"Oxidex." [188] Trademark for a proprietary anti-oxidant for soap fats and oils. It can be added to the oils before saponification or it can be added by milling into the finished soap in the same manner that perfume is incorporated. The correct proportion for solid soaps is 0.1%.

oxidized asphalts. See blown asphalts.

oxidized oils. See blown oils; also blown asphalts.

oxidized microcrystalline waxes (OMC). Waxes made from tank-bottom residues obtained in petroleum refining. This sludge is extracted with methyl ethyl ketone and the extract chilled, filtered, and refined. The product is ordinary microcrystalline waxes. These are then oxidized with air in the presence of manganese, cobalt, or other catalysts. The resulting waxes are emulsifiable, and are especially useful in floor polishes. See also waxes, microcrystalline.

oxine. Synonym for 8-hydroxyquinoline.

oxirane. A synonym for ethylene oxide, H_2COCH_2. Hence an oxirane group is one having the structure =COC=, and is one kind of an epoxy group.

"Oxiron 2000" Resin. [55] Brand name of an epoxy polyoleofin.

Properties: Sp. gr. 1.010 (20°C); epoxy equivalent 177; per cent epoxy 9.0; viscosity (25°C) 1800 poise. Has epoxy, hydroxy and double bond functionality and can be cross linked to a thermoset by agents like amines, anhydrides and acids, and combination of anhydride-glycol, anhydride-peroxide and anhydride-glycol-glycol-peroxides to give high temperature properties.

Also available: Low viscosity version (Oxiron 2001), viscosity (25°C) 160 poise; and enhanced double bond reactivity version (Oxiron 2002), viscosity (25°C) 15 poise.

Uses: Adhesives, electrical potting, encapsulating, laminating, tooling, prepregs, castings, coatings.

"Oxone." [28] Trademark for an acidic, white, granular, free-flowing solid containing the active ingredient potassium peroxymonosulfate; readily soluble in water; 1% solution has pH of 2-3. Minimum active oxygen content 4.5%. Strong oxidizing agent.

Containers: 50-lb bags.

Uses: For manufacture of dry laundry bleaches, detergent-bleach washing compounds, scouring powders, plastic dishware cleaners, and metal cleaners; preparation of hair wave neutralizers and pharmaceuticals; general oxidizing reactions.

2-oxopentanedioic acid. See alpha-ketoglutaric acid.

oxophenarsine hydrochloride. See 2-amino-4-arsenosophenol hydrochloride.

Oxo process. Production of alcohols, aldehydes and other oxygenated organic compounds by passage of olefin hydrocarbon vapors over cobalt catalysts in the presence of carbon monoxide and hydrogen gases. Aldehydes are produced as products, but in most cases these are hydrogenated at once to produce the corresponding alcohol. Propylene produces normal and isobutyraldehyde; higher olefins produce a mixture of aldehydes containing one more carbon atom than the olefins. n-Butyl, isobutyl, amyl, isooctyl, decyl and tridecyl alcohols are produced in large quantities.

oxosilanes. See siloxanes.

oxtriphylline (choline theophyllinate; theophylline cholinate) $C_7H_6N_4O_2 \cdot (CH_3)_3NC_2H_4OH$.

Properties: White crystalline solid (contains about 60% anhydrous theophylline). Extremely soluble in water.

Grade: N.N.D.

Use: Medicine.

oxy-. The –O– radical. Sometimes represents the hydroxy radical, –OH, but this is not considered good usage in the United States.

oxyacanthine (vinetine) $C_{37}H_{40}N_2O_6$. An alkaloid.

Properties: White crystalline powder, needles from alcohol or ether; m.p. 202-214°C; soluble in water, chloroform, benzene, alcohol, and ether.

Derivation: Occurs in the root of Berberis vulgaris.

Use: Medicine.

oxybenzoic acids. See hydroxybenzoic acids.

para, para'-**oxybis(benzenesulfonylhydrazide).** See "Celogen."

oxybutyric aldehyde. See aldol.

oxycellulose. See cellulose, oxidized.

oxyconiine. See conhydrine.

N-**oxydiethylenebenzothiazole-2-sulfenamide.**

Properties: Yellow color with sweet odor; masked by morpholine after sealed storage; sp. gr. 1.37; m.p. 75-80°C; insoluble in water; soluble in benzene, acetone, methanol.

Use: Rubber accelerator.

beta, beta'-**oxydipropionitrile** (ODPN) $NCCH_2CH_2OCH_2CH_2CN$.
Properties: Colorless liquid; m. p. −26. 3°C; b. p. 120°C (1 mm); b. p. 155°C (5 mm); sp. gr. 1. 0405 (30°C); viscosity (30°C) 8. 00 cps; refractive index (n 25/D) 1. 4392; flash point, Tag (closed cup) greater than 176°F; soluble in water. It is thermally unstable, yielding acrylonitrile and water above 175°C. Hydrolyzed by strong acids and bases; quite immiscible with paraffin hydrocarbons, but dissolves aromatics.
Derivation: From acrylonitrile.
Use: Solvent in fractional extraction.

oxyethylene oxypropylene polymer. See poloxalkol.

oxygen O. Element of atomic number 8; group VI of the periodic table.
Properties: Colorless, odorless, tasteless, diatomic gas, liquefiable at −183°C into a slightly bluish liquid, which is solidifiable at −227°C. Atomic weight 16. It constitutes roughly one-fifth (by volume) of the air. (Gas) Sp. gr. 1. 10535, referred to air. Soluble in molten silver; slightly soluble in water.
Derivation: (a) From liquid air by fractionation to remove the other gases of the air; (b) by electrolysis of water.
Impurities: Nitrogen, carbon dioxide, water vapor, ammonia, argon, helium and other rare gases.
Grades: Technical; pure; U. S. P. XVI.
Containers: As a compressed gas: in steel cylinders or "gas-bottles;" as a liquid: in vacuum-jacketed containers which range in size up to an entire truck load.
Uses: To increase capacity of steel and iron furnaces; with hydrogen or acetylene for production of exceedingly hot flames for cutting and welding metals, including platinum; for resuscitation in asphyxia and stimulation in various diseases; in combustion to promote better utilization of fuel; as a constituent of explosives produced by allowing liquid oxygen to be adsorbed on charcoal or similar material; anesthesia; for production of carbon monoxide-hydrogen gas mixtures by partial combustion of natural gas or coal; and thus the production of ammonia, methanol, etc.; as oxidizer for liquid rocket propellants.
Fire hazard: Dangerous.
Shipping regulations: Nonflammable gas. Green gas label. *

oxygen 18 (heavy oxygen). Oxygen isotope (nonradioactive) of atomic weight 18 (instead of the usual 16). Occurs in proportion of 8 parts to 10, 000 of ordinary oxygen in water, air, rocks, etc. The proportion may be increased by passing carbon dioxide gas repeatedly through a packed column down which water is passed. The carbon dioxide leaving the top of tower is enriched in heavy oxygen and the water leaving the bottom is depleted. Like other isotopes, the heavy oxygen 18 is useful in tracer experimentation. See also heavy water.

oxygen consumed (C. O. D.) (O. C.) (D. O. C.). A measure of the quantity of oxidizable components present in water. Since the carbon and hydrogen, and not the nitrogen, in organic matter are oxidized by chemical oxidants, the oxygen consumed is a measure only of the chemically oxidizable components and is dependent upon the oxidant, structure of the organic compound and manipulative procedure. Since this value does not differentiate stable from unstable organic matter, it does not necessarily correlate with the biochemical oxygen demand value. It is also known variously as chemical oxygen demand (C.O.D.); oxygen consumed (O.C.) and dichromate oxygen consumed (D.O.C.). See also biochemical oxygen demand; dissolved oxygen.

oxyhemoglobin. See hemoglobin.

oxymethandrolone. See oxymetholone.

oxymetholone (oxymethandrolone) $C_{21}H_{34}O_3$. 17-beta-Hydroxy-2-hydroxymethylene-17-alpha-methyl-3-androstanone.
Properties: White, odorless crystalline powder. M. p. 173-179°C.
Use: Medicine.

oxymethylene. See formaldehyde.

oxymorphone $C_{17}H_{19}NO_4$. 14-Hydroxydihydromorphinone.
Properties: M. p. 248-249°C (decomposes). Soluble in boiling acetone-chloroform mixture, boiling ethanol, aqueous alkalies; somewhat soluble in benzene.
Use: Medicine.

beta-**oxynaphthoic acid.** See 3-hydroxy-2-naphthoic acid.

oxyneurine. See betaine.

oxyphenbutazone. 1-Phenyl-2-(para-hydroxyphenyl)-3, 5-dioxo-4-n-butylpyrazolidine monohydrate. An analog of phenylbutazone and one of its metabolites.
Use: Medicine.

oxyphencyclimine hydrochloride $C_{20}H_{29}ClN_2O_3 \cdot HCl$. 1-Methyl-1, 4, 5, 6-tetrahydro-2-pyrimidylmethyl-alpha-cyclohexyl phenylglycollate hydrochloride.
Properties: White crystals; bitter taste; decomposes at 231-232°C; soluble in water.
Use: Medicine.

oxyphenonium bromide $C_{19}H_{30}NO_3Br$. Diethyl-(2-hydroxyethyl)methylammonium bromide alpha-phenyl-alpha-cyclohexylglycolate. A synthetic quaternary ammonium compound.
Grade: N. N. D.
Use: Medicine.

oxyphosphoranes. A class of compounds derived from trialkyl phosphites and orthoquinones. Their molecules have a five atom ring, OCCOP(OR)$_3$ in which the two carbon atoms are part of an aromatic ring. They react by liberating a phosphate ester.

oxypolygelatin (OPG). A purified gelatin treated with glyoxal, followed by oxidation

*See "I. C. C. Shipping Regulations," page xiii.
Reference numbers refer to name of manufacturer. See "List of Manufacturers," page v.

with hydrogen peroxide. A possible plasma substitute.

oxyquinoline. See 8-hydroxyquinoline.

oxyquinoline sulfate. See 8-hydroxyquinoline sulfate.

oxytetracycline $C_{22}H_{24}N_2O_9 \cdot 2H_2O$. An antibiotic obtained from Streptomyces rimosus, an actinomycete. It is relatively nontoxic. Its chemical structure is that of a modified naphthacene molecule.

Properties: Dull yellow, odorless, slightly bitter crystalline powder. M. p. 179-182° (dec). Soluble in acids and alkalies; very slightly soluble in acetone, alcohol, chloroform and water; practically insoluble in ether. Stable in air; affected by sunlight. Deteriorates in solution with pH below 2; destroyed rapidly by alkali hydroxide solutions; pH (saturated solution) about 6.5.

Grade: N. F. XI.

Use: Medicine; feed supplement; other agricultural and industrial uses.

oxytetracycline hydrochloride $C_{22}H_{24}N_2O_9 \cdot HCl$. The hydrochloride salt of oxytetracycline (q. v.).

Properties: Yellow, crystalline, odorless, powder with bitter taste. M. p. 180° (dec). Very soluble in water; soluble in alcohol; sparingly soluble in acetone; slightly soluble in chloroform; very slightly soluble in benzene and ether; pH (1% solution) about 2.5. Hygroscopic and affected by sunlight. Potency affected in solutions with pH below 2 and rapidly destroyed in alkali hydroxide solutions.

Grade: N. F. XI.

Use: Medicine.

oxytocin (alpha-hypophamine). One of the hormones secreted by the posterior lobe of the pituitary gland. Its chief action is the stimulation of the contraction of the smooth muscle of the uterus. It is an octapeptide containing eight different amino acids and has been purified and synthesized. It is available as a solution for injection (oxytocin injection, U. S. P. XVI.)

oxytoluene, meta-, ortho-, para-. See corresponding cresol.

oyster shells. Shells of Ostrea virginica, taken principally from the Gulf of Mexico coast in Texas and Louisiana; also from Chesapeake Bay. Average analysis: $CaCO_3$ 93-97%; $MgCO_3$ 1%; silica 0.5-2.0%; SO_4 (as $CaSO_4$) 0.3-0.4%; also miscellaneous substances.

Uses: Source of lime; drilling muds, road beds; poultry and cattle feeds.

"Ozark." [141] Trade name for zinc sulfate.

"Ozene." [292] Trademark for a chlorinated benzene composition.

"Ozide." [141] Trade name for zinc oxide.

"Ozlo." [141] Trade name for leaded zinc oxide produced by the co-fumed process.

Composition: Zinc oxide 50-88%; $PbSO_4 \cdot PbO$ 12-55%.

Grades: "Ozlo" 55: 55% $PbSO_4 \cdot PbO$; "Ozlo" 50: 50% $PbSO_4 \cdot PbO$; "Ozlo" 35: 35% $PbSO_4 \cdot PbO$; "Ozlo" 12: 12% $PbSO_4 \cdot PbO$; "Ozlo" 18M: 18% monobasic lead sulfate, more efficient because of higher available lead content.

Uses: Paint pigment; zinc salts; activator in rubber manufacture; mold growth inhibitor in paints.

ozocerite (mineral wax; fossil wax; native paraffin).

Properties: A native, waxlike hydrocarbon mixture, yellow-brown to black or green in color, translucent when pure and having a greasy feel. Soluble in light petroleum hydrocarbons, benzene, turpentine, kerosene, ether, carbon disulfide; slightly soluble in alcohol; insoluble in water.

Constants: Sp. gr. 0.85-0.95; m. p. 55-110°C, usually about 70°C.

Derivation: Found in nature in Utah, Austria, near the Caspian Sea, and Galicia.

Method of purification: Filtration.

Grades: Technical.

Containers: 80- to 100-lb stands; bags.

Uses: Electric insulation; rubber products; paints; leather polish; sealing wax; candles; lithographic and printing inks; electrotypers' wax; carbon paper; source of ceresin; floor polishes; impregnating furniture and parquet floor lumber; lubricating compositions; grease crayons; sizing and glossing paper; waxed paper; cosmetics; ointments; matrices for galvano-plastic work; textile sizings; waxed cloth; substitute for carnauba and beeswax.

Shipping regulations: None. *

ozocerite, purified. See ceresin wax.

ozokerite. Same as ozocerite.

ozone. An allotropic form of oxygen corresponding to the formula O_3.

Properties: Unstable blue gas with a pungent odor; decomposes to ordinary oxygen; powerful bleaching action; oxidizes more rapidly than oxygen and promotes spontaneous ignition of many substances. Its presence in air in known to contribute to the characteristic properties of smog. Embrittlement of rubber is accelerated by the presence of traces of ozone in the air.

One can detect the odor of ozone when present in the air to the extent of only one part in 500,000,000. In concentrations of about one part in 50,000,000 the odor is pleasant, resembling that of clover. When the concentration reaches one part in 1,000,000 the more characteristic sulfur-like odor becomes apparent. B. p. −112°C; m. p. −250°C; sp. gr. (liq.) about 1.6. Ozone is more soluble in water than is oxygen, a few milligrans per liter dissolving at ordinary temperatures. Its solubility is greater in acetic acid, acetic anhydride, propionic acid, chloroform, carbon tetrachloride, and dichloroacetic acid than in water.

*See "I. C. C. Shipping Regulations," page xiii.
Reference numbers refer to name of manufacturer. See "List of Manufacturers," page v.

Derivation: Commercial mixtures containing 1-2% ozone are produced by electronic irradiation of air or oxygen. Ozone is also formed when air or oxygen is exposed to ultraviolet light and when alkaline perchlorate solutions are electrolyzed. Since it is too expensive to ship because so diluted with air or oxygen, it is usually manufactured in place. Tonnage lots are being used.

Uses: Oxidizing agent; purification of drinking water; treatment of industrial wastes; deodorization of air and sewage gases; preservation of foods in cold storage; bleaching waxes, oils, textiles; promoting production of peroxides; bactericide. Oxidizing agent in several chemical processes (acids, aldehydes, ketones from unsaturated fatty acids); steroid derivatives; removal of chlorine from nitric acid; oxidation of phenols and cyanides.

Caution: Dangerous to breathe even in low concentrations for a protracted period! Liquid ozone is easily exploded as are concentrated ozone-oxygen mixtures (above about 3% ozone) in either the liquid or the vapor state.

Not shipped. See Derivation.

P

P. Symbol for phosphorus.

p-. Abbreviation for para-.

"P-10." [202] Trademark for a technical grade of castor oil fatty acids containing approximately 90% ricinoleic acid. Used to enhance dye colors in inks; to impart lubricity and rust-proofing qualities to soluble cutting oils; in germicidal soaps and pharmaceuticals; and in glycerine-free soaps for foam rubber stabilization.

"P-33." [69] Trademark for a proprietary thermatomic carbon. Available as a soft type of carbon black used in inner tubes, belting, hose, and molded rubber compounds. Also available in pellet form. It imparts high tensile, stretch, and resilience, with good tear and fair abrasion resistance.

Pa. Symbol for protactinium.

PA. Abbreviation for phthalic anhydride; also used for polyamide.

"PA." [241] Brand name for pre-attrited silica gel available in several mesh sizes. Marketed in a "PA 100" and a "PA 400" series. Used for liquid and vapor phase dehydration of refrigerants in either the high pressure side or low pressure side of the system; also for drying air, gases, and other organic liquids.

PABA. Abbreviation for para-aminobenzoic acid.

PABA sodium. See sodium para-aminobenzoate.

"Pabst Brewers Yeast." [114] Brand name for a proprietary product. Bottom-type, debittered yeast derived from the brewing process. Thiamin, 50 International Units per gram; riboflavin, 45 gammas. Contains nicotinic acid, pantothenic acid, filtrate factor, vitamin B_6.

Use: For mixing with pharmaceutical products requiring vitamin potency. Also finds use in the food industry as a reinforcement for its vitamin and protein values in infant and geriatric food products.

"Pabst Industrial Malt Extracts." [114] Brand name for a proprietary product. Maltose value: to specifications, 55-60%; total solids: to specifications, 79.5-82%; color: to specifications.

Uses: For all food industries; candy manufacturers; flavoring; malt and milk mixtures; ice-cream manufacturers; cake and cookie bakers.

"Pachkote." [323] Trademark for a heavy duty coal tar mastic.

"Pacific Crystals." [177] Trademark for pure, crystalline sodium sesquicarbonate, stable, non-caking and free-flowing.

pack fong. Chinese alloy of copper, nickel and zinc and resembling German silver.

packing-house pitch. See stearin and fatty acid pitches.

"Padutin." [162] Trademark for vascormone.

"Pagitane Hydrochloride." [100] Trade name for cycrimine hydrochloride (alpha-cyclopentyl-alpha-phenyl-1-piperidinepropanol hydrochloride or cyclopentyl-phenyl-3-(1-piperidyl)-1-propanol hydrochloride).

Properties: A white, odorless, bitter solid; m.p. 241-244°C; practically insoluble in benzene and ether; soluble in alcohol, chloroform and water; pH (0.5% solution) 4.9-5.4.

Use: Medicine.

PAHA. See para-aminohippuric acid.

paint. A liquid mixture which may be applied to surfaces to form a dry, thin, protective or decorative film. A paint is composed of a solid (the pigment) and the liquid "vehicle." The latter consists of a binder which forms the film and, usually, a volatile solvent or thinner to improve the ease of application. The solvent may be omitted from paints that are to be applied by hot melt or flame spraying techniques. In other cases the solvent may be such that it will change into a resin and become part of the film.

Paints may be either water-base or oil-base according to whether the thinner is water or an organic liquid such as turpentine, petroleum ether or naphtha, benzene, acetate esters, acetone, or an alcohol. Water-base paints have the advantage of a cheap, nonflammable thinner and suitability for use on damp or porous surfaces. A recent development is the heavy-bodied (thixotropic) paint. See gel paint.

Paint binders are classified or selected according to the manner in which the film is formed. (1) Binders that form films by oxidation or polymerization. Examples are drying oils (q.v.) (including drying oil-modified alkyd or epoxy resins, etc.); formaldehyde condensation resins (phenolic, urea, and triazine resins); allyl resins; and polyurethanes. (2) Binders that form films by evaporation of the thinner or by

congealing from a melt. Examples are nitrocellulose and most other cellulose esters and ethers, vinyl resins, styrene resins, many polyacrylates and polymethacrylates, rubber derivatives, some polyamide resins, and polyethylene.
(3) Binders that form a film when particles coagulate from a latex or dispersion of synthetic rubbers and other resins such as "Teflon" and high molecular weight vinyl resins. For these resins, subsequent heat treatment to fuse the particles into a continuous film is required. (4) Water-soluble binders such as casein or glue may be used in water-base paints although these have largely been supplanted by the latex type or by emulsions (emulsion paints) of typical oil-base vehicles.

The pigments (q. v.) may be various organic or inorganic materials used principally to impart color, opacity, and body to the paint film.

Paints may also contain as minor constituents plasticizers, driers, and extenders (q. v.). In addition, water-base paints usually contain emulsifiers, stabilizers, and anti-foam agents.

Special uses: Imparting resistance to corrosin, to fire, and to marine, mildew, or fungus growths; providing electrical insulation or conduction; protection against radiation; indication of temperature; reduction of frictional resistance (in aeronautical applications).

Shipping regulations: May be classified as flammable liquid. Red label. *

"Paint Deodorant No. 5." [188] Brand name for a deodorant for paints.

paint driers. See driers.

paint extenders. Primarily cost-reducing materials used in paint formulation. They have, however, secondary functions which vary in importance with the kind of paint in which they may be formulated.

The following is a list of the materials commonly used for this purpose:
barium sulfate (natural; barytes, precipitated)
calcium sulfate (gypsum, terra alba, plaster of Paris)
calcium carbonate (whiting, chalk)
magnesium carbonate
silica (quartz, diatomaceous earth)
magnesium silicates and related minerals (talc, soapstone)
kaolin
miscellaneous (mica, pumice).

"Paintodors." [12] Trademark for a series of specific masking agents designed for use in finished paint products to cover odors at application, during and after drying. No adverse properties with vehicles, pigments, preservatives or driers.

paint pigments. See pigments.

paint remover (varnish remover; finish remover). A preparation in liquid or paste form intended to be applied to objects coated with a dried film of paint, varnish, lacquer or enamel for the purpose of removing the finish. A uniform coating of the remover is generally applied to the surface under treatment, allowed to act for a longer or shorter time, and the softened or dissolved finish scraped, rubbed or washed away. A satisfactory remover will not affect the coated object and will leave it in good condition for refinishing. Removers containing volatile solvents usually contain nonvolatile constituents intended to retard the rate of evaporation of the liquid, thereby prolonging its action. Typical solvents are: methanol, denatured ethyl alcohol, toluene, benzene, and ethyl acetate. Paraffin (120°) is most often used as the retarder. Caustic removers are made of sodium phosphate, sodium silicate, caustic soda, or the like.

paint vehicle. The liquid portion of paint, composed of the film-forming component (binder) and volatile solvent or thinner. See paint.

"Palacet." [440] Trademark for a series of organic pigments used for dyeing and printing on acetate, polyamide and polyester fibers.

"Palanil." [440] Trademark for a series of organic pigments used for dyeing and printing on polyester fibers.

"Palanthrene." [440] Trademark for a series of vat dyestuffs, mainly anthraquinonoid types.

"Palatine." [307] Metallized acid dyestuffs approaching the fastness of chrome colors.

"Palatone." [233] Proprietary brand of maltol, 3-hydroxy-2-methyl-4-pyrone. A white crystalline solid, soluble in ethanol, propylene glycol, water, phenethyl alcohol and other synthetic aromatics; solid material has caramel-butterscotch odor, whereas alcohol solution has a definite pineapple character with a slight indication of strawberry.

palau. A palladium-gold alloy sometimes used by laboratory workers as a platinum substitute.

"Palconate." [126] Trade name for the sodium salt of phenolic acids extracted from redwood bark.

Properties: Brown-black powder or fine granules; density 33-35 lbs/cu ft; soluble in water and 1% caustic soda; insoluble in alcohol and ether; ash 31-32%; pH 10-11.

Uses: Drilling mud conditioner, ore flotation; battery expander; dispersing agent; replacement for phenol; ceramic binder.

"Palcotan." [126] Trade name for sodium salt of sulfonated complex organic acids extracted from redwood bark.

Properties: Dark brown powder or fine granules; density 37-39/lbs/cu ft; soluble in water and 1% caustic soda; insoluble in alcohol and ether; ash 31-33%; pH 7-8.

Containers: 50-lb paper bags.

Uses: Drilling mud conditioner; water

*See "I.C.C. Shipping Regulations," page xiii.
Reference numbers refer to name of manufacturer. See "List of Manufacturers," page v.

treatment; ore flotation; battery expander; leather tanning; dispersing agent; ceramic binder; rubber industry.

pale chrome. See chrome yellows.

pale oxide of iron. Red pigment consisting essentially of ferric oxide. See iron oxide reds.

palladium Pd. Element of atomic number 46, one of the triad ruthenium, rhodium, palladium.
Properties: Silver-white ductile metal. Does not tarnish in air, but on heating to about 600°C, becomes coated with an oxide which decomposes about 800°C, leaving the metal bright again. Absorbs large volumes of hydrogen. Insoluble in cold sulfuric acid and hydrochloric acid but is attacked somewhat by the hot acids; soluble in nitric acid; insoluble in organic acids. Readily soluble in aqua regia and fused alkalies.
Constants: Sp. gr. 12.0; m.p. 1554°C; Brinell hardness, hard 109, annealed 46 (i.e., slightly harder than platinum).
Derivation: Occurs in nature along with platinum in gold, nickel, and copper ores. The final refining of the platinum metals is done by chemical methods.
Grades: Chemically pure (99.99%); technical (99.0%).
Containers: Wooden or plastic boxes.
Uses: Catalyst, especially in hydrogenation processes; resistance wires; electroplating; component of alloys used for electrical contacts, jewelry, dentistry, aircraft spark plugs, non-magnetic watch parts; hydrogen "valves."
Shipping regulations: None. *

palladium chloride (palladous chloride; palladium dichloride) (a) $PdCl_2$ (b) $PdCl_2 \cdot 2H_2O$.
Properties: Dark brown, deliquescent powder or crystals; soluble in water, hydrochloric acid, alcohol and acetone. M.p. (a) 501°C (decomposes).
Derivation: By solution of palladium in aqua regia and evaporation.
Grades: Technical; reagent.
Containers: Glass bottles; tins.
Uses: Medicine; analytical chemistry; photography on porcelain; manufacture of indelible inks; electroplating with palladium; detecting carbon monoxide gas; ingredient of metal scouring compositions; starting point in the manufacture of photographic toning agents; textile mordant; manufacture of stearin, porcelains, germicides.
Shipping regulations: None. *

palladium dichloride. See palladium chloride.

palladium iodide (palladous iodide) PdI_2.
Properties: Black powder; sp. gr. 6.003 (18°C); soluble in a solution of potassium iodide; insoluble in alcohol and water. Decomposes above 100°C.

palladium monoxide. See palladium oxide.

palladium nitrate (palladous nitrate) $Pd(NO_3)_2$.
Properties: Brown salt, deliquescent; decomposed by heat; soluble in water with turbidity; soluble in dilute nitric acid.
Grades: Technical.
Use: Analytical reagent.

palladium oxide (palladium monoxide) PdO.
Properties: Black-green or amber solid; sp. gr. 8.31; m.p. (dec) 750°C; soluble in dilute acids.
Derivation: Careful ignition of the nitrate or prolonged heating of the finely divided metal at about 800°C.
Use: Reduction catalyst in organic synthesis.

palladium-potassium chloride (palladous-potassium chloride; potassium-palladium chloride) $PdCl_2 \cdot 2KCl$.
Properties: Reddish-brown crystals; soluble in water.
Uses: Reagent for carbon monoxide determination.

palladium-silver alloy. Used as means of purifying hydrogen to a very high degree of purity.

palladium-sodium chloride (palladous-sodium chloride; sodium-palladium chloride) $PdCl_2 \cdot 2NaCl \cdot 3H_2O$.
Properties: Brown salt; hygroscopic; soluble in alcohol and water.
Grades: Technical.
Use: Analysis (testing for carbon monoxide, ethylene, illuminating gas, iodine).

"Palladon." [169] Trademark for para-nitrosodimethylaniline used in the colorimetric determination of palladium and platinum.

palladous chloride. See palladium chloride.

palladous iodide. See palladium iodide.

palladous nitrate. See palladium nitrate.

palladous-potassium chloride. See palladium-potassium chloride.

palladous-sodium chloride. See palladium-sodium chloride.

"Pall Ring." [326] Trade name for tower filling material. Similar in design to a Raschig ring but with sections of the wall stamped and bent inward. Made in metals, plastics and ceramics.

palma christi. See ricinus.

palmarosa oil (Indian grass oil; Rusa oil; East Indian geranium oil; geranium oil, Turkish).
Properties: Colorless or light yellow, volatile oil, occasionally colored green by copper. Pleasant rose-like odor; sp. gr. 0.885-0.897 (15°C); optical rotation −2° to +3°C; refractive index (n 20/D) 1.4730-1.4775; soluble in 2 volumes of 70% alcohol, benzyl benzoate, fixed oils, propylene glycol, mineral oil; insoluble in glycerine.
Chief known constituent: Geraniol.
Derivation: By distilling Cymbopogon martini, var. motia, found in the East Indies and Java. The name Turkish geranium oil was formerly applied to this oil because it was usually shipped from India to Red Sea ports. Then it was sent by land to Constantinople

where it was distributed to other markets. This practice has now been replaced by direct shipments from India.
Adulterants: Kerosene, coconut oil, gurjun balsam oil, cedar oil, and turpentine. Their presence is indicated by the insolubility of the oil in 70% alcohol.
Containers: Cans.
Uses: Perfumery; manufacture of geraniol.
Shipping regulations: None. *

palm butter (palm oil; palm grease).
Properties: A fixed, reddish-yellow fatty oil of butter-like consistency, faint violet odor, which is conveyed to the soap made from the oil. Soluble in alcohol, ether, chloroform, and carbon disulfide.
Chief constituents: Free palmitic acid (12% in fresh oil to 55% in older oil), stearic acid, and glycerides of palmitic and oleic acids.
Constants: Sp. gr. 0.920-0.927; m.p. 27-42.5°C; iodine number 51.5; saponification number 202; Reichert number 0.5.
Derivation: By expression from the putrefied or fermented pulp of the fruit of a palm, Elaesis guineensis.
Occurrence: West coast of Africa; West Indies; South America.
Containers: Wooden barrels; casks; steel drums.
Uses: Manufacture of soaps and candles; emollient; coating iron plates; used in tinplate industry; lubricants; coloring butter substitutes.
Shipping regulations: None. *
See also palm-nut oil.

palm cake. See palm-nut cake.

"Palmex." [152] Trade name for a series of metal processing oils used for cold rolling of steel and aluminum, hot dip tinning and pickling.
Containers: Drums; tank trucks; tank cars.

palm grease. See palm butter.

palmitic acid (hexadecanoic acid; palmitinic acid; cetylic acid) $CH_3(CH_2)_{14}COOH$. One of the more common fatty acids. It occurs in natural fats and oils and in tall oil, and in large amounts in most commercial grade stearic acid.
Properties: White crystals; soluble in alcohol and ether; insoluble in water; sp. gr. 0.8414 (80/4°C); m.p. 62.9°C; b.p. 351.5°C (760 mm), 271.5°C (100 mm), 139.0°C (1 mm); refractive index 1.4309 (70°C).
Derivation: From spermaceti by saponification; hydrolysis of natural fats.
Method of purification: Crystallization.
Grades: Technical; 99.8%.
Containers: Wooden barrels; bags; boxes.
Uses: Starting point in manufacture of various metallic palmitates; soaps; lube oils; waterproofing.
Shipping regulations: None. *

palmitic acid cetyl ester. See cetin.

palmitin. See tripalmitin.

palmitinic acid. See palmitic acid.

palmitoleic acid (cis-9-hexadecenoic acid) $CH_3(CH_2)_5CH{:}CH(CH_2)_7COOH$. An unsaturated fatty acid found in nearly every fat, but largest amount in marine oils (15-20%).
Properties: Colorless liquid; m.p. 1.0°C; b.p. 140-141°C (5 mm). Insoluble in water; soluble in alcohol and ether.
Grade: Purified product 99%.
Uses: Organic synthesis; medical research; chromatographic standard.

palmitoyl chloride (hexadecanoyl chloride; palmityl chloride, so-called) $CH_3(CH_2)_{14}COCl$.
Properties: Colorless liquid; very soluble in ether, decomposes in water or alcohol; m.p. 11-12°C; b.p. 194.5 (17 mm).

palmityl alcohol. See cetyl alcohol.

palmityl chloride. Used, confusingly, as a synonym for palmitoyl chloride, the acid chloride $C_{15}H_{31}COCl$. It might also mean the alkyl chloride, $C_{15}H_{31}CH_2Cl$.

palm-kernel cake. See palm-nut cake.

palm-kernel oil. See palm-nut oil.

palm-nut cake (palm-kernel cake; palm cake). The cakes formed in the press when the seeds are subjected to hydraulic pressure in order to express the palm-nut oil. Contains various useful constituents, such as unexpressed oil, carbohydrates, proteins, and salts. Typical analysis: proteins 30.4%; fats 8.4%; fiber 41.0%; water 9.5%; ash 10.6%.
Containers: Bags; bulk.
Uses: Cattle-food; fertilizer ingredient.

palm-nut meal. The mealy form assumed by palm nuts after the crushing and heating operations preparatory to the expression of the oil in either the hydraulic presses or the expeller. If the oil cake be ground the product again is in this mealy form. Uses are similar to those of palm-cake (q.v.).

palm-nut oil (palm-kernel oil; palm oil).
Properties: A yellowish, fatty oil, free of fatty acids when fresh; rapidly becoming rancid in air. Soluble in alcohol, ether, chloroform, and carbon disulfide.
Chief constituents: Triolein (15-25%), triglycerides of stearic, palmitic and myristic acids (33%) and triglycerides of lauric, capric, caprylic and caproic acids (45-55%).
Constants: Sp. gr. 0.952; m.p. 26-30°C; iodine number 13.4-13.6; saponification number 247.6.
Derivation: By crushing the nuts of Elaesis guineensis and pressing, or extracting with solvents.
Method of purification: Filtration.
Grades: Crude; refined.
Containers: Wooden kegs; steel drums; tank cars.
Uses: Manufacture of soaps and chocolate products; pharmacy; margarine manufacture, coloring butter substitutes; candles; illuminant; cutting tool lubricant; tin plating;

*See "I.C.C. Shipping Regulations," page xiii.
Reference numbers refer to name of manufacturer. See "List of Manufacturers," page v.

cosmetics; softening and finishing cotton goods.
Shipping regulations: None. *
See also palm butter.

palm oil. See palm butter and palm-nut oil.

"Paludrine." [207] Trademark for proguanil or chloroguanide hydrochloride (q. v.). A synthetic antimalarial drug.

PAM. Abbreviation for 2-pyridine aldoxine methiodide; stated to be an antidote for nerve gas.

"Pamak." [266] Trademark for a series of fatty acids derived from tall oil.
Properties: Very light-colored liquids; having rosin content ranging from 1.3% to 25%.
Uses: Production of alkyd resins for protective coatings; soaps and detergents; core oils; flotation of nonmetallic and metallic minerals; metallic driers; floor coverings; rubber compounding; vinyl stabilizer-plasticizers.

pamaquine naphthoate $C_{42}H_{45}N_3O_7$. 6-Methoxy-8-(1-methyl-4-diethylaminobutylamino)-quinoline 1, 1'-methylene-bis(2-hydroxy-3-naphthoate).
Properties: Yellow to orange yellow odorless almost tasteless powder. Insoluble in water; soluble in alcohol and acetone.
Use: Medicine.

"Pamine" Bromide. [327] Trademark for methscopolamine bromide; scopolamine methyl bromide ($C_{18}H_{24}BrNO_4$).
Properties: Colorless crystals; m. p. 218°C; very soluble in water, less so in alcohol and acetone; insoluble in ether and benzene.
Derivation: Synthetic.
Use: Medicine.

"Pamisyl." [330] Trademark for para-aminosalicylic acid.

"Panaflex." [216] Trademark for liquid hydrocarbon plasticizers used as secondary plasticizers in compounding polyvinyl chloride; in vinyl plastisols; and in molded and extruded products.

Panama bark. See quillaja.

"Panapol." [216] Trademark for hydrocarbon liquid polymers derived from petroleum sources, produced by the polymerization of olefins and diolefins; available in a wide color range with different physical properties.
Containers: Tank cars or 55-gal drums.
Uses: As a partial replacement or extender for drying oils, or where a low cost resinous product is desirable.

"Panarez." [216] Trademark for hydrocarbon resins derived from petroleum sources, produced by the polymerization of olefins and diolefins; available in color grades ranging from pale lemon to dark brown with a normal softening point of 200-220°F. Higher softening grades can also be produced. Available as solid resin, solution, or flaked material.
Uses: Paints and varnishes; rubber compounding; floor tile; etc.

"Panasol." [216] Trademark for petroleum aromatic solvents available in a variety of boiling ranges. Used as solvents in paint and varnish applications, and in the formulation of insecticides.

pancreatic deoxyribonuclease. See pancreatic dornase.

pancreatic dornase (pancreatic deoxyribonuclease; deoxyribonuclease). Stabilized preparation of deoxyribonuclease.
Derivation: Prepared by fractional precipitation of aqueous acid extracts of beef pancrease followed by dialysis, sterilization by filtration and lyophilization.
Grade: N. N. D.
Use: Medicine.

pancreatin. A substance containing enzymes, principally pancreatic amylase, trypsin, and pancreatic lipase. Obtained from the pancreas of hog or ox.
Properties: Cream-colored amorphous powder; characteristic odor; acts upon starch and proteins. Soluble in water; insoluble in alcohol.
Derivation: Pancreas gland is extracted by macerating with chloroform, water, dilute boric acid, glycerol, or alcohol, filtered and evaporated.
Grades: N. F. XI.
Containers: Glass bottles; fiber cans.
Use: Medicine, as an emulsifying agent, and as a ferment.
Shipping regulations: None. *

"Pandrinox." [401] Trade name for a combination liquid insecticide-fungicide seed treatment material containing methylmercury dicyandiamide (0.75%) and heptachlor (24.5%).
Containers: 160-oz bottles; 6-, 30-gal drums.
Uses: For treatment of cereal grain and sorghum seeds for protection against soil insects such as wireworms and seed corn maggots and for control of fungi.
Warning! Poisonous when inhaled, swallowed, or absorbed through skin.

"Panobrome." [401] Trademark for a fumigant consisting of methyl bromide.
Containers: 1-lb cans; 10-, 100-, 175- and 375-lb cylinders.
Uses: For fumigation of grain or other stored products to control insects. For soil fumigation to control fungi, weeds and soil insects.
Danger! Vapor extremely hazardous. Highly volatile. Causes burns.

"Pano-drench." [401] Trademark for a soil treatment concentrate containing 0.6% cyano(methylmercuri)guanidine.
Containers: 4 and 16-oz bottles.
Use: As a soil drench in nurseries and greenhouses to control damping-off fungi.
Warning! This liquid is poisonous when inhaled, swallowed or absorbed through the skin. Do not breathe vapors. Do not get in eyes, on skin or on clothing. Handle carefully.

*See "I.C.C. Shipping Regulations," page xiii.
Reference numbers refer to name of manufacturer. See "List of Manufacturers," page v.

"Panogen 15." [401] Trademark for a liquid seed treatment material containing methylmercury dicyandiamide.
Containers: 1-qt, 1/2 gal and 160-oz. bottles; 6-, 30-, 54-gal drums.
Uses: Seed treatment of cereal grain seeds, flax, cotton, sorghum, rice, peanuts and safflower.
Warning! Poisonous when inhaled, swallowed or absorbed through the skin. Do not breathe vapors. Do not get in eyes, on skin or on clothing. Handle carefully.

"Panogen Turf Spray." [401] Trade name for a liquid containing methylmercury dicyandiamide (2.2%).
Containers: 8-, 16-, 32- and 160-oz bottles.
Uses: Diluted with water for spraying turf areas for control of fungus diseases such as melting out, fading out, dollar spot, copper spot, etc.

"Panoram 75." [401] Trade name for a seed treatment material containing 75% thiram. Sold as a wettable powder in 25- or 100-lb fiber drums.
Use: For seed treatment of corn, legumes, grasses and vegetables for control of fungi.
Caution! May cause irritation of eyes, nose, throat, and skin. May be harmful if inhaled or swallowed. Avoid contact with eyes, skin or clothing.

"Panoram D-31." [401] Trade name for a combination fungicide and insecticide consisting mainly of thiram (56%) and dieldrin (18%). Sold as a wettable powder in 25- and 100-lb fiber drums.
Use: As a seed treatment material for control of fungi and insects on corn, peas and beans.
Caution! Hazardous if swallowed, inhaled or absorbed through the skin. Avoid breathing dust. Avoid contact with eyes, skin or clothing, feed and foodstuffs.

"Panotectant." [401] Trade name for a liquid containing aldrin (about 42%).
Containers: 6-, 30-, 54-gal drums.
Uses: As a seed treatment insecticide on small grains, sorghum, cotton, rice and corn. For protection against soil insects such as seed corn maggots, wireworms, rice water weevil, etc.
Warning! Hazardous if swallowed, inhaled, or absorbed through skin.

pantethine $C_{22}H_{42}N_4O_8S_2$. The disulfide form of N-pantothenylthioethanolamine. Lactobacillus bulgaricus growth factor (LBF). A fragment of coenzyme A, a pantothenic acid derivative.
Use: Biochemical research.

panthenol. See pantothenol.

"Pantholin." [100] Trademark for racemic calcium pantothenate, U. S. P.

pantolactone $\underline{HC(OH)C(CH_3)_2CH_2OCO}$.
Properties: Crystals; sp. gr. (20/20°C), 1.180; m.p. 79.2°C; soluble in water.
Grades: 80% aqueous solution.
Use: Preparation of pantothenic acid.

pantothenic acid [N-(alpha, gamma-dihydroxy-beta, beta-dimethyl butyryl)-beta-alanine] $HOCH_2C(CH_3)_2CHOHCONH(CH_2)_2COOH$. A member of the vitamin B complex; it is a component of coenzyme A (q. v.), and may be considered a beta-alanine derivative with a peptide linkage. It is involved in the release of energy from carbohydrate utilization, and is necessary for synthesis and degradation of fatty acids, sterols and steroid hormones; it also functions in the formation of porphyrins. It is distributed in all living cells and tissues.
Properties: Yellow viscous oil; soluble in water and most organic solvents; insoluble in benzene and chloroform; dextrorotatory; stable to moist heat, oxidizing and reducing agents; unstable to dry heat, acid or alkaline media; hydrophilic.
Units: Amounts are usually expressed in milligrams or micrograms of pantothenic acid.
Sources: Food sources: liver, kidney, yeast, crude molasses, milk, whole grain cereals, rice. Commercial sources: produced synthetically from alpha, gamma-dihydroxy-beta, beta-dimethyl butyric acid and beta-alanine. Sold as the calcium salt.
Containers: Glass vials and bottles.
Uses: Medicine; nutrition.
See also calcium pantothenate.

pantothenol (D[+]-pantothenyl alcohol; panthenol) $HOCH_2C(CH_3)_2CHOHCONH(CH_2)_2CH_2OH$. alpha, gamma-Dihydroxy-N-(3-hydroxypropyl)-beta, beta-dimethylbutyramide. The alcohol corresponding to pantothenic acid, with vitamin activity.
Properties: A viscous liquid; soluble in water, ethanol, methanol; specific rotation +28.36° to 30.7° in water (c = 5); refractive index 1.497 (20°C).
Use: Biochemical research; vitamin supplement.

pantothenyl alcohol. See pantothenol.

papain (papayotin; vegetable pepsin).
Properties: White or gray slightly hygroscopic powder; soluble in water and glycerine; insoluble in other common organic solvents. The most thermostable enzyme known.
Derivation: Obtained as dried and purified latex of Carica papaya; source: tropical regions.
Grades: Technical; purified. Technical grade is susceptible to decomposition in storage.
Containers: Glass bottles; lined fiber drums.
Uses: Meat tenderizer; other food industries (mainly beer stabilizer); tobacco, pharmaceutical, cosmetic, leather and textile industries.

papaver (poppy heads; poppy capsules).
Derivation: Capsules and seeds of Papaver somniferum.
Occurrence: Europe and Asia.
Grade: Technical.

Containers: Bags.
Use: Medicine.
Shipping regulations: None. *

papaverine $(CH_3O)_2C_6H_3CH_2C_9H_4N(OCH_3)_2$. 6, 7-Dimethoxy-1-veratrylisoquinoline. An alkaloid.
Properties: White crystalline powder; poisonous! Soluble in chloroform, hot benzene, aniline, glacial acetic acid, and acetone; slightly soluble in alcohol and ether; insoluble in water. M. p. 147°C.
Derivation: From opium.
Method of purification: Crystallization.
Grades: Technical.
Containers: Glass bottles.
Use: Medicine, as such or as the hydrochloride which is soluble in water.
Shipping regulations: None. *

papaverine hydrochloride $C_{20}H_{21}NO_4 \cdot HCl$.
Properties: White crystals or white crystalline powder; odorless; slightly bitter taste; soluble in water, alcohol, and chloroform; practically insoluble in ether. Solutions acid to litmus.
Derivation: The hydrochloride of an alkaloid obtained from opium or prepared synthetically.
Grade: U. S. P. XVI.
Containers: Cans.
Use: Medicine.

papaw. Same as pawpaw.

papaya.
Derivation: Fruit of Carica papaya, a tropical tree. Not to be confused with the pawpaw of central United States, which is Asimina triloba.
Grades: Technical.
Containers: Bags.
Use: Manufacture of carpaine and papain.
Shipping regulations: None. *

papayotin. See papain.

paper board. A general term used to designate papers which range from 0.006" to 0.012" or more in thickness. Can be made from either chemical or mechanical fibrous pulps. See wood pulp.

paper chromatography. See chromatography.

paper clay. One suitable for mixing with paper pulp in order to give body, weight and finish to some grades of paper; also used in coating paper. Whiteness of color, plasticity, and freedom from sand are essential characteristics of these clays. The best paper clays come from England and Georgia.

paper pulp. The main raw material of paper, made from wood, rags, straw, bagasse. See wood pulp, and cellulose.

PAPI. Abbreviation for polymethylene polyphenylisocyanate.

papier-maché. A tough, plastic material made:
(a) from paper pulp admixed with size, paste, oil, resin or other substances;
(b) from sheets of paper glued and pressed together.
Use: Molded into boxes, trays, utensils, architectural ornaments and the like.
Shipping regulations: None. *

"Papricol." [342] Trademark for a brand of oleoresin of paprika for food coloring and flavoring.

paprika. A sweetish condiment made from the fruit of the common red pepper, Capsicum annum; much used by Hungarians.
Types: Bulgarian; Hungarian; Spanish; Yugoslav.
Containers: Bags.

para-. A prefix. For definition of para-compounds, see under ortho-. For para-compounds, see specific compound; thus para-cresol will be found under cresol.

para-acetaldehyde. See paraldehyde.

paracasein. See casein.

paracetaldehyde. See paraldehyde.

"Paracol." [266] Trademark for a series of stabilized water dispersions of petroleum waxes and wax-resin mixtures used in surface and internal sizing of paper and paperboard.

"Paracort." [330] Trademark for prednisone.

"Paracortol." [330] Trademark for prednisolone.

"Paracril." [248] Trademark for a group of synthetic rubbers of the Buna-N or nitrile type. They are produced by the copolymerization of butadiene and acrylonitrile. The "Paracril" rubbers resist deterioration by aliphatic hydrocarbon, mineral and vegetable oils, and animal fats and oils and are particularly suited for applications involving contact with petroleum products. "Paracril" is also used as a plasticizer for vinyls as well as other thermoplastic and thermosetting resins.

"Paracril OZO." [248] Trademark for a nitrile rubber and vinyl blend with excellent ozone resistance and good color stability. Used in wire and cable insulation, molded and extruded mechanicals and oil resistant shoe soles.

"Paradene." [21] Brand name for low-priced, dark, thermoplastic, coal-tar resins available in low to high softening point ranges; used in natural and synthetic rubber compounding.

"Paradi." [306] Trademark for para-dichlorobenzene.

paradichlorobenzene. See dichlorobenzene, para isomer.

"Paradione." [3] Trademark for paramethadione (q. v.).

"Paradone." [232] Brand name for a series of vat dyestuffs.

"Paradors." [12] Trademark for a series of masking agents for use in different rubber formulations. Soluble, stable and compatible under conditions of use providing masking or odor modification. Mixtures of synthetic, heat-resistant aromatic chemicals.

"Paradow." [233] Trade name for paradichlorobenzene.

"Paradyne." [29] Trademark for a group of chemical additives designed to improve the qualities of fuels. Includes additives for improvement of anti-icing, carburetor detergency, and deposit modification qualities of gasoline; storage ability and pour qualities of heating oils. Available in bulk and drum quantities.

paraffin. See paraffin wax.

paraffin, chlorinated. See chlorinated paraffin.

paraffin, fluorinated. See fluorinated paraffin.

paraffin hydrocarbons. See paraffins.

paraffin, liquid. See petrolatum, liquid.

paraffin, native. See ozocerite.

paraffin oils. Lubricating oils made by the dry distillation method.

Properties: The color range is from pale yellow, through yellowish-brown to reddish-brown to dark green but it is customary to decolorize them. In certain countries and localities kerosene is known as paraffin oil. Petrolatum, liquid (q. v.) is also known as paraffin oil.

Constants: Gr. 20° to 30° Bé; flash point (Cleveland open cup) 300° to 450°F; cold test 0° to 40°F; Saybolt viscosity 40 to 600 (70°F).

Method of purification: Filtration.

Containers: Metal cans, tank cars.

Uses: Lubricant; leather dressing; medicine.

Shipping reguations: None. *

paraffin oil, white. See petrolatum, liquid.

paraffins (paraffin hydrocarbons). A group of compounds with the empirical formula C_nH_{2n+2} varying from colorless gases through water-white liquids to low-melting point solids.

paraffin scale. See paraffin wax.

paraffin wax (a) Hard; (b) Soft (paraffin scale; ceresin wax; paraffin).

Properties: White translucent, waxy, tasteless; odorless solid; consisting of a mixture of solid hydrocarbons chiefly of the methane series obtained from petroleum. Before purification it is known as paraffin scale. Soluble in benzene, ligroin, warm alcohol, chloroform, turpentine, carbon disulfide and olive oil; insoluble in water and acids; sp. gr. 0.880-0.915; m. p. 42-60°C.

Derivation: (a) Paraffin oil is chilled and filter-pressed to remove the heavy oil; the remaining solid is paraffin wax. It is also made from paraffin oil by solvent extraction. (b) Treatment of ozocerite with sulfuric acid and bleaching.

Grades: Yellow crude scale; white scale; refined wax; N. F. XI.

Containers: 100-, 200-lb kegs; 200-lb cases (slabs); tank cars.

Uses: Manufacture of paraffin candles, waxed paper, etc; waterproofing wood, cork, etc; impregnating matches; stearin candles; lubricants; preserving eggs; oil crayons; pharmacy (to stiffen ointments); surgery; stoppers for acid bottles; preservative covers for food products; electrical insulation; laundering; preservative for railroad ties; phonograph records; floor polishes; extraction of perfumes from flowers; cosmetics; photography; antifrothing agent in sugar refining; packing tobacco products; coating interior of wine casks; protecting rubber products from sun-cracking.

Shipping regulations: None. *

"Paraflow." [29] Proprietary name for a mixture of synthetic organic compounds of high molecular weight, which are added to lubricating oils to reduce their pour points.

"Para-Flux." [94] Trademark for an asphaltic resin.

Properties: Sp. gr. < 1.06; carbon disulfide insolubles < 1%; flash point above 350°F; viscosity (Saybolt Furol at 122°F) 625-875 seconds; moisture < 0.5%; free sulfur < 0.1%; volatile content 1.5%; no effect on cure.

Containers: Drums and tank cars.

Uses: Rubber plasticizer to be used to replace pine tar, mineral oil, mineral rubber etc., in tires, mechanicals, footwear, heels and soles, tapes, etc.

paraform. See paraformaldehyde.

paraformaldehyde (paraform; polyoxymethylene; polyformaldehyde) $(HCHO)_n$ or $HO(CH_2O)_nH$. A polymer of formaldehyde in which n equals 6 plus. Not to be confused with the tripolymer, sym-trioxane (q. v.).

Properties: White solid with slight odor of formaldehyde; insoluble in alcohol and ether. Soluble in strong alkali solution. The higher polymers are insoluble in water.

Derivation: By evaporating an aqueous solution of formaldehyde.

Forms: Granular, powder, ground.

Grade: 91%, 96%.

Containers: Fiber drums; multiwall paper sacks (carloads).

Uses: Disinfectant; fumigant; fungicide; resins, including artificial horn and ivory; waterproofing animal and casein glues.

Warning: Causes irritation of skin, eyes, nose, and throat. MCA warning label.

"Paragon" Clay. [285] Proprietary brand name for a group of hydrous aluminum silicates (sedimentary kaolins) from South Carolina.

Properties: Sp. gr. 2.60; bulk density aerated 18-20 lbs/cu ft, packed 35-40 lbs/cu ft; creamy white; pH 4.5-5.5; airfloated; particle size 50-60% minus 2 microns.

Containers: 50-lb multiwall bags or bulk.

Uses: As a "soft" clay in rubber compounding to produce low modulus and tensile and a soft uncured compound; paper filler; in sodium silicate and other adhesives to impart body, retard silica migration, speed up setting and produce firmer glue lines; to condition (prevents caking) crushed or extruded resins. As filler in paint and ink. See "Suprex" Clay for a "hard" rubber clay.

"Paralac." [206] Brand name of proprietary line of synthetic resins for use in durable air drying and stoving finishes, paint, varnish and printing ink industries.

"Paralate." [244] Trademark for a product consisting of highly alkaline sodium silicates, complex phosphates, soda ash and soap. Also sold in "Paralate S" modification.
Properties: A white powder, soluble in water. Total alkali as Na_2O 50.7%.
Containers: 300-lb net drums.
Uses: Laundry break compound and builder; high pH type.

paraldehyde (para-acetaldehyde; paracetaldehyde; 2,4,6-trimethyl-1,3,5-trioxane) $C_6H_{12}O_3$. A polymer (trimer) of acetaldehyde.
Properties: Colorless liquid. Disagreeable taste; agreeable odor. Not so flammable as acetaldehyde. Decomposes on standing. Paraldehyde is not as reactive as acetaldehyde. It does not react with alkali disulfite, will not reduce silver solutions to form a mirror, nor does it unite with ammonia or hydroxylamine. It is stable toward alkalies, but is slowly decomposed to acetaldehyde when treated with a trace of mineral acid. Miscible with most organic solvents, volatile oils; soluble in water.
Constants: Sp. gr. 0.9960 at 20°C/20°C; b.p. 124.5°C; m.p. 12.6°C; vapor pressure 25.3 mm (20°C); flash point 111.2°F; specific heat 0.434; refractive index 1.40 to 1.42 (20°C); wt 8.27 lbs/gal (20°C).
Typical specifications, technical grade: Paraldehyde 98% min; acetaldehyde 2% max; acidity (as acetic) 0.5% max; boiling range 100 to 127°C; water-white; sp. gr. 0.991-0.993 (20°C); wt 8.27 lbs/gal (20°C).
Typical specifications, U.S.P. XVI grade: Paraldehyde 99% min; acetaldehyde 0.4% max; acidity (as acetic) 0.5% max; water white; boiling range 120-126°C; freezing point 11.0°C min; residue 0.06% max; dilution test, passes test; sulfates, chlorides, none; sp. gr. 0.99 at 25°C; wt 8.25 lbs/gal (25°C).
Derivation: Action of hydrochloric or sulfuric acid upon acetaldehyde.
Grades: Technical; U.S.P. XVI.
Containers: Technical: 7-lb jugs; 35-lb cans; 435-lb drums; tank cars. U.S.P. XVI grade: 50-lb cans; 875-lb drums.
Uses: Substitute for acetaldehyde; rubber accelerators; rubber antioxidants; making synthetic organic chemicals; dyestuff intermediates; medicine; solvent for fats, oils, waxes, gums, resins; leather; solvent mixtures for cellulose derivatives.
Caution: Vapor harmful, use with adequate ventilation. MCA warning label.
Shipping regulations: None.*

"Para Lube." [94] Trademark for a proprietary petroleum-base softener for rubber compounding.
Properties: Sp. gr. 1.00 ± 0.05; viscosity (Saybolt Univ.) 85 ± 10; flash point (open cup) 235°F min.
Containers: 50-gal steel drums; tank cars.
Use: In manufacture of GR-S molded goods, extruded goods, and hard rubber products.

paramethadione $(C_2H_5)(CH_3)\overline{COC(O)N(CH_3)CO}$. 3,5-Dimethyl-5-ethyloxazolidine-2,4 dione.
Properties: A clear, colorless liquid with esterlike odor; refractive index 1.449-1.501; freely soluble in alcohol, benzene, chloroform and ether; sparingly soluble in water.
Grade: N.F. XI.
Use: Medicine.

"Paramine." [232] Brand name for a series of direct cotton dyestuffs.

"Paramine." [300] Trademark for a cationic finishing agent which imparts softness and surface finish to textile fabrics. Amino-fatty condensation product solubilized with a volatile acid.

"Paramins." [29] General designation for the chemical additives marketed by Enjay Chemical Company.

"Paramul 115." [57] Trademark. A water-repellent of the retreatable or non-durable type, an emulsion of wax and aluminum salts.

"Paramul" Water Repellent DC-1. [57] Part of a two-package water repellent finish to be used with "Paramul" Water Repellent DC-2. These two compounds when mixed produce a water repellent finish on various fabrics. This finish is semidurable to washing and dry cleaning. These products can be applied by padding or exhaustion.

paranitraniline. See nitroaniline, para-.

paranitraniline red. See para red.

"Paranox." [29] Trademark for a series of additives compounded for use in lubricating oils. Several types composed of metallo-organic compounds characterized by the presence of one or more of the following: barium, calcium, zinc, sulfur and phosphorus. Imparts high temperature detergency, low temperature sludge inhibition and anti-wear characteristics to oils compounded for lubrication of automobile and diesel engines and for automatic transmission fluids. Several grades available in bulk and drum quantities.

para-oxon (diethyl para-nitrophenyl phosphate) $(C_2H_5O)_2P(O)OC_6H_4NO_2$. The oxygen analog of parathion. The name has been accepted as generic by the Ent. Soc.
Properties: Odorless, reddish-yellow oil; b.p. 148-151°C (1 mm); sp. gr. 1.269 (25/25°C); refractive index 1.5060 (25°C). Slightly soluble in water, soluble in most organic solvents. Decomposes rapidly in alkaline solutions.
Uses: Insecticide; nerve poison.

"Parapastels." [188] Brand name for a group of perfume and color combinations designed to impart, in one operation, agreeable fragrance and pleasing color to products made of para-dichlorobenzene or naphthalene.

"Paraplex." [23]
"Paraplex" Plasticizers for Coatings. Trademark for polymeric plasticizers for resinous coatings. Primarily polyester, but some grades also epoxides which impart heat and light stability. Supplied as viscous liquids, 100% solids or solutions in petroleum hydrocarbons. Most grades compatible with nitrocellulose, ethyl cellulose, polyvinyl butyral, and other high polymer film formers to produce tough, flexible, durable, abrasion- and water-resistant surface coatings.
Use: Coatings for wood, metal, fabrics, paper, and rubber. Also used in formulation of free films, caulking and sealing compounds.
"Paraplex" Plasticizers for PVC. Trademark for high- and intermediate-molecular-weight polymeric plasticizers for poly vinyl chloride and other high polymers. Supplied as viscous liquids. Primarily polyesters, but some grades also epoxides which impart heat and light stability. Excellent permanence and resistance to extraction and migration.
Use: Manufacture of film, sheeting, insulation, tape, and foam products. Also used in the formulation of plastisols and organisols for slush molding and coating.

"Paraplex" P resins. [23] Trademark for unsaturated polyesters that cure to a cross-linked structure. Supplied as light-colored, 100%-reactive solutions in styrene; some grades acrylic modified. Various grades differ in flexibility, toughness or hardness of final product. Produce tough chemically resistant plastics by laminating, casting or molding processes.
Use: Manufacture of reinforced plastic table tops, trays, boat bodies, awnings, automobile ducts, buttons, and other commercial and industrial products; potting of coils.

"Parapoid." [29] Proprietary name of extreme-pressure additives, especially for inclusion in automotive gear oils and the like.

"Parapol." [29] Trademark for a synthetic plastic prepared by copolymerization of isobutylene and styrene. This material is useful as a free film, as a supported film, and as a modifier for paraffin.

"Parapon." [300] Trademark for a highly sulfated fatty ester dyeing assistant and leveling agent for dyed fabrics. Also imparts softening and is a dispersing agent for acetate dyes.

para red (paranitraniline red)
$C_{10}H_6(OH)NNC_6H_4NO_2$. A pigment color formed by coupling diazotized para-nitroaniline with beta-naphthol. The term is also used to refer to a group of lakes based on this dye. See para toners.

"Para-Resin" 2457. [94] Trademark for a dark colored petroleum base resin.
Properties: Sp. gr. 1.09; softening point 200-220°F; carbon disulfide insolubles < 0.5%; carbon tetrachloride insolubles < 0.5%.
Containers: Drums.
Use: Rubber compounding.

para-**rosolic acid.** See rosolic acid.

"Parasepts." [138] Trademark for group of neutral esters of para-hydroxybenzoic acid; includes the methyl, ethyl, propyl, benzyl, and butyl esters.
Properties: White powder.
Grade: Purified; technical.
Uses: In parenteral solutions, pharmaceuticals and cosmetics for inhibiting the growth of mold and bacteria. Technical grade for use in glues and other industrial applications.

"Paratac." [29] Trademark for a high molecular weight isobutylene polymer used as an additive for both lubricating oils and greases. Particularly applicable in producing the so-called "non-drip," "non-spatter" oils.

"Paratex." [300] A water softener used in textile processing.

parathion (O, O-diethyl-para-nitrophenylthiophosphate) $(C_2H_5O)_2PSOC_6H_4NO_2$.
Properties: Deep brown to yellow liquid; often but not always with characteristic odor. Refractive index (n 25/D) 1.5367; sp. gr. 1.26; b. p. 157-162°C; vapor pressure 0.003 mm (24°C); very slightly soluble in water (20 ppm); completely soluble in esters, alcohols, ketones, ethers, aromatic hydrocarbons, animal and vegetable oils; insoluble in petroleum, ether, kerosene, spray oils. Stable in distilled water and in acid solution. Hydrolyzed in the presence of alkaline materials; slowly decomposes in open air.
Purity: Technical grade is about 95% pure. Also supplied diluted with inert carriers of various types, and in various proportions.
Derivation: From sodium ethylate, thiophosphoryl chloride and sodium para-nitrophenate.
Containers: 1-, 5-, and 10-gal tins; 55-gal steel drums.
Uses: Insecticide and acaricide.
Danger! Poisonous by skin contact, inhalation or swallowing. Rapidly absorbed through skin. Repeated exposure may, without symptoms, be increasingly hazardous. MCA warning label.
Shipping regulations: (liquid and dry): Poison, class B. Poison label. *

"Paratone." [29] Trademark for additives designed to improve viscosity index of lubricating oils. All grades are organic compounds. Some grades are multifunctional, with sludge dispersant and pour improvement qualities. One grade designed specifically for automatic transmission fluids and other oils requiring high shear stability. Available in bulk and drum quantities.

para toners. Insoluble red pigments derived from beta-naphthol and para-nitroaniline. The former is sometimes partly replaced by mono-acid F, 2-naphthol-7-sulfonic acid. Through varying the conditions of temperature and acid concentration, different shades may be obtained.
Containers: Barrels.
Uses: Pigments in paint trade, printing industry; making of para lakes.

"Par Clay." 69 Trademark for a proprietary product, a hard kaolin.
Properties: Cream; sp. gr. 2.62 ± .03; fineness (through 325 mesh) 99.5%.
Uses: Filler for rubber.

"Parco Lubrizing." 343 Trademark for an antifriction coating process.

"Paredrine." 71 Trademark for hydroxyamphetamine (q. v.).

pareira (pareira brava).
Derivation: Dried root of Chondodendron tomentosum or C. platphyllum.
Occurrence: Brazil and Peru.
Grades: Technical.
Containers: Bags.
Use: Medicine. Source of curare and curine.
Shipping regulations: None. *
See also Chondodendron tomentosum extract.

pareira brava. See pareira.

parent element. See daughter element.

parethoxycaine hydrochloride. See beta-diethylaminoethyl-para-ethoxybenzoate hydrochloride.

"Parez" Resins. 57 Trademark for a series of melamine-formaldehyde and urea-formaldehyde paper resins.
Uses: Used to provide wet and dry strength to paper, improved foldability and printability. May be used as pulp additives or incorporated into paper coating. They are also used to anchor top coats to cellophane.

Parian cement. See gypsum cements.

"Paricin." 202 Trademark for various alkyl hydroxystearates (soft, low melting point waxes useful as firming agents in cosmetics and specialty inks and as coupling agents for incompatible mixtures of polar and non-polar materials in hot melt applications) and acetoxystearates (plasticizers for nitrocellulose, cellulose acetate butyrate, ethylcellulose, and vinyl resins.)

"Pariflux." 250 Trade name for custom sized high purity ground calcium fluoride for use in welding electrode coatings.

"Parigran." 147 Brand name for a granular product containing 10% Paris green.
Containers: 35-lb bags.
Uses: As a larvicide in mosquito control.

"Paris Black." 133 Trademark for a regular color impingement carbon black for paper and phonograph records. Available as:
"Paris Black." 21 millimicron carbon used for photograph paper, wrappers, boards and other paper stocks. Powdered.
"Paris II." 25 millimicron carbon. A paper black showing unusual ease of dispersion and wetting with good color value and high strength. Available in powdered and semi-compressed forms.
"Paris Beads." 25 millimicron carbon. A paper and phonograph black in the form of a very soft bead showing unusual dispersion properties.
Containers: 25-lb bags.

Paris blue. A name applied loosely to any of a number of the varieties of iron blue pigments. See iron blues. Also used to designate a coal tar dye.

Paris green. See copper acetoarsenite.

Paris white. See whiting.

Paris white cliffstone. See cliffstone Paris white.

Paris yellow. See chrome yellows.

parity. A quantitative characterization of the manner in which a system differs from its mirror image. The concept is important in the quantum-mechanical interpretation of nuclei and their reactions, but does not have a simple analogy in ordinary mechanics. Loosely it is the "handedness" of a nucleus or a nuclear process, and is in general conserved, as is momentum. It has just recently been discovered that parity may not be conserved in some nuclear interactions, upsetting an assumption that had been taken as law of physics for three decades.

"Parkerizing." 343 Trademark. Inhibition of rust formation on iron or steel by coating surface with a phosphate layer, achieved by dipping in acid phosphate solution.

Parkes process. A standard process for the separation of silver from lead. From 1-2% molten zinc is added to the lead-silver mix, heated to above the melting point of zinc. A scum containing most of the silver and zinc forms on the surface. This is separated and the silver recovered. The separation of silver is not complete and the process is repeated several times.

park lily. See convallaria.

"Parlodion." 329 Trademark for a shredded form of pure, concentrated collodion.

"Parlon." 266 Trademark for a white, odorless, nonflammable granular powder; a chlorinated natural rubber; 4 viscosity grades; used as a film former in paints for alkaline surfaces and corrosion-resistant industrial and maintenance paints.

"Parmo." 51 Trademark for petrolatums meeting U.S.P. or N.F. requirements and having melting points, consistencies and colors suitable for pharmaceuticals and cosmetics.

"Parnol." 78 Trademark for an alkyl aryl sulfonate used in textile scouring and dyeing operations, and in synthetic detergents.

*See "I.C.C. Shipping Regulations," page xiii.
Reference numbers refer to name of manufacturer. See "List of Manufacturers," page v.

"Parolite." [159] $Zn(HSO_2 \cdot CH_2O)_2$. Trademark for soluble zinc sulfoxylate formaldehyde.
Use: As a reducing agent, particularly for removing color from dyed woolen materials. (Used at boiling temperature with addition of acetic or formic acid).

paromomycin sulfate. Antibiotic from a strain of Streptomyces.
Properties: Creamy white, odorless, hygroscopic powder. Soluble in water; insoluble in chloroform and ether.
Use: Medicine (antibiotic).

parrot green. See copper acetoarsenite.

parsley camphor. See apiole.

parsley oil.
Properties: A colorless or pale greenish-yellow liquid; strong odor of parsley. Soluble in alcohol, ether, and chloroform. Sp. gr. 1.07.
Derivation: Distilled from the fruit of Petroselinum sativum.
Method of purification: Rectification.
Grades: Technical.
Containers: 1-, 5-lb bottles.
Use: Medicine.
Shipping regulations: None. *

partial pressure. The pressure due to one of the several components of a gaseous or vapor mixture. In general this pressure cannot be measured directly but is obtained by analysis of the gas or vapor and calculation by use of Dalton's law. See also Raoult's law.

partial accelerator. Any device used for increasing the speed of charged particles such as electrons, protons, deuterons, alpha particles, etc. The high speed particles are used as projectiles in nuclear bombardment and other similar processes.

"Parvex." [327] Trademark for a betaine of 1-piperazine carbodithioic acid ($C_5H_{10}N_2S_2$).
Properties: Pale greenish-yellow powder; almost insoluble in water, alcohol, ether, and chloroform; soluble in caustic soda.
Derivation: Synthetic.
Use: Veterinary medicine.

"Parvo." [57] Trademark for folic acid feed supplement.

"Parzate." [28] Trademark for agricultural and horticultural fungicides based on zineb (dry formulation) and nabam (liquid formulation used in combination with zinc sulfate).
Use: For control of various fungous diseases on potatoes, celery, tomatoes, tobacco, cucurbits, peaches, and ornamentals.

PAS. Abbreviation for para-aminosalicylic acid. See 4-aminosalicylic acid.

PASA. Another abbreviation for para-aminosalicylic acid. See 4-aminosalicylic acid.

passiflora. See passion flower.

passion flower (passiflora).
Derivation: Dried flowering and fruiting top of Passiflora incarnata.
Occurrence: Southern United States.
Containers: Barrels.
Use: Medicine.

passivity. Term applied to the property shown by iron, chromium and related metals, in that they lose their normal chemical activity after treatment with strong oxidizing agents like nitric acid, and when oxygen is evolved upon them during electrolysis.

paste.
1. An adhesive composition of semisolid consistency, usually water dispersible. The common pastes are based upon starch, dextrin, or latex, often with the addition of gums, glue, and antioxidants. They are widely employed for the adhesion of paper and paperboard.
2. More generally, a soft, viscous mass.

paste resins. Finely divided resin mixed with plasticizer to form fluid or semifluid mixtures, without use of low boiling solvents or water emulsions.

paste solder. A paste containing flux, cleaner, tinning agent and powdered metallic solder.

pasteurization. Partial sterilization of organic liquids, particularly milk and fruit juices, by heating at 65°C for not less than 30 minutes.

PAT. Abbreviation for polyaminotriazoles.

patchouli oil.
Properties: Yellow-greenish brown or brown liquid having a strong penetrating camphoraceous odor; sp. gr. (15°C) 0.950-0.995; refractive index (n 20/D) 1.5070 to 1.5200; optical rotation −48° to −68°; soluble in 10 volumes of 90% alcohol, usually with opalescence. Acid value, not more than 5; saponification number not more than 18; soluble in ether, chloroform, benzyl benzoate, fixed oils, mineral oil; partially soluble in propylene glycol; insoluble in glycerine.
Derivation: Direct steam distillation of the dried leaves of Pogostemon patchouly, purified by rectification.
Chief constituents: Patchouly alcohol, eugenol, cinnamic aldehyde, cadinene.
Containers: Glass bottles; tins.
Use: Perfuming of toilet preparations.
Shipping regulations: None. *

patent alum. See aluminum sulfate.

patgreen. See copper acetoarsenite.

pathfinder elements. Elements, generally metallic in nature, associated with ore deposits at the time of formation. Mapping of the concentration variation of the selected element serves to locate the main ore deposit. Examples are zinc as the "pathfinder" for lead, copper and silver ores and molybdenum associated with porphyry copper deposits. As such elements are present in small proportions (less than 1%), variations in the ore bodies are more easily identified.

"Pathibamate." [57] Trademark for tridihexethyl chloride and meprobamate.

"Pathilon" Chloride. [315] Trademark for tridihexethyl chloride (q. v.).

patina. Variously used to refer to an ornamental and/or corrosion-resisting film on the surface of copper, copper alloys, including bronzes especially, and also sometimes of iron and other metals. Such a film is formed in some cases by exposure to the air, and in other cases by a suitable chemical treatment.
See also verdigris.

patronite. A mixture of vanadium-bearing substances with the approximate formula VS_4, found in Peru.

Pattinson process. Process for the removal of silver from lead. The silver-lead mixture is melted in one of a series of pots and allowed to cool slowly. The lead which is free from silver or poorer in silver separates out as crystals which are removed, leaving the silver-rich lead in the molten state. From a number of such operations in series a lead rich in silver is obtained, collected, and the silver recovered. See also Parkes process.

"Paveril Phosphate." [100] Trademark for dioxyline phosphate (q. v.).

paving brick clay. Usually impure shales and fire clays are used for this purpose. Desirable qualities in a paving brick clay are a fair degree of plasticity, good tensile strength, and suitable temperature viscosity characteristics.
Occurrence: United States (Ohio, Pennsylvania, Indiana, Illinois, New York, Maryland, Colorado).

pawpaw. See explanation under papaya.

Pb. Symbol for lead.

PBAA. Abbreviation for polybutadiene-acrylic acid copolymer.

PBD. See 1, 3, 4-phenylbiphenylyloxadiazole.

PBPB. Abbreviation for pyridinium bromide perbromide.

PC. Abbreviation for paper chromatography. See chromatography.

"PC-1244." [58] Trademark for a defoamer for non-aqueous systems. Soluble in benzene, toluene, kerosene, petroleum ether, carbon tetrachloride, isopropanol, tertiary amyl alcohol, butyl "Cellosolve" and ethyl acetate; insoluble in water, alcohol, methanol and methyl "Cellosolve."

PCNB. Abbreviation for pentachloronitrobenzene.

"PCON." [233] Trademark for a chloronitroaniline, $C_6H_3ClNO_2NH_2$, a dye and pigment intermediate.
Properties: Orange, crystalline powder; m. p. 116.3°C; insoluble in water; soluble in methanol and ether.

pcu. Abbreviation for pound centigrade unit, the amount of heat needed to raise one pound of water from 15 to 16°C. See also chu.

Pd. Symbol for palladium.

PDB. Abbreviation for para-dichlorobenzene.

PE. Abbreviation for pentaerythritol; also for polyethylene.

peach aldehyde. See gamma-undecalactone.

peach kernel oil. See persic oil.

peacock blue. A blue organic pigment used especially in inks for multicolor printing. It is the blue pigment most extensively used for this purpose and is also the organic pigment manufactured in greatest quantity in flushed form. It is a lake of acid glaucine blue dye (Color Index No. 671) on alumina hydrate. Structurally, the dye is alpha, alpha-bis[N-ethyl-N-(4-sulfobenzyl)-aminophenyl]-alpha-hydroxy-ortho-toluene-sulfonic acid sodium salt, $HSO_3C_6H_4COH[C_6H_4N(C_2H_5)CH_2C_6H_4SO_3Na]_2$, and is prepared from aniline, ethanol, benzyl chloride, ortho-chlorobenzaldehyde, sulfuric acid, and sodium bisulfite.
Containers: 250-lb barrels.
The term, peacock blue, is sometimes applied to other pigments of similar color, such as Prussian blue which has been treated with phosphotungstic acid.

peanut (ground nut; earth nut).
Varieties: White Spanish, most common in commercial use in America. Usually two well-rounded seeds to a pod. Seed coats pink to red. Also Red Spanish, Tennessee Red, Virginia Bunch, Jumbo Spanish, Carolina Runner; Virginia Runner.
The hulls: The fruit of the peanut plant which contains the seeds.
Properties: Cream-white to gray to dark brown depending upon variety. Composed mostly of cellulose. Contains pentose and some tannin and oils. Odorless, no taste, crisp. Contains whitish lining and considerable mucilaginous substances. Several varieties contain different amounts of carbohydrates, etc.
Grades: One grade, different varieties.
Containers: Shipped in box cars in bulk.
Uses: Preparation of cellulose for paper stock and other cellulose products; fertilizer; cattle feed; manufacture of furfural and xylose; base or synthetic plastics.
Shipping regulations: None. *
The skins:
Properties: Pink to brownish in small Spanish; red, in red Spanish; dull red, in Tennessee red; light brown in jumbo. Bitter, acrid, alkaloidal taste. Little oil, small percentage carbohydrates, alkaloids, large amount of protein.
Chief constitutents: Protein and pigment.
Grades: Technical.
Containers: Cloth bags.
Uses: Cattle feed; source of protein; source of dyes; pigments.

Shipping regulations: None.*

The seeds:

Properties: White to cream color; sweet taste, degrees of sweetness dependent upon varieties of peanuts; oily. Small white Spanish has sweetest taste and most oil. Contain the embryo or peanut hearts.

Chief constituents: Protein; 40% oil average; small amount of carbohydrates.

Grades: No. 1, best; No. 2, etc.

Containers: Bags; sacks.

Uses: Confections; roasted as a food; to make peanut butter by grinding in own oil; manufacture of oil by pressing.

Shipping regulations: None.*

The oil (arachis oil; peanut oil; earth-nut oil; ground-nut oil).

Properties: Yellow to greenish yellow. Typical non-drying oil of olive oil type. Soluble in ether, petroleum ether, carbon disulfide and chloroform; insoluble in alkalies, but saponified by alkali hydroxides with formation of soaps; insoluble in water; slightly soluble in alcohol.

Chief known constituents: Principally glycerides of oleic and linoleic acids, with lesser amounts of the glycerides of palmitic, stearic, arachidic, behenic, and lignoceric acids.

Constants: Sp. gr. 0.917-0.926; solidifying point -5 to +3°C; saponification value 186-194; iodine number 88-98; Hehner's number 95; refractive index (n 40/D) 1.463.

Derivation: By pressing ground peanut meats or by extraction with hot or cold solvents. Latter method more prominent.

Method of purification: Bleaching with fuller's earth or carbon. Hot pressed oil is frequently allowed to stand to deposit stearin (which it will do even at ordinary temperatures) and then filtered.

Grades: U. S. P. XVI; crude; refined; edible.

Containers: 5-gal tins; 50-gal drums; tank cars.

Uses: Substitute for olive oil; edible oils, both hydrogenated and unhydrogenated; soaps; vehicle for medicines.

Shipping regulations: None.*

peanut cake (ground nut cake; goober cake; earth nut cake). The press cake resulting from the extraction of oil from the peanut. See peanut oil meal.

peanut oil. See peanut.

peanut oil meal. The crushed form of peanut cake resulting from the extraction of oil from the peanut seed. This is prepared with or without the shells and the oil meal of commerce contains between 39-45% crude protein and is sold on that basis. A typical analysis of the 39% protein meal is 39.1% crude protein; 5.3% crude fiber; 34.3% nitrogen free extract; 6.2% ether soluble (fats); 5.3% ash and with the total digestible nutrient of approximately 80%.

Containers: Bulk or bags.

Uses: Animal feeds; fertilizer ingredient.

Shipping regulations: None.*

pearl alum. See aluminum sulfate.

pearl ash. See potassium carbonate.

"Pearlescent Pigment." [304] Trade name for a lead monohydrogen phosphate ($PbHPO_4$) pigment.

Properties: Soft white powder, sp. gr. 5.6, refractive index 1.85.

Containers: 50-, 300-lb fiber board drums.

Uses: Imparts pearlescence in a wide range of plastic products. Resists heat and finds direct use in flexible or rigid applications.

pearl essence. See nacreous pigment.

pearlite (perlite). During the process of slowly cooling steel from higher temperatures cementite (q. v.) and ferrite (q. v.) are liberated and form, at about 700°C, a mechanical mixture made up of definite amounts of each and in the proportion of about 7 parts of ferrite to 1 part of cementite, so that the resulting conglomerate contains approximately 0.85% carbon. It exists in the form of interstratified layers or bands of ferrite and cementite and is called pearlite. While pearlite commonly occurs in slowly cooled steels in the lamellar formation, composed of alternate layers of ferrite and cementite, it may, under different rates of cooling and dependent on the relative amounts of ferrite and cementite present, exist in other formations, or phases, of which some authorities have recognized at least four, making five modifications in all. Normal pearlite has a maximum tensile strength of about 105,000 lbs, and an elongation of about 10% in two inches. It is regarded as a separate and distinct constituent of steel because it forms masses or "grains," always contains this same percentage of carbon and, in slowly cooled carbon steel, is always formed at a definite temperature.

See also carbon, combined and carbon, graphite.

pearl moss. See chondrus.

pearl pigment. See nacreous pigment.

pearl white. See bismuth oxychloride; bismuth subnitrate.

pear oil. See amyl acetate.

peat.

Derivation: Partly decayed vegetable matter which has accumulated in marshes. Essentially the first stage in the development of coal from vegetable matter.

Typical analysis of partly dried material:

	Per Cent
Water	25
Ash	3
Woody fiber	50
Humus acids	22
Ammonia	2-3
Phosphoric acid and potash	0.10-0.20

Uses: Dried in blocks or briquetted and used as fuel; fertilizer material.

Shipping regulations: None.*

peat moss. See sphagnum.

pectic acid. Made from pectin by treating the latter with sodium hydroxide solution, washing with isopropyl alcohol, adding alcoholic hydrogen chloride and finally washing again with isopropyl alcohol and drying. Mentioned for use as an acidulant in pharmaceuticals.

pectinase. An enzyme present in most plants. It catalyzes the hydrolysis of pectin to sugar and galacturonic acid.
Use: Biochemical research; juice and jelly industry.

Pectinase "Takamine." [212] Trademark for a standardized combination of pectic enzymes which hydrolyze and depolymerize pectins.
Properties: Dry, off-white powder containing filter aid, non-hazardous; non-flammable. Optimum pH 3.5-4.5; optimum temperature range 80-125°F.
Grade: For food products.
Containers: All quantities; bulk orders, 100- and 150-lb fibre drums.
Uses: For the hydrolysis of pectins in fruit and fruit juices to speed processing and to make fruit juices sparkling clear for consumer appeal.

"Pectinol." [23] Trademark for formulated enzyme concentrates, of fungal origin, with varying degrees of pectinase activity which hydrolyze pectic substances.
Use: Clarification of wines and fruit juices and processing of jellies.

pectinose. See arabinose.

pectins. A group of high-molecular-weight substances, polyuronides, related to carbohydrates and found in varying quantities in fruits and plants. Pectin consists chiefly of partially methoxylated galacturonic acids joined in long chains.
Properties: White powder or syrupy concentrates. Commonest characteristic of pectins is their property of jelling or "setting" under suitable conditions, as on addition of sugar to fruit juices in the preparation of jams or jellies. Soluble in water; insoluble in organic solvents.
Derivation: By dilute-acid extraction of the inner portion of the rind of citrus fruits, or of fruit pomaces, usually apple.
Method of purification: Following decolorization, the extracts are concentrated by evaporation or the pectins precipitated with alcohol or acetone.
Grades: Pure (N. F. XI) containing not less than 7% methoxyl groups and not less than 78% galacturonic acid; 150-, 200-, 250-jelly grades, containing various diluents.
Containers: Bottles, plastic bags, 30-lb tins, 100-lb drums, 200-lb barrels.
Uses: Jellies, foods, cosmetics, drugs; protective colloids; emulsifying agents, dehydrating agents.

pectin sugar. See arabinose.

"Peerless." [133] Standard "flow" black for the printing ink, carbon paper, and typewriter ribbon industries. Available in four types:
"Standard Peerless." A long flow black with low oil absorption, good mass tone, high strength and loading capacity, and easy grinding characteristics. Used in lithographic, halftone and letter press inks, carbon papers and typewriter ribbons.
"Peerless Mark II." Equal to "Standard Peerless" in strength and shows even greater loading capacity and flow; for use where high concentrations of carbon black are required.
"Peerless Mark IIA." Same as "Peerless Mark II" except in denser form. Less dust; easier mixing. Containers 12.5-lb bags.
"Peerless Beads." Beaded form of the long flow "Peerless" type carbon black. Containers 12.5-lb bags; 50- and 75-lb cartons.

"Peetal." [51] Trademark for black greases made in a wide range of consistencies for use on steel mill roll necks whose housings are cooled externally with water.

"Pee Vee Cee." [41] Trade name for synthetic-resin, high-impact, rigid sheet and also pipe and fittings of the Type II unplasticized polyvinyl chloride type used to fabricate structures where optimum resistance to impact as well as corrosion is required.

PEG. Abbreviation for polyethylene glycol.

"Peg 42." [73] Trademark product. Polyoxyethylene stearate.
Properties: White soft-solid; sp. gr. 1.00; m. p. 28-31°C; iodine value <2; free fatty acid $<4\%$; soluble in alcohol and hydrocarbons; dispersible in hot water.
Containers: 425-lb drums; 40-, 8-lb containers.
Uses: Anti-staling and softening agent in baking.

pegmatite (giant granite). A rock consisting of the same constituents as ordinary granite, namely, quartz, feldspar and mica, but irregular in texture and composed of the constituent minerals in such large size that they can be differentiated. In addition to its use as a building stone, pegmatite is often a source of lithia, zircon, tin, tungsten, tantalum, tourmaline, uranium, etc.

"Peladow." [233] Trademark for calcium chloride, 94-97%, in pellet form.

"Pelargol." [188] Trademark for a perfume base for soap perfumes of the geranium and rose types.

pelargonic acid (n-nonoic acid; n-nonanoic acid; n-nonylic acid) $CH_3(CH_2)_7COOH$.
Properties: Colorless or yellowish oil with slight characteristic odor; sp. gr. 0.9052 (20/4°C); m. p. 12.5°C; b. p. 255.6°C (760 mm), 162.4°C (32 mm); refractive index 1.4322 (20°C); soluble in alcohol, ether, and organic solvents; insoluble in water.
Derivation: By the oxidation of nonyl alcohol or nonyl aldehyde; the oxidation of oleic acid, especially by ozone.
Grades: Technical; 99%.
Containers: Hardwood barrels; kegs; drums;

tank cars.
Uses: Organic synthesis; lacquers; plastics; production of hydrotropic salts; pharmaceuticals; synthetic flavors and odors; as a flotation agent; esters for turbojet lubricants; vinyl plasticizer.
Shipping regulations: None.*
See also nonoic acids.

pelargonic aldehyde. See nonanal.

pelargonyl chloride (n-nonanoyl chloride) $CH_3(CH_2)_7COCl$.
Properties: B.p. 80-85°C (5 mm); min assay 97%; soluble in hydrocarbons and ethers; decomposes in water.
Containers: Bottles, carboys, drums.
Uses: Intermediate in organic synthesis.

"Pelaspan." [233] Trademark for an expandable polystyrene in bead or pellet form. Each bead contains its own expanding agent, which is activated by heat.

Peligot's salt. See potassium chlorochromate.

"Pelletex." [275] Trade name for semi-reinforcing furnace (SRF) carbon blacks for use in rubber goods. Pelleted form of "Gastex."

pelletierine $C_5H_{10}N(CH_2)_2CHO$. beta-(2-Piperidyl) propionaldehyde.
Properties: Liquid alkaloid from the root of the pomegranate. Soluble in water, alcohol, ether, chloroform, benzene. Sp. gr. 20/4°C 0.988; b.p. 195°C.
Use: Medicine (in form of its salts, sulfate, tannate, valerate).
Shipping regulations: None.*

pellitory. See pyrethrum root.

"PEM." [195] Trade name for a finely-ground, portland cement-base compound for application over manufactured block, brick, concrete, stucco, or artificial stone.

"Penchlor." [204] Trademark for cold setting silicate-type acid-resistant cements. Completely stable in dry storage.
Grades: Acid-Proof, S-25, Fireproof, and FCC.
Containers: Cement powder available in 100-lb bags, solution in 50-lb and 600-lb steel drums.
Use: Quick setting cement mortar used with acidproof brick to resist acids and strong oxidizing agents, except hydrofluoric acid, up to 2000°F.

"Penco." [204] Trademark for an entire line of agricultural chemical products, including many insecticide formulations in addition to the following:
"Penco" De-Fol-Ate. A herbicide-defoliant containing 41.5% sodium chlorate and 27.5% magnesium chloride.
"Penco" Dimite E-2. An emulsifiable insecticide concentrate containing 2-lb DMC per gal.
"Penco" D-Phos 38-6. A wettable insecticide spray powder containing 38% DDT and 6.25% parathion.
"Penco" Endothal Harvest Aid. A defoliant-desiccant containing 6.3% disodium endothal.
"Penco" Endothal Weed Killer. A herbicide containing 19.2% disodium endothal.
"Penco" Pencal. An insecticide dust containing tricalcium arsenate (low lime).
"Penco" Kryocide. An insecticide dust containing 90% sodium fluoaluminate.
"Penco" Kryocide Super-Seventy. An insecticide dust containing 72% sodium fluoaluminate.
"Penco" Penite-6x. A herbicide containing 9 1/2 lb arsenic trioxide per gal.
"Penco" Penite-35. A herbicide containing 4 1/2 lb aresenic trioxide per gal.
"Penco" Sytam. A systemic emulsifiable insecticide concentrate containing 42% OMPA.
"Penco" Pentrete. A fungicide liquid seed treater containing 7% phenyl mercuric ammonium acetate.

"Pencogel." [342] Trademark for extract of Irish gums.

"Pendane." [342] Trademark for lindane insecticide concentrates.

"Penetone." [33] Proprietary, mildly alkaline detergent composed of neutral soaps and synthetic water softening compounds.
Containers: 5-, 15-, 30-, and 55-gal drums.
Uses: General maintenance and industrial cleaning.

penetrants. Any compounds used to increase the speed and ease with which a bath or liquid permeates a material being processed by effectively reducing the interfacial tension between the solid and liquid. Penetrants are widely used in the textile, tanning and paper industries for improving dyeing, finishing, etc., operations. Sulfonated oils, soluble pine oils and soaps are popular among the older penetrants while the salts of sulfated higher alcohols are typical of the synthetic organics developed for this purpose. See also wetting agents.

"Penex" Process. [416] Patented process for isomerization of normal pentane and normal hexane to corresponding branched-chain hydrocarbons in an environment of hydrogen over a platinum-containing catalyst of high activity. Preferably, the charge should be either predominantly normal pentane or normal hexane, but mixtures can be processed. Process is used in preparing high octane gasoline blending stock from the lower quality normal paraffin stocks. Commercial installations of both C_5 and C_6 Penex are in operation.

"Pen-Gleam." [204] Trademark for a mildly alkaline powdered cleaning compound used for manual washing of painted surfaces. Commercially available in 125-lb fiber drums; 125-, and 350-lb steel drums.

"Pen-Glo." [204] Trademark for a powdered, inhibited acidic material used for the removal of soil and oxides from railroad car bodies without damaging paint.

Commercially available in 400-lb fiber drums with polyethylene liners.

"Penglo 65." [79] Trade name for a pale maleic modified pentaerythritol ester of a special tall oil in mineral spirits.

Properties: Acid value (on solids) 9; concentration (total solids) 65%; color (Hellige) 6; viscosity (Gardner Holdt) 0; ash (on solids) 0.07%.

Containers: 55-gal drums; tank cars.

Uses: Paint and varnish.

"Penglo A." [79] Trade name for a pale pentaerythritol ester of a special tall oil in mineral spirits.

Properties: Acid value (on solids) 9; concentration (total solids) 70%; color (Hellige) 6; viscosity (Gardner Holdt) H; ash (on solids) 0.07%.

Containers: 55-gal drums; tank cars.

Uses: Paint and varnish.

penicillin $(CH_3)_2C_5H_3NSO(COOH)NHCOR$ (bicyclic). A group of isomeric and closely related antibiotic compounds with outstanding antibacterial activity, obtained from the liquid filtrate of the molds, Penicillium notatum and Penicillium chrysogenum, or also, more recently, by a synthetic process which includes fermentation.

The principal penicillins include:

G. R = benzyl, $C_6H_5CH_2-$.

O. R = allylmercaptomethyl, $CH_2{:}CHCH_2SCH_2-$.

V. R = phenoxymethyl, $C_6H_5OCH_2-$.

alpha-phenoxyethyl. R = $C_6H_5OCH(CH_3)-$.

Other varieties exist.

Derivation: The mold is grown in a nutrient solution such as corn steep liquor and lactose. Beginning in the second or third week of cultivation the mold excretes penicillin into its liquid culture medium. This liquid is then filtered off and the penicillin extracted and purified by countercurrent extraction with amyl acetate, adsorption on carbon, or other methods. Different varieties of penicillin are produced biosynthetically by adding the proper precursors to the nutrient solution. dℓ-alpha-Phenoxyethylpenicillin is produced synthetically from alpha-phenoxypropionic acid and 6-aminopenicillic acid, the latter derived from a fermentation process. The following varieties are in common use.

aluminum penicillin G $(C_{16}H_{17}N_2O_4S)_3Al$. The trivalent aluminum salt of penicillin G.

Properties: A light yellow powder with a characteristic odor and taste. It is only slightly soluble in water.

Grade: N. N. D.

benzathine penicillin G (N, N'-dibenzylethylenediamine dipenicillin G). $2C_{16}H_{18}N_2O_4S \cdot C_{16}H_{20}N_2 \cdot 4H_2O$.

Properties: White, odorless, crystalline powder; slightly soluble in alcohol; practically insoluble in water; pH of a saturated solution is 4.5-7.5.

Grade: U. S. P. XVI.

potassium penicillin G (benzylpenicillin potassium) $C_{16}H_{17}KN_2O_4S$.

Properties: Colorless or white crystals, or powder; odorless; moderately hygroscopic solutions dextrorotatory; relatively stable to air and to light; very soluble in water, in saline, and in dextrose solutions; moderately soluble in alcohol; pH of solution (30 mg/ml) is 5.0-7.5; m.p. 214-217°C (dec.).

Grade: U. S. P. XVI.

procaine penicillin G $C_{16}H_{18}N_2O_4S \cdot C_{13}H_{20}N_2O_2 \cdot H_2O$.

Properties: White, fine crystals or powder; odorless; relatively stable to air and light; solutions dextrorotatory; sparingly soluble in water; slightly soluble in alcohol; fairly soluble in chloroform; pH of saturated solution is 5.0-7.5.

Grade: U. S. P. XVI.

sodium penicillin G (benzylpenicillin sodium) $C_{16}H_{17}NaN_2O_4S$.

Properties: Identical to potassium penicillin G (q. v.).

Grade: U. S. P. XVI.

chloroprocaine penicillin O (crystalline salt of 2-chloroprocaine and penicillin O) $C_{13}H_{18}N_2O_4S_2 \cdot C_{13}H_{19}ClN_2O_2$.

Properties: White, crystalline powder; stable at room temperature for three years; practically insoluble in water.

Grade: N. N. R.

sodium penicillin O (sodium allylmercaptomethyl penicillin) $C_{13}H_{17}NaN_2O_4S_2$.

Properties: White, crystalline powder; onion-like taste and odor; freely soluble in water; stable for three years in dry form at room temperature; solutions stable for three days with refrigeration.

Derivation: Biosynthetically by growing the mold in a medium containing allylmercaptoacetic acid.

Grade: N. N. D.

penicillin V (phenoxymethylpenicillin) $C_{16}H_{18}N_2O_5S$.

Properties: White, odorless, crystalline powder; very slightly soluble in water; soluble in alcohol and acetone; insoluble in fixed oils; pH of saturated solution is 2.5-4.0.

Grade: U. S. P. XVI.

potassium penicillin V (potassium phenoxymethylpenicillin) $C_{16}H_{17}KN_2O_5S$.

Properties: White, odorless, crystalline powder; very soluble in water; soluble at all pH levels; slightly soluble in alcohol; insoluble in acetone; pH of solution (30 mg/ml) is 4.0-7.5.

Grade: U. S. P. XVI.

hydrabamine penicillin V (hydrabamine phenoxymethylpenicillin) $2C_{16}H_{18}N_2O_5S \cdot C_{42}H_{64}N_2$.

Properties: A water-insoluble mixture of crystalline phenoxymethylpenicillin salts consisting chiefly of the salt of N, N'-bis-(dehydroabietyl)ethylenediamine, with smaller amounts of the salts of the dihydro and tetrahydro derivatives.

Grade: N.N.D.

potassium alpha-phenoxyethylpenicillin (potassium penicillin 152; phenethicillin) $C_{17}H_{19}KN_2O_5S$. Synthetically prepared;

a mixture of two stereoisomers.

Properties: White, crystalline solid; moderately hygroscopic; decomposes above 220°C. Very soluble in water; resistant to acid decomposition.

Preparation: By N-acylation of alpha-phenoxypropionic acid and 6-amino-penicillanic acid (produced by fermentation using Penicillium chrysogenum).

dimethoxyphenylpenicillin sodium. Synthetic penicillin which resists inactivation by Staphylococci penicillinase; active against strains of staphylococci resistant to other penicillins.

penicillinase. A term applied to enzymes which antagonize the anti-bacterial action of penicillin. Such enzymes are found in many bacteria.

Grade: N. N. D.

Use: Pharmaceutical; biological research.

Penicillinase "Takamine." [212] Trademark for a special enzyme system for destroying penicillin.

penicillin G. See penicillin.

penicillin O. See penicillin.

penicillin V. See penicillin.

"Penicklor." [342] Trademark for chlordane insecticide concentrates.

"Pennclean." [204] Trademark for a liquid milkstone remover containing an inhibitor and wetting agent.

Use: For the softening and removal of milkstone and dried milk solids from farm dairy utensils, flash pasteurizers and other dairy plant equipment.

"Penncoat No. 101." [204] Trademark for an oxidized asphalt, hot-applied compound used as a sealing membrane beneath acid-proof brickwork on concrete floors and tanks. Applied with or without glass cloth membrane to thickness of $\frac{1}{4}$" or $\frac{3}{8}$". Resists acids and mild alkalies. Shipped in 100-lb drums.

"Pennex." [51] Trademark for dark colored cutting oils, including sulfurized fatty oil-mineral oil blends, sulfurized mineral oils and high film strength oils for difficult machining operations.

"Pennolox." [204] Trademark for a specially prepared solution for use in bleaching cellulose and synthetic fibers. Bleaching solutions can be used safely and are noncorrosive in the equipment customarily used. They are effective at temperatures below the "set" point of the synthetics. It is shipped in rubber or polyethylene lined drums and can be handled in polyethylene or rubber containers. It has also found a valuable use in the removal of metallic impurities from wet twisted cotton yarns prior to their bleaching in package machines.

"Pennply No. 101." [204] Trademark for an asphalt-coated glass cloth for use with hot-applied asphalt to form and strengthen the membranes. Shipped in 450 square foot rolls.

"Pennprime No. 101." [204] Trademark for a cut-back asphalt primer for priming concrete surfaces prior to application of hot-applied asphaltic membranes. Shipped in 5-gal drums.

"Pennsalt Asplit Cement." [204] Trademark for a cold-setting phenolic resin cement used with acidproof brick or tile for dilute and moderately concentrated acid service, except for strong oxidizing agents, up to 375°F.

"Pennsalt Cleaner A-27." [204] Trademark for a mildly alkaline powdered cleaning compound used for removing marking inks, oils, and soils from aluminum. Highly inhibited to prevent attack on the base metal. Commercially available in 250-lb steel or fiber drums.

"Pennsalt Cleaner AE-16." [204] Trademark for a highly alkaline powdered compound used to provide an attractive etched finish on aluminum. Strongly inhibited against sludging. Commercially available in 300-lb steel drums.

"Pennsalt Cleaner EC-10." [204] Trademark for a concentrated emulsifiable solvent used alone or in conjunction with alkalies to soften and remove heavy organic soils. Commercially available in 346-lb steel drums.

"Pennsalt Cleaner EC-51." [204] Trademark for an emulsifiable solvent used alone or in conjunction with alkalies to remove heavy organic soils from metal. Inhibited to provide temporary rust protection. Commercially available in 342-lb steel drums.

"Pennsalt Cleaner EC-54." [204] Trademark for an emulsifiable solvent used alone or in conjunction with alkalies for the removal of heavy organic soils. Very high flash point and low volatility as well as temporary rust protection. Commercially available in 345-lb steel drums.

"Pennsalt Cleaner K-8." [204] Trademark for a strongly alkaline powdered cleaning compound used with direct current for the cleaning of copper and steel prior to electroplating. Commercially available in 350-lb steel or fiber drums.

"Pennsalt Cleaner MC-1." [204] Trademark for a mildly alkaline powdered cleaning compound of many applications. Widely used as a pre-etch cleaner for aluminum. Commercially available in 125-lb and 350-lb steel or fiber drums.

"Pennsalt Cleaner PB-1." [204] Trademark for an alkaline powder used in water wall paint spray booth to avoid sticking of the paint. Commercially available in 400-lb steel drums.

"Pennsalt Corlok Cement." [204] Trademark for a cold-setting potassium silicate cement for use with acid-proof brick and tile to resist all acids, except hydrofluoric, up to 2000°F. For use especially with high temperature chlorinations and sulfuric acid processing. Exceptional bond strength to ceramic surfaces. Powder shipped in 100-lb bags, solution in 50-lb drums.

"Pennsalt Furan Cement." [204] Trademark for a cold setting furfural ketone resin cement, for use with acid-proof brick or tile to resist acid, alkaline, and solvent conditions. Powder "S" grade, with a siliceous filler, resists dilute and moderately concentrated acids, except strong oxidizing agents and hydrofluoric acid, also alkalies up to 50% caustic and most solvents. Powder "C" grade, with a carbon filler, resists hydrofluoric and phosphoric acids and other acids, except strong oxidizing agents, also strong alkalies and most solvents. Powder "S" shipped in 150-lb drums, Powder "C" in 100-lb drums, and Furan solution in 45- and 50-lb drums.

"Pennsalt HF Cement." [204] Trademark for a cold setting carbon-filled, phenolic resin cement for use with acid-proof brick or tile to resist acids, especially phosphoric, hydrofluoric and organic chlorinators, up to 375°F.

"Pennsalt HFK Cement." [204] Trademark for a HFK grade used for quicker setting requirements. Carbon powder shipped in 20- and 100-lb drums; resin solution in 10- and 50-lb drums.

"Pennsalt LockPrime." [204] Trademark for a synthetic resin primer for use with most synthetic coating materials, including neoprenes, vinyls, styrenes, alkyds, and furans. Exceptional bond strength and resistance to solvents. Shipped in 1- and 5-gal cans.

"Pennsalt PRF Cement." [204] Trademark for a cold setting modified phenolic resin cement, carbon filled, for use with acid-proof brick or tile to resist dilute and concentrated acids, including strong oxidizing agents, alkalies and most solvents, up to 375°C. Has exceptionally low shrinkage and high bond strengths.

"Pennsalt PRFK Cement." [204] Trademark for a grade of "PRF" used for quicker setting requirements. Carbon powder shipped in 20- and 100-lb drums; resin solution in 10- and 50-lb drums.

"Pennsalt Sulfur Cements." [204] Trademark for hot-pouring, plasticized sulfur cements for use in acid service, including strong oxidizing agents. Available with siliceous filler for normal use; Carbon-Sulfur grade for hydrofluoric service; and Double-Plasticized Sulfur grade for bell and spigot pipe connections.

"Pennsalt Thick-Coat." [204] Trademark for a high-build chlorinated hydrocarbon resin coating for application to steel, concrete, and wood surfaces as a general purpose maintenance type protective coating. Resistant to atmospheric corrosive conditions. Shipped in 1- and 5-gal cans.

pennyroyal oil, American. See hedeoma oil.

pennyroyal oil, European.
Properties: Yellowish to reddish-yellow essential oil, sometimes with bluish or greenish fluorescence; aromatic mint-like odor. Soluble in 4 to 7 vols. and more of 60% alcohol; in 1.5 to 2.5 vols. and more of 70%.
Chief constituents: Pulegone, limonene, dipentene, menthol, menthone.
Constants: Sp. gr. 0.930-0.950 (15°C); optical rotation + 15° to + 25°; refractive index 1.483-1.486.
Derivation: By distillation of Mentha pulegium.
Adulteration: Turpentine; sometimes eucalyptus oil.
Containers: 5-, 10-lb bottles; 25-, 50-lb tins.
Uses: Medicine; insectifuge; perfumery; manufacture of pulegone and its derivatives.
Shipping regulations: None. *

"Pen-o-led." [51] Trademark for extreme-pressure lubricants for gears and bearings operated under heavily loaded or overloaded conditions. Used extensively in metal working mills.

"Penolene 643." [33] Trade name for a purified grade of 1,1,1-trichloroethane (methyl chloroform). See also "Inhibisol."
Properties: Sp. gr. 1.324; boiling range 72-76°C; flash point none; fire point none.
Containers: 55-gal drums.
Uses: Electrical instrument cleaner; general degreasing; often as replacement for carbon tetrachloride.
Hazard: One of least toxic chlorinated solvents. Maximum allowable concentration: 500 ppm.

"Penozone." [204] Trademark for a hydrogen peroxide type bleach recommended for laundries and dry cleaners.

"Pensal." [204] Trademark for a detergent and soap builder.

"Pensuds." [204] Trademark for a powdered synthetic detergent promoted with complex phosphates for use in laundries and wet cleaning departments of dry cleaning establishments.

pentaborane B_5H_9 (boron hydride).
Properties: Liquid; m.p. −46.6°C; b.p. 48°C; sp. gr. 0.61; vapor pressure (0°C) 66 mm; decomposes at 150°C; ignites spontaneously in air. Toxic!
Derivation: Hydrogenation of diborane.
Uses: Fuel for air-breathing engines; propellant.
Shipping regulations: Flammable liquid. Red label. Not accepted by express. *

6,13-pentacenequinone. See 2,3,6,7-dibenzanthraquinone.

*See "I.C.C. Shipping Regulations," page xiii.
Reference numbers refer to name of manufacturer. See "List of Manufacturers," page v.

"Pent-Acetate." [204] Trademark for synthetic amyl acetate available in two grades based on acetate content (85% and 92% min).
Uses: Lacquer solvent; extractant for penicillin and other antibiotics; solvent.

pentachloroethane (pentalin) $CHCl_2CCl_3$.
Properties: Dense, high-boiling colorless liquid; sp. gr. 1.685 (15/4°C); b. p. 159.1°C; f. p. −22°C; refractive index 1.503 (24°C). Insoluble in water.
Derivation: By chlorination of trichloroethylene, which is obtained by a two-step process involving chlorination of acetylene to obtain tetrachloroethane, and removal of hydrochloric acid from this by action of alkali.
Method of purification: Distillation.
Containers: 700-lb drums; tank cars.
Uses: As solvent for oil and grease in metal cleaning. See tetrachloroethane for other uses. Also used for separation of coal from impurities by density difference.
Shipping regulations: None. *

pentachloronitrobenzene (PCNB) $C_6Cl_5NO_2$.
Properties: Cream crystals; musty odor; sp. gr. 1.718 (25/4°C); m. p. 142-145°C; b. p. 328°C (some decomposition). Practically insoluble in water; slightly soluble in alcohols; somewhat soluble in carbon disulfide, benzene, chloroform.
Derivation: Prepared by treating pentachlorobenzene with fuming nitric acid.
Grades: Dust; emulsion concentrate; wettable powder.
Containers: Casks or drums.
Uses: Intermediate; soil fungicide; slime preventative in industrial waters; herbicide.
Precautions: Avoid inhaling dust and spray. Wash hands and face thoroughly after using.

pentachlorophenol C_6Cl_5OH.
Properties: White powder or crystals; m. p. 190°C; b. p. 310°C with decomposition; sp. gr. 1.978 (22/4°C). Insoluble in water; soluble in dilute alkali, alcohol, acetone, ether, pine oil, benzene, "Carbitol," and "Cellosolve"; slightly soluble in hydrocarbons.
Derivation: Chlorination of phenol.
Containers: Fiber drums; multiwall paper sacks.
Uses: Fungicide; bactericide; algicide; herbicide; starting material for synthesis of such compounds; wood preservative.
Warning! Harmful dust. MCA warning label.
Shipping regulations: None. *

pentacites. Alkyd resins formed by use of pentaerythritol as the polyhydric alcohol. Used in protective coatings and printing inks.

"Pentacresol." [327] Trademark for a mixture of secondary amyl tricresols ($C_{12}H_{18}O$).
Derivation: Synthetic.
Use: Medicine.

pentadecane $C_{15}H_{32}$ and $CH_3(CH_2)_{13}CH_3$.
Properties: Colorless liquid; soluble in alcohol; insoluble in water. Sp. gr. 0.776; b. p. 270.5°C; m. p. 10°C.
Grades: Technical.
Use: Organic synthesis.

n-pentadecanoic acid (pentadecylic acid) $CH_3(CH_2)_{13}COOH$. A saturated fatty acid normally not found in vegetable fats, but produced synthetically.
Properties: Colorless crystals; sp. gr. 0.8423 (80/4°C); m. p. 51.8-52.8°C; b. p. 339.1°C (760 mm), 212°C (16 mm); refractive index 1.4529 (60°C). Insoluble in water; soluble in alcohols and ethers.
Available as 99% pure product for organic synthesis, medical research and reference standard in gas chromotography.

pentadecanolide. See "Thibetolide."

pentadecylic acid. See n-pentadecanoic acid.

1,3-pentadiene. See piperylene.

pentaerythrite. See pentaerythritol.

pentaerythritol (PE, pentaerythrite) $C(CH_2OH)_4$.
Properties: White, crystalline powder. Can be readily esterified by the common organic acids. Unaffected when boiled with dilute caustic alkali. Soluble in water; slightly soluble in alcohol; insoluble in benzene, carbon tetrachloride, ether, and petroleum ether.
Constants: B. p. 276°C (30 mm); m. p. 262°C; refractive index (20°C) 1.54 to 1.56. Sp. gr. 1.35.
Typical specifications (technical grade): Approximately 88% monopentaerythritol and 12% dipentaerythritol; sp. gr. 1.38 (25/4°C); m. p. 240°C; purity 99.5%.
Typical specifications (C. P. grade): M. p. 250°C min; color creamy white; chlorides trace; sulfates none; ash 0.05% max; solution 1 g in 12 cc water, clear.
Derivation: By reaction of acetaldehyde with an excess of formaldehyde in an alkaline medium.
Grades: Technical; nitration; C. P.
Containers: 5-, 25-, 50-lb fiber cartons; 50-, 80-lb multiwall bags; 25-, 100-, 125-, 250-, 300-lb drums.
Uses: Alkyd resins; rosin and tall oil esters; molding resins; special varnishes; organic synthesis such as pharmaceuticals; plasticizers; insecticides; synthetic drying oils; explosives.
Shipping regulations: None. *

pentaerythritol chloral. See petrichloral.

pentaerythritol tetraacetate $C(CH_2OOCCH_3)_4$.
Properties: White, crystalline powder. Nonflammable; extremely stable to sunlight. Soluble in water, alcohol and ether.
Constants: M. p. 84°C; b. p. 225°C (30 mm).
Typical specifications: Pentaerythritol tetraacetate 99% min; free acidity (as acetic) 0.4% min; m. p. 74°C min; color white; chlorides, sulfates, metals, none.
Derivation: By the esterification of pentaerythritol with acetic acid.
Grades: Technical.

pentaerythritol tetranitrate (PETN, penthrite) $C(CH_2ONO_2)_4$.
Properties: White crystalline material; sp. gr. 1.75; m.p. 138-140°C; decomposes above 150°C; explodes at 205-215°C. Very soluble in acetone; slightly soluble in alcohol and ether; insoluble in water.
Derivation: Esterification of pentaerythritol with nitric acid.
Containers: Specially lined 300-lb steel drums.
Use: Explosive; medicine.
Shipping regulations: Explosive, class A. Initiating explosive label. Not accepted by express. *

pentaglycerine. See trimethylolethane.

pentahydroxycyclohexane. See quercitol.

3, 5, 7, 3', 4'-pentahydroxyflavone-3-rutinoside. See rutin.

"Pentalarm No. 86." [204] Trademark for a warning agent for fuel gas.
Properties: Color, red; sp. gr. (20/20°C), 0.83-0.84; distillation not less than 95% below 266°F; amyl mercaptan content, 90% (min); flash point (open cup), 50°F; soluble in most organic solvents; insoluble in water.
Shipping regulations: Flammable liquid. Red label. *

"Pentalarm No. 1004." [204] Trademark for a fuel gas warning agent for liquid-injection odorization.
Properties: Color, blue; mercaptan sulfur, 34% (min); sp. gr. (68°F), 0.835-0.845; distillation 93% between 122-260°F; flash point < 0°F; average molecular weight 88; viscosity (77°F), 0.57 centistokes.
Shipping regulations: Flammable liquid. Red label. *

"Pentalene." [204] Trademark for mixed amylnaphthalenes, available in distilled (No. 195) and crude (No. 95) grades. Used as intermediates for anionic surface-active agents.

pentalin. See pentachloroethane.

"Pentalyn." [266] Trademark for a series of pentaerythritol esters of rosin and modified rosin.
Uses: Varnishes and enamels, primarily, where they contribute adhesion, hardness, flexibility, high gloss, and exterior durability. Also, they are used in printing inks, adhesives, and modifiers for waxes, and floor coverings.

pentamethonium salts. Similar to hexamethonium salts; see hexamethonium bromide, chloride.

1, 1, 3, 3, 5-pentamethyl-4, 6-dinitroindane. See "Moskene."

pentamethylene. See cyclopentane.

pentamethyleneamine. See piperidine.

pentamethylenediamine. See cadaverine.

pentamethylene dibromide (1,5-dibromopentane) $BrCH_2(CH_2)_3CH_2Br$.
Properties: Colorless, aromatic liquid; m.p. −35°C; b.p. 224°C; insoluble in water.
Containers: Bottles.

pentamethylene glycol. See 1, 5-pentanediol.

pentamethylene tetrazole. See pentylenetetrazol.

pentanal. See n-valeraldehyde.

n-pentane (amyl hydride) $CH_3(CH_2)_3CH_3$.
Properties: A colorless, mobile, flammable liquid; pleasant odor; freezing point −129.723°C; b.p. 36.074°C; refractive index (n 20/D) 1.35748; sp. gr. (20°C) 0.62624; soluble in hydrocarbons, oils, and ether; very slightly soluble in alcohol; insoluble in water. Flash point −45°F.
Typical specifications for a technical grade: Distillation range 94-100°F; sp. gr. of liquid (60/60°F) 0.631; A.P.I. gravity (60°F) 92.8; density of liquid (60°F) 5.25 lbs/gal; vapor pressure (70°F) 8.8 psi absolute; refractive index (n 20/D) 1.357; principal impurity approx. 4% isopentane; sulfur content 0.005% max; contains not less than 95 mole % normal pentane.
Derivation: Fractional distillation from petroleum; purified by rectification.
Grades: Pure; technical; commercial.
Containers: Pure: 1-qt bottle; gal bottle; 55-gal drum. Technical and commercial grade: 55-gal drums; tank cars.
Uses: Anesthetic; artificial ice manufacture; low-temperature thermometers; solvent for use in solvent extraction processes; general solvent.
Fire hazard: Dangerous.
Shipping regulations: Flammable liquid. Red label. *

pentanedinitrile. See glutaronitrile.

pentanedioic acid anhydride. See glutaric anhydride.

1, 5-pentanediol (pentamethylene glycol) $HOCH_2(CH_2)_3CH_2OH$.
Properties: Colorless liquid; b.p. 242.5°C; f.p. −15.6°C; sp. gr. 0.9921 (20/20°C); wt/gal 8.2 lbs (20°C); flash point (open cup) 265°F. Miscible with water.
Grades: Technical.
Containers: 5-, 55-gal drums; tank cars.
Uses: Polyester and urethane resins; hydraulic fluid; lube oil additive; anti-freeze.

2, 3-pentanedione. See acetyl propionyl.

2, 4-pentanedione. See acetylacetone.

pentanepentol. See xylitol.

pentanethiol. See amyl mercaptan.

pentanoic acid. See n-valeric acid.

1-pentanol. See n-amyl alcohol, primary.

2-pentanol (sec-n-amyl alcohol; sec-amyl alcohol, active; methyl propyl carbinol) $CH_3CH_2CH_2CHOHCH_3$.
Properties (racemic form): Colorless liquid; b.p. 119.3°C; sp. gr. (20/20°C) 0.811; wt/gal (20°C) 6.75 lbs; refractive index

*See "I.C.C. Shipping Regulations," page xiii.
Reference numbers refer to name of manufacturer. See "List of Manufacturers," page v.

(n 25/D) 1.4041; flash point (open cup) 105°F. Slightly soluble in water; miscible with alcohol and ether.
Derivation: Fractional distillation of the mixed alcohols resulting from the chlorination and hydrolysis of pentanes.
Grades: Technical.
Containers: 1-, 5-, 55-gal drums; tank cars.
Uses: Solvent for paints and lacquers; pharmaceutical intermediate.
Shipping regulations: None. *

3-pentanol (sec-n-amyl alcohol; 1-ethyl-1-propanol; diethyl carbinol) $CH_3CH_2CHOHCH_2CH_3$.
Properties: Colorless liquid; sp. gr. 0.82 (20°C); freezing point less than -75°C; b.p. 115.6°C; wt/gal 6.81 lbs; refractive index 1.41 (20°C). Soluble in alcohol and ether; slightly soluble in water.
Containers: 1-gal cans; (approx net contents 6.5 lbs); 5-gal cans (approx net contents 34 lbs); 55-gal drums (approx net contents 353 lbs); tank-car capacity (approx net contents 8000 gals).
Uses: Solvent; flotation agent; pharmaceuticals.
Fire hazard: Fire point (open cup) 100°F.
Shipping regulations: None. *

tert-**pentanol.** See tert-amyl alcohol.

2-pentanone. See methyl propyl ketone.

3-pentanone. See diethyl ketone.

penta resin. Ester gum made from rosin and pentaerythritol.

"Pentaryl A." [204] Trade name for amyl biphenyl, $C_5H_{11}C_6H_4C_6H_5$.
Properties: Light straw color; boiling range, 305-337°C; sp. gr. (20/20°C) 0.952-0.965; refractive index 1.555-1.567; viscosity 21 cps (25°C).
Use: Raw material for the manufacture of surface-active agents.

"Pentaryl B." [204] Trademark for diamyl biphenyl $C_5H_{11}C_6H_4C_6H_4C_5H_{11}$.
Properties: Straw color; boiling range, 355-385°C; sp. gr. (20/20°C) 0.932-0.945; refractive index, 1.540-1.552.
Uses: Raw material for the manufacture of surface active agents.

pentasodium triphosphate. See sodium tripolyphosphate.

"Pentasol." [204] Trademark for synthetic amyl alcohols.
"Pentasol" #26. A special esterification grade amyl alcohol for organic synthesis.
"Pentasol" #27. A mixture of synthetic isomeric amyl alcohols, primaries, secondaries and some tertiaries. Used as a solvent for nitrocellulose lacquers, urea-formaldehyde resins, and many organic compounds; flotation agent for concentration of non-ferrous ores; intermediate; hydraulic fluid.
"Pentasol" #258. A special blend of alcohols for lacquer solvents.
"Pentasol" Frother #124. A special amyl alcohol blend for the mining industry.

n-pentatriacontane $C_{35}H_{72}$ or $CH_3(CH_2)_{33}CH_3$.
Properties: Crystals; sp. gr. 0.782 at 75°C; b.p. 331°C at 15 mm; m.p. 75°C.
Grades: Technical.
Containers: Fiber cans; glass bottles.
Use: Organic synthesis.

"Pentecat L." [230] Trademark for a 50% aqueous solution of lithium naphthenate.
Use: As alcoholysis catalyst in alkyd varnish cooking.

"Pentek." [138] Trade name for pentaerythritol, technical.
Containers: 50-lb multiwall bags.
Uses: Alkyd resins, rosin esters, in situ varnishes, synthetic drying oils, and tall oil esters. Particularly useful in medium and long oil alkyds for architectural enamels, flat wall paints, trim paints, metal decorating paints and primers, marine paints.

1-pentene. See alpha-n-amylene.

cis-**2-pentene** $H_3CCH{:}CHCH_2CH_3$.
Properties: B.p. 37°C; sp. gr. 0.656 (20/4°C); refractive index 1.3820 (n 20/D); flash point -45°C.
Grade: 95%.
Containers: Bottles.
Shipping regulations: Flammable liquid. Red label. *
Note: The properties of the trans isomer are given under beta-n-amylene.

2-pentene (mixed isomers) $H_3CCH{:}CHCH_2CH_3$.
Properties: Flash point -45°C.
Grades: 95%, 99%.
Containers: Bottles and drums.
Shipping regulations: Flammable liquid. Red label. *

"Pentex." [248] Trademark for tetrabutylthiuram monosulfide.
Properties: Brown, free-flowing liquid; sp. gr. 0.99; soluble in acetone, benzol, gasoline, and ethylene dichloride; insoluble in water.
Uses: Accelerator for mechanicals and inner tubes. When mixed with 87.5% clay, it is used for sponge rubbers and called "Pentex Flour."

penthienate bromide $C_{18}H_{30}BrNO_3S$. Diethyl (2-hydroxyethyl)methylammonium bromide alpha-cyclopentyl-2-thiopheneglycolate; 2-diethylaminoethyl-2-cyclopentyl-2-(2-thienyl)-hydroxyacetate methobromide.
Properties: White, odorless, crystalline powder; soluble in water and in alcohol; slightly soluble in chloroform; practically insoluble in acetone and in ether; melting range 122-128°C.
Grade: N. F. XI.
Use: Medicine.

penthrite. See pentaerythritol tetranitrate.

"Pentite." [306] Trademark for $[(C_6H_5O)_2POCH_2]_4C$. A low melting (30-60°C) white, waxy solid

*See "I.C.C. Shipping Regulations," page xiii.
Reference numbers refer to name of manufacturer. See "List of Manufacturers," page v.

with slight phenolic odor. Sp. gr. 1.24 (25/15.5°C); refractive index (n 25/D) 1.5823.
Use: Resin stabilizer.

pentlandite (Fe, Ni)S.
Properties: Light bronze-yellow mineral; metallic luster. Contains 35.57% of nickel. Soluble in nitric acid. Sp. gr. 4.6-5. Hardness 3.5-4.
Occurrence: Canada (Ontario), Norway.
Use: Important nickel ore.

pentobarbital $C_{11}H_{18}O_3N_2$. 5-Ethyl-5-(1-methylbutyl)barbituric acid.
Properties: White, granular powder; m.p. 126-130°C; freely soluble in alcohol, chloroform, ether and in solutions of alkali hydroxides; slightly soluble in water.
Grade: N.N.D.
Containers: Drums.
Use: Medicine.

pentobarbital calcium $(C_{11}H_{17}N_2O_3)_2Ca$. Calcium 5-ethyl-5-(1-methylbutyl)barbiturate.
Properties: Very fine white powder; sparingly soluble in alcohol and water; practically insoluble in ether.
Grade: N.N.D.
Use: Medicine.

pentobarbital sodium $C_{11}H_{17}N_2O_3Na$. Sodium 5-ethyl-5-(1-methylbutyl)barbiturate.
Properties: White, crystalline granules or white powder; odorless with slightly bitter taste. Very soluble in water, freely soluble in alcohol but practically insoluble in ether. Solutions decompose on standing.
Grades: U.S.P. XVI.
Containers: 100-lb drums.
Uses: Same as barbiturates.

pentolinium tartrate $C_{23}H_{42}N_2O_{12}$. Pentamethylene-1,1-bis-(1-methylpyrrolidinium bitartrate).
Properties: White to light cream-colored, crystalline powder; slightly soluble in alcohol; insoluble in ether, chloroform; very soluble in water; pH of 1% solution in water is 3.0-4.0. Decomposes 203°C.
Grade: N.N.D.
Use: Medicine.

pentolite. A mixture of equal parts of pentaerythritol tetranitrate and trinitrotoluene (q.v.).
Shipping regulations: Explosives, Class A. High explosive label.*

"Penton." [266] Trademark for a thermoplastic resin. Derived from 3,3-bis(chloromethyl)oxetane $(CH_2Cl)_2CCH_2OCH_2$.
Properties: A linear polymer, of crystalline character, extremely resistant to chemicals, and to thermal degradation at molding and extrusion temperatures; sp. gr. 1.4; rated self-extinguishing in flammability tests; dimensionably stable; has very low water absorption, outstanding chemical resistance. Natural, black, or olive green molding powder. Finely divided powder for coatings.
Uses: "Penton" can be fabricated on conventional plastics equipment by injection molding or extrusion, or machined from tube, rod, or block stock to precision parts held to close tolerances. "Penton" is used both in solid form or as a metal liner in valves, pipe and fittings, pumps, meters and other components in chemical processing equipment. Tank linings are applied by the use of adhered sheet and conventional gas welding techniques. A variety of dry coating, water suspension, and organic dispersion spray coating systems are employed to provide tough, pinhole free protective coatings to metal substrates.

pentosans. Complex carbohydrates (hemicelluloses) present with the cellulose in many woody plant tissues, particularly cereal straws and brans; characterized by hydrolysis to give five-carbon-atom sugars (pentoses). Thus the pentosan xylan yields the sugar xylose ($HOH_2C \cdot CHOH \cdot CHOH \cdot CHOH \cdot CHO$) which is dehydrated with sulfuric acid to yield furfural ($C_5H_4O_2$).

pentoses. General term for sugars with five carbon atoms per molecule.

"Pentothal" Sodium. [3] Trademark for thiopental sodium.

"Pent-Oxol." [125] Trademark for a high boiling (IBP 163.8°C) glycol-ether type solvent, 4-methoxy-4-methylpentanol-2, $(CH_3)_2C(OCH_3)CH_2CHOHCH_3$.
Properties: Liquid; sp. gr. 0.894 (20/20°C); ASTM distillation, IBP 163.8°C, 50% v 166.7°, DP 167.0°C; flash point (Tag open cup) 140°F.
Containers: 1-, 5-, 55-gal pails and drums; 4500 to 10,000-gal tank cars; tank trucks.
Caution! May cause skin and eye irritation.
Uses: Solvent for nitrocellulose, acrylic, epoxy resins.

"Pent-Oxone." [125] Trademark for a high boiling (IBP 147.0°C) keto-ether solvent, 4-methoxy-4-methylpentanone-2. $(CH_3)_2C(OCH_3)CH_2COCH_3$.
Properties: Liquid; sp. gr. 0.910 (20/20°C); ASTM distillation, IBP 147°C, 50% v 159°, DP 163°C; flash point (Tag open cup) 141°F.
Containers: 1-, 5-, 55-gal pails or dums; 4500-10,000-gal tank cars; tank trucks.
Caution! May cause skin and eye irritation.
Uses: Solvent for nitrocellulose, acrylic, vinyl and epoxy resins.
Shipping regulations: None.*

"Pentrox." [204] Trademark for an emulsifier for solvents and greasy soils in industrial laundry washing.

pentyl. Synonym for the amyl radical, C_5H_{11}-.

pentyl acetate. See amyl acetate.

pentylamine. See n-amylamine.

alpha-**pentylcinnamaldehyde.** See "Buxine."

pentylenetetrazol (pentamethylenetetrazole) $C_6H_{10}N_4$.
Properties: White, odorless crystals with

slightly pungent, bitter taste; m. p. 57-59°C; freely soluble in water and alcohol; soluble in ether, chloroform, and carbon tetrachloride.
Grade: N. F. XI.
Use: Medicine.

pepo (pumpkin seed; cucurbita).
Derivation: Ripe seed of Cucurbita pepo.
Occurrence: Southern Asia, Europe, and North America.
Grades: Technical.
Containers: Bags.
Uses: Medicine; extraction of oil.
Shipping regulations: None. *

pepper, African. See capsicum.

pepper, bird. See capsicum.

pepperidge bush. See barberry.

pepper, Jamaica. See pimenta.

peppermint (brandy mint; lamb mint). Dried leaves and flowering tops of Mentha piperita.
Occurrence: Asia, Europe and North America, especially Michigan, Oregon and Washington (Yakima valley).
Constituents: Volatile oil, tannin, resin, gum.
Grades: Technical; U. S. P. XVI.
Containers: Bales; drums.
Uses: Medicine; peppermint oil; menthol.
Shipping regulations: None. *

peppermint-camphor. See menthol.

peppermint oil.
Properties: Colorless or slightly yellowish, volatile, essential oil; darkening in color and thickening in consistency on being exposed to the air for some time; very strong, aromatic, minty odor and taste, the latter being followed by a sensation of coolness. Soluble in alcohol, ether and chloroform.
Chief constituent: Menthol, varying in amount from 45 to 91% according to place of origin. Other constituents: esters of menthol, cineole, menthone, pinene, limonene, etc.
Constants: Sp. gr. 0.895-0.921; refractive index 1.4590-1.4650 (20°C); optical rotation −6° to −43° (constants differ according to species).
Derivation: By distilling the leaves and flowering tops of the peppermint plant, Mentha piperita, and from other Mentha species.
Method of purification: Rectification.
Grades: Technical; U. S. P. XVI.
Containers: Cans; bottles; drums;
Uses: Medicine (similar to menthol); tooth powders and pastes; mouth washes; manufacture of liqueurs; raw material for production of menthol; confectionery; flavoring; perfumery (soap).
Shipping regulations: None. *

pepsin (pepsinum).
Properties: White or yellowish-white powder or lustrous transparent or translucent scales; should have no odor; converts proteins into albumoses and peptones. Stable, when dry, to 100°C; soluble in water; insoluble in alcohol, chloroform and ether.
Derivation: Proteolytic ferment or enzyme from the glandular layer of fresh hogs' stomachs.
Grades: Technical; N. F. XI.
Containers: 1-, 5-lb bottles; 10-, 25-lb tins.
Uses: Medicine (as a digestive ferment); substitute for rennet in cheese making.

pepsinum. See pepsin.

peptidase. See protease.

"Pepton" 22 Plasticizer. [57] Proprietary name for 2,2'-dibenzamidodiphenyl disulfide. A chemical peptizer for use in the breakdown of natural rubber or butadiene-styrene copolymer.

peptone.
Properties: (a) From albumin: White or pale yellow, amorphous powder. (b) From meat: Light brown, amorphous powder. (c) From milk: Light brown powder. Soluble in water; insoluble in alcohol or ether.
Derivation: (a) By digestion of egg albumin by pepsin and a small quantity of dilute hydrochloric acid at 38-40°C (body temperature). (b) By digestion of red meat with pancreatin at body temperature. (c) By digestion of casein.
Grades: Technical; reagent.
Containers: Boxes; glass bottles.
Use: Preparation of nutrient media in bacteriology; nutrient.
Shipping regulations: None. *

para-**peptone.** See syntonin.

per-. A prefix signifying complete or extreme, and specifically denoting: (1) a compound containing an element in its highest state of oxidation, as perchloric acid; (2) presence of the peroxy group, -O-O-, as peracetic and perchromic acids; (3) exhaustive substitution or addition, as perchloroethylene.

"per." Slang for perchloroethylene.

peracetic acid (peroxyacetic acid) CH_3COOOH.
Properties: Available as a 40% solution in acetic acid, containing also water, hydrogen peroxide, and sulfuric acid. Colorless liquid; strong odor; b. p. 105°C; f. p. (approx.) −30°C; sp. gr. (20°C) 1.15; flash point (open cup) 105°F. Solubility similar to acetic acid. Decomposes with evolution of oxygen at elevated temperatures. Pure peracetic acid is reported to decompose explosively at about 110°C.
Derivation: (a) By reaction of acetic acid and hydrogen peroxide in the presence of sulfuric acid catalyst; (b) acetaldehyde, air and ethyl acetate react to form acetaldehyde monoperacetate (AMP). AMP is subjected to pyrolysis that produces peracetic acid, acetic acid, and acetaldehyde.
Grade: Technical; 40%.
Containers: 65-lb glass carboys; 250-lb aluminum drums.
Uses: Bleaching textiles, paper, oils,

*See "I.C.C. Shipping Regulations," page xiii.
Reference numbers refer to name of manufacturer. See "List of Manufacturers," page v.

waxes, starch; polymerization catalyst; bactericide and fungicide, especially in food processing; epoxidation of fatty acid esters and epoxy resin precursors; oxidant in organic synthesis.
Shipping regulations: Oxidizing material. Yellow label.*

per-acids. Derivatives of hydrogen peroxide, the molecules of which contain one or more directly linked pairs of oxygen atoms, -O-O- (see per-, second meaning). Examples are persulfuric, perchromic, peracetic, or perboric acids. Permanganic, perchloric, and periodic acids are not peracids in this sense (see per-, first meaning).

"Perandren." [305] Trademark for testosterone.
Use: Medicine.

"Perazil." [301] Trademark for chlorcyclizine hydrochloride; an antihistamine.

perchloric acid (Fraude's reagent) $HClO_4$.
Properties: Clear, colorless, hygroscopic liquid. The pure concentrated acid is not stable. In dilute solution the acid is stable. The more concentrated solutions are dangerously explosive if allowed to come in contact with oxidizable materials. Sp. gr. 1.764; m.p. −112°C; b.p. 16°C (8 mm) Soluble in water; forms a maximum boiling point mixture containing 71.6% acid. This is quite stable when other materials are absent.
Derivation: By distilling potassium perchlorate with strong sulfuric acid (96%), under reduced pressure in an oil bath at 140-190°C.
Method of purification: Rectification.
Grades: Technical; C.P., strength of solution 6 to 20%; 60%; 70-72%.
Containers: 1-, 5-lb glass bottles; 2-, 5-, 10-gal carboys.
Uses: Medicine; analytical chemistry; catalyst; manufacture of various esters; ingredient of the electrolytic bath in the deposition of lead; electro-polishing; explosives.
Danger! Strong oxidant. Contact with other material may cause fire or explosion, expecially if heated. Causes severe burns. MCA warning label.
Shipping regulations: Corrosive liquid. White label. (Not in excess of 72%. In excess of 72% must not be shipped.) Special regulations.*

perchlorobenzene. See hexachlorobenzene.

perchlorocyclopentadiene. See hexachlorocyclopentadiene.

perchloroethane. See hexachloroethane.

perchloroethylene ("per"; tetrachloroethylene) $Cl_2C{:}CCl_2$.
Properties: Colorless liquid; ether-like odor. Non-flammable. Nonexplosive. Toxicity is lower than that of most organic solvents. Extremely stable. Resists hydrolysis. Sp.gr. (20/20°C) 1.625; b.p. 121°C; f.p. −22.4°C; weight 13.46 lbs/gal (26°C); refractive index 1.5029 (25°C); flash point, none; fire point, none; heat of vaporization 50.1 cals/g (b.p.); specific heat 0.21 cal/g/°C; dielectric constant 2.20 (1000-cycle); dielectric strength >30,000 volts (0.1 in. gap); power factor 0.02% (1000-cycle); specific resistivity 1.8×10^{13} ohms/cm; viscosity 0.84 cps (25°C). Miscible with alcohol, ether, and oils, in all proportions. Very slightly soluble in water.
Derivation: (a) Carbon tetrachloride is vaporized and passed into a carbon-loaded furnace (800-900°C) where it decomposes into hexachloroethane and perchloroethylene. (Chlorine and unreacted carbon tetrachloride are recycled). (b) Thermal decomposition of pentachloroethane in the presence of aluminum chloride catalyst. (c) Treatment of pentachloroethane with milk of lime. (d) Chlorination of propane (LPG).
Method of purification: Distillation.
Grades: Purified; technical; U.S.P. XVI.
Containers: 5-, 10-, 55-gal drums; 8000- and 10,000-gal tank cars.
Uses: Dry cleaning solvent; vapor-degreasing solvent; drying agent for metals and certain other solids; vermifuge; solvent for rubber, waxes, tar, paraffin, gums; heat transfer medium; chemical synthesis.
Warning: Vapor harmful. MCA warning label.
Shipping regulations: None.*

perchloromethane. See carbon tetrachloride.

perchloromethyl mercaptan (trichloromethylsulfenyl chloride) $ClSCCl_3$.
Properties: Yellow, oily liquid. Disagreeable odor. Mildly decomposed by moist air. Subject to the action of chemical agents, such as oxidizing agents, reducing agents, chlorine, etc. Caution! Irritant!
Constants: Sp.gr. 1.722 (0°C); b.p. 148-149°C (decomposes); vapor density 6.414; volatility 18,000 mg/cu m (20°C).
Derivation: Chlorination of any of the following: carbon disulfide, thiophosgene, or methyl thiocyanate.
Grades: Technical.
Containers: Steel bottles.
Uses: Organic synthesis; dye intermediate; military poison gas; fumigant for granary weevils, lady-bird beetles; suggested as diesel fuel additive.
Shipping regulations: Poison, Class B. Poison label.*

"Perchloron." [204] Trademark for a high-test calcium hypochlorite. Contains not less than 70% available chlorine.
Containers: 55½-lb cases (9 cans/case). 109-lb drums (net wt 100 lbs).

perchloropropylene. See hexachloropropylene.

perchloryl fluoride $ClFO_3$.
Properties: Oxidizing agent; fluorinating and perchlorylating agent; m.p. −146°C; b.p. −46.8°C; sp.gr. (liq.) 1.434 (20°C).
Containers: Usually stored in cylinders as liquid under pressure. Is stable in storage

*See "I.C.C. Shipping Regulations," page xiii.
Reference numbers refer to name of manufacturer. See "List of Manufacturers," page v.

under anhydrous conditions.
Uses: In organic synthesis to introduce fluorine into the molecule; as oxidant in rocket fuels.
Caution: Very strong oxidant, especially dangerous with benzene and other materials whose resulting products can explode.

perchromic acid. Probably $(HO)_4Cr(OOH)_3$ or $H_3CrO_8 \cdot 2H_2O$.
Properties: A very unstable acid formed when a solution of chromic acid is added to hydrogen peroxide. Below −15°C, forms deep blue crystals; the blue color can be extracted from solutions by ether. Decomposes in acid solution to form chromic salts and in alkaline solution to form chromates. The blue color of perchromic acid can be used as a sensitive test for chloride or chromate.

"Perclene." [28] Trademark for dry cleaning grade and technical grade perchloroethylene. Boiling range 120.5°-122.0°C; wt/gal. 13.55 lb at 20°C.
Containers: 700-lb (55-gal) tank trucks; tank cars.

"Percorten." [305] Trademark for deoxycorticosterone U.S.P.
Use: Medicine.

"Perdox." [28] Trademark for sodium borate perhydrate $(NaBO_2 \cdot H_2O_2)$ (q.v.).

"Peregal." [307] Trademark for a series of textile chemicals.

"Peregal O." Polyoxyethylated fatty alcohol; nonionic.
Properties: Clear, yellow liquid; sp.gr. 1.02; soluble in water, ethanol, ethylene glycol, butyl "Cellosolve"; stable to acid, alkali, and metallic ions.
Uses: Dyeing assistant for use with basic and direct colors; assistant for dyeing, leveling or stripping of vat dyes; leveling agent in printing trade.

"Peregal OK." Methyl polyethanol quaternary amine; cationic.
Properties: Cloudy, amber liquid; sp.gr. 1.03; soluble in water, ethanol, and ethylene glycol.
Uses: Retarding, leveling and stripping agent for vat dyestuffs; dyeing assistant for vat blues.

"Peregal ST." Aqueous solution of polyvinyl pyrrolidone; cationic.
Properties: Pale yellow liquid; sp.gr. 1.00; soluble in water; stable to dilute acid and dilute alkali.
Uses: Stripping assistant for vat, sulfur, and direct color dyed or printed cotton and rayon; suitable for use in long liquors, package machines or jigs. Rag stripping assistant; used by the printing industry to prevent tinting of white grounds during soaping.

"Peregal TW." Alkyl polyoxyethylene glycol amine; cationic.
Properties: Cloudy, amber liquid; sp.gr. 1.005-1.015; soluble in water; stable to acid, alkali and metallic ions.
Uses: Leveling agent for wool raw stock, yarn and piece goods in dyeing with acid or chrome colors; recommended especially to overcome tippy dyeing effects. Dispersing agent for basic and acid dyestuffs used in the same bath for dyeing blends of wool and polyacrylonitrile fiber.

"Perfecta." [45] Trademark for petrolatum, U.S.P., of high melting point and medium consistency.
Properties: M.p. 128-135°F; Saybolt viscosity 55-75 (210°F); odorless.
Use: Pharmaceutical and cosmetic preparations.

"Perfection." [172] Brand name for sodium acid pyrophosphate, $Na_2H_2P_2O_7$.
Properties: White, crystalline material. Purity complies with all Food and Drug laws.
Containers: 100-lb paper bags.
Uses: Baking acid for prepared mixes and commercial baking powders; manufacture of instant puddings; formulation of acid-type metal cleaners; conditioning agent for oil well drilling muds.

"Perflow." [134] Trademark for brightener for nickel plating solutions.

perfluorobutyric acid. See heptafluorobutyric acid.

"Perfluoro Carboxylic Acids." [158] A group of acid products (perfluoropropionic acid, perfluorobutyric acid and perfluoroöctanoic acid) that are of commercial significance because of their unusual acid strength, chemical stability, surface activity and salt solubility characteristics.
Containers: 6½-gal glass carboy in wooden drum and perfluoroöctanoic in 3-pt glass jar in wooden box.

perfluoroethylene. See tetrafluoroethylene.

perfluoropropane (octafluoropropane) C_3F_8. A gaseous electrical insulator.
Typical specifications: Purity 99.0% min; specific volume (70°F) 2.9 cu ft/lb; moisture 20 ppm max.; non-condensible gases in vapor phase 2.0% max.
Containers: Cylinders.

performic acid (peroxyformic acid; permethanoic acid; formyl hydroperoxide) HCOOOH.
Properties: Colorless liquid. Explodes on contact with metals, their oxides, reducing substances or on distillation. Miscible with water, alcohol, ether; soluble in benzene, chloroform. Solutions are unstable.
Derivation: Mixture of formic acid, peroxide and sulfuric acid is allowed to interact for 2 hours and then distilled.
Grade: 90% solution.
Use: Oxidation, epoxidation and hydroxylation reactions.

perfumes. Pleasant smelling substances, usually in the form of blends of natural odor concentrates and components from synthetic sources. Due to their esthetic appeal they are used as alcoholic solutions (perfumes, toilet waters) or as additions to produce an agreeable aroma (soaps,

cosmetics). Sources of components are from steam distillation of flowers (flower oils, such as rose oil, geranium oil); steam distillation of plant tissue (essential oils, such as lemon oil, pine oil); extracts of animal tissue (musk); and synthetic organic chemicals not occurring in nature but desirable in perfumes (heptaldehyde; gamma-undecalactone). For perfumes too volatile or delicate to stand distillation, solvent extraction is sometimes used.
See concrete (2).

"Perglow." [134] Trademark for brightener for nickel plating solutions.

perhydrosqualene. See squalane.

peri acid. See 1-naphthylamine-8-sulfonic acid.

periclase MgO. Natural magnesium oxide, found in some marbles. Easily alters to brucite.
Properties: Colorless to grayish white, yellow, brown, green to black, luster vitreous, hardness 5.5; sp.gr. 3.56.
Occurrence: California, New Mexico; Europe.
Use: Refractories (specially prepared grade).

peridot. See olivine.

perilla oil.
Properties: Light yellow liquid. Soluble in alcohol, ether, chloroform, benzine, and carbon disulfide. Sp.gr. 0.932-0.945; saponification value 191-193; iodine value 187-202; refractive index 1.4841.
Derivation: From the seeds of Perilla ocimoides, a mint grown commonly in Japan and Korea.
Impurities: Sometimes adulterated with cottonseed oil.
Grades: Technical.
Containers: 75-lb cases (2 tins); 375-lb barrels.
Uses: Substitute for linseed oil in printer's ink, varnish, etc; edible oil in Japan, China, India, etc; manufacture of cheap varnishes; artificial leather.
Shipping regulations: None.*

periodic acid $HIO_4 \cdot 2H_2O$.
Properties: White crystals; loses $2H_2O$ at about 100°C. Soluble in water, alcohol; slightly soluble in ether.
Derivation: By the interaction of iodine and concentrated perchloric acid; by low temperature electrolysis of concentrated iodic acid.
Method of purification: Crystallization.
Grades: Technical.
Containers: Glass bottles; tins.
Use: Oxidizing material.

periodic table. A systematic classification of the elements according to atomic number (nearly the same order as by atomic weights) and by physical and chemical properties. Credited to Mendeleev and summarized as a periodic law: The properties of the chemical elements and their compounds are a periodic function of their atomic weights.

peristerite. A variety of feldspar showing a play of colors and used as a gem.

perlite (rock). A form of glassy rock similar to obsidian. Usually contains 65-75% SiO_2, 10-20% Al_2O_3, 2-5% H_2O, and smaller amounts of soda, potash, and lime. When perlite is heated to the softening point, it expands to form a light fluffy material similar to pumice (q.v.).
Occurrence: California, Colorado, New Mexico, Nevada, Oregon.
Use: Light weight concrete aggregate, plaster aggregate, acoustic and heat insulation; as insulation for liquid fuels; catalyst support, filtering. "Popped" perlite is also used as a packing.

perlite (eutectic). See pearlite for the iron-carbon eutectic present in certain carbon steels.

"Perluxe." [56] Trademark for a dry-cleaning solvent consisting of perchloroethylene or tetrachloroethylene.

"Permachlor Red." [141] Trade name for azo pigments made from para-chlor-ortho-nitraniline and beta-naphthol.
Properties: Good durability; especially in tints.
Uses: In bright, light red and pink house paints.

"Permachrom Red." [141] Trade name for permanent red azo pigments derived from beta-hydroxynaphthoic acid.
Properties: Excellent light resistance, good heat resistance, non-bleeding in water and organic solvents. May be used with light reds to give intermediate shades which are bright and permanent.
Grades: Medium red; dark red; maroon.
Uses: Printing inks, paints, enamels, lacquers, plastics.

permafils. Mixtures in which the liquid completely undergoes polymerization and hardens without the necessity of any evaporation. Anerobic permafils harden out of contact with air.

"Permal." [188] Brand name for a special type of aldehyde top note used in perfuming. This material retains the light, original character throughout the perfume's full range of evaporation.

permalloys. Nickel-iron alloys containing from 40-80% nickel and having high magnetic permeability and electrical resistivity.

"Permalume." [72] Trade name for a semi-bright (or dull) nickel plating process with good ductility and low tensile stress characteristics. Excellent mechanical properties, corrosion resistance when used as an undercoating in a two-deposit system. Plating bath composed of nickel sulfate, nickel chloride, boric acid and nickel acetate; sulfur free organic addition agents produce a deposit with <0.003% sulfur.

"Permalux." [28] Trademark for a di-ortho-tolylguanidine salt of dicatechol borate, $(HOC_6H_4O)_2BN(NHC_6H_4CH_3)_2$.
Properties: Light grayish brown powder; sp.gr. 1.27; f.p. not lower than 165°C.
Containers: 125-lb drums.
Use: To accelerate and improve the vulcanization of neoprene. A non-discoloring antioxidant for natural and synthetic rubbers.

permanent white. See barium sulfate.

permanent yellow. See chrome yellows.

"Permanickel." [283] Trademark for a wrought, age-hardenable nickel alloy containing approximately 98.5% nickel.

"Permanite." [250] Trademark for purified natural magnetic iron ore for use in radio cores, shields, etc.

"Permansa." [141] Trade name for organic color pigments.
Grades:
"Permansa" Yellow: Hansa yellows.
"Permansa" Orange: Dinitraniline orange.
"Permansa" Red: Chlorinated para red.
"Permansa" Green: Pigment green B.
Uses: Printing inks, paints, wallpaper, textiles.

"Permel Plus." [57] Trademark for composition for treating textiles.

"Permel" Resin B. [57] A water-repellent aqueous resin dispersion that can be used on cotton, rayon, wool and synthetics. It gives a soft finish and increases the tear strength and sewability of fabrics.

permenorm. Nickel iron alloy produced by magnetic annealing and drastic cold reduction, and used for mechanical rectifiers and low frequency amplifiers. This alloy has a rectangular hysteresis loop that eliminates arcing at the contacts of mechanical rectifiers, as well as other desirable properties.

permethanoic acid. See performic acid.

"Permigels." [57] Trademark based on U.S. Bureau of Mines standards. A line of permissible explosives approved for mining of coal. This line recommended where wet conditions are encountered.

"Permolite." [300] Trademark for a solubilized alkyd resin used as a slip-proof finish for low-count viscose and acetate fabrics.

"Permolith." [141] Trade name for lithopone.
Grades:
"Permolith" 10N: Low paint consistency.
"Permolith" 20L: Low paint consistency.
"Permolith" 40M: Medium paint consistency.
"Permolith" 60H: High paint consistency.
"Permolith" 60J: Quick dispersion in oils.
"Permolith" W-320: Water wetting, high dry hiding.
"Permolith" W-420: Water dispersible, medium water absorption.
"Permolith" W-421: Water dispersible, low water absorption.
"Permolith" W-560: Water dispersible, medium water absorption.
Uses: Paint pigment, printing ink, rubber, leather, plastics, wallpaper, textiles, floor coverings, face powders, cosmetics, paper coatings, linoleum, oil cloth and window shade cloth.

"Permutit." [184] Trademark for a line of cation and anion exchangers.
"Permutit" T. Treated precipitated potassium aluminosilicate cation exchanger.
"Permutit" Q. High capacity, sulfonated, acid resistant, styrene type resin cation exchanger.
"Permutit" Z. High capacity ammonium cation exchanger.
"Permutit" H-70. Weakly acidic carboxylic acid type resin cation exchanger.
"Permutit" A. Tertiary amine, moderately basic type anion exchanger.
"Permutit" W. Weakly basic, polystyrene base, polyamine type resin anion exchanger.
"Permutit" S-1. Trimethyl amine, strongly basic polystyrene quaternary amine type resin anion exchanger.
"Permutit" S-2. Strongly basic, dimethyl ethanol amine type resin anion exchanger.
"Permutit" SK. Highly porous, medium basic, high capacity resin type anion exchanger.
"Permutit" SM. Strongly basic, high capacity resin type anion exchanger.
"Manganese Permutit." Permanganate-regenerated glauconite cation exchanger.
"Permutit" DR (ASMIT). Specially prepared decolorizing resin; prepared from bone char.

"Permutit Wisprofloc-20 Coagulant Aid." [210] Coagulant aid used in conjunction with coagulation in the clarification of both potable and industrial waters.

pernambuco extract. See brasilin.

"Perone." [28] Trademark for hydrogen peroxide, 35 and 50% solutions.

peronine (morphine benzylether hydrochloride) $C_{17}H_{17}NO(OH)OC_7H_7 \cdot HCl$.
Properties: White prismatic crystalline powder, odorless, bitter taste; poisonous! Precipitated from its solutions by usual alkaloid reagents but differs from morphine and codeine in its solubilities and reactions toward special reagents. Soluble in boiling water, less so in alcohol (95%), sparingly soluble in hot water.
Containers: Glass bottles.
Use: Medicine.
Shipping regulations: None.*

perosmic acid, anhydride or **oxide.** See osmic acid, anhydride.

peroxidase. An enzyme found in nearly all plant cells which acts upon a large number of phenols, aromatic amines, etc., in the presence of hydrogen peroxide to produce water and an oxidation product of the substrate.
Use: Biochemical research.

peroxide. See hydrogen peroxide.

peroxyacetic acid. See peracetic acid.

peroxydol. See sodium perborate.

peroxyformic acid. See performic acid.

peroxysulfuric acid. See Caro's acid.

perphenazine $C_{21}H_{26}ClN_3OS$. 2-Chloro-10-(3-[4-(beta-hydroxyethyl)piperazinyl]propyl)-phenothiazine.
Properties: Crystals; sensitive to light; m.p. 97-100°C; insoluble in water, soluble in ethanol and acetone.
Grade: N.N.D.
Use: Medicine.

Persian bark. See cascara sagrada bark.

Persian insect flowers. See pyrethrum flowers.

Persian pellitory. See pyrethrum flowers.

Persian red. The term sometimes refers to red pigments derived from basic lead chromate, and at other times to red pigments whose chief compound is ferric oxide. See iron oxide reds.

persic oil (peach kernel oil; apricot kernel oil)
Properties: A pale yellow to colorless liquid; bland odor and taste similar to almond oil. Soluble in ether, chloroform and carbon disulfide; partly soluble in alcohol.
Constants: Sp.gr. 0.915; saponification value 191; iodine value 93-109.
Derivation: By expressing the blanched seeds of the peach, Prunus persica, or apricot, Prunus armenica.
Grades: Technical; N.F.XI.
Containers: Tins; drums.
Uses: Nutrient similar to almond and olive oils; flavoring; medicine.
Shipping regulations: None.*

persicol. See gamma-undecalactone.

persio. See cudbear.

Persoz's reagent. A reagent for the detection of silk in presence of wool. 10 g zinc chloride is dissolved in 10 cc water, 2 g zinc oxide is added and the whole shaken. If this solution is warned to 45°C, it will dissolve silk, but not wool.

persulfate of iron. See ferric sulfate.

persulfuric acid. See Caro's acid.

"Perthane." [23] Trademark for an agricultural insecticide based on diethyl diphenyl dichloroethane (see 1,1-dichloro-2,2-bis-(para-ethylphenyl)dichloroethane) and supplied as a wettable powder or emulsifiable concentrate.
Use: Controls insects on plants and livestock. Also for moth protection of textiles.

Peru apple. See stramonium.

Peru balsam. (Peruvian balsam; Indian balsam; China oil; Chinese oil; black balsam).
Properties: Dark, molasses-like liquid; pleasant aromatic odor; warm bitter taste. Soluble in alcohol and ether; miscible in acetone, glacial acetic acid, chloroform, and benzene; nearly insoluble in water.
Constants: Sp.gr. 1.150-1.170.
Derivation: Obtained from Myroxylon pereirae.
Constituents: Esters of cinnamic and benzoic acid; also resins; and vanillin.
Occurrence: San Salvador.
Method of purification: Rectification.
Grades: Technical; N.F. XI.
Containers: Iron drums.
Uses: Medicine; perfumery; chocolate manufacture.
Shipping regulations: None.*

Peruvian balsam. See Peru balsam.

Peruvian bark. See cinchona bark, calisaya.

"Pest-B-Gon." [253] Brand name for a type of insecticide liquid containing DDT.

pesticide. A descriptive term used to encompass all materials used for the control of animal or plant pests. These include insecticides, fungicides, herbicides and rodenticides.

"Pestmaster" Fumigant 37. [426] Trademark for 30% ethylene dibromide, 70% methyl bromide.
Containers: 1-lb cans, cylinders to 400 lbs.
Uses: Space and grain fumigant.
Shipping regulations: Class B poison. Poison label.*

"Pestmaster" Fumigant 73. [426] Trademark for 70% ethylene dibromide, 30% methyl bromide.
Properties: Liquid.
Containers: 6-oz cans.
Uses: Bulk grain fumigant.
Shipping regulations: Poison, Class B. Poison label.*

"Pestmaster" Methyl Bromide. [426] Trademark for methyl bromide, CH_3Br.
Properties: A colorless liquified gas, b.p. 3.2°C. Sp.gr. (0°/0°C) 1.732.
Grades: Pure.
Containers: 1-lb cans, steel cylinders to 400 lbs net.
Uses: Space and grain fumigant.
Shipping regulations: Class B poison. Poison label.*

"Pestmaster" Soil Fumigant-1. [426] Trademark for methyl bromide 98%, chloropicrin 2%.
Properties: A colorless liquified gas with a sharp pungent odor.
Containers: 1-lb cans; steel cylinders to 400-lb net.
Uses: Control insects, weeds, nematodes, and certain fungi in the soil.
Shipping regulations: Class B poison. Poison label.*

"Pestmaster" Spot Shot. [426] Trademark for 70% ethylene dibromide, 30% methyl bromide.
Containers: 1½-fl oz cans.
Uses: Fumigant of food handling equipment.
Shipping regulations: Class B poison. Poison label.*

petalite $LiAl(Si_2O_5)_2$ or $Li_2O \cdot Al_2O_3 \cdot 8SiO_2$.
Properties: Colorless, white, gray or

occasionally pink mineral, white streak, vitreous luster. Resembles spodumene (q.v.) in appearance. Contains 4.9% lithia, sometimes with partial replacement by sodium or, less often, by potassium. Insoluble in acids.
Constants: Sp.gr. 2.39-2.46; hardness 6-6.5.
Occurrence: United States (Massachusetts, Maine); Sweden.
Use: As a source of lithium salts; in ceramics and glass.

pethidine hydrochloride. See meperidine hydrochloride.

petitgrain citronier oil. See petitgrain oil.

petitgrain oil (petitgrain citronier oil).
Properties: Yellowish liquid; odor similar to neroli oil. Soluble in 70% alcohol, ether, chloroform and carbon disulfide. Sp.gr. 0.887-0.900.
Derivation: From various citrus species especially in Paraguay.
Method of purification: Rectification.
Grades: Technical.
Containers: 1-, 5-, 10-lb bottles; 20-, 25-lb cans; drums.
Use: Perfumery (soaps; synthetic neroli; skin creams).
Shipping regulations: None.*

PETN. Abbreviation for pentaerythritol tetranitrate.

"Petrex." [266] Trademark for a series of terpene-derived alkyd resins.
Properties: They vary from soft and tacky to hard and tack-free. They show high grease resistance and are practically insoluble in petroleum solvents. Acid number range, 9-500; color range (U.S.D.A. rosin scale) WW to WG; softening point range (Hercules drop method °C) 50-105; viscosity range (Gardnes-Holdt, 70-75% in ethanol or toluene) V to Z.
Uses: Lacquers, inks, adhesives and other products.

petrichloral (pentaerythritol chloral) $C_{13}H_{16}Cl_{12}O_8$.
Properties: Crystals.
Grade: N.N.D.
Use: Medicine.

"Petrobase." [25] Brand name for a series of proprietary, emulsifying and rust preventive bases. On dilution with suitable petroleum oils or light distillates the proper concentrate provides soluble cutting and grinding oils; solvent emulsion cleaners; preservative, slushing and household oils; and water displacing fluids.

"Petro Bond." [236] Trademark for a bonding agent in the preparation of waterless foundry sands used with oil and catalyst. Fine sands giving excellent reproduction of detail can be used since high permeability of the foundry sand is not required with this binder.
Containers: 50-lb multiwall paper bags.

petrochemicals. Chemicals for which petroleum or natural gas has served as the ultimate raw material. Thus, cracking of petroleum produces ethylene which is converted to ethylene glycol, the latter being a typical petrochemical. The term is also applied to substances such as ammonia, because the hydrogen used to form the ammonia is a product of petroleum refining. At least 175 substances are designed as petrochemicals. The common paraffin, olefin, naphthene, and aromatic hydrocarbons (methane, ethane, propane, ethylene, propylene, butenes, cyclohexane, benzene, toluene, naphthalene, etc) and their derivatives are referred to as petrochemicals. The term is often applied to such chemicals even though some of their commercial production is from sources other than petroleum.

petrol. A name used in the British Commonwealth (except Canada) to designate gasoline.

petrolatum (mineral fat; petroleum jelly; petroleum ointment; mineral jelly).
Properties: Almost colorless to amber-colored, gelatinous, oily, translucent, semisolid, amorphous mass whose consistency varies with the temperature. Soluble in chloroform, ether, benzine, carbon disulfide, benzene and oils; very slightly soluble in alcohol; insoluble in water.
Chief constituents: Hydrocarbons of the methane series, ($C_{16}H_{34}$ up to $C_{32}H_{66}$), and of the olefin series, ($C_{16}H_{32}$, etc.).
Constants: Sp.gr. 0.815-0.880 at 60°C; m.p. 38-60°C.
Derivation: By fractional distillation of still residues from the steam distillation of paraffin-base petroleum, or from steam-reduced amber crude oils (oils from which the light fractions have been removed).
Grades: Natural petrolatum produced as above; artificial petrolatums made by mixing heavy petroleum lubricating oil with a low m.p. paraffin wax; U.S.P. XVI (white petrolatum); N.F. XI (yellow petrolatum).
White petrolatum is made from the ordinary material by bleaching or additional treatment with decolorizing carbon.
Containers: Glass bottles; tins; drums.
Uses: Medicine and pharmacy (as a protective dressing and as a substitute for fats in ointments); modeling clay; shoe polishes; lubricating greases; metal polishes; leather grease; rust preventive; perfume extractor; insect repellents.
Shipping regulations: None.*

petrolatum, liquid (white mineral oil; paraffin, liquid; paraffin oil, white).
Properties: A colorless transparent oily liquid, almost tasteless and odorless even when warm; sp.gr. 0.828-0.880 (light), 0.860-0.905 (heavy); soluble in ether, chloroform, carbon disulfide, benzine, benzene, boiling alcohol, and fixed or volatile oils; insoluble in water, cold alcohol, glycerine.
Derivation: Distillation of high boiling (330-390°C) petroleum fractions.
Method of purification: Treatment with sulfuric acid, caustic soda, filtration through decolorizing carbon, and crystallization to

*See "I.C.C. Shipping Regulations," page xiii.
Reference numbers refer to name of manufacturer. See "List of Manufacturers," page v.

remove waxes.
Grades: Technical; U.S.P. XVI (heavy); N.F. XI (light).
Containers: Wooden barrels; tins; glass bottles; tank cars.
Use: Medicine, cosmetics; dispersants, diluents, etc. in plastics manufacture; compressor and textile lubricants; dispersants for reactive compounds such as metal hydrides; catalyst carriers.

"Petrolene." 200 Trademark for a petroleum solvent prepared by straight-run distillation.
Properties: Water-white; initial boiling point 140-145°F, 95% distills between 195 and 200°F; sp.gr. 0.701 (60°F); flash point (TCC) −16°F; mild; non-residual odor.
Use: In rubber cements; sealers; fast-drying lacquers; lacquer dopes; roto inks used on high speed presses.
Caution! Fire hazard; keep lights and fire away.
Shipping regulations: Flammable liquid. Red label.*

petroleum (crude oil, earth oil, Lima oil, Seneca oil, rock oil). All petroleums are complex mixtures of paraffin, naphthene, and aromatic hydrocarbons with small amounts of organic sulfur and very small amounts of nitrogen and oxygen compounds. The terms paraffin base crude, naphthene or asphalt base crude, and aromatic base crude are used to indicate the most prevalent constituents of crudes from various localities.
Properties: A thick, heavy, flammable liquid, varying in color from yellow to dark reddish-brown or black according to its place of origin. It has a peculiar distinct heavy odor also varying with its place of origin and composition. It usually shows a distinct greenish fluorescence. Sp.gr. 0.780-0.970, usually 0.85-0.95. The major petroleum fractions in approximate order of decreasing volatility are natural gas, ligroin, gasoline, naphtha, kerosine, fuel oil, gas oil, lubricating oil, paraffin wax, road oil, asphalt, and coke. See these individual topics for further products and uses. Petroleum and certain of its fractions are also a major raw material for chemicals of wide variety. See petrochemicals.
Shipping regulations: May be flammable liquid. Red label.*

petroleum asphalt. Asphalt obtained from petroleum.

petroleum benzin. A special grade of petroleum ether or ligroin (q.v.).
Properties: A clear, colorless, nonfluorescent, volatile liquid, with an ethereal or petroleum-like odor; insoluble in water; soluble or miscible with most organic solvents and most oils. Sp.gr. 0.634-0.660 (25/25°C); distillation range 35-80°C.
Derivation: By distillation from petroleum.
Use: Solvent or extraction medium, especially for drugs.
Caution: Flammable. Vapor may be explosive if mixed with air and ignited.
Shipping regulations: Flammable liquid. Red label.*

petroleum chemicals. Those used in making gasoline, lubricants and other products of the petroleum industry. They include tetraethyl lead, oil-soluble dyes, metal deactivators, catalysts, antioxidants, etc. See for example, under "Ethyl." They are distinct from petrochemicals, which are made from petroleum.

petroleum coke. See coke, petroleum.

petroleum dyes. Hydrocarbon soluble dyes for coloring gasoline, etc. See Oil Blue A, Oil Bronze, Oil Orange, Oil Red, and Oil Yellow N.
Containers: 100-lb fiber drums.

petroleum ether. See ligroin.
Note: Term is misleading and should not be used.

petroleum jelly. See petrolatum.

petroleum naphtha. See naphtha, petroleum.

petroleum ointment. See petrolatum.

petroleum resins. Obtained as a byproduct of petroleum refining. Used mainly in "asphalt" floor tile.

petroleum spirits. See naphtha, painters'. In Great Britain the term "petroleum spirits" refers to a very volatile hydrocarbon mixture having a flash point below 32°F.

petroleum sulfonates. Compounds usually made as by-products of white oil manufacturing and lube stock refining, usually by addition of sulfur trioxide to oils. They are often encountered as the metallic salts. Molecular weights vary from 475-600. Uses are lubricating oil additives; cutting oil emulsifiers; rust preventives; fat-splitting agents; antifreeze agents; and emulsifiers in textile processing oils.
Containers: Drums, tank cars.

petroleum tailings. See residual oils.

petroleum thinner. See naphtha, painters'.

petroleum waxes. Waxes derived from petroleum and usually divided into three groups: the paraffin waxes, the micro-crystalline waxes, and the petrolatums. All three are made mostly by solvent dewaxing, although pressing and sweating processes are still used in some instances. Paraffin waxes are obtained by dewaxing light lubricating oil stocks, petrolatum waxes by dewaxing heavy lubricating oil stocks, and microcrystalline waxes by deoiling and cutting the petrolatums or from pipe still bottoms. The use for petrolatums is limited to cosmetics and pharmaceuticals, and occasionally as plasticizers in wax formulations. The largest use of petroleum waxes is in the production of paper, candles, electrical items, chlorinated paraffins, textiles,

and polishes.
See also waxes, microcrystalline.

"Petrolite." [128] Trademark for a line of microcrystalline petroleum waxes.

"Petromix." [45] Trademark for a soluble oil base, made from petroleum sulfonates.
Uses: Emulsifying agent for oils and solvents.

"Petronates." [45] Trademark for salts of petroleum sulfonic acids varying in molecular weight and color.
Uses: Emulsifying agent, dispersing agent, wetting agents, corrosion-preventive.

"Petronauba." [128] Brand name for a series of emulsifiable petroleum microcrystalline waxes.
Properties: Color yellow, amber or cream; m.p. 180-185°F; saponification no. 50-60; emulsifiable.
Containers: 10-lb slabs, 8/carton or 168/pallet.
Uses: Water-emulsion type floor polishes; carbon paper.

"Petrosene." [45] Trademark for a series of fuel oil additives available as:
"Petrosene A": Sludge dispersant; rust inhibitor; soot destroyer and combustion improver for distillates and residual fuel oils.
"Petrosene B": Refinery fuel oil additive to improve storage stability of distillate fuels.
"Petrosene C": Sludge dispersant; peptizing agent and rust inhibitor for use primarily with residual fuels.
"Petrosene D": Corrosion inhibitor for residual fuels containing a high content of vanadium, sodium and sulfur, and can also be used to reduce super-heater slag deposits in marine boilers.

"Petrosul." [25] Brand name for a proprietary line of highly purified natural petroleum sulfonate products available in high, medium and low molecular weight ranges. Useful in applications requiring the surface active functions of foaming, detergency, emulsibility, dispersion, solubilization, spreading and rust protection.

"Petrotect." [25] Brand name for a proprietary series of rust preventive and hydraulic fluids in general meeting military specifications. The rust preventives are classified as solvent cut backs, petrolatum barriers, general purpose preservatives, and engine preservative lubricants. Hydraulic fluids include both preservative and operational types.

"Petrothene." [192] Trademark for a line of polyethylene resins available in various grades. The 100 series are film-grade resins for tubular and sheet film extrusion and blow molding; the 200 series are for general purpose extrusion, injection, compression molding and paper coating; the 300 series are especially adapted to wire covering and electrical insulation applications.

"Petrowet." [28] Trademark for wetting and penetrating agents used in acid solutions.
"Petrowet" R. Saturated hydrocarbon sodium sulfonates.
Properties: Light yellow liquid.
Use: A wetting and penetrating agent effective in high concentrations of electrolytes and especially for use in 15% hydrochloric acid for oil-well acidizing.
"Petrowet" WN. Fatty alcohol sodium sulfate.
Properties: Light yellow liquid.
Use: For wetting and penetrating in acid solution.

pettymorrel. See aralia.

petzite Ag_3AuTe_2. A natural telluride of silver and gold.
Properties: Color steel gray to iron black; luster metallic; hardness 2.5-3; sp.gr. 8.7-9.0.
Occurrence: Colorado, California, Australia, Canada.

pewter. Tin alloys with 5-15% antimony, 0-3% copper, and 0-15% lead. White metal (q.v.) and britannia metal are also of this general composition.

"Pexol." [266] Trademark for fortified rosin sizes based on modified rosins; available as pastes and dry products in both pale and dark types; paste products are at 70 and 77% solids.

PF resins. Abbreviation for phenol-formaldehyde resins (q.v.).

"PG"-16. [202] Trademark for butyl acetyl polyricinoleate; used in vinyls and synthetic rubbers to impart low temperature flexibility.

PGA. Abbreviation for pteroylglutamic acid. See folic acid.

pH. A means of expressing the degree of acidity or basicity of a solution. Thus at normal temperature a neutral solution such as pure distilled water has a pH of about 7, a tenth-normal solution of hydrochloric acid (approximately 3.65 g HCl/liter) has a pH near 1 and a normal solution of a strong alkali such as sodium hydroxide has a pH of nearly 14. Orignally pH was defined as the logarithm of the reciprocal of the hydrogen ion concentration in gram equivalents per liter of solution,

$$pH = \log \frac{1}{(H^+)}$$

and this is in many cases approximately correct. In some cases it is seriously in error. Actually pH values are obtained by measuring the potentials E of galvanic cells of the type (Pt)H_2; solution x, saturated KCl; reference electrode; and using the value of E in the equation

$$pH = \frac{E - E_0}{2.306\ RT/F}$$

In this equation, E_0 is a constant depending upon the nature of the reference electrode, and R, T and F are respectively the fundamental gas constant, the absolute temperature and the faraday. Commercial pH meters, however, are calibrated to read pH directly and no calculation is necessary.

phagocyte. See bacteriophage.

phalaris. Canary seed.

"Phaltan." [253] Brand name for fungicide formulations containing n-trichloromethylthiophthalimide.

"Phanadorn" Calcium. [162] Trademark for cyclobarbital calcium.

"Phantolid." [105] Trademark for a synthetic aromatic ketone, $C_{17}H_{24}O$. (1, 1, 2, 3, 3, 5-Hexamethylindan-6-methyl ketone).
Properties: Waxy, white solid; m.p. 35-40°C; odor between natural and synthetic musk.
Uses: Perfumery, as musk fixative.

"Pharmasorb." [99] Trademark for pharmaceutical grades (regular and colloidal) of activated attapulgites. Have high adsorptive capacities as a result of specific thermal treatment of the natural mineral. For typical analysis, see "Attasorb."
Properties: Tamped bulk density (regular grade) 12-18 lbs/cu ft,(colloidal grade) 25-37 lbs/cu ft; sp.gr. (regular grade) 2.47, (colloidal grade) 2.36; average particle size (regular grade) 2.9 microns, (colloidal grade) 0.14 microns.
Containers: 25-lb polyethylene bags.

"Phemerol." [330] Trademark for benzethonium chloride U.S.P.

phenacaine hydrochloride $C_2H_5OC_6H_4NHC(CH_3)NC_6H_4OC_2H_5 \cdot HCl \cdot H_2O$. N, N'-bis(para-ethoxyphenyl) acetamidine hydrochloride.
Properties: Small, white crystals; odorless; faintly bitter taste. Incompatible with alkalies. M.p. 190°C. Soluble in alcohol, boiling water and chloroform; less so in cold water, insoluble in ether.
Grade: N.F. XI; technical.
Use: Medicine.

phenacemide (phenylacetyl urea) $C_6H_5CH_2CONHCONH_2$.
Properties: White to creamy white, odorless, tasteless crystalline solid; m.p. 212-216°C; slightly soluble in alcohol, benzene, chloroform and ether; very slightly soluble in water.
Grade: N.N.D.
Use: Medicine.

phenacetin. See acetophenetidin.

phenacite Be_2SiO_4. A natural beryllium silicate.
Properties: Colorless or yellow, red or brown in color. White streak, vitreous luster.
Constants: Sp.gr. 2.5-2.8; hardness 5-6.5.
Occurrence: United States (Colorado, Montana, New Hampshire); Russia; France; Norway; Mexico; Brazil.
Use: Gem stones.

phenacyl chloride. See chloroacetophenone.

phenacyl fluoride. See fluoroacetophenone.

phenaglycodol $(CH_3)_2C(OH)C(CH_3)(OH)C_6H_4Cl$ (2-para-chlorophenyl-3-methyl-2, 3-butanediol).
Properties: White to cream colored, crystalline powder; m.p. 77.5-80.5°C; soluble in alcohol, chloroform and ether; insoluble in water, and dilute acids or alkalies.
Grade: N.N.D.
Use: Medicine.

"Phenamine." [307] Brand name of line of direct dyestuffs. Used for the dyeing of cotton and paper.

phenanthraquinone. See phenanthrenequinone.

phenanthrene (phenanthrin) $C_{14}H_{10}$. A tricyclic hydrocarbon.
Properties: Small, colorless, shining crystals. Soluble in alcohol, ether, benzene, carbon disulfide and acetic acid; insoluble in water.
Constants: Sp.gr. 1.063; m.p. 100.35°C; b.p. 340°C.
Derivation: Fractional distillation of high-boiling coal-tar oils, with subsequent recrystallization from alcohol.
Method of purification: Fractional oxidation (chromic or nitric acid) to remove anthracene.
Impurities: Anthracene.
Grades: Technical (90%).
Containers: Tins; glass bottles; fiber drums.
Uses: Dyestuffs; explosives; synthetis of drugs; phenanthrenequinone.
Shipping regulations: None.*

phenanthrenequinone. (Erroneously: phenanthraquinone) $C_{14}H_8O_2$.
Properties: Yellow-orange, needle-like crystals. Soluble in sulfuric acid, benzene, glacial acetic acid and hot alcohol; slightly soluble in ether; insoluble in water.
Constants: Sp.gr. 1.4045; m.p. 206-207°C; b.p. sublimes above 360°C.
Derivation: By oxidation of a boiling solution of phenanthrene of glacial acetic acid with chromic acid, solution in sodium disulfite, precipitation by means of hydrochloric acid and recrystallization.
Grades: Technical.
Containers: Wooden kegs; fiber drums.
Uses: Organic synthesis; dyes.

phenanthrin. See phenanthrene.

1, 10-phenanthroline (4, 5-phenanthroline; ortho-phenanthroline) $C_{12}H_8N_2 \cdot H_2O$.
Properties: White crystalline powder; m.p. 93-94°C, anhydrous 117°C. Slightly soluble in water; soluble in alcohol, benzene.
Derivation: Made by heating ortho-phenylenediamine with glycerin, nitrobenzene and concentrated sulfuric acid; or in like manner from 8-aminoquinoline.
Uses: Forms a complex compound with ferrous ions which is used as an indicator; used as a drier catalyst in coatings industry.

phenarsazine chloride (adamsite; diphenylaminechlorarsine; DM)

*See "I.C.C. Shipping Regulations," page xiii.
Reference numbers refer to name of manufacturer. See "List of Manufacturers," page v.

$C_6H_4(AsCl)(NH)C_6H_4$.
Properties: Canary-yellow crystals. Very poisonous; irritant to skin and respiratory tract. Sublimes readily. Sp.gr. 1.65; m.p. 195°C; b.p. 410°C (dec); insoluble in water; soluble in benzene, xylene, carbon tetrachloride.
Derivation: By heating diphenylamine with arsenic trichloride.
Uses: Dispersed in air as a poison gas; tear gas; wood treating.
Shipping regulations: Gas, liquid or solid form: Poison class C. Tear gas label.* Legal label name, diphenylaminechlorarsine.

phenarsone sulfoxylate. Sodium 3-amino-4-hydroxyphenylarsonate-N-formaldehyde sulfoxylate.
Properties: White, odorless, amorphous powder; soluble in water, dilute acids, alkalis and alkali carbonates; slightly soluble in methanol; insoluble in alcohol and ether; 5% solution has pH of 7.0-7.4.
Derivation: From 3-amino-4-hydroxybenzenearsonic acid and sodium formaldehyde sulfoxylate.
Use: Medicine.

phenazine (azophenylene) $C_6H_4N_2C_6H_4$. A tricyclic ring.
Properties: Yellow crystals; m.p. 170-171°C; b.p. >360°C; very slightly soluble in water; soluble in alcohol and ether.
Use: Organic synthesis; manufacture of dyes.

phenazocine (2'-hydroxy-5,9-dimethyl-2-(2-phenethyl)-6,7-benzomorphan) hydrobromide $C_{22}H_{27}NO \cdot HBr$. Colorless crystals, m.p. 166-170°C, used in medicine.

phenazone. See antipyrine.

phenelzine dihydrogen sulfate (beta-phenylethylhydrazine dihydrogen sulfate) $C_6H_5(CH_2)_2NHNH_2 \cdot H_2SO_4$.
Use: Medicine.

"Phenergan" Hydrochloride. [24] Trademark for promethazine hydrochloride [N-(2'-dimethylamino-2'-methyl) ethyl phenothiazine hydrochloride]; an antihistamine.

phenethicillin. See potassium alpha-phenoxyethyl penicillin, under penicillin.

phenethyl alcohol (phenylethyl alcohol; 2-phenylethanol; benzyl carbinol) $C_6H_5CH_2CH_2OH$.
Properties: Colorless liquid; light rose odor; sharp burning taste; sp.gr. 1.017-1.020 (25°C); refractive index (n 20/D) 1.5310-1.5330; m.p. −27°C; b.p. 219°C. Soluble in 50% alcohol; soluble 1 part in 50 parts of water; soluble in fixed oils, alcohol, and glycerol; slightly soluble in mineral oil.
Derivation: (a) By reduction of phenylacetic ethyl ester by sodium in absolute alcohol. (b) By the action of ethylene oxide on phenylmagnesium bromide and subsequent hydrolysis.
Method of purification: Rectification.
Grades: Technical; N.F. XI.
Containers: Tin cans and glass bottles; drums.
Uses: Organic synthesis; perfumery; synthetic rose oil; cosmetics; soaps; flavors; antibacterial; preservative; medicine.
Shipping regulations: None.*

sec-**phenethyl alcohol.** See styralyl alcohol.

phenethylamine. See beta-phenylethylamine.

ortho-**phenetidine** (2-aminophenetole) $NH_2C_6H_4OC_2H_5$.
Properties: Oily liquid; rapidly becomes brown on exposure to light or air. Solidifies below −20°C; b.p. 228-230°C. Soluble in alcohol and ether; insoluble in water.
Derivation: Prepared by reduction of orthonitrophenetole with iron filings and hydrochloric acid.
Containers: 475-lb drums.
Use: Manufacture of dyes.
Warning! Hazardous liquid and vapor; absorbed through skin! MCA warning label.

para-**phenetidine** (4-aminophenetole) $NH_2C_6H_4OC_2H_5$.
Properties: Colorless oily liquid; becomes red to brown on exposure to air and light. Sp.gr. 1.0613 (15°C); m.p. 2-4°C; b.p. 253-255°C. Insoluble in water; soluble in alcohol.
Derivation: Prepared by ethylating paranitrophenol with ethyl sulfate or chloride in presence of sodium hydroxide followed by reduction with iron filings and hydrochloric acid.
Grades: Technical (98%).
Containers: 5-, 10-, 55-gal drums.
Uses: Dyestuffs intermediate; pharmaceuticals; medicine.
Warning! Hazardous liquid and vapor; absorbed through skin! MCA warning label.

phenetidine amygdalate. See amygdophenine.

phenetidine citrate. See citrophen.

phenetole (phenyl ethyl ether) $C_6H_5OC_2H_5$.
Properties: A colorless liquid; b.p. 172°C; sp.gr. 0.967 (20/4°C); insoluble in water; soluble in alcohol and ether.

para-**phenetolecarbamide.** See dulcin.

phenetsal. See para-acetylaminophenyl salicylate.

"Phenex." [94] Trademark for a rubber accelerator, composed of selected salts of alpha-ethyl-beta-propylacrylaniline.
Properties: A clear, dark amber colored liquid; sp.gr. 1.01; stable in storage; insoluble in water; soluble in acetone, benzol, and gasoline; disperses readily.
Containers: 400-lb drums.
Uses: Recommended for molded mechanical goods, heels and soles, hose, wire insulation, hard rubber, tubes, tires, tire repair stocks and cement; effective with reclaimed rubbers; can be used in bright colored stocks, but not in white stocks.
Hazards: No health hazards when used in rubber in the amounts recommended.

phenindamine tartrate (2-methyl-9-phenyl-tetrahydro-1-pyridindene bitartrate) $C_{19}H_{19}N \cdot C_4H_6O_6$.
Properties: Creamy white powder with very faint odor. Solutions are acid to litmus. Soluble in water; slightly soluble in alcohol; practically insoluble in chloroform, ether and benzene. M.p. 160-162°C; re-solidifies at 163°C, melts with decomposition at 168°C.
Grades: U.S.P. XVI.
Use: Medicine.

phenindione (2-phenyl-1, 3-indanedione) $C_{15}H_{10}O_2$. A synthetic anticoagulant.
Properties: Pale yellow crystalline material; practically odorless; insoluble in water; soluble in methanol, alcohol, ether, acetone, benzene. Solutions in alkalis are red; in concentrated sulfuric acid blue.
Grade: N.N.D.
Use: Medicine.

pheniramine (prophenpyridamine) $C_6H_5CH(C_5NH_4)CH_2CH_2N(CH_3)_2$. 1-Phenyl-1-(2-pyridyl)-3-dimethylaminopropane.
Properties: Oily liquid; slightly yellow color; insoluble in water; soluble in dilute acids, alcohol, benzene, chloroform and ether; b.p. 135°C (0.5 mm), 181°C (13 mm); sp.gr. 1.008.
Use: Medicine.

pheniramine maleate (prophenpyridamine maleate) $C_{16}H_{20}N_2 \cdot C_4H_4O_4$. 1-Phenyl-1-(2-pyridyl)-3-dimethylaminopropane maleate).
Properties: White crystalline powder with faint amine-like odor. M.p. 104-108°. Very soluble in alcohol and water, only slightly soluble in benzene and ether. 1% solution has pH between 4.5-5.5.
Grade: N.F.XI.
Use: Medicine.

phenmetrazine hydrochloride $C_{11}H_{15}NO \cdot HCl$. (2-Phenyl-3-methyltetrahydro-1, 4-oxazine hydrochloride; 3-methyl-2-phenylmorpholine hydrochloride).
Grade: N.N.D.
Use: Medicine.

phenobarbital (phenylbarbital; 5-ethyl-5-phenylbarbituric acid; phenobarbitone) $C_{12}H_{12}N_2O_3$.
Properties: White, shining, crystalline powder, odorless, stable. Toxic unless properly used. M.p. 174°-178°C. Soluble in alcohol, ether, chloroform, benzene, alkali hydroxides, alkali carbonate solutions; sparingly soluble in water.
Derivation: Condensation of phenylethylmalonic acid derivatives and urea.
Grade: U.S.P. XVI.
Containers: Glass bottles; fiber cans, drums.
Use: Medicine.
Shipping regulations: None.*

phenobarbital sodium (phenobarbital, soluble; phenobarbitone, soluble) $C_{12}H_{11}N_2O_3Na$.
Properties: Flaky crystals, white, crystalline granules, or white powder; odorless, with a bitter taste; hygroscopic. It is very soluble in water; soluble in alcohol; but practically insoluble in ether and chloroform.
Grade: U.S.P. XVI.
Use: Medicine.

phenobarbital, soluble. See phenobarbital sodium.

phenobarbitone. See phenobarbital.

phenobarbitone, soluble. See phenobarbital sodium.

phenocoll acetate (aminoacetyl-para-phenetidine acetate; aminophenacetin acetate) $C_2H_5OC_6H_4NHCOCH_2NH_2 \cdot CH_3COOH$.
Properties: White powder; soluble in water.
Derivation: By the action of glycocoll upon phenetidine and acidifying.
Use: Medicine.

phenocoll hydrochloride (aminoacetyl-para-phenetidine hydrochloride; aminophenacetin hydrochloride) $C_2H_5OC_6H_4NHCOCH_2NH_2 \cdot HCl$.
Properties: Fine, white crystalline powder; soluble in water and warm alcohol; slightly soluble in chloroform, ether and benzene. M.p. 95°C.
Derivation: By the action of glycocoll upon phenetidine and acidifying.
Containers: Glass bottles; fiber cans.
Use: Medicine.
Shipping regulations: None.*

"Pheno" Direct Color Dyes. [57] Trademark for a group of direct dyes used for coloring paper.

"Phenoform." [307] Trademark for a line of dyestuffs and pigments. Used for the coloring of phenol-formaldehyde resins.

phenol.
1. A class of aromatic organic compounds in which one or more hydroxy groups are attached directly to the benzene ring. Examples are phenol itself (benzophenol), the cresols, xylenols, resorcinol, naphthols.
2. Phenol (carbolic acid; phenylic acid; benzophenol; hydroxybenzene) C_6H_5OH.

Properties: White, crystalline mass which turns pink or red if not perfectly pure or if under influence of light; absorbs water from the air and liquefies; distinctive odor; sharp burning taste; poisonous! When in very weak solution it has a sweetish taste; sp.gr. 1.07; m.p. 42.5-43°C; b.p. 182°C; flash point 172.4°F. Soluble in alcohol, water, ether, chloroform, glycerol, carbon disulfide, petrolatum, fixed or volatile oils and alkalies.
Derivation: (a) Classical method not now in much use: Treating the coal-tar oil fraction boiling 170-230°C with caustic soda to form phenolate, adding acid and distilling. (b) Converting benzene into the sulfonic acid and fusing the latter with caustic soda. On treating the sulfonate with acid, phenol is liberated and may be distilled off. (c) Raschig process (q.v.) from benzene and hydrogen chloride. (d) Chlorination of benzene and subsequent heating under pressure with caustic soda solution. The

*See "I.C.C. Shipping Regulations," page xiii.
Reference numbers refer to name of manufacturer. See "List of Manufacturers," page v.

phenol is set free with acid. (e) Cumene is oxidized to form phenol and acetone. (f) Solvent extraction of refinery wastes recovers phenol. (g) Most recent: oxidation of toluene to benzoic acid, its conversion by copper catalyst to phenylbenzoate, and hydrolysis to phenol.

Method of purification: Dissolved in water, crystallized out, centrifuged and redistilled.

Grades: Fused, crystals or liquid, all as technical (82%, 90%, 95%; other components mostly cresols), C.P. and U.S.P. XVI.

Containers: Bottles; tins; 25- and 55-gal drums; tank cars; tank trucks.

Uses: (in approximate order of volume): Phenolic resins; epoxy resins (bisphenol-A); nylon-6 (caprolactam); 2,4-D weed killers; selective solvent for refining lubricating oils; salicylic acid; pentachlorophenol; acetophenetidine; picric acid; germicidal paints; pharmaceuticals.

Danger! Rapidly absorbed through skin. Causes severe burns. MCA warning label.

Shipping regulations: Solid, or liquid if containing over 50% benzophenol: Poison, class B. Poison label.*

phenolate process. A process for removing hydrogen sulfide from gas by the use of sodium phenolate, which reacts with the hydrogen sulfide to give sodium hydrosulfide and phenol. The reaction can be reversed by steam heat in order to regenerate the sodium phenolate.

phenolbismuth. See bismuth phenate.

phenol coefficient. Method used to determine the effectiveness of a disinfectant using phenol as a standard of comparison. The phenol coefficient is obtained by dividing the highest dilution of the test disinfectant by the highest dilution of phenol which sterilizes a given culture of bacteria under standard conditions of time and temperature.

See also disinfectants.

phenoldisulfonic acid $C_6H_3OH(SO_3H)_2$.

Properties: Deliquescent, colorless needles. Soluble in water and alcohol.

Derivation: By the interaction of phenol and sulfuric acid with sulfurous acid.

Method of purification: Crystallization.

Grades: Technical.

Containers: Iron drums.

Uses: Intermediates; synthesis of drugs.

Shipping regulations: None.*

phenol-formaldehyde resins (PF resins). Synthetic resins formed from the compounds named. They were the first synthetic thermosetting plastics, then known as "Bakelite." Limited amounts of cresols, xylenols and other substituted phenols are used in such resins, but phenol itself, C_6H_5OH, is by far the most important starting material. Water is eliminated by the reaction of the carbonyl oxygen of formaldehyde with the active hydrogens the phenols.

In the presence of a basic catalyst such as sodium carbonate or ammonia, and excess formaldehyde, the condensation goes through three more or less distinct stages designated as A, B, and C or resol, resitol, and resite, respectively. The A-stage resin is thermoplastic and completely soluble in alcohol; the B-stage resin softens but does not melt on heating and swells without dissolving in alcohol. The fully cured C-stage resin is thermosetting and completely insoluble in all solvents. In the presence of an acid catalyst and excess phenol, the condensation proceeds much more rapidly and yields a thermoplastic product known as novolak. Novolak and B-stage resins can be cured to the thermosetting form by the addition of formaldehyde and an alkaline catalyst or with hexamethylenetetramine, which furnishes both the formaldehyde and ammonia to serve as a basic catalyst. The insoluble C-stage resin is cross-linked and can be formed only from phenolic compounds which have three active hydrogens, such as phenol, meta-cresol, and 3,5-xylenol. The 2,4- and 2,6-xylenols have only one active hydrogen and do not polymerize. The other cresols and xylenols form only thermoplastic polymers with formaldehyde. Thus, the properties of the resins depend largely on the starting materials and processing conditions used. The properties are further modified by the addition of fillers, plasticizers, and other monomers or polymers. Strength and shock resistance of the cured resin are greatly increased by the incorporation of fillers such as wood flour, cotton linters, canvas, asbestos, and mica. Electrical properties depend upon how thoroughly water and catalyst are removed from the product and upon the nature of the filler.

Phenol-formaldehyde resins are noted for good resistance to moisture, acids, solvents, and heat (up to 205°C); they are nonflammable and dimensionally stable over a wide temperature range. They have poor color stability, fair resistance to alkali, and are decomposed by oxidizing acids. An interesting property is sound-deadening, because the resins do not transmit and amplify sound.

Uses: Molded and cast plastic articles; ion exchange resins; laminating and impregnating resins; adhesives for sandpaper, plywood, etc; paints and baked enamel coatings. The resins have good ablative properties, so that when suitably compounded and filled with glass fiber they can withstand temperatures in the thousands of degrees for short exposure, and so are used for missile nose cones. More specifically, important uses include chemical equipment, machine and instrument housings, bottle closures, machine parts, pump impellers, electrical devices, handles, knobs, photographic equipment. See also resins, synthetic.

phenol-furfural resins. See phenolic resins.

phenolic resins. Synthetic thermosetting resins obtained by the condensation of phenol or substituted phenols with aldehydes such as formaldehyde, acetaldehyde, and furfural. Phenol-formaldehyde resins (q.v.) are typical and constitute the chief class of phenolics. Phenol-furfural resins exhibit a somewhat sharper transition from the soft, thermoplastic stage to the cured, infusible state and can be fabricated by injection molding since they have little tendency to harden before the actual curing conditions are reached.

1,2-phenolmethylol. See salicyl alcohol.

phenolphthalein $(C_6H_4OH)_2C_2O_2C_6H_4$ (an approximation). 3,3-bis(para-Hydroxyphenyl)-phthalide.
Properties: A pale yellow, crystalline powder; forms an almost colorless solution in neutral or acid solution and a bright purple-carmine solution in presence of alkali, but colorless in the presence of large amounts of alkali. Soluble in alcohol, ether, and alkalies; insoluble in water. Sp.gr. 1.2765; m.p. 261°C.
Derivation: Fused phenol is added to phthalic acid anhydride dissolved in concentrated sulfuric acid (cooled) and the whole heated 10 to 12 hours, then poured hot into boiling water and boiled with repeated changes of water. The residue is dissolved in warm dilute caustic soda and precipitated with acetic acid.
Method of purification: Recrystallization from absolute alcohol after filtering through animal charcoal.
Impurities: Phenol; phthalic acid.
Grades: Technical; pure reagent; N.F. XI.
Containers: 1-lb bottles; 1-lb tins; 5-, 25-lb cans; 50-, 100-, 250-lb drums.
Uses: Dyes; acid-base indicator in volumetric analysis (see indicators); medicine; proprietary laxatives.

phenol red. See phenolsulfonphthalein.

phenols, high boiling. See high boiling phenols.

phenolsulfonic acid (sulfocarbolic acid) $HOC_6H_4SO_3H$.
Properties: Yellowish liquid, becoming brown on exposure to air. A mixture of ortho- and para-phenolsulfonic acids. Soluble in water and in alcohol.
Derivation: By the action of sulfuric acid on phenol.
Grades: Technical; reagent.
Containers: Iron drums.
Uses: Water analysis; starting point in manufacture of intermediates and dyes; pharmaceutical (disinfectant, drugs, starting point in manufacture of various products).

phenolsulfonphthalein (phenol red) $(C_6H_4OH)_2COSO_2C_6H_4$ (an approximation). 3,3-bis(para-Hydroxyphenyl)-2,1,3H-benzoxathiole 1-dioxide. The names are also applied to the water-soluble sodium salt.
Properties: Bright to dark red crystalline powder. Stable in air. Slightly soluble in water, alcohol, and acetone; almost insoluble in chloroform and ether; soluble in alkali hydroxides and carbonates.
Derivation: Reaction of phenol with sulfobenzoic acid anhydride. Differs from phenolphthalein in containing an SO_2 group in place of CO.
Grades: U.S.P. XVI; technical; reagent.
Uses: Acid-base indicator in chemical analysis; diagnostic reagent in medicine.
See indicators.

phenol trinitrate. See picric acid.

"Phenoplast." [289] A cold-setting liquid phenolic resin-coating.

phenothiazine (thiodiphenylamine) $C_{12}H_9NS$ (tricyclic).
Properties: Grayish-green to greenish-yellow powder, granules or flakes. Tasteless; slight odor. Soluble in ether, benzene; slightly soluble in alcohol; insoluble in water. M.p. 175-185°C; b.p. 371°C; sublimes 130°C (1 mm).
Derivation: By reaction of diphenylamine and sulfur in presence of an oxidizing catalyst.
Grades: Technical; N.F. XI.
Containers: 175-lb wooden barrels; fiber drums.
Uses: Insecticide; vermifuge in livestock; manufacture of dyes.
Caution! May cause skin irritation. MCA warning label.
Shipping regulations: None.*

phenoxyacetic acid $C_6H_5OCH_2COOH$.
Properties: Light tan powder; b.p. 285°C; m.p. 98°C; soluble in ether, water, methanol, carbon disulfide, glacial acetic acid.
Uses: Intermediate for dyes, pharmaceuticals, pesticides, other organics; fungicides.

phenoxybenzamine hydrochloride (N-(2-chloroethyl)-N-(1-methyl-2-phenoxyethyl)benzylamine hydrochloride) $C_6H_5OCH_2CH(CH_3)N(CH_2C_6H_5)(CH_2)_2Cl \cdot HCl$.
Properties: Crystals; m.p. 137.5-140°C; soluble in alcohol, propylene glycol; slightly soluble in water.
Grade: N.N.D.
Use: Medicine.

phenoxydihydroxypropane. See phenoxypropanediol.

alpha-**phenoxyethylpenicillin.** See penicillin.

phenoxymethyl penicillin. See penicillin.

phenoxypropanediol (1-phenoxypropanediol-2,3; glyceryl alpha-monophenyl ether) $C_6H_5OCH_2CHOHCH_2OH$.
Properties: White crystalline solid; m.p. 53°C; b.p. 150-155°C (4 mm); soluble in water, alcohol, glycerine, carbon tetrachloride, warm benzene; insoluble in gasoline.
Derivation: Phenol and glycerol.
Uses: Medicine; plasticizer; resins; lacquers.

phenoxypropylene oxide $C_6H_5OCH_2\overline{CHCH_2O}$.
Properties: Practically colorless liquid with characteristic odor. Very slightly soluble

*See "I.C.C. Shipping Regulations," page xiii.
Reference numbers refer to name of manufacturer. See "List of Manufacturers," page v.

in water; sp.gr. 1.1110 (20/20°C); b.p. 244.2°C (760 mm); vapor pressure less than 0.1 mm (20°C); f.p. 2.8°C; viscosity 6.93 cps (20°C).

2-phenoxyquinizarin-3,4'-disulfonic acid. Reagent for the spectrophotometric determination of beryllium. Forms a stable violet complex. The reaction has been applied to the spectophotometric determination of beryllium in beryl, copper alloys, aluminum metal, and spiked samples of bronze and steel.

phensuximide (N-methyl-2-phenylsuccinimide) $C_6H_5C_4H_3O_2NCH_3$.
Properties: White crystalline solid; m.p. 71-73°C. Slightly soluble in water; readily soluble in methanol, ethanol.
Derivation: Prepared by the action of methylamine on phenylsuccinic acid.
Grade: N.N.D.
Use: Medicine.

phentolamine hydrochloride (tolylhydroxyphenylaminomethylimidazoline hydrochloride) $C_{17}H_{19}N_3O \cdot HCl$. 2-[N-(meta-Hydroxyphenyl)-para-toluidino methyl]-2-imidazoline hydrochloride.
Properties: White or slightly off-white, odorless, bitter, crystalline powder; m.p. 238-242°C; slightly soluble in alcohol and water; very slightly soluble in chloroform and ether; practically insoluble in acetone and ethyl acetate; aqueous solutions unstable.
Grades: N.F. XI.
Use: Medicine.

phentolamine methanesulfonate $C_{17}H_{19}N_3O \cdot CH_4SO_3$.
Properties: White, odorless, bitter, crystalline powder; m.p. 175-180°C; freely soluble in water; soluble in alcohol; very slightly soluble in acetone and chloroform; practically insoluble in ethyl acetate; stable when protected from moisture.
Grade: U.S.P. XVI.
Use: Medicine.

"Phenurone." [3] Trademark for phenacemide (q.v.).

phenylacetaldehyde (ethylalbenzene; hyacinthin; alpha-toluic aldehyde) $C_6H_5CH_2CHO$.
Properties: Colorless liquid; very strong hyacinth-like odor. Soluble in 2 parts of 80% alcohol; soluble in ether; very slightly soluble in water. Sp.gr. 1.023 to 1.030; m.p. below −10°C; b.p. 193 to 194°C; refractive index 1.520-1.530. Due to the ease with which this product polymerizes, these constants may not hold true after a period of shelf life following manufacture and purchase.
Derivation: From phenyl-alpha-chloroacetic acid, by action of alkalies; by heating phenyl lactic acid with dilute sulfuric acid; by oxidation of phenylethyl alcohol.
Method of purification: Rectification.
Grades: Technical; 50% solution in benzyl alcohol.
Containers: 1-, 5-lb bottles.
Use: Perfumes.

phenylacetaldehyde dimethylacetal. See "Viridine."

phenylacetamide (alpha-phenylacetamide) $C_6H_5CH_2CONH_2$.
Properties: White crystals; b.p. 280-290°C (decomposes); m.p. 156-160°C; soluble in hot water and alcohol; very slightly soluble in cold water and ether.
Derivation: Partial hydrolysis of benzyl cyanide; dehydration of ammonium phenyl acetate; Willgerodt reaction with acetophenone or styrene.
Method of purification: Crystallization.
Containers: Fiber drums.
Use: Organic synthesis; pharmaceuticals; penicillin precursors.

N-phenylacetamide. See acetanilide.

phenyl acetate (acetyl phenol) $C_6H_5OOCCH_3$.
Properties: Water-white liquid; infinitely soluble in alcohol and ether; very slightly soluble in water. Sp. gr. 1.073 (25/25°C); boiling point 195-196°C.
Derivation: (a) From phenol and acetyl chloride. (b) By heating triphenyl phosphate with potassium acetate and alcohol. (c) By heating lead acetate and phenol with carbon disulfide.
Method of purification: Fractional distillation.
Containers: Glass bottles; carboys.
Uses: Solvent; organic synthesis.
Shipping regulations: None.*

phenylacetic acid (alpha-toluic acid) $C_6H_5CH_2CO_2H$.
Properties: Shiny, white plate crystals; sp.gr. 1.0809; congealing point 74.5-76.5°C; b.p. 262°C; soluble in alcohol and ether.
Derivation: From benzyl cyanide.
Method of purification: Crystallization.
Grades: Technical.
Containers: 1-, 5-lb bottles; 5-, 25-lb tins.
Uses: Perfume; medicine; manufacture of penicillin; plant hormones.
Shipping regulations: None.*

phenylacetic acid nitrile. See benzyl cyanide.

phenyl acetylsalicylate (acetylsalol) $CH_3COOC_6H_4COOC_6H_5$.
Properties: Tasteless, odorless, white powder; decomposed by alkalies; insoluble in water; soluble in alcohol and ether. M.p. 97°C.
Use: Medicine.
Shipping regulations: None.*

phenylacetylurea. See phenacemide.

beta-**phenylacrylic acid.** See cinnamic acid.

phenylalanine (alpha-amino-beta-phenyl-propionic acid) $C_6H_5CH_2CH(NH_2)COOH$. An essential amino acid. Occurs as a racemic mixture of its optical isomers.
Properties: L(-)-phenylalanine: Plates and leaflets from concentrated aqueous solutions; hydrated needles from dilute aqueous solutions; decomposes 283°C; soluble in water; slightly soluble in methanol and ethanol.

*See "I.C.C. Shipping Regulations," page xiii.
Reference numbers refer to name of manufacturer. See "List of Manufacturers," page v.

D(+)-phenylalanine: Leaflets from water; decomposes 285°C; soluble in water; slightly soluble in methanol.
DL-phenylalanine: Leaflets or prisms from water or alcohol; sweet tasting; decomposes 318-320°C; soluble in water.
Sources: L(-)-Phenylalanine is isolated commercially from proteins (ovalbumin, lactalbumin, zein, and fibrin). DL-Phenylalanine is synthesized from the azlactone of alpha-benzoylaminocinnamic acid or from alpha-acetaminocinnamic acid.
Containers: Glass bottles; fiber cans.
Use: Biochemical research.
Available commercially as DL-dihydroxyphenylalanine and as DL-phenylalanine.

phenylallylic alcohol. See cinnamic alcohol.

phenylamine. See aniline.

phenylamine acetate. See aniline acetate.

phenylamino cadmium dilactate.
Use: Seed disinfectant.

phenyl-2-amino-5-naphthol-7-sulfonic acid (phenyl J acid; 6-anilino-1-naphthol-3-sulfonic acid) $HOC_{10}H_5(NHC_6H_5)(SO_3H)$.
Properties: Slate-colored crystals; soluble in alkali.
Derivation: Is prepared from H acid and aniline (condensation with heat).
Method of purification: Recrystallization.
Grades: Technical.
Containers: Wooden barrels.
Use: Dyes.
Shipping regulations: None.*

phenyl-2-amino-8-naphthol-6-sulfonic acid (phenyl gamma acid; 7-anilino-1-naphthol-3-sulfonic acid) $HOC_{10}H_5(NHC_6H_5)(SO_3H)$.
Properties: Gray crystals; soluble in alkali.
Derivation: Is prepared from gamma acid and aniline (condensation with heat).
Method of purification: Recrystallization.
Grades: Technical.
Containers: Wooden barrels.
Use: Dyes.
Shipping regulations: None.*

phenyl-2-aminopropane. See amphetamine.

phenyl para-**aminosalicylate** $C_6H_5CO_2C_6H_3OHNH_2$. Colorless crystals; m.p. 153°C.

N-phenylaniline. See diphenylamine.

ortho-**phenylaniline.** See ortho-aminobiphenyl.

phenylarsonic acid $C_6H_5AsO(OH)_2$.
Properties: Crystalline powder; m.p. 160°C with decomposition; soluble in water and alcohol; insoluble in chloroform.
Preparation: The Bart reaction between the diazonium salt and sodium arsenite.
Use: Analytical reagent for tin.

phenylazoaniline. See aminoazobenzene.

phenylbarbital. See phenobarbital.

phenylbenzamide. See benzanilide.

phenylbenzoylcarbinol. See benzoin.

phenylbiphenyloxadiazole. See 1,3,4-phenylbiphenylyloxadiazole.

1,3,4-phenylbiphenylyloxadiazole (PBD; phenylbiphenyloxadiazole) $C_{20}H_{14}N_2O$.
Properties: Crystals; m.p. 166-168°C.
Grade: Purified.
Uses: As primary fluors or as wave length shifters in solution scintillators.

phenylbis[1-(2-methyl)aziridinyl]-phosphine oxide (phenyl "MAPO") $C_6H_5(C_3H_6N)_2PO$.
Properties (technical): Straw-colored liquid; limited solubility in water; soluble in most organic solvents. The pure material is a low-melting solid.
Containers: 5- and 55-gal containers.
Uses: Polymerization additive.

phenylboric acid $C_6H_5B(OH)_2$.
Properties: White solid; m.p. 214-216°C.
Derivation: Reaction of phenylmagnesium bromide with boron esters.

phenyl bromide. See bromobenzene.

2-phenylbutane. See sec-butylbenzene.

phenylbutazone (4-butyl-1,2-diphenyl-3,5-pyrazolidinedione; butazolidine) $C_{19}H_{20}N_2O_2$. A synthetic pyrazolone derivative.
Properties: White or very light yellow powder; slightly bitter taste and very slight aromatic odor; m.p. 103-106°C; freely soluble in acetone, ether and ethyl acetate; very slightly soluble in water; stable if stored at room temperature in closed containers in absence of moisture.
Grade: N.N.D.
Use: Medicine.
Also available as sodium salt.

1-phenylbutene-2 $C_6H_5CH_2CH{:}CHCH_3$.
Properties: Boiling range 174-176°C; sp.gr. 0.888 (60/60°F); refractive index 1.511 (n 20/D); flash point 71°C.
Grades: 95%.
Containers: Bottles.
Shipping regulations: Red label not required.*

phenylbutynol (3-phenyl-1-butyn-3-ol) $HC{:}CC(C_6H_5)(OH)CH_3$.
Properties: Crystals; camphor odor; m.p. 51-52°C; b.p. 217-218°C; sp.gr. 1.0924 (20/20°C); slightly soluble in water; soluble in acetone, benzene, most organic solvents.
Uses: Acid inhibitor; organic synthesis.

2-phenylbutyric acid. See phenylethylacetic acid.

1-phenyl-3-carbethoxy-pyrazolone-5 $C_{12}H_{12}N_2O_3$.
Properties: White to light buff powder, stable in aqueous solution. M.p. 182-188°C.
Containers: Drums.
Use: Dyestuff intermediate.

phenyl carbimide. See phenylisocyanate.

phenylcarbinol. See benzyl alcohol.

phenylcarbylamine chloride $C_6H_5NCCl_2$.
Properties: Pale-yellow, oily liquid; onion-like odor. Mildly volatile. Caution! Very irritant! Soluble in alcohol, benzene,

*See "I.C.C. Shipping Regulations," page xiii.
Reference numbers refer to name of manufacturer. See "List of Manufacturers," page v.

ether; insoluble in water. Sp.gr. 1.30 at 15°C; b.p. 208-210°C; vapor density 6.03; coefficient of thermal expansion 0.000895; volatility 2100 mg/cu m (20°C).

Derivation: Chlorination of phenylisothiocyanate.

Grades: Technical.

Uses: Organic synthesis; military poison gas.

Shipping regulations: Poison, class A. Poison gas label. Not accepted by express.*

phenyl "Cellosolve." [214] Trademark for ethylene glycol monophenyl ether, $C_6H_5OC_2H_4OH$ (q.v.).

phenyl chloride. See chlorobenzene.

phenyl chloroform. See benzotrichloride.

phenyl chloromethylketone. See chloroacetophenone.

phenylcinchoninic acid. See cinchophen.

phenyl cyanide. See benzonitrile.

phenylcyclohexane (cyclohexylbenzene) $C_6H_5C_6H_{11}$. 1,2,3,4,5,6-Hexahydrobiphenyl.

Properties: Colorless, oily liquid; slight pleasant odor; sp.gr. 0.938 (25/15°C); m.p. 5°C; b.p. 237.5°C; refractive index (n 25/D) 1.523; flash point 210°C; insoluble in water, glycerine; very soluble in alcohol, acetone, benzene, carbon tetrachloride, castor oil, hexane, xylene.

Uses: High boiling solvent; penetrating agent; intermediate.

2-phenylcyclohexanol $HO\overline{CH(CH_2)_4C}H(C_6H_5)$.

Properties: Colorless to pale, straw-colored liquid; pour point −18°C; b.p. 276-281°C; sp.gr. 1.033 (25/25°C); refractive index 1.536 (n 25/D); flash point 280°F; very slightly soluble in water; soluble in methanol, ether.

Uses: Solvent; intermediate.

phenyldichloroarsine $C_6H_5AsCl_2$.

Properties: Liquid; microcrystalline mass at the m.p. Decomposed by water. Caution! Very irritant! Soluble in alcohol, benzene, and ether; insoluble in water. Sp.gr. 1.654 (20°C); b.p. 255-257°C; m.p. −20°C; vapor tension 0.014 mm (15°C); volatility 404 mg/cu m (20°C); coefficient of thermal expansion 0.00073.

Derivation: Arsenic trichloride and phenylmercuric chloride are heated together for 4 to 5 hours at 100°C.

Grades: Technical.

Containers: Steel bottles.

Uses: Military poison gas; solvent for diphenylcyanoarsine.

Shipping regulations: Poison, class B. Poison label.*

phenyl diethanolamine $C_6H_5N(C_2H_4OH)_2$.

Constants: M.p. 58°C; b.p. 190°C (1 mm); vapor pressure < 0.01 mm (20°C); wt 10.0 lbs/gal (20°C); sp.gr. 1.1203 at 60/20°C; viscosity 1.19 poise (20°C). Slightly soluble in water; soluble in ethyl alcohol and acetone.

Grades: Technical.

Containers: 1-gal cans; 5-, 55-gal drums; tank cars.

Uses: Organic synthesis; dyestuffs.

1-phenyl-2-diethylamino-1-propanone hydrochloride. See diethyl propion.

phenyl diglycol carbonate [diethylene glycol bis(phenyl carbonate)] $C_{18}H_{18}O_7$.

Properties: Colorless solid; sp.gr. 1.23 (20/4°C); m.p. 40°C; b.p. 225-229°C (2 mm); viscosity 1810 cps (supercooled) at 20°C; refractive index (n 20/D) 1.525; evaporation rate 0.026 mg/sq cm/hr at 100°C; insoluble in water (very stable to hydrolysis); widely soluble in organic solvents; compatible with many resins and plastics.

phenyl dihydrogen phosphate, disodium salt (disodium phenyl phosphate) $C_6H_5OPO_3Na_2$. This will release phenol quantitatively in the presence of phosphatase, so can be used to analyze for the latter.

phenyldimethylisopyrazolone. See antipyrine.

1-phenyl-2,3-dimethyl-5-pyrazolone-4-methylaminosulfonate sodium. See dipyrone.

3-phenyl-1,1-dimethylurea (fenuron) $C_6H_5NHCON(CH_3)_2$.

Properties: White crystalline solid; almost insoluble in water (3850 ppm at 25°C); sparingly soluble in hydrocarbon solvents. Stable towards oxidation and moisture. M.p. 127-129°C.

Use: Weed and brush killer.

phenylenediamine (diaminobenzene) $C_6H_4(NH_2)_2$.

(a) ortho- : (orthamine, ortho-diaminobenzene).

Properties: Colorless monoclinic crystals; darkens in air; m.p. range 102-104°C; b.p. 252-258°C; soluble in alcohol, ether, water, and chloroform; somewhat toxic.

Uses: Manufacture of dyes; photographic developing agent; organic synthesis.

(b) meta-:

Properties: Colorless needles; unstable in air; usually in the form of the stable hydrochloride; sp.gr. 1.1389; m.p. 63°C; b.p. 282-287°C; soluble in alcohol, ether, and water.

Uses: Dyestuff manufacture; reagent for detecting nitrous acid; textile developing agent.

(c) para-: (para-diaminobenzene).

Properties: White to light purple crystals (oxidizes on standing in air to purple and black); m.p. about 147°C; b.p. 267°C; soluble in alcohol, ether; slightly soluble in cold water and chloroform; affected by light.

Uses: Azo dyestuff intermediate; photographic developing agent; used in hair and fur dyes and in photochemical measurements; accelerator for vulcanization; chemical analysis.

Caution! May cause skin irritation. MCA warning label. (meta and para isomers).

Derivation: Reduction of ortho-, meta-, or

para-dinitrobenzenes or nitroanilines with iron and hydrochloric acid. Purified by crystallization.
Grades: Technical.
Containers: Drums (steel or fiber) (100 lbs net; 114 lbs gross); bottles.
Shipping regulations: None.*

phenylephrine hydrochloride (ℓ-1-(meta-hydroxyphenyl-2-methylaminoethanol hydrochloride)
$HOC_6H_4CH(OH)CH_2NHCH_3 \cdot HCl$.
Properties: White or nearly white crystals; odorless; has bitter taste; solutions are acid to litmus paper; freely soluble in water and in alcohol; m.p. 140-145°C; levorotatory in solution.
Grade: U.S.P. XVI.
Use: Medicine.

phenylephrine tartrate (para-methylamino-ethanolphenol tartrate).
Properties: White crystals; freely soluble in water; m.p. 182-185°C.
Use: Medicine.

phenylethane. See ethylbenzene.

2-phenylethanol. See phenethyl alcohol.

phenylethanolamine $C_6H_5NHCH_2CH_2OH$.
Constants: Sp.gr. 1.0970 (20/20°C); b.p. (760 mm) 285.2°C; vapor pressure < 0.01 mm (20°C); wt 9.1 lbs/gal (20°C); f.p. 10.6°C; viscosity 1.01 poise (20°C).
Typical specifications: Sp.gr. 1.094-1.099 (20/20°C); boiling range 280-290°C (760 mm).
Grades: Technical.
Containers: 1-gal cans; 5-, 55-gal drums. Net content 9, 45, 490 lbs.
Uses: Organic synthesis; dyestuffs.

phenyl ether. See diphenyl oxide.

phenylethyl acetate $C_6H_5CH_2CH_2OOCCH_3$.
Not the same as sec-phenylethyl acetate.
Properties: Colorless liquid; peach-like odor. Soluble in alcohol and ether. Sp.gr. 1.030-1.033; refractive index 1.497-1.498; b.p. 226°C.
Derivation: (a) Interaction of ethyl acetate and aluminum phenyl ethylate. (b) Interaction of acetic anhydride and phenylethyl alcohol in the presence of sodium acetate.
Grades: Technical.
Containers: Glass bottles.
Use: Perfumery.

sec-**phenylethyl acetate.** See styralyl acetate.

phenylethylacetic acid (2-phenylbutyric acid) $C_2H_5CHC_6H_5COOH$.
Properties: White crystals with aromatic odor; m.p. 41.0°C (min); insoluble in water; soluble in alcohol, ketones, and esters.
Use: Synthesis.

phenylethyl alcohol. See phenethyl alcohol.

beta-**phenylethylamine** (phenethylamine; 1-amino-2-phenylethane) $C_6H_5C_2H_4NH_2$.
Properties: Liquid with a fishy odor; absorbs carbon dioxide from the air; strong base; sp.gr. 0.9640; b.p. 194.5°C; soluble in water, alcohol, and ether.
Derivation: From phenylethyl alcohol and ammonia under pressure.
Containers: Drums.

phenylethyl anthranilate $H_2NC_6H_4COOC_2H_4C_6H_5$.
Properties: A colorless liquid which yellows with age, and has an odor of grape and orange; sp.gr. 1.14 (25/25°C).
Uses: Perfume; flavoring.

phenylethyl carbinol. See phenylpropyl alcohol.

phenylethylene. See styrene.

N-phenylethylethanolamine $C_6H_5N(C_2H_5)C_2H_4OH$.
Properties: A solid with m.p. 37.2°C; b.p. 268°C (740 mm); sp.gr. 1.04 (20/20°C); very slightly soluble in water. Flash point 270°F (Cleveland open cup). Soluble in alcohol, acetone, benzene.
Containers: Drums.
Uses: Solvents; chemical intermediates; preparation of dyes for acetate rayons.

phenyl ethyl ether. See phenetole.

5-phenyl-5-ethylhydantoin
$(C_6H_5)(C_2H_5)\underline{CNHCONHCO}$.
Properties: Colorless, odorless crystalline powder; m.p. 199°C; insoluble in water.
Use: Medicine.

beta-**phenylethylhydrazine dihydrogen sulfate.** See phenelzine dihydrogen sulfate.

phenylethyl isobutyrate $(CH_3)_2CHCOOC_2H_4C_6H_5$.
Properties: A colorless liquid; pleasant fragrance, resembling a somewhat fruity tearose odor; sp.gr. 0.988 (25/25°C); refractive index (n 20/D) 1.488; soluble in alcohol and ether.
Use: Perfumes.

phenylethylmalonylurea. See phenobarbital.

alpha-**phenylethyl mercaptan** $C_6H_5CH_2CH_2SH$.
Properties: Boiling range 193-225°C; sp.gr. 1.0264 (60/60°F); refractive index 1.5582 (n 20/D); flash point 71°C.
Containers: Bottles.
Shipping regulations: Flammable liquid. Red label not required.*

phenylethyl phenylacetate
$C_6H_5(CH_2)_2OOCCH_2C_6H_5$.
Properties: White crystals; hyacinth odor. Sp.gr. 1.080-1.082; congealing point 27°C.
Containers: Bottles.
Uses: Perfumery; flavors.

phenylethyl propionate $C_2H_5COOC_2H_4C_6H_5$.
Properties: A synthetic colorless liquid having a flower-fruit odor; miscible with alcohols and ether; sp.gr. 1.012 (25/25°C).
Uses: Perfumes; flavors.

phenylethyl salicylate.
Properties: Snow-white crystals; very faint aromatic odor. Soluble in 14 parts of 95% alcohol. Congealing point 41.5°C.

phenyl fluoromethylketone. See fluoroacetophenone.

phenylformamide. See formanilide.

*See "I.C.C. Shipping Regulations," page xiii.
Reference numbers refer to name of manufacturer. See "List of Manufacturers," page v.

phenylformic acid. See benzoic acid.

phenyl gamma acid. See phenyl-2-amino-8-naphthol-6-sulfonic acid.

phenylglucosazone $C_{18}H_{22}N_4O_4$.
Properties: Thin, yellow, needle-like crystals; soluble in alcohol; very slightly soluble in water. M.p. 217°C.
Derivation: By condensation of phenylhydrazine hydrochloride and glucose with subsequent crystallization.

phenylglycolic acid. See mandelic acid.

phenylhydrazine (hydrazinobenzene) $C_6H_5NHNH_2$.
Properties: Pale yellow crystals or oily liquid; becomes red-brown on exposure to air; poisonous! Soluble in alcohol, ether, chloroform, benzene, and dilute acids. Very slightly soluble in water. Sp.gr. 1.0978; m.p. 19.35°C; b.p. 243.5°C, with decomposition.
Derivation: Aniline is diazotized and then reduced producing diazobenzene hydrochloride, which is decomposed by caustic soda and dissolved in ether. The ethereal solution is dried, and the ether evaporated.
Method of purification: Vacuum distillation.
Grades: Commercial; C.P.; reagent.
Containers: Glass bottles; tins; drums.
Uses: Analytical chemistry (reagent for detecting aldehydes, sugars, etc.); organic synthesis (intermediates, dyestuffs, pharmaceuticals).
Shipping regulations: None.*

phenylhydrazine hydrochloride $C_6H_5NHNH_2 \cdot HCl$.
Properties: Colorless, crystalline scales; sublimable, if cautiously heated. Soluble in water, alcohol, and ether. M.p. 243-246°C with slight brown coloration.
Derivation: Reduction of benzene diazonium chloride with stannous chloride and hydrochloric acid.
Method of purification: Recrystallization.
Grades: Technical; reagent.
Containers: Glass bottles.
Uses: Analytical chemistry for differentiation of sugars; organic synthesis.
Caution! Avoid inhalation and exposure of skin.
Shipping regulations: None.*

alpha-**phenylhydroxyacetic acid.** See mandelic acid.

phenylic acid. See phenol.

2-phenyl-1,3-indanedione. See phenindione.

phenyl isocyanate (phenyl carbimide; carbanil) C_6H_5NCO.
Properties: Liquid; b.p. 165°C; density 1.095 (20/4°C); refractive index (n 19.6/D) 1.53684; decomposes in water and alcohol; very soluble in ether.
Use: Test reagent for identifying alcohols and amines.

phenyl isothiocyanate. See phenyl mustard oil.

phenyl J acid. See phenyl-2-amino-5-naphthol-7-sulfonic acid.

phenylmagnesium bromide C_6H_5MgBr. A Grignard reagent available as a solution in ether; sp.gr. 1.14.
Derivation: From magnesium and bromobenzene.
Containers: Glass bottles; 5-, 55-gal drums.
Uses: Arylating agent in organic synthesis.
Shipping regulations: Flammable liquid. Red label.*

phenyl "MAPO." [293] Trademark for phenylbis-[1-(2-methyl)aziridinyl]phosphine oxide (q.v.).

phenylmercuric acetate $C_6H_5HgOCOCH_3$.
Properties: White to cream prisms; m.p. 148-150°C. Slightly soluble in water; soluble in alcohol, benzene, and glacial acetic acid. Slightly volatile at ordinary temperatures.
Derivation: Action of heat on benzene and mercuric acetate.
Grades: C.P.; technical; commercial.
Containers: Fiber drums, 50 lbs net, 58 lbs gross; bottles.
Uses: Antiseptic, fungicide, herbicide.
Warning! Poisonous if inhaled or swallowed. May cause skin irritation. MCA warning label.
Shipping regulations: Poison, class B. Poison label.*

phenylmercuric borate $(C_6H_5Hg)_2HBO_3$.
Properties: White crystalline powder; m.p. 120-130°C. Slightly soluble in water; soluble in alcohol.
Derivation: Reaction of phenylmercuric acetate with boric acid.
Uses: Antiseptic, mildewproofing agent; fungicide.
Danger! Poisonous by inhalation or swallowing. May cause skin irritation.

phenylmercuric chloride C_6H_5HgCl.
Properties: White satiny crystals; m.p. 251°C. Insoluble in water; slightly soluble in hot alcohol; soluble in benzene, ether, pyridine.
Derivation: Reaction of phenylmercuric acetate and sodium chloride.
Uses: Antiseptic; fungicide; germicide.
Warning! Poisonous by inhalation or swallowing. May cause skin irritation.

phenylmercuric hydroxide C_6H_5HgOH.
Properties: Fine white to cream crystals; m.p. 197-205°C; slightly soluble in water; soluble in acetic acid; alcohol.
Typical specifications: Mercury content 68.71% (theory, 68.1%); ash 0.5% max; moisture content 5%.
Grades: Technically pure.
Containers: Bottles; fiber drums.
Uses: Manufacture of phenylmercuric salts; fungicide and germicide. Principal compound in manufacturing organic mercury derivatives; denaturant for alcohol.
Caution! Dust or strong solution causes blistering of skin unless washed off immediately.
Shipping regulations: Poison, class B. Poison label.*

phenylmercuric naphthenate
Prepared by interaction of phenylmercuric acetate and naphthenic acid, producing a colored solution. Used as a wood preservative and as a mildewproofing agent for paints and adhesives.
Danger! (Oil solution or oil-water emulsions 1% and over). Poisonous by swallowing or skin contact. May cause skin irritation or delayed chemical burns. Absorbed through skin. MCA warning label.

phenylmercuric nitrate (basic)
$C_6H_5HgNO_3 \cdot C_6H_5HgOH$.
Properties: Fine white crystals, or grayish powder; mercury content 63-65% (theory, 63.2%); melting range 175-185°C, with decomposition; ash 0.1% max; very slightly soluble in water; slightly soluble in alcohol; insoluble in ether; moderately soluble in glycerin.
Grades: N.F. XI.
Containers: Bottles; fiber drums.
Uses: Germicide, fungicide, antiseptic; denaturant for alcohol; mildew-proofing agent; preservative.
Caution! Dust or strong solution causes blistering of skin unless washed off immediately.
Shipping regulations: Poison, class B. Poison label.*

phenylmercuric oleate
$C_6H_5HgOOC(CH_2)_7CH{:}CHC_8H_{17}$.
Properties: White crystalline powder; m.p. 45°C; insoluble in water; soluble in organic solvents and some oils.
Derivation: Prepared by reaction of phenyl mercuric acetate with oleic acid.
Uses: Principally as mildewproofing agent for paints; generally as fungicide and germicide.
Danger! (Oil solution or oil-water emulsions 1% and over). Poisonous by swallowing or skin contact. May cause skin irritation or delayed chemical burns. Absorbed through skin. MCA warning label.

phenylmercuric propionate $C_6H_5HgOCOCH_2CH_3$.
Properties: Technical grades: White to off-white waxlike free flowing powder; m.p. 149-158°F; stable to 392°F for short periods; 57% min Hg content.
Containers: ½-lb container to 50-lb fiber drums.
Caution! Handle with care associated with all organic mercury compounds (dust mask, goggles, gloves).
Uses: Fungicide and bactericide for paints and industrial finishes.
Shipping regulations: Poison, class B. Poison label.*

phenylmercuric salicylate
$C_6H_4(OH)(COOHgC_6H_5)$. Toxic!
Uses: Seed disinfectant.

phenylmercuriethanolammonium acetate
$[(HOC_2H_4)NH_2(C_6H_5Hg)]OOCCH_3$.
Properties: White crystalline solid; soluble in water.
Derivation: Reaction of phenylmercuric acetate with monoethanolamine.
Use: As eradicant fungicide on fruit. Formulated as liquid, dust or soluble powder.
Warning: Poisonous if swallowed. May cause skin irritation.

phenylmercuritriethanolammonium lactate
[tris(2-hydroxyethyl)(phenylmercuri)ammonium lactate]
$[(HOC_2H_4)_3NHgC_6H_5]OOCCHOHCH_3$.
Properties: White crystalline solid; soluble in water.
Derivation: Reaction of phenylmercuric acetate with triethanolamine and lactic acid.
Uses: As turf fungicide and as eradicant fungicide for fruit trees.
Warning: Poisonous if swallowed. May cause skin irritation.

phenylmercury formamide $HCONHHgC_6H_5$. Toxic!
Uses: Seed disinfectant.

phenylmercury urea $C_6H_5HgNHCONH_2$. Toxic!
Uses: Disinfectant and fungicide for seed treatment.

phenylmethane. See toluene.

phenylmethyl acetate. See benzyl acetate.

1-phenyl-2-methylaminopropane hydrochloride. See methamphetamine hydrochloride.

1-phenyl-2-methylaminopropanol. See ephedrine.

1-phenyl-2-methylaminopropanol hydrochloride. See ephedrine hydrochloride.

1-phenyl-2-methylaminopropanol sulfate. See ephedrine sulfate.

phenylmethyl carbinol. See styralyl alcohol.

phenylmethyl carbinyl acetate. See styralyl acetate.

N-phenylmethylethanolamine
$C_6H_5N(CH_3)C_2H_4OH$.
Properties: Liquid which sets to a glass at −30°C; b.p. 192°C (100 mm); sp.gr. 1.0661 (20/20°C); slightly soluble in water; flash point 280°F (Cleveland open cup).
Uses: Solvents; chemical intermediate; preparation of dyes for acetate fibers.

phenyl methyl ketone. See acetophenone.

1-phenyl-3-methyl-5-pyrazolone (3-methyl-1-phenyl-5-pyrazolone)
$C_6H_5NN{:}C(CH_3)CH_2CO$.
Properties: White powder or crystals; soluble in water; slightly soluble in alcohol or benzene; insoluble in ether.
Constants: B.p. 287°C (205 mm); m.p. 127°C; vapor pressure < 0.01 mm (20°C).
Derivation: By condensation of phenylhydrazine with ethylacetoacetate.
Method of purification: Crystallization.
Grades: Technical; C.P.
Containers: Fiber drums.
Uses: Intermediate for drugs and dyes.
Shipping regulations: None.*

phenyl alpha-methylstyryl ketone. See dypnone.

*See "I.C.C. Shipping Regulations," page xiii.
Reference numbers refer to name of manufacturer. See "List of Manufacturers," page v.

N-phenylmorpholine $C_6H_5NCH_2CH_2OCH_2CH_2$.
Properties: White solid. Soluble in water.
Constants: B.p. (760 mm) 268°C; m.p. 57°C; vapor pressure < 0.1 mm (20°C).
Grades: Technical.
Use: Chemical intermediate in manufacture of dyestuffs, rubber accelerators, corrosion inhibitors and photographic developers.

phenyl mustard oil (thiocarbanil; phenyl isothiocyanate; phenylthiocarbonimide) C_6H_5NCS.
Properties: A pale yellow or colorless liquid; penetrating, irritating odor; readily volatilized with steam. Soluble in alcohol and ether; insoluble in water.
Constants: Sp.gr. 1.1382; m.p. −21°C; b.p. 221°C.
Derivation: (a) By action of concentrated hydrochloric acid on sulfocarbanilide; (b) by reaction of thiophosgene with aniline.
Method of purification: Rectification.
Grades: Technical.
Containers: Tins; glass bottles.
Uses: Medicine; organic synthesis.
Shipping regulations: None.*

N-phenyl-alpha-naphthylamine $C_{10}H_7NHC_6H_5$.
Properties: Crystallizes in prisms; white to slightly yellowish. Soluble in alcohol, ether, and benzene.
Constants: M.p. 62°C; b.p. 335°C (260 mm).
Derivation: From alpha-naphthylamine and aniline.
Method of purification: Distillation.
Grades: Technical.
Containers: Wooden barrels; fiber cans.
Uses: Dyes and other organic chemicals; rubber antioxidant.
Shipping regulations: None.*

phenyl-beta-naphthylamine $C_{10}H_7NHC_6H_5$.
Properties: "Pure grade": Light gray powder; set point 107°C (min); purity 99.25% min; sp.gr. 1.20; ash 0.25% max. Insoluble in water; soluble in alcohol, acetone, benzene, monochlorobenzene, isopropyl acetate and gasoline.

phenyl-1-naphthylamine-8-sulfonic acid $C_{16}H_{13}NO_3S$.
Properties: Greenish-gray needles. Rather insoluble in water; soluble in alcohol.
Derivation: Arylation of 1-naphthylamine-8-sulfonic acid with aniline.
Grades: Technical; mostly as sodium salt.
Containers: Barrels or steel drums.
Use: Azo colors.
Shipping regulations: None.*

ortho-phenylphenol (ortho-hydroxydiphenyl; ortho-xenol) $C_6H_5C_6H_4OH$.
Properties: Nearly white or light buff crystals; m.p. 56-58°C; b.p. 280-284°C. Soluble in alcohol, sodium hydroxide solution; insoluble in water.
Derivation: From reaction of chlorobenzene and caustic soda solution at elevated temperatures and pressure.
Method of purification: Distillation.
Grades: Technical; C.P.
Containers: Barrels; drums.
Uses: Intermediate for dyes; germicide; fungicide; rubber industry; food packaging.
Shipping regulations: None.*

para-phenylphenol (para-hydroxydiphenyl; para-xenol) $C_6H_5C_6H_4OH$.
Properties: Nearly white crystals; m.p. 164-165°C; b.p. 308°C. Soluble in alcohol, also in alkalies and most organic solvents; insoluble in water.
Derivation: From reaction of chlorobenzene and caustic soda solution at elevated temperatures and pressure.
Method of purification: Crystallization.
Grades: Technical; C.P.
Containers: Bags, 100-lb barrels and drums.
Uses: Intermediate for dyes; resins; rubber chemicals; fungicide.
Shipping regulations: None.*

N-phenyl-para-phenylenediamine. See para-aminodiphenylamine.

phenylphosphinic acid. See benzenephosphinic acid.

phenylphosphonic acid. See benzenephosphonic acid.

N-phenylpiperazine $C_6H_5NCH_2CH_2NHCH_2CH_2$.
Properties: Pale yellow oil; insoluble in water; soluble in alcohol and ether; sp.gr. 1.0621 (20/4°C); b.p. 286.5°C (760 mm), 156-7°C (10 mm); m.p. 18.8°C; flash point 285°F.
Uses: Intermediate for pharmaceuticals, anthelmintics, surface active agents, synthetic fibers.
Shipping regulations: Poison class B. Poison label.*

phenylpropanolamine hydrochloride (dℓ-norephedine hydrochloride; alpha-(1-aminoethyl)-benzyl alcohol hydrochloride) $C_6H_5CH(OH)CH(CH_3)(NH_2)\cdot HCl$.
Properties: White crystalline powder with odor similar to benzoic acid; m.p. 190-194°C; freely soluble in alcohol and water; insoluble in benzene, chloroform, and ether; aqueous solution neutral to litmus.
Use: Medicine.

1-phenylpropanone-1. See propiophenone.

3-phenylpropenal. See cinnamic aldehyde.

3-phenylpropenol. See cinnamic alcohol.

phenylpropiolic acid $C_6H_5C{:}CCOOH$.
Properties: Colorless needles. Keep dark. (Used in the form of sodium phenylpropiolate). Soluble in alcohol and ether; insoluble in water.
Constants: M.p. 136-137°C; b.p. sublimes.
Derivation: By heating ethylcinnamic dibromide with alcoholic potash.
Method of purification: Crystallization.

phenylpropyl acetate (hydrocinnamyl acetate) $C_6H_5CH_2CH_2CH_2OOCCH_3$.
Properties: Soluble in 70% alcohol.
Constants: Sp.gr. 1.012-1.016; refractive index 1.497.
Grades: Technical.

Containers: Glass bottles.
Use: Perfumery.
Shipping regulations: None.*

phenylpropyl alcohol (hydrocinnamic alcohol; phenylethyl carbinol) $C_6H_5CH_2CH_2CH_2OH$.
Properties: Colorless liquid with sweet floral odor; b.p. 219°C. Soluble in 70% alcohol; insoluble in water.
Constants: Sp.gr. 0.998-1.000; refractive index 1.524-1.528.
Containers: Glass bottles.
Use: Perfumery.
Shipping regulations: None.*

phenylpropyl aldehyde (hydrocinnamic aldehyde) $C_6H_5CH_2CH_2CHO$.
Properties: Colorless liquid, with floral odor of hyacinth. Soluble in 50% alcohol.
Constants: Sp.gr. 1.010-1.020; refractive index 1.520-1.532.
Grade: Chlorine-free.
Containers: Glass bottles; copper flasks.
Uses: Perfumery; flavors.
Shipping regulations: None.*

phenylpropylmethylamine (N, beta-dimethylphenylethylamine) $C_6H_5CH(CH_3)CH_2NHCH_3$.
Properties: A colorless to pale yellow liquid; 98% distils between 205-210°C; very soluble in alcohol, benzene, and ether; 1.2 parts dissolve in 100-parts water; aqueous solutions alkaline to litmus.
Grade: N.N.D.
Use: Medicine.

phenylpropylmethylamine hydrochloride $C_6H_5CH(CH_3)CH_2NHCH_3 \cdot HCl$. N, beta-Dimethylphenylethylamine hydrochloride.
Properties: Not available in dry state. Solution is clear, colorless and nearly odorless; pH between 5.5 and 6.5.
Grade: N.N.D.
Use: Medicine.

1-phenyl-3-pyrazolidone $C_6H_5\overline{NNHC(O)CH_2C}H_2$.
Properties: Crystals; m.p. 121°C.
Use: Photographic developer.

2-phenylquinoline-4-carboxylic acid. See cinchophen.

phenyl salicylate. See salol.

phenylsulfonic acid. See benzenesulfonic acid.

4-phenyl-1,4-thiazane $\underline{SCH_2CH_2N(C_6H_5)CH_2CH_2}$.
Properties: White powder. Soluble in hot toluene. M.p. 108-111°C.
Derivation: Interaction of dichlorodiethyl sulfide and an aliphatic amine in the presence of alcohol and sodium carbonate.

phenylthiocarbonimide. See phenyl mustard oil.

phenyltrichlorosilane $C_6H_5SiCl_3$.
Properties: Colorless liquid. B.p. 201°C; sp.gr. 1.321 (25/25°C); refractive index (n 25/D) 1.5240; flash point (Cleveland open cup) 185°F. Readily hydrolyzed by moisture, with the liberation of hydrochloric acid.
Derivation: By Grignard reaction of silicon tetrachloride and phenylmagnesium chloride; reaction of benzene with trichlorosilane; of chlorobenzene, silicon and copper.
Grades: Technical.
Containers: ½-, 1-, 10-lb bottles; 100-lb drums.
Use: Intermediate for silicones.
Shipping regulations: Corrosive liquid. White label.*

phenylurethane. See ethyl phenylcarbamate.

phenyl valerate $C_4H_9COOC_6H_5$. Colorless liquid; slightly soluble in water; soluble in alcohol and ether. Used in flavors and odorants.

phenytoin, soluble. See diphenylhydantoin sodium.

"Philback A." [303] Trademark for fast extrusion oil furnace carbon black, a reinforcing ingredient in natural and synthetic rubbers, contributing to processing of unvulcanized rubbers and to abrasion resistance, tensile and tear strength. Disperses heat and is non-staining.
Containers: 50-lb paper bags or hopper cars. Pelleted.
Uses: In inner tubes, molded and extruded goods, tire treads, sidewalls, carcass and tread rubber.

"Philback E." [303] Trademark for super abrasion oil furnace carbon black.
Containers: 50-lb paper bags or hopper cars. Pelleted.
Uses: In synthetic or natural rubber goods requiring high tensile strength and extreme abrasion resistance, such as tires, tread rubber, conveyor belts, and ability to withstand cracking, cutting and chipping. Good electrical conductivity.

"Philback I." [303] Trademark for intermediate super abrasion oil furnace carbon black.
Containers: 50-lb bags or hopper cars. Pelleted.
Uses: In natural and synthetic rubber goods for high tensile, tear and abrasion resistance properties. For tire treads, conveyor belts and mechanical goods. Good electrical conductivity. Excellent flex life and hot tensile. Easy processing.

"Philback O." [303] Trademark for high abrasion oil furnace carbon black. Pelleted.
Uses: In natural and synthetic rubber tires, conveyor belts, industrial hoses.

Philippine copal. A class of soft copal (q.v.) of semi-recent origin, obtained from the Philippines.
Containers: Bags.

Philippine physic nut oil (physic nut oil).
Properties: Sp.gr. 0.9820 (30/4°C); refractive index 1.4665; iodine value (Hanus) 94.8; saponification value 192.4; unsaponifiable matter 0.45; acid value 5.1; saturated acids (corr), 16.82%; unsaturated acids (corr), 78.0%; iodine value, unsaturated acids 110.0; the composition of the mixed fatty acid is oleic, 61.86; linoleic, 18.65; myristic, 0.45; palmitic, 11.84; stearic, 5.07%.
Derivation: From the kernels of the nut of

the small tree (Jatropha curcas) which is grown in the Philippines as hedge plant.
Containers: Steel drums.
Uses: Soap making; medicine.

"Philprene." [303] Trademark for a series of styrene-butadiene type synthetic rubbers, hot and cold, oil-extended, pigmented and non-pigmented, including special types for specific needs. Produced by emulsion polymerization of butadiene and styrene. Staining and non-staining types.
Containers: 70-90-lb bales.
Uses: Tire carcasses and treads, molded and extruded goods, sporting goods, footwear, coated fabrics, wire and cable jackets, hospital goods, floor tile, insulation.

"Phi-O-Sol WA." [328] Trademark for a light amber liquid compatible with mildly acid or alkaline solutions and hard water up to 75 ppm, and readily soluble in water in all proportions. It is the sodium salt of the sulfonated ester of butyl oleate and is a very effective wetting, penetrating and rewetting agent. Shipped in 55-gal drums.

phlogopite (magnesium mica, amber mica) $KMg_3AlSi_3O_{10}(OH)_2$. A silicate mineral of the mica group (q.v.).
Properties: Color yellow to dark brown; luster pearly; hardness 2.5-3; sp.gr. 2.86.
Occurrence: New York, New Jersey, Canada, Europe.
Use: Insulators in electrical apparatus; as heat insulation.

phloridzin (phlorizin; phlorrhizin) $C_{21}H_{23}O_9 \cdot 2H_2O$. A glycoside.
Properties: Light, white, small, silky needles; sweet taste and a bitter after-taste; poisonous! Soluble in alcohol and hot water; very slightly soluble in ether.
Constants: Sp.gr. 1.4298; m.p. 109°C; solidifies and then does not melt until a temperature of 170°C is reached.
Derivation: By extraction of the glycoside from the root-bark of apple, pear, plum, and cherry trees.
Grades: Technical.
Containers: Glass bottles.
Uses: Medicine; biochemical experimentation.
Shipping regulations: None.*

phlorizin. See phloridzin.

phloroglucine. See phloroglucinol.

phloroglucinol (phloroglucine; 1,3,5-trihydroxybenzene) $C_6H_3(OH)_3 \cdot 2H_2O$.
Properties: White to yellowish crystals, odorless; m.p. 212-217°C if rapidly heated; 200-209°C, if slowly heated; b.p., sublimes with decomposition. Soluble in alcohol, ether, and pyridine; slightly soluble in water.
Derivation: By fusion of resorcinol with caustic soda; by reduction of trinitrobenzene.
Method of purification: Crystallization.
Impurities: Resorcinol; diresorcinol.
Grades: Technical; C.P.
Containers: Tins; glass bottles; fiber drums.
Uses: Analytical chemistry (reagent for pentoses and with vanillin for determining the presence of free hydrochloric acid); medicine; decalcifying agent for bones; preparation of pharmaceuticals and dyes, resins; preservative for cut flowers.

phlorrhizin. See phloridzin.

phonochemistry. Chemistry dealing with reactions influenced by sound waves.

"Phoresin." [172] Trade name for diallyl benzenephosphonate $C_6H_5PO(OCH_2CHCH_2)_2$.
Properties: Clear, mobile liquid, soluble in common inert solvents. Polymerized by heating out of contact with air. Polymer is hard, clear, flameproof, and thermosetting.
Containers: Carboys.
Uses: Flameproofing resins; cross-linking agent.

phorone $(CH_3)_2CCHCOCHC(CH_3)_2$. Diisopropylidene acetone.
Properties: Yellow liquid or yellowish green prisms.
Constants: Sp.gr. 0.8791 at 20/20°C; b.p. (760 mm) 197.9°C; freezing point 28.0°C; vapor pressure 0.38 mm (20°C); flash point 185°F; wt 7.3 lbs/gal (20°C).
Grades: Technical.
Containers: 1-gal cans; 5-, 55-gal drums. Net content 7-, 35-, 400-lbs.
Uses: Solvent for nitrocellulose; lacquers; coating compositions; stains; intermediate (organic synthesis).
Shipping regulations: None.*

"Phorwite BUP." [422] Trade name for a stilbene derivative used for brightening papers and other fibrous materials.

"Phorwite K 2002." [422] Trade name for an optical brightener for thermoplastic and thermosetting materials.

"Phosdrin." [125] Trademark for an insecticide which contains not less than 60% w of the alpha isomer of 2-carbomethoxy-1-methylvinyl dimethyl phosphate, $(CH_3O)_2P(O)OC(CH_3){:}CHCOOCH_3$, and not more than 40% w of insecticidally active related compounds; it is 100% active.
Properties: Yellow to orange liquid; b.p. 210-218°F (0.03 mm); miscible with water, alcohols and aromatic and chlorinated hydrocarbons. Slightly soluble in aliphatic hydrocarbons.
Containers: 18-gauge, lined 30-gal steel drums (290-lbs).
Danger! (technical "Phosdrin" insecticide and formulations above 2%): Poisonous if swallowed, inhaled, or absorbed through skin.
Warning! (formulations 2% or less): Poisonous if swallowed, inhaled, or absorbed through skin.
Shipping regulations: (solids over 10%, all liquids): Poison, Class B. Poison label.*

"Phos-Feed." [196] Trademark for a brand of dicalcium phosphate used as a mineral supplement for animal and poultry feeds.

"Phosflake." [177] Trademark for a uniform blend of caustic soda and trisodium phosphate prepared in flake form, especially for bottle-washing use. Characterized by rapid solubility, sanitizing and rinsing properties, and ease of handling. Available in 450-lb drums.

"Phosfon." [40] Proprietary name for tributyl-2,4-dichlorobenzylphosphonium chloride.
Properties: White, crystalline solid with a mild aromatic odor; inversely soluble in water; soluble in acetone, ethanol, isopropanol and hot benzene; insoluble in hexane and ether; technical grade material melts 114-120°C.
Uses: Active ingredient for "Phosfon-D"; a chemical height retardant for potted and garden chrysanthemums, for potted Ace, Croft and Georgia Easter Lilies and certain other ornamental plants.

phosgene (carbonyl chloride; carbon oxychloride; chloroformyl chloride) $COCl_2$.
Properties: Colorless, very volatile liquid or easily liquefied gas; extremely poisonous! Sp.gr. 1.392; m.p. −104°C; b.p. 8.2°C. Slightly soluble in water and slowly hydrolyzed by it; soluble in benzene and toluene; specific volume (70°F) 3.9 cu ft/lb.
Derivation: By passing a mixture of carbon monoxide and chlorine over activated carbon.
Containers: Steel cylinders; special one-ton containers.
Uses: Lethal gas for warfare; bleaching sand for glass manufacture; chlorinating agent; dye manufacture (methyl violet); organic synthesis, especially of isocyanates, polyurethane and polycarbonate resins, carbamates, organic carbonates and chloroformates.
Shipping regulations: Poison, class A. Poison gas label. Not accepted by express.*

"Phos Kil." [55] Trade name for parathion-based insecticidal dusts and sprays.

phosphatase, alkaline. An enzyme which is excreted into the bile by the normal liver and found in the blood. It is concerned with bone formation, probably being produced by osteoblasts. It hydrolyzes phosphoric acid esters at an optimum pH 9.0, liberating phosphate ions.
Use: Biochemical research.

phosphate glass. A type of glass containing phosphorus pentoxide as a major ingredient. Aluminum metaphosphate is frequently the basic material. Many of these glasses have properties not attainable in silicate glasses; e.g., resistance to hydrofluoric acid.

phosphate of lime. See calcium phosphate.

phosphate rock (phosphorite). A natural rock consisting largely of calcium phosphate and used chiefly as a raw material for manufacture of phosphate fertilizers, phosphoric acid, and phosphorus, and therefore indirectly for practically all commercial phosphorus chemicals. A large amount is ground and applied directly to the soil. It is also a primary source of superphosphate for use in fertilizers. This is prepared by treatment of the pulverized rock with sulfuric acid (superphosphate having 16-18% P_2O_5) or by acidifying with phosphoric acid (triple superphosphate having 40-48% P_2O_5). Important deposits are in Florida, Tennessee, Wyoming, Utah, Idaho and North Africa.
See also apatite, land pebble, brown rock.

phosphate slag. Glassy calcium silicate by-product of electric furnace phosphorus manufacture.
Properties: Lumps; loose bulk density approximately 85 lbs/cu ft.
Containers: Hopper cars, gondolas.

phosphatic feed solution.
Properties: Clear, colorless, mobile liquid; density 13.1 lbs/gal.
Containers: 8000-gal tank cars, tank wagons.
Use: Fortification of animal feeds.

phosphatic fertilizer solution.
Grades: 75% and 78% grades of clear, colorless, mobile liquid miscible with water in all proportions.
Use: Manufacture of mixed liquid fertilizers.

phosphatide. See phospholipid.

phosphatidyl choline. See lecithin.

"Phosphen." [233] Trademark for certain aryl phosphates.

phosphine (hydrogen phosphide; phosphuretted hydrogen; phosphoretted hydrogen) PH_3.
Properties: Colorless; spontaneously flammable gas; disagreeable, garlic-like odor; exceedingly poisonous! Soluble in alcohol, ether and cuprous chloride; slightly soluble in cold water; insoluble in hot water.
Constants: Sp.gr. 1.185; m.p. −133.5°C; b.p. −85°C.
Derivation: By action of nascent hydrogen or of caustic potash on phosphorus.
Use: Organic preparations.
Fire hazard: Dangerous.
Shipping regulations: Poison gas. Not usually shipped.*
Note: There is also a synthetic dye, chrysaniline yellow, which is sometimes called phosphine.

phosphocozymase. See nicotinamide adenine dinucleotide phosphate.

phospholeum (superphosphoric acid)
Properties: A clear liquid containing approximately 76% P_2O_5 and equivalent to 105% H_3PO_4. It is an azeotropic mixture of orthophosphoric and polyphosphoric acids which on dilution with water hydrolyze to orthophosphoric acid. Sp.gr. 1.90 (approx) (75°F); viscosity 800 cps (75°F); temperature rise, 60°F on dilution to 75% H_3PO_4; f.p. less than −50°C.
Grade: Technical.
Containers: Tank cars and tank trucks.
Uses: High analysis liquid fertilizer

formulation; metal phosphatizing and aluminum bright dip bath component; desiccant; sequestrant for common trace minerals.

phospholipid. A group of lipid compounds that yield on hydrolysis phosphoric acid, an alcohol, fatty acid, and a nitrogenous base. They are widely distributed in nature and include such substances as lecithin, cephalin, and sphingomyelin.

phospholutein. See lecithin.

phosphomolybdic acid (PMA) $H_3PO_4 \cdot 12MoO_3 \cdot xH_2O$.
Properties: Yellowish crystals. Soluble in water, alcohol, and ether.
Derivation: By heating ammonium phosphomolybdate with aqua regia.
Method of purification: Crystallization.
Grades: Technical; C.P.
Containers: 1-lb bottles.
Uses: Reagent for alkaloids; pigments.
See phosphomolybdic pigments.

phosphomolybdic methyl violet complex.
Properties: Dark-colored powder.
Derivation: Reaction of methyl violet with phosphomolybdic acid.
Uses: Color lakes and toners for inks, water colors, crayons etc.

phosphomolybdic pigments (molybdenum lakes). Basic organic dyes, precipitated by phosphomolybdic acid or a mixture of phosphomolybdic and phosphotungstic acids. See also phosphotungstic pigments.

phosphonium iodide (iodophosphonium) PH_4I.
Properties: Colorless or slightly yellowish crystals.
Constants: Sp.gr. 2.86; m.p. 61.8°C (sublimes); b.p. 80°C; decomposed by water or alcohol.
Derivation: Action of phosphine upon hydrogen iodide.
Grades: Technical.
Use: Chemical synthesis.

phosphor. Any material which has been prepared artifically and has the property of luminescence is called a phosphor, regardless of whether it exhibits phosphorescence. Phosphors are either of the sulfide type (by far the most common is highly purified zinc sulfide, with or without admixed cadmium sulfide, but always with a trace of activator such as silver, copper, or manganese) or of the oxygen-dominated type. The latter group are used in common fluorescent light tubes, and most of these materials are excited by the mercury radiation at 2537A characteristic of a mercury arc. Silicates, borates, phosphates, and tungstates of zinc, beryllium and other metals are common examples. Phosphates and halophosphates and magnesium tungstate are widely used in fluorescent tubes. Television and radar tubes use the sulfide-type phosphors.

phosphor bronzes. Tin bronzes (see brass and bronze) which have been deoxidized by the addition of up to 0.5% phosphorus. They are relatively hard, strong, and corrosion resistant.
Grades: Grade A (5% tin), grade C (8% tin), grade D (10% tin), grade E (1.25% tin).
Uses: Springs, electrical switches, contact fingers, chains, etc.

phosphorescence. Fluorescence that continues for more than a very short time (10^{-6} seconds) after the exciting radiation is stopped.

phosphoretted hydrogen. See phosphine.

phosphoric acid. See the items which follow. See also pyrophosphoric acid; polyphosphoric acid; phospholeum.

phosphoric acid (orthophosphoric acid) H_3PO_4.
Properties: Phosphoric acid is a clear colorless, sparkling liquid or a transparent crystalline solid depending on the concentration and the temperature. At ordinary atmospheric temperature (20°C) the 50% and 75% strengths are mobile liquids, the 85% is of a syrupy consistency while the 100% acid is in the form of crystals; sp.gr. 1.884; m.p. 42.35°C; b.p. 260°C; soluble in water; very corrosive to ferrous metals and alloys.
Derivation: (a) Action of sulfuric acid on pulverized phosphate rock (Dorr strong-acid process; wet process); (b) by heating phosphate rock, coke, and silica in an electric or blast furnace, burning the elemental phosphorus produced, and then hydrating the phosphoric oxide (furnace acid); (c) action of aqueous hydrochloric acid on phosphate rock followed by solvent extraction. Acid purified by distillation.
Grades: Commercial (64 and 72%); technical (50, 75, 85, 90, 100%); food (50, 75, 85%); N.F. XI (85-88%) (Polyphosphoric acid is sometimes called 115% phosphoric acid).
Containers: 1-, 5-lb bottles; 5-, 6½-, 13-gal carboys; 15-, 55-gal drums and barrels; tank trucks; tank cars.
Uses: Fertilizers; inorganic phosphates; picking and rust-proofing metals; soft drinks and flavoring syrups; pharmaceuticals; sugar refining; gelatin manufacture; water treatment; animal feeds; electropolishing; lakes in cotton dyeing; yeasts; soil stabilizer; waxes and polishes; binder for ceramics; activated carbon production.
Caution: Causes skin irritation; avoid contact with skin or eyes. MCA warning label.
Shipping regulations: None.*

phosphoric acid, anhydrous. See phosphoric anhydride.

phosphoric acid, glacial. See phosphoric acid, meta-.

phosphoric acid, meta- (phosphoric acid, glacial) HPO_3.
Properties: Transparent, highly deliquescent, glassy mass; sp.gr. 2.2-2.488. Soluble in water, slowly forming the ortho-acid; soluble in alcohol.
Derivation: By heating phosphoric acid to redness; by treating phosphorus pentoxide

*See "I.C.C. Shipping Regulations," page xiii.
Reference numbers refer to name of manufacturer. See "List of Manufacturers," page v.

with the calculated quantity of water; by heating diammonium phosphate.
Grades: Technical; C.P.
Containers: 2-lb glass bottles; 13-gal carboys.
Use: Phosphorylating agent; dehydrating agent; manufacture of dental cements; analytical chemistry.
Shipping regulations: None.*

phosphoric acid, ortho-. See phosphoric acid.

phosphoric acid, reverted. A term used in connection with the solubility and availability to plants of calcium acid phosphate fertilizers.

The phosphoric acid, i.e. the phosphorus content, in calcium acid phosphate may be divided into three parts: (1) water-soluble, (2) insoluble in water but soluble in neutral ammonium citrate, and (3) insoluble in both water and neutral ammonium citrate.

When acid phosphate is first made nearly all of the phosphoric acid except that insoluble in neutral ammonium citrate solution is water-soluble. As the acid phosphate cures, the water-soluble usually decreases, the citrate-soluble portion increases and the citrate insoluble decreases.

This is referred to as reversion, and the phosphorus content of the fertilizer is referred to as reverted phosphoric acid.

phosphoric anhydride (phosphorus pentoxide; phosphoric oxide; phosphoric acid, anhydrous) P_2O_5.
Properties: Soft, white powder. Caution! Keep well stoppered! Phosphoric anhydride absorbs moisture from the air with avidity thus forming meta-, pyro-, or orthophosphoric acid, depending upon the amount of water absorbed and upon conditions under which absorption takes place; sp.gr. 2.387.
Derivation: By burning phosphorus in a current of dry air.
Grades: Technical.
Containers: Glass bottles; 10-lb cans; 60-, 375-lb drums.
Uses: Preparation of phosphoric acids; dehydrating agent; organic synthesis; medicine; sugar refining; analysis (dehydrating agent).
Warning! Causes burns. Avoid contact with skin or eyes. MCA warning label.
Shipping regulations: Flammable solid. Yellow label.*

phosphoric bromide. See phosphorus pentabromide.

phosphoric chloride. See phosphorus pentachloride.

phosphoric oxide. See phosphoric anhydride.

phosphoric perbromide. See phosphorus pentabromide.

phosphoric perchloride. See phosphorus pentachloride.

phosphoric sulfide. See phosphorus pentasulfide.

phosphorite. See phosphate rock.

phosphorodifluoridic acid. See fluophosphoric acids.

phosphorofluoridic acid. See fluophosphoric acids.

phosphorous acid, ortho- H_3PO_3.
Properties: White, or yellowish, crystalline mass. Very hygroscopic. Absorbs oxygen very readily with formation of orthophosphoric acid. Caution! Keep well stoppered! Soluble in alcohol, water.
Constants: Sp.gr. 1.651; b.p. 200°C (dec); m.p. 70°C (approx).
Grades: Reagent; technical; 70%.
Containers: Casks.
Uses: Analysis (testing for mercury); chemical (reducing agent); phosphite salts.
Shipping regulations: None.*

phosphorous bromide. See phosphorus tribromide.

phosphorous chloride. See phosphorus trichloride.

phosphorous iodide. See phosphorus triiodide.

phosphorous sulfide. See phosphorus trisulfide.

phosphorus P. Element of atomic number 15, group V of periodic system. A non-metallic element that exists in several allotropic forms (white, red and black).
Properties: White phosphorus (yellow phosphorus): White or yellow, soft, waxy solid. Darkens on exposure to light by conversion to the red form. B.p. 280°C; vapor density corresponds to formula P_4; m.p. 44.1°C; sp.gr. (solid, 20°C) 1.82, (liquid, 44.5°C) 1.745. Insoluble in water and alcohol; moderately soluble in chloroform and benzene; very soluble in carbon disulfide. At room temperature it exhibits phosphorescence (slow, luminous oxidation) in air; it ignites spontaneously in moist air at about 30°C. Stored and shipped beneath water to avoid ignition. It is very poisonous and causes severe burns.
Red phosphorus: Violet-red, amorphous powder obtained from white phosphorus by heating at 240-250°C in the presence of a catalyst, such as iodine. Sublimes 416°C; sp.gr. 2.34. Insoluble in all solvents. Non-poisonous and much less reactive than the white form. Ignites in air at about 260°C.
Black phosphorus: Black, lustrous crystals resembling graphite. Obtained by heating white phosphorus under high pressure. Insoluble in all solvents.
Occurrence and derivation: Phosphorus is present in nature in phosphate rock [impure $Ca_3(PO_4)_2$], in apatite [$Ca_5(PO_4)_3F$], in bones, teeth, and in organic compounds of living tissue. Phosphorus is produced by an electric furnace containing phosphate rock, sand and coke. The phosphorus vapor is driven off and condensed beneath water.
Grades: Technical; electronic grade, 99.9999%.

Containers: 380-lb drums; tank cars.
Uses: Manufacture of phosphoric acid and its derivatives (about 50% of this goes into detergents, sanitizers, soaps; 17% into fertilizers; remainder into gasoline additives, animal feeds and miscellaneous chemicals); phosphor bronzes and metallic phosphides; additive to semiconductors, and in electroluminescent coatings; incendiaries, pyrotechnics, and smoke bombs. White phosphorus is used to a small extent in rat poisons and red phosphorus is used in the manufacture of matches.
Danger! Must be kept under water. Causes severe burns; do not get on skin or in eyes. MCA warning label.
Shipping regulations: White and yellow phosphorus: Flammable solid. Yellow label. Not accepted by express. White or yellow phosphorus, in water: Flammable solid. Yellow label. Red phosphorus: Flammable solid. Yellow label.*

phosphorus 32. Radioactive phosphorus of mass number 32.
Properties: Half-life, 14.3 days; radiation, beta; radiotoxicity, moderately hazardous.
Derivation: Pile irradiation of potassium dihydrogen phosphate or sulfur and sulfur compounds.
Forms available: Phosphate ion in weak hydrochloric acid solution; solid potassium dihydrogen phosphate; P-32 sterile solution; in tagged compounds such as hexaethyltetraphosphate, ribonucleic acid, triphenylphosphine, etc.
Grades: N.N.D. (as sodium radio-phosphate, $NaH_2P^{32}O_4$ and $Na_2HP^{32}O_4$ in solution).
Uses: Medical treatment of polycythemia vera, leukemia, skin lesions; for the measurement of coverage and thickness of printing inks and paint films; for microradiography; as a tracer in assessing the effectiveness and utilization of fertilizers, in determining the amount of and location of phosphorus in steel, in determining the efficiency of phosphorus removal by slag, in tire tread wear tests, in studying the diffusion of plasticizers in rubber production, in locating water leaks, in detecting dye migration in textile dyeing, in studying phosphorus metabolism by plants and animals, in chemical analysis, in studying chemical reaction mechanisms, in locating brain tumors, in studying mosquito flight patterns, etc.
Shipping regulations: Poison, class D radioactive material. Red or blue label.*

phosphorus nitride P_3N_5.
Properties: Amorphous white solid; insoluble in cold water; decomposes in hot water; soluble in common organic solvents.

phosphorus oxychloride (phosphoryl chloride) $POCl_3$.
Properties: Colorless, fuming liquid; pungent odor. Sp.gr. 1.86 (20/20°C); m.p. 1.25°C; b.p. 107.2°C. Decomposed by water and alcohol.
Derivation: (a) From phosphorus trichloride and chlorates; (b) by distilling phosphorus pentoxide with phosphorus pentachloride.
Grades: Technical.
Containers: 200-, 650-lb steel-jacketed lead cylinders; 40-, 85-, 175-lb carboys; 15-, 55-gal barrels and drums.
Uses: Manufacture of organic phosphates for use as plasticizers, gasoline additives, hydraulic or similar fluids; also as a chlorinating agent and catalyst in organic synthesis.

phosphorus pentabromide (phosphoric bromide; phosphoric perbromide) PBr_5.
Properties: Yellow, crystalline mass. Caution! Keep hermetically sealed! Soluble in water (dec). B.p. 106°C (dec).
Grades: Technical.
Use: Organic synthesis.

phosphorus pentachloride (phosphoric chloride; phosphoric perchloride) PCl_5.
Properties: Slightly yellow, crystalline mass; irritating odor; fuming in moist air; strong irritating effect on the eyes.
Constants: Sp.gr. 3.60; m.p. (under pressure) 148°C. Ordinarily sublimes without melting; b.p. 160-165°C. Soluble in carbon disulfide; decomposed by water.
Derivation: By action of chlorine on phosphorus or phosphorus trichloride.
Grades: Technical; reagent.
Containers: 25-, 50-lb crocks; bottles; 500-lb drums.
Uses: Chlorinating agent in organic chemistry; catalyst.
Warning! Hazardous dust. Causes burns. MCA warning label.
Shipping regulations: Flammable solid. Yellow label.*

phosphorus pentasulfide (phosphoric sulfide; phosphorus persulfide; thiophosphoric anhydride) P_2S_5.
Properties: Light-yellow or greenish-yellow crystalline masses. Peculiar odor; similar to hydrogen sulfide. Caution! Keep in sealed containers. Very hygroscopic. Burns in air forming P_2O_5 and SO_2. Decomposed by moist air. Ignites by friction. Soluble in solutions of alkali hydroxides; slightly soluble in carbon disulfide. M.p. 280-283°C; b.p. 515°C (ignites in air at about 300°C); sp.gr. 2.03; vapor pressure 1 mm (300°C).
Derivation: By reaction of phosphorus and sulfur.
Grades: Technical; distilled.
Containers: 100-, 150-, 200-lb drums.
Uses: Intermediate for lube oil additives, insecticides, flotation agents, rubber additives.
Warning! Harmful dust. Contact with water or acids slowly liberates poisonous and flammable hydrogen sulfide gas. MCA warning label.
Shipping regulations: Flammable solid. Yellow label.*

phosphorus pentoxide. See phosphoric anhydride.

phosphorus persulfide. See phosphorus pentasulfide.

phosphorus salt. See sodium ammonium phosphate.

phosphorus sesquisulfide (tetraphosphorus trisulfide) P_4S_3.
Properties: Yellow, crystalline mass; very flammable. Soluble in carbon disulfide; insoluble in cold water; decomposed by hot water. Sp.gr. 2.00; m.p. 172°C; b.p. 407.8°C.
Derivation: By gently heating phosphorus and sulfur.
Grades: Technical.
Containers: 105-lb cases; drums.
Uses: Organic synthesis; manufacture of matches.
Fire hazard: Dangerous.
Shipping regulations: Flammable solid. Yellow label.*

phosphorus sulfide. See phosphorus trisulfide.

phosphorus tribromide (phosphorous bromide) PBr_3.
Properties: Fuming, colorless liquid; very penetrating odor; soluble in acetone, alcohol, carbon disulfide, hydrogen sulfide, water (decomposes). Sp.gr. 2.925 at 0°C; b.p. 175°C; m.p. −40°C.
Grades: Technical.
Containers: Glass bottles.
Use: Analysis (testing for sugar and oxygen); catalyst; synthesis.
Shipping regulations: Corrosive liquid. White label.*

phosphorus trichloride (phosphorous chloride) PCl_3.
Properties: Clear, colorless fuming liquid; decomposes rapidly in moist air. Soluble in ether, benzene, carbon disulfide and carbon tetrachloride; decomposed by water. Sp.gr. 1.574; m.p. −111.8°C; b.p. 76°C.
Derivation: By passing a current of dry chlorine over gently heated phosphorus which ignites. The trichloride, admixed with some pentachloride, distills over. A small amount of phosphorus is added and the whole distilled.
Grades: Technical; 99.9%.
Containers: 85-, 175-lb carboys; 200-, 650-lb cylinders; tank cars.
Uses: Intermediate for surfactants, phosphites (reaction with alcohols and phenols), gasoline additives, plasticizers, dyestuffs; chlorinating agent.
Danger! Causes severe burns. Vapor extremely irritating. Contact with water may cause flash fire. MCA warning label.
Shipping regulations: Corrosive liquid. White label.*

phosphorus triiodide (phosphorous iodide) PI_3.
Properties: Red crystals; hygroscopic. Soluble in alcohol, carbon disulfide, water (dec). M.p. 61°C (dec); sp.gr. 4.18.
Grades: Technical; reagent.
Use: Organic synthesis.

phosphorus trisulfide (phosphorous sulfide; phosphorus sulfide; tetraphosphorus hexasulfide; thiophosphorous anhydride) P_2S_3, or P_4S_6.
Properties: Grayish-yellow masses; tasteless; odorless. Caution! Keep well stoppered! Burns in air. Decomposed in moist air. Soluble in alcohol, carbon disulfide, ether.
Constants: B.p. 490°C; m.p. 290°C.
Grades: Technical.
Use: Organic chemistry (reagent).

phosphoryl chloride. See phosphorus oxychloride.

phosphotungstic acid (phosphowolframic acid; PTA) $H_3PO_4 \cdot 12WO_3 \cdot xH_2O$.
Properties: Heavy, greenish crystals. Soluble in water, alcohol and ether.
Derivation: By heating ammonium phosphotungstate with aqua regia.
Method of purification: Crystallization.
Grades: Technical; C.P.
Containers: Glass bottles.
Uses: Reagent for alkaloids; phosphotungstic pigments.
Shipping regulations: None.*

phosphotungstic pigments (tungsten lakes). Usually greens or blues which are manufactured by precipitating basic dyestuffs such as malachite green or Victoria blue with solutions of phosphotungstic acid, or phosphomolybdic acid, or mixtures of both. The pigments made from phosphomolybdic acid are also known as molybdenum lakes or phosphomolybdic pigments.
Uses: Chiefly in printing inks, the whitening of white paper, and in interior paints and enamels.

phosphowolframic acid. See phosphotungstic acid.

phosphuretted hydrogen. See phosphine.

"Phos-Trode." 407 Trademark for a phosphor bronze Grade C electrode and filler rod for joining like and dissimilar metals and overlaying surfaces resistant to wear and corrosion.

photochemistry. The branch of chemistry that deals with the effect of light in causing or modifying chemical changes. The most important examples are in natural photosynthesis, in the production of a photographic image, and in the reaction of chlorine on hydrocarbons and other organic compounds.

photo-glycin. See para-hydroxyphenylglycine.

photolysis. Breaking down of molecules into simpler units by use of light.

photon. A unit (quantum) of electromagnetic radiation. Light waves, gamma rays, x-rays, etc. consist of photons.
See fundamental particle.

photophor. See calcium phosphide.

photopolymer. A polymer or plastic that is made so that it is sensitive to and undergoes some kind of change on exposure to light. Such materials can be used for printing and lithography plates, for photographic prints and microfilm copying.

The effect of the light may be to cause further polymerization or crosslinking, or may cause degradation. One application involves the use of esters of polyvinyl alcohol which crosslink and so become insoluble, whereas unexposed portions of the material remain soluble.

photosynthesis. The natural process by means of which carbon dioxide and water are converted into carbohydrates in growing plants in sunlight. Chlorophyll (q.v.) is necessary as a catalyst. In the presence of light certain electrons in the chlorophyll molecule are displaced and the resultant unbalanced positive charge (hole) causes water to liberate its hydrogen and oxygen to other molecules present. In a subsequent reaction part of the oxygen is released to the air, the corresponding hydrogen is used in the synthetic cycle, while the remaining hydrogen and oxygen recombine with the liberation of energy for the synthesis cycle. It is now thought that the carbohydrate synthesis proceeds through addition of carbon dioxide in a complex cycle involving ribulose diphosphate (a five-carbon compound), phosphoglyceraldehyde (three carbons), sedoheptulose phosphate (seven carbons) and finally glucose phosphate (six carbons). The recent synthesis of chlorophyll is a long step forward in settling the details of photosynthesis.

photoxylin. See celloidin.

"PH-Plus." [84] Trademark for a special moisture-free alkali for water treatment. Fused at 2200°C and cast in ½-lb conical cakes. For treatment of swimming pool water and for general industrial water treatment.

"Phthalamaquin." [342] Trademark for a brand of aureoquin preparation.

phthalamide $C_6H_4(CONH_2)_2$. The double acid amide of phthalic acid.
Properties: Colorless crystals; m.p. 200-210°C (decomposes into phthalimide and ammonia). Very slightly soluble in water and alcohol; insoluble in ether.
Derivation: By stirring phthalimide with cold concentrated ammonia solution; by the reaction of phthalyl chloride and ammonia; or from the addition of ammonia to phthalic anhydride under pressure.
Containers: Barrels.
Use: Intermediate in organic synthesis.

phthalic acid (ortho-phthalic acid; naphthalic acid; ortho-benzene dicarboxylic acid) $C_6H_4(CO_2H)_2$.
Properties: Colorless crystals; soluble in alcohol; sparingly soluble in water and ether. Sp.gr. 1.585; m.p., decomposes at 191°C.
Derivation: From phthalic anhydride; or by direct oxidation of mixed alkyl aromatics, with heavy metal salts and bromine as catalysts.
Method of purification: Crystallization.
Grades: Technical; reagent.
Containers: 1-lb bottles; 50-lb cases; 100-lb barrels.
Uses: Dyes (synthesis of indigo, manufacture of phthaleins, various fluorescein and eosin dyes, rhodamines and pyronine dyes); medicine; phenolphthalein; phthalimide; anthranilic acid; synthetic perfumes.
Shipping regulations: None.*

meta-**phthalic acid.** See isophthalic acid.

ortho-**phthalic acid.** See phthalic acid.

para-**phthalic acid.** See terephthalic acid.

phthalic anhydride $C_6H_4(CO)_2O$ (acid phthalic anhydride).
Properties: White, crystalline needles; sublimes below b.p.; characteristic mild odor. Sp.gr. 1.527 (4°C); m.p. 130.8°C; b.p. 285°C. Soluble in alcohol; slightly soluble in ether and hot water.
Derivation: By air oxidation of naphthalene using vanadium pentoxide as a catalyst; also from ortho-xylene by a somewhat similar process.
Method of purification: Sublimation.
Grades: Pure.
Containers: 80-, 175-, 200-lb barrels; tank trucks; tank cars.
Uses: Alkyd resins; plasticizers; polyesters; synthesis of phenolphthalein and other phthaleins, many other dyes; chlorinated products; pharmaceutical intermediates; insecticides; diethyl phthalate; dimethyl phthalate.
Caution! May cause skin irritation. MCA warning label.

ortho-**phthalimide** $C_6H_4(CO)_2NH$.
Properties: White, crystalline leaflets. Slightly soluble in ether; insoluble in benzene; soluble in boiling benzene and in aqueous alkalies.
Constants: M.p. 233-238°C; b.p., sublimes.
Derivation: By dissolving phthalic anhydride in ammonium hydroxide, evaporating to dryness and using the residue.
Method of purification: Sublimation.
Grades: Technical; 97-98%.
Containers: Barrels; tins.
Uses: Production of indigo, via anthranilic acid; fungicide; organic synthesis.
Shipping regulations: None.*

phthalocyanine blue. See phthalocyanine pigments.

phthalocyanine green. See phthalocyanine pigments.

phthalocyanine pigments. A series of organic pigments having as a structural unit four isoindole groups $(C_6H_4)C_2N$, linked by four nitrogen atoms so as to form a conjugated chain. There are four commercially important modifications, including the basic compound: (1) phthalocyanine (metal-free), $(C_6H_4C_2N)_4N_4$, blue-green; (2) copper phthalocyanine $(C_6H_4C_2N)_4N_4Cu$, in which a copper atom is held by secondary valences of the isoindole nitrogen atoms; sp.gr. 1.59; (3) chlorinated copper phthalocyanine, green, in which 14 to 16 hydrogen atoms are

replaced by chlorine; and (4) sulfonated copper phthalocyanine, water-soluble, green, in which two hydrogen atoms are replaced by sulfonic acid (HSO_3) groups.
Properties: Extreme lightfastness and stability to acids, alkalies, and heat. The pigments are non-bleeding in the usual paint vehicles and have high tinctorial strength but are relatively expensive.
Derivation: (copper phthalocyanine, blue) (a) By reaction of phthalic anhydride, urea, and cuprous chloride at about 200°C; (b) reaction of ortho-chlorocyanobenzene with cuprous cyanide and copper powder.
Containers: Barrels.
Uses: Decorative enamels, automotive finishes, linoleum, plastics, roofing granules, printing inks, wallpaper, rubber goods, and similar applications where light fastness and chemical stability are required.

phthalonitrile $C_6H_4(CN)_2$.
Properties: Buff-colored crystals; m.p. 138°C; insoluble in water; soluble in acetone and benzene.
Derivation: Vapor phase reaction of ammonia and phthalic anhydride over alumina catalyst at high temperature.
Grades: Technical; purified.
Containers: Bags; bottles.
Uses: Intermediate in organic synthesis; especially for pigments and dyes; base material for high temperature lubricants and coatings.

meta-**phthalyl dichloride.** See isophthaloyl chloride.

phthalylsulfacetamide (N^1-acetyl-N^4-phthalylsulfanilamide) $C_{16}H_{14}N_2O_6S$ or $C_6H_4(COOH)CONHC_6H_4SO_2NHCOCH_3$.
Properties: White or creamy white crystals or crystalline powder. Slight odor; decomposes with liquefaction between 186-202°C. Very slightly soluble in water; soluble in acetone; freely soluble in solutions of alkali hydroxides.
Grade: N.F. XI.
Containers: Drums.
Use: Medicine.

phthalylsulfathiazole $C_{17}H_{13}N_3O_5S_2$. 4'-(2-Thiazolylsulfamyl) phthalanilic acid.
Properties: White or faintly yellowish-white crystalline powder; bitter taste; odorless. May darken slowly on long exposure to light. Practically insoluble in water and chloroform; slightly soluble in alcohol; very slightly soluble in ether; readily soluble in solutions of alkali hydroxides and their carbonates, and in hydrochloric acid. M.p. 272-277°C (dec).
Derivation: By condensing sulfathiazole with phthalic anhydride.
Grade: U.S.P. XVI.
Use: Medicine.

phthiocol ($C_{11}H_8O_3$). The principal pigment isolated from the lipids of Mycobacterium tuberculosis. It was the first identified form of vitamin K and is an antibiotic substance.
Properties: Yellow prisms from ether-petroleum ether; m.p. 173-174°C; sublimes; steam volatile; slightly soluble in water; soluble in organic solvents except petroleum ether.
Derivation: By treating 2-methylnaphthoquinone with bleaching powder and acidifying with sulfuric acid.
Use: Medicine.

"Phygon." [248] Trademark for a line of fungicides and seed protectants based on dichlone.
Uses: Fungicide for apple scab, brown rot of stone fruits, potato and tomato early blight; blue-green algae control in industrial water systems, ponds, irrigation canals. Formulated as "Phygon XL," a 50% wettable powder.

"Phyllicin." [9] Trademark for theophylline-calcium salicylate.

physic nut oil. See Philippine physic nut oil.

physostigma (calabar bean; ordeal bean; chop nut; split nut).
Derivation: Seed of Physostigma venenosum.
Occurrence: West Africa; India; Brazil.
Grades: Technical.
Containers: Bags.
Uses: Medicine; source of the alkaloid physostigmine or eserine.
Shipping regulations: None.*

physostigmine (eserine; calabarine) $C_{15}H_{21}O_2N_3$. An alkaloid.
Properties: Colorless or pinkish crystals. Poisonous! Slightly soluble in water; soluble in alcohol and diluted acids. M.p. 86-87°C and 105-106°C (unstable and stable forms).
Derivation: By solvent extraction from the seeds of Physostigma venenosum.
Uses: Medicine.
Available as the salicylate and sulfate.

physostigmine salicylate $C_{15}H_{21}O_2N_3 \cdot C_7H_6O_3$.
Properties: Colorless crystals; m.p. 182-183°C; soluble in water and alcohol. Very poisonous!
Grade: U.S.P. XVI.
Use: Medicine.

physostigmine sulfate $(C_{15}H_{21}O_2N_3)_2 \cdot H_2SO_4$.
Properties: White, deliquescent crystals; m.p. 150-151°C. Poisonous! Soluble in water and alcohol.
Use: Medicine.

phytic acid (inisitolhexaphosphoric acid) $C_6H_6[OPO(OH)_2]_6$. Occurs in nature in the seeds of many cereal grains, generally as the insoluble calcium-magnesium salt (see "Phytin.") In the body, phytic acid inhibits the absorption of calcium in the intestine.
Properties: A typical product is a white to pale yellow liquid; odorless with acid taste; pH less than 1.0 (in 1% solution); soluble in water and alcohol; sp.gr. 1.58; wt/gal 13.1 lbs.
Derivation: From corn steep liquor.
Grades: Technical (as a 70% solution).
Containers: (solution) 1-, 5-, 10-lb glass

*See "I.C.C. Shipping Regulations," page xiii.
Reference numbers refer to name of manufacturer. See "List of Manufacturers," page v.

containers; 70-lb carboys.
Uses: Chelation of heavy metals in processing of animal fats and vegetable oils; rust inhibitor; preparation of phytate salts; metal cleaning; treatment of hard water.

"Phytin." 305 Trademark for calcium-magnesium salt of inositol hexaphosphoric acid (phytic acid).
Use: Medicine (dietary supplement).

phytol $C_{20}H_{40}O$ or $CH_3[CH(CH_3)CH_2CH_2CH_2]_3C(CH_3){:}CHCH_2OH$. An alcohol obtained by the decomposition of chlorophyll.
Properties: Odorless liquid; b.p. 202-204°C (10 mm); sp.gr. 0.8497 (25/4°C); soluble in the common organic solvents; insoluble in water.
Use: Synthesis of vitamins E and K.

phytolacca (poke root; garget).
Derivation: Dried root of Phytolacca americana or decandra.
Occurrence: North America; southern Europe.
Grades: Technical.
Containers: Bags; bales.
Use: Medicine.
Shipping regulations: None.*

phytonadione (2-methyl-3-phytyl-1,4-naphthoquinone; vitamin K_1) $CH_3C_{10}H_4O_2C_{20}H_{39}$.
Properties: Clear yellow, viscous, odorless liquid; sp.gr. 0.967 (25/25°C); refractive index 1.5230-1.5252 (25°C); stable in air. Protect from sunlight! Insoluble in water; soluble in benzene, chloroform and vegetable oils; slightly soluble in alcohol.
Derivation: Synthetically, from 2-methyl-1,4-naphthoquinone and phytol.
Uses: Medicine; food supplement.

phytosterols. See sterols.

"Picco Aromatic Plasticizers." 140 Trade name for a series of alkylated, aromatic plasticizers.
Properties: Semi-liquids with softening point of 10° and 25°C; color coal tar 2-5; ash 0.2% max; sp.gr. 0.97-0.99.
Containers: 18 gauge, oil-type steel drums.
Uses: Plasticizers and tackifiers. Improve processing and permit higher filler loading. Good electrical properties make them useful in friction tapes.

"Picco Aromatic Plasticizing Oils." 140 Trade name for a series of alkylated, aromatic hydrocarbons useful as plasticizers and softeners.
Piccocizer R: Liquid with color coal tar 2 1/2; sp.gr. 0.930-0.960.
Piccocizer 30: Liquid with color coal tar 4; sp.gr. 0.995-1.010.
Dipolymer Oil: Liquid with color coal tar 4+; sp.gr. 0.990-1.005.
Pictar: Dark liquid; sp.gr. 0.980-1.000.
Containers: 18 gauge, oil-type, steel drums.
Uses: Plasticizers and softeners; impart low temperature flexibility.

"Piccoflex." 140 Trade name for a series of thermoplastic modified styrene copolymers with good grease resistance and color stability, flexibility and toughness. Available in ball and ring softening points of 100°, 115°, 120°C.
Properties: Color coal tar 3 max; sp.gr. 1.05; refractive index 1.58.
Containers: All grades shipped as solid in light gauge rust resistant, metal coated steel drums. All, except 100°C softening point material, also available in flaked form and shipped in multiwall paper bags.
Uses: Floor tile; molded goods; mats; traffic strips.

"Piccolastics." 140 Trade name for five series of thermoplastic polystyrene resins. "A" series includes the softer, more fluid materials; "C" series is intermediate in properties; "D" and "E" series are higher molecular weight resins; the "F" series, being highest in molecular weight, is available only in aromatic petroleum solvents. Solutions of some grades are also available.
Properties (depending on grade): Color 2 to 6 Gardner; softening point 5°-135°C; sp.gr. 1.020-1.065; refractive index 1.562-1.600; solvents include aromatic hydrocarbons, chlorinated solvents, methyl ethyl ketone, carbon disulfide, ethyl acetate, and turpentine. Compatible with "Chlorowax," "Arochlors," ester gums, some "Pentalyns," rosin, rosin oil and high styrene-butadiene polymers.
Containers: Grades above 50°C softening points in light gauge, rust resistant, metal coated steel drums. Grades below 50°C and Piccolastic solutions in 18 gauge, oil-type steel drums. Grades above 100°C softening points are also available in flaked form and are shipped in multiwall paper bags.
Uses: Adhesives; coatings; saturants for paper and fabrics; waterproofing.

"Piccolyte." 140 Trade name for a hydrocarbon thermoplastic terpene resin, composed essentially of polymers of pinene, especially beta-pinene. Piccolyte emulsions and solutions are also commercially available.
Properties: Ring and ball softening point varies from 10° to 135°C depending upon grade. Color very pale amber; sp.gr. 0.93-0.99; refractive index 1.507-1.533; ash 0.1% max. Soluble in low cost petroleum solvents, coal tar solvents, mineral oil, chlorinated hydrocarbons, long chain alcohols, long chain ketones. Compatible with waxes, oils, bituminous materials, resins such as rosin, ester gums, coumarone-indene, and phenolics.
Containers: Resins above 55°C softening point are shipped in light rust resistant, metal coated steel drums. Softening point grades above 100°C are also shipped in multiwalled bags. Piccolyte solutions are shipped in 18 gauge, oil-type steel drums. Emulsion grades are packed in lined fiber drums. Solutions and emulsions can also be shipped in tank cars or trucks.
Uses: Paint and varnish, printing inks, rubber compounding, paper coating,

adhesives, leather, textiles. Emulsified grades can be used as latex extenders.

"Piccopale." [140] Trade name for a thermoplastic hydrocarbon resin produced by the polymerization of unsaturates derived from the deep cracking of petroleum. Considerable cyclic but no aromatic structure is present. Available in ring and ball softening points of 70°, 85°, 100° and 110°C. Piccopale 100 is available in mineral spirits solution at 60 or 70% solids. A variety of cationic, anionic and nonionic emulsions is also produced.

Properties: Color - Gardner 11; sp.gr. 0.96 to 0.97; refractive index 1.53; soluble in aromatic and chlorinated solvents and in C_7 and higher aliphatic hydrocarbons; ash 0.1% max.

Containers: Solid grades are packed in light gauge, rust resistant, metal coated steel drums, in tank trucks, or in tank cars. Flaked Piccopale 100 or 110 can also be shipped in multiwall paper bags. Solutions are shipped in 18 gauge, oil type steel drums; also in tank trucks or tank cars.

Uses: Emulsion paints, hot-melt saturation of felt and paper, floor coverings, carpet backing, textile sizings, protective coatings, rubber compounding and fiberboard.

"Picco Piccopale Emulsions." [140] Trade name for a series of anionic, cationic, and nonionic emulsions. The emulsified resin is Piccopale, a petroleum hydrocarbon derivative.

Properties (depending upon emulsion): Total solids 45.50%; particle size 1 micron; viscosity 30-1400 cps; sp.gr. 0.97-1.00; good storage and mechanical stability. Degree of ion tolerance is dependent upon specific emulsion.

Uses: Water based paints, carpet backsizings, adhesives, pressure sensitive adhesives, binders for fiber boards.

"Picco Reclaiming Agents." [140] Trade name for a series of resinous aromatic oils identified as follows:

Identification	Color-Coal Tar	Specific Gravity
D-4	4 - 4½	0.88-0.92
D-12	2½ - 4	0.89-0.91
C-42	23	1.04-1.07
C-33	---	0.96-0.98
76-56	3½ - 4½	0.96-0.97

Containers: 18 gauge, oil type, steel drums having a gross weight of 450 pounds and net of 400 pounds.

Uses: Used primarily in the rubber industry as reclaim oils for whole tires, either by the pan or digestion process.

"Picco Resins." [140] Trade name for a series of thermoplastic aromatic hydrocarbons of coal tar or petroleum origin.

Properties: Coal Tar Type: Ball and ring softening points of 10°-120°C; color, coal tar ½-3; sp.gr. 1.05-1.10; refractive index 1.617-1.636. Soluble in aromatic hydrocarbon solvents, higher ketones, chlorinated solvents, aniline, diethyl aniline, morpholine, ethyl ether, ethyl acetate, and diethyl "Carbitol."

Aromatic Petroleum Type: Ball and ring softening points of 10°-130°C; color, coal tar - 1½-9. Soluble in aromatic hydrocarbon solvents, benzyl alcohol, higher ketones, chlorinated solvents, aniline, diethylaniline, morpholine, ethyl ether, ethyl acetate, diethyl "Carbitol."

Picco Resins Solutions: Gardner viscosity D-Z2. Color, coal tar 1½-9.

Containers: Resins melting above 50°C are shipped in light gauge, rust resistant, metal coated steel drums. Softer grades and solutions packed in 18 gauge, oil-type steel drums.

Uses: Protective coatings, rubber compounding, floor coverings, adhesives. Low softening point materials used as plasticizers, tackifiers, and softeners.

"Picco 480 Resins." [140] Trade name for a series of styrenated, thermoplastic hydrocarbon resins.

Properties: Color 2½ to 4 coal tar; softening point 115° to 145°C; sp.gr. 1.032 to 1.040; benzene insolubles 0.5% max; ash 0.5%max.

Containers: Solid resins in light gauge, rust resistant, metal coated steel drums. Material is also available in flake form and is shipped in multiwall bags.

Uses: Floor tile and rubber compounding. Used extensively for their reinforcing value and as a method of incorporating sytrene.

"Piccotex." [140] Trade name for a hard, color stable, substituted styrene copolymer resin available in ring and ball softening points of 100° and 120°C. Both grades are available in low odor mineral spirits solution at 50% solids; the 120°C grade is also available at 55% solids in the same solvent.

Properties: Color < 1 Gardner; sp.gr. 1.04 to 1.06; refractive index 1.58; ash < 0.1%. Soluble in mineral spirits, aromatic hydrocarbon solvents, ethyl and higher ethers, ethyl acetate and higher acetate esters, aniline, diethylamine, chlorinated hydrocarbons, carbon disulfide, and methyl ethyl ketone. Compatible with many chemical plasticizers such as DBP, DOP, TCP; with china wood oil, acrylics, some waxes, cellulose derivatives; with Piccolytes, Piccolastics, Piccoflexes, and other Picco resins.

Containers: Steel drums.

Uses: Textile applications, rubber compounding, and coatings.

pickle alum. See aluminum sulfate.

pickling acid. Sulfuric acid, usually 60° Bé. Used for treating iron and steel wire, plates, etc. in order to remove scale and rust.

"Pickling Compound No. 53." [175] Brand name for an acid solution of inhibitive coal-tar bases; dark-colored liquid; approximate wt/gal 10.16 lb.

Containers: 50-55-gal nonreturnable steel drums.

Uses: Retards excessive consumption of metal and evolution of fumes during removal of rust and scale from iron and steel products in acid pickling baths.

pico-. Prefix meaning 10^{-12} unit (symbol = p). 1 pg = 1 picogram = 10^{-12} gram.

alpha-**picoline** (2-methyl pyridine; 2-picoline) $C_5H_4N(CH_3)$, or $\underline{NC(CH_3)CHCHCHCH}$.
Properties: Colorless liquid; odor resembling pyridine; sp.gr. 0.952; b.p. 129°C; m.p. −69.9°C; miscible with water and alcohol.
Derivation: Dry distillation of bones or coal.
Containers: Drums; tank cars.
Uses: Organic intermediate for pharmaceuticals, dyes, rubber chemicals; solvent; source for vinyl pyridine.
Caution: Flammable.
Shipping regulations: None.*

beta-**picoline** (3-methyl pyridine; 3-picoline) $C_5H_4N(CH_3)$.
Properties: A colorless liquid; b.p. 143.5°C; m.p. −18.3°C; sp.gr. 0.9613 (15/4°C); refractive index 1.5060 (n 20/D); soluble in water, alcohol, and ether.
Derivation: Dry distillation of bones and coal.
Containers: Drums; tank cars.
Uses: Solvent in synthesis of pharmaceuticals, resins, dyestuffs, rubber accelerators, insecticides; in preparation of nicotinic acid, and nicotinic acid amide; waterproofing agents for fabrics.

gamma-**picoline** (4-methyl pyridine; 4-picoline) $C_5H_4N(CH_3)$.
Properties: Liquid; sp.gr. 0.957 (15/4°C); b.p. 144.9°C; refractive index 1.5050 (n 20/D); m.p. 3.8°C; soluble in water, alcohol, and ether.
Derivation: Dry distillation of bones or coal.
Containers: Drums; tank cars.
Uses: Solvent in synthesis of pharmaceuticals, resins, dyestuffs, rubber accelerators, insecticides; waterproofing agents for fabrics.
Shipping regulations: None.*

picoline-N-oxide $\underline{N(O)C(CH_3)CHCHCHCH}$. (2-picoline-N-oxide).
Properties: Crystals; very soluble in water. M.p. (2-isomer) 49.5°C; (3-isomer) 40.5°C; (4-isomer) 186.3°C.
Use: Chemical synthesis.

"Picragol." [24] Trademark for silver picrate (q.v.), for medicinal use.

picramic acid (picraminic acid; 2-amino-4,6-dinitrophenol; dinitroaminophenol) $C_6H_2(NO_2)_2(NH_2)OH$.
Properties: Red crystals. Soluble in alcohol, benzene, glacial acetic acid, aniline, and ether; sparingly soluble in water. M.p. 168°C.
Derivation: By partial reduction of picric acid.
Method of purification: Crystallization.
Grades: Technical.
Containers: 100-lb kegs; 300-lb barrels.
Use: Azo dyes; indicator; reagent for albumin.
Fire hazard: Dangerous.
Shipping regulations: None.*

picraminic acid. See picramic acid.

picric acid (picronitric acid; trinitrophenol; nitroxanthic acid; carbazotic acid; phenoltrinitrate) $C_6H_2(NO_2)_3OH$.
Properties: Yellow crystals; very poisonous! Explosive especially in contact with metals or metallic oxides. Soluble in water, alcohol, chloroform, benzene, and ether. Very bitter taste. Sp.gr. 1.767; m.p. 122°C.
Derivation: By the nitration of phenolsulfonic acid, obtained by heating phenol with concentrated sulfuric acid.
Grades: Technical; reagent.
Containers: 1-, 5-lb bottles; 25-lb boxes; 100-lb kegs; 300-lb barrels.
Uses: Explosives; medicine (external); dyes (starting point in manufacture of nigrosine and induline dyes, as well as substantive cotton dyes); textiles (dye producing bright yellow shades on the animal fibers, silk and wool); matches; electric batteries; etching copper; dyeing and printing textile fabrics with compound dyes which contain also such dyes as benzaldehyde green, methyl violet and indigo carmine; picrates.
Fire hazard: Dangerous; explosive; oxidizing material.
Shipping regulations: Various, according to whether dry or wet, type of packaging, and amounts.*

picrolonic acid $NO_2C_6H_4\overline{NNC(CH_3)C(NO_2)C}OH$. 3-Methyl-4-nitro-1-(para-nitrophenyl)-5-pyrazolone.
Properties: Yellow leaflets; m.p. 116-117°C; decomposes 125°C; slightly soluble in water; soluble in alcohol.
Use: Reagent for alkaloid identifications, for tryptophan and phenylalanine; for the detection and estimation of calcium.

picronitric acid. See picric acid.

picrotoxin (cocculin) $C_{30}H_{34}O_{13}$. A glucoside.
Properties: Flexible shining, prismatic crystals or microcrystalline powder; odorless; very bitter taste; stable in air; affected by light; m.p. 200°C. Soluble in boiling water, boiling alcohol, diluted acids and alkalies; sparingly soluble in ether and chloroform.
Derivation: Derived from the fruit of Anamirta paniculata or cocculus indicus, fishberries.
Grades: N.F.XI.
Use: Medicine.

"Pictol." [329] Trademark for monomethyl para-aminophenol sulfate, photo-developer.

"Pictone." [329] Trademark for a universal, "Pictol" hydroquinone developer, suitable for paper, press-type films and lantern slides. It produces neutral and cold-tone images on cold-tone types of papers.

Pidgeon process (ferrosilicon process; silicothermic process). Process for the production of high purity magnesium metal from dolomite by reduction with ferrosilicon at 1150°C under high vacuum.

piezochemistry. Study of reactions occurring at very high pressure, e.g., in interior of the earth's crust.

piezoelectricity. The property exhibited by certain crystals, of acquiring opposite electrical charges on different surfaces when subjected to mechanical stresses. Conversely, the property of expansion along one axis and contraction along another when subjected to an electrical field.

"Pigmentar." 296 Trademark for tar products derived from the distillation and decomposition of oleoresinous southern pine. Produced to various viscosity grades including thin, medium, heavy, extra heavy, and tar oil. Used primarily in rubber compounding and reclaiming, also in marine paints and roof coatings.

pigment E. See barium potassium chromate pigment.

pigments. A general term used for the various inorganic and organic, natural and synthetic chemical substances and mixtures that are used to confer color in manufacturing paints, printing inks, floor coverings of various types, rubber and many plastic compositions, leather, wax, chalk, crayons, cosmetics, etc. Pigments ordinarily impart color by the absorption of light but luminescent pigments, which emit colored light, are increasingly important. Examples are mentioned below and details are given under the various individual entries. Organic pigments generally are inferior to inorganic types as to lightfastness, heat resistance, and tendency to dissolve or "bleed" in oils and solvents but are available in brighter and more varied colors. Of the luminescent pigments the organic colors are much more intense than the inorganic types but are rapidly degraded by exposure to light. In paint applications, red lead and the chromate pigments, especially chrome yellow, are important for protection against corrosion of metal surfaces. Most commercial pigments are manufactured products. See under name of individual pigment.

A pigment is always a finely divided solid powder, insoluble but wettable under the circumstances of use. The pigment usually imparts opacity and body or consistency to the medium in which it is used. The distinction between pigments and dyes is not sharp, but pigments are almost without exception insoluble, while dyes are almost always organic substances for use in coloring textiles or other fibrous and plastic substances.

White pigments: Titanium dioxide is used where its whiteness and hiding power are required, especially in paints and printing inks; zinc oxide is used in compounding of rubbers where sulfur is a problem; calcium carbonate is used as an extender or pigment where a grey white is acceptable; barite, silica and china clay are used as extenders to control the cost of the final product. Lead white is used as a white pigment where sulfur is not present.

Black pigments: All of the black pigments used, with the exception of manganese black and iron oxide black, are carbonaceous in nature. These are seven distinct classes of such pigments, namely, lampblack, impingement carbon black, furnace black, thermal decomposition black, animal or bone black, vegetable or vine black and the miscellaneous carbon pigments of which graphite and mineral blacks are outstanding examples. As each of these vary somewhat in physical characteristics, their use is directed by the properties required for the final formulation.

Colored pigments: The natural and manufactured mineral pigments and the organic pigments have almost an unlimited range of hues. Many have been known and used for a great many years. Typical examples are the hydrated ferric oxide and manganese dioxide mixtures with clays (yellowish brown or sienna to dark brown or umber are obtained by heat treatment); Venetian red, Prussian blue, chromic oxide, chrome green, and copper or cobalt blues and blue greens. The organic pigments comprise a long list of insoluble colors derived from natural and synthetic dyestuffs. The natural products such as carmine (insect extract), alizarine and purpurin from extracted madder root and the like have become too costly to be largely used and synthetic products have replaced them. The synthetic product is used as a pure pigment or in precipitated form as a lake. Typical examples are the beta-hydroxynaphthoic pigments; the benzidine yellows; the triphenylmethane group; the metal-organic pigments using calcium, strontium, magnesium and manganese lakes with beta-naphthol; and the phosphotungstic - phosphomolybdic lake of methyl violet known as permanent violet.

Other types of pigments which produce color by luminescence rather than absorption are becoming more widely used. These are of the "daylight fluorescent" type based on zinc sulfide and cadmium sulfide with a metallic activator or organic compounds which emit radiation at a wave length in the visible range to produce the desired color such as the rhodamines and auramines.

pig-wrack. See chondrus.

pilchard oil. An oil expressed from the pickled fish, a member of the herring family.

Properties: Pale yellow oil; deposits stearin on long standing; sp.gr. 0.931-0.933; saponification value 186-189.6; refractive index 1.4751 (40°C).

Use: Making potash soft soap, paints.

pile. See nuclear reactor.

pill-bearing spurge. See euphorbia.

pilocarpine $C_{11}H_{16}N_2O_2$.
Properties: Colorless or yellow, hygroscopic, needle-like crystals or oil; very poisonous! Soluble in water, alcohol and chloroform; slightly soluble in ether. M.p. 34°C; b.p. 260° (5 mm).
Derivation: Alkaloid from the leaves of Pilocarpus jaborandi or Pilocarpus microphyllus.
Method of purification: Crystallization.
Grades: Technical.
Containers: Vials.
Uses: Medicine; hair pomades and tonics. Usually used in the form of the hydrochloride, nitrate or other salt.
Shipping regulations: None.*

pilocarpine hydrochloride $C_{11}H_{16}N_2O_2 \cdot HCl$. The hydrochloride of the alkaloid pilocarpine.
Properties: Colorless, translucent, odorless, faintly bitter crystals; hygroscopic; affected by light; solutions acid to litmus; very soluble in water; soluble in alcohol; nearly insoluble in chloroform; insoluble in ether. M.p. 200-203°C (?). Various authorities give values ranging from 195-205°C.
Grade: U.S.P. XVI.
Use: Medicine.

pilocarpine nitrate $C_{11}H_{16}N_2O_2 \cdot HNO_3$.
Properties: Shining, white crystals; stable to air; affected by light; m.p. 170-173°C. Soluble in water and alcohol; insoluble in chloroform and ether. Solutions acid to litmus.
Grades: U.S.P. XVI.
Use: Medicine.

pilocarpus (jaborandi).
Properties: Yellowish-green leaflets; bitterish, slightly salty, aromatic taste.
Chief constituent: Pilocarpine, a volatile oil, jaborine, and pilocarpidine.
Derivation: The dried leaflets of Pilocarpus jaborandi or P. microphyllus.
Occurrence: Brazil and Paraguay.
Grades: Pernambuco; Maranham.
Containers: Boxes; bales.
Uses: Medicine; source of pilocarpine.
Shipping regulations: None.*

pilocarpus oil. See jaborandi oil.

pimelic acid (1,7-heptanedioic acid) $HOOC[CH_2]_5COOH$. Is found in castor oil.
Properties: Crystals; m.p. 105-106°C; slightly soluble in water; soluble in alcohol and ether; nearly insoluble in cold benzene.
Use: Biochemical research; polymers; plasticizers.

pimelic ketone. See cyclohexanone.

pimenta (pimento; Jamaica pepper; allspice).
Properties: Dark brown berries or powder. The odor is throught to resemble that of a mixture of cinnamon, cloves and nutmeg, hence the name allspice.
Derivation: The dried, nearly ripe fruit of Pimenta officinalis.
Occurrence: East Indies; West Indies; Central America; South America.
Grades: Jamaican; Mexican.
Containers: Bags.
Uses: Perfumery; condiment.
Shipping regulations: None.*

pimenta oil (pimento oil, allspice oil).
Properties: Yellow to brownish-colored essential oil; agreeable, spicy, somewhat clove-like odor; pungent taste. Darkens with age; is affected by light. Phenol content: 65-80%. Soluble in 1 to 2 vols. and more of 70% alcohol (sometimes with opalescence and even turbidity upon dilution).
Chief known constituents: Eugenol, cineol, phellandrene, caryophyllene, eugenol methyl ether, palmitic acid.
Constants: Sp.gr. 1.018-1.048; optical rotation −0° to −4°; refractive index 1.5270-1.5400.
Derivation: By the distillation of the fruit of Pimenta officinalis, L.
Grade: N.F.XI.
Containers: Drums; cans.
Uses: Medicine; flavoring.
Shipping regulations: None.*

pimento. See pimenta.

pimento oil. See pimenta oil.

pi meson. See fundamental particle.

pinang. See areca nut.

alpha-**pinene** $C_{10}H_{16}$. A terpene hydrocarbon derived from sulfate wood turpentine in which it is the chief constituent.
Properties: Colorless, transparent liquid of characteristic terpene odor; sp.gr. 0.8620-0.8645 (15.5/15.5°C); refractive index (n 20/D) 1.4655-1.4670; boiling range 95% between 156-160°C; occurs in d-, ℓ-, and racemic forms.
Containers: Tank cars and galvanized drums.
Uses: Solvent for protective coatings, polishes, and waxes; synthesis of camphene, camphor, terpin hydrate, terpineol, synthetic pine oil, terpene esters and ethers, lube oil additives, synthetic resins, and their derivatives.

beta-**pinene** (nopinene) $C_{10}H_{16}$. A terpene hydrocarbon derived from sulfate wood turpentine in which it is a lesser constituent.
Properties: Colorless, transparent liquid of characteristic terpene odor; sp.gr. 0.8740-0.8770 (15.5/15.5°C); refractive index (n 20/D) 1.4775-1.4790; boiling range 95% between 164-169°C, laevo rotary.
Containers: Tank cars and galvanized drums.
Uses: Used primarily in polymer (Friedel-Crafts) resins, may be substituted for alpha-pinene.

pine-needle oils (firwood oil; douglas-fir oil). Derived from Douglas fir.
Properties: Greenish-yellow essential oil; limonene-like odor.
Chief known constituents: Terpenes.
Constants: Sp.gr. 0.8680 (23°C); optical rotation −62.5°; acid value 0; saponification value 86.6 (=30.3% bornyl acetate); after acetylation, 92.1 (=27.2% borneol).

The Douglas fir is found in the coniferous forests of North America.
Derived from Pinus mugo (dwarf pine needle oil).
Properties: Colorless to yellow liquid; pleasant aromatic odor; bitter pungent taste; soluble in alcohol; sp.gr. 0.853-0.871; optical rotation −5° to −15.5°; refractive index 1.4750-1.4800; boiling range, less than 10%, 165°C.
Grades: N.F. XI.
Containers: Glass bottles; drums.
Uses: Perfumery; medicine.
Shipping regulations: None.*

pinene hydrochloride. See bornyl chloride.

"Pine Oil 150 and 230." [296] Trade name for synthetic pine oils of various gravities consisting largely of terpineol and other monocyclic terpene alcohols. Used principally in disinfectants and soaps though is particularly well adapted for other uses where high alcohol content, light color, and odor are of prime importance.

pine oil, N.F.XI.
Properties: Colorless to light amber liquid having a characteristic pinaceous odor; miscible with alcohol in all proportions; sp.gr. 0.927-0.940; refractive index 1.4780-1.4820 (20°C) distilling range 200-225°C.
Derivation: From Pinus palustris by extraction and fractionation, or by steam distillation.
Grade: N.F. XI.
Uses: Deodorant; disinfectant.

"Pine Oil No. 220." [79] Trade name for a white pine oil.
Properties: Sp.gr. (15.5°C) 0.921; refractive index (20°C) 1.481; polymerization residue 0.6%; flash point (open cup) 150°F; Engler distillation 5%, 194°C; 50%, 207°C; 95%, 218°C.
Containers: 55-gal drums, tank cars.
Uses: Mining flotation; textile dyeing and cleaning; laundries; deodorants; cleaning compounds; paint and varnish.

pine oils. A somewhat loosely-defined term which covers a variety of volatile oils with characteristic pinaceous odors, consisting principally of isomeric tertiary and cyclic terpene alcohols, with variable quantities of terpene hydrocarbons, ethers, ketones, phenols and phenolic ethers. For example, some are obtained by the distillation of the cones and needles of the various species of pines; others are obtained from the stumps of the longleaf yellow pine trees; steam-distilled pine oils are obtained by the extraction of chipped stumps by means of solvents and hot steam. The pine stumps yield a pine oil known as destructively distilled pine oil or tar oil, wood (q.v.). Other commercial forms are steam distilled pine oil (obtained from pine wood by steam distillation or by solvent extraction followed by steam distillation) and synthetic pine oil (obtained by chemical hydration of terpene hydrocarbons to form terpene alcohols, or by dehydration of terpin hydrate).
Uses: Solvent for gums, resins, oils, cellulose ethers, other products; emulsifying agent; deodorizer; germicide; insecticide; chemicals (source of terpineol, terpene hydrate, fenchyl alcohol); alcohol denaturant; flotation agent; metal polishes; cutting oils; liquid soaps; emulsions of fats; greases and oils; paints and varnishes; fungicides; wax preparations; textile processing; other solvents, in the rubber industry; dyes; as antifoaming agents, wetting agents.
Shipping regulations: None.*

pine oleoresin. A blend of rosin and turpentine, usually encountered as a fused mass similar to rosin.

pine resin. See rosin.

pine tar. See tar, pine.

pine-tar oil. See tar oil, wood.

pine-tar pitch. The residue after distillation of practically all the volatile oils from pine tar. Similar to coal-tar pitch.

pinite. A variety of muscovite (q.v.) used in ceramics.

pink root. See spigelia.

pink salts. See rare earth salts.

pion. A pi meson, one of the fundamental particles (q.v.).

"Pipanol" Hydrochloride. [162] Trademark for trihexiphenidyl hydrochloride.

pipe clay. Usually of a ball clay nature but the term is rather wide in its application and may embrace almost any fine-grained, plastic clay.

2-pipecoline. See 2-methylpiperidine.

alpha-**pipecoline.** See 2-methylpiperidine.

3-(alpha-pipecolino)-1-propanol $CH_3C_5H_9N(CH_2)_3OH$.
Properties: A clear, colorless liquid. Refractive index 1.4740-1.4775.

pipenzolate methylbromide (1-ethyl-3-piperidylbenzilate methyl bromide) $C_{22}H_{28}BrNO_3$.
Grade: N.N.D.
Use: Medicine.

piperazidine. See piperazine.

piperazine (diethylenediamine; pyrazine hexahydride; piperazidine; ethyleneamine) $\underline{NHCH_2CH_2NHCH_2CH_2}$.
Properties: Colorless, deliquescent, transparent, needle-like crystals, which absorb carbon dioxide from the air. Keep well stoppered. Soluble in water, alcohol, glycerol, and glycols.
Constants: M.p. 104-107°C; b.p. 145°C.
Derivation: Treatment of ethylene bromide or chloride with alcoholic ammonia at 100°C.
Method of purification: Crystallization.
Grades: Technical.
Containers: Glass bottles; drums.

*See "I.C.C. Shipping Regulations," page xiii.
Reference numbers refer to name of manufacturer. See "List of Manufacturers," page v.

Uses: Medicine; corrosion inhibitor; anthelmintic; insecticide. Piperazine derivatives are suggested for surfactants, synthetic fibers, agricultural chemicals, stabilizing agents, rubber chemicals.
Shipping regulations: None.*

piperazine calcium edathamil $C_{14}H_{24}N_4O_8Ca$. A chelated compound produced by reacting edathamil with calcium carbonate and piperazine.
Grade: N.N.D.
Use: Medicine.

piperazine citrate $(C_4H_{10}N_2)_3 \cdot 2C_6H_8O_7 \cdot xH_2O$.
Properties: White crystalline powder; slight odor at most. Solutions acid to litmus. Soluble in water. Insoluble in alcohol and ether.
Grade: U.S.P. XVI.
Use: Medicine.

piperazine dihydrochloride $C_4H_{10}N_2 \cdot 2HCl$.
Properties: White needles. Soluble in water.
Uses: Fibers; insecticides; pharmaceuticals. The monochloride, $C_4H_{10}N_2 \cdot HCl$, is also commercially available.

piperazine estrone sulfate $C_4H_{10}N_2 \cdot C_{18}H_{22}O_5S$. When marketed for use as a drug it is stabilized with a small amount of free piperazine.
Properties: Fine, white to creamy white, odorless, crystalline powder; m.p. 185-195°C (dec. 240-250°C). Slightly soluble in water and alcohol.
Grade: N.F. XI.
Use: Medicine.

piperazine hexahydrate $C_4H_{10}N_2 \cdot 6H_2O$.
Properties: White crystals; m.p. 44°C. Soluble in water and alcohol.
Uses: Fibers; insecticides; pharmaceuticals.

piperidine (hexahydropyridine; pentamethyleneamine) $\overline{CH_2CH_2CH_2CH_2CH_2NH}$.
Properties: Colorless liquid with odor of pepper; sp.gr. 0.862; b.p. 106°C; m.p. -7° to -9°C; solidifies at -13° to -17°C. Soluble in water, alcohol, and ether. It is a strong base.
Derivation: By the reduction of pyridine with acid and a metal such as tin, or by heating piperine with alkali.
Grades: 95% and 98% pure.
Containers: Drums.
Uses: Solvent and organic chemical intermediate; curing agent in epoxy resins; agent in the manufacture of rubber; ingredient in oils and fuels; pharmaceutical intermediate.
Caution: Toxic; avoid breathing the vapors or allowing it come in contact with the skin.
Shipping regulations: Flammable liquid. Red label.*

2-piperidinoethanol (N-2-hydroxyethylpiperidine) $C_5H_{10}NCH_2CH_2OH$.
Properties: Sp.gr. 0.972-0.974 (20/4°C); b.p. 115-117°C (45 mm); refractive index (n 20/D) 1.478-1.480. Miscible with water and most organic solvents in all proportions.
Use: Intermediate.

3-piperidinopropiophenone hydrochloride $C_6H_5CO(CH_2)_2NC_5H_{10} \cdot HCl$.
Properties: A white to pale yellow, crystalline powder. M.p. not below 187°C.

piperidolate hydrochloride (1-ethyl-3-piperidyl diphenylacetate hydrochloride) $C_{21}H_{25}NO_2 \cdot HCl$.
Grade: N.N.D.
Use: Medicine.

piperidyl-1-cyclohexyl-1-phenyl-1-propanol hydrochloride. See trihexyphenidyl hydrochloride.

piperocaine hydrochloride $C_{16}H_{23}NO_2 \cdot HCl$. 3-(2-Methyl-1-piperidyl) propyl benzoate hydrochloride.
Properties: Occurs as small white crystals or as white crystalline powder; odorless and stable in air; m.p. 172-175°C; bitter taste plus sensation of numbness when placed on tongue. Solution (1 in 10) acid to litmus. Soluble in water, alcohol, and chloroform; almost insoluble in ether and fixed oils.
Grades: U.S.P. XVI.
Uses: Medicine.

piperonal. See heliotropin.

piperonyl aldehyde. See heliotropin.

piperonyl butoxide. Technical grade of 6-propylpiperonyl butyl diethylene glycol ether. Consists of 80% of the chemical and 20% of related compounds.
Properties: Light-brown liquid; mild odor; soluble in alcohol, benzene, petroleum hydrocarbons, "Freons." Sp.gr. 1.06 (25°C); refractive index 1.50 (20°C).
Containers: Glass bottles; tins; drums.
Use: Chiefly as a synergist in insecticides in combination with pyrethrins in oil solutions, emulsions, powders or aerosols.

piperonyl cyclonene. Technical grade of 3-isoamyl-5-(methylenedioxyphenyl)-2-cyclohexenone and its 6-carbethoxy derivative. Consists of 80% of the chemical and 20% of related compounds.
Use: As a synergist in insecticides in combination with rotenone, pyrethrins or rotenone-pyrethrins mixtures in oil solutions, emulsions, or powders.

piperoxan (benzodioxine) $C_{14}H_{19}NO_2$. 2-(1-Piperidylmethyl)-1,4-benzodioxan.
Properties: Usually encountered as the hydrochloride which melts at 232-234°C but darkens at slightly lower temperatures; it is also easily soluble in water, but the free base is not. The hydrochloride is soluble in isopropanol.
Use: Medicine.

piperylene (1,3-pentadiene) $CH_2{:}CHCH{:}CHCH_3$. Cis- and trans- forms.
Properties: Liquid; sp.gr. 0.693 (60/60°F); m.p. cis -141°C, trans -87°C; b.p. cis -44°C, trans -42°C; refractive index (n 20/D) cis 1.43634; trans 1.43008. Insoluble in water; soluble in alcohol and ether.
Uses: Polymers, maleic anhydride adducts; resins; intermediate.

pipestone. See catlinite.

pipradol hydrochloride (alpha, alpha-diphenyl-2-piperidine-methanol hydrochloride; alpha-2-piperidylbenzhydrol hydrochloride) $(C_6H_5)_2C(OH)C_5H_{10}N \cdot HCl$.

Properties: Small white crystals or crystalline powder. Odorless and stable in air. Soluble in water and in ethanol. Melting range 285-295°C; pH of 1% solution is between 5 and 7.

Grade: N.F. XI.

Use: Medicine.

pistachia galls. See mastic gum.

pitayine. See quinidine.

pitch. Thick, tenacious, dark-colored bituminous substances, secured either as the result of industrial destructive distillation processes, or as deposits on the earth's surface.

Properties: They are usually insoluble in water; miscible with carbon disulfide and benzene; have characteristic "tarry" odors.

Shipping regulations: None.*

See pitches, artificial, and also asphalt; asphaltite; and asphaltic pyrobitumen.

pitchblende. A massive variety of uraninite (q.v.), or uranium oxide, found in metallic veins. Contains 55-75% UO_2; up to 30% UO_3; usually a little water, and varying amounts of other elements. Thorium and the rare earths are generally absent.

Properties: Color black; streak brownish black; luster pitchy to dull; hardness 5.5; sp.gr. 6.5-8.5; radioactive.

Occurrence: Belgian Congo; Canada; Colorado; Europe.

Use: Most important ore of uranium; historical source of radium.

pitches, artificial. These pitches are usually named by the source from which derived. They may be divided into four general groups:

1. Residues from the distillation or oxidation of mineral oils. Blown asphalt (q.v.) is a representative of this class.
2. Residues from the distillation of tars. Coal-tar pitch, brown coal-tar pitch, coke-oven-tar pitch, blast-furnace-tar pitch, water-gas-tar pitch, generator-gas-tar pitch, wood-tar pitch, pine-tar pitch are representatives of this class.
3. Residues from the distillation of fusible organic substances, the process having been terminated before the actual formation of coke. Stearin and fatty acid pitches (q.v.) are a group which is representative of this class.
4. Various artificial mixtures having the general properties given in the definition of pitch (q.v.). Roofing pitch, brewer's pitch, and insulating pitch are representative of this class.

Shipping regulations: None.*

pitch, Jew's. See asphalt.

pitch, Judean. See asphalt.

pitch, mineral. See asphalt.

pitch, stearin. See stearin and fatty acid pitches.

"Pitocin." [330] Trademark for alpha-hypophamine.

"Pitressin." [330] Trademark for beta-hypophamine.

"Pitt Chem." [323] Trademark for a series of tar base protective coatings.

"Pitt Chem Hotline." [323] Trademark for a coal tar pipe enamel for high temperature conditions.

"Pittchlor." [177] $Ca(OCl)_2$. Trademark for a stable, high-test (70% available chlorine) calcium hypochlorite, granular in form; white; non-caking; water-soluble. Available in 3¾- and 5-lb cans; 100- and 130-lb drums. Used as a sanitizer, germicide, and bleach.

"Pittsburgh PX." [323] Trademark for a line of plasticizers.

pituitary (posterior pituitary). The dried, cleaned, powdered posterior lobe obtained from the pituitary body of domesticated animals which are used for food by man.

Properties: A yellow or grayish amorphous powder with characteristic odor; partially soluble in water.

Grade: U.S.P. XVI.

Use: In medicine.

pivalic acid. See trimethylacetic acid.

pix. See pitch.

pix pini. See pine tar.

pK. A measurement of the completeness of an incomplete chemical reaction. It is defined as the negative logarithm (to the base 10) of the equilibrium constant, K, for the reaction in question. The pK is most frequently used to express the extent of dissociation or the strength of weak acids, particularly fatty acids, and amino acids, and also complex ions, or similar substances. The weaker an electrolyte the larger its pK. Thus, at 25°C for sulfuric acid (strong acid), pK is about -3.0; acetic acid (weak acid), pK = 4.76; boric acid (very weak acid), pK = 9.24. The pK of a weak acid equals the pH of an equimolar solution of the acid and one of its salts. See pH.

"Placidyl." [3] Trademark for ethchlorvynol (q.v.).

plantago seed. The cleaned, dried ripe seed of Plantago psyllium or of Plantago indica, known in commerce as Spanish or French psyllium seed; or of Plantago ovata, known in commerce as blond psyllium or Indian plantago seed.

Properties: All varieties are nearly odorless and have a bland mucilaginous taste.

Grade: N.F. XI.

Use: Medicine.

*See "I.C.C. Shipping Regulations," page xiii.
Reference numbers refer to name of manufacturer. See "List of Manufacturers," page v.

plant hormones. Organic compounds, other than nutrients, which in small amounts promote, inhibit or modify any physiological process within the plant. Strict usage limits the term plant hormone to materials produced by growing plants. Similar substances of synthetic origin are usually referred to as growth regulators or growth substances. In many discussions the several terms are used synonymously and the term auxin includes all substances of this kind. An example which occurs naturally in the plant is 3-indoleacetic acid. This is responsible for cell elongation in the stem. Gibberellin, originally found in the fungus Gibberella, is known to exist in, and affect stem elongation in biennials. Other effects are to regulate root growth (indolebutyric acid, alpha-naphthalene acetic acid); fruit development in plants (indolebutyric acid); loss of leaves (alpha-naphthalene acetic acid); prevention of fall of fruit (2,4-D); killing of plants (2,4-D).

"Plaskon." [175] See "Halon."

plasma.
1. The liquid part of the blood separated by centrifugation.
2. The mixture of electrons and gaseous ions (electrically charged atoms or parts of atoms), with or without neutral atoms, which forms when any substance is heated to very high temperatures (i.e., 10,000 to 50,000°F or higher). A plasma may be formed by an electric arc of sufficient power, by sonic shock waves, or by other very sudden releases of very large quantities of energy, as in nuclear processes of fission or fusion. Uses are for spraying heat resistant coatings on missile cone surfaces and rocket nozzles, and to study high temperature chemistry and physics. In some cases the term plasma is used for materials that are very hot but still below the temperature at which electrons and ions are formed in appreciable quantities.

plaster.
1. A paste made by mixing together varying proportions of lime, sand and water, together with hair or other binding material. Used as a surface coating for walls, ceilings and partitions in buildings.
2. Plaster of paris, usually used for ornamental or intricate parts of plaster work.

plaster, hard-finished. Plaster made from over-burnt gypsum, which is dipped in alum solution and calcined a second time. Keene's cement and Parian cement are examples. See gypsum cements.

plaster of Paris. See gypsum cements.

plaster retarders. Substances used to slow up the setting of plaster. Blood, glue, dextrin, and hair are among those used.

"Plastex." [160] Trademark of wires and cables with oilproof and flameproof polyvinyl chloride insulation which resists the action of oxygen, ozone, and sunlight; has high dielectric strength, high resistance to water, acids, and alkalies. It is firm, dense, and has a smooth finish. Supplied in several colors.

plastic. A material that contains as an essential ingredient an organic substance of large molecular weight, is solid in its finished state, and, at some stage in its manufacture, or in its processing into finished articles, can be shaped by flow (definition from ASTM D883-54T). The term is sometimes used to include inorganic materials of similar character. The terms plastic and resin (q.v.) are used in overlapping senses but resin applies more specifically to the more or less chemically homogeneous polymers used as starting materials in the production of molded articles while plastic signifies the final solid product, which may contain fillers, plasticizers, stabilizers, pigments, etc.

"Plasticizer 136." [175] Brand name for an aryl alkyl hydrocarbon, used as a secondary plasticizer in vinyl formulations.

Containers: 55-gal steel drums.

plasticizers. Materials added to a plastic to facilitate compounding and improve flexibility and other properties of the finished product. The first industrial plasticizer was camphor, used to make "Celluloid" from nitrocellulose. At present, the important plasticizers are nonvolatile organic liquids or low-melting solids, especially the phthalate, adipate and sebacate esters and aryl phosphate esters. Plasticizers are used principally in the vinyl and cellulosic resins.

"Plasticone Red." [141] Trade name for pyrazolone red color pigments.

Properties: Good light resistance and heat resistance. Good resistance to acid and alkali. Non-bleeding in water and organic vehicles.

Uses: Paints, enamels, lacquers, plastics, rubber, printing inks, textiles and floor coverings.

"Plastic Steel." [445] Trade name for a mixture of steel and plastic materials used for repairing broken machinery, and for making plastic and rubber molds, metal forming dies, etc.

"Plastimer." [282] Trade name for a castor oil residue.

Typical properties: Dark brown color; acid value 108; iodine value 102; saponification value, 195; sp.gr. 0.94; refractive index 1.4795 (25°C).

Containers: Drums.

Uses: As plasticizer in dark colored compounds; can be sulfonated to form detergents.

plastisol. Liquid dispersion of finely divided resin in a plasticizer. It is usually 100% solid with no volatiles; when volatile content exceeds 5% of the total weight it is called an organosol. When the plastisol is heated, the plasticizer solvates the resin particles, and the mass gels. With continued application of heat the mass fuses to become a conventional thermoplastic material.

Plastisols are useful for molding, casting films, or coating, or printing with synthetic resins, often without the use of volatile solvents or high processing temperatures.

"Plasto." [243] Trademark for a line of solvent-soluble dyes used for coloring plastics.

"Plastogen." [69] Trademark for a plasticizing agent.
Properties: Liquid, amber to mahogany; sp.gr. 0.81-0.84; acid number 1.0-1.1.
Uses: Plasticizer and softener in all elastomers; effective in sponge rubber.

"Plastolein." [242] Trademark for a line of plasticizers for vinyls, cellulosics and synthetic rubbers, Primarily composed of esters and polyesters of azelaic and pelargonic acids.

plate glass. See glass.

"Platformate." [416] Trademark for the catalytic reformate product produced by the Platforming process; i.e., an aromatic-rich hydrocarbon mixture. See "Platforming process."

"Platforming" Process. [416] Patented process using special platinum-containing catalyst for making high octane gasoline and/or a highly aromatic fraction for subsequent recovery of pure aromatics. Reactions include aromatization, dehydrogenation, cyclization, isomerization and hydrocracking. Reaction product may contain up to 60% aromatics. By-product hydrogen also is produced.

platinic-ammonium chloride (ammonium chloroplatinate; platinic sal ammoniac; platinum-ammonium chloride) $(NH_4)_2PtCl_6$.
Properties: Orange-red crystals, or yellow powder. Slightly soluble in water; insoluble in alcohol.
Constants: Sp.gr. 3.06; m.p., decomposes.
Grades: Technical; C.P.
Containers: Glass bottles.
Uses: Plating; platinum sponge.
Shipping regulations: None.*

platinic chloride. See chloroplatinic acid; platinum chloride.

platinic sal ammoniac. See platinic-ammonium chloride.

platinic-sodium chloride (platinum-sodium chloride; sodium chloroplatinate; sodium platinichloride) $Na_2PtCl_6 \cdot 4H_2O$.
Properties: Yellow powder. Soluble in alcohol, water.
Grades: Technical; C.P.
Containers: Glass bottles.
Uses: Etching on zinc; ink (indelible); microscopy; mirrors; medicine; photography; plating; catalyst; determination of potassium.

platinic sulfate. See platinum sulfate.

platinous-ammonium chloride (ammonium chloroplatinite; platinous sal ammoniac; platinum-ammonium chloride) $PtCl_2 \cdot 2NH_4Cl$.
Properties: Dark ruby-red crystals. Soluble in water; insoluble in alcohol.
Constants: M.p., decomposes; sp.gr. 2.94.
Grades: Technical.
Containers: Glass bottles.
Use: Photography.
Shipping regulations: None.*

platinous chloride. See platinum dichloride.

platinous iodide. See platinum iodide.

platinous-potassium chloride. See potassium chloroplatinite.

platinous sal ammoniac. See platinous-ammonium chloride.

platinous-sodium chloride (platinum-sodium chloride; sodium chloroplatinite; sodium platinochloride) $Na_2PtCl_4 \cdot 4H_2O$.
Properties: Dark red crystals. Soluble in water.

platinum Pt. Element of atomic number 78, group VIII of the periodic system. See also platinum black.
Properties: Silvery-white ductile metal. Does not tarnish at any temperature. Insoluble in mineral and organic acids; soluble in aqua regia. Attacked by fused alkalies.
Constants: Sp.gr. 21.45; m.p. 1773.5°C; Brinell hardness, hard 97, annealed 42 (i.e., harder than silver or gold).
Derivation: Occurs alluvially in Russia, Colombia, Alaska; mined like gold in South Africa; main source is now as a by-product in electrolytic recovery of nickel from Canadian nickel ores. The natural material is generally admixed with the other platinum group metals and with gold, iron, etc. The pure metal is obtained by dissolving the crude material in aqua regia, precipitating the platinum by ammonium chloride as ammonium platinum chloride, igniting the precipitate to form platinum sponge. This is then melted in the oxyhydrogen flame or in an electric furnace.
Grades: Physically pure (99.99%); chemically pure (99.9%); crucible platinum (99.5%); commercial (99.0%).
Uses: Catalyst (nitric acid, sulfuric acid, etc.); laboratory ware of all kinds (dishes, crucibles, electrodes, wire, etc.); industrial equipment; spinnerets for rayon manufacture; jewelry; dentistry; electrical contacts; thermocouples; electroplating; high temperature furnace lining.

platinum-ammonium chloride. See platinic-ammonium chloride and platinous-ammonium chloride.

platinum (ous) barium cyanide (barium platinocyanide; barium cyanoplatinite) $BaPt(CN)_4 \cdot 4H_2O$.
Properties: Yellow or green crystals; m.p.

decomposes.
Grades: C.P.
Use: X-ray screens.
Shipping regulations: Poison, class B. Poison label.*

platinum black (platinum Mohr). Finely divided metallic platinum.
Properties: Black powder; exhibits a metallic luster when rubbed. Soluble in aqua regia. Sp.gr. 15.8-17.6 (apparent).
Derivation: Reduction of solution of a platinum salt with zinc or magnesium.
Grades: Technical.
Containers: Glass bottles.
Uses: Catalyst; absorbent of gases (hydrogen, oxygen, etc.) which it again liberates at red-heat; gas ignition apparatus.
Shipping regulations: None.*

platinum chloride (platinum tetrachloride; platinic chloride) (a) $PtCl_4$; (b) $PtCl_4 \cdot 5H_2O$. The platinum (ic) chloride of commerce is usually chloroplatinic acid (q.v.).
Properties: (a) Brown solid; (b) red crystals. Soluble in water and alcohol.
Constants: (a) M.p., decomposes; (b) sp. gr. 2.43; m.p., loses $4H_2O$ at 100°C.
Derivation: By solution of platinum in aqua regia and evaporation.
Method of purification: Crystallization.
Grades: Technical.
Containers: Glass bottles.
Use: Chemical reagent.
Shipping regulations: None.*

platinum dichloride (platinous chloride) $PtCl_2$.
Properties: Greenish-gray powder which forms double salts with the chlorides of the alkali metals. Soluble in hydrochloric acid and ammonium hydroxide; insoluble in water.
Constants: Sp.gr. 5.87; m.p., is decomposed at red-heat, yielding platinum.
Derivation: (a) By heating platinum sponge in presence of dry chlorine; (b) by heating chloroplatinic acid to 200°C.
Grades: Technical.
Containers: Glass bottles.
Use: Platinum salts.
Shipping regulations: None.*

platinum diiodide. See platinum iodide.

platinum iodide (platinous iodide; platinum diiodide) PtI_2.
Properties: Heavy, black powder. Slightly soluble in hydriodic acid; insoluble in alkalies, water.
Constants: Sp.gr. 6.4; m.p. 300-350°C (dec).

platinum-iridium alloys. These are the most important of the platinum alloys. Commercial alloys contain 1-30% iridium. As the iridium is increased the hardness of the alloy increases, as does the resistance to chemical attack. The m.p. of platinum is raised by the addition of iridium.
Uses: Jewelry ("medium" platinum is 95% Pt, 5% Ir and "hard" platinum is 90% Pt, 10% Ir); electrical contacts (10-25% Ir); fuse wire (10-20% Ir), hypodermic needles (20-30% Ir), and in general where hard noncorrodible material is needed.

platinum, liquid bright. See Liquid Bright Platinum.

platinum-lithium $LiPt_2$. Brittle, metallic-looking solid, nonreactive with water, made by direct combination at 540°C. If the lithium and platinum are combined at 200°C, the product can be decomposed by water, hydrolyzing and dissolving the lithium, and leaving behind unusually active platinum catalyst.

platinum metals. A group of six metals, all members of group VIII of the periodic system. They include ruthenium, rhodium, palladium, osmium, iridium, and platinum.

platinum, Mohr. See platinum black.

platinum-potassium chloride. See potassium chloroplatinate; also potassium chloroplatinite.

platinum-rhodium alloys. These alloys are in commercial use up to 40% rhodium. They are harder than platinum, but not as hard as the corresponding platinum-iridium alloys. The addition of rhodium to platinum increases the resistance to attack by aqua regia. The melting points of the alloys are higher than that of platinum.
Uses: Catalyst in nitric acid production; high temperature vessels; furnace resistors; thermocouples; spinnerets in the rayon industry.

platinum-sodium chloride. See platinic-sodium chloride and platinous-sodium chloride.

platinum sponge.
Properties: A grayish-black, porous mass of finely divided platinum. Soluble in aqua regia.
Derivation: By the ignition of platinum-ammonium chloride.
Grades: Technical.
Containers: Glass bottles.
Uses: Catalyst; ignition of hydrogen in Doebereiner lamps, illuminating gas, etc.
Shipping regulations: None.*
See also platinum black.

platinum sulfate (platinic sulfate) $Pt(SO_4)_2$.
Properties: Greenish-black mass. Hygroscopic. Soluble in acids (dilute), alcohol, ether, water.
Grades: Technical.
Use: Analysis (microtesting for bromine, chlorine, iodine).

platinum tetrachloride. See platinum chloride.

Plessy's green (chromium phosphate). Impure $CrPO_4$ with 2 to 6 H_2O.
Properties: A deep green pigment which is both stable toward chemical reagents and durable toward atmospheric influences. It consists of chromium phosphate mixed with variable amounts of chromium oxide and calcium phosphate.
Derivation: By boiling a solution of 1 part potassium bichromate in 10 parts water

with 3 parts of a solution of acid calcium phosphate and 1 part sugar.

pleurisy root. See asclepias.

"Plexiglas" Molding Powders.[23] Trademark for thermoplastic poly(methyl methacrylate)-type polymers in bead or granule form. Injection and compression molding grades, in a wide range of colors. Outstanding brilliance, optical properties, resistance to weathering, good dimensional stability.

Uses: Manufacture of lenses of automobile tail and stop lights, back-up lights, medallions, ornaments; molded parts for appliances, camera lenses, letters for signs; extruded sheet.

"Plexiglas" Sheet.[23] Trademark for thermoplastic poly(methyl methacrylate)-type polymers in cast sheet. Solid sheets in wide range of sizes, thicknesses. Colorless, colored, transparent, translucent, fluorescent, pearlescent grades. Strong, lightweight, impact-resistant plastic. Readily formable to complex shapes; machineable. Certain grades flame-retardant, craze-resistant, or ultraviolet filtering.

Use: Aircraft canopies and windows, outdoor signs, light diffusers, industrial and architectural glazing, chalk-boards, boat windshields, and other industrial and commercial products.

"Plexol."[23] Trademark for synthetic lubricants and additives for petroleum oils. Liquids. Most grades are diesters of dibasic acids, some are polyesters or polyether alcohols. The ester lubricants have very low freezing points, high flash points, little change of viscosity with temperature.

Uses: Aircraft engine lubricants, hydraulic systems, instrument oils, petroleum-base lubricant formulations.

"Plimasin."[305] Trademark for a compound containing tripelennamine hydrochloride U.S.P. and methylphenidate hydrochloride N.N.D.

"Pliobond."[265] Trademark for a general purpose adhesive which forms an excellent strong bond with all sorts of like and unlike surfaces. Adheres well to wood, glass, ceramics, metals, plastics, leather, rubber, concrete, and plaster. May be employed either as a wet adhesive, by solvent reactivation, or through the use of heat.

"Pliofilm."[265] Trademark product. It is said to be a rubber hydrochloride.

Properties: It is available in sheet or roll form in thicknesses of 0.001 inch or more. In most usual forms it is transparent. Inherently resistant to moisture vapor and water. Also resistant to oils, greases and most solvents. Is heat sealable and very flexible. Used as packaging film.

"Plioflex."[265] Trademark for a series of staining and nonstaining vulcanizable synthetic rubbers produced by the continuous polymerization of butadiene and styrene. They include hot, cold and oil-extended types, and lightcolored nonstaining polymers. Antioxidants are added to insure protection of the polymer during production and storage.

Properties: 23.5% styrene content, plus good stability in the raw polymer; good tensile strength, abrasion resistance, flexibility, and recovery in the cured polymer.

Uses: Tires, tubes, miscellaneous mechanical goods, proofed goods, hospital sheeting; wringer rolls; molded and extruded goods; shoe soles and heels; housewares, etc.

"Plioform."[265] Trademark for cyclized rubber, made by action of acid catalysts such as chlorostannic acid on natural rubber.

Properties: Tasteless, odorless, thermoplastic, resistant to solvents; requires no vulcanization.

Uses: In forms of lacquers, powder, and molded forms which may be machined.

"Pliolite."[265] Trademark for a series of butadiene-styrene copolymers or cyclized natural rubber resins produced primarily for coating and rubber compounding applications.

"Pliolite NR." A cyclized derivative of natural rubber. It is a hard, resinous material that has good solubility and produces films having excellent chemical resistance and excellent resistance to the transmission of water vapor.

Uses: Paper coatings, hot melts, vehicle for silk screen printing inks, rubber reinforcing, and paint vehicles in specialized coatings.

"Pliolite Resinous Latices." A series of styrene-butadiene emulsion polymers recommended for paint, paper, and textile applications. They possess a uniform particle size, good film clarity, excellent mechanical stability, and alkali resistance.

Uses: Latex paints, paper impregnation, paper coating, wire insulation, dipped goods, adhesives, masonry cement additive, textile printing and dyeing, textile binders, sizes, finishes and saturants.

"Pliolite Rubber Latices." A series of butadiene-styrene synthetic copolymers supplied in latex form. They vary in solids content from 24-60%, contain no residual styrene, and are essentially odorless.

Uses: Foam rubber, carpet backings, textile sizes, adhesives, tire cord dips, miscellaneous uses as textile binder and finishing agent.

"Pliolite S-3." A synthetic copolymer of butadiene and styrene. The resin has a specific gravity of 1.05, and is produced in the form of white friable granules. In a highly aromatic solvent system, the resin forms a gel at low concentrations. Small proportions of the resin impart desirable viscosity properties and control pigment settling in paints based on "Pliolite S-5."

Uses: With "Pliolite S-5" in masonry and

stucco paint formulations.

"Pliolite S-5." A thermoplastic resin available in the form of white, porous granules, odorless and tasteless. The pure resin produces films of unusual clarity, strength, hardness, and exceptional chemical resistance. The resin is particularly resistant to acids and alkalis making it extremely well suited for use as a paint vehicle in masonry paints.

Uses: Masonry and stucco paints, concrete floor enamels, sprayable industrial finishes, traffic marking paints, wall sealers, and aluminum paints.

"Pliolite S-7." A solution of a synthetic styrene copolymer resin in toluene. Specific gravity is 0.915, with a solids content of 30%. This solution grade resin is rigidly controlled in production to meet specific requirements of the moisture-proof packaging field.

Uses: Coatings for films, foils, glassine, bleached krafts, and paperboard.

"Pliolite S-6B." A styrene-butadiene copolymer of high styrene content. In rubber stocks, "Pliolite S-6B" upgrades physical properties of the rubber stock and aids processing.

Uses: Reinforcing resin for hot, cold and oil-extended rubbers; range of products includes shoe soles and heels, inner soles, leather-like goods, sporting equipment, rubber flooring, molded and extruded goods, and automotive parts.

"Plio-Tuf." [265] Trademark for a relatively new group of internally reinforced, high styrene resins. The resins are characteristically pale in color and light in weight. From these resins are produced rigid, stiff thermoplastic sheets with high impact strength even at low temperatures. Sheets made from "Plio-Tuf" resins are hard, tough, flexible, light, and unusually resistant to heat distortions. These sheets can be postformed economically by various methods into hundreds of useful products.

Uses: Tote boxes, advertising displays, lamp housings, serving tray, business machine housings, television masks, hoods, vents, luggage, toys, automotive parts, and athletic equipment.

"Pliovic." [265] Trademark for a group of vinyl chloride polymers, characterized by uniform particle size and porosity. The "Pliovic" resins are highly resistant to chemical attack and to the effects of heat and light. Supplied in the form of fine white powders, the resins are easily compounded for various forming operations including calendering, extruding, and injection molding. They are commonly available as dry blending resins. An electrical grade resin is available for specialized applications. "Pliovic AO" is a vinyl chloride copolymer dispersion resin available for organosols, plastisols, and plastigels. "Pliovic Latex 300" is a vinyl chloride copolymer, high solids latex, available for fabric sizing, fabric and paper coating, and impregnating.

Uses: Film, sheeting, flooring, electrical insulation, hose, gaskets, tubing, toys, foams, paper and fabric coating, rug backing, felt impregnation, and fabric sizing.

"Plioweld." [265] Trademark for a process for adhering rubber to metal.

Properties: The process is used for building rubberlined tanks and similar installations where rubber, in all its modifications, can be bonded to metal.

plumbago. See graphite.

plumbic acid, anhydrous. See lead dioxide.

plumbo-plumbic oxide. See lead oxide, red.

"Plumb-O-Sil B & C." [304] Trademarks for coprecipitates of lead orthosilicate vinyl stabilizers.

Properties: Soft white powders; sp. gr. (B) 3.3, (C) 2.96; refractive index (B & C) 1.58.

Containers: Fiberboard drums containing 25-, 135- and 50-lbs.

Uses: Heat and light stabilizers for translucent and brightly colored vinyl film, sheeting, and upholstery stocks. Imparts resistance to plate-out.

plumbous oxide. See litharge.

plumbous sulfide. See lead sulfide.

plumbum. The Latin name for lead, hence the symbol Pb and the names plumbic and plumbous.

"Pluracol." [203] Trademark for a series of polyoxyalkylene glycols of relatively high molecular weights. Includes polyethylene and polypropylene glycols; also includes a 3000-molecular weight polyoxypropylene derivative of glycerol. Various members of the series are useful as raw materials for nonionic detergents, textile specialties, and other surface active agents. Used as lubricants for rubber molds and textile yarns, and as carriers for various thermoplastic agents for external preparations. Also used in the production of pre-polymer and one-shot flexible urethane foams.

"Pluronic." [203] Trademark for a series of nonionic surface-active agents prepared by the addition of ethylene oxide to polypropylene glycols. They are available in liquid, paste, flake, and powder form and all are 100% active agent. Major uses are as low foaming detergents, lime soap and pigment dispersing agents, emulsifying and demulsifying agents, wetting agents and plasticizers of resins.

"Pluto." [307] Trademark for direct dyestuffs. Used for the dyeing of cotton, rayon, leather, paper. Characterized by good fastness to acid.

plutonium Pu. A synthetic element with atomic number 94 first prepared as the 238 isotope by bombarding uranium with deuterons. Several isotopes have since been made ranging in half-lives from 26 minutes to about 500,000 years and in mass numbers from 232-243. Plutonium 239 is of major importance since it is fissionable with slow neutrons. It is produced on a large-scale basis by bombarding uranium 238 with slow neutrons in a nuclear reactor. Plutonium 239 is used as a nuclear reactor fuel and as an explosive ingredient for nuclear weapons. Plutonium exists in nature in uranium-containing ores in very small concentrations. Pitchblende contains one part plutonium for 10^{14} parts uranium. Naturally occuring plutonium seems to consist only of the 239 isotope. For this reason it is not thought to be primeval but to form by the action of stray neutrons on uranium. Plutonium is similar chemically to uranium and neptunium, forming analogous compounds such as PuO_2, PuF_3, PuF_4, PuOCl, etc.

plutonium 239. See plutonium.

"Plyac." [50] Trademark for polyethylene spreader-sticker; non-oily and non-ionic.

"Plyacien." [36] A protein base dust-free adhesive for use with the cold press no-clamp gluing process.
Use: Manufacture of interior grade plywood.

"Plyamine." [36] A group of liquid water-soluble urea-formaldehyde adhesive resins used as binders in the manufacture of plywood, furniture, wood particle products, etc. Physical properties: nonvolatile 45-65%; volatile matter, water; stable in storage at 25°C, over 6 months; color, water-white.

"Plyophen." [36] Trademark for a water-soluble impregnating resin. Penetrates deeply and quickly into wood, canvas, asbestos, paper and other laminating and molding stocks. Confers water and chemical resistance. Can be diluted as much as 8-10 parts water to 1 part resin for spraying glass fiber or rock wool.

Pm. Symbol for promethium.

P.M. Abbreviation for Pensky-Martens, a type of flash point test procedure and apparatus.

PMA. Abbreviation for phosphomolybdic acid; and for pyromellitic acid.

PMAC. Abbreviation for polymethoxyacetal.

"PMAS." [49] Trademark for a colorless, odorless, stable water solution containing phenylmercury derivatives, used as a fungicide, germicide and herbicide.

PMHP. Abbreviation for para-menthane hydroperoxide. A polymerization catalyst.

PMP. Abbreviation for 1-phenyl-3-methyl-5-pyrazolone.

PMTA. Abbreviation for a mixture of phosphomolybdic and phosphotungstic acids, used in making pigments. See phosphotungstic pigments.

pn. An abbreviation for propylenediamine, as used in formulas for coordination compounds. See also dien; en; py.

"P.N." [57] Trademark for phenothiazine-nicotine.

Po. Symbol for polonium.

podophyllin (podophyllum resin).
Properties: Light yellow, green yellow, or light brown powder or small yellow, bulky, fragile lumps; bitter acrid taste; affected by heat and light. The dust is very irritating to the eyes and the mucous membranes. Soluble in alcohol, ether, ammonium hydroxide, chloroform, potassium hydroxide solution and sodium hydroxide solution; insoluble in water.
Chief known constituents: Podophyllotoxin, picropodophyllin, etc.
Derivation: Extraction of the rhizome or roots of Podophyllum peltatum with alcohol and subsequent precipitation from the extract with acidified water.
Grades: Technical; U.S.P. XVI.
Containers: 1-, 5-, 10-lb glass bottles; fiber cans; drums.
Use: Medicine.
Shipping regulations: None.*

podophyllum (May-apple; mandrake; devil's apple; Indian apple; vegetable calomel).
Derivation: Dried rhizome and roots of Podophyllum peltatum. Should yield not less than 5% podophyllum resin.
Occurrence: North America; China; Himalayas.
Grades: Technical; U.S.P. XVI.
Containers: Various size bags.
Use: Medicine.
Shipping regulations: None.*

podophyllum resin. See podophyllin.

POEMS. Abbreviation for polyoxyethylene monostearate (q.v.); see also polyoxyl 40 stearate.

POF. See dℓ-alpha-lipoic acid.

pogy oil. See menhaden oil.

poise. See centipoise.

poison.

1. In nuclear technology a poison is any material with a high capture probability for neutrons that may divert an undesirable number of neutrons from the fission chain reaction. A burnable poison is a material intentionally introduced into a nuclear reactor core intended to capture a determined fraction of the neutrons and be burned out by the nuclear reaction at the same rate that other poisons are formed in the fission process, thus maintaining a constant reactivity of the reactor core.
2. A substance that reduces or destroys the activity of a catalyst. Thus carbon monoxide, or phosphorus, arsenic, or sulfur compounds have this effect on the formation of ammonia from hydrogen and nitrogen gases, and the gases must be

*See "I.C.C. Shipping Regulations," page xiii.
Reference numbers refer to name of manufacturer. See "List of Manufacturers," page v.

highly purified to carry on a satisfactory plant production operation.

poison black cherry. See belladonna.

poison hemlock. See conium.

poison nut. See nux vomica.

poison parsley. See conium.

poison tobacco. See hyoscyamus.

poke root. See phytolacca.

"Polar." [175] Trade name for naphthalene.

"Polaris Red." [141] Trade name for precipitated azo pigments derived from beta-hydroxynaphthoic acid.
Properties: Good resistance to light, good heat resistance, non-bleeding in water and organic solvents.
Grades: Medium and dark shades of red.
Uses: Printing inks, paints, enamels, lacquers, rubber, plastics, and floor coverings.

"Polectron." [307] Trademark for a group of dielectric compounds.
"Polectron" monomer: N-vinylcarbazole.
Properties: Colorless to yellowish crystalline or flake material; m.p. 61-65°C; soluble in methyl alcohol, ethyl alcohol, pentane, hexane, cyclohexane, carbon tetrachloride, ethyl acetate, tetralin, dioxane, chloroform, acetone, methylene dichloride and chlorbenzene.
Uses: A dielectric impregnant for stationary electrical assemblies, such as rolled and stacked condensers.
"Polectron" polymer:
Properties: Grey to light brown solid varying in form from a powder to broken lumps. This material is not suitable for direct use in injection molding or for fibering purposes. Its excellent electrical properties are retained even at elevated temperatures and over a broad frequency range. May be molded by either compression or injection methods.

polianite MnO_2. A crystalline variety of pyrolusite (q.v.).

"Polidase." [91] Trademark for cultured vegetable enzyme product.
Properties: Shows activity through a range from pH 4-8.
Grades: "Polidase-S" (reagent grade); "Polidase-C" (crude).
Uses: As a laboratory reagent; in experimental nutrition; for the recovery of silver from photographic film; as a component of leather bates; for the conversion of starch in mashes used for alcohol production; as a low cost industrial enzyme.

polishing acids. See oxalic, hydrochloric, nitric or sulfuric acid.

polishing crocus. A form or grade of finely divided ferric oxide used for polishing. See iron oxide reds.

"Politol S," "Politol N." [229] Trademarks for combinations of "Indulin" and phosphates.
Properties: Brown, free-flowing powders. "S" is soluble in water; "N" is insoluble. Bulk densities 25-30 lbs/cu ft. Moisture 2-10%.
Containers: Packed in 50-lb MWP bags.
Uses: Degumming, clarifying and refining animal, vegetable, and fish oils and fats in accordance with U.S. Patent 2,654,766.

pollucite $Cs_4Al_4Si_9O_{26} \cdot H_2O$. A natural cesium aluminum silicate found in pegmatites. Colorless; hardness 6.5; sp.gr. 2.9. Source of cesium.

polonium Po. Element of atomic number 84, a member of group VI of the periodic system. Polonium is the first element to be discovered by means of its radioactivity. It occurs naturally as a decay product of radium in pitchblende, 25,000 tons of pitchblende containing about 1 gram. Polonium has been prepared synthetically by bombarding bismuth with neutrons. No stable isotopes have been found. The properties of polonium, which have been studied essentially by tracer techniques, are similar to those of tellurium, and, to a lesser extent, to those of bismuth. Elementary polonium may be obtained by preparing polonium chloride and reducing it with zinc. It is volatile at about 1000°C. Compounds similar to those formed by other members of the family have been prepared.

poloxalkol (oxyethylene oxypropylene polymer) $HO(C_2H_4O)_a(C_3H_6O)_b(C_2H_4O)_cH$. Relatively tasteless, non-ionic, surface-active agent.
Grade: N.N.D.
Use: Medicine.

poly-. A prefix signifying many. For example, a polymer is an aggregate formed by combination of a number of single molecules.

"Polyac." [28] Trademark for a butyl rubber conditioner containing 25% poly-para-dinitrosobenzene $[C_6H_4(NO)_2]_x$ with an inert wax.
Properties: Dark brown waxy pellets; sp.gr. 0.96.
Containers: 100-lb drums.
Uses: A processing aid and accelerator of vulcanization for butyl rubber.

polyacetylenes. One type, a dark polymer made from acetylene by use of Ziegler-Natta catalysts, has a conjugated structure, -CH:CH-CH:CH-, and has been suggested as a semiconductor. Non-conjugated types have been synthesized, such as H-C:C[$(CH_2)_4$C:C]$_8$-H, or $C_{50}H_{66}$, which is a solid, and have been suggested as high-energy binders, and plasticizers for solid propellants.

polyacrylamide $(CH_2CHCONH_2)_x$. White solid; water-soluble high polymer.
Uses: Thickening agent; suspending agent; production of uranium; additive to adhesives.
See also acrylate resins.

polyacrylic acid. A polymer of acrylic acid (q.v.) used as a sizing agent for nylon and

other synthetic fabrics.
See also acrylate resins.

polyalcohols (polyhydric alcohols; polyols). See alcohol.

polyamide resins. These are polymers made by condensation of diamines with dibasic acids, or by polymerization of lactams or amino acids. Polyamines or polybasic acids may also be used. The various kinds of nylon (q.v.) are the most important examples. See also "Versamids."

polyamine-methylene resin. A polyethylene polyamine methylene substituted resin of diphenylol dimethylmethane and formaldehyde in basic form.
Properties: Light amber, granular, freely flowing powder with appreciable odor. Insoluble in dilute acids and alkalies, alcohol, ether, and water.
Grade: N.N.D.
Use: Medicine, as an ion-exchange resin.

polyaminotriazoles (PAT). Synthetic polymers made from sebacic acid and hydrazine with small amounts of acetamide. Fibers have high tenacity, good elasticity and good dyeability. Polyoctamethylene-aminotriazole is a specific example.

polybasite $Ag_{16}Sb_2S_{11}$. A natural silver sulfantimonide.
Properties: Color steel gray to iron black; streak black; luster metallic; hardness 2-3; sp.gr. 6.0-6.2.
Occurrence: Nevada, Colorado, Idaho, Mexico, Chile, Europe.
Use: Ore of silver.

polybutadiene. Synthetic rubber made from butadiene, $H_2C{=}CH{-}CH{=}CH_2$. A number of different stereospecific polymer forms are possible and have been made and properties studied. Some of these are or will be in substantial production because of superior properties to older synthetic rubbers, particularly in those cases where synthetics were inferior to natural rubber, as in heat dissipation under heavy loads, and abrasion resistance. Cis-1,4-polybutadiene has the same stereospecific structure as natural rubber, and is referred to as a natural-synthetic type. It is stated to have improved properties but is harder to process. Commercial products of this type are known as "Budene," "Diene," and "Cis-4." Ziegler, Alfin and other stereospecific catalysts are used to produce the desired stereospecific polymers; the exact nature of the catalyst and conditions determine whether cis-1,4-, trans-1,4- or other polymers are produced. Trans-1,4-polybutadiene has also been made on a semi-commercial basis, being designated by the trademark "Trans-4," and resembling natural trans-polyisoprene rubbers such as gutta-percha and balata, and also being similar to synthetic trans-polyisoprene rubbers (q.v.). Trans-1,4-polybutadiene is hard, crystalline, and resin-like at room temperature but is thermoplastic at 180-280°F.

The designation 1,4- means that when the polymer molecules grow by union of simple monomer molecules of butadiene:

$$\underset{(1)\ (2)\ \ (3)\ \ (4)}{H_2C{=}CH{-}CH{=}CH_2} + \underset{(1)\ (2)\ \ (3)\ \ (4)}{H_2C{=}CH{-}CH{=}CH_2} + \underset{(1)\ (2)\ \ (3)\ \ (4)}{H_2C{=}CH{-}CH{=}CH_2} + \text{etc.}$$

the No. 1 carbon atom of any one butadiene molecule is joined to the No. 4 carbon of another butadiene molecule so that in the large polymer molecule, all carbon atoms are in one long chain.

It is also possible to cause butadiene to polymerize 1,2- so that the No. 1 carbon of each butadiene molecule becomes attached to the No. 2 carbon of another molecule. When this occurs the main backbone of the resulting polymer contains only the No. 1 and No. 2 carbons, while all the No. 3 and No. 4 carbons are in vinyl side chains.

```
 (1)  (2)  (1)  (2)  (1)  (2)  (1)  (2)  (1)
—C — C — C — C — C — C — C — C —C —etc.
      |         |         |         |
  (3) C     (3) C     (3) C     (3) C
      ||        ||        ||        ||
  (4) C     (4) C     (4) C     (4) C
```

These 1,2-polybutadienes exist in isotactic, syndiotactic and atactic forms, but cannot have cis and trans forms.
See polymer, stereospecific.

cis-**4-polybutadiene.** A synthetic rubber which promises to be an economical and stable alternate and/or substitute for natural rubber in many of its uses. Has low heat build-up, superior abrasion resistance, high resilience. Blends with natural rubber. Produced by solution polymerization from butadiene.
Containers: 75-lb film-wrapped bales in cardboard cartons.
Uses: In heavy-duty truck and passenger car tire treads and other items usually requiring natural rubber.

trans-**4-polybutadiene.** A synthetic polybutadiene of high trans configuration having a high melting point, high hardness and superior abrasion resistance. Blends with natural and synthetic rubbers or gutta percha.
Containers: Shipped in crumb form in cardboard cartons.
Uses: Golf ball covering; belting; footwear; floor tile; mechanical goods.

polybutadiene-acrylic acid copolymer. Binder now used in solid fuels for rockets.

polybutenes (polybutylenes). This term covers the polymers of isobutene of varying molecular weights that have been known since 1955 and before, and also the more recently developed polymers of butene-1 and butene-2, some of which are of the tactic type.

Butyl rubber is a polyisobutene, of relatively high molecular weight, having some isoprene units present in the polymer chain. Isobutene alone can be polymerized to various degrees, so that as the molecular weight changes from about 10 isobutene units to 1000 units, the polymer materials change from oils to tacky waxes to crystalline waxes to rubbery solids. These materials have had some use as lubricant supplements, viscosity index improvers, and binding agents in calking compounds, sealing tapes, and impregnated paper, but utilization has been limited.

Polymers of butene-1 can form films and plastic pipe, and by use of Ziegler catalysts isotactic polybutene-1 can be produced, with properties somewhat similar to those of polypropylene.

polybutylenes. See polybutenes.

polycarbonate resin. The commercial variety of this resin is derived from bisphenol A and phosgene, and has the following structure and composition: $(-COOC_6H_5C(CH_3)_2C_6H_5O-)_n$. In general, a polycarbonate resin might be formed from any dihydroxy compound and any carbonate diester.

Properties: The available resin is transparent, noncorrosive, nontoxic, stain resistant, self-extinguishing, has low water absorption, high impact strength, good heat resistance, dimensional stabilities, and electrical properties.

Uses: Molded products that range from dentures to ball bearings. Metals can be replaced by these resins in numerous operations.

polycarboxylic acids. Organic acids with two or more -COOH groups on a molecule.

polychloronaphthalenes. See chloronaphthalene oils and waxes.

polychlorotrifluoroethylene. See chlorotrifluoroethylene resins.

"Polycin." [202] Trademark for (1) an elastic, tacky, gel-like solid resulting from the complete oxidation of castor oil; used in rubber compounding, asphalt tile manufacture, and as a polymeric plasticizer; (2) a series of castor oil-derived polyols used in the preparation and curing of urethane polymers for protective coatings, foamed insulation and elastomers.

"Polyco." [65] Trademark for series of thermoplastic polymers in the form of water emulsions or solvent solutions. This name is applied to vinyl acetate polymers and copolymers, butadiene-styrene copolymer latices, polystyrenes, vinyl and vinylidene chloride copolymers, acrylic copolymers, and water-soluble polyacrylates.

Uses: Adhesives and coatings, in paint, leather, textiles, paper, cosmetics and construction fields.

polycrase $(Y, Ca)(Nb, Ta, Ti)_2O_6$. A natural oxide of rare earth metals, niobium, tantalum, titanium, calcium, and other metals. May contain uranium and thorium.

Properties: Color black; luster submetallic to greasy or vitreous; streak yellowish to reddish brown; hardness 5.5-6.5; sp.gr. 5.0-5.9.

Occurrence: South Carolina, North Carolina, Texas, Europe.

"Polycryl." [65] Trademark for a series of acrylic polymers and copolymers polymerized in solvent medium and supplied as solutions in toluene, methyl ethyl ketone or other solvent.

poly-1,1-dihydroperfluorobutyl acrylate.

Properties: A white, elastomeric gum rubber; sp.gr. 1.5; begins to degrade at 300°F; non-flammable; retains strength and elastomeric properties in contact with synthetic lubricants, solvents, hydraulic fluids, oils, etc., at temperatures in the range of 300-400°F; has limited flexibility at sub-zero temperatures.

Grades: Available in two grades; as a raw gum stock having the properties outlined above and as an aqueous latex containing 30% solids.

Containers: 1-, 2-, 5-, and 10-lb polyethylene bags inside fibreboard cartons; 25-lb polyethylene bags inside steel pails; larger quantities available.

Uses: In the fabrication of various rubber products, such as O-rings, seals, gaskets, diaphragms, hose, sheets, etc., as coating for fabric and other surfaces.

polyelectrolytes. High molecular weight electrolytes of either natural origin (proteins, polysaccharides) or of a synthetic nature (example - alkyl halide addition products to polyvinyl pyridine). They may be either weak or strong electrolytes. Since the polyelectrolytes in solution do not dissociate to give a uniform distribution of positive and negative ions, as do simple electrolytes, the ions of one sign are bound to the polymer chain. Thus for instance, the positive charges may be in the polymer chain, and only negative ions will be free to diffuse through the solvent.

polyester fiber. Generic name for a manufactured fiber in which the fiber-forming substance is any long chain synthetic polymer composed of at least 85% by weight of an ester of a dihydric alcohol and terephthalic acid (Federal Trade Commission). See, for example, "Dacron."

polyester resins. A large group of synthetic resins, almost all produced by reaction of dibasic acids with dihydric alcohols. In a few cases trifunctional monomers such as glycerol or citric acid are used.

The term polyester resins applies especially to the products made from unsaturated dibasic acids such as maleic. Other types are discussed at the end of this article. The unsaturated polyester resins can be further polymerized through cross linking because they are unsaturated. Often another unsaturated monomer such as styrene is added during this second stage of the polymerization, which can occur at ordinary temperature with suitable peroxide catalysts. In general the first stage of polymerization, i.e., the ester formation stage, is carried out at a chemical plant, while the second or vinyl polymerization can be carried out conveniently in the field. The most common applications involve the use of the unsaturated polyester resins plus a co-monomer to impregnate fabric, often glass fabric, which is then formed to the desired shape on a pattern or mold, followed by hardening of the resin to form a fabric-reinforced plastic item.

Properties: Reinforced polyester items are often lighter than aluminum and sometimes stronger than steel. They are resistant to corrosion, erosion, rusting, chemicals, solvents, detergents, salt water, weather, rot, and exhaust fumes. Addition of fireproofing agents is necessary to achieve non-flammable characteristics.

Derivation: Maleic anhydride and fumaric acid are the usual unsaturated acid components, while phthalic anhydride, or adipic or azelaic acid are the corresponding saturated materials. Commonly used glycols are ethylene, propylene, diethylene, dipropylene, and certain butylene glycols. The added polymerizable monomer is styrene, vinyltoluene, diallyl phthalate or methyl methacrylate.

Water is eliminated during the first stage combination of acid and glycol, but the second stage produces no gas or liquid products. This curing stage is exothermic but fast and relatively simple for field use.

Uses: Boat hulls; swimming pools; industrial glazing, awnings, partitions, panels, etc.; automotive parts, heater housings, ducts, plates, moldings and frames, trays, boxes, luggage; radomes and other airplane structural parts. Also used as a pipe joint sealer, and as adhesive.

In addition to the unsaturated polyester resins, there are other important types. One large group are the alkyd resins. These are made from saturated acid and alcohol monomers with many types of modifications, usually the inclusion of an unsaturated fatty acid. Special groups of polyester resins are used in producing films and fibers, foams, and plasticizers. "Dacron" and "Mylar" are the outstanding examples. Some polyurethane resins incorporate polyesters.

polyester rubber. See polyurethane rubber.

polyether foams. This term refers to polyurethane foams that have been made by use of a polyether as distinct from a polyester or other resin component.

polyethylene $(C_2H_4)_n$ Polymerized ethylene, available in various forms, but the white leathery resinous form is by far the most common.

Description: In general it is light weight, tasteless, odorless, and nontoxic. The low molecular weight polymers are high grade lubricating oils or oil additives (See "A-C" polyethylenes). The medium weight polymers are waxy materials miscible with paraffin. The high molecular weight materials (molecular weight greater than 6000) are tough white, leathery, resinous materials. The term polyethylene usually refers to the latter. Copolymers of polyethylene are also widely used and are sometimes referred to as polyethylene even though it may comprise only 50% of the total material.

Properties: These resins have outstanding electrical characteristics and impermeability to water, as well as being generally resistant to organic solvents and chemicals (particularly acids, alkalies and oxygenated solvents). They are translucent and remain tough and flexible even at low temperatures. The molded material is essentially non-breakable, flexible, easily processed and colored.

Typical properties of film: M.P. 85 to 110°C; sp.gr. 0.92; tensile strength, 1400-2600 psi; elongation, 200-800%; tearing strength (Elmendorf), 100-1000g; water absorption, 24 hours, less than 0.001%; low rate of transmission of water vapor. Resistance, good toward acids, alkalies, grease and oils, organic solvents (at 60°F), water, dampness, sunlight, heat (212°F), extreme cold (−150°F), fungus growth. Flammability – slow burning. The film is permeable to gases and vapors.

Derivation: Three density grades of high molecular weight material are produced: (a) low density, (b) medium density, (c) high density. This latter is a linear isotactic polymer.

(a) Low density material uses polymerization of ethylene at 1,000 to 2,500 atm and temperatures of 100-300°C in the presence of a peroxide catalyst. An alternative process employs the presence of an aromatic hydrocarbon, which acts as a diluent in preventing cross-linking.

(b) Intermediate pressures (500-1,500 psig) and temperatures (350-500°F), with a metallic oxide catalyst (molybdenum trioxide or chromic oxide, deposited on alumina) produce medium density material.

(c) A low temperature (140-175°F), low pressure (100 psig) process which employs a highly active catalyst gives high density polyethylene. The usual catalyst is an alkyl metal derivative (such as triethyl aluminum) activated with titanium tetrachloride or another heavy-metal derivative.

Forms: Available in milled or unmilled flake or cubed form with or without antioxidant and modifiers; extruded, cast, and calendered sheeting; extrusion and injection molding compounds. Also available as filaments, foam, and powder.

Uses: Films and sheets for packaging and covers of all kinds (thick sheets are used as radiation shields); molded materials such as all types of containers ranging from chemical and other equipment to kitchenware, pipe and tubing; electric wire insulation and various electrical parts and fixtures; coatings on metal, paper, and other surfaces; bottles; tank and pipe linings; textiles (filler cloth) and bristles; upholstery; cordage; window glazing.

polyethylene glycol. Name for polymers of ethylene glycol with the general formula $HOCH_2(CH_2OCH_2)_nCH_2OH$ or $H(OCH_2CH_2)_nOH$. Average molecular weights range from about 200 to at least 6000. Properties vary with molecular weight.
Properties: Clear, colorless, odorless, viscous liquids to waxy solids; soluble or miscible with water and for the most part with alcohol and other organic solvents; heat-stable; inert to many chemical agents; do not hydrolyze or deteriorate; have low vapor pressure.
Derivation: By condensation of ethylene glycol, or of ethylene oxide and water.
Uses: Chemical intermediates (lower molecular weight varieties); plasticizers; softeners and humectants; lubricants; bases for cosmetics and pharmaceuticals.
Descriptions of some typical polymers follow. The numbers indicate average molecular weight.

polyethylene glycol 300
$HOCH_2(CH_2OCH_2)_nCH_2OH$ (n varies from 5 to 5.75; average molecular weight from 285 to 315).
Properties: Clear, colorless, viscous liquid; m.p. −15 to +8°C; sp.gr. 1.124-1.130 (20°C); slightly hygroscopic; soluble in water; pH (1 in 20 solution) 4.0-7.0. Also soluble in alcohol, acetone and other glycols; insoluble in ether and aliphatic hydrocarbons.
Grade: N.F. XI; technical.
Uses: See general article.

polyethylene glycol 400. Similar to polyethylene glycol 300; n varies from 8 to 10; average molecular weight 400.
Properties: Liquid; sp.gr. 1.110-1.140 (20°C); m.p. 4-10°C; flash point 224°C.
Grades: Technical; U.S.P. XVI.

polyethylene glycol 1000.
Properties: White, waxy solid; m.p. 35-40°C; average molecular weight 1000; flash point above 232°C; solubility in water, about 70% by weight.
Uses: Cosmetics; pharmaceuticals; plasticizer, lubricants in textile processing; metal extrusion and molding of rubber articles; softeners and humectants for textiles, paper, cork, etc.

polyethylene glycol 1540. Average molecular weight 1300-1600.
Properties: White, waxy, plastic material, having a consistency similar to beeswax, and having a slight, characteristic odor. Melting range 42-46°C; pH of 5% solution between 4.0 and 7.0. Very soluble in water and chloroform; moderately soluble in alcohol; insoluble in ether.
Grade: N.F. XI.
Use: Cosmetics; pharmaceuticals; plasticizer.

polyethylene glycol 4000 $H(OCH_2CH_2)_nOH$. (n varies from 70 to 85.)
Properties: Pale creamy-white, waxy solid or flakes; m.p. 53-56°C; practically odorless and tasteless; soluble in water, alcohol, and chloroform; insoluble in ether. pH (1 in 20 solution) 4.5-7.5.
Grade: U.S.P. XVI.

polyethylene glycol 6000 $HO(CH_2CH_2O)_nH$.
Properties: White waxy solid; m.p. 58-62°C; flash point above 246°C; soluble in water.
Grade: Technical.

polyethylene glycol chlorides $H(OCH_2CH_2)_nCl$.
A group of polymers, usually colorless liquids with very low vapor pressures at room temperature. Molecular weights from about 100 to about 600 have been reported. Miscible with water; sp.gr. (20°C) for a low molecular weight polymer is 1.18, for a high molecular weight polymer 1.14. The former sets to a glass at −90°C, the latter sets to a wax-like solid at 20°C. Used as solvents for cleaning, extracting, and dewaxing.

polyethylene glycol tert-dodecylthioether.
Properties: Straw-colored liquid; sp.gr. 1.04 (20/20°C).
Containers: 5-gal cans, 55-gal drums, tank trucks, tank cars.
Uses: Non-ionic surface-active agent; textile detergent.

polyethylene glycol esters. See polyoxyethylene esters.

polyethylene glycol stearate. See polyglycol distearate; polyoxyl 40 stearate.

polyethylene oxide sorbitan fatty acid esters. See polyoxyethylene sorbitan fatty acid esters.

polyethylene terephthalate. A polyester resin (q.v.) formed from ethylene glycol and terephthalic acid. The product has a melting point of 265°C and a second order transition temperature of 70 to 80°C. It is produced as oriented films or fibers characterized by high strength, good electrical properties, and resistance to moisture.

A typical film ("Mylar") is made in thicknesses from 0.00025 to 0.01 inch and is useful from −60 to 150°C. It is used as electrical insulation for capacitors, motors, generators, transformers, and as a barrier tape for wire and cable. Its unusual balance of properties also permits its use in many non-electrical fields, such as decorative laminations, vapor barrier materials, as a printed covering for acoustical tile and various types of industrial tapes and magnetic recording tapes.
See also "Dacron", "Mylar", "Terylene" and "Cronar."

polyethylene thiuram sulfide.
Derivation: From oxidation of diammonium ethylene bisdithiocarbamate with calcium hypochlorite.

Grades: 50% vegetable powder; 95% technical powder.
Containers: 50-lb multiwall paper bags.
Use: Fungicide.

"Poly-FBA." [158] Trade name for poly-1, 1-dihydroperfluorobutyl acrylate (q.v.).

"Polyfil" Clay. [285] Proprietary brand name for chemically treated hydrous aluminum silicate (sedimentary kaolin) from South Carolina.
Properties: Sp.gr. 2.60; brightness 78-79 G.E.; airfloated; contains color stabilizer.
Containers: 50-lb multiwall bags or bulk.
Uses: A specially prepared and treated clay to impart color stability in PVC resin compounding for floor tile and electrical compositions. Imparts abrasion, scratch and scuff resistance.

"Polyfilm." [233] Trademark for virgin polyethylene film specially formulated for converting and packaging.

"Polyflo." [416] Trademark for a series of all-organic inhibitor-dispersants designed for treatment of all fuel and diesel oils. Depending on precise inhibitor needs, this series contains products (1) to improve jet fuel thermal stability, (2) to stabilize No. 2 heating oils and diesel fuels against color deterioration and sludge deposition, (3) to eliminate fouling in feed exchangers and reboilers, (4) to eliminate tank bottoms in crude and residual oil storage tanks. Each member of the series exhibits a different degree of effectiveness in different types of stocks.

"Polyfon H, O, T, R & F." [229] Trademarks for sodium lignosulfonates derived from "Indulin."
Properties: Brown, free-flowing powders; soluble in water, acids, and alkalies; sintering above 360°C; bulk density 25-30 lbs/cu ft; moisture 2-8%.
"Polyfon H": Sodium sulfonate groups 5.8%; sp.gr. 1.42; pH 8.0-8.2; surface tension (5% solution) 47 dynes/cm.
"Polyfon O": Sodium sulfonate groups 10.9%; sp.gr. 1.46; pH 9.0-9.5; surface tension (5% solution) 44 dynes/cm.
"Polyfon T": Sodium sulfonate groups 19.7%; sp.gr. 1.58; pH 9.7-10.2; surface tension (5% solution) 40-41 dynes/cm.
"Polyfon R": Sodium sulfonate groups 26.9%; sp.gr. 1.68; pH 10.0-10.5; surface tension (5% solution) 43-47 dynes/cm.
"Polyfon F": Sodium sulfonate groups 32.8%; sp.gr. 1.76; pH 10.2-10.7; surface tension (5% solution) 44-45 dynes/cm.
Containers: Packed in 50-lb MWP bags.
Uses: In dyestuff dispersion and levelling; boiler water conditioning; foam stabilization; insecticide compounding; emulsion thinning; concrete plasticizing; ceramics plasticizing and binding; drilling muds; and other dispersing applications.

polyform. A process for producing gasoline by the thermal non-catalytic conversion of naphthas and gas oils at high temperatures and pressures in the presence of recirculated C_3 and C_4 hydrocarbon gases. The gasoline produced is designated polyform distillate.

polyformaldehyde. See paraformaldehyde.

polyformaldehyde resins. Better known as acetal resins (q.v.).

"Poly G." [84] Trademark for a series of polyethylene glycols, polypropylene glycols and polyoxypropylene adducts of glycerol.
G200, 300, 400, and 600 are liquid polyethylene glycols; G1000, 1500, GB-1530 and GB-2000 are waxy polyethylene glycols. The number indicates the approximate molecular weight. See polyethylene glycols.
G420P, 1020P, 2020P are propylene oxide condensation polymers of propylene glycol. See polypropylene glycols.
G1030PG, 3030PG, 4030PG are propylene oxide condensation polymers of glycerol. See polyoxypropylene-glycerol adducts.

"Polygard." [248] Trade name for a mixture of alkylated aryl phosphites.
Properties: Liquid; clear amber; sp.gr. 0.99; soluble in acetone, alcohol, benzene, carbon tetrachloride, solvent naphtha and ligroin; insoluble in water but can hydrolyze.
Use: Nondiscoloring and nonstaining stabilizer for synthetic rubber and plastics.

polyglycol amine H-163 $HO[C_2H_4O]_2C_3H_6NH_2$.
Properties: Colorless liquid; sp.gr. 1.0556; b.p. 278°C; f.p. 14.5°C; soluble in water; wt/gal 8.8 lbs; flash point 295°F.
Containers: Drums.

polyglycol distearate (polyethylene glycol distearate) $C_{17}H_{35}COO(CH_2CH_2O)_xOCC_{17}H_{35}$. Distearate ester of polyglycol.
Properties: A soft, off-white solid; density 1.04 (50°C); m.p. 43°C; pH of 10% dispersion 7.26; saponification number variable; soluble in chlorinated solvents, light esters and acetone. Slightly soluble in alcohols, insoluble in glycols, hydrocarbons and vegetable oils.
Use: Used as plasticizer for various resins; as a component of grinding and polishing pastes to promote easy removal in water.

polyglycols. See polyoxyethylene esters.

polyhalite $2CaSO_4 \cdot MgSO_4 \cdot K_2SO_4 \cdot 2H_2O$. A naturally occurring potash salt found in Germany, Texas, and New Mexico.
Use: Source of potash for fertilizer.

polyhexamethyleneadipamide $[-NH(CH_2)_6NHCO(CH_2)_4CO-]_n$. A nylon obtained by the condensation of hexamethylenediamine with adipic acid. Used in food-processing equipment; for fibers. See also nylon 66.

polyhexamethylene sebacamide. See nylon 610.

polyhydric alcohols. See alcohol.

"Poly-Ionic." [300] Trademark for a series of polyethylene emulsions. Used as softeners and lubricating agents for textiles. Grades are A(anionic), C(cationic), N(nonionic).

polyisobutylenes. See polybutenes.

polyisoprene. Polymer found as the major component of natural rubber, and also made synthetically. Can exist in several forms including stereospecific cis-1, 4-, and trans-1, 4-polyisoprene. Both of these can be produced synthetically by the effect of heat and pressure on isoprene in the presence of stereospecific catalysts. Natural rubber is cis-1, 4-; the synthetic cis-1, 4- is therefore referred to by terms such as synthetic natural rubber. The latter is proposed for use in large size rubber tires as a complete or partial replacement for natural rubber. See "Natsyn" and "Coral." Trans-1, 4-polyisoprene resembles gutta-percha and is tough because its molecules are able to partially crystallize. It is therefore proposed for use as a replacement and extender for natural balata or gutta-percha in golf ball covers, shoe soles, floor tile, and cable covering. The corresponding polybutadiene derivatives may turn out to be more economic than the polyisoprenes.

polyisoprene, deutero. Polyisoprene in which heavy hydrogen atoms have replaced the ordinary hydrogen of atomic weight one. Cis-1, 4-deuteropolyisoprene is stated to be more elastic than natural rubber.

"Polylite." [36] A group of 100% reactive alkyd resins, dissolved in styrene and other monomers. Highly diversified applications both alone and in combination with such materials as fibrous glass. This group also includes resins for use with diisocyanate to form rigid or flexible polyurethane foams.

"Polymax." [332] Trademark for woven fabrics of polyolefin fibers for liquid and pneumatic filtration.

polymer. A substance, often synthetic, composed of giant molecules that have been formed by the union of a considerable number of simple molecules with one another. The number of simple molecules that unite to form a polymer molecule varies from two to hundreds or thousands. The simple molecules that will undergo such a change are known as monomers and their union is called polymerization. The monomer molecules may be all alike, or there may be two or more varieties of monomer involved in the formation of a single polymer. Thus ethylene molecules can be united with themselves to form polyethylene resin which is a monopolymer. On the other hand, SBR synthetic rubber (the most common type) is a copolymer, since two different kinds of monomers (styrene and butadiene) are required. See also dimer and trimer.

Condensation polymers are those such as the phenol-formaldehyde resins, in which the union of the simple molecules involves incidental formation of water or some simple substance. Many natural polymers are known, for example, natural rubber.

See also polymer, stereospecific.

polymerization. See polymer.

polymerized oils. See blown oils.

polymer, stereospecific (See also polymer structure.) A polymer whose molecular chains have a definite and specific spatial arrangement in the sense that there is a fixed position in three dimensional space for the various atoms and the various parts of the molecular chain with respect to one another. By way of contrast, the atoms and parts of molecules of a nonstereospecific polymer would have a random and varying spatial arrangement with respect to one another.

Synthetic stereospecific polymers and methods of making them have been known only recently. Natural rubber and its synthetic counterpart, cis-polyisoprene (see "Natsyn," "Coral") are examples of commercially useful polymers. Most older synthetic polymers were not stereospecific.

Five types of stereospecific structures are recognized. These are trans, cis, isotactic, syndiotactic (also spelled syndyotactic) and tritactic. Nonstereospecific polymers are designated atactic.

The isotactic and syndiotactic structures most commonly arise when double bonds are absent in the carbon chain but there are atoms other than hydrogen, or substituting groups, attached to the carbon atoms of the backbone chain. This type of polymer molecular structure is very common and is found in:

polystyrene $\left[\begin{array}{cc} H & C_6H_5 \\ | & | \\ -C & -\ C- \\ | & | \\ H & H \end{array}\right]_n$

polyvinyl chloride $\left[\begin{array}{cc} H & Cl \\ | & | \\ -C & -\ C- \\ | & | \\ H & H \end{array}\right]_n$

polypropylene $\left[\begin{array}{cc} H & CH_3 \\ | & | \\ -C & -\ C- \\ | & | \\ H & H \end{array}\right]_n$

and 1, 2-polybutadiene $\left[\begin{array}{cc} H & HC{=}CH_2 \\ | & | \\ -C & -\ C- \\ | & | \\ H & H \end{array}\right]_n$.

Whenever molecules of this type are formed from their simple monomer building blocks, the substituting atoms or groups may be arranged so that (1) all are above, (2) all below, or (3) all on the same side of the backbone chain when it is arranged so that all the backbone atoms are in a single plane. The three arrangements are actually completely identical from a three-dimensional point of view. This is an isotactic arrangement. In somewhat oversimplified form, this may be represented in two dimensions by the two equivalent structures following, in which R might be

C_6H_5, Cl, CH_3, or $-HC{=}CH_2$.

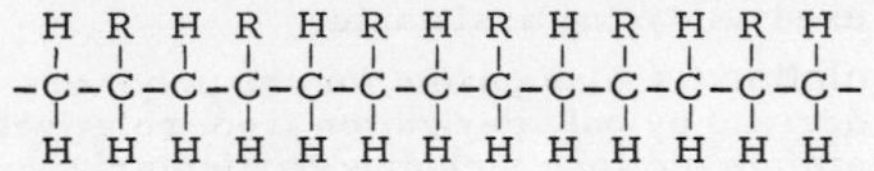

```
 H R H R H R H R H R H R H
 | | | | | | | | | | | | |
-C-C-C-C-C-C-C-C-C-C-C-C-C-
 | | | | | | | | | | | | |
 H H H H H H H H H H H H H
```

or

```
 H H H H H H H H H H H H H
 | | | | | | | | | | | | |
-C-C-C-C-C-C-C-C-C-C-C-C-C-
 | | | | | | | | | | | | |
 R H R H R H R H R H R H R
```

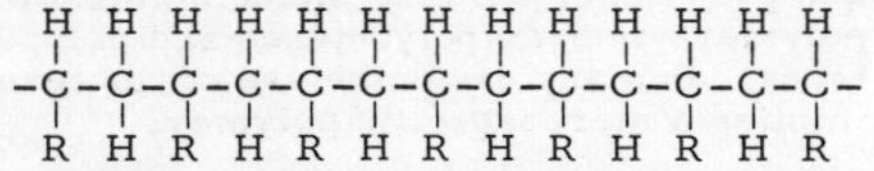

The syndiotactic arrangement arises when some of the substituting groups are above the plane of the backbone chain, and the remainder are below, but there is a definite and regular pattern of arrangement of these substituting groups such as alternate groups lying on each side of the backbone, represented as:

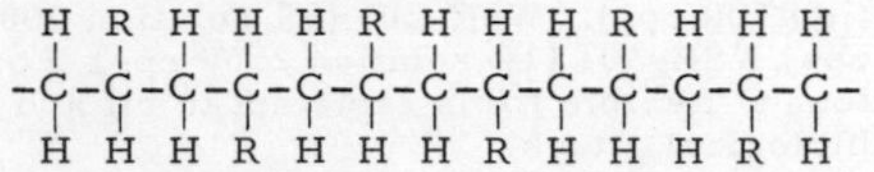

```
 H R H H H R H H H R H H H
 | | | | | | | | | | | | |
-C-C-C-C-C-C-C-C-C-C-C-C-C-
 | | | | | | | | | | | | |
 H H H R H H H R H H H R H
```

An atactic arrangement, in which there is no definite pattern or relationship with respect to the groups above and below the carbon chain, may be represented as:

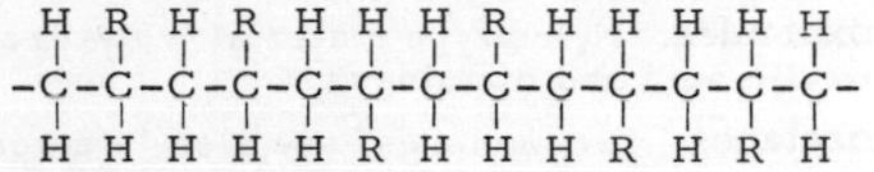

```
 H R H R H H H R H H H H H
 | | | | | | | | | | | | |
-C-C-C-C-C-C-C-C-C-C-C-C-C-
 | | | | | | | | | | | | |
 H H H H H R H H H R H R H
```

The cis and trans stereospecific forms differ from one another in the manner in which the atoms in the backbone chain are positioned with respect to adjacent atoms in the chain. These types depend on the presence of unsaturated (double) bonds in the molecule (see geometric isomerism). Thus cis-polybutadiene has the type of structure indicated (for the simplified case of two dimensions) by the following formula:

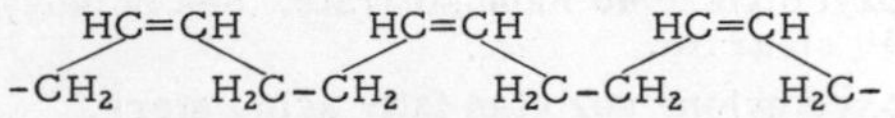

```
   HC=CH          HC=CH          HC=CH
-CH2     H2C-CH2       H2C-CH2        H2C-
```

The carbon atoms in such a structure can be thought of as having a backbone structure thus:

In trans-butadiene the backbone structure is:

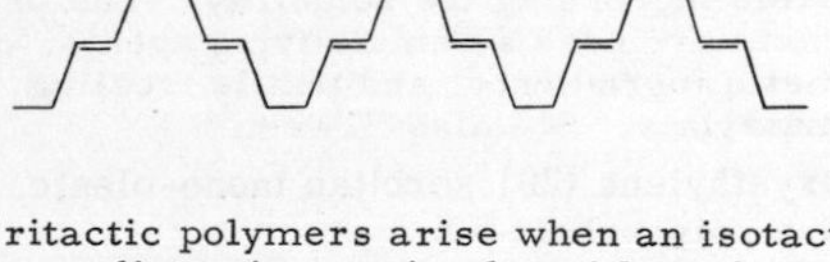

Tritactic polymers arise when an isotactic or syndiotactic type is also either cis or trans because the molecules are unsaturated and have double bonds. This can happen in:

1, 4-polyisoprene

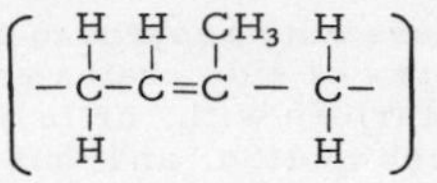

```
 ⎛ H  H CH3  H ⎞
 ⎜ |  |  |   | ⎟
 ⎜-C--C==C --C-⎟
 ⎜ |         | ⎟
 ⎝ H         H ⎠n
```

which has double bonds in the backbone chain which give rise to cis and trans forms, either of which can be iso- or syndiotactic according to the positions of the methyl groups relative to one another and the atoms in the backbone chain.

As with many phenomena, the above concepts are idealized and real polymers are usually mixtures of the several types, with one type being predominant in some cases. It is common to refer to the degree of tacticity, i.e., the degree to which the idealized state is achieved. It is also common for actual polymer molecules to be composed of blocks or sections of one type interspersed with blocks or sections of another type. These are referred to as stereoblock polymers.

polymer structure. This refers to the relative positions, arrangement in space, and the freedom of motion of the atoms in a polymer molecule, which in turn determine many polymer properties. Stereospecific polymers (q.v.) have regularity of some kind in the arrangement of the atoms and groups that make up the large molecule. This regularity permits or causes parts of molecules that happen to be near one another to line up somewhat like the atoms and molecules in crystals of ordinary low molecular weight substances, and consequently these stereospecific polymers behave somewhat like actual crystals. The lining up of polymer chains is never complete, so that resemblance to real crystals is limited. Stereospecific polymers have relatively higher softening temperatures than nonstereospecific, so that such polymers can be used at higher temperatures.

An important property of polymers that is determined to a large extent by polymer structure is the second order transition temperature, or glass temperature. Each kind of polymer has a characteristic second order temperature point, analogous and similar in some ways to a melting point. This in turn determines the flexibility and elasticity of a polymer at a particular temperature, and the temperature at which softening occurs. Below the second order transition temperature the polymer is relatively rigid; above this temperature the polymer becomes flexible and rubbery. The rate of increase of volume with temperature becomes greater above the second order transition temperature, and there are also changes in other characteristics such as tensile strength and elongation to break point. All these properties and characteristics are inherently dependent upon freedom of movement of the polymer chains. As temperature rises rotation of parts of the polymer chains about carbon-to-carbon bonds becomes possible and the polymer molecules can slip and otherwise move with respect to one another. Various types of side chains and substituent groups interfere with, or in some cases promote, such motion, and thus modify properties. Nonstereospecific polymers will behave entirely differently from those with definite stereospecific structure.

polymethoxy acetal (PMAC) $CH_3(CHOCH_3CH_2)_nCH(OCH_3)_2$. (3,5,7,x-polymethoxy dimethyl acetal).

Properties: High boiling yellow liquids with an ethereal odor, slightly hygroscopic. Slightly soluble in water, widely soluble in organic solvents. Excellent compatibility with water and insoluble plasticizers and resins.

PMAC-5: Sp.gr. 0.978; refractive index 1.440; pour point −55°F; initial b.p. 140°C; flash point 255°F.

PMAC-10: Sp.gr. 1.015; refractive index 1.453; pour point 5°F; initial b.p. 220°C; flash point 290°F.

PMAC-15: Sp.gr. 1.021; refractive index 1.456; pour point 15°F; initial b.p. 230°C; flash point 310°F.

Grades: Available in several molecular weight ranges.

Containers: Steel drums or tank cars.

Uses: Shell molding lubricants; phenolic resin modifiers; high boiling solvents; core binders; demulsifying agent; plasticizers and softeners.

Shipping regulations: None.*

polymyxin. Generic term for a series of antibiotic substances produced by strains of Bacillus polymyxa. Various polymyxins are differentiated by the letters A, B, C, D and E. All are active against certain gram-negative bacteria. Polymyxin B (q.v.) is the least toxic and most used.

Properties: All the polymyxins are basic polypeptides, soluble in water; the hydrochlorides are soluble in water and methanol, insoluble in ether, acetone, chlorinated solvents and hydrocarbons.

Use: Medicine (antibiotic).

polymyxin B. The least toxic and most used of the polymyxin series of antibiotics. It is a basic polypeptide containing leucine; threonine; phenylalanine; alpha, gamma-diaminobutyric acid and 6-methyloctanoic acid, $C_9H_{18}O_2$. It is stable in the dry state as the acid salt, polymyxin B sulfate (q.v.).

Uses: Medicine; also in beer production to kill bacteria which contaminate brewer's yeast.

polymyxin B sulfate. The commonly-used form of polymyxin B (q.v.).

Properties: White to buff-colored powder; odorless or faint odor; decomposes about 230°; soluble in water and isotonic saline solution; slightly soluble in alcohol. Solutions are slightly acid or neutral to litmus, having pH 5-7.5.

Grade: U.S.P. XVI.

Use: Medicine (antibiotic).

"Polynycel." [78] Trademark for compositions used as dyeing assistants.

polyolefins. A class name for the polymers derived by polymerization from relatively simple olefins, including particularly polyethylene and polypropylene from ethylene and propylene, but also including higher polymers such as polybutenes and polypentenes. In many cases the use of the term implies a stereospecific polymer.

polyols. See alcohol.

"Polyotic." [57] Trademark for tetracycline.

"Polyox." [214] Trademark for a series of water-soluble resins consisting of ethylene oxide polymers with molecular weights in the 100,000 to several million range.

Properties: White powder; water soluble to form viscous solutions. WSR-35 (5% solution 700 cps). WSR-205 (5% solution 6000 cps). WSR-301 (1% solution 2500 cps). Forms tough, flexible films resistant to oil and biological attack.

Containers: 5-, 20-, 50- and 150-lb fiber drums.

Uses: Textile warp size, paper coatings, detergents, hair spray, toothpastes, water-soluble packaging film, adhesives.

polyoxamides. Nylon type material made from oxalic acid and diamines.

polyoxetanes. See oxetanes; see also "Penton."

polyoxyethylene esters. The naming of these polymers is varied, so that entries may also be found under polyethylene glycol or polyglycol. Similarly polyoxypropylene and polypropylene glycol are used interchangeably for the propylene esters. Some coined generic names are also used to avoid the awkward chemical names.

polyoxyethylene ethers. Synonym for polyethylene glycol ethers.

polyoxyethylene 40 monostearate. See polyoxyl 40 stearate.

polyoxyethylene sorbitan fatty acid esters. (polyethylene oxide sorbitan fatty acid esters; sorbitan polyoxyethylene fatty acid esters). Nonionic surfactants of rather indefinite composition, obtained by esterification of sorbitol with one or three molecules of stearic acid or a similar fatty acid under conditions which result in the splitting out of water from the sorbitol. Varying amounts of ethylene oxide are added to the ester to control the water solubility of the product, additional ethylene oxide increasing the solubility. The products are used as emulsifying agents, cosmetic ingredients, and textile treating materials. See also "Tween."

polyoxyethylene (20) sorbitan mono-oleate. See polysorbate 80.

polyoxyl 40 stearate(polyoxyethylene 40 monostearate). The monostearate ester of a condensation polymer,

$H(OCH_2CH_2)_nOCOC_{17}H_{35}$ (n is approximately 40).
Properties: Waxy, light tan, nearly odorless solid; soluble in water, alcohol, ether, and acetone; insoluble in mineral oil and vegetable oils.
Grade: U.S.P. XVI.
Use: Ointments.

polyoxymethylene. Used as a name for paraformaldehyde (q.v.), and also for polyformaldehyde resins. For the latter, see acetal resins.

polyoxypropylene esters. See polyoxyethylene esters.

polyoxypropylene-glycerol adducts. Condensation polymers of propylene oxide and glycerol, with molecular weights in the range 1000 to 4000. Clear, stable, almost colorless non-corrosive liquids. Uses similar to those of polypropylene glycols (q.v.).

polyoxypropylene glycols $HO(C_3H_6O)_xH$. Polyethers derived from propylene oxide, with molecular weights in the range 3000 to 4000, and used as intermediates for polyurethane foams, elastomers and coatings.

"Poly-pale." [266] Trademark for a thermoplastic resin and several glycol and glycerol esters made from it.
Properties: These resins are characterized by pale color, oxidation resistance, and high softening points when compared to rosin and comparable esters.
Typical specifications: Acid number 148 for "Poly-pale" resin and 6 to 8 for its esters; sp.gr. (20°C) 1.072 for the resin and 1.062 to 1.085 for its esters; lbs/gal 8.95 for the resin and 8.86-8.97 for its esters; color (U.S.D.A. rosin scale) N to WG for the resin, and M to WG for the esters.
Uses: "Poly-pale" Resin is used in the production of ester gums, gloss oils, spirit varnishes, adhesives, and modified maleic and phenolic resins. The esters are used in the production of adhesives, lacquers, emulsion paints, varnishes, wax compositions, linoleum, floor tiles, and printing inks.

polyphosphoric acid $H_6P_4O_{13}$ (approx).
Properties: Water-white, hygroscopic, viscous liquid. Does not crystallize on standing. Soluble in water (with hydration to ortho-phosphoric acid).
Grade: Technical.
Containers: Carboys.
Use: Wherever a concentrated phosphoric acid is desired.

polypropylene $(C_3H_6)_n$ Polymerized propylene, a synthetic thermoplastic resin of relatively recent development but major importance. The lightest plastic. Highly crystalline in nature. The commercial material has a molecular weight of 40,000 or more (n = 1000 for such a molecular weight) and is isotactic to the extent of 60%. Syndiotactic polypropylene has also been made and has a higher melting point. Low molecular weight propylene polymers are also known. The tri- and tetrapropylenes are used in large quantities as gasoline additives and detergent intermediates. Somewhat higher molecular weight branched chain atactic propylene polymers have been used as greases or lube oil additives.
Properties: Sp.gr. 0.90; m.p. 168-171°C (isotactic); heat distortion temperature 220°F at 66 psi; color white or yellow; greater tensile strength (3000-5000 psi) and rigidity than high density polyethylene; remains tough and flexible at low temperatures; good clarity and freedom from haze; high gloss; low vapor transmission; dielectric strength 660 volts/mil; low creep; good abrasion resistance; lustrous surface; resistant to attack by chemical agents; withstands steam sterilization; resists moisture, oils, and greases; water absorption less than 0.01%. Somewhat sensitive to light and oxidation, so that antioxidants and inhibitors must be used. Relatively difficult to process. Polypropylene is also available in the form of a hard high-melting wax.
Typical properties of film: Sp.gr. 0.884-0.901; tensile strength 3500-6000 psi; elongation over 300; tensile modulus 9000-118,000; moisture vapor transmission 0.48-0.72 g/100 sq in/24 hrs/mil; resistance excellent to acids and alkalies; very good to oils and greases. Films can be coated, printed, or metallized, can be biaxially oriented to increase clarity, stiffness, and gloss.
Flammability: Slow burning.
Derivation: Aluminum alkyl and titanium tetrachloride or equivalent catalysts are mixed with solvents and propylene is introduced at 15 to 600 psi pressure, with temperatures up to about 212°F. The product varies widely with conditions and with the catalysts used.
Forms: Available as molding powder pellets with and without modifiers and pigments; in extruded sheet (10 to 250 mils thickness); in cast film (1 to 10 mils in thickness); in extruded monofilament; in textile staple and continuous filament yarn.
Uses: Film for packaging; molded parts for use in automobiles, utensils and appliances, housewares, closures; pipe fittings; wire and cable coating; coated and laminated products; bottles, pipe and tubing; filaments and fibers for upholstery webbing, cordage, ropes, bristles, and olefin yarns.

polypropylenebenzene. See dodecylbenzene.

polypropylene glycol esters. See polyoxyethylene esters.

polypropylene glycols $CH_3CHOH(CH_2OCHCH_3)_nCH_2OH$ (general formula). Comparable to the polyethylene glycols, but more oil-soluble and substantially less water-soluble. Classified by approximate molecular weight, as 425, 1025, and 2025. Nonvolatile, noncorrosive liquids; lower molecular weight members are soluble in water. All are solvents for vegetable oils,

waxes, resins.
Uses: Hydraulic fluids; rubber lubricants; antifoam agents; intermediates in urethane foams, adhesives, coatings, elastomers; plasticizers; paint formulations.
See polyethylene glycols.

"Polyrad." [266] Trademark for a series of reaction products of "Amine D" and ethylene oxide; used as corrosion inhibitors in the production and processing of petroleum products; also for inhibiting hydrochloric acid.

polysiloxanes. See siloxanes.

"Poly-Solv." [84] Trademark for a series of glycol ether solvents. See under individual names.
"Poly-Solv" EM. Ethylene glycol monomethyl ether, $CH_3OCH_2CH_2OH$.
"Poly-Solv" EE. Ethylene glycol monoethyl ether, $C_2H_5OCH_2CH_2OH$.
"Poly-Solv" EB. Ethylene glycol monobutyl ether, $C_4H_9OCH_2CH_2OH$.
"Poly-Solv" DM. Diethylene glycol monomethyl ether, $CH_3(OCH_2CH_2)_2OH$.
"Poly-Solv" DE. Diethylene glycol monoethyl ether, $C_2H_5(OCH_2CH_2)_2OH$.
"Poly-Solv" DB. Diethylene glycol monobutyl ether, $C_4H_9(OCH_2CH_2)_2OH$.
Used as solvents in the production of paints, varnishes, cleaners, soluble oils, dry-cleaning soaps, insecticides and cutting oils.
"Poly-Solv" EE Acetate. Ethylene glycol monoethyl ether acetate, $C_2H_5OCH_2CH_2$-OOC_2H_3. Used as solvent for paints and lacquers to give blush resistance and flow out. Has numerous other solvent uses requiring slow rate of evaporation.
"Poly-Solv" D2M. Diethylene glycol dimethyl ether, $(CH_3OCH_2CH_2)_2O$. Used as anhydrous reaction medium for organo-metallic syntheses.

polysorbate 80 (polyoxyethylene (20) sorbitan mono-oleate; sorbitan mono-oleate polyoxyethylene). An oleate ester of sorbitol and its anhydrides condensed with polymers of ethylene oxide consisting of approximately 20 oxyethylene units. See also polyoxyethylene sorbitan fatty acid esters.
Properties: Lemon-to-amber colored, oily liquid; sp.gr. 1.06-1.10; faint characteristic odor; warm, somewhat bitter taste; very soluble in water; soluble in alcohol, cottonseed oil, corn oil, ethyl acetate, methanol, and toluene; insoluble in mineral oil.
Grade: U.S.P. XVI.
Use: Emulsifier and dispersing agent.

"Polysperm 300." [221] Trademark for an industrial sperm oil product containing a mixture of oxidized, polymerized, hydroxylated, and possibly epoxidized higher fatty alcohol esters of fatty acids, and some triglycerides. Used in leather and textile chemicals, rust preventives, drawing compounds, lubricants and cordage manufacture.

polystyrene $(C_6H_5CHCH_2)_n$. Polymerized styrene, an important synthetic resin or plastic. When molded is a transparent glass-like material. Also available as rigid foam.
Properties: Water-white in color; very tough; has highest insulating power of the more common synthetic resins; soluble in aromatic hydrocarbons, chlorinated hydrocarbons and esters, very low water absorption; high mechanical strength.
Forms: Unmodified polystyrene contains only lubricants, stabilizers, plasticizers, and fillers to aid processing and properties. Heat resistant polystyrene includes also some alpha-methylstyrene copolymer. Impact resistant types contain butadiene copolymer; solvent and chemically resistant types are made with acrylonitrile as copolymer. See also ABS resins.
Uses: Major use is for refrigerators and air conditioners. Other uses are for packaging and foams, for containers or lids, dishes, tumblers, handles, fan blades, wall tile, signs, cabinets, and housings, sound records, impregnation of electric coils, lamination of fabrics, bonding of abrasives.

"Polysulfide." [28] A sodium sulfuret similar to liver of sulfur, having a sodium base instead of a potassium base. Yellow to yellow-green powder; sodium polysulfide content, 56% min; total sulfur, 50.7% min.
Containers: 10-lb tins (6/case); 100-lb drums.
Uses: For coloring copper and brass; for stripping copperplated deposits; for purifying cyanide plating solutions.

polysulfide rubbers. Synthetic polymers obtained by the reaction of sodium polysulfide with organic dichlorides such as dichlorodiethyl formal, alone or mixed with ethylene dichloride. The polysulfide rubbers are outstanding for their resistance to light, oxygen, oils, and solvents and for their impermeability to gases. In general, they have poor tensile strength and abrasion resistance but are resilient and have excellent low temperature flexibility. A recent use is as binder and fuel in solid propellants.
See also "Thiokol."

"Poly Tergent" B, G. [84] Trademark for a series of nonionic surface active agents consisting of ethoxylated nonyl or octyl phenols.
Properties: Pale yellow liquids; sp.gr. 1.02-1.06 (25/25°C); cloud point of 1% aqueous solution ranging from below 0° to 90°C; soluble or dispersible in water.
Uses: Emulsifiers, dispersing, wetting and foaming agents for petroleum oils, paraffins, chlorinated derivatives, aromatics, vegetable and pine oils.

"Poly Tergent" J. [84] Trademark for a series of polyethoxylated higher alcohols.
Properties: Yellow or amber liquids; mild pleasant odor; sp.gr. 1.01-1.06 (25/25°C); cloud point from 0° to over 100°C (1% aqueous solution). Soluble or dispersible

in water.
Uses: Emulsifiers; co-emulsifiers; wetting agents; detergents and stabilizers for corn oil, kerosene, xylene, fatty acids and systems containing other synthetic detergents, soaps, detergent builders, acids, alkalies and polyvalent metallic salts.

"Poly Tergent" LF.[84] Trademark for polyethoxy ether with low foaming properties.

polyterpene resins. A class of thermoplastic resins or viscous liquids of amber color, obtained by polymerization of turpentine in the presence of catalysts such as aluminum chloride or mineral acids. Unpolymerized material is removed by distillation at reduced pressure. The resins consist essentially of polymers of alpha- or beta-pinene and are soluble in most organic solvents.
Uses: Paints; wax polishes; rubber plasticizers; curing concrete; impregnating paper.

polytetrafluoroethylene (PTFE) $(C_2F_4)_n$. A polymer, plastic or resin, abbreviated PTFE, derived from tetrafluoroethylene, C_2F_4. Sometimes called TFE fluorocarbon resin. Available as resin powder, as sheets, rods, tape, and as aqueous dispersions and films. The material has a waxy texture, is opaque with a milk-white color. It can be molded by special techniques involving mixing with a diluent that is subsequently removed. Sintering is then used to consolidate the molded object. The PTFE is nonflammable, highly resistant to oxidation and action of chemicals, including strong acids, alkalies, oxidizing agents. It retains useful properties from −450° to +550°F and is strong and tough. It has an almost uniquely low coefficient of friction, and anti-stick properties.

Uses include gaskets, seals, electrical component insulation, coatings, linings for drums and containers, valve seats, bearings and packings, spacers for coaxial cable, laminates, diaphragms, molded parts of pumps and fittings, tubing and hose, coverings where stickiness must be avoided as on heat sealing guides and plates, bakery rolls, candy molds.

polythene. Generic name for polyethylene (q.v.) used as plastics. The word is no longer current in the United States, but is still used in England.

polyurethane foams. The foam form of polyurethane resins. May be either rigid or flexible, hard and abrasive, or soft and resilient, depending upon components, fillers, and method of foaming.

A polyether such as polypropylene glycol, or a propylene oxide derivative of glycerol or sorbitol, a polyester, or similar material is treated with a diisocyanate in the presence of some water and a catalyst (amines, tin soaps, organic tin compounds) as well as fillers, dispersing and emulsifying agents, and other substances. Simultaneously with the polymer-forming reactions, the water reacts with the isocyanate groups to cause cross linking and curing, and also produces carbon dioxide which causes foaming. In other cases trifluoromethane or similar volatile material is incorporated to serve as blowing agent, and to reduce the thermal conductivity of the finished foam.

The rigid foams when formed inside thin shells of metal, fabric, etc., add great strength and rigidity at little increase in weight. The flexible foams are unusual in having high strength, good heat insulating properties, resistance to water, oil, solvents, and abrasion.
Uses: Airplane construction; padding for mattresses and upholstery; interlinings for overcoats and sleeping bags; soundproof walls; insulation against heat loss; life preservers; fish net floats; foam rubber applications; cigarette and air filters; packaging; bone surgery; have been suggested for tires.

polyurethane resins (isocyanate resins). Synthetic polymers that may be either thermoplastic or thermosetting, and that range from soft and rubberlike to hard and brittle, usually made by action of tolylene diisocyanate or another diamine with polyols, polyethers, polyesters or other materials containing hydroxyl (OH) groups. The polyethers such as propylene or ethylene oxide derivatives are of increasing importance.
Properties: The polymers are made as flexible or rigid foams, flexible and stiff fibers (see spandex), coatings and linings, and as elastomers. These serve as foams of all kinds, insulation, reinforcing fillers, as adhesives, binders and sealers. The foams can be formed in place without heat or pressure, but in other applications heat is used. Density ranges from 2 to 30 lbs/cu ft.
Derivation: The basic polymer unit is formed as follows:

$$R_1NCO + R_2OH \rightarrow R_1NHCOOR_2.$$

If R_1 has a second CNO group and R_2 has a second OH group, further chain growth occurs. Cross linking takes place if R_2 has three or more OH groups. Many variations are possible and in use.
See also polyurethane rubber; polyurethane foams.

polyurethane rubber (polyester rubber). A non-foamed or solid but flexible type of polyurethane resins.
Properties: High abrasion resistance, strength, hardness, solvent resistance, resistance to oxygen aging, good flexibility, elasticity and shock absorption. Somewhat affected by high and low temperatures and by hot water or steam. Easily cast into complex shapes.
Uses: Development items.

polyvinyl acetal resins. General name for resins formed by the condensation of acetaldehyde or any other aldehyde with polyvinyl alcohol, obtained in turn by partial hydrolysis of the polyvinyl acetate. The physical and chemical properties of the polyvinyl acetals can be varied greatly by: (1) the molecular weight of the starting polyvinyl acetate, (2) the degree of hydrolysis, (3) the degree of acetal formation. The polymers contain varying percentages of acetate, hydroxyl and acetal groups. The aldehydes commonly used are formaldehyde, acetaldehyde and butyraldehyde, producing polyvinyl formals, acetals, and butyrals, respectively. See descriptions following:

Polyvinyl formals: Usually almost completely hydrolyzed and completely reacted with formaldehyde. Soluble in water, dioxane, chlorinated hydrocarbons, acetic acid, mixtures of polar and non-polar substances; insoluble in alcohol, benzene, toluene, etc.

Uses: Mainly in lacquers; coatings; impregnations; but may be used for molding and casting. They have high softening temperatures and poor water resistance. Available in several grades.

Polyvinyl acetals: Slightly yellow; soluble in lacquer solvents and aromatic hydrocarbons; insoluble in water and aliphatic hydrocarbons.

Uses: In lacquers, coatings, films, and adhesives. May be compression and injection molded. May be used as a photographic film base and for decorative articles. Available in several grades, and in rod, sheet and tube forms.

Polyvinyl butyral: White granules generally high in acetal content; soluble in esters, alcohols, ketones, and chlorinated hydrocarbons; insoluble in water and hydrocarbons; stable in dilute alkali, but not in acid. The butyral resin with high hydroxyl content is the best safety-glass interlayer available. Its extreme toughness, adhesiveness, clarity and water resistance make it ideal for the manufacture of safety glass and bulletproof glass, and also making shatterproof laminates of methyl methacrylate resins. It retains its toughness and flexibility at very low temperatures. It is also used in strippable coatings; antifouling paints; waterproofing fabrics; adhesives.

All the resins are also used in molding compounds and for textile coatings and waterproofing, adhesives, laminations. They have good compatibility with plasticizers and can be modified by the addition of thermosetting materials, such as phenolics, to raise water resistance and softening temperature. Unusual characteristics of toughness, resiliency, and thermal and abrasion resistance have been produced in this way.

Available in various grades and forms.

polyvinyl acetate (PVAc) $(H_2C{:}CHOOCCH_3)_x$.
A synthetic resin.

Properties: Colorless, odorless, tasteless, nontoxic, transparent, thermoplastic solid; sp.gr. 1.19 at 15°C; insoluble in water, gasoline, oils and fats; soluble in low molecular weight alcohols, esters, benzene, chlorinated hydrocarbons, ketones.

Derivation: Polymerization of vinyl acetate with peroxide catalysts. Produced by emulsion polymerization, solution polymerization, suspension and bulk polymerization.

Grades: Available as beads or pearls, powder, solutions, emulsions and latex.

Uses: Major use is in latex water paints because of low cost, stability to weathering, self-priming character, quick drying and easy recoatability. Also used in large quantities for hot melt and solution adhesives, as an intermediate for conversion to polyvinyl alcohol and acetals, for sealing, for coating and finishing fabrics, as a binder for nonwoven fabrics, as a component of lacquers, inks, and plastic wood, and as a strengthening agent for cements.

polyvinyl alcohol (PVA; PVOH) $(\text{-}CH_2CHOH\text{-})_x$.
A water-soluble synthetic resin made by hydrolysis of polyvinyl acetate. As recovered from the polymerization process the resin is a white to cream-colored powder; density 1.21-1.31 g/cc at 20°C. It is produced in a number of grades whose properties depend on degree of hydrolysis and on molecular weight or viscosity.

Forms and Properties: The resin powder is used to a large extent in the form of a water solution; or may also be converted into molded objects and into cast and extruded sheets, rods, films and thread. The water solutions are stable over considerable periods of time. Films and coatings made from these solutions are colorless and clear, tough and abrasion-resistant, unaffected by oils, greases, fats, hydrocarbons, and most organic solvents. They possess a high degree of impermeability to gases and air, and have good adhesive properties.

Composition: Even in "completely hydrolyzed" polyvinyl alcohol only about 95% of the acetate groups have been replaced by OH groups.

Typical properties of plastic film: Sp.gr. 1.21-1.31; refractive index approx. 1.50 (n 25/D); tensile strength 7800-8000 psi; elongation 180-250%; good resistance to sunlight; poor resistance to acids, alkalies, water and dampness; darkens slowly if heated above 100°C.

Flammability: Moderate.

Containers (resin powder): 10-, 25-lb cartons; 50-lb multiwall bags; 100-lb drums.

Uses: Textile warp size; size for nylon and rayon knitting yarns; binder for pigmented paper coatings; greaseproofing paper; paper size; size on paperboard for high gloss printing; base materials for water-resistant laminating adhesives; non-blocking, remoistenable adhesives; adhesives and binders for leather, cloth, nonwoven fabrics, and paper; pigment binder; temporary protective coatings; molded products; emulsifying agents; emulsion stabilizer and thickener; photosensitive films;

intermediate for polyvinyl butyral and other polyvinyl acetals.

polyvinyl butyral (polyvinyl butyral acetal). See polyvinyl acetal resins.

polyvinyl butyral resin sheeting. Soft, pliable sheeting composed essentially of polyvinyl butyral resin and plasticizer. Several types are manufactured, differing in identity and proportion of plasticizer. The sheeting is made in thicknesses from 0.015 to 0.030 inch and is supplied in continuous lengths of various widths. The surface has embossed pattern, and is dusted with powdered sodium bicarbonate to prevent sticking to itself in the roll. As interlayer in safety glass, the sheeting is transparent and colorless, and highly resistant to moisture and to the light and heat of the sun. Since it remains extremely flexible, tough and stretchable over a wide range of temperatures, it yields without breaking when the glass is broken by impact, and, since it forms a strong bond with the glass, it holds the broken pieces together.

Uses: The commercial products are designed primarily for use as interlayer in safety glass, in bullet-resistant glass, and in shatter-resistant acrylic sheeting for aircraft, and in implosion shields for television. Some sheeting is used as a die-covering in metal-shaping.

polyvinyl carbazole. A brown, thermoplastic resin obtained by the reaction of acetylene with carbazole. It softens about 150°C, has excellent electrical properties, good heat and chemical stability, but poor mechanical strength. It is used principally as a substitute for mica in electrical equipment and as an impregnant for paper capacitors.

polyvinyl chloride (PVC) $(-H_2CCHCl-)_n$. A common and widely used synthetic thermoplastic resin.

Properties: White powder or colorless granules or pearls which can be converted by heat and pressure into colorless sheets or films. These are tasteless, odorless, resistant to action of chemicals including moisture and air, flame retarding, dimensionally stable. They do not absorb or dissolve either oil or water, are insoluble in water and most organic solvents, but somewhat soluble in methyl ethyl ketone and phorone. The properties vary widely with the method of polymerization. Small amounts of stabilizers (lead, zinc, barium, calcium, tin compounds, epoxy compounds) are added during manufacture to prevent slow decomposition involving liberation of hydrogen chloride. The finished product is relatively hard, brittle and horn-like at 0°C unless plasticizers are added. These are usually nonvolatile esters, low molecular weight resins, or certain rubbers.

Derivation: By polymerization of vinyl chloride monomer with peroxide catalysts.

Uses: The resin is converted into final form by calendering, extrusion, molding, and by coating from solutions, dispersion or latexes, and by fluid bed coating.

In rubber-like final form it appears as electric wire insulation and sleeving, garden hose and other tubing, gaskets, welting, and flexible piping; in film form it appears in rainwear, aprons, and other garments, curtains and draperies, garment bags, hospital sheeting, upholstery, handbags and belts; it is coated on fabrics to be used as auto upholstery, clothing and wall coverings; one of the largest uses of rigid shapes is as floor tile and for related floor coverings; rigid shapes and sheets are also used as decorative panels, luminous ceilings, and for phonograph records, electrotype molds, display items, and packaging; as pipe in chemical plants, breweries and similar processing plants, and for pipe lines for gas, oil, water, and electrical conduits because of corrosion resistance, smooth inside surface and weight savings. Extruded rigid shapes are used as window frames and tracks, baffles, decorative moldings, relief maps, luggage, book bindings, etc. In foam or expanded form it is used for padding and cushions and for fishing floats; the latex is also used to coat or impregnate paper, fabric and leather to improve abrasion resistance, strength, chemical and flame resistance. Films and sheets are sometimes laminated to metal.

polyvinyl chloride-acetate. A synthetic resin made by copolymerizing vinyl chloride and vinyl acetate, to obtain a polymer that is inherently more flexible than polyvinyl chloride. The copolymer resin usually contains 85 to 97% of the chloride. It is produced as sheets, rods, tubes, granules, powder, and dispersions, and is generally similar in properties and uses to polyvinyl chloride.

polyvinyl dichloride. A modified vinyl resin, possibly chlorinated polyvinyl chloride, with improved high temperature and chemical stability properties.

See "Geon, Hi-temp."

polyvinyl fluoride $(-H_2CCHF-)_n$. Polymer of vinyl fluoride. In film form it is characterized by superior resistance to outdoor weather, toughness, and chemical resistance, as well as flexibility. Such film is of potential use as a replacement for paint and as an outdoors protective coating for siding and roofs, and for glazing, packaging, and electrical uses. The film has a high dielectric constant and is resistant to thermal and hydrolytic decomposition. The film has low permeability to air and water.

Properties: (1.5 mil film) Tensile strength 13,000 psi at 25°C, 6000 psi at 100°C; elongation, 150% at 25°C, 165% at 100°C; impact strength, 5 kg-cm/mil at 25°C; flex life 70,000 cycles at 25°C, 40,000 at

−17°C; dielectric constant 7.5 at 60 cps. See, for example, "Tedlar."

polyvinyl formals. See polyvinyl acetal resins.

polyvinylidene fluoride $(-CH_2CF_2-)_n$. A thermoplastic resin suitable for compression and injection molding and extrusion. Also can be applied from a dispersion. Crystalline melting point is 340°F; thermally stable at 650°F. Useful for packaging films, extruded jackets, insulation for high temperature wire, chemical tanks and autoclave linings. Tensile strength at 77°F, 7000 psi; yield stress 5500 psi; elongation 300%; compressive strength 10,000 psi; thermal conductivity 1.05 Btu/hr/sq ft/°F/in.; water absorption 0.04% in 24 hours; sp.gr. 1.76; refractive index 1.42.

polyvinyl isobutyl ether (PVI) $[-CH_2CHOCH_2CH(CH_3)_2-]_n$.
Properties: White, opaque elastomer or viscous liquid depending on molecular weight range; almost odorless. Soluble in hydrocarbons, esters, ethers, higher alcohols, and ketones. Insoluble in water and lower alcohols. Contains trialkylphenol stabilizer.
Typical specifications:
(1) High molecular weight PVI: white, rubbery sheets dusted with talc, density 0.93; brittle point −19° to +17°C; elongation at break approx. 500%.
(2) Medium molecular weight PVI: very viscous liquid; density 0.91; refractive index 1.46; softening point 44-46°C.
(3) Low molecular weight PVI: viscous liquid; density 0.85; refractive index 1.45.
Derivation: Polymerization of vinyl isobutyl ether.
Grades: As 100% material in three molecular weight ranges.
Containers: Fiber drums or open head steel drums.
Uses: Because of its adhesiveness, electrical properties and plasticity, for adhesives, waxes, tackifiers, plasticizers, surface coatings, laminating agents, cable filling compositions, lubricating oils.
Shipping regulations: None.*

polyvinyl methyl ether (PVM) $[-CH_2CHOCH_3-]_n$.
Properties: Light yellow to amber colored balsam-like tacky liquid; soluble in cold water, insoluble in hot water, soluble in most organic solvents except noncyclic hydrocarbons. Compatible with a wide range of synthetic and natural latices. Sp.gr. 1.05; refractive index 1.47; brittle, non-tacky solid at about 0°C. Very low toxicity.
Derivation: Polymerization of vinyl methyl ether.
Grades: Available as 100%, 50% aqueous solution, 50% toluene solution, 50% isopropanol solution.
Containers: Open head steel drums.
Uses: Component in pressure-sensitive and hot-melt adhesives for paper, polyethylene, polymer emulsions and rubbers, surface coating tackifier and plasticizer, heat sensitizer for natural and synthetic rubbers.
Shipping regulations: PVM 100% and aqueous solutions: None; toluene and isopropanol solutions: Flammable liquid. Red label.*

polyvinyl methyl ether–maleic anyhdride. See PVM/MA.

polyvinylpyrrolidone (PVP) $(C_6H_9NO)_n$.
Properties: A freely-flowing white amorphous powder; soluble in water, chlorinated hydrocarbons, alcohols, amines, nitroparaffins, and lower molecular weight fatty acids. Sp.gr. 1.23-1.29; viscosity range K20-90 (Fikentscher).
Derivation: Prepared by a Reppe synthesis; butynediol is formed by the action of acetylene on formaldehyde at high pressure. Butynediol is hydrogenated to form butanediol which is then dehydrogenated over copper at 200°C to form gamma-butyrolactone. This is then ammoniated to form alpha-pyrrolidone. Vinylation and polymerization complete the process.
Uses: Pharmaceuticals; medicine (synthetic blood plasma); cosmetics (hair spray, etc.); dye stripping agent; textile finish; protective colloid; detergent; food stabilizer or thickener; substitute for natural gums; beer and wine clarification.

polyvinyl resins. See vinyl plastics.

"Polywood." [448] Trade name for polyester coatings for wood furniture and other wood products.

"Polyzime" P. [212] Trademark for a product containing diastatic and proteolytic enzymes.
Properties: Dry, fine white powder; fully water soluble; non-hazardous; nonflammable; optimum pH for diastatic reaction, 7.0-7.2; for proteolytic reaction 7.5-8.0; optimum temperature 45°C.
Grades: Technical.
Packages: 1-, 5-, 10-, and 25-lb containers.
Uses: Desizing of textile fabrics preparatory to dyeing, bleaching, mercerizing, printing and finishing.

pomade. A perfumed oil or fat.

pomegranate bark. See granatum.

Pompeian red. Red pigment consisting, essentially, of ferric oxide. See iron oxide reds.

Pompey red. See iron oxide reds.

Ponceau 3R $(CH_3)_3C_6H_2NNC_{10}H_4(OH)(SO_3Na)_2$. Sodium cumeneazo-beta-naphthol disulfonate. A water-soluble red dye, C.I. No. 80.
Properties: Dark-red powder. Soluble in water and acids to form a cherry-red solution; slightly soluble in alcohol; insoluble in alkalies.
Use: Dyeing wool.

"Ponsol." [28] Trademark for a line of vat dyes derived from anthraquinone. Used for dyeing and printing of cotton, rayon, and silk and characterized by excellent

*See "I.C.C. Shipping Regulations," page xiii.
Reference numbers refer to name of manufacturer. See "List of Manufacturers," page v.

fastness properties. Under controlled conditions, they can be applied to wool. Also used to a limited extent on certain classes of paper.

"Pontachrome." [28] Trademark for a line of colors used on wool and nylon. Dyed by forming lakes on the fiber. Used largely on materials such as mens' suitings and overcoatings.

"Pontacyl." [28] Trademark for a line of colors for wool, nylon, silk and "Orlon" acrylic fiber. Dyed in an acid bath. Produce bright shades and are used largely on materials for women's wear, bathing suits, sweater yarns, upholstery fabrics, etc.

"Pontamine." [28] Trademark for a line of direct dyeing colors for all fibers, except acetate. Also applied to paper and leather. Certain types, after dyeing, are diazotized and developed with various "Pontamine" developers with resultant improvement in fastness, particularly to washing.

"Pontamine" Whites. [28] Trademark for fluorescent whitening agents that produce brilliant optical bleaches having outstanding wetfastness properties when applied to cotton and rayon. Also used in soaps, detergents, etc.

pontianak gum (jelutong). A copal.
Derivation: From species of Dyera indigenous to Malacca and Borneo and from the guayule from Parthenium, a shrub indigenous to the Chihuahua Desert of Mexico.
Properties: Grayish-white mass similar to burned lime, contains about 60% water; up to 25% rubberlike material.
Grades: Chips; nubs.
Containers: Bags.
Uses: Rubber manufacture; chewing gum; adhesives; lacquers; linoleum; varnishes and paints.
Shipping regulations: None.*

"Pontocaine" Hydrochloride. [162] Trademark for tetracaine hydrochloride.

poplar bud. The air-dried, closed winter leaf bud of Populus candicans, known in commerce as balm of Gilead buds, or of Populus tacamahacca, known in commerce as balsam poplar buds.
Properties: Pleasant and balsamic odor and aromatic, bitter taste.
Grade: N.F. XI.
Use: Medicine.

POPOP. Abbreviation for phenyl-oxazolyl-phenyl-oxazolyl-phenyl. See 1,4-bis[2-(5-phenyloxazolyl)]-benzene.

poppy capsules. See papaver.

poppy heads. See papaver.

poppy oil (poppy-seed oil).
Properties: Very pale golden yellow liquid; pleasant taste and odor. Soluble in ether, chloroform, petroleum ether, and carbon disulfide. Sp.gr. 0.924-0.928; saponification value 189-196; iodine value 133-157; refractive index 1.4751-1.4773.
Derivation: By expressing the seed of the poppy, Papaver somniferum, P. album and P. nigrum.
Method of purification: Filtration.
Grades: Crude; red; white.
Containers: Barrels.
Uses: Food oil; artist's colors; adulterant for olive oil; soap stock; varnishes; lubricant.
Shipping regulations: None.*

poppy-seed oil. See poppy oil.

porcelain. Ceramic ware made largely of baked clay (kaolin) coated or glazed with a fusible substance.

porcelain clay. See kaolin.

Porcelain Enamels for Aluminum. [28] Vitreous frits which fire at 521-538°C (970-1000°F) for use on aluminum alloys. Enamels have excellent resistance to impact, chemical attack, thermal shock, and salt-water corrosion.
Containers: 25-, 50-, 100-, and 200-lb fiber drums.
Use: To impart chemical resistance, dielectric strength, and color to aluminum alloys.

"Porocel." [99] Trademark for a hard, highly adsorbent, high surface area, activated bauxite utilizing either foreign or domestic ores. Supplied in various standard meshes, moisture contents, and in regular, low iron and low silica grades.
Uses: Recommended generally as an adsorbent, catalyst, and catalyst carrier. Used as a lubricating oil and wax percolation medium; vapor phase desulfurization catalyst; naphtha reforming catalyst; purification medium for H_2SO_4 alkylate, aluminum chloride, ferric chloride, copper chloride, nickel, molybdena, or chromia; refining medium for sugar liquors and syrups. Used to recover elemental sulfur from refinery, smelter, or natural gas, and as an agent for the removal of fluoride contaminants from hydrocarbon streams.

porphine. The basic molecule for porphyrins. Consists of four pyrrole rings with four carbons in a single large ring.

porphyrins. Organic compounds of extreme importance in the maintenance of both animal and plant life. Porphyrins are colored pyrrole derivatives which occur universally in protoplasm, where they form the basis of the respiratory pigments. They have the ability to combine with metals, especially iron and magnesium, to form metalloporphyrins. For example, protoporphyrin bound to reduced iron forms the heme of hemoglobin. Hemoglobin and the cytochromes contain the iron-porphyrins while the chlorophylls contain the magnesium-porphyrins.

porpoise oil (dolphin oil).
Properties: Pale yellow liquid; sp.gr. 0.926-0.929; saponification value about 290; the

iodine value varies with the part of the porpoise body used; jaw blubber 27, face blubber 22, body blubber 103. Soluble in ether, chloroform, benzene and carbon disulfide.
Derivation: By boiling the blubber of the brown porpoise.
Grades: Technical; also sold as body oil, jaw oil and junk oil (from the face blubber).
Containers: Wooden barrels; steel drums; tins.
Uses: Lubricant; soap stock; leather dressing; illuminating oil. The oil from the jaw blubber is used as a lubricant for watches and chronometers and is known as watch oil.

"Portite." [326] Trade name for corrosion-resistant masonry sulfur-base cement.

Portland cement. See cement, Portland.

Portugal orange. See orange peel, sweet.

positron. A fundamental particle identical in all aspects with the ordinary electron except that it carries a positive rather than a negative electrical charge. It is not stable in matter since it is annihilated by a negative electron to produce two quanta of radiant energy. See fundamental particle.

positronium. A positron associated with an electron during a very short period before the two interact to annihilate one another with the production of equivalent energy.

posterior pituitary. See pituitary.

potash. The word potash originally was applied to potassium carbonate recovered from wood ashes. The word is often used in trade in connection with any material containing the element potassium. The potash value is expressed as the equivalent amount of the oxide K_2O.
Sources of potassium: Stassfurt and Alsatian deposits of carnallite, kainite, and sylvinite; the Carlsbad, New Mexico, deposits of sylvinite and polyhalite; the Searles Lake, California, brines (4.7% KCl); Utah alunite. Other sources are: dust from cement kilns; dust from blast furnaces; ashes from molasses, sugar beet pulp residues, kelp and wood; flotation processes.

Since potassium is essential for the growth of plants, a large potash source is needed to supply the fertilizer industry.

potash alum. See aluminum-potassium sulfate.

potash blue. A pigment related to Prussian blue. Prepared by mixing potassium ferrocyanide and ferrous sulfate and oxidizing the resulting ferrous ferrocyanide.
Grade: C.P.
Containers: Barrels.
Use: In making carbon paper.

potash, caustic. See potassium hydroxide.

potash chrome alum. See chromium-potassium sulfate.

potash feldspar. See feldspar.

potash magnesia double salt. A material containing potassium carbonate, magnesium sulfate and a low proportion of chloride, containing 20-30% K_2O and used as a fertilizer.

potash muriate. Old term for potassium chloride.

potash, sulfurated (liver of sulfur; potassium, sulfurated; sulfurated potash). A mixture composed chiefly of potassium polysulfides and potassium thiosulfate, containing not less than 12.8% sulfur in combination as sulfide.
Properties: Liver brown when freshly made, changing to a greenish yellow; odor of hydrogen sulfide; bitter, acrid, and alkaline taste; decomposes upon exposure to air; soluble in water leaving a residue.
Grade: U.S.P. XVI.
Use: Medicine; production of decorative color effects on brass, bronze and nickel.

potassa. See potassium hydroxide.

potassio-cupric chloride. See copper-potassium chloride.

potassium (kalium) K. Element of atomic number 19, group Ia of the periodic system; one of the alkali metals. Potassium 40 is a naturally occurring radioactive isotope. The synthetic isotope, potassium 42, is used in tracer studies, primarily in medicine.
Properties: Soft, silvery metal; rapidly oxidized in moist air; igniting spontaneously if warm enough; reacts vigorously with water, acids, halogens. Must be kept submerged in liquids containing no oxygen, e.g., kerosine. Sp.gr. 0.862; m.p. 63°C; b.p. 770°C. Soluble in liquid ammonia, aniline, mercury, and sodium.
Source: See potash.
Derivation: Thermochemically by distillation of potassium chloride with sodium.
Grades: Technical.
Containers: 1-lb boxes and bottles; 50-, 250-lb drums.
Uses: Preparation of potassium peroxide; heat exchange alloys (see NaK).
Fire hazard: Dangerous.
Shipping regulations: Flammable solid. Yellow label.*

potassium abietate $KO_2CC_{19}H_{29}$. Water soluble soap resulting from action of rosin and potassium hydroxide. Used as a pesticide.

potassium acetate $KC_2H_3O_2$.
Properties: White, crystalline deliquescent powder; saline slightly alkaline taste. Keep well stoppered. Soluble in water and in alcohol; insoluble in ether. Solutions alkaline to litmus but not to phenolphthalein. M.p. 292°C.
Derivation: By the action of acetic acid on potassium carbonate.
Impurities: Chlorides; sulfates; heavy metals.
Grades: Pure; pure fused; C.P.; N.F. XI; reagent.

Containers: Bottles; drums.
Uses: Dehydrating agent; textile conditioner; reagent in analytical chemistry; medicine; acetone; cacodylic derivatives; crystal glass.
Shipping regulations: None.*

potassium acid carbonate. See potassium bicarbonate.

potassium acid fluoride. See potassium bifluoride.

potassium acid oxalate. See potassium binoxalate.

potassium acid phosphate. See potassium phosphate, monobasic.

potassium acid saccharate $HOOC(CHOH)_4COOK$.
Properties: Light, off-white powder; pH of solution, 3.5; soluble in hot water, acid, or alkaline solutions.
Uses: Chelating agent; in rubber formulations; in metal plating; in soaps and detergents.

potassium acid sulfate. See potassium bisulfate.

potassium acid sulfate, anhydrous. See potassium pyrosulfate.

potassium acid sulfite. See potassium bisulfite.

potassium acid tartrate. See potassium bitartrate.

potassium allylmercaptomethyl penicillin. See penicillin.

potassium alum. See aluminum-potassium sulfate.

potassium aluminate $K_2Al_2O_4 \cdot 3H_2O$.
Properties: Hard crystals, lustrous. Soluble in water with hydrolysis to form strongly alkaline solution; insoluble in alcohol.
Derivation: By fusing potassium hydroxide with aluminum oxide.
Grades: Technical.
Uses: Dyeing, printing (mordant); lakes; paper sizing.

potassium aluminosilicate. See feldspar.

potassium aluminum fluoride K_3AlF_6.
Properties: White powder, slightly soluble in water; poisonous!
Derivation: Aluminum fluoride, ammonium fluoride, and potassium chloride.
Containers: Fiber cans.
Use: Insecticide.
Shipping regulations: None.*

potassium-aluminum sulfate. See aluminum-potassium sulfate.

potassium anhydrosulfate. See potassium pyrosulfate.

potassium antimonyl tartrate. See antimony-potassium tartrate.

potassium arsenate (Macquer's salt) KH_2AsO_4.
Properties: Poisonous! Colorless crystals; sp.gr. 2.867; soluble in water; insoluble in alcohol.
Uses: Manufacture of fly paper; insecticidal preparations; as laboratory reagent; preserving hides; printing textiles.
Shipping regulations: Class B poison. Poison label.*

potassium arsenite (potassium metarsenite). Approx. $KH(AsO_2)_2$.
Properties: White powder; hygroscopic; decomposes slowly in air. Variable composition. Caution! Very poisonous!, Keep well stoppered! Soluble in alcohol, water.
Grades: Technical; reagent.
Uses: Analysis; medicine; mirrors.
Shipping regulations: Class B poison. Poison label.*

potassium aurate $KAuO_2$.
Properties: Yellow crystals; soluble in water.
Derivation: Gold oxide dissolved in potassium hydroxide solution.
Uses: To prepare other gold compounds.

potassium auribromide. See gold-potassium bromide.

potassium aurichloride. See gold-potassium chloride.

potassium auric iodide. See gold-potassium iodide.

potassium-beryllium fluoride. See beryllium-potassium fluoride.

potassium bicarbonate (potassium acid carbonate; known as baking soda in some countries) $KHCO_3$.
Properties: Colorless, odorless, transparent crystals or white powder; slightly alkaline, salty taste. Soluble in water and potassium carbonate solution; insoluble in alcohol. Sp.gr. 2.17; m.p., dec. between 100° and 120°C.
Derivation: By passing carbon dioxide into a solution of potassium carbonate in water.
Grades: Commercial; highest purity; U.S.P. XVI; reagent; technical.
Containers: 25 to 250-lb drums; 220-lb barrels; 700-, 784-lb casks.
Uses: In baking instead of yeast or baking powder; medicine; manufacture of pure potassium carbonate; fire-extinguishing agent for jet, petroleum and chemical fires.
Shipping regulations: None.*

potassium bichromate. See potassium dichromate.

potassium bifluoride (potassium acid fluoride; Fremy's salt) KHF_2.
Properties: Colorless crystals; decomposed by heat; corrosive and poisonous! Soluble in alcohol (dilute), water; insoluble in alcohol (absolute).
Grades: Technical.
Uses: Etching glass; production of fluorine; flux in metallurgy.

potassium binoxalate (potassium acid oxalate; acid potassium oxalate; salt of sorrel; salt of lemon; sorrel salt) $KHC_2O_4 \cdot 1/2\ H_2O$.
Properties: White crystals; bitter, sharp taste; somewhat hygroscopic; poisonous! Soluble in water; sp.gr. of the anhydrous

salt 2.088; m.p., decomposes when heated.
Derivation: Neutral potassium oxalate and oxalic acid are dissolved in water and crystallized.
Method of purification: Recrystallization.
Grades: Technical; C.P.
Containers: 1-, 5-lb bottles; 25-lb boxes; 100-lb kegs; 300-, 700-lb barrels.
Uses: Removing ink stains; scouring metals; cleaning wood; photography; bleaching straw.
Shipping regulations: None.*

potassium biphosphate. See potassium phosphate, monobasic.

potassium-bismuth tartrate (bismuth-potassium tartrate).
Properties: White, granular powder. Odorless; sweet taste; variable composition. Contains 60 to 64% bismuth. Decomposed by dilute mineral acid; darkens when exposed to light. Soluble in water; insoluble in alcohol, chloroform, ether.
Use: Medicine.

potassium bisulfate (acid potassium sulfate; potassium acid sulfate) $KHSO_4$.
Properties: Colorless crystals; the fused salt is deliquescent. Soluble in water, yielding a solution with acid reaction; decomposes in alcohol. Sp. gr. 2.245; m.p. 200°C; b.p., decomposes.
Derivation: (a) By heating potassium sulfate with sulfuric acid; (b) by heating potassium chloride with sulfuric acid to a moderate heat.
Method of purification: Crystallization.
Impurities: Heavy metals; chlorine; arsenic.
Grades: Commercial; technical; reagent; fused; highest purity; medicinal.
Containers: Glass bottles; barrels; multiwall paper sacks.
Uses: Conversion of tartrate of lime, wine lees and tartrates into potassium bitartrate; food preservative; flux; manufacture of mixed fertilizers by double decomposition with tricalcium phosphate; anisole; methyl acetate; ethyl acetate.
Shipping regulations: None.*

potassium bisulfide. See potassium hydrosulfide.

potassium bisulfite (potassium acid sulfite) $KHSO_3$.
Properties: White, crystalline powder; sulfur dioxide odor. Soluble in water; insoluble in alcohol. M.p., decomposes when heated.
Derivation: Sulfur dioxide is passed through a solution of potassium carbonate until no more carbon dioxide is given off; the solution is concentrated and allowed to crystallize.
Method of purification: Recrystallization.
Impurities: Heavy metals; arsenic.
Grades: Commercial; reagent; highest purity; medicinal.
Containers: Wooden barrels; glass bottles; multiwall paper bags.
Uses: Chemical (reduction of various organic compounds; purification of aldehydes and ketones, iodine, sodium hydrosulfite); antiseptic; source of sulfurous acid, particularly in brewing; analytical chemistry; tanning; bleaching straw and textile fibers; chemical preservative in foods.
Shipping regulations: None.*

potassium bitartrate (cream of tartar; potassium acid tartrate) $KHC_4H_4O_6$.
Properties: White crystals or powder; soluble in water; pleasant slightly acid taste; slightly soluble in alcohol. Sp. gr. 1.956.
Derivation: From argols by extraction with water and crystallization.
Method of purification: Recrystallization.
Grades: Technical; C.P.; N.F. XI; reagent.
Containers: 25- to 250-lb barrels; multiwall paper bags.
Uses: Baking powder; preparation of other tartrates; medicine; galvanic tinning of metals.
Shipping regulations: None.*

potassium borate. See potassium metaborate and potassium tetraborate.

potassium borohydride KBH_4.
Properties: White crystalline powder; soluble in water, ammonia; insoluble in ethers and hydrocarbons; sp. gr. 1.18; stable in moist and dry air; stable in vacuum to 500°C. Decomposed by acids with evolution of hydrogen.
Derivation: By metathetical reaction of sodium borohydride and potassium hydroxide.
Grades: Technical; powder; pellets.
Containers: Glass bottles and polyethylene bags in 1- to 55-gal metal containers.
Use: Source of hydrogen and other borohydrides; reducing agent for aldehydes, ketones and acid chlorides.
Shipping regulations: Flammable solid. Yellow label.*

potassium borotartrate (cream of tartar, borated; cream of tartar, soluble; potassium-sodium borotartrate).
Properties: White, crystalline powder. Odorless; soluble in water; sp. gr. 1.832.
Derivation: By evaporating a solution containing potassium bitartrate and borax in a proportion of 7:2.
Grades: Technical.
Containers: Glass bottles; fiber cans.
Uses: Medicine; photography.
Shipping regulations: None.*

potassium bromate $KBrO_3$.
Properties: White crystals or crystalline powder. Soluble in water; insoluble in alcohol; sp. gr. 3.27; m.p. 434°C, with decomposition above 370°C.
Derivation: By passing bromine into a solution of potassium hydroxide, potassium bromide and bromate being formed, which are separated by crystallization.
Method of purification: Recrystallization.

Impurities: Potassium bromide.
Grades: Pure; C.P.
Containers: 1-, 5-lb bottles; 200-lb drums.
Use: Reagent in analytical chemistry; strong oxidizing agent; additive for permanent wave compounds and for flour.
Fire hazard: Dangerous.
Shipping regulations: Oxidizing material. Yellow label.*

potassium bromide KBr.
Properties: White, crystalline granules or powder; pungent, strong, bitter saline taste; somewhat hygroscopic. Soluble in water and glycerin; slightly soluble in alcohol and ether; sp.gr. 2.749; m.p. 730°C; b.p. 1435°C.
Derivation: Solutions of iron bromide and potassium carbonate are mixed and heated, the solution filtered and concentrated and the bromide crystallized out.
Method of purification: Recrystallization.
Grades: Technical; C.P.; N.F. XI; reagent; single pure crystals.
Containers: 25-, 50-, 100-, 200-lb drums; 400-, 500-lb barrels; kegs.
Uses: Medicine; photography (gelatin bromide papers and plates); process engraving and lithography; special soaps; single crystals for spectroscopy; infrared transmission.
Shipping regulations: None.*

potassium carbonate (potash; pearl ash; American ashes; salt of tartar) (a) K_2CO_3; (b) $2K_2CO_3 \cdot H_2O$; (c) $K_2CO_3 \cdot H_2O$.
Properties: White, deliquescent, granular translucent powder; alkaline reaction. Soluble in water; insoluble in alcohol. (a) Sp.gr. 2.3312; m.p. 909°C; b.p. volatile at white heat.
Derivation: (1) From Stassfurt potassium beds by saturating a solution of magnesium and potassium chlorides with carbon dioxide, heating under pressure and evaporating the solution. (2) Lixiviation of wood and plant ashes with water, concentration of the solution and calcination of residue, which is extracted with water. (3) Aqueous residue of beet-sugar molasses after fermentation and distillation to remove alcohol is concentrated and treated as above. (4) Electrolysis of potassium chloride. (5) Recovered from wool washing. (6) Recovered in an impure form, by electrical precipitation from the fumes of cement factories, metallurgical furnaces.
Grades: Crystals; technical; reagent; N.F. XI; calcined 80-85%, 85-90%, 90-95%; 96-98%; hydrated 80-85%.
Containers: Drums; bags; kegs; barrels.
Uses: Chemical (dehydrating agent, potassium salts, potassium cyanide from potassium ferrocyanide, oxalic acid from sawdust, chromates, chlorates); brewing; ceramics; explosives; fertilizers; mineral waters; Bohemian glass; difficultly fusible glasses; tanning; electroplating; shampoo preparations; process engraving and lithography; soft soaps; textiles (dyeing, bleaching, wool washing); finishing oils and sizes; chocolate preparations.
Shipping regulations: None.*

potassium chlorate $KClO_3$.
Properties: Transparent, colorless crystals or white powder; cooling, saline taste; poisonous! Must not be ground with sugar, sulfur or other combustible substances as may cause explosions. Soluble in water and alkalies; very slightly soluble in alcohol. Sp.gr. 2.337; m.p. 368°C; b.p., decomposes at about 400°C, giving off oxygen.
Derivation: (a) By electrolyzing a hot concentrated alkaline solution of potassium chloride. Preferably (b) by interaction of solutions of potassium chloride and sodium chlorate or calcium chlorate.
Method of purification: Recrystallization.
Grades: Highest purity; pure granulated; C.P.; commercial.
Containers: 25-, 110-, 200-, 225-, 550-lb drums.
Uses: Oxidizing agent; explosives; matches; source of oxygen; printing textile fabrics; pyrotechnics; percussion caps; medicine; dyes; paper manufacture; disinfectant; bleaching.
Warning! Contact with combustible material may cause fire. MCA warning label.
Shipping regulations: Oxidizing material. Yellow label.*

potassium chloride (potassium muriate) KCl.
Properties: Colorless or white crystals or powder; strong saline taste. Occurs naturally as sylvite. Soluble in water; slightly soluble in alcohol; insoluble in absolute alcohol. Sp.gr. 1.987; m.p. 776°C; b.p. 1500°C.
Derivation: (a) Extracted from certain lake brines; purified from accompanying borax by recrystallization; (b) by fusion or extraction of carnallite, $MgCl_2 \cdot KCl \cdot 6H_2O$, with a solution of magnesium chloride; (c) from sylvinite by fractional crystallization, or by flotation. Most important U.S. process.
Grades: Refined (99.5 and 99.9%); technical (95%); agricultural grades sold as 60-62%; 48-52% and 22% K_2O; single pure crystals.
Containers: Bags; drums; bulk.
Uses: Fertilizer; source of potassium salts; pharmaceutical preparations; photography; single pure crystals for spectroscopy.
Shipping regulations: None.*

potassium chlorochromate (Péligot's salt) $KClCrO_3$.
Properties: Red crystals; on heating liberates chlorine. Soluble in water (dec). Sp.gr. 2.497.
Grades: Technical.
Use: Oxidizing agent.

potassium chloroiridate. See iridium potassium chloride.

potassium chloroplatinate (platinum-potassium chloride; potassium platinichloride) K_2PtCl_6.
Properties: Small, orange-yellow crystals

or powder. Insoluble in alcohol; very slightly soluble in water; m.p. decomposes when heated.
Derivation: By adding platinic chloride to a solution of a potassium salt and crystallizing.
Method of purification: Recrystallization.
Grades: Technical.
Containers: Glass bottles.
Use: Photography.
Shipping regulations: None.*

potassium chloroplatinite (platinous-potassium chloride) K_2PtCl_4.
Properties: Ruby-red crystals. Soluble in water; insoluble in alcohol. Sp.gr. 3.291.
Derivation: By adding potassium carbonate to a solution of chloroplatinous acid.
Method of purification: Crystallization.
Grades: Technical.
Containers: Glass bottles.
Use: Photography.
Shipping regulations: None.*

potassium chromate (potassium chromate, yellow; neutral potassium chromate) K_2CrO_4.
Properties: Yellow crystals; soluble in water; insoluble in alcohol; sp.gr. 2.7319; m.p. 971°C.
Derivation: Roasting powdered chromite with potash and limestone, treating the cinder with hot potassium sulfate solution and leaching.
Method of purification: Recrystallization.
Impurities: Free alkali; sulfates; aluminum; alkaline earths.
Grades: Highest purity; technical; reagent; commercial; crude.
Containers: 1-, 5-lb bottles; drums.
Uses: Reagent in analytical chemistry; aniline black; textile mordant; enamels; chromate pigments; inks; medicine; fungicide; leather finishing; making chromium compounds.
Shipping regulations: None.*

potassium chromate, neutral. See potassium chromate.

potassium chromate, red. See potassium dichromate.

potassium chromate, yellow. See potassium chromate.

potassium citrate $K_3C_6H_5O_7 \cdot H_2O$.
Properties: Colorless or white crystals or powder; deliquescent; cooling saline taste; odorless. Soluble in water and glycerol; almost insoluble in alcohol. Sp.gr. 1.98; m.p., decomposes when heated to about 230°C.
Derivation: By the action of citric acid on potassium carbonate.
Method of purification: Crystallization.
Grades: Technical; C.P.; N.F. XI.
Containers: 25- to 250-lb drums.
Use: Medicine.
Shipping regulations: None.*

potassium cobaltinitrite. See cobalt-potassium nitrite.

potassium columbate. Obsolete name for potassium niobate.

potassium-copper chloride. See copper-potassium chloride.

potassium-copper cyanide (copper-potassium cyanide).
Properties: White, crystalline, double salt of copper cyanide and potassium cyanide. Copper content (Cu) min. 25.8%; free KCN 1.25-3.0%.
Containers: 100-lb drums.
Uses: For preparing and maintaining cyanide copper plating baths based on potassium cyanide.

potassium-copper ferrocyanide. See copper-potassium ferrocyanide.

potassium-cupric ferrocyanide. See copper-potassium ferrocyanide.

potassium cyanate KOCN.
Properties: Colorless crystals; sp.gr. 2.05; decomposes 700-900°C. Soluble in water; insoluble in alcohol.
Derivation: Heating potassium cyanide with lead oxide.
Caution! Avoid breathing dust or spray mist. Avoid prolonged or repeated contact with skin. MCA warning label.
Use: Herbicide; manufacture of organic chemicals and drugs.

potassium cyanaurite. See gold-potassium cyanide.

potassium cyanide KCN.
Properties: White, amorphous, deliquescent lumps or crystalline mass; faint odor of bitter almonds; extremely poisonous! Do not handle with bare hands! Soluble in water, alcohol and glycerol. Sp.gr. 1.52; m.p. 634°C.
Derivation: (a) Potassium carbonate and carbon are heated in a current of ammonia. The fused mass is extracted with alcohol, the latter distilled off and the cyanide fused. (b) Calcium cyanamide is prepared from calcium carbide and nitrogen and is fused with caustic potash. (c) From by-products of beet-sugar manufacture.
Grades: Commercial; pure; C.P.; solution; reagent.
Containers: 25-, 100-lb drums.
Uses: Extraction of gold and silver from ores; electroplating; heat treatment of steel; reagent in analytical chemistry; insecticide; fumigant; reagent in manufacture of various intermediate organic cyanogen derivatives; paper manufacture; pharmaceutical preparations; fixative in photography; process engraving and lithography; fumigant for raw cotton; fumigant for grain elevators; fumigant for citrus fruits.
Caution! Not flammable, but evolves hydrocyanic acid on contact with acids or moisture. Toxic by ingestion, absorption in skin lesions, and inhalation.
Shipping regulations: Class B poison. Poison label (both solid and in solution).*

potassium cyclamate. Used as a non-nutritive sweetener. See sodium cyclamate.

potassium dichloroisocyanurate $Cl_2K(NCO)_3$. See also dichloroisocyanuric acid.
Properties: White; slightly hygroscopic; crystalline powder or granules; loose bulk density (approx) powder 37 lbs/cu ft, granular 64 lbs/cu ft. Active ingredient approx 59% available chlorine.
Containers: 200-lb fiber drums.
Uses: Active ingredient in household dry bleaches, dishwashing compounds, scouring powders, detergent-sanitizers, replacement for calcium hypochlorite.

potassium dichromate (potassium bichromate; red potassium chromate) $K_2Cr_2O_7$.
Properties: Bright, yellowish-red transparent crystals; bitter, metallic taste; poisonous! Soluble in water; insoluble in alcohol. Sp.gr. 2.692; m.p. 396°C; b.p., decomposes at 500°C.
Derivation: (a) By adding sulfuric acid to crude potassium chromate solution. (b) By heating an aqueous solution of sodium dichromate with potassium chloride, concentrating the solution, whereupon sodium chloride is deposited in the vessel. Lead rods are suspended in the solution and the dichromate crystallizes on these.
Methods of purification: Recrystallization.
Grades: Commercial; highest purity; highest purity fused; reagent.
Containers: 100-lb bags; 400-lb drums.
Uses: Oxidizing agent (chemicals, dyes, intermediates); analytical reagent; brass pickling compositions; electroplating; pyrotechnics; explosives; safety matches; textiles; dyeing and printing; glass; chrome glues and adhesives; milk and milk products preservative; chrome tanning leather; wood stains; histology; poison fly paper; process engraving and lithography; bleaching tallow, palm oil, etc.; photography; pharmaceutical preparations; synthetic perfumes; chrome alum manufacture; pigments; alloys; ceramic products.
Warning! Harmful dust may cause rash or external ulcers. MCA warning label.
Shipping regulations: None.*

potassium dihydrogen phosphate. See potassium phosphate, monobasic.

potassium diphosphate. See potassium phosphate, monobasic.

potassium dithionate (potassium hyposulfate) $K_2S_2O_6$.
Properties: Colorless crystals; soluble in water; insoluble in alcohol. Sp.gr. 2.278.
Grades: Technical.
Use: Reagent.

potassium ethyldithiocarbonate. See potassium xanthate.

potassium ethylxanthogenate. See potassium xanthate.

potassium ferric oxalate $K_3Fe(C_2O_4)_3 \cdot 3H_2O$.
Properties: Monoclinic green crystals; loses 3 molecules water 100°C; decomposes 230°C; soluble in water, acetic acid; incompatible with alkali and ammonia, since these react to precipitate ferric hydroxide.
Grades: Technical.
Use: Photography and blue-printing.

potassium ferricyanide (red prussiate of potash; red potassium prussiate) $K_3Fe(CN)_6$.
Properties: Bright red, lustrous crystals or powder; poisonous! Soluble in water; slightly soluble in alcohol. Sp.gr. 1.8109; m.p., decomposes when heated.
Derivation: Chlorine is passed into a solution of potassium ferrocyanide, the ferricyanide separating out.
Method of purification: Recrystallization.
Impurities: Ferrous salts; potassium chloride.
Grades: Pure crystals; pure powder; commercial; crude; highest purity reagent; technical.
Containers: 25-, 100-, 350-lb drums.
Uses: Calico printing; wood dyeing; tempering steel; etching liquid; production of pigments; electro-plating; leather; paper manufacture; ingredient of composition used to produce sensitive coatings on blue print paper; fertilizer compositions.
Shipping regulations: None.*

potassium ferrocyanide (yellow prussiate of potash) $K_4Fe(CN)_6 \cdot 3H_2O$.
Properties: Lemon-yellow crystals or powder; mild saline taste; effloresces on exposure to air. Soluble in water; insoluble in alcohol. Sp.gr. 1.853 (17°C); m.p., loses its water of crystallization when heated to 60°C; b.p., decomposes when heated to red heat.
Derivation: Produced from gas plant by-products and alkali metal or alkaline earth metal cyanides; or nitrogenous waste products, iron filings and potassium carbonate.
Method of purification: Recrystallization.
Impurities: Potassium carbonate, sulfate and chloride; calcium salts.
Grades: Technical; C.P.
Containers: 100-, 370-lb drums; multiwall paper sacks.
Uses: Medicine; potassium cyanide and ferricyanide; reagent in analytical chemistry; dry colors; tempering steel; dyeing; explosives; process engraving and lithography.
Shipping regulations: None.*

potassium fluoborate KBF_4.
Properties: A gritty white powder (when dried) or gelatinous crystals; a weak but bitter taste; not acid to litmus, forms 6-sided prisms when crystallized from an aqueous solution; sp.gr. 2.5 (20°C); wet crystals give a green and finally a violet flame test; m.p. 530°C; decomposes at a high temperature. Slightly soluble in water and alcohol (hot).
Derivation: By mixing fluoboric acid with a solution of a potassium salt forming a gelatinous precipitate that is washed and dried; by heating boric acid with potassium

silicofluoride and adding potassium carbonate.
Containers: Fiber drums.
Uses: Sand agents in casting of aluminum and magnesium; grinding aid in resinoid grinding wheels; flux for soldering and brazing; in electro-chemical processes and chemical research.
Shipping regulations: None.*

potassium fluophosphate. See fluophosphoric acids.

potassium fluoride (a) KF; (b) $KF \cdot 2H_2O$.
Properties: White, crystalline, deliquescent powder; sharp saline taste; poisonous! Soluble in water and hydrofluoric acid; insoluble in alcohol. Sp.gr. (a) 2.454. M.p. (a) about 800°C; (b) 41°C.
Derivation: By saturation of hydrofluoric acid with potassium carbonate.
Method of purification: Crystallization.
Impurities: Arsenic.
Grades: Technical; pure; purified; free of arsenic; C.P.
Containers: 1-, 5-lb waxed bottles; wooden kegs; tins; drums.
Uses: Etching glass; preservative; insecticide.

potassium fluosilicate (potassium silicofluoride) K_2SiF_6.
Properties: White, odorless, fine crystalline powder; sp.gr. 3.0; slightly soluble in water.
Containers: 100-, 200-, 400-lb and in bulk.
Uses: Vitreous enamel frits; synthetic mica; metallurgy of aluminum and magnesium; ceramics; insecticide.
Shipping regulations: None.*

potassium fluotantalate. See tantalum-potassium fluoride.

potassium fluozirconate. See zirconium potassium fluoride.

potassium germanium fluoride. See germanium potassium fluoride.

potassium gibberellate. Salt of gibberellic acid (q.v.). Used to promote and control development of malt in grain.

potassium gluconate $KC_6H_{11}O_7$.
Properties: Odorless, salty tasting, fine, white crystalline powder. Soluble in water; insoluble in alcohol and benzene.
Derivation: Reaction of potassium hydroxide or carbonate with gluconic acid.
Method of purification: Crystallization.
Grades: Pharmaceutical.
Containers: Cans; fiber drums.
Use: Medicine.
Shipping regulations: None.*

potassium glutamate Similar to sodium glutamate (q.v.).

potassium glycerinophosphate. See potassium glycerophosphate.

potassium glycerophosphate (potassium glycerinophosphate) $K_2C_3H_5O_2 \cdot H_2PO_4 \cdot 3H_2O$.
Properties: Pale yellow syrupy liquid; acid taste. Soluble in alcohol; miscible with water in all proportions.
Derivation: Glycerol and phosphorus pentoxide or metaphosphoric acid are mixed, warmed and exactly neutralized with potassium carbonate, warmed and concentrated.
Grades: Technical, 50 or 75% solution.
Containers: 5-, 10-lb bottles; 5-, 25-lb cans; 70-lb carboys.
Use: Medicine; dietary supplement.
Shipping regulations: None.*

potassium guaiacol sulfonate (orthocoll) $C_6H_3OCH_3OHSO_3K$.
Properties: Fine, white powder or crystals; gradually turns pink on exposure to air and light; bitter taste, afterward becoming sweetish; odorless. Contains approx. 60% guaiacol. Nonirritant; soluble in water; sparingly soluble in alcohol.
Grades: N.F. XI.
Containers: 25- to 300-lb drums.
Use: Medicine.

potassium 2,4-hexadienoate. See potassium sorbate.

potassium hexafluophosphate. See fluophosphoric acids.

potassium hexanitrocobaltate III. See cobalt potassium nitrite.

potassium hydrate. See potassium hydroxide.

potassium hydrogen phosphate. See potassium phosphate, dibasic.

potassium hydrogen phthalate. $HOOCC_6H_4COOK$.
Properties: Colorless crystals. Soluble in water; sp.gr. 1.636.
Derivation: Potassium hydroxide and phthalic anhydride.
Method of purification: Recrystallization.
Grades: C.P.; analytical.
Containers: Glass bottles.
Use: Alkalimetric standard.
Shipping regulations: None.*

potassium hydrosulfide (potassium sulfhydrate; potassium bisulfide) $(KSH)_2 \cdot H_2O$.
Properties: Colorless crystals; hydrogen sulfide odor. Turns yellow when exposed to air and forms the polysulfide. Hygroscopic. Soluble in alcohol and water.
Grades: Technical.
Use: Separation of heavy metals.

potassium hydroxide (caustic potash; potassium hydrate; potassa; lye) KOH.
Properties: White, deliquescent pieces, lumps, sticks, pellets, or flakes having a crystalline fracture. Keep well stoppered; absorbs water and carbon dioxide from the air. Soluble in water, alcohol, glycerin; slightly soluble in ether.
Constants: Sp.gr. 2.044; m.p. 360°C, but varies with water content.
Derivation: (a) Electrolysis of concentrated potassium chloride solution. (b) Boiling potassium carbonate with milk of lime.
Method of purification: Sulfur compounds are removed by the addition of potassium

nitrate to the fused caustic. The purest form is obtained by solution in alcohol, filtration and evaporation.
Impurities: Alumina; silica; sulfuric acid.
Grades: Commercial; ground; flake; fused (88 to 92%); purified by alcohol (sticks, lumps and drops); reagent; highest purity; U.S.P. XVI; liquid (45%).
Containers: Bottles; boxes; kegs; 100-, 225-, 275-, 700- to 730-lb drums; tank cars.
Uses: Soap manufacture; bleaching; manufacture of oxalic acid; manufacture of potassium compounds; reagent in analytical chemistry; medicine (caustic); matches; process engraving and lithography.
Caution! Generates heat on contact with water.
Warning! Causes severe burns to skin and eyes. MCA warning label.
Shipping regulations: Solution: corrosive liquid. White label.*

potassium hyperchlorate. See potassium perchlorate.

potassium hypophosphite (potassium hypophosphite, monobasic) KH_2PO_2.
Properties: White opaque crystals or powder with pungent saline taste, very deliquescent. Explosive if ground with nitrates, chlorates or other oxidizing agents. Soluble in water and alcohol; decomposed by heat.
Derivation: Interaction of calcium hypophosphite and potassium carbonate.
Containers: Drums.
Use: Medicine.
Fire hazard: Dangerous!

potassium hypophosphite, monobasic. See potassium hypophosphite.

potassium hyposulfate. See potassium dithionate.

potassium hyposulfite. See potassium thiosulfate.

potassium iodate KIO_3.
Properties: White, crystalline powder. Odorless; soluble in water, sulfuric acid (dilute); insoluble in alcohol. Sp.gr. 3.9; m.p. 560°C (partial decomposition).
Grades: Technical; C.P.
Uses: Analysis (testing for zinc and arsenic); iodometry; medicine; reagent; feed additive.

potassium iodide KI.
Properties: White crystals, granules or powder; strong bitter saline taste. Soluble in water, alcohol, and glycerol.
Constants: Sp.gr. 3.123; m.p. 723°C; b.p. 1420°C.
Method of purification: Recrystallization.
Impurities: Potassium carbonate; metals; sulfuric acid; potassium bromide; potassium chloride.
Grades: Crystals; granulated; powder; highest-purity; reagent; U.S.P. XVI; single pure crystals.
Containers: 1-, 5-lb bottles; 25- to 300-lb drums.
Uses: Medicine; reagent in analytical chemistry; photography (precipitating silver); feed additive; single crystals for spectroscopy, infrared transmission, scintillation.
Shipping regulations: None.*

potassium-iridium chloride. See iridium-potassium chloride.

potassium iridochloride. See iridium-potassium chloride.

potassium laurate $KOOCC_{11}H_{23}$.
Properties: Light tan paste; soluble in water.
Uses: Emulsifying agent; base for liquid soaps and shampoos.

potassium linoleate $KOOCC_{17}H_{31}$.
Properties: Light tan paste; soluble in water.
Use: Emulsifying agent.

potassium magnesium sulfate. $K_2SO_4 \cdot 2MgSO_4$.
Properties: White tetragonal crystals; sp.gr. 2.829; m.p. 927.
Use: Fertilizer.

potassium manganate K_2MnO_4.
Properties: Dark-green powder or crystals; soluble in potassium hydroxide solution and water; decomposes in acid solution. M.p. 190°C (decomposes).
Grades: Technical.
Uses: Bleaching chamois skins; bleaching fibers; disinfectants; mineral waters; mordant (wool); batteries; photography; printing; purifying and bleaching oils; source of oxygen (dyeing); water purification; oxidizing agent.

potassium manure salt. See manure salts.

potassium mercuric iodide. See mercuric potassium iodide.

potassium metabisulfite (potassium pyrosulfite) $K_2S_2O_5$.
Properties: White granules or powder; sulfur dioxide odor; sp.gr. 2.3; slightly soluble in water and alcohol.
Derivation: By heating potassium bisulfite until it loses water.
Grades: Technical; reagent.
Containers: 1-, 5-lb bottles; 100-, 300-, 350-lb drums.
Uses: Antiseptic; reagent in analytical chemistry; source of sulfurous acid; brewing (antiseptic and preservative, cleaning and sweetening casks and vats); food preservative; developing agent for photographic and motion picture films; process engraving and lithography; dyeing.
Shipping regulations: None.*

potassium metaborate (potassium borate) KBO_2.
Properties: White, crystalline powder. Alkaline reaction; caustic, alkaline taste. Soluble in water.

potassium metaphosphate (monopotassium metaphosphate) KPO_3.

*See "I.C.C. Shipping Regulations," page xiii.
Reference numbers refer to name of manufacturer. See "List of Manufacturers," page v.

Properties: White powder; soluble in acids (dilute); slightly soluble in water.

potassium metarsenite. See potassium arsenite.

potassium mica. See muscovite.

potassium molybdate $K_2MoO_4 \cdot 5H_2O$.
Properties: White, microcrystalline powder; deliquescent; soluble in water; insoluble in alcohol; sp.gr. 2.3; m.p. 919°C.
Grades: Technical; C.P.
Use: Reagent.

potassium monophosphate. See potassium phosphate, dibasic.

potassium muriate. See potassium chloride.

potassium naphthenate.
Properties: Gray paste; soluble in water.
Derivation: From naphthenic acids.
Uses: Driers; emulsifying agents.

potassium nickel sulfate. See nickel potassium sulfate.

potassium niobate (potassium columbate) $4K_2O \cdot 3Nb_2O_5 \cdot 16H_2O$. Potassium niobate is of importance technologically in the final purification of niobium material. Crystals of potassium niobate which are soluble in water form when niobic acid solutions are treated with concentrated potassium hydroxide.

potassium niobium oxyfluoride. See niobium-potassium oxyfluoride.

potassium nitrate (niter; nitre; saltpeter) KNO_3.
Properties: Transparent, colorless or white crystalline powder or crystals; slightly hygroscopic; cooling, pungent, saline taste. Sp.gr. 2.1062; m.p. 337°C; b.p., decomposes at about 400°C. Soluble in water; slightly soluble in alcohol and glycerin.
Derivation: Interaction of a solution of sodium nitrate and potassium chloride, or of nitric acid and potassium chloride.
Grades: Commercial; C.P.; N.F. XI.
Containers: Bottles; kegs; bags; barrels; drums.
Uses: Pyrotechnics; explosives; matches; pickling meat; fertilizer; reagent in analytic chemistry; tobacco; glass manufacture; metallurgy; medicine; oxidizer in solid rocket propellants.
Caution! Fire hazard; dangerous; in contact with organic materials causes violent combustion or ignition.
Shipping regulations: Oxidizing material. Yellow label.*

potassium nitrite KNO_2.
Properties: White or slightly yellowish prisms or sticks; deliquescent. Caution! Keep well stoppered! Sp.gr. 1.915; b.p., decomposes; m.p., various values from 297 to 450°C are recorded, the variation presumably being due in part to the decomposition of the material, also to the difficulty of obtaining a pure sample. Soluble in water; insoluble in alcohol.
Grades: C.P.; technical; reagent.
Uses: Analysis (testing for amino acids, cobalt, iodine, urea); medicine; organic synthesis.
Shipping regulations: Oxidizing material. Yellow label.*

potassium oleate $C_{17}H_{33}COOK$.
Properties: Gray-tan paste; soluble in water.
Use: Textile soaps; emulsifying agent.

potassium orthophosphate. See potassium phosphate, monobasic, dibasic or tribasic.

potassium orthotungstate. See potassium tungstate.

potassium osmate (potassium perosmate) $K_2OsO_4 \cdot 2H_2O$.
Properties: Violet crystals. Hygroscopic; Caution! Poisonous! Keep well stoppered! Soluble in water; insoluble in alcohol and ether.
Grades: Technical.
Containers: Ground glass stoppered bottles.
Use: Analysis (testing for nitrogenous matter in water).
Shipping regulations: None.*

potassium osmic chloride. See osmium potassium chloride.

potassium oxalate $K_2C_2O_4 \cdot H_2O$.
Properties: Colorless transparent crystals. Odorless; soluble in water; efflorescent in warm dry air; sp.gr. 2.08; m.p., decomposes when heated.
Derivation: Potassium formate or carbonate mixed with a small quantity of oxalate and a slight excess of alkali is heated, the oxalate extracted with water and crystallized.
Method of purification: Recrystallization.
Impurities: Heavy metals; chlorine.
Grades: Technical; C.P.
Containers: Bottles; 250-, 300-lb barrels; 25-, 50-, 100-lb drums.
Uses: Medicine; reagent in analytical chemistry; source of oxalic acid; bleaching and cleaning straw hats; removing stains from textiles; photography.
Shipping regulations: None.*

potassium oxide K_2O.
Properties: Gray, crystalline mass. Soluble in water; forms potassium hydroxide. Sp.gr. 2.32; m.p. red heat.
Derivation: By heating potassium nitrate and metallic potassium.
Grades: Technical.
Containers: Tins; iron barrels.

potassium oxyfluoniobate. See niobium-potassium oxyfluoride.

potassium palladium chloride. See palladium potassium chloride.

potassium penicillin 152. See penicillin.

potassium penicillin G. See penicillin.

potassium penicillin O. See penicillin.

potassium penicillin V. See penicillin.

*See "I.C.C. Shipping Regulations," page xiii.
Reference numbers refer to name of manufacturer. See "List of Manufacturers," page v.

potassium pentaborate $K_2B_{10}O_{16} \cdot 8H_2O$.
Properties: Colorless crystals; m.p. (anhydrous) 180°C. Very slightly soluble in cold water.
Containers: Drums.

potassium percarbonate $K_2C_2O_6 \cdot H_2O$.
Properties: Granular, white mass. Caution! Keep away from light and moisture! Soluble in water (liberates oxygen).
Grades: Technical.
Uses: Analysis (testing for cerium, chromium, vanadium, titanium); microscopy; oxidizing agent; photography; textile printing.

potassium perchlorate (potassium hyperchlorate) $KClO_4$.
Properties: Colorless crystals or white, crystalline powder. Decomposed by concussion, organic matter, and agents subject to oxidation. More stable than potassium chlorate. Soluble in water; insoluble in alcohol. Sp.gr. 2.524; decomposes at 400°C.
Grades: Technical; reagent.
Containers: Drums.
Uses: Explosives; medicine; oxidizing agent; photography; pyrotechnics; reagent; analysis; oxidizer in solid rocket propellants.
Caution! Oxidizing material, combustible in contact with organic materials.
Shipping regulations: Oxidizing material. Yellow label.*

potassium periodate KIO_4.
Properties: Small, colorless crystals or white, granular powder; slightly soluble in water. Sp.gr. 3.168; m.p. 582°C (decomposes at higher temperatures).
Grades: Technical; C.P.; reagent.
Use: Analysis (oxidizing agent).
Shipping regulations: Oxidizing material. Yellow label.*

potassium permanganate (purple salt) $KMnO_4$. A powerful oxidizing agent.
Properties: Dark purple crystals having a blue metallic sheen; sweetish, astringent taste; odorless. Soluble in water; decomposed by alcohol. Sp.gr. 2.7032; m.p., decomposes at 240°C.
Derivation: (a) By the oxidation of the manganate (prepared by the fusion of pyrolusite with caustic potash) in an alkaline electrolytic cell. (b) A hot solution of the manganate obtained as in (a), is treated with carbon dioxide gas. On cooling, the solution deposits crystals of the permanganate.
Method of purification: Recrystallization.
Grades: Technical; C.P.; U.S.P. XVI.
Containers: Bottles; cans; boxes; kegs; 50-, 75-, 135-, 500-lb drums.
Uses: Oxidizer; disinfectant; deodorizer; bleach; dye; reagent in analytical chemistry; medicine; manufacture of organic chemicals; absorbent for poison gases in military gas-masks; purification of carbon dioxide used in the manufacture of effervescent drinks; water purification.
Fire hazard: Dangerous. Explosions are liable to occur when brought in contact with organic or readily oxidizable materials either in solution or in the dry state.
Shipping regulations: Oxidizing material. Yellow label.*

potassium perosmate. See potassium osmate.

potassium peroxide K_2O_2.
Properties: Yellow, amorphous mass; decomposes in water, evolving oxygen.
Derivation: By the oxidation of potassium in oxygen.
Grades: Technical.
Containers: Tins.
Uses: Oxidizing agent; bleaching agent; oxygen-generating gas masks.
Caution! Fire hazard; dangerous! Does not burn or explode itself but mixtures of potassium peroxide and combustible substances are explosive and ignite easily.
Shipping regulations: Oxidizing material. Yellow label.*

potassium peroxydisulfate. See potassium persulfate.

potassium peroxymonosulfate. See "Oxone."

potassium persulfate (anthion; potassium peroxydisulfate) $K_2S_2O_8$.
Properties: White crystals; soluble in water; insoluble in alcohol; m.p., decomposes below 100°C.
Derivation: By electrolysis of a saturated solution of potassium sulfate.
Grades: Technical; C.P.
Containers: Glass bottles; 300-lb drums; 100-lb polyethylene-lined paper bags.
Uses: Bleaching; oxidizing agent; reducing agent in photography; antiseptic; soap manufacture; analytical reagent; polymerization promoter; pharmaceuticals; modification of starch; flour-maturing agent; defiberizing wet-strength paper; desizing of textiles.
Fire hazard: Dangerous; may cause explosion in a fire.

potassium phenethicillin. See potassium alpha-phenoxyethylpenicillin, under penicillin.

potassium phenoxymethylpenicillin. See penicillin.

potassium alpha-**phenoxymethylpenicillin.** See penicillin.

potassium phosphate. See potassium phosphate, dibasic; potassium phosphate, monobasic; or potassium phosphate, tribasic.

potassium phosphate, dibasic (DKP; potassium hydrogen phosphate; potassium monophosphate; dipotassium orthophosphate) K_2HPO_4.
Properties: Deliquescent white crystals or powder. Very soluble in water; slightly soluble in alcohol.
Derivation: By action of phosphoric acid on potassium carbonate.
Method of purification: Crystallization.
Impurities: Chlorine; potassium sulfate.
Grades: Commercial; pure; highest purity; N.F. XI.

Containers: Glass bottles; 100-lb bags; 100-lb fiber drums.
Uses: Buffer in antifreezes; ingredient of "instant" fertilizers; nutrient for penicillin culturing; humectant; pharmaceuticals.

potassium phosphate, monobasic (MKP; potassium acid phosphate; potassium diphosphate; potassium biphosphate; potassium dihydrogen phosphate) KH_2PO_4.
Properties: Colorless crystals. Acid in reaction. Soluble in water; insoluble in alcohol; sp.gr. 2.338; m.p. 96°C.
Derivation: By the action of phosphoric acid on potassium carbonate.
Method of purification: Crystallization.
Grades: Technical; C.P.
Containers: Wooden kegs; glass bottles; barrels; multiwall paper bags.
Uses: Medicine; baking powders; nutrient solution; yeast foods.
Shipping regulations: None.*

potassium phosphate, neutral. See potassium phosphate, tribasic.

potassium phosphate, normal. See potassium phosphate, tribasic.

potassium phosphate, tertiary. See potassium phosphate, tribasic.

potassium phosphate, tribasic (potassium phosphate, neutral; potassium phosphate normal; tripotassium orthophosphate; potassium phosphate, tertiary; tripotassium phosphate) $K_3PO_4 \cdot 3H_2O$, or K_3PO_4.
Properties: Granular white powder. Hygroscopic. Caution! Keep well stoppered! Soluble in water giving strongly basic solution; insoluble in alcohol. M.p. 1340°C.
Grades: Reagent; technical.
Containers: 275-, 300-, 400-lb drums.
Uses: Purification of gasoline; water-softening; liquid soaps; fertilizer.

potassium phosphite K_2HPO_3.
Properties: White powder; hygroscopic. Caution! Keep well stoppered. Soluble in water; insoluble in alcohol. Slowly oxidized by air to phosphate.

potassium platinichloride. See potassium chloroplatinate.

"Potassium Polymetaphosphate HV." [172]
Trade name for a grade of potassium metaphosphate having a high molecular weight and capable of yielding high viscosity solutions. $(KPO_3)_n$.
Properties: Insoluble in water. Soluble in sodium salt solutions; shows molecular weight of about 500,000.
Derivation: Dehydration of KH_2PO_4.
Containers: Bags.
Uses: As a fat emulsifier and moisture retaining agent in sausage.

potassium polysulfide K_2S_x.
Properties: Crystals; soluble in water and alcohol.
Use: Fungicide.

potassium prussiate, red. See potassium ferricyanide.

potassium prussiate, yellow. See potassium ferrocyanide.

potassium pyroantimonate $K_2H_2SbO_7 \cdot 4H_2O$ (approx).
Properties: White crystalline powder or granules; slightly soluble in cold water; readily soluble in hot water. Insoluble in alcohol.
Grades: Reagent; technical.
Uses: Used in starch sizes and flame retarding compounds.

potassium pyroborate. See potassium tetraborate.

potassium pyrophosphate (TKPP; tetrapotassium pyrophosphate; potassium pyrophosphate, normal) $K_4P_2O_7 \cdot 3H_2O$.
Properties: Colorless crystals or white powder; somewhat hygroscopic in air (deliquesces above a relative humidity of 40-45%). Similar to tetrasodium pyrophosphate except for greater solubility; sp.gr. 2.33; dehydrated below 300°C; m.p. 1090°C. Soluble in water; insoluble in alcohol.
Grades: Technical; 99.4%; 60% solution.
Containers: Fiber drums; multiwall paper bags.
Uses: In tin plating; stabilizing hydrogen peroxide baths; purification of china clay, oil drilling muds, dyeing; in soaps and detergents it sequesters magnesium, peptizes curds of calcium soaps and has solvent or dispersing action on gums, waxes or dirt; builder and clarifier of liquid soaps; in synthetic rubber production.

potassium pyrophosphate, normal. See potassium pyrophosphate.

potassium pyrosulfate (potassium anhydrosulfate; potassium acid sulfate, anhydrous) $K_2S_2O_7$.
Properties: Colorless needles or white, crystalline powder, or fused pieces. Soluble in water; converted to potassium bisulfate. Sp.gr. 2.27; m.p. (approx) 325°C.
Use: Acid flux in analysis.

potassium pyrosulfite. See potassium metabisulfite.

potassium rhodanide. See potassium thiocyanate.

potassium ricinoleate $C_{17}H_{32}OHCOOK$.
Properties: White paste; soluble in water.
Use: Emulsifying agent.

potassium selenate K_2SeO_4.
Properties: White powder or colorless crystals. Soluble in water; sp.gr. about 3.
Grades: Technical; C.P.
Use: Reagent.

potassium silicate (glass)
Properties: Weight ratio $SiO_2:K_2O = 2.5$; molar ratio $SiO_2:K_2O = 3.87$ (approx);

*See "I.C.C. Shipping Regulations," page xiii.
Reference numbers refer to name of manufacturer. See "List of Manufacturers," page v.

colorless, anhydrous lump, shattered or granular material; soluble in water only at elevated temperatures and pressure.
Derivation: Supercooled melt of potassium carbonate and pure silica sand.
Containers: Drums.
Uses: In the manufacture of glass and refractory material; in the manufacture of potassium silicate solutions; for dyeing and bleaching.
Shipping regulations: None.*

potassium silicate (solution).
Water solutions with various ratios of SiO_2:K_2O from 2.5 to 3.3.
Properties: Colorless to turbid liquid; gravity 29-41° Bé; sp.gr. 1.25-1.39; 10.4-11.6 lbs/gal.
Containers: 1- and 5-gal cans; 55-gal drums; tank cars and tank trucks.
Uses: As non-efflorescing base for inorganic paints and protective coatings; as a coating for roofing granules and welding rods; as a binder in the manufacture of carbon arc-light electrodes and for phosphors on television tubes; in detergents; as a catalyst; adhesives.
Shipping regulations: None.*

potassium silicofluoride. See potassium fluosilicate.

potassium-sodium borotartrate. See potassium borotartrate.

potassium-sodium carbonate. See sodium-potassium carbonate.

potassium-sodium ferricyanide $K_2NaFe(CN)_6$.
Constants: Red crystals, over 99% pure; m.p., decomposes; non-hygroscopic and stable; easily soluble in water.
Derivation: From ferrocyanides.
Containers: Fiber drums.
Uses: In making blueprint paper; in photographic work.
Shipping regulations: None.*

potassium-sodium phosphate. See sodium-potassium phosphate.

potassium-sodium tartrate (Rochelle salt; seignette salt; sodium-potassium tartrate) $KNaC_4H_4O_6 \cdot 4H_2O$.
Properties: Colorless transparent efflorescent crystals or white powder, having cool, saline taste. Soluble in water; insoluble in alcohol. Loses water of crystallization at 140°C; unstable above about 225°C. Sp.gr. 1.77; m.p. 70-80°C.
Derivation: Potassium acid tartrate is dissolved in water, the solution saturated with sodium carbonate, concentrated after purification and crystallized.
Method of purification: Hydrogen sulfide is passed in to remove copper and iron, the solution heated with animal charcoal and filtered.
Impurities: Copper; iron.
Grades: Highest purity; reagent; commercial crystals or powder; N.F. XI.
Containers: 25-, 225-, 250-lb drums.
Uses: Medicine; manufacture of Seidlitz powders; baking powders.
Shipping regulations: None.*

potassium sorbate (potassium 2,4 hexadienoate) $CH_3CH{:}CHCH{:}CHCOOK$.
Properties: White powder; m.p. 270°C; 58.5% sol in water (25°C); sp.gr. 1.36 (25/20°C).
Containers: 10-, 50-lb cartons; 150-lb fiber drums.
Uses: Water soluble form of sorbic acid for control of molds and yeasts in food.

potassium stannate $K_2SnO_3 \cdot 3H_2O$.
Properties: White to light tan crystals. Soluble in water; insoluble in alcohol. Sp. gr. 3.197.
Grades: Technical.
Containers: 100-, 350-lb drums.
Uses: Textiles (dyeing and printing); alkaline tinplating bath.

potassium stearate $C_{17}H_{35}COOK$.
Properties: White crystal powder; soluble in hot water; slight odor of fat.
Grades: Commercial, contains considerable palmitate.
Use: Base for textile softener.

potassium-strontium chlorate. See strontium-potassium chlorate.

potassium sulfantimonate (potassium thioantimonate) $(K_3SbS_4)_2 \cdot 9H_2O$.
Properties: Colorless to yellowish crystals; soluble in water; insoluble in alcohol.

potassium sulfate (salt of Lemery) K_2SO_4.
Properties: Colorless or white, hard crystals or powder; bitter, saline taste. Soluble in water; insoluble in alcohol. Sp.gr. 2.66; m.p. 1072°C.
Derivation: (a) By treatment of potassium chloride with sulfuric acid. (b) By fractional crystallization of kainite.
Method of purification: Recrystallization.
Grades: Highest purity medicinal; commercial; crude; C.P.; agricultural.
Containers: 100-lb bags; 25-, 100-, 350-, 375-lb drums.
Uses: Reagent in analytical chemistry; medicine; fertilizer; alum manufacture; glass manufacture.
Shipping regulations: None.*

potassium sulfhydrate. See potassium hydrosulfide.

potassium sulfide, fused or concentrated. (potassium sulfuret; hepar sulfuric) K_2S.
Properties: Red or yellow-red crystalline mass or fused solid; deliquescent in air. Keep well stoppered. Soluble in water, alcohol, and glycerin; insoluble in ether. Sp.gr. 1.805 (20/4°); m.p. 471°C.
Derivation: Potassium sulfate and carbon are heated in a tightly closed crucible to a moderate temperature.
Grades: Technical.
Containers: Cans; glass bottles; metal drums.
Uses: Reagent in analytical chemistry; depilatory; medicine.
Fire hazard: Moderately flammable, yields flammable hydrogen sulfide on contact

with mineral acids and sulfur dioxide when burning.
Shipping regulations: Flammable solid. Yellow label.*

potassium sulfite $K_2SO_3 \cdot 2H_2O$.
Properties: White crystals or powder. Soluble in water; sparingly soluble in alcohol.
Grades: Technical; C.P.
Containers: 1-, 5-lb bottles.
Use: Medicine; photography.
Shipping regulations: None.*

potassium sulfocarbonate (potassium trithiocarbonate K_2CS_3.
Properties: Yellowish-red crystals. Very hygroscopic. Soluble in alcohol and water.
Grades: Technical.
Uses: Analysis (testing for cobalt, nickel); medicine.

potassium sulfocyanate. See potassium thiocyanate.

potassium sulfocyanide. See potassium thiocyanate.

potassium, sulfurated. See potash, sulfurated.

potassium sulfuret. See potassium sulfide.

potassium-tantalum fluoride. See tantalum-potassium fluoride.

potassium tartrate $K_2C_4H_4O_6 \cdot \frac{1}{2}H_2O$.
Properties: Colorless, crystalline; soluble in water; decomposed by heat. Sp.gr. 1.98; insoluble in alcohol.
Grades: C.P.; technical.
Uses: Manufacture of potassium salts; medicine; laboratory reagent.

potassium tellurate $K_2TeO_4 \cdot 3H_2O$.
Properties: Colorless crystals; soluble in water.
Use: Medicine.

potassium tellurite K_2TeO_3.
Properties: Granular, white powder. Hygroscopic. Soluble in water.
Grades: Technical.
Use: Analysis (testing for bacteria).

potassium tetraborate (potassium pyroborate; potassium borate; potassium biborate) $K_2B_4O_7 \cdot 5H_2O$.
Properties: White powder; alkaline taste. Soluble in water.

potassium tetrathiocyanodiammono-chromate. See Reinecke salt.

potassium thioantimonate. See potassium sulfantimonate.

potassium thiocyanate (potassium rhodanide; potassium sulfocyanate; potassium sulfocyanide) KCNS.
Properties: Colorless, transparent, hygroscopic, odorless crystals; soluble in water, alcohol and acetone. Saline, cooling taste. Sp.gr. 1.88; m.p. 173°C; turns brown, green, blue when fused, white again on cooling. B.p., decomposes at 500°C.
Derivation: By heating potassium cyanide with sulfur.
Method of purification: Crystallization.
Impurities: Heavy metals, sulfates.
Grades: Commercial; pure; purified; reagent.
Containers: Glass bottles; drums.
Use: In freezing mixtures; reagent in analytical chemistry; manufacture of sulfocyanides; thioureas; in admixture with allyl bromide for making allyl mustard oil; printing and dyeing textiles; photographic restrainer and intensifier; manufacture of synthetic dyestuffs.
Shipping regulations: None.*

potassium thiosulfate (potassium hyposulfite) $K_2S_2O_3$ with varying proportions of water of crystallization.
Properties: Colorless crystals. Hygroscopic. Caution! Keep well stoppered! Soluble in water.
Grades: Technical; C.P.
Use: Analysis.

potassium titanate K_2TiO_3 (approx.).
Properties: White salt; hydrolyzes in water to give a strongly alkaline solution.
Derivation: From titanic acid and potassium hydroxide.
Containers: Cartons.
Use: See potassium titanate fibers.

potassium titanate fibers. Approximate composition $K_2O \cdot (TiO_2)_n$ where n is 4 to 7.
Properties: Crystalline fibers of small diameter; m.p. 2500°F; high refractive index; can diffuse and reflect infrared radiation.
Uses: Can be felted for use in rockets, missiles, nuclear-powered applications as an insulator, especially for the range 1300-2100°F.

potassium-titanium fluoride. See titanium-potassium fluoride.

potassium-titanium oxalate. See titanium-potassium oxalate.

potassium tripolyphosphate (KTPP) $K_5P_3O_{10}$.
Properties: White crystalline solid; m.p. 620-640°C; density 2.54; loose bulk density (approx) 67 lbs/cu ft; solubility in water (26°C), over 140 g/100 ml water.
Containers: 100-lb bags; 400-lb fiber drums.
Uses: Water emulsion plants; water-treating compounds; cleaners; specialty fertilizers; sequestrant (including use in food preparation).

potassium trithiocarbonate. See potassium sulfocarbonate.

potassium tungstate (potassium orthotungstate; potassium wolframate; potassium wolframate, normal) K_2WO_4.
Properties: Heavy, crystalline powder; sp. gr. 3.1; m.p. 921°C. Deliquescent. Caution! Keep dry! Soluble in water; insoluble in alcohol.
Grades: Technical.
Use: Magenta bronze.

potassium undecylenate $CH_2CH(CH_2)_8COOK$.
Properties: Finely divided white powder; decomposes above 250°C; limited solubility in most organic solvents; soluble in water.

Uses: Bacteriostat and fungistat in cosmetics and pharmaceuticals.

potassium wolframate. See potassium tungstate.

potassium wolframate, normal. See potassium tungstate.

potassium xanthate (potassium ethyldithiocarbonate; potassium xanthogenate; potassium ethylxanthogenate) $KS_2COC_2H_5$.
Properties: Colorless or light yellow crystals; soluble in water and alcohol; sp.gr. 1.558 (21.5°C).
Derivation: Prepared by reaction of potassium ethylate and carbon disulfide.
Containers: Glass bottles; fiber cans.
Uses: Fungicide for soil treatment; reagent in analytical chemistry.
Shipping regulations: None.*

potassium xanthogenate. See potassium xanthate.

potassium zinc iodide (zinc potassium iodide) $ZnI_2 \cdot KI$.
Properties: Colorless crystals; very hygroscopic.
Grades: Technical.
Use: Analysis (testing for alkaloids).

potassium-zinc sulfate. See zinc-potassium sulfate.

potassium zirconifluoride. See zirconium-potassium fluoride.

potassium zirconium chloride $KCl, ZrCl_4$. A source of zirconium for magnesium alloys, to remove iron in an insoluble form.

potassium-zirconium sulfate. See zirconium-potassium sulfate.

potato oil. See fusel oil.

potato spirit. See fusel oil.

pot clays. Refractory clays used for making the pots in which glass is produced.

"Pot N Pan." [108] A light blue, free-flowing granular composition containing special soaps, mild alkaline cleaners, corrosion inhibitors and "Calgon" water conditioner.
Containers: 325-, 100-lb drums.

potter's clay. See ball clay.

pottery body stains. Calcined-oxide finely ground pigments for coloring ceramic bodies.
Use: As colors or designs for tile, terra cotta, chinaware, etc., where the pigment becomes part of the ceramic body.

potting. A process in which an uncured liquid resin or its solution is caused to penetrate an assemblage of electrical wire and insulation, such as a solenoid or resistor coil, and the penetrated resin is then cured to produce a unit with improved electrical, handling, and deterioration characteristics. Epoxy resins are frequently used.

pound centigrade unit (pcu). See chu.

pour point.
1. The lowest temperature at which a liquid will flow when a test container is inverted.
2. The temperature at which an alloy is cast.

povidone-iodine. A water-soluble complex produced by reacting iodine with the polymer polyvinylpyrrolidone. Used in external medicine, for the slow release of its iodine.

powder, black. See gunpowder.

"Powdered Acid Cleaner." [244] Trade name for an acid cleaner compound of an acid salt and wetting agent.
Properties: Light gray granular material readily soluble in water.
Containers: 100-lb net drum.

powder metallurgy. The arts of producing metal powder and of the utilization of metal powders for the production of massive materials and shaped objects (MPA definition, MPA Standard 9-50T).

The production of finished metal shapes and products by pressing metal powder in suitably shaped molds and then sintering the briquettes at an elevated temperature to consolidate the structure, reduce porosity and impart useful strength is a major metal-fabricating operation. Since it need involve no liquid phase, it is the only commercial method for the manufacture of ductile tungsten, tantalum and similar high-melting metals, and also of tungsten-copper contacts, copper-lead bearings and such parts whose component metals do not readily mix when molten. Cemented tungsten carbides used for tools and dies, and porous self-lubricating bearings are also manufactured only from powdered metals. A fast-growing application is the use of powder metallurgy for making small metal parts such as gears, magnets, bushings, etc. that were formerly made by casting and machining, bar-machining, stamping, forging and machining, etc. In such cases, molding to shape by powder metallurgy may save machining and material cost without sacrificing quality or time.
See also metals, powdered.

powellite $CaMoO_4$ or $Ca(Mo, W)O_4$. Natural calcium molybdate in which a portion of the molybdenum is replaced by tungsten.
Properties: Bluish-green crystals contain 1.65 to 10.28% WO_3; sp.gr. 4.35-4.52. Pearly-gray scales are more common and contain only traces of tungsten; sp.gr. 4.25.
Occurrence: United States (Idaho, Michigan) Texas, Nevada, California); Siberia.
Uses: Minor ore of molybdenum.

"Power-Pak." [204] Trademark for a liquid dry cleaning detergent, recommended for use in charged systems at 1-4% concentrations.

power reactor. See nuclear reactor.

"Poxeal." [116] Trademark for an insulation system for dynamoelectric machines whereby the electrical circuits are sealed in a tough flexible material (such as epoxy

*See "I.C.C. Shipping Regulations," page xiii.
Reference numbers refer to name of manufacturer. See "List of Manufacturers," page v.

resin) capable of withstanding severe environmental and thermal shock conditions.

pozzolana cement (puzzolana cement; Santorin cement; gaize cement; silikat-cement; tarras cement; trass cement). A cement produced by grinding together Portland cement clinker and a pozzolana, or by mixing together a hydrated lime and a pozzolana. A pozzolana is defined as a material which is capable of reacting with lime in the presence of water at ordinary temperature to produce a cementitious compound. Natural pozzolanas are silicious material of volcanic origin. They include trass and Santorin earth. Blast furnace slag is used to produce artificial pozzolanas.

PPO. See 2,5-diphenyloxazole.

Pr. Symbol for praseodymium.

Prague red. A red pigment consisting, essentially, of red iron oxide.
See also iron oxide reds and hematite.

pramoxine hydrochloride
$C_4H_9OC_6H_4OC_3H_6C_4H_8NO \cdot HCl$.
Properties: White, practically odorless, crystalline powder; numbing effect when tasted; melts between 169° and 172°C. Freely soluble in water and alcohol; practically insoluble in chloroform and ether.
Grade: N.F. XI.
Use: Medicine.

"Pranone." [321] Brand name for ethisterone.

"Prantal." [321] Brand name of diphemanil methylsulfate.

praseodymia. See praseodymium oxide; see also rare earths.

praseodymium Pr. Element of atomic number 59; Group III of the periodic table; one of the rare earth elements of the cerium subgroup.
Properties: Yellowish metal, tarnishes easily (color of salts green); sp.gr. 6.78-6.81; m.p. 940°C; b.p. 3000°C (approx); ignites to oxide (200-400°C); liberates hydrogen from water; soluble in dilute acids.
Source: See rare earth minerals.
Derivation: Reduction of the chloride or fluoride with calcium powder.
Use: Praseodymium salts.
Shipping regulations: None.*
See also didymium.

praseodymium salts.
praseodymium ammonium nitrate $Pr(NO_3)_3 \cdot 2NH_4NO_3 \cdot 4H_2O$. Colored crystals; soluble in water. Purity 50% praseodymium salt.
praseodymium carbonate $Pr_2(CO_3)_3 \cdot xH_2O$. Green powder; insoluble in water; soluble in acids. Grades 99% Pr salts, available as 70% Pr_6O_{11}. Containers: Glass bottles, fiber drums.
praseodymium chloride $PrCl_3 \cdot xH_2O$. Green crystals, soluble in water. Purities to 99% praseodymium salt; color varies with the purity. Available as 45% Pr_6O_{11}. Containers: Glass bottles, fiber drums.
praseodymium fluoride $PrF_3 \cdot 2H_2O$. Green crystals. Grades up to 99% Pr salts. Available as 77% Pr_6O_{11}. Containers: Glass bottles, fiber drums.
praseodymium nitrate $Pr(NO_3)_3 \cdot xH_2O$. Green crystals; soluble in water. Color varies with the purity. Available as 40% Pr_6O_{11}. Containers: Glass bottles, fiber drums. Shipping regulations: Oxidizing material. Yellow label.*
praseodymium oxalate $Pr_2(C_2O_4)_3 \cdot xH_2O$. Green powder; insoluble in water; slightly soluble in acids. Grades 80%, 90%, 99%, 99.9% Pr salts. Available as 50% Pr_6O_{11}. Containers: Glass bottles, fiber drums.
praseodymium oxide (praseodymia) Pr_6O_{11}. Brown-black powder; insoluble in water; soluble in acids. Hygroscopic, absorbs carbon dioxide from the air. Purities to 99.8% oxide. Containers: Glass bottles, fiber drums.
praseodymium sulfate $Pr_2(SO_4)_3 \cdot xH_2O$. Green crystals; slightly soluble in hot water; more soluble in cold. Color varies with purity. Available as 45% Pr_6O_{11}. Containers: Glass bottles, fiber drums.
Uses: Coloring glass; decolorizing glass. The oxide is used to impart green color to synthetic emeralds.

precipitation. The formation of solid particles in a solution. Also, the settling out of small particles in either a liquid or gaseous medium.

precision. The extent to which a set of measurements or observations conform to their own mean, as frequently measured by the standard deviation.
See also accuracy.

prednisolone $C_{21}H_{28}O_5$. delta^{1,4}-Pregnadiene-11 beta, 17 alpha, 21-triol-3,20-dione. Generic name for an analog of hydrocortisone.
Properties: White to practically white, odorless, crystalline powder. Very slightly soluble in water; soluble in alcohol, chloroform, acetone, methanol, dioxane. M.p. about 235°C with some decomposition.
Grade: U.S.P. XVI.
Use: Medicine.

prednisolone acetate $C_{23}H_{30}O_6$.
Properties: White to practically white, odorless, crystalline powder. Practically insoluble in water; slightly soluble in alcohol, chloroform, acetone. M.p. about 235°C, with some decomposition.
Grade: U.S.P. XVI.
Use: Medicine.

prednisolone phosphate sodium $C_{21}H_{27}Na_2O_8P$.
Properties: White powder; slightly hygroscopic. Stable at room temperatures. Soluble in water, methanol, ethanol.
Grade: N.N.D.
Use: Medicine.

*See "I.C.C. Shipping Regulations," page xiii.
Reference numbers refer to name of manufacturer. See "List of Manufacturers," page v.

prednisone $C_{21}H_{26}O_5$. delta^{1,4}-Pregnadiene-17 alpha, 21-diol-3, 11, 20-trione. Generic name for an analog of cortisone.
Properties: White to practically white, odorless, crystalline powder. Very slightly soluble in water; slightly soluble in alcohol, chloroform, methanol, and dioxane. M.p. about 225°C with some decomposition.
Grade: U.S.P. XVI.
Use: Medicine.

pregnanediol $C_{21}H_{36}O_2$. 5 beta-Pregnane-3-alpha, 20 alpha-diol. A steroid; the metabolic product of progesterone.
Properties: Crystallizes in plates from acetone; m.p. 238°C; dextrorotatory in solutions; sparingly soluble in organic solvents; not precipitated by digitonin.
Derivation: Isolation from pregnancy urine of women, cows, mares, and chimpanzees; by reduction of pregnanedione.
Use: In the synthesis of progesterone; medically as a pregnancy test.

pregnenedione. See progesterone.

pregneninolone. See ethisterone.

4-pregnen-21-ol-3, 20-dione. See deoxycorticosterone.

pregnenolone $C_{21}H_{32}O_2$. delta^5-Pregnene-3-beta-ol-20-one. A steroid which is believed to be a precursor of progesterone and the adrenal steroid hormones.
Properties: Crystallizes in needles from dilute alcohol; m.p. 193°C; freely soluble in acetone, petroleum ether, benzene, and carbon tetrachloride.
Derivation: From stigmasterol or other steroids.
Containers: Bottles.
Use: Medicine; biochemical research.
Also available as acetate salt.

prehnite $Ca_2Al_2Si_3O_{10}(OH)_2$. A natural hydrous silicate of calcium and aluminum, related to the zeolites (q.v.).
Properties: Color, light green to white; luster vitreous; sp.gr. 2.8-2.95; hardness 6-6.5.
Occurrence: Europe; United States.

prehnitene (1, 2, 3, 4-tetramethylbenzene; prenitol) $(CH_3)_4C_6H_2$.
Properties: Colorless liquid; soluble in alcohol; insoluble in water. Sp.gr. 0.901; b.p. 204°C; m.p. −7.7°C.
Grades: Technical.
Use: Organic synthesis.

"Premose." [114] Trademark for a high-maltose content syrup made from malted and unmalted cereal grains.

"Prenderol." [412] Trademark for 2, 2-diethyl-1, 3-propanediol (q.v.).

prenitol. See prehnitene.

preparing salt. See sodium stannate.

"Prestabit Oil V." [307] Brand name for an anionic textile chemical consisting of purified sulfated castor-oil fatty acids.
Properties: Clear, yellow, oily liquid; soluble in water; stable to acid and alkali.
Uses: Dyeing assistant for both cotton and wool to give wetting, penetration, leveling as well as softness to the dyed fiber; in viscose manufacture, clarifying agent to prevent milkiness of the yarn; antistatic agent for acetate and polyacrylonitrile fibers.

"Prestone." [214] Trademark for a group of automotive service products which include an ethylene-glycol base antifreeze, containing special inhibitors for prevention of corrosion, foaming, creepage and rust-loosening.

"Preventol." [307] Trademark for a series of fungicides and bactericides.
"Preventol" GD. Dihydroxydichlorodiphenyl methane; 96% active.
Properties: Fine, sandy, off-white powder; density 0.60-0.65; insoluble in water; soluble in caustic solutions, butanol, isopropanol, petroleum naphtha, and acetone.
Uses: Fungicide and bactericide which is non-toxic and non-irritating in normal use concentrations; mildewproofing agent for cotton yarn, thread and fabrics; preservative for liquid and paste products such as glues and adhesives; bactericide in cutting oils; used for slime control in the paper industry.
"Preventol" GDC. A 38-40% active alkaline solution of "Preventol" GD.
"Preventol" I. Sodium trichlorophenate; 56% solids; 51% active.
Properties: Clear, thin liquid; soluble in water; sp.gr. 1.40.
Uses: Fungicide and bactericide for use in the leather trade; disinfectant and mold preventative for skins; preservative for print pastes, finishes, sizes, and adhesives.

priceite $Ca_4B_{10}O_{19}\cdot 7H_2O$. A hydrated calcium borate of variable formula, mined in Turkey for borax. Color white; luster earthy; hardness 3-3.5; sp.gr. 2.2-2.5.

prills. Material in small bead form.

"Prim." [204] Trademark for a solvent-soluble fabric size and water-repellent for dry cleaning use.

"Primal." [23] Trademark for aqueous dispersions of acrylic resins, supplied in various grades that differ in hardness and flexibility. Produce finishes which are water-insoluble, require no plasticizer for flexibility, unimpaired by aging, and adhere tenaciously to leather and lacquer coats.
Use: Base coat for leather finishes.
"Primal" Binders. Dispersions or solutions of waxes and auxiliary materials.
Use: Binders for water finishes on all types of leather.
"Primal" Colors. Aqueous dispersions of pigments.

primaquine $H_2N(CH_2)_3CHCH_3NHC_9H_5NOCH_3$. 6-Methoxy-8-(4-amino-1-methylbutylamino) quinoline.

Properties: Viscous liquid; b.p. 175-179°C (0.2 mm); moderately soluble in water.
Use: Medicine (base).

primaquine phosphate $C_{15}H_{21}N_3O \cdot 2H_3PO_4$. 8-(4-Amino-1-methylbutylamino)-6-methoxyquinoline phosphate.
Properties: Orange-red, crystalline powder; odorless; bitter taste; m.p. 200-205°C; solutions acid to litmus; soluble in water; insoluble in chloroform and ether.
Grade: U.S.P. XVI.
Use: Medicine.

primary. Term used to characterize certain types of compounds and distinguish them from other similar or isomeric substances. Thus a primary alcohol is one whose molecular structure may be written RCH_2OH rather than R_1R_2CHOH (secondary alcohol) or $R_1R_2R_3COH$ (tertiary alcohol). A primary amine is characterized by the formula RNH_2, a secondary amine by R_1R_2NH, and a tertiary amine by $R_1R_2R_3N$. In all these cases R_1, R_2, and R_3 may designate either identical or different hydrocarbon groups or substituted hydrocarbon groups. See, e.g., amyl alcohol. See tert- for another use of tertiary.

The terms primary, secondary, and tertiary are also used to name salts of orthophosphoric acid (H_3PO_4) in which one, two, or three of the hydrogen atoms have been replaced by metal or radicals. Thus NaH_2PO_4 is primary sodium phosphate, while Na_2HPO_4 is secondary sodium phosphate. The same system of names is used for salts of other acids containing three replaceable hydrogen atoms.

primary azo dyes. Azo dyes derived from primary amines.

primary calcium phosphate. See calcium phosphate, monobasic.

"Primene." [23] Trademark for tertiary alkyl amines (RNH_2, where R is C_{12}–C_{24} t-alkyl). Supplied as oil-soluble liquids.
Use: Fuel oil stabilizers, corrosion inhibitors; intermediates for oil additives, surfactants, and other chemical products.

primidone $C_{12}H_{14}N_2O_2$. 5-Ethyldihydro-5-phenyl-4, 6(1H, 5H)-pyrimidinedione.
Properties: White, crystalline powder. Odorless. Slightly bitter taste. Very slightly soluble in water and most organic solvents; slightly soluble in alcohol; pH of saturated aqueous solutions 5.6-6.0; m.p. 279°-284°C.
Grade: U S.P. XVI.
Use: Medicine.

"Priminox." [23] Trademark for ethylene oxide derivatives of "Primene". Most grades are oil-soluble liquids, some water-soluble.
Use: Formulation of corrosion inhibitors, surfactants, and fuel oil additives.

"Primol." [51] Trademark for high viscosity white mineral oils conforming with U.S.P. specifications for internal use; in the class known as "Russian type."

primrose chrome. See chrome yellows.

primrose yellow. See chrome yellows.

primuline. A yellow synthetic dye.

primuline dyes. See thiazole dyes.

printer's acetate. See mordant rouge.

printer's iron liquor. See iron acetate liquor.

printing ink. Usually a mixture of finely divided pigment such as carbon black suspended in a drying oil such as heat-bodied linseed oil. Alkyds, phenol-formaldehyde or other synthetic resins are frequently also present, and cobalt, manganese, and lead soaps are added to achieve rapid drying by oxidation and polymerization. Mineral oils are sometimes present and some types of inks dry by evaporation of a volatile solvent rather than by oxidation and polymerization of a drying oil or resin. For colored inks, pigments such as chrome yellows, benzidine yellows or lithol reds are used.
Shipping regulations: May be classified as flammable liquids. Red label.*

"Priodax." [321] Brand name for iodoalphionic acid.

"Priscoline." [305] Trademark for tolazoline.
Use: Medicine.

prisilidene hydrochloride. See alphaprodine hydrochloride.

"Prismlac." [333] Trade name for cellulose-base crystallizing lacquers which are used to produce a clear or colored, opaque finish.

"Pristane." [415] (norphytane; 2, 6, 10, 14-tetramethylpentadecane) $C_{19}H_{40}$.
Properties: A colorless, odorless, liquid; b.p. 290°C; sp.gr. 0.780-0.790 (20°C); refractive index 1.4382-1.4392 (20°C); flash point ca 183°C. Soluble in ether, petroleum ether, benzene, chloroform, carbon tetrachloride.
Grade: 90% min.
Containers: 1-lb bottles; 1-gal metal tins; 5-gal metal pails; 100-lb metal drums.
Uses: Suggested as a precision lubricant, and as heat transfer fluid.

"Privine." [305] Trademark for naphazoline.

"Pro-Banthine." [70] Trademark for propantheline bromide. (beta-Diisopropylaminoethyl xanthene-9-carboxylate methobromide).
Use: Medicine.

probarbital sodium $C_9H_{13}O_3N_2Na$. Sodium 5-ethyl-5-isopropylbarbiturate.
Properties: White, hygroscopic powder; insoluble in water; slightly soluble in alcohol; practically insoluble in ether and chloroform. Aqueous solutions are alkaline to litmus; bitter taste.
Use: Medicine.

probenecid $HOOCC_6H_4SO_2N(CH_2CH_2CH_3)_2$. (Dipropylsulfamyl) benzoic acid.
Properties: White, odorless crystalline powder; m.p. 198-200°C; soluble in acetone, alcohol, dilute alkalies, and dilute sodium

bicarbonate; insoluble in dilute acids and water.
Grade: U.S.P. XVI.
Use: Medicine.

procainamide hydrochloride
$H_2NC_6H_4CONHCH_2CH_2N(C_2H_5)_2 \cdot HCl$. para-Amino-N-(2-diethylaminoethyl)benzamide hydrochloride.
Properties: White to tan, odorless, crystalline powder; m.p. 165-169°C; very soluble in water; soluble in alcohol; slightly soluble in chloroform; very slightly soluble in benzene and ether.
Grade: U.S.P. XVI.
Use: Medicine.

procaine. See procaine hydrochloride.

procaine base (para-aminobenzoyldiethylaminoethanol base; 2-diethylaminoethyl-para-aminobenzoate)
$C_6H_4NH_2COOCH_2CH_2N(C_2H_5)_2$. Crystallizes with $2H_2O$ from aqueous alcohol.
Properties: White, stable, granular powder; odorless; melts near 60°C when anhydrous. Sensitive to light and air. Soluble in alcohol, ether, chloroform, benzene, fixed and volatile oils; insoluble in water.
Use: Medicine.
Shipping regulations: None.*

procaine benzylpenicillin. See under penicillin, as procaine penicillin G.

procaine hydrochloride (para-aminobenzoyldiethylaminoethanol hydrochloride; ethocaine; kerocaine; procaine)
$C_6H_4NH_2COOCH_2CH_2N(C_2H_5)_2 \cdot HCl$.
Properties: Small, colorless crystals or white, crystalline powder; odorless; stable in air. Soluble in water and in alcohol at 25°C; slightly soluble in chloroform; almost insoluble in ether; solutions acid to litmus. M.p. 153-156°C.
Derivation: (a) By heating chloroethyl-para-nitrobenzoic ester with diethylamine for 24 hours under pressure at 120°C. The product is then reduced with tin and hydrochloric acid. (b) By condensation of ethylene chlorohydrin with diethylamine. The chloroethyldiethylamine formed is heated with sodium para-aminobenzoate.
Containers: Drums.
Grade: U.S.P. XVI.
Use: Medicine.
Shipping regulations: None.*

procaine nitrate (para-aminobenzoyldiethylaminoethanol nitrate; beta-diethylaminoethyl-para-aminobenzoate nitrate)
$C_6H_4NH_2COOC_2H_4N(C_2H_5)_2 \cdot HNO_3$.
Properties: Small, colorless crystals; odorless. Soluble in water, alcohol. M.p. 100-102°C.
Use: Medicine.
Shipping regulations: None.*

procaine penicillin G. See penicillin.

"Processing Stiffiner 710." [248] Trademark for a hydrazine salt (26.4% active material) compound.
Properties: A gray-white powder; sp.gr. 2.50; the active constituent is slightly soluble in water and insoluble in gasoline, benzol, ethylene dichloride, and acetone.
Uses: Used to prevent excessive softness and low plasticity in rubber stocks in process. It is used in all types of unvulcanized natural, SBR, neoprene and nitrile rubbers.

prochlorperazine ethanedisulfonate
$C_{20}H_{24}ClN_3S \cdot C_2H_6O_6S_2$. 2-Chloro-10-[3-(4-methylpiperazinyl)propyl]phenothiazine ethanedisulfonate.
Properties: White to very light yellow, crystalline powder. Odorless. Solutions acid to litmus. Soluble in water; very slightly soluble in alcohol; insoluble in ether and chloroform.
Grade: U.S.P. XVI.
Use: Medicine.

prochlorperazine maleate $C_{20}H_{24}ClN_3S \cdot 2C_4H_4O_4$. Prochlorperazine dimaleate.
Properties: White to pale yellow, crystalline powder. Odorless. Saturated solution is acid to litmus. Practically insoluble in water and alcohol. Slightly soluble in warm chloroform.
Grade: U.S.P. XVI.
Use: Medicine.

"Procinyl" Dyes. [325] Brand name for reactive disperse dyes for nylon. Cover irregular-dyeing yarns, and produce dyeings having excellent wet fastness.

"Procion" Dyes. [325] Brand name for a range of dyes which form a chemical linkage with cellulosic fiber molecules. Applied to cotton and rayon by dyeing and printing.

"Procote." [428] Trademark for clear, fast drying, vinyl type lacquer and viscous, nonflammable, water soluble acrylic emulsion coatings for metals.

procyclidine hydrochloride (1-cyclohexyl-1-phenyl-3-pyrrolidino-1-propanol hydrochloride)
$C_6H_{11}(C_6H_5)C(OH)CH_2CH_2C_4H_8N \cdot HCl$.
Grade: N.N.D.
Use: Medicine.

"Prodag." [46] Trademark for a concentrated semi-colloidal dispersion of pure electric-furnace graphite in water.
Properties: Paste consistency; solids content 30%; sp.gr. 1.18; b.p. 100°C; freezing point 0°C; completely miscible with water.
Uses: General industrial mold-release and parting agent; "stopoff" coating; wire-drawing lubricant.

producer gas. A gas obtained by burning solid fuel with a restricted supply of air or by passing air and steam through a bed of incandescent fuel under such conditions that any carbon dioxide formed is, as far as possible, converted into carbon monoxide before it leaves the producer. The water vapor is split up chemically with the formation of carbon monoxide and hydrogen.
Uses: Producer gas is cheap but low Btu. It is used especially as an industrial fuel where it does not need to be transported,

as in coke-ovens and other furnaces, and in gas engines run for power purposes. Producer gas is also a source material for synthetic ammonia manufacture.

A typical gas will analyze:

	From Coal	From Coke
	Per Cent	
Illuminants	0.2	0.0
Carbon monoxide	17.6	25.3
Hydrogen	10.4	13.2
Methane	6.3	0.4
Ethane	0.0	0.0
Carbon dioxide	7.3	5.4
Oxygen	0.7	0.6
Nitrogen	58.1	65.2

The available Btu from a coal source is 161, and from a coke source is 137.

See also water gas; and synthesis gas.

"Product BCO." [28] A surface-active agent based on cetyl betaine. It is unique in that it is cationic when in acid media and anionic when in basic. Used as a wetting agent, detergent, and dyeing assistant.

production reactor. See nuclear reactor.

"Pro-fax." [266] Trademark for several grades of polypropylene. Available as natural or colored molding powder pellets.

Properties: Lightest of all plastics; high heat resistance; resistance to greases, oils, and many chemicals.

Uses: Manufacture of monofilament for rope, outdoor furniture webbing, replacement auto seat covers and bristles. Multifilaments serve in fabrics and rug yarn. Yields tough, clear films for packaging and industrial use, and is FDA approved for direct contact with foods. Injection molded uses include housewares, pipe fittings, automotive parts and a variety of industrial moldings.

proflavine sulfate (proflavine; 3,6-diaminoacridinium hydrogen sulfate) $C_{13}H_{11}N_3 \cdot H_2SO_4$.

Properties: Reddish-brown, odorless, crystalline powder. Soluble in water and alcohol forming brownish solutions which fluoresce green on dilution; nearly insoluble in ether and chloroform.

Derivation: Synthetic.

Method of purification: Crystallization.

Containers: Amber glass bottles.

Uses: Medicine.

Shipping regulations: None.*

"Profume." [233] Trademark for an odorized methyl bromide product, used as a soil fumigant and insecticide.

progesterone (delta^4-pregnene-3,20-dione) $C_{21}H_{30}O_2$. The female sex hormone secreted in the body by the corpus luteum, by the adrenal cortex, or by the placenta during pregnancy. It is important in the preparation of the uterus for pregnancy, and for the maintenance of pregnancy. It exists in two crystalline forms (alpha- and beta-) of equal physiological activity. Progesterone is believed to be the precursor of the adrenal steroid hormones.

Properties: White crystalline powder; odorless and stable in air but sensitive to light. M.p., alpha form 128-133°C, beta form, about 121°C. Practically insoluble in water; soluble in alcohol, acetone, and dioxane; sparingly soluble in vegetable oils.

Units: The international unit (IU) of progestational activity is expressed as 1 mg of progesterone.

Derivation: Isolation from corpus luteum of pregnant sows; synthesis from other steroids such as stigmasterol (q.v.).

Grade: U.S.P. XVI.

Containers: Bottles.

Use: Medicine.

"Progynon." [321] Brand name for estradiol.

"Progynon Benzoate." [321] Brand name for estradiol benzoate.

prolactin. See luteotropin.

prolamin. See gliadin.

"Prolase." [173] Trademark for a fungal protease system in powder form. It is useful for processing wheat gluten and other proteins in a pH range of 3.5-7.0.

proline (2-pyrrolidinecarboxylic acid) C_4H_8NCOOH. A nonessential amino acid found naturally in the L(−) form.

Properties: Colorless crystals; soluble in water and alcohol; insoluble in ether; optically active;

DL-proline: m.p. 205°C with decomposition.

D(+)-proline: m.p. 215-220°C with decomposition.

L(−)-proline: m.p. 220-222°C with decomposition.

Derivation: Hydrolysis of protein; organic synthesis.

Uses: Biochemical and nutritional studies; microbiological tests; culture media.

Available commercially as the L(−)-proline.

prolipin. A compound sterile solution of protein obtained from nonpathogenic bacteria, various animal fats and lipoids derived from bile.

"Prolixin." [412] Trademark for fluphenazine dihydrochloride (q.v.).

"Proluton." [321] Brand name for progesterone.

"Promacetin." [330] Trademark for acetosulfone (sodium 4,4'-diaminodiphenylsulfone-2-N-acetylsulfonamide).

Use: Medicine.

"Promat." [428] Trademark for a line of cadmium, zinc and copper plating brighteners.

promazine hydrochloride (10-(3-dimethylaminopropyl)phenothiazine hydrochloride) $C_{17}H_{20}N_2S \cdot HCl$. Isomeric with promethazine hydrochloride.

Properties: Crystals; hygroscopic. Soluble in water, methanol, ethanol, chloroform; practically insoluble in ether, benzene. Incompatible with alkalies, oxidizing agents, heavy metals.

Grade: N.N.D.
Use: Medicine.

promethazine hydrochloride $C_{17}H_{20}N_2S \cdot HCl$. N-(2'-Dimethylamino-2'-methyl)-ethylphenothiazine hydrochloride; 10-(2-dimethylaminopropyl)phenothiazine hydrochloride.
Properties: White to faint yellow, practically odorless powder. Slowly oxidized, particularly when moistened, on prolonged exposure to air, becoming blue in color. Melts within a 3° range between 215 and 225°. Very soluble in water, in hot absolute alcohol, and in chloroform; practically insoluble in ether, acetone, and ethyl acetate. 1 in 10 solutions in water or chloroform are colorless and clear. pH (5% solution) 4.5-5.5.
Grade: U.S.P. XVI.
Use: Medicine.

promethium Pm. A rare earth element with atomic number 61, first discovered in the fission products of uranium. It is prepared in the cyclotron by bombarding neodymium with protons and from nuclear reactor fuels by ion exchange separations. Five isotopes have been made, the longest lived having a half-life of 3.7 years, thus making the existence of natural stable isotopes unlikely.

"Promin." [330] Trademark for glucosulfone sodium (disodium para, para'-sulfonyldianiline-N,N'-diglucoside disulfonate).
Use: Medicine.

"Promizole." [330] Trademark for thiazolsulfone (2-amino-5-sulfanilylthiazole).
Use: Medicine.

promoters, in catalysts. Substances which, when added in relatively small quantities to a catalyst, increase its activity. Small amounts of aluminum and potassium oxide are commonly added as promotors to the iron catalyst used in facilitating combination of hydrogen and nitrogen to form ammonia.

promoters, in ore flotation. Those reagents which provide the minerals which are to be floated with a water-repellent surface that will adhere to air bubbles. Such reagents are generally more or less selective towards minerals of certain classes.

"Pronestyl." [412] Trademark for procainamide hydrochloride (q.v.).

"Pro-Noxfish." [342] Trademark for fish-toxicant composition containing rotenone and sulfoxide in emulsifiable vehicles.

proof, degree of. A method of designating the strength of ethyl alcohol-water mixtures, most frequently for purposes of taxation, according to the number of gallons of "proof spirit" or "100 proof" alcohol that can be made from 100 gallons of the alcohol-water mixture under discussion.
"Proof spirit" is defined by the regulation (in the U.S.) that it "contains one half its volume of alcohol of sp.gr. 0.7939 (60°F)." This latter is absolute alcohol, and pure absolute alcohol is, therefore, 200 proof. The degree of proof is twice the per cent by volume of alcohol. For example, one gallon of 95% alcohol is equivalent to 1.9 gallons proof alcohol. Due to volume contraction upon mixing alcohol and water, it is necessary to add 53.73 volumes of water to 50 volumes of alcohol in order to produce 100 volumes of 100 proof alcohol.

proof gallon. See proof, degree of.

propadiene. See allene.

propanal. See propionaldehyde.

propane (dimethylmethane) C_3H_8.
Properties: Colorless gas. Characteristic natural gas odor. Heavier than air. Has no corrosive action on metals.
Constants: B.p. −42.5°C; m.p. −189.9°C; density of liquid at 0°C 0.531; density of vapor at 0°C (760 mm) (air = 1) 1.56; explosive limits in air per cent by volume, lower 2.4, upper 9.5; flash point −104°C.
Grades: Technical; research (99.9%).
Containers: Cylinders; tank cars.
Uses: Organic synthesis; fuel for household and for many industrial purposes, either alone or in admixture with butane or air; extractant; solvent; refrigerant; gas enrichener; standby gas; aerosol propellant; mixture for bubble chambers.
Shipping regulations: Flammable gas. Red gas label.*

propanedinitrile. See malonic dinitrile.

1,2-propanediol. See propylene glycol.

1,3-propanediol. See trimethylene glycol.

propane hydrate. See gas hydrates.

propanenitrile. See ethyl cyanide.

1-propanethiol. See n-propyl mercaptan.

propanoic acid. See propionic acid.

1-propanol. See propyl alcohol.

2-propanol. See isopropyl alcohol.

propanolamine. See 2-aminopropanol; 3-aminopropanol.

2-propanolpyridine $C_5NH_4C_3H_6OH$.
Properties: B.p. (760 mm) 260.2°C; sp.gr. (25°C) 1.060; refractive index 1.5298 (n 20/D); miscible with water in all proportions at 20°C.

4-propanolpyridine $C_5NH_4C_3H_6OH$.
Properties: B.p. (760 mm) 289.0°C; f.p. 36.7°C; sp.gr. (40°C) 1.053; soluble in water.

2-propanone. See acetone.

2-propanone oxime. See acetoxime.

propanoyl chloride. See propionyl chloride.

propantheline bromide $C_{23}H_{30}BrNO_3$. (2-Hydroxyethyl)diisopropylmethylammonium bromide 9-xanthenecarboxylate.
Properties: White or nearly white crystals, or powder. Odorless. Bitter taste. Very

*See "I.C.C. Shipping Regulations," page xiii.
Reference numbers refer to name of manufacturer. See "List of Manufacturers," page v.

soluble in water, alcohol, and chloroform; practically insoluble in ether and benzene. M.p. 155°-160°C with decomposition.
Grade: U.S.P. XVI.
Use: Medicine.

proparacaine hydrochloride $C_3H_7OC_6H_3NH_2COOC_2H_4N(C_2H_5)_2 \cdot HCl$. 2-Diethylaminoethyl-3-amino-4-propoxybenzoate hydrochloride.
Properties: White crystalline substance; soluble in water; m.p. 178-179°C. (uncorrected).
Grade: N.N.D.
Use: Medicine.

propargyl alcohol (2-propyn-1-ol) $HC \vdots CCH_2OH$.
Properties: Colorless liquid; sp.gr. 0.9215; m.p. −17°C; b.p. 114°C; soluble in water, alcohol, and ether.
Derivation: From acetylene by high pressure synthesis.
Grades: Technical; 75% solution.
Containers: Up to tank cars.
Uses: Chemical intermediate; corrosion inhibitor; stabilizer.

propargyl bromide (3-bromo-1-propyne) $HC \vdots CCH_2Br$.
Properties: Liquid; sp.gr. 1.520; b.p. 88-90°C.
Derivation: From acetylene by high pressure synthesis.
Containers: Up to tank cars.
Uses: Chemical intermediate; soil fumigants.

propargyl chloride (3-chloro-1-propyne) $HC \vdots CCH_2Cl$.
Properties: Liquid; freezing point −76.9°C; b.p. range 56.0-57.1°C; refractive index (n 25/D) 1.4310; flash point 90-95°F. Has three centers of chemical reactivity, acetylenic hydrogen, triple bond and chlorine atom.
Derivation: From acetylene by high pressure synthesis.
Containers: Up to tank cars.
Uses: Chemical intermediate; soil fumigant.

propellants. Substances which contain the necessary ingredients (both a fuel and an oxidizer) for combustion in order to impart motion to an object. The term covers a wide range of materials, including gunpowder (q.v.) but now is usually associated with rockets and missiles. See rocket propellants.

"Pro-Pen." [123] Trademark for antibiotic and vitamin supplements for animal and poultry feed.

propenal. See acrolein.

propene. See propylene.

propenenitrile. See acrylonitrile.

2-propene-1-thiol. See allyl mercaptan.

propene-1, 2, 3-tricarboxylic acid. See aconitic acid.

propenoic acid. See acrylic acid.

2-propen-1-ol. See allyl alcohol.

propenyl alcohol. See allyl alcohol.

2-propenylamine. See allyl amine.

para-**propenylanisole.** See anethole.

alpha-**propenyldichlorohydrin.** See alpha-dichlorohydrin.

propenyl guaiacol. See methyl isoeugenol.

2-propenyl hexanoate. See allyl caproate.

2-propenyl isothiocyanate. See allyl isothiocyanate.

properdin. A protein which is a normal constituent of human blood serum and is important in natural immunity to infectious diseases.

prophenpyridamine. See pheniramine.

prophenpyridamine maleate. See pheniramine maleate.

propiodal (1,3-bis(trimethylamino)-2-propanol diiodide; iodisan; hexamethyldiaminoisopropanol diiodide) $[CH_2N(CH_3)_3I]_2CHOH$.
Properties: White, crystalline; m.p. about 275°C (dec). Turns brown at 240°C. Freely soluble in water, slightly in alcohol; insoluble in ether, acetone.
Use: Medicine, iodine therapy.

beta-**propiolactone** (BPL) $\underline{OCH_2CH_2CO}$.
Properties: A colorless liquid; pungent odor; b.p. 155°C (760 mm) with rapid decomposition; m.p. −33.4°C; refractive index (n 20/D) 1.4131; sp.gr. (20/4°C) 1.1460; soluble in water; miscible with ethyl alcohol, acetone, ether, and chloroform at 25°C. Reacts with alcohol. Flash point (open cup) 74°C; fire point (open cup) 74°C; quite stable when stored in glass at refrigeration temperature (+5° to +10°C).
Derivation: By the direct combination of ketene and formaldehyde.
Containers: Up to tank cars; tank trucks.
Use: Organic synthesis; vapor sterilant. Handle with caution!

propiomazine hydrochloride 1-[10(2-dimethylaminopropyl)phenothiazine-2-yl]-1-propanone hydrochloride.
Derivation: Synthetic compound.
Use: Medicine.

propionaldehyde (propanal; propyl aldehyde; propionic aldehyde) C_2H_5CHO.
Properties: Water-white liquid with suffocating odor; soluble in water.
Typical specifications: Flash point (open cup) 15°C; b.p. 48.8°C; sp.gr. 0.807 (20/4°C); refractive index (n 20/D) 1.364.
Derivation: (a) Oxidation of propyl alcohol with dichromate; (b) by passing propyl alcohol over copper at elevated temperatures.
Containers: 370-lb drums.
Uses: Manufacture of polyvinyl acetals and other types of plastics; synthesis of rubber chemicals; disinfectant; preservative.
Shipping regulations: Flammable liquid. Red label.*

propionamide nitrile. See cyanoacetamide.

*See "I.C.C. Shipping Regulations," page xiii.
Reference numbers refer to name of manufacturer. See "List of Manufacturers," page v.

propione. See diethyl ketone.

propionic acid (methylacetic acid; propanoic acid) $CH_3CH_2CO_2H$.
Properties: Clear, colorless liquid; pungent odor; sp.gr. 0.9942 (20/4°C); m.p. −20.8°C; refractive index 1.3862 (20°C); b.p. 140.7°C; soluble in water, alcohol, chloroform and ether.
Derivation: By reaction of ethyl alcohol with carbon monoxide, using a boron trifluoride catalyst; also from the mixtures of oxygenated products produced by the catalytic (Fischer-Tropsch and Hydrocol processes) reaction of carbon monoxide with hydrogen, olefins, or alcohols.
Method of purification: Rectification.
Grades: Technical.
Containers: Iron drums; carboys; tank cars.
Uses: Propionates, some of which are used as mold inhibitors in bread; emulsifying agents; solutions for electroplating nickel; perfume esters; artificial fruit flavors; pharmaceuticals; solvent mixtures for cellulose derivatives.
Shipping regulations: None.*

propionic aldehyde. See propionaldehyde.

propionic anhydride $(CH_3CH_2CO)_2O$.
Properties: Colorless liquid; pungent odor; m.p. −45°C; b.p. 167-169°C; sp.gr. 1.0119 at 20°C; vapor pressure 1 mm at 20°C; flash point 165°F; wt 8.4 lbs/gal at 20°C; coefficient of expansion 0.00107 at 20°C; viscosity 0.011 poise at 20°C; soluble in alcohol, ether, chloroform, and alkalies; insoluble in water.
Uses: Esterifying agent for fats, oils, cellulose; dehydrating medium for nitrations and sulfonations; production of alkyd resins, dyestuffs and pharmaceuticals.

propionic ester. See ethyl propionate.

propionitrile. See ethyl cyanide.

propionylbenzene. See propiophenone.

propionyl chloride (propanoyl chloride) CH_3CH_2COCl.
Properties: Colorless liquid with pungent odor; m.p. −94°C; b.p. 80°C; sp.gr. 1.065 (20/4°C). Decomposes in water and alcohol.
Uses: Chemical intermediates.

propionyl erythromycin ester lauryl sulfate See erythromycin propionate lauryl sulfate.

propiophenone (ethyl phenyl ketone; propionylbenzene; 1-phenyl propanone-1) $C_6H_5COC_2H_5$.
Properties: Water-white to light amber liquid with strong persistent odor; sp.gr. 1.012 (20/20°C); refractive index (n 20/D) 1.527; congealing temperature 17.5-21°C; b.p. 218°C; flash point (TOC) 210°F. Insoluble in water, ethylene glycol, glycerine; miscible with ethyl alcohol, ethyl ether, benzene, and toluene.
Containers: 55-gal drums containing 450 lbs net each; 30-gal drums containing 250 lbs net each.
Uses: As a fixative in perfumes; as the starting material for the synthesis of ephedrine and several other important pharmaceuticals; and as a raw material for the preparation of numerous synthetic organic chemicals.
Shipping regulations: None.*

propoxycaine hydrochloride. $H_2NC_6H_3(OC_3H_7)COO(CH_2)_2N(C_2H_5)_2 \cdot HCl$. 2-Diethylaminoethyl 4-amino-2-propoxybenzoate hydrochloride.
Properties: White, to pale yellow, odorless, crystalline solid; bitter taste. Discolors on prolonged exposure to light and to air. Very soluble in water. Slightly soluble in alcohol and in chloroform. Practically insoluble in acetone and in ether. Melting range 146°-151°C; pH of solution (about 1%) is about 5.4.
Grade: N.F. XI.
Use: Medicine.

propoxyphene hydrochloride (dextro propoxyphene hydrochloride; alpha-d-4-dimethylamino-1,2-diphenyl-3-methyl-2-butanol propionate hydrochloride) $C_6H_5CH_2C(C_6H_5)(OOCC_2H_5)CH(CH_3)CH_2N(CH_3)_2 \cdot HCl$.
Properties: Crystalline solid; m.p. 170°C.
Grade: N.N.D.
Use: Medicine.

n-propyl acetate $C_3H_7OOCCH_3$.
Properties: Clear, colorless liquid having pleasant odor. Miscible with alcohols, ketones, esters, oils, hydrocarbons.
Constants: Sp.gr. 0.887; flash point 68°F; boiling range 96.0-102.0°C; wt 7.36 lbs/gal.
Derivation: Interaction of acetic acid and n-propyl alcohol in the presence of sulfuric acid.
Grades: Technical.
Containers: 55-gal drums; tank cars.
Uses: Flavoring agents; perfumery; solvent for nitrocellulose and wide range of cellulose derivatives, natural and synthetic resins; lacquers; plastics; organic synthesis.
Shipping regulations: Flammable liquid. Red label.*

propyl acetone. See methyl butyl ketone.

propyl alcohol (1-propanol; n-propyl alcohol) $CH_3CH_2CH_2OH$.
Properties: Clear, colorless liquid; odor similar to ethyl alcohol; b.p. 97.2°C; m.p. −127.0°C; sp.gr. (20/4°C) 0.804; flash point (open cup) 96°F; lower explosive limit 2.6% by vol in air; autoignition temperature 540°C; refractive index (20°C) 1.385; viscosity 2.256 cps (20°C); soluble in water, alcohol, and ether.
Derivation: From oxidation of natural gas hydrocarbons, from Fischer-Tropsch process; also from fusel oil.
Containers: Up to 10,000 gal tank cars.
Uses: Organic synthesis; propionaldehyde; lacquers; cosmetics; solvent for dopes, waxes, vegetable oils, natural and synthetic resins, cellulose esters and ethers; solvent mixtures; polishing compositions.
Shipping regulations: None.*

n-propyl alcohol. See propyl alcohol.

sec-**propyl alcohol.** See isopropyl alcohol.

propyl aldehyde. See propionaldehyde.

n-propylamine $C_3H_7NH_2$.
Properties: Colorless liquid; sp.gr. at 20°C 0.7182; b.p. 47.8°C; vapor pressure 248 mm (20°C); f.p. −83°C; odor, amine; soluble in water, alcohol and ether.
Containers: Drums; tank cars.
Use: Intermediate.
Fire hazard: Flash point −35°F.
Shipping regulations: Flammable liquid. Red label.*

propyl butyrate $C_3H_7OOCC_3H_7$.
Properties: Colorless liquid.
Constants: Sp.gr. 0.8789 (15°C); b.p. 142.7°C; m.p. −95.2°C.
Grades: Technical.
Containers: Tin cans.
Use: Solvent mixtures for cellulose ethers.
Shipping regulations: None.*

propyl chlorosulfonate $CH_3CH_2CH_2OSO_2Cl$.
Properties: Caution! Very irritant!
Constants: B.p. 70-72°C (20 mm).
Derivation: Interaction of n-propyl alcohol and sulfuryl chloride.
Grades: Technical.
Uses: Organic synthesis; military poison gas (lachrymator).

n-propyl cyanide See n-butyronitrile.

n-propyl diethyl malonate $C_3H_7CH(COOC_2H_5)_2$.
Properties: Colorless liquid with an aromatic odor; insoluble in water; soluble in alcohols, ethers, esters and ketones.
Uses: Intermediate; tobacco flavoring agent.

propyl 3,5-diiodo-4-oxo-1(4H)pyridineacetate. See propyliodone.

propylene (propene) $CH_3CH{:}CH_2$. Now available in large quantities. Source of polypropylene.
Properties: Colorless gas; b.p. −47.7°C; m.p. −185.2°C; sp.gr. (liquid) 0.5139 (20/4°C); density of vapor at 0°C (760 mm; air 1) 1.46; critical temperature 92.1°C; critical pressure (absolute) 666 psi; explosive limits in air, % by volume, lower 2.2, upper 9.7; specific volume (70°F) 8.6 cu ft/lb. Flash point −108°C.
Derivation: From refinery off-gases or from cracking hydrocarbons during the production of ethylene.
Grades: 95%, 99% and research.
Containers: Cylinders; tank cars; or by direct pipeline.
Uses: (in approximate order of volume): Production of isopropyl alcohol, propylene trimer and dimer, propylene oxide, cumene, synthetic glycerol, polypropylene and synthesis of isoprene (this use is growing fast), Oxo alcohols.
Danger! Extremely flammable. MCA warning label.
Shipping regulations: Flammable liquid. Red label.*

propylene aldehyde. See crotonaldehyde.

propylene carbonate $C_3H_6CO_3$, possibly $CH_3\underline{CHOCOOC}H_2$.
Properties: Odorless, colorless mobile liquid. Freezing point −49.2°C (easily supercooled); b.p. 241.7°C; sp.gr. (20/4°C) 1.2057; wt/gal (20°C) 10 lbs; refractive index (n 20/D) 1.4209; flash point (Tag open cup) 270°F. Miscible with acetone, benzene, chloroform, ether, ethyl acetate. Moderately soluble in water and carbon tetrachloride.
Uses: Solvent extraction; plasticizer; organic synthesis; natural gas purification aid; synthetic fiber spinning solvent.

propylene chloride. See propylene dichloride.

propylene chlorohydrin (chloro-isopropyl alcohol; 1-chloro-2-propanol) $CH_2ClCHOHCH_3$.
Properties: Colorless liquid. Mild odor, nonresidual.
Constants: B.p. 127.5°C; vapor pressure 4.9 mm (20°C); flash point 125°F; wt 9.3 lbs/gal (20°C); sp.gr. 1.1128 at 20/20°C; coefficient of expansion 0.00094 (20°C); viscosity 0.00467 poise (20°C). Soluble in water (in all proportions).
Typical specifications: Acidity not more than 0.02% (as hydrochloric); purity 46-54% propylene chlorohydrin by weight; sp.gr. 1.1270 at 20/20°C; color water-white (if shipped in glass).
Grades: Technical.
Containers: 1-gal glass jugs; 5-, 12-gal returnable glass carboys; 55-gal returnable steel drums.
Use: Organic synthesis (introducing hydroxypropyl group).
Danger! Extremely hazardous liquid and vapor. May be fatal if inhaled, swallowed, or absorbed through skin. MCA warning label.

propylenediamine (1,2-diaminopropane) $NH_2CHCH_3CH_2NH_2$.
Properties: Colorless, very hygroscopic strongly alkaline liquid. Very soluble in water. Ammoniacal odor. Closely resembles ethylene diamine in its behavior, but yields products with greater oil solubility.
Constants: Sp.gr. 0.8732 at 20/20°C; vapor pressure 9.4 mm (20°C); flash point 160°F; wt 7.3 lbs/gal (20°C); b.p. 119.7°C, coefficient of expansion 0.00107 (20°C); viscosity 0.0170 poise (20°C).
Grades: Technical; 75% solution; 90% solution; 98% solution.
Containers: Tin-lined, nonreturnable containers, 5-, 10-, 55-gals.
Use: Synthesis of certain medicinals, dyes, rubber accelerators; electroplating; analytical test reagent.
Shipping regulations: None.*
See also 1,3-diaminopropane.

propylene dichloride (1,2-dichloropropane; propylene chloride) $CH_3CHClCH_2Cl$.
Properties: Colorless, stable liquid. Low (moderate) flammability. Chloroform-like

odor.

Constants: B.p. 96.3°C; sp.gr. 1.1583 at 20/20°C; wt/gal 9.6 lbs (20°C); refractive index 1.4068 (20°C); surface tension 31.4 dynes/cm (25°C); viscosity 0.0088 poise (20°C); vapor pressure 40.0 mm (20°C); flash point (ASTM open cup) 21°C (70°F); solubility in water 0.26% by wt. (20°C); solubility of water in propylene dichloride 0.07% by wt (20°C); explosive limits in air, lower 3.4% by vol (25°C), upper 14.5% by vol (100°C); dielectric constant (85.8 kilocycles) 8.925 recip ohms (26°C); ignition temperature in air 557°C; coefficient of expansion (per °C) 0.00113 to 20°C, 0.001153 to 55°C; freezing point −80°C (−112°F); fire point 38°C; heat of vaporization 68.1 cal/g (b.p.); specific resistivity 2.5×10^8 ohms/cm. Miscible with most common solvents; insoluble in water.

Derivation: Action of chlorine on propylene.

Method of purification: Distillation.

Grades: Refined.

Containers: 50-gal drums; tank cars.

Uses: Solvent for fats, oils, waxes, gums, and resins; solvent mixtures for cellulose esters and ethers; organic synthesis; dry-cleaning fluids; scouring compounds; spotting agents; metal degreasing agents; fumigant.

Caution: Do not store in aluminum.

Shipping regulations: Flammable liquid. Red label.*

1,2-propylene glycol (1,2-dihydroxypropane; methylethylene glycol; methyl glycol; 1,2-propanediol) $CH_3CHOHCH_2OH$.

Properties: Colorless, viscous, stable, hygroscopic liquid; slight odor. Very slight acrid taste. Miscible with water, alcohols, and many organic solvents.

Constants: B.p. 188.2°C; sp.gr. 1.0381 at 20/20°C; wt/gal 8.6 lbs (20°C); refractive index 1.4293 (27°C); surface tension 40.1 dynes/cm (25°C); viscosity 0.581 poise (20°C); vapor pressure 0.07 mm (20°C); specific heat 0.590 cal/g (20°C); latent heat of evaporation 168.6 cals/g at b.p.; flash point (ASTM open cup) 99°C (210°F); heat of combustion 431.0 kg cal/mole; coefficient of expansion (per °C) 0.000695 to 20°C, 0.000743 to 55°C.

Derivation: By hydration of propylene oxide.

Method of purification: By distillation.

Grades: Refined; U.S.P. XVI; technical.

Containers: Glass bottles; 5-, 55-gal drums; tank cars.

Uses: Organic synthesis; antifreeze solution; solvent for fats, oils, waxes, resins, dyes, flavoring extracts, perfumes; hygroscopic agent; lubricant in refrigeration machines; plasticizers, hydraulic fluids, bactericide, polyester resins; cosmetics; textile conditioners.

Shipping regulations: None.*

See also polypropylene glycols.

1,3-propylene glycol. See trimethylene glycol.

propylene glycol diricinoleate $[C_{17}H_{32}(OH)COO]_2(CH_2CHCH_3)$.

Properties: Solid; sp.gr. 0.938; m.p. 49-51°C; insoluble in water.

propylene glycol methyl ether $CH_3OCH_2CHOHCH_3$.

Properties: Colorless liquid; pour point −142°F; b.p. 120.1°C (760 mm); sp.gr. 0.919 (25/25°C); lbs/gal 7.65 (25°C); refractive index 1.402 (n 25/D); flash point 100°F. Soluble in water, methanol, ether.

Containers: Drums; tank cars.

Uses: Solvent; intermediate.

propylene glycol monoricinoleate $C_{17}H_{32}(OH)COOCH_2CHOHCH_3$.

Properties: A clear, pale yellow, moderately viscous oily liquid; mild odor; sp.gr. 0.960 (25/25°C); saponification value 160; hydroxyl value 285; solidifies at −26°C. Soluble in most organic solvents; insoluble in water.

Derivation: Castor oil and propylene glycol.

Grade: Technical.

Containers: 5-gal cans; 55-gal drums.

Uses: Plasticizer; dye solvent; lubricant; in cosmetics; urethane polymers and hydraulic fluids.

"Propylene Glycol Monostearate 70." [260] Proprietary brand of the stearic acid ester of propylene glycol. Monoester content 70%. Balance essentially diester with small amounts of free fatty acid and free glycol. Waxy solid; white; mild fatty odor; soluble in water; flash point (open cup) 390°F; acid value 3.0 max; iodine value 0.5 max; m.p. 34.5-39.5°C.

Uses: Emulsifier in cosmetic, food, pharmaceutical, textile and other emulsion systems.

propylene glycol phenyl ether $C_6H_5OCH_2CHOHCH_3$.

Properties: Almost colorless liquid; sp.gr. 1.060-1.070 (25/25°C); boiling range (760 mm) 5-95%, 237-242°C.

Use: High-boiling solvent; bactericidal agent; fixative for soaps and perfumes; intermediate for plasticizers.

"Propylene Glycol Stearate 8615." [260] Proprietary brand of modified ester of propylene glycol and stearic acid. Waxy solid; white; mild fatty odor; acid value 20.0 max; iodine value 3.0 max; m.p. 57-62°C; dispersible in water; anionic; flash point (open cup) 379°C.

Uses: Emulsifier and dispersing agent in cosmetic, pharmaceutical, textile and other emulsion systems.

propylene imine (2-methylethylene imine; 2-methylaziridine) $CH_3CHNHCH_2$.

Properties: A water-white liquid; soluble in water and most organic solvents.

Containers: Available in 5- and 30-gal drums.

Uses: Yields addition products when reacted with active hydrogen compounds, and polymers when reacted with other imine molecules.

Warning! Highly toxic and flammable. Observe handling instructions.

*See "I.C.C. Shipping Regulations," page xiii.
Reference numbers refer to name of manufacturer. See "List of Manufacturers," page v.

propylene laurate, edible. This product is of particular interest as an emulsifying agent to manufacturers of cosmetics, pharmaceuticals and foodstuffs, because of its nontoxicity and great degree of purity. It is physiologically harmless and definitely edible.
Properties: Amber oil, self-emulsifiable in water to form milky-white emulsions; non-hygroscopic and high-boiling; miscible with alcohol, glycerin, glycol, hydrocarbons, etc.; pH of 5% aqueous dispersion 8.0; practically odorless; non-toxic and edible, physiologically harmless, low surface tension and viscosity.
Containers: 2-, 10-lb cans; 50-, 225-, 417-lb drums.
Uses: Food and pharmaceutical stabilizers; food packaging.
Shipping regulations: None.*

propylene oxide $CH_3\overline{CHCH_2O}$.
Properties: Colorless liquid, ethereal odor. Sp.gr. 0.8304 at 20°C/20°C; b.p. (760 mm) 33.9°C; vapor pressure 445 mm (20°C); flash point −35°F; wt 6.9 lbs/gal (20°C); coefficient of expansion 0.00151 (20°C); freezing point −104.4°C; viscosity 0.0038 poise (20°C). Soluble in water, alcohol, ether.
Typical specifications: Acidity not more than 0.05% (as acetic); sp.gr. 0.829-0.831 at 20°C/20°C; boiling range (760 mm) 33-37°C; water, substantially anhydrous; non-volatile not more than 0.005 g/100 cc; color water-white; average wt 6.92 lbs/gal (20°C).
Derivation: (a) Action of alkalies on propylene glycol. (b) Action of alkalies on propylene chlorohydrin.
Methods of purification: By distillation.
Grades: Refined.
Containers: 1-, 5-gal cans; 55-gal drums; tank cars.
Uses:(in approximate order of volume): Propylene glycol; adducts for urethane foams; glycols for brake fluids; surfactants and detergents; isopropanol amines, dipropylene glycol; propylene compounds.
Shipping regulations: Flammable liquid. Red label.*

propylene tetramer. See tetrapropylene.

propylene trimer. See tripropylene.

propylethylene. See alpha-n-amylene.

propylformic acid. See butyric acid.

n-propyl furoate $C_4H_3OCO_2C_3H_7$.
Properties: Colorless, aromatic smelling liquid, becomes yellow in light. Practically insoluble in water; soluble in alcohol and ether. Sp.gr. 1.0745 (25.9°C/4°C); b.p. 210.9°C (corr.); refractive index 1.4737 (25.9°C/D).

propyl gallate $C_3H_7OOCC_6H_2(OH)_3$. Colorless crystals; m.p. 150°C. Almost insoluble in water; somewhat soluble in oils. Used in minute proportions (up to 0.01%) as an antioxidant or synergist to retard or prevent rancidity in lard and other edible fats and oils.
Containers: Canisters; fiber drums.

propylhexedrine (1-cyclohexyl-2-methylaminopropane; N-dimethylcyclohexaneethylamine) $C_6H_{11}CH_2CH(CH_3)NHCH_3$.
Properties: Clear, colorless liquid with amine odor; b.p. 202-206°C; very slightly soluble in water; soluble in dilute acids; miscible with alcohol, chloroform, and ether. Sp.gr. 0.848-0.852. Volatilizes slowly at room temperature; absorbs carbon dioxide from the air. Solutions are alkaline to litmus.
Grade: U.S.P. XVI.
Use: Medicine.

propyl hydride. See propane.

propyl para-**hydroxybenzoate.** See propylparaben.

propyliodone. (propyl 3,5-diiodo-4-oxo-1(4H)-pyridineacetate) $I_2(O)C_5H_2NCH_2COOC_3H_7$.
Properties: White, or almost white, crystalline powder. Odorless or nearly so. Practically insoluble in water. Soluble in acetone, alcohol, and ether. M.p. 187-190°C.
Grades: U.S.P. XVI.
Use: Medicine (radiopaque medium).

propylmagnesium bromide C_3H_7MgBr. A solution in ether. A Grignard reagent.
Use: Alkylating agent in organic synthesis.
Shipping regulations: Flammable liquid. Red label.*

n-propyl mercaptan (1-propanethiol).
Properties: Boiling range 65-80°C; sp.gr. (20/4°C) 0.8408; refractive index (20/D) 1.4380; flash point −20°C.
Grades: 95%.
Containers: Drums; tank cars.
Uses: Chemical intermediate.
Shipping regulations: Flammable liquid. Red label.*

n-propyl nitrate (NPN) $C_3H_7NO_3$.
Properties: Liquid, white to straw-colored; ethereal odor; density 1.057 g/ml (20°C); boiling range 104-127°C (760 mm); flash point (open cup) 75°F; freezing point <-100°C; refractive index 1.3975 (n 20/D). Insoluble in water.
Grade: 96-98% pure.
Chief impurities: sec-Butyl and other nitrates.
Containers: 55-gal steel drums.
Use: Monopropellant.
Shipping regulations: Flammable liquid. Red label.*

propylparaben (propyl para-hydroxybenzoate) $C_{10}H_{12}O_3$.
Properties: Small, colorless crystals or white powder; very slightly soluble in water; soluble in alcohol, ether and acetone; m.p. 95-98°C.
Grade: U.S.P. XVI.
Containers: Drums.
Use: Medicine; food additive as preservative; fungicide.

propylpiperidine. See coniine.

*See "I.C.C. Shipping Regulations," page xiii.
Reference numbers refer to name of manufacturer. See "List of Manufacturers," page v.

6-propylpiperonyl butyl diethylene glycol ether. See piperonyl butoxide.

n-propyl propionate $CH_3CH_2COOCH_2CH_2CH_3$.
Properties: Colorless liquid. Soluble in most organic solvents; very slightly soluble in water. Boiling range 122°-124°C; m.p. -76°C; flash point 79°C; wt 7.31 lbs/gal.
Grades: Technical.
Containers: 55-gal drums; tank cars.
Uses: Solvent for nitrocellulose; paints, varnishes, lacquers; coating agents.
Shipping regulations: None.*

4-n-propylpyridine $C_8H_{11}N$ or $C_3H_7C_5H_4N$. B.p. 188°C. Used as an intermediate. Not to be confused with coniine (propylpiperidine).

propylthiouracil (6-propyl-2-thiouracil) $C_7H_{10}N_2OS$.
Properties: White powdery crystalline substance; starchlike in appearance and to touch; has bitter taste; m.p. 218-221°C. Sensitive to light. Very slightly soluble in water; sparingly soluble in alcohol; slightly soluble in chloroform and ether; insoluble in benzene; soluble in ammonia and alkali hydroxides.
Derivation: By the condensation of beta-oxocaproate with thiourea.
Grade: U.S.P. XVI.
Containers: Bottles.
Use: Medicine.

n-propyltrichlorosilane $C_3H_7SiCl_3$.
Properties: Colorless liquid. B.p. 123.5°C; sp.gr. 1.195 (25/25°C); refractive index (n 25/D) 1.4292; flash point (Cleveland open cup) 100°F. Readily hydrolyzed by moisture, with the liberation of hydrochloric acid.
Derivation: By Grignard reaction of silicon tetrachloride and propylmagnesium chloride.
Grades: Technical.
Use: Intermediate for silicones.
Shipping regulations: Corrosive liquid. White label.*

propyl xanthate. See xanthic acids.

propyne. See methylacetylene.

2-propyn-1-ol. See propargyl alcohol.

"Proquel." [428] Trademark for viscous liquid containing foam depressants for developing a controlled foam blanket on chromic acid plating solutions to limit loss of spray, fumes, mist and "drag-out."

"Prorex Oils." [331] Trademark for a brand of highly-refined, light-colored, light-bodied mineral oils.

"Proseal." [428] Trademark for powdered acid conversion coating for electrodeposited zinc and cadmium.

prosthetic group. Low molecular weight proteins or metallo-proteins which can attach themselves to, and thus supplement, specific proteins to form active enzyme systems. The term coenzyme is considered to be synonymous with the term prosthetic group when used in connection with conjugated proteins which have enzyme activity.

"Prostigmin." [190] Trademark for neostigmine U.S.P.

"Pro-Strep." [123] Trade name for an antibiotic feed supplement.

protactinium Pa. A radioactive element of atomic number 91, a member of the actinide series of elements. Protactinium is found in nature as a constituent of all uranium ores, about 70 mg being extracted from one ton of pitchblende. Protactinium may also be produced by the pile irradiation of thorium-230. Purification is carried out by ion exchange and solvent extraction techniques. The longest lived isotope, Pa-231, decays by alpha emission and has a half-life of about 34,300 years. Protactinium may be precipitated as the double potassium fluoride K_2PaF_7 or the oxide Pa_2O_5. The metal may be prepared by reducing PaF_4 with barium or by heating PaI_4 in a vacuum.

protamine zinc insulin suspension. See insulin.

"Protan." [172] Trademark for sodium formate, HCOONa.
Properties: White, odorless material; soluble in water. Not less than 90% sodium formate.
Containers: 100-lb paper bags.
Uses: Masking agent in chrome tanning; neutralizer in wool dyeing; preparation of washable wallpapers; acidizing of oil wells; solubilizer and buffering agent for manufacture of chemicals.

protargin, mild. See silver protein, mild.

protargin, strong. See silver protein, strong.

"Protargol." [162] Trademark for strong silver protein for use as antibacterial. A different grade also available for use in staining nerve tissue.

protease. A class of enzymes which hydrolyze peptide linkages into peptides and proteins. These are subgrouped into two further classes: those which hydrolyze peptides into alpha-amino acids (peptidases), and those which hydrolyze proteins to polypeptides (proteinases). The latter group consists of many of the more widely known enzymes such as pepsin, trypsin, ficin, bromelin, papain and rennin. Proteases are water-soluble products used to solubilize proteins and commercially used for meat tenderizers, beer chill proofing, bread baking and as digestive aids.

"Protectol 3." [307] Trademark for a protective colloid, sulfite cellulose liquor; 46% active.
Properties: Brown liquid; soluble in water.
Uses: Protective colloid, primarily used as a retarding agent in vat color dyeing; fiber protective agent in wool dyeing and stripping.

proteinase. See protease.

*See "I.C.C. Shipping Regulations," page xiii.
Reference numbers refer to name of manufacturer. See "List of Manufacturers," page v.

protein fibers. Synthetic fibers produced by converting natural proteins to a fluid form and extruding or spinning through an opening in a die, followed by coagulation or hardening to a filament or bristle. Fibers have also been made from the corn protein (zein), milk casein, soybean protein, and peanut protein.

protein hydrolysate. Solution of protein hydrolyzed into its constituent amino acids. Used extensively in medicine and surgery. Usually administered by a stomach tube or intravenous injection.
Grades: N.N.D., U.S.P. XVI.

proteins. A large group of compounds of great importance in the structure and functioning of living matter. Proteins may serve as structural elements, enzymes, hormones, oxygen carriers, antibodies, and many other important types of compounds. They are compounds of large molecular weight and contain carbon, hydrogen, oxygen, nitrogen, and with few exceptions, sulfur also.

Amino acids (q.v.) are the fundamental structural units of proteins and are the end result of the complete hydrolysis of proteins. Proteins form colloidal solutions, and can behave chemically as both acids and bases because they contain both acidic (carboxyl) and basic (amino) groups. Proteins are easily modified (denatured) by changes in pH, by heating in aqueous solution, by ultraviolet radiation and by many organic solvents; denaturation causes alteration in such properties as solubility and specific activity. Proteins may be classified as simple, which contain only amino acids, or conjugated, which contain amino acids and non-amino acid substances such as nucleic acids, carbohydrates, lipids, metals, or phosphoric acid. Proteins are also classified by their solubilites and other characteristics. They are difficult to purify but some have been crystallized.

There has recently been increasing success in deciphering the structure of proteins (their amino acid makeup), and even in synthesizing them.

"Protek-Sorb." [241] Trademark for silica gel.
Properties: A hard, chemically inert material resembling rock salt in appearance. Will not swell, cake, or become sticky, and may be reactivated by heating at 205°F. Density, about 45 lb/cu ft. Will adsorb approximately 24% of its weight in moisture at 40% relative humidity and 12% at 20% relative humidity. Meets military specification MIL-D-3464 (latest revision).
Derivation: Silica gel is a synthetic desiccant made from sodium silicate and sulfuric acid.
Grades: "Protek-Sorb" 121 and 122.
Containers: Moisture-permeable bags ranging in size from $\frac{1}{6}$ unit to 80 units (a unit is about 1 oz) available in cans and drums ranging from 1 gal to 20 gal.
Uses: For protecting packages which contain materials that may rust, corrode, or mildew due to moisture during storage; for dehydrated packaging of military supplies, tools, electronic equipment, etc.

"Protexol" Fireproofed Wood. [171] Proprietary process consisting of the injection of ammonium salts into wood by the vacuum-pressure process in accordance with standard methods of the American Wood Preservers Association. Treatments vary in depth of penetration. Class A (doors, trim, finish) is total impregnation 6 lbs. per cu ft (dry salt); B (structural timbers) is partially impregnated 3 to 4 lbs. per cu ft; C (temporary scaffolding) is impregnated 2 to 3 lbs per cu ft.

Recommended for indoor use on wood or plywood. The impregnated salts form permanent, insoluble compounds, effective for the life of the structure.

"Protobore." [342] Trademark for protoveratrines for medicine.

protocatechuic aldehyde, methyl ether. See vanillin.

protogen. See thioctic acid.

"Protol." [45] Trademark for white mineral oil, U.S.P.
Properties: Sp.gr. 0.870-0.880; Saybolt viscosity 180-190 (100°F); odorless and tasteless.
Uses: Pharmaceutical and cosmetic formulations; plastics; tobacco; paper; animal husbandry.

"Protolin." [23] Trademark for reducing agents based on zinc sulfoxylate and zinc formaldehyde sulfoxylate. Supplied as water-soluble white powder.
Use: Stripping of colors from textile fabrics; chemical synthesis.

proton. A positively charged particle that comprises the nucleus of the ordinary hydrogen atom and having almost the same mass but very much smaller dimensions than the atom. Its charge is equal but opposite to that of the electron. The proton is similar in mass and dimensions to the neutron except that the latter lacks an electric charge.
See fundamental particle.

"Protopet." [45] Trademark for petrolatum of medium consistency and ranging in color from pure white to amber, but meeting U.S.P. and N.F. purity requirements for petrolatum.
Uses: Pharmaceutical and cosmetic bases; various industrial applications.

"Protovac." [65] Trademark for a series of solubilized caseins, which include modified caseins, zinc casein or sodium caseinates.
Uses: Binders, emulsifiers, stabilizers and coatings in the paint, paper, leather, printing and textile industries.

protoveratrine. A substance isolated from the Veratrum album plant. This substance is a mixture of two alkaloids, designated

protoveratrine A and protoveratrine B. The chemical formula has been determined as $C_{32}H_{51}NO_{11}$, but the structural formulas of the two alkaloids are unknown.
Properties: White, odorless, slightly bitter, crystalline powder; causes sneezing; m.p. 256-260°C (dec). Freely soluble in chloroform; very slightly soluble in ether; practically insoluble in petroleum ether and water. Stable to light and air. pH (saturated solution) 6.5-7.3.
Grade: N.N.D.
Use: Medicine.

protoveratrine A and B maleates. The maleate salts of protoveratrine A and B.
Properties: White to buff colored powder with faint characteristic odor and strong sternutatory action; m.p. 210-222°C (dec); freely soluble in alcohol, chloroform, and water; very slightly soluble in ether and petroleum ether. pH of 0.2% solution is 4.1-4.7.
Grade: N.N.D.
Use: Medicine.

"Protox" Zinc Oxide. [268] Brand name for a propionic-acid-treated zinc oxide having incorporation and dispersion characteristics particularly desired by compounders of rubber.
Grade: "Protox" 166.
Containers: 50-lb paper bags.
Use: In both natural and synthetic rubber where easier mixing and better dispersion is desired.

"Protozyme." [78] Trademark for a series of diastatic, proteolytic, and pectolytic enzymes of fungal origin on a cereal base carrier for application where subsequent filtration permits the removal of the inactive portion.

proustite (light ruby silver ore; light red silver ore) Ag_3AsS_3 or $3Ag_2S \cdot As_2S_3$, sometimes some Sb.
Properties: Scarlet or vermilion mineral with scarlet to aurora red streak. Adamantine luster. Contains 65.4% silver, 15.2% arsenic, 19.4% sulfur. Usually occurs disseminated through the gangue or as a stain or crust. Blackens when exposed to sunlight. Decomposed by nitric acid with separation of sulfur. Differs from pyrargyrite in scarlet streak and from cuprite and cinnabar by garlic odor when heated.
Constants: Sp.gr. 5.57-5.64 (5.57 if pure); hardness 2-2.5.
Occurrence: United States (Idaho, Nevada, Colorado, Arizona); Germany; France; Czechoslovakia; Spain; Sardinia; Mexico; Peru; Chile.
Use: Ore of silver.

"Provinite." [304] Trademark for a two part barium-cadmium organic vinyl stabilizer.
Properties: (Provinite A) soft white powder; sp.gr. 1.31; refractive index 1.56; (Provinite B) clear straw-colored liquid; sp.gr. 0.91; refractive index 1.45.
Containers: (Provinite A) fiber board drums containing 75 lbs; (Provinite B) metal drums containing 35 lbs.
Uses: Heat and light stabilizers for vinyl film, sheeting and dispersion resin systems.

provitamin. A precursor of a vitamin. It assumes vitamin activity upon activation within the animal body. No differentiation customarily is made between the free vitamin and the provitamin when speaking of the vitamin content of a food.

provitamin A. The chief dietary source of vitamin A. See carotene and cryptoxanthin.

provitamin D_2. See ergosterol and vitamin D.

provitamin D_3. See 7-dehydrocholesterol.

prunolide. See gamma-nonyl lactone.

prunus virginiana (wild cherry; wild black cherry bark).
Derivation: Bark of Prunus serotina.
Occurrence: North America.
Grades: Technical; U.S.P. XVI.
Containers: Bags; bales.
Use: Medicine.
Shipping regulations: None.*

Prussian blue. The most common and best known name for blue iron ferrocyanide (iron blue) pigments made by a variety of procedures. See iron blue; also peacock blue.

Prussian red. A name sometimes given to red varieties of ferric oxide (see iron oxides red), and at other times to potassium ferricyanide (red prussiate of potash).

prussiate of potash, red. See potassium ferricyanide.

prussiate of potash, yellow. See potassium ferrocyanide.

prussiate of soda, red. See sodium ferricyanide.

prussiate of soda, yellow. See sodium ferrocyanide.

prussic acid. See hydrocyanic acid.

"Prym." [42] Proprietary products. A group of thermosetting carbamide fiber reactants.
Properties: Colorless syrups; dissolve readily in water at 25°C.
Containers: 55-gal steel drums.
Use: Textile fabric finishes on cotton, rayon, acetate and blends. Impart crush resistance and shrinkage stabilization.

"Prym" CR. [42] Proprietary compound. Modified carbamide cellulose reactant.
Properties: Colorless clear liquid. Dispersible in water at 25°C.
Containers: 55-gal steel drums.
Use: Chlorine resistant resin finish for use on cotton or viscose fabrics in textile finishing where crush resistance and fabric stabilization are desirable.

pseudobutylene glycol. See 2,3-butylene glycol.

pseudocumene (1,2,4-trimethylbenzene; unstrimethylbenzene) $C_6H_3(CH_3)_3$.
Properties: Liquid, f.p. −43.91°C; b.p. 168.89°C; sp.gr. (20/4°C) 0.8758;

*See "I.C.C. Shipping Regulations," page xiii.
Reference numbers refer to name of manufacturer. See "List of Manufacturers," page v.

refractive index (20/D) 1.5045; flash point 54°C. Insoluble in water; soluble in alcohol, benzene, and ether.
Source: Coal tar and petroleum.
Grades: 95%, 99%, and research.
Containers: Bottles; drums.
Uses: Manufacture of dyes and perfumes; sterilizing catgut.
Shipping regulations: None.*

pseudocumidine (2,4,5-trimethylaniline; 1,2,4-trimethyl-5-aminobenzene) $C_6H_2(CH_3)_3NH_2$.
Properties: White crystals; sp.gr. 0.957; m.p. 62°C; b.p. 236°C. Soluble in alcohol and ether; insoluble in water.
Derivation: By heating crude xylidine with methyl alcohol and hydrochloric acid in an autoclave. The pseudocumidine is separated from the product by means of its sparingly soluble, crystalline nitrate.
Grades: Technical.
Containers: Wooden barrels.
Uses: Manufacture of dyes; organic synthesis.
Shipping regulations: None.*

d-pseudoephedrine hydrochloride. $C_6H_5CHOHCH(NHCH_3)CH_3 \cdot HCl$.
Properties: Needles; m.p. 181-182°C. Soluble in water, alcohol, chloroform.
Uses: Pharmaceuticals.

pseudoplasticity. See thixotropy.

psi. Abbreviation for pounds per square inch.

psia. Abbreviation for pounds per square inch absolute.

psilomelane $BaMn_9O_{16}(OH)_4$. A natural oxide of variable composition. Calcium, nickel, cobalt and copper frequently are present. The name sometimes refers to mixtures of manganese minerals.
Properties: Color black; streak brownish black; luster submetallic; hardness 5-6; sp.gr. 3.7-4.7.
Occurrence: U.S.S.R., India; South Africa; Cuba; Arkansas, Virginia, Georgia.
Use: Important ore of manganese.

psyllium (fleawort; fleaseed).
Properties: Dark brown, boat-shaped, shiny seeds, containing a mucilaginous compound.
Derivation: Seeds of Plantago psyllium.
Occurrence: Southern Europe.
Grades: Technical blonde, black.
Containers: Boxes; bags.
Uses: Sizing silk; printing fabrics; paper manufacture; medicine.
Shipping regulations: None.*

Pt. Symbol for platinum.

PTA. Abbreviation for phosphotungstic acid.

PTA pigments. See phosphotungstic pigments.

pteroylmonoglutamic acid. See folic acid.

PTFE. Abbreviation for polytetrafluoroethylene (q.v.).

PTMA. Abbreviation for a mixture of phosphotungstic and phosphomolybdic acids, used in making pigments. See phosphotungstic pigments.

ptomaines. Basic bodies resulting from the putrefaction or metabolic decomposition of animal proteins. Many of them are highly poisonous. Examples of ptomaines which have been isolated and prepared synthetically are cadaverine (1,5-diaminopentane) (q.v.), muscarine (hydroxyethyltrimethylammonium hydroxide), putrescine (tetramethylenediamine), and neurine (trimethylvinylammonium hydroxide) (q.v.).

PTSA. Abbreviation for para-toluenesulfonamide.

ptyalin. A salivary amylase which acts upon alpha-1,4-glycosidic linkages converting starch to various dextrins and maltose. It can act over a pH range of 4.0-9.0; optimum pH 5.6-6.5. It requires the presence of certain negative ions for activation; chlorides and bromides are the most effective.
Use: Biochemical research.

ptychotis oil. See ajowan oil.

p-type crystal. See transistor.

Pu. Symbol for plutonium.

pudding pipe. See cassia fistula.

pudding stick. See cassia fistula.

pulegium oil. See hedeoma oil.

pulegone $C_{10}H_{16}O$. A ketone found in pennyroyal oil.
Properties: Oily liquid with pleasant odor; sp.gr. 0.9323 (20°C); b.p. 221°C; dextrorotatory; refractive index (n 20/D) 1.4894.

pulsatilla camphor. See anemonin.

pumice. A highly porous igneous rock, usually containing 65-75% SiO_2 and 10-20% Al_2O_3; with a glassy texture. Potassium, sodium, and calcium are generally present in small amounts. Insoluble in water; not attacked by acids.
Occurrence: New Mexico, California, Idaho, Oregon, Nebraska; Italy; New Zealand.
Grades: Lump; powdered; ground; N.F. XI; technical.
Containers: Bags of various sizes.
Uses: Abrasive; light weight concrete aggregate; heat and sound insulation; insecticides; bricks; filtration; plastics; paint fillers; absorbents; solvents; support for catalysts; dental abrasive.
Shipping regulations: None.*

pumpkin seed. See pepo.

"Puratized" B-2. [307] Brand name for a mildewproofing agent consisting of phenylmercuricarbonate; 17% active. Used in lakes, rubber, paints, fabric and paper coatings.

"Puratized" N5DS. [307] Brand name for a mildewproofing agent consisting of phenylmercuritriethanolammonium lactate, 22% active. Used with emulsion-type water repellents.

*See "I.C.C. Shipping Regulations," page xiii.
Reference numbers refer to name of manufacturer. See "List of Manufacturers," page v.

"Puratized" Sprays. [55] Brand name for agricultural sprays containing phenylmercuri mono- or tri-ethanol ammonium acetate. Used on apples, pears and other fruits.

"Purayonier." [37] Trademark for a wood cellulose developed for applications requiring a highly refined and absorbent form of cellulose.
Properties: Cellulose–alpha 96%, beta 2.2%, gamma 1.8%, 10% KOH solubility, 5.3%.
Use: Manufacture of cellulose derivatives, saturating papers and plastics. It has also been used for diversifed special applications where high whiteness or cleanliness are of importance.

"Purecal." [203] Trademark for precipitated calcium carbonate.
Properties: Fine, white, microcrystalline powder; insoluble in water; soluble in dilute mineral acids.
Grades: M, T, and U, which meet U.S.P. requirements, having particle sizes 0.12-0.32, 0.045-0.055, and 0.033-0.040 microns respectively; "Purecal SC" (0.050-0.060 microns and surface treated with 2% modified fatty acid coating).
Containers: Bags and hopper cars.
Uses: Primary - paper, paint, rubber, ink, and plastic. Secondary - adhesives, ceramics, chemicals, cosmetics, crayons, dentifrices, filtering media; foods; animal and poultry feeds; baking powder, candy, ice cream concentrates; glass; lacquer; leather tanning; linoleum; neutralizing agent; pharmaceuticals; polishes; putty and caulking compounds; soaps; textiles; varnishes; water treatment; window shades; wood fillers.

purging cassia. See cassia fistula.

purging croton. See tiglium.

purified ozocerite. See ceresin wax.

purine [imidazo (4,5-d) pyrimidine] CHNCHNCCNHCHN.
Properties: Colorless crystals; m.p. 217°C; soluble in water, alcohol, toluene; slightly soluble in ether.
Derivation: Prepared from uric acid and regarded as the parent substance for compounds of the uric acid group, many of which occur naturally in animal waste products.
Uses: Organic synthesis; metabolism studies.

purines. Basic compounds found in living matter and having a purine-type molecular structure. For specific purines, see adenine, guanine, hypoxanthine, xanthine, uric acid, caffeine, and theobromine.

"Purinethol." [301] Trademark for 6-mercaptopurine (6MP), used in the medical treatment of leukemia.

"Purite." [84] Trademark for a specially prepared fused soda ash furnished in the form of two-pound cast pigs and stated to contain over 98% sodium carbonate.
Containers: 200-lb bags; 350-lb barrels; bulk.
Uses: Cupola flux; refining and desulfuring iron, steel and other metals.

"Purodigin." [24] Trademark for digitoxin (q.v.).

puromycin $C_{22}H_{29}N_7O_5$. Crystals; m.p. 176°C. An antibiotic which inhibits protein synthesis, prevents transfer of amino acid from its carrier to the growing protein.
Puromycin is produced by Streptomyces alboniger; it is effective against bacteria, protozoa, parasitic worms, an alga, and cancerous tumors. Toxic to living cells of all kinds.

purple cone flower. See echinacea.

purple copper ore. See bornite.

purple fox-glove. See digitalis.

purple lakes. A class of lakes derived from combination of such compounds as 2-diazonaphthalene-1-sulfonic acid and beta-hydroxynaphthoic acid. Used in printing inks.

purple of Cassius. See gold-tin purple.

purple oxide. Name given to certain varieties of ferric oxide.
See iron oxide reds.

purple salt. See potassium permanganate.

purpurin (1,2,4-trihydroxyanthraquinone) $C_{14}H_5O_2(OH)_3$.
Properties: Reddish needles; m.p. 256°C; slightly soluble in hot water; soluble in alcohol and ether.
Derivation: Occurs as a glucoside in madder root. Made synthetically by oxidation of alizarin.
Uses: Dye for cotton; stain for microscopy; reagent for calcium.

purpurin red. See anthrapurpurin.

"Purzaust." [416] Trademark name of catalytic converter developed for the removal of air pollutants normally present in raw exhaust gases from internal combustion engines. The device is self-contained and replaces the conventional acoustic muffler for both automotive and stationary engines. A fixed bed of solid catalyst promotes removal by oxidation of combustible pollutants to carbon dioxide and water. The device has no moving parts and does not require supplemental fuel or ignition systems; no additional operating or maintenance costs are involved. The device is designed to meet the 1960 California Motor Vehicle Emissions Standards without adversely affecting engine operation or materially increasing temperatures of any part of the vehicle.

"PūTrol." [188] Trademark for a powerful aromatic compound used for masking the odors of putrefaction associated with the decomposition of protein materials, as in fat rendering operations, sewage disposal, and certain problems involving industrial wastes.

putty. A mixture of whiting (chalk) with 18% of linseed oil, with or without white lead or other pigment. Containers must be air-tight.

putty oil. A petroleum fraction added to putty as a lubricant and softener so the putty will not become too hard after the linseed oil dries.

puzzolana cement. See pozzolana cement.

PVA. Abbreviation for polyvinyl alcohol or polyvinyl acetate.

PVAc. Abbreciation for polyvinyl acetate.

PVC. Abbreviation for polyvinyl chloride.

PVI. Abbreviation for polyvinylisobutyl ether.

PVM. Abbreviation for polyvinyl methyl ether.

PVM/MA. Common name for copolymers of vinyl methyl ether and maleic anhydride. $[-CH_2CHOCH_3CHCOOCOCH-]_n$.

Properties: White, amorphous powder available in a wide range of viscosities. Soluble with chemical reaction in water (over entire pH range) and in alcohols. Also soluble in ketones and esters, etc. Insoluble in aliphatic and aromatic hydrocarbons. Compatible with a wide variety of water-soluble resins and plasticizers. Gives typical anhydride reactions. Sp.gr. 1.3-1.4; bulk density 16 lbs/cu ft. Very low toxicity.

Derivation: Reaction of vinyl methyl ether and maleic anhydride.

Grades: Several viscosity grades available.

Containers: Fiber drums.

Uses: Protective colloid, dispersing agent, thickener, binder, adhesive and size in coatings, detergents, emulsions, paper, textiles, leather, latex, etc.

Shipping regulations: None.*

PVM/MA half amide. Common name for the copolymer of vinyl methyl ether and ammonium maleamate. $[-CHOCH_3CH_2CHCONH_2CHCOONH_4-]$.

Properties: White, amorphous, hygroscopic powder soluble in water and polar organic solvents. Sp.gr. (dry film) 1.2-1.3 g/ml, bulk density 13 lbs/cu ft; pH of aqueous sol. 6.8-7.2.

Derivation: Treatment of PVM/MA (vinyl methyl ether – maleic anhydride copolymer) with ammonia. Approximately 75% of the anhydride group is converted to the ammonium maleamate.

Grades: Several viscosity grades available.

Containers: Fiber drums.

Uses: Thickener, binder, adhesive and size for coatings, dispersions, latex, paper, textiles.

Shipping regulations: None.*

PVOH. Abbreviation for polyvinyl alcohol.

PVP. Abbreviation for polyvinylpyrrolidone.

"PX." [28] Trademark for vinyl or pyroxylin book cloth.

py. An abbreviation for pyridine, used as in formulas for coordination compounds. See also dien; en; pn.

"Pycal." [89] Trademark for a series of plasticizers used for cellulose acetate, cellulose acetate butyrate, nitrocellulose, vinyls, and synthetic rubber.

pycnometer. A standard vessel for measuring and comparing densities of liquids and solids. Most often a device for determining the volume of a known weight of material for density calculations. Volume determination is made by liquid displacement in standard vessels, or by gas displacement in air comparison pycnometers.

"Pydraul." [58] Trademark for a series of hydraulic fluids.

Properties: Oily fire resistant liquids with lubricating and non-corrosive characteristics.

Containers: 54-gal and 55-gal steel drums; 5-gal cans.

Uses: Used in high precision and low temperature equipment; diecasting machines; forging and extrusion presses; automatic welding machines; hydraulic presses and air compressers.

pyolipic acid. An antibiotic produced by certain strains of Pseudomonas pyocanea when grown on broth containing glucose. An oily liquid soluble in alcohol or ether. Composed of several fatty acids and the sugar, levo-rhamnose. Of possible use in treating tuberculosis infections. Differs from streptomycin in that the latter suppresses growth of the tuberculosis germs whereas pyolipic acid is fatal to these organisms.

"Pyramidon." [162] Trademark for aminopyrine.

"Py-Ran." [58] Trademark for anhydrous monocalcium phosphate $CaH_4P_2O_8$ or $CaH_4(PO_4)_2$. Conforms with federal and state pure food laws.

Use: As an acid component for leavening agents in self-rising flours and corn meal, also baking powders and prepared mixes.

Containers: 100-lb bags.

"Pyranol." [245] Trademark for dielectric material, principally of the askarel type (q.v.); also for capacitors.

pyrargyrite (dark ruby-silver ore) Ag_3SbS_3 or $3Ag_2S \cdot Sb_2S_3$, sometimes with some As.

Properties: Black to grayish-black mineral, light deep red by transmitted light. Purplish-red streak, metallic to adamantine luster. A natural silver sulfantimonite. Contains 59.9% silver, 22.3% antimony, 17.8% sulfur. Distinguished from proustite by color of streak and from cuprite, cinnabar and realgar by color of streak and silver reaction. Soluble in nitric acid.

Constants: Sp.gr. 5.77-5.86; hardness 2.5.

Occurrence: United States (Idaho, Nevada, Arizona, Colorado, Utah); Mexico; Chile; Germany; Czechoslovakia; Hungary; Norway; Spain; England; Canada.

Use: Ore of silver.

"Pyratex." [248] Trademark for a vinyl pyridine latex.

Properties: Total solids 40-42%, pH 10.5-11.5; sp.gr. 0.96.
Uses: Used in adhesive compounds to enhance the adhesion between rayon or nylon fibers and rubber. Main uses are in tire cord, belting, hose and other mechanical goods.

pyrathiazine hydrochloride (10-[2-[1-pyrrolidyl) ethyl] phenothiazine hydrochloride) $C_{18}H_{20}N_2S\cdot HCl$.
Properties: White or grayish white powder becoming darker upon exposure to light; odorless; solutions neutral to litmus. Soluble in water, chloroform, and alcohol; insoluble in ether and benzene.
Use: Medicine.

"Pyrax." [69] Brand name for a proprietary ground pyrophyllite-aluminum silicate used as a filler in paints, plastics, and for dusting unvulcanized rubber.

pyrazinamide (pyrazinoic acid amide; pyrazine carboxamide) $C_5H_5N_3O$.
Properties: Crystals; m.p. 189-191°C; begins to sublime at 159°C. Soluble in water.
Grade: N.N.D.
Use: Medicine.

pyrazine hexahydride. See piperazine.

pyrazoline $HNNCHCH_2CH_2$. B.p. 144°C.
Use: Organic synthesis.

pyrazolone $HNNCHCH_2CO$ (3-ketopyrazoline).
Properties: Solid; m.p. 165°C.
Use: Organic synthesis.

pyrazolone dyes. Dyes whose molecules contain both the –N=N– and the =C=C= chromophore groups in their structure and whose color index ranges from 636 to 654. These are acid dyes most used for silk and wool, and to some extent for lakes. Tartrazene, C.I. 640, is a very important member of this group. See dyes.

"Pyrefume." [342] Trademark for pyrethrin-extract insecticidal concentrates.

pyrene $C_{16}H_{10}$. A condensed ring hydrocarbon.
Properties: Colorless solid (tetracene impurities give a yellow color); solutions have a slight blue fluorescence. M.p. 156°C; density 1.271 (23°C); b.p. 404°C. Insoluble in water; fairly soluble in organic solvents.
Derivation: From coal tar. Also obtained by destructive hydrogenation of hard coal.

"Pyrene." [341] A proprietary name for liquid fire-extinguisher stated to consist of carbon tetrachloride.

"Pyrenone." [55] Proprietary insecticide products based on mixtures of blends of either piperonyl butoxide or piperonyl cyclonene and pyrethrins. Available in form of oil solutions, impregnated dusts, wettable powder, etc.
"Pyrenone" 20 New: A deodorized base oil concentrate standardized to contain 4.0 g technical piperonyl butoxide plus 0.5 g pyrethrins per 100 cc. Used in manufacture of household sprays and livestock sprays.
"Pyrenone" Dust Base 100: An impregnated dust base standardized to contain 2.5% technical piperonyl cyclonene plus 0.2% pyrethrins. Used in the manufacture of horticultural dusts and household dusts.

"Pyresote." [171] See "Protexol." Combines high fire resistance with a standard wood preservative.

pyrethrin I $C_{21}H_{28}O_3$. Pyrethrolone ester of chrysanthemummonocarboxylic acid. Most potent insecticidal ingredient of pyrethrum flowers. See also cinerin I and II and pyrethrin II.
Properties: Viscous liquid, oxidizes readily in air. Insoluble in water; soluble in other common solvents. Incompatible with alkalies.
Source: Pyrethrum extract (q.v.).

pyrethrin II ($C_{22}H_{28}O_5$). Pyrethrolone ester of chrysanthemumdicarboxylic acid. One of the four primary active insecticidal ingredients of pyrethrum flowers. See also pyrethrin I, cinerin I and II. Properties similar to those of pyrethrin I; less toxic than pyrethrin I.
Source: Pyrethrum extract (q.v.).

pyrethrosin $C_{17}H_{22}O_5$. A sesquiterpenoid.
Properties: White crystals; melts 188-189°C; insoluble in water; soluble in hot alcohol, chloroform; slightly soluble in ether or petrol.
Derivation: Obtained from pyrethrum flowers.

pyrethrum extenders (synergists). Substances not powerful insecticides when used alone, but which increase the potency of pyrethrum when used in combination with it. Among them are piperonyl butoxide, and piperonyl cyclonene (q.v.).

pyrethrum extract. An extract obtained from powdered pyrethrum flowers by using a hydrocarbon of the kerosene type. Not compatible with alkaline material. The chief constituents of the extract are called pyrethrins I and II and cinerins I and II (see also allethrin). These compounds are non-volatile and very slightly soluble in water. Used for medicine and insecticides.

pyrethrum flowers (Persian insect flowers; Dalmatian insect powder; Persian pellitory).
Derivation: Dried flowers of Chrysanthemum cinerariaefolium, C. coccineum and C. marshallii found in Africa, India, and southern Europe.
Grades: Technical. Sold by pyrethrin content.
Containers: Bags.
Use: Insecticide.
Shipping regulations: None.*

pyrethrum root (pellitory; Spanish pellitory; Spanish chamomile; bertram; longwort).
Derivation: Root of Anacyclus pyrethrum, D.C.
Occurrence: North Africa.

*See "I.C.C. Shipping Regulations," page xiii.
Reference numbers refer to name of manufacturer. See "List of Manufacturers," page v.

Grades: Technical.
Containers: Bags.
Use: Medicine.
Shipping regulations: None.*

"Pyretox." [55] Brand name for insecticidal product containing 1.0% by wt. of pyrethrin.

"Pyrex." [20] Trademark for heat and chemical resistant glassware of various compositions and physical properties, and accessories used therewith.

"Pyrex Glass Brand No. 7740." [20] Trademark. Borosilicate glass of American manufacture, widely used in the fabrication of laboratory and pharmaceutical glassware, domestic cooking utensils and industrial items; noted for its low expansion coefficient and high mechanical strength.
Physical characteristics: Linear coefficient of expansion 0.0000032 between 19° and 350°C; elasticity coefficient 6.230 kg/sq mm; hardness-scleroscope 120; sp.gr. 2.25; specific heat 0.20; refractive index 1.4754; dispersion 0.00738; light and heat transmission higher than the best plate glass.
Softening temperature: Softening does not commence at below 600°C, and for a limited time, these glasses can be used safely at temperatures somewhat higher.
Uses:
Chemical manufacture: Laboratory and pharmaceutical glassware and apparatus; tubes for horizontal and vertical coolers; heaters and condensers, and for Hart nitric acid condensers, tubes and return bends for atmospheric condensers; socket pipes for conveying gases and liquids; socket tower sections; "S" bends and "U" bends; plates and bonnets for fractionating and reaction columns; pulsometer tubes and "monkey" pumps; sight cylinders for pipe lines; sight glasses; sheets for lookboxes; ballon flasks and retorts; cascade dishes and pans; drying trays; pots and jars; bottles, carboys.
Electrochemical manufacture: Cell thimbles; brine wells; liquid and gas connectors; cooler pipes for sodium hypochlorite; intercell coolers; cyanide plating jars.
Silk and rayon manufacture: Corrugated reels and rollers; Godet wheels; guides; bleach coolers; piping.
Miscellaneous: Dielectric tubes for ozone generators; fruit-juice sterilizer tubes; motion-picture developing tanks (tubular type); rollers for plating machines; gage tubes for hot solutions; instrument tubing; vacuum-bottle blanks; battery jars.
Domestic: Cooking utensils; teapots; coffee makers.

"Pyrexcel." [342] Trademark for synergized pyrethrin extracts for insecticides.

"Pyribenzamine." [305] Trademark for tripelennamine, an antihistamine.

pyridine $\underline{N(CH)_4C}H$. An important organic base.
Properties: Slightly yellow or colorless liquid; sharp penetrating empyreumatic odor; burning taste; slightly alkaline in reaction. Soluble in water, alcohol, ether, benzene, ligroin and fatty oils. Sp.gr. 0.978; m.p. −42.0°C; b.p. 115.5°C; flash point (closed cup) 68°F.
Derivation: (a) Distillation of organic compounds containing nitrogen, or of gas liquor or light coal-tar oil. (b) Also synthetically from acetaldehyde and ammonia.
Method of purification: Fractional distillation.
Grades: Technical; medicinal; C.P.
Containers: Bottles; drums; barrels; tank cars.
Uses (in approximate order of volume): Synthesis of vitamins and drugs; waterproofing; solvent; rubber chemicals; denaturant for alcohol; dyeing assistant in textiles.
Warning! Flammable. Vapor harmful. MCA warning label.
Shipping regulations: Flammable liquid. Red label.*

2-pyridine aldoxime methiodide (PAM) $C_5NH_4CHNOH \cdot ICH_3$. Stated to be an antidote for nerve gas, because of its property of reactivating the enzyme cholesterase by removal of phosphoryl groups introduced by the nerve gas.

beta- or meta-**pyridinecarboxylic acid.** See nicotinic acid.

pyridine-3-carboxylic acid, diethylamide. See nikethamide.

2,5-pyridinedicarboxylic acid. See isocinchomeronic acid.

pyridine-N-oxide C_5H_5NO. F.p. 67.0°C. Soluble in water. Used as an intermediate.

pyridinium bromide perbromide (PBPB) $C_5H_6NBr \cdot Br_2$.
Properties: Red prismatic crystals; m.p. 135-137°C (dec) with preliminary softening. The salt is stable in the dry state and can be used in glacial acetic acid, ethanol and related solvents. This compound has 45% to 50% available bromine.
Uses: Brominating agent for phenols, and for the addition to double bond; agent for mono and polybromination of ketones, including aliphatic, alicyclic, steroid and amino carbonyls. Used in micro or semimicro quantitative analysis.

pyridinium hexafluophosphate. See fluophosphoric acids.

"Pyridose." [329] Trademark for a brand of pyridylmercuric acetate, an organo-mercurial fungicide for positive control of paper-mill slime. Available in 1-oz packets for use in paper mills.

pyridostigmine bromide $C_5H_4N(CH_3)OOCN(CH_3)_2Br$. 3-Hydroxy-1-methylpyridinium bromide dimethylcarbamate.
Properties: White or practically white, crystalline powder. Agreeable odor;

*See "I.C.C. Shipping Regulations," page xiii.
Reference numbers refer to name of manufacturer. See "List of Manufacturers," page v.

hygroscopic. Freely soluble in water, alcohol, and chloroform. Slightly soluble in solvent naphtha; insoluble in ether. M.p. 154-157°C.
Grade: U.S.P. XVI.
Use: Medicine.

pyridoxal hydrochloride $C_8H_9NO_3 \cdot HCl$. An aldehyde derivative of pyridoxine, with vitamin B_6 activity.
Properties: Rhombic crystals; m.p. 165°C (dec); soluble in water and 95% ethyl alcohol.
Use: Nutrition.

pyridoxal phosphate (2-methyl-3-hydroxy-4-formyl-5-pyridylmethylphosphoric acid) $CH_3C_5HN(OH)(CHO)(CH_2PO_4H_2)$. The active coenzyme form of pyridoxine; functions as a coenzyme for transaminases, decarboxylases, desulfurases and racemases. It is involved in the conversion of tryptophan to the nicotinamide portion of the pyridine coenzymes; it also participates in energy transformations in brain and nervous tissue.
Properties: Colorless in acid solution; bright yellow in alkaline solution.
Commercial derivation: (a) action of adenosine triphosphate on pyridoxal; (b) action of phosphorus oxychloride on pyridoxal in aqueous solution; (c) phosphorylation of pyridoxamine with 100% phosphoric acid followed by oxidation.
Uses: Medicine; nutrition.
Also described under codecarboxylase (a synonym) q.v.

pyridoxamine dihydrochloride $C_8H_{12}N_2O_2 \cdot 2HCl$. The dihydrochloride of the amino derivative of pyridoxine, exhibiting vitamin B_6 activity.
Properties: Platelets; m.p. 226-227°C (dec); soluble in water and 95% ethyl alcohol.
Uses: Medicine; nutrition.

pyridoxine (vitamin B_6) $C_8H_{11}O_3N$. 3-Hydroxy-4,5-dihydroxymethyl-2-methylpyridine. Pyridoxine is a group name to designate the naturally occurring pyridine derivatives with vitamin B_6 activity. It is essential for the dehydration and desulfhydration of amino acids and for the normal metabolism of tryptophan. It also appears to be related to fat metabolism. Pyridoxine is required in the nutrition of all species of animals.
Sources: Food source: vegetable fats, whole grain cereals, legumes, yeast, muscle meats, liver, and fish. Commercial source: synthetic pyridoxine, pyridoxal, and pyridoxamine are produced by a complex series of reactions from isoquinoline.
Units: Amounts are expressed in micrograms.
Uses: Medicine; nutrition.
Commercially available as pyridoxine hydrochloride (q.v.).

pyridoxine hydrochloride $C_8H_{11}O_3N \cdot HCl$. Usual commercial form of pyridoxine. Same biological function.
Properties: Colorless, white platelets; m.p. 204-206°C (dec); salty taste; stable in air and slowly affected by sunlight; stable to heat, concentrated acid and alkali; soluble in water, alcohol and acetone; slightly soluble in other organic solvents.
Grades: U.S.P. XVI.
Containers: Fiber drums, bottles.
Uses: Medicine; nutrition.

alpha-**pyridylamine.** See 2-aminopyridine.

beta-**pyridylamine.** See 3-aminopyridine.

3-pyridylcarbinol $C_5H_4NCH_2OH$.
Properties: B.p. (760 mm) 266°C; f.p. −6.5°C; density (20°C) 1.131 g/ml; refractive index (n 20/D) 1.5455. Soluble in water at 20°C in all proportions.
Use: Intermediate.

beta-**pyridylcarbinol tartrate** (3-pyridinemethanol tartrate) $C_6H_7NO \cdot C_4H_6O_6$.
Properties: M.p. 146.5-148.5°C. Soluble in water; sparingly soluble in alcohol; insoluble in chloroform and ether.
Use: Vasodilator.

2-(3-pyridyl)piperidine. See anabasine.

pyrilamine maleate $C_{17}H_{23}N_3O \cdot C_4H_4O_4$. 2-[(2-Dimethylaminoethyl)(para-methoxybenzyl)amino]-pyridine maleate; $CH_3OC_6H_4CH_2N(C_5H_4N)CH_2CH_2N(CH_3)_2 \cdot C_4H_4O_4$.
Properties: White crystalline powder with faint odor and bitter, saline taste. M.p. 99-101°C. Very soluble in chloroform and water; freely soluble in alcohol; slightly soluble in benzene and ether. 5% solution is clear and colorless (or nearly so) and has pH 4.5-5.5.
Grades: U.S.P. XVI.
Use: Medicine.

pyrimethamine $(NH_2)_2C_4N_2(C_2H_5)C_6H_4Cl$. 2,4-Diamino-5-(para-chlorophenyl)-6-ethylpyrimidine.
Properties: White, odorless, crystalline powder. Practically insoluble in water. Slightly soluble in alcohol, chloroform, and acetone. M.p. 238-241°C.
Grade: U.S.P. XVI.
Use: Medicine.

pyrimidine (1,3-diazine; miazine) $\overline{CHN(CH)_3N}$.
Properties: Liquid and crystalline mass with a penetrating odor; melting point 20-22°C; b.p. 123-124°C; soluble in water, alcohol, and ether.

pyrimidines. Basic compounds found in living matter and having a pyrimidine-type molecular structure. They may be isolated following complete hydrolysis of nucleic acids. Such pyrimidines include uracil, thymine, cytosine, and methylcytosine. Thiamine is also a pyrimidine derivative. Other pyrimidines such as alloxan and thiouracil are important for use in medicine and biochemical research.

pyrimidinetrione. See barbituric acid.

pyrite (iron pyrites; fool's gold) FeS_2, often with small amounts of Cu, As, Ni, Co, Au, Se.

Properties: Brass-yellow or brown tarnished mineral, greenish or brownish-black streak, metallic luster. Contains 46.7% iron, 53.3% sulfur. Differs from chalcopyrite and pyrrhotite in hardness, from marcasite in that its powder is completely soluble in strong nitric acid, and from gold in color, streak and brittleness. Is very widely distributed, being the most common sulfide mineral.
Constants: Sp.gr. 4.9-5.2; hardness 6-6.5.
Occurrence: United States (Virginia, New York, Massachusetts, Connecticut, New Jersey, Colorado, Utah); Norway; Germany; France; Italy; Spain; Portugal; England; Hungary; Sweden; Canada; North Africa.
Uses: Ore of iron; sulfuric acid manufacture; paper manufacture (sulfur dioxide); sulfur manufacture (by distillation); copperas; radio (detectors); cheap jewelry; recovery of gold, silver, copper.
See also ferrous sulfide.

pyrite, white. See marcasite.

pyrites, arsenical. See arsenopyrites.

pyrites, cinder.
Properties: Dark red lumps; composed mainly of iron oxide but contains, usually, from 1-4% sulfur as sulfate of the metals present.
Derivation: The residue from the burning of pyrite.
Use: Sometimes sold for the recovery of the iron or copper and silver.
Shipping regulations: None.*

pyrites, cobalt. See linnaeite.

pyrites, copper. See copper sulfide, also chalcopyrite.

pyrites, iron. See pyrite.

pyrites, magnetic. See pyrrhotite.

pyrites, tin. See stannite.

pyrithiamine (neopyrithiamine) $C_{14}H_{20}Br_2N_4O$. A thiamine antagonist.
Properties: Crystallizes from acetone; m.p. 219°C (dec); soluble in water.
Derivation: Synthetically from the condensation of 2-methyl-3-(beta-hydroxyethyl) pyridine with the pyrimidine moiety of thiamine.
Uses: Biochemical research.

pyro-. A prefix indicating formation by heat; specifically, an inorganic acid derived by loss of one molecule of water from two molecules of an ortho acid, as pyrophosphoric acid.

pyroacetic ether. See acetone.

pyrobitumen. A native, dark-colored, solid material composed of hydrocarbons, which may be partly oxygenated. Inorganic material may be present. The hydrocarbon portion is fusible and relatively insoluble in carbon disulfide. Pyrobitumen includes the asphaltic pyrobitumens (q.v.) and peat, coal, and bituminous shales.

"Pyrobor." [88] Trademark for dehydrated borax, $Na_2B_4O_7$.

pyrocatechol (ortho-dihydroxybenzene; 1,2-benzenediol; catechol) $C_6H_4(OH)_2$.
Properties: Colorless crystals; discolors to brown on exposure to air and light, especially when moist; sp.gr. 1.371; m.p. 104°C; b.p. 245°C, sublimes; soluble in water, alcohol, ether, benzene and chloroform, also in pyridine and aqueous alkaline solutions.
Derivation: (a) By fusion of ortho-phenolsulfonic acid with caustic potash at 350°C. (b) By heating guaiacol with hydriodic acid.
Method of purification: Crystallization.
Grades: Technical; C.P.; resublimed.
Containers: 25- to 200-lb drums.
Uses: Antiseptic; photography; dyestuffs; for dyeing; electroplating; specialty inks; antioxidants and light stabilizers.
Shipping regulations: None.*

pyrocatechol dimethyl ether. See veratrole.

pyrocatechol methyl ester. See guaiacol.

pyrocatechol monomethyl ether. See guaiacol.

"Pyroceram." [20] Trademark for crystalline ceramic materials made from glass of various compositions and physical properties and accessories used therewith.

"Pyroceram Brand Cement." [20] Trademark for powdered glasses which are thermosetting and utilized for sealing inorganic materials. The resultant seals are crystalline and have service temperatures in excess of the sealing temperatures.

"Pyroceram Brand Ceramic Code No. 9606." [20] Trademark for crystalline ceramic material made from glass having favorable electrical properties.
Properties: Flexural strength of abraded articles is 20,000 psi. Dielectric constant varies only slightly with temperature at frequencies on the order of 8.5 kmc and above.
Uses: High temperature, high frequency applications in the electronics field.

"Pyroceram Brand Ceramic Code No. 9608." [20] Trademark for crystalline ceramic material made from glass having high thermal shock resistance.
Properties: Thermal expansion coefficient of $3.9\text{-}11 \times 10^{-7}/°F$. Flexural strength of 14,000-23,000 psi.
Uses: Cooking ware, telescope mirrors.

pyrochlore $NaCaNb_2O_6F$. A complex oxide of sodium, calcium and niobium. Tantalum, rare earth metals, and other elements may be present. Color brown to black, streak light brown; hardness 5-5.5; sp.gr. 4.2-6.4. Forms a series with microlite. Koppite is a variety of pyrochlore.
Occurrence: Maine, California, Colorado; Africa; Europe.
Use: Ore of niobium.

pyrochroite. See manganous hydroxide.

"Pyrofax." [214] Brand name of a proprietary fuel gas.

Constants: Heat of combustion 2509 Btu/cu ft, 21,500 Btu/lb; vapor density (air = 1.0) 1.56; vapor pressure at 70°F approx. 117 lbs/sq in; vol/lb at 15.6°C and 760 mm 8.56 cu ft.

Containers: 50-, 100-lb cylinders (net wt) 120- to 1000-gal tanks; approx. 8000-gal tank cars.

Uses: As fuel for domestic and heavy-duty cooking; water heating; refrigeration; space heating; and for schools and laboratories not served with city gas. Fuel and atmosphere for heat-treating and carburizing metals; for annealing; tool dressing; brazing; soldering and other industrial applications. Also used for singeing and drying and for heating calender rolls.

pyroforic (also spelled pyrophoric). Spontaneously combustible upon exposure to air.

pyroforic fuel, or solutions.

Shipping regulations: Flammable liquid. Red label.*

pyrogallic acid (pyrogallol; 1,2,3-trihydroxybenzene) $C_6H_3(OH)_3$.

Properties: White, lustrous crystals; turn gray on exposure to light; sp.gr. 1.463; m.p. 132.5°C; b.p. 309°C; soluble in water, alcohol, and ether. A solution of pyrogallic acid acquires a brown color on exposure to air. This absorption of oxygen and change of color take place rapidly when the solution is made alkaline.

Derivation: By heating gallic acid with three times its weight of water, in an autoclave.

Grades: Technical; C.P.

Containers: 1-lb bottles; 1-, 5-lb cans; wooden barrels; drums.

Uses: Protective colloid in preparation of metallic colloidal solutions; photography; dyes; intermediates; synthetic drugs; medicine; process engraving; analysis of free oxygen in air and other gas mixtures; antioxidant in lubricating oils.

pyrogallol. See pyrogallic acid.

pyrogallol monoacetate (eugallol) $C_6H_3(OH)_2OOCCH_3$.

Derivation: By heating pyrogallol with glacial acetic acid or acetic anhydride.

Grades: Medicinal; 17% solution in acetone.

Use: Medicine.

pyrogallol triacetate (lenigallol) $C_6H_3(OOCCH_3)_3$.

Properties: White crystalline powder. Soluble in alcohol; almost insoluble in water; decomposed by alkali hydroxide solutions. M.p. 165°C.

Derivation: By heating pyrogallol with acetic anhydride.

Method of purification: Crystallization in alcohol.

Use: Medicine.

pyrographite. See graphite, pyrolytic.

pyroligneous acid (wood vinegar). Crude yellow to red liquid, a mixture of materials from wood distillation. Crude product contains methanol, acetic acid, acetone, furfural and various tars and related products; sp.gr. 1.018-1.030; miscible with water and alcohol.

Use: Formerly an important source of acetic acid and acetone, as well as certain medical products.

pyroligneous liquor. See pyroligneous acid.

pyroligneous vinegar. See pyroligneous acid.

pyrolignite of iron. See iron acetate liquor.

pyrolignite of lime. See calcium acetate.

pyrolusite (manganese dioxide, black) MnO_2.

Properties: Iron-black to dark steel-gray or bluish mineral; streak, black or bluish-black; luster, metallic or dull. Sufficiently soft to soil the fingers. Found with psilomelane and other black manganese oxides. Is thought to be an alteration product of manganite or polianite, etc. Contains 63.2% manganese. Soluble in hydrochloric acid.

Constants: Sp.gr. 4.73-4.86; hardness 2-2.5.

Occurrence: United States (Virginia, Georgia, Arkansas, Lake Superior region, Massachusetts, Vermont, New Mexico); Germany; Czechoslovakia; Australia; India; Canada.

Use: An important manganese ore. See also manganese dioxide. For many uses the ore and synthetic material are interchangeable.

"Pyrolux Maroon." [141] Trade name for organic maroon pigments used in plastics because of exceptional resistance to discoloration due to heat and excellent permanence.

pyrolysis. Breaking down of complex materials into simpler units by use of heat.

pyromellitic acid (PMA; 1,2,4,5-benzenetetracarboxylic acid) $C_6H_2(COOH)_4$.

Properties: White powder; sp.gr. 1.79; m.p. 257-265°C; b.p. converts to dianhydride; bulk density 32 lb/cu ft. Picks up moisture slowly if exposed to atmosphere. Keep sealed.

Grade: 99% purity.

Containers: 5-, 25-, 100-lb fiber drums.

Uses: Intermediate for polyesters and polyamides used in electrical and nonfogging plasticizers, lubricants and waxes.

Hazard: A primary irritant.

Shipping regulations: None.*

pyromellitic dianhydride $C_6H_2(C_2O_3)_2$.

Properties: White powder; sp.gr. 1.68; m.p. 286°C; b.p. 397-400°C (760 mm), 305-310°C (30 mm); bulk density 21 lb/cu ft. Soluble in some organic solvents. Hydrolyzes to the acid when exposed to moisture.

Grade: 98+% purity.

Containers: 5-, 25-, 100-lb fiber drums.

Uses: Curing agent for epoxy resins used in high temperature laminates, molds, and coatings; crosslinking agent for epoxy plasticizers in vinyls, in alkyd resins; as intermediates for esters of pyromellitic

acid for use as electrical grade and non-fogging plasticizers and high temperature lubricants.
Hazard: Strong primary irritant.
Shipping regulations: None.*

pyrometric cones (Seger cones). Small pyramids composed of mixtures of oxides which melt at known temperatures and are used to measure temperatures in the 1100-3700°F range.

pyromucamide. See furoamide.

pyromucic acid. See furoic acid.

pyromucic aldehyde. See furfural.

"Pyronate." [45] Trade name for water-soluble petroleum sulfonates.

"Pyronil." [100] Trademark for pyrrobutamine phosphate (q.v.).

pyrope. See garnet.

pyrophoric. See pyroforic.

pyrophoric alloy (sparking metal; ferrocerium; Auer metal). An alloy of misch metal with about 30% of other metals, chiefly iron.
Containers: Wooden or plastic boxes.
Uses: Pyrophoric alloy produces sparks on gentle friction and is used as the tip in various kinds of lighters, e.g., gas lighters, pocket lighters, etc.
Shipping regulations: None.*

pyrophosphoric acid $H_4P_2O_7$.
Properties: A viscous, syrupy liquid which tends to solidify on long standing at ordinary temperature. When diluted with water it is rapidly converted into orthophosphoric acid. Soluble in water. M.p. 61°C.
Derivation: By heating disodium phosphate, dissolving in water, and precipitating with a soluble lead compound. The lead in turn is precipitated with H_2S, the solution filtered off and concentrated in vacuo.
Grades: Technical.
Containers: Carboys; casks.
Use: Chemical catalyst; manufacture of organic phosphate esters.

pyrophyllite (agalmatolite) $Al_2Si_4O_{10}(OH)$. A natural hydrous aluminum silicate, found in metamorphic rocks.
Properties: Color white, green, gray, brown; luster pearly to greasy; good micaceous cleavage; sp.gr. 2.8-2.9; hardness 1-2. Similar to talc.
Occurrence: North Carolina, California; Newfoundland; Japan.
Use: Ceramics, insecticides, slate pencils, substitute for talc.

pyroracemic acid. See pyruvic acid.

"Pyroset" DO Fire Retardant." [57] This product is a water-soluble organic compound that is applied with phosphoric acid to provide a high degree of fire retardancy to cellulosic fabrics. The finish is durable to repeated commercial dry cleanings.

"Pyroset" Fire Retardant N-2. [57]
Resin used principally to produce a durably stiff, fire retardant finish on nylon.

pyrosulfuric acid. See sulfuric acid, fuming.

pyrosulfuryl chloride (disulfuryl chloride) $S_2O_5Cl_2$.
Properties: Colorless, mobile; very refractive, fuming liquid; sp.gr. 1.819; b.p. 146°C; m.p. −38°C; refractive index (n 19/D) 1.449.
Containers: Carboys.
Use: Synthetic processes.
Caution: Decomposes violently with water to sulfuric and hydrochloric acids.
Shipping regulations: Corrosive liquid. White label.*

pyrotartaric acid (methylsuccinic acid) $HOOCCH(CH_3)CH_2COOH$.
Properties: White or yellowish crystals. Soluble in water, alcohol, and ether.
Constants: Sp.gr. 1.4105; m.p. 111-112°C.
Derivation: By distilling tartaric acid.
Method of purification: Crystallization.
Grades: Technical.
Containers: Glass bottles; wooden kegs.
Use: Organic synthesis.
Shipping regulations: None.*

pyrotartaric acid, normal. See glutaric acid.

pyrovanadic acid. See vanadic acid.

pyroxenes. A group of silicate minerals with similar physical properties, chemical composition, and atomic structure. The group is characterized by prismatic cleavage and a silica:oxygen ratio of 1:3. Pyroxenes are common in igneous and metamorphic rocks.

pyroxylin (soluble guncotton). A nitrocellulose (q.v.) consisting chiefly of cellulose tetranitrate $[C_{12}H_{16}O_6(NO_3)_4]_n$.
Properties: Light yellow matted filaments; decomposed by light when kept in closed containers. Exceedingly flammable! Soluble in a mixture of ether-alcohol (3:1) (U.S.P. grade), in acetone and glacial acetic acid, and is precipitated from solution by water.
Grades: U.S.P. XVI; technical.
Containers: Cartons, protected from light. Must be loosely packed. Also bottles.
Uses: In making collodion.
Shipping regulations: (solid) Flammable solid, yellow label; (solution) flammable liquid, red label.*

pyrrhotine. See pyrrhotite.

pyrrhotite (magnetic pyrites, pyrrhotine) FeS. A natural iron sulfide. Frequently has a deficiency in iron. May contain small amounts of nickel, cobalt, manganese and copper.
Properties: Color brownish bronze; streak black; luster metallic; slightly magnetic; hardness 4; sp.gr. 4.6.
Occurrence: Tennessee, Pennsylvania; Europe; Canada.
Uses: Ore of iron; manufacture of sulfuric acid; sometimes an ore of nickel; possible source of sulfur.

pyrrobutamine phosphate
$ClC_6H_4CH_2C(C_6H_5):CHCH_2NC_4H_8 \cdot 2H_3PO_4$. (1-[4-(para-Chlorophenyl)-3-phenyl-2-butenyl] pyrrolidine diphosphate).
Properties: A light cream to off-white powder with a slight odor and a bitter taste. It is soluble in water, slightly soluble in alcohol and almost insoluble in chloroform and ether. The melting range (U.S.P.) is 127° to 131°C.
Grade: N.N.D.
Use: Medicine.

"Pyrrolazote." [327] Trademark for pyrathiazine-10-[2-(1-pyrrolidyl) ethyl] phenothiazine hydrochloride ($C_{18}H_{21}ClN_2S$).
Properties: M.p. 200°C; soluble in water, alcohol; insoluble in ether and benzene.
Derivation: Synthetic.
Use: Medicine; antihistamine.

pyrrole $\underline{CHNH(CH)_2CH}$.
Properties: Yellowish or brown liquid oil with a burning, pungent taste; odor similar to chloroform; readily polymerizes by the action of light and turns brown. The pure material is colorless if every trace of oxygen is kept away. Soluble in alcohol, ether, and dilute acids; insoluble in water and dilute alkalies.
Constants: B.p. 130-131°C; m.p. −24°C; sp.gr. 0.968 (20/4°C); refractive index (n 20/D) 1.5091; flash point (Tag closed cup) 39°C; slightly soluble in water; soluble in most organic chemicals.
Derivation: By the fractional distillation of bone-tar or bone-oil with sulfuric acid.
Method of purification: Conversion into the potassium compound (C_4H_4NK), washing with ether and treatment with water, followed by drying and distillation.
Grades: Technical.
Containers: Iron drums.
Use: Manufacture of drugs.
Shipping regulations: None.*

pyrrole tetraiodide. See iodole.

pyrrolidine C_4H_9N.
Properties: Colorless to pale yellow liquid; penetrating amine-like odor; poisonous. Sp.gr. 0.8660 (20/20°C); m.p. −60°C; b.p. 87°C (760 mm); refractive index 1.4425 (n 20/D).
Grades: 95% min purity.
Uses: Intermediate for pharmaceuticals, fungicides, insecticides, rubber accelerators, citrus decay control; cure for epoxy resins; inhibitor.
Containers: 1-, 5-gal cans; 55-gal drums.
Shipping regulations: Flammable liquid. Red label.*

2-pyrrolidinecarboxylic acid. See proline.

3-pyrrolidinopropiophenone hydrochloride
$C_6H_5COCH_2CH_2NC_4H_8 \cdot HCl$. A white to yellow crystalline powder; m.p. 160-166°C.

2-pyrrolidone $\underline{CH_2CH_2CH_2C(O)NH}$. Made from acetylene and formaldehyde by high pressure synthesis.

10-[2-(1-pyrrolidyl) ethyl] phenothiazine hydrochloride. See pyrathiazine hydrochloride.

pyruvic acid (alpha-ketopropionic acid; acetylformic acid; pyroracemic acid) $CH_3COCOOH$. An acid which is a fundamental intermediate in protein and carbohydrate metabolism in the cell. As the sodium salt, it is known to activate nitrogen fixation at least under artificial conditions.
Properties: Liquid with odor resembling acetic acid; m.p. 11.8°C; miscible with water, ether, and alcohol.
Derivation: Dehydration of tartaric acid by distilling with potassium acid sulfate.
Use: Biochemical research.

pyruvic alcohol. See acetol.

pyruvic aldehyde (methyl glyoxal) CH_3COCHO.
Properties: Supplied commercially as approximately 30% aqueous solution; sp.gr. 1.20 (20/20°C); 10 lbs/gal (20°C).
Containers: 1-gal cans; 5-, 55-gal drums.
Uses: Organic synthesis, as of complex chemical compounds such as pyrethrins; for the tanning of glove leathers.

pyrvinium chloride $C_{26}H_{28}ClN_3 \cdot 2H_2O$. 6-Dimethylamino-2-[2-(2,5-dimethyl-1-phenyl-3-pyrryl)vinyl]-1-methylquinolinium chloride dihydrate. A red powder; m.p. 250°C (dec). Slightly soluble.
Grade: N.N.D.
Use: Medicine.

*See "I.C.C. Shipping Regulations," page xiii.
Reference numbers refer to name of manufacturer. See "List of Manufacturers," page v.

"Q302.4." [233] Trademark for commercial divinylbenzene and related compounds.
Derivation: Diethylbenzene by-products of styrene manufacture.
Use: Polymerization monomer.

Q-lure. 4-(para-acetoxyphenyl)-2-butanone. A lure for the male melon fly.

"QO." [224] Trademark for a line of furan chemicals and derivatives.

"Quadrafos." [164] Trademark for a sodium phosphate glass commonly called sodium tetraphosphate (q.v.), a linear polymer containing a minimum of 63.5% P_2O_5. Soluble in water; insoluble in organic solvents.
Grades: Beads; granular; powder.
Containers: Bags; fiber drums. Beads are also packed in 20-lb cloth "dissolver" bags.
Uses: In general, as a sequestering, dispersing and deflocculating agent; specifically, to sequester alkaline earth and heavy metal ions; to prevent scaling and corrosion of pipes; to soften water; to disperse pigments and clays in paper making and oil well drilling.

"Quadrex." [309] Trademark for high potency vitamin-containing solids for fortifying foods and feeds with vitamins.

"Quadrol." [203] Trademark for N, N, N', N'-tetrakis (2-hydroxypropyl)-ethylenediamine. It is a colorless viscous liquid with a boiling point of 190°C at 1 mm. It has a low order of toxicity and is used as a humectant, chelating agent, plasticizer, and intermediate in resin and adhesive manufacture.

"Quakeral." [224] Trademark for furfural (q.v.).

"Quaker Blue." [204] Trademark. Highly concentrated sour liquid blue for finished work.

quaker buttons. See nux vomica.

"Quaker Improved Chlorinated Lime." [204] Trademark. A chloride of lime with 30-33% available chlorine content.

"Quakersol." [319] Trademark for a proprietary alcohol-type solvent used as a substitute for denatured alcohols in the manufacture of nitrocellulose and other lacquers, polishes, soldering fluxes, disinfectants, paint removers, cements, waterproofing materials, insecticides and cleaning compounds.

quartering, in sampling. Form of hand sampling used to obtain a representative sample of a granular or powdered solid. It is carried out by thoroughly mixing the solid and separating into four piles, rejecting the two opposite piles, mixing the remaining two parts again and repeating the procedure until a sample of the desired size is obtained.

quartz SiO_2. Crystallized silicon dioxide. It is the main constituent of sandstone, one of the two principal constituents of granite, pegmatite, and some other igneous and metamorphic rocks, and is the most frequent gangue mineral in mineral veins. It is very resistant to weathering and is left as quartz sand when rocks containing quartz are disintegrated.
Properties: Color variable; luster vitreous; hardness 7; sp.gr. 2.65; insoluble in acids except hydrofluoric; only slightly attacked by solutions of caustic alkali; piezoelectric and pyroelectric.
Varieties: The most important varieties are agate; chalcedony; chert; flint; opal; and rock crystal. See under individual entries for uses.
Derivation: Synthetic crystals of good size and purity are now grown by mass production methods under very carefully regulated conditions of temperature and concentration.
Uses of synthetic crystals: Electronic components, for piezoelectric control in filters, oscillators, frequency standards, wave filters, radio and TV components.

quartz, fused. Pure silica that has been fused so as to yield a glass-like material on cooling. Used for apparatus and equipment where its high melting point, ability to withstand large and rapid temperature changes, chemical inertness and transparency (including ultra violet light) are valuable and useful.

quartzite (quartz rock). A metamorphic or sedimentary rock composed almost wholly of quartz grains cemented by silica. Color white, gray, or brownish.
Occurrence: Pennsylvania, Maryland, Virginia, Wisconsin.
Use: Road material; tube mill linings; tower packings; glass sand; silica brick.

quartz rock. See quartzite.

quassia (bitter ash; bitterwood).
Derivation: The wood or bark of Picrasma excelsa or Quassia amara; very bitter taste. White to bright yellow chips or shavings.
Occurrence: Warmer regions of the world.
Chief constituents: Quassin, quassol,

picrasmin.
Grades: Technical, as chips.
Containers: Bags.
Uses: Decoction or tincture as a fly poison; surrogate for hops; medicine (powerful bitter); hair lotion.
Shipping regulations: None.*

quassin $C_{22}H_{30}O_6$. Extract of quassia.
Properties: Colorless crystals; m.p. 205°C; odorless but very bitter in taste. Slightly soluble in water; soluble in organic solvents.
Use: To denature alcohol.

quat. Slang for quaternary ammonium salt.

quaternary ammonium salt. A type of organic nitrogen compound in which the molecular structure includes a central nitrogen atom joined to four organic groups as well as to an acid radical of some sort. Octadecyldimethylbenzyl ammonium chloride and hexamethonium chloride are examples of this type of compound. In addition to the ordinary substituted ammonium compounds, pentavalent nitrogen ring compounds, such as lauryl pyridinium chloride, are also considered quaternary ammonium compounds. These are all cationic surface-active compounds and tend to be adsorbed on surfaces.
Uses: Disinfectant, cleanser and sterilizer; cosmetics (deodorants, dandruff removers, emulsion stabilizers); fungicides; mildew control; to increase affinity of dyes for film in photography; coating of pigment particles to improve dispersibility; to increase adhesion of road dressings and paints.
See also detergents, synthetic.

"Quaternary O." [219] Trade name for alkyl imidazolinium chloride ($C_{17}H_{35}C_3H_4N_2R^1R^2Cl$).
Properties: 80% active viscous oil, soluble in water and most organic solvents; unstable at pHs over 7.
Uses: Cationic wetting agent, foamer, penetrant for strong acids, salt solutions, solvents; germicide. Corrosion inhibitor in acids; emulsion breaking.

para-**quaterphenyl** $C_6H_5C_6H_4C_6H_4C_6H_5$.
Properties: Crystals; m.p. 316-318°C; b.p. 428°C (18 mm).
Grade: Purified.
Use: As primary fluor or as wave length shifter in solution scintillators.

quebrachine. See yohimbine.

quebracho (aspidosperma).
Derivation: The bark of Aspidosperma quebrachoblanco and quebracho-colorado from Argentina.
Chief constituents: Aspidospermine, tannin, quebrachine.
Grades: Technical.
Containers: Bags; multiwall paper sacks.
Uses: Quebracho-blanco in medicine for its tannin and alkaloid content; quebracho-colorado in tanning and dyeing for its tannin content; has also been used as an additive to well drilling muds.

quebracho extract.
Derivation: From the wood of Aspidosperma quebracho and Quebracho lorentzi, which is imported into the United States in logs.
Grades: Liquid: 35-37% tannin. Solid: 65% tannin.
Containers: Extract: 400-, 500-lb wooden barrels; tank-cars. Extract (powdered): 110-lb bales.
Uses: Tanning industry; dyeing; ore flotation; oil well drilling muds.
Shipping regulations: None.*

Queensland asthma weed. See euphorbia.

queen's metal. A jeweler's alloy composed approximately of 9 parts tin and 1 part of antimony, and small proportions of copper, zinc, lead, or bismuth.

queen's root. See stillingia.

"Quelicin Chloride." [3] Trademark for succinylcholine chloride (q.v.).

quercetin (meletin, quercetinic acid, tetrahydroxyflavanol, flavin) $C_{15}H_{10}O_7 \cdot 2H_2O$.
Properties: Brown or yellow crystalline powder. Soluble in alkalies; slightly soluble in water.
Constants: Dihydrate loses water at 95-97°C; anhydrous form melts with decomposition at 313-315°C.
Derivation: Action of dilute sulfuric acid on quercitrin.
Grades: Technical.
Containers: Glass bottles; tins.
Use: Dyeing.
Shipping regulations: None.*

quercetinic acid. See quercetin.

quercetin-3-rutinoside. See rutin.

quercimetin. See quercitrin.

quercite. See quercitol.

quercitol (acorn sugar; quercite; pentahydroxycyclohexane) $C_6H_7(OH)_5$.
Properties: Colorless crystals; sweet taste. Soluble in water; very slightly soluble in alcohol; insoluble in ether.
Constants: Sp.gr. 1.5845; m.p. 234°C.
Derivation: By extraction of acorn meal with water.
Method of purification: Crystallization.
Grades: Technical.
Containers: Glass bottles.
Use: Medicine.
Shipping regulations: None.*

quercitrin (quercimetin, quercitrinic acid) $C_{21}H_{20}O_{11}$.
Properties: Yellow glucosidal, crystalline powder. Soluble in acids, alkalies and amyl alcohol; slightly soluble in alcohol, ether and water.
Constants: M.p. 182-185°C.
Derivation: By extraction of the bark of the black-oak, Quercus tinctoria.
Grades: Technical.

*See "I.C.C. Shipping Regulations," page xiii.
Reference numbers refer to name of manufacturer. See "List of Manufacturers," page v.

Containers: Wooden kegs.
Use: Manufacture of quercetin.
Shipping regulations: None.*

quercitrinic acid. See quercitrin.

"Questex." [334] Trademark for ethylenediamine tetra-acetic acid (EDTA) and derivatives, a group of polyamino-acid based organic sequestering agents which complex or chelate multivalent, metallic cations (such as calcium, magnesium, copper and iron) into stable, coordinated anionic complexes.
"Questex" 4H: Purified form of anhydrous EDTA.
"Questex" 25W: Disodium dihydrogen EDTA dihydrate, offered in purified fine crystal form.
"Questex" 45W Crystal: Tetrahydrate of tetra-sodium-EDTA. This granular, crystalline form is unusually free flowing, non-dusting and readily soluble.
"Questex" 4S: Anhydrous granular form of tetrasodium-EDTA.

quicklime. See calcium oxide.

quicksilver. See mercury.

quicksilver, horn. See calomel, native.

"Quik-Gel." [236] Trademark for a peptized high-swelling bentonite. Used as thickener and suspending agent in fluids for core hole drilling.
Containers: 50-lb multiwall paper bags.

quillaia. See quillaja.

quillaja (soap bark; quillaia; panama bark; china bark; murillo bark).
Derivation: The dried bark of Quillaja saponaria from Bolivia, Peru, Chile, etc.
Chief constituents: Quillajac acid; quillaja sapotoxin and tannin.
Grades: Technical.
Containers: Whole: Bags and bales of variable size. Cut, crushed: 175-, 200-lb barrels; bags of variable size.
Uses: Medicine; manufacture of sapotoxin, saponin and quillajac acid; in the soft-drink and shampoo liquid industries as a foam producer; emulsifiant for oils; balsams and resins.
Shipping regulations: None.*

"Quilon." [28] Trademark for a Werner type chromium complex in isopropanol.
Properties: Dark green liquid with an alcoholic odor, completely soluble in water. and many short-chain aliphatic alcohols; approximately 30% solution of stearato chromic chloride in isopropanol; sp.gr. 0.94; b.p. 82.4°C.
Containers: 6 1/2-gal carboys; 55-gal drums.
Uses: As a water repellent and sizing treatment of paper and other cellulosic materials; for treatment of felt hats, leather, wool fabrics, hydrophobic fibers, and siliceous and other negatively charged surfaces; as an anti-blocking or release agent; for insolubilizing various water-soluble or swellable coatings.

quinacrine $C_{23}H_{30}ClN_3O$. The antimalarial drug brought to prominence during World War II. An alkaloid used in the form of one of its acid combinations. See quinacrine hydrochloride.

quinacrine hydrochloride
$C_{23}H_{30}ClN_3O \cdot 2HCl \cdot 2H_2O$. 3-Chloro-7-methoxy-9-(1-methyl-4-diethylaminobutylamino) acridine dihydrochloride).
Properties: Bright yellow, crystalline powder; odorless and with a bitter taste. Decomposes at 248-250°C. Soluble in water and alcohol; pH of 1% water solution 4.5.
Derivation: Organic synthesis.
Grade: U.S.P. XVI.
Use: Medicine.

quinaldine (chinaldine; alpha-methylquinoline) $C_9H_6NCH_3$.
Properties: Colorless oily liquid; odor of quinoline; darkens to reddish-brown in air. Soluble in alcohol, ether, and chloroform; soluble in water. B.p. 246-247°C.
Derivation: (a) By the treatment of aniline and paraldehyde with hydrochloric acid and heat. (b) From coal-tar.
Grades: Technical.
Containers: Iron drums.
Use: Manufacture of dyes, pharmaceuticals, fine organic chemicals.
Shipping regulations: None.*

quinaphthol (chinaphthol; quinine-beta-naphthol-alpha-sulfonate)
$C_{20}H_{24}N_2O_2 \cdot (OHC_{10}H_6 \cdot SO_3H)_2$.
Properties: Yellow, crystalline powder; containing 42% of quinine; butter taste. Moderately soluble in hot water or alcohol; insoluble in cold water.
Constants: M.p. 185-186°C.
Derivation: By the interaction of quinine and beta-naphtholsulfonic acid.
Grades: Technical.
Containers: Glass bottles.
Use: Medicine.
Shipping regulations: None.*

quince seed. See cydonia.

"Quindex." [74] Trademark for solubilized form of copper 8-quinolinolate. Contains 1.8% copper. Used when non-mercurial fungicide is required.

"Quindo." [438] Trademark for quinacridone pigments. Used in paints, printing inks, and plastics.

quinhydrone $C_6H_4O_2C_6H_4(OH)_2$.
Properties: Dark green crystals. Slightly soluble in water; soluble in alcohol, ether, hot water, ammonia.
Constants: M.p. 171°C; sp.gr. 1.40.
Derivation: Oxidation of hydroquinone with sodium dichromate.
Method of purification: Recrystallization.
Grades: Reagent; technical.
Containers: Glass bottles.
Use: Quinhydrone electrode for pH determination.
Shipping regulations: None.*

*See "I.C.C. Shipping Regulations," page xiii.
Reference numbers refer to name of manufacturer. See "List of Manufacturers," page v.

quinic acid (chinic acid)
$C_6H_7(OH)_4COOH \cdot H_2O$. Hexahydro-1,3,4,5-tetrahydroxybenzoic acid.
Properties: White, transparent crystals; very acid taste. Soluble in water, alcohol, and glacial acetic acid; insoluble in ether.
Constants: Sp.gr. 1.637; m.p. 162°C; b.p. decomposes.
Derivation: From cinchona bark.
Method of purification: Crystallization.
Grades: Technical.
Containers: Glass bottles.
Use: Medicine.
Shipping regulations: None.*

quinidine (chinidine; conchinine; beta-quinine; pitayine) $C_{20}H_{24}N_2O_2$.
Properties: Colorless, lustrous, crystalline alkaloid; efflorescing on exposure to air. Soluble in chloroform, alcohol, and ether; very slightly soluble in water. M.p. 171.5°C.
Derivation: By the extraction of certain species of cinchona bark; also by synthetic methods.
Method of purification: Recrystallization.
Grades: Technical.
Containers: Vials; 1-, 5-, 25-, 50-, 100-oz tins.
Use: Medicine (as the alkaloid or as the various salts).
Shipping regulations: None.*

quinidine gluconate $C_{20}H_{24}N_2O_2 \cdot C_6H_{12}O_7$.
Properties: White, odorless powder with very bitter taste. Freely soluble in water; only slightly soluble in alcohol.
Derivation: It is the gluconate of an alkaloid obtained from various species of Cinchona and their hybrids or from Remijia pedunculata, or prepared from quinine.
Grade: N.F. XI.
Use: Medicine.

quinidine sulfate $(C_{20}H_{24}N_2O_2)_2 \cdot H_2SO_4 \cdot 2H_2O$.
Properties: Fine, needle-like, white crystals, frequently cohering in masses; odorless with a very bitter taste and darkens on exposure to light. Soluble in water, alcohol, and chloroform; almost insoluble in ether.
Grade: U.S.P. XVI.
Containers: Tins; drums.
Use: Medicine.

quinine $C_{20}H_{24}N_2O_2 \cdot 3H_2O$. One of the most important natural alkaloids.
Properties: Bulky, white, amorphous powder or crystalline alkaloid; very bitter taste; odorless and levo-rotatory. Soluble in alcohol, ether, chloroform, carbon disulfide, ligroin, oils, glycerol, alkalies and acids (with formation of salts); very slightly soluble in water.
alkaloid, anhydrous $C_{20}H_{24}N_2O_2$.
M.p. 174.9°C.
alkaloid, hydrated $C_{20}H_{24}N_2O_2 \cdot 3H_2O$.
M.p. 57.0°C.
bisulfate $C_{20}H_{24}N_2O_2 \cdot H_2SO_4 \cdot 7H_2O$.
M.p. about 160.0°C.
hydrobromide $C_{20}H_{24}N_2O_2 \cdot HBr \cdot H_2O$
M.p. (anhydrous form) 152-200°C.
hydrochloride $C_{20}H_{24}N_2O_2 \cdot HCl \cdot 2H_2O$.
M.p. (anhydrous form) 156-160°C.
salicylate $C_{20}H_{24}N_2O_2 \cdot C_7H_6O_3 \cdot H_2O$.
M.p. 183-187°C.
Derivation: Finely ground cinchona bark mixed with lime is extracted with hot high-boiling paraffin oil. The solution is filtered, shaken with dilute sulfuric acid, the latter neutralized hot with sodium carbonate and on cooling quinine sulfate crystallizes out. The sulfate is treated with ammonia, the alkaloid being obtained.
Method of purification: Precipitation as tartrate from its solution by means of Rochelle salt.
Impurities: Other cinchona alkaloids.
Grades: Technical; N.F. XI (several of the salts).
Containers: Vials; up to 1000-oz drums.
Uses: Medicine (antimalarial), as the alkaloid or as the acetate, albuminate, arsenate, benzoate, camphorate, citrate, gallate, glycerophosphate, lactate, phenolate, salicylate, tannate, tartrate, etc. N.F. XI salts are the dihydrochloride, hydrochloride, phosphate, and sulfate. Most of these salts are soluble in water.
Shipping regulations: None.*

beta-**quinine**. See quinidine.

quinine acid sulfate. See quinine.

quinine bisulfate. See quinine.

quinine carbacrylic resin. The quinine salt of a polyacrylic carboxylic acid resin. Contains 1.85% of quininium ion.
Properties: Buff, odorless, tasteless, free-flowing amorphous, granular solid. Practically insoluble in dilute acids and alkalies, alcohol, ether and water.
Use: Medicine.

quinine carbonate. See aristoquin.

quinine dichloride. See quinine dihydrochloride.

quinine dihydrochloride (quinine dichloride) $C_{20}H_{24}N_2O_2 \cdot 2HCl$.
Properties: White powder, intensely bitter taste; odorless; affected by light. Contains 81.63% quinine. Very soluble in water (1 g in 0.6 cc) and alcohol (1 g in 12 cc); slightly soluble in chloroform and ether.
Derivation: (a) By passing hydrochloric acid gas over dry quinine. (b) Decomposition of quinine bisulfate with barium chloride.
Grades: N.F. XI.
Shipping regulations: None.*

quinine ethylcarbonate $C_{23}H_{28}O_4N_2$.
Properties: White, odorless needles, darken on exposure to light; m.p. 89-91°C; slightly soluble in water, soluble in alcohol, ether or chloroform.
Use: Medicine.

quinine hydrobromide. See quinine.

quinine hydrochloride. $C_{20}H_{24}N_2O_2 \cdot HCl \cdot 2H_2O$.
Properties: White silky needles; odorless; bitter taste; effloresces in warm air. Soluble in water; more soluble in alcohol

and chloroform.
Grade: N.F. XI.
Use: Medicine.

quinine-beta-**naphthol**-alpha-**sulfonate.** See quinaphthol.

quinine phosphate $(C_{20}H_{24}N_2O_2)_3\cdot 2H_3PO_4\cdot 5H_2O$.
Properties: Small, white crystals or as white crystalline powder. Odorless with bitter taste. Affected by light. Saturated solution is acid to litmus. Very slightly soluble in water; slightly soluble in boiling alcohol.
Grade: N.F. XI.
Use: Medicine.

quinine salicylate. See quinine.

quinine salts. See quinine.

quinine sulfate $(C_{20}H_{24}N_2O_2)_2\cdot H_2SO_4\cdot 2H_2O$.
Properties: White needle-like crystals; odorless; very bitter taste; turns brown in the light; a saturated solution is neutral or slightly alkaline to litmus paper; m.p. 205°C; if exposed to damp air it may take up 5 mols of water. Loses water of crystallization at about 100°C. Soluble in hot water and in alcohol. Slightly soluble in chloroform and ether.
Derivation: The sulfate of an alkaloid usually obtained from the cinchona tree (see quinine).
Grade: N.F. XI.
Containers: See quinine.
Use: Medicine.

quinine-urea hydrochloride (urea-quinine) $C_{20}H_{24}N_2O_2\cdot HCl\cdot CO(NH_2)_2\cdot HCl\cdot 5H_2O$.
Properties: White, granular powder, or colorless; translucent prisms. Bitter taste, odorless. Contains not less than 58% anhydrous quinine. Soluble in water and strong alcohol.
Derivation: By dissolving quinine hydrochloride in hydrochloric acid, filtering, adding urea, heating and crystallizing.
Purification: Recrystallization.
Containers: 1-oz bottles.
Grade: N.F. XI.
Use: Medicine.
Shipping regulations: None.*

quininic acid $CH_3OC_9H_5NCO_2H$.
Properties: Yellow prisms. Slightly soluble in alcohol; very slightly soluble in water and ether. M.p. decomposes at 280°C; b.p. sublimes in part.
Grades: Technical.
Containers: Glass bottles.
Use: Organic synthesis.
Shipping regulations: None.*

quinoidine (chinoidine).
Properties: Brownish-black lustrous mass; resinous appearance; conchoidal fracture; very bitter taste. It is a mixture of the amorphous alkaloids remaining in the solution from the extraction of cinchona bark, after the crystallizable alkaloids have been removed. Soluble in dilute acids, alcohol and chloroform. M.p. softens below 100°C.
Grades: Technical.
Containers: 70-, 140-lb drums.
Uses: Medicine, either as such, or as the borate, citrate, hydrochloride, sulfate or tannate.
Shipping regulations: None.*

quinol. See hydroquinone.

quinoline (chinoline; leucoline)
CHCHCHCHCCNCHCHCH.
Properties: An aromatic nitrogen compound, occurring in coal-tar and obtained from it, but more frequently by synthesis; highly refractive, colorless liquid; darkens with age; absorbs water from the air; peculiar, characteristic odor. Keep well stoppered. Soluble in water, alcohol, ether and carbon disulfide. Sp.gr. 1.0899; m.p. −15°C; b.p. 238°C.
Derivation: By treatment of aniline and nitrobenzene with glycerol and sulfuric acid and heat.
Method of purification: Rectification.
Grades: Pure, from cinchomine; synthetic; C.P.
Containers: Glass bottles; 5- and 55-gal drums; tank cars.
Uses: Medicine; preserving anatomical specimens; manufacture of quinosol, niacin, and copper-8-quinolinolate.
Shipping regulations: None.*

8-quinolinol. See 8-hydroxyquinoline.

8-quinolinol sulfate. See 8-hydroxyquinoline sulfate.

quinone (1,4-benzoquinone; chinone) $C_6H_4O_2$.
Properties: Yellow crystals; characteristic, irritating odor. Soluble in alcohol, ether and alkalies; slightly soluble in hot water. Sp.gr. 1.307; m.p. 115.7°C; b.p. sublimes; volatile with steam, being in part decomposed.
Derivation: By the oxidation of aniline with chromic acid, extraction with ether and distillation of the latter.
Method of purification: Steam distillation.
Grades: Technical.
Containers: 1-lb bottles; 25-lb boxes; 100-lb barrels.
Use: Manufacture of dyes and hydroquinone.
Shipping regulations: None.*

quinone oxime dyes. See nitroso dyes.

quinophthalone $NC_{10}H_6CCOHC_6H_4CO$.
Properties: Yellow powder.
Grade: Technical.
Use: Intermediate.

quinosol (chinosol) $C_9H_6NOSO_3K\cdot H_2O$.
Properties: Yellow, crystalline powder, weak saffron odor, burning taste, incompatible with alkalies, iron and mercury compounds. Does not coagulate albumin. M.p. 175-177.5°C. Soluble in water; difficultly soluble in alcohol; insoluble in ether.
Use: Medicine.
Shipping regulations: None.*

*See "I.C.C. Shipping Regulations," page xiii.
Reference numbers refer to name of manufacturer. See "List of Manufacturers," page v.

quinoxaline (1,4-benzodiazine; benzo-para-diazine) $C_8H_6N_2$ (bicyclic). An organic base.
Properties: Colorless crystalline powder; m.p. 30°C; b.p. 229°C; soluble in water and organic solvents.
Use: Organic synthesis.

N-(2-quinoxalinyl) sulfanilamide. See sulfaquinoxaline.

"Quixol." [319] Trademark for a proprietary alcohol-type solvent used as a solvent, cleaner, drying agent, shellac thinner, and as a lamp and stove fuel.

"Quotane." [71] Trademark for dimethisoquin.

q.v. Abbreviation for which see.

r. Abbreviation for roentgen.

"R-2" Crystals. [58] Trademark for a reaction product of carbon disulfide with 1, 1'-methylenedipiperidine.
Properties: Light brown, coarse crystalline solid; sp.gr. 1.11; m.p. 59°C min; soluble in acetone, benzene, and naphtha.
Containers: 50-lb fiber drums.
Uses: An ultra accelerator for latex and fast-curing cements.

Ra. Symbol for radium.

racemic acid (para-tartaric acid; tartaric acid, inactive; DL-tartaric acid) $HOOC(CHOH)_2COOH$. Not a mixture of D- and L- acids, but a true crystalline variety.
Properties: Transparent, colorless crystals. Soluble in water; slightly soluble in alcohol. Sp.gr. 1.697; m.p. 205-206°C.
Derivation: A by-product of the manufacture of tartaric acid.
Method of purification: Crystallization.
Grades: Technical.
Containers: Glass bottles; boxes; wooden kegs.
Use: Organic synthesis.
Shipping regulations: None.*

racemic ephedrine. See racephedrine.

racemic ephedrine hydrochloride. See racephedrine hydrochloride.

racemic ephedrine sulfate. See racephedrine sulfate.

racephedrine (racemic ephedrine; dℓ-ephedrine) $C_{10}H_{15}NO$. For optically active form, see ephedrine.
Properties: Crystals; m.p. 79°C; soluble in water, alcohol, chloroform, and oils.
Derivation: Synthetic.
Use: Medicine.

racephedrine hydrochloride (ephetonon; racemic ephedrine hydrochloride) $C_{10}H_{15}NO \cdot HCl$. For optically active form, see ephedrine hydrochloride.
Properties: White crystals; m.p. 187-188°C; soluble in water; slightly soluble in 95% alcohol; insoluble in ether; neutral to litmus; pH about 6; affected by light.
Derivation: Synthetic.
Grade: N.F. XI.
Use: Medicine.

racephedrine sulfate (racemic ephedrine sulfate) $(C_{10}H_{15}NO)_2 \cdot H_2SO_4$.
Properties: Crystals; m.p. 247°C; soluble in water and alcohol; neutral to litmus; pH about 6.
Use: Medicine.

R acid. See 2-amino-3-naphthol-6-sulfonic acid; 2-naphthol-3, 6-disulfonic acid.

2R acid. See 2-amino-8-naphthol-3, 6-disulfonic acid.

rad. The unit of absorbed dose of energy from ionizing radiations. It is equal to 100 ergs/gram. See roentgen.

"Radapon." [233] Trademark for a grass killer, based on dalapon.

raddle. See reddle.

radiation.
1. The process by which energy in forms such as light, heat, x-rays or electricity is transmitted through space without the presence of movement of matter in or through this space. The term is also applied to these forms of energy themselves, i.e., one speaks of light radiation, heat radiation, or electrical radiations as in transmission of electrical energy in radio and television.
2. The rapidly moving atomic or subatomic particles (alpha particles, protons, electrons, neutrons) encountered in nuclear processes such as radioactivity. The words soft and hard are often used to describe in a general way the ability of radiation to pass through matter, hard radiation being penetrating and soft radiation being easily absorbed.

radiation biology. A general name for the broad fields of study of the effects of radiation, especially ionizing radiation, on living systems. In particular it includes the study of the production of mutations and their genetics, radiation damage and repair of living systems, inhibitions of growth, radiation therapy, and the treatment of radiation sickness.

radiation catalysis. The activation or speeding up of chemical or physical changes by exposure to the proper amount of radiation. For example, polymerization can be brought about by radiation without using heat, pressure or chemical catalysts. Thus the irradiation of ordinary polyethylene plastic results in a product which is more heat resistant, more elastic, less soluble, harder, more dense and stronger than the original material (see "Irrathene"). The advantages of radiation catalysis are that no foreign material need be added to the reaction mixture and that the catalyst can not be poisoned and made inactive.

radiation chemistry. The study of chemical changes which are brought about by ionizing radiations such as gamma rays, neutrons, protons, electrons, etc.

radiation damage. The general term applied to changes in physical and chemical properties of materials caused by irradiation (bombardment) with neutrons, gamma rays, etc. Radiation exposure can cause compounds to decompose, plastics to become soft or brittle, metals to change in mechanical properties such as strength and toughness and in their ability to conduct heat and electricity, etc. A study of radiation damage to structural materials is of especial importance since the development of nuclear power plants, where very intense radiations are encountered.

radical. A group of atoms which occurs in the molecules of a number of compounds, and which remains unchanged through many chemical reactions. Ethyl, C_2H_5 (valence 1), and sulfate, SO_4 (valence 2), are typical organic and inorganic radicals, respectively.

radio-. A prefix denoting radiation or radioactivity. It is used in designating radioactive substances, as radioisotopes; radioactive elements, as radiocarbon (q.v.), radiogold, radiobarium, etc; and substances containing them, as radiocompounds, radiocolloids, radiochemicals, etc.

radioactive decay. Same as radioactivity. For the radioactive decay series, see actinium decay series, neptunium decay series, thorium decay series, uranium decay series.

radioactivity. The phenomenon of spontaneous nuclear transformation. The energy of the process is carried away by particulate or electromagnetic radiations, alpha, beta, or gamma radiation. Two fundamental laws apply to the process. One states that the rate of emission of radiation from a radioactive element is proportional to the amount of the element, and the other states that when particles are emitted the nuclear charge and mass are changed by the charge and mass carried off by the particle. Thus radium-226 undergoes radioactive decay by the emission of an alpha particle, and the new product is radon-222. Radioactivity is not affected by the physical state or chemical combination of the element.

The radioactivity of a particular nuclide is characterized by the nature of the radiations, their energy, and the half-life of the process, i.e., the time required for the activity to decrease to one-half of the original. Half-lives in different cases vary from microseconds to billions of years. Substances which are radioactive as found in nature are said to exhibit natural radioactivity. Induced, or artificial radioactivity may be brought about with almost every element by irradiation with neutrons in a nuclear reactor, or by charged particles from an accelerator.

Amounts of radioactive material are usually expressed in units of activity, the rate of the radioactive decay. The accepted unit is the curie and its metric fractions, the milli- and microcurie. A curie is 3.7×10^{10} disintegrations per second. Specific activities may be quoted either as a percentage abundance of the radioactive isotope, or in units of curies per unit weight or curies per mole. A common unit is millicuries per millimole.

radiobiology. A general name for the many areas in the biological sciences in which radioactive tracers are used. In particular, studies on intermediate metabolism in plant and animal forms of life, fertilizer uptake studies, trace nutrients, dispersion of insect hatches, blood flow patterns, transmission of materials through cell walls, and some diagnostic procedures are included.

radiocarbon. See carbon 14.

radiochemistry. That phase of chemistry which deals with the study of radioactive substances. Radiochemistry includes the investigation of the properties of radioactive materials, the study of the behavior of very small quantities of radioactive materials by means of their radioactivity and the use of radioactive materials in the study of chemical problems.

radiocyanocobalamin solution. A solution containing cobalt-60 labeled cyanocobalamin in which a portion of the molecules contain radioactive cobalt in the molecular structure.

Grade: U.S.P. XVI.

Use: Medicine (diagnosis).

radiogold. A radioactive colloidal concentrate of gold-198 (q.v.) in the particle range 0.003 to 0.007μ diameter. Stable to heat but not to autoclaving under pressure; miscible with many solvents and solutions but flocculated by metallic ions.

Grade: N.N.D.

Use: Medicine.

radiography. Determination of the internal characteristics and behavior of ordinarily opaque objects, particularly metals, machinery, and the human body, by passing x-rays or gamma rays into the object, and obtaining shadowgraphs of the transmitted x-rays on a photographic film or fluorescent screen. By this method, irregularities and abnormalities from normal conditions are distinguished. Hidden flaws in metallic castings are detected in this fashion.

radio-iodinated serum albumin. See iodine 131.

radioisotopes. Isotope forms of an element that exhibit radioactivity.

radionuclide. See nuclide.

radiopurity. The term is used in several different ways in connection with the proportion of radioactive material of a particular kind compared with the total radioactivity of a sample. Thus 90% of the carbon 14 in a particular sample of acetic acid may be in acetic acid molecules, while the remaining 10% of the carbon 14 is in an impurity such as propionic acid. In certain cases however the impurity or its radioactivity will not interfere with or enter into the reactions in which the material is used. When such is the case the radiopurity of the sample cited is given as 100%. Various other circumstances arise and must be known in order to be clear as to the meaning of any particular statement regarding radiopurity.

radium Ra. Naturally radioactive element of atomic number 88; group II of the periodic table. The best known radioactive element, until development of nuclear piles. Radium was discovered by Marie and Pierre Curie in France in 1898.

Properties: Brilliant white metal; m.p. about 700°C; b.p. 1140°C; sp.gr. nearly 6.0 (?); half-life 1600 years. Metal shows luminescence; turns black on exposure to air; soluble in water with evolution of hydrogen; causes serious flesh burns.

Derivation: Occurs in minute concentrations (one part in three million) in uranium ores such as pitchblende and carnotite, and has been found more recently in richer ores. Radium-bearing material is now produced as a byproduct in the treatment of uranium ores, when formerly it was the primary product. The general method used for isolating radium is similar to that used by Mme. Curie, and involves extensive fractional crystallization.

Containers: Dry salts are stored in sealed glass tubes opened regularly by experienced workers to relieve pressure. Glass tubes are kept in lead shields.

Caution: Radium and its salts are very poisonous to all life, due to the property of radioactivity (q.v.) which results in its spontaneous breakdown into decomposition products, one of which is radon (gaseous).

Uses: In medical treatments for malignant growths (either radium or radon); industrial radiography (inspecting metal castings); luminous paints; ionization agent in static elimination; as a source of neutrons in searching for oil deposits.

Shipping regulations: Poison, radioactive material. Blue or red label.*

radium bromide $RaBr_2$.

Properties: White crystals, becoming yellow or pink with age; radioactive; powerful corrosive effect on skin and flesh. Do not handle. M.p. 728°C; sublimes at 900°C; soluble in water.

Derivation: Freed from the ores as a bromide mixed with barium bromide.

Method of purification: Fractional crystallization.

Impurities: Barium salts.

Grades: Technical; pure. The purity is determined by the strength of the ionizing power of the salt, i.e., the extent to which it causes air to conduct electricity.

Containers: Glass bottles; sealed tubes enclosed in sheet lead.

Uses: Medicine (chiefly in the treatment of cancer); mixed with calcium sulfide to produce luminescent paint; physical research.

Shipping regulations: Poison, radioactive material. Blue or red label.*

radium carbonate $RaCO_3$.

Properties: Amorphous powder, white when pure, but sometimes yellow, orange, or pink due to impurities. Insoluble in water. Marketed as a mixture with barium carbonate.

Use: Medicine.

Caution: Usual precautions with radioactive material.

Shipping regulations: Poison, radioactive material. Blue or red label.*

radium chloride $RaCl_2$.

Properties: Yellowish-white crystals, becoming yellow or pink on standing; radioactive; powerful corrosive effect on skin and flesh. Do not handle. Soluble in water and alcohol. M.p. 1000°C; sp.gr. 4.91.

Derivation: Freed from the ores as a chloride mixed with barium chloride.

Method of purification: Fractional crystallization.

Impurities: Barium salts.

Grades: Technical; pure. The purity of radium salts is determined by the strength of their ionizing power, i.e., the extent to which they cause air to conduct electricity.

Containers: Glass bottles; sealed tubes enclosed in sheet lead.

Uses: Medicine (chiefly in the treatment of cancer); mixed with calcium sulfide to produce luminescent paint; physical research.

Shipping regulations: Poison, radioactive material. Blue or red label.*

radium emanation. See radon.

radium sulfate $RaSO_4$.

Properties: White crystals when pure, but sometimes yellow, orange, or pink due to impurities. Insoluble in acids and water.

Use: Medicine.

Caution: Care in handling as for all radioactive materials.

Shipping regulations: Poison, radioactive material. Blue or red label.*

radon (earlier names: radium emanation; niton; emanon) Rn. Radioactive element. Atomic number 86; helium group of periodic table.

Properties: Colorless gas; radioactive; density 9.72 g/liter (0°C, 760 mm); soluble in water; can be condensed to colorless transparent liquid (b.p. −61.8°C) and to an opaque, glowing solid. The heaviest gas known.

Derivation: Formed by radioactive decomposition of radium. Radon is obtained by bubbling air through a radium salt solution and collecting the gas plus air. Thoron and actinon are isotopes.
Caution: See radium.
Use: Medical treatment.
Shipping regulations: Poison, radioactive material. Blue or red label.*

raffinate. The portion of an oil that is not dissolved in solvent refining of lubricating oil.

raffinose (melitose; melitriose; gossypose) $C_{18}H_{32}O_{16}\cdot 5H_2O$. A trisaccharide composed of one molecule each of D(+)-galactose, D(+)-glucose, and D(−)-fructose.
Properties: White, crystalline powder; sweet taste. Soluble in water; very slightly soluble in alcohol. Sp.gr. 1.465; m.p. (anhydrous) 118° to 119°C; b.p. decomposes at about 130°C; optical rotation +104.5°C.
Derivation: By hydrolysis from cotton seed meal; from sugar beet concentrates.
Method of purification: Recrystallization.
Grades: Technical; pure.
Containers: Wooden barrels; glass bottles.
Uses: Medicine; preparation of melibiose; bacteriology.
Shipping regulations: None.*

raisin-seed oil. See grape-seed oil.

"Raldeine." [227] Trademark for methyl ionones, available in several grades; used in perfume materials, especially in cosmetics, to give a violet character.

Raman spectra. See spectroscopy.

"Ramapo." [28] Trademark for resinated forms of phthalocyanine blues and greens.
Uses: Paint, printing ink plastics, and other applications requiring excellent lightfastness and a high degree of chemical stability.

ramie. A natural fiber of vegetable origin, obtained from the stems of Boehmeria nivea, of the hemp family. Lustrous, strong fiber of high wet strength, highly absorbent but dries quickly; is light but can be spun or woven. Wears well and has great rot and mildew resistance; tensile strength four times that of flax; elasticity 50% greater than flax.
Sources: China; Formosa; Egypt; Brazil; Florida.
Uses: High grade paper (Europe); fabrics (wearing apparel and car seat covers); stern-tube packing in ships; patching water-mains (Great Britain.)

Raney nickel. A spongy form of nickel. It is pyrophoric and hence is shipped under water.
Derivation: By leaching the aluminum from an alloy of 50% Al-50% Ni with 25% caustic soda solution.
Use: Catalyst for hydrogenation.

Rankine. A scale of absolute temperature based on Fahrenheit degrees. See absolute temperature.

Raoult's law. The vapor pressure of a substance, in equilibrium with a solution containing the substance, is equal to the product of the mole fraction of the substance in the solution, and the vapor pressure of the pure substance at the temperature of the solution. The law is not applicable to most solutions, but is often approximately applicable for mixtures of closely similar substances, particularly for substances present in high concentration.

rape oil. See rape-seed oil.

rape-seed oil (colza oil; rape oil).
Properties: Dark brown when crude; pale yellow when refined. A very viscous, liquid oil; unpleasant taste and odor; may deposit stearin on standing. Soluble in ether, chloroform and carbon disulfide.
Constants: Sp.gr. 0.9132-0.9168; solidifying point −2° to −10°C; m.p. 17-22°C; refractive index 1.4720-1.4752; saponification value 167-179; iodine value 96 to 104; Hehner value 95.1; acid value 1.4-13.2.
Derivation: By the expression of rape-seed, Brassica napus, followed by treatment of the cake with solvents, and evaporation of the solvent. The oil is frequently oxidized or "blown" to increase its density and viscosity. It is refined by treatment with fuller's earth or sulfuric acid.
Grades: Crude; refined.
Containers: Tins; steel drums; tank cars.
Uses: Refined and "blown" rape-seed oil is used as a lubricant; illuminant; manufacture of rubber substitutes; heat treatment of steel. The refined cold-drawn oil is also used for edible purposes.
Shipping regulations: None.*

"Rapidase." [173] Proprietary name for an enzyme preparation containing starch-liquefying and proteolytic enzymes.
Properties: Liquid of low specific gravity and viscosity; active in slightly acid, neutral, and weakly alkaline solutions; can withstand temperatures from 25-75°C.
Derivation: Produced by growing pure cultures of certain microorganisms in sterilized wort.
Use: Desizing agent.
Shipping regulations: None.*

"Rapidogen." [307] Trademark of stabilized azoic-naphthol compounds soluble in caustic soda which are developed by means of acid aftertreatment. Used for printing on cotton rayon. Characterized by bright shades of very good fastness to washing, chlorine and light.

rare earth elements. See rare earth metals.

"Rare Earth Fluoride." [337] Trade designation for arc carbon grade materials of formula "RE"F_3.
Contains typically: rare earth elements, 54.7%, fluorine, 26.1%, carbon, balance. Black powder, bulk density 100-lb/cu ft; mesh size 80; approximate melting point 1350°C (2462°F). The anhydrous product

typically consists of 30.6% cerium, 15.6% lanthanum, 5.8% neodymium, 2.2% praseodymium, and minor amounts of samarium and gadolinium to make up the rare earth total of 54.7%. Used as source of rare earth elements suitable for addition to cored arc carbons.

rare earth metals (rare earth elements; rare earth series; lanthanide series). A group of elements, with names and atomic numbers thus:

lanthanum	La	57
cerium	Ce	58
praseodymium	Pr	59
neodymium	Nd	60
promethium	Pm	61
samarium	Sm	62
europium	Eu	63
gadolinium	Gd	64
terbium	Tb	65
dysprosium	Dy	66
holmium	Ho	67
erbium	Er	68
thulium	Tm	69
ytterbium	Yb	70
lutetium	Lu	71

The elements 57-62 are known as the cerium subgroup, and 63-71 as the yttrium subgroup. Yttrium, atomic number 39, although not a rare earth element, is found associated with the rare earths and is only separated with difficulty. See also didymium.

Properties: Soft and malleable metals becoming harder as the atomic number increases; para-magnetic; good heat conductors; active reducing agents; the freshly cut metals have a silver white to gray color and tarnish immediately in air; react slowly with water; are readily soluble in dilute acids.

Source: The rare earth elements are found in many minerals but only monazite and bastnaesite are important ores (see rare earth minerals). These elements are also found in the fission products of uranium and plutonium, this being the only source of promethium.

Derivation: Separation of these elements is very difficult because they have nearly identical chemical properties. Separation was originally made by fractional crystallization of the salts but more recently and successfully by selective absorption on synthetic resins. In the elementary state these elements are strongly electro-positive metals, prepared with difficulty by (1) displacement with sodium, calcium, or magnesium, (2) electrolysis of the fused halide, (3) as an amalgam by electrolysis of alcoholic solutions.

Uses: Misch metal (q.v.); pyrophoric material for gas lighters, etc; alloys.

See also rare earths and individual elements.

rare-earth minerals. Sources of the rare-earth elements. Monazite and bastnaesite are the most important. Less important are gadolinite, fergusonite, samarskite, xenotime, cerite, and allanite.

rare earths.

1. A term commonly used to designate the rare earth metals.
2. Since "earths" are usually considered to be metal oxides, the term may also be used in the latter sense, the rare earth oxides being known under the following names: lanthana La_2O_3, ceria CeO_2, neodymia Nd_2O_3, praseodymia Pr_2O_3, promethia Pm_2O_3, samaria Sm_2O_3, europia Eu_2O_3, gadolinia Gd_2O_3, terbia Tb_2O_3, dysprosia Dy_2O_3, holmia Ho_2O_3, erbia Er_2O_3, thulia Tm_2O_3, ytterbia Yb_2O_3, and lutetia Lu_2O_3. These basic oxides, very similar to one another in physical and chemical properties, are usually found associated in certain rare minerals of complex composition. They all absorb moisture and carbon dioxide from the air.

rare earth salts. Salts prepared from monazite and containing the rare earths in essentially the same ratio as found in monazite. The salts have a rare earth oxide composition of approximately 24% lanthana (La_2O_3), 48% ceric oxide (CeO_2), 6% praseodymium oxide (Pr_6O_{11}), 19% neodymia (Nd_2O_3), 2% samaria (Sm_2O_3), 0.5% gadolinia (Gd_2O_3), 0.2% yttria and yttrium earth oxides, 0.3% others. (The symbol RE as used here represents this mixture of rare earth elements. Note that the so-called didymium salts (q.v.) are also prepared from monazite, but do not contain ceria).

rare earth acetate $RE(C_2H_3O_2)_3 \cdot xH_2O$.
Fine light pink powder; soluble in water and acids. Used in textile water-proofing.

rare earth carbonate $(RE)_2(CO_3)_3 \cdot xH_2O$.
Fine, light pink powder; soluble in acids; insoluble in water. Used as a coloring agent.

rare earth chloride $RECl_3 \cdot xH_2O$.
Light pink crystals. Very soluble in water. Used in textile water-proofing and in embalming. The anhydrous chloride is used as a source for making rare earth metal (misch metal), used in lighter flints.

rare earth fluoride REF_3.
Pink powder; insoluble in water and acids. Used in manufacture of cored carbon for arc lighting; for nodular steel; special alloy steels.

rare earth hydrate.
Yellow hydrous rare earth oxide containing cerium in the tetravalent form. Soluble in acids; insoluble in water. Used in glass as a decolorizer and as an ultraviolet absorber.

rare earth nitrate $RE(NO_3)_3 \cdot xH_2O$.
Pink crystals; very soluble in water.
Shipping regulations: Oxidizing material. Yellow label.*

rare earth oxalate $(RE)_2(C_2O_4)_3 \cdot xH_2O$.
Pink powder; insoluble in water; slightly soluble in acids. Used in pharmaceuticals.

rare earth oxide.
Brown powder. Contains cerium in the tetravalent state. Soluble in acids;

*See "I.C.C. Shipping Regulations," page xiii.

insoluble in water. Readily absorbs moisture and carbon dioxide from the air. Used in glass coloring, glass polishing and in cores of arc carbons.

rare earth sodium sulfate $(RE)_2(SO_4)_3 \cdot Na_2SO_4 \cdot 2H_2O$.
Fine pink crystals; sparingly soluble in water and acids. Common name: pink salts. Used in the manufacture of ultraviolet absorbing glass and as an intermediate in the making of rare earth and cerium materials.

rare earth sulfate $(RE)_2(SO_4)_3 \cdot xH_2O$.
Fine pink crystals; soluble in cold water; sparingly soluble in hot water. Source material for conversion to other compounds.

Containers: Glass bottles; fiber and steel drums.

Other uses of rare earth salts: coloring, decolorizing glass; catalysts in esterification and hydrocarbon dehydrogenation; oxygen carrying catalyst; blood anticoagulant.

rare earth series. See rare earth metals.

rare earth sulfides. Cerium and samarium sulfides are being used as high temperature thermoelectric materials. Both have stability and thermoelectric efficiency at 2000°F. They are made by fusing samarium or cerium with sulfur at high temperatures in an induction furnace. The sulfide is crushed, made into pellets, and sintered in a vacuum furnace. Its composition is determined by x-ray diffraction analysis.

rare gases. Helium, neon, argon, krypton, xenon, and radon. Also known as inert or noble gases. They are elements, monatomic, and completely inert chemically.

rare metals. A loose term for the less common and more expensive metallic elements. Commonly included are the alkaline earth metals barium and strontium, beryllium, bismuth, cadmium, cobalt, gallium, germanium, hafnium, indium, lithium, molybdenum, rhenium, selenium, tantalum, niobium, tellurium, thallium, thorium, titanium, tungsten, uranium, vanadium, and zirconium. Also sometimes included are boron and silicon, the platinum metals, the rare earths, manganese and calcium.

"Rareox." [241] Brand name for a finely divided pink powder consisting of cerium oxide and other rare earth oxides.

Raschig process

1. A process for the production of phenol. Benzene in the vapor phase reacts with hydrogen chloride and air (only the oxygen reacts) in the presence of iron chloride and copper chloride catalysts at 200-300°C to produce chlorobenzenes. Monochlorobenzene is separated from the polychlorobenzenes and is then hydrolyzed with a silica or tricalcium phosphate catalyst at 350°C to produce phenol. The latter is washed with water, then with benzene, and finally distilled.
2. A commercial process for making hydrazine hydrate (q.v.) from ammonia, chlorine, and caustic soda.

"Rasorite." [441] Trademark for mineral kernite.

rattlesnake root. See senega.

"Raudixin." [412] Trademark for rauwolfia (q.v.).

"Raunormine." [342] Trademark for brand of deserpidine hydrochloride.

"Rau-Sed." [412] Trademark for reserpine (q.v.).

"Rauserfia." [342] Trademark for an extract of weakly basic alkaloids from Rauwolfia.

rauwolfia (so-called snake root). The powdered whole root of Rauwolfia serpentina. The plant is of value as a source of alkaloids, especially reserpine.

Properties: Light tan to light brown, bitter, fine, amorphous powder; slight aromatic odor. Sparingly soluble in alcohol and very slightly soluble in water.

Grade: N.F. XI.

Containers: Barrels; drums.

Use: Medicine.

"Raven." [133] Trademark for a series of general utility impingement carbon blacks for paint, paper and printing ink. Available as:

"Raven 11." Low oil absorption, good color, high tinting strength.

"Raven 15." Lowest oil absorption of all impingement blacks, which permits high loadings.

"Raven Beads." A low-viscosity black in dustless free-flowing form. Particularly suited for ball mill grinding of news and rotogravure inks.

Containers: 25-lb bags.

"Rayaceta." [37] Trademark for a highly refined wood cellulose.

Properties: Cellulose-alpha 96.0%, beta 2.2%, gamma 1.8%. 10% KOH solubility, 5.3%.

Uses: Production of cellulose acetate rayon; manufacture of acetate films, sheets, and molding powders for varied uses.

"Rayamo." [37] Trademark for a wood cellulose of low viscosity which was developed to meet the increasing need for greater economies in the manufacture of cellophane. It permits savings both in equipment and operating costs in cases where full advantage is taken of its special properties.

Properties: Cellulose-alpha 89.1%, beta 6.5%, gamma 4.4%. 10% KOH solubility, 15.2%.

"Raydex." [204] Trademark for a high wettability general cleaner for medium hard water, using either brush or spray method. Safe on all equipment and workers' hands. Packed in 125-lb and 350-lb drums.

"Rayocord." [37] Trademark for a highly purified wood cellulose developed specifically for the production of continuous filament rayon yarns of high tenacity.

Properties: Cellulose-alpha 95%, beta 2.5%, gamma 2.5%. 10% KOH solubility 12.9%.

rayon. Generic name for a manufactured fiber composed of regenerated cellulose, as well as manufactured fibers composed of regenerated cellulose in which substituents have replaced not more than 15% of the hydrogens of the hydroxyl groups (Federal Trade Commission).

Rayon was first made by denitration of cellulose nitrate fibers (Chardonnet process) but this process is now obsolete. Most rayon is now made by the viscose process. See also cuprammonium rayon and saponified acetate fiber.

rayon coning oils. Oils used to lubricate and reduce the static of yarns wound on cones by the coning machine. Usually composed of mineral oils of low viscosity or body, compounded to emulsify in water.

rayon, modified. Rayon composed principally of regenerated cellulose, and containing amounts of non-regenerated cellulose fiber-forming material; for example, a fiber spun from viscose containing casein or other protein. (ASTM definition; ASTM D123-54).

Rb. Symbol for rubidium.

RBE. Abbreviation for relative biological effectiveness (q.v.).

RDGE. See resorcinol diglycidyl ether.

RDX. See cyclonite.

Re. Symbol for rhenium.

RE. Abbreviation for rare earth elements in general or for various mixtures of these. See didymium salts and rare earth salts.

reagent. Any substance used in a reaction for the purpose of detecting, measuring, examining or producing other substances.

realgar As_2S_2 (arsenic disulfide).

Properties: Soft aurora-red mineral becoming orange-yellow on long exposure. Orange-red streak, resinous luster. Contains 70.1% arsenic, 29.9% sulfur. Usually occurs in veins associated with silver and lead ores. Has also been found in volcanic regions as a sublimation product and as a deposit from hot spring waters. Is often associated with orpiment and is frequently noted as a sublimation product from furnaces roasting ores of arsenic. Soluble in nitric acid and potassium hydroxide. Sp.gr. 3.4-3.6; hardness 1.5-2.

Occurrence: United States (Utah, Washington, Wyoming); Hungary; Germany; Iran; Switzerland; China.

Uses: Pyrotechnics (artificial now more used); calico printing; tanning (depilatory); shot manufacture (hardening); pigment.

Réaumur. Name of a temperature scale invented by Réaumur in which the freezing point of water is set at 0 and its boiling point at 80. Temperature is then measured in °R. Not widely used.

"Reax" 23 and "Reax" 31. [229] Trademark for lignin phenolic resin coreactants produced from the sulfate pulping process.

Properties: Brown, free-flowing powder; insoluble in water and mineral acids, soluble in phenol and aqueous alkaline solutions. Solvents include glycols, "Cellosolves," amines, dioxanes, and alcohol-ketone mixtures. Sp.gr. 1.3; bulk density 25-30 lbs/cu ft; moisture "Reax" 23: 2.0-7.0%, "Reax" 31: 2.0-5.0%; pH "Reax" 23: 6.5-7.5, "Reax" 31: 2.3-3.5. Composition dry basis: ash, "Reax" 23: 0.8% maximum, "Reax" 31: 0.5% maximum.

Containers: 50-lb MWP bags.

Uses: Phenolic and cresylic resin coreactants.

rectification. The enrichment or purification of the vapor during the distillation process by contact and interaction with a countercurrent stream of liquid condensed from the vapor.

red acetate. See mordant rouge.

red arsenic. See arsenic disulfide.

red bark. See cinchona bark, succirubra.

red bole. See iron oxide reds.

red brass. Copper-zinc (brass) alloys characterized by their red color and high copper content. The term is used for several different types of brass. One ASTM classification permits 2-8% zinc, tin less than zinc, and lead less than 0.5%. Other alloys referred to as red brass include those with 75-85% copper, up to 20% zinc, and usually very small amounts of lead and tin. In one alloy possessing good machining qualities, the lead content may be as high as 10%, the tin as high as 5%.

Red brasses are widely used for decorative purposes and in plumbing and piping because of their resistance to atmospheric corrosion and dezincification.

red chalk. See iron oxide reds.

red cinchona. See cinchona bark, succirubra.

"Red Cross Extra." [28] Trademark for a high density general purpose ammonia dynamite of 20 to 60% strength. Used for quarrying, stripping, agricultural and general construction work.

"Red Cross" (F.R.). [28] Trademark for five grades of ammonia dynamite (based upon strengths) with free-running characteristics (F.R.). Designed for use in dry holes.

reddle (raddle). A name for a naturally occurring mixture of red hematite and clay. See also iron oxide reds.

red earth. See iron oxide reds.

red glass. About 1% selenium may be added to the melt of soda-zinc glass containing a small amount of cadmium. Red may also be obtained by using cuprous oxide or gold chloride, the latter usually in the form of purple of Cassius.

red hematite. See hematite, red.

red iron ore. See hematite, red.

red iron trioxide. See ferric oxide.

"Redisol # 4." 170 Trademark for refined pre-cooked pulverized tapioca starch derived from cassava, for room temperature method of thickening and stabilizing foods.

Properties: Maximum moisture 8%, maximum ash 0.35%, pH: 5.0-7.0.

Containers: 80-lb multiwall poly-lined paper bags.

red lake C pigments. A family of organic acid azo pigments prepared by coupling the diazonium salt of orthochloro-meta-toluidine-para-sulfonic acid with beta-naphthol. Both sodium and barium salts of the parent dye are used as pigments.

sodium salt. Known as "bronze orange" because of its very yellowish tone and high print bronze. Since it has noticeable bleed tendencies in both aqueous and organic media and discolors under heat, uses are limited.

barium salt. Known as "red lake C" and represents toners having varying degrees of fulltone color, bronze, strength and transparency depending on the amount of resination and method of processing. Exhibits excellent bleed resistance on aqueous media, alkalies and acids, together with good ease of dispersion. They are superior to lithol reds in permanency and color stability on baking.

Uses: Printing ink, plastic, rubber, metal decorating.

red lead. See lead oxide, red.

red liquor. See mordant rouge.

red ocher. A red earthy impure hematite, due to admixture with clay which is often partly removed by treatment (see ocher) of the original materials (also known as reddle, raddle, ruddle, red chalk, or red iron froth). The term red ocher is used both for these raw materials and the treated or finished product.

See also iron oxide reds.

red oil. Used to describe the commercial grade of oleic acid.

red orpiment. See arsenic disulfide.

redox. Handy version of the term oxidation-reduction, as in redox reactions, redox conditions, redox equations.

redox fuel cells. See fuel cells.

red oxide. See iron oxide reds.

"Redoxyvat." 328 Brand product consisting of an antioxidant dissolved in sulfonated oil. It is employed in the textile industry as an oxidation moderator to prevent premature oxidation of vat dyes.

red pepper. See capsicum.

red Peruvian bark. See cinchona bark, succirubra.

red precipitate. See mercuric oxide, red.

red prussiate of potash. See potassium ferricyanide.

red prussiate of soda. See sodium ferricyanide.

red puccoon. See sanguinaria.

red root. See sanguinaria.

red rudd. See iron oxide reds.

red sandalwood. See santalum rubrum.

red saunders. See santalum rubrum.

red stone. See iron oxide reds.

reduced states. See corresponding states.

red zinc ore. See zincite.

refiner's syrup. See molasses.

refinery gas. The mixture of hydrocarbon gases (and often some sulfur compounds) produced in cracking and distilling petroleum in the course of refinery operations. The usual components are hydrogen, methane, ethane, propane, butanes, pentanes, ethylene, propylene, butenes, pentenes, and small amounts of other components such as butadiene.

Use: As a source of raw material for components of high octane fuel; butadiene for synthetic rubber; and organic synthesis of alcohols.

reflux. Term used in distillation with a fractionating column for the liquid condensed from the rising vapor and allowed to flow down the column toward the still.

Reformatzky reaction. The interaction of an alpha-halo ester, an aldehyde or ketone and zinc in the presence of anhydrous ether or ether-benzene, followed by hydrolysis, to yield a beta-hydroxy ester.

reforming. The use of controlled heat and pressure, with or without catalysts, to cause cracking and isomerization of the hydrocarbon molecules in low octane petroleum fractions. The hydrocarbons formed are lower in molecular weight, somewhat more branched and somewhat unsaturated, and so have higher octane numbers. Several kinds of reforming are in use, such as hydroforming, catalytic reforming, "Platforming," etc.

"Refractaloy." 308 Trademark for a nickel-cobalt-chromium-molybdenum-iron alloy. Type 26 is a precipitation-hardenable material using titanium as the hardening agent, and having high strength, high ductility, and corrosion resistance at high temperatures – up to 1450°F. Gas turbine discs, bolting and blading are typical applications.

refraction. When a light ray passes from one medium to another of different density it is bent from its original path. The ratio of the sine of the angle of incidence to the sine of the angle of refraction is the index of refraction of the second medium. Index of refraction of a substance may also be expressed as the ratio of the velocity of light in a vacuum to its velocity in the substance. It varies with the wave length of the incident light, temperature and pressure. The usual light source is the D line of sodium at 20°C; the expression for refractive index is n 20/D.

refractive index. See refraction.

refractories. (See also brick). Nonmetallic structural materials for use at high temperatures. In addition to their primary function, they may also be called upon to resist abrasion, corrosion, pressure, and rapid changes in temperature. No single refractory will stand up equally well under all of these conditions so that a large variety of materials have been developed. These include fire clay, kaolinite, bauxite, magnesite, dolomite, chromite, carbon, silicon carbide, zirconia, spinel, mullite, sillimanite, forsterite, olivine, electrocast and unburned and insulating refractories. In each case, the refractory consists of an infusible portion (the refractory proper) and a relatively small proportion of binder. Refractories are classified as acid, basic, or neutral according to the character of the oxide present in largest proportion. Refractories are graded according to fusion temperature, porosity, spalling, strength, resistance to rapid temperature change, thermal conductivity and heat capacity.

refractories, basic. Include both magnesia and chrome refractories, which are commonly used together.

refractories, electrocast. Consist of relatively pure aluminum silicate of mullite composition ($3Al_2O_3 \cdot 2SiO_2$) that has been fused in an electric furnace, and cast into blocks. The product is vitreous, nonporous, hard, and has a low coefficient of expansion.

refractories, insulating. Are of a porous nature, resulting from the use of porous diatomaceous earth as raw material, or expanded perlite or vermiculite, diaspore clay or similar light weight materials. In some cases sawdust or other combustible is used.

refractories, specialties. These are castables, ramming mixes and plastics that form monolithic structures and high temperature bonding mortars. Castables are ground refractory materials containing added bonding agents. Supplied in dry form but when mixed with water they form a strong hydraulic set at room temperatures and a ceramic bond at high temperatures.

Ramming mixes are made of ground refractory materials, carefully graded as to relative proportions of different particle sizes and other materials added for workability and self bonding. Ramming mixtures are supplied in plastic form, or in dry form, needing water addition. They are rammed into place and set upon drying.

refractories, unburned. Are shaped without burning, by use of high pressures after deaeration to reduce voids between grains. Chemical bonding and metal encasement are also used.

refractory. As an adjective, means resistant to heat. For noun, see refractories.

refractory clays. Clays that withstand high temperature, i.e., fire clays.

"Refrasil." [161] Trademark for fibrous silica of high purity SiO_2; available as bulk fibers, batts, cloth, tape, sleeving, cordage and yarn; useful as extreme high temperature thermal insulation, as in jet aircraft tail cones and pipes. Batted fibers resemble cotton in appearance and feel. Fiber diameters are 0.0002 to 0.0005 inches while lengths are ½ to 2 inches. No binders are present. Blankets made from batts are resistant to high frequency vibrations encountered on jet aircraft. Fibrous silica cloth resembles cloth made of spun glass. The fibers used are continuous, and no lubricant is used. Textile products are widely used as electrical space separators at high temperatures.

"Refrax." [280] Trademark for silicon nitride-bonded silicon carbide refractories. Available in brick and precision-formed shapes and parts.

Properties: High thermal conductivity, high dimensional stability, high resistance to spalling, cracking, mechanical wear, heat shock and extreme temperatures. Thermal conductivity is 113.5 Btu/sq ft/inch thickness/°F/hr. Porosity ranges from 7 to 10%. Modulus of rupture at 2462°F is 5000-7000 psi.

Uses: Brazing and furnace fixtures; pumps and pump parts handling corrosive, abrasive slurries; rocket motor components; spray nozzles; burners; pyrometer protection tubes; sinker assemblies in wire aluminizing; bolts and nuts; valve parts; aluminum melting furnace linings and parts; conveyor parts.

"Refrex." [214] Trademark for a blend of alcohols, for refrigerator car-heating fuels.

"Regal." [275] Trade name for a series of oil furnace carbon blacks for use in rubber. Available as:

"Regal 600." Highly reinforcing oil furnace black.

"Regal 300." Fully reinforcing oil furnace black.

"Regal SRF." Oil semi-reinforcing furnace carbon black. (SRF).

"Regent 12XX." [172] Trademark for a hydrated monocalcium phosphate.

Properties: Brilliant white, free-flowing, crystalline material. Its purity complies with all Food and Drug laws.

Containers: 100-lb paper bags.

Uses: Manufacture of household baking powders, prepared mixes, and phosphated flours; ingredient in commercial bread improvers; yeast manufacture; manufacture of frits for vitreous enamels.

"Regent Yellows." [141] Trade name for lightfast chrome yellow pigments.
Properties: Considerably better light resistance and alkali resistance than regular chrome yellow pigments.
Grades: Primrose, light or lemon and medium shades.
Uses: In paints for all exterior uses; in vinyl, particularly automotive or furniture upholstery.

"Regithane" Foaming Resins. [27] Urethane resin systems which can be formulated to densities ranging from less than 2 to 40 pounds per cubic foot and are adaptable to many end uses.
Properties of cured foams: Low thermal conductivity; good energy absorption; good strength; light weight; excellent adhesion to most materials; room temperature cures; compatibility with colorants; nonfriable; self-extinguishing; impervious to vermin, mildew, and rot; very good "K" factors.
Uses: Thermal insulation for trucks, railroad cars, refrigerators, etc.; molded toys, shipping containers; encapsulation; structural reinforcement; void filling; vibration dampening; sound absorption; packaging; life preservers, rafts, etc.; submarines.

"Regitine." [305] Trademark for phentolamine.

regulus of antimony. See antimony.

Reichert-Meissl number. A measure of volatile soluble fatty acids derived under arbitrary conditions.

Reich process. A method for purifying carbon dioxide produced in fermentation. The small amounts of organic impurities are oxidized and absorbed and the gas is then dehydrated with chemicals.

Reimer-Tiemann reaction. Reaction for the formation of phenolic aldehydes by heating a phenol with chloroform in the presence of alkali.

Reinecke salt (ammonium tetrathiocyanodiammonochromate; ammonium reineckate) $NH_4[Cr(NH_3)_2(SCN)_4] \cdot H_2O$.
Properties: Dark-red crystals or crystalline powder; moderately soluble in cold water; soluble in hot water and alcohol; decomposes in aqueous solutions.
Derivation: From fusion of ammonium thiocyanate with ammonium dichromate.
Grade: Reagent; technical.
Use: As a reagent for organic bases, such as choline, amines, and for certain amino acids, etc.; as a reagent for mercury.

reinforced concrete. See concrete.

relative biological effectiveness (RBE). A factor used in health physics to adjust for the different biological effects of the various radiations, and is defined to be the ratio of dose in rads of 200 kv x-rays that produces some particular effect to the dose in rads of the radiation in question that produces the same effect. Thus since fast neutrons are about ten times as effective as x-rays in their deleterious effects on tissue, the RBE for fast neutrons is ten.

relaxin. A polypeptide hormone.
Properties: Amorphous powder. Slightly soluble in water and alcohol; soluble in acid or alkaline solutions. Insoluble in absolute alcohol, ether, acetone, benzene. Decomposes above pH 9.0.
Derivation: Obtained commercially from pregnant sows' ovaries. Isolation procedure and purification by resin chromatography.
Grade: N.N.D.
Use: Medicine.

rem. A rem (roentgen equivalent man) is a radiation dose of any ionizing radiation estimated to produce a biological effect equivalent to that produced by one roentgen of x-rays.

"Renex." [89] Trademark for a group of nonionic synthetic organic detergents, chiefly ethylene oxide esters and ethers.

renierite $(Ca, Fe)_3(Fe, Ge, Zn, Sn)(S, As)$. A mineral reported to contain up to 8% germanium, found in the Congo.
Use: Possible source of germanium.

rennase. See rennin.

rennin (rennase; chymosin). The enzyme secreted by the glands of the stomach which causes curdling of milk. It has the power of coagulating 25,000 times its own weight of milk.
Properties: Yellowish-white powder or as yellow grains or scales. Characteristic and slightly salty taste and peculiar, not unpleasant odor. Slightly hygroscopic. Partially soluble in water and diluted alcohol.
Derivation: From the glandular layer (inner lining) of the true stomach of the calf.
Grade: N.F. XI. Rennet is the dried commercial extraction containing the rennin.
Containers: Glass bottles; fiber cans and drums.
Uses: Medicine; pharmacy; cheese-making; coagulation of casein for plastics.
Shipping regulations: None.*

"Renografin." [412] Trademark for methylglucamine diatrizoate (q.v.).

"Reogen." [69] Trademark for a plasticizing agent.
Properties: Liquid, clear mahogany; sp.gr. 0.82-0.85.
Uses: Plasticizer and processing aid in all elastomers.

rep. Abbreviation for roentgen equivalent physical. See roentgen.

"Repel-O-Flame." [328] Trademarked product consisting of ammonium salts and designed to serve as a fire retardant for textiles.

*See "I.C.C. Shipping Regulations," page xiii.
Reference numbers refer to name of manufacturer. See "List of Manufacturers," page v.

"Repel-O-Tex D3 and D4." [328] Trademark for a combination that produces a wax emulsion containing heavy metal formates. It serves as a durable noncuring water repellent for rayon, acetate, and cotton-rayon mixtures.

"Repel-O-Tex ST." [328] A silicone-type milk-white dispersion, 30% active, and instantly dispersible in cold water; used in water repellent finishes.

replacement. See substitution.

Reppe processes (Reppe chemistry). Various reactions or processes by which acetylene is converted to other products. Included are reaction of acetylene (a) with formaldehyde to produce 2-butyne-1,4-diol, which can be converted to butadiene, (b) with formaldehyde under different conditions to product propargyl alcohol and from this, allyl alcohol, (c) with hydrocyanic acid to yield acrylonitrile, (d) with alcohols to give vinyl ethers, (e) with amines or phenols to give vinyl derivatives, (f) with carbon monoxide, and (g) by polymerization to produce cyclooctatetraene. The use of catalysts, of pressures up to 30 atmospheres, and of special techniques to avoid or contain explosions, is an important factor in these processes.

"Republic '65'" High Strength Alloy Steel. [251] Trademark for a high strength steel with corrosion resistance higher than ordinary copper-bearing steels.
Typical Analysis: 0.12% carbon; 0.45% manganese; 0.04% phosphorus; 0.04% sulfur; 0.20% silicon; 1.00% copper; 1.30% nickel; 0.20% molybdenum.
Properties: Yield point 67,000 psi; tensile strength 89,000 psi; elongation in 2 inches, 18%.
Forms: Wide ranges of sizes, sheets, plates and bars.
Uses: For lightweight construction.

"Republic '70'" High Strength Alloy Steel. [251] Trademark for a high strength steel with corrosion resistance higher than ordinary copper-bearing steels.
Typical Analysis: 0.14% carbon; 0.75% manganese; 0.04% phosphorus; 0.04% sulfur; 0.25% silicon; 1.15% copper; 1.50% nickel; 0.22% molybdenum.
Properties: Yield point 73,000 psi; tensile strength 96,000 psi; elongation in 2 inches, 17%.
Forms: Wide ranges of sizes, sheets, plates and bars.
Uses: For lightweight construction.

"Republic '50'" High Strength Steel. [251] Trademark for a low alloy steel with 50% higher yield point and four to six times the corrosion resistance of carbon steel.
Composition: 0.12% carbon (max); 0.50-1.00% manganese; 0.04% phosphorus (max); 0.05% sulfur (max); 0.50-1.00% copper; 0.50-1.10% nickel; 0.10% molybdenum (min).
Properties: Yield point, min. 50,000 psi; tensile strength min 70,000 psi; elongation in 2 inches, ASTM Standard Flat Specimen, 22% min.
Forms: Available in sheet and plate.
Use: Used for lightweight construction.

rescinnamine $C_{35}H_{42}N_2O_9$. Alkaloid from certain species of Rauwolfia.
Properties: White or pale buff to cream colored, odorless, crystalline powder. Darkens slowly on exposure to light, more rapidly when in solution. Soluble in chloroform and acetic acid. Slightly soluble in alcohol. Practically insoluble in water. M.p. 238°C (in vacuum).
Grade: N.F. XI.
Use: Medicine.

research reactor. See nuclear reactor.

reserpine $C_{33}H_{40}N_2O_9$. An important alkaloid.
Properties: White or pale buff to slightly yellowish, odorless powder. Darkens slowly on exposure to light, and darkens more rapidly in solution. Insoluble in water; very slightly soluble in alcohol. Soluble in chloroform and acetic acid. M.p. 264-265°C (decomposes).
Derivation: From Rauwolfia serpentina (q.v.).
Grade: U.S.P. XVI.
Containers: Bottles.
Use: Medicine, often in the form of its salts.

"Resi-Chem." [231] Trademark for urea-formaldehyde resins, polyester resins, polyamide resins; liquid resin syrups supplied in bulk and in drums. Used for wet strength in paper products, furniture and plywood adhesives, as textile and paper size, and in structural plastics.

residual oils. Liquid or semi-liquid products obtained as residues from the distillation of petroleum. They contain the asphaltic hydrocarbons. Residual oils are also known as asphaltum oil, liquid asphalt, black oil, petroleum tailings, and residuum.
Shipping regulations: None.*

residuum. See residual oils.

"Resilon." [326] Trademark for a bituminous base lining for chemical process equipment.
Use: To resist acids and alkalies up to temperatures of 150°F.

"Resimene." [58] Trademark for a series of resins based on melamine. Used for baked enamel and industrial finishes for appliances, automobiles and other fabricated metal items; for mar-resistant laminates for furniture; for electrical, electronic and radar components.

"Resin 731." [266] Trademark for a disproportionated, pale-colored rosin, unusually stable to the effects of light and heat.
Uses: In adhesives; coatings; rubber compounding.

resinated lithols. See lithol reds.

resinols. A coal tar distillation fraction containing phenols. It is the fraction soluble

*See "I.C.C. Shipping Regulations," page xiii.
Reference numbers refer to name of manufacturer. See "List of Manufacturers," page v.

in benzene but insoluble in light petroleum, obtained by solvent extraction of low temperature tars or similar materials. Resinols are very sensitive to heat and oxidation.

"Resinox." [58] Trademark for a series of phenolic resins, supplied in various forms suitable for applications as bonding agents for shell molding and as core binders for metal casting; impregnants or bonding materials for grinding wheels, brake linings, insulation and similar industrial uses; as pipe linings, air conditioning equipment coatings, special primers; as laminating, bonding, impregnating resins for paper, fibers; for use in can, drum, and tank car linings requiring a high degree of chemical and solvent resistance; and for heavy-duty products, such as equipment housings.

Resin S. [175] Brand name for a neutral synthetic coal tar resin of high styrene content.

Properties: Light color; m.p. 115-123°C; sp.gr., approx 1.05; mineral oil cloud point, 130-150°C.

Containers: 55-gal destructible steel drums.

Uses: Principally to impart alkali- and grease-resistance to floor tile.

resins, natural (resins, true). Solid or semisolid viscous materials derived mostly from secretions of certain plants and trees. Common examples are rosin, amber, copal, Kauri, pine tar, pitch, and Canada balsam. The latter three are oleoresins (q.v.). Amber is a fossil resin, found as a deposit rather than taken from a living tree. Hard resins are those that occur naturally as hard solids. Most of these will soften on warming. All these materials are compounds of carbon, hydrogen and oxygen. They are usually clear or translucent yellow or brown materials that burn easily with a yellow and smoky flame. They are insoluble in water, soluble in ether, alcohol, benzene, and carbon disulfide. Spirit-soluble resins are those that are easily soluble in alcohol or other volatile solvents. Oil-soluble resins are those that are more easily soluble in fatty oils.

Natural resins are used in lacquers and varnishes; as modifiers of synthetic plastics; and in inks and adhesives.

Some gums (q.v.) are natural resins.

See also shellac.

resins, synthetic (See also plastic.) Amorphous, organic, semi-solid or solid materials produced by union (polymerization or condensation) of a large number of molecules of one, two, or less frequently three relatively simple compounds. Properties vary widely with the raw materials, their proportions, and the conditions of formation of the resin. The term synthetic resin is also sometimes applied to chemically modified natural resins. Resins are broadly classified as thermoplastic or thermosetting according as they soften or harden with application of heat. In the following table the principal types of synthetic resins are classified according to their derivation by (1) modification of natural polymers, (2) polymerization, or (3) condensation. Many commercially important plastics represent combinations of these types, resulting from co-polymerization, crosslinking, etc.

Typical Resins Derived from Natural Polymers
- Hydrocarbons
  - rubber, chlorinated
  - rubber hydrochloride
- Cellulose
  - cellulose nitrate
  - cellulose acetate
  - methylcellulose
- Protein
  - casein
  - zein
- Miscellaneous
  - ester gums
  - lignin

Typical Resins Derived by Polymerization (union of small molecules without formation of water or some other simple molecule as a byproduct)
- acetal resins
- acrylate resins
- allyl resins
- coumarone-indene resins
- fluorocarbon resins
- furan resins
- polyethylene
- polypropylene
- polystyrene
- vinyl plastics
- vinylidene resins

Typical Resins Derived by Condensation (union of small molecules with formation of water or some other simple molecule as a byproduct)
- alkyd resins
- epoxy resins
- melamine resins
- phenolic resins
- polyamide resins
- polycarbonate resins
- polyester resins
- polyurethane resins
- polyether foams
- silicone resins
- urea-formaldehyde resins

resins, true. See resins, natural.

"Resipon." [300] Trademark covering a series of textile resin finishes. Includes thermosetting types for crush resistance and dimensional control, polyvinyl acetate emulsions and alkyd types.

"Resipon" BC: Methylated urea-formaldehyde type.

"Resipon" C: Standard urea-formaldehyde type.

"Resipon" NC: Ethylene urea type.

"Resipon" NDC: Cyclic urea type, for wash and wear finishes.

"Resipon" ST: Triazone type.

"Resipon" V: Polyvinyl acetate emulsion.
"Resipon" RF: Solubilized alkyd type.

resist. A term used in textile printing to describe a material which will prevent the fixation of dye on the fiber, in order to obtain patterned prints. The resist may act mechanically, as a wax, resin or gel which may prevent the absorption of the dye, or it may act chemically to neutralize either the dye or its accompanying mordant. Citric acid, oxalic acid, and various alkalies are among the more common resists of the chemical type.

"Resistac." [111] Trade name for an aluminum bronze alloy available in three special grades containing 89% copper, 10% aluminum, 1% iron (Resistac No. 1); 89.5% copper, 8% aluminum, 2.5% iron (Resistac No. 2); 78% copper, 10-11.5% aluminum, 3-5% iron, 3-5.5% nickel (Resistac No. 3).
Properties: Gold color; resistant to dilute sulfuric acid solutions, salt solutions, and vapors; heat resistant; good fatigue resisting qualities.
Forms available: Castings; rods and bars; forgings.
Uses: Applications requiring resistance to chemical corrosion; hot mill guides; valve guides in internal combustion engines; valve seats for superheated steam; oscillating equipment.

Resistor Compositions. [28] A new type of fired-on resistor composition of specially treated semiconducting metal powders compounded with glass binders and temporary organic carriers. Can be applied to glass or ceramic surfaces by stenciling, spraying, brushing or dipping; firing range 1300° to 1400°F. Resistance of 500 to 10,000 ohms/square obtained with 1-mil film. Fired resistors have good reproducibility, low temperature and voltage coefficients, and stability to abrasion, moisture, and relatively high (125°C) ambient temperature.
Containers: 1- to 32-oz jars.
Uses: To produce fired-on resistor components for electronic circuits.

"Resistox" Copper Powders. [294, 296] Trademark for stabilized grades of metallic copper powder assaying a minimum of 99% copper with a specific gravity of 8.9 and apparent density range of 2.0-3.5 grams per cubic centimeter. Marketed in several grades of various particle sizes.
Containers: 500-lb pails. (Also available in smaller containers).
Uses: Fabrication of porous bearings; infiltration of sintered ferrous machine parts; chemical catalysts; additive to magnesium chloride cements; metal friction surfaces; electric brushes; electrical contacts; metallic paints.

resite. See C-stage resin.

resitol. See B-stage resin.

"Resloom." [58] Trademark for a line of synthetic plastic resins based on melamine. Used to impart shrinkage control, wrinkle-resistance and durability to woolens, rayon, synthetics and cottons.

"Resmetal." [65] Trademark for a resin-metal composition which when catalysed converts to metal-like solid. Recommended for mold making, patching, forming and general repair of metal surfaces and objects.

"Resodors." [12] Trademark for a group of heat stable, residual odor modifiers and re-odorants for application to resins and latices. Compatible, soluble and non-deteriorating to plastic.

"Resoform." [307] Brand name of a line of dyestuffs used for the coloring of plastics.

resol. See A-stage resins.

resonance. In chemistry, a property of certain types of molecular structure in which the atoms forming it remain substantially in a fixed spatial arrangement with their electrons arranged so as to simultaneously satisfy two or more classical structural formulas. (See formula, chemical). Resonance usually is responsible for high molecular stability. The structure of benzene is one of the best-known examples of this property.

resorcin. See resorcinol.

resorcinol (resorcin; meta-dihydroxybenzene; 3-hydroxyphenol) $C_6H_4(OH)_2$.
Properties: Very white crystals, becoming pink on exposure to light when not perfectly pure; unpleasant sweet taste; sp.gr. 1.2717; m.p. 110.7°C; b.p. 281°C; soluble in water, alcohol, ether, glycerol, benzene and amyl alcohol; slightly soluble in chloroform.
Typical specifications: 98.5-99.5% pure; freezing point 108.5-109.1°C; fine crystals or flakes; flash point (open cup) 339°F.
Derivation: By fusing benzene-meta-disulfonic acid with sodium hydroxide, dissolving the melt in water, acidifying the solution with hydrochloric acid and extracting the resorcinol with a volatile solvent, followed by evaporation of the latter.
Method of purification: Redistillation.
Impurities: Diresorcinol; phenol; salicylic acid.
Grades: U.S.P. XVI; powder; resublimed; pure; reagent; technical; crude.
Containers: 25- to 350-lb fiber drums; multiwall paper sacks.
Uses: Manufacture of dyes; medicine; reagent in analytical chemistry; celluloid (camphor substitute); manufacture of styphnic acid, tanning agents, synthetic resin adhesives, pharmaceuticals, rubber tackifiers.
Caution: Avoid inhalation of dust or vapor and contact with skin (except under medical care).
Shipping regulations: None.*

resorcinol acetate (resorcinol monoacetate) $HOC_6H_4OCOCH_3$.
Properties: Viscous, pale yellow or amber liquid with a faint characteristic odor and a burning taste. B.p. about 283°C (dec);

*See "I.C.C. Shipping Regulations," page xiii.

Reference numbers refer to name of manufacturer. See "List of Manufacturers," page v.

boiling range (10 mm) 150-153°C. Saturated solution is acid to litmus. Soluble in alcohol and most organic solvents; sparingly soluble in water. Sp.gr. 1.203-1.207.
Derivation: Action of acetic anhydride on resorcinol.
Method of purification: Fractional distillation.
Grades: C.P.; N.F. XI.
Containers: Glass bottles; drums.
Uses: Medicine; cosmetics.

resorcinol blue. See lacmoid.

resorcinol diglycidyl ether (RDGE; 1,3-diglycidyloxybenzene; meta-bis(2,3-epoxypropoxybenzene) $C_6H_4(OCH_2CHOCH_2)_2$.
Properties: Straw-yellow liquid; sp.gr. 1.21 (25°C); b.p. 172°C (0.8 mm); refractive index 1.541 (n 25/D); viscosity 500 cps (25°C); flash point 350°F (Cleveland open cup). Miscible with most organic resins.
Use: Epoxy resins.

resorcinol monoacetate. See resorcinol acetate.

resorcinol monobenzoate $C_6H_5COOC_6H_4OH$.
Properties: A white, crystalline solid; m.p. 132-135°C; b.p. 140°C (0.15 mm).
Uses: Non-coloring ultraviolet inhibitor for various plastics; color stabilizer in cosmetic compositions.

resorcinolphthalein. See fluorescein.

resorcinolphthalein sodium. See uranine.

alpha-**resorcylic acid** (3,5-dihydroxybenzoic acid) $(OH)_2C_6H_3COOH$.
Properties: White crystals, m.p. 237°C. Soluble in water, ethanol and ether.
Grade: C.P.
Uses: Intermediate for dyes; pharmaceuticals; light stabilizers; resins.

beta-**resorcylic acid** (BRA; 2,4-dihydroxybenzene carboxylic acid; 2,4-dihydroxybenzoic acid; 4-hydroxysalicylic acid; 4-carboxyresorcinol) $(OH)_2C_6H_3COOH$.
Properties: White needles; m.p. (with decomposition) 219-220°C; b.p., decomposes; almost insoluble in water and benzene; soluble in alcohol, ethyl ether. The sodium, potassium, ammonium, calcium and barium salts are soluble in water; the silver, lead and copper salts are only slightly soluble.
Typical specifications: White or light buff crystalline powder; moisture 2.5% max; m.p. (decomposition point) 215°C min; assay not less than 97.0% pure.
Containers: 100-lb net nonreturnable fiber drums.
Use: As a dyestuff and pharmaceutical intermediate and as a chemical intermediate in the synthesis of fine organic chemicals.

"Reswax." [65] Trademark for a series of wax-resin blends used as coatings and hot melt adhesives in paper conversion. The resins or polymers used include butyl rubber, polyisobutylene, chlorinated rubber, polyethylene, and styrene co-polymers.

"Resyn." [53] Trademark for a line of synthetic resin-base adhesives and coatings.

"Resyn 3600." [53] Trademark for aqueous dispersion of modified polyvinylidene chloride, of special value for paper coating, resulting in a film characterized by excellent barrier properties.

ret. To reduce or digest fibers, especially linen, by enzymatic action.

retarder. Designation of a group of cationic surface-active agents operative in acid and electrolyte solutions. Used as dye retarders, stripping agents, and as oil well acidizing assistants.

"Retarder E-S-E-N." [248] Trademark for phthalic anhydride.
Properties: White crystalline powder; sp.gr. 1.52; melting range 127-132°C; soluble in acetone; moderately soluble in ethylene dichloride; slightly soluble in benzol and water. Insoluble in gasoline.
Uses: Retarder or antiscorch for rubber compounding.
Hazards: Excessive dustiness is to be avoided as a source of possible irritation to mucous membranes.

"Retarder J." [248] Trademark for N-nitrosodiphenylamine. (q.v.).

"Retarder PD." [57] A proprietary name for a modified phthalic anhydride. Effective as a retardant of premature cure or set-up during processing of rubber compounds. Normally used in combination with thiazole or mixtures of thiazole-guanidine accelerators.

"Retarder W." [28] Trade name for salicylic acid with a dispersing agent.
Properties: Buff-colored, non-dusty powder; sp.gr. 1.43.
Containers: 300-lb drums.
Uses: To prevent premature curing of natural and synthetic rubbers; to activate blowing agents.

"Retardex." [94] Trademark for a vulcanization retarder.
Properties: A light colored, finely divided solid; easily dispersed; no staining characteristics; may safely be used in white or pastel colored rubber; stable in storage; sp.gr. 1.14.
Uses: In low temperature curing to elevate "critical" temperature without affecting the time-tensile curing curve.

"Retardion." [233] Trademark for an ion-exchange resin used in ion retardation, a separation process in which the flow of ions through a column is retarded by the ion-absorbing properties of the resin.

retinene (vitamin A aldehyde) $C_{20}H_{28}O$. A necessary component of rhodopsin, the light-sensitive pigment of the eye. Retinene is the aldehyde form of vitamin A, which is an alcohol.

retinol. Vitamin A_1 or axerophthol. See vitamin A.

retort carbon. See carbon, gas.

retort graphite. See carbon, gas.

"Retrol." [217] Brand name of acid activated clay used as decolorizing adsorbent for refining used crankcase or other industrial oils. See also "Filtrol."

"Rexforming" Process. [416] Patented process combining certain of the elements of the "Platforming" process (q.v.) and "Udex" extraction (q.v.) to convert a naphtha fraction into a highly aromatic motor fuel blending component of high octane rating. A comparatively much smaller volume of an essentially paraffinic product also is produced, which has desirable characteristics as a component of jet fuels but which may, if preferred, be recycled to extinction. By-product hydrogen also is produced.

Reynold's number. The function DUP/μ used in fluid flow calculations to estimate whether flow through a pipe or conduit is streamline or turbulent in nature. D is the inside pipe diameter, U is the average velocity of flow, P is density and μ is the viscosity of the fluid. Different systems of units give identical values of the Reynold's number, and values much below about 2100 correspond to streamline flow, while values above 3000 correspond to turbulent flow.

"Rezgard." [58] Trademark for a fire retardant.
Properties: A white, water-soluble crystalline powder. An alkaline mixture of flame-retarding chemicals to give proper balance to flame extinguishing, elimination of afterglow and retention of tensile strength.
Containers: 100-lb bags; 300-lb fiber drums.
Uses: Fire retardant for cellulosic fabrics.

"Rezyl" Resins. [57] Proprietary products. Reaction products of polybasic acids and polyhydric alcohols modified with various oils and oil acids. Some grades further modified with resins of other types. Those based on drying oils are used in air drying surface coatings and alone, or in combination with amino resins, for baking finishes. The nondrying oil modifications are used as plasticizers for amino resins in baking finishes and for nitrocellulose and other high polymer film-forming materials. Available in numerous grades which vary in oil content and are designed for specific applications. Usually supplied in solution in organic solvents.

RF black. Abbreviation for reinforcing furnace black. See furnace black.

R film. Name used for "Tedlar" polyvinylfluoride film during development stages.

RFNA. Abbreviation for red fuming nitric acid. See nitric acid, fuming.

Rh. Symbol for rhodium.

rhabarberone. See aloe-emodin.

rhenium Re. Element, atomic number 75; group VII of the periodic table.
Properties: Silver white metal or gray to black powder; sp.gr. 20.53; m.p. about 3100°C; tensile strength 80,000 psi; soluble in concentrated nitric acid; slowly soluble in sulfuric acid; practically insoluble in hydrochloric acid; can be hot forged or rolled.
Sources: Principally molybdenite.
Derivation: Solutions from refinery residues (molybdenum ore flue dust, copper ore treatment) are (a) concentrated by salting-out processes and reduced by hydrogen gas under pressure to give the metal or (b) passed through an anionic resin from which pure rhenium can be extracted by a strong mineral acid.
Forms: Metal powder techniques are used to make rods, wires and strips.
Uses: As a catalyst for dehydrogenation; thermocouples; electric contact points, spark-plug points, electrodes, high temperature and corrosion resistant alloys.

rhenium compounds. Rhenium is multivalent, but its more stable compounds are heptavalent. Rhenium heptoxide, Re_2O_7, is a yellow crystalline material, sp.gr. 8.2; m.p. 220°C; dissolves in water to form perrhenic acid, $HReO_4$. Rhenium trioxychloride, ReO_3Cl, is a colorless liquid, m.p. 4°C; b.p. 131°C (760 mm); reacts readily with organic substances.

rhenium heptoxide. See rhenium compounds.

rhenium trioxychloride. See rhenium compounds.

rheology. Study of the deformation and flow of matter in terms of stress, strain and time. See Newtonian flow, consistency, dilatancy, thixotropy, and viscosity.

rhesus factor. A substance present in the red blood cells of the rhesus monkey, and of approximately 85% of an average white American population. Those whose red cells contain this factor are termed Rh-positive; others, Rh-negative. A negative individual may develop anti-Rh antibodies, if Rh-positive red cells enter his blood; such antibodies can then agglutinate Rh-positive cells. Hemolytic reactions may thus occur following transfusion of Rh-positive bloods cells into a recipient previously sensitized and having Rh antibodies in the serum. Likewise, an Rh-positive fetus may give rise to antibodies in the blood of an Rh-negative mother; the antibodies, returning into the fetus, may then produce the disease erythroblastosis fetalis. There are many subtypes of the Rh factor; these can be distinguished by serologic tests, and the laws of their inheritance have been determined.

Rh factor. See rhesus factor.

rhm. The abbreviation for one roentgen per hour at one meter.

rho acid. See anthraquinone-1,5-disulfonic acid.

rhodamine. A class of organic pigments which fluoresces bright orange to red when excited by ultraviolet radiation.

rhodamine B $C_{28}H_{31}ClN_2O_3$. A basic red fluorescent dye, C.I. No. 749, structurally related to xanthene.
Properties: Green crystals or reddish-violet powder. Very soluble in water and alcohol, forming bluish-red, fluorescent solution; slightly soluble in acids or alkalies.
Derivation: By fusion of meta-diethylaminophenol and phthalic anhydride followed by acidification with hydrochloric acid.
Uses: Principally, as a red dye for paper; also used for wool and silk where brilliant fluorescent effects are desired and lightfastness is of secondary importance; analytical reagent for certain heavy metals; biological stain.

rhodamine toners. Red to maroon lakes of rhodamine dyes and phosphotungstic or phosphomolybdic acid. They have good lightfastness and are used principally in printing inks. See also phosphomolybdic or phosphotungstic pigments.

rhodanates. The same as sulfocyanates or thiocyanates (preferred). See under the individual metals which form thiocyanates.

rhodanides. The same as rhodanates.

rhodanine (2-thio-4-keto-thiazolidine)
$\overline{SCH_2C(O)NHCS}$.
Typical specifications: Finely crystalline, light yellow color; sp.gr. 0.868; bulk density 0.617; m.p. decomposes, often violently, 166°C; pure material, 167-168.5°C; soluble in methanol, ethyl ether and hot water.
Use: Organic synthesis (phenylalanine).
Handle with caution!

rhodinol. A mixture of citronellol and geraniol.
Properties: Colorless liquid; very fine, heavy rosy odor.
Constants: Sp.gr. 0.850-0.870; refractive index 1.457-1.467.
Derivation: From oil of geranium.
Containers: 5-lb cans.
Use: Perfumery, for rose and other floral notes.

rhodinyl acetate.
Properties: Colorless liquid consisting of a mixture of citronellyl and geranyl acetates.
Derivation: Action of acetic anhydride on rhodinol in the presence of sodium acetate.
Grades: Technical.
Use: Perfumery.

rhodium Rh. Element having atomic number 45, group VIII of the periodic system.
Properties: The whitest of the platinum group of metals; somewhat limited ductility but can be obtained in the form of foil and fine wire. Insoluble in acids and practically so in aqua regia; soluble in fused $KHSO_4$. Insoluble in fused alkalies.
Constants: Sp.gr. 12.44; m.p. 1950-2000°C. Brinell hardness, hard 390, annealed 135.
Derivation: Occurs with platinum, from which it is recovered during the purification process.
Uses: Catalyst (in alloy with platinum); electroplating; mirrors; thermocouples; printed electrical circuits; alloyed with platinum, it is used for spinnerets in the rayon industry. The sulfate is used in electroplating.

rhodium chloride (rhodium trichloride) $RhCl_3$.
Properties: Brownish-red powder. Soluble in solutions of alkalies and cyanides; insoluble in acids, water.
Constants: B.p. 800°C (sublimes); m.p. 450-500°C (dec).

rhodium trichloride. See rhodium chloride.

rhodochrosite $MnCO_3$ with partial replacement by Fe, Ca, Mg, Zn.
Properties: Light pink, rose-red, brownish-red or brown mineral; white streak; vitreous to pearly luster; photoluminescent. Contains 61.7% MnO, 38.3% CO_2. Found in veins with ores of silver, lead, copper, manganese.
Constants: Sp.gr. 3.3-3.6; hardness 3.5-4.5.
Occurrence: United States (Connecticut, New Jersey, Colorado, Montana, Nevada); Hungary; Rumania; Germany; Belgium.
Use: Manganese ore.

rhodolite. See garnet.

rhodopsin. The red-light-sensitive pigment of the eye, visual purple, consisting of the protein, opsin, and retinene (vitamin A aldehyde). It is found in land and marine vertebrates.

"Rhoduline." [307] Trademark of basic dyestuffs. Used for the dyeing and printing of silk, cotton, rayon, for the dyeing of paper, leather, lacquers, and plastics. Also used as phosphotungstic toners or lakes for printing inks, wall paper and coated paper. Characterized by exceptional brilliancy.

"Rhonite." [23] Trademark for thermosetting modified and unmodified urea-formaldehyde condensates. Supplied as water-clear solutions and aqueous pastes. Reactive with cotton, various grades producing shrink-resistance, crease-proofing or modification of hand.
Use: Textile finishing of natural and synthetic fiber fabrics.

"Rhoplex." [23] Trademark for aqueous dispersions of acrylic copolymers. White opaque emulsions. Various grades differing in hardness, flexibility, adhesion and tack of film; some thermosetting. Produce colorless transparent films with outstanding permanence properties, durability, adhesion, and pigment-binding capacity.
Use: Emulsion paints; paper coatings and saturation of paper; floor sealers and wax emulsions; textile backing and textile finishing; bonding of textile fibers and pigments; clear and pigmented coatings on wood and metals.

"Rhothane." [23] Trademark for an agricultrual insecticide based on 1,1-bis (chlorophenyl) -2,2-dichloroethane and supplied as a

wettable powder or emulsion concentrate.
Use: Controls insects on plants and livestock. Also used for mosquito larvacide and adulticide.

"Rhozyme." 23 Trademark for enzyme concentrates with diastatic or proteolytic activity. Buff-colored powders or liquids of fungal or bacterial origin which hydrolyze and solubilize proteins and starches, depending upon type.
Use: Desizing of textile fabrics; dry-cleaning spotting; liquefaction of starch paste; fermentation processes; manufacture of corn syrup, fish solubles, septic tank formulations; medicinal, animal feed, and meat tenderizer applications.

rhubarb. The dried rhizome and root of certain species of Rheum. The official rhubarb of N.F. XI is a Chinese rhubarb: Rheum officinale, Rheum palmatum or certain other species grown in China. Indian rhubarb, N.F. XI (Himalayan rhubarb) is Rheum webbianum, Rheum emodi, and certain other species native to India, Pakistan or Nepal.
Chief constituents: Chrysophanic acid and other derivatives of methylanthracene.
Grades: Technical; N.F. XI.
Containers: Bags.
Use: Medicine.

"Rhulitol." 57 Trademark for tannic acid solution.

riboflavin. Alternate spelling of riboflavine. The latter is preferred by the IUPAC.

riboflavine (vitamin B_2) $C_{17}H_{20}N_4O_6$. 7, 8-Dimethyl-10-(1^1-D-ribityl) isoalloxazine. A crystalline pigment, the principal growth promoting factor of the vitamin B_2 complex. It functions as a flavoprotein in tissue respiration. A syndrome resembling pellagra is thought to be due to riboflavine deficiency.
Properties: Orange yellow crystals; bitter taste; m.p. 282°C (dec). Slightly soluble in water and alcohols, insoluble in lipid solvents, stable to heat in dry form and in acid solution; stable to ordinary oxidation; unstable in alkaline solution, and quite sensitive to light. In solution, riboflavine has an intense greenish-yellow fluorescence.
Units: Amounts are expressed in milligrams or micrograms of riboflavine.
Sources: Food sources: milk, green leafy vegetables, egg yolk, liver, enriched flour, yeast. Commercial sources: distiller's residues, fermentation solubles; synthetic production (indirectly from dextrose).
Grades: U.S.P. XVI.
Uses: Medicine; nutrition; animal feed supplement; enriched flours.

riboflavine 5'-phosphate (FMN, flavine mononucleotide). The phosphate ester of riboflavine (q.v.) in which the phosphate is esterified to the ribityl portion of riboflavine.

It functions as a coenzyme for many flavine enzymes (q.v.). The riboflavine group has the ability to take up hydrogen atoms, thus oxidizing the substrate.
Properties (sodium salt): Yellow crystals. Much more soluble than riboflavine in water. Quite sensitive to ultraviolet light.
Derivation: By treating riboflavine with chlorophosphoric acid.
Containers (sodium salt): Fiber drums.
Use: Biochemical research.

9-beta-D-ribofuranosyladenine. See adenosine.

ribonucleic acid (ribose nucleic acid; yeast nucleic acid; RNA) $C_{38}H_{49}O_{29}N_{15}P_4$ (for the tetranucleotide). Contains the sugar D-ribose, the bases adenine, guanine, cytosine and uracil, and phosphoric acid. Ribonucleic acid is universally present in the cytoplasm (cellular material exclusive of the cell nucleus) of cells, and some evidence indicates that it is also present in the cell nucleus. The acid is usually bound to proteins to form ribonucleoproteins. See deoxyribonucleic acid for recent ideas about the structure of the nucleic acids.
Properties: White to buff powder; insoluble in water and dilute acid; soluble in alkali.
Derivation: From yeast.
Uses: Biochemical research.

D-ribose $CH_2OH(CHOH)_3CHO$ D(-)-ribose. A five-carbon-atom sugar which is a constituent of components in living cells, or their decomposition products, such as deoxyribonucleic acid, ribonucleic acid, nucleotides, nucleosides, and many coezymes.
Properties: White, crystalline, slightly sweet odor; m.p. 86-87°C; soluble in water, ethyl alcohol, and methyl alcohol. Insoluble in ether, acetone, benzene, and chloroform.
Use: Physiological and biochemical studies of cell function.

ribose nucleic acid. See ribonucleic acid.

D-ribose-5-phosphoric acid $C_5H_9O_4 \cdot H_2PO_4$. A constituent of nucleotides and nucleic acids.
Properties: The barium salt ($5\frac{1}{2} H_2O$) is sparingly soluble in cold water and crystallizes in hexagonal plates.
Derivation: From inosinic acid.
Use: Biochemical research.

D-ribosyl uracil. See uridine.

Rice's bromine solution. A reagent used for the quantitative determination of urea; depends upon the oxidizing action of bromine.

ricin. White powder. The tox-albumin of the castor oil bean, being the poisonous principle. Extracted from the pressed seeds with 10% solution of sodium chloride followed by precipitation with magnesium sulfate.
Uses: As a reagent for pepsin and trypsin.
Caution! Extremely poisonous; handle with care; small particle in cut or abrasion, eye or nose, may prove fatal. Poison label.

ricinine $C_8H_8O_2N_2$. 1, 2-Dihydro-4-methoxy-1-methyl-2-oxonicotinonitrile.
Properties: White crystalline alkaloid; bitter;

soluble in water, chloroform and ether. M.p. 201.5°C; sublimes at 170-180° under 20 mm pressure.
Derivation: From castor oil seeds and leaves.
Shipping regulations: None.*

ricinola oil. See castor oil.

ricinoleic acid (cis-12-hydroxyoctadec-9-enoic acid; 12-hydroxyoleic acid; castor oil acid) $CH_3(CH_2)_5CH(OH)CH_2CH{:}CH(CH_2)_7COOH$. A C_{18} unsaturated fatty acid which comprises approximately 80% of the fatty acid content of castor oil. The presence of the hydroxyl group imparts a number of properties (following sulfonation or dehydration) which widens its utility.
Properties: Colorless to yellow viscous liquid; soluble in most organic solvents; insoluble in water. Sp.gr. 0.940 (27.4/4°C); m.p. 5.5°C; b.p. 226°C (10 mm); refractive index 1.4697 (20°C).
Derivation: Saponification of castor oil.
Method of purification: Rectification.
Grades: Technical; purified (99+%).
Containers: 5-gal cans; 55-gal drums.
Uses: Soaps, Turkey red oils; textile finishing; source of sebacic acid and heptanal; ricinoleate salts; 12-hydroxystearic acid.
Shipping regulations: None.*

ricinoleyl alcohol $CH_3(CH_2)_5CH(OH)CH_2CH{:}CH(CH_2)_7CH_2OH$. The fatty alcohol derived from ricinoleic acid. It has a long straight chain with one double bond and one hydroxyl (OH) group in a secondary position on this chain besides the primary group on the end. Available as a 90% product. A colorless non-drying liquid at room temperature.
Typical specifications: Iodine value 91.8; cloud point below 10°F; boiling range 170-328°C; viscosity 51 (SSU/21°C); fire point 435°F.
Derivation: Reduction of acid made from castor oil.
Impurities: Oleyl and linoleyl alcohols.
Uses: Protective coatings; polyesters; plasticizers; organic synthesis, pharmaceuticals, lubricants, surface active agents.

ricinus (castor oil plant; palma christi; Mexico seed; oil plant; castor bean).
Derivation: The seeds of Ricinus communis.
Occurrence: United States; East Indies; West Indies; southern Europe and Africa.
Grades: Technical.
Containers: Bags.
Use: Source of castor oil.
Shipping regulations: None.*

ricinus oil. See castor oil.

Riegler's test. A reagent composed of sodium naphthionate, beta-naphthol, and concentrated hydrochloric acid for the detection of nitrites and nitrous acid in very small amounts.

"Rila." [4] Trademark for a series of phosphorescent and fluorescent pigments (ZnS-CdS, CaS-SrS), compatible with a variety of vehicles.
Uses: Lacquers, paints, printing inks, paper, tape and plastics.

"Rimex." [250] Trademark for a chemically compounded material used in the production of rimming steels. It is an effective ingot mold addition agent which reduces fuming characteristics and improves rimming action.
Containers: 4- and 8-oz polyethylene bags; bulk drums.

"Rimifon." [190] Trademark for isoniazid (isonicotinylhydrazine).

"Ringots." [329] Trademark for uncoated, doughnut-shaped pieces of misch metal. They are available in 1, 2, and 5-lb sizes with the center hole large enough to permit a standard 1" pipe or rod to be passed through. This hole is convenient for mounting, plunging and effectively mixing misch metal with molten metal.
Use: Special steels.

"Ringwood Repellent." [401] Trade name for a rodent repellent containing trinitrobenzene-aniline complex (5%).
Containers: 1 pint, 1 quart and 1 gallon containers.
Use: To protect dormant deciduous trees and shrubs against rabbit damage.
Caution: Harmful if swallowed, inhaled or absorbed through skin.

"Rio Resin." [69] Trademark for proprietary blend of resinous and protective materials.
Properties: Resinous; orange to dark red; sp.gr. 1.13 ± .03; softening point 54°C min.
Uses: In compounding heat resistant copolymer and corona resistant neoprene.

"RISA." [3] Radioactive iodinated blood serum albumin (human) used as research and diagnostic tool.

ristocetin. An antibiotic produced by the fermentation of Nocardia lurida, a species of Actinomycetes. The antibiotic has two components, ristocetin A and ristocetin B, the chemistry of which is not completely known. The commercial product is a lyophilized preparation representing a mixture of A and B.
Grade: N.N.D.
Use: Medicine.

"Ritalin." [305] Trademark for methylphenidate.
Use: Medicine.

Rittinger's law. The energy required for reduction in particle size of a solid is directly proportional to the increase in surface area. See also Kick's law.

"R Monel." [283] Trademark for a wrought alloy containing approximately 66% nickel and 31% copper which has had its machining qualities enhanced by a controlled addition of sulfur.

Rn. Symbol for radon.

*See "I.C.C. Shipping Regulations," page xiii.
Reference numbers refer to name of manufacturer. See "List of Manufacturers," page v.

RNA. Abbreviation for ribonucleic acid.

road oil. 1. Asphaltic-, see asphalt. 2. Non-asphaltic-, a non-hardening petroleum residual oil used as a dust-laying oil. It has sufficiently low viscosity to be applied without preheating.

roasting. Heating in the presence of air or oxygen. Most commonly used in converting natural metal sulfide ores to oxides as a first step in recovery of metals such as zinc, lead, copper, etc.

"Robane." [415] Trademark for squalane (q.v.) Purity 90% min.

Containers: 1-lb bottles; 1-gal metal tins; 5-gal metal pails; 100-lb metal drums.

Uses: Since it is a derivative or possible component of the sebum hydrocarbons, it is especially suited as a vehicle for skin medicines and cosmetics.

Robison ester. See glucose 6-phosphate.

"Roccal." [162] Trademark for a mixture of high molecular weight alkyl benzyl dimethyl ammonium chlorides in 10 and 50% aqueous solution. The alkyl groups are straight chain C_8 to C_{18} from coconut fatty acids. Used as a germicide in the food and beverage industry, and for similar applications.

rochelle salt. See potassium-sodium tartrate.

rock candy. Semi-transparent crystals of hydrated sugar, the purest form of sucrose commercially obtainable.

Derivation: Gently heating concentrated sugar syrup, adding alcohol, and cooling slowly.

rock crystal. 1. Colorless and very pure quartz (q.v.) occurring in crystals found in Brazil and Madagascar; used for oscillator plates in electronic equipment; optical equipment; raw material for fused quartz; gemstone.

2. Rock crystal is also used to refer to highly polished blown glassware that has been handcut or engraved.

rocket fuels. See rocket propellants.

"Rocketon." [349] Trademark for a wet or dry phenolic-abestos molding compound used when good heat and ablative properties are needed.

rocket propellants (rocket fuels). Solid and liquid propellants of various classifications. The homogeneous solid propellant consists of a base–usually nitrocellulose–which acts both as oxidizer and fuel. Stabilizers, of which the most common are diphenylamine and diethyldiphenylurea, remove oxides of nitrogen which are formed in nitrocellulose decomposition during storage. Plasticizers, added to improve mechanical qualities, may be nonexplosive, such as various phthalates, or explosive, such as nitroglycerine. The latter is not only a plasticizer, but a propellant in itself. Darkening agents, such as carbon black, are added to stop radiation transfer which results in premature ignition. The composite solid propellant, on the other hand, is a mixture of a finely ground oxidizer and a plastic-consistency fuel, which acts as the binder. The oxidizers are usually alkali nitrates or perchlorates. The fuels are of several types: synthetic resins, elastomers (such as polysulfide rubber), polyesters, and formerly cellulose derivatives and asphalt. Inert plastics are added to the molded propellant along certain surfaces to produce burning in a particular section.

Liquid propellants fall into two classifications: the mono- and bipropellants. The liquid monopropellant is analogous to the homogeneous solid propellant in that it is basically a single substance which is both oxidizer and fuel. Typical monopropellants are nitroglycerin, nitromethane, diethylene glycol dinitrate (the German DEGN of World War II). The bipropellant consists of fuel and oxidizer stored in separate chambers, then mixed for combustion. Oxidizers are most commonly liquid oxygen or nitric acid, but other oxygen compounds or halogens and their compounds (and in special cases, water) are also used. The fuels include hydrogen and wide range of hydrogen compounds-ammonia, hydrazine, metallic hydrides, such as those of boron, hydrocarbons, amines, alcohols. Bipropellants are far more commonly used than monopropellants. The following table gives common systems:

Fuel	Oxidizer
ammonia	liquid oxygen
ethyl alcohol	liquid oxygen
gasoline or kerosine	liquid oxygen
gasoline	red fuming nitric acid
aniline	red fuming nitric acid
hydrogen	fluorine
methanol	chlorine trifluoride
hydrazine	hydrogen peroxide
furfuryl alcohol	white fuming nitric acid
boron hydride	water
uns-dimethyl-hydrazine and hydrazine	nitrogen tetroxide
hydrazine	red fuming nitric acid
ammonia	red fuming nitric acid
hydrogen	ozone
hydrogen	oxygen
hydrogen	white fuming nitric acid
hydrogen	nitrogen tetroxide

The hybrid type of propellant combines solid and liquid propellant characteristics. It is also called a slurry. There are three varieties:

(a) Solid oxidizer and liquid fuel. Example: kerosene, a gelling agent, perchlorate oxidizer.

(b) Solid binder containing part of the oxidizer with the rest of the propellant in the liquid state.

(c) Solid fuel binder with a liquid oxidizer.

Rocket propellants are usually compared

on the basis of specific impulse which means the pounds of thrust produced per pound of propellant burned per second. High specific gravity is also desirable, to reduce the size of the containing rocket or missile.

Generally, the liquid propellants are more easily controlled during operation, and certain liquid systems have the highest performance potential. However, solid propellants have been gaining favor due to their simplicity of storage, handling, and the fact that they can be fired almost immediately, with no complex mixing, and with great reliability.

Future propellant possibilities, which are being studied due to the definite limits on performance of chemical fuels, include use of the recombination energy of free radicals, use of nuclear propulsion, ion flow, solar energy, and (most ultimately) propulsion by high energy photons.

rock oil. A synonym for petroleum (q.v.).

rock salt. See halite.

rock salt moss. See chondrus.

rock wool. See mineral wool.

"Rodar." [155] Trademark for an iron-nickel-cobalt alloy.

Properties: Produces a permanent vacuum-tight seal with simple oxidation procedure; resists mercury corrosion; readily machined and fabricated; can be welded, soldered, or brazed.

Forms: Wire; strip; bar.

Use: For sealing metal to hard glass.

rodenticide. Chemicals used to kill rats and other undesirable rodents. Many older rodenticides (arsenic, strychnine, ANTU) were toxic to all animals. Newer materials, such as the indandione compounds and warfarin (a coumarin derivative), are more selective, and owe their activity to their anticoagulating effect on the blood and subsequent hemorrhages in the animal.

"Rodform." [69] Trademark for rubber accelerators supplied in the form of extruded rods or pellets.

"Rodine." [342] Trademark for red squill liquid-extract rodenticides.

"Rodo No. 0." [69] Trademark for a blend of essential oils.

Properties: Sweet odor; clear; nontoxic; sp.gr. 0.96.

Use: To offset the usual rubber odor, leaving the finished product practically odorless.

"Rodo No. 4." [69] Trademark for a product similar to "Rodo No. 0." Imparts a pleasing, distinctive odor to finished rubber.

"Rodo No. 10." [69] Trademark. Same as "Rodo No. 0." Imparts to rubber stocks a pleasing and lasting odor.

roentgen (r). The international unit of quantity or dose for both x-rays and gamma rays. It is defined as the quantity of x- or gamma rays which will produce as a result of ionization one electrostatic unit of electricity of either sign in 1 cc (0.001293 g) of dry air as measured at 0°C and standard atmospheric pressure.

The use of the roentgen unit has been extended to include particle radiation such as alpha and beta particles and protons and neutrons. The roentgen equivalent physical, abbreviated rep, is defined as the quantity of particle radiation which upon absorption in 1 gram of body tissue is accompanied by the gain of 93 ergs of energy. One rep of ionizing radiation produces essentially the same physical effect in soft tissue as does one roentgen of x-rays. One mrep is one thousandth of a rep.

Roentgen rays. See x-rays.

"Roetinic." [299] Trademark for a vitamin and mineral preparation, used in medicine.

Rohrbach solution.

Properties: Clear, yellow liquid. Very refractive; sp.gr. 3.5.

Derivation: An aqueous solution of mercuric-barium iodide.

Uses: Separating minerals by their specific gravity; microchemical detection of alkaloids.

Caution: Poison; keep well closed.

"Rolandshuette." [447] An aluminous cement of the fused type which originated in Germany. See cement, aluminous.

"Romark." [448] Trade name for alkyd and chlorinated rubber type road marking paints.

"Romilar" Hydrobromide. [190] Trademark for a brand of dextromethorphan hydrobromide (q.v.).

"Rongalite CX." [307] Trademark for a stripping and reducing agent, sodium sulfoxylate formaldehyde; 99% pure.

Properties: Hard, white lumps; decomposes slowly in moist, warm air.

Uses: Textile stripping agent; reducing agent for discharge and vat prints; reducing agent in GR-S latex low temperature redox polymerization systems.

"Roniacol" Tartrate. [190] Trademark for a brand of beta-pyridylcarbinol tartrate (q.v.).

ronnel (O,O-dimethyl O-(2,4,5-trichlorophenyl) phosphorothioate). Accepted as generic name by the Ent. Soc.

Use: Systemic insecticide; used for the elimination of external parasites on animals by oral treatment.

"Ronopole" Oil. [165] Trademark for a highly oxidized sulfonated castor oil.

"Roofmate." [233] Trademark for a "Styrofoam" product used in insulation.

"Rootone." Trade name for indolebutyric acid.

"Roracyl." [28] Trademark for a group of soluble dyes that have good affinity and fastness properties on "Orlon" acrylic fiber. They are used with selected acid dyes in many instances.

"Rosaldehyde." [188] Brand name for a synthetic floral perfume base.

"Rosanlik." [188] Trademark for a synthetic replacement for otto of rose.

roscoelite $K_2V_4Al_2Si_6O_{20}(OH)_4$. A vanadium-bearing species of mica (q.v.). Formula variable, with V_2O_3 up to 28%. Occurs as minute scales with micaceous cleavage; dark green to brown in color; pearly luster. Hardness 2.5; sp.gr. 3.0.
Occurrence: Colorado, California, Australia.
Use: Minor source of vanadium.

rose absolute. Pure oil of rose. The first filtrate obtained on separation of waxes from the cooled alcohol solution of rose concrete in perfume manufacture.

rose concrete. Semi-solid residue, a mixture of essential oils and waxes, resulting from extraction of rose flower petals, leaves, seeds, fruit, roots, gums or bark by means of a volatile solvent. Term used in perfume manufacture. See rose absolute.

rose flower oil. See rose oil.

rose geranium oil. See geranium oil, rose.

rosemary (garden rosemary).
Derivation: Flowers and leaves of Rosmarinus officinalis.
Occurrence: Mediterranean basin; cultivated in gardens.
Containers: Boxes.
Uses: Perfumery; medicine.
Shipping regulations: None.*

rosemary oil.
Properties: Colorless or pale yellow, limpid liquid, volatile oil; warm, somewhat camphoraceous taste; pungent; rosemary odor; sp.gr. 0.894-0.912; optical rotation −5° to +10°; saponification value 12-20; refractive index (n 20/D) 1.4640-1.4760. Soluble in alcohol, ether, and glacial acetic acid.
Chief known constituents: Pinene, borneol, camphene, cineole, camphor and bornyl acetate.
Derivation: By steam distillation from the leaves of Rosmarinus officinalis.
Occurrence: Southern Europe.
Grades: Technical; N.F. XI.
Containers: Cans; drums.
Uses: Perfumery; medicine.
Shipping regulations: None.*

rose oil (otto of rose oil; attar of roses).
Properties: Pale yellow, pale green, or pale red, transparent, volatile, liquid oil; mild, sweet taste; strong, fragrant odor; semi-solid at ordinary temperature; keep containers well stoppered. Sp.gr. 0.845-0.865; solidifying point 18-37°C; saponification value 10-17; acid value 0.5-3; refractive index (n 30/D) 1.457-1.463.
Chief constituents: Geraniol, citronellol and phenylethyl alcohol.
Derivation: By steam distillation of the fresh flowers of Rosa damascena, Rosa centifolia, Rosa gallica and Rosa alba.
Grades: Bulgarian; French; Turkish.
Containers: 1-, 8-, 16-oz bottles; 1-, 8-, 16-, 32-oz packages; 1-, 2-kilo packages.
Uses: Perfumes; flavoring.
Shipping regulations: None.*

Rose's alloy. See table under fusible alloys.

"Rosetone." [19] Trademark for trichloromethyl phenyl carbinyl acetate (q.v.).

rose water. A water solution of the odorous materials of Rosa centifolia, prepared by steam distillation of the fresh flowers.
Grade: U.S.P. XVI.

rosewood oil. See oil bois de rose Brazilian.

rosin (gum rosin; colophony; pine resin; wood rosin; common rosin)
Properties: Angular, translucent, amber-colored fragments; sp.gr. 1.08; m.p. 100-150°C; acid no. not less than 150. Insoluble in water; freely soluble in alcohol, benzene, ether, glacial acetic acid, oils, carbon disulfide, dilute solutions of fixed alkali hydroxides.
Chief constituents: Resin acids of the abietic and pimaric types, having the general formula $C_{19}H_{29}COOH$, and having a phenanthrene nucleus. They are unsaturated and hence reactive. An unsaponifiable portion of rosin (3-10%) contains hydrocarbons and high molecular weight alcohols. These constituents vary greatly according to the source of the rosin.
Derivation: From pine trees, chiefly Pinus palustris and Pinus caribaea. (a) Gum rosin is the residue obtained after the distillation of turpentine oil from the crude turpentine oleoresin. (b) Wood rosin is obtained by extracting pine stumps with naphtha and distilling off the volatile fraction. (c) Tall oil rosin is a byproduct in the fractionation of tall oil (q.v.).
Grades: Virgin; yellow dip; hard; N.F. XI. Rosin is graded B, C, D, E, F, G, H, I, K, L, M, N, W-G (window-glass), W-W (water-white). The grading is done by color, B being the darkest and W-W the lightest rosin. Ordinarily the first three grades, B, C, and D, are not separated. Occasionally factors other than color are considered in the grading such as the acidity and the melting point.
Containers: Multiwall paper bags; drums.
Uses: Linoleum; soldering compounds; core oils; insulating compounds; molding compounds; sealing waxes. Metallic salts (soaps) of rosin are used in soaps, emulsifiers for SBR manufacture, paper sizing, printing inks, varnishes. Rosin esters are used in protective coatings, as are polymers and copolymers.
Caution: Combustible; gives off flammable vapors when heated.

rosin essence. That portion of rosin distilling below 360°C.

rosin ester gums. See ester gums.

rosin esters. See ester gums.

rosin oil (rosinol; codoil).
Properties: Water-white to brown liquid; viscous; odorless; strong, peculiar taste. Soluble in ether, chloroform, fatty oils and carbon disulfide; slightly soluble in alcohol; insoluble in water. Essentially decarboxylated rosin acids.
Constants: Sp.gr. 0.980-1.110; iodine number 112-115.
Derivation: By fractional distillation of rosin in a retort, that portion distilling above 360°C being rosin oil.
Method of purification: Second distillation.
Grades: Technical (first, second, third run).
Containers: 50-gal wooden barrels.
Uses: Lubricant; adulterant for boiled linseed oil; printing inks; impregnating paper for wrapping electric cables; rubber reclaimers; core compounds; linoleum.
Shipping regulations: None.*

rosinol. See rosin oil.

rosin size. A material compounded of rosin, soda ash and alum used for paper sizing to impart water and ink resistance to the paper.

rosin soap. See sodium abietate, and soap.

rosolic acid. The names rosolic acid, pararosolic acid and aurin are used confusingly either as direct synonyms for each other or to mean similar compounds or groups of compounds. The preponderance of opinion seems to lean toward the use of rosolic acid as a direct synonym for aurin (q.v.).

"Rosottone." [342] Trademark for a composition from natural and synthetic sources reproducing Rose de Mai absolute.

"Rossville." [319] Brand name for a line of ethyl alcohols available under the following proprietary names:
"Rossville Hexagon Cologne Spirits." Pure or specially denatured, highly refined grain spirit, 190 proof, pure, clear, and free from foreign odor, used in manufacture of finest perfumes, colognes, and toilet waters.
"Rossville Algrain Alcohol." Crystal clear, 190 proof, grain alcohol free from foreign odor or flavor for use in manufacture of medicinals, perfumes, toilet waters, flavoring extracts, and other high-grade products.
"Rossville Gold Shield Alcohol." Pure or specially denatured, 190 or 200 proof, high quality alcohol for industrial, scientific, pharmaceutical, and cosmetic purposes.

"Rotalin." [28] Trademark for a line of spirit-soluble printing colors.
Uses: Principally in the printing of glassine and other types of paper and cellophane.

"Rotax." [69] Trademark for a purified grade of 2-mercaptobenzothiazole.
Properties: Pale yellow powder; sp.gr. 1.52 ± .03; melting range 169-180°C; very soluble in dilute caustic, benzene, carbon disulfide, chloroform, alcohol; insoluble in water.
Uses: Primary accelerator for natural and nitrile rubber and SBR. Used in tires and tubes, wire and cable, molded and extruded goods, coated fabrics, footwear.

rotenone (tubatoxin) $C_{23}H_{22}O_6$. A pentacyclic compound.
Properties: White, odorless crystals. Soluble in ether, alcohol, acetone, carbon tetrachloride, chloroform and other organic solvents; insoluble in water. Not compatible with alkalies.
Constants: Sp.gr. 1.27 at 20°C; m.p. 163°C; strongly levorotatory in solution, specific rotation for D line, 230° in benzene, 62° in ethylene dichloride.
Derivation: By solvent extraction of derris and cube root.
Method of purification: Recrystallization from alcohol.
Grades: C.P. crystals; technical; also as extracts of derris and cube root.
Containers: Fiber drums; tins; multiwall paper sacks.
Uses: A very powerful insecticide (harmless to mammals and birds) claimed to be also very toxic to fish; flea powders; fly sprays; mothproofing agents.

"Rotessenol." [55] Trademark for an insecticide concentrate.
Typical analysis: Rotenone, 6%; cube extractives (other than rotenone), 9%; mono- and di-isopropyl cresols, 26%; dipentene, 29%; petroleum oil, 30%.
Uses: In insecticide formulations; in water emulsion sprays for truck garden crops; and for emulsion concentrates for warbles and cattle grub.
Caution: Harmful if swallowed. Do not leave on skin, in eyes or on clothing. Avoid contamination of feed and foodstuffs.
Fire hazard: Combustible mixture.

"Rotoclor." [56] Trademark for ferric chloride solution with other additives specially prepared for rotogravure work. Contains 43% $FeCl_3$. Sold only in 155-lb carboys.

rotten-stone. See tripoli.

rouge (synthetic iron oxide). A high-grade red pigment used as a polishing agent. It is similar to Venetian red (q.v.). The finer grades of rouge are known as jeweler's rouge. See also iron oxide reds. The term rouge is also applied to a cosmetic prepared from dried flowers of the safflower.

rouge black. See iron oxide, black.

rouge de mars (red iron oxide). A form of ferric oxide. See iron oxide reds.

"Rovana." [233] Trademark for a thermoplastic vinylidene chloride copolymer filament in

*See "I.C.C. Shipping Regulations," page xiii.
Reference numbers refer to name of manufacturer. See "List of Manufacturers," page v.

the form of a folded flat tape, offered in 300, 400 and 550 deniers.

"Roxtone." [51] Trademark for low cold test oils for journal bearings. Typical uses are on construction equipment, crushers and mine equipment. Heavier grades are applied to rough bearings and gears.

"Royal Methyl Violet." [141] Trade name for violet pigment produced by precipitation of the basic methyl violet dyestuff with phosphomolybdic acid.
Properties: Brilliant shade, very high strength, fair lightfastness.
Uses: Printing inks.

"Royal Spectra." [133] Trademark for an impingement carbon black. Used in specialty application requiring highest blackness.

"Royal Victoria Blue." [141] Trade name for blue pigment produced by precipitation of basic victoria blue dye with phosphomolybdic acid.
Properties: Excellent strength; good permanence.
Uses: Printing inks and some paints.

royal yellow. See orpiment.

RPA No. 2 Rubber Peptizing Agent. [28] ($33\frac{1}{3}$% 2-naphthalene thiol $C_{10}H_7SH$, and inert hydrocarbon $66\frac{2}{3}$%).
Properties: White, waxy material or slightly cream colored waxy flakes; sp.gr. 0.92.
Containers: 50-lb boxes.
Use: As a plasticizing and peptizing agent for crude and synthetic rubbers.

RPA No. 3 Rubber Peptizing Agent. [28] (36.5% mixed xylene thiols and 63.5% petroleum oil).
Properties: Amber or light yellow liquid; sp.gr. 0.91.
Containers: 400-lb drums.
Uses: As a plasticizing and peptizing agent for crude and synthetic rubbers; as a rubber reclaiming agent.

RPA No. 6 Rubber Peptizing Agent. [28] (90% pentachlorothiophenol C_6Cl_5SH, and 10% inert hydrocarbon wax.)
Properties: White or light gray powder; sp.gr. 1.79.
Containers: 125-lb drums.
Use: As a plasticizing and peptizing agent for natural and synthetic rubber.

RR-10. [28] Trade name for a rubber reclaiming chemical. Mixed dixylyl disulfides $[(CH_3)_2C_6H_3]_2S_2$.
Properties: Dark amber liquid; sp.gr. 1.05.
Containers: 450-lb drums.
Use: As agent for accelerating devulcanization of natural rubber, SBR (styrene butadiene rubber) and mixed scrap during the reclaiming process.

RR acid. See 2-amino-8-naphthol-3, 6-disulfonic acid.

R salt. The disodium salt of 2-naphthol-3, 6-disulfonic acid (q.v.).

"RSR." [173] A proteolytic enzyme preparation for removal of stains available in powder form specifically designed for removing albuminous spots and stains from garments.

"RTV" Silicone Rubber. [245] Trade name for a family of silicone rubber compounds. RTV (room temperature vulcanizing) rubbers have good physical properties and possess electrical properties similar to silicone rubber. These compounds are available in forms from easily pourable liquids to stiff pastes and cure at room temperature, after the addition of a catalyst, to the consistency of rubber. They are operative over the temperature range −65 to 600°F. RTV's possess good release characteristics and excellent bonding ability and exhibit shrinkage of less than 0.02%.
Uses: The material can be applied to objects of virtually any size or shape. Used in sealing, caulking, potting, encapsulating and flexible mold-making applications. Useful in electronic, aircraft, missile, and building industries and wherever a long term protective and thermal coating or sealant is required. Provides protection against heat, cold, moisture, ozone, arc, corona, and corrosive atmospheres of many types.

Ru. Symbol for ruthenium.

"Rubanox Red." [141] Trade name for lithol rubine pigments.
Composition: Calcium salts of the azo pigments formed when 4-aminotoluene-3-sulfonic acid is coupled with beta-hydroxynaphthoic acid.
Properties: Bright bluish shade reds, clean in masstone, clean and strong in tint. Good resistance to light, good heat resistance, non-bleeding in water and organic solvents. May be blended with light reds to give clean intermediate shades.
Grades: Resinated and non-resinated.
Uses: Printing inks, paints, enamels, lacquers, rubber, plastics, wallpaper, floor coverings.

rubber accelerators. See accelerator (1).

rubber, chlorinated. See chlorinated rubbers. See also rubber hydrochloride.

rubber, cold. See cold rubber.

rubber, crepe brown. An inferior grade of raw natural rubber.

rubber fiber. Generic name for a manufactured fiber in which the fiber-forming substance is comprised of natural or synthetic rubber (Federal Trade Commission). Often the rubber is a core around which yarns of cotton or other fibers are wrapped to make an elastic yarn. Such yarns are used for girdles, swimwear, elastic bands and tapes.

rubber, hard. Rubber compounded with 30-50% sulfur. Usually contains some lime or magnesia as a filler. Used for acid and alkali resistant tank linings, battery boxes, etc.

rubber hydrochloride. A hydrochloride derivative rather than a chlorine derivative.

Properties: A thermoplastic white powder or clear film. Odorless, tasteless, nonflammable. Chlorine content 29-30.5%. Soluble in aromatic hydrocarbons. Softens at 110-120°C. Films are highly resistant to moisture, oils, acids, and alkalis but tend to become brittle on exposure to sunlight. The life of such films is greatly extended by the incorporation of suitable stabilizers and plasticizers.

Derivation: A solution of rubber is treated with anhydrous hydrogen chloride under pressure and at low temperature. After neutralization of excess hydrogen chloride the product is precipitated by the addition of ethyl alcohol.

Uses: Protective coverings for machinery, rainclothing, shower curtains, food packaging.

See also "Pliofilm."

rubber, natural (India rubber; caoutchouc) $(C_5H_8)_x$. Essentially cis-1,4-polyisoprene, a stereospecific polymer produced by natural growth processes in rubber trees and plants. The elastic solid obtained from the sap (latex) of the rubber tree (Hevea brasiliensis) by coagulation and drying, or from other similar sources.

Properties: Light cream to dark amber, amorphous, elastic, dry loaves, sheets, or slabs, consisting of caoutchouc, resins, and proteins. Sp.gr. about 0.9. Soluble in carbon disulfide, petroleum and coal-tar hydrocarbons, particularly solvent naphtha, chlorinated hydrocarbons and essential oils. Vulcanization (q.v.) with sulfur, etc., improves the properties markedly for practical application.

Grades: Rubber is graded according to the localities from which it is obtained, the best being first latex pale crepe and smoked sheets.

Occurrence: Brazil, British Malaya, Sumatra.

Containers: Bales, wooden boxes.

Uses: Electric insulation; elastic bands and webbing; combs; pen-holders; footwear; brush and other handles; toys; gas, air, and water hose; containers; vehicle tires; belting; etc.

Shipping regulations: None.*

rubber, pale crepe. A yellowish-white raw rubber obtained by addition of sodium bisulfite to latex, and then drying the washed coagulum without smoking.

rubber-plated metal. Metal with a coating of rubber applied by "electrodeposition" or "ionic coagulation" from latex or artificial dispersions of rubber. In forming a rubber coating by electrodeposition the surface to be covered is made the positive electrode and the layer is built up by the migration of the rubber particles toward this positive electrode and their coagulation thereon by positively charged ions formed at that electrode by the passage of the current.

In ionic coagulation no electric current is employed. In this case the positively charged coagulating ions necessary to build up the rubber layer from the latex or rubber dispersion are supplied entirely by diffusion from a surface layer of salts or acids, which has been applied mechanically to the surface to be rubber covered. In either case the freshly deposited layer of rubber contains a large percentage of water and some soluble materials and hence is washed and dried before vulcanization.

The properties of the vulcanized coating may be varied from those of a soft, highly resilient rubber to a hard ebonite.

In general, rubber plating is applied to metal for the following reasons: (1) Protection against corrosion; (2) electrical insulation; (3) protection against abrasion or to furnish a cushioned surface. Hard rubber coatings furnish the greatest resistance to corrosion, and soft the greatest resistance to abrasion. Either hard or soft rubber coatings are satisfactory for electrical insulation.

Specific applications in each of the above three fields are as follows: (1) Corrosion resistance (dipping baskets, fan and pump rotors, centrifuge baskets, conveyor pins, spinnerette tubes, and bobbins; (2) electrical insulation (plating racks; pliers, screw drivers, lamp guards); (3) abrasion resistance (screen, perforated metal, conveyor buckets, and dishwashing racks).

rubber, reclaimed. A term used to describe replasticized vulcanized rubber. It is made from old tires, ground up and digested with caustic soda, and from pulverized inner tubes by long heating with various hydrocarbon oils.

Properties:

Unvulcanized material: Somewhat rubbery. Vulcanizes with sulfur to become nonthermoplastic.

Vulcanized material: Rubbery. Said to have fair tensile strength, fair elongation at break.

Uses: Reclaimed rubber is widely used in mechanical rubber goods, footwear, etc. It may be used alone or in combination with crude rubber, depending on the quality of product desired. Its chief uses are in baby carriage tires, heels, jar rings, friction tape, garden hose and lower grades of belting.

Shipping regulations: Flammable solid. Yellow label.*

rubber resin. See rubber, synthetic.

rubber, smoked sheets. Made by sheeting coagulated natural rubber on even-speed rolls and drying in an atmosphere of smoke which acts as a preservative and turns the rubber brown. Pressed into bales of 200-220 lbs for shipment.

*See "I.C.C. Shipping Regulations," page xiii.

rubber softener. See softening agents.

rubber solvent. A petroleum distillate used in making rubber cements and in tire manufacture. The volatility is usually similar to that of gasoline (boiling range 150-300°F) but may vary according to application.

rubber sponge (foam rubber) A cellular rubber containing 4-5 times the volume of air ordinarily found in rubber. May be produced by beating air into soapy latex with subsequent curing, or by incorporating ammoniun carbonate or stearic acid and sodium bicarbonate in a very soft rubber stock. The heat of vulcanization releases NH_3 and CO_2 which inflate the rubber to a spongy mass. Certain organic nitrogen compounds (diazoaminobenzene, etc.) are often substituted for the above chemicals, since they also decompose with heat to yield a gas, and thus function as blowing agents.

rubber substitute. See factice.

rubber, synthetic. Synthetic elastic products whose properties resemble those of natural rubber. Synthetic rubbers are inferior to natural rubber in some respects, but definitely superior in other properties. Each variety of rubber tends to become a specialty product for the uses requiring a particular set of properties. See styrene-butadiene rubber (SBR, formerly GR-S), acrylonitrile rubber, butyl rubber, cold rubber (a variety of SBR), neoprene, polysulfide rubbers, polyurethane foams and rubber, and the new stereoregular polymers, polybutadiene, cis- and trans-polybutadiene, and polyisoprene. In addition to these commercially important synthetic rubbers, a great variety of experimental and special synthetic rubbers are known.

rubber, synthetic natural. An awkward term for the stereospecific synthetic rubbers which closely resemble natural rubber. These include cis-polybutadiene and cis-polyisoprene, which are like ordinary natural rubber, and trans-polybutadiene, which is like balata and gutta percha.

rubeanic acid. See dithiooxamide.

rubellite. See tourmaline.

rubene. See naphthacene.

rubidium Rb. Element of atomic number 37; group I of the periodic table; radioactive isotopes known.
Properties: Soft, silvery-white metal; very easily oxidized in air. Must be kept immersed in naphtha, kerosine or the like. Soluble in acids and alcohol; decomposes water.
Constants: Sp.gr. 1.532; m.p. 39°C; b.p. 680°C.
Derivation: (a) By thermochemical reduction of rubidium chloride with calcium; (b) electrolysis of fused cyanide or chloride.
Source: Lepidolite.
Method of purification: Redistillation.
Grades: Technical.
Containers: Glass bottles under naphtha or kerosine.
Uses: Photocells; radio vacuum tubes; catalyst or catalyst promotor. The metal has been suggested as working fluid for vapor turbines in space vehicles; plasmas in thermoelectric devices; heat transfer fluids in power generators; grain-refining agents in metals.
Fire hazard: Dangerous.
Shipping regulations: Flammable solid. Yellow label.*

rubidium alum. See aluminum-rubidium sulfate.

rubidium bromide (rubidium monobromide) RbBr.
Properties: Lustrous, colorless, crystalline powder. Soluble in water; insoluble in alcohol.
Constants: Sp.gr. 3.35; m.p. 682°C.
Grade: C.P.
Use: Medicine.

rubidium carbonate (rubidium carbonate, normal) Rb_2CO_3.
Properties: White powder; m.p. 837°C. deliquescent. Soluble in water.
Grades: C.P.
Caution: Keep well stoppered!

rubidium carbonate, normal. See rubidium carbonate.

rubidium-cesium-ammonium bromide. See cesium-rubidium-ammonium bromide.

rubidium-cesium bromide. See cesium-rubidium bromide.

rubidium-cesium chloride. See cesium-rubidium chloride.

rubidium chloride RbCl.
Properties: White, crystalline powder. Lustrous. When heated it decrepitates, melts and volatilizes. Soluble in water; very slightly in alcohol.
Constants: Sp.gr. 2.76; m.p. 715°C.
Grades: Technical; C.P.
Use: Analysis (testing for perchloric acid).

rubidium hydrate. See rubidium hydroxide.

rubidium hydroxide (rubidium hydrate) RbOH.
Properties: Grayish-white mass; deliquescent, strong base. Soluble in alcohol, water.
Constants: Sp.gr. 3.2; m.p. 300°C.

rubidium iodide RbI.
Properties: White crystals. Soluble in water.
Constants: Sp.gr. 3.65; m.p. 642°C; b.p. 1300°C. Discolors in air and light.
Use: Medicine.

rubidium monobromide. See rubidium bromide.

rubidium sulfate Rb_2SO_4.
Properties: White rhombic crystals. Stable. Soluble in water.
Constants: Sp.gr. 3.613; m.p. 1060°C. Insoluble in alcohol.
Grades: C.P.
Use: Medicine.

rubigo. See iron oxide reds.

*See "I.C.C. Shipping Regulations," page xiii.
Reference numbers refer to name of manufacturer. See "List of Manufacturers," page v.

"Rubinol." [307] Fast alizarine direct red dyestuff.

"Rubramin." [412] Trademark for vitamin B_{12} (q.v.).

"Rub-Sol." [200] Trademark for a petroleum solvent prepared by straight-run overhead distillation of crude petroleum.
Properties: Water-white; boiling range 105-204°F; sp.gr. 0.687 (60°F); wt/gal 5.72 lbs at 60°F; flash point, Tag closed cup, -40°F.
Caution: Fire hazard — keep lights and fire away.
Use: In rubber industry for rubber cement and swabbing.
Shipping regulations: Flammable liquid. Red label.*

rubus (blackberry bark).
Derivation: The dried root-bark of the blackberry, Eubatus rubus.
Occurrence: Eastern United States.
Grades: Technical.
Containers: Barrels.
Use: Medicine.
Shipping regulations: None.*

ruby. See corundum for the true ruby. Synthetic rubies are made from aluminum oxide (containing small amounts of other metals) by the single crystal-growing technique. These have found some use in masers.

ruby wood. See santalum rubrum.

"Ruelene." [233] Trademark for organophosphorus products and formulations for use as parasiticides, insecticides, veterinary drugs and animal health products.

rue oil.
Properties: Essential oil; colorless to yellow; usually fluoresces; intense, persistent, characteristic odor.
Chief known constituents: Methyl n-nonyl ketone, methyl n-heptyl ketone; methyl anthranilate.
Constants: Variable, according to botanical origin and vegetative condition of plant at time of distillation.
Range of properties: Sp.gr. 0.8328-0.847 (15°C); optical rotation -5° to +2°; soluble in 1½-4 vols 70% alcohol.
Derivation: Distilled from various species of the genus Ruta.
Adulteration: Mineral oil; turpentine oil.
Containers: Bottles.
Uses: Veterinary medicine; organic synthesis; perfumery, and for isolation of methyl nonyl ketone.
Shipping regulations: None.*

"Rufert." [134] Trademark for nickel catalyst for hardening margarine.

"Ruflux." [337] Trade name for a line of titanium and zirconium products.
"Ruflux" C: See "Weldopax."
"Ruflux" I: Ilmenite (granular and milled) 60% TiO_2 min, black to brown color; sp.gr. 4.7; bulk density 180 lb/cu ft; m.p. 2800°F; particle size (granular) +100 mesh 3%, +200 mesh 67%; (milled) +200 mesh 2%. Used in weld rod coatings, ceramic glaze speck material, ceramic body and glaze colorant.
"Ruflux" P: Potassium titanate (K_2TiO_3 + TiO_2). 72% min TiO_2. Tan powder; sp.gr. 4.1; bulk density 91 lbs/cu ft; m.p. 2940°F. Used in weld rod coatings.
"Ruflux" S: Sodium titanate, 78% min TiO_2. Used in weld rods to help stabilize the arc.
"Ruflux" T: See "Treopax." Used in weld rod coatings to increase slag viscosity.
"Ruflux" Z: See Zircon Milled "G". Used in weld rod coatings to increase slag viscosity.
"Ruflux" 61: Rutile milled; 92% min TiO_2; gray brown powder; sp.gr. 4.2; bulk density 120 lbs/cu ft; m.p. 3250°F. Used in weld rod coatings to aid fluxing, control fusion, stabilize the arc, and contribute to welding speed.
"Ruflux" 84: Rutile granular, 92% min TiO_2; gray-brown powder; sp.gr. 4.2; bulk density 171 lbs/cu ft; m.p. 3250°F; fineness +140 mesh 57.2%; +180 mesh 37.5%; -180 mesh 5.4%. Used in coatings for mild and alloy steel welding rods to provide arc stabilizing flux and slag forming material; minimizes cracking from excessive shrinkage of the weld rod coating; also used as source of TiO_2, as a speckling agent in ceramic glazes, and as a raw material for titanium metal.

"Rulan." [28] Trademark for flame-retardant plastic, comprised essentially of "Alathon" polyethylene resin, a flame-retardant, and an antioxidant.
Form: "Cube-cut," in 50-lb multiwall paper bags.
Properties: "Rulan" has most of the good properties of polyethylene, particularly low electrical loss, high resistivity, and moisture-resistance. It does not support combustion, and does not drip when heated. It has passed the Underwriters' Laboratories vertical flame test.
Constants: Tensile strength 1400 psi; yield point 1400 psi; elongation 300%; water-absorption 0.02%; power factor 0.0015; dielectric constant 2.5.
Uses: As a high-frequency dielectric with flame-retardant characteristics, particularly as insulation on wire.

rumex (yellow dock; curl dock).
Derivation: Dried root of Rumex crispus or obtusifolius.
Occurrence: United States; Europe.
Chief constituent: Tannins.
Use: Medicine.

rusa oil. See palmarosa oil.

Russian blistering flies. See cantharides.

Russian flies. See cantharides.

"Rust-Ban." [51] Trademark for a series of products designed to prevent corrosion of metals. Group comprises hard-drying and bituminous types with thinners and

*See "I.C.C. Shipping Regulations," page xiii.
Reference numbers refer to name of manufacturer. See "List of Manufacturers," page v.

primers. Also grades suitable for aviation and automotive crank-case use.

"Rustmaster." [448] Trade name for alkyd type primers and top coatings for metal, wood, and masonry.

ruthenic chloride. See ruthenium chloride.

ruthenium Ru. Element of atomic number 44, group VIII of the periodic system.
Properties: Silvery-white, nonductile metal of the platinum group. Sp.gr. 12.2; m.p. 2450°C; b.p. 3700°C; Brinell hardness 220 (cast). Insoluble in acids and in aqua regia; is attacked by fused alkalies.
Derivation: Occurs with platinum, from which it is recovered.
Uses: Hardener for platinum and palladium in jewelry; electrical contact alloys.

ruthenium chloride (ruthenic chloride; ruthenium sesquichloride) $RuCl_3$.
Properties: Brownish-red crystalline mass; deliquescent. Soluble in water.
Grades: Technical; C.P.
Use: Analysis (testing for sulfur trioxide).

ruthenium red (ammoniated ruthenium oxychloride) $Ru_2(OH)_2Cl_4 \cdot 7NH_3 \cdot 3H_2O$.
Properties: A brownish-red powder. Soluble in water.
Grade: C.P.
Uses: Microscopic stain and reagent for pectin, plant mucin and gum.

ruthenium sesquichloride. See ruthenium chloride.

ruthenocene $(C_5H_5)_2Ru$. Dicyclopentadienylruthenium. A coordination compound. See dicyclopentadienyl metal compounds.

rutile TiO_2. Natural titanium dioxide. May contain up to 10% iron.
Properties: Color red, reddish brown, to black; luster adamantine to submetallic; streak pale brown; hardness 6-6.5; sp.gr. 4.18-4.25. May occur included in other minerals in hair or needle-like penetrations. Insoluble in acids.
Occurrence: Virginia, Florida, North Carolina, Arkansas, Maryland, Nevada; Europe, Brazil.
Use: Source of titanium and titanium compounds; ceramics; steel deoxidizer; pigment for paints, enamels and tiles. See also titanium dioxide.

"Rutile, Ceramic." [337] Trade designation for 92% TiO_2. Light brown powder; sp.gr. 4.2, bulk density 98 lbs/cu ft; m.p. 3250°F; fineness (average) 44 microns max. Insoluble in water and alkalies, slightly soluble in dilute mineral acids, soluble in hot concentrated sulfuric acid.
Uses: Glazes, floor tile, art ware and dinner ware bodies.
Containers: 100-lb paper bags; 600-lb barrels; 40,000-lbs min carload.

rutin (melin; quercetin-3-rutinoside) $C_{27}H_{30}O_{16} \cdot 3H_2O$. The 3-rhamnoglucoside of 5,7,3',4'-tetrahydroxyflavonol. A bioflavonoid.
Properties: Tasteless, bright yellow or greenish yellow powder; nontoxic; m.p. about 190°C; dec. 215°C. Very slightly soluble in cold water, more soluble in hot water and hot alcohol; soluble in isopropyl alcohol, pyridine and solutions of alkali hydroxides.
Derivation: By extraction from buckwheat.
Grades: N.F. XI; technical.
Containers: Fiber drums.
Use: Medicine; vitamins.

rutinsol.
Properties: A water-soluble rutin obtained from the reaction of rutin with magnesia. Reddish-brown amorphous powder; odorless, bitter taste. Water solutions slightly alkaline, pH about 9. Insoluble in organic solvents.
Grades: 83-88% pure rutin.
Use: Liquid medical preparations.

ryania. Ground stem wood or Ryania Speciosa. The active principles are alkaloids.
Toxicity: Low to animals.
Containers: Bags.
Use: Insecticide.

"Ryanicide." [342] Trademark for powdered ryania insecticide concentrates.

rye ergot. See ergot.

"RZ-50-A." [58] Trade name for 50% solution of N,N-dimethylcyclohexylamine salt of dibutyldithiocarbamic acid. Also available as "RZ-50-B" with emulsifying agents for use in latex.
Properties: Clear, dark liquid; sp.gr. 0.964.
Containers: 40-lb cans.
Uses: Rubber accelerators.

"RZ-50-B." [58] Trademark for 50% solution in "Cellosolve" of the N,N-dimethylcyclohexylamine salt of dibutyl-dithiocarbamic acid, with emulsifying agents added.
Specifications: A clear dark-brown liquid; sp.gr. (25°C) 0.98; turbid point 10°C max.
Use: Stabilizer for latex emulsions.

S

S. Symbol for sulfur.

"SA-326." [58] Trade name for ortho-biphenyl biguanide (q.v.).

sabadilla (cevadilla).
Derivation: Seeds of Sabadilla officinalis.
Occurrence: Mexico to Guatemala and Venezuela.
Grades: Technical.
Containers: Bags; barrels.
Uses: Medicine; source of veratrine, insecticides.
Shipping regulations: None.*

sabal. See serenoa.

sabina. See savin.

D-saccharic acid (2, 3, 4, 5-tetrahydroxyhexanedioic acid; tetrahydroxyadipic acid) $COOH(CHOH)_4COOH$. The 1, 6-dicarboxylic acid formed by the oxidation of D-glucose.
Properties: White needles or syrup; very soluble in water, alcohol, or ether; deliquescent; m.p. 125-126°C with decomposition.
Derivation: Oxidation of cane sugar, glucose, starch by nitric acid.

saccharin (ortho-benzosulfimide; gluside; benzoylsulfonic imide) The anhydride of ortho-sulfimide benzoic acid.

CHCHCHCHCCC(O)NHSO$_2$.

Properties: White, crystalline powder; exceedingly sweet taste (500 times that of cane sugar). M.p. 226°-230°C. Soluble in amyl acetate, ethyl acetate, benzene and alcohol; slightly soluble in water, chloroform and ether.
Derivation: A mixture of toluenesulfonic acids is converted into the sodium salt, then distilled with phosphorus trichloride and chlorine to obtain the ortho-toluene sulfonyl chloride, which by means of ammonia is converted into ortho-toluenesulfamide. This is oxidized with permanganate, treated with acid and saccharin crystallized out.
Method of purification: Recrystallization.
Grades: Commercial; C.P.; U.S.P. XVI.
Containers: Bottles; drums.
Uses: Manufacture of syrups; medicine (substitute for sugar); sweetening champagne, oils, soft drinks, foods, etc.
Shipping regulations: None.*

saccharin calcium (calcium ortho-benzosulfimide) $C_{14}H_8CaN_2O_6S_2 \cdot 3\frac{1}{2}H_2O$.
Properties: White crystalline powder or white crystals. Odorless, or with faint aromatic odor. Intensely sweet taste (500 times that of cane sugar). Soluble in water.
Grade: U.S.P. XVI.
Containers: Fiber drums.
Use: Medicine; sweetening agent.

saccharin, sodium (sodium benzosulfimide; soluble gluside; soluble saccharin) $C_7H_4NNaO_3S \cdot 2H_2O$. The sodium salt of saccharin (q.v.). A sweetening agent. (In dilute solutions, 500 times as sweet as sucrose).
Properties: White crystals or crystalline powder; odorless; intensely sweet taste; very soluble in water; slightly soluble in alcohol.
Grade: U.S.P. XVI.
Containers: Drums.
Use: Medicine; sweetening agent.

saccharin, soluble. See saccharin, sodium.

saccharolactic acid. See mucic acid.

saccharose. See sugar, cane.

saccharum lactis. See lactose.

saccharum, U.S.P. See sugar, cane.

Sachsse process. See BASF process.

S-acid. See 1-amino-8-naphthol-4-sulfonic acid; 1-naphthylamine-8-sulfonic acid.

"Sacon." [173] Trademark for a concentrated water-soluble material used to restore sizing in dry cleaned garments. It is a thermoplastic resin type product which restores sizing lost through wear and in the course of dry cleaning.

sacred bark. See cascara sagrada bark.

SAE. Abbreviation for Society of Automotive Engineers. The initials are applied to its specifications and tests for motor fuels, oils and steels.

SAF black. Abbreviation for super abrasion furnace black. See furnace black.

"Safemark." [99] Trade name for a specially processed natural earth mineral of exceptional whiteness, which is completely free flowing even in humid weather. Will go into water suspensions quickly and easily with minimum stirring, making it an excellent wet marking material.
Containers: 50-lb multiwall paper bags.

safety glass. See polyvinyl butyral resin sheeting.

"Safety-Linen Sour." [244] A proprietary product consisting chiefly of acid fluorides.
Properties: White granular solid; soluble in water; neutralizing value 21.7 oz sodium bicarbonate per lb.

*See "I.C.C. Shipping Regulations," page xiii.
Reference numbers refer to name of manufacturer. See "List of Manufacturers," page v.

Containers: 150-lb, 300-lb net fiber drums.
Uses: Laundry sour, iron removing type.

safety solvent. See Stoddard solvent.

safflor carmine. See carthamin.

safflor red. See carthamin.

safflower. See carthamus.

safflower oil. Drying oil from safflower (carthamus) seed, somewhat similar to linseed oil. It is non-yellowing. Rich in linoleic acid (unsaturated fatty acid).
Properties: Sp.gr. (25/25°) 0.923-0.927; refractive index (25°C) 1.4740-1.4745; acid value 0.6-1.5; iodine no. 140-152; saponification no. 186-193; unsaponifiable 0.3-1.0%.
Derivation: Oil removed from seeds by press or solvent extraction.
Grade: N.N D.
Containers: Drums; tank cars.
Uses: Paints; varnishes; linoleum; medicine; special diets; foods.

saffron. See crocus.

"Saflex." [58] Trademark for vinyl butyral adhesive film for manufacture of automotive or TV safety glass.

safranine $C_{18}H_{14}N_4$. A dye prepared by oxidizing para-toluylenediamine, aniline and ortho-toluidine.
Uses: Stain in microscopy; textile industry.

"Safrella." [188] Trademark for a proprietary product. It is claimed to be useful for the replacement of artificial sassafras in practically all technical processes, but should not be used for the replacement of the artificial product in flavor compounds. Its strength is about the same and its odor remarkably close.
Uses: Laundry soaps; insecticides; germicides; other products calling for masking agents.

safrole (4-allyl-1,2-methylenedioxybenzene) $C_3H_5C_6H_3O_2CH_2$.
Properties: A colorless or pale yellow liquid oil; odor of sassafras; is the odor-giving constituent of sassafras, camphorwood and other oils; sp.gr. (15°C) 1.100-1.107; m.p. 11°C; b.p. 233°C; optical rotation (15°C) −0°30' to +0°30'; refractive index (n 20/D) 1.5363-1.5385; soluble in alcohol, ether, benzyl benzoate, fixed oils, mineral oil, chloroform; slightly soluble in propylene glycol; insoluble in water and glycerine.
Derivation: From oil of sassafras or camphor oil.
Method of purification: Rectification or freezing.
Grades: Technical.
Containers: Bottles; drums.
Uses: Perfumery and soaps; manufacture of heliotropin; medicine; insecticides; flavors; denaturing fats in soap making. May not be used in foods or beverages (FDA).
Shipping regulations: None.*

sage. See salvia.

sage oil (salvia oil).
Properties: Yellowish or greenish-yellow liquid, penetrating sage odor. Soluble in alcohol.
Chief known constituents: Cineol; thujone; pinene.
Constants: Sp.gr. 0.915-0.925; optical rotation +10° to +25°; saponification value 107.
Derivation: By distillation from the leaves of Salvia officinalis.
Grades: Dalmatian; Spanish.
Containers: Glass bottles; copper flasks; cans.
Uses: Medicine; perfumery; flavors.
Shipping regulations: None.*

sage oil, clary. See clary sage oil.

sagger clay. Clays used in the making of saggers where high-grade pottery is burned. These clays may be of two types: (a) Open-burning and (b) semi-open or tight-burning. Saggers are containers in which whiteware and pottery are burned to prevent fire damage.
Occurrence: United States (Tennessee, Kentucky, New Jersey, Pennsylvania, Missouri, California).

SAIB. See sucrose acetate isobutyrate.

sal ammoniac. See ammonium chloride.

sal chalybis. See ferrous sulfate.

salcomine (cobalt disalicylal ethylene diamine) $Co(C_6H_4)_2O_2N_2(CH)_2(CH_2)_2$. A compound which will absorb oxygen from air and liberate the oxygen later under modified conditions of temperature and pressure.
Use: Suggested for producing oxygen pure enough for cutting and welding without the usual liquefication and distillation operations. Apparently not economical under ordinary circumstances.

"Sal-Ethyl" Carbonate. [330] Trademark for carbethyl salicylate. Used in medicine.

salicin (salicyl alcohol glucoside) $CH_2OHC_6H_4OC_6H_{11}O_5$. A glucoside obtained from several species of Salix and Populus.
Properties: Colorless crystals or white powder. Soluble in water, alcohol, alkalies, glacial acetic acid; insoluble in ether. M.p. 199-202°C; sp.gr. 1.43.
Containers: Cartons; 1-, 5-, 25-lb cans.
Uses: Medicine; reagent for nitric acid.

salicitrin. See methylenecitrosalicylic acid.

salicylacetal (salicylacetol) $C_6H_4(OH)COOCH_2COCH_3$.
Properties: White or pinkish needles. Slightly soluble in alcohol, cold water; soluble in ether, chloroform, petroleum benzine. Fusing point 71°C.

salicylacetol. See salicylacetal.

salicylal. See salicylaldehyde.

salicyl alcohol (ortho-hydroxybenzyl alcohol; saligenin; saligenol) $HOC_6H_4CH_2OH$.
Properties: White crystals; m.p. 86-87°C; sublimes at 100°C; somewhat soluble in

*See "I.C.C. Shipping Regulations," page xiii.
Reference numbers refer to name of manufacturer. See "List of Manufacturers," page v.

water; very soluble in alcohol, chloroform, ether; soluble in propylene glycol, benzene and fixed oils; sp.gr. 1.16.
Derivation: Hydrolysis of salicin; heating phenol and methylene chloride with caustic.
Use: Medicine (external).

salicylaldehyde (salicylal; salicylic aldehyde; ortho-hydroxybenzaldehyde) C_6H_4OHCHO.
Properties: Clear, colorless, oily liquid or dark red oil; bitter almond-like odor; burning taste; sp.gr. 1.165-1.172; m.p. -7°C; b.p. 196°C. Soluble in alcohol, ether and benzene; very slightly soluble in water.
Derivation: By the interaction of phenol and chloroform in presence of aqueous alkali.
Method of purification: Distillation.
Grades: Technical; reagent.
Containers: Glass bottles; 10-, 55-gallon drums.
Uses: Analytical chemistry; perfumery (violet); synthesis of coumarin.
Shipping regulations: None.*

salicylamide (ortho-hydroxybenzamide) $C_6H_4(OH)CONH_2$.
Properties: White or slightly pink; tasteless, gritty, crystals; m.p. 139-142°C; b.p. decomposes at 270°C; soluble in hot water, alcohol, ether, chloroform; slightly soluble in cold water, naphtha, and carbon tetrachloride.
Derivation: Treatment of methyl salicylate with dry ammonia gas.
Method of purification: Crystallization.
Grades: Technical; N.F. XI.
Containers: Glass bottles; drums.
Use: Medicine.
Shipping regulations: None.*

salicylanilide $HOC_6H_4CONHC_6H_5$.
Properties: Odorless, white or slightly pink crystals. M.p. 136-138°C; stable in air; slightly soluble in water; freely soluble in alcohol, ether, chloroform, and benzene.
Grade: N.F. XI.
Uses: Medicine; fungicide; anti-mildew agent; intermediate.

salicylated mercury. See mercuric salicylate.

salicylazosulfapyridine
$HOOCC_6H_3(OH)NNC_6H_4SO_2NHC_5H_4N$.
5-[para-(2-Pyridylsulfamyl)phenylazo] salicylic acid.
Properties: Brownish yellow, odorless powder. M.p. 220-240° (dec). Slightly soluble in alcohol; practically insoluble in benzene, chloroform, ether and water.
Grade: N.N.D.
Use: Medicine.

salicylic acid (ortho-hydroxybenzoic acid) $C_6H_4(OH)(COOH)$.
Properties: White needle crystals, or powder; sweetish, afterward acrid taste; stable in air but gradually discolored by light; sp.gr. 1.483; m.p. 158-161°C; b.p. about 211°C at 20 mm; sublimes at 76°C; soluble in acetone, oil or turpentine, alcohol, ether, benzene; very slightly soluble in water.
Derivation: By treatment of a hot solution of sodium phenolate with carbon dioxide, and acidifying the sodium salt thus formed.
Method of purification: Sublimation.
Grades: Technical; U.S.P. XVI; crude.
Containers: 25 to 250 lb fiber drums.
Uses: Manufacture of aspirin and other medicinals; preservative; dyes; perfumes; to prevent precuring in rubber compounds; organic intermediate; fungicide.
Caution: Salicylic acid dust forms explosive mixtures with air.
Shipping regulations: None.*

salicylic acid dipropylene glycol monoester. See dipropylene glycol monosalicylate.

salicylic acid methoxymethyl ester (methoxymethyl salicylate) $C_6H_4OHCOOCH_2OCH_3$.
Properties: Clear yellow liquid; miscible with oils in all proportions; aromatic odor. Burns the skin unless diluted with vegetable oils or petrolatum. Slightly soluble in water, readily soluble in usual organic solvents. Sp.gr. 1.2; b.p. 162°C.
Use: Medicine.
Shipping regulations: None.*

salicylic acid phenyl ester. See salol.

salicylic acid, salicylic ester (salicylosalicylic acid; salicylsalicylate) $OHC_6H_4COOC_6H_4COOH$.
Properties: White crystalline powder. Insoluble in water and dilute acids; m.p. 148°C.
Use: Medicine.
Shipping regulations: None.*

salicylic aldehyde. See salicylaldehyde.

salicylic naphthyl ester. See beta-naphthyl salicylate.

salicylosalicylic acid. See salicylic acid, salicylic ester.

salicyl-para-phenetidine (malakin) $C_2H_5OC_6H_4NHCOC_6H_4OH \cdot H_2O$.
Properties: Small, fine, bright yellow needles; insoluble in water; soluble in sodium hydroxide, hot alcohol, benzene and ether. M.p. 92°C.
Use: Medicine.
Shipping regulations: None.*

salicylsalicylate. See salicylic acid, salicylic ester.

saligenin. See salicyl alcohol.

saligenol. See salicyl alcohol.

salmon oil.
Properties: Pale golden-yellow liquid; mild, sweet tolerably pleasant taste. Soluble in ether, alcohol, chloroform, benzine and carbon disulfide. Sp.gr. 0.9258; saponification value 182-188; iodine value 161.
Derivation: A by-product of the salmon canning industry.
Method of purification: Filtration.
Grades: Crude; refined.
Container: Wooden barrels; steel drums.
Uses: Soap; leather dressing.
Shipping regulations: None.*

salol (phenyl salicylate; salicylic acid phenyl ester) $C_6H_4OHCOOC_6H_5$.
Properties: White crystalline powder; faint aromatic odor and taste. Soluble in alcohol, ether, chloroform, benzene and fixed or volatile oils; sparingly soluble in water. Sp.gr. 1.2614; m.p. 41.9°C; b.p. 172°-173°C. Absorbs light, especially at 290-330μ.
Derivation: By heating salicylic acid and phenol with phosphorus pentachloride or other dehydrating agent.
Method of purification: Recrystallization.
Grades: Technical; N.F. XI (as phenyl salicylate).
Containers: 100-lb kegs; barrels.
Use: Medicine; preservative; as light absorber in plastics, waxes, polishes, suntan oils.
Shipping regulations: None.*

sal soda (washing soda; sodium carbonate decahydrate) $Na_2CO_3 \cdot 10H_2O$.
Properties: White crystals; sp.gr. 1.44; m.p. 32.5-34.5°C, loses water at this temperature. Easily soluble in water; insoluble in alcohol. A pure but relatively expensive form of sodium carbonate (soda ash).
Containers: Steel drums; multiwall paper sacks; wooden barrels; fiber cans.
Uses: Washing textiles; bleaching linen and cotton; general cleanser.
Shipping regulations: None.*

salt.
1. The compound, other than water, which is formed by the reaction of an acid with a base; e.g., copper sulfate is the salt resulting from the reaction between sulfuric acid and copper hydroxide.
2. Specifically, sodium chloride.

sal tartar. See sodium tartrate.

salt, black. A monohydrated sodium carbonate containing some caustic soda, as was obtained in the Leblanc soda process.

salt cake Impure sodium sulfate (90-99%) obtained, usually as a byproduct, from (a) production of hydrochloric acid; (b) by crystallization from natural brines, such as the Searles Lake brine; (c) from the coagulating bath for viscose rayon; (d) by the Hargreaves process. Impurities vary with the source.
Grades: Technical; glassmakers' (iron-free).
Containers: 200-lb bags; 500-lb barrels; carloads.
Uses: Paper pulp; detergents and soaps; plate and window glass; sodium salts; ceramic glazes; dyes.

salt, common. See sodium chloride.

salt, Glauber's. See Glauber's salt.

salting out process. The addition of salts to a mixture to recover proteins, soaps, or simpler organic substances. The salting out or precipitation of the desired component results from the dehydrating action of the added salt.

salt of Lemery. See potassium sulfate.

salt of lemon. See potassium binoxalate.

salt of phosphorus. See sodium ammonium phosphate.

salt of sorrel. See potassium binoxalate.

saltpeter. See niter, and potassium nitrate.

saltpeter, Chile. See caliche.

saltpeter, soda. See caliche.

salt, preparing. See sodium stannate.

salt, rock. See halite.

salt, sea. See sodium chloride.

salts, Epsom. See magnesium sulfate.

salt, table. See sodium chloride.

salt, tin. See stannous chloride.

salufer. See sodium fluosilicate.

"Salvarsan." See arsphenamine.

salvia (sage; garden sage). Dried leaves of perennial shrub, Salvia officinalis.
Occurrence: Southern Europe; cultivated in United States; England; France and Italy.
Grades: Dalmatian; Greek, Italian.
Containers: Bags; cans.
Use: Condiment.
Shipping regulations: None.*

salvia oil. See sage oil.

"Salyrgan." [162] Trademark for mersalyl.

samarium. Sm. One of the rare earth metals (Group III) of the cerium subgroup; atomic number 62; valence +2 and +3.
Properties: Hard brittle metal which quickly tarnishes in air. Found in small quantities in such minerals as cerite, gadolinite and others. Ignites to form oxide at 200-400°C; liberates hydrogen from water. Sp.gr. about 7.50; m.p. 1052°C; b.p. 1600°C (approx); about as hard as iron.
Derivation: Reduction of the oxide with barium or lanthanum.
Grades: Ingots; lumps; turnings (high purity).
Uses: Nuclear reactor control rods; neutron shields; metallurgy.

samarium salts.
samarium chloride $SmCl_3 \cdot xH_2O$. Faintly yellow crystals; soluble in water. Hygroscopic. Obtained by treating the carbonate or oxide with hydrochloric acid. Grades: Up to 99.9% samarium salts. Available as 45% Sm_2O_3.
samarium fluoride $SnF_3 \cdot 2H_2O$. Grades: Up to 99.9% samarium salts. Available as 77% Sm_2O_3.
samarium nitrate $Sm(NO_3)_3 \cdot xH_2O$. Faintly yellow crystals; soluble in water. Grades: Up to 99.9% samarium salts. Available as 40% Sm_2O_3. Shipping regulations: Oxidizing material. Yellow label.*
samarium oxalate $Sm_2(C_2O_4)_3 \cdot xH_2O$. White powder, insoluble in water, slightly soluble in acids. Grades: Up to 99.9% samarium salts. Available as 50% Sm_2O_3.

*See "I.C.C. Shipping Regulations," page xiii.
Reference numbers refer to name of manufacturer. See "List of Manufacturers," page v.

samarium oxide Sm_2O_3. Cream-colored powder; insoluble in water; soluble in acids. Absorbs moisture and carbon dioxide from the air. M.p. 2300°C; sp.gr. 7.43.
Grades: Up to 99.9% samarium oxide.

samarium sulfate $Sm_2(SO_4)_3 \cdot xH_2O$. Faintly yellow crystals, slightly soluble in hot water, more soluble in cold. Grades: Up to 99.9% samarium salts. Available as 45% Sm_2O_3.
Containers: Glass bottles; fiber drums.
Uses: In red and infrared sensitive phosphors; (oxide) infrared-absorbing glass; catalysis.

samarskite $(Y, Er, Fe, U)(Ta, Nb)_2O_6$.
A natural oxide of rare earth metals, iron, uranium, niobium, and tantalum.
Properties: Color black to brown, streak reddish brown, black; luster vitreous to resinous; hardness 5-6; sp.gr. 5.7.
Occurrence: North Carolina, Colorado, Idaho; Europe; Brazil; Madagascar.

sand. A natural material, usually composed largely of quartz, with grain sizes ranging from 0.05-5 mm. It is found in rivers and lakes, on seashores, in soils, and elsewhere. Varieties include furnace sand (q.v.), glass sand (q.v.), and also filter sand and molding sand.
Uses: Building and construction work; glassmaking; filtration; foundry work; by railroads; as a filler; dusting agent; abrasive.

sand acid. See fluosilicic acid.

sandal oil. See sandalwood oil.

sandalwood. See santalum album.

sandalwood oil (santalwood oil; santal oil; sandal oil; East Indies sandalwood oil).
Properties: Pale yellow, somewhat viscous essential oil; faint, aromatic, persistent odor; harsh, resinous taste, somewhat pleasant. Sp.gr. 0.974-0.985 (15°C); optical rotation −15° to −20° in 100 mm tube at 25°C; refractive index 1.504-1.508; acid value 0.5-8.0; ester value 3-17, after acetylation, not less than 196; soluble in 3-5 vols of 70% alcohol; in 5-6 vols of 69% alcohol; in 6-7 vols of 68% alcohol.
Derivation: By the steam distillation of the wood of Santalum album of India.
Adulteration: Cedarwood oil; copaiba and gurjun balsam oils; castor oil; sesame oil; liquid paraffin; linseed oil.
Containers: Cans.
Uses: Medicine; perfumery.
Shipping regulations: None.*

sandalwood oil, Australian.
Properties: Differs slightly from Indian sandalwood oil and is cheaper.
Derivation: By distillation of wood of Eucarya spicata in West Australia.
Use: Perfumery.

sandalwood oil, East Indies. See sandalwood oil.

sandalwood oil, West Indies (amyris oil).
Properties: Viscous essential oil. Odor, faint, somewhat unpleasant. Soluble in 1 vol or less of 90% alcohol; in 2-10 vols of 80% alcohol (occasionally).
Chief known constituents: Amyrol; amyrolin; cadinene; caryophyllene; etc.
Constants: Sp.gr. 0.950-0.970 at 15°C; optical rotation +19° to +29°; refractive index 1.508-1.513; acid value up to 3.0; ester value up to 6.0, after acetylation, 66-125.
Derivation: By steam distillation of the wood of Amyris balsamifera, L.
Containers: 24-oz bottles; 25-, 40-lb tins; 80-lb cases.
Uses: Medicine; perfumery.
Shipping regulations: None.*

sandalwood, red. See santalum rubrum.

sandarac gum (juniper gum).
Properties: Yellow, brittle, translucent, amorphous lumps or powder. Soluble in alcohol, ether, acetone, amyl alcohol and hot caustic alkali; partially soluble in volatile oils, carbon disulfide, chloroform and oil of turpentine; insoluble in light petroleum hydrocarbons, benzene and water.
Derivation: The resin from Callitris quadrivalvis, indigenous to Morocco.
Grades: Technical.
Containers: Bags.
Uses: Incense; varnishes, lacquers, dental cements.
Shipping regulations: None.*

sand, glass. A sand of medium grain consisting of 98 to 100% of silica (SiO_2) and containing less than 1% of iron oxides. Found in many parts of the country.

sand, molding. A sand used for making the molds for casting metal.

sand, placing. A sand used in the pottery industry for the placing of bisque ware. It must be free of iron and fluxes so that it will not flux with the body with which it comes in contact.

"Sandril." [100] Trade name for reserpine (purified reserpine alkaloid of Rauwolfia species) (3, 4, 5-trimethoxybenzoic acid plus an amino acid called reserpic acid).
Properties: M.p. 264-265°C; optically active, $[a]^{26}/D$ −115° (in chloroform).
Use: Medicine.

sandstone. A variety of sedimentary rock consisting essentially of grains of quartz, sometimes with feldspar, mica, and other materials. The cohesion of the grains may be caused by pressure or by a binder of silica, iron oxide, calcite, or clay. The corresponding rocks are siliceous, ferruginous, calcareous or argillaceous sandstones.
Use: Building stone.

sandwich molecule. See molecular sandwich.

"Sanforized." [7] Trademark used on fabrics treated by a special process to limit shrinkage to not over 1% by the government standard wash test. Used extensively on cotton goods and to lesser extent on others.

The process mechanically rearranges the yarns of a fabric by the amount the yarns would be shortened if laundered. The shrinkage characteristics are first determined in separate tests. The desired shrinkage is obtained by water spray and steam heat while the cloth is in a special machine that regulates the width and length of the cloth. The machine regulating the length consists of a large steam heated cylinder, an endless thick woolen felt blanket in close contact with the cylinder for most of its perimeter, and an electrically heated shoe which presses the cloth against the blanket while the former is in a stretched condition as it curves around the feed-in roll.

sanguinaria (bloodroot; blood geranium; red puccoon; red-root; tetterwort).
Derivation: Dried rhizome and roots of Sanguinaria canadenis.
Occurrence: North America.
Grades: Technical; N.F. XI.
Containers: Bags.
Use: Medicine.
Shipping regulations: None.*

sanitizers. A special class of disinfectants designed for use on food processing equipment; dairy utensils; dishes and glassware in restaurants. Among them are the hypochlorites; chloramines and other organic chlorine-liberating compounds, and quaternary ammonium compounds, many of which are proprietary. See antiseptics; disinfectants.

santal oil. See sandalwood oil.

santalol $C_{15}H_{24}O$. A sesquiterpene alcohol.
Properties: Colorless liquid, with a characteristic odor of oil of sandalwood. Soluble in 3 parts of 70% alcohol. Insoluble in water; sp.gr. 0.971-0.973; refractive index 1.504-1.508; b.p. about 300°C.
Derivation: From sandalwood oil.
Use: Perfumery.

santalum album (sandalwood; yellow saunders). Heartwood of Santalum album.
Occurrence: India, Malaya.
Use: Source of sandalwood oil; incense; fumigant.

santalum rubrum (red saunders; ruby wood; sandalwood). Heartwood of tree Petrocarpus santalinus.
Properties: (Unground) purplish, reddish-orange or reddish-brown powder; dusky red to dark reddish-orange chips; nearly odorless; slightly astringent taste.
Occurrence: East Indies.
Containers: (Chips) bags; (powder) fiber drums.
Use: Coloring extract.
Shipping regulations: None.*

santalyl acetate. Acetic acid ester of a mixture of alpha- and beta-santalols.
Properties: Colorless liquid, with a light odor of sandalwood. Sp.gr. 0.982-0.985; refractive index 1.487-1.492.
Derivation: By the treatment of sandalwood oil or santalol with acetic anhydride.
Use: Perfumery.

"Santicizer." 58 Trademark for a line of resin plasticizers including:

"Santicizer 1-H": N-Cyclohexyl-para-toluenesulfonamide, $C_{13}H_{19}O_2NS$.
Properties: Yellowish brown fused mass. M.p. 83.5°C min; moisture 0.25% max; color of melt APHA 250 max; soluble in alcohols, esters, ketones, aromatic hydrocarbons, and vegetable oils; slightly soluble in gasoline hydrocarbons; insoluble in water.
Use: In paper coatings; textile coatings and adhesives.

"Santicizer 3": N-Ethyl-para-toluenesulfonamide.
Properties: Practically white crystalline solid; m.p. 58.0°C min; acidity (as acetic acid) 0.10% max; sp.gr. (65/65°C) 1.166-1.176; compatible with cellulose acetate up to about 50 parts per 100 parts of resin.
Uses: Especially in combination with other plasticizers.

"Santicizer 8": Mixture of N-ethyl-ortho- and para-toluenesulfonamides.
Properties:; Light yellow, viscous liquid; odor slight, characteristic; free amide 9-13%; acidity (as acetic acid) 0.1% max.
Solubility: Readily miscible with all common organic solvents except the petroleum hydrocarbons in which it is only very slightly soluble. In water, soluble to approx. 0.13% at 23°C, 0.21% at 48°C; soluble in castor oil; slightly soluble in china wood oil; insoluble in linseed oil. Its solubility in water decreased by presence of ammonia but increased by other alkalies.
Uses: Plasticizer for cellulose acetate, for coatings designed to withstand gasoline and related products. Some plasticizing action on casein, glue, soybean protein and corn protein. Increases the flexibility of shellac with considerable softening effect; also useful with polyvinyl acetate adhesives; and synthetic polyamides such as nylon.

"Santicizer 9": Mixture of ortho- and para-toluenesulfonamides.
Properties: Fine granular particles; m.p. 105°C min; melt clear and light yellow; acidity, pH 4.6 min in dilute acetone (2 g sample in 15 ml A.R. acetone and 35 ml of CO_2-free distilled water).
Solubility: Quite low in coal-tar hydrocarbons and negligible in petroleum hydrocarbons; readily soluble in most other solvents, and in hot linseed, china wood and castor oils; soluble in water only about 1.0% at 34°C, but this is increased by the presence of alkali; solubility in ethylene dichloride approx. 4%, in benzene approx. 1.5%, each at 30°C.
Uses: Relatively small amounts impart a gloss to polished articles, give smooth working and uniform curing of molded thermosetting products; blends readily with most synthetic resins, does not impart the softening action of liquid plasticizers. Valuable for molding compositions. In

some phenol-aldehyde resins it appears to act as a promoter, giving more uniform and thorough curing.

"Santicizer B-16": Butyl phthalyl butyl glycolate.

Properties: A liquid; sp.gr. (25/25°C) 1.098 ± 0.005; refractive index 1.490 ± 0.002 at 25°C; acidity (as acetic acid) 0.02% max; essentially odorless; sulfur (copper strip test) negative; b.p. (5 mm) 219°C; does not crystallize at temperatures as low as −35°C; water solubility 0.0012% at 30°C; extremely light-stable; darkens on heating above 290°C.

Uses: A polyvinyl-chloride plasticizer where a tasteless and non-toxic composition is required; may be used in vinyl food wrappings. Solvent for nitrocellulose; compatible with most resins, imparting flexibility. Miscible with castor, linseed and china wood oils; somewhat retards gel formation in the latter on heating. Swells crepe rubber but appears to have no effect on vulcanized rubber; plasticizer for chlorinated rubber and polystyrene.

"Santicizer E-15": Ethyl phthalyl ethyl glycolate.

Properties: A liquid; color (APHA), 20 max; sp.gr. (25°/25°C) 1.182 ± 0.003; refractive index 1.498 ± 0.002 (25°C); acidity (as acetic acid) 0.02% max; odor slight; sulfur (copper-strip test) negative.

Solubility: Miscible with the common organic solvents except the petroleum hydrocarbons; when heated, miscible with vegetable oils but partially separates on cooling; solubility in water 0.0175% at 30°C.

Use: Solvent for nitrocellulose and cellulose acetate and many other resins, giving clear, tough flexible films which have increased resistance to moisture penetration; lightfast; does not evaporate even from the thinnest films. May be used as plasticizer in composition for food wrapping and packaging.

"Santicizer M-17": Methyl phthalyl ethyl glycolate.

Properties: A liquid; color (APHA) 20 max; sp.gr. (25/25°C) 1.220 ± 0.005; refractive index 1.504 ± 0.002 (25°C); acidity (as acetic acid) 0.02% max; slight characteristic odor; sulfur (copper strip test) negative.

Solubility: Not miscible with castor, linseed, or china wood oils or with petroleum hydrocarbons; readily miscible with other organic solvents; solubility in water 0.09% at 30°C.

Uses: A solvent for nitrocellulose and cellulose acetate; imparts plasticity which is undiminished even in thin films exposed to air currents for a long time. Light-stable. Dissolves "Santolite" resins to produce gasoline-resistant coatings.

"Santicizer 140": Mixed cresyl diphenyl phosphates.

Properties: Clear mobile liquid; color (APHA) 40 max; sp.gr.(25/25°C) 1.208 ± 0.005; refractive index 1.56 ± 0.01 (25°C); acidity (as H_3PO_4) 0.01% max; odorless; b.p. (10 mm) 258°C; pour point −37°C approx.

Uses: Imparts flame retardance, permanence, chemical stability; and heat and light stability in vinyl sheeting, lacquers, paper coating, cellulose acetate, cellulose nitrate, ethyl cellulose, cellulose acetate-propionate, polyvinyl formal and polyvinyl butyral.

"Santicizer 141": Alkyl aryl phosphate. Used as a non-flammable plasticizer for vinyl resins.

"Santicizer 160": Butyl benzyl phthalate.

Properties: Nearly water-white, oily liquid with faint characteristic odor; sp.gr. about 1.113-1.121 (25/25°C); acidity (as phthalic acid) 0.02% max.

Uses: A relatively nonvolatile plasticizer for polyvinyl chloride, polyvinyl chloride-acetate, polyvinyl butyral, ethyl cellulose and nitrocellulose. Compatible with cellulose triacetate, polyvinyl formal, cellulose acetate-butyrate, cellulose propionate, polyvinyl acetate and acrylics. Effective in cellulosic lacquers imparting toughness, flexibility, low-moisture vapor transfer, heat and light stability. Also useful for low molecular weight polystyrene.

"Santicizer 165": Mixed-alcohol phthalate used as plasticizer for vinyl resins and nitrocellulose.

"Santicizer 602":

Properties: Clear oily liquid.

Uses: A primary plasticizer for polyvinyl chloride with lower volatility than dioctyl phthalate (DOP); resistant to soapy water extraction and alkali. Used in vinyl floor tile, plastisols, film and sheeting.

"Santicizer 603":

Properties: Clear oily liquid free of sediment or turbidity; odor, characteristically mild.

Uses: Low cost primary plasticizer for polyvinyl chloride and copolymer type resins, in calendered vinyl products such as film and sheeting for shower curtains, drapes and upholstery.

"Santicizer 630": Modified-phthalate plasticizer more volatile than dioctyl phthalate.

"Santicizer 636": 2-Ethylhexyl isodecyl phthalate. Nearly colorless liquid miscible with common solvents, thinners, and oils. Used as plasticizer.

"Santobane." [58] Trademark for DDT.

"Santobrite." [58] Trademark for sodium pentachlorophenate, technical.

"Santocel." [58] Trademark for silica aerogel.

Properties: Colors, white, transparent in vehicles; absolute density 17.1 lb/gal, 0.0585 gal/lb; refractive index 1.464; oil absorption 2.5 g oil/g.

Typical analysis: SiO_2, 90% min; Na_2SO_4, 2.7% max; Al_2O_3 and Fe_2O_3, 1.0%; volatile 4.0-6.0%; pH (4.0 g in 100 cc H_2O) 3.5-4.0. Volatile portion is water, alcohol and acetaldehyde.

Grades: "Santocel" 54, "Santocel" C, "Santocel" CX, "Santocel" CS, "Santocel" DCS, "Santocel" CF (unground, crude).

Containers: Multiwall bags, net 20-lbs for 54; 25-lbs for C, CS and DCS; 15-lbs for CX.

Uses: Flatting agent to produce matte finishes in coatings of all types (lacquers, varnishes, vinyl resin coatings and resin dispersions); bodying agent in printing inks; bulking agent for dry powders; free-flowing and anticaking agent for dry powders; insecticide for stored industrial grains; application aid for rubber cements; reinforcing agent for silicone rubbers; bodying agent to make lubricating greases; dry grinding aid for DDT and other sticky powders; to raise head distortion point, as in molded brake linings; mold lubricants for hard rubber; bodying agent for paint removers; antisettling agent for pigments; antislip agent in paste floor waxes; flatting agent for free film and textile coatings; thixotropic agent for polyester and epoxy resins, dispersions and solutions.

"Santochlor." 58 Trademark for para-dichlorobenzene.

"Santocure." 58 Trademark for N-cyclohexyl-2-benzothiazole-sulfenamide.
Properties: Tan or buff powder; m.p. 94°C; sp.gr. 1.27; soluble in ethyl ether.
Containers: 50-lb bags, 150-lb drums.
Use: Rubber accelerator.

"Santocure" NS. 58 Trademark for N-tert-butyl-2-benzothiazole-sulfenamide.
Properties: Light tan or buff powder; m.p. 105°C min; sp.gr. 1.29; ether insolubles 0.5% max.
Use: Rubber accelerator.

"Santoflex" 75. 58 Trademark for a mixture of N,N'-diphenyl-para-phenylenediamine (DPPD), 75%, and 6-dodecyl-1,2-dihydro-2,2,4-trimethylquinoline ("Santoflex" DD), 25%. Dark flakes; used for reducing aging and oxidation effects of rubber.

"Santoflex" AW. 58 Trademark for 6-ethoxy-1,2-dihydro-2,2,4-trimethylquinoline.
Properties: Dark, viscous liquid; sp.gr. 1.04 at 45°C; benzene insolubles, trace max. Causes discoloration and staining.
Uses: Antioxidant and flex-cracking inhibitor for rubber.

"Santoflex" DD. 58 Trademark for 6-dodecyl-1,2-dihydro-2,2,4-trimethylquinoline. A dark, viscous liquid. Used for inhibiting flex-cracking and oxidation of rubber.

"Santolene." 58 Trademark for a line of fuel oil additives of which the following are examples.
"Santolene" C:
Properties: Oily liquid; viscosity (SU) 250 sec. at 100°F; flash point 150°F min; neutralization number 84-100.
Containers: 55-gal steel drums; tank cars.
Uses: Corrosion and rust inhibitor for use in light petroleum products.
"Santolene" F:
Properties: Oily liquid; viscosity (SU) 75 sec (210°F); sp.gr. 0.98 (60/60°F); sulfur 1.9%; barium 7.5%.
Containers: 55-gal steel drums; tank cars.
Uses: Sludge inhibitor for fuel oils to prevent burner screen clogging and reduce rusting.

"Santolite" MS. 58 Trademark for light colored viscous resin made by condensation of formaldehyde with aromatic sulfonamides. Free-flowing above 20°C. Used for spraying, dipping and brushing lacquer. Compatible with nitrocellulose and cellulose acetate; miscible with common organic solvents except hydrocarbons; not miscible with varnish and paint oils.
Containers: 5-, 55-gal drums.

"Santolube." 58 Trademark for a series of oil additives.
"Santolube" 70: A rust inhibitor for turbine oils. Oily viscous liquid; sp.gr. 0.926 (60/60°F); Saybolt Universal viscosity 240 seconds at 210°F.
Containers: 55-gal steel drums.
"Santolube" 203-C: A detergent-dispersant additive for motor oils and heavy duty oils of the MIL-L-2104A type. Barium concentration 10%; sp.gr. 0.981 (60/60°F); Saybolt Universal viscosity 450 seconds at 100°F.
Containers: 55-gal drums; tank cars.
"Santolube" 333: A detergent-dispersant additive for heavy duty motor oils, MIL-L-2104A, Supplement I and Series III oils; a phenate. Barium concentration 10%; sp.gr. 1.01 (60/60°F); viscosity at 210°F 11 centistokes.
Containers: 55-gal drums; tank cars.
"Santolube" 392: An inhibitor and motor oil antioxidant for upgrading the quality of straight base oils when used alone or in combination with detergent-dispersant additives. Flash point 330°F; phosphorus 7.3%, sulfur 15.2% zinc 6.3%; viscosity at 100°F 1260 SUS, at 210°F 60 SUS; sp.gr. 1.104 (60/60°F).
Containers: 55-gal drums; tank cars.
"Santolube" 393: Zinc dialkyl dithiophosphate.
Properties: Liquid; viscosity at 100°F 150 centistokes; flash point 260°F; sp.gr. 1.120 (60/60°F).
Containers: 55-gal black iron drums; tank cars.
Uses: Antioxidant and bearing corrosion inhibitor for lubricating oils.
"Santolube" 394-C: An antioxidant and bearing corrosion inhibitor for engine lubricating oils. Amber, viscous liquid; viscosity 15-25 cts. (210°F); sp.gr. 1.02 (60/60°F); phosphorus 4.7%; sulfur 13.1%.
"Santolube" 395-X: An oxidation and bearing corrosion inhibitor for crankcase oils. Amber liquid; viscosity 12-18 cts (210°F); sp.gr. 1.005 (60/60°F); phosphorus 3.7%; sulfur 11% and barium 1.1%.
"Santolube" 401: A detergent-dispersant-inhibitor additive which when used alone or in combination with inhibitor ("Santolube" 392 or 393) permits the compounding of single and cross-graded motor oils of the MIL-L-2104A and Supplement I types. Liquid; viscosity at 100°F 1153.2 SUS, at 210°F 113.3 SUS; sp.gr. 1.0151 (60/60°F); barium 8.4%; sulfur 3.35%; zinc 0.80%.
"Santolube" 410: A detergent-dispersant additive used in varying concentrations to formulate motor oils meeting performance

requirements of the military and automotive industries. Liquid; viscosity at 100°F 1350 SUS, at 210°F 148 SUS; flash point 380°F; barium 9.5% and sulfur 1.82%.

"Santomask" II. [58] Trademark for an odor-masking agent.
Properties: Clear, light yellow liquid; pleasant characteristic odor; flash point approx 210°F (Cleveland open cup method); flow point below 0°C; wt/gal approx. 8.7 lbs.
Containers: 33- and 5-gal drums.
Uses: Odor-masking ingredient for interior paints, printing inks, and similar products having objectionable natural odors. Will not affect drying time, color durability and other characteristics of paints and inks.

"Santomerse." [58] Trademark for allkyl aryl sulfonates available in the following types:
"Santomerse" No. 1. Neutral flakes and granular form, 40% active.
"Santomerse" 80. Neutral flakes, 80% (min) active.
"Santomerse" No. 3. Neutral powder, 100% active.
"Santomerse" No. 3. Paste (75% "Santomerse" No. 3 in water).
Containers: "Santomerse" No. 1: 75-lb bags; 175-, 220- and 225-lb drums. "Santomerse" 80: 200-lb drums. "Santomerse" No. 3: (powder) 115-lb drums. "Santomerse No. 3: (paste) 175-lb steel drums.
Uses: Surface-active wetting, spreading, penetrating, emulsifying agents and detergents used in the following industries: textile, paper, leather, adhesive, rubber, insecticide, dust collection; soap, cosmetic, ceramic, detergent, metal cleaning, plating, emulsions, dyes, pharmaceutical, embalming fluid, laundry, steel, paint, cement, glue, felt, color, etc.

santonica (levant worm-seed; cina; worm-seed).
Derivation: Dried, unexpanded flower heads of Artesmisia maritima.
Occurrence: Iran, Turkestan and Russia.
Grades: Technical.
Containers: Bags.
Use: Medicine.
Shipping regulations: None.*

santonin $C_{15}H_{18}O_3$. The inner anhydride of santoninic acid.
Properties: Glossy, colorless crystals or white powder, turning yellow on exposure to light; odorless; tasteless at first, then bitter; poisonous! Soluble in chloroform, alcohol, alkalies and most volatile and fatty oils; very slightly soluble in water. Solutions are levorotatory. Sp.gr. 1.187; m.p. 170-173°C; b.p. sublimes; specific rotation −170 to −175° (2 g/100 ml alcohol).
Derivation: By extraction from Artemisia cina, Artemisia maritima, or other species of Artemisia.
Grades: Technical; N.F. XI.
Containers: Bottles; cans.
Use: Medicine.

"Santonox." [58] Trademark for antioxidant for polyethylene. Light gray to light tan powder; m.p. 150°C min.; readily soluble in methanol, ethanol, acetone; insoluble in benzene, ether, carbon tetrachloride and water.
Containers: 150-lb fiber containers.

"Santophen" 1 Germicide. [58] Trademark for ortho-benzyl-para-chlorophenol.

"Santophen" 1 Solution. [58] Trademark for 75% "Santophen" 1 and 25% isopropanol. Used as the germicidal active principle, or as an enhancing agent for disinfectant.

"Santopoid." [58] Trademark for a series of oil additives for compounding gear lubricants meeting Military Specification MIL-L-2105.
Properties: Dark oily liquids, some of which are foam-inhibited, some have rust prevention properties. Flash points 250-300°F; typical viscosities 435 and 750 SUS at 100°F, 60 and 87 SUS at 210°F; sp.gr. (60/60°F) average 1.13, 1.15, 1.29.
Typical analyses:
"Santopoid" 22: S 12.0%; Cl 16.5%; P 3.3%; Zn 3.5%.
"Santopoid" 22-RI: S 13.3%; P 4.0%; Cl 14.5%; Zn 3.7%.
"Santopoid" 23-RI: S 13.3%; Cl 14.0%; P 4.1%; Zn 3.7%.
"Santopoid" 33: S 8.5%; Cl 26.0%; P 0.56%.
"Santopoid" 35: S 8.4%; Cl 26%; P 0.54%.
Containers: 55-gal drums; tank cars.

"Santopoid" S. [58] Trademark for a chlornaphtha xanthate meeting Federal Specification VV-L-761.
Properties: Dark, oily, viscous liquid; sp. gr. approx. 1.19 (60°F); flash point 250°F; viscosity 63 centistokes at 100°F.
Containers: 55-gal steel drums; tank cars.
Uses: Additive for lubricants for hypoid and other gears; for extreme pressure lubricants when compounded with mineral oil.

"Santopour" C. [58] Trademark for pour point depressant for wax-bearing oils.
Properties: A dark liquid; flash point 360°F; viscosity 1020 SUS at 210°F; sp.gr. 0.910 (60/60°F).
Containers: 55-gal drums; tank cars.

"Santoquin." [58] Trademark for 6-ethoxy-1,2-dihydro-2,2,4-trimethyl quinoline (q.v.).

Santorin cement. See pozzolana cement.

Santorin earth. A variety of pumice mined on Santorin Island (Greece) and used in making pozzolana cement.

"Santosite." [58] Trademark for sodium sulfite anhydrous, technical.

"Santotan" KR. [58] Trademark for basic chrome sulfate containing approximately 50% $Cr_2(SO_4)_2(OH)_2$.
Properties: Dark green crystals; basicity (Cr_2O_3/SO_3) 101-105; basicity (Schorlemmer) 37.3-39.7; chromium as Cr_2O_3 23.9% min; clear solution in water (5:500).
Uses: In chrome tanning of calf, cow and horse hides and sheepskins.
Containers: 100-lb paper bags.

*See "I.C.C. Shipping Regulations," page xiii.
Reference numbers refer to name of manufacturer. See "List of Manufacturers," page v.

"Santovar" A. [58] Trade name for 2,5-di-tert-amylhydroquinone.
Properties: Buff powder; m.p. 172°C min; sp.gr. 1.05; soluble in acetone.
Uses: Antioxidant for rubber and adhesives.

"Santowax" R. [58] Trademark for mixed terphenyls. Yellowish-white, non-crystalline flaked solid. Used as extender for polystyrene.
Containers: 75-lb bags; 225-lb fiber drums.

"Santowhite" Crystals. [58] Trademark for (essentially) 4,4'-thiobis-(6-tert-butyl-meta-cresol).
Properties: Light gray to tan powder; m.p. 140°C min; sp.gr. 1.097; soluble in acetone and alcohol.
Uses: Rubber antioxidant.

"Santowhite" L. [58] Trademark for thiobis-(di-sec-amylphenol).
Properties: Dark, viscous liquid; softening point 0°C max; sp.gr. 0.98.
Use: Rubber antioxidant.

"Santowhite" MK. [58] Trademark for a reaction product of 6-tert-butyl-meta-cresol and SCl_2.
Properties: Dark viscous liquid; sp.gr. 1.06; soluble in acetone, alcohol, benzene, and carbon tetrachloride; softening point 20-30°C.
Use: Rubber antioxidant, especially for uncured GRS latex compounds.

"Santowhite" Powder. [58] Trademark for 4,4'-butylidene-bis(6-tert-butyl-meta-cresol).
Properties: White powder; m.p. 109°C; sp. gr. 1.02.
Containers: 125-lb drums.
Uses: Antioxidant, non-staining, in rubber compounding.

saponaria (soapwort; soaproot; fuller's herb bruisewort).
Derivation: Rhizome and roots of perennial herb, Saponaria officinalis.
Occurrence: United States; Europe; middle Asia.
Uses: Source of saponin which constitutes about one-third of the root; medicine; detergent.

saponification. The chemical reaction or process in which an ester is heated with aqueous alkali such as sodium hydroxide, in order to form an alcohol and the sodium salt of the acid corresponding to the ester. The process is most frequently carried out on fats, which are glyceryl esters of fatty acids. The sodium salts formed in this case are soaps, which circumstance leads to the use of the term saponification.

saponification number. The number of milligrams of potassium hydroxide required to saponify 1 gram of a sample of an ester (glyceride; fat) or mixture.

saponified acetate fiber. Regenerated cellulose fibers obtained by complete saponification of highly oriented cellulose acetate fibers. Available in continuous filament form having a high degree of crystallinity and great strength.
Properties: Tensile strength (psi) 136,000-155,000; elongation 6%; sp.gr. 1.5-1.6; moisture regain 9.6-10.7%; decomposes about 300°F; burns readily. Similar to cotton in chemical resistance, dyeing, and resistance to insects and mildew.
Uses: Balloon fabric; cargo-parachutes; typewriter ribbons; fire hose; belts; webbing; tapes; carpet backing.

saponin
1. A general term applied to two groups of plant glucosides that on shaking with water form colloidal solutions giving soapy lathers. They also have the ability to hemolyze red blood corpuscles at very great dilutions. The two groups are:
   A. Triterpenoid saponins. This group of saponins is derived from soap wort, soaproots and soap barks. These noncardiac-active saponin preparations find use as detergents, medicinals, foaming agents in fire extinguishers and fish poisons. Hydrolysis yields sugars and naphthalene and picene derivatives. Structure is not fully known.
   B. Steroid saponins. These cardiac-active saponins were isolated originally from plants of the digitalis series, have also been separated from California soaproot, Mexican sarsparilla root and a number of Mexican plants of the lily family. They are used for the synthesis of sex hormones. Hydrolysis yields various sugars and steroids.
2. Specific term. Saponin derived from saponaria or quillaja.

Properties: White, amorphous glucoside; pungent, disagreeable taste and odor; poisonous! It foams very strongly when shaken with water. Soluble in water.
Chief constituents: Sapotoxin, lactosin, quillaic acid.
Grades: Crude; purified; highest purity.
Containers: 1-, 5-lb bottles; 1-, 5-, 10-lb cans; 25-, 50-lb boxes; 100-lb kegs.
Uses: Foam producer in beverages; detergent in the textile industries; sizing; substitute for soap; fire extinguishers; emulsification agent for fats and oils.
Shipping regulations: None.*

SAPP. Abbreviation for sodium acid pyrophosphate. See sodium pyrophosphate, acid.

sappan wood. Heartwood of the tree, Caesalpinia sappan.
Occurrence: China, Japan and India.
Grades: Technical.
Containers: Bags.
Uses: Textile dyeing, as a substitute for logwood; medicine.
Shipping regulations: None.*

sapphire. For the naturally occurring material, see corundum. Synthetic sapphires are now made from aluminum oxide by the single crystal growing technique for use as gems, as support rods in fine apparatus, as windows and domes for microwave and infrared systems. Their advantages are flexural

*See "I.C.C. Shipping Regulations," page xiii.
Reference numbers refer to name of manufacturer. See "List of Manufacturers," page v.

strength at elevated temperatures, good dielectric properties, small-diameter rigidity, low loss characteristics, and zero porosity.

"SAPP No. 4." [172] Trade name for a grade of sodium acid pyrophosphate possessing an unusually slow rate of reaction necessary when manufacturing canned biscuit doughs.
Derivation: Dehydration of NaH_2PO_4.
Containers: Bags.
Uses: Leavening agent in refrigerated biscuits.

"Saraloy." [233] Trademark for a thermoplastic flashing material based on a saran copolymer. It is both flexible and elastic.

saran. Generic term for thermoplastic resins obtained by the polymerization of vinylidene chloride or copolymerization of vinylidene chloride with lesser amounts of other unsaturated compounds. Sarans are available with softening points from 70° to 175°C. They have excellent chemical resistance to most acids and alkalies (except ammonium hydroxide) and are unaffected by most solvents. They are softened by some chlorinated hydrocarbons and dissolve in cyclohexanone and dioxane. Available as oriented fibers, films, and extruded or molded forms.
Uses (nonfiber): Packaging film; pipes and fittings for chemical processing equipment; bristles; latex coatings.

saran fiber. Generic name for a manufactured fiber in which the fiber-forming substance is any long chain synthetic polymer composed of at least 80% by weight of vinylidene chloride units ($-CH_2CCl_2-$) (Federal Trade Commission).
Properties (of fiber): Tensile strength (psi) 15,-45,000; elongation 15-30%; moisture regain, none; sp.gr. 1.65-1.75; softens at 240-280°F; nonflammable. Highly resistant to most chemicals and solvents, to weather, moths and mildew; flameproof.
Uses: Screens; upholstery, curtain and drapery fabrics; rugs and carpets; awnings; filter cloth.

sarcolysin (para-di(2-chloroethyl)aminophenylalanine) $(ClCH_2CH_2)_2NC_6H_4CH_2CH(NH_2)COOH$. A nitrogen mustard (q.v.). Crystals; m.p. 180°C. Used in medicine.

sarcosine (methyl glycocoll, methyl aminoacetic acid) CH_3NHCH_2COOH.
Properties: Deliquescent crystals with sweet taste; m.p. 210-215°C (dec). Very soluble in water; slightly soluble in alcohol.
Derication: Decomposition of creatine or caffeine.
Containers: Up to tank cars.
Grades: Technical.
Use: Synthesis of foaming antienzyme compounds for toothpaste.

sard. A variety of quartz (q.v.).
Color: Brownish red, similar to but darker than carnelian, classed by some as carnelian.
Use: Gem stone.

sardine meal. Dry fish scrap, approx. 11.5% nitrogen, used for animal and poultry feed.

sardine oil.
Properties: Pale golden-yellow liquid. Soluble in alcohol, ether, chloroform, benzene and carbon disulfide.
Constants: Sp.gr. 0.9274-0.9330; m.p. 28-36°C; acid value 4-25; Hehner value 95-97; saponification value 189-193; iodine value 181-193; refractive index 1.4802-1.4808.
Derivation: By chopping up sardines and subjecting them to boiling and pressing.
Grades: Crude; refined.
Containers: Wooden barrels; steel drums.
Uses: Soap; leather stuffing; lubricant.
Shipping regulations: None.*

sardonyx. A natural crystalline variety of silica, with rather uniform layers of red, white, brown and black; waxlike luster; and may be transparent or translucent.

Sarin (methylisopropoxyfluorophosphine oxide) $(C_3H_7O)(CH_3)FPO$. A nerve gas.

"Sarkosyl." [219] Trademark for a series of surface active acylated sarcosines available in several types:
NL-30: 30% aqueous sodium lauroylsarcosinate; water white, odorless, practically tasteless, low in toxicity, anti-corrosive, anti-enzyme.
L: lauroyl sarcosine free acid.
LC: cocoyl (whole coconut) sarcosine free acid.
O: oleoyl sarcosine free acid.
S: stearoyl sarcosine free acid.
Uses: Tooth pastes, cosmetics, and pharmaceuticals (anti-enzyme); corrosion inhibitors; lubricants and greases; anti-oxidants; inks.

sarsaparilla. The dried root of Smilax aristolochiaefolia (Mexican sarsaparilla) and other Smilax.
Occurrence: Southern United States; Honduras; Jamaica; Mexico; Guatemala; Brazil.
Grades: N.F. XI.
Containers: Honduras, 160-lb bales; Mexican, 200-lb bales; bags and bales of variable size.
Uses: Medicine; soft drinks.
Shipping regulations: None.*

"Sartorius No. 1." [188] Trademark for a grade of ylang ylang oil.

SAS. Abbreviation for sodium aluminum sulfate. See aluminum sodium sulfate.

S.A.S. baking powder. A baking powder containing sodium aluminum sulfate (S.A.S.), as well as sodium bicarbonate and starch.

sassafras bark (saxifrax; ague tree; cinnamon wood) The dried bark of the root of Sassafras albidum.
Occurrence: North America.
Grades: Technical; N.F. XI.

Containers: 150-lb bales; bags and bales of variable size.
Uses: Medicine; flavoring; perfumery.
Shipping regulations: None.*

sassafras oil.
Properties: Yellowish or reddish-yellow volatile liquid oil; pungent, aromatic odor and warm, aromatic taste. Sp.gr. 1.065-1.077 (25/25°C); optical rotation +2 to +4°; refractive index 1.5250-1.5350 (20°C). Soluble in alcohol, ether, chloroform, glacial acetic acid and carbon disulfide.
Chief constituents: Safrol, eugenol, camphor, pinene, phellandrene.
Derivation: Steam distillation of the root of Sassafras albidum.
Method of purification: Rectification.
Impurities: Terpenes.
Grades: Technical; N.F. XI.
Containers: Glass bottles; cans; drums.
Uses: Flavoring; perfumery; medicine. May not be used in foods or beverages (FDA).
Shipping regulations: None.*

"Sassafrol." [188] Trademark for a substitute for oil of sassafras.

sassolite H_3BO_3. A natural boric acid found as pearly yellowish-white scale on native sulfur in the crater of Vulcano (Lipari Isles) and as a sublimation around Vesuvian fumaroles. Also forms a small proportion of the boron compounds in Californian borax localities. See boric acid.

satin spars. See calcite and gypsum.

"Satintone." [99] Trademark for anhydrous aluminum silicate pigments derived from water-washed kaolin by calcination and grinding. Available in three particle size grades: "Satintone" Special, "Satintone" #1, and "Satintone" #3. Used in coatings for roofing granules and as a hiding extender in paints. Also used as a polishing and cleaning agent.
Typical analysis: Silica 52.1-52.3%; alumina 44.4-44.6%; L.O.I. 0.5-0.9%; iron oxide trace; titanium dioxide 2.0%; manganese none.

satin white. See calcium sulfate.

saturation.
1. The state in which all available valence bonds of an atom (esp. carbon) are attached to other atoms. The straight-chain paraffins are typical saturated compounds.
2. The state of a solution when it holds the maximum equilibrium quantity of dissolved matter at a given temperature.

satureja. See savory.

"Savelite." [141] Trade name for an interior white finish for ceilings, walls and the interior of warehouses, storehouses and factories.

savin (sabina).
Derivation: Tops of evergreen shrub Juniperus sabina.
Occurrence: Europe; northern Asia; North America south to New York and Montana.
Grades: Technical.
Containers: Bags.
Uses: Medicine; savin oil.
Shipping regulations: None.*

savin oil.
Properties: Colorless to pale yellow liquid; soluble in alcohol, ether, and chloroform.
Chief known constituents: Sabinol; cadinene; pinene.
Constants: Sp.gr. 0.910-0.930; optical rotation +40 to +60°.
Derivation: Distilled from the fresh leaves and twigs of Juniperus sabina.
Method of purification: Rectification.
Grades: Technical.
Containers: 1-, 5-lb bottles; 11-, 25-lb tins.
Use: Medicine.
Shipping regulations: None.*

"Savonafix." [188] Trademark for a perfume specialty with fixation properties used in the perfuming of soaps.

savory (summer savory; satureja).
Derivation: Dried leaves of Satureia hortensis, an aromatic mint of the United States and Europe. Also applied to the essential oil derived from these leaves.
Uses: Medicine; flavors; perfumes.

savory, summer. See savory.

saw palmetto. See serenoa.

saw palmetto berries. See serenoa.

saxifrax. See sassafras bark.

Sb. Symbol for antimony.

SBA. Abbreviation for sec-butyl alcohol (q.v.).

SBR. Abbreviation for styrene-butadiene rubber (q.v.).

"S-B Zinc." [72] Trade name for a semi-bright (dull) zinc plating process, composed of zinc cyanide, sodium cyanide, sodium hydroxide and addition agents. High surface activity for subsequent surface conversion coatings.

Sc. Symbol for scandium.

"SC-50." [245] Aqueous solutions of alkali-metal salts of hydrocarbon-substituted silanols used for treating various surfaces to render them water-repellent and to improve other properties of such surfaces.

scammony resin. A mixture of resins obtained by extraction from scammony root.
Properties: Soluble in alcohol.
Use: Medicine.

scammony root. Root of Convolvulus scammonia.
Occurrence: Asia Minor and Greece.
Containers: Bags.
Use: Source of scammony resin.
See also ipomea.

scandia. See scandium oxide.

scandium Sc. Element of atomic number 21; Group III of the periodic table.

Properties: Metal; m.p. 1550-1600°C; b.p. 2470°C; density 3.02g/cc; colorless salts.
Source: Occurs in certain uncommon minerals as wolframite, thortveitite, euxenite. It is not a member of the rare earth elements, though it is rare in the sense of not being abundant, and occurs associated with the rare earth elements.
Derivation: From the oxide, by way of the fluoride, which is reduced to the metal. Distillation gives 99+% pure scandium.

scandium antimonide ScSb. Used as a high-purity semiconductor.

scandium arsenide ScAs. Used as a high-purity semiconductor.

scandium borate $ScBO_3$.
Properties: White powder; soluble in dilute acid.
Derivation: By fusing scandia and boric acid.
Shipping regulations: None.*

scandium oxide (scandia) Sc_2O_3.
Properties: White amorphous powder resembling magnesia. Soluble in hot acids, less so in cold acids. Sp.gr. 3.864; specific heat 0.153 (0-100°C).
Derivation: From thortveitite, an ore.
Uses: Manufacture of scandium; ceramics.
Shipping regulations: None.*

scandium phosphide ScP. Used as a high-purity semiconductor.

scandium sulfide Sc_2S_3.
Properties: Light yellow powder decomposed by boiling water and by dilute acids with evolution of hydrogen sulfide.
Derivation: By heating the sulfate in hydrogen sulfide.
Shipping regulations: None.*

scapolite, common. See wernerite.

scarlet corns. See kermes.

scarlet red (Biebrich red; Sudan IV) $C_{24}H_{20}ON_4$. ortho-Tolueneazo-ortho-tolueneazo-beta-naphthol. A red dye.
Properties: Dark brown odorless powder; m.p. 181-188°C; decomposes before boiling; insoluble in water; soluble in oils; slightly soluble in alcohol; somewhat soluble in chloroform.
Derivation: From beta-naphthol and ortho-amino-azo-toluene.
Uses: Medicine (in ointments); biological stain for fats.
Also refers to red varieties of ferric oxide. See iron oxide reds.

scavengers. From a chemical point of view this term may be used for any chemical which is added to a system or mixture in order to consume or convert to inactive form small quantities of impurities or undesired materials. In metallurgical operations active metals are added to combine with oxygen and nitrogen in the molten metal, and so cause removal into the slag. A more recent use is the addition of oxygen, iodine, or more complex materials which combine with free radicals in a mixture, and make possible the measurement of these radicals.

Schaeffer's acid. See 2-naphthol-6-sulfonic acid.

Schaeffer's salt $C_{10}H_6OHSO_3Na$. Sodium salt of 2-naphthol-6-sulfonic acid (q.v.).
Uses: Intermediate for organic chemicals.
Containers: (Paste) drums; (powder) bags.

Schaffer's salt. See Schaeffer's salt.

Scheele's green. See copper arsenite.

scheelite $CaWO_4$. A natural calcium tungstate, found in igneous rocks, usually with granite. Some molybdenum may replace tungsten.
Properties: Color white, yellow, green, brown; luster vitreous to adamantine; hardness 4.5-5; sp.gr. 5.9-6.1; usually fluorescent in ultraviolet light.
Occurrence: Nevada, California, Arizona, Utah, Colorado; New Zealand; Europe.
Use: Ore of tungsten; phosphor.

Schiff bases. The class of compounds derived by chemical reaction (condensation) of aldehydes or ketones with primary amines. The general formula is RR'C=NR".
Properties: Usually colorless crystalline solids, although some are dyes. Very weakly basic and hydrolyzed by water and strong acids to carbonyl compounds and amines.
Uses: Accelerators for vulcanization of rubber; dyes, for example phenylene blue, toluylene blue, and naphthol blue; intermediates in the preparation of many organic compounds. Those prepared from ortho-hydroxyacetophenone and diamines such as ethylenediamine are oil-soluble sequestering agents which are used to inhibit the undesirable effects of trace metal ions in mineral oils and gasoline.

Schlippe's salt. See livers of antimony.

Schoch process. A process for making acetylene by passing hydrocarbons through an electric arc.

Schoelkopf's acid. See 1-naphthylamine-8-sulfonic acid; and 1-naphthol-4,8-disulfonic acid.

schoenite $MgSO_4 \cdot K_2SO_4 \cdot 6H_2O$. Natural salt obtained from Stassfurt, Germany, salt deposits, and an important source of potassium compounds.

schorlite. A black iron-bearing variety of tourmaline (q.v.).

schradan. Generic name for octamethyl pyrophosphoramide.

schroeckingerite $NaCa_3(UO_2)(CO_3)_3(SO_4)F \cdot 10H_2O$. A natural hydrated fluocarbonate-sulfate of sodium, calcium and uranium.
Properties: Color greenish yellow; luster vitreous; hardness 2.5; sp.gr. 2.51; fluorescent in ultraviolet light; radioactive.
Occurrence: Wyoming, Utah, Arizona; Europe.
Use: Ore of uranium.

schwatzite. A mercury-bearing variety of tetrahedrite (q.v.).

Schweinfurth green. See copper acetoarsenite.

Schweitzer's reagent. A reagent used in analytical chemistry as a test for wool. It dissolves cotton, silk and linen and consists of a solution of copper hydroxide in strong ammonia.

scilla. See squill.

scintillation counter. A form of detector of ionizing radiations much used in work with nuclear radiations. It depends on the property possessed by many transparent substances of emitting visible or near ultraviolet light when traversed by an ionizing particle. By mounting the scintillating material on the face of a photomultiplier tube, the pulse of light produced by the radiation is converted to an electrical pulse which in turn may be amplified and counted. For work with short range, densely ionizing radiations like alpha particles, the scintillator may be finely crystalline zinc sulfide coated in a thin layer on the face of the photomultiplier tube. For more penetrating radiations like gamma rays the scintillator is usually a large single crystal of sodium iodide activated with a trace of thallium iodide. Recent developments in the field include the discovery of clear plastic materials with this scintillating property, which may be molded into required complex shapes, and the use of solutions of substituted oxazole compounds dissolved in toluene into which radioactive material may be dissolved to measure their radioactivity. Various complex solutions are used, adjusted to have a scintillation spectrum matched to the spectral sensitivity of the photomultiplier by a proper choice of scintillating compounds in the solvent.

The oxazole compounds in such use are commonly known by their abbreviations, as DPO, BBO, POPOP, NOPON, BOPOB. Some of the other materials used in scintillation counting are terphenyl, quaterphenyl, tetraphenylbutadiene, diphenylacetylene, diphenylstilbene, and of the older materials, anthracene and naphthalene.

scleroproteins (albuminoids). A large class of proteins that have a supporting or protective function in tendons, bones, cartilages, ligaments and other hard or tough parts of the animal body. They include the collagens of skin, tendons and bones, as well as the elastic proteins known as elastins, and the keratins. Specific examples are the keratin of hair, hoofs and horns and fibroin from silk. Collagens are the material from which gelatin is produced when protein materials are boiled with water.

scoparius (broom; green broom; Scotch broom; Irish broom; hogweed).
Derivation: Dried tops of shrub Cytisus scoparius.
Occurrence: Western Asia; southern and western Europe; cultivated in the United States.
Grades: Technical.
Containers: Bags.
Uses: Medicine; perfumery.
Shipping regulations: None.*

scopola (Japanese belladonna).
Derivation: Dried rhizome of Scopola carniolica.
Occurrence: Japan; Germany; Austria; Hungary; Russia.
Grades: Technical.
Containers: Bags.
Use: Medicine; source of alkaloids.
Shipping regulations: None.*

scopolamine. See hyoscine.

scopolamine hydrobromide. The U.S.P. XVI name for hyoscine hydrobromide (q.v.).

scopolamine methylbromide. See methscopolamine bromide.

scopolamine methylnitrate. See methscopolamine nitrate.

scopoline (oscine) $C_8H_{13}O_2N$.
Properties: White crystals; m.p. 108-109°C when crystallized from ether; b.p. 248°C; soluble in water, acetone, ether, and alcohol.
Derivation: An alkaloid obtained by hydrolysis of scopolamine with dilute alkali.
Use: Medicine.

"Scorbord." [233] Trademark for an expanded polystyrene plastic product used as insulating material.

scorodite. Natural ferric arsenate, $Fe_2O_3 \cdot As_2O_5 \cdot 4H_2O$.
Properties: Color, pale leek-green or liver-brown.
Occurrence: Utah and Washington.
Use: A lesser ore of arsenic.

Scotch broom. See scoparius.

"Scotch-em Aerosol Insecticide." [204] Trademark for a mixture containing allethrin, DDT, and methoxychlor. For space spraying to control flying insects.

"Scotchgard" Leather Protector. [158] Brand name for fluorochemical applied to leather.
Properties: Colored liquid with a specific gravity of 1.0 is 30 percent solids in isopropyl alcohol. When applied to leather, the fibers react with the fluorochemicals and become permanently bound. Soluble in water, the chemical develops qualities of repellency upon drying. Does not affect color, hand or porosity of the leather.
Containers: 5-gal pails; 15- and 55-gal drums.
Uses: Imparts repellency to water and oil and resistance to chemicals (including acids and alkalis) and soil resistance to brushed pigskins, suedes and side leathers. Applied after the fatliquoring run (in the same equipment) at the tannery. Also makes garment leather drycleanable.

"Scotchgard" Stain Repeller. [158] Brand name for a group of textile chemicals (available in various formulations for different types of fabric applications including cationic and non-ionic stabilized emulsion of fluorochemical resins and solvent soluble solutions of fluorochemical resins).

Properties: The fluorochemicals are liquids, generally composed of about 28 per cent solids.

Containers: 5-gal pails and 30-gal drums.

Uses: Imparts oil and water repellency to a wide range of fabrics and textile blends as well as resistance to dust and dry soil.

scotophors. Materials which are sensitive to cathode-ray beams and respond by a change in opacity or light-transmission properties. Are useful in screens for television. Crystals of alkali halides are used.

scouring agents. Compounds used to remove the natural oils and fats from raw wool; also used to remove lubricants applied to rayon yarns or fabrics during such operations as throwing, winding, weaving, knitting, etc.

scram. A colloquialism that has come into general use to mean the rapid shut-down of a nuclear reactor, used both as a verb and as a noun. An accidental scram is a shut-down by the safety circuits, though it may also be scrammed by intentionally dropping in the control rods.

"Scriptite" 33. [58] Trademark for a dry, finely divided melamine formaldehyde powder paper treating resin; used to provide wet strength in paper.

Containers: 50-lb polyethylene-lined multiwall bags.

"Scriptite" 40. [58] Trademark for a liquid urea-formaldehyde resin used for coating paper to achieve wet strength and printability.

Containers: 525-lb drums.

"Scriptite" Series 50-52-54. [58] Trademark for styrene copolymers (dry powders), used as paper coating resins with or without clay to improve surface printability and wax hold-out properties.

Containers: Bags and drums.

scrubbing. Process for removing one or more components from a mixture of gases and vapors by its passage upward and usually countercurrent to and in intimate contact with a stream of descending liquid, the latter being chosen so as to dissolve the desired components and not others. The gas or vapor may be broken into fine bubbles upon entering a tower filled with liquid, but more frequently the tower is filled with coke, broken stone or other packing, over which the liquid flows while exposing a relatively large surface to the rising gas or vapor. See also absorption.

scrubbing oil. See absorption oils.

"Scutamol." [206] Brand name for organic mercury used for control of slime in paper mills.

"SD-75." [430] Trade name for alkenyl dimethyl ethyl ammonium bromide. Used primarily as an algaecide in water towers and swimming pools.

SDA No. 1. Specially denatured alcohol (government regulation formula). It consists of 5 gal wood alcohol added to 100 gal 95% ethyl alcohol. See denaturants for alcohol; denatured alcohol.

Se. Symbol for selenium.

Seaboard process. Method of removing hydrogen sulfide from a gas by absorption in sodium carbonate solution. Sodium bicarbonate and sodium hydrosulfide are formed. By blowing air through this solution, hydrogen sulfide is released and carried off, and the sodium carbonate is regenerated.

"SeaKem." [124] Trade name for a line of carrageenan extractives, hydrocolloids which may be derived from a number of sea plants in the class of Rhodophyceae, order of Gigartinales, principally, Irish moss sea plants. Used in the food, pharmaceutical and cosmetic industries as gelling agents; stabilizers for emulsions, suspensions, foams; binding agents; viscosity producers; as emollients.

"Sealstix." [18] Trademark for a cement of the deKhotinsky type (q.v.) which adheres to almost any surface. Insoluble in all common reagents except alcohol, strong caustics and chromic acid cleaning solution.

Uses: Cementing glass, in air, vacuum, or under water; for sealing cracks and pinholes in vacuum systems; corrosion protection; insulation for exposed conductors.

"Sealz." [248] Trademark for a line of thermoplastic rubber compounds which are used as additives for asphalt and tar products to raise softening point, improve low temperature flexibility, enhance adhesive qualities and increase elasticity.

"Seam-It." [65] Trademark for a material comprising, in part, natural latex, stabilizers and vulcanizing ingredients for binding carpets.

sea onion. See squill.

Searles Lake brine. Source of trona (q.v.). In California.

sea salt. The actual content of various salts in sea water is as follows, the values being grams per liter, i.e., approximately grams per thousand grams: NaCl (sodium chloride) 27.319; $MgCl_2$ (magnesium chloride) 4.176; $MgSO_4$ (magnesium sulfate) 1.668; $MgBr_2$ (magnesium bromide) 0.076; $CaSO_4$ (calcium sulfate) 1.268; $Ca(HCO_3)_2$ (calcium bicarbonate) 0.178; K_2SO_4 (potassium sulfate) 0.869; B_2O_3 (boron oxide) 0.029; SiO_2 (silica) 0.008; iron and alumina (R_2O_3) (iron and aluminum oxide) 0.022.

Salt recovered from sea water by the usual evaporation methods contains 96-99% NaCl and small amounts of $CaSO_4$, $MgCl_2$

and $MgSO_4$, varying with the degree of evaporation and methods of handling and washing.

"Sea Sorb." [55] Brand name for patented grades of magnesia, MgO, with specific adsorption characteristics.
Properties: White powder or granules; iodine number 80-90; surface area 120 sq m/g; bulk density 35-40 lbs/cu ft; ignition loss 6-7%; particle size 12% +200 mesh, max.
Grades: 43 (powder); 53 (granules).
Containers: 200-lb fiber drums (43); 250-lb steel drums (53).
Uses: Wide applications involving catalysts, chromatography, purification, decolorization, and dehumidification.

sea weed. See kelp.

sebacic acid (octanedicarboxylic acid; sebacylic acid; decanedioic acid) $COOH(CH_2)_8COOH$.
Properties: White leaflets; m.p. 133°C; b.p. 295.0°C (100 mm); refractive index 1.422 (133.3°C); slightly soluble in water; soluble in alcohol and ether.
Derivation: From butadiene via dichlorobutene and its nitrile derivatives; dry distillation of castor oil with alkali.
Containers: Fiber drums; bags.
Grades: C.P. and purified.
Uses: In plasticizers; alkyd resins; fibers; paint products; candles and perfumes; low temperature lubricants and hydraulic fluids; manufacture of certain kinds of nylon.
Shipping regulations: None.*

"Sebacol." [173] Trademark for a tanners' depilatory containing sodium hydrosulfite suitably stabilized for use in alkaline unhairing systems.

sebaconitrile $NC(CH_2)_8CN$.
Properties: Straw-colored oily liquid; b.p. 199°C (15 mm).
Use: As chemical intermediate for drugs, dyes and high polymers.

sebacoyl chloride (n-octane-1,8-dicarboxylic acid dichloride) $ClOC(CH_2)_8COCl$.
Properties: Liquid; b.p. 137-140°C (3 mm); 97% pure. Decomposes slowly in cold water; soluble in hydrocarbons and ethers.
Containers: Bottles; carboys; 55-gal drums.
Use: Organic intermediate.

sebacylic acid. See sebacic acid.

sec-. Abbreviation for secondary, as applied to names of organic compounds. See primary. In this book the prefix sec- is ignored in the alphabetical arrangement.

secale cornutum. See ergot.

secobarbital $C_{12}H_{18}N_2O_3$. 5-Allyl-5(1-methylbutyl) barbituric acid.
Properties: White, amorphous, odorless powder with slightly bitter taste. M.p. about 82°C. Very soluble in alcohol and ether; very slightly soluble in cold water.
Grade: U.S.P. XVI.
Use: Medicine.

secobarbital sodium $C_{12}H_{17}N_2NaO_3$. Sodium 5-allyl-5-(1-methylbutyl)barbiturate.
Properties: White, hygroscopic odorless powder with bitter taste. Very soluble in water; soluble in alcohol; practically insoluble in ether; pH (5% solution) 9.8-10.1.
Grade: U.S.P. XVI.
Use: Medicine.

"Seconal." [100] Trade name for secobarbital [5-allyl-5-(1-methylbutyl)barbituric acid].

"Seconal Sodium." [100] Trademark for secobarbital sodium, U.S.P.

secondary. Term used to characterize certain types of compounds. See primary; see also sec-.

secondary azo dyes. Azo dyes derived from secondary amines.

secondary calcium phosphate. See calcium phosphate, dibasic.

sedimentary rocks. Rocks derived from the disintegration and weathering of older rocks, and deposited in layers by water, wind or ice. Common sedimentary rocks are sandstone, limestone, shale, conglomerate.

sedimentation. The settling of finely divided suspended solid particles from a liquid.

"Sedulon." [190] Trademark for a brand of dihyprylone (q.v.).

seed oil. See cottonseed oil.

seed oil cake. See cottonseed cake.

"Seedrin." [147] Brand name for an insecticidal seed treatment containing aldrin.
Containers: "Seedrin W75" - 50-lb bags; "Seedrin Liquid" 55-gal drums.
Uses: "Seedrin Liquid" is used for seed treatment of rice; "Seedrin W75" for rice, wheat, oats, barley, corn, milo and sorghum.

Seger cones. See pyrometric cones.

Seidlitz powders (Seidlitz mixtures). Two powders to be mixed in water and drunk while effervescing.
In the blue paper: 3 parts Rochelle salt with 1 part sodium bicarbonate.
In the white paper: Tartaric acid.
Containers: Fiber drums.
Use: Medicine.

seignette salt. See potassium-sodium tartrate.

"Seismex." [28] Trademark for an ammonia dynamite for seismic prospecting.

"Seismogel." [28] Trademark for a straight gelatin dynamite for seismic prospecting.

"Selectacel." [231] Trademark for ion-exchange cellulose products which are derived from purified wood pulp by chemical reaction and substitution. White fibrous powders packed in glass, plastic and fiber drums. Used for the isolation and purification of proteins.

selenious acid. See selenous acid.

selenite. Single crystals of gypsum (q.v.).

selenium Se. A nonmetallic element, atomic number 34, in Group VI of the periodic table. It is similar to sulfur and occurs

*See "I.C.C. Shipping Regulations," page xiii.
Reference numbers refer to name of manufacturer. See "List of Manufacturers," page v.

with it in sulfide ores as well as in certain soils. It seems to be an essential nutrient in minute amounts, but is toxic in larger amounts.

Properties: Steel-gray, nonmetallic rods or buttons; very high luster; crystalline surface on being broken. Also occurs in the form of dark red crystals or powder; soluble in carbon disulfide, concentrated sulfuric acid; insoluble in water and alcohol. Selenium burns in air with a bluish-red flame, forming selenium dioxide. Its electrical conductivity increases with increasing brightness of any light irradiating it. Selenium and its compounds are quite poisonous, like arsenic. Sp.gr. 4.26-4.28; m.p. 217°C; b.p. 690°C.

Derivation: From anode mud of electrolytic copper refining process, or from flues of pyrites burners.

Method of purification: By sublimation and reduction by means of an aqueous solution of sulfur dioxide.

Grades: Selenium is available as a powder and in the form of small cast sticks or cakes. Various grades of selenium may be obtained but the usual quality is 99% Se. A high purity grade (impurities less than 10 ppm) is now available. There is also a lower grade 97.5-99%. The powdered selenium may be obtained in various grades of fineness, the usual classifications being 40 mesh (through 40 on 120), 120 mesh (through 120 on 200) and 200 mesh (through 200).

Containers: Glass bottles; drums.

Uses: The allotropic, red powder form is used in microscopy as an imbedding material. Selenium is used in the glass industry for making red glass; in the rubber industry as a vulcanizing agent; in electrical rectifiers, photoelectric cells, as a stabilizer in lubricating oils; stainless steels; xerography (q.v.), an inkless printing process based on the electrical properties of this element.

Shipping regulations: None.*

selenium diethyldithiocarbamate $Se[SC(S)N(C_2H_5)_2]_4$.

Properties: Orange-yellow color; sp.gr. (20/20°C) 1.32; melting range 63-71°C (commercial grade); characteristic odor. Soluble in carbon disulfide, benzol, chloroform; insoluble in water.

Containers: 5-, 50-, 100-lb drums.

Use: Vulcanization agent without added sulfur or as a primary or secondary accelerator with sulfur. Functions at relatively low temperatures.

selenium dimethyldithiocarbamate. See "Methyl Selenac."

selenium dioxide (selenous acid anhydride; selenous anhydride) SeO_2.

Properties: White or yellowish white to slightly reddish, lustrous, crystalline powder or needles; sp.gr. 3.954 (15/15°C); m.p. 340°C; b.p. 317°C (sublimes); soluble in alcohol, water.

Grades: Technical.

Uses: Analysis (testing for alkaloids); medicine; an oxidizing agent; as an antioxidant in lubricating oils; catalyst.

selenium disulfide. See selenium sulfide.

selenium sulfide (selenium disulfide) SeS_2.

Properties: Bright orange powder; faint odor at most. Practically insoluble in water and organic solvents.

Grade: U.S.P. XVI.

Use: Medicine.

1,4-selenothiane $\overline{SeCH_2CH_2SCH_2CH_2}$.

Properties: Thin, colorless leaflets; b.p. 86.5°C (97 mm); m.p. 107°C.

Derivation: By boiling an aqueous solution of sodium selenide with dichloroethyl sulfide.

selenous acid (selenious acid) H_2SeO_3.

Properties: Transparent, colorless deliquescent crystals; soluble in water and alcohol; insoluble in ammonia. Sp.gr. 3.0066; m.p. decomposes.

Derivation: By the action of hot nitric acid on selenium.

Method of purification: Sublimation.

Grades: Technical.

Containers: Glass bottles.

Use: Chemical reagent.

Shipping regulations: None.*

selenous acid anhydride. See selenium dioxide.

selenous anhydride. See selenium dioxide.

"Sellogen." [78] Trademark for a series of detergent, emulsifying, and wetting agents used in the textile industry and in the manufacture of cosmetics.

"Selsun." [3] Trademark for selenium sulfide preparation used to control dandruff. Not a cure.

semecarpus nut (acajou nut; bhilawan nut; malacca nut; marany nut; marking nut; mangle). The fruit of semecarpus anacardium; oriental cashew nut. Properties and uses similar to cashew nut. See cashew nutshell liquid.

"Semesan." [28] Trademark for seed disinfectant and turf fungicide containing hydroxymercurichlorophenol.

Containers: 4-oz cans, 12-oz cans and 25-lb pails.

Use: For dust or dip treatment to control seed-borne diseases, reduce seed decay, and check damping-off of seedlings of vegetable, flower, and other crop seeds, bulbs, corms, roots and tubers. Also for prevention and control of brown patch and other fungus diseases on fine turf of lawns and golf greens.

"Semesan Bel." [28] Trademark for dip disinfectant containing hydroxymercurinitrophenol and hydroxymercurichlorophenol.

Containers: 3-oz, 1- and 4-lb cans; 40-lb pails.

Use: For dip treatment of Irish and sweet potatoes (seed pieces or slips) to protect against decay and control certain surface-borne diseases.

semicarbazide hydrochloride (amidourea hydrochloride; carbamylhydrazine hydrochloride; aminourea hydrochloride) $H_2NCONHNH_2 \cdot HCl$.

Properties: Snow-white crystals; m.p. 172-175°C (dec); soluble in water; insoluble in absolute alcohol and ether; soluble in diluted alcohol.

Derivation: From hydrazine sulfate, potassium or sodium cyanate and sodium carbonate, or electrolytically by the reduction of nitro-urea.

Grades: C.P.; technical.

Containers: Bottles; fiber drums of all sizes.

Uses: Reagent for aldehydes and ketones; isolation of hormones and isolation of certain fractions from essential oils.

Shipping regulations: None.*

semiconductor. A material with an electrical conductivity greater than that of insulators or dielectrics, but far less than the conductivity of true conductors such as metals. The range of conductivity of a semiconductor is usually 10^{-6} to 10^{-3} mhos/cm, but a wider range is sometimes included. The most common semiconductors are highly purified silicon and germanium, but slight traces (parts per million or billion) of selected impurities and/or crystal imperfections must be present to produce semiconductor properties. Without the impurities the materials are insulators.

These impurities cause either (1) loosely bound electrons that can move and carry some current, or (2) the impurities remove electrons from their normal place in the lattice and so form a "hole" which can be filled by an adjacent electron whose movement creates a new hole which is in turn filled. The resulting movement of the hole is the equivalent of electrical conduction in a direction opposite to that occurring when electrons move. The conductivity of a semiconductor usually varies rapidly with temperature. The most important semiconductors are silicon, germanium, selenium, cuprous oxide (Cu_2O), lead sulfide, silicon carbide, lead telluride and other intermetallic compounds. The production of these materials with suitable ultra high purity and with appropriate controlled impurities has become a significant segment of industrial chemistry. Applications are as rectifiers, modulators, detectors, thermistors, photocells and transistors in electrical circuits.

See also transistor.

semidrying oils. See drying oils.

senarmontite (antimony trioxide, octahedral) Sb_2O_3.

Properties: White or gray mineral with white streak and resinous, almost adamantine luster. Yellow on melting. Insoluble in water; slightly soluble in sulfuric or nitric acids; soluble in hydrochloric acid, alkali hydroxides or sulfides. Sp.gr. 5.2-5.3; hardness 2-2.5. See valentinite.

Occurrence: Europe and northern Africa.

Uses: As a paint pigment and in flameproofing.

Seneca oil. A term sometimes applied to petroleum in colonial days, when it was used in medicine.

Seneca root. See senega.

senega (senega snakeroot; Seneca root; rattlesnake root).

Derivation: Dried root of Polygala senega.

Occurrence: North America (Canada to South Carolina, west to Wisconsin).

Grades: Technical.

Containers: Bales; bags of variable size.

Use: Medicine.

Shipping regulations: None.*

Senegal gum. See arabic gum.

senega snakeroot. See senega.

senna.

Derivation: Dried leaflets of Cassia acutifolia (Alexandria senna) or of Cassia angustofolia (Tinnevelly senna).

Occurrence: Nubia; Barbary; Abyssinia; Egypt; southern India.

Grades: Technical; N.F. XI.

Containers: Bales; powder in boxes and barrels.

Use: Medicine.

Shipping regulations: None.*

"Sentry." [214] Trademark for sorbic acid and potassium sorbate (q.v.). Used in fungistats for the control of certain molds and yeasts in foods.

"Separan." [233] Trademark for synthetic flocculating agents.

"Separan" NP10: Synthetic nonionic, water soluble polymer of acrylamide.

Uses: Thickening and filtration of ore slurries; clay slurries and flotation concentrates; flocculation of solids in the lime-sulfur process; clarification of potash brines; removal of ferric hydroxide in manganese purification; filtration of coal-washery slimes; disposal of tailings; filtration of gypsum in wet-process phosphoric acid plants.

"Separan" NP20: Nonionic polyacrylamide polymer.

Uses: As for "Separan" NP10.

"Separan" AP30: Synthetic anionic polymer.

Uses: Clarification of coal tar.

sepia.

1. A reddish-brown pigment prepared from the ink of the cuttlefish. It is a mixture of calcium carbonate, magnesium carbonate, melanin, and an organic black coloring matter.
2. Pulverized bones of the cuttlefish, used as a polishing agent, as an ingredient of tooth powders, and for jewelry models.

sepiolite. See meerschaum.

"Septo-Sour." [244] A proprietary product consisting chiefly of zinc salts of fluorine compounds.

Properties: White, dustless crystals; readily

*See "I.C.C. Shipping Regulations," page xiii.
Reference numbers refer to name of manufacturer. See "List of Manufacturers," page v.

soluble in water; neutralizing value 25.6 oz sodium bicarbonate per lb.
Containers: 10-lb net corrugated paper drums; 60-lb net plywood drums; 150- and 300-lb net fiber drums.
Uses: Laundry sour, especially for goods washed at low temperatures.

"Seqonyx." [328] Trademark for a heavy crystalline sodium phosphoborate product that is most soluble in water at 90°F. Used as a processing assistant in the manufacture of light weight papers such as towelling tissue, crepe papers, bleached and unbleached sulfite newsprint, and bleached bond papers; also a dispersant in pitch control.

sequestration. (See also chelate). Usually the coordination complex of certain phosphates with metallic ions in solution so that the usual precipitation reactions of the latter are no longer possible. Thus, calcium soap precipitates are not produced from hard water treated with hexametaphosphates. The term sequestration may be used for any instance in which an ion is prevented from exhibiting its usual properties due to close combination with an added material.

Phosphates (see sodium metaphosphate) are the most widely used inorganic sequestering agents. Two groups of organic sequestering agents (chelates in these examples) of economic importance are the aminopolycarboxylic acids such as ethylenediaminetetraacetic acid and the hydroxycarboxylic acids such as gluconic, citric, and tartaric acids.

"Sequet." [244] Trademark for a compound consisting of complex phosphate base and an anionic surface active agent.
Properties: Soluble in water; stable, high sudsing, mildly alkaline compound; white, dedusted mechanical mixed powder.
Containers: 120-lb plywood drums; 300-lb wooden barrels.
Uses: Bottle washing additive for free rinsing operations in all hard water areas.

"Serene Green." [141] Trade name for color pigments made from benzidine yellow and phthalocyanine blue.

"Serenium." [412] Trademark for ethoxazene (q.v.).

serenoa (saw palmetto; sabal).
Derivation: Partly dried ripe berries of Serenoa repens or S. serrulata.
Occurrence: South Carolina to Florida and West Indies.
Grades: Technical.
Containers: Bags.
Use: Medicine.
Shipping regulations: None.*

"Serfin." [330] Trademark for reserpine.

"Serichrome." [243] Trademark for mordant acid dyes for wool.

serine (beta-hydroxyalanine; alpha-amino-beta-hydroxypropionic acid) $HOCH_2CH(NH_2)COOH$. A nonessential amino acid occurring naturally in the L(−) form.
Properties: Colorless crystals; soluble in water; insoluble in alcohol and ether; optically active; DL-serine, m.p. 246°C with decomposition; D(+)-serine, m.p. 228°C with decomposition; L(−)-serine, m.p. 228°C with decomposition.
Derivation: Hydrolysis of protein (especially silk protein); organic synthesis.
Containers: Bottles.
Use: Biochemical and nutritional studies; culture media; microbiological tests; feed additive. Available commercially in all three forms.

"Serizyme." [173] Trademark for a standardized bacterial proteolytic enzyme preparation intended for removal of gelatin and similar protein type sizes from fabrics. Also used to assist the degumming of natural silk. Available as a liquid or as a highly concentrated powder.

"Seromycin." [100] Trademark for cycloserine (q.v.).

serotonin (5-hydroxytryptamine; 5-hydroxy-3-(beta-aminoethyl)indole) $C_{10}H_{12}N_2O$. A powerful vasoconstrictor found in the blood serum and the platelets of bone marrow in mammals. It also causes other smooth muscle to contract. It has been isolated from beef serum and may be synthesized from 5-benzyloxyindole. It is recovered as a complex with creatinine sulfate, $C_{14}H_{23}N_5O_7S$, m.p. (decomposes) 215°C; soluble in water, glacial acetic acid; sparingly soluble in methanol; insoluble in acetone, ether, benzene, absolute alcohol, chloroform, and ethyl acetate.

"Serpasil." [305] Trademark for reserpine, U.S.P.

"Serpatilin." [305] Trademark for a compound containing reserpine U.S.P. and methylphenidate hydrochloride N.N.D.
Use: Medicine.

serpentine. A natural hydrated magnesium silicate corresponding to the formula $3MgO \cdot 2SiO_2 \cdot 2H_2O$, sometimes containing small amounts of nickel and with partial replacement of magnesium by ferrous oxide.
Properties: Dark to blackish-green, brownish-red or yellow. White streak; greasy; waxy, subresinous luster. Sp.gr. 2.5-2.65; hardness 2.5-4.
Varieties:
Antigorite: Platy or massive variety.
Chrysotile: Fibrous variety. See asbestos.
Occurrence: United States; Austria; Russia; Norway; Sweden; Scotland; England; Germany; Siberia.
Use: Serpentine has been suggested as a source of magnesium compounds.

serpentine asbestos. See asbestos.

SES (sesone; sodium 2,4-dichlorophenoxyethyl sulfate) $C_6H_3Cl_2OCH_2CH_2SO_4Na$. Used in sprays for pre-emergence weed control; is

*See "I.C.C. Shipping Regulations," page xiii.
Reference numbers refer to name of manufacturer. See "List of Manufacturers," page v.

toxic to germinating seeds but harmless to many established plants.

sesame oil (benne oil; beni oil; teal oil; teel oil; til oil; gingelly oil; gingily oil; gingili oil; gigily oil).

Properties: A fixed, bland, yellow, almost odorless liquid oil. Does not readily become rancid; sp.gr. 0.9187 (25/25°C); solidifying point −5°C; m.p. 26-32°C (for free fatty acids); Hehner value 95.7; saponification value 188-193; iodine value 103-114; refractive index 1.4748-1.4762. Soluble in chloroform, carbon disulfide, ether, and benzine; slightly soluble in alcohol.

Chief known constituents: Olein (75%); stearin; palmitin; myristin; linolein, and sesamin.

Derivation: By pressing from Sesamum indicum (orientale).

Occurrence: China; Japan; East Indies; South America.

Method of purification: Filtration.

Grades: Edible, should contain less than 1% free fatty acids; semi-refined; coast; U.S.P. XVI.

Containers: Drums.

Uses: Manufacture of oleomargarine, soap, cosmetics, etc; general use similar to olive and almond oils which are frequently adulterated with sesame oil; medicine; insecticides.

Shipping regulations: None.*

"Sesamex." (Sesoxane). Is 2-(2-ethoxyethoxy)-ethyl-3, 4-methylenedioxyphenyl acetal of acetaldehyde. A synergist for pyrethrins and methoxychlor.

sesamolin $C_{22}H_{18}O_7$ [2-(3, 4-methylene-dioxyphenoxy)-6-(3, 4-methylenedioxy-phenyl)cis-3, 7-dioxobicyclo(3.3.0)octane]. A constituent of sesame oil which is the most powerful pyrethrum synergist known. A 1:1 mixture of pyrethrum and sesamolin has 31 times the insecticidal activity of pyrethrum alone.

sesone. See SES.

Sesoxane. See "Sesamex".

"Sesquilate." [244] Trademark for a buffered, inhibited, alkaline cleaning compound with low sudsing characteristics.

Uses: Mechanical washing of dairy cans; spray washing of equipment in dairies and food plants.

"Setit." [337] Tradename for an aluminum hydrate containing 48.2% Al_2O_3. Buff colored powder, sp.gr. 2.5; particles are of colloidal dimensions; insoluble in water and alkalies; slightly soluble in dilute mineral acids and hot concentrated sulfuric acid; soluble in hydrofluoric acid. Used to aid in suspension of nonplastic porcelain enamels, glazes, and polishing slurries, often as a partial replacement for colloidal clays in suspending titania frits.

Containers: 50-lb paper bags; 40,000-lb carloads.

"Setole." [42] Proprietary products. Thermoplastic resin type emulsions.

Properties: White emulsion; dispersible in water at 20°C.

Containers: 55-gal steel drums.

Use: Finishing agents for all types of textile fabrics imparting "firm to medium hand" as desired by amount used.

"Setrete." [49] Trademark for a seed disinfectant containing phenyl mercuric ammonium acetate. Used for cereal grains and cotton seed.

"Setsit 5." [69] Trademark for an activated dithiocarbamate liquid accelerator; 100% active.

Properties: Reddish brown liquid miscible with water.

Uses: Ultra accelerator for natural and synthetic latex films and adhesives. Fast precure rate.

"Setsit 9." [69] Trademark for an activated dithiocarbamate liquid accelerator; 100% active.

Properties: Amber to brown liquid miscible with water.

Uses: Ultra accelerator for natural and synthetic latex films, adhesives, and foams. Slow precure rate.

"Sevin." [214] Trademark for 1-naphthyl N-methyl carbamate (q.v.); an insecticide.

"Sevron." [28] Trademark for a line of cationic dyes especially suited for dyeing "Orlon" acrylic fiber, having outstanding fastness properties on this fiber.

sewage pitch. See stearin and fatty acid pitches.

sewage sludge. An organic material obtained by purifying city sewage. Obtained in two varieties:

(a) Imhoff sludge: A low-grade sludge containing from 2 to 3% ammonia and about 1% phosphoric acid.

(b) Activated sludge: A high-grade sludge containing from 5.0 to 7.5% ammonia and from 2.5 to 4.0% phosphoric acid.

Derivation: (a) By running sewage through settling tanks without the access of air. The sludge, or solid matter, is settled by the aid of anaerobic bacteria. (b) By running sewage through settling tanks and forcing air in through porous plates at the bottom of the tanks. 20% of the current "make" is also added. The action of the bacteria causes the solid organic matter to coagulate and settle. It is subsequently filtered and dried.

Use: Fertilizer.

Shipping regulations: None.*

See also milorganite.

sewer pipe clay. These clays are similar in characteristics to those used for paving brick. They must contain enough silica and stand sufficient heat to combine with the salt fumes to produce the required glaze. A high percentage of iron is also necessary since it aids in the formation of the salt glaze. A high percentage of soluble salts is desirable.

SFS. Abbreviation for sodium formaldehyde sulfoxylate.

shaddock oil. See grapefruit oil.

shale oil. A crude oil obtained from oil shale (q.v.) by destructive distillation.

shark liver oil (shark oil; dog fish oil).
Properties: Yellow to red-brown liquid; characteristic odor, not disagreeable if oil is refined; sp.gr. 0.917-0.928. Soluble in ether, chloroform, benzene, and carbon disulfide.
Derivation: By expression from shark and dogfish livers.
Method of purification: Chilling and filtration.
Grades: Crude; refined.
Containers: Wooden barrels.
Uses: Leather dressing; oil tannage; paints and varnishes; soaps; source of vitamins A and D.

shark oil. See shark liver oil.

sharps. A term used in flour-milling. It is synonymous with middlings (q.v.).

shatterproof glass. See glass, safety.

"Shed-A-Leaf." [147] Trademark for a chlorate-borate defoliant and desiccant.
Containers: 5-, 30-, 55-gal drums (liquid); dust and powder: 50-lb bags.
Uses: To hasten maturity of crops and reduce late insect infestation especially for cotton and beans.

sheep berry. See viburnum prunifolium.

sheep dip. See tar acids.

"Shell A-A Weed Seed Killer." [125] A preplant herbicide containing not less than 98 wt % allyl alcohol (propen-1-ol-3) and not more than 2 wt % inert ingredients.
Properties: Colorless, mobile liquid; boiling range 95-98°C; characteristic, irritating odor; lachrymal vapor at, and above, a concentration of 10 ppm; miscible with water and with most common organic solvents.
Containers: 55-gal unlined metal drums (389 lb net).
Hazards: Warning! Poisonous by skin contact, inhalation or swallowing. Rapidly absorbed through skin. Wash thoroughly with soap and water after handling and before eating or smoking. Keep away from heat and open flame.
Shipping regulations: Class B poison label; package in metal drums.*

shellac (lac; garnet lac; gum lac; stick lac).
Derivation: A resin secreted by the insect Laccifer lacca (coccus lacca) and deposited on the twigs of various species of trees in India, Siam, and Indo-China. When collected and dried it is known as stick lac. This is ground and washed to remove an adherent red dye (lac dye), after which it is known as seed lac. The seed lac is refined by melting and straining and is then poured in thin films over cylinders or plates and allowed to cool. When it hardens and scales off in thin flakes, it is known as shellac, or it is poured into molds to form "button" or garnet lac. This is the orange shellac of commerce. White shellac is made by bleaching orange shellac with sodium hypochlorite.
Properties: Insoluble in water; soluble in alcohol.
Grades (partial list): Garnet; button; orange lemon; lemon; bleached bonedry; bleached refined; bonedry.
Containers: Bags; kegs; barrels.
Uses: Paints; stains; varnishes; leather dressing; polishes; phonograph records; dielectric compositions; general binder; engraving and lithography; sealing wax; inks; rubber; match heads; paper; felt and crepe stiffeners.
Shipping regulations: Liquid solutions: Flammable liquid. Red label.*

"Shellacol." [319] Trademark for a proprietary alcohol-type solvent, available in anhydrous form.
Uses: Solvent for resins, shellac, dyes, inks, and oils; as a substitute for denatured alcohols.

"Shell Catalysts 105 and 205." [125] Trademark for iron oxide catalysts promoted with potassium carbonate and chromium oxide; they are dehydrogenation catalysts used in the selective dehydrogenation of olefins to the corresponding diolefins and of ethyl-substituted aromatic compounds to the corresponding vinyl compounds. The basic difference between these two catalysts is the greater potassium carbonate content of the Shell 205 which makes it possible for the dehydrogenation process to operate at lower steam dilution ratios.
Containers: 41-gal fiber drums.
Caution! Dust may cause irritation of eyes and respiratory passages.
Shipping regulations: None.*

"Shell Isoprene Rubber." [125] A synthetic elastomer possessing a close molecular similarity to natural rubber. Chemically, it is a stereospecific isoprene polymer which has been reacted to a high structural purity of about 92% cis-1,4-polyisoprene. The weight average molecular weight of the polymer is about 2.75×10^6, and the number average molecular weight is about 2.5×10^6. Compatible with natural and synthetic polymers (SBR, nitrile, neoprene, etc.) and may be blended with them in any proportion. The specific gravity of the polymer is 0.92 or about the same as that of natural rubber.

"Shell-S-Polymers." [125] Manufactured by copolymerizing the two monomers, styrene and butadiene, under suitable conditions in a soap emulsion solution in which small amounts of other chemicals are present. Hot polymers, reacted at about 120°F, and cold polymers, reacted at about 40°F, can be prepared. Choice of reaction temperature, type of soap, chemical additives and

*See "I.C.C. Shipping Regulations," page xiii.
Reference numbers refer to name of manufacturer. See "List of Manufacturers," page v.

reaction time determine many of the properties of the final copolymer. After polymerization to a predetermined point, the reaction is terminated, and unreacted monomers are removed from the emulsion. At this stage the emulsion is called latex and can be used in this form. To produce dry rubber, a stabilizer is added to the latex which is then coagulated in crumb form. The rubber crumb is filtered, washed, dried and compressed into bales for shipment.

sherardizing. The process by which steel is coated with zinc powder. The metal forms an alloy with the steel surface and produces a thin tightly adherent coating.

"Sherbelizer." [322] A trademark for an algin derivative-vegetable gum composition.
Properties: A cream colored, highly refined powder passing essentially through 20 mesh and having a moisture content of about 13%; dissolves in hot or cold water to give viscous solutions. Soluble in milk or ice cream products.
Grades: Refined (complies with N.F. requirements.
Containers: 10-, 50-, 100-, and 300-lb drums.
Uses: Stabilizer for sherbets, water ices, frozen fruits, syrups, purees, chocolate ice cream.

"Sherdye." [141] Trade name for textile pigment emulsions used for impregnating fabrics by means of pad dyeing and by printing on said fabrics.

shield fern. See aspidium.

"Shirlan." [28] Trademark for a group of products which have bacteriostatic or fungicidal properties. Included in this group are products based on salicylanilide which are used to impart mildew resistance to textiles, paints, and varnish.

"Shirlan." [206] Brand name for a powerful fungicide for use in sizing and finishing recipes to prevent the development of mildew. It is neither volatile nor injurious to the fabric, and is claimed to be superior to zinc chloride in this respect and in fungicidal action. Specially recommended for the treatment of tent cloth. It is marketed in paste and powder form.

"Shirlan" NA. [206] Brand name for a water-soluble form of "Shirlan."

shorts. A term used in flour-milling. Shorts are that part of the husk or outer coat of the wheat finer than bran (q.v.).

shot metal. See lead shot metal.

"Showertex." [307] Trademark for a water repellent, an emulsion of wax and metal salts; 28% solids content.
Properties: Creamy white emulsion; dispersible in water; sp.gr. 1.03-1.04.
Uses: Non-durable water repellent for long liquor or pad application on cotton, wool, rayon and mixed fiber fabrics.

Si. Symbol for silicon.

"SiC, H-W." [446] See "Harbide."

siderite (chalybite; spathic iron ore) $FeCO_3$, usually with some Ca, Mg, or Mn. Spherosiderite is a secondary siderite occurring in basalt cavities. The term siderite is also used for an iron alloy found in meteorites.
Properties: Gray, yellow, brown, green, white or brownish-red mineral, vitreous inclining to pearly luster, white streak; sp.gr. 3.83-3.88; hardness 3.5-4.
Occurrence: United States (Vermont, Massachusetts, Connecticut, New York, North Carolina, Pennsylvania, Ohio); England; Wales; Germany; Austria; Switzerland; France.
Use: An ore of iron; when high in manganese, used in the manufacture of spiegeleisen.

sienna. A yellowish clay which is colored due to the presence of the oxides of iron and manganese. Raw sienna is a brown-tinted ferruginous manganiferous yellow ocher occurring in Alabama, California, Pennsylvania; Cyprus and Italy. Burnt sienna is an orange-brown pigment made by carefully calcining raw sienna. See also ocher.
Containers (raw or burnt): Paper bags.

"Sierralite." [38] Trade name for hydrous magnesium-aluminum silicate derived from the mineral prochlorite. Used in ceramics in the preparation of synthetic cordierite for high resistance to heat shock. Supplied as 200 mesh powder.

"Sight Savers." [149] Trademark for silicone treated polishing tissues used for the polishing of glass and other vitreous surfaces.

sigma particle. See fundamental particle.

"Signemycin." [299] Trademark for an antibiotic combination containing tetracycline hydrochloride and triacetyloleandomycin.

silane SiH_4, the simplest silicon hydride.
Properties: A gas with repulsive odor. Solidifies at about −200°C, b.p. −112°C; stable at ordinary temperatures. Slowly decomposes in water; insoluble in alcohol, ether, benzene, chloroform, and silicon tetrachloride.

"Silaneal." [149] Trademark for organo polysiloxanes used for making materials water-repellent.

silanes. Compounds of silicon and hydrogen of the formula Si_nH_{2n+2} analogous to the alkanes or saturated hydrocarbons. SiH_3- is called silyl (analogous to methyl), and Si_2H_5- is disilanyl (analogous to ethyl). A cyclic silicon and hydrogen compound having the formula $(SiH_2)_n$ is called a cyclosilane. See also silicone.

"Silastic." [149] Trademark for compositions in physical character comparable to milled and compounded rubber prior to vulcanization but containing organosilicon polymers. Parts fabricated of "Silastic" are

serviceable from −100 to 500°F; retain good physical and dielectric properties in such service; show excellent resistance to compression set, weathering, and corona. Thermal conductivity is high; water absorption low.
Uses: Diaphragms, gaskets and seals, O-rings, hose, coated fabrics, wire and cable, and insulating components for electrical and electronic parts.

"Silastic" Gums. [149] Trade name for unvulcanized organopolysiloxane elastomers, for compounding silicone rubber. Reinforced gum with low shrink and low compression set characteristics, partly filled with silica to minimize compounding time; and sulfur-vulcanizable gum for blending with sulfur-vulcanizable organic rubber.

"Silastic" R Tape. [149] Trade name for electrical insulating tape of glass coth coated with semi-vulcanized silicone rubber. It is easily applied by wrapping, and vulcanization is completed by heating for a short time, to form a resilient, void-free, moisture-proof and oil resistant jacket with excellent dielectric properties. It is a Class H insulating material designated for hottest spot temperatures of 180°C.

silberol (silver phenolsulfonate; silver sulfophenylate; silver sulfocarbolate) $C_6H_4(OH)SO_3Ag$.
Properties: Crystalline powder. Contains 28% silver. Undergoes spontaneous decomposition. Has antiseptic properties of silver without its corrosive action. Soluble in water and alcohol.
Derivation: By the interaction of silver oxide and phenolsulfonic acids.
Containers: Blue glass bottles.
Use: Medicine.
Shipping regulations: None.*

"Silco-Flex." [116] Trademark for a silicone-rubber insulation for use on large motors and generators.
Properties: Good dielectric properties at operating temperatures; resistance to thermal degradation; good flexibility, strength and resilience; high thermal conductivity; low moisture absorption; high resistance to ionic discharge; abrasive resistant; chemically inert.

"Silene." [177] Trademark for a white, highly absorptive, finely divided, precipitated, hydrated calcium silicate.
Properties: (typical analysis) SiO_2 64%; CaO 18%; loss on ignition 15%; pH in water suspension 10; sp.gr. 2.1; refractive index 1.47; density (lbs/cu ft) 12; oil absorption (rub-in method) 120 g; average particle size 0.03 micron.
Grades: "Silene" EF; "Silene" JA.
Containers: 50-lb paper bags.
Uses: In rubber compounding to improve physical properties of non-black products; as partial stabilizer and reinforcing filler in vinyl floor tile and highly loaded rubber; in printing inks; to prevent caking in table salt and other granular materials. "Silene" JA is used in paper manufacture to impart opacity and to improve formation and finish.

"Silflake." [31] Trademark for silver flakes made by milling a chemical precipitate in a lubricant to impart a leaf-like quality to the silver particle. Available as:
"Silflake" 131. A dry flake used as a pigment for silver conductive coatings.
"Silflake" 135. A dry flake for the same purpose. It has appreciably higher conductivity in air dry silver paints than "Silflake" 131.
"Silflake" 850. An 85% silver paste with denatured alcohol or solvent of choice. Used as a pigment in air dry silver conductive coatings.
Containers: 100- and 750-troy ounce cans.

"Silfrax." [280] Trademark for bonded refractories containing from 40% to 78% silicon carbide.
Properties: High refractoriness; great strength; high thermal conductivity; freedom from spalling; resistance to clinker adhesion; and resistance to mechanical and flame abrasion.
Uses: Bricks for boiler and furnace installations; kiln furniture in ceramic kilns; shapes for boiler furnaces, air-cooled furnace linings, glass lehrs, pit furnaces, and enameling furnace ware supports.

silica (silicon dioxide) SiO_2. Occurs in nature as quartz, sand, flint, chalcedony, opal, agate, diatomite, and numberous less common modifications. May also be made from a soluble silicate such as water glass by acidification, thorough washing and ignition to drive off water. Silicon dioxide combines chemically with most metallic oxides in many different proportions. See also silicic acid and quartz.
Properties: Colorless, transparent crystals, or white tasteless powder; sp.gr. 2.2-2.6; insoluble in water and acids except hydrofluoric; soluble in molten alkali when finely divided and amorphous. Silica melts to a glass with the lowest known coefficient of thermal expansion. M.p. 1710°C (cristobalite modification); b.p. 2230°C.
Containers: Paper bags.
Uses: Manufacture of glass, water glass, ceramic products and enamel ware; abrasives; foundry molds; carborundum; ferrosilicon and elemental silicon; filler; ingredient of concrete and mortar. Fused silica is used as a protectant in rockets. Powdered silica is used as a filler or bodying agent in cosmetics, pharmaceuticals, paper, insecticides; as a plasticizer in resins; flatting agent; thermal insulator. See also the minerals listed above (especially quartz), silica gel, and silicic acid.

silica gel. A regenerative adsorbent, consisting of amorphous silica.
Grades: Commercial grades capable of withstanding temperatures up to 500-600°F are supplied in the following mesh sizes: 3-8, 6-16, 14-20, 14-42, 28-200 and through 325.

Containers: 1-, 5-, 10-, 25-, 100-lb, and 120-lb, air tight metal containers.
Uses: Dehumidifying and dehydrating agent; in air-conditioning and in the drying of compressed air and other gases, liquids, such as refrigerants, and oils containing water in solution and in suspension; recovery of natural gasoline from natural gas; also used as a carrier for active catalysts, in some cases as a catalyst itself; in chromatography; anticaking agent in cosmetics and pharmaceuticals; in waxes to prevent slipping.
See silicic acid.

silicate. Any member of the very widely occurring compounds characterized by the presence of the elements silicon, oxygen, and one or more metals with or without hydrogen. Infrequently, the silicon and oxygen are combined with organic radicals to form silicate esters. Most common rocks (except limestone and dolomite) and many minerals are silicates or mixtures thereof, as are ordinary glass, water glass, and common bricks. The silicates may be considered as salts or esters of one or another of the many hypothetical silicic acids derived from varying proportions of silica and water. An alternative is to consider them as resulting from the union of silica and various metallic oxides, sometimes with water also present.

silicate cotton. See mineral wool.

silicate of soda. See sodium silicate.

siliceous earth. See diatomite.

siliceous earth, purified. See diatomite.

siliceous limestone. See limestone.

silicic acid (hydrated silica). This term is applied to the jelly-like precipitate obtained when sodium silicate solution is acidified. The formula H_2SiO_3 is often used for convenience but no such compound has been isolated, and $SiO_2 \cdot nH_2O$ is the proper formula. The proportion of water varies with the conditions of preparation and decreases gradually during drying and ignition, until relatively pure silica, SiO_2, remains. During drying the jelly is converted to a white amorphous powder or lumps. In this form the material has great surface area and adsorbing power and is therefore used for decolorizing (bleaching) oils, fats and waxes, as a catalyst, carrier or base for other catalysts and for chromatographic adsorption, and recovery of gases and vapors including moisture from air or gas streams. The adsorbent properties may be regenerated by heating to 300°F. See silica gel.

The term silicic acid is also sometimes applied to various hypothetical acids of silica such as H_4SiO_4, $H_2Si_2O_5$ etc.

silicochloroform. See trichlorosilane.

silicofluoric acid. See fluosilicic acid.

silicomanganese. Alloys consisting principally of manganese, silicon and carbon. Three forms are usually marketed having approximately the following analyses:
(1) C 1.00 (max), Mn 65-70, Si 20-25%;
(2) C 2.00 (max), Mn 65-70, Si 16-20%;
(3) C 2.5 (max), Mn 65-70, Si 16-20%.
Forms available: In crushed form or in lumps up to 75 lbs.
Uses: For making low-carbon steel in which silicon is not objectionable, silicomanganese may be used effectively for the introduction of manganese because of its low carbon content. It is also used effectively for 13% manganese steel when large quantities of returned scrap are in the charge. Makers of low carbon-chromium and manganese steels favor it because of the low carbon to manganese ratio. Silicon manganese steels are used for springs and high strength structural steels.
See also manganese steels and ferromanganese.

silicon. Nonmetallic element with atomic number 14; group IV of the periodic table. In trade, the element is commonly referred to as silicon metal. Silicon in organic compounds is the basis for a considerable industry; see silicone; silanes.
Properties: (a) Dark brown, amorphous, nonmetallic powder, which burns in air when ignited. (b) Also obtained as hard, lustrous, crystalline leaflets which do not burn in air. This latter form is a conductor of electricity. Both forms are soluble in hydrofluoric acid and alkalies; insoluble in water, nitric acid and hydrochloric acid.
Constants: (a) Sp.gr. 2.00; (b) 2.49; m.p. 1410°C; b.p., 2480°C; hardness 7 on Mohs scale; thermal conductivity 0.39 cal/cm/sec/°C; thermal expansion 4.15×10^{-6}/°C; specific heat 0.168 at 25°C; dielectric constant 13 at 9.37×10^9 cps; elastic modulus 19×10^6 psi; flexural strength 20,000 psi.
Occurrence: Second to oxygen in abundance. Never found in free state in nature, but composes a major portion of silicate rocks and of quartz, sand, sandstone, clay, granite, feldspar, mica, and many other common minerals.
Derivation: (a) By heating sand with coke in an electric furnace. (b) Heating sand with powdered magnesium and treating the mixture with water.
Method of purification: Treatment with hydrochloric and hydrofluoric acids. For semiconductor transistor and rectifier grades very high purity is achieved by operations such as distillation of silicon tetrachloride, tetraiodide or silane, reduction to silicon by use of ultrapure zinc, and vacuum or argon zone refining and special single crystal-growing techniques.
Purity: Total impurities are estimated to be as low as 10 parts per billion in some semiconductor grades, boron content being reduced to 1 part per billion.
Impurities of technical grades: Iron; carbon.
Grades: Ferrosilicon (50% Si), regular (97% Si), and semiconductor or hyperpure (for

electronic devices). Available in crystalline or powdered form.
Containers: 500-lb barrels; drums.
Uses: Manufacture of silicon tetrachloride, silicon-bronze, silicon-copper and ferrosilicon; production of halogenated silanes; organo-silicon compounds; rectifiers and transistors; photocell elements; electrodes; solar battery. Silicon is combined with certain high temperature refractory ceramic materials to give cermets and other special refractories that are advantageous because the silicon confers higher thermal conductivity, better resistance to thermal shock, and resistance to oxidation.
Shipping regulations: None.*

silicon bromide. See silicon tetrabromide.

silicon-bronze. An alloy of copper, tin and silicon used for telephone and telegraph wires.

silicon carbide SiC. One of the hardest substances known.
Properties: Bluish black iridescent crystals; very hard (Mohs' scale hardness 9); sp.gr. 3.17; sublimes with decomposition at about 2210°C. Insoluble in water and alcohol; soluble in fused alkalies.
Derivation: Heating of carbon and silica sand in a horizontal resistance furnace.
Containers: Bags, barrels, drums.
Uses: Abrasive; heat refractory material. Single high-purity crystals are now used as semiconductors, specially suitable for high temperatures.

silicon chloride. See silicon tetrachloride.

silicon-copper (copper silicide).
Properties: A hard, tough, bronze-like alloy containing 10-30% silicon.
Derivation: From silicon and copper electrolytically.
Grades: Technical.
Containers: Boxes.
Use: Manufacture of silicon-bronze.
Shipping regulations: None.*

silicon dioxide. See silica.

silicone. Group name for semi-inorganic polymers made up of a skeleton structure of alternate silicon and oxygen atoms with various organic groups attached to the silicon. (See, for example, dimethyl silicone.) Silicones always include organic groups; siloxanes may be completely inorganic. Silicones range from low molecular weight volatile materials to cyclic, linear and cross-linked high molecular weight polymers. Produced in the basic forms of fluids, resins and elastomers, they are also further compounded to yield greases, rubbers, protective coatings and foamable powders. In general, silicone products are characterized by an unique combination of properties. They are heat-stable, serviceable over a wide temperature span, water repellent, resistant to oxidation and weathering. They retain good physical and dielectric properties in severe operating conditions.
Uses: As adhesives, compressible fluids, cosmetic ingredients, damping media, dielectric fluids and compounds, diffusion pump fluids, defoamers, heat transfer media, protective coatings, surface treatments for glass and ceramics, textile finishes, leather tanning, lubricants, release agents, water repellents, electrical insulating materials and extreme temperature rubber.
See also siloxanes.

silicone, carbon-functional. Designation for silicon compounds containing at least one hetero atom or nonbenzenoid unsaturated linkage attached to silicon through carbon, in contrast to better known silicon compounds in which reactive groups are attached directly to silicon (see silicones). Examples are dichlorophenyltrichlorosilane ($Cl_2C_6H_3SiCl_3$), which is an intermediate for superior high temperature lubricants, and vinyltrichlorosilane ($CH_2CHSiCl_3$), used in novel gums and finishes for glass cloth. Carbon-functional groups modify the physical properties of silicones and increase their reactivity, permitting new syntheses.

silicone fluids. Organosiloxane polymers, commercially available in a range of viscosities from 0.65 to over 1,000,000 centistokes; include methyl and phenyl polysiloxanes. Characterized by heat stability, water repellency, good dielectric properties, and incompatibility with many organic polymers which makes them effective release agents. See also silicone.

silicone resins. Organosiloxane polymers, generally supplied as solvent solutions, used in formulating protective coatings, in making glass cloth laminates, and as electrical insulating varnishes. Are characterized by unusual heat stability; resistance to oxidation, weathering and corrosive chemicals; and retention of dielectric properties in high temperature service.

silicone rubber. A silicone or siloxane of such structure and molecular weight as to have many properties characteristic of rubber. Vulcanized and cured, the silicones produce rubbery products serviceable from −150 to 600°F; retain good electrical properties at elevated temperatures, maintain resistance to compression set over a wide temperature range, and resist the effect of ozone, aging, sunlight, weathering, oil and water.
Available as pure gum, reinforced gum, and compounded in various ways.
Used in molded, extruded, and calendered form for high-temperature electrical and electronic insulation, and in the formation of O-rings, gaskets, seals, cushions, hot air ducts and hose, and insulating tapes.

silicon, ferro-. See ferrosilicon.

silicon fluoride. See silicon tetrafluoride.

silicon-gold alloy. See gold-silicon alloy.

silicon nitride Si_3N_4. An important refractory.
Properties: Greyish powder (can be prepared

*See "I.C.C. Shipping Regulations," page xiii.
Reference numbers refer to name of manufacturer. See "List of Manufacturers," page v.

as crystals); sublimes at 1900°C; density 3.44 g/cu m; bulk density 70-75 lb/cu ft, depending on mesh; Mohs' hardness 9+; modulus of rupture (room temperature) 10,000 psi; modulus of elasticity (room temperature) 13.4×10^6 psi; thermal conductivity 10.83 Btu/in/sq ft/hr/°F at 400-2400°F. Resistant to oxidation, various corrosive media, molten aluminum, zinc lead and tin. A beta-phase variety has unusual thermal shock resistance.

Derivation: One method involves the reaction of powdered silicon and nitrogen gas in an electric furnace.

Uses: Coatings; bonding of silicon carbide; mortars; abrasives; thermocouple tubes in molten aluminum; crucibles for zone-refining germanium; rocket nozzles.

silicon steel. Fourteen to sixteen per cent silicon steels are used for corrosion resistance to some inorganic acids and other chemicals. These contain 1% or less of carbon. Such alloys are not machinable except for limited turning with tungsten carbide tools, generally finished by grinding. Used for erosion resistance as well as corrosion resistance.

Silicon to the extent of 0.5-5% in iron makes good electrical iron. It has low hysteresis and eddy current losses, and satisfactory permeability. Used as armatures and transformer coils.

Silicon is also used as a general deoxidizer in steel manufacture; its presence strengthens low alloy steels, increases hardenability and improves oxidation resistance.

See also ferrosilicon.

silicon tetrabromide (silicon bromide) $SiBr_4$.

Properties: Fuming, colorless liquid. Turns yellow in air. Disagreeable odor. Decomposed by water with evolution of heat. Sp.gr. 2.82 (0°C); b.p. 153°C; m.p. 5°C.

silicon tetrachloride (silicon chloride) $SiCl_4$.

Properties: A clear, colorless, exceedingly mobile, fuming liquid; suffocating odor. Since it is decomposed by water to yield silicic acid and hydrochloric acid, it is quite corrosive to most metals when water is present with it. In the absence of water it has practically no action on iron, steel, or the common metals and alloys, and can be stored and handled in metal equipment without danger. Sp.gr. 1.48 (68/60°F); wt 12.4 lbs/gal; m.p. −70°C; b.p. 57.6°C; refractive index (n 20/D) 1.412. Miscible in all proportions with carbon tetrachloride, tin tetrachloride, titanium tetrachloride, and sulfur mono- and di-chlorides. Decomposed by water and alcohol.

Derivation: Silicon carbide is packed around a resistor and electrically heated, whereupon chlorine is passed through the mass and the silicon tetrachloride condensed from the escaping gas.

Impurities: Silicon hexachloride.

Grades: Technical; 99.5%; C.P. (99.8%).

Containers: Iron drums; bottles; tank cars.

Uses: Smoke screens in warfare; raw material in making ethyl silicate and similar compounds which result from the reaction between anhydrous alcohols and silicon tetrachloride; production of silicones; source of pure silicon and silica; convenient source of hydrogen chloride.

Shipping regulations: Corrosive liquid. White label.*

silicon tetrafluoride (silicon fluoride) SiF_4.

Properties: Colorless gas; suffocating odor similar to hydrogen chloride; fumes strongly in air; m.p. −95.7°C; b.p. −65°C (181 mm). Absorbed readily in large quantities by water with decomposition.

Derivation: (a) Action of hydrofluoric acid or concentrated sulfuric acid and a metallic fluoride on silica or silicates. (b) Direct synthesis.

Grades: Pure, 99.5% min.

Containers: Gas cylinders.

Uses: Manufacture of fluosilicic acid; chemical analysis; to seal water out of oil wells during drilling.

Shipping regulations: Nonflammable gas. Green gas label.*

silicothermic process. See Pidgeon process.

silicotungstic acid (silicowolframic acid) $SiO_2 \cdot 12WO_3 \cdot 26H_2O$ (approximately). Composition is probably 12 molecules of tungsten trioxide to one molecule of silicon dioxide with varying amounts of water.

Properties: White to yellowish-white crystals; soluble in water and alcohol.

Derivation: By heating ammonium silicotungstate with aqua regia.

Method of purification: Crystallization.

Grades: Technical; C.P.

Containers: Kegs; glass bottles.

Uses: Chemical reagent for alkaloids; mordant for basic dyes.

Shipping regulations: None.*

silicowolframic acid. See silicotungstic acid.

"Silicure." [74] Trademark for a series of catalysts for curing silicone resins and rubbers. They consist of iron, lead, tin, zirconium and zinc octoates, and zinc stearate.

silikatcement. See pozzolana cement.

"Silimite." [166] Trademark for a high magnesium dolomitic lime used for silica reduction in hot process water softening equipment.

silk. An animal fiber secreted as a continuous filament by the silkworm, Bombyx mori. Silk consists essentially of the protein fibroin and, in the raw state, is coated with a gum, which is usually removed before spinning the filaments into threads. The important characteristics of silk are high strength, luster, and moisture absorbency, and fineness of the filaments. The principal sources are Japan, Italy, France, India, and China.

Uses: Apparel; parachute fabrics; bolting cloth.

silk, artificial. See fibers, synthetic.

"Silk-O-Fast FF." [328] A brand product consisting of urea-polypeptide-formaldehyde condensate. It provides a durable body to silk and other textile fabrics.

"Silk-O-Ray." [328] Trademark for a textile finishing and scrooping agent consisting of fatty esters in dispersible form.

sillimanite [The refractory material.] An aluminum silicate. A high heat-resisting material containing a maximum amount of mullite, developed from the alteration of andalusite during firing. This necessitates firing above 1550°C for the development of a suitable crystalline structure.
Uses: Spark plugs; chemical laboratory ware; pyrometer tubes; special porcelain shapes; furnace patch and refractories.

sillimanite [The mineral.] (fibrolite) Al_2OSiO_4. A silicate of aluminum with the same formula as andalusite and cyanite (q.v.). See also mullite. Used in refractories.
Properties: Color white, gray, brown, or greenish; vitreous luster; hardness 6-7; sp.gr. 3.23; usually found as fine fibrous masses.
Occurrence: Massachusetts, Connecticut, New Hampshire, Pennsylvania, South Carolina, Georgia; India; Brazil; Australia.

"Sil-O-Cel." [247] Trademark for a type of diatomaceous earth products used in filtration and insulation.

siloxanes (oxosilanes; polysiloxanes). Compounds of silicon, oxygen, usually also containing carbon and hydrogen, and containing in their molecules the structural unit R_2SiO in which R is usually CH_3 but may be H, C_2H_5, C_6H_5 or more complex substituents. Disiloxane ($H_3Si\text{-}O\text{-}SiH_3$) and trisiloxane ($H_3Si\text{-}O\text{-}SiH_2\text{-}O\text{-}SiH_3$) are the simplest examples, but the most interesting are those of higher molecular weight and having the composition $(R_2SiO)_n$. These are polyorganosiloxanes or silicones (q.v.) whose molecules consist of chains of alternate silicon and oxygen atoms

$$(-O-\overset{|}{\underset{|}{Si}}-O-\overset{|}{\underset{|}{Si}}-O-\overset{|}{\underset{|}{Si}}-)$$

with the free valences of the silicon atoms joined usually to hydrocarbon (R) groups but also to some extent to oxygen atoms that are joined to (cross-linked) silicon atoms in a second chain. The properties of the resulting materials vary from relatively mobile fluids through oils, greases, rubbers, to resins or plastics, depending on the length of the chain, the nature of the R groups, and the extent of crosslinking. In commercial silicones R is usually CH_3, i.e., they are methyl siloxanes. See also silicones.

"Silpaints." [31] Trademark for a series of conductive coatings pigmented with metallic silver together with bonding agents. Available in two classes: (a) Fire-on-types for base material that can withstand heating in range 750-1700°F. (b) Air dry types for organic base materials that are usually force-dried in range of room temperature to 800°F.
Containers: 25- and 50-oz jars.
Use: To make conductive surfaces on nonconductive materials such as ceramics, glass, quartz, mica, plastics and paper.

"Silpowder." [31] Trademark for series of silver powders, available as:

"Silpowder" No.	Average Particle Size (microns)	% Silver
120	5.0-10.0	99.9
130	0.6- 2.0	99.6
131	1.0- 5.0	99.6
150		99.9

Containers: 100- to 150- troy ounce metal containers.
Uses: Powder metallurgy; electrical contacts; battery plates.

"Sil-Temp." [349] Trademark for a substantially pure fibrous SiO_2 for use in rocket and missile constructions and for high temperature insulation of motor components and similar applications. As a construction material it is used in laminate form impregnated and bonded with high temperature resin systems.

"Silvacel." [129] Trademark for a series of wood fibers prepared from western softwood species by use of a defibrating process that separates individual fibers of the wood rather than cutting or grinding. Various grades are prepared according to degree of fiber separation and the chemical treatment that is used during manufacture. Fiber and fiber bundle lengths range up to ¾ in. long and down to less than ⅛ in. Usually shipped in tightly compressed bales. When fluffed, the fibers form a woolly mass resistant to settling or compression.
Properties: Natural wood color; fineness variable according to grade; moisture content as desired; grades treated with chemicals may have special properties including fire resistance.
Uses: Insulation; pulp molding; special papers and boards where high bulk and absorbent qualities are desirable; filter aid and filter media; treatment of oil well drilling muds.

"Silvacon." [129] Trademark for a series of products made from Douglas fir (Pseudotsuga taxifolia) bark. The basic constituents of the bark are pliable spongy flakes (cork), tough needle-like fibers and a fine amorphous powder. The "Silvacon" products are purified constituents or blends and are available in different particle sizes. The basic constituents vary in properties, the fibers being largely cellulose (65%) and the cork and powder constituents largely lignin-like polyphenols with tannins, natural waxes, and resins present in fair quantity.

Properties: Color from buff to dark brown; sp.gr. (ultimate) 1.32-1.47; particle size of powder is 95% minus 200 mesh and cork particles are up to ⅜ in. diameter; acidic (pH about 3.6); low solubility in water or dilute acids (15% or less); soluble in caustic solution up to 90% of cork but only 25% of fiber.

Uses: Phenolic adhesive extender; molding compound component; conditioner for fertilizer, insecticides, etc; burnout material in foundry sands; rubber sponge manufacture; other vinyl and rubber compounds; treatment of oil well drilling fluids; replacement for European cork in various applications.

silver Ag. A metallic element, atomic number 47, in Group I of the periodic table.

Properties: Soft, ductile and malleable, lustrous white metal. Best metallic conductor of heat and electricity. Resists oxidation, but tarnishes slightly in ordinary air through reaction with atmospheric sulfur compounds. Sp.gr. 10.53; m.p. 961°C; b.p. 1950°C. Soluble in nitric acid, hot sulfuric acid, and alkali cyanide solutions; insoluble in water and alkalies.

Derivation: Most silver production comes as a by-product of operations on copper, zinc, lead, or gold ores, but some smelters still operate on silver ores. The recovery ranges from 166 ounces per ton to a few thousandths of an ounce per ton. The chief silver ores are native silver, argentite (silver sulfide) and cerargyrite or horn silver (silver chloride).

Forms available: Pure ("fine"), sterling (7.5% Cu), various alloys, silver plate; ingot, bullion, coins, moss, sheet, including porous silver sheets, wire, tubing, castings; powder; high purity (impurities less than 10 ppm).

Uses: Pure silver is used for the manufacture of silver bromide, photographic chemicals, etc.; for lining vats, tanks barrels, cans, pipes, autoclaves, stills, condensers and other equipment for distilling water, processing organic acids, aldehydes and food products. Other major uses of the pure metal are heavy electric conductors, such as bus bars, and silver plating. Alloys containing a relatively high proportion of silver are used for low temperature brazing alloys, table cutlery, decorative and ornamental objects, jewelry, dental, medical and scientific equipment, coinage, electrical contacts as in relays, and for bearing metal. High purity silver is used in electronics.

silver acetate CH_3COOAg.

Properties: White crystals or powder; sp.gr. 3.26. Moderately soluble in hot water; soluble in nitric acid.

Use: Medicine (external).

silver arsenite Ag_3AsO_3.

Properties: Fine, yellow powder. Sensitive to light. Caution! Poisonous! Soluble in acetic acid, ammonium hydroxide, and nitric acid; insoluble in alcohol and water. M.p. 150°C (decomposes).

Use: Medicine.

Shipping regulations: Poison, class B. Poison label.*

silver arsphenamine (silver diaminodihydroxyarsenobenzene, sodium salt; silver diarsenol).

Properties: Brownish-black powder (containing approximately 20% arsenic, 15% silver) for which the exact molecular formula has not been established. Unstable in air. Soluble in water.

Derivation: By the action of silver salts on arsphenamine, converting the product to the disodium salt and precipitating with alcohol, acetone, or ether.

Grades: Medicinal.

Containers: Ampules.

Use: Medicine.

Shipping regulations: None.*

silver bichromate. See silver dichromate.

silver bromate $AgBrO_3$.

Properties: White powder. Caution! Keep in amber bottle! Soluble in ammonium hydroxide; slightly soluble in water (hot). Decomposed by heat. Sp.gr. 5.2.

Caution: Poison label. Protect from contact with organic matter.

silver bromide AgBr.

Properties: Pale yellow crystals or powder, darkening on exposure to light, finally turning black. Sp.gr. 6.473; m.p. 432°C; b.p. decomposes at 700°C. Soluble in potassium bromide, potassium cyanide and sodium thiosulfate solutions; very slightly soluble in ammonia water; insoluble in water.

Derivation: Silver nitrate is dissolved in water and a solution of alkali bromide added slowly. The precipitated silver bromide is washed repeatedly with hot water; the whole operation must be carried on in a darkroom under a ruby-red light.

Impurities: Silver nitrate; alkali bromide.

Grade: Technical.

Containers: Amber or black glass bottles. Poison label.

Use: Silver bromide is the light sensitive material on ordinary photograph film and plates.

Shipping regulations: None.*

silver carbonate Ag_2CO_3.

Properties: Yellow to yellowish-gray, crystalline powder. Contains 78% (approx.) silver. Caution! Keep away from light! Soluble in ammonium hydroxide, nitric acid; insoluble in alcohol and water. Sp. gr. 6.077; m.p. 220°C (decomposes).

silver chloride AgCl.

Properties: White granular powder, which darkens on exposure to light, finally turning black. Exists in several modifications differing in conduct toward light and also in their solubility in various solvents. Soluble in ammonium hydroxide, concentrated sulfuric acid and sodium thiosulfate and potassium bromide solutions; very slightly

soluble in water. Sp.gr. 5.56; m.p. 445°C; b.p. 1550°C.
Derivation: Silver nitrate solution is heated and hydrochloric acid or salt solution added. The whole is boiled, then filtered, all in the dark or under a ruby-red light.
Method of purification: Re-solution in ammonium hydroxide and precipitation by hydrochloric acid.
Impurities: Silver nitrate.
Grades: Technical; C.P.; single pure crystals.
Containers: Amber or black glass bottles. Poison label.
Uses: Photography; photometry and optics; silver plating; production of pure silver; medicine. Single crystals are used for infrared absorption cells and lens elements.
Shipping regulations: None.*

silver chromate Ag_2CrO_4.
Properties: Dark, brownish-red, crystalline powder. Soluble in acids, ammonium hydroxide, potassium cyanide, solutions of alkali chromates; insoluble in water. Sp. gr. 5.625.
Use: Reagent.
Shipping regulations: None.*

silver, colloidal. Metallic silver, insoluble silver salts and silver protein precipitates in sufficiently finely divided form to assume the colloidal state and form permanent suspensions in water. They are prepared either by means of chemical reactions or electrolytically and are used in medicine. See, for example, silver protein, mild; and silver protein, strong.

silver cyanide AgCN.
Properties: White, odorless, tasteless powder which darkens on exposure to light. Soluble in ammonium hydroxide, dilute boiling nitric acid and potassium cyanide and sodium thiosulfate solutions; insoluble in water. Sp.gr. 3.95; decomposes at 320°C.
Derivation: By adding sodium or potassium cyanide to a solution of silver nitrate.
Grade: Technical.
Containers: Amber or black glass bottles. Poison label.
Uses: Medicine; silver plating.
Shipping regulations: None.*

silver diaminodihydroxyarsenobenzene sodium salt. See silver arsphenamine.

silver diarsenol. See silver arsphenamine.

silver dichromate (silver bichromate) $Ag_2Cr_2O_7$.
Properties: Dark red, almost black, crystalline powder; sp.gr. 4.770; soluble in ammonium hydroxide and nitric acid; slightly soluble in water.

silver fluoride $AgF \cdot H_2O$.
Properties: Yellow, or brownish, crystalline masses. Very hygroscopic. Becomes dark on exposure to light. Caution! Keep away from light! Soluble in water. Sp.gr. 5.852 (dehydrated); m.p. 435°C (dehydrated).
Use: Medicine.

silver glance. See argentite.

silver haloids, natural. A series of natural isomorphous haloids of silver forming a class of secondary minerals commonly found in the upper parts of silver deposits and occurring massive or in crusts with the consistency and luster of horn or wax, hence the term "horn silver." They may consist of the simple chloride, bromide or iodide, or may be mixed in varying proportions. The principal members of this series are cerargyrite, bromyrite, embolite, iodargyrite, iodembolite.

silver iodate $AgIO_3$.
Properties: White powder; decomposed by sulfuric acid. Soluble in ammonium hydroxide, nitric acid, solution of potassium iodide (conc.); slightly soluble in water.
Constants: Sp.gr. 5.65; m.p. above 200°C.
Use: Medicine.

silver iodide AgI.
Properties: Pale yellow, odorless, tasteless powder, darkening on exposure to light. Soluble in hydriodic acid, potassium iodide, potassium cyanide, ammonium hydroxide, sodium chloride and sodium thiosulfate solutions; insoluble in water. Sp.gr. 5.675; m.p. 556°C.
Derivation: Silver nitrate solution is heated, alkali iodide solution added and the precipitate washed with boiling water in the dark or under ruby-red illumination.
Impurities: Silver nitrate and alkali iodide.
Grades: Technical; pure.
Containers: Amber or black glass bottles. Poison label.
Uses: Medicine; photography; artificial rainmaking.
Shipping regulations: None.*

silver lactate $CH_3CHOHCOOAg \cdot H_2O$.
Properties: White to gray crystals; affected by light. Slightly soluble in water and alcohol.
Use: Medicine.

silver leaf. See stillingia.

silver, light ruby. See proustite.

"Silver-Lume." [72] Trade name for a bright silver plating process. Prepared from silver cyanide, potassium cyanide, potassium carbonate and organic addition agents. Solution has exceptional throwing power and deposits are hard, wear resistant with high protective value and tarnish resistance. Produces bright deposits directly.

silver-mercury iodide. See mercuric-silver iodide.

silver methylarsonate (methanearsonic acid, disilver salt) $CH_3AsO_3Ag_2$.
Derivation: Reaction of disodium methylarsonate with silver salts.
Use: Algicides.

silver nitrate $AgNO_3$.
Properties: Colorless, transparent, tabular,

*See "I.C.C. Shipping Regulations," page xiii.
Reference numbers refer to name of manufacturer. See "List of Manufacturers," page v.

rhombic crystals, becoming gray or grayish-black on exposure to light in the presence of organic matter; odorless; bitter, caustic metallic taste; corrosive and poisonous! Soluble in cold water; more soluble in hot water, glycerol and hot alcohol; slightly soluble in ether. Sp.gr. 4.328; m.p. 212°C; b.p. decomposes.

Derivation: Silver is dissolved in dilute nitric acid and the solution evaporated. The residue is heated to a dull red-heat to decompose any copper nitrate, dissolved in water, filtered and recrystallized.

Grades: Technical; C.P.; U.S.P. XVI.

Containers: 16-, 50-, 80-, 100-, 160-, 200-oz amber or black glass bottles.

Uses: Photography; hair dyeing; mother of pearl; reagent in chemical laboratories; silver plating; indelible ink; silver salts; glass manufacture; silvering mirrors; medicine (external); germicide (as a wall spray).

Warning: May cause burns. Poison label. MCA warning label.

Shipping regulations: Oxidizing material. Yellow label.*

silver nitrate, fused (lunar caustic; silver nitrate, toughened).

Properties: White, hard pencils or cones containing not less than 94.5% silver nitrate. Odorless; bitter, caustic, and strongly metallic taste. Becomes gray or grayish-black on exposure to light in the presence of organic matter. Solubilities similar to silver nitrate (q.v.).

Derivation: (a) By adding hydrochloric acid (4 parts) to silver nitrate (100 parts), melting and pouring into molds. (b) By melting silver nitrate (95 parts) and potassium nitrate (5 parts) together and pouring into molds.

Grades: U.S.P. XVI.

Containers: Amber or glass bottles.

Use: Medicine (external).

Warning: May cause burns. Poison! MCA warning label.

Shipping regulations: Oxidizing material. Yellow label.*

silver nitrate, toughened. See silver nitrate, fused.

silver nitrite $AgNO_2$.

Properties: Small, yellow or grayish-yellow needles. Become gray on exposure to light. Contain 70% (approx.) silver. Decomposed by acids. Soluble in (hot) water; insoluble in alcohol. Decomposes at 140°C. Sp.gr. 4.4.

Grade: Technical.

Uses: Organic synthesis; standardizing potassium permanganate solutions; water analysis; analysis (testing for alcohols).

silver ore, brittle. See stephanite.

silver ore, dark ruby. See pyrargyrite.

silver ore, light ruby. See proustite.

silver oxide (argentous oxide) Ag_2O.

Properties: Dark brown odorless powder; metallic taste; must not be triturated with organic matter; may cause explosions. Soluble in ammonium hydroxide, potassium cyanide solution, nitric acid and sodium thiosulfate solution; very slightly soluble in water; insoluble in alcohol. Sp.gr. 7.14; m.p. decomposes when heated above 300°C.

Derivation: Silver nitrate and alkali hydroxide solutions are mixed, the precipitate filtered and washed.

Grades: Technical; 92.5% silver, particle size 2-3 micron.

Containers: Glass bottles; 100-, 500-oz cans.

Uses: Medicine; polishing glass; coloring glass yellow; catalyst; purifying drinking water; battery plates; silver paint.

Caution: A powerful oxidizing agent. Caution should be exerted when silver oxide is brought into contact with ammonia or combustible solvents. An ammoniacal solution of a silver salt is potentially explosive and this hazard increases with standing time.

Shipping regulations: None.*

silver oxide battery. See zinc-silver oxide battery.

silver oxide, divalent (argentic oxide) AgO. Made by chemical precipitation. Average particle size, 1.5-3.0 microns. Fineness 730.0 Ag pts/1000 (min) (73% Ag).

Containers: 80-oz jars in metal outer-containers.

Caution: Do not mix with any organic material.

silver permanganate $AgMnO_4$.

Properties: Violet, crystalline powder. Contains 47.5% (approx.) silver. Decomposed by alcohol. Caution! Keep in dark-colored bottle! Soluble in water.

Grade: Technical.

Use: Gas masks; medicine.

silver phenolsulfonate. See silberol.

silver phosphate (silver orthophosphate) Ag_3PO_4.

Properties: A yellow powder; turns brown when heated or on exposure to light. Soluble in acids, potassium cyanide solution, ammonium hydroxide, ammonium carbonate, and acetic acid; very slightly soluble in water. Sp.gr. 6.37; m.p. 849°C.

Derivation: By the interaction of silver nitrate and sodium phosphate.

Grade: Technical.

Containers: Amber or black glass bottles. Poison label.

Uses: In photographic emulsions; catalyst; pharmaceuticals.

Shipping regulations: None.*

silver picrate $C_6H_2O(NO_2)_3Ag \cdot H_2O$.

Properties: Yellow crystals containing 30% silver; soluble in water; slightly soluble in alcohol and acetone; insoluble in ether and chloroform.

Use: Medicine.

silver-potassium cyanide $KAg(CN)_2$.

Properties: White crystals; sensitive to light; exceedingly poisonous! Soluble in

water and alcohol; insoluble in acids.
Derivation: By adding silver chloride to a solution of potassium cyanide.
Impurities: Silver.
Grade: Technical.
Containers: Glass bottles. Poison label.
Uses: Silver plating; bactericide; antiseptic.
Shipping regulations: Poison, class B. Poison label.*

silver protein, mild (protargin, mild). See also silver protein, strong. Silver rendered colloidal by the presence of, or combination with, protein. Contains 19-23% silver.
Properties: Brown or black shining scales or granules; odorless; hygroscopic; affected by light. Must be freshly prepared. Soluble in water but almost insoluble in alcohol, chloroform, or ether.
Containers: 16-oz bottles.
Grade: N.F. XI.
Use: Medicine.

silver protein, strong (protargin, strong) See also silver protein, mild. The strong variety contains 7.5-8.5% silver, but since this is principally in ionic form, it is more irritating and a stronger germicide than the mild.
Properties: Orange to black, odorless powder; somewhat hygroscopic; affected by light. Must be freshly prepared. Soluble in water but almost insoluble in alcohol, chloroform, and ether.
Use: Medicine.

silver salt. See anthraquinone-2-sodium sulfonate.

silver-sodium chloride (sodium-silver chloride) $AgCl \cdot NaCl$.
Properties: Hard, white crystals. Decomposed by water. Soluble in solution of sodium chloride (conc.).

silver-sodium thiosulfate (sodium-silver thiosulfate) $Ag_2S_2O_3 \cdot 2Na_2S_2O_3 \cdot 2H_2O$.
Properties: White or gray, crystalline powder. Sweet taste; soluble in water.

silver sulfate (silver sulfate, normal) Ag_2SO_4.
Properties: Small, colorless, lustrous crystals or crystalline powder. Contains 69% (approx.) silver. Turns gray on exposure to light. Soluble in ammonium hydroxide, nitric acid, sulfuric acid, (hot) water; insoluble in alcohol. Sp.gr. 5.45; b.p. 1085°C (decomposes); m.p. 652°C.
Grades: Technical; C.P.
Use: Analysis.

silver sulfate, normal. See silver sulfate.

silver sulfide Ag_2S.
Properties: A grayish-black, heavy powder. Soluble in concentrated sulfuric and nitric acids; insoluble in water. Sp.gr. 6.85-7.32; b.p. decomposes; m.p. 825°C.
Derivation: By passing hydrogen sulfide gas into silver nitrate solution, washing and drying.
Grade: Technical.
Containers: Glass bottles.
Uses: Inlaying in niello metal-work; ceramics.
Shipping regulations: None.*

silver sulfocarbolate. See silberol.

silver sulfophenylate. See silberol.

silvichemicals. Chemicals made from wood. They include: alcohol (ethyl), lignins, lignosulfonates (from spent sulfite liquor); vanillin; yeast (from fermentation of wood sugars); tall oil; sulfate turpentine; bark extracts; phenolic materials.

"Silvol." 330 Trademark for silver protein, mild, N.F.

silyl. See silane.

simaroubidin. See simarubidin.

simaroubin. See simarubin.

simarubidin (simaroubidin) $C_{22}H_{32}O_9$.
Properties: Tasteless, crystallizes as needles. M.p. 260-261°C. Soluble in glacial acetic acid, pyridine.
Derivation: Bark and wood of Simaruba amara.
Use: Medicine.

simarubin (simaroubin) $C_{22}H_{30}O_9$.
Properties: Needle-like crystals; m.p. 230-231°C; slightly soluble in water; soluble in alcohol, acetone, pyridine, glacial acetic acid, methanol. Insoluble in benzene, ether, and chloroform.

Simons process. An electrochemical fluorination process which makes fluorocarbons by passing an electric current through a mixture of the organic starting compound and liquid anhydrous hydrogen fluoride. The products are hydrogen and the desired fluorocarbon.

simple distillation. Distillation in which no appreciable rectification of the vapor occurs i.e., the vapor formed from the liquid in the still is completely condensed in the distillate receiver and does not undergo change in composition due to partial condensation or contact with previously condensed vapor.

sinapis alba. See mustard, white.

sinapis nigra. See mustard, black.

sinapis oil. See mustard oil, volatile.

single nickel salt. See nickel sulfate.

"Singoserp." 305 Trademark for syrosingopine N.N.D.
Use: Medicine.

sinopis. A variety of red hematite (q.v.) used as a pigment.

sintered carbides. See cemented carbides.

sintering. The partial welding together of powder particles at temperatures below the melting point. Occurs in both powder metallurgy and ceramic firing with or without the presence of liquid. While heat and pressure encourage the process, the driving force results from the decrease in surface area

*See "I.C.C. Shipping Regulations," page xiii.
Reference numbers refer to name of manufacturer. See "List of Manufacturers," page v.

(see Rittinger's law). Sintering produces greater strength, conductivity, and density.

"Si-O-Lite." [329] Trademark for a pure, bulky, precipitated silica hydrogel.

SIPP. Abbreviation for sodium iron pyrophosphate.

sisal. Hard, strong, light-yellow to reddish fibers obtained from the leaves of Agave sisilana. It is inferior to abaca in strength and water resistance.
Source: Africa; Java; Haiti; Bahama.
Grades: Based on country of origin, color, and length of fibers.
Containers: 400-lb bales.
Uses: Binder twine; rope; sacking; upholstery.

siserskite. A natural alloy of osmium and iridium, with osmium predominating.

"Sitol." [28] Trademark for sodium salt of nitrobenzene-meta-sulfonic acid.
Properties: Cream-colored powder.
Use: As a protective oxidizing agent in discharge printing.

beta-**sitosterol** $C_{29}H_{50}O$. 22, 23-Dihydrostigmasterol.
Properties: Waxy white solid; almost odorless and tasteless; insoluble in water, soluble in benzene, chloroform, carbon disulfide and ether. Can be crystallized from ether as anhydrous needles, or from aqueous alcohol as leaflets with one molecule of water.
Derivation: See sitosterols.
Uses: Sitosterol differs from cholesterol by a C_2H_5 group at the 24 position and has some properties and uses in common.
See sitosterols.

sitosterols. The N.F. XI grade is a mixture of beta-sitosterol and certain saturated sterols. It contains not less than 95% total sterols nor less than 85% unsaturated sterols calculated as beta-sitosterol (q.v.).
Properties: White, nearly odorless, tasteless powder or waxy white solid. Insoluble in water; soluble in chloroform, benzene and carbon disulfide. Specific rotation (2% solution) −25° to −38°. Melting range 136-142°C.
Derivation: From soybeans.
Uses: Medicine; additive in cosmetics and emulsions; intermediate.

"Six-Twelve." [214] See "6-12."

size oils. The same as throwing oils (q.v.).

sizing compounds. Materials applied to yarns, fabrics, paper, leather and other products to improve or increase their stiffness, strength, smoothness or weight; materials used to modify the cooked starch solutions applied to warp ends prior to weaving.

skatole (3-methylindole; beta-methylindole) C_9H_9N.
Properties: A white crystalline substance, browning upon aging, and having an extremely fecal odor. M.p. 93° to 95°C; b.p. 265°C. Soluble in hot water, alcohol, benzene. Gives violet color in potassium ferrocyanide and sulfuric acid.
Use: Sparingly in perfumery as a fixative and for artificial civet.

"Skellysolve." [409] Proprietary name applied to any one of a series of highly refined petroleum hydrocarbon fractions manufactured in various grades having closely specified and controlled boiling ranges, gravity, evaporation rates, etc. for uses in the oil, fat, rubber, paint, polish, ink, insecticide, and other chemical industries. The various grades are designated "Skellysolve"-B, -C, -F, etc. as follows.

"Skellysolve" B. Normal hexane-type solvent with low evaporation residue and freedom from greasy ends and odor. Used in rubber compounding, vegetable extraction, manufacture of cans, and as a petroleum ether. Produced by exhaustive chemical treatment.
Typical specifications:
Sp.gr. (60°F) 0.683.
Reid vapor pressure, psi (100°F) 5.1.
Evaporation rate at 68°F: 50%, 1.80 minutes; 100%, 4.30 minutes.
Evaporation residue % by weight 0.0003.
Flash point (Tag closed cup) −25°F.
Aniline point 142.5°F.
Kauri butanol value 30.5.
% sulfur by weight 0.0040.
A.S.T.M. distillation I.B.P., 146°F; 20%, 149°F; 40%, 149°F; 60%, 150°F; 80%, 151°F; 95%, 153°F; dry point, 156°F.
Acidity of residue from distillation, neutral.

"Skellysolve" C. Low evaporation residue and slower evaporation rate than "Skellysolve"-B. Used in rubber compounding, meat scrap extraction, manufacture of cans, adhesive tape.
Typical specifications:
Sp.gr. (60°F) 0.726.
Reid vapor pressure, psi (100°F) 2.2.
Evaporation rate at 68°F: 50%, 4.00 minutes; 100%, 9.80 minutes.
Evaporation residue % by weight 0.0005.
Flash point (Tag closed cup) + 13°F.
Aniline point 130.2°F.
Kauri butanol value 36.2.
% sulfur by weight 0.0060.
A.S.T.M. distillation I.B.P., 190°F; 20%, 196°F; 40%, 197°F; 60%, 199°F; 80%, 200°F; 95%, 203°F; dry point, 208°F.
Acidity of residue from distillation, neutral.

"Skellysolve" F. A petroleum ether naphtha. Used as a laboratory reagent, in aerosol-type insecticides, herbicide formulations. Meets petroleum ether specifications for the American Chemical Society, for the American Oil Chemists' Society and for Federal Specification O-E-751b.
Typical specifications:
Sp.gr. (60°F) 0.643.
Reid vapor pressure, psi (100°F) 13.5.
Evaporation rate at 68°F:
50%, 0.80 minutes; 100%, 2.40 minutes.
Evaporation residue % by weight 0.0003.
Flash point (Tag closed cup) −70°F.
Kauri butanol value 26.4.
% sulfur by weight 0.0040.

A.S.T.M. distillation I.B.P., 96°F; 20%, 100°F; 50%, 103°F; 80%, 110°F; 90%, 116°F; 95%, 126°F; dry point, 136°F.
Acidity of residue from distillation, neutral.

"Skellysolve" H. Has complete absence of foreign taste or odor. Used in rubber compounding, ink manufacture, pharmaceutical, edible, and inedible oil extraction, and in the manufacture of tape.
Typical specifications:
Sp.gr. (60°F) 0.704.
Reid vapor pressure, psi (100°F) 4.0.
Evaporation rate at 68°F:
50%, 1.80 minutes; 100%, 5.50 minutes.
Evaporation residue % by weight 0.0005.
Flash point (Tag closed cup) -20°F.
Aniline point, 136.0°F.
% sulfur by weight 0.0040.
A.S.T.M. distillation, I.B.P. 156°F; 20%, 168°F; 50%, 175°F; 80%, 188°F; 90%, 194°F; 95%, 200°F; dry point 205°F.
Acidity of residue from distillation, neutral.

"Skellysolve" H, Special. Has extremely low evaporation residue. Used for paint formulations, rubber cements, tapes, ink diluent, edible and inedible oil extraction.
Typical specifications:
Sp.gr. (60°F) 0.685.
Reid vapor pressure, psi (100°F) 5.42.
Evaporation rate at 68°F:
50%, 1.70 minutes; 100%, 4.10 minutes.
Evaporation residue % by weight 0.0004.
Flash point (Tag closed cup) -24°F.
Aniline point, 141.4°F..
% sulfur by weight 0.0018.
A.S.T.M. distillation, I.B.P. 147°F; 20% 150°F; 50%, 152°F; 80%, 155°F; 95%, 164°F; dry point, 174°F.
Acidity of residue from distillation, neutral.

"Skellysolve" L. Meets U.S. Government Paint Specification TT-N-95a, Type I.
Typical specifications:
Sp.gr. (60°F) 0.733.
Reid vapor pressure, psi (100°F) 2.0.
Evaporation rate at 68°F:
50%, 4.40 minutes; 100%, 12.50 minutes.
Evaporation residue % by weight 0.0007.
Flash point (Tag closed cup) +17°F.
Aniline point, 130.4°F.
% sulfur by weight 0.005.
A.S.T.M. distillation, I.B.P. 202°F; 20%, 206°F; 50%, 210°F; 60%, 211°F; 90%, 217°F; 95%, 221°F; dry point 229°F.
Acidity of residue from distallation, neutral.

"Skellysolve" R. A rubber naptha with uniform evaporation rate and freedom from heavy greasy ends. High initial boiling point helps eliminate excessive evaporation losses.
Typical specifications:
Sp.gr. (60°F) 0.711.
Reid vapor pressure, psi (100°F) 4.4.
Evaporation rate at 68°F:
50%, 2.30 minutes; 100%, 9.80 minutes.
Evaporation residue % by weight 0.0013.
Flash point (Tag closed cup) -18°F.
Aniline point 137.0°F.
% sulfur by weight 0.0080.
A.S.T.M. distillation, I.B.P., 132°F; 20%, 174°F; 50%, 196°F; 80%, 218°F; 90%, 228°F; 95%, 235°F; dry point, 244°F.
Acidity of residue from distillation, neutral.

"Skellysolve" S. Meets specifications for both mineral spirits and Stoddard solvent. Used in the paint and protective coatings industry, the dry cleaning industry, in wood treating, in wax formulations, in metal and tool cleaning and in the chemical industry. Meets U.S. Government Paint Specification TT-29la and also Federal Specification PS-66lb, Type I. Also meets Stoddard Solvent ASTM Specification D484-52.
Typical specifications:
Sp.gr. (60°F) 0.780.
Reid vapor pressure, psi (100°F) 0.1.
Evaporation rate at 68°F:
50%, 70.0 minutes; 100%, 300.0 minutes.
Evaporation residue % by weight 0.0050.
Flash point (Tag closed cup) 105°F.
Aniline point 138.0°F.
% sulfur by weight 0.0200.
A.S.T.M. distillation, I.B.P., 305°F; 20%, 321°F; 50%, 334°F; 80%, 351°F; 95%, 368°F; dry point, 375°F.
Acidity of residue from distillation, neutral.

"Skellysolve" S1. Narrow boiling range and controlled front end distillation points. Meets U.S. Government Paint Specification TT-29la and Federal Specification PS-66lb, Type I.
Typical specifications:
Sp.gr. (60°F) 0.785.
Reid vapor pressure, psi (100°F) 0.1.
Evaporation rate at 68°F:
50%, 110.0 minutes; 100%, 340.0 minutes.
Evaporation residue % by weight 0.0060.
Flash point (Tag closed cup) 118°F.
Aniline point, 138.0°F.
% sulfur by weight 0.0180.
A.S.T.M. distillation, I.B.P. 330°F; 20%, 340°F; 50%, 344°F; 80%, 351°F; 95%, 365°F; dry point, 370°F.
Acidity of residue from distillation, neutral.

"Skellysolve" S2. Mineral spirits type with extremely short boiling range and minimum flash of 100°F. Used in dry cleaning, in manufacture of waxes, in paint formulations and in cloth impregnation. Meets U.S. Government Paint Specification TT-29la, Federal Specification PS-66lb, Type I, and Stoddard Solvent ASTM Specification D484-52.
Typical specifications:
Sp.gr. (60°F) 0.775.
Reid vapor pressure, psi (100°F) 0.2.
Evaporation rate at 68°F:
50%, 80.0 minutes; 100%, 210.0 minutes.
Evaporation residue % by weight 0.0015.
Flash point (Tag closed cup) 101°F.
Aniline point, 137.0°F.
% sulfur by weight 0.0160.
A.S.T.M. distillation, I.B.P. 305°F; 20%, 311°F; 50%, 314°F; 80%, 318°F; 95%, 323°F; dry point, 327°F.
Acidity of residue from distillation, neutral.

"Skellysolve" T. A 140°F flash-type naphtha. Gives a longer wet edge in paint and protective formulations. Meets Federal Specification PS-66lb, Type II.

Typical specifications:
Sp. gr. (60°F) 0.796.
Reid vapor pressure, psi (100°F) 0.05.
Evaporation rate at 68°F:
50%, 299.0 minutes; 100%, 1040.0 minutes.
Evaporation residue % by weight 0.015.
Flash point (Tag closed cup) 141°F.
Aniline point 143.0°F.
% sulfur by weight 0.0360.
A.S.T.M. distillation, I.B.P., 360°F; 20%, 369°F; 50%, 374°F; 80%, 381°F; 95%, 389°F; dry point, 400°F.
Acidity of residue from distillation, neutral.

"Skellysolve" V. A VMP naphtha with sweet naphtha-like odor, free from oiliness or heavy greasy ends. Used in the paint and protective coatings industry as a diluent, in rubber compounding, tape manufacture, and as a base for cigar and cigarette lighter fluid.

Typical specifications:
Sp. gr. (60°F) 0.750.
Reid vapor pressure, psi (100°F) 0.8.
Evaporation rate at 68°F:
50%, 13.00 minutes; 100%, 38.00 minutes.
Evaporation residue % by weight 0.0010.
Flash point (Tag closed cup) 54°F.
Aniline point, 137.5°F.
% sulfur by weight 0.0090.
A.S.T.M. distillation, I.B.P., 242°F; 20%, 255°F; 60%, 265°F; 90%, 280°F; 95%, 285°F; dry point, 290°F.
Acidity of residue from distillation, neutral.

"Skellysolve" X. A high-flash, long wet-edge mineral spirits solvent, used as an industrial cleaning solvent, insecticide formulation diluent and in paint formulations.

Typical specifications:
Sp. gr. (60°F) 0.813.
Flash point (Tag closed cup) 141°F.
Aniline point 151.9°F.
% sulfur by weight 0.0700.
A.S.T.M. distillation, I.B.P., 345°F; 20%, 408°F; 50%, 436°F; 80%, 461°F; 95%, 485°F; end point, 508°F.
Acidity of residue from distillation, neutral.

"Skiabaryt." [123] Trademark for a chemical preparation used in x-ray examinations.

"Skiodan." [162] Trademark for methiodal sodium.

Skraup synthesis. Synthesis of quinoline or its derivatives by heating aniline or an aniline derivative, glycerol and nitrobenzene in the presence of sulfuric acid.

skutterudite (Co, Ni)As_3. A natural cobalt-nickel arsenide with cobalt in excess of nickel. Chloanthite and smaltite (q.v.) are similar species.

Properties: Color tin white to silver gray; luster metallic; hardness 5.5-6; sp. gr. 6.5.

Occurrence: Colorado; Canada; Norway.

Use: Minor ore of cobalt and nickel.

"Skydrol." [58] Trademark for fire resistant jet aircraft hydraulic fluid.

"Skydrol" 500-A. A clear, purple liquid; operational at -65°F; moisture 0.20% max; sp. gr. (25°/25°C) 1.060-1.066.

"Skydrol" 7000. A clear green liquid; moisture 0.25% max; sp. gr. (25°/25°C) 1.080-1.086.

"S. L." [329] (Standard Luminescent) Trademark for chemicals specially developed and standardized to meet the exacting requirements of phosphor manufacturers.

slag. The fused product which separates in metal smelting and floats on the bath of metal. Formed by combination of flux with gangue of ore, ash of fuel, and perhaps furnace lining. The slag is often the medium by means of which impurities may be separated from metal. A nonreactive slag or cover, such as glass, may be used to protect the melt from the furnace atmosphere in melting brass. Cinder is a term used interchangeably with slag. Contains compounds derived from silica (SiO_2), lime (CaO), alumina (Al_2O_3), magnesia (MgO), manganese (Mn), phosphorus (P), sulfur (S) as well as the major elements of the ore or metal being refined.

Uses: Railroad ballast; material for highway construction; concrete aggregate; raw material for Portland cement; raw material for glass fibers (see mineral wool); fertilizer.

slag cement (lime slag cement). Cement prepared either by mixing granulated and finely ground blast-furnace slag with slaked lime or by spraying the granulated slag with a solution of alum, magnesium sulfate or other salt, and grinding the product with lime and gypsum. Slag cements contain less lime and much more alumina than Portland cement and are usually blue gray in color but may be brown due to the presence of manganese. They are also slow setting but attain maximum hardness faster than Portland cement. They are well suited for underwater work but poor when exposed to the atmosphere. The setting time may be decreased by adding caustic soda, potash, or soda ash.

slag wool. See mineral wool.

slaked lime. See calcium hydroxide; lime hydrated.

slate. A fine grained metamorphic rock which breaks into thin slabs or sheets. Color usually gray to black, sometimes green, yellow, brown, or red. Slates are composed of micas, chlorite, quartz, hematite, clays, and other minerals.

Occurrence: Pennsylvania, Vermont, Maine, Virginia, California, Colorado; Europe.

Uses: Roofing; blackboards; decorative stone; various building applications such as stair treads, shower stalls, walks; in crushed form on shingles; filler in paint, linoleum, rubber; abrasive; pigment.

slate black. See mineral black.

slip clay. A type of clay containing such a high percentage of fluxing impurities and of such a texture that it melts at a relatively low temperature to a greenish or brown glass, thus forming a natural glaze. It must be fine-grained, free from lumps or concretions, show a low air-shrinkage and mature in burning at as little above 1300°F as possible.

"Slipicone." [149] Trademark for fluid silicone compositions to prevent the adhesion of materials to one another. Used on food processing and packaging equipment.

"Slipkote." [51] Trademark for a smooth, buttery lubricant for application to the hardened base coat on ship launching ways. It has a low coefficient of fraction, is water-repellent and adhesive to the base coat, and may be easily applied in cold weather.

slippery elm. See ulmus.

slip stains. See pottery body stains.

sloe. See viburnum prunifolium.

sludge. A soft mud, slush, or mire; for example, the solid product of a filtration process before drying (filter cake). See also sewage sludge.

sludge acid. Waste or spent sulfuric acid, usually that from refining petroleum oils or crude benzenes.
Shipping regulations: Corrosive liquid. White label.*

sludge asphalt. Asphalt-like products obtained by separation from the acid sludge produced in the refining of petroleum.
Shipping regulations: None.*

slurry. A thin watery suspension; for example, the feed to a filter press or other filtration equipment.

slushing compound. A nondrying oil, grease or similar organic compound which, when coated over a metal, affords at least temporary protection against corrosion.

Sm. Symbol for samarium.

smalt.
Properties: Blue powder.
Derivation: A potash-cobalt glass made by fusing pure sand and potash with cobalt oxide, grinding and powdering.
Grades: Technical.
Containers: Wooden kegs; multiwall paper sacks.
Uses: Paint pigments; ceramic industries (pigment); coloring glass; bluing paper, starch and textiles; coloring rubber.
Shipping regulations: None.*

smaltite $(Co, Ni)As_2$. A natural cobalt-nickel arsenide, with cobalt in excess of nickel. Skutterudite and chloanthite (q.v.) are similar species.
Properties: Color tin white to silver gray; luster metallic; hardness 5.5-6; sp.gr. 6.5.
Occurrence: Canada; Chile; Europe.
Use: Minor ore of cobalt and nickel.

smelting. Melting or fusing of an ore to separate and refine the metal. Roasting and subsequent reduction are usually involved.

"Smentox." [236] Brand name for a chemical compound for reconditioning cement-contaminated drilling mud, or for preventing mud from becoming cement-contaminated, since contact with cement flocculates untreated drilling mud, rendering it unfit for use by causing the solids to settle out and raising the water loss to an impractical level.

smithsonite (dry-bone) $ZnCO_3$.
Properties: A white mineral often colored yellow or brown by iron. Vitreous to dull luster. Contains 64.8% of zinc oxide. Distinguished from other carbonates by its hardness. Soluble in acids. Many zinc deposits consist in their upper portions of smithsonite and calamine. Sp.gr. 4.3-4.5; hardness 5.
Occurrence: United States (Mississippi valley, New Mexico); Poland.
Use: Source of zinc.

smog. An aerosol (q.v.). Household rubbish burners and automobile exhaust gases are known to be contributing factors to eye-irritating city smogs.

smokeless powder. Nitrocellulose containing about 13.1% nitrogen, produced by blending material of somewhat lower (12.6%) and slightly higher (13.2%) nitrogen content, converting to a dough with alcohol-ether mixture, extruding, cutting and drying to a hard horny product. Small amounts of stabilizers (amines) and plasticizers are usually present, as well as various modifying agents (nitrotoluene, nitroglycerine salts).

"S Monel." [283] Trademark for a cast age-hardenable alloy containing approximately 65% nickel, 30% copper, and 4% silicon.
Use: Valve seats, pump impellers, and other sliding or moving elements.

smudge oil. An oil burned in fruit orchards to prevent frost from injuring the trees. No. 3 fuel oil is typical of oils used.
Shipping regulations: None.*

Sn. Symbol for tin.

snake root. See rauwolfia.

snake-root oil. See asarum oils.

snake-root oil, Canada. See asarum canadense oil.

snake-weed. See euphorbia.

snapping hazel. See hamamelis.

SN numbers (survey numbers). A designation adopted by the antimalarial commission of the Office of Scientific Research and Development for identification of the compounds tested by that body during World War II.

"Snodotte." [221] Trademark for a mixture of fatty acids (C_{14} to C_{22}) which are 97.6% saturated. Its snow-white color makes

*See "I.C.C. Shipping Regulations," page xiii.
Reference numbers refer to name of manufacturer. See "List of Manufacturers," page v.

it very useful in candles, shaving cream and cream shampoos as well as metallic stearates, buffing compounds, lubricating greases, and plastic molding.

"Sno-Gon." [108] A chloride-free ice-melting compound, colorless, odorless, non-toxic, non-flammable, non-corrosive to aluminum and ferrous metals.
Containers: 100-lb bags.

"S-1 Non-Ionic Surfactant." [108] A non-ionic surfactant in liquid form.
Containers: 40-lb cans; 425-lb drums.
Uses: To increase flood-water injectivity rate in secondary oil recovery.

snowball bush. See viburnum opulus.

"Snowdros." [232] Brand name for a proprietary product of the hydrosulfite class.

snuff bean. See tonka.

"SOA." [214] Trademark for sucrose octaacetate (q.v.).

"Soak-Eze." [108] Concentrated highly active crystalline compound specifically designed for pre-soaking of wares prior to machine washing.
Containers: 4-lb boxes.

soap. Ordinary soap is a mixture of the sodium salts of various fatty acids of natural oils and fats. Thus common soap is largely a mixture of the sodium salts of palmitic, stearic, and oleic acids. The term soap is also applied to the individual components such as sodium palmitate, sodium stearate, etc. In case some other metal or basic radical is present instead of sodium a modified term such as potash soap, calcium soap or amine soap is used. See also soaps, metallic.

A great variety of special soaps are produced, which are for the most part variations of ordinary sodium soaps. Hard soap contains a relatively large proportion of sodium stearate.

Rosin soaps as used for laundry purposes are made by adding a soap made from rosin, or rosin itself, to an ordinary soap. Castile or Marseilles soaps are made from olive oil. Mottled soaps are produced by the addition of small amounts of ferrous sulfate, ferric oxide or ultramarine. Transparent soaps are made from decolorized fats with the addition of glycerol or sugar, or both. Liquid soap is usually a potash soap dissolved in water, containing from 8 to 30% soap; the solutions of 30% and higher contain alcohol.
See also detergents, synthetic.

soap bark. See quillaja.

soap builder. Any material mixed with soap to improve the cleaning properties, modify the alkali content, or confer water-softening characteristics.

soap, green. See soap, soft.

soap, hard (neutral soap). An N.F. XI grade of soap containing no heavy metal soap or alkaline salts. Generally it contains not more than 0.05% of free caustic alkali to prevent its turning rancid due to free fatty acids being formed.

soaproot. See saponaria.

soaps, metallic. A term usually limited to insoluble soaps of such easily available fatty acids as stearic, naphthenic, octoic or 2-ethylhexoic, rosin (resinates), or tall oil (tallates) with the heavier metals such as aluminum, calcium, cadmium, copper, iron, lead, tin, or zinc. A pure compound is seldom required. For descriptions and properties, see names of individual metallic soaps.
Derivation: (a) The fusion method, by heating a fatty acid with a metallic oxide, carbonate, etc. (b) The precipitation process, by the reaction of soluble sodium or potassium soaps with solutions of heavy metal salts.
Uses: Waterproofing; gels; conditioning agents in cements, paints, plastics; fungicides; lubricants; driers.

soap, soft (green soap).
Properties: A yellowish-green or brownish slippery soft mass with a slight characteristic odor, and an alkaline taste; made from vegetable oil and potassium hydroxide; soluble in hot water and hot alcohol.
Containers: 5-, 25-lb tins; 150-lb kegs.
Grades: U.S.P. XVI specifies vegetable oil excluding coconut and palm kernel oil.
Uses: In medicine as a detergent, antiseptic and disinfectant; also as a lubricant.

soapstock. See foots.

soapstone. See talc.

soapwort. See saponaria.

soda. Any one of the forms of sodium carbonate (q.v.); also used loosely as equivalent to the word sodium in compounds.

soda alum. See aluminum sodium sulfate.

soda amatol.
Shipping regulations: Explosive, Class A. High explosive label.*

soda ash. Na_2CO_3 (soda, calcined; sodium carbonate, anhydrous). The crude sodium carbonate of commerce.
Properties: A grayish-white powder or lumps containing up to 99% sodium carbonate. Soluble in water; insoluble in alcohol.
Derivation: By the Solvay ammonia soda process (q.v.). Over 90% of the world's production is made by this process. Appreciable amounts are recovered from natural deposits or brines particularly in California and Wyoming (see Trona process) and by passing carbon dioxide into the negative electrode chamber of the diaphragm-type cells used for the electrolytic production of chlorine (see diaphragm cell).
Impurities: Sodium chloride, sodium sulfate, calcium carbonate and magnesium carbonate, sodium bicarbonate.
Grades: Dense 58%; light 58%; extra light; natural; refined.

Containers: Bags; barrels; drums; bulk.
Uses: Glass; ceramics; soap; detergents; cleaners; water softening; petroleum refining; aluminum production; textiles; pulp and paper; metals processing; caustic soda; sodium bicarbonate; sodium nitrate; and miscellaneous other uses.
Shipping regulations: None.*

"Soda Ash, Foundry Grade." [244] Available as Foundry Grade #1, #2, and Briquettes. Grades #1 and #2 are hard, dustless, and free flowing, readily adaptable to automatic feeding equipment. Briquettes are adaptable to automatic feeding; can also be used directly in a cupola. They are dependable desulfurizing agents, being 100% active materials.
Containers: 100-lb Kraft multiwall bags; bulk shipments.

soda, baking. See sodium bicarbonate. Potassium bicarbonate is sometimes used in baking but is not, strictly speaking, baking soda.

soda Bordeaux (Burgundy mixture). A fungicide mixture made by mixing copper sulfate and sodium carbonate solutions, and used similarly to ordinary Bordeaux mixture.

"Soda Briquettes." [177] Trademark for soda ash held in walnut-size pellet form by a hydrocarbon bonding material. Packed in 100-lb paper bags. Used extensively in desulfurizing and cleansing pig iron; speeds up these operations and brings about sulfur reduction of 30 to 70%, as well as physically cleaning the iron.

soda, calcined. See soda ash.

soda, caustic. See sodium hydroxide.

soda crystals. See sodium carbonate monohydrate.

soda lime. A mixture of calcium hydroxide with sodium or potassium hydroxide intended for the absorption of carbon dioxide gas and water vapor.
Properties: White or grayish-white granules unless colored by a specified indicator. Must be kept in air tight containers.
Grades: Technical; U.S.P. XVI. Usually % moisture and mesh size are stated.
Uses: Drying agent and carbon dioxide absorbent for technical laboratory and medical work.

soda lime glass. See glass.

sodalite $3NaAlSiO_4 \cdot NaCl$. A mineral found in igneous rocks.
Properties: Color blue, white, gray; luster vitreous; hardness 5.5-6; sp.gr. 2.1-2.3.
Occurrence: Montana, Maine; Canada; Europe.
Use: Gem stone.

sodamide. See sodium amide.

soda monohydrate. See sodium carbonate, monohydrate.

soda niter. See sodium nitrate.

"Sodaphos." [55] Brand name for glassy sodium tetraphosphate (q.v.).

soda pulp. See wood pulp.

soda saltpeter. See caliche.

sodas, modified. Combinations of soda ash and bicarbonate of soda in definite proportions marketed for purposes where an alkali is needed ranging in causticity between bicarbonate of soda and soda ash. Modified sodas are white crystalline powders, readily soluble and possessing valuable cleansing and purifying properties. They are prepared in various strengths both as to alkalinity and causticity.
Containers: See soda ash.
Uses: Washing powders; laundry sodas; wool scouring powders; bottle cleansers; textile sodas; mild detergents.
Shipping regulations: None.*

sodas, natural. See trona; thermonatrite.

soda, washing. See sal soda.

sodio-cupric chloride. See copper-sodium chloride.

alpha-**sodio-sodium acetate** (sodium alpha-sodioacetate) $NaCH_2COONa$.
Properties: Free flowing powder; stable in dry air; decomposes slowly in moist air. Decomposes 280°C without melting. Insoluble in ethers and hydrocarbons; reacts mildly with water. It is a strong base and should be handled with caution like sodium hydroxide.
Grades: 80-85% pure. Impurities are sodium acetate, sodium amide and sodium hydroxide.
Uses: Organic intermediate; drying agent for organic solvents.
Caution! Exposure to alpha-sodio sodium acetate dust should be avoided. Wash affected areas of skin with plenty of water.

"Sodite." [253] Brand name for an insecticide product containing arsenic trioxide.

sodium (natrium) Na. A metallic element; atomic number 11, group I of the periodic table. Long handicapped by its reactivity and the danger involved in handling it, sodium is now sold in tonnage lots and is finding many uses.
Properties: Light, soft, ductile, malleable, silver-white metal, oxidizing rapidly in air; of wax-like consistency at ordinary temperatures, but brittle at low temperatures; must be kept immersed in naphtha or other similar liquid which does not contain water or free oxygen. Sp.gr. 0.9712; m.p. 97.6°C; b.p. 892°C. Decomposes water on contact with vigorous evolution of hydrogen and forms sodium hydroxide; insoluble in benzene, kerosine, and naphtha.
Derivation: (a) Electrolysis of a fused mixture of sodium chloride and calcium chloride (see also Downs cell); (b) sodium amalgams are produced by electrolysis of sodium hydroxide or sodium chloride solutions in mercury cells (see Castner cell;

*See "I.C.C. Shipping Regulations," page xiii.
Reference numbers refer to name of manufacturer. See "List of Manufacturers," page v.

chlorine); (c) thermochemical reduction of sodium carbonate or sodium hydroxide with carbon or iron-carbon mixtures.
Method of purification: Distillation.
Grades: Technical; brick; amalgam; coated powders; dispersions (in oils, etc).
Containers: 1-, 5-lb tins in cases; 290-lb steel barrels; 80,000-lb tank cars.
Uses (in approximate order of volume): Tetraethyl lead; chemical reductions, particularly of fats for detergent production; sodium cyanide; sodium peroxide; sodium hydride; pharmaceuticals; petroleum refining; metallurgy of titanium; heat transfer medium; as a component in non-glare street lights; in fluidized beds for desulfurization of benzene and toluene; in small nuclear reactors as a cooling agent.
Danger! Reacts violently with water liberating and igniting hydrogen. May cause burns. MCA warning label.
Shipping regulations: Flammable solid. Yellow label.*
See also sodium dispersions.

sodium abietate (rosin soap; sodium resinate) $C_{19}H_{29}COONa$.
Properties: White powder. Soluble in water.
Derivation: By leaching rosin with sodium hydroxide solution.
Uses: Medicine; soap making; paper coating. (See also soap.)

sodium acetate (a) $NaC_2H_3O_2$; (b) $NaC_2H_3O_2 \cdot 3H_2O$.
Properties: Colorless, odorless crystals; efflorescent; very soluble in water; slightly soluble in alcohol; soluble in ether.
Constants: Sp.gr. 1.4; m.p. (a) 324°C; (b) 58°C.
Derivation: (1) Neutralization of acetic acid with sodium carbonate, concentration of the solution, crystallization and centrifuging; (2) calcium acetate is treated with sodium sulfate and a little soda, the solution filtered, evaporated to dryness, the residue heated to about 250°C, dissolved in water, filtered, concentrated and crystallized.
Method of purification: Recrystallization.
Grades: Highest purity; pure fused; C.P.; N.F. XI (the trihydrate); technical.
Containers: Bags; drums.
Uses: Intermediates; dyes; manufacture of dry colors; pharmaceuticals; cinnamic acid; acetic anhydride; copper acetate; mordant; soaps; photography; purification of glucose; meat preservation; medicine; Schweinfurth green; textile printing; separation of the opium alkaloids; rubber industry; electroplating reagent; preserving meats; tanning; dehydrating agent.
Shipping regulations: None.*

sodium N-acetoacetyl-para-**sulfanilate** $CH_3COCH_2CONHC_6H_4SO_3Na$.
Properties: Available in the form of a red colored 40% aqueous solution. Sp.gr. 1.203.

sodium acetone bisulfite (acetone-sodium bisulfite) $(CH_3)_2CONaHSO_3$.
Properties: Crystalline material; soluble in water; decomposed by acids; slightly soluble in alcohol.
Derivation: Interaction of sodium bisulfite and acetone.
Grades: Technical.
Uses: Chemical (pure acetone); photography; textile (dyeing and printing).

sodium acetrizoate $NaOOCC_6HI_3NHCOCH_3$. Sodium 3-acetylamino-2,4,6-triiodobenzoate. Marketed as a solution.
Properties: Aqueous solutions are clear and practically colorless. U.S.P. XVI solution has pH 7.0-7.5. (The free acid is a white powder which decomposes at 278-283°C and is soluble in alcohol, slightly soluble in water and ether.)
Grade: U.S.P. XVI.
Use: Medicinal.

sodium-para-**acetylaminophenylantimoniate** (stibenyl) $CH_3CONHC_6H_4SbO_3HNa \cdot H_2O$.
Properties: Light yellow powder; antimony content is 35%; soluble in water.
Use: Medicine.

sodium para-**acetylaminophenylarsonate.** See arsacetin.

sodium acetylarsanilate. See arsacetin.

sodium acetylformate. See sodium pyruvate.

sodium acetylide (disodium acetylide) NaC⋮CNa.
Use: Intermediate.

sodium acetylsulfanilate $C_6H_4(NHCH_3CO)SO_3Na$.
Properties: White crystals. Soluble in water; slightly soluble in alcohol; insoluble in ether.
Derivation: By heating together glacial acetic acid and sulfanilic acid.
Use: Medicine.

sodium acid carbonate. See sodium bicarbonate.

sodium acid fluoride. See sodium bifluoride.

sodium acid phosphate. See sodium phosphate, monobasic.

sodium acid pyrophosphate. See sodium pyrophosphate, acid.

sodium acid sulfate. See sodium bisulfate.

sodium acid sulfite. See sodium bisulfite.

sodium acid tartrate. See sodium bitartrate.

Sodium "Aerofloat" Promoters. [57] Trademark for dry, water-soluble reagents containing the active promoting constituent of liquid "Aerofloat" promoters. Strong zinc sulfide promoters are also extensively used in flotation of gold, silver, and copper sulfide minerals in the presence of pyrite, which they do not actively promote.

sodium alginate. See algin; algin fibers.

sodium allylisopropylbarbiturate. See 5-allyl-5-isopropylbarbituric sodium.

sodium allylmercaptomethyl penicillin. See sodium penicillin O, under penicillin.

sodium 5-allyl-5(1-methylbutyl)-2-thiobarbiturate. See thiamylal sodium.

sodium alum. See aluminum sodium sulfate.

sodium aluminate $Na_2Al_2O_4$ or $NaAlO_2$.
Properties: White powder. Soluble in water; insoluble in alcohol; aqueous solution strongly alkaline. M.p. 1650°C.
Derivation: By heating bauxite with sodium carbonate and extracting the sodium aluminate with water.
Grades: Technical; reagent; also 27° Bé solution.
Containers: Wooden kegs; bags; multiwall paper sacks.
Uses: Mordant; zeolites; water purification; sizing paper; manufacture of milk-glass, soap and cleaning compounds; hardening building stones.
Shipping regulations (solution): Corrosive liquid. White label.*

sodium-aluminum chloride. See aluminum-sodium chloride.

sodium aluminum fluoride. See cryolite, synthetic.

sodium aluminum hydride $NaAlH_4$.
Properties: White crystalline material, sp.gr. 1.24 g/cc. Stable in dry air at room temperature but very sensitive to moisture. Begins to melt at 183°C with decomposition to evolve hydrogen. Soluble in tetrahydrofuran, dimethyl "Cellosolve."
Derivation: By reaction of aluminum chloride with sodium hydride.
Containers: Glass bottles.
Uses: As a reducing agent similar to lithium aluminum hydride.
Shipping regulations: Flammable solid. Yellow label.*

sodium aluminum phosphate
$NaH_{14}Al_3(PO_4)_8 \cdot 4H_2O$.
Properties: Dry, white powder having a neutralizing strength or baking value of 100% expressed in units of bicarbonate of soda.
Containers: Drums and bags.
Uses: As a slow acting baking acid in the manufacture of preleavened flour mixtures, in baking powder and in commercial bakery cake operations.
Shipping regulations: None.*

sodium aluminum silicofluoride $Na_5Al(SiF_6)_4$.
Sodium aluminum fluosilicate.
Properties: White powder; somewhat soluble in cold water; corrosive to galvanized iron.
Uses: Moth proofing and insecticides; also to obtain acid medium in dyebath.

sodium aluminum sulfate. See aluminum sodium sulfate.

sodium amalgam Na_xHg_x.
Properties: A silver-white, porous, crystalline mass, containing 2-10% of metallic sodium. Decomposes water.
Derivation: Mercury is heated to about 200°C and sodium, in small pieces, added slowly. Also formed at one stage of process for making chlorine and sodium hydroxide by mercury cell process.
Grades: 2, 3, 4, 5, 6, 7, 8, 9, 10, and 20%.
Containers: Glass bottles.
Uses: Preparation of hydrogen; reduction of metal halogen compounds and organic compounds; reagent in analytical chemistry.
Fire hazard: Dangerous.
Shipping regulations: Flammable solid. Yellow label.*

sodium amide (sodamide) $NaNH_2$.
Properties: White crystalline powder. Decomposes in water.
Constants: M.p. 210°C; b.p. 400°C.
Derivation: Dry ammonia gas is passed over metallic sodium at 350°C.
Grades: Technical.
Containers: Wooden kegs.
Use: Manufacture of sodium cyanide; organic synthesis.
Fire hazard: Dangerous.
Shipping regulations: Flammable solid. Yellow label.*

sodium para-**aminobenzenesulfonate.** See sodium sulfanilate.

sodium para-**aminobenzoate** (PABA sodium) $NH_2C_6H_4COONa$. Sodium salt of para-aminobenzoic acid.
Properties: Crystals; soluble in water; slightly soluble in alcohol; nearly insoluble in ether.
Use: Medicine.

sodium para-**aminohippurate**
$NH_2C_6H_4CONHCH_2COONa$. Water soluble salt.
Grade: N.N.D.
Use: Clinical test (medicine).

sodium aminophenylarsonate. See sodium arsanilate.

sodium para-**aminosalicylate** (PAS sodium) $NaOOCC_6H_3(OH)NH_2$.
Properties: White to pale yellow crystalline powder. Practically odorless, with sweet saline taste. Freely soluble in water; sparingly soluble in alcohol; very slightly soluble in ether and chloroform; 2% solution is clear and colorless and has pH 7.0-7.5. Solutions decompose slowly and darken in color.
Grade: U.S.P. XVI.
Containers: Drums.
Use: Medicine.

sodium-ammonium hydrogen phosphate. See sodium-ammonium phosphate.

sodium-ammonium phosphate (microcosmic salt; sodium-ammonium hydrogen phosphate; salt of phosphorus; phosphorus salt) $NaNH_4HPO_4 \cdot 4H_2O$.
Properties: Transparent, colorless, odorless, efflorescent, monoclinic crystals. Gives off water and ammonia on heating, leaving $NaPO_3$. Soluble in water; insoluble in alcohol.
Constants: Sp.gr. 1.57; m.p. about 79°C with decomposition.
Derivation: Mixing solutions of sodium phosphate and ammonium chloride.
Grades: Granular; C.P.; technical.
Containers: 1-, 5-lb bottles; 1-, 5-lb cans;

25-, 50-, 150-lb kegs; 300-lb barrels.
Use: Analytical reagent.

sodium-ammonium sulfate (ammonium-sodium sulfate) $Na_2SO_4 \cdot (NH_4)_2SO_4 \cdot 4H_2O$.
Properties: White powder; soluble in water.

sodium anilinearsonate. See sodium arsanilate.

sodium anilinesulfonate. See sodium sulfanilate.

sodium antimonate (antimony sodiate) $NaSbO_3$. Other forms are sodium metaantimonate $2NaSbO_3 \cdot 7H_2O$ and sodium pyroantimonate $Na_2H_2Sb_2O_7 \cdot H_2O$.
Properties: White, granular powder; slightly soluble in water and alcohol. Insoluble in dilute alkalies, mineral acids, but soluble in tartaric acid.
Grades: Technical.
Containers: 500-lb barrels; 100-lb paper lined burlap bags.
Uses: Opacifier in enamels, for cast iron and glass; ingredient of acid-resisting sheet steel enamels.

sodium antimony bis-pyrocatechol-2, 4-disulfonate. See stibophen.

sodium antimonyl tartrate. See antimony sodium tartrate.

sodium arsanilate (sodium anilinearsonate; sodium aminophenylarsonate) $C_6H_4NH_2(AsO \cdot OH \cdot ONa)$, often with 1 or more H_2O.
Properties: White, crystalline, odorless powder; faint salty taste; poisonous! May produce optic atrophy. Soluble in water.
Derivation: By dissolving arsanilic acid in sodium carbonate solution and crystallizing.
Method of purification: Recrystallization.
Grades: Technical; medicinal.
Containers: Tins; glass bottles.
Uses: Medicine; veterinary medicine; organic synthesis.

sodium arsenate $Na_3AsO_4 \cdot 12H_2O$.
Properties: Clear, colorless crystals; mild alkaline taste; poisonous! Soluble in water; slightly soluble in alcohol; insoluble in ether.
Constants: Sp.gr. 1.7593; m.p. 86°C.
Derivation: (a) Arsenic trioxide is heated with sodium nitrate, dissolved in water and crystallized. (b) Arsenic trioxide is dissolved in sodium carbonate solution, sodium nitrate is added, the solution evaporated to dryness, the residue calcined, dissolved in water and crystallized.
Method of purification: Recrystallization.
Impurities: Sodium binarsenate.
Grades: Highest purity; pure crystals; pure dry; C.P.; technical (60% arsenic pentoxide).
Containers: 1-lb bottles; wooden kegs; multiwall paper sacks; 100-lb drums.
Uses: Medicine; insecticides; dry colors; textiles (mordant and assist in dyeing and printing); making other arsenates; germicide.
Shipping regulations: Poison, class B. Poison label.*

sodium arsenite (sodium metaarsenite) $NaAsO_2$.
Properties: Grayish-white powder, which absorbs carbon dioxide from the air; poisonous! Soluble in water. Sp.gr. 1.87.
Derivation: Arsenic trioxide is dissolved in a solution of sodium carbonate or hydroxide and boiled for some time.
Grades: Crude; pure; 75% arsenious oxide.
Containers: 1-, 5-lb bottles; drums.
Uses: Manufacture of arsenical soaps for taxidermists; antiseptic; dyeing; insecticides; hide preservation; herbicide.
Shipping regulations: Solution: Poison, class B. Poison label.*

sodium arsphenamine (sodium salt of 3-diamino-4-dihydroxy-1-arsenobenzene) $C_{12}H_{10}O_2N_2As_2Na_2$.
Properties: Bright yellow powder containing not less than 19% arsenic; very unstable in air; poisonous! Soluble in water.
Containers: Ampules.
Use: Medicine.

sodium ascorbate
$CH_2OH(CHOH)_2COHCOHCOONa$. The sodium salt of ascorbic acid.
Properties: White odorless crystals or powder; m.p. 218°C (dec); stable under ordinary conditions; soluble in water; insoluble in alcohol.
Derivation: By reacting ascorbic acid with hot methyl alcohol and a warm solution of sodium methylate and stirring until the stable crystalline salt is formed.
Containers: Bottles; drums.
Grade: N.N.D.
Use: Vitamin therapy; antioxidant for foods.

sodium auribromide. See gold-sodium bromide.

sodium aurichloride. See gold-sodium chloride.

sodium aurothiosulfate. See gold-sodium thiosulfate.

sodium azide NaN_3.
Properties: Highly poisonous, colorless, hexagonal crystals; decomposes at about 300°C; sp.gr. 1.846; soluble in water and in liquid ammonia; slightly soluble in alcohol; insoluble in ether.
Use: Preparation of lead azide explosives, medicinals.
Shipping regulations: Poison, class B. Poison label.*

sodium benzenesulfonchloramine. See chloramine-B.

sodium benzoate C_6H_5COONa.
Properties: White, amorphous, crystalline or granular, odorless powder; sweetish, astringent taste. Soluble in water and alcohol.
Derivation: Benzoic acid is neutralized with sodium bicarbonate solution, the solution filtered, concentrated and allowed to crystallize.
Method of purification: Recrystallization.
Grades: U.S.P. XVI; technical.
Containers: Cartons; bottles; boxes; kegs; 5 to 175-lb drums.

Uses: Food preservative (its use for this purpose being limited by law in most countries); antiseptic; medicine; tobacco; pharmaceutical preparations; intermediate for manufacture of dyes; rust and mildew inhibitor.
Shipping regulations: None.*

sodium benzosulfimide. See saccharin, sodium.

sodium benzyl succinate $C_6H_5CH_2O_2C(CH_2)_2COONa$.
Properties: White amorphous or crystalline powder having a slight benzyl odor and cool salty taste. Soluble in hot and cold water.

sodium-beryllium fluoride. See beryllium-sodium fluoride.

sodium bicarbonate (baking soda; sodium acid carbonate) $NaHCO_3$.
Properties: White powder or crystalline lumps; cooling, slightly alkaline taste. Soluble in water; insoluble in alcohol. Stable in dry air, but slowly decomposes in moist air.
Constants: Sp.gr. 2.20; m.p., loses carbon dioxide at 270°C.
Derivation: Principally, by treating a saturated solution of soda ash with carbon dioxide to precipitate the less soluble bicarbonate; also by purifying the crude product from the Solvay process.
Method of purification: Recrystallization.
Impurities: Sulfuric acid; chlorine; silica; heavy metals; sodium thiosulfate; sodium carbonate; potassium salts; ammonium salts.
Grades: Commercial; pure; reagent; highest purity; C.P.; U.S.P. XVI.
Containers: Bottles and cartons; 100-lb bags; drums.
Uses: Manufacture of effervescent salts and beverages, artificial mineral waters, baking powder; reagent in analytical chemistry; gold and platinum plating; tanning industry; treating wool and silk; fire extinguishers; medicine; ceramics; preserving butter; prevention of timber mold.
Shipping regulations: None.*

sodium bichromate. See sodium dichromate.

sodium bifluoride (sodium acid fluoride) $NaHF_2$.
Properties: White, crystalline powder; poisonous! Soluble in water. Decomposed on heating.
Grades: Technical.
Containers: Tins; 125-, 375-lb drums.
Uses: For neutralizing last traces of alkali in laundry rinsing operations; preservative, for zoological and anatomical specimens; etching glass; antiseptic; production of tinplate.
Shipping regulations: None.*

sodium binoxide. See sodium peroxide.

sodium biphosphate. See sodium phosphate, monobasic.

sodium bismuthate $NaBiO_3$.
Properties: Yellow or brown, amorphous powder. Slightly hygroscopic.
Grades: Technical.
Uses: Analysis (testing for manganese in iron, steel and ores); reagent; pharmaceuticals.

sodium bisulfate (sodium acid sulfate; niter cake; nitre cake) $NaHSO_4$ or $NaHSO_4 \cdot H_2O$.
Properties: Colorless crystals or white, fused lumps; aqueous solution is strongly acid. Soluble in water.
Constants: Sp.gr. 2.742; m.p. 315°C with decomposition.
Derivation: A by-product in the manufacture of hydrochloric and nitric acids.
Method of purification: Recrystallization.
Impurities: Heavy metals; chlorides; arsenic; potassium bisulfate; sulfuric acid.
Grades: Pure crystals; pure fused; pure dry; reagent; crude; C.P.; technical; cakes; ground.
Containers: 1-, 5-lb bottles; 100-lb kegs; various sizes of drums to 500-lb barrels; bulk in cars.
Uses: Flux for decomposing minerals; substitute for sulfuric acid in dyeing; disinfectant; dyeing; manufacture of sodium sulfate and soda alum; liberating carbon dioxide in CO_2 baths; in thermophores; carbonizing wool; manufacture of magnesia cements, paper, soap, perfumes, brick, glue; in general, as a strong solid acid.
Shipping regulations: None.*

sodium bisulfide. See sodium hydrosulfide.

sodium bisulfite (sodium acid sulfite; leucogen) $NaHSO_3$.
Properties: White crystals or crystalline powder; slight sulfurous odor; disagreeable taste; soluble in water; insoluble in alcohol. Unstable in air.
Constants: Sp.gr. 1.48; m.p., decomposes.
Derivation: Sodium carbonate solution is saturated with sulfur dioxide gas and the solution crystallized.
Method of purification: Recrystallization.
Grades: Crystals; pure dry; commercial dry (this consists chiefly of sodium meta-bisulfite); reagent; commercial solution 35°Bé; powder (67% SO_2); C.P.; U.S.P. XVI.
Containers: Commercial, dry: 25-lb boxes; 100-lb kegs; 500-lb barrels; 500-, 600-lb drums; solution: 500-lb barrels. C.P: 1-, 5-lb cans and bottles.
Uses: Chemicals (sodium salts, cream of tartar); dyes; intermediates; organic chemicals; perfumery; wood pulp (digestion); leather (depilatory, tanning); textiles (antichlor, mordant, discharge); food preservative; bleaching straw and cork; photographic reducing agent; fermentation industries; medicine; glucose and sugar syrups; brewing (cask sterilization); copper and brass plating; rubber (coagulating latex); general antiseptic; analytical reagent; pesticides; source of sulfur dioxide.
Shipping regulations: None.*

sodium bitartrate (acid sodium tartrate) $NaHC_4H_4O_6 \cdot H_2O$.
Properties: White, crystalline powder.

Soluble in water (aqueous solution is acid); slightly soluble in alcohol.
Grades: C.P.; technical; reagent.
Uses: Analysis (testing for potassium); effervescing mixtures.

sodium borate (sodium tetraborate; sodium borate (2, 4, 7); sodium pyroborate; borax) $Na_2B_4O_7 \cdot 10H_2O$. See also borax, dehydrated; borax, pentahydrate.
Properties: White crystals or powder; odorless; sp.gr. 1.73 (20/4°C); loss of water of crystallization when heated, with melting, between 75 and 200°C; fuses to a glassy mass at red heat (borax glass); effloresces slightly in warm, dry air. Soluble in water and glycerol; insoluble in alcohol.
Derivation: In California, by fractional crystallization from Searles Lake brine; or by solution of kernite ore followed by crystallization. Also obtained from colemanite, natural borax, ulexite, and other borates.
Grades: Crystal; granulated; powdered (refined, U.S.P. XVI); C.P.; technical.
Containers: 1-, 5-lb bottles; 100-lb kegs; 100-lb sacks; 340-, 390-lb barrels; bulk (granulated).
Uses (in approximate order of volume): Glass; porcelain enamel; starch and adhesives; detergents; herbicides; fertilizers; rust inhibitors; pharmaceuticals; leather; photography; bleaches; paint; boron compounds; smelters; insulation materials.

sodium borate perhydrate $NaBO_2 \cdot H_2O_2$.
Properties: Free-flowing, white powder. Dissolves readily in water to produce mildly alkaline peroxide solution. Active oxygen content, 15.5% min.
Containers: 125-lb nonreturnable fiber drums.
Uses: In dentrifices; hair wave kits; bleaching; oxidation of dyestuffs; stain remover for plastic dishes.
See also sodium perborate.

sodium boroformate
$NaH_2BO_3 \cdot 2HCOOH \cdot 2H_2O$.
Properties: White, crystalline product. Soluble in water; m.p. 110°C.
Grades: Technical.
Containers: Barrels; bags; kegs.
Uses: An excellent buffering agent toward both acid and alkali in the range of pH 8.5. Useful in textile treating and tanning baths.

sodium borohydride $NaBH_4$.
Properties: White crystalline powder; soluble in water, ammonia, amines, and dimethyl ether of diethylene glycol; insoluble in other ethers, hydrocarbons and alkyl chlorides. Sp.gr. 1.07; hygroscopic; stable in dry air to 300°C; decomposes slowly in moist air or in vacuum at 400°C.
Derivation: By reaction of sodium hydride and trimethyl borate.
Grades: Technical; powdered; pellets.
Containers: Glass bottles; and in polyethylene bags in from 1-gal to 55-gal metal containers.
Uses: Source of hydrogen and other borohydrides. Reduces aldehydes, ketones and acid chlorides; blowing agent for plastics.
Shipping regulations: Flammable solid. Yellow label.*

sodium borohydride SWS. A stabilized water solution of sodium borohydride.
Properties: Sp.gr. 1.4; viscosity (23°C) 79.0 cps; approx. composition 12% $NaBH_4$, 40% NaOH, 48% H_2O; decomposes very slowly with evolution of hydrogen.
Containers: 1-gal polyethylene containers and 55-gal drums with polyethylene liners.
Use: Reducing agent for aldehydes and ketones.
Shipping regulations: Corrosive liquid. White label.*

sodium bromate $NaBrO_3$.
Properties: White odorless crystals or crystalline powder. Soluble in water; insoluble in alcohol.
Constants: Sp.gr. 3.339; m.p. 381°C (dec).
Derivation: By passing bromine into a solution of sodium carbonate, sodium bromide and sodium bromate being formed which are separated by crystallization.
Method of purification: Recrystallization.
Grades: Pure; reagent; C.P.
Containers: Glass bottles; 25-lb tin boxes.
Use: Analytical reagent.
Fire hazard: Dangerous.
Shipping regulations: Oxidizing material. Yellow label.*

sodium bromide (a) NaBr, (b) $NaBr \cdot 2H_2O$.
Properties: White, crystalline powder or granules; saline and somewhat bitter taste; absorbs moisture from the air, becoming very hard. Keep well stoppered. Soluble in water; moderately soluble in alcohol.
Constants: Sp.gr. (a) 3.208, (b) 2.176; m.p. (a) 757.7°C; b.p. (a) 1455°C.
Derivation: Occurs naturally in some salt deposits. Made synthetically by first causing iron to react with bromine and water. The resulting ferroso-ferric bromide is dissolved in water, sodium carbonate added, the solution filtered and evaporated.
Method of purification: Recrystallization.
Grades: C.P.; crystalline; powdered; commercial; pure; highest purity.
Containers: 1-, 5-lb bottles; 25-, 50-, 100-lb boxes; 100-lb kegs; 500-lb barrels; drums.
Uses: Photography; medicine; preparation of bromides; organic chemicals; source of bromine.
Shipping regulations: None.*

sodium cacodylate (sodium dimethylarsenate) $(CH_3)_2AsOONa \cdot 3H_2O$.
Properties: White, amorphous crystals or powder; deliquescent; poisonous! Melts about 60°C. Soluble in water and alcohol.
Derivation: Oxidation and neutralization of cacodyl oxide.
Grades: Technical.
Containers: Vials; bottles.
Use: Medicine.

Shipping regulations: Poison, class B. Poison label.*

sodium caprylate $CH_3CH_2(CH_2)_4CH_2COONa$.
Properties: Cream-colored granules. Freely soluble in water; sparingly soluble in alcohol.
Grade: N.N.D.
Use: Medicine (topical).

sodium carbolate. See sodium phenate.

sodium carbonate (soda). See soda ash; sal soda; sodium bicarbonate; sodium carbonate, monohydrate; sodium sesquicarbonate.

sodium carbonate, anhydrous. See soda ash.

sodium carbonate, decahydrate. See sal soda.

sodium carbonate, monohydrate (crystal carbonate; soda monohydrate; soda crystals).
Properties: A crystalline sodium carbonate containing one molecule of water ($Na_2CO_3 \cdot H_2O$). White; odorless; small crystals or crystalline powder; alkaline taste; sp.gr. 1.55. Soluble in water. M.p. 109°C (loses water), 851°C.
Grades: U.S.P. XVI; technical.
Containers: 100-lb bags.
Uses: Medicine; photography; cleaning and boiler compounds; pH control of water.
Shipping regulations: None.*

sodium carbonate peroxide
$2Na_2CO_3 \cdot 3H_2O_2$ or $Na_2CO_3 \cdot H_2O_2 \cdot \frac{1}{2}H_2O$.
Properties: White crystalline powder. Soluble in water. Stable at room temperature when dry; decomposes rapidly at 100°C, evolving oxygen. Active oxygen content 14% min. Soluble in water; pH of saturated solution 10.2.
Derivation: Crystallization from solution of soda ash and hydrogen peroxide.
Grade: Technical.
Containers: 100-lb fiber drums.
Uses: Source of hydrogen peroxide; household detergents; dental cleansers; bleaching and dyeing; modification of starch.

sodium carboxymethylcellulose (CMC; sodium cellulose glycolate; cellulose gum). A synthetic cellulose gum containing 0.4 to 1.5 sodium carboxymethyl groups ($-CH_2COONa$) per glucose unit of the cellulose. These added groups have each replaced an hydroxyl hydrogen to form ether linkages along the cellulose chain.
Properties: Colorless, odorless, tasteless, non-toxic, hygroscopic powder or granules, readily dispersible in hot or cold water; pH (1% solution) 6.5-8.0; stable in pH range 2-10. Viscosity (1% solution) 5-2000 cps, depending upon the number of hydroxyl radicals of the cellulose that have been etherified. Insoluble in most organic liquids. In water solution it is a colloidal electrolyte which reacts with heavy metal salts to give films insoluble in water, colorless, transparent, relatively tough, and unaffected by common organic solvents, oils, fats, and greases. Many of the colloidal properties are similar or superior to those of water-soluble starches, gelatins, gums, and other hydrophilic colloids in thickening and stabilizing solutions and dispersing and suspending particles. It is also thixotropic.
Derivation: By reaction of alkali cellulose and sodium chloroacetate.
Grades: Crude; technical (about 75% pure); high viscosity; low viscosity; semirefined; refined (99.5+% pure); U.S.P. XVI.
Containers: 13-, 51-gal drums; 50-lb bags.
Uses: As water binder, thickener, suspending agent, film-former, and emulsion stabilizer for detergents and soaps; textile and paper sizing; latex paints; drilling muds; medicine; foods (especially ice cream); pharmaceuticals; adhesive; as a protective colloid in general, replacing or in combination with natural water-soluble gums.

sodium caseinate. See casein-sodium.

sodium catechol disulfonate. See disodium 1,2-dihydroxybenzene-3,5-disulfonate.

sodium cellulose glycolate. See sodium carboxymethylcellulose.

sodium chlorate $NaClO_3$.
Properties: Colorless, odorless crystals; cooling, saline taste; must not be triturated with any combustible substance. Soluble in water and alcohol. Sp.gr. 2.490; m.p. 255°C; b.p., decomposes.
Derivation: A concentrated acid solution of sodium chloride is heated and electrolyzed, the chlorate crystallizing out.
Method of purification: Recrystallization.
Grades: Technical; C.P.; crystals; powder.
Containers: Bottles; cartons; boxes; 100 to 600-lb drums.
Uses: Oxidizing agent and bleach (especially to make chlorine dioxide) for paper pulps; ore processing; herbicide and defoliant; medicine; substitute for potassium chlorate, being more soluble in water; matches; explosives; recovery of bromine from natural brines; leather tanning and finishing; textile mordant; to make perchlorates.
Warning: Strong oxidant, contact with other material may cause fire. MCA warning label.
Shipping regulations: Oxidizing material. Yellow label.*

sodium chloraurate. See gold-sodium chloride.

sodium chloride (table salt; sea salt; common salt; rock salt) NaCl.
Properties: Colorless, transparent crystals or white, crystalline powder; occurs in nature as the mineral halite (q.v.); somewhat hygroscopic; soluble in water and glycerol; very slightly soluble in alcohol. Sp.gr. 2.161; m.p. 804°C; b.p. 1490°C.
Derivation: (a) By solution of rock salt in water, filtration, and crystallization. (b) By evaporation and crystallization of naturally occurring brines. (c) By evaporation of sea water by the heat of the sun and crystallization.
Method of purification: Recrystallization.
Impurities: Sulfates; heavy metals; alkaline earths; magnesium salts; ammonium salts.

*See "I.C.C. Shipping Regulations," page xiii.
Reference numbers refer to name of manufacturer. See "List of Manufacturers," page v.

Grades: Highest purity medicinal, crystals; highest purity, dried; highest purity, fine powder; highest purity, fused; reagent; reagent, fused; sea evaporated; ground; powdered; table salt; rock salt; C.P.; U.S.P. XVI; single pure crystals.
Containers: Bottles; cartons; kegs; 100-, 200-lb bags; 280-lb barrels; bulk in cars.
Uses: Chemical (sodium salts, especially soda ash, and hydrochloric acid, chlorine, metallic sodium); dyes; ceramic glazes; metallurgy (silver, gold, copper, zinc); refrigeration; glass; leather; food preservative; mineral waters; soap (salting out); cattle foods; fertilizers; production of sodium light for polariscopic, spectroscopic and other similar work; medicine; analytical chemistry; photography; paper; seasoning of foods; source of chlorates; butter; salting out dyestuffs; freezing mixtures. Single crystals are used for spectroscopy, ultraviolet and infrared transmission.
Shipping regulations: None.*

sodium chlorite $NaClO_2$.
Properties: White crystals or crystalline powder; slightly hygroscopic; soluble in water.
Grades: Technical; reagent.
Containers: Drums.
Uses: For improving taste and odor of potable water (as an oxidizing agent); bleaching agent for textiles, paper pulp, edible and inedible oils, shellacs, varnishes, waxes and straw products; oxidizing agent; reagent.
Caution! Strong, oxidizing material; decomposes with evolution of heat at about 175°C. Extremely explosive in contact with combustible material.
Shipping regulations: Oxidizing material. Yellow label.*

sodium chloroacetate $ClCH_2COONa$.
Properties: White nonhygroscopic powder, odorless, easier to handle than chloroacetic acid. Soluble in water; slightly soluble in methanol; insoluble in acetone, benzene, ether, and carbon tetrachloride.
Grades: Technical.
Containers: 300-lb drums.
Use: Manufacture of weed killers, dyes, vitamins, pharmaceuticals, other organics; also a defoliant.

sodium chloroaurate. See gold-sodium chloride.

sodium para-**chloro**-meta-**cresolate.** A water-soluble preservative for cutting oils.

sodium chloroplatinate. See platinic-sodium chloride.

sodium chloroplatinite. See plantinous-sodium chloride.

sodium ortho-**chlorotoluene**-para-**sulfonate** $NaSO_3C_6H_3(CH_3)Cl$.
Properties: Gray to light tan powder; soluble in water and organic solvents.
Derivation: Made by sulfonation of orthochlorotoluene, neutralized to form the sodium salt.
Containers: 250-lb drums.
Uses: In the synthesis of dyes, intermediates and drugs.

sodium chromate $Na_2CrO_4 \cdot 10H_2O$.
Properties: Yellow, translucent, efflorescent crystals. Soluble in water; slightly soluble in alcohol.
Constants: Sp.gr. 1.483; m.p. 19.92°C.
Derivation: Chrome iron ore is melted in a reverberatory furnace with lime and soda, in presence of air. The melt is dissolved in water, a small amount of sodium carbonate added, the solution decanted, acidified with acetic acid, concentrated and crystallized.
Method of purification: Recrystallization.
Grades: Pure neutral; highest purity; technical; C.P.; reagent.
Containers: Bottles; 100-lb bags; 100-, 400-lb drums; barrels.
Uses: Inks; dyeing; paint pigment; leather tanning; other chromates; protection of iron against corrosion and rusting; wood preservative; starting material for making chromium metal.
Shipping regulations: None.*
Anhydrous sodium chromate is also available commercially.

sodium chromate, tetrahydrate (chromate of soda) $Na_2CrO_4 \cdot 4H_2O$.
Properties: Yellow crystals; deliquescent; soluble in water.
Grade: Technical.
Containers: Multiwall paper bags, 100 lb net; fiber drums, 400 lbs net; special packages upon request.
Uses: Pigments manufacture; corrosion inhibition; leather tanning; raw material for other chromium chemicals.
Hazards: Harmful dust; avoid breathing or prolonged contact; wash thoroughly after contact; avoid ingestion.

sodium cinchophenate $C_6H_5C_9H_5NCOONa \cdot H_2O$.
Properties: Yellowish white powder; solubility in water, 1 part in 3 parts water.
Derivation: From cinchophen by neutralizing with sodium hydroxide.
Grade: Pharmaceutical.
Containers: Drums.
Use: In medicine.
Shipping regulations: None.*

sodium citrate $C_6H_5O_7Na_3 \cdot 2H_2O$.
Properties: White crystals or granular powder; odorless; stable in air; pleasant acid taste. Soluble in water; insoluble in alcohol.
Constants: M.p., loses $2H_2O$ at 150°C; b.p. decomposes at red heat.
Derivation: Sodium sulfate solution is treated with calcium citrate, filtered, concentrated and crystallized.
Method of purification: Recrystallization.
Impurities: Calcium citrate; sodium sulfate.
Grades: Highest purity, medicinal; pure; commercial; C.P.; U.S.P. XVI; technical.
Containers: Bottles; 50-lb cans; 100-, 250-lb drums; bags.

Uses: Medicine; soft drinks; photography; in special cheeses; in electroplating.
Shipping regulations: None.*

sodium coerulin sulfate. See indigo carmine.

sodium columbate. Obsolete name for sodium niobate.

sodium copper chloride. See copper sodium chloride.

sodium-copper cyanide (copper-sodium cyanide).
Properties: White, crystalline, double salt of copper cyanide and sodium cyanide. Copper content (as Cu) 28.7%; free NaCN 0.4-2.0%.
Containers: 100-lb drums.
Uses: For preparing and maintaining cyanide copper plating baths based on sodium cyanide.

sodium cyanate NaOCN.
Properties: White crystalline powder; soluble in water; insoluble in alcohol and ether.
Special specifications: 0.5% max moisture content; purity 94.5% min.
Containers: Fiber drums; bottles.
Uses: Organic synthesis; heat treating of steel; intermediate for manufacture of medicinals.
Shipping regulations: None.*

sodium cyanide NaCN.
Properties: White deliquescent, crystalline powder; exceedingly poisonous! Soluble in water; slightly soluble in alcohol; m.p. 563°C. The aqueous solution is strongly alkaline and decomposes rapidly on standing.
Derivation: (a) Sodamide is produced from sodium and ammonia. The sodamide is heated with charcoal and the resultant soda cyanamide is then heated with an excess of charcoal resulting in the formation of sodium cyanide. (b) By the fusion of calcium cyanamide, common salt and a small amount of calcium carbide. (c) By absorption of hydrocyanic acid in a solution of sodium hydroxide.
Method of purification: Recrystallization.
Impurities: Sodium cyanate; sodium carbonate.
Grades: 30% solution; 73 to 75%; 96 to 98%; reagent; technical; briquettes; granular.
Containers: 25-lb packages; 100-, 160-, 200-lb drums.
Uses: Extraction of gold and silver from ores; electroplating; heat treatment of metals; making hydrocyanic acid; insecticide; cleaning metals; fumigation; manufacture of dyes and pigments; nylon intermediates; chelating compounds.
Danger: Contact with acid liberates poisonous gas. (MCA warning label) Keep dry.
Shipping regulations: Poison, class B. Poison label (both in solid and liquid form).*

sodium cyclamate (sodium cyclohexylsulfamate) $C_6H_{11}NHSO_3Na$.
Properties: White, crystalline, practically odorless powder with very sweet taste. Freely soluble in water; practically insoluble in alcohol, benzene, chloroform and ether; pH (10% solution) 5.5 - 7.5. Sweetening power approximately 30 times that of sucrose.
Grade: N.F. XI.
Containers: 100-lb drums.
Uses: Sweetening agent in certain soft drinks and in low-calorie and diabetic diets.

sodium cyclohexylsulfamate. See sodium cyclamate.

sodium decaphosphate. A sodium phosphate glass similar in a general way to sodium hexametaphosphate, but having a higher ratio of Na_2O to P_2O_5. There is considerable confusion and uncertainty as to the precise meaning of this term. See sodium metaphosphate.

sodium dehydroacetate $C_8H_7NaO_4 \cdot H_2O$.
Properties: Tasteless white powder. Soluble in water and propylene glycol. Insoluble in most organic solvents. See also dehydroacetic acid.
Uses: Fungicide; plasticizer; antienzyme toothpaste; pharmaceutical.

sodium deoxycholate. See deoxycholic acid.

sodium 3,5-diacetamido-2,4,6-triiodobenzoate. See sodium diatrizoate.

sodium diacetate $CH_3COONa \cdot x(CH_3COOH)$, anhydrous, or $CH_3COONa \cdot x(CH_3COOH) \cdot yH_2O$, technical.
Properties: White crystals, soluble in water; slightly soluble in alcohol; insoluble in ether.
Containers: 5- to 250-lb drums.
Uses: Buffer; mold inhibitor; souring agent; intermediate for acid salts; mordants, varnish hardeners; antitarnishing agents.

sodium diatrizoate (sodium 3,5-diacetamido-2,4,6-triiodobenzoate) $C_6I_3(COONa)(NHCOCH_3)_2$.
Properties: White crystals; soluble in water. Solutions are radio-opaque.
Grade: U.S.P. XVI (as solution for injection).
Use: Radiopaque medium.

sodium alpha, beta-**dichloroisobutyrate.** Used as a plant growth regulator.

sodium dichloroisocyanurate (sodium salt of dichloro-s-triazine-2,4,6-trione) NaNC(O)NClC(O)NClCO.
Properties: White, slightly hygroscopic, crystalline powder; loose bulk density (approx.) powder 37 lb/cu.ft., granular 57 lb/cu.ft. Active ingredient: approx 60% available chlorine.
Containers: 200-lb fiber drums.
Uses: Active ingredient in household dry bleaches, dishwashing compounds, scouring powders, detergent-sanitizers, swimming pool disinfectants, water treatment, replacement of calcium hypochlorite.

sodium 2,4-dichlorophenoxyacetate (2,4-D, sodium salt) $C_6H_3(OCH_2COONa)Cl_2$.
Properties: Crystalline solid. Decomposes at 215°C. Slightly soluble in water.
Use: Herbicide.

sodium 2,4-dichlorophenoxyethyl sulfate. See SES.

sodium 2,2-dichloropropionate (alpha, alpha-dichloropropionic acid sodium salt) CH_3CCl_2COONa.
Properties: Crystals; decompose 174-176°C; salty taste. Corrosive to iron. Soluble in water; aquous solutions hydrolyze above 70°C.
Caution: Causes skin irritation. MCA warning label.
Use: Herbicide, narrow-leafed grasses.

sodium dichromate (sodium bichromate; bichromate of soda) $Na_2Cr_2O_7 \cdot 2H_2O$.
Properties: Red or red-orange deliquescent crystals. Sp.gr. 2.52; m.p. 320°C; decomposes at 400°C; loses $2H_2O$ on prolonged heating at 105°C. Very soluble in water; insoluble in alcohol.
Derivation: (a) Manufactured from chromite ore by alkaline roasting and subsequent leaching; (b) action of sulfuric acid on sodium chromate.
Method of purification: Recrystallization.
Grades: Technical crystalline; technical liquor containing 69-70% $Na_2Cr_2O_7 \cdot 2H_2O$.
Containers: Crystals; multiwall paper bags, 100 lbs net; fiber drums, 400 lbs net. Liquor: tank cars or tank trucks.
Uses: Chemical reactant for oxidation reactions; raw material for other chromium chemicals; corrosion inhibitor; manufacture of pigments; tanning of leather; electroplating; mordanting.
Warning! Harmful dust. May cause rash or external ulcers. MCA warning label.

sodium dihydrogen phosphate. See sodium phosphate (monobasic).

sodium dihydroxyethylglycine. A chelating agent similar to EDTA.

sodium dimethylarsenate. See sodium cacodylate.

sodium dimethyldithiocarbamate $(CH_3)_2NCS_2Na$.
Properties: 40% solution is amber to light green; sp.gr. 1.17-1.20 (25/25°C).
Derivation: Reaction of dimethylamine, carbon disulfide and caustic.
Containers: 5-gal cans; 55-gal drums; tanks (solution).
Uses: Fungicide; corrosion inhibitor; rubber accelerator; intermediate.

sodium dinitro-ortho-cresylate.
Properties: Brilliant orange-yellow dye which may stain clothing and wood.
Derivation: By treating 4,6-dinitro-ortho-cresol with sodium hydroxide.
Caution: Continued inhalation of sprays containing the compound may cause toxic effects; dust masks or respirators should be worn as a safety precaution when spray is used on large scale.
Uses: Herbicide (control of mustard and other susceptible weeds); fungicide.
See also 4,6-dinitro-ortho-cresol.

sodium dioxide. See sodium peroxide.

sodium dipropionamido-2,4,6-triiodobenzoate. See sodium diprotrizoate.

sodium diprotrizoate (sodium 3,5-dipropionamido-2,4,6-triiodobenzoate) $(C_2H_5CONH)_2C_6I_3COONa$. Marketed as a solution. Aqueous solutions are clear and colorless; solution has a pH of 7.0-7.4.
Grade: U.S.P. XVI.
Use: Medicinal.

sodium dispersions. Stable suspensions of microscopic sodium particles in inert hydrocarbon or other media which boil at temperatures above the melting point of sodium (97.5°C), e.g., heptane, n-octane, toluene, xylene, naphtha, kerosene, mineral oil, n-butyl ether, etc. Particles range in size from submicron to 30 microns depending on the method of preparation. Dispersions contain up to 50% (by weight) of sodium metal.
Uses: For chemical reactions where advantages of controlled reaction rate, lower reaction temperature, increased yields, or substitution for more expensive reagents can be achieved.

sodium dithionate (sodium hyposulfate) $Na_2S_2O_6 \cdot 2H_2O$.
Properties: Large transparent crystals; bitter taste. Soluble in water; insoluble in alcohol and concentrated hydrochloric acid. Sp.gr. 2.175.
Grade: Technical.
Containers: Glass bottles.
Use: Chemical reagent.
Shipping regulations: None.*

sodium dithionite. See sodium hydrosulfite.

sodium diuranate (uranium yellow) $Na_2U_2O_7 \cdot 6H_2O$.
Properties: Yellow-orange solid, insoluble in water; soluble in dilute acids.
Derivation: By treating a solution of a uranyl salt with sodium hydroxide.
Uses: Ceramics, to produce colored glazes; manufacture of fluorescent uranium glass.

sodium dodecylbenzene sulfonate $C_{12}H_{25}C_6H_4SO_3Na$.
Properties: White to light yellow flakes, granules, or powder.
Derivation: Benzene is alkylated with a 12-carbon straight- or branched-chain olefin, to yield dodecylbenzene, which is then sulfonated with oleum and neutralized with caustic soda.
Grades: 30%, 40%, 65%, 85%, 96% active ingredient.
Uses: Detergent.

sodium dodecyldiphenyl oxide disulfonate $C_{24}H_{34}O(SO_3)_2Na_2$. M.p. 150°C (dec); dry form 90% min active; solution (45%), sp.gr. 1.1. Very soluble in water, strong acids, bases and electrolytes; stable to oxidation. See "Dowfax" 2A1.

sodium erythorbate (sodium isoascorbate) $NaC_6H_7O_6$. White, free-flowing crystals. Used as an antioxidant.

sodium ethoxide. See sodium ethylate.

sodium ethylate (sodium ethoxide; caustic alcohol) C_2H_5ONa.
Properties: White powder, sometimes having brownish tinge. Decomposed by moisture or moist living tissue into alcohol and sodium hydroxide.
Derivation: By carefully adding small amounts of sodium to absolute alcohol kept at a temperature of 10°C, heating carefully to 37.7°C, again carefully adding sodium, cooling to 10°C, and adding the same quantity of absolute alcohol as was used originally.
Uses: Organic synthesis.

sodium ethylene bisdithiocarbamate. See nabam.

sodium 2-ethylhexyl sulfoacetate $C_8H_{17}OOCCH_2SO_3Na$.
Properties: Light cream colored flakes, highly soluble. Good foaming properties and excellent resistance to hard water. Solutions practically neutral and stable to mineral acids.
Uses: Solubilizing agent, particularly adapted for soapless shampoo compositions; as an electroplating assistant by virtue of its ability to lower the surface tension of plating solutions.

sodium ethylmercurithiosalicylate. See thimerosal.

sodium 5-ethyl-5(1-methyl-1-butenyl) barbiturate. See vinbarbital sodium.

sodium ethyl oxalacetate. See ethyl sodium oxalacetate.

sodium ferric EDTA. See ethylenediaminetetraacetic acid salts.

sodium ferricyanide (red prussiate of soda) $Na_3Fe(CN)_6 \cdot H_2O$.
Properties: Ruby-red, deliquescent crystals; poisonous! Soluble in water; insoluble in alcohol.
Derivation: Chlorine is passed into sodium ferrocyanide solution, crystals of the ferricyanide separating out.
Method of purification: Recrystallization.
Impurities: Sodium ferrocyanide; sodium chloride.
Grades: Technical; C.P.
Containers: 1-, 5-lb bottles; wooden kegs.
Uses: Production of pigments; dyeing; printing.
Shipping regulations: None.*

sodium ferrocyanide (yellow prussiate of soda) $Na_4Fe(CN)_6 \cdot 10H_2O$.
Properties: Yellow, semi-transparent crystals. Soluble in water; insoluble in alcohol. Sp.gr. 1.458.
Derivation: "Spent oxide" from illuminating gas manufacture is treated first with water and then with carbon disulfide to remove ammonium and other soluble salts and sulfur. The residue is mixed with lime and heated in closed pans yielding ammonia and calcium ferrocyanide solution. A boiling solution of sodium chloride is added and the precipitate formed is heated with a solution of sodium carbonate. The solution is filtered, concentrated and crystallized.
Method of purification: Recrystallization.
Grade: Technical.
Containers: Bags.
Uses: Photography; manufacture of sodium ferricyanide; blue pigments; blueprint paper; metallurgy; tanning; dyes.
Shipping regulations: None.*

sodium fluoaluminate. See cryolite, synthetic.

sodium fluoborate $NaBF_4$.
Properties: White powder; bitter acid taste; slowly decomposed by heat; m.p. 384°C; transparent, rectangular prisms. Soluble in water; slightly soluble in alcohol.
Derivation: By heating sodium fluoride and hydrofluoboric acid and cooling slowly to form crystals of sodium fluoborate.
Uses: Sand agents in casting of aluminum and magnesium; in electrochemical processes; oxidation inhibitor; fluxes for nonferrous metals; fluorinating agent.

sodium fluophosphate. See fluophosphoric acids.

sodium fluoride NaF.
Properties: Clear, lustrous crystals or white powder or balls; poisonous, avoid breathing dust! Insecticide grade frequently dyed blue. Soluble in water; slightly soluble in alcohol.
Constants: Sp.gr. 2.766; m.p. 993°C.
Derivation: By adding sodium carbonate to hydrofluoric acid.
Grades: Pure; C.P.; U.S.P. XVI; technical; single pure crystals.
Containers: 100-lb multiwall paper bags; 125-, 400-lb drums.
Uses: Fluoridation of municipal water supplies; insecticide (not to be used on living plants and animals), fungicide, and rodenticide; metallurgy (rimmed steel); chemical cleaning; electroplating; glass manufacture; vitreous enamels; preservative for adhesives and wood; dentistry; single crystals used as windows in ultraviolet and infrared radiation detecting systems.
Warning! May be fatal if swallowed. MCA warning label.
Shipping regulations: None.*

sodium fluoroacetate (also known as 1080) FCH_2COONa.
Properties: Fine, white, odorless powder; soluble in water. Nonvolatile and not irritating to the skin. Decomposes at 200°C; soluble in water; insoluble in most organic solvents.
Derivation: Ethyl chloroacetate and potassium fluoride react to form ethyl fluoroacetate, which is then treated with a methanol solution of sodium hydroxide.
Containers: 8-oz cans, 2, 25, or 50 per case; 1-lb cans, 16 or 64 per case.
Uses: Rodenticide.
Warning: For use by trained operators only! Poisonous if swallowed! U.S. Pesticides Regulations.

sodium fluosilicate (sodium silicofluoride; salufer) Na_2SiF_6.
Properties: White, odorless, tasteless, granular, free-flowing, amorphous powder; sp.gr. 2.7; m.p., decomposes at red heat; very slightly soluble in cold water; insoluble in alcohol.
Derivation: By neutralization of fluosilicic acid with sodium carbonate, or addition of sodium chloride to the acid.
Grades: Technical; C.P.
Containers: 100-lb multiwall bags; 100-, 400-lb fiber drums; various sized barrels.
Uses: Fluoridation of drinking water; laundry sours; opalescent glass; vitreous enamel frits; latex foam rubber; metallurgy of aluminum and beryllium; insecticides and rodenticides; chemical intermediate; glue, leather and wood preservative.
Warning! May be fatal if swallowed. MCA warning label.
Shipping regulations: None.*

sodium formaldehyde bisulfite $HOCH_2SO_3Na$.
Properties: White water-soluble solid.
Derivation: Action of sodium bisulfite, formaldehyde and water.
Use: Textile stripping agent.
See also hydrosulfite-formaldehyde compounds.

sodium formaldehyde sulfoxylate (SFS) $HCHO \cdot HSO_2Na$.
Properties: White solid; m.p. 64°C; soluble in water.
Purity: Usually admixed with a sulfite.
Containers: Drums.
Use: Stripping and discharge agent for textiles; bleaching agent for molasses, soap.
See also hydrosulfite-formaldehyde compounds.

sodium formate HCOONa.
Properties: White, slightly hygroscopic, crystalline powder, slight odor of formic acid; soluble in water; slightly soluble in alcohol; insoluble in ether.
Constants: Sp.gr. 1.919; m.p. (dec) 245°C.
Derivation: Carbon monoxide and sodium hydroxide are heated under pressure.
Method of purification: Recrystallization.
Grades: Technical; C.P.
Containers: Bags.
Uses: Reducing agent; medicine; manufacture of formic acid and oxalic acid; intermediates; organic chemicals; mordant; tanning of leather; wallpaper printing; plating.
Shipping regulations: None.*

sodium gentisate $C_7H_5O_4Na$. The sodium salt of gentisic acid.
Properties: Crystallizes with 5.5 H_2O from water; loses $3H_2O$ rapidly upon exposure to air; soluble in water.
Containers: Drums.
Use: Medicine.

sodium glucoheptonate $HOCH_2(CHOH)_5COONa$. A sequestering agent for polyvalent metals. Used in metal cleaning; bottle washing; kier boiling; mercerizing; caustic boiloff.

sodium gluconate $NaC_6H_{11}O_7$.
Properties: White to yellowish, crystalline powder; readily soluble in water.
Derivation: From glucose by fermentation.
Method of purification: Crystallization.
Grades: Purified and technical.
Containers: Fiber drums; bags.
Uses: Foods and pharmaceuticals; sequestering agent; to prevent precipitation by alkalies of iron, aluminum, etc., from solution.
Shipping regulations: None.*

sodium glucosulfone $C_{24}H_{34}N_2Na_2O_{18}S_3$. para, para'-Diaminodiphenylsulfone-N,N'-di-(dextrose sodium sulfonate). The U.S.P. XVI grade is an aqueous solution (for injection), clear and pale yellow; pH 5.5-6.5. Used in medicine.

sodium glutamate (monosodium glutamate; MSG) $COOH(CH_2)_2CH(NH_2)COONa$. Sodium salt of glutamic acid, one of the common naturally occurring amino acids. Has meat-like taste.
Properties: White crystalline powder; m.p., decomposes; soluble in water and alcohol. Shows optical activity. Enhances food tastes without contributing any noticeable odor or taste of its own. Most effective between pH 6 and 8.
Derivation: (a) Alkaline hydrolysis of the waste liquor from beet sugar refining; (b) a similar hydrolysis of wheat or corn gluten; (c) organic synthesis based on acrylonitrile.
Grade: Technical, 99%; N.N.D.
Containers: Bottles, cans, fiber drums; retail shaker cans; paper bags.
Use: Valuable as a flavor intensifier, especially for meats.
See also glutamic acid.

sodium glycerinophosphate. See sodium glycerophosphate.

sodium glycerophosphate (sodium glycerinophosphate) $Na_2C_3H_5(OH)_2PO_4$.
Properties: White crystals or powder; saline taste. Soluble in water; insoluble in alcohol.
Derivation: By neutralizing glycerophosphoric acid with sodium carbonate.
Grade: Technical.
Containers: Solution: Bottles, 50-lb carboys; crystals: cans, 25 to 250-lb barrels and drums.
Use: Medicine.
Shipping regulations: None.*

sodium-gold chloride. See gold-sodium chloride.

sodium gynocardate (sodium hydnocarpate). Sodium salt of the fatty acids in chaulmoogra oil.
Properties: Yellow powder. Soluble in alcohol, water. Aqueous solution slightly alkaline.
Use: Medicine.

sodium heptaphosphate. A sodium phosphate glass similar in a general way to sodium hexametaphosphate, but having a higher ratio of Na_2O to P_2O_5. There is considerable confusion and uncertainty as to the precise meaning of this term. See sodium metaphosphate.

sodium hexametaphosphate (Graham's salt; sodium (1:1) phosphate glass; sodium polyphosphate; glassy sodium phosphate).
Derivation: Soluble Na_2O (1:1) P_2O_5 glass derived from monobasic sodium phosphate by prolonged fusion and rapid cooling of the melt. The degree of association is probably greater than indicated by the subscript 6 in the formula $(NaPO_3)_6$ commonly used in the past for this glass. The molecule is now considered to be a linear polymer.
Forms and Uses: See "Metafos."
For more about structure, see sodium metaphosphate.

sodium hexylene glycol monoborate $C_6H_{12}O_3BNa$.
Properties: Amorphous white solid; bulk density 0.25 g/cc; m.p. 426°C. Very soluble in nonpolar solvents up to 35% by weight.
Purity: Minimum 98%.
Uses: Corrosion inhibitor to organic systems; additive to lubricating oils; flame retardant additive; siloxane resin additive.

sodium hydnocarpate. See sodium gynocardate.

sodium hydrate. See sodium hydroxide.

sodium hydride NaH.
Properties: Microcrystalline dispersion of gray powder in oil. Particle size range 5-50μ. Concentration approximately 50% by weight. The sodium hydride starts to decompose with evolution of hydrogen at about 255°C. The hydride is insoluble in all inert organic solvents.
Preparation: By reaction of sodium metal with hydrogen.
Containers: The dispersion is packed in polyethylene bags in a metal can. Packaging ranges from 1 pint to 55-gallon drums.
Uses: As a basic condensing or alkylating agent. Particularly useful for alkylation of amines.
Hazards: Reacts vigorously with water, liberating hydrogen. Should not be separated from oil.
Shipping regulations: Flammable solid. Yellow label.*

sodium hydride descaling. Sodium and hydrogen are reacted in situ forming sodium hydride which first dissolves in a nonelectrolytic carrier bath of fused sodium hydroxide, and then reacts to reduce scale to lower oxides or free metal.
Uses: Removing oxide scale from steel, copper, and other ferrous and nonferrous metals and alloys.

sodium hydrosulfide (sodium sulfhydrate; sodium bisulfide) $NaSH \cdot 2H_2O$.
Properties: Colorless needles to lemon colored flakes. Soluble in water, alcohol, and ether.
Typical specifications: 70-72% NaSH; m.p. 55°C; water of crystallization 26-28%.
Derivation: Obtained from calcium sulfide by treating it in the cold with sodium bisulfate.
Grades: Technical; flake, 70-72%; solution, 40-44%.
Containers: 90-, 350-lb drums; tank cars.
Uses: Chemical intermediate for dyestuffs and various organic chemicals; leather (depilatory); desulfurizing viscose rayon; bleaching agent.
Danger! Contact with acid liberates poisonous gas. Burns skin and eyes. MCA warning label.

sodium hydrosulfite (sodium hyposulfite; hydrosulfite; sodium dithionite) $Na_2S_2O_4$ or $Na_2S_2O_4 \cdot 2H_2O$.
Properties: Light lemon-colored solid in powder or flake form or white to grayish-white crystalline powder; m.p. 55°C. Soluble in water; insoluble in alcohol.
Derivation: Zinc is dissolved in a solution of sodium bisulfite, the zinc-sodium sulfite precipitated by milk of lime leaving the hydrosulfite in solution. On adding salt, the hydrosulfite, containing water of crystallization, is precipitated. The latter is removed by treatment with hot alcohol. An equivalent simpler process is employed for making solutions to be used in situ.
Grades: Technical; reagent.
Containers: 50-lb pails and 250-lb drums.
Uses: As a strong reducing agent, in stripping dyes and discharge printing of textiles; bleaching sugar, soap, oils, minerals, straw, etc.
Caution: Subject to fire if allowed to become damp.
Shipping regulations: Flammable solid. Yellow label.*

sodium hydroxide (caustic soda; sodium hydrate; lye; white caustic) NaOH. The most important commercial caustic.
Properties: White, deliquescent pieces, lumps or sticks; crystalline fracture. Keep well stoppered, absorbs water and carbon dioxide from the air. Handle with care; it destroys organic tissue. Sp.gr. 2.13; m.p. 318°C; b.p. 1390°C. Soluble in water, alcohol, and glycerol.
Derivation: (a) By the electrolysis of sodium chloride; (b) by treating a solution of soda ash with a solution of lime commonly called "milk of lime." The lime and soda ash combine to form caustic soda and calcium carbonate. The former is soluble and remains in the solution while the latter is insoluble and is precipitated out of the solution. The precipitated calcium carbonate (lime mud) is removed from the caustic soda solution by filtration or decantation, thus leaving a clear solution of caustic soda.
Method of purification: Solution in alcohol, in which the carbonate, chloride and sulfate are practically insoluble, evaporation of the alcohol, followed by fusion.
Grades: Commercial; ground; flake;

granulated (60% and 76% Na_2O); rayon (low in iron, copper, and manganese); solution (50% and 73% NaOH); purified by alcohol (sticks, lumps, and drops); reagent; highest purity; C.P.; U.S.P. XVI.
Containers: U.S.P., C.P.: 1-lb bottles; 5-, 10-lb cans; commercial: drums and barrels; solution: drums; 8000-gal tank cars; barges.
Uses (in approximate order of volume): Manufacture of other chemicals; rayon and film; petroleum refining; pulp and paper; detergents; soap; textile processing; vegetable oil refining; reclaiming rubber.
Caution: Causes severe burns to skin and eyes. MCA warning label.
Shipping regulations: Solution: Corrosive liquid. White label.*

sodium hypochlorite NaOCl.
Properties: This salt is unstable in air unless mixed with sodium hydroxide. It is usually stored and used in solution, known as Labarraque's solution (q.v.) having a disagreeable, sweetish odor and a pale greenish color. Soluble in cold water; decomposed by hot water.
Constants: M.p., decomposes.
Derivation: By electrolyzing a cold dilute solution of salt, sometimes seawater.
Grades: Technical.
Containers: 5-, 13-gal carboys; 30-gal drums.
Uses: Bleaching paper pulp, textiles, etc.; intermediate; organic chemicals; water purification; medicine; manufacture of indigo; fungicides.
Shipping regulations: None.*
See also Javelle water.

sodium hypophosphite $NaH_2PO_2 \cdot H_2O$.
Properties: Colorless, pearly, crystalline plates or white, granular powder; saline taste. Keep well stoppered. May explode if heated, or if mixed with oxidizing agents. Deliquescent. Soluble in water, alcohol, and glycerol.
Derivation: By neutralizing hypophosphoric acid with sodium carbonate.
Grades: Technical; C.P.
Containers: Glass bottles; 25-, 50-, 100-lb drums.
Use: Medicine.
Shipping regulations: None.*

sodium hyposulfate. See sodium dithionate.

sodium hyposulfite. See sodium thiosulfate, or sodium hydrosulfite.

sodium indigotindisulfonate. The U.S.P. XVI name for indigo carmine (q.v.).

sodium iodate $NaIO_3$.
Properties: White crystals; sp.gr. 4.28. Soluble in water and acetone; insoluble in alcohol.
Derivation: Interaction of sodium chlorate and iodine in presence of nitric acid.
Grades: C.P.; reagent; technical.
Containers: 1-lb bottles.
Use: Medicine; disinfectant.

sodium iodide (a) NaI (b) $NaI \cdot 2H_2O$.
Properties: White cubical crystals or powder, or colorless, odorless, crystals; slowly becomes brown in air; deliquescent; saline, somewhat bitter taste. Soluble in water, alcohol, and glycerin.
Constants: Sp.gr. (a) 3.665; (b) 2.67. M.p. (a) 653°C; b.p. (a) 1350°C.
Derivation: Ferroso-ferric iodide, prepared from iron, iodine and water, is treated with pure sodium carbonate, filtered, the solution concentrated and crystallized.
Method of purification: Recrystallization.
Impurities: Sodium iodate.
Grades: Technical; C.P.; U.S.P. XVI.
Containers: 25-, 100-, 300-lb drums.
Uses: Photography; solvent for iodine; organic chemicals; reagent in analytical chemistry; medicine.
Shipping regulations: None.*

sodium iodide, thallium activated.
Grades: Single pure crystals.
Uses: Scintillation.

sodium iodipamide $C_{20}H_{12}I_6N_2Na_2O_6$. N,N'-adipoyl-bis(3-amino-2,4,6-triiodobenzoic acid), disodium salt. See iodipamide for formula.
Properties: Radiopaque; water soluble. Available as a 20% solution for injection as a clear colorless to pale yellow, slightly viscous liquid.
Derivation: Prepared by dissolving the free acid in dilute sodium hydroxide and buffering to pH 6.5-7.7.
Grade: N.F. XI; N.N.D.
Use: Roentgenographic contrast medium.

sodium iodomethamate $C_8H_3I_2NNa_2O_5$. Disodium 1-methyl-3,5-diiodo-4-pyridone-2,6-dicarboxylate.
Properties: White, odorless, powder or crystals; soluble in water; slightly soluble in alcohol; insoluble in ether and chloroform.
Grade: N.F. XI.
Use: Medicine.

sodium iodomethanesulfonate. See methiodal sodium.

sodium 5-iodo-2-thiouracil. See iothiouracil sodium.

sodium-iridium chloride. See iridium-sodium chloride.

sodium iron pyrophosphate (SIPP) $Na_8Fe_4(P_2O_7)_5 \cdot xH_2O$.
Properties: 300 mesh powder, tan in color, insoluble in water but soluble in dilute acid. Minimum 14.5% iron. Iron is in complex form and will not catalize oxidation reactions.
Derivation: By reacting tetrasodium pyrophosphate with a soluble iron salt.
Containers: Drums.
Uses: For iron enrichment; particularly in flours and cereals where there is a possibility of rancidity development.

sodium isoascorbate. See sodium erythorbate.

sodium isopropyl xanthate.
Properties: Light yellow crystals soluble in water.
Uses: Chemical weed killer of seeded crops; fortifying agent for certain oils; flotation reagent for base and precious metal ores.

sodium isovalerate. See sodium valerate.

sodium lactate $CH_3CHOHCOONa$.
Properties: Colorless or yellowish syrupy liquid, very hygroscopic. Soluble in water. M.p. 17°C.
Derivation: Interaction of lactic acid with sodium carbonate.
Grades: Technical; U.S.P. XVI (solution with pH 6.0 to 7.3).
Uses: Medicine; hygroscopic agent; glycerine substitute; plasticizer; corrosion inhibitor in alcohol antifreeze.
Shipping regulations: None.*

sodium N-lauroyl sarcosinate $C_{15}H_{28}NO_4Na$.
Used in oral dentifrices.

sodium lauryl sulfate $NaC_{12}H_{25}SO_4$.
Properties: Small, white or light yellow crystals; slight characteristic odor. Soluble in water, forming an opalescent solution.
Grades: U.S.P. XVI; technical.
Containers: 55-gal drums; tank cars.
Use: Wetting agent in textiles; detergent in toothpaste; food additive.

sodium lead alloys. An alloy, usually containing 10% sodium and 90% lead, was used in the manufacture of lead tetraethyl. An alloy containing 2% sodium is used as a deoxidizer and homogenizer in nonferrous metals where lead is a constituent. A sodium lead alloy is used as a stabilizer and deoxidizer for lead in cable sheathing.

sodium-lead hyposulfite. See lead-sodium thiosulfate.

sodium-lead thiosulfate. See lead-sodium thiosulfate.

sodium levothyroxine (sodium thyroxine) $HOC_6H_2I_2OC_6H_2I_2CH_2CH(NH_2)COONa \cdot 5H_2O$.
Sodium salt of levo isomer of thyroxine.
Properties: Light yellow to buff, odorless, tasteless powder. Very slightly soluble in chloroform and ether; slightly soluble in alcohol and water. Hygroscopic; stable in dry air and at room temperature; pH (saturated solution) 8.35-9.35.
Grade: U.S.P. XVI.
Use: Medicine.

sodium lignosulfonate.
Properties: Available as a tan, free-flowing spray-dried powder, containing wood sugars.
Containers: 50-lb bags.
Uses: Dispersant, emulsion stabilizer, chelating agent.
See also lignin sulfonates.

sodium liothyronine
$NaOOCCH(NH_2)CH_2C_6H_2I_2OC_6H_3IOH$.
Sodium L-3[4-(4-hydroxy-3-iodophenoxy)-3,5-diiodophenyl]alanine.
Properties: Light tan, odorless, crystalline powder. Very slightly soluble in water. Slightly soluble in alcohol, practically insoluble in most other organic solvents.
Grade: U.S.P. XVI.
Use: Medicine (a thyroid hormone).

sodium-magnesium sulfate (magnesium-sodium sulfate) $Na_2SO_4 \cdot MgSO_4 \cdot 4H_2O$.
Properties: White crystals. Soluble in water.

Sodium MBT. [28] A 50% water solution of sodium mercaptobenzothiazole of high purity.
Use: Corrosion inhibitor for anti-freeze.

sodium mercaptoacetate. See sodium thioglycolate.

sodium meta-antimonate. See sodium antimonate.

sodium meta-arsenite. See sodium arsenite.

sodium metabisulfite (sodium pyrosulfite) $Na_2S_2O_5$. Chief constituent of commercial dry sodium bisulfite, with which most of its properties and uses are practically identical.

sodium metaborate $NaBO_2$.
Properties: White lumps. Soluble in water. Solution alkaline. M.p. 966°C.
Derivation: By fusing sodium carbonate and borax.
Containers: Bags.
Use: Herbicide.
Also available commercially as octahydrate and tetrahydrate.

sodium metaborate peroxyhydrate. See sodium perborate.

sodium metanilate $NaSO_3C_6H_4NH_2$.
Derivation: The meta-sodium sulfonate of aniline sold as a solid or 20% aqueous solution prepared by neutralizing metanilic acid.
Grades: Technical; 99%; also 20% solution.
Containers: Barrels.
Use: In manufacture of synthetic dyestuffs and drugs.

sodium metaperiodate. See sodium periodate.

sodium metaphosphate $NaPO_3$, or possibly some multiple thereof such as $(NaPO_3)_3$. Sodium metaphosphate exists in a large number of forms and varieties, and in addition certain mixtures are known by this name or some derivation of it. As a result there is considerable confusion and difficulty in properly associating names, compositions and properties. The terms monometaphosphate, di-, tri-, tetra-, penta-, hexa-, hepta-, octa-, deca-, dodeca- and tetradecametaphosphate have all been used, and unfortunately are not clearly and uniquely associated with particular compounds. Based chiefly on x-ray diagrams, six different crystalline sodium metaphosphates have been shown to exist, and in addition there is a sodium phosphate glass of the same composition which is one of the series of glasses with varying ratios of Na_2O to P_2O_5. As a result, the following

*See "I.C.C. Shipping Regulations," page xiii.
Reference numbers refer to name of manufacturer. See "List of Manufacturers," page v.

nomenclature has been proposed:
(a) For sodium hexametaphosphate (Graham's salt), commonly assigned the formula $(NaPO_3)_6$, the name and formula sodium (1:1) phosphate glass [$Na_2O(1:1)P_2O_5$ glass] are suggested.
(b) For sodium trimetaphosphate, commonly assigned the formula $(NaPO_3)_3$, the names and formulas sodium metaphosphate I ($NaPO_3$ I), sodium metaphosphate I' ($NaPO_3$ I'), and sodium metaphosphate I'' ($NaPO_3$ I'') are suggested.
(c) For Maddrell's salt, commonly assigned the formula $NaPO_3$, the names and formulas sodium metaphosphate II ($NaPO_3$ II) and sodium metaphosphate III ($NaPO_3$ III) are suggested.
(d) For Kurrol's salt, commonly assigned the formula $NaPO_3$, the name and formula sodium metaphosphate IV ($NaPO_3$ IV) are suggested.
Uses: Principally as sequestering agents. See sequestration.

sodium metasilicate, anhydrous Na_2SiO_3. A crystalline silicate.
Properties: Dustless white granules; m.p. 1089°C; total Na_2O content 51.5%; percent of total Na_2O in active form 48.6%; density 2.61 or 75 lb/cu ft. Soluble in water; precipitated by acids and by alkaline earth and heavy metal ions; pH of 1% solution 12.6.
Derivation: Crystallized from a melt of Na_2O and SiO_2 below 1089°C.
Containers: 100-lb bags; 125-lb fiber drums; 400-lb fiber drums.
Uses: Laundry, dairy and metal cleaning; floor cleaning; base for detergent formulations.
Shipping regulations: None.*

sodium metasilicate, pentahydrate $Na_2SiO_3 \cdot 5H_2O$.
Properties: Dustless white granular crystals; m.p. 72.2°C; total Na_2O content 29.3%; percent of total Na_2O in active form 27.8%; density 1.75, or 55 lb/cu ft. Soluble in water; pH of 1% solution 12.5.
Derivation: Crystallization from aqueous solution.
Containers: 100-lb bags; 125-, 300-, 400-lb fiber drums.
Uses: Laundry and dairy cleaning; metal cleaning; paper mills; textile finishing; base for cleaning compounds; bleaching aid; deinking paper.

sodium metavanadate $NaVO_3$, often with $4H_2O$.
Properties: Colorless, monoclinic, prismatic crystals, or pale green crystalline powder. Soluble in hot water. M.p. 630°C.
Derivation: Sodium hydrate and vanadium pentoxide in water solution.
Method of purification: Recrystallization.
Grades: Technical; C.P.
Containers: Glass bottles; compressed paper drums.
Uses: Inks; fur dyeing; therapeutics; in photography to impart a red tone to films and plates; inoculation of plant life; mordants and fixers; corrosion inhibitor in gas-scrubbing systems.
Shipping regulations: None.*

sodium methacrylate $CH_2{:}C(CH_3)COONa$.
Containers: Drums.
Uses: Water-soluble polymerizable monomer; chemical intermediate.

sodium methanearsonate. See disodium methylarsonate.

sodium methoxide. See sodium methylate.

sodium methylate (sodium methoxide) CH_3ONa.
Properties: White, amorphous, free-flowing powder, sensitive to oxygen; decomposed by water. Soluble in methyl and ethyl alcohol. Decomposes in air above 260°F.
Containers: 10-lb pails; 25 to 200-lb steel drums; also 25% solution in methanol, 425-lb drums.
Uses: Condensation reactions in general; catalyst for treatment of edible fats and oils, especially lard; intermediate for sulfa drugs and other pharmaceuticals.
Shipping regulations: Solid: Flammable solid. Yellow label. Alcohol mixture: Flammable liquid. Red label.*

sodium methyl carbonate $CH_3OCOONa$.
Properties: White powder; m.p. 330°C (decomposes); density 1.66; purity 90% min.

sodium N-methyl-N-oleoyl taurate (oleyl methyl tauride)
$(CH_3)(CH_2)_7CHCH(CH_2)_7CON(CH_3)CH_2CH_2SO_2ONa$.
Properties: Fine white powder. Pleasant, slight, sweet odor. Nontoxic.
Grades: Technical, 32% purity (remainder is mainly sodium sulfate).
Use: Detergent; pesticide adjuvant.

sodium N-methyltaurate. See N-methyltaurine.

sodium molybdate. As an article of commerce, usually the normal molybdate Na_2MoO_4 or its dihydrate (called sodium molybdate crystals). Chemically, however, a wide variety of complex molybdates with sodium are known.
Properties: Small, lustrous, crystalline plates;* soluble in water; m.p. 687°C; sp.gr. 3.28.
Derivation: By the action of sodium hydroxide on molybdenum trioxide. Complex molybdates are prepared by dissolving large amounts of molybdenum trioxide in solutions of normal molybdates.
Grades: Anhydrous; crystals.
Containers: 100-, 150-, 200-, 375-lb drums.
Uses: Reagent in analytical chemistry; medicine; casein glues; paint pigment; production of molybdate orange, molybdated toners and other colors; metal finishing; fertilizer; brightening agent for zinc plating; corrosion inhibitor; catalyst in dye and pigment production.
Shipping regulations: None.*

sodium molybdophosphate. See sodium phosphomolybdate.

sodium monoxide (sodium oxide) Na_2O.
Properties: A white powder, soluble in molten caustic soda or potash. Converted to sodium hydroxide by water.
Containers: 60-lb pails; 300-lb drums.
Uses: As a condensing or polymerizing agent in organic reactions; as a dehydrating agent; strong base.

sodium naphthalenesulfonate $C_{10}H_7SO_3Na$.
Properties: Yellowish, crystalline plates, or white, odorless scales. Soluble in water; insoluble in alcohol.
Derivation: Naphthalene sulfonic acid is prepared by sulfonating naphthalene with sulfuric acid. The solution is treated with sodium carbonate, filtered, concentrated and crystallized.
Method of purification: Recrystallization.
Grades: Technical.
Containers: 360-, 500-lb barrels.
Uses: Organic preparations; liquefying agent in animal glue preparations; naphthols.
Shipping regulations: None.*

sodium naphthenates. A white paste. The most important of the naphthenic acid salts. Commercial soaps have consistency of grease, but this will vary with source and manner of processing. Combine excellent emulsifying and foam producing properties with a low hydrolytic dissociation, and are of value as mild detergents. Very powerful emulsifying agents for water-mineral oil systems. Possess marked disinfectant properties. Also used in manufacture of driers. See naphthenic acid.

sodium naphthionate (sodium alpha-naphthylaminesulfonate) $NH_2C_{10}H_6SO_3Na \cdot 4H_2O$.
Properties: White crystals, become violet on exposure to light. Soluble in water; insoluble in ether.
Derivation: Alpha-naphthylamine is fused and poured into concentrated sulfuric acid, heated to 180°C, then oxalic acid is added. The molten mass is poured on lead plates and baked for 8 hours to 180°C. When cold the porous mass is neutralized with hot caustic soda solution, filtered, concentrated and crystallized.
Method of purification: Recrystallization.
Grades: Technical (paste, crystals).
Containers: Paste: 235-, 300-, 500-lb barrels; crystals: 340-lb barrels.
Uses: For Riegler's reagent for nitrous acid; manufacture of dyestuffs.
Shipping regulations: None.*

sodium beta-**naphtholate** (beta-naphthol sodium; microcidine) $C_{10}H_7ONa$.
Properties: White powder, non-caustic, nontoxic. More powerful than phenol. Soluble in water.
Use: Medicine.

sodium naphthylaminesulfonate. See sodium naphthionate.

sodium niobate (sodium columbate) $7Na_2O \cdot 6Nb_2O_5 \cdot 31H_2O$. Sodium niobate is of importance technologically in the purification of niobium materials. The crystalline compound forms when a niobium compound is treated with hot concentrated sodium hydroxide. It is sparingly soluble in water.

sodium nitrate (soda niter) $NaNO_3$. Chile saltpeter (caliche) is impure natural sodium nitrate.
Properties: Colorless, transparent, odorless crystals; saline, slightly bitter taste; sp.gr. 2.267; m.p. 308°C; b.p., decomposes. Soluble in water and glycerol; slightly soluble in alcohol.
Derivation: From nitric acid and sodium carbonate; and from Chile saltpeter.
Method of purification: Recrystallization.
Grades: Granular, sticks, powder; crude; 95%; double refined; recrystallized; C.P.; technical; reagent; diuretic.
Containers: Tins; glass bottles; multiwall paper sacks; bags up to 100-lb; bulk.
Uses: Manufacturing sulfuric and nitric acids and potassium nitrate; oxidizing agent; oxidizer in solid rocket propellants; fertilizer; flux; glass manufacture; pyrotechnics; reagent in analytical chemistry; medicine; refrigerant; matches; dynamites; military explosives and gases; manufacturing sodium salts, dyes; pharmaceuticals; food preservative; enamel for pottery; modifying burning properties of tobacco.
Caution: Fire hazard: dangerous! In contact with organic or other readily oxidizable (combustible) substances it will cause violent combustion or ignition.
Shipping regulations: Oxidizing material. Yellow label.*
See also caliche.

sodium nitrite (diazotizing salts) $NaNO_2$.
Properties: Slightly yellowish or white crystals, pellets, sticks or powder. Oxidizes on exposure to air. Soluble in water; slightly soluble in alcohol and ether. Sp.gr. 2.157; m.p. 271°C; b.p., decomposes at red heat above 320°C.
Grades: Reagent; technical; U.S.P. XVI.
Containers: 25-, 100-lb drums; 150-lb kegs; 400-lb barrels.
Uses: Dyestuff manufacture, for diazotizing; organic synthesis; preparation of nitric oxide; reagent in analytical chemistry; pharmaceuticals; photographic reagent; pickling meat; medicine; dyeing and printing textile fabrics; bleaching flax, silk and linen; rustproofing; prevention of corrosion; metal cleaner; in cutting oils.
Fire hazard: Dangerous.
Shipping regulations: Oxidizing material. Yellow label.*

sodium nitroferricyanide (sodium nitroprussiate; sodium nitroprusside) $Na_2Fe(CN)_5NO \cdot 2H_2O$.
Properties: Red, transparent crystals; sp.gr. 1.72. Soluble in water, with slow decomposition; slightly soluble in alcohol.
Grades: Reagent; technical.
Use: Analytical reagent.

sodium nitroprussiate. See sodium nitroferricyanide.

*See "I.C.C. Shipping Regulations," page xiii.
Reference numbers refer to name of manufacturer. See "List of Manufacturers," page v.

sodium nitroprusside. See sodium nitroferricyanide.

sodium nucleate. See sodium nucleinate.

sodium nucleinate (sodium nucleate).
Properties: Yellowish-white almost odorless powder containing approximately 4.5% sodium. Soluble in water; insoluble in alcohol.
Derivation: From yeast.
Use: Medicine.
Shipping regulations: None.*

sodium oleate $C_{17}H_{33}COONa$.
Properties: White powder; slight tallow-like odor. Soluble in water with partial decomposition; soluble in alcohol.
Derivation: Action of alcoholic sodium hydroxide on oleic acid.
Containers: 100-lb bags; 180-lb barrels; 200-lb drums.
Use: Medicine; ore flotation; waterproofing textiles.
Shipping regulations: None.*

sodium orthophosphate, primary, secondary and tertiary. See sodium phosphate, mono-, di-, and tribasic.

sodium orthosilicate $Na_2SiO_3 \cdot 2NaOH$ or other proportions such as $2Na_2O \cdot SiO_2$ (anhydrous) or $2Na_2O \cdot SiO_2 \cdot 5.4H_2O$.
Properties: (Composition $2Na_2O \cdot SiO_2$) Dustless white, flaked product; density 75 lbs/cu ft; total Na_2O content 60.8%; percent of total Na_2O in active form 59.0%. Soluble in water; pH of a 1% solution 13.0.
Derivation: Crystallized from anhydrous melts. Commercial products are mixtures of anhydrous metasilicate and sodium hydroxide, (either crystallized integrally or compounded) or an integral mixture of sodium hydroxide with the metasilicate pentahydrate.
Containers: 100-lb bags; 100-, 275-, 400-lb fiber drums.
Uses: Commercial laundries; metal cleaning; heavy duty cleaning.
Warning: Causes severe burns to skin and eyes. MCA warning label.

sodium orthovanadate Na_3VO_4.
Properties: Colorless hexagonal prisms. Soluble in water; insoluble in alcohol. M.p. 866°C.
Derivation: Fusion of vanadium pentoxide and sodium carbonate.
Method of purification: Recrystallization.
Containers: Glass bottles.

sodium-osmic chloride. See osmium-sodium chloride.

sodium oxalate $Na_2C_2O_4$.
Properties: White, crystalline powder; poisonous! Soluble in water; insoluble in alcohol.
Derivation: Oxalic acid is dissolved in water, neutralized with sodium carbonate, the solution filtered, concentrated, and crystallized.
Method of purification: Recrystallization.
Impurities: Sodium carbonate; sodium binoxalate; chlorine; sulfates; iron salts; potassium salts; organic impurities.
Grades: Reagent; technical, 88%, 99%.
Containers: 100-lb bags; 225-lb barrels.
Uses: Reagent in analytical chemistry; textile finishing; pyrotechnics; leather tanning and finishing; blue printing.
Shipping regulations: None.*

sodium oxide. See sodium monoxide.

sodium palconate. The sodium salt of an acid that may be extracted with alkali from redwood dust. The dark reddish brown material consists mainly of a partially methylated phenolic acid containing aliphatic hydroxyls, phenolic hydroxyls and carboxyl groups in the ratio 2:4:3. The viscosity of aqueous solutions rises rapidly with concentration.
Uses: It is used to control viscosity and water loss in drilling muds, and is also useful as a dispersing agent.

sodium-palladium chloride. See palladium-sodium chloride.

sodium penicillin G. See penicillin.

sodium penicillin O. See penicillin.

sodium pentaborate $Na_2B_{10}O_{16} \cdot 10H_2O$.
Properties: White crystals; free-flowing; stable under ordinary conditions; solubility in water, 15.40% (20°C), increasing with temperature; sp.gr. 1.72; pH of solution approx 7.5.
Containers: 100-lb paper bags.
Uses: Weed killer; in cotton defoliant and fireproofing compositions; in glass manufacture; boron supplement for tree fruit and truck crops.

sodium pentachlorophenate C_6Cl_5ONa.
Properties: Tan powder; soluble in water, ethanol, and acetone; insoluble in benzene.
Containers: 50-lb bags; 60-, 70-, 100-, 200-lb drums.
Grades: Technical; powder, pellets, or briquettes.
Caution: Avoid skin contact and inhalation of dust.
Uses: Used as a fungicide, herbicide, for control of algae and slime; and as a fermentation disinfectant, especially in finishes and papers, including food packaging.

sodium perborate (peroxydol; sodium metaborate peroxyhydrate) $NaBO_2 \cdot H_2O_2 \cdot 3H_2O$. Often described as sodium perborate tetrahydrate, $NaBO_3 \cdot 4H_2O$; see also sodium borate perhydrate.
Properties: White, odorless crystals or powder; salty taste. Stable in cool, dry air but decomposes with the evolution of oxygen in warm or moist air. Moderately soluble in water (with decomposition) and glycerol. pH of aqueous solutions 10.0 to 10.3. Active oxygen content 10% min.
Derivation: (a) Electrolysis of a solution of borax and soda ash; (b) crystallization from solution of borax or boric acid, sodium peroxide, and hydrogen peroxide.
Method of purification: Recrystallization.
Grades: Technical; C.P.; N.F. XI (as

$NaBO_3 \cdot 4H_2O$).
Containers: Cartons; boxes; bags.
Uses: Developing vat dyes; textile bleaching; household bleaches and detergents; neutralizing cold wave preparations; dental compositions; electroplating; laboratory reagent; germicide; deodorant; general applications as a mild, alkaline oxidizing agent.
Shipping regulations: None.*

sodium perchlorate $NaClO_4$, sometimes with $1H_2O$.
Properties: White deliquescent crystals; must not be triturated with organic or combustible substances, may cause explosions; explosive in contact with concentrated sulfuric acid! Soluble in water and alcohol. M.p. 482°C; b.p., decomposes; sp.gr. 2.02.
Derivation: (a) Sodium chlorate and sodium chloride are mixed and heated until fused. The unchanged chloride is leached out. (b) A cold solution of sodium chlorate is electrolyzed, the solution concentrated and crystallized.
Method of purification: Recrystallization.
Impurities: Sodium chloride; sodium chlorate.
Grades: Technical.
Containers: Glass bottles; wooden kegs.
Use: Explosives; jet fuel; analytical reagent.
Fire hazard: Dangerous.
Shipping regulations: Oxidizing material. Yellow label.*

sodium periodate (sodium metaperiodate) (a) $NaIO_4$; (b) $NaIO_4 \cdot 3H_2O$.
Properties: Colorless crystals; sp.gr. (a) 4.172 (25°C), (b) 3.219 (18°C); m.p. (a) 300°C (dec), (b) 175°C (dec). Very soluble in water.
Containers: Glass bottles; steel pails.
Uses: Source of periodic acid; analytical reagent.
Shipping regulations: Oxidizing material. Yellow label.*

sodium permanganate $NaMnO_4 \cdot 3H_2O$.
Properties: Purple to reddish-black crystals or powder; soluble in water. M.p., decomposes.
Derivation: Sodium manganate is dissolved in water and a current of chlorine or ozone passed in. The solution is concentrated and crystallized.
Method of purification: Recrystallization.
Impurities: Sodium hydroxide; sodium manganate.
Grades: Technical; sold commercially in solution.
Containers: Wooden barrels; steel drums; 1-, 5-, and 10-lb glass bottles.
Uses: Oxidizing agent; disinfectant; bactericide; manufacture of saccharin; antidote for poisoning by morphine, curare and phosphorus.
Fire hazard: Dangerous.
Shipping regulations: Oxidizing material. Yellow label.*

sodium peroxide (sodium dioxide; sodium superoxide, q.v., a misnomer) Na_2O_2.
Properties: Yellowish-white powder, turning yellow when heated. Keep away from alcohol and other similar flammable liquids, as it will cause ignition, particularly in presence of water. Absorbs water and carbon dioxide from air. Active oxygen content approximately 20% by weight; sp.gr. 2.805; m.p. 660°C (approx) with decomposition. Soluble in cold water developing great heat; decomposed in water.
Derivation: Metallic sodium is heated at 300°C in aluminum trays in a retort in a current of dry air, from which the carbon dioxide has been removed.
Grades: Technical; reagent.
Containers: 20-, 75-lb pails; 280-, 400-lb drums.
Uses: Oxidizing agent; bleaching agent in paper and textile industries; deodorant; germicide; antiseptic; disinfectant; medicine; purifying air in sick rooms; bleaching fats, oils, resins, waxes, bones, gelatin, bristles, straw, ivory, sponges, feathers; medicinal soaps; organic chemicals; peroxides; water purification; pharmaceuticals; oxygen generation for diving bells, submarines, etc; textile dyeing and printing; ore processing; analytical reagent; calorimetry.
Caution: Fire hazard: Dangerous, does not burn or explode itself but mixtures of sodium peroxide and combustible substances are explosive and ignite very easily, particularly if moisture is present.
Shipping regulations: Oxidizing material. Yellow label.*

sodium peroxydisulfate. See sodium persulfate.

sodium persulfate (sodium peroxydisulfate) $Na_2S_2O_8$.
Properties: White, crystalline powder; soluble in water; decomposed by alcohol.
Uses: Analytical; bleaching agent (fats, oils, soap); battery depolariziers; medicine.

sodium phenate (sodium phenolate; sodium carbolate; phenol sodium) C_6H_5ONa.
Properties: White, deliquescent crystals. Keep well stoppered. Soluble in water and alcohol; decomposed by carbon dioxide of the air.
Derivation: Phenol is dissolved in caustic soda solution, concentrated and crystallized.
Method of purification: Recrystallization.
Grades: Technical.
Containers: Glass bottles.
Uses: Antiseptic; in military gas-masks with charcoal and hexamethylenetetramine as absorbent for phosgene; salicylic acid; organic synthesis.
Shipping regulations: None.*

sodium phenobarbital. See phenobarbital sodium.

sodium phenolate. See sodium phenate.

sodium phenolsulfonate (sodium sulfocarbolate) $HOC_6H_4SO_3Na \cdot 2H_2O$.
Properties: Colorless crystals or granules, slightly efflorescent; chars at high

temperature, evolving phenol. Soluble in water, hot alcohol, and glycerol.
Containers: 25-, 100-, 200-, 275-lb drums.

sodium phenoneacetate. See guaiacetin.

sodium N-phenylglycinamide-para-**arsonate.** See tryparsamide.

sodium ortho-**phenylphenate** (sodium ortho-phenylphenolate) $C_6H_4(C_6H_5)ONa \cdot 4H_2O$.
Properties: Practically white flakes. Bulk density 38-43 lb/cu ft; pH of saturated solution in water 12.0-13.5. Soluble in water, methanol, acetone.
Uses: Industrial preservative (as a bactericide and fungicide.)

sodium ortho-**phenylphenolate.** See sodium ortho-phenylphenate.

sodium phenylphosphinate $C_6H_5PH(O)(ONa)$.
Properties: Crystals; m.p. 355°C (decomposes to give phenylphosphine); stable, non-hygroscopic at room temperature; soluble in water.
Uses: Antioxidant; heat and light stabilizer.

sodium phosphate. See sodium hexametaphosphate; sodium metaphosphate; sodium phosphate, dibasic; sodium phosphate, hemibasic; sodium phosphate, monobasic; sodium phosphate, tribasic; sodium pyrophosphate; sodium pyrophosphate, acid.

sodium phosphate, dibasic (DSP; disodium phosphate; hydrodisodium phosphate; sodium orthophosphate, secondary; disodium orthophosphate) (a) Na_2HPO_4; (b) $Na_2HPO_4 \cdot 2H_2O$; (c) $Na_2HPO_4 \cdot 7H_2O$; (d) $Na_2HPO_4 \cdot 12H_2O$.
Properties: Colorless, translucent crystals or white powder; cooling, saline taste. Soluble in water; very slightly soluble in alcohol.
Constants: (a) Hygroscopic; converted to sodium pyrophosphate at about 240°C. (b) M.p. loses H_2O at 92.5°C; sp.gr. (15°C) 2.066; (d) M.p. 35°C; sp.gr. 1.5235. Readily loses $5H_2O$ on exposure to air at ordinary temperature; loses $12H_2O$ at 100°C.
Derivation: (1) By treating phosphoric acid with a slight excess of soda ash, boiling the solution to drive off carbon dioxide, and cooling to permit the dodecahydrate to crystallize; (2) by precipitating calcium carbonate from a solution of dicalcium phosphate with soda ash.
Method of purification: Recrystallization.
Grades: Commercial; food; N.F. XI (c).
Containers: 100-lb paper bags; 100-, 300-, 400-lb fiber drums; 300-lb barrels.
Uses: Chemicals; dyes; fertilizers; pharmaceuticals; medicine; textiles (weighting silk, dyeing and printing, fire-proofing); fire-proofing wood, paper and other products; ceramic glazes; tanning; paint pigments; baking powders; galvano-plastics; soldering enamels; analytical reagent; cheese; detergents; water treatment.
Shipping regulations: None.*

sodium (1:1) phosphate glass. See sodium hexametaphosphate.

sodium phosphate, glassy. See sodium tetraphosphate.

sodium phosphate, hemibasic $NaH_5(PO_4)_2$.
Properties: A hygroscopic, strongly acid salt. Soluble in water.
Grades: Technical.
Containers: 350-, 500-lb barrels (moisture-proof liner).
Uses: Treating silage; contact tinning of brass; boiler-water treatment. Useful wherever a strong acid is desired in solid form.

sodium phosphate, monobasic (sodium acid phosphate; sodium biphosphate; sodium orthophosphate, primary) (a) NaH_2PO_4, (b) $NaH_2PO_4 \cdot H_2O$.
Properties: (a) White crystalline powder; slightly hygroscopic; very soluble in water; has acid reaction; forms sodium acid pyrophosphate at 225-250°C and sodium metaphosphate at 350-400°C; (b) large, transparent crystals; m.p., loses H_2O at 100°C; sp.gr. 2.040; very soluble in water; insoluble in alcohol.
Derivation: By treating disodium phosphate with proper proportion of phosphoric acid.
Method of purification: Recrystallization.
Grades: Commercial; food; (b) U.S.P. XVI.
Containers: 100-lb paper bags; 125-lb drums; 350-lb barrels.
Uses: Boiler water treatment; electroplating; dyeing; acid cleansers; baking powders; cattle food supplement.
Shipping regulations: None.*

sodium phosphate, tribasic (TSP; trisodium orthophosphate; trisodium phosphate; tertiary sodium phosphate; sodium orthophosphate, tertiary) $Na_3PO_4 \cdot 12H_2O$.
Properties: Colorless crystals; soluble in water. Sp.gr. 1.618-1.645; m.p. 77°C; b.p., loses $11H_2O$ at 100°C.
Derivation: By mixing soda ash and phosphoric acid in proper proportions to form disodium phosphate, and then adding caustic soda.
Method of purification: Recrystallization.
Grades: Commercial; highest purity; C.P. Anhydrous salt, Na_3PO_4, also available.
Containers: 1-, 5-lb bottles; 350-, 500-lb barrels; bags.
Uses: Water softeners; boiler compounds; detergent; metal cleaner; textiles; manufacture of paper; laundering; tanning industry; sugar purification; photographic developers; medicine; paint removers; industrial cleaners.

sodium phosphate, tribasic, monohydrate (trisodium phosphate monohydrate) $Na_3PO_4 \cdot H_2O$.
Properties: White powder. Absorbs water as water of crystallization. Soluble in water. Density 65 lb/cu ft.
Containers: 100-lb bags; 125-, 400-lb drums.

*See "I.C.C. Shipping Regulations," page xiii.
Reference numbers refer to name of manufacturer. See "List of Manufacturers," page v.

Uses: Water softening; detergents; industrial cleaners. See preceding article.

sodium phosphite $Na_2HPO_3 \cdot 5H_2O$.
Properties: White, crystalline powder. Hygroscopic. Caution! Keep well stoppered! Soluble in water; insoluble in alcohol. M.p. 53°C.
Use: Medicine; antidote in mercuric chloride poisoning.

sodium phosphoaluminate. White powder composed primarily of sodium aluminate (hydrated), sodium phosphate (ortho) and small amounts of sodium carbonate and sodium silicate. Used primarily in the paper industry as a sizing adjunct, as an aid in retention of filler and fiber and in control of the pH of the stock. Also used in boiler feed water treatment.

sodium phosphomolybdate (sodium molybdophosphate) $Na_3PO_4 \cdot 12MoO_3$.
Properties: White crystals; soluble in water.
Grades: Technical.
Uses: Analysis; neuromicroscopy.

sodium phosphotungstate (sodium phosphowolframate) $Na_3PO_4 \cdot 12WO_3 \cdot 18H_2O$.
Properties: White, granular powder. Soluble in water.
Derivation: By neutralizing phosphotungstic acid with sodium carbonate.
Method of purification: Crystallization.
Grades: Technical.
Containers: Glass bottles.
Use: Reagent for detecting and determining alkaloids, uric acid, potassium.
Shipping regulations: None.*

sodium phosphowolframate. See sodium phosphotungstate.

sodium picramate $NaOC_6H_2(NO_2)_2NH_2$.
Derivation: Yellow, water-soluble salt resulting from neutralization of picramic acid with caustic soda.
Grades: Technical.
Containers: Drums.
Uses: Manufacture of dye intermediates; organic chemicals; synthesis.
Shipping regulations: Wet with 20% of water: Flammable solid. Yellow label.*

sodium platinichloride. See platinic-sodium chloride.

sodium platinochloride. See platinous-sodium chloride.

sodium plumbate $Na_2PbO_3 \cdot 3H_2O$.
Properties: Fused, light-yellow lumps. Hygroscopic. Decomposed by water. Caution! Keep well stoppered!
Grades: Technical.

sodium plumbite. See doctor treatment.

"Sodium Polyphos." [84] Brand name for a water-soluble glassy sodium phosphate of standardized composition, $(Na_{12}P_{10}O_{31})$ analyzing 63.5% P_2O_5 (ratio of $Na_2O:P_2O_5$ is 1.2:1). It is closely similar to sodium hexametaphosphate and sodium tetraphosphate; frequently the three names are used interchangeably.

Grades: Ground; walnut-size lumps; and pea-size lumps.
Containers: 100-lb bags; 100-, and 350-lb drums.
Uses: Boiler water compounds; detergents; textiles; leather tanning; photographic film developing; deflocculation of clays; flotation and desliming of minerals; dispersion of pigments; paper processing; conditioning agent for oil well drilling muds; industrial and municipal water treatment.

sodium polyphosphate. See sodium hexametaphosphate; sodium tetraphosphate.

sodium polysulfide Na_2S_x.
Properties: Yellow-brown granular free-flowing material; density 56 lbs/cu ft.
Containers: 100-lb, 400-lb drums.
Uses: Manufacture of sulfur dyes and colors, insecticides, synthetic rubber, petroleum additives; electroplating.

sodium-potassium alloy. See NaK.

sodium potassium carbonate (potassium-sodium carbonate) $NaKCO_3 \cdot 6H_2O$.
Properties: Colorless crystals. The double salt fuses more readily than the single salts; sp.gr. 1.6344; m.p. 135°C (dec). Soluble in water.
Derivation: Mixture of potassium and sodium carbonates.
Grades: Technical.
Use: Analysis (flux).

sodium-potassium phosphate (potassium-sodium phosphate) $NaKHPO_4 \cdot 7H_2O$.
Properties: White powder; stable in air. Soluble in water.

sodium-potassium tartrate. See potassium-sodium tartrate.

sodium propionate CH_3CH_2COONa or $C_2H_5COONa \cdot xH_2O$.
Properties: Transparent crystals or granules; almost odorless; deliquescent in moist air; very soluble in water; slightly soluble in alcohol.
Grades: N.F. XI.
Containers: 25-, 100-, 250-lb drums.
Uses: Fungicide; mold preventative, widely used in foods.

sodium prussiate, red. See sodium ferricyanide.

sodium prussiate, yellow. See sodium ferrocyanide.

sodium pyroantimonate. See sodium antimonate.

sodium pyroborate. See sodium borate.

sodium pyrocatechin acetate. See guaiacetin.

sodium pyrophosphate (tetrasodium pyrophosphate; sodium pyrophosphate, normal; TSPP) (a) $Na_4P_2O_7$; (b) $Na_4P_2O_7 \cdot 10H_2O$.
Properties: Colorless, transparent crystals or white powder; mild on the hands (has a pH of only 10.2 in 1% solutions). (a) M.p. 988°C; sp.gr. 2.45; soluble in water; decomposes in alcohol; (b) m.p. 94°C; sp.gr. 1.82; soluble in water; insoluble in alcohol and ammonia.

Derivation: By fusing disodium phosphate.
Grades: Pure crystals; dried; fused; C.P.
Containers: Bottles; 100-lb paper bags; drums; bulk cars; 350-lb barrels.
Uses: Water softener; soap and synthetic detergent builder; dispersing and emulsifying agent; metal cleaner; boiler water treatment; viscosity control of drilling muds; de-inking news print; synthetic rubber manufacture; textile dyeing; scouring of wool.
Shipping regulations: None.*

sodium pyrophosphate, acid (disodium pyrophosphate; sodium acid pyrophosphate; SAPP) $Na_2H_2P_2O_7 \cdot 6H_2O$.
Properties: White crystalline powder; m.p. (dec) 220°C; sp.gr. 1.862; soluble in water.
Derivation: Incomplete decomposition of monobasic sodium phosphate.
Grades: Technical; food.
Containers: 100-lb bags; 125-, 350-lb drums.
Uses: Electroplating; metal cleaning and phosphatizing; conditioning agent for drilling muds; baking powders and leavening agent.
Shipping regulations: None.*

sodium pyrophosphate, normal. See sodium pyrophosphate.

sodium pyrophosphate peroxide $Na_4P_2O_7 \cdot 2H_2O_2$.
Properties: White powder, apparent bulk density 73 lb/cu ft. Active oxygen minimum 9.0% by wt. Water soluble; mildly alkaline.
Containers: Fiber drums.
Uses: Denture cleaners, dentifrices, household and laundry detergents.

sodium pyroracemate. See sodium pyruvate.

sodium pyrosulfite. See sodium metabisulfite.

sodium pyrovanadate $Na_2V_2O_7 \cdot 8H_2O$.
Properties: Colorless six-sided plates. Soluble in water; insoluble in alcohol; m.p. (anhydrous) 654°C.
Derivation: Sodium hydroxide and vanadium pentoxide in water solution.
Method of purification: Recrystallization.

sodium pyruvate (sodium pyroracemate; sodium acetylformate) $NaOOCCOCH_3$. White powder; apparent melting point 205°C. Very soluble in water.
Use: Biochemical research.

sodium radio-chromate $Na_2Cr^{51}O_4$. A radioactive form of sodium chromate which uses chromium 51 as a biological tracer. See chromium 51.
Grade: U.S.P. XVI (as injection).

sodium radio-iodide NaI^{131}. A radioactive form of sodium iodide containing iodine 131 which can be used as a tracer. See iodine 131.
Grade: U.S.P. XVI (as capsules or solution).

sodium radio-phosphate. N.N.D. 1960 solution contains sodium phosphate monobasic, $NaH_2P^{32}O_4$, and sodium phosphate dibasic, $Na_2HP^{32}O_4$, and has a pH of 5.0-6.0. See phosphorus 32.
Grades: N.N.D; U.S.P. XVI.
Use: Medicine.

sodium resinate. See sodium abietate.

sodium rhodanate. See sodium thiocyanate.

sodium rhodanide. See sodium thiocyanate.

sodium ricinoleate $C_{17}H_{32}OHCOONa$.
Properties: White or slightly yellow, odorless or nearly odorless powder; soluble in water or alcohol; the aqueous solution being alkaline and the alcoholic solution not alkaline.
Derivation: Sodium salt of the fatty acids from castor oil.
Uses: Emulsifying agent in making special soaps. Also used in medicine.

sodium salicylate HOC_6H_4COONa.
Properties: Lustrous, white, crystalline scales or amorphous powder; sweetish, saline taste. The salt prepared from natural salicylic acid has a faint, aromatic odor. Soluble in water, alcohol and glycerol.
Derivation: By heating sodium phenate in an autoclave with carbon dioxide, dissolving and crystallizing.
Method of purification: Recrystallization.
Grades: Technical; C.P.; U.S.P. XVI.
Containers: 25-, 50-, 100-, 200-, 300-lb drums; barrels.
Uses: Medicine; production of salicylic acid; preservative for paste, mucilage, glue and hides.
Shipping regulations: None.*

sodium sarcosinate (sodium sarcosine) CH_3NHCH_2COONa.
Grade: 33% aqueous solution.
Use: Intermediate; stabilizer for diazonium salts; chelating agent.

sodium sarcosine. See sodium sarcosinate.

sodium selenate $Na_2SeO_4 \cdot 10H_2O$.
Properties: White crystals. Soluble in water. Poisonous!
Grades: Technical; reagent.
Uses: Medicine; reagent; as an insecticide for nonedible plants.

sodium selenite $Na_2SeO_3 \cdot 5H_2O$.
Properties: White crystals. Soluble in water; insoluble in alcohol; poisonous!
Derivation: By neutralizing selenious acid with sodium carbonate and crystallizing.
Method of purification: Recrystallization.
Grades: Reagent; technical.
Containers: Glass bottles; wooden kegs.
Uses: Glass manufacture; reagent in bacteriology; testing germination of seeds; decorating porcelain.
Shipping regulations: None.*

sodium sesquicarbonate $Na_2CO_3 \cdot NaHCO_3 \cdot 2H_2O$.
Properties: White needle-shaped crystals; sp.gr. 2.112; m.p., decomposes; composition Na_2CO_3 46.90%; $NaHCO_3$ 37.17%; H_2O 15.93%; Na_2O 41.15%. Soluble in water. Less alkaline than sodium carbonate.

Derivation: By crystallization of a solution containing equimolar quantities of sodium carbonate and sodium bicarbonate.
Grades: Technical.
Containers: Wooden barrels and kegs; paper or burlap bags; fiber drums.
Uses: Detergent and soap builder in laundries; mild alkaline agent for general cleaning and water softening; bath crystals; alkaline agent in leather tanning; cream neutralizer in butter making.

sodium sesquicarbonate, native. Soda occurring in nature as the sesquicarbonate.
Occurrence: Extensive deposits of this mineral, known as natrona, trona, or urao, are found in California; Wyoming; Hungary; Egypt; and the deserts of Africa; Asia; South America.

sodium sesquisilicate. Formula given variously as $Na_4SiO_4 \cdot Na_2SiO_3 \cdot H_2O$; $Na_3HSiO_4 \cdot 5H_2O$; $3Na_2O \cdot 2SiO_2 \cdot 11H_2O$; $Na_2SiO_3 \cdot NaOH \cdot 5H_2O$.
Properties: White, granular powder; soluble in water; pH of 1% solution 12.7.
Derivation: Obtained by crystallization from solutions obtained by heating silica or sodium metasilicate with sodium hydroxide. Intermediate in composition between ortho- and metasilicates; less alkaline than sodium orthosilicate.
Containers: Bags; barrels and fiber drums.
Uses: Heavy duty cleaning (metals, laundries), ingredient in cleaning compounds; textile processing.
Shipping regulations: None.*

sodium silicate (soluble glass; silicate of soda; liquid glass; water glass). See also the other soluble sodium silicates; sodium metasilicate anhydrous, sodium metasilicate pentahydrate, sodium sesquisilicate, sodium orthosilicate.
Formulas: Products varying in ratio from $Na_2O \cdot 3.75\ SiO_2$ to $2Na_2O \cdot SiO_2$ and with various proportions of water.
Properties: Lumps of greenish glass soluble in steam under pressure, white powders of varying degrees of solubility, or liquids cloudy or clear and varying from highly fluid to extreme viscosity; viscosity range from 0.4 to 600,000 poises; f.p. slightly lower than water; miscible with some polyhydric alcohols; partially miscible with primary alcohols and ketones. Gels form with acids between pH 3 and 9; coagulated by brine; precipitated by alkaline earth and heavy metal ions.
Containers: Paper bags; barrels; drums; tank trucks.
Uses: Textiles (fire-proofing; weighting silk; resist in dyeing and printing; boiling of cotton); manufacture of corrugated paperboard, mailing tubes, veneer products, etc.; greaseproofing paper containers, etc.; manufacture of cements; concrete hardeners, etc.; manufacture of cold water paints; filling for soap; cementing stones; waterproofing in hydraulic and acidproof mortars; dyeing and bleaching; cottonseed-oil refining; cementing-pipe insulations; preservative for eggs; in medicine for fastening splints; manufacture of abrasive wheels, stones, etc; adhesive preparations; refining petroleum; ore flotation; lining Bessemer converters; stainproofing wood; sizing paper; boiler compounds; binder; digester linings; acid concentrator linings; ceramic cements; artificial stones; purification of fats and oils; sizing fertilizer bags; inks; paint removers; tanning; drilling mud; manufacture of silica gel.
Shipping regulations: None.*

Sodium Silicate "B-W." [201] Trade name for a specific grade of sodium silicate, 58.5° Bé liquid; SiO_2: Na_2O = 1.60. Other grades in the series are:
"C." 59.3° Bé liquid; SiO_2:Na_2O, 2.00.
"D." 50.5° Bé liquid; SiO_2:Na_2O, 2.00.
"E." 40.0° Bé liquid; SiO_2:Na_2O, 3.22.
"G." Hydrated powder, 18.5% H_2O; SiO_2:Na_2O, 3.22.
"GD." Hydrated powder, 18.5% H_2O; SiO_2:Na_2O, 2.00.
"K." 47.0° Bé liquid; SiO_2:Na_2O, 2.90.
"N." 41° Bé liquid; SiO_2:Na_2O, 3.22.
"O." 42.2° Bé liquid; SiO_2:Na_2O, 3.22.
"RU." 52° Bé liquid; SiO_2:Na_2O, 2.40.
"S 35." 35.0° Bé liquid; SiO_2:Na_2O, 3.75.
"SS." Anhydrous glass and powders; SiO_2:Na_2O, 3.22.

sodium silicofluoride. See sodium fluosilicate.

sodium-silver chloride. See silver-sodium chloride.

sodium-silver thiosulfate. See silver-sodium thiosulfate.

sodium alpha-sodioacetate. See alpha-sodio-sodium acetate.

sodium sorbate $CH_3CH{:}CHCH{:}CHCOONa$. Used as a food preservative. See sorbic acid.

sodium stannate (preparing salt) $Na_2SnO_3 \cdot 3H_2O$.
Properties: White to light tan crystals; soluble in water; insoluble in alcohol; decomposes in air. Aqueous solution slightly alkaline.
Derivation: (a) By fusion of metastannic acid and sodium hydroxide. (b) By boiling tin scrap and sodium plumbate solution.
Grades: Technical; C.P.
Containers: 1-lb bottles; 100-, and 350-lb drums.
Uses: Mordant in dyeing; ceramics; glass; source of tin for electroplating and immersion plating; textile fireproofing.
Shipping regulations: None.*

sodium stearate $NaOOCC_{17}H_{35}$.
Properties: White powder with fatty odor. Soluble in hot water and hot alcohol; slowly soluble in cold water and cold alcohol; insoluble in many organic solvents.
Impurities: The commercial article contains varying quantities of sodium palmitate.
Grades: U.S.P. XVI; technical.
Containers: 150-lb drums; 200-lb barrels.
Uses: Medicine; waterproofing and gelling agent; in toothpaste and cosmetics; as stabilizer in plastics.
Shipping regulations: None.*

sodium subsulfite. See sodium thiosulfate.

Sodium "Sucaryl." [3] Trademark for sodium cyclamate (q.v.).

sodium succinate $Na_2C_4H_4O_4 \cdot 6H_2O$.
Properties: White crystals or odorless granules; soluble in water.
Use: Medicine.
Shipping regulations: None.*

"Sodium Sulamyd." [321] Brand name for sulfacetamide sodium.

sodium 2-sulfanilamidothiazole. See sulfathiazole sodium.

sodium sulfanilate (sodium anilinesulfonate; sodium para-aminobenzenesulfonate) $NaC_6H_4(NH_2)SO_3 \cdot 2H_2O$.
Properties: White, lustrous, crystalline leaflets. Soluble in water.
Derivation: Sulfanilic acid is dissolved in a solution of sodium hydroxide, or carbonate, concentrated and crystallized.
Method of purification: Recrystallization.
Grades: Technical.
Containers: Glass bottles; barrels.
Uses: Medicine; organic synthesis.
Shipping regulation: None.*

sodium sulfantimonate. See livers of antimony.

sodium sulfate. See sodium sulfate, anhydrous; sodium sulfate, decahydrate; thenardite; mirabilite; salt cake.

sodium sulfate, anhydrous (sodium sulfate, exsiccated) Na_2SO_4. See also salt cake.
Properties: White crystals or powder; odorless; bitter saline taste; sp.gr. 2.671; m.p. 888°C; soluble in water and glycerol; insoluble in alcohol.
Derivation: (a) As a byproduct of hydrochloric acid production from salt and sulfuric acid. (b) By sintering a mixture of soda ash and sulfur (product used for kraft paper process only). (c) By passing hot sulfur dioxide and air over salt (Hargreaves process), and (d) By purification of natural sodium sulfate from deposits or brines.
Grades: Technical; C.P.; detergent; rayon; glass makers.
Containers: 100-, 200-lb bags; 100-, 350-lb drums.
Uses: In manufacturing of kraft paper, paperboard, and glass; detergents; sodium salts; ceramic glazes; processing textile fibers, dyes; tanning; stock tonic; pharmaceuticals; freezing mixtures; miscellaneous.
Shipping regulations: None.*

sodium sulfate, crystals. See sodium sulfate decahydrate.

sodium sulfate decahydrate (sodium sulfate, crystals; Glauber's salt) $Na_2SO_4 \cdot 10H_2O$.
Properties: Large transparent crystals, small needles, or granular powder; sp.gr. 1.464 (crystals); m.p. 33°C (liquefies); loses water of hydration at 100°C. Soluble in water and glycerin; insoluble in alcohol.
Derivation: Crystallization of sodium sulfate from water solutions. The name Glauber's salt carries the connotation of the manufactured salt; the natural variety is mirabilite.
Grades: Technical; N.F. XI.
Uses: Textile dyeing; many uses same as for anhydrous variety.

sodium sulfate, exsiccated. See sodium sulfate, anhydrous.

sodium sulfhydrate. See sodium hydrosulfide.

sodium sulfide (sodium sulfuret) Na_2S or $Na_2S \cdot 5H_2O$ or $9H_2O$.
Properties: Yellow or brick-red lumps or flakes or deliquescent crystals; sp.gr. 1.856; m.p. 920°C; soluble in water; slightly soluble in alcohol; insoluble in ether; largely hydrolyzed to sodium acid sulfide and sodium hydroxide.
Derivation: By heating sodium acid sulfate with salt and coal to above 950°C; extraction with water and crystallization.
Method of purification: Recrystallization.
Impurities: Ammonium salts; sodium sulfite; ferrous sulfide.
Grades: Flake; fused; chip sulfide (60% Na_2S), 60% fused and broken; 30% crystals; liquid.
Containers: Barrels; drums.
Uses: Organic chemicals; dyes (sulfur); intermediates; rayon (denitrating); leather (depilatory); dyeing; paper pulp; solvent for gold in hydrometallurgy of gold ores; sulfiding oxidized lead and copper ores preparatory to flotation; flotation processes for lead and copper ores; calico printing; sheep dips; photographic reagent; engraving and lithography; analytical reagent; soap; rubber.
Danger: Contact with acid liberates poisonous gas. Burns skin and eyes. MCA warning label.
Shipping regulations: Flammable solid. Yellow label, except when crystallized or fused solid in metal container.*

sodium sulfite (a) Na_2SO_3 (b) $Na_2SO_3 \cdot 7H_2O$.
Properties: White crystals or powder; saline, sulfurous taste. Soluble in water; sparingly soluble in alcohol.
Constants: Sp.gr.: (a) 2.6334; (b) 1.5939. M.p.: (a) decomposes; (b) loses $7H_2O$ at 150°C.
Derivation: Large sodium carbonate crystals are placed in a lead-lined vat on a perforated false bottom, a current of sulfur dioxide is passed up through the crystals, a solution of sodium bisulfite collecting at the bottom of the vat. This is saturated with sodium carbonate, concentrated and allowed to crystallize.
Method of purification: Recrystallization.
Impurities: Heavy metals; arsenic.
Grades: Reagent; technical.
Containers: 100-, 200-lb bags; 234-, 450-lb barrels; 100-, 375-, 400-lb drums.
Uses: Dyes; intermediates; organic chemicals; sodium thiosulfate; textiles (bleaching delicate fabrics, antichlor); chemical reducing agent; preservative, especially in foods; photography (developer); engraving and lithography; medicine; silvering mirrors; treating rubber latex; sterilizing beer kegs;

permanent wave solutions.
Shipping regulations: None.*

sodium sulfocarbolate. See sodium phenolsulfonate.

sodium sulfocyanate. See sodium thiocyanate.

sodium sulfocyanide. See sodium thiocyanate.

sodium sulfonates. Class name for various sulfonates derived from petroleum. Some specific sulfonates are sodium tert-butylbenzenesulfonate; sodium hexylbenzenesulfonate; sodium toluenesulfonate.
Typical specifications: Equivalent weight, min. 450; inorganic salts (as Na_2SO_4) max 0.8% by weight; sulfonate content (water-free basis) min. 62% by weight; water content 3.0-5.0% by weight.
Uses: In textile processing oils; oils for metal-working (emulsifying and antirust agents); lubricating oils; as emulsifiers for insecticides, herbicides, fungicides; as a preparation of dyes and intermediates; a hydrotropic solvent; in some cases for coatings in food packaging.

sodium sulforicinoleate.
Derivation: Product of successive sulfonation (partial) and saponification of castor oil. Composition indefinite.
Use: As emulsifying and wetting agent.

sodium sulfoxone $(NaO_2SCH_2NHC_6H_4)_2SO_2$. Disodium sulfonylbis(para-phenyleneimino) di(methanesulfinate).
Properties (of U.S.P. XVI mixture containing buffers, inert ingredients, and 73-81% sodium sulfoxone): White to pale yellow powder, with characteristic odor. Very soluble in water, yielding a clear, pale yellow solution; slightly soluble in alcohol.
Grade: U.S.P. XVI.
Use: Medicine.

sodium sulfoxylate. See sodium formaldehyde sulfoxylate.

sodium sulfoxylate formaldehyde. See sodium formaldehyde sulfoxylate.

sodium sulfuret. See sodium sulfide.

sodium superoxide NaO_2. Contains O_2^- group.

sodium suramin (suramin) $C_{51}H_{34}N_6O_{23}S_6Na_6$.
Properties: White or slightly pink powder; odorless; slightly bitter taste; very hygroscopic and affected by light. Soluble in water; slightly soluble in alcohol; insoluble in ether, chloroform, and benzene.
Grade: U.S.P. XVI.
Use: Medicine.

sodium tartrate (sal tartar) $Na_2C_4H_4O_6 \cdot 2H_2O$.
Properties: White crystals or granules. Soluble in water; insoluble in alcohol. Sp. gr. 1.794.
Derivation: Neutralization of tartaric acid with sodium carbonate, concentration and crystallization.
Method of purification: Recrystallization.
Grades: Technical; C.P.; reagent.
Containers: 1-lb bottles; wooden kegs.
Use: Medicine; chemical reagent.
Shipping regulations: None.*

sodium tartrate, acid. See sodium bitartrate.

sodium TCA. See sodium trichloroacetate.

sodium tellurate Na_2TeO_4.
Properties: White powder; soluble in water.
Use: Medicine.

sodium tellurite (sodium tellurite, normal) Na_2TeO_3.
Properties: White powder; soluble in water.
Grade: Technical.
Use: Bacteriology.

sodium tellurite, normal. See sodium tellurite.

sodium tetraborate. See sodium borate.

sodium 2, 3, 4, 6-tetrachlorophenate $C_6HCl_4ONa \cdot H_2O$.
Properties: Buff to light brown flakes; bulk density 26-29 lb/cu ft; pH of water-saturated solution 9.0-13.0. Soluble in water, methanol, acetone.
Use: Industrial preservative (as a bactericide and fungicide).

sodium tetradecyl sulfate $(CH_3)_2CHCH_2$-$CH(OSO_3Na)CH_2CH_2CH(C_2H_5)CH_2(CH_2)_2CH_3$. Sodium 7-ethyl-2-methyl-4-hendecanol sulfate.
Properties: White, waxy, odorless solid. Soluble in alcohol, ether, and water. 5% solution is clear and colorless. pH (5% solution) 6.5-9.0.
Grade: N.N.D.
Use: Medicine; wetting agent.

sodium tetraiodophenolphthalein. See iodophthalein sodium.

sodium tetraphosphate (sodium polyphosphate; glassy sodium phosphate). A sodium phosphate glass similar in a general way to sodium hexametaphosphate, but having a higher ratio of sodium oxide (Na_2O) to phosphorus pentoxide (P_2O_5). There is a considerable confusion and uncertainty as to the precise meaning of this term. For more about structure, see sodium metaphosphate.
Forms and uses: See "Quadrafos."

sodium tetrasulfide Na_2S_4.
Properties: Yellow, hygroscopic crystals or clear dark red liquid; m.p. of crystals 275°C.
Grade: Aqueous solution containing 40% by weight of compound.
Containers: Glass bottles; carboys; 700-lb drums.
Uses: Reducing organic nitro bodies; manufacture of sulfur dyes; insecticides and fungicides; ore flotation agent; soaking hides and skins; preparation of metal sulfide finishes.
Shipping regulations: None.*

sodium theophylline glycinate. See theophylline sodium glycinate.

sodium thiocyanate (sodium sulfocyanate; sodium sulfocyanide; sodium rhodanate; sodium rhodanide) NaSCN.
Properties: Colorless, deliquescent crystals

or white powder; m.p. 287°C, poisonous! Soluble in water and alcohol. Hygroscopic and affected by light.
Derivation: By boiling sodium cyanide with sulfur.
Method of purification: Crystallization.
Grades: Technical; pure, crystal or dried; C.P.
Containers: 1-, 5-lb bottles; tins; drums.
Uses: Analytical reagent; dyeing and printing textiles; medicine; black nickel plating; in manufacturing other thiocyanate salts; in manufacturing of artificial mustard oil, in treatment of rubber; solvent for polyacrylates.

sodium thioglycolate (sodium mercaptoacetate) $HSCH_2COONa$. The sodium salt of thioglycolic acid.
Properties: Crystals; characteristic odor; hygroscopic; discolors on exposure to air or iron; soluble in water; slightly soluble in alcohol.
Containers: Bottles; 6-, 13-gal carboys.
Uses: Bacteriology; cold waving of hair; depilatory; analytical reagent.

sodium thiosulfate (sodium hyposulfite; antichlor; sodium subsulfite; hypo) $Na_2S_2O_3 \cdot 5H_2O$. The anhydrous salt is also commercially available.
Properties: White, translucent crystals or coarse, crystalline powder; cooling taste and bitter after-taste. Soluble in water and oil of turpentine; insoluble in alcohol; deliquescent in moist air; efflorescent above 33°C in dry air. Sp.gr. 1.69; m.p. 48°C; b.p., decomposes.
Derivation: (a) Combination of sulfur and sodium sulfite in aqueous solution; (b) as a by-product in the manufacture of sodium sulfide; a sulfide-carbonate liquor is converted to sodium thiosulfate by reaction with sulfur dioxide; (c) as a by-product in sulfur dye manufacture; (d) drom the reaction of sulfur and sodium bisulfite, which is obtained from reaction of soda ash and sulfur dioxide.
Method of purification: Recrystallization.
Impurities: Sulfates; sulfites; free alkali; calcium sulfides.
Grades: Technical; crystals; granulated; photographic; C.P.; pure; U.S.P. XVI.
Containers: 1-, 5-lb bottles; 100-lb bags; 175-lb drums; 375-lb barrels.
Uses: Photography (fixing agent to dissolve unchanged silver salts from exposed negatives); chrome tanning; removing chlorine in bleaching and papermaking; extraction of silver from its ores; dechlorination of water; mordant in dyeing; reagent in analytical chemistry; medicine; bleaching bone, straw, ivory; reducing agent in chrome dyeing.
Shipping regulations: None.*

sodium thyroxine. See sodium levothyroxine.

sodium titanate (sodium trititanate) $Na_2Ti_3O_7$.
Properties: White crystals. Insoluble in water.
Containers: Cartons, drums.
Use: Welding.

sodium toluenesulfonate (para-toluenesulfonic acid, sodium salt) $CH_3C_6H_4SO_3Na$.
Properties: Crystals; very soluble in water.
Use: Dye chemistry; hydrotropic solvent.

sodium para-**toluenesulfonchloramine.** See chloramine T.

sodium trichloroacetate (sodium TCA) CCl_3COONa.
Containers: 10-lb cans; 50- and 100-lb drums.
Uses: Grass killer, pesticide.
Warning: Irritating to skin and eyes. May cause burns. MCA warning label.

sodium 2,4,5-trichlorophenate $C_6H_2Cl_3ONa \cdot 1\frac{1}{2}H_2O$.
Properties: Buff to light brown flakes; bulk density 28-33 lb/cu ft; pH of water-saturated solution 11.0-13.0. Soluble in water, methanol, acetone.
Use: Industrial preservative (as a bactericide and fungicide).

sodium tridecylbenzenesulfonate. A synthetic detergent. See sodium dodecylbenzenesulfonate.

sodium trimetaphosphate $(NaPO_3)_3$, or better $NaPO_3$ I, I' or I'' according to the particular form involved. There is some evidence to justify the subscript 3 for the molecular formula, but the names sodium metaphosphate I, I', and I'' are preferable. $NaPO_3$ I is the crystalline form resulting from experiments in which the preparation is heated to temperatures just below the melting point (627.6°C), while $NaPO_3$ I' and I'' result from carefully controlled cooling of molten sodium metaphosphate. All three forms are crystalline, soluble in water, and yield $NaPO_3 \cdot 2H_2O$ when the solution crystallizes at room temperature.
See sodium metaphosphate.

sodium triphosphate. See sodium tripolyphosphate.

sodium tripolyphosphate (STPP; sodium triphosphate; pentasodium triphosphate) $Na_5P_3O_{10}$.
Properties: Powdered or granular form; soluble in water; pH (1% solution) 9.75; is chelating agent for certain metals in solution.
Derivation: Slow crystallization from a heated mixture of mono- and disodium phosphates, made from proper amounts and concentrations of phosphoric acid and soda ash.
Containers: Bulk cars; 100-lb paper bags.
Uses: Soap builder; manufacture of detergents, water softeners; purification of china clay; conditioning oil drilling muds; particularly effective in bar soaps-will not crystallize or bloom; disperses soap curds in hard water and eliminates scum; clay dispersant; antipitch agent in paper making; textile processing; dispersant in cements.

sodium trititanate. See sodium titanate.

sodium tungstate (sodium wolframate) $Na_2WO_4 \cdot 2H_2O$.
Properties: Colorless crystals; soluble in

water; insoluble in alcohol and acids. Sp. gr. 3.245; m.p., loses $2H_2O$ at 100°C and then melts at 692°C.
Derivation: By dissolving tungsten trioxide in caustic soda solution, concentration and crystallization.
Method of purification: Recrystallization.
Impurities: Chlorine; sulfates.
Grades: Technical; C.P.; crystalline.
Containers: 1-, 5-lb bottles; 25-, 50-lb cans; 100-lb kegs.
Uses: Intermediate in manufacture of tungsten, tungstic acid and oxides; reagent in analytical chemistry; production of phosphotungstates and borotungstates; fireproofing fabrics and cellulose.
Shipping regulations: None.*

sodium undecylenate $CH_2{:}CH(CH_2)_8COONa$.
Properties: A fine, white powder; decomposes above 200°C; limited solubility in most organic solvents; soluble in water.
Uses: Bacteriostat and fungistat in cosmetics and pharmaceuticals.

sodium uranate. See sodium diuranate, the compound which is considered better to represent the pigment, uranium yellow.

sodium uranium acetate. See uranyl sodium acetate.

sodium valerate (sodium isovalerate) C_4H_9COONa.
Properties: White deliquescent mass. Soluble in water or alcohol; m.p. 140°C.
Use: Medicine.
Shipping regulations: None.*

sodium vanadate. See sodium orthovanadate; sodium metavanadate.

sodium wolframate. See sodium tungstate.

sodium xylene sulfonate (dimethylbenzenesulfonic acid, sodium salt) $(CH_3)_2C_6H_3SO_3Na \cdot H_2O$.
Use: Hydrotropic solvent, used in detergents, etc.

sodium zirconium glycolate
$NaH_3ZrO(H_2COCOO)_3$. Available as a clear, light straw-colored solution, sp.gr. 1.28-1.30, containing 35.7-38.6% solids; 12.5-13.5% ZrO_2.
Containers: 5-gal pails; 50-gal drums.
Uses: Personal deodorant; astringent; germicide; sequestrant; fire retardant.

sodium zirconium lactate
$NaH_3ZrO(CH_3CHOCOO)_3$. Available as a clear, straw colored solution, sp.gr. 1.28-1.30, containing 12.5-13.5% ZrO_2, equivalent to 42.5-45.9% sodium zirconium lactate; pH 7.5-8.0.
Containers: 5-gal pails; 50-gal drums.
Uses: Personal deodorant and antiperspirant.

sod oil. See degras.

soft coal. See bituminous coal.

"Soft-Cote." [173] Trademark for a soft finish water repellent. A clear light amber liquid; may be used with either petroleum or synthetic solvent to convey soft flexible water repellent finish to any garment or textile.

softening agents. Substances used (in many industries) for the purposes of promoting and increasing softness of products such as textiles, leather, paper and rubber. They may be added during the processing for facilitating operations or they may be present in the finished product.

Rubber softeners are usually added to lubricate the rubber mass on the compounding mills, thus effecting reduction in power consumption in the milling operation; to increase the dispersion of pigments; or to prevent premature vulcanization during processing. Greater smoothness of finish may also result from their use. Rubber softeners are composed chiefly of the following single materials or their mixtures; coal-tar resins (indene-coumarone types); mixtures of mineral and vegetable oils, asphalts and pitches, petroleum, derivatives, etc.

Textile softeners are used primarily to counteract the stiffness and harshness which may be imparted to yarns and fabrics by other finishing materials or operations. Such softeners are based on solubilized castor oil (ricinoleates) or on emulsions or dispersions of synthetic fatty bodies, or stearic acid; on sulfonated oils, sulfated fatty alcohols or quaternary ammonium salts, usually referred to as cationic agents. Silicones and polyoxyethylenes are also used.

"Soilfume." [55] Brand name for highly effective soil fumigant against root-knot nematodes, wireworms and seed-corn maggots. Active ingredient is ethylene dibromide.
Caution: Vapor harmful; avoid breathing vapor.

soil stabilizers. Chemicals which can alter an engineering property of a natural soil to suit an intended use of the soil. Usually refers to chemicals which convert sandy soils to satisfactory traffic-bearing materials.

soja-bean oil. See soy-bean oil.

sol. A liquid colloidal suspension or solution; a colloidal dispersion that is a liquid.

"Solacet." [206] Brand name for a line of soluble dyestuffs for acetate materials.

"Solantine." [243] Trademark of fast to liquid direct dyes for cellulose.

"Solar." [152] Trade name for a series of surfactants and emulsifying agents that range from anionic to nonionic and are available in solid and liquid forms. Activity of these products ranges from 20% active to 100% active. Used as base materials in the manufacture of various types of detergents and cleaning compounds and in a large variety of industrial applications.
Containers: 25-lb packages; 50-lb bags; 100-, 200-lb drums; 30-, 55-gal drums for liquids.

"Solar." [307] Brand name of proprietary line of phosphotungstic and phosphomolybdic lakes

*See "I.C.C. Shipping Regulations," page xiii.
Reference numbers refer to name of manufacturer. See "List of Manufacturers," page v.

in dispersed powder form.
Use: For the coloring of various grades of paper.

solar cell. A device which converts light energy directly into electrical energy by use of a p-type semiconductor. An example is a pure silicon crystal in the form of a thin wafer, which has been treated with small amounts of impurities which cause positive and negative charges to collect on opposite surfaces of the wafer. Radiant energy, such as sunlight, impinging on the positive surface disturbs the electrical balance causing a current flow through the wafer and then to any circuit desired.

"Solastral." 243 Trademark for phthalocyanine pigments.

solder. An alloy of relatively low melting point, used for joining other metals of higher melting point. The term applies particularly to lead tin alloys.

soldering acid. See hydrochloric acid.

"Soledon." 206 Brand name for solubilized vat dyestuffs and accessory materials.

"Solfast Blue." 141 Trade name for phthalocyanine blue pigments.
Properties: Excellent light resistance; excellent heat resistance; excellent resistance to acids, alkalies and organic solvents. Non-bleeding in water and organic solvents.
Grades: Red shade, medium shade and green shade blues; non-crystallizing and non-flocculating types.
Uses: Paints; enamels; lacquer; printing inks; rubber, plastic, wallpaper, textiles, floor covering and paper coatings.

"Solfast Green." 141 Trade name for chlorinated copper phthalocyanine pigments.
Properties: Bright green shades with excellent resistance to light, heat, acid and alkali. Non-bleeding in water, alcohol, lacquer or petroleum solvents. Good resistance to crystallization and flocculation.
Grades: Yellow shade green and blue shade green.
Uses: Paints, enamels, lacquer, printing inks, rubber, plastic, wallpaper, textiles, floor coverings and paper coatings.

"Solfast Methyl Violet." 141 Trade name for violet pigment produced by precipitation of the basic methyl violet dyestuff with phosphotungstic acid.
Properties: Brilliant shade; very high strength; good lightfastness. Produces very clean tints. Non-bleeding in oil, water, and paraffin.
Uses: Printing inks.

"Solfast Red." 141 Trade name for precipitated azo pigments derived from beta-hydroxynaphthoic acid.
Properties: Good resistance to light, good heat resistance, non-bleeding in water and organic solvents.
Grades: Medium shade red and dark shade red.
Uses: Printing inks, paints, enamels, lacquers, rubber, plastics, and floor coverings.

"Solfast Victoria Blue." 141 Trade name for blue pigment produced by precipitation of basic Victoria Blue dye with phosphotungstic acid.
Properties: Brilliant shade; very high strength; fairly good lightfastness; good bleed resistance.
Uses: Printing inks.

"Solganal." 321 Brand name for aurothioglucose.

"Solidogen." 307 Trademark for a group of dye fixing agents.
"Solidogen LT-13." Resinous type compound; 35% active; cationic.
Properties: Aqueous solution; sp.gr. 1.13-1.15.
"Solidogen WF." Resinous type compound: 100% active; cationic.
Properties: Fine, white powder; soluble in water; compatible with urea-formaldehyde resin finishes.
Uses: Fixing agent for direct dyestuffs; improves their water fastness on cotton, rayon and other cellulosic fibers; can be applied in long liquor ratios or by padding; can be used with copper salts to improve wash fastness of direct cellulosic fibers.

"Solid Phosphoric Acid Condensation Process." 416 Patented process for polymerization or alkylation of hydrocarbons employing a solid phosphoric acid catalyst (specially treated mixtures of phosphoric acid and a diatomaceous earth such as kieselguhr). Mixtures of olefins may be polymerized to form so-called "polymer gasoline" or propylene may be polymerized to form propylene dimer, trimer, tetramer, etc., as examples of the polymerization application of the process. Under other process conditions and with additional equipment, benzene may be alkylated with propylene tetramer to form dodecylbenzene, or with ethylene to form ethylbenzene, or with propylene to form cumene.

solid propellants. See rocket propellants.

solid solution. A homogeneous crystalline material containing two or more substances in variable proportions.

solid state physics. The explanation and understanding of the physical properties of solids through knowledge of the arrangement and behavior of the atoms and molecules, of ions and electrons in the individual crystals composing the solid, and particularly the effect of imperfections in the arrangements of the atoms, etc., in these crystals. There are important applications in metallurgy, ceramics, glass technology, high polymer properties, semiconductors, magnetic materials, and corrosion. See crystal.

"Solinox." 64 Trademark for a line of modified linseed and soybean oil plasticizers used in lacquers and coatings.

solions. Circuit control elements in which the charge carriers are ions in solution rather than electrons as in vacuum tubes, or electrons and "holes" as in transistors.

"Solithane 113." [27] An extremely versatile liquid urethane prepolymer that can be cured to produce solid materials for many applications. Depending upon the selection of curing system, polymerized urethane adducts will range from very soft rubbery compounds to products of a rigid nature.
Properties: Abrasion and impact resistance; negligible shrinkage at room temperature; good dielectric properties; good adhesion; good resistance to most chemicals; complete compatibility with colorants.
Uses: Protective coatings for metallic and non-metallic surfaces; fabric coatings; encapsulating; potting; printing rolls; casting of components.

"Solivap." [206] Brand name for a proprietary green dye for use in solar evaporation of sea water.

"Solka." [231] Trademark for a purified wood cellulose.

"Solka-Floc." [231] Trademark for flock, alpha flock, or powdered cellulose products which are derived by mechanical comminution of purified wood pulp. Available in various fiber lengths including dense free-flowing powders.
Properties: White in color; brightness up to 92; particle size 40 to 165 microns; moisture 5-7%; sp.gr. 1.58; apparent density 9-34 lbs/cu ft; relatively inert to acids, alkalies and solvents; practically ashless; and when bone dry is 99.5+ % cellulose. Soft, adsorbent and non-abrasive.
Containers: Multiwall bags.
Uses: Filter aid; raw material for cellulose derivatives; processing aid in rubber; component in welding rod coatings; inert bulking agent in food products.

"Solochromate." [206] Brand name of proprietary line of mordant dyestuffs particularly suitable for application to wool, by the one bath method; yielding results of excellent fastness to light, washing, milling, etc.

"Solochrome." [206] Brand name for mordant dyestuffs for wool.

"Solox." [192] Proprietary product said to consist of 100 parts of specially denatured alcohol No. 1, to which is added 1 part of ethyl acetate, 1 part of aviation gasoline and 2 parts denaturing grade wood alcohol. Also available in anhydrous form, characterized by mild odor.
Regular 190°.
Constants: Sp.gr. 0.8158 (60/60°F); acidity, free acid as acetic, not more than 0.02 g/100 cc; odor mild, non-residual; purity not less than 87.1% ethyl alcohol by volume; color water-white; non-volatile matter not more than 0.003 g/100 cc. Apparent proof at 60/60°F 190.0; flash point approx 71°F; water solubility at 25°C: 100 cc of "Solox" yields a clear solution with 70 cc of water, above these limits turbidity develops. Wt/gal (60°F) 6.79 lbs.
Anhydrous 200°.
Constants: Sp.gr. 0.7962 (60°F); acidity, free acid as acetic not more than 0.015 g/100 cc; odor mild, non-residual; purity not less than 91.5% ethyl alcohol by volume; dryness miscible without turbidity with 20 vols. 60° Bé. gasoline at 20°C; non-volatile matter not more than 0.003 g/100 cc; color water-white. Apparent proof 199.0 (60/60°F); flash point approx. 71°F; water solubility: 100 cc (25°C) of anhydrous "Solox" yields a clear solution with 80 cc of water, above these limits turbidity develops; wt/gal (60°F) 6.63 lbs.
Containers: 1-gal cans; 5-, 55-gal steel drums; 30,000-lb drum cars; 6000- and 8000-gal tank cars.
Uses: Solvent (household and industrial); solvent mixtures with ethylene dichloride for cellulose acetobutyrate lacquers; solvent mixtures with toluene ("Solox" 20%, toluol 80%) for ethyl cellulose; resin solvent; lacquers, paints, varnishes, coating compositions; ethyl cellulose adhesives; solvent-cleaner for watches, jewelry, electrical equipment, delicate parts, etc.; dry-cleaning fluids; aviation ice-formation; aniline.
Shipping regulations: Flammable liquid. Red label.*

"Solozone." [28] Trademark for sodium peroxide. Pale yellow beadlike particles; stable when dry; readily dissolves in cold water to form alkaline hydrogen peroxide solution. Na_2O_2 content min 96%; active oxygen content approx 20% by weight. Dustless and free-flowing; bulk density approx. 95 lb/cu ft; melting point above 450°C.
Containers: 75-lb pails; 400-lb drums.
Uses: Bleaching textiles and wood pulp; deinking waste papers; purification of metal salt solutions; manufacture of hydrogen peroxide and organic peroxides; general oxidizing agent.
Shipping regulations: Oxidizing material. Yellow label. *

"Solricin" 135. [202] Trademark for an aqueous solution of potassium ricinoleate containing approximately 32% of the soap and 3% glycerine.
Derivation: From castor oil.
Containers: 5-gal cans; 55-gal drums.
Uses: Emulsifier and stabilizer in foamed rubber; cleaning compounds; germicides.

"Solricin" 285. [202] Trademark for an 85% aqueous solution of ammonium ricinoleate; used as a rust proofing agent.

"Solros." [79] Trademark for a special "FF" wood rosin, containing no lime or other inorganic chemicals, which is distinguished by its excellent solubility in a wide range of solvents.
Constants: M.p. (capillary tube) 53°C; m.p. (ball and ring) 72°C; acid number 125; unsaponifiable matter 22.0%; color "FF."

*See "I.C.C. Shipping Regulations," page xiii.
Reference numbers refer to name of manufacturer. See "List of Manufacturers," page v.

Uses: Adhesive tape; artificial Burgundy pitch; belt dressings; branding paint; core oil; electric insulating compounds; pitch; printing ink; rock wool; roofing cement; rubber cement; shellac diluent; smoking molds; spirit varnishes; sticky fly paper; synthetic rosin oil; tree banding; Venice turpentine; wire coating compounds.

"Soltrols." [303] Trademark for complex mixtures of saturated branch-chain hydrocarbons, having extremely low odors.
Properties: (Soltrol 130): Boiling range 168-210°C; sp.gr. (60/60°F) 0.7539; flash point 57°C; (Soltrol 170): Boiling range 215-246°C; sp.gr. (60/60°F) 0.7745; flash point 85°C.
Containers: Drums and tank cars.
Uses: Odorless solvents in paint and insecticide formulations.

soluble blue.
1. Soluble Prussian blue, made by precipitating a ferric salt with potassium or sodium ferrocyanide or by pressing Prussian blue to a firm pulp and working in oxalic acid or ammonium oxalate. The pulp becomes runny and is then dried on lead or lead-lined pans.
2. An organic dye, C.I. 707, based on triphenylmethane.

soluble glass. See sodium silicate.

soluble gluside. See saccharin sodium.

soluble indigo. See indigo carmine.

soluble guncotton. See pyroxylin.

soluble iodophthalein. See iodophthalein sodium.

soluble oils. These oils are known as emulsifying oils, since they are normally bright, clear oils which, when mixed with water, produce milky emulsions. In some soluble oils the emulsion is so fine that instead of milky solutions in water, amber colored transparent solutions are formed. Typical examples are sodium and potassium petroleum sulfonates. See detergents, synthetic (1).
Uses: The soluble oils, when mixed in appropriate proportions with water are used as metal cutting lubricants; textile lubricants; metal boring lubricants; emulsifying agents.

soluble saccharin. See saccharin, sodium.

soluble starch (thin boiling starch). Starch that has been modified by oxidizing agents, acids, glycerin, enzymes or other agents or treatments so that the product is soluble or dispersible in hot water. Such starches have relatively little adhesive character.
Uses: For sizing paper and textiles; chemical indicator.

"Solu-Rez." [170] Trademark for a modified polyvinyl resin emulsion designed as a multi-purpose packaging adhesive.
Properties: White liquid, dries colorless, of varied viscosities with excellent mechanical stability. Will not build up or change viscosity on prolonged machining. Skin formulations will redissolve on agitation. Can be pumped through circulating systems.
Uses: High speed carton sealing, carton forming, tube winding, bag seaming, sealing, and tight-wrapping machines.

"Solutax" Poster Paste. [170] Trademark for a dehydrated vegetable cellulose compound designed as a cold water poster gum for outdoor advertising poster pasting. Material remains uniform even upon freezing and thawing, therefore affords adhesion through all types of weather.

solute. A dissolved substance, or a substance that is to be dissolved in another substance. Usually the solution components present in relatively small concentration are called solutes. See solvent, solution.

solution. A true solution is a homogeneous mixture of two or more substances that has the following characteristics: (1) spontaneous formation; (2) subdivision down to molecular magnitudes; (3) absence of settling; and (4) no fixed proportions of the component substances. The best known examples of solutions are cases in which solids are dissolved in liquids (salt or sugar in water) but solutions of liquids in liquids, gases in liquids, gases in solids and solids in solids are also known.

solutrope. Ternary mixtures having two liquid phases between which one component is distributed in an apparent ratio varying with concentration from less than one to more than one. In other words, the solute may be selectively dissolved in one or the other of the phases or solvents depending on the concentration. This phenomenon has been compared to azeotropic behavior.

"Solvat." [243] Trademark of leuco esters of vat dyes used for wool, cellulose and synthetic fibers.

"Solvay Nitrox." [292] Trademark for a product designed to provide in a single operation the cleansing and neutralizing action of caustic soda and the corrosion protection properties of sodium nitrite. It is produced in a fused, flake form to insure a uniform, free-flowing, non-caking product. Used in cleaning, protection and neutralizing of iron and steel during processing and storage.

Solvay process (ammonia soda process). Manufacture of sodium carbonate (soda ash, Na_2CO_3) from salt, ammonia, carbon dioxide and limestone by an ingenious sequence of reactions involving recovery and reuse of practically all the ammonia and part of the carbon dioxide. Limestone is heated to produce lime and carbon dioxide. The latter is dissolved in water containing the ammonia and salt, with resultant precipitation of sodium bicarbonate. This is separated by filtration, dried and heated to form the normal sodium carbonate. The liquor from the bicarbonate filtration is heated and treated with lime to regenerate the ammonia. Calcium chloride is a major byproduct.
See also soda ash.

"Solvay Snowflake Crystals." [292] ($Na_2CO_3 \cdot NaHCO_3 \cdot 2H_2O$). Trademark for small white sparkling needle-like crystals uniform in size which will not cake under ordinary conditions. Used in general cleaning where mild alkaline action is desired; as a laundry detergent and as a base for bath crystals, both foaming and nonfoaming types.

Also applied to cleansers containing the crystals.

"Solvenol No. 1." [266] Trademark for monocyclic terpene hydrocarbons with minor amounts of terpene alcohols and ketones; clear, colorless liquid; sp.gr. 0.8560 (15.6/15.6°C); ASTM distillation range, 5-95%, 174-184°C.

"Solvenol No. 2." [266] Trademark for monocyclic terpene hydrocarbons with minor amounts of terpene alcohols and ketones; clear colorless liquid.

solvent. That component, usually liquid, that is present in excess in a solution. The term may however by also applied to a substance frequently used as a solvent, such as water or alcohol, even though the proportion is less than indicated above. When no one substance is present in excess, the choice of solvent is arbitrary and unnecessary. In principle, one of the gases in a gaseous mixture, or one of the solids comprising a solid solution may be designated as the solvent.

solvent extraction. See extraction, liquid-liquid.

solvent naphtha. See naphtha, solvent.

"Solvesso." [51] Trademark for aromatic solvents prepared from petroleum. Grades available include "Solvesso" Toluol, "Solvesso" Xylol, "Solvesso" 100 and "Solvesso" 150.

"Solvotone." [214] Trademark for mixture of low boiling alcohols and ketones, including isopropanol and acetone.

"Solway." [206] Brand name for anthraquinone acid dyestuffs for wool, and animal fibers.

Soman (methyl-1, 2, 2-trimethylpropoxy-fluorophosphine oxide) $(CH_3)(C_6H_{13}O)FPO$. A German nerve gas.

somatotropic hormone (STH; somatotropin; growth hormone). One of the hormones secreted by the anterior lobe of the pituitary gland. It causes an increase in general body growth and also has an effect on carbohydrate and lipid metabolism. It is a protein which has been crystallized in pure form.

somatotropin. See somatotropic hormone.

"Sonaquol." [45] Trademark for a water-miscible mineral oil which is emulsifiable in either cold or warm water, without the aid of alkali or sulfonated oils.

Uses: Cosmetics.

"Sono-Jell." [45] Trademark for a balanced blend of white mineral oils and waxes of U.S.P. purity.

Properties: Smooth, white, non-sweating cream base.

Uses: Waterless type cleaning creams; pharmaceutical ointments.

soot. A finely divided powder produced during the combustion of coal or wood, consisting essentially of carbon but containing some tar, ash, ammonium salts, etc.

Uses: Used in agriculture as a fertilizer, and as a slug and snail repellent.

Shipping regulations: None.*

"Sopanox." [58] Trademark for ortho-tolyl biguanide (q.v.).

"Sorapon" SF-78. [307] Trademark for an anionic surfactant, sodium alkyl aryl sulfonate; 85% active.

Properties: Flaky, off-white powder; density 0.43; soluble in water; stable to acid, alkali and hard water.

Uses: Wetting, emulsifying and dispersing agent used in textile processing as a detergent, dyeing and bleaching assistant, and lime soap dispersant; detergent base for industrial cleaning compounds.

sorbic acid (2, 4-hexadienoic acid) $CH_3CH{:}CHCH{:}CHCOOH$.

Properties: White, crystalline solid. M.p. 134.5°C; b.p. 228°C (dec), 153°C (50 mm); flash point (open cup) 260°F. Slightly soluble in water; soluble in many organic solvents.

Derivation: Trimerization of acetaldehyde and catalytic air oxidation of the resulting hexadienal. Found in berries of mountain ash, Sorbus aucuparia L.

Containers: Glass bottles; fiber cans.

Uses: Fungicide and food preservative; copolymerization; upgrading of drying oils; cold rubber additive; intermediate for plasticizers and lubricants.

Shipping regulations: None.*

sorbide (dianhydrosorbitol) $C_6H_8O_2(OH)_2$. Generic name for anhydrides (dicyclic ether dihydric alcohols) derivable from sorbitol by the removal of two molecules of water. The name is also applied to specific commercial varieties. One of these is crystalline, m.p. 62°C. Soluble in water, the lower alcohols, and in ethylene glycol; insoluble in most other solvents. Used as conditioning agent and in medicine.

"Sorbistat." [299] Trademark for sorbic acid.

"Sorbistat-K." [299] Trademark for potassium sorbate.

"Sorbit" AC. [219] Trademark for sodium dibutyl naphthalene sulfonate $(C_4H_9)_2C_{10}H_5SO_3Na$, 65% active light tan paste. Highly soluble in water, polar organic solvents, strong electrolytes. Stable to acid and alkali. Wetting agent and penetrant; hydrotrope; dispersant and thinning agent. Used in detergents.

sorbitan (monoanhydrosorbitol) $C_6H_8O(OH)_4$. Generic name for anhydrides (cyclic ether tetrahydric alcohols) derivable from sorbitol by the removal of one molecule of water. (Some derivatives are marketed as "Spans" and "Tweens.")

The name is also applied to specific commercial varieties. One of these is crystalline, m.p. 110°C; b.p. 225-250°C; soluble in water and acetic acid.

sorbitan mono-oleate polyoxyethylene. See polysorbate 80.

sorbitan monostearate. Sold in large amounts, as is the tristearate, as an emulsifying agent. Can be used in foods.

Containers: Drums.

sorbitan polyoxyethylene fatty acid esters. See polyoxyethylene sorbitan fatty acid esters.

d-**sorbite.** See sorbitol.

sorbitol (d-sorbite; d-sorbitol; sorbol; hexahydric alcohol) $C_6H_8(OH)_6$.

Properties: White, odorless, crystalline powder of faint, sweet, cooling taste. Soluble in water; slightly soluble in methanol, ethanol, acetic acid, phenol and acetamide. Almost insoluble in most other organic solvents. M.p. 93-97.7°C (hydrate); 110°C (anhydrous). Sp.gr. 1.47 (−5°C).

Derivation: By pressure hydrogenation of dextrose with nickel catalyst. Occurs in small amounts in various fruits and berries.

Grades: Crystals; technical; 70% aqueous solution (U.S.P. XVI); resin; powder.

Containers: Powder, pellets, 150- to 325-lb drums; tank cars; tank trucks.

Uses: Explosive manufacture; ascorbic acid (vitamin C) fermentation. In solution form, for moisture-conditioning and otherwise improving quality of cosmetic creams and lotions, toothpaste, tobacco, gelatin, glue specialites; as bodying agent for paper, textiles, and liquid pharmaceuticals, such as elixirs, syrups, etc; softener for candy, shredded sweetened coconut and other confections; sugar crystallization inhibitor. Also used for synthesis of resins, surface active agents, varnishes.

"Sorbo." [89] Trademark for 70% sorbitol solution. Meets the Toilet Goods Association and U.S.P. XVI standards.

"Sorbo-Cel." [247] Trademark for a special chemical coated diatomite filter aid used for selectively removing traces of oil from oil-in-water emulsions. Effective for removing other trace components from free or emulsified systems.

Use: In conditioning of boiler feed water.

sorbol. See sorbitol.

L(−)-**sorbose** $HOCH_2CO(CHOH)_3CH_2OH$.

Properties: White crystalline powder; sweet taste; m.p. 159-161°C; soluble in water; slightly soluble in ethyl or isopropyl alcohol, insoluble in ether, acetone, benzene, chloroform.

Derivation: Made from sorbitol by submerged culture aerobic fermentation.

Grades: Technical; reagent.

Uses: In the manufacture of ascorbic acid (vitamin C) and for preparation of special diets and media for the study of metabolism in animals and microorganisms.

Sorel cement. See magnesium oxychloride cement.

sorghum syrup. Syrup produced from a cane-like grass (sorghum, kafir, Indian millet) resembling maize in appearance and cultivated in the United States. The grass yields a sugar-containing juice from which the syrup is made. It is used for food.

"Sorolene G Paste." [328] A dark amber paste with green fluorescence—the sodium salt of an alkyl-naphthalene sulfonic acid, of approximately 28% activity.

Uses: Agricultural dispersions; cleaning compounds; wall paper removers; pigment dispersions; flotation agent in refining operations; and foaming agent for rubber, insulation, and concrete.

"Soromine." [307] Trademark for a series of softening agents.

"Soromine AT." Complex fatty amido compound; 20% active; amphoteric.

Properties: White paste; density 0.91; readily dispersible in warm water; functions as a cationic softening agent in acid liquors and a nonionic and/or anionic softener in neutral of alkaline liquors.

Uses: Softener for animal, cellulosic and synthetic fibers; applied on quetch or pad, in package machines, jigs, overhead reel machines, etc.; does not discolor whites nor adversely affect dyed fabrics.

"Soromine BSA." Alkyl biguanidine; 73% active; cationic.

Properties: Brown paste; density 0.90; soluble in water; compatible with urea formaldehyde resins.

Uses: Highly substantive softening agent for cellulosic fibers which is applied in long liquors or by padding; improves water and perspiration fastness of direct dyeings; anti-static agent for polyester fibers and polyacrylonitrile fibers.

"Soromine FW." Sodium salt of a fatty amide complex; anionic.

Properties: Off-white, gritty paste; density 1.19; soluble in water; compatible with starch, gelatine, sulfonated oils.

Uses: A cellulosic fiber softening agent which causes no discoloration of bleached white or dyed yarns, stock or piece goods; does not yellow during high temperature drying; applied by padding, in package machines and in conventional long liquor dyeing equipment.

sorrel salt. See potassium binoxalate.

"Sotex." [83] Trade name for a series of dispersing agents based on long chain fatty acid esters.

Soudan coffee. See cola.

sour. Any substance used in textile or laundry operations to neutralize residual alkali or decompose residual hypochlorite bleach. The commonly used sours are sodium bifluoride and sodium fluosilicate.

sour gas. Slang for either natural gas or a gasoline contaminated with odor-causing sulfur compounds. In natural gas the contaminant is usually hydrogen sulfide, which can be removed by passing the gas mixture through carbonate solutions containing special metal or organic activators. In gasolines, the sour contaminents are usually mercaptans, which are removed in the doctor treatment (q.v.) or by ethylene oxide with a phenolic catalyst. The improved gas or gasoline is known as sweet gas.

sour-spine. See barberry.

sowberry. See barberry.

"S.O. Wetting Agent S." [151] Trademark for paste type sulfonate derived from "Alkane" detergent intermediate and used for the preparation of commercial and industrial detergents.

soya bean oil. See soybean oil.

soybean cake. The press cake resulting from the extraction of soybeans for their oil. The crushed press cake is termed soybean oil meal (q.v.).

soybean oil (soja-bean oil; soya bean oil; Chinese bean oil; soy oil).

Properties: Pale, yellow, fixed oil. Soluble in alcohol, ether, chloroform and carbon disulfide.

Constants: Sp.gr. 0.924-0.929; m.p. 22-31°C; refractive index 1.4760-1.4775; solidifying point −15 to −8°C; Hehner value 94-96; saponification value 190-193; iodine value 137-143.

Derivation: Soya-beans (Soja hispida, S. japonica or Phaseolus hispida) are crushed, packed into jute bags, heated over jets of steam and pressed. Solvent extraction is now extensively used both alone and after pressing.

Method of purification: Oil to be used for edible purposes is bleached with fuller's earth; oil for technical use is purified with chemicals.

Grades: Coast; refined (salad); crude; foots (for soapstock); clarified.

Containers: 75-lb cases; 375-, 400-lb barrels; 8000-gal tank cars.

Uses: Soap manufacture; foods (this oil has always been one of the chief articles of diet in China, Japan, etc.); inks; adulterant and substitute for linseed oil in paints and varnishes; cattle feeds; butter substitutes and salad dressings; resins; linoleum.

Shipping regulations: None.*

soybean oil meal (soybean cake). The crushed residue from the extraction of soybeans. Extraction by the hydraulic or expeller process produces normally a meal with approximately 6% residual oils while from the solvent process approximately 1% residual oil. Typical analyses show crude protein 43%; crude fiber 5.5%; nitrogen-free extract 30%; ash 6% and oil content between 1 and 6%. The total digestible nutrients approximate 75%.

Containers: Bulk or bags.

Uses: Animal feeds; manufacture of plastics; meat substitutes; medium for bacitracin production.

Shipping regulations: None.*

soy oil. See soybean oil.

sozoiodolic acid (2,6-diiodophenol-4-sulfonic acid) $C_6H_2I_2(OH)(SO_3H)\cdot 3H_2O$.

Properties: White odorless crystals; m.p. 120° (anhydrous). Soluble in water, alcohol and ether.

Use: Medicine (radiopaque medium).

space velocity. The volume of gas or liquid, measured at specified temperature and pressure, (usually standard conditions) passing through unit volume in unit time. Used in comparing flow processes involving different conditions, rates of flow, and sizes or shape of containers.

spalling. Chipping an ore for crushing; or the cracking, breaking or splintering of materials due to heat.

"Span." [89] Trademark for each member of a series of general purpose emulsifiers and surface active agents. They are fatty acid partial esters of hexitol anhydrides (or sorbitan). Generally insoluble or dispersible in water and soluble in most organic solvents.

spandex. Generic name for a manufactured fiber in which the fiber-forming substance is a long chain synthetic polymer comprised of at least 85% of a segmented polyurethane (Federal Trade Commission). Adds elasticity to garments. See, for example, "Lycra."

Spanish blistering flies. See cantharides.

Spanish chamomile. See pyrethrum root.

Spanish flies. See cantharides.

Spanish grass. See esparto.

Spanish lavender oil. See lavender-spike oil.

Spanish oxide. See iron oxide reds.

Spanish pellitory. See pyrethrum root.

Spanish saffron. See crocus.

Spanish spike oil. See lavender-spike oil.

Spanish white. See bismuth subnitrate.

spar, adamantine. See corundum.

spar, dogtooth. See calcite.

spar, Greenland. See cryolite.

spar, heavy. See barite.

spar, ice. See feldspar.

spar, Iceland. See calcite.

"Sparine" Hydrochloride. [24] Trademark for promazine hydrochloride (q.v.).

sparking metal. See pyrophoric alloy.

spar, nailhead. See calcite.

spar, satin. See calcite; also gypsum.

sparteine $C_{15}H_{26}N_2$ (lupinidine).
Properties: Colorless, thick, oily, liquid alkaloid; bitter taste; distinctive peculiar odor; usually used in the form of the sulfate, hydriodide, hydrochloride and triiodide; poisonous! Soluble in alcohol and ether; very slightly soluble in water.
Constants: B.p. 173°C; sp.gr. 1.027.
Derivation: By extraction of the tops of Spartium scoparium (broom) with alcohol and evaporation of the latter.
Grades: Technical.
Containers: Glass bottles; cans (sparteine sulfate).
Use: Medicine.
Shipping regulations: None.*

spar varnish. A very durable, water resistant varnish for severe service on exterior exposure. It consists of one or more drying oils: for example, linseed, tung or dehydrated castor oil; one or more resins: rosin, ester gum, phenolic resin or modified phenolic resin; one or more volatile thinners: turpentine or petroleum spirits; and driers: linoleates, resinates or naphthenates of lead, manganese and cobalt. It is classed as a long-oil varnish and generally consists of 45-50 gallons of oil to each 100 lb of resin.
See also varnish.

spathic iron ore. See siderite.

"SPDX-GH." [94] Trade name for a rubber accelerator, composed of the lead salt of dithiocarbamate mixed with a diphenyl ethylenediamine radical, blended with selected oils.
Properties: Light grey soft granules; sp.gr. 1.51; m.p., decomposes before melting; insoluble in water, acetone, benzol, and gasoline; odorless; non-toxic; disperses well; excellent storage stability.
Use: Mechanicals, heels and soles, hose, insulation, tires, tubes, and tire repair stock. Also available as "SPDX GL" for use in latex.

spearmint.
Derivation: Dried leaves and tops of the herb Mentha spicata or cardiaca.
Occurrence: Europe; Asia; eastern and central United States, and the Yakima valley, Washington.
Grade: N.F. XI.
Containers: Bales.
Uses: Flavoring material; source of spearmint oil.

spearmint oil.
Properties: Colorless to pale yellowish liquid; characteristic odor and taste. Soluble in alcohol, ether, and chloroform.
Chief known constituents: Carvone (40-60%); linalool; pinene.
Constants: Sp.gr. 0.930-0.940; optical rotation −48° to −59°; refractive index about 1.4910.
Derivation: By distillation of spearmint leaves.
Method of purification: Rectification.
Grades: Technical; N.F. XI.
Containers: 5-, 10-lb bottles; 20-, 25-lb tins; drums.
Use: Flavoring.
Shipping regulations: None.*

"Special Dynamite." [413] Brand name applies to a series of ammonia type dynamites. They are more economical, less flammable, and relatively safer than nitroglycerine dynamites. Made in strengths from 15 to 60%.
Containers: Packaged in cartridges 7/8" diameter and up, in 50-lb shipping cases.
Uses: Open pit mining; quarrying; construction projects and general blasting.
Fire hazard: Dangerous.
Shipping regulations: Explosives. Red label.*

"Special Gelatin." [84] Brand name applied to a series of an ammonia gelatin type dynamites. Economical and relatively safe. Made in strengths from 30 to 90%.
Containers: Packaged in cartridges 7/8" in diameter and up, in 50-lb shipping cases.
Uses: Open pit mining; underground metal mining; quarrying; construction projects; primers for non-sensitive explosives compositions.
Fire hazards: Dangerous.
Shipping regulations: Explosives. Red label.*

specific activity. See radioactivity.

specific gravity. The weight of a particular volume of any substance, compared with the weight of an equal volume of water. Since these weights will vary differently with the temperature, it is necessary to specify both temperatures involved, except for rough or approximate values. Thus the specific gravity of alcohol should be given as 0.7893 at 20/4°C, the first temperature referring to the alcohol and the latter to the water.

specific heat. The ratio of the heat capacity of a substance to the heat capacity of water; or the quantity of heat required for a one degree temperature change in a unit weight of material. Commonly expressed in Btu/lb/°F or in cal/g/°C. For a gas the specific heat at constant pressure is greater than that at constant volume by the amount of heat needed for expansion.

specific impulse. See rocket propellants.

specific volume. The volume of unit weight of a substance, as cubic feet per pound, or gallons per pound, but more frequently milliliters per gram. The reciprocal of density.

spectroscopy. Observation of the wave length and intensity of light, or other electromagnetic waves, absorbed or emitted by various materials. When excited by an arc or spark each element emits light of certain well defined wave lengths. Even in very minute quantities, the presence of any element may often by determined spectroscopically. These spectra may be used to obtain information on the structure of the atom. In the visible and infra-red regions, absorption spectra and Raman spectra, while more complex, serve as analytical tools for substances not otherwise readily determined. Theoretical interpretation of spectra in these regions leads to a knowledge of molecular structure. X-ray spectroscopy first established the atomic numbers of the elements. A more recent technique is nuclear magnetic resonance spectroscopy (q.v.).

specular iron ore. A variety of hematite with brilliant black color and metallic luster.

"Speedenamel." [448] Trade name for alkyd gloss general purpose enamels for interior or exterior, brush, dip or spray.

spelter. Term applied loosely to relatively pure zinc as encountered in industrial operations such as galvanizing. Lead and/or iron are common impurities.

spent acid. Mixed acid (q.v.) which has given up part of its nitric acid.
Fire hazard: Dangerous.
Shipping regulations: Corrosive liquid. White label.*

spent oxide. The residue resulting from passage of impure coal gas over iron oxide, as one of the final steps in purification of the gas before storage or domestic and industrial use. Contains unchanged ferric oxide, ferrous sulfide, sulfur as well as nitrogen compounds and other impurities from the gas. The sulfur comes from hydrogen sulfide which must be removed from the gas. Sometimes referred to as gashouse tankage.
Use: As fertilizer for its nitrogen content.
Shipping regulations: Flammable solid. Yellow label. Not accepted by express.*

"Spergon." [248] Trademark for a series of fungicides and seed protectants based on tetrachloro-para-benzoquinone. Seed protectants are in powder form, with or without DDT for dust or slurry treatment of various field, flower and vegetable seeds before planting or storage; fungicidal formulations are available for spray or dust applications on cabbage downy mildew, flowering bulbs, lawns, seed beds, and plant beds.
See chloranil.

"Sperm 42." [403] Trade name for a self-emulsifiable 45 NW sperm oil.
Properties: Amber liquid; 100% active. Contains no water, added emulsifiers, ethylene oxide, or added non-ionic compounds. It is made up largely of fatty alcohol esters of fatty acids in the C_{14} to C_{20} range, plus a small percentage of a chemically reacted nitrogen compound which acts as an emulsifier.
Typical specifications: Cloud point 55°F; pour point 49°F; saponification number 105-125; iodine no. 70-80; flash point 420-440°F; ; unsaponifiable matter 30-40%. Soluble in mineral spirits, kerosine, petroleum oils and most organic solvents; insoluble in alcohol and water.
Containers: Steel drums or tank cars.
Uses: Cutting oils; wire drawing compounds, lubricants, wetting agents; corrosion inhibitor; gear oils; rubber processing aid; leather oils; metal working compounds; oil and grease additive.

spermaceti (cetaceum).
Properties: Pearly-white, unctuous, semi-transparent, concrete, fatty substance; almost odorless and tasteless; becomes rancid on exposure. Soluble in ether, chloroform, carbon disulfide and hot alcohol; insoluble in water and cold alcohol. Sp.gr. 0.945; m.p. 42-50°C; refractive index about 1.4330; saponification no. 120-136; iodine no. 3-4.4.
Chief constituents: Cetyl palmitate, cetyl alcohol, esters of lauric, myristic and stearic acids.
Derivation: Found in the head of the sperm-whale or floating in the ocean; is filtered under pressure to remove stearin, boiled with water and a small amount of caustic soda, followed by repeated washing with water.
Grades: Technical; U.S.P. XVI; as blocks or cakes.
Containers: 50-, 60-lb cases.
Uses: Base for ointments, cerates and emulsions; manufacture of candles, soaps, cosmetics, laundry wax; finishing and lustering linens.
Shipping regulations: None.*

"Spermafol." [221] Trademark for a hydrogenated sperm oil derivative with applications in cosmetics, hand soaps, wax compounds, leather and textile chemicals, lubricants, drawing compounds, waterproof packing, buffing and polishing compounds.

sperm oil (sperm whale oil).
Properties: Light yellow liquid; sp.gr. 0.8781-0.8835; saponification number 123-147; iodine number 79.5-84; acid number 13.2. Soluble in chloroform, ether, and benzene.
Grades: Bleached winter; natural winter.
Containers: Drums, tank cars.
Use: High-grade lubricating oil for light machinery such as watches, clocks and scientific instruments; heat treating; rust proofing.

sperm whale oil. See sperm oil.

sperrylite $PtAs_2$.
Properties: Tin-white mineral, black streak, metallic luster. Of very rare occurrence but of interest as the only native compound of platinum. Contains 52.57% Pt., 43.5% As, with some replacement of platinum by rhodium and palladium.
Constants: Sp.gr. 10.60; hardness 6-7.
Occurrence: United States (Wyoming, North Carolina, Nevada); Canada.

*See "I.C.C. Shipping Regulations," page xiii.
Reference numbers refer to name of manufacturer. See "List of Manufacturers," page v.

Sperry process. An electrolytic process for the manufacture of lead carbonate, basic (white lead) from desilverized lead containing some bismuth. The impure lead forms the anode. A diaphragm separates anode and cathode compartments, and carbon dioxide is passed into the solution. Impurities, including bismuth, remain on the anode as a slime blanket.

sp. gr. Abbreviation for specific gravity (q.v.).

sphagnum (peat moss; bog moss). The moss found in marshes and wet places, which when it decays and dies forms peat. It is a paler green in color than the true mosses and forms taller growths. Its stems and leaves possess colorless cells, which are connected by small perforations that have the property of sucking up, and the capacity to retain, large quantities of liquid.
Shipping regulations: None.*

sphalerite (blende; zinc blende, black jack) ZnS. Natural zinc sulfide, usually containing some cadmium, iron, and manganese.
Properties: Color yellow, brown, black, or red; luster resinous; hardness 3.5-4; sp.gr. 3.9-4.1; good cleavage; soluble in hydrochloric acid.
Occurrence: Missouri, Kansas, Oklahoma, Colorado, Montana, Wisconsin, Idaho; Australia; Canada; Mexico.
Use: Most important ore of zinc; also a source of cadmium; phosphor; source of sulfur dioxide for production of sulfuric acid and other sulfur compounds.

sphene. See titanite.

"Spheron." 275 Trade name for a series of channel carbon blacks for rubber.
Available as:
"Spheron 9." Easy processing channel black (EPC).
"Spheron 6." Medium processing channel black (MPC).

spherosiderite. See siderite.

sphingomyelin. Diaminophosphatides occurring primarily in nervous tissue and containing a fatty acid, phosphoric acid, choline, and sphingosine. They are soluble in hot absolute alcohol, and insoluble in ether, acetone, and water.

"Spiceal." 342 Trademark for water-miscible flavor concentrates from essential oils and oleoresin bases.

spice berry. See aralia.

spiegel. See spiegeleisen.

spiegeleisen (spiegel; spiegel iron). An alloy of manganese (15% to 30%) and iron containing approximately 5% carbon and small amounts of silicon, sulfur and phosphorus. In Bessemer practice molten spiegel is added to steel after the blow and an after-blow applied. The manganese serves to reduce oxides of iron and to furnish manganese to the steel. At the same time the carbon content is adjusted.

spiegel iron. See spiegeleisen.

spigelia (pinkroot; Indian pink; Maryland pink; wormgrass).
Derivation: Dried rhizome and roots of Spigelia marilandica.
Occurrence: North America.
Grades: Technical.
Containers: Bags; bales.
Use: Medicine.
Shipping regulations: None.*

spignet. See aralia.

spike lavender oil. See lavender-spike oil.

spikenard. See aralia.

spike oil. See lavender-spike oil.

spin. An important concept in nuclear physics, which includes both the total angular momentum of a system, which may be a nucleus or a nuclear reaction, and the intrinsic angular momentum of the nucleons which make up the nucleus. Since a single nucleon may spin in only one of two directions, the intrinsic angular momentum may have only one of two possible values, but it will be coupled to the orbital angular momentum, and the manner in which the spin and orbital angular momenta are coupled together to produce the net value is important in an understanding of nuclear processes.

spinacane. See squalane.

spinacene. See squalene.

spindle oil. Low viscosity lubricating oil for lubrication of relatively high speed machinery.

spindle tree. See euonymus.

spinel $MgAl_2O_4$. A natural oxide of magnesium and aluminum, with replacement of magnesium by iron, zinc and manganese, and of aluminum by iron and chromium. There are also synthetic spinels, as magnesia-alumina or magnesia-chromia.
Properties: Color, various shades of red, grading to green, brown, and black; luster vitreous; hardness 8. There are many varieties.
Occurrence: New York, New Jersey, Massachusetts, Virginia, North Carolina; Ceylon; Burma; Thailand; Madagascar.
Uses: Gem stone (for example, balas ruby); synthetic spinel is used as a refractory.

"Spinesso." 51 Trademark for lubricating oils for textile spindles and other high-speed shafts. Viscosity grades cover the entire range now in use, and all grades contain oxidation inhibitor, oiliness agent and a corrosion preventive.

Spinning Lubricant L. 28
Properties: A pale yellow liquid.
Use: As a textile spinning lubricant, with antistatic properties for yarns and fibers.

"Spiralloy." 266 Trademark for continuous filament-wound, resin-bonded glass fiber structures. Used in rocket and missile

*See "I.C.C. Shipping Regulations," page xiii.
Reference numbers refer to name of manufacturer. See "List of Manufacturers," page v.

cases and commercial applications. Possesses best strength-to-weight property of any yet-known structural material. Nonconductive to electricity and resistant to corrosion.

spirit, potato. See fusel oil.

spirits. In medicine, alcoholic solutions of volatile principles, procured either by distillation or by simple dilution. Spirits as a rule are prepared by solution of the active substance in alcohol, and are employed as therapeutic agents or merely as flavoring agents. See tincture.

spirits, cologne. See ethyl alcohol.

spirits of wine. See ethyl alcohol.

spirit soluble dyes. Those that are soluble in oils and organic solvents as contrasted with solubility in water. Azo and triaryl methane dyes are most frequently used for this purpose.

spirit varnish. See varnish.

spirocyclane. See spiropentane.

spiropentane (spirocyclane; cyclopropane-spirocyclopropane) $H_2CH_2CCCH_2CH_2$.

Properties: Colorless liquid; refractive index (n 20/D) 1.41220; density 0.7551 (20/4°C); freezing point −107.05°C; b.p. 39.03°C (760 mm).

Derivation: Heating pentaerythrityl tetrabromide in ethanol with zinc dust.

spiro system. A structural formula consisting of two rings having one atom in common. Most bicyclic compounds, such as naphthalene, have two atoms in common. See, for example, spiropentane.

"Spl-Ash." [244] Trade name for soda ash (Na_2CO_3).

Properties: Solid, pillow-shaped block; bulk density 60-65 lb/cu ft; specific gravity 2.4; dust free, slowly soluble in water. Minimum Na_2O content 58%.

Containers: 100-lb multiwall bags; bulk shipments.

Uses: Water treatment; alkalinity control in swimming pools.

"Splic-It." [65] Trademark for a material comprising natural latex, stabilizers, antioxidants, for use by textile and carpet manufacturers.

splint coal. A variety of bituminous or subbituminous coal, commonly having a dull luster and grayish-black color, of compact structure, often containing a few thin irregular bands with vitreous luster. When struck, it is resonant. It is hard and tough and breaks with an irregular, rough, sometimes splintery fracture. It is free burning and does not swell on heating. (ASTM definition, ASTM D493-39).

Splint coal consists principally of the durain structure.

split nut. See physostigma.

spodumene $LiAl(SiO_3)_2$.

Properties: White, pale green, emerald green, pink or purple mineral; white streak; vitreous luster. Contains 8.4% lithium oxide with some replacement by sodium. Insoluble in acids. Hiddenite and kunzite are gem varieties.

Constants: Sp.gr. 3.13-3.20; hardness 6.5-7.

Occurrence: United States (North Carolina, California, Massachusetts, South Dakota); Brazil; Madagascar.

Uses: Source of lithium; gemstone; in ceramics and glass as a source of lithia and alumina.

sponge iron. See iron sponge.

sponge, platinum. See platinum black.

"Spotleak." [204] Trademark for mercaptan type, fuel gas odorants.

"Spotrete." [49] Trade name for a 75% thiuram fungicide used as a seed disinfectant and turf fungicide.

spotted cowbane. See conium.

spotted hemlock. See conium.

"Spraycop" 340. [50] Trademark for a neutral copper fungicide which contains 34% metallic copper and a built-in spreader adhesive.

"Spray Flo." [204] Trademark for a 100% active cleaner with high active alkali content. For in-place lines and spray cleaning of tanks. Effective over wide temperature range, low foaming, easy rinsing. Packed in 25-, 125-, and 350-lb drums.

spreader-sticker. A substance which will reduce surface tension and increase spray adherence to surfaces, especially for agricultural and herbicidal use.

spruce oil (hemlock oil; hemlock needle oil).

Properties: Colorless to light yellow oil; characteristic agreeable odor; sp.gr. (15°C) 0.907-0.920; optical rotation −18° to −25°; refractive index (n 20/D) 1.4675-1.4700; usually clearly soluble in 0.5 and more volumes of 90% alcohol, benzyl benzoate, ether, chloroform, carbon disulfide, fixed oils, mineral oil; slightly soluble in propylene glycol; insoluble in glycerol.

Chief known constituents: Bornyl acetate, cadinene and pinene.

Derivation: Direct steam distillation of needles and branches of Tsuga canadensis, Picea alba, Picea nigra.

Method of purification: Rectification.

Containers: Cans; drums.

Uses: Medicine; veterinary liniments; for odor value in soaps and cosmetic preparations.

spruce sulfite extract. A by-product of the paper industry used in tanning; as a core binder in foundries; and as a road binder.

spunk. See agaric.

"S.Q." [123] Trademark for sulfaquinoxaline preparation for the treatment of coccidiosis in poultry.

"SQ" Phosphate. [58] Trademark for a glassy polyphosphate with high calcium sequestering properties.
Typical analysis: P_2O_5 63.0%; Na_2O 36.0%; pH (1% soln) 7.9.
Use: Formulating industrial cleaning compounds such as dishwashing and dairy cleaners.
Containers: 100-lb net paper bags or triotex bags.

squalane (perhydrosqualene; 2,6,10,15,19,23-hexamethyltetracosane; spinacane) $C_{30}H_{62}$. See also "Robane."
Properties: Colorless, odorless, tasteless liquid; miscible with vegetable and mineral oils, organic solvents and lipophilic substances. Is nonrancid, nondrying, nonoxidizing, noncongealing. Sp.gr. 0.805-0.812 (20°C); b.p. approx. 350°C; f.p. approx. −38°C; refractive index 1.4520-1.4525 (n 20/D).
Derivation: Hydrogenation of squalene; may occur naturally in sebum.
Containers: Bottles; drums.
Uses: High grade lubricating oil; vehicle for externally applied pharmaceuticals and cosmetics.

squalene (spinacene; 2,6,10,15,19,23-hexamethyl-2,6,10,14,18,22-tetracosahexene) $C_{30}H_{50}$. A natural raw material found in human sebum (5%), as well as in other fatty and waxy deposits. An unsaturated aliphatic hydrocarbon with six unconjugated double bonds.
Properties: Oil with faint agreeable odor; sp.gr. 0.858-0.860 (20°C); b.p. c. 225°C; f.p. −60°C; refractive index 1.49-1.50; iodine no. 360-380; saponification value 0-5. Insoluble in water; slightly soluble in alcohol; soluble in lipids and organic solvents.
Grade: 90% min.
Containers: 100-lb metal drums; 5-gal metal pails; 1-gal metal tins; 1-lb bottles.
Uses: Biochemical and pharmaceutical research; a precursor of cholesterol in biosynthesis; possible fungicide.

squaw bush. See viburnum opulus.

Squeegee Mediums. [28] (Squeegee Oils). Specially compounded organic vehicles for paste colors applied by stencil screen.
Containers: 55-gal drums.
Uses: As vehicles for paste colors for stencil application on glassware, tumblers, bottles, porcelain enamel, pottery.

Squeegee Oils. [28] See Squeegee Mediums.

squill (scilla; sea onion; squill white; Mediterranean squill; Indian squill).
Derivation: Bulb of Urginea maritima or Urginea indica, deprived of its dry, membranous, outer scales, cut into thin slices and carefully dried.
Occurrence: Mediterranean basin near the sea (Spain, France, Italy, Morocco and Algeria).
Grades: Crude; powdered.
Containers: Bags; barrels.
Uses: Medicine; rat poison.
Shipping regulations: None.*

squill, red. Similar to white squill and may be derived from the same sources.
Derivation: Also obtained from Urginea burkei. Sun dried or oven dried.
Use: Relatively nontoxic to humans, fowls, or domestic animals. Used as rat poison.

squill white. See squill.

Sr. Symbol for strontium.

"SR-406." [51] Trade name for captan.

SRF black. Abbreviation for semi-reinforcing furnace black. See furnace black.

"SRHS." [288] Trademark for chemical compounds and compositions in dry form for use in chromium plating baths and also in the form of aqueous solutions.

SS acid. See 8-amino-1-naphthol-5,7-disulfonic acid.

"S-1 Surfactant." [108]
"S-2 Surfactant." [108] Water-soluble setting agents.
Containers: 40-lb cans; 425-lb drums.
Uses: Emulsifiers; lower surface tension, increase "injectivity" in oilfield waterflood surfaces.

"ST-115." [256] Trade name for denaturant.
Containers: 55-gal non-returnable steel drums; tank cars.
Shipping regulations: Flammable liquid. Red label.*

"Stab-a-dry." [123] Trademark for vitamin supplements for animal and poultry feeds.

"Stabilide." [329] Trademark for potassium iodide stabilized with calcium stearate.

"Stabilite." [94] Trademark for N,N-diphenylethylenediamine (q.v.).

"Stabilite Alba." [94] Trademark for N,N-di-ortho-tolylethylenediamine, a rubber antioxidant.

"Stabilite L." [94] Trademark for N,N-diphenylpropylenediamine (q.v.), a rubber antioxidant.

"Stabilite White." [94] Trademark for a polyalkyl substituted monohydric phenol; used as a rubber antioxidant.

"Stabilizer D-22." [214] $(CH_3CH_2CH_2CH_2)_2Sn[OCO(CH_2)_{10}CH_3]_2$.
Properties: Sp.gr. 1.0525 (20/20°C); f.p. 8°C; insoluble in water; viscosity 45.2 cps (20°C); wt/gal 8.9 lbs (20°C); flash point 440°F.
Uses: Outstanding heat and light stabilizer for vinyl chloride resins.

stabilizers. In general, any substance which makes a solution, mixture, suspension, or state, etc. more stable. Specifically, there are stabilizers which may retard a reaction rate or preserve a chemical equilibrium, act as antioxidants, keep pigments and other components in emulsion form in an emulsion paint, or prevent the particles in a colloidal suspension from precipitating.

"Sta-fast." [50] Trademark for a naphthalene-acetic acid product (q.v.). Used for thinning apples and reducing pre-harvest drop of apples and pears.

"Sta-fresh." [50] Trademark for a sodium bisulfite powder for keeping grass silage fresh, green and sweet-smelling.

stagbush. See viburnum prunifolium.

"Stamere." [406] Trade name for a series of edible hydrocolloids extracted from Irish Moss. Several members of the series are:

Stamere Type DG: A highly refined extract of Irish Moss, commonly known as carrageenan, which is standardized for use as a suspending agent for cocoa.

Stamere Type CB: A blend of carrageenan and other vegetable gums, used for thickening such things as soups, toothpastes, and ice cream.

"Stanco." [110] Brand name for ordinary grade channel black for use in inks, plastics and some types of paints.

"Standard" Leaded Zinc Oxide. [268] Brand name for a low consistency zinc oxide containing 3 to 5.5% lead, calculated as $PbSO_4$. Shipped in 50-lb bags.

Use: Extensively used in exterior paints and primers.

standard wood spirits. A purified methanol (q.v.).

"Standard" Zinc Dust. [268] Trademark for a finely divided gray metallic zinc powder. Contains a minimum of 96% metallic zinc, the remainder being mainly zinc oxide. Shipped in 100-lb sealed containers.

Use: Metal primers and metal protective paints because of its rust-preventive and adhesion properties.

"Standfast Metal." [60] Trade mark for a bismuth, lead, tin, and cadmium alloy. M.p. 158°F.

Use: As liquid heat and pressure transfer medium.

stand oil. This is a drying oil which has been subjected to a heat-treatment process under conditions of minimum oxidation. It may be prepared by this means from any drying oil, such as linseed, tung, perilla, soya.

stannic acid. See stannic oxide.

stannic anhydride. See stannic oxide.

stannic bromide (tin bromide; tin tetrabromide) $SnBr_4$.

Properties: White, crystalline mass. Fumes when exposed to air. Soluble in water, alcohol, and carbon tetrachloride. Sp.gr. 3.3; b.p. 203°C; m.p. 31°C.

Containers: Glass bottles.

Use: Mineral separations.

Shipping regulations: None.*

stannic chloride (tin chloride; tin tetrachloride; sometimes erroneously called tin bichloride; butter of tin) $SnCl_4$. Often sold in the form of the double salt with sodium chloride, $Na_2SnCl_6 \cdot H_2O$.

Properties: Colorless, fuming, caustic liquid, which water converts into the crystalline butter of tin, $SnCl_4 \cdot 5H_2O$. Keep well stoppered! Sp.gr. 2.2788; m.p. −33°C; b.p. 114°C. Soluble in cold water, alcohol, carbon disulfide and oil of turpentine; decomposed by hot water.

Derivation: Treatment of tin or stannous chloride with chlorine.

Grades: Technical; C.P.

Containers: Various bottles; 560-, 1000-lb drums.

Uses: Electroconductive and electroluminescent coatings; textiles (mordant, weighting silk, brightening colors on wool, addition to sizing compounds for cotton warps to retard decomposition of the constituents of the size); perfume stabilization; manufacture of fuchsin; color lakes; ceramics; bleaching agent in the sugar industry; drug; stabilizer for certain resins; manufacture of blue print and other sensitized papers; other tin salts.

Shipping regulations: Corrosive liquid. White label. Legal label name, tin tetrachloride.*

stannic chloride pentahydrate $SnCl_4 \cdot 5H_2O$.

Properties: White solid, m.p. 56°C; soluble in water or alcohol.

Uses: An easily handled solid used in place of stannic chloride anhydrous where the presence of water is not objectionable.

stannic chromate (tin chromate) $Sn(CrO_4)_2$.

Properties: Brownish-yellow, crystalline powder. Partially soluble in water.

Derivation: By the action of chromic acid on stannic hydroxide.

Method of purification: Crystallization.

Grades: Technical.

Containers: Glass bottles.

Use: Decorating porcelain and china in rose and violet colors.

Shipping regulations: None.*

stannic iodide (tin iodide; tin tetraiodide) SnI_4.

Properties: Yellow to reddish crystals; sp.gr. 4.46; b.p. 341°C; m.p. 143.5°C. Soluble in alcohol, benzene, carbon disulfide, chloroform, and ether; soluble in water. Sublimes at 180°. Decomposed by hot water.

stannic oxide (stannic anhydride; tin peroxide; tin dioxide; stannic acid; flowers of tin; tin ash; tin anhydride) SnO_2, or $SnO_2 \cdot nH_2O$.

Properties: White powder, anhydrous or containing variable amounts of water; sp.gr. 6.6-6.9; m.p. 1127°C. Soluble in concentrated sulfuric acid, hydrochloric acid and alkalies; insoluble in water.

Derivation: (a) Found in nature as in the mineral cassiterite (q.v.). (b) Tin is melted and heated in air. (c) Tin ash, resulting from the oxidation of the baths of molten tin used in making tin plate, is refined. (d) Precipitated from stannic

chloride solution by ammonium hydroxide. (e) Acidifying a solution of an alkali stannate.
Grades: White, pure; white; gray; C.P.
Containers: 1-, 5-lb bottles; drums.
Uses: Tin salts; catalyst; ceramics (glazes); production of ceramic colors; putty; perfume preparations and cosmetics; textiles (mordant, weighting); polishing powder for steel, glass, etc; manufacture of milk-glass, alabaster glass, enamel and opaque glass.
Shipping regulations: None.*

stannic phosphide (tin phosphide) Sn_2P_2.
Properties: Silver-white, hard mass or lumps. Soluble in acids. Sp.gr. 6.56; m.p., forms Sn_4P_3 at 415°C.
Derivation: By heating tin and phosphorus.

stannic sulfide (artificial gold; mosaic gold; tin bisulfide; tin bronze; tin disulfide) SnS_2.
Properties: Yellow to brown powder. Soluble in concentrated hydrochloric acid and alkaline sulfides; insoluble in water. Sp.gr. 4.42-4.60; m.p., decomposes at red-heat.
Derivation: (a) By the action of sulfide on a solution of stannic chloride. (b) By heating tin amalgam with sulfur and ammonium chloride, distilling off the mercury sulfide and ammonium chloride.
Grades: Technical; reagent.
Containers: Glass bottles; boxes; 100-lb drums.
Uses: Imitation gilding; pigment.
Shipping regulations: None.*

"Stannine." [206] Brand name for a proprietary line of restrainers for use in the acid pickling of iron, steel and ferrous alloys. Recommended especially to be used with hot sulfuric acid.

stannite (tin pyrites) $Cu_2S \cdot FeS \cdot SnS_2$ or Cu_2FeSnS_4.
Properties: Steel-gray to nearly black mineral, metallic luster; often intermixed with chalcopyrite (q.v.). Soluble in nitric acid; sp.gr. 4.3-4.52; hardness 4.
Occurrence: United States (South Dakota); England; Germany; Bolivia.

"Stannochlor." [288] Trademark for anhydrous stannous chloride. Soluble in a number of non-aqueous solvents.
Uses: In the plating industry; as a catalyst in a variety of organic reactions; as a bleaching agent in white soaps; fuel additive; lubricant additive; as a stabilizer in frozen pineapple juice to prevent off-flavors during frozen storage.

stannous bromide (tin bromide; tin dibromide) $SnBr_2$.
Properties: Yellow powder. Soluble in hydrochloric acid (dilute); soluble in water, alcohol, ether, and acetone. Oxidizes and turns brown in air. Sp.gr. 5.1; b.p. 619°C; m.p. 215°C.

stannous chloride (tin crystals; tin salt; tin dichloride; tin protochloride) (a) $SnCl_2$ (b) $SnCl_2 \cdot 2H_2O$.
Properties: White, crystalline mass, which absorbs oxygen from the air, being converted into the insoluble oxychloride. Soluble in water, alkalies, tartaric acid, and alcohol.
Constants: (a) M.p. 246.8°C; b.p. 603-628°C. (b) Sp.gr. 2.71; m.p. 37.7°C; b.p., decomposes.
Derivation: By dissolving tin in hydrochloric acid.
Grades: Technical; C.P.; reagent; anhydrous; hydrated.
Containers: 1-, 5-lb bottles; drums.
Uses: Reducing agent in manufacture of chemicals, intermediates, dyes, etc., manufacture of lakes; textiles (reducing agent in dyeing, mordant in cochineal dyeing; discharge in printing); tin galvanizing; reagent in analytical chemistry; medicine; removing ink stains; bleaching sugar; silvering mirors; revivification of yeast sown in must (accelerator); antisludging agent for lubricating oils; chemical preservative, stabilizer for perfume in soaps.
Shipping regulations: None.*

stannous chromate (tin chromate) $SnCrO_4$.
Properties: Brown powder. Almost insoluble in water.
Derivation: By the interaction of stannous chloride and sodium chromate.
Use: Decorating porcelain.

stannous 2-ethylhexoate (stannous octoate) $Sn(C_8H_{15}O_2)_2$.
Properties: Light yellow liquid; insoluble in water, methanol; soluble in benzene, toluene, petroleum ether; hydrolyzed by acids and bases. Sp.gr. 1.25; Gardner color, 3 (max).
Uses: Polymerization catalyst for urethane foams; lubricant; addition agent.

stannous fluoride (tin difluoride; tin fluoride) SnF_2.
Properties: White, lustrous, crystalline powder. Slightly soluble in water.
Use: In toothpastes as fluoride source.

stannous octoate. See stannous 2-ethylhexoate.

stannous oleate $Sn(C_{18}H_{33}O_2)_2$.
Properties: Light yellow liquid; insoluble in water and methanol; soluble in benzene, toluene, petroleum ether; hydrolyzed by acids and bases.
Uses: Polymerization catalyst; inhibitor.

stannous oxalate (tin oxalate) SnC_2O_4.
Properties: Heavy, white, crystalline powder; sp.gr. 3.56; soluble in acids; insoluble in water.
Derivation: By the action of oxalic acid on stannous oxide.
Grades: Technical; C.P.; reagent.
Containers: 1-lb bottles; wooden kegs.
Uses: Dyeing and printing textiles.
Shipping regulations: None.*

stannous oxide (tin oxide; tin protoxide) SnO.
Properties: Brownish-black powder; unstable in air. Soluble in acids and strong bases; insoluble in water. Sp.gr. 6.3; m.p., decomposes with combustion.
Derivation: By heating stannous hydroxide in

a current of carbon dioxide.
Grades: Technical; C.P.
Containers: 1-, 5-lb bottles; wooden kegs.
Uses: Reducing agent; intermediate in preparation of stannous salts as used in plating and glass industries; pharmaceuticals.
Shipping regulations: None.*

stannous pyrophosphate $Sn_2P_2O_7$.
Properties: White, free-flowing crystals insoluble in water.
Uses: Toothpaste additive.

stannous sulfate (tin sulfate) $SnSO_4$.
Properties: Heavy, white or yellowish, crystals; soluble in water and sulfuric acid. Water solution decomposes rapidly. M.p. loses sulfur dioxide at 360°C.
Derivation: By the action of sulfuric acid on stannous oxide.
Method of purification: Crystallization.
Grades: Technical; pure.
Containers: Fiber drums.
Uses: Dyeing; tin-plating; particularly for plating automobile pistons and steel wire.
Shipping regulations: None.*

stannous sulfide (tin monosulfide; tin protosulfide; tin sulfide) SnS.
Properties: Dark-gray or black crystalline powder; sp.gr. 5.080; b.p. 1230°C; m.p. 880°C. Soluble in concentrated hydrochloric acid (decomposes); insoluble in dilute acids and water.
Containers: 100-, 400-lb wooden barrels and kegs.
Uses: Making bearing material; as catalyst in polymerization of hydrocarbons; chemical reagent.

stannous tartrate (tin tartrate) $SnC_4H_4O_6$.
Properties: Heavy, white, crystalline powder; soluble in water, dilute HCl.
Derivation: By the action of tartaric acid on stannous oxide.
Grades: Technical.
Containers: Wooden kegs.
Uses: Dyeing and printing fabrics.
Shipping regulations: None.*

St. Ann's bark. See cinchona bark, succirubra.

stannum. The Latin name for tin, hence the symbol Sn in chemical nomenclature.

"Star." [201] Trade name for sodium silicate, 42° Bé. Liquid, $SiO_2:Na_2O = 2.50$. Used with peroxide in textile bleaching.

starch (amylum) $(C_6H_{10}O_5)_x$.
Properties: White, amorphous, tasteless powder; irregular lumps, or fine powder. Insoluble in cold water, alcohol and ether; forms a jelly with hot water. See also soluble starch. Sp.gr. 1.499-1.513; m.p., does not melt; b.p. decomposes (burns) when heated.
Derivation: From corn (maize), arrow-root, potatoes and the like; the material is rasped or ground with water; the resulting pulp is ground in various types of mills; the milky liquid is strained through sieves, allowed to stand a short time to permit sand and the like to settle. The milk is then removed to another tank and the starch allowed to settle, the supernatant liquid removed and the starch washed a number of times and then dried.
Grades: Commercial; powdered; pearl; laundry; technical; reagent; edible; U.S.P. XVI (corn starch).
Varieties: Corn, wheat, rice, potato, tapicoa, cassava or arrow-root, alant or inulin.
Containers: 140-, 200-, 280-lb bags; barrels of various sizes; cardboard cartons.
Uses: Manufacture of adhesives; sizing and finishing textiles; foods (e.g., cocoas, chocolates, confectionery, ice cream, sausages, etc.); sizing paper; explosives; dextrin; reagent in analytical chemistry, indicator in iodometric analysis; medicine; face powders, cosmetics, etc.; bookbinding; distilled liquors; glucose; malt sugar; caramel; colloidal preparations; cattle foods.
Shipping regulations: None.*

starch, chlorinated. Derivatives of starch produced by the treatment with alkaline hypochlorites. A variety of products are obtained depending upon the exact procedure. Chlorinated starches are water-soluble and the sols are highly fluid and are important in the proper sizing of paper and textiles.

starch gum. See dextrin.

starch, liquid. A water suspension of starch usually containing a small proportion of stabilizing agent so that the mixture remains homogeneous.

starch nitrate. See nitrostarch.

starch, oxidized. See starch, chlorinated.

starch, permanent. Term frequently applied to materials that are not starch, but instead consist of a water emulsion of a synthetic resin or plastic. These may be applied to a fabric and upon ironing produce stiffness similar to starch but of a relatively more permanent character.

starch phosphate. An ester made from the reaction of a mixture of orthophosphate salts (sodium dihydrogen phosphate and disodium hydrogen phosphate) with starch.
Properties: Soluble in cold water (unlike regular starch), and has high thickening power. Can be frozen and thawed repeatedly without change in physical properties.
Suggested uses: Thickener for frozen foods; taconite ore binder; in adhesives, drugs, cosmetics; low-cost substitute for arabic gum, locust bean gum, and carboxymethyl cellulose.

starch, potato. Starch manufactured from potato flour. Used as a substitute for grain starches. Important commercially in continental Europe. See starch.

starch syrup. See glucose.

"Starwax." [128] Trade name for a microcrystalline wax available in brown or amber colors; m.p. 180°F.
Containers: 10-lb slabs, 8/carton or 168/pallet; 350-lb drums; tank cars.

Uses: Electrical insulation; paper coatings; wax compounds; precision casting, etc.

starwort. See helonias.

"Sta-set." [50] Trademark for a trichlorophenoxyproprionic acid product for use on apples to reduce pre-harvest drop.

"Statex." [133] Trademark for furnace carbon blacks used in rubber, printing inks and protective coatings. Available in 25- and 50-lb bats and hopper cars as:

"Statex B." A fine furnace black (FF). High reinforcing carbon, producing rubber compounds with low heat development, good tear resistance, high tensile and high resilience. Usage: Tire carcass, undertread and sidewall, wire and cable jacket compounds, and tank blocks. Neoprene stocks particularly.

"Statex B-12 & B-12A." Medium color furnace blacks used in printing inks and protective coatings. B-12 uncompressed, B-12A densed.

"Statex F-12." High color furnace black, for printing inks.

"Statex G." General purpose furnace black (GPF).

"Statex M." Fast extruding furnace black (FEF). A high structure carbon, producing high modulus, high hardness, dimensional stability and smooth extrusion. Used in inner tubes, molded and extruded goods; in blends in tire treads, sidewalls, carcass and camel backs. Excellent uncured stock retension of dimension.

"Statex R." High abrasion furnace black (HAF) for tires, conveyor belts, and miscellaneous molded and extruded items.

"Statex 93." High modulus furnace black (HMF). Intermediate reinforcement with a minimum heat build-up. Used in tire carcasses and sidewall, footwear, inner tubes, molded and extruded goods and V-belts. Also available non-staining.

"Statex 125." Intermediate super abrasion furnace black (ISAF). Used in rubber goods for high tensile, tear and abrasion resistance, also for products with electrical conductivity, including plastics, paints, etc.

"Statex 160." Super abrasion furnace black (SAF). Used for high tensile tear and abrasion resistance, also for products developing electrical conductivity.

"Statex 930." High modulus furnace black (HMF) which has been treated with 20% oil. Used where freedom from dust is desirable. Primary use in reclaiming.

"Staybelite." [266] Trademark for pale, medium hard, thermoplastic resin made from hydrogenated wood rosin; acid number 162; softening point 76°C; density 1.045 at 20°C. Characterized by resistance to oxidation and discoloration as compared to rosin. Used for adhesives and protective coatings.

"Staybelite Esters." [266] Trademark for a series of synthetic resins that are esters of hydrogenated rosin, including:

"Staybelite Ester 1": Pale, soft resin; ethylene glycol ester of hydrogenated rosin; acid number 12 max; softening point 55-65°C; sp.gr. 1.07 (25/25°C).

"Staybelite Ester 2": Pale, soft resin; diethylene glycol ester of hydrogenated rosin; acid number 10 max; sp.gr. 1.05 (25/25°C).

"Staybelite Ester 3": Medium pale, viscous resin; triethylene glycol ester of hydrogenated rosin; acid number 10 max; sp.gr. 1.08 (25/25°C).

"Staybelite Ester 5": Pale, hard resin; glycerol ester of hydrogenated rosin; acid number 6 max; softening point 78-86°C; sp.gr. 1.06 (25/25°C).

"Staybelite Ester 10": Pale, hard resin; glycerol ester of hydrogenated rosin; acid number 10 max; softening point 80-88°C; sp.gr. 1.07 (25/25°C).

"Staybelite Ester 10-75X": "Staybelite Ester 10"; 75% solids in xylene; acid number 8; sp. gr. 1.01 (25/25°C).

"Staybelite Ester 2043-66": Medium pale, hard resin; mixed ester of hydrogenated rosin; acid number 5 max; softening point 93-101°C; sp.gr. 1.07 (25/25°C).

Uses: Adhesives; protective coatings; printing inks; chewing gum.

"Stayrite." [104] Trademark for stabilizers used to provide increased heat and light stability during processing of vinyl resins. Available in range of products, both liquid and solid, including non-toxic products.

steapsin. A lipase in the pancreatic juice. See enzymes.

stearamide $CH_3(CH_2)_{16}CONH_2$. Octadecanamide.

Properties: Colorless leaflets; m.p. 109°C; b.p. 251°C (12 mm); insoluble in water; slightly soluble in alcohol and ether.

Use: Corrosion inhibitor in oil wells.

stearato chromic chloride. A polynuclear complex in the form of a six-membered ring. The two chromium atoms are bridged on one side by a hydroxyl group, and on the other side by the carboxyl oxygens of the stearic acid. The water-soluble complex results from the neutralization of stearic acid with basic chromic chloride. It acts as a water repellent and non-adhesive.

stearic acid (n-octadecanoic acid) $CH_3(CH_2)_{16}CO_2H$. The most common fatty acid occurring in natural animal and vegetable fats. Most commercial stearic acid is about 45% palmitic acid, 50% stearic acid, and 5% oleic acid.

Properties: Colorless, odorless, wax-like material. Odor and taste slight suggesting tallow. Soluble in alcohol, ether, chloroform, carbon disulfide, carbon tetrachloride; sparingly soluble in water. Sp.gr. 0.8390 (80/4°C); m.p. 69.6°C; b.p. 361.1°C (760 mm), 160°C (1 mm); refractive index 1.4299 (80°C).

Derivation: From high-grade tallows and yellow grease stearin by washing, saponification with the Twitchell or similar reagent, sometimes under pressure, boiling, distilling, cooling and pressing; or from oleic acid by hydrogenation.

Method of purification: Crystallization with

pressing; solvent crystallization, hydrogenation, and distilling.

Grades: Saponified; distilled; single-pressed; double-pressed; triple-pressed; U.S.P. XVI; 90% stearic with low oleic; grade free from chick edema factor; 99.8% pure.

Containers: 1-, 5-lb cans; 1-lb cartons; barrels; bags; tank cars and trucks.

Uses: Chemicals, especially stearates and stearate driers; lubricants; soaps; candles; pharmaceuticals and cosmetics; rubber compounding; shoe and metal polishes; coatings; food packaging.

Shipping regulations: None.*

stearin (tristearin; glyceryl tristearate) $C_3H_5(C_{18}H_{35}O_2)_3$.

Properties: Colorless crystals or powder; odorless; tasteless; m.p. 71.6°C; sp.gr. 0.943 (65°C); insoluble in water; soluble in alcohol, chloroform, carbon disulfide; insoluble in ligroin and ether.

Derivation: Constituent of most fats.

Grades: Technical; also graded as to source.

Containers: Barrels.

Use: Soap; candles; candies; adhesive pastes; artificial ivory; metal polishes; waterproofing paper; artificial stone; textile sizes; leather stuffing.

Shipping regulations: None.*

stearin and fatty acid pitches. A series of pitches obtained as by-product residue from the following operations: (a) Soap stock and candle stock manufacture. (b) Refining of vegetable oils. (c) Refining of refuse greases. (d) Refining of wool grease.

Properties: Dark brown to black; properties analogous to complex hydrocarbons; contains fixed carbon (5% to 35%). Soluble 80-100% in 88° naphtha; 95-100% in carbon disulfide; sp.gr. 0.90-1.10 (77°F).

Derivation: These pitches are named by the source from which derived, e.g., stearin; candle-tar; candle; fat; bone-fat; cottonseed-oil; cotton-stearin; cottonseed-foots; corn-oil; corn-oil-foots; packing-house; garbage; sewage; wool; wool-fat; cholesterol.

Uses: Manufacture of black paints and varnishes; tarred papers; printers' rolls; rubber filling agent; impregnating agent; ingredient of electrical insulations; marine caulking; waterproofing.

Shipping regulations: None.*

stearin pitch. See stearin and fatty acid pitches.

"Stearite." [104] Trademark for stearic acid produced by selective hydrogenation process. Used in rubber vulcanization and other processing where a purer stearic acid is desired.

stearoyl chloride (n-octadecanoyl chloride) $CH_3(CH_2)_{16}COCl$.

Properties: M.p. 23°C; b.p. 174-178°C (2 mm); soluble in hydrocarbons and ethers.

Containers: Bottles, carboys, steel drums.

Uses: Suggested for preparation of substituted amines and amides, acid anhydrides, for esterification of alcohols, synthesis of other organic compounds.

stearyl alcohol (1-octadecanol; octadecyl alcohol) $CH_3(CH_2)_{16}CH_2OH$.

Properties: Occurs as unctuous white flakes or granules with faint characteristic odor and bland taste; sp.gr. 0.8124 (59/4°C); b.p. 210.5°C (15 mm); m.p. 59°C. Soluble in alcohol, acetone, and ether; insoluble in water.

Typical specifications: M.p. 57.5-59°C; b.p. 191-194°C (7 mm).

Derivation: Reduction of stearic acid.

Grades: Commercial; technical; U.S.P. XVI.

Containers: 50-lb bags; 200-lb fiber drums; steam-coiled tank cars.

Uses: Perfumery; cosmetics; intermediate; surface active agents; detergents; lubricants; plastics and resins.

stearyl mercaptan $CH_3(CH_2)_{17}SH$. Octadecyl mercaptan.

Properties: M.p. 25°C; b.p. 205-209°C (11 mm); sp.gr. 0.8420 (25°C/4); refractive index (n 34/D) 1.4591.

Grades: 95% (min) purity.

Use: Organic intermediate; synthetic rubber processing.

stearyl methacrylate. In this case, a group name for $CH_2{:}C(CH_3)COO(CH_2)_nCH_3$ ($n = 13$ to 17).

Containers: Drums.

Uses: Polymerizable monomers for plastics, molding powders, solvent coatings, adhesives, oil additives; emulsions for textile, leather, paper finishing.

steatite. See talc.

"Steclin." [412] Trademark for tetracycline hydrochloride (q.v.).

"Sted." [428] Trademark for low sudsing, synthetic detergent. Contains non-ionic synthetic detergents, silicates, carbonates, complex phosphates, carboxymethylcellulose and optical bleach.

"Stedbac." [430] Trade name for stearyl dimethyl benzyl ammonium chloride, a hair conditioner used primarily in after-shampoo hair rinses.

steel. An alloy of iron and 0.1-1.5% carbon, which is produced from pig iron (q.v.) by oxidizing out the excess carbon and other impurities (phosphorus, sulfur, silicon) with air and hematite (q.v.). In the open-hearth process the hot air blows through a furnace over a mixture of molten pig iron, scrap iron, and hematite. In the Bessemer process the air blows directly through the mixture, contained in a large pot or "converter." The open-hearth process, although more time consuming and expensive, permits accurate addition of desired metal constituents such as manganese, nickel, and chromium (see ferro-alloys). Over 90% of the steel in the United States comes from open-hearth mills. Recent modifications include the use of an oxygen-air mixture to hasten the combustion of impurities; and the "vacuum degassing" of the molten steel in a vacuum chamber to remove excess oxygen. About 90% of the steel produced in the U.S. is plain carbon steel with the remainder divided into various alloy steels.

steel, alloy: Steels containing certain varying proportions of special elements which provide specific properties in the steels. Alloy steels are made by adding one or more of the following alloying metals: manganese, nickel, copper, chromium, molybdenum, vanadium, tungsten and cobalt.

steel, acid: Steel melted in a furnace the inner bottom and lining of which is composed of materials that have an acid reaction in the melting process and under a slag that is dominantly silicates.

steel, basic: Steel melted in a furnace that has a basic (crushed, burned dolomite, magnesite, magnesite bricks or basic slag) bottom and lining, and under a slag that is dominantly basic.

steel, high-carbon: Steel containing more than 0.85% carbon.

steel, killed: Steel deoxidized with a strong deoxidizing agent such as silicon or aluminum in order to reduce the oxygen content to a minimum so that no reaction occurs between carbon and oxygen during solidification.

steel, medium carbon: Steel having more than 0.3% carbon, but less than 0.85%.

steel, low carbon: Steel containing from 0.03 to 0.3% carbon.

steel, stainless: A group of steels, of three different classes (see below), which have in common the fact that they resist corrosion and oxidation much more strongly than ordinary steels and most alloy steels. All steels of the stainless group have high percentages of chromium, ranging from 10% up to as high as 25%. For typical commercial products, see "Enduro."

Austenitic stainless steels contain both chromium and nickel with a minimum chromium content of 16% and minimum nickel 7%. AISI Types 302, 303, 304 and 316 are the most extensively used alloys in this group, which is often known by the name 18:8, referring to the percentages of principal alloying materials. Austenitic stainless steels work harden readily; are shock resistant and unless they contain sulfur or selenium are difficult to machine. They are are readily hot worked, welded, and surpass other stainless steels in strength and resistance to scaling at high temperatures, as well as having better corrosion resistance. Austenitic stainless steels, unless stabilized, are subject to intercrystalline corrosion at the temperature range of 800 to 1600°F. These steels are used for a wide range of parts under most temperature and corrosive conditions.

ferritic stainless steels: Steels in this group contain chromium and are characterized as not being hardenable through heat treatment. They can be hot or cold worked, and often show excessive grain growth upon extended exposure to high temperatures. Likewise, these ferro-magnetic steels are likely to become brittle after welding by most commonly used methods. Resistance to scaling at high temperatures is better than for martensitic stainless steels, although strength is somewhat lower. Although machinability is rated as good, it can be improved if sulfur or selenium is included in the composition. Most steels in this group are easily formed.

martensitic stainless steels: Stainless steels which can be hardened by heat treatment and are straight chromium grades fall in the martensitic group. Martensitic stainless steels are ferro-magnetic, can be cold worked; resist the effects of water, weather and some chemicals; machine readily; are tough, but easily forged and hot worked. Hardening usually increases corrosion resistance in these steels, which tend to harden when air cooled from above 1500°F. The martensitic stainless steels are used for many mechanical and stressed parts.

SAE steels: A number system devised to indicate principal ingredients of alloy steels: The first figure indicates the general class of the steel: the second figure the amount of the alloy and the last figures carbon content in hundredths of 1%. 1 stands for carbon; 2, nickel; 3, nickel-chromium; 4, molybdenum; 5, chromium; 6, chromium-vanadium; 9, silicon-manganese. Thus, SAE 2515 is a 5% nickel steel of 0.15% carbon.

self-hardening steel (air-hardening steel). Alloy steel which may be hardened by cooling in air from above the critical temperature, without special quenching. High speed tool steels are commonly of this type.

steel, high speed. Steel which does not soften at elevated temperatures, and is therefore used to make cutting tools which will retain their cutting edges at very high temperatures, even to a red heat, and hence can be operated at much higher speeds than ordinary tool steels.

The majority of machine tools are now made of high speed steel. There are at least 15 major types based on varying compositions of molybdenum, vanadium, tungsten, and cobalt. Special heat treatments, hardening, annealing and tempering are applied to develop desired characteristics in these steels. Surface treatments may also be used. The steels can all be forged. The four major types are: molybdenum steels; molybdenum-cobalt steels; tungsten steels; tungsten-cobalt steels.

Steffens molasses. See molasses.

Steffens process, for sugar. A process used in beet sugar manufacture to separate residual sugar from molasses. Based on the formation of insoluble tricalcium saccharate and its subsequent decomposition to sugar in the presence of a weak acid such as carbonic acid.

Steinbuhl yellow. See calcium chromate; also barium chromate.

"Stellite." [214] Trademark for a series of cobalt-chromium-tungsten alloys for metal-cutting tools.
Forms: Rods; castings.
Uses: For hard-facing wearing surfaces; for applications requiring high resistance to abrasion and corrosion, such as centerless grinder rests, wear strips, values for homogenizing equipment and refrigeration, reflectors, bushings, burnishing rollers, rubber-cutting knives, and dies.

Stengel process. Process for making ammonium nitrate fertilizer from anhydrous ammonia and nitric acid. The fertilizer particles can be made in different sizes according to the use.

"Stepan." [451] Trade name for a series of methyl esters of C_6 to C_{18} saturated fatty acids.
C25 (caprylate, caprate) 55% C_8, 40% C_{10}; m.p. −30°C.
C40 (laurate) 96% C_{12}; m.p. 5°C.
C41 (laurate) 90% C_{12}; m.p. 1°C.
C42 (laurate) 74% C_{12}; 24% C_{14}.
C43 (laurate) 69% C_{12}, 29% C_{14}.
C49 (whole coco esters) 9.7% C_8, 7.4% C_{10}, 48.7% C_{12}, 17.9% C_{14}, 7.9% C_{16}, 7.6% C_{18}.
C50 (myristate) 93% C_{14}; m.p. 15°C.
C60 (palmitate) 80% C_{16}; m.p. 28°C.
C65 (palmitate-oleate) 50%-50%; m.p. 14°C; iodine no. 45.
C66 (stearate) 50% palmitate, 50% stearate; m.p. 30°C; iodine no. 0.5.
C68 (oleate) 16% palmitate, 82% oleate-stearate; m.p. 16°C; iodine no. 75.

stephanite (brittle silver ore) Ag_5SbS_4 or $5Ag_2S \cdot Sb_2S_3$.
Properties: Soft, but brittle, iron-black mineral; streak same color; metallic luster. A natural silver sulfantimonite. Contains 68.5% silver, 15.2% antimony, 16.3% sulfur. Found with argentite, pyrargyrite and polybasite (q.v.).
Constants: Sp.gr. 6.2-6.3; hardness 2-2.5.
Use: Extraction of silver.

"Sterane." [299] Trademark for prednisolone.

sterculia gum. See karaya gum.

stereoblock polymer. A polymer whose molecule is made of comparatively long sections of identical stereospecific structure, these sections being separated from one another by segments of different structure, as, for example, blocks of an isotactic polymer interspersed with blocks of the same polymer with a syndiotactic structure. See polymer, stereospecific.

stereochemistry. The chemistry of stereoisomers.

stereoisomers. Compounds whose molecules contain the same number and kind of atoms arranged in the molecule in an identical manner except for their relative positions in space. Stereoisomers are three-dimensional. See geometric isomerism; optical isomerism.

stereoregular polymers. Isotactic and syndiotactic polymers. See polymer, stereospecific.

stereospecific catalysts. Catalysts that cause formation of stereospecific polymers. Widely used examples are Ziegler and Natta catalysts derived from a transition metal halide and a metal alkyl, or similar substances. There are many patented catalysts of this general type, most of them developed in connection with the production of polypropylene or other polyolefins.

stereospecific polymer. See polymer, stereospecific.

steric hindrance. A characteristic of molecular structure in which the molecules have a spatial arrangement of their atoms such that a given reaction with another molecule is prevented or retarded in rate.

"Sterimine." [282] Trade name for trichloromelamine (TCM) compounded with salt, wetting agent, and acid phosphate buffer. Available chlorine: 50%.
Properties: White, free-flowing powder; slight chlorine odor.
Containers: 10-, 25-, 100-lb fiber containers.
Uses: Sterilization of ion exchange beds, and of surfaces of equipment and utensils.

"Sterling." [275] Trade name for a series of furnace and thermal carbon blacks for rubber, inks, paints, plastics and paper.
Available as:
"Sterling FT." Fine thermal black (FT).
"Sterling L and LL." High modulus furnace blacks (HMF).
"Sterling MT." Medium thermal black (MT).
"Sterling R and S." Semi-reinforcing furnace black (SRF).
"Sterling SO." Fast extruding furnace black (FEF).
"Sterling 10." Furnace carbon black.
"Sterling 99." Fine furnace black (FF).
See furnace black and thermal black.

"Sterofat." [228] A fatty acid ester of sterols and higher alcohols found in tall oil.
Properties are similar to degras.
Uses: In leather; non-drying film applications; rust preventatives; and similar uses common to degras.

steroids. A group of compounds including many important hormones, precursors of certain vitamins, bile acids, alcohols (sterols), and certain natural drugs and poisons (such as the digitalis compounds). Steroids have as a common "nucleus" a fused reduced 17-carbon-atom ring system, cyclopentanoperhydrophenanthrene. Most steroids also have two methyl groups and an aliphatic side-chain attached to the "nucleus." The length of the side-chain varies and generally contains 8, 9 or 10 carbon atoms in the sterols, 5 carbon atoms in the bile acids, 2 in the adrenal cortical steroids, and none in the estrogens and androgens. Steroids are classed as lipids because of their solubility in organic solvents and insolubility in water.

Most of the naturally-occurring steroids have been synthesized and many new steroids unknown in nature have been synthesized for use in medicine, such as the fluorosteroids like dexamethasone.

sterols. The steroid alcohols. They contain the common steroid "nucleus", plus an 8 to 10 carbon-atom side-chain and an alcoholic group. Sterols are widely distributed in plants and animals, both in the free form and esterified to fatty acids. Cholesterol is the most important animal sterol; ergosterol is an important plant sterol, or phytosterol.

"Sterox." [58] Trademark for a line of nonionic surface-active liquids, including:

"Sterox" AJ and AJ-100: Tridecanolethylene oxide condensates (polyoxyethylene ether); used for powdered detergent formulations, emulsifier; textile, paper and leather processing.

"Sterox" AP and AP-100: Tridecanolethylene oxide condensates (polyoxyethylene ether); used in liquid detergents.

"Sterox" CD: Polyoxyethylene ester; used in insecticide and herbicide formulations; detergent compositions; rewetting agent for paper towels; dispersing agent for powdered materials.

"Sterox" DF: Dodecyl phenol-ethylene oxide condensate (alkyl aryl polyoxyethylene ether); used as an intermediate for sulfation; emulsifier.

"Sterox" DJ: Dodecyl phenol-ethylene oxide condensate (alkyl aryl polyoxyethylene ether); used in powdered and liquid detergent formulation.

"Sterox" NJ: Nonyl phenol-ethylene oxide condensate (alkyl aryl polyoxyethylene ether); used in powdered and liquid detergents; textile, leather and paper processing; metal cleaning; toxicant formulations; deduster.

"Sterox" NL: Nonyl phenol-ethylene oxide condensate (alkyl aryl polyoxyethylene ether); used in liquid detergents.

"Sterox" SK: Polyoxyethylene thioether; used as a wetting agent and penetrant for insecticidal and herbicidal formulations.

"Sterox" No. 6: Polyoxyethylene thioether; used for textile scouring operations; rewetting agent for textiles.

Containers: 450- and 470-lb drums; tank cars and tank trucks.

"Sterozol." [173] Trademark for a non-coloring, non-interfering germicide and preservative of low relative toxicity used to inhibit bacterial and mold development on hides, skins and leathers in tannery wet work operations. Also used to preserve leather finish preparations.

STH. See somatotropic hormone.

stibamine glucoside $C_{36}H_{49}N_3NaO_{22}Sb_3$. A nitrogen glucoside of sodium para-aminobenzenestibonate. Incompletely defined structure.

Properties: Odorless, pale cream to light buff, amorphous powder. Unstable if warmed or exposed to air. Soluble in water. pH (6% solution) 8.5-9.0.

Derivation: Condensation of para-aminobenzenestibonic acid and glucose in slightly basic solution followed by precipitation with absolute alcohol and final drying.

Grade: N.N.D.

Use: Medicine.

stibenyl. See sodium-para-acetylamino-phenylantimonate.

stibic anhydride. See antimony pentoxide.

stibium. The Latin name for the elementary metal antimony, hence the symbol Sb in chemical nomenclature.

stibnite (gray antimony; antimony glance; antimonite) Sb_2S_3, sometimes contains silver or gold.

Properties: Lead-gray mineral inclining to steel-gray; subject to blackish tarnish; metallic luster. Differs from galenite and other sulfides by ease of fusion. Contains 71.8% antimony, 28.2% sulfur. Soluble in concentrated boiling hydrochloric acid with evolution of H_2S. Sp.gr. 4.52-4.62; hardness 2.

Occurrence: United States (Arkansas, Alaska, California, Nevada); Germany; Hungary; France; Japan; China; Mexico; Chile.

Use: The most important ore of antimony.

stibophen $C_{12}H_4Na_5O_{16}S_4Sb \cdot 7H_2O$. Sodium antimony III bis-catechol-2,4-disulfonate heptahydrate.

Properties: White, crystalline, odorless powder. Affected by light. Freely soluble in water; nearly insoluble in alcohol, ether, and chloroform.

Derivation: Reaction of sodium pyrocatechol-3,5-disulfonate with antimony trioxide and precipitating with alcohol.

Use: Medicine.

stick lac. See shellac.

stigmasterol $C_{29}H_{48}O \cdot H_2O$. A plant sterol.

Properties: Anhydrous form has m.p. of 170°C. Insoluble in water; soluble in usual organic solvents.

Derivation: From soy or calabar beans.

Uses: In the preparation of progesterone and other important steroids.

stilbamidine isethionate

$H_2NC(NH)C_6H_4CH{:}CHC_6H_4C(NH)NH_2 \cdot$-$2HOCH_2CH_2SO_3H$. 4,4'-Stilbenedicarboxamidine di-(beta-hydroxyethanesulfonate).

Properties: White, odorless, crystalline powder. Darkens with heating above 250°C. M.p. 325-335°C (dec); stable in air; decomposed by light. Freely soluble in water; very slightly soluble in alcohol; practically insoluble in ether. pH (1% solution) 5.0-7.0.

Use: Medicine.

stilbene (toluylene; trans form of alpha, beta-diphenylethylene) $C_6H_5CH{:}CHC_6H_5$.

Properties: Colorless or slightly yellow crystals; sp.gr. 0.9707; m.p. 124-125°C;

b.p. 306-307°C. Soluble in benzene and ether; slightly soluble in alcohol; insoluble in water.

Derivation: By passing toluene over hot lead oxide.

Method of purification: Crystallization; zone melting used for very pure crystals.

Grades: Technical; pure.

Containers: Wooden casks; fiber drums.

Uses: Manufacture of dyes and optical bleaches; crystals are used as phosphors and scintillators.

Note: The cis form of alpha, beta-diphenylethylene (isostilbene), is a yellow oil; b.p. 145°C (13 mm); m.p. 1°C.

Shipping regulations: None.*

stilbene dyes. Dyes whose molecules contain both the –N=N– and the >C=C< chromophore groups in their structure and whose color index ranges from 620 to 635. A common example is Direct Yellow R, color index 620. These are direct cotton dyes.

stilbestrol. See diethylstilbestrol.

"Stilbetin." [412] Trademark for diethylstilbestrol (q.v.).

stilbite $(CaNa_2)Al_2Si_7O_{18}\cdot 7H_2O$. A mineral, one of the zeolites (q.v.).

Properties: Color white, yellow; luster vitreous to pearly; hardness 3.5-4; sp.gr. 2.1-2.2.

Occurrence: New Jersey, Michigan, Colorado; Europe.

stillage. The grain residue from alcohol production, used in feeds and feed supplements.

stillingia (queen's root; yaw root; silver leaf).

Derivation: Root of Stillingia sylvatica.

Occurrence: Southeastern United States.

Grades: Technical.

Containers: Bags.

Use: Medicine.

Shipping regulations: None.*

stillingia oil (tallow-seed oil).

Properties: Pale yellow, limpid, drying oil; peculiar odor like linseed oil; mustard-like taste. Slightly soluble in alcohol. Sp.gr. 0.9432-0.9458; iodine number 160; saponification number 210.

Derivation: From the seeds of the tallow tree, Stillingia sebifera, by pressing.

Grades: Technical; regular; acidless.

Containers: 375-lb barrels; 8000-gal tank cars.

Uses: Candles; soap; dressing textiles; manufacturing lubricating compositions.

Shipping regulations: None.*

"Stimplants." [299] Trademark for diethylstilbestrol ear implants. Used in agriculture.

stinkweed. See stramonium.

St. John's bread. See carob seed.

Stoddard solvent. A widely used dry-cleaning solvent. See also naphtha, cleaners'. Specifications of the U.S. Bureau of Standards for this product are as follows:

Material: A petroleum distillate conforming to the requirements given herein.

Appearance: Clear and free from suspended matter and undissolved water.

Color: Water-white or not darker than 21 by Saybolt chromometer.

Odor: Sweet.

Flash point: Not lower than 100°F (Tag closed cup).

Corrosion test: A clean copper strip shall show not more than extremely slight discoloration when submerged in the solvent for 3 hours at 212°F.

Distillation range: Not less than 50% shall be recovered in the receiver when the thermometer reads 350°F. The dry or end point shall be not higher than 410°F. No tolerance shall be allowed above 410°F.

Acidity: The residue remaining in the flask after the distillation is completed shall not show an acid reaction.

Doctor test: Negative.

Sulfuric acid absorption test: Not more than 5% of the solvent shall be absorbed by concentrated "c.p." sulfuric acid (sp.gr. 1.835) (approximately 93.2%).

Containers: 1-, 5- and 10-gal cans; 55-gal steel drums; tank cars; tank trucks.

Fire hazard: Caution. MCA warning label.

Shipping regulations: None.*

"Stod-Sol." [200] Trademark for a petroleum solvent prepared by straight-run distillation.

Properties: Water-white; initial boiling point 308-316°F, 95% distills between 363-373°F; sp.gr. 0.780 (60°F); flash point (TCC) 103°F; mild, nonresidual odor.

Use: In professional dry cleaning.

stoichiometry. The branch of chemistry and chemical engineering that deals with the quantities of chemical substances that enter into and are produced by chemical reactions. Thus when the hydrocarbon gas methane undergoes union with oxygen in complete combustion, sixteen grams of methane will require 64 grams of oxygen. At the same time 44 grams of carbon dioxide and 36 grams of water will be produced. Every other chemical reaction has its own characteristic proportions, and stoichiometry deals with the methods of obtaining these from chemical formulas, equations, atomic weights and molecular weights, and with the practical use of this information to determine quickly and easily what and how much is used and produced in chemical processes.

stone flax. See asbestos.

stone red. A fine red pigment consisting essentially of red iron oxide. See also iron oxide reds and hematite, red.

stoneware clay. Although an inferior material may sometimes be used, these clays usually range from semi-refractory up to the refractoriness of firebrick. In the latter case they differ from fire-brick in burning to a very dense body at a low temperature. Desirable characteristics of a good stoneware clay are: (a) tensile strength of 125 lbs per sq. in. or over; (b) low fire-shrinkage; (c) sufficient plasticity and toughness to permit its being turned on the potter's wheel; (d) absence of concretionary minerals such as lime or iron; (e) low coarse sand content.

Average of ten analyses of stoneware clays used in Ohio potteries: Clay base, 56.65%; sandy matter, 37.45%; fluxing matter, 4.44%; moisture, 1.57%. Total silica, 65.09%.

Occurrence: United States (Ohio, New York, New Jersey, Missouri, Texas, Minnesota, Illinois, Indiana).

"Stonex." [318] Trademark for a white powder, free from nitric and chromic acids, used to remove beerstone; develops no gases.

"Stonite." [194] Trademark applied to a series of protective and decorative coatings for a wide range of surfaces and range of service conditions.

storage battery (storage cell). An electrolytic cell (q.v.) which can be recharged after the electrode materials are partly consumed during generation of electricity. Recharging is achieved by forcing a current to flow into the cell in the opposite direction to the flow during the generating period. This reverse current flow regenerates the electrode materials. In practice several cells are usually connected in series, the result being a battery. Automobile batteries are a common example, the electrolyte being sulfuric acid in distilled water (see battery acid), with one electrode of lead and the other of lead dioxide, both of these being converted into lead sulfate when current is generated.

storage cell. See storage battery.

storax. The U.S.P. XVI name for styrax (q.v.).

storax oil. See styrax oil.

"Storcavite." [299] Trademark for a preparation containing vitamins and minerals; used in medicine.

storksbill. See geranium.

STPP. Abbreviation for sodium tripolyphosphate.

stramonium (Jamestown weed; jimson weed; thorn apple; stinkweed; devil's apple; apple of peru).

Derivation: Dried leaves and flowering tops of Datura stramonium or its variety tutula.

Constituents: Atropine, scopolamine, proteins, albumin.

Occurrence: Europe; Asia; America.

Grades: Technical; N.F. XI.

Containers: 200-lb bales; bags of various sizes.

Uses: Medicine; source of its alkaloids.

Shipping regulations: None.*

strange particle. A kind of nuclear particle, not well understood, differing in subtle ways from similar more common nuclear particles.

strawberry aldehyde. See ethyl methylphenylglycidate.

strawberry tree. See euonymus.

"Stren." [28] Trademark for a high strength, limp and low elongation monofilament for use in spin fishing.

"Strep-Combiotic." [299] Trademark for a combination drug containing penicillin and streptomycin.

"Strep-Dust." [299] Trademark for an agricultural product containing streptomycin sulfate. Used in veterinary medicine.

strepogenin. A peptide constituent of many pure proteins shown to be essential to the growth of several microorganisms, mice, and rats. Contains glutamic acid and glycine residues, but pure strepogenin has not been prepared. In insulin or trypsinogen, richest known sources of strepogenin, some of the free amino groups apparently constitute the amino groups of strepogenin.

streptidine $C_8H_{18}N_6O_4$. 1,3-Diguanidino-2,4,5,6-tetrahydroxycyclohexane. A substituted cyclohexane found in the antibiotic, streptomycin.

Properties: Optically inactive, diacidic base. Isolated as dipicrate dihydrate ($C_{20}H_{24}N_2O_{18}\cdot 2H_2O$), which crystallizes in yellow needles from water and decomposes at 283-284°C.

Derivation: From streptomycin by acid hydrolysis.

streptobiosamine $C_{13}H_{23}NO_9$. A disaccharide, composed of streptose and N-methyl-L-glucosamine.

Derivation: From streptomycin by acid hydrolysis.

streptodornase. An enzyme, a deoxyribonuclease, obtained from hemolytic streptococci. It helps to split up the deoxyribonucleoproteins and nucleic acids of purulent exudates.

Grade: N.N.D. (as streptokinase-streptodornase).

streptoduocin. White powder consisting of a mixture of approximately equal parts of dihydrostreptomycin sulfate and streptomycin sulfate.

Grade: U.S.P. XVI.

Use: Medicine.

streptokinase. An enzyme, or specifically, an extracellular enzyme activator, obtained from hemolytic streptococci, which helps to dissolve blood clots and fibrinous exudates.

Grade: N.N.D. (as streptokinase-streptodornase).

streptolin.

Properties: An antibiotic, isolated as the hydrochloride. Gummy mass; soluble in water; most stable at pH 3.0-3.5.

Derivation: Produced by Streptomyces no. 11.

Use: Antibiotic; possible rodenticide.

streptomycin $C_{21}H_{39}N_7O_{12}$. The name of a specific antibiotic, but the word is also used loosely to designate several chemically related antibiotics produced by actinomycetes belonging to the genus Streptomyces or related genera.

Streptomycin is produced by Streptomyces griseus and consists of streptidine attached in glycosidic linkage to the disaccharide, streptobiosamine. It is active against gram-negative bacteria and the tubercle bacillus. It is now used mainly in the treatment of tuberculosis.

Properties: A base which readily forms salts with anions. Quite stable but must be kept from moisture as it is very hygroscopic.

Units: One unit equals one microgram of pure crystalline streptomycin base.

Derivation: From Streptomyces griseus by submerged culture. The streptomycin is then concentrated by adsorption on activated carbon and purified.

Use: Medicine (usually as sulfate salt).

See also streptomycin sulfate and dihydrostreptomycin.

streptomycin sulfate $(C_{21}H_{39}N_7O_{12})_2 \cdot 3H_2SO_4$.

Properties: White or practically white powder. Odorless or very faint odor; slightly bitter taste; hygroscopic, stable toward air or light. Solutions are acid or nearly neutral to litmus, and are levorotatory. Freely soluble in water; very slightly soluble in alcohol; practically insoluble in chloroform.

Grade: U.S.P. XVI.

Use: Medicine.

streptonigrin $C_{24}H_{20}O_8N_4$. Derived from Streptomyces flocculus. Dark brown rectangular crystals. Suggested as an antitumor antibiotic.

L-streptose $C_6H_{10}O_5$ or $CH_3CHOHC(OH)(CHO)CHOHCHO$. A peculiar sugar which is found as the disaccharide, streptobiosamine, in streptomycin. It is an unstable compound and has not been isolated.

streptothricin. An antibiotic that combats gram-negative bacteria. Its use is restricted by toxic effects.

striped alder. See hamamelis.

"Strip-Gard." [194] Trade name used to identify a series of strippable vinyl films intended to protect metal surfaces from marring during fabrication, handling, and assembly.

"Stripkolex." [28] Trademark for a dynamite for coal stripping operations.

"Stripolite." [159] Trade name for sodium formaldehyde sulfoxylate used as a reducing agent (stripper) for textiles.

stripping.

1. Removal of relatively volatile components from a gasoline or other liquid mixture by distillation, evaporation, or by passage of steam, air or other gas through the liquid mixture.
2. Rapid removal of color from a dyed fabric or fiber by chemical action, as is sometimes required in the course of dyeing operations. See also stripping agents.

stripping agents. Compounds used to remove dyes from improperly dyed fabric so it can be redyed. When used in vat dyeing or in discharge printing the products are termed "discharge agents." Such compounds should possess good penetration, should strip evenly, rinse freely, and should not attack the fiber. Substances commonly used as strippers are sodium hydrosulfite, titanous sulfate, sodium and zinc sulfoxylate formaldehydes (q.v.).

"Strobane." [138] Trademark for a versatile pesticide. Active ingredients are terpene polychlorinates (camphene, pinene and related terpene polychlorinates) with 65% chlorine value.

strontia. See strontium oxide.

strontianite $SrCO_3$. Natural strontium carbonate.

Properties: Color white, gray, yellow, green; luster vitreous; hardness 3.5-4; sp.gr. 3.7.

Occurrence: California, New York, Washington; Germany; Mexico.

Use: Source of strontium chemicals.

strontium Sr. A metallic element, of atomic number 38, classed in group II on the periodic chart.

Properties: Pale yellow, soft metal, similar to sodium. Must be kept immersed in naphtha. Soluble in alcohol and acids; decomposes water on contact. Occurs in nature in the minerals, strontianite (carbonate), and celestite (sulfate).

Constants: Sp.gr. 2.54; m.p. 770 ± 10°C; b.p. 1380°C; burns when heated in air above m.p.

Derivation: By electrolysis of molten strontium chloride in a graphite crucible with cooling of the upper, cathodic space; by the thermal reduction of the oxide with metallic aluminum (strontium aluminum alloy formed).

Grades: Technical.

Containers: Glass bottles containing sufficient naphtha to cover the rods or lumps of metals.

Uses: Alloys; "getter" in electronic tubes.

Fire Hazard: Dangerous.

strontium 90. Radioactive strontium of mass number 90.

Properties: Half-life, 25 years; radiation, beta; radiotoxicity, very hazardous.

Derivation: From the fission products of nuclear reactor fuels.

Forms available: A mixture containing strontium 90, yttrium 90, and strontium 89 chlorides in hydrochloric acid solution.

Uses: The radiation source in thickness gauges for measuring sheet paper, steel, flooring, rubber, etc. Strontium 90 is also used for the elimination of static charge; for the treatment of eye diseases; in radioautography to determine the uniformity of material distribution; in electronics for studying strontium oxide in vacuum tubes; for the activation of phosphors; as a source of ionizing radiation in luminous paint; for cigarette density control; for measuring silt density; in atomic batteries, etc.

In fallout: This isotope, a member of the mass 90 fission product chain which is produced in high yield from fission, is the longest lived member of that chain and consequently is itself produced in high yield. Because of its half-life, which is comparable to times of importance in meteorological and biological processes, this isotope is one of the principal radioactive species, in terms of activity, in fission products that are a few months to a few years old. In addition the element is classed as a "bone seeker", is metabolized in much the same manner as calcium, and is thus generally considered the most hazardous component of fall-out and the source of much of the concern for the fall-out problem. It is also the one generally isolated and measured when fall-out measurements are made.

Shipping regulations: Class D poison. Radioactive material. Blue label.*

strontium acetate $Sr(C_2H_3O_2)_2 \cdot \frac{1}{2}H_2O$.

Properties: White, crystalline powder; soluble in water.

Derivation: Interaction of strontium hydroxide and acetic acid; followed by crystallization.

Grades: C.P.

Containers: 1-lb bottles.

Use: Medicine.

Shipping regulations: None.*

strontium arsenite $Sr(AsO_2)_2 \cdot 4H_2O$ (approx.)

Properties: White powder; slightly soluble in water. Odorless; soluble in dilute acids.

Use: Medicine.

Shipping regulations: Class B poison. Poison label.*

strontium bromate $Sr(BrO_3)_2 \cdot H_2O$.

Properties: Colorless or yellowish, crystalline, lustrous powder. Hygroscopic. Caution! Keep well stoppered! Soluble in water. Sp.gr. 3.8.

strontium bromide $SrBr_2 \cdot 6H_2O$.

Properties: White, hygroscopic crystals or powder. Keep well stoppered. Soluble in water, alcohol, and amyl alcohol. Insoluble in ether. Sp.gr. 2.358. Loses all its water by 180°C; m.p. anhydrous salt 643°C.

Derivation: A strontium carbonate is treated with bromine or hydrobromic acid in presence of a reducing agent.

Method of purification: Recrystallization.

Grades: Anhydrous powder; crystal; technical; C.P.

Containers: 1-, 5-lb bottles; 150-lb drums.

Use: Medicine.

Shipping regulations: None.*

strontium carbonate $SrCO_3$.

Properties: White, impalpable powder. Soluble in acids, carbonated water and solutions of ammonium salts; very slightly soluble in water.

Constants: Sp.gr. 3.62; m.p. decomposes at about 1100°C.

Derivation: Celestite is boiled with a solution of ammonium carbonate or is fused with sodium carbonate.

Method of purification: Ignition to pale red heat.

Grades: Pure; precipitated; technical; C.P.; natural.

Containers: 50-lb bags; drums.

Uses: Pyrotechnics; manufacture of iridescent glass; strontium salts; medicine.

Shipping regulations: None.*

strontium chlorate (a) $Sr(ClO_3)_2$; (b) $Sr(ClO_3)_2 \cdot 8H_2O$.

Properties: White, crystalline powder. Must not be triturated with organic materials, liable to cause explosions. Soluble in water; slightly soluble in alcohol. (a) Sp.gr. 3.152; m.p. decomposes at 120°C.

Derivation: Strontium hydroxide solution is warmed and chlorine passed in, with subsequent crystallization.

Method of purification: Recrystallization.

Grades: Technical.

Containers: Tins; glass bottles.

Use: Manufacture of red-fire and other pyrotechnics.

Fire hazard: Dangerous.

Shipping regulations: Oxidizing material. Yellow label (applies to both wet and dry salt).*

strontium chloride (a) $SrCl_2$; (b) $SrCl_2 \cdot 6H_2O$.

Properties: White, crystalline needles; odorless; sharp, bitter taste; soluble in water and alcohol.

Constants: (a) Sp.gr. 3.054; m.p. 872°C. (b) Sp.gr. 1.964; m.p. loses $6H_2O$ at 112°C.

Derivation: Strontium carbonate is fused with calcium chloride, the melt extracted with water, the solution concentrated and crystallized.

Method of purification: Recrystallization.

Grades: Reagent; technical; anhydrous.

Containers: 1-, 5-lb bottles; wooden kegs; fiber drums.

Uses: Strontium salts; pyrotechnics; medicine.

Shipping regulations: None.*

strontium chromate $SrCrO_4$.

Properties: Clean, light yellow shade pigment; sp.gr. 3.84; with rust inhibiting and corrosion resistant properties. Has good heat and light resistance and low reactivity in high acid vehicles.

Containers: Fiber drums.

Uses: In metal protective coatings to prevent corrosion; in polyvinyl chloride resins to produce pastel primrose yellows.

strontium dioxide. See strontium peroxide.

strontium fluoride SrF_2.

Properties: White powder; soluble in hydrochloric acid and hydrofluoric acid. Insoluble in water. Sp.gr. 2.4; m.p. 1190°C.

Uses: Medicine; substitute for other fluorides.

strontium hydrate. See strontium hydroxide.

strontium hydroxide (strontium hydrate) (a) $Sr(OH)_2$; (b) $Sr(OH)_2 \cdot 8H_2O$.

Properties: Colorless, deliquescent crystals. Soluble in acids and hot water; slightly soluble in cold water. Absorbs carbon dioxide from air. Sp.gr. (a) 3.625;

(b) 1.396; m.p. (a) 375°C.
Derivation: Heating the carbonate in steam.
Grades: Technical; C.P.
Containers: 1-, 5-lb bottles; wooden kegs.
Use: Sugar industry; lubricants from soap; stabilizer for plastics.
Shipping regulations: None.*

strontium hyposulfite. See strontium thiosulfate.

strontium iodide (a) SrI_2 (b) $SrI_2 \cdot 6H_2O$.
Properties: (a) White, crystalline plates; decomposes in moist air. Keep well stoppered! Becomes yellow on exposure to air or light. (b) White crystals; soluble in water and alcohol.
Constants: (a) Sp.gr. 4.549; m.p. 515°C; b.p., decomposes. (b) Sp.gr. 2.76.
Derivation: By treating strontium carbonate with hydriodic acid.
Method of purification: Crystallization.
Grades: Technical.
Containers: 1-, 5-lb bottles; 25-lb jars.
Use: Medicine.
Shipping regulations: None.*

strontium lactate $Sr(C_3H_5O_3)_2 \cdot 3H_2O$.
Properties: White crystals or granular powder. Soluble in water, alcohol, ether. Odorless.
Containers: 1-lb bottles.
Use: Medicine.
Shipping regulations: None.*

strontium monosulfide. See strontium sulfide.

strontium nitrate (a) $Sr(NO_3)_2$ (b) $Sr(NO_3)_2 \cdot 4H_2O$.
Properties: White powder. Soluble in water; very slightly soluble in absolute alcohol. Sp.gr. (a) 2.98, (b) 2.249; m.p. (a) 570°C.
Derivation: A concentrated solution of strontium chloride is precipitated by means of sodium nitrate.
Method of purification: Recrystallization.
Grades: Technical; reagent.
Containers: 200-lb bags.
Uses: Pyrotechnics; marine signals, railroad flares; matches; medicine.
Shipping regulations: Oxidizing material. Yellow label.*

strontium nitrite (strontium nitrite, anhydrous) $Sr(NO_2)_2$.
Properties: White or yellowish powder, or hygroscopic needles. Soluble in water; insoluble in alcohol. Sp.gr. 2.8.

strontium nitrite, anhydrous. See strontium nitrite.

strontium oxide (strontia) SrO.
Properties: Grayish-white powder; sp.gr. 4.7; m.p. 2430°C; converted to hydroxide by water.
Derivation: Decomposition of strontium carbonate or hydroxide.
Uses: Manufacture of strontium salts; pyrotechnics; pigments; medicine.

strontium peroxide (strontium dioxide) (a) SrO_2 (b) $SrO_2 \cdot 8H_2O$.
Properties: White powder; odorless and tasteless; sp.gr. (a) 4.56; m.p. (a) decomposes, (b) loses $8H_2O$ at 100°C and decomposes when heated to a higher temperature. Soluble in ammonium chloride solution; decomposes in hot water. Very slightly soluble in cold water. Soluble in alcohol.
Derivation: (a) By passing oxygen over heated strontium oxide. (b) Reaction of strontium hydroxide and hydrogen peroxide.
Containers: 100-, 500-lb drums.
Uses: Bleaching; medicinal; fireworks.
Caution: Fire hazard: dangerous; oxidizing material, does not burn or explode by itself but mixtures of strontium peroxide and combustible substances are explosive and ignite easily.
Shipping regulations: Oxidizing material. Yellow label.*

strontium phosphate $Sr_3(PO_4)_2$.
Properties: White tasteless powder. Insoluble in water; soluble in acids.
Grades: C.P.
Containers: 1-lb bottles.
Use: Medicine.
Shipping regulations: None.*

strontium-potassium chlorate (potassium-strontium chlorate) $Sr(ClO_3)_2 \cdot 2KClO_3$.
Properties: White, crystalline powder; explosive on contact with reducing materials. Soluble in water.
Grades: Technical.
Use: Pyrotechnics.
Shipping regulations: Oxidizing material. Yellow label.*

strontium salicylate $Sr(C_7H_5O_3)_2 \cdot 2H_2O$.
Properties: White crystals or powder; odorless with sweet saline taste. Soluble in water and alcohol; decomposes when heated. Protect from light.
Derivation: By the interaction of strontium carbonate and salicylic acid.
Method of purification: Crystallization.
Grades: Technical.
Containers: 5-, 25-lb cartons; drums.
Uses: Medicine; pharmaceutical and fine chemical manufacture.
Shipping regulations: None.*

strontium stearate $Sr(OOCC_{17}H_{35})_2$.
Properties: White powder; m.p. 130-140°C. Insoluble in alcohol; soluble (forms gel) in aliphatic and aromatic hydrocarbons.
Uses: In grease and wax compounding.

strontium sulfate $SrSO_4$.
Properties: White precipitate or crystals of the mineral celestite. Odorless. Slightly soluble in concentrated acids; very slightly soluble in water; insoluble in alcohol and dilute sulfuric acid. Sp.gr. 3.71-3.97; m.p. 1605°C.
Derivation: (a) Celestite is ground. (b) Precipitation of any soluble strontium salt by means of sodium sulfate.
Method of purification: Washing with water.
Impurities: Sodium sulfate.
Grades: Reagent; technical; air floated, 90%, 325 mesh; free from sodium salts; C.P.
Containers: 1-, 5-lb bottles; bags; fiber drums.

Uses: Pyrotechnics; ceramics; paper manufacture.
Shipping regulations: None.*

strontium sulfide (strontium monosulfide) SrS.
Properties: Gray powder. Has hydrogen sulfide odor when in presence of moist air. Soluble in acids (decomposes); slightly soluble in water. Sp.gr. 3.7.
Grades: Technical.
Containers: Glass bottles.
Uses: Depilatory; luminous paints.

strontium thiosulfate (strontium hyposulfite) $SrS_2O_3 \cdot 5H_2O$.
Properties: Fine needles. Soluble in water; insoluble in alcohol; sp.gr. 2.17; m.p. loses $4H_2O$ at 100°C.

strophanthin. A glycoside or a mixture of glycosides obtained from Strophanthus kombé. This is K-strophanthin; for G-strophanthin see ouabain.
Properties: White or yellowish powder containing varying proportions of water, which it does not lose entirely without decomposition. Stable in air but affected by light. Solutions are neutral to litmus. Soluble in water and diluted alcohol, less soluble in dehydrated alcohol; nearly insoluble in chloroform, ether and benzene.
Containers: Bottles.
Use: Medicine.
Caution: Extremely poisonous!

strophanthin thoms. See ouabain.

strophanthus. Poisonous.
Derivation: Ripe seeds of Strophanthus kombé or hispidus.
Occurrence: Central Africa, Asia, Philippines.
Grades: Technical.
Containers: Bags.
Use: Medicine.
Shipping regulations: None.*

structural formula. See formula, chemical.

strychnidine $C_{21}H_{24}ON_2$.
Properties: Colorless crystals; m.p. 256°C.
Use: Reagent for nitrate determination.

strychnine $C_{21}H_{22}N_2O_2$. An alkaloid.
Properties: Hard, white, crystals or powder; very bitter taste; very poisonous! M.p. 268-290°C; b.p. 270°C (5 mm). Soluble in chloroform; slightly soluble in alcohol and benzene; very slightly soluble in water and ether.
Derivation: By extraction of the seeds of Nux vomica with acetic acid, filtration, precipitation by alkali and filtration.
Method of purification: Recrystallization.
Grades: Crystal; powder; technical.
Containers: Vials; tins.
Uses: In medicine as such, or as the hydrochloride, disulfate, lactate, nitrate, sulfate or other salt, which are more soluble in water; for destroying rodents and predatory animals and for trapping fur animals. See also strychnine salts.
Danger: Poisonous if swallowed. MCA warning label.
Shipping regulations: Class B poison. Poison label.*

strychnine salts. Under this heading the following salts with their corresponding formulas are described:
strychnine acetate: $C_{21}H_{22}N_2O_2 \cdot HC_2H_3O_2$.
strychnine arsenate: $C_{21}H_{22}N_2O_2 \cdot H_3AsO_4 \cdot 2H_2O$.
strychnine hydrochloride: $C_{21}H_{22}N_2O_2 \cdot HCl \cdot 2H_2O$.
strychnine hypophosphite: $C_{21}H_{22}N_2O_2 \cdot H_3PO_2 \cdot 2H_2O$.
strychnine nitrate: $C_{21}H_{22}N_2O_2 \cdot HNO_3$.
strychnine phsophate: $C_{21}H_{22}N_2O_2 \cdot H_3PO_4 \cdot 2H_2O$. (N.F. XI.)
strychnine sulfate: $(C_{21}H_{22}N_2O_2)_2 \cdot H_2SO_4 \cdot 5H_2O$. (N.F. XI).
Properties: All above salts are fine white crystalline powders; odorless; stable in air. In general, the salts are slightly to moderately soluble in water.
Containers: Cans.
Uses: In medicine. In industry as a poison for animals and rodents.
Danger: Poisonous if swallowed.
Shipping regulations: Class B poison. Poison label.*

stuff. Term used by paper makers to refer to mixtures containing the constituents of paper as supplied to the papermaking machine.

"Stymer." [58] Trademark for vinyl and styrene resins used for sizes for acetate, rayon, and other synthetic fibers.

S-type synthetic elastomer. See styrene-butadiene rubber.

"Styphen." [233] Trademark for a mixture of styrenated phenols.
Properties: Light amber liquid; pour point 13°C; b.p. (reflux) 353°C; sp.gr. 1.080 (25/25°C); flash point 450°F. Insoluble in water; soluble in methanol and ether.
Use: Antioxidant.

styphnic acid (2,4,6-trinitroresorcinol) $C_6H(OH)_2(NO_2)_3$.
Properties: Yellow crystals from ethyl acetate; astringent taste; m.p. 179-180°C; explodes on rapid heating; forms additive compounds with many hydrocarbons. Soluble in ethyl alcohol and ether; slightly soluble in water.
Derivation: Nitration of resorcinol.
Use: A constituent of priming agents in the explosive industry.
Shipping regulations: Explosive, class A. High explosive label.*

styracin. See cinnamyl cinnamate.

styralyl acetate (methyl phenyl carbinyl acetate; "Gardenol"; phenyl methyl carbinyl acetate; sec-phenylethyl acetate) $C_6H_5CH(CH_3)OOCCH_3$.
Properties: Colorless liquid with very strong odor suggesting gardenia. Soluble in 70% alcohol. Sp.gr. 1.023-1.026; refractive index 1.493-1.496.
Containers: Bottles.
Uses: Perfumery.

styralyl alcohol (methyl phenyl carbinol; phenyl methyl carbinol; alpha-methylbenzyl alcohol; sec-phenethyl alcohol) $C_6H_5CH(CH_3)OH$.
Properties: Colorless liquid with a "green" floral odor. Soluble in 18 parts 30% alcohol; slightly soluble in water; sp.gr. 1.008-1.015; m.p. 21.4°C.
Use: Perfumery; dyes.

styramate $C_6H_5CHOHCH_2OOCNH_2$. 2-Phenyl-2-hydroxyethyl carbamate. Used in medicine.

styrax (storax; oriental sweet gum; liquid amber orientalis).
Properties: (a) American. Amber-colored droplets or powder. (b) Levant. Thick, tough, gray, semi-liquid mass. Characteristic odor and taste. Soluble in ether, acetone, carbon disulfide, and warm alcohol, some residue usually remaining; insoluble in water.
Constituents: Storesin; cinnamic acid and esters; small amounts of styrene.
Constants: (a) Sp.gr. 0.890-1.100; b.p. 150-300°C; optical rotation −3° to −38°.
Derivation: A balsam obtained from the inner bark of Liquidambar orientalis or L. styraciflua.
Occurrence: (a) Honduras; (b) Asia Minor.
Method of purification: Solution in ether and treatment with fused calcium chloride.
Grades: Technical; U.S.P. XVI.
Containers: 1-, 5-, 10-lb bottles; 90-lb cases.
Uses: Medicine; microscopy; fumigating powders and tablets; perfumery.
Shipping regulations: None.*

styrax oil (storax oil).
Properties: Pale yellow, liquid, volatile oil.
Chief known constituents: Styrol; cinnamic acid esters.
Constants: Sp.gr. 0.890-0.900; b.p. 150-300°C, with decomposition; optical rotation −15°.
Derivation: By distillation from styrax.
Grades: Technical.
Containers: Glass bottles.
Use: Perfumery.
Shipping regulations: None.*

styrenated oils. Drying oils whose drying and hardening characteristics have been modified by incorporation with styrene or a similar easily polymerized monomer.

styrene (styrene monomer; vinylbenzene; phenylethylene; cinnamene; cinnamol) $C_6H_5CH{:}CH_2$.
Properties: Highly refractive; colorless liquid; aromatic odor. F.p. −30.63°C; b.p. 145.2°C; sp.gr. (25/25°C) 0.9045; wt/gal (20°C) 7.55 lbs; flash point (Tag open cup) 100°F. Insoluble in water; soluble in alcohol and ether. Readily undergoes polymerization when heated or exposed to light or a peroxide catalyst, becoming increasingly viscous until a clear solid is produced. The polymerization releases heat and may become explosive. Inhibitors, such as hydroquinone or para-tert-butylcatechol, must be added to prevent polymerization during storage or shipping.
Derivation: (a) From ethylene and benzene in the presence of aluminum chloride to yield ethylbenzene, which is catalytically dehydrogenated at about 630°C to form styrene; (b) from ethylbenzene, which is oxidized to acetophenone, hydrogenated to form methyl phenyl carbinol, and finally dehydrated to styrene. Large amounts of ethylbenzene are obtained by superfractionation of petroleum streams.
Purification: Vacuum and azeotropic distillation.
Grades: Technical 99.2%; polymer 99.6%.
Containers: Glass bottles; carboys; steel drums; in bulk in tank cars and tank trucks.
Uses: Polystyrene plastics; SBR and ABS resins; protective coatings (styrene-butadiene latex; alkyds); intermediate; impregnation of magnesium castings.
Caution: Vapor harmful. Use with adequate ventilation. Avoid prolonged or repeated contact with skin. MCA warning label.
Shipping regulations: None.*

styrene-butadiene rubber (SBR, GR-S, S-type synthetic elastomer). The most common type of synthetic rubber, amounting to over 80% of the total produced in recent years. Manufacture involves co-polymerization of about 3 parts butadiene with 1 part styrene. These materials are suspended in finely divided emulsion form in a large proportion of water, in the presence of some kind of a soap, or detergent. Also present in small amounts are an initiator or catalyst which is usually a peroxide, and a chain-modifying agent (usually a mercaptan such as dodecyl mercaptan).

A redox activator (ferrous sulfate, dextrose, and a complex phosphate) is also used for cold-rubber production. The latter differs from ordinary synthetic production only in that the activator causes the polymerizations to occur at about 40°F which in turn gives polymer molecules that have properties superior to those produced at higher temperatures. The earlier U.S. processes for SBR operated at about 120°F in the absence of the redox activator. Redox refers to the reactions between the reducing agents (dextrose, etc) and the oxidized form of the ferrous sulfate, which is instrumental in providing a controlled supply of free radicals at the 40°F temperature.

The polymerization process usually requires 5 to 15 hours, at the end of which the water contains suspended finely divided globules of synthetic rubber, i.e., a rubber latex. Usually only 60 to 80% of the butadiene and styrene are allowed to react because further reaction reduces quality and proceeds too slowly to be practical. The reaction is stopped at the desired time by adding an active reducing agent such as hydroquinone, referred to as a short stop.

The unreacted starting materials are vaporized to remove them from the latex and recover them for recycling. A rubber stabilizer (antioxidant), and other materials are often added at this point, and the latex then coagulated by adding sodium chloride and various acidic materials. Crumbs of synthetic rubber form and are filtered, washed and dried to produce a product with physical appearance and properties similar to natural gum rubber. The material is white at first but quickly turns brown due to slight oxidation.

A great many special varieties of SBR are produced in addition to the regular and cold types. The most important special types are oil extended, cold oil black, and regular black masterbatch. The word oil signifies a rubber to which 20 to 35% of a petroleum oil has been added to improve properties and reduce cost. Regular black masterbatch is a synthetic rubber made at elevated temperature (120°F usually), and to which carbon black and other processing and compounding agents have been added prior to coagulation. Variations of many kinds are achieved for special purpose rubbers by changing the proportions of butadiene and styrene, by varying the percentage of completion of the reaction, by the nature and amounts of additives, etc. A considerable volume of the synthetic rubber latex is used as such.

SBR is used mainly for automobile and other tires, but molded and extruded industrial items (gaskets, shock absorbers, etc.) belts, hose, packing, footwear, flooring, heels and soles are also major uses.

styrene monomer. Same as styrene.

styrene nitrosite. A compound resulting from the reaction between styrene and nitrogen dioxide and used as a qualitative or quantitative specific test for monomeric styrene in mixtures with other hydrocarbons.

styrene oxide $C_6H_5\overline{CHOCH}_2$.

Properties: Colorless to pale straw-colored liquid. Boiling range (5 to 95%) (760 mm) 194.2-195°C; f.p. −36.6°C; flash point 165°F; refractive index (n 25/D) 1.5328; sp.gr. (25/4°C) 1.0469; miscible in benzene, acetone, ether, and methanol.

Use: Highly reactive organic intermediate.

"Styresol." 36 A group of styrenated alkyd resins with air-drying and baking properties and high resistance to gasoline, alkalies, acids, and water.

Physical properties: Non-volatile 44-51%; viscosity (Gardner-Holdt) Q-Z^5; color (Gardner 1933) 2-9; acid number 3-15.

"Styrofoam." 233 Trademark for a brand of expanded, cellular polystyrene.

"Styron." 233 Trademark for polystyrene resin thermoplastic molding granules. Available in colors and a number of forms for compression or injection molding, extrusion, and transfer molding.

Properties: Colorless (unless pigmented); odorless; tasteless; resistant to weather, age, sunlight. Unaffected by water, strong and weak acids, strong and weak alkalies, and alcohols. Solvated by ketones, esters, aromatic and aliphatic hydrocarbons, and some mineral, animal and vegetable oils.

Uses: Electrical insulators; high-frequency coaxial cables; decorative items; ornaments; refrigerator moldings; medical and chemical ware; combs; panels; optical parts; edge-lighted dials; reflector buttons; cosmetic containers; dishes; teaspoons.

styryl alcohol. See cinnamic alcohol.

styrylic alcohol. See cinnamic alcohol.

subbituminous coal. A rank of coal between lignite and bituminous coal. Usually has a glossy black color and pitchy luster. As mined, it may contain 12-40% moisture. The principal U.S. deposits are in Wyoming, Montana, New Mexico, and Washington.

subcritical. See chain reaction.

suberane. See cycloheptane.

suberic acid (octanedioic acid) $HOOC(CH_2)_6COOH$.

Properties: Colorless crystals from water; m.p. 140°C; b.p. 279°C at 100 mm. Sparingly soluble in ether; soluble in alcohol; insoluble in chloroform.

Use: As intermediate for the synthesis of drugs, dyes and high polymers.

suberone. See cycloheptanone.

sublimation. The direct passage of a substance from the solid state to the gaseous state and back into the solid form without at any time appearing in the liquid state. Also applied to the conversion of solid to vapor without the later return to solid state, and to a conversion directly from the vapor phase to the solid state.

sublimed blue lead. See lead sulfate, blue basic.

sublimed white lead. See lead sulfate, basic.

subnuclear particle. Particles either found in the nucleus or observed coming from the nucleus as the result of nuclear reaction or rearrangement, i.e., neutrons, mesons, etc. See fundamental particle.

substantive dyes. Direct dyes.

substitution. Any chemical reaction in which one element replaces another in a compound. Chlorination of benzene to produce chlorobenzene is a typical example; in this case chlorine replaces hydrogen in the benzene molecule.

substrate. A substance upon which an enzyme or ferment acts.

subtilin. An antibiotic produced by the metabolic processes of a strain of Bacillus subtilis. It is a cyclic polypeptide similar to bacitracin in chemical structure and antibiotic activity, but not as important clinically. Subtilin is active against many gram-positive bacteria, some gram-negative cocci, and some species of fungi. It is a surface tension depressant, and its antibiotic action is increased by use of wetting agents.

Properties: Soluble in water in pH range 2.0-6.0; soluble in methanol and ethanol (up to

80%); insoluble in dry ethanol or other common organic solvents. Relatively stable in acid solutions. Inactivated by pepsin and trypsin, and destroyed by light.
Uses: Medicine; seed disinfectant.

"Sucaryl." [3] Trademark for sodium, potassium, and calcium salts of cyclohexylsulfamic acid. These are referred to as "Sucaryl" sodium, "Sucaryl" calcium, etc. Also known as cyclamates (q.v.).

succinaldehyde (butanedial) $OHCCH_2CH_2CHO$.
Properties: Liquid; sp.gr. 1.064 (20/4°C); b.p. 169-170°C. Refractive index 1.4254. Soluble in water, alcohol, and ether.
The name succinaldehyde is also often incorrectly used in commerce as a synonym for succinic anhydride.

succinic acid (butanedioic acid; ethylenedicarboxylic acid) $CO_2H(CH_2)_2CO_2H$.
Properties: Colorless crystals; soluble in water; sparingly soluble in alcohol and ether; odorless; acid taste. Sp.gr. 1.552; m.p. 185°C; b.p. 235°C.
Derivation: By the fermentation of ammonium tartrate.
Method of purification: Crystallization.
Grades: Technical; C.P.
Containers: 1-lb bottles; wooden barrels; kegs; fiber drums.
Uses: Medicine; organic synthesis; manufacture of lacquers, dyes, esters for perfumes, succinates; photography.
Shipping regulations: None.*

succinic acid peroxide.
Shipping regulations: Oxidizing material. Yellow label.*

succinic anhydride (2,5-diketotetrahydrofurane; succinyl oxide; butanedioic anhydride) $H_2\overline{C}C(O)OC(O)\overline{C}H_2$. Note: The term succinaldehyde is incorrectly used as a synonym for succinic anhydride, but more correctly is a name for a different compound.
Properties: Colorless or lightly colored needles or flakes; sp.gr. 1.104 (20/4°C); m.p. 120°C; b.p. 261°C. Soluble in water (with conversion into succinic acid), alcohol, and chloroform; very slightly soluble in ether. Sublimes at 115°C at 5 mm pressure.
Grade: Distilled.
Containers: 250-lb drums.
Use: Manufacture of chemicals, pharmaceuticals, esters, and resins.

succinimide (2,5-diketopyrrolidine) $H_2\overline{C}C(O)NHC(O)\overline{C}H_2$ or $C_4H_5O_2N \cdot H_2O$.
Properties: Colorless crystals or thin light tan flakes; nearly odorless; sweet taste; m.p. 125-127°C; b.p. 287-288°C; sp.gr. 1.41. Very soluble in hot water; completely soluble in cold water or sodium hydroxide solution; slightly soluble in alcohol; insoluble in ether and chloroform.
Grades: Purified; technical.
Use: Certain derivatives are useful as growth stimulants for plants; also used in organic synthesis.

succiniodimide. See N-iodosuccinimide.

succinite. See amber.

succinonitrile. See ethylene cyanide.

succinylcholine chloride (choline succinate dichloride dihydrate) $[Cl(CH_3)_3N(CH_2)_2OOCCH_2]_2 \cdot 2H_2O$.
Properties: White, odorless, slightly bitter, crystalline powder. Very soluble in water; slightly soluble in alcohol; very slightly soluble in benzene and chloroform; practically insoluble in ether. Aqueous solutions are unstable at room temperature. pH (2% solution) 3.0-4.5; m.p. 160-164°.
Grade: U.S.P. XVI.
Use: Medicine.

succinyl oxide. See succinic anhydride.

succinylsulfathiazole $C_{13}H_{13}N_3O_5S_2 \cdot H_2O$.
Properties: White or yellowish white, crystalline powder; odorless and stable in air, slowly darkening on exposure to light. Almost insoluble in water; soluble in solutions of alkali hydroxides and in solutions of sodium bicarbonate with evolution of carbon dioxide; sparingly soluble in alcohol and acetone; insoluble in chloroform and ether.
Derivation: From sulfathiazole and succinic anhydride.
Grade: N.F. XI.
Use: Medicine.

"Suconox." [212] Trademark for a series of amides of para-aminophenol, having antioxidant properties.
Suconox-4: N-n butyryl-para-aminophenol.
Suconox-9: N-n pelargonyl-para-aminophenol.
Suconox-12: Essentially N-lauroyl-para-aminophenol.
Suconox-18: Essentially N-stearoyl-para-aminophenol.
Uses: In synthetic rubber and polyethylene.

"Sucostrin." [412] Trademark for succinylcholine chloride (q.v.).

sucrase. See invertase.

sucroblanc. A mixture used to defecate and bleach sugar solution in one operation. Contains high-test calcium hypochlorite, calcium superphosphate, lime and Filtercel.

sucrol. See dulcin.

sucrose. Pure sugar. See sugar, cane and beet.

sucrose acetate isobutyrate (SAIB) $(CH_3COO)_2C_{12}H_{14}O_3[OOCCH(CH_3)_2]_6$. Clear sucrose derivative available either as a semi-solid (100%), or as a 90% solution in ethyl alcohol.
Properties: Sp.gr. 1.146 (25/25°C); molecular weight 847; color (APHA) max 100.
Containers: 55-gal drums.
Uses: As a modifier for lacquers and hot-melt coating formulations.

sucrose monostearate.
Properties: Odorless, tasteless white powder.
Derivation: By reaction of sugar and methyl stearate in a suitable solvent and with potassium carbonate catalyst.

*See "I.C.C. Shipping Regulations," page xiii.
Reference numbers refer to name of manufacturer. See "List of Manufacturers," page v.

Use: Low-foam non-ionic detergent; surfactants.

sucrose octaacetate $C_{12}H_{14}O_3(OOCCH_3)_8$.
Properties: White, hygroscopic, crystalline material. Bitter taste. When once melted does not recrystallize on cooling but becomes a transparent film. Ultra-violet light increases transparency. Molten film is very adhesive but this property is lost when the film is cooled below its melting point. Rate of hydrolysis, practically nil. Gives no action with Fehling's solution. Soluble in acetic acid, acetone, benzene, ethylene dichloride, methyl acetate, toluene; slightly soluble in water.
Constants: Sp.gr. 1.28 (fused) (20/20°C); b.p. 260°C (0.1 mm); m.p. 79-84°C; refractive index 1.4660 (20°C); decomposes above 285°C; viscosity (100°C) 29.5 poises; specific rotation (CCl_4) + 54.96°.
Typical specification: Sucrose octaacetate 99.0% min; free acidity (as acetic) 0.01-0.10%; m.p. 78-84°C; insolubles (in alcohol) 0.05% max; sp.gr. (fused) 1.28; color creamy white.
Grades: Technical; reagent; denaturing.
Containers: Bags; plywood drums; 200-lb slack barrels.
Uses: Plasticizer for cellulose esters and synthetic resins; adhesive compositions; coating compositions; insecticide; termite repellent; denaturant in rubbing alcohol formulas; paper; plastics; lacquers.
Shipping regulations: None.*

"Sudafed." [301] Trademark for pseudoephedrine hydrochloride (q.v.).

"Sudan." [307] Trademark of a line of dyestuffs.
Properties: Soluble in hydrocarbons.
Use: For the coloring of fats, oils, waxes, etc.

"Sudan III." [307] Trademark for aminoazobenzene-beta-naphthol (q.v.).

"Sudan IV." [307] Trademark for scarlet red (q.v.).

"Suganilla." [342] Trademark for concentrated extractives of vanilla beans adsorbed on sugar for food flavoring purposes.

sugar, acorn. See quercitol.

sugar, beet. See sugar, cane and beet.

sugar, burnt. See caramel.

sugar, cane and beet (saccharose; saccharum; sucrose) $C_{12}H_{22}O_{11}$.
Properties: Hard, white, dry crystals, lumps or powder; sweet taste; odorless. Soluble in water; very slightly soluble in alcohol. Solutions are neutral to litmus.
Constants: Sp.gr. 1.5877; decomposes 160-186°C.
Derivation: By crushing and extraction of sugar cane (Saccharum officinarum) with water or extraction of the sugar-beet (Beta vulgaris) with water, evaporating and purifying with lime, absorbent carbon and various liquids.
Grades: Reagent; U.S.P. XVI; technical; refined.
Containers: Multiwall paper sacks and in bulk in box cars.
Uses: Food; sweetening; manufacture of syrups; confectionery; preserves and jams; demulcent and lenitive; soap; pharmaceutical products; caramel; as chemical intermediate for detergents; emulsifying agents and other sucrose derivatives, including plasticizers, resins, explosives, glues, insecticides.
Shipping regulations: None.*

sugar cane wax. A hard wax varying from dark green to tan and brown; produced by solvent extraction. M.p. 76-79°C.
Grades: Domestic refined, in slabs.
Containers: Cartons.
Uses: Polishes, pigment disperser, castings, lubricant for plastics in food wrappers.

sugar coloring. See caramel.

sugar, corn. See dextrose.

sugar, grape. See dextrose.

sugar, invert. See invert sugar.

sugar of lead. See lead acetate.

sugar of milk. See lactose.

sugar, reducing. Sugars that will reduce Fehling's solution or similar test liquids, with conversion of the blue soluble copper salt to a red, orange or yellow precipitate of cuprous oxide. Glucose and maltose are typical examples of reducing sugars, their molecules containing an aldehyde group that is the basis for this type of reaction.

Suida process. A process for the separation of acetic acid from pyroligneous acid vapor by absorption using a high-boiling wood-oil fraction as the absorption oil.

"Sulamyd." [321] Brand name for sulfacetamide.

sulfacetamide (N^1-acetylsulfanilamide) $NH_2C_6H_4SO_2NHCOCH_3$.
Properties: Odorless, white crystalline powder. Characteristic sour taste. M.p. 181-184°C; decomposes with evolution of gas at 190-195°C. Soluble in alcohol; slightly soluble in water, ether and chloroform; practically insoluble in benzene.
Grade: N.F. XI.
Containers: Fiber drums.
Use: Medicine.

sulfacetamide sodium (N^1-acetylsulfanilamide sodium) $NH_2C_6H_4SO_2NNaCOCH_3 \cdot H_2O$.
Properties: White, odorless, bitter, crystalline powder. Soluble in water; sparingly soluble in alcohol, and acetone; practically insoluble in benzene, chloroform and ether. pH (1 in 20 solution) 8.0-9.5. Aqueous solutions must be refrigerated and protected from light.
Grade: U.S.P. XVI.
Containers: Fiber drums.
Use: Medicine.

sulfadiazine (2-sulfanilamidopyrimidine) $H_2NC_6H_4SO_2NHC_4H_3N_2$.
Properties: White or slightly yellow powder

or crystals nearly odorless; stable in air, slowly darkening on exposure to light; m.p. 252-256°C. Very slightly soluble in water; sparingly soluble in alcohol, acetone, and chloroform. Freely soluble in dilute mineral acids and in solutions of potassium and sodium hydroxides.
Derivation: Made from sulfanilamide by the action of pyrimidine.
Grade: U.S.P. XVI.
Containers: Drums.
Use: Medicine.

sulfadiazine sodium (sulfadiazine, soluble) $C_{10}H_9N_4O_2SNa$.
Properties: White powder. Protect from air and light! Soluble in water; slightly soluble in alcohol.
Grade: U.S.P. XVI.
Containers: Fiber drums.
Use: Medicine.

sulfadiazine, soluble. See sulfadiazine sodium.

sulfadimethoxine $H_2NC_6H_4SO_2NHC_4HN_2(OCH_3)_2$. N^1-(2, 6-Dimethoxy-4-pyrimidinyl)sulfanilamide.
Properties: M.p. 197-202°C. Insoluble in water; slightly soluble in alcohol; very slightly soluble in ether and chloroform.
Use: Medicine.

"Sulfads." [69] Trademark for proprietary preparation of dipentamethylene thiuram tetrasulfide $[(CH_2)_5NC(S)S]_2$.
Properties: Light yellow to buff powder (also supplied in "rodform"); sp.gr. 1.50 ± 0.03; melting range 126-135°C; soluble in chloroform, benzene, acetone; slightly soluble in gasoline, carbon tetrachloride; insoluble in water, dilute caustic.
Uses: Accelerator in natural, nitrile and butyl rubbers and SBR.

sulfaethidole $H_2NC_6H_4SO_2NHC_2N_2S(C_2H_5)$. N^1-(Ethyl-1, 3, 4-thiadiazole-2-yl)sulfanilamide. White crystalline powder.
Grade: N.N.D.
Use: Medicine.

sulfaguanidine (sulfanilylguanidine) $H_2NC_6H_4SO_2NHC(NH_2){:}NH \cdot H_2O$.
Properties: White needle-like crystalline powder; m.p. 190-193°C; nearly odorless; stable in air, darkening on exposure to light. Soluble in dilute mineral acids; sparingly soluble in alcohol and acetone; insoluble in solutions of sodium hydroxide at room temperature; soluble in hot water.
Derivation: Condensation of guanidine with N-acetylsulfanilyl chloride.
Grade: N.F. XI.
Containers: Drums.
Use: Medicine.

"Sulfamate Nickel." [72] Proprietary nickel sulfamate process used to produce heavy, smooth, nickel deposits with good mechanical properties. Prepared from nickel sulfamate, nickel chloride, boric acid and organic addition agents. The addition agents function to control the stress of the deposit.
Uses: Electroforming and electrotyping.

sulfamerazine $H_2NC_6H_4SO_2NHC_4H_2N_2CH_3$. N-(4-Methyl-2-pyrimidyl)sulfanilamide.
Properties: White or faintly yellowish-white crystals or powder; m.p. 234-238°C; slightly bitter taste; odorless or nearly so; stable in air but slowly darkens on exposure to light. Readily soluble in dilute mineral acids and in solutions of potassium, ammonium and sodium hydroxides; practically insoluble in water; sparingly soluble in acetone; slightly soluble in alcohol.
Grade: U.S.P. XVI.
Use: Medicine.

sulfamerazine sodium (sulfamerazine, soluble) $C_{11}H_{11}N_4O_2SNa$.
Properties: White, or faintly yellowish-white crystals, or a crystalline powder, which slowly darkens on exposure to light, odorless or nearly so with a bitter taste. Soluble in water; slightly soluble in alcohol; insoluble in ether and chloroform.
Containers: Drums.
Use: Medicine.

sulfamethazine $NH_2C_6H_4SO_2NHC_4N_2H(CH_3)_2$. N-(4, 6-Dimethyl-2-pyrimidyl)sulfanilamide.
Properties: White to yellow-white powder. Almost odorless; slightly bitter taste. Affected by light. M.p. 197-200°C. Soluble in acetone; slightly soluble in alcohol; very slightly soluble in water and ether.
Grade: U.S.P. XVI.
Containers: Drums.
Use: Medicine.

sulfamethizole $H_2NC_6H_4SO_2NHC_2N_2S(CH_3)$. N^1-(5-Methyl-1, 3, 4-thiadiazol-2-yl)sulfanilamide.
Properties: White crystals or powder; slightly bitter taste; almost odorless. Very slightly soluble in water; soluble in solutions of alkalies; soluble in dilute mineral acids; sparingly soluble in alcohol. Melting range 207°-211°.
Grade: N.F. XI.
Use: Medicine.

sulfamethoxypyridazine $C_{11}H_{12}N_4O_3S$. N^1-(6-methoxy-3-pyridazinyl)sulfanilamide.
Properties: White or yellowish white, crystalline powder. Odorless or nearly so; bitter aftertaste. Stable in air, but slowly darkens on exposure to light; m.p. 180-183°C. Very slightly soluble in water. Sparingly soluble in alcohol and acetone. Freely soluble in dilute mineral acids and in solutions of alkali hydroxides.
Derivation: Prepared from 6-chloro-3-sulfanilamidopyridazine.
Grade: U.S.P. XVI.
Use: Medicine.
Also available as the hydrochloride.

sulfamic acid HSO_3NH_2.
Properties: White crystalline solid; nonvolatile; non-hygroscopic; sp.gr. 2.1; m.p. 205°C (decomposes). Moderately soluble in water; slightly soluble in organic solvents. Odorless. Aqueous solutions are highly ionized, giving pH values lower than solutions of formic, citric, tartaric, phosphoric, and oxalic acids. All of the common salts

(including calcium, barium, and lead salts) are extremely soluble in water.
Grades: Reagent; crystalline; granular.
Containers: 50- and 100-lb fiber drums.
Uses: Metal and ceramic cleaning; metal pickling; nitrite removal in azo dye operations; gas liberating compositions; electroplating and refining of metals; organic synthesis; analytical acidimetric standard; preparation of odorless formaldehyde; preparation of amine sulfamates used as plasticizers, softening agents, and fire retardants for paper and other cellulosic materials.
Shipping regulations: None.*

sulfamidyl. See sulfanilamide.

"Sulfamylon" Hydrochloride. [162] Trademark for maphenide hydrochloride.

sulfanilamide (p-aminobenzenesulfonamide; sulfamidyl) $H_2NC_6H_4SO_2NH_2$.
Properties: White; odorless crystals, or a crystalline powder; m.p. 164.5-166.5°C. Soluble in acetone, glycerin, hydrochloric acid, alcohol, and boiling water; slightly soluble in cold water; insoluble in chloroform, ether and benzene.
Derivation: Condensation of acetyl sulfanilyl chloride with ammonia and subsequent hydrolysis.
Method of purification: Recrystallization.
Grades: N.F. XI.
Containers: Ampoules; glass vials and bottles; fiber drums.
Uses: Medicine; chemotherapeutics.
Shipping regulations: None.*

2-sulfanilamidopyridine. See sulfapyridine.

2-sulfanilamidopyrimidine. See sulfadiazine.

sulfanilic acid (para-aminobenzenesulfonic acid; para-anilinesulfonic acid) $H_2NC_6H_4SO_3H \cdot H_2O$.
Properties: Grayish-white, flat crystals. Soluble in fuming hydrochloric acid; slightly soluble in water; very slightly soluble in alcohol and ether.
Constants: M.p., chars at 280-300°C.
Derivation: By heating aniline with weak fuming sulfuric acid and pouring the reaction product into water.
Method of purification: By boiling a solution of the sodium salt with animal charcoal.
Grades: Technical; pure; reagent.
Containers: 1-, 5-lb bottles; drums.
Uses: Dyestuffs; organic synthesis; medicine.
Shipping regulations: None.*

meta-**sulfanilic acid.** See metanilic acid.

sulfanilylbutylurea. See 1-butyl-3-sulfanilylurea.

sulfanilylguanidine. See sulfaguanidine.

"Sulfanole." [42] Proprietary products. Various types.
Properties: Amber solutions and gels, pastes and powders. Disperse readily in water at temperatures of 60°C.
Containers: 55-gal steel and fiber drums.
Uses: Detergents in all types of textile fabric preparation for scouring and dyeing.

"Sulfanole" NO-9. [42] Proprietary product. Non-ionic surfactant.
Properties: Colorless thick syrup at 25°C. Disperses readily in water at 20°C.
Containers: 55-gal steel drums.
Use: Penetrating agent and detergent for all types of textile fabrics during preparation in scouring, dyeing and in resin application for finishing.

"Sulfanthrene." [28] Trademark for a line of thioindigoid vat dyes characterized by good fastness properties.
Use: Principally for the dyeing and printing of cotton, rayon and silk.

sulfapyridine (2-sulfanilamidopyridine) $H_2NC_6H_4SO_2NHC_5H_4N$.
Properties: White, or faintly yellowish-white crystals or crystalline powder. Soluble in dilute mineral acids and aqueous solutions of potassium or sodium hydroxide; slightly soluble in water and alcohol. M.p. 190-193°C.
Derivation: Condensation of acetyl sulfanilyl chloride with 2-aminopyridine and subsequent hydrolysis.
Method of purification: Crystallization.
Grades: U.S.P. XVI.
Containers: Ampoules; glass vials and bottles; tins.
Uses: Medicine; chemotherapeutics.
Shipping regulations: None.*
Also available as the sodium monohydrate salt.

sulfaquinoxaline [N-(2-quinoxalinyl)sulfanilamide] $H_2NC_6H_4SO_2NHC_8H_5N_2$.
Properties: Crystals; m.p. 247°C; almost insoluble in water; soluble in alkaline solutions.
Containers: Drums.
Use: Veterinary medicine.

sulfarsphenamine (3,3'-disodium-4,4'-diaminodihydroxyarsenobenzene-N-dimethylenesulfonate) $NaOSO_2CH_2NHOHC_6H_3As{:}AsC_6H_3OHNHCH_2OSO_2Na$.
Properties: Orange-yellow powder; almost odorless. Must not contain less than 19% of arsenic. Readily soluble in water yielding a yellow solution which is acid to litmus paper; slightly soluble in alcohol; insoluble in ether.
Derivation: Synthetic.
Containers: Ampoules.
Use: Medicine.
Shipping regulations: None.*

"Sulfasan" R. [58] Trade name for 4,4'-dithiodimorpholine.
Properties: Coarse, gray powder; m.p. 122°C min; flash point 250°F; sp.gr. 1.29.
Containers: 150-lb fiber drums.
Uses: Rubber compounding.

sulfate pulp. See wood pulp.

sulfate resisting cement. See cement, chemical-resisting.

sulfathiazole $C_9H_9N_3O_2S_2$.
Properties: White, or faintly yellowish-white

*See "I.C.C. Shipping Regulations," page xiii.
Reference numbers refer to name of manufacturer. See "List of Manufacturers," page v.

crystals or powder; odorless or nearly so; stable in air; slowly darkens on exposure to light. Soluble in acetone, dilute mineral acids, or aqueous solutions of potassium and sodium hydroxide; slightly soluble in water and alcohol. M.p. 200-204°C.
Derivation: Condensation of 2-aminothiazole with acetylsulfanilyl chloride and a subsequent hydrolysis.
Method of purification: Crystallization.
Grades: N.F. XI; technical.
Containers: Ampoules; glass vials and bottles; drums.
Uses: Medicine; chemotherapeutics.
Fire hazard: None.
Shipping regulations: None.*

sulfathiazole sodium (sulfathiazole, soluble; sodium 2-sulfanilamidothiazole) $C_9H_8N_3O_2S_2Na \cdot 1\frac{1}{2}H_2O$.
Properties: White to faintly yellowish-white powder or granules, affected by light. Soluble in water and alcohol. Absorbs CO_2 on prolonged exposure to humid air, becoming incompletely soluble in water.
Grades: N.F. XI.
Containers: Drums.
Use: Medicine.

sulfathiazole, soluble. See sulfathiazole, sodium.

ortho-**sulfhydrylbenzoic acid.** See thiosalicylic acid.

"Sulfidal." [138] Trademark for a dried colloidal sulfur powder.
Properties: A pale yellow powder, containing about 80% sulfur, which disperses readily in water up to about 10% to give a milky fluid.
Use: Medicine (external).

sulfide dyes (sulfur dyes). A group of dyes produced by heating various organic compounds with sulfur. The characteristic chromophore groupings are ≡C—S—C≡ and ≡C—S—S—C≡. Color index ranges from 933-1012. Application is usually to cotton from a sodium sulfide bath. Sulfur black (Color Index 978) is an important example.

"Sulfindone." [243] Trademark of sulfur dyes for cotton.

sulfinpyrazone $(C_6H_5)_2C_3N_2H(O)_2C_2H_4S(O)C_6H_5$. 1,2-Diphenyl-4-(2'-phenylsulfinylethyl)-3,5-pyrazolidinedione. Used in medicine.

sulfisomidine $H_2NC_6H_4SO_2NHC_4N_2H(CH_3)_2$. N^1-(2,6-Dimethyl-4-pyrimidinyl)sulfanilamide.
Grade: N.N.D.
Use: Medicine.

sulfisoxazole (N-3,4-dimethyl-5-isoxazolyl-sulfanilamide) $H_2NC_6H_4SO_2NHC_3NO(CH_3)_2$.
Properties: White to slightly yellowish, odorless, slightly bitter, crystalline powder. M.p. 192-195°C. Freely soluble in dilute hydrochloric acid; soluble in boiling alcohol; almost insoluble in water.
Grade: U.S.P. XVI.
Use: Medicine.

sulfisoxazole diethanolamine $C_{11}H_{13}N_3O_3S \cdot HN(CH_2CH_2OH)_2$.
Properties: Slightly yellowish, crystalline powder. Very soluble in water; soluble in alcohol; very slightly soluble in chloroform; practically insoluble in ether. M.p. 120-123°C.
Grade: U.S.P. XVI.
Use: Medicine.

sulfite acid liquor. An aqueous solution of calcium bisulfite or calcium and magnesium bisulfites containing a large amount of free sulfur dioxide. It is prepared from sulfur dioxide and limestone or dolomite or lime by passing sulfur dioxide gas up towers, packed with limestone, down which water flows.
Use: In the manufacture of sulfite pulp in the paper industry.

sulfite pulp. See wood pulp.

sulfite waste liquor. A waste liquor produced in the sulfite process of making paper. Sold in three forms, (a) Dilute, (b) Concentrated, (c) Solid.
Properties: (a) Light brown liquid. (b) Dark brown viscous liquid. (c) Brown granular solid.
Constants: (a) 10% total solids. (b) 50-52% solids; gr. 30-35°Bé.
Derivation: (a) Neutralized with lime or soda. (b) Neutralized and evaporated to 30° Bé. (c) Neutralized and evaporated to dryness.
Containers: (a) Drums; carboys. (b) Drums; barrels. (c) Bags; barrels.
Uses: Foam producer; emulsifier; adhesive; tanning agent; binder for briquets, cores, unpaved roads; source of torula yeast.
Shipping regulations: None.*
See also lignin sulfonates.

sulfobenzeneazodimethylaniline. See dimethylaminoazobenzene sulfonate.

meta-**sulfobenzoic acid** $HO_3SC_6H_4COOH \cdot 2H_2O$.
Properties: Grayish-white solid; stable but hygroscopic in air; m.p. 98°C, anhydrous form melts at 141°C. Soluble in water, alcohol; very soluble in ether; insoluble in benzene, carbon tetrachloride, petroleum ether.
Derivation: Direct sulfonation of benzoic acid with sulfur trioxide.
Typical specifications: Ground powder or ½-1 in. lumps. Less than 0.2% insoluble in water; neutralization equivalent 99-101; m.p. range 130-141°C; para isomer, small amount.
Uses: Derivative for surface active agents.
Shipping regulations: None.*

ortho-**sulfobenzoic acid** $HO_3SC_6H_4COOH$.
Properties: White needles. Soluble in water and alcohol; insoluble in ether.
Constants: M.p. 68-69°C (with $3H_2O$ of crystallization); m.p. 134°C (dry).
Derivation: (a) From saccharin and concentrated hydrochloric acid; (b) by the oxidation of thiosalicylic acid, with potassium permanganate in alkaline solution.

Method of purification: Crystallization from water.
Containers: Glass bottles.
Uses: Manufacture of sulfonaphthalein indicators; dyes.
Shipping regulations: None.*

sulfobromophthalein sodium $C_{20}H_8Br_4O_{10}S_2Na_2$.
Properties: White, crystalline powder; odorless with a bitter taste; hygroscopic. Soluble in water; insoluble in alcohol and acetone.
Derivation: From phenol and tetrabromophthalic acid or anhydride.
Grades: U.S.P. XVI; technical.
Use: Medicine (diagnostic aid).

sulfocarbanilide. See thiocarbanilide.

sulfocarbolic acid. See phenolsulfonic acid.

1-(4-sulfo-2, 3-dichlorophenyl)-3-methylpyrazolone $(Cl_2C_6H_2SO_3H)NNC(CH_3)CH_2CO$.
Properties: White or yellowish powder or crystals. Very soluble in water; soluble in alkalies.
Derivation: By condensation of dichlorophenylhydrazine sulfonic acid with ethylacetoacetate.
Method of purification: Crystallization.
Grades: Technical; C.P.
Use: Intermediate for dyes.
Shipping regulations: None.*

"Sulfogene." [28] Trademark for a line of sulfur colors; dyed from a solution of sodium sulfide.
Use: Used extensively on cotton workclothing and similar fabrics and to a limited extent on rayon and other materials.

"Sulfole." [303] Trademark for tert-dodecyl mercaptan.
Properties: Boiling range (5 mm) 85-106°C; sp.gr. (60/60°F) 0.862; refractive index (20/D) 1.464; flash point 99°C.
Containers: Drums; tank cars.
Uses: Polymer modification and manufacturing of detergents.
Hazards: Flammable liquid.
Shipping regulations: Red label not required.*

"Sulfo-Merthiolate." [100] Trademark for the sodium salt of para-ethylmercurithiobenzene sulfonate (q.v.).

"Sul-fon-ate." [93] Trade name for a line of derivatives of organic sulfonic acids produced by a special manufacturing process and intended for use generally in applications utilizing the surface active properties of these derivatives.

"Sul-fon-ate AA-9." [93] Trade name for a 90-92% active, flake dodecylbenzene sulfonate, sodium salt.
Uses: Used as a wetting agent and detergent.

"Sul-fon-ate AA-10." [93] Trade name for a 96% plus active, flake dodecylbenzene sulfonate, sodium salt.
Uses: Used as emulsification aid, wetting agent, detergent.

"Sul-fon-ate AA-T." [93] Trade name for an amber liquid dodecylbenzene sulfonate, triethanolamine salt.
Uses: Similar to "Sul-fon-ate AA-9."

sulfonated castor oil. See Turkey red oil.

sulfonated oils. Vegetable or animal oils which have been treated with sulfuric acid and the excess of acid washed out and the oil neutralized with a small amount of caustic soda or ammonia. Sometimes paraffin oil is also included in the formula. Chemically this trade term, sulfonated, is incorrect since the oils are sulfated (contain the $-OSO_2OH$ group and not the $-SO_2OH$ group). These oils are soluble (emulsifiable) in water and are claimed to be capable of: (1) supplying lubrication for various purposes; (2) emulsifying (dispersing) in water various other materials; (3) plasticizing; (4) defoaming; (5) dispersing; (6) penetrating; (7) scouring; (8) wetting out; and (9) softening. For some of the foregoing operations specially formulated oils may be necessary.
Properties:
Solubility in water: From coarse emulsions to actual solutions, according to the degree of sulfation.
Color: Vary from pale yellow to dark brown.
Viscosity: Mobile liquids to semi-solids.
Emulsification: Emulsify mineral oils, solvents, fatty oils, fats, and waxes.
Penetration: Proportionate to sulfation.
Lubrication: In proportion to unchanged oil present.
Containers: Glass bottles; carboys; 1-, 5-, and 10-lb tins; steel drums.
Uses: They are very widely used in industrial processing; e.g., textiles, leather, paper, glue, cosmetics, metal working, cleansing, inks, disinfectants, petroleum products, ceramic clays, agricultural sprays, cutting oils, laundering, and many others.
Shipping regulations: None.*
See also soluble oils.

"Sul-fon-ate OA-5." [93] Trade name for a sulfonate of oleic acid, sodium salt.
Properties: An amber liquid, which is a true sulfonate containing the $-SO_2ONa$ group, i.e., with a carbon-to-sulfur rather than carbon-to-oxygen linkage as in sulfated oils.
Uses: Used as wetting agent, defoamer, particularly effective in acid media.

sulfonation. The formation of a sulfonic acid, i.e., a compound containing the $-SO_2OH$ group in its molecular structure. Thus the conversion of benzene (C_6H_6) into benzene sulfonic acid ($C_6H_5HSO_3$) is an example. Common sulfonating agents are: concentrated sulfuric acid, fuming sulfuric acid (oleum), sulfur trioxide, alkali disulfates, pyrosulfates, chlorosulfonic acid and a mixture of manganese dioxide and sulfurous acid.

para-**sulfondichloraminobenzoic acid.** See halazone.

sulfonethylmethane (methylsulfonal; diethylsulfonmethylethylmethane) $CH_3(C_2H_5)C(SO_2C_2H_5)_2$.
Properties: Colorless, lustrous, crystals or powder; odorless; characteristic, slightly bitter taste; m.p. 76°C; decomposes on heating with evolution of sulfur dioxide. Soluble in hot water, alcohol, and ether; moderately soluble in cold water.
Derivation: By passing dry hydrochloric acid gas into a mixture of anhydrous ethyl mercaptan and methyl ethyl ketone and oxidizing the product with potassium permanganate.
Method of purification: Crystallization.
Grades: Technical.
Containers: 1-lb bottles; 5-, 25-lb boxes; 50-lb kegs.
Use: Medicine.
Shipping regulations: None.*

sulfonmethane (diethylsulfondimethylmethane) $(CH_3)_2C(SO_2C_2H_5)_2$.
Properties: Colorless, crystalline powder; odorless; nearly tasteless; m.p. 125-126°C; b.p. decomposes at 300°C. Soluble in alcohol, ether, chloroform, benzene; slightly soluble in water.
Derivation: Anhydrous acetone and anhydrous ethylmercaptan are combined by means of a stream of anhydrous hydrochloric acid, and oxidized with potassium permanganate.
Method of purification: Recrystallization.
Grades: Technical.
Containers: Glass bottles; 1-lb cartons.
Use: Medicine.
Shipping regulations: None.*

sulfonyl chloride. See sulfuryl chloride.

sulfonyldianiline (DDS; 4,4'-diaminodiphenyl sulfone) $(C_6H_4NH_2)_2SO_2$. Bis(4-aminophenyl)sulfone.
Properties: Colorless crystals. M.p. 175-176°C. Insoluble in water; soluble in alcohol.
Use: Medicine; curing agent for epoxy resins.

sulfonyl diphenol. See dihydroxy diphenyl sulfone.

para-**1-sulfophenyl-3-methyl-5-pyrazolone** $(C_6H_4SO_3H)\underline{NNC(CH_3)CH_2C}O$.
Properties: White or yellowish powder. Only very slightly soluble in water; soluble in alkalies.
Derivation: By condensation of phenylhydrazine sulfonic acid with ethylacetoacetate.

4-sulfophthalic anhydride $HO_3SC_6H_3(CO)_2O$.
Properties: Reddish-brown nonflammable syrup; crystallizes partially on long standing; hygroscopic; fluorescent in solution under ultraviolet radiation; sp.gr. 1.62 (25°C), 1.56 (90°C); density 97.5 lbs/cu ft. Very soluble in water, alcohol; insoluble in ether, benzene.
Grades: Technical.
Uses: Esters of 4-sulfophthalic acid used in wetting, cleansing, emulsifying, softening and equalizing agents with textiles. Derivatives suggest application as surface active agents.
Shipping regulations: None.*

"Sulforon." [28] Trademark for very finely divided sulfur; readily dispersible in water; not less than 97% sulfur; contains a wetting agent.
Containers: 6- and 50-lb bags.
Use: As a fruit fungicide for control of scab, brown rot, etc.

"Sulforon" X. [28] Trademark for extremely finely divided wettable sulfur of 5 microns average particle size.
Containers: 5- and 50-lb bags.
Use: For control of brown rot, scab and certain other fungus diseases of fruits.

5-sulfosalicylic acid $C_6H_3OH(SO_2OH)COOH \cdot 2H_2O$.
Properties: Colorless crystals; colored pink by traces of iron. Very soluble in water. M.p. 120°C; decomposes at higher temperatures.
Derivation: Action of sulfuric acid on salicylic acid.
Method of purification: Recrystallization.
Grades: C.P.; analytical; reagent.
Containers: Glass bottles.
Uses: Reagent for albumin; colorimetric reagent for ferric iron.
Shipping regulations: None.*

sulfotepp. See tetraethyldithiophosphate.

sulfovinic acid. See ethylsulfuric acid.

sulfovinous acid. See ethylsulfurous acid.

"Sulfoxide." [342] Trademark for N-octyl sulfoxide of isosafrole.

sulfoxone sodium. See sodium sulfoxone.

"Sulframins." [449] Trademark for a series of alkyl aryl sulfonates, surface active agents used for general industrial applications and for household and toilet use, including dishwashing, cleaning of walls, woodwork, linoleum, rubber, tile. Used as detergent, wetting and dispersing agent for textile fabrics.

"Sulfricin." [202] Trademark for a modified castor oil of increased hydroxyl content used for sulfonation.

sulfur (brimstone; flowers of sulfur; sulfur flour; sulfur flowers) S. Nonmetallic element with atomic number 16, group VI of periodic system.
Properties: Sulfur exists in two stable crystalline forms and at least two amorphous and two liquid forms.
(a) Alpha-sulfur, rhombic, octahedral yellow crystals stable at room temperature. Sp.gr. 2.06; transition to beta form 94.5°C; m.p. (rapid heating) 112.8°C; refractive index 1.957.
(b) Beta-sulfur, monoclinic, prismatic pale yellow crystals slowly changing to alpha form below 94.5°C. Sp.gr. 1.96; m.p. 119.3°C; b.p. 444.6°C; index of refraction 2.038. Both forms are insoluble in water; slightly soluble in alcohol and ether; soluble in carbon disulfide, carbon tetrachloride, and benzene.
Occurrence: Texas, Louisiana: Sicily.
Derivation: Mined as such or by Frasch

*See "I.C.C. Shipping Regulations," page xiii.
Reference numbers refer to name of manufacturer. See "List of Manufacturers," page v.

process (q.v.) in which it is melted underground by superheated water, pumped up and allowed to solidify in bins from which it is shipped; recovered from waste hydrogen sulfide and from natural gas, by oxidation.

Method of purification: Filtration of liquid or sublimation.

Grades: Technical (lumps, roll, flour); rubber makers; N.F. XI (sublimed); crude; refined; Mexican; high purity (impurities less than 10 ppm).

Containers: Bags; barrels, bulk in carloads and vessels, including molten sulfur.

Uses (in approximate order of volume): Sulfuric acid; pulp and paper; agricultural fungicide; carbon disulfide; other chemicals and dyes; rubber vulcanization; medicine.

Shipping regulations: None.*

sulfur 35. Radioactive sulfur of mass number 35.

Properties: Half-life, 87.1 days; radiation, beta; radiotoxicity, moderately hazardous.

Derivation: By pile irradiation of elemental sulfur or of various chlorides.

Forms available: Solid elemental sulfur; sulfate in weak hydrochloric acid; barium sulfide in barium hydroxide solution; elemental sulfur in benzene solution; in tagged compounds such as carbon disulfide, chlorosulfonic acid, thiourea, sulfanilamide, thiamine, heparin, insulin, "Sucaryl," etc.

Uses: A research tool in studying the mechanism of rubber vulcanization; the mechanism of the polymerization of synthetic rubber; the role of sulfur in the coking process and in steel; the effect of sulfur on engine wear; sulfur removal in the viscose process for the manufacture of rayon; the behavior of detergents during washing; sulfur deposition in diesel engines; the action of sulfur in silver plating solutions; protein metabolism; surface active agents and surface phenomena; drug actions; etc.

Shipping regulations: Class D poison, radioactive material. Blue label.*

sulfurated lime. See lime, sulfurated.

sulfurated potash. See potash, sulfurated.

sulfur bichloride. See sulfur dichloride.

sulfur bromide (sulfur monobromide) S_2Br_2.

Properties: Yellow liquid. Becomes red when exposed to air. Decomposed by water. Soluble in carbon disulfide. Sp.gr. 2.6 (15°C); b.p. 54°C.

sulfur cement Cement used for iron vessels and consisting of equal parts of sulfur and pitch. Acid resisting.

sulfur chloride (sulfur subchloride; sulfur monochloride) S_2Cl_2.

Properties: Amber to yellowish-red, oily, fuming liquid; penetrating odor; irritating effect on the eyes, lungs and mucous membranes. Keep well stoppered. Reacts violently with water when contained in a closed vessel. Soluble in alcohol, ether, benzene, carbon disulfide and amyl acetate; decomposes on contact with water. Sp.gr. 1.690 (15.5°C); m.p. −80°C; b.p. 138°C; flash point 266°F.

Derivation: By passing chlorine over molten sulfur.

Method of purification: Distillation.

Grades: Technical.

Containers: 25-, 50-, 75-lb jugs; 150-lb carboys; 700-lb drums; tank cars.

Uses: Chemicals (sulfur solvent, acetic anhydride, thionyl chloride, carbon tetrachloride from carbon disulfide, various chlorohydrins from glycerol, glycol, etc.); analytical reagent; rubber industry for vulcanizing; manufacturing vulcanized oils; rubber substitutes and cements; purifying sugar juices; manufacture of military poison gas; insecticide; hardening soft woods (by treatment with sulfur chloride dissolved in carbon disulfide); pharmaceuticals; textile finishing and dyeing; extraction of gold from its ores.

Warning: Causes burns. Vapor irritating. M.C.A. warning label.

Shipping regulations: Corrosive liquid. White label.*

sulfur dichloride (sulfur bichloride) SCl_2.

Properties: Reddish-brown fuming liquid with pungent chlorine odor. Sp.gr. 1.638 (15.5°C); m.p. −78°C; b.p. decomposes above 40°C; on rapid heating, boils near 60°C; decomposes in water, and alcohol; refractive index 1.567 (n 20/D); flash point, none; fire point, none. Very corrosive.

Derivation: Chlorine is passed into sulfur monochloride to saturation, at 6° to 10°C, followed by carbon dioxide to drive off the excess of chlorine.

Grades: Technical.

Containers: Drums; tank cars.

Uses: In general, as a chlorine carrier or chlorinating agent; rubber industry for vulcanizing; manufacturing vulcanized oils; rubber substitutes and cements; purifying sugar juices; sulfur solvent; as chloridizing agent in metallurgy; in manufacture of organic chemicals and insecticides.

Warning: Causes burns. Vapor irritating. MCA warning label.

Shipping regulations: Corrosive liquid. White label.*

sulfur dioxide (sulfurous acid anhydride) SO_2.

Properties: Colorless gas or liquid. Caution: extremely irritating gas and liquid. Soluble in water, alcohol, and ether. Forms sulfurous acid H_2SO_3. Sp.gr. 1.4337, liquid at 0°; m.p. −76.1°C; b.p. −10°C; vapor pressure 3.2 atmospheres at 68°F; refractive index (liquid) 1.410 (n 24/D). An outstanding oxidizing and reducing agent.

Derivation: (a) By roasting pyrites in special furnaces. The gas is readily liquefied by cooling it with ice and salt, or at a pressure of three atmospheres. (b) By purifying and compressing sulfur dioxide gas from smelting operations, and (c) by burning sulfur.

Grades: Commercial; U.S.P. XVI; technical; refrigeration; anhydrous 99.98% min.

Containers: 2- to 300-lb cylinders; ton drums on multi-unit cars; 40,000-lbs tank cars.
Uses: Chemicals (sulfuric acid, salt cake, sulfites, hydrosulfites of potassium and sodium, thiosulfates, alum from shale, recovery of volatile substances); intermediates; solvent extraction of lubricating oils; general bleaching agent of oils, foods; preservative for beer, wine, and meats; restoring the yellow color of the new grain to old barley and oats; cellulose and paper industries; artificial ice industry; refrigeration; disinfecting and fumigating; tanning; field mouse destruction; agricultural fumigant; extraction of bituminous matters in lignite coal; sulfite pulp manufacture; annealing of glass.
Shipping regulations: Nonflammable gas. Green gas label.*

sulfur dyes. See sulfide dyes.

sulfuret of antimony. See antimony trisulfide.

sulfuretted hydrogen. See hydrogen sulfide.

sulfur flour. See sulfur.

sulfur flowers. See sulfur.

sulfur hexafluoride SF_6.
Properties: Colorless gas; boiling point −63.8°C (sublimes); m.p. −56°C; density of gas 6.5 grams per liter, of liquid 1.91; specific volume 2.5 cu ft/lb (70°F); slightly soluble in water; soluble in alcohol and ether.
Derivation: Sulfur and fluorine.
Grade: 98% pure.
Containers: 60-lb cylinders.
Use: Dielectric (gaseous insulator for electrical equipment).
Shipping regulations: Nonflammable gas. Green label.*

sulfuric acid (hydrogen sulfate; oil of vitriol; battery acid; dipping acid) H_2SO_4. See also sulfuric acid, fuming.
Properties: Strongly corrosive, dense, oily liquid; colorless to dark brown depending on purity. Miscible with water in all proportions but great caution is necessary in mixing due to evolution of much heat that can cause explosive spattering. Very reactive, dissolves most metals; concentrated acid oxidizes, dehydrates, or sulfonates most organic compounds, often causes charring. Sp.gr. of pure material 1.84; m.p. 10.4°C; b.p. varies over range 315-338°C due to loss of sulfur trioxide during heating to 300°C or higher.
Derivation: From sulfur, pyrites, sulfide smelter gases, hydrogen sulfide recovery processes, by the contact and chamber processes, q.v.
Grades: Technical; 50° Baumé (sp.gr. 1.53, 62.2% H_2SO_4); 60°Baumé (sp.gr. 1.71, 77.7% H_2SO_4); 66° Baumé (sp.gr. 1.84, 93.2% H_2SO_4); 98% (sp.gr. 1.84); 99%; 100%; reagent A.C.S.; battery acid (q.v.); C.P.
Containers: Bottles; 5-, 13-gal carboys; 55-, 110-gal drums; tank trucks; tank cars.
Uses (in approximate order of volume; includes fuming sulfuric acid): Fertilizer; chemicals; petroleum refining; paints and pigments; iron and steel; rayon and cellulose film; industrial explosives; nonferrous metallurgy; textile finishing. Generally considered the most important industrial chemical.
Danger: Causes severe burns. MCA warning label.*
Shipping regulations: Corrosive liquid. White label.*

sulfuric acid, aromatic (elixir of vitriol).
Properties: Clear, reddish-brown liquid; peculiar, aromatic odor; pleasant acid taste when diluted. Soluble in water.
Derivation: A mixture of sulfuric acid with alcohol, tincture of ginger, and oil of cinnamon.
Containers: Glass bottles.
Use: Medicine.

sulfuric acid, fuming (oleum; pyrosulfuric acid; disulfuric acid). A solution of sulfur trioxide in sulfuric acid as $H_2S_2O_7$.
Properties: A heavy, oily liquid. Colorless to dark brown depending on purity. Fumes strongly in moist air. Extremely corrosive and hygroscopic.
Derivation: Sulfur trioxide produced by the contact process (q.v.) is absorbed in concentrated sulfuric acid.
Grades: Technical (20-, 40-, 60-, 66% SO_3); C.P.
Containers: Bottles; 5-, 13-gal carboys; 55-, 110-gal drums; tank cars.
Uses: Sulfating and sulfonating agent; dehydrating agent in nitrations; dyes; explosives; petroleum refining.
Danger! Causes severe burns. MCA warning label. Reacts violently with water.
Shipping regulations: Corrosive liquid. White label.*

sulfuric anhydride. See sulfur trioxide.

sulfuric chloride. See sulfuryl chloride.

sulfuric ether. See ether.

sulfuric oxychloride. See sulfuryl chloride.

sulfur iodide. See sulfur iodine.

sulfur iodine (sulfur iodide; iodine disulfide; iodine bisulfide) I_2S_2.
Properties: Brittle masses; grayish-black; iodine odor; metallic luster. Soluble in carbon disulfide; slightly soluble in glycerol; insoluble in water.
Derivation: Heating iodine mixed with sulfur to fusion in a closed vessel.
Containers: 1-lb bottles.
Use: Medicine.
Shipping regulations: None.*

sulfurized asphalts (Dubb's asphalt). Products obtained by heating residual oil or residual asphalt with sulfur at high temperatures.
Shipping regulations: None.*

sulfur, lac (milk of sulfur; precipitated sulfur). Powdered sulfur itself.

Properties: A fine, light-colored, amorphous powder, without taste or odor. For other properties, see sulfur.
Derivation: Obtained by the action of hydrochloric acid upon a solution prepared by boiling together sulfur and lime in water.
Grade: U.S.P. XVI (as precipitated sulfur).
Use: Medicine.
Shipping regulations: None.*

sulfur lotum (washed sulfur). Fine yellow crystalline powder without odor or taste. Sublimed sulfur which has been treated for 3 days with dilute ammonia water, then thoroughly washed with water.
Use: Medicine.

sulfur monobromide. See sulfur bromide.

sulfur monochloride. See sulfur chloride.

sulfur olive oil. See olive oil (grades).

sulfurous acid. A solution of sulfur dioxide in water. The formula H_2SO_3 is used, but the acid is known only through its salts.
Properties: Colorless liquid; suffocating sulfur odor. Sp.gr. about 1.03; unstable. Soluble in water.
Derivation: Absorption of sulfur dioxide in water.
Grades: Technical; C.P.
Containers: Carboys; drums.
Uses: Organic synthesis; bleaching straw hats, wickerware, textiles, etc.; paper manufacture; wine manufacture; brewing; metallurgy; ore flotation; medicine; reagent in analytical chemistry; sulfites; as preservative for fruits, nuts, foods, wines; disinfecting ships; refining crude oils and paraffins.
Shipping regulations: None.*

sulfurous acid, anhydride. See sulfur dioxide.

sulfurous oxychloride. See thionyl chloride.

sulfur oxychloride. See thionyl chloride.

sulfur, precipitated. See sulfur, lac.

sulfur subchloride. See sulfur chloride.

sulfur tetrafluoride SF_4. Sold in cylinders as a selective fluorinating agent.

sulfur trioxide (sulfuric anhydride) SO_3; $(SO_3)_n$.
Properties: Exists in three solid modifications; alpha, m.p. 62°C; beta, m.p. 32.5°C; gamma, m.p. 16.8°C. The alpha form appears to be the stable form but the solid transitions are commonly slow; a given sample may be a mixture of the various forms, and its m.p. not constant. The solids sublime easily. The gamma form boils at 45°C. An explosive increase in vapor pressure occurs when the alpha form melts. The anhydride combines with water, forming sulfuric acid and evolving a large amount of heat. It is strongly corrosive and an active oxidizing agent. Will produce severe burns.
Containers: (Stabilized, liquid) 750-lb drums; tank cars.
Uses: Sulfonation of organic compounds.
Shipping regulations: Corrosive liquid. White label.* It is usually generated in the plant where it is to be used.

sulfuryl chloride (chlorosulfuric acid; sulfonyl chloride; sulfuric chloride; sulfuric oxychloride) SO_2Cl_2.
Properties: Colorless liquid. Pungent odor. Rapidly decomposed by alkalies and by hot water. Soluble in glacial acetic acid.
Constants: Sp.gr. 1.667 at 20°C; b.p. 69.2°C; m.p. −64.1°C; vapor density 4.6.
Derivation: (a) By heating chlorosulfonic acid in the presence of catalysts. (b) From sulfur dioxide and chlorine in the presence of either activated carbon or camphor.
Grades: Technical.
Containers: 5-gal carboys; 55-gal drums; 725-lb drums.
Uses: Organic synthesis (chlorinating agent, dehydrating agent; acylating agent); making pharmaceuticals; dyestuffs; rubber-base plastics; rayon; military poison gas mixtures and making certain poison gases; as a solvent; as a catalyst.
Warning: Causes burns. Vapor irritating. MCA warning label.
Shipping regulations: Corrosive liquid. White label.*

"Sulmet." [57] Trademark for sulfamethazine.

"Sulphon." [307] Trademark of acid dyestuffs used on wool and silk. Characterized by fairly good fastness to light, good fastness to washing, etc. Can be used on leather and paper.

sultam acid. See 1,8-naphthosultam-2,4-disulfonic acid.

sumac wax. See Japan wax.

sumbul (musk root).
Derivation: Dried rhizome and root of Ferula sumbul, or closely related species possessing the characteristic musklike odor.
Occurrence: Central Asia, East Indies.
Grades: Technical.
Containers: Bags.
Use: Medicine.
Shipping regulations: None.*

sumbul oil.
Properties: Dark-colored viscous essential oil; musk-like odor. Soluble in 1 vol or less of 90% alcohol.
Constants: Sp.gr. 0.941-0.964; optical rotation, −6°20'; acid value, 7.0; saponification value, 24-92.
Derivation: Distilled from the root of Ferula sumbul.
Shipping regulations: None.*

"Sumstar." [212] Trademark for a dialdehyde starch made by a periodate oxidant (electrolytic technique) from starch.
Properties: Fine powder, nonvolatile, odorless. Its acute toxicity is claimed to be only a small fraction of that of other commercially available aldehydes. Available as: Sumstar-S: Over 90% oxidized; Sumstar-R: 75-80% oxidized; Sumstar-J: 50% oxidized. Used in adhesives, leather, tobacco, plastics, paper.

"Sumycin." [412] Trademark for tetracycline phosphate complex (q.v.).

sunflower cake. The cakes formed in the press when the seeds are subjected to hydraulic pressure in order to express the sunflower oil.
Constants: Contains various useful constituents, such as unexpressed oil, carbohydrates, proteins and salts. Typical analysis: proteins 21.0%; fats 8.5%; fiber 48.9%; water 10.2%; ash 11.4%.
Containers: Bags; bulk.
Uses: Cattle food; fertilizer ingredient.
Shipping regulations: None.*

sunflower meal. The mealy form assumed by sunflower seeds after the crushing and heating operations, preparatory to the expression of the oil in either hydraulic presses or expeller. If the oil cake is ground the product again is in this mealy form. Uses are similar to those of sunflower cake (q.v.).

sunflower oil.
Properties: Pale yellow liquid; mild taste; pleasant odor. Soluble in alcohol, ether, chloroform and carbon disulfide.
Constants: Sp.gr. 0.924-0.926; iodine value 125; refractive index 1.4611.
Derivation: By expression from the seeds of Helianthus annus.
Method of purification: Filtration.
Grades: Crude; refined.
Containers: Wooden barrels; steel drums; tank cars.
Uses: Varnishes; soap; illuminant; edible oil, particularly in Russia.
Shipping regulations: None.*

"Sunny South" Rosin Oils. [296] Trademark for a line of rosin oils obtained from processing and blending of oils of terpenic origin. Available in three standard grades which differ in viscosity and acid number. Special grades produced to specifications.

"Sunolite." [104] Trademark for anti-sun-checking wax used to prevent disintegration of rubber stocks which are constantly exposed to sunlight.

"Sunolith." [296] Brand name for a proprietary product consisting of various types of lithopone.

"Sunolox." [204] Trademark for a proprietary formulation designed to promote the stability of hydrogen peroxide bleach solutions.
Properties: White, free-flowing powder; bulk density, 56.6 lbs/cu ft; solubility, 48.1 g/100 ml of water at 25°C; insoluble in organic solvents. pH (0.1% sol), 7.5.
Use: Bleaching of textiles near the neutral point; softness of fabric is improved by low silicate content.

"Sunproofs." [248] Trademark for a series of protective waxes to prevent static atmospheric cracking in all types of synthetic and natural rubbers. There are five types, namely, "Sunproof 713", "Sunproof Improved," "Sunproof Junior," "Sunproof Regular" and "Sunproof Super."

"Super Ad-It." [74] Trademark for di-phenylmercuric dodecenylsuccinate (10% mercury as metal). For mildew resistance in pigmented and clear coatings. Easily handled. Does not affect film hardness, drying, color, or gloss.

superalloys (wrought heat-resisting alloys). Steels with oxidative resistance and tensile strength at temperatures up to about 2500°F. All heat-resisting steels contain chromium to combat oxidation; while nickel, manganese, nitrogen, molybdenum, or tungsten make the metal "austenitic" (provides strength). The addition of cobalt and sometimes titanium, aluminum, or silicon will produce a "superstrength wrought heat-resisting alloy." Uses include high pressure equipment, jet-engines, missiles, etc.

"Super-amides: L-9; GR; GC." [328] Alkanolamides claimed to contain up to 50% more amide than the alkanolamide fatty acid condensates of highest concentration hitherto available. Used in textile applications based on their synergistic action with conventional textile processing assistants, to reinforce detergency, foam building, and foam stability. They provide versatility in combining wetting, penetrating, foaming, and dispersing properties.

"Superba." [133] Trademark for a medium high-color carbon black for automotive finishes, locomotive paints, high grade industrial work. Available in two types:
"Standard Superba." Standard loose uncompressed black. Packed in 12½-lb bags.
"Superba Beads." Dustless form. Especially suited for ball mills. Low viscosity mill pastes are produced. Packed in 25-lb bags.

"Super-Beckacite." [36] Oil-soluble, oil-reactive, heat-hardening pure phenolic synthetic resins. Available in both foaming and nonfoaming types.
Physical properties: Color grades range from X to D (U.S. Department of Agriculture rosin standards); acid number various types range from 2-105; melting range 158°F to the theoretical equivalent of 650°F, capillary tube method.
Chemical properties: Imparts fast drying, quick hardening and strong resistance to wear, water, weather and other reagents to varnishes and oleoresinous vehicles.
Uses: Varnishes and enamels of the spar and marine type.

"Super-Beckosol." [36] A group of isophthalic acid alkyd resins for fast drying, flexibility, and high exterior durability when used in the manufacture of paints and printing inks.
Physical properties: Non-volatile 49-100%; viscosity (Gardner-Holdt) L-Z^1; color (Gardner 1933) 5-8; acid number 4-11.
See also "Beckosol."

"Supercarbovar." [275] Trade name for channel carbon black for paints and plastics.

*See "I.C.C. Shipping Regulations," page xiii.
Reference numbers refer to name of manufacturer. See "List of Manufacturers," page v.

"Super-CE." [88] Trademark for a cerium polish, consisting of cerium oxide and other ingredients.

"Super-Cel." See "Celite" Filter Aids.

"Superchrome." [243] Trademark of acid mordant dyes for wool and synthetics.

"Superclear." [78] Trademark for a chemical preparation used for thickening solutions of coloring matters, and mordants used in textile printing and for finishing textiles in order to bind, strengthen, stiffen, or otherwise desirably affect the handle of the treated textile materials.

superconductivity. The property which causes certain metals, alloys, and compounds near absolute zero to lose both electrical resistance and magnetic permeability. For instance, an "isolated" wire loop at superconductive temperatures has been known to maintain a constant current for 18 months. Depending upon the substance, the maximum temperature (transition temperature) for the behavior is 0.5-18°K. Superconductivity is absent in alkali metals, noble metals, ferro- and antiferromagnetic metals. The property often occurs, however, in elements having 3, 5, or 7 valence electrons per atom; and is associated with high room-temperature resistivity.

"Super Cordura." [28] Trademark for high tenacity viscose process yarns having special properties of fatigue and heat resistance, and durability. Available denier sizes from 1100-4400.
Containers: Cones and beams.
Uses: In pneumatic tires and other industrial and mechanical goods uses.

supercritical. See chain reaction.

"Super Dylan." [11] Trademark for polyethylene for use where strength and rigidity are required. Heat resistant to 250°F, solvent resistant, has low temperature strength, electrical insulation properties. Developed by the low pressure Ziegler process.

"Super Exsize." [114] See "Exsize."

"Superfloc 16." [57] Trademark for a flocculant.

"Super Floss." See "Celite" Mineral Fillers.

"Superglo." [108] Trademark for a completely neutral, liquid synthetic detergent. Combination of highly active wetting agents and foam stabilizers. Soluble in hot and cold water.
Containers: 5-gal cans; 15-gal drums.

"Super Glo Gloss Oils." [296] Trade name for limed rosin in petroleum solvent processed in three standard grades differing only in color.
Use: Primarily in paint and varnish industry.

"Supergum." [78] Trademark for compositions used as thickeners and dyestuff carriers in textile printing.

"Superior Red." [141] Trade name for azo red pigments.
Properties: Bright, yellow-red pigments with excellent stability in moisture set inks. Produce inks having excellent flow and body.
Grades: Resinated and non-resinated.
Uses: In all types of inks; especially recommended for moisture set inks.

"Superkel." [244] Trademark for a bottle washing product. Highly alkaline composition with scale and foam control agents.
Uses: Bottle washing, dairies, breweries and soft drink plants.
Hazards: Same as those for sodium hydroxide.

"Superlan." [232] Brand name for a series of level-dyeing acid dyestuffs of good fastness properties.

"Superlith." [223] Trade name for imported zinc sulfide pigments.
Grades: Available as pure zinc sulfide, 60% zinc sulfide.

"Superloid." [322] Trademark for ammonium alginate.
Properties: A tan colored, refined, granular product passing essentially through 20-mesh and having a moisture content of about 13%; dissolves in hot or cold water to give a high viscosity solution (1% by weight about 1200 cps) of slightly acid pH.
Grades: Technical.
Uses: As a hydrophilic colloid for suspending, thickening, emulsifying and stabilizing agent in creaming and bodying of natural and synthetic rubber latex products; protective colloid in resin emulsion paints; adhesives; fire-retarding compositions; ceramics, etc.
Shipping regulations: None.*

"Superlume." [72] Trade name for a bright nickel plating process. Prepared from nickel sulfate, nickel chloride, boric acid and organic addition agents. Used for high speed reproduction where leveling and brightness are desired. Applications include plating steel stampings, forgings, die castings; brass, copper and aluminum parts.

"Super-Multifex." [244] Brand name for an ultra fine calcium carbonate—surface coated.
Properties: Weight per solid gal 22.07 lbs; color, white; particle size 0.03 microns.
Derivation: Precipitated calcium carbonate.
Containers: Multi-wall paper bags, 50-lb net.
Uses: Rubber and plastics.

"Supernilla." [342] Trademark for natural vanilla concentrate in liquid form.

"Superoxol." [123] Trademark for hydrogen peroxide (q.v.).

"Superpax." [337] Trade name for 92 to 94.5% zircon with bulk density 68 lbs/cu ft; average particle size 5 microns max. "Superpax A" has average particle size 4 microns max. Used as opacifiers and for texture control in ceramic products. See "Zircopax."

superphosphate (acid phosphate). The most important phosphorus fertilizer. It is made by the action of sulfuric acid on insoluble phosphate rock (essentially calcium

phosphate, tribasic) to form a mixture of gypsum and calcium phosphate, monobasic. A typical composition is $CaH_4(PO_4)_2 \cdot H_2O$ 30%; $CaHPO_4$ 10%; $CaSO_4$ 45%; iron oxide, alumina, silica 10%; water 5%. This mixture is the superphosphate marketed by the fertilizer industry.
Typical analysis: Moisture 10-15%; available phosphoric acid (as P_2O_5) 18-21%; insoluble phosphoric acid 0.3-2%; total phosphoric acid (as P_2O_5) 19-23%.
Grades: Based on available P_2O_5.
Containers: Bags; bulk; multi-wall paper sacks; carloads.
Use: Fertilizer.
Shipping regulations: None.*
See also triple superphosphate and nitrophosphate.

superphosphoric acid. See phospholeum.

"Super Prime." 84 Brand name for a specially constructed explosive primer for detonating non-cap sensitive blasting agents and ammonium nitrate-fuel oil mixtures.
Containers: 1-lb and 2-lb units in 50-lb shipping cans.
Fire hazard: Dangerous.
Shipping regulations: Explosives. Red label.*

"Superset." 57 Trademark for a resin finish which renders cotton wrinkle-resistant with little loss in tensile strength. Also gives shrinkage control, remains durable after washing and dry cleaning, makes starching unnecessary. Based on "Aerotex Resin M-3" and "Aerotex Resin 133."

"Supersil." 436 Trade name for powdered silica, 99.9% pure.
Grades: Available in a complete range of grades from 80 mesh to material finer than 325 mesh.
Containers: 100-lb multi-wall paper bags and bulk carloads.
Uses: Manufacture of ceramics, porcelain enamel, scouring powders and buffing compounds, fiber glass, autoclave concrete products, chemicals, asbestos products and for cementing deep oil wells.

"Supersilicate." 244 Trademark for a compound consisting essentially of the formula, $1.5\ Na_2O \cdot SiO_2 \cdot 5.5\ H_2O$.
Properties: Dustless, white, granular product containing water of crystallization; soluble in water; m.p. 80-85°C; total Na_2O content 36.6%; loose bulk density 60 lb/cu ft.
Containers: 100-lb multi-wall bags; 125-lb fiber drums; 350-lb fiber drums.
Uses: Laundry and metal cleaning; paint remover; concrete floor cleaner; base for cleaning compounds.

"Super-sol." 25 Brand name for an odorless petroleum naphtha; a rapid-drying highly purified solvent.
Properties: Flash point 130°F, distillation range 330-430°F, water-white color, 98% unsulfonatable residue.
Uses: As a carrier for insecticides and mothicide, in the preparation of odorless paints, and in cleaning compositions.

"Super Spectra." 133 Trademark for high color impingement carbon black used for jet black enamel and lacquers requiring satin finish. Containers $6\frac{1}{4}$-lb bags.

"Super Stod-Sol." 200 Trade name for a petroleum solvent.
Properties: Water-white color; boiling range 310-353°F; sp.gr. 0.779 (60°F); wt/gal 6.49 lbs (60°F); flash point 102°F.
Containers: Drums, tank wagon, tank cars.
Uses: Dry cleaning solvent.

"Super X." 84 Trademark applied to each of a series of permissible ammonia gelatin type dynamites developed for use in underground coal mining operations.
Containers: Packaged in cartridges 1" to $2\frac{1}{2}$" in diameter in 50-lb shipping cases.
Fire hazard: Dangerous.
Shipping regulations: Explosives. Red label.*

"Suplex." 244 Trademark for a formulation for detergent aid. Mildly alkaline water conditioner, detergent, and foam control agent. Uniform, rapidly soluble granules.
Uses: Bottle washing.

"Supracet." 232 Brand name for a series of disperse dyestuffs for acetate rayon and synthetic fibers.

"Suprak." 57 Trade name for a line of tanning agents classified as synthetic phenolic resins. Used in the leather industry as a tanning agent.

"Supralan." 307 Metallized acid colors of good fastness and level dyeing properties.

"Supramine" XA. 307 Trademark for a leather chemical.
Composition: Solubilized sulfur phenol condensate; 75% active.
Properties: Fine tan powder; soluble in water. Density 0.90-1.00.
Uses: Penetrating, leveling and toning agent in the application of acid, direct and chrome dyestuffs to leather.

"Supranol." 307 Trademark. Acid dyestuffs used on wool and silk. Characterized by good fastness to light, washing, and sea water. Can also be used on leather.

"Suprarenin." 162 Trademark for synthetic epinephrine.

"Suprexcel." 232 Brand name for a series of light fast direct cotton dyestuffs.

"Suprex" Clay. 285 Proprietary brand name for a group of hydrous aluminum silicates (sedimentary kaolins) from South Carolina.
Properties: Sp.gr. 2.60; bulk density, aerated 18-20 lbs/cu ft., packed 35-40 lbs/cu ft; creamy white; pH 4.5-5.5; airfloated; particle size 90% minus 2 microns.
Containers: 50-lb multiwall bags or bulk.
Uses: As a "hard" clay in rubber compounding to produce high modulus and tensile, good abrasion resistance and a stiff uncured

compound; as a carrier in pesticides where its fine particle size, plate-like shape and good wetting provide high adsorption in concentrates and excellent suspension behavior in wettable powders. See "Paragon" Clay for a "soft" rubber clay.

"Supronyx." [328] Modified sodium lauryl sulfate, synergically blended for superior wetting, foaming, emulsifying, and detergency properties. It is claimed to be an excellent scouring compound and dye assistant for wool, synthetics, and silk, and with alkaline builders, outstanding on cotton. Although neutral and mild in character it is said to possess outstanding soil-carrying properties.

suramin. See sodium suramin.

"Surett." [51] Trademark for tacky, adhesive lubricants which afford protection for open gears and wire ropes. They resist water and prevent rusting.

"Surfacaine." [100] Trade name for cyclomethycaine.

surface active agent. Any compound that affects (usually reduces) surface tension when dissolved in water or water solutions, or which similarly affects interfacial tension between two liquids. Soap is such a material, but the term is most frequently applied to organic derivatives such as sodium salts of high molecular weight alkyl sulfates, or sulfonates.

Uses: As detergents; wetting agents; penetrants; spreaders; dispersing agents; and foaming agents.

See detergents, synthetic.

surface combustion. Combustion of fuel-air mixtures at incandescent surfaces that catalyze the process to give rapid, flameless and complete combustion.

surfactant. Abbreviated term for surface active agent.

"Surfactol." [202] Trademark for a series of castor oil-derived non-ionic surfactants including:

"Surfactol" 13: A water dispersible grade of glyceryl monoricinoleate.

Uses: Emulsifier; foam inhibitor; deflocculant for colored pigments in water-based pigment dispersions and latex emulsion paints.

"Surfactol" 318, 340, 365, 380: Alkoxy adducts of castor oil listed in the order of their increasing tolerance for water (from moderately self-emulsifiable to completely water soluble).

Uses: Emulsifiers; defoamers; plasticizers; solubilizers for oils, dyes; lubricants; in emulsion paints, pigment dispersions, cosmetics and polishes.

"Surfaseptic." [233] Trademark for synthetic molding resins containing germicides.

"Surfex." [244] Brand name for a high purity precipitated calcium carbonate (surface coated).

Properties: Oil absorption, 18-20; density as shipped, 50-55 lbs/cu ft; wt/solid gal, 22.07 lbs; color, white; particle size, 1-5 microns.

Derivation: Precipitated calcium carbonate.

Containers: Multiwall paper bags, 50 lbs net.

Uses: Paint, plastics, rubber, inks.

"Surfex MM." [244] Brand name for a high purity precipitated calcium carbonate (surface coated, micro milled).

Properties: Oil absorption, 18-20; density as shipped, 50-55 lbs/cu ft; wt/solid gal, 22.07 lbs; color, white; particle size, 1-5 microns.

Derivation: Precipitated calcium carbonate.

Containers: Multiwall paper bags, 50 lbs net.

Uses: Paints, plastics, rubber, inks.

"Surfonic." [137] Trademark for a group of nonionic surface active agents, each with 100% active ingredient.

"Surfonic" N-10 (ethylene oxide nonyl phenol).

Properties: Color 500 Pt-Co max; completely soluble in acetone, methanol, xylene, carbon tetrachloride, and mineral oils; insoluble in water; flash point, 355°F; sp.gr. 0.980 (20/20°C); refractive index 1.5090 (20°C).

Uses: Emulsifier; anti-foaming agent; detergent; penetrant; solubilizing agent; dispersant.

"Surfonic" N-40 (polyoxyethylene nonylphenol, 4 to 1 mole ratio).

Properties: Clear liquid; color, 200 Pt-Co max.; soluble in Stoddard solvent, acetone, methanol, xylene, carbon tetrachloride; gel in water, not dispersible; sp.gr. 1.026 at 20/20°C; refractive index 1.4979 (20°C); flash point 435°F.

Uses: Metal cleaning; degreasing; high-foaming liquid detergents; oil soluble emulsifier; dispersant; penetrant; solubilizing agent.

"Surfonic" N-60 (polyoxyethylene nonylphenol, 6 to 1 mole ratio).

Properties: Color, 200 Pt-Co max.; clear liquid; sp.gr. 1.041 (20/20°C); refractive index 1.4938 (20°C); flash point 475°F; f.p. less than 0°C. Soluble in Stoddard solvent, acetone, methanol, xylene, carbon tetrachloride; gel in water, dispersible.

Uses: Agricultural chemical concentrates; metal cleaning; degreasing; dry cleaning; emulsifier for styrene-butadiene latices; dispersant.

"Surfonic" N-95 (polyoxyethylene nonylphenol, 9.5 to 1 mole ratio).

Properties: Color, 100 Pt-Co max.; clear liquid; completely soluble in water, acetone, methanol, xylene, carbon tetrachloride; sp.gr. 1.061 (20/20°C); refractive index 1.4893 (20°C); freezing point 5°C; flash point 500°F.

Uses: Agricultural chemical concentrates; household and industrial cleaning compounds; textile processing; removal of wall-paper; fire fighting compounds; fruit and vegetable washing; caking prevention.

"Surfonic" N-120 (polyoxyethylene nonylphenol, 12 to 1 mole ratio).

Properties: Color, 200 Pt-Co max; viscous

clear liquid; completely soluble in water, acetone, methanol, xylene, carbon tetrachloride; insoluble in kerosine and Stoddard solvent; sp.gr. 1.070 (20/20°C); refractive index 1.4869 (20°C); freezing point 14°C; flash point 525°F.

Uses: Agricultural chemical concentrates; textile processing; wetting agent; degreasing; wool scouring; emulsification; removal of wallpaper; dispersant.

"Surfonic" N-150 (polyoxyethylene nonylphenol, 15 to 1 mole ratio).

Properties: Color, 200 Pt-Co max., white semi-solid; completely soluble in water, acetone, methanol, xylene, carbon tetrachloride; sp.gr. 1.065 (30/4°C); refractive index, 1.4815 (30°C); f.p. 23°C; flash point 500°F.

Uses: High temperature emulsification; dispersion; high foaming detergents; wetting agents; penetrants.

"Surfonic" N-200 (polyoxyethylene nonylphenol, 20 to 1 mole ratio)

Properties: Color, 200 Pt-Co max; white waxy solid; completely soluble in water, acetone, methanol, xylene, carbon tetrachloride; insoluble in kerosine and Stoddard solvent; refractive index 1.4720 (50°C); f.p. 34°C; flash point 550°F.

Uses: High temperature emulsification; dispersion; high foaming detergents; wetting agents; penetrants.

"Surfonic" N-300 (polyoxyethylene nonylphenol, 30 to 1 mole ratio).

Properties: Color, 200 Pt-Co max; white waxy solid; completely soluble in water, acetone, methanol, xylene, carbon tetrachloride; insoluble in kerosine and Stoddard solvent; refractive index 1.4690 (50°C); f.p. 44°C.

Uses: High temperature emulsification; dispersion; wetting agent; very high foaming detergents; penetrants.

All the above "Surfonic" products available in 55-gal steel drums and in tank cars.

"Surfȳnol." [144] Trade name for ditertiary acetylenic glycols, $R_1R_2C(OH)C{:}CC(OH)R_1R_2$.

Properties: White solids, non-foaming, nonionic surface active agents, soluble in a wide variety of organic solvents, available in the following types:

"Surfȳnol" 82, m.p. 49°C; b.p. 222°C.
"Surfȳnol" 102, m.p. 61°C; b.p. 253°C.
"Surfȳnol" 104, m.p. 37°C; b.p. 260°C.

Uses: Pigment dispersion and defoaming in emulsion paints; rinsing agents; viscosity reductions; foam suppression agents; wetting agents; components of descaling compounds.

"Surital" Sodium. [330] Trademark for thiamylal sodium.

S. U. S. Abbreviation for Saybolt Universal Seconds, a method of expressing viscosity.

suspension. A liquid medium having small solid particles more or less uniformly dispersed through it. If the particles are small enough to pass through ordinary filters and do not settle out on standing, the suspension is called a colloidal suspension or colloid. See also sol.

"Suspenso." [244] Brand name for a high purity precipitated calcium carbonate.

Properties: Oil absorption, 20-22; density as shipped, 50-55 lbs/cu ft; wt/solid gal 22.07 lbs; color, white; particle size, 1-5 microns.

Derivation: Precipitated calcium carbonate.

Containers: Multiwall paper bags, 50 lbs net.

Uses: Paint, plastics, rubber.

"Suspensoil." [244] Trademark for a rapidly soluble alkaline powder laundry detergent.

Uses: Complete detergent and builder for commercial laundry applications.

"Sustane." [416] Trademark for a group of nontoxic antioxidant formulations based on UOP butylated hydroxyanisole; variations may include other approved antioxidants.

Properties: "Sustane BHA" is in the form of white tablets and "Sustane 1-F" in the form of solid white flakes; m.p. 135°F; b.p. 516°F (745 mm Hg), 270°F (5 mm Hg); viscosity (kinematic) 3.3 cs at 100°F; soluble to slightly soluble in wide variety of oils, fats, various organic solvents.

Other variations: "Sustane 3-F" and "Sustane 5-F" are flake form mixtures. "Sustane 3, 6 and 8" are liquid formulations.

Uses: Antioxidants for food products, especially for processors of animal fats and vegetable oils; in cooked food products such as potato chips, baked goods, nuts, etc; stabilizer for petroleum wax coatings for food packaging; antioxidants for petroleum products.

sweet basil oil. See basil oils.

sweet bay. See laurus.

sweet bay oil. See laurel oil, volatile.

sweet cane. See calamus.

sweet fennel. See fennel.

sweet flag. See calamus.

sweet gas. See sour gas.

sweet grass. See calamus.

sweet gum, oriental. See styrax.

sweet oil.
1. See olive oil.
2. A petroleum oil, free of sulfur compounds.

sweet orange. See orange peel, sweet.

sweet spirits of niter.

Properties: Clear mobile liquid of pale yellow or greenish-yellow tint; sp.gr. not above 0.823 (25°C). Unstable in light and air.

Derivation: Alcoholic solution of about 4% ethyl nitrite.

Containers: Well-stoppered amber bottles, kept in cool, dark place.

Use: Medicine.

Fire hazard: Dangerous; keep tightly closed, remote from fire.

Shipping regulations: Flammable liquid. Red label.*

*See "I.C.C. Shipping Regulations," page xiii.

sweet viburnum. See viburnum prunifolium.

sweet water. The glycerin and water mixture obtained when fats are split (or hydrolyzed) with water to give fatty acid and glycerin. Also, the washings from char used in sugar refining.

sweet wood bark. See cascarilla.

"Sylfat." [296] Trade name for high-purity tall oil fatty acids used in protective coatings, soaps, detergents, disinfectants, chemical intermediates, and flotation chemicals.

"Sylflex." [149] Trademark for compositions used in the preparation of leather to impart permanent water repellence to the leather.

"Sylkyd." [149] Trademark for organosiloxy compositions for use as intermediates in the production of resins and formulating varnishes, paints, and enamels of general utility.

"Sylmer." [149] Trademark for silicone compositions used in the finishing of knitted, netted and textile fabrics, and also used in the preparation and treatment of leather to impart permanent water repellence to the leather.

"Syloid." [241] Brand name for a line of finely divided silica gels.

Properties: White powder, transparent in vehicles; refractive index 1.46; oil adsorption 0.9-3.0 g oil/g; liquid bulking value 0.06 solid gal/lb; bulk density 4-30 lb/cu ft.

Typical analysis: Untreated grades, SiO_2, 99.7% ignited basis; volatile portion (6%) is water. Several grades surface-treated with organic and inorganic compounds.

Containers: 50-, 100-lb drums; 15-, 30- and 50-lb bags.

Grades and Uses:

"Syloids" 308, 978: Flatting agents for lacquers.

"Syloids" 161, 162: Flatting agents for heat set finishes.

"Syloid" 404: For air dry varnishes producing matte surfaces.

"Syloid" 72: Anti-blocking agent plastic film, specialty paper coatings.

"Syloid" 73: Anti-blocking agent for plastic film and surfaces.

"Syloid" 75: Reinforcement of adhesives.

"Syloid" AL-1: Gassing preventative in metallic paints.

"Syloid" 244: Anti-caking agent, thickening organic liquids, bodying agent for paste inks.

"Sylox." [241] Brand name for an antiseptic body dusting powder. Principal component is silica gel.

sylvanite (Au, Ag)Te_2.

Properties: One of the gold telluride group of minerals. Corresponds to the same general formula as calaverite (q.v.) and krennerite (q.v.). Steel-gray to silver-white color. Brillant metallic luster becoming dull on exposure. Contains 24.2% Au, 13.3% Ag.

Constants: Sp.gr. 7.9-8.3; hardness 1.5-2.

Occurrence: United States (California, Colorado); Australia.

Use: Source of gold.

"Sylvanol." [342] Trademark for ketonic aromatic having an intense heavy woody note for use in perfumery.

"Sylvaros." [296] Trade name for high-purity tall oil rosins used in sizing of paper and manufacture of resins for use in protective coatings and food containers.

"Sylvatal." [296] Trade name for refined tall oil.

sylvic acid. See abietic acid.

sylvine. See sylvite.

sylvinite. Granular masses of sylvite (q.v.) found naturally intermixed with rock salt and kieserite (q.v.). Used as a fertilizer.

"Sylviola." [342] Trademark for woody cedarlike ketonic concentrate for perfumery.

sylvite (sylvine) KCl. A natural potassium chloride. Contains about 43% potassium chloride, 57% sodium chloride, sometimes with up to 0.26% bromine.

Properties: Colorless or white, bluish or reddish in color; streak, white; vitreous luster. Resembles rock salt in appearance but is easily identified by difference in sp.gr.

Constants: Sp.gr. 1.97-1.99; hardness 2.

Occurrence: Italy; Germany; West Texas; New Mexico.

Use: Major source of potassium compounds in the United States; used for fertilizers.

sym-. Abbreviation for symmetrical. A prefix denoting the structure of organic compounds in which substituents are disposed symmetrically with respect to the carbon skeleton or to a functional group, such as a double bond. For example, sym-dichloroethane is ClH_2CCH_2Cl. In this dictionary, it is disregarded in alphabetizing. Uns- and as- are used for unsymmetrical or asymmetrical.

"Sympatol" Tartrate. [162] Trademark for para-methylaminoethanol-phenol tartrate. $(HOC_6H_4CHOHCH_2NHCH_3)_2C_4H_6O_6$.

Properties: White crystals; freely soluble in water and alcohol; m.p. 182-185°C.

Use: Medicine.

Note: These properties have been incorrectly ascribed in this book to phenylephrine tartrate, which is the meta-isomer.

synaptase. See emulsin.

"Synasol." [214] Trademark for proprietary solvent. Composed of 100 gal of S.D. 1 ethanol denatured with 1 gal of methyl isobutyl ketone, 1 gal of ethyl acetate (87%), and 1 gal of aviation gasoline.

Properties of anhydrous grade: Sp.gr. 0.7900-0.7940 at 20/20°C; b.p. (760mm) 74.5°-79.5°C; flash point 55°F.

Grades: Anhydrous and 190 proof.

Containers: 1-gal can; 5- and 55-gal drums; tank cars up to 10,000 gal.

Uses: Shellac thinner and aniline ink solvent;

coupler or latent solvent in nitro-cellulose lacquers; recommended for general industrial use wherever an alcohol-type solvent is required, except for anti-freeze manufacture where its use is prohibited by the Bureau of Internal Revenue.
Shipping regulations: Flammable liquid. Red label.*

synchrocyclotron. See cyclotron.

synchrotron. See cyclotron.

"Syncurine." 301 Trademark for decamethonium bromide, used in anesthesia.

syndets. Abbreviated form for "synthetic detergents." See detergents, synthetic.

syndiotactic (syndyotactic). A type of polymer molecule in which groups of atoms that are not part of the backbone structure are located in some symmetrical and recurring fashion above and below the atoms in the backbone chain, when the latter are arranged so as to be in a single plane. See polymer, stereospecific.

syndyotactic. See syndiotactic.

"Synektan." 78 Trademark for a line of synthetic tanning materials.

syneresis. The contraction of a gel on standing, with exudation of liquid. The separation of serum from a blood clot, or of the whey in milk souring or cheese making are examples.

synergist. A material which enhances the remedial efficiency of a therapeutically active agent, or the effectiveness of an insecticide, fungicide, or similar biological agent.

"Synkamin." 330 Trademark for 4-amino-2-methyl-1-naphthol.

"Synkayvite" Sodium Diphosphate. 190 Trademark for a brand of menadiol sodium diphosphate (q.v.).

"Synoca." 108 Trademark for a granular, alkaline hexametaphosphate compound containing a wetting agent.
Containers: 100- and 300-lb drums.
Uses: Washing paper machine felts; washing motor vehicles.

"Synpex." 333 Trade name for metal enamels composed of a combination of synthetic resin and cellulose bases.

syntans. Synthetic organic tanning materials made from phenolsulfonic acids and formaldehyde. Used in making chrome and vegetable tanned leathers.

"Syntergent." 309 Trademark for a series of liquid, nonionic detergents especially useful in the textile industry.

"Syntharol." 325 Brand name for blends of lubricants and binders used as warp size additions.

"Synthe-Copal." 36 Pure rosin ester gum.
Physical properties: Color WG to N (U.S. Department of Agriculture rosin standards); acid number 6-8; melting range 149-162°F, capillary tube method.
Chemical properties: Soluble in acetates, coaltar solvents, turpentine and drying oils. Imparts fair resistance to water, weather, abrasion, and many reagents.
Use: General application in the formulation of varnishes and oleoresinous vehicles.

"Synthenol." 64 Trademark for a series of dehydrated castor oils with fast, controllable rate of bodying, excellent color retention, and water resistance.
Uses: Varnishes, enamels, house paints.

synthesis gas. Any mixture of carbon monoxide and hydrogen, usually intended to be used for catalytic conversion to hydrocarbons, alcohols or other organic compounds. The hydrogen and carbon monoxide may be in various proportions, and production may be by high temperature action of steam on carbon or natural gas, by partial oxidation of natural gas, or by other processes. See water gas, and Oxo process.

synthetic indigo blue. See indigo.

synthetic muguet. See hydroxycitronellal.

synthetic oil of bitter almonds. See benzaldehyde.

Synthine process. A name sometimes applied to the more highly developed forms of the Fischer-Tropsch process.

Synthol. A name originally used for a catalytic process developed by Fischer and Tropsch for obtaining mixtures of alcohols and other oxygenated organic compounds from carbon monoxide and hydrogen. The name has also been applied to more recent varieties of such processes and to the products of such processes. See Fischer-Tropsch process.

"Synthrapols." 325 Series of ethylene oxide condensation products. Used as nonionic all-purpose surface-active agents.

"Synthravon." 325 Brand name for a series of non-substantive softeners used as textile napping assistants, lubricants, and resin finishing plasticizers.

syntonin (para-peptone; muscle fibrin).
Properties: Yellow powder. Soluble in dilute hydrochloric acid and alkaline carbonates.
Derivation: By conversion of albumose by means of dilute hydrochloric acid.

"Syntropan" Phosphate. 190 Trademark for amprotropine phosphate (q.v.).

Syrian asphalt. A glance pitch (q.v.) found in Syria, containing mineral matter up to 5% and having a fusing point (B & R) of 275°F.
Shipping regulations: None.*

Syrian gum. See carmania gum.

syrosingopine. An analog of reserpine, prepared from it by hydrolysis and reesterification.
Grade: N.N.D.
Use: Medicine.

syrup. Trade name for solution of cane or beet sugar sold in tank car lots to manufacturers of candy, soft drinks, soda-fountain goods, etc.

syrup, U.S.P. XVI. Aqueous solution of cane sugar (85 g/100 ml). A viscous liquid with sp.gr. about 1.313.

systemics. Pesticides which are absorbed without harm by the host but which are toxic to parasites which feed upon it.

"Systox." [181] Trademark for O,O-diethyl O-(and S)-2-(ethylthio)ethyl phosphorothioates (q.v.).

"Syton." [58] Trademark for colloidal silicas, dispersed in water; available in various grades:

"Syton" AS-200 is used as an anti-soilant for pile fabrics.

"Syton" C-30 and 200 arc used as anti-blocking agents in water finishes; gloss control agents in water finishes; increase bond strength of rubber latex adhesives; increase modulus of latex film and thread; surface harden water-base laminating resins; suspending agents for pigments in water systems; mold lubricant; binders for precision casting molds; flame retarding agents for water-based paints.

"Syton" DS and DS-200 are used in spinning, wet processing and finishing of textiles.

"Syton" P and P-200 are used on paper containers to make slip-resistant coatings.

"Syton" W-200 is used as an anti-slip agent in wax emulsion polishing compounds.

T

T. Symbol for tritium (q.v.).

2,4,5-T. Abbreviation for 2,4,5-trichlorophenoxyacetic acid.

2,4,6-T. Abbreviation for 2,4,6-trichlorophenol.

Ta. Symbol for tantalum.

tabbyite. A mineral hydrocarbon found in Utah and elsewhere.
Uses: As a filler in rubber and roofing materials.

table salt. See sodium chloride.

tabun (dimethylphosphoramidocyanidic acid, ethyl ester) $(CH_3)_2NP(O)(C_2H_5O)(CN)$.
Properties: Liquid, m.p. −50°C; b.p. 240° (760 mm); sp.gr. 1.4250 (20/4°C); readily soluble in organic solvents; miscible with water but readily hydrolyzed. Destroyed by bleaching powder, which, however, generates cyanogen chloride.
Use: Proposed as military nerve gas and experimental cholinesterase inhibitor. Toxic symptoms similar to parathion.

tachysterol $C_{28}H_{44}O$. A sterol.
Properties: Oil; levorotatory; insoluble in water, soluble in most organic solvents. Protect from air.
Use: Medicine, as the dihydrotachysterol.

tack. See tackiness.

tackiness (tack). Property of being sticky or adhesive.

taconite. A low grade iron ore consisting essentially of a mixture of hematite and chert (q.v.). It contains about 25% iron. Found in the Lake Superior district.

tactic. Refers to definite regularity or symmetry of some kind in the molecular arrangement or structure of a polymer molecule. Contrasts with random positioning of substituent groups along the polymer backbone, or random position with respect to one another of successive atoms in the backbone chain of the polymer molecule.
See polymer, stereospecific.

"Tag." [253] Brand name for a type of fungicide containing phenyl mercuric acetate.

"Tagathen." [315] Trademark for chlorothen citrate (q.v.).

tagged compound. A compound whose molecules or formula units contain one or more radioactive atoms. See tracer.

tailed pepper. See cubeba.

tailings. In flour-milling, the product left after grinding and bolting middlings (q.v.); impurities remaining after the extraction of useful minerals from an ore. In general, any residue from a mechanical refining or separation process.

"Takalab TLM." [212] Trademark for a product containing diastatic and proteolytic enzymes.
Properties: Dry, fine, white powder, fully water-soluble, nonhazardous, non-flammable; optimum pH for diastatic reaction 6.5-7.0; for proteolytic reaction 7.0-8.4; optimum temperature 45°C.
Containers: 1-lb packages, each containing two ½-lb sifter-top bottles.
Use: For the digestion of albumen and starch-containing stains in commercial dry-cleaning plants.
Fire hazard: None.

"Take-Hold." [172] Brand name for a proprietary product.
Properties: Completely water-soluble mixture of ammonium and potassium phosphates giving a substantially neutral solution.
Containers: 50-lb paper bags.
Use: An agricultural starter solution mix for transplanting set-outs.

"Takimerse." [212] Trademark for a product containing diastatic and proteolytic enzymes.
Properties: Dry, fine white powder, water-soluble, nonhazardous, nonflammable. Optimum pH 7.0-8.0; optimum temperature 40-45°C.
Grade: Technical.
Containers: 2- and 4-lb cans.
Uses: For the digestion of albumen and starch-containing stains by immersion, in commercial dry-cleaning plants.

"Talase." [212] Trademark for product containing diastatic and proteolytic enzymes.
Properties: Dry, fine white powder, water-soluble, nonhazardous, nonflammable. Optimum pH for diastatic reaction 7.0-7.2, for proteolytic reaction 7.5-8.0; optimum temperature 45°C.
Grade: Technical.
Containers: 50-, 100-, and 300-lb drums.
Uses: Desizing of textile fabrics preparatory to dyeing, bleaching, mercerizing, printing and finishing.

talbutal (5-allyl-5-sec-butylbarbituric acid) $C_{11}H_{18}N_2O_3$.
Properties: White crystals; m.p. 103.4-106.4°C.
Use: Medicine.

talc (talcum; mineral graphite; steatite) $Mg_3Si_4O_{10}(OH)_2$. A natural hydrous magnesium silicate, usually occurring as a natural alteration of magnesium silicate rocks or in metamorphosed dolomites. Compact massive varieties may be called steatite in distinction to the foliated varieties which are called talc. Soapstone is an impure variety of steatite.

Properties: Color white, apple green, gray; luster pearly or greasy; feel greasy; can be cut with a knife; hardness 1-1.5 (may be harder when impure). High resistance to acids, alkalies and heat. Sp.gr. 2.7-2.8.

Occurrence: New York, North Carolina, California, Vermont, Georgia, Maryland, Virginia, Nevada, Montana, Texas, Washington.

Grades: Crude; washed; air floated; U.S.P. XVI; fibrous (99.5%, 99.95%).

Containers: Railroad cars; 50-lb paper bags; 200-lb burlap bags.

Uses: In paint as an extender and as a pigment; in ceramics; in tar paper, asphalt shingles, and roll roofing; in cosmetics and pharmaceuticals; as a filler in rubber, insecticides, soap, putty, plaster, linoleum, oilcloth, rope, fabrics; as a dusting agent for linoleum, oilcloth, leather; as a lubricant; in paper; in slate pencils and crayons; in gas burner tips and electrical insulation.

See also magnesium silicate.

talcum. See talc.

"Tallene." [228] Trademark for tall oil pitch, the residue obtained from distillation and fractionation of whole tall oil.

Properties: 35-50% fatty acids, 24-32% rosin acids, and 22-32% sterols, higher alcohols, etc. It is dark brown in color and a semi-solid at room temperature.

Uses: Protective coatings, adhesives, linoleum, asphalt tile, plasticizers, and rust preventatives. "Tallene" will undergo reactions common to fatty and rosin acids such as esterfication, saponification, and liming.

tall oil (tallol; liquid rosin). The oily mixture of rosin acids, fatty acids, and other materials obtained by acid treatment of the alkaline liquors from the digesting (pulping) of pine wood.

Derivation: The spent black liquor from the pulping process is concentrated until the sodium salts (soaps) of the various acids separate out and are skimmed off. These are acidified by sulfuric acid to obtain the crude tall oil.

Constituents: Rosin acids, including abietic and its isomers; unsaturated fatty acids, including oleic, linoleic and linolenic; and phytosterols and higher alcohols and hydrocarbons, many still unidentified. The composition and properties of tall oil vary widely. The crude fatty acid derivatives, especially metal compounds, are known as tallates.

Grades: Crude; refined.

Containers: Drums; tank cars; tank trucks.

Uses: Drying oils; alkyd resins; linoleum; soaps, cutting oils and emulsifiers; driers; flotation agents; oil well drilling muds; core oils; lubricants and greases; asphalt derivatives; rubber; synthesis of cortisone and sex hormones.

tallol. See tall oil.

tallow.

Derivation: The fat extracted from the solid fat or "suet" of cattle, sheep or horses. The quality varies depending on the season, the food and age of the animal.

Chief constituents: Stearin, palmitin and olein.

Properties: The solidifying points of the different tallows are as follows: from 20-45°C for horse fat; 27-38°C for beef tallow; 54-56°C for stearin and oleo; 32-41°C for mutton tallow.

Grades: Edible; inedible; beef tallow; mutton tallow; horse fats, acidless; edible, extra.

Containers: 50-lb tierces; 375-lb wooden barrels; tank trucks; tank cars.

Uses: Soap stock; leather dressing; candles; food; railway axle grease; manufacture of stearin and oleo oil.

Shipping regulations: None.*

tallow-seed oil. See stillingia oil.

tallow shrub. See myrica.

talmi gold. See Abyssinian gold.

"Tamarax." [342] Trademark for extract of tamarind rind for flavoring.

"Tamax." [408] Trade name for a premium grade mullite refractory made from the best grade of calcined Indian kyanite to which a mineralizer is added. The mineralizer increases the mullite content of the bond. Al_2O_3 68.0%, PCE Cone 38-39. Available in bonded brick and special shapes.

Uses: Recommended for use under extreme conditions of temperature, load, spalling and attack by most slags; specific uses are glass melting superstructure and feeder parts, ferrous and non-ferrous melting refractories, crowns and linings for all types of furnaces and kilns, car top blocks, muffles and piers.

TAM Color "A." [337] Trademark for aluminum iron silicate containing Al_2O_3 47%, SiO_2 32%, Fe_2O_3 13.5%; available as granular +50 mesh 6.6%, +70 mesh 32.1%, +100 mesh 45.0%, +140 mesh 12.0% and milled +200 mesh 3.8%. Granular is used as a speckle ingredient for glazes; milled is used as a colorant for ceramic glazes and bodies.

Containers: 100-lb drums or bags; 50,000-lb carloads.

"Tamol." [23] Trademark for anionic, polymer-type dispersing agents. Supplied as light-colored powders or aqueous solutions. Effective dispersant for aqueous suspensions of insoluble dyestuffs, polymers, clays, tanning agents and pigments.

Use: Manufacture of dyestuff pastes; textile printing and dyeing; pigment dispersion in textile backings, latex paints and paper coatings; retanning and bleaching of leather; dye resist in leather dyeing; dispersion of pitch in paper manufacture; pre-floc

*See "I.C.C. Shipping Regulations," page xiii.

Reference numbers refer to name of manufacturer. See "List of Manufacturers," page v.

prevention in the manufacture of synthetic rubber.

"Tamul." [408] Trade name for a refractory made of sintered synthetic mullite grain; Al_2O_3 68.21%, PCE Cone 39. Available in bonded brick and special shapes; also refractory heat setting cements, air setting cements, hydraulic setting cements and ramming mixes.
Uses: Refractories for glass melting superstructure, ferrous and non-ferrous metal melting, construction of high temperature furnaces and kilns using all types of fuel.

"TAM" Zircon. [337] See Zircon Granular "TAM" and Zircon Milled "TAM."

"Tanak." [57] Brand name for a group of highly concentrated synthetic tanning agents in liquid and dry form.
Uses: In leather manufacture for bleaching dye leveling, for penetration of vegetable tans, for tanning. Used for pitch control in paper manufacture.

"Tanak" MRX. [57] Trademark for melamine-formaldehyde resin tanning agent used to make pure white leather and for bleaching and filling chrome leather.

"Tanamer." [57] Trademark for sodium polyacrylate adhesive for use during the drying of leather.

"Tanarc." [250] Trademark for a closely controlled rutile substitute specifically processed for welding electrode coatings. The product is essentially titanium dioxide plus iron oxides with minor impurities of silica and alumina.
Grades: Granular (80 mesh) and airfloated (325 mesh).

"Tanasol." [78] Trademark for a series of synthetic, organic sulfonic acid condensation products used in tanning of hides and skins.

tangerine oil. See mandarin oil.

"Tanigan" DLNA. [307] Trademark for a leather chemical.
Composition: Sulfonated condensation product of a dihydroxy diaryl sulfone and diphenylol propane; 19% tannin content.
Properties: Dark brown, opaque liquid; sp.gr. 1.22; soluble in water.
Uses: In the leather industry as a replacement for vegetable extracts in combination tanned grain leather; bleaching agent for chrome tanned leather; leveling agent for pastel shades; mordant for basic dyestuffs.

"Taninol" BMN. [206] Brand name for a mordant for basic dyestuffs on cotton and viscose.

"Taninol" WR. [206] Brand name for a wool resist.

tankage (animal tankage). The product obtained in abattoir by-product plants from meat scraps and bones. These are boiled under pressure and allowed to settle. The grease is removed from the top and the liquor drawn off. The scrap is then pressed, dried and sold for fertilizer.
Grades: Based on per cent of ammonia and bone phosphate. A medium grade has 10% ammonia and 20% bone phosphate. Concentrated tankage has had the boiled down tank liquor and press water added to it before drying, and runs about 15-17% ammonia.
Shipping regulations: Flammable solid. Yellow label. Not accepted by freight.*

tankage, garbage. Garbage treated with steam under pressure, the water and some of the grease removed by pressing and further grease removed by solvent extraction. Contains from 3-4% ammonia, 2-5% phosphoric acid and 0.50-1.00% potash.
Use: Fertilizer.
Shipping regulations: Less than 8% moisture; flammable solid. Yellow label. Not accepted by express.*

tankage, gashouse. A misnomer for spent oxide (q.v.).

"Tannex." [236] Brand name for a full substitute for quebracho in the thinning of drilling muds. Used in mud thinning when exposed to the high bottom-hole temperatures of deep wells. May be added directly to the mud system, but gives better results when dissolved in caustic soda solution before adding to the mud system.

tannic acid (gallotannic acid; tannin; digallic acid) $C_{14}H_{10}O_9$. Naturally occurring substance widely found, probably as a glucoside, in nutgalls, tree barks such as sumac, oak, and hemlock, and in other plant parts. The molecular formula is only an approximation, although the tannins are known to be gallic acid derivatives. They have been known in the crude form for centuries for their ability to tan skins. A solution of tannic acid will precipitate albumen.
Properties: Lustrous, faintly yellowish, amorphous powder, glistening scales or spongy mass, odorless; strong astringent taste; m.p., decomposes at 210°C; soluble in water, alcohol, and acetone; almost insoluble in benzene, chloroform, ether, and petroleum ether.
Derivation: Extraction of powdered nutgalls with water and alcohol.
Grades: Technical; C.P.; N.F. XI; fluffy.
Containers: Barrels; drums.
Uses: Chemicals (tannates, gallic acid, pyrogallic acid, hydrosols of the noble metals); alcohol denaturant; tanning; textiles (mordant and fixative); electroplating; galvanoplastics (gelatin precipitant); clarification agent in wine manufacture and brewing; inks; pharmaceuticals; deodorization of crude oil; rubber substitutes; photography; paper (sizing, mordant for colored papers); stove polishing compounds; medicine.
Shipping regulations: None.*

tannin. See tannic acid.

tannin albuminate. See albumin tannate.

tannin-formaldehyde. See tannoform.

tanning extracts. These are now prepared from nearly all of the tannin substances

*See "I.C.C. Shipping Regulations," page xiii.
Reference numbers refer to name of manufacturer. See "List of Manufacturers," page v.

by extraction with water in specially designed extracting equipment and then evaporating the tannin solution to a thick syrup or even to dryness, generally by aid of vacuum.

tanning grease. See degras.

tannoform (methylene ditannin; tannin-formaldehyde) $CH_2(C_{14}H_9O_9)_2$.
Properties: Reddish, odorless, tasteless, bulky powder. Soluble in alcohol, dilute ammonia, sodium hydroxide, potassium hydroxide; insoluble in organic solvents and water.
Constants: Decomposes at 230°C.
Derivation: Reaction of tannin and formaldehyde in hot aqueous solution and precipitating with concentrated hydrochloric acid.
Use: Medicine.
Shipping regulations: None.*

tannyl acetate. See acetyltannic acid.

"Tanolin." 244 Trademark for basic chromium sulfate, specifically adjusted to varying basicities.
Properties: Green crystalline powder; soluble in water.
Derivation: From sodium bichromate.
Grades: Available in 6 basicities, ranging from 34-60%; available in solid or liquid.
Brands: "R"; "T"; "KXD"; "225"; "W2XD"; "S".
Containers: Multiwall paper bags, 80-lb net; fiber drums, 200-lb net. Liquid shipped in tank cars or tank trucks.
Uses: Chrome tanning of leather; colors; compounding.

"Tanoyl." 309 Trademark for sulfonated oils with or without raw oils blended therewith, for use in the fat-liquoring, oiling-off and stuffing of leather.

tansy oil.
Properties: Yellowish liquid; strong odor; becomes brown or exposure to air and light; poisonous! Soluble in alcohol, ether, carbon disulfide and chloroform; almost insoluble in water.
Chief known constituents: Thujone; camphor; borneol.
Constants: Sp.gr. 0.925-0.955.
Derivation: Distilled from the herb, Tanacetum vulgare.
Method of purification: Rectification.
Grades: Technical.
Containers: 1-, 5-lb bottles; 20-lb tins; drums.
Use: Medicine.
Shipping regulations: None.*

tantalic acid anhydride. See tantalum oxide.

tantalic chloride. See tantalum chloride.

tantalite. See columbite.

tantalum Ta. Element of atomic number 73 in group V of the periodic system.
Properties: (a) Black powder. (b) Steel-blue-colored metal when unpolished; nearly a platinum-white color when polished. Soluble in fused alkalies; insoluble in acids except hydrofluoric and fuming sulfuric acids.
Constants: Sp.gr. (a) 14.491; (b) 16.6 (worked metal); m.p. 2996 ± 50°C. Tensile strength of drawn wire may be as high as 130,000 lbs/sq in. Linear coefficient of expansion is only slightly less than platinum and more than molybdenum or tungsten. Electrical resistance about 8 times that of copper and 3 times that of tungsten. With 10% tungsten it has great strength at very high temperatures.
Source: Columbite.
Derivation: From tantalum-potassium fluoride, by heating in an electric furnace, by sodium reduction, or by electrolysis. The powdered metal is converted to the massive metal by sintering in a vacuum. Foot-long crystals can be grown by an arc-fusion process.
Corrosion resistance: 99.5% pure tantalum is resistant to all concentrations of hot and cold sulfuric (except concentrated boiling), hydrochloric, nitric and acetic acids, hot and cold dilute sodium hydroxide, all dilutions of hot and cold ammonium hydroxide, mine and sea waters, moist sulfurous atmospheres, aqueous solutions of chlorine.
Forms and grades available: Powder; sheet; rods; wire; ultrapure; single crystals.
Uses: Dental instruments; surgical tools; container material in nuclear reactors; pen points; filament wire, plates, and support wire for incandescent lamps of thermionic tubes; hypodermic needles; cathodes for use in electrochemical analysis; analytical weights; laboratory ware; parts of scientific instruments; acidproof pumps and parts of chemical equipment; electrical devices; radio; sutures; capacitors; alloys with tungsten used in missiles.

tantalum carbide TaC. A very hard, heavy, high-melting, brown crystalline solid; m.p. 3700°C; b.p. 5500°C; sp.gr. 14.5; hardness 1800 kg/sq min; resistivity 30 micro-ohm cm (room temp); extremely resistant to chemical action except at elevated temperatures.
Derivation: Tantalum oxide and carbon heated at high temperatures.
Use: In cutting tools and dies although alloy does not have "seizing" effect of pure Ta; in cemented carbide tools.

tantalum chloride (tantalic chloride; tantalum pentachloride) $TaCl_5$.
Properties: Pale yellow, crystalline powder. Decomposed by moist air. Caution! Keep well stoppered! Sp.gr. 3.7; b.p. 242°C; m.p. 221°C. Soluble in alcohol and potassium hydroxide.
Grades: Technical.
Uses: Chlorination of organic substances; medicine; production of pure metal.

tantalum nitride TaN.
Properties: Hexagonal brown, bronze or black crystals; sp.gr. 16.3; m.p. 3090°C; insoluble in water; slightly soluble in aqua regia, nitric acid, hydrofluoric acid.

*See "I.C.C. Shipping Regulations," page xiii.
Reference numbers refer to name of manufacturer. See "List of Manufacturers," page v.

Grade: Technical, powder.
Shipping regulations: None.*

tantalum ore. See columbite.

tantalum oxide (tantalic acid anhydride; tantalum pentoxide) Ta_2O_5.
Properties: Rhombic, crystalline prisms; sp.gr. 7.6; m.p. infusible; insoluble in water and acids except hydrofluoric.
Derivation: From tantalite, by removal of other metals.
Grades: Technical.
Containers: Wooden kegs.
Uses: Production of tantalum; "rare-element" optical glass; intermediate in preparation of tantalum carbide; electronics.
Shipping regulations: None.*

tantalum pentachloride. See tantalum chloride.

tantalum pentoxide. See tantalum oxide.

tantalum potassium fluoride (potassium tantalum fluoride; potassium fluotantalate) K_2TaF_7.
Properties: White, silky needles. Slightly soluble in cold water, quite soluble in hot water.
Use: Intermediate in preparation of pure tantalum.

tantiron. A ferrous alloy containing 84.87% iron, 13.5% silicon, 1% carbon, 0.4% manganese, 0.18% phosphorus, 0.05% sulfur. It is resistant to acids; used for chemical equipment. Very brittle.

"Tanz." [48] Trademark for an ammonium lignin sulfonate used as an extender for vegetable tannin extracts.

"Tao." [299] Trademark for triacetyloleandomycin.

"Tao-AC." [299] Trademark for combination drug containing triacetyloleandomycin, phenacetin, caffeine, salicylamide, and buclizine hydrochloride. Used in medicine.

"Taomid." [299] Trademark for a combination drug containing triacetyloleandomycin, sulfadiazine, sulfamerazine, and sulfamethazole.

"Tapazole." [100] Trademark for methimazole, U.S.P.

tapioca. See cassava starch.

tapioca dextrin. See dextrin.

tar. A dark-colored, bituminous substance, liquid or semi-liquid at room temperature, obtained by the destructive distillation of coal, wood, peat, or other carbonaceous or vegetable materials. If often possesses a characteristic "tarry" odor; usually insoluble in water, but miscible with carbon disulfide, benzene, etc. On further distillation, oxidation, etc., forms a pitch. Its composition and origin are variable.
Shipping regulations: May be classed as flammable liquid. Red label.*

tar acid oil. Same as tar acids or a blend of same with neutral oils.

tar acids. Mixtures of phenols present in tars or tar distillates and extractable by caustic soda solutions. The term is generally recognized as applying particularly to tar acids from coal tar; when applied to the products from other tars it should be qualified by the appropriate prefix, e.g., wood-tar acids, lignite-tar acids, etc.
Properties: Soluble in alcohol and coal-tar hydrocarbons.
Grades: 15-18%, 25-28% and 50-53% phenol.
Containers: Drums; tank cars.
Uses: Wood preservative; as an insecticide for cattle and sheep dipping, called dip oils or sheep dip; and in the manufacture of disinfectants.
Danger: Rapidly absorbed through skin. Causes severe burns. MCA warning label.

taraxacum (dandelion; lion's tooth).
Properties: Blackish-brown roots; odorless; bitter taste.
Derivation: The dried rhizome and root of the dandelion, Taraxacum officinale or of Taraxacum levigatum.
Components: Taraxasterol, choline, levulin, inulin.
Occurrence: North America and Europe.
Grades: Technical; N.F. XI.
Containers: Bags; barrels.
Use: Medicine.
Shipping regulations: None.*

tar bases. Basic nitrogen compounds from coal tar, such as pyridine, picoline, lutidine, and quinoline.

tar camphor. See naphthalene.

tar camphor, chlorinated. See chloronaphthalenes.

tar, dehydrated (tar, refined).
Properties: Dark brown, thick, viscid liquid; poisonous!
Derivation: Tar from which the water has been driven off.
Grades: Technical.
Containers: Barrels; tank cars.
Uses: Water-proofing compounds; roads; medicine.
Shipping regulations: None.*

tar, hardwood. See tar, wood.

"106 Tarmastic." [323] Trademark for a series of tar base protective coatings.

tar oil. See creosote, coal tar.

tar oil, rectified. The volatile oil from pine tar rectified by steam distillation.
Properties: Thin liquid with dark reddish-brown color and strong, burning odor and taste. Alcohol solution is acid to moistened blue litmus. Miscible with alcohol. Sp.gr. 0.960-0.990.
Use: Medicine.

tar oil, wood (pine-tar oil).
Properties: Almost colorless liquid when freshly distilled; turns dark reddish-brown; strong tarry odor and taste; poisonous! Sp.gr. 0.862-0.872. Soluble in ether, chloroform, alcohol and carbon disulfide.
Chief constituents: Phenols.
Derivation: Obtained by the destructive

*See "I.C.C. Shipping Regulations," page xiii.
Reference numbers refer to name of manufacturer. See "List of Manufacturers," page v.

distillation of the wood of Pinus palustris.
Method of purification: Rectification.
Grades: Technical; rectified; refined.
Containers: Iron drums; glass bottles; tank cars.
Uses: Renders paper proof against moisture and insects. Used in paints and stains. A heatquenching oil for steel and iron castings and as an insecticide in cattle dips. Used in the manufacture of reclaimed rubber due to its softening action and its mild antioxidant property. Used by the mining industry as a reagent in ore flotation; by the paint industry in roofing compounds; as the vehicle in insulating varnishes; as a wood stain; and as a solvent for various compounds and in emulsions. Used as a spreader for nicotine; as a medicine.

tar, pine.
Properties: Very viscous, dark brown to black liquid or semi-solid with strong characteristic odor and sharp taste. Translucent in thin layers; hardens with aging. Sp.gr. 1.03-1.07; boiling range 240-400°C. Soluble in alcohol, ether, chloroform, acetone, glacial acetic acid, fixed and volatile oils and in sodium hydroxide solution; insoluble in water. Chief constituents are complex phenols; also present are turpentine, rosin, toluene, xylene and other hydrocarbons.
Derivation: By destructive distillation of pine wood, especially Pinus palustris.
Grades: Kiln burnt; retort; N.F. XI.
Containers: Tanks, drums, and barrels.
Uses: Medicine; ore flotation; roofing compositions; paints and varnishes; plastics; tar soaps; linoleum; asphaltic compositions; general preservative; marine preservative.
Shipping regulations: None.*

tarragon oil. See estragon oil.

tarras cement. See pozzolana cement.

tar, refined. See tar, dehydrated.

"Tarset." 323 Trademark for a coal tar-epoxy coating.

tartar, chalybeated. See iron-potassium tartrate.

tartar, cream of. See potassium bitartrate.

tartar, crude. See argols.

tartar emetic. See antimony-potassium tartrate.

tartaric acid (dihydroxysuccinic acid) $HOOC(CHOH)_2COOH$.
Properties: Colorless, transparent crystals, or white, fine to granular, crystalline powder; odorless, has acid taste, stable in air. Soluble in water, alcohol, and ether.
Constants: Sp.gr. 1.7598; m.p. 170°C.
Derivation: (a) Wine-lees containing cream of tartar and calcium tartrate are treated with sufficient milk of lime to convert the cream of tartar into calcium tartrate. The calcium tartrate is treated with sulfuric acid, the solution filtered and the tartaric acid obtained by crystallization. (b) From maleic anhydride and hydrogen peroxide.
Method of purification: Recrystallization.
Grades: Technical; C.P.; crystalline; powder; granular; N.F. XI.
Containers: Bags; drums.
Uses: Chemicals (tartrates, e.g., cream of tartar, tartar emetic, acetaldehyde); as a sequestrant; tanning; effervescent beverages; baking powder preparations; fruit esters; ceramics; galvano-plastics; effervescent medicinal preparations; photography (printing and developing, light-sensitive iron salts); textile industry; silvering glass mirrors; coloring metals.
Shipping regulations: None.*

tartaric acid, inactive. See racemic acid.

para-**tartaric acid.** See racemic acid.

tartar, salt of. See potassium carbonate.

tartrated antimony. See antimony-potassium tartrate.

tartrated iron. See iron-potassium tartrate.

tar, water-gas.
Properties: Dark brown, thick, viscid liquid; poisonous! Sp.gr. 1.005-1.15.
Derivation: A by-product from the manufacture of illuminating gas (carburetted water gas).
Grades: Technical.
Containers: Iron drums; tank cars.
Uses: Distillation for benzene, phenol, etc.; in flame projectors.
Shipping regulations: None.*

tar weed. See grindelia.

tar, wood (tar, hardwood).
Properties: Black, syrup-like, viscous fluid.
Derivation: A by-product of the destructive distillation of wood.
Grades: Technical.
Containers: Tank cars.
Use: Hardwood pitch; wood creosote; heavy, high boiling wood oils; wood preserving oils; paint thinners; medicine.
Shipping regulations: None.*

"Tasil." 408 Trade name for a mullite super refractory made from the best grade of calcined Indian kyanite, selected for low content of impurities including iron, titania and alkalies. Al_2O_3 59.0%, PCE Cone 37-38. Available in bonded brick and special shapes; also refractory heat, air setting and hydraulic setting cements, and ramming mixes.
Uses: Refractories for glass melting superstructure and feeder parts, ferrous and nonferrous metal melting and heat treating, linings and crowns for high temperature furnaces and kilns firing all kinds of fuel, kiln furniture, car top blocks, muffles and piers.

taurine (2-aminoethanesulfonic acid) $NH_2CH_2CH_2SO_3H$. A crystallizable amino acid found in combination with bile acids; its combination with cholic acid is called

taurocholic acid.
Properties: Rods; decompose 300°C; soluble in water; insoluble in alcohol.
Derivation: Isolated from ox bile; organic synthesis.
Use: Biochemical research; pharmaceuticals; organic synthesis; wetting agents.

taurocholic acid (cholaic acid; cholyltaurine) $C_{26}H_{45}NO_7S$. Occurs as sodium salt in bile. It is formed by the combination of the sulfur-containing amino acid, taurine, and cholic acid (q.v.) as the sodium salt. It aids in digestion and absorption of fats.
Properties: Crystals; stable to air. M.p. 125°C (dec). Freely soluble in water; soluble in alcohol; almost insoluble in ether and ethyl acetate.
Derivation: Isolation from bile.
Use: Biochemical research; medicine. The acid and its salts are emulsifying agents.

"Taycor." [408] Trade name for a corundum base super-refractory made of sintered high purity alumina. Al_2O_3 88.0-90.0%, PCE Cone 41. Available in bonded brick and shapes and also heat, air setting and hydraulic setting cements, patches and ramming mixes. Outstanding resistance to abrasion, load and iron oxide slag attack at high temperatures.
Uses: Rails in billet heating furnaces and other service where iron oxide slag is a problem; lining for ferrous melting in electric furnaces, direct arc, indirect arc and induction types; high temperature furnaces of all types including crowns, linings, kiln furniture, muffles and piers.

"Taylor Zircon." [408] Trade name for a refractory made of selected grades of refined zirconium silicate. PCE above Cone 42. Available in bonded bricks and special shapes; also high temperature cements and ramming mixes of all types.
Uses: Paving superstructure and feeder parts for glass melting furnaces; linings for aluminum melting and holding furnaces, gas, oil and electrically fired; refractories for melting ferrous alloys in induction and arc electric furnaces; refractories for phosphate, sodium silicate and frit melting; kiln furniture for firing of certain porcelain and other ceramic bodies.

Tb. Symbol for terbium.

TBH. Abbreviation for technical benzene hexachloride. See 1,2,3,4,5,6-hexachlorocyclohexane. Used as an insecticide.

TBT. Abbreviation for tetrabutyl titanate.

"TBTO." [288] Trademark for bis(tributyltin) oxide.
Properties: Colorless to slightly yellow liquid; b.p. 180°C/2 mm; f.p. lower than −45°C; sp.gr. 1.17 (25°C); flash point above 212°F (Tag closed cup); viscosity 4.8 centistokes at 25°C; practically insoluble in water.
Containers: 6-gal steel pails; 55-gal steel drums.
Uses: Potential application as a fungicide in the fields of textile, wood, leather, paper and plastic preservation.

Tc. Symbol for technetium.

TCA. Abbreviation for trichloroacetic acid or its sodium salt, a herbicide.

TCA cycle (tricarboxylic acid cycle; Krebs cycle; citric acid cycle) A special mechanism in the normal metabolism of living cells for the final degradation of 2-, 3-, or possibly 4-carbon fragments of metabolites by a combination of decarboxylation and dehydrogenation.

TCB. Abbreviation for tetracarboxybutane (q.v.).

T.C.C. Abbreviation for Tag (actually Tagliabue) closed cup, a type of flash point test.

TCP. Abbreviation for tricresyl phosphate.

TDE (2,2-bis(para-chlorophenyl)-1,1-dichloroethane; dichlorodiphenyldichloroethane; DDD) $(ClC_6H_4)_2CHCHCl_2$. Insecticide similar to DDT but considered less toxic to mammals. Name accepted as generic by Ent. Soc. TDE is the abbreviation for tetrachlorodiphenylethane.
Properties: Colorless crystals; m.p. 109-110°C; soluble in organic solvents; insoluble in water; not compatible with alkalies.
Derivation: Chlorination of ethanol and condensation with chlorobenzene.
Grades: Technical.
Containers: Fiber drums.
Caution! Harmful if swallowed. Absorbed through skin when in solution. MCA warning label.
Uses: Similar to DDT; as dusts, emulsions and wettable powders for contact control of leaf rollers and other insects not readily controlled by DDT.

TDI. Abbreviation for toluene diisocyanate.

TDQP. Abbreviation for trimethyldihydroquinoline polymer.

Te. Symbol for tellurium.

TEA. Abbreviation for triethanolamine.

teaberry. See gaultheria.

TEAC. Abbreviation for tetraethylammonium chloride.

TEA chloride. Another abbreviation for tetraethylammonium chloride.

teal oil. See sesame oil.

"TEC-Anti-Freeze." [256] Trade name for methyl alcohol solution.
Use: Freezing point depressant.

technetium Tc. Element with atomic number 43 of group VII of the periodic system. Technetium was first obtained by the deuteron bombardment of molybdenum but since has been found in the fission products of uranium and plutonium. Apparently all the known isotopic forms of technetium are radio-active. Several isotopes have been prepared ranging in mass numbers from 93 to 99. Technetium 99 is the most important because of its long half-life, 9.4×10^5 years.

The chemistry of technetium has been studied by tracer techniques and is similar to that of rhenium and manganese. The free metal is obtained from reactor fission products by solvent extraction followed by crystallization as an ammonium salt of technetium, which is reduced with hydrogen. The metal is silver gray in appearance, and melts at 2200°C (4000°F). The specific gravity is approximately the same as that of silver. It is slightly magnetic. Compounds of the types TcO_2, Tc_2O_7, NH_4TcO_4, etc, have been prepared. The principal chemical interest in technetium derives from the remarkably strong corrosion-inhibiting properties of pertechnate compounds.

"Technoscents." [188] Trademark for odor modifiers designed to cover a disagreeable odor associated with products of a technical nature: solvents, cutting oils, paints, lubricating oils, detergents, scrub soaps and powders, fuel oils.

"Tecmangam." [256] Trade name for free flowing granular solid containing 70% manganese sulfate.
Containers: 5-ply paper bags (50-lb net); bulk.
Use: Source of manganese in fertilizers and animal feeds.

"Tecquinol." [256] Trade name for technical hydroquinone (98.5% min).
Properties: White to cream colored crystals; melting point 169°C min.
Containers: 100- and 110-lb net fiber drums.
Use: As a polymerization inhibitor for many monomers; as an arrester for peroxide-catalyzed polymerizations; and as a raw material for organic synthesis.

"Tecsol." [256] Trade name for proprietary solvents consisting of denatured anhydrous and denatured 95% ethyl alcohol.
Properties: Anhydrous grade: Sp.gr. (15.6°/15.6°C) 0.794-0.798; boiling range 74°-80°C; miscible without turbidity with 19 vols 60° A.P.I. gasoline at 20°C; flash point (Tag open cup) 51°F.
Properties: 95% grade: Sp.gr. (15.6°/15.6°C) 0.813-0.817; boiling range 74°-80°C; miscible without turbidity with 19 vols 60° A.P.I. gasoline at 20°C; flash point (Tag open cup) 54°F.
Containers: 55-gal non-returnable steel drums; 4000-, 6000-, 8000-, and 10,000-gal tank cars.
Use: A basic raw material and solvent interchangeable with most denatured ethyl alcohol formulations.
Shipping regulations: Flammable liquid. Red label.*

"Tedion." Trademark for 2,4,4',5-tetrachlorodiphenylsulfone, $Cl_3C_6H_2SO_2C_6H_4Cl$. White crystalline powder, m.p. 147°C. Used as a miticide, especially for fruit trees, including citrus fruit trees.

"Tedlar." [28] Trademark for polyvinylfluoride film.

teel oil. See sesame oil.

teeming. Metallurgical term for the process of pouring molten metal (steel or iron) from a ladle into molds.

"Teepol" 610. [125] Trademark for a detergent and wetting agent in liquid form based on the sodium salt of secondary alkyl sulfates, of the type $R_1R_2CHOSO_2ONa$, where R_1 is a large alkyl group and R_2 is a relatively small alkyl group.
Properties: A clear amber liquid, pH 8.5-9.0; sp.gr. 1.08 (60/60°F); soluble even in hard water; gives clear solutions. The calcium and magnesium salts are extremely soluble in water and are surface-active also. The surface-active properties of "Teepol" are apparent even in low concentrations.
Containers: Specially lined drums.
Shipping regulations: None.*

"Teflon." [28] Trademark for tetrafluoroethylene (TFE) fluorocarbon resins available as molding powder, extrusion powder, aqueous dispersion and fiber. The trademark also applies to fluorinated ethylene-propylene resins designated as 100 FEP. See polytetrafluoroethylene.

TEG. Abbreviation for tetraethylene glycol, or triethylene glycol.

"Tego." [23] Trademark for thin tissue impregnated with heat-convertible phenol-formaldehyde resin, supplied in rolls. Produces waterproof bond with plywood veneers.
Use: Hot press bonding of furniture veneers, premium wall panelling.

Teichmann's crystals. See hemin.

TEL. Abbreviation for tetraethyl lead.

"Telar." [28] Trademark for an anti-freeze and summer coolant. Concentrated ethylene glycol with rust inhibitors that protect cooling system metals, including aluminum, against rust and corrosion in winter and summer. Can be mixed with any potable water and never needs to be drained from a properly operating cooling system. The color of "Telar," bright red, changes to yellow when serious cooling system failure occurs and the coolant turns corrosive.

"Teldrin." [71] Trademark for chloroprophenpyridamine maleate, an antihistamine drug.

"Telepaque." [162] Trademark for iopanoic acid.

"Telloy." [69] Trademark for a proprietary product. Finely ground tellurium. A secondary vulcanizing agent for rubber products.

"Tellurac." [69] Trademark for proprietary product, tellurium diethyldithiocarbamate $[(C_2H_5)_2NC(S)S]_4Te$.
Properties: Orange-yellow powder (also supplied in "rodform"); sp.gr. 1.44 ± .03; melting range 108-118°C; soluble in benzene, carbon disulfide, chloroform; slightly soluble in alcohol, gasoline; insoluble in water.
Uses: Primary or secondary (with thiazoles)

accelerator in natural, nitrile and butyl rubber and SBR; for curing bags, inner tubes, hose, molded and extruded goods.

telluric acid (trihydrated telluric oxide; hydrogen tellurate) H_2TeO_4.
Properties: White heavy crystals. Soluble in hot water and alkalies; slightly soluble in cold water.
Constants: Sp.gr. 3.07; m.p., decomposes at 160°C.
Derivation: By the action of sulfuric acid on barium tellurate.
Method of purification: Crystallization.
Grades: Technical.
Containers: Glass bottles.
Use: Chemical reagent.
Shipping regulations: None.*

telluric bismuth. See tetradymite.

telluric bromide. See tellurium tetrabromide.

tellurium Te. Element having atomic number 52; in Group VI of the periodic system.
Properties: Silvery white, lustrous solid with metal characteristics. Density 6.24 g/cc (30°C); Mohs hardness 2.3; m.p. 450°C ± 10°C; b.p. 990°C. Soluble in sulfuric acid, nitric acid, potassium hydroxide and potassium cyanide solution; insoluble in water. Imparts garlic-like odor to breath of workers; can be depilatory.
Source: From anode slime produced in the electrolytic refining of copper and lead.
Derivation: By the reduction of telluric oxide with sulfur dioxide; by dissolving the oxide in a caustic soda solution and plating out the metal.
Grades: Powder, sticks, slabs and tablets, 99.5% pure; crystals, 99.99% and 99.999% pure.
Uses: Alloys (tellurium lead, stainless steel, etc.); secondary vulcanizing agent in the rubber industry; in manufacture of iron and stainless steel castings; coloring agent in glass and ceramics; thermoelectric devices.
Shipping regulations: None.*

tellurium bromide. See tellurium dibromide and tellurium tetrabromide.

tellurium chloride. See tellurium dichloride.

tellurium dibromide (tellurium bromide; tellurous bromide) $TeBr_2$.
Properties: Blackish-green, crystalline mass, or gray to black needles. Very hygroscopic. Decomposed by water. Violet vapor.
Constants: B.p. 340°C; m.p. 280°C.

tellurium dichloride (tellurium chloride; tellurous chloride) $TeCl_2$.
Properties: Amorphous, black mass. Greenish yellow when powdered. Decomposed by water.
Constants: Sp.gr. 6.9; b.p. 327-377°C; m.p. 175°C.
Grades: Technical.

tellurium dioxide (tellurous acid anhydride) TeO_2.
Properties: Heavy, white, crystalline powder, odorless. Soluble in acids (conc.), alkalies; slightly soluble in acids (dilute), water.
Constants: Sp.gr. 5.89; m.p. 700°C.
Grades: Technical; C.P.
Uses: Antiseptic; detecting bacteria in vaccines.

tellurium disulfide (tellurium sulfide) TeS_2.
Properties: Red powder. Turns in time to a dark brown, amorphous powder. Fuses to gray, lustrous mass. Soluble in alkali sulfides; insoluble in acids, water.

tellurium lead. See lead, tellurium.

tellurium sulfide. See tellurium disulfide.

tellurium tetrabromide (telluric bromide; tellurium bromide) $TeBr_4$.
Properties: Red crystals when hot, orange when cold. Soluble in a little water (decomposes in excess water).
Constants: Sp.gr. 4.3; m.p. 380°C; b.p. approx 420°C, with decomposition into bromine and dibromide.

tellurous acid H_2TeO_3.
Properties: White, crystalline powder. Soluble in acids (dilute), alkalies; slightly soluble in water, alcohol.
Constants: Sp.gr. 3.053; m.p. 40°C (decomposes).
Grades: Technical.

tellurous acid anhydride. See tellurium dioxide.

tellurous bromide. See tellurium dibromide.

tellurous chloride. See tellurium dichloride.

"Tel-Tale." [241] Trademark for 6-16 mesh silica gel which is impregnated with cobalt chloride and turns from blue to pink as the relative humidity increases.

"Telura." [51] Trademark for pale, filtered lubricating oils for textile mills. Compounded grades are supplied for knitting machines and shear oil requirements. Low viscosity grades are suitable for process use where white mineral oil quality is not required.

"Telvar." [28] Trademark for a monuron weed killer on sites where bare ground is desired; and for selective control of weed seedlings in certain crops.

TEM. Abbreviation for triethylene melamine.

"Temasept." [430] Trade name for polybrominated salicylanilide, an active germicidal agent.

"Temex." [304] Trademark for a series of organic vinyl stabilizers. Available as:
"Temex 3." Barium-zinc organic compound.
Properties: Fine white powder, sp.gr. 1.15.
Containers: 30- and 175-lb fiberboard drums.
Uses: Heat and light stabilizer for asbestos-filled vinyl flooring with outstanding resistance to moisture and curling, and freedom from sulfide staining.
"Temex 3A." Barium-zinc organic compound.
Properties: Fine white powder, sp.gr. 1.37.
Containers: 30- and 125-lb fiberboard drums.
Uses: Stabilizer for vinyl asbestos-filled

flooring compounds. Provides heat stability and color retention properties in these systems.

"Temex 4." Barium-zinc organic compound.
Properties: Fine white powder, sp.gr. 1.17.
Containers: 15- and 75-lb fiberboard drums.
Uses: Stabilizer for all types of vinyl flooring. Imparts heat and light stability, and excellent retention of color shades.

"Temex 5." Metal salt-organic complex.
Properties: Fine white powder, sp.gr. 1.5.
Containers: 200-lb fiberboard drums.
Uses: Stabilizer for utility grade vinyl asbestos flooring. Imparts good heat and light stability and resistance to sulfide stain.

temperature. The thermal state of a body considered with reference to its ability to communicate heat to other bodies (J. C. Maxwell). There is a distinction between temperature and heat, as is evidenced by Helmholtz' definition of heat as "energy that is transferred from one body to another by a thermal process," where, by a thermal process, is meant radiation, conduction, and/or convection.

Temperature is measured by such instruments as thermometers, pyrometers, thermocouples, etc.; and by scales such as Centigrade, Fahrenheit, Réaumur, and absolute (Kelvin). See also absolute temperature.

"Tenamene 1." [256] Trade name for a tan to red liquid composed of N-butylated-p-aminophenol in isopropanol.
Properties: Sp.gr. (60/60°F) 0.91; flash point (Tag closed cup) 52°F; soluble in dilute acid and 10% sodium hydroxide; miscible with benzene.
Containers: 30-, 55-gal non-returnable steel drums.
Use: An antioxidant and gum inhibitor for gasoline.
Shipping regulations: Flammable liquid. Red label.*

"Tenamene 2." [256] Trade name for a tan to red liquid consisting of N,N'-di-sec-butyl-p-phenylenediamine.
Properties: Sp.gr. (60/60°F) 0.94; viscosity at 100°F (Saybolt Universal) 63 secs; flash point (Cleveland open cup) 285°F; soluble in dilute acid; insoluble in water and 10% sodium hydroxide; miscible with benzene.
Containers: 30-, 55-gal non-returnable steel drums; 6000-gal tank cars.
Use: A gum inhibitor and sweetening agent for motor fuels and aviation gasolines.
Shipping regulations: Poison label.*

"Tenamene 3." [256] Trade name for a white to light yellow solid consisting of 2,6-di-tert-butyl p-cresol, or a pale yellow liquid composed of this compound in toluene (⅓ solids, ⅔ solvent).
Properties: Sp.gr. of liquid (60/60°F) 0.89; viscosity at 100°F (Saybolt Universal) 28 secs; flash point (Tag closed cup) 44°F; insoluble in water, dilute acid or 10% sodium hydroxide; soluble in benzene.
Containers: Solid, 100-lb fiber drums; liquid, 55-gal non-returnable steel drums, 6000- and 8000-gal tank cars.
Use: An antioxidant and gum inhibitor for gasoline, transformer, turbine and lubricating oils. Also as a stabilizer for paraffin wax, polyethylene and other plastic resins.
Shipping regulations: Liquid: flammable liquid. Red label.*

"Tenamene 30." [256] Trademark for a dioctyl derivative of para-phenylenediamine.
Properties: Dark reddish brown liquid; sp.gr. 0.896 (80°F); flash point (C.O.C.) 420°F; soluble in dilute acid; insoluble in 10% sodium hydroxide; miscible with benzene.
Containers: 55-gal drums and 5-gal cans.
Uses: Sweetening agent and antioxidant for gasoline and related products.

"Tenamene 31." [256] Trademark for a dioctyl derivative of para-phenylenediamine.
Properties: Dark reddish-brown liquid; sp.gr. 0.903 (80°F); flash point (C.O.C.) 410°F. Soluble in dilute acid; insoluble in 10% sodium hydroxide; miscible with benzene.
Containers: 55-gal drums; 5-gal cans.
Uses: Sweetening agent and antioxidant for gasoline and related products.

"Tenamene 60." [256] Trade name for a red-brown liquid composed of disalicylal propylenedi-imine, 80%, in toluene.
Properties: Sp.gr. (60/60°F) 1.07; viscosity at 100°F (Saybolt Universal) 80 secs; flash point (Tag closed cup) 20°F; decomposes in dilute acid; soluble in 10% sodium hydroxide; miscible with benzene.
Containers: 30-gal non-returnable steel drums and 5-gal steel cans.
Use: A copper deactivator for gasoline.
Shipping regulations: Flammable liquid. Red label.*

"Tenamene MD 50." [256] Trademark for a reddish-brown liquid composed of disalicylal propylenedi-imine, 50% in mixed xylene.
Properties: Sp.gr. 1.014 (60/60°F); flash point, Tag open cup, 106°F. Decomposes in dilute acid; miscible with benzene.
Containers: 30-gal lined drums.
Use: A copper deactivator for gasoline.

"Tenex." [79] Trademark of an extremely pale, "all purpose" wood rosin claimed to show excellent solubility and freedom from crystallization.
Constants: M.p. (capillary tube) 53°C; m.p. (ball and ring) 72°C; acid number 148; unsaponifiable matter 17.0%; color "X."
Containers: Non-returnable, light-weight galvanized drums of about 500 lbs gross wt. Tare 14-16 lbs.
Uses: Adhesive tape; artificial Burgundy pitch; core oil; cosmetics; disinfectants; electric insulating compounds; emulsions; ester gum; printing ink; rubber cement; shellac diluent; solder; spirit varnishes; synthetic resins; varnishes; Venice turpentine.

"Tenite." [256] Trademark. "Tenite" Acetate—cellulose acetate molding composition. "Tenite" Butyrate—cellulose acetate butyrate molding composition. "Tenite" Polyethylene—polyethylene molding composition. Thermoplastic.

"Tenite" Acetate and "Tenite" Butyrate:

Colors: The molded products are made in all shades of color and any degree of translucence, including clear transparent.

Forms: Granulation and pellets for injection molding and continuous extrusion.

Physical properties: Vary with formula as follows: sp.gr. (20/20°C) 1.13-1.34; elongation 3-94%; tensile strength at yield 900-8000 lbs/sq in; flexural strength at yield 1000-13,500 lbs/sq in; compressive strength at yield 650-9400 lbs/sq in; impact strength (Izod) 0.2-2.9 (−40°F) ft lbs/in of notch; 0.3-9.6 (77°F) ft lbs/in of notch; modulus of elasticity 0.49-8.14 x 10^5 lbs/sq in; water absorption (24-hr immersion) wt gained 1.1-5.7%, soluble matter lost 0.1-2.8%; refractive index 1.46-1.50; Rockwell hardness 32-120 (R scale); softening temperature 140-250°F; max recommended service temperature 160°F.

Chemical properties: All cellulose ester compositions are somewhat hygroscopic, absorbing moisture to some extent, depending upon the relative humidity and temperature as well as the formula used. "Tenite" Butyrate absorbs on immersion about the same amount of water as cellulose nitrate plastic, or about half as much as cellulose acetate plastic. Articles molded or extruded of "Tenite" cellulosic plastics should not be subjected to concentrated acids or alkalies. Alcohol or some essential oils of an aromatic type will spot the surface of a molded piece. "Tenite" cellulosic plastics are not damaged by contact with most vegetable and mineral oils but are soluble in ketones and certain esters.

Uses: Miscellaneous articles, wherever appearance and strength in resistance to impact is desired without exposure to excessive heat. Continuously extruded in the form of film, sheets, strips, rods, tubing, pipe, and various profile sections.

"Tenite" Polyethylene:

Properties: White, translucent waxy solid; tasteless, odorless, and non-toxic in natural formulation; it is waterproof, has low rate of water-vapor permeability; excellent electrical insulation properties; insoluble in all ordinary solvents at room temperature; slightly soluble in vegetable oils at 200°F and above; highly resistant to chemicals including concentrated sulfuric, nitric, hydrochloric and acetic acids at room temperature; fairly resistant to strong oxidizing agents such as chromic acid or acid permanganate solutions.

Grades: It is available in melt index ranges from 0.3-200, ranging in molecular weights from 10,000-38,000, and in mechanical properties from tough, flexible plastics to wax-like materials.

Uses: Housewares, containers, closures, packaging materials, flexible pipe, sheathing for wire and cable, coatings for paper and metal foil to give heat-sealing packaging materials.

"Tenlo." [309] Trademark for polyhydroxy alcohol fatty acid esters sold to the paint making trade as an oil soluble, non-ionic surface active agent adapted for use as an aid in grinding and mixing pigments.

tennantite $(Cu, Fe)_{12}As_4S_{13}$. A natural sulfarsenide of copper and iron, found in metallic veins. A variety of fahlore (q.v.).

Properties: Color flint gray to iron black; luster metallic; streak black to brown; hardness 3.5-4.5; sp.gr. 4.6.

Occurrence: Colorado, Idaho, Utah, Montana; Canada; Europe.

Use: Minor ore of copper.

tenorite (melaconite) CuO. For the synthetic material, see copper oxide, black.

Properties: Copper mineral occurring as dull black earthy masses, black powder, or shining black scales; streak, black; luster, metallic in scales, dull in masses.

Constants: Sp.gr. 5.82-6.25; hardness 3.

Occurrence: United States (Utah, Arizona, Wyoming, Oregon, Tennessee), Italy.

"Tenox." [256] Trade name for a line of food-grade antioxidants containing one or more of the following ingredients: butylated hydroxyanisole; butylated hydroxytoluene and/or propyl gallate with or without citric acid. Some formulas are supplied in solvents such as propylene glycol.

Containers: Solids: 1-lb bottle, 5-, 10- and 100-lb fiber drums (net); liquids: 1-lb bottle, 1-gal bottle, 15- and 30-gal nonreturnable steel drums.

Use: Antioxidants for edible fats and oils.

"Tenox BHT." [256] Trademark for a form of butylated hydroxytoluene especially prepared for feed use.

Properties: White free flowing, nondusting powder; melting range 69-70°C; molecular weight 220.

Grade: Agricultural.

Containers: 50-, 100-lb fiber drums.

Uses: Stabilizer for fish meals and scraps, and antioxidant for poultry and animal feeds.

"Tensilon" Chloride. [190] Trademark for a brand of edrophonium chloride (q.v.).

"Tensol." [83] Trade name for a dispersing agent, emulsifier, and colloidal film breaker used in the textile, leather, and rubber industries, and for mineral flotation.

"Ten-Ten." [233] Trademark for 2,4-D compositions.

"Tentone." [315] Trademark for methoxypromazine maleate (q.v.).

"Teox 120." [84] Trademark for a nonionic surface active agent composed of polyethenoxy tallate.

Properties: Light straw-colored liquid; bland odor; sp.gr. 1.065-1.070 (77°F);

refractive index 1.4762 (77°F); viscosity 190 cps (77°F); heat stable to 644°F; very slightly hygroscopic; pH (0.2% soln) 7.2; soluble in water, acetone, benzene, ethyl ether, carbon tetrachloride, ethanol, methanol, and xylene.
Containers: 450-lb drums; tank trucks; tank cars.
Uses: Detergents; surface cleaners; textile cleaning and dyeing; emulsion formulations.

"Teox" Compounds. [84] Detergent products consisting of "Teox 120" compounded with alkaline builders to form an all-purpose controlled sudsing nonionic detergent, "Teox Compound 3," or formulated with soda ash and sodium tripolyphosphate to form an anti-dusting additive for alkaline detergents, "Teox Compound DD."

TEPA. Abbreviation for triethylene phosphoramide. See also nitrogen mustards.

tephroite. See olivine.

"Tepidone." [28] Trademark for 47% water solution of sodium dibutyldithiocarbamate.
Properties: Clear, amber liquid.
Containers: Drums (125 lb, net).
Use: To accelerate and improve the vulcanization of natural and synthetic rubber and latex compounds.

TEPP. Abbreviation for tetraethylpyrophosphate.

tera-. Prefix meaning 10^{12} units, i.e., 1 Tg = 1 teragram = 10^{12} grams.

terbia. See terbium oxide.

terbium Tb. Atomic number 65; Group III of the periodic table; one of the rare-earth elements of the yttrium sub-group. See also rare earth metals.
Properties: Metallic luster. Reacts slowly with water, and is soluble in dilute acids. Sp.gr. 8.332; m.p. 1360°C; b.p. 2500°C (approx); salts colorless.
Source: See rare-earth minerals.
Derivation: Reduction of fluoride with calcium.
Grades: Regular, high purity (ingots, lumps).
Uses: Phosphor activator.

terbium chloride $TbCl_3 \cdot 6H_2O$.
Properties: Transparent, colorless, prismatic crystals; readily soluble in water or alcohol; density 4.35; m.p. (anhydrous) 588°C; very hygroscopic.
Derivation: By treatment of carbonate or oxide with hydrochloric acid in an atmosphere of dry hydrogen chloride.

terbium fluoride $TbF_3 \cdot 2H_2O$.
Grades: Up to 99.9% terbium salts. Available as 77% Tb_4O_7.
Containers: Glass bottles.
Use: To make the element.

terbium nitrate $Tb(NO_3)_3 \cdot 6H_2O$.
Properties: Colorless monoclinic needles or white powder, soluble in water; m.p. 89.3°C.
Derivation: By treatment of oxide, carbonate or hydroxide with nitric acid.
Containers: Glass bottles.
Shipping regulations: Oxidizing material. Yellow label.*

terbium oxide (terbia) Tb_2O_3.
Properties: Dark brown powder; soluble in dilute acids; slightly hygroscopic; absorbs carbon dioxide from air.
Derivation: By ignition of hydroxides or salts of oxyacids.
Grades: 98-99%.
Containers: Glass bottles.
See also rare earths.

terbium sulfate $Tb_2(SO_4)_3 \cdot 8H_2O$.
Properties: Colorless crystals which lose $8H_2O$ at 360°C. Soluble in water.
Grades: Up to 99.9%.
Containers: Glass bottles.

terebene. A mixture of terpenes, chiefly dipentene and terpinene.
Properties: Colorless liquid; sp.gr. 0.862-0.866; optical rotation, inactive; b.p. 160-172°C. Soluble in alcohol; insoluble in water.
Derivation: Prepared from oil of turpentine.
Use: Medicine.
Shipping regulations: None.*

terebinthina canadensis. Canada turpentine. See Canada balsam.

terephthalic acid (para-phthalic acid; benzene-para-dicarboxylic acid) $C_6H_4(COOH)_2$.
Properties: White crystals or powder; insoluble in water, chloroform, ether, acetic acid; slightly soluble in alcohol; soluble in alkalies.
Derivation: Oxidation of para-xylene or of mixed xylenes and other alkyl aromatics, with heavy metal salts and bromine as catalyst. Also by reacting benzene and potassium carbonate over a cadmium catalyst.
Uses: Production of synthetic resins, fibers and films by combination with glycols, e.g., "Dacron," "Mylar," "Terylene." Also used as a reagent for alkali in wool.

terephthaloyl chloride $C_6H_4(COCl)_2$. 1,4-Benzenedicarbonyl chloride.
Properties: Colorless needles; m.p. 78°C; b.p. 259°C; decomposes in water and alcohol; soluble in ether.
Uses: Dye manufacture; synthetic fibers, resins, films; ultraviolet adsorption; pharmaceuticals; rubber chemicals; cross-linking agent for polyurethanes and polysulfides.

"Teresso." [51] Trademark for high quality lubricants for purposes ranging from applications requiring light bearing oils through turbine oils and diesel engine oils to gear lubricants. Made from paraffin-base stocks by solvent refining methods to impart high viscosity index, low carbon content, high flash point, resistance to sludging, and rapid separation from moisture contamination.

"Tergitol." [214] Trademark for a group of surface active agents including:
Nonionic NPX (alkyl phenyl ether of

*See "I.C.C. Shipping Regulations," page xiii.
Reference numbers refer to name of manufacturer. See "List of Manufacturers," page v.

polyethylene glycol).
Properties: Active ingredient 100% by wt; colorless liquid; solubility in water, complete at 20°C; cloud point (0.5% aqueous soln), 60-65°C; freezing point 5-7°C; sp. gr. 1.063; wt/gal 8.9 lb; viscosity, 373.8 cks at 20°C, 11.9 cks at 100°C; flash point 550°F.
Uses: Detergent, wetting agent, emulsifier in acid, alkaline, neutral or hard water systems.

Nonionic NP-14 (alkyl phenyl ether of polyethylene glycol).
Properties: Active ingredient 100% by wt; colorless liquid; solubility in water, less than 0.01% by wt at 20°C; freezing point, sets to a glass below -40°C; sp.gr. 1.031 (20/20°C); wt/gal 8.6 lb at 20°C; viscosity, 240 cks at 20°C, 6.7 cks at 100°C; flash point, 480°F.
Uses: Oil-soluble emulsifier in acid, alkaline, neutral or hard water systems.

Nonionic NP-27 (alkyl phenyl ether of polyethylene glycol).
Properties: Active ingredient 100% by wt; colorless liquid; solubility in water, 20% by wt at 20°C; cloud point (0.5% aqueous solution), 8°C; freezing point, -5 to -7°C; sp.gr. 1.055 at 20/20°C; wt/gal 8.8 lb at 20°C; viscosity, 224 cks at 20°C, 8.7 at 100°C; flash point 510°F.
Uses: Aromatic-soluble emulsifier in acid, alkaline, neutral or hard water systems.

Nonionic NP-35 (alkyl phenyl ether of polyethylene glycol).
Properties: Active ingredient 100% by wt; white semi-solid; solubility in water, complete at 20°C; cloud point (0.5% aqueous solution) 95°C; freezing point, 25-27°C; sp.gr. 1.065 at 40/20°C; wt/gal 8.8 lb at 40°C; viscosity, 39 cks at 65°C, 15.6 at 100°C; flash point, 490°F.
Uses: Detergent and wetting agent at elevated temperatures.

Nonionic NP-40 (alkyl phenyl ether of polyethylene glycol).
Properties: Active ingredient 100% by wt; white solid; solubility in water, complete at 20°C; cloud point (0.5% aqueous solution), above 100°C; freezing point, 36-38°C; sp. gr. 1.077 at 40/20°C; wt/gal 8.9 lb at 40°C; viscosity, 55.1 cks at 65°C, 21.7 cks at 100°C; flash point, 505°F.
Uses: Detergent and wetting agent at elevated temperatures.

Nonionic 12P-6 (dodecyl phenol adduct with 6 mols of ethylene oxide).
Properties: Active ingredient 100% by wt.
Uses: General purpose nonionic; sulfation.

Nonionic 12P-12 (dodecyl phenol adduct with 12 mols of ethylene oxide).
Properties: Active ingredient 100% by wt; cloud point (0.5% aqueous solution) 60°C; sp.gr. 1.0555; f.p. 15.1°C.
Uses: General purpose nonionic.

Nonionic TMN-3 (trimethyl nonyl ether of polyethylene glycol).
Properties: Active ingredient 100% by wt; sp.gr. 0.9355; f.p. sets to glass at -40°C.
Uses: Aromatic-soluble nonionic surfactant.

Nonionic TMN-10 (trimethyl nonyl ether of polyethylene glycol).
Properties: Active ingredient 90% by wt; solubility in water, complete at 20°C; cloud point (0.5% aqueous solution), 36°C; freezing point, 15-17°C; sp.gr. 1.024 at 20/20°C; wt/gal 8.5 lb at 20°C; viscosity 118 cks at 20°C, 6.5 cks at 100°C; no flash point.
Uses: Wetting, leveling and spreading agent in dilute solutions of acids, bases, salts, and hard water.

Nonionic TP-9 (alkyl phenyl ether of polyethylene glycol).
Properties: Active ingredient 100% by wt; colorless liquid; soluble in water; cloud point (0.5% aqueous solution) 51-56°C; f.p. 5°C; sp.gr. 1.057 (20/20°C); 8.8 lbs/gal; flash point 550°F.
Uses: Detergent; wetting agent; emulsifier in acid, alkaline, neutral or hard water systems.

Nonionic XD (polyalkylene glycol ether).
Properties: Active ingredient 100% by wt; white, soft-solid; solubility in water, complete at 20°C; cloud point (0.5% aqueous solution), 62°C; f.p. 28-38°C; sp.gr. 1.053 at 40/20°C; wt/gal 8.7 lb at 40°C; viscosity 177 cks at 55°C, 54.4 cks at 100°C; flash point, 480°F.
Uses: Detergent and emulsifier in acid, alkaline, neutral or hard water systems.

Nonionic XH (polyalkylene glycol ether).
Properties: Active ingredient 100% by wt; white semi-solid; solubility in water, complete at 20°C; cloud point (0.5% aqueous solution), 85-95°C; f.p. 35-40°C; viscosity 258.8 cks at 55°C, 58.1 cks at 100°C.
Uses: Detergents and emulsifier at elevated temperatures.

Anionic 7 $C_4H_9CH(C_2H_5)C_2H_4CH(SO_4Na)C_2H_4CH(C_2H_5)_2$.
Properties: Active ingredient 25-27% by wt; colorless liquid; solubility in water, complete at 20°C; pH (0.1% aqueous solution), 5.3; sp.gr. 1.046 at 20/20°C; wt/gal 8.7 lb at 20°C.
Uses: Wetting agent in aqueous solutions containing less than 1% dissolved solids.

Anionic P-28 $[C_4H_9CH(C_2H_5)CH_2]_2NaPO_4$.
Properties: Active ingredient 24-26% by wt; amber liquid; solubility in water, complete at 20°C; pH (0.1% aqueous solution) 6.7; sp.gr. 1.034 at 20/20°C; wt/gal 8.6 lb at 20°C.
Uses: Wetting agent and penetrant in aqueous solutions containing 1-2% dissolved alkali or 2-5% dissolved salts.

Anionic 4 $C_4H_9CH(C_2H_5)C_2H_4CH(SO_4Na)CH_2CH(CH_3)_2$.
Properties: Active ingredient 26-28% by wt; colorless liquid; solubility in water, complete at 20°C; pH (0.1% aqueous solution), 8.5; sp.gr. 1.056 at 20/20°C; wt/gal 8.8 lb at 20°C.
Uses: Wetting agent and penetrant in aqueous solutions containing 1-10% dissolved acid, 2-5% dissolved alkali or 1-10% dissolved salts.

Anionic 08 $C_4H_9CH(C_2H_5)CH_2SO_4Na$.
Properties: Active ingredient 38-40% by wt;

colorless liquid; solubility in water, complete at 20°C; pH (0.1% aqueous solution), 7.3; sp.gr. 1.144 at 20/20°C; wt/gal 9.5 lb at 20°C.
Uses: Wetting agent and penetrant in aqueous solutions containing 15-30% dissolved acid, 10-20% dissolved alkali, or 10-25% dissolved salts.

"Teridax." [321] Brand name for iophenoxic acid.

terneplate. Terneplate is a product made by coating steel sheet with an alloy of lead and tin.

"Terola." [51] Trademark for naphthenic lubricating oils especially recommended for small, two-stroke cycle, crankcase scavenging engines in marine service.

1,4(8)-terpadiene. See terpinolene.

1,8(9)-terpadiene. See dipentene.

terpene alcohol. An alcohol directly related to or derived from a terpene hydrocarbon; the following are common examples: terpineol (tertiary cyclic), borneol (secondary cyclic), geraniol (primary, acyclic), linalool (tertiary, acyclic).

terpene (hydrocarbon). An unsaturated organic compound having the empirical chemical formula $C_{10}H_{16}$, occurring in most essential oils and oleo-resins of plants. The terpenes may be considered as polymers of isoprene, C_5H_8, and may be either open-chain or cyclic with one or more benzenoid groups. They are classified as monocyclic (dipentene), dicyclic (pinene), or acyclic (myrcene), according to the molecular structure.

terpene hydrochloride. See bornyl chloride.

terphenyl (1,4-diphenylbenzene) $(C_6H_5)_2C_6H_4$.
Properties: Solid; sp.gr. 1.234 (0°C); m.p. 213°C; sublimes at 427°C.
Derivation: From para-dibromobenzene, or bromobenzene and sodium.
Method of purification: Zone-melting.
Grades: Technical; scintillation.
Containers: Glass bottles; fiber drums.
Uses: Polymerized with styrene to make a plastic phosphor. Single crystals used as scintillation counters.

terpilenol. See terpineol.

terpineol (alpha-terpineol; beta-terpineol; gamma-terpineol; terpilenol) $C_{10}H_{17}OH$.
Properties: Colorless liquid or low melting transparent crystals; lilac odor; sp.gr. (alpha) 0.933-0.936, (beta) 0.923-0.924, (gamma) 0.936; m.p. (alpha) 35°C, (beta) 32°C, (gamma) 69-70°C; b.p. (alpha) 218°C, (beta) 210°C, (gamma) 218°C.
Usually sold commercially as a mixture of the three isomers.
Typical specifications: Sp.gr. (15°C) 0.936-0.941; optical rotation between −0° 10' and +0° 10'; boiling range between 214 and 224°C, 90% within 5°C; refractive index (n 20/D) 1.4825-1.4850; soluble in 2 vols 70%, 4 vols 60% alcohol, 8 vols of 50% alcohol; soluble in diethyl phthalate, benzyl benzoate, mineral oil; slightly soluble in water, glycerin.
Derivation: By heating terpin hydrate with phosphoric acid and distilling, or with dilute sulfuric, using an azeotropic separation; fractional distillation of pine oil. Occurs naturally in several essential oils.
Method of purification: Distillation.
Grades: Technical; perfumery; extra; prime.
Containers: 1-, 5-, 10-lb bottles; 50-lb tins; 1000-lb drums.
Uses: Solvent for resins, gums, waxes, oils, other products; mutual solvent for resins and cellulose esters and ethers; perfumes; soap; disinfectant; antioxidant; medicine.
Shipping regulations: None.*

alpha-**terpineol.** See terpineol.

beta-**terpineol.** See terpineol.

gamma-**terpineol.** See terpineol.

"Terpineol 318." [266] Trade name for a highly refined mixture of alpha- and beta-terpineols; 96.5% tertiary terpene alcohols; colorless liquid; freezing point less than −10°C; sp.gr. 0.9374 at 15.6/15.6°C; ASTM distillation range, 10-95%, 217-220°C.

terpin hydrate (dipentene glycol) A cyclohexane derivative. $CH_3(OH)C_6H_9C(CH_3)_2OH \cdot H_2O$.
Properties: Colorless, lustrous, rhombic crystalline prisms, or white powder; slight characteristic odor; slightly bitter taste; efflorescent; m.p. 115-117°C; anhydrous, m.p. 102-105°C; b.p. 258°C; soluble in alcohol and ether; slightly soluble in water.
Derivation: Action of nitric acid and dilute sulfuric acid, or dilute sulfuric alone, on pine oil boiling between 200-220°C.
Grades: Technical; N.F. XI.
Containers: 1-lb cans and cartons; 100-lb net fiber drums.
Uses: Pharmaceuticals; raw material for terpineol.
Shipping regulations: None.*

terpinolene (1,4(8)-terpadiene) $C_{10}H_{16}$.
Properties: Water-white to pale amber liquid. Insoluble in water; soluble in alcohol, ether, glycol.
Constants: Sp.gr. (15.5/15.5°C) 0.864; b.p. (760 mm) 183-185°C; flash point (closed cup) 38°C; wt/gal 7.2 lbs (15.5°C).
Typical specifications: Sp.gr. at 15.5/15.5°C 0.864-0.868; boiling range 180-200°C; acidity none; color water-white; moisture trace.
Derivation: By fractionation from crude wood turpentine.
Grades: Technical.
Containers: 55-gal non-returnable galvanized drums.
Uses: Solvent for resins; essential oils; manufacture of synthetic resins, chemical derivatives.
Fire hazard: Flammable.
Shipping regulations: None.*

terpinyl acetate $C_{10}H_{17}OOCCH_3$.
Properties: Colorless liquid; characteristic odor suggestive of bergamot and lavender;

sp.gr. 0.958-0.968 (15°C); refractive index (n 20/D) 1.4640-1.4660; optical rotation −0° 30' and +0° 30'; m.p. −50°C; b.p. 220°C. Soluble in 5 or more vols of 70% alcohol; slightly soluble in water and glycerol.
Derivation: By heating terpineol with acetic acid or anhydride in the presence of sulfuric acid, and subsequent distillation.
Method of purification: Rectification.
Grades: Technical; prime; extra.
Containers: Cans; drums.
Use: Perfumes.
Shipping regulations: None.*

terpinyl propionate $C_{10}H_{17}OOCC_2H_5$.
Containers: Drums.
Use: In perfumes.

"Terposol 8." [266] Trademark for a technical grade of terpinyl ethyleneglycol ether.
Properties: Colorless to pale yellow liquid; sp.gr. 0.9813 (15.6/15.6°C); distillation range 248-284°C.
Uses: Powerful solvent for resins, waxes, natural oils, fats, polymerized oils, and rubber; in enamels, printing inks and perfume; special grade used in insecticides.

terra alba $CaSO_4 \cdot 2H_2O$. Finely pulverized powder made from gypsum and used in the manufacture of paper, paints, artificial marble, and composition plastics. See also kaolin.

"Terraclor." [84] Trade name for pentachloronitrobenzene (q.v.). Available as technical grade and formulations of 75% wettable powder, 2 lb concentrate, 20% dust, and 10% granular. Useful as soil fungicide.
Caution! Avoid skin exposure, and inhaling dust or mist.

"Terra-Cortril." [299] Trademark for a combination drug containing oxytetracycline and hydrocortisone.

terra-cotta clay. Terra-cotta clays differ quite widely, but semi-fire clays or a mixture of these with a more impure clay or shale are mostly used. Buff-burning clays are preferred because of the hard body produced on burning. Desirable qualities in the terra-cotta clay: (a) dense burning character and strong bonding, (b) low shrinkage and freedom from warping, (c) absence of soluble salts.
Occurrence: United States (New Jersey, Pennsylvania, Indiana, Missouri).

terra di siena. A raw sienna (q.v.) found in Cyprus and various parts of Italy. See also iron oxide reds.
Use: Pigment.

"Terramix ABD-25." [299] Trade name for vitamin-fortified antibiotic mix for agricultural feeds.

"Terramycin." [299] Trademark for oxytetracycline hydrochloride.

"Terran A, B and E." [349] Trademark for a high temperature epoxy cement used in the rocket and missile industry.

"Terran S." [349] Trademark for a phenolic varnish with special high heat resistant properties for use as an impregnant for various reinforcing materials, as fabric or rovings, nylon, asbestos, glass, or graphite.

terra ponderosa. See barium sulfate.

terra rosa. A variety of hematite, red (q.v.) used as a pigment.

terre verte. See earths, green.

"Tersan" 75. [28] Trademark for a turf fungicide containing tetramethylthiuramdisulfide (thiram). It is used to control brown patch and dollar spot; for control of stem rot of sweet potatoes and basal rot and decay of gladiolus bulbs.
Containers: 8-oz cans and 3-lb bags.

"Tersan" OM. [28] Trademark for turf fungicide containing thiram and hydroxymercurichlorophenol.
Containers: 3-lb cans.
Use: For the prevention and control of large brown patch, dollar spot, copper spot, and certain other diseases on golf greens, grass tennis courts, lawns and other fine turf.

tert-. Abbreviation for tertiary. See primary. May also mean a trisubstituted methyl radical, $R_1R_2R_3C-$, in which the central carbon is attached to three other carbons.

tertiary calcium phosphate. See calcium phosphate, tribasic.

tertiary sodium phosphate. See sodium phosphate, tribasic.

"Tervan." [51] Trademark for a group of petroleum wax specialties. The 2800 grade is a blend of "Vistanex"-polybutene and crystalline paraffin, applicable as a modifier for both crystalline and microcrystalline petroleum waxes. Tervan 2850, 2865, and 2875 are especially suitable for coating paperboard milk containers.

"Terylene." [56, 206] Trademark for a synthetic polyester textile fiber based on terephthalic acid.

"Tessalon." [305] Trademark for benzonatate N.N.D.
Use: Medicine.

testosterone $C_{19}H_{28}O_2$. An androgenic steroid; the male sex hormone produced by the testis. It has six times the androgenic activity of its metabolic product, androsterone.
Properties: White or slightly creamy-white crystals or as crystalline powder; odorless and stable in air; m.p. 153-157°C; dextrorotatory in dioxane solution. Very soluble in chloroform; soluble in alcohol, dioxane and vegetable oils; slightly soluble in ether; insoluble in water.
Derivation: Isolation from extract of testis; synthesis from cholesterol or from the plant steroid diosgenin.
Grade: N.F. XI.
Containers: Bottles.
Use: Medicine; biochemical research.
See also methyltestosterone.

*See "I.C.C. Shipping Regulations," page xiii.
Reference numbers refer to name of manufacturer. See "List of Manufacturers," page v.

testosterone cyclopentylpropionate $C_{19}H_{27}O \cdot OOC(CH_2)_2C_5H_9$.
Properties: Off-white, odorless, tasteless, crystalline powder; m.p. 98-101°C. Freely soluble in alcohol, chloroform, and ether; soluble in vegetable oils; slightly soluble in water.
Grade: U.S.P. XVI.
Use: Medicine.

testosterone enanthate $C_{19}H_{27}O \cdot OOCC_6H_{13}$.
Properties: White or creamy-white crystalline powder or a viscous, amber-colored liquid. Odorless or faint odor resembling enanthic acid; m.p. 34-39°C. Insoluble in water; soluble in ether and vegetable oils.
Grade: U.S.P. XVI.
Use: Medicine.

testosterone propionate $C_{19}H_{27}O \cdot OOCC_2H_5$.
The propionate ester of testosterone; has androgenic activity.
Properties: White or creamy white crystals or crystalline powder; odorless and stable in air; m.p. 118-123°C. Freely soluble in alcohol, dioxane, ether, and other organic solvents; soluble in vegetable oils; insoluble in water.
Derivation: By organic synthesis.
Grade: U.S.P. XVI.
Containers: Bottles.
Use: Medicine.

"Testryl." [412] Trademark for testosterone (q.v.).

TETD. Abbreviation for tetraethylthiuram disulfide.

"Tetmosol." [207] Trademark for tetraethylthiuram sulfide.
Uses: In human and animal medicine.

tetraamylbenzene $(C_5H_{11})_4C_6H_2$.
Properties: Sp.gr. at 20°C 0.89; boiling range 320-350°C; color light straw; odor faintly aromatic; flash point 295°F.
Shipping regulations: None.*

tetra base. See tetramethyldiaminodiphenylmethane.

tetrabromobisphenol A. A flame retardant. See "Firemaster" BP4A.

sym-**tetrabromoethane.** See acetylene tetrabromide.

tetrabromoethylene C_2Br_4.
Properties: Colorless crystals; m.p. 55-56°C; b.p. 227°C.
Derivation: Bromination of dibromoacetylene.
Use: Organic synthesis.

tetrabromofluorescein. See eosin.

tetrabromofluorescein, disodium salt. See eosin, yellowish.

tetrabutylthiuram disulfide $[(C_4H_9)_2NCS_2]_2$.
Properties: Amber color; sp.gr. (20/20°C) 1.03-1.06°C; solidifies approx. −30°C; slight sweet odor. Soluble in carbon disulfide, benzene, chloroform, and gasoline; insoluble in water and 10% caustic.
Uses: Vulcanizing and accelerating agents.

tetrabutyltin $(C_4H_9)_4Sn$.
Properties: Colorless or slightly yellow oily liquid; b.p. 145°C (10 mm); decomposes at 265°C. Insoluble in water; soluble in most common organic solvents.
Derivation: Reaction of tin tetrachloride with butyl magnesium chloride.
Uses: Stabilizing and rust-inhibiting agent for silicones; lubricant and fuel additive; polymerization catalyst; hydrochloric acid scavenger.

tetrabutyl titanate (TBT; butyl titanate; titanium butylate) $Ti(OC_4H_9)_4$.
Properties: Colorless to light yellow liquid. B.p. 310-314°C; forms a glass below −55°C; sp.gr. 0.996; refractive index 1.486; flash point 170°F; decomposes in water; soluble in most organic solvents except ketones.
Derivation: Reaction of titanium tetrachloride with butyl alcohol.
Uses: Ester exchange reactions; heat resistant paints (up to 500°C); improving adhesion of paints, rubber, and plastics to metal surfaces; cross-linking agent; condensation catalyst.

tetrabutyl urea $(C_4H_9)_2NCON(C_4H_9)_2$.
Properties: Liquid; sp.gr. 0.880; refractive index 1.4535; vapor pressure less than 0.01 mm; b.p. 305°C; m.p. less than −60°C; flash point 146°C; insoluble in water.
Use: Plasticizer.

tetrabutyl zirconate $(C_4H_9O)_4Zr$. White solid from reaction of zirconium tetrachloride with butyl alcohol. Used as condensation catalyst and cross-linking agent.

tetracaine
$CH_3(CH_2)_3NHC_6H_4COOCH_2CH_2N(CH_3)_2$.
2-Dimethylaminoethyl para-butylaminobenzoate.
Properties: White or light yellow, waxy solid. Very slightly soluble in water; soluble in alcohol, ether, benzene, chloroform. M.p. 41-46°C.
Grade: U.S.P. XVI.
Use: Medicine.

tetracaine hydrochloride $C_{15}H_{24}N_2O_2 \cdot HCl$.
Properties: Fine, white crystalline, odorless powder with a bitter taste and a melting range of 147-150°C. Soluble in water and alcohol; insoluble in ether and benzene.
Derivation: By reaction of para-butylaminobenzoyl chloride with dimethylaminoethanol.
Grade: U.S.P. XVI.
Use: Medicine.

tetracalcium aluminoferrate. An ingredient of cement. See cement, Portland, and other cement articles.

1,2,3,4-tetracarboxybutane (TCB). A substance used for making alkyd resins; as an epoxy curing agent; and as a sequestrant.

tetracene. See naphthacene.

1,2,3,4-tetrachlorobenzene $C_6H_2Cl_4$.
Properties: White crystals; m.p. 46.6°C; b.p. (760 mm) 254°C; flash point 161°C; insoluble in water.

Use: Component of dielectric fluids; synthesis.

1,2,4,5-tetrachlorobenzene $C_6H_2Cl_4$.
Properties: White flakes; m.p. 137.5-140°C; distillation range 240-246°C.
Uses: Intermediate; insecticide; impregnant for fire and moisture resistance, for electrical insulation, for temporary protection in packing.

tetrachloro-para-benzoquinone. See chloranil.

tetrachlorobisphenol A $C_{15}H_{12}Cl_4O_2$. A monomer for flame-retardant epoxy, polyester and polycarbonate resins.

sym-**tetrachlorodifluoroethane** CCl_2FCCl_2F.
Properties: White solid or colorless liquid with slightly camphor-like odor when concentrated. B.p. 92.8°C; f.p. 26°C; critical temperature 278°C; sp.gr. (25°C) 1.6447; refractive index (25°C) 1.413; wt/gal 13.8 lbs.
Grades: Purified; solvent.
Use: Nonflammable, nonexplosive degreasing solvent.

tetrachlorodiphenylethane. See TDE.

tetrachlorodiphenyl sulfone. See "Tedion."

sym-**tetrachloroethane** (acetylene tetrachloride) $CHCl_2CHCl_2$.
Properties: Heavy, colorless, mobile, nonflammable, corrosive, toxic liquid. Chloroform-like odor, but more toxic than the latter. Soluble in alcohol and ether; insoluble in water.
Constants: Sp.gr. 1.593 (25/25°C); b.p. 146.5°C; f.p. −43°C; weight 13.25 lbs/gal (25°C); refractive index 1.4918 (25°C); flash point none; fire point none; heat of vaporization 55.1 cals/gm (b.p.); specific heat 0.27 cals/gm/°C; specific resistivity 4.2×10^7 ohms/cm; viscosity 1.59 centipoises at 25°C.
Typical specifications: Colorless.
Derivation: By the interaction of acetylene and chlorine, and subsequent distillation.
Method of purification: Rectification.
Grades: Technical.
Containers: 55-gal drums.
Uses: Solvent; cleansing and degreasing metals; paint removers, varnishes, lacquers, photographic film; resins and waxes; extraction of oils and fats; ethyl alcohol denaturant; organic synthesis; insecticides; as a weed killer; fumigant.
Danger! Vapor extremely hazardous; do not breathe vapor or get on skin or clothing. MCA warning label.
Shipping regulations: None.*

tetrachloroethylene. See perchloroethylene.

tetrachloromethane. See carbon tetrachloride.

tetrachloronaphthalene. See chloronaphthalenes.

2,3,4,6-tetrachlorophenol C_6HCl_4OH.
Properties: Brown flakes or sublimed mass with a strong characteristic odor; m.p. 69-70°C; b.p. 164°C (23 mm); no flash or fire point; sp.gr. 25°/4°C 1.839; soluble in acetone, benzene, ether, and alcohol.
Use: Fungicide.
Warning! Harmful dust. MCA warning label.

2,4,5,6-tetrachlorophenol C_6HCl_4OH.
Properties: Brown solid; phenol odor. Sp.gr. 1.65 at 60/4°C; m.p. >50°C. Soluble in sodium hydroxide solutions and most organic solvents; insoluble in water.
Uses: Fungicide; wood preservative.
Warning! Harmful dust. MCA warning label.

tetrachlorophthalic acid $C_6Cl_4(CO_2H)_2$.
Properties: Colorless, crystalline plates. Soluble in hot water; sparingly soluble in cold water.
Derivation: By passing a stream of chlorine through a mixture of phthalic anhydride and antimony pentachloride.
Method of purification: Crystallization.
Grades: Technical.
Containers: Wooden kegs.
Uses: Dyes; intermediates.
Shipping regulations: None.*

tetrachlorophthalic anhydride $C_6Cl_4(CO)_2O$.
Properties: White, odorless, free-flowing, non-hygroscopic powder; slightly soluble in water; m.p. 254-255°C; b.p. 371°C.
Containers: 60-, 120-lb fiber drums; 400-lb barrels.
Uses: Intermediate in dyes, pharmaceuticals, plasticizers, and other organic materials.

tetrachloroquinone. See chloranil.

tetrachlorothiophene CClCClCClCClS.
Properties: M.p. 29-30°C; b.p. 104°C (10 mm). Soluble in benzene, hexane, alcohols, chlorocarbon.
Uses: Agricultural chemicals; lubricants.

tetracopper calcium oxychloride. A complex compound containing not less than 45% copper; finely divided bluish powder.
Containers: 6- and 50-lb bags.
Uses: Fungicide; manufacture of agricultural copper dust.

tetracosane $C_{24}H_{50}$ and $CH_3(CH_2)_{22}CH_3$.
Properties: Crystals; soluble in alcohol; insoluble in water. Sp.gr. 0.779 (51°/4°C); b.p. 324.1°C; m.p. 51.5°C.
Grades: Technical.
Use: Organic synthesis.

n-tetracosanoic acid. See lignoceric acid.

tetracyanoethylene $(CN)_2C{:}C(CN)_2$. The first member of a new class of compounds called cyanocarbons.
Properties: Colorless crystals; sublimes above 120°C; m.p. 198-200°C; b.p. 223°C. Has high thermal stability; burns in oxygen with a hotter flame than acetylene.
Caution: Hydrolyzes with water or in moist air to give hydrogen cyanide.
Use: Organic synthesis; dyes; makes colored solutions with aromatics.

tetracycline $C_{22}H_{24}N_2O_8 \cdot 0$ to $6H_2O$. An antibiotic obtained from certain Streptomyces species. It can also be prepared by catalytic hydrogenation of chlortetracycline or oxytetracycline, which it resembles in its actions

*See "I.C.C. Shipping Regulations," page xiii.
Reference numbers refer to name of manufacturer. See "List of Manufacturers," page v.

and uses. Its chemical structure is that of a modified naphthacene molecule.
Properties: Yellow, odorless, crystalline powder. Stable in air; affected by strong sunlight. Potency affected in solutions with pH below 2 and destroyed in alkali hydroxide solutions. Practically insoluble in chloroform and ether; very slightly soluble in water; slightly soluble in alcohol; very soluble in dilute hydrochloric acid and alkali hydroxide solutions; pH (saturated solution) 3.0-7.0. Induces fluorescence in mitochondria of living cells in tissue culture or in fresh preparations from various organs.
Derivation: From para-chloro-meta-cresol in 24 steps.
Grade: U.S.P. XVI.
Use: Medicine.

tetracycline hydrochloride $C_{22}H_{24}N_2O_8 \cdot HCl$. The hydrochloride salt of tetracycline.
Properties: Yellow, odorless, crystalline powder. Stable in dry air; affected by strong sunlight in moist air. Potency affected in solutions with pH below 2, and destroyed by alkali hydroxide solutions; pH (1% solution) 1.8-2.8. Slowly hydrolyzes in aqueous solution. Soluble in water; slightly soluble in alcohol; soluble in solutions of alkali hydroxides and carbonates; practically insoluble in chloroform and ether.
Grade: U.S.P. XVI.
Use: Medicine.

tetracycline phosphate complex.
Properties: Yellow, odorless, hydrated, fine crystalline powder. Insoluble in water.
Derivation: Prepared by adding a solution of sodium metaphosphate to a solution of tetracycline or tetracycline hydrochloride.
Grade: N.N.D.
Use: Medicine.

"Tetracyn." [299] Trademark for tetracycline hydrochloride.

n-**tetradecane** $C_{14}H_{30}$ and $CH_3(CH_2)_{12}CH_3$.
Properties: Colorless liquid; sp.gr. (20/4°C) 0.7653; m.p. 5.5°C; b.p. 253.5°C; refractive index (n 20/D) 1.4302; flash point 121°C. Soluble in alcohol; insoluble in water.
Grades: 95%, 99%.
Containers: Bottles; small drums.
Uses: Organic synthesis; solvent; standardized hydrocarbon.

tetradecanoic acid. See myristic acid.

tetradecanol. See myristyl alcohol; see also 7-ethyl-2-methyl-4-undecanol.

1-tetradecene (alpha-tetradecylene) $CH_2{:}CH(CH_2)_{11}CH_3$.
Properties: Colorless liquid; density 0.775 g/ml (20°/4°C); m.p. −12°C; b.p. 256°C (760 mm); insoluble in water; very slightly soluble in alcohol and ether.
Uses: Perfumes, flavors, medicines, dyes, oils, resins, plastics.

tetradecylamine $C_{14}H_{29}NH_2$.
Properties: A white solid with odor of ammonia; m.p. 37°C; b.p. 291.2°C; insoluble in water; soluble in alcohol and ether.
Grade: 90% purity.
Containers: Drums; tank cars.
Uses: Intermediate for manufacture of cationic surface-active agents; germicides.

tetradecyl chloride (myristyl chloride) $CH_3(CH_2)_{13}Cl$.
Properties: Water-white distilled liquid, mild odor. Sp.gr. 0.8590; f.p. −0.2°C; b.p. 154-155°C (15 mm); 15.2% chloride; subject to mild hydrolysis on standing.
Grade: 97% min.

alpha-**tetradecylene.** See 1-tetradecene.

tetradecyl mercaptan (myristyl mercaptan) $CH_3(CH_2)_{13}SH$.
Properties: Liquid; m.p. 6.5°C; b.p. 176-180°C (22 mm); sp.gr. 0.8398 (25/4°C); refractive index 1.4612 (n 20/D).
Grade: 95% (min) purity.
Uses: Organic intermediate; synthetic rubber processing.

tert-**tetradecyl mercaptan** $C_{14}H_{29}SH$.
Properties: Flammable liquid; boiling range (5 mm) 104-129°C; sp.gr. (60/60°F) 0.865; refractive index (20/D) 1.467; flash point 113°C.
Containers: 1-, 5-, 54-gallon drums; tank cars.
Uses: Polymer modifications.
Shipping regulations: No label required.*

tetradymite (bismuth telluride; telluric bismuth; bismuth tritelluride) Bi_2Te_3. A natural telluride of bismuth, frequently containing sulfur and selenium.
Properties: Color and streak pale steel gray; luster metallic; hardness 1.5-2; sp.gr. 7.3.
Occurrence: California, Colorado, Arizona, Montana, New Mexico, Virginia; Canada; Europe.
Use: Ore of bismuth.

tetraethanolammonium hydroxide $(HOCH_2CH_2)_4NOH$.
Properties: A white, crystalline solid; m.p. 123°C; vapor pressure less than 0.01 mm (20°C); completely soluble in water. A strong base, approaching sodium hydroxide in alkalinity. Its aqueous solutions are stable at ordinary temperatures but decompose on heating to form weakly basic polyethanolamines.
Grades: Commercial grade is a 40% water solution.
Containers: 1-gal cans; 5-, 55-gal drums.
Uses: As an alkaline catalyst; solvent for certain types of dyes; in screen printing with "Rapidogen" dyestuffs; in the application of rubber latices; in metal-plating solutions.

tetraethylammonium chloride (TEAC; TEA chloride) $(C_2H_5)_4NCl$.
Properties: Anhydrous: Colorless, odorless, hygroscopic crystals; sp.gr. 1.080; freely soluble in water, alcohol, chloroform, acetone; slightly soluble in benzene and ether. Tetrahydrate: $(C_2H_5)_4NCl \cdot 4H_2O$, crystals;

*See "I.C.C. Shipping Regulations," page xiii.
Reference numbers refer to name of manufacturer. See "List of Manufacturers," page v.

m.p. 37.5°C; sp.gr. 1.084.
Grade: N.N.D.
Use: Medicine.

tetraethylammonium hexafluophosphate. See fluophosphoric acids.

tetra-(2-ethylbutyl) silicate $[(C_2H_5)C_4H_8O]_4Si$.
Properties: Sp.gr. (20/20°C) 0.8920-0.9018; m.p. below −100°C; b.p. 238°C (50 mm); insoluble in water; slightly soluble in methanol; miscible with most organic solvents.
Uses: Heat transfer medium; hydraulic fluid; wide-temperature-range lubricant.

tetraethyl dithiopyrophosphate (ethylthiopyrophosphate; sulfotepp) $(C_2H_5)_4P_2S_2O_5$. Used as an insecticide.
Shipping regulations: Liquid or any mixture containing sulfotepp: Poison, class B. Poison label. Compressed gas mixture: Poison, class A. Poison gas label. Not accepted by express.*

tetraethylene glycol (TEG) $HO(C_2H_4O)_3C_2H_4OH$.
Properties: Colorless liquid; hygroscopic. Soluble in water; insoluble in benzene, toluene, or gasoline.
Constants: Sp.gr. 1.1248 at 20/20°C; b.p. (760 mm) 327.3°C; vapor pressure < 0.001 mm (20°C); flash point 345°F; wt/gal 9.4 lbs (20°C).
Grades: Technical.
Containers: 1-gal cans; 5-gal (tin-lined) drums; 55-gal drums. Net content 9.0, 45, 510 lbs.
Uses: Solvent for nitrocellulose; plasticizer; lacquers; coating compositions; suggested as a resin intermediate, heat transfer medium.

tetraethylene glycol dimethacrylate
Properties: Water-white to pale straw liquid; b.p. (1 mm) 200°C; sp.gr. (20/20°C) 1.075; refractive index (20°C) 1.4620; viscosity 12 cps; insoluble in water; soluble in styrene, many esters and aromatics; limited solubility in aliphatic hydrocarbons.
Use: Plasticizers.
Caution: Avoid contact with skin or eyes!

tetraethylene glycol dimethyl ether. See dimethoxytetraglycol.

tetraethylene glycol distearate
$(C_{17}H_{35}COOCH_2CH_2OCH_2CH_2)_2O$.
Properties: M.p. 32-33°C; insoluble in water.
Use: Plasticizer.

tetraethylene glycol monostearate
$C_{17}H_{35}COO(CH_2CH_2O)_4H$.
Properties: Sp.gr. 0.971; m.p. 30-31°C; insoluble in water.
Use: Plasticizer.

tetraethylenepentamine
$NH_2(CH_2CH_2NH)_3CH_2CH_2NH_2$.
Properties: Somewhat viscous, hygroscopic liquid. Sp.gr. 0.9980 at 20/20°C; b.p. (760 mm) 333°C; vapor pressure < 0.01 mm (20°C); wt/gal 8.3 lbs (20°C). Coefficient of expansion 0.00076 (20°C); viscosity 0.962 poise (20°C).
Typical specifications: Sp.gr. 0.990-1.000 at 20/20°C; boiling range 280-360°C (760 mm). Soluble in most organic solvents and water.
Grades: Technical.
Containers: 1-gal cans; 5-, 55-gal drums; tank cars.
Uses: Solvent for sulfur, acid gases, various resins and dyes; saponifying agent for acidic materials; manufacture of synthetic rubber.
Danger! Causes severe eye and skin burns. MCA warning label.

tetra-(2-ethylhexyl) silicate
$[C_4H_9CH(C_2H_5)CH_2O]_4Si$.
Properties: Sp.gr. 0.8838; b.p. 350-370°C; f.p. −90°C; solubility in water < 0.01; pounds/gallon 7.4; flash point 390°F.
Shipping regulations: No label required.*

tetra-(2-ethylhexyl) titanate
$[C_4H_9CH(C_2H_5)CH_2O]_4Ti$. A light-yellow, viscous liquid from the transesterification of isopropyl titanate with 2-ethylhexanol. Used as cross-linking agent; condensation catalyst; adhesion promotor. Available in commercial quantities.

tetraethyllead (TEL) $Pb(C_2H_5)_4$.
Properties: Colorless, oily liquid; pleasant characteristic odor. Caution! Fairly strong poison. Absorbed by the skin. Soluble in all organic solvents. Insoluble in water and dilute acids or alkalies.
Constants: Sp.gr. 1.65; b.p. 198-202°C (at 760 mm, calculated), 75-85°C (13-14 mm); f.p. −136°C; decomposes slowly at room temperature, rapidly at 125-150°C.
Derivation: (a) By treating lead-sodium alloy with ethyl chloride; (b) by the Ziegler process based on the reaction of metallic lead and a triethylaluminum complex in an electrolytic cell. Raw materials are lead, ethylene and hydrogen.
Method of purification: Careful and repeated steam distillation and drying over $CaCl_2$; vacuum distillation.
Grades: One grade only–approximately 98% pure.
Containers: May be shipped in tight metal containers.
Uses: Used for preventing knocking in internal combustion engines; certain ethylation operations.
Shipping regulations: Class B poison. Poison label.*

O,O,O',O'-tetraethyl-S,S'-methylenediphosphorodithioate. Used in commercial form as insecticide and acaricide. See ethion.

tetraethyl orthosilicate. See ethyl silicate.

tetraethyl pyrophosphate (TEPP) $(C_2H_5)_4P_2O_7$.
Properties: Water-white to amber liquid depending on purity; hygroscopic; miscible in all proportions with water and all organic solvents except aliphatic hydrocarbons; hydrolyzed in water with formation of mono-, di- and triethyl ortho-phosphates; water solutions attack metals. Commercial material contains 40% TEPP; sp.gr. approx. 1.20; refractive index 1.420.
Derivation: From phosphorus oxychloride

and ethanol or phosphorus oxychloride and triethyl phosphate.
Grades: 40%.
Containers: 5-gal cans; 50- and 55-gal drums.
Uses: To formulate insecticides for aphids and mites; as a rodenticide.
Danger: Poisonous by skin contact, inhalation or swallowing. Rapidly absorbed through skin. Repeated exposure may, without symptoms, be increasingly hazardous. MCA warning label.
Shipping regulations: Tetraethyl pyrophosphate and compressed gas mixture: Poison, class A. Poison gas label. Not accepted by express. Liquid or mixture, dry, or liquid: Poison, class B. Poison label.*

tetraethylthiuram disulfide [disulfiram; TTD; TETD; bis (diethylthiocarbamyl) disulfide] $[(C_2H_5)_2NCS]_2S_2$.
Properties: Cream color; sp.gr. (20/20°C) 1.17; freezing range 65-70°C; slight odor. Soluble in carbon disulfide, benzene and chloroform; insoluble in water.
Grades: N.N.D.; technical.
Containers: 5-, 50-, 100-, 150-lb drums.
Uses: May be used without sulfur as vulcanizing agents or with sulfur as ultra accelerators and as activators for thiazole-type accelerators. Also causes the vulcanizates to be nonstaining, non-discoloring and exceptionally resistant to heat aging; medicine; fungicide and insecticide.

tetraethylthiuram sulfide [bis-(diethylthiocarbamyl) sulfide] $[(C_2H_5)_2NCS]_2S$.
Properties: Dark brown; slight odor; sp.gr. (20/20°C) 1.12; boiling range 225-240°C (3 mm).
Uses: Pharmaceutical ointments; fungicide; insecticide.

tetraethyltin $Sn(C_2H_5)_4$.
Properties: Colorless liquid; sp.gr. (23°C) 1.187; b.p. 181°C; m.p. −112°C; insoluble in water; soluble in alcohol and ether.

tetrafluorethylene. See tetrafluoroethylene.

tetrafluorodichloroethane. See dichlorotetrafluoroethane.

tetrafluoroethylene (perfluorethylene; tetrafluorethylene) $F_2C{:}CF_2$. Raw material for polytetrafluoroethylene polymers.
Properties: Colorless gas; m.p. −142.5°C; b.p. −78.4°C; insoluble in water.
Derivation: By passing chlorodifluoromethane through a hot tube.
Shipping regulations: Inhibited: Flammable gas. Red gas label.*

tetrafluorohydrazine F_2NNF_2.
Properties: Colorless mobile liquid and colorless gas; b.p. (calc) −73°C; heat of vaporization 3170 cal/mole; critical temperature 36°C.
Use: Organic synthesis; oxidizer in fuels for rockets, missiles, etc.

tetrafluoromethane CF_4.
Properties: A colorless gas; m.p. −184°C; b.p. −128°C; slightly soluble in water. Density of liquid 1.96 g/ml at −184°C; sp. vol. (70°F) 4.4 cu ft/lb.
See "Freon 14."

tetrafluoro-1-propanol. See fluoroalcohols.

tetraglycine hydroperiodide. See "Globaline."

tetraglycol dichloride $(ClCH_2CH_2OCH_2CH_2)_2O$.
Properties: Colorless liquid. Slightly miscible with water. Sp.gr. 1.186; b.p. (2 mm) 114°C.
Grade: Technical.
Uses: High-boiling solvent and extractant for oils, fats, waxes and greases; chemical intermediate.

tetrahedrite $(CuFe)_{12}Sb_4S_{13}$. A natural sulfantimonide of copper and iron. Silver, zinc, lead and mercury may be present. A variety of fahlore (q.v.).
Properties: Color grayish black to black; luster metallic; streak black to brown; hardness 3-4; sp.gr. 4.6-5.1.
Occurrence: Colorado, Montana, Nevada, Arizona, Utah; Mexico; Europe; South America.
Use: Ore of silver and copper.

1,2,3,6-tetrahydrobenzaldehyde. See 3-cyclohexene-1-carboxaldehyde.

1,2,3,4-tetrahydrobenzene. See cyclohexene.

tetrahydrofuran (THF) C_4H_8O.
Properties: Water-white liquid with ethereal odor; sp.gr. (20°C) 0.888; refractive index (n 20/D) 1.4070; f.p. −65°C; b.p. 66°C; flash point (open cup) 5°F. Soluble in water and organic solvents.
Derivation: From furfural, as an intermediate in the production of adiponitrile.
Containers: 7-lb (1 gal) containers; 35- and 375-lb drums; 29,000-lb tank cars.
Uses: Solvent for natural and synthetic resins, particularly vinyls, in topcoating solutions, polymer coating cellophane, protective coatings, adhesives, printing inks, etc. Useful reaction solvent, e.g., in Grignard reactions, $LiAlH_4$ reductions, and polymerizations. Versatile chemical intermediate and monomer.
Shipping regulations: Flammable liquid. Red label.*

tetrahydrofurfuryl acetate $C_4H_7OCH_2OOCCH_3$.
Properties: Colorless liquid. Soluble in water, alcohol, ether, and chloroform. Sp. gr. 1.061 (20/0°C); b.p. 194-195°C/753 mm.
Derivation: By treatment of tetrahydrofurfuryl alcohol with acetic anhydride.
Grade: Refined.
Containers: Bottles.
Shipping regulations: None.*

tetrahydrofurfuryl alcohol (tetrahydrofuryl carbinol) $C_4H_7OCH_2OH$.
Properties: Water-white liquid with a mild odor; sp.gr. 1.0543 (20/20°C); b.p. 178°C (760 mm); refractive index 1.4520 (20/D); flash point (open cup) 183°F; viscosity 5.49 cps (25°C).
Derivation: Catalytic hydrogenation of furfural.
Grades: Commercial; industrial (about 80%).

Containers: 1-, 5-, 10-gal cans; 55-gal drums; tank cars; tank trucks.
Uses: Solvent for resins, dyes, chlorinated rubber; preparation of esters.
Shipping regulations: None.*

tetrahydrofurfuryl benzoate $C_4H_7OCH_2OOCC_6H_5$.
Properties: Colorless liquid; insoluble in water; soluble in alcohol, ether, and chloroform. Sp.gr. 1.137 (20/0°C); b.p. 300-302°C/750 mm, 138-140°C/2 mm.
Derivation: Tetrahydrofurfuryl alcohol and benzoic acid by usual esterification procedure.
Grade: Refined.
Containers: Bottles.
Shipping regulations: None.*

tetrahydrofurfuryl laurate $C_{17}H_{32}O_3$.
Properties: Sp.gr. (25°C) 0.930; m.p. −76 to −9°C; insoluble in water.
Use: Plasticizer.

tetrahydrofurfuryl levulinate $CH_3CO(CH_2)_2COOCH_2C_4H_7O$.
Properties: M.p. 59-61°C; soluble in water.
Use: Plasticizer.

tetrahydrofurfuryl oleate $C_{17}H_{33}COOCH_2C_4H_7O$.
Properties: Colorless liquid; sp.gr. (25°C) 0.923; b.p. (5 mm) 240°C; m.p. −30°C; insoluble in water.
Containers: 1-, 5-, 55-gal drums.
Uses: Plasticizer.

tetrahydrofurfuryl phthalate $C_6H_4(COOCH_2C_4H_7O)_2$.
Properties: Sp.gr. (25°C) 1.194; m.p. less than 15°C; insoluble in water.
Use: Plasticizer.

tetrahydrofuryl carbinol. See tetrahydrofurfuryl alcohol.

tetrahydrolinalool $C_{10}H_{21}OH$.
Properties: A colorless liquid with a delicate floral odor. Sp.gr. 0.832-0.837; optically inactive.
Use: Perfumery.

tetrahydromethylnicotinic acid. See arecaidine.

tetrahydro-para-methyloxyquinoline. See thalline.

1,2,3,4-tetrahydro-6-methylquinoline. See "Tetraquinone."

tetrahydronaphthalene $C_{10}H_{12}$.
Properties: Colorless liquid; pungent odor. Fairly stable but will, however, polymerize and oxidize giving rise to discoloration and resinous material. Miscible with most solvents, expecially the resinous and petroleum thinners, and compatible with natural and synthetic vehicles; insoluble in water.
Constants: Sp.gr. 0.981 at 13°C; b.p. 206°C; refractive index 1.540-1.547; flash point 78°C; freezing point −25°C; moisture content none; residue on evaporation none; acidity neutral; wt/gal approx. 8 lbs.
Derivation: Hydrogenation of naphthalene in the presence of a catalyst at 150°C.
Grade: Technical.
Containers: 1-, 2-, 5-, 10-, 50-gal drums.
Uses: Solvent for waxes, resins, rubber, gums; oils, resinates, cellulose ethers, asphalt, linoxyn, liquid driers, metallic soaps, greases, benzene, toluene, naphthalene, casinghead gasoline, other products; removing printing ink from paper; turpentine substitute; purifying coal gas; extracting sulfur from spent oxide in gas purification; paint and varnish; bituminous emulsions; water-proofings; solvent mixtures; motor fuels; paint and varnish removers; textile processing; agricultural sprays; shoe polishes, floor polishes, etc.; extracting casinghead gasoline from natural gas; other purposes.

2-tetrahydronaphthyl-2-imidazoline hydrochloride. See tetrahydrozoline hydrochloride.

tetrahydro-1,4-oxazine. See morpholine.

tetrahydrophenobarbital. See cyclobarbital.

tetrahydrophthalic anhydride $C_6H_8(CO)_2O$.
Properties: White crystalline powder, solidification point 99-101°C; sp.gr. (105°C) 1.20; slightly soluble in petroleum ether and ethyl ether; soluble in benzene.
Derivation: Diels-Alder reaction of butadiene and maleic anhydride.
Containers: Fiber drums.
Uses: Chemical intermediate for manufacture of light colored alkyds, polyesters, plasticizers and adhesives; intermediate for pesticides.

tetrahydropyran-2-methanol $\overline{OCH_2CH_2CH_2CH_2C}HCH_2OH$.
Properties: Liquid; sp.gr. (20°C) 1.0272; b.p. b.p. 187.2°C; f.p., sets to glass below −70°C; miscible with water.
Use: Chemical intermediate.

1,2,5,6-tetrahydropyridine C_5H_9N.
Properties: Sp.gr. (20/4°C) 0.912-0.914; m.p. −44°C; b.p. 115.5-120.0°C.
Purity: 96% min.
Use: Organic intermediate.

tetrahydrothiophene $\overline{CH_2CH_2CH_2CH_2S}$.
Properties: Water-white liquid; sp.gr. 1.00 (15.6/15.6°C); boiling range 115-124.4°C.
Uses: Solvent; intermediate. Stable fuel gas odorant.

tetrahydroxyadipic acid. For two of fourteen possible isomers, see mucic acid; saccharic acid.

tetrahydroxybutane. See erythritol.

tetrahydroxydiphenyl. See diresorcinol.

tetrahydroxyethylethylenediamine [N,N,N',N'-tetrakis(2-hydroxyethyl)ethylenediamine] $(HOCH_2CH_2)_2NCH_2CH_2N(CH_2CH_2OH)_2$.
Properties: Clear, viscous liquid; good heat stability; low toxicity.
Uses: Organic intermediate; cross-linking of rigid polyurethane foams; chelating agent; humectant; gas absorbent; resin formation; detergent processing.

tetrahydroxyflavanol. See quercetin.

2,3,4,5-tetrahydroxyhexanedioic acid. See D-saccharic acid.

*See "I.C.C. Shipping Regulations," page xiii.
Reference numbers refer to name of **manufacturer**. See "List of Manufacturers," page v.

tetrahydrozoline hydrochloride (2-(1,2,3,4-tetrahydro-1-naphthyl)-2-imidazoline hydrochloride) $C_{13}H_{16}N_2 \cdot HCl$.
Properties: White, odorless solid; soluble in water and alcohol; slightly soluble in chloroform; insoluble in ether; melting range 253-259°C.
Grade: N.F. XI.
Use: Medicine.

tetraiodoethylene (diiodoform; ethylene periodide; ethylene tetraiodide; iodoethylene) $I_2C:Cl_2$.
Properties: Light-yellow, heavy, small odorless crystals; on exposure to light turns brown; m.p. 187°C; sp.gr. 2.98. Insoluble in water; soluble in most organic solvents.
Derivation: Action of iodine on diiodoacetylene obtained from calcium carbide and iodine.
Uses: Surgical dusting powder; antiseptic ointment; fungicide.
Warning: Protect from light.

tetraiodofluorescein. See iodeosin.

tetraiodophenolphthalein. See iodophthalein.

tetraiodophenolphthalein, sodium salt. See iodophthalein sodium.

tetraiodopyrrole. See iodole.

tetraisopropylthiuram disulfide $[(CH_3CH_3CH)_2NCS]_2S_2$.
Properties: Tan color; sp.gr. (20/20°C) 1.12; melting range 95-99°C; amine odor. Soluble in benzene, chloroform, gasoline; insoluble in water, 10% caustic, carbon disulfide.
Uses: Vulcanizing and accelerating agents.

tetraisopropyl titanate (TPT; titanium isopropylate; isopropyl titanate) $Ti[OCH(CH_3)_2]_4$.
Properties: Light yellow liquid which fumes in moist air. B.p. (10 mm) 102-104°C; m.p. 14.8°C; sp.gr. 0.954; index of refraction 1.46; apparent viscosity (25°C) 2.11 cps. Decomposes rapidly in water; soluble in most organic solvents.
Derivation: Reaction of titanium tetrachloride with isopropanol.
Uses: Ester exchange reactions; improving adhesion of paints, rubber, and plastics to metal surfaces; condensation catalyst.

tetraisopropyl zirconate $(C_3H_7O)_4Zr$.
Properties: White solid; decomposes before melting.
Derivation: By reaction of zirconium tetrachloride with isopropanol.
Uses: Condensation catalyst; cross-linking agent.

N,N,N',N'-tetrakis(2-hydroxyethyl)ethylenediamine. See tetrahydroxyethylethylenediamine.

tetrakishydroxymethylphosphonium chloride $(HOCH_2)_4PCl$. A crystalline compound made by the reaction of phosphine, formaldehyde and hydrochloric acid. Used to render cotton fabrics flame resistant. It is more effective when used with bromoform allyl phosphate or with other resin-forming materials.

"Tetralin." [28] Trademark for tetrahydronaphthalene $(C_{10}H_{12})$.
Use: As a solvent in the textile, soap, and manufactured gas industries; chemical intermediate.

tetralite. See tetryl.

tetramer. A molecule formed by union of four identical simpler molecules. Also applied to the substances composed of such quadruple molecules. Thus C_8H_8 is a tetramer of C_2H_2. See polymer.

tetramethylammonium chloride $(CH_3)_4NCl$. A quaternary ammonium compound.
Properties: White crystalline solid; sp.gr. 1.1690 (20/4°C); m.p., decomposes. Soluble in water and alcohol; insoluble in ether.

tetramethylammonium chlorodibromide $(CH_3)_4NClBr_2$.
Properties: M.p. 118-126°C. Soluble in water and other polar solvents. Liberates elemental bromine on contact with water.
Uses: Dry brominating agent; ingredient in formulation of sanitizers.

1,2,3,4-tetramethylbenzene. See prehnitene.

1,2,3,5-tetramethylbenzene. See isodurene.

sym-**tetramethylbenzene.** See durene.

2,2,3,3-tetramethylbutane. See hexamethylethane.

N,N,N',N'-tetramethyl-1,3-butanediamine $CH_3CHN(CH_3)_2CH_2CH_2N(CH_3)_2$.
Properties: Colorless, stable liquid; f.p. below −100°C; b.p. 165.0°C (760 mm); sp. gr. 0.8020 (20/20°C); vapor pressure 1.64 mm (20°C); miscible in all proportions with water; viscosity 1.0 cps (20°C).
Uses: Amine catalyst for polyurethane foams; catalyst for epoxy resins; high energy fuels.

tetramethyldiamidophosphoric fluoride. See dimefox.

tetramethyldiaminobenzhydrol (tetramethyldiaminodiphenylcarbinol; Michler's hydrol; hydrol) $(CH_3)_2NC_6H_4CH(OH)C_6H_4N(CH_3)_2$.
Properties: Colorless prisms; forms a colorless solution in ether or benzene and a blue solution in alcohol or acetic acid. Soluble in alcohol, ether, benzene, and acetic acid. M.p. 96°C.
Derivation: By the reaction of tetramethyldiaminodiphenylmethane, hydrochloric acid and glacial acetic acid; oxidized with lead peroxide.
Grade: Technical.
Containers: Wooden kegs or fiber drums; tank trucks.
Uses: Dye intermediate; organic synthesis.
Shipping regulations: None.*

tetramethyldiaminobenzophenone (Michler's ketone) $CO[C_6H_4N(CH_3)_2]_2$. (4,4'-bis(dimethylamino) benzophenone).
Properties: Crystalline leaflets; m.p. 172°C;

b.p., decomposes at 360°C; soluble in alcohol, ether, and water.
Derivation: From dimethylaniline by reaction with phosgene.
Method of purification: Crystallization.
Grade: Technical.
Containers: Wooden barrels or fiber containers.
Use: Synthesis of dyestuffs and auramine derivatives.
Shipping regulations: None.*

tetramethyldiaminodiphenylcarbinol. See tetramethyldiaminobenzhydrol.

tetramethyldiaminodiphenylmethane (tetra base) $H_2C[C_6H_4N(CH_3)_2]_2$.
Properties: Yellowish leaflets or glistening plates; m.p. 90-91°C; sublimes with decomposition; b.p. 390°C; insoluble in water; soluble in benzene, ether, carbon disulfide and acids.
Derivation: By heating dimethylaniline with hydrochloric acid and formaldehyde.
Method of purification: Crystallization.
Grade: Technical.
Containers: Wooden barrels.
Use: Dye intermediate.
Shipping regulations: None.*

tetramethyldiaminodiphenylsulfone $[(CH_3)_2NC_6H_4]_2SO_2$ (4,4'-bis(dimethylamino) diphenylsulfone).
Constants: M.p. 259-260°C.
Grades: Technical; reagent.
Use: Intermediate in making dyestuffs and medicinal chemicals; analytical reagent for lead.

tetramethylene. See cyclobutane.

tetramethylene bismethanesulfonate. See busulfan.

tetramethylenediamine $H_2N(CH_2)_4NH_2$.
Properties: Colorless crystals with strong odor; m.p. 27°C; b.p. 158-159°C. Soluble in water with strongly basic reaction.
Use: As a chemical intermediate.

tetramethylene dichloride. See 1,4-dichlorobutane.

tetramethylene glycol. See 1,4-butylene glycol.

1,1,4,4-tetramethyl-6-ethyl-7-acetyl-1,2,3,4-tetrahydronaphthalene $C_{18}H_{26}O$.
Properties: Colorless crystals; m.p. 45°C; b.p. (2 mm) 130°C; insoluble in water.
Uses: Perfumes; cosmetics; soaps.

tetramethylethylenediamine (TMEDA; N,N,N',N'-tetramethylethylenediamine) $(CH_3)_2NCH_2CH_2N(CH_3)_2$.
Properties: Colorless, anhydrous liquid with slight ammoniacal odor. Soluble in water and most organic solvents. B.p. 121-122°C; sp.gr. 0.7765 (20/4°C); refractive index (n 25/D) 1.4170; f.p. −55.1°C.
Grade: Anhydrous.
Containers: Up to 55-gal steel drums.
Uses: Preparation of epoxy curing agents, and polyurethane formation; corrosion inhibitor; textile treating compounds; intermediate for quaternary ammonium compounds.

tetramethylguanidine $(CH_3)_2NC(NH)N(CH_3)_2$.
Properties: Liquid with slight ammoniacal odor; b.p. 159-160°C; soluble in both water and organic solvents.
Containers: Drums.
Use: A strong, all-organic base.

tetramethyllead (TML) $(CH_3)_4Pb$.
Properties: Colorless liquid. Sp.gr. 1.995; m.p. −27.5°C; b.p. 110°C (10 mm). Insoluble in water; slightly soluble in benzene, petroleum ether, alcohol.
Derivation: Lead-sodium alloy and methyl chloride.
Use: Gasoline additive (antiknock compound).

1,2,6,6-tetramethyl-4-mandeloxypiperidine hydrochloride. See eucatropine hydrochloride.

tetramethylmethane. See neopentane.

3,3'-tetramethylnonyl thiodipropionate. See ditridecyl thiodipropionate.

N,2,3,3-tetramethyl-2-norcamphanamine hydrochloride. See mecamylamine hydrochloride.

2,6,10,14-tetramethylpentadecane. See "Pristane."

tetramethylsilane $(CH_3)_4Si$.
Properties: Colorless volatile liquid. B.p. 26.5°C; sp.gr. 0.646 (20/4°C). Insoluble in water and cold, concentrated sulfuric acid; soluble in most organic solvents. Ignites in air.
Derivation: By Grignard reaction of silicon tetrachloride and methylmagnesium chloride.
Grades: Technical; purified.
Containers: Bottles; 50-lb drums.
Use: Liquid in high-speed aircraft.
Shipping regulations: Flammable liquid. Red label.*

tetramethylthiuram disulfide (thiram; thiuram; bis(dimethylthiocarbamyl) disulfide; TMTD) $[(CH_3)_2NCS]_2S_2$.
Properties: White crystalline powder with a characteristic odor; soluble in alcohol, benzene, chloroform, carbon disulfide; insoluble in water, dilute alkali, gasoline; sp. gr. 1.29 (20°C); melting range 146-148°C.
Grades: 75% wettable powder; 95% technical powder.
Containers: 50-lb multiwall paper bags.
Uses: As a rubber accelerator to impart heat-resistance; fungicide; insecticide; seed disinfectant; rat repellent in vinyl film; lube oil additive; bacteriostat.
Caution! May cause irritation; harmful if inhaled or swallowed. MCA warning label.

tetramethylthiuram monosulfide (bis(dimethylthiocarbamyl) sulfide) $[(CH_3)_2NCS]_2S$. Yellow powder. A powerful accelerator of vulcanization for rubber. It is usually used in combination with some other accelerator. It has also been used as a fungicide and insecticide.

tetraminoditolylmethane $[CH_3C_6H_2(NH_2)_2]_2CH_2$.
Properties: White crystalline powder. Insoluble in water. M.p. 195°C.
Derivation: Condensation of meta-tolylenediamine and formaldehyde.

"Tetramix." [28] Trademark for a liquid antiknock compound, the active component of which is a redistribution mixture of tetraethyl and tetramethyl lead; sp.gr. 1.591.
Use: In concentrations up to 2.52 cc active ingredient per gal as an antiknock agent for motor gasolines.
Containers: Liter cans, 10- and 55-gal drums; 3,000- and 6,000-gal tank cars.

tetranitroaniline (TNA) $C_6H(NO_2)_4NH_2$. A nitration product of aniline which melts at 170°C and explodes at 237°C.
Use: In the manufacture of detonators and primers.
Fire hazard: Dangerous.
Shipping regulations: Explosive, class A. High explosive label.*

tetranitrol. See erythrityl tetranitrate.

tetranitromethane $C(NO_2)_4$.
Properties: Colorless liquid. Very toxic; irritating to the eyes and respiratory passages. B.p. 125.7°C; m.p. 12.5°C; sp.gr. (13°C) 1.650. Miscible with alcohol and ether; insoluble in water. Decomposed by alcoholic solution of potassium hydroxide.
Derivation: By action of fuming nitric acid on benzene, acetic anhydride, or acetylene.
Uses: Rocket fuel, as an oxidant or monopropellant; qualitative test for unsaturated compounds.
Caution: Poison; fire hazard.
Shipping regulations: Oxidizing material. Yellow label.*

"Tetranol." [300] Trademark for a highly sulfated fatty ester of the oleic type, with wetting and dye-leveling properties.

tetra(octylene glycol)titanate. See octylene glycol titanate.

1,1,4,4-tetraphenylbutadiene $(C_6H_5)_2C:CHCH:C(C_6H_5)_2$.
Properties: White crystals; m.p. 205°C; insoluble in water; soluble in most organic solvents.
Grade: Purified.
Use: As primary fluor or as wave length shifter in solution scintillators.

tetraphenylsilane $(C_6H_5)_4Si$.
Properties: White solid; m.p. 237°C; b.p. 428°C. Very stable and inert.
Derivation: By Grignard reaction of silicon tetrachloride and phenylmagnesium chloride; or by chlorination of phenyltrichlorosilane.
Grade: Technical.
Use: Heat transfer medium; polymers.

tetraphenyltin $(C_6H_5)_4Sn$.
Properties: White powder. Sp.gr. 1.490; m.p. 225-228°C; b.p. above 420°C. Insoluble in water; soluble in hot benzene, toluene, xylene.
Derivation: Reaction of tin tetrachloride with phenylmagnesium bromide.
Uses: Stabilizer in chlorinated transformer oils; mothproofing agent; scavenger in dielectric fluids; intermediate.

tetraphosphorus hexasulfide. See phosphorus trisulfide.

tetraphosphorus trisulfide. See phosphorus sesquisulfide.

tetrapotassium EDTA. See ethylenediaminetetraacetic acid salts.

tetrapotassium pyrophosphate (TKPP). See potassium pyrophosphate.

tetrapropenylsuccinic anhydride (TPSA) $CH_3CH_2CH_2CH(CH_3)CH_2C(CH_3):CHC(CH_3)_2$-$CHC_2O_3CH_2$. Molecule with both a hydrophobic and a hydrophilic end-section. Used as curing agent for epoxy resins; in drying oils and lacquers; in waterproofing resins and polyester resins; and as organic intermediate.

tetrapropylene (dodecene; propylene tetramer) $C_{12}H_{24}$. Mixture of C_{12} monoolefins.
Properties: Liquid; sp.gr. (20/20°C) 0.770; boiling range 183°-202°C; weight 6.44 lb/gal (60°F).
Derivation: Olefin fraction obtained from catalytic polymerization of propylene.
Containers: Barge or tank car.
Uses: Detergents (dodecylbenzene); lubricant additives; plasticizers.

tetrapropylthiuram disulfide $[(C_3H_7)_2NCS]_2S_2$.
Properties: Light cream color; sp.gr. (20/20°C) 1.13; melting range 49-51.5°C; musty odor. Soluble in carbon disulfide, benzene, chloroform and gasoline; insoluble in water and 10% caustic.
Uses: Vulcanizing and accelerating agents.

"Tetraquinone." [227] Trademark for 1,2,3,4-tetrahydro-6-methylquinoline, $C_{10}H_{13}N$. (Civettal).
Properties: Yellowish crystals; characteristic civet odor; stable; may cause discoloration; congealing point, minimum 32.8°C; soluble in 2 parts of 80% alcohol.
Uses: A contributive note to the odor of civet; an effective fixative.

tetrasodium EDTA. See ethylenediaminetetraacetic acid salts.

tetrasodium 2-methyl-1,4-naphthalenediol diphosphate. See menadiol sodium diphosphate.

tetrasodium monopotassium tripolyphosphate $Na_4KP_3O_{10}$.
Properties: White crystalline solid; m.p. 580-600°C; density, 2.55; solubility in water (26°C) 30 g/100 ml.
Use: Sequestrant.

tetrasodium pyrophosphate. See sodium pyrophosphate.

tetrastearyl titanate. A substance used as organic intermediate, adhesion promoter, pigment dispersant.

tetrazene (guanyl nitrosaminoguanyl tetrazene). An initiating explosive.
Shipping regulations: Class A explosive. Not accepted by express.*

tetrazolium chloride (tetrazolium salts; TTC; 2, 3, 5-triphenyltetrazolium chloride) $CN_4Cl(C_6H_5)_3$.
Properties: White to pale-yellow crystalline powder which darkens on exposure to light. Readily soluble in water. M.p. (with decomposition) 245°C.
Uses: In germination and viability tests. Viable parts of seed are stained red by deposition of red insoluble triphenyl formazan. Also used in medicine.

tetrazolium salts. See tetrazolium chloride.

"Tetrine." [73] Trademark for a series of organic chelating, sequestering, and complexing agents consisting of ethylene-diaminetetraacetic acid and its sodium salts.

tetrol. See furan.

"Tetron." [88] Trademark for tetraethyl pyrophosphate (TEPP), technical, and in liquid formulations for insecticidal use.

"Tetrone" A. [28] Trademark for dipentamethylenethiuramtetrasulfide rubber accelerator.
Properties: Grayish-yellow powder.
Containers: 50-lb bags.
Use: To accelerate and improve the vulcanization of natural and synthetic rubber and latex compounds.

"Tetronic." [203] Trademark for a series of nonionic surface active agents prepared by the sequential addition of propylene and ethylene oxides to ethylenediamine. They are available in liquid, paste, and flake form and all are 100% active agents. Major uses are as low foaming detergents, lime soap and pigment dispersing agents, emulsifying and demulsifying agents, wetting agents, and as plasticizers for a variety of resins.

"Tetrosan 60%." [328] Trademark for an alkyl dimethyl 3, 4-dichlorobenzyl ammonium chloride and alkenyl dimethyl ethyl ammonium bromide in the ratio of 5:1.
Properties: Clear, mobile liquid, light yellow to amber with mild characteristic odor and a bitter taste; miscible in any proportion with water, methanol, ethanol, isopropanol, and acetone. Insoluble in hydrocarbons, halides, vegetable, animal and mineral oils. Use dilutions are colorless and virtually without odor.
Containers: 5- and 13-gal returnable glass carboys; 15- and 55-gal lined drums.
Use: Disinfectant and antiseptic in the veterinary and pharmaceutical fields.

tetryl (tetralite; nitramine). Common commercial names for trinitrophenyl-methylnitramine.
Properties: Yellow crystals; m.p. 130-132°C. explodes at about 180-190°C. Insoluble in water; soluble in alcohol, ether, benzene, glacial acetic acid.
Use: As explosive used in detonators; also as a "priming" or intermediary detonating agent for less sensitive high explosives.
Fire hazard: Dangerous.
Shipping regulations: Explosive, class A. High explosive label.*

tetterwort. See sanguinaria.

"Texavon." [57] Trademark for textile penetrant or scouring agent.

"Texicote" Emulsions. [263] Proprietary product. Vinyl acetate polymer and copolymer emulsions, plasticized and unplasticized. Several grades: Solids content 35-63%; pH less than 5. Hard to very flexible films.
Uses: Emulsion paints; finishes for textiles, leather, and paper; adhesives; floor coatings.

"Texicryl" Emulsions. [263] Proprietary product. Acrylic polymer and copolymer emulsions. Several grades. Solids content 30-50%; pH less than 10. Hard to very flexible films.
Uses: Emulsion paints; textile, leather and paper finishes.

"Texigel" Thickening Agent. [263] Proprietary product. Water-soluble, anionic colloid; solids content less than 20%; viscosities 15,000-50,000 cps, or over 100,000 cps. (Ferranti).
Uses: Thickening agent for natural and synthetic latex; protective colloid for emulsions and dispersions; textile and leather finishes.

"Texilac" Solutions. [263] Proprietary product. Vinyl polymer and copolymer solutions in various solvents, plasticized and unplasticized. Solids content 30-65%.
Uses: Adhesives; impregnants and surface coatings; printing inks.

"Texiprint" Color Pastes. [263] Proprietary product. Synthetic resin pigment pastes and thickeners for screen and roller printing on textiles.

textile clays. China clays low in grit and silica, and used as fillers for textiles.

textile oils. This term includes the various specially compounded oils used to condition all types of raw fibers, yarns or fabric for manufacturing, bleaching, dyeing and finishing operations. While the primary purpose of these oils is to lubricate or soften the fibers, they also may add any or all of the following properties or conditions: (1) increase tenacity and tensile strength; (2) diminish static electricity; (3) prevent breakage and abrasion during spinning and twisting; (4) coat and bind the filaments together; (5) make the yarn less sensitive to atmospheric conditions; (6) serve as a carrier for any special agents; for example, tints applied to mixed rayon and silk yarns so each can be readily distinguished by its code color.

"Textolite." [245] Trademark for industrial laminated plastic sheets, tubes, and rods used as insulating materials. Includes laminations of various combinations of phenolic, melamine, silicone, and epoxy resins with such base materials as paper, asbestos, cotton, linen, and nylon. Also some of the above materials with copper foil cladding for printed circuit applications.

*See "I.C.C. Shipping Regulations," page xiii.
Reference numbers refer to name of manufacturer. See "List of Manufacturers," page v.

"Textone." [84] Brand name for a proprietary product. A sodium chlorite product that bleaches textile fabrics (cotton, linen, rayon, nylon) to high whiteness without degradation of the fibers.

textryl. A generic name for non-woven structures which may be manufactured by wet-processing from staple fibers and fibrid binder.

"Texzyme." [114] Trademark for a liquid enzyme preparation, principally proteolytic enzyme, produced by growing pure cultures of micro-organisms on select media.
Uses: Removal of protein type sizes such as gelatine type coatings from photographic film.
Shipping regulations: None.*

TFE. Abbreviation for tetrafluoroethylene. See also polytetrafluoroethylene.

"TG-8." [52] Trade name for triethylene glycol dicaprylate (q.v.).
Uses: Low temperature plasticizer for synthetic resins and rubbers.

"T-Glo-8-210." [79] Trade name for a tall oil gloss oil, i.e., solution of a limed special tall oil in mineral spirits.
Properties: Acid value (on solution) 28; concentration (total solids) 62%; viscosity (Gardner-Holdt) K; color (Hellige) 9; percent lime 8.
Containers: 55 gal drums; tank cars.
Uses: Paint and varnish; caulking compounds.

"T-Glo-8Y." [79] Trademark for a tall oil gloss oil, i.e., solution of limed special tall oil in mineral spirits.
Properties: Acid value (on solution) 33; concentration (total solids) 65%; viscosity (Gardner-Holdt) Y; color (Hellige) 7-8; percent lime 9.
Containers: 55-gal drums; tank cars.
Uses: Paint and varnish; caulking compounds.

"T-Glo-8Y-210." [79] Trade name for a tall oil gloss oil, i.e., solution of a limed special tall oil in mineral spirits.
Properties: Acid value (on solution) 33; concentration (total solids) 65%; viscosity (Gardner-Holdt) Y; color (Hellige) 9-10; percent lime 9.
Containers: 55-gal drums; tank cars.
Uses: Paint and varnish; caulking compounds.

Th. Symbol for thorium.

"Thalamyd." [321] Brand name for phthalylsulfacetamide.

thallic oxide. See thallium peroxide.

thalline (tetrahydro-para-methyloxyquinoline) $C_9H_{10}N(OCH_3)$.
Properties: White crystalline powder, darkens on exposure to air and light; m.p. 42-43°C; b.p. 283-285°C; soluble in water, petroleum ether, alcohol, ether.
Derivation: Reaction of para-aminoanisole, para-nitroanisole, glycerol and sulfuric acid, to form para-quinanisole, followed by reduction with tin and hydrochloric acid.
Use: Medicine.
Shipping regulations: None.*

thallium Tl. Element of atomic number 81, group III of the periodic system.
Properties: Bluish-white, lead-like metal. Soluble in nitric and sulfuric acids; insoluble in water. Its compounds are poisonous.
Constants: Sp.gr. 11.85; m.p. 302°C; b.p. 1650°C.
Source: Cadmium-bearing flue dust from zinc smelting.
Purification: By heating thallium iodide with metallic sodium.
Grades: Technical.
Containers: Glass bottles.
Use: Thallium salts; alloys.
Shipping regulations: None.*

thallium acetate (thallous acetate) $TlOCOCH_3$.
Properties: White, deliquescent crystals. Poisonous! Soluble in water, alcohol.
Constants: M.p. 110°C; sp.gr. 3.68.
Derivation: Interaction of acetic acid and thallium carbonate.
Grades: Technical.
Containers: Bottles.
Uses: Medicine; high specific gravity solutions used to separate ore constituents by flotation.
Shipping regulations: Poison, class B. Poison label.*

thallium bromide (thallous bromide) TlBr.
Properties: Yellowish-white, crystalline powder. Very poisonous! Soluble in alcohol; slightly soluble in water; insoluble in acetone.
Constants: Sp.gr. 7.557; b.p. 815°C; m.p. (approx) 460°C.
Uses: Mixed crystals with thallium iodide for infrared radiation transmitters.
Shipping regulations: Poison, class B. Poison label.*

thallium carbonate (thallous carbonate) Tl_2CO_3.
Properties: Heavy shiny, colorless or white crystals. Highly refractive. Melts to dark-gray mass. Slightly alkaline taste. Soluble in water; insoluble in alcohol. Poisonous!
Constants: Sp.gr. 7; m.p. 272°C.
Grades: Technical.
Uses: Analysis (testing for carbon disulfide); artificial diamonds.
Shipping regulations: Poison, class B. Poison label.*

thallium chloride (thallous chloride) TlCl.
Properties: White, crystalline powder. Becomes violet on exposure to light. Slightly soluble in water; insoluble in alcohol, ammonium hydroxide. Very poisonous!
Constants: Sp.gr. 7; m.p. 430°C.
Containers: Bottles.
Uses: Catalyst (chlorination); medicine; tungsten lamps.
Shipping regulations: Poison, class B. Poison label.*

thallium hydroxide (thallous hydroxide) $TlOH \cdot H_2O$.
Properties: Yellow needles. Soluble in

alcohol, water. Poisonous!
Constants: B.p. (dehydrated) 100°C (dec).
Grades: Technical.
Use: Analysis (testing for ozone; indicator).
Shipping regulations: Poison, class B. Poison label.*

thallium iodide (thallous iodide) TlI.
Properties: Yellow powder. Insoluble in alcohol; slightly soluble in water; soluble in aqua regia.
Constants: Sp.gr. 7.09; b.p. 824°C; m.p. 440°C. Becomes red at 170°C.
Uses: Medicine; mixed crystals with thallium bromide for infrared radiation transmitters.
Shipping regulations: Poison, class B. Poison label.*

thallium monoxide (thallium oxide; thallous oxide) Tl_2O.
Properties: Black powder. Oxidizes when exposed to air. Caution! Keep well stoppered! Soluble in alcohol, water (decomposes). Poisonous! M.p. 300°C.
Grades: Technical.
Uses: Analysis (testing for ozone); artificial gems; optical glass of high refractive index.
Shipping regulations: Poison, class B. Poison label.*

thallium nitrate (thallous nitrate) $TlNO_3$.
Properties: Colorless crystals. Soluble in hot water; insoluble in alcohol.
Constants: Sp.gr. 5.5; m.p. 206°C (solidifies to a glass-like solid; decomposes at 450°C).
Grades: Technical.
Uses: Analysis; pyrotechnics, green fire.
Shipping regulations: Poison, class B. Poison label.*

thallium oxide. See thallium monoxide; thallium peroxide.

thallium peroxide (thallic oxide; thallium oxide; thallium sesquioxide; thallium trioxide) Tl_2O_3.
Properties: Brown or dark-red powder. Soluble in acids; insoluble in basic solutions, water.
Constants: Sp.gr. 9.6; m.p. 759°C.
Grades: Technical.
Use: Matches.
Shipping regulations: Poison, class B. Poison label.*

thallium sesquichloride (thallo-thallic chloride) $TlCl_3 \cdot 3TlCl$ or Tl_2Cl_3.
Properties: Yellow, crystalline powder. Slightly soluble in water.
Constants: Sp.gr. 5.9; m.p. 400-500°C.
Shipping regulations: Poison, class B. Poison label.*

thallium sesquioxide. See thallium peroxide.

thallium sulfate (thallous sulfate) Tl_2SO_4.
Properties: Colorless crystals. Soluble in water.
Constants: Sp.gr. 6.77; m.p. 632°C.
Grades: Technical, 99%.
Containers: Bottles.
Uses: Analysis (testing for iodine in the presence of chlorine); medicine; ozonometry; rodent and insect poisons.
Danger! Cumulative poison. Absorbed through skin. MCA warning label.
Shipping regulations: Poison, class B. Poison label.*

thallium sulfide (thallous sulfide) Tl_2S.
Properties: Blue-black, lustrous, microscopic crystals or amorphous powder. Soluble in mineral acids; insoluble in water, alcohol or ether.
Constants: Sp.gr. 8.0; m.p. 448°C.
Uses: Infrared-sensitive photocells.
Shipping regulations: Poison, class B. Poison label.*

thallium trioxide. See thallium peroxide.

thallo-thallic chloride. See thallium sesquichloride.

thallous acetate. See thallium acetate.

thallous bromide. See thallium bromide.

thallous carbonate. See thallium carbonate.

thallous chloride. See thallium chloride.

thallous hydroxide. See thallium hydroxide.

thallous iodide. See thallium iodide.

thallous nitrate. See thallium nitrate.

thallous oxide. See thallium monoxide.

thallous sulfate. See thallium sulfate.

thallous sulfide. See thallium sulfide.

THAM. See tris(hydroxymethyl)aminomethane.

"Thanite." [266] Trademark for a technical grade of isobornyl thiocyanoacetate.
Properties: Amber liquid; sp.gr. 1.107 (15.6/15.6°C).
Uses: Insecticides.

thebaine (para-morphine) $C_{19}H_{21}NO_3$.
Properties: White, crystalline alkaloid; very poisonous! Slightly soluble in water; soluble in alcohol and ether. M.p. 193°C.
Derivation: From opium.
Method of purification: Crystallization.
Grades: Technical.
Containers: Glass bottles.
Use: Medicine.
Shipping regulations: None.*

thebaine hydrochloride $C_{19}H_{21}NO_3 \cdot HCl \cdot H_2O$.
Properties: Large rhombic prisms, white to slightly yellow. Soluble in water, alcohol and ether. Poisonous!
Derivation: By the action of hydrochloric acid on thebaine.
Method of purification: Crystallization.
Grades: Technical.
Containers: Glass bottles.
Use: Medicine.
Shipping regulations: None.*

"Thedane Blue." [243] Trade name for biological staining solution useful for the rapid and precise identification of blood parasites.

theine. See caffeine.

thenardite Na_2SO_4 (verde salt). A natural anhydrous sulfate of sodium. White to brownish in color.

Occurrence: United States (Arizona, California), Central Asia, Chile, Spain, Germany.

Thénard's blue. See cobalt blue.

"Thenfadil" Hydrochloride. [162] Trademark for thenyldiamine hydrochloride.

thenyl. The radical $C_4H_3SCH_2-$ or CHCHSCHCCH₂–, based on methylthiophene. Thenyl alcohol is a synonym for thiophenemethanol.

thenyldiamine (2-[2-dimethylaminoethyl)-3-thenylamino]pyridine) $(C_4H_3SCH_2)N(C_5H_4N)CH_2CH_2N(CH_3)_2$. Thenyldiame is a coined name, not a true chemical name.

Properties: Liquid; b.p. (1.0 mm) 169-172°C.

Derivation: Prepared by condensing N,N-dimethylaminoethyl-alpha-amino-pyridine with 3-thenyl bromide.

Use: Medicine (as base for various salts, especially the hydrochloride).

thenyldiamine hydrochloride (N,N-dimethyl-N'-(3-thenyl)-N'-(2-pyridyl)ethylenediamine hydrochloride).

Properties: White, practically odorless, crystalline powder; solutions neutral to litmus; soluble in water, alcohol, and chloroform; nearly insoluble in ether and benzene; m.p. 167-171°C.

Grade: N.F. XI.

Use: Medicine.

"Thenylene" Fumarate. [3] Trademark for methapyrilene fumarate (q.v.).

"Thenylene" Hydrochloride. [3] Trademark for methapyrilene hydrochloride (q.v.).

theobroma oil. See cacao butter.

theobromine (3,7-dimethylxanthine) $C_7H_8N_4O_2$. The alkaloid found in cocoa and chocolate products.

Properties: White crystalline powder; poisonous! Insoluble in water, alcohol and ether; sublimes at 260°C.

Derivation: By extraction from the seeds of the Theobroma cacao.

Method of purification: Crystallization.

Grades: Technical; N.F. XI.

Containers: 1-lb cans; 1-lb bottles; fiber drums.

Use: Medicine.

Shipping regulations: None.*

theobromine calcium salicylate (calcium theobromine salicylate). A complex or double salt of equimolecular proportions of theobromine calcium $(C_{14}H_{14}CaN_8O_4)$ and calcium salicylate $(C_{14}H_{10}CaO_6)$.

Properties: White, odorless powder with saline taste. Solutions are alkaline to litmus and phenolphthalein. Slightly soluble in water; insoluble in alcohol.

Grade: N.F. XI.

Use: Medicine.

theobromine-sodium acetate $C_7H_7O_2N_4Na \cdot NaC_2H_3O_2$.

Properties: White crystalline powder; odorless, bitter taste. Contains not less than 55% theobromine. Incompatible with carbonated beverages, acids, saccharin, mucilaginous liquids, alkaloids. Soluble in cold water; slightly soluble in cold alcohol, more soluble in hot alcohol; hygroscopic; absosrbs carbon dioxide from the air and liberates theobromine.

Grades: N.F. XI.

Containers: Fiber drums.

Use: Medicine.

Shipping regulations: None.*

theobromine sodium salicylate. A mixture of sodium theobromine and sodium salicylate in equimolecular proportions.

Properties: White powder; odorless, sweet, saline alkaline taste. Contains not less than 46.5% theobromine and 35% salicylic acid when dried to constant weight at 105°C. Slowly absorbs carbon dioxide from the air, liberating theobromine; develops characteristic odor. Soluble in water; slightly soluble in alcohol.

Grades: N.F. XI.

Containers: 1-lb bottles; 25-, 100-lb fiber drums.

Use: Medicine.

Shipping regulations: None.*

"Theocalcin." [9] Trademark for theobromine-calcium salicylate.

theophylline (1,3-dimethylxanthine) $C_7H_8N_4O_2 \cdot H_2O$.

Properties: White, crystalline alkaloid; poisonous! Odorless with a bitter taste; m.p. 270-274°C; slightly soluble in water and alcohol, more soluble in hot water, sparingly soluble in ether or chloroform.

Derivation: (a) By extraction from tea leaves; (b) synthetically from ethyl cyanoacetate.

Method of purification: Crystallization.

Grades: Technical; N.F. XI.

Containers: Glass bottles; 1-, 5-lb tins; 25-, 100-lb drums.

Use: Medicine.

Shipping regulations: None.*

theophylline cholinate. See oxtriphylline.

theophylline ethylenediamine. See aminophylline.

theophylline-methylglucamine. An equimolecular mixture of theophylline $(C_7H_8N_4O_2 \cdot H_2O)$ and N-methyl-L-glucosamine $(C_7H_{17}NO_5)$.

Use: Medicine.

theophylline sodium acetate. A hydrated mixture of theophylline sodium $(C_7H_7N_4O_2Na)$ and sodium acetate $(C_2H_3O_2Na)$ in approximately equimolecular proportions.

Properties: White, crystalline powder; odorless; bitter, salty taste; gradually absorbs CO_2 from the air liberating free theophylline. Soluble in water to give solution basic to phenolphthalein; insoluble in alcohol, ether, and chloroform.

Grade: N.F. XI.
Use: Medicine.

theophylline sodium glycinate (sodium theophylline glycinate). A mixture containing theophylline sodium ($C_7H_7N_4NaO_2$) and glycine (NH_2CH_2COOH) in approximately equimolecular proportions buffered with an additional mole of glycine.
Properties: White crystalline powder with slight ammoniacal odor and a bitter taste. Freely soluble in water; very slightly soluble in alcohol; practically insoluble in chloroform. pH (saturated solution) 8.5-9.5.
Grade: N.F. XI.
Use: Medicine.

"Thephorin" Tartrate. [190] Trademark for a brand of phenindamine tartrate (q.v.).

therm. A unit of heat equal to 100,000 Btu. It has also been used to mean 1 Btu or 1 small calorie, but these alternative uses have been abandoned in the U.S.

"Thermaflow." [89] Trademark for a series of reinforced polyester molding compounds.

thermal black. Carbon black made by the thermatomic process. It consists of coarser particles than channel black and is less suitable for rubber reinforcement but is used as a pigment.
Grades: Fine thermal black (FT); medium thermal black (MT).

thermal conductivity. The capacity for conducting heat, usually expressed as the number of calories which pass per second through a plate one square centimeter in area and one centimeter thick having its opposite faces differ in temperature by 1°C. In engineering applications thermal conductivity is expressed as Btu/hr/sq ft/°F/ft of thickness of the material under consideration.

thermal expansion coefficient. The change in volume per unit volume per degree change in temperature (cubical coefficient). For isotropic solids the expansion is equal in all directions; and the cubical coefficient is approximately three times the linear coefficient of expansion. These coefficients vary with temperature; but for gases at constant pressure the coefficient of volume expansion is nearly constant and equals 0.00367 for 1°C at any temperature.

thermal glass. A glass in which boron oxide replaces the calcium oxide in ordinary lime soda glass. Has low coefficient of expansion, can be heated and cooled rapidly without breakage. See "Pyrex."

thermal neutrons. See neutron.

"Therma-Tite." [170] Trademark for a specially formulated adhesive for high speed bottle label machines to withstand the dangers of thermoshock. Material will adhere to cold glass without crystallizing.
Properties: Light tan liquid; pH 3.0-3.5. Humidity resistant and fast-setting.

thermatomic process. Methane is cracked over hot bricks at a temperature of 1600°F to form amorphous carbon (carbon black) and hydrogen.

"Thermax." [69] Trademark for thermatomic carbon; a medium thermal carbon black, available in powder and pellet form.
Use: In rubber and synthetic rubber compounds. See "P-33."

thermionics. Study of the process by which electrons are produced and caused to escape from a material by use of high temperatures. There are many applications, as in the vacuum tubes used in radio and other electronic devices. Recent studies relate to the use of this phenomenon as a means for direct conversion of heat to electricity.

"Thermit." [288] Trademark for a mixture of iron oxide and finely divided aluminum used in welding iron and steel, and for incendiary bombs. It is said to develop a temperature of from 2000 to 5000°F.

thermodynamics. Originally defined as the study of the conversion of heat into work and vice versa. Now defined as the study of the laws of transformation of energy from any one form to another. The original kind of thermodynamics is now often called engineering thermodynamics to distinguish it from chemical thermodynamics, which is chiefly concerned with the relationship of heat and work and other forms of energy to equilibrium in chemical reactions and changes of state.

thermoelectric alloy. An alloy used as a junction element in a thermocouple or other thermoelectric device. New alloys of silver, antimony, and tellurium are expected to replace semiconductors such as lead telluride and bismuth telluride. See thermoelectricity.

thermoelectricity. Electricity produced directly by applying a temperature difference to different parts of electrically conducting or semiconducting materials. Usually two different materials are used and the points of contact are kept at different temperatures (Peltier effect). Many temperature measuring devices (thermocouples, thermopiles) work on this principle, since the voltage is proportional to the temperature difference. Metallic conductors are usually used for these "thermometers," which produce a rather small current. A newer use for the effect is as a source of electrical energy, i.e., a means of direct conversion of heat into electricity (or vice-versa) without the use of a generator (or motor). The materials used for these new thermoelectric couples are semiconductors (i.e., zinc antimonide; lead, bismuth, germanium tellurides; samarium sulfide) or thermoelectric alloys (q.v.), either of which produce relatively large currents. Several of these "cells" are then hooked in series much like the cells of a battery.

"Thermoflex" A. [28] Trademark for a rubber antioxidant containing 25% di-para-methoxydiphenylamine $(CH_3OC_6H_4)_2NH$, 25% diphenyl-para-phenylenediamine $C_6H_4(NHC_6H_5)_2$, and 50% phenyl-beta-naphthylamine $C_{10}H_7NHC_6H_5$.
Properties: Dark gray pellets; sp.gr. 1.21; f.p. not lower than 67°C.
Containers: 250-lb drums.
Use: Promote outstanding heat aging and flex life of natural and synthetic rubbers.

Thermofor process. Catalytic cracking process in which petroleum vapor is passed up through a reactor and countercurrent to a flow of small beads of aluminum silicate catalyst.

"Thermoguard H." [288] Trade name for antimony oxide, Sb_2O_3, of high tinctorial strength.
Properties: Very fine white powder; fineness 99.5% minimum through 325 wet sieving; refractive index 2.087; m.p. 656°C; sp.gr. 5.7 (25°C); insoluble in common organic solvents; very slightly soluble in water; soluble in aqueous hydrochloric acid, potassium hydroxide and tartaric acid.
Containers: 50-lb moisture-proof 3-ply paper bags.
Uses: Flame retardant pigment for vinyl chloride, polyesters and polyethylene compounds used in the manufacture of films, sheets, textiles, paper and paints. For special applications where high tinctorial strength is desired.

"Thermoguard L." [288] Trade name for antimony trioxide, Sb_2O_3, of low tinctorial strength.
Properties: Very fine white powder; fineness of 99.5% minimum through 325 wet sieving; refractive index 2.087 (25°C); m.p. 656°C; sp.gr. 5.7 (25°C). Insoluble in common organic solvents; very slightly soluble in water, soluble in aqueous hydrochloric acid, potassium hydroxide and tartaric acid.
Containers: 50-lb moisture-proof 3-ply paper bags.
Uses: Flame retardant pigment for vinyl chloride, polyesters, and polyethylene compounds used in the manufacture of films, sheets, textiles, paper, paints, and special applications where low tinctorial strength is desired.

"Thermolite." [288] Trademark for a series of heat and light stabilizers for polyvinyl chloride resins and other chlorinated organic compounds.

"Thermolite 12" A high boiling, slightly yellow, oily liquid; b.p. above 400°F (10 mm); bulk density 8.75 lbs/gal; sp.gr. 1.05 (25°C); refractive index 1.470 (25°C); viscosity 43 cps (25°C); completely miscible in all commonly used PVC plasticizers; soluble in petroleum ether, octane, benzene, toluene, carbon tetrachloride, ethyl ether, esters; insoluble in water and methyl alcohol.

"Thermolite 13" A powerful organotin stabilizer; white powder; insoluble in water; soluble in benzene, ethyl alcohol and acetone.

"Thermolite 17" A complex organotin composition; light yellow liquid of faint odor; sp.gr. 1.15 (25°C); 9.6 lbs/gal; refractive index 1.48; pour point (ASTM) −17°C; viscosity 203 cps (25°C); completely miscible in all commonly used PVC plasticizers; soluble in petroleum ether, octane, benzene, toluene, carbon tetrachloride, ethyl ether, esters.

"Thermolite 20" A sulfur-containing organotin product; a pale amber liquid; f.p. −10°C; b.p., non-distillable at 10 mm; sp.gr. 0.995 (25°C); refractive index 1.496 (25°C); viscosity 19 centistokes (25°C); lbs/gal 8.3; insoluble in water; sparingly soluble in lower alcohols; soluble in higher alcohols and other organic solvents; miscible with most PVC plasticizers.

"Thermolite 25" A complex organotin composition; a light yellow liquid; sp.gr. 1.15 (25°C); decomposition temperature above 200°C; f.p. below 0°C.

"Thermolite 26" A complex organotin composition; slightly viscous, light yellow liquid; sp.gr. 1.035 (25°C); decomposition temperature above 200°C; f.p. below 0°C.

"Thermolite 31" A sulfur-containing organotin compound; clear liquid of slightly characteristic odor; f.p. below −35°F; sp.gr. 1.11 (25°C); viscosity 33.2 cps (25°C); refractive index 1.504 (25°C); miscible with all common PVC plasticizers; highly soluble in esters, ethers, ketones, alcohols, aliphatic and aromatic hydrocarbons, chlorinated hydrocarbons and other organic types; insoluble in water.

"Thermolite 35" A sulfur-containing organotin compound; white, free-flowing crystals; bulk density approx. 45 lbs/cu ft; slight characteristic odor; soluble in most organic solvents but insoluble in water; stable in air but should be kept in dry, tightly-closed containers.

"Thermolite 45" A complex organotin composition; pale amber liquid; freezing point −10°C; density 0.995 (25°C); refractive index 1.496 (25°C); viscosity 19 centistokes (25°C); insoluble in water, sparingly soluble in lower alcohols; soluble in higher alcohols and other organic solvents; miscible with most PVC plasticizers.

"Thermolite 112" A liquid barium-cadmium stabilizer completely free of any fatty acids such as octoates; an amber colored liquid; sp.gr. 1.05 (25°C); flash point 100°C; pour point below −20°C (ASTM).

"Thermolite 166" An organic zinc vinyl resin stabilizer; a light yellow liquid; sp.gr. 1.02 (25°C); flash point 275°F; Gardner color 9 max; pour point −36°C; viscosity 238-333 centistokes (25°C); soluble in ester plasticizers and the usual solvents employed in the vinyl industry such as ketones, aromatic hydrocarbons, esters, etc; insoluble in water.

"Thermolite 180" An organic vinyl stabilizer

containing no diluents, plasticizers or epoxy groups; a yellow liquid; sp.gr. 0.992 (25°C); flash point 175°C; pour point −7°C.

thermoluminescence. A term used to denote phosphorescence which occurs at a particular temperature, but not at other temperatures. This, however, does not imply luminescence excited by heat.

thermonuclear reactions. See nuclear fusion.

thermoplastic. Term applied particularly to synthetic resins that may be softened by heat, and then regain their original properties upon cooling. Polyvinyl, polystyrene, and acrylate resins are of this type. See thermosetting.

"Thermoscents." [188] Trademark. Aromatic designed for use in plastics; available in various forms for incorporation into latex, milled plastics, and for use as a dressing. Unusual features of the "Thermoscents" are their stability and resistance to high temperatures.

thermosetting. Term applied to synthetic resins which solidify or set on heating and cannot be remelted. The thermosetting property is usually associated with a cross-linking reaction of the constituents to form a three-dimensional network of polymer molecules. Phenol-formaldehyde and urea-formaldehyde resins are of this type. Products made of thermosetting resins cannot be reshaped once they have been fully cured.

"THFA." [244] Trademark for tetrahydrofurfuryl alcohol (q.v.).

thiacetic acid. See thioacetic acid.

thiamine (vitamin B_1) $C_{12}H_{17}ClN_4OS$. 3-(4-Amino-2-methylpyrimidyl-5-methyl)-4-methyl-5, beta-hydroxyethylthiazolium chloride. The antineuritic vitamin which is essential for growth and the prevention of beriberi. It functions in intermediate carbohydrate metabolism in coenzyme form in the decarboxylation of alpha-keto acids. Deficiency symptoms: emotional hypersensitivity, loss of appetite, susceptibility to fatigue, muscular weakness and polyneuritis.

Sources: Enriched and whole grain cereals, milk, legumes, meats, yeast. Most of the thiamine commercially available is synthetic.

Uses: Medicine; nutrition; enriched flours.

Available as thiamine hydrochloride and thiamine mononitrate.

thiamine hydrochloride $C_{12}H_{17}ClN_4OS \cdot HCl$.

Properties: Small white crystals or crystalline powder; hygroscopic; m.p. 248°C (dec); nut-like odor; bitter taste; soluble in water and glycerol; slightly soluble in alcohol; insoluble in ether and benzene; pH (1% solution) 2.7-3.4.

Units: One USP Unit or IU is the activity possessed by 3.0 micrograms.

Grades: U.S.P. XVI.

Containers: Fiber drums.

Uses: Medicine; nutrition.

thiamine mononitrate $C_{12}H_{17}N_5O_4S$.

Properties: White crystals or crystalline powder; nonhygroscopic; m.p. 196-200°C (dec.); slightly soluble in water, alcohol, and chloroform; more stable than the chloride; pH (2% solution) 6.0-7.5.

Units: One USP Unit or IU is the activity possessed by 2.92 micrograms.

Grades: U.S.P. XVI.

Containers: Fiber drums.

Uses: Medicine; nutrition.

thiamine pyrophosphate chloride. See cocarboxylase.

thiamylal sodium $C_{12}H_{17}N_2NaO_2S$. Sodium 5-allyl-5-(1-methyl butyl)-2-thiobarbiturate.

Properties: Pale yellow, hygroscopic liquid, with disagreeable odor. Solutions are alkaline to litmus. Soluble in water.

Grade: U.S.P. XVI (as an injection solution, buffered with anhydrous soldium carbonate).

Use: Medicine.

"Thiate A." [69] Trademark for a proprietary thiohydropyrimidine.

Properties: White, crystalline powder; sp.gr. 1.12 ± .03; melts at 250°C min; moderately soluble in chloroform, ethyl alcohol; insoluble in water; soluble in caustic, carbon disulfide, gasoline, benzene).

Uses: Accelerator for neoprenes to be cured in open steam. Used alone or with thiurams or guanidines plus sulfur for neoprene Type W accelerator.

"Thiate B." [69] Trademark for a proprietary trialkyl thiourea.

Properties: Reddish brown liquid; sp.gr. 1.05 ± .02; very soluble in benzene, carbon disulfide, chloroform, acetone, ethanol; soluble in water; insoluble in dilute caustic, gasoline.

Uses: Accelerator for press-cured neoprene, especially where low compression set is required.

1,4-thiazane $\underline{CH_2SCH_2CH_2NHCH_2}$.

Properties: Colorless liquid. Pyridine-like odor. Fumes in air. Absorbs carbon dioxide from the air. Soluble in alcohol, benzene, ether, water. B.p. 169°C (758 mm).

Derivation: Interaction of alcoholic ammonia and dichlorodiethyl sulfide.

Grades: Technical.

Use: Organic synthesis.

thiazole $\overline{SCH:NCH:CH}$.

Properties: Colorless or pale yellow liquid; sp.gr. 1.198; b.p. 116.8°C; soluble in alcohol and ether; slightly soluble in water; odor resembles that of pyridine.

Use: Organic synthesis of fungicides, dyes.

thiazole dyes. Dyes whose molecular structure contains the thiazole ring (see thiazole). The chromophore groups are =C=N−, -S-C=, but the conjugated double bonds are also of importance. The members of the class are mainly used as direct or developed dyes for cotton, though some find use as union dyes. One example is Direct Fast Yellow (Color Index 814).

thiazolsulfone $NH_2C_6H_4SO_2C_3HNS(NH_2)$. 2-Amino-5-sulfanilylthiazole.
Properties: White to off-white, odorless, crystalline powder; slightly soluble in alcohol and water.
Grade: N.N.D.
Use: Medicine.

2-thiazylamine. See 2-aminothiazole.

"Thibetolide." 227 Trademark for pentadecanolide; 15-hydroxypentadecanoic acid lactone. $(CH_2)_{14}CO_2$ (cyclic).
Properties: Colorless liquid, congealing to white crystals at cool room temperature; has an extremely strong odor of musk; stable; not known to cause discoloration; clearly soluble in 6 parts of 80% alcohol. Congealing point, minimum 36.0°C.
Occurrence: Found in angelica root oil.
Uses: Musky note; and fixative for perfumes.

thickened oils. See blown oils.

2-thienylalanine $C_4H_3SCH_2CHNH_2COOH$. White, microcrystalline powder of characteristic sweetish taste; m.p. 243-245°C.

thimerosal (sodium ethylmercurithiosalicylate) $NaOOCC_6H_4SHgC_2H_5$.
Properties: Light cream-colored crystalline powder with slight characteristic odor. pH (1% solution) about 6.7. Affected by light. Soluble in water and alcohol; almost insoluble in ether and benzene.
Derivation: Reaction between ethylmercuric chloride and thiosalicylic acid in alcoholic sodium hydroxide.
Grade: N.F. XI.
Use: Medicine; bacteriostat, fungistat.

"Thimet." 57 Trademark for a systemic insecticide based on O,O-diethyl-S-(ethylthiomethyl)phosphorodithioate, $(CH_3CH_2O)_2SPSCH_2SCH_2CH_3$. It can be applied to cottonseed to protect the young plants against insect attacks, by absorption through the root system. Also used on alfalfa, potatoes and sugar beets.

thin boiling starch. See soluble starch.

thioacetamide CH_3CSNH_2.
Properties: Colorless leaflets; stable in solution; m.p. 109°C; soluble in water, alcohol, ether, benzene.
Use: To replace gaseous hydrogen sulfide in the qualitative analysis of Group II and III cations.

thioacetic acid (thiacetic acid; ethanethiolic acid) CH_3COSH.
Properties: Clear yellow liquid; pungent acetic and hydrogen sulfide odor; sp.gr. 1.05 (25°C); m.p. −17°C; b.p. 81.8°C (630 mm); soluble in water, alcohol and ether.
Derivation: By heating glacial acetic acid and phosphorus pentasulfide, with subsequent distillation.
Method of purification: Rectification.
Grades: Technical.
Containers: Glass bottles.
Use: Chemical reagent; lachrymator.
Shipping regulations: None.*

thioallyl ether. See allyl sulfide.

thiobenzoic acid (benzenecarbothioic acid) C_6H_5COSH.
Properties: Yellow oil or crystals. Sp.gr. (20/4°C) 1.1825-1.1835; m.p. 24°C; b.p. 77.5°C (5 mm), 122°C (30 mm); refractive index (n 20/D) 1.602-1.604. Insoluble in water; miscible in all proportions with organic solvents.
Grade: 95% min.
Uses: Organic intermediate; in zinc thiobenzoate.

thiocarbamide. See thiourea.

thiocarbanil. See phenyl mustard oil.

thiocarbanilide (N,N'diphenylthiourea; sulfocarbanilide) $CS(NHC_6H_5)_2$.
Properties: Gray powder; m.p. 146°C (min); sp.gr. 1.32. Soluble in alcohol and ether; insoluble in water.
Derivation: By the interaction of aniline and carbon disulfide and alcohol in the presence of sulfur.
Grades: Technical.
Containers: Bags, drums.
Uses: Intermediates; dyes (sulfur colors, indigo, methyl indigo); vulcanization accelerator; synthetic organic pharmaceuticals; flotation agent; acid inhibitor.
Shipping regulations: None.*

thiocarbonyl chloride. See thiophosgene.

para-**thiocresol.** See toluene-para-thiol.

thioctic acid. See dl-alpha-lipoic acid.

thiodiethylene glycol. See thiodiglycol.

thiodiglycol (thiodiethylene glycol; beta-bishydroxyethyl sulfide) $(CH_2CH_2OH)_2S$.
Properties: Syrupy colorless liquid. Characteristic odor; nontoxic. Sp.gr. 1.1847 at 20°C; b.p. 282°C; freezing point −10°C; viscosity 0.652 poises (20°C); flash point 320°F; weight per gallon 9.86 lbs; refractive index (20/D) 1.5217. Soluble in acetone, alcohol, chloroform, water; slightly soluble in benzene, carbon tetrachloride, and ether.
Derivation: Hydrolysis of dichloroethyl sulfide; interaction of ethylene chlorhydrin and sodium sulfide.
Grades: Technical.
Containers: 1-gal cans; 5- and 55-gal drums; tank cars.
Use: Organic synthesis; solvent for dyes in textile printing; antioxidant.
Caution: Do not use with hydrochloric acid.

thiodiglycolic acid $HOOCCH_2SCH_2COOH$. Colorless crystals; m.p. 128°C. Soluble in water and alcohol. Used as an analytical reagent.

thiodiphenylamine. See phenothiazine.

thiodipropionic acid $S(CH_2CH_2CO_2H)_2$. Used as a food preservative.

thiodipropionic acid, dilauryl ester. See dilauryl thiodipropionate.

thiodipropionic acid, dioctyl ester. See dioctyl thiodipropionate.

thiodipropionic acid, distearyl ester. See distearyl thiodipropionate.

thiodipropionic acid, ditridecyl ester. See ditridecyl thiodipropionate.

beta, beta-**thiodipropionitrile** $S(CH_2CH_2CN)_2$.
Properties: White crystals; sp.gr. (30°C) 1.1095; m.p. 28.65°C; slightly soluble in water and alcohol; soluble in acetone, chloroform and benzene.
Suggested uses: Preservative; selective solvent.

thioethanolamine. See cysteamine.

"Thiofast." [438] Trademark for a thio-indigo deep maroon pigment with a red-violet undertone. Used in paints, printing inks, and plastics.

"Thiofide." [58] Trademark for 2-2'-dithiobis-(benzothiazole) (MBTS).
Properties: Cream powder; sp.gr. 1.50; m.p. 168°C min; petroleum ether extractable 1.5-3.0%. Available in seed form as "Thiofide S."
Use: Rubber accelerator.

Thioflavin T $CH_3C_6H_3N(HCl)SCC_6H_4N(CH_3)_2$.
Properties: A yellow basic dye of the thiazole class. Color Index No. 815. Fluoresces yellow to yellowish-green when excited by ultraviolet.
Derivation: By heating para-toluidine with sulfur in the presence of lead oxide.
Uses: Textile dyeing; fluorescent sign paints; in combination with green or blue pigments to produce brilliant greens; preparation of phosphotungstic pigments.

thiofuran. See thiophene.

1-thioglycerol $CH_2(OH)CH(OH)CH_2SH$.
Properties: Water white; b.p. 118°C at 5 mm; sp.gr. 1.295 (14.4°C). Soluble in water, alcohol and ether.
Uses: Reducing agent for cystine molecule in human hair and wool; for stabilization of acrylonitrile polymers; medicine.

thioglycolic acid (mercaptoacetic acid) $HSCH_2COOH$.
Properties: A colorless liquid or white crystals with a strong, unpleasant odor; sp.gr. 1.325; m.p. −16.5°C; b.p. 123°C (29 mm). Miscible with water, alcohol or ether.
Derivation: Heating chloracetic acid with potassium hydrogen sulfide.
Containers: Carboys; drums.
Uses: Reagent for iron; manufacture of thioglycolates, permanent wave solutions and depilatories; making complex antimony derivatives and esters used as drugs.

2-thiohydantoin $C_3H_4N_2OS$.
Properties: Crystals. M.p. 230°C. Slightly soluble in water; insoluble in alcohols and ethers.
Purity: 99% min.
Uses: Intermediate, pharmaceuticals, rubber accelerators, copper plating brighteners and dyestuffs.

2-thio-4-keto-thiazolidine. See rhodanine.

"Thiokol." [27] Trademark for a line of products including the polymers produced by the chemical reaction between dichlorodiethylformal and an alkali polysulfide. These are distributed as liquid polymers, water-dispersed latices, and dry rubber (crude).
"Thiokol" Liquid Polymers: LP-2, LP-32, LP-3, LP-8. Liquids of varying viscosity capable of being converted to tough resilient rubbers at room temperature, without appreciable shrinkage, by the addition of curing agents.
Properties of cured polymers: Grease-, oil-, solvent-, and water-swell resistant; good electrical resistivity; superior aging characteristics; better than adequate flexibility at temperatures as low as −65°F; good plastic flow under stress; highly impermeable to gases and moisture; resistant to ozone and sunlight; strong adhesiveness to many materials.
Uses: As sealants for fuel cells; sealer adhesives for machine components; for potting and sealing electrical parts; for caulking ship decks and buildings; as the flexibilizing constituents of resin-based adhesives and potting compounds.
"Thiokol" Organic Polysulfide Rubbers: A, FA, ST. Synthetic rubbers made from dihalogenated organic compounds and inorganic polysulfides or combinations and modifications of these.
Properties: Low swell in oils and solvents; excellent resistance to deterioration in oils, greases, solvents; good resistance to ozone, sunlight, corona discharge, and natural aging; more resistant to aromatic blended fuels than natural or other synthetic rubbers; tensile strengths lower than those of natural rubbers, but good resilience, elongation, low-temperature flexibility, abrasion resistance, and resistance to cold flow.
Uses: Paint spray hose; oil suction and discharge hose; extruded and molded goods; printers' rolls; tank linings; cable covers; tubing; gaskets; putties; cements; and other permanent oil- and weather-resistant putties.
"Thiokol" Water Dispersions: MX, WD-6, WD-2. Aqueous dispersions of polysulfide polymers developed for special applications in the protective coating field.
Properties: Stability to chemical and physical action; excellent resistance to oils and solvents (except the chlorinated type) and to aromatic and aliphatic fuels; good resistance to water; strong adhesion to clean metal, wood and concrete surfaces; excellent flexibility at low temperatures; better than average resistance to aging, adequate resistance to dilute acids and alkalies.

thiomalic acid (mercaptosuccinic acid) $HOOCCH(SH)CH_2COOH$.
Properties: White crystals or powder; sulfuric odor; m.p. 149-150°C; soluble in water, alcohol, acetone, and ether; slightly soluble in benzene.
Use: Biochemical research; intermediate; suggested as antidote to heavy metal poisoning; rust inhibitor aid; antidarkening agent for crepe rubber; tackifier for synthetic rubber.

"Thionex." [28] Trademark for tetramethylthiuram monosulfide. $[(CH_3)_2NCS]_2S$.
Properties: Lemon yellow powder or grains; sp.gr. 1.39; m.p. not lower than 110°C.
Containers: 50-lb boxes.
Use: To accelerate and improve the vulcanization of natural and synthetic rubber and latex compounds.

"Thionone." [232] Brand name for a series of sulfur dyestuffs.

thionyl chloride (sulfurous oxychloride; sulfur oxychloride) $SOCl_2$.
Properties: Clear, pale yellow to red liquid; b.p. 78.8°C; sp.gr. 1.638; m.p. −105°C; decomposes in water.
Grades: 93%, 97.5%.
Containers: Glass carboys; drums.
Use: Organic synthesis; preparation of acid chlorides and anhydrides; catalyst.
Warning! Causes burns. Vapor irritating. MCA warning label.
Shipping regulations: Corrosive liquid. White label.*

2-thio-4-oxypyrimidine. See thiouracil.

thiopental sodium (thiopentone soluble) $C_{11}H_{17}N_2O_2SNa$. Sodium 5-ethyl-5-(1-methylbutyl)-2-thiobarbiturate.
Properties: Yellowish-white, hygroscopic powder with a disagreeable odor. Soluble in water and alcohol; insoluble in absolute ether, benzene and ligroin. Solution decomposes on standing. Precipitation occurs on boiling of solution.
Grade: U.S.P. XVI.
Use: Medicine.

thiopentone soluble. See thiopental sodium.

thiophene (thiofuran) CHCHCHCHS.
Properties: Colorless mobile liquid; refractive index (n 20/D) 1.5285; sp.gr. 1.0644 (20°/4°C); f.p. −38.5°C; b.p. 84°C. Soluble in alcohol and ether; insoluble in water.
Derivation: Found in coal tar (benzene fraction) and petroleum; synthetically from heating sodium succinate with phosphorus trisulfide.
Use: Organic synthesis.

alpha-thiophenealdehyde C_4H_3SCHO. 2-Thiophenecarboxaldehyde.
Properties: Oily liquid with almond-like odor; b.p. 198°C (760 mm), 90°C (20 mm); sp. gr. 1.210-1.220; very soluble in alcohol, benzene, ether; slightly soluble in water.
Grades: 95%.
Containers: Drums.
Uses: For thiophene derivatives; for introducing thenyl group into organic compounds.

thiophenol C_6H_5SH.
Properties: Water white liquid, repulsive odor; b.p. (760 mm) 168.3°C; b.p. (15 mm) 71°C; refractive index 1.5891; sp.gr. (25/25°C) 1.075; insoluble in water; soluble in alcohol and ether.
Derivation: Reduction of benzenesulfonyl chloride with zinc dust in sulfuric acid.
Grades: 99%.
Uses: Pharmaceutical; synthesis.

"Thiophos." [57] Trademark for parathion technical.

thiophosgene (thiocarbonyl chloride) $CSCl_2$. A reddish liquid; sp.gr. 1.5085 (15°C); b.p. 73.5°C. Decomposes in water and alcohol. Poisonous!
Use: Organic synthesis.
Shipping regulations: Poison, class B. Poison label.*

thiophosphoric anhydride. See phosphorus pentasulfide.

thiophosphorous anhydride. See phosphorus trisulfide.

thiophosphoryl chloride $PSCl_3$.
Properties: Colorless liquid; sp.gr. 1.68; b.p. 126°C; m.p. −35°C; decomposed by water; soluble in carbon disulfide, carbon tetrachloride.
Shipping regulations: Corrosive liquid. White label.*

thiopropazate hydrochloride $C_{23}H_{28}ClN_3O_2S\cdot 2HCl$. 2-Chloro-10-{3-[1-(2-acetoxyethyl)-4-piperazinyl] propyl}-phenothiazine dihydrochloride.
Properties: Crystals, m.p. 223° (decomposes). Soluble in water; slightly soluble in organic solvents.
Grades: N.N.D.
Use: Medicine.

thiopropionic acid, dioctyl ester. See 3,3'-(2-ethylhexyl) thiodipropionate.

thioridazine hydrochloride $C_{21}H_{26}N_2S_2$. 2-Methylthio-10-[2-(1-methyl-2-piperidyl)-ethyl]phenothiazine hydrochloride.
Properties: Colorless crystals; m.p. 158°C; soluble in water and alcohol.
Use: Medicine.

"Thiosafast." [438] Similar to "Thiofast." Trademark for a thio-indigo deep maroon pigment with a red-violet undertone. Used in paints, printing inks, and plastics.

thiosalicylic acid (2-mercaptobenzoic acid; ortho-sulfhydrylbenzoic acid) $HOOC(C_6H_4)SH$.
Properties: Yellow solid; m.p. 164-165°C; sublimes; slightly soluble in hot water; soluble in alcohol, ether and acetic acid.
Grades: Reagent, technical, 80%.
Containers: Drums.
Uses: Dyes; reagent for iron determination.

thiosemicarbazide (aminothiourea) $NH_2CSNHNH_2$.

*See "I.C.C. Shipping Regulations," page xiii.
Reference numbers refer to name of manufacturer. See "List of Manufacturers," page v.

Properties: Melting point 180-184°C; white crystalline powder, no odor, soluble in water and alcohol.
Derivation: From potassium thiocyanate and hydrazine salts.
Grades: Technical and pure.
Containers: 1-, 5-lb bottles; 25-, 100-lb drums.
Uses: Reagent for ketones and certain metals; photographic; rodenticide relatively non-toxic to humans.

thiosemicarbazide hydrochloride $NH_2CSNHNH_2 \cdot HCl$.
Properties: White crystalline powder; m.p. 184°C (min); no odor; soluble in water.
Grades: Pure.
Uses: See thiosemicarbazide.

thiosemicarbazone (para-acetylaminobenzaldehyde thiosemicarbazone) $CH_3CONHC_6H_4CH{:}NNHCSNH_2$.
Properties: Yellow solid; decomposes at 230°C; m.p. 207°C; insoluble in water and organic solvents except glycols.
Use: Medicine.

thiosinamine. See allyl thiourea.

"Thiostop K." [248] Trade name for a 50% aqueous solution of potassium dimethyl dithiocarbamate.
Properties: Clear yellow to amber liquid. Sp.gr. 1.23; good storage stability. The salt will start to crystallize out at 20°F. Avoid long storage in partially filled containers.
Uses: Used as a shortstop in SBR latices to arrest the catalytic action which induces polymerization of monomer to polymer.

"Thiostop N." [248] Trade name for a 40% aqueous solution of sodium dimethyl dithiocarbamate.
Properties: Clear yellow to amber liquid; sp.gr. 1.18; good storage stability; salt will start to crystallize out at 32°F. Avoid long storage in partially filled containers.
Uses: Used as a shortstop in SBR polymerization to arrest the catalytic action which induces polymerization of monomer to polymer.

thiostrepton. An antibiotic.
Properties: White to off-white powder; m.p. 246-256°C. Insoluble in water and lower alcohols; soluble in chloroform, dioxane, glacial acetic acid and dimethylformamide.
Derivation: Produced by fermentation with Streptomyces azureus.
Use: Medicine.

"Thiotax." [58] Trade name for 2-mercaptobenzothiazole.

Thio-TEPA. See triethylenethiophosphoramide.

thiouracil (2-thio-4-oxypyrimidine; 2-mercapto-4-hydroxypyrimidine) OCNHC(S)NHCHCH.
Properties: Bulky, white, odorless, crystalline powder of intense bitter taste; melts with decomposition at about 340°C; has a characteristic ultraviolet absorption spectrum $A_{260} = 11.0 \times 10^3$ at 260 millimicrons (pH 7.0); slightly soluble in water, insoluble in alcohol.
Use: Experimental work in nutrition and physiology; medicine.

thiourea (thiocarbamide) $(NH_2)_2CS$.
Properties: White, lustrous crystals; bitter taste; sp.gr. 1.406; m.p. 180-182°C; b.p. sublimes in vacuo at 150-160°C; soluble in cold water, ammonium thiocyanate solution, and alcohol; nearly insoluble in ether.
Derivation: By heating dry ammonium thiocyanate, extraction with a concentrated solution of ammonium thiocyanate with subsequent crystallization.
Method of purification: Sublimation in vacuo.
Grades: Technical; reagent.
Containers: Bottles; bags.
Uses: Photography; organic synthesis (intermediates, dyes, drugs); rubber accelerator; analytical reagent; medicine (external).

thiourea resins. A type of amino resins (q.v.).

"Thiovanic Acid." [312] Trademark for thioglycolic acid (q.v.).

"Thiovanol." [312] Trademark for vacuum-distilled alpha-monothioglycerol. See 1-thioglycerol.

thiram. See tetramethylthiuram disulfide.

"Thiram-75W." [248] Trade name for a tetramethylthiuram disulfide formulation especially prepared for slurry seed treating to reduce seed decay and "damping-off" of beans, corn, peas, and certain other vegetable and field crops. Also available as "Thiram 50 Dust" for dry seed treatment and "Thiram Technical" for manufacturing.
Hazards: May cause skin irritation. Do not inhale dust. Do not get in eyes or on skin. Wash thoroughly after handling.

thistle saffron. See carthamus.

"Thiurad." [58] Trademark for tetramethylthiuram disulfide.

thiuram.
1. In general, the radical R_2NCS. The most important derivatives are the disulfides $[R_2NCSS]_2$ or $(R_2NCS)_2S_2$. Some of them are tetraethylthiuram disulfide, tetramethylthiuram disulfide, tetrabutylthiuram disulfide, tetraisopropylthiuram disulfide and tetrapropylthiuram disulfide (q.v.).
2. Specifically, the tetraethyl- and tetramethylthiuram disulfides, which are often referred to by the name thiuram alone.

Use: To accelerate and improve the vulcanization of synthetic and natural rubber compounds; as insecticides and fungicides.

Thiuram E. [28] Trade name for tetraethylthiuram disulfide.

Thiuram M. [28] Trade name for tetramethylthiuram disulfide.

"Thixcin." [202] Trademark for a fine, white nondiscoloring, non-toxic powder used as multipurpose thixotropic agent in paints, inks, calk compounds, plastisols, non-drip lubricants, greases and similar compositions.

thixotropic paint. See gel paint.

thixotropy. The property enabling certain colloidal gels to liquefy when agitated (as by shaking or ultrasonic sound) and then to return to the jelly-like form when at rest. This is often observed in paints or printing inks which flow freely only when force (brushing or rolling pressure) is applied.

Thomas & Gilchrist process for steel. A basic Bessemer process used to remove phosphorus from pig iron, involving the use of a basic liner of dolomite brick for the converter and the addition of lump lime during the process.

Thomas balsam. See tolu balsam.

Thomas metal. See basic slag.

Thomas phosphate. See basic slag.

Thomas slag. See basic slag.

thomsonite $NaCa_2Al_5(SiO_4)_5 \cdot 6H_2O$. A mineral; one of the zeolites (q.v.).
Properties: White, reddish or green; streak uncolored; vitreous luster; sp.gr. 2.3-2.4; hardness 5-5.5.
Occurrence: Europe, Iceland, United States, Nova Scotia.

thonzylamine hydrochloride $C_{16}H_{22}N_4O \cdot HCl$; 2-[(2-dimethylaminoethyl)(para-methoxybenzyl)amino]pyrimidine hydrochloride. $CH_3OC_6H_4CH_2N(C_4H_3N_2)CH_2CH_2N(CH_3)_2 \cdot HCl$.
Properties: White, crystalline powder with faint odor. M.p. 173-176°C. Very soluble in water; freely soluble in alcohol and chloroform; practically insoluble in ether and benzene; pH (2% solution) 5.0-6.0.
Grade: N.F. XI.
Containers: Bottles (syrup form); also tablet form.
Use: Medicine.

"Thor." [65] Trademark for phenolic or urea resins in powder or liquid form.
Uses: Foundry sand for foundry applications, core binders, shell mold resins, core pastes, shell pastes and parting adhesives.

"Thorazine." [71] Trademark for a brand of chlorpromazine (q.v.).

thoria. See thorium dioxide.

thorin. See ortho-(2-hydroxy-3, 6-disulfo-1-naphthylazo)-benzenearsonic acid.

"Thorite." [195] Trade name for a nonshrink grouting compound used as a patching mortar to fill holes and blisters in masonry surfaces and to prevent further destruction of steel reinforcing.

thorite $ThSiO_4$. A natural thorium silicate, usually impure, found in pegmatites.
Properties: Color black to orange; luster vitreous to resinous; hardness 4.5-5; sp.gr. 4.4-5.2; radioactive.
Occurrence: Norway, Ceylon.
Use: Source of thorium.

thorium Th. Metallic element having atomic number 90; classified as a member of the actinide series.
Properties: Soft metal with bright silvery luster when freshly cut; similar to lead in hardness when pure. Sp.gr. about 11.7; m.p. about 1520°C; b.p. about 4500°C. Soluble in acids; insoluble in alkalies and water. Radioactive. Also encountered in the powdered form (see under derivation).
Source: Monazite, thorite.
Derivation: Reduction of thorium dioxide with calcium; fused salt electrolysis of the double fluoride $ThF_4 \cdot KF$. The product of either process is thorium powder. The powder is fabricated into the metal by powder-metallurgy techniques. Hot-surface decomposition of the iodide produces crystal-bar thorium.
Forms available: Powder, unsintered bars, sintered bars, sheets.
Uses: In special lamps such as sun lamps; in magnesium alloys. The radioactive 228 isotope can be used in dissipating static electricity. Thorium is a source of nuclear energy, the naturally occurring thorium 232 being converted to the nuclear fuel uranium 233 when bombarded by neutrons.
Shipping regulations: (For powdered thorium) Flammable solid. Yellow label.*

thorium acetate $Th(OH)_2(OOCCH_3)_2 \cdot H_2O$.
Properties: Soluble in formic acid; insoluble in water.
Grades: Technical.

thorium acetylacetonate $Th(C_5H_7O_2)_4$. Crystalline powder. Slightly soluble in water. Resistant to hydrolysis. A chelating, nonionizing compound.

thorium anhydride. See thorium dioxide.

thorium chloride (thorium tetrachloride) $ThCl_4$.
Properties: Colorless or white, lustrous needles (light-yellow color caused by iron trace). Hygroscopic. Partially volatile. Crystallizes with variable water of crystallization. Soluble in alcohol, water. Sp.gr. 4.59; b.p. 1100°C (dec); m.p. 820°C.
Grades: Technical; as 50% ThO_2.
Containers: Glass bottles; fiber drums.
Use: Incandescent lighting.

thorium decay series. The series of radioactive elements produced as successive intermediate products when the element thorium undergoes its spontaneous natural radioactive disintegration into lead.

thorium dioxide (thorium anhydride; thorium oxide; thoria) ThO_2.
Properties: Heavy, white powder; sp.gr. 9.7; m.p. about 3300°C; b.p. 4400°C; hardness 6.5 (Mohs). Very refractory. Soluble in sulfuric acid; insoluble in water.
Derivation: By the reduction of thorium nitrate.
Grades: Technical and purities to 99.8% ThO_2; granular particles, crystals.
Containers: Lined drums.
Uses: Ceramics (high temperature); nuclear fuel; flame spraying; crucibles; medicine; non-silica optical glass; catalyst; thoriated tungsten; electronics.

thorium fluoride ThF_4.
Properties: White powder having approximate formula $ThF_4 \cdot 1.4H_2O$. Dehydrated between 200 and 300°C; m.p. 1111°C. Above 500°C reacts with atmospheric moisture to form thorium oxyfluoride, $ThOF_2$, and finally the oxide ThO_2. Forms a series of compounds with other metallic fluorides such as NaF and KF.
Uses: For production of thorium metal and magnesium-thorium alloys; used in high temperature ceramics. $ThOF_2$ is used as a protective coating on reflective surfaces.
Grades: 79-80% ThO_2.
Containers: Glass bottles; fiber drums.

thorium nitrate $Th(NO_3)_4 \cdot 4H_2O$.
Properties: White, crystalline mass. Soluble in water and alcohol.
Derivation: By extraction from monazite sand.
Method of purification: Crystallization.
Grades: Technical; C.P.
Containers: Glass bottles; fiber drums.
Uses: Medicine; reagent for determination of fluorine; thoriated tungsten.
Caution: Oxidizing material, in contact with organic or other readily oxidizable (combustible) substances it will cause violent combustion on ignition.
Shipping regulations: Oxidizing material. Yellow label.*

thorium ore. See monazite.

thorium oxalate $Th(C_2O_4)_2 \cdot 2H_2O$.
Properties: White powder, insoluble in water and most acids. Soluble in solutions of alkali and of ammonium oxalates. Above 300-400°C decomposes to thorium oxide, ThO_2.
Grades: Purities to 99.9%; as 59% ThO_2.
Containers: Glass bottles; fiber drums.
Use: Ceramics.

thorium oxide. See thorium dioxide.

thorium sulfate (thorium sulfate, normal) $Th(SO_4)_2 \cdot 8H_2O$.
Properties: White, crystalline powder; sp. gr. 2.8; m.p. loses water at 400°C. Slightly soluble in water; soluble in ice water.
Grades: As 43% ThO_2.
Containers: Glass bottles; fiber drums.

thorium sulfate, normal. See thorium sulfate.

thorium tetrachloride. See thorium chloride.

thorn apple. See stramonium.

"Thoroclear." [195] Trade name for a clear, water-repellent silicone resin for porous brick, stone, concrete stucco, asbestos siding, plaster and for all masonry surfaces.

thorogummite. A natural hydrated silicate of uranium and thorium approximately $UO_3 \cdot ThO_2 \cdot 3SiO_3 \cdot 6H_2O$.
Occurrence: Texas.

thoron. An isotope of radon (q.v.).

"Thoroseal." [195] Trade name for a cementious product used to fill and seal pores and voids in masonry surfaces.

"Thoroset." [195] Trade name for a metallic nonshrink grouting compound for hardening mortars, grouting and bedding, accelerates strength and set.

thortveitite. An ore containing 37-42% scandium oxide. A basic material in production of scandium.

"THPC." [306] Trademark for tetrakis (hydroxymethyl) phosphonium chloride (q.v.).

"Thram." [342] Trademark for bird-repellent composition.

threonine (alpha-amino-beta-hydroxybutyric acid) $CH_3CH(OH)CH(NH_2)COOH$. An essential amino acid.
Properties: Colorless crystals, soluble in water; optically active.
DL-threonine, m.p. 228-229°C with decomposition;
L(−)-threonine (natural occurring), m.p. 255-257°C with decomposition;
DL-allo-threonine, m.p. 250-252°C.
Derivation: Hydrolysis of protein (casein); organic synthesis.
Containers: Bottles; drums.
Uses: Nutrition and biochemical studies.
Commercially available as D-, DL-, L-, and DL-allo-threonine.

throwing oils. Oils applied to prepare raw silk and filament rayon for "throwing," the operation by which strands are twisted and wound into proper size threads. Applied by a bath, the oils improve the strength, elasticity and suppleness of the yarns, properly conditioning them for subsequent weaving or knitting operations. The oils are usually compounded to be self-emulsifying and may contain a sizing agent such as dextrin, gelatin, etc.

throwing power. A term denoting the effectiveness of an electrolytic cell for depositing metal at the cathode. The throwing power is the weight of deposition per unit distance between the electrodes.

thuja (arbor vitae; white cedar; tree of life). Dried leafy twigs of Thuja occidentalis.
Occurrence: Northern and central United States.
Uses: Source of thuja oil; medicine.

thuja oil (arbor vitae oil). Sometimes called cedar leaf oil, but different from the cedar leaf oil listed in this book.
Properties: Pale yellow liquid; characteristic, rather agreeable odor; sp.gr. 0.910-0.920; refractive index (n 20/D) 1.459; optical rotation −10° to −13° in 100 mm tube. Soluble in alcohol, ether, chloroform and carbon disulfide.
Chief known constituents: Dextro-pinene; levo-fenchone; thujone; should contain not less than 60% ketones calculated as thujone ($C_{10}H_{16}O$).

*See "I.C.C. Shipping Regulations," page xiii.
Reference numbers refer to name of manufacturer. See "List of Manufacturers," page v.

Derivation: Distilled from the leaves of the white cedar, Thuja occidentalis.
Method of purification: Rectification.
Grades: Technical.
Containers: Glass bottles.
Uses: Medicine; perfumery.
Shipping regulations: None.*

thujone $C_{10}H_{16}O$. A terpene-type ketone contained in thuja oil and the oils of sage, tansy and wormwood.
Properties: Colorless liquid; sp.gr. 0.915-0.919 (20°/20°C); b.p. 203°C. Insoluble in water; soluble in alcohol.

thulia. See thulium oxide.

thulite. See zoisite.

thulium Tm. Atomic number 69; Group III of the periodic table; one of the rare-earth elements of the yttrium sub-group.
Properties: Metallic luster; reacts slowly with water; soluble in dilute acids; salts colored green; sp.gr. 9.346; m.p. 1550-1600°C; b.p. (approx) 2100°C.
Derivation: For source see rare-earth minerals. Isolated by reduction of the fluoride with calcium. See rare-earth metals.
Grades: Regular high purity (ingots, lumps).

thulium 170. Radioactive thulium of mass number 170. Thulium 170 is used as the x-ray source in portable x-ray units.
Shipping regulations: Poison, class D, radioactive material. Red or blue label.*

thulium oxalate $Tm_2(C_2O_4)_3 \cdot 6H_2O$.
Properties: Greenish-white precipitate. Soluble in aqueous alkali oxalates.
Derivation: Precipitation from a solution containing a thulium salt and a mineral acid by addition of oxalic acid.
Uses: For analytical separation of thulium (and other rare-earth metals) from the common metals.
Shipping regulations: None.*

thulium oxide (thulia) Tm_2O_3.
Properties: Dense white powder with greenish tinge; slightly hygroscopic; absorbs water and CO_2 from the air; sp.gr. 8.6. Exhibits a reddish incandescence on heating, changing to yellow and then white on prolonged heating. Slowly soluble in strong acids.
Derivation: By ignition of thulium oxalate, salt of other oxyacids or hydroxide.
Grades: 99-99.9%.
Containers: Glass bottles; fiber drums.
Shipping regulations: None.*
See also rare earths.

thulium salts. Other thulium salts are available in 99% or 99.9% purity; in glass bottles or fiber drums. They include:
thulium chloride, $TmCl_3 \cdot xH_2O$.
Available as 45% Tm_2O_3.
thulium fluoride, $TmF_3 \cdot xH_2O$.
Available as 77% Tm_2O_3.
thulium nitrate, $Tm(NO_3)_3 \cdot 6H_2O$.
Available as 42% Tm_2O_3.
thulium sulfate, $Tm_2(SO_4)_3 \cdot 8H_2O$.
Available as 47% Tm_2O_3.

"Thuricide." A microbial insecticide inducing disease in certain insects, but apparently harmless to plants, fish, and warmblooded animals. It is the first microbial insecticide to get full tolerance exemption from FDA. An example of a "biological" poison, "Thuricide" consists of the spores of Bacillus thuringiensis Berliner.

thus, American. See turpentine.

thus gum. See olibanum, or turpentine.

"Thylate." [28] Trademark for a wettable off-white powder containing thiram (tetramethylthiuramdisulfide) used to control scab and other apple diseases.
Containers: 5- and 50-lb bags.

Thylox process. A process whereby hydrogen sulfide and organically combined sulfur are absorbed from gases by a solution of arsenious oxide and soda ash (or sodium thioarsenate) in water.

thyme.
Deviation: The dried leaves and flowering tops of Thymus vulgaris.
Constituents: Volatile oil, tannin, gum.
Occurrence: Southern Europe; cultivated in England and United States.
Grades: Technical; French; Spanish.
Containers: Bags; bales.
Uses: Source of thyme oil; flavoring; medicine.
Shipping regulations: None.*

thyme camphor. See thymol.

thyme oil. Essential oil distilled from flowering plant Thymus vulgaris or Thymus zygis and its variety gracilis.
Properties: A colorless to reddish brown liquid with a pleasant odor and sharp taste; sp.gr. 0.910-0.935 (25/25°C); refractive index 1.4950-1.5050 (20°C); very slightly soluble in water; soluble in alcohol.
Chief constituents: Not less than 40% by volume of phenols, and including thymol, carvacrol, cymene, pinene, linolool, and bornyl acetate.
Grades: N.F. XI; red; white.
Containers: Cans; drums.
Use: Medicine, perfumery, cosmetics, toilet soaps, flavoring.

thyme oil, Cyprian. This is an origanum oil (q.v.).

thymic acid. See thymol.

thymidine (thymine-2-desoxyriboside) $C_{10}H_{14}N_2O_5$. The nucleoside (deoxyriboside) of thymine. It is found in deoxyribonucleic acid.
Properties: Crystalline needles; m.p. 185°C; dextrorotatory in solution; soluble in water, methanol, hot alcohol, hot acetone and hot ethyl acetate; sparingly soluble in hot chloroform; soluble in pyridine and glacial acetic acid. $A_m = 9.65 \times 10^3$ at 267 millimicrons and pH 7.2.
Use: Biochemical research.
Also available as trityl thymidine, and as tritiated thymidine in a radioactive form.

thymidylic acid. The nucleotide of thymine; i.e. the phosphate ester of thymidine.

thymine (5-methyluracil) $CH_3CC(O)NHC(O)NHCH$. 5-Methyl-2,4-dioxypyrimidine.
Properties: White crystalline powder; decomposes at 335-337°C; slightly soluble in hot water; insoluble in cold water, alcohol; sparingly soluble in ether; readily soluble in alkalies. $A_M = 7.89 \times 10^3$ at 264.5 millimicrons and pH 7.0.
Derivation: Hydrolysis of deoxyribonucleic acids; from methylcyanacetylurea by catalytic reduction.
Use: Biochemical research.

thymine-2-desoxyriboside. See thymidine.

thyminic acid (nucleotinphosphoric acid) $C_{30}H_{46}N_4O_{15} \cdot 2P_2O_5$.
Properties: Yellow amorphous powder. Soluble in water.
Use: Medicine.
Shipping regulations: None.*

thymol (isopropyl-meta-cresol; 5-methyl-2-isopropylphenol; 3-hydroxy-para-cymene; thyme camphor; thymic acid) $(CH_3)_2CHC_6H_3(CH_3)OH$.
Properties: White crystals with aromatic odor and taste. Soluble in alcohol, carbon disulfide, chloroform, glacial acetic acid, ether and fixed or volatile oils; slightly soluble in water and glycerol. Sp.gr. 0.979; m.p. 48-51°C; b.p. 233°C.
Derivation: From thyme oil or other oils; synthetically from meta-cresol and isopropyl chloride by the Friedel-Crafts method at −10°C.
Grades: Technical; N.F. XI; reagent.
Containers: Bottles; tins; 25-, 50-, 100-lb drums.
Uses: Medicine; perfumery; thymol compounds; microscopy; preservative; embalming; antioxidant.
Shipping regulations: None.*

thymol blue (thymolsulfonephthalein) $C_{27}H_{30}O_5S$.
Properties: Brown green powder or crystals; insoluble in water, soluble in alcohol or dilute alkali.
Use: Acid base indicator. See indicators.

thymol iodide. Principally the dithymoldiiodide, $[C_6H_2(CH_3)(OI)(C_3H_7)]_2$.
Properties: Red-brown powder or crystals, slight aromatic odor; affected by light. Soluble in ether, chloroform and fixed or volatile oils; slightly soluble in alcohol, insoluble in water.
Derivation: By the interaction of thymol and potassium iodide in alkaline solution.
Grades: Technical.
Containers: 1-lb bottles; 5-, 25-lb boxes; drums.
Use: Medicine; feed additive.
Shipping regulations: None.*

thymolphthalein $(C_{10}H_{13}O)_2CC_6H_4COO$.
Properties: White powder; m.p. 245°C; insoluble in water, soluble in alcohol and acetone and in dilute alkali and acids.
Use: Used in medicine and as acid base indicator in pH range 9.3 (colorless) to 10.5 (blue). See indicators.

thymolsulfonephthalein. See thymol blue.

para-thymoquinone (2-isopropyl-5-methylbenzoquinone) $C_{10}H_{12}O_2$.
Properties: Bright yellow crystals; penetrating odor. M.p. 45.5°C; b.p. 232°C. Slightly soluble in water; soluble in alcohol and ether.
Derivation: From diazonium salt of aminothymol and nitrous acid.
Use: Fungicide.

thymus nucleic acid. See deoxyribonucleic acid.

"Thyrite." 245 Trademark for composite dielectric or resistance material in molded form applied primarily to silicon carbide type resistance material having a negative resistance-voltage characteristic.

thyroid.
Properties: Yellow, amorphous powder with slight, characteristic odor, saline taste.
Derivation: Dried, cleansed, powdered thyroid gland obtained from domesticated animals.
Grade: U.S.P. XVI.
Use: Medicine.

thyroid-stimulating hormone. See thyrotropic hormone.

thyronine (desiodothyroxine) $HOC_6H_4OC_6H_4CH_2CH(NH_2)COOH$. The para-hydroxyphenyl ether of tyrosine. Thyronine and its iodinated derivatives are used for physiological and biochemical research regarding the thyroid gland and its activity.

thyrotropic hormone (TSH; thyroid-stimulating hormone; thyrotropin). One of the hormones secreted by the anterior lobe of the pituitary gland. It increases the rate of the removal of iodine from the blood by the thyroid gland, of the synthesis of the thyroid hormone, and of the release of thyroid hormone into the bloodstream. The thyrotropic hormone has not been completely purified but is known to be a protein which has a low molecular weight (approximately 10,000) and which contains some carbohydrate.

thyrotropin. See thyrotropic hormone.

thyroxine $C_{15}H_{11}I_4NO_4$ or $HOC_6H_2I_2OC_6H_2I_2CH_2CH(NH_2)COOH$. 3,5,3',5'-Tetraiodothyronine. The hormone produced by the thyroid gland. (See also triiodothyronine). It is an amino acid and a derivative of tyrosine. It increases the metabolic rate and oxygen consumption of animal tissues.
Properties: Optically active; the L-isomer is the natural and physiologically active form.
DL-thyroxine: Needles; decompose 231-233°C; insoluble in water, alcohol, and the common organic solvents; soluble in alcohol in the presence of mineral acids or alkalies.
L-thyroxine: Crystals; decompose 235-

236°C.
D-thyroxine: Crystals; decompose 237°C.
Derivation: Obtained from the thyroid glands of animals; organic synthesis.
Uses: Medicine; physiological research.

Ti. Symbol for titanium.

TIBAL. Abbreviation for triisobutylaluminum.

"Ti-Cal." [28] Trademark for titanium-calcium pigment containing approximately 30% titanium dioxide (TiO_2) and 70% calcium sulfate ($CaSO_4$).
Properties: Fine, white powder; sp.gr. 3.25.
Containers: 50-lb bags.
Uses: In flat paints, architectural finishes, gloss paints and enamels, fade-resistant tinted exterior finishes, paper, and rubber.

"Ticon." [337] Trade name for a line of metallic stannates, titanates and zirconates chiefly used as additives in ferroelectric and piezoelectric devices; also used for capacitors, and in some cases for electroluminescent devices. The group includes the following:
"Ticon B": Barium titanate, $BaTiO_3$. Insoluble in water and alkalies; soluble in acids; high dielectric constant; K approx. 1200-1600. Used in electronics industries; as electroluminescent devices.
"Ticon Bi": Bismuth titanate, $Bi_2Ti_2O_7$. Insoluble in water; dielectric K approx. 100-115, P.F. = 0.5%. Used as piezoelectric ceramics and capacitors.
"Ticon C": Calcium titanate, $CaTiO_3$. Insoluble in water and alkalies; soluble in acids; dielectric K approx. 150, P.F. = 0.05%.
"Ticon Ce": Cerium titanate, $CeTiO_4$. Insoluble in water; dielectric properties.
"Ticon Co": Cobalt titanate, $CoTiO_3$. Insoluble in water; dielectric K approx. 19-20, P.F. = 0.2%.
"Ticon Cu": Copper titanate, $CuTiO_3$. Insoluble in water; dielectric.
"Ticon M": Magnesium titanate, $MgTiO_3$. Insoluble in water and alkalies; soluble in acids; dielectric; K approx. 15, P.F. = 0.1%.
"Ticon N": Nickel titanate, $NiTiO_3$. Insoluble in water; dielectric K approx. 15, P.F. = 1.6%.
"Ticon P": Lead titanate, $PbTiO_3$. Insoluble in water; dielectric.
"Ticon S": Strontium titanate, $SrTiO_3$. Insoluble in water and alkalies; soluble in acids; dielectric K = 225-250, P.F. = 0.06%.
"Ticon T": Titanium oxide, TiO_2. Insoluble in water and alkalies; slightly soluble in dilute acids; soluble in hot concentrated H_2SO_4; dielectric. Used as capacitors and as a thread guide body.
"Ticon ZB": Zinc titanate, Zn_2TiO_4. Insoluble in water; dielectric.
"Ticon BT": Barium stannate, $BaSnO_3$. Insoluble in water and alkalies; soluble in dilute acids; dielectric.
"Ticon CT: Calcium stannate, $CaSnO_3$. Insoluble in water and alkalies; soluble in dilute acids; dielectric.
"Ticon CeT": Cerium stannate, $CeSnO_4$. Insoluble in water; dielectric.
"Ticon MT": Magnesium stannate, $MgSnO_3$. Insoluble in water and alkalies; soluble in dilute acids; dielectric.
"Ticon PT": Lead stannate, $PbSnO_3$. Insoluble in water; dielectric.
"Ticon ST": Strontium stannate, $SrSnO_3$. Insoluble in water and alkalies; soluble in dilute acids; dielectric.
"Ticon BZ": Barium zirconate, $BaZrO_3$. Insoluble in water and alkalies; slightly soluble in dilute acids and hot concentrated H_2SO_4; soluble in HF. Used as a filler in plastics and resins.
"Ticon BiZ": Bismuth zirconate, $Bi_4Zr_3O_{12}$. Insoluble in water; dielectric.
"Ticon CZ": Calcium zirconate, $CaZrO_3$. Insoluble in water and alkalies; slightly soluble in dilute acids and hot concentrated H_2SO_4; soluble in HF; dielectric.
"Ticon CdZ": Cadmium zirconate, $CdZrO_3$. Insoluble in water; dielectric.
"Ticon CeZ": Cerium zirconate, $CeZrO_4$. Insoluble in water; dielectric.
"Ticon MZ": Magnesium zirconate, $MgZrO_3$. Insoluble in water; dielectric.
"Ticon PZ": Lead zirconate, $PbZrO_3$. Insoluble in water.
"Ticon SZ": Strontium zirconate, $SrZrO_3$. Insoluble in water and alkalies; slightly soluble in dilute acids and hot concentrated H_2SO_4; soluble in HF; dielectric.
"Ticon ZZ": Zinc zirconate, $ZnZrO_3$. Insoluble in water; dielectric.

tiff. See barite.

"Tigan" Hydrochloride. [190] Trademark for a brand of trimethobenzamide hydrochloride (q.v.).

tiger's eye. A pseudomorph of quartz (q.v.) or silicified crocidolite (q.v.).
Properties: Golden or tawny yellow; luster, silky.
Use: Ornamental stone.

tiglic acid (methylcrotonic acid; crotonolic acid; trans-2-methylbutenoic acid; alpha, beta-dimethylacrylic acid) $CH_3CH:C(CH_3)COOH$. The trans isomer of angelic acid.
Properties: Thick, syrupy liquid or colorless crystals; spicy odor; very poisonous! Soluble in alcohol and ether; slightly soluble in water. Sp.gr. 0.9641; m.p. 65°C; b.p. 198.5°C.
Derivation: Obtained from croton oil, which is extracted from seeds of Croton tiglium. Also occurs in Roman chamomile oil.
Method of purification: Rectification.
Grades: Technical.
Containers: Glass bottles.
Use: Medicine.
Shipping regulations: None.*

tiglium (croton; purging croton; molucca grains; grana tilli).
Properties: Ovoid seed, reddish-brown when fresh, turning grayish-brown with age.
Chief constituents: Croton oil; tiglic acid; crotonol.

Derivation: The seed of Croton tiglium.
Occurrence: East Indies and Philippines.
Grades: Technical.
Containers: Bags; boxes.
Uses: Medicine; source of croton oil.
Shipping regulations: None.*

tiglium oil. See croton oil.

til oil. See sesame oil.

timbo root. See cube root.

tin (stannum) Sn. Element of atomic number 50, member of group IV of the periodic system.
Properties: Silver-white, ductile metal; density (20°C) 7.29; m.p. 232°C; b.p. 2260°C; changes to brittle grey (alpha) tin at temperatures below 18°C but the transition is normally very slow. Soluble in acids, and hot potassium hydroxide solution; insoluble in water.
Derivation: By roasting the ore (cassiterite) to oxidize sulfates and to remove arsine, then reducing with coal in a reverberatory furnace, or by smelting in an electric furnace. Also recovered from tin plate.
Occurrence: The tin of commerce comes chiefly from Malaya (called Straits tin), Bolivia and Nigeria.
Grades: Tin is available in 6 grades, A, B, C, D, E, F (99.8% or higher to below 99%). Block tin is a common designation for pure tin.
Forms: Anodes (for plating); wire; tape; pipe; sheet; bar; ingot; pig form.
Uses: Tin plate; terneplate; solder; Babbitt metal; brass and bronze; foil; collapsible tubes; tinning and retinning; white, type and casting metal; manufacture of chemicals; tinned wire (all copper wire which is to be rubber covered). Block tin is used in coating copper vessels for culinary purposes. It is also used for coating lead sheet or lining lead pipe for distilled water and some chemicals.
Shipping regulations: None.*
Note: In speaking of fabricated articles "tin" is often incorrectly used when tinplate (thin sheets of steel coated with tin) is meant; e.g. "a tin can." To distinguish, articles (such as condenser coils) actually made of solid tin are said to be made of "block tin." See also block tin lining.

tin alloys. See fusible tin alloys in the tables under fusible alloys.

tin anhydride. See stannic oxide.

tin ash. See stannic oxide.

tin base Babbitt. A bearing metal with 3-8% copper, 7.5-8.3% antimony and balance tin, with or without lead. Soft, with relatively low load strength but good antiseize properties and corrosion resistance. See Babbitt metal.

tin bisulfide. See stannic sulfide.

tin bromide. See stannic bromide; and stannous bromide.

tin bronze. See stannic sulfide.

tin, butter of. See stannic chloride.

tincal. See borax.

tin chloride. See stannic chloride.

tin chromate. See stannic chromate and stannous chromate.

tin crystals. See stannous chloride.

tincture. An alcoholic or water-alcoholic solution of an animal or vegetable drug or a chemical substance. The tincture of potent drugs is essentially a 10% solution. Tinctures are more dilute than fluid extracts and less volatile than spirits (q.v.).

tincture of opium. See laudanum.

tinder. See agaric.

tin dibromide. See stannous bromide.

tin dichloride. See stannous chloride.

tin difluoride. See stannous fluoride.

tin dioxide. See stannic oxide.

tin disulfide. See stannic sulfide.

tin, flowers of. See stannic oxide.

tin fluoride. See stannous fluoride.

tin iodide. See stannic iodide.

tin monosulfide. See stannous sulfide.

tin naphthenate. See soaps, metallic.

tin octoate $Sn(C_7H_{15}COO)_2$.
Properties: Pale amber viscous liquid; soluble in aliphatic and aromatic hydrocarbon solvents.
Uses: Catalyst; antioxidant, stabilizer for transformer oils.

"Tinopal." [219] Trademark for a group of optical brighteners, which absorb ultraviolet light in the near visible range and remit the energy as visible light.
Uses: In heavy duty detergents, and detergent specialties to whiten fabrics; starch products; plastics (for example, molded grade nylon); soap bars as product brighteners.

tin ore. See cassiterite.

tin oxalate. See stannous oxalate.

tin oxides. See stannic oxide and stannous oxide.

tin peroxide. See stannic oxide.

tin phosphide. See stannic phosphide.

tin plate. Sheet steel coated with pure tin. The function of the tin coating is both to protect and beautify the steel base sheet. Also it facilitates soldering.

tin plating. The processes of covering steel, iron or other metal with a layer of tin by dipping in the molten metal, by electroplating, or by immersion in solutions which deposit tin by chemical action of their components. In the molten tin process it is first necessary thoroughly to clean the surface of the steel by pickling in sulfuric acid, annealing, dipping in zinc chloride flux, and subsequent to the dip, the sheet steel is passed through hot palm oil. Electrolytically deposited tin is dull in color, and a great variety of solutions and procedures have been proposed. The usual ingredients for the chemical process are cream of tartar and stannous chloride. The metal being plated also takes part in the process. The objective of all the types of tin plating is to take advantage of the superior corrosion resistance of tin and in some cases improve appearance.

tin protochloride. See stannous chloride.

tin protosulfide. See stannous sulfide.

tin protoxide. See stannous oxide.

tin pyrites. See stannite.

tin resinate. See soaps, metallic.

tin salt. See stannous chloride.

tin spirits. Solutions of tin salts used in dyeing.

tin stearate. See soaps, metallic.

tin-stone. See cassiterite.

tin, stream. See cassiterite.

tin sulfate. See stannous sulfate.

tin sulfide. See stannous sulfide.

tin tallate. See soaps, metallic.

tin tartrate. See stannous tartrate.

tin tetrabromide. See stannic bromide.

tin tetrachloride. See stannic chloride.

tin tetraiodide. See stannic iodide.

"Tinuvin P." [219] Trademark for ultraviolet absorber. The product is a substituted benzotriazole.

Properties: Off-white, crystalline powder; m.p. 131-132°C. Soluble in acetone, styrene, methyl methacrylate, ethyl acetate; insoluble in water. Absorbs ultraviolet light from 200 millimicrons up to 380 millimicrons. Transmits 100% light at 410 millimicrons, therefore contributes no color to the systems it is used in. Stable to strong acid or alkali; to peroxide catalysts; to reducing agents; to temperatures normally encountered in production of plastics.

Uses: Protection of plastics, cosmetics, oils, and waxes, pigments and dyes, lacquers, etc.

tin, wood. See cassiterite.

"Tipagon." [328] Trademarked product consisting of a blend of nonionic agents designed to serve as a dyeing assistant for tippy wool. It is a light amber colored liquid with a faint alcoholic odor.

"Tipersul." [28] Trademark for fibrous potassium titanate. Crystalline fibers 1 micron in diameter melting at 2500°F; useful to 2200°F. Used for high temperature thermal, acoustical and electrical insulation; also for filter media. Available as lumps, blocks, and loose fibers.

TIPPS. Abbreviation for tetraiodophenolphthalein sodium. See iodophthalein sodium.

"Ti-Pure." [28] Trademark for titanium dioxide (TiO_2) pigment, available in two different crystalline forms.

Properties: Fine, dry, white powder; pH 7.0-9.5. Anatase form: Sp.gr. 3.88; index of refraction 2.53. Rutile form: Sp.gr. 4.2; index of refraction 2.71.

Containers: 50-lb paper bags.

Uses: Both anatase and rutile forms as pigments in paints, linoleum, lacquers, paper, leather, inks, and rubber.

"Ti-Pure" VG. [28] Trademark for titanium dioxide. Free flowing dry powder especially for vitreous enamels for iron and steel, but also for other ceramic applications. Not a pigment. Particle size 80 to 325 mesh.

"Tiron." [169] Trademark for disodium-1, 2-dihydroxybenzene-3, 5-disulfonate used in the colorimetric determination of ferric iron, titanium, or molybdenum.

Tischenko reaction. Reaction for the formation of esters by the condensation of two molecules of aldehyde catalyzed by aluminum alcoholate in the presence of a halide.

titanellow. See titanium trioxide. (Not to be confused with titan yellow, an organic dye containing no titanium).

titania. See titanium dioxide.

titanic acid (titanic hydroxide; metatitanic acid) H_2TiO_3 or $Ti(OH)_4$. Water content variable.

Properties: White powder; insoluble in mineral acids and alkalies except when freshly precipitated; insoluble in water.

Derivation: From hydrochloric acid solution of titanates by treating with ammonia and then drying over concentrated sulfuric acid or by boiling titanium sulfate solution.

Grades: Technical.

Containers: 1-, 5-lb bottles; fiber containers; multiwall paper sacks.

Use: Mordant.

Shipping regulations: None.*

titanic acid, meta-. See titanic acid.

titanic acid, anhydride. See titanium dioxide.

titanic anhydride. See titanium dioxide.

titanic chloride. See titanium tetrachloride.

titanic hydroxide. See titanic acid.

titanic iron ore. See ilmenite.

titanic oxide. See titanium dioxide.

titanic sulfate. See titanium sulfate.

titanite (sphene) $CaTiSiO_5$. A natural calcium titanium silicate. Contains variable amounts of iron and sometimes small amounts of yttrium and cerium earths. Hardness 5-5.5; sp.gr. 3.4-3.55.

*See "I.C.C. Shipping Regulations," page xiii.
Reference numbers refer to name of manufacturer. See "List of Manufacturers," page v.

Properties: Yellow, green, brown, black, gray, rose-red; white streak; adamantine or resinous luster.

Occurrence: United States (New York, Pennsylvania, Arkansas, Maine, Massachusetts, New Jersey, North Carolina); Canada; Switzerland; France; Italy; Austria; England; Wales; Norway; Sweden; Russia.

Use: Transparent crystals of good color are sometimes used as gem stones; a source of titanium for paint pigments.

titanium Ti. Ninth most abundant element in earth's crust; very light and strong. Atomic number 22, Group IV of periodic system.

Properties: Silvery metal or dark gray amorphous powder; density 4.5 (20°C); m.p. 1730°C; b.p. greater than 3000°C; linear coefficient of thermal expansion 5.0×10^{-6}/°F; specific heat 0.13 Btu/lb/°F; thermal conductivity 105 Btu/ft²/in/°F/hour; tensile strength up to 125,000 psi at room temperature, 96,000 psi (400°F), 20,000 psi (1000°F); very hard (scratches steel); has excellent resistance to atmospheric and seawater corrosion and to numerous chemicals when cold; reactive when hot or molten. Insoluble in cold water, decomposes hot water.

Sources: Ilmenite, rutile, titanium slag from certain iron ores.

Derivation: From titanium carbide by electrolysis; from ores by treatment with chlorine. The titanium tetrachloride is then reduced with magnesium or sodium (Kroll process) in an inert atmosphere of helium or argon. The titanium sponge is consolidated by melting.

Grades: Technical (powder); commercially pure (sheets, bars, tubes, rods, wire and sponge).

Containers: Wet powder in cans/box or 10-, 100-lb kegs; sponge in steel pails.

Uses: As metal or alloy (especially ferrotitanium) as structural material in aircraft, jet engines (replacing steel in missile frames); marine equipment, textile machinery, chemical equipment, surgical instruments, orthopedic appliances, sporting equipment, foodhandling equipment; also in x-ray tube targets; abrasives; cermets, metal-ceramic brazing, especially in nickel-cadmium batteries for space vehicles.

Uses of titanium compounds: Pigments (titanium dioxide); electronics; smoke clouds; porcelain enamels; fire retardants; waterproofing agents; gems.

Caution: The dry powder ignites in air above 250°C. It can be ignited by static sparks and by grinding.

Shipping regulations: Powder, wet (with not less than 20% water) or dry: flammable solid. Yellow label.*

titanium acetylacetonate. See titanyl acetylacetonate.

titanium acylates. Compounds whose general formula is

R'O[RCO-O(-TiO-)OR']$_x$R'

where RCO-O- may be a saturated or unsaturated organic acid radical, and where R' is a hydrocarbon radical; e.g., isopropoxytitanium stearate.

Uses: Surface-active agents in non-polar solvents; water repellents for masonry, wood, and paper.

titanium boride (titanium diboride) TiB_2.

Properties: Solid with oxidative resistance up to 1400°C. M.p. 2480°C; sp.gr. 4.50; hardness 9+ (Mohs); low electrical resistivity.

Uses: Metallurgical additive, high temperature electrical conductor, refractory, cermet component; coatings resistant to attack by molten metals; aluminum manufacture; super alloys; nuclear steels.

titanium butylate. See tetrabutyl titanate.

titanium carbide TiC.

Properties: Crystalline solid with gray metallic color; hardness 3200 kg/sq mm; m.p. 2700°C; b.p. 4300°C; sp.gr. 4.93; resistivity 60 micro-ohm-cm (room temperature). Insoluble in water; soluble in nitric acid and aqua regia.

Uses: Additive with tungsten carbide in making cutting tools and other parts submitted to thermal shock; arc-melting electrodes; cermets.

titanium chelates. Compounds whose general formula is $(HOYO)_2Ti(OR)_2$ or $(H_2NYO)_2Ti(OR)_2$ where Y and R may be hydrocarbon radicals, e.g., octylene glycol titanate; triethanol amine titanate.

Uses: Surface-active agents; corrosion inhibitors; cross-linking agents.

titanium diboride. See titanium boride.

titanium dichloride $TiCl_2$.

Properties: Black powder; burns like tinder in air. Decomposed by water. Hygroscopic; soluble in alcohol; insoluble in chloroform, ether, carbon disulfide. Keep under water or inert gas.

Caution: Keep away from air!

titanium dioxide (titanic anhydride; titanic acid anhydride; titanic oxide; titanium white; titania) TiO_2.

Properties: White to black powder, depending on purity; is also prepared in two crystalline forms:

anatase, sp.gr. 3.8; index of refraction 2.5; m.p. 1560°C;

rutile, sp.gr. 4.3; index of refraction 2.7; m.p. 1640°C (dec). Rutile occurs naturally and when pure is a light yellow. Both decompose at 1640°C; insoluble in water and cold dilute acids; soluble in hot concentrated sulfuric acid and alkalies. Titanium dioxide possesses the greatest hiding power of all the white pigments.

Derivation: From ilmenite or rutile (q.v.). Ilmenite is treated with sulfuric acid and the titanium sulfate further processed.

Grades: Technical, of many variations;

*See "I.C.C. Shipping Regulations," page xiii.

pure; U.S.P. XVI.
Containers: Fiber drums; multiwall paper sacks.
Uses (in order of volume): Paint pigments; opacifying agent in paper; white rubber and plastics; floor coverings (linoleum, etc); rubber manufacture; glassware and ceramics; enamel frits; delustering synthetic fibers; printing inks; welding rods. Single crystals are high temperature transducers.
Shipping regulations: None.*

titanium disilicide Ti, Si. Used for special alloy applications, as a flame or blast impingement resistant coating material.

titanium esters. Compounds whose general formula is $Ti(OR)_4$ where R is a hydrocarbon radical, e.g., tetraisopropyl titanate. See also tetrabutyl titanate, tetra(2-ethylhexyl) titanate.
Uses: Adhesion promotors, ester exchange catalysts, cross-linking agents, and in heat-resistant paints.

titanium ferrocene. See dicyclopentadienyl-titanium chloride.

titanium hydride TiH_2.
Properties: A grey-black metallic powder, which dissociates above 550°F. The evolution of hydrogen is gradual and practically complete at 1200°F. The hydride is inert at room temperature; can be handled in air without the hazard of explosion associated with the powdered metal. Sp.gr. 3.8. Attacked by strong oxidizing agents.
Derivation: Direct combination of titanium with hydrogen; reduction of titanium oxide with calcium hydride in the presence of hydrogen above 600°C.
Containers: Polyethylene bags in metal drums. (100-lb net).
Uses: Powder metallurgy; production of pure hydrogen (contains approx 1800 cc (STP) hydrogen per cc of hydride); production of foamed metals; solder for metal-glass; electronic getter; reducing atmosphere for furnaces; hydrogenation agent; refractories.

titanium isopropylate. See tetraisopropyl titanate.

titanium nitride TiN.
Properties: Golden-brown, hard, brittle plates; m.p. 2927°C; sp.gr. 5.24; specific heat 8.86 cal/mole at 25°C; electrical resistivity 21.7 micro-ohm-cm.
Uses: High temperature bodies, cermets, alloys, rectifiers, semiconductor devices.

titanium ore. See rutile and ilmenite.

titanium oxalate (titanous oxalate) $Ti_2(C_2O_4)_3 \cdot 10H_2O$.
Properties: Yellow prisms. Soluble in water; insoluble in alcohol and ether.
Derivation: By the action of oxalic acid on titanous chloride.
Method of purification: Crystallization.

titanium oxide. See titanium dioxide or trioxide.

titanium peroxide. See titanium trioxide.

titanium-potassium fluoride (potassium-titanium fluoride) TiK_2F_6.
Properties: White leaflets. Soluble in water (hot).
Grades: Technical.
Uses: Titanic acid; titanium.

titanium-potassium oxalate $TiO(CO_2 \cdot CO_2K)_2 \cdot 2H_2O$.
Properties: Colorless, lustrous crystals. Soluble in water.
Derivation: By treating titanium hydroxide with potassium oxalate and oxalic acid.
Grades: Technical; pure; 22% TiO_2 (min)
Containers: 112-lb kegs; 200-, 300-, 350-lb barrels.
Use: Mordant in cotton and leather dyeing; sensitization of aluminum for photography.
Shipping regulations: None.*

titanium sesquisulfate. See titanous sulfate.

titanium sulfate (titanium sulfate cake; titanic sulfate; basic titanium sulfate; titanyl sulfate) $(TiSO_4)_2 \cdot 9H_2O$; $TiOSO_4 \cdot H_2SO_4 \cdot 8H_2O$. A commercial material, possibly a mixture of both formulas.
Properties: White cake-like solid; highly acidic, similar to 50% sulfuric acid; typical composition 20% TiO_2, 50% H_2SO_4, 30% H_2O; hygroscopic; sp.gr. about 1.47; soluble in water; solutions hydrolyze readily unless protected from heat and dilution.
Derivation: By the action of sulfuric acid on ilmenite ore.
Containers: 300-lb net fiber drums.
Uses: Treatment of chrome yellow and other colors; production of titanous sulfate used as reducing agent or stripper for dyes; also a laundry chemical.

titanium sulfate, basic. See titanium sulfate.

titanium sulfate cake. See titanium sulfate.

titanium sulfates. See titanium sulfate and titanous sulfate.

titanium tetrachloride (titanic chloride) $TiCl_4$.
Properties: Colorless liquid. Fumes strongly when exposed to moist air forming a dense and persistent white cloud. Pure: Sp.gr. 1.7609 at 0°C; b.p. 136.4°C (760 mm); freezing point −30°C; specific heat, liquid 0.188 between 13 and 99°C, vapor at constant pressure, 0.12897 between 152° and 272°C; average cubical coefficient of expansion 0.001086 from 0-100°C; critical temperature 358°C; dielectric constant at 24°C, 2.73; heat of solution 57,870 cals at 17°C for 1 mole $TiCl_4$ in 1600 moles water; heat of formation 185 kg cals; vapor pressure log p = 7.64433 − (1947.6/T). Commercial: Density 14.5 lbs/gal (approx); b.p. between 132° and 137°C; vapor pressure 8 mm (20°C); analysis (a typical analysis is as follows): free chlorine 0.0 to 0.05%; dissolved gases 0.0 to

0.20%; vanadium and zirconium chlorides small amounts; silicon tetrachloride 1.0 to 6.00%; titanium tetrachloride 94-99%. Soluble in dilute hydrochloric acid; soluble in water with evolution of heat; concentrated aqueous solutions are stable and corrosive; dilute solutions precipitate insoluble basic chlorides.
Derivation: By heating titanium dioxide or the ores and carbon to redness in a current of chlorine.
Grades: Technical; C.P.
Containers: Glass bottles; steel drums; tank cars.
Uses: Pure titanium and titanium salts; the textile industry as a mordant; iridescent effects in glass; artificial pearls; smoke screens; titanium pigments; polymerization catalyst.
Caution: Avoid breathing fumes and vapor.
Shipping regulations: Corrosive liquid. White label.*

titanium trichloride (titanous chloride) $TiCl_3$.
Properties: (Titanium trichloride-1). Dark violet anhydrous deliquescent crystals. Sp.gr. 2.6; decomposes above 440°C; decomposes in air and water. Soluble in alcohol, acetonitrile, certain amines; slightly soluble in chloroform. Insoluble in ether and hydrocarbons.
(Titanium trichloride-2). Light red powder decomposing above 178°C; slightly soluble in trichloroethylene, ketones, chloroform; insoluble in benzene, carbon tetrachloride; other properties similar.
Uses: Reducing agent; organic synthesis; co-catalyst for polyolefin polymerization; organometallic synthesis involving titanium.

titanium trioxide (titanium peroxide; titanellow) TiO_3. (Not to be confused with titan yellow, an organic dye containing no titanium.)
Properties: Yellow powder; soluble in acids.
Containers: 1-, 5-lb bottles; fiber containers.
Uses: Dental porcelain and cements; yellow tile.

titanium white. See titanium dioxide.

titanous chloride. See titanium trichloride.

titanous oxalate. See titanium oxalate.

titanous sulfate (titanium sesquisulfate) $Ti_2(SO_4)_3$.
Properties: Green crystalline powder; insoluble in water, alcohol, concentrated sulfuric acid; but soluble in dilute hydrochloric or sulfuric acids giving violet solutions.
Grades: Commercial grade made and supplied as a dark purple solution containing about 15% $Ti_2(SO_4)_3$.
Containers: Glass bottles; carboys.
Use: Textile industry as reducing agent for stripping or discharging colors.
Shipping regulations: (solution): Corrosive liquid. White label.*

"Titanox." [336] Trademark of an extensive series of white titanium pigments comprising titanium dioxide (TiO_2) in both anatase and rutile crystal forms, and titanium dioxide extended with calcium sulfate (titanium-calcium pigments). The titanium dioxide pigments run 94-99% TiO_2 depending upon type of pigment and application. The titanium-calcium pigments contain the rutile form either 30% TiO_2 or 50% TiO_2.
Properties: Fine, dry, white powders; relatively inert. Titanium dioxide pigments, anatase form: sp.gr. 3.9; refractive index 2.55. Rutile form: sp.gr. 4.2; refractive index 2.7. Rutile TiO_2 has about 25% greater tinting strength and opacity than anatase TiO_2. Titanium-calcium pigments, 30% TiO_2 type: sp.gr. 3.25. 50% TiO_2 type: sp.gr. 3.47.
Containers: 50-lb paper bags.
Uses: Paints, enamels and lacquers; paper; plastics and rubber; floor coverings; coated fabrics; inks; delustering synthetic fibers; leather; porcelain enamels; welding rod coatings, etc.

titanyl acetylacetonate (titanium acetylacetonate) $TiO[OC(CH_3):CHCOCH_3]_2$.
Properties: Crystalline powder; slightly soluble in water. Resistant to hydrolysis. A chelating, non-ionizing compound.
Derivation: Reaction of titanium oxychloride with acetylacetone and sodium carbonate.
Uses: Cross-linking agent for cellulosic lacquers.

titanyl sulfate. See titanium sulfate.

titer. In solutions (1) the concentration of a dissolved substance as determined by titration, (2) the minimum amount or volume needed to bring about a given result in titration; or (3) the solidification point of the fatty acids which have been liberated from the fat by hydrolysis.

titration. A method for determining volumetrically the concentration of a desired substance in solution by adding a standard solution of known volume and strength until the reaction is completed, usually as indicated by a change in color due to an indicator or by electrical measurements.

TKP. Abbreviation for tripotassium phosphate. See potassium phosphate, tribasic.

TKPP. Abbreviation for tetrapotassium pyrophosphate. See potassium pyrophosphate.

Tl. Symbol for thallium.

Tm. Symbol for thulium.

TMA. Abbreviation for trimethylamine.

"TME." [138] Trade name for trimethylolethane.
Properties: Non-hygroscopic, non-corrosive fine, white crystals.
Containers: 50-lb multi-wall bags.
Uses: For use in the manufacture of alkyd resins, drying oils, polyesters and isocyanate resins; also of interest as an intermediate for producing plasticizers and a variety of specialty chemicals, surface active agents, and explosives.

TMEDA. Abbreviation for tetramethylethylenediamine.

*See "I.C.C. Shipping Regulations," page xiii.
Reference numbers refer to name of manufacturer. See "List of Manufacturers," page v.

TML. Abbreviation for tetramethyllead.

TMTD. Abbreviation for tetramethylthiuram disulfide.

TNA.
1. Abbreviation for tetranitroaniline (q.v.). It is a high explosive made by nitrating aniline and is used in detonators.
2. Abbreviation for thymus nucleic acid. See deoxyribonucleic acid.

TNB. Abbreviation for trinitrobenzene.

TNT. Abbreviation for trinitrotoluene.

tobacco. See nicotine.

tobacco mosaic virus. The first virus to be obtained in crystalline form. Recently the protein portion of this virus (95% of each particle is protein) was found to contain about 2200 protein molecules, each with a molecular weight of about 18,000. The complete sequence of the 158 amino acids for this virus protein molecule has recently been determined.

tobacco stems. Tobacco stems and stalks contain from 1.2-3.3% nitrogen and 4-9% potash (K_2O).
Use: They are ground and used as a fertilizer material.
Shipping regulations: None.*

tobacco wood. See hamamelis.

Tobias acid. See 2-naphthylamine-1-sulfonic acid.

"Tobin Bronze-452." [324] A high-strength, corrosion-resistant rod alloy, developed originally for marine use. Nominal composition is copper 60%, zinc 39.25%, tin 0.75%. Produced under special procedure as propeller shafting in diameters up to 6 in. Also available as die pressed forgings. With a slight adjustment of composition used as rod and wire for oxyacetylene braze welding of steel, cast iron, and copper alloys.

"Toclase." [299] Trademark for carbetapentane citrate.

"Tocopherex." [412] Trademark for d-alpha-tocopheryl acetate (q.v.).

tocopherols (Vitamin E). A group of related substances: alpha-, beta-, gamma-, and delta-tocopherol, which constitute vitamin E. The alpha-form, $C_{29}H_{50}O_2$, (which occurs naturally as the d-isomer), is the most potent. All are derivatives of dihydrobenzo-gamma-pyran, and differ from each other only in the number and position of methyl groups. Vitamin E is required by certain rodents (but possibly not by humans) for normal reproduction. Muscular and central nervous system depletion along with generalized edema are deficiency symptoms in all animals. The tocopherols function as antioxidants; and are used, thus, as preservatives. They are marketed as the acetate, since this is stable to oxidation. See dℓ-alpha-tocopheryl acetate.
Properties: These vitamins are viscous oils; soluble in lipid solvents; insoluble in water; stable to heat in the absence of oxygen, to strong acids, and to visible light; unstable to ultraviolet light, alkalies and oxidation.
Units: One international unit is the vitamin E activity of 0.1 g of the International Standard, containing 1 mg of pure synthetic racemic alpha-tocopherol acetate.
Grades: N.F. XI (dℓ-alpha-tocopherol).
Uses: Medicine; nutrition; antioxidants.

d-alpha **tocopheryl acetate** (d-alpha tocopherol acetate) $C_{29}H_{49}O \cdot OOCCH_3$.
Properties: Yellow, nearly odorless, clear, viscous oil; unstable in the presence of alkalies; affected by light; m.p. about 25°. Insoluble in water; freely soluble in alcohol; miscible with ether, chloroform, acetone, and vegetable oils. Angular rotation: 10% solution in $CHCl_3$ = +0.25° in a 200 mm tube. Refractive index 1.4940-1.4985; sp.gr. 0.950-0.964.
Grade: N.F. XI.
Use: Medicine; antioxidant.

dℓ-alpha **tocopheryl acetate** (dℓ-alpha tocopherol acetate) $C_{29}H_{49}O \cdot OOCCH_3$.
Properties: Yellow, nearly odorless, clear, viscous oil; unstable in the presence of alkalies; affected by light. Insoluble in water; freely soluble in alcohol; miscible with ether, chloroform, acetone, and vegetable oils. 10% $CHCl_3$ solution shows no appreciable angular rotation in a 200 mm tube. Sp.gr. 0.950-0.964; refractive index 1.4940-1.4985.
Grade: N.F. XI; powdered 25%, 33%.
Containers: Bottles.
Use: Medicine.

d-alpha **tocopheryl acid succinate** (d-alpha tocopherol acid succinate) $C_{29}H_{49}O \cdot OOC(CH_2)_2COOH$.
Properties: White, crystalline powder; little or no taste or odor; stable to air; unstable to alkali and to heat. Insoluble in water; slightly soluble in aqueous alkali; soluble in alcohol, ether, acetone, and vegetable oils; very soluble in chloroform. Melting range 73°-78°.
Grade: N.F. XI.
Use: Medicine; antioxidant.

"Tofaxin." [162] Trademark for tocopherol.

tolan (diphenylacetylene) $C_6H_5C{:}CC_6H_5$.
Properties: Monoclinic crystals; m.p. 59-61°C; b.p. 300°C (760 mm); 170°C (19 mm); sp.gr. 0.966 (100/4°C). Insoluble in water; soluble in ether or hot alcohol.
Grades: Technical; purified.
Use: Organic synthesis; purified grade as primary fluor or as wave length shifter in solution scintillators.
Shipping regulations: None.*

tolazoline hydrochloride (2-benzyl-2-imidazoline hydrochloride; benzazoline hydrochloride) $C_{10}H_{12}N_2 \cdot HCl$ or
$NHCH_2CH_2NCCH_2C_6H_5 \cdot HCl$.
Properties: White or creamy-white, bitter, crystalline powder with slight aromatic odor. M.p. 172-176°C. Freely soluble

in alcohol, chloroform and water; very slightly soluble in ether and ethyl acetate. pH (2.5% solution) 4.9-5.3.
Grade: U.S.P. XVI.
Use: Medicine.

tolbutamide (1-butyl-3-para-tolylsulfonylurea) $H_3CC_6H_4SO_2NHCONH(CH_2)_3CH_3$.
Properties: White, or practically white, crystalline powder; m.p. 126-132°C. Tasteless and practically odorless. Insoluble in water; soluble in alcohol and chloroform.
Grade: U.S.P. XVI.
Use: Medicine.

"Toleron." [329] Trademark for ferrous fumarate, an anhydrous salt of a combination of ferrous iron and fumaric acid. Used in medicine.

ortho-**tolidine** (dimethylbenzidine; diaminoditolyl) $[C_6H_3(CH_3)NH_2]_2$.
Properties: Glistening plates, white to reddish; m.p. 129-131°C. Soluble in alcohol and ether; sparingly soluble in water.
Derivation: By the reduction of ortho-nitrotoluene with zinc dust and caustic soda and conversion of the hydrazotoluene by boiling with hydrochloric acid.
Method of purification: Crystallization.
Grades: Technical, dry or paste.
Containers: 200-, 350-lb kegs.
Uses: Dyes; sensitive reagent for gold (1:10 million detectable); and for free chlorine in water.
Shipping regulations: None.*

meta-**tolidine dihydrochloride** $[C_6H_3(CH_3)NH_2]_2 \cdot 2HCl$.
Properties: M.p. (of free amine) 107-108°C; soluble in hot water.
Use: Synthesis.

ortho-**tolidine dihydrochloride** (dimethylbenzidine hydrochloride) $C_{14}H_{16}N_2 \cdot 2HCl$.
Properties: White crystals. Soluble in water and in dilute hydrochloric acid solutions.
Containers: 350-lb drums (available as a paste).
Use: Determination of small amounts of chlorine in water.
Caution: Store reagent in dark or amber bottles in cool place. Do not use rubber stoppers.

tolonium chloride (3-amino-7-dimethylamino-2-methyl-phenazothionium chloride) $C_{15}H_{16}SN_3Cl$.
Properties: Green, crystalline powder with bronze luster. Soluble in water; slightly soluble in alcohol; very slightly soluble in chloroform; practically insoluble in ether.
Grade: N.N.D.
Use: Medicine.

3-ortho-**toloxy-1,2-propanediol.** See mephenesin.

3-ortho-**toloxy-1,2-propanediol-1-carbamate.** See mephenesin carbamate.

"Tolrez." [79] Trade name for tall oil pitch.
Properties: Acid number 69; flash pt (open cup) 410°F; saponification number 113; color (Hellige) (10% n.v. in benzol) 16; viscosity (100°C) Brookfield #2 spindle (60 rpm) 118 cps., Sayboldt Fural 65 secs; specific quantity (25°/25°C) 0.99.
Containers: 55-gal drums; tank cars.
Uses: Roofing compounds, asphalt emulsions and anti-stripping compounds, etc.

"Tolseram." [412] Trademark for mephenesin carbamate.

"Tolserol." [412] Trademark for mephenesin (q.v.).

tolualdehyde. See tolyl aldehydes.

toluazotoluidine. See ortho-aminoazotoluene.

tolu balsam (Thomas balsam; tolu resin).
Properties: A brown or yellowish-brown plastic solid with a pleasant, aromatic odor resembling that of vanilla, and a mild, aromatic taste. Brittle when old or cold, soft when fresh. Soluble in alcohol, chloroform and ether; nearly insoluble in water. Soluble in alkalies.
Derivation: By incision into the wood of Toluifera balsamum, indigenous to Colombia.
Grades: Natural; U.S.P. XVI; cleaned.
Containers: Natural: 50-lb tins; 90-, 100-lb cases; cleaned: 1-lb bottles; 5-, 25-, 50-lb tins.
Uses: Medicine; perfumery (hyacinth); confectionery (glaze); fumigating compositions; chewing gum.

tolu balsam oil. See tolu oil.

toluene (toluol; methylbenzene; methylbenzol; phenylmethane) $CH_3C_6H_5$.
Properties: Colorless, refractory, flammable liquid; benzene-like odor. As compared with benzene its vapors are less dangerously toxic, less flammable, and it has a slower rate of evaporation; sp.gr. 0.866 (20/4°C); m.p. −94.5°C; b.p. 110.7°; aniline equivalent 15; flash point 6-10°C. Soluble in alcohol, benzene, and ether; insoluble in water.
Derivation: (a) By catalytic reforming of petroleum. (b) By fractional distillation of coal-tar light oil. (c) By extraction from coal gas.
Method of purification: Rectification.
Grades: Pure; commercial; straw-colored; nitration; industrial. These are usually defined in terms of boiling ranges.
Containers: 5-gal can; 55-, 110-gal drums; 8000-gal tank cars.
Uses: Aviation gasoline and high-octane blending stock; solvent, for paints and coatings, gums, resins, most oils, rubber cement, vinyl organosols; chemicals, including benzoic acid, benzyl and benzoyl derivatives, saccharin, medicines, dyes, perfumes; source of toluenediisocyanates (polyurethane resins); explosives (TNT) toluene sulfonates (detergents).
Warning! Flammable; vapor harmful. MCA warning label.
Shipping regulations: Flammable liquid. Red label.*

*See "I.C.C. Shipping Regulations," page xiii.
Reference numbers refer to name of manufacturer. See "List of Manufacturers," page v.

ortho-**tolueneazonaphthylamine.** See yellow OB.

toluene-2,4-diamine (meta-tolylenediamine; meta-toluylenediamine; diaminotoluene) $CH_3C_6H_3(NH_2)_2$, (CH_3 = 1).
Properties: Colorless crystals. Soluble in water, alcohol and ether. M.p. 99°C; b.p. 280°C.
Derivation: By the reduction of meta-dinitrotoluene with iron and hydrochloric acid.
Method of purification: Crystallization.
Grades: Technical.
Containers: 175-lb barrels; 200-, 250-lb drums.
Uses: Dye intermediate; direct oxidation black for furs and hair.
Shipping regulations: None.*

toluene-2,4-diisocyanate (2,4-tolylene diisocyanate; meta-tolylene diisocyanate; TDI) $CH_3C_6H_3(NCO)_2$
Properties: Water-white to pale-yellow liquid; sharp, pungent odor; b.p. 251°C; flash point 270°F; m.p. (pure isomer) 19.5-21.5°C; sp.gr. 1.22 (25°/15.5°C); vapor pressure (approx) 0.01 mm at 20°C. Relatively non-corrosive; reacts with water producing CO_2; reacts with compounds containing active hydrogen (may be violent). Soluble in ether, acetone, and other organic solvents. Irritating to eyes and nose.
Derivation: Reaction of 2,4-diaminotoluene with phosgene.
Method of purification: Distillation to remove hydrochloric acid.
Grade: 99% (min).
Containers: Drums; tank cars.
Uses: Polyurethane foams; elastomers and resins.
Danger! Hazardous liquid and vapor. Causes burns. MCA warning label.
Availability: 100% 2,4-isomer; 80% and 65% 2,4-isomer both mixed with 2,6-isomer.

para-**toluenesulfamine.** See para-toluenesulfonamide.

para-**toluenesulfanilide** $CH_3C_6H_4SO_2C_6H_4NH_2$.
Properties: White to pink crystalline solid. M.p. 103°C. Soluble in most lacquer solvents.
Derivation: Para-toluene sulfonchloride treated with aniline in presence of lime or carefully regulated amounts of alkalies.
Grades: Technical.
Use: Softener for acetylcellulose in quantities up to 50%; dyestuff intermediate.

toluenesulfochloride (toluene sulfonchloride) $CH_3C_6H_4SO_2Cl$.
Properties: (a) Ortho-: oily liquid. (b) Para-: rhombic crystals. Soluble in alcohol and ether; insoluble in water. M.p. (b) 69°C; b.p. (b) 145°-146°C.
Derivation: By the action of chlorosulfonic acid on toluene.
Grades: Technical.
Containers: Wooden kegs.
Use: Organic synthesis; manufacture of dyestuffs, saccharin.
Shipping regulations: None.*

para-**toluenesulfonamide** (para-toluenesulfamine; PTSA). $CH_3C_6H_4SO_2NH_2$.
Properties: White leaflets. Soluble in alcohol; very slightly soluble in water. M.p. 137°C.
Derivation: By amination of para-toluenesulfochloride.
Method of purification: Crystallization.
Grades: Technical.
Containers: Wooden kegs, barrels.
Uses: Organic synthesis; plasticizers and resins; fungicide and mildewicide in paints and coatings.
Shipping regulations: None.*

ortho-**toluenesulfonate.** See ortho-toluenesulfonic acid.

para-**toluenesulfonate.** See para-toluenesulfonic acid.

toluenesulfonchloride. See toluenesulfochloride.

para-**toluenesulfondichloroamide.** See dichloramine-T.

ortho-**toluenesulfonic acid** (orthotoluenesulfonate) $C_6H_4(SO_3H)(CH_3)$.
Properties: Colorless crystals; m.p. 67.5°C; b.p. 129°C; soluble in alcohol, water, and ether.
Derivation: By sulfonating toluene with concentrated sulfuric acid below 100°C.
Method of purification: Crystallization.
Availability: Anhydrous; monohydrate; 40% aqueous solution.
Containers: 55-gal drums.
Uses: Dyes; organic synthesis; acid catalyst.
Shipping regulations: None.*

para-**toluenesulfonic acid** (para-toluenesulfonate) $C_6H_4(SO_3H)(CH_3)$.
Properties: Colorless leaflets; m.p. 107°C; b.p. 140°C (20 mm). Soluble in alcohol, ether, and water.
Derivation: By action of chlorosulfonic acid on toluene at a low temperature.
Method of purification: Crystallization.
Availability: Anhydrous; monohydrate; 40% aqueous solution.
Containers: 55-gal drums; monohydrate, 125-lb drums.
Uses: Dyes; organic synthesis; organic catalyst.
Shipping regulations: None.*

para-**toluenesulfonic acid, sodium salt.** See sodium toluenesulfonate.

toluene-para-thiol (para-thiocresol) $CH_3C_6H_4SH$.
Properties: Cream to white moist crystals; m.p. 43-44°C; b.p. about 195°C. Insoluble in water; soluble in alcohol or ether.
Containers: Tinned steel drums.
Uses: Medicine; intermediate.

toluene trichloride. See benzotrichloride.

toluene trifluoride. See benzotrifluoride.

toluhydroquinone $CH_3C_6H_3(OH)_2$.
Properties: Pink to white; m.p. 126-127°C;

ash 0.01% max; assay 99% min.
Grade: Technical.
Containers: Fiber drums.
Uses: Antioxidant; polymerization inhibitor.

alpha-**toluic acid.** See phenylacetic acid.

meta-**toluic acid** (meta-toluylic acid; 3-methylbenzoic acid) $C_6H_4CH_3COOH$.
Properties: White to yellowish crystals; slightly soluble in water; soluble in alcohol and ether. Sp.gr. 1.0543; m.p. 109°C; b.p. 263°C; ionization constant 5.3×10^{-5}.
Derivation: Oxidation of meta-xylene with nitric acid.
Method of purification: Crystallization.
Grade: Technical.
Containers: Wooden kegs; 250-lb fiber drums.
Use: Organic synthesis; to form N,N-diethyl-meta-toluamide, an important broad-spectrum insect repellent.
Shipping regulations: None.*

ortho-**toluic acid** (ortho-toluylic acid; 2-methylbenzoic acid) $C_6H_4CH_3COOH$.
Properties: White crystals; slightly soluble in water; soluble in alcohol and chloroform. Sp.gr. 1.0621; m.p. 103.5-104°C; b.p. 259°C; refractive index (114.6°C) 1.512; ionization constant 1.2×10^{-5}.
Derivation: Oxidation of ortho-xylene with dilute nitric acid.
Method of purification: Crystallization.
Grade: Technical.
Containers: Wooden kegs; 250-lb fiber drums.
Uses: Organic synthesis; alkyd and polyester resins.
Shipping regulations: None.*

para-**toluic acid** (para-toluylic acid; 4-methylbenzoic acid) $C_6H_4CH_3COOH$.
Properties: Transparent crystals; slightly soluble in water; soluble in alcohol and ether. M.p. 180°C; b.p. 275°C; ionization constant 4.3×10^{-5}
Derivation: By treating cymene or turpentine with nitric acid.
Method of purification: Crystallization.
Grade: Technical.
Containers: Wooden kegs; 175-lb fiber drums.
Use: Organic synthesis; plasticizers, dye carriers.
Shipping regulations: None.*

alpha-**toluic aldehyde.** See phenylacetaldehyde.

meta-**toluidine** (meta-aminotoluene) $CH_3C_6H_4NH_2$.
Properties: Colorless liquid; sp.gr. 0.980; m.p. −31.5°C; b.p. 203.3°C; slightly soluble in water; soluble in alcohol or ether.
Derivation: By the reduction of meta-nitrobenzylidine chloride with zinc at a low temperature.
Grade: Technical.
Containers: 450-, 900-lb iron drums; tank cars.
Uses: Dyes; manufacture of organic chemicals.
Danger: Hazardous liquid and vapor rapidly absorbed through skin. MCA warning label.
Shipping regulations: None.*

ortho-**toluidine** (ortho-aminotoluene) $CH_3C_6H_4NH_2$.
Properties: Light yellow liquid; becomes reddish-brown on exposure to air and light; volatile with steam. Sp.gr. 1.008 (20/20°C); m.p. −21°C; b.p. 200-202°C; flash point 87°C. Soluble in alcohol and ether; very slightly soluble in water.
Derivation: By the reduction of ortho-nitrotoluene or obtained mixed with para-toluidine by the reduction of crude nitrotoluene.
Grade: Technical.
Containers: 450-, 900-lb iron drums; tank cars.
Uses: Dyes; saccharin; printing textiles blue-black and making various colors fast to acids; vulcanization accelerators; organic synthesis.
Danger: Hazardous liquid and vapor rapidly absorbed through skin. MCA warning label.
Shipping regulations: None.*

para-**toluidine** (para-aminotoluene) $CH_3C_6H_4NH_2$.
Properties: White, lustrous plates or leaflets. Sp.gr. 1.046 (20/4°C); m.p. 45°C; b.p. 200.3°C. Soluble in alcohol and ether; very slightly soluble in water.
Derivation: By the reduction of para-nitrotoluene with iron and hydrochloric acid.
Grades: Technical, flake or cast.
Containers: 450-, 900-lb iron drums.
Uses: Dyes; organic synthesis; test reagent for lignin, nitrite, phloroglucinol.
Danger: Hazardous solid and vapor rapidly absorbed through skin. MCA warning label.
Shipping regulations: None.*

toluidine maroon
$CH_3C_6H_3NO_2N_2C_{10}H_5OHCONHC_6H_4NO_2$. An organic azo pigment obtained by the azo coupling of meta-nitro-para-toluidine with the meta-nitroanilide of beta-hydroxynaphthoic acid.
Properties: Good lightfastness and resistance to bleeding in oils.
Uses: Automotive finishes, sign enamels, printing inks.

ortho-**toluidine**-meta-**sulfonic acid** (CH_3= 1). [2-aminotoluene-5-sulfonic acid (CH_3=1); 4-amino-meta-toluenesulfonic acid (SO_3H=1)] $C_6H_3(CH_3)(NH_2)SO_3H$.
Properties: Colorless crystals; soluble in hot water; insoluble in alcohol and ether.
Derivation: By heating acid ortho-toluidine sulfate.
Method of purification: Crystallization.
Grade: Technical.
Containers: Wooden kegs or fiber drums.
Use: Dye intermediate.
Shipping regulations: None.*

para-**toluidine**-ortho-**sulfonic acid** (CH_3=1) [4-aminotoluene-2-sulfonic acid (CH_3=1); 5-amino-ortho-toluenesulfonic acid (SO_3H=1)] $C_6H_3(CH_3)(NH_2)SO_3H$.
Properties: Monoclinic crystals; soluble in water; insoluble in ether and alcohol.

Derivation: From para-toluidine sulfate by heating in oven (baking process).
Method of purification: Recrystallization as sodium salt.
Grade: Technical.
Containers: Wooden barrels or fiber drums.
Use: Dye intermediate.
Shipping regulations: None.*

6-(para-toluidino)-metanilic acid. See 4-amino-4'-methyldiphenylamine-2-sulfonic acid.

tolu oil (tolu balsam oil; albahaca oil).
Properties: Yellow liquid; hyacinth-like odor. Soluble in alcohol, ether, chloroform and carbon disulfide.
Chief known constituents: A terpene, $C_{10}H_{16}$, and esters of cinnamic and benzoic acid.
Constants: Sp.gr. 0.945-1.09.
Derivation: From tolu balsam by distillation.
Method of purification: Rectification.
Grade: Technical.

toluol. See toluene.

toluquinone (2-methylquinone; para-toluquinone) $CH_3C_6H_3O_2$.
Properties: Yellow leaflets or needles; m.p. 65-67°C; soluble in hot water; very soluble in alcohol, ether, acetone, ethyl acetate, and benzene.
Containers: 50-lb fiber drums.

tolu resin. See tolu balsam.

toluyl aldehydes. See tolyl aldehydes.

toluylene. See stilbene.

meta-**toluylenediamine.** See toluene-2, 4-diamine.

toluylene red. See neutral red.

meta-, ortho-, and para-**toluylic acid.** See corresponding toluic acid.

meta-**tolylaldehyde** (meta-toluyl aldehyde; meta-tolualdehyde, meta-methylbenzaldehyde) $CH_3C_6H_4CHO$.
Properties: Colorless liquid; refractive index (n 21.4/D) 1.54068; sp.gr. 1.019 (20/4°C); b.p. 199°C; slightly soluble in water; soluble in alcohol and ether.

ortho-**tolylaldehyde** (ortho-toluyl aldehyde; ortho-tolualdehyde; ortho-methylbenzaldehyde) $CH_3C_6H_4CHO$.
Properties: Colorless liquid; refractive index (n 19/D) 1.54852; sp.gr. 1.039 (20/4°C); b.p. 195.5°C; slightly soluble in water; soluble in alcohol and ether.

para-**tolylaldehyde** (para-toluyl aldehyde; paratolualdehyde; para-methylbenzaldehyde) $CH_3C_6H_4CHO$.
Properties: Colorless liquid; refractive index (n 16.6/D) 1.54693; sp.gr. 1.020; b.p. 204°C; slightly soluble in water; soluble in alcohol and ether.
Grades: Technical; pure.
Containers: Tins; drums.
Uses: Flavors; perfumes; synthetic aromatic; pharmaceutical and dyestuff intermediate.

alpha-**tolylaldehyde dimethylacetal.** See "Viridine."

4-ortho-**tolylazo**-ortho-**diacetotoluide.** See diacetylaminoazotoluene.

ortho-**tolyl biguanide** $NH_2(CNHNH)_2C_6H_4CH_3$.
Properties: White to off-white powder; melting point 138°C (min).
Containers: 140-lb drums.
Use: Antioxidant for soaps produced from animal or vegetable oil.

meta-**tolyldiethanolamine** $(HOC_2H_4)_2NC_6H_4CH_3$.
Properties: Its soaps form stable emulsions that are distinguished by their mild alkalinity, noncorrosiveness, ease of preparation, and flexibility in formulation. M.p. 62°C; b.p. 297.1°C (760 mm); vapor pressure < 0.1 (20°C); sp.gr. 1.0723 (20/20°C); solubility in water 1.67% by weight (20°C); viscosity 155 cps (20°C). Very soluble in acetone, ethanol, ethyl acetate, benzene.
Containers: 500-lb drums.
Uses: Emulsifier, dyestuff intermediate.

meta-**tolylenediamine.** See toluene-2, 4-diamine.

meta-**tolylenediaminesulfonic acid** [4, 6-diamino-meta-toluenesulfonic acid (SO_3H=1)] $CH_3C_6H_2(NH_2)_2SO_3H$.
Properties: White crystalline product, soluble in alkalies.
Derivation: By addition of meta-toluylenediamine sulfate to oleum and heating.
Grade: Technical.
Containers: Wooden barrels or fiber drums.
Use: Dyes.
Shipping regulations: None.*

meta-**tolylenediisocyanate.** See toluene-2, 4-diisocyanate.

ortho-**tolylethanolamine** $(HOC_2H_4)NHC_6H_4CH_3$.
Properties: Its soaps form stable emulsions that are distinguished by their mild alkalinity, noncorrosiveness, ease of preparation, and flexibility in formulation; mol wt 151; m.p. 63.2°C; flash point 385°F; lbs/gal 8.93 (80°C).
Use: Emulsifier.

(tolylhydroxyphenylaminomethyl)imidazoline hydrochloride. See phentolamine hydrochloride.

para-**tolyl**-alpha-**naphthylamine** $C_{10}H_7NHC_6H_4(CH_3)$.
Properties: Colorless, short prisms. Soluble in alcohol and ether. M.p. 79°C; b.p. 236°C (15 mm).
Derivation: By heating alpha-naphthyl-amine hydrochloride with para-toluidine.

para-**tolyl**-beta-**naphthylamine** $C_{10}H_7NHC_6H_4(CH_3)$.
Properties: Short, colorless, crystalline plates, sparingly soluble in alcohol. M.p. 103°C.
Derivation: From beta-naphthol and para-toluidine by heating.

para-**tolyl-1-naphthylamine-8-sulfonic acid** (tolylperi acid) $C_{17}H_{15}NO_3S$.
Properties: Greenish-gray needles. Soluble

in alcohol; rather insoluble in water.
Derivation: Arylation of 1-naphthylamine-8-sulfonic acid with para-toluidine.
Method of purification: Recrystallization.
Grades: Technical; mostly as sodium salt.
Containers: Barrels or steel drums.
Use: Azo colors.
Shipping regulations: None.*

tolyl-peri acid. See para-tolyl-1-naphthylamine-8-sulfonic acid.

ortho-**tolylpropanolamine** $CH_3C_6H_4NH(CH_2CHOHCH_3)$.
Properties: Slightly soluble in water; completely miscible with acetone, ethanol, ethyl acetate, benzene.
Containers: 500-lb drums.

tolyl-para-toluene sulfonate. See cresyl-para-toluene sulfonate.

tomatidine. A steroid secondary amine; the nitrogenous aglycone of tomatine. Isolated from the roots of the Rutgers tomato plant as the hydrochloride, $C_{27}H_{45}NO_2 \cdot HCl$. Crystals decompose at 275-280°C.

tomatine. A glycosidal alkaloid prepared from the dried leaves and stems of the tomato plant. White crystals, used as plant fungicide and as a specific precipitating agent for cholesterol. The crude extract is referred to as tomatin.

"Tona." [173] Trademark for a proteolytic enzyme meat-tenderizer. It is supplied as a powder and used as a liquid.

"Tonalid." [105] Trademark for 1,1,2,2,3,3,5-heptamethyl indan-6-methyl ketone ($C_{18}H_{26}O$), a synthetic aromatic ketone.
Properties: White crystalline powder; m.p. 55-60°C; odor similar to natural macrocyclic musks.
Uses: Perfumery, as musk odorant.

toner. An organic pigment which does not contain inorganic pigment or inorganic carrying base. (ASTM definition, ASTM D16-52). See also lake.

tonka (tonka bean; coumarouna bean; snuff bean; English bean; dipteryx).
Properties: Black-brownish seeds with wrinkled surface and brittle shining or fatty skins; aromatic, bitterish taste; balsamic, vanilla-like odor; efflorescences of coumarin are often observed on the surface.
Derivation: Bean of Dipteryx oppositifolia and other species of the Dipteryx.
Occurrence: Tropical America, Guiana and Angostura.
Grades: Angostura; Brazilian.
Containers: Casks; cases.
Uses: Production of natural coumarin; medicine; flavoring extracts; toilet powders.
Shipping regulations: None.*

tonka bean. See tonka.

tonka bean camphor. See coumarin.

"Tonox." [248] Trademark for p,p'-diaminodiphenylmethane.
Properties: Brown waxy lumps; sp.gr. 1.15; m.p. above 73°C; soluble in acetone and ethylene dichloride; moderately soluble in benzene; insoluble in water and gasoline.
Use: A general toner for improving the properties of vulcanized rubber; curing agent for epoxy resins.

"Tontine." [28] Trademark for pyroxylin or vinyl coated fabric used as a window shade cloth.

topaz $Al_2SiO_4(F,OH)_2$. A natural fluosilicate of aluminum, found in igneous rocks.
Properties: Colorless, yellow, pink, bluish, or greenish; luster vitreous; hardness 8; sp.gr. 3.4-3.6; one good cleavage.
Varieties:
Brazil rubies. Red or pink. "Burnt topaz" is obtained by heating yellow topaz until it turns pink.
Brazil sapphires. Blue.
Spanish topaz, Scotch topaz, Occidental topaz are actually yellow quartz or cirtine.
Oriental topaz. See corundum.
Occurrence: Colorado, California, Maine, Utah; Brazil; U.S.S.R.; Japan; Mexico.
Use: Gem stone.

topaz, false. See cirtine.

tops. A distillate obtained from crude petroleum.

"Toranil." [168] Trademark for desugared extract of coniferous woods, consisting of 96% calcium salt of lignosulfonic acid and 1.2% glucose.
Grades: "Toranil A," a viscous coffee-colored 50% solution with a characteristic tart odor; b.p. 107-108°C; sp.gr. 1.24 (60/60°F); available in tank-car lots.
"Toranil B," a free-flowing light tan, odorless, non-hygroscopic, water-soluble powder available in 50-lb multiwall kraft bags.
Uses: Dispersants; tanning agent; adhesive base; binder; chemical raw material.

torbanite. A variety of oil shale (q.v.).

torbernite (copper uranite) $Cu(UO_2)_2(PO_4)_2 \cdot 12H_2O$. A natural hydrated copper uranium phosphate, found in the oxidized parts of uranium deposits.
Properties: Color emerald green, grass green; luster vitreous to pearly; good micaceous cleavage; hardness 2-2.5; sp.gr. 3.22; radioactive.
Occurrence: Utah, Colorado, South Dakota, New Mexico; Congo; Europe; Australia.
Use: Minor ore of uranium.

torula yeast. A yeast that utilizes fermentable sugar in industrial wastes, such as fruit cannery refuse and sulfite liquor from pulp mills. The dried yeast is high in protein and vitamin content, enabling it to be used for enriching animal feeds. The enzymes present are destroyed during drying.

tosyl (Ts). The para-toluenesulfonyl radical, $CH_3C_6H_4SO_2$-. Esters of para-toluenesulfonic acid are known as tosylates.

totaquine. A mixture containing 7-12% anhydrous quinine and 70-80% total anhydrous crystallizable cinchona alkaloids.

*See "I.C.C. Shipping Regulations," page xiii.
Reference numbers refer to name of manufacturer. See "List of Manufacturers," page v.

Properties: White, to grayish white, or slightly yellowish white powder which darkens with exposure to light. It is nearly odorless, with a bitter taste. Soluble in alcohol and chloroform; partly soluble in ether; almost insoluble in water.
Containers: Cans.
Use: Medicine.

touchstone. See lydian stone.

touchwood. See agaric.

tourmaline
$(Na, Ca)(Al, Fe)B_3Al_3(AlSi_2O_9)(O, OH, F)_4$.
A complex borosilicate of aluminum. Varying amounts of lithium, sodium, potassium, calcium, iron, magnesium, and manganese may be present. Found in igneous rocks.
Properties: Color variable, black, brown, yellow, pink, blue; luster vitreous to resinous; hardness 7-7.5; sp.gr. 3.0-3.2; piezoelectric and pyroelectric.
Varieties:
Schorlite. A black iron-bearing variety.
Rubellite. A pink lithium-bearing variety used as a gem.
Brazilian emerald. Green gem stone.
Indicolite. Dark blue gem stone.
Occurrence: Ceylon; Madagascar; Brazil; U.S.S.R.; Island of Elba; California, Maine, Connecticut, New York.
Uses: Gem stone; pressure gauges; optical equipment; oscillator plates; source of boric acid.

"Tovex." [28] Trademark for a non-nitroglycerin, water-compatible explosive slurry that gives very high loading density. For use where rock or ore is massive and hard to break.

toxaphene. Technical chlorinated camphene with the approximate formula $C_{10}H_{10}Cl_8$. Contains 67-69% chlorine.
Properties: Amber, waxy solid with a mild odor of chlorine and camphor; melting range 65-90°C; density 1.66 (27°C); good residual toxicity; may be stored for a year as a solid or in solution without deterioration; is attacked by bases. Soluble in common organic solvents.
Containers: 50-lb bags; 5-, 55-gal drums.
Use: As an agricultural insecticide against insects and grasshoppers.
Warning: May be fatal if swallowed. May be absorbed through skin. MCA warning label.

TPA. Abbreviation for terephthalic acid.

TPG. Abbreviation for triphenylguanidine.

TPN. Abbreviation for triphosphopyridine nucleotide. See nicotinamide adenine dinucleotide phosphate.

TPP.
1. Abbreviation for triphenyl phosphate.
2. Abbreviation for thiamine pyrophosphate. See cocarboxylase.

TPT.
1. Abbreviation for triphenyltetrazolium chloride. See tetrazolium chloride.
2. Abbreviation for tetraisopropyl titanate.

trace elements. Elements present in soils, foods, water, etc., in extremely small amounts. Trace elements necessary in plant and animal nutrition are molybdenum, copper, iron, cobalt. Traces of zinc are necessary in plant nutrition, as are boron and manganese.

tracer. An isotopic form of an element used to allow the tracing of the element through a process in the presence of material containing the ordinary form of the element. Often a radioactive isotope is used. Usually the tracer element is incorporated into a suitable compound which is then referred to as tagged or labelled. The isotopically labelled form is for all practical purposes chemically and physically identical with the ordinary form. The two forms will follow identical paths through a complex process, and the tracer can be distinguished at any time by its radioactivity or atomic weight.

For instance, carbon 14 is widely used in studying organic reactions. A particular carbon 12 atom in a given product is replaced by carbon 14, as in adenine-8-C^{14}, so that the carbon atom in the 8 position can be followed throughout the particular reaction. In other situations, a known proportion of all the carbons is replaced by carbon 14 and their behavior followed. Such a compound might be written, for example, as C^{14}-adenine, 1 μc/mg.

Radioactive carbon dioxide has been used to follow the complex sequence of steps in the photosynthesis of compounds in plants. Other examples include the use of sodium-24 to determine blood circulation patterns in surgical diagnosis, the injection of radioactive oil-soluble material into the interface between two materials in a pipe-line to signal the necessity of directing the stream into a different storage container, and the use of radioactive satellite ablation coatings, along with a radiation detector and telemetering, to determine the resistance of such coatings during re-entry.

tragacanth gum.
Properties: Dull white, translucent plates or spirally twisted, yellowish powder. Soluble in alkaline solutions, aqueous hydrógen peroxide solution; swells up with water; insoluble in alcohol.
Derivation: An exudation from Astragalus gummifer.
Occurrence: Native to southwestern Europe, Greece, Turkey, Asia Minor, Iran.
Grades: U.S.P. XVI; Nos. 1, 2, 3.
Containers: Barrels; cases.
Uses: Pharmacy for making emulsions and trochees; adhesives; leather dressing; calico printing; emulsifying agent; food preservative; cosmetics; dyes.
Shipping regulations: None.*

tragacanthin. See bassorin.

"Tragtex" R, Number 1. [325] Gum tragacanth. Textile printing vehicle; sizing gum.

"Tral." [3] Trademark for hexocyclium methylsulfate (q.v.).

trans-. A prefix denoting that one of two geometrical isomers (q.v.) in which certain atoms or groups are on opposite sides of a plane. In this dictionary, it is disregarded in alphabetizing. See also cis-.

"Trans-4." [303] Trademark for trans-1,4-polybutadiene synthetic rubber.

transferase. An enzyme whose activity causes a transfer of a radical from one molecule to another. Examples are transaminases, transacetylases, and transmethylases, which effect the transfer of amino, acetyl, and methyl groups respectively.

transformer compound. A compound which acts as a cooling and insulating medium and for coating or impregnating electrical apparatus subject to contact with oil. The essential ingredient is a specially treated gum. The flowing point may be around 212°F.

Shipping regulations: None.*

transformer oil. Any refined petroleum fraction suitable for use in surrounding the coils of transformers. Such oils have two purposes, (1) to provide electrical insulation and (2) to conduct heat away from the coils. The usual product is nonviscous (usually less than 100 SUS at 100°F) and is refined to maintain oxidation, moisture, acid, soap, salts, and suspended matter at a minimum. A typical product is a nonviscous neutral oil with gravity 34° Bé; flash point 340°F, fire test 400°F, cold test 20°F, and Saybolt viscosity 80.

"TransistAR." [329] Trademark for chemicals specifically controlled and standardized to meet the critical requirements of semiconductor device manufacturers.

transistor. A device for electrical rectification and amplification. It consists of a semiconductor material to which contact is made by two or more electrodes, usually by metal points or soldered connections.

A diode type transistor consists of an n-type semiconductor crystal (n meaning negative, having extra electrons in lattice) in contact with a p-type crystal (p meaning positive, having holes in lattice due to deficiency of electrons). At the area in contact an n-p boundary is formed. Commonly the crystals are high purity germanium or silicon, but with traces of appropriate impurities. Thus, germanium with arsenic as an impurity produces an n-type crystal, while with boron as the trace impurity a p-type crystal is formed. A diode transistor functions as a rectifier of alternating current because it permits current flow across the n-p boundry only from the n- to the p-crystal and not from p to n. When the transistor is connected into an electrical circuit so that the n-crystal terminal is negative and the p-crystal terminal is positive, a current will flow. This takes place because the extra electrons in the n-crystal move across the n-p boundary, and displace lattice electrons of the p-crystal so that some of these flow into the external circuit, or, these electrons displace holes in the direction of the n-crystal and so cause further movement of extra electrons in the n-crystal to the n-p boundary, and consequent entry of electrons from the external circuit into the n-type crystal. With the opposite electrical polarity in the external circuit, current cannot flow through the transistor because the extra electrons in the n-crystal, and also the holes in the p-crystal, both start to move away from the n-p boundary and it becomes an electrical insulator.

A triode type transistor consists of an n-p-n arrangement of semiconductor crystals, with electrical connections to each part. This functions as an amplifier in a manner analogous to a vacuum tube device.

transmission oil (gear case oil). Steam refined cylinder oil with a gravity of about 25° Bé; flash point 600°F; cold test 30°F; Saybolt viscosity of 240 at 210°F.

transmutation. The transformation of atoms of one element into atoms of a different element as the result of a nuclear reaction. The reaction may be one in which two nuclei interact, as in the formation of oxygen from nitrogen and helium nuclei (alpha particles), or one in which a nucleus reacts with an elementary particle such as a neutron or a proton. Thus a sodium atom and a proton form a magnesium atom.

The term is also used in a more general sense to include reactions in which an atom is transformed into one of its isotopes.

"Transphalt." [140] Trade name for a series of dark thermoplastic resins which are polymeric polynuclear hydrocarbons. Available as 100°C softening point material, as a liquid whose viscosity is 45-55 S.S.U. (210°F), and as a 60% solution in aromatic solvent.

Properties: Color, coal tar 22; sp.gr. 1.01-1.14 depending upon grade; ash < 0.1%; benzene insoluble - nil. Soluble in aromatic, chlorinated, and terpene solvents; only partially soluble in aliphatic hydrocarbons.

Containers: Liquids or solutions in 18 gauge, oil type, steel drums, tank trucks or tank cars. Higher softening point materials in light gauge, rust resistant, metal coated steel drums.

Uses: Paint, saturants for paper and wall board, pipe coatings, joint sealers, floor tile, extenders in epoxy systems, in the rubber industry under the trade name of "Resinex" in calendered and extruded goods, and secondary plasticizers.

transuranic elements. Elements of higher atomic number than uranium, not found naturally, and produced by nuclear bombardment. See actinide elements.

"Trapex." [401] Trade name for a soil fumigant containing methylisothiocyanate.

Use: For injection or drench treatment of

soil to control fungi, weeds, nematodes and soil insects.
Warning! Hazardous vapor and liquid. Irritating to eyes, nose, throat and skin. Do not inhale or swallow. Avoid skin contact.

"Trasentine." [305] Trademark for adiphenine.

trass. A pale yellow, or gray-colored, metamorphosed, volcanic ash.
Use: In the preparation of hydraulic cements.

trass cement. See pozzolana cement.

Trauzl test. Test used to determine the strength of an explosion. Measured by exploding a known weight of substance in the cavity of a standard test block of lead and measuring its increase in volume resulting from explosion of the charge.

travertine (tufa). A porous, cellular variety of limestone in banded layers, formed by precipitation of calcium carbonate from calcareous springs and rivers. Travertine forms the deposits at Mammoth Hot Springs, Yellowstone National Park.

"Treadsure." [205] An anti-skid floor coating containing fine abrasive aggregates. Applied with a brush.

treble superphosphate. See triple superphosphate.

"Trebo-Phos." [57] Trademark for triple superphosphate.

tree of life. See thuja.

"Trek." [214] Trademark for a proprietary concentrated synthetic methanol base antifreeze containing special corrosion inhibitors.
Properties: Deep violet (dye added) liquid; practically odorless. B.p. 149°F; sp.gr. (60/60°F) 0.800; vol concentration of antifreeze to lower freezing point of water solution to 0°F, 27%; -20°F, 37%; wt/gal at 68°F, 6.6 lbs.
Containers: 1-qt, 1-gal tamper-proof cans; 54-gal drums (all nonreturnable).
Use: Antifreeze for use with water in automotive cooling systems.

tremolite $Ca_2Mg_5Si_8O_{22}(OH)_2$. A variety of amphibole. Some tremolite is sold as "fibrous talc."
Properties: Color white to light green; luster vitreous to silky; hardness 5-6; sp.gr. 3.0-3.3. Resistant to acids.
Occurrence: New York, California, Maryland; South Africa.
Use: As asbestos, particularly in acid-resisting applications; ceramics; paint.

tremorine dihydrochloride
$C_4H_8NCH_2C{:}CCH_2NC_4H_8 \cdot 2HCl$.
1,4-Dipyrrolidino-2-butyne dihydrochloride.
Properties: Odorless, white crystalline solid; melts at about 225°C with decomposition; soluble in water, alcohol, and chloroform.
Use: Medicine.

"Treopax." [337] Trade name for zirconium oxide containing 91.5% ZrO_2. White cream powder, with sp.gr. 5.2; average particle size 15 microns. Used as a mill addition opacifier in antimony and zirconium sheet iron and cast iron enamels, to increase reflectance and stabilize colors. See "Opax."

"Trepidone." [315] Trademark for mephenoxalone (q.v.).

tri. Chemical slang for trichloroethylene.

triacetin (glyceryl triacetate) $C_3H_5(CO_2CH_3)_3$.
Properties: Colorless liquid with slight fatty odor and a bitter taste; sp.gr. 1.160 (20°C); b.p. 258-260°C; m.p. -78°C; flash point 300°F; wt/gal 9.7 lbs. Slightly soluble in water; very soluble in alcohol, ether, and other organic solvents.
Typical specifications: B.p. 258-259°C (760 mm); sp.gr. 1.159 (20/20°C); refractive index n 25/D 1.4288-1.4296; viscosity 15.1 cps (25°C).
Derivation: By the action of acetic acid on glycerol.
Method of purification: Vacuum distillation followed by neutralization and filtration.
Grades: Technical; C.P.; N.N.D.
Containers: Tins; 500-lb drums; tank cars.
Uses: Camphor substitute in pyroxylin industries; plasticizer; fixative in perfumery; manufacture of cosmetics; specialty solvent; to remove carbon dioxide from natural gas; medicine (external).
Shipping regulations: None.*

triacetyloleandomycin. Triacetyl ester of an antibacterial substance produced by the growth of species of Streptomyces antibioticus.
Properties: White, odorless, crystalline powder; pH of a solution in diluted alcohol is between 7.5 and 9.0. Soluble in alcohol; slightly soluble in ether and in water. Specific rotation in trichloroethylene solution containing 200 mg of triacetyloleandomycin in 10 ml is -16° to -22°.
Grade: N.F. XI.
Use: Medicine.

triacontanoic acid. See melissic acid.

1-triacontanol (formerly confused with 1-hentriacontanol under the names myricyl or melissyl alcohol) $CH_3(CH_2)_{28}CH_2OH$. A long-chain fatty alcohol.
Properties: Colorless needles from ether; m.p. about 85-88°C; soluble in most organic solvents; insoluble in water.
Derivation: The triacontanyl palmitate is one of the chief constituents of beeswax.
Purification: Crystallization from benzene.
Use: Biochemical research.

trialkyl boranes R_3B.
Properties: Stable in absence of air or oxidizing agents; do not react with water; soluble in hydrocarbons and other organic solvents; insoluble in water; act as reducing agents at high temperatures. The trialkyl boranes isomerize internal olefins to terminal olefins by exchange reactions. Oxidation of the intermediate boro-compounds leads to primary alcohols. (Mixed

*See "I.C.C. Shipping Regulations," page xiii.
Reference numbers refer to name of manufacturer. See "List of Manufacturers," page v.

hexenes give a ninety per cent yield of 1-hexanol by this process.) Also, the trialkyl boranes can polymerize vinyl-type monomers at lower temperatures than other catalysts.
Uses: Petro-chemical, pharmaceutical, fatty acid, and essential oil industries.
See tributylborane; triethylborane.

triallyl cyanurate $(CH_2{:}CHCH_2OC)_3N_3$. Cyclic.
Properties: Colorless liquid or solid; m.p. 27.32°C; flash point greater than 176°F, Tag open cup; sp.gr. 1.1133 (30°C); refractive index n 25/D 1.5049. Miscible with acetone, benzene, chloroform, dioxane, ethyl acetate, ethyl alcohol, and xylene.
Uses: Polymers and organic intermediates.

triamcinolone (9-alpha-fluoro-16-alpha-hydroxyprednisolone) $C_{21}H_{27}FO_6$.
Properties: White crystalline powder; m.p. 264-268°C; insoluble in water; slightly soluble in usual organic solvents; soluble in dimethylformamide.
Grade: N.N.D.
Use: Medicine.

triamcinolone acetonide. 9-alpha-Fluoro-16-alpha, 17 alpha-isopropylidenedioxyprednisolone, $C_{24}H_{31}FO_6$.
Properties: Crystals; m.p. 276-279°C. Optical rotation (25°/D) + 124.9° (in dimethylformamide). Soluble in organic solvents; insoluble in water.
Derivation: Prepared by stirring a suspension of triamcinolone in acetone in the presence of a trace of perchloric acid.
Grades: N.N.D.
Use: Medicine.

1, 3, 5-triaminobenzene $C_6H_3(NH_2)_3$.
Properties: M.p., anhydrous, 129°C; hydrate, 84-86°C (1.5 moles water). Soluble in water, acetone, and alcohol. Insoluble in ether, cold benzene, carbon tetrachloride, and petroleum ether.
Containers: Bottles; fiber drums. Supplied as hydrochloride.
Uses: Possible ion exchange resin intermediate, possible wetting and frothing agent component, in photographic developers and organic reactions.
Shipping regulations: None.*

2, 4, 6-triaminotoluene trihydrochloride $C_6H_2(NH_2HCl)_3CH_3 \cdot H_2O$.
Properties: Fine light tan to cream crystals; very soluble in water; soluble in alcohol and acetone; insoluble in benzene. Melting point 119°C (free base).
Grades: Technical.
Containers: Bottles; fiber drums.
Uses: In nongelatin photographic emulsion with ethylenediamine for fixation; possible ion exchange resin component; possible wetting and frothing agent component; in photographic developers; possible intermediate in making various organic chemicals and pharmaceuticals; and as bases for varnishes and rubber chemicals.
Shipping regulations: None.*

2, 4, 6-triamino-sym-triazine. See melamine.

triamylamine $(C_5H_{11})_3N$.
Properties: Color yellow; sp.gr. (20°C) 0.79-0.80; triamylamine content at least 98.0%; initial b.p. not below 215°C, 95% boils between 225 and 260°C; wt/gal 6.60 lbs; flash point 174°F. Viscosity (20°C) 0.02421 poise; refractive index (18°C) 1.4374; surface tension (13°C) 24.4 dynes/cm; specific heat at room temperature 0.51 cal/gm; coefficient of expansion 0.00091 at 20-60°C; vapor pressure (26°C) 7 mm; heat of vaporization 79 cals/gm; insoluble in water; soluble in gasoline.
Derivation: From the reaction of amyl chloride and ammonia.
Containers: 1-gal, 5-gal cans; 55-gal drums.
Uses: Corrosion inhibitor, insecticidal preparations.

triamylbenzene $(C_5H_{11})_3C_6H_3$.
Properties: Sp.gr. (20°C) 0.87; boiling range 300-320°C; color water-white; odor faintly aromatic. Flash point 270°F.

triamyl borate $(C_5H_{11})_3BO_3$.
Properties: Sp.gr. (20°C) 0.845; boiling range 220-280°C; flash point 180°F; color water-white; odor faintly alcoholic. Soluble in alcohol and ether.
Derivation: Direct heating of boric acid and amyl alcohol.
Use: Varnish.

tri-para-tert-amylphenyl phosphate $(C_5H_{11}C_6H_4)_3PO_4$.
Properties: Boiling range 305-345°C at 5mm; m.p. 62-63°C; white solid; odorless; insoluble in water.
Use: Plasticizer.

"Triangle." [110] Brand name for general purpose channel black for use in paints, inks and plastics.

tri-para-anisylchloroethylene. See chlorotrianisene.

triarylmethane dyes. Dyes whose molecular structure involves a central carbon atom joined to three aromatic nuclei. The Colour Index ranges from 657 to 738. The color of these dyes is due in part to the aromatic rings and to the groups =C=NH and =C=N-. The members of this class function as basic dyes for cotton, using tannin as a mordant, or, if they contain sulfonic acid groups, as acid dyes for wool and silk. Examples are malachite green and methyl violet.

as-triazine-3, 5(2H, 4H)dione riboside. See 6-azauridine.

s-triazine-2, 4, 6-triol. See cyanuric acid.

s-triazine-2, 4, 6-trione. See isocyanuric acid.

"Tribase." [304] Trade name for hydrous tribasic lead sulfate $(3PbO \cdot PbSO_4 \cdot H_2O)$ vinyl stabilizer.
Properties: Fine white powder, sp.gr. 7.1, refractive index 2.1.
Containers: Fiberboard drums containing 75 and 400 lbs.

*See "I.C.C. Shipping Regulations," page xiii.
Reference numbers refer to name of manufacturer. See "List of Manufacturers," page v.

Uses: For electrical and other vinyl compounds requiring high heat stability. Special "XL" grade available for vinyl electrical insulation.

"Tribase-E." [304] Trade name for basic lead silicate sulfate vinyl stabilizer.
Properties: Fine white powder, sp.gr. 5.55, refractive index 2.1.
Containers: Multiwall paper bags (50 lbs. net).
Uses: A low specific gravity stabilizer with good electrical properties and moisture repellency for vinyl insulation. Special "XL" grade available for vinyl electrical insulation.

tribasic copper sulfate $CuSO_4 \cdot 3Cu(OH)_2 \cdot H_2O$.
Properties: Aqua colored powder of extremely fine particle size; water insoluble; stable in storage; forms essentially neutral water dispersion.
Containers: 50-lb bags and 48-lb cases (8 x 6 lb bags).
Uses: A fixed copper fungicide. Also nutritional trace element for plants. Compatible with DDT, arsenicals, organic insecticides, sulfur and cryolite. Used as spray or dust. Does not inhibit photosynthesis.

tribenzoin. See glyceryl benzoate.

triboluminescence. The emission of energy as light as a result of the fracture of crystals by impact or friction. It is believed to be due to an imbalance of electric charges on the newly formed crystal cleavage planes, and a consequent transfer of electrons. Illustrations are the pulverizing of sugar crystals in a mortar or the shaking of certain liquid or solid particles in vessels of different electron affinities.

tribromoacetaldehyde (bromal) CBr_3CHO.
Properties: An oily yellowish liquid; sp.gr. 2.66; b.p. 174°C. Soluble in water, alcohol, or ether.
Derivation: (a) By adding bromine to a solution of paraldehyde in ethylacetate. (b) By adding bromine to absolute alcohol, fractionating, treating the fraction boiling at 165° to 180°C with water and distilling.
Uses: Medicine; organic synthesis.
Shipping regulations: None.*

tribromoacetic acid CBr_3COOH.
Properties: Colorless crystals; soluble in water, alcohol, or ether. M.p. 135°C; b.p. 245° to 250°C.
Derivation: By oxidizing bromal with nitric acid.
Method of purification: Crystallization.
Grades: Technical.
Containers: Glass bottles; kegs.
Use: Organic synthesis.
Shipping regulations: None.*

tribromo-tert-butyl alcohol (acetone-bromoform) $CBr_3C(CH_3)_2OH$.
Properties: Fine white prismatic crystals; camphor odor and taste; m.p. 176°C. Slightly soluble in water; soluble in alcohol and ether.
Derivation: Reaction of acetone and bromoform with solid potassium hydroxide.
Use: Medicine.
Shipping regulations: None.*

tribromoethanol (1,1,1-tribromoethyl alcohol) CBr_3CH_2OH.
Properties: White crystals or powder with slight aromatic odor and taste; m.p. 79-82°C; b.p. 94°C (11 mm); unstable in air and light; slightly soluble in water; soluble in alcohol, ether, benzene, and amylene hydrate; aqueous and alcoholic solutions decompose on exposure to light.
Grade: U.S.P. XVI.
Derivation: By reduction of tribromoacetaldehyde with aluminum isopropylate.
Use: Medicine.

1,1,1-tribromoethyl alcohol. See tribromoethanol.

tribromomethane. See bromoform.

1,1,1-tribromo-2-methyl-2-propanol $CBr_3C(CH_3)_2OH$.
Properties: Fine white crystals; m.p. 176-177°C; soluble in water, methanol, ether.
Use: Organic synthesis.

tribromonitromethane. See bromopicrin.

tribromophenol. See bromol.

tribromophenol-bismuth. See bismuth tribromophenate.

"Triburon" Chloride. [190] Trademark for a brand of triclobisonium chloride (q.v.).

tributoxyethyl phosphate.
Properties: Slightly yellow oily liquid. Insoluble or limited solubility in glycerin, glycols and certain amines. Soluble in most other organic liquids.
Typical specifications: Sp.gr. 1.020 (20°C); f.p. < -70°C (viscous liquid); boiling range 215-228°C (4 mm); acidity 0.06% (max) as acetic acid; flash point 435°F; fire point 485°F; vapor pressure < 0.1 mm Hg (150°C); refractive index 1.434 (25°C); viscosity 12.2 cps (20°C); coefficient of thermal expansion 0.00081 from 10-40°C; wt/gal 8 lbs.
Containers: 5-gal cans (40-lb net); 55-gal steel drums (460-lbs net).
Uses: Primary plasticizer for most resins imparting low temperature flexibility, flame retardance and permanent flexibility.

tri-n-butyl aconitate $C_3H_3(COOC_4H_9)_3$.
Properties: Colorless, odorless liquid; sp.gr. 1.018 (20°C); refractive index 1.4500-1.4530 (25°C); b.p. 190°C (3 mm); insoluble in water; soluble in organic solvents; free acidity (max) 0.1%; residue on ignition (max) 0.05%; color (max) 150 APHA; water (Karl Fischer) 0.12% (max).
Containers: 8-oz sample; 450-lb drums; tank cars and wagons.
Uses: Plasticizer-stabilizer for vinylidene chloride polymers, nitrile and Buna-S rubbers, and cellulose-type lacquers. Insecticides.
Shipping regulations: None.*

tri-n-butylaluminum $(CH_3CH_2CH_2CH_2)_3Al$.
Properties: Colorless, pyrophoric liquid.
Derivation: Exchange reaction of butene-1 and isobutyl aluminum.
Uses: Production of organo-tin compounds.

tri-n-butylamine $(C_4H_9)_3N$.
Properties: Pale yellow liquid with amine odor. B.p. 214°C; f.p. below −70°C; sp. gr. (20/20°C) 0.7782; wt/gal 6.5 lbs; flash point (open cup) 175°F. Insoluble in water; soluble in most organic solvents.
Derivation: By reaction of butanol or butyl chloride with ammonia.
Grade: Technical.
Containers: 5-, 55-gal drums; tank cars.
Uses: Solvent; inhibitor in hydraulic fluids; intermediate.
Shipping regulations: None.*

tri-n-butylborane (tri-n-butylborine) $(CH_3CH_2CH_2CH_2)_3B$.
Properties: Colorless pyrophoric (spontaneously flammable) liquid; m.p. −34°C; b.p. +170°C (222 mm); density (25°C) 0.747 g/ml; vapor pressure (20°C) 0.1 mm; refractive index (n 20/D) 1.4285; insoluble in water; soluble in most organic solvents. Flash point −32°F.
See trialkyl boranes for uses and further properties.
Caution: Must be stored, transferred, or used in an inert atmosphere such as dry nitrogen or argon. Should be stored in dry, ventilated rooms at normal temperature.
Shipping regulations: Flammable liquid. Red label. (up to 150 lbs in cylinders).*

tributyl borate (butyl borate) $(C_4H_9)_3BO_3$.
Properties: Water-white liquid; sp.gr. 0.8550-0.8570; b.p. 232.4°C (760 mm); distillation range, 85% distills between 135°C and 140°C (40 mm); refractive index (n 25/D) 1.4071; m.p. less than −70°C; viscosity 1.601 cps (25°C); flash point (open cup) 185°F; hydrolyzes rapidly in presence of water; miscible with common organic liquids such as aliphatic alcohols, esters, diacetone, chloroform, carbon tetrachloride, and naphtha.
Derivation: From butyl alcohol and boric acid.
Uses: Agent for impregnating with crystalline boric acid to render textiles fire-resistant and to prevent plastic sheets and fibers from sticking together. Improves adhesion of lacquers and inks to metal surfaces; inhibits formation of wax crystals in oil at low temperatures; as gas welding flux leaves smoother, cleaner surface; drying agent to remove water from nonaqueous systems; antigelling agent.

tri-n-butylborine. See tri-n-butylborane.

tributyl citrate (butyl citrate) $C_3H_5O(COOC_4H_9)_3$.
Properties: Colorless or pale yellow, stable, odorless nonvolatile liquid. Practically insoluble in water.
Constants: M.p. −20°C; b.p. approximately 233.5°C at 22.5 mm; flash point 185°C (365°F); refractive index 1.4453 at 20°C; sp.gr. (25/25°C) 1.042; wt/gal 8.7 lbs at 68°F; pour point −80°F; viscosity (25°C) 31.9 cps; evaporation rate at 105°C 0.000065 g/sq cm/hr.
Typical specifications: Purity not less than 99% ester by weight; sp.gr. 1.043-1.049 (20/20°C); acidity not more than 0.2%, calculated as citric acid; water content, no turbidity when one volume is mixed with 19 vols. of 60° Bé gasoline at 20°C; color, water-white.
Grade: Technical.
Containers: 1-gal, 25-lb cans; 55-gal, 87-lb drums; tank cars.
Uses: Plasticizer; antifoam agent; solvent for cellulose nitrate.
Shipping regulations: None.*

tri-para-tert-butylphenyl phosphate $[(CH_3)_3CC_6H_4O]_3PO$.
Properties: Solid; b.p. (5 mm) 320°C; m.p. 102-105°C; flash point 275°C; insoluble in water.
Use: Plasticizer.

tributyl phosphate $(C_4H_9)_3PO_4$.
Properties: Stable, colorless, odorless, light-fast liquid. Miscible with most solvents and diluents.
Constants: Refractive index 1.4226 at 20°C; b.p. 177-178°C at 27 mm; latent heat of vaporization 55.1 cals/gm at 289°C; m.p. below −80°C; flash point 380°F; Saybolt viscosity 38.6 seconds at 85°F; wt/gal 8.19 lbs.
Typical specifications: Sp.gr. 0.973-0.983 at 20/20°C; acidity not more than 0.05%, calculated as phosphoric acid; water no turbidity when 1 vol. is mixed with 19 vols. of 60° Bé gasoline at 20°C; color water-white.
Grade: Technical.
Containers: 1-gal cans; 5-, 55-gal steel drums; tank cars.
Uses: Heat exchange medium; solvent extraction of metal ions from solution of reactor products; solvent for nitrocellulose, cellulose acetate; plasticizer; lacquers; plastics; pigment grinding assistant; solvent in inks; antifoam agent; dielectric; blending agent.
Shipping regulations: None.*

tri-n-butyl phosphine $(CH_3CH_2CH_2CH_2)_3P$.
Properties: Sp.gr. 0.8100 (min at 25/4°C); f.p. −60 to −65°C; b.p. 249°C (max); flash point 40°C; fire point 43°C; auto ignition point 260°C; refractive index (25°C) 1.4588; almost insoluble in water; miscible with ether, methanol, ethanol, benzene.
Uses: Fuel additive; epoxy resin curing catalyst; vinyl and isocyanate polymerization; organic intermediate.

tributyl phosphite $(C_4H_9O)_3P$.
Properties: Water-white liquid; b.p. 120°C (8 mm); sp.gr. 0.911 (25°C); refractive index (n 25/D) 1.4301. Soluble in common organic solvents.
Containers: Carboys.
Uses: Additive for greases and extreme-pressure lubricants; stabilizer for fuel oils and polyamides; gasoline additive.

O, O, O-tributyl phosphorothioate (tributyl thiophosphate) $(C_4H_9O)_3PS$.
Properties: Colorless liquid with characteristic odor. B.p. (4.5 mm) 142-145°C; sp.gr. 0.987; flash point (Cleveland open cup) 295°F. Insoluble in water; soluble in most organic solvents.
Containers: 1-, 4-lb bottles; 5-, 55-gal drums.
Uses: Plasticizer; lubricant additive; antifoam agent; hydraulic fluid; intermediate.

S, S, S-tributyl phosphorotrithioate ("DEF") $(C_4H_9S)_3PO$.
Properties: A liquid with boiling point 150°C (0.3 mm). Insoluble in water; soluble in aliphatic, aromatic, and chlorinated hydrocarbons.
Use: Cotton defoliant.

tributyl thiophosphate. See tributyl phosphorothioate.

tributyltin acetate $(C_4H_9)_3SnOOCCH_3$.
Properties: White crystalline solid.
Derivation: Reaction of sodium acetate with tributyltin chloride.
Uses: Fungicide and bactericide.

tributyltin chloride $(C_4H_9)_3SnCl$.
Derivation: Reaction of tetrabutyltin with dibutyltin chloride.
Use: Rodenticide.

tributyltin oxide $(C_4H_9)_3SnOSn(C_4H_9)_3$.
Properties: Colorless to pale yellow liquid; soluble in many organic solvents; practically insoluble in water; b.p. 180°C (2 mm).
Uses: Bactericide, fungicide.

tri-n-butyl tricarballylate $(C_4H_9OCOCH_2)_2CHCOOC_4H_9$.
Properties: Sp.gr. (24°C) 1.004; refractive index (26.5°C) 1.4388; b.p. 305°C; insoluble in water.
Use: Plasticizer.

tributyrin. See glyceryl tributyrate.

tricalcic phosphate. See calcium phosphate, tribasic.

tricalcium aluminate. See calcium aluminate.

tricalcium orthoarsenate. See calcium arsenate.

tricalcium orthophosphate. See calcium phosphate, tribasic.

tricalcium phosphate. See calcium phosphate, tribasic.

tricalcium silicate. See cement, Portland, and other cement articles. Also used as an anticaking agent in foods.

tricaprin (glyceryl tricaprinate) $C_3H_5(C_{10}H_{19}O_2)_3$. Triclinic crystals; insoluble in water; sp.gr. 0.921; m.p. 31°C.

tricarbimide. See cyanuric acid.

tricarboxylic acid cycle. See TCA cycle.

trichloroacetaldehyde. See chloral.

trichloroacetic acid (TCA) CCl_3COOH.
Properties: Deliquescent colorless crystals; sharp pungent odor; strongly corrosive; sp. gr. 1.6298; m.p. 57.5°C; b.p. 197.5°C; soluble in water, alcohol, and ether.
Derivation: (1) Treating chloral hydrate with fuming nitric acid; (2) from glacial acetic acid by the action of chlorine in presence of sunlight, ultraviolet radiation or catalysts.
Method of purification: Crystallization.
Grades: Technical; C.P.; U.S.P. XVI.
Containers: Tightly stoppered glass bottles; drums.
Uses: Organic synthesis; reagent for detection of albumin; medicine; pharmacy; herbicides.

trichloroacetic aldehyde. See chloral.

trichloroacetic aldehyde, hydrated. See chloral hydrate.

1, 2, 3-trichlorobenzene $C_6H_3Cl_3$.
Properties: White crystals; sp.gr. (solid) 1.69; refractive index (19°C) 1.5776; b.p. (760 mm) 221°C; m.p. 52.6°C; insoluble in water; slightly soluble in alcohol; soluble in ether.
Use: Synthesis.

1, 2, 4-trichlorobenzene $C_6H_3Cl_3$.
Properties: Colorless, stable, refractive liquid. Odor similar to that of orthodichlorobenzene. Miscible with most organic solvents and oils. Insoluble in water. Sp. gr. 1.4634 (25°C); b.p. 213°C; m.p. 17°C; flash point 100°C.
Derivation: Further chlorination of monochlorobenzene.
Grades: Technical; 99%; mixture of 1, 2, 4- and 1, 2, 3- isomers distilling at 213-219°C.
Containers: 1-, 2-gal cans; 60- 650-lb drums; tank cars.
Uses: Solvent in chemical manufacturing; dyes and intermediates; dielectric fluid; synthetic transformer oils; lubricants; heat transfer medium; insecticides.

1, 1, 1-trichloro-2, 2-bis(para-chlorophenyl)-ethane. Correct chemical name for DDT.

B-trichloroborazole $\overline{BClNHBClNHBClNH}$.
Properties: White crystalline solid; m.p. 84.5-85.5°C; b.p. 96.5-98°C/37 mm. Soluble in many organic solvents. Highly reactive.
Uses: Intermediate; gelling agent; catalyst; complexing agent.

trichlorobromomethane. See bromotrichloromethane.

2, 2, 3-trichlorobutanal. See butyl chloral.

trichloro-tert-butyl alcohol. See chlorobutanol.

trichlorobutyraldehyde. See butyl chloral.

trichlorobutyraldehyde hydrate. See butyl chloral hydrate.

3, 4, 4'-trichlorocarbanilide $C_6H_3Cl_2NHCONHC_6H_4Cl$. Colorless, heat resistant, highly insoluble bacteriostat, useful in soaps and detergents; plastics.

1, 1, 1-trichloroethane (methyl chloroform) CH_3CCl_3.
Properties: A colorless liquid; sp.gr. 1.325;

b.p. 75°C. Insoluble in water; soluble in alcohol and ether.
Containers: Drums; tank cars.
Use: Medicine; solvent.
See "Chlorothene."

1,1,2-trichloroethane (vinyl trichloride; beta-trichloroethane) $CHCl_2CH_2Cl$.
Properties: Clear, colorless liquid. Characteristic sweet odor; nonflammable. B.p. 113.7°C; latent heat of vaporization 68.7 cals/g, 123.5 Btu/lb; specific heat 0.270 (20°C) cal/g/°C; sp.gr. 1.4432 (20°C/4°C); refractive index 1.4458; vapor pressure 16.7 mm (20°C); wt/gal 12.0 lbs (20°C); f.p. −36.4°C; flash point, none; fire point, none; specific resistivity 5.2×10^8 ohms/cm; viscosity 1.20 cps (20°C). Miscible with alcohols, ethers, esters, and ketones; insoluble in water.
Grade: Technical.
Containers: 55-gal drums; tank cars.
Uses: Solvent for fats, oils, waxes, resins, other products; organic synthesis.

beta-trichloroethane. See 1,1,2-trichloroethane.

trichloroethanol CCl_3CH_2OH.
Properties: Viscous liquid; ether-like odor. Slightly soluble in water; miscible with alcohol, ether, and carbon tetrachloride. B.p. 150°C; f.p. (approx) 13°C; sp.gr. (25/4°C) 1.541.
Use: Intermediate.

trichloroethylene (tri) $CHCl{:}CCl_2$.
Properties: Stable, low-boiling, colorless, heavy, mobile, toxic liquid. Use with adequate ventilation. Chloroform-like odor. Nonflammable, nonexplosive, and noncombustible. Will not attack the common metals, even in the presence of moisture; b.p. 86.7°C; m.p. −73°C; sp.gr. 1.456-1.462 (25/25°C); refractive index 1.4735 (27°C); surface tension 32.0 dynes/cm (25°C); vapor pressure 60.0 mm (20°C); specific heat 0.229 cal/gm (23°C); flash point (ASTM open cup) none at b.p.; latent heat of evaporation 57.3 cals/ g at b.p.; coefficient of expansion (per °C) 0.00115 to 20°C; vapor density 0.277 lbs/ cu ft (90°C), 3.6 cu ft/lb (90°C); viscosity at 25°C 0.550 cps; heat of vaporization 57 kcal/kg, 104.5 Btu/lb; thermal conductivity (liquid) (64°F) 0.0672 Btu/sq ft/ft/°F/hr, wt/gal 12.16 lbs (25°C); fire point none; dielectric constant 3.27 (1000 cycle); power factor 2.2% (1000 cycle); specific resistivity 6.6×10^9 ohms/cm.
Typical specifications: Acidity not more than 0.001% (as hydrochloric); color water-white; sp.gr. 1.47-1.48 (15°C/15°C); boiling range, 95% or better distills from 86.0 to 87.5°C (760 mm); free chlorine none; residue none from filtered sample; average wt 12.20 lbs/gal (20°C). Miscible with all common organic solvents; practically insoluble in water.
Derivation: (a) From tetrachloroethane by treatment with lime or alkali in the presence of water, or by thermal decomposition, followed by steam distillation. Tetrachloroethane is obtained by chlorination of acetylene. (b) From ethylene by chlorination followed by fractional distillation.
Grades: U.S.P. XVI; technical; high purity; electronic.
Containers: Cans; drums; tank trucks; tank cars.
Uses (in approximate order of importance): Metal degreasing; extraction solvent for oils, fats, waxes; dry cleaning; refrigerant and heat exchange liquid; organic syntheses; fumigant. The electronic grade is used for cleaning and drying electronic parts.
Warning: Vapor harmful. MCA warning label.
Shipping regulations: None.*

trichlorofluoromethane (fluorotrichloromethane; fluorocarbon-11) CCl_3F.
Properties: Colorless, nearly odorless, volatile liquid. B.p. 23.7°C; f.p. −111°C; sp.gr. 1.494 (17.2°C); critical pressure 43.2 atm.
Derivation: From carbon tetrachloride and hydrogen fluoride, in the presence of fluorinating agents such as antimony tri- and penta- fluorides.
Grades: Technical; 99.9% min.
Containers: Drums; cylinders.
Uses: Solvent; fire extinguishers; refrigerant; aerosol propellants; air conditioning.
Shipping regulations: Nonflammable gas. Green label.*

trichloroisocyanuric acid (1,3,5-trichloro-s-triazine-2,4,6-trione) OCNClCONClCONCl.
Properties: White, slightly hygroscopic, crystalline powder or granules; loose bulk density (approx) powder 31 lbs/cu ft, granular 60 lbs/cu ft.
Active ingredient: Approx. 90% available chlorine.
Containers: 200-lb fiber drums.
Uses: Active ingredient in household dry bleaches, dishwashing compounds, scouring powders, detergent-sanitizers, and commercial laundry bleaches.

trichloroisopropyl alcohol. See isopral.

trichloromelamine (N,N',N''-trichloro-2,4,6-triamine-1,3,5-triazine) NC(NHCl)NC(NHCl)NC(NHCl).
Properties: Fine white powder; slightly soluble in water and glacial acetic acid; insoluble in carbon tetrachloride and benzene; pH saturated aqueous solution 4.
Derivation: By chlorination of melamine.
Grades: 89% available chlorine; see also "Sterimine."
Containers: Polyethylene bags in fiber drums.
Hazard: Flammable; can ignite spontaneously with oils and other reactive organic materials.
Uses: As a chlorine bleach and bactericide.
Shipping regulations: None.*

trichloromethane. See chloroform.

trichloromethyl chloroformate (diphosgene) $ClCOOCCl_3$.
Properties: Colorless, mobile liquid. Odor

*See "I.C.C. Shipping Regulations," page xiii.
Reference numbers refer to name of manufacturer. See "List of Manufacturers," page v.

is somewhat like that of phosgene (new-mown hay). Decomposed by heat, porous substances, activated carbons (with evolution of phosgene). Also decomposed by alkalies, hot water. Caution! Not so irritant as the mono- and di- compounds but more toxic and asphyxiating! Soluble in alcohol, benzene, and ether.

Constants: Sp.gr. 1.65 (15°C); b.p. 127-128°C; m.p. −57°C; vapor density 6.9 (air = 1); refractive index 1.45664 (22°C).

Derivation: (a) By chlorinating methyl formate. (b) By chlorinating methyl chloroformate. In both methods the mixture of chloro-derivatives is then separated by fractionation.

Grade: Technical.

Uses: Organic synthesis; military poison gas.

Shipping regulations: Poison, class A. Poison gas label. Legal label name: diphosgene.*

trichloromethyl ether $CHCl_2OCH_2Cl$.

Properties: Liquid. Pungent odor. Caution! Very irritant! Lachrymatory. Sp.gr. 1.5066 (10°C); b.p. 130-132°C. Soluble in alcohol, benzene, and ether; insoluble in water.

Shipping regulations: Poison, class A. Poison gas label.*

N-trichloromethylmercapto-4-cyclohexene-1, 2-dicarboximide. See captan.

trichloromethyl phenyl carbinyl acetate.

Properties: White crystalline substance; intense rose odor; m.p. 86-88°C. Clearly soluble in 18 parts of 95% alcohol.

Containers: Fibre drums.

Uses: In rose perfumes.

Shipping regulations: None.*

trichloromethylphosphonic acid $CCl_3PO(OH)_2$.

Strong dibasic acid, soluble in water and alcohol; insoluble in benzene and hexane. Used as an acid catalyst and condensation agent.

1, 1, 1, -trichloro-2-methyl-2-propanol. See chlorobutanol.

trichloromethylsulfenyl chloride. See perchloromethyl mercaptan.

N-trichloromethylthiotetrahydrophthalimide. See captan.

trichloronaphthalene. See chloronaphthalenes.

trichloronitromethane. See chloropicrin.

trichloronitrosomethane CCl_3NO.

Properties: Dark blue liquid. Unpleasant odor. Slowly decomposes, but is more stable in solution. Caution! Very irritant! Soluble in alcohol, benzene, ether; insoluble in water. Sp.gr. 1.5 (20°C); b.p. 5°C (70 mm).

Derivation: Interaction of sulfuric acid, sodium trichloromethylsulfonate, potassium nitrate, and sodium nitrate.

Grade: Technical.

Uses: Organic synthesis; military poison gas (lachrymator).

Shipping regulations: Poison, class A. Poison gas label.*

2, 4, 5-trichlorophenol $C_6H_2Cl_3OH$.

Properties: Gray flakes in sublimed mass with a strong phenolic odor; sp.gr. (25°/4°C) 1.678; b.p. 252°C; m.p. 61-63°C; no flash or fire point. Soluble in alcohol, ether, and acetone.

Use: Fungicide, bactericide.

Caution: May cause skin irritation. MCA warning label.

2, 4, 6-trichlorophenol $C_6H_2Cl_3OH$. (2, 4, 6-T).

Properties: Yellow flakes with strong phenolic odor; sp.gr. (25/4°C) 1.675; f.p. 61°C; b.p. 248-249°C; no flash or fire point. Soluble in acetone, alcohol, and ether.

Use: Fungicide.

Caution: May cause skin irritation. MCA warning label.

2, 4, 5-trichlorophenoxyacetic acid (2, 4, 5-T) $C_6H_2Cl_3OCH_2CO_2H$.

Properties: Light tan solid; m.p. 151-153°C; soluble in alcohol; insoluble in water; available as sodium and amine salts.

Containers: 50-, 200-lb drums.

Uses: Plant hormone; herbicide.

Warning: Irritating to eyes, nose and throat. MCA warning label.

Also available in form of esters, as, 2, 4, 5-trichlorophenoxyacetic acid, isopropyl ester.

2, 4, 5-trichlorophenyl acetate $C_6H_2Cl_3OOCCH_3$.

Use: As fungicide, especially on cotton seed.

1, 2, 3-trichloropropane $CH_2ClCHClCH_2Cl$.

Properties: Colorless liquid; sp.gr. 1.3888 (20/4°C); b.p. 156.17°C; refractive index (n 20/D) 1.4841. Slightly soluble in water; dissolves oils, fats, waxes, chlorinated rubber and numerous resins.

Derivation: Chlorination of propylene.

Uses: Paint and varnish remover; nonflammable solvent; degreasing agent.

trichlorosilane.

1. $SiHCl_3$ (silicochloroform).

Properties: Colorless very volatile liquid; b.p. 31.8°C; m.p. −127°C; density 1.35 (0°C).

Use: Since this silane is relatively easy to prepare, it has been much used as a source material in laboratory synthesis of organic silanes such as trimethyl silane and triphenyl silane.

Shipping regulations: Flammable liquid. Red label.*

2. Generic name for compounds of the formula $RSiCl_3$ of which methyl trichlorsilane, CH_3SiCl_3, is most important.

N, N', N" -trichloro-2, 4, 6-triamine-1, 3, 5-triazine. See trichloromelamine.

2, 4, 6-trichloro-1, 3, 5-triazine. See cyanuric chloride.

1, 3, 5-trichloro-s-triazine-2, 4, 6-trione. See trichloroisocyanuric acid.

trichlorotrifluoroacetone (1, 1, 3-trichloro-1, 3, 3-trifluoroacetone) $CCl_2FCOCClF_2$.

Properties: Colorless lachrymatory liquid; b.p. 84.5°C. Soluble in all proportions with water and most organic solvents. Stable to acid but not alkalies.
Containers: 1-lb bottles.
Uses: Solvent in acid media; complexing agent.

1,1,2-trichloro-1,2,2-trifluoroethane (trifluorotrichloroethane; fluorocarbon 113) CCl_2FCClF_2.
Properties: Colorless, nearly odorless, volatile liquid. B.p. 47.6°C; f.p. −35°C; critical pressure 33.7 atm; sp.gr. 1.42 (25°C).
Grade: Technical.
Containers: Drums.
Uses: Solvent; fire extinguishers; refrigerant; to make chlorotrifluoroethylene.

tricholine citrate (tris(2-hydroxyethyl)trimethylammonium citrate).
Containers: Carboys (65% solution).
Use: Medicine, nutritional factor.

"Triclene." [28] Trademark for trichloroethylene. Available in five grades.
Technical and Dry Cleaning Grade. Boiling range 86.6°-88.0°C. Wt/gal 12.19 lb at 20°C.
Uses: Dry cleaning and spotting; textile cleaning; as solvent for adhesives and waxes; asphaltic paint remover; for coating paper; organic reagent.
Metal Degreasing Grade. Boiling range 86.4°-87.9°C. Wt/gal 12.16 at 20°C.
Uses: Specifically stabilized for vapor degreasing of metals; liquid flushing of liquid oxygen and missile fuel systems and their component parts; low temperature heat transfer liquid; vapor drying of metals.
Paint Grade. Boiling range 85.4°-87.9°C. Wt/gal 12.14 lb at 20°C.
Uses: As a thinner for paints based on several types of alkyd resins; on asphaltic materials, including gilsonite and on some specialty resins, such as certain acrylics, epoxy esters, chlorinated rubbers, ethyl cellulose, etc.
Extraction Grade. Boiling range 86.5°-87.5°C. Wt/gal 12.23 lb at 20°C.
Uses: Extraction solvent for many waxes, resins, and alkaloids; flushing solvent; cold cleaner.
Freezing-point Depressant Grade. Wt/gal 12.21 lb at 20°C.
Containers: All grades, 55-gal (660-lb) drums; tank trucks; tank cars.
Uses: Freezing-point depressant for fire extinguisher fluids.

triclobisonium chloride $C_{36}H_{74}Cl_2N$. N,N'-Bis-[1-methyl-3-(2,2,6-trimethylcyclohexyl)-propyl]-N,N'-dimethyl-1,6-hexanediamine bis(methochloride). A crystalline powder; soluble in water, alcohol, chloroform; insoluble in ether. Used in medicine (external).

tricobalt tetraoxide. See cobalto-cobaltic oxide.

"Tricoloid." [301] Trademark for tricyclamol, an anticholinergic, used in medicine.

tricosane (n-tricosane) $CH_3(CH_2)_{21}CH_3$.
Properties: Glittering leaflets. Soluble in alcohol; insoluble in water. Sp.gr. 0.779 (48°C); b.p. 234°C (15 mm); m.p. 48°C.
Grade: Technical.
Containers: 1- and 5-lb glass bottles; fiber containers.
Use: Organic synthesis.

n-tricosanoic acid $CH_3(CH_2)_{21}COOH$. A saturated fatty acid not normally found in natural fats or oils. Synthetic compound is a white crystalline solid; m.p. 79.1°C. Purified product is used in medical research and as reference standard for gas chromatography.

tricresyl phosphate (tritolyl phosphate; TCP) $(CH_3C_6H_4O)_3PO$. A mixture of isomers.
Properties: Practically colorless, odorless liquid. Stable, nonvolatile. The ortho isomer is reputed to be the toxic element when present in isomeric mixtures. B.p. 420°C; dilution ratio with toluene 3.2; evaporative residue at atmospheric pressure not weighable; ignition temperature, nonflammable; refractive index 1.556 (25°C); sp.gr. 1.162 (25/25°C); wt/gal 9.7 lbs; crystallizing point below −35°C. Typical specifications: (of "ortho-free" grade) Form, clear oily liquid; essentially colorless; sp.gr. 1.166±0.007 (25/25°C); refractive index 1.556±0.001 (20°C); acidity (as H_3PO_4) 0.01% max; free phenols (permanganate test) a distinct purple color to be present after 30 minutes. (10 gram sample−40 cc N/100 $KMnO_4$). Miscible with all the common solvents and thinners, also with vegetable oils; insoluble in water.
Derivation: From cresylic acid and phosphorus oxychloride.
Grades: Coal tar; petroleum; to meet custom specifications.
Containers: 1-, 5-, 55-gal drums; tank cars.
Uses: Plasticizer for polyvinyl chloride, polystyrene, nitrocellulose; fire retardant for nitrocellulose; solvent mixtures; gasoline additive as lead scavenger; waterproofing and fireproofing compositions; additive to extreme pressure lubricants; hydraulic fluid and heat exchange medium.

tricresyl phosphite $(CH_3C_6H_4O)_3P$. (Not to be confused with tricresyl phosphate.)
Properties: Colorless liquid; slight phenolic odor. B.p. (0.11 mm) 191°C; sp.gr. (20/4°C) 1.115; flash point (open cup) 440°F. Insoluble in water; miscible with acetone, alcohol, benzene, ether, and kerosine.
Grades: Technical.
Uses: Stabilizer, plasticizer, and flame retardant for plastics and resins.

tricyanic acid. See cyanuric acid.

tricyclamol chloride
$C_6H_{11}C(C_6H_5)(OH)CH_2CH_2C_4H_8N \cdot CH_3Cl$.
1-Cyclohexyl-1-phenyl-3-pyrrolidino-1-propanol methylchloride.
Properties: White, extremely bitter, crystalline powder; faint characteristic odor; soluble in alcohol and in water; stable.
Grade: N.N.D.
Use: Medicine.

n-tridecane $CH_3(CH_2)_{11}CH_3$.
Properties: Colorless liquid. Soluble in alcohol; insoluble in water. Sp.gr. 0.755 (20/4°C); b.p. 236°C; f.p. −5.45°C; refractive index 1.4250 (20/D); flash point 80°C.
Grades: 95%; 99%; research.
Containers: Glass bottles; 1-, 5-gal drums.
Use: Organic synthesis.
Shipping regulations: None.*

n-tridecanoic acid (tridecylic acid; tridecoic acid) $CH_3(CH_2)_{11}COOH$. A saturated fatty acid normally not found in vegetable fats but prepared synthetically.
Properties: Colorless crystals; m.p. 44.5°C; sp.gr. 0.8458 (80/4°C); b.p. 312.4°C (760 mm), 192.2°C (16 mm); refractive index 1.4328 (50°C). Slightly soluble in water; soluble in alcohol and ether.
Available as 99% pure product for organic synthesis; medical research.

tridecanol. See tridecyl alcohol.

tridecoic acid. See n-tridecanoic acid.

tridecyl alcohol (tridecanol). General term for a commercial mixture of isomers of the formula $C_{12}H_{25}CH_2OH$.
Properties: Water-white liquid with pleasant odor; boiling range 252-272°C; sp.gr. (20/20°C) 0.845; wt/gal 7.0 lbs; flash point (Tag open cup) 180°F.
Derivation: By Oxo process (q.v.) from C_{12} hydrocarbons.
Grade: Technical.
Containers: 55-gal drums; tank cars.
Uses: Esters for synthetic lubricants; detergents; antifoam agent; tridecyl mercaptan; perfumes; tridecyl phthalate (plasticizer).

tri-n-decylaluminum $(C_{10}H_{21})_3Al$. A colorless liquid.
Derivation: From n-decene and isobutylaluminum.
Use: Polyolefin catalyst.

tridecyl bromide $C_{13}H_{27}Br$.
Properties: Colorless liquid; density 1.025 (20°C); b.p. 158-160°C (15 mm); insoluble in water.

tridecylic acid. See n-tridecanoic acid.

tridecyl phosphite $(C_{10}H_{21}O)_3P$.
Properties: Water-white liquid; decyl alcohol odor; sp.gr. 0.892 (25/15.5°C); m.p. less than 0°C; refractive index 1.4565 (25°C).
Containers: 55-gal drums.
Uses: Chemical intermediate; stabilizer for polyvinyl and polyolefin resins.

tridihexethyl chloride
$C_6H_{11}C(C_6H_5)(OH)CH_2CH_2N(C_2H_5)_2 \cdot C_2H_5Cl$. 3-Diethylamino-1-phenyl-1-cyclohexylpropanol ethochloride; triethyl(3-hydroxy-3-cyclohexyl-3-phenyl propyl) ammonium chloride.
Properties: White, odorless, crystalline powder, bitter taste. Freely soluble in water, in methanol, in chloroform, and in alcohol. Practically insoluble in ether and in acetone. Melting range 198-202°C.
Grade: N.F. XI.
Use: Medicine.

tri(dimethylphenyl) phosphate (trixylenyl phosphate) $[(CH_3)_2C_6H_3O]_3PO$.
Properties: Sp.gr. 1.155; refractive index 1.5535; b.p. (10 mm), 243-265°C; flash point, 233°C; solubility in water (85°C), 0.002% by weight.
Use: Plasticizer.

"Tridione." [3] Trademark for trimethadione.

tridodecyl amine. See trilauryl amine.

tridymite SiO_2. A vitreous, colorless or white, native form of pure silica. Found variously but not so commonly as quartz (q.v.). Quartz will change into tridymite with a 16.2% increase in volume at 870°C. Unlike quartz, it is soluble in boiling sodium carbonate solution. Sp.gr. 2.28-2.3; hardness 7.

triethanolamine (TEA; tri(2-hydroxyethyl)-amine) $(HOCH_2CH_2)_3N$.
Properties: Colorless, viscous, hygroscopic liquid with slight ammoniacal odor; m.p. 21.2°C; b.p. 360°C; vapor pressure < 0.01 mm (20°C); sp.gr. 1.126; flash point (open cup) 375°F; wt/gal 9.4 lbs; miscible with water, alcohol; soluble in chloroform; slightly soluble in benzene and ether; slightly less alkaline than ammonia. Commercial product contains up to 25% diethanolamine and up to 5% monoethanolamine.
Derivation: Reaction of ethylene oxide and ammonia.
Grades: Technical; regular; 98%; N.F. XI.
Containers: 5-, 10-, 55-, 110-gal drums; 6000-, 8000-gal tank cars.
Uses: Fatty acid soaps used in drycleaning, cosmetics, household detergents, and a wide variety of oil emulsions (cutting oils, lubricants, textile processing, leather fatliquoring, etc.); wool scouring; textile antifume agent and water-repellent; dispersion of dyes, casein, shellac, rubber latex; corrosion inhibitor; softening agent, humectant, and plasticizer for textiles, glues, leather coatings, waxes, polishes; insecticide.

triethanolamine lauryl sulfate. A liquid or paste.
Containers: Drums, tank cars; tank trucks.
Uses: Detergent; wetting, foaming and dispersing agent for industrial, cosmetic and pharmaceutical applications.

triethanolamine oleate. See trihydroxyethylamine oleate.

triethanolamine stearate. See trihydroxyethylamine stearate.

triethanolamine titanate. See titanium chelates.

triethanolamine trinitrate phosphate. See trolnitrate phosphate.

1,1,3-triethoxyhexane
$CH(OC_2H_5)_2CH_2CH(OC_2H_5)C_3H_7$.
Properties: Liquid; sp.gr. 0.8746 (20/20°C); b.p. 133°C (50 mm); f.p. −100°C; wt/gal

*See "I.C.C. Shipping Regulations," page xiii.
Reference numbers refer to name of manufacturer. See "List of Manufacturers," page v.

7.3 lb; flash point 210°F. Insoluble in water.
Use: Starting point for synthesis of aldehydes, acids, esters, chloride, amines, etc.

triethoxymethane. See triethyl orthoformate.

1, 1, 3-triethoxy-3-methoxypropane (triethylmethyl malonaldehyde diacetal) $(CH_3O)(C_2H_5O)CHCH_2CH(OC_2H_5)_2$.
Properties: Sp.gr. (25/4°C), 0.9300; b.p. (6 mm) 86°C.
Grade: 99%.
Use: Intermediate; crosslinking and insolubilizing agent.

triethyl aconitate $C_2H_5OOCCHC(COOC_2H_5)CH_2COOC_2H_5$.
Properties: Sp.gr. (25°C) 1.096; refractive index (26°C) 1.4517; b.p. (5 mm) 154-156°C.
Use: Plasticizer.

triethylaluminum (ATE) $(C_2H_5)_3Al$.
Properties: Clear colorless liquid; sp.gr. 0.837; m.p. −52.5°C; b.p. 194°C; specific heat 0.527 (91.4°F). Miscible with saturated hydrocarbons. Reacts violently with water; ignites on exposure to air. Also reacts vigorously with acids, halogens, alcohols and amines.
Derivation: Synthesized by introduction of ethylene and hydrogen into an autoclave containing aluminum. The reaction proceeds under moderate temperature and varying pressures.
Grade: 88-94%.
Containers: Cylinders.
Uses: Catalyst intermediate for polymerization of olefins, especially ethylene; pyrophoric fuels; production of alpha-olefins and long-chain alcohols; gas plating of aluminum.
Shipping regulations: Flammable liquid. Red label.*

triethylamine $(C_2H_5)_3N$.
Properties: Colorless liquid; strong ammoniacal odor. B.p. 89.7°C; f.p. −115.3°C; sp.gr. (20/20°C) 0.7293; wt/gal (20°C) 6.1 lbs; flash point (open cup) 25°F. Slightly soluble in water above 20°C; miscible with water below 18°C, alcohol, and ether.
Derivation: From ethyl chloride and ammonia under heat and pressure.
Containers: 1-, 5-gal cans; 55-gal drums; tank cars.
Uses: Catalytic solvent in chemical synthesis; manufacture of accelerator activators for rubber, and wetting, penetrating and waterproofing agents of quaternary ammonium types; solvent; corrosion inhibitor; propellant.
Shipping regulations: Flammable liquid. Red label.*

triethylbenzylammonium chloride $(C_2H_5)_3C_6H_5CH_2NCl$.
Properties: White crystals; odorless; decomposes on heating; soluble in water and alcohol.

triethylborane (triethylborine) $(C_2H_5)_3B$.
Properties: Colorless, spontaneously flammable liquid. Density 0.68 (25°C); flash point −32°F; m.p. −93°C; b.p. 95°C; refractive index 1.3971; heat of combustion 20,000 Btu/lb. Miscible with most organic solvents; immiscible with water.
Derivation: Reaction of triethylaluminum and boron halide, or of diborane and ethylene.
Use: Igniter or fuel for jet and rocket engines.
Shipping regulations: Flammable liquid. Red label (up to 140 lb in cylinders).*

triethylborine. See triethylborane.

triethyl citrate $C_3H_5O(COOC_2H_5)_3$.
Properties: Colorless, mobile liquid. Bitter taste; b.p. (760 mm) 294°C; b.p. (1 mm) 126-127°C; sp.gr. 1.136 (25°C); refractive index 1.4405 (24.5°C); evaporation rate 0.000676 g/sq cm/hr (105°C); pour point −50°F; viscosity (25°C) 35.2 cps; solubility in water 6.5 g/100 cc; solubility in oil 0.8 g/100 cc.
Derivation: Esterification of citric acid.
Grades: Technical; refined.
Containers: Metal drums and cans; tank cars.
Uses: Solvent for cellulose nitrate, acetate and ethers, natural resins such as dammar and ester gums, starch ethers; plasticizer for cellulose nitrate and acetate and vinyl resins; softener; paint removers; agglutinant; perfume base.
Shipping regulations: None.*

triethylene glycol (TEG) $HO(C_2H_4O)_3H$.
Properties: Colorless, hygroscopic, practically odorless liquid. Similar in properties to diethylene glycol. Sp.gr. 1.1254 (20/20°C); b.p. 287.4°C (760 mm); vapor pressure less than 0.01 mm (20°C); flash point 330°F; wt/gal 9.4 lbs (20°C); coefficient of expansion 0.00069 (20°C); freezing point −7.2°C; viscosity 0.478 poise (20°C). Soluble in water; immiscible with benzene, toluene and gasoline.
Typical specifications: Acidity not more than 0.02% (as acetic); color (500 mm tube) not more than 5 yellow Lovibond; sp.gr. 1.122-1.127 (20/20°C); boiling range (760 mm) below 270°C none, below 280°C not more than 20%, below 290°C not less than 85%, below 300°C not less than 95%; average wt/gal 9.36 lbs (20°C).
Grade: Technical.
Containers: 1-, 5-gal cans; 55-gal drums; tank cars.
Uses: Solvent for nitrocellulose, various gums and resins; lacquers; organic synthesis; in air-conditioning units; bactericide (in vapor form); humectant in printing inks; textile conditioner.
Shipping regulations: None.*

triethylene glycol diacetate $CH_3COOCH_2CH_2OCH_2CH_2OCH_2CH_2OOCCH_3$.
Properties: Sp.gr. (25°C) 1.112; refractive index n(25°C) 1.437; b.p. 300°C; m.p. less than −60°C.
Use: Plasticizer.

triethylene glycol dicaprylate (triethylene glycol dioctoate).
Properties: (typical specification) Clear

*See "I.C.C. Shipping Regulations," page xiii.
Reference numbers refer to name of manufacturer. See "List of Manufacturers," page v.

liquid; sp.gr. 0.973 (20°C); acidity 0.3% max. (caprylic); moisture 0.05% max; m.p. –3°C; b.p. 243°C (5 mm).
Uses: Low temperature plasticizer for synthetic resins and rubbers.

triethylene glycol dichloride. See triglycol dichloride.

triethylene glycol didecanoate
$C_9H_{19}COO(C_2H_4O)_3OCC_9H_{19}$.
Properties: B.p. 237°C at 2.0 mm Hg; sp.gr. 0.9584 (20/20°C); viscosity 28.6 cps (20°C).
Use: Plasticizer.

triethylene glycol di(2-ethylbutyrate)
$C_5H_{11}OCOCH_2(CH_2OCH_2)_2CH_2OCOC_5H_{11}$.
Properties: A light-colored liquid; sp.gr. 0.9946 (20/20°C); 8.3 lb/gal (20°C); b.p. 196°C (5 mm); vapor pressure 5.8 mm Hg (200°C); solubility in water 0.02% by wt (20°C); viscosity 10.3 cps (20°C).
Use: Plasticizer.

triethylene glycol di(2-ethylhexoate)
$C_7H_{15}OCOCH_2(CH_2OCH_2)_2CH_2OCOC_7H_{15}$.
Properties: A light-colored liquid; sp.gr. 0.9679 (20/20°C); 8.1 lb/gal (20°C); b.p. 219°C (5 mm); vapor pressure 1.8 mm Hg (200°C); insoluble in water; viscosity 15.8 cps (20°C).
Use: Plasticizer.

triethylene glycol dihydroabietate
$C_{19}H_{31}COOCH_2CH_2OCH_2CH_2OCH_2$-$CH_2OOCC_{19}H_{31}$.
Properties: Sp.gr. (25°C) 1.080-1.090; refractive index (20°C) 1.5180; vapor pressure (225°C) 2.5; flash point 226°C; insoluble in water.
Use: Plasticizer.

triethylene glycol dimethyl ether
$CH_3(OCH_2CH_2)_3OCH_3$.
Properties: Water-white liquid with mild ether odor; sp.gr. (20/20°C) 0.9862; refractive index 1.4233 (n 20/D); flash point 232°F; b.p. (760 mm) 216.0°C; b.p. (100 mm) 153.6°C; m.p. –46°C. Competely soluble in water and hydrocarbons at 20°C. May contain dangerous peroxides.
Containers: Glass bottles; 1-, 5-gal cans; 55-gal drums.
Uses: Solvent, for gases; for coupling immiscible liquids.

triethylene glycol dioctoate. See triethylene glycol dicaprylate.

triethylene glycol dipropionate
$C_2H_5CO(OCH_2CH_2)_3OOCC_2H_5$.
Properties: Sp.gr. (25°C) 1.066; refractive index (25°C) 1.436; b.p. (2 mm) 138-142°C; m.p. less than –60°C; solubility in water, 6.70% by weight.
Use: Plasticizer.

triethylenemelamine
(TEM; 2,4,6-tris(1-aziridinyl)-s-triazine; 2,4,6-tris(ethyleneimino)-s-triazine)
$NC[N(CH_2)_2]NC[N(CH_2)_2]NC[N(CH_2)_2]$.
Properties: White, crystalline, odorless powder; m.p. 160°C (polymerizes); polymerizes readily with heat or moisture; soluble in alcohol, water, methanol, chloroform, and acetone.
Grade: U.S.P. XVI.
Hazard: Very poisonous!
Use: Medicine. See nitrogen mustards.

triethylenephosphoramide
(TEPA; tris-(1-aziridinyl)phosphine oxide; APO) $(NCH_2CH_2)_3PO$.
Properties: Colorless crystals; m.p. 41°C; soluble in water, alcohol and ether.
Use: Medicine. See nitrogen mustards. Also used with tetrakis(hydroxymethyl)phosphonium chloride (THPC) to form a condensation polymer suitable for flameproofing cotton. See also tris[1-(2-methyl)aziridinyl]phosphine oxide.
Caution! Toxic.

triethylenetetramine $NH_2(C_2H_4NH)_2C_2H_4NH_2$.
Properties: Moderately viscous yellowish liquid. It is less volatile than diethylenetriamine but resembles it in many other properties. Soluble in water.
Constants: B.p. 277.5°C; sp.gr. 0.9818 (20/20°C); vapor pressure $<$ 0.01 mm (20°C); flash point 260°F; wt 8.2 lbs/gal (20°C); coefficient of expansion 0.00081 (20°C); viscosity 0.267 poise (20°C).
Typical specifications: Sp.gr. 0.980-0.985 (20/20°C); boiling range 260-290°C (760 mm).
Grades: Technical; anhydrous.
Containers: 1-gal cans; 5-, 10-, 55-gal drums; tank cars.
Uses: Making detergents and softening agents; synthesis of dyestuffs, pharmaceuticals and rubber accelerators.
Danger! Causes severe eye and skin burns. MCA warning label.

triethylenethiophosphoramide
(TSPA; thio-TEPA; tris(1-aziridinyl)phosphine sulfide).
Properties: White crystals; m.p. 51°C; moderately hygroscopic; soluble in benzene, acetone, water, warm petroleum ether, and warm diethyl ether. Slight turbidity in aqueous solution. Polymerizes in aqueous solution or in presence of moisture, expecially at an acid pH.
Use: Medicine. See nitrogen mustards.
Caution! Toxic.

tri(2-ethylhexyl)phosphate
$[C_4H_9CH(C_2H_5)CH_2]_3PO_4$.
Properties: A light-colored liquid; sp.gr. 0.9260 (20/20°C); 7.7 lb/gal (20°C); b.p. 220°C (5 mm); vapor pressure 2.0 mm Hg (200°C); insoluble in water; viscosity 14.1 cps (20°C). Used as a plasticizer.

triethylmethane. See 3-ethylpentane.

triethylmethyl malonaldehyde diacetal. See 1,1,3-triethoxy-3-methoxypropane.

triethyl orthoformate (orthoformic ester; triethoxy methane) $CH(OC_2H_5)_3$.
Properties: Colorless liquid; pungent odor; b.p. 145.9°C (760 mm); refractive index 1.39218 (18.8°C); sp.gr. 0.895 (20/20°C). Soluble in alcohol, ether, and water.

Derivation: Reaction of sodium ethylate on chloroform or reaction of hydrochloric acid on hydrogen cyanide in ethyl alcohol solution.
Method of purification: Fractional distillation.
Containers: 55-gal steel drums.
Use: Organic synthesis; pharmaceuticals.

triethyl phosphate (TEP) $(C_2H_5)_3PO_4$.
Properties: Colorless, high-boiling liquid. Mild odor; very stable at ordinary temperatures. Compatible with many gums and resins. Is very difficultly flammable and contributes fireproofing characteristics to some products in which it is used. Soluble in most organic solvents; is completely miscible in water. When mixed with water is quite stable at ordinary temperatures, but at elevated temperatures it hydrolyzes slowly.
Constants: M.p. −56.4°C; b.p. 216°C (760 mm); flash point 115.6°C (240°F); refractive index 1.4055 (20°C); wt/gal 8.90 lbs (68°F).
Typical specifications: Purity not less than 97% ester by weight; sp.gr. 1.068-1.072 at 20/20°C; acidity not more than 0.02%, calculated as phosphoric acid; water no turbidity when 1 vol. is mixed with 19 vols. of 60° Bé gasoline at 20°C; b.p. 215-216°C; wt/gal 8.9 lbs.
Grades: Technical.
Containers: 1-, 5-, 55-gal drums; tank cars; tank trucks.
Uses: High-boiling solvent; plasticizer for resins, plastics, gums; in manufacture of pesticides; catalyst; lacquer remover.

triethyl phosphite $(C_2H_5)_3PO_3$.
Properties: Colorless liquid; sp.gr. 0.9687 (20°C); b.p. 156.6°C; insoluble in water; soluble in alcohol and ether.
Containers: Glass bottles; 5-, 55-gal drums.
Uses: Synthesis; plasticizers; stabilizers; lube and grease additives; flameproofing composition.

O, O, O-triethyl phosphorothioate (triethyl thiophosphate) $(C_2H_5O)_3PS$.
Properties: Colorless liquid with characteristic odor. B.p. (10 mm) 93.5-94°C; sp.gr. 1.074; flash point (Cleveland open cup) 225°F.
Containers: 1-, 4-lb bottles; 5-, 55-gal drums.
Uses: Plasticizer; lubricant additive; antifoam agent; hydraulic fluid; intermediate.

triethyl thiophosphate. See triethyl phosphorothioate.

triethyl tricarballylate
$(C_2H_5OCOCH_2)_2CHCOOC_2H_5$.
Properties: Sp.gr. (20°C) 1.087; refractive index (26°C) 1.4234; b.p. (5mm) 158-160°C; solubility in water (20°C) 0.62% by weight.
Uses: Plasticizer.

"Triexcel." [342] Trademark for rotenone and isome-synergized pyrethrin extracts in concentrate form for insecticidal formulations.

trifluoroacetic acid CF_3COOH.
Properties: Colorless fuming liquid; hygroscopic; pungent odor. A strong acid. B.p. 72.4°C; sp.gr. 1.535; m.p. −15.25°C; index of refraction (n 20/D) 1.2850; very soluble in water.
Containers: Custom packed.
Use: Very strong non-oxidizing acid; reaction medium; solvent; catalyst.

trifluorochloroethylene. See chlorotrifluoroethylene.

trifluorochloromethane. See chlorotrifluoromethane.

trifluoromethylbenzene. See benzotrifluoride.

trifluoromethylhydrothiazide. See hydroflumethiazide.

3-trifluoromethyl-4-nitrophenol (alpha, alpha, alpha-trifluoro-4-nitro-meta-cresol) $CF_3C_6H_3(NO_2)OH$.
Properties: Crystals; m.p. 74-76°C.
Use: To exterminate lampreys, especially in the Great Lakes. It is placed in tributary streams where it kills the lamprey larvae.

alpha, alpha, alpha-**trifluoro-4-nitro**-meta-**cresol.** See 3-trifluoromethyl-4-nitrophenol.

trifluoronitrosomethane CF_3NO.
Properties: Bright blue, fairly stable gas. Disagreeable odor. Caution! Very irritant. B.p. −84°C; m.p. −150°C.
Derivation: (a) Interaction of fluorine and silver cyanide in the presence of silver nitrate; (b) from nitric oxide and iodotrifluoromethane or bromotrifluoromethane in the presence of ultraviolet light.
Use: Monomer for nitroso rubber.

trifluoropentachloropropane $CF_3CCl_2CCl_3$.
Properties: Flash point 109°C; b.p. 153°C.

trifluorotrichloroethane. See trichlorotrifluoroethane.

trifluorovinylchloride. See chlorotrifluoroethylene.

triflupromazine hydrochloride
(10-(3-dimethylaminopropyl)-2-(trifluoromethyl)phenothiazine hydrochloride) $C_{18}H_{19}F_3N_2S \cdot HCl$.
Properties: White crystalline powder; decomposes 173-174°C. Soluble in water, alcohol, acetone; solutions are sensitive to air and light.
Grade: N.N.D.
Use: Medicine.

triformol. See sym-trioxane.

"Trigamine." [73] Trademark for buffered aliphatic amine.
Properties: Water-white to pale-yellow viscous liquid with a pleasant odor. Soluble in water, ethyl alcohol (50%), glycerin and diethylene glycol. Insoluble in oils and hydrocarbon solvents.
Constants: Sp.gr. (25°C) 1.17; pH (10% solution) 9.5; neutralization value 208-210.
Containers: 1-gal cans (10 lbs); 5-gal cans (48 lbs); 55-gal drums (525 lbs).
Uses: Solvent and plasticizer for aqueous solutions of casein and shellac in place of borax,

ammonia, etc. Emulsifying agent for the manufacture of "soluble" waxes for sizing and finishing of textiles, leather, paper, etc. Emulsifying agent for the manufacture of automobile, furniture and floor polishes.

triglycerides. Triglycerides, the chief constituents of fats and oils, are naturally occurring esters of normal acids (fatty acids) and glycerol. They have the general formula $CH_2(OOCR_1)CH(OOCR_2)CH_2(OOCR_3)$, where R_1, R_2, and R_3 are usually of different chain length. Refining processes will often yield a commercial product in which R_1, R_2, and R_3 are the same chain length.
See stearin, olein, etc.
Derivation: Extraction from animal, vegetable and marine matter.
Uses: Fatty acids and derivatives; manufacture of edible oils and fats (such as cooking oil and margarine); manufacture of monoglycerides.

triglycol dichloride (triethylene glycol dichloride) $Cl(C_2H_4O)_2C_2H_4Cl$.
Properties: Colorless liquid; sp.gr. 1.1974 (20/20°C); b.p. 241.3°C (760 mm); vapor pressure 0.03 mm (20°C); flash point 250°F; wt/gal 10.0 lbs (20°C); freezing point −31.5°C; viscosity 0.0493 poise (20°C); coefficient of expansion 0.00092 (20°C). Insoluble in water.
Typical specifications: Sp.gr. 1.1950-1.2000 (20/20°C); boiling range 230-245°C (760 mm); acidity not more than 0.01% (as hydrochloric acid).
Grades: Technical.
Containers: 1-gal cans; 5-gal (tin-lined) drums; 55-gal (galvanized) drums.
Uses: Solvent for hydrocarbons, oils, other products; extractant; intermediate for making dyes, resins and insecticides; organic synthesis.

triglycollamic acid (nitrilotriacetic acid) $N(CH_2COOH)_3$.
Properties: Odorless, white to grayish white crystals; m.p. 246-249°C; insoluble in water and most organic solvents; soluble in caustic alkalies.
Use: Synthesis; sequestering agent.

trigonella foenum graecum. See fenugreek.

trigonelline (coffearine; caffearine; gynesine) $CH_3N^+C_5H_4COO^- \cdot H_2O$. N-Methylnicotinic acid betaine. A base formed in the seeds of many plants.
Properties: Colorless prisms; m.p. 218°C (dec); very soluble in water; slightly soluble in alcohol; nearly insoluble in ether and benzene.
Use: Biochemical research.

tri-n-hexylaluminum $(C_6H_{13})_3Al$.
Properties: Colorless liquid; b.p. (0.001 mm) 105°C.
Derivation: Exchange reaction between hexene and isobutyl aluminum.
Use: Polyolefin catalyst.

trihexylene glycol diborate $(C_6H_{12}O_2)_3B_2$.
A colorless liquid.
Derivation: Reaction of hexyleneglycol with boric oxide.
Use: Gasoline additive.

trihexyl phosphite $(C_6H_{13}O)_3P$.
Properties: Mobile, colorless liquid with characteristic odor; sp.gr. 0.897 (20/4°C); b.p. 135-141°C (0.2 mm); flash point 320°F (C.O.C.). Miscible with most common organic solvents; insoluble in water, hydrolyzes very slowly in water. High degree of thermal stability. Exposure to air should be kept to a minimum.
Containers: 5-gal, 55-gal steel drums.
Uses: Intermediate for insecticides; component of vinyl stabilizers; lubricant additive; specialty solvent.

trihexyphenidyl hydrochloride (cyclohexylphenyl-1-piperidinepropanol hydrochloride; 3-(1-piperidyl)-1-cyclohexyl-1-phenyl-1-propanol hydrochloride) $C_6H_{11}C(OH)(C_6H_5)CH_2CH_2C_5H_{10}N \cdot HCl$.
Properties: White, odorless solid. M.p. 249.0-249.5°C (dec); freely soluble in methanol; soluble in alcohol and chloroform; slightly soluble in water; very slightly soluble in ether and benzene. pH (1% solution) 5.5-6.0.
Grade: U.S.P. XVI.
Use: Medicine.

trihydrated telluric oxide. See telluric acid.

1,2,3-trihydroxyanthraquinone. See anthragallol.

1,2,4-trihydroxyanthraquinone. See purpurin.

1,2,7-trihydroxyanthraquinone. See anthrapurpurin.

1,2,3-trihydroxybenzene. See pyrogallic acid.

1,3,5-trihydroxybenzene. See phloroglucinol.

3,4,5-trihydroxybenzoic acid. See gallic acid.

2,4,5-trihydroxybutyrophenone $C_6H_2(OH)_3COC_3H_7$.
Properties: Yellow-tan crystals; m.p. 149-153°C. Very slightly soluble in water; soluble in alcohol and propylene glycol.
Use: Solid antioxidant for polyolefins and paraffin waxes.

tri(2-hydroxyethyl)amine. See triethanolamine.

trihydroxyethylamine oleate (triethanolamine oleate) $(HOCH_2CH_2)_3N \cdot HOOCC_{17}H_{33}$. A surface active agent made by reaction of triethanolamine with oleic acid.
Use: Emulsifying agent.

trihydroxyethylamine stearate (triethanolamine stearate) $(HOCH_2CH_2)_3N \cdot HOOC \cdot C_{17}H_{35}$.
Properties: Cream colored, wax-like solid. Faint fatty odor. Soluble in methyl alcohol, ethyl alcohol, mineral spirits, mineral oil, vegetable oil. Dispersible in hot water.
Constants: Titer 42°C; sp.gr. 0.968; pH (25°C) 8.8-9.2 (5% aqueous dispersion); m.p. 42-44°C.
Containers: 1-, 5-gal cans; 50-gal drums.
Uses: Emulsifying agent for the manufacture of fluid oil emulsions for the cosmetic and

pharmaceutical industries; in general a surface active agent.

tri(hydroxymethyl)aminomethane. See tris(hydroxymethyl) aminomethane.

2, 4, 6-trihydroxytoluene. See methylphloroglucinol.

triiodomethane. See iodoform.

triiodothyronine
$HOC_6H_3IOC_6H_2I_2CH_2CH(NH_2)COOH$. 3, 5, 3'-Triiodothyronine. Either a derivative or precursor of thyroxine (q.v.). Triiodothyronine may be the true hormone of the thyroid gland; some evidence indicates that the physiological activity of triiodothyronine is at least three times that of thyroxine. It increases the metabolic rate and oxygen consumption of animal tissues.
Use: Biochemical and physiological research.

triisobutylaluminum (TIBAL)
$[(CH_3)_2CHCH_2]_3Al$.
Properties: Clear, colorless liquid; sp.gr. 0.7876 (20°C); f.p. 1.0°C; b.p. 114°C (30 mm); reacts violently with water; fumes violently or ignites with air. Also reacts vigorously with acids, halogens, alcohols and amines.
Derivation: Reaction of isobutylene and hydrogen with aluminum under moderate temperature and varying pressures.
Uses: Polyolefin catalyst; manufacture of primary alcohols and olefins; pyrophoric fuel.
Shipping regulations: Flammable liquid. Red label.*

triisobutylene. A mixture of isomers of the formula $(C_4H_8)_3$ readily prepared by polymerizing isobutylene under suitable conditions of temperature and pressure and usually in the presence of a catalyst. A typical mixture is 2, 2, 4, 6, 6-pentamethylheptane-3 and 2-neopentyl-4, 4-dimethylpentene-1. May be depolymerized to simpler isobutylene derivatives.
Properties: Sp.gr. 0.764 (60°F); boiling range 348-354°F.
Containers: 55-gal drums; tank cars.
Uses: For synthesis of resins, rubbers, and intermediate organic compounds; lubricating oil additive, raw material for alkylation in producing high octane motor fuels.

triisooctyl phosphite.
Properties: Mobile, colorless liquid with characteristic odor. Miscible with most common organic solvents; insoluble in water. Hydrolyses very slowly in water. Exposure to air should be kept to a minimum. High thermal stability. Sp.gr. 0.891 (20/4°C); b.p. 161-164°C (0.3 mm); flash point 385°F (C.O.C.).
Containers: 5-gal, 55-gal steel drums.
Uses: Intermediate for insecticides; component of vinyl stabilizers; lubricant additive; specialty solvent.

O, O, O-triisooctyl phosphorothioate (triisooctyl thiophosphate) $(C_8H_{17}O)_3PS$.
Properties: Colorless liquid with characteristic odor. B.p. (0.2 mm) 160-170°C; sp. gr. 0.933; flash point (Cleveland open cup) 410°F. Insoluble in water; soluble in most organic solvents.
Containers: 1-, 4-lb bottles; 5-, 55-gal drums.
Use: Plasticizer; lubricant additive; hydraulic fluid; intermediate.

triisooctyl thiophosphate. See triisooctyl phosphorothioate.

triisopropanolamine $N(C_3H_6OH)_3$.
Properties: Crystalline pure-white solid. Mild base. (A mixture of isopropanolamines which has sp.gr. of 1.004-1.010 and is liquid at room temperature is also marketed). Sp.gr. 0.9996 (50/20°C); m.p. 45°C; b.p. 305°C; vapor pressure < 0.01 mm (20°C); freezing point 58°C; viscosity 1.38 poise (60°C). Soluble in water.
Typical specifications: Melts 46-50°C.
Grade: Technical.
Containers: Drums; tank cars.
Use: Making emulsifying agents.

triisopropyl borate $[(CH_3)_2CH]_3BO_3$.
Properties: Colorless liquid; b.p. 138-140°C.
Derivation: Reaction of isopropyl alcohol with boric oxide.

triisopropyl phosphite $[(CH_3)_2CH]_3PO_3$.
Properties: Mobile, colorless liquid with a characteristic odor; sp.gr. 0.914 (20/4°C); b.p. 94-96°C (50 mm); flash point 165°F (C.O.C.). Miscible with most common organic solvents; insoluble in water. Hydrolyzes slowly in water. Exposure to air should be held to a minimum. High thermal stability.
Containers: 55-gal, 5-gal steel drums.
Uses: Intermediate for insecticides; component of vinyl stabilizers; lubricant additive; specialty solvent.

triketohydrindene hydrate $C_9H_4O_3 \cdot H_2O$. 1, 2, 3-Indantrione hydrate.
Properties: White crystals; becomes red at 125°C, swells at 193°C, melts at 239-240°C; freely soluble in water.
Use: Biological test reagent.

trilaurin. The glyceride of lauric acid; glyceryl trilaurate.

trilauryl amine (tridodecyl amine) $(C_{12}H_{25})_3N$. A liquid; sp.gr. 0.82; m.p. 14°C; soluble in organic solvents; insoluble in water.
Uses: Chemical intermediate; metal complexes.

"Trilene." [207] Trademark for trichloroethylene. A volatile liquid used by inhalation to produce anesthesia and analgesia.

"Triluxe." [56] Trademark for dry cleaning solvent consisting of trichloroethylene.

trimagnesium phosphate. See magnesium phosphate, tribasic.

"Tri-Mal." [304] Trademark for tribasic lead maleate vinyl stabilizer.
Properties: Soft, yellowish-white crystalline powder; sp.gr. 6.0; refractive index 2.08.
Containers: 75-, 400-lb fiberboard drums.

*See "I.C.C. Shipping Regulations," page xiii.
Reference numbers refer to name of manufacturer. See "List of Manufacturers," page v.

Uses: Stabilizer in vinyl plastics; vulcanizing agent for chlorosulfonated polyethylene.

"Trimene Base." [248] Trademark for a reaction product of ethyl chloride, formaldehyde, and ammonia. Rubber accelerator.
Properties: Dark-brown, viscous liquid; sp. gr. 1.10; soluble in water and acetone; insoluble in gasoline and benzene.

trimer. A molecule formed by union of three identical simpler molecules. Also applied to the substances composed of such triple molecules; thus C_6H_6 is a trimer of C_2H_2.
See polymer.

trimer acid. See dimer acid.

trimercuric orthophosphate. See mercuric phosphate.

trimercurous orthophosphate. See mercurous phosphate.

trimethadione (3,5,5-trimethyl-2,4-oxazolidinedione) $C_6H_9NO_3$ or $\overline{OC(O)N(CH_3)C(O)C}(CH_3)_2$.
Properties: White, granular, crystalline substance. Camphorlike odor. M.p. 45-47°C; soluble in water; freely soluble in alcohol, chloroform, and ether; pH (5% solution) about 6.0.
Grade: U.S.P. XVI.
Use: Medicine.

trimethaphan camphorsulfonate $C_{22}H_{25}N_2OS \cdot C_{10}H_{15}O_4S$. d-3,4-(1',3'-Dibenzyl-2'-ketoimidazolido)-1,2-trimethylenethiophanium d-camphorsulfonate.
Properties: White crystals; slight odor. M.p. 230-235°C. Soluble in water, alcohol and chloroform; insoluble in ether.
Grade: U.S.P. XVI.
Use: Medicine.

trimethobenzamide hydrochloride $C_{21}H_{28}N_2O_5 \cdot HCl$. 4-(2-Dimethyl aminoethoxy)-N-(3,4,5-trimethoxybenzoyl) benzyl amine hydrochloride.
Properties: Crystals; m.p. 187-190°C; soluble in water and alcohol; insoluble in ether.
Use: Medicine.

trimethoxyborine. See trimethyl borate.

trimethoxyboroxine (methyl metaborate) $(CH_3O)_3B_3O_3$.
Properties: Colorless liquid; m.p. 10-11°C; b.p., dissociates; density 1.216 (25°C); refractive index 1.3986.
Derivation: Reaction of methyl borate with boric oxide.
Grade: 99%.
Use: Metal-fire extinguishing fluid.

3,4,5-trimethoxyphenethylamine. See mescaline.

trimethylacethydrazide ammonium chloride. See Girard's T Reagent.

trimethylacetic acid (pivalic acid) $(CH_3)_3CCOOH$.
Properties: Colored crystals; sp.gr. 0.905 (50°C); refractive index 1.3931 (36.5°C); m.p. 35.5°C; b.p. 163.8°C; soluble in water, alcohol and ether.
Use: Intermediate.

trimethylaluminum (ATM) $(CH_3)_3Al$.
Properties: A clear colorless pyrophoric liquid; b.p. 126°C; f.p. 15.4°C; sp.gr. 0.752. Flames instantly on contact with air; reacts violently with water, vigorously with acids, halogens, alcohols and amines.
Derivation: By sodium reduction of dimethylaluminum chloride.
Use: Catalyst for olefin polymerization; pyrophoric fuel; manufacture of straight-chain primary alcohols and olefins.
Shipping regulations: Flammable liquid. Red label.*

trimethylamine (TMA) $(CH_3)_3N$.
Properties: Colorless, liquefied gas; fishy, ammoniacal odor; flammable; sp.gr. 0.662 (−5°C); b.p. 2.87°C; m.p. −117.1°C; flash point of 25% solution (Tag open cup) 38°F. Soluble in water, alcohol, and ether.
Derivation: By the interaction of methanol and ammonia over a catalyst at high temperature. The mono-, di-, and trimethylamines are all produced, and yields are regulated by conditions.
Method of separation: Azeotropic distillation.
Grades: Anhydrous 99% min; aqueous solution 25, 30, 40%.
Containers: Solution: 1-gal glass bottles; 5-, 55-gal drums; tank cars. Anhydrous: 25-, 50-, 100-, 1400-lb cylinders; tank cars.
Uses: Organic synthesis; warning agent for gas; manufacture of disinfectants; flotation agent; insect attractant; quaternary ammonium compounds; synthetic resins.
Danger! Extremely flammable. Hazardous liquid and vapor under pressure. Liquid causes burns. Vapor extremely irritating. MCA warning label for anhydrous form.
Shipping regulations: Anhydrous: Flammable gas. Red gas label. Aqueous solution: Flammable liquid. Red label.*

1,2,4-trimethyl-5-aminobenzene. See pseudocumidine.

2,4,5-trimethylaniline. See pseudocumidine.

1,2,4-trimethylbenzene. See pseudocumene.

1,3,5-trimethylbenzene. See mesitylene.

sym-**trimethylbenzene.** See mesitylene.

uns-**trimethylbenzene.** See pseudocumene.

trimethyl borate (methyl borate; trimethoxyborine) $(CH_3O)_3B$.
Properties: Water-white liquid. B.p. 67-68°C; sp.gr. 0.915; f.p. −29°C. Miscible with ether, methanol, hexane, tetrahydrofuran; decomposes in presence of water.
Derivation: Reaction of boric acid and methanol.
Containers: Truck and car lots.
Uses: Solvent; dehydrating agent; flame retardant for plastics, paints, and lacquers; fungicide for citrus fruit; neutron

scintillation counters; brazing flux; boron compounds intermediate.

2,2,3-trimethylbutane (isopropyltrimethylmethane; triptane) C_7H_{16}; and $CH_3C(CH_3)_2C(CH_3)CH_3$.
Properties: Colorless liquid. Soluble in alcohol; insoluble in water.
Constants: Sp.gr. 0.691; b.p. 81.0°C; f.p. −24.96°C; refractive index 1.3895 (20°C).
Grade: Technical.
Uses: Organic synthesis; aviation fuel.

trimethyl carbinol. See tert-butyl alcohol.

trimethylchlorosilane $(CH_3)_3SiCl$.
Properties: Colorless liquid. B.p. 57°C; sp.gr. 0.854 (25/25°C); refractive index (n 25/D) 1.3893; flash point (Cleveland open cup) 0°F. Readily hydrolyzed by moisture, with the liberation of hydrochloric acid.
Derivation: By Grignard reaction of silicon tetrachloride and methylmagnesium chloride.
Grade: Technical.
Use: Intermediate for silicone fluids, as a chain terminating agent.
Shipping regulations: Flammable liquid. Red label.*

3,3,5-trimethylcyclohexanol-1 $C_6H_8(CH_3)_3OH$.
Properties: Sp.gr. 0.878 (40/20°C); m.p. 35.7°C; b.p. 198°C (760 mm). Soluble in most organic solvents, hydrocarbons, oils; insoluble in water.
Uses: Menthol and camphor substitute; antifoaming agent; manufacture of hydraulic fluids and textile soaps; odor masking.
Caution: Vapor harmful. MCA warning label.

4-(2,6,6-trimethyl-1-cyclohexenyl)-buten-3-one-2 (beta-ionone). See ionone.

3,3,5-trimethylcyclohexyl mandelate. See cyclandelate.

trimethyl dihydroquinoline polymer (TDQP) $(C_{12}H_{15}N)_n$ (n approx 3). Consists of polymer containing probably three or more quinoline groups.
Properties: Amber pellets; sp.gr. 1.08; softening point 75°C. Insoluble in water; miscible with ethanol, acetone, benzene, monochlorobenzene, isopropyl acetate and gasoline.
Use: Antioxidant; stabilizer or polymerization inhibitor.
Handle with caution. Toxic!

trimethylene. See cyclopropane.

trimethylene bromide (1,3-dibromopropane) $CH_2BrCH_2CH_2Br$.
Properties: Colorless liquid; sweet odor; sp.gr. 1.979 (20/4°C); b.p. 166°C; insoluble in water; soluble in organic solvents; m.p. −34.4°C.
Derivation: Synthetic.
Method of purification: Distillation after washing with concentrated sulfuric acid.
Grades: Technical; C.P.
Containers: Carboys, drums.
Use: Intermediate for dyestuff and pharmaceutical industries; cyclopropane manufacture.
Shipping regulations: None.*

trimethylene chlorobromide. See 1-bromo-3-chloropropane.

trimethylenedicyanide. See glutaronitrile.

trimethylene glycol (1,3-propylene glycol; 1,3-propanediol) $CH_2OHCH_2CH_2OH$.
Properties: Colorless, odorless liquid; sp.gr. 1.0537 (25°C); b.p. 210-211°C; soluble in water, alcohol and ether.
Derivation: From acrolein.
Grades: Technical, 95%; pure, 99%.
Use: Intermediate.

sym-**trimethylene trinitramine.** See cyclonite.

trimethylethylene. See 3-methyl-2-butene.

trimethylglycine. See betaine.

trimethylheptanoic acid. See isodecanoic acid.

2,2,5-trimethylhexane $(CH_3)_3CCH_2CH_2CH(CH_3)_2$.
Properties: Liquid; f.p. −105.84°C; b.p. 124.06°C; sp.gr. 0.711 (60/60°F); refractive index 1.399 (20/D); flash point 15°C.
Grades: 95%, 99%; research.
Containers: Bottles; drums.
Use: Synthesis.
Shipping regulations: Flammable liquid. Red label.*

3,5,5-trimethylhexan-1-ol $C_9H_{20}O$.
Properties: A colorless, mobile liquid of mild odor; b.p. 194°C; sp.gr. 0.8236 (25/4°C); wt/gal 6.86 lbs (25°C); refractive index (n 25/D) 1.4300; flash point (open cup) 200°F. Insoluble in water.
Derivation: High-pressure synthesis.
Uses: Suggested for synthetic lubricants; additives to lubricating oils; wetting agent; softener in manufacture of various plastics; disinfectants and germicides.

trimethylmethane. See isobutane.

trimethylnonanol. See 2,6,8-trimethylnonyl-4 alcohol.

2,6,8-trimethyl-4-nonanone $(CH_3)_2CHCH_2COCH_2CH(CH_3)CH_2CH(CH_3)_2$.
Properties: Practically water-white liquid with pleasant odor; possesses a high solvent power for vinyl resins and certain other synthetic resins, the cellulose esters and ethers, and many substances soluble with difficulty in other solvents; insoluble in water; sp.gr. 0.8165 (20/20°C); wt/gal 6.8 lbs (20°C); f.p. −75°C; b.p. 211-219°C (760 mm); viscosity 1.91 cps (20°C).
Uses: Solvent; dispersant; intermediate; lube oil dewaxing.

2,6,8-trimethylnonyl-4 alcohol (trimethylnonanol) $CH_3CH(CH_3)CH_2CHOHCH_2CH(CH_3)CH_2$-$CH(CH_3)CH_3$.
Properties: Colorless liquid with characteristic odor; sp.gr. 0.8913 (20/20°C); wt/gal 6.9 lbs (20°C); b.p. 225.2°C (760 mm); f.p. −60°C; viscosity 21.4 cps. Insoluble in water.
Uses: Surface-active and flotation agents; lube additives; rubber chemicals.

trimethylolethane (pentaglycerine; methyltrimethylol methane) $CH_3C(CH_2OH)_3$.
Properties: Colorless hygroscopic crystals. Soluble in water and alcohol.
Containers: 200-lb drums.
Uses: Conditioning agent; manufacture of varnishes, alkyd and hard resins, synthetic drying oils.

trimethylolpropane (hexaglycerine) $C_2H_5C(CH_2OH)_3$.
Properties: Colorless hygroscopic crystals. Soluble in water and alcohol.
Containers: 200-lb drums.
Uses: Conditioning agent; manufacture of varnishes, alkyd and hard resins, synthetic drying oils.

trimethylolpropane monooleate. Theoretically $C_2H_5C(CH_2OH)_2CH_2OOCC_{17}H_{33}$. The commercial product is a mixture of mono-, di-, and tri- esters, free polyol and free oleic acid.
Properties: Oily liquid; sp.gr. 0.954 (25°C); f.p. less than −20°C. Insoluble in water; soluble in most organic solvents.
Uses: Water-in-oil emulsifier; corrosion inhibitor; low-temperature plasticizer; deicing agent for gasoline.

3,5,5-trimethyl-2,4-oxazolidinedione. See trimethadione.

2,2,4-trimethylpentane. See isooctane.

2,3,4-trimethylpentane $(CH_3)_2CHCH(CH_3)CH(CH_3)CH_3$.
Properties: Liquid; f.p. −109.43°C; b.p. 113°C; sp.gr. 0.723 (60/60°F); refractive index 1.4042 (20/D); flash point 5°C.
Grades: 95%, 99%; research.
Containers: Bottles; drums.
Use: Intermediate.
Shipping regulations: Flammable liquid. Red label.*

2,4,4-trimethylpentene-1 (alpha-diisobutylene) $H_2C{:}C(CH_3)CH_2C(CH_3)_3$.
Properties: Colorless liquid; b.p. 101.44°C; f.p. −93.5°C; refractive index 1.4086 (20°C); density 0.7150 (20°C); flash point 0°C.
Derivation: Polymerization of isobutene.
Grades: 95%; 99%; research.
Containers: Bottles; drums.
Use: Organic synthesis; motor fuel synthesis, particularly isooctane.
Shipping regulations: Flammable liquid. Red label.*
See diisobutylene.

2,4,4-trimethylpentene-2 (beta-diisobutylene) $H_3CC(CH_3){:}CHC(CH_3)_3$.
Properties: Colorless liquid; b.p. 104.55°C; sp.gr. 0.724 (60/60°F); f.p. −106.4°C; refractive index 1.416 (20/D); flash point 0°C.
Grade: 95%.
Containers: Bottles; drums.
Use: Organic synthesis.
Shipping regulations: Flammable liquid. Red label.*
See diisobutylene.

tri-2-methylpentylaluminum $[(CH_3)_2CH(CH_2)_3]_3Al$.
Properties: Colorless liquid.
Derivation: Reaction of 2-methylpentene and isobutylaluminum.
Use: Polyolefin catalyst.

N, alpha, alpha-**trimethylphenethylamine.** See mephentermine.

N-alpha, alpha-**trimethylphenethylamine sulfate.** See mephentermine sulfate.

trimethyl phosphate $(CH_3O)_3PO$.
Typical properties: Colorless liquid; 22.1% phosphorus; density 1.210 g/ml at 68°F; flash point greater than 305°F; boiling point 379°F; refractive index (n 20/D) 1.397; pour point −51°F. Soluble in both gasoline and water.
Uses: For controlling spark plug fouling, surface ignition and rumble in gasoline engines.
Warning: Do not take internally. Avoid contact with skin and eyes; flush immediately with much water if contact occurs.

trimethyl phosphite $(CH_3O)_3P$.
Properties: Colorless liquid; b.p. 108-108.5°C; pour point, less than −75°F; sp. gr. (20/4°C) 1.046; flash point (Cleveland open cup) 100°F. Insoluble in water; soluble in hexane, benzene, acetone, alcohol, ether, carbon tetrachloride, and kerosine.
Uses: Chemical intermediate, especially for insecticides.

trimethylpropylmethane. See 2,2-dimethylpentane.

2,4,6-trimethylpyridine. See 2,4,6-collidine.

2,4,6-trimethyl-1,3,5-trioxane. See paraldehyde.

trimethylvinylammonium hydroxide. See neurine.

trimethylxanthine. See caffeine.

"Trimeton." [321] Brand name for pheniramine maleate.

"Trimulso." [236] Trademark for a liquid synthetic surfactant used in the preparation of oil-in-water emulsion drilling muds; effective in both fresh water and brine muds.
Containers: 5-gal cans and 55-gal drums.

trimyristin. The glyceride of myristic acid; glyceryl trimyristate.

trinickelous orthophosphate. See nickel phosphate.

Trinidad pitch. See asphalt.

1,3,5-trinitrobenzene (TNB) $C_6H_3(NO_2)_3$.
Properties: Yellow crystals with a sp.gr. of 1.688 (20/4°C); m.p. 122°C; soluble in alcohol and ether; insoluble in water.
Derivation: From TNT by removal of methyl group.
Use: Explosive.
Shipping regulations: (Dry) explosive, class A. High explosive label. (Wet) (not to exceed 16 ozs) flammable solid. Yellow label.*

2,4,6-trinitrobenzoic acid (trinitrobenzoic acid) $C_6H_2(NO_2)_3COOH$.
Properties: Orthorhombic crystals; m.p. 228.7°C; sublimes with decomposition forming carbon dioxide and trinitrobenzene; slightly soluble in water and benzene; soluble in alcohol, ether, and acetone.
Derivation: Oxidation of 2,4,6-trinitrotoluene with chromic acid.
Use: Synthesis.
Shipping regulations: Dry: high explosive, class A. High explosive label. Wet: Label varies.*

trinitrocellulose. A nitrocellulose (q.v.).

2,4,6-trinitro-1,3-dimethyl-5-tert-butylbenzene. See musk xylol.

trinitroglycerin. See nitroglycerin.

trinitrophenol. See picric acid.

2,4,6-trinitroresorcinol. See styphnic acid.

2,4,6-trinitrotoluene (TNT; trinitrotoluol; methyltrinitrobenzene) $CH_3C_6H_2(NO_2)_3$.
Properties: Yellow, monoclinic needles; sp. gr. 1.654; m.p. 80.9°C; soluble in alcohol and ether; insoluble in water.
Derivation: By the nitration of toluene with mixed acid. Small amounts of the 2,3,4- and 2,4,5-isomers are produced which may be removed by washing with aqueous sodium sulfite solution.
Grade: Technical.
Containers: Wooden cases or kegs; multiwall paper sacks.
Uses: Explosive; intermediate in dyestuffs and photographic chemicals.
Fire hazard: Dangerous.
Shipping regulations: Dry: explosive, class A. High explosive label. Wet (not to exceed 16 ozs): flammable solid. Yellow label.*

trinitrotoluol. See trinitrotoluene.

trinitrotrimethylenetriamine. See cyclonite.

tri-n-octylaluminum $(C_8H_{17})_3Al$.
Properties: Colorless liquid.
Derivation: Reaction between octene and isobutylaluminum.
Use: Polyolefin catalyst.

trioctyl phosphate (octyl phosphate) $(C_8H_{17})_3PO_4$.
Properties: Sp.gr. 0.924 (26°C); b.p. 220-30 (8 mm); soluble in alcohol, acetone, and ether.
Uses: Solvent, antifoaming agent; plasticizer.

"Triodine." [284] Trademark for a cleaner-sanitizer-disinfectant particularly formulated for use in the bottling industry. Contains nonionic-iodine complexes. Claimed to be non-toxic, non-irritating, non-staining when used as directed.

"Tri-Ol." [45] Trade name for a combination of concentrated sulfonated solubilized oils, fortified with a white mineral oil of U.S.P. purity.
Uses: Soapless shampoos; brushless shaving creams; non-lathering soaps.

triolein. See olein.

"Triosul." [58] Trademark for sulfur trioxide, stabilized liquid. Colorless to pale yellow liquid; b.p. 44.8°C; freezing point 16.8°C.

"Triox." [253] Brand name for a weed killer containing sodium arsenite.

sym-**trioxane** (triformol; trioxin; metaformaldehyde) $(CH_2O)_3$ or $\underline{CH_2OCH_2OCH_2O}$. A trimer of formaldehyde; not to be confused with paraformaldehyde (q.v.) which may consist of many more formaldehyde units.
Properties: White crystals with formaldehyde odor; m.p. 62°C; b.p. 115°C; flash point (open cup) 113°F. Slightly soluble in water; soluble in alcohol and ether.
Derivation: By distillation of formaldehyde.
Containers: Drums.
Uses: Organic synthesis; disinfectant; non-luminous, odorless fuel.
See also formaldehyde.

trioxin. See sym-trioxane.

2,6,8-trioxypurine. See uric acid.

tripalmitin (palmitin; glyceryl tripalmitate) $C_3H_5(OC_{16}H_{31}O)_3$.
Properties: White, crystalline powder. Soluble in ether and chloroform. Insoluble in water.
Constants: M.p. 65.5°C; sp.gr. 0.866 (80/4°C).
Derivation: From glycerine and palmitic acid.
Grade: Technical.
Containers: Tins; fiber containers.
Uses: Medicine; soap; leather dressing.
Shipping regulations: None.*

triparanol. 1-(para-beta-Diethylaminoethoxyphenyl)-1-(para-tolyl)-2-(para-chlorophenyl) ethanol. Said to act as a cholesterol depressant in the body.

tripelennamine citrate $C_{16}H_{21}N_3 \cdot C_6H_8O_7$.
N-Benzyl-N',N'-dimethyl-N-2-pyridylethylenediamine citrate.
Properties: White, bitter crystalline powder. Solutions are acid to litmus. M.p. 107°C. Soluble in water and alcohol; very slightly soluble in ether; practically insoluble in chloroform and benzene. 1% solution in water has a pH of about 4.3.
Grade: U.S.P. XVI.
Use: Medicine.

tripelennamine hydrochloride $C_{16}H_{21}N_3 \cdot HCl$.
2-[Benzyl(2-dimethylaminoethyl)amino]pyridine hydrochloride; $C_6H_5CH_2N(C_5H_4N)$-$CH_2CH_2N(CH_3)_2 \cdot HCl$.
Properties: White, crystalline powder, bitter taste. Affected by light. M.p. 188-193°C. Freely soluble in water; soluble in alcohol and chloroform; very slightly soluble in acetone; insoluble in benzene, ether, and ethyl acetate. Solutions neutral to litmus.
Grade: U.S.P. XVI.
Use: Medicine.

tripentaerythritol
$(CH_2OH)_3CCH_2OCH_2C(CH_2OH)_2$-$CH_2OCH_2C(CH_2OH)_3$. See "Tripentek" for properties and uses.

"Tripentek." [138] Trade name for tripentaerythritol, technical.
Properties: A non-hygroscopic, white to ivory powder; has eight primary hydroxyl groups, all of which are esterifiable.
Containers: 50-lb multiwall bags.
Uses: For hard resins, varnishes, and fast drying tall oil vehicles.

triphenylantimony (triphenylstibine) $Sb(C_6H_5)_3$.
Properties: White crystalline solid; sp.gr. 1.434 (25°C); m.p. 46-53°C; b.p. less than 360°C. Insoluble in water; slightly soluble in alcohol; soluble in most organic solvents.
Derivation: Reaction of antimony trichloride with phenyl magnesium bromide or phenyl sodium.
Containers: 10-, 50-, 200-lb drums.
Uses: Forms stibonium salts; co-catalyst in converting trienes to aromatics and hydroaromatics; reacts with nitric-sulfuric acid to give trinitro derivatives; polymerization inhibitor catalyst; lubricating oil additive.

triphenylbismuth $(C_6H_5)_3Bi$.
Properties: White crystalline solid; sp.gr. 1.585; m.p. 77.6°C. Slightly soluble in alcohol. Soluble in chloroform, ether and acetone.
Derivation: Reaction of bismuth chloride with phenylmagnesium bromide.

triphenylfluorosilane $(C_6H_5)_3SiF$.
Properties: White solid; m.p. 63°C.
Uses: Silicone modifier.

triphenyl formazan $CN_4H(C_6H_5)_3$. Red insoluble derivative of tetrazolium chloride; formed when the latter comes into contact with viable portions of a seed. Used in germination and viability tests.

triphenylguanidine (TPG) $C_6H_5NC(C_6H_5NH)_2$.
Properties: White crystalline powder; soluble in alcohol.
Constants: Sp.gr. 1.10; m.p. 144°C.
Derivation: Desulfurization of thiocarbanilide in presence of aniline.
Method of purification: Crystallization.
Grade: Technical.
Containers: 100-lb barrels.
Use: Accelerator for vulcanization of rubber.
Shipping regulations: None.*

triphenylmethane dyes. Dyes whose molecular structure is basically derived from $(C_6H_5)_3CH$, usually by substitution of NH_2, OH, HSO_3, or other groups or atoms for some of the hydrogen of the C_6H_5 groups. A very large number of the coal tar and other synthetic dyes are of this class, including rosaniline, fuchsine, malachite green, fast green and crystal violet.

triphenylmethyl $(C_6H_5)_3C$. The first free radical to be isolated (by Gomberg in 1900). Exists only in solution in inert solvents, producing a deep yellow color. Combines with itself to form hexaphenylethane $(C_6H_5)_3CC(C_6H_5)_3$ when solvent is removed, and reacts vigorously with oxygen, halogen elements and metallic sodium.
See free radical and carbonium ion.

triphenyl phosphate (TPP) $PO(OC_6H_5)_3$.
Properties: Colorless, odorless, nonflammable, crystalline powder. Soluble in most lacquers, solvents, thinners, oils.
Constants: M.p. 48.5°C; b.p. 370°C; sp.gr. 1.268 (60°C); wt/gal 10.5 lbs; fire retardation excellent; refractive index 1.550 (60°C).
Typical specifications: Appearance white flakes; odor very faintly aromatic; m.p. 48.5°C min; not more than faintly opalescent; permanganate test, distinct purple color to be present after 30 minutes; purity 99% min; phenol 0.1% max; acidity 0.003% max (as phosphoric acid).
Derivation: Phenol and phosphorus oxychloride are boiled in presence of a little zinc chloride, until no more hydrogen chloride is given off. The product is shaken with caustic soda solution, filtered and the residue dissolved in ether. The ethereal solution is dehydrated and the ether evaporated.
Method of purification: Crystallization.
Grade: Technical.
Containers: 200-lb fiber drums; barrels; multiwall paper sacks.
Uses: Fire-retarding agent; plasticizer for cellulose acetate and nitrocellulose.

triphenylphosphine. See triphenylphosphorus.

triphenyl phosphite $(C_6H_5O)_3P$.
Properties: Water white to pale yellow solid or oily liquid; clean pleasant odor; sp.gr. 1.184 (25/25°C); m.p. 22-25°C; refractive index 1.589 (25°C); color APHA 50 max.
Containers: 55-gal drums.
Uses: Chemical intermediate; ingredient in stabilizer systems for resins; acts as a metal scavenger by chelating.

triphenylphosphorus (triphenylphosphine) $(C_6H_5)_3P$.
Properties: White, crystalline solid; m.p. 79-82°C; b.p. above 360°C; sp.gr. 1.132 (25°C). Insoluble in water; slightly soluble in alcohol; soluble in benzene, acetone, carbon tetrachloride.
Derivation: By a modified Grignard synthesis.
Containers: 10-, 50-, 150-lb drums.
Uses: Synthesis of organic compounds, phosphonium salts, other phosphorus compounds.

triphenylsilanol $(C_6H_5)_3SiOH$.
Properties: White solid; m.p. 155°C.
Derivation: Reaction of triphenylchlorosilane with ammonium hydroxide.

triphenylstibine. See triphenylantimony.

triphenyltetrazolium chloride. See tetrazolium chloride.

triphenyltin acetate $(C_6H_5)_3SnOOCCH_3$. An agricultural biocide. White crystalline solid, made by reaction of sodium acetate with triphenyltin chloride.

triphenyltin chloride $(C_6H_5)_3SnCl$.
Properties: White crystalline solid; m.p. 106°C; b.p. 240°C (13.5 mm). Insoluble in water; soluble in organic solvents.
Derivation: Reaction of tin tetrachloride with phenylmagnesium bromide.
Use: Biocidal intermediate.

"Triphosaden." [91] Trademark for a brand of adenosine triphosphate used for biochemical and clinical research, and medicine.

triphosgene. See hexachloromethyl carbonate.

triphosphopyridine nucleotide. See nicotinamide adenine dinucleotide phosphate.

"Triple Mix." [55] Trade name for insect repellents which contain dimethyl phthalate, butyl dimethyl-dihydro-gamma-pyronecarboxylate (see "Indalone"), and 2-ethylhexanediol-1, 3 (Rutgers 612). Known in World War II as "622."

triple point. The temperature and pressure at which the solid, liquid, and vapor of a substance are in equilibrium with one another. The term can also be applied to the similar equilibrium between any three phases, i.e., two solids and a liquid, etc. The triple point of water is of especial importance because it is a basic fixed point (273.16°K) for the absolute scale of temperature.

triple superphosphate. A dry, granular, free-flowing product, gray in color. It is produced by the addition of phosphoric acid to phosphate rock, thus avoiding the formation of insoluble gypsum as in superphosphate and achieving about three times the amount of available phosphate (as P_2O_5).
Typical chemical analysis: Moisture 2%; available P_2O_5 50%; water soluble P_2O_5 45%; free phosphoric acid 1%; also minor ingredients.
Containers: Bags; bulk; carloads.
Use: Fertilizer.
See also superphosphate and nitrophosphate.

tripoli (rotten stone).
Derivation: A porous, siliceous rock resulting from the natural decomposition of siliceous sandstone.
Grades: Various grades according to fineness for polishing; rose; cream; white.
Containers: 150-, 200-, 220-lb bags; 500-lb barrels.
Uses: Abrasive; polishing powder; filtering material; absorbent for insecticidal chemicals; paints (inert filler, wood filler); rubber filler; base for scouring soaps and powders.

tripolite. See diatomite.

tripoly. See sodium tripolyphosphate.

tripolyphosphate. See sodium tripolyphosphate.

tripotassium orthophosphate. See potassium phosphate, tribasic.

tripotassium phosphate. See potassium phosphate, tribasic.

triprolidine hydrochloride
$CH_3C_6H_4C(C_5H_4N){:}CHCH_2C_4H_8N \cdot HCl$.
trans-1-(4-Methylphenyl)-1-(2 pyridyl)-3-pyrrolidinoprop-1-ene hydrochloride.
Properties: Crystals; m.p. 116.5-118°C; moderately soluble in water, ethanol and methanol.
Use: Medicine.

tri-n-propylaluminum $(C_3H_7)_3Al$.
Properties: Colorless pyrophoric liquid.
Derivation: Reaction of propylene and isobutylaluminum.
Uses: Polyolefin catalyst.

tripropylamine $(CH_3CH_2CH_2)_3N$.
Properties: Water white; amine odor; boiling range 150-156°C; sp.gr. (20/20°C) 0.754; refractive index 1.417 (20°C); flash point 105°F.
Containers: 5-gal cans; 55-gal drums; tank cars.

tripropylene (propylene trimer) C_9H_{18}. Mixture of C_9 monoolefins.
Properties: Liquid; sp.gr. (20/20°C) 0.738; boiling range 133.3°-141.7°C. Weight 6.17 lb/gal (60°F); flash point 75°F (TOC).
Derivation: Catalytic polymerization of propylene.
Containers: Drums; barges or tank cars.
Uses: Oxo feed stock; lubricant additive; plasticizers; nonyl phenol.
Shipping regulations: Flammable liquid. Red label.*

tripropylene glycol $HO(C_3H_6O)_2C_3H_6OH$.
Properties: Colorless liquid; supercools instead of freezing; b.p. 268°C (760 mm); sp.gr. 1.019 (25/25°C); lbs/gal 8.51; refractive index 1.442 (n 25/D); flash point 285°F. Soluble in water, methanol, ether.
Containers: Drums; tank cars; tank trucks.
Uses: Synthesis; intermediate in resins, plasticizers, pharmaceuticals; insecticides; dyestuffs; mold lubricants.

tripropylene glycol, methyl ether
$HO(C_3H_6O)_2C_3H_6OCH_3$.
Properties: Sp.gr. 0.961 (25°C); b.p. 242°C (760 mm); 116°C (10 mm); viscosity 5.5 cps (25°C); refractive index 1.427 (25°C); fire point 127°C; completely miscible with water, VMP naphtha, acetone, ethanol, benzene, carbon tetrachloride, ether, methanol, monochlorobenzene, and petroleum ether.
Use: Ingredient in hydraulic fluids.

triptane. See 2, 2, 3-trimethylbutane.

"Triptide." [91] Trademark for a brand of lyophilized monosodium glutathione used for biochemical and clinical research, and medicine.

tris amine buffer. See tris(hydroxymethyl)-aminomethane.

tris(1-aziridinyl)phosphine oxide. See triethylenephosphoramide.

tris(1-aziridinyl)phosphine sulfide. See triethylenethiophosphoramide.

2, 4, 6-tris(1-aziridinyl)-s-triazine. See triethylenemelamine.

trisazo dyes. A subdivision in the chemical classification of dyes; one of the four kinds of azo dyes. They are characterized by the presence of three azo couplings (-N=N-) in each molecule. The Colour Index number of this class of dyes is C.I. 531-605.

tris-(beta-chloroethyl) phosphate $(ClC_2H_4O)_3PO$.
Properties: A clear transparent liquid; sp. gr. 1.425 (20/20°C).
Use: A flame-retardant plasticizer.

tris(2-chloroethyl) phosphite $(ClC_2H_4O)_3P$.
Properties: Mobile, colorless liquid with characteristic odor. Sp.gr. 1.353 (20/4°C); b.p. 119°C (0.15 mm); flash point 375°F (C.O.C.). Miscible with most common organic solvents. Undergoes intramolecular isomerization at higher temperatures. Exposure to air should be kept at a minimum. Insoluble in water, and hydrolyzes in water.
Containers: 5-gal, 55-gal steel drums.
Uses: Intermediate; component of vinyl stabilizers; grease additives; flameproofing compositions; color inhibitor.

tris(beta-chloroisopropyl)thionophosphate $[CH_3(CH_2Cl)CHO]_3PS$.
Typical properties: Phosphorus content 9.0%; density 1.282 g/ml at 68°F; flash point (open cup) greater than 347°F; pour point less than −58°F. Readily soluble in gasoline; insoluble in water.
Uses: To extend spark plug life, to control deposit-induced knocking and to reduce rumble in gasoline engines.

tris(2,3-dibromopropyl) phosphate $(CH_2BrCHBrCH_2O)_3PO$. Used as a flame retardant for plastics.

tris(2,3-dichloropropyl) phosphate $(CH_2ClCHClCH_2O)_3PO$. Used as a flame retardant in plastics.

tris(diethylene glycol monoethyl ether) citrate $HOC[CH_2COO(CH_2CH_2O)_2C_2H_5]_2$-$COO(CH_2CH_2O)_2C_2H_5$. Plasticizer.
Properties: Sp.gr. (25°C) 1.28; m.p. 16-19°C; soluble in water.

2,4,6-tris-(ethyleneimino)-s-triazine. See triethylene melamine.

tris-2-ethylhexyl phosphite $P(OC_8H_{17})_3$.
Properties: Colorless liquid; sp.gr. 0.902; b.p. (0.3 mm) 163-164°C; insoluble in water; soluble in alcohol and ether.
Containers: Glass bottles; 5-, 55-gal drums.
Uses: Synthesis; plasticizers; stabilizers; lube and grease additives; flameproofing compositions.

tris(2-hydroxyethyl) (phenylmercuri)ammonium lactate. See phenylmercuritriethanolammonium lactate.

tris(2-hydroxyethyl)trimethylammonium citrate. See tricholine citrate.

tris(hydroxymethyl)aminomethane [tri(hydroxymethyl)aminomethane; THAM; tris amine buffer] $(CH_2OH)_3CNH_2$.
Properties: White crystalline solid. Solubility in water, 80 g/100 cc at 20°C. M.p. 171-172°C; b.p. (10 mm) 219-220°C; pH 0.1M aqueous solution 10.36. Corrosive to copper, brass, aluminum.
Containers: Fiberpak boxes; drums.
Uses: Emulsifying agent (in soap form) for oils, fats, and waxes; absorbent for acidic gases; chemical synthesis; buffer.
Hazard: Avoid repeated exposure to skin.
Shipping regulations: None.*

tris(hydroxymethyl)nitromethane $(CH_2OH)_3CNO_2$.
Properties: White crystals or amorphous solid; m.p. 175°C; b.p. decomposes. Soluble in water and alcohol. Non-flammable; non-irritating to eyes and skin.
Uses: Bactericide and slimicide for aqueous systems, cutting oil emulsions, pulp and paper industry, industrial water systems, drilling muds.

tris[1-(2-methyl) aziridinyl]phosphine oxide $(C_3H_6N)_3PO$. ("MAPO").
Properties: Amber-colored liquid; high boiling amine odor; completely soluble in water and all common organic solvents.
Derivation: Reactive, tri-functional imine derivative.
Containers: 5-gal and 55-gal steel containers.
Uses: Addition products for textile treatments, adhesives, paper and rubber processing; crosslinking agent in polymer systems which contain active hydrogens; as monomer for polymers.

trisodium dipotassium tripolyphosphate $Na_3K_2P_3O_{10}$.
Properties: White crystalline solid; m.p. 620-640°C; density 2.48; solubility in water (26°C) 80 g/100 ml.
Use: Sequestrant.

trisodium EDTA, monohydrate. See ethylenediaminetetraacetic acid salts.

trisodium EDTA, trihydrate. See ethylenediaminetetraacetic acid salts.

trisodium orthophosphate. See sodium phosphate, tribasic.

trisodium phosphate. See sodium phosphate, tribasic.

trisodium phosphate monohydrate. See sodium phosphate, tribasic, monohydrate.

tristearin. See stearin.

tritium T. Radioactive hydrogen of mass number 3.
Properties: Half-life 12.5 years; radiation, beta; radiotoxicity, very low.
Derivation: Pile irradiation of lithium. If tritium occurs naturally it is present in amounts which are less than one part in 10^{17} parts of ordinary hydrogen.
Forms available: As a gas packaged in ampules and in tagged compounds such as water, streptomycin, cortisone, epinephrine, octadecane, stearic acid, etc.
Uses: In the hydrogen bomb. The fusion of tritium nuclei to form helium releases about seven times the energy per hydrogen nucleus which is obtained from nuclear fission. Also used as a bombarding particle in cyclotrons; an activator in self-luminous phosphors; in cold cathode tubes; as a tracer in studying hormone metabolism, the solubility of water in various organic compounds, the bulk and surface diffusion of hydrogen

on metals, the distribution of water in pre-cooked foods, the moisture gradient in attached protective coatings, petroleum products and fuels, etc.

Tritium is the significant exception to the general rule that the radioactive isotopes of the elements behave chemically and physically the same as the usual form. This is because the mass of tritium is quite different from that of hydrogen. However when tritium is incorporated into heavy molecules, this mass difference becomes insignificant, and tritium is used extensively to label organic molecules to follow their behavior.

Shipping regulations: Class D poison, radioactive material. Blue label.*

tritolyl phosphate. See tricresyl phosphate.

triton. Nucleus of tritium or hydrogen 3.

"Triton." [23] Trademark for surfactants based on alkylaryl polyether alcohols, sulfonates and sulfates; nonionic, cationic and anionic types; oil-soluble and water-soluble types. Supplied as viscous liquids, pastes, or aqueous solutions. Surface activity includes detergency, emulsification, wetting, spreading, dispersing action.

Use: Processing of textile fabrics and yarns; leather tannery operations; emulsification of pesticides; sanitizing and cleansing formulations; cosmetic preparations; medicinal soaps; paper manufacture; oil well flooding; polymer manufacture; petroleum oil additives.

"Triton B-1956." [23] Trademark for a free-flowing emulsifier, spreader, sticker, and depositing agent for insecticides.

tritopine. See laudanidine.

"Tri-Una-Sol." [58] Trademark for nitrogen fertilizer solution.

Properties: Clear water solution; varying analysis of 28, 30, and 32% nitrogen containing urea, ammonium nitrate and water.

Containers: Insulated and uninsulated tank cars; tank trucks.

Uses: Direct application to soil; formulation of mixed fertilizers.

triuranium octoxide (uranous-uranic oxide; uranyl uranate) U_3O_8.

Properties: Olive green to black solid; crystals or granules; insoluble in water; soluble in nitric acid and sulfuric acid. Sp.gr. 8.39; decomposes when heated to 1450°C.

Source: It is the naturally occurring uranium oxide found in pitchblende.

Derivation: (a) As one of the forms of uranium produced from the ores, often by a solvent extraction process. The solvent used is dodecylphosphoric acid. (b) A common form of triuranium octoxide is yellow cake, the powder obtained by evaporating an ammonia solution of the oxide.

Uses: Nuclear technology; preparation of other uranium compounds.

trixylenyl phosphate. See tri-dimethylphenyl phosphate.

"Tro-Grees." [25] Brand name for a proprietary, odorless and tasteless mineral type grease for lubricating troughs, chutes, etc. in the baking industry.

troilite. A variety of pyrrhotite (q.v.) found in meteorites.

trolnitrate phosphate (triethanolamine trinitrate phosphate; aminotrate phosphate) $N(CH_2CH_2ONO_2)_3 \cdot 2H_3PO_4$.

Properties: Crystals; banana-like odor; m.p. 107-109°C. Pure compound is potentially explosive.

Derivation: Prepared by the nitration of triethanolamine followed by precipitation with phosphoric acid.

Grade: N.N.D. (is dispensed as a 1:1 trituration with lactose).

Use: Medicine.

"Troluoil." [200] Trademark for a petroleum solvent prepared by straight-run distillation.

Properties: Water-white, initial boiling point 194-205°F, 95% distills between 235 and 247°F; sp.gr. 0.741 (60°F); flash point (TCC) 25°F; mild, nonresidual odor.

Uses: In lacquer formulations, rubber cements, and roto inks.

Shipping regulations: Flammable liquid. Red label.*

trona (urao) $Na_2CO_3 \cdot NaHCO_3 \cdot 2H_2O$. A natural sodium sesquicarbonate and the most important of the natural sodas.

Properties: White, gray or yellow with vitreous, glistening luster; contains 41.2% Na_2O, 38.9% CO_2, 19.9% H_2O with some impurities.

Occurrence: Extensive deposits in Hungary, Egypt, Africa, Venezuela, and United States(Wyoming; California, especially Searles Lake, Owens Lake).

Use: Source of sodium compounds.

"Tronabor." [88] Trademark for crude borax pentahydrate, from Searles Lake brines.

Trona process. The method used for separation and purification of soda ash, anhydrous sodium sulfate, boric acid, borax, potassium sulfate, bromine and potassium chloride from the Searles Lake (California) brine.

"Tronothane." [3] Trademark for pramoxine hydrochloride (q.v.).

troostite. A transition substance occurring in the heat treatment of steel. It is a special mixture of iron and Fe_3C.

troostite (mineral). See willemite.

tropacocaine hydrochloride $C_{15}H_{19}NO_2 \cdot HCl$. An alkaloidal salt.

Properties: White crystals; poisonous! Soluble in water, alcohol, and ether. M.p. 271°C.

Derivation: From a variety of Erythroxylon coca.

Method of purification: Crystallization.

Grade: Technical.

Containers: Glass bottles.

Use: Medicine.

3-tropanol. See tropine.

tropeolin D. See methyl orange.

tropeolin OO $NaSO_3C_6H_4NNC_6H_4NHC_6H_5$. para-Diphenylamino-azobenzene-sodium sulfonate. A biological stain and acid-base indicator, red at pH 1.4, yellow at pH 2.6. See indicators.

"Tropicel." [248] Trademark for a family of translucent and highly styled sandwich panels of reinforced polyester resin. Two flat panels are bonded to geometric cores forming a rigid panel for architectural applications.

tropine (3-tropanol) $C_8H_{15}NO$.
Properties: White, crystalline alkaloid; very hygroscopic; poisonous! Soluble in water; alcohol, ether, and chloroform.
Constants: M.p. 61.2-63°C; b.p. 233°C.
Derivation: By heating atropine or hyoscyamine with barium hydroxide.
Method of purification: Crystallization.
Grade: Technical.
Containers: Glass bottles, boxes.
Use: Medicine.
Shipping regulations: None.*

Trouton's rule. States that the molal heat of vaporization of normal liquids, at the boiling point and under atmospheric pressure, divided by the absolute boiling temperature is a constant, approximately 22.

true lavender. See lavender.

"Truline." [266] Trademark for a dry, pulverized resin used to bind core and molding sands. Made from a high-melting resin extracted from pine wood.

"Trycite." [233] Trademark for an oriented polystyrene film.

trypaflavine. See acriflavine hydrochloride.

trypaflavine neutral. See acriflavine.

tryparsamide (sodium N-phenylglycineamide-para-arsonate) $NaOAs(OOH)C_6H_4$-$NHCH_2CONH_2 \cdot \frac{1}{2}H_2O$.
Properties: White crystalline powder, odorless. Contains 24.6% arsenic. May affect eyes. Soluble in water; almost insoluble in alcohol, ether, chloroform, benzene.
Grades: Medicinal; U.S.P. XVI.
Use: Medicine.
Shipping regulations: None.*

trypsin. The proteolytic enzyme of the pancreatic juice. Yellow to grayish-powder; soluble in water; insoluble in alcohol or glycerin. It acts on the albuminoid material producing amino acids. The maximum result is obtained in a neutral or slightly alkaline medium. Trypsins or similar materials are found not only in the pancreas but also in the spleen, leucocytes and urine and also in beer yeast, molds and bacteria. Used in medicine.
Grade: N.F. XI.

trypsinogen. An inactive precursor of trypsin (q.v.).

tryptophan (indole-alpha-aminopropionic acid; 1-alpha-amino-3-indolepropionic acid) $C_6H_4NHCHCCH_2CHNH_2COOH$. One of the essential amino acids occuring naturally in the L(-)-form.
Properties:
DL-: White crystals; slightly soluble in water; stable in alkaline solution; decomposed by strong acids.
D(+)-: Characteristic sweet taste; m.p. 275-290°C (dec); soluble in water, hot alcohol, alkali hydroxides; insoluble in chloroform.
L(-)-: Flat taste (other properties identical with D(+)-tryptophan).
Derivation: Synthetic tryptophan can be made by the conversion of indole to gramine, followed by methylation, interaction with acetylaminomalonic ester and hydrolysis; hydrolysis of proteins.
Grades: Reagent; technical.
Containers: 1-, 5-lb bottles; drums.
Uses: Nutrition and research; medicine.
Available commercially in all three forms, as well as acetyl-DL-tryptophan.

"Trysben" 200. [28] Trademark for a weed killer based on an aqueous solution of the dimethylamine salt of trichlorobenzoic acid, containing 2 lb of acid equivalent per gallon.
Use: Control of broadleaf weeds (especially deep-rooted perennials) and woody vines and brush.
Containers: 1-gal cans, 5- and 30-gal drums.

Ts. Abbreviation for tosyl.

TSA. Abbreviation for toluene sulfonic acid.

TSH. See thyrotropic hormone.

TSP. Abbreviation for trisodium phosphate. See sodium phosphate, tribasic.

TSPA. Abbreviation for triethylenethiophosphoramide.

TSPP. Abbreviation for tetrasodium pyrophosphate. See sodium pyrophosphate.

TTC. Abbreviation for tetrazolium chloride.

TTD. Abbreviation for tetraethylthiuram disulfide.

"Tuamine." [100] Trademark for tuaminoheptane sulfate, N.F.

tuaminoheptane. See 2-aminoheptane.

tuaminoheptane sulfate
$[CH_3(CH_2)_4CHNH_2CH_3]_2 \cdot H_2SO_4$.
Properties: White odorless powder; readily soluble in water and alcohol.
Grade: N.F. XI.
Use: Medicine.

"Tubarine." [301] Trademark for d-tubocurarine chloride (q.v.).

tubatoxin. See rotenone.

tuberose oil.
Properties: Colorless to very light colored oil; sp.gr. 1.007-1.035 at 15°C; taken from Polianthes tuberosa, by enfleurage.
Use: Perfume.

*See "I.C.C. Shipping Regulations," page xiii.
Reference numbers refer to name of manufacturer. See "List of Manufacturers," page v.

tubocurare. A tube form of curare.

d-tubocurarine chloride $C_{38}H_{44}Cl_2N_2O_6 \cdot 5H_2O$.
Properties: White to light tan, odorless, crystalline powder. M.p. (anhydrous) 274-275°C with decomposition; soluble in water and alcohol; insoluble in acetone, chloroform, and ether. Aqueous solution is strongly dextrorotatory. (Specific rotation for 1% solution of anhydrous + 208° to + 218°.)
Grade: U.S.P. XVI.
Use: Medicine.

"Tuex." [248] Trademark for tetramethylthiuram disulfide.
Properties: Grayish white powder; sp.gr. 1.29; m.p. not less than 135°C; soluble in benzene and ethylene dichloride; moderately soluble in acetone; insoluble in water and gasoline.
Uses: Rubber accelerator for natural rubber wire insulation; inner tubes; druggist sundries; mechanicals; proofing; footwear; sponge rubber and transparent pure gum stocks. In SBR and butyl rubber, a general-purpose accelerator.

tufa. See travertine.

tuff. See volcanic ash.

"Tuftop." [329] Trademark for a sensitized cold top enamel.
Use: Photoengraving.

"Tumerol." [342] Trademark for oleoresin of tumeric for food coloring and flavoring.

"Tumescal." [206] Brand name for proprietary swelling agents for "Terylene" and other polyester fibers.

tuna oil (tunny-fish oil).
Properties: Pale yellow to red-brown liquid; characteristic odor. Soluble in alcohol, ether, chloroform, carbon disulfide and ligroin.
Constants: Iodine value 156.
Derivation: By expressing the livers of Thynnus vulgaris.
Method of purification: Filtration.
Grades: Crude; refined.
Containers: Wooden barrels and steel drums.
Uses: Paints; source of vitamins A and D.
Shipping regulations: None.*

tung oil (Chinese-wood oil; China-wood oil; wood oil).
Properties: Yellow, drying oil. Jellies or solidifies when kept. Reputed to be poisonous. Soluble in chloroform, ether, carbon disulfide, and oils. Sp.gr. 0.9360-0.9432; saponification value 193; iodine value 150-165; refractive index 1.5030.
Derivation: From the seeds of Aleurites cordata, a tree indigenous to China and Japan, by roasting, grinding and pressing. The tree is now being grown on a commercial scale in Florida and Mississippi.
Impurities: Tung oil, being expensive, is frequently adulterated with cottonseed oil, soya bean oil, etc.
Grades: White; black; cold-pressed, yellow; hot-pressed, dark. Japanese tung oil is an inferior grade with notably poorer drying qualities. Also domestic; imported.
Containers: 75-lb cases; drums; 8000-gal tank cars.
Uses: Varnishes; linoleum; making varnish dryers; india rubber substitutes; insulating masses; waterproofing paper and other tissues.
Shipping regulations: None.*

tungstated green. See phosphotungstic pigments.

tungstated pigments. See phosphotungstic pigments.

tungstate white. See barium tungstate.

tungsten (wolfram) W. Element with atomic number 74, group VI of the periodic system. An official ruling of the commission on Inorganic Nomenclature, International Union of Pure and Applied Chemistry, recognizes the use of tungsten for English-speaking countries. Wolfram is also official, however.
Properties: Hard, brittle, gray metal; very heavy. Not found native. The ores are scheelite and wolframite. Sp.gr. 19.3 (20°C); m.p. about 3400°C (the highest melting point of all the metals). Soluble in a mixture of nitric acid and hydrofluoric acid.
Derivation: By the aluminothermic reduction of tungstic oxide; hydrogen reduction of tungstic acid or its anhydride. The metal can be plated onto objects by vapor deposition from tungsten hexafluoride or hexacarbonyl. Tungsten powder is converted into solid metal by powder metallurgical techniques. Large single crystals now grown by an arc-fusion process.
Occurrence: Found in Korea, China, Mexico, Spain and in the United States in Arizona, California, Colorado, Nebraska, Nevada, New Mexico, and Texas.
Grades: Technical; ultrapure.
Container: 5-, 10-, 25-lb cans, barrels, drums (powder).
Uses: High-speed tool steel; nonferrous alloys; electric lamp industry; contact points in automotive and telegraph industries (wireless apparatus); dentistry; pen points; targets of x-ray tubes; phonograph needles; shell steel; chemical apparatus; high speed rotors as used in gyroscopes; counterweights; vibration damping devices; shielding radioactive materials and missile surfaces.
Shipping regulations: None.*

tungsten carbide WC.
Properties: A fine gray powder; sp.gr. 15.6; m.p. 2780°C; b.p. 6000°C; hardness approaches that of diamond (Mohs hardness 9+); insoluble in water but readily attacked by a nitric acid-hydrofluoric acid mixture. Stable to 400°C with chlorine; burns with fluorine at room temperature; oxidizes on heating with air to tungstic oxide.
Derivation: Chemical combination of tungsten metal powder and lamp black at 1500-1600°C.
Uses: Cemented carbide tools; dies; wear-resistant parts; cermets; electrical resistors.

tungsten carbide, cemented. A mixture consisting of tungsten carbide, 85-95% and cobalt, 5-15%.
Properties: Sp.gr. 12-16; hardness about that of corundum, and not affected by severe high industrial temperatures.
Derivation: Ball milling of powdered tunsten carbide with metallic cobalt, followed by sintering.
Uses: Machine tools and abrasives for machining and grinding metals, rocks, molded products, porcelain and glass; in gages, knife edges, blast nozzles.

tungsten hexacarbonyl $W(CO)_6$.
Properties: White, volatile, highly refractive, crystalline solid; decomposes without melting at 150°C. One of the more stable carbonyls. Sp.gr. 2.65; vapor pressure 0.1 mm (20°C). Insoluble in water; soluble in organic solvents.
Derivation: By the reaction of tungsten with carbon monoxide at high pressures; reduction of tungsten hexachloride with iron alloy powders in CO atmosphere.
Use: To obtain tungsten coatings on base metals through deposition and decomposition of the carbonyl.

tungsten hexachloride WCl_6.
Properties: Dark blue or violet hexagonal crystals; volatile; m.p. 275°C; b.p. 347°C; sp.gr. 3.52; vapor pressure 43 mm (215°C); electrical conductivity (fused state) poor. Soluble in organic solvents including ligroin and ethanol. Decomposed by moist air and water; reduced by hydrogen to the metal.
Derivation: By treating tungsten metal with dry chlorine at red heat.
Uses: Formation of tungsten coatings on base metals; formation of single crystal tungsten wire; additive to tin oxide to produce electrically conducting coating for glass.

tungsten hexafluoride WF_6.
Properties: Colorless gas or light yellow liquid. M.p. 2.5°C; b.p. 19.5°C. Similar to fluorine in toxicity.
Derivation: Direct fluorination of powdered tungsten. Purified by distillation under pressure.
Uses: Vapor phase deposition of tungsten; fluorinating agent.

tungsten lakes. See phosphotungstic pigments.

tungsten oxychloride $WOCl_4$.
Properties: Dark-red, acicular crystals. Decomposed by water and moist air. Caution! Keep in sealed glass containers! B.p. (approx) 227.5°C; m.p. (approx) 211°C; sp.gr. 11.92; soluble in carbon disulfide.
Derivation: By the action of chlorine on tungsten or tungstic oxide at elevated temperatures.
Purification: Vacuum distillation.
Grade: Technical.
Use: Incandescent lamps.

tungsten silicide. A ceramic. Probably WSi_2.
Grades: Cylindrical shapes, lumps, standard sieve sizes.
Uses: Oxidation resistant coatings; electrical resistance applications.

tungsten steel. In many of its alloying effects, tungsten is similar to molybdenum, and therefore is not used for any standard SAE or AISI alloy steels, since the cheaper molybdenum will give the same properties. Tungsten increases the density of alloys to which it is added. It is used to obtain steels with great wear resistance and special resistance to tempering, as indicated by the following examples:
high speed steels: The outstanding application. Typical composition 18% tungsten, 4% chromium, 1% vanadium, 0.70% carbon. Tungsten makes steel very resistant to tempering and the carbide has great wear resistance, making an ideal high speed cutting tool.
hot work steels: 10% tungsten, 3% chromium, 0.30% vanadium, 0.35% carbon. This can be used even where surface working temperatures exceed tempering temperature due to very slow loss of hardness occurring in these conditions.
finishing steels: 3.5% tungsten, 1.35% carbon; maintains keen cutting edge and has great wear resistance.
creep resisting steels: 3% tungsten, 0.12% carbon, 12% chromium, 2% nickel; useful in resisting creep at temperatures up to 1100°F.
oxidation resistant, high temperature, high strength alloys: Used in stainless steels for high temperature work, i.e., in exhaust valves. Most promising alloys—iron, cobalt, nickel, chromium with tungsten, molybdenum, titanium and niobium.
magnet steels: Tungsten was used for magnets but is now completely superseded by the iron, nickel, aluminum, and iron, nickel, aluminum, cobalt alloys.

tungsten trioxide. See tungstic oxide.

tungstic acid (wolframic acid; orthotungstic acid) H_2WO_4.
Properties: Yellow powder; sp.gr. 5.5. Insoluble in water; soluble in hydrofluoric acid and alkalies. A white form of tungstic acid exists, having the formula $H_2WO_4 \cdot H_2O$. This is formed by acidifying tungstate solutions in the cold.
Derivation: Decomposition of sodium tungstate with hot sulfuric acid.
Method of purification: Crystallization.
Grades: Technical; C.P.; reagent.
Containers: Drums; in tonnage lots.
Uses: Textiles (mordant, color resist); plastics; tungsten metal, wire, etc.
Shipping regulations: None.*

tungstic acid anhydride. See tungstic oxide.

tungstic anhydride. See tungstic oxide.

tungstic oxide (tungstic acid anhydride; tungstic anhydride; tungsten trioxide; wolframic acid, anhydrous) WO_3.
Properties: Canary yellow, heavy powder; dark orange when heated and regains

original color on cooling; m.p. 1473°C; sp. gr. 7.16. Insoluble in water; soluble in caustic alkalies; only soluble with difficulty in acids.
Derivation: Scheelite ore is treated with hydrochloric acid and the resulting product dissolved out with ammonia. The complex ammonium tungstate can then be ignited to tungstic oxide.
Uses: To form metal by reduction; alloys; preparation of tungstates for x-ray screens and for fire-proofing fabrics; yellow coloring agent in ceramics.

"Tungstide." [289] Brand name for a line of coatings consisting of metallic tungsten of near-colloidal particle-size suspended in a liquid plastic and incorporated with a form of "Liqui-Moly", (a molybdenum disulfide product) to produce a hard, abrasion-resistant self-lubricating coating.

tunicine (animal cellulose). A variety of cellulose found in the tunic of the ascidians (e.g., sea-squirts).

tunny-fish oil. See tuna oil.

turbine oil. The desirable characteristics of an oil for use with heavy horizontal turbines are bright yellow color, high flash test, low viscosity, the ability to maintain its body and efficiency under high temperatures and low moisture-absorptive capacity. One good steam-turbine oil is said to have the following characteristics: Gravity 30° Bé; flash point, 420°F; Saybolt viscosity, 150 at 70°F.
Shipping regulations: None.*

turbith mineral. See mercuric subsulfate.

turkey brown (turkey umber). Natural earth which serves as permanent pigment. Contains iron and manganese oxides with some clay.
Use: In paint.

turkey galls. See galls.

Turkey red. See iron oxide reds.

Turkey red oil (castor oil, sulfonated; castor oil, soluble). It is also known as alizarin assistant and alizarin oil because of its use in dyeing with alizarin.
Properties: Sp.gr. 0.95; iodine no. 82.1; acid no. 174.3; saponification no. 189.3. Soluble in water.
Derivation: By sulfonating castor oil with sulfuric acid and washing.
Grades: Sulfonated castor oil graded as to moisture and color.
Containers: 55-gal barrels and drums; 500-lb barrels.
Uses: Textiles; leather; manufacture of soaps; alizarin dye assistant.
Shipping regulations: None.*

"Turkey Red Oil PO." [206] Brand name for product combining emulsifying action of turkey red oil with improved wetting properties.

turkey umber. See turkey brown.

Turkish geranium oil. See palmarosa oil.

turmeric. See curcuma.

turmeric oil. See curcuma oil.

turmeric root. See hydrastis.

turmeric yellow. See curcumin.

Turnbull's blue. Blue precipitate or pigment resulting from reaction of a ferrous salt and potassium ferricyanide. See iron blues.

turpentine (gum turpentine; thus, American; thus, gum). The gum or oleoresin from which spirits of turpentine (turpentine oil, popularly called turpentine) is produced. It is an exceedingly sticky, viscid, liquid balsam, a mixture of rosin and volatile oil, obtained from coniferous trees. Yellowish, opaque masses, stocky and more or less glossy with characteristic odor and taste. Soluble in alcohol, ether, chloroform, and glacial acetic acid.
Chief constituents: A volatile oil (turpentine oil) and a resin.
Derivation: By incision into the wood of coniferous trees in the spring of the year or by solvent extraction from stumps of coniferous trees.
Containers: Tins; bottles; barrels.
Use: Source of spirits of turpentine and rosin.

turpentine camphor. See bornyl chloride.

turpentine, Canadian. See Canada balsam.

turpentine, destructively-distilled wood. See turpentine, spirits of.

turpentine gum. See turpentine.

turpentine, gum spirits. See turpentine, spirits of.

turpentine, oil of. See turpentine, spirits of.

turpentine, spirits of (turps; turpentine, oil of). General formula $C_{10}H_{16}$. A volatile oil obtained by distilling the oleoresin exuded by or contained in the wood of certain species of pine trees. See turpentine. The principal countries producing the oil are America, France, Russia, Spain, Austria, Greece and (recently) Mexico.
Chief known constitutents: Pinene; dipentene. Four kinds of turpentine oil are now recognized:
(a) Gum turpentine or gum spirits of turpentine, made from the gum (oleoresin) collected from living trees.
(b) Steam-distilled wood turpentine, obtained from the oleoresin within the wood by steam distillation of the wood itself or of an extract therefrom.
(c) Sulfate wood turpentine, recovered during the conversion of wood to paper pulp by the sulfate process.
(d) Destructively-distilled wood turpentine, obtained by fractionation of certain oils recovered by condensing the vapors formed during the destructive distillation of pine wood.
American turpentine oil.
Species from which crude is derived:

Pinus palustris, Mill.; Pinus heterophylla, Sudw.; Pinus echinata, Mill.
Properties: Colorless, mobile liquid; rosin-like odor.
Constants: Considerable variation appears in constants reported. The following are based on tests made by the U.S. Department of Agriculture, Forest Products Laboratory: Sp.gr. (15°C) 0.860-0.875; moisture content-trace; refractive index 1.463-1.483 (20°C); Kauri-Butanol Solvency Test 80.2; flash point (closed cup) 102.5°F; acidity none; unpolymerized residue (28 N H_2SO_4) 2%; color water-white (not below Saybolt plus 16); wt in lbs/gal 7.18; representative distillation 10% (161.2°C), 50% (165.2°C), 80% (171°C), 90% (175.5°C), end (181°C).
Grades: Technical; N.F. XI.
Containers: Bottles; cans; drums.
Uses: Medicine; solvent; thinner for paints, varnishes and lacquers; rubber solvent and reclaiming agent.
Shipping regulations: None.*

turpentine, steam-distilled wood. See turpentine, spirits of.

turpentine substitute. See naphtha, painters'. The term turpentine substitute is misleading and should not be used.

turpentine, sulfate wood. See turpentine, spirits of.

turpentine, Venice (larch turpentine).
Properties: Yellowish to green oleoresin; pleasant aromatic odor; hot, pungent, bitter taste; becomes hard and brittle on exposure to air; sp.gr. 1.09-1.19; soluble in most organic solvents.
Chief constituents: Volatile oil and resin.
Derivation: Distilled from Larix europaea.
Occurrence: Middle and southern Europe.
Uses: Varnish; sealing wax; lithography.
Shipping regulations: None.*

turpentine, Venice, artificial. A mixture of rosin and oil of turpentine and similar to pine oleoresin.
Properties: Soluble in ether, acetone, acetic acid, aqueous alkalies and slowly soluble in alcohol.
Uses: In making varnishes, sealing wax and plasters; lithographic work.

turpeth mineral. See mercuric subsulfate.

"Turpol" Brand Rubber Plasticizer NC-1200.[158]
A plasticizer and processing aid that has a wide range of compatibility with various synthetic rubbers. A non-extractable plasticizer used for the compounding of low durameter stocks such as in printing rolls. Exerts softening effect upon harder, tougher, drier rubbers and it imparts a cohesiveness to all rubbers during the mixing process. Mold flow, particularly in transfer molding, is aided by the addition of the plasticizer.
Properties: Amber colored rubbery terpene-derived synthetic polymer. Sp.gr. 1.20 at 25°C., resists all common solvents, Shore "A" hardness, 5-10 with nil ash percentage.
Containers: 50-lb cartons.

"Turpol" Brand Terpene Based Resin. [158] A reactive diluent for epoxy resins which improves adhesion and peel strength.
Containers: Quart, 1-gal, 5-gal and 55-gal drums.

turps. See turpentine, spirits of.

turquoise $CuAl_6(PO_4)(OH)_3 \cdot 2H_2O$. A natural basic hydrated phosphate of copper and aluminum.
Properties: Color blue, blue-green, green; luster waxy; streak white or green; hardness 6; sp.gr. 2.6-2.9.
Occurrence: New Mexico, Arizona, Nevada, California, Colorado; Iran.
Use: Gem stone.

turtle oil.
Properties: A yellow semi-solid oil with an odor of beef-drippings.
Derivation: In South America from turtle eggs; in Seychelle Islands and Jamaica from turtle fat.
Use: Similar to cod-liver oil.

tuscan oxide. See tuscan red.

tuscan red (tuscan oxide). A red iron oxide pigment.

tutia. See zinc carbonate, precipitated.

tutocaine (butamin) $C_{14}H_{22}O_2N_2 \cdot HCl$ (para-aminobenzoyldimethylamino-1, 2-dimethylpropanol hydrochloride).
Properties: Light, ivory-colored, crystalline powder, virtually odorless; faintly bitter taste. More toxic than procaine. M.p. 212-215°C. Soluble in water; with difficulty in alcohol.
Use: Medicine.
Shipping regulations: None.*

"TVM." [194] Trademark applied to a series of protective and decorative thermosetting vinyl color metal coatings.

Tw. Abbreviation for Twaddell, used in reporting specific gravities for densities greater than water, as °Tw. A Twaddell reading, multiplied by five and added to 1000, gives specific gravity with reference to water as 1000.

Twaddell. See Tw.

"Twecotan." [57] Brand name for a group of proprietary products. Combination of natural and synthetic tanning materials used for retanning leather where a fine grained full leather is required.

"Tween." [89] Trademark for each member of a series of general purpose emulsifiers and surface active agents. They are polyoxyethylene derivatives of fatty acid partial esters of hexitol anhydrides. Generally soluble or dispersible in water , and differ widely in organic solubilities.

twist setting agents. Following twisting operations in the manufacture of silk and rayon yarns, the threads are likely to be too lively or "springy" to process easily. Consequently they are subjected to a steaming operation to render the yarns limp. To avoid loss of twist during steaming, especially in yarns for crepes, they are treated with compounds which assist in holding the twist. Modified casein and caseinates, highly sulfated oils, and various other sulfonated penetrants or a combination of a melamine–formaldehyde resin and 2-mercaptoethanol are among the products used for twist setting.

"Twitchell." [242] Trademark for a group of textile processing oils, fat-splitting reagents, and bases for soluble oils.

Twitchell process. Acid hydrolysis of fats in the presence of the Twitchell reagent to produce fatty acids and glycerine.

Twitchell reagent. A catalyst for the Twitchell process. Consists of a sulfonated addition product of naphthalene and oleic acid, a naphthalenestearosulfonic acid.

"Two-Sixty-Two." [28] Trademark for a urea feed compound.
Properties: Free-flowing, non-caking granular product having a nitrogen content of 42%, all in the form of urea, equivalent to 262% protein from nonprotein nitrogen.
Containers: 80-lb multiwall moisture-resistant paper bags.
Use: As ingredient of feeds for ruminants in an amount not more than one-third of the total equivalent protein in the mixture. Since too large a content of urea in such feeds is toxic, the product is recommended for use by commercial feed-manufacturers only.

"Ty-Bond." [428] Trademark for a line of phosphate coatings for dip or spray application to steel, iron and zinc base die castings.

"Tybrene." [233] Trademark for tert-butylstyrene and similar materials.

"Tygobond 30." [326] Trade name for a vinyl rubber resin adhesive for cementing porous and semiporous substances to each other.

"Tygofil." [326] Trade name for a modified epoxy base metal filler.

"Tygoflex." [326] Trade name for a plastisol-vinyl compound derived from "Tygon." For specialty corrosion protection.

"Tygon." [326] Trademark for a series of vinyl compounds used as linings, coatings, adhesives, tubing, and extruded shapes applied to chemical process equipment as corrosion protection.

"Tygonite." [326] Trade name for a group of general utility cements or adhesives derived from vinyls and rubber. Available in several consistencies.

"Tygorust." [326] Trade name for a vinyl-based primer for application to damp or dry rusted steel.

"Tygoweld." [326] Trade name for a modified epoxy base adhesive for bonding both similar and dissimilar materials.

"Tylose." [450] Trademark for a wide range of water-soluble cellulose ethers.
"Tylose" UM, UMK and UH are grades of methylcellulose, prepared by treating alkali cellulose with methyl chloride. The sodium chloride formed during this etherification is washed out with boiling water in which methylcellulose itself is not soluble.
"Tylose" UC, UCB, UCBR and UCR are grades of sodium carboxymethylcellulose (q.v.). The UC and UCB grades are highly purified; the UCR and UCBR grades contain some of the sodium chloride from the original reaction.
Properties: The "Tylose" products differ from one another in the type of substituent and also in the degree of substitution and polymerization. The numerical nomenclature refers to the viscosity, for example "Tylose" UM 4000 (high viscosity) and "Tylose" UM 70 (low viscosity). They also differ in purity, in the active ingredient content and in appearance (granular or powder form). (The types assigned the letter K after the numeral must not be used for foodstuffs, pharmaceuticals or cosmetics.) They are not affected by fats, oils or most organic solvents.
Uses: As thickeners, binders, dispersing agents, emulsifiers, protective colloids, lubricants and film forming materials. They are used as drilling muds; in detergents; in textile, paper, print, and varnish industries; in ceramics; and in cosmetics and pharmaceuticals.

"Tynex." [28] Trademark for nylon filaments. Available tapered with an essentially uniform taper from butt to tip, and also level, i.e., in a wide range of constant diameters. The tapered form used primarily in paint brushes; the level form in other brushes and wigs, surgical sutures and racquet strings.

type metal. Alloy of 75-95% lead, 2½-18% antimony with a little tin and sometimes copper which expands slightly upon solidification and produces sharp castings.

tyramine (tyrosamine; para-beta-amino-ethylphenol) $HOC_6H_4CH_2CH_2NH_2$. A base which is found in mistletoe, putrefied animal tissue, certain cheeses, and ergot. It is usually made synthetically.
Properties: Colorless crystals; m.p. 164-165°C; soluble in boiling alcohol; slightly soluble in water, benzene, and xylene.
Use: Medicine.

tyramine hydrochloride $C_8H_{11}NO \cdot HCl$. A solid with m.p. 269°C. Soluble in water with neutral reaction.
Use: Medicine.

"Tyril." [233] Trademark for a rigid resin which is a copolymer of styrene and acrylonitrile.

tyrocidine. An antibiotic produced by the metabolic processess of the bacteria, Bacillus brevis. It is a cyclic polypeptide which is active against most grampositive pathogenic (disease-causing) bacteria. It is one of the

two antibiotic components of tyrothricin (q.v.) but has been isolated and used alone.
Properties (probably the hydrochloride): Fine crystalline needles which decompose at 240°C. Soluble in 95% alcohol, acetic acid and pyridine; slightly soluble in water, acetone and absolute alcohol; insoluble in ether, chloroform and hydrocarbons. Depresses surface tension; forms fairly stable colloidal emulsion in distilled water.
Use: Medicine (usually as component of tyrothricin); possible fungistat and bacteriostat.

tyrosamine. See tyramine.

tyrosinase. An enzyme containing copper which occurs in plant and animal tissue and is responsible for turning peeled potatoes black when exposed to air.
Use: Medicine.

tyrosine (beta-para-hydroxyphenylalanine; alpha-amino-beta-para-hydroxyphenylpropionic acid) $C_6H_4OHCH_2CHNH_2COOH$. A nonessential amino acid.
Properties: White crystals, readily oxidized by the animal organism; soluble in water; slightly soluble in alcohol; insoluble in ether; optically active.
DL-tyrosine m.p. 316°C;
D(+)-tyrosine m.p. 310-314°C;
L(-)-tyrosine m.p. 295°C with decomposition; sp.gr. 1.456 (20/4°C).
Derivation: Hydrolysis of protein (casein); organic synthesis.
Uses: A growth factor in nutrition; biochemical studies.
Available commercially as DL-tyrosine.

tyrothricin. An antibiotic produced by growth of Bacillus brevis. It consists of a mixture of antibiotics, principally gramicidin and tyrocidine (the latter usually present as the hydrochloride). Gramicidin is the more active component. Use is generally limited to local external applications. It is active against some gram-positive bacteria; including species of pneumococci, streptococci and staphylococci.
Properties: White to buff-colored powder; nearly odorless and tasteless; soluble in alcohol, acetone and dioxane; insoluble in water, chloroform and ether. Resistant to action of pepsin and trypsin.
Grade: N.F. XI.
Use: Medicine.

"Tysonite." [69] Trademark for an organic rubber-like plastic for use in rubber products.
Properties: Dark brown; no odor in rubber compounds.
Constants: Sp.gr. 1.04.
Uses: To resist the deteriorating effect of ozone and improve electrical properties of rubber insulation compounds.

tyuyamunite $Ca(UO_2)_2(VO_4) \cdot nH_2O$. A natural hydrated vanadate of calcium and uranium, similar to carnotite (q.v.).
Properties: Color yellow; luster adamantine to pearly; usually occurs as fine crystals or as a powder; radioactive.
Occurrence: New Mexico, Utah, Colorado, U.S.S.R.
Use: Ore of uranium.

"Tyzine." [299] Trademark for tetrahydrozoline hydrochloride.

"Tyzor" Organic Titanates. [28] Trademark for a series of esters of orthotitanic acid.
Use: As adhesion primers, binders for high temperature paints, pigment dispersants and moisture scavengers.

U

U. Symbol for uranium.

U-233. See uranium 233.

U-235. See uranium 235.

U-238. See uranium 238.

"Ubatol." [22] Trademark for a series of fine particle size styrene- and acrylic-based polymer and co-polymer emulsions, some less than .01 microns. Films deposited from compounded polymer emulsions exhibit high gloss, increased water resistance and durability. Film characteristics of straight polymers range from hard non-film forming to soft-flexible. U-2000 series - modified styrene homo-polymers; U-3000, U-4000, and U-7000 series - acrylic and acrylic-styrene homo-polymers, co-polymers and inter-polymers.

Uses: Coatings for paper, leather, textiles, tapes; vehicle for gloss latex paints; self polishing floor wax, latex compounding, detackifying agents, detergent opacifying agent.

"Ucar." [214] Trademark for various types of synthetic organic chemicals.

"Ucar" butylene oxide 12 (1,2-epoxy-butane). $\overline{CH_2OCHCH_2CH_3}$.

Properties: Water white reactive liquid; b.p. 63.2°C; f.p. −150°C; sp.gr. 0.8312 (20/20°C); 6.9 lb/gal; 5.9% soluble in water; flash point less than 20°F.

Uses: Intermediate for detergents; oil additives; lubricants; stabilizer chlorinated solvents.

"Ucar" butylphenol 4T $(CH_3)_3CC_6H_4OH$.

Properties: Flaked white solid; melting point 97°C; b.p. 237°C; sp.gr. 0.9081 (114/4°C).

Uses: Intermediate oil soluble phenolic resins; oil additives; paint driers; rubber vulcanizers; insecticides; antioxidant.

"Ucar" nonylphenol $C_9H_{19}C_6H_4OH$.

Properties: Clear liquid; b.p. 297.6°C; f.p. −20°C; sp.gr. 0.9385 (20/20°C); viscosity 563 cps at 20°C.

Uses: Intermediate surface active agents; oil soluble phenolic resins; oil additives; plasticizers.

"Ucar" triphenol P [1,1,3-tris(hydroxyphenyl)-propane] $(C_6H_4OH)_2CHCH_2CH_2C_6H_4OH$.

Properties: White solid; m.p. 84°C; sp.gr. 1.226 (25/20°C).

Uses: Antioxidant; intermediate polyester and alkyd type resins.

"Ucet." [214] Trademark for epoxy resin type of textile wrinkle resistant finishes for cotton and rayon.

"Ucilon." [288] Trademark for corrosion resisting coating materials (paint) and thinners.

"Ucon." [214] Trademark for various types of synthetic organic chemicals.

"Ucon" fluids and lubricants: polyalkylene glycols and diesters. Available as both water-soluble or water-insoluble products. Non-corrosive to metals; little effect on rubber. Numerous grades available; viscosities range from 50 to 90,000 SUS; pour points as low as −85°F. Used as high-temperature lubricants, low temperature fluids, compressor lubricants, hydraulic brake fluids, leather and paper-treating compounds, rubber lubricants, plasticizers and solvents, chemical intermediate.

"Ucon" fluorocarbons, for refrigerant and propellant use, include the following:

"Ucon" 11: trichlorofluoromethane (q.v.).
"Ucon" 12: dichlorodifluoromethane (q.v.).
"Ucon" 22: chlorodifluoromethane (q.v.).
"Ucon" 113: 1,1,2-trichloro-1,2,2-trifluoroethane (q.v.).
"Ucon" 114: 1,2-dichloro-1,1,2,2-tetrafluoroethane (q.v.).

"Ucon" JL-6, a diester: sp.gr. 0.9265; (20/20°C); pour point lower than −85°F; viscosity index 151.

"Udex." [233] Trademark for glycols, such as diethylene glycol (q.v.), specially prepared for use in liquid-liquid extraction processes, such as the "Udex" process.

"Udex" Process. [233,416] Patented process for extracting aromatic hydrocarbons from mixed hydrocarbons by using a glycol-water mixture as solvent. The aromatic extract is distilled from the solvent and extremely high purity individual aromatics separated from one another by further distillation.

UDMH. Abbreviation for uns-dimethylhydrazine.

UDP. Abbreviation for uridine diphosphate. See uridine phosphates.

UDPG. Abbreviation for uridine diphosphate glucose.

"U. F. Concentrate-85." [197] Trade name for a solution of formaldehyde and urea.

Typical analysis: Formaldehyde, 60%; urea, 25%; water, 15%; methanol < 0.3%; salts < 0.2%; free formic acid, nil.

Properties: Clear, colorless, viscous solution; sp.gr. 1.33 at 25/15°C; refractive index 1.472 at 25°C; b.p. about 100°C; f.p. −20 to −30°C; wt/gal 11.1 lbs; viscosity, 260 cps (25°C); flash point (Cleveland open cup) none; pH, approx. 8.

Containers: 55-gal lined drums; 8000-, 10,000-gal tank cars; 3500-gal tank trucks.
Hazards: Toxicity: similar to 50% formaldehyde.
Uses: Adhesives; textile finishing agents; treating paper products to impart wet strength; industrial finishes; molding powders.

"Uformite." [23] Trademark for synthetic resins based on urea-formaldehyde, melamine-formaldehyde, and triazine condensates. Supplied as colorless or light-colored aqueous solutions or solutions in volatile solvents. Solvent type produces hard, alkali-resistant, colorless coatings on curing, with adhesion to a variety of surfaces.
Use: With alkyd resins in coatings; industrial finishes on appliances, automobiles and other commercial products; adhesive for paperboard boxes; paper coatings; manufacture of wet-strength paper; textile pigment binding.

uintahite. See gilsonite.

uintaite. See gilsonite.

ulexine. See cytisine.

ulexite (cotton balls) $NaCaB_5O_9 \cdot 8H_2O$. A natural hydrated borate of sodium and calcium.
Properties: Color white; luster silky; hardness 1-2.5; sp.gr. 1.96; usually found as rounded, loose-textured masses of fine crystals.
Occurrence: Chile; Argentina; California, Nevada.
Use: Source of borax.

Ullmann reaction. A modification of the Fittig synthesis (q.v.) in which copper powder is used instead of the sodium.

ulmin brown. See Van Dyke brown.

ulmins. Class of amorphous substances resulting from the decomposition of the cellulose and lignite tissues of plants. Ulmins represent one of the initial changes by which vegetable matter is converted into coal.

ulmus (elm; slippery elm). Flat pieces, externally light brown with dark brown patches; mucilaginous taste.
Derivation: Dried bark of Ulmus fulva, deprived of its periderm.
Occurrence: Eastern and central North America.
Grade: Technical.
Containers: Burlap bags; boxes; barrels.
Use: Medicine.
Shipping regulations: None.*

"Ultandren." [305] Trademark for fluoxymesterone N.N.D.
Use: Medicine.

"Ultex." [94] Trade name for a rubber accelerator; composed of a selected organic salt of dithiocarbamic acid, containing no metallic radicals.
Properties: White powder; sp.gr. 1.14; m.p. 121-124°C; stable in storage; soluble in acetone and ethylene dichloride; insoluble in water, gasoline, and carbon disulfide.
Containers: 100-lb fiber drums.
Use: In rubber drug sundries, sponge rubber, pure gum translucent products, patching rubber, cements, footwear, mechanical goods, molded goods, heels and soles.
Hazards: No health hazards when used in rubber, GR-S, and Buna N in amounts recommended.

ultra-accelerator. An unusually powerful accelerator of rubber vulcanization, typified by thiuram sulfides and dithiocarbamates.

"Ultraflex." [128] Brand name for a grade of petroleum microcrystalline wax.
Properties: Color, amber or white; m.p. 140-145°F.
Containers: 10-lb slabs, 8/carton or 168/pallet; 350-lb drums; tank cars.
Uses: Coating and laminating paper, foil, and board; impregnating and waterproofing fabrics.

"Ultralan." [206] Brand name of proprietary line of metal complex acid dyestuffs for wool, leather, and synthetic fibers.

ultramarine. See ultramarine blue.

ultramarine blue (imitation ultramarine blue; ultramarine). Blue pigment of variable composition; probably a double silicate of sodium and aluminum, with some sodium sulfide, the latter appearing to influence the color.
Properties: Light blue powder or lumps with a reddish hue that is transparent in oils and enamels and comparatively weak tinctorially. Better in tints than in dark blue shades. Does not retain color on exterior exposure, but reverts toward the dry appearance of the pigment when the enamel or paint film disintegrates. (This is not a true fading of pigment.) Color-fast to light, soap and alkalies, but sensitive to even weak acids.
Derivation: (a) Found in nature as the mineral lapis lazuli (q.v.). (b) Artificial product: (1) Direct method, by heating in muffles or crucibles a charge of soda ash, kaolin (or zeolites), charcoal, and sulfur; sometimes with the addition of powdered quartz, sand, or kieselguhr in order to make the product less sensitive to acids. (2) Indirect method, by first producing green ultramarine by heating in crucibles a mixture of kaolin or zeolites, anhydrous sodium sulfate, and charcoal or rosin. The green ultramarine is then powdered and "colored" by heating in contact with powdered sulfur. The product is ground, washed, and then boiled with sodium sulfide solution.
Containers: Kegs, barrels, or fiber drums, or multiwall paper sacks.
Uses: Paint pigment; paper manufacture (neutralizing yellow color); calico printing; laundry blue; printing inks; coloring mottled soaps, rubber, linoleum, plastics, etc.; feed additive in salt for animals. The artificial product has now superseded the ground mineral for these uses.

*See "I.C.C. Shipping Regulations," page xiii.
Reference numbers refer to name of manufacturer. See "List of Manufacturers," page v.

Shipping regulations: None.*
For cobalt ultramarine, see cobalt blue.

ultramarine, green. A green pigment formed as an intermediate product in the production of artificial ultramarine blue (q.v.).

ultramarine, red. A red pigment prepared by heating ultramarine blue to not over 145°C in the presence of nitric acid vapors.
Shipping regulations: None.*

ultramarine, violet. A violet pigment prepared by heating a silica-rich ultramarine blue to 175°C, in an atmosphere of chlorine and steam. It has poor tinctorial properties.
Shipping regulations: None.*

ultramarine, yellow. See barium chromate; also calcium chromate and chrome yellows.

"Ultramins." [449] Trademark for a series of amine condensates used in the softening and finishing of textile fabrics.

"Ultran." [100] Trademark for phenaglycodol (q.v.).

"Ultrapoles." [449] Trademark for a series of alkanolamine condensates, detergent base and surface active materials used as detergents and ingredients of detergents, wetting and foaming agents, foam stabilizers, ingredients for cosmetic preparations, emollients, thickening agents.

ultraquinine. See cupreine.

ultrasonics. The study of effects of sound vibrations at and beyond the limit of audible range of frequencies. Suggested or used for dust, smoke and mist precipitation; preparation of colloidal dispersions or emulsions such as homogenizing of milk; formation of catalysts, degassing and solidification of molten metals; extracting flavor oils in brewing; speeding electroplating; drilling of hard or brittle materials; fluxless soldering; and in nondestructive testing of metals. Also used for investigation of physical properties, determination of molecular weights of liquid polymers, degree of association of water, and for causing chemical reactions to occur. Biological effects are also under study.

ultraviolet. The region of the electromagnetic spectrum including wave lengths from 100 to 3900A. (i.e., longer than x-rays and shorter than visible light).

ultraviolet absorber. A substance which absorbs radiant energy of wave length in the ultraviolet range (100-3900 Angstrom units-just below the range of visible radiation). The energy absorbed is dissipated in some harmless form other than visible light. Double bonds, triple bonds, and (to a less degree) unbonded electron pairs all contribute to the ultraviolet absorbing capacity of molecules. Ultraviolet absorbers are added to unsaturated substances (plastics, rubbers, etc.) to decrease light sensitivity.

"Ultrawets." [136] Trade name for alkyl aryl sulfonate anionic detergents or surface active materials (dodecyl benzene cyclic type).
Use: Light and heavy duty household detergent formulations; liquid dishwashing formulations; penetrating; wetting and emulsifying agents.

"Ultron." [58] Trademark for vinyl chloride flexible films or rigid sheets.

"Ultrox." [288] Trademark for zirconium silicate opacifiers.
Grades: Ultrox - fineness all through 400 mesh, all below 7 microns; average particle size 2 microns. Ultrox 500 W - fineness all through 400 mesh, all below 5 microns; average particle size 1 micron.

umber. A naturally occurring brown earth containing ferric oxide together with silica, alumina, manganese oxides and lime. See limonite and ferric oxide, yellow. Raw umber is umber which is ground and then levigated. Burnt umber is umber calcined at low heat.
Grades: Based on tinctorial power and iron content; the best come from Cyprus or Turkey.
Containers: Bags.
Uses: Paint pigment; lithographic inks; wall paper (pigment).
Shipping regulations: None.*

UMP. Abbreviation for uridine monophosphate. See uridine phosphates and also uridylic acid.

"UN-32." [266] Brand name for a urea-ammonium nitrate liquid fertilizer containing 32% nitrogen.

"Unadol." [221] A line of unsaturated fatty alcohols used in protective coatings, resins, polymers, defoamers and as chemical intermediates.

"Unads." [69] Trademark for tetramethylthiuram monosulfide $[(CH_3)_2NC(S)]_2S$.
Properties: Yellow powder (also supplied as rods); sp.gr. 1.37 ± .03; melting range 103-114°C; slightly soluble in water and gasoline; very soluble in acetone, benzol, chloroform.
Uses: In natural and nitrile rubber and in SBR, primary accelerator; secondary accelerator (with thiazoles). In sponge, tile, soles.

gamma-**undecalactone** (aldehyde C-14 "so-called"; peach aldehyde; gamma-undecyl lactone; persicol). $CH_3(CH_2)_6\underline{CHCH_2CH_2COO}$.
Properties: Colorless to light yellow liquid with a fruity odor like that of peach. Sp.gr. 0.941-0.944; refractive index 1.450-1.454. Soluble in 4 to 5 vols. of 60% alcohol; soluble in benzyl alcohol and benzyl benzoate.
Derivation: By heating undecylenic acid in the presence of sulfuric acid.
Grades: Chlorine-free.
Uses: Perfumery; flavors.

n-undecane (hendecane) $CH_3(CH_2)_9CH_3$.
Properties: Colorless liquid; sp.gr. 0.7402 (20/4°C); f.p. −25.75°C; b.p. 195.6°C;

refractive index 1.41725 (n 20/D); flash point 65°C.
Grades: 95%; 99%; research.
Containers: Bottles and drums.
Use: Petroleum research; organic synthesis.

undecanoic acid (n-undecylic acid; hendecanoic acid) $CH_3(CH_2)_9COOH$. Small amounts occur in castor oil. It is best derived from undecylenic acid by hydrogenation.
Properties: Colorless crystals. Sp.gr. 0.8505 (80/4°C); m.p. 28.5°C; b.p. 284.0°C (760 mm), 222.2°C (128 mm); refractive index 1.4319 (40°C); insoluble in water; soluble in alcohol and ether.
Grades: Technical; 99%.
Uses: Organic synthesis.

1-undecanol (n-undecyl alcohol; decyl carbinol; 1-hendecanol; alcohol, C-11) $CH_3(CH_2)_9CH_2OH$.
Properties: Colorless liquid with a citrus odor. Sp.gr. 0.829-0.834; refractive index 1.435-1.443; m.p. 15°C. Soluble in 60% alcohol.
Use: Perfumery.

2-undecanol (2-hendecanol) $CH_3(CH_2)_8CHOHCH_3$.
Properties: Colorless liquid; sp.gr. 0.8363 (20°C); m.p. 12°C; b.p. 228-229°C; insoluble in water; soluble in alcohol and ether.
Containers: Drums.
Uses: Anti-foaming agent; intermediate; perfume fixatives; plasticizer.

2-undecanone. See methyl nonyl ketone.

10-undecenoic acid. See undecylenic acid.

10-undecen-1-ol. See undecylenic alcohol.

undecoylium chloride-iodine. Iodine complex of acylcolaminoformylmethylpyridinium chloride.
Grade: N.N.D.
Use: Medicine (topical).

n-undecyl alcohol. See 1-undecanol.

undecylenic acid (10-undecenoic acid) $CH_2{:}CH(CH_2)_8COOH$.
Properties: Light colored liquid with characteristic fruity-rosy odor. Almost insoluble in water; miscible with alcohol, chloroform, ether, benzene, and with fixed and volatile oils. Congealing point 21°C; sp. gr. (25/25°C) 0.910-0.913; refractive index (25°C) 1.4475-1.4485.
Derivation: Destructive distillation of castor oil.
Grades: Technical; N.F. XI.
Containers: 6-, 13-gal carboys; 5-, 10-, 50-gal drums.
Uses: Perfumery; flavoring materials; medicinals; plastics; modifying agent (plasticizer, lubricant additive, etc.).

undecylenic alcohol (n-undecylenic alcohol; 10-undecen-1-ol; 10-11-undecylenic alcohol; alcohol C-11) $CH_2{:}CH(CH_2)_8CH_2OH$.
Properties: Colorless liquid with fatty, somewhat citrus odor; sp.gr. 0.842-0.847; refractive index n 20 1.449-1.454; m.p. -3.0°C; soluble in 70% alcohol.
Use: Perfumes.

undecylenic aldehyde (aldehyde C-11; 10-hendecen-1-al) $CH_2{:}CH(CH_2)_8CHO$.
Properties: Colorless liquid; strong odor suggesting rose. Sp.gr. 0.842-0.850; refractive index 1.442-1.447. Soluble in 80% alcohol.
Use: Perfumery.

undecylenyl acetate (acetate C-11; 10-hendecenyl acetate) $CH_3COO(CH_2)_9CH{:}CH_2$.
Properties: Colorless liquid with a floral-fruity type odor. Sp.gr. 0.876-0.883; refractive index 1.438-1.442. Soluble in 80% alcohol.
Use: Perfumery.

n-undecylic acid. See undecanoic acid.

gamma-**undecyl lactone.** See gamma-undecalactone.

underglaze colors. Finely ground calcined oxides for colored designs beneath the glaze on ceramic surfaces.
Use: For coloring or decorating pottery, tile, terra cotta and similar glazed ceramic surfaces.

underground gasification. A process for decomposing coal in place. Two or more wells are drilled to the vein of coal. At one of these the coal is ignited and supplied with a forced draft of air or oxygen. The products of combustion are drawn off from the other wells and used as fuel or chemical raw materials.

unguentum. Ointment, a fatty base in which a drug or mixture of drugs is incorporated.

UNH. Abbreviation for uranyl nitrate hydrated. See uranyl nitrate.

"Unicel" ND. [28] Trademark for rubber chemical comprising 40% dinitrosopentamethylenetetramine and 60% inert organic filler.
Properties: Light cream-colored powder.
Containers: Drums (100 lbs net).
Use: A non-discoloring blowing agent for the manufacture of natural and synthetic rubber sponge.

"Unicel" NDX. [28] Trademark for a mixture of rubber blowing agent 80% di-N-nitrosopentamethylene-tetramine (see "Unicel" ND) and 20% inert filler. Cream-colored spongy crumbs; sp.gr. 1.40.
Containers: 100-lb drums.
Use: A non-discoloring blowing agent for natural or synthetic rubber sponge.

"Unicel" S. [28] Trademark for rubber chemical comprising a 50% dispersion of finely divided sodium bicarbonate in oil.
Properties: A cream-colored, free-flowing liquid.
Containers: Drums (500 lbs net).
Use: Blowing agent for the manufacture of natural and synthetic rubber sponge.

"Unichrome." [288] Trademark for (1) synthetic resinous materials in the form of solutions for use to form coatings on electroplating apparatus; (2) organic solvents used as thinners for solutions of synthetic resinous materials; (3) papers impregnated with dyestuffs of the class known as "Indicators," for testing hydrogen ion concentration (pH) of solutions; (4) cleaning compositions for use in making up solutions used for cleaning metals preparatory to electroplating; (5) cements in liquid and solid (plastic) form, used for sealing joints of containers of chemical solutions.

"Unicor." [416] Trademark for ash free, oil-soluble, surface active film-former.
Properties: Oily liquid; density, 7.3 lbs/gal at 60°F; API, 29.4 at 60°F; flash point (Pensky-Martin), 118°F; pour point, −5°F, cold point, −10°F; viscosity (kinematic) 103.8 cs at 100°F, (universal) 481 sec; completely combustible; insoluble in water and LPG.
Uses: Corrosion inhibitor in various refinery operations in crude units, thermal cracking units, Platformers, etc.; it is also effective in run-down and storage tanks, pipelines and tankers.

"Unicor LHS." [416] Trademark for film-forming, organic phosphate base anti-icer, corrosion inhibitor and carburetor detergent.
Properties: Liquid; sp.gr. 0.909 at 68°F; pour point, −30°F; PM flash point (open cup), 110°F; viscosity (kinematic) 11.82 cs at 100°F, (universal) 65.4 sec; soluble in liquid propane and light hydrocarbons.
Containers: 55-gal steel drums; tank wagons.
Uses: Corrosion inhibitor in refinery environments where acidic gases and moisture collect on condenser tubes and exchangers, particularly in areas condensing propanes and butanes; detergent and anti-icer to prevent engine-stalling caused by ice formation in carburetors and induction systems.

"Unicor V." [416] Trademark for an oil-soluble, film-forming corrosion inhibitor for the petroleum industry.
Properties: Oily liquid; density, 0.854; b.p. 680-698°F; pour point, below −30°F; flash point (Pensky-Martin), 325°F; viscosity (kinematic) 9.426 cs at 100°F, 2.228 cs at 200°F; will evaporate at atmospheric pressure without decomposition; low order of reactivity.
Uses: Corrosion inhibitor in process units, where its non-fouling and volatility properties give it an advantage over more reactive and less stable inhibitors.

"Unifining" Process. [416] Patented process for the removal of sulfur, nitrogen and metals; for the saturation of olefins and for the decomposition of oxygen compounds from all petroleum derived distillate stocks. The process is also applied to the purification of coal tar distillates. The process operates at moderate pressures in an environment of hydrogen, employing a catalyst containing cobalt and molybdenum.

"Uniflo." [51] Trademark for a multi-grade motor oil, an all-weather lubricant which reduces the amount of engine deposits. Available in two grades: 5W-20 and 10W-30.

union dyes. Those that are suitable for dyeing fabrics containing both cotton and wool.

"Uniphats." [259] Trade name for the monohydric alcohol esters of a series of fatty acids ranging from technically pure fatty acids to natural and synthetic mixtures.
Uses: Chemical raw material; detergents; gasoline additives.

"Unisol." [136] A process for removing mercaptans from gasoline and burning oils by use of caustic methanol solutions.

"Unisol" Process. [416] Patented process for removing mercaptans from gasoline and burning oils by use of caustic methanol solutions.

"Unitane." [57] Trademark for titanium dioxide (TiO_2) pigment available in both anatase and rutile crystal forms.
Anatase form: Sp.gr. 3.8-3.9; index of refraction 2.52.
Rutile form: Sp.gr. 4.2; index of refraction 2.76. Rutile TiO_2 has approximately 25% greater tinting strength and opacity than the anatase type.
Containers: 50-lb paper bags.
Uses: Both anatase and rutile forms are used in paints, lacquers, enamels, printing inks, paper, rubber, leather, linoleum, ceramics, welding rods, etc.

"Unitol CMT." [420] Trade name for tall oil fatty acids.
Composition:

	Range min	Range max
Rosin acids, %	0.2	0.6
Unsaponifiables, %	0.1	0.6
Fatty acids, %	98.8	99.7
Saturated fatty acids, %	2.0	2.8
Acid number	198	201
Saponification number	198	202
Color, Gardner	1	2
Iodine number	128	133
Titre	-1.0	+1.0

Containers: 55-gal resin-lined steel drums; resin-lined tank cars.

unit operation. A particular kind of a physical change that is repeatedly and frequently encountered as a step in the processes for industrial production of various chemicals and related materials. Filtration, evaporation, distillation, fluid flow and heat transfer are examples.

unit process. A process characterized by a particular kind of chemical reaction and equipment, of which many specific examples are encountered, to which the same basic principles of designs and operation may be applied. Oxidation, hydrolysis, esterification, and nitration are examples.

"Univis." [51] Trademark for a series of power transmission or hydraulic oils. They have viscosity indexes of 150 or higher and pour points of −50°F or lower, permitting wide temperature ranges in operation.

*See "I.C.C. Shipping Regulations," page xiii.
Reference numbers refer to name of manufacturer. See "List of Manufacturers," page v.

"Univolt."[51] Trademark for oils used as electrical insulating mediums in transformers, switches and some electrical cables. Suitable for transformers, whether used indoors or outdoors or under low-temperature conditions.

"Unox."[214] Trademark for epoxides and wetting agent.
"Unox" epoxide 101 - vinyl cyclohexene monoxide. Viscosity 1.69 cps (20°C); sp.gr. 0.9598 (20/20°C).
"Unox" epoxide 201 - 3,4-epoxy-6-methylcyclohexylmethyl 3,4-epoxy-6-methylcyclohexanecarboxylate. Viscosity 1810 cps (25°C); sp.gr. 1.121 (20/20°C).
"Unox" epoxide 206 - vinyl cyclohexene dioxide. Viscosity 7.77 cps (20°C); sp.gr. 1.098 (20/20°C).
"Unox" epoxide 207 - dicyclopentadiene dioxide. Density 1.331 (25°C).
"Unox" epoxide 269 - dipentene dioxide. Sp. gr. 1.032 (20/20°C).
Uses: Intermediate for epoxy resins, plasticizers, pharmaceuticals; polymer cross linking agents.
"Unox" fire-fighting penetrant. Trademark for a wetting agent. Increases the effectiveness of water in extinguishing fires. It helps water to penetrate dense burning materials, putting out the fire quickly-with less smoke, less water damage, and less overhaul.

uns- (unsym). Abbreviation for unsymmetrical. A prefix denoting the structure of organic compounds in which substituents are disposed unsymmetrically with respect to the carbon skeleton or to a functional group, such as a double bond. For example, uns-dichloroethane is CH_3CHCl_2. In this dictionary, it is disregarded in alphabetizing. See also sym-.

unsaturation. The property of an organic compound that causes or allows its ready combination with hydrogen, chlorine, oxygen and various other substances. An unsaturated compound (as ethylene, C_2H_4, butadiene, C_4H_6, acetylene, C_2H_2) has therefore fewer hydrogen atoms or equivalent groups than the corresponding saturated compound (ethane, C_2H_6, butane, C_4H_{10}, ethane respectively). This phenomenon can be conveniently visualized by the assumption of the presence in a molecule of one or more double or triple bonds (unsaturated bonds). Thus ethylene can be represented by the line formula $H_2C{=}CH_2$, butadiene by $H_2C{=}CHCH{=}CH_2$, and acetylene, $HC{\equiv}CH$. (The double bond may also be represented by a colon, as $H_2C{:}CH_2$.) Each of these multiple bonds connects just two atoms.

The unsaturated character results from the ability of the second and third bonds to detach and connect with additional atoms taken up by the molecule. A cyclic structure, however, which contains double bonds alternating with single bonds (benzene, naphthalene, etc., and derivatives) is said to have aromatic character. Such a molecule does not participate in addition reactions. This lack of unsaturation in aromatic substances is explained by resonance (q.v.).

unsym-. See uns-.

"UOP #5."[416] Trademark for N,N'-di-secondary butyl-para-phenylenediamine.
Properties: Normally exists as a supercooled liquid below 64°F; sp.gr. (60/60) 0.94; pour point ASTM, below 0°F; flash point, 285°F; viscosity (kinematic) 10.1 cs at 100°F; miscible in all proportions with absolute alcohol and benzene; insoluble in water or caustic solutions.
Uses: Oxidation inhibitor and stabilizer in both aviation and motor gasoline; catalyzes air-oxidation of mercaptans; prevents gum formation and decomposition of tetraethyllead in gasoline.

"UOP 88."[416] Trademark for an N,N'-dioctyl-para-phenylenediamine inhibitor and antiozonant.
Properties: Liquid; b.p. about 735°F at 760 mm; sp.gr. 0.912 at 60°F; pour point, 25°F; flash point (Pensky-Martin), 395°F; viscosity (kinematic) 43.85 cs at 100°F, (universal) 204.0 sec; refractive index, (n 20/D) 1.5129; completely miscible in methanol, pentane and benzene; vapor pressure (absolute) 0.33 mm Hg at 302°F.
Containers: Steel drums, tank cars, tank trucks.
Uses: Antioxidant, antiozonant. Prevents gum formation; stabilizes tetraethyllead in both motor and aviation gasoline; catalyst in air-oxidation of mercaptans; reduces engine deposits; retards cracking of synthetic rubbers due to ozone attack.

"UOP 288."[416] Trademark for an N,N'-dioctyl-para-phenylenediamine; isomeric with "UOP 88."
Properties: Liquid; m.p. (solid isomer), 99.5°F; b.p. 788°F; flash point (Pensky-Martin), 325°F; viscosity (kinematic) 25.9 cs at 212°F; vapor pressure (absolute), 0.18 mm Hg at 302°F; density, 0.901 (20/4°C); refractive index, 1.5098 (n 20/D).
Uses: Antiozonant for protection against cracking in natural rubber formulations, in SBR synthetic rubber, buna-N and oil-extended polymers.

"UOP Copper Deactivator."[416] Trademark for formulations based primarily on disalicylalaminopropane.
Properties: Liquid; sp.gr. (60/60) 1.08; density, 9.0 lbs/gal; pour point, 0°F; flash point (Tag closed cup), above 70°F; viscosity 25 cs at 100°F; insoluble in water; miscible with benzene.
Grades: Regular (properties as above); "UOP AW(50)" is 50% active and gasoline soluble.
Uses: Prevents traces of copper from harming civilian and "combat grade" gasolines; synergist for fuel oil inhibitors and dispersants.

"UOP #1 Inhibitor."[416] Trademark for a phenolic fraction of hardwood tar.
Properties: Liquid; sp.gr. (60/60) 1.08; pour point ASTM, 0-5°F; b.p. above 400°F;

viscosity 17.6 cs at 100°F; slightly soluble in water.

Uses: Oxidation inhibitor for gasolines which are completely free of water and caustic contamination.

"UOP #4 Inhibitor." [416] Trademark for an alcoholic solution of N-n-butyl-para-aminophenol.

Properties: Liquid; sp.gr. (60/60) 0.90; pour point ASTM, below −35°F; flash point (Cleveland Open Cup Fire), 60°F; viscosity (kinematic) 4.4 cs at 100°F; miscible in all proportions with absolute alcohol and benzene; soluble in caustic solutions; very slightly soluble in water.

Uses: Oxidation inhibitor to protect unstable gasolines; particularly effective in reducing copper dish gum.

"UOP #7 Inhibitor." [416] Trademark for 2, 6-di-tert-butyl-4-methylphenol.

Properties: F.p., 156°F; flash point, 260°F; soluble in aromatic hydrocarbons; insoluble in either acid or alkali.

Grades: Crystalline (No. 7); 33% solution in toluene (No. 7-S).

Uses: Oxidation inhibitor for relatively stable gasolines; prevents tetraethyllead decomposition. Approved for use in military fuels.

"UOP # 5-S." [416] Trademark for a modified phenylenediamine type antioxidant.

Properties: Liquid (supercools); sp.gr. (60/60) 0.94; pour point ASTM, below 0°F; viscosity (kinematic) 11.9 cs at 100°F; miscible with absolute alcohol and benzene; insoluble in caustic.

Uses: Oxidation inhibitor for gasoline; more rapid sweetener than UOP #5. Approved for addition to "combat grade" gasoline; prevents gum formation and decomposition of tetraethyllead in gasoline.

"Urab." [50] Trademark for a complex of fenuron and TCA, available in liquid concentrate, granular, and pelleted formulations. A brush and weed killer for control of woody plants and deep-rooted weeds in non-crop land.

uracil $\overline{HNC(O)NHC(O)CHCH}$. 2, 4-Dioxypyrimidine. A pyrimidine that is a constituent of ribonucleic acids and the coenzyme, uridine diphosphate glucose (q.v.).

Properties: Crystalline needles; m.p. 335°C (dec). Soluble in hot water, ammonium hydroxide and other alkalies; insoluble in alcohol and ether; $A_M = 8.2 \times 10^3$ at 260 millimicrons and pH 7.0.

Derivation: Hydrolysis of nucleic acids; precipitation from urea and ethyl formylacetate. Radioactive forms available.

Use: Biochemical research.

uracil-6-carboxylic acid. See orotic acid.

uracil, D-ribosyl. See uridine.

"Urac" Resins. [57] Trademark for proprietary products based on urea-formaldehyde condensates used mainly as adhesives for the production of moisture-proof bonds in plywood manufacture, plywood assembly, and furniture manufacture.

"Uramite." [28] Trademark for an odorless yellow or gray white granular solid containing nitrogen principally in the form of methylene ureas; remains free-flowing under all conditions; contains 38% nitrogen, only slightly soluble in water.

Containers: 25-, 50-, 80-lb paper bags.

Use: As a direct application fertilizer for turf grasses and ornamental plants, or other crops requiring a long period of a continuous and uniform supply of nitrogen.

"Uramon" Ammonia Liquors. [28] Trademark for solutions of crude urea and ammonium sulfate in aqueous ammonia.

	Nitrogen Content	Vapor Pressure (lb/sq in, 70°F)
UAL-B	45.5%	22
UAL-K	40.5%	1.0
UAL-37	37.0%	2
UAL-S	43.5%	20

UAL-37 also supplies water-insoluble organic nitrogen.

Containers: Tank cars.

Use: In manufacture of fertilizers.

urania-thoria. Crystals of the mixed oxides of uranium and thorium are available. Used as nuclear fuel. The crystals are denser and cheaper than the old pellet form.

uranic chloride. See uranium tetrachloride.

uranic oxide. See uranium dioxide.

uranine (uranine yellow; fluorescein-sodium; resorcinolphthalein sodium) $Na_2C_{20}H_{10}O_5$. Colour Index No. 766.

Properties: Orange red, odorless powder; hygroscopic; soluble in water and sparingly soluble in alcohol.

Derivation: By treatment of fluorescein with sodium carbonate solution and crystallizing.

Method of purification: Recrystallization.

Grades: Technical; U.S.P. XVI (as fluorescein sodium).

Containers: Custom packed.

Uses: Dyeing silk and wool yellow; following the course of subterranean waters; marking water for air-sea rescues; clinical test solution.

Shipping regulations: None.*

uranine yellow. See uranine.

uraninite UO_2. A natural oxide of uranium, usually partly oxidized to UO_3, with variable amounts of lead, radium, thorium, rare earth metals, helium, argon, and nitrogen.

Varieties:

Broggerite. A thorium-bearing variety.

Nivenite and Cleveite. Contain rare earth metals.

Pitchblende (q.v.). A very finely crystalline variety.

The name uraninite is also used to refer to a well-crystallized variety of uranium dioxide found in pegmatites, while the term pitchblende has been used both for uraninite found in metallic veins and for a distinct

*See "I.C.C. Shipping Regulations," page xiii.

species.

Occurrence: Colorado, Utah, Connecticut, North Carolina, Texas, Canada, Czechoslovakia, Norway, Congo, Germany, England.

Use: Source of uranium and radium.

uranite. A mineral consisting either of autunite (lime uranite) or torbernite (copper uranite) (q.v.).

uranium U. Element number 92, a member of the actinide series, and the heaviest naturally occurring element. More important, it is the only primeval element that can readily undergo direct, spontaneous nuclear fission. Natural uranium is a mixture of three radioactive isotopes: U-234 (0.006%), U-235 (0.7%), and U-238 (99%). U-235 is the isotope which can capture slow neutrons to undergo fission with large energy release in nuclear reactors (q.v.) or in uranium bomb explosions.
U-238 isotope will not directly support fission, it will also absorb neutrons to eventually form plutonium-239, which has fission properties similar to U-235. Natural uranium can serve as fuel for an atomic pile, but uranium fuel enriched in U-235 is easier to regulate. This enrichment is usually achieved by gas diffusion; but other gas methods, including gas centrifugation, have been proposed. (For the diffusion separation, which is the production method in use at Oak Ridge, Tenn., uranium hexafluoride is forced through a series of barriers, which permit the lighter isotope to pass through slightly more rapidly than the heavy one.)

Properties: Very dense, silvery metal; strongly electropositive; ductile and malleable; poor conductor of electricity; sp.gr. 18.685; m.p. 1132°C; b.p. 3818°C; heat of fusion 4.7 kcal/mole; heat capacity 6.6 cal/mole -°C. Powdered uranium is spontaneously flammable. Forms solid solutions (for nuclear reactors) with molybdenum, niobium, titanium and zirconium. The metal reacts with nearly all nonmetals. It is attacked by water, acids, and peroxides; but is inert towards alkalies. Green tetravalent uranium and yellow uranyl ion (UO_2^{++}) are the only species which are stable in solution.

Occurrence: Pitchblende (essentially UO_2), a variety of uraninite, coffinite ($USiO_4$) and carnotite are the most important sources. Other ores include autunite, torbernite, tyuyamunite, thorianite, uranophane, samarskite, davidite, schroeckingerite, and various rare earth minerals. Low-grade ore occurs in phosphate deposits, bituminous shales, and lignite. Principal locations are the Congo, Canada, the Colorado Plateau, U.S.S.R., and North Carolina (samarskite).

Derivation: Finely ground ore is leached under oxidizing conditions to give uranyl nitrate solution. The uranyl nitrate, purified by solvent extraction (ether, alkyl phosphate esters), is then reduced with hydrogen to uranium dioxide. This is treated with hydrogen fluoride to obtain uranium tetrafluoride. Winning of the free metal proceeds by either electrolysis of the tetrafluoride in fused salts or, more often, by reduction with calcium or magnesium. Fluorination of uranium tetrafluoride yields the hexafluoride, the form used in gas diffusion and centrifugation techniques for uranium isotope separation.

Uses: As nuclear fuel when alloyed with zirconium, molybdenum, etc. (the oxide and carbide are more recent nuclear fuels); as atomic bomb explosive. Uranium compounds are coloring agents in ceramics.

See also uses for uranium 238.

uranium 233 (U-233). A fissionable isotope of uranium produced artificially by bombarding thorium with neutrons.

uranium 235 (U-235). The readily fissionable isotope of uranium used in one type of atomic bomb. Concentrated from natural uranium by gaseous diffusion, centrifugation, or electromagnetic methods.

uranium 238 (U-238). The abundant isotope of uranium; 140 times as plentiful as U-235. It is nonfissionable, but will capture neutrons in a nuclear reactor to eventually produce Pu-239, a nuclide which can substitute for U-235 as a fuel or explosive. This production of Pu-239 is called "breeding." U-238 (obtained as natural uranium from which U-235 has been removed) can also be used as a coloring agent; analytical reagent; "getter" for vacuum tubes; for cathodic protection; alloys; catalysis; ion exchange systems, etc.

uranium acetate. See uranyl acetate.

uranium ammonium carbonate. See uranyl ammonium carbonate.

uranium ammonium fluoride. See uranyl ammonium fluoride.

uranium barium oxide (barium diuranate; barium uranium oxide) BaU_2O_7.

Properties: Yellow or orange powder. Caution! Poison! Soluble in acids.

Grade: Technical.

Use: Ceramics (coloring porcelain).

uranium-bismuth. A liquid metal alloy suggested as a liquid metal fuel for nuclear reactors.

uranium carbides. See uranium dicarbide; uranium monocarbide.

uranium decay series (uranium-radium series). The series of elements produced as successive intermediate products when the element uranium undergoes its spontaneous natural radioactive disintegration into lead. Radium and radon are members of this series.

uranium, depleted. Uranium from which most of the U-235 isotope has been removed. See uranium-238.

uranium dicarbide (uranium carbide) UC_2.

Properties: Gray crystals; sp.gr. 11.28 (18°C); m.p. 2260°C; b.p. 4100°C.

Decomposes in water; slightly soluble in alcohol.
Use: As crystals, pellets, or microspheres for nuclear reactor fuel.

uranium dioxide (uranium oxide; uranic oxide) UO_2.
Properties: Black crystals, insoluble in water, soluble in nitric acid and concentrated sulfuric acid; frequently pyrophoric in finely divided form; sp.gr. 10.9; m.p. 3000 ± 200°C.
Derivation of pure oxide: The powdered uranium ore is digested with hot nitric-sulfuric acid mixture and filtered to remove the insoluble portion. Sulfate is precipitated from the solution with barium carbonate and uranyl nitrate is extracted with ether. The uranyl nitrate, after re-extraction into water, is heated to drive off nitric acid, leaving uranium trioxide. The latter is reduced with hydrogen to the dioxide. More recently prepared from uranium hexafluoride by treating with ammonia and subsequent heating of the ammonium diuranate to get the dioxide.
Uses: A crystalline (or pellet) form is used to pack nuclear fuel rods. Used also in ceramics; pigments; photographic chemicals; catalyst; a source of uranium for the fluorides used for isotope separation.

uranium, enriched. Uranium containing more than the normal proportion of uranium-235 isotope.

uranium hexafluoride UF_6.
Properties: Colorless volatile crystals; sublimes; triple point 64.0°C (1134 mm); m.p. 64.5°C (2 atm); sp.gr. 5.06 (25°C); soluble in liquid bromine, chlorine, carbon tetrachloride, sym-tetrachloroethane and fluorocarbons. Reacts vigorously with water, alcohol, ether, and most metals. Vapor behaves as nearly perfect gas.
Derivation: Fluorination of uranium tetrafluoride, which is obtained by hydrofluorination of uranium dioxide. The dioxide, in turn, results from fluid bed reduction of higher oxides by hydrogen.
Use: In gas diffusion process for separating isotopes of uranium.

uranium hydride UH_3.
Properties: Brown gray to black powder; sp. gr. 10.92; conductor of electricity.
Derivation: Action of hydrogen on hot uranium.
Uses: Preparation of finely divided uranium metal by decomposition; separation of hydrogen isotopes; reducing agent; laboratory source of pure hydrogen.
Hazards: Pyrophoric; should be handled in inert gas atmosphere; protective clothing should be worn.

uranium monocarbide (uranium carbide) UC.
Properties: Lumps or powder that can be formed into desired shapes by powder metallurgy or arc-melt casting; m.p. 2375°C, density 13.63 g/cc; thermal conductivity 0.08 cal/sec/cm²/°C/cm; must be stored in inert atmosphere.
Use: Nuclear reactor fuel.

uranium nitrate. See uranyl nitrate.

uranium nitride U_3N_4. Dark brown crystals; sp. gr. 10.09; decomposed by water. Said to be used as a nuclear fuel.

uranium oxides. Uranium forms several oxides, including U_2O_3 (uranous oxide), UO_2 (uranium dioxide, uranic oxide), UO_3 (uranium trioxide), U_3O_8 (triuranium octoxide), $UO_4 \cdot xH_2O$ (uranium peroxide).

uranium oxychloride. See uranyl chloride.

uranium peroxide (uranium oxide; uranium tetroxide). $UO_4 \cdot xH_2O$. (The water of hydration varies according to the conditions under which oxide is made.)
Properties: Yellow crystals, hygroscopic; sp.gr. (15°C) 2.5; decomposes at 115°C; insoluble in water; decomposes in hydrochloric acid.
Derivation: Precipitation from solutions of uranyl salts by hydrogen peroxide.
Uses: Ceramics, pigments.

uranium-radium series. See uranium decay series.

uranium sodium acetate. See uranyl sodium acetate.

uranium sulfate. See uranyl sulfate.

uranium tetrachloride (uranic chloride) UCl_4.
Properties: Dark green volatile crystals; soluble in water and in alcohol; hygroscopic; sp.gr. 4.98; m.p. 590°C, b.p. 792°C.
Derivation: By the reaction of uranium dioxide with carbon tetrachloride or phosphorus pentachloride or other strong chlorinating agents.
Purification: Sublimation or fractional distillation.

uranium tetrafluoride (green salt) UF_4.
Properties: Green, non-volatile crystalline powder; sp.gr. 6.70; m.p. 1036°C; insoluble in water.
Derivation: Treatment of uranium dioxide with hydrogen fluoride. See uranium hexafluoride.
Uses: Preparation of uranium metal and uranium hexafluoride.

uranium tetroxide. See uranium peroxide.

uranium trioxide (uranium oxide) UO_3.
Properties: Red or yellow powder; insoluble in water, soluble in nitric acid; sp.gr. 8.34; decomposes with heating.
Derivation: Thermal decomposition of uranyl nitrate or ammonium diuranate. See uranium dioxide.
Uses: Ceramics and pigments.

uranium yellow. See sodium diuranate.

uranocircite $Ba(UO_2)_2(PO_4)_2 \cdot 8H_2O$. A natural phosphate of barium and uranium, found in the oxidized portions of some uranium deposits.
Properties: Color yellow-green; luster pearly; one good cleavage; hardness 2-2.5;

sp.gr. 3.5; radioactive.
Occurrence: South Dakota, Europe.

"Uranon." [169] Trademark for dibenzoylmethane used in the colorimetric determination of uranium.

uranophane $Ca(UO_2)_2Si_2O_7 \cdot 6H_2O$. A natural hydrated silicate of calcium and uranium. Color green; sp.gr. 3.8-3.9; radioactive.
Occurrence: Wyoming, North Carolina.

uranospinite $Ca(UO_2)_2(AsO_4)_2 \cdot 8H_2O$. A natural hydrated arsenate of uranium and calcium.
Properties: Color yellow to green; good micaceous cleavage; hardness 2-3; sp.gr. 3.5; radioactive.
Occurrence: Utah, Europe.

uranous oxide U_2O_3. The least important of the several oxides of uranium.

uranous-uranic oxide. See triuranium octoxide.

uranyl acetate (uranium acetate) $UO_2(C_2H_3O_2)_2 \cdot 2H_2O$.
Properties: Small, yellow crystals; decomposed by light; poisonous! M.p., loses $2H_2O$ at 110°C; decomposes at 275°C; sp. gr. 2.893 at 15°C. Soluble in cold water and alcohol; decomposes in hot water.
Derivation: By the action of acetic acid on uranium oxide.
Method of purification: Crystallization.
Grades: Technical; C.P.
Containers: Amber glass bottles.
Uses: Medicine; analytical chemistry; bacterial oxidations.

uranyl ammonium carbonate (ammonium uranium carbonate; uranium ammonium carbonate) $UO_2CO_3 \cdot 2(NH_4)_2CO_3 \cdot 2H_2O$.
Properties: Monoclinic yellow crystals. M.p. 100°C (dec); sp.gr. 2.773. Decomposes in air; soluble in cold water.
Grade: Technical.
Use: Uranium-yellow glazes.

uranyl ammonium fluoride (uranium ammonium fluoride; ammonium uranium fluoride) $UO_2F_2 \cdot 3NH_4F$.
Properties: Greenish-yellow, crystalline powder. Soluble in water; slightly soluble in hydrofluoric acid; insoluble in alcohol.
Grade: Technical.
Use: In x-ray work because of its fluorescence.

uranyl chloride (uranium oxychloride) $UO_2Cl_2 \cdot H_2O$.
Properties: Yellow, deliquescent crystals; decomposed on heating; soluble in water, alcohol, and ether.

uranyl nitrate (uranium nitrate; UNH; yellow salt) $UO_2(NO_3)_2 \cdot 6H_2O$.
Properties: Yellow, rhombic crystals. Sp. gr. 2.807; m.p. 60.2°C; b.p. 118°C. Soluble in water, alcohol, and ether.
Derivation: By the action of nitric acid on uranium oxide.
Method of purification: Crystallization.
Grades: Technical; C.P.
Containers: Glass bottles; boxes.
Uses: Source of uranium dioxide; photography; uranium glaze; medicine; extraction of uranium into non-aqueous solvents.
Caution! Fire hazard; dangerous in contact with organic or other readily oxidizable (combustible) substances. It will cause violent combustion on ignition.
Shipping regulations: Oxidizing material. Yellow label.*

uranyl sodium acetate (uranium sodium acetate) $UO_2(C_2H_3O_2)_2 \cdot 2NaC_2H_3O_2$.
Properties: Yellow crystals. Soluble in water and alcohol.
Derivation: By mixing sodium acetate and uranium acetate solutions and crystallizing.
Impurities: Sodium acetate.
Grades: Technical.
Containers: Wooden kegs.
Use: Uranium compounds.

uranyl sulfate (uranium sulfate) $UO_2SO_4 \cdot 3H_2O$.
Properties: Yellow crystals; soluble in water and concentrated hydrochloric acid.
Grades: Technical; purified.
Use: Chemical analysis.

uranyl uranate. See triuranium octoxide.

urao. See trona.

urea (carbamide) $CO(NH_2)_2$. Occurs in urine and other body fluids; basis of urea-formaldehyde resins. Urea was the first organic compound to be synthesized (Wöhler, 1828).
Properties: White crystals or white powder, almost odorless; cool saline taste; sp.gr. 1.335; m.p. 132.7°C; decomposes before boiling. Soluble in water, alcohol and benzene; slightly soluble in ether, almost insoluble in chloroform.
Derivation: (a) Liquid ammonia and liquid carbon dioxide at 1750-3000 psi and 160-200°C react to form ammonium carbamate, which decomposes at lower pressure (about 80 psi) to urea and water. Several variations of the process have been devised to improve yields and reduce corrosion. (b) Hydrolysis of calcium cyanamid with a solution of carbon dioxide.
Method of purification: Crystallization.
Grades: Technical; C.P.; N.F. XI; fertilizer (45-46% nitrogen); feed grade (about 42% nitrogen).
Containers: 80-, 100-lb bags; 100-, 225-lb drums. Solution in tank cars and tank trucks.
Uses (in approximate order of volume): Fertilizer; animal feed; resins; miscellaneous, including chemical intermediate, stabilizer in explosives, medicine, adhesives; separation of hydrocarbons (as urea adducts); sulfamic acid production.
Shipping regulations: None.*

urea adducts. See inclusion complexes.

urea ammonia liquor. A solution of crude urea in aqueous ammonia containing ammonium carbamate.
Containers: Tank cars.
Use: Reaction with superphosphate in preparation of fertilizers, furnishing combined nitrogen.

urea-form. A urea formaldehyde reaction product that contains more than one molecule of urea per molecule of formaldehyde. It can be used as a fertilizer because of its high nitrogen content, its insolubility in water and its gradual decomposition in the soil during the growing season to produce soluble nitrogen.

urea-formaldehyde resins. Urea and formaldehyde are united in a two-stage process in the presence of pyridine, ammonia, or certain alcohols with heat and control of pH to form intermediates (methylolurea, dimethylolurea) that are mixed with fillers to produce molding powders. These are converted to the thermosetting insoluble infusible resin by further controlled heating and pressure in the presence of catalysts. The resins are strong and rigid, free of odor and taste, and have excellent light diffusion characteristics. See also dimethylol ethylene urea and dimethylol ethyltriazone, which are cyclic examples. Melamine resins (q.v.) are similar.
Uses: Buttons, baking enamels, tableware, light reflectors, housings for apparatus and equipment such as scales; also in textile finishes for wrinkle resistance.

urea half-chloride $(NH_2CONH_2)_2 \cdot HCl$.
Properties: Practically white, odorless powder. Very soluble in water.
Use: Catalyst.

urea hydrogen peroxide. See urea peroxide.

urea nitrate (acidogen nitrate) $CO(NH_2)_2 \cdot HNO_3$.
Properties: Colorless crystals. Decomposes 152°C. Slightly soluble in water, soluble in alcohol.
Shipping regulations: Variable according to weight and moisture content, sometimes calling for high explosives label, sometimes yellow label (flammable solid).*

urea peroxide (urea hydrogen peroxide; carbamide peroxide) $CO(NH_2)_2 \cdot H_2O_2$.
Properties: White crystals or crystalline powder; m.p. (dec) 75-85°C. Decomposed by moisture and temperatures above 40°C. Soluble in water, alcohol, and ethylene glycol. Solvents such as ether and acetone extract the hydrogen peroxide and may form explosive solutions. Active oxygen (min) 16%.
Grade: Technical.
Containers: 100-lb fiber drums.
Uses: Source of water-free hydrogen peroxide; bleaching; disinfectant; cosmetics; pharmaceuticals; blue print developer; modification of starches.
Fire hazard: Dangerous.
Shipping regulations: Oxidizing material. Yellow label.*

urea phosphoric acid. See carbamide phosphoric acid.

urea-quinine. See quinine-urea hydrochloride.

urease. Enzyme present in the soy bean and in the jack bean and possibly in other legumes. Also present in blood and urine, and secreted by certain microorganisms. Its principal use is in the determination of urea in urine and in blood. It splits urea into ammonia and carbon dioxide or ammonium carbonate.

"Urecholine." [123] Trademark for bethanechol chloride, a choline ester used in medicine.

para-**ureidobenzene arsonic acid.** See carbarsone.

5-ureidohydantoin. See allantoin.

urethan. See urethane.

urethane (urethan; ethyl carbamate; ethyl urethane) $CO(NH_2)OC_2H_5$. Not used directly in urethane (polyurethane) resins and foams, but its structure is typical of the repeating unit in such polymers. See polyurethane resins.
Properties: Colorless crystals or white, granular powder; odorless; saltpeter-like taste; solutions are neutral to litmus. Soluble in water, alcohol, ether, glycerol and chloroform; slightly soluble in olive oil. Sp. gr. 0.9862; m.p. 49°C; b.p. 180°C.
Derivation: (a) By heating ethyl alcohol and urea nitrate at 120-130°C; (b) by action of ammonia on ethyl carbonate or ethyl chloroformate.
Method of purification: Crystallization.
Grades: Technical; U.S.P. XVI.
Containers: 55-gal steel drums; 250-lb fiber drums.
Uses: Medicine; intermediate or solvent for pharmaceuticals, pesticides and fungicides; biochemical research.

urethane foams. See polyurethane foams.

6,6'-ureylenebis-1-naphthol-3-sulfonic acid. Preferred name for 5,5'-dihydroxy-7,7'-disulfonic-2,2'-dinaphthylurea.

uric acid (lithic acid; uric oxide; 2,6,8-trioxypurine) OCNHC(O)NHCCNHC(O)NH (keto form). May also be written in phenolic form. The end-product of purine metabolism in man and other primates, the Dalmatian dog, birds, and some reptiles.
Properties: Odorless, tasteless, white crystals. Soluble in hot concentrated sulfuric acid; very slightly soluble in water; insoluble in alcohol and ether; soluble in glycerol, solutions of alkali hydroxides, sodium acetate, and sodium phosphate. Sp.gr. 1.855-1.893; m.p., decomposes on heating with evolution of hydrogen cyanide. Caution!
Derivation: From urine or bird excrement.
Method of purification: Crystallization.
Grades: Technical; reagent.
Use: Organic synthesis.
Shipping regulations: None.*

uric oxide. See uric acid.

uridine (D-ribosyl uracil) $C_9H_{12}N_2O_6$. The nucleoside of uracil. It is a constituent of ribonucleic acid and some coenzymes (such as uridine diphosphate glucose).
Properties: White, odorless powder, of slightly acrid and faintly sweet taste; m.p. 165°C (uncorrected); soluble in water, acid,

and base; slightly soluble in dilute alcohol; insoluble in strong alcohol.
AM = 10.1×10^3 at 262 millimicrons and pH 7.0.
Derivation: From nucleic acid hydrolyzates, from yeast.
Use: Experimental biochemical studies. Radioactive forms available.

uridine diphosphate glucose (UDPG). A coenzyme which acts in the transfer of glucose from the coenzyme to another chemical compound during the process for which coenzyme is a catalyst. Important in biochemical research.

uridine phosphates. Nucleotides used by the body in growth processes; important in biochemical and physiological research. Those isolated and commercially available (as sodium salts) are the monophosphate (UMP), the diphosphate (UDP), and the triphosphate (UTP).
See also uridine diphosphate glucose (UDPG).

uridine-phosphoric acid. See uridylic acid.

uridylic acid (uridine-phosphoric acid; UMP) $C_9H_{13}N_2O_9P$. The nucleotide of uracil.
Properties: Crystallizes in prisms from methanol. M.p. 202°C (dec). Freely soluble in water and alcohol. Dextrorotatory in solution. $A_M = 10.1 \times 10^3$ at 262 millimicrons and pH 7.0.
Derivation: From yeast ribonucleic acid. Radioactive forms available.
Use: Biochemical research.

urine salt, fusible. See sodium ammonium phosphate.

"Uritone." [330] Trademark for hexamethylenetetramine.

Urner's liquid. See dichloroacetic acid.

"Urokon Sodium." [329] Trademark for sodium acetrizoate, a water-soluble, x-ray contrast medium.

uronic acids. A class of compounds similar to sugars but differing from them in that the terminal carbon has been oxidized from an alcohol to a carboxyl group. The most commonly occurring are galacturonic acid and glucuronic acid.

"Urox." [50] Trademark for a complex of monuron and TCA, available as 11% and 22% granular formulations, and as liquid oil concentrate. Used on non-crop lands for control of most annual and perennial grasses and broadleaved weeds.

ursin. See arbutin.

usnic acid (usninic acid) $C_{18}H_{16}O_7$. A constituent of many lichens. Known in d-, ℓ-, and dℓ- forms.
Properties: Crystalline yellow solid; melting range 193-203°C. Insoluble in water; slightly soluble in alcohol and ether.
Derivation: From Usnea barbata, a lichen growing on trees.
Use: Medicine.

usninic acid. See usnic acid.

U.S.P. Abbreviation for United States Pharmacopoeia, the official United States book of standard drugs. The latest edition at the time that this dictionary was written was the 16th, noted as U.S.P. XVI.

"U.S.P.-12" pharmaceutical zinc oxide. [268] Trademark for a pharmaceutical zinc oxide of extreme fineness and whiteness. Meets the specifications of the United States Pharmacopoeia, including the tests prescribed for heavy metals, lead and arsenic.
Containers: 50-lb cartons and 100-lb drums.
Uses: Especially suitable for cosmetics, ointments and other medical purposes. For chemical purposes where high purity is required.

ustilagic acid. An antibiotic from corn smut.

UTP. See uridine phosphates.

uvanite $U_2V_6O_{21} \cdot 15H_2O$. A natural hydrated uranium vanadate. Color brownish-yellow. Found in Utah.

uvarovite. See garnet.

"Uverite." [134] Trademark for opaquing agent for vitreous enamels.

"Uversoft." [134] Trade name for softening agents for fabrics and paper products. Based on a quaternary ammonium salt containing two long straight chain hydrocarbon chains attached to the nitrogen atom. Are cationic and substantive to cellulose.

"Uversol." [134] Trademark for metal salts of naphthenic acids. Available as solids and liquids of most of the common metals. Used as paint driers, wetting agents, catalysts, etc.

"Uvinul." [307] Trademark for a series of ultraviolet absorbers, designated as "Uvinul" 400, 490, D-49, M-40, and D-50. These are nearly pure substituted benzophenones, containing only traces of inorganic salts.
Properties: Cream or tan powders; somewhat soluble in alcohols, ethyl acetate, methanol, methyl ethyl ketone; insoluble in toluene and water.
Uses: Organic ultraviolet light absorbers effective in the range from 200 millimicrons (2000 Angstrom units). They do not darken or decompose upon prolonged exposure to an intense ultraviolet source. The absorbed energy is not reemitted in the visible spectrum.
Typical applications include protection of plastics, oils, cosmetics, paper, wood, and leather; for most protective uses, 0.05 to 2.0% of "Uvinul" is recommended.

"UVO-Cryst." [309] Trademark for vitamin D concentrates in milk constituents for use in milk and foods where milk is used as an ingredient.

V. Symbol for vanadium.

"V-90." [172] Trademark for a proprietary product, an anhydrous, heat-treated monocalcium phosphate.

Properties: Finely granulated, white, free-flowing particles having thin coatings of relatively insoluble phosphate which delay solution in aqueous liquids.

Containers: 100-lb paper bags.

Uses: Acid-reacting leavening ingredient in self-rising flour, pancake flour, self-rising cornmeal, baking powder, and prepared flours. In the production of commercial angel food cakes.

"Vabar." [195] Trade name for a cementitious product used as a vapor barrier and plaster bond to seal the surface and to provide a bond for gypsum plaster.

vacuum deposition of metals. Aluminum and other metals can be caused to deposit in the form of a smooth reflective film on other metals by heating a sample of the aluminum or other metal in an evacuated space which also contains the metal on which the deposit is desired. This surface must be kept relatively cool. Deposition of metals is also made by this means on plastics, paper, glass, and fabrics or yarns. The process is essentially one of molecular distillation.

vacuum distillation. Distillation at a pressure less than atmospheric but not so low that it would be classed as molecular distillation. Since lowering the pressure also lowers the boiling point, vacuum distillation is useful for distilling high boiling and heat-sensitive materials such as heavy distillates in petroleum, fatty acids, etc.

valence. In the simplest sense, valence is an integer representing the number of hydrogen or chlorine atoms which one atom of an element can hold in combination. Thus, silver, sodium and bromine have a valence of one in silver chloride, AgCl, sodium chloride, NaCl, and hydrogen bromide, HBr, while aluminum and nitrogen have a valence of three in aluminum chloride, $AlCl_3$, and ammonia, NH_3. Groups of atoms, i.e., radicals, also have valences. Thus SO_4 has a valence of two in H_2SO_4, and PO_4 of three in H_3PO_4. Each element usually enters into compounds on the basis of one or two characteristic valences, but there are many exceptions and complexities, and the term valence has been and is used in many special ways, some of which are subject to controversy.

Valence varies in a rather regular way according to position in the periodic system, the underlying reasons being similarities and variations in the number and energies of the electrons in the outermost parts of atoms of the different elements. Helium and other inert gases which form no compounds occupy a position in the Periodic System corresponding to a valence of zero.

valentinite (antimony trioxide, ortho-rhombic; white antimony) Sb_2O_3. White or gray mineral, sometimes pale red. White streak and adamantine or silky luster. An alteration product of stibnite and other antimony minerals. Contains 83.3% antimony.

Constants: Sp.gr. 5.57-5.76; hardness 2-3.

Occurrence: Algeria; Yugoslavia; Italy; Germany.

Use: Ore of antimony.

valeral. See n-valeraldehyde.

n-valeraldehyde (valeric aldehyde; valeral; amyl aldehyde; pentanal) $CH_3(CH_2)_3CHO$.

Properties: Colorless liquid; sp.gr. 0.8095; (20/4°C); f.p. −91°C; b.p. 102-103°C; refractive index (n 20/D) 1.3944; flash point (open cup) 54°F. Slightly soluble in water; soluble in alcohol and ether. See isovaleraldehyde.

Derivation: Oxidation of amyl alcohol, also by the Oxo process.

Shipping regulations: Flammable liquid. Red label.*

valerian. Dried rhizome and roots of Valeriana officinalis.

Occurrence: Europe, northern Asia and eastern United States.

Grades: Belgian; Indian.

Containers: Bags.

Uses: Medicine; valerian oil.

Shipping regulations: None.*

valerian, American. See cypripedium.

valerianic acid. See n-valeric acid.

valerian oil.

Properties: Yellowish or brownish liquid; characteristic, penetrating odor. Soluble in alcohol, ether, chloroform, acetone, benzene, and carbon disulfide.

Chief known constituents: Pinene, camphene, borneol and esters of borneol and valeric acid.

Constants: Sp.gr. 0.930-0.960; refractive index (n 20/D) about 1.486.

Derivation: Distilled from roots and rhizome of Valeriana officinalis.

Method of purification: Rectification.

Grade: Technical.
Containers: Iron drums; glass bottles.
Uses: Medicine; tobacco perfume; industrial odorant; flavors.

valerian oil, Japanese (kesso oil).
Properties: Green, thick liquid, essential oil. Cannot be distinguished from the ordinary valerian oil. Soluble in alcohol, ether, chloroform, benzene, acetone, and carbon disulfide.
Constants: Sp.gr. 0.960-1.004; refractive index (n 20/D) 1.47-1.48.
Derivation: Distilled from the rhizome and roots of Valeriana officinalis, var. angustifolia.
Method of purification: Rectification.
Grades: Technical.
Containers: Iron drums; glass bottles.
Uses: Medicine; industrial odorants.

valeric acid (valerianic acid; n-pentanoic acid) $CH_3(CH_2)_3COOH$.
Properties: Colorless liquid; penetrating odor and taste; sp.gr. 0.9394 (20/4°C); b.p. 185.4°C; refractive index 1.4081 (20°C); vapor pressure 0.08 mm (20°C); f.p. −34°C; slightly soluble in water; soluble in alcohol and ether. Undergoes reactions typical of normal monobasic organic acids.
Derivation: With other C_5 acids by distillation from valerian; by oxidation of n-amyl alcohol.
Grades: Technical; reagent.
Containers: Drums; tank cars.
Uses: Intermediate for flavors and perfumes; ester-type lubricants; plasticizers; pharmaceuticals; vinyl stabilizers.

valeric aldehyde. See n-valeraldehyde.

gamma-**valerolactone** ($C_5H_8O_2$). A solvent miscible with water and most organic solvents, resins, waxes, etc. Slightly miscible with zein, beeswax, petrolatum. Not miscible with anhydrous glycerin, glue, casein, arabic gum, and soybean protein.
Properties: Sp.gr. (25/25°C) 1.0518; b.p. 205-206.5°C; crystallizing point −37°C; flash point (Cleveland open cup) 205°F; fire point (Cleveland open cup) 220°F; refractive index (25°C) 1.4301. Surface tension (25°C) 39 dynes/cm; viscosity (25°C) 2.18 cps; pH, anhydrous 7.0; pH 10% solution in distilled water 4.2.
Uses: In dye baths (coupling agent), brake fluids, cutting oils, and as solvent for adhesives, insecticides and lacquers.

valeryl diethylamide (diethylvaleramide; isovaleryl diethylamide) $C_4H_9CON(C_2H_5)_2$.
Properties: Colorless liquid; burning taste; characteristic odor. Produces convulsions of cerebral origin. B.p. 210°C; slightly soluble in water; soluble in alcohol and ether. Used as medicine.

valethamate bromide
$C_6H_5CH(C_4H_9)COOC_2H_4N(C_2H_5)_2 \cdot CH_3Br$. 2-Diethylaminoethyl 3-methyl-2-phenylvalerate methylbromide.
Properties: Crystals. Freely soluble in water. Aqueous solutions are stable to storage.
Grade: N.N.D.
Use: Medicine.

valine (alpha-aminoisolvaleric acid) $(CH_3)_2CHCH(NH_2)COOH$. An essential amino acid.
Properties: White crystalline solid; soluble in water; very slightly soluble in alcohol; insoluble in ether. Shows the following optical activity:
DL-valine: M.p. 298°C with decomposition.
D-valine (natural isomer): M.p. 315°C with decomposition.
L-valine: M.p. 293°C with decomposition.
Derivation: Hydrolysis of proteins; synthesized by the reaction of ammonia with alpha-chloroisovaleric acid. Available commercially as D-, L-, or DL-valine.
Containers: Drums.
Uses: Medicine; food; culture media; biochemical and nutritional investigations.

"Vallestril." [70] Trademark for a brand of methallenestril, 3-(6-methoxy-2-naphthyl)-2,2-dimethylpentanoic acid, used in medicine.

"Valmid." [100] Trademark for ethinamate (q.v.).

valonia.
Derivation: The acorn cups of an oak Quercus aegilops, native of Greece, Asia Minor and France. The cups are very large and are covered with coarse hair or "beard" which is very rich in tannin. Good valonia contains 30-40% tannin.
Containers: Cups: 100- to 200-lb (average about 150-lb) burlap bags. Extract: Wooden barrels.
Use: Tanning industry.

vanadic acid (a) meta-HVO_3; (b) ortho-H_3VO_4; (c) pyro-$H_4V_2O_7$. These acids apparently do not exist in the pure state, but are represented in the various alkali and other metal vanadates. Ordinarily, when vanadic acid is mentioned, vanadium pentoxide (vanadic acid anhydride) is meant.

vanadic acid anhydride. See vanadium pentoxide.

vanadic sulfate. See vanadyl sulfate.

vanadic sulfide. See vanadium sulfide.

vanadinite $Pb_5Cl(VO_4)_3$. A natural chlorovanadate of lead. Grades into mimetite, and endlichite (q.v.).
Properties: Color ruby red, orange red, brown, yellow; luster resinous to adamantine; hardness 3; sp.gr. 6.7-7.1. Soluble in strong nitric acid.
Occurrence: New Mexico, Arizona; Africa; Scotland; U.S.S.R.
Use: Ore of vanadium and lead.

vanadium V. Element having atomic number 23, of group V of the periodic system.
Properties: Silvery-white ductile metal. Insoluble in water; soluble in nitric, hydrofluoric, and concentrated sulfuric acids;

attacked by alkali, forming water-soluble vanadates. Sp.gr. 6.11; m.p. 1900 ± 25°C. Vanadium exists in five states of valence, behaves chemically as either a metal or a nonmetal, and forms a variety of complex compounds.
Source: Not found native. More than sixty-five vanadium minerals have been identified. The principal ores are patronite, roscoelite, carnotite and vanadinite.
Occurrence: Found in Colorado, Utah, New Mexico, Arizona; Mexico and Peru.
Derivation: Calcium iodide reduction of vanadium pentoxide yields 99.8+ % pure ductile vanadium. Aluminum, cerium, etc. reduction produces a less pure product. Large, single crystals of vanadium can now be produced by an arc-fusion process.
Uses: Target material for x-rays; manufacture of steel (see ferrovanadium); vanadium compounds, especially catalysts for sulfuric acid manufacture; vanadium alloys.

vanadium acetylacetonate.
Properties: Blue to blue-green crystals. Decomposes before melting.
Derivation: Reaction of vanadyl sulfate with acetyl acetone and sodium carbonate.

vanadium carbide.
Properties: Crystals with hardness 2800 kg/sq mm; sp.gr. 5.77; m.p. 2800°C; resistivity 150 micro-ohm cm (room temp.).
Use: Alloys for cutting tools.

vanadium catalyst.
Properties: Alkali salts of vanadium within a porous siliceous carrier. Density 42-48 lb/cu ft; activation temperature 750-850°F.
Uses: Conversion of sulfur dioxide to the trioxide.

vanadium dichloride (vanadous chloride) VCl_2.
Properties: Apple green hexagonal plates. Soluble in alcohol and ether; decomposes in hot water. Sp.gr. 3.23 (18°C).
Derivation: From vanadium trichloride by heating in atmosphere of nitrogen.
Method of purification: Sublimation in nitrogen.
Grade: C.P.
Containers: Sealed glass containers.
Uses: As strong reducing agent; purification of hydrogen chloride from arsenic.

vanadium ethylate $(C_2H_5O)_4V$.
Properties: Dark reddish-brown solid.
Derivation: Reaction of vanadium chloride with sodium ethylate.
Uses: Polymerization catalyst.

vanadium ore. See vanadinite.

vanadium oxides. See vanadium pentoxide, vanadium tetraoxide and vanadium trioxide.

vanadium oxydichloride. See vanadyl chloride.

vanadium oxytrichloride $VOCl_3$.
Properties: Sp.gr. 1.811 (32°C); m.p. −78.9°C; b.p. 125-127°C; nonionizing solvent; dissolves most nonmetals; dissolves and/or reacts with many organic compounds.
Uses: Catalyst in olefin polymerization; organovanadium synthesis.

vanadium pentasulfide. See vanadium sulfide.

vanadium pentoxide (vanadic acid anhydride) V_2O_5.
Properties: Yellow to red crystalline powder; sp.gr. 3.357 (18°C); m.p. 690°C; b.p., decomposes at 1750°C. Soluble in acids and alkalies; slightly soluble in water.
Derivation: (a) Alkali or acid extraction from vanadium minerals. (b) By igniting ammonium meta-vanadate. (c) From concentrated ferrophosphorus slag by roasting with sodium chloride, leaching with water, and purification by solvent extraction followed by precipitation and heating.
Method of purification: Alkali solution, precipitation as ammonium meta-vanadate and ignition to V_2O_5.
Grades: Commercial air dried; commercial fused; C.P. air dried; C.P. fused.
Containers: Compressed paper drums; multiwall paper sacks.
Uses: Starting material for other vanadium salts; catalyst for oxidation of sulfur dioxide; ferrovanadium (q.v.); gasoline catalyst; catalyst for organic reactions, including automobile exhaust hydrocarbon elimination; ceramic coloring material; inoculation of plant life; inhibiting ultraviolet transmission in glass; black inks; photographic developer; dyeing textiles; medicine; nuclear energy uses.
Shipping regulations: None.*

vanadium sesquioxide. See vanadium trioxide.

vanadium steel.
Vanadium in steel (1) elevates coarsening temperature of austenite (promotes fine grain), (2) increases hardenability (when dissolved), (3) resists tempering and causes marked secondary hardening.

It increases tensile strength and elastic limit without decreasing ductility. Prevents hardening and embrittlement in welding and castings. Refinement of grain size permits excellent ductility and impact resistance during the development of high tensile strength and yield strength. Vanadium retards rate of softening in tempering condition, hence vanadium steels retain hardness at elevated temperatures. Also improves creep strength at elevated temperatures. Used in tool steels to improve serviceability, refine carbide structure, improve "red hardness" and abrasion resistance. Vanadium raises the endurance limit and ratio of endurance limit to tensile strength in constructional steels.
See also ferrovanadium.

vanadium sulfate. See vanadyl sulfate.

vanadium sulfide (vanadium pentasulfide; vanadic sulfide) V_2S_5.
Properties: Black-green powder. Soluble in acids, alkali-metal sulfides and alkalies; insoluble in water. Sp.gr. 3.0.
Derivation: By the action of hydrogen sulfide on vanadium chloride solution.

Grade: Technical.
Containers: Wooden kegs or fiber drums.
Use: Vanadium compounds.
Shipping regulations: None.*

vanadium tetrachloride VCl_4.
Properties: Red liquid. Soluble in absolute alcohol and ether; decomposes slowly to vanadium trichloride and chlorine at temperatures below 63°C. Sp.gr. 1.8584 (0°C); b.p. 154°C.
Derivation: Chlorination of ferrovanadium.
Method of purification: Distillation and fractionation.
Grades: Technical; C.P.
Containers: Sealed glass bottles.
Uses: Medicinal; starting material for preparation of vanadium trichloride and vanadium dichloride.

vanadium tetraoxide V_2O_4.
Properties: Blue-black powder; sp.gr. 4.339; m.p. 1967°C; insoluble in water; soluble in alkalies and acids.
Derivation: (1) From vanadium pentoxide by oxalic acid reduction. (2) From vanadium pentoxide by carbon reduction.
Grades: Technical; C.P.
Containers: Glass bottles.
Use: Catalyst at high temperature.

vanadium trichloride VCl_3.
Properties: Pink deliquescent crystals; soluble in absolute alcohol and ether. Decomposes in water. Sp.gr. 3.0 (18°C); decomposes on heating.
Derivation: From vanadium tetrachloride boiling under reflux condenser.
Grade: C.P.
Containers: Glass bottles.
Use: Starting material for preparation of vanadium dichloride.

vanadium trioxide (vanadium sesquioxide) V_2O_3.
Properties: Black crystals; soluble in alkalies and HF; slightly soluble in water. Sp.gr. 4.87 (18°C); m.p. 1970°C.
Derivation: From vanadium pentoxide by either hydrogen or carbon reduction.
Grades: Technical; C.P.
Containers: Glass bottles.
Uses: Catalyst of ethylene to ethyl alcohol.

vanadous chloride. See vanadium dichloride.

vanadyl chloride (vanadium oxydichloride; vanadyl dichloride; divanadyl tetrachloride) $V_2O_2Cl_4 \cdot 5H_2O$.
Properties: Green, very deliquescent crystals. Slowly decomposed by water. Usual technical product is a dark green syrupy mass 76-82% pure, or a solution. Poisonous. Soluble in water, alcohol, and acetic acid.
Grade: Technical.
Use: Mordanting textiles.
Shipping regulations: None.*

vanadyl dichloride. See vanadyl chloride.

vanadyl sulfate (vanadic sulfate; vanadium sulfate) $VOSO_4 \cdot 2H_2O$.
Properties: Blue crystals; soluble in water.
Derivation: Reduction of cold solution of concentrated sulfuric acid and vanadium pentoxide by sulfur dioxide gas.
Method of purification: (a) Recrystallization; (b) wash with alcohol.
Grades: Technical; C.P.
Containers: Glass bottles; kegs.
Uses: Mordant; catalyst; aniline black preparation; reducing agent; blue and green colors in glasses and ceramics.
Shipping regulations: None.*

"Vanaldol." [19] Brand name for a proprietary product. Ethyl vanillin. Claimed to be 3½ times stronger than vanillin.

"Van Caloria." [51] Trademark for lubricants consisting of "Caloria" to which colloidal graphite has been added. When high operating temperatures cause the lubricant base to vaporize, the graphite film remains to reduce solid friction.

"Vancide 89." [69] Trademark for a proprietary product, N-(trichloromethylmercapto)-4-cyclohexene 1,2-dicarboximide.
Use: Fungicide for vinyl compositions.

"Vancide 26EC." [69] Trademark for the lauryl pyridinium 5-chloro-2-mercaptobenzothiazole.
Use: Preservative for cotton fabrics used in rubber structures. Applied directly to cotton from an aqueous solution.

"Vancide 51Z." [69] Trademark for a proprietary product, zinc dimethyldithiocarbamate $[(CH_3)_2NC(S)S]_2Zn$, with a small proportion of zinc 2-mercaptobenzothiazole.
Use: Fungicide for neoprene rubber compositions.

"Vancocin." [100] Trademark for vancomycin hydrochloride (q.v.).

vancomycin hydrochloride. An antibiotic substance.
Properties: White solid. Soluble in water; moderately soluble in dilute methanol; insoluble in higher alcohols, acetone, ether.
Derivation: Produced by Streptomyces orientalis from Indonesian and Indian soil. Isolated as the hydrochloride.
Grade: N.N.D.
Use: Medicine.

Van der Waal's equation. A modified equation of state for gases to compensate for actual volume of molecules and for attractive forces existing between the molecules. The latter are known as Van der Waal's forces.

"Vandex." [69] Trademark for a proprietary product. Finely ground selenium. A secondary vulcanizing agent for rubber products.

"Vandrynilla." [342] Trademark for concentrated extract of vanilla adsorbed on non-caloric adsorbents for dietetic flavoring.

Van Dyke brown (Cassel brown; Cassel earth; Cologne brown; Cologne earth; ulmin brown). A naturally occurring pigment.
Derivation: Indefinite mixtures of iron oxide and organic matter. Obtained from bog-earth, peat deposits or from ochers containing bituminous matter.

Grades: Based on iron oxide content and tinctorial value.
Containers: Wooden barrels or fiber drums.
Use: Pigment for artists' color and stains.
Shipping regulations: None.*

Van Dyke red. A brownish-red pigment consisting of copper ferrocyanide; sometimes used to refer to red varieties of ferric oxide. See iron oxide reds.
Use: Pigments.

"Van Estan." [51] Trademark for lime-base lubricants that consist of "Estan" and graphite. This makes them suitable for use on plungers in water pumps, stuffing boxes, elevator plungers and slides, etc.

"Vanfre." [69] Trademark for a series of proprietary release agents.
"Vanfre." Milky liquid; total solids 18-20%; sp. gr. 0.87 ± .02; flash point, 52°F min; vehicle, isopropanol. Release agent for lead press cured items.
"Vanfre Clear." Clear amber liquid; total solids 15-17%; sp. gr. 0.88 ± .02; vehicle, water.
"Vanfre No. 3." Clear amber liquid, total solids 14-16%; sp. gr. 1.00 ± .02; vehicle, water. General purpose; for molded goods and frame cured sponge.
"Vanfre No. 4." Milky liquid; total solids 9-11%; sp. gr. 1.00 ± .02; flash point, 118°F min.; vehicle, water plus isopropanol. For molded goods and frame cured sponge.

vanilla bean. Cured, full grown, but immature fruit of Vanilla planifolia.
Chief known constituents: Coniferin and two ferments which later change coniferin to vanillin during the process of curing. Resins and considerable sugar are also present.
Grades: Whole; cuts; Bourbon; Mexican; South American; N.F. XI.
Containers: Bags; boxes; tins.
Occurrence: Mexico; West Indies; Madagascar.
Uses: Confectionery; flavoring (source of vanillin); perfumery; pharmaceuticals.
Shipping regulations: None.*

vanillal. See ethyl vanillin.

vanilla plant. See liatris.

vanillic aldehyde. See vanillin.

vanillin (3-methoxy-4-hydroxybenzaldehyde; vanillic aldehyde) $(CH_3O)(OH)C_6H_3CHO$. The methyl ether of protocatechuic aldehyde. It occurs in vanilla bean extract and in many balsams and resins.
Properties: White crystalline needles; pleasant aroma; vanilla taste. M.p. 80-82°C; b.p. 285°C. Soluble in 125 parts water; in 20 parts glycerol and in 2 parts 95% alcohol.
Derivation: By dichromate oxidation of isoeugenol; treatment of guaiacol with formaldehyde, hydrochloric acid, and para-nitrosodimethylaniline; or by extraction of the vanilla bean. Also from sulfite waste liquor upon heating with alkali at 160°C, and from lignin, or wood itself (especially spruce).
Method of purification: Crystallization.
Grades: Technical; U.S.P. XVI.
Containers: 1-, 5-lb cartons; 5-lb bottles; 5-, 25-lb cans; 25-, 100-lb drums.
Uses: Perfumes; flavoring; pharmaceuticals; a reagent.

vanillin, ethyl. See ethyl vanillin.

vanirom. See ethyl vanillin.

"Vanitrope." [85] Trademark for propenyl guaethol (1-ethoxy-2-hydroxy-4-propenylbenzene) $C_{11}H_{14}O_2$.
Properties: Free-flowing white powder; odor and taste strongly similar to vanilla; 16-25 times stronger flavor than vanillin, depending on use; m.p. 85-86°C; very soluble in fats, edible solvents, and essential oils; very slightly soluble in water; appears to stimulate the sense of taste, thus intensifies many other flavors.
Containers: 1-, 5-, and 25-lb packages.
Use: Artificial vanilla flavoring (F.D.A. approved).

"Van Nakta." [51] Trademark for lubricant for uses where the grease must contain graphite.

"Vanstay." [69] Trademark for a series of heat and light stabilizers for vinyl resins.

"Vantoc." [206] Brand name of proprietary bactericides for the food industries.

"Vanwax." [69] Trademark for proprietary wax emulsions.
"Clear" Total solids, 12-13%; sp. gr. 1.00 ± .02; pH 9.5 - 10.5.
"Black" Total solids 13-15%; sp. gr. 1.00 ± .02; pH 9.5 - 10.5.

"Vapam." [1] Soil fumigant composed of sodium N-methyldithiocarbamate dihydrate $CH_3NHC(S)SNa \cdot 2H_2O$.
Properties: White crystalline solid; readily soluble in water; moderately soluble in alcohol; sparingly soluble in other common organic solvents; stable in concentrated aqueous solution but decomposed in dilute aqueous solution; acids and heavy metal salts promote decomposition; compound is unstable in all types of moist soil.
Uses: Fungicide, nematocide, weed killer, and insecticide.
Hazards: Irritating to the eyes, skin and mucous membrane. May be toxic to living plants.

"Vapona." [125] Trademark for an insecticide which contains not less than 93%w 2,2-dichlorovinyl dimethyl phosphate (see DDVP) and not more than 7%w active, related compounds.
Properties: Colorless to amber liquid with a mild, fruity odor; b.p. approx. 183°F (1 mm). Miscible with aromatic and chlorinated hydrocarbons, solvents and alcohols. Moderately soluble in diesel oil, kerosene, isoparaffinic hydrocarbons and

mineral oil; slightly soluble in water (about 1%) but undergoes hydrolysis in the presence of water and is readily decomposed by strong acids and bases.

Containers: 55-gal steel drums with molded polyethylene insert containing 590 lbs; 5-gal steel pails with molded polyethylene insert containing 55 lbs.

Warning! (technical insecticide and formulations containing over 15%): Poisonous if swallowed, inhaled, or absorbed through skin.

vapor-liquid chromatography. See gas chromatography.

vapor pressure. The pressure (usually expressed in millimeters of mercury) characteristic at any given temperature of a vapor in equilibrium with its liquid or solid form.

vapor tension. Same as vapor pressure.

"Vapotone." [253] Brand name for a line of insecticides containing tetraethylpyrophosphate.

"Varjamine." [307] An azoic composition for printing blues on cotton and rayon.

"Varidase." [57] Trade name for streptokinase-streptodornase derived from the fermentation of a group C streptococcus. After purification for removal of toxic components, it is prepared for clinical use as a sterile frozen-dried powder containing buffer and a preservative. The mixture contains streptokinase, the activator of human plasminogen, which digests fibrin; and streptodornase, a group of enzymes attacking deoxyribonucleo protein, the main constituent of pus.

Properties: Water and saline soluble; may be destroyed in presence of organic solvents at room temperature; compatible with major antibiotics at pH 7.4. Denatured by heat at 56°C; powder stable at refrigeration temperatures approximately 18 months; solutions stable for 1 week at refrigeration temperautes.

Use: Medicine.

"Varkon." [300] Trademark for sequestering agents of the ethylenediaminetetraacetate type. Used in kier boiling, peroxide bleaching, dyeing and stripping operations. Available in both powdered and liquid forms.

varnish. An unpigmented, oil-base paint (see paint) composed of a solvent and either of two types of binders: (1) those which form a film by oxidation or polymerization, such as drying oils (q.v.), alone or in combination with natural or synthetic resins, chlorinated rubber, etc. (see resins, natural; resins, synthetic; rubber, chlorinated), and (2) those which form films by evaporation of the solvent, such as the shellac, cellulose ester or ether, alkyd and phenolic resin varnishes. The first type of binders characterizes oil varnishes, which contain such solvents as turpentine and petroleum naphtha. A "long" oil varnish contains a larger proportion of oil to resin than a "short" varnish and forms a more elastic film. Spar varnish is "long;" rubbing and furniture varnishes are "short." The second type of binders listed above characterize spirit varnishes, which employ such solvents as methanol, methyl isobutyl ketone, butyl acetate, toluene, etc.

Bituminous varnish contains asphalt or bituminous materials and may be either of the oil or spirit type.

Shipping regulations: May be classified as flammable liquid. Red label.*

varnish-makers' naphtha. See naphtha, painters'.

varnish oil. An oil obtained by the distillation of a gum resin and used in the manufacture of varnishes.

varnish remover. See paint remover.

"Varnon." [446] Trade name for hard fired super duty fireclay brick which resists the destructive action of carbon monoxide and other reducing gases. Dense, strong, good volume stability at high temperatures and has high rigidity under soaking heat conditions. Used as glass tank regenerator checkers; also to line various metallurgical furnaces, rotary kilns, shaft kilns, carbon baking furnaces, and incinerators.

"Varox." [69] Trademark for proprietary product, a 50% active blend of 2,5-bis(tert-butylperoxy)-2,5-dimethylhexane with an inert mineral carrier.

Properties: White powder; sp.gr. (calc) 1.50.

Uses: Cross-linking agent for polymers such as polyethylene.

Shipping regulations: Oxidizing material. Yellow label.*

"Varox Liquid." [69] Trademark for proprietary product, 2,5-bis(tert-butylperoxy)-2,5-dimethylhexane.

Properties: Water-white to light yellow liquid; sp.gr. 0.87 ± .02; min. assay 92%; active O_2 10.1% min.

Uses: Cross-linking agent for polymers such as polyethylene.

Shipping regulations: Oxidizing material. Yellow label.*

"Varsol." [51] Trademark for straight petroleum aliphatic solvents used as paint and varnish thinners, for dry cleaning and for general plant machinery cleaning. Conform to CS3-40, the U.S. Dep't. of Commerce commercial standard for Stoddard Solvent and have minimum Tag closed cup flash points of 100°F.

"Vaseline" Petroleum Jelly. [97] Trademark for a familiar brand of petrolatum. A commercial product of petroleum, largely employed in pharmacy, alone and as a vehicle for external applications of medicinal agents, especially when local action rather than absorption is desired; as a protective coating for metallic surfaces, and for other purposes.

"Vaseline" petroleum jelly consists of a semisolid mixture of hydrocarbons, having a m.p. usually ranging between 38° and 45°C. It is colorless, or of a pale yellow color, translucent, fluorescent, and amorphous. It does not readily oxidize on exposure to the air and is not readily acted on by chemical reagents. It is soluble in chloroform, benzene, carbon disulfide, and oil of turpentine. It also dissolves in warm ether and is slightly soluble in hot alcohol, but separates from the latter in flakes on cooling.

vasopressin (betahypophamine; antidiuretic hormone). One of the hormones secreted by the posterior lobe of the pituitary gland. It causes an increase in blood pressure and an increase in water retention by the kidney. Vasopressin is an octapeptide consisting of eight different amino acids. It is available as the tannate in solution for injection (vasopressin injection, U.S.P. XVI).

"Vasoxyl." [301] Trademark for methoxamine hydrochloride, used in medicine.

vat dyes. Vat dyes are those that can be easily reduced to a soluble and usually colorless leuco or vat form in which they can readily impregnate fibers. Subsequent oxidation then produces the insoluble colored dyestuff in a form that is remarkably fast to washing, light and chemicals. Examples are indigo (C.I. 1177), Indanthrene Blue GCD (C.I. 1113), and Anthraquinone Vat Yellow GC (C.I. 1095). The reducing agents are usually an alkaline solution of sodium hydrosulfite ($Na_2S_2O_4$) or some derivative of the latter. Oxidation is by air, perborate, dichromate or similar materials.

vat printing assistants. Compounds of gums, reducing and wetting agents used to carry the dye in printing fabrics with vat dyes. They assist in securing penetration of the fabric and help to convert the dyes from a semi-leuco state to a leuco state.

"Vatro Gum." [159] Trademark for fully prepared gums. Contain everything necessary for printing except glycerin, dyestuff and necessary amount of water to bring them to printing consistency.

Use: For printing vat colors in both screen and roller printing.

"Vatrolite." [159] Trademark for a proprietary product. Sodium hydrosulfite.

Uses: As a reducing agent. For bleaching soaps, removing color from textile fabrics. As a reducing agent in vat dyeing of textiles or fast color dyeing.

"Vatrolite" 58 Series. [159] A series of buffered sodium hydrosulfites used as bleaching agents. Available as "Vatrolite" 58-E, 58-ELD, 58-EAC.

"Vatsol." [57] Brand name for a proprietary product. Wetting agents consisting of several different grades for use with insecticide dusts and sprays.

Grades: Made in several different types and grades: OS, sodium isopropyl naphthalene sulfonic acid; OT, sodium dioctyl sulfosuccinate.

Containers: Fiber and stainless steel drums.

Uses: When used in insecticide dusts, brings the insecticide into more intimate contact with bodies of insects, thus increasing insect kill; when used in insecticide sprays, reduces the surface tension thus giving better contact and coverage.

Fire hazard: None.

Shipping regulations: None.*

"V-Bor." [88] Trademark for refined borax pentahydrate $Na_2B_4O_7 \cdot 5H_2O$; 99.8% purity. Obtained from Searles Lake brines.

"VBR Ester Gums." [296] Brand name for a series of esters of rosin acids prepared by reaction with polyhydric alcohols under conditions to yield substantially neutral esters of pale color and high softening point. Used extensively in paints, varnishes, lacquers, and printing inks.

"VBR-700 Modified Maleic Resins." [296] Brand name for a series of partial esters of maleic anhydride and rosin acids with polyhydric alcohols. Thermoplastic and oil soluble resins used in lacquers, printing inks, and paints requiring good color retention.

"VBR-900 Modified Phenolic Resins." [296] Brand name for a series of condensation products of phenol and/or substituted phenols with formaldehyde modified with rosin acids and subsequently esterfied with polyhydric alcohols to produce a series of resins of various softening points and solubility characteristics.

Use: Principally in varnishes and printing inks.

"VBR-800 Pure Phenolic Resins." [296] Brand name for a series of unmodified condensation products of substituted phenols and formaldehyde. These resins are produced from both acid and alkaline catalyzed systems. All resins are of the oil soluble type and are used in chemical resistant, fast drying, protective coatings.

VC. Abbreviation for vinyl chloride or vinylidene chloride.

"V-Cillin." [100] Trademark for phenoxymethyl penicillin, U.S.P.

"V-Cillin K." [100] Trademark for potassium phenoxymethyl penicillin, U.S.P.

"V-C 13 Nemacide." [40] Trademark for a 75% emulsifiable concentrate used for control of nematodes and as an insecticide for lawn chinch bugs and onion maggots. The active ingredient is O-2,4-dichlorophenyl-O,O-diethylphosphorothioate.

Properties (typical): Dark, straw-colored liquid with a mildly unpleasant odor; insoluble in water; readily soluble in most organic solvents. B.p. (0.1 mm) 106-127°C;

*See "I.C.C. Shipping Regulations," page xiii.

Reference numbers refer to name of manufacturer. See "List of Manufacturers," page v.

sp.gr. (20°C) 1.30; refractive index (25°) 1.532.

vegetable black (Frankfort black). In general any form of more or less pure carbon produced by incomplete combustion or destructive distillation of vegetable matter, wood, vines, wine lees. See active carbon; charcoal.

vegetable calomel. See podophyllum.

vegetable char. See vegetable black.

vegetable dyes. Vegetable substances which yield coloring matter. In general, the majority of the natural dyes belong to the mordant class, though a few will combine with wool directly after the manner of the acid dyes, and a limited number act in the same manner as substantive dyes. Typical dyes of this class are logwood, natural indigo, fustic, hypernic, cochineal, madder, cutch, camwood, brazil-wood, archil, quercitron, safflower, Persian berries, turmeric, etc. Logwood is still used to a considerable extent, and small amounts of cochineal, fustic and cutch are still employed.

vegetable gelatin. Erroneous name for agar-agar.

vegetable glue. Made by treating starch with caustic soda. See also glue.

vegetable gum. See dextrin.

vegetable ivory. A mannose polysaccharide obtained from the hardened albumin of the seeds of the ivory-nut (corozo-nut) tree (Phytelephas macrocarpa). The latter must must be distinguished from the cohune-palm (Attalea cohume) whose seeds are also sometimes referred to as corozo nuts.
Uses: Button manufacture; mannose production.

vegetable oils. Edible oils extracted from the seeds, fruit or leaves of plants and generally considered to be mixtures of mixed glycerides (e.g., cottonseed, flaxseed, peanut, perilla, oiticica, etc.).

vegetable parchment. A product made by immersing unsized paper in 75-84% sulfuric acid for a short period of time followed by immediate washing and drying. The process results in a layer of semi-transparent gelatinous amyloid or cellulose hydrate being deposited on the surface. The paper becomes tough and resembles natural parchment in appearance. Thick sheets can be made by pressing together layers of paper which have been processed.
Uses: Semi-permeable membranes, packing greasy materials, substitute for natural parchment.
Shipping regulations: None.*

vegetable pepsin. See papain.

vegetable spermaceti. See Chinese wax.

vegetable sulfur. See lycopodium.

vegetable tanning. The tanning of leather by plant extracts. See tannic acid; tanning extracts.

vegetable wax of Japan. See Japan wax.

"Vegimal." [170] Trademark for a vegetable base adhesive.
Properties: Viscous, light tan liquid; difficult to spread at 20°C, but spreads easily at 52°C.
Derivation: By conversion of domestic starches with plasticizing chemicals to produce various drying and setting speeds.
Uses: To glue fabric to wood and paper to fabric; as sealer coats, fillers, stiffener for fabrics.

"Velban." [100] $C_{46}H_{58}O_9N_4$.
Properties: A dimeric alkaloid containing both indole and dihydroindole. Extracted from the common garden shrub periwinkle.
Use: Medicine.

"Velvapex." [165] Trade name for a series of cation-active softeners which are compatible over a pH range of 4 to 10. Used in the textile, leather, and cosmetic industries for substantive action on natural and synthetic fibers with unusually wide compatibilities.

"Velvasil." [245] Trademark for a group of drug grade silicone fluids available in viscosities ranging from 50 to 100,000 centistokes. These materials provide resistance to water-borne irritants; are non-sticky. They are physiologically inert, have excellent water repellency, low surface tension, excellent chemical, thermal and physical stability, are essentially colorless, odorless and tasteless and readily emulsify.
Uses: In protective skin creams and lotions, sun tan lotions, lip pomades, lipstick bases, hair dressings, ointment bases, waterless hand cleaners and medicinal bases.

"Velvoray." [159] Trademark for a high-grade finishing oil made from a blend of sulfonated vegetable oil combined with specially selected fats.
Properties: Compatible with all commonly used textile finishing materials; no foaming, no smoking, no oxidizing, no rancid odor; adds "body" and has emulsion stability and uniformity.
Uses: To impart a silky softness to fabrics of all kinds of fibers.

"Velvosheen." [328] A blend of sulfonated vegetable oils used as a softening agent for textile fabrics, particularly rayon.

"Velvo Softener." [159] Trade name for sulfonated tallow. A creamy white paste containing total fats amounting to 25% and pH of 9.3 - 9.5.
Uses: General finishing of all types of textile fabrics.

veneer. A thin layer or sheet of wood.

Venetian red. A high-grade ferric oxide pigment of a purer red hue than either light red (q.v.) or Indian red (q.v.). It is obtained either native as a variety of hematite red (q.v.) or more often artificially, by

*See "I.C.C. Shipping Regulations," page xiii.
Reference numbers refer to name of manufacturer. See "List of Manufacturers," page v.

calcining copperas in the presence of lime. The composition ranges from 15 to 40% ferric oxide and from 60 to 80% calcium sulfate. The 40% ferric oxide is the "pure" grade and has a sp.gr. of 3.45.
Grades: 20 to 40% ferric oxide.
Containers: Bags.
See also iron oxide reds.

Venice turpentine. See turpentine, Venice.

Venice turpentine, artificial. See turpentine, Venice, artificial.

"Veon." [233] Trademark for herbicides consisting of amine salts of 2,4-D and 2,4,5-T.

"Vera Blanc." [148] Brand name for a proprietary calcium carbonate product.
Properties: White; inert; water-ground, water-floated, and silk-bolted; sp.gr. 2.72; wt/gal 22.66 lbs; one pound bulks 0.04413 gal; oil absorption, 13.5; calcium carbonate, 99.5%; 100% through 325 mesh; average particle size, 4 microns.
Containers: 50-lb paper bags (net).
Uses: Inert extender pigment for oil and water paints, colors, rubber compounds, adhesives, oil-cloth, linoleum, textile fabrics, fur dressing, putty, plastics, ceramics, etc.

"Verabore." [342] Trademark for extract of antihypertensive alkaloids from Veratrum species.

"Verafleurs." [188] Brand name for a series of synthetic replacements for natural floral absolutes.

veratria. See veratrine.

veratridine. See veratrine.

veratrine (veratria).
Properties: A mixture of colorless, crystalline alkaloids; very poisonous! M.p. 145-155°C. Probably includes cevadine (also called veratrine), veratridine, cevadilline, sabadine, cevine. Soluble in alcohol, chloroform, and ether; very slightly soluble in water.
Derivation: By extraction from the seeds of Asagraea officinalis (sabadilla).
Method of purification: Crystallization.
Grade: Technical.
Containers: ⅛-, 1-, 5-oz vials and bottles.
Use: Medicine.
Shipping regulations: None.*

veratrole (ortho-dimethoxybenzene; pyrocatechol dimethyl ether) $C_6H_4(OCH_3)_2$.
Properties: Colorless crystals or liquid; m.p. 21-22°C; b.p. 206-207°C; sp.gr. 1.084 (25/25°C). Soluble in alcohol and ether; slightly soluble in water.
Derivation: Treatment of catechol in methyl alcohol with dimethyl sulfate and caustic.
Use: Medicine (antiseptic).

veratrum viride (American hellebore; green hellebore; Indian poke). Dried rhizome and roots of Veratrum viride.
Occurrence: North America.
Constituents: Several alkaloids, including jervine, cevadine, and protoveratrine.
Grade: Technical.
Containers: Boxes; burlap bags.
Use: Medicine.
Shipping regulations: None.*

"Verban." [57] Trademark for piperazine.

verbena oil. Oil distilled from the leaves of Verbena triphylla L. which is cultivated as an ornamental plant in Spain, northern France and Central America. Very pleasant, lemon-like odor, resembling that of lemon-grass oil. Principal component is the aldehyde citral. Not a regular article of commerce due to its scarcity and high price. Used in perfumes.

verbena oil, East Indian. See lemon-grass oil.

verde antique. A dark-green rock composed essentially of serpentine (hydrous magnesium silicate). Usually criss-crossed with white veinlets of marble. Used as an ornamental stone. In commerce often classed as a marble.
Occurrence: California, Georgia, Maryland, Massachusetts, New York, and Virginia.
Not to be confused with verte antique (q.v.).

verde salt. See thenardite.

verdigris. True verdigris is basic copper acetate, also called blue or green verdigris according to variety. A false verdigris formed on uncleaned copper vessels may consist of basic copper carbonate, while the green patina which coats old copper and bronze statues is a basic copper sulfate, or if on copper exposed to sea air or water, a basic copper chloride.

verdigris, blue. See copper acetate, basic.

verdigris, crystallized. See copper acetate.

verdigris green. See copper acetate, basic.

"Verelite." [233] Trademark for light stabilized polystyrene resins.

"Vergitryl." [412] Trademark for veratrum viride (q.v.).

vermiculite. A mineral of the mica group but hydrated, and with the property of expanding six to twenty times the volume of the unexpanded mineral when heated to about 2000°F.
Composition: It is a hydrated magnesium-aluminum-iron silicate containing approximately 39%SiO_2, 21%MgO, 15%Al_2O_3, 9% Fe_2O_3, 5-7%K_2O, 1%CaO, 5-9%H_2O and small quantities of Cr, Mn, P, S, Cl. Free oxides as such do not exist in the vermiculite crystals.
Occurrence: Montana, North Carolina, South Carolina, Wyoming, Colorado; South Africa.
Properties: Platelet-type crystalline structure; high porosity; high void volume to surface area ratio; low density; relative chemical inertness; large range of particle size; insoluble in water and organic solvents; water vapor adsorption capacity of expanded vermiculite less than 1%, liquid adsorption dependent on conditions and particle size, ranges 200-500%.

Grades: Unexpanded (ore concentrate); expanded (also called exfoliated); flake; activated.
Containers: Multiwall paper bags, 4 cu ft.
Uses: Lightweight concrete aggregate; insulation; sound conditioning; fireproofing; plaster; soil conditioner; additive for fertilizers; seed bed for plants; refractory; lubricant; oil well drilling mud; filler in rubber, paint plastics; in wall paper printing; extender in gold and bronze printing ink or paint; for packing; carrier for more active materials, such as insecticides; catalyst and catalyst support; litter for hatcheries; adsorbent.

vermilion. See mercuric sulfide, red.

vermilion, natural. See cinnabar.

vermilion, permanent. This is usually orange mineral (q.v.) tinted with paranitraniline. It has great tinctorial strength.

vermilion, quicksilver. See mercuric sulfide, red.

Verneuil process. Production of artificial corundum for jewels and bearings by fusing pure alumina with a hydrogen-air flame.

"Veronal." [162] Trademark for barbital.

"Versacaine." [57] Trademark for chlorprocaine hydrochloride.

"Versalide." [227] Trademark for 1,1,4,4-tetramethyl-6-ethyl-7-acetyl-1,2,3,4-tetrahydronaphthalene, a polycyclic musk. It is a white crystalline solid, stable both in color and odor over a wide range of pH.

"Versamid." [259] Trademark for a series of polyamide resins.
Derivation: Condensation product of dimeric fatty acid with polyamines.
Grades: Thermoplastic resins with softening points from 90 to 190°C. Reactive resins for copolymerization with epoxy resins. Amine values from 83 to 400. Solutions of the reactive resins are also available.
Containers: Thermoplastics: multiwall bags. Reactive resins: 7-, 40-, 100- and 400-lb net pails; drums.
Uses: Thermoplastic resins: flexographic inks, overprint varnish, heat seal and hot melt adhesives. Reactive resins: copolymerized with epoxy resins for coatings, adhesives, concrete topping, and patching compounds; castings; laminates; potting and encapsulating.

"Versenate." [233] Trademark for disodium salt of ethylenediaminetetraacetic acid and related compounds.

"Versene." [233] Trademark for sodium salts of ethylenediaminetetraacetic acid useful as complexing, chelating, or sequestering agents for hard water salts and other polyvalent metals. Recommended as general complexing agents for all polyvalent metals. Iron is complexed efficiently only in acid pH range. In addition to their water-softening properties, these products are recommended as cleaning and degreasing agents, as textile processing assistants, as decontaminating agents for radioactive materials, and as dissolving agents for proteins.

"Versene Fe-3 Specific." [233] Trademark for an organic complexing agent that exhibits preferential complexing action towards iron. Formula of active ingredient given as $C_6H_{12}O_4NNa$. It forms extremely stable complexes with ferric iron in the pH range of 3.5 to 12.5. It is also capable of complexing cobalt, nickel, copper, and zinc although its action on these divalent metals is not as strong as that of "Versene."

"Versenex." [233] Trademark for pentasodium salt of diethylenetriaminepentaacetic acid.

"Versenol." [233] Trademark for the trisodium salt of N-hydroxyethylethylene-diamine-triacetic acid, useful as complexing, chelating, or sequestering agent for hard water salts and other polyvalent metals. This product is very soluble in water and is particularly useful in alkaline media.

"Versilad." [244] Trademark for a fortified sodium silicate admixture, containing basically sodium silicate, clay, and urea.
Properties: Suspension of solids; thixotropic.
Uses: Adhesives for bonding corrugated boxboard. Made up in users' plant.

"Versilate." [244] Trademark for a composition of fortified sodium silicate.
Properties: Viscous liquid; opalescent.
Containers: Tank trucks and tank cars.
Uses: Base for "Versilad" adhesives.

"Versilube." [245] Trademark for silicone lubricating and hydraulic fluids. "Versilube F-50" is a lubricating fluid with an operating range from −100 to 400°F, and is unusually inert chemically. F-44 is an improved silicone fluid which possesses lubricating properties superior to other types of silicones. It is designed for use at −100 to 500°F. The bulk of the fluid has such a high boiling point that it has essentially no vapor pressure until some degradation or rearrangement occurs at 600°F. This defines the maximum temperature to which the fluids can be subjected without thermal degradation.
Uses: "Versilube F-50" is useful in all kinds of antifriction or rolling equipment and has greatly improved ability over other types of silicones to lubricate many metal combinations under conditions of sliding friction, except where both bearing surfaces are soft metals. "Versilube F-44" is designed for use as an engine oil or hydraulic fluid.

"Versimine." [233] Trademark for hydroxyiminodiacetic acid for use as a chelating agent.

verte antique (copper green). A paint pigment, essentially bicarbonate of copper, used for producing a corroded copper effect. Note: Not to be confused with the mineral known as verde antique (q.v.).

"Vertifume." [233] Trademark for fumigants containing carbon tetrachloride and carbon bisulfide.

"Vesperin." [412] Trademark for triflupromazine hydrochloride (q.v.).

vesuvianite (idocrase) $Ca_{10}Al_4(MgFe)_2(SiO_4)_5(Si_2O_7)_2(OH)_4$. Natural hydrous silicate of aluminum, calcium, magnesium and iron; found in metamorphic rocks. May contain boron, fluorine, or beryllium.
Properties: Color brown or green; luster vitreous to resinous; hardness 6.5; sp.gr. 3.4.
Occurrence: Maine, New York, New Jersey, California, Europe, U.S.S.R.
Use: Gem stone; possible source of beryllium.

"Vetamox." [57] Trademark for sulfonamide.

"Veticillin." [57] Trademark for penicillin.

vetivenol. See vetiverol.

vetiver oil (cuscus oil; vetivert).
Properties: Viscid, very aromatic oil; violet-like odor; sp.gr. 0.990-1.040 (15°C); optical rotation +15 to +45°; refractive index (n 20/D) 1.5200-1.5280; saponification value 14-45; soluble in 1-3 volumes of 80% alcohol, in fixed oils, diethyl phthalate, benzyl benzoate, mineral oil, ether, chloroform, acetone, carbon disulfide; insoluble in glycerin and propylene glycol.
Derivation: Steam distillation of partially dried roots of East Indian grass, Vetiveria zizanioides. Purified by rectification.
Grades: The oil is characterized by its geographical origin (Java, Haiti, East Indian, Bourbon Reunion, French).
Containers: Bottles; cans.
Use: Perfumery.

vetiverol (vetivenol; vetivol) $C_{15}H_{24}O$. An alcohol or a mixture of alcohols, structure not certain, obtained from vetiver oil.
Properties: Straw-colored, viscous liquid with pleasant odor; sp.gr. 0.980-1.002 (25/25°C); refractive index (n 20/D) 1.510-1.517.
Use: Perfumery.

vetivert. See vetiver oil.

vetivert acetate $C_{15}H_{23}OOCCH_3$ (an approximation).
Properties: Viscous, yellow liquid with pleasant odor; sp.gr. 0.979-0.999 (25/25°C); soluble in 80% alcohol.
Derivation: Treatment of vetiver oil or vetiverol with acetic anhydride.
Grade: Technical (about 50% pure).
Use: Perfumery.

vetivol. See vetiverol.

"Vetstrep." [123] Trademark for a veterinary preparation containing streptomycin sulfate.

"V-G-B." [248] Trademark for reaction product of acetaldehyde and aniline.
Properties: Brown resinous powder; sp.gr. 1.152; m.p. 60-80°C; soluble in acetone, benzene, and ethylene dichloride; insoluble in water and gasoline.
Uses: Rubber antioxidant for pure gum, mechanicals, solid tires, rubber-covered rolls, and specialties.

"Viadril." [299] Trademark for hydroxydione sodium.

"Vialon." [440] Trademark for a series of metal complex dyestuffs used in dyeing and printing on polyamide fibers.

"Viandarome." [188] Trademark for a flavor base for meat products.

"Vibrathane." [248] Trademark for a group of polyurethane raw materials for manufacture of foam and elastomers. Includes isocyanates, polyesters, polyether glycols, polyester and polyether prepolymers, liquid casting resins and gums and catalysts.

"Vibrin." [248] Trademark for resin compositions of polyesters and cross-linking monomers which, when catalyzed, will polymerize to infusible solid resins without evolving water or other by-products. Consistencies range from low viscosity liquids to waxlike solids, and the cured resins are of many types from soft, flexible to hard and rigid.
Properties: Resistant to aging, abrasion, weathering, and chemical reagents; good electrical properties; good compressive, tensile, and flexural strength. Self-extinguishing types are available.
Uses: Molding; laminating; impregnating; casting; automotive and aircraft structural parts; wall panels, table tops; coating for paper; boat hulls; chemically inert tanks; large-diameter pipe.

viburnum opulus (cramp bark; high cranberry; cranberry tree; water elder; squaw bush; snowball bush). Dried bark of Viburnum opulus.
Occurrence: Europe, Asia and northern North America south to Pennsylvania.
Chief constituents: Valeric acid, an enzyme and viburnin.
Grade: Technical.
Containers: Burlap bags; bales.
Use: Medicine.
Shipping regulations: None.*

viburnum prunifolium (black haw; sweet viburnum; sheep-berry; stag bush; sloe). Dried bark of root of Viburnum prunifolium or Viburnum rufidulum.
Occurrence: United States.
Grade: Technical.
Containers: Burlap bags.
Use: Medicine.
Shipping regulations: None.*

vic-. Prefix meaning vicinal (q.v.).

vicinal (abbreviated as vic-). Neighboring or adjoining positions on a carbon ring or chain; the term is used in naming derivatives with substituting groups in such locations in the structural formula or molecule.

"Victalube 5810." [172] Trade name for a brown viscous oil, not volatile at ordinary temperature.
Properties: Insoluble in water. Soluble in hydrocarbons.
Containers: 5-, 55-gal steel drums.
Use: Rust-inhibiting additive for industrial oils.

"Victamide." [172] Trade name for a proprietary product. Ammonium salt of an amido polyphosphate.
Properties: Exceedingly fine particles, practically all less than five microns. Slowly soluble in cold water; more rapidly soluble in hot water.
Uses: Sequestering agent for metallic ions; flameproofing agent of non-crystalline type; deflocculating agent for oil drilling muds, paint pigments, and clay slips.

"Victamines." [172] Trade name for proprietary products. Cationic surface-active phosphorus compounds.
Properties: Tan, waxy solids which disperse in water.
Containers: 90-lb fiber drums.
Uses: Softening agents for textiles and leather; oil additives.

"Victamuls." [172] Trade name for nonionic, surface-active phosphorus compounds.
Containers: Carboys and drums.
Uses: Wherever properties of spreading, emulsifying, penetrating, solubilizing, or dispersing are indicated.

"Victawets." [172] Trademark for surface-active phosphorus compounds.
Properties: Anionic and nonionic wetting agents.
Uses: Penetrants, dye carriers, and dispersing agents.

"Victor Cream." [172] Trade name for sodium acid pyrophosphate. $Na_2H_2P_2O_7$.
Properties: White, crystalline material. Purity meets all requirements of Federal and State Pure Food laws.
Containers: 100-lb paper bags.
Uses: Baking acid in doughnut and prepared flours. Manufacture of commercial baking powders and instant puddings. For conditioning the mud used in drilling oil wells. Formulation of acid-type metal cleaners.

Victoria blue $C_{33}H_{31}N_3 \cdot HCl$.
Properties: Crystalline powder, bronze colored. Soluble in hot water, alcohol or ether.
Derivation: Michler's ketone is condensed with phenyl-alpha-naphthylamine.
Uses: In textile industry for dyeing silk, wool, and cotton; biological stain; dye intermediate for producing complex acid pigment toners.

Victoria blue toners. See phosphotungstic pigments.

Victoria green. See malachite green.

Victoria red. See chrome red.

"Victory." [128] Brand name for a series of petroleum microcrystalline waxes.
Properties: Colors, brown, amber or white; m.p. 155 or 165°F min.
Containers: 10-lb slabs, 8/carton or 168/pallet; 350-lb drums; tank cars.
Uses: Coating and laminating paper, foil, and board; impregnating and waterproofing fabrics.

"Vi-Delta." [57] Trademark for vitamins A, D.

"Videne." [265] Trademark for thermoplastic polyester resin film.
Properties: Available in roll form in both glass clear and satin finish. It can also be reverse printed with any grain a camera can capture. It is resistant to water, oils, greases and most solvents. It is heat sealable and can be readily laminated to various surfaces by heat and pressure.

Vienna caustic. See caustic, Vienna.

Vienna green. See copper acetoarsenite.

Vienna paste. See caustic, Vienna.

"Vigofac." [299] Trademark for an animal feed supplement containing a growth ingredient.

"Vinac." [144] Trademark for polyvinyl acetate homopolymer emulsions, beads, powders.
Emulsions Properties: Water-white; 55-57% solids; viscosity varies according to grade; range 500-4200 cps (60 RPM-RVO Brookfield); pH 4.0-6.0. Excellent adhesive characteristics.
Grades: XX-210, XX-220, HF-300, WR-20, WR-50, CE 1-P, AA-63.
Containers: 5-, 10-, 30-gal drums; lined 55-gal drums.
Uses: Bases for compounding adhesives, binders, paints, coatings, textile finishes, paper coatings.
Bead Properties: Spherical granules; odorless, tasteless, nontoxic, light-stable; 98.5%-100% resin solids; softening point varies according to grade: range 105-155°C; molecular weight varies according to grade. Soluble in alcohols, chlorinated solvents; esters, hydrocarbons, ketones, dimethyl furane, nitromethane, nitropropane.
Grades: B-7, B-15, B-25, B-100, B-800, ASB-10.
Containers: 250-lb fiber drums; 50-lb multiwall paper bags.
Uses: Adhesives; coatings; textile finishes.
Powder Properties: White polyvinyl acetate powder which redisperses readily in cold water. Dispersions resemble polyvinyl acetate homopolymer emulsions, with adhesive and binding characteristics. Particle size of powder -98.8 through 100 mesh; bulk density 0.53 grams/cc; in dispersion mean particle size 2-6 microns; pH 4-6; viscosity (50% dispersion, No. 3 spindle, 60 RPM) 1,000 cps; viscosity (60% dispersion, No. 4 spindle, 60 RPM) 9,500 cps.
Grades: RP-250.
Containers: 55-gal fiber drums.
Uses: Formulation of dry-mix products, such as joint cements for dry-wall construction; spackling compounds; cement and cinder

block fillers; powder paints. Also compounding adhesives, textile binders and stiffening agents, heat-seal coatings.

"Vinactane." [305] Trademark for viomycin.
Use: Medicine.

vinal. Generic name for a manufactured fiber in which the fiber-forming substance is any long chain synthetic polymer composed of at least 50% by weight of vinyl alcohol units, $-CH_2CHOH-$, and in which the total of the vinyl alcohol units and any one or more of the various acetal units is at least 85% by weight of the fiber (Federal Trade Commission). It has been developed and first used in Japan. It has good chemical resistance, low affinity for water; good resistance to mildew and fungi.
Uses (suggested): Fishing nets; stockings, gloves; hats; rainwear; swimsuits.

vinasse. The residue obtained from beet sugar molasses fermentation, containing mineral salts.

vinbarbital (5-ethyl-5(1-methyl-1-butenyl) barbituric acid) $C_{11}H_{16}N_2O_3$.
Properties: White powder with characteristic odor and bitter taste. Very slightly soluble in water; sparingly soluble in ether; soluble in alcohol. M.p. 160-163°C.
Grade: N.F. XI.
Use: Medicine.

vinbarbital sodium (sodium 5-ethyl-5(1-methyl-1-butenyl)barbiturate) $C_{11}H_{15}N_2NaO_3$.
Properties: White odorless powder with bitter taste. Soluble in alcohol and water; slightly soluble in chloroform and ether. Unbuffered aqueous solutions unstable. Powder is hygroscopic and is affected by moisture and carbon dioxide. pH (1% solution) 8.5-9.5.
Grade: N.N.D.
Use: Medicine.

"Vincel." [332] Trademark for a line of woven fabrics of polyvinyl chloride fibers for liquid and pneumatic filtration.

vine black. The charcoal resulting from incomplete combustion (destructive distillation) of grape vines.

vinegar (cider vinegar). A dilute impure solution of acetic acid. Unless otherwise stated, it is assumed to be made from apple cider and to contain not less than 4% acetic acid. It can be made by the traditional process of slow fermentation through the alcohol stage to the acid stage, or by the quick vinegar process of trickling the fermented liquid mixed with some vinegar over shavings. See also vinegar, malt; vinegar, distilled; vinegar, wine.

vinegar acid. See acetic acid.

vinegar, cider. See vinegar.

vinegar, distilled. A colorless vinegar which has been distilled after its formation. It is usually stronger in acetic acid than cider vinegar.

vinegar, malt. A vinegar made from fermented malted barley liquor.

vinegar naphtha. See ethyl acetate.

vinegar salts. See calcium acetate.

vinegar, wine. A vinegar made from wine and having a characteristic flavor. It contains about 8% acetic acid.

"Vinethene." [123] Trademark for an anesthetic preparation consisting of vinyl ether.

vinetine. See oxyacanthine.

"Vinol." [144] Trademark for polyvinyl alcohol resins.
Properties: White powders; soluble in water; form tough, flexible, transparent films which are impermeable to oxygen and other gases, abrasion resistant, grease and solvent resistant. Films range from water soluble to water resistant. Viscosity and hydrolysis vary according to grade; hydrolysis ranges from 99.85+% to 87%. Excellent adhesive and binding properties.
Grades: 125, 260, 230, 205, 350, 325, 540, 523, 505.
Containers: 50-lb multiwall bags.
Uses: Adhesive compounding; textile sizing and finishing; paper sizing and coating; packaging films; water soluble films; protective coatings; molded and foamed plastics.

"Vinsol." [266] Trademark for a dark brittle thermoplastic resin; ruby-red by transmitted light, dark-brown by reflected light; sp.gr. 1.218; acid number 94; softening point 116°C.
Uses: Asphalt emulsions; in Portland and mortar cements; thermoplastic stiffener in rubber and paper board; phonograph records; oil resistant compounds; resins; lacquers; varnishes and plastics.

"Vinsol NVX." [266] Brand name for sodium salt of Vinsol resin.
Properties: Dark dry powder, soluble in water.
Uses: Emulsifying asphalt; agent for air-entraining cement.

"Vinycol." [233] Proprietary products consisting of pigments dispersed in "Vinylite" resins for coatings.
Grades and Uses:
"Vinycol" No. 100 White: Titanium dioxide dispersed in "Vinylite" resin and plasticizer. Used in high-gloss vinyl finishes.
"Vinycol" No. 200 Black: High-color carbon black dispersed in "Vinylite" resin and plasticizer. Used in high-gloss vinyl finishes.

"Vinylac." [65] Trademark for a series of resinous tackifier dispersions for modification of polyvinyl acetate adhesive systems.

vinyl acetate $CH_3COOCH{:}CH_2$. A major raw material for vinyl plastics (q.v.). See also polyvinyl acetal resins; polyvinyl acetate.
Properties: Colorless liquid, which is stabilized with either hydroquinone or diphenylamine inhibitors. The HQ-stabilized

material can be polymerized without redistillation. The DPA-stabilized material must be distilled prior to polymerization.

Typical specifications: Vinyl acetate 99.8% min; acetaldehyde 0.013% max; acid (as acetic acid) 0.007% max; moisture 0.04% max; colorless to light yellow; b.p. 72.3-73.0°C; m.p.—100.2°C; sp.gr. 0.9335-0.9345 (20/20); refractive index 1.3941; flash point 22°F (open cup); wt/gal 7.79 lbs; vapor pressure 3.7 lbs/sq in; hydroquinone 14-17 ppm; diphenylamine 275-325 ppm. Soluble in most organic solvents including chlorinated solvents; insoluble in water.

Derivation: Obtained by the reaction of acetylene and acetic acid in the presence of a mercuric oxide catalyst. Purified by distillation.

Grade: Technical.

Containers: 1-, 5-gal cans; 55-gal steel drums. All nonreturnable. 8000-gal tank cars.

Uses: Polyvinyl acetate, polyvinyl alcohol, polyvinyl butyral, and polyvinyl chloride-acetate resins. Also in latex paints; adhesives; textile finishing; safety glass interlayers.

Danger! Extremely flammable. MCA warning label.

Shipping regulations: Flammable liquid. Red label.* (For vinyl acetate, inhibited.)

vinylacetonitrile. See allyl cyanide.

vinylacetylene C_4H_4 or $H_2C:CHC:CH$. The dimer of acetylene, formed by passing it into a solution of cuprous and ammonium chlorides in hydrochloric acid.

Properties: Colorless gas or liquid; sp.gr. 0.6867 (0/20°C); b.p. 5°C.

Use: Intermediate in manufacture of neoprene synthetic rubber and for various organic syntheses.

See also divinyl acetylene.

vinyl alcohol (ethenol) $CH_2:CHOH$. Isolated only in form of its esters, or the polymer, polyvinyl alcohol (q.v.).

vinylation. The formation of a vinyl derivative by reaction with acetylene. Thus vinylation of alcohols yields vinyl ethers such as vinyl ethyl ether, $C_2H_5OC_2H_3$. The process was used in Germany during World War II (Reppe chemistry), and requires catalysts as well as some heat, pressure, and dilution of acetylene with nitrogen. Amines, mercaptans, and fatty acids can be vinylated. The products are useful as intermediates for further synthesis, especially polymerization.

vinylbenzene. See styrene.

vinyl n-butyl ether (n-butyl vinyl ether; BVE) $CH_2:CHOC_4H_9$.

Properties: Liquid; sp.gr. 0.7803 (20°C); b.p. 94.1°C (760 mm); f.p. —92°C; refractive index 1.3997; flash point 30°F; wt/gal 7.45 lbs (20°C). Slightly soluble in water; soluble in alcohol and ether.

Derivation: Reaction of acetylene with n-butyl alcohol.

Method of purification: Washing with water; drying in presence of alkali, and distillation from metallic sodium.

Grade: Technical (95%).

Containers: Glass bottles; drums; tank cars.

Uses: Synthesis; copolymerization.

Shipping regulations: Flammable liquid. Red label.*

vinyl butyrate $CH_2:CHOOCC_3H_7$.

Properties: Liquid; sp.gr. 0.9022 (20/20°C); b.p. 116.7°C; f.p. —86.8°C. Very slightly soluble in water. Flash point 68°F.

Containers: 55-gal drums.

Uses: Polymers; emulsion paints.

Shipping regulations: Flammable liquid. Red label.*

N-vinylcarbazole $C_{12}H_8NCH:CH_2$.

Derivation: From acetylene and carbazole.

Use: Polymerizes to form heat resistant and insulating resins somewhat similar to mica in dielectric properties. See polyvinyl carbazole.

vinyl chloride (VC; chloroethene; chloroethylene) $CH_2:CHCl$. The most important of the vinyl monomers.

Properties: Easily liquefied gas, with a pleasant ethereal odor. Usually handled as the liquid. A small amount of phenol is added as a polymerization inhibitor. Sp.gr. 0.9121 (liquid, at 20/20°C); b.p. —13.9°C (760 mm); f.p. —159.7°C; vapor pressure 2300 mm (20°C); flash point —108°F. Slightly soluble in water; soluble in alcohol and ether.

Typical specifications: Color water-white and clear; boiling range (760 mm) not less than 95% distils over before the temperature of the liquid reaches 10°C; water, no water layer present in the cylinder when sampled; acetaldehyde not more than 0.5% by weight; residue not more than 0.5% by vol; average wt/gal (liquid) 7.59 lbs (20°C).

Derivation: (a) Reaction of acetylene and hydrochloric acid, either as liquids or gases. Mercuric chloride is a catalyst for the dry process; cupric chloride for the liquid process. (b) Cracking of ethylene dichloride (made from ethylene and chlorine) at high temperatures. (c) Reaction of ethylene dichloride and caustic soda.

Method of purification: Distillation.

Grades: Technical; pure, 99.9%.

Containers: Cylinders; tank cars.

Danger! Extremely flammable liquid and gas under pressure. MCA warning label.

Uses: Mainly for polyvinyl chloride (q.v.), and copolymers (see also vinyl plastics); organic synthesis.

Shipping regulations: (For vinyl chloride, inhibited). Flammable gas. Red gas label.*

vinyl 2-chloroethyl ether $CH_2:CHOCH_2CH_2Cl$.

Properties: Liquid having sp.gr. 1.0498 (20°C); b.p. 109.1°C; m.p. —69.7°C; very slightly soluble in water; flash point (open cup) 80°F.

Use: Copolymerization to produce plastics.

Shipping regulations: Flammable liquid. Red label.*

vinyl beta-**chloroethyl sulfide**
$CH_2:CHSCH_2CH_2Cl$.
Properties: Liquid. B.p. 71-72°C (50 mm).
Grade: Technical.
Use: Organic synthesis.

vinyl compounds. Compounds having the vinyl grouping (CH_2=CH—), specifically vinyl chloride, vinyl acetate and similar esters, but also referring more generally to other types of compounds such as styrene $C_6H_5CH:CH_2$, methyl methacrylate $CH_2:C(CH_3)COOCH_3$ and acrylonitrile $CH_2:CHCN$. The vinyl compounds are highly reactive and polymerize easily, hence are basic materials for plastics. For typical inhibitors used to avoid polymerization, see vinyl acetate. See vinyl plastics.

vinyl cyanide. See acrylonitrile.

vinylcyclohexene (1-vinylcyclohexene-3; 4-vinylcyclohexene-1; cyclohexenylethylene)
$CH_2:CHCHCH_2CH:CHCH_2CH_2$.
Properties: Liquid; sp.gr. 0.8303 (20/4°C); f.p. −108.9°C; b.p. 128°C; refractive index 1.464 (n20/D); flash point 70°F. Temperatures above 80°F and prolonged exposure to oxygen-containing gases should be avoided as these conditions lead to discoloration and gum formation.
Typical specifications: Boiling range 126.5°C min; 133.5°C max; sp.gr. 0.830-0.835 (15.5/15.5°C); color, not less than 10 Saybolt; appearance, clear and free of suspended matter; flash point (open cup) 74°F.
Grades: Technical, 95%; pure, 99%; research.
Containers: 1-, 5-gal cans; 55-gal drums; tank trucks; tank cars.
Uses: Polymers; organic synthesis.
Shipping regulations: Flammable liquid. Red label.*

vinyl cyclohexene monoxide $CH:CHC_6H_9O$.
Properties: Liquid; sp.gr. 0.9598 (20/20°C); b.p. 169°C; f.p. −100°C; flash point 126°F. Very slightly soluble in water.
Uses: Polymers; organic synthesis.

vinyl ether (divinyl ether; divinyl oxide)
$CH_2:CHOCH:CH_2$.
Properties: Clear colorless liquid with characteristic odor; sp.gr. 0.773; b.p. 28-31°C; refractive index (20/D) 1.3989. Slightly soluble in water; miscible with alcohol, acetone, chloroform, and ether. Must be protected from light.
Derivation: Treatment of dichloroethyl ether with alkali.
Grade: U.S.P. XVI required about 96% vinyl ether, 4% dehydrated alcohol and allows 0.025% harmless preservative which may impart purple fluorescence.
Containers: 50- and 75-cc bottles.
Caution: Highly volatile flammable liquid. Flash point below −22°F. Gives off even at comparatively low temperatures vapors which form flammable mixtures with air or oxygen.
Use: Medicine (anesthetic).

vinyl beta-**ethoxyethyl sulfide**
$CH_2:CHSCH_2CH_2OC_2H_5$.
Properties: Colorless, mobile liquid. Pungent, camphor-like odor.
Constants: Sp.gr. 0.9532 (15°C); b.p. 65°C (8 mm).
Grade: Technical.
Use: Organic synthesis.

vinylethylene. See butadiene.

vinyl ethyl ether (ethyl vinyl ether; EVE)
$CH_2:CHOC_2H_5$.
Properties: Colorless liquid. Extremely reactive and can be polymerized in either the liquid or vapor phase; slightly soluble in water: 0.9% by wt; sp.gr. 0.754 (20/20°C); 6.28 lbs/gal (20°C); b.p. 35.5°C (760 mm); vapor pressure 428 mm (20°C); f.p. −115.0°C; viscosity 0.22 cps (20°C); refractive index 1.3739; flash point, less than 0°F. Commercial material contains inhibitor to prevent premature polymerization.
Derivation: Reaction of acetylene with ethyl alcohol.
Method of purification: Washing with water, drying in presence of alkali, distillation from metallic sodium.
Grade: Technical.
Containers: Drums; tank cars.
Uses: Copolymerization; intermediate.
Shipping regulations: Flammable liquid. Red label.*

vinyl 2-ethyl hexoate $CH_2:CHOOCCH(C_2H_5)C_4H_9$.
Properties: Liquid; sp.gr. 0.8751 (20/20°C); b.p. 185.2°C; f.p. −90°C; flash point 165°F. Insoluble in water.
Containers: 55-gal drums.
Uses: Polymers; emulsion paints.

vinyl 2-ethylhexyl ether
$CH_2:CHOCH_2CH(C_2H_5)C_4H_9$.
Properties: Liquid; sp.gr. 0.8102 (20/20°C); b.p. 177.7°C; f.p. −100°C; flash point 135°F. Insoluble in water.
Containers: 55-gal drums.
Uses: Intermediate for pharmaceuticals, insecticides, adhesives; viscosity index improver.
Shipping regulations: None.*

2-vinyl-5-ethylpyridine $(CH_2:CH)C_5H_3N(C_2H_5)$.
Properties: Sp.gr. (20/20°C) 0.9449; b.p. (100 mm) 138°C; vapor pressure (20°C) 0.2 mm; f.p. −50.9°C; flash point 200°F (Cleveland open cup); solubility in water (20°C) less than 0.01% by weight.
Use: Copolymer; synthesis.

vinyl fluoride (fluoroethylene) $CH_2:CHF$.
Properties: Colored gas; b.p. −51°C; insoluble in water; soluble in alcohol and ether.
Containers: Cylinders.
Use: Monomer for resins. See polyvinyl fluoride.
Shipping regulations: (Inhibited) flammable gas. Red gas label.*

vinylformic acid. See acrylic acid.

vinylidene chloride (VC) $CH_2:CCl_2$.
Properties: Colorless volatile liquid;

f.p. −122.53°C; b.p. 32°C; flash point (open cup) −10°C. The material readily undergoes polymerization, and in commerce is always encountered with a small proportion of polymerization inhibitor present.

Uses: Vinylidene chloride is copolymerized with vinyl chloride (principal use) or acrylonitrile to form various kinds of saran. Other copolymers are also used.

Shipping regulations: Flammable liquid. Red label.*

vinylidene fluoride (1, 1-difluoroethylene) $CH_2:CF_2$. A monomer of growing importance.

Properties: Colorless gas; b.p. less than −70°C; insoluble in water; soluble in alcohol, and ether.

Use: To make elastomers with such copolymers as chlorotrifluoroethylene and hexafluoropropylene. "Genetron" 1132A and "Viton" A are examples. See also polyvinylidene fluoride.

vinylidene resins (polyvinylidene resins). These are resins in which the unit structure in the polymer molecule is $(-H_2CCX_2-)$, in which X is usually chlorine, fluorine, or cyanide radical. Examples are saran, "Viton" A, "Genetron" 1132 A.

vinyl isobutyl ether (isobutyl vinyl ether; IVE) $CH_2:CHOCH_2CH(CH_3)_2$.

Properties: Colorless liquid; sp.gr. 0.7706 (20/20°C); b.p. 83.3°C; vapor pressure 68 mm (20°C); f.p. −112°C; refractive index 1.3938; flash point 20°F. Very slightly soluble in water; soluble in alcohol and ether; easily polymerized.

Derivation: Catalytic union of acetylene and isobutyl alcohol.

Method of purification: Washing with water, drying in presence of alkali, and distillation from metallic sodium.

Grade: Technical.

Containers: Drums; tank cars.

Uses: Polymer and copolymers used in surgical adhesives, coatings and lacquers, modifier for alkyd and polystyrene resins, plasticizer for nitrocellulose and other plastics; chemical intermediate.

Shipping regulations: Flammable liquid. Red label.*

"Vinylite." [214] Trademark for a series of synthetic thermoplastic resins and plastics. Available in four series—A, Q, V, and X. Data on these are as follows:

Series A:

Type: Polymerized vinyl acetate, vinyl alcohol-acetate.

Grades: AYAA, AYAF, AYAT, AYAC, A-35, A-70, MA-28-14, MA-28-18, T-24-9, W-125.

Properties: White granular powder, solutions, dispersions. Colorless, odorless, tasteless, nontoxic, resistant to dilute salt and acid solutions. Stable to both heat and light. Soluble in most alcohols, ketones, esters, glycol-ethers, chlorinated hydrocarbons, and lower-boiling coal-tar hydrocarbons, or mixtures of these solvents. Insoluble in petroleum naphtha, glycols, turpentine, and vegetable and mineral oils. Insoluble in water, although from 5-20% water, mixed with lower alcohols, improves their solubility characteristics. Some nonsolvents, such as xylol, can be used as diluents in the presence of active solvents.

Uses: Solvent-type and thermoplastic adhesives for cloth, paper, cardboard, porcelain, metal, mica, stone, glass, wood, leather, and plastic sheets and film. Also binding agent for artificial wood.

Series Q:

Type: Polymerized vinyl chloride.

Grade: QYNA.

Properties: White, granular powder. Soluble in dioxane, ethylene dichloride, chlorobenzene and mesityl oxide. Partially soluble in acetone. Insoluble in acids, alcohols, water, and most other solvents.

Uses: In plasticized compositions for the production of calendered products, cloth coatings with the chemical resistance, permanence and performance depending on the type of plasticizer used.

Series V:

Type: Copolymerized vinyl chloride and vinyl acetate.

Grades: VYHH, VYLF, VYNS, VYNW, VYDR, VYNY, VYNV, VYCM, VYCC, VMCH, and VAGH.

Properties: White granular powders. Colorless, tasteless, odorless, nontoxic. Nonburning except on direct exposure to flame. Thermoplastic, having high internal plasticity. Can be plasticized to any degree of flexibility. Soluble in lower-boiling ketones, chlorinated hydrocarbons, dioxane, propylene oxide and mesityl oxide. Softens in higher-boiling ketones, aldehydes, esters, ethers, ether-alcohols, carbon disulfide and aromatic hydrocarbons. Insoluble in most other solvents.

Uses: Injection and compression molding compounds, flexible film and sheeting, rigid sheets, coated paper, extrusion compounds for electrical insulation and other uses. Base resins for coatings, formulations for paper, cloth, metal and wood as supplied by leading lacquer manufacturers.

Series X:

Type: Polyvinyl butyral.

Grades: XYSG, XYSL, XYSG-XYHL.

Properties: Colorless granular powder. Very resistant to sunlight and ultra-violet radiation, also very heat-stable. Soluble in methanol, ethanol, isopropanol, butanol, methyl "Cellosolve" methyl ether, "Cellosolve" ethyl ether, "Cellosolve" butyl ether, dioxane, and dioxolane. Swelled by ketones, aromatic hydrocarbons, chlorinated hydrocarbons, and many esters. Insoluble in water and aliphatic hydrocarbons.

Uses: Compounded as interlayer for safety glass for automotive and aircraft use. Adhesive and leather finishing base. As the base resin, compounded by lacquer manufacturers into wash primer of cohesive and protective properties for metals, wood.

*See "I.C.C. Shipping Regulations," page xiii.
Reference numbers refer to name of manufacturer. See "List of Manufacturers," page v.

vinyl methyl ether (methyl vinyl ether; MVE) $CH_2:CHOCH_3$.
Properties: Colorless, easily liquefied gas, colorless liquid; sp.gr. 0.7500 (20/20°C); b.p. 6.0°C; vapor pressure 1052 mm (20°C); flash point −60°F; f.p. −121.6°C. Slightly soluble in water; soluble in alcohol and ether; easily polymerized, and commercial material contains a small proportion of polymerization inhibitor to avoid reaction during shipment or storage.
Derivation: Catalytic union of acetylene and methyl alcohol.
Method of purification: Washing with water, drying in presence of alkali, distillation from metallic sodium.
Grades: Technical (95% min); pure.
Containers: 150-lb cylinders; tank cars.
Uses: Copolymers used in coatings and lacquers, modifier for alkyl and polystyrene resins, plasticizer for nitrocellulose and other plastics. Also potentially useful as a starting point in synthesis. See polyvinyl methyl ether.
Shipping regulations: (Inhibited) flammable gas. Red gas label.*

vinyl methyl ether copolymers. See PVM/MA.

vinyl methyl ketone (3-butene-2-one) $CH_3OCH:CH_2$.
Shipping regulations: (Inhibited) flammable liquid. Red label.*

vinyl plastics. Polymers and resins derived by polymerization or copolymerization of vinyl monomers (vinyl compounds) including vinyl chloride and acetate, vinylidene chloride, methyl acrylate and methacrylate, acrylonitrile, styrene, the vinyl ethers, and numerous others characterized by presence of a carbon double bond in the monomer molecule, which opens during polymerization to make possible the carbon chain of the polymer. A simple case is the conversion of vinyl chloride $H_2C:CHCl$ to polyvinyl chloride $(-CH_2-CHCl-)_n$. In a narrower sense the term vinyl plastics refers to polyvinyl chloride, acetate, alcohol, etc., and copolymers or closely related materials. See under both vinyl and polyvinyl.
Uses: Adhesives; protective coatings (lacquers, etc.); inks; films; sheets; solid molded objects; textile sizing; stiffening and coating; paper sizing; as a binder-fuel in solid rocket propellants; and a variety of other applications. See especially polyvinyl chloride.

vinyl propionate $CH_2:CHOOCC_2H_5$.
Properties: Liquid; sp.gr. 0.9173 (20/20°C); b.p. 95.0°C; f.p. −81.1°C; flash point 34°F. Almost insoluble in water.
Containers: 55-gal drums; tank cars.
Use: Polymer; emulsion paints.
Shipping regulations: Flammable liquid. Red label.*

2-vinylpyridine $C_5H_4NCH:CH_2$.
Properties: Colorless liquid, boils with resinification at about 159°C (760 mm); sp.gr. 0.9746 (20°C); refractive index 1.5509 (n 20/D). Dissolves in water to extent of 2.5%; water dissolves in it to about 15%; soluble in dilute acids, hydrocarbons, alcohols, ketones, esters. Commercial material contains inhibitor to prevent premature polymerization.
Containers: Drums; tank cars.
Uses: Production of synthetic rubber, polymers, pharmaceuticals.

4-vinylpyridine $CH_2:CHC_5H_4N$.
Properties: Liquid; b.p. (150 mm) 121°C; sp. gr. (20°C) 0.988; refractive index (n 20/D) 1.5525. Slightly soluble in water.
Grade: 95% min.
Uses: Polymers; synthesis of pyridine derivatives.
Caution! Both the liquid and its vapor are irritants to the skin, eyes, and respiratory tract. Avoid contact with the liquid and prolonged exposure to high vapor concentration.

N-vinyl-2-pyrrolidone $CH_2CH:\overline{NCH_2CH_2CH_2CO}$.
Made from acetylene and formaldehyde by high pressure synthesis.
Use: See polyvinylpyrrolidone.

vinyl resins. See vinyl plastics.

vinyl stabilizers. Substances added to vinyl chloride resins during compounding, to retard the rate of deterioration due to formation of hydrogen chloride from the polyvinyl chloride. Many of these substances are of such a chemical nature as to combine readily with hydrogen chloride but do not otherwise interfere with the properties and uses of the final plastic product. Amines, basic oxides, and metallic soaps are commonly used.

vinyl stearate $CH_3(CH_2)_{16}COOCH:CH_2$.
Properties: White, waxy solid. M.p. 28-30°C; b.p. 175°C (3 mm); sp.gr. 0.9037 (20/20); refractive index 1.4355-1.4362 (n 55/D); iodine no. 80-82. Insoluble in water and alcohol; moderately soluble in ketones and vegetable oils; soluble in most hydrocarbon and chlorinated solvents.
Containers: 35-, 210-, and 390-lb drums.
Use: Plasticizer (copolymerizer).

vinylstyrene. See divinyl benzene.

vinyltoluene $CH_2:CHC_6H_4CH_3$.
Properties: Colorless liquid; f.p. −76.8°C; b.p. 170-171°C (760 mm); sp.gr. 0.890 (25/25°C); lb/gal 7.41; refractive index 1.534 (n 35/D); flash point 140°F. Very slightly soluble in water. Soluble in methanol, ether.
Containers: Drums; tank cars.
Use: Solvent; intermediate.

vinyl trichloride. See 1,1,2-trichloroethane.

vinyltrichlorosilane $CH_2CHSiCl_3$.
Properties: Colorless or pale yellow liquid. B.p. 90.6°C; sp.gr. 1.265 (25/25°C); refractive index (n 20/D) 1.432; flash point (Cleveland open cup) 70°F. Readily hydrolyzed by moisture, with the liberation of hydrochloric acid; polymerizes easily.
Derivation: By the reaction of acetylene and

trichlorosilane in the presence of a peroxide catalyst; reaction of trichlorosilane with vinyl chloride.
Grade: Technical.
Use: Intermediate for silicones.
Shipping regulations: Flammable liquid. Red label.*

"Vinymul." [170] Trademark for a textile finisher for all fabrics or fibers where flexibility is desired at low temperatures, such as back-coating or immersion finishing of automobile fabrics.
Properties: White, fluid stable synthetic vinyl resin copolymer, giving tough, nonbrittle film which is oil, fat, solvent and water resistant. Material does not require curing temperatures and has no plasticizer migration. Has good specific adhesion to vinyl surfaces without adverse effect on chemical or dimensional properties of the vinyl.
Uses: As primer with adhesives on vinyl pressure sensitive tapes or as a prime coat on metal for adhesion to metal and improved adhesion for second protective coating.

vinyon. Generic name for a manufactured fiber in which the fiber-forming substance is any long chain synethetic polymer composed of at least 85% by weight of vinyl chloride units, $-CH_2CHCl-$ (Federal Trade Commission). It has good resistance to chemicals, bacteria, moths; is unaffected by water and sunlight; and has a low softening point.
Uses: Mixing with other fibers for heat bonding; for fishing nets and lines; industrial filters.

"Vinyzene." [8] Trade name for a series of fungicides and bactericides formulated as additives to vinyl films.

viobin process. Process for desiccating and defatting tissues such as liver, pancreas, thyroid, etc., for pharmaceutical use. The process is a combination of azeotropic distillation and solvent extraction.

"Viocin." [299] Trademark for viomycin sulfate.

"Vioform." [305] Trademark for iodochlorhydroxyquin, U.S.P.

"Violamine." [307] Brand name of proprietary line of acid dyestuffs. Used on wool and silk. Characterized by fairly good fastness to light, washing, fulling, etc. Can also be used on paper and leather.

violanthrone. See dibenzanthrone.

"Violite." [4] Trademark for phosphorescent and fluorescent pigments (ZnS-CdS, CaS-SrS); compatible with a variety of vehicles.
Uses: Lacquers, paints, printing inks, paper, tape, and plastics.

viomycin. An antibiotic produced by Streptomyces puniceus. It is unique among antibiotics in that it is more active against acid-fast organisms than it is against other groups of bacteria. Mycobacteria are most sensitive to viomycin, and the antibiotic is active against strains of Mycobacterium tuberculosis which are resistant to other antibiotics.
Properties: It is a strongly basic polypeptide, which readily forms neutral salts with acids. The sulfate and the hydrochloride are very soluble in water. Moderately stable at pH 5-6 at room temperature but loses activity at high temperatures.
Units: Unit established arbitrarily as 1 microgram of the pure base.
Use: Medicine.
Available commercially as sulfate.

viomycin sulfate.
Properties: White, odorless, hygroscopic powder; soluble in water; only slightly soluble in alcohol; pH (1% solution) 4.5-7.0.
Grade: N.N.D.
Use: Medicine.

viosterol. Vitamin D_2. See vitamin D.

"Vircol-82." [40] Trademark for a neutral phosphorus polyol used as co-reactant flame retardant for urethane foams.
Properties: Nearly colorless, pleasant-smelling, high-boiling fluid; insoluble in water up to 45% but miscible above this level; soluble in a variety of organic solvents; insoluble in ether and aliphatic hydrocarbons. Typical product has sp.gr. (20°C) 1.10; refractive index (25°C) 1.445; flash point (open cup) 260°F.
Uses: Chemically bound, permanent flame retardency, nonmigrating and not extracted by solvents; useful with both polyesters and polyethers with prepolymer and one-shot techniques in flexible and rigid foams.

"Vircol-189." [40] Trademark for a neutral phosphorus polyol with outstanding properties as a plasticizer for poly(vinyl alcohol).
Properties: Colorless, pleasant-smelling, high-boiling fluid; soluble in water below 2% and above 20%, but forms two phases between these levels; soluble in a variety of organic solvents; insoluble in ether and aliphatic hydrocarbons. Typical product has sp.gr. (20°C) 1.130; refractive index (20°C) 1.448; viscosity (100°F) 27.2 centistokes; flash point (open cup) 205°F.
Uses: Plasticizer for poly(vinyl alcohol); films prepared therefrom are less water sensitive and retain properties over a wide range of conditions; useful in films, moldings, paper sizing, textile sizing and other applications.

"Virco-Pet 20." [40] Trademark for a neutral organic phosphorus chemical for use as a corrosion inhibitor for steel and aluminum.
Properties: A tan, viscous liquid; forms an emulsion with water; miscible with a variety of organic solvents; sp.gr. (20°C) 0.96-1.00.
Uses: Protection of petroleum-handling facilities; in protective coatings such as paints and waxes; additive to antifreeze compounds (forms stable emulsions with glycol-water mixture), cleaners and polishes; and

wherever a liquid is in contact with a metal under corrosive conditions.

"Virco-Pet 30." [40] Trademark for a neutral organic phosphorus chemical which gives outstanding corrosion protection to aluminum and its alloys.

Properties: A tan, viscous liquid; soluble in water and a variety of organic solvents; insoluble in ether and aliphatic hydrocarbons; sp.gr. (20°C) 1.06; refractive index (20°C) 1.447.

Uses: Complements and extends protection offered by Virco-Pet 20; protects aluminum and its alloys even when the aluminum is galvanically coupled with other metals; affords protection also for steel, iron, copper, bronze and brass.

"Virflux." [288] Trademark for titanium ore and titanium oxide.

Grades: "Virflux" W (Standard M and T Virginia rutile). Mineral consisting principally of titanium dioxide. "Virflux" C (Ceramic Grade M and T Virginia rutile). Rutile having a minimum chromium and iron content.

Uses: In the metallurgical and ceramic industries.

virginium. See francium.

"Virgo." [306] Trademark for a line of descaling, desanding, degraphitizing and deenameling salts used in molten baths. The descaling operation is primarily applied to various steels and some alloys.

"Viridine." [227] Trademark for phenylacetaldehyde dimethylacetal. $C_6H_5CH_2CH(OCH_3)_2$.

Properties: Colorless liquid; a very potent green odor; more stable than phenylacetaldehyde; not known to cause discoloration; sp.gr. (25/25°C) 1.000-1.004; refractive index (20°C) 1.493-1.496; soluble in 2 parts of 70% alcohol.

Uses: Substitute for phenylacetaldehyde, adding freshness to the lighter bouquets, and an attractive note to lilac, rose, muguet and geranium bouquets.

"Viscarin." [124] Trade name for a carrageenan extractive, a hydrocolloid which may be derived from a number of sea plants in the class of Rhodophyceae, order of Gigartinales, principally, Irish moss sea plants. Used in food, pharmaceutical and cosmetic industries as a viscosity producing agent, a stabilizer for emulsions, suspensions, foams; a binding agent, as in toothpaste, and for its emollient properties as in hand lotions.

"Viscasil." [245] Trademark for a series of polish grade silicone fluids. These fluids are a group of dimethyl silicone fluids characterized by small change of viscosity with temperature change, a high order of thermal and oxidative stability, excellent water repellency, and a low order of physiological activity. They also have a high order of chemical inertness and offer excellent release properties between many kinds of surfaces. Available in viscosities from 50 to 1,000 centistokes at 25°C (77°F). Low viscosities provide easier rubout whereas high viscosities yield superior gloss.

Uses: In wax and nonwax polishes for furniture and enameled metal surfaces. Are not compatible with waxes and act as lubricants, producing a polish which is more easily rubbed out and which gives a higher gloss.

viscose process. A process for making rayon by converting cellulose to the soluble xanthate, which can be spun into fibers and then reconverted to cellulose by treatment with acid. Wood pulp is steeped with 17-20% caustic soda; the resulting alkali cellulose is pressed to remove excess liquor and the soluble beta- and gamma-cellulose, and then shredded and aged. During this period, absorption of atmospheric oxygen degrades the cellulose polymer, producing shorter units and resulting in decreased viscosity at later stages of the process. When the material has aged sufficiently it is treated with carbon disulfide and sodium hydroxide to form an orange, viscous solution of cellulose xanthate. After filtration and deaeration, this solution (viscose) is forced through minute spinneret openings (or long slit dies in the case of cellophane film) into a bath containing sulfuric acid and various salts as sodium and zinc sulfate. The salts cause the viscose to gel immediately, forming a fiber of sufficient strength to permit it to be drawn through the bath under tension. At the same time the sulfuric acid decomposes the xanthate, converting the fibers to cellulose, in which form they are washed and dried.

High tenacity viscose rayon is made by modifications of the process which lead to greater crystallinity of the fiber and longer polymer units. These modifications include reduced aging of the alkali cellulose, additional carbon disulfide in the xanthate solution, a high concentration of salts in the spinning bath (which permits greater tension), and reduced acid concentration in the bath. The latter factor slows the regeneration process so that the fiber can be stretched 50-150% while in the bath and the molecules become highly oriented before they are completely converted to cellulose.

viscose rayon. Regenerated cellulose fibers made by the viscose process. Available in staple and filament forms.

Properties: Tensile strength (psi) 29,000-46,000 (regular), 46,000-58,000 (medium), and 58,000-88,000 (high tenacity); elongation 15-30%; sp.gr. 1.50-1.52; moisture regain 11-13% (70°F, 65% relative humidity); loses strength at 300°F; does not melt; burns readily. Similar to cotton in chemical resistance, dyeing, and resistance to insects and mildew.

Uses: All types of woven and knitted fabrics, alone or blended with cotton, wool, or other fibers. High tenacity rayon is used especially in tires.

viscosity. The resistance to flow exhibited by a liquid resulting from the combined effects of cohesion and adhesion. The units of measurement are the poise and the stoke. A liquid has a viscosity of one poise if a force of one dyne per square centimeter causes two parallel liquid surfaces one square centimeter in area and one centimeter apart to move past one another at a rate of one centimeter per second. One poise equals 100 centipoises (cp). A viscosity in centipoises divided by liquid density gives viscosity in centistokes (cs). One hundred centistokes equal 1 stoke. Viscosity is most frequently measured by noting the time required for a definite quantity of liquid to flow through a standard capillary. The time is directly proportional to stokes or centistokes. Water is the universal standard for viscosity and has a viscosity of almost exactly one centistoke (1.0038) at 20°C. Gasoline hydrocarbons such as hexane are less viscous (around 0.4 cs). Molasses may have a viscosity of several hundred centistokes, while for a very heavy lubrication oil the viscosity may be one thousand centistokes. An asphalt, although not a true liquid, may have its viscosity expressed as several hundred thousand centistokes. There are a great many crude and empirical methods for measuring viscosity which generally involve measurement of the time of flow or movement of a ball, ring or other object in a specially shaped or sized apparatus.

"Viscotone." [271] Trademark for a mineral by-product used for oil well drilling muds, tanning agents, ore dressing, and as a dispersant and emulsifier.

viscous flow (streamline flow; laminar flow). Flow of fluids usually at low velocities in which fluid elements flow in a straight line parallel to the axis of the conduit with little or no bulk mixing. Flow characterized by absence of turbidity.

"Viscrome." [206] Brand name for dyestuffs for wool-spun viscose unions.

"Visqueen." [80] Trademark for an unsupported, polyethylene film.

Uses: Packaging; moisture vapor barrier; waterproofing membrane; protective covering.

"Vistac." [230] Trademark for a series of synthetic hydrocarbon polymers, highly compatible with rubber, used in compounding rubber base adhesives and cements, and as softeners and modifiers for waxes and asphalts. Also used in paints and caulking compounds, and in latex and asphalt emulsions.

"Vistanex." [29] Trademark for polyisobutylene, a completely saturated polymer unvulcanizable by itself, but frequently added to impart special properties of vulcanized compounds to other elastomers and natural rubber. Also used in unvulcanized compositions, alone and in combinations. Outstanding properties include exceptional electrical properties, chemical inertness, high ozone resistance, low gas permeability, excellent aging and high degree of tackiness. Soluble in hydrocarbon solvents. Principal uses include cable insulation compounds, pressure sensitive tapes and adhesives, caulking and sealing compounds, additives to wax, asphalt, and polyethylene.

"Vistaril." [299] Trademark for hydroxyzine pamoate.

"Vitallium." [404] Trademark for a cobalt-chromium alloy used for cast full and partial dentures, and for surgical appliances, prostheses and instruments.

vitamins. Relatively complex organic compounds present in small and variable amounts in natural products, and essential in small amounts in the diet for life and growth. Earlier practice led to the use of letters of the alphabet in naming the vitamins; the recent trend is to use chemical names where possible. The vitamins listed in the following articles are those needed in human nutrition; others are being found essential for microorganisms and other forms of life. For individual vitamins see: para-aminobenzoic acid, ascorbic acid (vitamin C), biotin, choline, folic acid, inositol, nicotinamide, nicotinic acid, pantothenic acid, pyridoxine (vitamin B_6), riboflavine (vitamin B_2), thiamine (vitamin B_1), tocopherol (vitamin E), vitamin A, vitamin B_{12}, vitamin D, vitamin K, and vitamin P.

vitamin A Usually considered synonymous with vitamin A_1, but may also mean collectively vitamin A_1 and A_2. Both are alcohols, A_1 being $C_{20}H_{29}OH$ or 3,7-dimethyl-9-(2,6,6-trimethyl-1-cyclohexen-1-yl)-2,4,6,8-nonatetraen-1-ol, $(CH_3)_3C_6H_6CH$:-$CHC(CH_3)$:CHCH:$CHC(CH_3)$:$CHCH_2OH$. A_2 is $C_{20}H_{27}OH$, identical except for an additional double bond in the cyclohexyl ring.

Vitamin A active substances function in night vision through their conversion to retinene and rhodopsin (visual pigment); they also function in the control of cell differentiation, in normal bone growth and in carbohydrate metabolism. Vitamin A active substances are: the provitamin A's, carotene and cryptoxanthin (q.v.); natural vitamins A_1 and A_2; and synthetic vitamin A. Vitamin A_1 is the most widely distributed of these in nature.

Properties: Vitamin A_1 (retinol; axerophthol). Yellow prisms; m.p. 62-64°C; optically active; soluble in fats and most organic solvents; insoluble in water; unstable in air (stabilized with antioxidants). Vitamin A_2 has not been crystallized.

Source: Vitamin A is found in animal tissues as the free alcohol or in the ester form. The provitamins carotene and cryptoxanthin are found in plant tissue. Food source: Green and yellow vegetables, apricots, yellow peaches, butter, cream, milk, cheddar, eggs, liver, fish liver oils.

Commercial source: Concentrates of fish

liver oils and cereal grass; synthetic.
Derivation: Synthetic crystalline vitamin A_1 esters are synthesized from citral, acetone, formaldehyde, and acetylene in a patented process.
Units: One U.S.P. or IU unit is the specific biologic activity of 0.3 mcg vitamin A alcohol.
Containers: Bottles and drums.
Grade: U.S.P. XVI (as oleovitamin A).
Uses: Medicine; nutrition.

vitamin A acetate $C_{20}H_{29}OOCCH_3$. Synthetic vitamin A acetic acid ester.
Properties: Finely divided, dry, light yellow crystalline powder; nearly odorless and tasteless; insensitive to oxidation and humidity.
Containers: Bottles; drums.
Use: See vitamin A.

vitamin A aldehyde. See retinene.

vitamin A palmitate $C_{20}H_{29}OOCC_{15}H_{31}$. Synthetic vitamin A palmitic acid ester.
Properties: Yellow liquid; nearly odorless and tasteless.
Containers: Under nitrogen in aluminum bottles.
Use: See vitamin A.

vitamin A, water-miscible. Water-miscible vitamin A is oleovitamin A (q.v.) rendered miscible by harmless dispersing agents.
Properties: A pale yellow to yellow liquid; viscous; neutral or slightly acid to litmus in its 1 to 10 solution; miscible with 10 volumes of water or alcohol.
Grades: U.S.P. XVI.
Uses: Medicine and nutrition.

vitamin B_1. See thiamine.

vitamin B_2. See riboflavine.

vitamin B_4. Abandoned name for choline.

vitamin B_6. See pyridoxine.

vitamin B_{12} (cobalamin) $C_{63}H_{90}N_{14}O_{14}PCo$. The antipernicious anemia vitamin. All vitamin B_{12} compounds contain the cobalt atom in its trivalent state. There are now thought to be three active forms: vitamin B_{12a} (cyanocobalamin), containing a cyano radical attached to the cobalt; B_{12b} (hydroxocobalamin), in which the cyano is replaced by hydroxyl; B_{12c} (nitrocobalamin), in which the replacement is the nitro radical. The exact metabolic mechanism has not been defined, although vitamin B_{12} is known to have an influence on nucleic acid synthesis, fat metabolism; conversion of carbohydrate to fat, and metabolism of glycine, serine, methionine and choline.
Properties: Dark-red crystals (B_{12a}); nearly black crystals (B_{12b}); hygroscopic; soluble in water and alcohol; insoluble in acetone, chloroform, ether. Aqueous solutions deteriorate in the presence of acacia, aldehydes, ascorbic acid, ferrous gluconate, ferrous sulfate. Destroyed by alkalies and strong acids.
Source: Food source: Liver, eggs, milk, meats, and fish. Commercial source: Produced by microbial action on various nutrients (spent antibiotic liquors, sugar beet molasses, whey, etc.).
Containers: Bottles.
Grades: U.S.P. XVI (cyanocobalamin).
Uses: Medicine; nutrition; animal feed supplements.

vitamin B complex. Term for a group of B vitamins, often found associated in foods. The group is not clear-cut, but has been thought to include: thiamine, riboflavine, nicotinic acid, pantothenic acid, biotin, pyridoxine, folic acid, vitamin B_{12}. See individual entries.

vitamin C. See ascorbic acid.

vitamin D. The anti-rachitic vitamin. It aids in the utilization of calcium and phosphorus, and is essential to the development and maintenance of strong bones and teeth. It prevents rickets in children and osteomalacia in adults. Four crystalline forms have been isolated: (1) vitamin D_2 ($C_{28}H_{44}O$), irradiated ergosterol (natural and synthetic), also called calciferol or viosterol; (2) vitamin D_3 ($C_{27}H_{44}O$), irradiated 7-dehydrocholesterol (natural and synthetic) (see the activated form); (3) vitamin D_4 ($C_{28}H_{46}O$), irradiated 22-dihydroergosterol (synthetic); (4) vitamin D_5 ($C_{29}H_{48}O$), irradiated 7-dihydrositosterol (synthetic). Vitamin D_2 and vitamin D_3 are the common forms.
Properties: The D vitamins are white, odorless crystals; soluble in fats and organic solvents; insoluble in water; stable to heat and aeration; have characteristic ring structure, but differ in side chain structure.
Vitamin D_2: Melting point 121°C (highly purified), 116°C (commercial); crystallizes from methanol in clusters of needles, from acetone in long prisms.
Vitamin D_3: M.p. 84-85°C (see 7-dehydrocholesterol, activated).
Vitamin D_4: M.p. 96-98°C; crystallized from acetone in long needles.
Sources: Food source: Milk, fish, eggs. Commercial source: Fish liver oils and irradiation of provitamins.
Units: One U.S.P. or IU unit is the vitamin D activity of one milligram International Standard solution of irradiated ergosterol found equal to 0.025 micrograms of crystalline vitamin D_2.
Grade: U.S.P. XVI (calciferol).
Uses: Medicine and nutrition.
See also ergosterol.

vitamin D_3. See 7-dehydrocholesterol, activated.

vitamin E. See tocopherols.

vitamin H. See biotin.

vitamin K. Vitamin K active substances aid in the clotting of blood and the prevention of hemorrhage through maintaining the prothrombin content of blood. The four most important forms are: (1) vitamin K_1 or phytonadione (2-methyl-3-phytyl-1,4-naphthoquinone) $C_{31}H_{46}O_2$ (see phytonadione); (2) vitamin K_2 (2-methyl-3-difarnesyl-1,4-naphthoquinone) $C_{41}H_{56}O_2$; (3) vitamin K_3

or menadione (q.v.); (4) phthiocol (q.v.).
Properties: K vitamins are fat soluble; fairly stable to heat; unstable to alkali and light.
Vitamin K_2: Yellow, crystalline solid; m.p. 53-54°C.
Sources: Food source: Green leafy vegetables, tomatoes, vegetable oils. Commercial source: Preparations of dried cereal grasses; synthetic 3-alkyl and 3-alkenyl derivatives of 2-methyl-1,4-naphthoquinone.
Grade: U.S.P. XVI (phytonadione).
Units: A unit of vitamin K is the minimum amount which will render the blood clotting of a K-depleted chick normal within six hours. One unit is equivalent to one microgram of menadione.
Uses: Medicine; nutrition.

vitamin P. A term no longer accepted that has been used to designate a group of plant pigments and related compounds with similar biological activities. See bioflavonoids.

vitamin P complex. See bioflavonoids.

vitamin PP. See nicotinamide.

"VitaStain." [109] Trademark for (2,3,5-triphenyltetrazolium chloride) $C_{19}H_{15}N_4Cl$.
Properties: White to pale yellow crystalline powder; m.p. 245°C (decomposes).
Uses: Reagent for the determination of the germinability of seeds, for staining the cambium layer of living twigs, for staining yeasts, bacteria and a wide variety of other living tissue.

"Vitel." [265] Trademark for polyester resins used in making fiber and film. Fibers are useful in various textile fabrics, particularly for clothing.

"Viterra." [299] Trademark for a vitamin supplement preparation.

"Viterra Therapeutic." [299] Trademark for a high potency vitamin supplement preparation.

"Vitex." [309] Trademark for fat-soluble vitamin concentrates.

"Viton." [28] Trademark for several fluoroelastomers.
Properties: White translucent solids; sp.gr. 1.81-1.86. "Viton" A is standard type, A-HV has high viscosity, B has superior heat and chemical resistance.
Containers: 50-lb drums.
Uses: For rubber products that must withstand unusually high temperatures up to 600°F in air, chemicals or most liquids.

"Vitrafix." [206] Brand name for proprietary bonding assistant for glass fiber laminates.

"Vitrafos." [172] Trademark for a clear, glassy, granular sodium phosphate.
Properties: Phosphorus pentoxide, 63% min; sieve size, 10- to 80-mesh screen; pH (1% soln), 7.8; bulk density 79 lbs/cu ft.
Containers: Moisture-proof bags, 100 lbs net.
Uses: Builder in household and industrial cleaning compounds; deflocculating agent in oil well drilling muds; a sequestering agent in the textile industry.

vitrain. One of the types of physical structure found in coal (see also clarain, durain, and fusain). Vitrain has a bright, glassy appearance, a conchoidal fracture and is usually free from striations. It is associated with good coking qualities. See also anthraxylon.

"Vitra-Tint." [194] Trade name used to cover an organic spray coat system of transparent and opaque color coatings for glass.

"Vitreosil." [220] Trademark for heat and acid-proof utensils fashioned from pure fused silica. Used in laboratory ware.

vitreous. Glass-like in luster, color, brittleness, or composition.

vitreous antimony. See antimony glass.

"Vitrex." [41] Trade name for an acid-proof silicate cement which sets by chemical action. Inert to acids, except hydrofluoric, at temperatures up to 1600°F.

"Vitric 10." [326] Trade name for a silicate base acid proof, chemical setting cement suitable for use with strong oxidizing acids such as nitric and chromic.

vitrification. The process of converting into glass or a glassy substance by heat and fusion.

vitriol, blue. See copper sulfate; also chalcanthite.

vitriol, green. See ferrous sulfate.

vitriol, iron. See ferrous sulfate.

vitriol, lead. See lead sulfate.

vitriol, oil of. See sulfuric acid.

vitriol, white. See zinc sulfate.

vitriol, zinc. See zinc sulfate, also goslarite.

"Vitrobond." [41] Trade name for a silica-filled, sulfur-based compound for use as a hot-pour cement for jointing acid brick, used where temperatures do not exceed 200°F.

"Vitroplast." [41] Trade name for a silica-filled, synthetic-resin, acid-proof cement of the vinyl type which sets in contact with concrete, metal, etc. Good up to 275°F. Excellent resistance to oxidizing materials such as chlorine dioxide.

"Vivana." [233] Trademark for fibers, filaments and yarn of polymers and interpolymers of vinylidene chloride.

vivianite $Fe_3(PO_4)_2 \cdot 8H_2O$. A natural hydrated ferrous phosphate. Colorless when unaltered but gradually changing to blue or bluish-green on exposure; colorless to bluish-white streak changing to indigo blue and liver brown; vitreous or pearly luster. Contains 43.0% FeO, 28.3% P_2O_5. Sp.gr. 2.58-2.69; hardness 1.5-2.
Occurrence: United States (New Jersey, Virginia, Colorado, Kentucky); Canada;

England; Germany; Japan; Russia; Austria; Guatemala; Bolivia; Australia; Greenland.

"Viz-Thin." [48] Trademark for a ferro-chrome lignin sulfonate, used as a thinner for gypsum base oil well drilling muds.

V.M.&P. naphtha. See naphtha, V.M.&P.

"Volan." [28] Trademark for a Werner type chromium complex (methacrylato chromic chloride) in isopropanol.

Properties: Dark green liquid with an alcoholic odor; miscible with water; sp.gr. 1.026; b.p. 180.3°F.

Containers: 6½-gal carboys; 55-gal drums.

Uses: As a bonding agent applied to glass fibers, it is used in reinforced plastic laminates to improve the adhesion between glass and resin, especially under moist conditions; also to improve adhesive bond between other hydrophilic surfaces such as paper and wood and polymeric coatings or impregnants.

volatility. In general, the tendency of a solid or liquid material to pass into the vapor state at ordinary temperature. Specifically the vapor pressure of a component divided by its mole fraction in the liquid or solid.

volcanic ash (volcanic tuff). A material made up of either loose or solidified small fragments and dust of lava, commonly glassy in character, blown from a volcano and deposited either on land or under water. If numerous large fragments are embedded in the ash, the deposit is called volcanic agglomerate or volcanic breccia.

Occurrence: Colorado, Montana, Nevada, Oklahoma and South Dakota.

Uses: Abrasive, building stone and in the manufacture of fire-brick, hydraulic mortar and concrete.

volcanic glass. See obsidian.

volcanic tuff. See volcanic ash.

Volhard's solution. A solution of potassium thiocyanate used in analytical chemistry.

"Voltex." [110] Brand name for medium color channel black for use in paints and plastics. Also classified as a conductive channel black - CC.

vomit nut. See nux vomica.

"Voranate." [233] Trademark for a series of urethane intermediates which are the reaction products of polyols and isocyanates. They are adducts or quasi-prepolymers to be used in combination with "Voranol" products to obtain rigid urethane foams.

"Vorane." [233] Trademark for a group of polyurethane chemicals, raw materials for polyurethane elastomers, coatings and foams.

"Voranol." [233] Trademark for a series of urethane intermediates which are polypropylene diols, polyether triols of alkylene oxides, or polyether glycols. Used as crosslinkers in conjunction with "Voranate" adducts to obtain rigid urethane foams, or to make prepolymers for elastomer or flexible foams.

"Voraspan." [233] Trademark for urethane resins.

Vorce cell. A diaphragm-type electrolytic cell (see diaphragm cell) for the production of chlorine and caustic soda. The Vorce cell is the most widely used cylindrical cell in the U.S. It consists of a steel pot in which the hydrogen and caustic are collected. Within the pot is mounted a vertical cylindrical cathode of wire mesh, on the inside of which a diaphragm of asbestos paper is clamped. Carbon anodes are arranged in a ring within the cathode. In some designs, a second, smaller cylindrical cathode and diaphragm are placed at the center of the ring of anodes. Brine is fed to the anode chamber, where chlorine is released, and trickles through the diaphragm to the cathode, where caustic forms with the evolution of hydrogen. The Vorce cell produces about 7 lbs of chlorine per day per square foot of floor space.

"Vorlex." [401] Trade name for a soil fumigant containing methylisothiocyanate and chlorinated C_3 hydrocarbons, including 1,3-dichloropropene.

Containers: 6 and 30 gallon drums.

Use: For injection or drench treatment of soil to control fungi, weeds, nematodes and soil insects.

Warning: Hazardous vapor and liquid. Irritating to eyes, nose, throat and skin. Do not inhale or swallow. Avoid skin contact.

"VPM." [28] Trade name for a soil fumigant solution containing sodium methyldithiocarbamate. Used for control of certain weeds, soilborne fungus disease, nematodes and garden centipedes in plant propagation beds, crop-land and in the preparation of new turf areas.

Containers: 1-gal bottles; 5-gal cans; 30-gal drums.

"Vuepak." [58] Trademark for cellulose acetate.

Properties: Crystal-clear rigid film available in sheets and rolls.

Uses: Protective film; "see through" pack.

"Vulca." [53] Trademark for a series of ether derivatives of ungelatinized starch. They are graded according to their resistance to swelling in boiling water. "Vulca" 100 is completely non-swelling, resistant to agents that gelatinize starch, retains stability during autoclaving and pressure cooking, and may be steam sterilized without any appreciable change in its powdery appearance.

Uses: Thickeners; cosmetics; dusting powders.

"Vulcabond." [206] Brand name for a proprietary line of adhesives based on rubber and for bonding to rubber, metals, cotton and textiles.

"Vulcacel." [206] Brand name for proprietary blowing agent for cellular rubber.

"Vulcaflex." [206] Brand name of line of proprietary anti-oxidant and anti-flex cracking products for use in rubber compounding.

"Vulcafor." [206] Brand name of line of proprietary vulcanization accelerators for use in rubber.

"Vulcamel." [206] Brand name for proprietary peptizing agents for natural and synthetic rubber.

"Vulcan." [307] Trademark for a series of organic pigments used for the coloring of rubber. Characterized by their fastness to vulcanizing.

"Vulcan." [275] Trade name for a series of oil furnace carbon blacks. Available as:
"Vulcan C." Conductive furnace black (CF).
"Vulcan SC." Super-conductive furnace black (SCF).
"Vulcan XC-72." Extra-conductive furnace black (ECF).

vulcanization. Process of combining rubber with sulfur or certain other additives under the influence of heat and pressure to eliminate tackiness when warm and brittleness when cool, and to otherwise improve the useful properties of rubber such as strength, elasticity and abrasion resistance.

vulcanization accelerators. See accelerator.

"Vulcastab." [206] Brand name for a proprietary line of surface active agents for use as auxiliaries in natural and synthetic latex.

"Vulcatard." [206] Brand name for proprietary retarders for natural and synthetic rubbers.

"Vulcatex." [205] Plastic caulking compound having unusual adhesion and elasticity between extremely high and low temperatures.

"Vulcoid." [281] Trademark for a resin-impregnated vulcanized fiber. It is a light-weight electrical insulator of great mechanical adaptability which has arc and moisture resistance and can be drawn and formed. Approved by Underwriters' Laboratories as Class A Insulation.
Forms: Sheets; tubes; rods; fabricated parts.
Uses: Contact and instrument panels; knife switch guides; arc deflector spacer bushings; electrical insulation; motor and transformer lead bushings and terminal blocks.

"Vulcosal." [233] Trademark for the industrial grade of salicyclic acid; used as a stabilizer.

"Vulklor." [248] Trade name for tetrachloro-para-benzoquinone.
Properties: Golden yellow powder; sp.gr. 1.97; m.p. 290°C; good storage stability; slightly soluble in acetone and benzene. Insoluble in water, gasoline, and ethylene dichloride.
Use: Rubber vulcanizing agent.

"Vultac." [204] Trademark for alkylphenol disulfides used as vulcanizing agents as total or partial replacement for sulfur; processing aid for improving pigment dispersion; resinous type softener; tack increaser for butadiene rubber.

"Vycor." [20] Trademark for heat and chemical resistant glassware of various compositions and physical properties, all characterized by extremely low coefficients of expansion, and accessories used therewith.

"Vycor Brand Glass No. 7900." [20] Trademark for a glass comprising approximately 96% silica made by a unique process in which an article fabricated by conventional methods is leached in hot chemical solutions to remove substantially all of the ingredients except silica. The silica residue, after being washed and dried, is fired at high temperatures, becoming a transparent, vitreous glass of simple chemical composition, exceptional chemical stability, high softening point and extremely low expansion coefficient.
Physical properties: Softening point approximately 1500°C. Linear coefficient of expansion per °C 0.0000008; sp.gr. 2.18; refractive index 1.458; loss in weight on heating and cooling, negligible; visible light transmission for 2 mm thickness, 92%; ultraviolet transmission at 254 millimicron line for 2 mm thickness, 2 to 4% (not controlled). (A similar glass, No. 7910, made under controlled conditions, will transmit over 60% radiation at 254 millimicrons in 2 mm section.) Temperature limit in service, 900°C.
Chemical durability: This glass is very stable and resists chemical attack as shown by the following:
Loss in weight, 100 lbs steam pressure, 96 hours, 0.0001 grams per sq cm.
Loss in weight, boiling 6 hours in 5% sodium hydroxide, 0.0015 grams per sq in.
Loss in weight, 5% hydrochloric acid for 72 hours at 80°C, negligible.
Durability at elevated temperatures:
No. 7900 glass is attacked at elevated temperatures by basic material, the rate of attack increasing as the temperature increases; therefore, this glass can not be recommended for alkaline fusions, or for ashing material which gives a basic ash, unless the condition of the glassware and its tare weight are unimportant after ashing. Some metals at high temperatures produce an attack on, or cause devitrification of No. 7900 glass. Accelerated tests at 1150°C for two hours in vacuum show that the following metals affect the glass: magnesium, aluminum, manganese, zinc, iron, thallium, cobalt, vanadium, and tungsten. These are listed in order of decreasing attack.
Strongly reducing gases tend to accelerate devitrification at high temperatures.
Uses: The manufacture of laboratory and industrial glassware, including beakers, crucibles, flasks, dishes, tubes, cylinders, containers, flat glass rods.

"Vycron." [287] Trademark for a polyester fiber spun from "Vitel." "Vycron" is a tough durable fiber wet or dry; produced in 1.5, 3 and higher denier for apparel and other uses. Also available as direct spun yarn and as tow.

"Vydax." [28] Trademark for fluorocarbon telomers.

Properties: Chemically and thermally stable; insoluble; nonflammable.

Uses: Dry lubricants and release agents.

"Vygen Resins." [179] Trademark for a complete series of polyvinyl chloride polymers. "Vygen" PVC resins are thermoplastic and used in calendering, extrusion and molding applications. They are characterized by good heat and light stability, toughness, chemical resistance, flame resistance, and electrical insulation properties.

Uses: Film, sheeting, coated fabrics, luggage, wall covering, flooring, extruded goods, gaskets, wire insulation, toys, and automotive parts.

W. Symbol for tungsten.

"W-545." [308] Trademark for an austenitic iron-base alloy containing nickel, chromium, and relatively small proportions of molybdenum, titanium, boron, silicon, and manganese. This alloy is precipitation-hardening and was developed primarily to meet the need for improved gas turbine discs, one of the most critical components of jet engines. It has exceptionally high creep strength combined with good ductility, resistance to notch sensitivity, and excellent oxidation resistance in the temperature range of 1000°F to 1350°F — the range in which gas turbine discs operate.

wad (bog manganese). Mixture of manganese oxides, often with oxides of other metals. Dark brown or black in color; often soft and loose, but sometimes hard and compact. Luster, dull. Contains 10-20% of water. Sp.gr. 3-4.26; hardness 0.5-6.
Occurrence: United States (large deposits in Montana, also found in sixteen other states); Canada (New Brunswick); Norway.
Uses: Manufacture of chlorine; paint.

wad clay. A low-grade fire-clay used in grouting the joints between saggers when they are set up in bungs in the kilns.

wahoo. See euonymus.

"Wallerstat." [173] Trademark for an antistatic, antilint agent for the dry cleaner. It eliminates static and lint and permits the dry cleaner to run mixed loads. It works in the wheel with regular brand and amount of soap.

"Wallkyd." [36] A group of alkyd resins used in the manufacture of interior flat paints where good soil-removal and odorless characteristics are desirable.
Physical properties: Non-volatile 34-41%; viscosity (Gardner-Holdt) Z-Z^3; color (Gardner 1933) 4-8; acid number 2-5.

"Wallpol." [36] Colloidal dispersions of solid polyvinyl acetate resin particles in water. Generally used for the manufacture of interior and exterior paints and primer sealers.
Physical properties: Non-volatile 54-56%; pH 3.5-6.8; color, white opaque.

walnut oil.
Properties: The cold-pressed oil is a colorless or pale yellowish-green liquid; pleasant odor; agreeable, nutty taste. The hot-pressed oil has a greenish tint and an acrid taste and odor. Sp.gr. (15°C) 0.925-0.927; saponification value 188-196; iodine value 143-148; refractive index (15°) 1.4808. Soluble in alcohol, ether, chloroform, and carbon disulfide.
Derivation: By expressing the kernels of Juglans regia, Persian or (commonly) English walnut.
Method of purification: Filtration.
Grade: Technical.
Containers: Wooden barrels.
Uses: Cold-pressed oil: edible products. Hot-pressed oil: paints, artists' colors, soap.
Shipping regulations: None.*

Warburg's yellow enzyme (old yellow enzyme; yellow oxidation enzyme). A yellow, oxidation enzyme found in yeast originally by Warburg and Christian. It is composed of a protein united to riboflavine through phosphoric acid. It acts as a dehydrogenase. The above workers have prepared the enzyme synthetically.

"Warcofix" Z. [42] Proprietary product.
Properties: White powder; disperses in water above 20°C.
Containers: 44-gal fiber container.
Use: Color fixative for application in textile and in leather dyeing operations as an aftertreatment to prevent dye bleed.

"Warco" GFI. [42] Proprietary product. High molecular weight amine compound.
Properties: Dark brown oily liquid; soluble in water above 20°C.
Containers: 55-gal open head steel drums.
Use: In textile dyeing of acetate fabrics as a "gas fading" inhibitor.

"Warcolene" D. [42] Proprietary product. A non-ionic product.
Properties: Creamy white emulsion; disperses readily in water at 25°C.
Containers: 55-gal steel drums.
Use: Textile lubricant specifically designed for difficult sewing problems.

"Warcolene" 362-M. [42] Proprietary product. A non-ionic surfactant.
Properties: White paste; disperses readily in water at 60°C.
Containers: 55-gal open head steel drums.
Use: Cotton textile softener with excellent wetting and rewetting properties.

"Warco" Silk Soaking Oils. [42] Proprietary products. Comprised of blends of sulfated vegetable oils and mineral oils, alkyl amine condensates, high molecular weight amines.
Properties: Dark amber liquids; dispersible

*See "I.C.C. Shipping Regulations," page xiii.
Reference numbers refer to name of manufacturer. See "List of Manufacturers," page v.

in water at 30°C.
Containers: 55-gal steel drums.
Uses: Textile products for silk soaking, throwing and finishing.

"Warcosol." [42] Proprietary products. Alkyl naphthalene sodium sulfonates.
Properties: Light yellow liquids and tan colored pastes. Disperse readily in water from 25 to 60°C.
Containers: 55-gal steel drums.
Uses: Wetting agents for wet textile processes in scouring and dyeing.

"Warcosol" 23-T. [42] Proprietary product. Sulfonated ester.
Properties: Colorless viscous liquid; soluble in water above 25°C.
Containers: 55-gal open head steel drums.
Use: Textile wet finishing agent. Extremely efficient as a rewetting agent for fabrics to be stored and then rewet. Used in dyehouse and in sanforizing operations.

ware clay. A term sometimes used synonymously with ball clay.

warfarin (3-(alpha-acetonylbenzyl)-4-hydroxycoumarin) $C_{19}H_{16}O_4$.
Properties: Colorless, odorless, tasteless crystals; m.p. 161°C. Soluble in alcohol and ether; insoluble in water.
Derivation: Condensation of benzylideneacetone and 4-hydroxycoumarin.
Method of purification: Recrystallizing from alcohol.
Grades: Technical.
Containers: Drums.
Use: Rodenticide.
Warning! Poison. Keep away from humans, domestic animals and pets (U.S.D.A. pesticides regulations; similar MCA warning label.)

warfarin sodium $C_{19}H_{15}NaO_4$. 3-(alpha-Acetonylbenzyl)-4-hydroxy-coumarin sodium.
Properties: Colorless, odorless, tasteless solid. Soluble in alcohol, acetone, and water; insoluble in chloroform, ether; pH of 1% solution 7.2-8.3.
Derivation: Neutralization of warfarin with aqueous sodium hydroxide.
Grades: U.S.P. XVI; clinical.
Forms: Tablets; ampules.
Use: Medicine.

washed clays. Purified clays, with low silica and grit. They result from mixing raw clay with water and allowing sedimentation to cause separation of the impurities from the clay.

washed sulfur. See sulfur lotum.

washing blue. A name applied loosely to any of a number of the varieties of iron blue pigments. See iron blues.

washing soda. See sal soda.

wash oil. An absorption oil.

Wassermann test. A diagnostic test used for the detection of syphilis by subjecting the blood sample under investigation to action of a specially prepared hemolytic serum.

waste disposal. See Zimmermann process.

watch oil. See porpoise oil.

"Watchung." [28] Trademark for a line of red and maroon pigments possessing excellent heat and grease resistance.
Uses: Principally in printing ink and plastics.

water (hydrogen oxide) H_2O.
Properties: Clear, colorless liquid which is practically tasteless and odorless. Is the standard of reference for many physical properties (centigrade temperature scale, density, calorie, etc.). Sp.gr. (4°C) 1.0000; m.p. 0°C (32°F); b.p. 100°C (212°F); vapor pressure (100°) 760 mm. Weight per gallon (15°) 8.337 lbs; wt. per cu ft 62.3 lbs. Water is the most common solvent.
Purification: Distillation; ion exchange systems (see zeolites).
For chemical use, water is usually classified as purified, soft or hard. Purified water of U.S.P. standards is either distilled or de-ionized and is neutral and free from contaminants according to a set of prescribed tests of ordinary precision. Hard water contains dissolved salts of calcium and magnesium. Degree of hardness is often expressed in terms of grains per gallon of calcium carbonate. (1 grain per gallon $CaCO_3$ is equivalent to 17.1 parts per million.) Water containing up to about 5 grains of hardness is considered soft; over 30 grains is very hard. For water of other than the common isotopic hydrogen or oxygen, see heavy water.

water-base paint. See water paint.

water demineralizers. See zeolites.

water elder. See viburnum opulus.

water gas (blue gas, blue water gas). A gas made by decomposing steam by passing it over a bed of incandescent coke, or coke and coal. In the operation of the water-gas generator, after the bed of incandescent coke is partly quenched by the "make" or "steam blow" it is revived by an "air blow." A water gas may also be made by high temperature reaction of steam with natural gas or similar hydrocarbons. Synthesis gas (see under Uses) may be made by either process, but its composition may vary considerably from the example below.
Typical composition:

Illuminants	0.0%
Carbon monoxide	40.9
Hydrogen	50.8
Methane	0.2
Carbon dioxide	3.4
Oxygen	0.9
Nitrogen	3.5
Btu per cubic foot, 299.0	

Since this gas is low in heating value and burns with a nonluminous flame, it is enriched for ordinary city gas purposes with oil gas and is then known as carburetted water gas (q.v.).
Uses: Welding; heat treatment; melting of

*See "I.C.C. Shipping Regulations," page xiii.
Reference numbers refer to name of manufacturer. See "List of Manufacturers," page v.

W

metals; glass making; other industrial uses in which its high flame temperature (up to 3000°F) is of advantage. As synthesis gas, also used as source material in synthesis, as for ammonia, methanol and in the Oxo and Fischer-Tropsch processes for hydrocarbons, oxygenated organic compounds, and synthetic fuels.

water glass. See sodium silicate.

water of crystallization. Water chemically combined in many crystallized substances which may be removed at 100°C, usually with loss of crystalline properties.

water paint (water-base paint). A paint that is a water emulsion or water dispersion. Water paints formerly included whitewash, cement suspensions, and calcimine, and were limited in use. They are now almost entirely emulsion paints (q.v.), having latex, casein or synthetic resins as binders, and are used in large volume. Gel paints (q.v.) are a fast-developing type of water paint.

"Waterplug." [195] Trade name for a quickset hydraulic cement used to stop water running or seeping through masonry walls.

waterproof cement. Hydraulic cement (Portland cement) that has been given a surface coating, or else made with the use of substances or a manufacturing process that results in greatly reduced moisture absorption and transmission. Common surface treatments are alternate treatments with soap and alum to produce insoluble aluminum soap in the pores. Water glass coatings function similarly, as do aluminum, magnesium, and zinc fluosilicates. Paints can be used only after a pre-treatment as, for example, with zinc sulfate to convert calcium hydroxide of the cement to materials that do not affect the paint film. Casein and formaldehyde are sometimes used to protect the paint from the calcium hydroxide. Paraffin and tar are frequently used to waterproof cement. In addition to surface coatings, cement is made more or less waterproof by properly incorporating hydrated lime or soap in the mixture.

waterproofing salts. See aluminum acetate. This is a specific synonym. There are many other salts which act as water-proofing agents.

water softeners. Substances, which, when added to water, lower its degree of hardness (see water). They include a variety of compounds such as sal soda, trisodium phosphate, sodium metaphosphate, sodium tetraphosphate, and the zeolites. All soften water by either removing calcium and magnesium ions from the water or by sequestering these ions into a form which does not exhibit their usual properties.

water-soluble gums. See gums.

water-soluble oils. (See also soluble oils.) Ammonia, potash, or sodium soaps of oleic, sulfo-fatty, rosin, or naphthenic acids dissolved in mineral oils. Sometimes the rosin oils are blown beforehand (see blown oils) and sometimes ammoniacal liquor, naphtha or alcohol is added. The final products form permanent emulsions or almost clear solutions with water.

Uses: Boring; lathe-cutting; milling; polishing lubricants; dressing textile fibers; dust laying.

wattle bark (Australian bark, mimosa bark).

Derivation: From the Australian wattles, Acacia pycnantha, Acacia mollissina and Acacia binervata, and other native Australian and South African acacias. Bark contains 25-35% tannin.

Grades: Based on tannin content.

Use: Source of wattle bark extract, used in tanning industry.

wattle gum (Australian black wattle gum). A variety of arabic gum.

Properties: Dark reddish, hard, glossy tears or lumps, strong astringent taste, contains tannin and more galactan, less araban than ordinary arabic gum.

Uses: Adhesives; polishes; printing textiles; paper size; inks.

wavellite $Al_3(OH)_3(PO_4)_2 \cdot 5H_2O$. A natural hydrated basic aluminum phosphate.

Properties: Color white, yellow, green; luster vitreous; hardness 3.5-4; sp.gr. 2.33; usually occurs in radiating aggregates.

Occurrence: Pennsylvania, Arkansas; Europe; Brazil.

Use: Has been used as a source of phosphorus.

wax-berry. See myrica.

wax distillate. Lubricating oil distillate from petroleum, which is dewaxed, earth treated, and filtered to produce neutral lubricating oils and wax.

waxes. Unctuous, fusible, variably viscous to solid substances, having a characteristic waxy luster, which are insoluble in water but soluble in most organic solvents. They are extremely susceptible to changes in temperature and their origin, composition and color are variable. They are usually composed of high molecular weight substances, and may be grouped according to their origin as follows: (a) Animal: spermaceti, beeswax, stearic acid, Chinese wax, etc. (b) Vegetable: carnauba, Japan, bayberry, candelilla, etc. The animal and vegetable waxes are mostly fatty acid esters of higher monohydroxy alcohols, as for example, ceryl cerotate (q.v.). (c) Mineral: ozocerite, montan, ceresin, paraffin, etc. (see also petroleum waxes). These mineral waxes are usually high molecular weight esters or hydrocarbons. (d) Synthetic (varied chemically), as the medium weight polyethylenes, polyethylene glycols and polyoxyethylene esters, chloronaphthalenes, sorbitols, chlorotrifluoroethylene resins.

See also specific waxes mentioned above.

waxes, microcrystalline. Waxes derived from petroleum and characterized by the fineness of their crystals in distinction to the larger crystals of paraffin wax. They consist of saturated aliphatic hydrocarbons such as $C_{48}H_{98}$.

Properties: White, amber or black solids; usually odorless, tasteless and chemically inert.

Typical specifications: Sp.gr. (60°F) 0.92-0.94; m.p. 190-195°F; penetration (ASTM D5-25) 2/7; saponification no. (max) 2. These waxes are often emulsifiable, but are impermeable to water vapor. They have useful electrical, heat-sealing and adhesive properties, and are coming into increasing use in the processing and packaging industries.

Uses: Paper and packaging; wire insulation compounds; printing inks; protective coatings and polishes; ski wax; leather treatment; rust preventives; cosmetics; waterproofing; lubricant manufacture; phonograph records.

See also oxidized microcrystalline waxes.

"Waxine" Sizes. [57] Trademark for acid-stable, aqueous emulsions of various petroleum waxes, combined with rosin. Rosin present entirely in free state (not saponified).

Use: Used widely in paper industry because they impart the desirable features of wax without excessive slipperiness in sheet.

See also "Alwax" sizes.

waxmyrtle. See myrica.

"Waxol." [206] Brand name for wax and tallow emulsions for finishing textiles.

"Waxoline." [206] Brand name of line of proprietary dyestuffs soluble in oils and waxes. "Waxoline" lacquers are dyestuffs for clear nitrocellulose lacquers.

wax tailings. A brown, sticky, semi-asphalt product obtained in the destructive distillation of petroleum tar just prior to the formation of the coke.

Uses: Wood preservative and in the manufacture of roofing paper.

websterite. See aluminite.

"Weedez." [147] Trademark for weed killer based on 2,4-D, in the form of impregnated wax bar and stick.

Containers: 3-lb bars (12 in case); sticks (12 in case).

Uses: To control broadleaved weeds in turf.

weed-killers. See herbicides.

weight. The force with which a body is drawn toward the earth by gravity.

weight, atomic. See atomic weight.

weighting agent. In soft drink technology, a weighting agent is an oil or oil-soluble compound of high specific gravity, such as a brominated olive oil, which is added to the citrus flavoring oils to raise the specific gravity of the mixed oils to about unity so that stable emulsions with water can be made for use in flavoring soft drinks.

"Weldal." [142] Trademark for chromate-containing composition designed for preparing aluminum for resistance welding. It is supplied as a yellow powder that is dissolved in water.

welding. "A metal-joining process by which coalescence is produced by heating to suitable temperatures with or without the application of pressure, and with or without the use of filler metal." (American Welding Society.) Brazing means welding above 800°C with a non-ferrous filler metal; soldering implies use of a lower melting filler (solder) with the joint depending on adhesion rather than alloying.

Most industrial welding involves the fusion welding of two molten edges or surfaces. The three principal heat sources are electric arc (most frequent), electric resistance, and flame.

Electric arc welding often involves the use of consumable electrodes of the same composition as the work, but nonconsumable electrodes of tungsten or carbon are also used. Covered or cored electrodes are designed with the electrode metal surrounded by or inserted through a casing which feeds flux, inert gases, or other additives to the working area.

Resistance welding is used on sheet metal assemblies, such as automobile bodies. The coalescence of the parts occurs from the alternate application to the work of pressure and electric current, which heats the work by resistance. Spot, seam and flash welding are common techniques.

Gas welding produces a relatively low temperature, useful for brazing, welding thin material, and avoiding cracks (as with cast iron). The most common fuel is an equal-volume mixture of oxygen and acetylene.

Forge welding involves pressure applied after furnace-heating. Aluminum can be cold welded by pressure alone. Ultrasonic welding is a new technique needing no heat and only light pressure.

"Weldopax." [337] ("Ruflux C"). Trade name for zirconium oxide used in weld rod coatings to increase slag viscosity.

"Weltone." [108] Trademark for a powdered composition of sodium hexametaphosphate, alkaline salts, a disinfecting agent and a wetting agent.

Containers: 15-lb cans; 40- and 100-lb drums.

Uses: For cleaning and developing water wells. Removes clay, silt, drilling mud and other mineral deposits from water wells.

Werner complex. See coordination compound.

wernerite (scapolite) $(Na,Ca)_4Al_3(Al,Si)_3Si_6O_{24}(Cl,CO_3,SO_4)$. A complex silicate mineral of variable formula. Marialite is a sodium-rich member of the series, and meionite is the calcium-rich member.

Properties: Color white, gray, greenish, bluish, reddish; luster vitreous; hardness 5-6; sp.gr. 2.65-2.74.

Occurrence: Massachusetts, New York, Canada, Madagascar.
Use: Gem stone.

"Wescodyne." [284] Trademark for a cleaner-disinfectant containing nonionic-iodine complexes. Claimed to be non-toxic, non-irritating, non-staining, when used as directed. Germicidally effective against bacteria, viruses, M. Tuberculosis, spores, etc.
Use: In hospitals.

"W.E.S. Oil." [175] Brand name for a semi-refined coal tar distillate.
Properties: Straw to dark amber colored liquid; distils 10% to not below 165°C; 70% to not above 190°C and 95% to not above 235°C; sp.gr. (15.5/15.5°C) 0.930-0.950; approx wt/gal 7.75 lb.
Use: A high-boiling, slow-drying solvent designed for use in wire enamels and metal coatings.

"Wes-X." [308] Trademark for proprietary addition agents and plating salts used in the deposition of bright and leveled deposits of tin, copper and nickel.

"Wetanol." [73] Trademark. Modified sodium lauryl sulfate.
Properties: White, practically odorless powder, pH of 1% aqueous dispersion at 25°C, 5. Soluble in water, insoluble in alcohol, hydrocarbons, mineral oil, vegetable oils.
Containers: 400-lb drums; 50-, 8-lb cont.
Uses: Wetting, penetrating, scouring agent.

"Wet-Ege Spirits." [200] Trademark for a petroleum solvent prepared by straight-run distillation.
Properties: Water-white; initial boiling point 320-328°F, 95% distills between 380 and 388°F; sp.gr. 0.784 (60°F); flash point (TCC) 118°F; mild, nonresidual odor, holds a wet edge longer than mineral spirits.
Uses: In brushing enamels, fly sprays, polishes, for cleaning garments, degreasing hides, washing metal before painting, and as a carrier for chemicals in application.

Wetherill process (American process). Method for making zinc oxide pigment directly from franklinite ore. The ore is roasted with anthracite coal and the zinc vapor oxidized with air.

"Wetsit." [78] Trademark for a wetting agent of the alkyl aryl type which also contains aromatic properties for added solvent action on pectins, waxes, grease, etc. Used for various cleaning compounds; for scouring and dyeing textiles; for removal of emulsions from photographic film.

"Wetting Agent F-126." [158] Trade name for a mixture of ammonium salts of completely fluorinated carboxylic acids, chiefly perfluorocaprylic acid.
Properties: Fine, white, free-flowing powder with a faintly musty odor; bulk density, 0.6-0.7 g/cc; surface tension (0.35% aqueous solution), 16 dynes; decomposes at 175°C (before melting); soluble in water, acetone, diethyl ether, methanol, methyl ethyl ketone, formamide; insoluble in benzene, toluene, xylene, Stoddard solvent, heptane, carbon tetrachloride, perchloroethylene; stable toward extreme oxidizing conditions.
Uses: Specialized wetting agent for low surface tension or non-sudsing detergency.

wetting agents. Any compounds that cause water to penetrate more easily into, or to spread over the surface of, another material. Used particularly with respect to textiles, paper, and leather, and similar materials, but also used more generally. Soaps, detergents, and surface-active agents are used as wetting agents, their effectiveness being related to their capacity for reducing surface tension or interfacial tension. See surface-active agents and detergents.

WFNA. Abbreviation for white fuming nitric acid. See nitric acid, fuming.

whale oil (body oil; blubber oil).
Properties: Yellowish-brown, non-drying, fixed oil; strong fishy odor. Soluble in alcohol, ether, benzene, chloroform, and carbon disulfide.
Constants: Sp.gr. 0.925-0.930; saponification value 188-193; iodine value 120.
Derivation: By boiling the blubber of the Greenland or other whales, and skimming off the oil.
Method of purification: Filtration.
Grades: Crude No. 1; crude No. 2; natural winter; bleached winter.
Containers: 375-lb barrels; 8000-gal tank cars.
Uses: Leather tanning and dressing; lubrication; tempering steel; soap-making; illumination; fat manufacture (by hydrogenation); oleomargarine; plant insecticide; sheep dips, edible oils.
Shipping regulations: None.*

wheat germ oil. A fat-soluble oil extracted from the wheat germ. Useful as a source of vitamin E.
Containers: 5-gal drums.

whey (milk serum). The liquid remaining after the fat and casein have been removed from milk. It is essentially a 5% (approx) water solution of lactose.

whiskers. Minute hair-like crystals of certain metals which have been obtained under special conditions in a very pure state. Iron whiskers, for example, are said to have remarkable tensile strength.

whisky. A distilled alcoholic beverage.
Properties: Light yellow to amber liquid; sp. gr. 0.923-0.935 (15.56°C); 47-53% alcohol by volume.
Derivation: Distillation of fermented malted grains, as corn, rye, or barley. After distillation the whisky is aged in wooden containers four years or more to improve its flavor. The characteristics of various brands and grades are due to slight "impurities", and also to the principal grain

*See "I.C.C. Shipping Regulations," page xiii.

used in the mash, i.e., rye whisky comes from a mash in which rye predominates.

white acid. A mixture of ammonium bifluoride and hydrofluoric acid used for etching glass.

white antimony. See valentinite.

white arsenic. See arsenic trioxide.

white bole. See kaolin.

white carbon black. A form of silica produced in the gas phase and having extremely fine particle size, and rubber reinforcing properties similar to carbon black.

white cedar. See thuja.

white cinnamon. See canella.

white clay. See kaolin.

white copperas. See zinc sulfate.

white dye. An optical bleach (q.v.) or, in general, any substance, such as bluing, which may be added to a white article to increase its apparent whiteness.

white flag. See orris.

white iron pyrites. See marcasite.

"Whitekote." [139] Trademark for a dolomitic, normal, hydrated lime having a neutralizing value reported as 166% in terms of calcium carbonate; used in neutralizing and soil stabilization and other processes requiring an air-floated lime high in magnesium.

white lead, basic silicate. A pigment made up of an adherent surface layer of basic lead silicate and basic lead sulfate cemented to silica.

Properties: Excellent film-forming properties with vegetable drying oils combined with low specific gravity.

Derivation: Fine silica is mixed with litharge and sulfuric acid. The mixture is then furnaced in a rotary kiln and ground to break up agglomerates.

Use: As white lead pigment in exterior mixed pigment house paints.

white leads. Among the most used white pigments. Name primarily applied to lead carbonate, basic (q.v.), but also used for lead sulfate, basic (white lead sulfate) and lead silicate, basic (white lead silicate).

white lead, sublimed. See lead sulfate, basic.

white metal.

1. A group of alloys having relatively low melting points. They usually contain mainly tin, lead, or antimony as the chief component. Type metal, Babbitt, pewter, and Britannia metal are of this group.

2. The term white metal is also applied to copper matte containing about 75% copper, as obtained in copper smelting operations. See copper.

white mineral oil. See petrolatum, liquid.

white petrolatum. See petrolatum.

white pine bark. The inner bark of the white pine, Pinus strobus of eastern North America. Used in making cough syrups.

Grade: N.F. XI.

Containers: Bales.

white Portland cement. A very pure Portland cement free of iron compounds. See cement, Portland.

white powder. See gunpowder, white.

white precipitate. See mercury, ammoniated.

white precipitate, fusible. See mercury, ammoniated.

white Senaar gum. See arabic gum.

"White Star." [67] Trademark for white arsenic.

"Whitetex." [432] Trademark for a clay used as a filler in rubber products.

white vitriol. See zinc sulfate.

white wax. See beeswax, bleached.

white wood bark. See canella.

whiting (Paris white; gilder's whiting). Finely ground, naturally occurring calcium carbonate, $CaCO_3$, about 98% pure, contaminated with silica, iron, aluminum, or magnesium. Not to be confused with chalk, prepared, or chalk, precipitated. See calcium carbonate for distinctions.

Properties: White amorphous powder; sp.gr. 2.7; insoluble in water, soluble in acids.

Derivation: Traditionally from chalk (q.v.), the crude product obtained from the chalk cliffs of England, France, and Belgium. A pure limestone or calcite is the principal commercial source of whiting. The crude chalk or limestone is ground dry or wet, air- or water-floated and sieved. Grades are based on particle size, softness and light reflectance. Dry ground, air-floated limestone whiting can be as fine as 99% through 300 mesh.

Grades: Various. Paris white is the finest grade; coarser grades are extra gilders whiting, gilders whiting and commercial, the last being quite coarse and of poor color. A putty grade is also sold. See also cliffstone Paris white.

Uses: As an inexpensive filler and extender wherever safe from acids, as pigment extender, rubber filler; putty (mixture with linseed oil); ceramics; glass; soaps; detergents; linoleum; whitewash; paper; inks; wood filler.

wild black cherry bark. See prunus virginiana.

wild cherry. See prunus virginiana.

wild ginger. See asarum.

wild ginger oil. See asarum canadense oil.

wild saffron. See colchicum.

wild tobacco. See lobelia.

wild vanilla. See liatris.

wilkinite. See bentonite.

willemite Zn_2SiO_4. Natural zinc orthosilicate. Troostite is a manganese-bearing variety.
Properties: Color yellow, green, red, brown, white; luster vitreous to resinous; sometimes fluoresces in ultraviolet light; hardness 5.5; sp.gr. 3.3.
Occurrence: New Jersey; New Mexico; Africa; Greenland.
Use: Ore of zinc; a phosphor.

Williamson's blue. A name applied loosely to any of a number of the varieties of iron blue pigments. See iron blues.

Williamson synthesis. An organic method for preparing ethers by the interaction of an alkyl halide with a sodium alcoholate (or phenolate).

"WIN 3000." [162] Trade name for ammonium polystyrene sulfonate. Also known as Permutit Z. Used in decalcification of bone for cytologic study.

window glass. See glass.

wine. The fermented juice of grapes or other fruits or plants. It usually contains about 7-20% alcohol (by volume). The higher percentages are obtained by the addition of pure alcohol (called fortifying). Coloring matter, sugars, and small amounts of acetic acid, salts (see wine lees), higher fatty acids, etc. give wines their distinctive appearance and flavor.

wine ether. See ethyl pelargonate.

wine gallon. Same as ordinary U.S. gallon.

wine lees. A deposit or sediment which collects in the bottom of wine casks during the fermentation of wine. Wine lees vary greatly in quality, but usually contain from 20-35% potassium acid tartrate and up to 20% calcium tartrate. They also contain yeast cells, proteins, and other solid matter which was suspended in the grape juice.
Use: Similar to argols (q.v.).
Shipping regulations: None.*

winestones oil. See grape-seed oil.

"Wing-Stay." [265] Trademark for the first of a series in rubber chemicals used as a nonstaining, nondiscoloring antioxidant.
Properties: Pale amber liquid; stable to heat; exhibits comparatively low volatility; will not hydrolyze; completely soluble in most organic solvents.
Uses: Protection against deterioration and discoloration in GR-S and natural rubber compounds, and in natural and synthetic rubber latices. Also suitable for a wide variety of light colored foam rubber products.

"Wingstop." [265] Trademark for a series of rubber chemicals used as polymerization terminating agents to stop polymerization reactions at a certain point.
"Wingstop B" - sodium dimethyldithiocarbamate.
"Wingstop K" - potassium dimethyldithiocarbamate.
Uses: As short-stopping agents in the preparation of synthetic rubbers, particularly for SBR latices and similar emulsion polymerization systems.

winter bloom. See hamamelis.

wintergreen. See gaultheria.

wintergreen oil (gaultheria oil). See also methyl salicylate.
Properties: Colorless, yellowish or reddish liquid; characteristic strongly aromatic odor quite distinct from that of betula oil; sweetish, warm and aromatic taste. Solubility in alcohol: in 6 to 8 vols of 70% alcohol.
Chief known constituents: Methyl salicylate; a paraffin; a ketone; an alcohol; an ester.
Constants: Sp.gr. 1.180-1.193 (15°C); optical rotation —0° 25' to —1°; refractive index 1.535-1.536; saponification value 354-365 (96-99% methyl salicylate).
Derivation: Distilled from the leaves of Gaultheria procumbens.
Adulteration: Mineral oil; betula oil; artificial methyl salicylate.
Containers: Cans.
Uses: Flavoring compounds; medicine; perfumery; confectionery.
Shipping regulations: None.*

winterize. A process of refrigerating edible and lubricating oils to crystallize the saturated glycerides, which are then removed by filter pressing.

wire glass. Made by rolling into a softened sheet of heavy glass a netting of wire, then annealing in the usual way. Such glasses have the important advantage that they may crack (as in a fire) but will, nevertheless, remain suspended.

"Witall." [104] Trademark for a line of driers manufactured from highly refined tall oil acids available in usual metallic salts including cobalt, lead and manganese. Equals naphthenate driers in action.

"Witcarb R." [104] Trademark for a finely precipitated calcium carbonate having a particle size of 0.03-0.05 micron.

witch hazel. See hamamelis.

"Witcizer." [104] Trademark for ester plasticizers used in the plastics, paint and lacquer industries.

"Witcoblak." [104] Trademark for pigment blacks including both channel and furnace grade blacks. Suitable for use in news inks, paints, plastics, phonograph records, paper, pigmentation of paperboard, caulking compounds, etc.

"Witcolite." [104] Trademark for a resinous hydrocarbon used as underground pipe insulation. Exhibits high thermal insulation and protects against corrosion and acids and alkalis found underground.

"Witcote." [104] Trademark for industrial asphalt paint; a bituminous mastic compound, provides protection against rust and corrosion, unaffected by water or chemicals.

withdrawing agent. See entrainer.

witherite $BaCO_3$. A natural barium carbonate usually found in veins with lead ores.
Properties: White, yellowish or grayish. Vitreous, inclining to resinous luster.
Constants: Sp.gr. 4.27-4.35; hardness 3-3.75.
Occurrence: United States (Kentucky, Lake Superior region), England (most important source), Germany, Austria, Japan.
Uses: Chemicals (barium dioxide, barium hydroxide, blanc fixe); plate glass and porcelain; brick making (prevents efflorescence and discoloration by soluble sulfates); rat poison.

"Wittox." [104] Trademark for copper and zinc naphthenate fungicides.

Wohlwill process. The official process of the U.S. mints for refining gold. It consists in subjecting gold anodes to electrolysis in a hot solution of hydrochloric acid containing gold chloride, the solution being continuously agitated with compressed air.

"Wolfco Solvent." [78] Trademark for a detergent used for scouring cotton and rayon goods.

wolfram. See tungsten. Wolfram is the official international alternate name for tungsten. The latter is preferred in the U.S.

wolframic acid. See tungstic acid.

wolframic acid, anhydrous. See tungstic oxide.

wolframite $(Fe, Mn)WO_4$. A natural tungstate of iron and manganese. Ferberite is the iron-rich member of the series, and huebnerite is the manganese-rich member. Usually found with granite and pegmatite.
Properties: Color black to brown; luster submetallic to resinous; streak black to brown; hardness 5-5.5; sp.gr. 7.0-7.5.
Occurrence: Colorado, South Dakota, Nevada, China, Burma, Australia, Bolivia, Europe.
Use: Chief ore of tungsten.

wolfram white. See barium tungstate.

wolfsbane. See aconite.

wollastonite $CaSiO_3$. A natural calcium silicate, found in metamorphic rocks.
Properties: Color white to brown, red, gray, yellow; luster vitreous to pearly; hardness 4.5-5; sp.gr. 2.8-2.9.
Occurrence: New York; California.
Grades: Fine, medium paint grades.
Uses: In ceramics; paint extender; welding rod coatings; rubber; alloying agent; in silica gels; as paper coating pigment; as reinforcing fiber in plastics, cements, and wallboard; mineral wool; soil conditioner.

"Wolman" Salts. [11] Trademark for a patented wood preservative containing sodium fluoride, sodium chromate, sodium arsenate and dinitrophenol.

wood. Wood is composed principally of 40-60% cellulose and 20-40% lignin, together with gums, resins, a variable amount of water, and inorganic matter left as ash when the wood is burned. Its fuel value varies widely around 3000-6000 Btu per lb according to variety, moisture, etc.
Uses: Pulp and paper; construction; containers, especially because of its cheapness and resistance to dilute acids; for destructive distillation (not so important as formerly), from which are obtained charcoal, acetic acid, methyl alcohol, turpentine, pine oil, etc.; for extraction, to give turpentine, rosin, pine oil, tars, etc.; source for carbohydrates, cattle food, ethyl alcohol; see also wood ashes, wood flour, wood pulp.

wood, agatized. See wood, petrified.

wood alcohol. Methanol (q.v.) (CH_3OH) from destructive distillation of wood. Also called natural methanol. When pure its properties are identical with those of synthetic methanol.

wood ashes.
Use: As a fertilizer for their potash content, which varies widely around 4% K_2O.

woodbine, wild. See gelsemium.

wood distillation gas. See wood gas.

wood ether. See dimethyl ether.

wood flour (wood meal). Pulverized dried wood from either soft or hardwood wastes but mostly from spruce, white pine, and poplar. Graded according to color and fineness.
Grades: Domestic standard, domestic fine, imported 40-60 mesh, 70-80 mesh, etc.
Uses: An absorbent for nitroglycerin in the manufacture of dynamite; filler for plastics, linoleums, paperboard; soaps; fur cleaning; polishing agents; Sorel cement.

wood gas (wood distillation gas). Gas produced during production of charcoal by heating wood in absence of air.
Typical composition:

	% by volume
Hydrogen	2.2
Methane	16.8
Hydrocarbons	1.2
Carbon monoxide	23.4
Carbon dioxide	37.9
Oxygen	2.4
Nitrogen	16.0
Btu's/cu ft, 290.	

Usually used as a fuel at the production site.

"Wood-Glu No. 2140." [170] Trademark. Cold run, fast setting thermoplastic resin emulsion which requires no heating and can be safely used winter and summer. Used as a general woodworking glue. Available in liquid form, weighing approx 9 lbs/gal.

wood, indurated. A wood hardened by impregnation with a phenol-formaldehyde product. Used in storage batteries.

wood meal. See wood flour.

wood oil. See tung oil.

wood, petrified (wood, silicified; wood, agatized). A natural material composed of opal or chalcedony (agate) and formed by the replacement of wood by silica. The replacement takes place in such a way that the original form and structure of the wood is preserved.

Occurrence: Arizona, California, Colorado, South Dakota, Utah, and Wyoming.

wood pulp (paper pulp). Produced for its cellulose content and used for the making of various kinds of paper, paperboard, rayon, and nitrocelluloses. The cellulose fibers from wood are used mainly in two forms: (1) as ground wood or mechanical pulp, which is merely finely divided wood without purification, and is made into newsprints, cheap manila papers, and nonpermanent tissues; (2) as chemical pulp, of which there are three kinds: (a) soda process pulp, obtained from the digestion of wood chips (mostly poplar) by caustic soda; (b) sulfite process pulp (mostly spruce and other coniferous woods) obtained by digestion with a solution of magnesium, ammonium or calcium disulfite containing free sulfur dioxide; and (c) sulfate process pulp, in which sodium sulfate is added to the caustic liquors but is reduced by the carbon present to the sulfide, which becomes a digesting agent. Sulfite and sulfate pulps (chiefly pine and other coniferous woods) make up the bulk of the paper pulps, and may be bleached or unbleached.

Bleaching is usually done with chlorine gas, chlorine dioxide, or alkaline hypochlorites. Sulfate pulps are known as kraft pulps because of their strength, and the unbleached varieties are used for wrapping papers and shipping containers. Bleached kraft pulp is used in print paper, but the bulk of such papers are made from mixtures of ground wood and bleached sulfite pulp. See also cellulose; lignin.

wood rosin. See rosin.

wood, silicified. See wood, petrified.

Wood's metal. A very low melting cadmium-bismuth alloy chiefly used for sprinkler systems. See table under fusible alloys.

wood-sour. See barberry.

wood sugar. See D(+)-xylose.

wood-tar. See tar, wood.

wood vinegar. See pyroligneous acid.

wool. Stable fibers, usually 2-8 in. long, obtained from the fleece of sheep (and also alpaca, vicuna, and certain goats). Physically, wool differs from hair in fineness and by the presence of prominent cortical scales and a natural crimp. The latter properties are responsible for the felting properties of wool and the ability of the fibers to cling together and be spun into stable yarns. Chemically, wool consists essentially of protein chains (keratin) bound together by disulfide cross linkages.

Properties: Tensile strength 20,000-29,000 psi; elongation 25-50%; sp.gr. 1.32; moisture regain 16% (70°F, 65% relative humidity); decomposes about 260°F, scorches at 400°F. Resistant to most acids except hot sulfuric; destroyed by alkalies and chlorine bleach; resistant to mildew but attacked by insects.

Sources: Australia, Argentina, U.S., New Zealand, Uruguay, Russia, England.

Uses: Outerwear; blankets; carpets; upholstery; felt.

wool fat. See lanolin, anhydrous.

wool-fat, hydrous. See lanolin.

wool-fat pitch. See stearin and fatty acid pitches.

wool grease. See degras.

wool pitch. See stearin and fatty acid pitches.

wool wastes. Wool from old woolen materials which have been cut up for re-making into cloth. Some of the wool is so cut up as to cause rejection as unfit for spinning. Used as a fertilizer. Wool waste usually contains from 4-7% ammonia. Pure wool shoddy may contain as much as 15% nitrogen and is particularly valued by hop growers.

Shipping regulations: None.*

worm-grass. See spigelia.

wormseed. See santonica.

wormseed oil, American. See chenopodium oil.

wormseed oil, Levant (chenopodium oil, Levant).

Properties: Yellow essential oil; penetrating, disagreeable odor. Soluble in 2-3 vols and more of 70% alcohol.

Chief known constituents: Cineol, pinene, terpenes.

Constants: Sp.gr. 0.915-0.940 (15°C); optical rotation −1° 50' to −7°; refractive index, 1.465-1.469 (20°C).

Derivation: By distillation of the unexpanded flower heads of Artemisia maritima, L. (var. Stechmanni).

Use: Medicine.

Shipping regulations: None*

wormwood. See absinthium.

wormwood oil (artemisia absinthium oil; absinthe oil).

Properties: The oil from the fresh herb has a dark green color, while that from the dry herb is yellowish-green or yellowish-brown; becomes dark brown with age; strong somewhat unpleasant odor characteristic of absinthe; bitter, scratching, lingering taste. Soluble in 1-2 vols of 80% alcohol (additional solvent may cause opalescence and turbidity); usually soluble in all proportions of 90% alcohol.

Chief known constituents: Thujone, thujyl alcohol, phellandrene, cadinene.

Constants: The condition of the herb at the time of distillation and the length of the distillation influence greatly the properties of the oil; consequently there is considerable variation in the properties of the different market samples. Sp.gr. 0.901-0.954;

refractive index 1.460-1.470; acid value, American, up to 2.2, French, up to 6.7; ester value, American, 46-89, French, 11-108.

Derivation: Distilled from the leaves of Artemisia absinthium, L.

Adulteration: Turpentine.

Containers: Bottles; cans.

Uses: Medicine; flavors.

Shipping regulations: None.*

wort. A clear infusion of plant materials (such as grains) obtained as a preliminary to fermentation; term used especially in brewing.

"Worthite." [269] Trademark for an austenitic stainless steel, containing 3% silicon, 20% chromium, 24% nickel, 2.75% molybdenum, 1.75% copper. Corrosion resisting properties are excellent for most concentrations and temperatures of sulfuric, sulfurous, phosphoric, nitric, acetic, fatty, mixed and most other acids, and to hot caustic soda and other alkalies. It is widely used in the chemical pump field for many varieties of corrosive environments, such as the paper industry, the citrous fruit canners, sea water, sulfuric acid plants, etc. It is available as castings and wrought bars.

"WR 1339." [162] Trade name for "Triton A-20", an alkyl aryl polyether alcohol. No diluent is present in "WR 1339". A non-ionic surface active agent of low toxicity. Viscous amber colored liquid, soluble in oil and water; chemically stable in the presence of acids, bases and salts at sterilizing temperatures. Less satisfactory as a wetting agent than other "Tritons", but is a good emulsifying agent and interfacial tension depressant.

"Wrinkle" Finish. [435] Trademark for an organic protective coating, applicable to any surface, which inherently develops a textured surface during the drying phase.

wulfenite $PbMoO_4$, sometimes with Ca, Cr, V. Yellow, orange or bright orange-red mineral of resinous luster. Found in veins with ores of lead.

Constants: Sp.gr. 6.7-7.0; hardness 2.75-3.

Occurrence: United States (Massachusetts, New York, Pennsylvania, Nevada, Utah, New Mexico, Arizona); Hungary; Austria; Germany; Australia.

Uses: Ore of molybdenum.

Wulff process. A process for producing acetylene by treating a hydrocarbon gas such as butane with superheated steam in a regenerative type refractory furnace that operates at 2100-2500°F. Contact times are very short and various other by-products such as ethylene, hydrogen, and carbon dioxide are formed and various distillation and solvent extraction processes are used to purify the acetylene.

wurtzite. A natural zinc sulfide of the same composition as sphalerite, ZnS, but hexagonal in its crystallization. Found in United States (Missouri, Montana), Czechoslovakia, France, Cornwall.

"Wyamine." [24] Trademark for mephentermine (N-methyl, phenyl-tert-butylamine).

"Wydase." [24] Trademark for a highly purified testicular hyaluronidase.

"Wyex" (EPC). [285] Proprietary brand name for easy processing channel carbon black.

Properties: Sp.gr. 1.77; free-flowing pellets; bulk density 25 lbs/cu ft; particle diameter 30 millimicrons; pH 4.1-4.5; ash 0.05% max; 99.9% thru 325 mesh screen; color (Nigrometer) 85-86.

Containers: 50-lb paper bags or bulk.

Uses: As a reinforcing ingredient for compounding in natural and most synthetic rubbers, contributing to abrasion resistance, good tensile and tear strength; as a black coloring agent in rubber, paper, plastics, paint and ink.

"X-12." [28] Trademark for a flame retardant based on ammonium sulfamate and modified to retain original fabric or paper properties, prevent afterglow, and insure penetration.
Properties: White, crystalline powder completely soluble in hot water; resistant to removal by dry cleaning solvents.
Containers: 50-lb fiber drums; 100-lb paper bags.
Use: One-step, renewable-type flame retardant treatment for cellulosic materials, including cotton, viscose rayon, and paper.

xanthates. See xanthic acids.

xanthene (diphenylenemethane oxide, tricyclic) $CH_2(C_6H_4)_2O$, the central structure of the fluorescein, eosin and rhodamine dyes.
Properties: Yellowish, crystalline leaflets.
Constants: M.p. 100.5°C; b.p. 315°C. Soluble in ether; slightly soluble in alcohol; very slightly soluble in water.
Derivation: By the condensation of phenol and ortho-cresol by means of aluminum chloride.
Grade: Technical.
Containers: Tins; glass bottles.
Use: Organic synthesis; fungicide.
Shipping regulations: None.*

xanthene-9-carboxylic acid $O(C_6H_4)_2CHCOOH$.
Properties: White crystals; m.p. 220°C; soluble in alkalies.
Use: Intermediate.

xanthene dyes. Dyes whose molecular structure is related to that of xanthene. The aromatic (C_6H_4) groups are the source of the color and thus constitute the chromophore group. The Colour Index number ranges from 739 to 784. The dyes are closely related structurally to diaryl methane dyes. Phenolphthalein (Colour Index 764), and eosin (Colour Index 768) are examples of this class.

xanthene ketone. See xanthone.

xanthenol. See xanthydrol.

xanthic acids (xanthogenic acids). Substituted dithiocarbonic acids of the type ROC(S)SH, in which R is ordinarily an alkyl radical. Unless otherwise designated, xanthic acid is understood to be the ethyl derivative $C_2H_5OC(S)SH$. This is a liquid melting at −53°C and undergoing decomposition at room temperature. Xanthic acid salts, in which a metal replaces the hydrogen, are called xanthates. The simpler salts, sodium, potassium ethyl, propyl, or butyl xanthates, are used as flotation collectors.

xanthine (dioxopurine) $C_5H_4N_4O_2$. A purine base found in the blood and urine and some plants. Theophylline and theobromine, the alkaloids found in tea and cocoa respectively, are both dimethyl xanthines.
Properties: Yellowish-white powder; sublimes with partial decomposition. Soluble in potassium hydroxide; insoluble in water and acid. $A_M = 9.3 \times 10^3$ at 277.5 millimicrons and pH 10.
Derivation: By the action of nitrous acid on guanine.
Grades: Technical, C.P.; monohydrate; sodium salt. Radioactive forms available.
Containers: Tins; glass bottles.
Use: Organic synthesis, medicine.
Shipping regulations: None.*

xanthine oxidase. An enzyme found in animal tissues which acts upon hypoxanthine, xanthine, aldehydes, reduced coenzyme I, etc., producing, respectively xanthine, uric acid, acids, oxidized coenzyme I, etc.
Use: Biochemical research.

xanthogenic acids. See xanthic acids.

xanthone (benzophenone oxide, dibenzopyrone, xanthene ketone) $CO(C_6H_4)_2O$. Occurs in some plant pigments.
Properties: White needles or crystalline powder; m.p. 173-4°C; b.p. 350°C; sublimes. Insoluble in water, soluble in alcohol, chloroform, and benzene, especially when hot.
Uses: Larvicide; intermediate for dyes, perfumes and pharmaceuticals.

xanthophyll $C_{40}H_{56}O_2$.
Properties: Yellow pigment which accompanies chlorophyll in green leaves of plants. Chemically it is allied to the hydrocarbon carotene, $C_{40}H_{56}$. M.p. 190-3°C; insoluble in water, slightly soluble in alcohol and ether.
Occurrence: It is especially found in autumn leaves and is believed to be the residue obtained by the fading of chlorophyll.
See also lutein.

xanthophyll oil. Obtained from corn gluten; used in poultry feeds.

xanthopterin $C_6H_5N_5O_2 \cdot H_2O$. 2-Amino-4,6-dihydroxypteridine. Pigment found in the wings of butterflies. Can be converted by yeast into folic acid.
Properties: Orange-yellow crystals; sinters around 360°C; decomposes above 410°C. Practically insoluble in water; freely soluble in dilute ammonium or sodium hydroxide giving yellow solutions, and in 2N

hydrochloric acid giving colorless solutions.
Use: Biochemical research.

xanthorrhea resin. See accroides gum.

xanthuremic acid (4,8-dihydroxyquinaldic acid) $C_{10}H_7NO_4$. A factor in vitamin B deficiency studies.
Properties: Sulfur-yellow crystals, m.p. 286°C. Insoluble in water, soluble in alkaline hydroxides or hot dilute hydrochloric acid.
Derivation: Excreted by pyridoxine-deficient animals after ingestion of tryptophan.
Use: Medicine.

xanthydrol (xanthenol) $HCOH(C_6H_4)_2O$. A derivative of xanthene.
Properties: White powder; m.p. 123°C. Insoluble in water; soluble in alcohol.
Derivation: Reduction of xanthone with alcohol and sodium.
Method of purification: Recrystallization.
Grades: C.P. (analytical).
Containers: Glass bottles.
Use: Determination of urea, and of DDT.

Xe. Symbol for xenon.

xenol. See phenylphenol.

xenon. Xe. Element of atomic number 54 of the zero group in the periodic table.
Properties: Colorless wholly inert gas—does not combine chemically with any element. Liquefies at −106.9°C; density 5.89 g/l.
Derivation: By fractional distillation of liquid air.
Grades: Highest purity.
Containers: Hermetically sealed glass flasks; cylinders.
Use: Filling electrical luminescent tubes; radio and television tubes.
Shipping regulations: Nonflammable gas. Green label.*

xenotime YPO_4. A natural phosphate of yttrium. Rare earth metals, thorium, and uranium, also may be present.
Properties: Color, yellowish brown, reddish brown, streak pale brown; luster, vitreous; hardness 4-5; sp.gr. 4.4-5.1.
Occurrence: North Carolina, Georgia, Alabama, New York, Colorado; Europe; Africa.

xenyl. The biphenyl radical, $C_6H_5C_6H_4$-.

xerography. A "dry" method of photography or photocopying employing a smooth metallic plate covered with a layer of photoconductive powder, such as selenium or anthracene. The surface of this specially coated plate is given an electric charge by passage under a series of charged wires. An image of the material to be photographed is projected onto the charged plate through a camera lens and the electric charges disappear in the areas exposed to light but elsewhere the surface retains its charge. A powder consisting of a coarse carrier and a fine developing resin is then spread over the surface of the plate. Adhesion between powder and plate occurs only where the plate retained its charge. Elsewhere developing resin and carrier are not retained on the plate, which thus has become a negative of the original image. A positive is obtained by placing a piece of paper against the plate, and applying an electric charge as in the first stages of the process. This causes adhesion of developing resin and its carrier to the paper. This positive print is fixed by heating in a press for a few seconds to melt the developing resin and fuse it to the paper. Colored prints are possible by use of suitable developing resins. Various materials other than paper can be printed. Applications are in copying letters, drawings, and charts, in lithography, photoengraving, and printing and in transferring designs to other materials.

"Xerols." [19] Brand name for a series of self-emulsifying waxes.

xi particle. See fundamental particle.

"X-OX." [241] Brand name for a red powder consisting primarily of rare earth oxides used in pitch polishing or precision optical elements.

"X-3 Oxygen Corrosion Inhibitor." [108] Specially formulated metaphosphate-type corrosion inhibitor. This granular product contains inorganic film formers and accelerators.
Containers: 100-lb drums.
Uses: In oilfield secondary waterflood operations to protect distribution lines, pumps, storage tanks, etc., from oxygen corrosion.

"X-2 Oxygen Inhibitor." [108] Highly polar, film-forming corrosion inhibitors in liquid form. Easily dispersible in fresh waters or concentrated brines.
Containers: Lined cans, 40 lbs. Lined drums, 425 lbs.
Uses: Control carbon dioxide and hydrogen sulfide corrosion in water flood systems.

x-rays (roentgen rays). Invisible radiation of the same nature as light radiation, but of extremely short wave length (0.06-120A). Will be emitted as the result of electron transitions in the inner orbits of heavy atoms which are being bombarded in a vacuum tube by cathode rays (electrons). The most notable properties of these waves are:
1. Penetration through various thicknesses of all solids.
2. Action on photographic plates, fluorescent screens.
3. Ionization of a gas through which they pass.

The applications are many. Used widely in analysis of crystalline structure, in treatment of cancer, surgery (location of fractures, foreign bodies), in metallurgy, determination of molecular structure, etc.

xylene (dimethylbenzene) $C_6H_4(CH_3)_2$. A commercial mixture of the three isomers, ortho-, meta-, and para-xylene (q.v.). The last two isomers predominate.
Properties: Clear liquid, toxic and flammable; soluble in alcohol and ether; insoluble in

water. Sp.gr. approximately 0.86; see under Grades for boiling range.
Derivation: Fractional distillation from petroleum (90%), coal tar or coal gas.
Grades: Nitration (b.p. range 137.2-140.5°C); 3° (b.p. range 138-141°C); 5° (b.p. range 137-142°C, high in meta isomer); 10° (b.p. range 135-145°C); industrial (b.p. 90% below 150°C, complete below 160°C). Also other grades depending upon use. In some cases one or another of the individual isomers are partially removed for use in chemical production.
Containers: Glass bottles, 5-, 55-, 110-gal drums; tank cars, tank trucks.
Uses: Aviation gasoline (largest use); protective coatings; solvent for alkyd resins, lacquers, enamels, rubber cements; synthesis of various organic chemicals (see individual isomers).
Warning! Flammable. Use with adequate ventilation. MCA warning label.
Shipping regulations: Flammable liquid. Red label.*

meta-**xylene** (1,3-dimethylbenzene) 1,3-$C_6H_4(CH_3)_2$.
Properties: Clear, colorless liquid, toxic and flammable, soluble in alcohol and ether; insoluble in water. Sp.gr. (15°C) 0.8684; m.p. −47.4°C; b.p. 138.8°C; refractive index (20°C) 1.4973.
Derivation: Selective crystallization or solvent extraction of meta-para mixtures.
Grade: 95%.
Uses: Solvent; intermediate for dyes and organic syntheses, especially isophthalic acid; insecticides.
Caution: Keep away from heat and open flame; use with adequate ventilation.
Shipping regulations: Flammable liquid. Red label.*

ortho-**xylene** (1,2-dimethylbenzene) 1,2-$C_6H_4(CH_3)_2$.
Properties: Clear, colorless liquid; toxic; flammable; soluble in alcohol and ether; insoluble in water. Sp.gr. (20/4°C) 0.8968; m.p. −25°C; b.p. 144°C; refractive index (20°C) 1.505; wt/gal 7.36 lb.
Grade: 99%, free of hydrogen sulfide and sulfur dioxide.
Uses: Manufacture of phthalic anhydride; vitamin and pharmaceutical syntheses; dyes; insecticides.
Warning: Flammable. Keep away from heat and open flame. Use with adequate ventilation. MCA warning label.
Shipping regulations: Flammable liquid. Red label.*

para-**xylene** (1,4-dimethylbenzene) 1,4-$C_6H_4(CH_3)_2$.
Properties: Colorless liquid; crystals at low temperature; toxic and flammable; soluble in alcohol and ether; insoluble in water. Sp.gr. (20°C) 0.8611; m.p. 13.2°C; b.p. 138.5°C; refractive index (21°C) 1.5004.
Derivation: Selective crystallization or solvent extraction of meta-para mixtures.
Uses: Synthesis of terephthalic acid for production of synthetic resins and fibers (for example "Dacron," "Mylar," "Terylene"); vitamin and pharmaceutical synthesis; insecticides.
Caution: Keep away from heat and open flame; use with adequate ventilation.
Shipping regulations: Flammable liquid. Red label.*

para-**xylene**-alpha, alpha'-**diol** $C_6H_4(CH_2OH)_2$.
Properties: White crystalline solid. M.p. 118°C; b.p. 138-144°C (0.8-1.0 mm). Slightly soluble in water (25°C).
Purity: Approximately 98%.
Uses: Crosslinker for polyurethanes; in esters, polyethers, polyesters, polycarbonates.

xylene substitute. A petroleum product having the following characteristics:
Constants: Sp.gr. 0.760; distillation range 130-170°C; refractive index 1.41 at 20°C; flash point 45°F.
Uses: Solvent and diluent.
Shipping regulations: Flammable liquid. Red label may be required.*

xylenol (dimethylphenol, hydroxydimethyl benzene; dimethylhydroxybenzene) $(CH_3)_2C_6H_3OH$. Found in the cresylic acid or tar acid fraction from coal tar. There are six isomers. See commonly used isomers following.
Properties: White, crystalline solids; sp.gr. (15°C) 1.02-1.03; m.p. varies, 20-76°C; b.p. 203-225°C. Only slightly soluble in water, soluble in most organic solvents and in caustic soda solution. Approximate wt/gal: 8.52 lb.
Derivation: By fusing the xylenesulfonic acids with potassium hydroxide.
Method of purification: Crystallization.
Grades: Technical.
Containers: Fiber drums.
Uses: Disinfectants; solvents, pharmaceuticals, insecticides and fungicides; plasticizers; rubber chemicals; additives to lubricants and gasolines, plastics; wetting agents; dyestuffs.
Shipping regulations: None.*

2,4-xylenol (1-hydroxy-2,4-dimethylbenzene; 1,3-dimethyl-4-hydroxybenzene; 1,3,4-xylenol) $1,2,4(OH)(CH_3)_2C_6H_3$.
Properties: Colorless needles; sp.gr. 1.036 (20°C); m.p. 26°C; b.p. 211.5°C; very slightly soluble in water, soluble in alcohol and ether.

2,5-xylenol (1-hydroxy-2,5-dimethyl benzene; 1,4-dimethyl-2-hydroxybenzene; 1,4,2-xylenol; 2,5-dimethylphenol).
Properties: Colorless liquid; sp.gr. 1.169 (15°C); m.p. 74.5°C; b.p. 211.5°C. Soluble in water and alcohol; very soluble in ether.
Grades: Up to 90%; other isomers are the chief impurity.
Containers: Drums; tank cars.

2,6-xylenol (1-hydroxy, 2,6-dimethyl benzene; 1,3,2-xylenol; 1,3-dimethyl-2-hydroxybenzene).
Properties: Colored solid; m.p. 45°C; b.p.

201°C; soluble in hot water and alcohol.
Grade: 95% pure.

3,4-xylenol (1-hydroxy-3,4-dimethyl benzene; 1,2-dimethyl-4-hydroxybenzene; 1,2,4-xylenol; 3,4-dimethyl phenol) $1,3,4(OH)(CH_3)_2C_6H_3$.
Properties: Solid; sp.gr. 1.023 (17°C); m.p. 65°C; b.p. 225°C; soluble in water, alcohol, and ether.
Grades: 95%, 62.5%, 56%, 51%, 46%; other xylenols are the chief impurity.
Containers: Drums; tank cars.

3,5-xylenol (1-hydroxy-3,5-dimethyl benzene; 1,3-dimethyl-5-hydroxybenzene; 1,3,5-xylenol; meta-xylenol; 3,5-dimethyl phenol) $1,3,5(OH)(CH_3)_2C_6H_3$.
Properties: Solid; m.p. 68°C; b.p. 219.5°C; sp.gr. 1.01 (approx.); slightly soluble in water, soluble in alcohol and NaOH.
Grades: 95%, 90%, 60.5%, 55.5%, 50%, 46%, 41%; other xylenols are the chief impurity.
Containers: Drums; tank cars.

1,2,4-xylenol. See 3,4-xylenol.

1,3,2-xylenol. See 2,6-xylenol.

1,3,4-xylenol. See 2,4-xylenol.

1,3,5-xylenol. See 3,5-xylenol.

1,4,2-xylenol. See 2,5-xylenol.

meta-**xylenol.** See 3,5-xylenol.

"Xylex 780." 277 Trademark for a chemical peptizing agent for rubber.
Containers: 1-, 5-, and 55-gal metal drums; tank cars.
Uses: Peptizing agent or softener to facilitate processing of natural and synthetic rubbers; devulcanizing agent for reclaiming natural and/or synthetic rubbers.
Shipping regulations: None.*

xylidine (aminodimethylbenzene; aminoxylene, dimethylaniline) $(CH_3)_2C_6H_3NH_2$. A varying mixture of the six isomers. See commonly used isomers following.
Properties: Liquid; sp.gr. 0.97-0.99; b.p. 213-226°C; insoluble in water; soluble in alcohol and ether.
Derivation: Nitration of xylene and subsequent reduction.
Grade: Technical.
Containers: 425- and 800-lb drums; tank cars.
Uses: Dye intermediates; organic syntheses; pharmaceuticals.
Danger: Hazardous liquid and vapor rapidly absorbed through skin. MCA warning label.

2,3-xylidine (ortho-xylidine; 1-amino-2,3-dimethylbenzene) $1,2,3(NH_2)(CH_3)_2C_6H_3$.
Properties: Liquid; sp.gr.(15°C) 0.991; m.p. less than −15°C; b.p. 224°C; slightly soluble in water; soluble in alcohol and ether.

2,4-xylidine (meta-xylidine; 1-amino-2,4-dimethylbenzene) $1,2,4(NH_2)(CH_3)_2C_6H_3$.
Properties: Liquid; sp.gr. (20°C) 0.974; b.p. 216°C; slightly soluble in water, soluble in alcohol, ether, and benzene.

2,5-xylidine (para-xylidine; 1-amino-2,5-dimethylbenzene) $1,2,5(NH_2)(CH_3)_2C_6H_3$.
Properties: Liquid; sp.gr. (15°C) 0.980; m.p. 15.5°C; b.p. 217°C; slightly soluble in water and alcohol, soluble in ether.

2,6-xylidine (2,6-dimethylaniline; 1-amino-2,6-dimethylbenzene) $(CH_3)_2C_6H_3NH_2$.
Properties: Colorless liquid; b.p. 216.9°C. Insoluble in water; slightly soluble in alcohol and ether.
Purity: 98.0% min.
Containers: Drums.
Use: Organic intermediate.

meta-**xylidine.** See 2,4-xylidine.

ortho-**xylidine.** See 2,3-xylidine.

para-**xylidine.** See 2,5-xylidine.

xylitol (pentanepentol) $CH_2OH(CHOH)_3CH_2OH$.
Properties: White, odorless, crystals. Sp. gr. (55% aq.sol.) 1.2; m.p. 95°C. Soluble in water.
Purity: 99%.
Uses: Softener, humidifier, sweetener, organic synthesis; pharmaceuticals, cosmetics, plastics, synthetic resins, paper, foodstuffs.

xylometazoline hydrochloride
$C_4H_9C_6H_2(CH_3)_2CH_2C_3H_5N_2 \cdot HCl$. 2(4-tert-Butyl-2,6-dimethylbenzyl)-2-imidazoline hydrochloride.
Properties: Crystals; m.p. 320°C (decompose). Somewhat soluble in water; soluble in alcohol.
Use: Medicine.

D(+)-xylose (wood sugar) $C_5H_{10}O_5$. (Not to be confused with phenylosazone of the same name.)
Properties: White crystalline dextrorotatory powder, sweet taste; sp.gr. (20°C) 1.525; m.p. 144°C (also given as 153°C); soluble in water and alcohol.
Derivation: Hydrolysis with hot dilute acids of wood, straw, corn cobs, etc.
Grades: Reagent, technical.
Uses: Dyeing, tanning; diabetic food.

xylyl bromides (alpha-bromoxylenes) $CH_3C_6H_4CH_2Br$. Mixed ortho-, meta-, and para-isomers.
Properties: Colorless liquid; pleasant aromatic odor. Decomposed slowly by water. Caution! Very irritant!
Constants: Sp.gr. 1.4; b.p. 210-220°C.
Derivation: Bromination of xylene.
Grade: Technical.
Containers: Lead-lined containers.
Uses: Organic synthesis; military poison gas.
Shipping regulations: Poison, class C. Tear gas label. 75-lb max lot.*

"Xyno Finish 9909." 328 A brand-named cationic-nonionic liquid employed as a softening agent that produces a minimum yellowing of white fabrics.

"Xynofix FL." 328 A brand-named dye-fixative of most direct colors on textile fabrics with no effect on the hand or color shade of dyed fabrics. It is a water-white viscous liquid practically odorless.

"Xynomine Paste." 328 Trademark for a product used as a detergent, wetting agent, dyeing assistant, emulsifier, bleaching assistant for peroxide, and scouring agent for wool yarn.

"Xyno Resin." 328 Proprietary name for a line of synthetic resins in emulsion and solution forms.

Grades: Vinyl and acrylic type emulsion polymers and copolymers include: "Xyno Resin" 362, X-99, AA-40, S-69, and AN-25.

Properties: Stable water emulsions of thermoplastic, filmforming polymers of various particle size and charge, that eliminate the use of expensive and flammable solvents. They combine high solid contents with low viscosities and may be compounded into a wide variety of binders, adhesives, sizing and coating compositions.

Uses: Textile sizes and finishes, including stiffening, bodying, dulling, slipproofing and other finishing applications; nylon hosiery finishing; paper coating for greaseproofing and heat sealing; binders and adhesive formulations; cosmetics.

"Xyno Resin" LVH. 328 Trade name for cationic vinyl acetate, 50% liquid.

Uses: Specialty paper coatings additive.

"Xyno Resin R." 328 Proprietary product consisting of modified water-soluble alkyd resin solution.

"Xynotaf HV; V; and LV." 328 Solutions of a thermoplastic, vinyl type polymer available in three different viscosities, and providing different degrees of stiffness.

Uses: Textile finishes to impart special stiffening and bodying effects and effective back-filling finish for a wide variety of fabrics.

Y. Symbol for yttrium.

yara-yara. See beta-naphthyl methyl ether.

"Yarmor." [266] Trademark for several grades of pine oil supplied to meet varying needs. They consist largely of terpineols with minor amounts of borneol, fenchyl alcohol, terpene ketones, and other terpene materials.

Grades and properties:

"Yarmor 302." High gravity pine oil; primarily a blend of tertiary and secondary terpene alcohols; colorless to pale yellow liquid; sp.gr. 0.9436 at 15.6/15.6°C; distillation range, 213-219°C.

"Yarmor 302W." Medium-gravity pine oil; a blend of terpenes and terpene alcohols; colorless to pale yellow liquid; sp.gr. 0.9246 at 15.6/15.6°C; distillation range 199-220°C.

"Yarmor 350." Low-gravity pine oil; colorless to pale yellow liquid; sp.gr. 0.9037 at 15.6/15.6°C; distillation range 187-222°C.

"Yarmor F." Flotation grade pine oil; primarily a blend of terpene alcohols and other oxygen-containing terpenes; amber colored liquid; distillation range 200-234°C.

Uses: Making disinfectants, insecticides, and essential oils; frothing agent in the flotation of ores; wetting and penetrating agent in the wet processing of textile fibers, liquid soaps, laundry detergents, metal cleaning compounds, water paints and paint emulsions, paint and varnish solvents, and other applications.

yarrow. See achillea.

yaw root. See stillingia.

Yb. Symbol for ytterbium.

yeast (barm). Living unicellular organisms known as saccharomycetaceae. The following description applies to the cultured commercial product, and not to various wild varieties.

Properties: Yellowish-white viscid liquid or soft mass, flakes, or granules, consisting of cells and spores of Saccharomyces cerevisiae.

Derivation: A ferment obtained in brewing. Yeasts induce fermentation through the agency of enzymes (zymases) which convert glucose and some other carbohydrates into carbon dioxide and water in the presence of oxygen or into alcohol and carbon dioxide (or lactic acid) in the absence of oxygen.

Grades: Technical; brewers'; cooking; compressed (contains about 74% moisture); dried yeast; N.F. XI (contains no starch or filler, not more than 7% moisture nor more than 8% ash). Also graded according to vitamin B_1 content.

Containers: Tins, boxes, drums, tank trucks.

Uses: Fermentation of sugars, molasses, and cereals for alcohol; brewing; medicine; baking bread and the like; important source of vitamin B complex.

Shipping regulations: None.*

yeast adenylic acid. See adenylic acid.

yeast nucleic acid. See ribonucleic acid.

yellow AB (1-benzeneazonaphthylamine-2; External D&C No. 9) $C_6H_5N_2C_{10}H_6NH_2$. Colour Index No. 22.

Properties: Orange or red platelets; m.p. 102-104°C; insoluble in water; soluble in alcohol and oils.

Use: Color for use in drugs and cosmetics, but not in foods.

yellow arsenic sulfide. See orpiment; arsenic trisulfide.

yellow cake. See triuranium octoxide.

yellow calisaya bark. See cinchona bark, calisaya.

yellow cinchona bark. See cinchona bark, calisaya.

yellow cobalt. See cobalt-potassium nitrite.

yellow copper. See chalcopyrite.

yellow dock. See rumex.

yellow gentian. See gentian.

yellow glass. Selenium is added to a soda lime glass.

yellow jasmin. See gelsemium.

yellow lakes. Pigments made by precipitating soluble yellow dyes on an aluminum hydrate base. The yellow lakes are transparent in oil and lacquer vehicles and are used for metal decorating finishes, such as cans, bottle caps, novelties, etc. They withstand the customary baking schedules for finishes of this type and in combination with orange lakes, produce gold shades. Aluminum hydrate has the desirable property of being transparent in oil and varnish vehicles, but unfortunately it also is reactive with some vehicles and care must be taken to be sure that the vehicle used is compatible with pigments containing aluminum hydrate. The yellow lakes do not have sufficient permanency for finishes that are to receive severe exterior exposure.

yellow moccasin flower. See cypripedium.

yellow OB (1-ortho-tolueneazonaphthyl-amine-2; External D & C No. 10) $CH_3C_6H_4N_2C_{10}H_6NH_2$. Colour Index No. 61.
Properties: Orange or yellow powder; m.p. 122-125°C; insoluble in water; soluble in alcohol and oils.
Use: Color for use in drugs and cosmetics, but not in foods.

yellow oxidation enzyme. See Warburg's yellow enzyme.

yellow precipitate. See mercuric oxide, yellow.

yellow prussiate of potash. See potassium ferrocyanide.

yellow prussiate of soda. See sodium ferrocyanide.

yellow puccoon. See hydrastis.

yellow root. See hydrastis.

yellow salt. See uranyl nitrate.

yellow saunders. See santalum album.

yellow ultramarine. See calcium chromate.

yellow wax. See beeswax.

yerba maté. The leaves of Ilex paraguayensis, a tree found in Paraguay. Used in South America in the same manner as tea, for a beverage.

yerba santa. See eriodictyon.

ylang ylang oil (mosoi flower oil; anona oil). A yellowish volatile oil distilled from the flowers of Cananga odorata, the principal sources of which are the Philippines and Réunion (Bourbon variety). The oil is often confused with cananga oil (q.v.), which comes from a Javanese variety of Cananga odorata. There is considerable difference in the constants, according to the quality and the fraction that is sold. The major components are linalool, geraniol and their esters, pinene, with small amounts of para-cresol methyl ether, and various other substances.
Constants:
Manila ylang ylang oil: Sp.gr. 0.911-0.958 (30/4°); optical rotation −27 to −49° 7' (mostly −32 to −45°); refractive index 1.4747-1.4940 (rarely over 1.4900); ester value 90-150 (usually 100 or more). This is the first fraction to distil over. Later fractions are sometimes called cananga oil.
Réunion ylang ylang oil: Sp.gr. 0.932-0.962 (15°C); optical rotation −34 to −64°; ester value 96-134.
Adulteration: Coconut fat, fatty oils, turpentine oil, alcohol, and petroleum.
Containers: Bottles.
Uses: Medicine; perfumery.
Shipping regulations: None.*

ylem. The original flux of neutrons that was the beginning of all matter.

yohimbine (aphrodine; corynine; quebrachine) $C_{21}H_{26}O_3N_2$. Poisonous! An alkaloid.
Properties: Glistening, needle-like alkaloid; m.p. 234°C; soluble in alcohol and ether; very slightly soluble in water.
Derivation: By extraction from the bark of Corynanthe yohimbé, found in the Cameroons.
Method of purification: Crystallization.
Grade: Technical.
Use: Medicine.
Shipping regulations: None.*

yohimbine hydrochloride $C_{21}H_{26}O_3N_2 \cdot HCl$.
Properties: White crystals or powder; m.p. 302°C; soluble in water and alcohol. Poisonous.
Containers: Bottles; tins.
Use: Medicine.

ysop. See hyssop.

ytterbia. See ytterbium oxide.

ytterbium Yb. One of the rare-earth metals (q.v.) of the yttrium group; atomic number 70; vanence +2 and +3; very scarce.
Properties: Metallic luster; quite malleable; m.p. 824°C; b.p. 1500°C (approx.); sp.gr. 7.01; reacts slowly with water; soluble in dilute acids and liquid ammonia. The solid dichloride when heated near the melting point decomposes into the trichloride and metallic ytterbium.
Derivation: Reduction of the oxide with lanthanum or misch metal.
Grades: Regular high purity (ingots and lumps).
Uses: Special alloys.

ytterbium oxide (ytterbia) Yb_2O_3.
Properties: Colorless mass when free of thulia but tinted brown or yellow when containing thulia. The weakest base of the yttrium group with the exception of scandia and lutetia. Slightly hygroscopic, absorbs water and carbon dioxide from the air; sp. gr. 9.2. Soluble in hot, dilute acids, less so in cold acids.
Grades: Purities to 99.9%.
Containers: Glass bottles; fiber drums.
Uses: Special alloys; dielectric ceramics.
Shipping regulations: None.*

ytterbium salts.
ytterbium chloride $YbCl_3 \cdot xH_2O$. Available as 45% Yb_2O_3.
ytterbium fluoride $YbF_3 \cdot xH_2O$. Available as 77% Yb_2O_3.
ytterbium nitrate $Yb(NO_3)_2 \cdot 6H_2O$. Available as 43% Yb_2O_3.
ytterbium sulfate $Yb_2(SO_4)_3 \cdot 8H_2O$. Colorless prisms; sp.gr. 3.29; soluble in cold water.
Grades: (all salts) 99%; 99.9% Yb salts.
Containers: Glass bottles; fiber drums.

yttria. See yttrium oxide.

yttrialite. One of the rare-earth minerals. It is a silicate of the yttrium metals (43 to 47%), thorium (10 to 20%), and cerium metals (5 to 8%). Color on the fresh fracture olive-green, changing to orange-yellow on surface.
Constants: Sp.gr. 4.575.
Occurrence: Texas.

yttrium Y. Element of atomic number 39, of group III of the periodic table. Historically considered one of the rare-earth metals, but strictly speaking, not a member of this group.

Properties: Dark gray metal; sp.gr. 4.47; m.p. 1500°C; b.p. 3200°C (approx.). Soluble in dilute acids and potassium hydroxide solution; decomposes water; known only in the tripositive state.

Derivation: Reduction of the fluoride with calcium.

Impurities: Rare earths.

Grades: Regular high purity (ingots, lumps, turnings); metallurgical; low-oxygen; crystal sponge.

Containers: Wooden kegs or fiber drums.

Uses: In nuclear technology because of its high neutron transparency; iron alloys; incandescent gas mantles.

Shipping regulations: None.*

yttrium acetate $Y(C_2H_3O_2)_3 \cdot 8H_2O$.

Properties: Colorless crystals. Soluble in water.

Derivation: By the action of acetic acid on yttrium oxide.

Method of purification: Crystallization.

Impurities: Rare earths.

Grade: Technical.

Containers: Glass bottles.

Use: Analytical chemistry.

Shipping regulations: None.*

yttrium antimonide YSb. Used as a high-purity binary semiconductor.

yttrium arsenide YAs. Used as a high-purity binary semiconductor.

yttrium bromide $YBr_3 \cdot 9H_2O$.

Properties: Colorless crystals. Hygroscopic. Soluble in water; slightly soluble in alcohol; insoluble in ether.

yttrium carbonate $Y_2(CO_3)_3 \cdot 3H_2O$.

Properties: Reddish-white to white powder. Soluble in acids; insoluble in water, alcohol and ether.

Derivation: By the interaction of solutions of yttrium chloride and sodium carbonate.

Impurities: Erbium salts.

Grades: Technical.

Containers: Wooden kegs or fiber drums.

Use: Incandescent gas mentles.

Shipping regulations: None.*

yttrium chloride $YCl_3 \cdot 6H_2O$.

Properties: Reddish-white, transparent, deliquescent prisms. Soluble in water and alcohol; insoluble in ether. Sp.gr. 2.18; decomposes at 100°C.

Derivation: By the action of hydrochloric acid on yttrium oxide.

Method of purification: Crystallization.

Impurities: Erbium salts.

Grades: Purities to 99%.

Containers: Glass bottles.

Use: Analytical chemistry.

Shipping regulations: None.*

yttrium fluoride $YF_3 \cdot xH_2O$.

Grade: Up to 99.9+% yttrium salts (available as 65% Y_2O_3).

Containers: Glass bottles.

yttrium-iron garnet. Used for microwave devices. See garnet.

yttrium nitrate $Y(NO_3)_3 \cdot 6H_2O$.

Properties: Reddish-white crystals; loses $3H_2O$ at 100°C; soluble in water, alcohol and nitric acid.

Derivation: By the action of nitric acid on monazite sand.

Method of purification: Fractional crystallization.

Impurities: Rare earths.

Grades: Purities to 99%.

Containers: Glass bottles.

Use: Production of yttrium oxide.

Fire hazard: Dangerous.

Shipping regulations: Oxidizing material. Yellow label.*

yttrium oxalate $Y_2(C_2O_4)_3 \cdot xH_2O$.

Properties: White powder, insoluble in water, slightly soluble in acid. Purities up to 75% yttrium salt.

yttrium oxide (yttria) Y_2O_3.

Properties: Yellowish-white powder. Soluble in dilute acids; insoluble in water. Sp.gr. 4.84; m.p. 2410°C.

Derivation: By the ignition of yttrium nitrate.

Impurities: Rare earths.

Grades: Purities to 99.8%.

Containers: Wooden kegs or fiber drums.

Use: Incandescent gas mantles; optical glasses; special ceramics; electronics; arc welding; organic synthesis.

Shipping regulations: None.*

yttrium phosphate. See xenotime.

yttrium phosphide YP. Used as a high-purity binary semiconductor.

yttrium sulfate $Y_2(SO_4)_3 \cdot 8H_2O$.

Properties: Small reddish-white, monosymmetric crystals. Soluble in concentrated sulfuric acid; sparingly soluble in water; insoluble in alkalies. Sp.gr. 2.558.

Derivation: By the action of sulfuric acid on monazite sand.

Method of purification: Fractional crystallization.

Grade: C.P.

Containers: Glass bottles.

Use: Reagent.

Shipping regulations: None.*

Z

"Zac." [20] Trademark for electrically melted and cast refractories containing zirconia, alumina, and silica for use in glass-melting units.

"Zactirin" Citrate. [24] Trademark for ethoheptazine citrate. See ethoheptazine.

"Zalba." [28] Trademark for a rubber antioxidant containing a hindered phenol.
Properties: Amber colored viscous liquid; sp.gr. 0.94.
Containers: 50-lb drums.
Use: Nondiscoloring antioxidant for natural and synthetic rubbers and latex. A stabilizer for the manufacture of SBR.
"Zalba" Special Rubber Antioxidant is a fortified hindered phenol; yellow cream colored powder; sp.gr. 1.27.

"Zamak." [268] See "Horse Head Zamak."

"Zanchol." [70] Trademark for a brand of florantyrone, gamma-oxo-gamma-(8-fluoranthene)butyric acid.
Use: Medicine.

Zanzibar gum. A hard, usually fossil type of copal.
Derivation: Found on the island of Zanzibar and the adjoining African mainland.
Properties: Sp.gr. 1.062-1.068; m.p. 240-250°C; insoluble in most solvents.
Grades: Bean, pea, sorts.
Containers: Bags.
Use: Varnishes.
Shipping regulations: None.*

"Zarontin." [330] Trade name for ethosuximide (q.v.).

"Z-B-X." [248] Trademark for zinc butyl xanthate.
Properties: White powder; sp.gr. 1.45; decomposes when heated; moderately soluble in benzol and ethylene dichloride; slightly soluble in acetone; insoluble in water and gasoline.
Uses: A low temperature nonstaining and nondiscoloring ultra accelerator used in rubber cements.

"Z-C" Spray. [55] Trademark for a fungicidal spray containing 70% ziram (q.v.).

ZDP. Abbreviation for zinc dithiophosphate. See "Ethyl."

"Zeecon." [48] Trademark for sodium lignin sulfonate with controlled wood sugar content. Used as a cement dispersant in the manufacture of concrete.

"Zefran." [233] Trademark of an acrylic fiber in white staple form, based on polyacrylonitrile and supplemented with a dye-receptive component; used in fibers, filaments, and yarns.

zein.
Properties: White to slightly yellow powder. An odorless, nontoxic protein of the prolamine class, derived from corn. Tasteless. It is free of cystine, lysine, and tryptophane. It is a resinous material dispersible in water with neutral sulfonated castor oil. Soluble in dilute alcohol; insoluble in water, dilute acids, anhydrous alcohols, turpentine, esters, oils, fats. Sp. gr. 1.226.
Derivation: A by-product of corn processing. The commercial production process consists of the extraction of gluten meal with 85% isopropyl alcohol, clarification of the extract, extraction of the zein from the extract with hexane, precipitation by water and spray drying.
Containers: Bags.
Uses: Plastics; paper coating; grease-resistant coatings; adhesives; laminated board; solid color prints; printing inks; films and fibers; coating formulations. Basis of imitation shellacs.

"Zelan." [28] Trademark for a line of durable water repellent textile finishes based on a long chain nitrogen complex. Used principally on rainwear fabrics.

"Zelcon" C. [28] Trademark for a cationic fabric softener and conditioner, 100% active, viscous amber fluid.
Use: As the active ingredient in compounded home laundry fabric softeners and conditioners.

"Zelec." [28] Trademark for a line of antistatic agents including both durable and non-durable ones. The non-durable ones are used principally as textile processing assistants, the durable ones as finishes for the newer non-cellulosic synthetic fiber fabrics.

"Zenite." [28] Trademark for a line of rubber accelerators based on zinc salt of 2-mercaptobenzothiazole, with or without various modifying agents.
Properties: Pale yellow powder.
Use: To accelerate and improve the vulcanization of natural and synthetic rubber and latex compounds.

"Zeo-Dur." [184] Trademark for a processed glauconite (naturally occurring greensand) cation exchanger.

"Zeogel." [236] Brand name for a special clay used in drilling fluids to give an equally high yield and stable viscosity and gel characteristics in either fresh or salt water, regardless of the concentration of salt in the latter.

"Zeo-Karb." [184] Trademark for a sulfonated coal type, acid resistant, cation exchanger.

"Zeolex." [285] Proprietary brand name for a series of precipitated, hydrated sodium silico-aluminates and sodium calcium silico-aluminates (microlitic zeolites).

Properties: Sp.gr. 2.1; bulk density 3 lbs/cu ft aerated, 15-20 lbs/cu ft packed; mean particle diameter ranges of 0.02-0.03 microns and 0.03-0.05 microns; pH 7-11.5; refractive index 1.52-1.55; various oil absorbencies; 325 mesh screen residue 0.1% max; in bright, white powder form.

Containers: 50-lb multiwall, moisture-barrier bags.

Uses: Has maximum reinforcing properties and contributes extra long life to rubber for footwear, flooring, rolls, etc.; as a high-concentrate, highly-sorptive carrier for pesticide dust bases and wettable powders; as a conditioner to promote free flow and non-caking in table salt, other food products, and various deliquescent or efflorescent products; as a bulking agent, bodying agent, and flatting agent; as a mild abrasive and polishing agent; in paperboard specialties, writing, offset and book stocks to develop brightness and opacity at minimum cost; to condition certain plastic molding powders.

zeolite process. Water softening process using zeolites (q.v.).

zeolites. A class of hydrated silicates of aluminum and either sodium or calcium or both, of the type $Na_2O \cdot Al_2O_3 \cdot nSiO_2 \cdot xH_2O$.

The term originally described a group of naturally occurring minerals, principally sodium or calcium aluminosilicates. Both natural and artificial zeolites are now used extensively for water softening. For this purpose the sodium or potassium compounds are required, since their usefulness depends on the cationic exchange of the sodium of the zeolite for the calcium or magnesium of the hard water. When the zeolite has become saturated with calcium or magnesium ion, it is flooded with strong salt solution, a reverse exchange of cations takes place, and the material is regenerated for use again. The natural zeolites are analcite, chabazite, heulandite, natrolite, stilbite, and thomsonite. See also glauconite, which is similar in use.

Artificial zeolites are made in a variety of forms ranging from gelatinous to porous and sandlike and are used as gas adsorbents and drying agents as well as water softeners. The term zeolite now includes such diverse groups of compounds as sulfonated organics or basic resins, which act in a similar manner to effect either cation or anion exchange. See ion exchange resins.

Containers: Bags; drums.

"Zeo-Rex." [184] Trademark for a sulfonated coal type, acid-resistant cation exchanger.

"Zeotone." [108] Trademark for a powdered, corrosion-inhibiting, hexametaphosphate compound specially formulated for cleaning and disinfecting domestic and industrial water softeners.

Containers: 2-lb cans; (12/case); 100-lb drums.

"Zepar" BP. [28] Trademark for a reducing agent. White free-flowing granular powder readily soluble in water.

Use: Bleaching agent for groundwood pulp.

"Zephfleur." [188] Trademark for identifying a series of perfume bases for colognes.

"Zephiran" Chloride. [162] Trademark for benzalkonium chloride.

Zerewitinoff reagent. Solution of methyl magnesium iodide in purified n-butyl ether. A clear, light colored liquid which reacts rapidly with moisture and oxygen.

Uses: As analytical reagent for active hydrogen atoms in organic compounds; also to determine water, alcohols, and amines in inert solvents.

"Zerex." [28] Trademark for anti-rust anti-freeze based on ethylene glycol and containing MR8 rust inhibitor that protects all metals including aluminum. A clear, odorless, slightly viscous liquid with a fluorescent color.

Containers: 1-gal cans, 6/carton; 1-qt cans, 24/carton.

Uses: As a high-boiling anti-freeze for combustion engines, particularly at high altitudes or very low temperatures or under severe driving conditions.

"Zerice." [51] Trademark for light lubricating oils suitable for many general machinery applications, particularly where low temperatures are encountered. Used in refrigerating machines having low evaporator temperatues.

"Zerlate." [28] Trademark for agricultural and horticultural fungicide based on ziram (zinc dimethyldithiocarbamate).

Containers: 3- and 50-lb bags.

Use: Control of certain vegetable diseases and brown rot of peaches.

"Zerlon." [233] Trademark for a methyl methacrylate-styrene copolymer used as a plastic molding grade material.

"Zerok." [41] Trade name for a synthetic resin coating of the vinyl type which is resistant to oxidizing acids and useful for protection against fumes and splashing.

"Zerone." [28] Trademark for anti-rust anti-freeze based on methanol and containing a dye and chemical inhibitors. A clear violet mobile liquid with odor of methanol.

Containers: 54-gal drums; 1-gal cans, 6/carton; 1-qt cans, 24/carton.

Uses: Anti-freeze for combustion engines,

recommended for use in cooling systems with normal thermostats operating below 160°F.

Precaution: Same as for methanol.

"Zeset." [28] Trademark for a reactive type finish for imparting durable crease resistance, dimensional stability, and hydrophobic characteristics to cellulosic textiles.

"Zetafin." [233] An ethylene-ethyl acrylate copolymer with properties similar to flexible polyvinyl polymers.

zeta potential (electrokinetic potential). The potential across the interface of all solids and liquids. More specifically, it is the potential across the diffuse layer of ions surrounding a charged colloidal particle; and is largely responsible for colloidal stability. Destruction of the zeta potential, accompanied by precipitation of the colloid, can occur by addition of polyvalent ions of sign opposite to that of the colloidal particles. (This principle is used in water purification.) Zeta potentials can be calculated from electrophoretic mobilities, i.e., the rates at which colloidal particles travel between charged electrodes placed in the solution.

"Zetax." [69] Trademark for zinc 2-mercaptobenzothiazole.

Properties: Pale yellow powder; sp.gr. 1.70 ± .03; melting point above 300°C; zinc content 15-18%.

Uses: In natural rubber, primary accelerators, activator of benzothiazyl disulfide and for ultras or combinations, and accelerator for low sulfur. In nitrile rubber and SBR, primary accelerator. In butyl, activator for ultras. Also as latex accelerator for foams and films.

zibeth. See civet.

Ziegler catalysts. Particular types of stereospecific catalysts which are usually chemical complexes derived from a transition metal halide and a metal hydride or a metal alkyl. The transition metal may be any of a large number of those in groups IV to VIII of the periodic table; the hydride or alkyl metals are those of groups I, II, and III. Typically, titanium chloride is added to aluminum alkyl in a hydrocarbon solvent to form a dispersion or precipitate of the catalyst complex. These catalysts usually operate at atmospheric pressure and are used to convert ethylene to linear polyethylene, and also in stereospecific polymerization of propylene to crystalline polypropylene.

Ziegler process. A process for polymerizing ethylene or propylene into linear polyethylene or polypropylene. Higher olefins may also be used. The process features low pressures and aluminum alkyls as catalysts.

Ziehl's stain. See carbolfuchsin.

"Zilloy." [268] See "Horse Head Zilloy."

Zimmermann process. A process for waste disposal used for sewage sludge and industrial wastes, such as sulfite pulp from paper mills. The organic material is oxidized with air in water at high temperature (up to 700°F) and pressure (up to 1800 psi). The water from the process may then be treated by conventional processes and released into streams.

"Zinar." [79] Trademark for a high melting, pale colored zinc resinate which has a negative acid number (slightly basic).

Constants: Sp.gr. (25°C) 1.150; m.p. (capillary tube) 160°C; metallic zinc (combined) 5.6%; color M.

Containers: Non-returnable, light gauge galvanized drums containing about 530 lbs net. Tare 14-16 lbs.

Uses: Paints; varnishes; adhesives; linoleum print paint; ethyl cellulose lacquers.

zinc Zn. Element of atomic number 30, Group II of periodic table. Referred to in metal trades as spelter.

Properties: Shining white metal with bluish gray luster. Not found native. Soluble in acids and alkalies. Insoluble in water. Sp. gr. 7.14; m.p. 419°C; b.p. 907°C.

Ores and minerals: See calamine, franklinite, hydrozincite, smithsonite, sphalerite, willemite, wurtzite, zincite.

Derivation: By roasting the ore, and then heating with carbon out of contact with air. Also obtained by electrolysis.

Grades: Technical; high purity (impurities less than 10 ppm).

Forms available: Slab, rolled (strip, sheet, rod, tubing), wire, mossy zinc, zinc dust (see separate entry), zinc anodes.

Uses: Alloys such as brass, bronze, Babbitt, German silver, and many die-casting alloys; galvanizing iron and other metals; electroplating; metal spraying; automotive parts; electrical fuses, anodes, meter cases; household articles, roofing, gutters; engravers' plates; cable wrappings, organ pipes, etc.

zinc 65. Radioactive zinc of mass number 65.

Properties: Half-life, 250 days; radiation, beta, gamma and K.

Derivation: Pile irradiation of zinc metal and, in the cyclotron, by bombarding copper 65 with deuterons.

Forms available: Zinc metal and zinc chloride in hydrochloric acid solution.

Uses: To study wear in alloys, the nature of phosphor activators, galvanizing, the function of traces of zinc in body metabolism, the functions of oil additives in lubricating oils, etc.

Shipping regulations: Class D poison, radioactive material. Red label.*

zinc acetate $Zn(C_2H_3O_2)_2 \cdot 2H_2O$.

Properties: White, monoclinic, crystalline plates; pearly luster; faint acetous odor; astringent taste. Soluble in water and alcohol. Sp.gr. 1.735; m.p. 237°C.

Derivation: By the action of acetic acid on

*See "I.C.C. Shipping Regulations," page xiii.

Reference numbers refer to name of manufacturer. See "List of Manufacturers," page v.

zinc oxide.
Method of purification: Crystallization.
Grades: Technical; C.P.
Containers: 1-, 5-lb bottles and cartons; fiber drums.
Uses: Medicine; preserving wood; textiles (mordant in dyeing with alizarin blue and similar dyes, resist in dyeing with aniline black, substitute for tartar emetic in dyeing with basic dyes, calico printing); zinc chromate; laboratory reagent; feed additive.
Shipping regulations: None.*

"Zincalume." [72] Trade name for a bright zinc plating process; composed of zinc cyanide, sodium cyanide, sodium hydroxide and addition agents.

zinc ammonium chloride $ZnCl_2 \cdot 2NH_4Cl$. A complex salt. Double salts with 3 or 6 molecules ammonium chloride have also been prepared.
Properties: White powder or crystals; soluble in water; sp.gr. 1.8.
Grade: Technical (foaming and non-foaming).
Containers: 100-lb paper bags; 400-lb barrels.
Uses: Welding; soldering flux; dry batteries; galvanizing.
Shipping regulations: None.*

zinc ammonium nitrite.
Shipping regulations: Oxidizing material. Yellow label.*

zinc ammonium sulfate (ammonium zinc sulfate) $ZnSO_4 \cdot (NH_4)_2SO_4 \cdot 6H_2O$.
Properties: White crystals. Soluble in water.

zinc antimonide. Used in thermoelectric devices.

zinc arsenate. Variable composition approximating $5ZnO, 2As_2O_5, 4H_2O$. Occurs in nature as mineral kottigite, $Zn_3(AsO_4)_2 \cdot 8H_2O$. Insoluble in water; soluble in acids or alkalies. White, odorless powder. Poisonous.
Preparation: Reaction of a solution of sodium arsenate and a soluble zinc salt.
Use: Insecticide.
Shipping regulations: Class B poison. Poison label.*

zinc arsenite (zinc meta-arsenite; ZMA) $Zn(AsO_2)_2$.
Properties: Colorless powder, soluble in acids, insoluble in water. Federal Specification TT-W-581 describes the composition of the solution used for wood preservation.
Uses: Timber preservative; insecticide.
Warning: Poisonous if swallowed. MCA warning label.
Shipping regulations: Class B poison. Poison label.*

zinc bacitracin.
Properties: Creamy-white powder; slightly soluble in water. May be used in formulations requiring heat processing. Usually has 50-60 units/mg of bacitracin activity.
Derivation: By the action of zinc salts on bacitracin broth.
Use: Preserving silage for feed.

zinc baryta white. See lithopone.

zinc benzoate $Zn(C_6H_5CO_2)_2$.
Properties: White powder; slightly soluble in water.
Use: Medicine.

zinc bichromate. See zinc dichromate.

zinc blende. See sphalerite.

zinc bloom. See hydrozincite.

zinc borate. Of indefinite composition, containing zinc oxide (ZnO) and boric oxide (B_2O_3) in various ratios. A typical specification is ZnO 45%, B_2O_3 34%. May have 20% water of hydration.
Properties: White, amorphous powder; soluble in dilute acids; slightly soluble in water.
Derivation: Interaction of the oxides at 500-1000°C, or of zinc oxide slurries with solutions of boric acid or borax.
Method of purification: Recrystallization.
Grade: Technical.
Containers: Tins, bags.
Use: Medicine; fireproofing textiles; fungistat and mildew inhibitor; flux in ceramics.

zinc bromide $ZnBr_2$.
Properties: White, hygroscopic, crystalline powder. Soluble in water, alcohol, and ether. Sp.gr. 4.219; m.p. 394°C; b.p. 650°C.
Derivation: By the interaction of solutions of barium bromide and zinc sulfate, with subsequent crystallization.
Method of purification: Crystallization.
Grades: Technical; C.P.
Containers: Glass bottles.
Uses: Medicine; photography (plates, papers); manufacture and finishing of rayon. A solution of 80% zinc bromide is used as a radiation viewing shield.
Shipping regulations: None.*

zinc, butter of. See zinc chloride.

zinc cadmium sulfide. A fluorescent pigment; a phosphor.

zinc calcium resinate. A mixed soap similar to zinc resinate, but containing varying proportions of calcium.

zinc caprylate $Zn(C_8H_{15}O_2)_2$.
Properties: Lustrous scales; slightly soluble in boiling water, fairly soluble in boiling alcohol. M.p. 136°C. Decomposes in moist atmosphere giving off caprylic acid.
Derivation: By precipitating from a solution of ammonium caprylate with zinc sulfate.
Use: Fungicide.

zinc carbolate. See zinc phenate.

zinc carbonate $ZnCO_3$.
Properties: White, crystalline powder. Soluble in acids, alkalies and ammonium salt solutions; insoluble in water. Sp.gr. 4.42-4.45; loses carbon dioxide at 300°C.
Derivation: (a) By grinding the mineral smithsonite; (b) By the action of sodium

bicarbonate on a solution of a zinc salt.
Method of purification: Crystallization.
Impurities: Zinc oxide; zinc hydroxide.
Grades: Technical; C.P.
Containers: 1-lb bottles; 25-, 50-lb boxes; 100-, 150-lb kegs; 250-, 300-lb barrels; multiwall paper sacks.
Uses: Pigment; ceramics; fire-proofing agent; cosmetics and lotions; pharmaceuticals (ointments, dusting powders); zinc salts; feed additive.
Shipping regulations: None.*

zinc carbonate, basic. See zinc carbonate, precipitated.

zinc carbonate, precipitated (zinc subcarbonate; zinc carbonate, basic; tutia). $2ZnCO_3 \cdot 3Zn(OH)_2$; composition variable.
Properties: Impalpable, white powder. Soluble in dilute acids, ammonium hydroxide and ammonium carbonate solution; insoluble in alcohol.
Derivation: By the action of sodium carbonate on a solution of a zinc salt.
Impurities: Zinc oxide.
Grade: Technical.
Containers: Wooden barrels, fiber drums, or tins.
Uses: Medicine (similar to zinc oxide); pigment.
Shipping regulations: None.*

zinc chlorate $Zn(ClO_3)_2 \cdot 4H_2O$.
Properties: Colorless to yellowish crystals; deliquescent; sp.gr. 2.15; decomposes at 60°C. Very soluble in water and alcohol; soluble in glycerin and ether.
Shipping regulations: Oxidizing material. Yellow label.*

zinc chloride (butter of zinc) $ZnCl_2$.
Properties: White granular deliquescent crystals or crystalline powder; poisonous; soluble in water, alcohol, glycerine and ether. Sp.gr. 2.91; m.p. 290°C; b.p. 732°C.
Derivation: By the action of hydrochloric acid on zinc or zinc oxide.
Method of purification: Recrystallization.
Grades: C.P.; technical. Fused; crystal; granulated; 50% solution; N.F. XI.
Containers: Bottles; drums. Solutions in tank cars and tank trucks.
Uses: Galvanizing iron; catalyst, dehydrating and condensing agent in organic synthesis; wood preservative; ingredient of soldering fluxes; burnishing and polishing compounds for steel; electroplating; disinfectant; antiseptic and deodorant preparations; textiles (mordant, carbonizing agent, mercerizing, sizing and weighting compositions, resist for sulfur colors, albumin colors and para red); cold water glues and other adhesives; special cements; glass etching compositions; petroleum refining; parchment paper; vulcanized fiber; dental cements; dentifrices; embalming and taxidermists fluids; candles; vulcanizing rubber; medicine; dyestuffs; pigments; antistatic; feed additive; some promise for fuel cells.
Warning: May cause severe skin irritation. MCA warning label.
Shipping regulations: None.*

zinc chloride, chromated. A mixture of zinc chloride and sodium dichromate used as a wood preservative. Federal Specification TT-W-551 requires that it contain not less than 77.5% zinc chloride and 17.5% sodium dichromate dihydrate. See also CZC Chromated Zinc Chloride.

zinc chloride, chromated, copperized. See Copperized CZC Chromated Zinc Chloride.

zinc chloroiodide. A mixture of zinc chloride and iodide.
Properties: White powder. Soluble in water.
Containers: Glass bottles.
Grades: Technical.
Uses: Disinfectant; pharmaceutical preparations.
Shipping regulations: None.*

zinc chromate. Of variable composition. The C.P. salt is alleged to be $ZnCrO_4 \cdot 7H_2O$, a yellow, crystalline powder. A series of compounds are known in which the ratio of ZnO to CrO_3 is 5:1, 4:1, 2:1, 1:1 and 1:2, usually with some water of crystallization. The basic compounds containing the highest zinc content are more insoluble and stable than the acidic compounds.
Derivation: By the action of chromic acid on slurries of zinc oxide, or on zinc hydroxide.
Grades: Technical; C.P.
Containers: 1-lb bottles; barrels.
Use: Pigments; artists' color; varnishes; linoleum.
See zinc yellow, which is principally zinc potassium chromate.

zinc chrome. See zinc yellow.

zinc citrate $Zn_3(C_6H_5O_7)_2 \cdot 2H_2O$.
Properties: White, amorphous powder. Slightly soluble in water.
Derivation: By the action of citric acid on zinc hydroxide.
Method of purification: Crystallization.
Grade: Technical.
Containers: Glass bottles.
Use: Medicine.
Shipping regulations: None.*

zinc cyanide $Zn(CN)_2$.
Properties: White powder; poisonous! Decomposes on heating; soluble in dilute mineral acids with production of hydrogen cyanide; insoluble in water and alcohol.
Derivation: By precipitation of a solution of zinc sulfate or chloride with potassium cyanide.
Grades: Technical.
Containers: Glass bottles; kegs; drums.
Uses: Medicine; metal plating; chemical reagent; insecticide; purifying illuminating gas.
Caution: Contact with acid liberates poisonous gas. Avoid breathing gas or dust and avoid contact with skin.

zinc dibutyldithiocarbamate $Zn[SC(S)N(C_4H_9)_2]_2$.
Properties: White powder; sp.gr. (20/20°C)

1.24; melting range 104-108°C; pleasant odor, nontoxic. Soluble in carbon disulfide, benzene, and chloroform; insoluble in water.
Use: Rubber vulcanization accelerator; lubricating oil additive.

zinc dichromate (zinc bichromate) $ZnCr_2O_7 \cdot 3H_2O$.
Properties: Orange-yellow powder; soluble in acids and hot water; insoluble in alcohol and ether.
Derivation: By the action of chromic acid on zinc hydroxide.
Grade: Technical.
Containers: Wooden kegs or fiber drums.
Use: Pigment.
Shipping regulations: None.*

zinc diethyl. See zinc ethyl.

zinc diethyldithiocarbamate $Zn[SC(S)N(C_2H_5)_2]_2$.
Properties: White powder; sp.gr. (20/20°C) 1.47; melting range 172-176°C; nontoxic. Soluble in carbon disulfide, benzene and chloroform; insoluble in water.
Containers: Drums.
Use: Rubber vulcanization accelerator.

zinc dimethyldithiocarbamate. See ziram.

zinc dimethyldithiocarbamate cyclohexyl amine complex (zinc dithioamine complex). White powder or slurry of low solubility, used as a fungicide and rat poison.

zinc dioxide. See zinc peroxide.

zinc dithioamine complex. See zinc dimethyldithiocarbamate cyclohexylamine complex.

zinc dithionite. See zinc hydrosulfite.

zinc dithiophosphate. Term used for commercial dialkyl zinc dithiophosphate. See "Ethyl."

zinc dust. A gray powder. May form explosive mixtures with air; in bulk when damp may heat and ignite spontaneously on exposure to air.
Grades: Commercial; pigment.
Containers: Barrels.
Uses: Zinc salts and other zinc compounds; as reducing agent, precipitating agent, purifier, catalyst, polymerizing agent; in rust-resistant paints; bleaches; pyrotechnics; soot-removal; pipe-thread compounds; sherardizing.

zinc ethyl (zinc diethyl; better, diethylzinc). $Zn(C_2H_5)_2$.
Properties: Colorless liquid; takes fire on contact with air; sp.gr. (20°C) 1.207; m.p. −28°C; b.p. 118°C; decomposes in water.
Derivation: By the action of ethyl iodide on zinc and sodium-zinc; or by reacting zinc chloride with triethyl aluminum.
Grade: Technical.
Containers: Sealed tubes; steel cylinders.
Use: Organic synthesis; polyolefin catalyst; production of ethyl mercuric chloride.
Fire hazard: Dangerous.
Shipping regulations: Flammable liquid. Red label.*

zinc ethylenebisdithiocarbamate. See zineb.

zinc 2-ethylhexoate. See zinc octoate.

zinc ethylsulfate $Zn(C_2H_5SO_4)_2 \cdot 2H_2O$.
Properties: Clear, colorless, hygroscopic, crystalline leaflets. Keep well stoppered. Soluble in water and alcohol.
Derivation: By the interaction of zinc hydroxide and diethyl sulfate.
Grade: Technical.
Containers: Glass bottles; tins.
Use: Organic synthesis.
Shipping regulations: None.*

zinc ferrocyanide $Zn_2Fe(CN)_6 \cdot 3H_2O$.
Properties: White powder; decomposes on heating. Soluble in ammonium hydroxide; insoluble in water and hydrochloric acid.
Derivation: By the interaction of zinc sulfate and potassium ferrocyanide.
Grade: Technical.
Containers: Glass bottles; boxes.
Use: Medicine.
Shipping regulations: None.*

zinc, flowers of. See zinc oxide.

zinc fluoborate $Zn(BF_4)_2$. Handled as 40 or 48% solution, a colorless liquid.
Uses: Plating and bonderizing; resin curing.

zinc fluoride ZnF_2.
Properties: White powder; poisonous! Soluble in hot acids; slightly soluble in water; insoluble in alcohol. Sp.gr. (15°C) 4.84; m.p. 872°C.
Derivation: (a) By the action of hydrofluoric acid on zinc hydroxide; (b) By the addition of sodium fluoride to a solution of zinc acetate.
Grade: Technical, about 95% pure.
Containers: Drums, barrels.
Uses: Ceramic glazes and enamels; impregnating lumber; galvanizing.
Shipping regulations: None.*

zinc fluosilicate. See zinc silicofluoride.

zinc formaldehyde sulfoxylate $Zn(HSO_2 \cdot CH_2O)_2$ (normal); $Zn(OH)(HSO_2 \cdot CH_2O)$ (basic).
Properties: Rhombic prisms. Very soluble in water (normal); insoluble in alcohol. Decomposes in acid.
Derivation: Reaction of formaldehyde and zinc sulfoxylate.
Grades: Basic; normal.
Containers: 250-lb and 300-lb drums.
Uses: Stripping and discharging agent for textiles.
See also hydrosulfite-formaldehyde compounds.

zinc formate $Zn(CHO_2)_2 \cdot 2H_2O$.
Properties: White crystals; sp.gr. (20°C) 2.207; decomposes on heating. Soluble in water; insoluble in alcohol.
Derivation: By the action of formic acid on zinc hydroxide.
Method of purification: Crystallization.
Grade: Technical.
Containers: Glass bottles.
Uses: Medicine; catalyst for production of methyl alcohol; waterproofing agent; textiles; wood preservative.

zinc glass. A glass in which zinc oxide (ZnO) replaces part of the calcium oxide of ordinary lime soda glass.

zinc gluconate. Dietary supplement accepted by F.D.A.

zinc glycerinophosphate. See zinc glycerophosphate.

zinc glycerophosphate (zinc glycerinophosphate) $C_3H_5(OH)_2OPO_3Zn$.
Properties: White, amorphous powder. Soluble in water; insoluble in alcohol and ether.
Derivation: By the action of glycerophosphoric acid on zinc hydroxide.
Method of purification: Crystallization.
Grade: Technical.
Containers: Glass bottles.
Use: Medicine.
Shipping regulations: None.*

zinc greens. Brilliant green pigments mostly consisting of mixtures of Prussian blue and zinc yellow. They are permanent to light but not to alkali or water. Mostly used for flat wall paints and interior work.

zinc hydrosulfite (zinc dithionite) (ZnS_2O_4).
Properties: White, amorphous solid; soluble in water.
Grade: Technical.
Containers: Cartons, 50-lb pails, 275-lb drums.
Uses: Brightening ground wood, kraft, and other paper pulps; treatment of beet and cane sugar juices; as a depressant in mining flotations; and for bleaching textiles, vegetable oils, straw, hemp, vegetable tannins, animal glues, etc.

zinc hypophosphite $Zn(H_2PO_2)_2 \cdot H_2O$.
Properties: White hygroscopic crystals. Keep well stoppered. Soluble in water and alkalies.
Derivation: By the action of hypophosphorous acid on zinc hydroxide.
Method of purification: Crystallization.
Grade: Technical.
Containers: Glass bottles.
Use: Medicine.
Shipping regulations: None.*

zinc iodate $Zn(IO_3)_2$. (May contain water of crystallization, 1- or $2H_2O$).
Properties: White, crystalline powder; sp.gr. (anhydrous) 5.06. Soluble in nitric acid and alkalies; very slightly soluble in water.
Derivation: By the interaction of barium iodate and zinc sulfate, with subsequent crystallization.
Method of purification: Recrystallization.
Grade: Technical.
Containers: Glass bottles.
Use: Medicine (topical).

zinc iodide ZnI_2.
Properties: Hygroscopic, white, crystalline powder; sharp, saline taste. Keep well stoppered. Turns brown on exposure to light or air. Soluble in water, alcohol and alkalies. Sp.gr. 4.67; m.p. 446°C; b.p. 625°C.
Derivation: By the interaction of barium iodide and zinc sulfate, with subsequent crystallization.
Method of purification: Recrystallization.
Grade: Technical.
Containers: Wooden kegs; glass bottles.
Use: Medicine; analytical reagent.

zinc isovalerate. See zinc valerate.

zincite (red zinc ore; zinc oxide, red) ZnO. Natural zinc oxide, usually with some manganese.
Properties: Color red to orange-yellow; luster subadamantine; streak orange-yellow; hardness 4-4.5; sp.gr. 5.6.
Occurrence: New Jersey.
Use: Ore of zinc.

zinckenite. See zinkenite.

zinc lactate $Zn(C_3H_5O_3)_2 \cdot 3H_2O$.
Properties: White crystals. Soluble in water.
Derivation: By the action of lactic acid on zinc hydroxide.
Method of purification: Crystallization.
Grade: Technical.
Containers: Glass bottles.
Use: Medicine.
Shipping regulations: None.*

zinc laurate $Zn(C_{12}H_{23}O_2)_2$.
Properties: White powder; m.p. 128°C; insoluble in water and alcohol.
Derivation: Precipitation of a soluble coconut oil soap with a solution of a zinc salt.
Containers: Barrels.
Uses: Paints; varnishes; rubber compounding.

zinc linoleate $Zn(C_{17}H_{31}COO)_2$. A firm, tan-colored solid containing 8.5 to 9.5% zinc.
Derivation: Precipitation from solutions of the sodium linoleate and soluble zinc salt, or by fusion of the fatty acid and zinc oxide.
Uses: As drier, especially with cobalt and manganese soaps.

zinc malate $Zn(OOCCH_2CHOHCOO) \cdot 3H_2O$.
Properties: White, crystalline powder. Soluble in water.
Derivation: By the action of malic acid on zinc hydroxide.
Grades: Technical.
Method of purification: Crystallization.
Containers: Glass bottles.
Use: Medicine.
Shipping regulations: None.*

zinc metaarsenite. See zinc arsenite.

zinc naphthenate.
Properties: Amber, viscous, basic liquid or basic solid. The liquid contains 8-10% Zn, the solid contains 16% Zn. Very soluble in acetone.
Derivation: Fusion of zinc oxide or hydroxide and naphthenic acid, or precipitation from mixture of soluble zinc salts and sodium naphthenate.
Containers: Small metal drums; fiber drums.
Uses: As drier and wetting agent in paints, varnishes and resins; insecticide and fungicide; wood preservative; waterproofing of textiles; insulating materials.

zinc nitrate $Zn(NO_3)_2 \cdot 6H_2O$.
Properties: Colorless lumps or crystals. Soluble in water and alcohol. Sp.gr. (13°C) 2.065; m.p. 36.4°C; b.p. 131°C; loses water of crystallization at 105°C.
Derivation: By the action of nitric acid on zinc or zinc oxide.
Method of purification: Crystallization.
Grades: Technical; C.P.
Uses: Acidic catalyst; resin setter, latex coagulant; medicine; chemical reagent and intermediate; mordant.
Fire hazard: Dangerous.
Shipping regulations: Oxidizing material. Yellow label.*

zinc octoate (zinc 2-ethylhexoate). Light straw colored viscous liquid.
Properties: Sp.gr. 1.16; insoluble in water, soluble in common organic hydrocarbon solvents.
Use: Catalyst.

zinc oleate $Zn(C_{17}H_{33}COO)_2$.
Properties: Dry, white to tan, greasy, granular powder, containing from 8.5 to 10.5% zinc; m.p. 85.5°C. Soluble in alcohol, ether, carbon disulfide and ligroin; insoluble in water.
Derivation: By the interaction of solutions of zinc acetate and sodium oleate, or by fusion of zinc oxide and oleic acid.
Grade: Technical.
Containers: Tins; wooden kegs; glass bottles; barrels.
Use: Medicine; paints, resins and varnishes.
Shipping regulations: None.*

"Zincon." [169] Trademark for 2-carboxy-2'-hydroxy-5'-sulfoformazylbenzene used in colorimetric determination of zinc and copper.

zinc ores. See under zinc.

zinc orthophosphate. See zinc phosphate.

zinc orthosilicate. See willemite.

zinc oxalate $ZnC_2O_4 \cdot 2H_2O$.
Properties: White powder. Soluble in acids and alkalies; slightly soluble in water. Sp.gr. (24°C) 2.562.
Derivation: By the interaction of zinc sulfate and sodium oxalate.
Grades: Technical; C.P.
Containers: 1-lb bottles; boxes.
Uses: Zinc oxide; organic synthesis.
Shipping regulations: None.*

zinc oxide (Chinese white; zinc white; flowers of zinc; nil alba; philosopher's wool) ZnO.
Properties: Amorphous, odorless, white or yellowish white powder; absorbs carbon dioxide from the air; sp.gr. 5.47; m.p. above 1800°C; soluble in acids; insoluble in water and alcohol.
Derivation: (a) Oxidation of vaporized pure zinc (French process) or previously roasted sulfide ore (American process). (b) By heating zinc carbonate. (c) Oxidation of vapor fractionated die castings. See also Wetherill process.
Impurities: Zinc carbonate.
Grades: American process, lead-free; French process, lead-free, green seal, red seal, white seal (according to fineness); leaded (with lead sulfate); U.S.P. XVI.
Containers: 50-lb kegs; 100-lb drums; multiwall paper sacks; glass bottles.
Uses: Paint pigment; as mold growth inhibitor in paints; zinc salts; accelerator-activator and reinforcing agent in rubber manufacture; ceramic glazes; matches; linoleum (pigment); dental cements; medicine; feed additive; seed treatment; chemical warfare; cosmetics; zinc soaps; opaque glass; white printing inks; candles; "Celluloid;" textile printing (resist); white glue and gelatin; in rayon manufacture.
Shipping regulations: None.*

zinc oxide, leaded. Contains lead sulfate.

zinc oxide, red. See zincite.

zinc palmitate $Zn(C_{16}H_{31}O_2)_2$.
Properties: White amorphous powder; sp.gr. 1.121; m.p. 100°C; insoluble in water and alcohol; slightly soluble in benzene and toluene.
Grade: Technical.
Containers: Cartons, bags.
Use: Flatting agent in lacquer; pigment suspending agent for paints; rubber compounding; lubricant in plastics.

zinc perborate $Zn(BO_3)_2$ with water of hydration.
Properties: Amorphous, white powder; insoluble in water but slowly decomposed by it, liberating hydrogen peroxide.
Derivation: Interaction of sodium peroxide, boric acid and zinc salt, or of boric acid and zinc peroxide.
Grade: Technical.
Containers: Tins, glass bottles.
Use: Medicine; oxidizing agent.
Shipping regulations: None.*

zinc permanganate $Zn(MnO_4)_2 \cdot 6H_2O$.
Properties: Violet-brown or black, hygroscopic crystals; sp.gr. 2.47; loses $5H_2O$ at 100°C. Decomposes on exposure to light and air. Soluble in water and acids; decomposes in alcohol.
Grades: Technical (about 95% pure).
Containers: Glass bottles, tins.
Use: Medicine; oxidizing agent.
Fire hazard: Dangerous.
Shipping regulations: Oxidizing material. Yellow label.*

zinc peroxide (zinc dioxide) ZnO_2.
Properties: White powder containing 45-60% ZnO_2, balance ZnO; sp.gr. 1.571; decomposes rapidly above 150°C. Decomposes in acids; insoluble in water but gradually decomposed by it.
Derivation: By the action of barium peroxide on zinc sulfate solution, followed by filtration.
Grades: U.S.P. XVI (mixture of peroxide, carbonate and hydroxide); technical, 50-60%.
Containers: Glass bottles; 100-, 200-lb drums.
Uses: Cosmetics; medicine; vulcanizing agent;

high temperature oxidation reactions.
Fire hazard: Dangerous.
Shipping regulations: Oxidizing material. Yellow label.*

zinc phenate (zinc carbolate; zinc phenolate) $Zn(C_6H_5O)_2$. (May be only a mixture of zinc oxide and phenol.)
Properties: White powder. Soluble in alcohol; slightly soluble in water.
Derivation: By heating zinc hydroxide with phenol and extracting with alcohol.
Method of purification: Recrystallization.
Grade: Technical.
Use: Medicine; insecticide.
Shipping regulations: None.*

zinc phenolate. See zinc phenate.

zinc phenolsulfonate (zinc sulfophenate; zinc sulfocarbolate) $Zn(SO_3C_6H_4OH)_2 \cdot 8H_2O$.
Properties: Colorless, transparent crystals or white granular powder; odorless; astringent metallic taste; effloresces in air; turns pink on exposure to air and light; loses water of crystallization at 120°C; soluble in water and alcohol.
Derivation: By heating zinc hydroxide with phenol-sulfonic acid.
Method of purification: Crystallization.
Grades: N.F. XI; technical.
Containers: Glass bottles; 200-lb drums.
Use: Medicine.

zinc phosphate (zinc orthophosphate; zinc phosphate, tribasic) $Zn_3(PO_4)_2 \cdot 4H_2O$.
Properties: White powder. Soluble in acids and ammonium hydroxide; insoluble in water. Sp.gr. 3.03.
Derivation: By the interaction of zinc sulfate and trisodium phosphate.
Grades: Technical, about 98% pure.
Containers: Boxes; wooden kegs; glass bottles.
Uses: Medicine; dental cements; phosphors.
Shipping regulations: None.*

zinc phosphate, tribasic. See zinc phosphate.

zinc phosphide Zn_3P_2.
Properties: Dark gray, gritty powder; sp.gr. 4.55; stable if kept dry; insoluble in water and alcohol; soluble in acids with production of flammable phosphine. Reacts violently with oxidizing agents.
Derivation: By passing phosphine into a solution of zinc sulfate.
Grades: Technical, about 80-85% pure.
Containers: Tins; glass bottles.
Uses: Medicine; rat poisons.
Warning! Poisonous if swallowed. MCA warning label.
Shipping regulations: None.*

zinc phosphite $ZnHPO_3 \cdot 2\frac{1}{2}H_2O$.
Properties: White granular, crystalline powder. Soluble in cold water; insoluble in hot water.
Derivation: The present commercial method is by reacting zinc powder and phosphorus. By the action of phosphorous acid on zinc hydroxide.
Method of purification: Crystallization.
Grades: Technical.
Containers: Tins; glass bottles.
Use: Medicine.
Shipping regulations: None.*

zinc potassium chromate. See zinc yellow.

zinc potassium iodide. See potassium zinc iodide.

zinc potassium sulfate (potassium zinc sulfate) $K_2SO_4 \cdot ZnSO_4 \cdot 6H_2O$.
Properties: White crystals; soluble in water.

zinc powder. Finely divided metallic zinc. See zinc dust.

zinc propionate $Zn(OOCC_2H_5)_2$.
Properties: Occurs as platelets, tablets, or needlelike crystals. Fairly soluble in water, slightly soluble in alcohol. Decomposes in moist atmosphere, liberating propionic acid.
Derivation: By dissolving zinc oxide in dilute propionic acid and concentrating the solution.
Use: Fungicide on adhesive tape.

zinc pyrophosphate $Zn_2P_2O_7$.
Properties: White powder; sp.gr. 3.756; soluble in acids and alkalies; insoluble in water.
Derivation: By heating a soluble zinc salt with ammonium phosphate.
Grade: Technical.
Containers: Wooden kegs or fiber drums.
Use: Pigment.
Shipping regulations: None.*

zinc resinate.
Properties: Powder; clear amber lumps, or clear yellowish liquid. May be acid, basic or neutral. Soluble in some organic solvents, as ether, amyl alcohol.
Derivation: By fusion of zinc oxide and rosin, or by precipitation from solutions of zinc salts and sodium resinate.
Containers: 55-gal drums.
Uses: Wetting, dispersing and hardening agent; drier, in paints, varnishes and resins.

zinc rhodanide. See zinc thiocyanate.

zinc ricinoleate
$Zn[CH_3(CH_2)_5CHOHCH_2CH:CH(CH_2)_7CO_2]_2$. Fine white powder with faint fatty acid odor.
Properties: M.p. 92-95°C; sp.gr. (25/25°C) 1.10.
Containers: 50-lb bags.
Uses: Fungicide; emulsifier; greases; lubricants; waterproofing; lubricating oil additive; stabilizer in vinyl compounds.

zinc salicylate $Zn[C_6H_4(OH)COO]_2 \cdot 3H_2O$.
Properties: White, crystalline needles or powder; soluble in water and alcohol.
Derivation: By heating zinc hydroxide and salicylic acid.
Method of purification: Crystallization.
Grades: Technical.
Containers: Tins; glass bottles.
Use: Medicine.
Shipping regulations: None.*

zinc silicate (zinc orthosilicate). See willemite.

zinc silicates. Used for phosphors; spray ingredients; to remove traces of copper from gasoline.

zinc silicofluoride (zinc fluosilicate) $ZnSiF_6 \cdot 6H_2O$.
Properties: White crystals; sp.gr. 2.104. Decomposes on heating. Soluble in water.
Derivation: Reaction of zinc oxide and fluosilicic acid.
Containers: Bulk; barrels, drums.
Grades: Technical.
Uses: Concrete hardener; laundry sour; preservative; mothproofing agents.

zinc-silver oxide battery. Primary or secondary battery used where space and weight are critical, i.e., in missiles. The battery has large energy output for its weight, but the components are expensive and the cycle life is short. To avoid deterioration, the potassium hydroxide electrolyte is added just before use.

zinc stearate $Zn(C_{18}H_{35}O_2)_2$. Percentage of zinc may vary according to intended use, some products being more basic than others.
Properties: (pure substance) White, agglutinating powder, free from grittiness and with faint characteristic odor; sp.gr. 1.095; m.p. about 120°C. Soluble in acids; soluble in common solvents when hot.
Derivation: By the action of sodium stearate on solution of zinc sulfate.
Grades: U.S.P. XVI; technical; available free from chick edema factor.
Containers: Glass bottles, bags, cartons.
Uses: Medicine; cosmetics; drying lubricant in rubber; waterproofing agent; lacquers; plastics; powder metallurgy; as a dietary supplement.
Shipping regulations: None.*

zinc subcarbonate. See zinc carbonate, precipitated.

zinc sulfate (white vitriol; white copperas; zinc vitriol) $ZnSO_4 \cdot 7H_2O$. See goslarite.
Properties: Colorless crystals, small needles or granular crystalline powder, without odor; astringent, metallic taste; efflorescent in air. Keep well stoppered. Solutions acid to litmus. Sp.gr. 1.9661; m.p. 50°C if rapidly heated; soluble in water and glycerol; insoluble in alcohol.
Derivation: (a) By roasting zinc blende in a reverberatory furnace and lixiviating, with subsequent purification; (b) by the action of sulfuric acid on zinc or zinc oxide.
Method of purification: Crystallization.
Grades: Technical; U.S.P. XVI; reagent.
Containers: Glass bottles; barrels; fiber drums; multiwall paper sacks.
Uses: Medicine; mordant in calico printing; preservative for skins and wood; bleaching paper; preparing zinc chemicals; clarifying glue; reagent in analytical chemistry; feed additive; pesticide adjuvant; fungicide.

zinc sulfate dried (zinc sulfate monohydrate) $ZnSO_4 \cdot H_2O$. Much less common than $ZnSO_4 \cdot 7H_2O$. Used in warmer climates because it is less likely to cake than the heptahydrate.
Properties: White, free flowing powder; soluble in water; insoluble in alcohol.
Use: Agricultural sprays; chemical intermediate; glue clarifying; dyestuffs; electroplating; manufacturing of rayon.

zinc sulfate monohydrate. See zinc sulfate dried.

zinc sulfide. ZnS or $ZnS \cdot H_2O$.
Properties: Yellowish, white powder; stable if free from water of crystallization and kept dry; sp.gr. greater than 4; m.p. 1020°C; sublimes at 1180°C; soluble in acids; insoluble in water.
Derivation: By passing hydrogen sulfide gas into a solution of a zinc salt. See also wurtzite and sphalerite.
Grades: Technical; C.P.; fluorescent or luminous.
Containers: 1-lb bottles; barrels; bags.
Uses: Pigment (paint, linoleum, artificial leather); white and opaque glass; white opaque glues and gelatins; base for color lakes; rubber; plastics; dyeing (hydrosulfite process); ingredient of lithopone, as a phosphor in x-ray and television screens and on luminous watch faces.
Shipping regulations: None.*

zinc sulfide white. See lithopone.

zinc sulfite $ZnSO_3 \cdot 2H_2O$.
Properties: White, crystalline powder; absorbs oxygen from the air to form sulfate. Soluble in sulfurous acid; insoluble in cold water and alcohol; decomposes in hot water.
Derivation: By the action of sulfurous acid on zinc hydroxide.
Method of purification: Crystallization.
Impurities: Zinc sulfate.
Grades: Technical; C.P.
Containers: Glass bottles; tins.
Uses: Medicine; preservative for anatomical specimens.
Shipping regulations: None.*

zinc sulfocarbolate. See zinc phenolsulfonate.

zinc sulfocyanate. See zinc thiocyanate.

zinc sulfophenate. See zinc phenolsulfonate.

zinc sulfoxylate $ZnSO_2$.
Properties: White crystalline material; decomposed by heat; salt of unstable sulfoxylic acid, H_2SO_2; zinc sulfoxylate is a strong reducing agent.
Derivation: By the action of zinc and sulfuryl chloride in ethereal solution, or by the action of sulfur dioxide on granulated zinc in absolute alcohol.
Containers: Fiber drums; multiwall shipping sacks.
Use: A stripping agent in dyeing.
Shipping regulations: None.*

zinc sulfoxylate formaldehyde. See zinc formaldehyde sulfoxylate.

zinc tallate. A mixture of zinc resinate and zinc salts of unsaturated fatty acids. Produced from a soluble tall oil soap and a solution

of a zinc salt. Of limited use as a drier in paints.
See also soaps, metallic.

zinc thiocyanate (zinc rhodanide; zinc sulfocyanate) $Zn(CNS)_2$.
Properties: White hygroscopic powder or crystals. Keep well stoppered. Soluble in water, alcohol and ammonium hydroxide.
Derivation: By the interaction of zinc hydroxide and ammonium thiocyanate.
Method of purification: Crystallization.
Grade: Technical. See also solution, next article.
Containers: Glass bottles.
Use: Analytical chemistry.
Shipping regulations: None.*

zinc thiocyanate solution.
Typical specifications: Colorless; 32.5-33.5% aqueous solution; sp.gr. 1.25-1.30.
Containers: Stainless steel drums; 550 lbs net.
Uses: Swelling agent for cellulose esters, dyeing assistant.

zinc-tin amalgam. Composed of zinc 25%, tin 25%, mercury 50%.
Grade: Technical.
Containers: Glass bottles.
Uses: Electrical machines; dental cement.
Shipping regulations: None.*

zinc undecylenate $[CH_2{:}CH(CH_2)_8COO]_2Zn$.
Properties: A fine, white amorphous powder; nearly insoluble in water and alcohol; m.p. 115-116°C.
Grade: N.F. XI.
Containers: Drums.
Use: Medicine; cosmetics; chemical intermediate.

zinc valerate (zinc isovalerate) $Zn[OOCCH_2CH(CH_3)_2] \cdot 2H_2O$.
Properties: White, pearly crystals or powder with disagreeable odor. Keep well stoppered. Soluble in hot water and alcohol; very slightly soluble in ether.
Derivation: Interaction of zinc sulfate and sodium isovalerate or zinc carbonate and isovaleric acid.
Containers: 1-lb bottles.
Use: Medicine.

zinc vitriol. See zinc sulfate; also goslarite.

zinc white. Zinc oxide pastes used by commercial artists. See also zinc oxide.

zinc yellow (citron yellow, buttercup yellow, zinc potassium chromate, zinc chrome).
Properties: A greenish yellow pigment of comparatively low tinting strength. Is partially water-soluble. Consists principally of zinc potassium chromate, about $4ZnO \cdot K_2O \cdot 4Cr_2O_3 \cdot 3H_2O$.
Derivation: Made by reaction of a solution of potassium dichromate with zinc oxide and sulfuric acid.
Containers: 250-lb bags.
Uses: Principally in the manufacture of rust-inhibitive paints, particularly primers such as are now being used by the aircraft industry. In making green trim paints in combination with hydrated chromium oxide, it should not be used where the soluble chromates may be objectionable. Not used alone for yellow enamels since severe exposure tends to turn it greenish and less brilliant. Occasionally used as an artists' color. See zinc chromate.

zinc zirconium silicate $ZnO \cdot ZrO_2 \cdot SiO_2$.
Properties: White powder; sp.gr. 4.8; density 115 lbs/cu ft; m.p. 3800°C; soluble in hydrofluoric acid; insoluble in water and alkalies; slightly soluble in mineral acids and hot conc. sulfuric acid.
Containers: Bags, barrels, carloads.
Uses: Opacifier for ceramic glazes.

zineb (zinc ethylenebisdithiocarbamate) $Zn(CS_2NHCH_2)_2$.
Properties: Light tan solid, insoluble in water.
Derivation: Reaction of sodium ethylenebisdithiocarbamate with zinc sulfate or other zinc salts. In practical application as a fungicide these reactants are mixed in the presence of lime, and the zineb is not formed until after reaction of the carbon dioxide of the air with the film of the other chemicals on the leaf or fruit.
Grades: Commercial dusts and wettable powders usually contain 65% active material.
Caution! May cause irritation of eyes, nose, throat, and skin. May be harmful if inhaled or swallowed. MCA warning label.
Uses: Insecticide and fungicide.

zingiber. See ginger.

zinkenite (zinckenite) $Pb_6Sb_{14}S_{27}$. A natural sulfantimonide of lead, found in metallic veins.
Properties: Color and streak steel gray; luster metallic; hardness 3-3.5; sp.gr. 5.3.
Occurrence: Colorado; Arkansas; Nevada; Europe.

"Zinol." [79] Trademark for a special zinc resinate in solution in mineral spirits.
Properties: Solids 64.0%; viscosity (Gardner-Holdt) "K" (25°C); acid number 12; color (Hellige) 7-8; metallic zinc (combined) 4.3%; wt/gal (25°C) 8.18 lbs. The resin contained in "Zinol" conforms quite closely to the following typical analysis: m.p. (capillary tube) 80°C; acid number 18; metallic zinc (combined) 6.7%.
Containers: 55-gal drums; tank cars.
Uses: Printing ink; paint and varnish.

"Zin-O-Lyte." [28] Trademark for a series of zinc plating compounds.
"Zin-O-Lyte" Salts. Fine, white powder containing sodium cyanide and other chemicals for the zinc plating bath.
Containers: 100-lb drums.
Addition Agent O is a heavy gray powder containing molybdenum; soluble in caustic soda solution; insoluble in water.
"Zin-O-Lyte" Brightener. Organic addition agent; fine powder; water-soluble. Used to increase brightness and current efficiencies and to widen the bright current density range.

"Z. I. P." [401] Trade name for a deer and rabbit taste repellent containing zinc dithiocarbamate-amine complex (20%).
Containers: 1-quart and 1-, 5-, 55-gal containers.
Uses: Is diluted with water and used as a spray on vegetation such as nursery stock, forestry seedlings and ornamentals for protection against deer and rabbit damage.
Caution! May be harmful if swallowed. May cause irritation of nose, throat and skin. Avoid breathing spray. Avoid contact with eyes, skin or clothing.

ziram (zinc dimethyldithiocarbamate) $Zn(SCSNCH_3CH_3)_2$.
Properties: White and odorless when pure; m.p. 246°C; almost insoluble in water; soluble in acetone, carbon disulfide, chloroform, in dilute alkalies, and in concentrated hydrochloric acid.
Derivation: Reaction of sodium dimethyldithiocarbamate with a soluble zinc salt in aqueous solution.
Grades: 76% wettable powder; 90% technical powder.
Containers: 3-, 50-lb multiwall bags; drums.
Caution! May cause irritation and be harmful if inhaled. MCA warning label.
Uses: Rubber accelerator; fungicide.

"Zirberk." [81] Trademark for zinc dimethyldithiocarbamate. See ziram.

zircite. See zirkite.

"Zirco." [230] Trademark for a zirconium organic complex in odorless mineral spirits. Not a paint drier; has synergistic action on metallic driers.

"Zircofrax." [280] Trademark for super-refractory products made from zirconium oxide and zirconium silicate.
Properties: High refractoriness; great strength; high thermal conductivity; high resistance to attack by acids and acid slags; porosity about 25%; permeability low.
Uses: Bricks and special shapes for ceramic kiln furniture and in chemical and metallurgical furnaces where severe slagging occurs.

zircon $ZrSiO_4$. A natural zirconium silicate found in igneous rocks. May contain uranium, hafnium, and other elements in minor quantities.
Properties: Color brown, gray, red, colorless; luster adamantine; hardness 7.5; sp.gr. 4.68. Insoluble in acids.
Occurrence: North Carolina; Maine; New York; Florida; India; Australia; Brazil.
Uses: Source of zirconium oxide and metallic zirconium; gem stone; abrasive; refractories; enamels; source of hafnium.

zircon, Australian. See Zircon Granular "TAM" and Zircon Milled "TAM."

zircon, Florida. See Zircon Milled "G" and Zircon Milled "TAM."

zircon flour. Finely milled zircon sand used as a mold wash.

Zircon Granular "G." [337] Trade name for 98% zircon ($ZrSiO_4$). Gray white powder, sp.gr. 4.6; bulk density 184 lbs/cu ft; fineness −70 mesh to +270 mesh. Used as coarse ingredient in zircon super-refractories, as a smelter addition in producing zirconium porcelain enamel frits, and as a sand seal in high temperature furnace insulations.
Containers: 100-lb paper bags, 500-lb barrels, 60,000-lb carloads.

Zircon Granular "TAM." [337] (Australian zircon) Trade designation for zirconium silicate ($ZrSiO_4$) with 98+% zircon, 0.05% Fe_2O_3, 0.1% TiO_2, 0.5% Al_2O_3 (max); white powder, sp.gr. 4.6; bulk density 179 lbs/cu ft; dissociates at 3200°F; screen analysis +80 mesh 3%, −200 mesh < 1%. Insoluble in water, dilute mineral acids, alkalies, hot concentrated sulfuric acid; slightly soluble in hydrofluoric acid. Used as the coarse ingredient of mixes for special crucibles and refractory items, and in electrical cements.
Containers: 100-lb paper bags; 500-lb barrels; 36,000-lb carloads.

"Zircon H-W." [446] Trade name for compound made from the purified mineral by impact pressing, or by air ramming or slip casting for more intricate shapes.
Properties: High density (230 lbs/cu ft) helps it to resist the wetting and penetration of molten glass. Resistant also to thermal spalling, fluxing conditions; has constancy of volume under soaking heat in excess of 2900°F.
Uses: To pave glass tank bottoms and floors; line sodium meta-phosphate and sodium silicate furnaces; tap hole blocks for aluminum and non-ferrous melting; and as nozzles for continuous steel casting operations.

zirconia. See zirconium oxide.

zirconia, stabilized, fused. Zirconium oxide which has been fused with small additions of other oxides to improve its thermal-mechanical properties.
Typical assay: 0.20% SiO_2; 0.52% Fe_2O_3; 0.22% TiO_2; 4.96% CaO; 94.1% ZrO_2.
Properties: Sp.gr. 5.6; fusion point 2540-2600°C; coefficient of thermal expansion 0.0000080.
Grades: Available in standard mesh sized grains.
Uses: High temperature furnace linings; rocket motors; setter-plates on which dielectrics of barium and strontium titanate are fired.

zirconic anhydride. See zirconium oxide.

zirconium Zr. Element of atomic number 40, group IV of the periodic table. Ninth most abundant metal in the earth's crust.
Properties: Hard, lustrous, grayish, crystalline scales or gray amorphous powder; sp. gr. 6.4; m.p. about 1850°C; soluble in hot, very concentrated acids; insoluble in water and cold acids.
Sources: Zircon; baddeleyite.

Derivation: The ore is converted to a carbonitride, which is chlorinated to obtain zirconium tetrachloride. This is reduced with magnesium (Kroll process) in an inert atmosphere. The metal can be prepared in a highly pure and ductile form by vapor-phase decomposition of the tetraiodide. Hafnium must be removed for uses in nuclear reactors (see hafnium).

Grades: Plate, strip, bars, wire; sponge and briquettes; powder; technical; pure (hafnium-free).

Containers: Cans/box.

Uses: Greatest consumption for atomic energy purposes because when free of hafnium, zirconium is corrosion-resistant and has low absorption for neutrons; hence used in nuclear reactor chambers, tubing, etc. Also used as a "getter" material in high-vacuum work; deoxidizer in metal castings; ingredient in flashlight powder; suggested antidote for plutonium poisoning; alloying agent in nickel-chromium and other nonferrous alloys; bonding agent for ceramic-to-metal seals; high intensity electric arc light. Radioactive zirconium is used in medicine. Zirconium compounds are used in three major fields: (1) chemical process industry, as in leather tanning, catalysts, water repellents for textiles, (2) ceramics and (3) metallurgy. See also zirconium ferro alloys.

Caution: Powder has comparatively low ignition temperature; highly flammable in dry state; explosive in contact with oxidizing agents.

Shipping regulations: Powder, sponge, or scrap, wet or dry: flammable solid. Yellow label. Liquid solutions or mixtures: flammable liquid. Red label.*

zirconium 95. Radioactive zirconium of mass number 95.

Properties: Half-life, 63 days; radiation, beta and gamma.

Derivation: Obtained in a mixture with niobium from the fission products of nuclear reactor fuels.

Forms available: Zirconium oxalate complex in oxalic acid solution.

Uses: To trace the flow of heavy petroleum products in pipelines; to measure the rate of catalyst circulation in petroleum cracking plants; to study the cracking and polymerization of hydrocarbons with various catalysts; etc.

Shipping regulations: Class D poison, radioactive material. Red label.*

zirconium acetate solution $H_2ZrO_2(C_2H_3O_2)_2$ in aqueous solution.

Properties: (a) Available as 22% ZrO_2. Clear to pale amber solution; sp.gr. 1.46 (approx); pH 3.8-4.2 (20°C); f.p. −7°C; stable at room temperature. (b) Available as 13% ZrO_2. Pale-amber liquid; sp.gr. 1.20 (approx); pH 3.3-4.0 (20°C); stable at room temperature, but temperature of hydrolysis decreases with pH; undergoes exchange with anion exchange resins, but not with cation exchangers.

Containers: 500-lb (polyethylene lined) drums; 30,000-lb carloads.

Uses: Marine fuel oil additive to combat vanadium deposits on boiler tubes; water-repellent ingredient for textiles; curing agent for silicone resins used for water repellency; precipitating agent for gelatin and starch on paper and fabrics.

zirconium acetylacetonate. See zirconium tetraacetylacetonate.

zirconium aluminum Zr, Al alloy. Used for grain refinement in magnet alloys and as a source of zirconium.

zirconium ammonium fluoride (ammonium zirconifluoride) $Zr(NH_4)_2F_6$.

Properties: White crystals. Soluble in water.

zirconium anhydride. See zirconium oxide.

zirconium boride ZrB_2.

Properties: Gray metallic crystals or powders; sp.gr. 6.085; m.p. 3000°C; Mohs hardness 8; electrical resistivity 9.2 microohm-cm at 20°C; excellent thermal shock resistance; poor oxidation resistance above 1100°C.

Uses: Refractory for special high temperature aircraft and rocket applications; electrodes in metal refining; thermocouple protection tubes; metallurgical additive; high temperature electrical conductor.

zirconium carbide ZrC.

Properties: Sp.gr. 6.78; hardness Mohs 8+; m.p. 3540°C; insoluble in water and hydrochloric acid; soluble in oxidizing acids and attacked by oxidizers; fine powder is pyrophoric.

Derivation: By heating zirconium oxide and coke in an electric furnace.

Grade: Technical.

Containers: Iron drums.

Uses: Incandescent filaments, abrasive; source for pure zirconium and zirconium compounds; cermet component; metallurgical additive; high temperature electrical conductor; refractory.

zirconium carbonate. See zirconium carbonate, basic.

zirconium carbonate, basic (zirconyl carbonate, zirconium carbonate) $ZrOCO_3$ or $ZrOCO_3 \cdot xH_2O$.

Properties: White, amorphous powder. Soluble in acids; insoluble in water.

Derivation: By adding sodium carbonate to a solution of zirconium salt.

Grade: Technical.

Containers: Glass bottles; wooden kegs.

Uses: Preparation of zirconium oxide.

Shipping regulations: None.*

zirconium carbonitride ZrCN. Zirconium cyanonitride. A solid used for the production of zirconium and its compounds including the the tetrachloride. Also, a deoxidizer and source of zirconium in steels.

zirconium chloride. See zirconium tetrachloride.

zirconium chloride, basic. See zirconium oxychloride.

zirconium dioxide. See zirconium oxide.

zirconium disilicide Zr, Si. A solid which forms coatings resistant to flame or blast impingement. Also has special alloy applications.

zirconium ferro alloys (ferrozirconium alloys). Alloys used in the manufacture of steel.
12 to 15 percent zirconium alloy: Approximate analysis: zirconium 12-15%, silicon 39-43%; iron 40-45%. Application: Steel of high silicon content.
35-40 percent zirconium alloy: Approximate analysis: zirconium 35-40%, silicon 47-52%, iron 8-12%. Application: Steel of low silicon content.
Uses: Zirconium is used as a deoxidizer and scavenger of steel in amounts between 0.05 and 0.10%. It acts first on the oxygen and nitrogen in the steel, tending to eliminate them as well as non-metallic inclusions. In amounts of 0.10 to 0.15%, it forms zirconium nitride. When present in steel above 0.15%, it combines with sulfur, forming zirconium sulfide. This prevents tearing in rolling and produces a better surface on high-sulfur steel.

zirconium glycolate $H_4ZrO(C_2H_2O_3)_3$.
Properties: Solid; decomposes without melting on heating to about 220°C; insoluble in water and organic solvents; soluble in alkali and sulfuric acid solutions. One or more of the acidic hydrogens may be replaced by alkali metals or ammonium to give water-soluble salts.
Containers: 250-lb (polyethylene lined) fiber drums (41 gals).
Uses: Cosmetics (deodorant); medicine; sequestrant; source of high purity zirconia.

zirconium hydride ZrH_2. Contains 1.7 to 2.1% combined hydrogen which can be driven off in a vacuum above 600°C. Because powdered zirconium is extremely hazardous, it is shipped as the hydride.
Properties: Gray-black metallic powder. Stable toward air and water. Sp.gr. 5.6.
Derivation: Reduction of zirconia with calcium hydride or magnesium in the presence of hydrogen; direct combination of hydrogen and zirconium metal.
Grades: Commercial (contains hafnium); reactor (hafnium free).
Uses: Vacuum tube getter; powder metallurgy; metal-foaming agent; combustible mixtures; nuclear moderators; reducing agent; hydrogenation catalyst.
Hazards! Reactive with oxidizing agents.

zirconium hydroxide $Zr(OH)_4$.
Properties: White, bulky, amorphous powder. Soluble in dilute mineral acids; insoluble in water and alkalies. Sp.gr. 3.25; decomposes to ZrO_2 at 550°C.
Derivation: By the action of a solution of sodium hydroxide on a solution of a zirconium salt.
Grade: Technical.
Containers: Wooden kegs; glass bottles.
Use: Zirconium compounds; pigments, dyes, glass.
Shipping regulations: None.*

zirconium lactate $H_4ZrO(CH_3CHOCO_2)_3$.
Properties: White, slightly moist pulp; decomposes without melting; very slightly soluble in water and the common organic solvents; soluble in aqueous alkalies with formation of salts; decomposes to hydrous zirconia above pH 10.5. Efficient odor absorber.
Grade: Zirconia 25% (min).
Containers: 250-lb (polyethylene lined) fiber drums (41 gals).
Uses: Body deodorants; source of zirconia.

zirconium naphthenate.
Properties: Amber-colored, heavy liquid. Completely transparent. Consistency equivalent to that of heavy lubricating oil. Very stable. Unlike other metallic naphthenates possesses no drying properties. Soluble in all common solvents. Sp.gr. 1.05.
Derivation: By heating a mixture of naphthenic acid and zirconium sulfate.
Grade: Technical.
Uses: Ceramics (enamels, glazes); lubricants; paints and varnish (anti-chalking agent, minimizer of moisture and solar radiation effects).

zirconium nitrate $Zr(NO_3)_4 \cdot 5H_2O$.
Properties: White hygroscopic crystals. Soluble in water and alcohol. Decomposes at 100°C.
Derivation: By the action of nitric acid on zirconium oxide.
Method of purification: Crystallization.
Grade: Technical.
Containers: Kegs, drums.
Use: Preservative.
Shipping regulations: Oxidizing material. Yellow label.*

zirconium nitride ZrN. A brassy-colored powder produced by heating the metal in nitrogen.
Properties: Sp.gr. 7.09; hardness Mohs 8+; m.p. 2930°C; slightly soluble in dilute hydrochloric or sulfuric acid; soluble in concentrated acid.
Uses: Special crucibles; cermets; refractories.

zirconium orthophosphate. See zirconium phosphate.

zirconium oxide (zirconia; zirconium dioxide; zirconic anhydride; zirconium anhydride) ZrO_2. Occurs in nature as baddeleyite and zirkite.
Properties: Heavy white amorphous powder; sp.gr. 5.73; m.p. 2700°C; hardness 6.5; refractive index 2.2; insoluble in water and most acids or alkalies at room temperature; soluble in nitric acid and hot concentrated hydrochloric, hydrofluoric, and sulfuric acids. Most heat resistant of commercial refractories.
Derivation: By heating zirconium hydroxide or zirconium carbonate.
Grades: Reagent; technical; crystals; fused;

*See "I.C.C. Shipping Regulations," page xiii.
Reference numbers refer to name of manufacturer. See "List of Manufacturers," page v.

C.P. (99% zirconia).
Containers: Bags; drums; carloads.
Uses: Pigment for paints, lacquers, resins and inks; component of dielectrics; pharmaceutical agent; catalyst; piezoelectric crystals; color stabilizer of organic dyes; to increase light efficiency of lacquers or resins used as light reflectors; crucibles; furnace linings; x-ray photography; substitute for calcium oxide in calcium lights; opacifier in white glass for indirect electric lighting; ceramics; acid-proof enamel; refractory utensils and cermets; metallurgy; source of zirconium; abrasive and polishing agent.
Shipping regulations: None.*
See also zirconia, fused, stabilized.

Zirconium Oxide 45006. [337] Trade designation for proprietary zirconium oxide containing 98+% ZrO_2. Salmon colored powder; sp. gr. 5.7; bulk density 89 lbs/cu ft; m.p. 4900°F. Available as −20 mesh with 0.5% SiO_2, 0.5% TiO_2, 0.5% Al_2O_3, (max); also as −325 mesh with 1.0% SiO_2, 0.5% TiO_2, 0.75% Al_2O_3, (max). Used as setter material in firing ceramic dielectrics; a commercial source of high purity ZrO_2; also used as a raw material for the manufacture of super-refractories.
Containers: 100-lb drums, 36,000-lb carloads.

"Zirconium oxide, E. F." [337] Trade name for electrically fused zirconium oxide available as salmon colored ½-inch lumps (98.5% ZrO_2) or milled material (95% ZrO_2) of -35, -80, -200, or -325 mesh; sp.gr. 5.7; bulk density 205 lbs/cu ft; m.p. 4900°F. The lumps are used in the massive packing of high temperature furnaces. The milled material is used in the manufacture of super-refractory crucibles, rods, tubes, and special shapes for use up to 4000°F under oxidizing conditions. Also used as a setter material in firing ceramic dielectrics.
Containers: 100-lb drums; 36,000-lb carloads.

zirconium oxychloride (zirconium chloride, basic; zirconyl chloride) $ZrOCl_2 \cdot 8H_2O$.
Properties: White, silky crystals; m.p. about 115°C with evolution of hydrochloric acid and water; density 44 lbs/cu ft. Soluble in water, methanol, and ethanol; insoluble in other organic solvents. Aqueous solutions are acidic.
Derivation: By the action of hydrochloric acid on zirconium oxide.
Method of purification: Crystallization.
Grades: Technical; 36% ZrO_2; H.P.
Containers: Barrels; 250-lb (plastic lined) fiber drums; 30,000-lb carloads.
Uses: Textile, cosmetic, and grease additive; chemical reagent; source of zirconium salts; in lakes and toners of acid and basic dyes.
Shipping regulations: None.*

zirconium phosphate (zirconium phosphate, basic; zirconium orthophosphate) $ZrO(H_2PO_4)_2 \cdot 3H_2O$.
Properties: White, dense, amorphous powder. Decomposes on heating. Soluble in acids; insoluble in water and organic solvents. Extensively hydrolysed in basic solution.
Derivation: By the action of phosphoric acid on zirconium hydroxide.
Grade: Technical.
Containers: Glass bottles; 250-lb fiber drums; 30,000-lb carloads.
Use: Chemical reagent; cation scavenger; coagulant; carrier for radioactive phosphorus.
Shipping regulations: None.*

zirconium phosphate, basic. See zirconium phosphate.

zirconium picramate.
Shipping regulations: Wet with 20 percent of water. Oxidizing material. Yellow label.*

zirconium potassium fluoride ZrK_2F_6. (potassium fluozirconate, potassium zirconifluoride).
Properties: White crystals. Soluble in water (hot).
Grade: Technical.
Use: Preparation of metallic zirconium.

zirconium potassium sulfate (potassium zirconium sulfate) $2K_2SO_4 \cdot Zr(SO_4)_2 \cdot 3H_2O$.
Properties: White, crystalline powder. Slightly soluble in water.

zirconium pyrophosphate ZrP_2O_7.
Properties: White solid; stable to about 1550°C; insoluble in water and dilute acids other than hydrofluoric acid; coefficient of thermal expansion 5×10^{-6} (approx) at 1000°C.
Containers: 250-lb fiber drums; 30,000-lb carloads.
Use: Suggested as refractory; olefin polymerization catalyst; phosphor.

zirconium silicate. See zircon.

zirconium sodium sulfate $2Na_2SO_4 \cdot Zr(SO_4)_2 \cdot 3H_2O$.
Properties: White solid.
Containers: Drums.
Use: Manufacture of zirconium.

"Zirconium Spinel." [337] Trade designation for a synthetic complex containing 40% ZrO_2, 20% SiO_2, 20% Al_2O_3, 19% ZnO. White powder, particle size 44 microns max; sp.gr. 4.7; bulk density 98 lb/cu ft; m.p. 3100°F. Insoluble in water, partly soluble in dilute acids, insoluble in alkalies, soluble in hot concentrated sulfuric acid. Used as an opacifier that widens the firing range of a glaze, and promotes high gloss and smooth texture.
Containers: 80-lb paper bags; 500-lb barrels; 30,000-lb carloads.

zirconium sulfate $Zr(SO_4)_2 \cdot 4H_2O$.
Properties: White, crystalline powder; bulk density 70 lbs/cu ft; decomposes to monohydrate at about 100°C. Soluble in water; slightly soluble in alcohol; insoluble in hydrocarbons. Aqueous solutions are strongly acidic; will precipitate potassium and amino

*See "I.C.C. Shipping Regulations," page xiii.
Reference numbers refer to name of manufacturer. See "List of Manufacturers," page v.

acids from solution; are decomposed by bases and heat.
Derivation: By the action of sulfuric acid on zirconium hydroxide.
Method of purification: Crystallization.
Grade: Technical.
Containers: Glass bottles; 500-lb (plastic lined) steel drums.
Uses: Chemical reagent; lubricants; catalyst support; protein precipitation; tanning.
Shipping regulations: None.*

zirconium sulfate, basic (zirconyl sulfate) $ZrOSO_4$ (approximately). Similar in properties to the oxychloride, and is prepared in a similar fashion, the end result being in cake form. Used in textile treatment. The material is also handled as an acidic water solution.

zirconium tetraacetylacetonate (zirconium acetylacetonate; zirconium tetrapentanedionate) $Zr(C_5H_7O_2)_4$.
Properties: A colorless, crystalline tetrachelate; density 1.415; m.p. 194-5°C (decomposition begins at 125°C). Soluble in pyridine, acetone, benzene, and other organic solvents having some polarity; slightly soluble in water.
Derivation: Reaction between zirconyl chloride, acetylacetone and sodium carbonate.
Uses: Cross-linking agent for polyol, polyester, and polyalkoxy resins; lubricant and grease additive; reagent; catalyst.

zirconium tetrachloride (zirconium chloride) $ZrCl_4$.
Properties: White, lustrous crystals. Soluble in alcohol; decomposes in water. Sp. gr. 2.8; sublimes above 300°C.
Derivation: By the action of hydrochloric acid on zirconium hydroxide.
Grade: Technical.
Containers: Glass bottles; drums.
Uses: Source of the pure metal (formed as intermediate in process); analytical chemistry; water repellents for textiles; pigments; tanning agent; catalysis.
Shipping regulations: None.*

zirconium tetrapentanedionate. See zirconium tetraacetylacetonate.

Zircon Milled, "G." [337] (Florida zircon). Trade name for zircon with 96+% $ZrSiO_4$. Gray white powder; sp.gr. 4.5; bulk density 95 lbs/cu ft; available as fine grind (0.1% +200 mesh, 1.0% +325 mesh) and coarse grind (1% +200 mesh, 10% +325 mesh). Used as fine ingredient in zircon refractories, and as a component of enamel frits, ceramic insulators for spark plugs, low-loss electrical porcelains and acid resistant chemical porcelain.

Zircon Milled "TAM." [337] (Australian and Florida zircon milled). Trade designation for 96% $ZrSiO_4$; white powder with average particle size less than 44 microns, sp.gr. 4.5; bulk density 109 lbs/cu ft. Used for slip casting of high grade zircon refractories, as an ingredient of electrical and chemical porcelains, and as an opacifier in some ceramics.
Containers: 80-lb paper bags; 550-lb barrels; 36,000-lb carloads.

zircon sand. Term applied to sand containing considerable zirconium, titanium and related metals. Used as a source of these elements, and also as a high heat resistant sand for casting of alloys.

zirconyl carbonate. See zirconium carbonate, basic.

zirconyl chloride. See zirconium oxychloride.

zirconyl hydroxychloride solution $ZrOOHCl \cdot nH_2O$.
Properties: Colorless or slightly acid solution; sp.gr. 1.26; forms a soluble glass on evaporation; pH of solution 0.8 (approx); reacts with alkalies to form hydrous zirconia. Solution assays 20% zirconia.
Containers: 500-lb (plastic lined) steel drums; 30,000-lb carloads.
Uses: Pharmaceuticals; deodorants; precipitation of acid dyes; synthesis of zirconium compounds; water-repellents for textiles.

zirconyl hydroxynitrate solution. See zirconyl nitrate solution (basic).

zirconyl nitrate solution (basic) (zirconyl hydroxynitrate solution) $ZrO(OH)NO_3$.
Properties: Sp.gr. (25°C) 1.35.
Uses: Gelations and improving lamination bonds of polyvinyl alcohol.

zirconyl sulfate. See zirconium sulfate, basic.

"Zircopax." [337] Trade name for 94-96.5% pure zircon; white powder, sp.gr. 4.5; bulk density 98 lbs/cu ft; average particle size 15 microns max. Used as an opacifier in glazes for wall tile, sanitary ware, electrical porcelain and art ware to impart craze resistance, whiteness, color stability, and reduction of chrome flashing.
Containers: 80-lb paper bags; 500-lb drums; 36,000-lb carloads.

"Zircotan." [23] Trademark for zirconium tanning agents which produce through-white leather.
Use: Production of white kid suede, glove leathers, retannage of chrome leather.

"Zirex." [79] Trademark of a special zinc resinate having a high melting point; high zinc content (twice as high as "Zitro") with a pale color and a negative acid number (slightly basic).
Constants: Sp.gr. (25°C) 1.162; m.p. (capillary tube) 132°C; metallic zinc (combined) 8.9%; color "N."
Containers: Non-returnable, light-weight galvanized drums of about 500 lbs gross wt. Tare 14-16 lbs.
Uses: Paints; varnishes; fiber coating compounds (with ethyl cellulose); adhesives; linoleum print paint; ethyl cellulose lacquers.

zirkite (zircite). Name applied to baddeleyite that contains appreciable zirconium oxide. Found in Brazil.

*See "I.C.C. Shipping Regulations," page xiii.
Reference numbers refer to name of manufacturer. See "List of Manufacturers," page v.

"Zirmet." [250] Trademark for ductile zirconium for vacuum tube getters; available as thin sheet and wire.

"Zirox B." [337] Trade name for zirconium oxide containing 90% ZrO_2 and 5-8% SiO_2. White cream powder, sp.gr. 5.4; bulk density 72 lbs/cu. ft.; maximum particle size 5 microns. Used as a polishing compound for opthalmic lenses, precision glass polishing, also for polishing glazes, marble and granite. "Zirox C" is closely similar with bulk density 70 lbs/cu. ft.
Containers: 50-lb drums, 100-lb drums, 36,000-lb carloads.

"Zirox D." [337] Trade name for a zirconium oxide similar to "Zirox B" but with improved suspension properties to permit use in recirculating automatic equipment.

"Zitro." [79] Trademark of a zinc resinate having a high melting point, high zinc content, pale color and low acid number.
Constants: Sp.gr. (25°C) 1.130; m.p. (capillary tube) 132°C; metallic zinc (combined) 4.9%; acid number 15; color "N."
Containers: Nonreturnable light-weight galvanized drums of about 500 lbs gross weight. Tare 14-16 lbs.
Uses: Paints; varnishes; fiber coating compounds (with ethyl cellulose); adhesives; linoleum print paint; ethyl cellulose lacquers.

ZMA. Abbreviation for zinc metaarsenite. See zinc arsenite.

Zn. Symbol for zinc.

"Zobar." [28] Trademark for a weed killer based on an aqueous solution of the dimethylamine salts of polychlorobenzoic acids, containing 4 lb of acid equivalent per gallon.
Use: See "Trysben" 200.

zoisite (thulite) $Ca_2(AlOH)Al_2(SiO_4)_3$. One of the epidote (q.v.) group of minerals.
Properties: Grayish-white, gray, peach-blossom to rose-red, green; white or uncolored streak; vitreous or pearly luster. Sp.gr. 3.25-3.37; hardness 6-6.5.
Occurrence: United States (Tennessee, Massachusetts, Pennsylvania), Austria, Switzerland, Norway, Italy.
Uses: Ornamental stone.

"Zonyl" E-7. [28] Trademark for a fluoroalcohol pyromellitate. Amber oil.
Use: Thermally stable fluid-lubricant and power fluid.

"Zonyl" E-91. [28] Trademark for fluoroalcohol camphorate. Amber oil.
Use: Thermally stable liquid-lubricant and power fluid.

"Zonyl" S-13. [28] Trademark for a fluorochemical surfactant, a free acid of fluoroalkyl phosphate. Brown waxy solid.
Use: Chemically stable surface active agent.

"Zopaque." [296] Trademark for pure titanium dioxide (q.v.), manufactured from ilmenite and specially processed to control crystal growth. It is used in nearly every industry requiring a white opacifying agent.

"Zoron" CR. [28] Trademark for a resin finish. A clear solution of thermosetting resin.
Use: Synthetic tanning assistant to tighten leather grain.

"Zoron" TR. [28] Trademark for an anionic alkaline dispersion of unplasticized thermoplastic resin. Light cream-colored liquid.
Use: As a base for seasoning to obtain a glossy top finish for leather and leather articles.

zoxazolamine $C_7H_5ClN_2O$.
Properties: White to creamy white, odorless powder or glistening crystals; freely soluble in alcohol; nearly insoluble in water; m.p. 183-188°C.
Grade: N.F. XI.
Use: Medicine.

Zr. Symbol for zirconium.

Zr "G." [250] Trademark for a highly pyrophoric zirconium metal powder of low oxide and impurity content. Particle size is 99% minus 200 mesh with average Fisher subsieve size 3±1 microns.
Uses: Production of pyrotechnics, ammunition, photo flash bulbs.
Containers: Packed under water in special I.C.C. approved containers.

zymase. The enzyme present in yeast which converts sugars to alcohol and carbon dioxide.

zymohexase. See aldolase.

"Zytel." [28] Trademark for nylon resin available as molding powders, extrusion powder and soluble resin.

"Zytron." [233] O-[2,4-dichlorophenyl]-O-methyl isopropylphosphoramidothioate.
Use: As active ingredient for crab grass control; either as dry material or liquid emulsion.

Nos.

"99." [233] Trademark for 2,4-D base compounds.

"#140." [133] Trademark for a general purpose black for inks and paints; low oil absorption. Made by impingement process. Available in uncompressed and densed form.
Container: 25-lb bags.

606. See arsphenamine. The name arose from the fact that this was the 606th and finally successful compound Ehrlich had tried against syphilis.

"6-12" ("Six-Twelve.") [214] Trademarks for an insect repellent. Active ingredient, ethylhexanediol, 100%.
Properties: Colorless, odorless liquid; contains no grease or oil.
Uses: Effective repellent against mosquitoes, chiggers, stable flies, black flies and gnats.

"#999." [133] Trademark for an intermediate color impingement carbon black made from natural gas. Used for industrial paints and plastics; excellent ultraviolet screening properties. Protects polyethylene from ultraviolet breakdown.
Containers: 12½-lb bags (powdered form); 25-lb bag (beads).

1080. See sodium fluoroacetate.